Geological Society books refereeing procedures

The Society makes every effort to ensure that the scientific and production quality of its books matches that of its journals. Since 1997, all book proposals have been refereed by specialist reviewers as well as by the Society's Books Editorial Committee. If the referees identify weaknesses in the proposal, these must be addressed before the proposal is accepted.

Once the book is accepted, the Society Book Editors ensure that the volume editors follow strict guidelines on refereeing and quality control. We insist that individual papers can only be accepted after satisfactory review by two independent referees. The questions on the review forms are similar to those for *Journal of the Geological Society*. The referees' forms and comments must be available to the Society's Book Editors on request.

Although many of the books result from meetings, the editors are expected to commission papers that were not presented at the meeting to ensure that the book provides a balanced coverage of the subject. Being accepted for presentation at the meeting does not guarantee inclusion in the book.

More information about submitting a proposal and producing a book for the Society can be found on its web site: www.geolsoc.org.uk.

It is recommended that reference to all or part of this book should be made in one of the following ways:

MCCANN, T. (ed.) 2008. *The Geology of Central Europe. Volume 1: Precambrian and Palaeozoic*. Geological Society, London.

BELKA, Z. & NARKIEWICZ, M. 2008. Devonian. *In*: MCCANN, T. (ed.) *The Geology of Central Europe. Volume 1: Precambrian and Palaeozoic*. Geological Society, London, 383–410.

The Geology of Central Europe
Volume 1: Precambrian and Palaeozoic

EDITED BY

T. McCANN

Bonn University, Germany

2008

Published by

The Geological Society

London

THE GEOLOGICAL SOCIETY

The Geological Society of London (GSL) was founded in 1807. It is the oldest national geological society in the world and the largest in Europe. It was incorporated under Royal Charter in 1825 and is Registered Charity 210161.

The Society is the UK national learned and professional society for geology with a worldwide Fellowship (FGS) of over 9000. The Society has the power to confer Chartered status on suitably qualified Fellows, and about 2000 of the Fellowship carry the title (CGeol). Chartered Geologists may also obtain the equivalent European title, European Geologist (EurGeol). One fifth of the Society's fellowship resides outside the UK. To find out more about the Society, log on to www.geolsoc.org.uk.

The Geological Society Publishing House (Bath, UK) produces the Society's international journals and books, and acts as European distributor for selected publications of the American Association of Petroleum Geologists (AAPG), the Indonesian Petroleum Association (IPA), the Geological Society of America (GSA), the Society for Sedimentary Geology (SEPM) and the Geologists' Association (GA). Joint marketing agreements ensure that GSL Fellows may purchase these societies' publications at a discount. The Society's online bookshop (accessible from www.geolsoc.org.uk) offers secure book purchasing with your credit or debit card.

To find out about joining the Society and benefiting from substantial discounts on publications of GSL and other societies worldwide, consult www.geolsoc.org.uk, or contact the Fellowship Department at: The Geological Society, Burlington House, Piccadilly, London W1J 0BG: Tel. +44 (0)20 7434 9944; Fax +44 (0)20 7439 8975; E-mail: enquiries@geolsoc.org.uk.

For information about the Society's meetings, consult *Events* on www.geolsoc.org.uk. To find out more about the Society's Corporate Affiliates Scheme, write to enquiries@geolsoc.org.uk

Published by The Geological Society from:
The Geological Society Publishing House, Unit 7, Brassmill Enterprise Centre, Brassmill Lane, Bath BA1 3JN, UK

(*Orders*: Tel. +44 (0)1225 445046
Fax +44 (0)1225 442836)
Online bookshop: www.geolsoc.org.uk/bookshop

British Library Cataloguing in Publication Data
A catalogue record for this book is available from the British Library.

ISBN 978-1-86239-245-8 (hardback)
ISBN 978-1-86239-246-5 (paperback)

Typeset by Keytec Tyesetting Ltd, Bridport, UK
Printed by MPG Books Ltd, Bodmin, UK

Distributors
North America
For trade and institutional orders:
The Geological Society, c/o AIDC, 82 Winter Sport Lane, Williston, VT 05495, USA
Orders: Tel +1 800-972-9892
Fax +1 802-864-7626
Email gsl.orders@aidcvt.com

For individual and corporate orders:
AAPG Bookstore, PO Box 979, Tulsa, OK 74101-0979, USA
Orders: Tel +1 918-584-2555
Fax +1 918-560-2652
Email *bookstore@aapg.org*
Website http://bookstore.aapg.org

India
Affiliated East-West Press Private Ltd, Marketing Division, G-1/16 Ansari Road, Darya Ganj, New Delhi 110 002, India
Orders: Tel. +91 11 2327-9113/2326-4180
Fax +91 11 2326-0538
E-mail affiliat@vsnl.com

Contents of Volume 1

Contents of Volume 2

Acknowledgements

The idea of publishing a reference work on Central European geology, written in English, by research-active experts from a range of European countries, was initially conceived in 2004. From this idea to final completion was a long road! Manuscripts began to arrive in 2005, were refereed and revised during 2006, with the completed work submitted to the Geological Society in May 2007. A mammoth work such as this involves the efforts of many, many people. I would, therefore, like to first thank the co-ordinators of the individual chapters for their work in steering to completion each of their manuscripts. These co-ordinators built up expert teams of contributing authors who wrote on specific areas or topics, and I would also like to thank each individual, in turn, for their excellent work.

A work such as this, however, could not have been completed without the published scientific papers of the many hundreds, if not thousands, of individual authors working over the last century or so. Many of these studies were carried out at a time when Central Europe, both politically and economically, was a very different region from the one we know today. Working in sometimes difficult circumstances, these authors have contributed to our knowledge and understanding of the evolution of this complex area. Without the detailed studies of these many scientists, any overview such as this would have been impossible, and I would, therefore, like to acknowledge our collective debt to these many workers.

Finally, I would like to acknowledge the support and guidance of Angharad Hills of the Geological Society Publishing House as well as the superlative editorial skills of Sally Oberst and her colleague Jessica Pollitt in bringing this work to fruition.

The editor gratefully acknowledges the work of the following referees:

Richard Aldridge
Marek Awdankiewicz
Thilo Bechstädt
David Blundell
Christoph Breitkreuz
Tim Brewer
Robin Cocks
Juergen Ehlers
Mary Ford
Sven-Oliver Franz
Hans-Joachim Franzke
Bernhard Fügenschuh
Andrew Gale
Reinhard Gast
Reinhard Gaupp
Alice Giannetti
Phil Gibbard
Ken Glennie
David Harper
Adrian Hartley
Jeremy Inglis
Ian Jarvis
Peter Konigshof
Phil Leat
Cherry Lewis
Ralf Littke
Jerzy Nawrocki
Franz Neubauer
Onno Oncken
Alan Owen
Christophe Pascal
Holger Paulick
Andreas Peterek
Niels Poulsen
Cecilio Quesada
Bettina Reitenbacher
David Rickard
Fred Rögl
Reinhard Schaeffer
Andreas Siemes
Randell Stephenson
Rob Strachan
Jaromir Ulrych
Mario Valdivia-Manchego
Cees van Staal
Geoff Warrington
Paul Wignall
John Winchester
Thomas Will
Gernold Zulauf

Contributing authors

Paweł Aleksandrowski
Institute of Geological Sciences, University of Wrocław, Wrocław, Poland (e-mail: palex@ing.uni.wroc.pl)
Olga Y. Anistratenko
Institute of Geological Sciences of NAS Ukraine, Kiev, Ukraine (e-mail: nistrat@rambler.ru or anistrat@izan.kiev.ua)
Vitaliy V. Anistratenko
Schmalhausen Institute of Zoology of NAS Ukraine, Kiev, Ukraine (e-mail: anistrat@izan.kiev.ua or nistrat@rambler.ru)
Didier Arbouille
IHS Energy 24, Geneva, Switzerland (e-mail: didier.arbouille@ihsenergy.com)
Janusz Badura
Polish Geological Institute, Wroclaw, Poland (e-mail: janusz.badura@pgi.gov.pl)
Davide Bassi
Dipartimento di Scienze della Terra, Università degli Studi di Ferrara, Ferrara, Italy (e-mail: bsd@unife.it)
Mirko Belak
Croatian Geological Survey, Zagreb, Croatia (e-mail: mbelak@hgi-cgs.hr)
Zdzislaw Bełka
Institute of Geology, Adam Mickiewicz University, Poznań, Poland (e-mail: zbelka@amu.edu.pl)
Jean-Pierre Berger
Institut de Géologie, Université de Fribourg, Fribourg, Switzerland (e-mail: jean-pierre.berger@unifr.ch)
Gianluca Bianchini
Dipartimento di Scienze della Terra, Università degli Studi di Ferrara, Ferrara, Italy (e-mail: bncglc@unife.it)
Ravi Borkhataria
Institute of Geosciences, University of Tübingen, Tübingen, Germany. Current address: Shell International Exploration and Production B.V., Rijswijk, The Netherlands (e-mail: ravi.borkhataria@shell.com)
Sibila Borojević-Šoštarić
Department of Geology, University of Zagreb, Zagreb, Croatia (e-mail: sborojsost@geol.pmf.hr)
Pavel Bosák
Institute of Geology, Academy of Sciences of the Czech Republic, Prague, Czech Republic (e-mail: bosak@gli.cas.cz)
Frédéric Boulvain
Département de Géologie, Université de Liège, Liège, Belgium (e-mail: fboulvain@eulg.ac.be)
Zbigniew Buła
Polish Geological Institute, Sosnowiec, Poland (e-mail: zbigniew.bula@pgi.gov.pl)
Stanislav Čech
Czech Geological Survey, Prague, Czech Republic (e-mail: stanislav.cech@geology.cz)
Safet Čičić
Mining & Civil Engineering Faculty, University of Tuzla, Tuzla, Bosnia Herzegovina.
L. Robin M. Cocks
Natural History Museum, London UK (e-mail: r.cocks@nhm.ac.uk)
Vlasta Ćosović
Department of Geology, University of Zagreb, Zagreb, Croatia (e-mail: vcosovic@geol.pmf.hr)
Harald G. Dill
Federal Institute for Geosciences and Natural Resources, Hanover, Germany (e-mail: h.dill@bgr.de)
Richard S. D'Lemos
Crustal Dynamics Group, Oxford Brookes University, Oxford, UK. Present address: Somerton, Oxfordshire, UK (e-mail: richard@dlemos2.wanadoo.co.uk)
Nela Doláková
Institute of Geological Sciences, Masaryk University, Brno, Czech Republic (e-mail: nela@sci.muni.cz)
Katica Drobne
Institute of Palaeontology ZRC-SAZU, Ljubljana, Slovenia (e-mail: katica@zrc-sazu.si)
Kerstin Drost
Museum für Mineralogie und Geologie, Dresden, Germany (e-mail: kerstin.drost@snsd.smwk.sachsen.de)
Michiel Dusar
Geological Survey of Belgium, Brussels, Belgium (e-mail: michiel.dusar@naturalsciences.be)
Jerzy Dzik
Instytut Paleobiologii PAN, Warsaw, Poland (e-mail: dzik@twarda.pan.pl)
Olak Elicki
Institut für Geologie, TU Bergakademi Freiberg, Freiberg/Sachsen, Germany (e-mail: elicki@geol.tu-freiberg.de)
Raymond Enay
UFR des Sciences de la Terre, Université Claude Bernard Lyon 1, Villeurbanne, France (e-mail: raymond.enay@univ-lyon1.fr)
Richard England
Department of Geology, University of Leicester, Leicester, UK (e-mail: rwe5@leicester.ac.uk)
Oldrich Fatka
Department of Paleontology, Faculty of Science, Charles University, Prague, Czech Republic (e-mail: fatka@natur.cuni.cz)
Susanne Feist-Burkhardt
Natural History Museum, London, UK (e-mail: s.feist-burkhardt@nhm.ac.uk)
Peter J. Felder
Université de Liège, Paléontologie, Liège, Belgium (e-mail: sjeuf.felder@wanadoo.nl)
Anna Feldman-Olszewska
Polish Geological Institute, Warsaw, Poland (e-mail: anna.feldman-olszewska@pgi.gov.pl)
Sorin Filipescu
Department of Geology, Babes-Bolyai University, Cluj-Napoca, Romania (e-mail: sorin@bioge.ubbcluj.ro)
Hans-Joachim Franzke
Institute for Geology and Paleontology, Technische Universität, Clausthal-Zellerfeld, Germany (e-mail: hans.joachim.franzke@tu-clausthal.de)
Manfred Frechen
Leibniz Institute for Applied Geoscience (GGA-Institut), Hanover, Germany (e-mail: manfred.frechen@gga-hannover.de)
Nikolaus Froitzheim
Geologisches Institut, University of Bonn, Bonn, Germany (e-mail: niko.froitzheim@uni-bonn.de)
Hanspeter Funk
Seminarstraße 26, Baden, Switzerland (e-mail: h.funk@hispeed.ch)

Giorgio Garuti
Department of Earth Sciences, University of Modena and Reggio Emilia, Modena, Italy (e-mail: garutig@unimore.it)
Thorsten Geisler
Institut für Mineralogie, Universität Münster, Münster, Germany (e-mail: tgeisler@nwz.uni-muenster.de)
Mark Geluk
Shell International Exploration and Production B.V., Rijswijk, The Netherlands (e-mail: mark.geluk@shell.com)
Axel Gerdes
Institut für Geowissenschaften, Mineralogie, Johann Wolfgang Goethe-Universität, Frankfurt/Main, Germany (e-mail: gerdes @em.uni-frankfurt.de)
Gerd Geyer
Institut für Paläontologie, Universität Würzburg, Würzburg, Germany (e-mail: gerd.geyer@mail.uni-wuerzburg.de)
Jan Golonka
University of Science and Technology, Kraków, Poland (e-mail: jan_golonka@yahoo.com)
Annette Götz
Institute for Applied Geosciences, Darmstadt University of Technology, Schnittspahnstrasse 9, D-64287 Darmstadt, Germany (e-mail: goetz@energycenter.tu-darmstadt.de)
P. Grecula
Geological Survey of Slovak Republic, Kosice, Slovakia (e-mail: grecula@gssr-ke.sk)
Karl Gürs
Landesamt für Natur und Umwelt Schleswig-Holstein, Flintbek, Germany (e-mail: kguers@lanu.landsh.de)
Jacek Gutowski
Polish Geological Institute, Warsaw, Poland (e-mail: jacek. gutowski@pgi.gov.pl)
János Haas
Geological, Geophysical & Space Science Research Group of the Hungarian Academy of Sciences, Eötvös Loránd University, Budapest, Hungary (e-mail: haas@ludens.elte.hu)
Horst Peter Hann
Institut für Geologie und Paläontologie, Universität Tübingen, Tübingen, Germany (e-mail: horst.hann@uni-tuebingen.de)
Martin Hansen
Hydro Oil & Energy, Research Centre Bergen, Norway (e-mail: martin.bak.hansen@hydro.com)
Mathias Harzhauser
Naturhistorisches Museum in Wien, Vienna, Austria (e-mail: mathias.harzhauser@nhm-wien.ac.at)
G.F.W. Herngreen
Kenniscentrum Biogeologie (UU/TNO), Utrecht University, Utrecht, The Netherlands (e-mail: g.f.w.herngreen@bio.uu.nl)
Thomas Heuse
Thüringer Landesanstalt für Umwelt und Geologie, Geologischer Landesdienst, Jena, Germany (e-mail: t.heuse@TLUGJena. Thueringen.de)
Šarka Hladilová
Institute of Geological Sciences, Masaryk University, Brno, Czech Republic (e-mail: sarka@sci.muni.cz)
Jens Hornung
Institute for Applied Geosciences, Darmstadt University of Technology, Darmstadt, Germany (e-mail: hornung@geo. tu-darmstadt.de)
Hazim Hrvatović
Geological Survey of Bosnia and Herzegovina, Iiidza-Sarajevo, Bosnia Herzegovina (e-mail: hharish@bih.net.ba)
Christian Hübscher
Institut für Geophysik, Universität Hamburg, Hamburg, Germany (e-mail: christian.huebscher@zmaw.de)
Hermann Huckriede
Thüringer Landesanstalt für Umwelt und Geologie (Außenstelle Weimar), Jena, Germany (e-mail: h.huckriede@tlugjena. thueringen.de)
John W.M. Jagt
Natuurhistorisch Museum Maastricht, Maastricht, The Netherlands (e-mail: john.jagt@maastricht.nl)
Marek Jarosiński
Polish Geological Institute, Warszawa, Poland (e-mail: mjar@pgi.gov.pl)
Teresa Jeffries
Natural History Museum, London, UK (e-mail: t.jeffries@nhm. ac.uk)
Bogomir Jelen
Geoloski zavod Slovenije, Ljubljana, Slovenia (e-mail: bogomir. jelen@guest.arnes.si)
Peter Jordan
Institute of Geology and Palaentology, University of Basel, Basel, Switzerland (e-mail: peter.jordan@tiscali.ch)
Václav Kachlík
Department of Geology and Paleontology, Faculty of Science, Charles University, Prague, Czech Republic (e-mail: kachlik@natur.cuni.cz)
Jacek Robert Kasiński
Polish Geological Insititute, Warsaw, Poland. (e-mail: jacek. kasinski@pgi.gov.pl)
Oliver Kempf
Landesgeologie, swisstopo, Seftigenstr. 264, CH-3084 Wabern Switzerland (e-mail: oliver.kempf@swisstopo.ch)
Hubert Kiersnowski
Polish Geological Institute, Warsaw, Poland (e-mail: hubert. kiersnowski@pgi.gov.pl)
Frank Körner
Geologisches Insitut, University of Bonn, Bonn, Germany (e-mail: koerner@uni-bonn.de)
Kristijan Kornpihl
Integrated Exploration Systems, Aachen, Germany (e-mail: k.kornpihl@ies.de)
Michal Kováč
Department of Geology and Paleontology, Comenius University, Bratislava, Slovakia (e-mail: kovacm@fns.uniba.sk)
Karl Krainer
Geologie und Paläontologie, Universität Innsbruck, Innsbruck, Austria (e-mail: karl.krainer@uibk.ac.at)
Polona Kralj
Geological Survey of Slovenia, Ljubljana, Slovenia (e-mail: polona.kralj@geo-zs.si)
Charlotte Krawczyk
Leibnitz Institute for Applied Geosciences (GGA), Hannover, Germany (e-mail: lotte@gga-hannover.de)
Michael Krings
Department für Geo- und Umweltwissenschaften an der LMU München, München, Germany (e-mail: m.krings@lrz-uni-muenchen)
Jíří Kříž
Czech Geological Survey, Prague, Czech Republic (e-mail: jiri.kriz@geology.cz)
Michał Krobicki
University of Science and Technology, Kraków, Poland (e-mail: krobicki@geol.agh.edu.pl)
Uwe Kroner
Institut für Geologie, TU Bergakademie Freiberg, Freiberg, Germany (e-mail: uwe.kroner@geo.tu-freiberg.de)
Ewa Krzemińska
Polish Geological Institute, Warsaw, Poland (e-mail: ewa. krzeminska@pgi.gov.pl)
Piotr Krzywiec
Polish Geological Institute, Warsaw, Poland (e-mail: piotr. krzywiec@pgi.gov.pl)

Frédéric Lacquement
Bureau de Recherches Géologiques et Minières, Orleáns, France (e-mail: f.lacquement@brgm.fr)

Juliette Lamarche
Centre de Sédimentologie-Paléontologie, Université de Provence Aix-Marseille 1, Marseille, France (e-mail: jula@up.univ-mrs.fr)

Bernard Lathuiliere
UMR CNRS 7566, Géologie et Gestion des Ressources Minérales et Energétiques (G2R), Université de Nancy I, Vandoeuvre-lès-Nancy, France (e-mail: bernard.lathuiliere@g2r.u-nancy.fr)

Patrick Ledru
BRGM/Research Division 3, Orléans, France (e-mail: p.ledru @brgm.fr)

Reinhold R. Leinfelder
Museum für Naturkunde, Humboldt-Universität Berlin, Berlin, Germany (e-mail: leinfelder@museum.hu-berlin.de)

Ulf Linnemann
Museum für Mineralogie und Geologie, Dresden, Germany (e-mail: ulf.linnemann@snsd.smwk.sachsen.de)

Thomas Litt
Institut für Paläontologie, Universität Bonn, Bonn, Germany (e-mail: t.litt@uni-bonn.de)

John McBride
Brigham Young University, Provo, Utah, UT, USA (e-mail: john_mcbride@byu.edu)

Tom McCann
Geologisches Insitut, University of Bonn, Bonn, Germany (e-mail: tmccann@uni-bonn.de)

Jörg Maletz
Department of Geology, University at Buffalo, SUNY, Buffalo, NY, USA (e-mail: jorgm@buffalo.edu)

Tihomir Marjanac
Department of Geology, University of Zagreb, Zagreb, Croatia (e-mail: marjanac@geol.pmf.hr)

Emö Márton
Eötvös Loránd Geophysical Institute of Hungary, Budapest, Hungary (e-mail: paleo@elgi.hu or h11000mar@helka.iif.hu)

Yuriy Maystrenko
GeoForschungsZentrum Potsdam, Potsdam, Germany (e-mail: yuram@gfz-potsdam.de)

Stanisław Mazur
Institute of Geological Sciences, University of Wrocław, Wrocław, Poland (e-mail: smazur@ing.uni.wroc.pl)

Jozef Michalík
Slovak Academy of Science, Bratislava, Slovakia (e-mail: geolmich@savba.sk)

Paolo Mietto
Dipartimento di Geoscienze, Università di Padova, Padova, Italy (e-mail: paolo.mietto@unipd.it)

Eckhard Mönnig
Naturkunde-Museum Coburg, Coburg, Germany (e-mail: e.moennig@naturkunde-museum-coburg.de)

Gediminas Motuza
Department of Geology and Mineralogy, Faculty of Natural Sciences, Vilnius, Lithuania (e-mail: gediminas.motuza@gf.vu.lt)

Alan Moro
Department of Geology, University of Zagreb, Zagreb, Croatia (e-mail: amoro@geol.pmf.hr)

Rainer Müller
Institute for Geology and Palaeontology, Technische Universität, Clausthal-Zellerfeld, Germany (e-mail: rainer.mueller@ tu-clausthal.de)

J. Brendan Murphy
Department of Earth Sciences, St. Francis Xavier University, Antigonish, NS, Canada (e-mail: bmurphy@stfx.ca)

András Nagymarosy
Department of Physical and Historical Geology, Eötvös Loránd University, Budapest, Hungary (e-mail: gtorfo@ludens.elte.hu)

R. Damian Nance
Department of Geological Sciences, Ohio University, Athens, OH, USA (e-mail: nance@ohio.edu)

Marek Narkiewicz
Polish Geological Institute, Warsaw, Poland (e-mail: marek. narkiewicz@pgi.gov.pl)

Jerzy Nawrocki
Paleomagnetic Laboratory, Polish Geological Institute, Warsaw, Poland (e-mail: jerzy.nawrocki@pgi.gov.pl)

James H. Nebelsick
Institute of Geosciences, University of Tübingen, Tübingen, Germany (e-mail: nebelsick@uni-tuebingen.de)

Slavomír Nehyba
Masaryk University, Institute of Geological Sciences, Brno, Czech Republic (e-mail: slavek@sci.muni.cz)

Birgit Niebuhr
GeoZentrum Nordbayern, Fachgruppe Paläoumwelt, FAU Universität Erlangen-Nürmberg, Erlangen, Germany (e-mail: niebuhr@pal.uni-erlangen.de)

Nanna Noe-Nygaard
Department of Geography and Geology, Copenhagen, Denmark (e-mail: nannan@geol.ku.dk)

Bojan Ogorelec
Geological Survey of Slovenia, Ljubljana, Slovenia (e-mail: bojan.ogorelec@geo-zs-si)

Stanislav Oplustil
Institute of Geology and Palaeontology, Charles University, Praha, Czech Republic (e-mail: oplustil@natur.cuni.cz)

Nestor Oszczypko
Institute of Geological Sciences, Jagiellonian University, Kraków, Poland (e-mail: nestor@geos.ing.uj.edu.pl)

József Pálfy
Hungarian Academy of Sciences, Hungarian Natural History Museum, Budapest, Hungary (e-mail: palfy@nhmus.hu)

Ladislav A. Palinkaš
Department of Geology, University of Zagreb, Zagreb, Croatia (e-mail: lpalinkas@geol.pmf.hr)

Florentin Paris
Géosciences Rennes, UPR 4661 CNRS, Université de Rennes, Renne, France (e-mail: florentin.paris@univ-rennes1.fr)

Davor Pavelić
Faculty of Mining, Geology and Petroleum Engineering, University of Zagreb, Zagreb, Croatia (e-mail: dpavelic@yahoo. com)

Rajko Pavlovec
Department of Geology, University of Ljubljana, Ljubljana, Slovenia (e-mail: rajko.pavlovec@ntf.uni-lj.si)

Jernej Pavšič
Department of Geology, University of Ljubljana, Ljubljana, Slovenia (e-mail: jernej.pavsic@ntf.uni-lj.si)

Tadeusz Peryt
Polish Geological Institute, Warsaw, Poland (e-mail: tadeusz. peryt@pgi.gov.pl)

Pavla Tomanová Petrová
Czech Geological Survey, Brno, Czech Republic (e-mail: pavla. petrova@geology.cz)

Timothy C. Pharaoh
British Geological Survey, Keyworth, UK (e-mail: tcp@bgs. ac.uk)

Grzegorz Pieńkowski
Polish Geological Institute, Warsaw, Poland (e-mail: grzegorz. pienkowski@pgi.gov.pl)

Christian Pin
Département des Sciences de la Terre, CNRS & Université Blaise Pascal, Clermont-Ferrand, France (e-mail: c.pin@opgc.univ-bpclermont.fr)

Anna Pint
Institut für Geologische Wissenschaften, Freie Universität Berlin, Berlin, Germany (e-mail: annapint@web.de)

Sergei A. Pisarevsky
Tectonics Special Research Centre, University of Western Australia, Crawley, WA, Australia (e-mail: spisarevsky@tsrc.uwa.edu.au)

Marcin Piwocki
Polish Geological Institute, Warsaw, Poland (e-mail: marcin.piwocki@pgi.gov.pl)

Dušan Plašienka
Department of Geology and Paleontology, Comenius University, Bratislava, Slovakia

Marijan Poljak
Geoloski zavod Slovenije, Ljubljana, Slovenia (e-mail: marijan.poljak@geo-zs.si)

Paweł Poprawa
Polish Geological Institute, Warszawa, Poland (e-mail: pawel.poprawa@pgi.gov.pl)

Edouard Poty
Département de Géologie, Université de Liège, Liège, Belgium (e-mail: e.poty@ulg.ac.be)

W. Prochaska
Department of Applied Geosciences and Geophysics, University of Leoben, Leoben, Austria (e-mail: walter.prochaska@unileoben.ac.at)

Nevio Pugliese
Dipartimento di Scienze Geologiche Ambientali e Marine, Università di Trieste, Trieste, Italy (e-mail: pugliese@univ.trieste.it)

Michael W. Rasser
Staatliches Museum für Naturkunde, Stuttgart, Germany (e-mail: rasser.smns@naturkundemuseum-bw.de)

Rejhana Redžepović
Mining Geology & Civil Engineering Faculty, Universtiy of Tuzla, Tuzla, Bosnia Herzegovina (e-mail: rredzepovic@yahoo.com)

Mike Reich
Geowissenschaftliches Zentrum der Universität Göttingen (GZG), Museum, Sammlungen & Geopark, Göttingen, Germany (e-mail: mreich@gwdg.de)

Klaus Reicherter
Institut für Neotektonik und Georisiken, RWTH Aachen, Aachen, Germany (e-mail: k.reicherter@nug.rwth-aachen.de)

John Reinecker
Geological Institute, Tübingen University, Tubingen, Germany (email: reinecker@uni-tuebingen.de)

Lutz Reinhardt
Federal Institute for Geosciences and Natural Resources, Hanover, Germany (e-mail: l.reinhardt@bgr.de)

Achim G. Reisdorf
Institute of Geology and Palaeontology, University of Basel, Basel, Switzerland (e-mail: achim.reisdorf@unibas.ch)

Werner Ricken
Department of Geology and Mineralogy, University of Cologne, Cologne, Germany (e-mail: wricken@uni-koeln.de)

Helena Rifelj
Geoloski zavod Slovenije, Ljubljana, Slovenia (e-mail: helena.rifelj@geo-zs.si)

Reinhard Roetzel
Geological Survey of Austria, Vienna, Austria (e-mail: reinhard.roetzel@geologie.ac.at)

Heinz-Gerd Röhling
State Authority for Mining, Energy and Geology, Geological Survey of Lower Saxony and Bremen, Hannover, Germany (e-mail: heinz-gerd.roehling@lbeg.niedersachsen.de)

Rolf L. Romer
GeoForschungsZentrum Potsdam, Potsdam, Germany (e-mail: romer@gfz-potsdam.de)

Thomas Rüffer
Halle (Saale), Germany (e-mail: thomas.rueffer@t-online.de)

Reinhard F. Sachsenhofer
Department of Applied Geosciences and Geophysics, University of Leoben, Leoben. Austria (e-mail: reinhard.sachsenhofer@mu-leoben.at)

Scott D. Samson
Department of Earth Sciences, Syracuse University, Syracuse, NY, USA (e-mail: sdsamson@syr.edu)

Tibor Sasvári
Faculty of Mining, Ecology, Process Control and Geotechnologies, Technical University of Košice, Košice, Slovakia (e-mail: tibor.sasvari@tuke.sk)

Andreas Schäfer
Geologisches Institut, University of Bonn, Bonn, Germany (e-mail: schaefer@uni-bonn.de)

Magdalena Scheck-Wenderoth
GeoForschungsZentrum Potsdam, Potsdam, Germany (e-mail: leni@gfz-potsdam.de)

Christian Schlüchter
Institut für Geologie der Universität Bern, Bern, Switzerland (e-mail: christian.schluechter@geo.unibe.ch)

Dieter U. Schmid
Department für Geo- und Umweltwissenschaften/GeoBio-Center an der Ludwig-Maximilians-Universität München, München, Germany (e-mail: d.schmid@lrz.uni-muenchen.de)

Hans-Ulrich Schmincke
Leibniz Institut für Meereswissenschaften IFM-GEOMAR, Kiel, Germany (e-mail: hschmincke@ifm-geomar.de)

Jörg Schneider
Department of Palaeontology, Universität Freiberg, Freiberg, Germany (e-mail: schneidj@geo.tu-freiberg.de)

Hans Peter Schönlaub
Kötschach 350, A9640 Kötschach-Mauthen, Austria (e-mail: schoenlaub@aon.at)

Michael E. Schudack
Institut für Geologische Wissenschaften, Freie Universität Berlin, Berlin, Germany (e-mail: schudack@zedat.fu-berlin.de)

Anne S. Schulp
Natuurhistorisch Museum Maastricht, Maastricht, The Netherlands (e-mail: mail@nhmmaastricht.nl)

Bernhard Schulz
Institut für Geologie und Mineralogie, Erlangen, Germany. (e-mail: bschulz@geol.uni-erlangen.de)

Hans-Martin Schulz
GeoForschungsZentrum Potsdam, Potsdam, Germany (e-mail: schulzhm@gfz-potsdam.de)

Ralf Schuster
Geologische Bundesanstalt, Vienna, Austria (e-mail: ralf.schuster@geologie.ac.at)

Günter Schweigert
Staatliches Museum für Naturkunde, Stuttgart, Germany (e-mail: schweigert.smns@naturkundemuseum-bw.de)

Thomas Servais
Géosystèmes, UMR8157 du CNRS, Université des Sciences et Technologies de Lille, Villeneuve d'Ascq, France (e-mail: thomas.servais@univ-lille1.fr)

Dragomir Skaberne
Geoloski zavod Slovenije, Ljubljana, Slovenia (e-mail: dragomir.skaberne@geo-zs.si)

Stanislaw Skompski
Institute of Geology, Warsaw University, Warsaw, Poland (e-mail: skompski@uw.edu.pl)
Ľubomír Sliva
Department of Geology and Paleontology, Comenius University, Bratislava, Slovakia (e-mail: sliva@fns.uniba.sk)
Werner Stackebrandt
Landesamt für Geowissenschaften und Rohstoffe Brandenburg, Kleinmachnow, Germany (e-mail: werner.stackebrandt@lbgr-brandenburg.de)
Gerda Standke
Sächsisches Landesamt für Umwelt und Geologie, Dresden, Germany (e-mail: gerda.standke@smul.sachsen.de)
Harald Stollhofen
Geologisches Institut der RWTH, Aachen, Germany (e-mail: stollhofen@geol.rwth-aachen.de)
Robin A. Strachan
School of Earth and Environmental Sciences, University of Portsmouth, Portsmouth, UK (e-mail: rob.strachan@port.ac.uk)
Sabina Strmić-Palinkaš
Department of Geology, University of Zagreb, Zagreb, Croatia (e-mail: sabina.strmic@inet.hr)
Finn Surlyk
Department of Geography and Geology, Copenhagen, Denmark (e-mail: finns@geol.ku.dk)
Joachim Szulc
Institute of Geological Science, Jagiellonian University, Kraków, Poland (e-mail: szulc@geos.ing.uj.edu.pl)
Jennifer Tait
School of Geosciences, University of Edinburgh, Edinburgh, UK (e-mail: jenny.tait@ed.ac.uk)
Martin Timmerman
Institut für Geowissenschaften, Universität Potsdam, Potsdam, Germany (e-mail: timmer@geo.uni-potsdam.de)
Ákos Török
Department of Construction Materials and Engineering Geology, Budapest University of Technology and Economics, Budapest, Hungary (e-mail: torokakos@mail.bme.hu)
Giorgio Tunis
Dipartimento di Scienze Geologische Ambientali e Marine, Università di Trieste, Trieste, Italy (e-mail: tunis@univ.trieste.it)
David Uličný
Geophysical Institute of the Academy of Sciences of the Czech Republic, Boční 11/1401, 14131 Prague 4, Czech Republic (e-mail: ulicny@ig.cas.cz)
Jens Ulrich
Museum für Mineralogie und Geologie, Dresden, Germany. (e-mail: jens.ulrich@snsd.smwk.sachsen.de)
Dionýz Vass
Dionyz Stur Institute of Geology, Bratislava, Slovakia
Marco Vecoli
Géosystèmes, UMR 8157 du CNRS, Université des Sciences et Technologies de Lille, Villeneuve d'Ascq, France (e-mail: marco-vecoli@univ-lille1.fr)
Jacques Verniers
Department of Geology and Pedology, Ghent University, Ghent, Belgium (e-mail: jacques.verniers@ugent.be)
Silke Voigt
IfM-Geomar, Germany (e-mail: svoigt@ifm-geomar.de)
Thomas Voigt
Institut für Geowissenschaften, Universität Jena, Jena, Germany (e-mail: thomas.voigt@uni-jena.de)
Hilmar von Eynatten
Geowissenschaftliches Zentrum der Universität Göttingen, Göttingen, Germany (e-mail: hilmar.von.eynatten@geo.uni-goettingen.de)
Anna Vozárová
Department of Mineralogy and Petrology, Comenius University Bratislava, Bratislava, Slovakia. (e-mail: vozarova@fns.uniba.sk)
Michael Wagreich
Centre for Earth Sciences, University of Vienna, Vienna, Austria (e-mail: michael.wagreich@univie.ac.at)
Ireneusz Walaszczyk
Institute of Geology, University of Warsaw, Warsaw, Poland (e-mail: i.walszczyk@uw.edu.pl)
Frank Wesselingh
Naturalis – National Museum of Natural History, Leiden, The Netherlands (e-mail: wesselingh@naturalis.nl)
Andreas Wetzel
Geologisch-Paläontologisches Institut, Universität Basel, Basel, Switzerland (e-mail: andreas.wetzel@unibas.ch)
Frank Wiese
Fachrichtung Paläontologie, FU Berlin, Berlin, Germany (e-mail: frwiese@snafu.de)
Markus Wilmsen
GeoZentrum Nordbayern, Fachgruppe Paläoumwelt, FAU Erlangen-Nürmberg, Erlangen, Germany (e-mail: markus.wilmsen@pal.uni-erlangen.de)
Janina Wiszniewska
Polish Geological Institute, Warsaw, Poland (e-mail: janina.wiszniewska@pgi.gov.pl)
Theo E. Wong
TNO-NITG, Utrecht, The Netherlands (e-mail: theoewong@yahoo.co.uk)
Ryszard Wrona
Instytut Paleobiologii PAN, Warsaw, Poland (e-mail: wrona@twarda.pan.pl)
Stanislaw Wybraniec
Polish Geological Institute, Warsaw, Poland (e-mail: swyb@pgi.waw.pl)
Jerzy Żaba
Faculty of Earth Sciences, University of Silesia, Sosnowiec, Poland (e-mail: jzaba@ultra.cto.us.edu.pl)
Federica Zaccarini
Department of Applied Geosciences and Geophysics, University of Leoben, Leoben, Austria (e-mail: fedezac@tsc4.com)
Hubert Zedler
Regierungspräsidium Freiburg, Landesamt für Geologie, Rohstoffe und Bergbau, Freiburg im Breisgau, Germany (e-mail: hubert.zedler@rpf.bwl.de)
Armin Zeh
Mineralogisches Institut der Universität Würzburg, Würzburg, Germany (e-mail: armin.zeh@mail.uni-wuerzburg.de)
Živilė Žigaitė
Department of Geology and Mineralogy, Vilnius University, Vilnius, Lithuania (e-mail: zivile.zigaite@gf.vu.lt)
Ulf Zimmermann
Staatliche Naturhistorische Sammlungen Dresden, Dresden, Germany (e-mail: ulf.linnemann@snsd.smwk.sachsen.de)
Witold Zuchiewicz
Faculty of Geology, Geophysics & Environmental Protection, University of Science and Technology, Krakow, Poland (e-mail: witoldzuchiewicz@geol.agh.edu.pl)
Rainer Zühlke
Institute of Geology and Paleontology, Ruprecht Karl University, Heidelberg, Germany (e-mail: zuehlke@uni-hd.de)
Gernold Zulauf
Institut für Geowissenschaften, Frankfurt, Germany (e-mail: g.zulauf@em.uni-frankfurt.de)
Anna Żylińska
Institute of Geology, University of Warsaw, Warsaw, Poland (e-mail: anna.zylinska@uw.edu.pl)

1 Introduction and overview

TOM McCANN

Go my Sons, buy stout shoes, climb the mountains, search the valleys, the deserts, the sea shores, and the deep recesses of the earth. Look for the various kinds of minerals, note their characters and mark their origin ... observe and experiment without ceasing, for in this way and in no other will you arrive at a knowledge of the nature and properties of things.

(Wallerius, *Systema Mineralogicum*, Preface, Vienna, 1778) (Adams 1954, p. 210).

Once out of nature I shall never take
My bodily form from any natural thing,
But such a form as Grecian goldsmiths make
Of hammered gold and gold enamelling
To keep a drowsy Emperor awake;
Or set upon a golden bough to sing
To lords and ladies of Byzantium
Of what is past, or passing, or to come.

(W.B. Yeats, from *Sailing to Byzantium*)

Central Europe is a complex mosaic of crustal elements, including Cadomian fragments, Caledonian and Variscan microcontinents and suture zones, as well as Mesozoic and Alpine tectonostratigraphic successions (Fig. 1.1). These disparate elements were assembled during the various Precambrian and Phanerozoic (Caledonian, Variscan and Alpine) orogenies, the end result being a geological history of over 3500 million years. Over this long and complex period of crustal evolution, older consolidated crustal elements were repeatedly remobilized and overprinted by later events.

Rather than a physical entity, the term 'Central Europe' is more a concept of shared history, and in some ways refers to a similar area to that which is defined by the German term 'Mitteleuropa' (or Middle Europe), referring to territories under German cultural hegemony until World War I. In terms of the current volume, the definition of Central Europe is a broad one and includes Austria, Belgium, the Czech Republic, eastern France, Germany, Hungary, the Netherlands, Lichtenstein, Luxembourg, Poland, Slovakia, Slovenia and Switzerland. However, this core area was expanded according to the needs of the particular geological event or period under review. For example, the chapter on Caledonian tectonics (Krawczyk *et al.* 2008) includes a discussion of the situation in Britain and Ireland, while the Silurian chapter (Verniers *et al.* 2008) includes information from the Baltic countries. These additions to the core area vary from chapter to chapter, and have been included, as noted, in order to provide a more thorough understanding of the geological situation at a particular time. The majority of the chapters, however, are restricted to the core region which in itself is of significant complexity and interest.

Geological research

The geology of Europe has been studied intensively for well over two hundred years and the surface geology is probably better known than anywhere else in the world. In this time an enormous amount of geological data have been amassed, ranging from surface mapping and analysis through to the more recent deep surveys (e.g. seismic, borehole). However, this work did not represent the beginnings of the geological sciences in Europe; this extends back to the voluminous literature of the Greeks, much of it now sadly lost (Oldroyd 1996). Therefore, any overview of the development of the geological sciences in Europe can only be an introduction, since a detailed history is beyond the scope of this chapter. For more detail, there is a wealth of information on offer both general (e.g. Gohau 1991; Oldroyd 1996) and more specific (e.g. Dean 1992; Klaver 1997; Laudan 1987).

Greeks and Romans

One of the earliest Greek writers to observe geological phenomena and try to explain them was Xenophanes of Colophon (540–510 BC) who wrote on the phenomenon of erosion, as well as noting the presence of marine fossils on mountain tops. Aristotle of Stagira (384–322 BC) commented in his treatise *De Respiratione* that 'a great many fishes live in the earth motionless and are found when excavations are made' (Adams 1954, p. 12). Similar fossil occurrences were also remarked upon much later by the Roman poet Ovid (43 BC–AD 17) who noted that 'Often sea-shells lie far from the beach, and men have found old anchors on mountain tops' (Gohau 1991, p. 10). Explaining the presence of these fossils on mountain tops was, however, more difficult. Xanthus of Lydia (*c.* 480 BC) suggested that these upland areas must have, at one time, been covered by the sea. Herodotus of Halicarnassus (*c.* 484–425 BC), meanwhile, describing the presence of abundant nummulites in the Eocene limestones used to build the pyramids in Egypt, suggested that they were the indurated remains of the lentils left over from the store of food used to feed the workers.

The Greeks and Romans were also interested in volcanic and tectonic processes. Indeed, there are extensive classical writings on volcanoes (see Oldroyd 1996, table 1.1, for details). One of the earliest Greek writers, Thales of Miletus (*c.* 624–548/545 BC), suggested that the Earth floated on a great ocean, and that earthquakes might be caused by occasional perturbations (Oldroyd 1996). Aristotle, however, believed that volcanic eruptions

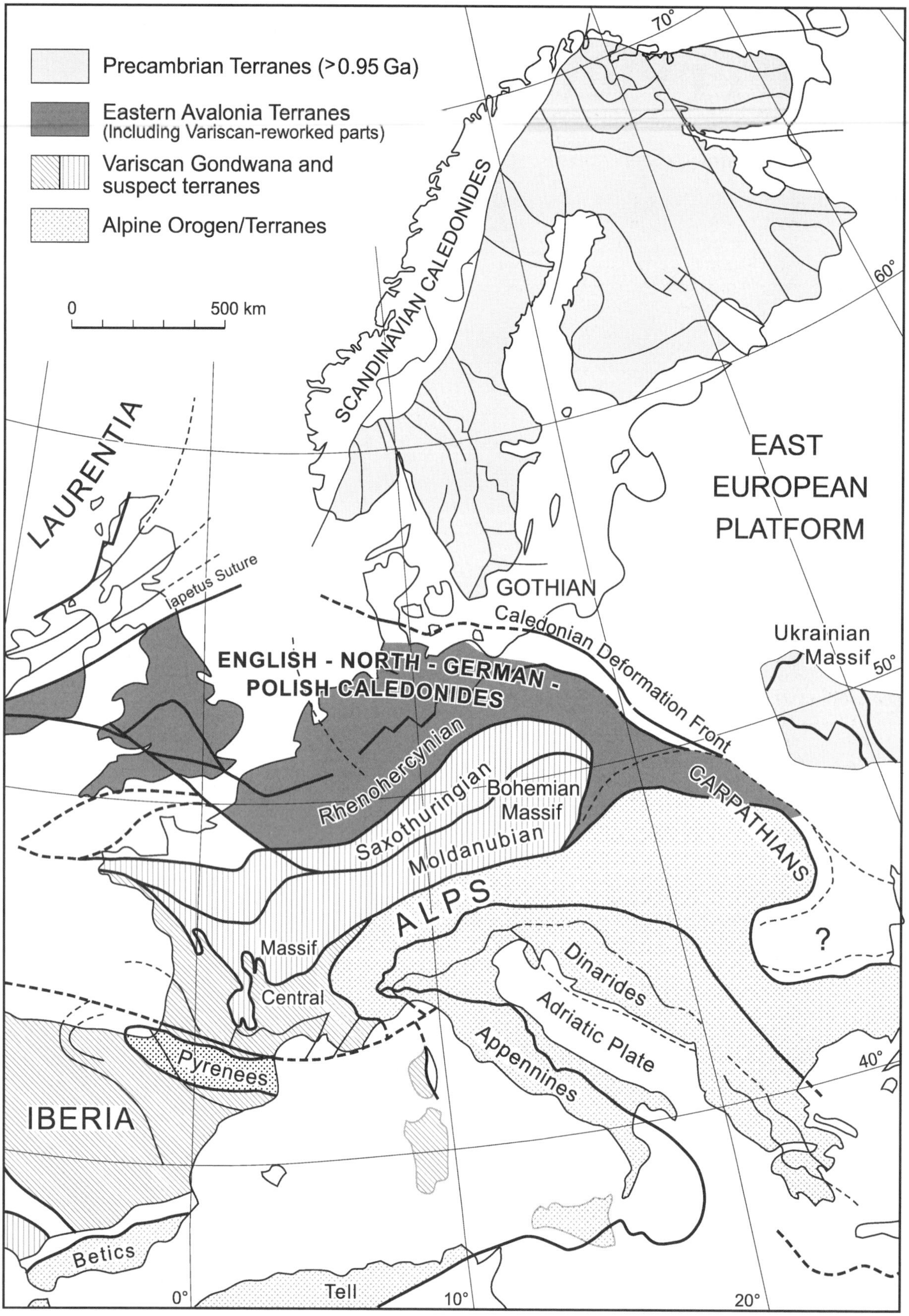

Fig. 1.1. Main tectonic units of Precambrian and Phanerozoic Europe (after Berthelsen 1992).

and earthquakes were the result of violent winds escaping from the Earth's interior. A different interpretation was proposed by the most influential of the early Greek writers, Strabo (*c.* 63 BC–after AD 21), writing in the early days of the Roman Empire. While only a fragment of his works survive, his *Geography* (in 17 volumes) is the most important work of antiquity on that subject, describing the world as it was known in Roman times and largely based on his own descriptions and observations. He suggested that variations in elevation in tectonically active areas were probably related to the so-called central fires (which also explained the origin of volcanoes). Furthermore, he believed that the effect of 'earthquakes, volcanic eruptions, uplifting of the submarine floor ... and sudden collapses or landslides' could raise or lower sea level (Gohau 1991, p. 11). This contrasted with the position of Eratosthenes, the first person to measure the circumference of the Earth, who believed that the level of the Mediterranean had lowered during the opening of the Strait of Gibraltar, and that proof of the original higher sea level was to be found in the occurrence of fossils and 'wreckage from seafaring ships' above the present level (Gohau 1991, p. 11).

In addition to tectonics and sea-level variations, the Greeks and Romans were also interested in the processes, and implications, of sediment transport and deposition. For example, Herodotus, writing about the Nile Valley and delta, noted that the valley had been gradually infilled by sediments transported by the Nile. Indeed, he suggested that if the Red Sea had a similar river flowing into it, it would take *c.* 20 000 years for it to fill with sediment (Oldroyd 1996). Other writers, however, were more concerned with the economic value of the Earth's resources, particularly in the uses of certain minerals in agriculture, mining or building construction. For example, Agatharchides (181–146 BC) in his treatise on the Red Sea mentioned rich gold deposits on the Nile. Diodorus Siculus (*c.* 44 BC) wrote a history of the world in 40 volumes (from creation to *c.* 60 BC) in which he included information concerning the occurrences of ore deposits in various countries, as well as noting the presence of amber in the area of the Baltic Sea, the volcanoes and volcanic deposits in central and southern Italy and the presence of alluvial gold in Gaul (i.e. France), and tin in Belerion (i.e. Cornwall, England). Similarly, the work of Strabo included information concerning the character and mode of occurrence of many celebrated ore deposits, as well as quarries for building stones and gem localities (Adams 1954).

The Romans were the successors to the Greeks in terms of European culture, although science was not accorded the same importance. The Romans were more concerned with the practical nature of science, and developed a great understanding of the use of building materials. Caius Plinius Secundus (i.e. Pliny the Elder, AD 23–79) made the most outstanding and important contribution to our knowledge of the mineral kingdom. His *Natural History* (*c.* AD 77) was divided into 37 books, the last five of which deal with the mineral kingdom.

Not satisfied with merely cataloguing and describing the Earth around them, the Greeks and Romans also wanted to understand the Earth as a dynamic system. Aristotle and Zeno (*c.* 336–264 BC) and their followers promulgated two opposing views of the Earth, with the debate resembling the controversies between the Catastrophists and Uniformitarians at the beginning of the nineteenth century. Aristotle's group (the Peripatetics), believed that the various cycles of change operating on and within the Earth overlapped with one another. The alterations wrought by these changes would eventually compensate for each other and, thus, the Earth as a whole remained in equilibrium. The opposing view was that of the Stoics (and their Roman disciples, including Seneca) who suggested that the Earth was a finite system. They believed that destructive forces would lead to the end of the current world, whereupon a new Earth would evolve (Gohau 1991). This philosophy was described by Seneca (*c.* 3 BC – AD 65) who noted that the world was in a phase of increasing tension, and that eventually the oceans would disappear. At that point there would be a great conflagration, and a new world would be formed from the ashes of the old (Oldroyd 1996). A more moderate approach than that of Seneca, however, was taken by Plato, who considered that the catastrophes affecting the Earth (e.g. floods, earthquakes) were surface phenomena only. The purpose of these divinely instigated catastrophes was to destroy all living things and, thus, to purify the Earth.

Early Christianity and the early Middle Ages

The philosophical debate between the Peripatetics and Stoics continued on into the early days of the Christian church. As noted above, Aristotle and his followers believed in an eternal world. Such a philosophy, however, could not be accepted by the early Church, which preferred the catastrophic purifications of Plato. Origen (?185–?254) combined Plato's concepts with Christian doctrine, where Plato's god, endowed with a personality and free will, compared closely to the Christian god (Gohau 1991). Thus, the Stoic philosophy influenced Christian cosmology. All the Christians needed to do was to replace the Stoic's notion of successive worlds with a unique act of creation and destruction by the Christian god (Gohau 1991).

This period can be broadly characterized as one where knowledge regressed, mainly due to the Church's disdain for science (Gohau 1991), and this was closely linked to the belief in a closed universe, i.e. that the world was centred around the Earth, and that the stars regulated terrestrial phenomena. Indeed, medieval science was permeated with astrology. The Italian scholar Ristoro d'Arezzo (thirteenth century), for example, imagined that the sky was responsible for keeping parts of the Earth free of water, and that there was a relationship between the pattern of stars in the sky and the pattern of mountain peaks and valleys in a landscape

The Arab world

While Early Christian Europe turned away from science and secular knowledge, the Arabs became the immediate heirs to the Greek intellectual inheritance. Indeed, by AD 1000 Arab and Byzantine scholars had already translated the major scientific works of the ancient world, Scholars such as Ibn Sina (also called Avicenna; 980–1037) and Ibn Ruschd (also called Averroes; 1126–1198) wrote important commentaries on Aristotle's writings which added considerably to their ideas about the Earth (Oldroyd 1996). Avicenna classified mineral substances, and wrote on the formation of mountains, fluvial erosion and diagenesis (a concept not commented upon in Europe until almost 200 years later).

In the tenth century, two opposing theories were formulated in the Arab world, one which favoured erosion and deposition as being the determining processes of Earth's evolution, while the other prioritized internal processes (Gohau 1991). The Brothers of Purity, a group of tenth century Muslim scholars (and authors of the *Encyclopedia*) taught that erosion destroys mountains continuously: 'Rain waters transport rocks, stones and sand to the bed of torrents and rivers; those in turn, carry these materials to swamps, lakes, seas ... where they are piled up ... in the form of superposed layers ... intermingled these deposits accumulate one on top of the other so that on the seafloor, mounds, hills, and mountains emerge as in the steppes and deserts where the wind blows and sand is piled up in dunes' (Gohau 1991, p. 25). The Brothers believed that an individual cycle of erosion and deposition would take *c.* 36 000 years, whereupon a new cycle would begin. However, their theory was flawed since peneplanation and ocean filling would ultimately result in the sea covering the smooth surface of the Earth. This was understood by an unknown Islamic author who wrote in the book *Treatise of Elements* that at this stage all evolution would cease. For the cycle to continue, it would be necessary for the ancient sea to rise or for the land to subside. Thus, the purely neptunistic theory of the Brothers of Purity necessitated additional processes which were capable of raising or lowering the Earth's surface. Such processes were favoured by Ibn Sina (Avicenna), who stated in his *De Mineralibus* that when 'a violent earthquake ... raises part of the ground ... a mountain is suddenly formed' (Gohau 1991, p. 26).

The late Middle Ages and the Renaissance

The diffusion of Arabic learning into Europe began *c.* 900 and was initially concentrated in the major schools of learning in Spain (e.g. Cordoba, Sevilla, Toledo, Granada), spreading from there throughout the continent. Treatises translated by the Arabs only became widely known in the Christian West in the twelfth century (and some not until the Renaissance) (Gohau 1991). At that time, translations from Arabic to Latin of original Greek works, as well as those of the various Arab commentators, began to appear. Thus, in the thirteenth century, Aristotle's ideas slowly began to enter the Christian world view. Important in this respect was the work of both Albertus Magnus (*c.* 1200–1280) and Roger Bacon (1214–1294) which represented a medieval return to Aristotle's ideas after centuries of Neoplatonism. Albertus Magnus, following on from the work of Ibn Sina (Avicenna), suggested that internal processes were responsible for seismic and orogenic activity. He believed that vapours released from the interior of the Earth would result in the uplift of mountainous areas.

This rediscovery of Aristotle in Europe had a sudden and profound effect on medieval thought. However, Aristotle's ideas penetrated very slowly and were very much counter to the religious tradition (Gohau 1991). In 1215, for example, the Fourth Lateran Council condemned the Aristotlean doctrine. Some years later, the Bishop of Paris attacked followers of Aristotle, including Thomas Aquinas (1225–1274) and Roger Bacon, for their denial that God could have created other worlds. Indeed, the Church was not averse to murder, as witnesssed by the death of Giordano Bruno in 1600 for believing in a plurality of worlds (Gohau 1991).

The onset of the Renaissance in Italy and elsewhere coincided with a renewed appreciation of the Greek and Roman teachings. In the geological sciences this period was marked by the work of the so-called Forefather of Geology Georgius Agricola (1494–1555) in Germany. He wrote seven books dealing with geological subjects which have come down to us. *Bermannus* (1530), was his first mineralogical work and covered the history of the mining industry in Germany as well as discussing various aspects of mineralogy and mining. These aspects were expanded in *De Veteribus et Novis Metiallis* (1546) and *De Re Metallica* (1556). In *De Natura Fossilium* (1546) he compiled a treatise on mineralogy with additional information on petrography and palaeontology. In this book he sought to modernize the mineralogies of the medieval scholars Ibn Sina (Avicenna) and Albertus Magnus (Laudan 1987). In Switzerland, the work of Conrad Gesner (1516–1565) *De rerum fossilum, lapidum et gemmarum* (1565) classified minerals according to their resemblances to geometric forms, celestial objects, animals, plants and artefacts. Similar works were produced by Ulysse Aldrovandi (1522–1605) in Italy.

Perhaps the greatest of the Renaissance scholars was Leonardo da Vinci (1452–1519) who made a great number of observations concerning the Earth and its processes. One significant point was his recognition that fossil shells were the remains of once-living organisms which had been deposited at the bottom of the sea, thus opposing the theory that water had gushed out through subterranean conduits. In this he rejected the popular theory of the day that suggested that fossils were related to the biblical flood. As Leonardo maintained, '... if the shells had been carried by the muddy deluge (i.e. flood) they would have been mixed up, and not in regular steps and layers, as we see them now in our time' (Gohau 1991, p. 34). He also recognized that there had been significant changes in the relationship between land and sea.

Chronological and biblical problems

At the beginning of the sixteenth century, there was increasing interest in tracing the history of humanity back to the Creation. The Christian concept of the Earth's history differed radically from that of the Greeks and Romans in that it was concerned with a linear pathway (from Creation to the return of the Messiah) and it was presumed that these events could be quantified by the interpretation of biblical texts. This was not a new pursuit, since Theophilus of Antioch (AD 115–181), the founder of Christian chronology, had already determined that 5529 BC was the year of the Creation. However, with the incorporation of Aristotelian philosophy in the Middle Ages, notably through the efforts of St Thomas Aquinas, such a relatively short timespan was questioned. This was exemplified by the work of Jean Buridan (*c.* 1328–1358), Rector of the University of Paris, who in arguing against the 36 000 year cycle (as proposed by the Brothers of Purity), envisaged a much slower cycle of 'a hundred thousand million years, perhaps' (Oldroyd 1996, p. 26).

Renewed attempts at precise dating of the age of the Earth continued with the attempts of scholars in the Early Renaissance. As noted above, Mosaic chronology (i.e. based on biblical accounts of creation and family lineages, e.g. Townsend 1813; Sumner 1816) suggested an age of approximately 6000 years. However, humanist scholars such as Joseph Scaliger (1540–1609) in France set historical chronology on a more scholarly basis. In his *De emendatione temporum* (1583) he revised the age of the Earth by using his skills in classical languages and history to compare the Mosaic chronology with other non-Christian chronologies. Orthodox defenders of the Mosaic account, however, maintained that non-Christian civilizations exaggerated their antiquity. Thus, the orthodox apologist Archbishop James Ussher of Ireland (1581–1656) ignored non-biblical chronologies and in *Annales veteris et novi testament* (1650–54) dated the creation as occurring on the night before October 23rd in the year 4004 BC. Ussher's research carried so much weight that this date was printed in the margin of the King James Bible in 1701, a decision which was criticized by Charles Lyell as late as 1850 (Klaver 1997). Indeed, Charles Darwin graduated from Cambridge University believing that the world was 6000 years old, give or take a few years (Baxter 2003), while these dates are still quoted and accepted by creationists in the USA and elsewhere.

Conceptualizing the possibly immense age of the Earth was a fundamental problem in the development of geology as a science. This was compounded by the fossil evidence that other strange worlds had existed. John Ray (1627–1705), for example, noted 'Whatever may be said for the Antiquity of the Earth itself and bodies lodged in it, the race of mankind is new' (Klaver 1997, p. 8). Reconciling such observations with the stories in the Bible became increasingly problematic. These difficulties also extended beyond the truth of the Mosaic chronology to question the veracity of the various events outlined in the Book of Genesis. For example, the French philsopher René Descartes (1596–1650) in his *Principia philosophiae* (1644) described the internal structure of the Earth which he envisaged as consisting of concentric layers, including the central fire, the ocean, the atmosphere, and a subterranean cavity (Oldroyd 1996). He suggested that cracking of the layers led to the formation of mountains and oceans and the mixing of certain layers. In France in the 1640s, these ideas were considered heretical. Despite this, Descartes made no effort to reconcile his system with the biblical Book of Genesis. Nevertheless, his idea of a crust broken by collapse was eagerly accepted by authors (e.g. Thomas Burnet, 1635–1656) who sought to defend the biblical Deluge in their theories of the Earth.

Steno and his contemporaries

Geology proper was a relatively recent science in the early nineteenth century, emerging slowly as a distinct field of study out of the collective eighteenth century denomination Natural Philosophy (Klaver 1997). Like other branches of Natural Philosophy, geological investigation was initially a search for facts which, as noted above, proved the creation in terms of the Mosaic account in Genesis. Despite some important discoveries, most eighteenth century studies in geology are concerned more with cosmogony and aprioristic reasoning rather than pure inductive science (Klaver 1997). However, early intimations of the separation of geology from cosmogony and mineralogy can be seen in the works of Nicolaus Steno (1638–1686) in Italy and Robert Hooke (1635–1703) in England, as well as a number of other writers, including Lehmann, Füchsel and Arduino (Oldroyd 1996).

In 1669 Nicholaus Steno published *De solido intra solidum naturaliter contento dissertationis prodromius* in Florence where he demonstrated the organic origin of fossils. In this work he recognized the similarity between shark's teeth and what had been considered to be the petrified tongues of reptiles. Additionally, he stated that fossils found in rocks resemble each other no matter what kind of rock forms the layer, and furthermore, that they are identical to the parts of animals which they resemble.

Steno reasoned that if fossils and other fragments are the remains of ancient living beings, then the rocks that contain them must have been deposited on the sea floor. While Leonardo da Vinci had also understood this, Steno went further in that he distinguished two depositional environments: 'If in a certain strata we discover traces of salt of the sea, the remains of marine animals, the timbers of ships, and a substance similar to the bottom of the sea, it is certain that the sea was at one time in that place' (Gohau 1991, p. 61). If, on the contrary, 'we find a great abundance of rush, grass, pine cones, trunks and branches of trees, and similar objects, we might rightly conclude that this matter was swept there by the flooding of a river or the inflowing of a torrent' (Gohau 1991, p. 61). Thus, Steno assigned to fossils the role of defining the facies of geological formations. He expanded this work by defining the principle of superposition (where successive layers are increasingly younger going up in the series) as well as that of original horizontality, where he noted that 'strata either perpendicular to the horizon, or inclined towards it, were at one time parallel to the horizon' (Gohau 1991, p. 62). Thus any resultant dip would have resulted from subsequent folding or tilting. With such observations, Steno laid the groundwork for reading the archives of nature. This was exemplified in his history of the geology of Tuscany, where he illustrated transgressions, regressions and the formation of angular unconformities.

In England, Robert Hooke went even further than Steno when he denounced the idea that fossils 'are generally believed to be real Stones form'd into these Shapes, either by some plastick Vertue inherent in these Parts of the Earth ... or else by some Celestial Influence or Aspect of the Planets' (Gohau 1991, p. 60) and 'that there may be divers new kinds now which have not been from the beginning' (Gohau 1991, p. 60). Furthermore, the German, Gottfried Wilhelm Leibnitz (1646–1716) suggested that 'A large number of forms were transformed during the great changes to which the Earth was subjected' (Gohau 1991, p. 60). In Renaissance times, Leonardo da Vinci had tried to convince his contemporaries that fossils were the remains of organisms; however, such proof was not valid when fossils and living organisms had different forms.

The increasing recognition that fossils were sometimes radically different from living organisms led some authors to try to explain these differences. For example, Benoit de Maillet (1656–1738), a French consul in Cairo, circulated a manuscript in the 1720s entitled *Nouveau système du monde ou entretien avec Telliamed* (his name spelled backwards; the work itself was only published in 1748 after his death). He noted how sea level is falling (a process which he suggested lasted 500 000 years), and suggested that all terrestrial species descended from marine ancestors underwent transformation during the lowering of sea level.

Classifying the Earth

The work of Steno and his contemporaries provided a new scientific framework for determining the history of the Earth. In the mid- to late eighteenth century, a number of scientists began to use this work to subdivide rocks into a series of distinct groups. It is, however, difficult to determine who was the first to describe rocks in this way, and it may be that it was a case of simultaneous, independent discovery (Laudan 1987). The Italian Abbé Anton Lazzaro Moro (1687–1764) distinguished primary and secondary mountains in his *De crostacei e degli altri corpi marini che si trovano su' monti* (1740). In this work, he noted that the primitive ocean did not rise more than *c.* 350 m above its present level and suggested that any mountains with strata and fossils higher than this elevation had been uplifted, a concept he explained by reference to volcanic activity. His primary mountains 'originated in the bowels of the Earth when the surface area where they were born was still covered by water and not burdened by terrestrial material' (Gohau 1991, p. 78). After their emergence, erosion produced materials that were deposited at the bottom of the ocean and were subsequently uplifted to form layered secondary mountains. This was similar to the scheme proposed by Benoit de Maillet (Oldroyd 1996). In contrast, the Swiss geologist Élie Bertrand proposed three different classes of rock in his *Mémoires sur la structure intérieure de la terre* (1752).

Johann Lehmann (1719–1767), working in Germany, distinguished three orders of mountains (those formed when the world was made, those formed from sediment deposited in sheets under water, and volcanic mountains). He characterized primitive mountains as having perpendicular or diagonal strata, rare fossils, and being rich in mineral deposits. Secondary, or stratified mountains were formed by horizontal, thin and regular layers. These mountains contain fossils, have few minerals, and are of low elevation with gentle slopes. The significance of Lehmann's work was that he had a clear idea of a sequence of stratified rocks, and he illustrated such a succession in a section between the areas of Ilfeld and Mansfeld in Thuringia (Oldroyd 1996, fig. 3.4). It was this particular aspect that marked a fresh approach to the study of the Earth since it intimated that a stratigraphical succession was an indicator of time. Furthermore, the representation of this concept by means of a stratigraphic section was a major advance. Lehmann specifically referred to himself as a historian of the Earth and he claimed that no-one before him had treated the layered mountains in this fashion.

However, the simple distinction between two (or three) classes of mountains was not good enough for miners. Within each of these classes, the order of superposition needed to be established. Lehmann was the first to realize the importance of the order of strata and to make use of the accumulated knowledge of quarrymen and miners; he subdivided the stratified rocks (*Flötz-Gebürge*) that surround the Harz Mountains, which always lean against primitive mountains (*Gang-Gebürge*), into red-beds consisting of coarse sand and pebbles, followed by beds of coal, shales, limestones and 'salted springs'. This was the first detailed description of the Permian.

This findings of Lehmann were amplified by Georg Christian Füchsel (1722–1773) who published two important geological works. The first of these (in two parts, entitled *Historia terrae et maris* and *Eiusdem usus historia suae terraae et maris*) contained what are arguably the first geological maps (as opposed to the mineralogical maps of Guettard, for example). In his book *A History of the Earth and the Sea, Based on a History of the Mountains of Thuringia* (1761) Füchsel recognized a stratigraphic succession for Thuringia, with more subdivisions than originally proposed by Lehmann. Additionally, he noted that certain fossils were associated with particular strata, although he actually distinguished the strata in terms of their lithology and colour. Furthermore, he recognized sandstones and fossiliferous limestones (Muschelkalk) above the salt and gypsum layers, thus providing the basis for the recognition of the Triassic. His superiority over Lehmann was in his understanding that layers were formed successively over long periods of time and not during the unique event of the biblical Deluge. He also noted that 'for formation of deposits, nature must have followed present-day laws: every deposit forms one stratum, and a series of strata of the same composition represents a formation or an epoch in the history of the globe' (Gohau 1991, pp. 101–102). Füchsel's work, however, did not attract much attention at the time, but his stratigraphic principles and some of his terminology were later popularized by Abraham Werner (Oldroyd 1996). Indeed, all three (Lehmann, Füchsel, Werner) were drawing on the discourse of the German mining fraternity, so that terms such as *Muschelkalk* and *rothe tode Lager* (i.e. Old Red Sandstone) entered the geologist's regular vocabulary (Oldroyd 1996).

Georges-Louis Leclerc, Compte de Buffon (1707–1788) in his *Histoire naturelle* (in 44 volumes) suggested that the oldest mountains were of igneous origin, rather than of aqueous origin as suggested, for example, by Lehmann. According to Leclerc the globe had gone through a molten stage and during its slow cooling the molten matter became solid, resulting in the formation of high mountains. He also subdivided the history of the Earth into various epochs, beginning with the formation of the Earth and planets and ending with that epoch defined by the power of humans. The timespan for these epochs was *c.* 75 000 years (although his manuscripts revealed that he himself believed the true figure to be closer to three million years) (Oldroyd 1996).

In Italy, Giovanni Arduino (1735–1795) produced a remarkable geological section up the valley of the Agno from Montecchio Maggiore to the high alpine peaks to the NW of Vicenza. Within the section he identified various divisions of the Earth's crust (Primary, Secondary, Tertiary, Quaternary) while also noting that fossils from the uppermost divisions were more like creatures still extant than those of the deeper and older rocks (Oldroyd 1996).

As noted above, the second half of the eighteenth century heralded the beginning of field mapping in several parts of Europe, as well as the drawing of geological sections and the first steps towards the establishment of the divisions of the stratigraphical column (Oldroyd 1996). The Earth was increasingly studied as an object in its own right, and not as a branch of cosmology. Biblical considerations, while not excluded, declined in importance for practical purposes.

Geological maps began to increase in terms of their complexity with the addition of colour (first used by Friedrich Gottlob Gläser, 1749–1804), and dips and strikes (as used in a map of the Pyrenees by Pierre Bernard Palassou, 1745–1830). In the eighteenth century, the French were the first to establish a scientifically based topographic survey of their nation, this being useful for military and commercial purposes. The naturalist Jean-Etienne Guettard (1715–1786) undertook the massive task of preparing a mineral map of the kingdom (1746), which was followed by a more detailed survey in 1780.

Werner and Hutton

Geology was first taught as a separate branch of science in 1775 by Abraham Gottlieb Werner (1750–1817) in Freiberg, Germany. (More than forty years earlier *The philosophical grammar* (1735) defined *Geology* as that 'which treats of the Nature, Make, Parts and Productions of the Globe of Earth on which we live'; and Samuel Johnson's *A dictionary of the English language* (1755), described the term *Geology*, as 'the doctrine of the earth; the knowledge of the state and nature of the earth'. However, the term did not fall into common use in the form we understand it today, until after 1800. Even in 1804 the English physician and palaeontologist James Parkinson was complaining: 'I have to ... treat of a science which has not yet acquired a peculiar name'.) While still a student, Werner had published *Von den äusserlichen Kennzeichen der Fossilien* (1774) which classified minerals based on their external characteristics. Following this publication, he became a professor and his next major work *Kurze Klassifikation und Beschreibung der verschiedenen Gebirgsarten* (1786) can be considered as the beginning of modern petrology (Dean 1992).

Werner suggested that the order of deposition of the principal rocks made it possible to distinguish four classes of rocks (or mountains): primitive (*Uranfängliche*), in layers (*Flötze*), volcanic and alluvial. Around 1796 he inserted a new group between the first two classes: the transitional mountains (*Übergangsgebirge*) which included tilted and fossiliferous late Palaeozoic units resting on primitive igneous and metamorphic rocks. Each of these classes was further subdivided into various categories. For example, stratified mountains were subdivided into 12 formations, beginning with the units today recognized as Devonian and extending through the Zechstein and Triassic to basalts and other volcanic rocks (which Werner believed were sedimentary) and followed by the Cretaceous chalk (Gohau 1991).

According to Werner, the origins of the Earth lay in the universal ocean which had formed around a primeval core of solid matter. This universal ocean contained, in solution, great masses of material which crystallized onto the irregular surface of the primeval core, beginning with granite and continuing through to siliceous schist (for this reason, he and his followers were called Neptunists) (Oldroyd 1996). Following the completion of this succession, it was supposed that the water level of the ocean had declined to such an extent that the higher parts of the core were exposed and thus weathering and erosion could begin, depositing sediments in the ocean basins. Werner believed that the various formations were arranged around the Earth like the layers of an onion, with each lithology being characteristic of a particular time, and his ambition was to reconstruct a single column that would include all of the Earth's formations and thus it would be possible to recognize this universal order. To this end, between 1790 and 1830 Wernerians surveyed many parts of Europe and further afield. Indeed, Alexander von Humboldt (1769–1859), a student of Werner, travelled all over South America trying to establish such an order of layers.

There were, however, a number of problems with Werner's philosophy. He believed that formations could be classified as both general and partial: the former were produced by regional effects while the latter were determined by local factors. However, Werner's belief in the law of superposition made it difficult to visualize lateral transitions between the formations. Thus partial formations (i.e. formations whose appearance was determined by local factors) of the same age were simply superposed in one way or another. Gradually, the original Wernerian concentration on the primary and transition rocks shifted and Wernerian gradualism was replaced by a theory of Earth history interrupted by catastrophic revolutions (Laudan 1987).

The position of Werner and his followers was challenged by James Hutton (1726–1797) who lived and worked in Edinburgh, where he regularly met with fellow scholars at the Royal Society of Edinburgh. It was there that Hutton first presented his geological theories (see below), based on his geological observations around Edinburgh, to a professional world which fully accepted Werner's Neptunism. Some years later, in response to criticism from Richard Kirwan and others of the Wernerian school, Hutton extended his original paper (1788) to two long volumes, the *Theory of the Earth, with Proofs and Illustrations* (1795). In the first chapters, he restated his original theory, and the remainder of the book was taken up with providing empirical support from his own observations and extensive reading of the geological literature (Laudan 1987). However, on the whole, Hutton's two-volume version of his geological theory attracted less attention than did the earlier version of 1788 (*Theory of the Earth; or an Investigation of the Laws Observable in the Composition, Dissolution, and Restoration of Land upon the Globe*) (Dean 1992).

Hutton postulated (as others had done before) that the Earth's interior was hot and that from time to time the internal material of the Earth would begin to expand, forcing its way into the crust above, bending the layers of the crust upwards in the process. Sometimes the molten rock would break through to the surface, giving rise to volcanoes (Oldroyd 1996). Additionally, Hutton believed that angular unconformities have the same age wherever they occur, i.e. the processes controlling them were global rather than regional. Because he was searching for a model of Earth processes he established a cyclic theory where the same phenomena recur indefinitely. This concept led him to state that 'We find no vestige of a beginning, no prospect of an end' (Gohau 1991, p. 116).

Unlike Werner, Hutton (and his Plutonist theories) did not attract a major following, partly due to his bad prose style (Playfair 1802), his unorthodox theories of erosion (Herries-Davies 1969) and heat (Gerstner 1971), and his theology (Howard 1797; Dean 1973). Additionally, Huttons's theory flew in the face of all conventional wisdom about geology (Laudan 1987). Kirwan (1793) noted that while there were problems with the Wernerian system, these were nothing when compared to those of the Huttonian system which appeared to contradict all known mineralogical and geological knowledge at that time. Indeed, Kirwan was even more vociferous when arguing against the idea of infinite time, which Hutton's concept of uniformity of geological phenomena implied. Hutton's desciption of the endless cycle of uplift and erosion implied that the Earth had to be infinitely older than was acceptable in biblical terms (Klaver 1997). Indeed, his friend John Playfair (1748–1819) commented, on their expedition to the unconformity at Siccar Point, that 'We found ourselves carried back to the time when the schistus on which we stood was yet at the bottom of the sea, and when the sandstone before us was only beginning to be deposited ... An epocha still more remote presented itself, when even the most ancient of these rocks, instead of standing upright in vertical beds, lay in horizontal planes at the bottom of the sea and was not yet disturbed by that immeasurable force which has burst asunder the solid pavement of the globe. Revolutions still more remote appeared in the distance of this extraordinary perspective. The mind seemed to grow giddy by looking so far into the abyss of time' (Dean 1992, p. 122).

John Playfair's *Illustrations of the Huttonian Theory of the Earth* (1802) was the most comprehensive defence of Hutton's theory. This, together with the experimental work of Sir James Hall (1761–1852) and others (Dean 1992) on melting limestones and basalts, supported the initial ideas of Hutton. However, it was increasingly clear that there was a major theoretical dispute within the new science of geology and that the Huttonian and Wernerian hypotheses were mutually contradictory. The need for new research and mapping was evident, and the initial sources came from France.

In 1779 Nicolas Demarest (1725–1815), in an article entitled *On the Definition of Some Epochs of Nature by the Products of Volcanoes and on the Use of These Epochs in the Study of Volcanoes*, noted that basalt is ancient lava. The publication of this article sparked increased debate between the Plutonists and the Neptunists, who continued to believe that basalt was a sedimentary rock. However, Demarest's work was supported by the mapping of J.L. Giraud Soulavie (1752–1813) who published a seven-volume study entitled the *Natural History of Southern France* (1780–84). In this he recognized the volcanic nature of some of the area as well as including a relatively detailed and precise stratigraphic study. The idea that basalt was a sedimentary rock was becoming increasingly untenable. However, the most famous Neptunist, Werner, remained adamant until his death, despite the increasing proof to the contrary (Gohau 1991).

Catastrophism and Uniformitarianism

More than any other science, geology was bound to create tension with regard to religiously accepted truths since its new discoveries inevitably tended to shed light on the age and origin of the Earth (Klaver 1997). Theories, such as those of Werner and his universal ocean, explained why fossils of salt-water animals could be found at high altitudes and thus supported the biblical account for a great flood (Klaver 1997). However, the increasing recognition of the variety and abundance of fossils seemed to provide evidence of mass extinctions, while the proof of a universal deluge seemed incomplete. Orthodox geologists often had to force new discoveries into a religious context, which was becoming increasingly untenable (Klaver 1997). The very idea that organic and inorganic material had existed on the Earth millions of years before the Creation undermined the Christian belief that god had created the world purposely for human use (Klaver 1997).

In Switzerland, Jean-André Deluc (1727–1817), published a series of works between 1778 and 1809 in which he noted that layers differed not only in their lithological nature but also in the fauna they contained. This was not a new discovery, but what Deluc did was to establish a superposition of layers, thus defining their relative age, as well as that of the fossil species contained within the layers. He thus determined the ages of the fossils based on the position of the layer within which they occurred. There were, however, two problems: a process was needed which would change the form of organisms in order for faunas of the same age in different locations to be similar; and secondly, it was necessary to determine the process responsible for regulating this variation. Deluc suggested that the process of 'transformism' resulted in the observed changes in the fossil record. This process was related to compaction of the Earth's interior where the resultant collapse of surface layers led to fluid escape. These fluids modified the oceanic composition and led to the disappearance of living organisms, some of which were destroyed while others were transformed (Gohau 1991).

Georges Cuvier (1769–1832), a French zoologist and comparative anatomist, and Alexandre Brongniart (1770–1847) published a memoir entitled *Essai sur la géographie minéralogique des environs de Paris* in 1808 (final form 1811). This work, which was concerned with the Chalk and the overlying Tertiary succession in the Paris Basin, included both a geological map of the region as well as a series of cross-sections. They noted that the superposed rocks differed in terms of their lithology and thickness and also in terms of the fossil content, and they suggested that it would be possible to correlate a known formation with an unknown one on the basis of the fossils present. However, a necessary precondition for correlating formations by fossils was the preparation of adequate classification and identification guides to fossil fauna and flora, and thus it was not until the 1820s that geologists could produce tables of correlation that covered much of Europe (Laudan 1987). One early example of such work was the article published by Brongniart in 1821 in which he noted that the Chalk of both France and Poland contained a similar fauna. The rocks in Poland had previously been classified by Wernerians as transition rocks (i.e. Palaeozoic). Brongniart's work reclassified them, and led him to note 'When petrographic and palaeontologic criteria differ, which one is more reliable?' (Gohau 1991, p. 130).

In his work, Deluc had originally recognized six major catastrophes (corresponding to the biblical days of creation, and ending with Noah's flood) and these ideas dominated European geology for several decades. Cuvier was strongly in favour of extinction, linking it to a view of geological change that stressed the effects of sudden physical events, or 'catastrophes', at the Earth's surface. While he agreed with Deluc that the evidence pointed to an earlier pre-human world, and considered it to have been destroyed by some kind of catastrophe, this was not specified as being in any way related to the biblical flood. Cuvier suggested in the foreword to his *Ossements fossiles* (1812) that there had been great catastrophes or revolutions from time to time, but they were not global. This foreword, later republished as *Discours sur les révolutions de la surface du globe*, commenced with Brongniart's observation of alternating marine and freshwater deposits which he interpreted as the products of revolutions which had dried up areas once covered with water. Cuvier suggested that violent phenomena could destroy fauna that cannot live in the newly created environments. However, he did not provide any solution (other than migration) for the derivation of new species. Thus, his system was directionalist, but non-evolutionary since he was ambivalent as to whether new creations occurred from time to time after each catastrophe. However, he did accept that there had been a major catastrophe (i.e. the biblical flood) less than five thousand years ago (Oldroyd 1996) and this message, that the findings of the new historicized geology did not necessarily conflict with Judaeo-Christian history, was the one that many people took from Cuvier's geological work.

Coincident with these debates on the mechanisms underlying the changing fossil assemblages, geological mapping and data-gathering proceeded apace. In England, for example, William Smith (1769–1839) published the first comprehensive geological map, *A map of the Strata of England and Wales with a part of Scotland, exhibiting the Collieries, Mines and Canals, the Marshes and fen lands, originally overflowed by the sea and the varieties of soils according to the variations in the substrata,* in 1815, while the detailed analysis of particular stratigraphic systems commenced with the work of Adam Sedgwick (1785–1873) on the Cambrian, and Roderick Impey Murchison (1792–1871) on the Silurian and the Permian. Together, Sedgwick and Murchison also worked on defining the Devonian.

Despite these advances, however, there was still no resolution to the ongoing debate between the Christian view of the world and the increasing body of scientific data. When Charles Lyell (1797–1875) attended Oxford University he came under the influence of the Reverend William Buckland (1784–1856). At that time Buckland was endeavouring to show how geology (essentially Cuvierian) and Anglican theology could be reconciled. However, Lyell was not convinced by the theoretical assumptions espoused by Buckland and was instead drawn to the ideas of Hutton and Playfair, as well as to some alternative ideas being developed in France by George Poulett Scrope (1797–1876). Scrope's work, which was published in two memoirs (in 1825 and 1827), was an overview of the volcanic history of the Auvergne district. Subsequently, Lyell visited the area himself and continued his journey on to the volcanic areas of Italy. It was there, with his observations of mollusc borings on Roman columns, that he concluded that significant geological changes could occur gradually and need not involve great catastrophes such as envisaged by Cuvier.

On returning from his journeys Lyell began work on a major geological treatise published in three volumes (1830–33) and entitled *Principles of Geology, an Attempt to Explain the Former Changes of the Earth's Surface, by Reference to Causes Now in Operation*. He envisaged a steady-state system where processes were constant both in kind and degree (Klaver 1997). Indeed, in 1829 Lyell told Murchison that 'no causes whatever have from the earliest time to which we can look back to the present ever acted but those now acting & that they never acted with different degrees of energy from that which they now exert' (Gohau 1991, p. 144). The apparently slow, and by extension uniformitarian, turnover of species within the stratigraphic record led Lyell to develop a palaeontological method for the subdivision of the Tertiary rocks. These he classified, based on the collections of Gérard Paul Deshayes (1795–1875) in Paris, according to the proportion of extant species they contained: Newer Pliocene (96%), Older Pliocene (35–50%), Miocene (17%) and Eocene (3.5%). He could not, however, extend it any further due to the lack of extant forms. Based on his investigations, Lyell concluded that the fossil record provided evidence of change at the species level, but he did not believe that classes had changed. Instead, he argued that the same classes of plants and animals had always existed, albeit not always in the same location since environmental conditions varied (Laudan 1987). Thus the constancy of the various plant and animal classes was maintained through a succession of extinctions and creations of individual species. Lyell thought that the various species appeared and disappeared from time to time, but he had no clear idea of the mechanism for the production of new species (Oldroyd 1996). With this uniformitarian approach to change in the fossil record, Lyell took the opposite view of Cuvier (although it should be noted that the intensity of the Uniformitarian–Catastrophist debate has often been exaggerated, and that leading Catastrophists were far from being dogmatic Mosaicists (e.g. Cannon 1976; Porter 1976; Gould 1987). However, the geological evidence seemed to indicate that the Earth's past had been more violent and catastrophic than the present, and of particular interest was the evidence of more recent catastrophes.

The suggestion that the Alpine glaciers might have extended significantly beyond the positions of their present terminal moraines was made in 1787 by Bernard Kuhn. Such early suggestions were presented by Jean de Charpentier in a book (1841) showing the former extent of the Rhone glacier into the central plains of Switzerland, based on the distribution of erratics. Subsequently, Louis Agassiz (1807–1873) took up the cause and promoted the idea of an Ice Age (*Eiszeit*). This theory of climate change was, in a sense, a kind of Catastrophist doctrine. While Agassiz could not explain how or why the Earth's temperature changed in such a peculiar manner (this would not begin until the work of Joseph-Alphonse Adhémar in his *Révolutions de la mer* (1842) which related climate change to astronomical causes; Oldroyd 1996), it was, however, clear was that the history of the Earth was the result of both uniformitarian principles as well as catastrophic processes.

The quarter of a century between Lyell's *Principles* and Charles Darwin's *Origin of Species* (1859) was a time in which there was a slowly but steadily growing theory of evolution. This had become possible because of Lyell's interpretation of the extent of geological time (Klaver 1997). Despite this, the publication of Darwin's *Origin of Species* caused (and continues to cause) a great furore. It did, however, provide an understanding of the various observations which had been made with regard to the stratigraphic distribution of fossils, and made it possible to read the record in a new way. Extrapolating from this, it can be said that a more complete understanding of the Earth's history requires a blend of two distinct scientific methods, namely *causal* scientific enquiry and *historical* analysis (Woodcock & Strachan 2002). While the former is concerned with understanding, for example, the fundamental processes of the Earth such as plate tectonics, the fluid mechanics of sediment transport or the chemical evolution of a magma chamber, the latter recognizes that the complexity of Earth processes is dependent on both when they occur as well as on the cumulative geological history of the place in which they occur. Thus any geological history is concerned not only with those events which can be recently observed, but also with detailing the complexity of the preceding geological events. The reason for this is that recent processes are often the cumulative outcome of preceding events (Woodcock & Strachan 2002). For example, faults and sutures in the Earth's crust lead to changes in the crustal rheology, possibly weakening the area and making it more susceptible to subsequent deformation. An additional point is that as we get closer to the present day it becomes relatively easier to reconstruct the geological history of a region. Thus, our understanding of the complexity of the evolution of the Alps (see Froitzheim *et al.* 2008; Reicherter *et al.* 2008) provides us with a greater appreciation for the task involved in trying to reconstruct those ancient mountain chains which resulted from the Caledonian and Variscan events (Krawczyk *et al.* 2008; Kroner *et al.* 2008).

As previously noted, the regional focus of this book is Central Europe, an area which has had a remarkably varied geology for a relatively small fragment of continent. This is largely the result of the complex geological evolution of the region, extending from the Precambrian through to the Quaternary and encompassing significant orogenic events (including Cadomian, Variscan and Alpine) and resulting in a record which is both fascinating and perplexing. Any attempt, such as the present work, to reconstruct a geological history for the region is, therefore, bound to be controversial to a certain degree. While there may be broad agreement about the main geological developments, the details are often open to discussion. This is partly due to different viewpoints and interpretations, and partly a result of the relative lack of evidence in some areas (e,g, Krawczyk *et al.* 2008). Despite these obstacles, the geology of Central Europe can be broadly summarized, and this summary provides us with an overview of the history of Central Europe, at least from the dawn of the Palaeozoic onwards.

Overview of the geological and tectonic structure of Europe

As previously noted, Europe comprises a complex mosaic of crustal elements which have been assembled over geological time (Fig. 1.1). The basement can be subdivided into two large and distinct regions: to the north and east there is a stable Precambrian craton known as the East European Platform (Craton), while in the south and west Europe comprises a series of crustal blocks that were successively attached to the ancient cratonic nucleus (Berthelsen 1992). The boundary between these two regions is marked by the NW–SE trending Trans-European Suture Zone (TESZ) which extends for approximately 2000 km from the North Sea to the Dobrogea region of the Black Sea. The TESZ is defined as a broad zone of NW–SE striking faults by subsurface geology, drilling results and geophysical methods, including deep seismic reflection data.

The Precambrian areas of Europe have a lithosphere which is typically 150–220 km thick (Artemieva *et al.* 2006). This varies across the region with thicknesses of *c.* 120–180 km recorded in the mid- and late Proterozoic provinces of the Baltic Shield. An exceptionally thick lithospheric root is revealed by seismic, thermal and xenolith data for the Karelian part of the shield (which locally reaches depths of *c.* 250–300 km) (Artemieva *et al.* 2006). The areas dominated by the older orogens (e.g. Caledonian, Variscan) were significantly overprinted by late Palaeozoic and Mesozoic–Cenozoic tectonic processes associated with the convergence of the Eurasian and the African plates (Ziegler & Dèzes 2006). Crustal thickness in these areas tends to be uniform (typically 28–32 km although values of up to 80–140 km are recorded from beneath the Proterozoic–early Palaeozoic terranes, e.g. Armorican, Bohemian and Anglo-Brabant massifs, northern part of the Massif Central) (Artemieva *et al.* 2006).

Closure of the Tethyan Ocean domains and subsequent continental subduction (e.g. Alps, Carpathians) in Cenozoic times was associated with the development of backarc basins (e.g. Pannonian region). In these orogenic areas, the crustal thickness ranges from 60–65 km up to 150–200 km in the convergence zones, while the backarc regions are characterized by thin crust (25–30 km) and thin lithosphere (60–80 km) (Artemieva *et al.* 2006).

The Cadomian Orogen

The Cadomian Orogeny comprises a series of complex sedimentary, magmatic and tectono-metamorphic events that commenced in the mid-Neoproterozoic (*c.* 750 Ma) and extended into the earliest Cambrian (*c.* 540–530 Ma). These events occurred along the periphery of the supercontinent of Gondwana and were the last in a series of events which formed the crystalline basement rocks of Europe (see Linnemann *et al.* 2008*a*, *b*). Earlier events include those forming the Laurentian Shield area (small outcrops of which are preserved in the Lewisian Complex of NW Scotland; Strachan 2002) and the Fenno-Scandian Shield (e.g. Kola-Karelian, Svecofennian, Gothian and Sveco-Norwegian orogens). The late Neoproterozoic was a period of major tectonic, climatic, biogeochemical and biological change (Fig. 1.2a). The collision of Laurentia, Baltica and Amazonia resulted in the formation of the supercontinent Rodinia (*c.* 1.1–1.0 Ga). Laurentia was located at its centre, surrounded by the cratons of East Gondwana (India, Antarctica, Australia), Siberia, Baltica and the separated parts of West Gondwana (South America, Africa). The reconstruction of Rodinia is based on matching the Grenville orogenic belts (*c.* 1300–1000 Ma) which are thought to have resulted from the assembly of the supercontinent. The SE margin of Laurentia was located against the proto-Andean margin of South America, while the SW margin was linked to East Antarctica (the so-called SWEAT hypothesis) (Hartz & Torsvik 2002).

The breakup of Rodinia took place in several stages, commencing with the opening of the Palaeo-Pacific Ocean separating East Gondwana from the western margin of Laurentia (800–700 Ma) (although this model has been challenged by a number of authors; see Krawczyk *et al.* 2008), with final disintegration (including the separation of Laurentia from Amazonia) occurring at some stage between 590 and 550 Ma. The onset of the breakup of Rodinia was accompanied by the subduction of the Mozambique and Brazilide oceans leading to the collision of East Gondwana and the several continental blocks forming West Gondwana, resulting in the Pan-African-Baikalian-Brasiliano orogens, at *c.* 620 Ma (Fig. 1.2b). These orogenic events led to the formation of a late Neoproterozoic supercontinent comprising Gondwana, Laurentia and Baltica, and termed the Vendian supercontinent (i.e. Vendia/Pannotia) (Strachan & Holdsworth 2002). During and after the Late Neoproterozoic amalgamation of this supercontinent, several Andean-type continental margins began to form around it (Fig. 1.2b; see also Linnemann *et al.* 2008*a*). These peripheral orogens, including the Cadomian, or Avalonian, belt were marked by the widespread subduction of oceanic lithosphere, the growth of island arcs and the formation of associated accretionary prisms and backarc basins. However, no continent–continent collisional episodes have been recorded (Linnemann & Romer 2002). Between 600 and 580 Ma, the Ediacaran-age supercontinent began to break up along the eastern margins of Laurentia. Baltica rifted from the Greenland margin, and Gondwana (western South America) rifted from the North American margin to initiate the formation of the Iapetus Ocean (Fig. 1.2c). Indeed, by 550 Ma Gondwana, Laurentia and Baltica were all separate continents (Cocks & Torsvik 2006).

The Caledonian and Variscan basement complexes of western and central Europe, as well as some basement inliers within the Alpine-Carpathian Orogen, contain several continental crustal blocks which date back to the Cadomian Orogeny. These include the Irish Sea Horst, the London Platform, the East Silesian Block and the Armorican, Bohemian and Malopolska massifs. Cadomian crustal elements are also recognized in the Alpine foldbelt (based on their geology and radiometric age dating).

The Caledonian Orogen

The Caledonian orogenic cycle occurred between Late Cambrian and earliest Devonian times, and involved the collision of the late Proterozoic continents of Baltica and Laurentia with Avalonia, a fragment of Gondwana (Figs 1.2d–f, 1.3a, b & 1.4; see Krawczyk *et al.* 2008). These events marked the complex closures of the Iapetus Ocean, between Avalonia, Baltica and Laurentia, and the Tornquist Sea, between Avalonia and Baltica. While Laurentia comprised much of present-day North America, and Scandinavia formed the main part of Baltica, there is much discussion concerning the precise boundaries and extent of Avalonia, particularly in terms of its central and eastern European parts as well as its southern margin (see Krawczyk *et al.* 2008). The Caledonian events coincided with a period of dramatic change in evolution commencing with the appearance of mineralized skeletons (Cambrian), and a massive expansion of life (Ordovician radiation, where the highest diversity levels for the Palaeozoic were attained) but also by extinctions (end of the Ordovician) and the colonization of land (Silurian).

Localized orogenic events in Cambrian (Elicki *et al.* 2008) to Early Ordovician times (Servais *et al.* 2008) are thought to reflect the collision of volcanic arcs with some of the continental margins of Iapetus (Fig. 1.2d-f). These events include the Finnmarkian Orogeny (540–490 Ma) on the Baltica margin and the Grampian Orogeny (470–460 Ma), which affected the Scottish and Irish sectors of the Laurentian margin. These events were, however, of wider significance as evidenced by the similarity between the geodynamic evolution of the Grampian Orogeny in Ireland and Scotland and the early middle Ordovician Taconic Orogeny in North America. The collision of volcanic arcs with continental margins resulted in the closure of the intervening marginal basins as well as the associated obduction of ophiolites.

Avalonia detached from Gondwana in late Tremadoc to early Arenig times, drifting northwards. Rifting coincided with the onset of closure of the Iapetus Ocean and the opening of a new ocean, the Rheic Ocean, to the south. Subduction also occurred at this time along the western margin of Baltica, which was rotating anticlockwise and converging with Laurentia. The earliest linkage of Laurentian and Gondwanan marginal fragments occurred in

Fig. 1.2. Global palaeocontinental reconstructions from the Neoproterozoic. (**a**) Rodinia at *c.* 750 Ma immediately prior to breakup. Arrows summarize plate motions leading to the Pan-African orogenesis. (**b**) Plate configuration as Laurentia began to rift away from the late Neoproterozoic (Ediacaran) supercontinent *c.* 580 Ma. (**c**) Late Ediacaran, *c.* 550 Ma. (**d**) Early Cambrian, *c.* 535 Ma. (**e**) Early Ordovician, 490 Ma. (**f**) Mid-Ordovician, 470 Ma (after Cocks & Torsvik 2006; Dalziel 1992; Torsvik *et al.* 1996; Holdsworth *et al.* 2002; Meert & Torsvik 2003).

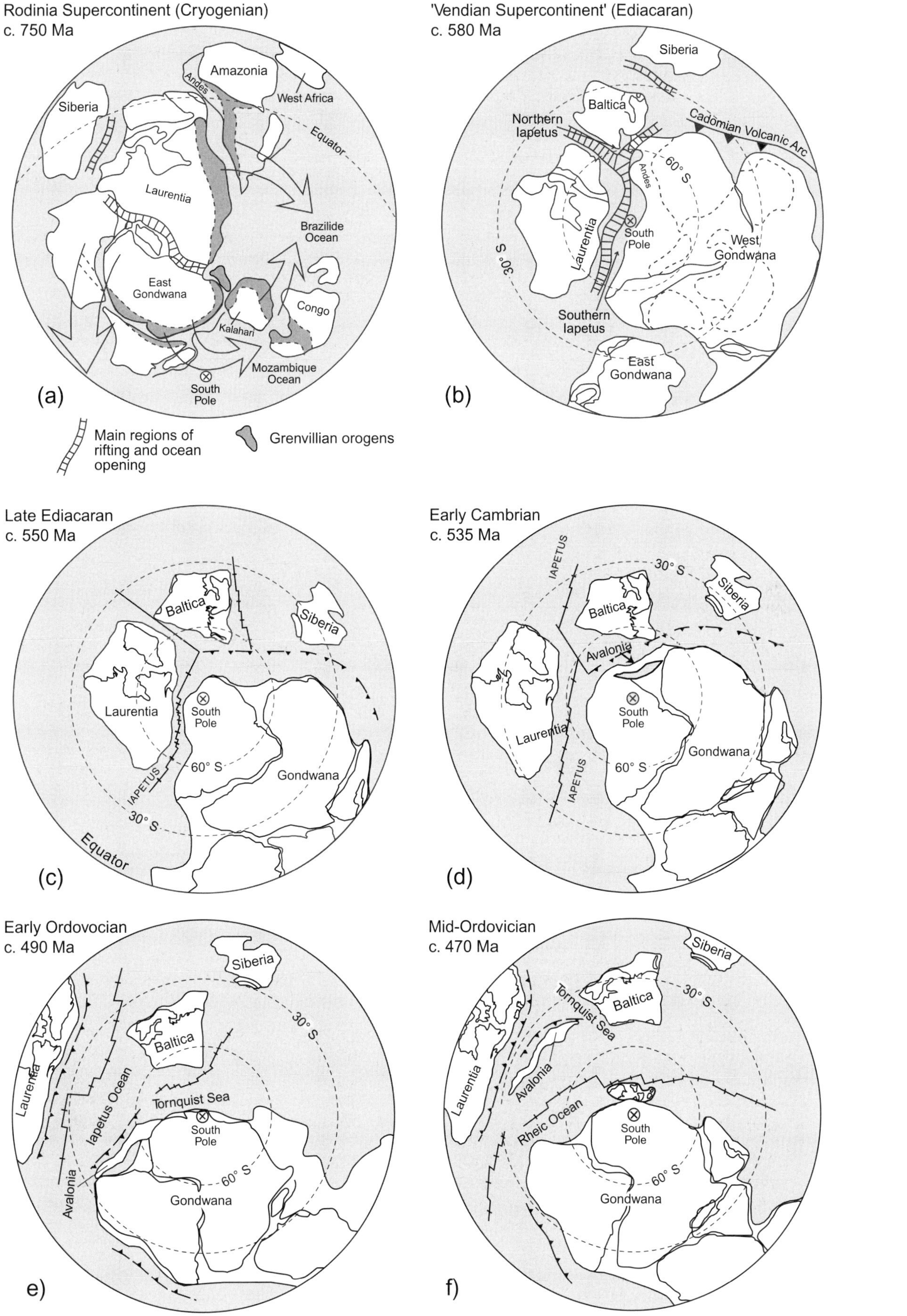

Rodinia Supercontinent (Cryogenian)
c. 750 Ma
Amazonia
Andes
West Africa
Siberia
Equator
Laurentia
Brazilide Ocean
East Gondwana
Congo
Kalahari
Mozambique Ocean
South Pole
(a)
Main regions of rifting and ocean opening
Grenvillian orogens
'Vendian Supercontinent' (Ediacaran)
c. 580 Ma
Siberia
Baltica
Northern Iapetus
Cadomian Volcanic Arc
60° S
Andes
Laurentia
South Pole
30° S
West Gondwana
Southern Iapetus
East Gondwana
(b)
Late Ediacaran
c. 550 Ma
Baltica
Siberia
Laurentia
South Pole
60° S
IAPETUS
Gondwana
30° S
Equator
(c)
Early Cambrian
c. 535 Ma
IAPETUS
30° S
Siberia
Baltica
Avalonia
South Pole
Laurentia
Gondwana
60° S
IAPETUS
(d)
Early Ordovician
c. 490 Ma
Siberia
30° S
Baltica
Laurentia
Iapetus Ocean
Tornquist Sea
South Pole
60° S
Avalonia
Gondwana
e)
Mid-Ordovician
c. 470 Ma
Siberia
Tornquist Sea
Baltica
30° S
Laurentia
Avalonia
Rheic Ocean
South Pole
60° S
Gondwana
f)

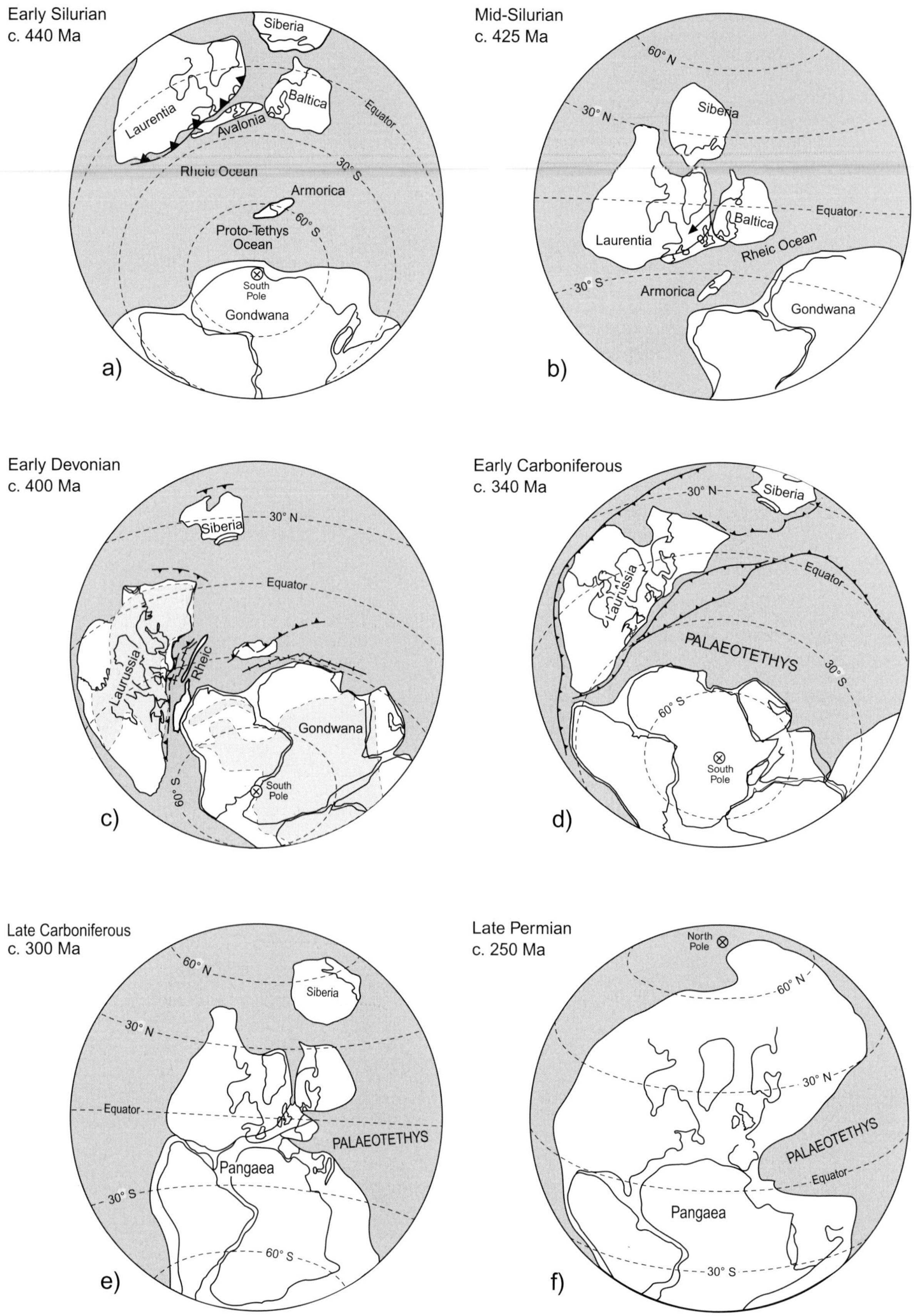

Fig. 1.3. Global palaeocontinental reconstructions for the Palaeozoic. (**a**) Early Silurian, *c*. 440 Ma. (**b**) Mid-Silurian, *c*. 425 Ma. (**c**) Early Devonian, *c*. 400 Ma. (**d**) Early Carboniferous, *c*. 340 Ma. (**e**) Late Carboniferous, *c*. 300 Ma. (**f**) Late Permian, *c*. 250 Ma (after Cocks & Torsvik 2006; Torsvik *et al.* 1993, 1996; Holdsworth *et al.* 2002; Ziegler 1990).

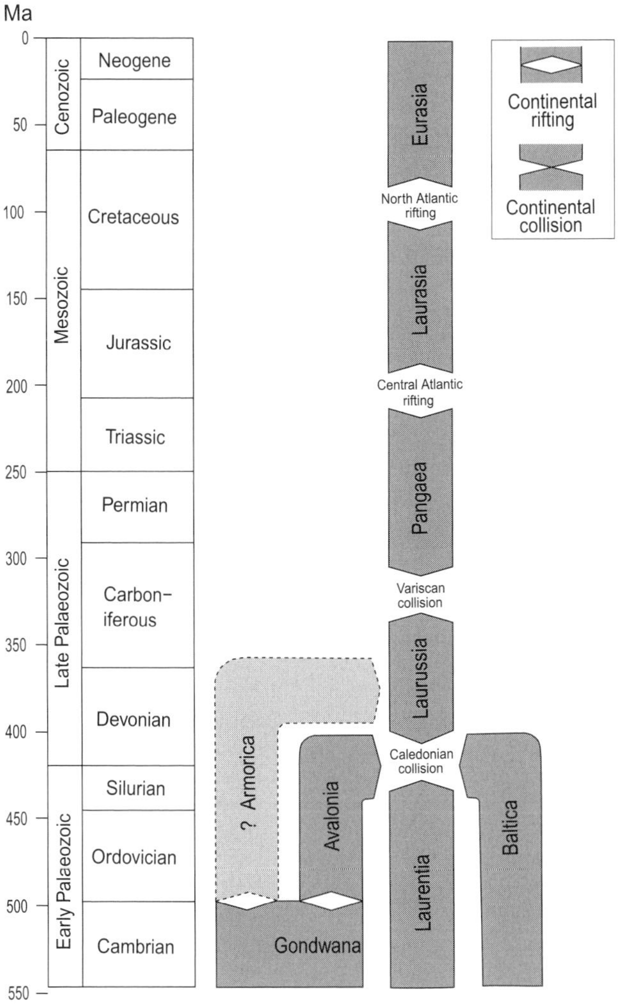

Fig. 1.4. Summary of the Phanerozoic collision and rift histories for central and western Europe (after Holdsworth *et al.* 2002).

Newfoundland in the mid-Ordovician (*c.* 450 Ma). However, the main closure events appear to be Silurian (*c.* 425 Ma), with Baltica and parts of Avalonia converging obliquely with the Laurentian margin (Fig. 1.3a, b; see Verniers *et al.* 2008).

The relative and actual timing of the closure of the various oceans (Tornquist Sea, Iapetus Ocean) as well as the collision of Laurentia, Baltica and Avalonia is still the subject of ongoing debate (for details see Krawczyk *et al.* 2008). Complex, collision-related deformation continued well into the Devonian (*c.* 390 Ma) in many areas, while significant strike-slip faulting associated with oblique closure disrupted the Laurentian continental margin (Fig. 1.3c; Holdsworth *et al.* 2002). Furthermore, the subsequent destruction of the crustal 'roots' of the Irish–Scottish–Scandinavian, North German–Polish and Mid-European Caledonides in the period of Early Devonian post-orogenic wrench faulting and backarc rifting, and subsequent Late Devonian and Carboniferous rifting (Ziegler & Dèzes 2006), makes palaeohistoric reconstructions even more difficult (see Krawczyk *et al.* 2008).

The Variscan Orogen

The Caledonian and Variscan orogenic cycles were, in many respects, part of a continuum involving the breakup of one supercontinent in the Late Neoproterozoic and the reassembly of another by the late Palaeozoic. The Variscan Orogeny of Europe is correlative to the final stages of the Appalachian orogenies along eastern North America, and marks the final amalgamation of Laurasia and Gondwana to form the supercontinent of Pangaea (Figs 1.3d, e & 1.4; see Kroner *et al.* 2008). The major oceans which had separated the various plates constituting Europe had closed by the Devonian (see Bełka & Narkiewicz 2008), although final consolidation into the present-day configuration of Variscan Europe did not occur until some time later. The formation of Pangaea resulted in major changes in oceanic circulation and was also coincident with the development of marked biogeographic differentiation and a high degree of bioprovincialism, including land plants on the newly created landmass (Fig. 1.3e, f). Also of note is the fact that the Late Carboniferous–Early Permian Kiaman Superchron is the longest known period of predominantly reversed polarity in the stratigraphic record (Davydov *et al.* 2004, 2008).

Compared with other orogens, the Variscan Orogeny had an unusually complex evolution and many aspects are the subject of active debate (Franke 2000). Despite this, its development and architecture have become much clearer with the identification and correlation of plate tectonic elements (see McCann *et al.* 2008*b*; Kroner *et al.* 2008) as well as the integration of palaeomagnetic and geophysical data with high precision dating (Timmerman 2004; 2008).

The Variscan area of Europe can be subdivided into a number of distinct zones. The most northerly of these, the Rhenohercynian Zone, extends from Cornwall through to the Moravo-Silesian Belt and possibly into southern Poland. The zone records two convergent episodes, namely the closure of the Rheic Ocean between Avalonia and Armorica (= Armorican Terrane Assemblage) and the closure of the Rhenohercynian Basin. This latter basin formed during the period of Variscan-related extension throughout western and central Europe in early to middle Devonian times. Basin formation led to the renewed separation of Avalonia from the Armorican Terrane Assemblage to the south and the resultant ocean, the Rhenohercynian Ocean, extended from southern England through Belgium and northern Germany. The onset of SE-directed subduction during the late Devonian–early Carboniferous led to the closure of the basin.

South of the Rhenohercynian Zone, the Saxothuringian Zone contains the Mid-German Crystalline High, which represents the SE active margin of the Rhenohercynian Zone. Southwards, the crystalline high is thrust over the Palaeozoic rift sequences of the Saxothuringian Basin, which are exposed in Bavaria, Thuringia and Saxony, as well as in the northern parts of the Black Forest and Vosges. These, in turn, were overthrust by nappes of virtually unmetamorphosed Palaeozoic deeper-water facies and crystalline rocks, derived from the NW margin of the Moldanubian region to the SE. The internal structure of the Moldanubian Zone is complex. In places, SE-directed nappe thrusts emplace high-grade metamorphic successions over medium-grade ones. To the south, the Moldanubian Zone probably continues into the Variscan basement of the Alps.

The Late Palaeozoic and Mesozoic development of Europe

Variscan deformation continued into the Late Carboniferous (McCann *et al.* 2008*b*), with widespread thrust and wrench faulting, folding, post-tectonic granite emplacement and the accumulation of thick sedimentary successions in the developing foreland basins. Following the Late Westphalian–Early Stephanian consolidation of the Variscan foldbelt, convergence between Gondwana and Laurussia apparently changed from an essentially north–south directed collision to an east–west orientation, leading to the development of continental-scale dextral shears (e.g. Tornquist-Teisseyre Zone) which were linked by secondary sinistral and dextral shear systems (Ziegler *et al.* 2006). The late stages of orogenic evolution were characterized by crustal extension, high heat flow, post-kinematic emplacement of granitoids and widespread late-orogenic volcanism (see Timmerman 2008). While this volcanic episode was centred in the Oslo Rift and NE Germany, magmatic activity was also recorded across much of Europe (McCann *et al.* 2006). Coincident with this tectonic and magmatic activity, the supercontinent of Pangaea moved northwards, resulting in a dramatic change in climate from the more humid conditions of the Late Carboniferous to the aridity, and related evaporitic formation, of the Permian (see McCann *et al.* 2008*a*, *b*).

Two major intracontinental basins, generally referred to as the Northern and Southern Permian basins, began to subside within western and central Europe in response to thermal contraction of the lithosphere (Figs 1.3f & 1.4; see McCann *et al.* 2008*a*; Scheck-Wenderoth *et al.* 2008). Both basins were initiated as broad crustal downwarps with little associated faulting. The Southern Permian Basin extended over a distance of 1700 km from England through northern Germany into Poland and the Baltic states. Flooding of this basin in late Permian times led to the development of thick Zechstein salt units. The latest Permian was also coincident with a major extinction event.

Late Permian and extensive Mesozoic rifting affected the area of the Atlantic shelves, the North Sea, the Tethys shelves and to a lesser degree the Sorgenfrei-Tornquist-Teisseyre Line (see Feist-Burkhardt *et al.* 2008; Scheck-Wenderoth *et al.* 2008). This resulted in significant crustal thinning that led to the breakup of Pangaea, commencing along the central Atlantic axis in Jurassic times. In the Middle and Late Jurassic, new ocean basins formed in the Mediterranean area as a result of the separation of Gondwana from Laurasia, and the western Tethys formed a shallow sea covering significant parts of Europe in Jurassic and Cretaceous times (Figs 1.5a, b; see Pieńkowski *et al.* 2008).

Crustal extension continued into the Early Cretaceous with the development of the North Atlantic rift system; spreading began in the South Atlantic and led to the separation of South America from Africa (see Voigt *et al.* 2008). The consequent anticlockwise rotation of Africa tended to narrow the Tethys Ocean, resulting in the development of a complex series of rifts and strike-slip faults which propagated westward towards the Atlantic. These rift systems demarcated the southern continents from Eurasia through the present Mediterranean region. By Late Cretaceous times, continued spreading of the North Atlantic led to the separation of Greenland from Eurasia, and the propagation of rifts into the Arctic Ocean (Fig. 1.5b). This period of final separation was coincident with a marked increase in volcanic activity and spreading-ridge formation which also enhanced greenhouse conditions in the middle Late Cretaceous. The Cretaceous ended dramatically with an asteroid impact (Ogg *et al.* 2004).

Late Cretaceous tectonic events heralded the Alpine cycle, marked by major east–west shortening and the onset of cordillera-building processes. However, in the Alps and elsewhere, remnant oceanic basins (Alpine Tethys) were finally closed only in Oligocene to Early Miocene times, a period corresponding to the first Alpine orogenic phase (Stampfli & Kozur 2006) (see Froitzheim *et al.* 2008; Reicherter *et al.* 2008).

The Palaeogene and Neogene (Alpine Orogeny) and beyond

Data from the Alps indicate that there were at least two distinct major episodes of tectonic activity, one during the Cretaceous and a later one during the Palaeogene and Neogene (Fig. 1.5c; see Froitzheim *et al.* 2008; Rasser *et al.* 2008, Reicherter *et al.* 2008). This second phase of convergence resulted from the north–south to NNE–SSW directed plate motions between Europe and Africa in the Eocene, and NW–SE directed motions in the Miocene. To the north of the Alps, a foreland basin developed in late Eocene times, its depocentre migrating northward in response to loading and flexure of the lithosphere caused by the continental collision of Europe and Africa.

To the south of the Alps, the formation of the Po Basin of Italy, a foredeep in which syntectonic sediments accumulated, was linked to south-vergent thrusting in the southern Alps. In the western and eastern Alps, the rotating WNW motion of the Adriatic microplate was associated with transcurrent faulting. The late stages of the collision produced spectacular large-scale backfolds in the Penninic nappes of the central Alps, related to the movement of an indentor of Adriatic crust pushed into the European crust. Shortening of the entire south Alpine fold-and-thrust belt is of the order of 70–115 km.

The Palaeogene notably contains evidence of a period of rapid and extreme global warming (the Palaeocene–Eocene Thermal Maximum) which upset oceanic and atmospheric circulation and led to the extinction of numerous deep-sea benthic foraminifera, and on land to a major turnover in mammals (see Rasser *et al.* 2008). The event may have been triggered by gas hydrates releasing methane into the ocean–atmosphere reservoir, upsetting the global carbon cycle (e.g. Sluijs *et al.* 2006) (similar events have also been recognized in the Mesozoic; see Jenkyns 2003).

The onset of the Neogene Period coincided with the gradual closing of Tethys and the formation of more distinctive latitudinal belts under the influence of dramatic climatic cooling phases (Lourens *et al.* 2004) (see Litt *et al.* 2008; Rasser *et al.* 2008). The period also coincided with a change in convergence direction between Eurasia and Africa–Arabia (dominated by dextral translations), and coupled with the development of intra-Alpine shear systems, and the concentration of crustal shortening in the western Alps and the Eastern Carpathians (Ziegler 1988, 1990). The Alps continued to be affected by a stress regime associated with crustal thickening, although in the central Alps rising topography was already counterbalanced by the onset of extensional collapse tectonics.

Frequent and marked climatic oscillations were characteristic of the Neogene. The late Neogene climatic changes coincided with the rapid spread of grasses around 8–10 Ma (Agustí *et al.* 1999). Additionally, climatic shifts and falling temperatures, due to the spread of the Antarctic ice, were followed by the Plio-Pleistocene Arctic glaciations in the northern hemisphere. Thick icesheets spread outwards from the Fennoscandian Shield region, as well as from other upland areas (e.g. Scotland, Switzerland). These icesheets locked away large amounts of water resulting in eustatic falls in sea level and the development of an extensive continental landmass. At the last glacial maximum, when the centre of the Weichselian icesheet was an estimated 2.5 km thick, sea level fell by over 120 m exposing the continental shelf (now *c.* 90 m below sea level).

The most recent stage in the development of Europe is one in which humans have played the central role. Anthropogenic changes to the environment became more marked and extensive as humans migrated northwards across Europe following climatic

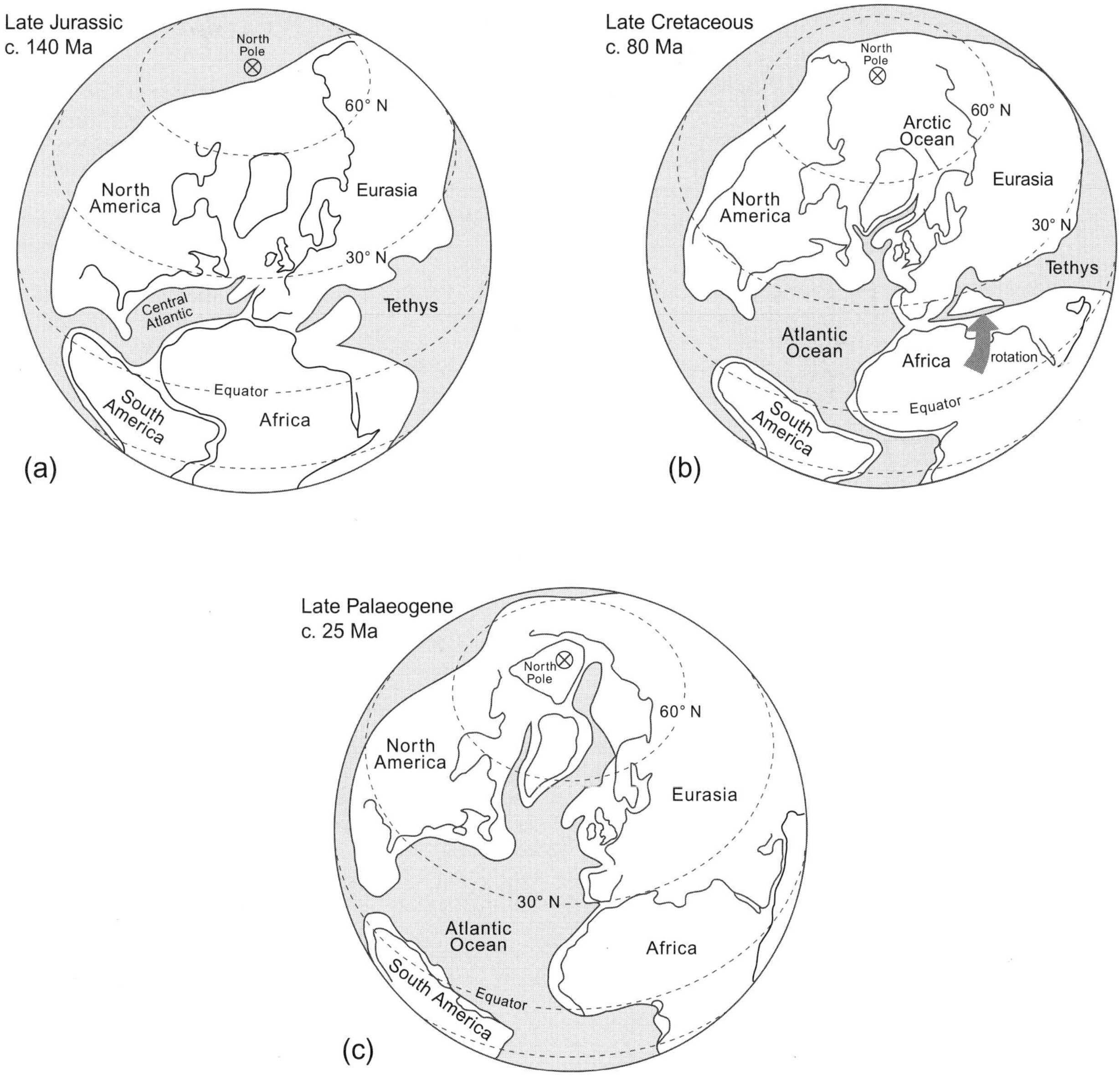

Fig. 1.5. Global palaeocontinental reconstructions for the Mesozoic and Cenozoic. (**a**) Late Jurassic, *c.* 140 Ma. (**b**) Late Cretaceous, *c.* 80 Ma. (**c**) Late Palaeogene, *c.* 25 Ma (after Holdsworth *et al.* 2002; Ziegler 1990).

amelioration. The onset of changes in the vegetation record can be traced across Europe in the pre- and post-Roman periods (for details see Litt *et al.* 2008). These changes, which commenced with the development of settled agriculture, became more marked with the onset of industrialization and urbanization and have resulted in increasing pollution of the environment. Indeed, in the five millennia that have followed our appearence in Central Europe, almost no part of the landscape or environment has been spared human influence (Woodcock 2002). Even if *Homo sapiens* is seen as just another species in the long geological record of the region, no other species has ever had such a major and profound impact.

Back to the future

What the future holds for us as a species is difficult to determine. As noted by Woodcock (2002), a book on historical geology such as this can be seen as an academic pursuit, something of intrinsic interest but of no practical use. However, this perception is changing as the scale of the human impact on regional and, indeed, global environments is increasingly recognized. Predicting future threats such as climatic change, sea-level rise or coastal erosion means extrapolating from our observations on the spatial organization of these variables, as well as our measurement of their present-day rates. However, to place this into a context of increasing change, it is necessary to understand the variability of 'natural' or pre-human environments, as well as to determine the controls which may have acted upon them (Fig. 1.6).

Climate change is one of the great environmental concerns facing humankind in the twenty-first century. Surface temperatures are expected to continue to rise globally, and major changes are likely to occur in the global hydrological and energy cycles (Watson *et al.* 2001). The greatest threat to humans will be due

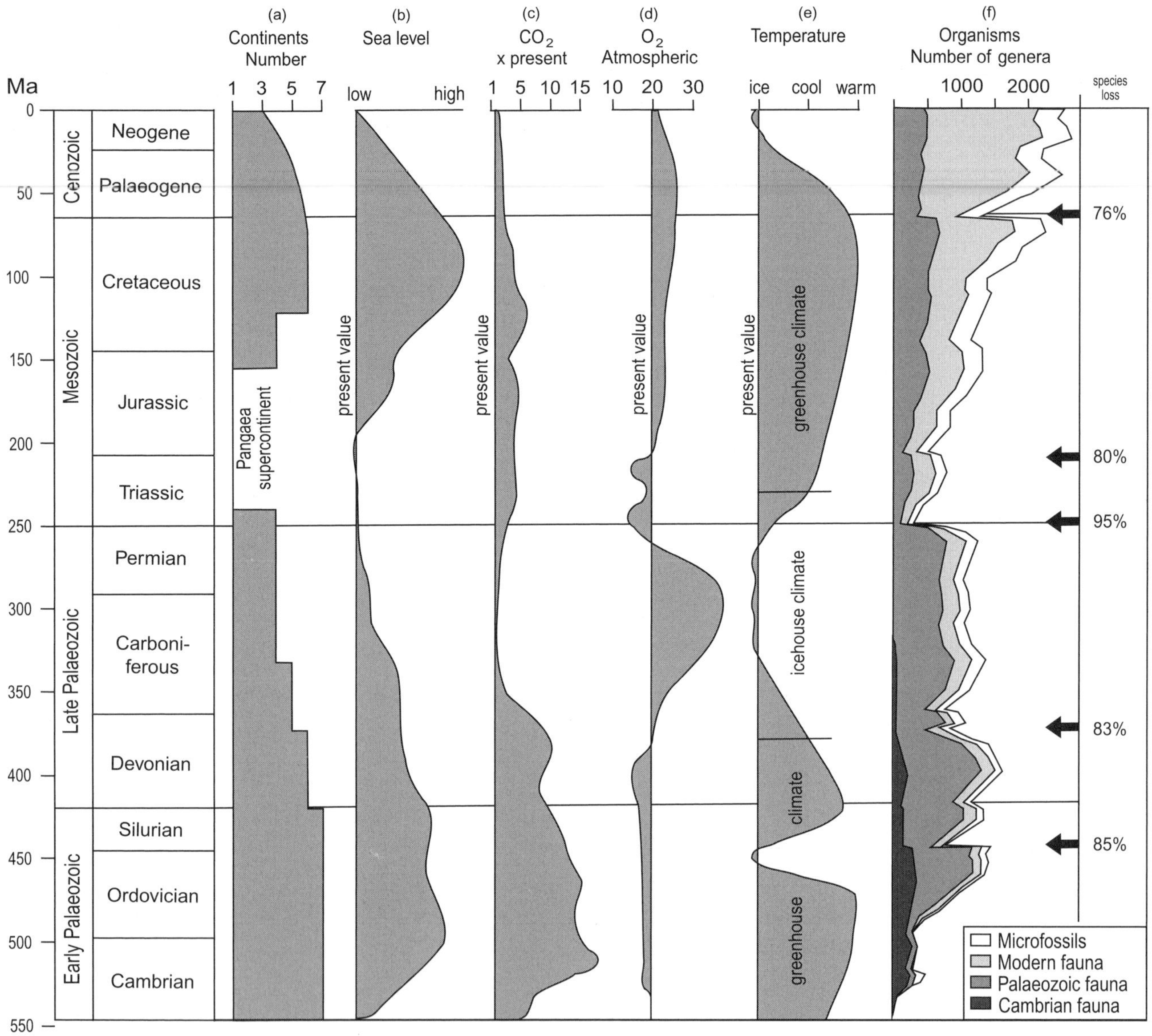

Fig. 1.6. Parameters of global change through Phanerozoic time. (**a**) Number of major continents (Veevers 1990). (**b**) Global sea level (Veevers 1990). (**c**) Atmospheric concentration of carbon dioxide (Berner 1991). (**d**) Atmospheric concentration of oxygen (Berner & Canfield 1989). (**e**) Global average surface temperature (Veevers 1990). (**f**) Number of genera in representative groups of organisms, subdivided into component faunas, and showing percentage species loss in major mass extinctions (Sepkoski 1998) (after Holdsworth *et al.* 2002).

to regional changes in extreme weather and climate events. Indeed, evidence from recent years has shown that European society is particularly vulnerable to changes in the frequency and intensity of such extreme events, for example heat waves, heavy precipitation, droughts and wind storms (Beniston *et al.* 2007).

With increasing concern over the effects that anthropogenic modification of future climates will bring, palaeoclimate studies of previous periods in Earth history are invaluable since they can provide us with possible analogues for such change. Much current research focuses on environmental change in the Quaternary, which is the most recent geological period for the examination of perturbations on a geologically short timescale (e.g. Braconnot *et al.* 2007*a*, *b*). However, Woodcock (2002) notes that some of the human-induced perturbations to the global environment may be on a scale unrepresented in the last two million years and we must, therefore, seek out analogues in the more distant past which would allow us to better understand those changes which may not, at first sight, appear to fall within a 'natural' pattern. The palaeoclimate of the European and Mediterranean region in middle Pliocene (*c.* 3 Ma) times, for example, was generally warmer (by 5°C), wetter (by 400–1000 mm/a), and less seasonal than present (Haywood *et al.* 2000, 2002, 2007; Haywood & Valdes 2004). Mechanisms proposed to account for this warming include enhanced thermohaline circulation and/or greater concentrations of CO_2 in the atmosphere. If future warming facilitates less ice over Greenland and the Arctic and warmer sea-surface temperatures in the North Atlantic, a climate scenario similar to that of the middle Pliocene is likely to occur (Haywood *et al.* 2000). Recent work has also focused on other periods of extreme climate, including the Jurassic (e.g.

Rees *et al.* 1999) and the Cretaceous (e.g. Haywood *et al.* 2004; Beerling *et al.* 1999; Price *et al.* 1997).

Information on possible changes in future climates can be obtained from numerical climate models (e.g. Watson *et al.* 2001; Claussen 2007) and only with such information is it possible to quantify the risks of a changing climate brought about by anthropogenic emissions of greenhouse gases (Christensen *et al.* 2007). The quantification of any possible changes is absolutely necessary in order to formulate and implement realistic adaptation and mitigation strategies (e.g. Arnell 1996). However, the various climatic models behave differently and there is no established way to determine which model represents the most probable version of the future (e.g. Kohfeld & Harrison 2000; Déqué *et al.* 2007). Evaluating the uncertainty, therefore, is as important as averaging the different possible scenarios to derive the best estimate.

As noted above current estimates of past and future climate change, as well as projections of future greenhouse gas emissions and their effects, are subject to various uncertainties (Wanner *et al.* 2006). According to Dessai & Hulme (2004) there are two different sources of uncertainty: the 'epistemic' (due to incomplete knowledge of processes that influence events) and 'stochastic' (sources that are considered 'unknowable', e.g. variability in the system, the chaotic nature of the climate system, and the indeterminacy of human systems). The sources of uncertainty in climate forecasting include: (a) uncertainty in anthropogenic forcing due to different emission paths, which is also a function of uncertain socio-economic and technological developments; (b) uncertainty due to natural variability, encompassing internal chaotic climate variability and externally driven (e.g. solar, volcanic) natural climate change; and (c) uncertainty in the climate system's response to external forcing due to incomplete knowledge of feedbacks and timescales in the system (Allen *et al.* 2004). Additionally, there are problems due to the inevitably incomplete and inadequate representation of the climate system in a weather or climate model (Schwierz *et al.* 2007) as well as the amounts of future atmospheric emissions (Christensen & Christensen 2007). The urgent need for improved numerical models and scenarios (e.g. Williams *et al.* 2007) becomes particularly apparent when considering extreme weather events (Christensen *et al.* 2007), the frequency and magnitude of which will be affected by any future climate changes as a result of an intensified hydrological cycle (Stocker *et al.* 2001). Therefore, a single realization of simulated climate is often insufficient to provide the information needed for a comprehensive assessment of potential climate change and its impacts.

Projections of future climate change from numerical models have existed for three decades, but these remain deficient both in regional detail and in the characterization of their uncertainty. The assessment of potential regional impacts of climate change has, to date, generally relied on data from global atmosphere–ocean general circulation models, which do not resolve spatial scales of less than *c.* 300 km (Mearns *et al.* 2001). Such models do not provide information on the spatial structure of temperature and precipitation in areas of complex topography and land-use distribution. Additionally, their depiction of regional and local atmospheric circulations and the representation of processes at high-frequency temporal scales are similarly insufficient (Christensen *et al.* 2007).

Numerous model-based studies have been conducted in Europe to investigate the potential impacts of climate change on natural and human systems. Many of these have been summarized in the European Commission-sponsored ACACIA report (Parry 2000) and in successive Intergovernmental Panel on Climate Change (IPCC) reports (e.g. Beniston *et al.* 1998; Kundzewicz *et al.* 2001; Solomon *et al.* 2007). The most important factors in shaping the present-day climate over the North Atlantic, Europe and the Mediterranean are the Icelandic low and Azores high-pressure systems combined with surface ocean currents and thermohaline circulation (Bjerknes 1964; Rodwell *et al.* 1999; Wood *et al.* 1999). However, regional climate processes and feedback (second-order trends) may locally override any global (first-order) trend (Haywood *et al.* 2000). Additionally, there are numerous factors that also play a sometimes indeterminate role, e.g. sea ice (Singarayer *et al.* 2006), behaviour of ocean currents (Stephenson & Pavan 2003), interannual to decadal variations (Collins *et al.* 2006) and latitudinal variations in warming (Kaplan & New 2006).

Observations and reconstructions for the late Holocene indicate that the warming since the 1960s is unprecedented over the last millennium. Modelling studies provide some evidence that the temperature changes of at least the second half of the twentieth century can only be explained by including anthropogenic forcing (Raible *et al.* 2006, and references therein). Indeed, the recent upward temperature trend for the Atlantic/European region is unprecedented over the last 500 years and will continue for the next 100 years under the constraint that greenhouse gas emissions will rise (Raible *et al.* 2006).

Although these extrapolated human-induced temperature changes are impressive, they are just within the natural range for these parameters experienced during Quaternary time (albeit on an accelerated timescale). Potentially more problematic could be the response of the ocean–atmosphere system. Changes in ocean circulation can occur on timescales of hundreds of years or less (e.g. Manabe & Stouffer 1994; Whorf & Keeling 1998). For instance, a global temperature rise of 4°C would result in substantial loss of Arctic and Greenland ice. Fresh water from these sources could possibly flow into the North Atlantic disrupting the North Atlantic thermohaline circulation. The slowing, or even stopping, of this deep-ocean circulation would lead to a dramatic reduction in winter temperatures and rainfall levels in western Europe (ACIA 2004). However, as with other modelling scenarios, reconstructing the strength and structure of the North Atlantic Ocean overturning circulation is difficult. Results from various climate models give very different glacial thermohaline circulations even with comparable circulations at present. Additionally, coupled modelling studies of the Last Glacial Maximum have shown widely varying changes to the Last Glacial Maximum Atlantic meridional overturning circulation, and the interpretation of these differences has been difficult due to the various forcings and boundary conditions used (Otto-Bliesner *et al.* 2007, and references therein).

In summary

From the above discussion it is apparent that our understanding of the Earth and its evolution is not as complete as it could be. Amid such uncertainties, it is increasingly clear that our present databases are insufficient to fully comprehend and model the world around us. It is, therefore, imperative that we continue to gather information both to increase our knowledge base, and also to apply this to understanding the complexity of our dynamic Earth. As editor of this unique volume on the geology of Central Europe, it is my wish that the work of more than 200 scientists, from various disciplines, contained within these pages will provide a basis for future discussion and co-operation across the region, as well as acting as a stimulant for more research. As noted by Charles Lyell in his *Principles of Geology,* the present is the key to the past. However, in terms of understanding the complexity of our science, and its implications for us as humans, it is important to remember that the past is also the key to the future.

I would like to thank A. Siemes and T. Beilfuß for the diagrams. The manuscript was greatly improved by the comments of S.-O. Franz, A. Giannetti, M. Valdiva-Manchego and particularly C. Lewis and P. Leat. I would also like to take this opportunity to thank my teachers over the years, especially J. Graham, P. Coxon, R. Pickerill, and P. Shannon. Additionally, I would like to thank my family – my parents Tommy (RIP) and Eileen, my sister Jackie, and my brothers Gary and Robert, for all their support over the years. And finally, Fionán, Josh and Bernd ... for the good times.

References

ACIA (Arctic Climate Impact Assessment) 2004. *Impacts of a Warming Arctic*. Cambridge University Press, Cambridge.

Adams, F. D. 1954. *The Birth and Development of the Geological Sciences*. Dover Publications, New York.

Agustí, J., Rook, L. & Andrews, P. 1999. Introduction. *In:* Agustí, J., Rook, L. & Andrews, P. (eds) *The Evolution of Neogene Terrestrial Ecosystems in Europe*. Hominoid Evolution and Climate Change in Europe, Volume 1. Cambridge University Press, Cambridge, 1–6.

Allen, M. R., Booth, B. B. B., Frame, D. J., Gregory, J. M., Kettleborough, J. A., Smith, L. A., Stainforth, D. A. & Stott, P. A. 2004. Observational constraints on future climate: distinguishing robust from model-dependent statements of uncertainty in climate forecasting. *In:* Manning *et al.* (eds) *Describing Scientific Uncertainties in Climate Change to Support Analysis of Risk and of Options*. May 2004 IPCC Workshop Report. IPCC Working Group I Technical Support Unit, Boulder, Colorado, 53–57. (World Wide Web address: http://www.ipcc.ch/)

Arnell, N. W. 1996. *Global Warming, River Flows and Water Resources*. Wiley, Chichester.

Artemieva, I. M., Thybo, H. & Kaban, M. K. 2006. Deep Europe today: geophysical synthesis of the upper mantle structure annd lithospheric processes over 3.5 Ma. *In:* Gee, D. G. & Stephenson, R. A. (eds) *European Lithosphere Dynamics*. Geological Society, London, Memoirs, **32**, 11–41.

Baxter, S. 2003. Revolutions in the Earth. James Hutton and the True Age of the World. Phoenix, London.

Beerling, D., Woodward, F. I. & Valdes, P .J. 1999. Global terrestrial productivity in the mid-Cretaceous (100 Ma): model simulations and data. *In:* Barrera, E. & Johnson, C. C. (eds) *Evolution of the Cretaceous Ocean-Climate System*. Geological Society of America, Special Paper, **332**, 385–390.

Bełka, Z.& Narkiewicz, M. 2008. Devonian. *In*: McCann, T. (ed.) *The Geology of Central Europe. Volume 1: Precambrian and Palaeozoic*. Geological Society, London, 383–410.

Beniston, M., Tol, R. S. J., *et al.* 1998. Europe. *In:* Watson, R. T. *et al.* (eds) *The Regional Impacts of Climate Change: An Assessment of Vulnerability*. A special report of Working Group II of the Intergovernmental Panel on Climate Change. Cambridge University Press, Cambridge, 149–185.

Beniston, M., Stephenson, D. B., *et al.* 2007. Future extreme events in European climate: an exploration of regional climate model projections. *Climatic Change*, **81**, 71–95.

Berner, R. A. 1991. A model for atmospheric CO_2 over Phanerozoic time. *American Journal of Science,* **291**, 339–376.

Berner, R. A. & Canfield, D. E. 1989. A new model for atmospheric oxygen over Phanerozoic time. *American Journal of Science*, **289**, 333–361.

Berthelsen, A. 1992. Mobile Europe. *In:* Blundell, D., Freeman, R. & Mueller, S. (eds). *A Continent Revealed - The European Geotraverse*. Cambridge University Press, Cambridge, 11–32.

Bjerknes, J. 1964. Atlantic air–sea interaction. *Advanced Geophysics*, **10**, 1–82.

Braconnot, P., Otto-Bliesner, B., *et al.* 2007*a*. Results of PMIP2 coupled simulataions of the Mid-Holocene and Last Glacial Maximum – Part 1: experiments and large-scale features. *Climate of the Past,* **3**, 261–277.

Braconnot, P., Otto-Bliesner, B., *et al.* 2007*b*. Results of PMIP2 coupled simulations of the Mid-Holocene and Last Glacial Maximum – Part 2: feedbacks with emphasis on the location of the ITCZ and mid- and high latitudes heat budget. *Climate of the Past*, **3**, 279–296.

Cannon, W. F. 1976. Charles Lyell, Radical Actualism, and Theory. *British Journal for the History of Science*, **9**, 104–120.

Christensen, J. H. & Christensen, O. B. 2007. A summary of the PRUDENCE model projections of changes in European climate by the end of the century. *Climatic Change*, **81**, 7–30.

Christensen, J. H., Carter, T. R., Rummukainen, M. & Amanatidis, G. 2007. Evaluating the performance and utility of regional climate models: the PRUDENCE project. *Climatic Change*, **81**, 1–6.

Claussen, M. 2007. Climate system models – A brief introduction. *In:* Sirocko, F., Claussen, M., Sanchez Goni, M. F. & Litt, T. (eds) *The Climate of Past Interglacials*. Developments in Quaternary Science, **7**. Elsevier, Amsterdam.

Cocks, L. R. M. & Torsvik, T. H. 2006. European geography in a global context from the Vendian to the end of the Palaeozoic. *In:* Gee, D. G. & Stephenson, R. A. (eds) *European Lithosphere Dynamics*. Geological Society, London, Memoirs, **32**, 83–95.

Collins, M., Botzet, M., *et al.* 2006. Interannual to decadal climate predictabililty in the North Atlantic: A multimodel-ensemble study. *Journal of Climate,* **19**, 1195–1203.

Dalziel, I. W. D. 1992. Antarctica: a tale of two supercontinents? *Annual Review of Earth and Planetary Science*, **20**, 501–526.

Davydov, V., Wardlaw, B. R. & Gradstein, F. M. 2004. The Carboniferous Period. *In:* Gradstein, F. M., Ogg, J. G. & Smith, A. G. (eds) *A Geologic Time Scale 2004*. Cambridge University Press, Cambridge, 222–248.

Dean, D. R. 1973. James Hutton and his public, 1785–1802. *Annals of Science*, **30**, 89–105.

Dean, D. R. 1992. *James Hutton and the History of Geology*. Cornell University Press, Ithaca.

Déqué, M., Rowell, D. P., *et al.* 2007. An intercomparison of regional climate simulations for Europe: assessing uncertainties in model projections. *Climatic Change*, **81**, 53–70.

Dessai, S. & Hulme, M. 2004. Does climate adaptation policy need probabilities? *Climate Policy*, **4**, 107–128.

Feist-Burkhardt, S., Götz, A. E. *et al.* 2008. Triassic. *In:* McCann, T. (ed.) *The Geology of Central Europe. Volume 2: Mesozoic and Cenozoic*. Geological Society, London, 749–821.

Franke, W. 2000. The mid-European segment of the Variscides: Tectono-stratigraphic units, terrane boundaries and plate tectonic evolution. *In:* Franke, W., Haak, V, Oncken, O. & Tanner, D. (eds) *Orogenic Processes: Quantification and Modelling in the Variscan Belt*. Geological Society, London, Special Publication **179**, 35–62.

Froitzheim, N., Plašienka, D. & Schuster, R. 2008. Alpine tectonics of the Alps and Western Carpathians. *In:* McCann, T. (ed.) *The Geology of Central Europe. Volume 2: Mesozoic and Cenozoic*. Geological Society, London, 1141–1232.

Gerstner, P. 1971. The reaction to James Hutton's use of heat as a geological agent. *British Journal for the History of Science*, **5**, 353–362.

Geyer, G., Elicki, O., Fatka, O. & Żylińska, A. 2008. Cambrian. *In:* McCann, T. (ed.) *The Geology of Central Europe. Volume 1: Precambrian and Palaeozoic*. Geological Society, London, 155–202.

Gohau, G. 1991. *A History of Geology* (revised & translated by A. V. Carozzi & M. Carozzi). Rutgers University Press, New Brunswick & London.

Gould, S. J. 1987. *Time's Arrow, Time's Circle: Myth and Metaphor in the Discovery of Geological Time*. Harvard University Press, Cambridge.

Hartz, E. H. & Torsvik, T. H. 2002. Baltica upside down: a new plate tectonic model for Rodinia and the Iapetus Ocean. *Geology*, **30**, 255–258.

Haywood, A. M. & Valdes, P. J. 2004. Modelling Pliocene warmth: contribution of atmosphere, oceans and cryosphere. *Earth and Planetary Science Letters*, **218**, 363–377.

Haywood, A. M., Sellwood, B. W. & Valdes, P. J. 2000. Regional warming: Pliocene (3 Ma) paleoclimate of Europe and the Mediterranean. *Geology*, **28**, 1063–1066.

Haywood, A. M., Valdes, P. J. & Sellwood, B. W. 2002. Magnitude of climate variability during middle Pliocene warmth: a palaeoclimate modelling study. *Paleogeography, Paleoclimatology, Paleoecology*, **188**, 1–24.

Haywood, A. M., Valdes, P. J. & Markewich, P. 2004. Cretaceous (Wealden) climates: a modelling perspective. *Cretaceous Research*, **25**, 303–311.

Haywood, A. M., Valdes, P. J. & Peck, V. L. 2007. A permanent El Nino-like state during the Pliocene. *Paleooceanography*, **22,** 1–21.

Herries-Davies, G. L. 1969. *The Earth in Decay: A History of British Geomorphology 1578–1878*. Macdonald Technical and Scientific, London.

Holdsworth, R. E., Woodcock, N. H. & Strachan, R. A. 2002. Geological framework of Britain and Ireland. *In:* Woodcock, N. H. & Strachan, R. (eds) *Geological History of Britain and Ireland*. Blackwell Publishing, Oxford, 19–37.

Howard, P. 1797. (Thoughts on the Structure of the Globe and) The Scriptural History of the Earth and Mankind. *Monthly Review*, **25** (1798), 241–254.

Jenkyns, H. 2003. Evidence for rapid climate change in the Mesozoic-Palaeogene greenhouse world. *Philosophical Transactions of the Royal Society : Mathematical, Physical & Engineering Sciences*, **361**, 1885–1916.

Kaplan, J. O. & New, M. 2006. Arctic climate change with a 2oC global warming: Timing, climate patterns and vegetation change. *Climatic Change*, **79**, 213–241.

Kirwan, R. 1793. Examination of the supposed igneous origin of stony substances. *Transactions of the Royal Irish Academy*, **5,** 51–87.

Klaver, J. M. I. 1997. *Geology and Religious Sentiment. The Effect of Geological Discoveries on English Society and Literature between 1829 and 1859*. Brill, Leiden.

Kohfeld, K. E. & Harrison, S. P. 2000. How well can we simulate past climates? Evaluating earth system models using global palaeoenvironmental datasets. *Quaternary Science Reviews*, **19,** 321–346.

Krawczyk, C. M., McCann, T., Cocks, L. R. M., England, R., McBride, J. & Wybraniec, S. 2008. Caledonian tectonics. *In:* McCann, T. (ed.) *The Geology of Central Europe. Volume 1: Precambrian and Palaeozoic*. Geological Society, London, 303–389.

Kroner, U., Mansy, J.-L. *et al.* 2008. Variscan tectonics. *In:* McCann, T. (ed.) *The Geology of Central Europe. Volume 1: Precambrian and Palaeozoic*. Geological Society, London, 599–664.

Kundezewicz, Z. W., Parry, M. L., *et al.* 2001. Europe. *In:* McCarthy, J. *et al.* (eds) *Climate Change 2001: Vulnerability, Impacts and Adaptation*. Contribution of Working Group II to the third assessment report of the Intergovernmental Panel on Climate Change. Cambridge University Press, Cambridge, 641–692.

Laudan, R. 1987. *From Mineralogy to Geology. The Foundations of a Science, 1650–1830*. University of Chicago Press, Chicago.

Linnemann, U. & Romer, R. L. 2002. The Cadomian orogeny in Saxo-Thuringia, Germany: geochemical and Nd-Sr-Pb isotopic characterization of marginal basins with constraints to geotectonic setting and provenance. *Tectonophysics*, **352**, 33–64.

Linnemann, U., D'Lemos, R. *et al.* 2008*a*. Cadomian tectonics. *In:* McCann, T. (ed.) *The Geology of Central Europe. Volume 1: Precambrian and Palaeozoic*. Geological Society, London, 103–154.

Linnemann, U., Romer, R. L. *et al.* 2008*b*. Precambrian. *In:* McCann, T. (ed.) *The Geology of Central Europe. Volume 1: Precambrian and Palaeozoic*. Geological Society, London, 21–101.

Litt, T., Schmincke, H.-U., Frechen, M. & Schlüchter, C. 2008. Quaternary. *In:* McCann, T. (ed.) *The Geology of Central Europe. Volume 2: Mesozoic and Cenozoic*. Geological Society, London, 1287–1340.

Lourens, L., Hilgen, F., Shackleton, N. J., Laskar, J. & Wilson, D. 2004. The Neogene Period. *In:* Gradstein, F. M., Ogg, J. G. & Smith, A. G. (eds) *A Geologic Time Scale 2004*. Cambridge University Press, Cambridge, 409–440.

McCann, T., Pascal, C., *et al.* 2006. Post-Variscan (end Carboniferous–early Permian) basin evolution in Western and Central Europe. *In:* Gee, D. G. & Stephenson, R. A. (eds) *European Lithosphere Dynamics*. Geological Society, London, Memoirs, **32**, 355–388.

McCann, T., Kiersnowski, H., *et al.* 2008*a*. Permian. *In:* McCann, T. (ed.) *The Geology of Central Europe. Volume 1: Precambrian and Palaeozoic*. Geological Society, London, 531–597.

McCann, T., Skompski, S., *et al.* 2008*b*. Carboniferous. *In:* McCann, T. (ed.) *The Geology of Central Europe. Volume 1: Precambrian and Palaeozoic*. Geological Society, London, **1**, 411–529.

Manabe, S. & Stouffer, R. J. 1994. Multiple-century response of a coupled ocean-atmosphere model to an increase of atmospheric carbon dioxide. *Journal of Climate*, **7**, 5–23.

Mearns, L. O., Hulme, M., Carter, T. R., Leemans, R., Lal, M. & Whetton, P. H. 2001. Climate scenario development. *In:* Houghton, J. *et al.* (eds) *Climate Change 2001: The Scientific Basis*. Contribution of Working Group I to the third assessment report of the Intergovernmental Panel on Climate Change. Cambridge University Press, Cambridge, 739–768.

Meert, J. G. & Torsvik, T. H. 2003. The making and unmaking of a supercontinent: Rodinia revisited. *Tectonophysics*, **375**, 261–268.

Ogg, J. G., Agterberg, F. P. & Gradstein, F. M. 2004. The Cretaceous Period. *In:* Gradstein, F. M., Ogg, J. G. & Smith, A. G. (eds) *A Geologic Time Scale*. Cambridge University Press, Cambridge, 344–383.

Oldroyd, D. 1996. *Thinking about the Earth: A History of Ideas in Geology*. Harvard University Press, Cambridge, Massachusetts.

Otto-Bliesner, B. L., Hewitt, C. D., *et al.* 2007. Late Glacial Maximum ocean thermohaline circulation: PMIP2 model intercomparisons and data constraints. *Geophysical Research Letters*, **34**, L12706, doi: 10.1029/2007 GL 029475

Parry, M. (ed.) 2000. *The Changing Climate of Europe. Assessment of potential effects and adaptations for climate change in Europe: the Europe ACACIA project*. Jackson Environmental Institute, University of East Anglia, Norwich.

Pieńkowski, G., Schudack, M. E., *et al.* 2008. Jurassic. *In:* McCann, T. (ed.) *The Geology of Central Europe. Volume 2: Mesozoic and Cenozoic*. Geological Society, London, 823–922.

Playfair, J. 1802. *Illustrations of the Huttonian Theory of the Earth*. Edinburgh. (Facsimile reprint, 1964, Dover Publications, New York.)

Porter, R. 1976. Charles Lyell and the principles of the history of Geology. *British Journal for the History of Science*, **9**, 93.103.

Price, G. D., Valdes, P. J. & Sellwood, B. W. 1997. Quantitative palaeoclimate GCM validation: Late Jurassic and Cretaceous case studies. *Journal of the Geological Society*, **154**, 769–772.

Raible, C. C., Casty, C., *et al.* 2006. Climate variability – observations, reconstructions, and model simulations for the Atlantic-European and Alpine region from 1500–2100 AD. *Climatic Change*, **79**, 9–29.

Rasser, M. W., Harzhauser, M., *et al.* 2008. Palaeogene and Neogene. *In:* McCann, T. (ed.) *The Geology of Central Europe. Volume 2: Mesozoic and Cenozoic*. Geological Society, London, 1031–1139.

Rees, P. A., Ziegler, A. M. & Valdes, P. J. 1999. Jurassic phytogeography and climates: new data and model comparisons. *In:* Huber, B., MacLeod, K. & Wing, S. (eds) *Warm Climates in Earth History*. Cambridge University Press, Cambridge, 297–318.

Reicherter, K., Froitzheim, N., *et al.* 2008. Alpine tectonics north of the Alps. *In:* McCann, T. (ed.) *The Geology of Central Europe. Volume 2: Mesozoic and Cenozoic*. Geological Society, London, 1233–1285.

Rodwell, M. J., Rowell, D. P. & Folland, C. K. 1999. Oceanic forcing of the wintertime North Atlantic Oscillation and European climate. *Nature*, **398**, 320–323.

Scheck-Wenderoth, M., Krzywiec, P., Zühlke, R., Maystrenko, Y. & Froitzheim, N. 2008. Permian to Cretaceous tectonics. *In:* McCann, T. (ed.) *The Geology of Central Europe. Volume 2: Mesozoic and Cenozoic*. Geological Society, London, 999–1030.

Schwierz, C., Appenzeller, C., Davies, H. C., Liniger, M. A., Müller, W., Stocker, T. F. & Yoshimori, M. 2006. Challenges posed by and approaches to the study of seasonal-to-decadal climate variability. *Climatic Change*, **79**, 31–63.

Sepkoski, J. J. 1998. Rates of speciation in the fossil record. *Philosophical Transactions of the Royal Society, London,* **B353**, 315–326.

Servais, T., Dzik, J., Fatka, O., Heuse, T., Vecoli, M. & Verniers, J. 2008. Ordovician. *In:* McCann, T. (ed.) *The Geology of Central Europe. Volume 1: Precambrian and Palaeozoic*. Geological Society, London, 203–248.

Singarayer, J. S., Bamber, J. L. & Valdes, P. J. 2006. Twenty-first-century climate impacts from a declining Arctic sea ice cover.

Journal of Climate, **19**, 1109–1125.

SLUIJS, A., SCHOUTEN, S., *et al.* 2006. Subtropical Arctic Ocean temperatures during the Palaeocene/Eocene thermal maximum. *Nature*, **441**, 610–613.

SOLOMON, S., QIN, D. *et al.* (eds) 2007. *Climate Change 2007.* The Physical Science Basis, Contribution of Working Group I to the Fourth Assessment Report of the Intergovernmental Panel on Climate Change. Cambridge University Press, Cambridge.

STAMPFLI, G. M. & KOZUR, H. W. 2006. Europe from the Variscan to the Alpine cycles. In: GEE, D. G. & STEPHENSON, R. A. (eds) *European Lithosphere Dynamics.* Geological Society, London, Memoirs, **32**, 57–82.

STEPHENSON, D. B. & PAVAN, V. 2003. The North Atlantic Oscillation in coupled climate models: a CMIP1 evaluation. *Climate Dynamics*, **20**, 381–399.

STOCKER, T. F., CLARKE, G. K. C., *et al.* 2001. Physical climate processes and feedbacks. *In:* HOUGHTON, J. *et al.* (eds) *Climate Change 2001: The Scientific Basis.* Contribution of Working Group I to the third assessment report of the Intergovernmental Panel on Climate Change. Cambridge University Press, Cambridge, 417–470.

STRACHAN, R. A. 2002. Early Earth history and development of the Archaean crust. *In:* WOODCOCK, N. H. & STRACHAN, R. (eds) *Geological History of Britain and Ireland.* Blackwell Publishing, Oxford, 41–51.

STRACHAN, R. A. & HOLDSWORTH, R. E. 2002. Proterozoic sedimentation, orogenesis and magmatism on the Laurentian Craton (2500–750 Ma). *In:* WOODCOCK, N. H. & STRACHAN, R. (eds) *Geological History of Britain and Ireland.* Blackwell Publishing, Oxford, 52–72.

SUMNER, J. B. 1816. *A Treatise on the Records of the Creation.* 2 vols, London.

TIMMERMAN, M. J. 2004. Timing, geodynamic setting and character of Permo-Carboniferous magmatism in the foreland of the Variscan Orogen, NW Europe. *In:* WILSON, M., NEUMANN, E.-R., DAVIES, G. R., TIMMERMAN, M. J., HERREMANS, M. & LARSEN, B. T. (eds) *Permo-Carboniferous Magmatism and Rifting in Europe.* Geological Society, London, Special Publications, **223**, 41–74.

TIMMERMAN, M. J. 2008. Palaeozoic magmatism. *In:* McCANN, T. (ed.) *The Geology of Central Europe. Volume 1: Precambrian and Palaeozoic.* Geological Society, London, 665–748.

TORSVIK, T. H., TRENCH, A., SVENSSON, I. & WALDERHAUG, H. J. 1993. Palaeogeographic significance of mid-Silurian palaeomagnetic results from southern Britain – a major revision of the apparent polar wander path for Eastern Avalonia. *Geophysical Journal International*, **113**, 651–668.

TORSVIK, T. H., SMETHURST, M. A., *et al.* 1996. Continental break-up and collision in the Neoproterozoic and Palaeozoic – a tale of Baltica and Laurentia. *Earth Science Reviews*, **40**, 229–258

TOWNSEND, J. 1813. The Character of Moses established for Veracity as an Historian, Recording Events from the Creation to the Deluge. *Monthly Review* (Bath), **74**(1814), 225–238.

VEEVERS, J. J. 1990. Tectonic-climatic supercycle in the billion-year plate tectonic eon: Permian Pangaean icehouse alternates with Cretaceous dispersed-continents greenhouse. *Sedimentary Geology*, **68**, 1–16.

VERNIERS, J., MALETZ, J., KŘÍŽ, J., ŽIGAITĖ, Ž., PARIS, F., SCHÖNLAUB, H. P. & WRONA, R. 2008. Silurian. *In:* McCANN, T. (ed.) *The Geology of Central Europe. Volume 1: Precambrian and Palaeozoic.* Geological Society, London, 249–302.

VOIGT, S., WAGREICH, M., *et al.* 2008. Cretaceous. *In:* McCANN, T. (ed.) *The Geology of Central Europe. Volume 2: Mesozoic and Cenozoic.* Geological Society, London, 923–997.

WANNER, H., GROSJEAN, M., RÖTHLISBERGER, R. & XOPLAKI, E. 2006. Climate variability, predictability and climate risks: a European perspective. *Climatic Change*, **79**, 1–7.

WATSON, R. T. & CORE WRITING TEAM (eds) 2001. *Intergovernmental Panel on Climate Change (IPCC): Climate Change 2001: Synthesis Report.* IPCC, Geneva.

WHORF, T. P. & KEELING, C. D. 1998. Rising carbon. *New Scientist*, **157**, 54.

WILLIAMS, M., HAYWOOD, A. M., GREGORY, J. & SCHMIDT, D. N. (eds) 2007. *Deep-Time Perspectives on Climate Change: Marrying the Signal from Computer Models and Biological Proxies.* The Micropalaeontological Society Special Publications **2**. The Geological Society, London.

WOOD, R. A., KEEN, A. B., MITCHELL, J. F. B. & GREGORY, J. M. 1999. Changing spatial structure of the thermohaline circulation in response to atmospheric CO2 forcing in a climate model. *Nature*, **399**, 572–575.

WOODCOCK, N. H. 2002. The Quaternary: history of an ice age. *In:* WOODCOCK, N. H. & STRACHAN, R. (eds) *Geological History of Britain and Ireland.* Blackwell Publishing, Oxford, 392–411.

WOODCOCK, N. H. & STRACHAN, R. A. 2002. Regional geological history: why and how? *In:* WOODCOCK, N. H. & STRACHAN, R. (eds) *Geological History of Britain and Ireland.* Blackwell Publishing, Oxford, 3–18.

ZIEGLER, P. A. 1988. Laurussia – the Old Red Continent. *In:* McMILLAN, N. J., IMBRY, A. F. & GLASS, D. J. (eds) *Devonian of the World.* Canadian Society of Petroleum Geologists, Memoir, **14(1)**, 15–48.

ZIEGLER, P. A. 1990. *Geological Atlas of Western and Central Europe.* Shell International Petroleum Maatschappij B.V..

ZIEGLER, P. A. & DÈZES, P. 2006. Crustal evolution of Western and Central Europe. *In:* GEE, D. G. & STEPHENSON, R. A. (eds) *European Lithosphere Dynamics.* Geological Society, London, Memoirs, **32**, 43–56.

ZIEGLER, P. A., SCHUMACHER, M. E., DÈZES, P., VAN WEES, J.-D. & CLOETINGH, S. 2006. Post-Variscan evolution of the lithosphere in the area of the European Cenozoic Rift System. *In:* GEE, D. G. & STEPHENSON, R. A. (eds) *European Lithosphere Dynamics.* Geological Society, London, Memoirs, **32**, 97–112.

2 Precambrian

ULF LINNEMANN, ROLF L. ROMER, CHRISTIAN PIN (co-ordinators), PAWEŁ ALEKSANDROWSKI, ZBIGNIEW BUŁA, THORSTEN GEISLER, VÁCLAV KACHLIK, EWA KRZEMIŃSKA, STANISŁAW MAZUR, GEDIMINAS MOTUZA, J. BRENDAN MURPHY, R. DAMIAN NANCE, SERGEI A. PISAREVSKY, BERNHARD SCHULZ, JENS ULRICH, JANINA WISZNIEWSKA, JERZY ŻABA & ARMIN ZEH

Around 88% of the history of the Earth occurred during the Precambrian period, which can be subdivided into the Archaean and the Proterozoic eons (Figs. 2.1 & 2.2). The Archaean eon (Greek *archaia* = ancient ones; 4.56–2.5 Ga) comprises the Eo-Palaeo-, Meso- and Neoarchaean eras. For the early Archaean the term Hadean is also used (Greek *hades* = unseen or hell; 4.56–3.8 Ga) (Fig. 2.1). The Proterozoic eon (Greek *proteros* = first, *zoon* = creature; 2.5–0.542 Ga) is composed of the Palaeo-, Meso- and Neoproterozoic eras (Fig. 2.2). The latter eras can be subdivided into different periods defined by the International Commission on Stratigraphy on the basis of geochronological data and characteristic features such as particular geotectonic settings and events (Gradstein *et al.* 2004). Palaeoproterozoic periods include the Siderian (Greek *sideros* = iron; 2.5–2.3 Ga), the Rhyacian (Greek *rhyax* = steam of lava; 2.3–2.05 Ga), the Orosirian (Greek *orosira* = mountain range; 2.05–1.8 Ga) and the Statherian (Greek *statheros* = stable; 1.8–1.6 Ga). The Calymmian (Greek *calymma* = cover; 1.6–1.4 Ga), Ectasian (Greek *ectasis* = extension; 1.4–1.2 Ga), and Stenian (Greek *stenos* = narrow; 1.2–1.0 Ga) are the Mesoproterozoic periods, while the Neoproterozoic is subdivided into the Tonian (Greek *tonas* = stretch; 1.0–0.85 Ga), Cryogenian (Greek *cryos* = ice, *genesis* = birth; 0.85–0.635 Ga), and finally Ediacaran (0.635–0.542 Ma). This latter is named after the Ediacara Hills (Flinders Ranges, Australia) and characteristically contains the Ediacara biota which represents the dawn of evolved life-forms. The Ediacaran period was ratified in 2004 by the International Union of Geological Sciences and replaces the Vendian period.

The lower limit of the Precambrian is assumed to be very close to the origin of the solar system. It was Clair Cameron Patterson (1956) who first produced a meteorite isochron by using lead isotopes and determined an age of *c.* 4.55 Ga which he interpreted as being indicative for the origin of the solar system and the early Earth. This isochron has since been modified slightly, although the 4.53–4.58 Ga age of the solar sytem has been repeatedly confirmed by all known chronometers. Recent calculations have provided an age for the solar system of *c.* 4.571 Ga and formation was coincident with a supernova event (Lugmair & Shukolyukov 2001; Gilmour & Saxton 2001). Within the next million years planetary embryos were formed and the accretion of these embryos was largely responsible for the formation of the terrestrial planets. Early Earth formed around 4.561 Ga (Allègre *et al.* 1995). The impact of the Mars-sized planetary embryo Theia with the early Earth at *c.* 4.550–4.540 Ma is widely accepted as having formed the moon (Steward 2000). Early Earth became differentiated into a metallic core, a silicate mantle and a chemically differentiated crust at *c.* 4.551 Ga (Eriksson *et al.* 2004). Intense meteor bombardment affected early Earth during planetary accretion. Later in the Earth's early Precambrian history, impact events continued to be important. Indeed, the impact rate in the Earth–Moon system exceeded the present-day rate by about 15 times at 3.8 Ga, eventually declining to about two times by 3.0 Ga (Ryder 2003).

Sensitive high-resolution ion microprobe (SHRIMP) U–Pb ages have been derived from different parts of the oldest known zircon grain from the Jack Hills metasedimentary belt (Yilgarn Craton, Western Australia). These ages range between 4.4 and 4.3 Ga (Wilde *et al.* 2001). The oldest rocks known on Earth are 4.03 Ga granitic gneisses of the Acasta Gneiss Complex of the NW Territories of Canada (Stern & Bleeker 1998; Bowring & Williams 1999). Pillow lavas within the oldest known and unique supracrustal rock complex of the Isua greenstone belt (West Greenland) provide clear evidence for submarine magmatism at *c.* 3.8–3.7 Ga (Myers 2001). This, in conjunction with the formation of the granite–greenstone belts, suggests that Phanerozoic-like plate tectonic processes were possibly operating as early as 3.8 Ga (e.g. Nutmann *et al.* 2002).

A major global peak in continental crustal growth occurred at 2.7 Ga as demonstrated by the development of voluminous juvenile crust with thermal overprinting of pre-existing crust on every continent. Prior to 3.0 Ga continental crust may have been only *c.* 20% of the present volume covering *c.* 7% of Earth. By 2.5 Ga the Archaean continents had attained *c.* 80% of their modern volume and occupied *c.* 27% of the Earth (the present-day value is about 41%). Regional peaks in the growth rate of continental crust are recognized at 2.5 Ga (China and India), 2.0–2.1 Ga (West Africa and South America), and 1.8–1.9 (North America, Baltica and Australia) (Fig. 2.1). The contentious issue of how large amounts of continental crust were formed may be directly related to the presence of mantle plumes (superplume events), subduction processes and continent–continent collisions. Mantle plumes provide both under- and overplating (e.g. flood basalts), whereas horizontal plate movements related to subduction and collision may generate much of the volcanoplutonic material at active continental margins (Eriksson *et al.* 2004).

The ongoing development of Precambrian plate tectonics processes, together with the increase in the rate of formation of continental crust, led to several cycles of supercontinent coalescence and dispersal driven by large-scale mantle convection (Fig. 2.1). It would appear that each of these cycles, from

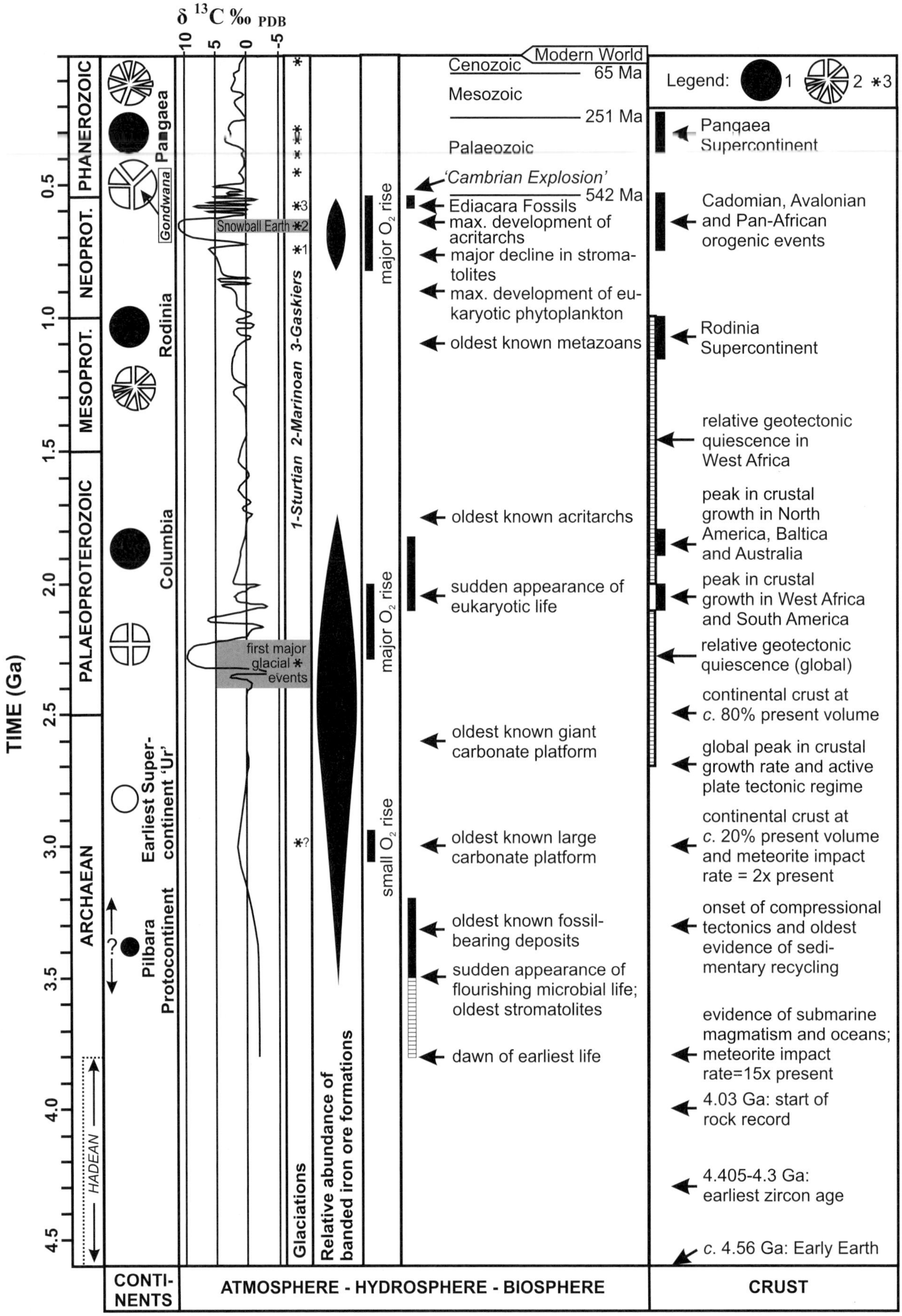

Fig. 2.1. Summary of geotectonic, isotopic, biospheric and environmental changes through time (modified after Lindsay & Brasier 2004; Eriksson *et al.* 2004). Key: 1, Supercontinent; 2, Continental dispersal; 3, Glaciations.

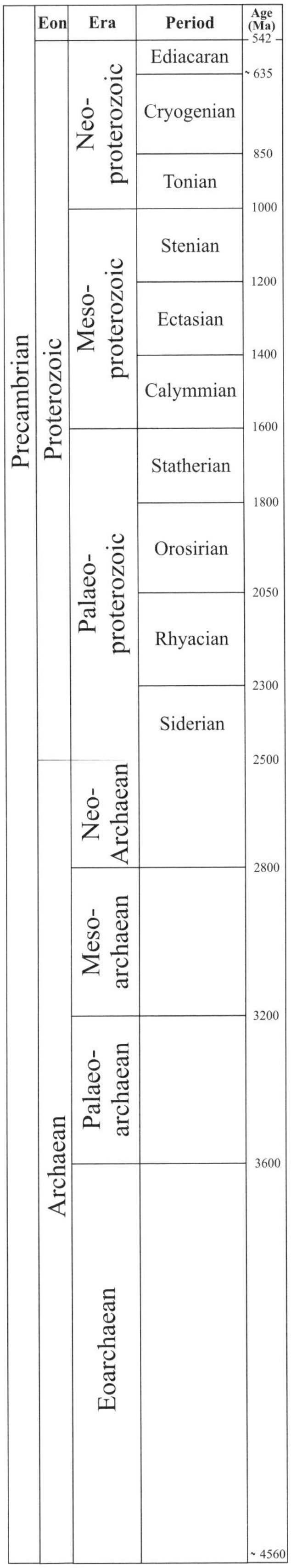

Fig. 2.2. Timescale of the Precambrian (from Gradstein *et al.* 2004).

origin to dispersal and the forming of a new supercontinent, is about 700–800 Ma in duration. The development and dispersal of one particular Neoproterozoic supercontinent (Rodinia, beginning at *c.* 1.0 Ga), while not as well documented as that of Pangaea, has been broadly outlined (e.g. McMenamin & McMenamin 1990; Dalziel 1991; Li *et al.* 1996). Rogers & Santosh (2002) have reconstructed a supercontinent, now called Columbia, which was assembled in the Palaeoproterozoic at *c.* 1.8 Ga.

The earliest continent-like assembly, around 2.5–3.0 Ga, called Ur (Fig. 2.1) formed from the integration of some of the oldest cratons in Africa, India, Australia and eastern Antarctica (Rogers & Santosh 2002). In the core of the Pilbara craton a well-preserved stratigraphic succession records the initiation of an early protocontinent beginning at some time before *c.* 3.5 Ga (Lindsay & Brasier 2004) (Fig. 2.1). Other published examples of possible continent assemblages include ones which formed at *c.* 2.5 Ga (*Arctica:* Siberian and Canadian cratons), *c.* 2.0 (*Atlantica:* West Africa, Congo, South America) and *c.* 1.8 Ga (*Nena:* Baltica, Greenland, Siberia, North America) (see compilation of Sankaran 2003 and references therein). The supercontinent of Gondwana, which is of great importance for the geological history of Central and western Europe, originated as a result of the dispersal of Rodinia and associated Late Neoproterozoic collisional events which culminated at *c.* 600–620 Ma in the Pan-African Orogeny. A short-lived supercontinent called Vendia or Pannotia (e.g. Dalziel 1997) was formed by the western part of Gondwana and Amazonia and Laurentia in late Neoproterozoic time. West Gondwana consisted of South America, Africa and Arabia, whereas Antarctica, Australia and India formed East Gondwana. The term 'peri-Gondwana' is used for Neoproterozoic magmatic arcs and related volcanosedimentary complexes which were accreted marginal to Gondwana and their Palaeozoic overstep sequence. Typical peri-Gondwanan fold belts were formed during the Avalonian and Cadomian orogenic events in Ediacaran and early Cambrian times (see Linnemann *et al.* 2008).

The global carbon isotopic curve is flat both for the NeoArchaean and on into the early Palaeoproterozoic (Fig. 2.1). The two major $\delta^{13}C_{carb}$ oscillatory excursions *c.* 2.2–2.3 Ga and *c.* 0.65 Ga are separated by essentially flat patterns. These $\delta^{13}C_{carb}$ patterns have been interpreted both in terms of changes in atmospheric composition and climate, and as a result of supercontinent assemblage. Major supercontinental assembly events at *c.* 2.8, 1.9 and 1.0 Ga were associated with mantle instability. Concomitantly, large intracratonic sag basins accumulated sediments over periods of 200–500 Ma, and these sediment piles included major carbonate platform successions. The large positive $\delta^{13}C_{carb}$ excursions at *c.* 2.3–1.9 Ga are inferred to reflect the significant burial of organic carbon which would imply that an increased flux of O_2 production was possibly responsible for the major O_2 rise at 2.0–2.3 Ga (Fig. 2.1) (e.g. Lindsay & Brasier 2004; Eriksson *et al.* 2004 and references therein).

Iron formations, especially the form characteristic to the early Precambrian known as banded iron formations, are unique among the sedimentary rocks in that there are no similar processes or products operating at the present time. Thus, many aspects concerning the origin of these banded iron deposits remain speculative. Iron formations are iron-rich sedimentary deposits occurring within PalaeoArchaean to Palaeoproterozoic and in Neoproterozoic rocks (Fig. 2.1). Most banded iron formations are older than *c.* 2.0 Ga. In most cases, younger deposits are represented by granular iron formations (Trendal & Blockley 2004). The oldest known banded iron formation occurs at Isua (West Greenland), and has an age of *c.* 3.8 Ga. In general, there would appear to be a major peak in the deposition of iron at *c.* 2.5 Ga (Fig. 2.1). After 1.7–1.8 Ga the deposition of iron formations ceased with a final Neoproterozoic scattering of predominantly small iron formations of various types (Trendal & Blockley 2004 and references therein).

From *c.* 3.8 Ga onwards iron formations were formed in a variety of basinal settings. It would appear that, at least in part, Neoproterozoic iron formations were linked to major glaciations. Hoffman *et al.* (1998) have developed the idea of Kirschvink (1992) concerning 'snowball Earth' (based on an earlier idea of Harland 1964), into an integrated hypothesis in which the development of iron formations is one possible geological outcome of rapid and extreme oscillations in global climate. During cold periods ice covered much of the land area, and the oceans were insulated from the atmosphere by the development of thick ice sheets. This situation resulted in a fall in the level of dissolved oxygen, leading to an increase in the amount of dissolved iron in seawater. Meanwhile, a rise in carbon dioxide levels in the atmosphere, associated with continuing volcanic activity, led to the onset of greenhouse warming and resulted in the rapid melting of both terrestrial and oceanic ice. The resultant melt-related deposition of glaciogene sediments at the start of the warm period was coeval with the precipitation of dissolved ferrous iron from the re-oxygenated oceans, leading to a global association of iron formations with glacial material. The snowball Earth hypothesis is still controversial, and at a stage of ongoing testing and discussions. The Snowball Earth event is linked to the Marinoan glaciation in Australia and the equivalent glacial deposits on the other continents (Fig. 2.1).

While there is little evidence of glaciation in Archaean times (Young *et al.* 1998) there is, in contrast, strong indication for an initial major (global?) glacial event near the beginning of the Proterozoic Eon (2.4–2.2 Ga), evidence of which is preserved in the Huronian stratigraphic succession (Lake Huron, North America), and in Finland, Western Australia and South Africa (Young 2004 and references therein). There have been several reports suggesting that glaciation in the Palaeoproterozoic, as in the Neoproterozoic, took place at low palaeolatitudes (Evans *et al.* 1997). Neoproterozoic glacial events, with at least two glacial-cap carbonate sequences, occurred in the interval of *c.* 760–670 Ma (Sturtian glacial events; Fanning & Link 2004), at *c.* 635 Ma (Marinoan glaciation; Condon *et al.* 2005; Hoffmann *et al.* 2004) and at *c.* 582 Ma (Gaskiers glaciation; Bowring *et al.* 2003).

The Precambrian record comprises *c.* 85% of the history of life and is dominated by prokaryotic (bacterial and cyanobacterial) microbes. By *c.* 3.5 Ga microbial life was flourishing and presumably widespread. There is evidence for sustained life at *c.* 3.9 Ga after the cessation of continual sterilizing meteor storms (Schopf 1999). The oldest known fossils are 11 kinds of cellular thread-like microbes petrified in the 3.465 Ma old Apex chert of northwestern Australia (Schopf 1993). These earliest fossils were filamentous prokaryotes. The oldest stromatolites, which are small and of limited lateral extent, are known from the *c.* 3.5–3.4 Ga old Warrawoona Group of the Pilbara craton. Unicellular eukaryotes appeared at *c.* 2.1–1.8 Ga and sexual eukaryotes at *c.* 1.1 Ga (Schopf 1999). Metazoan fossils appeared in the Mesoproterozoic, the oldest being preserved as trace fossils in *c.* 1.1 Ga old sandstones in India (Seilacher *et al.* 1998). Acritarchs (eukaryotes of unknown biological affinity) appeared at *c.* 1.75 Ga, and are the most widespread fossils in Meso- and Neoproterozoic rocks. Stabilization of siliciclastic sediments by microbial mats ('elephant skin') was apparently important in the Proterozoic, and rare examples are recognized from the Neoarchaean (Altermann 2002). At *c.* 1.1 Ga, a rapid diversification of eukaryotic phytoplankton occurred, reaching a maximum at *c.* 0.9 Ga. A major decline in stromatolite diversity and abundance at *c.* 0.8–0.7 Ga was related to a decrease in atmospheric CO_2 and an increase of O_2 (Holland 1984) with concomitant glaciation (Sturtian glacial events), followed by the Marinoan glaciation (snowball Earth) at about 635 Ma.

Following the short-lived Gaskiers glaciation (at *c.* 582 Ma), the oldest known Ediacaran fossils (Fig. 2.3) first appear within 4 Ma of deglaciation. The Ediacaran biota marks a critical stage in the evolution of life on Earth, the late Neoproterozoic appearance of large, soft-bodied organisms including the oldest definitve animals. Ediacaran fossils first appear significantly above the last Neoproterozoic glacial deposits (Gaskiers glaciation) and disappear near the level of the first Cambrian-style shelly fossils (Narbonne 1998; Narbonne & Gehling 2003). These temporal relationships have led some workers to hypothesize that the Neoproterozoic glaciations functioned as a bottleneck in eukaryote evolution and/or as an impetus for the subsequent radiation of animals (Hoffman *et al.* 1998; Runnegar 2000). Ediacaran fossils and the classical occurrence in the Ediacara Hills (Flinders Ranges, Australia) provide the name for the final Neoproterozoic period, the Ediacaran (Knoll *et al.* 2004).

At *c.* 560–551 Ma, the global carbon cycle underwent a major reorganization consistent with the progressive oxidation and remineralization of the organic reservoir. At about the same time the first complex ichnofossils (trace fossils) as well as the stem group mollusc, *Kimberella,* occurs (Martin *et al.* 2000; Knoll *et al.* 2004). Weakly calcified metazoans, such as *Cloudina* and *Namacalathus*, appear at *c.* 548 Ma and continue up to the Ediacaran/Cambrian boundary where they become extinct (Amthor *et al.* 2003). The spectacular first animal fossils, megascopic eggs, embryos and segmented skeletal tubes, preserved in the Doushantou Formation of southern China (Xiao & Knoll 2000), occurred at *c.* 599–551 Ma (Barfod *et al.* 2002).

U–Pb zircon dates on ash beds from Oman and Namibia constrain the top of the Ediacaran and the base Cambrian to be *c.* 542 ± 1 Ma (Amthor *et al.* 2003; Bowring & Schmitz 2003). The type locality for the Global Stratotype Section and Point

Fig. 2.3. Ediacara fossil (probably *Kullingia sp.*, large arrow) from the Ediacaran Period (Bradgate Formation, East Avalonia, Charnwood Forest near Leicester, UK). Small arrow indicates pale remnants of the volcanic ash-fall that buried the organism. Diameter of the coin is 2 cm.

(GSSP) for the top of the Ediacaran and the base of the Cambrian is situated at Fortune Head (southeastern Newfoundland). The onset of Cambrian strata is defined by the earliest occurrence of the trace fossil *Trichophycus pedum* (formerly *Phycodes pedum*).

Central Europe and its neighbouring crustal segments belong to (i) Baltica and (ii) peri-Gondwana (including Avalonia and Cadomia, Fig. 2.4). The latter is an association of crustal segments which originated at the northern margin of the Gondwana Supercontinent. Both of these main crustal provinces are today divided by the Tornquist–Teissere Line and the Thor–Tornquist Suture (Caledonian Deformation Front) (Fig. 2.4). In most reconstructions Baltica also formed part of the Gondwana Supercontinent in Neoproterozoic times (Unrug 1996; Torsvik 2003; Murphy *et al.* 2004; Linnemann *et al.* 2004). From 635 to 590 Ma there was a close linkage of Baltica to Amazonia, West Africa and Laurentia, with the former separating off at *c.* 590–540 Ma (Murphy *et al.* 2004; Nawrocki *et al.* 2007).

Avalonia rifted off from the mainland of Gondwana during late Cambrian and early Ordovician times (*c.* 500–490 Ma) while Cadomia remained at the Gondwanan margin (Linnemann *et al.* 2004, 2007). A consequence of the separation of Avalonia from Gondwana was the opening of the Rheic Ocean. After its drift, Avalonia collided first with Baltica at *c.* 440–435 Ma which led to the closure of the Tornquist Sea and the formation of the Thor–Tornquist Suture (see also Krawczyk *et al.* 2008). At

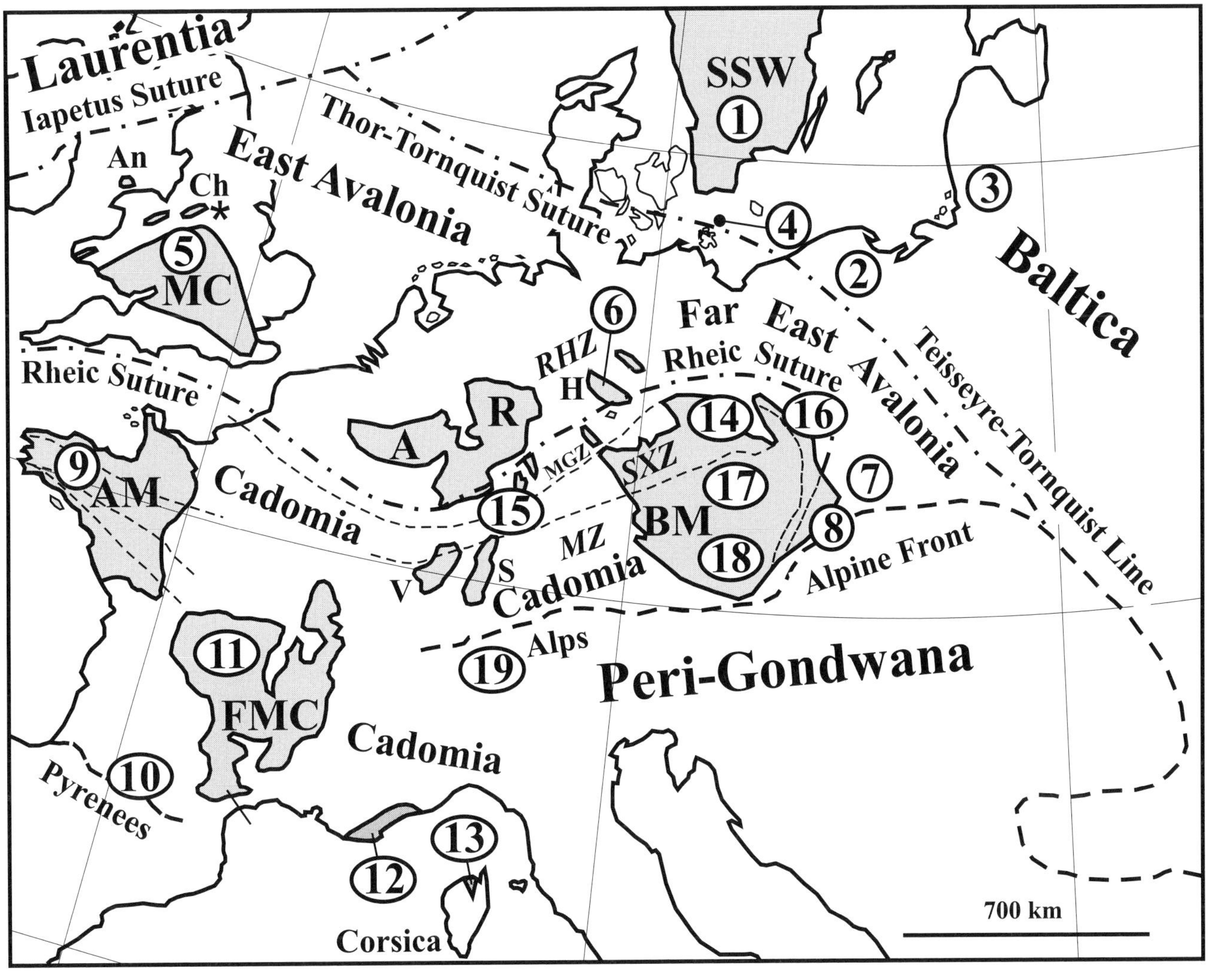

Fig. 2.4. Plate tectonic structure and distribution of Precambrian rock complexes described in this chapter. Grey shading represents areas dominated by Precambrian and Palaeozoic basement rocks outside of the Alpine orocline. Asterisks represent occurrence of Ediacara fossils. Abbreviations: A, Ardennes; AM, Armorican Massif; An, Anglesey; BM, Bohemian Massif; Ch, Charnwood Forest; FMC, French Massif Central; H, Harz Mountains; MC, Midlands Microcraton; MGZ, Mid-German Crystalline Zone; MZ, Moldanubian Zone; R, Rheinisches Schiefergebirge (Rhenish Massif); RHZ, Rhenohercynian Zone; S, Schwarzwald (Black Forest); SSW, southern Sweden; V, Vosges; SXZ, Saxothuringian Zone; Precambrian of Baltica (Fennoscandia): 1, Southern Sweden; 2, NE, Poland; 3, Lithuania; 4, the G14 borehole in the Baltic Sea. Precambrian of East Avalonia: 5, Southern Britain and Ireland. Precambrian of Far East Avalonia and related terranes: 6, The Eckergneiss Complex (Harz Mountains); 7, Upper Silesian and Małopolska Blocks (SE Poland); 8, Thaya/Slavkov Terranes and the Metabasite Belt (E Bohemian Massif). Precambrian of Cadomia and related terranes: 9, Armorican Massif; 10, Pyrénées; 11, French Massif Central; 12, Maures Massif; 13, Corsica; 14, Saxothuringian Zone; 15, Mid-German Crystalline High; 16, Sudetes; 17, Teplá-Barrandian Unit (Moldanubian Zone); 18, Moldanubicum *sensu stricto* (Moldanubian Zone); 19, basement of the Alps.

c. 415–400 Ma Laurentia and Baltica + Avalonia collided to form Laurussia. This was coincident with the closure of the Iapetus Ocean. Subsequently in Devono-Carboniferous times, as Pangaea was slowly assembled, the Rheic Ocean also closed. Therefore, Avalonia and Cadomia, in terms of the present-day geological situation, are very closely connected in Central and western Europe and are only divided by the Rheic suture as represented by the Mid-German Crystalline Zone (High) (Fig. 2.4).

In this chapter we represent an overview of the complex history of the various Precambrian rock units in Central Europe and adjoining areas including a summary of the most important data and a palaeogeography. The Cadomian Orogeny is the predominant Precambrian orogenic event in Central Europe and is described in Linnemann *et al.* 2008.

Palaeogeography (J.B.M. & S.A.P.)

Over the past 20 years, there has been an emerging consensus that the Late Palaeozoic supercontinent, Pangaea, is the youngest of several supercontinents that have formed, dispersed and subsequently reformed (Nance *et al.* 1986). Although there is general agreement that these processes have probably dominated the Earth's evolution for at least the last 2 billion years, there are still many uncertainties such as the configuration of the supercontinents, the timing of major events, and the paths of drifting continental blocks.

Neoproterozoic tectonics is dominated by the amalgamation of the supercontinent Rodinia by *c.* 1.0 Ga, its breakup beginning at *c.* 0.75 Ga, and the collision between East and West Gondwana between 0.6 and 0.5 Ga (e.g. Hoffman 1991; Wingate & Giddings 2000; Pisarevsky *et al.* 2003). Amalgamation is recorded by a number of continental collisions that occurred between 1.2 and 1.0 Ga, most notably the Grenville Belt of the eastern margin of North America (Laurentia) and its correlatives in NW Europe (Baltica), Mexico (Oaxaquia) and South America (Amazonia). Unlike Pangaea, Rodinia did not consist of 'all lands'. Recent palaeomagnetic data show that while much of Rodinia occupied a low latitude position, India was located near the North Pole (Pisarevsky *et al.* 2003).

The breakup of Rodinia, beginning at *c.* 0.75 Ga, is recorded by the abundance of passive margin sedimentary successions that surround Laurentia, implying that large continents rifted away and new oceans formed (Hoffman 1991). In the earliest stages, South China, Australia and Antarctica rifted from the western portion of Laurentia to form the Pacific Ocean (Li & Powell 2001) (Fig. 2.5). From *c.* 610 to 550 Ma, Baltica, and Amazonia and West Africa (together) rifted from eastern Laurentia to form the Iapetus Ocean (Cawood *et al.* 2001). In the late Neoproterozoic, however, a series of continental collisions, commonly referred to as 'Pan-African', amalgamated several continental blocks into a large continental land mass called Gondwana between *c.* 700 to 500 Ma. This event included the collision of the eastern portion of Africa with some of the continental blocks that previously rifted from western Laurentia to form the East African Orogen. In the earliest stages of these collisions, the western portion of Gondwana (Amazonia and West Africa) was still connected to Laurentia, resulting in a short-lived supercontinent variously known as Vendia or Pannotia (e.g. Dalziel 1997). However, it appears that not all of Gondwana had assembled prior to the formation of the Iapetus Ocean, so it is probable, as noted above, that the late Neoproterozoic supercontinent did not consist of 'all lands'.

Owing to the lack of reliable palaeomagnetic data, there is considerable disagreement on the location and configuration of the continents between 800 and 500 Ma (Figs 2.6–2.8). For example, the position of Laurentia (high latitude or low latitude) in the interval between 630 and 550 Ma is controversial, and resolution of this controversy has important implications for the understanding of Neoproterozoic tectonics. Two reliable, but mutually contradictory, palaeomagnetic poles from Laurentia exist for this time interval. One is from the 577 Ma Callander Complex (Symons & Chaisson 1991), the other from the 565 Ma Sept Iles intrusion (Tanczyk *et al.* 1987; Higgins & van Breeman 1998). Such data have led some authors to propose a significant component of true polar wander during this time period, in which case all palaeomagnetic data would be invalidated (Evans 2003). Additionally, as Amazonia and West Africa are palaeomagnetically unconstrained during this interval, their location in most tectonic syntheses is inferred from an interpreted connection with Laurentia. Such a connection, although widely accepted, has not been demonstrated and some problems exist with interpreting the geological connection.

Resolution of these controversies has fundamental implications for Laurentia–Gondwana connections. It is widely believed that Laurentia occupied a low-latitude position from *c.* 800 to 615 Ma, and was at low latitudes again by 550 Ma. Therefore, in the high latitude model, Laurentia–Amazonia–West Africa would have drifted rapidly south between 625 and 550 Ma. Such motion would provide a geodynamic explanation for the voluminous subduction-related magmatism along the leading edge of this plate (the so-called Avalonian–Cadomian Belt) at that time, in a manner analogous to the Mesozoic–Cenozoic westward drift of North America and South America and the voluminous subduction-related magmatism along their leading edges (i.e. eastern margin of the Pacific Ocean). The low latitude position of Laurentia by the Early Cambrian would require rapid northward drift of Laurentia as it separated from Amazonia–West Africa to form the Iapetus Ocean.

On the other hand, if Laurentia–Amazonia remained at low latitudes during this interval, a different explanation is required for Avalonian–Cadomian magmatism, such as the re-establishment of subduction zones along the Amazonian–West African margin following a terrane accretionary event at *c.* 650 Ma. Additionally, the origin of the Iapetus Ocean would be coeval with the southward drift of Amazonia–West Africa from Laurentia.

Clearly more palaeomagnetic data are required to document the movement histories of Laurentia, Amazonia and West Africa, and to test the connections between them. Although most reconstructions juxtapose Amazonia and the eastern margin of Laurentia during this time period, this reconstruction has several problems. According to Forsythe *et al.* (1993), the Grenville-age Arequipa–Antofalla terrane of Peru and Chile moved independently in the Early Palaeozoic. Loewy *et al.* (2000) have shown that this terrane cannot be correlated with either the Grenville Belt of eastern Laurentia or the Grenville-age Sunsas orogen of Bolivia and may represent a microcontinent bordered by oceans. Such an interpretation is consistent with the presence of a Neoproterozoic orogen (the Maranón Belt and its continuation into the Tucavaca Belt; Ramos & Alemán 2000) and would call the Laurentia-Amazonia connection into question.

The position of Baltica with respect to Laurentia is also controversial. Most reconstructions juxtapose the Scandinavian margin of Baltica and some part of the Labradorian–Greenland margin of Laurentia (e.g. Winchester 1988; Hoffman 1991; Park 1992; Dalziel 1992, 1997; Starmer 1993; Weil *et al.* 1998; Grelling & Smith, 2000; Pisarevsky *et al.* 2003). However, Torsvik *et al.* (1996) and Hartz & Torsvik (2002) proposed a

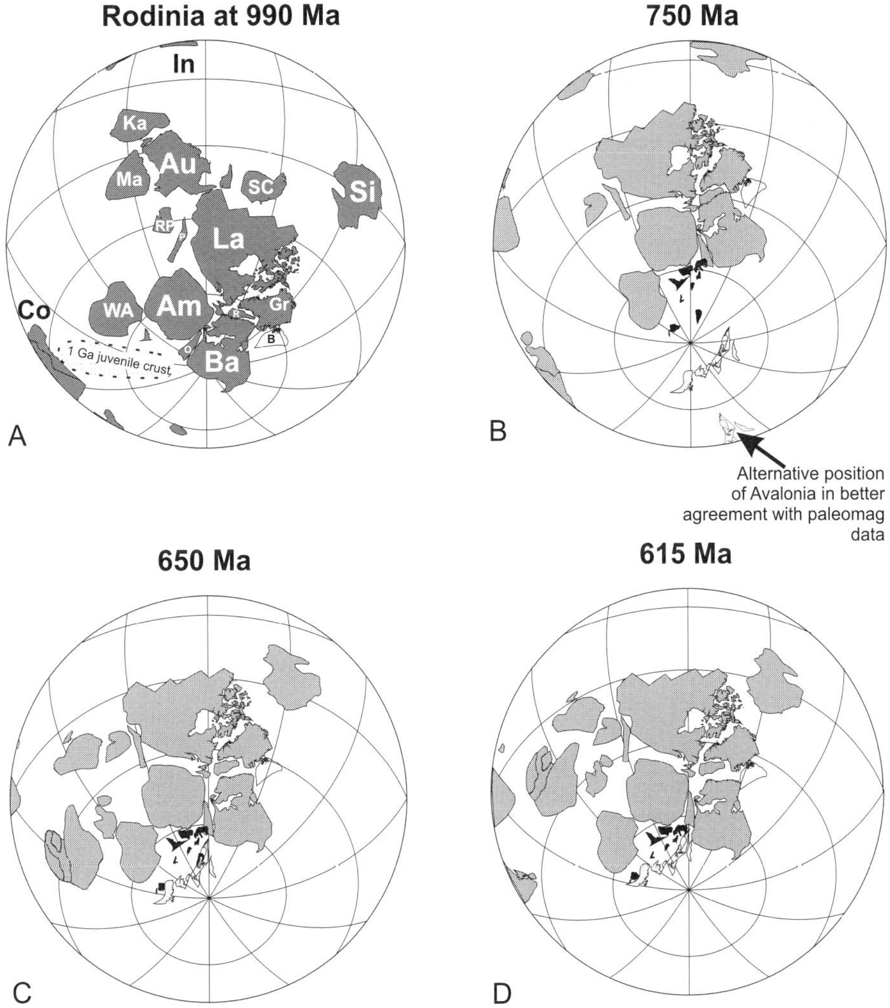

Fig. 2.5. (**A**) Rodinia at 990 Ma (modified after Pisarevsky *et al.* 2003); Am, Amazonia; B, Barentsia; Ba, Baltica; Ch, Chortis; Gr, Greenland; La, Laurentia; O, Oaxaquia; P, Pampean terrane; R, Rockall; RP, Rio de La Plata; WA, West Africa; In, India; Ka, Kalahari; Si, Siberia; (**B**, **C**, **D**) Reconstructions for 750–615 Ma of Laurentia–Gondwana–Baltica, emphasizing the history of the peri-Gondwanan terranes (modified from Murphy *et al.* 2004). This reconstruction assigns the minimum movement to the continents required to satisfy the palaeomagnetic data. Euler rotation parameters are given in Murphy *et al.* (2004). Shading: grey, cratons and continents; black, Cadomia and related terranes; light grey, Avalonia and related terranes.

'Baltica upside-down' model which juxtaposes the eastern (Uralian) margin of Baltica to East Greenland. This model is based on palaeomagnetic data from the Komagnes dyke in the southern part of the Varanger peninsula, two dykes in the Sredny peninsula (see references in Torsvik *et al.* 1996), and the subset of Laurentian palaeomagnetic poles corresponding to the high latitude model of Laurentia (see above). The 580 Ma assigned age for these Baltica dykes was based on the old K–Ar determinations and varied from 360 to 580 Ma. However, recently Guise & Roberts (2002) reported a high quality 378 ± 2 Ma $^{40}Ar/^{39}Ar$ plateau age for the Komagnes dyke, probably ruling out the possibility of the mentioned poles being Ediacaran or Cambrian. Popov *et al.* (2002) published the 555 Ma high-quality pole from the Winter Coast sediments, which supports the conventional fit between Baltica and Laurentia. More recent palaeomagnetic data from western Ukraine latest Neoproterozoic basalts (Nawrocki *et al.* 2004*a*) and sediments (Iosifidi *et al.* 2005) are in good agreement with the Winter Coast result. Early Cambrian palaeomagnetic data from the Brunovistulian terrane (Nawrocki *et al.* 2004*b*) also better fit geological data with the Apparent Polar Wander Path (APWP) for Baltica proposed by Popov *et al.* (2002), rather than with the APWP of Torsvik & Rehnström (2001). Popov *et al.* (2002) note the similarity of the Komagnes/Sredny poles with the Jurassic part of the European APWP, suggesting a Jurassic remagnetization for these dykes. Geological data such as the correlation of the Precambrian crustal blocks and orogens in Laurentia and Baltica (e.g. Winchester 1988; Hoffman 1989; Gorbatschev & Bogdanova 1993; Park *et al.* 1994; Karlstrom *et al.* 1999), and the lack of evidence for the Grenvillian Orogeny along the eastern margin of Baltica, support the conventional fit. Moreover, the presence of thick (up to 15 km) Uralian Riphean passive margin successions with a change to active margin conditions after Early Ediacaran times (*c.* 620 Ma) (e.g. Nikishin *et al.* 1996; Willner *et al.* 2001, 2003; Maslov & Isherskaya 2002; Puchkov 2003 and references therein) are inconsistent with the 'upside-down' model which implies intracontinental basin history until *c.* 550 Ma, followed by rifting and passive margin development until *c.* 500 Ma. Although the 'upside-down' model would predict a thick Cambrian passive margin succession in the Urals,

Fig. 2.6. Reconstruction incorporating the high latitude option for Laurentia between 600 and 550 Ma (modified from Murphy *et al.* 2004). Euler rotation parameters are given in Murphy *et al.* (2004); Shading: see caption of Figure 2.5.

such deposits are virtually absent (Maslov *et al.* 1997 and references therein). For these reasons our reconstructions (both high and low latitude) show Baltica in its traditional fit.

The reconstructions presented for the crucial time interval between 750 and 500 Ma permit an examination of the potential geodynamic linkages between the tectonothermal evolution of Laurentia, Gondwana and Baltica. There are several simplifying assumptions and some critical uncertainties. However, these serve to focus attention on potential geodynamic linkages between widespread tectonothermal events. The most critical uncertainty is the palaeolatitude of Laurentia and, by implication, Amazonia, at *c.* 570 Ma. The high latitude and low latitude options for Laurentia at this time allow for very different geodynamic interpretations of the evolution of peri-Gondwanan terranes and the opening of the Iapetus Ocean.

Baltica

The Baltica part of the East European Craton (EEC) consists of three major elements, namely the Fennoscandia, Sarmatia and Volga–Ural blocks, which are characterized by their different structure and evolution over a period extending from Archaean to Palaeoproterozoic (Bogdanova 1999; Bogdanova *et al.* 2001). Palaeomagnetic data reveal that Fennoscandia and Sarmatia were separated up until *c.* 1.85 Ga, i.e. until the time of their amalgamation into the Baltica palaeocontinent (Elming *et al.* 1993). It is assumed that Fennoscandia comprises a mosaic of several Archaean-age blocks surrounded by accreted crust, generally younging westwards, i.e. in the direction of the Trans-European Suture Zone (TESZ) (Gaal & Gorbatschev 1987). A recent tectonic model of Gorbatschev & Bogdanova (1993; see also Bogdanova *et al.* 1994) suggests the presence of Svecofennian terranes extending across the Baltic and into the Black Sea area. All parts of Baltica in Central Europe and adjoining areas belong to Fennoscandia (southern Sweden, NE Poland, Lithuania, parts of the basement in the Baltic Sea).

Southwestern Fennoscandia (southern Sweden) (R.L.R.)

The Fennoscandian part of the Baltica Shield represents the largest area of exposed Precambrian rocks in Europe. Baltica shares most of its geological evolution with Laurentia and came into contact with the terranes that make up Central Europe during the Caledonian and Variscan orogenies. Fennoscandia is deeply eroded and was, for extended periods, a source area for detritus, some of which was shed to the areas covered by the Baltic Sea and the East European Platform (Winterhalter *et al.* 1981) and – subsequent to the docking of Baltica with Avalonia and the Variscan terranes – also to areas to the south of the Caledonian Deformation Front. The southern margin of Baltica was reworked during Late Variscan wrenching and the Late Carboniferous to Early Permian development of the Sorgenfrei Tornquist Zone (STZ) and the Teisseyre Tornquist Zone (TTZ). Those parts of the Baltic Shield to the south of the TTZ subsided and formed the basement of the subsequent Permian, Mesozoic and Caenozoic basins (cf. Berthelsen 1992). The age pattern of detrital micas and zircons in clastic sediments of Variscan Europe thus provides important constraints on the source region of the sediments and the Caledonian and Variscan docking history of Baltica with the Gondwana-derived terranes of Central Europe (e.g. Belka *et al.* 2000).

Based on structural and lithologic arguments, the Precambrian rocks of Fennoscandia have been grouped into a relative age

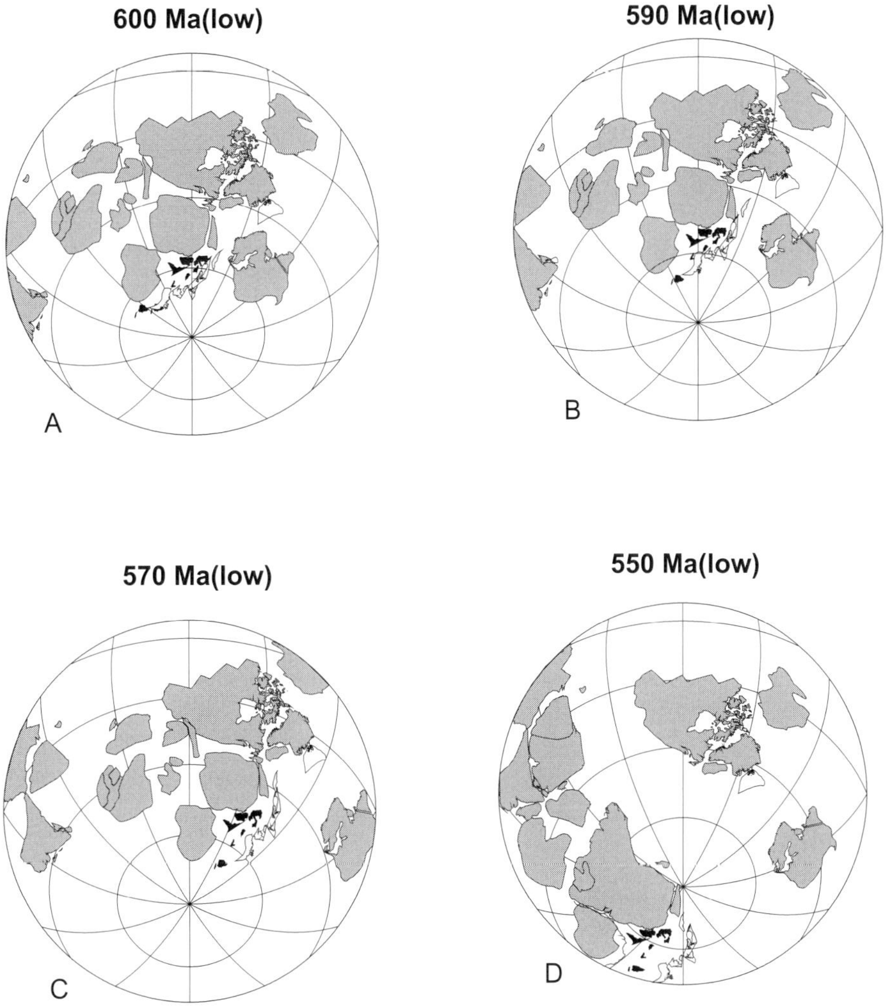

Fig. 2.7. Reconstruction incorporating the low latitude option for Laurentia between 600 and 550 Ma Euler rotation parameters are given in Murphy *et al.* (2004); Shading: see caption of Figure 2.5.

sequence, with the presumably oldest rocks in the NE (Fig. 2.9). The first of these subdivisions of Fennoscandia were proposed by Ramsay (1909) and Högbom (1913). They used regional names for the various deformation events (e.g. Karelian, Svecofennian, Gothian, Sveconorwegian). In a similar way, sediments deposited during these events received partly regional and partly event names. However, it later became apparent that in some regions the same event may have been given different names. Thus the use of a nomenclature that was based on events, but was also a time indicator and geographical, eventually led to changes in the use of terms such as Svecofennian, Svecokarelian and Karelian (see discussion in Gaal & Gorbatschev 1987). Later, some groups started to use Svecofennian to describe the distribution and the early arc evolution, and Svecokarelian for the deformation and metamorphism of rocks in the same region and time range. Although early subdivisions of the Precambrian of Fennoscandia relied on the then widespread belief that more deformed and higher grade metamorphic rocks are older than less deformed and less metamorphosed rocks, these original assumptions were broadly confirmed by isotopic ages (exceptions include the high-grade metamorphic regions of west and south Norway and SW Sweden) that became increasingly available after the 1960s. Early geochronological work on Precambrian rocks of Scandinavia relied mainly on K–Ar whole-rock and regionally sampled Rb–Sr whole-rock dating (Magnusson 1960; Welin 1980), whereas geochronological work on rocks from Finland and Canada predominantly employed the U–Pb method. The major result of early geochronological work was the belief that the geological evolution in old orogens lasted for several hundred million years and was not necessarily synchronous on Baltica and Laurentia.

After the 1970s, several new data sets and the interpretation of the Baltic Shield in a plate-tectonic context drastically changed the understanding of its geological evolution.

(1) Low-altitude aeromagnetic data became available from large parts of the shield (Fig. 2.10), allowing for (i) regional lithologic correlation and (ii) structural analysis (e.g. Henkel & Eriksson 1987). The aeromagnetic data, in particular, allowed mapping of the distribution of granites, gabbros and mafic dyke swarms, the latter related to brittle deformation events affecting the interior of the shield. These dyke swarms, intruded in extensional stress regimes, are excellent time markers and allow us to correlate the tectonic evolution of Fennoscandia with Laurentia (e.g. Romer 1996; Åhäll & Connelly 1998; Söderlund *et al.* 2005). Sedimentary basins may show banded magnetic patterns due to the occurrence of intermediate and mafic

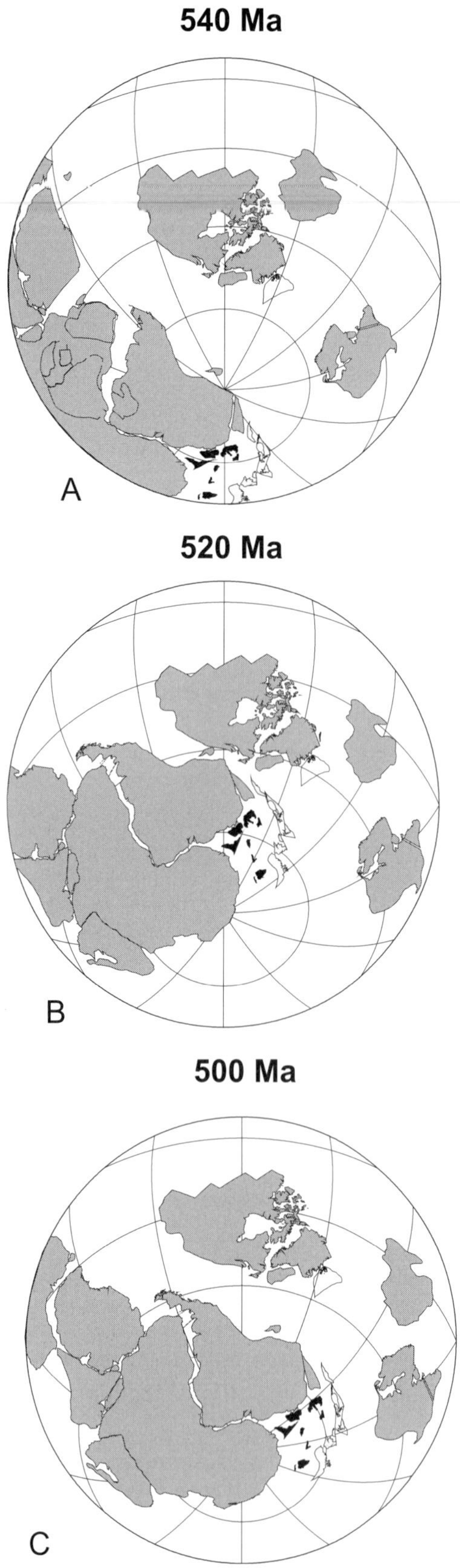

Fig. 2.8. Reconstructions for 540–500 Ma of Laurentia–Gondwana–Baltica, emphasizing the history of the peri-Gondwanan terranes. This reconstruction assigns the minimum movement to the continents required to satisfy the palaeomagnetic data. Euler rotation parameters are given in Murphy *et al.* (2004). Shading: see caption of Figure 2.5.

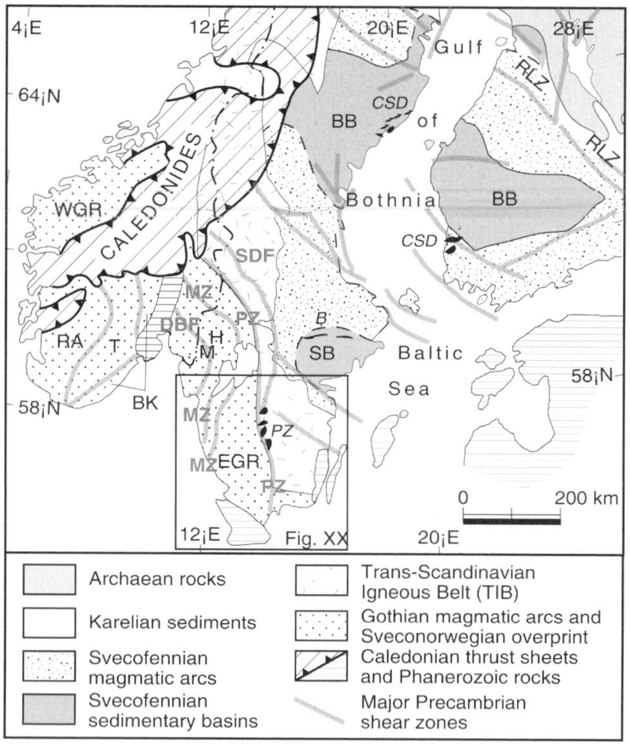

Fig. 2.9. Distribution of the major Precambrian tectonic provinces on Fennoscandia. Modified from Gaal & Gorbatschev (1987) and Mellqvist *et al.* (1999). Note that the original separation of tectonic provinces was based on structural and cross-cutting relations, and received regional names that later were used for orogenic cycles and specific time ranges. Because of this multiple use, the use of several names has changed through time (cf. Gaal & Gorbatschev 1987). Tectonic zones (black labels): ÅH, Åmål-Horred; BB, Bothnian Basin; BK, Bamble-Kongsberg; EGR, Eastern Gneiss Region; ÖM, Östfold-Marstrand; SB, Sörmland Basin; RA, Rogaland-Agder; T, Telemark; WGR, Western Gneiss Region. Major deformation zones (grey labels): DBF, Dalsland Boundary Fault; MZ, Mylonite Zone; PZ, Protogine Zone; RLZ, Raahe-Ladoga Zone; SDF, Sveconorwegian Deformation Front. Mid-Proterozoic major mafic intrusions (black italic labels): B, Breven Dolerite; CSD, Central Scandinavian Dolerites; PZD, Protogine Zone Dolerites.

volcanic rocks, which have a stronger magnetization than pelitic and arenitic rocks of the basin. The pattern of banded magnetic anomalies may reveal the deformation style of the sedimentary basin and the fading of the magnetic anomalies may indirectly reflect migmatization and metamorphic grade. Truncating relations among different magnetic patterns permits the reconstruction of a relative chronology (e.g. Henkel & Eriksson 1987; Romer & Nisca 1995).

(2) Precise and accurate U–Pb ages from Precambrian magmatic rocks of Sweden, which became increasingly available after the 1990s, corrected several misconceptions originating from early age determinations. For instance, (i) processes previously thought to have lasted several hundred million years now appear to have been completed in short periods of 30–70 Ma, i.e. they have durations similar to those operating in modern orogens. (ii) Distorted timescales originating from imprecise and, in part, incorrect (rotated mixing lines) age determinations were replaced by highly resolved time–process relations, that allowed the comparison and correlation of magmatic and tectonic processes on a large scale. (iii) U–Pb age data

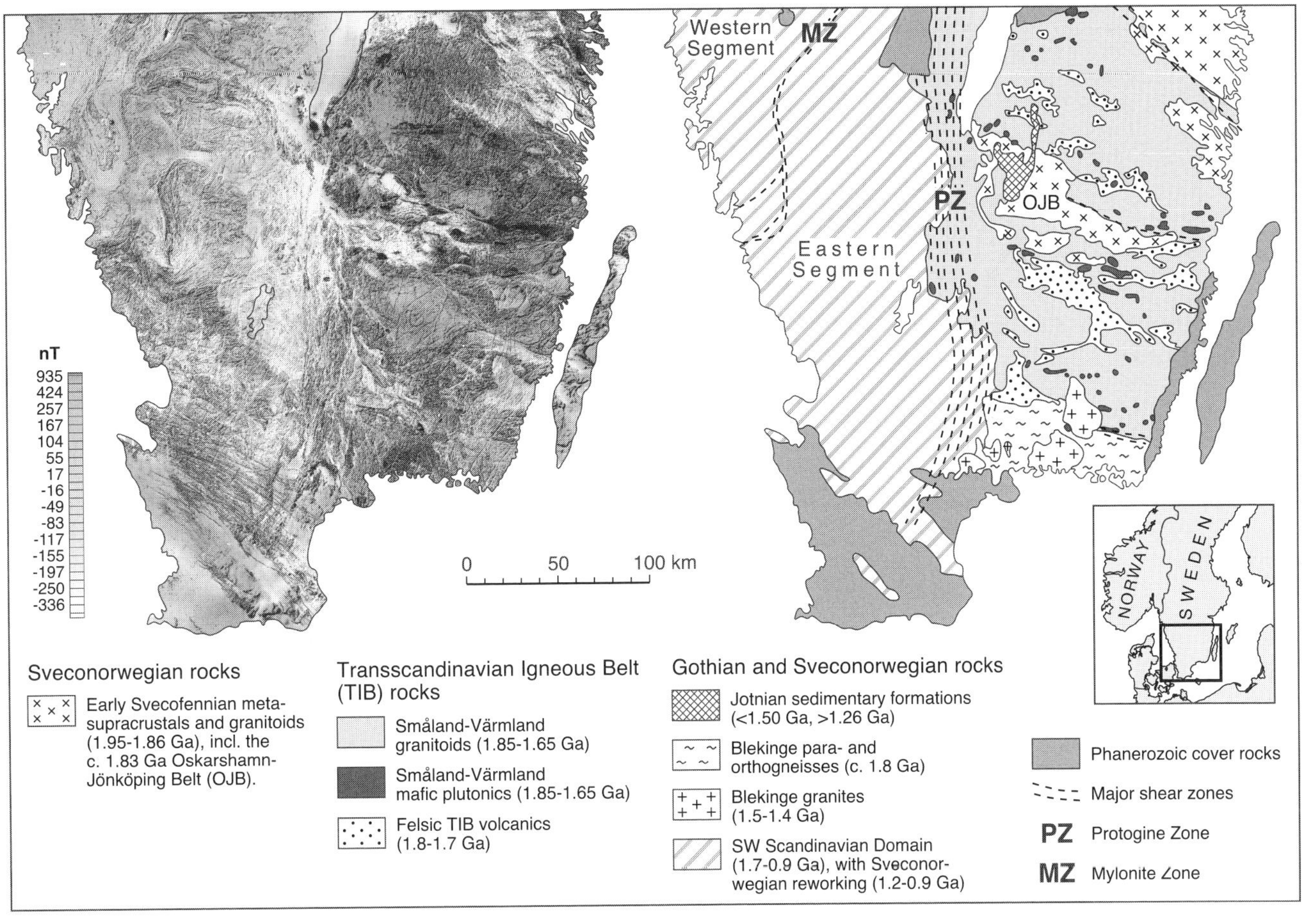

Fig. 2.10. Left: low-altitude aeromagnetic map of southern Sweden (from the SGU's aeromagnetic database. © Swedish Geological Survey, Uppsala). Data were collected typically 30 m above ground with a line spacing of 200 m and a sampling distance of 16 m along the lines. Right: simplified interpretation map showing the major tectonic units and elements. Modified from Lundqvist *et al.* (1994). Note that the magnetic pattern allows the continuation of the Precambrian basement under the Phanerozoic cover to be traced.

show no age differences between extension-related magmatic rocks in Sweden and those in Finland and Laurentia. Instead, different extensional events show close temporal coincidence (cf. Romer 1996; Åhäll & Connelly 1998; Söderlund *et al.* 2005)

(3) The importance of dating old structural elements to constrain the assemblage of the Baltic Shield was recognized and resulted in the dating of the most important of these structures (e.g. Lindroos *et al.* 1996; Högdahl & Sjöström 2001; Högdahl *et al.* 2001). The recognition that these old lineaments are associated with the seismically most active zones on the Baltic Shield (e.g. Talbot & Slunga 1989) and that major offsets in old erosional surfaces (Jotnian peneplain, Cambrian peneplain, Jurassic land surface) are spatially associated with them (cf. morphologic models of Rudberg 1970; Lidmar-Bergström 1988) demonstrated that these old structural elements were not only repeatedly reactivated during stress build-up at the margins of the craton, but also influenced the positioning of younger structural elements and the distribution of sedimentary basins on passive margins (e.g. Talbot & Slunga 1989; Romer *et al.* 1994; Gabrielsen *et al.* 1999).

Growth of the Baltic Shield

The western and northern part of the Baltic Shield has grown over the course of several major orogenies, during which older crustal material became reworked and new material was added. The oldest rocks are found in the NE and east of the Fennoscandian Shield and include granodiorite–trondhjemite–tonalite terranes, some of which include material as old as 3.1 Ga, and granite–greenstone and high-grade gneiss terranes aged 2.7–2.8 Ga. Palaeoproterozoic rocks belonging to the Svecofennian Zone were accreted to this cratonized Archaean core along the Raahe–Ladoga Zone (Fig. 2.10), an up to 50 km wide, repeatedly reactivated dextral strike-slip fault (e.g. Gaal & Gorbatschev 1987). The Svecofennian crust includes the remains of magmatic arcs aged 1.92–1.84 Ga, with the most voluminous magmatic activity occurring from 1.90 to 1.88 Ga (Skiöld 1988; Skiöld *et al.* 1993; Vaasjoki *et al.* 2003), which are separated from each other by former sedimentary basins filled with extensive deposits of clastic sediments with locally important volcanic intercalations (e.g. Claesson 1987). The age of arc accretion is bracketed by the ages of the youngest subduction-related magmatic rocks and the oldest post-collisional rocks. Post-accretion crustal thickening gave rise to regionally widespread migmatization events and the formation of major strike-slip shear zones at the border between the magmatic arc and the sediment piles deposited in the basins. These shear zones acted as magma conduits for K-feldspar porphyric granites (Ehlers *et al.* 1993; Lindroos *et al.* 1996; Högdahl & Sjöström 2001). Post-accretion crustal shortening in the Svecofennian region was accomplished by (1) deformation of the Archaean Craton along shear zones and the lateral extrusion of terranes by dextral strike-slip movements along terrane boundaries, and (2) thickening of the Svecofennian region by thrusting and terrane-internal deformation, especially in conjunction with regional migmatization and emplacement of voluminous granites (cf. Romer *et al.* 1992; Ehlers *et al.* 1993; Högdahl & Sjöström 2001).

To the west and SW, the Svecofennian domain is bordered by the Trans-Scandinavian Igneous Belt (TIB). The occurrence of the oldest TIB rocks (1.86–1.83 Ga; Persson & Wikström 1993; Wikström & Andersson 2004) temporally coincides with the end of arc magmatism and the onset of major strike-slip movement in the Svecofennian region. Some of the TIB-type granites seem to be concentrated along these major crustal-scale shear zones (Högdahl & Sjöström 2001; Wikström & Andersson 2004). The onset of TIB magmatism reflects the shift of crust-consuming processes from the interior to the western margin of the Svecofennian region. Westward growth of the Baltic Shield includes TIB magmatism (1.81–1.65 Ga) and subsequent accretion of magmatic arcs (1.66–1.50 Ga) in southern Sweden and Norway (e.g. Åhäll *et al.* 2000; Andersen *et al.* 2004). Parts of the TIB, and the entire region of post-TIB accretion of magmatic arcs (in southern Sweden also referred to as Gothian Orogeny), were overprinted during the 1.1.0–0.95 Ga Sveconorwegian Orogeny that reactivated old terrane boundaries and overprinted the old structural grain of the various terranes.

The Svecofennian rocks

In southern Sweden, there are only two small areas of Svecofennian crust: (1) the southernmost part of the Sörmland Basin (Fig. 2.9) and (2) the Oskarshamn–Jönköping Belt (Fig. 2.10). The Sörmland Basin is bounded to the north by the 1.90–1.86 Ga volcanic arc and to the south it is truncated by the TIB along a major strike-slip zone. The metasedimentary basin fill is predominantly of argillaceous to greywacke composition, with subordinate carbonates (Lundqvist 1979). A minor portion of the detrital zircon grains in quartzites from the supracrustal rocks in southern Sweden yield Archaean U–Pb ion-probe ages and were possibly derived from a craton to the SW (Claesson *et al.* 1993), in line with the involvement of old crustal lead in sulphide deposits syngenetic with the supracrustal rocks (Romer & Wright 1993). These magmatic arcs and supracrustal rocks became variably deformed and metamorphosed at 1.84–1.82 Ga (Andersson *et al.* 2000) and at 1.80 to 1.77 Ga were intruded by post- and late-kinematic granitoids These granitoids include mainly true granites and can be divided into two major types: (i) large massifs of coarse-grained granite with large megacrysts of K-feldspar with the geochemical characteristics of 'within-plate' granitoids derived from deep crustal sources; and (ii) geochemically and mineralogically very heterogeneous small intrusions of migmatite-type, minimum-melt granites that generally contain a significant contribution of enclaves from local metasedimentary country-rocks (Öhlander & Zuber 1988).

The Oskarshamn–Jönköping Belt includes the remains of a magmatic arc that can be distinguished from all of the other Svecofennian magmatic arcs by its slightly younger age (1.84–1.82 Ga) of subduction-related calc-alkaline magmatism (Mansfeld 1996). The belt consists of volcanosedimentary rocks deposited along an active plate margin and related calc-alkaline intrusions. The entire belt is bordered by TIB rocks and is, in part, intruded by slightly younger TIB granites (Mansfeld 2004). Magmatic evolution in the Oskarshamn–Jönköping Belt was synchronous with the oldest TIB magmatism at the western margin of the Svecofennian domain, but differs from the TIB with regard to the chemical character of the magmatic rocks, which closely correspond to rocks from other Svecofennian arcs. The Oskarshamn–Jönköping Belt is considered to be a terrane that was accreted to the Svecofennian margin around 1.81 Ga and was subsequently intruded by TIB-1 rocks.

The Trans-Scandinavian Igneous Belt (TIB)

The Trans-Scandinavian Igneous Belt (= Småland-Värmland Granite-Porphyry Belt (SVB); Trans-Scandinavian Granite-Porphyry Belt (TGPB)) is a more than 1200 km long magmatic belt along the western margin of the Svecofennian domain, extending from SE Sweden to NW Norway. It is distinctive on aeromagnetic maps, because of its abundant mafic rocks and magnetite-bearing granites, and also on gravity maps (Eriksson & Henkel 1983). The TIB consists of large massifs of granitoid rocks and associated mafic rocks and commonly has a geochemical character ranging from I- to A-type alkali-calcic rocks. The granitoids are typically coarse to coarsely medium grained with K-feldspar megacrysts that may, or may not, show plagioclase mantles. The rocks show textural evidence of magma mingling and magma mixing between felsic and mantle-derived mafic melts. TIB magmatism has been divided into several age groups ranging from 1.81–1.76 Ga (TIB-1) to 1.71–1.69 Ga (TIB-2) and 1.67–1.65 Ga (TIB-3; Larson & Berglund 1992). Several granitoids with TIB characteristics and ages of around 1.86–1.84 Ga have been grouped as TIB-0 (e.g. Persson & Wikström 1993; Högdahl & Sjöström 2001; Andersson *et al.* 2004). TIB-1 rocks tend to occur farther to the east than TIB-2 rocks. The distinction between TIB-2 and TIB-3 is controversial (cf. Gorbatschev 2004).

The Sveconorwegian Province and its precursors

The Sveconorwegian Province consists of a series of terranes that are separated by crustal-scale shear zones (Fig. 2.9). The contrasting nature and age of protoliths, metamorphism and deformation in the various segments suggests that the shear zones separating the various terranes were repeatedly reactivated (e.g. Åhäll *et al.* 1995). As subsequent reactivations only approximately coincide with the former terrane boundaries, this repeated reactivation smoothed the originally present lithological contrasts. Therefore, the precise positions of some Sveconorwegian terrane boundaries are not well defined locally.

The intensely reworked protoliths of each belt consist of lithologically distinct assemblages (Åhäll *et al.* 1990, 1995; Park *et al.* 1991). The Eastern Gneiss Region is a polymetamorphic terrane, bound by the Mylonite Zone and the Protogine Zone (Fig. 2.9), dominated by 1650–1800 Ma old granodioritic to granitic gneisses, veined migmatites, and locally mafic intrusions that vary from gabbro to anorthosite (Larson *et al.* 1990). The coincidence in age range of the gneisses with the TIB rocks suggests that there is a lithologic correspondence across the Protogine Zone and the Sveconorwegian Deformation Front. This is in agreement with geochemical and geophysical data (e.g. Zuber & Öhlander 1990). The Protogine Zone represents a zone in which tectonic style changes from isolated shear zones in the east to more penetrative deformation to the west, where increasingly deeper levels are exposed (e.g. Wahlgren *et al.* 1994).

The oldest basement to the west of the Eastern Gneiss Region comprises calc-alkaline volcanic rocks, aged *c.* 1.64–1.61 Ga, in the Åmål-Horred Belt (Fig. 2.10; Åhäll *et al.* 1995), which together with the sediment-dominated Östfold-Marstrand Belt

(Fig. 2.10) formed consolidated continental crust by *c.* 1580 Ma (Åhäll *et al.* 1990, 1995). The Åmål Formation consists of volcanic and sedimentary rocks that were metamorphosed under greenschist and amphibolite facies conditions and were subsequently regionally deformed, albeit to differing degrees. The volcanic rocks are predominantly feldspar and quartz porphyric rhyolites and dacites which are intercalated with arkosic sandstones. The Horred Formation includes discontinuous sheets of mafic to felsic volcanic rocks that were metamorphosed under amphibolite facies conditions. Geochronological analysis work indicates that the Horred Formation may be older than the Åmål Formation (1660 versus 1615 Ma; Brewer *et al.* 1998). The Östfold-Marstrand Belt predominantly consists of migmatized metasedimentary rocks with locally occuring mafic volcanic rocks (pillows and dykes) that were deposited after 1.60 Ga (Brewer *et al.* 1998). The Åmål-Horred Belt and the Östfold-Marstrand Belt were subsequently joined together by 1.59 Ga old granitoids (Åhäll *et al.* 1995).

The Sveconorwegian Province is differentiated – analogous to studies in the Grenville Province – into one internal and two external zones (Falkum 1985; Gower *et al.* 1991). This subdivision is based on the contrasting structural styles, protolith ages, level of Grenville metamorphism, and the distribution of post-kinematic granitoids across the three zones. The internal zone includes allochthonous terranes (Bamble-Kongsberg, Telemark, and Rogaland-Agder belts, all in southern Norway) which are intruded by voluminous post-kinematic plutons of monzonite, quartz syenite, and granite compositions. The external zones consist of (1) allochthonous terranes which have been thrust over large distances onto the former continental margin, and (2) para-autochthonous terranes, i.e. areas that were reworked during the Sveconorwegian Orogeny and represent slices of the former continental margin that were thrust, in a piggy-back style, on the craton. The border between the allochthonous and para-autochthonous terranes corresponds to the Mylonite Zone. The allochthonous External Thrust Belt includes the Åmål-Horred Belt and the Östfold-Marstrand Belt. Both of these belts were affected by a regional Sveconorwegian deformation under greenschist to amphibolite facies conditions (Åhäll *et al.* 1995). Field relations demonstrate that locally penetrative Sveconorwegian deformation is older than the deposition of the *c.* 1050–1100 Ma Dal Group (Åhäll *et al.* 1995). The para-autochthonous External Thrust Belt, which includes the Eastern Gneiss Region, is bordered by the Sveconorwegian Deformation Front to the east and the Mylonite Zone to the west (Fig. 2.10). The southern part of the Eastern Gneiss Region would appear to have undergone a higher grade of metamorphism (eclogite facies) than the northern part. Since metamorphism is younger in the southern segment (cf. Johansson *et al.* 2001), it is likely that the high-grade areas had a Sveconorwegian tectonic and metamorphic evolution that was uncoupled from the northern part of the Eastern Gneiss Region.

Compressional and extensional events

The temporal variation of compressional and extensional tectonic settings in the Precambrian of southern Sweden is shown schematically on Figure 2.11. Compressional events in southern Sweden included the accretion of the Svecofennian magmatic arcs and the development of dextral strike-slip shear zones subsequent to the cessation of subduction across the Svecofennian domain and the onset of TIB magmatism along the western margin of the Svecofennian region. Early TIB magmatism seems to have been concentrated along the major strike-slip zones in the Svecofennian domain. The major phase of TIB-1 magmatism (1.81–1.76 Ga) was coeval with migmatization and metamorphism in the Svecofennian region and the subsequent orogenic collapse and emplacement of post-kinematic granites (e.g. Öhlander & Zuber 1988; Andersson *et al.* 2000). With regard to the possible position of the TIB along the margin of the Svecofennian region, two different interpretations focus either on the presence of compressional features, comparing the region to an Andino-type orogen, or on the extensional features, concentrating on the interaction of mantle and crustal-derived melts in

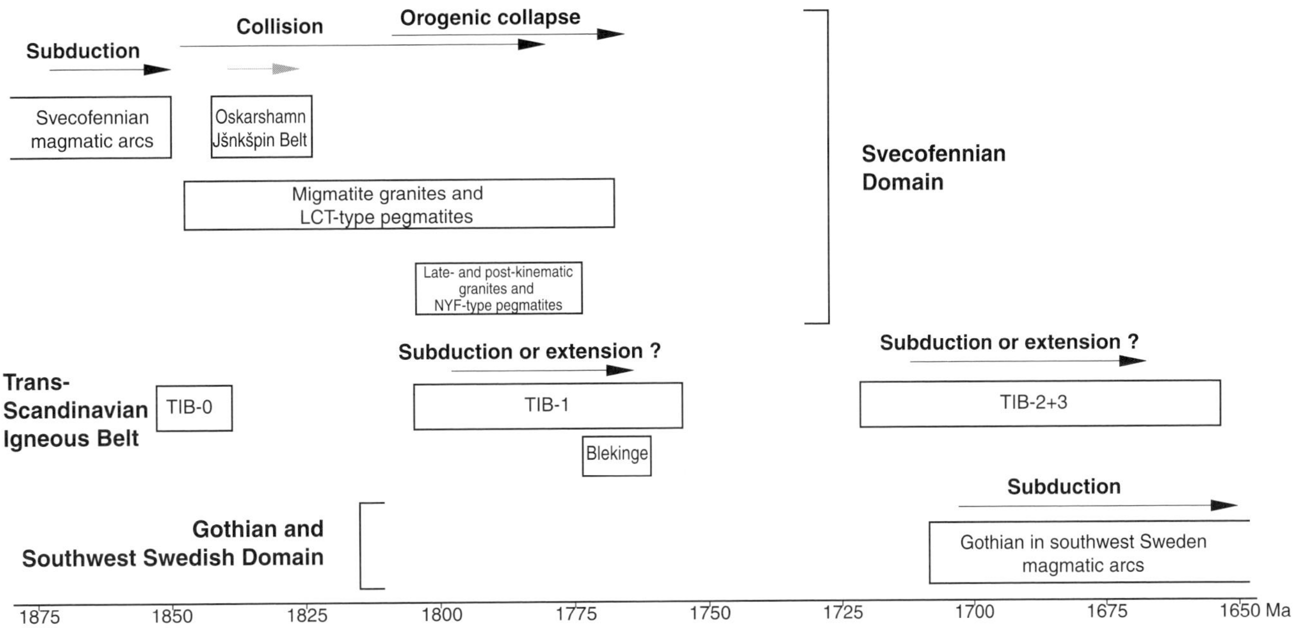

Fig. 2.11. Time-evolution diagram showing the duration of crust-forming processes in southern Sweden. Simplified from Andersson & Wikström (2004).

scenarios involving backarc crustal thinning or asthenospheric upwelling (cf. discussion in Gorbatschev 2004). The westward growth of the Baltic Shield included the accretion of the Gothian terranes (that were later involved in the Sveconorwegian Orogeny). Coeval to this stepwise growth was the onset of bimodal rapakivi magmatism and the emplacement of mafic dyke swarms in the Svecofennian domain (Åhäll *et al.* 2000; Karlstrom *et al.* 2001; Söderlund *et al.* 2005). Subsequent to the accretion of the Gothian terranes, the western Baltic Shield was deeply eroded and emplaced by several swarms of mafic dykes (cf. Suominen 1991; Romer 1996; Söderlund *et al.* 2005), possibly reflecting events at the craton margin (Åhäll *et al.* 2000; Karlstrom *et al.* 2001). Particularly prominent are the Breven-Hällefors dykes, the Värmland dolerites, the Central Scandinavian Dolerite Group, and the Protogine Zone dolerites (Figs 2.9 & 2.10) all of which appear to have been emplaced within relatively short time spans at 1595–1600 Ma, 1460–1462 Ma, 1272–1264 Ma and 1222–1215 Ma, respectively (data from Söderlund *et al.* 2005).

The Sveconorwegian Orogeny involved the collision of an unknown craton (presumably Amazonia) with the combined Laurentia and Baltica (cf. Hoffman 1991; Åhäll & Connelly 1998). Crustal thickening in southern Sweden seems to have occurred no later than 1040 Ma, which coincides with the age of the post-kinematic granitic pegmatites derived from extensive crustal melting (Romer & Smeds 1996) and the various metamorphic minerals (Johansson & Johansson 1993; Page *et al.* 1996*a*), although there is a tendency to younger ages in the external zones (Page *et al.* 1996*b*). Extension-related magmatism, subsequent to crustal thickening in the Sveconorwegian overprinted region, may have commenced as early as 960 to 950 Ma (Romer & Smeds 1996; Möller & Söderlund 1997; Söderlund *et al.* 2005) with rare-element pegmatites showing a distinct mantle input and abundant mafic dykes in the foreland of the Sveconorwegian Deformation Front (Söderlund *et al.* 2005). Extension-related magmatism became especially prominent with the emplacement of the 920–930 Ma old anorthosites in southern Norway (Schärer *et al.* 1996) and 920 Ma old granites in southern Sweden (e.g. Eliasson & Schöberg 1991) and slightly younger but voluminous granite magmatism in southern Norway. This extension is thought to reflect the collapse of the orogenically overthickened crust and the subsequent separation of the unknown craton from the combined Laurentia and Baltica (cf. Söderlund *et al.* 2005). Crustal thinning and reactivation of older crustal-scale deformation zones led to the development of graben systems, that became filled with sandstones (including arkoses) and which were also locally intruded by mafic dykes. The remains of these sediments are preserved in the Vättern Graben (Andréasson & Rodhe 1990) and in the lowermost Caledonian nappes (cf. Nystuen & Siedlecka 1988). Continued extension eventually resulted in the opening of the Iapetus Ocean at around 600 Ma (cf. Kamo *et al.* 1989) (see also Krawczyk *et al.* 2008).

Southern Fennoscandia I (NE Poland) (J.W. & E.K.)

The Precambrian crystalline basement of the Polish part of Baltica (also East European Craton, EEC) is concealed beneath a thick cover succession comprising unmetamorphosed Upper Precambrian–Carboniferous, Permo-Mesozoic and Cenozoic sedimentary and volcanogenic rocks of variable thickness, ranging from *c.* 0.5 km in the NE to 5–8 km along the SW margin of the EEC (Ryka 1993). Thus is it only possible to garner a true picture of the deep basement structures via geophysical data and from direct petrological studies of drill cores from the *c.* 250 deep boreholes, distributed unevenly across the area.

Systematic geophysical and geological studies of the Precambrian crystalline basement in NE Poland has been carried out since the 1950s (Karaczun 1953; Dąbrowski & Karaczun 1956, 1958). The early results were interpreted in terms of a mosaic composed of Katarchaic, Archaic and Svecofenno-Karelian structures, showing an approximately north–south trend, and paralleled by younger, Karelian and Gothian-age structures (Semenenko 1965). The observed platform structure was believed to have resulted from the superimposition of multiple tectonic cycles, the number and timing of which was the subject of much discussion (Ryka 1984). The crystalline basement of NE Poland was originally considered to be the youngest accreted component of the EEC. The first geochronological Ar/Ar data on biotites yielded ages of *c.* 1600 Ma and 1382–1368 Ma from the Krynki and the Sokółka gneisses, respectively (Łaszkiewicz 1960). Of interest also is the fact that the region is typified by particularly intense magnetic anomalies (Figs 2.12 & 2.13).

A tripartite Precambrian lithostratigraphic scheme for the crystalline basement of the region was originally proposed by Znosko (1960). This comprised: (1) gneisses and migmatites (older Precambrian); (2) greenshists, biotite schists, phyllites and granites (younger Precambrian); and (3) Jotnian quartzites (the youngest Precambrian). This scheme was subsequently extended by Ryka (1961, 1964), who classified the basement rocks into:

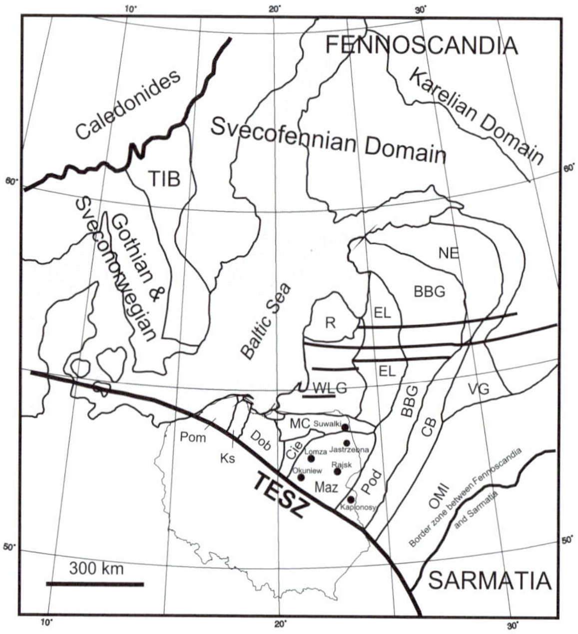

Fig. 2.12. Sketch map of the East European Craton, after Bogdanova *et al.* (1997) and Bogdanova *et al.* (1999) showing the principal structural units. Abbreviations: BBG, Baltic-Belarus Granulite Belt (BBG + Pod = PBG); CB, Central Belarus Belt; EL, East Lithuanian Belt; EN, North Estonian Belt; OMI, Osnitsk-Mikashevichi Igneous Belt; R, rapakivi-like granites and related rocks; TIB, Trans-Scandinavian Igneous Belt; VG, Vitebsk Granulite Domain; WLG, West Lithuanian Granulite Belt. Previous structural units, within the Polish section of the EEC (after Ryka 1998). Massifs: Dob, Dobrzyn. Maz, Mazowsze, Pom, Pomorze. Metamorphic complexes: Cie, Ciechanow; Ks, Kaszuby, MC, Mazury Complex with rapakivi-like granitoids and anorthosite massifs (Suwałki & Ketrzyn); Pod, Podlasie; TESZ, Trans-European Suture Zone. The locations of boreholes described in the text are also shown.

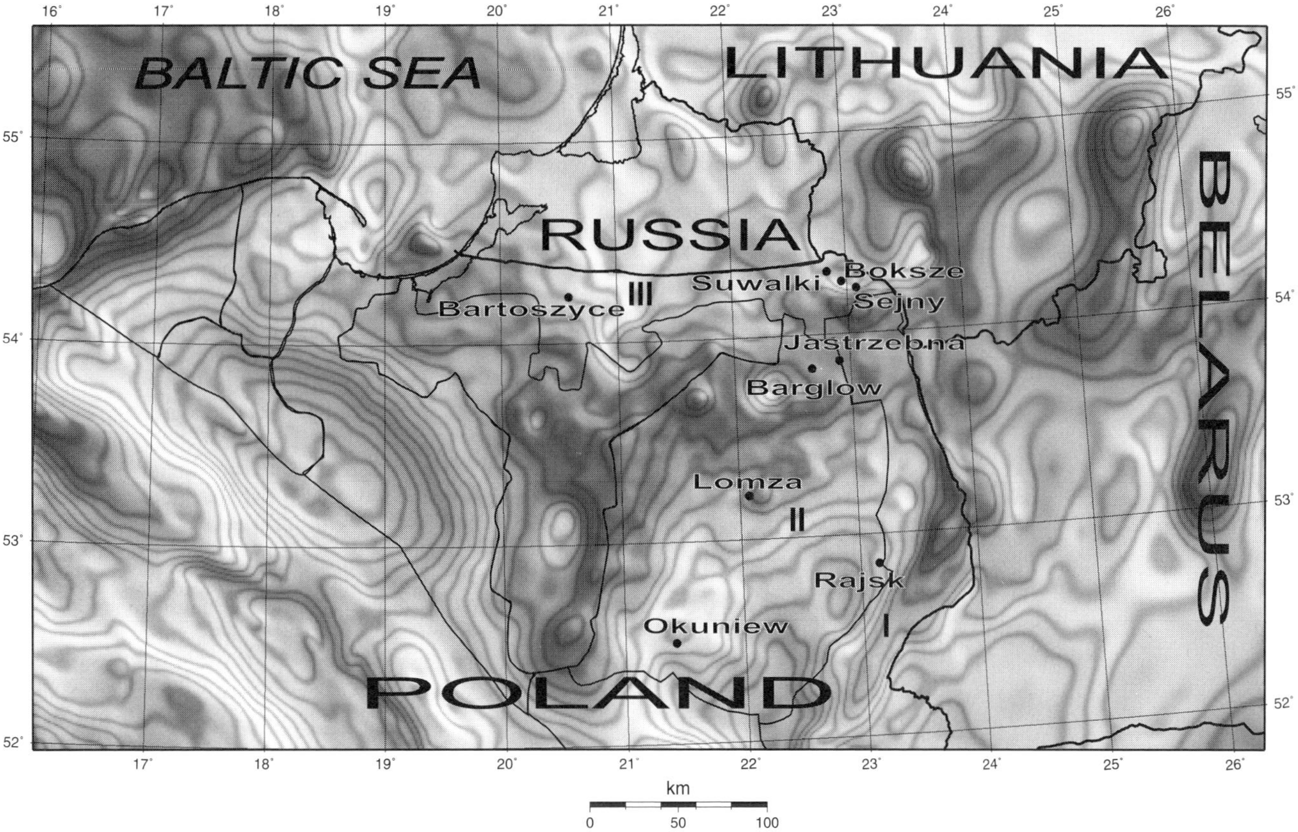

Fig. 2.13. Gravity map of Baltic region in grey scale (after Wybraniec 1999) showing the major tectonic features with higher fields and contrast. The locations of boreholes examined by SHRIMP U-Pb are also shown. The borders of structural units of Precambrian basement of NE Poland after Ryka (1998) are marked: I, part of the Palaeoproterozoic Podlasie-Belarus Granulite Belt (PBG); II, Mazowsze Massif; III, Mesoproterozoic Suwałki Anorthosite Massif and related Mazury granitoids. For other names see Figure 2.12.

(1) oldest Precambrian: Archaean, or lower Proterozoic (granulites, enderbites, hornfelses and pyroxenite gneisses); (2) lower and middle Proterozoic (gneisses, amphibolites, schists, phyllites); (3) Mezoproterozoic (anorthosites, norites, gabbrogranites); and (4) Neoproterozoic (quartzites, diabases, porphyries, syenites). The key element of this complex interpretation was the presence of a granitoid massif and the extensive belts of strongly, NE–SW orientated magnetic anomalies, which were interpreted to represent the trends of fold structures preserved in tectonic depressions (Pożaryski 1963; Ryka 1964). Additionally, the zones with weakly contrasting magnetic properties, occupying almost half of the crystalline basement subcrop area (and with little borehole control), were considered to represent dome-shaped granitoid massifs comparable to the central massifs in the Svecofenno–Karelian domain. These were considered to have undergone a tectonothermal overprint during the Svecofenno–Karelian orogenic stage (Ryka 1973); this event influenced the geometry and style of the structures developed at that time.

A more recent lithostratigraphic scheme (Kubicki & Ryka 1982) distinguished six main units in the Precambrian basement of NE Poland: (1) Pre-Karelian granitoid massifs, e.g. Mazowsze (or Mazovia), Dobrzyn and Pomorze (or Pomerania); (2) Pre-Karelian metamorphic fold zones, e.g. Podlasie Complex (including Lublin Zone), Ciechanow Complex (divided into Warmia and Zuławy parts) and Kaszuby Complex; (3) the Karelian metamorphic–magmatic Kampinos Complex; (4) the Gothian magmatic zone, represented by the Mazury rapakivi Granitoid Complex and by Suwałki and Ketrzyn anorthosite intrusions; (5) the Sub-Jotnic–Jotnic Biebrza complex; (6) the Palaeozoic Elk, Tajno and Pisz intrusions.

Due to the lack of geochronological data, the mutual relationship between the Pre-Karelian granitoid massifs and the metamorphic complexes remains hypothetical. The available biotite and amphibole K–Ar ages usually suggest middle and upper Proterozoic cooling (Kubicki *et al.* 1973). It has been generally assumed that the granitoid massifs represented either relatively young domes, emplaced into the earlier metamorphosed fold structures, or fragments of the dismembered continental crust incorporated into a younger fold belt (Ryka 1993).

A later modification of the Precambrian lithostratigraphy of the region (Ryka 1998) introduced the concept of Palaeoproterozoic metamorphic complexes (e.g. Podlasie, Ciechanow and Kaszuby) and granitoid massifs (e.g. Mazowsze, Dobrzyn and Pomorze), thus replacing the 'Pre-Karelian' fold belts and granitoid domains. Rapakivi-like granitoids and anorthosite massifs were assigned to the Mezoproterozoic, including the Mazury Igneous Complex and the Kampinos Metamorphic Complex. The Neoproterozoic consisted of the Biebrza Complex with a platform metamorphosed cover.

The evolution of the Precambrian basement was previously subdivided into four stages: (1) granulitization, (2) gneissification, (3) migmatization and (4) granitization. Several events, not always well documented, have also been described (Ryka 1984; Kubicki & Ryka 1982). The collision of the protocontinents, following the Pre-Karelian stage, contributed to the folding of the granulite and gneissic rock complexes. The Karelian Orogeny commenced with the deposition of the volcanosedimentary successions found in the Kampinos Complex. Gneisses and granulites were transformed into metasomatic granitoids and charnockites. Thick mylonite zones recording multiple displacements, and rapakivi-like granite intrusions associated with anorthosite massifs (Suwałki, Ketrzyn) also developed. The period of post-Gothian erosion was followed by the deposition of a thin, presently greenschist-grade, platform cover (Ryka 1984, 1998). At the same time, magmatic activity led to the formation of a series of intrusions (e.g. Lomza, Tajno, Pisz) as well as lamprophyre bodies. Volcanic activity at the end of the Precambrian produced basalt flows that are associated with clastic units (Juskowiakowa 1971; Szczepanowski 1977).

Tectonometamorphic evolution

The Precambrian basement of NE Poland represents an integral part of Fennoscandia and is spatially and petrogenetically related to the Baltic–Belarus area. A major feature of this area is the occurrence of lens-shaped zones of high-grade metamorphic rocks yielding Palaeoproterozoic ages (Bibikova *et al.* 1995; Taran & Bogdanova 2001; Skridlaite & Motuza 2001). All of the reliably dated rocks, however, were formed during Palaeoproterozoic times and there is no evidence of the presence of an Archaean-age crust in the area between the Baltic and Ukrainian shields (Bogdanova *et al.* 1994; Krzeminska *et al.* 2006).

The Belarus-Podlasie Granulite Belt (BPG, Fig. 2.12), 100–200 km wide, extends in a SW–NE direction for a distance of over 600 km, from SE Poland to western Belarus, and represents a zone of magnetic and gravity highs. The belt consists of small lens-like granulite bodies, separated by steep faults and blastomylonitic zones (Taran 2001). These rocks were subjected to a variable tectonothermal evolution extending from 1.89 to 1.50 Ga (Taran & Bogdanova 2001). The granulites, with geochemical characteristics typical of matured island arc tholeites, are located only in the marginal parts of the BPG. Their protolith was emplaced due to subduction-related magmatism at 1.9–1.85 Ga. Granulite grade metamorphism occurred under temperatures of 750–760 °C (grt-bt thermometer) and pressures of 7.5–8.5 kbar, and was related to the emplacement of bimodal (gabbro, enderbites, charnockites) intrusions at *c.* 1.79 Ga (Bibikova *et al.* 1996).

The Precambrian basement of the central part of NE Poland (previously called the Mazowsze Massif) consists of calc-alkaline orthogneisses of granodioritic composition (Okuniew), calc-alkaline monzonites (Rajsk) and calc-alkaline orthoamphibolites (Lomza). Their geochemical features, particularly their depletion in Th–U, Nb, P and Ti relative to primitive mantle, and the fact that the data plot in the volcanic arc field on Y versus Nb, Y + Nb versus Rb, Hf–Rb–Ta and La–Y–Nb discrimination diagrams, are all suggestive of a volcanic arc palaeoenvironment. The results of new U–Pb zircon SHRIMP analysis suggest a phase of magmatic activity at 1.826–1.800 Ga (Wiszniewska *et al.* 2007; Krzeminska *et al.* 2005). Compilation of age determinations and new geochemical data provide evidence for the presence of a Late Palaeoproterozoic arc system (1.82–1.80 Ga) as well as Late Svecofennian crust on the Mazowsze Massif (Wiszniewska *et al.* 2005).

The TDM_{Nd} model ages from 1.99 to 2.25 Ga for the main tectonostructural units of NE Poland (including granite massifs and metamorphic complexes) are similar to those obtained for the Svecofennian massifs in Scandinavia and Lithuania (Claesson & Ryka 1999). From these results, it is clear that the crystalline basement of NE Poland forms part of an extensive region with essentially juvenile Palaeoproterozoic crust. Indeed, a continuation of the main crustal domains as imaged on the Eurobridge'95 reflection seismic profile (Eurobridge'95 Seismic Working Group 2001) in Lithuania could also be expected in NE Poland. These major structures include the East Lithuanian Block (EL), the West Lithuanian Granulite Block (WL) and, separating them, the Middle Lithuanian Suture Zone (MLSZ) (see below).

Palaeoproterozoic metamorphic complexes composed of granulite and amphibolite facies rocks define the structure of the crystalline basement in NE Poland. The main phase of deformation and metamorphism of the Polish part of the EEC can be ascribed to the Svecofennian Orogeny (Claesson *et al.* 2001). The orientation of the regional metamorphic foliation S_1 reported by Cymerman (2004) is almost perpendicular to the trend of regional structures and lithostratigraphic units. According to Cymerman (2004), the evolution of the Precambrian crust in the Polish part of the EEC was stimulated by two tectonic processes: (1) deformation related to subduction–collision events under a compressional to transpressional regime during the Svecofennian Orogeny (1.95–1.85 Ga) resulting in the development of ductile thrusts with a NW and northward polarity; (2) extensional deformation related to either the Gothian or the Danopolonian orogenies (Bogdanova 2005) due to post-collisional collapse of an overthickened crust. The collapse was associated with the emplacement of post-collisional/anorogenic (rapakivi-type) intrusions (AMC magmatism) in a within-plate setting in the Mazury Complex at *c.* 1.5 Ga (Lorenc & Wiszniewska 1999). Magmatic rocks related to the first phase were dated at 1.835 Ga (U–Pb, Barglow felsic gneiss), 1.826 Ga (U–Pb, Jastrzębna pegmatite, Rajsk granitoids), 1.802 Ga (U–Pb, Lomza mafic suite) and 1.8 Ga (U–Pb, Okuniew granitoids), whereas magmatic rocks related to the Gothian crustal evolution and orogenic events were intruded at 1.59 Ga (Re–Os, Suwałki anorthosite), 1.54–1.51 Ga (U–Pb, Sejny jotunite), 1.525 Ga (U–Pb, Krasnopol granites), 1.522 Ga (U–Pb, Bartoszyce Q-monzonites), 1.513 Ga (U–Pb, Boksze monzonites), 1.436 Ga (Ar/Ar cooling age Bartoszyce granites) and 1.425 Ga (Ar/Ar cooling age Krasnopol granites) (Compston *et al.* 1995; Bogdanova *et al.* 2001; Dörr *et al.* 2001; Skridlaite *et al.* 2003; Wiszniewska 2002; Krzeminska *et al.* 2005, 2006; Wiszniewska *et al.* 2007). Furthermore, there was a period of synrift volcanism at 0.551 Ga as demonstrated by the presence of continental flood basalts in the Lublin region (Compston *et al.* 1995).

Mesoproterozoic post-collisional magmatism: Mazury Complex

The Palaeoproterozoic domains of NE Poland were later affected by post-collisional, anorogenic magmatism of bimodal, silicic and basic compositions. The magmatic activity at 1.65–1.5 Ga (Sundblad *et al.* 1994) involved the emplacement of shallow rapakivi-like granite intrusions, such as the Mosty Complex in NW Belarus and the Kabelai Complex in SE Lithuania. A series of east–west trending shear zones controlled the occurrence of the various magmatic units comprising the Mazury Complex (Duchesne *et al.* 1998; Wiszniewska *et al.* 2002*a*). The complex represents a polyphase pluton composed of anorthosite-norite intrusions (e.g. Ketrzyn Massif, Suwałki Massif and Sejny intrusion; Wiszniewska *et al.* 2002*b*) and A-type granitoids (Skridlaite *et al.* 2003; Bagiński *et al.* 2001). Granites of similar age, but with I-type geochemistry, have been encountered in Lithuania (e.g. Kabeliai, Gardasiai) and in Belarus. The Kabeliai

Complex was dated at 1505 ± 11 Ma and is probably an eastward extension of the Mazury Complex. Geochronologically, the Mazury Complex correlates with the Ragunda (Sweden) rapakivi complex and several smaller complexes in central Sweden which are dated at between 1530 and 1470 Ma (Persson 1999), as well as the 1547–1530 Ma Salmi (Russia) rapakivi granite-anorthosite complex (Amelin *et al.* 1997). This latter complex also occurs in a peripheral position with respect to the Palaeoproterozoic (Svecofennian) metamorphic crust of the Fennoscandian Shield.

The Mazury Complex is a 200 km long and 40 km wide zone of magmatic rocks, comprising the Suwałki anorthosite–norite intrusion with Fe-Ti-V-bearing deposits and silicic and intermediate rocks of variable compositions, including leucogranites, quartz-monzonites, monzonites, granodiorites and monzodiorites (Bagiński *et al.* 2001). The Suwałki Anorthosite Massif (SAM) and associated rocks belong to the widespread, magmatic anorthosite–mangerite–charnockite–granite (AMCG) 'rapakivi' suite, related to Proterozoic deep crustal structures. The Re–Os chronometer allowed for direct dating of magmatic Fe-Ti-V ores and the Fe-Cu-Co-Ni sulphide mineralization dispersed in the Suwałki anorthosites and norites. The isochron ages of the main Suwałki Fe-Ti-V deposits (Krzemianka + Jezioro Okrągłe and Udryn) are 1559 ± 37 Ma and 1556 ± 94 Ma, respectively (Stein *et al.* 1999; Wiszniewska *et al.* 2002*b*). The obtained initial isotope $^{187}Os/^{188}Os$ ratios (1.16 ± 0.06 for Krzemianka and 0.87 ± 0.20 for Udryń) and ε_{Nd} values (−1.56 to −6.3) indicate a lower crustal source of the parental Suwałki anorthosite magma (Morgan *et al.* 2000). Rocks of gabbronorite composition, remelted at a depth interval corresponding to a pressure of 10–13 kbar and a temperature of *c.* 1300°C, are believed to be the most probable protholith for the anorthosites (Wiszniewska 2002).

The geochemical and petrological investigations of jotunites (hypersthene monzodiorites) from the Suwałki and Sejny massifs and deformed plagioclase and pyroxene phenocrysts in the medium-grained anorthosite matrix highlight the fact that the parental magma was channelled to the surface as a plagioclase crystal mush, lubricated by a minor amount of interstitial liquids. The diapiric emplacement of the plagioclase mush took place under a pressure of *c.* 3–5 kbar. Fe-Ti-V ore mineralization was concentrated in the marginal parts of the massif (Krzemianka, Jezioro Okrągłe, Jeleniewo ore fields), as a result of granulation and polygenization processes of anorthosites from the central part of the massif, as well as filter-pressing and squeezing out of the ore-mineral mush into tectonic cracks and faults (Wiszniewska 2002).

Petrographically, the studied granitoid rocks of the Mazury Complex represent two groups, namely monzodiorites and granodiorites from the Krasnopol and Boksze area in the southern part of the SAM, and quartz monzonites from the Gołdap area in the central part of the complex. In Bartoszyce, situated in the western part of the Kętrzyn anorthosite massif, coarse-grained, porphyritic quartz monzonites are present, and these are petrographically and geochemically similar to the Fennoscandian rapakivi granites, and commonly contain K-feldspar megacrysts (microcline and microcline microperthite), frequently mantled by plagioclase (Skridlaite *et al.* 2003). The dioritic-granodioritic rocks of the Krasnopol area and the monzodioritic-monzonitic rocks of the Boksze area are typically medium-grained, greyish and show some evidence of magmatic foliation (Baginski *et al.* 2001). This group of rocks generally does not display the textural features of the typical Finnish rapakivi granites. The geochemical characteristics of the above-mentioned rocks are typical of rapakivi granites, with low MgO, CaO, Al_2O_3 and Sr values, but high values of K_2O (4–7.5 wt%), Rb (64–192 ppm), Zr (412–1316 ppm), Hf, Th, U and rare earth elements (REE) (Rämö & Haapala 1996). The SiO_2 content in the monzodiorite-granodiorite suite ranges from 50.3 to 53.1 wt% for the Boksze rocks and from 51.6 to 61.1 wt% for those from the Krasnopol area, whereas the total alkalis for both localities range from 5 to 9 wt%. The quartz monzonite suite from the Gołdap and Bartoszyce boreholes contains 60 to 65 wt% of SiO_2 and 6 to 9 wt% $Na_2O + K_2O$. These two suites, including the basic samples, show the major element characteristics of the potassic (shoshonitic) series: $Na_2O + K_2O > 5$ and $K_2O/Na_2O > 1$. (Baginski *et al.* 2001; Dörr *et al.* 2001; Skridlaite *et al.* 2003). All the rocks of the Mazury Complex are subalkaline. In the agpaitic index versus 10 000*Ga/Al diagram, they generally plot in the A-type granite field and partly overlap the field of the typical Finnish rapakivi granites.

The U–Pb age determinations, performed on single zircons and on titanite fractions from the region yield ages of 1525 ± 4 Ma (Krasnopol), 1522 ± 2 Ma (Bartoszyce), 1512 ± 1.1 Ma (Boksze) as well as 1499 ± 4 Ma and 1502 ± 2 Ma for two other quartz monzonites (Claesson *et al.* 1995; Dörr *et al.* 2001). These dates imply a *c.* 25 Ma emplacement history for the felsic and intermediate rocks of the Mazury AMCG complex. Of interest is the fact that there is no apparent correlation between the ages and the geochemical and petrographic compositions of the granitoids. Apparently the diorite, granodiorite, quartz monzonite and monzonite plutons were intruded at different times and in various localities into the Palaeoproterozoic crust, forming the resultant 200 km long east–west trending belt. A comparison of recently obtained U–Pb data for the Mazury Complex with several occurrences of anorogenic magmatism in Fennoscandia linked the Mazury Complex to the Riga-Aland subprovince, the Salmi subprovince and also to the Swedish magmatic complexes in Ragunda and Nordingra (Dörr *et al.* 2001; Persson 1999).

Late Neoproterozoic volcanic activity

In the eastern part of Poland, Late Precambrian volcanic rocks are noted exclusively from boreholes within the area extending from Kruszyniany (near Białowieża) to Łopiennik. This area is defined as the Lublin–Podlasie (L-P) Basin, which developed at the crossing of the Orsha–Volhyn aulacogen (AOV), a supraregional tectonic structure, and the eastern part of the Trans-European Suture Zone (TESZ). These two mutually perpendicular structures formed on the borders of the megablocks of Sarmatia, Fennoscandia (Fig. 2.12) and probably Amazonia (Sadowski & Bettencourt 1996). The rifting zones of the AOV and the TESZ are distinguished in terms of their tectonic activity in the Mesoproterozoic and Neoproterozoic. It has been suggested that rifting occurred in both directions, leading to the creation of a triple-junction (Pożaryski & Kotański 1979; Poprawa & Pacześna 2000, 2002; Żelaźniewicz *et al.* 2001). The latest Neoproterozoic phase of tectonic reactivation is recorded by the presence of continental flood basalts (CFB) in the SE part of the AOV (Bakun-Czubarow *et al.* 2000). A similar volcanogenic succession occurs in the area of the western slope of the Ukrainian Shield (Ukraina and Moldova), where it is known as the Volhynian series. The Volhynian series has been lithologically correlated with the Slawatycze series in eastern Poland. The uppermost tuff horizon of the Slawatycze series from the Lublin region (eastern Poland) was dated at 551 ± 4 Ma using the U–Pb method (Compston *et al.* 1995).

The flood basalts of eastern Poland are underlain almost everywhere in the region by continental to shallow-marine siliciclastics of the Polesie Formation, which represent the Early to Middle Neoproterozoic synrift phase (Pacześna & Poprawa 2005). The upper part of the Neoproterozoic section is also represented by synrift volcanics. The lava, tuff, and agglomerate/breccia volanigenic successions were originally between 70 m and 350 m thick (the latter value representing the agglomerate/breccia). The tuff layers, which are predominant in the north, range in thickness up to 30 m, whereas the flood basalts, which are more common in the south are up to 300 m. The subalkaline volcanics include picrobasalts, basalts, trachybasalts and basaltic trachyandesites. An important feature of the rocks of the Slawatycze series is that they fall into two groups, namely low-Ti and high-Ti tholeiites. These groups are also geographically separated, with the high-Ti basalts (>1 wt% TiO_2) dominant in the northern part of the L-P Basin, and the low-Ti basalts (<1 wt% TiO_2) in the southern part of the basin.

In general, the lava floods are dominated by relatively evolved tholeiitic basalts but with minor amounts of picrobasalts (>12 wt% MgO and higher trace compatible Cr and Ni contents: 1998–2107 ppm, and 364–389 ppm, respectively) restricted to part of the southern region (i.e. Kaplonosy–Wisznice–Busówno area). The tholeiitic basalts exhibit a wide compositional range in terms of Ni (50–120 ppm), Sr (128–605 ppm), Ba (359–869 ppm) and Rb (14–63 ppm). Tholeiites are enriched in REE, total REE ranging from 70 to 190 ppm. The primitive mantle-normalized incompatible element profiles reveal that the rocks are enriched in K and Pb, but exhibit moderate depletion in Nb, Sr, P and Ti. The lavas show a strong negative Nb anomaly, which is characteristic for the continental crust and can be an indicator of the crust's involvement in magmatic processes. Tholeiites and picrites are typically enriched in LREE and LILE, but have relatively low concentrations of HFSE (e.g. Nb, Zr) with respect to LILE and LREE. The Slawatycze series basalts are enriched in incompatible elements in relation to N-MORB, resembling E-MORB and OIB. The diagnostic ratios (K/P > 7, La/Ta > 22 and La/Nb > 1.5), as well as the image of the Nb- and Pb-anomalies in all of the volcanics of the Slawatycze series, suggest that there was marked crustal contamination. Volhynian series basalts from the Ratno beds (Białowolska *et al.* 2002) show similar LREE enrichment and LREE/HREE fractionation as the rocks of the Slawatycze series. Flood basalts from eastern Poland display the typical geochemical characteristics of within-plate lavas and show similar incompatible element patterns as basalts from the CFB provinces of Keweenawan, Ferrar, Deccan and Siberia. The Late Neoproterozoic (Ediacaran) volcanic activity in the western part of the EEC was probably related to the formation of continental rifts during the latest stages of the breakup of the Rodinia supercontinent (Białowolska *et al.* 2002; Shumlyanskyy & Shumlyanskyy 2004).

Southern Fennoscandia II (Lithuania) (G.M.)

The Precambrian crystalline basement in the SW part of the East European Craton (former Baltica palaeocontinent) is entirely covered by Phanerozoic and, in places, also by Neoproterozoic unmetamorphosed sediments. The thickness of this cover is variable, ranging from a few hundred metres on the Belarus–Mazurian High to more than 3000 m in the central part of the Baltic Basin. The first direct data on the geology of the Precambrian crystalline basement were provided by wells drilled in the 1930s, with systematic drilling beginning in the 1960s. A total of *c.* 520 wells have been drilled into the crystalline basement in Lithuania with a number of wells also drilled in the areas of Belarus and Latvia.

Structure of the lithosphere

The seismic structure of the upper lithosphere in the Lithuanian part of Southern Fennoscandia has been revealed by deep seismic sounding data acquired along the Eurobridge, Polonaise P4, P5 and Kohtla-Jarve-Sovietsk profiles as well as from other geophysical offshore experiments on the Baltic Sea. P- and S-wave velocity models were compiled by the Eurobridge'95 Seismic Working Group (2001), the Polonaise Seismic Working group (Grad *et al.* 2003) and individual researchers (e.g. Giese 1998). Summarizing the available data, a number of structural and compositional blocks can be distinguished in the upper lithosphere along the Eurobridge DSS profile (Fig. 2.14), including (1) the Vestervik-Gotland (VG) block, (2) the West Lithuanian Granulite Domain (WLGD) block, (3) the East Lithuanian Domain (ELD) block, and (4) the Podlasie–Belarus Granulite Belt (PBGB) block.

The Moho depth and thickness of the crust (Fig. 2.15), as well as its internal structure and composition, vary significantly along the Eurobridge DSS profile. The thickness of the crust is 47–51 km in the westernmost VG block, 42–44 km in the WLGD block, and in the EL and PBGB blocks it ranges between 50 and 57 km (i.e. the zone of maximum variability). Deepening of the Moho towards the boundary between the WLGD and ELD blocks takes place over a distance of 35–40 km. The seismic velocities in the uppermost part of the mantle vary between 8.65 and 8.9 km/s and the corresponding density is 3374–3406 kg/m^3, increasing from west to east along the profile. In the central part of the WLGD, the seismic velocities (8.1–8.5 km/s) and the density of the mantle (3358 kg/m^3) are essentially lower.

The thickness of the lower crust in the VG and ELD blocks is up to 22–25 km, whereas in the WLGD block it is only 10 km. The density of the lower crust along the Eurobridge profile varies between 2900 and 3000 kg/m^3, with the exception of the WLGD block. The latter has a density of *c.* 2850 kg/m^3, i.e. below the average density of the lower continental crust (2900 kg/m^3), and this is presumably due to the presence of felsic or intermediate granulites along with mafic rocks or/and migmatites. The lower density also implies an absence of rocks rich in garnet and pyroxene.

The seismic middle crust is characterized by seismic velocity values varying between 6.5 and 6.75 km/s and the corresponding densities vary between 2820 and 2845 kg/m^3. In the WLGD block the density is lower, at 2750–2800 kg/m^3. Taking into account the increase in density with depth, such an average density would correspond with a predominantly felsic lithology, similar to that of the upper crust. This presumption implies that the geological middle crust is absent in the WLGD block, whereas the felsic upper crust is much thicker. Such a crustal structure model is indicated by the presence of a regional negative gravity anomaly located roughly in the area of the WLGD block, as well as being supported by the results of geophysical modelling.

The density of the felsic upper crust varies between 2650 kg/m^3 in the uppermost part up to 2750–2760 kg/m^3 in the lower part. The thickness of the upper crust in the WLGD block is up to 35 km, while in the adjacent blocks it varies between 13 and 20 km. Granitic and charnockitic intrusives, with relict felsic and intermediate supracrustal rocks, predominate in the upper crust of the WLGD block. In other blocks, particularly the ELD, migmatites are predominant. A seismic velocity model for the region indicates a low velocity layer in the upper crust, about 7.5 km thick at a depth of *c.* 10 km. This low velocity layer is situated mainly in the area of the WLGD block and in the

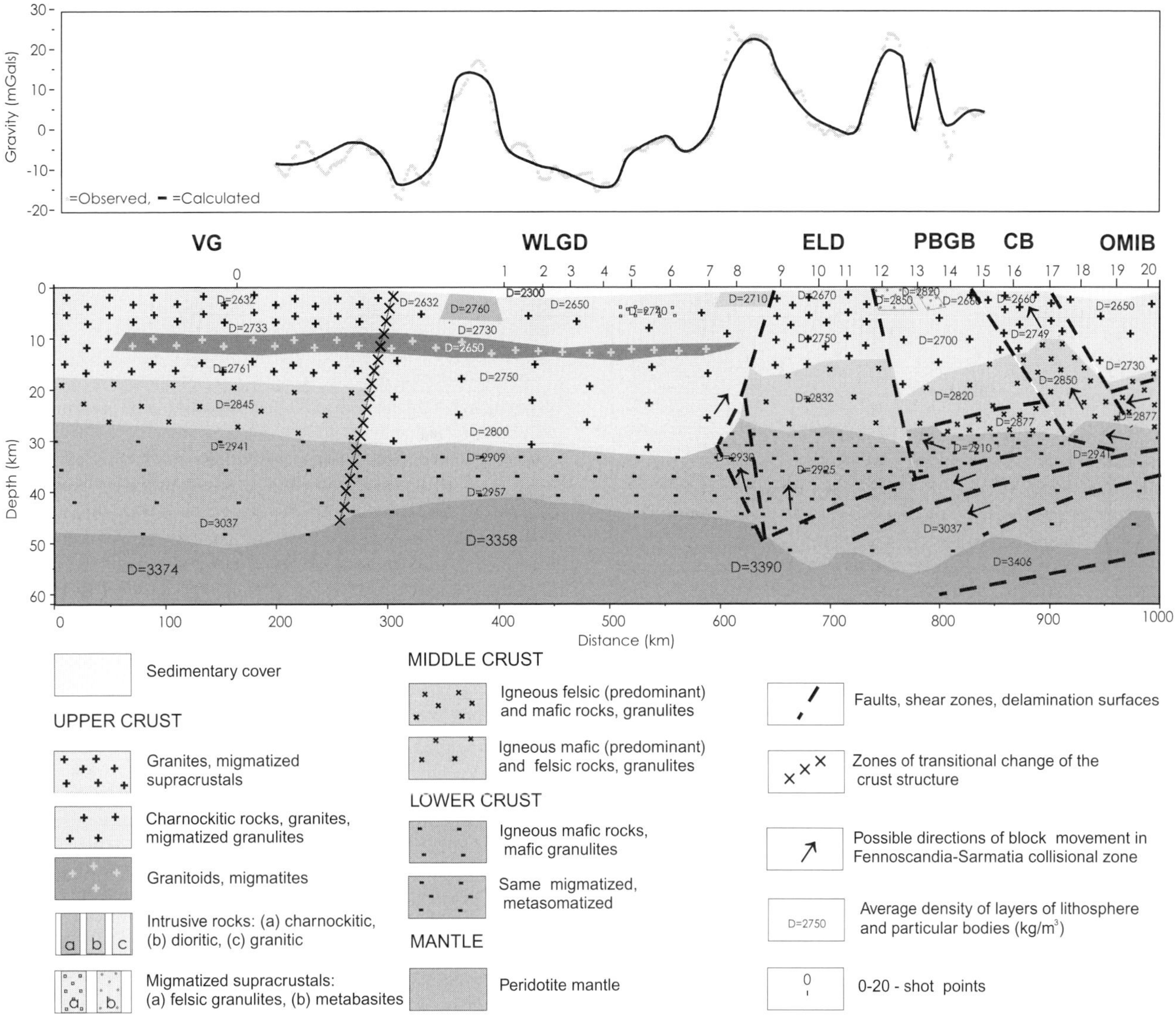

Fig. 2.14. Deep structure of the lithosphere of Lithuania and adjacent areas along the deep seismic sounding profile Eurobridge (Motuza *et al.* 2000). For position of the profile see Figure 2.15. Domains and blocks of crystalline crust: CB, Central Belarus; ELD, East Lithuanian Domain; OMIB, Osnitsk-Mikashevichi Igneous Belt; PBGB, Podliasie-Belarus Granulite Belt; VG, Vestervik-Gotland; WLGD, West-Lithuanian Granulite Domain.

eastern part of the VG block. The interpreted density of this layer (2650–2700 kg/m^3) is slightly lower when compared to the surrounding rocks and may indicate the presence of granite.

The Central Lithuanian Suture Zone (CLSZ) is considered to represent a boundary or transitional zone between the WLGD and ELD blocks. In this zone, there is marked thickening of the crust and there are also abrupt changes in the physical parameters of the crust and mantle. The width of the transitional zone is *c.* 35–40 km. The contact between the ELD and PBGB blocks follows the border of granulites exposed at the palaeosurface of the crystalline basement.

Lithologies

The major crustal domains outlined above also differ in terms of their lithostratigraphic and tectonic evolution. Specific lithological complexes are formed in the junction zone between the WLGD and the ELD (i.e. the CLSZ). The principal lithotectonic and genetic units are presented on a generalized geological map of the crystalline basement of the region (Fig. 2.16) as well as on the chronostratigraphic chart (Fig. 2.17).

Supracrustal rocks

The Precambrian crystalline basement of Lithuania, and adjacent areas, was formed under amphibolite and granulite facies metamorphic conditions. The supracrustal succession is heavily migmatized and the original metasediments are preserved mainly in the form of the palaeosomes of migmatites and relict bodies. The lower age limit of the supracrustal rocks in the ELD is estimated from the youngest U–Pb age of detrital zircon at *c.* 1.91 Ga, whereas the upper possible limit is indicated by the age of the oldest migmatite neosome (1.884 Ga) (Mansfeld 2001; Claesson *et al.* 2001).

Metasedimentary rocks of the ELD are predominantly greywackes and arkoses, in places with an admixture of volcaniclastic material, that have been metamorphosed to biotite–plagioclase–

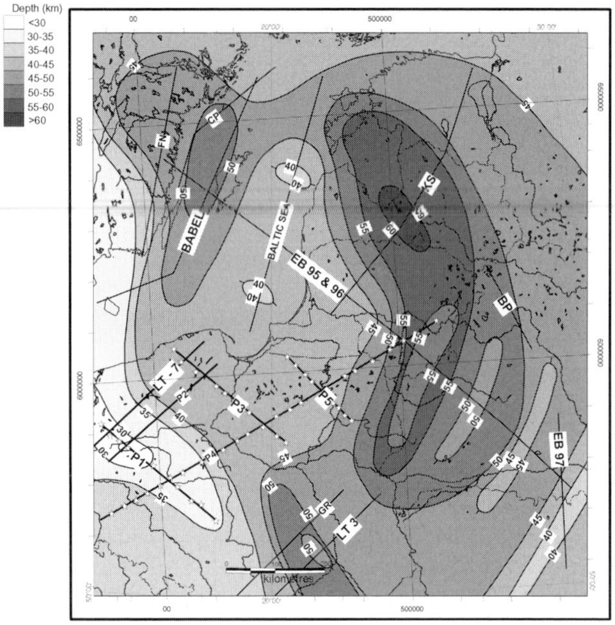

Fig. 2.15. Crust thickness (Moho depth) map of Lithuania and adjacent areas. Compiled by G. Motuza using data of the deep sesimc sounding experiments: BABEL (Lund *et al.* 2001); Baltic Sea, Baltijskoje More (Ostrovsky *et al.* 1994); B–P, Braslav–Pleščenicy (Girin 1990; Gareckij 2001); CP, Coastal profile (Lund *et al.* 2001); EB, Eurobridge'95 (Giese 1998, Eurobridge'95 Seismic Working Group 2001); FN, FENNOLORA (Lund *et al.* 2001); GR, gravity model of the crust structure (Grabowska & Bojdys 2001); KS, Kochtla Jarve–Sovetskas (Ankundinov *et al.* 1994); LT-3 and LT-7 (Guterch *et al.* 1986; Królikowski *et al.* 1999; Petecki 2002); P2, P3, P4, P5, POLONAISE (Środa *et al.* 1999; Czuba *et al.* 2001, 2002).

quartz gneisses, in places with garnet and sillimanite. There are subordinate mafic gneisses, marbles and metapelites. The latter are characterized by a fine lepidogranoblastic texture and a banded structure, regarded as relicts of primary bedding and caused mainly by variations in biotite content. The geochemical data suggest that there were two main sources for the sedimentary rocks: one with mafic rocks, probably mainly local, and the other with felsic rocks, which might be both local volcanics and granitoids from a distant block of older, Proterozoic and/or Archaean continental crust. The $\varepsilon Nd_{1.9\ Ga}$ ratio in biotite–plagioclase–quartz gneisses from the Šalčia 403 borehole is intermediate between that typical of Archaean ($\varepsilon Nd = -9$) and of Palaeoproterozoic ($\varepsilon Nd = +4$) rocks and can be interpreted as resulting from the mixing of different age source rocks. Higher values (e.g. Lazdijai 32 borehole) might be due to the addition of volcanic material. The chemical index of alteration indicates that primary clastic material of the ELD metasediments was not weathered. Mafic metavolcanics are represented by amphibolites and pyroxene plagioclase gneisses with relicts of primary porphyritic and ophitic texture. The geochemical characteristics of the supracrustal mafic rocks mostly correspond to tholeiitic basalts, although some of them demonstrate features similar to a calc-alkaline series. Three groups of mafic metavolcanics can be distinguished, differing in the abundance and the pattern of the REE. On a discrimination diagram the majority of the amphibolite analyses fall in the subduction-related volcanic island arc field.

In the WLGD, the predominant supracrustal rocks are metapelitic and felsic gneisses (metapsammites), while marbles and mafic rocks are practically absent. In comparison with the ELD metasediments, the WLGD rocks contain less Na and K and more Fe, Mg, and Ti. This may indicate that the source rocks of the WLGD metasediments were more mafic in character. The high chemical index of alteration implies that there was weathering of the primary clastic material of the metasediments. The possible time interval for the formation of the WLGD supracrustal rocks is between 2.145 and 1.849 Ga. The first figure is the age for the youngest detrital zircon (Claesson *et al.* 2001) and the second that of the oldest plutonic rocks (i.e. charnockites). The most probable time interval for deposition is considered to be 1.88–1.86 Ga, which is roughly the same as the age of deposition in the Västervik Basin (SE Sweden, i.e. the nearest part of the Fennoscandian Shield) (Sultan *et al.* 2005)

In the western part of the CLSZ a belt of pyroxene–biotite–hornblende–feldspar–quartz gneisses with well-developed porphyritic texture, interpreted as metavolcanics of andesitic-dacitic composition, occurs. In the eastern part of the CLSZ (western margin of the ELD), amphibolites predominate. These show geochemical features of subduction-related magmatic arc volcanics as illustrated by the well differentiated REE patterns, the negative Nb anomaly or an affinity to calk-alkaline series. The age of the CLSZ rocks is constrained by single datings of andesitic-dacitic metavolcanics and granitoids at 1.842 and 1.836 Ga, respectively.

Migmatites

Migmatites are the most common rock type in the basement of Lithuania. They comprise different textural varieties, including nebulites (diatexite), and are classified according to the composition of the palaeosome and the leucosome. The palaeosome consists of supracrustal rocks typical of particular crustal domains and its composition mirrors the distribution of supracrustal sequences before the migmatization, as shown on the map (Fig. 2.16). The radiometric age of the migmatites (U–Pb on monazite) is constrained only in the WLGD to 1800 Ma (Skridlaite *et al.* 2004). This age indicates that migmatization in the WLGD may have been induced by the emplacement of large charnockitic intrusions.

Plutonic rocks

Major plutonic suites recognized in Lithuania, in terms of their tectonic setting are classified as pre- to synorogenic and intracratonic. In the ELD, ultramafics found in several boreholes as serpentinized peridotites, serpentinites or serpentine–actinolite–phlogopite–talc rocks are considered to be pre-orogenic intrusives. Synorogenic intrusions are represented by gabbroic, dioritic to granodioritic plutons occurring in the MLSZ and the eastern part of the WLGD. The gabbros of the Randamonys Complex are geochemically similar to the tholeitic series, while the diorites and granodiorites are of calc-alkaline character. The age of the Randamonys Complex is constrained by rare age determinations (U–Pb method on zircon) falling in the range between 1.837 Ga (Rimsa *et al.* 2001) and 1850 Ma (Skridlaite *et al.* 2004). This age interval overlaps with the age range of the felsic metavolcanic rocks in the CLSZ.

Large plutons of synorogenic charnockitic and granitic rocks are also recognized in the WLGD. These rocks are mainly ferroan, subalkaline, and peraluminous S-type granites. Two age determinations of charnockitic and two of granitic rocks fall within a time span between 1.815 and 1.846 Ga (Claesson *et al.* 2001).

Intracratonic intrusions occur in all of the domains of the Precambrian basement in Lithuania. The largest is the Ryga

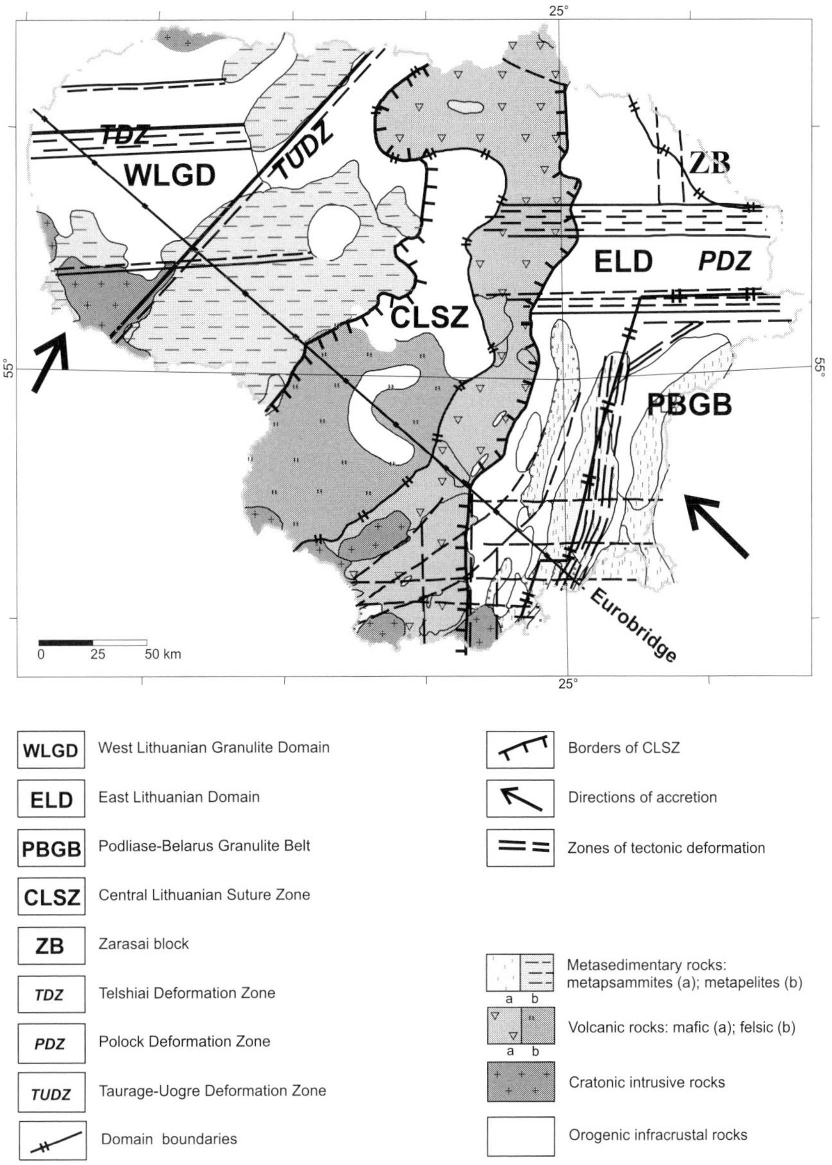

Fig. 2.16. Tectonic sketch map showing the structure of crystalline crust of Lithuania (compiled by G. Motuza).

Batholith which consists of rapakivi granites–gabbros–anorthosites (Bogatikov & Birkis 1973), and is dated at 1.58 Ga (Rämö *et al.* 1996). Other intrusions of this type are present in western Lithuania, NE Poland, southern Lithuania, NW Belarus and probably in the Kaliningrad district. The predominant rocks are granitoids, monzodiorites and monzonites, although in NE Poland (Suwałki-Kętrzyn) the AMCG suite is also documented (Bagiński *et al.* 2001; Sundblad *et al.* 1994; Wiszniewska *et al.* 2004). The granitoids are mostly ferroan, metaluminous, I-types, enriched in REE and other incompatible elements. Geochronological data from southern Lithuania and Poland (U–Pb on zircons; Re–Os) indicate that the time of their emplacement was between 1.56 and 1.478 Ga (Sundblad *et al.* 1994; Dörr *et al.* 2002*b*; Wiszniewska *et al.* 2002*a*; Stein *et al.* 1998). In western Lithuania similar rocks have been dated at 1.46 Ga (Motuza *et al.* 2004). Rare mafic dykes of unmetamorphosed diabases or basalts up to a few metres thick are reported from a few boreholes in both the WLGD and the ELD; their age is presumably Meso- or Neoproterozoic.

Metasomatic rocks

A magnesium and calcium skarn association is developed within a supracrustal dolomite marble in the Varena area of southern Lithuania. The skarns are composed of olivine (serpentine), spinel, clino- and orthopyroxene, magnetite, hornblende, diopside-plagioclase, phlogopite and microcline. Magnetite-bearing rocks in the region form iron ore deposits with reserves estimated at hundreds of millions of tonnes.

Impact rocks

The Misarai impact crater, which is 5 km in diameter, is known from southern Lithuania. Various impact rocks are identified in the crater, including shocked rocks, impact breccias and suevite (Puura *et al.* 1994). The age of the impact is estimated as

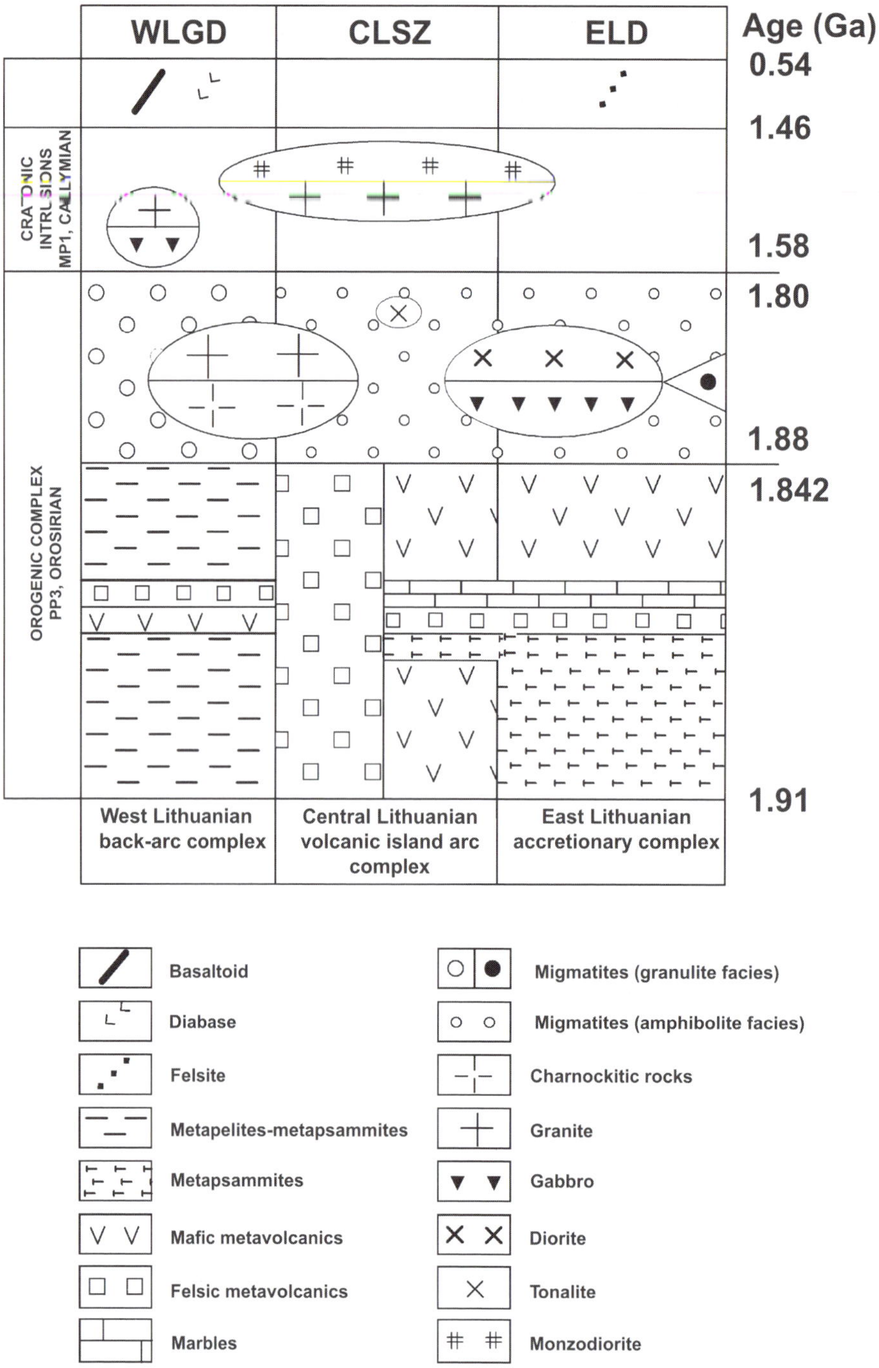

Fig. 2.17. Schematic chronology of the rock complexes in the crystalline basement of Lithuania.

Vendian–Cambrian, based on acritarch fossils in the rocks filling the crater.

Tectonic evolution

The most recent investigations of the western portion of the EEC including Lithuania and the surrounding areas did not reveal the presence of Archaean crust (Claesson *et al.* 2001). Therefore, it is presumed that the formation of the crust in this region took place in the Palaeoproterozoic Orosirian period and may thus be regarded as part of the Svecofennian Orogeny *senso lato*. The WLGD contains evidence for the development of NW–SE striking island arcs which were accreted towards the NNE (Nironen 1996). The formation of continental crust in the eastern part of Lithuania, NE Poland and NW Belarus (ELD, PBGB) was complicated by the distinct movement of a separate plate in a WNW direction. The subduction of the ELD block towards the WNW took place along the CLSZ, corresponding to an active plate margin and associated volcanic island arc.

The supracrustal successions of the WLGD, ELD and CLSZ

form distinct lithotectonic units differing in terms of their lithological composition, provenance and tectonic setting. The ELD block, composed of metapsammites with non-weathered plagioclases, suggests that it was deposited under a regime of rapid sedimentation in an active tectonic environment. The geochemical data clearly support the idea of a subduction-related volcanic island arc as its tectonic setting. Calc-alkaline plutons of mafic-intermediate composition are typical of such an environment. The ELD block comprises successions with contrasting compositions which may correspond to the structure of an accretionary prism. In contrast, the WLGD block, composed of predominantly metasedimentary felsic and metapelitic gneisses, contains intensely weathered plagioclases, while mafic metavolcanics are absent. These data suggest that the tectonic setting was a backarc one. The MLSZ comprises mostly metavolcanics of mafic and felsic to intermediate composition with a geochemical signature corresponding to that of a volcanic arc setting. The metavolcanic rocks form belts between the WLGD and ELD blocks, an area where essential changes in composition and structure of the entire crust are noted.

The protolith ages of the supracrustal rocks in the region are indirectly estimated at 1.91–1.88 Ga and 1.945–1.85 Ga for the ELD and the WLGD, respectively. The age of the MLSZ volcanics is estimated at *c.* 1.84 Ga. Intense plutonic magmatism and the emplacement of synorogenic S-type charnockitic and granitic plutons in the WLGD took place over a timespan of 1.85–1.836 Ga and may have been related to the same orogenic event that resulted in the cratonization of the WLGD crust. The intense deformation of the cratonized ELD crust took place due to collision with the older Sarmatian palaeocontinent (Bogdanova *et al.* 2001). This collisional event resulted in large-scale deformation of the lithosphere, including the delamination of the crust and the upper mantle, the formation of thrust systems, the uplift of granulite blocks and the development of shear zones along their margins, as well as the resultant exhumation of a granulitic lower crust. The strike direction of thrusts and of uplifted granulite bodies is mainly NE–SW. The timing of this tectonic event is constrained by palaeomagnetic (Elming *et al.* 2001) and $^{40}Ar/^{39}Ar$ data to a period between 1.7 and 1.65 Ga (Bogdanova *et al.* 2001). Intracratonic plutons of predominantly granitoid, monzodiorite and in places anorthositic and gabbroic compositions intruded the western margin of the EEC from southern Sweden up to NW Belarus between 1.58 and 1.42 Ga. The tectonic context of this magmatism, however, is still unknown.

Southern Fennoscandia III (the offshore G14–1 borehole, Baltic Sea) (U.L.)

Around 40 km to the NE of the island of Rügen in the Baltic Sea the basement of the Fennoscandian part of the EEC (Baltica) was found in the core of the offshore G14 borehole at a depth of between 1941.7 and 1997.0 m (final depth). The basement comprises a monzogranite, that proably represents a restite-poor melting product of a granodioritic protolith. The G14–1 granite is unconformably overlain by Cambrian to Silurian shelf sediments (McCann 1996; Maletz *et al.* 1997; Beier *et al.* 2000). The Pb/Pb age of 1460 ± 3 Ma is interpreted to be the age of intrusion for the granite (Obst *et al.* 2004). The drilled granite from the G14–1 borehole is correlated with the Mesoproterozoic intrusions from Bornholm and southeastern Sweden.

Avalonia and related terranes

Avalonia is a microcontinent or terrane derived from the margin of West Gondwana (peri-Gondwana). Avalonia is subdivided into West Avalonia (eastern Nordamerica) and East Avalonia (southern Britain, parts of Ireland, Brabant Massif). Other parts of Avalonia situated in Central Europe share a diagnostic Cambro-Ordovician overstep sequence with Acado-Baltic (Avalonian) fauna (Theokritoff 1979) and comprised a continous tectonostratigtraphic belt that was sundered by the opening of the North Atlantic (Murphy *et al.* 2004; Winchester *et al.* 2006; see also Krawczyk *et al.* 2008).

The main events in the evolution of Avalonia are (see Nance *et al.* 2002): (1) the development of juvenile crust at *c.* 1.3–1.0 Ga, (2) an early arc phase (pre-650 Ma), (3) its accretion to Gondwana at *c.* 650 Ma, (4) a main arc phase (640–570 Ma), (5) its transition to a platform (570–540 Ma), (6) the rifting of Avalonia during the Cambro-Ordovician. During the rift/drift of Avalonia the Rheic Ocean opened as the Tornquist Sea became closed by the collision of Avalonia (7) with Baltica during the late Ordovician and early Silurian (*c.* 440 Ma). In the late Silurian, Baltica plus Avalonia collided with Laurentia by the closure of the Iapetus Ocean to form the short-lived continent Laurussia. During its drift and after amalgamation with Baltica and Laurentia, Avalonia formed the northern margin of the Rheic Ocean (e.g. Linnemann *et al.* 2007).

The Neoproterozoic basement of Avalonia formed and was reworked during Avalonian orogenic processes and related tectonomagmatic events. Neoproterozoic and Early Palaeozoic sediments of Avalonia contain detrital zircon grains of Mesoproterozoic age, which are assumed to have formed in a juvenile crust between 1.0 and 1.3 Ga (Murphy *et al.* 2004). There is evidence that large parts of Avalonia formed at the margin of Amazonia (Nance & Murphy 1994).

Avalonia forms the poorly exposed basement of the Rhenohercynian Zone, extending from the Brabant Massif of Belgium to the eastern margin of the Bohemian Massif. However, the Precambrian–Early Palaeozoic geology of Avalonia is best known from southern Britain and Ireland where the tectonostratigraphy of this crustal block is locally well constrained. That part of Avalonia (East Avalonia) is therefore included for the purposes of comparison and because models for the tectonothermal evolution of Avalonia are based on this record. Other more easterly parts of Avalonia and related terranes, as well as terranes of uncertain origin at the recent SW margin of Baltica, are summarized as Far East Avalonia *sensu* Winchester *et al.* (2006). We relate the Eckergneiss Complex in the Harz Mountains (Rhenohercynian Zone), the Upper Silesian and Małopolska Blocks (SE Poland) and the Thaya and Slavkov terranes, including the so-called Metabasite Belt, to Far East Avalonia, although that interpretation is uncertain and under discussion.

East Avalonia (southern Britain and Ireland) (R.D.N.)

Precambrian rocks of Avalonia underlie most of southern Britain but are largely concealed by younger strata. Outcrops are restricted to small, structurally controlled inliers (Fig. 2.18) in the English Midlands (Charnian, Caldecote volcanics, Malverns Complex/Warren House Formation), the Welsh Borderlands (Uriconian, Longmyndian, Stanner-Hanter Complex), South Wales (Pebidian, Johnston Complex, Llangynog), and North Wales (Coedana Complex, Monian, Arfon Group, Sarn Complex). Additionally, Precambrian rocks resembling those of the Coedana Complex/Monian crop out in SE Ireland (Rosslare

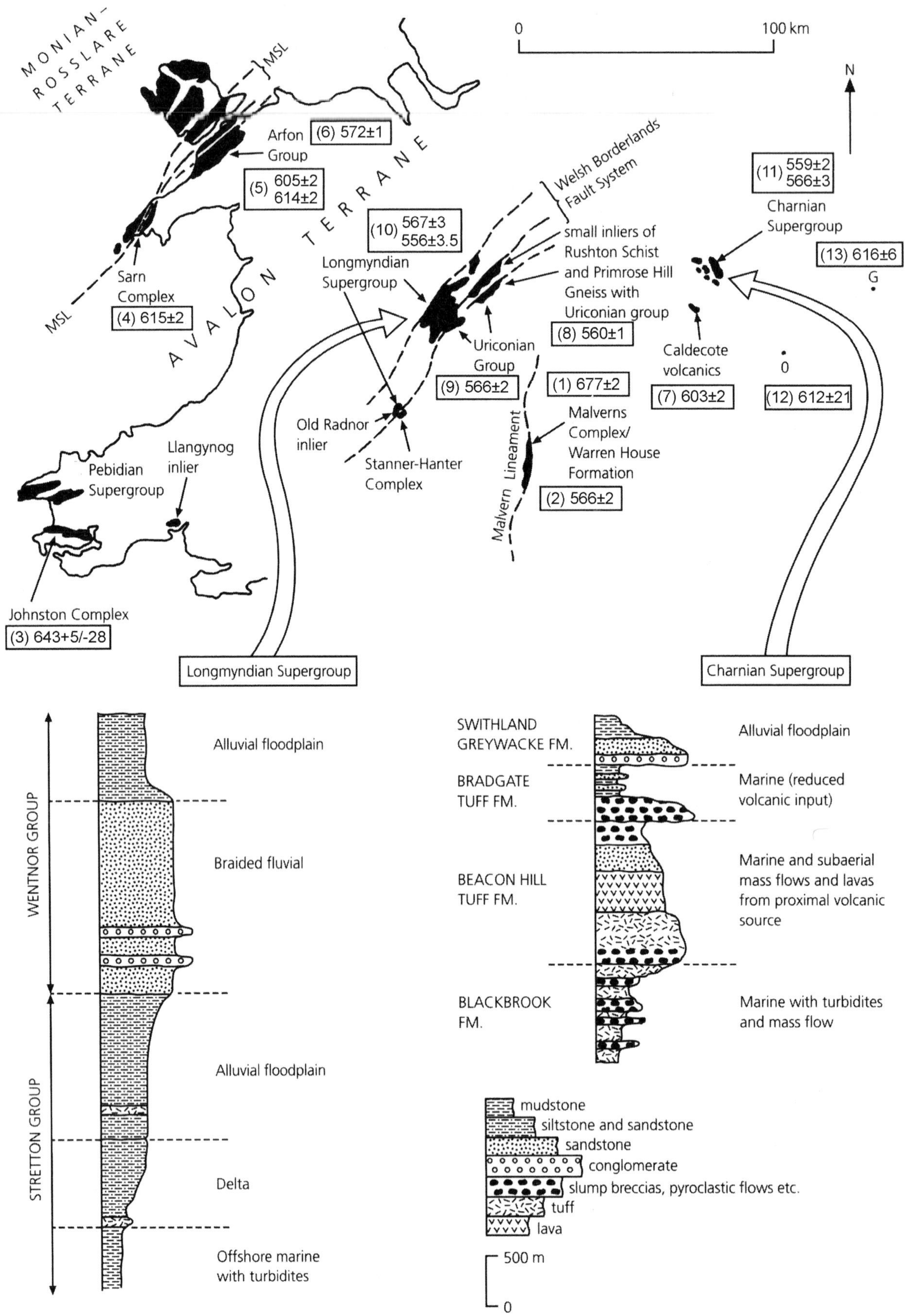
MONIAN-ROSSLARE TERRANE
MSL
0 100 km
N
Arfon Group
(6) 572±1
(5) 605±2 614±2
AVALON TERRANE
Welsh Borderlands Fault System
(11) 559±2 566±3
Charnian Supergroup
(10) 567±3 556±3.5
Longmyndian Supergroup
small inliers of Rushton Schist and Primrose Hill Gneiss with Uriconian group
(13) 616±6
G
Sarn Complex
(4) 615±2
(8) 560±1
Uriconian Group
Caldecote volcanics
O
(9) 566±2
(1) 677±2
(7) 603±2
(12) 612±21
Old Radnor inlier
Malverns Complex/ Warren House Formation
Llangynog inlier
Pebidian Supergroup
Stanner-Hanter Complex
Malvern Lineament
(2) 566±2
Johnston Complex
(3) 643+5/-28
Longmyndian Supergroup
Charnian Supergroup
WENTNOR GROUP
STRETTON GROUP
Alluvial floodplain
Braided fluvial
Alluvial floodplain
Delta
Offshore marine with turbidites
SWITHLAND GREYWACKE FM.
BRADGATE TUFF FM.
BEACON HILL TUFF FM.
BLACKBROOK FM.
Alluvial floodplain
Marine (reduced volcanic input)
Marine and subaerial mass flows and lavas from proximal volcanic source
Marine with turbidites and mass flow
mudstone
siltstone and sandstone
sandstone
conglomerate
slump breccias, pyroclastic flows etc.
tuff
lava
500 m
0

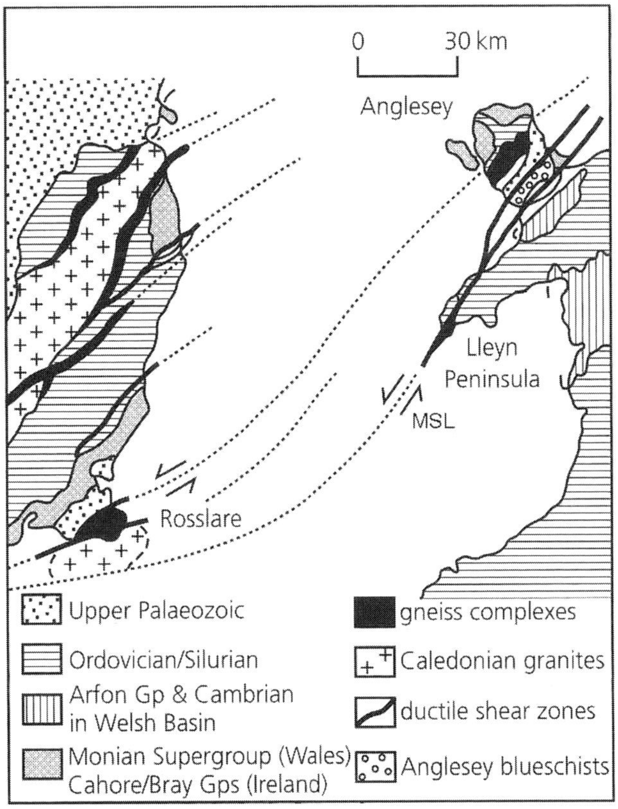

Fig. 2.19. Simplified geological map of the Irish Sea region showing continuation of Monian–Rosslare Terrane into southeastern Ireland (modified from Strachan 2000). MSL, Menai Strait Line.

Complex, Cahore Group; Fig. 2.19). Together, the geology of these areas records the development of volcanic arcs and arc-related sedimentary basins along the active northern margin of Neoproterozoic Gondwana during the interval *c.* 700–555 Ma (Nance *et al.* 1991).

Within this Precambrian crustal block, two terranes are identified (Strachan 2000): (i) the Avalon Terrane of central England and Wales, and (ii) the Monian–Rosslare Terrane of NW Wales and SE Ireland. The tectonic boundary separating these two terranes is a prominent structural lineament known as the Menai Strait Line (Fig. 2.18).

The oldest components of the Avalon Terrane are calc-alkaline plutonic complexes that comprise sheared and variably banded granites, granodiorites, diorites and gabbros. They include the *Stanner-Hanter Complex* (702 ± 8 Ma, Rb–Sr whole-rock; Patchett *et al.* 1980), the *Malverns Complex* (670 ± 10 Ma, U–Pb monazite; Thorpe *et al.* 1984; 677 ± 2 Ma, U–Pb zircon; Tucker & Pharaoh 1991), and the *Johnston Complex* (643 +5/ −28 Ma; U–Pb zircon; Patchett & Jocelyn 1979). The Malverns Complex was deformed and metamorphosed under amphibolite facies conditions at *c.* 650 Ma ($^{40}Ar/^{39}Ar$ hornblende age; Strachan *et al.* 1996*a*), an event that may also be recorded in the garnetiferous Rushton schist to the north, which has yielded an imprecise Rb–Sr whole-rock age of 667 ± 20 Ma (Thorpe *et al.* 1984). Other granitoid plutons are less deformed and yield significantly younger ages. They include the granites and gabbros of the *Sarn Complex* (615 ± 2 Ma, U–Pb zircon; Horák *et al.* 1996), and a suite of diorites that intrude the Caldecote volcanics (603 ± 2 Ma; U–Pb zircon; Tucker & Pharaoh 1991) and the Charnian Supergroup. The subvolcanic *Ercall Granophyre*, which is intrusive into the Uriconian Group, has yielded a younger U–Pb zircon age of 566 ± 2 Ma (Tucker & Pharoah 1991).

Other outcrops of the Avalon Terrane expose major volcanic-volcaniclastic assemblages. These include the Arfon Group and Pebidian Supergroup of Wales, the Uriconian Group and Longmyndian Supergroup of the Welsh Borderlands, and the Caldecote volcanics and Charnian Supergroup of central England (Fig. 2.18). Available age data suggest that these record two main periods of volcanic activity (at *c.* 620–590 Ma and *c.* 575–555 Ma) that match the ages of the younger granitoids. The *c.* 4 km thick *Arfon Group* is made up of two units separated by an unconformity: a lower series of ignimbrites (Padarn Tuff) that have yielded U–Pb zircon ages of 614 ± 2 Ma (Tucker & Pharoah 1991) and 605 ± 2 Ma (Compston *et al.* 2002), and an upper series of interbedded tuffs and sandstones that pass upwards into slates. Compston *et al.* (2002) reported a U–Pb zircon age of 572.5 ± 2 Ma for a tuff interbedded with the slates. The *c.* 2 km thick *Pebidian Supergroup* comprises felsic-intermediate tuffs and lavas interbedded with thin conglomerates. The succession is of uncertain age, but is intruded by the St. Davids granophyre, from which Patchett & Jocelyn (1979) obtained an imprecise U–Pb zircon age of 587 +25/−14 Ma, and was deformed under greenschist facies conditions prior to the Early Cambrian. A succession of felsic-intermediate tuffs and lavas exposed at Llangynog (Fig. 2.18) are interbedded with sandstones and siltstones that preserve the Ediacaran fauna *Cyclomedusa* and *Medusinites* (Cope & Bevins 1993).

The *Uriconian Group* comprises a thick sequence of subaerial volcanic rocks that include pyroclastic breccias, agglomerates, lavas and tuffs of felsic, intermediate and mafic composition. They show both arc and within-plate geochemical characteristics and are interpreted to have erupted in an ensialic marginal basin (Pharaoh *et al.* 1987). The timing of volcanism is given by a U–Pb zircon age of 566 ± 2 Ma from a flow-banded rhyolite (Tucker & Pharaoh 1991). A similar sequence of basalts, basaltic andesites and rhyolitic tuff in the Malvern Hills (Warren House Formation) has yielded an identical U–Pb zircon age of 566 ± 2 Ma (Tucker & Pharaoh 1991), but shows geochemical affinities more consistent with an oceanic marginal basin (Pharaoh *et al.* 1987).

Tectonically juxtaposed against the Uriconian Group along the Welsh Borderlands Fault System (Fig. 2.18), the *Longmyndian Supergroup* is a *c.* 6 km thick, upward-coarsening sequence of turbidites, deltaic sandstones and alluvial floodplain siltstones (Stretton Group) overlain by braided alluvial sandstones and conglomerates (Wentnor Group) (Pauley 1990). Minor pyroclastic deposits occur within the Stretton Group, and the succession

Fig. 2.18. Inliers of Avalonia in southern Britain with simplified stratigraphic columns for the Longmyndian and Charnian supergroups (modified from Strachan 2000). Boxes give U-Pb zircon ages for: 1, Malverns Complex granodiorite (Tucker & Pharaoh 1991); 2, Warren House Formation rhyolitic tuff (Tucker & Pharaoh 1991); 3, Johnston Complex diorite (Patchett & Jocelyn 1979); 4, Sarn Complex diorite (Horák *et al.* 1996); 5, Arfon Group ignimbrite (Tucker & Pharaoh 1991; Compston *et al.* 2002); 6, Arfon Group tuff (Compston *et al.* 2002); 7, diorite intrusive into Caldecote volcanics (Tucker & Pharaoh 1991); 8, Ercall Granophyre (Tucker & Pharaoh 1991); 9, Uriconian Group rhyolite (Tucker & Pharaoh 1991); 10, Longmyndian Supergroup tuff and bentonite (Compston *et al.* 2002); 11, Charnian Supergroup tuffs (Compston *et al.* 2002); 12, 13, ignimbrite from Orton and Glinton boreholes (Noble *et al.* 1993). MSL, Menai Strait Line.

is interpreted to have been deposited in a forearc or interarc basin adjacent to the Uriconian volcanic arc. This is supported by U–Pb zircon ages of 567 ± 3 Ma and 556 ± 3.5 Ma for bentonite and tuff in the Stretton Group (Compston *et al.* 2002).

The *c.* 3 km thick *Charnian Supergroup* is dominated by volcaniclastic rocks that were most likely deposited within an interarc basin. They comprise marine turbiditic pyroclastic deposits that are locally associated with volcanic breccias and tuffs, and pass upwards into fluvial conglomerates and sandstones. Tuffs within the marine succession preserve the Ediacaran fauna *Charnia masoni* and *Charniadiscus concentricus* (Ford 1980) and have yielded U–Pb zircon ages of 566 ± 3 Ma and 559 ± 2 Ma (Compston *et al.* 2002). Charnian volcanic rocks plot in the arc lava field (Thorpe *et al.* 1984) and are interpreted to record derivation from an immature volcanic arc floored by either oceanic or juvenile continental crust (Pharaoh *et al.* 1987). The *Caldecote volcanics*, which crop out to the south (Fig. 2.18), have been correlated with the Charnian Supergroup, although Tucker & Pharaoh (1991) have recorded a U–Pb zircon age of 603 ± 2 Ma from a diorite that cuts these volcanic rocks, suggesting they belong to the older period of volcanic activity. Further to the east, U–Pb zircon ages of 612 ± 21 Ma and 616 ± 6 Ma have been obtained from ignimbrites encountered in the Orton and Glinton boreholes (Noble *et al.* 1993).

West of the Menai Strait Line, the Monian–Rosslare Terrane of NW Wales (Fig. 2.18) includes the Coedana Complex, a belt of blueschists and the Monian Supergroup, the history and relationships of which are not well constrained. The *Coedana Complex* comprises high-grade paragneisses, migmatites and amphibolites that are cut by the unmetamorphosed calc-alkaline Coedana granite (614 ± 4 Ma, U–Pb zircon; Tucker & Pharaoh 1991). The blueschists occur within a belt of quartz-phengite schists and barroisite/crossite-bearing metabasites derived from an earlier greenschist facies protolith. $^{40}Ar/^{39}Ar$ ages of *c.* 580–590 Ma obtained from actinolite may record sub-seafloor metamorphism and pre-date those obtained from the blue amphiboles (*c.* 550–560 Ma), which are considered to date the high-pressure blueschist event (Dallmeyer & Gibbons 1987). The assemblage likely represents oceanic basalt and accretionary prism sediment exhumed from a late Neoproterozoic subduction zone.

The *Monian Supergroup* is a *c.* 5–7 km thick assemblage of low-grade, metasedimentary rocks made up of three major units (Phillips 1991): (i) a basal succession of quartz arenites and turbiditic metasandstones deposited in a sand-rich continental margin fan system (South Stack Group), (ii) a pelitic sequence of continental island-arc-derived, turbiditic phyllites with volcanic arc metabasites and serpentinites (New Harbour Group), and (iii) a chaotic mélange (Gwna Group) with clast lithologies that include quartzite, stromatolitic limestone, pillow basalt, chert, granite and serpentinite. The Monian Supergroup has long been considered part of Neoproterozoic–Cambrian Avalonia although it is not thought to represent an accretionary prism and its depositional age was, until recently, poorly constrained. However, a detrital zircon age of 501 ± 10 Ma recently reported from the South Stack Group by Collins & Buchan (2004) requires the Monian Supergoup to have been deposited in the Late Cambrian–Early Ordovician. Detrital age maxima at 543–552 Ma and 604–627 Ma suggest that Avalonia was the major sediment source.

In southeasternmost Ireland, the *Rosslare Complex* shows many similarities with the Monian geology of North Wales (Gibbons & Horák 1996) and is considered to represent a continuation of the Monian–Rosslare Terrane along strike (Fig. 2.19). Although blueschists have not been reported, the Rosslare Complex includes amphibolite facies paragneisses and amphibolites like those of the Coedana Complex, faulted against low-grade metasedimentary rocks (Cahore Group) like those of the Monian Supergroup. Gneisses from both complexes show identical Sm–Nd model ages (1.8–1.7 Ma from the paragneisses, 1.1–1.0 Ma from amphibolites; Davies *et al.* 1985; Horák 1993) and the amphibolites share a continental tholeiitic geochemistry. $^{40}Ar/^{39}Ar$ hornblende ages of 626 ± 6 Ma from the amphibolites and 618 ± 6 Ma from a calc-alkaline mafic pluton (St Helens gabbro) that intrudes the gneisses are thought to date post-metamorphic cooling and pluton emplacement, respectively (Max & Roddick 1989). The *Cahore Group* is separated from the Rosslare gneisses by a steep zone of mylonitic gneisses and metasediments, but comprises marine siliciclastics that closely resemble those of the Monian Supergroup, the quartz-rich turbidites of the Cullenstown Formation having been directly correlated with those of the South Stack Group (Phillips 1991). Such a correlation would be consistent with the new depositional age constraints for the Monian Supergroup since the Cahore Group has long been considered Cambro-Ordovician in age (Gibbons *et al.* 1994).

The evolution of Avalonia in Britain is interpreted in terms of a three-stage model (Fig. 2.20) that describes the episodic construction of a long-lived magmatic arc (Gibbons & Horák 1996). An early phase of arc construction at *c.* 700–640 Ma (Fig. 2.20a) is represented by the Malverns, Stanner-Hanter and Johnston complexes, and was followed by metamorphism, deformation and uplift. The continental crust upon which this early arc was constructed is not exposed, but isotopic evidence suggests that it is of Mesoproterozoic age (Tucker & Pharaoh 1991; Murphy *et al.* 2000). The main phase of arc magmatism at *c.* 620–585 Ma (Fig. 2.20b) is represented by the Coedana granite and St Helens gabbro, the lower Arfon Group, the Sarn Complex, the Pebidian Supergroup, the Caldecote volcanics, and the volcanic rocks exposed in the Orton and Glinton boreholes. A late phase of magmatism and marginal basin sedimentation at *c.* 570–550 Ma (Fig. 2.20c) is represented by the upper Arfon Group, the Llangynog succession, the Longmyndian Supergroup, the Uriconian Group and Ercall Granophyre, the Warren House Formation, and the Charnian Supergroup. Subduction was towards the present SE and may have been initially orthogonal. However, it is thought to have become increasingly oblique in the latest Neoproterozoic, leading to basin inversion and the dispersal of terranes along the plate margin by way of sinistral strike slip. Deformation of the Monian–Rosslare Terrane in response to movement along the Menai Strait Line continues into the Palaeozoic. Within the Avalon Terrane, however, deformation had ceased by the Cambrian, at which time the accretionary history of Avalonia was essentially complete.

Far East Avalonia

The Eckergneiss Complex (Harz Mountains) (T.G.)

The Eckergneiss Complex (EGC) is a geologically unique medium- to high-grade metamorphic unit within the Harz Mountains of the Rhenohercynian Zone of the Mid-European Variscides (Fig. 2.21a). The Harz area essentially comprises diagenetic and low-grade metamorphosed sedimentary rocks (Friedel *et al.* 1995) of Ordovician to Upper Carboniferous age (Fig. 2.21b). Major basaltic–spilitic intercalations occur within Middle Devonian and Lower Carboniferous age shales and slates, frequently as piles of pillow lavas. Granitoid and gabbroic plutonic rocks, aged *c.* 295 Ma (Baumann *et al.* 1991), form

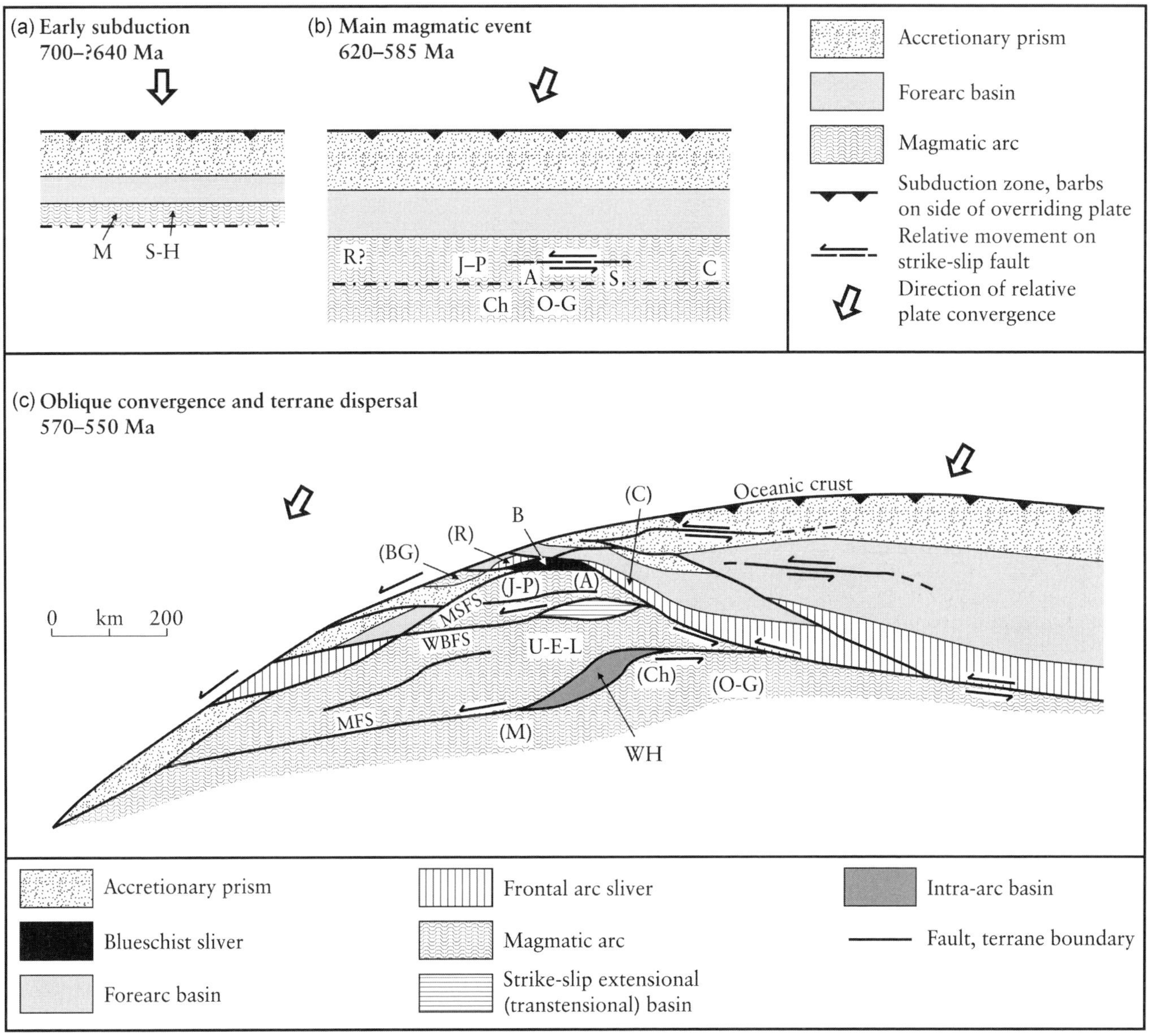

Fig. 2.20. Model for the late Neoproterozoic evolution of the Avalonian subduction system in southern Britain as modified from Gibbons & Horák (1996) by Pharaoh & Carney (2000). (**a**) Early arc construction at *c.* 700–640 Ma; (**b**) main arc magmatism at *c.* 620–585 Ma; (**c**) oblique convergence and terrane dispersal at *c.* 570–550 Ma. Abbreviations: A, Arfon Group; B, Anglesey blueschists; BG, Bwlch Gwyn Tuff and related strata (Anglesey); C, Coedana Complex; Ch, Charnian Supergroup; J-P, Johnston Complex and Pebidian Supergroup; M, Malverns Complex; MFS, Malverns lineament; MSFS, Menai Strait fault system; O-G, volcanics in Orton and Glinton boreholes; R, Rosslare Complex; S, Sarn Complex; S-H, Stanner-Hanter Complex; U-E-L, Uriconian Group, Ercall Granophyre, Longmyndian Supergroup; WBFS, Welsh Borderland fault system; WH, Warren House Formation. Letters in parentheses refer to relative positions of volcanic belts that were by then extinct.

high-level intrusions into Devonian and Lower Carboniferous country-rocks. The EGC, which represents a 7 km^2 polymetamorphic crustal segment, is enclosed by the Harzburg Gabbronorite and the Brocken Granite (Fig. 2.21c). Petrographic and petrological investigations revealed that the EGC underwent multiple metamorphic events, ranging from contact metamorphism through to amphibolite and granulite-facies conditions (Franz *et al.* 1997; Müller & Strauss 1985; Schlüter 1983). The complex consists of pelitic and siliciclastic metasediments (quartzites, quartz-(cordierite)-feldspar granofelses, gneisses, hornfelses, and mica schists) with subordinate metabasaltic intercalations.

A conventional lower intercept U–Pb age of 560 ± 100 Ma from detrital zircon grains of biotite-corderite-bearing schistose hornfelses of the EGC was interpreted to reflect the timing of the main phase of metamorphism (Baumann *et al.* 1991), suggesting that the EGC represents the oldest geological unit of the Rhenohercynian Zone. This age has led to controversial speculation about the tectonic setting of the EGC (Bankwitz 1995; Baumann *et al.* 1991; Franz *et al.* 1997; Franzke 2001). However, a recent SHRIMP U–Pb provenance study on detrital zircons from various EGC quartzites revealed the occurrence of Lower Devonian and Silurian ages in the zircon population, implying that sedimentation and metamorphism of the EGC protolith must have taken place after 410 ± 10 Ma, as indicated

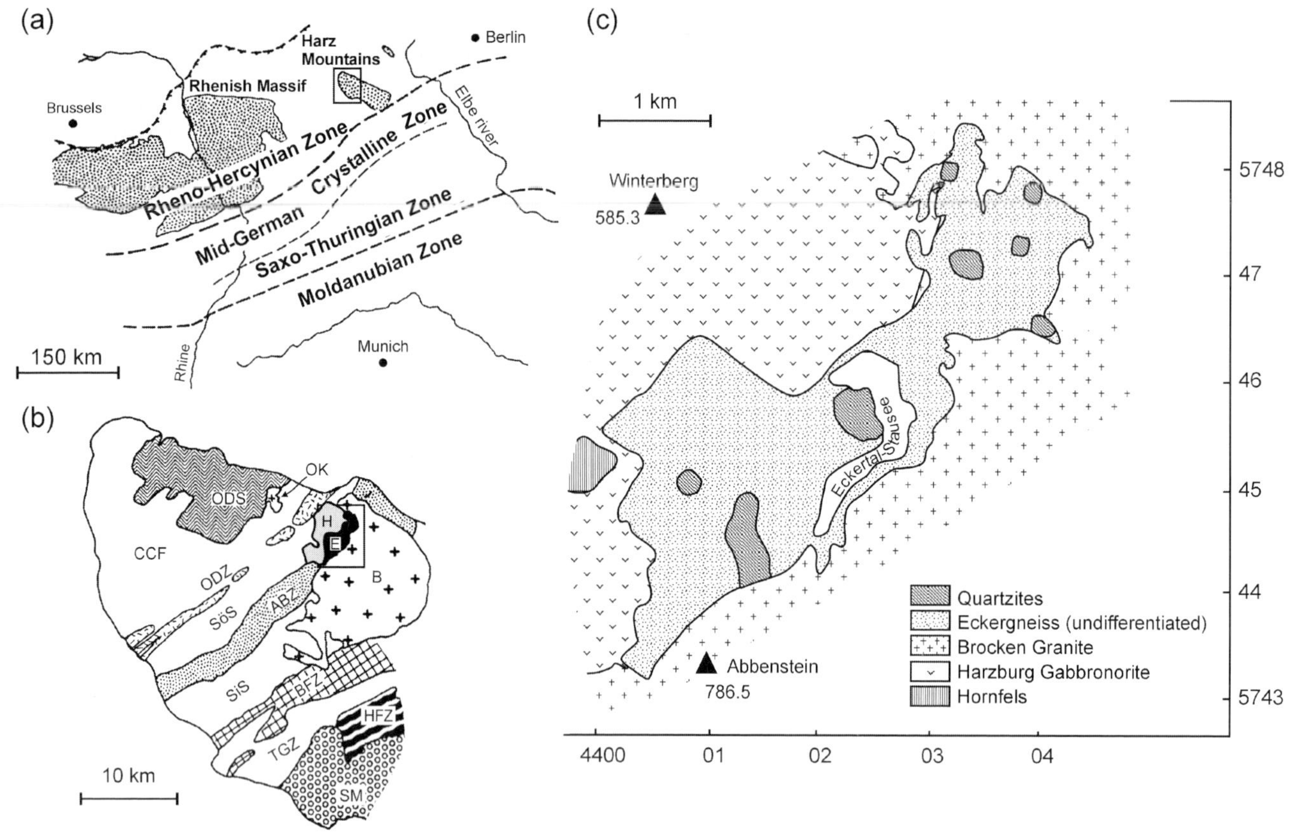

Fig. 2.21. Geology of the Eckergneiss Complex (Harz Mountains). (**a**) subdivision of the Mid-European Variscides after Franke (1989). (**b**) Geological subdivision of the Upper and Middle Harz: ABZ, Acker-Bruchberg zone; B, Brocken granite; BFZ, Blankenburger fold zone; CCF, Clausthaler Culm fold zone; E, Eckergneiss Complex; H, Harzburg gabbronorite; HFZ, Harzgeröder fold zone; ODS, Devonian anticline of the Upper Harz; ODZ, Upper Harz diabase range; OK, Ocker granite; SiS, Sieber syncline; SöS, Söse syncline; TGZ, Tanne greywacke zone. (**c**) Simplified geological map of the Eckergneiss Complex with quartzite occurrences and sample localities (from Geisler *et al.* 2005 and references therein).

by the youngest zircon dated (Geisler *et al.* 2005). The age population was, nevertheless, dominated by Proterozoic ages, ranging from *c.* 900 to *c.* 1800 Ma. Archaean age components were not found in any of the samples. Sveconorwegian ages of *c.* 0.9 to 1.2 Ga are particularly well represented by analyses from metamorphic recrystallization/alteration zones penetrating into igneous zircon, suggesting a SW Baltica provenance for the detritus.

The lower limit for high-grade metamorphism is constrained by the intrusion of the Harzburg gabbronorite and the Brocken granite at *c.* 295 Ma, which thermally affected the EGC as shown by mineral reactions (Franz *et al.* 1997) and a reset concordant titanite U–Pb age of *c.* 295 (Baumann *et al.* 1991). It is, therefore, clear that any Cadomian metamorphism in the EGC can be unambiguously excluded. The sedimentation of the EGC protolith thus occurred contemporaneously with the deposition of the Devonian to Carbonifereous siliciclastic sediments, which make up the largest portion of the Palaeozoic outcrop area in the Harz Mountains (Fig. 2.22). Consequently, sedimentation and metamorphism must have taken place between *c.* 410 and 295 Ma (Fig. 2.22). This timespan coincides with most of the sedimentation within the virtually non-metamorphosed (lowest grade) Rhenohercynian Zone rocks in the Harz Mountains, including those in the direct vicinity of the EGC, as well as being coeval with the phase of high-grade metamorphism (Fig. 2.22). The EGC can thus be interpreted as forming part of a nappe rather than uplifted local basement. A possible candidate for the root region is an easterly, concealed marginal segment of the Rhenohercynian Zone, which was tectonically overridden by the Mid-German Crystalline Zone (High) during continent–continent collision. However, according to the concept of the post-accretion intraplate deformation resulting in the strike-slip movement of Variscan terranes with different P-T-t histories (Krohe 1996), the EGC may also represent a fault-bounded complex with an origin located far to the east or SE of the present location.

Upper Silesian and Małopolska Blocks (SE Poland) (Z.B. & J.Ż.)

Precambrian igneous and clastic sedimentary rocks, some of which have been metamorphosed (middle and low grade), have been recognized in *c.* 590 boreholes in south and SE Poland (Buła *et al.* 1997*a*; Moryc & Heflik 1998; Żaba 1999; Buła 2000; Moryc & Jachowicz 2000; Moryc & Łydka 2000 and references therein). The Precambrian has been drilled at depths of *c.* 200 to more than 7000 m, beneath unmetamorphosed Palaeozoic and younger sedimentary successions. Its subcrops occur directly to the north of the Carpathian Orogenic Front and SW of the TTZ and, thus, are located in the area of the Trans-European Suture Zone (TESZ), composed of terranes that were accreted to the SW edge of Baltica during Palaeozoic times (e.g. Pharaoh 1999).

The Precambrian to Palaeozoic basement of south and SE

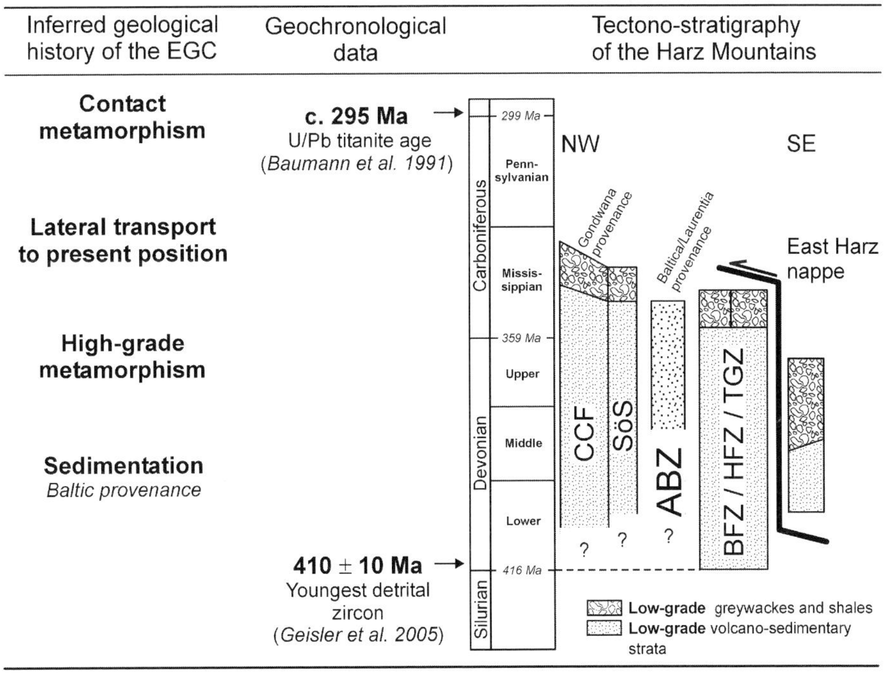

Fig. 2.22. Schematic summary of geochronological data and the inferred geological history of the Eckergneiss Complex in comparison with the tectonostratigraphy of the Harz Mountains. For abbreviations see Figure 2.1 (from Geisler *et al.* 2005 and references therein).

Poland is divided into two major tectonic units, the Upper Silesian Block (USB) and the Małopolska Block (MB). Both blocks were overlain by the Carpathian Foreland Basin during Neogene times and their southern portions were covered by nappes of the Outer Carpathian fold-and-thrust belt (Figs 2.23 & 2.24). The USB and MB have been distinguished as separate terranes taking into account the markedly different development of Palaeozoic (especially Early Palaeozoic) sequences within them and their different structural histories (Buła *et al.* 1997*a*; Żaba 1999; Buła 2000). The Precambrian rocks of both units also show some different characteristics. Within the USB the rocks are represented by genetically diverse metamorphic and igneous complexes as well as by anchimetamorphic clastics. The Precambrian of the MB corresponds exclusively to anchimetamorphic or non-metamorphic clastic rocks similar to those in the USB area.

The USB is a unit overprinted by the deformation of the Cadomian Orogeny (see Linnemann *et al.* 2008) and which is composed of Precambrian metamorphic and igneous rocks and anchimetamorphic clastic rocks (which lack fossils) (Fig. 2.24), unconformably overlain by non-metamorphosed Early Cambrian clastics (Buła & Jachowicz 1996; Buła 2000). In the nearby Brno Massif in the Czech Republic, metamorphic and igneous rocks occur in an analogous position (Buła *et al.* 1997*b*), and, therefore, it has been suggested the Brno Massif and the USB are elements of a common tectonic unit, called the Brunovistulian Block (formerly termed Brunovistulicum; Dudek 1980; Kotas 1982 1985; see below). This unit is delimited by major fault/shear zones in the west along the Moravo-Silesian Zone (MSZ), the Peri-Pieninian Zone (PPZ) along its eastern boundary (following the Outer/Inner Carpathian dividing line) and the Kraków-Lubliniec Fault Zone (KLFZ) along the NE boundary with the MB.

The age of basement consolidation in the MB has not yet been determined. Additionally, the precise locations of its NW, NE and southern boundaries are not clear due to their deep burial below the Palaeozoic–Cenozoic cover. At its NE periphery the MB includes the Kielce Zone of the Holy Cross Mountains (Pożaryski 1990; Pożaryski & Tomczyk 1993), with the precise position of the NE boundary located along the Holy Cross Fault (HCF). The continuation of this fault, to the east or SE of Sandomierz, is not known (Fig. 2.23). The relative position of the metamorphosed clay–mudstone–sandstone succession of Precambrian (Ediacaran) through to Cambrian rocks that crop out in the Kielce area of the Holy Cross Mountains and continue further to the SE is also undetermined, thus leading to controversies regarding the age of deformation in the Precambrian basement of the MB.

The boundary line between the USB and MB is the narrow KLFZ (Fig. 2.23) which shows evidence of both sinistral and dextral movement (Buła 1994, 2000; Żaba 1999). This zone was active from late Precambrian/earliest Cambrian to Permian times (Brochwicz-Lewiński *et al.* 1983; Oberc 1993; Franke & Hoffmann 1999). The extension of the KLFZ to the SE of Kraków can only be traced using geophysics.

The Precambrian metamorphic and igneous rocks of the USB have been cored in the southern part of the block (Figs 2.23 & 2.24). The top surface has been reached at depths ranging from 833.4 m (Rzeszotary 1 borehole) to 4136 m (Lachowice 1 borehole). This variability is the result of dip-slip displacements on the two dominant sets of faults, trending west–east and NNW–SSE. These movements led to the formation of two distinct highs within the Precambrian basement, namely the Bielsko-Biała–Andrychów High and the Rzeszotary Horst (Figs 2.23 & 2.24). The former has an east–west elongation and its top surface is at a depth ranging from *c.* 1400 to 2000 m. The latter trends NNE–SSW and the top surface is at a depth of *c.* 800 to 2400 m.

The metamorphic rocks of the USB constitute two distinct lithostratigraphic complexes: the older one is represented by medium-grade metamorphic rocks, and the younger (Ediacaran)

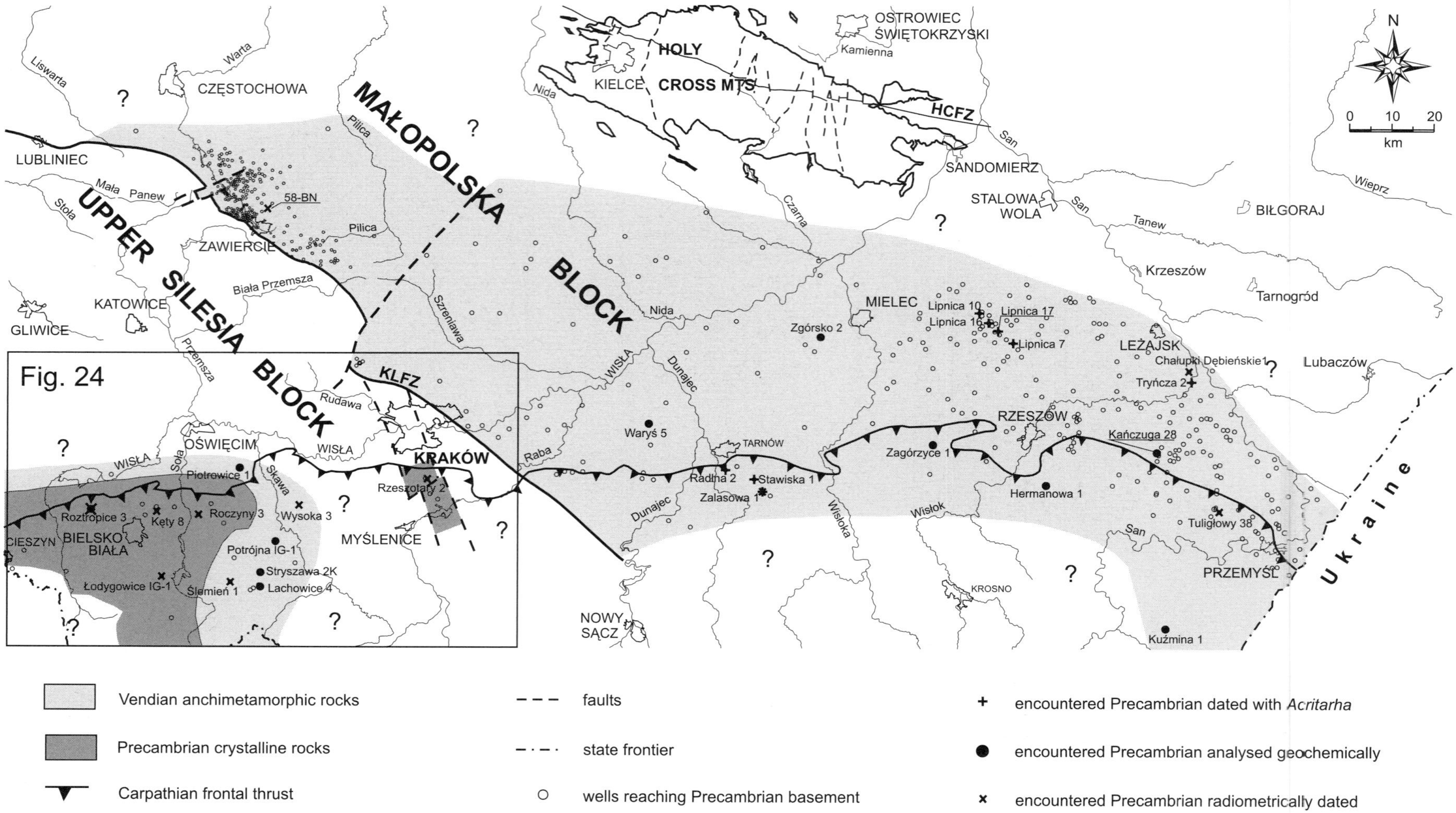

Fig. 2.23. Extent of the Precambrian basement in the Upper Silesia and Małopolska blocks. HCFZ, Holy Cross Fault Zone; KLFZ, Kraków-Lubliniec Fault Zone.

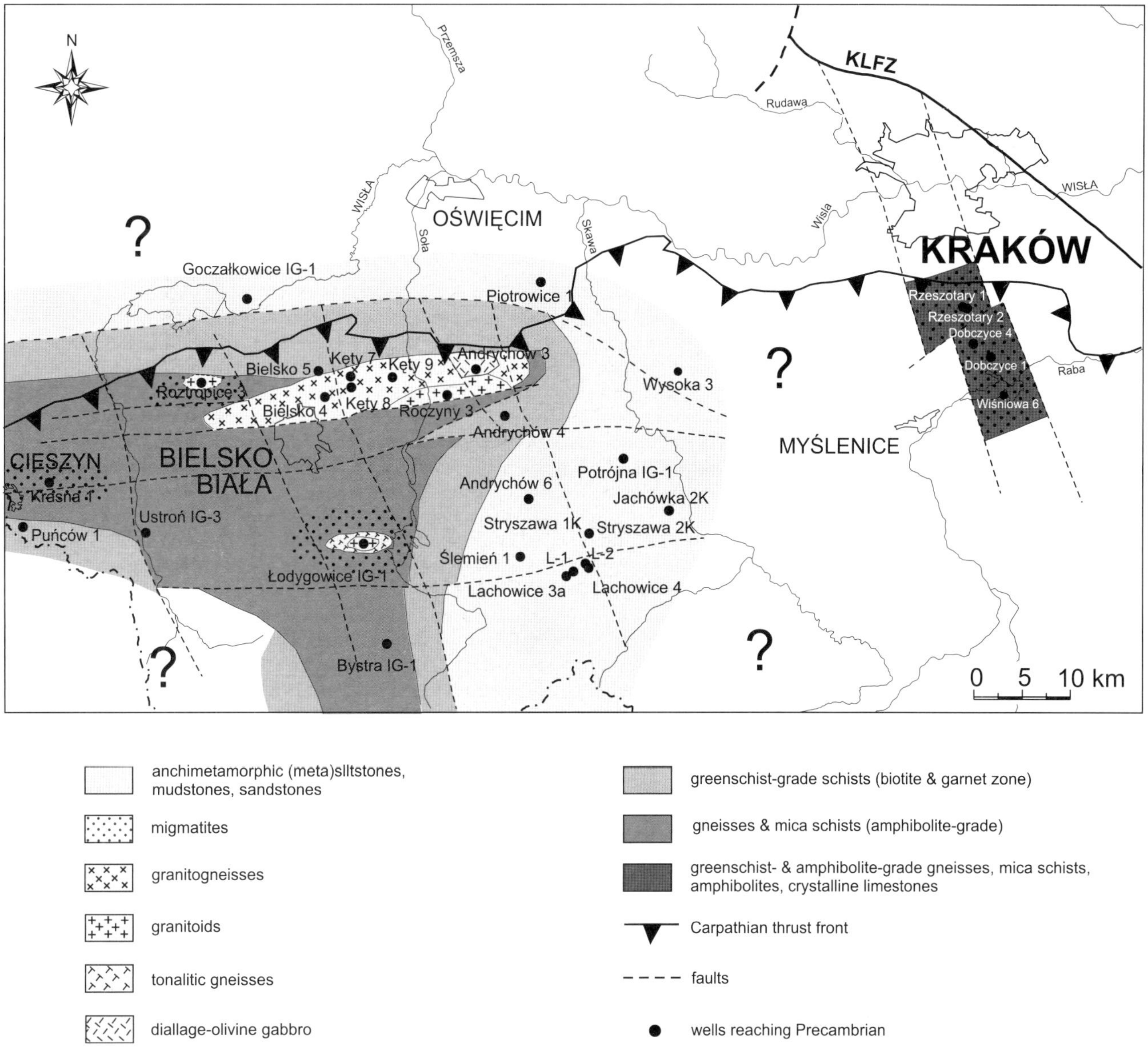

Fig. 2.24. Precambrian of the Upper Silesia Block. KLFZ, Kraków-Lubliniec Fault Zone.

is composed of anchimetamorphic (locally epizonal) rocks. The older rocks are present within the Bielsko-Biała–Andrychów High and the Rzeszotary Horst (e.g. Moryc & Heflik 1998 and references therein). These are mainly gneisses (predominantly paragneisses) and mica schists altered under amphibolite facies conditions and which are locally migmatized. On the southern slope of the Bielsko-Biała–Andrychów High, hornfels and tonalitic gneisses of trondhjemitic composition occur; these are accompanied by leucogneisses and leptynites (Wieser 1974). The metamorphic grade in the rocks of the Bielsko-Biała–Andrychów High decreases to that of greenschist facies in its peripheral parts. In the Rzeszotary Horst, the older metamorphic complex comprises various gneisses and schists, accompanied by amphibolites and carbonate intercalations. Both the gneisses and amphibolites are migmatized.

Early radiometric determinations (K–Ar method) yielded an age of 837 Ma for the crystalline rocks at the Rzeszotary Horst (Burchart 1971). More recently the amphibolites of this area were dated at 2.7 Ga using conventional U–Pb methods (Bylina *et al.* 2000). Age determinations (U–Pb SHRIMP) of zircons derived from various parts of the migmatized amphibolites indicate (Żelaźniewicz *et al.* 2004) that these rocks represent an Archaic (2.8–2.6 Ga) mafic intrusion, metamorphosed and subsequently deformed during Proterozoic times (2 Ga).

Precambrian igneous rocks have been found only in the central, axial part of the Bielsko-Biała–Andrychów High (e.g. Heflik & Konior 1974; Moryc & Heflik 1998; Żelaźniewicz & Żaba 2002; Żelaźniewicz *et al.* 2002). These include granites and diallage–olivine gabbros. U–Pb SHRIMP analyses of zircon samples from the granites (Żelaźniewicz *et al.* 2002) yielded intrusion ages of 573 ± 3 Ma (Kęty 8 borehole), 589 ± 35 Ma (Roztropice 3 borehole), 544 ± 4 Ma (Roczyny 3 borehole) and *c.* 620–580 Ma (Łodygowice IG-1 borehole). Thus the granitoids were intruded at *c.* 590–545 Ma, having originated from the melting of rocks that had crystallized at 640–620 Ma. These data suggest that there was a thermal event which included both the gneisses and migmatites of the Bielsko-Biała–Andrychów High at *c.* 625–555 Ma (Żelaźniewicz *et al.* 2004).

Both the Upper Silesian and Małopolska blocks contain lithologically similar, anchimetamorphic, locally phyllitic, clastic rocks represented by (meta)claystones, mudstones and sandstones with minor conglomerates. The Ediacaran age of these rocks was confirmed palaeontologically using acritarchs (Moryc & Jachowicz 2000; Jachowicz *et al.* 2002).

The proportions of particular clastic lithologies present in the Ediacaran profiles drilled from the USB and the MB are variable. The sedimentary succession of the USB is characterized by sandstones with fine-grained intercalations and, locally, with conglomerate beds containing clasts of metamorphic and igneous rocks up to 15 cm in diameter (Kotas 1982). This contrasts with the succession from the MB where the Ediacaran deposits are predominantly fine-grained clay-mud-rich rocks with rare sandy intercalations. However, the western and SW parts of the MB are more sandstone-rich, and even include thick conglomerate beds (Moryc & Jachowicz 2000).

The Ediacaran anchimetamorphic rocks which overlie the USB and the MB show common lithological, sedimentological and biofacies features, including: (a) distinct rhythmicity (cyclicity) of deposition defined by the alternation of coarse- to fine-grained sediments; (b) conglomerates and sandstones are generally poorly sorted; (c) in coarse-grained clastics, normal graded bedding is common; (d) horizontal lamination, often tectonically deformed, is frequent in the mudstones and claystones; (e) variegated colours in the sediments, due to haematite admixture, are typical, particularly in the rocks which overlie the MB; (f) sandstone beds sometimes contain single lithoclasts, mostly represented by dark, poorly rounded claystones and mudstones, from one to tens of centimetres in diameter; (g) thin tuffs and tuffites are present, mostly overlying the MB (Jurkiewicz 1975; Kowalczewski 1981; Compston *et al.* 1995).

The lithological and sedimentological characteristics of the Ediacaran rocks suggest that they were deposited in a marine setting, where turbidity currents played a significant role. The minimum thickness of the Ediacaran deep-marine succession of the USB and the MB is >1000 m and probably up to *c.* 3000 m. The sediments contain detrital fragments of igneous (mostly acid volcanic and plutonic), metamorphic (mylonites, paragneisses, mica schists, quartzites) and anchimetamorphic or strongly lithified clastic sedimentary rocks (e.g. Jurkiewicz 1975; Moryc & Heflik 1998; Moryc & Łydka 2000; Jachowicz *et al.* 2002; Żelaźniewicz *et al.* 2004). The source areas for the protolith of these rocks were similar to those of the sandstones and conglomerates, and must have been located on an active continental margin composed of quartzo-feldspathic rocks or on a magmatic arc developed on continental crust (Żelaźniewicz *et al.* 2004). U–Pb geochronological studies of detrital zircons from the coarse-grained Ediacaran rocks of the USB and the MB indicated that they were eroded from rocks that underwent a tectonothermal event at *c.* 620–545 Ma (Jachowicz *et al.* 2002; Żelaźniewicz *et al.* 2004), typical of the crust produced by the Cadomian Orogeny.

The Ediacaran anchimetamorphic rocks of the USB and the MB record at least three tectonic deformation events (D_1–D_3) associated with pre-Variscan folding. The dominant structural style of these rocks was mostly determined by the D_2 buckle folding, which produced recumbent, tight to open macrofolds (less frequently mesofolds). The earlier F_1 folding resulted in asymmetric, micro- to mesofolds (less frequently macrofolds) and a penetrative S_1 axial-plane cleavage. The third-generation folding F_3 was mainly controlled by low-angle thrust faults.

The Thaya/Slavkov Terranes and the Metabasite Belt (east Bohemia) (V.K.)

The Thaya and Slavkov terranes and a related Metabasite Belt form part of the Brunovistulian Block (Dudek 1980; Finger *et al.* 2000*a*) which is incorporated into the Moravo-Silesian Zone (MSZ) in the eastern part of the Bohemian Massif (Fig. 2.25A). These terranes show a strong relation to Avalonia (e.g. Finger *et al.* 2000*a*) and not to Cadomia and are, therefore, described herein separately from the rest of the Bohemian Massif.

The western boundary of the MSZ, comprising the Brunovistulian Block, is represented by the Lugodanubian Fault along which the Brunovistulian Block is underthrust top-to-the-west below the Moldanubian Zone (see below). The eastern limit of the MSZ is formed by the Peri-Pieninian Fault in the region of the Carpathians. To the south, the MSZ extends to the Danube Fault. The northern border of the MSZ is located in Poland and is defined by the system of NW–SE trending faults of the Odra Fault Zone (Buła *et al.* 1997*b*; Belka *et al.* 2002). It has been suggested that the USB in SE Poland (see above) represents the northernmost part of the Brunovistulian Block (Dudek 1980; Buła *et al.* 1997*b*).

The Brunovistulian Block was affected by a late Neoproterozoic orogenic event (Dudek 1980) and is covered by a Palaeozoic overstep sequence (Fig. 2.26). The block can be subdivided into three separate parts represented by the Thaya Terrane (Finger & Pin 1997; Finger *et al.* 2000*b*), the Metabasite Belt (Štelcl & Weiss 1986; Hanžl & Melichar 1997; Leichmann 1996; Finger *et al.* 2000*a*), and the Slavkov Terrane (Finger & Pin 1997).

Rocks of the *Thaya Terrane* are exposed in the eastern part of the Thaya Massif and in small relicts along the Diendorf Fault (Fig. 2.25B). The Thaya Terrane consists chiefly of potassium-rich biotite and two-mica granodiorites and granites close to the contact with the Moravian nappes, which are highly deformed. More basic amphibole–biotite quartz diorites, abundant in the western sector of the Metabasite Belt, cover less than 10% of the area of the Thaya pluton.

The metamorphic rocks in the mantle of the Thaya Massif in Austria are represented by phyllites and mica schists of the Therasburg Formation (Höck & Vetters 1975; Höck *et al.* 1990) formed by alternating initially argillaceous and sandy sediments with interbeds of limestones and marl slates. These low-to-medium grade metamorphic rocks with bodies of basic rocks are exposed near Krhovice, east of Znojmo, and probably also in the lower parts of the Miroslav Horst (Dudek 1963).

The *Metabasite Belt* is up to 5 km wide, trends roughly north–south and consists mainly of basic intrusive and volcanic rocks (Fig. 2.25B). Ultrabasic rocks and some acid dykes, mainly of rhyolite composition, are also present (Hanžl *et al.* 1995;

Fig. 2.25. Geology of the Moravo-Silesian Zone. (**A**) Position of the Brunovistulian Block with the Slavkov and Thaya Terranes in the tectonic framework of the Central European Variscides. (**B**) Geological setting of the Moravo-Silesian Zone based on surface outcrops and from drillings (modified according to Dudek 1980; Finger *et al.* 1995). DF, Diendorf Fault; MB, Metabasite Belt; ST, Slavkov Terrane; TT, Thaya Terrane; Orthogneisses of the Moravo-Silesicum: BO, Bíteš orthogneiss; DO, Desná orthogneiss; KO, Keprník orthogneiss; NO, Nectava orthogneiss.

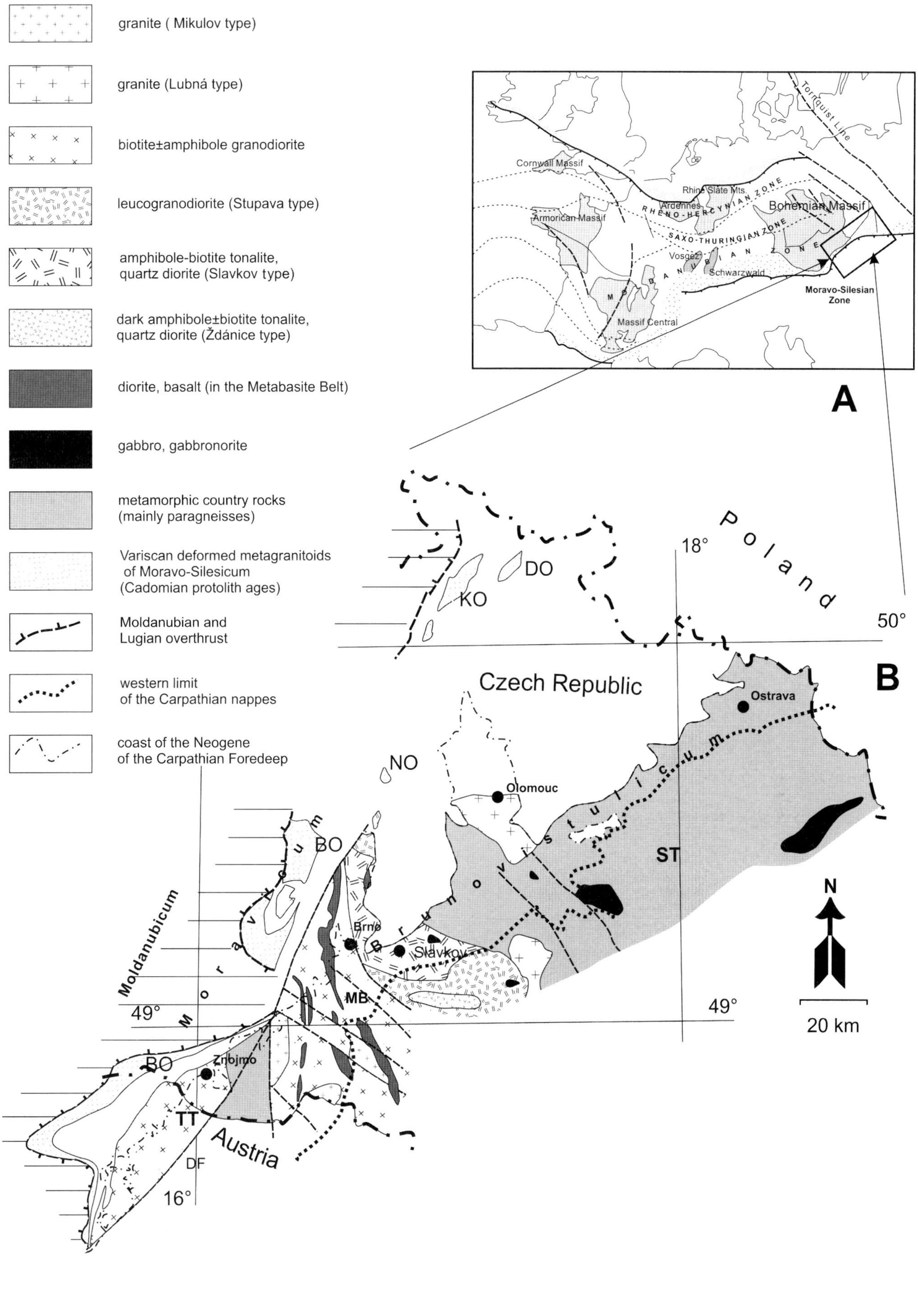
granite (Mikulov type)
granite (Lubná type)
biotite±amphibole granodiorite
leucogranodiorite (Stupava type)
amphibole-biotite tonalite, quartz diorite (Slavkov type)
dark amphibole±biotite tonalite, quartz diorite (Ždánice type)
diorite, basalt (in the Metabasite Belt)
gabbro, gabbronorite
metamorphic country rocks (mainly paragneisses)
Variscan deformed metagranitoids of Moravo-Silesicum (Cadomian protolith ages)
Moldanubian and Lugian overthrust
western limit of the Carpathian nappes
coast of the Neogene of the Carpathian Foredeep
Tornquist Line
Cornwall Massif
Rhine Slate Mts.
Ardennes
RHENO-HERCYNIAN ZONE
Bohemian Massif
Armorican Massif
SAXO-THURINGIAN ZONE
Vosges
MOLDANUBIAN ZONE
Schwarzwald
Massif Central
Moravo-Silesian Zone
A
Poland
18°
DO
KO
50°
Czech Republic
B
Ostrava
NO
Olomouc
Silesicum
BO
ST
Moldanubicum
Moravicum
Brunovistulicum
Brno
Slavkov
N
MB
49°
49°
20 km
BO
Znojmo
TT
Austria
DF
16°

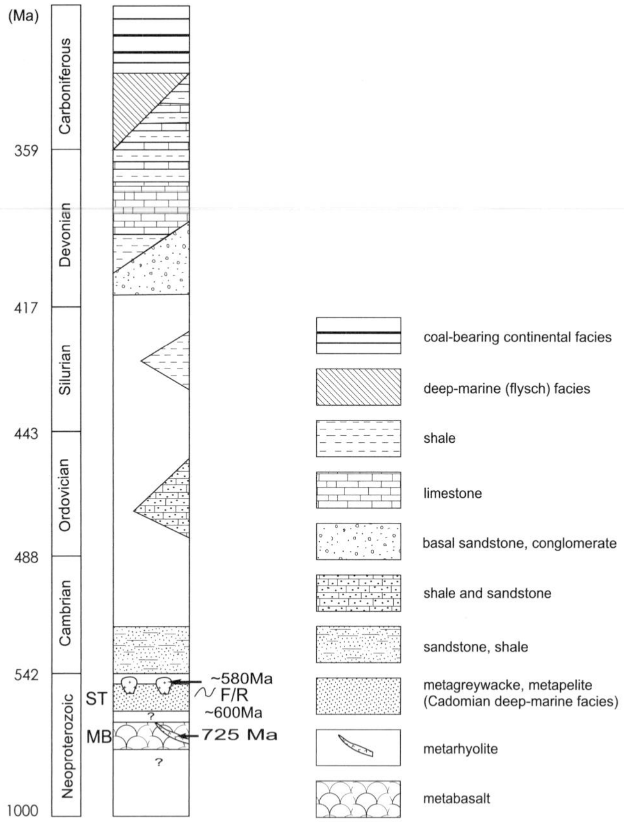

Fig. 2.26. Lithostratigraphic scheme of the Neoproterozoic basement and Palaeozoic cover of the Brunovistulian Block (according to Kalvoda *et al.* 2002, slightly modified). F/R, Peak of folding and regional metamorphism during a Late Neoproterozoic orogeny related to Avalonian–Cadomian orogenic processes; MB, Metabasite Belt; ST, Slavkov Terrane.

Finger *et al.* 2000*a*). Basic intrusives (metadiorites, metagabbros) predominate in the western part of the zone, whereas volcanic members – chiefly tholeiitic metabasites, locally with preserved textures of pillow lavas (Štelcl & Weiss 1986; Hanžl *et al.* 1995; Leichmann 1996) and penetrated by dykes of metarhyolite (725 ± 15 Ma; Finger *et al.* 2000*b*) and metadolerite – are more abundant in the eastern part of the zone.

The *Slavkov Terrane* includes the remaining part of the Brunovistulian Block located to the east of the Metabasite Belt. It is exposed only in the eastern part of the Brno Massif and as uplifted blocks in the Neogene-age sediments of the Upper Moravian valley. The remainder is covered by younger sediments ranging from Lower Cambrian to Neogene in age. NW–SE trending fault zones divide this terrane into the Central Moravian and North Moravian blocks, and these differ in terms of the proportions of intrusive and metasedimentary rocks. Granitoids predominate in the Central Moravian Block, SE of Brno. These granites are mainly represented by biotite and amphibole–biotite granodiorites and leucotonalites and minor quartz diorites or gabbronorites. Granites are also present along the Haná Fault Zone (SE of Olomouc). Metasediments are rare in the NE part of this block (Fig. 2.25B). Varied types of phyllites, mica schists and paragneisses predominate in the North Moravian Block (Fig. 2.25B) (Dudek 1980). These were mostly derived from greywacke protoliths, with the more pelitic lithologies being subordinate. Intercalations of basic rocks (chlorite–actinolite schists, epidote amphibolites) enclosed in chlorite and graphite phyllites are rare. The metamorphic grade ranges from the greenschist through to the amphibolite zone. Banded garnet-bearing paragneisses were often migmatized in the region of diorite intrusions. The low contents of Al, together with higher proportions of plagioclase and low initial $^{87}Sr/^{86}Sr$ ratios (0.704–0.706) and εNd_{580} ratios (−1 to +2), similar to the adjacent granitoids, indicate that the greywackes were derived from a primitive crustal source such as an island arc (Finger *et al.* 2000*a*). Lower values of εNd (−3 to −7) in the paragneisses of the Thaya Terrane indicate the presence of a more mature crust in its source area.

Jelínek & Dudek (1993) and Hanžl & Melichar (1997) assumed that the Brunovistulian Block was formed at the Avalonian–Cadomian active margin of peri-Gondwana in Neoproterozoic times. Additionally, numerous authors have correlated the Brunovistulian Block with Avalonian-type basement units of the Rhenohercynian Zone based on the overlying Palaeozoic cover (Dvořák 1995; Tait *et al.* 1997; Friedl *et al.* 2000). Possible equivalents have also been identified in the Zonguldak Terrane in Turkey by Kalvoda *et al.* (2002).

The SHRIMP dating of inherited zircons from the Bíteš (Friedl *et al.* 2000), Keprník and Desná orthogneisses (Hegner & Kröner 2000) and the Nd model ages of granitoids and metasediments of the Desná and Keprník domes (Hegner & Kröner 2000) suggest that the Brunovistulian Block was most probably an Avalonia-type terrane according to Murphy *et al.* (2004) and that it was derived from an area related to the Amazonian Craton. This idea has been supported by magmatic ages in the Brunovistulian Block (630–580 Ma) which are similar to ages of Avalonian magmatic events and differ from the ages representing the climax of magmatic activity in those units of the Bohemian Massif which are related to Cadomia (570–540 Ma; Linneman *et al.* 2004, 2007). There are also differences between the regional metamorphism age (*c.* 600 Ma) in the Brunovistolian Block and the Cadomia-related Teplá-Barrandian Unit of the Bohemian Massif (*c.* 540 Ma; Zulauf *et al.* 1999). Additionally, the age of basaltic volcanism in the Metabasite Belt (*c.* 725 Ma; Finger *et al.* 2000*b*) is close to that of similar basaltic volcanites from Newfoundland *c.* 780 Ma; O'Brien *et al.* 1996), which represent the type region for Avalonia.

The measured difference between the ages of the Metabasite Belt and the younger granitoids would appear to be in accordance with the idea of prolonged subduction processes taking place along the northern edge of the Gondwana Supercontinent during the Neoproterozoic. At this time, sedimentation of deep-marine arc-derived sediments (preserved in the Slavkov Terrane) occurred in an island arc/backarc setting between *c.* 700 and 600 Ma (Finger *et al.* 2000*a*). The onset of regional metamorphism and associated deformation at *c.* 600 Ma may have reflected the collision of the Gondwana margin with the volcanic arc. During this event, the Thaya and Slavkov terranes collided and the Metabasite Belt was obducted onto the Thaya Terrane (Finger *et al.* 2000*a*). This major collisional event was followed by the intrusion of post-collisional granitoids at *c.* 590–580 Ma (Dudek & Melková 1975; Friedl *et al.* 2000; Finger *et al.* 2000*a*). An unmetamorphosed Lower Cambrian sedimentary succession unconformably overlies the Brunovistulian Block (Jachowicz & Přichystal 1997) and post-dates the orogenic event in the Late Neoproterozoic (which was most probably related to the Avalonian–Cadomian active margin). There is evidence from the Brunovistulian Block that the rift–drift transition and subsequent separation from the Gondwana margin commenced in the Cambrian (Winchester *et al.* 2002) (see also Krawczyk *et al.* 2008).

Cadomia and related terranes

Cadomia is a belt of Neoproterozoic rock units formed by Cadomian orogenic processes at an active margin at the periphery of the West African Craton of Gondwana. The main events in the evolution of Cadomia characterized by the Cadomian Orogen are as follows (see Linnemann *et al.* 2008): (1) development of a juvenile island arc from *c.* 750 Ma onwards, which resulted in the formation of a number of juvenile felsic island arcs at *c.* 750–650 Ma; (2) transition to continental magmatic arc and significant crustal thickening at *c.* 620 Ma and contamination by cratonic basement aged *c.* 2 Ga; (3) development of the early marginal basins by *c.* 610 Ma to *c.* 585 Ma; (4) a major tectonomagmatic pulse at 580–570 Ma in the continental magmatic arc leading to the inversion and deformation of early basins which were intruded by *c.* 580–570 Ma old plutons; (5) origin of late marginal basins and cessation of calc-alkaline magmatism and increasing oblique subduction and a gradual transition to a transform margin (*c.* 570–540 Ma); several backarc basins were originated and passive margin sequences witness the continental margin location (West African Craton) of some of these basins; and (6) collisional processes resulted in the formation of a fold-and-thrust belt and the development of retro-arc basins in which the debris of the magmatic arc and the backarc basins accumulated and final transpressive amalgamation of the magmatic arc and possibly other terranes led to transpressional thickening, magmatic underplating and slab break-off to generate the thermal energy to generate widespread anatexis and voluminous intracrustal plutonism during latest Ediacaran and earliest Cambrian time, represented by the final tectonomagmatic pulse at *c.* 540–530 Ma.

Rock units of Cadomia contain few or no zircons with ages in the range 0.75–1.6 Ga and are dominated by detritus from the West African Craton (2.05 Ga and older). Crust aged *c.* 2.0 Ga was recycled (Murphy *et al.* 2004). Basement slivers of rocks such as the Icartian basement, *c.* 2 Ga, occur in places. The characteristic Cadomian basement was formed by the Cadomian Orogeny (*c.* 750–530 Ma; see above).

Most probably, Cadomia never left the Gondwana mainland and did not drift off like Avalonia during the Palaeozoic (Linnemann *et al.* 2004). The overstep sequence covering the Cadomian basement includes Cambro-Ordovician, Silurian and Devono-Carboniferous sediments that were derived from the West African Craton (Linnemann *et al.* 2004, 2007) and a distinct peri-Gondwanan fauna (Fortey & Cocks 2003). After the separation of Avalonia and related terranes from the Gondwana mainland during the Cambrian–Ordovician, Cadomia formed the southern margin of the Rheic Ocean.

During the assemblage of the Pangaea Supercontinent by Variscan orogenic events, the Rheic Ocean closed as a result of continent–continent collision between Laurussia and Gondwana in Late Devonian–Early Carboniferous times. The Mid-German Crystalline High at the northern rim of the Bohemian Massif is believed to be the suture of the Rheic Ocean. To Cadomia (formerly 'Armorica' or 'Armorican Terrane Assemblage'; see also Robardet 2003) we include the Armorican Massif, Pyrénées, French Massif Central, Maures Massif, Corsica, and much of the Bohemian Massif, including parts of the Saxo-Thuringian Zone, the Sudetes, and the Moldanubian Zone. The latter contains the Tepla-Barrandian Unit and the Moldanubicum *sensu stricto*.

Armorican Massif (C.P.)

The northern and NE parts of the Armorican Massif contain the majority of the documented Precambrian outcrops in France. This is because, in marked contrast with the Southern Armorican Massif and the rest of the pre-Mesozoic basement in France, Variscan tectonic overprinting was non-penetrative in these regions, and was limited to the reactivation of major fault zones, accompanied by relatively minor granite magmatism. In addition, the Armorican Massif includes the type locality for the Late Proterozoic Cadomian Orogen (from *Cadomus*, the Latin name of the town Caen), part of the worldwide pan-African belt.

For these reasons, and despite the fact that the Armorican Massif is located outside of the area of Central Europe as defined in this book (Fig. 2.27), its major pre-Palaeozoic features will be briefly reviewed herein. The northern Armorican Massif comprises a collage of different tectonostratigraphic domains bounded by major shear zones of Cadomian age and subsequently reactivated to varying degrees during Variscan times. To circumvent this level of complexity, the following description is a geographical one, extending from the Trégor–Channel Islands–North Cotentin Peninsula Domain in the north, to the Mancellian and Central Armorican Domains in the south, via the intermediate units of St Brieuc, Guingamp, and St Malo (Chantraine *et al.* 2001). Information is also provided on possible correlatives in Vendée (SE Armorican Massif) and beneath the Paris Basin.

Small fragments of ancient Palaeoproterozoic basement have been documented in several areas of the northernmost Cadomian Block (Trégor–North Cotentin Domain): (a) in the Channel Islands, the Icart meta-igneous gneisses of Guernsey which were emplaced at *c.* 2.06 Ga (Samson & D'Lemos 1998); (b) along the shore of Trégor, the Trebeurden granodioritic orthogneiss with igneous emplacement ages of 2031 +36/−28 Ma and 2000 +35/−30 (Auvray *et al.* 1980); and (c) at the northern tip of the Cotentin Peninsula (La Hague), where the Culeron orthogneisses have an igneous crystallization age of 2061 ± 3 Ma (Inglis *et al.* 2004). In contrast to previous hypotheses which suggested that there was a Palaeoproterozoic metamorphic Icartian event, more recent studies indicate that these granitoids underwent high-grade metamorphism only in the Neoproterozoic (620–610 Ma) during the Cadomian Orogeny, as inferred from the precise U–Pb analyses of syntectonic intrusions (Miller *et al.* 1999), as well as SHRIMP dating of zircon overgrowths in the Culeron gneisses (Inglis *et al.* 2004).

A very large gap (*c.* 1.3 Ga) in the geological record is a major feature of the chronostratigraphy of this region, since there are no radiometric dates before the subsequent phase of magmatism, which occurred at *c.* 750 Ma, with the emplacement of the igneous protoliths of plagioclase gneisses from St Brieuc Bay (Jospinet and Port-Morvan orthogneisses). These have been dated by the U–Pb zircon method at 755 ± 1 Ma and 747 ± 1 Ma, respectively (Samson *et al.* 2003). These orthogneisses occur as a basement slice (referred to as 'Pentevrian') forming the base of the composite St Brieuc Unit. The next documented igneous event occurred *c.* 90 Ma later, and is evidenced by trondhjemite boulders from the Cesson conglomerate in the St Brieuc Unit, and dated at between 666 ± 1 Ma and 656 ± 5 Ma by the U–Pb zircon method (Guerrot & Peucat 1990; Samson *et al.* 2003). Younger detrital zircons from sandstones of the same sedimentary unit yielded concordant U–Pb ages ranging from 650 to 624 Ma (Samson *et al.* 2003). The most recent of these dates coincides with the emplacement of the Jospinet granodiorite (626 +4/−2 Ma; Nagy *et al.* 2002), which itself is unconformably overlain by Brioverian-age metasediments.

The subsequent emplacement of syntectonic quartz-dioritic intrusions in the Channel Islands and on northern Cotentin coincided with an important episode of deformation, magmatism

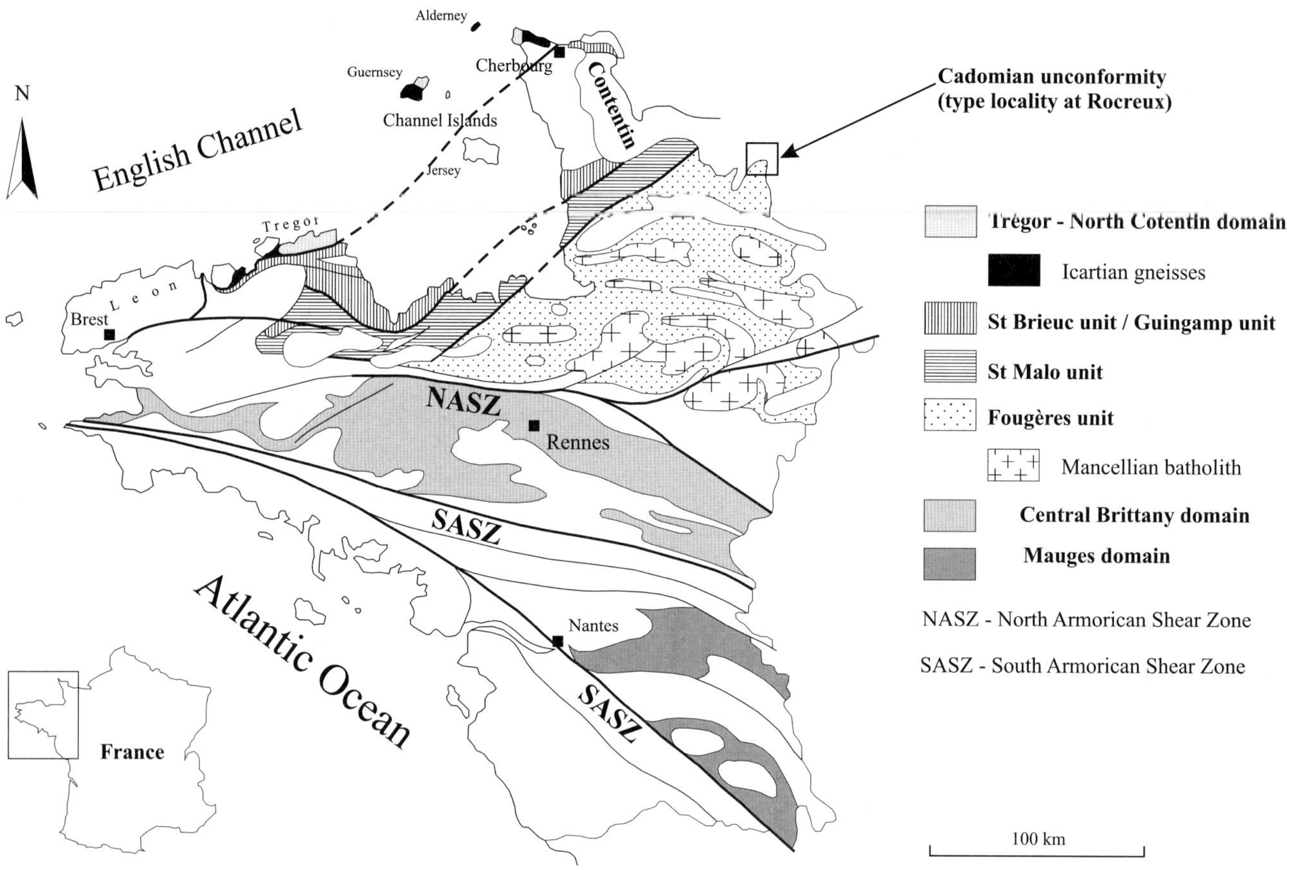

Fig. 2.27. Geological and structural domains and units of the Armorican Massif.

and metamorphism which has been dated at between 616 and 608 Ma (see references in Samson *et al.* 2003). The North Trégor Batholith was emplaced at the same time (615 +13/−7 Ma, U–Pb zircon; Graviou 1984) along with coeval felsic volcanics (Lézardrieux ignimbrite, *c.* 615 Ma; Chantraine *et al.* 2001). Ar/Ar ages indicate that post-metamorphic cooling occurred in these regions at *c.* 600–610 Ma (Dallmeyer *et al.* 1991*a*; Ruffet *et al.* 1991).

Further to the south, in the St Brieuc Unit, a pre-tectonic tonalite (Port-Moguer) has been dated at 600 ± 1 Ma by the U–Pb zircon method (Nagy *et al.* 2002). This was followed by a younger generation of mafic to intermediate intrusions which were emplaced syn- to late-kinematically at *c.* 580 Ma, as exemplified by the U–Pb zircon ages measured for the St Quay-Portrieux gabbrodiorite (575 ± 2 Ma) and the Fort-la-Latte quartz diorite (576 ± 2 Ma) by Nagy *et al.* (2002) in St Brieuc Bay, and for the Moulinet (585 ± 12 Ma) and Coutances (584 ± 4 Ma) quartz diorites in Normandy (Guerrot & Peucat 1990). Ar/Ar ages suggesting cooling through *c.* 500°C cluster at *c.* 560–565 Ma (Dallmeyer *et al.* 1991*b*). These data imply that the major tectonothermal event occurred shortly before 575 Ma in this part of the Cadomian Belt. Similar igneous emplacement ages have been obtained for a rhyolite (583 ± 3 Ma) and a post-tectonic granite (580 ± 2 Ma) from Jersey in the Channel Islands (Miller *et al.* 2001).

Significant variations in magma sources within the region are indicated by the contrasting Nd isotope signatures which have been measured (Guerrot 1989; Samson *et al.* 2003). The 620–610 Ma syntectonic granitoids of the Trégor–North Cotentin Unit are characterized by strongly negative values (*c.* −8/−10) suggesting that there was a major contribution of ancient sialic crust enriched in LREE; such crust may be represented by the 2 Ga gneisses which locally occur as the country-rocks of the intrusions. These granitoids were possibly generated in a continental arc setting during oblique subduction (e.g. Gibbons & Moreno 2002), or alternatively as the result of arc–continent collisional processes. In marked contrast, the pre-tectonic 755 to 625 Ma rocks from the St Brieuc Unit show strongly positive εNd values (from +5 to +7; Samson *et al.* 2003) reflecting time-integrated depleted mantle sources, probably in an intra-oceanic arc setting. The syn- to late-kinematic intrusions of the St Brieuc Unit display relatively unradiogenic Sr isotope signatures ($^{87}Sr/^{86}Sr$ *c.* 0.704–0.705) and positive εNd values from +2 to +5 in the *c.* 580 Ma diorites of Fort-la-Latte and Coutances (Guerrot 1989). These results would suggest that the Icartian basement played a minor role, or even was absent at depth, in the area outside of the Trégor–North Cotentin Domain.

In addition to calc-alkaline intrusives, the tectonically composite St Brieuc Unit contains volcanic and sedimentary units. The Paimpol Formation displays weak, heterogeneous deformation and metamorphism. It comprises basaltic lavas, sometimes pillowed (the Paimpol spilites), mafic and acid pyroclastics, and doleritic and keratophyric dykes (Egal *et al.* 1996). The metabasalts are chemically comparable to present-day arc tholeiites (Egal *et al.* 1996), with high εNd values (+5 to +6; Dabard *et al.* 1996). The mafic volcanics of the Paimpol Formation contain minor rhyolites with a 610 ± 9 Ma zircon Pb/Pb-evaporation age (Egal *et al.* 1996). Possible correlatives in Normandy may be the

Montsurvent spilite–keratophyre volcanics. This latter formation, post-dated by the 584 Ma Coutances diorite, displays a wide range of compositions, from basalt to rhyolite through andesites and voluminous pyroclastic rocks, and shows strong affinities with modern intra-oceanic island arcs (Dupret *et al.* 1990).

The Paimpol Formation spilites are conformably overlain by detrital sediments (La Roche Derrien Formation) made up of alternating greywackes, siltstones and mudstones with some volcanogenic intercalations. These sediments are interpreted as turbidites which were deposited in an active marginal basin (Denis 1988). Rare andesitic and rhyolitic lavas and tuffs also occur (Egal *et al.* 1996).

The Paimpol and La Roche Derrien formations are thrust over a bimodal volcanic suite (Lanvollon Formation) which was strongly deformed and metamorphosed under greenschist facies (in the north) to amphibolite facies (in the south). This formation contains rare sedimentary intercalations, including the basal Cesson conglomerate deposited later than 625 Ma, as indicated by the detrital zircon data of Samson *et al.* (2003). A 588 ± 11 Ma Pb/Pb-evaporation age was measured on zircon from a felsic gneiss (Egal *et al.* 1996). The interpretation of the geochemical affinities of metavolcanic rocks, which have mildly positive εNd values (+2 to +3; Dabard *et al.* 1996), has proved to be equivocal, with inferred geodynamic settings ranging from backarc basin (Cabanis *et al.* 1987) to intraplate rift (Lees *et al.* 1987; Egal *et al.* 1996).

The *c.* 2 km thick Erquy Mafic Volcanic Formation, broadly equivalent to the Lanvollon Formation and initially used as a stratotype for the Lower Brioverian, was originally believed by some authors to be Ordovician in age on the basis of Rb–Sr isotopic (Vidal *et al.* 1971; Clauer *et al.* 1985) and micropalaeontological (Deunff *et al.* 1973) data, although other scientists did not accept this (e.g. Brown & Roach 1972; Lees *et al.* 1987). More recently, SHRIMP U–Pb dating of zircons extracted from interbedded cinerites gave an age of 608 ± 7 Ma (Cocherie *et al.* 2001), thereby confirming the initial assignment to the Brioverian.

The Lanvollon Formation shows a progressive transition to the overlying Binic Formation, which is composed of rhythmically bedded immature sediments of volcaniclastic origin that have been interpreted as turbidites (Denis 1988). These sediments are weakly deformed and metamorphosed, apart from the area of the contact aureole developed around the St Quay Portrieux gabbro-diorite. The age of deposition of the Binic Formation is younger than that of the underlying Jospinet granodiorite (i.e. 626 +4/−2 Ma), and older than that of the cross-cutting St Quay intrusion (i.e. 575 ± 2 Ma) (Nagy *et al.* 2002).

The Trégor–Channel Islands–North Cotentin and St Brieuc–Coutances regions (corresponding to the so-called Domnonean Domain; Cogné 1972) contain mainly igneous rocks (including *c.* 2 Ga relict in the northern Trégor–North Cotentin Unit) with subordinate basins filled by immature sediments. Further to the south, the broad Mancellian–Central Armorican Domain exposes mainly terrigenous sediments, which were intruded by a large granite batholith emplaced at *c.* 540 Ma (Pasteels & Doré 1982). These two contrasting domains are separated by an intervening high-metamorphic-grade domain (the Guingamp Unit) which is bounded by two major Cadomian faults, subsequently reactivated during Variscan times (Chantraine *et al.* 2001).

In addition to the Cadomian-age anatexites and late-stage (540–530 Ma) leucogranites, the Guingamp Unit includes, at its southern and basal part, a tectonic slice composed mainly of gabbroic rocks with ultrabasic cumulates. The two main gabbroic bodies (Yffiniac and Belle-Isle-en-Terre) were dated at 602 ± 8 Ma and 602 ± 4 Ma, respectively (Guerrot & Peucat 1990; Peucat *et al.* 1981). Certain amphibolites from the unit are similar to oceanic tholeiites in composition (Chantraine *et al.* 2001). The Yffiniac-Belle-Isle Formation underwent relatively high-pressure metamorphism (9–11 kbar, *c.* 700°C; Ballèvre *et al.* 2001), and is interpreted as representing a major contact, possibly a suture (Graviou 1992; Chantraine *et al.* 2001), of the Cadomian Belt.

On the southern side of this major contact zone, the St Malo Unit, and its eastern prolongation in the St Lô region, occupy an external position with respect to the Cadomian Belt. The St Malo Unit is predominantly composed of the weakly metamorphosed monotonous terrigenous sediments (including greywackes rich in lithic fragments, sandstones, siltstones and mudstones) of the Lamballe Formation. These can be correlated laterally with the St Lô Formation in Normandy. Hummocky cross-stratification, observed within these terrigenous sediments, indicates a depositional environment around storm wave base (Dabard & Loi 1998). The presence of interbedded siliceous carbonaceous beds (i.e. black cherts, formerly referred to as 'phtanites'; see Roblot (1968) and references therein) is a typical feature of these formations, and was classically considered to be characteristic of the Lower Brioverian succession in general (e.g. Cogné 1972). However, recent studies would appear to preclude a primary origin for the black cherts, but suggesting that their origin is related to diagenetic processes involving the silicification of terrigenous and mixed terrigenous evaporitic facies (Dabard 2000). Such an interpretation would thus question the use of these black cherts as a lithostratigraphic marker. Silicified dolomitic limestones also occur within the units in Normandy, but fault contacts generally prevent clear correlation of the closely associated facies of black cherts (Dupret *et al.* 1990). The only recorded fossil evidence consists of microspheres of cyanophytan origin (Chauvel & Schopf 1978). Clastic supply was derived from broadly coeval acid to intermediate volcanics as well as from reworked basement, *c.* 2 Ga old (Dabard 1990). Basaltic rocks associated with these two formations are either of within-plate alkaline affinity (Cabanis *et al.* 1987) in the Lamballe Formation (Chateau-Serein), or similar to N-MORB (Dupret *et al.* 1990) in the St Lô Formation (La Terrette).

In summary, the formations containing the interbedded cherts may have been deposited in a shallow, coastal shelf environment (Dabard & Loi 1998) adjacent to an incipient intra-arc or backarc basin. This presumably occurred during an early stage of magmatic arc activity and may, thus, have been coeval with the formation of the volcanic and volcaniclastic rocks of the St Brieuc Unit. Alternatively, the deposition of these sediments may have occurred in a tectonic accretionary prism associated with a subducting slab dipping to the north (Graviou 1992). The metasediments of the Lamballe Formation are generally affected by low-grade metamorphism, but locally, they suffered *in situ* partial melting. This melting resulted in the formation of the St Malo migmatite dome (541 ± 5 Ma; Peucat 1986), and was possibly connected with local extension in a broad sinistral transpressive regime (Egal *et al.* 1996).

Further to the SW, beyond the Cancale-Granville Fault, the Fougères Unit exposes a monotonous deep-marine clastic series of alternating greywackes, siltstones and mudstones, with intercalations of conglomerates. These detrital sediments rework earlier deposits, particularly volcanic rocks and black cherts. Although such reworked 'phtanites' are classically considered to be a characteristic feature of the Upper Brioverian Group, no unconformity can be directly observed between the two groups of the Brioverian, which crop out on either side of the major

Cadomian shear fault marking the boundary between the St Malo and Fougères units (Chantraine *et al.* 1994). It is, therefore, not possible to determine whether the two groups which appear to follow each other in time were deposited in the same basin, or whether they were accumulated in two distinct basins or sub-basins (Chantraine *et al.* 1994).

A microflora found within the turbidites is comparable with similar records from the Late Proterozoic series of northern Europe and North America suggesting an Ediacaran age for these Upper Brioverian sediments (Choubert & Faure-Muret 1980; Chauvel & Schopf 1978). Such a conclusion would be in agreement with the biostratigraphic material found in the strata just above the unconformably overlying conglomerates and the subsequent Cambrian beds. This material includes trace fossils and microfossils (hyolithids, heteractinid sponges) which have been ascribed to the Tommotian (Doré 1994). The Vassy pillowed basalts are the only volcanic rocks documented in the Upper Brioverian Group. They are geochemically similar to N-MORB (Dupret *et al.* 1990), and provide evidence for an extensional tectonic regime and marked crustal thinning. The Upper Brioverian succession is interpreted to have been deposited in a turbidite-rich foreland basin, on an unknown substratum (Chantraine *et al.* 2001). Diamictites occur locally (Granville); their former interpretation in terms of tillites has since been refuted, and they are instead considered to be debris flows and other mass flow deposits (Doré *et al.* 1985).

The Upper Brioverian sediments were weakly deformed by open folds with steeply dipping cleavages, and metamorphosed before the intrusion of the Mancellian Batholith, at 540 ± 10 Ma (monazite U–Pb age; Pasteels & Doré 1982). The batholith is composed of several high-level plutons with narrow contact aureoles. Apart from the volumetrically minor basic bodies, it contains mainly peraluminous cordierite- or biotite-bearing granodiorites. These were probably formed by partial melting of metasediments in the lower crust, as suggested by their distinctly negative εNd values (from −4 to −7 on three samples analysed by D'Lemos & Brown (1991), and from −3 to −6 on nine samples measured by Le Fèvre (1997)), and model ages (T_{DM}; DePaolo 1981) ranging from 1.55 to 1.85 Ga. However, the geodynamic explanation for the copious crustal melting in this external domain of the Cadomian belt (i.e. thermal relaxation following crustal thickening, or crustal extension, possibly in a transcurrent regime, with associated underplating by mafic magmas) are still poorly understood (see discussion in Ballèvre *et al.* 2001).

The Central Brittany Domain is isolated from the Cadomian Belt by the major North Armorican Shear Zone. This domain contains widespread exposures of a monotonous, at least 1000 m thick, terrigenous succession composed of greywackes, siltstones, mudstones, conglomerates with chert clasts (Upper Brioverian Group), and rare limestones (Chantraine *et al.* 1994). These sediments are unconformably overlain by red beds of Early Ordovician age, with the stratigraphic gap interpreted as reflecting block tilting activity during an episode of Ordovician extension (Ballard *et al.* 1986; Brun *et al.* 1991). Despite the generally unconformable nature of the overlying Palaeozoic strata, the main deformation and slaty cleavage event observed in the underlying succession is considered to be of Variscan age (Le Corre 1977). The pre-Palaeozoic (Cadomian) deformation in the area was very mild, associated with the development of north–south orientated folds without any associated cleavage, and possibly of synsedimentary origin (Le Corre 1977). Therefore, cleavage cannot be used to differentiate between Cadomian-age basement and Palaeozoic cover in central Brittany, and thus the upper part of the Neoproterozoic Brioverian succession could well have been deposited during the Cambrian (Chantraine *et al.* 1994). LREE-depleted amphibolites, which are geochemically similar to N-MORB, were described in the NW part of the Brioverian outcrops of central Brittany (Saunier & Chauvel 1986). As with the Vassy spilite from lower Normandy, these indicate that there was a marked extensional tectonic regime during the deposition of the Upper Brioverian Group.

Close to the SE end of Armorican Massif, Precambrian rocks are well documented in the Mauges Domain. This domain is separated from the rest of the Armorican Massif by the Ligerian branch of the Variscan Belt. It is characterized by marked deformation and metamorphism – and possibly major displacements – during mid- to late Palaeozoic times. In this domain, which was generally shielded from the effects of pervasive Variscan overprinting, the rocks are slightly metamorphosed but undeformed. The sediments, including siltstones and conglomerates containing trilobites of Middle Cambrian age (Cavet *et al.* 1966), unconformably overlie a basement unit referred to as the Mauges Group. This basement is strongly deformed with a composite flat-lying S_{1-2} foliation, and metamorphosed under deep epizonal to mesozonal (5–7 kbar, 620–650°C) conditions (Wyns *et al.* 1998). The Mauges Group, which is several kilometres thick, is composed of monotonous clastic rocks ranging from quartz-feldspar metagreywackes to metapelites, together with black cherts and graphitic schists. Mafic meta-igneous rocks ranging from gabbros to cinerites through to lava flows, breccias and dykes also occur (Wyns *et al.* 1998). The major and trace element characteristics of these magmatic rocks reflect their affinities with transitional tholeiitic basalts (Cabanis & Wyns 1986), emplaced in a within-plate extensional setting.

To the east, the Cadomian Belt is concealed by the Mesozoic sediments of the Paris Basin. An extension of the Cadomian domain beneath this cover has been suggested by the occurrence, 1040 m below the surface, of a mylonitized granitoid in a borehole drilled along the Pays de Bray Fault, *c.* 100 km NW of Paris. A U–Pb zircon age of 570 ± 15 Ma was determined for the igneous emplacement of this rock, while biotite Ar/Ar ages of *c.* 320 Ma were ascribed to Variscan ductile deformation in a dextral strike-slip regime (Matte *et al.* 1986).

In summary, the existence of a very large gap (from 2.06 to 0.76 Ga) in the igneous rock record from the Armorican Massif is a key feature of the succession in this area. The existence of this gap is further corroborated by analyses of detrital zircons from late Neoproterozoic sediments of the Binic Formation (Fernández-Suárez *et al.* 2002). A similar gap is characteristic of the geological evolution of the NW African realm (e.g. in the Anti-Atlas of Morocco), and strongly suggests that there was a close palaeogeographic connection between the two areas. Indeed, a broadly similar Neoproterozoic evolution, characterized by the formation of intra-oceanic arcs and their accretion to active continental margins, has been described in the pan-African belt of West Africa (e.g. Caby *et al.* 1989).

The Late Proterozoic evolution of the Armorican region was clearly dominated by subduction-related processes, as documented by the distinct affinities of the majority of igneous and sedimentary rocks which include evidence of island arc, backarc and active continental margin settings. It is, however, not yet possible to clearly determine the Cadomian development of the region. Furthermore, whether the Cadomian Orogeny resulted from oblique collisional processes causing southwestward thrusting of a series of volcanic arc and backarc units onto a sedimentary margin (Brun & Balé 1990; Brun *et al.* 2001), or involved the amalgamation of different domains through dominantly strike-slip tectonics (Strachan *et al.* 1996*b* and references therein) is still controversial. Similarly, the number, location and polarity of subduction zones in the region is still a matter of great debate, with no agreed plate tectonic model for the area as yet.

Pyrénées, French Massif Central, Maures Massif, Corsica (C.P.)

Due to the marked effects of Variscan tectonometamorphic overprinting, there is little direct evidence of Precambrian rocks in France outside of the region of the Armorican Massif. Indeed, the Cadomian Belt cannot be recognized as such in any of the other occurrences of pre-Mesozoic basement domains, namely the Pyrenean Variscan Massifs, the Massif Central, the External Crystalline Massifs in the Western Alps, and the Maures–Corsica–Sardinia Block.

Pyrenees

In the Axial Zone of the Eastern and Central Pyrenees (Canigou, Albères and related Aston and Hospitalet massifs), the oldest palaeontologically documented formations are of Late Ordovician age. At a deeper level, orthogneisses and surrounding metasediments were, for a long time, considered to correspond to a Cadomian basement and its transgressive Cambrian–Lower Palaeozoic cover, respectively. In the well-exposed Canigou Massif these meta-igneous and metasedimentary series have been interpreted as having formed in a large recumbent fold nappe of Penninic style during the Variscan Orogeny (Guitard *et al.* 1996 and references therein). This model was initially based on detailed mapping which shows a remarkably concordant contact over several tens of kilometres between the orthogneisses and paragneisses on the one hand, and the metasediments of the Canaveilles series, including a marker layer of marble, on the other. This pattern was interpreted to reflect a cartographic unconformity between a sedimentary cover (Canaveilles series with its basal limestone, believed to be of Early Cambrian age) and an older basement made of paragneisses and porphyritic granitoids (Autran *et al.* 1966). U–Pb zircon dating of the Canigou orthogneisses (580 ± 20 Ma; Vitrac-Michard & Allègre 1975) apparently corroborated this concept, as did a Rb–Sr whole-rock isochron (553 ± 25 Ma; Vitrac-Michard 1975) measured from high-grade paragneisses and norites in the neighbouring North Pyrenean Agly Massif (Vitrac-Michard & Allègre 1971). These radiometric dates were interpreted to reflect the presence of pan-African metamorphic basement. However, the results were far from unequivocal because of the poor alignment of the zircon data points and the disagreement with a 524 ± 20 Ma Rb–Sr whole-rock isochron obtained from the same rocks in the Canigou Massif (Vitrac-Michard 1975), as well as the doubtful synchroneity and cogenetism of samples from the Agly area. Nevertheless, the idea that Late Proterozoic (Cadomian) basement was present in the Pyrenees was widely accepted for two decades (see Zwart 1979 for a different opinion), although the Penninic-style nappe model of the Canigou structure was challenged by two structural studies (Lagarde 1978; Soliva *et al.* 1989).

All of the subsequent geochronological investigations have failed to reproduce the Precambrian dates but instead suggest an Ordovician age for the succession (see references in Laumonier *et al.* 2004). Furthermore, new field observations of contact metamorphism documented, at least locally, the intrusive nature of the igneous protolith of the Canigou gneisses (Barbey *et al.* 2001). The most recent study, using SHRIMP measurements to circumvent the adverse effects of zircon inheritance, gave clear Early Ordovician ages (from 477 ± 4 Ma to 471 ± 8 Ma) for the igneous emplacement of the protoliths of various rock types from the Canigou body (Cocherie *et al.* 2004). Thus, the concept of metamorphic and granitic basement of Late Proterozoic (Cadomian) age, unconformably overlain by Palaeozoic units is clearly false.

The Canaveilles Group, as redefined by Laumonier *et al.* (2004), comprises a 2000–3000 m thick succession of shales and greywackes with rhyodacitic volcanic rocks and carbonate beds. Metarhyodacite interlayered in the lower part of the Canaveilles series yield a SHRIMP age of 581 ± 10 Ma (Cocherie *et al.* 2005). Thus, the age of the Canaveilles Group is interpreted as Late Ediacaran to possibly earliest Cambrian, while the overlying Evol Formation is consigned to the Early Cambrian based on an inferred correlation with Archaeocyata limestones of Atdabanian–Botomian age on the Spanish side of the eastern Pyrenees. The orthogneiss bodies were derived from Ordovician granitoids that intruded at different levels of a broadly continuous Late Proterozoic (Ediacaran) to Early Cambrian sedimentary series, without any evidence for an unconformity, nor for any Precambrian metamorphism or deformation.

French Massif Central

The southernmost part of the Massif Central (Montagne Noire) shows distinct similarities with the Variscan basement of the Pyrenees, although Cambrian rocks are palaeontologically well-documented in the region. Some authors (e.g. Demange 1975; Demange *et al.* 1996) have applied the Canigou model (see above) to the orthogneisses and metasediments of the Axial Zone of the Montagne Noire, interpreting them to represent Cadomian granitic basement and associated lower Palaeozoic sedimentary cover, respectively; both were subsequently deformed in Penninic-style nappe structures. This model, however, has not been unanimously accepted (e.g. Bard & Loueyit 1978), both for structural reasons and also because all of the geochronological studies have failed to obtain Precambrian ages for the orthogneisses. Recent U–Pb zircon work on the orthogneisses provides Ordovician ages ranging from 450 ± 6 Ma to 456 ± 3 Ma (by isotope dilution thermal ionization mass spectrometry; Roger *et al.* 2004) to 471 ± 4 Ma (by SHRIMP; Cocherie *et al.* 2005). These dates are interpreted to provide the age of igneous emplacement of granite magmas into the sedimentary country-rocks of the La Salvetat and St Pons–Cabardès series. A dacitic layer, interbedded in the upper part of the St Pons–Cabardès series, was dated at 545 ± 15 Ma (Pb/Pb; Lescuyer & Cocherie 1992), implying a probable Precambrian age for the underlying clastic series. Based on this age constraint together with a comparison of the lithological characteristics, Laumonier *et al.* (2004) considered these two series to be – as with the Canaveilles Group – of Late Ediacaran age. A *c.* 600 Ma U–Pb zircon age (upper intersection) was obtained for a small mafic–ultramafic body (Airette eclogite) from the SE part of the Zone Axiale (Gebauer & Grünenfelder 1982).

In the NE part of the Montagne Noire, the Neoproterozoic–Cambrian transition has been recently recognized within the lower part of the thick (>1000 m, base unknown; Guérangé-Lozes *et al.* 1991) siliciclastic Marcory Formation. The age dating is based on the presence of the metazoan *Cloudina* in carbonate and phosphate beds (Clausen & Alvaro 2004).

With the exception of the Montagne Noire, there is as yet no direct radiometric evidence for Precambrian igneous rocks in the remainder of the Massif Central area. A *c.* 600 Ma U–Pb zircon age has been published for the Palanges metagranite (Rouergue area), but the analytical data were not given in this report (Lévêque 1985). Recent ion probe measurements document the presence of old inherited zircon components in this granitoid, and favour an igneous emplacement age of *c.* 500 Ma (*c.* Pin unpublished data). Two occurrences of metaquartz diorites with U–Pb zircon igneous emplacement ages close to the Precambrian–Cambrian boundary have also been reported, from the medium-grade Variscan metamorphic basement of the Rouergue (Caplongue body: 557 +12/−10 Ma; Lafon 1984) and Aubrac (Picades body 540 ± 15 Ma; Pin & Lancelot 1978) areas. Based on these data, a probable Neoproterozoic age can be ascribed to the deposition of the detrital sedimentary protoliths of the amphibolite facies mica schists forming the country-rocks of these small plutons. In general, however, the absence of direct or circumstantial evidence for the existence of pre-Variscan cratonic basement in the Massif Central at any crustal level has been emphasized by Roques *et al.* (1995). These authors concluded that the thick accumulations of detrital sediments of Late Proterozoic and Early Palaeozoic age were deposited in an extensional setting, on extremely attenuated continental crust and/or oceanic crust which was subsequently consumed by subduction during middle and late Palaeozoic convergence.

Maures Massif

In the pre-Mesozoic basement of Provence (SE France), porphyritic metagranites (Bormes orthogneiss) resembling the Canigou gneisses from the eastern Pyrenees have been considered to be Late Proterozoic in age (*c.* 550 Ma), based on fairly scattered Rb–Sr whole-rock data and the assumption of an initial $^{87}Sr/^{86}Sr$ ratio equal to 0.705 (Maluski & Allègre 1970). This age was subsequently supported by $^{39}Ar/^{40}Ar$ dating of 578 ± 10 Ma from biotites present within an undeformed body (Barral granite) which occurs as a preserved relict within the Bormes orthogneiss (Maluski & Gueirard 1978). The muscovites from the same rock, interpreted as secondary, gave a much younger age (*c.* 340 Ma). This led several authors to infer the occurrence, prior to the extremely marked tectonometamorphic Variscan overprinting, of Cadomian basement–Palaeozoic cover relationships (see references in Crevola & Pupin 1994). However, more recent U–Pb investigations (Moussavou 1998) do not confirm the Late Precambrian ages, but rather suggest that the Bormes orthogneiss is a Variscan syntectonic intrusion emplaced at 345 ± 3 Ma (U–Pb zircon lower intercept and monazite). Because of the importance of inherited Pb components in zircon in this S-type granitoid, further studies on single grains and/or *in situ* analyses are needed to clearly establish the igneous emplacement age of the Bormes Massif.

In addition to metagranitoids of crustal origin, the Maures basement also contains two igneous bimodal suites (referred to as leptyno-amphibolitic groups in the French geological literature). Igneous emplacement ages close to the Cambrian–Ordovician boundary have been measured on some of the felsic members (alkaline metarhyolites) of these associations (Ducrot *et al.* 1982). However, a significantly older U–Pb age of 548 +15/−7 Ma has been published recently on zircon extracted from a leptynite from the leptyno-amphibolitic group of Central Maures (Innocent *et al.* 2003). This result shows that the bimodal volcanic formations from the Maures are composite, possibly due to tectonic telescoping. Whatever its precise origin, this felsic igneous sample, *c.* 550 Ma old, is the oldest rock reported so far in SE France. The presence of significant Precambrian units in this region is still, however, an open question

Corsica

The Variscan basement of Corsica is almost entirely made up of Carboniferous and Permian granitoids. However, rare and very small-scale outcrops of country-rocks are preserved locally. Among these occurrences, the Argentella region near Galeria (NE Corsica) includes detrital Palaeozoic formations, with probable Hirnantian glaciomarine diamictites and graptolite-bearing black cherts of Llandoverian age in their upper part (Barca *et al.* 1996). These formations lie unconformably upon sericite–chlorite mica schists with banded amphibolites. Although the age of the basal clastic rocks is not precisely known, and is tentatively ascribed to the Cambro-Ordovician, the metasediments and amphibolites below the unconformity are reasonably interpreted to reflect the existence of low-grade metamorphic basement of pan-African (Cadomian) age (Barca *et al.* 1996). Further evidence for Precambrian metamorphic basement in Corsica may be provided by a poorly defined 747 ± 120 Ma Sm–Nd age based on an ilmenite–amphibole–whole rock three-point isochron, measured on a sample of amphibolite found together with metaquartzites and metagreywackes in the Tenda Massif (Rossi *et al.* 1995). However, the geological significance of this date does not appear to be firmly established, in view of the absence of clear evidence of isotope equilibration in the sample.

Bohemian Massif

The Bohemian Massif is the most important inlier of basement rocks in Central Europe and forms the inner part of the European Variscides. It is subdivided into two major and principal zones, the Saxo-Thuringian Zone to the north and the Moldanubian Zone to the south (Fig. 2.28; Kossmat 1927). The Moldanubian Zone includes the Teplá-Barrandian Unit and the Moldanubicum *sensu stricto*. To the north, the Sudetes include diverse parts of the Saxo-Thuringian and the Moldanudian zones. The Moravo-Silesian Zone, in the eastern part of the Bohemian Massif, comprises a late Neoproterozoic basement unit named the Brunovistulian Block and its Palaeozoic cover. To the north and NW, the massif is borded by the Mid-German Crystalline Zone (High), which is assumed to represents an important Variscan suture zone, perhaps the Rheic Suture (Kroner *et al.* 2007; Linnemann *et al.* 2007; Zeh *et al.* 2003, 2005). The suture was formed by oblique collision between the Bohemian (Cadomia, Gondwana) and the Rhenish massifs (Avalonia, Laurussia) during the Late Devonian to Early Carboniferous (Kroner *et al.* 2007). To the south and SE, the Bohemian Massif is overthrust by Mesozoic and Cenozoic rocks of the Alps and the Carpathians. The oldest units of the Bohemian Massif are remnants of Palaeo- to Mesoproterozoic cratonic basement slivers such as the Dobra gneiss (1.38 Ga) and the Světlik gneiss (2.05–2.1 Ga).

The Bohemian Massif records a complex Neoproterozoic to Palaeozoic history that includes the Cadomian and Variscan orogenies. Some rock units (e.g. the Erzgebirge, Saxo-Thuringian Zone) locally experienced ultra-high pressure metamorphic conditions during the Variscan Orogeny with the formation of diamond-bearing rocks (Massonne 1998). More marginal rock units and inliers, however, were essentially unaffected by the Variscan tectonometamorphic overprint. These rock units comprise the Neoproterozoic to Palaeozoic successions at the northern margin of the Saxo-Thuringian Zone and the Teplá-Barrandian Unit.

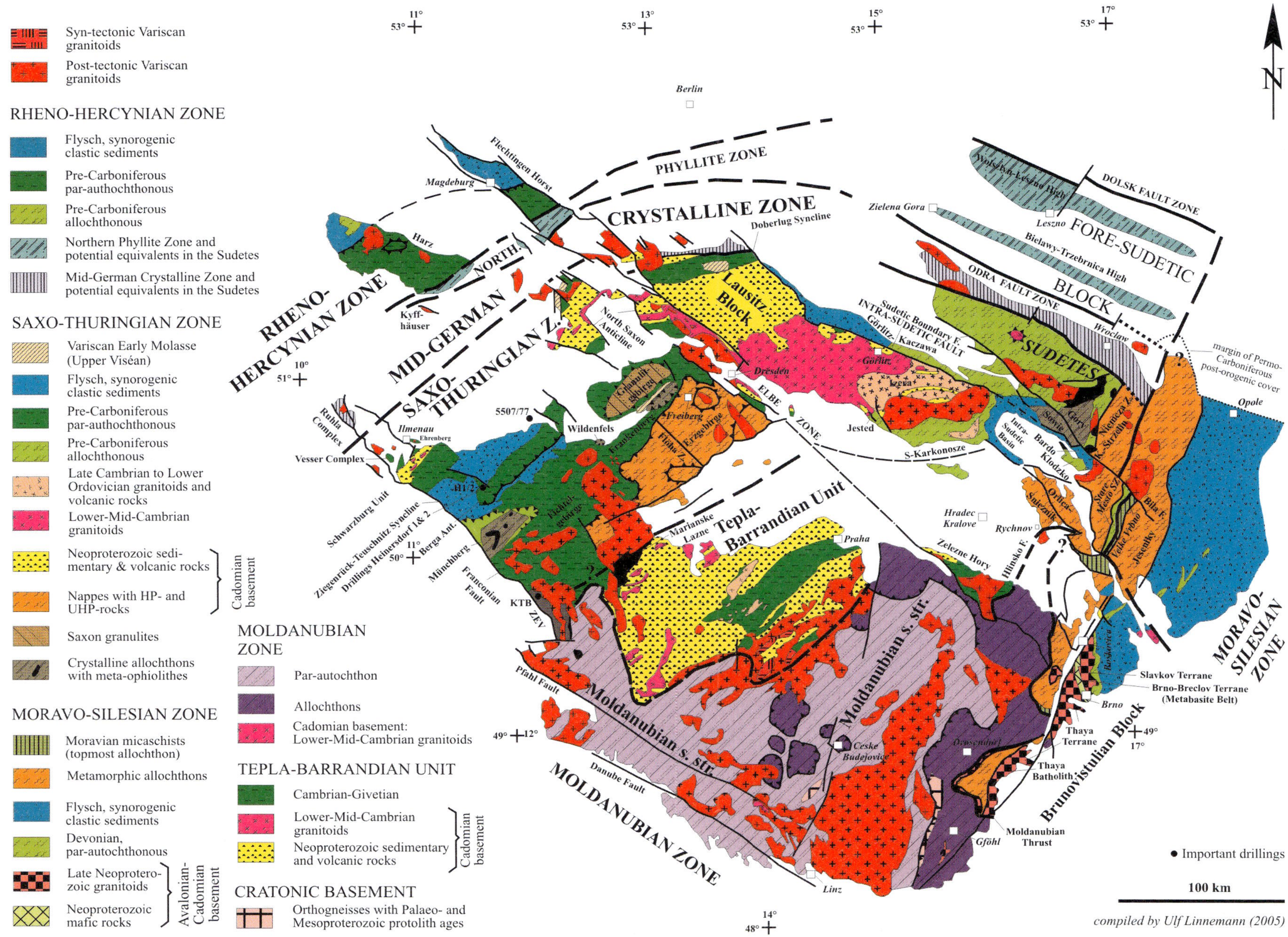

Fig. 2.28. Geology of the Bohemian Massif (map compiled after Franke & Żelaźniewicz 2002; Kozdrój *et al.* 2001; Linnemann & Schauer 1999; Linnemann & Romer 2002).

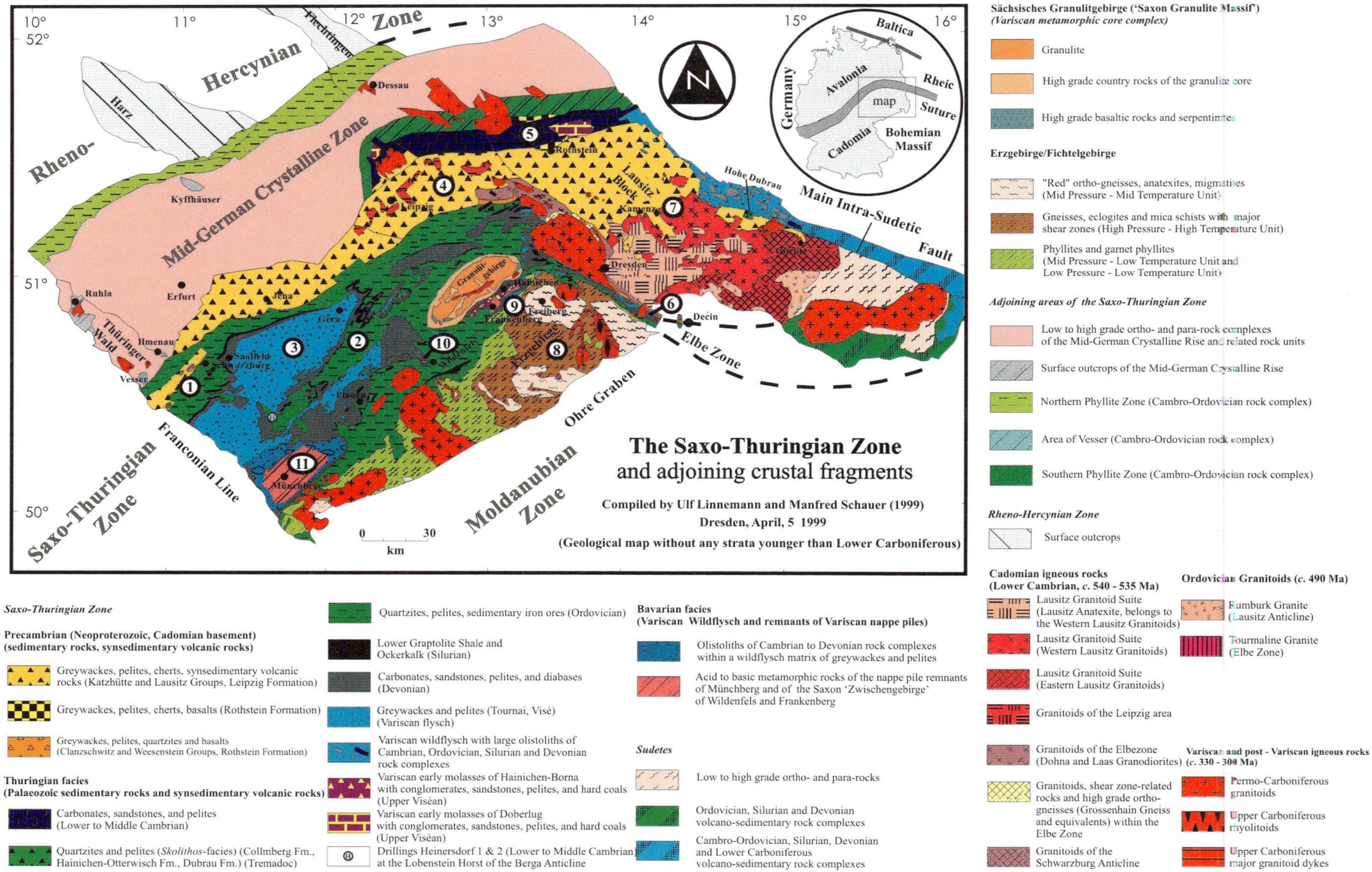

Fig. 2.29. Geological map of the Saxo-Thuringian Zone showing only Lower Carboniferous and older strata (after Linnemann & Schauer 1999; Linnemann & Romer 2002). Key: 1, Schwarzburg Unit; 2, Berga Anticline; 3, Ziegenrück-Teuschnitz Syncline; 4, North Saxon Anticline; 5, Torgau-Doberlug Syncline; 6, Elbe Zone; 7, Lausitz Block; 8, Erzgebirge; 9, Frankenberg Zwischengebirge; 10, Wildenfels Zwischengebirge; 11, Münchberg Nappe Pile.

We relate most parts of the Bohemian Massif to Cadomia *sensu* Nance & Murphy (1994, 1996). Exceptions are the Brunovistulian Block of the Moravo-Silesian Zone (eastern Bohemia) and the Upper Silesian and Małopolska blocks (SE Poland), which are related to Far East Avalonia (Finger *et al.* 2000*a*; Winchester *et al.* 2006).

The Saxo-Thuringian Zone (U.L., R.L.R. & J.U.)

The Saxo-Thuringian Zone, together with the area of the Mid-German Crystalline Zone (High), represents the northwestern part of the Bohemian Massif (Fig. 2.28) and was first defined by Kossmat (1927) (see also McCann *et al.* 2008; Kroner *et al.* 2008). The structural complexity of the Precambrian to Lower Carboniferous units of the Saxo-Thuringian Zone is shown in a geological map (Fig. 2.29) that was constructed using data from abundant deep wells and available subsurface maps from the former East German mining company SDAG WISMUT (Linnemann & Schauer 1999; Linnemann & Romer 2002). Around 25% of the basement is occupied by Precambrian (Ediacaran) rocks.

The Saxo-Thuringian Zone is fault-bounded by the Mid-German Crystalline Zone to the NW and north. The SW boundary is represented by the Franconian Line which is a Meso-Cenozoic structural element. The Ohre Graben and the Sudetes form the southeastern and western limits, respectively (Fig. 2.29).

The Precambrian of the Saxo-Thuringian Zone is dominated by Ediacaran clastic sediments and subordinate, locally occurring, mafic rocks (Linnemann *et al.* 2000, 2004, 2007; Linnemann & Romer 2002). U–Pb and Pb/Pb datings of detrital and magmatic zircon has revealed that within the Saxo-Thuringian Zone, the Precambrian (Ediacaran) rock units are not older than *c.* 570–580 Ma (Linnemann *et al.* 2000, 2004, 2007; Buschmann *et al.* 2001; Gehmlich 2003; Tichomirowa *et al.* 2001; Tichomirowa 2003). All of the Ediacaran rocks were affected by Cadomian orogenic processes and were later intruded at *c.* 540–535 Ma by granitoid intrusions (Linnemann *et al.* 2000, 2008; Linnemann & Romer 2002).

The very low grade metamorphosed Ediacaran succession is exposed in six areas, all of which belong to the para-autochthonous domain of the Saxo-Thuringian Zone (i.e. Schwarzburg Unit, Torgau-Doberlug Syncline, North Saxon Anticline, Lausitz Block, and Elbe Zone) (Figs 2.30 & 2.31). Ediacaran protoliths, with metamorphic overprinting to differing degrees, are also present in the allochthonous parts of the Variscan slices and nappes of the Saxo-Thuringian Zone (e.g. Erzgebirge, Frankenberg Zwischengebirge, Münchberg Massif).

Pioneering work in Precambrian units of the Schwarzburg Unit was carried out by von Gaertner (1931, 1932, 1944), Deubel (1959), Bankwitz (1967) and Bankwitz & Bankwitz (1975, 1995). An early synthesis of the Precambrian in the Saxo-Thuringian Zone was presented by Hoth (1968). First compilations and lithostratigraphic subdivisions of Precambrian units in the Erzgebirge were published by Hoth *et al.* (1979, 1985) and Lorenz & Hoth (1990). Hirschmann (1966), Brause *et al.* (1981) and Kemnitz & Budzinski (1994) presented early lithostratigraphic and geodynamic concepts of the Lausitz Block. Earliest geochronological K–Ar datings of the Schwarzburg Unit and the Lausitz Block were published by Watznauer (1966) and Haake *et al.* (1973), respectively. Late Neoproterozoic microfossil assemblages from the Cadomian Basement of Schwarzburg Unit were described by Heuse (1990) and the Lausitz Block by Burmann (1966) and Lorenz & Burmann (1974). From the Frankenberg Zwischengebirge and the Münchberg Massif Late Neoproterozic microfossils were published by Heuse *et al.* (1994).

The Neoproterozoic units and their Palaeozoic overstep successions of the Saxo-Thuringian Zone were originally formed at the northern margin of Gondwana (Linnemann *et al.* 2004). Based on SHRIMP U–Pb dating of detrital zircon grains from Neoproterozoic and Cambro-Ordovician sediments, Linnemann *et al.* (2000, 2004, 2007) concluded that the provenance of the rocks of the Saxo-Thuringian Zone is clearly related to the West African Craton.

Ediacaran sediments in the Saxo-Thuringian Zone are dominated by siliciclastic rocks such as conglomerates, greywackes, mudstones and shales. The geotectonic setting of these sediments is interpreted to be within a backarc basin (Linnemann *et al.* 2000, 2004, 2007, 2008). The input of old cratonic crust is demonstrated by Nd model ages for the sediments ranging from *c.* 1.4 to 1.9 Ga (Linnemann & Romer 2002) and detrital zircon grains with Palaeoproterozoic and Archaean ages ranging from *c.* 1.8 to 2.5 Ga and from 2.6 to 3.45 Ga (Linnemann *et al.* 2007). The majority of the younger zircon grains are in the range of *c.* 543 to 750 Ma with a maximum peak at 570 Ma (Linnemann *et al.* 2004, 2007). This population most probably marks the main activity of the Cadomian magmatic arc (Linnemann *et al.* 2007, 2008).

The backarc sediments are typically pyroclastic turbidites and debris flows which are locally intercalated with black cherts, mafic lavas and tuffs. In places the Ediacaran sediments were intruded by mafic sills and dykes. The geochemical and clastic composition of these sediments suggests an active margin provenance and derivation from the Cadomian-age magmatic arc (Buschmann 1995; Linnemann & Romer 2002). In addition, hydrothermal cherts and related effusive volcanic rocks formed during backarc spreading (Buschmann 1995). Backarc sediments are represented by the Rothstein and Altenfeld Formations in the Torgau-Doberlug Syncline and the Schwarzburg Unit, respectively (Fig. 2.30; see Fig. 3.18).

Successions related to the passive margin of the back-basin are rarer and are represented by the Weesenstein Group (Elbe Zone; see Fig. 3.21) and the Clanzschwitz Group (North Saxon Anticline) (Fig. 2.31). These units are dominated by sediments derived directly from the weathered West African Craton (Linnemann *et al.* 2007). Typical sediments, represented by high mature quartzites, quartz-rich shale and conglomerates, were deposited during a sea level lowstand possibly related to glacio-eustatic effects (Linnemann 1991). These lowstand deposits are overlain by diamictites in the upper parts of the Weesenstein and Clanzschwitz Groups (Fig. 2.31). Some units within the diamictites are interpreted as being glaciomarine in origin (Linnemann & Romer 2002). Furthermore, in the Lausitz Block area (a roadcut near Kunnersdorf) a diamictite was recently discovered that may also be glaciomarine (J. Ulrich, unpublished data). Pb/Pb and U–Pb ages of granitoid pebbles from the diamictites cluster at *c.* 570 Ma and represent a maximum age of sedimentation (Linnemann *et al.* 2000; Gehmlich 2003). A post-570 Ma Ediacaran glaciation of local extent in the Cadomian Orogen would agree with a proposed position for peri-Gondwanan terranes at high southern latitudes (Scotese *et al.* 1999; Murphy *et al.* 2004).

The collision of the Cadomian magmatic arc with the West African Craton during the latest Ediacaran resulted in the closure of the backarc basin and the formation of a retro-arc basin in latest Ediacaran times (see Fig. 3.27). Igneous and sedimentary rocks, particularly black cherts, became recycled within a retro-arc wedge and redeposited in a retro-arc or foreland basin. The

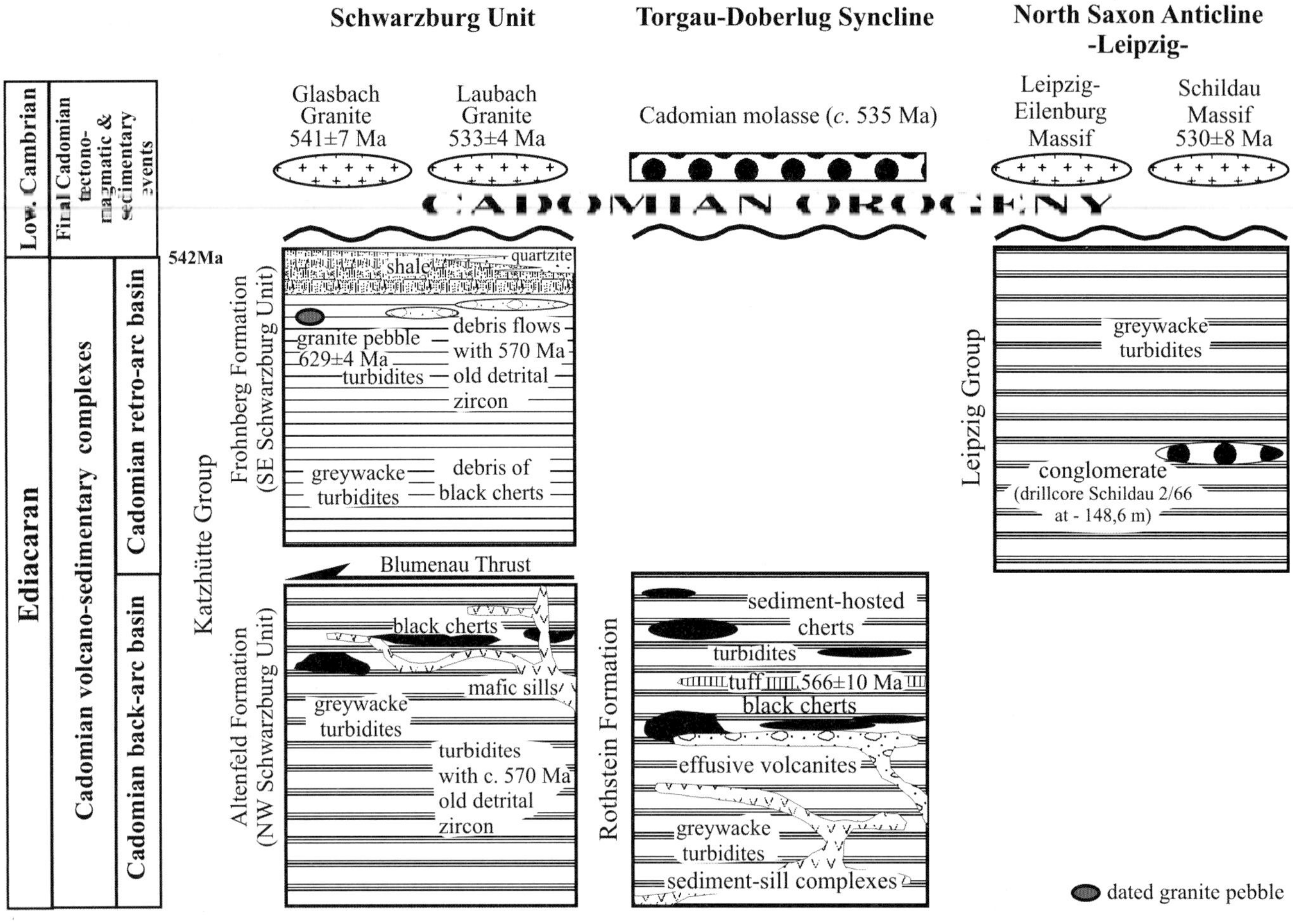

Fig. 2.30. Tectonostratigraphic columns for Neoproterozoic volcanosedimentary complexes and early Cambrian intrusions (Cadomian Basement) of the Schwarzburg Unit, the Torgau-Doberlug Syncline, and the North Saxon Anticline in the Leipzig area (geochronology by Linnemann *et al.* (2000) with exception of the Schildau Massif, that was carried out by Hammer *et al.* (1999); figure after Linnemann & Romer 2002; Linnemann *et al.* 2007).

youngest detrital zircon grains from chert-bearing microconglomerates suggests that the maximum age of sedimentation was *c.* 543 ± 4 Ma (Linnemann *et al.* 2007). Since this age overlaps slightly with the Precambrian–Cambrian boundary (542 ± 1 Ma), the Cadomian retro-arc deposits could also be earliest Cambrian in age (see Linnemann *et al.* 2008). Cadomian retro-arc successions from the Saxo-Thuringian Zone are mainly composed of monotonous dark grey greywacke–mudstone turbidites and rare intercalations of microconglomerates containing fragments of black chert and volcanic rocks derived from the eroded and redeposited material of the Cadomian backarc basin. The Lausitz Group in the Lausitz Block and the Leipzig Group in the North Saxon Anticline form the Lausitz–Leipzig Greywacke Complex (see Fig. 3.19), which is the most representative part of the Cadomian retro-arc basin in the Saxo-Thuringian Zone. The lowermost part of the Lausitz Group may belong to the older Cadomian backarc basin (Fig. 2.31). There are also Cadomian retro-arc deposits in the Frohnberg Formation in the SE part of the Schwarzburg Unit (see Fig. 3.17).

In the *c.* 540–530 Ma (Early Cambrian) time interval voluminous granitoid plutons (granites, granodiorites, tonalites) were intruded into the Cadomian sediments (Linnemann *et al.* 2000, 2007; Tichomirowa *et al.* 2001; Gehmlich 2003). These melts were mainly crustal derived. Xenoliths of mafic rock, calc-silicate rock, and fragments of contact-metamorphic greywacke are found in the 539 ± 6 Ma anatexites of the Lausitz Granitoid Complex (Hammer 1996; Linnemann *et al.* 2000; Linnemann & Romer, 2002) (see Fig. 3.19). Other units of the Saxo-Thuringian Zone contain granitoid intrusions aged *c.* 540–535 Ma such as the Glasberg and Lauchbach granites (541 ± 7 and 533 ± 4 Ma, Schwarzburg Unit), the Laas Granodiorite (531 ± 7 Ma, North Saxon Anticline), the Dohna Granodiorite (537 ± 7 Ma, Elbe Zone), and the granitoids of the Leipzig-Eilenburg and Schildau massifs (530 ± 8 Ma) (Linnemann *et al.* 2000; Linnemann & Romer 2002) (Figs 2.30 & 2.31). The magmatic event at *c.* 540–530 Ma was of a widespread and penetrative nature.

Abundant inherited zircon grains from the granitoids show an age pattern dominated by Palaeoproterozoic and Archaean zircon grains very similar to those of the West African Craton (Linnemann *et al.* 2004; Gehmlich 2003). The granitic melts obviously recycled both the old cratonic crust and parts of the young *c.* 570 Ma Cadomian magmatic arc and related sediments (Linnemann & Romer 2002). Locally, the influence of a mantle source during the melt processes is apparent in the geochemical signature (Hammer 1996). The cause of the intrusion of plutons and batholiths at 540–530 Ma is still under consideration. The possibility of slab break-off of the subducted oceanic plate beneath the magmatic arc resulting in an enormous shock-like heat flow may have produced the voluminous melts that intruded at *c.* 540–535 Ma. This model would also explain the presence

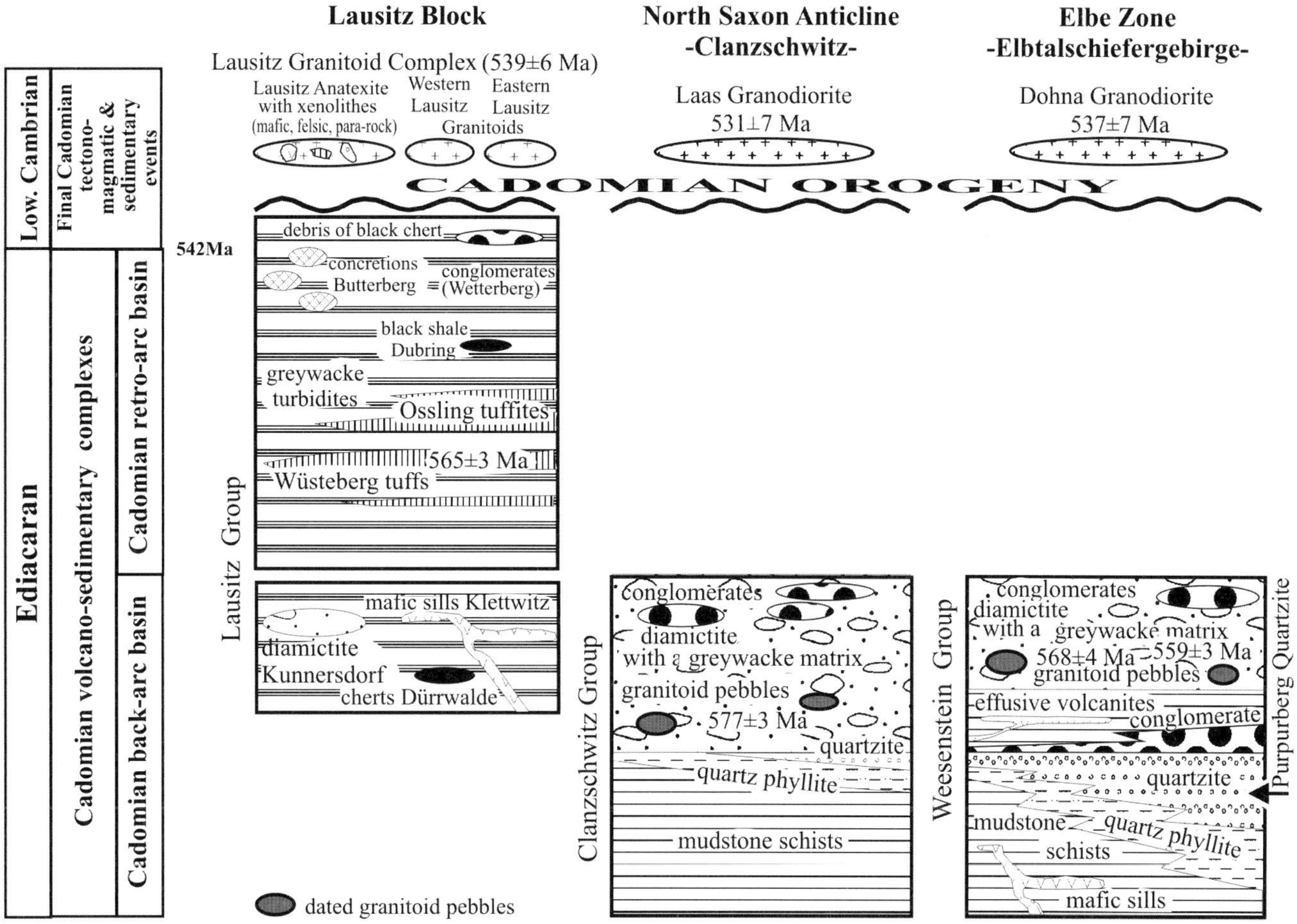

Fig. 2.31. Tectonostratigraphic columns for Neoproterozoic volcanosedimentary complexes and early Cambrian intrusions (Cadomian Basement) of the Lausitz Block, the North Saxon Anticline in the Clanzschwitz area, and the Elbe Zone in the Elbtalschiefergebirge at Dresden (geochronology from Linnemann *et al.* 2000; figure after Linnemann & Romer 2002; Linnemann *et al.* 2007).

of slices and blocks of MORB-like basalts in the *c.* 540–530 Ma granitoids of the Lausitz Block, which could be possible remnants of the subducted late Neoproterozoic oceanic plate (see Linnemann *et al.* 2008).

A typical area containing metamorphosed Ediacaran protoliths of Cadomian basement is the Erzgebirge region in SE Germany (Figs 2.29 & 2.32). This is a NE–SW striking Variscan structure which occupies an area of *c.* 5600 km^2 along the southern rim of the Saxo-Thuringian Zone. It is mainly composed of low-grade to very high-grade ortho- and parametamorphic rocks, which form several mantled gneiss dome structures with a periclinal trend to the main mylonitic foliation (Fig. 2.32). The Freiberg gneiss dome in the eastern Erzgebirge is the most pronounced of the dome structures.

Petrological (Schmädicke *et al.* 1992; Willner *et al.* 1997; Reich 1997; Rötzler *et al.* 1998; Massonne 1998, 2001; Košler *et al.* 2004), geochemical (Kröner *et al.* 1995; Mingram 1998; Tichomirowa *et al.* 2001; Mingram *et al.* 2004), and structural (Sebastian 1995; Nega *et al.* 1997, 1998; Kroner & Sebastian 1997; Kroner *et al.* 1998; Konopásek *et al.* 2001) data have provided clear evidence for crustal stacking of nappes with differing P-T-t evolutions and deformation histories in the regiom during the Variscan Orogeny (Fig. 2.32). The lowermost nappe units crop out mainly in the eastern Erzgebirge, and subordinately in the western Erzgebirge. These footwall nappes consist of the widespread metagreywackes of the Lower Detachment and the metagranitoids of the basal mid-pressure–mid-temperature (MP-MT) unit. The Lower Shear Zone is considered to be an antithetical branch detachment to the SE-dipping master detachment between the Erzgebirge and the metamorphic core complex of the Saxonian Granulites caused by the Variscan post-collisional extension (Sebastian 1995; Kroner & Sebastian 1997). Metamorphic rocks with Cadomian protolith ages occupy nearly half of the total area of the Erzgebirge (Fig. 2.32).

Cadomian metagranitoids are essential parts of the lowermost MP-MT unit. They comprise the Inner Freiberg Gneiss including the adjacent gneisses and migmatites of the easternmost Erzgebirge as well as the periclinal Wolkenstein-Bärenstein Augengneiss belt and the augengneisses in the cores of the Mědĕnec Structure and the Klínovec Antiform in the western Erzgebirge. U–Pb and Pb/Pb zircon ages of between *c.* 540 and 525 Ma are interpreted as representing the intrusion ages of the igneous protoliths of these rocks (Nasdala *et al.* 1996; Tichomirowa 2003; Košler *et al.* 2004). Tectonically overlying metagreywackes of the Lower Detachment comprise the widespread Outer Freiberg Gneiss in the eastern Erzgebirge, the migmatites of the Flöha Zone, and the Marienberg-Annaberg gneisses in the western Erzgebirge. Pb/Pb zircon data of the Outer Freiberg Gneisses yielded a formation age for the magmatic source rocks of the metagreywackes of *c.* 570 Ma, and thus a maximum age

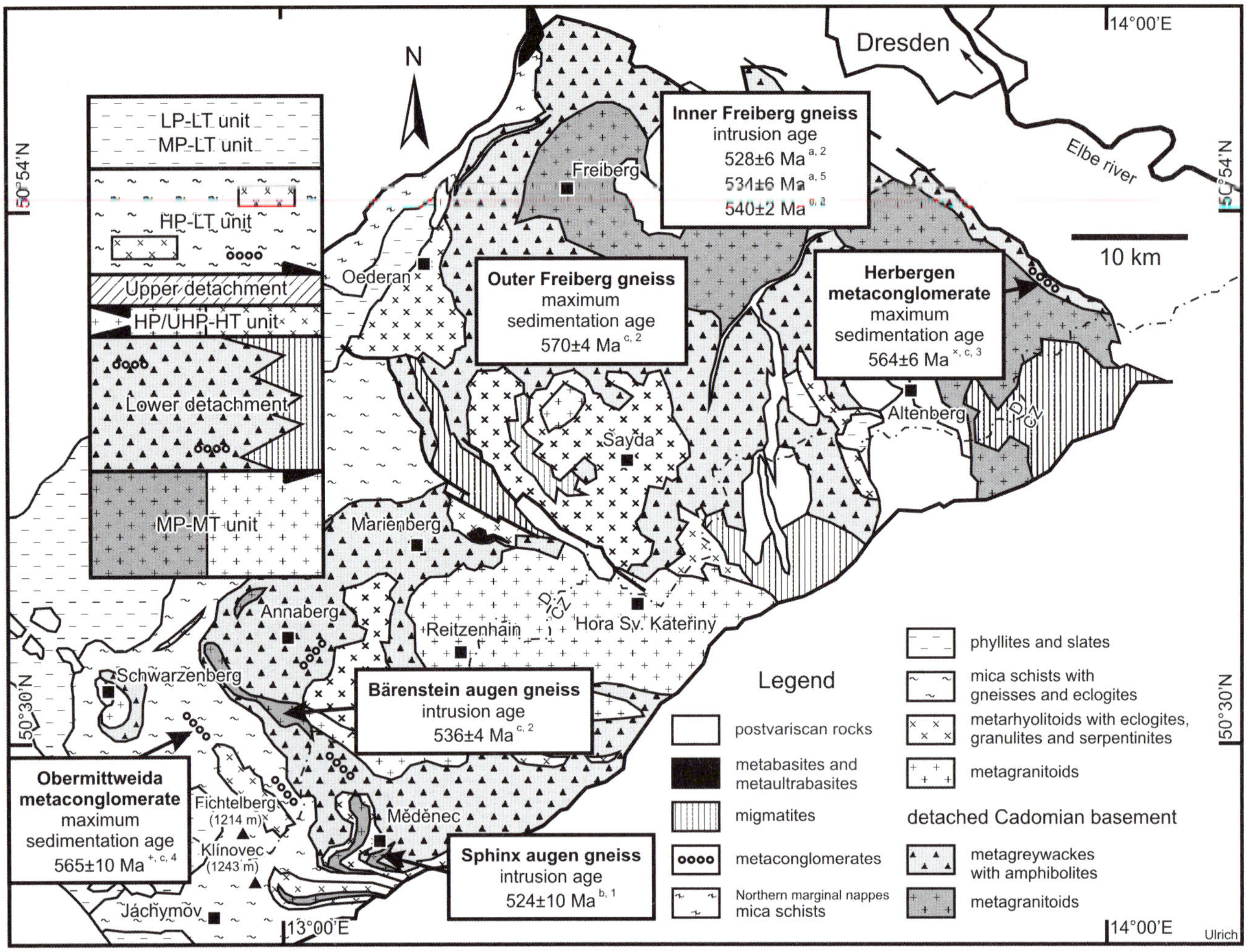

Fig. 2.32. Simplified tectonostratigraphic map and crustal section of the Erzgebirge nappe pile (map basics after Pietzsch 1962; tectonometamorphic units compiled after Sebastian 1995 and references therein; Willner *et al.* 1997; Reich 1997; Rötzler *et al.* 1998; Massonne 1998, 2001; Tichomirowa 2003; Košler *et al.* 2004). Zircon age data are given with analytical uncertainties at 2σ level. Analytical methods; a, U–Pb SHRIMP; b, U–Pb LA-ICP-MS; c, Pb/Pb evaporation. Sources of age data: 1, Košler *et al.* (2004); 2, Tichomirowa (2003); 3, Hammer *et al.* (1999); 4, Palen & Werner (1998); 5, Nasdala *et al.* (1996). Metaconglomerates: +, granitoid pebble, ×, rhyolitic pebble.

for sedimentation of the metagreywackes. Geochronological data from the feldspathoblastic Marienberg-Annaberg gneisses indicated an anatectic event at about 540–520 Ma (Tichomirowa 2003). Metaconglomeratic intercalations (e.g. the Herbergen metaconglomerate in the eastern Erzgebirge) are known from different tectonic levels within the metagreywackes of the Lower Detachment. These also occur as large shear bodies, which are tectonically embedded in the various high-pressure units of the western Erzgebirge (e.g. the famous Obermittweida metaconglomerate; Mehnert 1938) within the high-pressure–low-temperature unit. Both of the metaconglomerates have maximum sedimentation ages of *c.* 565 Ma (Palen & Werner 1998; Hammer *et al.* 1999). Nd–Sr whole-rock isotope data from the metagreywackes and metagranitoids suggest that there was varying but significant input of old crustal material into the protoliths of the para-rocks (Kröner *et al.* 1995; Tichomirowa *et al.* 2001; Mingram *et al.* 2004) (see Fig. 3.28). Geochronological, geochemical and isotopic data from the medium-grade Cadomian basement rocks in the Erzgebirge are comparable with corresponding data from surrounding very low-grade Cadomian units of the Lausitz Block, the Elbe Zone and the Schwarzburg Unit (Linnemann *et al.* 2000, 2004, 2007; Linnemann & Romer 2002; Tichomirowa 2003; Gehmlich 2003).

Diverse allochthonous rock complexes from the Saxo-Thuringian Zone and the Granulitgebirge were formed during the Variscan Orogeny in the Devonian–Carboniferous. These units contain, in places, tectonic slices and nappes with Ediacaran protoliths derived from the Cadomian basement and overprinted by Variscan metamorphism ranging from lower greenschist facies up to ultra-high pressure/temperature conditions. In the nappe pile of the Frankenberg Zwischengebirge, detrital zircon grains of the Paragneiss Subcomplex indicate ages in the range of *c.* 530–600 Ma. Therefore these paragneisses were interpreted in terms of a metasediment derived from the Cadomian basement (Gehmlich 2003). The prasinites in the Frankenberg Zwischengebirge could also be remnants of Neoproterozoic or early Palaeozoic oceanic crust due to the close field relations observed in the Paragneiss Subcomplex. The Wildenfels Zwischengebirge is a small remnant of a Variscan nappe located *c.* 50 km SW of the Frankenberg Zwischengebirge (Fig. 2.29). During his geochronological investigations Gehmlich (2003) found only Ordovician ages and could not identify any Cadomian protolith ages.

The sedimentation of the Prasinit-Phyllit-Serie in the Münchberg Massif most probably occurred in the Ediacaran according to the evidence of acritarchs (Pflug & Reitz 1987; Reitz & Höll 1988). An Ediacaran age for the Hangend-Serie and the Liegend-Serie is also possible, but can only be inferred from Sr evolution diagrams (Söllner *et al.* 1981*a*, *b*). The age of the Saxonian granulites (Granulitgebirge) has often been estimated as Precambrian because of the high metamorphic overprint. New U–Pb datings have shown that lower Palaeozoic protoliths dominate the Granulitegebirge (Romer & Rötzler 2003; Kröner *et al.* 1998). Although the presence of Precambrian protoliths cannot be excluded, the classification of the Granulitgebirge as a Precambrian unit as a whole is thus completely revised. The peak of the granulite facies metamorphism, formerly suggested to be Precambrian, is now dated at *c.* 340–341 Ma (Kröner *et al.* 1998; Romer & Rötzler 2001).

The Mid-German Crystalline Zone (A.Z.)

Based on lithostratigraphic correlations with other basement units of Central Europe (e.g. the Moldanubian basement of Bohemia) it has long been assumed that the area of the Mid-German Crystalline Zone (MGCZ or Mid-German Crystalline High) contained abundant Precambrian ortho- and paragneiss units (Fig. 2.33). This, was postulated, in particular, for the Böllsteiner gneiss, forming the NE part of the Odenwald Crystalline Complex, the Rotgneiss and Haibach-Schweinheimer Gneiss of the Spessart Crystalline Complex, and the migmatitic gneisses of the Liebensteiner Group, which form the central part of the Ruhla Crystalline Complex (see Neumann 1974; Hirschmann & Okrusch 1988). In addition, high-grade (ortho)gneisses of the

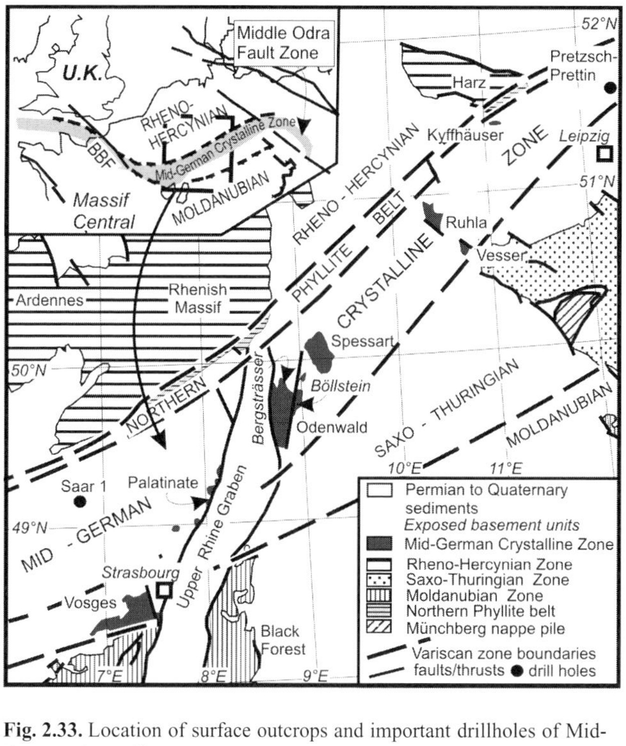

Fig. 2.33. Location of surface outcrops and important drillholes of Mid-German Crystalline Zone.

Kyffhäuser Crystalline Complex were also considered to have formed during the Cadomian Orogeny (Schüller 1957).

Comprehensive geochronological dating carried out over the last ten years, however, has indicated that the protoliths of most of these orthogneisses were intruded predominantly during the Devonian and, to a minor extent, at the Silurian–Devonian boundary, and were subsequently metamorphosed during the Variscan Orogeny, e.g. the Böllsteiner gneiss of the Odenwald Crystalline Complex (405 Ma; Reischmann *et al.* 2001), the Haibach-Schweinheimer Gneiss and the Rotgneiss of the Spessart Crystalline Complex (418–407 Ma; Lippolt 1986; Dombrowski *et al.* 1995), and orthogneisses of the Ruhla Crystalline Complex (Liebenstein Group, 413–400 Ma; Ruhla Formation, 426–422 Ma; see summary in Brätz 2000; Zeh & Wunderlich 2003). Furthermore, some rocks formerly classified as orthogneisses were identified as high-grade paragneisses. These commonly contain abundant detrital zircon grains, which often have metamorphic overgrowths of Variscan age (Zeh *et al.* 2003, 2005). U–Pb data of the youngest zircon cores indicate that the protolith of the high-grade metasediments of the Kyffhäuser Crystalline Complex is younger than 470 Ma (Zeh *et al.* 2005) and that the Brotterode metasediments of the Ruhla Crystalline Complex were deposited after 500–460 Ma (Zeh *et al.* 2001, 2003; Gerdes & Zeh 2006). Thus, the zircon ages suggest that abundant (meta)sediments of the MGCZ were deposited during and after the breakup of the northern Gondwanan margin in the Ordovician. This interpretation is consistent with the finding of *c.* 500 Ma old MOR-basalts in the Vesser Rift, *c.* 30 km SW of the Ruhla Crystalline Complex (Kemnitz *et al.* 2002), with meta-MOR-basalts intercalated in several metasedimentary units of the MGCZ (Hansch & Zeh 2000), and with palynological data (Reitz 1987), indicating that garnet-grade metamorphic rocks of the Geiselbach Formation of the Spessart Crystalline Complex were deposited during the Silurian.

The age spectra of the detrital zircons, in combination with Sm–Nd and new Lu–Hf isotope investigations, reveals that the metasediments of the MGCZ contains material that originates from three distinct crustal sources: juvenile and evolved rocks of the Avalonian–Cadomian Belt (550, 590, 670 and 720 Ma); the West African Craton and/or Amazonian cratons (1.75, 2.05, 2.65, 2.75 Ga); and evolved rocks of the Grenville Belt (1.0 Ga) (Zeh *et al.* 2001; Gerdes & Zeh 2006). In the light of these data, zircon xenocrysts of Precambrian ages, as dated in some Silurian/Devonian orthogneisses and Rotliegend volcanic rocks of the MGCZ (e.g. Todt *et al.* 1995; Reischmann *et al.* 2002; Zeh & Brätz 2000, 2002), can be interpreted as resulting from partial anatexis and/or assimilation of Gondwana-derived sediments, rather than from remelting of a consolidated Precambrian crust beneath the MGCZ. In fact, all of the data are consistent with a model suggesting that the sediments of the MGCZ were deposited after 500–470 Ma and subsequently intruded and metamorphosed during metamorphic–magmatic events in the Silurian/Devonian (425–400 Ma), Carboniferous (350–335 Ma) and Permian (300–280 Ma) (see summary in Zeh & Wunderlich 2003).

The Sudetes (S.M. & P.A.)

The pre-Permian crystalline basement exposed in the Sudetes along the NE margin of the Bohemian Massif (in SW Poland and the northern Czech Republic) contains inliers of Precambrian rocks representing Neoproterozoic crust that developed during, or immediately prior to, the Cadomian Orogeny in suprasubduction and/or syn- to post-collisional tectonic settings (e.g. Mazur *et al.* 2006). The Precambrian domains of the Sudetes were variously affected by Late Devonian to Early Carboniferous Variscan tectonothermal events. Sudetic rocks with a proven Precambrian age occur within the Lausitz Block (also called the Lusatian Massif), in the Strzelin Massif and its equivalent Silesian domain in the East Sudetes (in the Czech Republic), as well as in the Kłodzko Metamorphic Complex (e.g. Kryza *et al.* 2004; Fig. 2.34). These blocks and massifs were predominantly formed or emplaced during the Cadomian Orogeny. It is also likely that other metasedimentary–metavolcanic complexes in the Sudetes, in particular those intruded by Lower Palaeozoic granitoids, have Neoproterozoic protoliths.

Within the Lausitz Block, the Cadomian rocks are represented by the Lausitz–Leipzig Greywacke Complex (*c.* 570–545 Ma) and the granodiorites and other granitoids of the Lausitz Granitoid Complex (*c.* 540–535 Ma) (Fig. 2.35; see also Fig. 3.19). In Poland, the continuation of the Lausitz–Leipzig Greywacke Complex is confined to a small area near the Polish western border, on the NE periphery of the Lausitz Granitoid Complex. It consists of a succession of anchimetamorphic medium- to coarse-grained greywackes, intercalated with slates and interpreted as hemipelagic turbidites (Kemnitz & Budzinski 1994). The composition of detrital clasts and the whole-rock geochemical signature suggest that the greywackes were deposited on an active continental margin (Kemnitz & Budzinski 1994) and derived from a magmatic arc dated at *c.* 575 Ma (SHRIMP U–Pb; Buschmann *et al.* 2001) to 562 ± 4 Ma (Pb/Pb; Gehmlich *et al.* 1997). The Lausitz–Leipzig Greywacke Complex was subjected to very low-grade regional metamorphism and folded into open east–west to SE–NW trending folds during the Cadomian Orogeny. After folding, the greywacke succession was intruded and thermally metamorphosed by granodiorite intrusions. The latter crop out near Zgorzelec and were dated at *c.* 540–530 Ma (Pb/Pb; Kröner *et al.* 1994) and by the SHRIMP U–Pb method (Tikhomirova 2002).

Cadomian basement occurs in the northern part of the Strzelin Massif (Figs 2.34 & 2.35), and comprises the fine- to medium-grained porphyritic biotite–muscovite Strzelin Gneiss, containing conformable, centimetre- to metre-thick intercalations of amphibolites, interpreted as former mafic dykes (Szczepański & Oberc-Dziedzic 1998). The granitic protolith of the Strzelin Gneiss has been dated at *c.* 600 Ma and *c.* 568 Ma by U–Pb zircon SHRIMP analysis (Oberc-Dziedzic *et al.* 2003). These two distinct mean values are interpreted as the age of the magmatic protolith and subsequent metamorphic dissolution and/or anatectic resorption, respectively. To the south of the massif, Neoproterozoic rocks are represented by the migmatitic sillimanite Nowolesie Gneiss. An age of *c.* 1020 Ma was obtained for these rocks by the single zircon evaporation method (Kröner & Mazur 2003). These Neoproterozoic dates indicate an affinity of the Strzelin and Nowolesie gneisses to the Brunovistulian Block, that is mainly exposed further south, within the Moravo-Silesian Zone of the Variscan Belt. The gneisses of the Strzelin Massif are predominantly peraluminous rocks, mostly medium- to highly potassic granites or granodiorites (Oberc-Dziedzic 1999). Their compositions and the proportions of major and trace elements suggest that greywacke-type rocks were the magma source for the granitic protolith (Szczepański 1999). The peraluminosity of the gneisses, their monotonous, fine- to medium-grained, porphyritic fabric, and the lack of mafic enclaves also suggest an S-type granite as a precursor of the gneisses (Oberc-Dziedzic *et al.* 2003).

The Neopretorozoic succession of the Kłodzko Metamorphic Massif (KMM, Fig. 2.34) comprises three separate tectonic elements, thrust over the Palaeozoic rocks within a Variscan nappe pile (Mazur 2003; Mazur *et al.* 2004). The Neoproterozoic components of this tectonic stack consist of two distinct lithological associations: (1) an amphibolite-grade, mostly meta-igneous suite composed of metagabbros and amphibolites accompanied by felsic metavolcanics; and (2) an epidote–amphibolite grade volcanosedimentary succession (Fig. 2.35). Zircon from a plagioclase gneiss, which is intimately associated with metagabbros, yielded an age of 590 ± 10 Ma, interpreted as the time of igneous crystallization for the entire plutonic suite (Mazur *et al.* 2004). Furthermore, a fraction of euhedral zircon crystals with magmatic zonation, derived from a rhyodacite/andesite tuffaceous rock, were dated as 590–600 Ma, which constrained a maximum deposition age for the volcanosedimentary succession. Although this age was expected to represent the time of volcanic activity (Mazur *et al.* 2004), subsequent SHRIMP analysis of similar rocks revealed younger ages clustering at 528.4 ± 5.2 and 543.3 ± 5 Ma. These are interpreted as indicating the time of protolith formation and the age of an inherited component, respectively (Turniak *et al.* 2005).

Bulk compositional and Nd-isotope results reveal that the Neoproterozoic succession of the KMM is characterized by an association of calc-alkaline, felsic metavolcanics, with metagabbros and cumulates of variable trace-element features, partly resembling N-MORB. The metagabbros range from slightly enriched to depleted rocks and their εNd_{560} values vary from +2.2 to +8.6 (Kryza *et al.* 2003). The intermediate and acidic rocks are peraluminous to metaluminous rhyolites, rhyodacites/dacites and andesites, with trace-element patterns generally rather flat to slightly enriched in most incompatible elements. Their εNd_{560} values range from +2.9 to +8.6, similar to the metagabbros and, together with the lack of alkaline rocks, suggest a subduction-related environment (Kryza *et al.* 2003).

The recognition of Neoproterozoic subduction-related magmatism in the KMM provides new evidence for a pan-African active margin in the Sudetes. In comparison with other peri-Gondwanan crustal blocks, this can be interpreted as evidence for subduction of an oceanic-type domain beneath the Gondwana active margin prior to the climax of the Cadomian Orogeny (Nance & Murphy 1996). Thus, the plutonic crustal fragment of the KMM may correspond to the floor of a backarc basin, while the volcanosedimentary succession was subsequently deposited in a magmatic arc setting.

Although the varied volcanosedimentary series of the Śnieżnik Massif (Fig. 2.34) are believed to be mostly Neoproterozoic (Don *et al.* 1990), available micropalaeontological and radiometric data reveal Early Cambrian ages for protoliths of some supractustal rocks (Gunia 1984, 1990; Kröner *et al.* 1997). As in the KMM, these age data suggest that deposition was continuous across the Proterozoic–Cambrian boundary. Similar age constraints have been postulated for the Góry Sowie Massif, largely composed of amphibolite facies paragneisses (Figs 2.34 & 2.35). The precursors to these rocks are thought to have been derived from Neoproterozoic to Early Cambrian sedimentary protoliths (Gunia 1985, 1999).

Teplá-Barrandian Unit (Moldanubian Zone) (V.K.)

The Teplá-Barrandian Unit is a separate crustal segment within the Moldanubian Zone of the Bohemian Massif and represents one of the best-preserved relicts of the Cadomian Orogeny incorporated into the structure of the European Variscides (Edel & Weber 1995). To the NW, it is fault-bounded against the Saxo-Thuringian Zone and the SE the Moldanubicum.

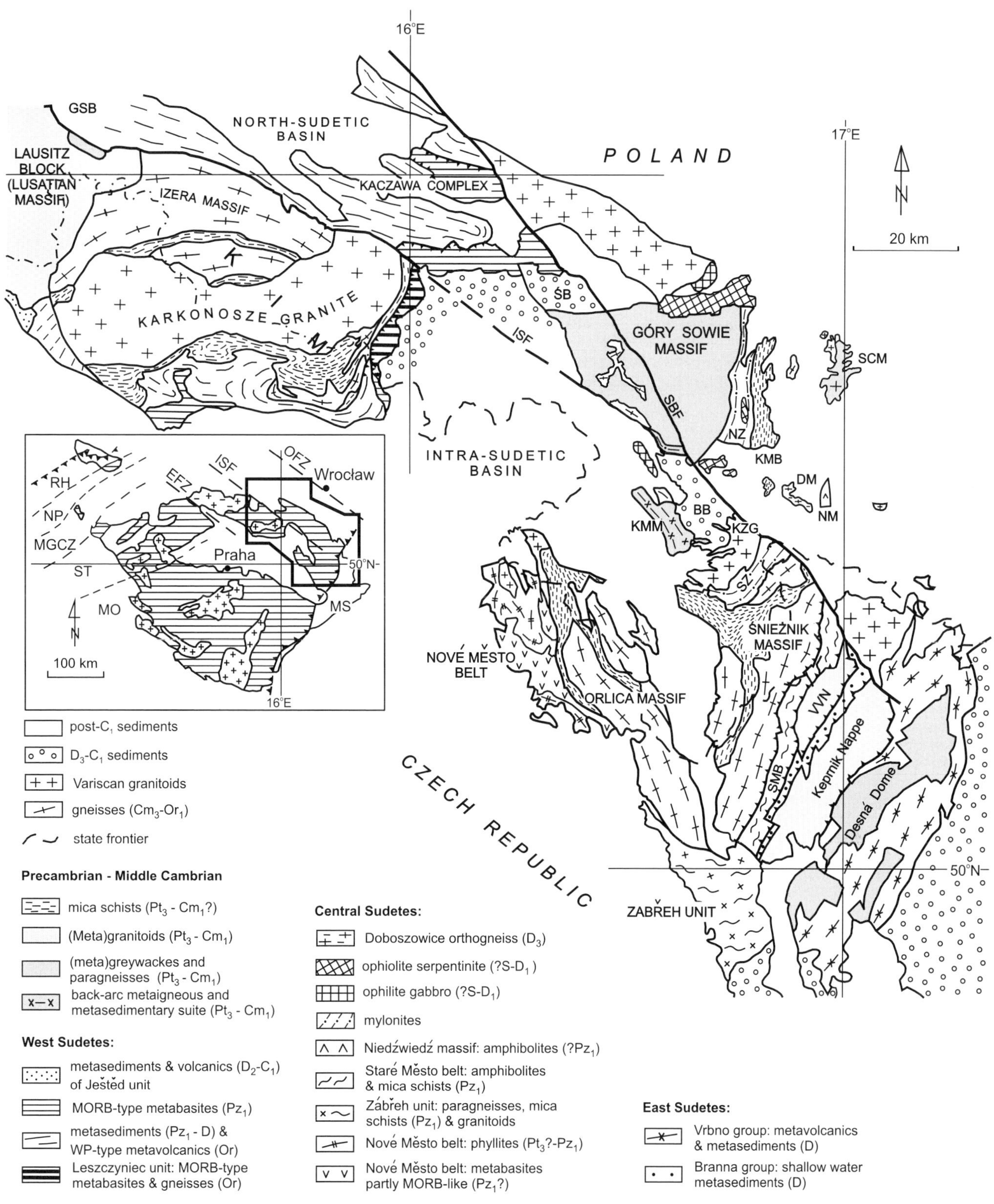

Fig. 2.34. Geological sketch map of the Sudetes. Abbreviations: BB, Bardo Basin; DM, Doboszowice Massif; GSB, Görlitz Slate Belt; ISF, Intra-Sudetic Fault; KIM, Karkonosze-Izeta Massif; KMB, Kamieniec Metamorphic Belt; KMM, Kłodzko Metamorphic Massif; KZG, Kłodzko–Złoty Stok Granite; NM, Niedźwiedź Massif; NZ, Niemcza Shear Zone; ŚB, Świebodzice Basin; SBF, Sudetic Boundary Fault; SCM, Strzelin Crystalline Massif; SMB, Staré Město Belt; SZ, Skrzynka Shear Zone; VVN, Velké Vbrno Nappe. Inset: EFZ, Upper Elbe Fault Zone; MGCZ, Mid-German Crystalline Zone; MO, Moldanubian Zone; OFZ, Middle Odra Fault Zone; NP, Northern Phyllite Zone; RH, Rheno-Hercynian Zone; ST, Saxo-Thuringian Zone; hatched area corresponds to the outcrop of crystalline basement. Age assignments: Pt, Proterozoic; Pz, Palaeozoic; Cm, Cambrian; Or, Ordovician; D, Devonian; C, Carboniferous; Subscripts: 1, early; 2, middle; 3, late.

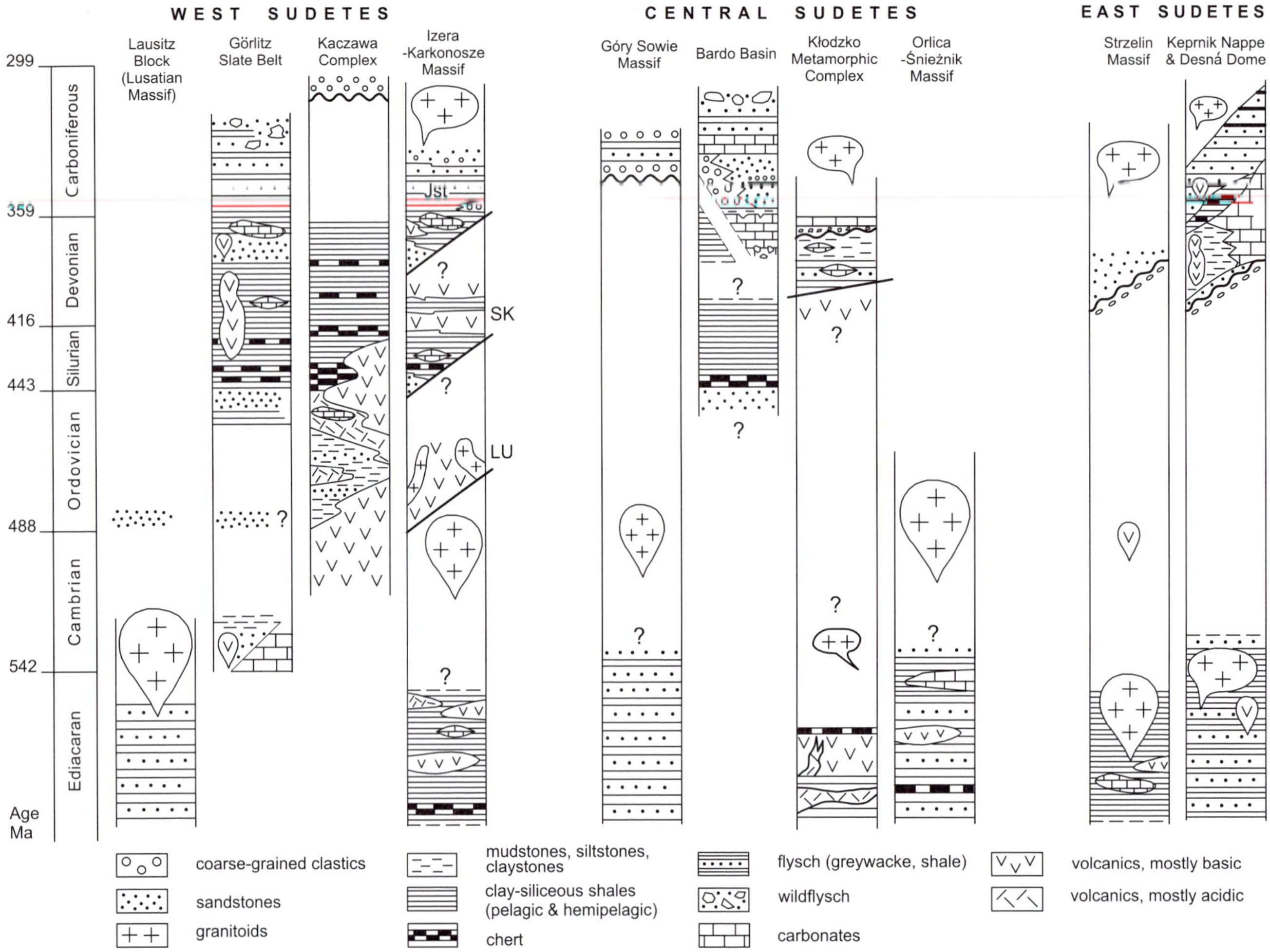

Fig. 2.35. Simplified stratigraphic columns of the main structural units of the Sudetes (modified from Aleksandrowski & Mazur 2002). Metamorphic rocks are represented by their sedimentary and igneous protoliths. Jst, Ještěd Unit; LU, Leszczyniec Unit; SK, South Karkonosze Unit.

With the exception of its marginal parts, the Teplá-Barrandian Unit was affected by weak Variscan tectonometamorphism, in contrast with the strongly metamorphosed and deeply eroded segments of the Moldanubicum to the south and SE, and the Saxo-Thuringian Zone to the NW and north (Fig. 2.36). The weak Variscan metamorphic overprint allows two structural levels to be distinguished: (1) Cadomian basement, slightly metamorphosed and deformed (with the exception of the SW and NW margins along the contact with neighbouring units); and (2) Early Palaeozoic cover, ranging from Cambrian up to Middle Devonian. These rocks rest unconformably on Ediacaran sequences which were folded during the Cadomian Orogeny. The cover rocks were, in turn, folded in the Variscan Orogeny.

The Cadomian basement of the region is similar to the Cadomian succession of the Armorican Massif (Zoubek *et al.* 1988*b*). The Teplá-Barrandian Unit can be distinguished from the adjacent units by its different geophysical and crustal parameters and in terms of its geological development during the Variscan Orogeny. The basement can be characterized by: (1) the higher density of the Cadomian crust, which is evidenced, on a regional scale, by a positive gravity anomaly typical of Cadomian blocks, already incorporated into the Variscan Orogen (Edel & Weber 1995); and (2) the earlier onset of Variscan deformation in the region. These commenced as early as Early Devonian (Emsian) times (Chlupáč & Kukal 1986, 1988) and climaxed during the Middle to Late Devonian (Kukal & Jäger 1988; Franke 2000). Available geochronological data (Dallmeyer & Urban 1998; Kreuzer *et al.* 1989) indicate that the crust of the Teplá-Barrandian Unit was already cooled below the closing value of the K–Ar system of muscovite at the Devonian–Early Carboniferous boundary, 20 to 30 Ma earlier than in the Moldanubicum (Kalt *et al.* 2000; Fritz *et al.* 1996) and the Saxo-Thuringian Zone (Werner & Lippolt 2000).

The Teplá-Barrandian Unit occupies a central position in the mosaic of units within the Bohemian Massif, and is mainly exposed in central and western Bohemia (Fig. 2.36). Its boundaries with the neighbouring units are mostly tectonic. Pre-existing discontinuities later served as conduits for the emplacement of Variscan plutons, which obscure the original character of the Teplá-Barrandian Unit boundaries.

Relicts of the Ediacaran basement of the Teplá-Barrandian Unit and the Early Palaeozoic cover are preserved as roof pendants in the Central Bohemian Pluton (Kettner 1930; Svoboda 1933; Kachlík 1992). Units of the Barrandian Palaeozoic and Ediacaran extend under the Bohemian Cretaceous Basin eastwards into the area of the Železné Hory Mountains where they are again exposed on the surface (Fig. 2.36).

The Ediacaran units of the Teplá-Barrandian Unit are up to 10 km thick, and mostly comprise deep-marine sediments (greywackes, shales, cherts) together with associated submarine basic

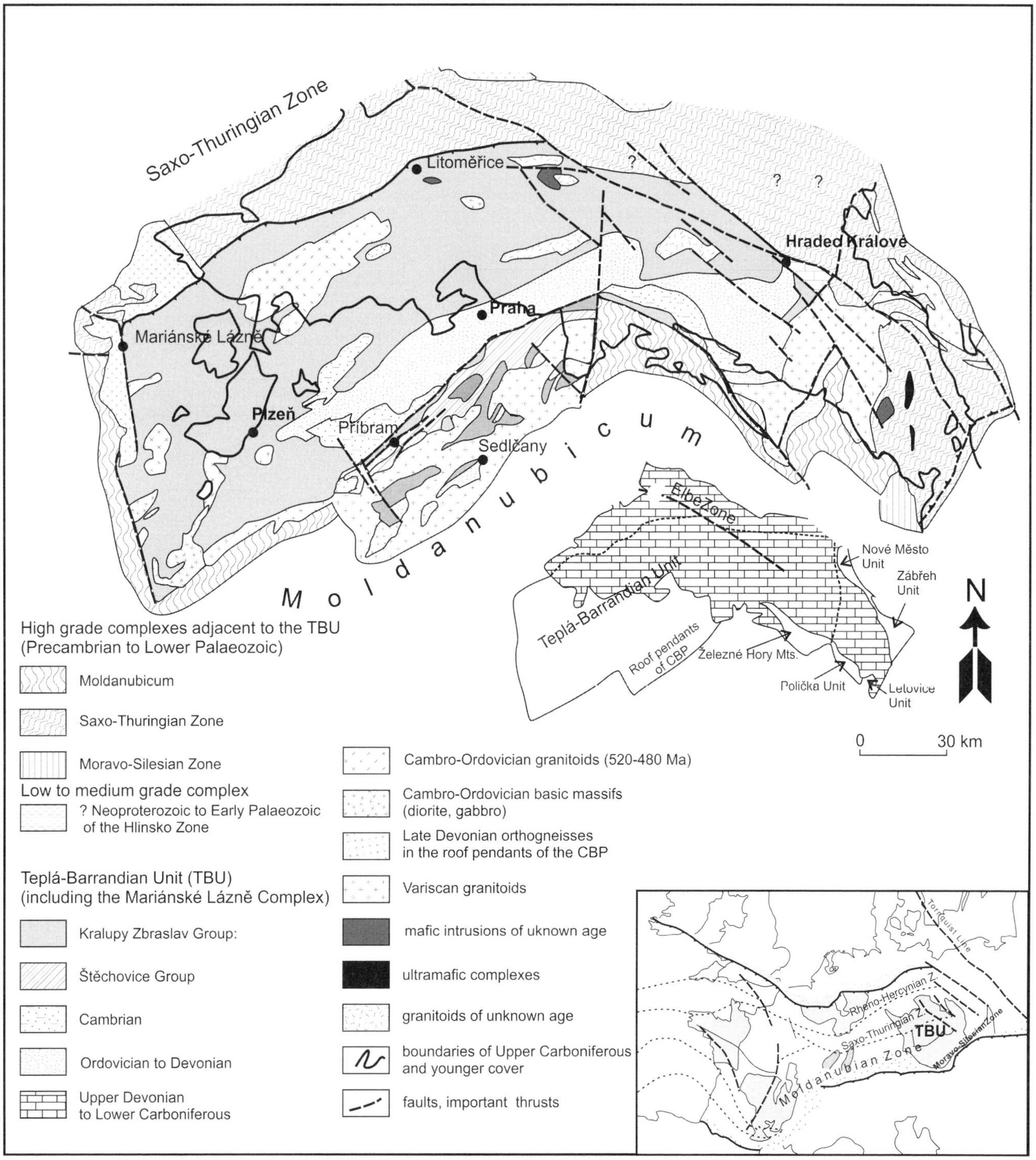

Fig. 2.36. Pre-Upper Carboniferous geology of the Teplá-Barrandian Unit (TBU) (modified according to Chaloupský *et al.* (1995).

and acid volcanics and rare conglomerates in the uppermost parts of the succession. The units were originally ascribed to the Precambrian, following the recognition of an unconformity between the Cambrian and its basement in Central Bohemia. Kettner (1917) provided the first significant lithostratigraphic classification of the Precambrian in Central Bohemia and subdivided the Ediacaran succession into three units according to the occurrence of volcanic rocks (Fig. 2.37). His original concept was later revised by Röhlich (1965), Cháb & Pelc (1968) and Mašek & Zoubek (1980). However, none of these classifications is applicable to the the entire area of the Barrandian Ediacaran. The classification introduced by Mašek & Zoubek (1980), for example, can only be applied along the ESE flank of the Barrandian Ediacaran, where the younger Štěchovice Group overlies the older Kralupy-Zbraslav Group. In the central, southern and western parts, where the prominent marker horizon

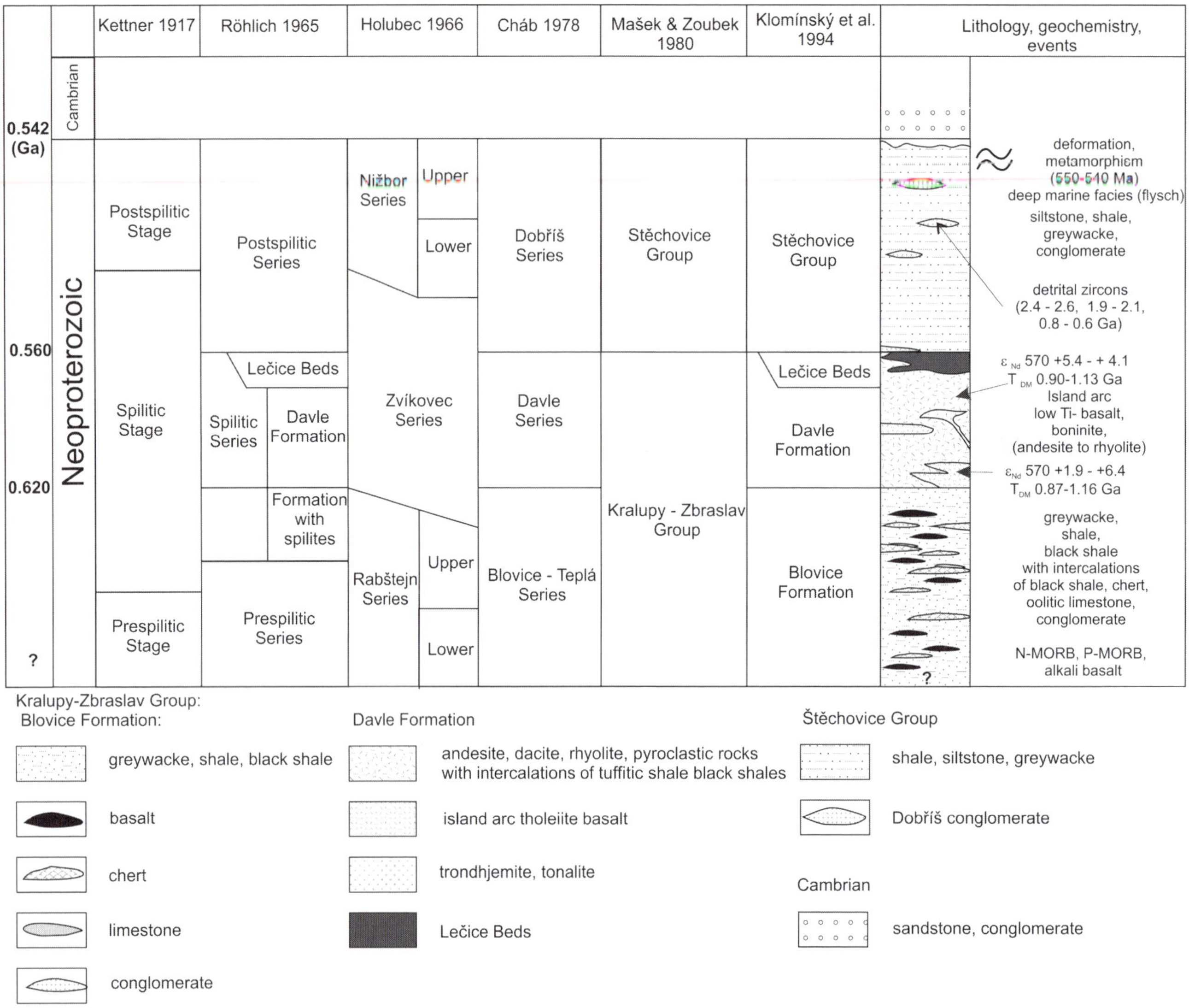

Fig. 2.37. Litostratigraphy and lithology of the Neoproterozoic in the Teplá-Barrandian Unit. Lithostratigraphic scheme modified according to Mašek (2000).

comprising the acid volcanics of the Davle Formation with the overlying black shales of the Lečice Member is missing, it is difficult to apply their scheme (see discussion in Röhlich 2000).

The succession of Ediacaran-age strata in the Železné Hory Mountains can be subdivided into the Chvaletice Group and the overlying Sovolusky Group. The Chvaletice Group, consisting of rhythmic alternations of dark shales and greywackes with rare intercalations of Telčice conglomerates, is thought to be an equivalent of the Kralupy-Zbraslav Group in the Barrandian area (although there is evidence of facies variations). The overlying Sovolusky Group consists of alternations of greywackes and shales with bimodal volcanics. The Litošice conglomerates were interpreted by Fiala & Svoboda (1956) as glaciomarine sediments. The conglomerates contain large numbers of pebbles including both acid and basic intrusives, as well as acidic and basic volcanics, greywackes and cherts. These are the characteristic rocks of the lower part of the Kralupy-Zbraslav Group. Due to the presence of both acidic and basic volcanics, the Sovolusky Group cannot be considered to be a direct equivalent of the Štěchovice Group, since volcanic activity in the latter was only of limited extent.

The Ediacaran rocks in the roof of the Central Bohemian Pluton consist of metamorphosed greywackes and siltstones, with subordinate intercalations of conglomerates, and basic volcanics (low Ti-basalts, boninites), which originated in an island arc setting (Vítková & Kachlík 2001). In addition to the abundant quartz pebbles, the polymictic Vletice conglomerates contain granite pebbles (*c.* 600–580 Ma), quartzites, limestones and rhyolites (Sláma *et al.* 2003), probably derived from the Davle Formation of the Jílové Belt (Fig. 2.38). Consequently, it is probable that the Ediacaran succession in the roof pendant of the Central Bohemian Pluton is partly the equivalent of the younger Štěchovice Group. Facies development of Ediacaran rocks in the roof of the Central Bohemian Pluton was more closely related to the Ediacaran units cropping out in the Železné Hory Mountains.

The Ediacaran successions of the Teplá-Barrandian Unit are represented by relatively monotonous units comprising rhythmically alternating sediments such as dark shales and siltstones,

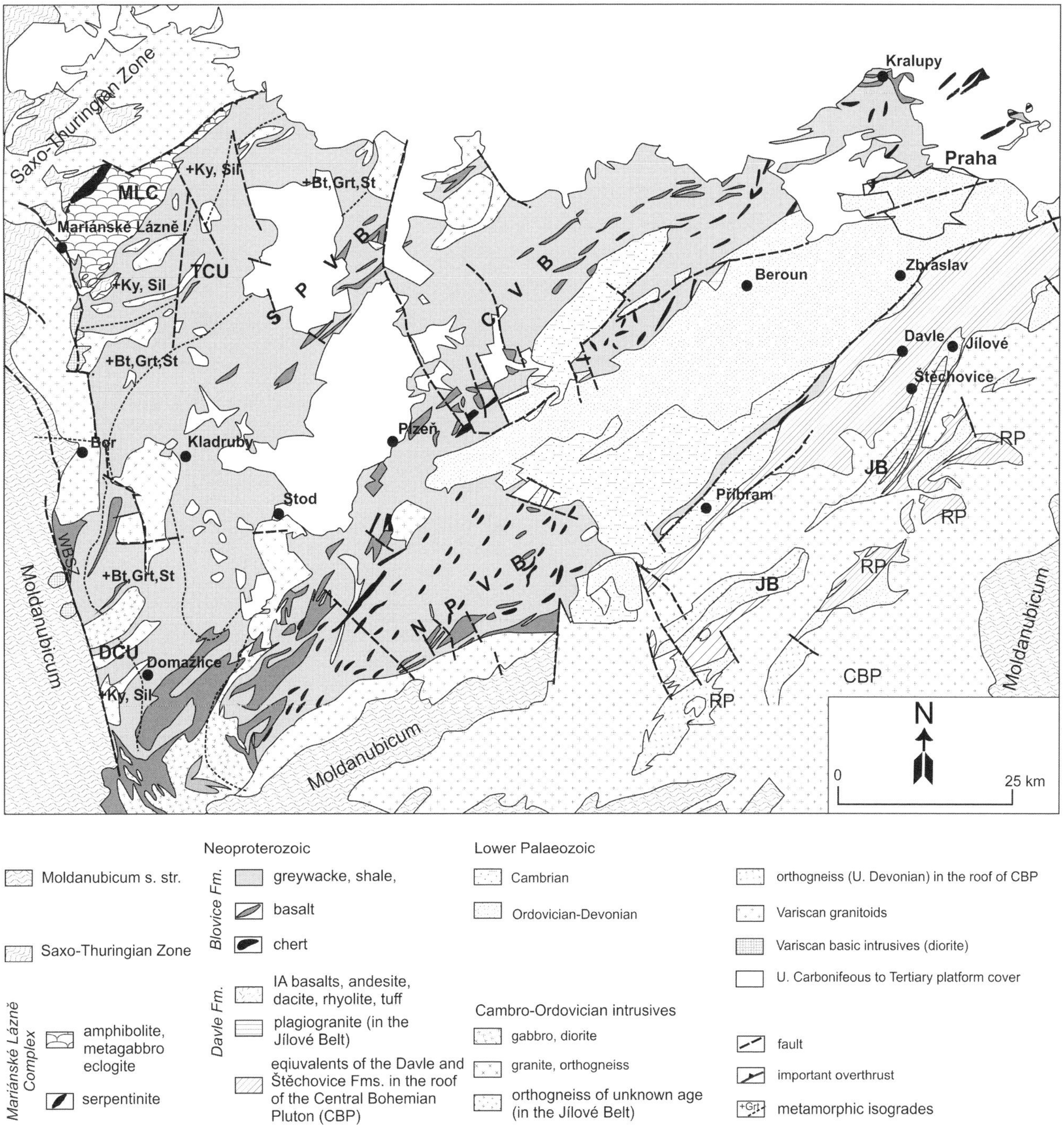

Fig. 2.38. Geological map of the Neoproterozoic in the Teplá-Barrandian Unit (simplified according to the geological map of the Czech Republic 1:500 000). Abbreviations: CBP, Central Bohemian Pluton; CVB, Central Volcanic Belt; DCU, Domažlice Crystalline Unit; JB, Jílové Belt; MLC, Marianské Lázně Unit; TCU, Teplá Crystalline Unit; NPVB, Nepomuk–Příbram Volcanic Belt; RP, roof pendants of the CBP; SPVB, Stříbro-Plasy Volcanic Belt; WBSZ, West Bohemian Shear Zone.

with layers of sulphide-bearing shales, cherts and conglomerates accompanied by the products of basic and acid volcanism. The succession can be subdivided into two lithostratigraphic units (Fig. 2.37): the Kralupy-Zbraslav Group and the Štěchovice Group. The former, with a minimum thickness of *c.* 7 km, crops out mainly in SW and western parts of the Barrandian Ediacaran (Fig. 2.38). Deep-marine units of the Štěchovice Group are confined to the SE flank of the Barrandian Ediacaran and to roof pendants of the Central Bohemian Pluton (Fig. 2.38).

Cháb & Pelc (1968) recognized three subfacies in the lower part of the Kralupy-Zbraslav Group (the Blovice Formation; Fig. 2.37). (1) A deep marine (flysch) facies which comprises rhythmically arranged, thick sequences of laminated shales, siltstones and greywackes which are mainly graded. Sediments were deposited from turbidity currents and, to a lesser extent, by mudflows and slumps. Intercalated volcanic rocks are rare. (2) A volcanogenic facies, which is rich in recycled sediments derived from submarine slumps, slides and mass flows, and redeposited from volcanic highs. The greywackes differ from those of the deep-marine facies in that they contain a higher proportion of lithoclasts derived from basic volcanics. The sediments are interbedded with numerous bodies of submarine tholeiite basalts, transitional and alkaline basalts (including pillow lavas and brecciated lavas) with rare tuffaceous rocks (Fiala 1978), forming several NE–SW trending belts within the Kralupy-Zbraslav Group. Occurrences of pisolitic and oolitic carbonates and micrites (Skoček & Pouba 2000), associated with black metal-bearing shales (Pašava 2000), are confined to volcanic highs. Cherts, which are characteristic components of both the volcanogenic and the monotonous facies (Zbiroh-Šárka belt), are abundant close to the volcanic belt (Fig. 2.38). (3) A monotonous facies, which comprises rhythmically alternating siltstones and greywackes without typical graded bedding. The contacts of the thicker greywacke beds with the siltstones are often very sharp and erosional.

The uppermost part of the Kralupy-Zbraslav Group comprises the 500–1000 m thick Davle Formation occurring at the SE flank of the Barrandian. This formation consists of basic, intermediate and acidic volcanics, tuffs, tuffites and tuffaceous shales with a typical island arc affinity (Waldhausrová 1984; Fediuk 1992; Vítková & Kachlík 2001). A horizon of carbon-rich sulphide-bearing shales (Lečice Member; Röhlich 1965) occurs at the top of calc-alkaline volcanics. These shales represent the uppermost part of the Kralupy-Zbraslav Group succession.

The Davle Formation is overlain by a sequence of deep-marine turbidites and mass flow deposits up to 3 km thick (Štěchovice Group) (Fig. 2.38) along the SE flank of the Barrandian Ediacaran. This succession commences with a 300–400 m thick unit of thinly laminated and graded siltstones and shales locally containing slumps breccias (Röhlich 1964). Interbedded greywackes and conglomerates increase in abundance up-section. The Dobříš conglomerates contain abundant greywacke and shale clasts derived from the older parts of the Ediacaran successions. These conglomerates have been interpreted as glaciomarine sediments (Svoboda; Fiala 1948) that were redeposited by debris flows. The conglomerates also inlcude a range of granitoids (leucocratic plagiogranites) similar to those of the Jílové Belt, as well as acid volcanics derived from the Davle Formation.

The Ediacaran age of the successions of the Teplá-Barrandian Unit is supported by micropalaeontological data. The majority of the findings come from the Blovice Formation of the Kralupy-Zbraslav Group. These include cyanophyta, green algae, acritarchs and micro-organisms of uncertain systematic classification. Unicellular or colonial organisms with more complex morphologies are reported from greywackes or redeposited silicified fragments of greywackes (Konzalová 2000). Associations of algae and blue-green algae were found in oolitic limestones, black lagoonal shales and in stromatolitic build-ups (Vavrdová 2000), where they often form algal mats. Stratigraphically important species are characteristic of the Neoproterozoic (Fatka & Gabriel 1991; Pacltová 1990).

Ediacaran sediments exhibit rather low chemical maturity, typical for sediments derived from rather primitive crustal sources, such as oceanic island/or continental island arcs. Limited hydraulic sorting and predominantly mechanical weathering are reflected in the high SiO_2/Al_2O_3 and Na_2O/K_2O, in comparison with the overlying Early Palaeozoic sediments (Kukal 1985; Kachlík 1992; Drost *et al.* 2004). The higher contents of Fe, Ti and Ca, Cr, Ni and a high Na_2O/K_2O ratio indicate that basic volcanics were an important source of detritus mainly for the older Kralupy-Zbraslav Group rocks (Matějka 1988; Jakeš *et al.* 1979). In the younger Štěchovice Group, the amount of material was derived from more evolved sources, for example from continental island arcs or active continental margins, and this is reflected in the characteristic depletion in Nb, Ta and HFSE elements and by the enrichment in incompatible elements such as Ba, Rb and K (Kachlík 1992).

Volcanic rocks are concentrated mainly in the Kralupy-Zbraslav Group (Figs 2.37 & 2.38) in the Barrandian area. In the overlying Stěchovice Group, the volcanic rocks are only of minor importance. Extensive volcanic activity has been documented from the Železné Hory Mountains and from the roof rocks of the Central Bohemian Pluton (Sovolusky & Svrchnice Formation; Fiala 1978; Vítková & Kachlík 2001).

Subalkaline tholeiite basalts corresponding to recent MORB predominate in the central volcanic belt (Fig. 2.38). Primitive tholeiite basalts are accompanied by transitional P-MORB basalts and andesites (Waldhausrová 1997*a*, *b*). Varieties of alkaline basalts to trachytes occur in the Pavlíkov volcanic belt and in the volcanic zone between Nepomuk and Příbram (Pelc; Walhausrová 1984; Fig. 2.38). A relatively high proportion of transitional and alkaline volcanics indicates that the volcanic rocks of the Blovice Formation represent basalts of a backarc or interarc basin rather than true MORBs.

Volcanic rocks of the Davle Formation from the SE flank of the Barrandian, the Jílové Belt and the Islet Zone of the Central Bohemian Pluton, and possibly even in the Železné Hory area, would appear to include older low-Ti basalts exhibiting an affinity with island arc tholeiites (Waldhausrová 1984; Fediuk 1992). These volcanics are usually accompanied by dykes or sills of boninites (Fediuk 1992; Kachlík 1992; Vítková & Kachlík 2001). However, intermediate and acid volcanics (dacites, rhyolites) and their tuffs begin to predominate upwards into the roof of the succession. These more highly evolved rocks correspond geochemically to the calc-alkaline series of a recent island arc (Waldhausrová 1984).

The Nd isotopic data of basalts and associated Ediacaran metaboninites of the roof pendants of the Central Bohemian Pluton Ediacaran indicate either an important mantle heterogeneity, different degrees of crustal contamination and/or variable mantle metasomatism (εNd_{570} = +1.9 to +6.4, T_{DM} = 0.87–1.16 Ga). Intermediate and acid volcanics have also relatively primitive features (εNd_{570} = +5.4 to +4.1, T_{DM} = 0.90–1.13 Ga) pointing to their unevolved, probably mantle-derived parental magma (Kachlík *et al.* 2001).

The limited amount of geochronological data and the doubtful reliability of earlier data do not permit more detailed reconstruction of the tectonomagmatic evolution and determination of the range of sedimentation of individual units of Teplá-Barrandian Unit in Ediacaran times. The radiometric ages obtained from basic volcanics of the Blovice Formation (see discussion in Mašek 2000) are scattered over a relatively large timespan. The age of basalt from the Darová locality in the central volcanic belt has been dated by the K–Ar method at about 625 Ma (Šmejkal 1964). The ages obtained by Sm–Nd isochron dating of tholeiite basalts from Zbečno, which are accompanied by substantial errors, vary between 790 and 900 Ma (von Drach *et al.* 1993).

The age of the acid volcanics of the Davle Formation has been more precisely established. U–Pb dating of zircons from pebbles of acid volcanics derived from the Davle Formation in the Dobříš conglomerates and basal Cambrian conglomerates (Dörr *et al.* 1992, 2002*b*) indicate that volcanic arc activity lasted over an interval of *c.* 620–560 Ma. U–Pb analyses of detrital zircons from the matrices of the Dobříš and Vletice conglomerates and zircons separated from granitoid pebbles (Sláma *et al.* 2003; Drost *et al.* 2004), whose age varies between 620 and 564 Ma, placed the deposition age of the Štěchovice Group in the interval of *c.* 560–540 Ma. The U–Pb dating of monazites from the Domažlice (Zulauf *et al.* 1999) and the Teplá Crystalline Complexes (Timmerman *et al.* 2003) indicate that the onset of the Cadomian tectonometamorphic processes, which completed the Štěchovice Group deposition, occurred in the timespan *c.* 550–540 Ma.

The age of detrital zircons from Ediacaran greywackes of the Teplá-Barrandian Unit (2.4–2.6, 2–2.2, 0.8–0.6 Ga; Sláma *et al.* 2003; Drost *et al.* 2004) with a prominent maximum at *c.* 580–620 Ma, and the absence of Mesoproterozoic zircons suggest that during the latest Ediacaran period the Teplá-Barrandian Unit was situated at the margin of the West African Craton. Lithological features, volcanic rock chemistry and ages indicate that the Ediacaran succession in the Teplá-Barrandian Unit has a close relationship to the Brioverian succession of the Armorican Massif (Zoubek *et al.* 1988*a*; Chaloupský 1989; Kříbek *et al.* 2000). Furthermore, the age of deposition of greywacke successions overlaps with the age of similar sedimentary Ediacaran rocks in the Saxo-Thuringian Zone (Linnemann *et al.* 2000, 2004, 2007).

The Ediacaran development of the Teplá-Barrandian Unit commenced with the deposition of a thick sequence of deep marine siliciclastic sediments (Kralupy-Zbraslav Group), initially accompanied by N-MORB, P-MORB and alkaline basalts (Fiala 1978; Waldhausrová 1997*a*, *b*). The timespan of sedimentation is not precisely dated but, according to limited biostratigraphic data, may include Ediacaran and older Neoproterozoic rocks. The sediment character and also that of volcanics indicates that these rocks were deposited in a basin with prominent chains of volcanoes, perhaps partly floored by spreading oceanic crust (Jakeš *et al.* 1979).

During sedimentation of the upper part of the Davle Formation, a system of island arcs began to develop, of which the most prominent is the Jílové Belt. The chemistry of the volcanism changed from low Ti basalts in the lower parts to much more evolved calc-alkaline rocks, such as dacites and rhyolites and their tuffs, at the top. Active magmatism in the island arc system continued from at least 620 Ma to 560 Ma (Sláma *et al.* 2003; Dörr *et al.* 2002*b*). Magmatic activity decreased during deposition of the Lečice Member and the overlying deep-marine succession, which contains few volcanic beds. Deposition of deep-marine sediments was accompanied by gradual deformation of the basin fill, during which the older sediments and volcanics were redeposited to form the Štěchovice Group.

At the end of the Ediacaran period, *c.* 550–540 Ma ago (Timmerman *et al.* 2003; Zulauf *et al.* 1999), all of the units were folded into NE–SW trending megafolds (Holubec 1968, 1995). According to Zulauf *et al.* (1999), top-to-north shearing also originated during this event. The grade of Cadomian metamorphism in the central part of the Barrandian Ediacaran reached the prehnite–pumpellyite facies (Cháb & Bernardová 1974). However, the metamorphic grade increases to the NW and SE up to the amphibolite facies grade (Vejnar 1982; Žáček & Cháb 1993). The onset of the Cadomian Orogeny at *c.* 550–540 Ma was probably caused by arc–continent or arc–arc collision close to the Gondwana mainland. The Lower Cambrian continental deposits unconformably overlying the Cadomian basement reflect transition of the geotectonic regime from convergence to transtensional deformations.

The Moldanubicum sensu stricto *(Moldanubian Zone) (V.K.)*

The Moldanubicum *sensu stricto* (Suess 1901) forms the easternmost part of the Moldanubian Zone and constitutes the southern and southwestern part of the Bohemian Massif (Fig. 2.39A). The Moldanubicum has tectonic boundaries with all of the major units which surround it. The Teplá-Barrandian Unit, the Saxo-Thuringian Zone in the area of the Bohemian Forest, and the Moravo-Silesian Zone (Fig. 2.39A). The present contacts between the individual units are the result of Variscan collision processes and post-orogenic extension (Rajlich *et al.* 1988; Zulauf 1994). The contact with the Teplá-Barrandian Unit, in particular, served as a pathway for the intrusion of Variscan granitoids which obscure the original character of the boundary between the two units. To the south, the units of the Moldanubicum units are covered by sediments of the Alpine Foredeep and flysch units of the Eastern Alps. To the SW in Bavaria, they are overlain by Permian and Mesozoic platform sediments, partly affected by repeated movements along the faults of the Bavarian Pfahl and the Franconian Line.

The marginal Kutná Hora and Svratka Crystalline units (Figs 2.39B & 2.40), are herein incorporated into the Moldanubicum (Mísař *et al.* 1983; Report of the Working Party for Regional Geological Classification of the Bohemian Massif 1994). These units consist of alternating nappes of middle and lower crustal rocks (Synek & Oliveriová 1993; Kachlík 1999; Schulmann *et al.* 2005; Tajčmanová *et al.* 2006). The internal structure of the units is influenced by strong NW–SE dextral shearing along the Elbe Zone, which cuts the NE margin of the Teplá-Barrandian Unit along the boundary with the Saxo-Thuringian Zone. The character and kinematics of the tectonic processes along the boundaries of the Moldanubicum and the Teplá-Barrandian Unit changed during the Devonian to Lower Carboniferous, and the structural evolution is very complex and not fully understood (see summary by Pitra *et al.* 1999). To the east, the Moldanubicum is thrust over the Moravo-Silesian Zone.

The Moldanubicum forms the thickest and most eroded part of the root of the Variscan Orogen. During thickening, crustal rocks close to the suture zones were often carried into the upper mantle, as reflected in the relicts of HP-HT rocks of crustal origin (Kotkova *et al.* 1997; Vrána & Frýda 2003). HP-HT garnet peridotites and associated eclogites were mostly enclosed in HP leucogranulites (Medaris *et al.* 1995*a*, *b*; Brueckner & Medaris 2000; Medaris *et al.* 2003). During the Early to Late Carboniferous exhumation, Moldanubian rocks were widely and intensely migmatized and intruded in several stages from *c.* 350 to 284 Ma (Finger *et al.* 1997 and references therein) by large Variscan late-kinematic and post-kinematic plutons.

The structure of the Moldanubicum is influenced by a changing stress regime along the tectonic contact with the Teplá-Barrandian Unit and, to the east, also by strong oveprinting associated with collision with the Moravo-Silesian Zone. These two superimposed events produced a rather complex foliation pattern (Urban & Synek 1995; Vrána 1979). During the older event, the internal lower crustal complexes of the Gföhl Unit were thrust over the para-autochthonous middle crustal complexes of the Varied and Monotonous units. The external belt of HP rocks of the Gföhl Unit straddle the tectonic boundary with the Teplá-Barrandian Unit, Moravo-Silesian Zone (Fritz *et al.* 1996; Schulmann *et al.* 2005) and probably the Saxo-Thuringian Zone (Polička Crystalline Unit). The resulting distribution of individual units became slightly modified by updoming of the

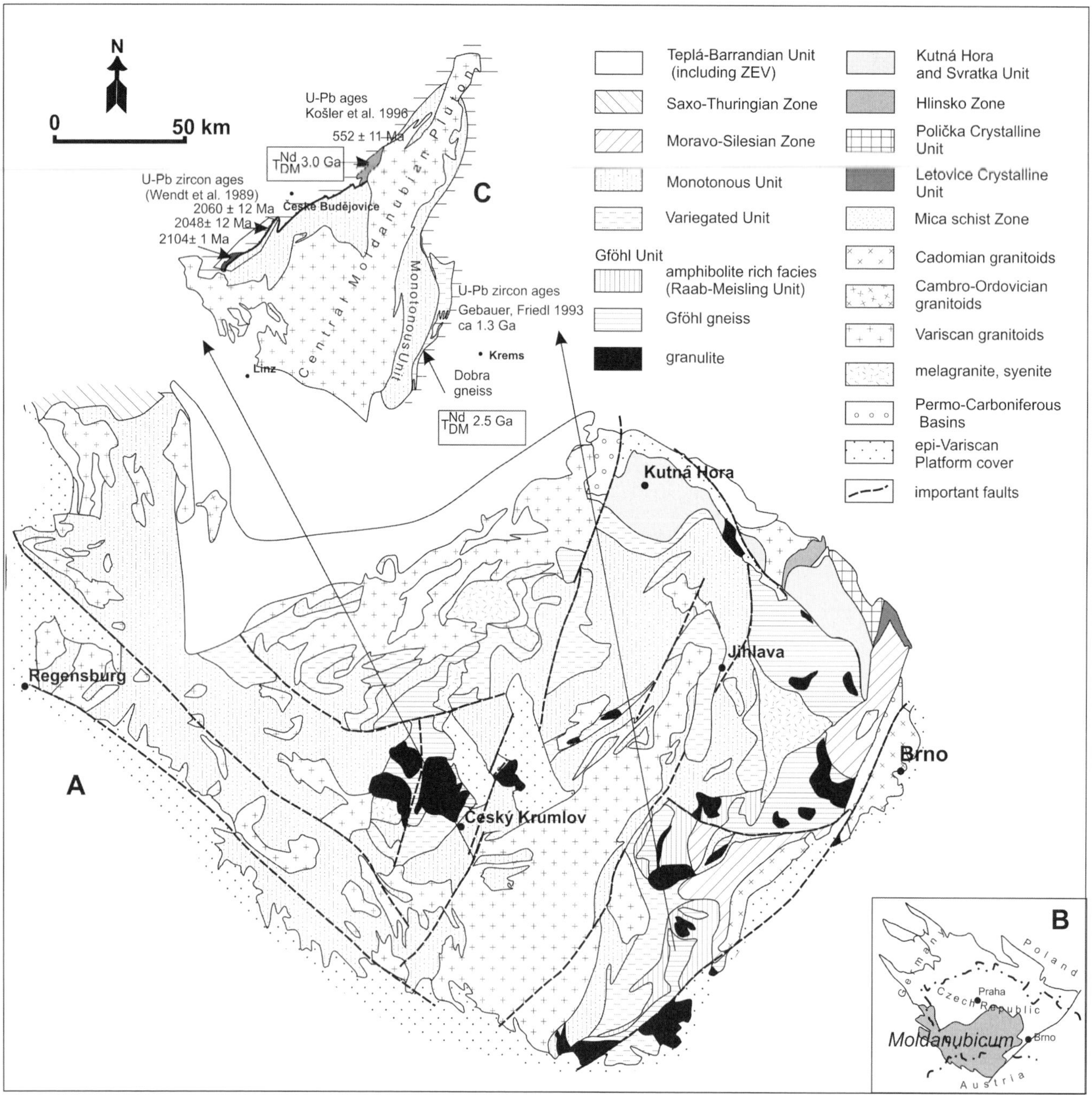

Fig. 2.39. (**A**) Lithology and extent of major lithotectonic units of Moldanubicum *sensu stricto* (modified according to Urban & Synek 1995). (**B**) Position of the Moldanubicum within a frame of Bohemian Massif. (**C**) Geological map showing distribution of Palaeoproterozoic and Mesoproterozoic orthogneisses cropping out at the tectonic boundary between the Monotonous and Variegated Units (modified according to Wendt *et al.* 1988).

central parts of the Moldanubicum during the intrusion of the Central Moldanubian Pluton.

The present consensus on the tectonostratigraphy of the Moldanubicum is that three lithotectonic units can be distinguished. From the bottom to the top, these are the Monotonous (Ostrong), Varied (Drosendorf) and Gföhl Units (Thiele 1976; Matura 1976; Fuchs 1986; Finger & Steyrer 1995; Fiala *et al.* 1995; Franke 2000). However, it is usually very difficult to distinguish between the first two units in the field. Consequently, the mapped extent of these units in different maps can vary considerably (Tollmann 1982; Fiala & Patočka 1994; Mísař *et al.* 1983).

Structurally, the lowest succession is the Monotonous Unit is characterized by its monotonous lithology. Its extent is shown in Figure 2.39C. The main rocks are K-feldspar–cordierite–biotite paragneisses and migmatites, which originated mainly from greywacke protoliths of presumably Late Proterozoic or Cambro-Ordovician ages. Less metamorphosed mica schist complexes, such as the Kaplice Unit and Královský hvozd (Künischer Wald) near Nýrsko in the Šumava Mountains also contain two-mica paragneisses and mica schists. Quartzites and calc-silicate rocks are rare intercalations in the host gneisses and migmatites. Bodies of orthogneisses and eclogites rarely occur in the Monotonous Unit (O'Brien & Vrána 1995).

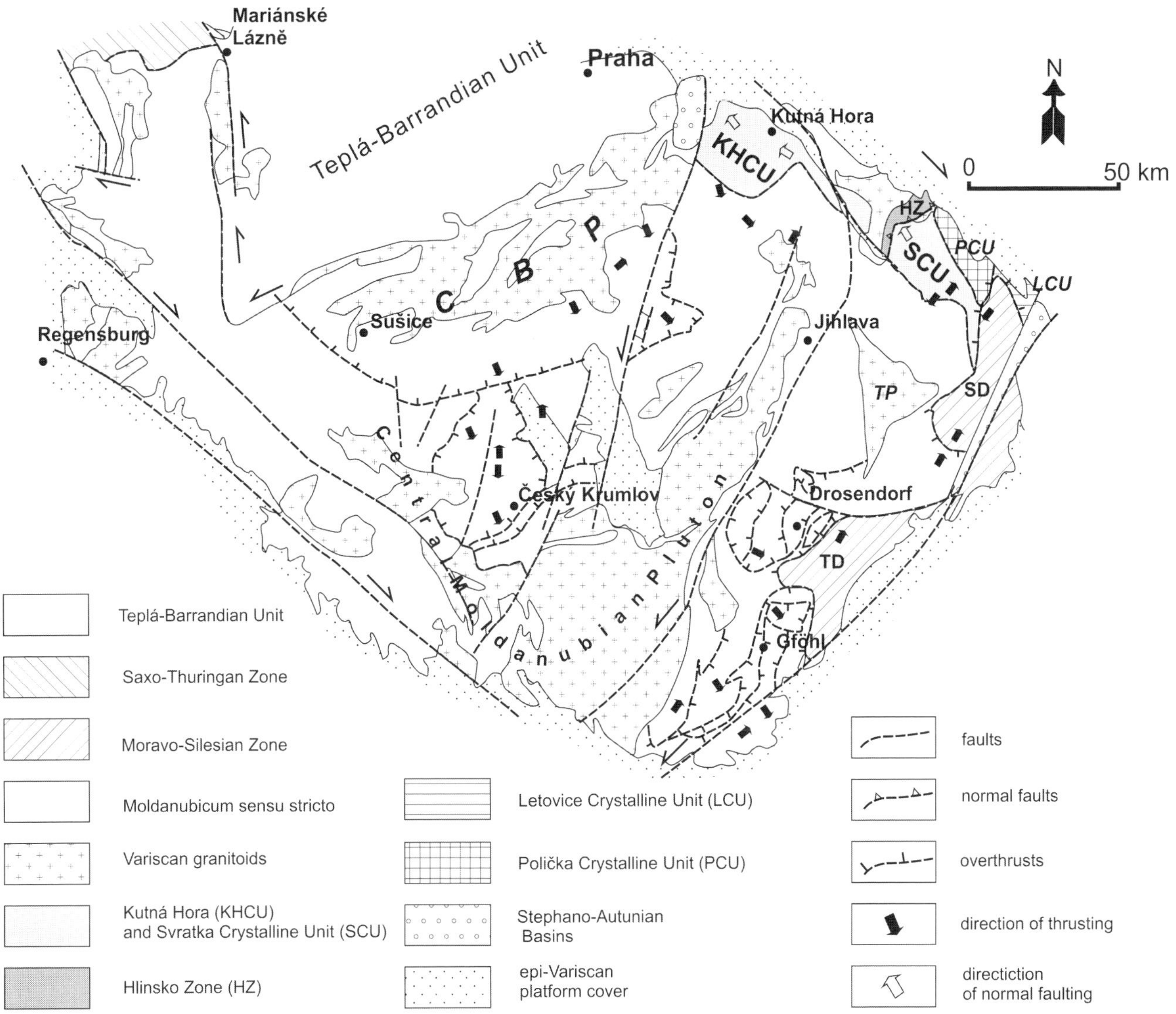

Fig. 2.40. Principal lithotectonic units of the Moldanubicum *sensu stricto* and adjacent units with emphasis on major tectonic boundaries and their kinematics (modified according to Urban & Synek 1995). CBP, Central Bohemian Pluton; SD, Svratka Dome; TD, Thaya Dome; TP, Třebíč Pluton.

Detrital zircons in the paragneisses of the Monotonous Unit in Bavaria indicate a heterogeneous source area in which zircons with an age of 3.8 Ga to 0.7 Ga occurred (Gebauer *et al.* 1989; Grauert *et al.* 1974; Kröner *et al.* 1988; Friedl *et al.* 2004). This indirectly implies a maximum age of Ediacaran for the deposition of the sediments of this unit.

SHRIMP U–Pb dating of zircons from metarhyolites belonging to the Monotonous Unit of the southern Bavarian Forest provides further evidence that Cadomian magmatism (*c.* 550–540 Ma) occurred (Teipel *et al.* 2004). Friedl *et al.* (2004) used SHRIMP methods on zircon from the Natschlag migmatite from the roof of the Central Bohemian Pluton and interpreted the rock as a Cadomian migmatite or S-type granitoid, which recycled the Late Archaean to Palaeoproterzoic crust, or as Early Palaeozoic metasediment with Cadomian detrital zircon only weakly recrystallized during Variscan processes.

As in other parts of the Bohemian Massif (except for the Moravo-Silesian Zone), the Cambro-Ordovician magmatism (*c.* 515–480 Ma) and anatexis were demonstrated by means of the SHRIMP method in the NE Bavarian part of the Moldanubicum (Teipel *et al.* 2004; Friedl *et al.* 2004; Schulmann *et al.* 2005).

The Varied Unit overlies the Monotonous Unit and differs from the underlying unit in having a much more varied lithology and a different P-T path. Biotite–sillimanite paragneisses represent the major rock type. Numerous bodies of quartzite, graphite-bearing gneisses, marbles, acid and basic metavolcanic rocks of tholeiitic and intraplate character are intercalated (Patočka 1991; Kachlík 1999).

The age of the Varied Unit is controversial. SHRIMP dating of zircons from the Spitz orthogneiss in Lower Austria, intercalated within the Varied Unit, indicated an Ediacaran age (610 Ma) of its granitic protolith. Parts of the Varied Unit could possibly be Palaeozoic in age. This is indicated by the similarity in facies development of the Silurian and Devonian of the Barrandian (Chlupáč 1988) and the Varied Unit metasediments and by sporadic finds of Silurian microfossils in Varied Units marbles associated with layers of graphite paragneisses (Andrusov & Čorná 1976; Pacltová 1994; Reitz 1992).

The Gföhl Unit forms the uppermost structural unit of the

Moldanubicum. It characteristically shows great lithological heterogeneity and includes bodies of HP-HT mantle rocks (garnet peridotites and garnet pyroxenites), eclogites and skarns which take the form of boudins or larger bodies within various types of crustal rocks such as migmatites, migmatized granitic gneisses, metagranites and granulites. Its extent is depicted in Figure 2.39C. An Ediacaran age of 551 ± 1 Ma for one sample of granitic gneiss from Biskupice was reported by Schulmann *et al.* (2005) from the South Moravian part of the Gföhl Unit. SHRIMP U–Pb and Pb/Pb evaporation protolith ages of zircons from felsic and mafic granulites and Gföhl gneisses are mostly Early Palaeozoic (Friedl 1997; Friedl *et al.* 1998; Kröner *et al.* 2000*a*; Schulmann *et al.* 2005; Janoušek *et al.* 2006). Cambro-Ordovician ages are more common between various types of Gföhl orthogneisses and migmatites (Finger & von Quadt 1995; Friedl *et al.* 2004; Schulmann *et al.* 2005). Protolith ages for both HP-HT and LP granulites (Lišov Massif) span the period between 390 and 360 Ma (Wendt *et al.* 1994; Schulmann *et al.* 2005; Janoušek *et al.* 2006), while Ordovician or Silurian ages are rare.

Precambrian intrusive rocks, metamorphosed to various types of orthogneisses, straddle the tectonic boundary between the Monotonous and Varied units (Fig. 2.39C) (Vrána 1979; Wendt *et al.* 1993). Little evidence exists for the occurrence of Precambrian igneous rocks in the Gföhl Unit (Schulmann *et al.* 2005) and inner parts of the Varied Unit, such as the Spitz orthogneiss (Friedl *et al.* 2004).

The oldest known Moldanubicum intrusives are represented by the 2.1 Ga Světlík orthogneisses (Wendt *et al.* 1993; Fig. 2.41), which occur together with small bodies of ultrabasic rocks and eclogites. The Světlík orthogneisses correspond geochemically to quartz diorites, tonalites and granodiorites, correlating with I-type granitoids (Patočka *et al.* 2003) of island arcs or active continental margins. Low values of εNd (2.1–2.5) and high T_{DM} of 3.0 Ga indicate that a primitive Archaean tonalite crust was involved in their origin (Wendt *et al.* 1993).

However, the Stráž orthogneiss, exhibiting a similar structural position NE of the evidently Palaeoproterozoic metagranitoids, revealed an Ediacaran to Early Cambrian age of *c.* 552 ± 11 Ma (Košler *et al.* 1996). On the other hand, the Dobra orthogneiss in the Austrian sector of the Moldanubicum yields a Mesoproterozoic protolith age (1.4 Ga, SHRIMP; Gebauer & Friedl 1994; Fig. 2.41). Based on a *c.* 600 Ma overgrowth on Mesoproterozoic zircons, Friedl *et al.* (2004) suggest that they may represent a restite-rich Caromian pluton, derived from the a 1.4 Ga igneous source. The tonalitic to granitic Spitz orthogneiss crops out structurally above the Dobra Gneiss inside the Varied Unit metasediments and is of Ediacaran age (611 ± 10 Ma, U–Pb SHRIMP; Friedl *et al.* 2004).

According to the classification of Wimmenauer (1984), the

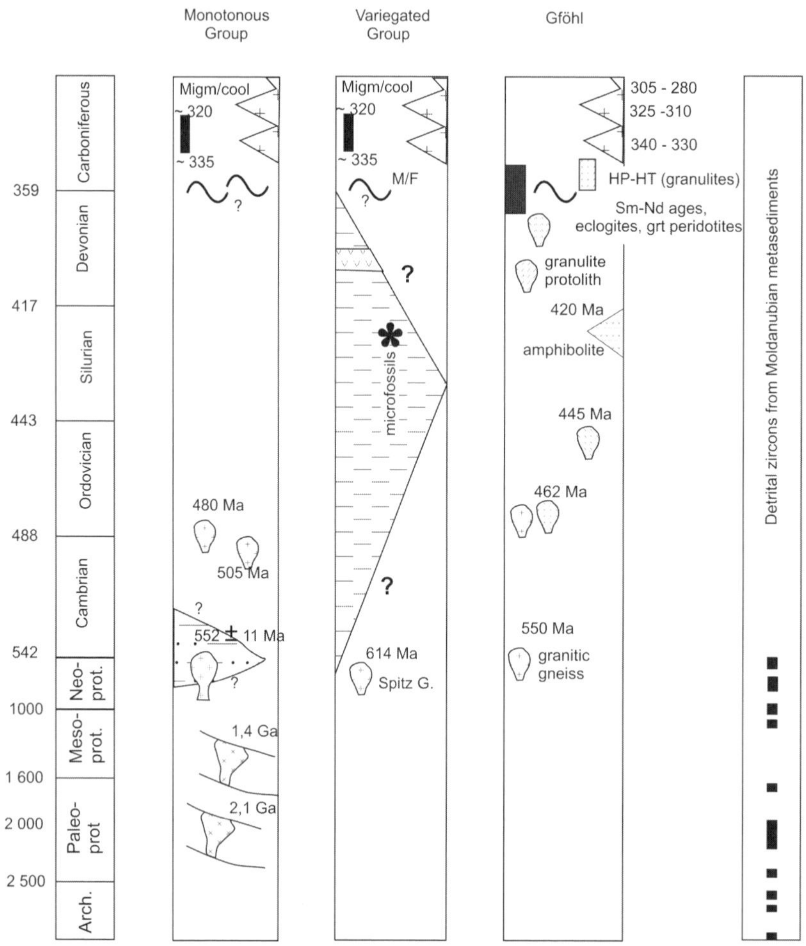

Fig. 2.41. Dynamic stratigraphy of the principal lithotectonic units of the Moldanubicum *sensu stricto* inferred from protolith ages of metagranitoids, mafic and felsic volcanic intercalations, scarce microfossil findings and ages of detrital zircons.

biotite–sillimanite paragneisses and migmatites of both the Monotonous and Varied units correspond mainly to greywackes and rarely to meta-arkoses. The chemistry of paragneisses from both of these units mostly overlaps (Vrána & Bártek 2005). Compared with the average composition of the continental crust at an interval of 0.8–0.5 Ga, these rocks are enriched in K_2O and P_2O_5 but depleted in CaO. The enrichment in lithophile elements (Rb, Ba, K, Th), and their relatively steep REE curves with a prominent Eu anomaly, distinguish the Moldanubian metasediments from the Late Proterozoic metapsammites of the SE flank of the Barrandian (Kachlík 1999).

The more evolved lithology of the Moldanubian metasediments compared to the Ediacaran rocks of the Teplá-Barrandian Unit can also be demonstrated by Rb–Sr and Sm–Nd isotopes (Gorokhov *et al.* 1983; Köhler & Muller-Sohnius 1986; Liew & Hofman 1988). The εNd_{330} values of the Moldanubian sediments vary between -5 and -12 and the initial ratio of $^{87}Sr/^{86}Sr_{330}$ varies between *c.* 0.710 and 0.718, which contrasts with the considerably lower values of the initial Sr ratios and, with the higher εNd values for the Ediacaran metagreywackes of the Teplá-Barrandian Unit. On average, the higher T_{DM} values of the Moldanubian metasediments (1.5–1.7 Ga), in comparison with the values of the Proterozoic metagreywackes of the Barrandian, indicate differences in the source areas of metasediments of the two units (Ihlenfeld *et al.* 1998; Henjes-Kunst *et al.* 1995; Köhler & Miller-Sohnius 1986). The distribution of trace elements and Zr/Th, Zr/Nb and Nb/Y ratios indicate that the source of material for the sediments was an active continental margin with contribution of synsedimentary basic volcanism.

Most of the Precambrian orthogneisses are former I-type calc-alcaline granitoids of tonalitic composition, which originated in a subduction-related environment or by remelting of older crust of calc-alkaline composition. The repeated subduction-related magmatic activity during the Precambrian suggests that the Moldanubian crust was formed during several episodes on the active continental margin of Gondwana or in an island arc setting, probably close to the West African Craton (Friedl *et al.* 2004).

Basement of the Alps (B.S.)

The Alps result from the closure of the Meliata and Penninic oceans and the collision of the northern European Plate and the southern Adriatic Plate, one of the Mediterranean microplates that were squeezed and rotated by the Cretaceous and Tertiary northward motion of Africa (Schmid *et al.* 2004). Previous to their Permian–Mesozoic sedimentation and the opening of the Meliata and Penninic oceans, both the European and Adriatic plates were part of the Variscan Belt, the central tectonic element of Pangaea. Uplift and erosion following Alpine collisional crustal thickening and nappe tectonics led to a nearly continuous outcrop representing the Variscan Orogen (Pfiffner *et al.* 1997; Lammerer & Weger 1998). Indeed the Alpine basement units provide a good place to study the pre-Mesozoic rocks despite the subsequent Alpine overprint (von Raumer & Neubauer 1993 and references therein).

Alpine sutures subdivide the pre-Mesozoic basement into (a) the Helvetic Domain which forms a part of the European Plate, (b) the Penninic Domain, assembled by the Mesozoic sediments and ophiolites of the Penninic Ocean, and some older continental fragments from the European Plate, (c) the Austro-Alpine, and (d) the South Alpine domains both of which are regarded as parts of the Adriatic Plate (Fig. 2.42). These domains belong to a Palaeozoic Intra-Alpine Terrane (Stampfli 1996; von Raumer *et al.* 2002) within the other former peri-Gondwanan microcontinents or terrane assemblages, namely Avalonia, Armorica, Cadomia and Bohemia. These terranes split off from the north Gondwanan margin at various times, moved and collided with Laurussia and constitute the Variscan Belt (von Raumer *et al.* 2003).

Pre-Mesozoic basement in the Alps

Late Variscan age granitoids are abundant in the Helvetic and Penninic basement and less frequent in the Austro-Alpine and South Alpine domains. Carboniferous and Permian (meta)sediments are locally preserved. Thus, the Alpine basement is comparable in terms of the pre-Variscan evolution to most of the Gondwana-derived microcontinents (von Raumer 1998), as it contains relicts of Cadomian continental crust, of an Early Palaeozoic active-margin setting, and the remnants of Early Palaeozoic sediments and magmatic rocks (Neubauer 2002). Consequently, most, if not all of the Alpine pre-Mesozoic basement units enclose pre-Variscan elements which are critical for the evaluation of the Precambrian evolution. On one hand, Precambrian rocks can be expected in pre-Variscan basement segments where Cambrian to Ordovician intrusives of an Early Palaeozoic active margin occur. In these tectonic settings, the Precambrian successions host ophiolites as in the Helvetic Domain (Menot *et al.* 1988; Pin & Carme 1987), oceanic-island or island-arc magmatic rock associations as in the Penninic Domain (Frisch *et al.* 1993; Eisele *et al.* 1997; von Quadt *et al.* 1997; Stille & Tatsumoto 1985) and a mature island arc or continental magmatic arc in the Austro-Alpine Domain (Miller & Thöni 1996; Schaltegger *et al.* 1997; Schulz *et al.* 2004). Their zonal distribution, while poorly studied, may serve as a baseline for tracing older structures and suture lines. On the other hand, an indirect record of Precambrian geodynamics is possible in the post-Early Ordovician phyllitic, weakly metamorphic and/or fossiliferous sequences of the Austro-Alpine and South Alpine domains.

In detail, Precambrian rocks occur in the Aar-Gotthard, Aiguilles Rouges-Mont Blanc, Belledonne, Haut Dauphiné and Argentera massifs of the Helvetic Domain (Fig. 2.42). Apart from the Alpine overprint, there is no difference between the Helvetic Domain and the extra-Alpine French Massif Central, Vosges or Black Forest units of the Variscan Moldanubian Zone.

In the Penninic Domain of the Western Alps, Precambrian rocks are suspected in the external units of the Briançonnais and in the Siviez-Mischabel-Bernard-Mont Fort units. In the internal zone of the Penninic Domain with its strong Alpine overprint of basement and cover, the Dora Maira and Gran Paradiso massifs, the stack of the Lepontine Nappes (with the Simano, Maggia, Leventina, Lebendun units), Antigorio and Adula/Cima Lungo may also contain Precambrian rocks. In the Eastern Alps, the Penninic Domain is exposed in the Tauern Window, a tectonic window beneath the Austro-Alpine nappes, and is subdivided into an upper part of Mesozoic metasediments and ophiolites, and a structurally lower part. In the lower part Permian and Carboniferous granitoid intrusions occur within an older basement (Tuxertal, Zillertal and Granatspitz; Hochalm and Sonnblick) which can be considered as the eastern prolongation of the basement exposed in the Helvetic Domain and thus as a part of the European Plate (Lammerer 1988; von Raumer 1998). Within these lower parts of the Penninic Domain in the Tauern Window (also called the Venediger nappe system) Precambrian rocks crop out in the Habach-Storz and Stubach groups which mantle the granitoids.

The klippes of the Dent Blanche Nappe, the Sesia-Lanzo Zone and the Seconda Zona Diorite-Kinzigite represent the Austro-

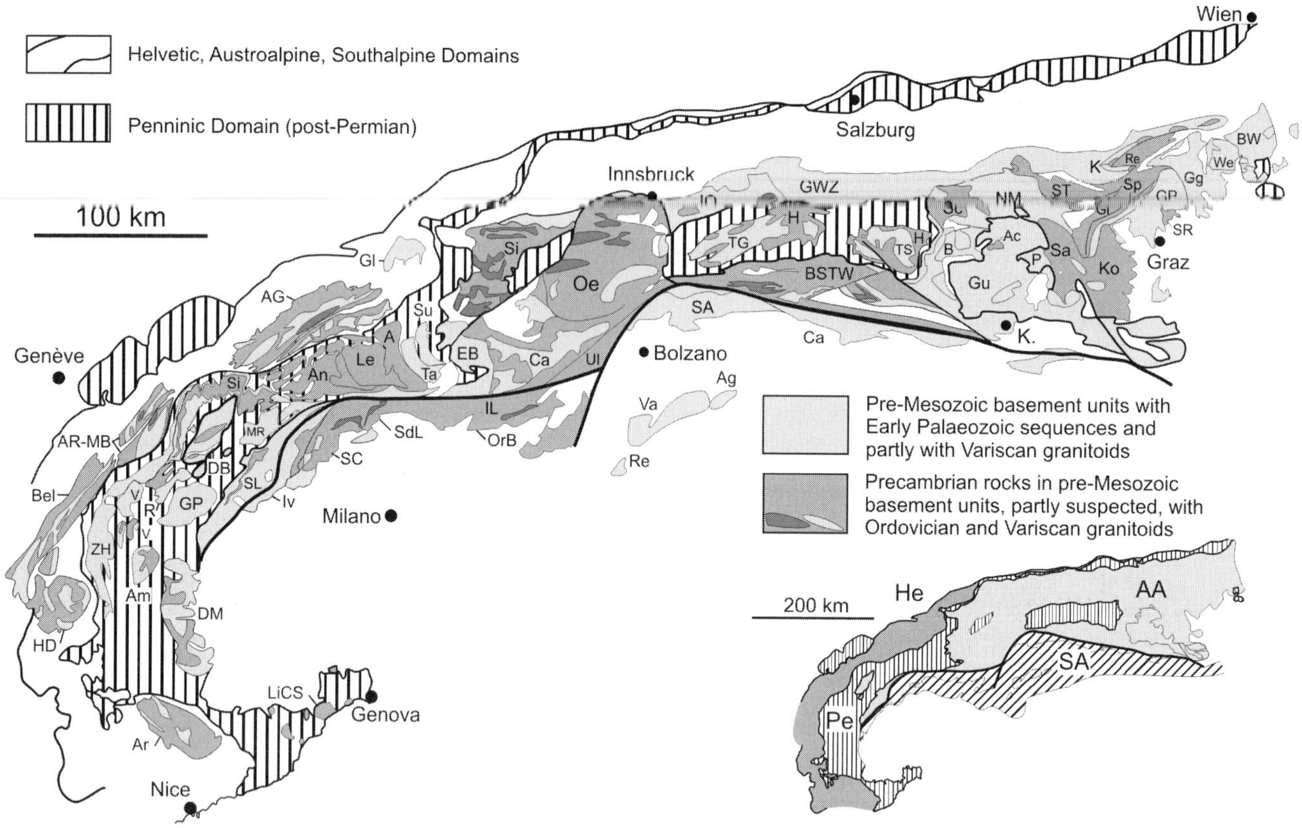

Fig. 2.42. Pre-Mesozoic basement units in the Alps and occurrence of segments with Precambrian rocks (compiled according to von Raumer & Neubauer 1993; Neubauer & Frisch 1993; von Raumer 1998; von Raumer *et al.* 2002; Neubauer 2002, and references therein). *Helvetic Domain*: AG, Aar-Gotthard; AR-MB, Aiguilles Rouges - Mont Blanc; Ar, Argentera; Bel, Belledonne; Gl, Glarus; HD Haute Dauphiné. *Penninic Domain*: A, Adula/Cima Lungo; Am, Ambin; An, Antigorio with Berisal and Monte Leone; DM, Dora Maira; GP, Gran Paradiso; H, Habach-Storz and Stubach Groups; Le, Lepontine nappes with Simano, Maggia, Leventina, Lebendun; LiCS, Ligurian Calizzano-Savona; MR, Monte Rosa; R, Ruitor; Si, Siviez-Mischabel-Bernard-Mont Fort; Su, Suretta & Ta Tambo; TG, Tauern Basement Zillertal-Granatspitz; TS, Tauern Basement Hochalm-Sonnblick; V, Northern and Southern Vanoise; ZH, Zone Houillère with Sapey Gneiss Zone. *Austro-Alpine Domain*: Ac, Ackerl Crystalline; B, Bundschuh; BSTW, Basement south of Tauern Window; BW, Bucklige Welt; Ca, Campo; DB, Dent Blanche; EB, Err-Bernina; Gg, Grobgneis complex; Gl, Gleinalm; GP, Palaeozoic of Graz; Gu, Gurktal nappes; GWZ, Greywacke Zone; IQ, Innsbruck Quartzphyllite; K, Kaintaleck slices (Frauenberg, Prieselbauer, Ritting); Ko, Koralpe; NM, Niedere Tauern and Micaschist Complexes; Oe, Oetztal; P, Plankogel; Re, Rennfeld; Sa, Saualpe; SL, Sesia-Lanzo Zone and Seconda Zona Diorite Kinzigite; Sc, Schladming-Wölz; Sil, Silvretta; Sp, Speik Complex with Kraubath ophiolite; SR, St. Radegund; ST, Seckauer Tauern; Ul, Ulten; We, Wechsel. *South Alpine Domain*: Ag, Agordo-Cereda; Ca, Carnic Alps; Iv, Ivrea Zone; OrB, Orobic Basement; SA, Southern Alps and Pusteria; SC, Strona-Ceneri Zone; SdL, Serie dei Laghi; Re, Recoaro; Va, Valsugana.

Alpine Domain in the Western Alps. In the Central and Eastern Alps, the Lower Austro-Alpine Bernina basement is situated below the Austro-Alpine Silvretta, Oetztal, Scarl-Campo and Ulten units. These units and the Austro-Alpine basement to the south and to the east of the Tauern Window consist of metasedimentary basement rocks which were intruded by igneous rocks of Early Cambrian to Ordovician ages, especially the Silvretta, Oetztal and Ulten basement and the basement south of the Tauern Window. This intrusion defines a Precambrian age of the metasedimentary host rocks. However, these Austro-Alpine basement units (also called Altkristallin in the Eastern Alps) underwent subsequent Variscan, Permian and Alpine metamorphism to variable degrees. Therefore, at least some parts could represent remnants of Early Palaeozoic and Palaeozoic sediments, masked by the metamorphic overprint. The occurrence of Ordovician metagranitoids, which are not observed in the Austro-Alpine Saualpe and Koralpe to the east, appears to be critical for the recognition of former Precambrian rocks. The Austro-Alpine units of the Eastern Alps have been assigned to various Palaeozoic terranes, partly with ophiolites and with Precambrian cores but also with sedimentary and volcanic sequences of Early Palaeozoic to Palaeozoic ages (Frisch & Neubauer 1989; Neubauer *et al.* 1989).

Within the western part of the South Alpine Domain, Precambrian rocks occur in the Ivrea and Strona-Ceneri zones, in the Serie dei Laghi and in the Orobic basement. In the eastern segment of the South Alpine Domain (South Alpine Basement), post-Early Ordovician phyllitic series such as those around Brixen and in the Agordo-Cereda, Recoaro and Valsugana exposures, and fossiliferous sequences as in the Carnic Alps predominate. These pre-Mesozoic units in the eastern South Alpine Domain are quite similar to the Early Palaeozoic sequences in the Austro-Alpine Domain and Precambrian age rocks have not yet been identified there (Fig. 2.42 & 2.43).

Rocks of sedimentary origin

Metasediments hosting Cambrian to Ordovician intrusives are strikingly similar in the polymetamorphic Helvetic, Austro-Alpine and South Alpine basement units. They are monotonous

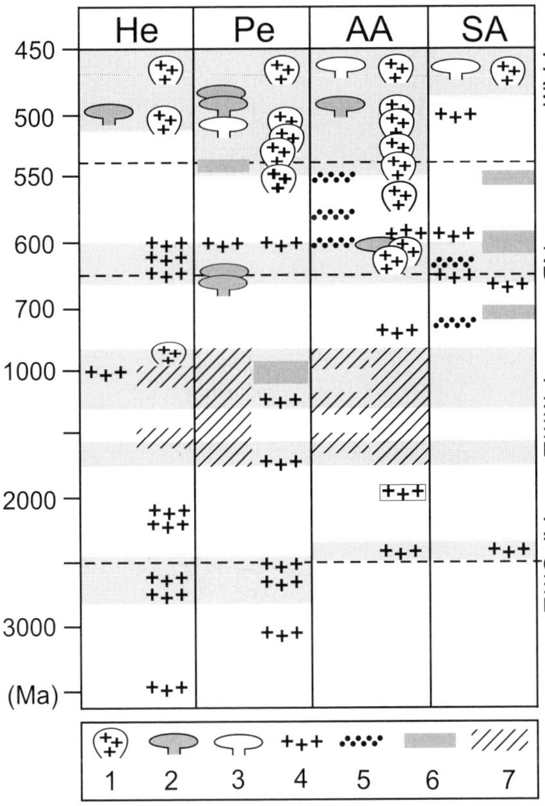

Fig. 2.43. Precambrian and Early Palaeozoic events in the Alpine basement domains: He, Helvetic; Pe, Penninic; AA, Austro-Alpine; SA, South Alpine. The major stages of the pre-Variscan magmatic evolution in the Alpine basement can be grouped into a Palaeozoic active margin (PAM), a Neoproterozoic breakup (NB), the Proterozoic mantle mobilization events (PMME), and the Archaean–Proterozoic crustal mobilization event (APCME) recorded by detrital and inherited zircons. Key: 1, Intrusion of granites, tonalites, diorites and gabbros (magmatic protolith zircon ages); 2, mafic extrusions and dykes (magmatic protolith zircon ages); 3, rhyolites; 4, ages of detrital and inherited zircons; 5, detrital minerals in Early Palaeozoic sequences (Ar/Ar mica ages); 6, Sm–Nd whole-rock isochron ages; 7, Lu–Hf and Sm–Nd T_{DM} (depleted mantle) model ages of igneous and/or sedimentary rocks.

psammopelitic sequences, several kilometres thick and with occasional intercalations of carbonate- and graphite-bearing horizons. Calc-silicate lenses and layers are characteristic and locally abundant. Metabasites also occur, but they are not necessarily of Precambrian age and could be related to the various Early Palaeozoic magmatic suites of mid-ocean ridge or alkaline within-plate basalt type geochemical character.

Precambrian sedimentation ages are constrained by the youngest detrital zircons which have been dated at between 0.62 and 0.5 Ga in the Swiss Alps (Grünenfelder *et al.* 1984; Gebauer & von Quadt 1991; Schaltegger 1992). In the Monte Rosa basement, detrital zircons record a magmatic event at 620 and/or 690 Ma. Metasediments of the Strona-Ceneri Zone were derived from a very heterogeneous Proterozoic source region as documented by detrital zircon ages ranging from 2.4 to 0.6 Ga (Gebauer 1993). Zircon age determinations from different units of the Swiss Alps (Gotthard, Lepontine nappes, Ivrea Zone) by SHRIMP revealed the existence of ancient components. The oldest zircon is 3.43 Ga, while the 2.8–2.6 Ga, 2.3–2.1 Ga, 1.0 Ga and 0.65–0.6 Ga time intervals are common (Gebauer 1993). This distribution of ages and its comparison with the pattern of igneous zircon age intervals observed from Africa, India and South America suggests that the provenance of the detritus was one or more ancient craton(s) with Gondwanan affinity. The presence or absence of the 1 Ga orogenic phase (Grenvillian) is a determining factor as to whether the west Gondwanan Brasilian and/or Guyana Shield may be the provenance area for the metasediments, rather than the West African or the East African/Indian Craton (Schaltegger & Gebauer 1999). Precambrian depositional ages of metasediments can also be estimated via their Rb–Sr whole-rock systematics and were observed from paragneisses of the Austro-Alpine Silvretta Unit (Grauert 1969) and the metapelites of the South Alpine Ivrea Zone (Hunziker & Zingg 1980). Further evidence for Late Precambrian (Cadomian) magmatic or metamorphic protoliths and source terrains is provided by $^{40}Ar/^{39}Ar$ ages (650–580 Ma) of detrital muscovite in the Early Palaeozoic sequences of the Carnic Alps and in the Greywacke Zone of the Austro- and South Alpine domains in the Eastern Alps (Dallmeyer & Neubauer 1994; Dallmeyer *et al.* 1996; Handler *et al.* 1997; Panwitz *et al.* 2000; Neubauer *et al.* 2001). These occurrences of Cadomian-age detritus in post-Early Ordovician deposits provide another link to Cadomia and other peri-Gondwanan terranes which were affected by the Cadomian Orogeny.

Precambrian microfossils are rare in the Alpine basement units. The Habach Group in the Penninic Tauern Window is a metasedimentary–volcanic island arc suite which is dominated by black phyllites intercalated with amphibolites and gneisses, the latter representing basic and intermediate volcanic lithologies of a magmatic arc (Höck 1993; Frisch *et al.* 1993). Reitz & Höll (1988) and Reitz *et al.* (1989) described organic-walled spheroidal acritarchs in the black phyllites, which they interpreted as having a Neoproterozoic and Lower Ediacaran depositional age. Fossiliferous rocks of Cambrian age have not yet been recognized in the Alps, and previous reports on such occurrences were misleading and not based on true fossils (Schönlaub 1993). Other work on microfossils mainly concerns the Early Palaeozoic sedimentary successions, where Lower Ordovician acritarchs have been found in the Greywacke Zone (Reitz & Höll 1989; Reitz 1994). Cambrian to Ordovician microfossils have been discovered in the black and green schists of the Huez Formation in the Helvetic Grandes Rousses Massif (Giorgi *et al.* 1979) and in the South Alpine Agordo-Caerda unit (Sassi *et al.* 1984; Kalvacheva *et al.* 1986). No palaeomagnetic data yet exist for the Precambrian in the Alps.

Igneous rocks

Igneous rock suites with zircons allowing protolith age determinations, which could serve as a backbone for the reconstruction of Precambrian geodynamics, are rare in the Alps. In the Gotthard Massif of the Helvetic Domain an eclogite with a 870 Ma protolith age may be the result of active-margin magmatism (Gebauer *et al.* 1988*b*). A 670 ± 36 Ma magmatic event, probably related to mantle melting, has been recognized from zircons of a metapyroxenite layer in the Ivrea Zone (Gebauer *et al.* 1992*b*). In the Penninic Domain exposed in the Tauern Window (Eastern Alps), a protolith age of 657 ± 15 Ma (U–Pb zircon) has been reported by von Quadt (1992) from an amphibolite of the Habach Group. Further zircon dating in this unit indicated a history of pervasive Early Cambrian to Early Ordovician intrusive magmatism (von Quadt *et al.* 1997; Eichhorn *et al.* 1995, 1999, 2001). In the Austro-Alpine Silvretta basement, the Val Lavinnoz diorite with an age of 609 ± 3 Ma (Schaltegger *et al.* 1997) intruded a volcanosedimentary succession. A U–Pb zircon age of 603 ± 18 Ma was reported from the alkaline Wismath orthogneiss from the easternmost Austro-Alpine region (Korokovsky *et al.* 1998). In the Austro-Alpine Domain to the south of the Tauern Window, an eclogitic amphibolite with N-MORB-type geochemical signature in the Schobergruppe mountains was dated at 590 ± 4 Ma by the Pb/Pb single zircon evaporation method (Schulz & Bombach 2003). As no meta-ultrabasic members or an ophiolite or ocean-floor magmatic suite are related to the eclogitic amphibolite, the magmatic protolith of the metabasite could have been a sill-like intrusion into a pre-existing crust. The εNd characteristics and evolution line of the eclogitic amphibolite almost match the pattern of a depleted mantle (Schulz *et al.* 2004). However, some minor contribution of a crustal or enriched mantle source is indicated by elevated $\delta^{18}O$, LILE and LREE. This can be interpreted as a subduction-related component (Schulz 1995) and has been attributed to a backarc magmatic setting.

A further source of data on the Precambrian are the inherited zircons in intrusive rocks. Discrete events of mantle mobilization may be recorded by zircons in mafic and ultramafic rocks that formed during partial melting of mantle material. Ages of 3.17, 2.67, 2.45 and 1.27 Ga were reported from an eclogite in the Gotthard Massif of the Helvetic Domain (Gebauer *et al.* 1988*b*) and an age of 1.72 Ga from the Alpe Arami peridotite in the Adula/Cima Lunga Nappe system of the Penninic Domain (Gebauer *et al.* 1992*a*). The Cambrian to Ordovician intrusives are widespread in the pre-Variscan basement units of the Alps. These rocks bear many inherited zircons which can serve as a record of Precambrian magmatic events. Several Ordovician granitoids such as the Modane porphyritic granite, the Péclet orthogneiss in the Sapey Gneiss Unit (eastern part of Zone Houillère) and further orthogneisses with protolith ages ranging between 500 and 450 Ma within the Briançonnais basement of the Penninic Domain (Western Alps) have inherited zircons with SHRIMP U–Pb ages of around 660–600 Ma (Bertrand *et al.* 2000). From the Austro-Alpine Oetztal crystalline basement, Klötzli-Chowanetz *et al.* (1997) reported a 2.4 Ga population of inherited zircons. The inherited zircons in the Ordovician orthogneisses and metaporphyroids in the Austro-Alpine Domain to the south of the Tauern Window were dated by thermal ionization mass spectrometry (TIMS) evaporation and SHRIMP and have ages up to 2.0 Ga, around 800 Ma and 600 Ma (Schulz & Bombach 2003; Siegesmund *et al.* 2007). All the data from detrital and inherited zircons indicate that the continental source areas were affected by several older orogenies, around 1.0, 1.7 and possibly >2.0 Ga (von Raumer *et al.* 2003).

For the Austro-Alpine Silvretta basement, Köppel *et al.* (1997) used the Pb isotopic technique to model source compositions and mantle/crust ratios. They demonstrated the presence of a continental mass that has been structured and mainly formed during the Cadomian Orogeny, a conclusion that can also be derived from Nd isotopes. A mantle mobilization event has been invoked to explain ages of around 0.9 Ga. The lead isotope data do not indicate the presence of remnants of an old Palaeoproterozoic or Archaean crust.

Additional information on provenance is provided by Sm–Nd whole-rock and Sm–Nd and Lu–Hf model ages from igneous rocks (Meisel *et al.* 1996). The Sm–Nd T_{DM} model ages refer to the depleted mantle and reflect the mean ages of a mantle differentiation (mean crustal residence ages) and only accidentally date geological events. From the Penninic Domain, Sm–Nd whole-rock ages of *c.* 1 Ga were reported from amphibolites in the Berisal Complex (1070 ± 43 Ma; Stille & Tatsumoto 1985) and the Siviez-Mischabel Nappe (978 ± 59 Ma; Zingg 1989). From the Ivrea Zone, Voshage *et al.* (1987) and Pin & Sills (1986) reported 607 ± 19 Ma and 596 ± 35 Ma Sm–Nd whole-rock protolith ages. Model T_{DM} ages of granites from the Helvetic basement domain scatter around a mean value of 1.5–1.6 Ga (Schaltegger & Gebauer 1999). Hf model ages of 1 Ga from highly evolved granites in the Aar Massif have been reported by Schaltegger & Corfu (1995). An addition of subcontinental mantle-derived magmas to a 1.7 Ga crust was postulated to produce T_{DM} model ages of around 0.9–1.3 Ga (Stille & Steiger 1991). Cambrian (520 Ma) metagabbros and metatonalites in the Silvretta yielded Sm–Nd T_{DM} ages of 0.9–1.1 Ga (Schaltegger *et al.* 1997). Recent attempts using Sm–Nd and Re–Os isotopes to date ultramafic rocks in the Austro-Alpine Kraubath harzburgite–dunite sequence suggest a Late Proterozoic age of *c.* 745 Ma whereas pyroxenites crystallized at 550 Ma (Melcher & Meisel 2004). Nd isotopic data from the numerous amphibolites and ultramafites within the Austro-Alpine basement are still rare; however, a Sm–Nd T_{DM} age of 0.9 Ga has been also recorded in within-plate basalt-type amphibolites to the south of the Tauern Window (Schulz *et al.* 2004). Mafic to ultramafic rocks often display Nd and Hf model ages of *c.* 0.9–1.0 Ga, which are interpreted as the signal of a mantle mobilization event, which could either be partial melting and/or a metasomatism in the subcontinental lithosphere (Stille & Schaltegger 1996).

Metamorphic evolution

Apart from detrital muscovite with a weak Variscan and/or Alpine overprint, observed in the Early Palaeozoic sedimentary successions, there is little information of pre-Ordovician metamorphism in the Alpine basement. An Early Ordovician metamorphic event has been reported from the Austro-Alpine Oetztal Winnebach migmatite (Klötzli-Chowanetz *et al.* 1997) while the Gotthard Streifengneisses in the Helvetic Domain (Schaltegger 1993) appear to be related to contemporaneous anatexis and pervasive granitoid intrusions, the latter being a common phenomenon in the Alpine basement (Schaltegger & Gebauer 1999). However, the possible existence of a Barrovian metamorphic overprint of Ordovician age is quite controversial (e.g. Boriani & Villa 1997; Zurbriggen *et al.* 1997; Romer & Franz 1998) in the Alpine basement where Variscan and Alpine metamorphism predominate.

Plate tectonic setting and evolution

Plate tectonic reconstructions for the Alpine basement units commence not earlier than the Early Ordovician and consider the Intra-Alpine Terrane as a lateral and eastward continuation of Avalonian-type crustal pieces along the north Gondwanan margin (Stampfli 1996; Stampfli & Borel 2002; von Raumer *et al.* 2003). The Precambrian evolution of the Alpine basement can be subdivided into two parts: (a) the pre-Ediacaran (>660 Ma) period, and (b) a peri-Gondwanan Ediacaran period (<660 Ma), which culminated in a Cambrian to Ordovician active-margin setting (Fig. 2.43).

The pre-Ediacaran period is characterized by a Gondwana-related age pattern for detrital and inherited zircon, which begin at 3.43 Ga and can be grouped around 2.8–2.6 Ga, 2.4–2.1 Ga, 2.0 Ga, 1.7 Ga and 1.0 Ga. The Grenvillian zircon ages (1 Ga) signal proximity to a west Gondwanan realm, possibly the Brasilian and/or Guyana Shield. From Sm–Nd whole-rock data and T_{DM} model ages from various granites, metasediments and metabasites within the Alpine basement, it is possible to identify major events of mantle differentiation at 1.7–1.5 Ga (Thöni 1999; Schaltegger & Gebauer 1999) and *c.* 1.1 0.9 Ga (Schaltegger *et al.* 1997; Schulz *et al.* 2004).

An important stage in the Precambrian evolution of the Alpine basement is marked by 0.65–0.6 Ga magmatic protoliths in various domains. An age of 0.6 Ga is typical of protoliths in juvenile European crust, found in a primitive island arc situated off the Gondwana continent. Acidic plutonic rocks as well as basalts resembling N-MORB tholeiites were emplaced in the pre-existing Gondwana-derived continental margin deposits. This magmatic period can be labelled as Ediacaran breakup. Detailed observations demonstrate that the Alpine basement units were involved in a peri-Gondwanan Cadomian active-margin setting, which continued throughout the Cambrian and Ordovician, before they underwent rifting and drifting to form parts of the Variscan basement.

Summary

The Precambrian of Central Europe and adjoining areas belongs to two principal geotectonic domains represented by (1) Baltica that forms a part of the East European Craton (EEC), and (2) peri-Gondwana. Both parts are separated by the Trans-European Suture Zone which is the boundary between the ancient Precambrian lithosphere of the craton and the younger lithosphere beneath the late Neoproterozoic and Palaeozoic peri-Gondwanan mobile belts of Central and western Europe. The peri-Gondwanan units of Central Europe are situated SW of the suture and subdivided into Avalonia and Cadomia, separated by the Rheic Ocean suture.

The Baltica part of the EEC consists of three major blocks, namely, Fennoscandia, Sarmatia and Volga–Ural which are characterized by different structures and evolution over a period extending from the Archaean to Palaeoproterozoic. Palaeomagnetic data reveal that Fennoscandia and Sarmatia were separate until *c.* 1.85 Ga, i.e. until the time of their amalgamation into Baltica. It is assumed that Fennoscandia comprises a mosaic of several Archaean blocks surrounded by accreted crust, generally younging westwards, i.e. in the direction of the Trans-European Suture Zone. These younger orogenic belts include the *c.* 1.95–1.82 Ga Svecofennian Domain, the *c.* 1.80–1.65 Ga Trans-Scandinavian Igneous Belt, and a series of 1.75–1.55 Ga magmatic arcs that were variably overprinted during the 1.04–0.97 Ga Sveconorwegian Orogeny. A recent tectonic model suggests the presence of Svecofennian terranes extending across the Baltic and into the Black Sea area. All parts of Baltica in Central Europe and adjoining areas belong to Fennoscandia (southern Sweden, NE Poland, Lithuania, parts of the basement in the Baltic Sea).

Avalonia is a microcontinent or terrane derived from the periphery of West Gondwana. The Neoproterozoic basement of Avalonia formed during Avalonian orogenic processes and related tectonomagmatic events (*c.* 750–540 Ma). Neoproterozoic and Early Palaeozoic sediments of Avalonia contain detrital zircon grains of Mesoproterozoic age, which are assumed to have formed in a juvenile crust between 1.0 and 1.3 Ga. There is evidence that a large part of Avalonia originated at the margin of Amazonia. Avalonia rifted off from the Gondwanan margin in the late Cambrian/early Ordovician. The Rheic Ocean opened as the Tornquist Sea closed due to the collision of Avalonia with Baltica during the late Ordovician and the early Silurian. In the late Silurian, Baltica plus Avalonia collided with Laurentia coincident with the closure of the Iapetus Ocean to form the short-lived continent of Laurussia. During its drift, and following amalgamation with Baltica and Laurentia, Avalonia represented the northern margin of the Rheic Ocean. Avalonia forms the poorly exposed basement of the Rhenohercynian Zone, extending from the Brabant Massif of Belgium to the eastern margin of the Bohemian Massif. The most important parts of East Avalonia are located in southern Britain and Ireland. Other more easterly parts of Avalonia and related terranes, as well as terranes of uncertain origin along the recent SW margin of Baltica, are summarized as Far East Avalonia. In this chapter, we include the Eckergneiss Complex in the Harz Mountains (Rhenohercynian Zone), the Upper Silesian and Małopolska blocks (SE Poland) and the Thaya and Slavkov terranes, including the so-called Metabasite Belt (eastern Bohemia), as parts of Far East Avalonia.

Cadomia is a belt of Neoproterozoic rock units formed by Cadomian orogenic processes at an active margin at the periphery of the West African Craton of Gondwana. Rock units of Cadomia show few or no zircon grains with ages in the range 0.75–1.6 Ga and are dominated by detritus from the West African Craton (2.05 Ga and older). A *c.* 2.0 Ga crust was recycled. Basement slivers of *c.* 2 Ga rocks such as the Icartian basement occur locally. The Cadomian basement was formed by Cadomian orogenic processes (*c.* 750–530 Ma). It includes Neoproterozoic volcanosedimentary complexes and intrusions and early Cambrian (*c.* 540–530 Ma) plutons intruded during a final intense tectonothermal event. Most probably, Cadomia never left the Gondwana mainland and did not drift off like Avalonia.

During the assemblage of the Pangaea Supercontinent by Variscan orogenic events, the Rheic Ocean closed as a result of the continent–continent collision between Laurussia and Gondwana in Late Devonian–Early Carboniferous times. The Mid-German Crystalline Zone at the northern rim of the Bohemian Massif is believed to represent the suture of the Rheic Ocean. The suture was formed by oblique collision between the Bohemian (Cadomia, Gondwana) and Rhenish massifs (Avalonia, Laurussia) in Variscan times during the Late Devonian to Early Carboniferous. During its origin Neoproterozoic crust became recyled and parts of Cadomia and Avalonia amalgamated.

Cadomia includes the Armorican Massif, Pyrénées, French Massif Central, Maures Massif, Corsica, and the largest part of the Bohemian Massif such as the Saxo-Thuringian Zone, the Sudetes, and the Moldanubian Zone. The latter contains the Teplá-Barrandian Unit and the Moldanubicum *sensu stricto*. The oldest units of the Bohemian Massif are remnants of Palaeo- to Mesoproterozoic cratonic basement slivers such as the Dobra gneiss (1.38 Ga) and the Svetlik gneiss (2.05–2.1 Ga). Parts of the Bohemian Massif are related to Far East Avalonia such as the Brunovistulian Block of the Moravo-Silesian Zone (eastern Bohemia) and the Upper Silesian and Małopolska blocks (SE Poland).

It is likely that the peri-Gondwanan basement of the Alps also was a part of Cadomia although the occurrence of Mesoproterozoic zircon points to the possibility of an origin west of the West African Craton.

The authors thank very much C. Quesada (Madrid) and R. Strachan (Portsmouth) for their constructive reviews which improved the manuscript. Many thanks to T. McCann (Bonn) and R. L. Romer (Potsdam) who corrected the English. J. B. M. is grateful for the support of the Natural Sciences and Engineering Research Council Canada, the University Council of Research, St. Francis Xavier University, and a Visiting Senior Gledden Fellowship, Tectonics Special Research Centre, University of Western Australia. S.A. P. thanks the University of Western Australia for a Visiting Senior Gledden Fellowship. Reconstructions were made in the Western Australian Geotectonic Mapping facility co-funded by the Government of Western Australia (Office of Industry and

Innovation) using the PLATES reconstruction program of the University of Texas at Austin, and Generic Mapping Tools of P. Wessel and W.H.F. Smith. B. Walker, D. Erler and E. Kühe (Dresden) are thanked very much for the compilation of the reference list. C. P. thanks J. Ulrich (Dresden) for the production of one figure. U. Chmiel (Dresden) is thanked by U. L. for drawing one of the figures. K. Drost (Dresden, Cape Town) is thanked very much for reading parts of the manuscript.

References

ÅHÄLL, K.-I. & CONNELLY, J. 1998. Intermittent 1.53–1.113 Ga magmatism in western Baltica age constraints and correlations within a postulated supercontinent. *Precambrian Research*, **92**, 1–20.

ÅHÄLL, K.-I., DALY, J. S. & SCHÖBERG, H. 1990. Geochronological constraints on Mid-Proterozoic magmatism in the Östfold-Marstrand belt: implications for crustal evolution in SW Sweden., *In*: RYAN, A. B., RIVERS, T. & GOWER, C. F. (eds) *Mid-Proterozoic Laurentia Baltica*. Geological Association of Canada, Special Papers, **38**, 97–115.

ÅHÄLL, K.-I., PERSSON, P.-O. & SKIÖLD, T. 1995. Westward accretion of the Baltic Shield: implications from the 1.6 Ga Åmål-Horred Belt, SW Sweden. *Precambrian Research*, **70**, 235–251.

ÅHÄLL, K.-I., CONNELLY, J. & BREWER, T. S. 2000. Episodic rapakivi magmatism due to distal orogenesis? Correlation of 1.69–1.50 Ga orogenic and inboard, 'anorogenic' events in the batic Shield. *Geology*, **28**, 823–826.

ALEKSANDROWSKI, P. & MAZUR, S. 2002. Collage tectonics in the northeasternmost part of the Variscan Belt: the Sudetes, Bohemian Massif. *In*: WINCHESTER, J. A., PHARAOH, T. C. & VERNIERS, J. (eds) *Palaeozoic Amalgamation of Central Europe*. Geological Society, London, Special Publications, **201**, 237–277.

ALLÈGRE, C.J., MANHÈS, G. & GÖPEL, C. 1995. The age of the Earth. *Geochimica et Cosmochimica Acta*, **59**(8), 1445–1456.

ALTERMANN, W. 2002. The evolution of life and its impact on sedimentation. *In*: ALTERMANN, W. & CORCORAN, P.L. (eds) *Precambrian Sedimentary Environments: A Modern Approach to Ancient Depositional Systems*. International Association of Sedimentologists, Special Publications, **33**, Blackwell, Oxford, 15–32.

AMELIN, Y. V., LARIN, A. M. & TUCKER, R. D. 1997. Chronology of multiphase emplacement of the Salmi rapakivi granite-anorthosite complex, Baltic Shield: implications for magmatic evolution. *Contributions to Mineralogy and Petrology*, **127**, 353–368.

AMTHOR, J.E., GROTZINGER, J.P., SCHRÖDER, S., BOWRING, S.A., RAMEZANI, J., MARTIN, M.W., MATTER, A. 2003. Extinction of *Claudina* and *Namacalathus* at the Precambrian–Cambrian boundary in Oman. *Geology*, **31**, 431–434.

ANDERSEN, T., GRIFFIN, W. L., JACKSON, S. E, KNUDSEN, T.-L. & PEARSON, N. J. 2004. Mid-Proterozoic magmatic arc evolution at the southwest margin of the Baltic Shield. *Lithos*, **73**, 289–318.

ANDERSSON, U.B. & WIKSTRÖM, A. 2004. The Småland-Värmland belt (SVB). *In:* HÖGDAL, K., ANDERSSON, U.B. & EKLUND, O. (eds) *The Transskandinavian Igneous Belt (TIB) in Sweden: A Review of its Character and Evolution.* Geological Survey of Finland, Special Paper, **37**, 15–20.

ANDERSSON, U. B., HÖGDAHL, K., SJÖSTRÖM, H. & BERGMAN, S. 2000. Geochronologic constraints on metamorphic, intrusive and detrital ages of rocks from the Bergslagen area, southern Sweden. *Sveriges Geologiska Undersökning, Rapporter och meddelanden*. Contract-no 03–1025/97, 1–12.

ANDERSSON, U. B., CLAESON, D. T., HÖGDAHL, K. & SJÖSTRÖM, H. 2004. Geological feartures of the Småland-Värmland belt along the Svecofennian margin, part II: the Nygården, Karlskoga and Filipstad areas. *In*: HÖGDAHL, K., ANDERSSON, U. B. & EKLUND, O. (eds) *The Trasscandinavian Igneous Belt (TIB) in Sweden: A Review of its Character and Evolution*. Geological Survey of Finland, Special Papers, 37, 39–47.

ANDRÉASSON, P. G. & RODHE, A. 1990. Geology of the Protogine Zone south of Lake Vättern, southern Sweden: a reinterpretation. *Geologiska Föreningens i Stockholm Förhandlingar*, **112**, 107–125.

ANDRUSOV, D. & ČORNÁ, O. 1976. Über das Alter des Moldanubikums nach mikrofloristischen Forschungen. *Geologicke Práce Zpravy*, **65**, 81–89.

ANKUNDINOV, S., SADOV, A. & BRIO, H. 1994. Crustal structure of Baltic countries on the basis of deep seismic sounding data. *Proceedings of the Estonian Academy of Sciences*, **43**, 129–136.

AUTRAN, A., FONTEILLES, M. & GUITARD, G. 1966. Discordance du Paléozoïque inférieur métamorphique sur un socle gneissique antéhercynien dans le Massif des Albères (Pyrénées orientales). *Comptes Rendus de l'Académie des Sciences Paris*, **263**, 317–320.

AUVRAY, B., CHARLOT, R. & VIDAL, P. 1980. Données nouvelles sur le Protérozoïque inférieur du domaine nord-armoricain (France). *Canadian Journal of Earth Siences*, **17**, 52–538.

BAGIŃSKI, B., DUCHESNE, J.-C., VANDER AUWERA, J., MARTIN, H. & WISZNIEWSKA, J. 2001. Petrology and geochemistry rapakivi-type granites from the crystalline basement of NE Poland. *Geological Quarterly*, **45**, 33–52.

BAKUN-CZUBAROW, N., BIAŁOWOLSKA, A. & FEDORYSHYN, Y. 2000. Petrology, geochemistry and geologic setting of the Precambrian basalts of Volyn. *Joint Meeting of EUROPROBE (TESZ) and PACE Project.* Abstract volume.

BALLARD, J. P., BRUN, J. P. & DURAND, J. 1986. La discordance Briovérien-Paléozoïque inférieur en Bretagne Centrale: signature d'un épisode de distension ordovicien. *Comptes Rendus de l'Académie des Sciences Paris*, **303**, 1327–1332.

BALLÈVRE, M., LE GOFF, E. & HÉBERT, R. 2001. The tectonothermal evolution of the Cadomian belt of northern Brittany, France: a Neoproterozoic volcanic arc. *Tectonophysics*, **331**, 19–43.

BANKWITZ, P. 1967. Probleme des Faltenbaus, besonders in den Katzhütter Schichten (Präkambrium) im Schwarzburger Sattel (Thüringen). *Geologie*, **16**, 1083–1102.

BANKWITZ, P. 1995. Die Erdkruste der östlichen Rhenoherzynischen Zone im Umfeld des Harzes. *Zentralblatt für Geologie und Paläontologie. T. 1, Allgemeine, angewandte, regionale und historische Geologie*, **9/10**, 1551–1557.

BANKWITZ, P. & BANKWITZ, E. 1975. Zur Sedimentation proterozoischer und kambrischer Gesteine im Südteil der DDR (Strainanalyse). *Zeitschrift für Geologische Wissenschaften*, **3**, 1279–1305.

BANKWITZ, P. & BANKWITZ, E. 1995. Proterozoikum/Schwarzburger Antiklinorium. *In*: SEIDEL, G. (ed.) *Geologie von Thüringen*. E. Schweitzerbart'sche Verlagsbuchhandlung, Stuttgart, 46–77.

BARBEY, P., CHEILLETZ, A. & LAUMONIER, B. 2001. The Canigou orthogneisses (eastern Pyrénées, France, Spain): an early Ordovician rapakivi granite laccolith and its contact aureole. *Comptes Rendus de l'Académie des Sciences Paris*, **332**, 129–136.

BARCA, S., DURAND-DELGA, M., ROSSI, P. & STORCH, P. 1996. Les micaschistes panafricains de Corse et leur couverture paléozoïque: leur interprétation au sein de l'orogène varisque sud-européen. *Comptes Rendus de l'Académie des Sciences Paris*, **322**, 981–989.

BARD, J. P. & LOUEYIT, J. 1978. Sur l'origine des gneiss oeillés de l'Espinouse dans la zone axiale de la Montagne Noire (Massif Central): consequences tectoniques. *Bulletin de la Société Géologique de France*, **20**, 751–772.

BARFOD, G.H., ALBAREDE, F., KNOLL, A.H., XIAO, S.H., TELOUK, P., FREI, R. & BAKER, J. 2002. New Lu–Hf and Pb/Pb age constraints on the earliest animal fossils. *Earth and Planetary Science Letters*, **201**, 203–212.

BAUMANN, A., GRAUERT, B., MECKLENBURG, S. & VINX, R. 1991. Isotopic age determination of crystalline rocks of the Upper Harz Mountains, Germany. *Geologische Rundschau*, **80**, 669–690.

BEIER, H., MALETZ, J. & BÖHNKE, A. 2000. Development of an Early Palaeozoic foreland basin at the SW margin of Baltica. *Neues Jahrbuch für Geologie und Paläontologie, Abhandlungen*, **218**, 129–152.

BELKA, Z., AHRENDT, H., FRANKE, W. & WEMMER, K. 2000. The Baltica-Gondwana suture in central Europe: evidence from K–Ar ages of detrital muscovites and biogeographical data. *In*: FRANKE, W., HAAK, V., ONCKEN, O. & TANNER, D. (eds) *Orogenic Processes: Quantification* and *Modelling in the Variscan Belt.* Geological Society of London, Special Publications, **179**, 87–102.

BELKA, Z., VALVERDE VAQUERO, P., DÖRR, W., AHRENDT, H., WEMMER, K., FRANKE, W. & SCHÄFER, J. 2002. Accretion of first Gondwana-derived terranes at the margin of Baltica. *In*: WINCHESTER, A., PHARAOH, T. C., & VERNIERS, J. (eds) *Palaeozoic Amalgamation of Central Europe*. Geological Society of London, Special Publications, **201**, 19–36.

BERTHELSEN, A. 1992. From Precambrian to Variscan Europe. *In*:

BLUNDELL, D., FREEMAN, R., MUELLER, ST. (eds) *A Continent Revealed – the European Geotraverse*. Cambridge University Press, Cambridge, 153–164.

BERTRAND, J. M., PIDGEON, R. T., LETERRIER, J., GUILLOT, F., GASQUET, D. & GATTIGLIO, M. 2000. SHRIMP and IDTIMS U-Pb zircon ages of the pre-Alpine basement in the Internal Western Alps (Savoy and Piemont). *Swiss Bulletin of Mineralogy and Petrology*, **80**, 225–248.

BIAŁOWOLSKA, A., BAKUN-CZUBAROW, N. & FEDORYSHYN, Y. 2002. Neoproterozoic flood basalts of the upper beds of the Volhynian Series (East European Craton). *Geological Quarterly*, **46**, 37–57.

BIBIKOVA, E., BOGDANOVA, S., GORBATSCHEV, R., CLAESSON, S. & KIRNOZOVA T. 1995. Stratigraphy and geological correlation. *Interperiodika*, **3**, 591–601.

BIBIKOVA, E., CLAESSON, S., MANSFELD, J., BOGDANOVA, S., GORBATSCHEV, R., SKRIDLAITE, G. & TARAN, I. 1996. Isotope geochronological evidence of Palaeoproterozoic accretion in the western part of the East European Craton. *EUROBRIDGE Workshop*, 8–15 June 1996, Oskarshamn, Sweden. Abstracts, 5–6.

BOGATIKOV, O. A. & BIRKIS, A. P. 1973. *Precambrian Magmatism of West Latvia*. Nauka, Moscow. [in Russian].

BOGDANOVA, S.V. 1999. The Palaeoproterozoic terrane pattern in the western part of the East European Craton. *Between EUROBRIDGE and TESZ, Seventh Eurobridge Workshop*, 26–30 May 1999, Szelment Poland. Abstacts, 11 –12.

BOGDANOVA, S.V. 2005. The East European Craton: some aspects of the Proterozoic evolution in its south-west. *Special Papers PTMin.*, **26**, 18–24.

BOGDANOVA, S. V., BIBIKOVA, E. V. & GORBATSCHEV, R. 1994. Paleoproteroterozoic U-Pb zircon ages from Belorussia: new tectonic implication for the East European Craton. *Precambrian Research*, **68**, 231–240.

BOGDANOVA, S., PASKEVICH, I. K., GORBATCHEV, R. & ORLYUK M. I. 1997. Riphean rifting and major Palaeoproterozoic crustal boundaries in the basement of the East European Craton: Geology and geophysics. *Tectonophysics*, **268**, 1–21.

BOGDANOVA, S. V., PAGE, L. M., SKRIDLAITE, G. & TARAN L. N. 2001. Proterozoic tectonothermal history in the western part of the East European Craton: $^{40}Ar/^{39}Ar$ geochronological constraints. *Tectonophysics*, **339**, 39–66.

BORIANI, A.C. & VILLA, I. M. 1997. Geochronology of regional metamorphism in the Ivrea-Verbano Zone and Serie dei Laghi, Italian Alps. *Swiss Bulletin of Mineralogy and Petrology*, **77**, 381–401.

BOWRING, S.A. & SCHMITZ, M.D. 2003. High-precision U-Pb zircon geochronology and the stratigraphic record. *In*: HANCHAR, J.M. & HOSKIN, P.W.O. (eds) Zircon. *Reviews in Mineralogy and Geochemistry*, **53**, 305–326.

BOWRING, S.A. & WILLIAMS, I.S. 1999. Priscoan (4.00–4.03 Ga) gneisses from northwestern Canada. *Contributions to Mineralogy and Petrology*, **134**, 3–16.

BOWRING, S.A., MYROW, P., LANDING, E., RAMEZANI, J. & GROTZINGER, J. 2003. Geochronological constraints on terminal Neoproterozoic events and the rise of metazoans. *Geophysical Research Abstracts*, **5**, 219.

BRÄTZ, H. 2000. *Radiometrische Altersdatierungen und geochemische Untersuchungen von Orthogneissen, Graniten und Granitporphyren aus dem Ruhlaer Kristallin, Mitteldeutsche Kristallinzone*. PhD thesis, Universität, Würzburg.

BRAUSE, H., SCHUBERT, G. & HORTENBACH, R. 1981. Beitrag zur präkambrischen und zur pleistozänen Tektonik im Gebiet von Kamenz. *Veröffentlichungen des Museums der Westlausitz Kamenz*, **5**, 9–27.

BREWER, T. S., DALY, J. S. & ÅHÄLL, K.-I. 1998. Contrasting magmatic arcs in the Palaeoproterozoic of the south-western Baltic Shield. *Precambrian Research*, **92**, 297–315.

BROCHWICZ-LEWIŃSKI, W., POŻARYSKI, W. & TOMCZYK, H. 1983. Paleozoic strike-slip movements in southern Poland. *Przegląd Geologiczny*, **31**, 651–658 [in Polish with English summary].

BROWN, M. & ROACH, R. A. 1972. Precambrian rocks south of Erquy and around St Cast. *Nature. Physical Science*, **236**, 77–79.

BRUECKNER, H. K. & MEDARIS, L. G. 2000. A general model for the intrusion and evolution of 'mantle' garnet peridotites in high-pressure and ultra-high-pressure metamorphic terranes. *Journal of Metamorphic Geology*, **18**, 123–133.

BRUN, J. P. & BALÉ, P. 1990. Cadomian tectonics in northern Brittany. *In*: D'LEMOS, R. S., STRACHAN, R. A. & TOPLEY, C. G. (eds) *The Cadomian Orogen*. Geological Society, London, Special Publications, **51**, 95–114.

BRUN, J. P., BALLARD, J. F. & LE CORRE, C. 1991. Identification of Ordovician block-tilting in the Hercynian fold belt of Central Brittany (France): field evidence and computer models. *Journal of Structural Geology*, **13**, 419–429.

BRUN, J. P., GUENNOC, P., TRUFFET, C., VAIRON, J. & THE ARMOR WORKING GROUP OF THE GEOFRANCE 3-D PROGRAM. 2001. Cadomian tectonics in northern Brittay: a contribution of 3-D crustal-scale modelling. *Tectonophysics*, **331**, 229–246.

BUŁA, Z. 1994. Problems of stratigraphy and development of Lower Paleozoic sediments of the NE margin of the Upper Silesian Coal Basin (in Polish with English summary). *In:* RÓŻKOWSKI, A., ŚLÓSARZ, J. & ŻABA, J. (eds) *65th Meeting of the Polish Geological Society, Sosnowiec, Guidebook*. Prace Naukowe Uniwersytetu Śląskiego wKatowicach, **1431**, 31–57.

BUŁA, Z. 2000. The Lower Paleozoic of Upper Silesia and West Małopolska. *Prace Państwowego Instytutu Geologicznego*, **171**, 1–89 [in Polish with English summary].

BUŁA, Z., JACHOWICZ, M. & ZABA, J. 1997*a*. Principal characteristics of the Upper Silenian Block and Małopolska Block border zone, southern Poland. *Geological Magazine*, **134**, 669–677.

BUŁA, Z., JACHOWICZ, M. & PRICHYSTAL, A. 1997*b*. Lower Paleozoic deposits of the Brunovistulicum. *Terra Nostra*, **97**(11), 32–38.

BURCHART, J. 1971. Absolute ages of rocks from Poland (A catalogue of geochronological determinations). *Rocznik Polskiego Towarzystwa Geologicznego*, **41**, 241–255.

BURMANN, G. 1966. Mikroreste aus der Lausitzer Grauwackenformation. *Monatsberichte der Deutschen Akademie der Wissenschaften*, **8**, 218–224.

BUSCHMANN, B. 1995. *Geotectonic facies analysis of the Rothstein Formation (Neoproterozoic, Saxothuringian Zone, Germany)*. PhD thesis, Bergakademie Freiberg.

BUSCHMANN, B., NASDALA, L., JONAS, P., LINNEMANN, U. & GEHMLICH, M. 2001. SHRIMP U-Pb dating of tuff-derived and detrital zircons from Cadomian marginal basin fragments (Neoproterozoic) in the northeastern Saxothuringian Zone (Germany). *Neues Jahrbuch für Geologie und Paläontologie, Abhandlungen*, **6**, 321–342.

BYLINA, P., ZELAŹNIEWICZ, A. & DÖRR, W. 2000. Archaean basement in the Upper Silesia Block: U-Pb zircon age from amphibolites of the Rzeszotary horst. *Joint Meeting EUROPROBE and PACE projects*, Zakopane 2000. Abstracts, 11–12.

CABANIS, B. & WYNS, R. 1986. Le volcanisme précambrien des Mauges (SE du Massif armoricain) et ses caractères géochimiques. *Hercynica II*, **1**, 71–78.

CABANIS, B., CHANTRAINE, J. & RABU, D. 1987. Geochemical study of the Brioverian (late Proterozoic) volcanic rocks in the northern Armorican Massif (France). Implications for geodynamic evolution during the Cadomian. *In*: PHARAOH, T. C, BECKNSALE, R. D. & RICKARD, D. (eds) *Geochemistry of Mineralization of Proterozoic Volcanic Suites*. Geological Society, London, Special Publications, **33**, 503–523.

CABY, R., ANDREOPOULOS-RENAUD, U. & PIN, C. 1989. Late Proterozoic arc–continent and continent–continent collision in the pan-African trans-Saharan belt of Mali. *Canadian Journal of Earth Sciences*, **26**, 1136–1146.

CAVET, P., GRUET, M. & PILLET, J. 1966. Sur la présence du Cambrien à Paradoxides à Cléré-sur-Layon dans le nord-est du Bocage Vendéen (massif armoricain). *Comptes Rendus de l'Académie des Sciences Paris*, **263**, 1685–1688.

CAWOOD, P. A., MCCAUSLAND, P. J. A & DUNNING, G. R. 2001. Opening Iapetus: constraints from the Laurential margin of Newfoundland. *Geological Society of America Bulletin*, **113**, 443–453.

ČEČYS, A. & BOGDANOVA, S. A. 2004. 1450 Ma late- to post- collisonal granitoids in southern Sweden: a last episode of Dano-Polonian orogeny. *The 26th Nordic Geological Winter Meeting.* Abstract volume, **126**, 1–19.

CHÁB, J. & BERNARDOVÁ, E. 1974. Prehnite and pumpellyite in Upper Proterozoic basalts of the NW part of the Barrandian. *Krystalinikum*, **10**, 53–65.

CHÁB, J. & PELC, Z. 1968. Lithology of the Upper Proterozoic in the NW limb of the Barrandian area. *Krystalinikum*, **6**, 141–167.

CHALOUPSKÝ, J. 1989. Major tectonostratigraphic units of the Bohemian Massif. *In*: DALLMEYER, R. D. (ed.) *Terranes in the Circum-Atlantic Palaeozoic Orogens*. Geological Society of America, Special Publications, **230**, 101–114.

CHALOUPSKÝ, J., CHLUPÁČ, I., MAŠEK, J., WALDHAUSROVÁ, J. & CHÁB, J. 1995. Moldanubian region: Teplá-Barrandian zone (Bohemicum): Stratigraphy (ch.VII.B.1). *In*: DALLMEYER, R. D., FRANKE, W. & WEBER, K. (eds) *Pre-Permian Geology of Central and Eastern Europe*. Springer-Verlag, Berlin, 379–391.

CHANTRAINE, J., CHAUVEL, J. J. & RABU, D. 1994. The Cadomian orogeny in the Armorican Massif. Lithostratigraphy. *In*: KEPPIE, J. D. (ed.) *Pre-Mesozoic Geology in France and Related Areas*. Springer-Verlag, Berlin, 81–95.

CHANTRAINE, J., EGAL, E., THIÉBLEMONT, D., LE GOFF, E., GUERROT, C., BALLÈVRE, M. & GUENNOC, P. 2001. The Cadomian active margin (North Armorican Massif, France) : a segment of the North Atlantic Panafrican belt. *Tectonophysics*, **331**, 1–18.

CHAUVEL, J. J. & SCHOPF, J. W. 1978. Late Precambrian mirofossils from Brioverian cherts and limestones of Brittany and Normandy, France. *Nature*, **275**, 640–642.

CHLUPÁČ, I. 1988. The metamorhic palaeozoic of the 'Islet Zone' as a possible connecting link between the Barrandian and the Moldanubian. *"Bohemian Massif" conference*, Prague, 26 September–1 October 1988. Abstracts, 1–16.

CHLUPÁČ, I. & KUKAL, Z. 1986. Reflection of possible global Devonian events in the Barrandian area, ČSSR. *Lecture Notes in Earth Sciences. Global Bio-Events*, **8**, 171–179.

CHLUPÁČ, I. & KUKAL, Z. 1988. Possible global events and the stratigraphy of the Barrandian Palaeozoic (Cambrian–Devonian). *Sbornik geologických Věd Řada Geologie*, **43**, 83–146.

CHOUBERT G. & FAURE-MURET, A. 1980. The Precambrian in North Peri-Atlantic and South Mediterranean mobile zones: general results. *Earth Sci Rev*, **16**, 85–219.

CLAESSON, S. 1987. Nd isotope data on 1.9–12. Ga old basic rocks and metasediments from the Bothnian Basin, Central Sweden. *Precambrian Research*, **35**, 115–126.

CLAESSON, S., 1996. U-Pb age determination from the Precambrian of Lithuania and Poland. *EUROBRIDGE Workshop*, 8–15 June 1996, Oskarshamn, Sweden. Abstracts, 11.

CLAESSON, S. & RYKA, W. 1999. Nd model ages of the Precambrian crystalline basement of the NE Poland. *Between EUROBRIDGE and TESZ, Seventh Eurobridge Workshop*, 26–30 May 1999, Szelment, Poland. Abstracts, 17.

CLAESSON, S., HUHMA, H., KINNY, P. D. & WILLIAMS, I. S. 1993. Provenance of Svecofennian detrital zircon ages – implications for the Precambrian evolution of the Baltic Shield. *Precambrian Research*, **64**, 109–130.

CLAESSON, S., SUNBLAD, K., RYKA, W. & MOTUZA, G. 1995. The Mazury Complex – an extension of the transcandinavian Igneous Belt (TIB) into the East European platform? *Terra Nova, EUG Abstracts* **8**, 107.

CLAESSON, S., BOGDANOVA, S., BIBIKOVA, E. V. & GORBATSCHEV, R. 2001. Isotopic evidence for Palaeoproterozoic accreation in the basement of the East European Craton. *Tectonophysics*, **339**, 1–18.

CLAUER, N., VIDAL, P. & AUVRAY, B. 1985. Differential behaviour of the Rb-Sr and K-Ar systems of spilitic flows and interbedded metasediments: the spilite group of Erquy (Brittany, France). Paleomagnetic implications. *Contributions to Mineralogy and Petrology*, **89**, 81–89.

CLAUSEN, S. & ALVARO, J. 2004. Implications biostratigraphiques et paléobiogéographiques de la découverte du métazoaire Cloudina dans le versant nord de la Montagne Noire (France). *Colloque de Paléogéographie*. Société Géologique de France, Program with Abstracts, 42–43.

COCHERIE, A., CHANTRAINE, J., FANNING, C.M., DABARD, M.P., PARIS, F., LE HERISSÉ, & EGAL, E. 2001. Datation U/Pb: âge Brivérien de la série d'Erquy (Massif armoricain, France). *Comptes Rendus de l'Accadémie des Sciences Paris* **333**, 427–434.

COCHERIE, A., BAUDIN, T., AUTRAN, A., GUERROT, C., FANNING, C. & LAUMONIER, B. 2005. U-Pb zicron (ID-TIMS and SHRIMP) evidence for the early Ordovician intrusion of metagranites in the Late Proterozoic Canaveilles Group of the Pyrenees and the Montagne Noire (France). *Bulletin de la Société Géologique de France*, **176**, 269–282.

COGNÉ, J. 1972. Le Briovérien et le cycle orogénique cadomien dans le cadre des orogènes fini-précambriens. *Notes et Mémoires Services Géolegiques Maroc*, Rabat, **236**, 193–218.

COLLINS, A. S. & BUCHAN, C. 2004. Provenance and age constraints of the South Stack Group, Anglesey, UK: U-Pb SIMS detrital zircon data. *Journal of the Geological Society, London*, **161**, 743–746.

COMPSTON, W., SAMBRIDGE, M. S., REINFRANK, R. F., MOCZYDŁOWSKA, M., VIDAL, G. & CLAESSON, S. 1995. Numerical ages of volcanic and the earliest faunal zone within the Late Precambrian of east Poland. *Journal of the Geological Society, London*, **152**, 599–611.

COMPSTON, W., WRIGHT, A. E. & TOGHILL, P. 2002. Dating the Late Precambrian volcanicity of England and Wales. *Journal of the Geological Society, London*, **159**, 323–339.

CONDON, D., ZHU, M.Y., BOWRING, S., WANG, W., YANG, A.H. & JIN, Y. G. 2005. U-Pb ages from the neoproterozoic Doushantuo Formation, China. *Science*, **308**, 95–98.

COPE, J. C. W. & BEVINS, R. E. 1993. The stratigraphy and setting of the Precambrian rocks of the Llangynog Inlier, Dyfed, South Wales. *Geological Magazine*, **130**, 101–111.

CYMERMAN, Z. 2004. *Prekambr platformy wschodnioeuropejskiej na obszarze Polski: tektonika i rozwój skorupy*. Prace Panstwowego Instytutu Geologicznego, Warsaw, Poland, 1–131.

CZUBA, W., GRAD, M., LUOSTO, U., MOTUZA, G., NASEDKIN, V. & POLONAISE P5 WORKING GROUP. 2001. Crustal structure of the East European Craton along the POLONAISE P5. *Acta Geophysica Polonica*, **XLIX**(2), 145–168.

CZUBA, W., GRAD, M., LUOSTO, U., MOTUZA, G., NASEDKIN, V. & POLONAISE P5 WORKING GROUP. 2002. Upper crustal seismic structure of the Mazery Complex and Mazowsze Massif within East European Craton in NE Poland. *Tectonophysics*, **360**, 115–128.

DABARD, M. P. 1990. Lower Brioverian formations (Upper Proterozoic) of the Armorican Massif (France): geodynamic evolution of source areas revealed by sandstone petrography and geochemistry. *Sedimentary Geology*, **69**, 45–58.

DABARD, M. P. 2000. Petrogenesis of graphitic cherts in the Armorican segment of the Cadomian orogenic belt (NW France). *Sedimentology*, **47**, 787–800.

DABARD, M. P. & LOI, A. 1998. Environnement de dépôt des formations à phtanites interstratifiés du Protérozoïque supérieur armoricain (France) : conséquences sur la genèse des phtanites. *Comptes Rendus de l'Académie des Sciences Paris*, **326**, 763–769.

DABARD, M. P., LOI, A. & PEUCAT, J. 1996. Zircon typology combined with Sm-Nd whole-rock isotope analysis to study Brioverian sediments from the Armorican Massif. *Sedimentary Geology*, **101**, 243–260.

DABROWSKI, A. & KARACZUN, K. 1956. Morfologia podłoża prekambryjskiego w północno-wschodniej Polsce. *Przeglad Geologiczny*, **4**, 341–344.

DABROWSKI, A. & KARACZUN, K. 1958. *Mapa magnetyczna Polski*. Biuletyn Instytut Geologiczny. Scale 1:2 000 000, 137.

DALLMEYER, R. D. & GIBBONS, W. 1987. The age of blueschist metamorphism in Anglesey, North Wales: evidence from $^{40}Ar/^{39}Ar$ mineral age dates of the Penmynydd schists. *Journal of the Geological Society, London*, **144**, 843–852.

DALLMEYER, R. D. & NEUBAUER, F. 1994. Cadomian $^{40}Ar/^{39}Ar$ apparent age spectra of detrital muscovites from the Eastern Alps. *Journal Geological Society, London*, **151**, 591–598.

DALLMEYER, R. D. & URBAN, M. 1998. Variscan vs Cadomian tectonothermal activity in northwestern sectors of the Tepla-Barrandian zone, Czech Republic: constraints from Ar^{-40}/Ar^{-39} ages. *Geologische Rundschau*, **87**, 94–106.

DALLMEYER, R. D., D'LEMOS, R. S., STRACHAN, R. A. & MUELLER, P.A. 1991*a*. Tectonothermal chronology of early Cadomian arc development in Guernsey and Sark, Channel Islands. *Journal of the Geological Society, London*, **148**, 691–702.

DALLMEYER, R. D., STRACHAN, R. A. & D'LEMOS, R. S. 1991*b*. Chronology of Cadomian tectonothermal activity in the baie de St Brieuc (north Brittany), France: evidence from $^{40}Ar/^{39}Ar$ mineral ages. *Canadian Journal of Earth Sciences*, **28**, 762–773.

DALLMEYER, R. D., NEUBAUER, F., HANDLER, R., FRITZ, H., MULLER, W., PANA, D. & PUTIS, M. 1996. Tectonothermal evolution of the Internal Alps and Carpathians: evidence from $^{40}Ar/^{39}Ar$ mineral and whole-rock data. *Eclogae Geologiae Helvetia*, **89**, 203–227.

DALZIEL, I.W.D. 1991. Pacific margins of Laurentia and east Antarctica-

Australia as a conjugate rift pair: evidence and implications for an Eocambrian supercontinent. *Geology*, **19**, 598–601.

Dalziel, I. W. D. 1992. On the organization of American plates in the Neoproterozoic and the breakout of Laurentia. *Geological Society of America, Today*, **2**, 238–241.

Dalziel, I. W. D. 1997. Overview: Neoproterozoic-Paleozoic geography and tectonics: review, hypotheses and environmental speculations. *Geological Society of America Bulletin*, **109**, 16–42.

Davies, G. R., Gledhill, A. & Hawkesworth, C. 1985. Upper crustal recycling in southern Britain: Evidence from Nd and Sr isotopes. *Earth and Planetary Science Letters*, **75**, 1–12.

Demange, M. 1975. Style pennique de la Zone Axiale de la Montagne Noire entre St Pons et Murat-sur-Vèbre (Massif Central). *Bulletin du Bureau de recherches géologique et miniéres*, **2**, 91–139.

Demange, M., Guérangé-Lozes, J., *et al.* 1996. *Notice explicative, Carte géologique France* 1 :50 000, feuille Lacaune (987). BRGM Orléans, 153.

Denis, E. 1988. *Les sédiments briovériens (Protérozoïque supérieur) de Bretagne septentrionale et occidentale : nature, mise en place et évolution.* PhD Thesis, University Rennes. Mémoires et Documents du CAESS, **18**.

DePaolo, D. J. 1981. Neodymium isotopes in the Colorado Front Range and crust-mantle evolution in the Proterozoic. *Nature*, **291**, 193–197.

Deubel, F. 1959. Über die Grenze Vorpaläozoikum/Paläozoikum in Böhmen und Thüringen. *Geologie*, **8**, 818–827.

Deunff, J., Auvray, B., Cogné, J., Hameurt, J., Jeannette, D. & Vidal, P. 1973. Confirmation micropaléontologique de l'âge radiométrique Ordovicien inférieur du groupe spilitique d'Erquy (Côtes du Nord). *Comptes Rendus de l'Académie des Sciences Paris*, **D276**, 935–937.

D'Lemos, R. S. & Brown, M. 1991. Sm–Nd isotope characterisics of late Cadomian granite magmatism in northern France and the Channel Islands. *The Geological Magazine*, **130**, 797–804.

Dombrowski, A., Henjes-Kunst, F., Höhndorf, A., Kröner, A., Okrusch, M. & Richter, P. 1995. Orthogneisses in the Spessart Crystalline Complex, Northwest Bavaria: Witnesses of Silurian granitoid magmatism at an active continental margin. *Geologische Rundschau*, **84**, 399–411.

Don, J., Dumicz, M., Wojciechowska, I. & Żelaźniewicz, A. 1990. Lithology and tectonics of the Orlica - Śnieżnik Dome, Sudetes: recent state of knowledge. *Neues Jahrbuch für Geologie und Paläontologie, Abhandlungen*, **179**, 159–188.

Doré, F. 1994. Cambrian of the Armorican Massif. *In*: Keppie, J. D. (ed.) *Pre-Mesozoic Geology in France and Related Areas*. Springer-Verlag, Berlin, 136–141.

Doré, F., Dupret, L. & Le Gall, J. 1985. Tillites et tilloïdes du massif armoricain. *Palaeogeography, Palaeoclimatology, Palaeoecology*, **51**, 85–96.

Dörr, W., Fišera, M. & Franke, W. 1992. Cadomian magmatic events in the Bohemian Massif – U-Pb data from felsic magmatic pebbles. *7th Geology Workshop, Styles of Superposed Variscan Nappe Tectonics*. Czechoslovak Academy of Science. Abstracts, 3.

Dörr, W., Valverde-Vaquero, P., Marhaine, D., Schastok, J. & Wiszniewska, J. 2001. U-Pb and Ar/Ar geochronology of rapakivi-type granites from Mazury Complex, Poland. Cambridge Publications, Cambridge. *Journal of Conference*, Abstracts, 4, (SS02) European Union of Geosciences XI, 772.

Dörr, W., Belka, Z., Marheine, D., Schastok, J., Valverde-Vaquero, P. & Wiszniewska, J. 2002*a*. U-Pb and Ar/Ar geochronology of anorogenic granite magmatism of the Mazury Complex, NE Poland. *Precambrian Research*, **119**(1–4), 101–120.

Dörr, W., Zulauf, G., Fiala, J., Franke, W. & Vejnar, Z. 2002*b*. Neoproterozoic to Early Cambrian history of an active plate margin in the Tepla-Barrandian unit – a correlation of U-Pb isotopic-dilution-TIMS ages (Bohemia, Czech Republic). *Tectonophysics*, **352**(1–2), 65–85.

Drost, K., Linnemann, U., *et al.* 2004. New data on the Neoproterozoic - Cambrian geotectonic setting of the Tepla-Barrandian volcano-sedimentary successions: geochemistry, U-Pb zircon ages, and provenance (Bohemian Massif, Czech Republic). *International Journal of Earth Sciences*, **93**, 742–757.

Duchesne, J. C., Vander Auwera, J. & Wiszniewska, J. 1998. Geochemical evidence of a crustal derivation of the Suwałki anorthosites (NE Poland). *EUROBRIDGE Workshop*, Tallin-Arbavere, 2–4 June 1998.

Ducrot, J., Allégret, A. & Lancelot J. R. 1982. Prevariscan evolution of the western Europe at the light of U/Pb zircon dating of ortogneisses and eclogites. *5th International Conference Geochronolog, Cosmochronology and Isotope Geology*, Nikko, Japan. 83–84.

Dudek, A. 1963. Beitrag zum Problem der Moldanubischen Überschiebung (Mislitzer Horst). *Sborník geologických Věd Řada Geologie*, **1**, 7–20.

Dudek, A. 1980. The crystalline basement block of the Outer Carpathians in Moravia – Brunovistulicum. *Rozpravy Československé Academie Věd*, **90**, 1–85.

Dudek, A. & Melková, J. 1975. Radiometric age determinantion in the crystalline basement of the Carpathian Foredeep and of the Moravian Flysh. *Věstnik Ústředniho ústavu geolgického*, **50**, 257–264.

Dupret, L., Dissler, E., Doré, F, Gresselin, F. & Le Gall, J. (1990) Cadomian geodynamic evolution of the northeastern Armorican Massif (Normandy and Maine). *In:* D'Lemos R.S., Strachan R.A. & Topley C.G. (eds) *The Cadomian Orogen.* Geological Society, London, Special Publications, **51**, 115–131.

Dvořák, J. 1995. Stratigraphy of the Moravo-Silesian Zone. *In*: Dallmeyer, R. D., Franke, W. & Weber, K. (eds) *Pre-Permian Geology of Central and Earstern Europe*. Springer Verlag, Berlin, 477–489.

Edel, J. B. & Weber, K. 1995. Cadomian terranes, wrench faulting and thrusting in the central-europe variscides – geophysical and geological evidence. *Geologische Rundschau*, **84**, 412–432.

Egal, E., Guerrot, C., Le Goff, E., Thiéblemont, D. & Chantraine, J. 1996. The Cadomian orogeny revisited in northern Brittany. Avalonian and related Peri-Gondwanan terranes of the Circum-North Atlantic. *In*: Nance, R.D. & Thompson, M.D. (eds) *Avalonian and related peri-Gondwanan terranes of the Circum-North Atlantic.* Geological Society of America, Special Papers, **304**, 281–318.

Ehlers, C., Lindroos, A. & Selonen, O. 1993. The late Svecofennian granite-migmatite zone of southern Finland–a belt of transpressive deformation and granite emplacement. *Precambrian Research*, **64**, 295–309.

Eichhorn. R., Schärer, U. & Höll, R. 1995. Age and evolution of scheelite-hosting rocks in the Felbertal deposit (Eastern Alps): U-Pb geochronology of zircon and titanite. *Contributions to Mineralogy and Petrology*, **119**, 377–386.

Eichhorn, R., Höll, R., Loth, G. & Kennedy, A. 1999. Implications of U-Pb SHRIMP zircon data on the age and evolution of the Felbertal tungsten deposit (Tauern Window, Austria). *International Journal of Earth Sciences*, **88**, 496–512.

Eichhorn, R., Loth, G. & Kennedy, A. 2001. Unravelling the pre-Variscan evolution of the Habach terrane (Tauern Window, Austria) by U-Pb SHRIMP zircon data. *Contributions to Mineralogy and Petrology*, **142**, 147–162.

Eisele, J., Geiger, S. & Rahn, M. 1997. Chemical characterization of metabasites from the Turtmann valley (Valais, Switzerland): implications for their protoliths and geotectonic origin. *Swiss Bulletin of Mineralogy and Petrology*, **77**, 403–417.

Eliasson, T. & Schöberg, H. 1991. U-Pb dating of the post-kinematic Sveconorwegian (Grenvillian) Bohus Granite, SW Sweden; evidence of restitic zircon. *Precambrian Research*, **51**, 337–350.

Elming, S.-Å., Pesonen, L.J., Leino, M.A.H., Khramov, A.N., Mikhailova, N.P., Krasnova, A.F., Mertanen, S., Bylund, G. & Terho, M. 1993. The drift of the Fennoscandian and Ukrainian shields during the Precambrian: A palaeomagnetic analysis. *Tectonophysics* **223**, 177–198.

Elming, S.Å. Mikhailova, N.P. & Kravchenko, S. 2001. Paleomagnetism of Proterozoic rocks from the Ukrainian Shield: new tectonic reconstruction of the Ukrainian and Fennoscandian shields. *Tectonophysics* **339**, 19–38.

Eriksson, L. & Henkel, H. 1983. Deep structures in the Precambrian interpreted from magnetic and gravity maps of Scandinavia. International Basement Tectonics Association publication, **4**, 351–358.

Eriksson, P.G., Catuneanu, O., Nelson, D.R., Mueller, W.U. & Altermann, W. 2004. Towards a synthesis. *In:* Eriksson, P.G., Altermann, W. Nelson, D.R., Mueller, W.U. & Catuneanu,

O. (eds) *The Precambrian Earth: Tempos and Events*. Developments in Precambrian Geology, Vol. 12, Elsevier, Amsterdam, 739–769.

EUROBRIDGE'95 SEISMIC WORKING GROUP 2001. EUROBRIDGE'95 deep seismic profiling within the East European Craton. *Tectonophysics*, **339**, 153–175.

EVANS, D.A., BEUKES, N.J. & KIRSCHVINK, J.L. 1997. Low latitude glaciation in the Palaeoproterozoic era. *Nature*, **386**, 262–266.

EVANS, D.A.D. 2003. True polar wander and supercontinents. *Tectonophysics*, **362**, 303–320.

FALKUM, T. 1985. Geotectonic evolution of southern Scandinavia in light of a late-Proterozoic plate-collision. *In*: TOBI, A. C. & TOURET, J. L. R. (eds) *The Deep Proterozoic Crust in the North Atlantic Provinces*. NATO Advanced Science Institute Series, **C158**, 309–322.

FANNING, C.M. & LINK, P.K. 2004. U-Pb-SHRIMP ages of Neoproterozoic (Sturtian) glaciogenic Pocatello Formation, southeastern Idaho. *Geology*, **32**, 881–884.

FATKA, O. & GABRIEL, Z. 1991. Microbiota from siliceous stromatolitic rocks of the Barrandian Proterozoic (Bohemian Massif). *Časopis pro mineralogii a geologii*, **36**, 143–148.

FEDIUK, F. 1992. Metavolcanites in the Proterozoic Jílové Belt, Central Bohemia. *Věstnik Ústředniho ústavu geolgického*, **67**, 297–310.

FERNANDEZ-SUAREZ, J., GUTTIÉRREZ ALONSO, G. & JEFFRIES, T. E. 2002. The importance of along-margin terrane transport in northern Gondwana: insights from detrital zircon parentage in Neoproterozoic rocks from Iberia and Brittany. *Earth and Planetary Science Letters*, **204**, 75–88.

FIALA, F. 1948. Algonkické slepence ve středních Čechách. *Sbornik Státniho geologického ústavu Československé Republiky*, **15**, 399–612.

FIALA, F. 1978. Proterozoic and Early Paleozoic volcanism of the Barrandian–Železné Hory zone. *Sbornik geologický Věd*, **30**, 7–47.

FIALA, F. & SVOBODA, J. 1956. Problém subkambria a subkambrického zalednění v Železných horách. *Sbornik Ústředniho ústavu geologického. Oddil geologický*, **22**, 257–309.

FIALA, J. & PATOČKA, F. 1994. The evolution of Variscan terranes of the Moldanubicum Region, Bohemian Massif. *Niedersächsisches Landesamt für Bodenforschung, Hannover*, **94**(3), 1–8.

FIALA, J., FUCHS, G. & WENDT, I. 1995. VII.C.1 Stratigraphy. *In:* DALLMEYER, R.D., FRANKE, W. & WEBER, K. (eds) *Pre-Permian Geology of Central and Eastern Europe*. Springer, Berlin, 417–428.

FINGER, F. & PIN, C. 1997. Arc-type crustal zoning in the Bruno-Vistulicum, Eastern Czech Republic: a trace of the Late-Proterozoic Euro-Gondwana margin. *Journal of the Czech Geological Society*, **42**(3/4), 53.

FINGER, F. & STEYRER, H. P. 1995. A tectonic model for the eastern variscides – indications from a chemical study of amphibolites in the south-eastern Bohemian Massif. *Geologica Carpathica*, **46**, 137–150.

FINGER, F., FRASL, G., DUDEK, A., JELÍNEK, E. & THÖNI, M. 1995. Igneous activity (Cadomian plutonism in the Moravo-Silesian basement). *In*: DALLMEYER, R. D., FRANKE, W. & WEBER, J. (eds) *Pre-Permian Geology of Central and Eastern Europe*. Springer Verlag, Berlin, 494–507.

FINGER, F., ROBERTS, M. P., HAUNSCHMIDT, B., SCHERMAYER, A. & STEYER, H. P. 1997. Variscan granitoids of Central Europe: their typology, potential sources and tectonothermal relations. *Mineralogy and Petrology*, **61**, 67–96.

FINGER, F., HANŽL, P., PIN, C., VON QUADT, A. & STEYRER, H. P. 2000*a*. The Brunovistulian: Avalonian Precambrian sequence at the eastern Bohemian Massif: speculations on palinsplastic reconstruction. *In*: FRANKE, W., HAAK, V., ONCKEN, O. & TANNER, D. (eds) *Orogenic Processes: Quantification and Modelling in the Variscan Belt*. Geological Society, London, 103–113.

FINGER, F., TICHOMIROWA, M., PIN, C. & HANZL, P. 2000*b*. Relics of an early-Panafrican metabasite-metarhyolite formation in the Brno Massif, Moravia, Czech Republic. *International Journal of Earth Sciences*, **89**(2), 328–335.

FORD, T. D. 1980. The Ediacaran fossils of Charnwood Forest. *Proceedings of the Geologists' Association*, **91**, 81–84.

FORSYTHE, R.D., DAVIDSON, J., MPODOZIS, C. & JESINKEY, C. 1993. Lower Paleozoic relative motion of the Arequipa block and Gondwana; paleomagnetic evidence from Sierra de Almeida of Northern Chile. *Tectonics*, **12**, 219–236.

FORTEY, R.A. & COCKS, L.R.M. 2003. Palaeontological evidence bearing on global Ordovician-Silurian continental reconstructions. *Earth-Science Reviews*, **61**, 245–307.

FRANKE, D. & HOFFMANN, N. 1999. Das Elbe-Lineament – bedeutende Geofraktur oder Phantomgebilde? – Teil 2: Regionale Zusammenhänge. *Zeitschrift für Geologische Wissenschaften*, **27**, 319–350.

FRANKE, W. 1989. Tectonostraphic units in the Variscan belt of central Europe. *In*: DALLMEYER, R. D. (ed.) *Terranes in the Circum-Atlantic Palaeozoic Orogens*. Geological Society of America, Special Publications, **230**, 67–90.

FRANKE, W. 2000. The mid-European segment of the Variscides: tectonostratigraphic units, terrane boundaries and plate tectonic evolution. *In*: FRANKE, W., HAAK, V., ONCKEN, O. & TANNER, D. (eds) *Orogenic Processes: Quantification and Modelling in the Variscan Belt*. Geological Society, London, Special Publications, **179**, 35–61.

FRANKE, W. & ZELAZNIEWICZ, A. 2000. The eastern termination of the Variscides: terrane correlation and kinematic evolution. *In*: FRANKE, W., HAAK, V., ONCKEN, O. & TANNER, D. (eds) *Orogenic Processes: Quantification and Modelling in the Variscan Belt*. Geological Society, London, Special Publications, **179**, 63–85.

FRANKE, W. & ZELAZNIEWICZ, A. 2002. Structure and evolution of the Bohemian arc. *In:* WINCHESTER, J.A., PHARAOH, T.C. & VERNIERS, *J.* (eds) *Palaeozoic Amalgamation of Central Europe*. Geological Society, London, Special Publications, **201**, 279–293.

FRANZ, L., SCHUSTER, A. K. & STRAUSS, K. W. 1997. Basement evolution in the Rhenohercynian Segment: Discontinuous exhumation history of the Eckergneis complex (Harz Mountains, Germany). *Chemie der Erde*, **57**, 105–135.

FRANZKE, H. J. 2001. Die strukturelle Einbindung des Eckergneises zu seinem variscisch geprägten Umfeld. *Braunschweiger Geowissenschaftliche Arbeiten*, **24**, 1–26.

FRIEDL, G. 1997. *U/Pb -Datierungen an Zirkonen uind Monaziten aus Gesteinen vom Österreichischen Anteil der Böhmischen Masse*. University of Salzburg, Salzburg.

FRIEDEL, C.H., HOTH, P., FRANZ, G. & STEDINGK, K. 1995. Niedriggradige Regionalmetamorphose im Harz. *Zentralblatt für Geologie und Paläontologie*, **1**, 1993, **9/10**, 1213–1235.

FRIEDL, G., MCNAUGHTON, N., FLETCHER, I. R. & FINGER, F. 1998. New SHRIMP-zircon ages for othogneiss from the south-eastern part of the Bohemian Massif (Lower Austria). *Acta Universitatis Carolinae. Geologica*, **42**, 251–252.

FRIEDL, G., FINGER, F., MCNAUGHTON, N. J. & FLETCHER, I. R. 2000. Deducing the ancestry of terranes: SHRIMP evidence for South America-derived Gondwana fragments in central Europe. *Geology*, **28**, 1035–1038.

FRIEDL, G., FINGER, F., PAQUETTE, J.-L., QUADT, A. V., MCNAUGHTON, N. J. & FLETCHER, I. R. 2004. Pre-Variscan geological events in the Austrian part of the Bohemian Massif deduced from U–Pb zircon ages. *International Journal of Earth Sciences*, **93**, 802–823.

FRISCH, W. & NEUBAUER, F. 1989. Pre-Alpine terranes and tectonic zoning in the eastern Alps. *In*: DALLMEYER, R. D. (ed.) *Terranes in the Circum-Atlantic Palaeozoic Orogens*.Geological Society of America. Special Publications, **230**, 91–99.

FRISCH, W., VAVRA, G. & WINKLER, M., 1993. Evolution of the Penninic Basement of the Eastern Alps. *In*: VON RAUMER, J. F. & NEUBAUER, F. (eds) *The Pre-Mesozoic Geology of the Alps*. Springer Verlag, Heidelberg, 349–360.

FRITZ, H., DALMEYER, D. R. & NEUBAUER, F. 1996. Thick-skinned versus thin-skinned thrusting: Rheology controlled thrust propagation in the Variscan collisional belt (The southeastern Bohemian Massif, Czech Republic, Austria). *Tectonics*, **15**, 1389–1413.

FUCHS, G. 1986. Zur Diskussion um den Deckenbau der Böhmischen Masse. *Jahrbuch der Geologischen Bundesanstalt*, **129**, 41–49.

GAAL, G. & GORBATSCHEV, R. 1987. An outline of Precambrian evolution of the Baltic Shield. *Precambrian Research*, **35**, 15–52

GABRIELSEN, R. H., ODINSEN, T. & GRUNNALEITE, I. 1999. Structuring of the northern Viking Graben and the Møre Basin; the influence of basement structural grain, and the particular role of the Møre-Trøndelag fault complex. *Marine and Petroleum Geology*, **16**, 443–465.

GARECKIJ, R. G. 2001. *Tektonosphere of Belarus: Deep Structure and Regularities of the Distribution of Mineral Resources*. Minsk [in Russian].

GEBAUER, D. 1993. The pre-Alpine evolution of the continental crust of the Central Alps – an overview. *In*: VON RAUMER, J. F. & NEUBAUER, F. (eds) *The Pre-Mesozoic Geology of the Alps*, Springer Verlag, Heidelberg, 93–117.

GEBAUER, D. & FRIEDL, G. 1994. A 1.38 Ga protolith age for the Dobra Orthogneiss (Moldanubian Zone of southern Bohemian Massif, NE Austria): evidence from ion-microprobe (SHRIMP) dating of zircon. *Journal of the Czech Geological Society*, **39**, 34–35.

GEBAUER, D. & GRÜNENFELDER, M. 1982. Geological development of the Hercynian Belt of Europe based on age and origin of high-grade and high-pressure mafic and ultramafic rocks. *5th International Conference, Geochronology, Cosmochronology and Isotope Geology*, Nikko, Japan, 111–112.

GEBAUER, D. & VON QUADT, A. 1991. Geochronological evidence for tectonic contacts between Caledonian meta-eclogite of the Gotthard Massif and ist country rocks (Central Alps, Switzerland). *Terra Nova*, **3**, 6.

GEBAUER, D., WILLIAMS, I. S., COMPSTON, W. & GRUNENFELDER, M. 1988*a*. The development of the Central European continental-crust since the Early Archaean. *Chemical Geology*, **70**, 68–68.

GEBAUER, D., VON QUADT, A., WILLIAMS, I. S., COMPSTON, W. & GRÜNENFELDER, M. 1988*b*. Archaean zircons in a retrograded, Caledonian eclogite of the Gotthard Massif. *Swiss Bulletin of Mineralogy and Petrology*, **68**, 485–490.

GEBAUER, D., WILLIAMS, I. S., COMPSTON, S. & GRÜNENFELDER, M. 1989. The development of the Central European continental crust since the Early Archaean based on conventional and ion-microprobe dating of up to 3.84 b.y. old detrital zircons. *Tectonophysics*, **157**, 81–96.

GEBAUER, D., GRÜNENFELDER, M., TILTON, G. R., TROMMSDORFF, V. & SCHMID, S. 1992*a*. The geodynamic evolution of garnet-peridotites, garnet-pyroxenites and eclogites of Alpe Arami and Cima di Gagnone (Central Alps) from Early Proterozoic to Oligocene. *Swiss Bulletin of Mineralogy and Petrology*, **72**, 107–111.

GEBAUER, D., SCHMID, R., VON QUADT, A. & ULMER, P. 1992*b*. Oligocene, Permian and Panafrican zircon ages from rocks of the Balmuccia Peridotite and of the Lower Layered Group in the Ivrea Zone. *Swiss Bulletin of Mineralogy and Petrology*, **72**, 113–122.

GEHMLICH, M. 2003. Die Cadomiden und Varisziden des Saxothuringischen Terranes–Geochronologie magmatischer Ereignisse. *Freiberger Forschungshefte*, C500, TU Bergakademie, Freiberg, 1–129.

GEHMLICH, M., LINNEMANN, U. G., TICHOMIROWA, M., LÜTZNER, H. & BOMBACH, K. 1997. Die Bestimmung des Sedimentation salters cadomischer Krustenfragmenteim Saxothuringikum durch die Einzelzircon–Evaporatios methode. *Terra Nostra*, **5**, 46–49.

GEISLER, T., VINX, R., MARTIN-GOMBOJAV, N. & PIDGEON, R. T. 2005. Ion microprobe (SHRIMP) dating of detrital zircon grains from quartzites of the Eckergneiss Complex, Harz Mountains (Germany): Implications for the provenance and geological history. *International Journal of Earth Sciences*, **94**, 369–384.

GERDES, A. & ZEH, A. 2006. Combined U-Pb and Hf isotope LA-(MC-) ICP-MS analyses of detrital zircons: Comparison with SHRIMP and new constraints for the provenance and age of an Armorican metasediment in Central Germany. *Earth and Planetary Science Letters*, **249**, 47–61.

GIBBONS, W. & HORÁK, J. M. 1996. The evolution of the Neoproterozoic Avalonian subduction system: Evidence from the British Isles, *In:* NANCE, R. D. & THOMPSON, M. D. (eds) *Avalonian and Related Peri-Gondwanan Terranes of the Circum-North Atlantic*. Geological Society of America, Special Publications, **304**, 269–280.

GIBBONS, W. & MORENO, T. 2002. Tectonomagmatism in continental arc: evidence from the Sark arc complex. *Tectonophysics*, **352**, 185–201.

GIBBONS, W., TIETZSCH-TYLER, D., HORÁK, J. M. & MURPHY, F. C. 1994. Precambrian rocks in Anglesey, southwest Llyn and southeast Ireland, *In:* GIBBONS, W. & HARRIS, A. L. (eds) *A Revised Correlation of Precambrian Rocks of the British Isles*. Geological Society, London, Special Reports, **22**, 75–84.

GIESE, R. 1998. Eine zweidimensionale Interpretation der Geschwindigkeitenstruktur der Erdkruste des sudwestlichen Teils der osteuropäischen Plattform (Project EUROBRIDGE). *Scientific Technical Report*, **16**, 1–190.

GIORGI, L., GIRAUD, P. & VACHARD, D. 1979. Sur la présence de microorganismes d'âge cambrien dans les schistes cristallins du versant occidental du massif cristallin externe des Grandes Rousses (Alpes occidentales). *Comptes Rendus de l'Académie des Sciences Paris*, **D267**, 1251–1254.

GILMOUR, J.D. & SAXTON, J.M. 2001. A time-scale of formation of the first solids. *Philosophical Transactions of the Royal Society London Series A*, **359**, 2037–2048.

GIRIN, R. J. 1990. Deep seismic section of the lithosphere along profile Braslav-Pleshczenicy. *Dokl. AN BSSR*, **35**(7), 640–644 [in Russian].

GORBATSCHEV, R. 2004 The Transscandinavian Igneous Belt – introduction and background. *In*: HÖGDAHL, K., ANDERSSON, U. B. & EKLUND, O. (eds) *The Trasscandinavian Igneous Belt (TIB) in Sweden: A Review of its Character and Evolution*. Geological Survey of Finland, Special Papers, **37**, 9–15.

GORBATSCHEV, R. & BOGDANOVA, S. 1993. Frontiers in the Baltic Shields. *Precambrian Research*, **64**(1), 3 –21.

GOROKHOV, I.M., VARSCHAVSKAYA, E.S. & KUTJAVIN, E.P. 1983. Rb-Sr dating of magmatic and metamorphic events in the eastern part of the Bohemian Massif. *Časopis pro mineralogii a geologii*, **28**, 349–361.

GOWER, C.F., HEAMAN, L.M., LOVERIDGE, W.D., SCHÄRER, U. & TUCKER, R.D. 1991. Grenvillian magmatism in the eastern Grenville Province, Canada. *Precambrian Research*, **51**, 315–336.

GRABOWSKA, T. & BOJDYS, G. 2001. The border of the East-European Craton in south eastern Poland based on gravity and magnetic data. *Terra Nova*, **13**, 92–98.

GRAD, M., JENSEN, L. *et al.* 2003. Crustal structure of the Trans-European suture zone region along POLONAISE'97 seismic profile P4. *Journal of Geophysical Research*, **108**(B11), 2541.

GRADSTEIN, F., OGG, J. & SMITH, A. (eds) 2004. *A Geologic Time Scale*. Cambridge University Press, Cambridge.

GRAUERT, B. 1969. *Die Entwicklungsgeschichte des Silvretta-Kristallins auf Grund radiometrischer Altersbestimmungen*. PhD Thesis, University, Bern.

GRAUERT, B., HÄNNY, R. & SOPTRAJANOVA, G. 1974. Geochronology of a polymetamorphic and anatectic gneiss region: The Moldanubicum of the area Lam-Deggendorf, Eastern Bavaria, Germany. *Contributions to Mineralogy and Petrology*, **45**, 37–63.

GRAVIOU, P. 1984. *Pétrogénèse des magmas calco-alcalins : exemple des granitoïdes cadomiens de la région trégorroise (Massif Armoricain)*. Thesis, University Rennes.

GRAVIOU, P. 1992. Reconnaissance d'une suture majeure au sein de la chaîne cadomienne. *Comptes Rendus de l'Académie des Sciences Paris*, **315**, 1799–1802.

GRELLING, R. O. & SMITH, A. G. 2000. The Dalradian of Scotland: missing link between the Vendian of northern and southern Scandinavia? *Physics and Chemistry of the Earth, Part A, Solid Earth and Geodesy*, **25**, 495–498.

GRÜNENFELDER, M., WILLIAMS, I. S. & COMPSTON, W. 1984. Use of the ion-microprobe in deciphering complex U-Th-Pb systems in zircons from the pre-Alpine basement, Switzerland. *Terra Cognita, Special Issue ECOG*, **VIII**, 26–27.

GUÉRANGÉ-LOZES, J., GUÉRANGÉ, B. *et al.* 1991. *Notice explicative, Carte géologique de France*, 1.50 000, feuille Camarès (961). BRGM, Orléans, 84.

GUERROT, C. 1989. *Archéen et Protérozoïque dans la chaîne hercynienne ouest-européenne : géochimie isotopique (Sr-Nd-Pb) et géochronologie U-Pb sur zircons*. PhD Thesis, University Rennes. Mémoires et documents du Centre Armoricain d'Etude Structurale des Socles, **25**.

GUERROT, C. & PEUCAT, J.-J. 1990. U–Pb geochronology of the Upper Proterozoic Cadomian orogeny orogeny in the northern Armorican Massif, France. *In*: D'LEMOS, R. S., STRACHAN, R. A. & TOPLEY, C. G. (eds) *The Cadomian Orogen*. Geological Society, London, Special Publications, **51**, 13–26.

GUISE, P. G. & ROBERTS, D. 2002. Devonian ages from ^{40}Ar-^{39}Ar dating of plagioclase in dolerite dykes, eastern Varanger Peninsula, North Norway. *Bulletin Werges Geologiske Undersøkalse: Miscellaneous Reseach Papers*, **440**, 27–36.

GUITARD, G., AUTRAN, A. & FONTEILLES, M. 1996. Le substratum précambrien du Paléozoïque. *In* : BARNOLAS, A. & CHIRON, J. C. (coord.) *Synthèse géologique et géophysique des Pyrénées. Tome I : cycle hercynien*. BRGM-ITGE, Orléans-Madrid, 37–156.

GUNIA, T. 1984. Microfossils from the quartzitic schists in vicinity of Goszów, Śnieżnik Kłodzki Massif, Central Sudetes (in Polish,

English summary). *Geologia Sudetica*, **18**, 47–57.

GUNIA, T. 1985. Geological position of the Sowie Góry block and its influence on the palaeography of the Palaeozoic complexes of Central Sudetes (in Polish, English summary). *Geologia Sudetica*, **20**, 83–119.

GUNIA, T. 1990. Acritarcha and microproblematica of the crystalline limestones from the vicinity of Romanowo Górne (Central Sudetes Mts, Krowiarki) (in Polish, English summary). *Geologia Sudetica*, **24**, 101–137.

GUNIA, T. 1999. Microfossils from the high-grade metamorphic rocks of the Góry Sowie Mts. (Sudetes area) and their stratigraphic importance. *Geological Quarterly*, **43**, 519–536.

GUTERCH, A., GRAD, M., MATERZOK, R. & PERCHUC, E. 1986. Deep structure of the Earth crust in the contact zone of Palaeozoic and Precambrian platforms in Poland (Tornquist–Teisseir Zone). *Tectonophysics*, **128**, 251–279.

HAAKE, R., HERRMANN, C., PÄLCHEN, W. & PILOT, J. (1973). Zur Altersstellung der Granodiorite der westlichen Lausitz und angrenzender Gebiete. *Zeitshrift für Geologische Wissenschaften*, **1**, 1669–1671.

HAMMER, J. 1996. Geochemie und Petrogenese der cadomischen und spätvariszischen Granitoide der Lausitz. *Freiberger Forschungshefte*, **C463**, 1–107.

HAMMER, J., EIDAM, J., RÖBER, B. & EHLING, B.-C. 1999. Prävaricischer granitoider Magmatismus am NE-Rand des Böhmischen Massifs – Geochemie und Petrogenese. *Zeitschrift für Geologische Wissenschaften*, **27**, 401–415.

HANDLER, R., DALLMEYER, R. D. & NEUBAUER, F. 1997. $^{40}Ar/^{39}Ar$ ages of detrital white mica from Upper Austroalpine units in the Eastern Alps, Austria: evidence for Cadomian and contrasting Variscan sources. *Geologische Rundschau*, **86**, 69–80.

HANSCH, R. & ZEH, A. 2000. Metabasites from the Ruhla Crystalline Complex: Evidence for distinct pre-Variscan, plate-tectonic environments within the Mid-German Crystalline Rise. *Chemie der Erde*, **60**, 1–25.

HANŽL, P. & MELICHAR, R. 1997. The Brno Massif: a section through the active continental margin or a composed terrane? *Krystalinikum*, **23**, 33–58.

HANŽL, P., PŘICHYSTAL, A. & MELICHAR, R. 1995. The Brno Massif: Volcanites of the northern part of the metabasite Zone. *Acta Universitatis Palackianae Olomucensis, Facultas Rerum Naturalium: Geologica*, **34**, 75–82.

HARLAND, W.B. 1964. Critical evidence for a great Infra-Cambrian glaciation. *Geologische Rundschau*, **54**, 45–61.

HARTZ, E. H. & TORSVIK, T. H. 2002. Baltica upside down: a new plate tectonic model for Rodinia and the Iapetus Ocean. *Geology*, **30**, 255–258.

HEFLIK, W. & KONIOR, K. 1974. The present state of knowledge concerning the crystalline basement in the Cieszyn-Rzeszotary area. *Biuletyn Instytutu Geologicznego*, **273**, 195–228 [in Polish with English summary].

HEGNER, E. & KRÖNER, A. 2000. A review of Nd izotopic data and xenocrystic and detrital zircon ages from the pre-Variscan basement in the eastern Bohemian Massif: speculations on palinsplastic reconstructions. *In*: FRANKE, W., HAAK, V., ONCKEN, O. & TANNER, D. (eds) *Orogenic Processes: Quantification and Modelling in the Variscan Belt*. Geological Society, London, Special Publications, **179**, 113–129.

HENJES-KUNST, F., MÜLLER-SIGMUND, H., VEJNAR, Z. & FIALA, J. 1995. Geochemical and Sr-Nd isotopic composition of Teplá-Barrandian Metasediments. *Terra Abstracts*, **7**, 339.

HENKEL, H. & ERIKSSON, L. 1987. Regional aeromagnetic and gravity studies in Scandinavia. *Precambrian Research*, **35**, 169–180.

HEUSE, T. 1990. Mikrofossilien aus den Altenfelder Schichten (Präkambrium) des Schwarzburger Antiklinoriums. *Freiberger Forschungsheft*, **C436**, 114–123.

HEUSE, T., KURZE, M. & REITZ, E. 1994. Paläontologischer Nachweis von Oberproterozoikum (Vendium) in der Zone Münchberg-Frankenberg. *Neues Jahrbuch für Geologie und Paläontologie, Monatshefte*, **2**, 95–107.

HIGGINS, M. D. & VAN BREEMEN, O. 1998. The age of the Sept Iles layered mafic intrusion, Canada: Implications for the Late Neoproterozoic /Cambrian history of southern Canada. *Journal of Geology*, **106**, 421–431.

HIRSCHMANN, G. 1966. Assyntische und variszische Baueinheiten im Grundgebirge der Oberlausitz (unter spezieller Berücksichtigung der Geologie des östlichen Görlitzer Schiefergebirges). *Freiberger Forschungsheft*, **C212**, 1–146.

HIRSCHMANN, G. & OKRUSCH, M. 1988. Spessart-Kristallin und Ruhlaer Kristallin als Bestandteile der Mitteldeutschen Kristallinzone – ein Vergleich. *Neues Jahrbuch für Geologie und Paläontologie, Abhandlungen*, **177**, 1–39.

HÖCK, V. 1993. The Habach-Formation and the Zentralgneis–a key in understanding the Palaeozoic evolution of the Tauern Window (Eastern Alps). *In*: VON RAUMER, J. F. & NEUBAUER, F. (eds) *The Pre-Mesozoic Geology of the Alps*. Springer Verlag, Heidelberg, 361–374.

HÖCK, V. & VETTERS, W. 1975. Bericht 1974 über geologischen Aufnahmen auf Blatt Horn (21). *Verhandlungen der Geologischen Bundesanstalt*, **1975**, A22–26.

HÖCK, V., MARSCHALLINGER, R. & TOPPA, D. 1990. Granat-biotit-Geothermometrie in Metapeliten der Moravischen Zone Österreich. *Österreichische Beiträge zu Meteorologie und Geophysik*, **3**, 149–167.

HOFFMAN, P.F., 1989. Precambrian geology and tectonic history of North America. *In*: BALLY, A.W. & PALMER, A.R. (eds) *The Geology of North America – An Overview*. Geological Society of America, Boulder, Colorado, 447–512.

HOFFMAN, P. F. 1991. Did the breakout of Laurentia turn Gondwanaland inside out? *Science*, **252**, 1409–1412.

HOFFMAN, P.F., KAUFMANN, A.J., HALVERSON, G.P. & SCHRAG, D.P. 1998. A Neoproterozoic Snowball Earth. *Science*, **281**, 1342–1346.

HOFFMANN, K.-H., CONDON, D.J., BOWERING, S.A. & CROWLEY, J.L. 2004. U-Pb zircon date from the Neoproterozoic Ghaub Formation, Namibia: Constraints on Marinoan glaciation. *Geology*, **32**, 817–820.

HÖGBOM, A. G. 1913. *Fennoskandia. Handbuch der regionalen Geologie*. Carl Winter's Universitätsbuchhandlung, Heidelberg, 1–197.

HÖGDAHL, K. & SJÖSTRÖM, H. 2001. Evidence for 1.82 Ga transpressive shearing in a 1.85 Ga granitoid in central Sweden: implications for the regional evolution. *Precambrian Research*, **105**, 37–56.

HÖGDAHL, H., GROMET, L. P. & BROMAN, C. 2001. Low P-T Caledonian resetting of U-rich Paleoproterozoic zircons, central Sweden. *American Mineralogist*, **86**, 534–546.

HOLLAND, H.D. 1984. *The Chemical Evolution of the Atmosphere and the Oceans*. Princeton University Press, Princeton, New Jersey.

HOLUBEC, J. 1968. Structural development of the geosynclinal Proterozoic and its relations to the deeper zones of the earth's crust (Central European domain). *Rozpravy Československé Academie Věd, Řada Matematických a přírodních Věd*, **78**, 1–77.

HOLUBEC, J. 1995. Structure. *In*: DALLMEYER, R. D., FRANKE, W. & WEBER, K. (eds) *Pre-Permian Geology of Central and Eastern Europe*. Springer Verlag, Heidelberg, 392–397.

HORÁK, J. M. 1993. *The late Precambrian Coedana and Sarn Complexes – A Geochemical and Petrological Study*. PhD thesis, University of Wales, Cardiff.

HORÁK, J. M., DOIG, R., EVANS, J. A. & GIBBONS, W. 1996. Avalonian magmatism and terrane linkage: new isotopic data from the Precambrian of North Wales. *Journal of the Geological Society, London*, **153**, 91–99.

HOTH, K. 1968. Präkambrium oder Kryptozoikum. *In: Grundriß der Geologie der Deutschen Demokratischen Republik*, Band 1. Akademie-Verlag, Berlin, 77–94.

HOTH, K., LORENZ, W., HIRSCHMANN, G., BERGER, H.-J. 1979. Lithostratigraphische Gliederungsmöglichkeiten regionalmetamorphen Jungproterozoikums am Beispiel des Erzgebirges. *Zeitschrift für geologische Wissenschaften*, **7**, 397–404.

HOTH, K., BRAUSE, H., FREYER, G., LORENZ, W., PÄLCHEN, W.,WAGNER, S. 1985. Neue Ergebnisse zur Gliederung des Proterozoikums im Erzgebirge–Z padné Sudety/Sudety Zachodnie – Antiklinorium sowie an seiner Nordflanke. *Wissenschaftliche Zeitschrift der Ernst Moritz Arndt-Universität Greifswald*, **34**, 5–13.

HUNZIKER, J. C. & ZINGG, A. 1980. Lower Palaeozoic amphibolite to granulite metanorphism in the Ivrea zone (Southern Alps, N. Italy). *Swiss Bulletin of Mineralogy and Petrology*, **60**, 181–214.

IHLENFELD, C., ROHMÜLLER, J. & KÖHLER, H., 1998. The Teplá-Barrandian /Moldanubian boundary: Isotopic investigations on rocks and minerals from the Rittsteig drilling (NE-Bavaria, Germany). *In*:

Novák, M. & Rosenbaum, J. (eds) *Challenges to Chemical Geology Referred Papers from MAEGS-10*. Czech Geology Survey, Prague, 69–92.

Inglis, J. D., Samson, S. D., D'Lemos, R. S. & Hamilton, M. 2004. U-Pb geochronological constraints on the tectonothermal evolution of the Paleoproterozoic basement of Cadomia, La Hague, NW France. *Precambrian Research*, **134**, 293–315.

Innocent, C., Michard, A., Guerrot, C. & Hamelin, B. 2003. Datation U-Pb sur zircons à 548 Ma de leptynites des Maures Centrales. Signification géodynamique des complexes leptyno-amphibolitiques de l'Europe Varisque. *Bulletin de la Société Géologique de France*, **174**, 585–594.

Iosifidi, A.G., Khramov, A.N. & Bachtadse, V. 2005. Multicomponent magnetization of Vendian sedimentary rocks in Podolia, Ukraine. *Russian Journal of Earth Sciences*, **7**, 1–14.

Jachowicz, M. & Přichystal, A. 1997. Lower Cambrian sediments in deep boreholes in South Moravia. *Věstník Českého geologického ústavu*, **72**, 329–332.

Jachowicz, M., Zelaźniewicz, A., Buła Z., Bibiński, W., Habryn, R., Markowiak, M. & Zaba, J. 2002. *The stratigraphy and origin of anchimetamorphic, pre-Cambro-Ordovician rocks in south Poland–deposits of a foreland of the Neoproterozoic orogen*. Archives of Polish Geological Institute, Sosnowiec-Warszawa (in Polish).

Jakeš, P., Zoubek, J., Zoubková, J. & Franke, W, 1979. Graywackes and metagraywackes of the Teplá-Barrandian Proterozoic area. *Sbornik geologických Věd Řada Geologie*, **33**, 83–122.

Janoušek, V., Gerdes, A., Vrána, S., Finger, F., Erban, V., Friedl, G. & Braithwaite, C.J.R. 2006. Low-pressure granulites of the Lišov Massif, Southern Bohemia: Visean metamorphism of Late Devonian plutonic arc rocks. *Journal of Petrology*, **47**, 705–744.

Jelínek, E. & Dudek, A. 1993. Geochemistry of Precambrian Plutonites of the covered part of the Brunovistulicum (Variscides, Czechoslovakia). *Precambrian Research*, **62**,103–125.

Johansson, Å. & Johansson, L. 1993. U-Pb age of titanite in the Mylonite Zone, southwestern Sweden. *Precambrian Research*, **48**, 395–414.

Johansson, L., Möller, C. & Söderlund, U. 2001. Geochronology of eclogite facies metamorphism in the Sveconorwegian province in SW Sweden. *Precambrian Research*, **106**, 261–275.

Jurkiewicz, H. 1975. The geological structure of the basement of the Mesozoic in the central part of the Miechów Trough. *Biuletyn Instytutu Geologicznego*, **283**, 5–100 (in Polish with English summary).

Juskowiakowa, M. 1971. Bazalty wschodniej Polski. *Biuletyn Instytutu Geologicznego*, **245**(VII), 173–252.

Kachlík, V. 1992. *Litostratigrafie, paleogeografický vyvoj a metamorfní postižení hornin pláště v severovychodní části ostrovní zóny stredočeského plutonu*. PhD thesis, Universita Karlova, Praha.

Kachlík, V. 1999. Relationship between Moldanubicum, the Kutná Hora Crystalline Unit and Bohemicum (Central Bohemia, Czech Republic) A result of the polyphase Variscan nappe tectonic. *Journal of the Czech Geological Society*, **44**, 201–292.

Kachlík, V., Janoušek, V., Heřmánek, R. & Vítková, P. 2001. Whole-rock geochemistry and Nd isotopic composition of metavolcanics from the Netvořice-Neveklov and Sedlčany Krásná Hora Islets: their petrogenesis and implication for geodynamic processes at the Teplá-Barrandian-Moldanubian boundary. *Geolines*, **13**, 72–73.

Kalt, A., Corfu, F. & Wijbrans, J. 2000. Time calibration of a P-T path from a Variscan high temperature low-pressure metamorphic complex (Bayerisches Wald, Germany), and the detection of inherited monazite. *Contributions to Mineralogy and Petrology*, **138**, 143–163.

Kalvacheva, R., Sassi, F. P. & Zanferrari, A. 1986. Acritarch evidence for the Cambrian age of phyllites in the Agordo area (South Alpine basement of the Eastern Alps, Italy). *Review of Palaeobotany and Palynology*, **48**, 311–336.

Kalvoda, J., Melichar, R., Bábek, O. & Leichmann, J. 2002. Late Proterozoic–Paleozoic tectonostratigraphic development and paleogeography of Brunovistulian Terrane and comparison with other terranes at the SE margins of Baltica-Laurussia. *Journal of the Czech Geological Society*, **47**, 81–102.

Kamo, S. L., Gower, C. F. & Krogh, T. E. 1989. Birthdate for the Iapetus ocean? A precise U-Pb zircon and baddeleyite age for the Long Range dikes, southeast Labrador. *Geology*, **17**, 602–605.

Karaczun, K. 1953. *Regionalne badania magnetyczne w regionie Mazowsza, Mazur i Bałostocczyzny*. Central Archiv Geology Instytutu Geologicznego, Warsawa.

Karlstrom, K. E., Williams, M. L., McLelland, J., Geissman, J. W. & Ahäll, K.-I. 1999. Refining Rodinia: geologic evidence for the Australia-western U.S. connection in the Proterozoic. *Geological Society of America, Today*, **9**, 1–7.

Karlstrom, K. E., Åhäll, K.-I., Harlan, S. S., Williams, M. L., McLelland, J. & Geissmann, J. A. 2001. Long-lived (1.8–1.0 Ga) convergent orogen in southern laurentia, its extensions to Australia and baltica, and implications for refining Rodinia. *Precambrian Research*, **111**, 5–30.

Kemnitz, H. & Budzinski, G. 1994. Die Grauwacken der Lausitz und ihre cadomische Prägung. *Abhandlungen des Staatlichen Museums für Mineralogie und Geologie Dresden*, **40**, 37–98.

Kemnitz, H., Romer, R. & Oncken, O. 2002. Gondwana breakup and the northern margin of the Saxothuringian belt (Variscides of Central Europe). *International Journal of Earth Sciences*, **91**, 246–259.

Kettner, R. 1917. Versuch einer stratigraphischen Einteilung des böhmischen Algonkium. *Geologische Rundschau*, **8**, 169–188.

Kettner, R. 1930. O postavení metamorfovaných ostrovů v oblasti středočeského žulového masivu. *Sbornik Státniho geologického ústavu Československé Republiky*, **9**, 301–332.

Kirschvink, J.L. 1992. Late Proterozoic low-latitude glaciation: The snowball Earth. *In*: Schopf, J.W. & Klein, C. (eds) *The Proterozoic Biosphere: A Multidisciplinary Study*. Cambridge University Press, Cambridge, 51–52.

Klötzli-Chowanetz, E., Klötzli, U. & Koller, F. 1997. Lower Ordovician migmatization in the Ötztal crystalline basement (Eastern Alps, Austria): linking U-Pb and Pb-Pb dating with zircon morphology. *Swiss Bulletin of Mineralogy and Petrology*, **77**, 315–324.

Knoll, A. H., Walter, M.R., Narbonne, G.M. & Christie-Blick, N. 2004. A new period fort he Geologic time scale. *Science*, **305**, 621–622.

Kodzdrói, W., Krentz, O. & Olpetal, M. (eds) 2001. Geological map Lausitz-Jizera-Karkonosze (without Cenozoic sediments). Scale 1:100 000, Warsaw.

Köhler, H. & Muller-Sohnius, D. 1986. Rb-Sr Altersbestimmungen un Sr-Isotopen-systematik an Gesteinen des Regensburger Waldes (Moldanubikum NE Bayerns). Teil I: Paraanatexite. *Neues Jahrbuch für Mineralogie, Abhandlungen*, **151**, 1–28.

Konopásek, J., Schulmann, K. & Lexa, O. 2001. Structural evolution of the central part of the Krušné Hory (Erzgebirge) Mountains in the Czech Republic – evidence for changing stress regime during Variscan compression. *Journal of Structural Geology*, **23**(9), 1373–1392.

Konzalová, M. 2000. Organic-walled microbiota from the greywackes and other siliciclastic sediments of the Barrandian Neoproterozoic (Bohemian Massif, Czech Republic). *Věstnik Ústředniho ústavu geolgického*, **75**, 319–331.

Köppel, V., Hansmann, W. & Maggetti, M. 1997. Pb isotope and trace element signatures of polymetamorphic rocks from the Silvretta nappe, a comparison. *Swiss Bulletin of Mineralogy and Petrology*, **77**, 325–335.

Korokovsky, S. P., Putis, M., Kotov, A. B., Sal'nikova, E. B. & Kovach, V. P. 1998. High-pressure metamorphism of Phengite Gneisses in the Lower Austro-Alpine Nappe Unit, Eastern Alps: mineral equilibria, *P-T* parameters, and age. *Petrology*, **6**, 655–672.

Košler, J. 1991. Geochemie metasedimentů ostrovní zóny. *Acta Universitatis Carolinae. Geologica*, **35**, 143–149.

Košler, J., Aftalion, M., Vokurka, K., Klečka, M. & Svojtka, M. 1996. Early Cambrian granitoid magmatism in the Moldanubian Zone: U-Pb zircon isotopic evidence from the Stráž orthogneiss. *Abstrakta semináře skupiny tektonických studií Jeseník 26.–29. 4. 1996*, 22–23.

Košler, J., Bowes, D. R., Konopásek, J. & Míková, J. 2004. Laser ablation ICPMS dating of zircons in Erzgebirge orthogneisses: evidence for Early Cambrian and Early Ordovician granitic plutonism in the western Bohemian Massif. *European Journal of Mineralogy*, **16**, 15–22.

Kossmat, F. 1927. Gliederung des varistischen Gebirgsbaues. *Abhandlun-*

gen des Sächsischen Geologischen Landesamtes, **1**, 1–39.

Kotas, A. 1982. Zarys budowy geologicznej Górnośląskiego Zagłębia Węglowego. *In*: Różkowski A. & Ślósarz J. (eds) *65th Meeting of the Polish Geological Society, Sosnowiec, Guidebook.* Wydawnictwa Geologiczne, Warszawa, 45–72.

Kotas, A. 1985. Structural evolution of the Upper Silesian Coal Basin (Poland). *X International Congress on Carboniferous Stratigraphy and Geology.* Compte Rendu, 3, Madrid, 459–469.

Kotkova, J., Harley, S. L. & Fišera, M. 1997. A vestige of very high-pressure (ca. 28 kbar) metamorphism in the Variscan Bohemian Massif, Czech Republic. *European Journal of Mineralogy*, **9**, 1017–1033.

Kowalczewski, Z. 1981. Litostratygrafia wendu w Górach Świętokrzyskich i niecce miechowskiej. *In*: Żakowa, H. (ed.) *53 Meeting of the Polish Geological Society, Kielce, Guidebook.* Wydawnictwa Geologiczne, Warszawa, 7–19.

Krawczyk, C. M., McCann, T., Cocks, L. R. M.. England, R., McBride, J. & Wybraniec, S. 2008. Caledonian Tectonics. *In*: McCann, T. (ed.) *The Geology of Central Europe. Volume 1: Precambrian and Palaeozoic.* Geological Society, London, 303–381.

Kreuzer, H., Seidel, E., Schussler, U., Okrush, M., Lentz, K.L. & Rashka, H. 1989. K-Ar geochronology of different tectonic units at the northwestern margin of the Bohemian Massif. *Tectonophysics*, **157**, 149–178.

Kříbek, B., Pouba, Z., Skoček, V. & Waldhausrová, J. 2000. Neoproterozoic of the Teplá-Barrandian Unit as a part of the Cadomian orogenic belt: A review and correlations aspects. *Věstník Ústředního ústavu geologického*, **75**,175–196.

Krohe, A. 1996. Variscan tectonics of central Europe: Post-accretionary intraplate deformation of weak continental lithosphere. *Tectonics*, **15**, 1364–1388

Kröner A., Hegner E. & Jaeckel P. 1997. Cambrian to Ordovician granitoid orthogneisses in the Polish and Czech West Sudetes Mts. and their geodynamic significance. *Terra Nostra*, **97**, 67–68.

Królikowski, C, Petecki, Z. & Żółtowski, Z. 1999. Main structural units of the Polish part of the East European Platform in the light of gravimetric data. *Biuletyn Państwowego Instytutu Geologicznego*, **386**, 5–58 (in Polish).

Kröner, A. & Mazur, S. 2003. Proterozoic a nd Palaeozoic crustal components accros the East/Central Sudetes boundary at the eastern margin of the Bohemian Massif: new U/Pb and Pb/Pb single zircon ages from the eastern Fore-Sudetic block (SW Poland). *Journal of the Czech Geological Society*, **48**(1–2), 83–84.

Kröner, A., Wendt, I., *et al.* 1988. U-Pb zircon and Sm-Nd model ages of high grade Moldanubian metasediments, Bohemian Massif, Czechoslovakia. *Contributions to Mineralogy and Petrology*, **99**, 257–266.

Kröner, A., Hegner, E., Hammer, J., Haase, G., Bielicki, K.-H., Krauss, M. & Eidam, J. 1994. Geochronology and Nd-Sr systematic of Lusatian granitoids: significance for the evolution of the Variscan orogen in east-central Europe. *Geologische Rundschau*, **83**, 357–376.

Kröner, A., Willner, A. P., Hegner, E., Frischbutter, A., Hofmann, J. & Bergner, R. 1995. Latest Precambrian (Cadomian) zircon ages, Nd isotopic systematics and P-T evolution of granitoid orthogneisses of the Erzgebirge, Saxony and Czech Republic. *Geologische Rundschau*, **84**, 437–456.

Kröner, A., Jaeckel, P., Reischmann, T. & Kroner, U. 1998. Further evidence for an early carboniferous (*c.* 340 Ma) of high-grade metamorphism in the Saxonian granulite complex. *Geologische Rundschau*, **86**, 751–766.

Kröner, A., O'Brien, P., Nemchin, A. & Pidgeon, R. T. 2000*a*. Zircon ages for high pressure granulites from South Bohemia, Czech Republic, and their connection to carboniferous high temperature processes. *Contributions to Mineralogy and Petrology*, **138**, 127–142.

Kröner, A., Štípská, P., Schulmann, K. & Jaeckel, P. 2000*b*. Chronological constraints on the pre-Variscan evolution of the northeastern margin of the Bohemian Massif, Czech Republic. *In*: Franke,W., Haak, V., Oncken, O. & Tanner, D. (eds) *Orogenic Processes: Quantification and Modelling in the Variscan Belt.* Geological Society, London, Special Publications, **179**, 175–197.

Kroner, U. & Sebastian, U. 1997. Der Deckenbau des östlichen Saxothuringikums. *Freiberger Forschungshefte*, **C470**, 35–52.

Kroner, U., Hahn, T., Romer, R.L. & Linnemann, U. 2007. The Variscan Orogeny in the Saxo-Thuringian Zone-heterogenous overprint of Cadomian/Palaeozoic peri-Gondwana crust. *In:* Linnemann, U., Nance, D., Kraft. P. & Zulauf, G. (eds) *The Evolution of the Rheic Ocean: From Avalonian-Cadomian Active Margin to Alleghenian-Variscan Collision.* Geological Society of America, Special Papers, Boulder, Colorado, **423**, 153–172.

Kroner, U., Mansy, J.-L., *et al.* 2008. Variscan tectonics. *In*: McCann, T. (ed.) *The Geology of Central Europe. Volume 1: Precambrian and Palaeozoic.* Geological Society, London, 599–664.

Kryza, R., Mazur, S. & Pin, C. 2003. Subduction- and non-subduction-related igneous rocks in the Central-European Variscides: geochemical and Nd isotope evidence for a composite origin of the Kłodzko Metamorphic Complex, Polish Sudetes. *Geodinamica Acta*, **16**, 39–57.

Kryza, R., Mazur, S. & Oberc-Dziedzic, T. 2004. The Sudetic geological mosaic: Insights into the root of the Variscan orogen. *Przegląd Geologiczny*, **52**(8/2), 761–773.

Krzeminska, E. 2005. *Outline of Geochemistry of the Neoproterozoic Flood Basalts in Lublin-Podlasie Basin, E Poland.* The Mineralogical Society of Poland, Special Papers, **26**, 48–52.

Krzeminska, E., Williams, I. & Wiszniewska, J. 2005. A Late Paleoproterozoic (1.80 Ga) subduction-related mafic igneous suite from Lomza, NE Poland. *Terra Nova*, **17**, 442–449.

Krzeminska E. , Wiszniewska, J. & Williams, I. S. 2006. Revision of the "oldest rock" age in the crystalline basement of NE Poland. *Prz. Geologiczny*, **58**, 967–973 (in Polish, English abstract).

Kubicki, S. & Ryka, W. 1982. *Atlas geologiczny podłoża krystalicznego polskiej części platformy wschodnioeuropejskiej.* Instytut Geologiczny Warszawa, Wydawnictwa Geologiczne, Warszawa.

Kubicki, S., Ryka, W. & Znosko J. 1973. Tektonika. *In*: Łaszkiewicz, A. (ed.) *Skały platformy prekambryjskiej w Polsce. cz. 12. Podłoże krystaliczne.* Prace Instytutu Geologicznego, **68**, 129–138.

Kukal, Z. 1985. Vývoj sedimentů Českého masivu. *Knihovna Ústředního ústavu geologického*, **61**(1), 221.

Kukal, Z. & Jäger, O. 1988. Siliciclastic signal of the Variscan orogenesis: the Devonian Srbsko Formation of Central Bohemia. *Věstnik Ústředniho ústavu geolgického*, **63**(2), 65–80.

Lafon, J. M. 1984. La granodiorite de Caplongue, nouveau témoin d'un magmatisme cambrien dans le Rouergue occidental. *Comptes Rendus de l'Académie des Sciences, Paris*, **298**, 595–600.

Lagarde, J. L. 1978. Analyse du sens de déversement d'une nappe gneissique. Application à la nappe hercynienne du Canigou. *Comptes Rendus de l'Académie des Sciences, Paris*, **286**, 937–940.

Lammerer, B. 1988. Thrust-regime and transpression-regime tectonics in the Tauern Window (Eastern Alps). *Geologische Rundschau*, **77**, 143–156.

Lammerer, B. & Weger, M. 1998. Footwall uplift in an orogenic wedge: the Tauern Window in the Eastern Alps of Europe. *Tectonophysics*, **285**, 213–230.

Larson, S.Å. & Berglund, J. 1992 A chronological subdivision of the Transscandinavian Igneous Belt – three magmatic episodes? *Geologiska Föreningens i Stockholm Förhandlingar*, **114**, 459–461.

Larson, S. Å, Berglund, J. Stigh, J. & Tullborg, E.-L. 1990. The Protogine Zone, southwest Sweden: a new model — an old issue. *In*: Gower, C. F., Rivers, T. & Ryan, A. B. (eds) *Mid-Proterozoic Laurentia-Baltic.* Geological Association of Canada, Special Paper, **38**, 317–333.

Łaszkiewicz, A. 1960. Skały i minerały krystalicznego podłoża Niżu Polskiego. *Kwartalnik Geologiczny*, **4**, 819–825.

Laumonier, B., Autran, A., Barbey, P., Cheilletz, A., Baudin, Cocherie, A. & Guerrot, C. 2004. Conséquences de l'absence de socle cadomien sur l'âge et la signification des séries pré-varisques (anté-Ordovicien supérieur) du sud de la France (Pyrénées, Montagne Noire). *Bulletin de la Société Géologique de France*, **175**, 643–655.

Le Corre, C. 1977. Le Briovérien de Bretagne centrale : essai de synthèse lithologique et structurale. *Bulletin BRGM, section I*, **3**, 21–254.

Le Fevre, B. 1997. *Géochimie isotopique Sm-Nd et Rb-Sr des granitoïdes cadomiens de Bretagne (France) et de Moravie (Rép. Tchèque/Autriche).* Memoire DEA University of Clermont-Ferrand, 1–50.

LEES, G. J., ROACH, R. A., SHUFFLEBOTHAM, M. M. & GRIFFITHS, N. H. 1987. Upper Proterozoic basaltic volcanism in the northern Massif Armoricain, France. *In*: PHARAOH, T. C., BECKINSALE, R. D. & RICKARD, D. (eds) *Geochemistry of Mineralization of Proterozoic Volcanic Suites*. Geological Society, London, Special Publications, **33**, 503–523.

LEICHMANN, J. 1996. *Geologie und Petrologie der Brünner Massiversus* Universität Salzburg, 1–118.

LESCUYER, J. L. & COCHERIE, A. 1992. Datation sur monozircons des métadacites de Sériès: argument pour un âge protérozoïque terminal des « schistes X » de la Montagne Noire (Massif Central français). *Comptes Rendus de l'Académie des Sciences Paris*, **314**, 1071–1077.

LÉVÊQUE M.H. 1985. Mise en évidence d'un socle Précambrien dans le Massif Central français: l'orthogneiss des Palanges (Aveyron). *Comptes Rendus de l'Académie des Sciences Paris*, **300**, 277–282.

LI, Z. X. & POWELL, C. MCA. 2001. An outline of the palaeogeographic evolution of the Australasian region since the beginning of the Neoproterozoic. *Earth Science Reviews*, **53**, 237–277.

LI, Z.X., ZHANG, L. & POWEL, C.M. 1996. Positions of the east Asian cratons in the Neoproterozoic supercontinent Rodinia. *Australian Journal of Earth Sciences*, **43**, 593–604.

LIDMAR-BERGSTRÖM, K. 1988. Denudation surfaces of a shield area in south Sweden. *Geografiska Annaler*, **70**, 337–350.

LIEW, T. C. & HOFMAN, A. W. 1988. Precambrian crustal components, plutonic association, plate environment o the Hercynian fold belt of central Europe: implication from a Nd and Sr isotopic study. *Contributions to Mineralogy and Petrology*, **98**, 129–138.

LINDROOS, A., ROMER, R. L. EHLERS, C. & ALVIOLA, R. 1996. Late-orogenic Svecofennian deformation in SW Finland constrained by pegmatite emplacement ages. *Terra Nova*, **8**, 567–574.

LINDSAY, J.F. & BRASIER, M.D. 2004. The evolution of the Precambrian atmosphere: carbon isotopic evidence from the Australian continent. *In*: ERIKSSON, P.G., ALTERMANN, W., NELSON, D.R., MUELLER, W.U. & CATUNEANU, O. (eds) *The Precambrian Earth: Tempos and Events*. Developments in Precambrian Geology, Vol. 12, Elsevier, Amsterdam, 388–403.

LINNEMANN, U. 1991. Glazioeustatisch kontrollierte Sedimentationsprozesse im Oberen Proterozoikum der Elbezone (Weesensteiner Gruppe/Sachsen). *Zentralblatt für Geologie und Paläontologie Teil I*, **12**, 2907–2934.

LINNEMANN, U. & ROMER, R. L. 2002. The Cadomian Orogeny in Saxo-Thuringia, Germany: geochemical and Nd-Sr-Pb isotopic characterization of marginal basins with constraints to geotectonic setting and provenance. *Tectonophysics*, **352**, 33–64.

LINNEMANN, U. & SCHAUER, M. 1999. Die Entstehung der Elbezone vor dem Hintergrund der cadomischen und variszischen Geschichte des Saxothuringischen Terranes – Konsequenzen aus einer abgedeckten geologischen Karte. *Zeitschrift für Geologische Wissenschaften*, **27**, 529–561.

LINNEMANN, U., GEHMLICH, M., *et al.* 2000. From cadomian subduction to Early Palaeozoic rifting: the evolution of Saxo-Thuringia at the margin of Gondwana in the light of single zircon geochronology and basin development (Central European Variscides, Germany). *In*: FRANKE, W., HAAK, V., ONCKEN, O. & TANNER, D. (eds) *Orogenic Processes: Quantification and Modelling in the Variscan Belt of Central Europe*. Geological Society, London, Special Publications, **179**, 131–153.

LINNEMANN, U., MCNAUGHTON, N. J., ROMER, R. L., GEHMLICH, M., DROST, K. & TONK, C. 2004. West African provenance for Saxo-Thuringia (Bohemian Massif): Did Armorica ever leave pre-Pangean Gondwana? U/Pb-SHRIMP zircon evidence and the Nd-isotopic record. *International Journal of Earth Sciences*, **93**(5), 683–705.

LINNEMANN, U., GERDES, A., DROST, K. & BUSCHMANN, B. 2007. The continuum between Cadomian Orogenesis and opening of the Rheic Ocean: Constraints from LA-ICP-MS U–Pb zircon dating and analysis of plate-tectonic setting (Saxo-Thuringian zone, NE Bohemian massif, Germany). *In:* LINNEMANN, U., NANCE, D., KRAFT. P. & ZULAUF, G. (eds). *The Evolution of the Rheic Ocean: From Avalonian-Cadomian Active Margin to Alleghenian-Variscan Collision*. Geological Society of America, Special Paper, Boulder, Colorado, **423**, 61–69.

LINNEMANN, U., D'LEMOS, R., *et al.* 2008. Cadomian tectonics. *In*: MCCANN, T. (ed.) *The Geology of Central Europe. Volume 1: Precambrian and Palaeozoic*. Geological Society, London, 103–154.

LIPPOLT, H. J. 1986. Nachweis altpaläozoischer Primäralter (Rb–Sr) und karbonischer Abkühlalter (K–Ar) der Muskovit-Biotit-Gneise des Spessarts und der Biotit-Gneise des Böllsteiner Odenwaldes. *Geologische Rundschau*, **75**, 569–583.

LOEWY, S., CONNELLY, J. N., DALZIEL, I. W. D., GOWER, C. F. & CAWOOD, P. A. 2000. Testing a proposed Rodinia reconstruction using Pb isotopes and U–Pb geochronology. *Geological Society of America*. Abstracts, **32**(7), A455.

LORENC, W. M. & WISZNIEWSKA, J. 1999. The Mazury Complex: one magmatic influx or more? *Between EUROBRIDGE and TESZ, Seventh EUROBRIDGE Workshop*, 26–30 May 1999, Suwałki-Szelment, Poland.

LORENZ, W. & BURMANN, G. 1974. Alterskriterien für das Präkambrium am Nordrand der Böhmischen Masse. *Geologie*, **21**(4/5), 405–418.

LORENZ, W. & HOTH, K. 1990. Lithostratigraphie im Erzgebirge – Konzeption, Entwicklung, Probleme und Perspektiven. *Abhandlungen des Staatlichen Museums für Mineralogie und Geologie zu Dresden*, **37**, 7–35.

LUGMAIR, G.W. & SHUKOLYUKOV, A. 2001. Early solar system events and timescales. *Meteoritics and Planetary Science*, **36**, 1017–1026.

LUND, C. E., GORBATSCHEW, R. & SMIRNOV, A. 2001. A seismic model of the Precambrian crust along the coast of southeastern Sweden: the coast profile wide-angle airgun experiment and the southern part of FENNOLORA revisited. *Tectonophysics*, **339**, 93–111.

LUNDQVIST, T. 1979. The Precambrian of Sweden. *Sveriges Geologiska Undersökning, Rapporter och meddelanden*, **C768**, 1–87.

LUNDQVIST, T., BYGGHAMMAR, B., STEPHENS, M. B., BECKHOLMEN, M. & NORLING, E. 1994. Sveriges berggrund, 1:250000. *In*: FREDEN, C. (ed.) *Berg och jord. Sverigs nationalatlay*. Sverigs Nationalatlas Förlag.

MCCANN, T. 1996. Silurian facies from the G-14 well, offshore Northern Germany. *Zeitschrift der geologischen Gesellschaft*, **147**, 209–219.

MCCANN, T. & KRAWCZYK, C. M. 2001. The Trans-European Fault: a critical reassessment. *Geological Magazine*, **138**, 19–29.

MCCANN, T., SKOMPSKI, S., *et al.* 2008. Carboniferous. *In*: MCCANN, T. (ed.) *The Geology of Central Europe. Volume 1: Precambrian and Palaeozoic*. Geological Society, London, 411–529.

MCMENAMIN, M. A. S. & MCMENAMIN, D. L. 1990. *The Emergence of Animals; The Cambrian Breakthrough*. Columbia University Press, New York.

MAGNUSSON, N. H. 1960. Age determinations of Swedish Precambrian rocks. *Geologiska Föreningens i Stockholm Förhandlingar*, **82**, 407–432.

MALETZ, J. 1997. Ordovician and Silurian strata of the G-14 well (Baltic Sea): Graptolithe faunas and biostratigraphy. *Zeitschrift für Geologische Wissenschaften*, **25**, 29–39.

MALUSKI, H. & ALLEGRE, C. J. 1970. Problème de la datation par le couple $^{87}Rb/^{86}Sr$ des socles gneissiques : exemple des gneiss de Bormes (massif hercynien des Maures, France). *Comptes Rendus de l'Académie des Sciences Paris*, **270**, 18–21.

MALUSKI, H. & GUEIRARD, S. 1978. Mise en évidence par la méthode $^{39}Ar/^{40}Ar$ de l'âge à 580 Ma du granite de Barral (Massif des Maures, Var, France). *Comptes Rendus de l'Académie des Sciences Paris*, **287**, 195–198.

MANSFELD, J. 1996. Geological, geochemical and geochronological evidence for a new Palaeoproterozoic terran in southeastern Sweden. *Precambrian Research*, **77**, 91–103.

MANSFELD, J. 2001. Age and ε_{Nd} constraints on the Paleoproterozoic tectonic evolution in the Baltic-Sea region. *Tectonophysics*, **339**, 135–151.

MANSFELD, J. 2004. The Småland-Värmland Belt in southeastern Sweden. *In*: HÖGDAHL, K., ANDERSSON, U. B. & EKLUND, O. (eds) *The Trasscandinavian Igneous Belt (TIB) in Sweden: A Review of its Character and Evolution*. Geological Survey of Finland, Special Papers, **37**, 20–21.

MARTIN, M.W., GRAZHDANKIN, D.V., BOWRING, S.A., EVANS, D.A.D., FEDONKIN, M.A. & KIRSCHVINK, J.L. 2000. Age of Neoproterozoic bilatarian body and trace fossils, White Sea, Russia: Implications for metazoan evolution. *Science*, **288**, 841–845.

MAŠEK, J. 2000 Stratigraphy of the Proterozoic of the Barrandian area. *Vestnik Ceskeho geologickeho ustavu*, **75**, 197–199.

MAŠEK, J. & ZOUBEK, J. 1980. Návrh vymezení a označování hlavních stratigrafických jednotek barrandienského proterozoika. *Věstnik Ústředniho ústavu geolgického*, **55**(2), 121–123.

MASLOV, A. V. & ISHERSKAYA, M. V. 2002. Riphean sedimentary sequences of the eastern and northeastern margins of the Eastern European craton. *Russian Journal of Earth Sciences*, **4**, 271–276.

MASLOV, A. V., ERDTMANN, B.-D., IVANOV, K. S., IVANOV, S. N. & KRUPENIN, M. T. 1997. The main tectonic events, depositional history, and the palaeogeography of the southern Urals during the Riphean – early Palaeozoic. *Tectonophysics*, **276**, 313–335.

MASSONNE, H.-J. 1998. A new occurrence of microdiamonds in quartzofeldspathic rocks of the Saxonian Erzgebirge, Germany, and their metamorphic evolution: *7th International Kimberlite Conference*, Cape Town, Abstracts, 552–554.

MASSONNE, H.-J. 2001. First find of coesite in the ultrahigh-pressure metamorphic area of the central Erzgebirge, Germany. *European Journal of Mineralogy*, **13**, 565–570.

MATĚJKA, D., 1988. Geochemistry of Upper Proterozoic rocks of the Svojšín volcanic strip (Western Bohemia) and problem of the source material of metasediments. *Geologický sborník Geologica Carpathica*, **39**, 109–122.

MATTE, PH., RESPAUT, J. P., MALUSKI, H., LANCELOT, J. R. & BRUNEL, M. 1986. La faille NW-SE du Pays de Bray, un décrochement ductile dextre hercynien: déformation à 330 Ma d'un granite à 570 Ma dans le sondage Pays de Bray 201. *Bulletin de la Société Géologique de France*, **II**, 69–77.

MATURA, A. 1976. Hypothesen zum Bau und zur geologischen Geschichte des kristallinen Grundgebirges von Südwestmahren und dem niederösterreichischen Waldviertel. *Jahrbuch der Geologischen Bundesanstalt*, **119**, 63–74.

MAX, M. D. & RODDICK, J. C. 1989. Age of metamorphism in the Rosslare Complex, SE Ireland. *Proceedings of the Geologists' Association*, **100**, 113–121.

MAZUR, S. 2003. Structural evolution of the Kłodzko Metamorphic Complex and the implications for the Variscan tectonics of the Sudetes. *Acta Universitatis Wratislaviensis no 2581, Prace Geologiczno-Mineralogiczne*, **74**, 1–197.

MAZUR, S., TURNIAK K. & BRÖCKER, M. 2004. Neoproterozoic and Cambro-Ordovician magmatism in the Variscan Kłodzko Metamorphic Complex (West Sudetes, Poland): new insights from U/Pb zircon dating. *International Journal of Earth Sciences*, **93**, 758–772.

MAZUR, S., ALEKSANDROWSKI, P., KRYZA, R. & OBERC-DZIEDZIC, T. 2006. The Variscan Orogen in Poland. *Geological Quarterly*, **50**, 89–118.

MEDARIS, L.G., JELÍNEK, E. & MÍSAŘ, Z. 1995*a*. Czech eclogites. Terrane settings, interpretation for Variscan tectonic evolution of the Bohemian Massif. *European Journal of Mineralogy*, **7**, 7–28.

MEDARIS, L. G., BEARD, B. L., JOHNSON, C. M., VALLEY, J. W., SPICUZZA, M. J., JELÍNEK, E. & MÍSAŘ, Z. 1995*b*. Garnet pyroxenite and eclogite in the Bohemian Massif – geochemical evidence for Variscan recycling of subducted lithosphere. *Geologische Rundschau*, **84**, 489–505.

MEDARIS, L.G. JR, WANG, H. F., JELÍNEK, E. & JAKEŠ, P. 2003. Garnet peridotite in the Moldanubian Zone in the Czech Republic – A heat source for Variscan metamorphism? *Journal of the Czech Geological Society*, **48**, 92–93.

MEHNERT, K.-R. 1938. Die Meta-Konglomerate des Wiesenthaler Gneiszuges im sächsischen Erzgebirge. *Mineralogische und Petrographische Mitteilungen*, **50**, 194–272

MEISEL, T., BIINO, G. G. & NÄGLER, TH. 1996. Re-Os, Sm-Nd and rare earth element evidence for Proterozoic oceanic and possible subcontinental lithosphere in tectonized ultramafic lenses from the Swiss Alps. *Geochimica et Cosmochimica Acta*, **60**, 2583–2593.

MELCHER, F. & MEISEL, T. 2004. A metamorphosed early Cambrian crust-mantle transition in the Eastern Alps, Austria. *Journal of Petrology*, **45**, 1689–1723.

MELLQVIST, C., ÖHLANDER, B., SKIÖLD, T. & WIKSTRÖM, A. 1999. The Archaean-Proterozoic palaeoboundary in the Lulea area, northern Sweden; field and isotope geochemical evidence for a sharp terrane boundary. *Precambrian Research*, **96**, 225–243.

MÉNOT, R. P., PEUCAT, J. J., SCARENZI, D. & PIBOULE, M. 1988. 496 Ma age of plagiogranites in the Chamrousse ophiolitic complex (external crystalline massifs in the French Alps): evidence of a Lower Paleozoic oceanization. *Earth and Planetary Science Letters*, **88**, 82–92.

MILLER, B., SAMSON, S. D. & D'LEMOS, R. S. 1999. Time span of plutonism, fabric development, and cooling in a Neoproterozoic magmatic arc segment: U-Pb age constraints from syntectonic plutons, Sark, Channel Islands, U.K. *Tectonophysics*, **312**, 79–95.

MILLER, B., SAMSON, S. D. & D'LEMOS, R. S. 2001. U-Pb geochronological constraints on the timing of plutonism, volcanism, and sedimentation, Jersey, Channel Islands, UK. *Journal of the Geological Society, London*, **158**, 243–252.

MILLER, CH. & THÖNI, M. 1996. Origin of eclogites from the Austroalpine Ötztal basement (Tirol, Austria): geochemistry and Sm-Nd vs Rb-Sr isotope systematics. *Chemical Geology*, **122**, 199–225.

MINGRAM, B. 1998. The Erzgebirge, Germany – a subducted part of northern Gondwana: geochemical evidence for repetition of early Palaeozoic metasedimentary sequences in metamorphic thrust units. *Geological Magazine*, **135**, 785–801.

MINGRAM, B., KRÖNER, A., HEGNER, E. & KRENTZ, O. 2004. Zircon ages, geochemistry, and Nd isotopic systematics of pre-Variscan orthogneisses from the Erzgebirge, Saxony (Germany), and geodynamic interpretation. *International Journal of Earth Sciences*, **93**, 706–727.

MÍSAŘ, Z., DUDEK , A., HAVLENA , V. & WEISS, J. 1983. *Geologie ČSSR I – Český masiv*. Státní pedagogické nakladatelství, Praha.

MÖLLER, C. & SÖDERLUND, U. 1997. Age constraints on the regional deformation within the Eastern Segment, S. Sweden: Late Sveconorwegian granite dyke intrusion and metamorphic-deformational relations. *Quarterly Journal of the Geological Society of Sweden*, **119**, 1–12.

MORGAN, J., STEIN, H. J., HANNAH, J. L., MARKEY, R. & WISZNIEWSKA, J. 2000. Re-Os study of Fe-Ti-V oxide and Fe-Cu-Ni sulfide deposits, Suwalki Anorthosite Massif, northeast Poland. *Mineralium Deposita*, **35**, 391–401.

MORYC, W. & HEFLIK, W. 1998. Metamorphic rocks in the basement of the Carpathians between Bielsko-Biała and Cracow. *Geological Quarterly*, **42**, 1–14.

MORYC, W. & JACHOWICZ, M. 2000. Precambrian deposits in the Bochnia-Tarnów-Dębica area, Southern Poland (in Polish with English summary). *Przegląd Geologiczny*, **48**, 601–606.

MORYC, W. & ŁYDKA, K. 2000. Sedimentation and tectonics of the Upper Proterozoic–Lower Cambrian deposits of the southern Małopolska Massif (SE Poland). *Geological Quarterly*, **44**, 47–58.

MOTUZA, G., KORABLIOVA, L. & NASEDKIN, V. 2000. Geological structure of the upper lithosphere beneath northwestern part of the DSS EUROBRIDGE profile. *Geophysics in the Baltic Region: Problems and Prospects for the New Millenium, International Conference, Abstracts and Excursions Guide*, Tallinn, September 26–30, 47–49.

MOTUZA, G., ČEČYS, A., SALNIKOVA, J. & KOTOV, A. 2004. Mezoproterozoic granitoid magmatism in Western Lithuania: new evidences. *In*: ČEČYS, A. (ed.) *Tectonic Implications of the ca. 1,45 Ga Granitoid Magmatism at the Southwestern Margin of the East European Craton.* LITHOLUND theses- No. 5, PhD thesis, Department of Geology, Lund University.

MOUSSAVOU, M. 1998. *Contribution à l'histoire thermo-tectonique varisque du Massif des Maures, par la typologie du zircon et la géochronologie U-Pb sur minéraux accessoires*. Thesis, University of Montpellier, II.

MURPHY, J. B., PISAREVSKY, S. A., NANCE, R. D. & KEPPIE, J. D. 2004. Neoproterozoic–early Paleozoic configuration of peri-Gondwanan terranes: implications for Laurentia-Gondwanan connections. *International Journal of Earth Sciences*, **93**, 659–682.

MYERS, J.S. 2001. Protoliths of the 3.8–3.7 Isua greenstone belt, west Greenland. *Precambrian Research*, **105**, 129–141.

NAGY, E. A., SAMSON, S. D. & D'LEMOS, R. S. 2002. U-Pb geochronological constraints on the timing of Brioverian sedimentation and regional deformation in the St Brieuc region of the Neoproterozoic Cadomian orogen. *Precambrian Research*, **116**, 1–17.

NANCE, R. D., & MURPHY, J. B. 1994. Contrasting basement isotopic signatures and the palinspastic restoration of peripheral orogens: Example from the Neoproterozoic Avalonian–Cadomian belt. *Geology*, **22**, 617–620.

NANCE, R. D. & MURPHY, J. B. 1996. Basement isotopic signatures and Neoproterozoic paleogeography of Avalonian–Cadomian and related terranes in the circum-North Atlantic. *In*: NANCE, R. D. & THOMPSON, M. D. (eds) *Avalonian and Related Peri-Gondwanan Terranes of the Circum-North Atlantic.* Geological Society of America. Special Publications, **304**, 333–346.

NANCE, R. D. MURPHY, J. B. & KEPPIE, J. D. 2002. A Cordilleran model for the evolution of Avalonia. *Tectonophysics*, **352**, 11–31.

NANCE, R. D., MURPHY, J. B., STRACHAN, R. A., D'LEMOS, R. S. & TAYLOR, G. K. 1991. Late Proterozoic tectonostratigraphic evolution of the Avalonian and Cadomian terranes. *Precambrian Research*, **53**, 41–78.

NANCE, R.D., WORSLEY, T.R., AND MOODY' J.B. 1986. Post-Archaean biogeochemical cycles and long-term episodicity in tectonic processes. *Geology*, **14**, 514–518.

NARBONNE, G.M. 1998. The Ediacara biota: a terminal Neoproterozoic experiment in the evolution of life. *GSA Today*, **8**, 1–6.

NARBONNE, G.M. & GEHLING, J.G. 2003. Life after snowball: The oldest complex Ediacaran fossils. *Geology*, **31**, 27–30.

NASDALA, L., GRUNER, T., NEMCHIN, A. A., PIDGEON, R. T. & TICHOMIROWA, M. 1996. New SHRIMP ion microprobe measurements on zircons from Saxonian magmatic and metamorphic rocks. *Freiberger Isotopenkolloquium, 1996, Proceedings*, 205–214.

NAWROCKI, J., BOGUCKIJ, A. & KATINAS, V. 2004*a*. New late Vendian palaeogeography of Baltica and TESZ. *Geological Quarterly*, **48**, 309–316.

NAWROCKI, J., ZYLINSKA, A., BULA, Z., GRABOWSKI, J., KRZYWIEC, P. & POPRAWA, P. 2004*b*. Early Cambrian location and affinities of the Brunovistulian terrane (Central Europe) in the light of palaeomagnetic data. *Journal of the Geological Society, London*, **161**, 513–522.

NAWROCKI, J., DUNLAP, J., *et al.* 2007. Late Neoproterozoic to Early Palaeozoic palaeogeography of the Holy Cross Mountains (Central Europe): an integrated approach. *Journal of the Geological Society, London*, **164**, 405–423.

NEGA, M., KRUHL, J. H., KRENTZ, O. & LEONHARDT, D. 1997. Gefügeprägung und Exhumierung durch Kontinent-Kontinent-Kollison: Die spätvariszische Kinematik im westlichen Erzgebirge. *Terra Nostra*, **97**, 109–115.

NEGA, M., WERNER, O., KRUHL, J. H., LIPPOLT, H. J., KRENTZ, O. & LEONHARDT, D. 1998. Exhumierung von HP-Gesteinen durch Kontinent-Kontinent-Kollision: Ein variszisches Szenario im westlichen Erzgebirge. *Freiberger Forschungshefte*, **C471**, 156–157.

NEUBAUER, F. 2002. Evolution of late Neoproterozoic to early Paleozoic tectonic elements in Central and Southeast European Alpine mountain belts: review and synthesis. *Tectonophysics*, **352**, 87–103.

NEUBAUER, F. & FRISCH, W. 1993. The Austro-Alpine metamorphic basement east of the Tauern Window. *In*: VON RAUMER, J. F. & NEUBAUER, F. (eds) *The Pre-Mesozoic Geology of the Alps.* Springer Verlag, Heidelberg, 515–536.

NEUBAUER, F., FRISCH, W., SCHMEROLD, R. & SCHLÖSER, H. 1989. Metamorphosed and dismembered ophiolite suites in the basement units of the Eastern Alps. *Tectonophysics*, **164**, 49–62.

NEUBAUER, F., KLÖTZLI, U. & POSCHESCHNIK, P. 2001. Cadomian magmatism in the Alps recorded in Late Ordovician sandstones of the Carnic Alps: preliminary results from zircon Pb/Pb evaporation dating. *Swiss Bulletin of Mineralogy and Petrology*, **81**, 175–179.

NEUMANN, W. 1974. Mitteldeutsche Kristallinzone. *In*: HOPPE, W. & SEIDEL, G. (eds) *Geologie von Thüringen*. Hermann Haack, Gotha-Leipzig, 77–118.

NIKISHIN, A. M., ZIEGLER, P.A., *et al.* 1996. Late Precambrian to Triassic history of the East European Craton: dynamics of sedimentary basin evolution. *Tectonophysics*, **268**, 23–63.

NIRONEN, M. 1997. The Svekofennian Orogen: a tectonic model. *Precambrian Research*, **86**, 21–44.

NOBLE, S. R., TUCKER, R. D. & PHARAOH, T. C. 1993. Lower Paleozoic and Precambrian igneous rocks from eastern England, and their bearing on late Ordovician closure of the Tornquist Sea: constraints from U-Pb and Nd isotopes. *Geological Magazine*, **130**, 835–846.

NUTMANN, A.P., FRIEND, C.R.L. & BENNET, V.C. 2002. Evidence for 3650–3600 Ma assembly of the northern end of the Itsaq Gneiss Complex, Geenland: implication for early Archaean tectonics. *Tectonics*, **21**, 1005–1029.

NYSTUEN, J. P. & SIEDLECKA, A. 1988. The 'Sparagmites' of Norway. *In*: WINCHESTER, J.A. (ed.) *Later Proterozoic Stratigraphy of the Northern Atlantic Regions*. Blackie, Glasgow, 237–252.

OBERC, J. 1993. The role of longitudinal dislocation zones and strike-slip transversal deep fracture of Silesia–Lubusza (Hamburg–Kraków) in formation of main zone of meridional folds on Silesia and Moravia areas). *Geological Quarterly*, **37**, 1–18.

OBERC-DZIEDZIC, T. 1999. The metamorphic and structural development of gneisses and older schist series in the Strzelin crystalline massif (Fore-Sudetic Block, SW Poland). *Mineralogical Society of Poland, Special Papers*, **14**, 10–21.

OBERC-DZIEDZIC, T., KLIMAS, K., KRYZA, R. & FANNING, C. M. 2003. SHRIMP zircon geochronology of the Strzelin gneiss, SW Poland: evidence for a Neoproterozoic thermal event in the Fore-Sudetic Block, Central European Variscides. *International Journal of Earth Sciences*, **92**, 701–711.

O'BRIEN, P. & VRÁNA, S. 1995. Eclogites with a short-lived granulite-facies overprint in the moldanubian zone, Czech Republic–petrology, geochemistry and diffusion modelling of garnet zoning. *Geologische Rundschau*, **84**, 473–488.

O'BRIEN, S. J., O'BRIEN, B. H., DUNNING, H. & TUCKER, R. D. 1996. Late Neoproterzoic Avalonian and related Peri-Gonwanan rocks of teh Newfoundland Appalachians. *In*: NANCE, R. D. & THOMPSON, D. (eds) *Avalonian and Related Peri-Gondwanian Terranes of the Circum-North Atlantic.* Geological Society of America, Special Publications, **304**, 9–28.

OBST, K., HAMMER, J., KATZUNG, G. & KORICH, D. 2004. The Mesoproterozoic basement in the southern Baltic Sea: insights from the G 14–1 off-shore borehole. *International Journal of Earth Sciences* (*Geologische Rundschau*), **93**, 1–12.

ÖHLANDER, B. & ZUBER, J. 1988. Genesis of the Fellingsbro-type granites: evidence from gravity measurements and geochemistry. *Geologiska Föreningens i Stockholm Förhandlingar*, **110**, 39–54.

OSTROVSKY, A. A., FLUEH, E. R. & LUOSTO, U. 1994. Deep seismic structure of the Earth's crust along the Baltic Sea Profile. *Tectonophysics*, **233**, 279–292.

PACLTOVÁ, B. 1990. Late Proterozoic organic remains from the Mítov Locallity. *Journal of Paleontology*, **91**, 261–276.

PACLTOVÁ, B. 1994. Silurian (Wenlockian) sporomorph in South Bohemian Moldanubicum. *Journal of the Czech Geological Society*, **39**, 81–82.

PACZEŚNA, J. & POPRAWA, P. 2005. Rola procesów tektonicznych oraz eustatycznych w rozwoju sekwencji stratygraficznych utworów neoproterozoiku i kambru basenu lubelsko-podlaskiego. *Przegląd Geologiczny*, **57**, 262–272.

PAGE, L., MÖLER, L. & JOHANSSON, L. 1996*a*. $^{40}Ar/^{39}Ar$ geochronology across the Mylonite Zone and the Sveconorwegian Granulite Province in the Sveconorwegian orogen of S. Sweden. *Precambrian Research*, **79**, 239–259.

PAGE, L. M., STEPHENS, M. B. & WAHLGREN, C.-H. 1996*b*. $^{40}Ar/^{39}Ar$ geochronological constraints on the tectonothermal evolution of the Eastern sement of the Sveconorwegian orogen, south-central Sweden. *In*: BREWER, T. S. (ed.) *Precambrian Crustal Evolution in the North Atalantic Region*. Geological Society, London, Special Publications, **112**, 315–330.

PALEN, R. & WERNER, C.-D. 1998. Zirkonalter von Granitgeröllen aus Obermittweida/Erzgebirge. *Terra Nostra*, **98**(2), 119–121.

PANWITZ, C., HANDLER, R., HEINISCH, H. & NEUBAUER, F. 2000. $^{40}Ar/^{39}Ar$-dating on detrital white micas from Austroalpine Palaeozoic of Northern Greywacke Zone (Tyrol/Salzburg – Austria): a pilot study. *Terra Nostra*, **2000**, 83.

PARK, R. G. 1992. Plate kinematic history of Baltica during the Middle to Late Proterozoic: a model. *Geology*, **20**, 725–728.

PARK, R.G., ÅHÄLL, K.-I. & BOLAND, M. P. 1991. The Sveconorwegian shear-zone network of SW Sweden in relation to mid-Proterozoic plate movements. *Precambrian Research*, **49**, 245–260.

PARK, R.G., CLIFF., R.A., FETTES, D.J. & STEWART, A.D. 1994. Precambrian rocks in northwest Scotland west of the Moine Thrust: the Lewisian Complex and the Torridonian. *In*: GIBBONS, W. & HARRIS, A. L. (eds) *A revised Correlation of Precambrian Rocks in the British Isles*. Geological Society London, Special Reports, **22**, 6–20.

PAŠAVA, J. 2000. Normal versus Metal-rich black shales in the Barrandian Neoproterozoic of the Teplá-Barrandian Unit: a summary with new data. *Vestník Ceského Geologického Ústavu*, **75**(3), 229–239.

PASTEELS, P. & DORÉ, F. 1982. Age of the Vire-Carolles granite. *In*:

ODIN, G. S. (ed.) *Numerical Dating in Stratigraphy.* J Wiley, New York, 784–790.

PATCHETT, P. J. & JOCELYN, J. 1979. U-Pb zircon ages for late Precambrian igneous rocks in South Wales. *Journal of the Geological Society, London*, **136**, 3–19.

PATCHETT, P. J., GALE, N. H., GOODWIN, R. & HUMM, M. J. 1980. Rb-Sr whole-rock isochron ages of late Precambrian to Cambrian igneous rocks from southern Britain. *Journal of the Geological Society, London*, **137**, 649–656.

PATTERSON, C. 1956. Age of meteorites and the earth. *Geochimica et Cosmochimica Acta*, **10**, 230–237.

PATOČKA, F. 1991. Geochemistry and primary tectonic environment of the amphibolites from the Český Krumlov Varied Group (Bohemian Massif, Moldanubicum). *Jahrbuch der Geologischen Bundesanstalt*, **134**, 117–133.

PATOČKA, F., KACHLÍK, V., DOSTAL, J. & FRÁNA, J. 2003. Granitoid gneisses with relict orbicular metagranitoids from the Varied Group of southern Bohemian Massif Moldanubicum: protolith derived from melting of Archaean crust? *Journal of the Czech Geological Society*, **48**, 100–101.

PAULEY, J. C. 1990. Sedimentology, structural evolution and tectonic setting of the late Precambrian Longmyndian Supergroup of the Welsh Borderland, UK. *In:* D'LEMOS, R. S., STRACHAN, R. A. & TOPLEY, C. G. (eds) *The Cadomian Orogeny.* Geological Society, London, Special Publications, **51**, 341–351.

PERSSON, A. I. 1999. Absolute (U-Pb) and relative age determinations of intrusive rocks in the Ragunda rapakivi complex, central Sweden. *Precambrian Research*, **95**, 109–127.

PERSSON, P.-O. & WIKSTRÖM, A. 1993. A U-Pb dating of the Askersund granite and its marginal augen gneiss. *Geologiska Föreningens i Stockholm Förhandlingar*, **115**, 321–329.

PETECKI, Z. 2002. Modelowanie grawimetryczne i magnetyczne wzdluž profilu LT-7. *Przegląd Geologiczny*, **50**, 630–633.

PEUCAT, J. J. 1986. Behaviour of Rb-Sr whole-rock and U-Pb zircon systems during partial melting as shown in migmatitic gneisses from the St Malo Massif, NE Brittany, France. *Journal Geological Society, London*, **143**, 875–886.

PEUCAT, J. J., HIRBEC, Y., AUVRAY, B., COGNÉ, J. & CORNICHET, J. 1981. Late Proterozoic zircon age from a basic-ultrabasic complex: a possible cadomian orogenic complex in the Hercynian Belt of Western Europe. *Geology*, **9**, 169–173.

PFIFFNER, O. A., LEHNER, P., HEITZMANN, P., MÜLLER, ST. & STECK, A. (eds) 1997. *Deep Structure of the Swiss Alps: Results of NRP 20.* Birkhäuser Verlag, Basel.

PFLUG, H. D. & REITZ, E. 1987. Palynology in metamorphic rocks: indication of early land plants. *Naturwissenschaften*, **74**, 386–387.

PHARAOH, T.C. 1999. Paleozoic terranes and their lithospheric boundaries within the Trans-European Suture Zone (TESZ): a review. *Tectonophysics*, **314**, 17–41.

PHARAOH, T. C., WEBB, P. C., THORPE, R. S. & BECKINSALE, R. D. 1987. Geochemical evidence for the tectonic setting of late Proterozoic volcanic suites in central England, *In:* PHARAOH, T. C., BECKINSALE, R. D. & RICKARD, D. (eds) *Geochemistry and Mineralization of Proterozoic Volcanic Suites.* Geological Society, London, Special Publications, **33**, 541–552.

PHARAOH, T. C. & CARNEY, J. 2000. Introduction. *In:* CARNEY, J., HORAK, J. M., PHARAOH, T. C., ET AL. (eds) *Precambrian rocks of England and Wales.* Geological Conservation Review, Joint Nature Conservation Committee, Chapman & Hall, 1–18.

PHILLIPS, E. 1991. The lithostratigraphy, sedimentology and tectonic setting of the Monian Supergroup, western Anglesey, North Wales. *Journal of the Geological Society, London*, **148**, 1079–1090.

PIETZSCH, K. 1962. *Geologie von Sachsen.* VEB Deutscher Verlag der Wissenschaften, Berlin.

PIN, C. & CARME, C. 1987. A Sm-Nd isotope study of 500 Ma old oceanic crust in the Variscan belt of Western europe: the Chamrousse ophiolite complex, Western Alps (France). *Contributions to Mineralogy and Petrology*, **96**, 406–413.

PIN, C. & LANCELOT, J. R. 1978. Un exemple de magmatisme Cambrien dans le Massif Central: les métadiorites quartziques intrusives dans la série du Lot. *Bulletin de la Société géoogique de France*, **7**, 203–208.

PIN, C. & SILLS, J. D. 1986. Petrogenesis of layered gabbros and ultramafic rocks from Val Sesia, the Ivrea Zone, NW Italy: trace element and isotope geochemistry. *In:* DAWSON, J. B., HALL, J. & WEDEPOHL, K. H. (eds) *The Nature of the Lower Continental Crust.* Geological Society, London, Special Publications, **24**, 231–249.

PISAREVSKY, S. A., WINGATE, M. T. D., POWELL, C. MCA., JOHNSON, S. & EVANS, D. A. D. 2003. Models of Rodinia assembly and fragmentation. *In:* YOSHIDA, M., WINDLEY, B. & DASGUPTA, S. (eds) *Proterozoic East Gondwana: Supercontinent Assembly and Breakup.* Geological Society, London, Special Publications, **206**, 35–55.

PITRA, P., BURG, J. P. & GUIRAUD, M. 1999. Late Variscan strike-slip tectonics between the Tepla-Barrandian and Moldanubian terranes (Czech Bohemian Massif): petrostructural evidence. *Journal of the Geological Society, London*, **156**, 1003–1020.

POPOV, V., IOSIFIDI, A., KHRAMOV, A., TAIT, J. & BACHTADSE, V. 2002. Paleomagnetism of Upper Vendian sediments from the Winter Coast, White Sea region, Russia: implications for the paleogeography of Baltica during Neoproterozoic times. *Journal of Geophysical Research*, **107**(B11), 2315. DOI: 10.1029/2001JB001607.

POPRAWA, P. & PACZEŚNA, J. 2000. Late Neoproterozoic to Early Paleozoic tectonic evolution of the Lublin-Podlasie slope of Eatr European Craton –subsidence analysis and facies development. *Joint Meeting of EUROPROBE (TESZ) and PACE Project.* Abstract volume.

POPRAWA, P. & PACZEŚNA, J. 2002. Rozwój ryftu w późnym neoproterozoiku – wczesnym paleozoiku na lubelsko–podlaskim skłonie kratonu wschodnioeuropejskiego –analiza subsydencji i zapisu facjalnego. *Przegląd Geologiczny*, **50**, 49–63.

POŻARYSKI, W. 1963. Jednostki geologiczne Polski. *Przegląd Geologiczny*, **1**, 4–10.

POŻARYSKI, W. 1990. The Middle Europe Caledonides – wrenching orogen composed of terranes (in Polish with English summary). *Przegląd Geologiczny*, **38**, 1–9.

POŻARYSKI, W. & KOTAŃSKI, Z. 1979. Rozwój tektoniczny bajkalski oraz kaledońsko-waryscyjski przedpola wschodnioeuropejskiej platformy w Polsce. *Kwartalnik Geologiczny*, **23**, 7–19.

POŻARYSKI, W. & TOMCZYK, H. 1993. Geological cross-section through SE Poland. *Przegląd Geologiczny*, **41**, 683–694 (in Polish with English summary).

PUCHKOV, V. N. 2003. *Magmatism on Passive Margins of the Rodinia. Materials of the International conference.* Geological Institute of the Ufimian Scientific Centre, Russian Academy of Science, Ufa, 1–40.

PUURA, V., LINDSTROOM, M., FLODEN, T., PIPPING, F. & MOTUZA, G. 1994. Structure and stratigraphy of meteorite craters in Fennoscandia and the Baltic region: a first outlook. *Proceedings of the Estonian Academy of Sciences*, **43**, 2.

RAJLICH, P., SCHULMANN, K. & SYNEK, J. 1988. Strain analysis on conglomerates from the Central Bohemian shear zone. *Krystalinikum*, **19**, 119–134.

RÄMÖ, O. T. & HAAPALA, I. 1996. Rapakivi granite magmatism: a global review with emphasis on petrogenesis. *In:* DEMAIFFE, D. (ed.) *Petrology and Geochemistry of Magmatic Suites of Rocks in Continental and Oceanic Crust.* Universite Libre de Bruxelles, Royal Museum of Central Africa (Tervuren), 177–200.

RÄMÖ, O. T., HUHMA, H. & KIRS, J. 1996. Radiogenic isotopes of the Estonian and Latvian rapakivi granite suite: new data from the concealed Precambrian of the East European Craton. *Precambrian Research*, **79**, 209–226.

RAMOS, V. A. & ALEMÁN, A. 2000. Tectonic evolution of the Andes. *In:* CORDANI, U. G., MILANI, E. J., THOMAZ, F., A. & CAMPOS, D. A. (eds) *Tectonic Evolution of South America, 31st International Geological Congress*, Rio de Janeiro, Brasil, 635–685.

RAMSAY, W. 1909. *Geologins grunder.* Hälsingfors, Finland.

REICH, S. 1997. Die metamorphe Entwicklung der Erzgebirgsnordrandzone. *Freiberger Forschungshefte*, **C470**, 65–85.

REISCHMANN, T., ANTHES, G., JAECKEL, P. & ALTENBERGER, U. 2001. Age and origin of the Böllsteiner Odenwald. *Mineralogy and Petrology*, **72**, 29–44.

REITZ, E. 1987. Palynologie in metamorphen Serien: I. Silurische Sporen aus einem granatführenden Glimmerschiefer des Vor-Spessart. *Neues Jahrbuch für Geologie und Paläontologie, Monatshefte*, **1987**, 699–704.

REITZ, E. 1992. Silurische Mikrosporen aus einem Biotit-Glimmerschiefer bei Rittsteig, nördlicher Bayerischer Wald. *Neues Jahrbuch für*

Geologie und Paläontologie, Monatshefte, **6**, 351–358.

REITZ, E. 1994. Unterordovizische Acritarchen aus der Magdalensberg-Gruppe in Ostkärnten und aus der Salzburger Grauwackenzone (Österreich). *Jahrbuch der Geologischen Bundesanstalt*, **137**, 661–668.

REITZ, E. & HÖLL, R. 1988. Jungproterozoische Mikrofossilien aus der Habachformation in den mittleren Hohen Tauern und dem nordostbayerischen Grundgebirge. *Jahrbuch der Geologischen Bundesanstalt*, **132**, 329–340.

REITZ, E. & HÖLL, R. 1989. Unterordovizische Acritarchen aus der Nördlichen Grauwackenzone (Ostalpen). *Jahrbuch der Geologischen Bundesanstalt*, **132**, 761–774.

REITZ, E., DANECK, T. & MILLER, H. 1989. Ein Nachweis jungproterozoischen Alters in Schwarzphylliten am Tauern-Nordrand (Salzburg, Österreoch) und seine Bedeutung für den Bau der Hohen Tauern. *Jahrbuch Geologischer Bundesanstalt*, **132**, 751–760.

Report of the Working Group for Regional Geological Classification of the Bohemian Massif at the former Czechoslovak Stratigraphic Commission. 1994. Regional geological subdivision of the Bohemian massif on the territory of the Czech Republic. *Journal of the Czech Geological Society*, **39**, 127–144.

RIMSA, A., BOGDANOVA, S., SKRIDLAITE, G. & BIBIKOVA, E. 2001. The Randamonys TTG-intrusion in Southern Lithuania: Evidence of an 1.84 Ga island arc. *EUG XI abstract volume*. Strasbourg, p. 358.

ROBARDET, M. 2003. The Armorica 'microplate': fact or fiction? Critical review of the concept and contradictory palaeobiogeographical data. *Palaeogeography, Palaeoclimatology, Palaeoecology*, **195**, 125–148.

ROBLOT, M.M. 1968. *Les roches silico-carbonées du Précambrien armoricain*. Dr Sc. Thesis, University of Paris.

ROGER, F., RESPAUT, J. P., BRUNEL, M., MATTE, P. & PAQUETTE, J. L. 2004. Première datation U-Pb des orthogneiss oeillés de la zone axiale de la Montagne Noire (sud du Massif Central) : nouveaux témoins du magmatisme ordovicien dans la chaîne Varisque. *Comptes Rendus de l'Académie des Sciences Paris*, **33**, 19–28.

ROGERS, J.J.W. & SANTOSH, M. 2002. Configuration of Columbia, a Mesoproterozoic supercontinent. *Gondwana Research*, **5**, 5–22.

RÖHLICH, P. 1964. Podmořské skluzy a bahnotoky v nejmladším středočeském algonkiu. *Sbornik geologických Věd Řada Geologie*, **6**, 89–121.

RÖHLICH, P. 1965. Geologische Probleme des mittelböhmischen Algonkiums. *Geologie*, **14**, 373–403.

RÖHLICH, P. 2000. Some stratigraphic problems of the Barrandian Neoproterozoic. *Vestnik Ceského Geologického Ústavu*, **75**, 201–204.

ROMER, R. L. 1996. Contiguous Laurentia and Baltica before the Grenvillian-Sveconorwegian orogeny? *Terra Nova*, **8**, 173–181.

ROMER, R. L. & FRANZ, L. 1998. Ordovician Barrow-type metamorphism in the Strona-Ceneri Zone (Northern Italy) dated by U-Pb on staurolite. *Swiss Bulletin of Mineralogy and Petrology*, **78**, 383–395.

ROMER, R. L. & NISCA, D. H. 1995. Svecofennian crustal deformation of the Baltic Shield and U-Pb age of late-kinematic tonalitic intrusions in the Burträsk Shear Zone, northern Sweden. *Precambrian Research*, **75**, 17–29.

ROMER, R. L. & RÖTZLER, A. J. 2001. P-T-t evolution of ultrahigh-temperature granulites from the Saxon Granulite Massif, Germany. Part II: Geochronology. *Journal of Petrology*, **42**, 2015–2032.

ROMER, R. L. & RÖTZLER, J. 2003. Effect of metamorphic reaction history on the U-Pb dating of titanite. *In*: VANCE, D., MÜLLER, A. & VILLA I. (eds) *Geochronology: Linking the Isotopic Record with Petrology and Textures*. Geological Society, London, Special Publications, **220**, 147–158.

ROMER, R. L. & SMEDS, S.-A. 1996. U-Pb columbite ages of pegmatites from Sveconorwegian terranes in southwestern Sweden. *Precambrian Research*, **76**, 15–30.

ROMER, R. L. & WRIGHT, J. E. 1993. Lead mobilization during tectonic reactivation of the western Baltic Shield. *Geochimica et Cosmochimica Acta*, **57**, 2555–2570.

ROMER, R. L., KJØSNES, B., KORNELIUSSEN, A., LINDAHL, I., SKYSETH, T., STENDAL, H. & SUNDVOLL, B. 1992. *The Archaean–Proterozoic boundary below the Caledonides of northern Norway and Sweden: U-Pb, Rb-Sr and εNd isotope data from the Rombak-Tysfjord area*. Norges Geologiske Undersøkelse Rapport, **91.255**, 1–67.

ROMER, R. L., BAX, G. & KATHOL, B. 1994. Basement control of the Caledonian orogen along the Torneträsk section, northern Sweden. *Schweizerische Mineralogische und Petrographische Mitteilungen*, **74**, 469–481.

ROQUES, M., PIN, C. & DUTHOU, J. L. 1995. Substratum antéhercynien et collision hercynienne dans le Massif Central français. *Comptes Rendus de l'Académie des Sciences Paris*, **320**, 817 821.

ROSSI, P., COCHERIE, A. & DURAND-DELGA, M. 1995. Arguments géochronologiques en faveur de la présence d'un socle panafricain (cadomien) en Corse; conséquences sur la paléogéographie de l'orogène varisque sud-européen. *Comptes Rendus de l'Académie des Sciences Paris*, **321**, 983–992.

RÖTZLER, K., SCHUMACHER, R., MARESCH, W. V. & WILLNER, A. P. 1998. Contrasts in the metamorphic evolution and exhumation history of different metapelite units in the Western Erzgebirge (Saxony, Germany). *European Journal of Mineralogy*, **10**, 261–280.

RUDBERG, S. 1970. The sub-Cambrian peneplain in Sweden and its slope gradient. *Zeitschrift für Geomorphologie, (Supplement)*, **9**, 156–167.

RUFFET, G., FÉRAUD, G. & AMOURIC, M. 1991. Comparison of ^{40}Ar-^{39}Ar conventional and laser dating of biotites from the North Trégor Batholith. *Geochimica et Cosmochimica Acta*, **55**, 1675–1688.

RUNNEGAR, B. 2000. Loophole for snowball Earth. *Nature*, **405**, 403–404.

RYDER, G. 2003. Bombardement of the Hadean Earth: wholesome or deleterious? *Astrobiology*, **3**, 3–6.

RYKA, W. 1961. O problemach podłoża prekambryjskiego północnowschodniej Polski w świetle badań petrograficznych skał metamorficznych z Sokółki, Kruszynian i Mielnika. *Kwartalnik Geologiczny*, **5**, 497- 525.

RYKA, W. 1964. O budowie i stratygrafii krystaliniku północno-wschodniej Polski. *Kwartalnik Geologiczny*, **8**, 42–57.

RYKA, W. 1973. Stratygrafia. *In*: ŁASZKIEWICZ, A. (ed.) *Skały platformy prekambryjskiej w Polsce. cz. 12. Podłoże krystaliczne*. Prace Instytutu Geologicznego, **68**, 138–148.

RYKA, W. 1984. Precambrian evolution of the East European Platform in Poland. *Biuletyn Instytutu Geologicznego*, **347**, 17–28.

RYKA, W. 1993. Fundament krystaliczny Suwalszczyzny. *Przegląd Gelogiczny*, **41**, 546 –551.

RYKA, W. 1998. Geologic position of the Suwałki Anorthosite Massif. *Prace Panstwowego Instytutu Geologicznego*, Warsawa, Poland, **169**, 19–26.

SADOWSKI, G. R. & BETTENCOURT, S. J. 1996. Mesoproterozoic tectonic cerrelations between eastern Laurentia and western border of the Amazon Craton. *Precambrian Research*, **76**, 213–227.

SAMSON, S. D. & D'LEMOS, R. S. 1998. U-Pb geochronology and Sm-Nd isotopic composition of Proterozoic gneisses, Channel Islands, U.K. *Journal of the Geological Society, London*, **155**, 609–618.

SAMSON, S. D., D'LEMOS, R. S., BLICHERT-TOFT, J. & VERVOORT, J. 2003. U-Pb geochronology and Hf-Nd isotope compositions of the oldest Neoproterozoic crust within the Cadomian orogen: new evidence for a unique juvenile terrane. *Earth and Planetary Science Letters*, **208**, 165–180.

SANKARAN, A.V. 2003. The supercontinent medley: Recent views. *Current Science*, **85**(8), 1121–1124.

SASSI, F. P., KALVACHEVA, R. & ZANFERRARI, A. 1984. New data on the age of deposition of the South Alpine phyllitic basement in the Eastern Alps. *Neues Jahrbuch für Mineralogie, Abhandlungen*, Monatsheft **1984**(12), 741–751.

SAUNIER, J. F. & CHAUVEL, J. J. 1986. Les amphibolites du domaine cristallophyllien de Plouguenast. N magmatisme tholéiitique dans le Protérozoïque de Bretagne Centrale Massif Armoricain. *Comptes Rendus de l'Académie des Sciences Paris*, **302**, 671–674.

SCHALTEGGER, U. 1992. Die polymetamorphe Geschichte des Aarmassivs: neue U/Pb-Resultate. *Swiss Bulletin of Mineralogy and Petrology*, **72**, 148–149.

SCHALTEGGER, U. 1993. The evolution of the polymetamorphic basement in the central Alps unravelled by precise U-Pb zircon dating. *Contributions to Mineralogy and Petrology*, **113**, 466–478.

SCHALTEGGER, U. & CORFU, F. 1995. Late Variscan "basin and range" magmatism and tectonics in the Central Alps: Evidence from U-Pb geochronology. *Geodinamica Acta*, **8**, 82–98.

SCHALTEGGER, U. & GEBAUER, D. 1999. Pre-Alpine geochronology of the Central, Western and Southern Alps. *Swiss Bulletin of Mineralogy and Petrology*, **79**, 79–87.

SCHALTEGGER, U., NÄGLER, TH. F., CORFU, F., MAGGETTI, M., GALETTI, G. & STOSCH, H. G. 1997. A Cambrian island arc in the Silvretta nappe: constraints from geochemistry and geochronology study. *Swiss Bulletin of Mineralogy and Petrology*, **77**, 337–350.

SCHÄRER, U., WILMART, E. & DUCHESNE, J.-C. 1996. The short duration and anorogenic character of anorthosite magmatism; U-Pb dating of the Rogaland Complex, Norway. *Earth and Planetary Science Letters*, **139**, 335–350.

SCHLÜTER, J. 1983. *Petrographische und Geochemische Untersuchungen am Eckergneis, Harz*. Unpublished doctoral thesis, University of Hamburg.

SCHMÄDICKE, E., OKRUSCH, M. & SCHMIDT, W. 1992. Eclogite facies rocks in the Saxonian Erzgebirge, Germany: high pressure metamorphism under contrasting P-T conditions. *Contributions to Mineralogy and Petrology*, **110**, 226–241.

SCHMID, S. M., FÜGENSCHUH, B., KISSLING, E. & SCHUSTER, R. 2004. Tectonic map and overall architecture of the alpine orogen. *Eclogae Geologiae Helvetia*, **97**, 93–117.

SCHÖNLAUB, H. P. 1993. Stratigraphy, biogeography and climatic relationships of the Alpine Palaeozoic. *In*: VON RAUMER, J. & NEUBAUER, F. (eds) *The Pre-Mesozoic Geology in the Alps*. Springer Verlag, Berlin, 65–92.

SCHOPF, J.W. 1993. Microfossils of the early Archaean Apex chert: new evidence of the antiquity of life. *Science*, **260**, 640–646.

SCHOPF, J.W. 1999. *Cradle of Life: The Discovery of Earth's Earliest Fossils*. Princeton University Press, Princeton, New Jersey.

SCHÜLLER, A. 1957. Zur Grundgebirgsgeologie des Kyffhäusers. *Geologica*, **11**, 107–114.

SCHULMANN, K., KRÖNER, A., HEGNER, E., WENDT, I., KONOPÁSEK, J., LEXA, O. & ŠTIPSKÁ, P. 2005. Chronological constraints on the pre-orogenic history, burial and exhumation of deep-seated rocks along the eastern margin of the Variscan Orogen, Bohemian Massif, Czech Republic. *American Journal of Science*, **305**, 407–448.

SCHULZ, B. 1995. Geochemistry and REE magmatic fractionation patterns in the Prijakt amphibolitized eclogites of the Schobergruppe, Austroalpine basement (Eastern Alps). *Swiss Bulletin of Mineralogy and Petrology*, **75**, 225–239.

SCHULZ, B. & BOMBACH, K. 2003. Single zircon Pb/Pb geochronology of the Early-Palaeozoic magmatic evolution in the Austroalpine basement to the south of the Tauern Window. *Jahrbuch der Geologischen Bundesanstalt*, **143**, 303–321.

SCHULZ, B., BOMBACH, K., PAWLIG, S. & BRÄTZ, H. 2004. Neoproterozoic to Early-Palaeozoic magmatic evolution in the Gondwana-derived Austroalpine basement to the south of the Tauern Window (Eastern Alps). *International Journal of Earth Sciences*, **93**, 824–843.

SCOTESE, C. R., BOUCOT, A. J. & MCKERROW, W. S. 1999. Gondwanan palaeogeography and palaeoclimatology. *Journal of African Earth Sciences*, **28**, 99–114.

SEBASTIAN, U. 1995. Die Strukturentwicklung des spätorogenen Erzgebirgsaufstiegs in der Flöhazone – Ein weiterer Beitrag zur postkollisionalen Extension am Nordrand der Böhmischen Masse. *Freiberger Forschungshefte*, **C461**, 114.

SEILACHER, A., POSE, P.K. & PFLUGER, F. 1998. Triploblastic animals more than 1 Billion years ago: trace fossil evidence from India. *Science*, **282**, 80–84.

SEMENENKO, S. P. 1965. *Geochronologija dokembrija Ukrainy*, Naukowa Dumka, Kijew (in Russian).

SHUMLYANSKYY, L.V. & SHUMLYANSKYY, V.O. 2004. About genesis and tectonic position of the Volynian Vend continental flood basalts. *Proceedings of the Institute of Fundamental Studies*, **7**, 38–44.

SIEGESMUND, S., HEINRICHS, T., ROMER, R. L. & DOMAN, D. 2007. Age constraints on the evolution of the Austroalpine basement to the south of the Tauern Window. *International Journal of Earth Sciences (Geologische Rundschau)*, **96**, 415–432.

SKIÖLD, T. 1988. Implications of new U-Pb zircon chronology to early Proterozoic crustal accretion in northern Sweden. *Precambrian Research*, **38**, 147–164.

SKIÖLD, T., ÖHLANDER, B., MARKKULA, H., WIDENFALK, L. & CLAESSON, L.-Å. 1993. Chronology of Proterozoic orogenic processes at the Archaean continental margin in northern Sweden. *Precambrian Research*, **64**, 225–238.

SKOČEK, V. & POUBA, Z. 2000. Neoproterozoic sedimentary carbonates and their silicified equivalents: Barrandian, Czech Republic. *Vestnik Ceského Geologického Ústavu*, **75**, 241–260.

SKRIDLAITE, G. & MOTUZA G. 2001. Precambrian domains in Lithuania: evidence of terrane tectonics. *Tectonophysics*, **339**, 113–133.

SKRIDLAITE, G., WISZNIEWSKA, J. & DUCHESNE, J.-C. 2003. Ferro-potassic A-type granites and related rocks in NE Poland and S Lithuania: west of the East European Craton. *Precambrian Research, Special Issue*, **124**, 305–326.

SKRIDLAITE, G., TARAN, L., BOGDANOVA, S., WHITEHOUS, M., RIMŠA, A. & SMILGIENĖ, Z. 2004. New evidence for the 1.8 Ga collisional/accretionary event in the west of the East European Craton. *32nd International Geological Congress*, Abstract volume, Pt. 1, abstract 241–23.

SLÁMA, J., KACHLÍK, V. & KOŠLER, J. 2003. Neoproterozoic metaconglomerates of the Sedlčany Krásná Hora Metamorphic Islet. *Geolines*, **16**, 96–97.

ŠMEJKAL, V. 1964. Absolutní stáří některých vyvřelých a metamorfovaných hornin Českého masivu stanovené kalium argonovou metodou. *Sbornik geologických Věd Řada Geologie*, **64**, 121–136.

SÖDERLUND, U., ISACHSEN, C.E., BYLUND, G., HEAMAN, L.M., PATCHETT, P.J., VERVOORT, J.D. & ANDERSSON, U.B. 2005. U-Pb baddeleyite ages and isotopic chemistry constraining repeated magmatism in the Fennoscandian Shield from 1.6 to 0.9 Ga. *Contributions to Mineralogy and Petrology*, **150**, 174–194.

SOLIVA, J., SALEL, J. F. & BRUNEL, M. 1989. Shear deformation and emplacement of the gneissic Canigou thrust nappe (eastern Pyrénées). *Geologie en Mijnbouw*, **68**, 357–366.

SÖLLNER, F., KOHLER, H. & MÜLLER-SOHNIUS, D. 1981*a*. Rb/Sr Alterbestimmungen am Gesteinen der Münchberger Gneissmasse, NE Bayern, Teil 1 Gesamtsteindatierungen. *Neues Jahrbuch für Mineralogie, Abhandlungen*, **141**, 90–112.

SÖLLNER, F., KÖHLER, H. & MÜLLER-SOHNIUS, D. 1981*b*. Rb/Sr-Altersbestimmungen an Gesteinen der Münchberger Gneismasse (MM), NE-Bayern – Teil 2, Mineraldatierungen. *Neues Jahrbuch für Mineralogie, Abhandlungen*, **142**, 178–198.

ŚRODA, P. & POLONAISE P3 WORKING GROUP 1999. P- and S-wave velocity model of the southwestern margin of the Precambrian East European Craton; POLONAISE'97, PROFILE P3. *Tectonophysics*, **314**, 175–192.

STAMPFLI, G. 1996. The intra-Alpine terrain: a palaeotethyan remnant in the Alpine Variscides. *Eclogae Geologicae Helveticae*, **89**, 12–42.

STAMPFLI, G. M. & BOREL, G. D. 2002. A plate tectonic model for the Paleozoic and Mesozoic. *Earth and Planetary Science Letters*, **196**,17–33.

STARMER, I.C. 1993. The Sveconorwegian Orogeny in southern Norway, relative to deep crustal structures and events in the North Atlantic Proterozoic supercontinent. *Norsk Geologisk Tidsskrift*, **73**, 109–132.

STEIN, H. J., MORGAN, J. W., MARKEY, R . J. & WISZNIEWSKA, J. 1998. A Re-Os study of the Suwałki anorthosite massif, northeast Poland. *Geophysical Journal*, **20**, 111–114.

STEIN, H.J., MORGAN, J.W., HANNAH, I.L., MARKEY, R.J. & WISZNIEWSKA, J. 1999. Crustal origin for oxide-sulphide ores and anorthosite: Re-Os evidence from Suwałki massif, northeast Poland. *Journal of Conference. Abstracts*, **4**(1), 688. EUG10, Strasbourg.

ŠTELCL, J. & WEISS, J. 1986. *Brněnský masiv*. Univerzita JEPurkyně. Brno.

STERN, R.A. & BLEEKER, W. 1998. Age of the world's oldest rocks refined using Canada's SHRIMP: The Acasta gneiss complex, Northwest Territories, Canada. *Geoscience Canada*, **25**, 27–31.

STEWARD, G.R. 2000. Outstanding questions for the Giant Impact Hypothesis. *In:* CANUP, R.M. & RIGHTER, K. (eds) *Origin of the Earth and the Moon*. University of Arizona, Tucson, Arizona, 217–233.

STILLE, P. & SCHALTEGGER, U. 1996. The lithospheric mantle beneath Central Europe: Nd isotopic constraints for Late Proterozoic enrichment and implications for early crustal evolution. *In*: BASU, A. & HART, S. R. (eds) *Earth Processes: Reading the Isotopic Code*. Geophysical Monograph Series, **95**, American Geophysical Union, Washington, 269–276.

STILLE, P. & STEIGER, H. R. 1991. Hf isotope systematics in granitoids from the Central and Southern Alps. *Contributions to Mineralogy and Petrology*, **107**, 273–278.

STILLE, P. & TATSUMOTO, M. 1985. Precambrian tholeiitic-dacitic rock-suites and Cambrian ultramafic rocks in the Penninic nappe systems of the Alps: evidence from Sm-Nd isototpes and rare earth elements. *Contributions to Mineralogy and Petrology*, **89**, 184–192.

STRACHAN, R. A. 2000. Late Neoproterozoic to Cambrian accretionary history of Eastern Avalonia and Armorica on the active margin of Gondwana. *In:* WOODCOCK, N. & STRACHAN, R. A. (eds) *Geological History of Britain and Ireland*. Blackwell, Oxford, 127–140.

STRACHAN, R. A., BROWN, M., D'LEMOS, R. S., ROACH, R. A. & TRELOAR, P. J. 1992. Discussion on the timing and kinematics of Cadomian deformation (Northern Brittany). *Journal of the Geological Society, London*, **149**, 151–155.

STRACHAN, R. A., NANCE, R. D., DALLMEYER, R. D., D'LEMOS, R. S., MURPHY, J. B. & WATTS, G. R. 1996*a*. Late Precambrian tectonothermal evolution of the Malverns Complex. *Journal of the Geological Society, London*, **153**, 589–600.

STRACHAN, R. A., D'LEMOS, R. S. & DALLMEYER, R. D. 1996*b*. Neoproterozoic evolution of an active plate margin: North Armorican Massif, France. *In*: NANCE, R. D. & THOMPSON, D. (eds) *Avalonian and Related Peri-Gondwanian Terranes of the Circum-North Atlantic*. Geological Society of America, Special Paper, **304**, 319–332.

SUESS, F. E. 1901. Zur Tektonik der Gneissgebiete am Ostrande der böhmischen Masse. *Verhandlungen der Kaiserlich-Königlichen Geologischen Reichsanstalt*, **1901**, 399–402.

SULTAN L., CLAESSON S. & PLINK-BJÖRKLUND, P. 2005. Proterozoic and Archaean ages of detrital zircon from the Palaeoproterozoic Västervik Basin, SE Sweden: Implications for provenance and timing of deposition. *GFF*, **127**, 17–24.

SUNDBLAD, K., MANSFELD, J., MOTUZA, G., AHL, M. & CLAESSON, S. 1994. Geology, geochemistry, and age of a Cu-Mo bearing granite at Kabeliai, southern Lithuania. *Mineralogy and Petrology*, **50**, 43–57.

SUOMINEN, V. 1991. The chronostatigraphy of southwestern Finland with special reference to Postjotnian and Subjotnian diabases. *Bulletin of the Geological Survey of Finland*, **356**, 1–100.

SVOBODA, J. 1933. *Metamorfovaný ostrov sedlčansko-krásnohorský*. Archiv pro prírodovědecký výzkum Čech, **18**(4), 1–62.

SYMONS, D. T. A. & CHAISSON, A. D. 1991. Paleomagnetism of the Callander Complex and the Cambrian apparent polar wander path for North America. *Canadian Journal of Earth Siences*, **28**, 355–363.

SYNEK, J. & OLIVERIOVÁ, D. 1993. Terrane character of the north-east margin of the Moldanubian Zone: the Kutná Hora Crystalline Complex, Bohemian Massif. *Geologische Rundschau*, **82**, 566–582.

SZCZEPANOWSKI, W. J. 1977. Badania geochemiczne bazaltów serii sławatyckiej. *Biuletyn Instytutu Geologicznego*, **278**, 35–69.

SZCZEPAŃSKI, J. 1999. Geochemistry of the orthogneisses from the Strzelin crystalline massif (SW Poland, Fore-Sudetic Block). *Geolines*, **8**, 68–69.

SZCZEPAŃSKI, J. & OBERC-DZIEDZIC, T. 1998. Geochemistry of amphibolites from the Strzelin crystalline massif, Fore-Sudetic Block, SW Poland. *Neues Jahrbuch für Mineralogie, Abhandlungen*, **173**, 23–40.

TAIT, J. A., BACHTADSE, V., FRANKE, W. & SOFFEL, H. C. 1997. Geodynamic evolution of the European Variscan fold belt: palaeomagnetic and geological constraints. *Geologische Rundschau*, **86**, 585–598.

TAJČMANOVÁ, L., KONOPÁSEK, J. & SCHULMANN, K. 2006. Thermal evolution of the orogenic lower crust during exhumation within a thickened Moldanubian root of the Variscan belt of Central Europe. *Journal of Metamorphic Geology*, **24**, 119–134.

TALBOT, C. J. & SLUNGA, R. S. 1989. Patterns of active shear in Fennoscandia. *In*: GREGERSEN, S. & BASHAM, P. W. (eds) *Earthquakes at North-Atlantic Passive Margins; Neotectonics and Postglacial Rebound*. NATO ASI Series, Series C: Mathematical and Physical Sciences, **266**, 441–466.

TANCZYK, E. I., LAPOINTE, P., MORRIS, W. A. & SCHMIDT P. W. 1987. A paleomagnetic study of the layered mafic intrusion at Sept-Iles, Quebec. *Canadian Journal of Earth Sciences*, **24**, 1431–1438.

TARAN, L., 2001. Distribution of the Okolovo unit in thew crystaline basement of Belarus. *Lithosphere*, **1**(14), 72–81.

TARAN, L. N. & BOGDANOVA, S. V. 2001. The Fennoscandia – Sarmatia junction in Belarus new inferences from a PT –study. *Tectonophysics*, **339**, 193 –214.

TEIPEL, U., EICHHORN, R., LOTH, G., ROHMÜLLER, J., HÖLL, R. & KENNEDY, A. 2004. U-Pb SHRIMP and Nd isotopic data from the western Bohemian Massif (Bayerischer Wald, Germany): Implications for Upper Vendian and Lower Ordovician magmatism. *International Journal of Earth Sciences*, **93**, 782–801.

THEOKRITOFF, G. 1979. Early Cambrian provincialism and biogeographic boundaries in the North Atlantic Region. *Lethaia*, **12**, 281–295.

THIELE, O. 1976. Zur Tektonik des Waldviertels in Niederösterreich (südliche Böhmische Masse). *Nova Acta Leopoldina*, **45**, 67–82.

THÖNI, M. 1999. A review of geochronological data from the Eastern Alps. *Swiss Bulletin of Mineralogy and Petrology*, **79**, 209–230.

THORPE, R. S., BECKINSALE, R. D., PATCHETT, P. J., PIPER, J. A., DAVIES, G. R. & EVANS, J. A. 1984. Crustal growth and late Precambrian-early Paleozoic plate tectonic evolution of England and Wales. *Journal of the Geological Society, London*, **141**, 521–536.

TICHOMIROWA, M. 2003. Die Gneise des Erzgebirges – hochmetamorphe Äquivalente von neoproterozoisch-frühpaläozoischen Grauwacken und Granitoiden der Cadomiden. *Freiberger Forschungshefte*, **C495**, 1–222.

TICHOMIROWA, M., BERGER, H.-J., KOCH, E. A., BELYATSKI, B. V., GÖTZE, J., KEMPE, U., NASDALA, L. & SCHALTEGGER, U. 2001. Zircon ages of high-grade gneisses in the Eastern Erzgebirge (Central European Variscides) – constraints on origin of the rocks and Precambrian to Ordovician magmatic events in the Variscan foldbelt. *Lithos*, **56**, 303–332.

TIKHOMIROVA, M. 2002. Zircon inheritance in diatexite granodiorites and its consequence on geochronology—a case study in Lusatia and the Erzgebirge (Saxo-Thuringia, eastern Germany. *Chemical Geology*, **191**, 209–224.

TIMMERMANN, H., KRENN, E., DORR, W. & FINGER, F. 2003. U-Pb geochronology revealing different monazite generations in the polyphase paragneisses of the Tepla Barrandian Unit, Northwestern Bohemian Massif: Implications for the processes involving monazite formation. *Journal of the Czech Geological Society*, **48**, 125–126.

TODT, W. A., ALTENBERGER, U. & RAUMER, J. F. 1995. U-Pb data on zircons for the thermal peak of metamorphism in the Variscan Odenwald, Germany. *Geologische Rundschau*, **84**, 466–472.

TOLLMANN, A. 1982. Grossraumiger variszischer Deckenbau im Moldanubikum und neue Gedanken zum Variszikum Europas. *Geotektonische Forschungen*, **64**, 1–91.

TORSVIK, T.H. 2003. The Rodinia jigsaw puzzle. *Science*, **300**, 1379–1381.

TORSVIK, T.H. & REHNSTRÖM, E.F. 2001. Cambrian palaeomagnetic data from Baltica: implications for true polar wander and Cambrian palaeogeography. *Journal of the Geological Society, London*, **158**, 321–329.

TORSVIK, T. H., SMETHURST, M. A., *et al.* 1996. Continental break-up and collision in the Neoproterozoic and Palaeozoic-a tale of Baltica and Laurentia. *Earth Science Reviews*, **40**, 229–258.

TRENDALL, A.F. & BLOCKLEY, J.G. 2004. Precambrian Iron-Formation. *In*: ERIKSSON, P.G., ALTERMANN, W. NELSON, D.R., MUELLER, W.U. & CATUNEANU, O. (eds) *The Precambrian Earth: Tempos and Events*. Developments in Precambrian Geology Vol. 12, Elsevier, Amsterdam, 403–421.

TUCKER, R. D. & PHARAOH, T. C. 1991. U-Pb zircon ages for Late Precambrian igneous rocks in southern Britain. *Journal of the Geological Society, London*, **148**, 435–443.

TURNIAK, K., MAZUR, S., FANNING, M. & BRÖCKER, M. 2005. A record of multi-phase Proterozoic and Palaeozoic magmatism in the Central Sudetes, N Bohemian Massif, revealed by shrimp zircon dating. *Mineralogical Society of Poland, Special Papers*, **26**, 272–275.

UNRUG, R. 1996. *Geodynamic Map of Gondwana Supercontinent Assembly, 1:10,000,000*. Council for Geosciences, Bureau de Recherches Géologiques et Minières, Orléans.

URBAN, M. & SYNEK, J. 1995. Structure (Moldanubian region: Moldanubian zone, ch. VII.C.2). *In*: DALLMEYER, R. D., FRANKE, W. & WEBER, K. (eds) *Pre-Permian Geology of Central and Eastern Europe*. Springer-Verlag, Berlin, 430–443.

VAASJOKI, M., HUHMA, H., LAHTINEN, R. & VESTIN, J. 2003. Sources of Svecofennian granitoids in the light of ion probe U-Pb measurements on their zircons. *Precambrian Research*, **121**, 251–262.

VAVRDOVÁ, M. 2000. Microfossils in carbonaceous cherts from Barrandian Neoproterozoic (Blovice Formation, Czech Republic). *Věstnik Ústředniho ústavu geolgického*, **75**, 351–360.

VEJNAR, Z. 1982. Regionální metamorfóza psamiticko-pelitickych hornin domažlické oblasti (Regional metamorphism of psammopelitic rocks in the Domažlice area). *Sbornik geologických Věd Řada Geologie*, **37**, 9–70.

Vidal, P., Auvray, B., Cogné, J., Hameurt, J. & Jeannette, D. 1971. Données géochronologiques sur la série spilitique d'Erquy : problèmes nouveaux à propos du Briovérien de Bretagne septentrionale. *Comptes Rendus de l'Académie des Sciences Paris*, **D273**, 132–135.

Vítková, P. & Kachlík, V. 2001. Petrology and geochemistry of high-Mg mafic metavolcanic rocks from the Neoproterozoic sequence of Sedlčany-Krásná Hora metamorphic "Islet", Central Bohemian Pluton, Czech Republic. *Krystalinikum*, **27**, 7–25.

Vitrac-Michard, A. 1975. *Chronologie et développement d'une orogenèse. Exemple des Pyrénées Orientales*. PhD Thesis, University of Paris VII.

Vitrac-Michard, A. & Allègre, C. J. 1971. Datation ^{87}Rb-^{86}Sr des gneiss du Canigou et de l'Agly (Pyrénées orientales, France). *Comptes Rendus de l'Académie des Sciences Paris*, **273**, 2411–2413.

Vitrac-Michard, A. & Allègre, C. J. 1975. ^{238}U-^{206}Pb, ^{235}U-^{207}Pb systematics on Pyrenean basement. *Contributions to Mineralogy and Petrology*, **95**, 255–268.

von Drach, V., Veress, C., Hofmann, B., Koehler, H., Vejnar, Z. & Waldhauserrova, J. 1993. Die isotopische Entwicklung des Oberen Erdmantels im Bereich der NW Boehmischen Masse. *Berichte der Deutschen Mineralogischen Gesellschaft*, **1**(132).

von Gaertner, H.R. 1931. Über den Nachweis von Algonkium in Thüringen. *Sitzungsberichte der Preußischen geologischen Landesanstalt*, **6**, 39–40.

von Gaertner, H.R. 1932. Algonkium und Tremadoc des Schwarzburger Sattels. *Zeitschrift der deutschen geologischen Gesellschaft*, **84**, 742–745.

von Gaertner, H. R. 1944. Die Schichtgliederung der Phyllitgebiete in Thüringen und Nordbayern und ihre Einordnung in das stratigraphische Schema. *Jahrbuch der Reichsanstalt für Bodenforschung*, **62**, 54–80.

von Quadt, A. 1992. U-Pb zircon and Sm-Nd geochronology of mafic and ultramafic rocks from the central part of the Tauern Window (eastern Alps). *Contributions to Mineralogy and Petrology*, **110**, 57–67.

von Quadt, A., Günther, D., Frischknecht, R., Zimmermann, R. & Franz, G., 1997. The evolution of pre-Variscan eclogites of the Tauern Window (Eastern Alps): A Sm-Nd-, conventional and Laser ICP-MS zircon U-Pb study. *Swiss Bulletin of Mineralogy and Petrology*, **77**, 265–279.

von Raumer, J. F. 1998. The Palaeozoic evolution in the Alps: from Gondwana to Pangea. *Geologische Rundschau*, **87**, 407–435.

von Raumer, J. F. & Neubauer, F. (eds) 1993. *The Pre-Mesozoic Geology in the Alps*. Springer Verlag, Berlin.

von Raumer, J. F., Stampfli, G. M., Borel, G. & Bussy, F. 2002. Organization of pre-Variscan basement areas at the north-Gondwanan margin. *International Journal of Earth Sciences*, **91**, 35–52.

von Raumer, J. F., Stampfli, G. M. & Bussy, F. 2003. Gondwana-derived microcontinents – the constituents of the Variscan and Alpine collisional orogens. *Tectonophysics*, **365**, 7–22.

Voshage, H., Hunziker, J. C., Hofmann, A. W. & Zingg, A. 1987. A Nd and Sr isotopic study of the Ivrea Zone, Southern Alps, N-Italy. *Contributions to Mineralogy and Petrology*, **97**, 31–42.

Vrána, S. 1979. Polyphase shear folding and thrusting in the Moldanubicum of southern Bohemia. *Věstnik Ústředniho ústavu geolgického*, **51**, 75–86.

Vrána, S. & Bártek, J. 2005. Retrograde metamorphism in a regional shear zone and related chemical changes: The Kaplice Unit of muscovite-biotite gneisses in the Moldanubian Zone of Southern Bohemia, Czech Republic. *Journal of the Czech Geological Society*, **50**, 43–57.

Vrána, S. & Frýda, J. 2003. Ultrahigh-pressure grossular-rich garnetite from the Moldanubian Zone, Czech Republic. *European Journal of Mineralogy*, **15**, 43–54.

Wahlgren, C.-H., Cruden, A. R. & Stephens, M. B. 1994. Kinematics of a fan-like structure in the eastern part of the Sveconorwegian orogen, Baltic Shield, south central Sweden. *Precambrian Research*, **70**, 67–91.

Waldhausrová, J. 1984. Proterozoic volcanics and intrusive rocks of the Jílové Zone in Central Bohemia. *Krystalinikum*, **17**, 77–97.

Waldhausrová, J. 1997*a*. Proterozoic volcanics geochemistry and mineral chemistry: a contribution to the Barrandian Upper Proterozoic stratigraphy (Bohemian Massif, Czech Republic). *Krystalinikum*, **23**, 151–180.

Waldhausrová, J. 1997*b*. Geochemistry of volcanites (metavolcanites) in the western part of the TBU Precambrian and their original tectonic setting. *In*: Vrána, S. & Štědrá, V. (eds) *Gological Model of Western Bohemia Related to KTB Borehole in Germany*. Czech Geological Survey, Prague, 85–90.

Watznauer, A. 1966. Zwei bemerkenswerte radiometrische Altersdatierungen aus dem Gebiet des Scharzburger Sattels (Thüringen), *Geologie*, **15**, 419–420.

Weil, A. B., Van der Voo, R., Mac Niocaill, C. & Meert, J.G. 1998. The Proterozoic supercontinent Rodinia: paleomagnetically derived reconstruction for 1100 to 800 Ma. *Earth and Planetary Science Letters*, **154**, 13–24.

Welin, E. 1980. Tabulation of recalculated radiometric ages published 1960–1979 for rocks and minerals in Sweden. *Geologiska Föreningens i Stockholm Förhandlingar*, **101**, 309–320.

Wendt, I., Höhndorf, A., Kreuzer, H., Müller, P., Stettner, G. 1988. Gesamtgesteins- und Mineraldatierungen der Steinwaldgranite (NE_Bayern). *Geologisches Jahrbuch*, **E 42,** 167–194.

Wendt, I., Kröner, A., Fiala, J. & Todt, W. 1993. Evidence from zircon dating for existence of approximately 2.1 Ga old crystalline basement in southern Bohemia, Czech Republic. *Geologische Rundschau*, **82**, 42–50.

Werner, O. & Lippolt, H. J. 2000. White mica ^{40}Ar/^{39}Ar ages of Erzgebirge metamorphic rocks: simulating the chronological results by a model of Variscan crustal imbrication. *In*: Franke, W., Haak, V., Oncken, O. & Tanner, O. (eds) *Orogenic Processes: Quantification and Modelling in the Variscan Belt*. Geological Society, London, Special Publications, **179**, 323–337.

Wieser, T. 1974. Fundament krystaliczny w otworze wiertniczym Łodygowice IG-1 *Kwartalnik Geologiczny*, **18**, 934–935 (in Polish).

Wikström, A. & Andersson, U. B. 2004. Geological feartures of the Småland-Värmland belt along the Svecofennian margin, part I: from the Loftahammar to the Tiveden-Askersund areas. *In*: Högdahl, K., Andersson, U. B. & Eklund, O. (eds) *The Trasscandinavian Igneous Belt (TIB) in Sweden: A Review of its Character and Evolution*. Geological Survey of Finland, Special Papers, **37**, 22–39.

Wilde, S.A., Valley, J.W., Peck, W.H. & Graham, C.M. 2001. Evidence from detrital zircons for the existence of continental crust and oceans on the Earth 4.4 Ga ago. *Nature*, **409**, 175–178.

Willner, A. P., Rötzler, K. & Maresch, W. V. 1997. Pressure-temperature and fluid evolution of quartzo-feldspathic metamorphic rocks with a relic high-pressure, granulite-facies history from the Central Erzgebirge (Saxony, Germany). *Journal of Petrology*, **38**(3), 307–336.

Willner, A.P., Ermolaeva, T., *et al.* 2001. Contrasting provenance signals in Riphean and Vendian sandstones in the SW Urals (Russia): constraints for a change from passive to active continental margin conditions in the Neoproterozoic. *Precambrian Research*, **110**, 215–239.

Willner, A.P., Sindern, S., Metzger, R., Ermolaeva, T., Kramm, U., Puchkov, V. & Kronz, A. 2003. Typology and single grain U/Pb ages of detrital zircons from Proterozoic sandstones in the SW Urals (Russia): early time marks at the eastern margin of Baltica. *Precambrian Research*, **124**, 1–20.

Wimmenauer, W. 1984. Das prävariszische Kristallin im Schwarzwald. *Fortschritte der Mineralogie: Beihefte*, **62**(2), 69–86.

Winchester, J. A. 1988. Later Proterozoic environments and tectonic evolution in the northern Atlantic lands. *In*: Winchester, J. A. (ed.) *Later Proterozoic Stratigraphy of the Northern Atlantic Regions*. Blackie, Glasgow, 253–270.

Winchester, J. A, Pharaoh, T. C. & Verniers, J. 2002. Palaeozoic amalgamation of Central Europe: An introduction and synthesis of new results from recent geological and geophysical investigations. *In*: Winchester, J.A., Pharaoh, T.C., & Verniers, J. (eds) *Palaeozoic Amalgamation of Central Europe*. Geological Society, London, Special Publications, **201**, 1–18.

Winchester, J.A., Pharaoh, T.C., Verniers, J., Ioane, D. & Seghedi, A. 2006. Palaeozoic accretion of Gondwana-derived terranes to the East European Craton: recognition of detached terrane fragments dispersed after collision with promontories. *In*: Gee, D.G. & Stephenson, R.A. (eds) *European Lithosphere Dynamics*. Geological Society, London, Memoirs, **32**, 323–332.

WINGATE, M. T. D. & GIDDINGS, J. W. 2000. Age and paleomagnetism of the Mundine Well dyke system, Western Australia: implications for Australia-Laurentia connection at 755 Ma. *Precambrian Research*, **100**, 335–357.

WINTERHALTER, B., FLODÉN, T., IGNATIUS, H., AXBERG, S. & NIEMISTÖ, L. 1981. Geology of the Baltic Sea. *In*: VOIPIO, A. (ed.) *The Baltic Sea.* Elsevier Oceanography Series, **30**, 1–122.

WISZNIEWSKA, J. 2002. Wiek i geneza rud Fe-Ti-V i skał towarzyszących w suwalskim masywie anortozytowym (NE Polska). *Biuletyn Instytutu Geologicznego*, **401**, 1–96.

WISZNIEWSKA, J. CLAESSON, S., STEIN, H. & VANDER AUWERA, J. 2002*a*. The north-east Polish anorthosite massifs: petrological, geochemical and isotopic evidence for a crustal derivation. *Terra Nova*, **14**, 451–460.

WISZNIEWSKA, J., DUCHESSE, J.-C. & DOERR, W. 2002*b*. Titanomagnetite and Cu-Co-Ni sulfide mineralization in the Suwalki and Sejny AMCG complexes (NE Poland). *11th Quadrennial IAGOD Symposium and Geocongress,* 22–26 July 2002, Windhoek.

WISZNIEWSKA, J., KRZEMINSKA, E. SKRIDLAITE, G., MOTUZA, G., WILLIAMS, I. & WHITEHOUSE, M. 2005. Metasedimentary and metavolcanic rocks from Poland and Lithuania: implications for Precambrian crustal evolution. *In:* KOZŁOWSKI, A., ILNICKI, S. & PAŃCZYK, M. (eds) *Crystalline Rocks of the East European Craton.* Mineralogical Society of Poland, Special Papers, **26**, 104–108.

WISZNIEWSKA J., KRZEMINSKA, E. & DÖRR, W. 2007. Evidence of arc-related Svecofennian magmatic activity in the southwestern margin of the East European Craton in Poland. *Gondwana Research*, **12**, 268–278.

WYBRANIEC, S. 1999. *Transformations and visulization of potential field data.* Polish Geological Institute, Special Papers, **1**, 1–88.

WYNS, R., LARDEUX, H. *et al.* 1998. *Notice explicative*, *Carte géologique France 1:50 000*, feuille Chemillé (483). BRGM, Orléans, 1–72.

XIAO, S. & KNOLL, A.H. 2000. Phosphatized animal embryos from the Neoproterozoic Doushantuo Formation at Weng'an, Guizhou Province, South China. *Journal of Paleontology*, **74**, 767–788.

YOUNG, G.M. 2004. Earth's two great Precambrian glaciations: aftermath of the "Snowball Earth" hypothesis. *In:* ERIKSSON, P.G., ALTERMANN, W. NELSON, D.R., MUELLER, W.U. & CATUNEANU, O. (eds) *The Precambrian Earth: Tempos and Events*. Developments in Precambrian Geology, Vol. 12, Elsevier, Amsterdam, 440–448.

YOUNG, G.M., VON BRUNN, V., GOLD, D.J.C. & MINTER, W.E.L. 1998. Earth's oldest reported glaciation: Physical and chemical evidence from the Archaean Mozaan Group ($\sim$2.9 Ga) of South Africa. *Journal of Geology*, **106**, 523–538.

ZABA, J. 1999. The structural evolution of Lower Paleozoic succession in the Upper Silesia Block and Małopolska Block border zone. *Prace Instytutu Geologicznego*, **166**, 1–162 (in Polish with English summary).

ŽÁČEK, V. & CHÁB, J. 1993. Metamorphism in the Teplá Upland, Bohemian Massif, Czech Republic. *Bulletin of the Czech Geological Survey*, **68**, 33–37.

ZEH, A. & BRÄTZ, H. 2000. Radiometrische und morphologische Untersuchungen an Zirkonen aus Granitporphyren, Rhyolithen und Granitgeröllen des nordwestlichen Thüringer Waldes. *Zeitschrift der deutschen geologischen Gesellschaft*, **151**, 187–206.

ZEH, A. & BRÄTZ, H. 2002. Timing of Late Carboniferous/Permian granite and granite porphyry intrusions in the Ruhla Crystalline Complex (central Germany), new constraints from SHRIMP and $^{207}Pb/^{206}Pb$ single zircon dating. *Chemie der Erde*, **62**, 303–316.

ZEH, A. & WUNDERLICH, J. 2003. Mitteldeutsche Kristallinzone *In*: SEIDEL, G. (ed.) *Geologie von Thüringen*. E.Schweitzerbart'sche Verlagsbuchhandlung, Stuttgart, 24–51.

ZEH, A., BRÄTZ, H., MILLAR, I. L. & WILLIAMS, I. S. 2001. A combined zircon SHRIMP and Sm-Nd isotope study on high-grade paragneisses from the Mid-German Crystalline Rise: Evidence for northern Gondwanan and Grenvillian provenance. *Journal of the Geological Society*, **158**, 983–994.

ZEH, A., WILLIAMS, I. S., BRÄTZ, H. & MILLAR, I. L. 2003. Different age response of zircon and monazite during the tectonometamorphic evolution of a high grade paragneiss from the Ruhla Crystalline Complex, Central Germany. *Contributions to Mineralogy and Petrology*, **145**, 691–706.

ZEH, A., GERDES, A., WILL, T. M. & MILLAR, I. L. 2005. Provenance and magmatic-metamorphic evolution of a Variscan island-arc complex: constraints from U-Pb dating, petrology, and geospeedometry of the Kyffhäuser Crystalline Complex, central Germany *Journal of Petrology*, **46**, 1393–1420.

ZELAŹNIEWICZ, A. & ZABA, J. 2002. The Upper Silesia and Małopolska Blocks: new data on their geodynamics at the Precambrian–Palaeozoic boundary *Przegląd Geologiczny*, **50**, 1227–1228 (in Polish).

ZELAŹNIEWICZ, A., GRAD, M. & GUTERCH, A. 2001. Late Proterozoic through Early Paleozoic crustal additions at the Teisseyre-Tornquist margin of Baltica: geological and geophysical data from Poland. *ESF EUROPROBE Meeting "Neoproterozoic- Early Paleozoic Time Slice Symposium*, Ankara, Abstracts, 96–97.

ZELAŹNIEWICZ, A., BUŁA, Z. & JACHOWICZ, M. 2002. Neoproterozoic granites in the Upper Silesia massif of Bruno-Vistulicum, S Poland: U-Pb SHRIMP evidence. *Schriftenreich der Deutschen Geologischen Gesellschaft*, **21**, 361–362.

ZELAŹNIEWICZ, A., BIERNACKA, J., *et al.* 2004. *Palaeozoic Accretion of Poland – Task 1. The Geological Structure of the Upper Silesia Block and of the Southern Part of the Małopolska Block – Model of a pre-Variscan Tectonogenesis.* Archives of Polish Geological Institute. Sosnowiec-Warszawa (in Polish).

ZINGG, M. A. 1989. Die Siviez-Mischabel-Decke: Entstehung und Entwicklung eines Altkristallins. *Swiss Bulletin of Mineralogy and Petrology*, **69**, 130.

ZNOSKO, J. 1960. Uwagi o stratygrafii podłoża krystalicznego północno-wschodniej Polski. *Kwartalnik Geologiczny*, **4**, 281–287.

ZOUBEK, V., COGNÉ, J., KOZHOUKHAROV, D. & KRÄUTNER, H.G. 1988*a*. *Precambrian in Younger Fold Belts: European Variscides, the Carpathians And Balkans*. International Geological Correlation Programme, **22**, John Wiley, Chichester.

ZOUBEK, V., FIALA, J., VAŇKOVÁ, V., MACHART, J. & STETTNER, G. 1988*b*. Moldanubian region. *In*: ZOUBEK, V., COGNÉ, J. & KOZHOUKHAROV, D. (eds) *Precambrian in Younger Fold Belts. European Variscides, the Carpathians and Balkans*. John Willey, Chichester, 83–268.

ZUBER, J. A. & ÖHLANDER, B. 1990. Geophysical and geochemical evidence of Proterozoic collision in the western marginal zone of the Baltic Shield. *Geologische Rundschau*, **79**, 1–11.

ZULAUF, G. 1994. Ductile normal faulting along the West Bohemian Shear Zone (Moldanubian/Tepla-Barrandian boundary): evidence for late Variscan extensional collapse in the Variscan Internides. *Geologische Rundschau*, **83**, 276–292.

ZULAUF, G., SCHITTER, F., RIGLER, G., FINGER, F., FIALA, J. & VEJNAR, Z. 1999. Age constraints on the Cadomian evolution of the Teplá Barrandian unit (Bohemian Massif) through electron microprobe dating of metamorphic monazite. *Zeitschrift der deutschen geologischen Gesellschaft*, **150**, 627–639.

ZURBRIGGEN, R., FRANZ, L. & HANDY, M. 1997. Pre-Variscan deformation, metamorphism and magmatism in the Strona-Ceneri Zone (southern Alps of northern Italy and southern Switzerland). *Swiss Bulletin of Mineralogy and Petrology*, **77**, 361–381.

ZWART, H. J. 1979. The geology of Central Pyrénées. *Leidse geologische mededelingen*, **50**, 1–74.

3 Cadomian tectonics

ULF LINNEMANN (co-ordinator), RICHARD D'LEMOS, KERSTIN DROST, TERESA JEFFRIES, AXEL GERDES, ROLF L. ROMER, SCOTT D. SAMSON & ROB A. STRACHAN

The Cadomian Orogeny comprises a series of complex sedimentary, magmatic and tectonometamorphic events that spanned the period from the mid-Neoproterozoic (*c.* 750 Ma) to the earliest Cambrian (*c.* 540–530 Ma) along the periphery of the supercontinent Gondwana (peri-Gondwana, Fig. 3.1). Modern data demonstrate broad continuity between Cadomian events and the later opening of the Rheic Ocean during Cambrian–Ordovician times (Linnemann *et al.* 2007). Due to very similar contemporaneous orogenic processes in the Avalonian microcontinent, the collective terms 'Avalonian–Cadomian' Orogeny and 'Avalonian–Cadomian' Active Margin have often been used in the modern literature (e.g. Nance & Murphy 1994; Fig. 3.1). Rock units formed during the Cadomian Orogeny are commonly referred to collectively as 'Cadomian Basement'. Peri-Gondwanan terranes, microcontinents and crustal units in Central, Western, Southern and Eastern Europe, in the Appalachians (eastern USA and Atlantic Canada), and in North Africa were affected by the Cadomian Orogeny. This orogenic event is also apparently present in Baltica because of the 'Cadomian affinity' of late Precambrian orogenic events in the Urals and in the Timanides on the margin of Baltica (Roberts & Siedlecka 2002).

The Cadomian Orogeny *sensu stricto* was first defined in the North Armorican Massif in France on the basis of the unconformity that separates deformed Precambrian rock units from their Early Palaeozoic (Cambro-Ordovician) overstep sequence (see below). This unconformity is commonly referred to as the 'Cadomian unconformity' (Fig. 3.2). However, it cannot be precluded that the youngest metasedimentary rocks affected by Cadomian deformation are in fact earliest Cambrian in age. Many geologists currently assume that the final stages of Cadomian orogenesis were diachronous, lasting from the latest Neoproterozoic (Ediacaran) to the earliest Cambrian. In this overview, the term 'Cadomian basement' includes all sedimentary, volcanosedimentary and igneous Neoproterozoic rock units as well as early Cambrian plutons that were emplaced during final Cadomian tectonomagmatic events at *c.* 540–530 Ma.

The Cadomian unconformity was first described from Rocreux near Caen (Normandy) by Bunel (1835) without any illustration. The first description of this unconformity in Normandy, however, is often attributed to Dufrenoy (1838) who published the first field sketch. The wider geographic extent of the unconformity in central Brittany was recognized by Dufrenoy (1838) and Barrois (1899). The first illustration of the unconformity from Brittany was printed in the paper by Kerforne (1901). 'Cadomus' and 'Cadomum' are old Latin terms for the modern city of Caen and are the source of the name of the orogeny. The term '*discordance cadomienne*' was first used by Bertrand (1921). The type locality of the Cadomian unconformity is located at the northern edge of the village of Jacob Mesnil (Rocreux) close to Bretteville sur Laize near Caen (Normandy). We present the original illustration of the unconformity by Graindor (1957) in Figure 3.2.

In some publications, the term 'pan-African Orogeny' is used in the same sense as Cadomian Orogeny, because both events were related to the Gondwana Supercontinent in the late Precambrian and occurred at more or less the same time. To further the confusion, some geologists are of the opinion that 'Cadomian' in the sense of an orogenic event is a local name, the major difference between the more or less synchronous Cadomian and pan-African orogenic events being their position within the configuration of the Gondwana Supercontinent in the Neoproterozoic. However, the crustal units affected by pan-African orogenic events are located *between* the cratons that were assembled to form Gondwana and in most cases reflect continent–continent collision (see compilation of Windley 1995). In contrast, the Cadomian Orogeny (or alternatively the Avalonian–Cadomian Belt) was a major peripheral orogeny at the margin of Gondwana (Fig. 3.1) that was characterized by orogenic processes similar to those at present observable in the Andes and the Cordilleran chains of the American continents (Nance *et al.* 2002) and in the western Pacific region (Buschmann 1995; Linnemann *et al.* 2007).

On the basis of provenance studies based on U–Pb ages of detrital zircon grains from sedimentary rocks and inherited zircon of igneous rocks in combination with palaeomagnetic and palaeobiogeographic data, most geologists accept that the Cadomian basement of Central Europe was located at the periphery of the West African Craton (Fig. 3.1). Remnants of the West African cratonic basement are represented by the Icartian Basement (2.01–2.06 Ga) in the Armorican Massif and the Svetlik Gneiss (2.05–2.1 Ga) in the Bohemian Massif. In view of the occurrence of these very old rock complexes and many Archaean and Palaeoproterozoic detrital zircon grains in Neoproterozoic sediments, it seems that most of the Cadomian basement was developed on a stretched and thinned old cratonic crust with eroded basement slices in horst structures, from which cratonic material was available for erosion and deposition into basins and for the incorporation of large blocks from the basement into the Cadomian Orogeny.

The aim of this chapter is to summarize the present state of knowledge concerning the geological processes related to the Cadomian Orogeny. The Variscan Orogeny in Devonian and Carboniferous times resulted in penetrative overprinting and recrystallization of Cadomian basement in many areas. Therefore, we have focused on those parts of the Cadomian Orogeny in Central and Western Europe which were least affected by younger orogenic events, namely the Armorican and the Bohemian massifs (Fig. 3.3). Our analysis is focused on the external parts of both basement inliers (North Armorican Massif and Saxo-Thuringian Zone). Both of these inliers belong to a group of peri-Gondwanan crustal units underlain, at least in part, by recycled *c.* 2–3 Ga old West African crust and known variously as Cadomia-type terranes or 'Cadomia' (Fig. 3.1). In

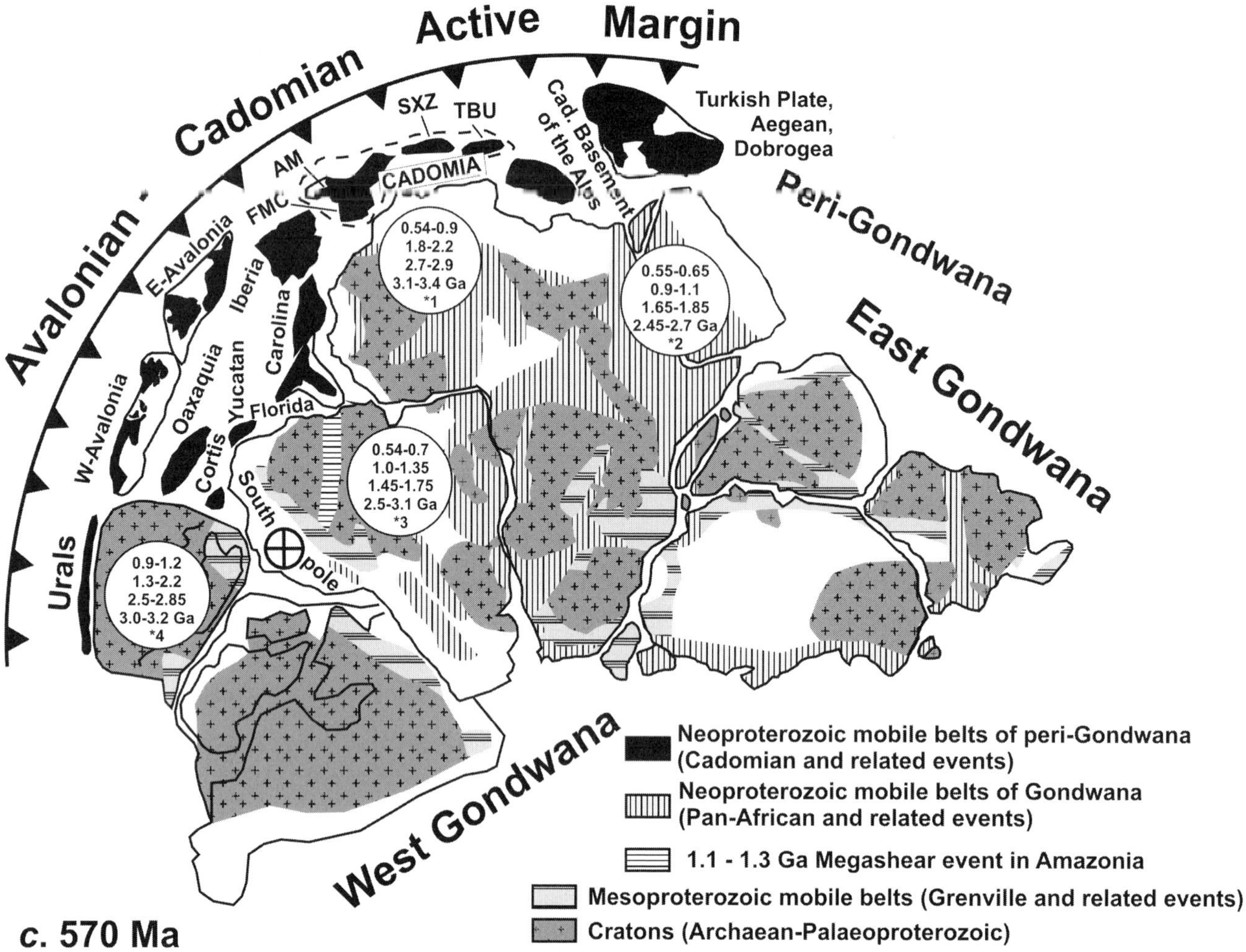

Fig. 3.1. Palaeogeography of the Cadomian–Avalonian active margin and related major peri-Gondwanan terranes at *c.* 570 Ma (modified after Nance & Murphy 1994, 1996; Linnemann *et al.* 2000, 2004; Murphy *et al.* 2000; Linnemann & Romer 2002; Nance *et al.* 2002), and palaeogeography of the Gondwanan continental plates after Unrug (1996). Abbreviations: AM, Armorican Massif; FMC, French Massif Central; SXZ, Saxo-Thuringian Zone (part of the Bohemian Massif); TBU, Teplá–Barrandian Unit (part of the Bohemian Massif). Numbers in circles are zircon ages from the cratons in Ga. *1, From the compilation of Nance & Murphy (1994 and references therein); *2, from Avigad *et al.* (2003); *3, from Schneider Santos *et al.* (2000); *4, from the compilation of Zeh *et al.* (2001).

contrast, the Bruno-Vistulian Block of the Moravo-Silesian Zone in the eastern Bohemian Massif belongs to the group of Avalonia-type terranes (Finger *et al.* 2000*a*) derived from recycled *c.* 1.0–1.3 Ga old juvenile crust within the Panthalassa-type ocean surrounding Rodinia and accreted to the northern Gondwanan margin by 650 Ma (Murphy *et al.* 2004).

The North Armorican Massif (R. S. D'L., S. D. S. & R. A. S.)

Introduction

The Armorican Massif is divided into three contrasting crustal blocks by the North and South Armorican shear zones (Fig. 3.4, inset). The well-documented dextral displacements along these structures during the Variscan Orogeny (Watts & Williams 1979; Gapais & Le Corre 1980; Jegouzo 1980) may have been associated with the reactivation of older structures. In all three blocks, Neoproterozoic volcanosedimentary successions are overlain unconformably by early Palaeozoic sedimentary successions. In contrast to central and southern parts of Armorica, Variscan orogenic activity in large parts of North Armorica was weak, with the result that here early Palaeozoic sedimentary rocks and intrusions are generally undeformed and unmetamorphosed. Thus, earlier, late Neoproterozoic geological relationships are often preserved and this makes the area exceptional for the study of Cadomian geology. The Cadomian Belt of the North Armorican Massif mainly comprises late Neoproterozoic supracrustal sequences of the Brioverian Supergroup (Graindor 1957; Cogné 1962). Small areas of Palaeoproterozoic basement occur in the northern part of the belt (e.g. Calvez & Vidal 1978; Samson & D'Lemos 1998) and a juvenile Neoproterozoic plutonic complex in the Baie de Saint-Brieuc region of Brittany forms the local basement to the Brioverian succession (Egal *et al.* 1996; Samson *et al.* 2003). The Brioverian succession and the basement units were variably deformed, metamorphosed and intruded by igneous complexes during the Cadomian Orogeny. Detailed geochronological studies, particularly over the last 15 years, have demon-

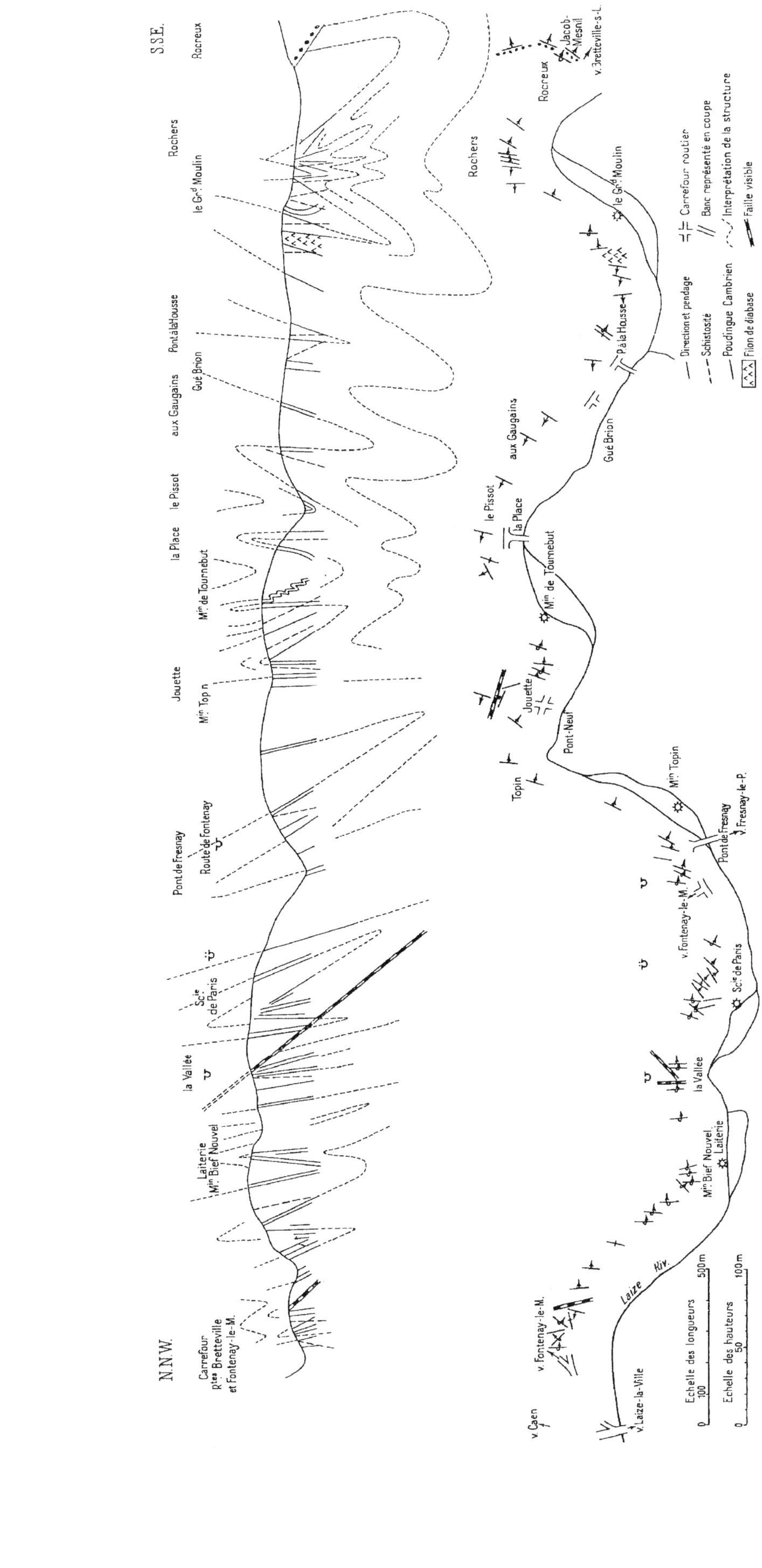

Fig. 3.2. Geological profile of the type locality of the Cadomian Unconformity close to Rocreux (Jacob-Mesnil, right in the figure), NW of the city of Bretteville-sur Laize in the region of Caen (Normandy, France), where Lower Cambrian conglomerates unconformable overlie folded Late Neoproterozoic (Ediacaran) greywacke turbidites (original figure by Graindor 1957).

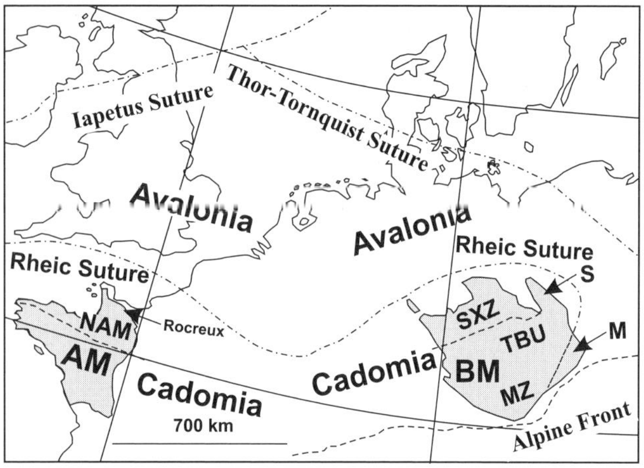

Fig. 3.3. Location of the Armorican and Bohemian Massifs and boundaries relevant for the reconstruction of the Cadomian Orogen in Central Europe. The position of the type locality of the Cadomian Unconformity is marked (Rocreux). Abbreviations: AM, Armorican Massif; BM, Bohemian Massif; M, Moravo-Silesian Unit; MZ, Moldanubian Zone; NAM, North Armorican Massif; S, Sudetes; SXZ, Saxo-Thuringian Zone; TBU, Teplá–Barrandian Unit.

strated that this was not a single orogenic event, but a series of possibly discrete accretionary events between *c.* 620 Ma and 540 Ma. In this overview, only the most reliable U–Pb (mainly zircon) and $^{40}Ar/^{39}Ar$ mineral studies are discussed. Earlier Rb–Sr dates, which have largely proven to be unreliable and/or lack the precision necessary to elucidate a detailed history of events, are neither summarized nor utilized.

There is broad agreement that the evolution of the Cadomian belt can be interpreted in terms of the amalgamation of calc-alkaline magmatic arc complexes and intra-arc/marginal basin sequences along an active plate margin (e.g. Auvray 1979; Graviou & Auvray 1985; Cabanis *et al.* 1987; Dissler *et al.* 1988; Strachan *et al.* 1989, 1996; Brown *et al.* 1990; Dupret *et al.* 1990; Rabu *et al.* 1990; Brown 1995; Egal *et al.* 1996; Ballèvre *et al.* 2001; Chantraine *et al.* 2001). Furthermore, it is also agreed that the belt is subdivided by steeply dipping shear zones and faults into a number of structural blocks with contrasting geological histories, although the palaeogeographic relationships of these tectonic blocks and the timing and mechanism of their tectonic juxtaposition continue to be debated. The structural blocks have been variously termed 'domains' (Cogné 1972; Egal *et al.* 1996), 'terranes' (Strachan *et al.* 1989, 1996), 'units' (e.g. Ballèvre *et al.* 2001; Chantraine *et al.* 2001) and 'segments' (Nagy *et al.* 2002). In the following account we use the terrane divisions of Strachan *et al.* (1989) as a framework for discussion. From north to south these are the Trégor-La Hague, Saint-Brieuc, Saint-Malo and Mancellian terranes (Fig. 3.4). We follow Coney *et al.* (1980) in defining a terrane as a fault-bounded crustal entity that may have evolved entirely separately from adjacent terranes prior to their amalgamation. The usefulness of this approach is that it encourages the objective description of each terrane followed by a rigorous assessment of its likely accretionary history.

Trégor–La Hague Terrane

The Trégor–La Hague Terrane is distinct as it includes the only exposed Palaeoproterozoic basement in the North Armorican Massif, and it lacks extensive metasedimentary successions that can be correlated with the Brioverian Supergroup. Small areas of basement are exposed on Guernsey, on the La Hague peninsula and as rafts in the North Trégor Batholith (Fig. 3.4). The most common lithology is granite gneiss, with subordinate metasedimentary and metabasic rocks. Early U–Pb zircon studies using multigrain fractions yielded ages of *c.* 2200–1800 Ma from the granite gneisses (Fig. 3.4; Calvez & Vidal 1978; Auvray *et al.* 1980; Piton 1985). The term 'Icartian' has been widely used to refer to these basement rocks because the Icart granite gneiss at Soldiers Bay on Guernsey (Fig. 3.5A) was the first to yield a *c.* 2 Ga isotopic age (Calvez & Vidal 1978). Recent, high-precision U–Pb zircon dating of other basement components has obtained protolith ages of 2061 +1.7/−1.5 Ma (Saints Bay Gneiss, Guernsey; Samson & D'Lemos 1998; Fig. 3.6A) and 2061 ± 2.7 Ma (Culeron orthogneiss, La Hague; Inglis *et al.* 2004; Fig. 3.7B). Strongly deformed granites, amphibolites and metasedimentary units on Sark (Fig. 3.6B) have also been assigned to the Icartian basement (Gibbons & Power 1975; Power & Gibbons 1994). However, at least some of these units can be shown on geochronological grounds to be late Neoproterozoic in age (Samson & D'Lemos 1998), demonstrating the difficulty of interpreting all gneissic lithologies in the region as 'basement' without robust chronometric data.

On Guernsey, Sark, Alderney and at La Hague variably deformed and metamorphosed calc-alkaline quartz diorite, tonalite, granodiorite and granite stocks occur (Sutton & Watson 1957; Roach 1966; Power *et al.* 1990a) (Figs 3.6 & 3.7). The results of integrated isotopic and structural studies indicate that these were emplaced during a significant tectonothermal event between *c.* 620 and 610 Ma. A U–Pb zircon date of 615.6 +4.2/−2.3 Ma obtained from a mylonitic granite on Sark (Tintageu gneiss; Fig. 3.6B), interpreted to date crystallization of the igneous protolith, places an older limit on the timing of strongly penetrative deformation (Samson & D'Lemos 1998). This is consistent with U–Pb zircon (SHRIMP) dates of *c.* 618 Ma obtained from rims of zircons from orthogneiss at La Hague considered to date peak metamorphism at 700–800°C and 8–9 kbar (Inglis *et al.* 2004). A number of younger syntectonic intrusions were emplaced between *c.* 615 and *c.* 608 Ma (Table 3.1), with the youngest intrusions showing the least amount of penetrative deformation (Tribe & D'Lemos 1996; Samson & D'Lemos 1999; Miller *et al.* 1999; D'Lemos *et al.* 2001; Inglis *et al.* 2005). $^{40}Ar/^{39}Ar$ hornblende dates define an upper limit of *c.* 600 Ma for regional cooling following deformation and associated amphibolite facies metamorphism in this part of the Trégor-La Hague Terrane (Dallmeyer *et al.* 1991*a*, 1994). The undeformed North Trégor Batholith has yielded a U–Pb zircon date of 615 +13/−7 Ma (Graviou *et al.* 1988; Fig. 3.4). Within analytical error, this could have been emplaced towards the end of the deformation event identified to the NE.

The lithological and chemical compositions of these intrusions are consistent with an origin in a suprasubduction environment as a result of volatile-enhanced partial melting of the mantle wedge above a subducting oceanic lithospheric plate (Graviou & Auvray 1985, 1990; Graviou *et al.* 1988; Brown *et al.* 1990). Isotopic studies of the North Trégor Batholith suggest contamination of the calc-alkaline magmas by continental crust (Guerrot 1989). Identical conclusions have been reached on Guernsey, Sark and at La Hague where Nd isotopic compositions imply the involvement of significant amounts of evolved continental crustal material (Samson *et al.* 2003). The abundance of xenocrystic zircon in the intrusions rules out derivation from subducted clastic sediment, as mantle temperatures are too high for detrital

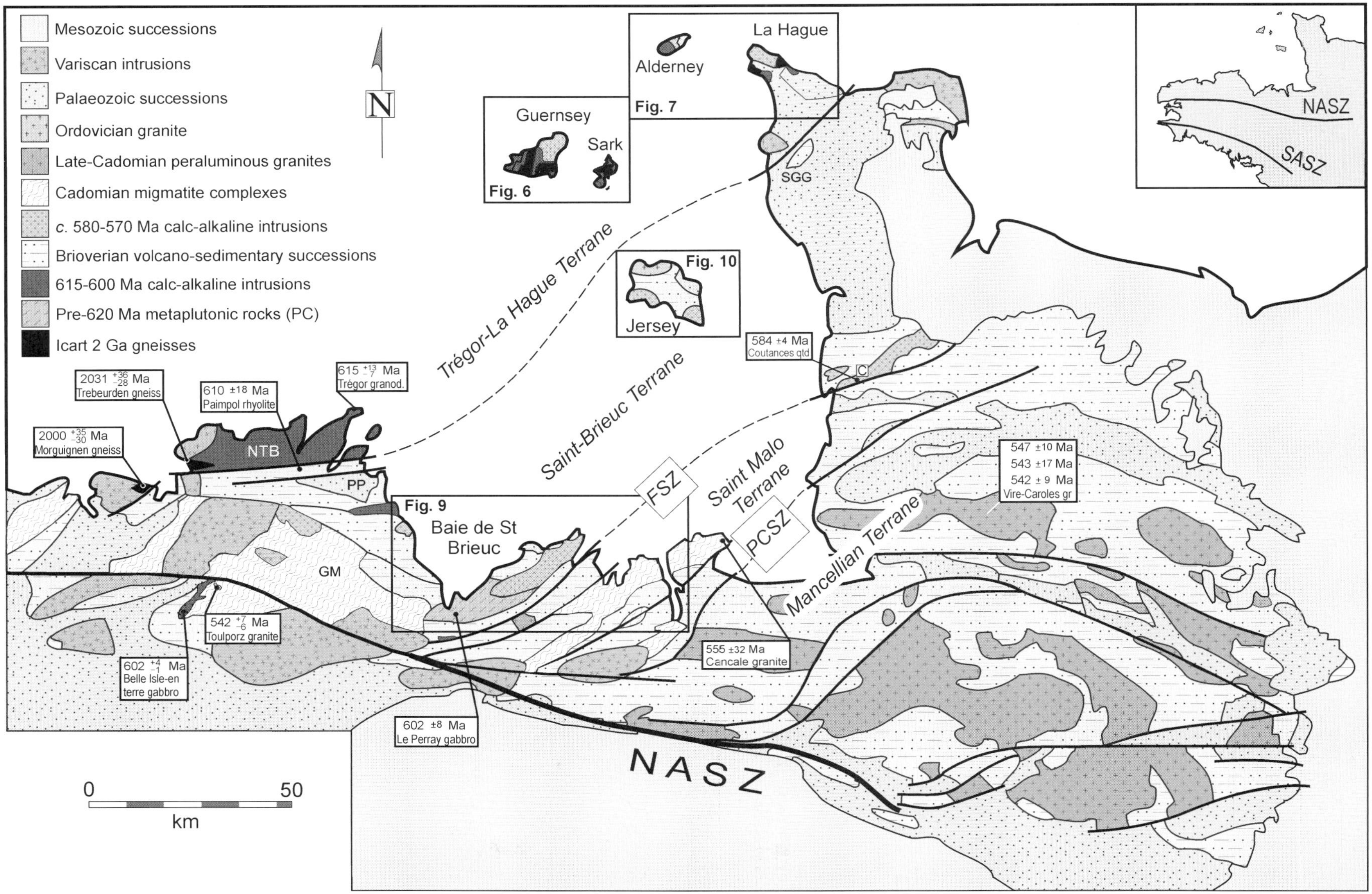

Fig. 3.4. Generalized geological map of the North Armorican Massif. For source of U–Pb zircon ages (upper intercepts only) and zircon evaporation ages (all errors quoted at 2σ) see Table 3.1. Rectangles outline the regions shown in detail in Figures 3.6, 3.7, 3.9 & 3.10. Abbreviations: FSZ, Fresnaye shear zone; PCSZ, Plouer-Cancale shear zone. Inset shows the North Armorican shear zone (NASZ) and South Armorican shear zone (SASZ). Other abbreviations: C, Coutances; GM, Guingamp migmatites; NTB, North Trégor Batholith; PC, Penthièvre Complex; PP, Plourivo-Plouézec; SGG, Saint Germain le-Gaillaird.

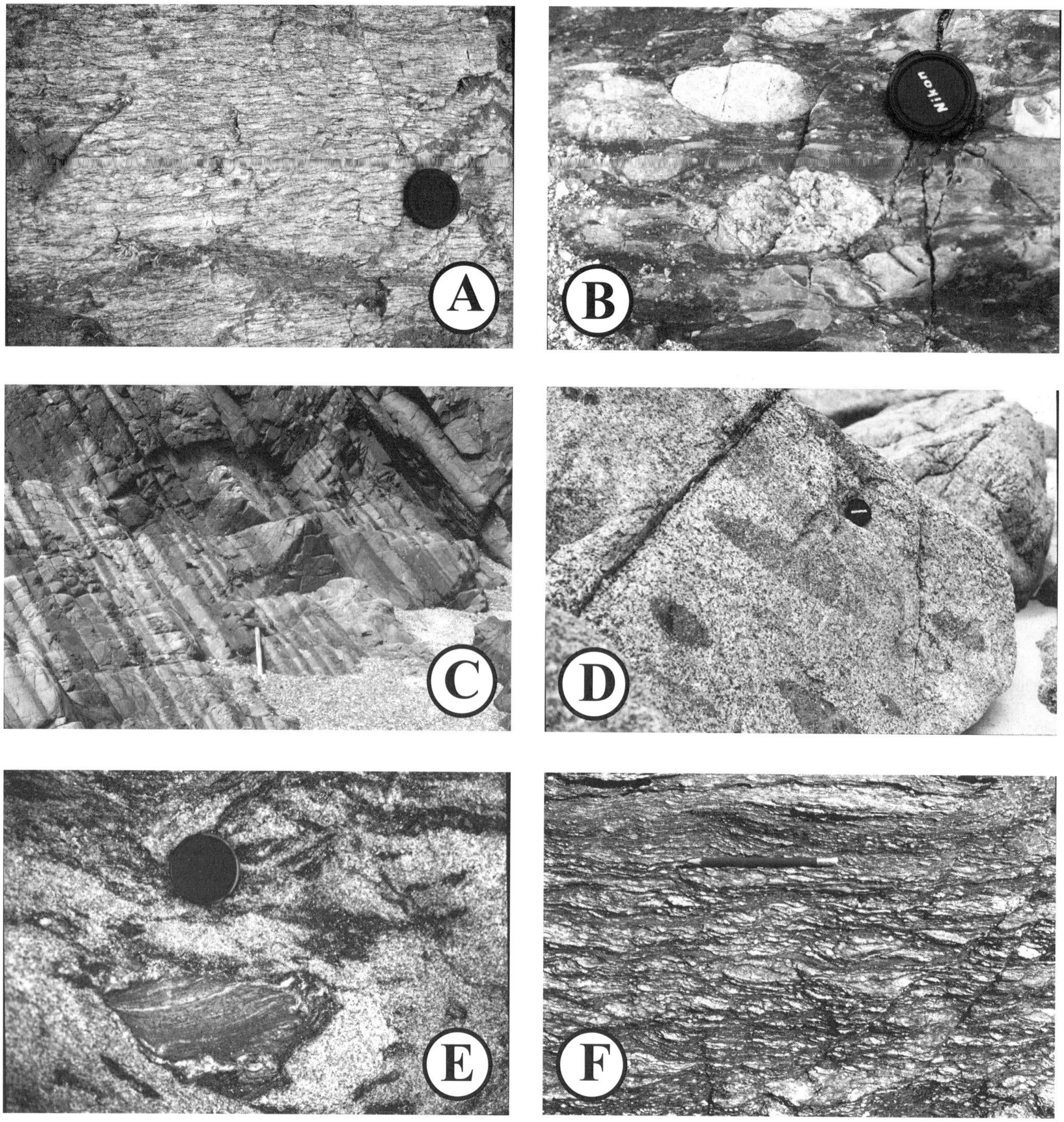

Fig. 3.5. (**A**) Icart Granite Gneiss at Saints Bay, Guernsey. (**B**) Granitic pebbles within the Brioverian Cesson Conglomerate, Baie de Saint-Brieuc. (**C**) Brioverian turbidites, Binic, Baie de Saint-Brieuc. (**D**) Flattened mafic enclaves within the Fort La Latte quartz diorite, La Fosse; (**E**) Relict pod of Brioverian metasediment engulfed by anatectic melt within the Saint-Malo migmatite belt, Port Briac; (**F**) S-C fabrics within highly deformed Brioverian semipelites within the Saint-Cast migmatite belt; the sinistral sense of displacement shown by these fabrics is parallel to a subhorizontal stretching and mineral lineation, Saint Cast.

zircon to survive complete melting, and is more consistent with interaction with Palaeoproterozoic basement (Samson *et al.* 2003).

On Guernsey, Sark and at La Hague, the syntectonic Cadomian intrusions and their country-rocks carry similarly orientated ductile fabrics and kinematic indicators (Tribe & D'Lemos 1996; Tribe *et al.* 1996; Inglis *et al.* 2005). This concordance of structure demonstrates that the dominant fabrics and metamorphic parageneses within the Icartian basement formed during the Cadomian Orogeny. The ductile fabrics trend mainly north–south: on Guernsey, steep planar fabrics with associated gently plunging lineations formed during oblique dextral shear, and on Sark the dominant structures are recumbent folds and gently dipping to subhorizontal planar fabrics and lineations that locally carry top-to-the-south kinematic indicators. This contrasting kinematic pattern is consistent with Sark representing a flat-lying

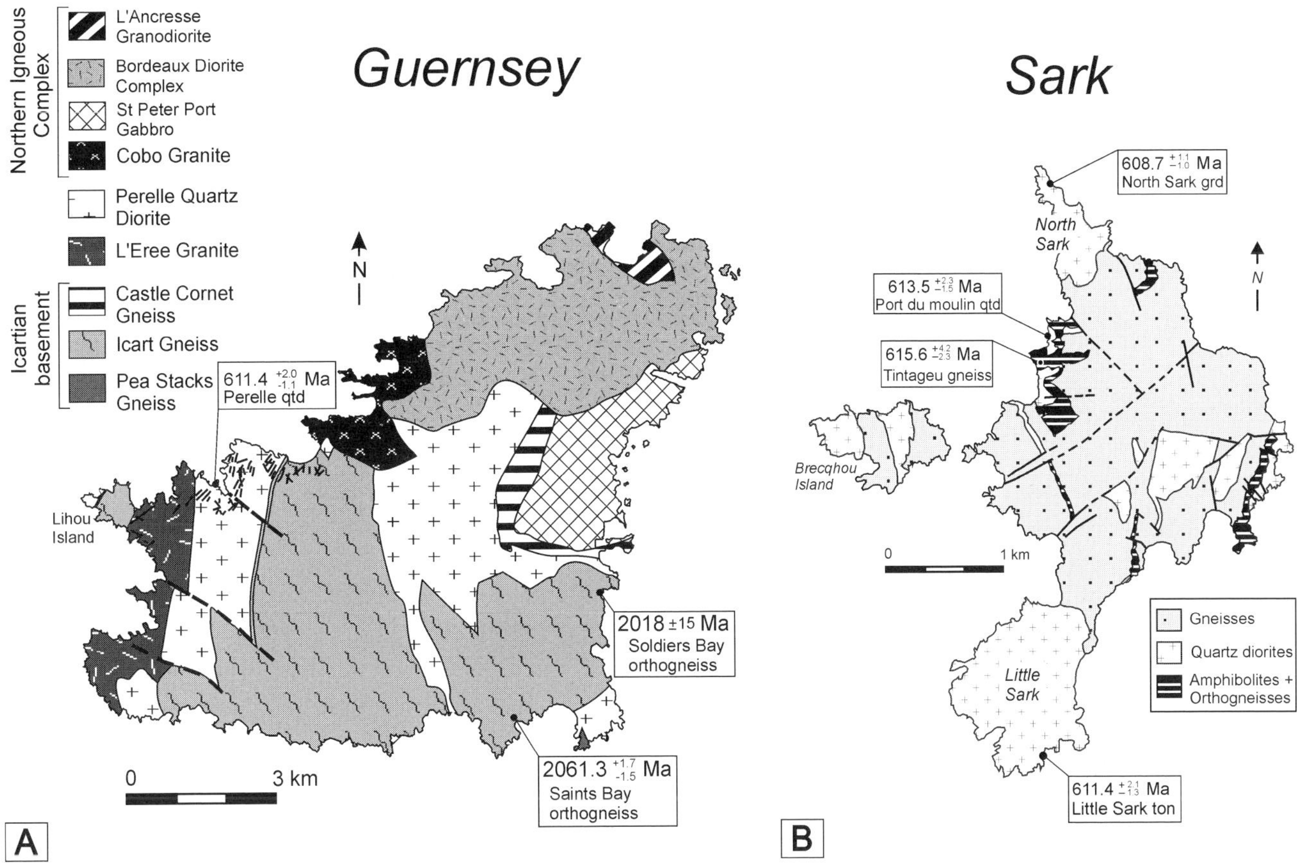

Fig. 3.6. Generalized geological maps of Guernsey and Sark. Dates listed are from published U–Pb zircon dates (see also Table 3.1). Errors are 2σ. Note differences in scale for the two islands.

shear zone that accommodated oblique dextral movement on Guernsey. The regional significance of these structures, and how they might relate kinematically to the steep north–south fabrics at La Hague, is uncertain due to the incomplete exposure.

A second, geochemically, isotopically and temporally distinct period of magmatism occurred at 580–570 Ma (Table 3.1). The intrusions include the Northern Igneous Complex of Guernsey (Topley *et al.* 1990; Dallmeyer *et al.* 1992), the Bibette Head granodiorite on Alderney (D'Lemos *et al.* 2001) and the Thiébot granodiorite and the St Martin monzonite at La Hague (Power *et al.* 1990*b*; Inglis *et al.* 2005) (Figs 3.6 & 3.7). The intrusions form mixed magma complexes with calc-alkaline to alkaline compositions: the majority of the intrusions are diorite to granite *sensu stricto* but also include layered gabbro (Brown *et al.* 1990; Power *et al.* 1990*b*). Most are undeformed and cross-cut ductile structures in their host rocks and, hence, have been regarded as post-tectonic. However, intrusion may have overlapped the development of localized shear zones and faults. On the La Hague peninsula, emplacement of the Thiebot granite at 574 +1.3/−1.0 Ma was contemporaneous with ductile deformation of its country rocks (Inglis *et al.* 2005). On Guernsey, Sark and Alderney, Tribe (1994) recorded a suite of greenschist facies retrogressive shear zones and widespread cataclasis, which both pre- and post-date the intrusions. This suggests that the plutons were emplaced at mid- to upper crustal levels during localized semi-brittle deformation. The $\varepsilon_{Nd}(t)$ values for these granitoids typically range from −6 to −2 (total range −10 to −0.6), indicative of derivation from, or interaction with, evolved crustal materials (D'Lemos & Brown 1993; Samson *et al.* 1997). However, the Nd isotopic ratios are more radiogenic than the earlier, quartz diorititic–tonalitic, syntectonic intrusions of the Trégor–La Hague Terrane which have $\varepsilon_{Nd}(t)$ values in the range of −10.4 to −8.3 (Fig. 3.8; Samson *et al.* 2003). Hence, paradoxically, the younger, chemically more evolved intrusions (granites *sensu stricto*) have more juvenile isotopic compositions than the older, chemically less evolved intrusions (quartz diorites).

Saint-Brieuc Terrane

The Saint-Brieuc Terrane is distinct from the Trégor–La Hague Terrane in that it exposes Brioverian metavolcanic and metasedimentary successions that, at least in part, were deposited on juvenile Neoproterozoic crust represented by the Penthièvre Complex (Fig. 3.9). A further significant difference is that the timing of the main tectonothermal event in this terrane is distinctly younger. Both the Brioverian and the Penthièvre Complex were intruded by a range of pre-, syn- and post-tectonic gabbroic, dioritic and granitic calc-alkaline plutons.

In the south of the Baie de Saint-Brieuc, the lowermost parts of the Brioverian succession unconformably overlie variably deformed and metamorphosed plutonic rocks of the Penthièvre Complex (Fig. 3.9; Cogné 1959; Rabu *et al.* 1983; Shufflebotham 1990; Strachan & Roach 1990). The complex itself comprises plutons with protolith ages that span at least 120 Ma. The oldest dated components are the Port Morvan orthogneisses that have yielded U–Pb zircon dates of 754.6 ± 0.8 Ma and 746.0 ± 0.9

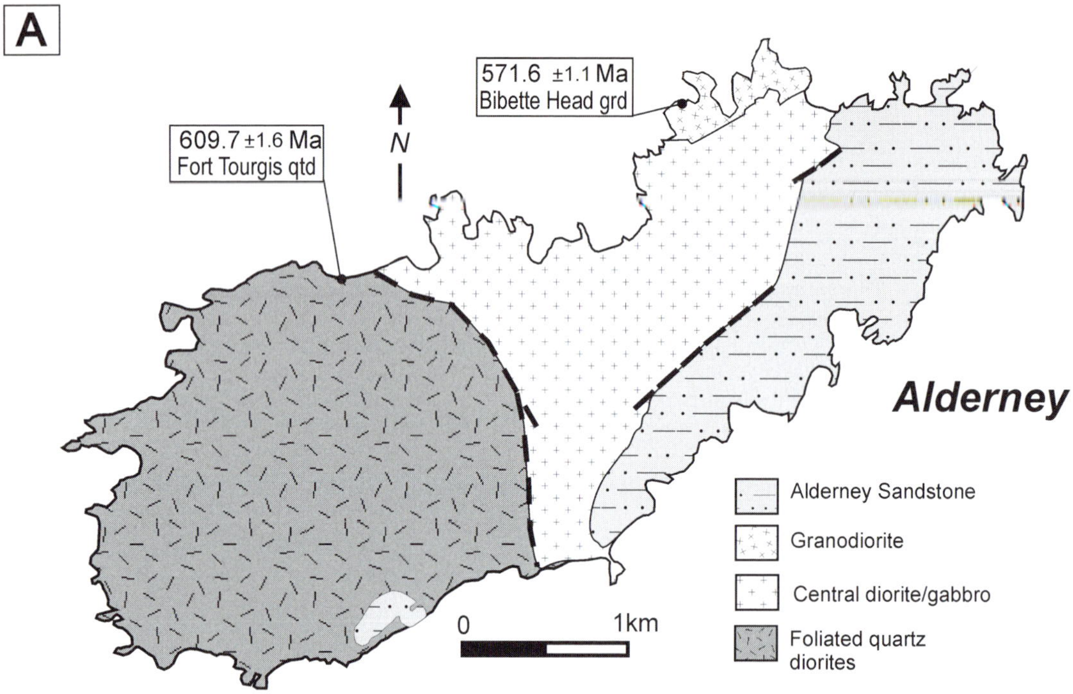

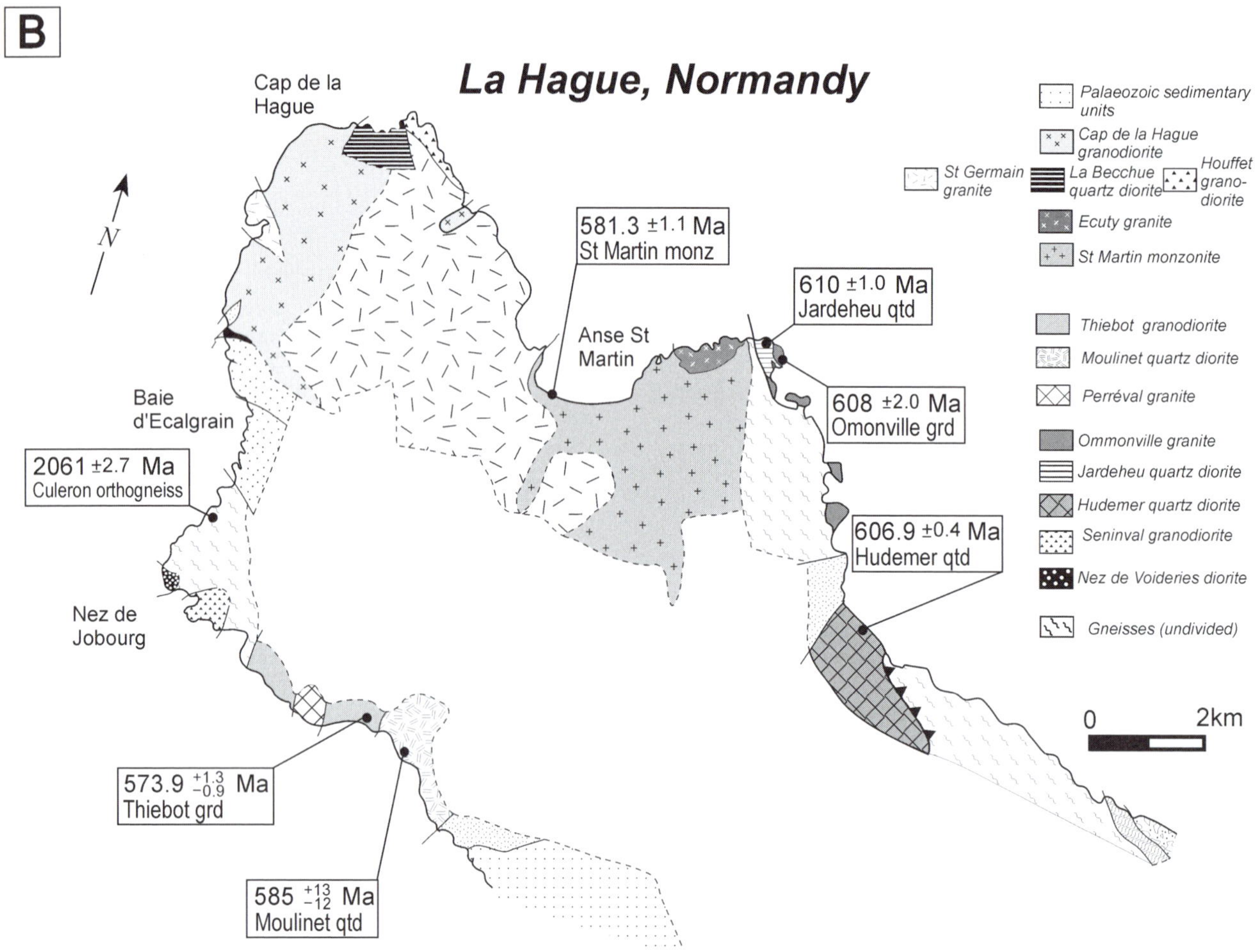

Fig. 3.7. Generalized geological maps of Alderney and the La Hague region. Dates listed are from published U–Pb zircon dates (see also Table 3.1). Errors are 2σ. Note differences in scale for the two regions.

Table 3.1. *Summary of isotopic data derived from meta-igneous units in North Armorica*

Unit	Age (± 2σ error)	Number of analyses (N), comments	Reference
Tregor–La Hague Terrane			
Protolith ages/Icartian basement			
Saint's Bay orthogneiss (Guernsey)	2061.3 + 1.7/−1.5 Ma MSWD = 1.45	N = 7, abr zircon UI age	Samson & D'Lemos (1998)
Culeron orthogneiss (La Hague)	2061 ± 2.7 Ma MSWD = 0.33	N = 4, abr zircon UI age	Inglis *et al.* (2004)
Culeron orthogneiss (metamorphic age)	*c.* 618 Ma MSWD not given	N = 3, SHRIMP analyses/zircon rims	Inglis *et al.* (2004)
Soldier's Bay orthogneiss (Guernsey)	2018 ± 15 (±68) Ma (recalculated error)	Bulk fraction, unabr zircon	Calvez & Vidal (1978)
Syntectonic intrusions			
Tintageu leucogneiss (Sark)	615.6 + 4.2/−2.3 Ma MSWD = 0.035	N = 4, abr zircon UI age	Samson & D'Lemos (1998)
(cooling age)	603 ± 2 Ma	$^{40}Ar/^{39}Ar$, hbl isotope corr age	Dallmeyer *et al* (1991*a*)
Trégor granodiorite	615 +13/-7 Ma MSWD not given		Graviou *et al.* (1988)
Port du Moulin quartz diorite (Sark)	613.5 + 2.3/−1.5 Ma MSWD = 0.38	N = 6, abr zircon UI age	Miller *et al.* (1999)
Little Sark quartz diorite (Sark)	611.4 + 2.1/−1.3 MSWD = 0.029	N = 6, abr zircon UI age	Miller *et al.* (1999)
Perelle quartz diorite (Guernsey)	611.4 + 2.0/−1.1 MSWD = 0.41	N = 10, abr zircon UI age	Samson & D'Lemos (1999)
Jardeheu quartz diorit (La Hague)	610.4 + 0.9/−0.7 Ma MSWD = 0.13	N = 6, abr zircon UI age	Inglis *et al.* (2005)
Fort Tourgis quartz diorite (Alderney)	609.7 ± 1.6 Ma MSWD = 0.47	N = 5, abr zircon UI age	D'Lemos *et al.* (2001)
North Sark granodiorite (Sark)	608.7 + 1.1/−1.0 Ma MSWD = 0.43	N = 7, abr zircon UI age	Miller *et al.* (1999)
Omonville granodiorite (La Hague)	608 + 2.8/−1.8 Ma MSWD = 0.83	N = 6, abr zircon UI age	Inglis *et al.* (2005)
Hudemer quartz diorite (La Hague)	606.9 ± 1.0 Ma MSWD = 0.53	N = 3, abr zircon concordant age	S.D. Samson (unpublished)
Post-tectonic intrusions			
Moulinet quartz diorite (La Hague)	585 + 13/−12 Ma MSWD = 0.04	N = 3, unabr zircon UI age	Guerrot & Peucat (1990)
St. Martin monzonite (La Hague)	581.3 ± 1.1 Ma MSWD = 0.83	N = 5, abr zircon UI age	Inglis *et al.* (2005)
Thiébot granodiorites (La Hague)	573.9 + 1.3/−1.0 Ma MSWD = 0.71	N = 4, abr zircon UI age	Inglis *et al.* (2005)
Bibette Head granodiorite (Alderney)	571.6 ± 1.1 Ma MSWD = 0.73	N = 9, abr zircon UI age	D'Lemos *et al.* (2001)
Saint-Brieuc Terrane			
Penthièvre Complex			
Jospinet orthogneiss	754.6 ± 0.8 Ma MSWD – 0.85	N = 7, abr zircon mean $^{207}Pb/^{206}Pb$ date	Samson *et al.* (2003)
Port Morvan orthogneiss	746.0 ± 0.9 Ma MSWD = 0.38	N = 5, abr zircon mean $^{207}Pb/^{206}Pb$ date	Samson *et al.* (2003)
Port Morvan orthogneiss	746.0 ± 34 Ma range: 763–739 Ma MSWD not given	N = 3, zircon evaporation, mean date from 3 crystals	Egal *et al.* (1996)
Cesson conglomerate Tonalite clast	667 ± 4 Ma MSWD not given	N = 4, unabr zircon UI age	Guerrot & Peucat (1990)
Cesson conglomerate Trondhjemite clast	665.5 ± 0.7 Ma MSWD = 0.63	N = 4, abr zircon conc $^{238}U–^{206}Pb$ dates	Samson *et al.* (2003)
Cesson conglomerate Trondhjemite clast	665.2 ± 0.5 Ma MSWD = 0.16	N = 3, abr zircon conc $^{238}U–^{206}Pb$ dates	Samson *et al.* (2003)
Cesson conglomerate Tonalite clast	656 ± 5 Ma MSWD not given	N = 4, unabr zircon UI age	Guerrot & Peucat (1990)
Post-Penthièvre igneous units			
Jospinet granodiorite	625.9 +3.6/−1.9 Ma MSWD = 0.32	N = 7, abr zircon UI age	Nagy *et al* (2002)
Paimpol Formation rhyolite	610 ± 18 Ma range: 613–608 Ma MSWD not given	N = 4, zircon evaporation, mean date from 4 crystals	Egal *et al.* (1996)
Erquy Series (acidic tuff)	608 ± 7 Ma MSWD = 1.5	N = 9, SIMS	Cocherie *et al.* (2001)
Le Perray metagabbro	602 ± 8 Ma MSWD not given	N = 4, unabr zircon mean $^{207}Pb/^{206}Pb$ date	Guerrot & Peucat (1990)
Port Moguer tonalite	600.4 ± 0.9 Ma MSWD = 0.51	N = 6, abr zircon mean $^{207}Pb/^{206}Pb$ date	Nagy *et al* (2002)
Lanvollon Fm (acidic volcanic unit)	588 ± 22 Ma range: 594–579 Ma MSWD not given	N = 7 zircon evaporation, mean date from 7 crystals	Egal *et al.* (1996)
Coutances quartz diorite	584 ± 4 Ma MSWD not given	N = 4, unabr zircon UI age	Guerrot & Peucat (1990)
Anne Port rhyolite	582.8 + 3.0/−2.7 Ma MSWD = 0.23	N = 6, abr zircon UI age	Miller *et al.* (2001)
SW Jersey granite	580 + 2.3/−1.6 Ma MSWD = 0.34	N = 7, abr zircon UI age	Miller *et al.* (2001)
Fort la Latte quartz diorite	576.3 + 1.5/−1.2 Ma MSWD = 0.50	N = 5, abr zircon UI age	Nagy *et al* (2002)
(cooling age)	564.7 ± 1.6 Ma MSWD = 0.56	$^{40}Ar/^{39}Ar$, hbl isotope corr date	Dallmeyer *et al.* (1991*b*)
St. Quay quartz diorite	574.6 + 1.8/−1.5 Ma MSWD = 0.08	N = 7, abr zircon UI age	Nagy *et al.* (2002)
(cooling age)	563.1 ± 1.9 Ma MSWD = 1.4	$^{40}Ar/^{39}Ar$, hbl isotope corr date	Dallmeyer *et al.* (1991*b*)
Paleozoic intrusions			
NW Jersey granite	482.7 ± 1.6 Ma MSWD = 0.33	N = 7, abr zircon UI age	Miller *et al.* (2001)
(cooling age)		$^{40}Ar/^{39}Ar$, hbl	D'Lemos et al. (1992*a*)
Saint-Malo Terrane			
Belle-Isle-en-Terre gabbro	602 + 4/−1 Ma	N = 4, unabr zircon UI age	Peucat *et al.* (1981)

(*continued*)

Table 1. (*continued*)

Unit	Age (± 2σ error)	Number of analyses (N), comments	Reference
Cancale granite	555 ± 32 Ma range: 556–554 Ma MSWD not given	N = 4 zircon evaporation, mean date from 4 crystals	Egal *et al.* (1996)
Toulporz granite	542 = 7/–6 Ma MSWD not given	N = 7, unabr zircon UI age	Andriamarofahatra & de la Boisse (1988)
Anatectic granite (2880)	555 ± 13 Ma MSWD not given	N = 3, unabr zircon UI age	Peucat (1986)
Granitic vein, La pointe de Garel	541 ± 5 Ma MSWD not given	N = 4, unabr zircon mean $^{207}Pb/^{206}Pb$ date	Peucat (1986)
Saint Brieuc pluton	533 ± 12 Ma MSWD not given	zircon evaporation	Hébert *et al.* (1993)
Mancellian Terrane			
Vire-Carolles granite			
St-Michel de Loups	547 ± 10 Ma MSWD not given	Monazite analyses	Pasteels & Doré (1982)
St-Michel de Montjoie	543 ± 17 Ma MSWD not given	Monazite analyses	Pasteels & Doré (1982)
Villedieu	542 ± 9 Ma MSWD not given	Monazite analyses	Pasteels & Doré (1982)

abr, abraded; conc, concordant; corr, correlation; hbl, hornblende; MSWD, mean squares weighted deviate; SIMS, secondary ion mass spectrometer; UI, upper intercept; unabr, unabraded.

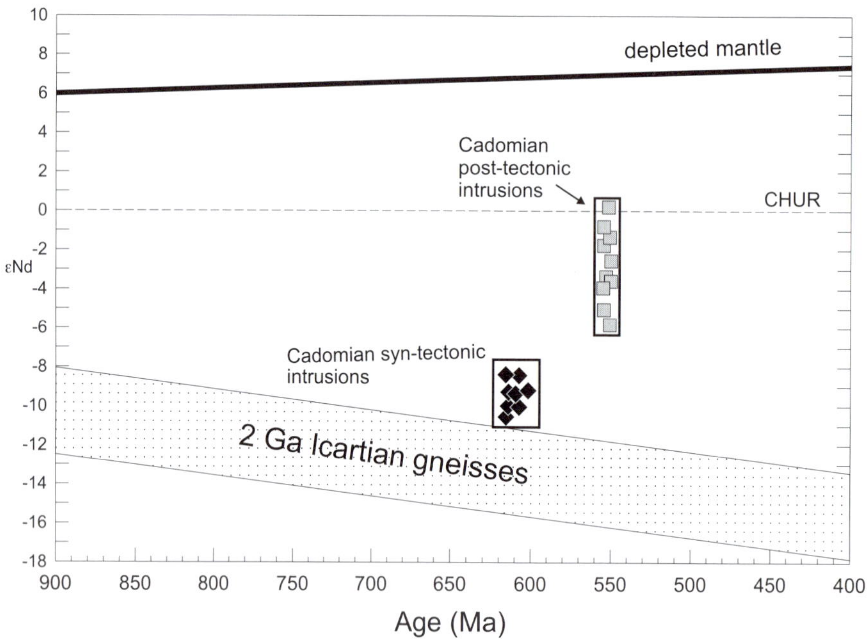

Fig. 3.8. Crystallization age versus initial εNd diagram for two major groups of granitoids within the North Armorican Massif (data from D'Lemos & Brown 1993; Samson *et al.* 2003; Samson unpublished data). Note that the syntectonic plutons have less evolved chemical compositions but more evolved isotopic compositions than the post-tectonic granitoids.

Ma (Egal *et al.* 1996; Samson *et al.* 2003) (Fig. 3.9). The Jospinet granodiorite has yielded a U–Pb zircon date of 625.9 +3.6/–1.9 Ma, which places a maximum limit on the depositional age of the non-conformably overlying Brioverian succession (Nagy *et al.* 2002). U–Pb zircon dates in the range 670–625 Ma have been obtained from detrital zircons of both cobble conglomerates and boulders of granite and trondhjemite within the basal Brioverian Cesson Conglomerate (Fig. 3.5B; Guerrot & Peucat 1990; Samson *et al.* 2003). Geochemical and isotopic data indicate that the *c.* 750–620 Ma igneous units are highly juvenile and not contaminated by *c.* 2 Ga basement (Thiéblemont *et al.* 1999; Samson *et al.* 2003). Both whole-rock Nd depleted-mantle model ages (800–690 Ma) and zircon Hf depleted-mantle model ages (750–650 Ma) are very similar to crystallization ages, consistent with the view that the original source magmas must have been almost completely mantle-derived. The high SiO_2 contents of some of the units (e.g. Cesson Conglomerate boulders) suggest either that the original mantle-derived material underwent marked fractionation, or, more likely, a rapid series of sequential partial melting events occurred which would preserve depleted mantle-like isotopic compositions, but produce felsic melts.

The Port Morvan and Jospinet plutons are the only rock units that have been shown conclusively to pre-date deposition of the Brioverian succession in the Saint-Brieuc Terrane. Although presently undated, ultramafic to mafic cumulates exposed inland and south of the Baie de Saint-Brieuc, record P-T conditions of 9 ± 1 kbar and 700 ± 50°C (Hébert 1993) and could possibly represent the oceanic basement to the Penthièvre Complex

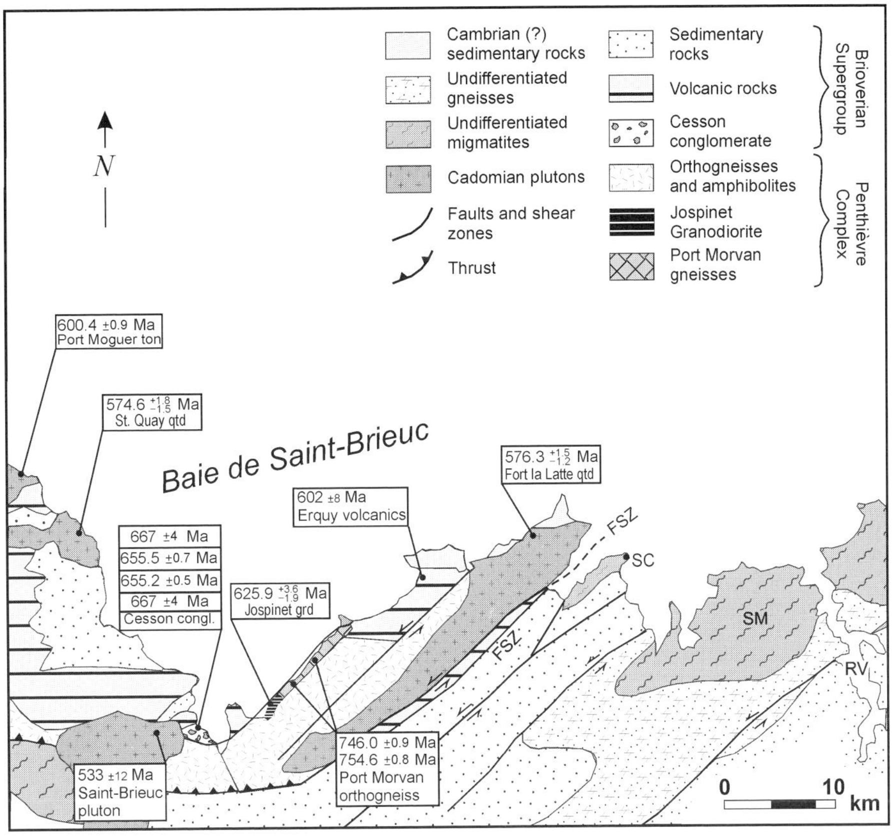

Fig. 3.9. Generalized geological map of Saint-Brieuc region, Brittany. Dates listed are from published U–Pb zircon dates/zircon evaporation dates (see also Table 3.1). Ages based on lower intercept U–Pb dates are not shown. Errors are 2σ. Abbreviations: FSZ, Fresnaye shear zone; RV, Rance Valley; SC, Saint-Cast; SM, Saint-Malo migmatite belt.

(Ballèvre *et al.* 2001). It is noteworthy that there are no exposures of Icartian-type basement within the Saint-Brieuc Terrane, nor is there any geochemical or isotopic evidence to suggest the presence of unexposed Icartian gneisses at depth within the terrane (see below).

Where local Brioverian successions can be established within the Saint-Brieuc Terrane, lowermost parts are represented by up to 2 km thick successions of submarine volcanics thought to have been erupted into marginal basins (Dupret *et al.* 1985; Cabanis *et al.* 1987; Lees *et al.* 1987; Roach *et al.* 1990; Strachan & Roach 1990). These are well exposed along the southern shores of the Baie de Saint-Brieuc where they occur mainly as pillowed, massive or composite basalt flows interdigitated with intrusive sheets (Roach *et al.* 1990). Whole- and broken-pillow breccias, hyaloclastites and peperites are locally common. Brioverian volcanics north of Coutances (Fig. 3.4) display a range of compositions from basalt to rhyolite, and voluminous pyroclastic rocks (Dupret *et al.* 1985). A rhyolitic sheet within the Spilites de Paimpol along the south side of the North Trégor Batholith (Fig. 3.4) has yielded a $^{207}Pb/^{206}Pb$ single zircon evaporation date of 610 ± 9 Ma (Egal *et al.* 1996), and a tuff within the Erquy Volcanic Formation on the east side of the Baie de Saint-Brieuc has provided a U–Pb zircon (SHRIMP) date of 608 ± 7 Ma (Fig. 3.9; Cocherie *et al.* 2001). The Brioverian succession in the Saint-Brieuc Terrane is probably, therefore, no older than *c.* 610 Ma. A metadolerite dyke suite, and possibly also the Le Perray metagabbro (Fig. 3.4; 602 ± 8 Ma, U–Pb zircon; Guerrot & Peucat 1990), cut the older orthogneisses of the Penthièvre Complex, and may have been co-magmatic with the Brioverian basic volcanics described above (see also Lees *et al.* 1987). Geochemical data (summarized in Egal *et al.* 1996) indicate that the basic volcanism around the Baie de Saint-Brieuc was dominated by continental tholeiites, consistent with an extensional environment, either a within-plate or backarc setting, while the chemistry of mafic lavas close to the Trégor batholith in the NW is more consistent with a suprasubduction setting.

On the west side of the Baie de Saint-Brieuc, the Brioverian volcanic successions are thought to pass conformably upwards into turbiditic sandstones and siltstones (Fig. 3.5C; Ryan 1973; Denis & Dabard 1988; Rabu *et al.* 1983; Strachan & Roach 1990). Petrographic and chemical analyses indicate a mixed source region, potentially comprising volcanic arc and metamorphic material (Denis & Dabard 1988). The island of Jersey (Figs 3.4 & 3.10) has been considered to be part of the Saint-Brieuc Terrane (Strachan *et al.* 1996) on the basis of the presence of lithologically similar turbidites of the Jersey Shale Formation (Squire 1974). Deposition of the Jersey Shale Formation is well constrained between the U–Pb zircon ages of the youngest detrital zircons (586.7 ± 3.0 Ma and 587.1 ± 2.7 Ma) and a rhyolite within the conformably overlying Jersey Volcanic Group (582.8 +3/−2.7 Ma; Miller *et al.* 2001). The latter is a succession of mainly intermediate to acidic rocks, over 2200 m thick, formed of lava flows, pyroclastic ash-flow deposits, shallow intrusive domes, and volcaniclastic deposits mostly of mass flow derivation (Fig. 3.10; Lees & Roach 1993). Their overall character is that of a high-K calc-alkaline (shoshonitic) suite with

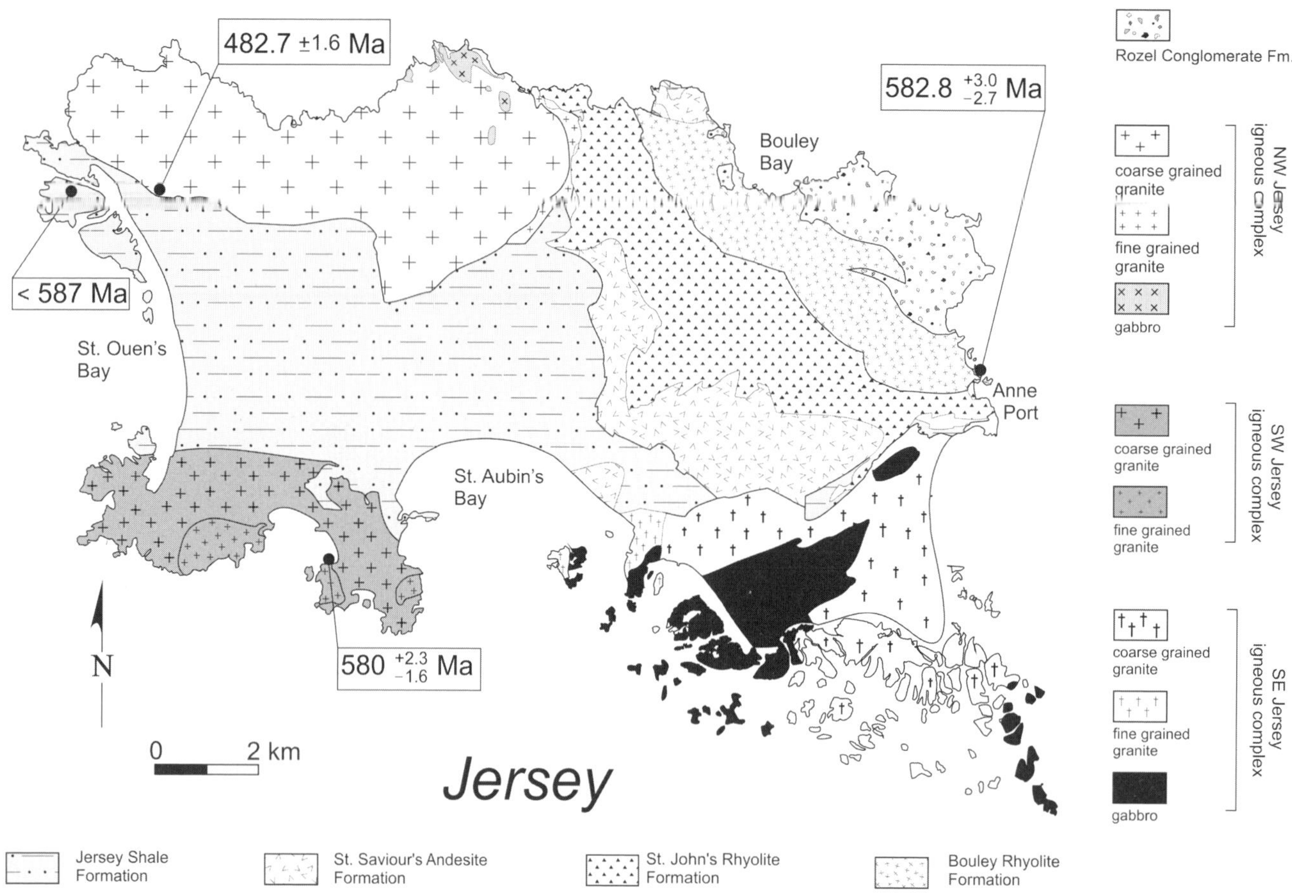

Fig. 3.10. Generalized geological map of Jersey. Dates listed are from published U–Pb zircon dates (see also Table 3.1). Errors are 2σ.

marked enrichment in alkalis and incompatible elements in the upper part of the succession (Lees & Roach 1993). Broad correlation has been commonly assumed between the Jersey Volcanic Group and currently undated ignimbrites exposed to the NE in Normany at Saint-Germain-le-Gaillard (Fig. 3.4; Dupret *et al.* 1990), but remains to be proven.

Around the Baie de Saint-Brieuc (Fig. 3.9), calc-alkaline intrusions were emplaced into the Brioverian rocks and the Penthièvre Complex prior to, and during, regional deformation. The earliest of these intrusions are deformed and include the Port Moguer (also known as Plouha) metatonalite emplaced at 600.4 ± 0.9 Ma (U–Pb zircon; Nagy *et al.* 2002). A younger group of less deformed and metamorphosed intrusions include the syntectonic Fort La Latte quartz diorite (Fig. 3.5D), and the late-tectonic St Quay quartz diorite which were emplaced, at 576.3 +1.5/−1.2 Ma and 574.6 +1.8/−1.5 Ma respectively, (U–Pb zircon; Nagy *et al.* 2002). These intrusions contrast sharply with the quartz diorites in the Trégor–La Hague Terrane as they have positive initial ε_{Nd} values (Guerrot 1989). In contrast, the Southwest Igneous Complex of Jersey (Fig. 3.7) which was emplaced at 580 +2.3/−1.6 Ma (U–Pb zircon; Miller *et al.* 2001) is entirely undeformed and unmetamorphosed, and is composed of granite *sensu stricto* and has negative $\varepsilon_{Nd}(t)$ values of −2.6 to −2.0 (D'Lemos & Brown 1993).

Deformation of the Brioverian rocks resulted in the development of upright, open-to-isoclinal folds with associated axial-planar foliations. Stretching and mineral lineations vary in attitude from subhorizontal to steeply plunging. Strain is markedly hetereogeneous and concentrated in mylonitic shear zones (e.g. Rabu *et al.* 1983; Balé & Brun 1983, 1989; Brun & Balé 1990; Strachan & Roach 1990). Metamorphic grade varies from prehnite-pumpellyite to low amphibolite facies (Ryan 1973; Strachan & Roach 1990; Hèbert & Ballèvre 1993; Le Goff *et al.* 1994; Ballèvre *et al.* 2001). On the west side of the Baie de Saint-Brieuc, the timing of deformation is bracketed by the *c.* 600 Ma Port Moguer tonalite and the *c.* 575 Ma St Quay quartz diorite (see above). $^{40}Ar/^{39}Ar$ hornblende dates of *c.* 575–565 Ma from intermediate plutons and metabasic rocks record post-metamorphic cooling of the region (Dallmeyer *et al.* 1991*b*). On Jersey, deformation of the Brioverian succession is constrained tightly between the *c.* 583 Ma eruption age of the Anne Port Rhyolite and the *c.* 580 Ma intrusion age of the undeformed SW Jersey granite that thermally metamorphoses the folded Jersey Shale Formation (Fig. 3.10; Miller *et al.* 2001). In summary, regional deformation of the Brioverian succession occurred in the interval 583–575 Ma, and was possibly slightly earlier on Jersey than on the west side of the Baie de Saint-Brieuc.

Saint-Malo Terrane

The Saint-Malo Terrane includes greenschist facies metasedimentary rocks that in north Brittany were partially melted at *c.* 540 Ma under low pressure–high temperature conditions to form migmatite belts (Figs 3.4 & 3.9). It is distinct from the Saint-Brieuc Terrane in recording this contrasting and significantly younger metamorphic event. However, the metasedimentary protolith of the migmatite belts has been considered to be essentially part of the same Brioverian succession. In north Brittany, the stratigraphically lowest Brioverian unit in the Saint-Malo Terrane is a sequence of metabasic volcanic rocks located to the SE of the Fort La Latte quartz diorite (Fig. 3.9). These have an alkaline chemistry but have been assigned to both passive-margin (Cabanis *et al.* 1987) and oceanic environments (Graviou 1992). The volcanics are overlain by interbedded sandstones and siltstones that also include thin black chert layers. Otherwise, the protolith sediments are petrographically and chemically similar to Brioverian sediments in parts of the Saint-Brieuc Terrane (Denis & Dabard 1988; Darlet *et al.* 1990). The depositional setting of the Brioverian sediments in the Saint-Malo Terrane is uncertain, and both coastal shelf zone (Dabard 1997) and accretionary prism (Graviou 1992) environments have been proposed.

There are three main migmatite belts, of which the largest is the Saint-Malo Belt (Brown 1974, 1979; Brun 1977; Brun & Martin 1978). These belts comprise metatexites and diatexites that grade into homogeneous anatectic granitoids (Fig. 3.5E). P-T estimates of 650–750°C and 3–4 kbar (see discussion in Ballèvre *et al.* 2001) are consistent with melting at relatively shallow crustal levels (*c.* 9–12 km). Contacts between the migmatites and the Brioverian metasediments are generally defined by steeply dipping sinistral shear zones that largely obscure their original relationships (Brown 1974, 1978; Brun & Balé 1990; Treloar & Strachan 1990). However, in the Rance Valley (Fig. 3.9) Brioverian phyllites have transitional contacts with migmatites of the Saint-Malo Belt, clearly indicating that the migmatites resulted from the partial melting of Brioverian sediments (Brun 1977; Brun & Martin 1978). U–Pb zircon (multigrain) analyses indicate that anatexis occurred at 541 ± 5 Ma (Peucat 1986).

Deformation within and along the margins of the Saint-Malo Terrane was associated with strike-slip displacements along the Fresnaye and Plouer-Cancale shear zones (Figs 3.4 & 3.9; Brun & Balé 1990; Treloar & Strachan 1990) and reverse displacements along dip-slip shear zones in the Rance Valley (Fig. 3.9; Treloar & Strachan 1990). Important evidence relating to the timing of deformation is provided by the observation that granitic sheets interpreted as fractions of the migmatite melt were emplaced syntectonically into sinistral shear zones on all scales (Brown 1978; Brun & Balé 1990; Gapais & Balé 1990; Treloar & Strachan 1990). Egal *et al.* (1996) reported a $^{207}Pb/^{206}Pb$ zircon evaporation date of 555 ± 32 Ma for the Cancale Granite, a syntectonic, anatectic granite emplaced within the Plouer-Cancale shear zone (Fig. 3.4). This evidence, in combination with that of Peucat (1986) (see above), constrains the main phase of Cadomian melting and deformation within the Saint-Malo Terrane to have occurred at 550–540 Ma.

Inland and to the west of the Baie de Saint-Brieuc the Guingamp migmatites and leucogranites (Fig. 3.4), described in detail by Ballèvre *et al.* (2001 and references therein) are exposed. Most authors have recognized a 'Guingamp-Saint-Malo Migmatite Belt' and assumed that migmatites from both areas have the same age and origin (e.g. Egal *et al.* 1996; Strachan *et al.* 1996). However, even though the Guingamp migmatites record similar temperatures to the Saint-Malo examples, they formed at distinctly higher pressures of 5–7 kbar (Ballèvre *et al.* 2001). The leucogranites formed at a late stage during migmatization and cross-cut major structures, including the major tectonic boundary that forms the southern margin of the Saint-Brieuc Terrane. Hèbert (1995) and Ballèvre *et al.* (2001) suggested the existence of two separate episodes of partial melting, both triggered by crustal thickening, the Guingamp migmatites being older than the St Malo migmatites, although this suggestion requires further, precise geochronological testing.

Mancellian Terrane

The Mancellian Terrane corresponds to the Fougères Unit of Ballèvre *et al.* (2001) and Chantraine *et al.* (2001) and mainly comprises Brioverian sediments that are intruded by voluminous granites of the Mancellian Batholith (Fig. 3.4; Dissler *et al.* 1988; Dupret *et al.* 1990). The Mancellian Terrane is distinguished from the Saint-Malo Terrane by the absence of migmatites, and the presence of predominantly undeformed granite batholiths that intrude relatively lower-grade Brioverian rocks. The Brioverian sediments are composed of sandstones, siltstones and mudstones, with minor conglomerate horizons, and were deposited as turbidite and submarine fan successions (Garlan 1985; Dupret *et al.* 1990). The supposed tillitic nature of some conglomerate horizons is now rejected (Winterer 1964; Doré *et al.* 1985). In addition to a variety of volcanic, plutonic and metamorphic clasts, the sediments also include clasts of reworked sediment, most notably black chert. Rare interbedded pillow basalts occur (Dupret *et al.* 1990). The Brioverian rocks were affected by low-grade metamorphism during the development of open-to-tight, upright, northeasterly trending folds.

The Mancellian Batholith comprises a series of peraluminous plutons, six of which exceed 50 km^2 at current exposure. Lithologically they range between biotite granodiorite and two-mica tourmaline leucogranite (Jonin 1981). The plutons are discordant to country-rock structures and contain locally stoped, pre-deformed blocks that, together with contact metamorphic aureole assemblages, indicate emplacement at upper crustal levels (maximum depth of about 6 km; Ballèvre *et al.* 2001). Minor basic complexes occur close to granites in the southern parts of the region (Le Gall & Mary 1983; Le Gall & Barrat 1987), but their temporal relationships are uncertain. Mafic magma involvement in granite genesis may be recorded by magma-mixed mafic enclaves present in some granites. More common, however, are enclaves of metasedimentary material, including spinel- and corundum-bearing restite, and, in the Athis Granite, garnet-cordierite restite (Brown 1995). There is general agreement that the chemical characteristics of the Mancellian Batholith are indicative of melting of a metasedimentary crustal source with a limited mantle input (Jonin 1981; Graviou & Auvray 1985; Le Gall & Barrat 1987; Graviou *et al.* 1988; Brown & D'Lemos 1991; D'Lemos & Brown 1993). There are no high precision zircon dates for any components of the Mancellian Batholith. Monazite U–Pb dates of 547 ± 10 Ma and 542 ± 9 Ma from the Vire Granite (Fig. 3.4, Table 3.1; Pasteels & Doré 1982) provide the best current estimate of crystallization of this pluton, and a broadly similar age has generally been assumed for the other Mancellian Batholith intrusions based upon Rb–Sr dating (Graviou *et al.* 1988).

Structural evolution

The *c.* 620–610 Ma structures preserved on Guernsey, Sark and on the La Hague peninsula are the earliest deformational features that might be assigned to Cadomian orogenic activity, and most probably pre-date accumulation of the Brioverian succession of the Saint-Brieuc Terrane. Because of their geographical restriction, their tectonic significance is uncertain, although it is notable that the north–south orientation of ductile fabrics is highly oblique to the trend of younger Cadomian structures to the SE. The metamorphic data obtained by Inglis *et al.* (2004) from La Hague indicate that deformation was accompanied by significant crustal thickening. The southern boundary of the Trégor–La Hague Terrane is exposed only in north Brittany where it is represented by a brittle fault to the south of the North Trégor Batholith (Fig. 3.4). It is therefore, difficult to establish with certainty the time of its juxtaposition with the Saint-Brieuc Terrane.

In northern Brittany, there is evidence for two younger and quite distinct phases of Cadomian deformation. The older of these at *c.* 580–575 Ma has been identified most clearly in the Saint-Brieuc Terrane, and the younger at *c.* 550–540 Ma is restricted to the Saint-Malo Terrane (e.g. Treloar & Strachan 1990; Egal *et al.* 1996). In the Saint-Brieuc Terrane, widespread folding and fabric development was synchronous with displacement along apparently contemporaneous oblique-slip and strike-slip shear zones (Balé & Brun 1983, 1989; Strachan & Roach 1990), a structural pattern that is consistent with oblique shortening (Brun & Balé 1990; Strachan & Roach 1990). Hèbert & Ballèvre (1993) show that deformation of the Brioverian succession was accompanied by low-pressure metamorphism (3–4 kbar), and, hence, there is no requirement for major crustal thickening during this event.

Deformation within and along the margins of the Saint-Malo Terrane was dominated by sinistral strike-slip displacements parallel to stretching and mineral lineations that plunge gently to the NE (Fig. 3.5F; Brun & Balé 1990; Gapais & Balé 1990; Treloar & Strachan 1990). Development of widespread upright, tight to isoclinal folds accompanied strike-slip displacements. These structures were interpreted by Treloar & Strachan (1990) and Egal *et al.* (1996) to result from sinistrally transpressive deformation in a regional strike-slip belt. Displacements along reverse shear zones in the Rance Valley (Fig. 3.9) are thought to have occurred during strike-slip movements as a result of strain partitioning (Treloar & Strachan 1990). Final amalgamation of these terranes, and the Mancellian Terrane, is considered to have occurred at *c.* 550–540 Ma as a result of regionally significant sinistral displacements along the Fresnaye and Plouer-Cancale shear zones. Egal *et al.* (1996) suggested that these shear zones linked into the North Armorican shear zone that was also probably active during this event as a major sinistral strike-slip structure. Seismic reflection studies carried out north of the Baie de Saint-Brieuc suggest that within the upper crust the Fresnaye and Plouer-Cancale shear zones are steeply dipping (*c.* 60°) towards the NW, consistent with oblique convergence (Fig. 3.11; Brun *et al.* 2001).

The relative importance of strike-slip versus thrust displacements has been debated (e.g. Brun 1992; Strachan *et al.* 1992). Contrasting interpretations have been placed on a zone of thrusting identified south and west of the Baie de Saint-Brieuc (Fig. 3.9; Balé & Brun 1983, 1989). One model views this structural feature as a localized constrictional bend within the regional sinistral strike-slip belt (Egal *et al.* 1996; Strachan *et al.* 1996). An alternative model stresses the importance of oblique thrusting, preferring to interpret this feature as a major thrust (the 'Main Cadomian Thrust'; e.g. Brun & Balé 1990; Ballèvre *et al.* 2001; Brun *et al.* 2001 and references therein). These workers envisage that early displacement at 580–570 Ma along this structure resulted in overthrusting of the Saint-Brieuc Terrane onto the Saint-Malo Terrane, resulting in crustal thickening and consequent formation of the migmatite belts that were reworked at 550–540 Ma by continued SW-directed overthrusting. In this model, the sinistral Fresnaye and Plouer-Cancale shear zones are viewed as lateral ramps or transfer zones within an overall oblique convergent system.

Post-Cadomian sedimentation and magmatism

Unfossiliferous sandstones and conglomerates lie unconformably on Cadomian plutons and Brioverian metasediments on Alderney, Jersey and around the Baie de Saint-Brieuc (Fig. 3.4). Facies analysis indicates that these were deposited in fluvial and marginal marine environments, and regional palaeoflow was broadly to the east (Went & Andrews 1990). These successions that been correlated by Went & Andrews (1990) with red beds which, in Normandy, unconformably overlie Brioverian sedimentary rocks and the Mancellian Batholith and pass transitionally up into fossiliferous Cambrian marine strata (Doré 1994 and references therein). Cambrian sandstones also unconformably overlie the Sary Granodiorite on the La Hague peninsula (Fig. 3.7B; Power *et al.* 1990*b*).

Early Palaeozoic, post-Cadomian magmatism in North Armorica is rare. Felsic volcanics are interbedded with the Plourivo-Plouézec red beds, south of the North Trégor Batholith (Fig. 3.4; Auvray *et al.* 1980) and with Cambrian–Ordovician(?) sediments in Lower Normandy and Maine (Le Gall 1993), but none has yielded reliable isotopic ages. One component of the NW Jersey igneous complex (Fig. 3.10) has a U–Pb zircon upper intercept date of 482.7 $\pm$ 1.6 Ma, taken as the best estimate of the timing of intrusion (Miller *et al.* 2001), and similar to a $^{40}Ar/^{39}Ar$ hornblende cooling date of *c.* 473 Ma obtained for another component of the complex (D'Lemos *et al.* 1992*a*). However, the tectonic significance of early Ordovician plutonism in North Armorica is uncertain at present.

Palaeogeographical constraints in the Late Neoproterozoic

Most palaeogeographic reconstructions rely on palaeomagnetic studies, with supporting information from geographically characteristic faunas and/or climatic indicators (e.g. van der Voo 1988). Application of these tools is limited for North Armorica since most primary magnetic inclinations were reset by younger geological events, and the rock record is mostly unfossiliferous. However, the ages of detrital zircons from Brioverian metasediments shed light on its latest Neoproterozoic position (Fernández-Suárez *et al.* 2002; Samson *et al.* 2005). Detrital zircons give late Neoproterozoic, Palaeoproterozoic and, much more rarely, Archaean ages (Fig. 3.12). The late Neoproterozoic grains provide no palaeogeographic information because all of these could have been locally derived from 750–580 Ma plutons. It is not clear if zircons with ages between 2.1 and 2.0 Ga could have been locally derived from the Saint-Brieuc Terrane as no known rocks of that age are currently exposed. However, the zircons could have been derived from Icartian gneisses within the Trégor–La Hague Terrane, which would require a geographical connection between the two terranes from about 610 Ma onwards. Alternatively, zircons of this age could also be derived from other continental sources, which is probably a requirement for the zircons with ages of $\geqslant$2.2 Ga since there are no known crustal sources anywhere in North Armorica with rocks of such antiquity.

Because the majority of non-Neoproterozoic detrital zircon is aged between 2.4 and 2.0 Ga it is considered very unlikely that either Laurentia or Baltica were sources because they do not

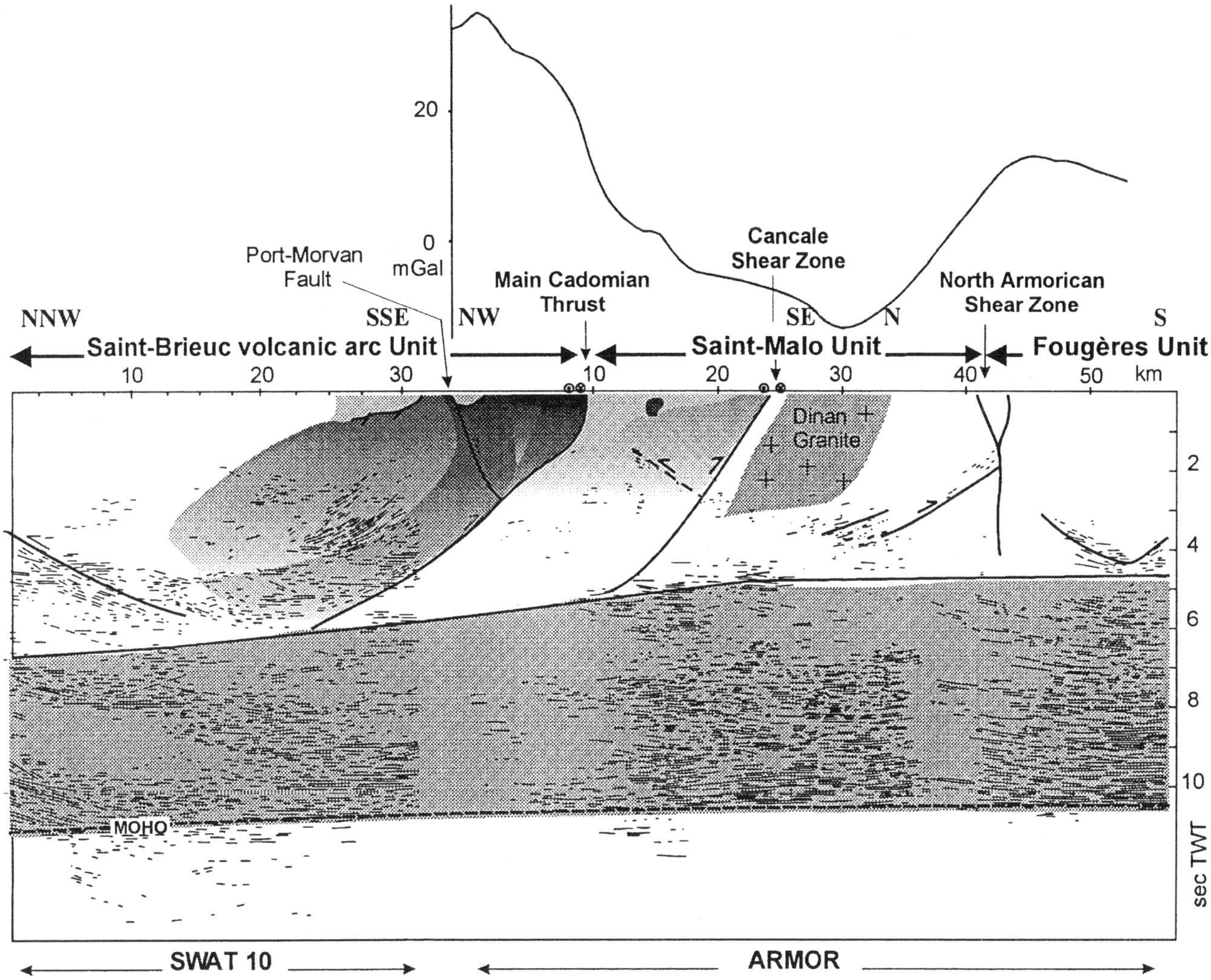

Fig. 3.11. Composite line drawing of the ARMOR–SWAT10 (southern part) seismic lines and gravity anomaly profile across the Cadomian belt in north Brittany (from Brun *et al.* 2001). The 'units' depicted correspond to the terranes described in the present text, with the Fougères Unit representing the Mancellian Terrane. The 'Main Cadomian Thrust' corresponds to the Fresnaye shear zone in Figures 3.4 abd 3.9. Interpretative bodies within the upper crust are those obtained from the gravity model, with the denser rocks within the Saint-Brieuc Unit/Terrane in dark grey and the less dense ones within the Saint-Malo and Mancellian units/terranes in light grey. The geometry of the main faults has been deduced from some shallow reflectors.

incorporate common igneous rocks of that age range (see discussion in Samson *et al.* 2005). Western Gondwana contains abundant rocks of this age range, making it a more likely source region. However, some regions within western Gondwana, such at the Amazon Craton, also contain extensive regions of Mesoproterozoic crust (e.g. Tassinari & Macambria 1999) that would therefore be expected to be present in the detrital zircon suites. However, no Mesoproterozoic zircons have yet been identified in North Armorica, thus making the Amazon Craton and similar regions unlikely candidates for providing the detritus. The West African Craton, in contrast, does contain extensive 2.3–2.0 Ga rocks and Archaean exposures, but has virtually no Mesoproterozoic record, and thus is the most likely non-local sediment source for Brioverian detritus (Fig. 3.9; Fernández-Suárez *et al.* 2002; Samson *et al.* 2005). However, it is difficult to constrain how far away the West African Craton may have been from the sampled Brioverian rocks at the time of their deposition, given that very long transport distances of zircon appear to be possible (e.g. Rainbird *et al.* 1992, 1997). Linnemann *et al.* (2004), for example, suggest that Armorica remained part of West Africa until at least the earliest Carboniferous, based on the similarity of detrital zircon dates from Neoproterozoic and Ordovician sandstones and the similarity in Nd isotopic compositions of Cambrian to Carboniferous shales.

Stratigraphic framework of the Brioverian

Correlation of the Brioverian volcanosedimentary successions in North Armorica is difficult because of the lack of distinctive or datable marker horizons and the tectonically segmented nature of the Cadomian belt. Early workers divided the Brioverian into three major groups: the Lower, Middle and Upper Brioverian (e.g. Graindor 1957; Cogné 1962). This was supplanted by the view that the Brioverian could be divided into an older (pre-570 Ma), more deformed sequence that contained interbedded black chert (essentially the Saint-Brieuc and Saint-Malo terranes) and a younger (post-570 Ma), less deformed sequence that contained detrital black chert (i.e. the Mancellian Terrane and the Brioverian of Jersey) and was thus viewed as having been derived from

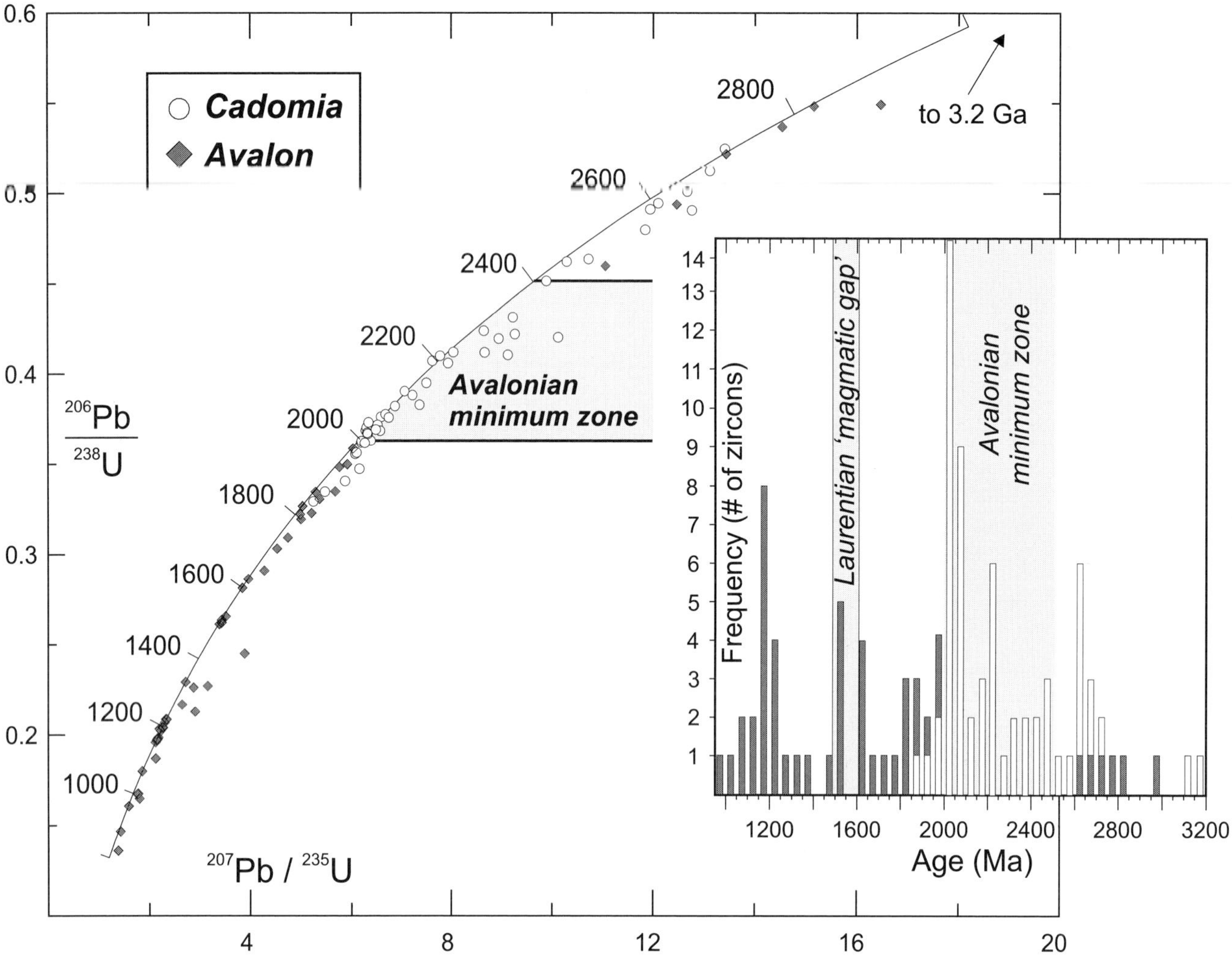

Fig. 3.12. Compilation by Samson *et al.* (2005) from published sources of pre-Neoproterozoic detrital zircon Cadomia and Avalonia. Inset shows histogram, using 50 Ma increments, for the combined data sets. The histogram emphasizes the high abundance of 2400–2050 Ma zircon crystals in Cadomia and their virtual absence in Avalon. The interval of 1610–1490 Ma is a period of magmatic quiescence in Laurentia, but zircons of this age are well represented in Avalonian sedimentary rocks.

the erosion of the older sequence (e.g. Dupret *et al.* 1990; Rabu *et al.* 1990 and references therein).

As indicated earlier, tight geochronological controls exist on the age of the Brioverian successions on Jersey (*c.* 587–583 Ma) and around the Baie de Saint-Brieuc (*c.* 610–575 Ma), but the Brioverian successions in the Saint-Malo and Mancellian terranes are less well constrained. The suggestion that the presence of detrital black chert could be used to distinguish between pre- and post-570 Ma successions is challenged by the geochronological evidence that the Brioverian on Jersey, containing chert clasts, was deformed at *c.* 583–580 Ma. Taking these considerations into account, as well as the structural evidence for potentially large lateral relative movements between the various terranes that make up the belt, it seems highly unlikely that it is possible to erect a coherent Brioverian stratigraphy, and indeed the 'Brioverian' may have been deposited in a number of entirely separate basins.

Terranes

Tectonic models for the Cadomian Orogeny need to take into account and explain the main characteristics of, and differences between, the various terranes in North Armorica. These may be summarized as follows (Table 3.2).

- On a regional scale, the boundary between the Saint-Brieuc and Saint-Malo terranes separates different Cadomian magmatic provinces: mantle-derived, calc-alkaline suites occur to the NW, and intracrustal granitoids and migmatites to the SE (Auvray 1979; Graviou & Auvray 1985).
- The Trégor–La Hague Terrane contains small areas of Palaeoproterozoic (*c.* 2 Ga) basement that were deformed and intruded at *c.* 615–610 Ma by syntectonic calc-alkaline igneous complexes. These plutons were contaminated significantly by the 2 Ga basement. A second pulse of magmatism at 580–570 Ma was possibly synchronous with semi-ductile to brittle faulting.
- The Saint-Brieuc Terrane incorporates a juvenile Neoproterozoic (750–620 Ma) magmatic arc in the Baie de Saint-Brieuc region, which shows no evidence for interaction with Palaeoproterozoic basement. It is unconformably overlain by the post-610 Ma Brioverian succession that underwent widespread deformation and metamorphism up to amphibo-

Table 3.2. *Tectonomagmatic events in the North Armorican Massif*

Approx. age	Trégor-La Hague Terrane	Saint-Brieuc Terrane	Saint-Malo Terrane	Mancellian Terrane
2 Ga	Formation of Icartian protoliths			
?	Rifting of Icartian block from West African Craton			
750 Ma		Inception of juvenile island arc (Port Morvan)		
650 Ma		Further growth/accretion of island arc material (Cesson provenance)		
620 Ma	**Crustal thickening and regional deformation** (700–800°C, 8–9 kbar)	Emplacement of island arc plutons (Jospinet granodiorite)		
620–610 Ma	Emplacement of syntectonic arc plutons during waning deformation. Exhumation, cooling below 500°C	Exhumation and erosion of island arc		
610–580 Ma	Magmatic quiescence. Erosion to provide detritus for Brioverian basin(s)	Extension, eruption of volcanics, followed by detrital accumulation. Intrusion of mafic – intermediate plutons	? Accumulation of detrital sequence	? Accumulation of detrital sequence
580–570 Ma	Emplacement of intermediate granitic plutons in mainly post-tectonic setting with localized deformation (mid–upper crust)	**Regional deformation and metamorphism** (locally to 650–750°C, 8–10 kbar), emplacement of syn- to late tectonic intermediate plutons. Regional cooling below 500°C	? Regional deformation	? Regional deformation
550–540 Ma			Anatexis (600–700°C, 3–4 kbar) **Transpressional deformation and terrane amalgamation**	Intracrustal melting and emplacement of peraluminous post-tectonic batholith at shallow levels. Erosion
540–530 Ma	Onset Cambrian deposition	Onset Cambrian deposition	Exhumation of migmatite terranes. Emplacement of post-tectonic plutons	Onset Cambrian deposition

lite facies grade at *c.* 580–570 Ma. Coeval calc-alkaline plutons also appear to be mostly juvenile.

- The Saint-Malo Terrane comprises Brioverian metasediments that underwent deformation and widespread anatexis at *c.* 550–540 Ma. The Guingamp Unit of this terrane may have undergone anatexis at a slightly earlier time.
- The Mancellian Terrane comprises low-grade Brioverian metasediments that were deformed (timing uncertain) and intruded by voluminous crustally derived, high-level granitoids at *c.* 540 Ma.

The degree to which these terranes evolved independently prior to amalgamation is uncertain. There is general agreement that the boundary between the Saint-Brieuc and Saint-Malo terranes is an important regional structure (e.g. Cogné 1972; Auvray 1979; Graviou & Auvray 1985; Cabanis *et al.* 1987; Dissler *et al.* 1988; Brun & Balé 1990; Dupret *et al.* 1990; Rabu *et al.* 1990), but less clear is the significance of differences to the NW between the Trégor–La Hague and Saint-Brieuc terranes, and to the SE between the Saint-Malo and Mancellian terranes.

The suggestion that the Trégor–La Hague Terrane might in part represent the basement to the Saint-Brieuc Terrane (Strachan *et al.* 1996) is now rejected as it is not supported by field or isotopic evidence. Some points of commonality might indicate that these terranes evolved in close proximity, although none are conclusive. The Palaeoproterozoic detrital zircons within the Saint-Brieuc Terrane could have been derived from the Trégor–La Hague Terrane, but a West African source to the south cannot be excluded. Both terranes were intruded by calc-alkaline plutons at *c.* 580–575 Ma, but this most likely reflects the location of both terranes along the same active plate margin. The available evidence does not, therefore, resolve whether or not these terranes were in their present relative positions prior to *c.* 580 Ma.

The contrasting nature of the metamorphic and deformation histories of the Saint-Malo and Mancellian terranes has been taken to indicate that they were unrelated prior to amalgamation at *c.* 540 Ma (e.g. Strachan *et al.* 1989; Ballèvre *et al.* 2001). Conversely, they could represent deeper and shallower exposure levels of essentially the same terrane, respectively (e.g. D'Lemos *et al.* 1992*b*; Brown 1995; Fig. 3.13). Geochemical and isotopic similarities between the migmatites and anatectic granites of the Saint-Malo Terrane and some components of the Mancellian Batholith indicate that both may have been derived from a similar, metasedimentary (Brioverian) source (Guerrot 1989; Brown & D'Lemos 1991; D'Lemos & Brown 1993). Clearly, the entire Mancellian Batholith was not derived directly from the Saint-Malo, or similar, migmatites. The mineralogies of the Saint-Malo migmatites and locally sourced muscovite-bearing leucogranites are indicative of dominantly vapour-present melting at mid to upper crustal levels, and there is little evidence for mass melt extraction from the exposed migmatite belt. Furthermore, while leucogranite is present in the Mancellian Batholith, biotite granodiorite is volumetrically dominant. However, it can be envisaged that melting at deeper, hotter, mid-lower crustal levels beneath a migmatite zone similar to that exposed in the Saint-Malo Belt would have resulted in greater dehydration melting and the production of larger volumes of biotite grano-

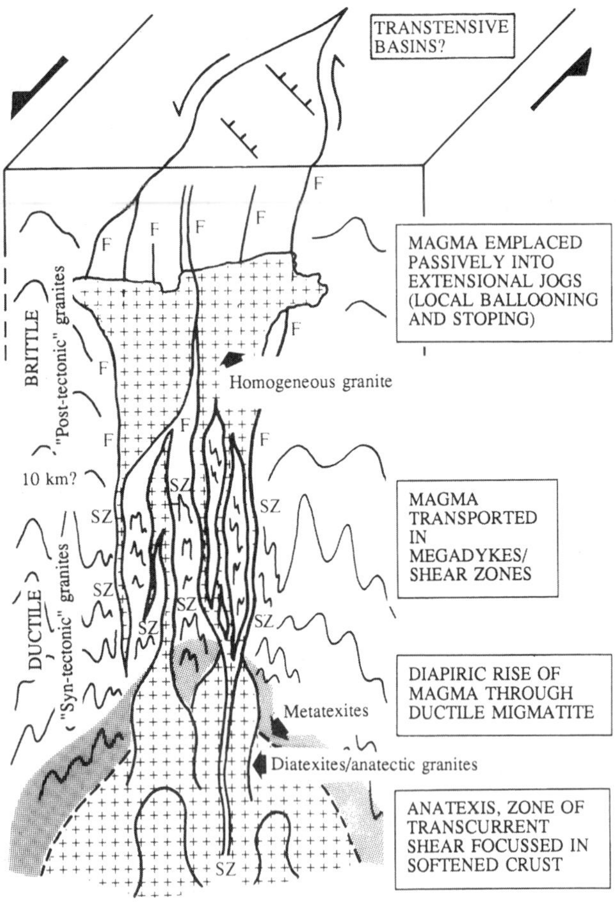

Fig. 3.13. Generalized model for the generation, ascent and emplacement of granite magma within a transpressional Cadomian orogen (F, fault; SZ, shear zone). The lower part of the diagram represents the Saint-Malo Terrane and the upper part the Mancellian Terrane (from D'Lemos *et al.* 1992*b*).

diorite that were transferred via transpressional shear zones and fault systems to upper crustal levels – essentially the model proposed by D'Lemos *et al.* (1992*b*). Small intrusions of this type are present as minor diapiric bodies of anatectic granitoids within the Saint-Malo Belt (Brown 1979). Restitic components found within the Mancellian batholith are consistent with a migmatitic source (Brown 1995). Nonetheless, Ballèvre *et al.* (2001) argue that whereas granite magmatism in the Saint-Malo region was driven by crustal thickening at intermediate crustal levels, the magmatism in the Mancellian region was induced by mafic underplating within the lower crust. Further isotopic studies are needed to determine and compare the precise timing of magmatism and deformation, to establish the sources of the various granitoid magmas, and to examine indirectly the nature of the underlying basement.

Tectonic models

Tectonic models for the Cadomian belt are necessarily speculative because of a lack of unambiguous evidence for the location and direction of dip of associated subduction zones and the palaeogeographic setting of constituent terranes. This is reflected, for example, in published models that have advocated either SE-dipping (e.g. Graviou & Auvray 1985; Cabanis *et al.* 1987; Brown *et al.* 1990; Dupret *et al.* 1990; Rabu *et al.* 1990; Brown 1995; Strachan *et al.* 1996) or NW-dipping (Brun & Balé 1990; Graviou 1992; Hébert 1993; Chantraine *et al.* 2001) directions of subduction. In either interpretation, it could be argued that the various terranes are in approximately their original palinspastic locations, and that the Cadomian belt resulted from the more or less orthogonal accretion of Icartian continental basement, Penthièvre-type juvenile arcs, and Brioverian volcanosedimentary successions that have been variously assigned to intra-arc, forearc and continental margin settings, depending upon the preferred tectonic model. Equally permissible, however, are more mobile models that stress the potential importance and magnitude of strike-slip displacements during accretion meaning that none of these terranes originated in their present relative positions (Strachan *et al.* 1989). In this context, we can attempt to summarize the general tectonic evolution of the Cadomian belt in terms of active plate boundary processes along the Gondwanan margin.

There is broad consensus that the Icartian basement of the Trégor–La Hague Terrane is probably a detached fragment of the West African Craton, that records significant crustal growth and emplacement of voluminous granitoids during the Eburnian orogenic event at around 2 Ga (e.g. Boher *et al.* 1992). We suggest that detachment most likely occurred during the main rifting event that ultimately led to the development of the passive margin of the West African Craton. Ages of 900–800 Ma are suggested for the depositional age of passive margin sedimentary rocks along the western margin of the West African Craton (e.g. Bouougri & Saquaque 2004) implying rifting prior to that time, but such ages are poorly constrained.

The earliest event that may be assigned to Cadomian-style tectonic processes corresponds to the formation of juvenile mid-Neoproterozoic felsic island arc crust, represented by the Penthièvre Complex of the Saint-Brieuc Terrane. This is assumed to define the onset of subduction along the Gondwanan margin from 750 Ma onwards. The development of a number of juvenile arcs is implied by the 750–650 Ma age span of the observed components and proximally eroded detritus in overlying Brioverian sediments. It has been assumed that these arcs developed broadly to the south (in current co-ordinates) of the Icartian crust of the Trégor–La Hague Terrane (e.g. Chantraine *et al.* 2001), although given the lack of evidence for linkage until 580 Ma or later, this may not have been the case.

The Trégor–La Hague Terrane experienced crustal thickening at *c.* 620 Ma, possibly as a result of the collision with an island arc, perhaps the Penthièvre or analogous arcs. Much of the pan-African Belt of North Africa (e.g. Anti-Atlas region of Morocco) exhibits a similar history as numerous arcs collided with the West African Craton during the period 750 to 600 Ma. Hence, while it is possible that a crustal block composed of Icartian basement had rifted from the West African Craton at an earlier stage, it may have remained near the edge of the craton at this time.

Continued subduction is recorded by numerous calc-alkaline plutons from *c.* 620 Ma onwards in the Trégor–La Hague Terrane. Although the magmas were likely mantle-derived they were significantly contaminated by 2 Ga basement. Magmatic activity of generally similar age is also recorded in the Saint-Brieuc Terrane (e.g. Jospinet Granodiorite, Port Moguer Tonalite) but here the lack of an ancient basement has preserved the depleted, mantle-like isotopic composition of the intrusions. By *c.* 610 Ma, the Brioverian mafic volcanics and sedimentary rocks began to accumulate in the Saint-Brieuc Terrane, although whether in a backarc (e.g. Dupret *et al.* 1990; Rabu *et al.* 1990) or a forearc setting (e.g. Graviou 1992; Chantraine *et al.* 2001) is uncertain, as are temporal and spatial relationships with the Brioverian successions of the Saint-Malo and Mancellian terranes. Early basic volcanism in the Baie de Saint-Brieuc was succeeded by clastic sedimentation derived from a source dominated by 2.4–2.0 Ga crust and 750–650 Ma juvenile arc material. One potential source is represented by the Trégor–La Hague Terrane but others cannot be precluded. Deposition continued until *c.* 585 Ma when the basin was inverted and deformed, and by 580–575 Ma intruded by syn- to post-tectonic plutons. Deformation and metamorphism at *c.* 583–575 Ma of the Saint-Brieuc Terrane was plausibly driven by sinistrally oblique collision with the Trégor–La Hague Terrane. In this context, we speculate that the *c.* 580–570 Ma magmatic pulse in the Channel Islands portion of the Trégor–La Hague Terrane could have resulted from slab break-off following collision. Upward flow of hot asthenospheric mantle into the slab-break window could have provided both the heat and isotopically distinct material necessary to generate the *c.* 580 Ma, more radiogenic magmas in this part of the belt (Samson *et al.* 1997).

The abrupt nature of the tectonic boundary that separates arc-related and intracrustal magmatic provinces (the Fresnaye Shear Zone) suggests that the amalgamated Trégor–La Hague/Saint-Brieuc terranes were not proximal to the Saint-Malo and Mancellian terranes at 580–570 Ma. The cessation of calc-alkaline magmatism shortly afterwards at 570–560 Ma is thought to reflect increasingly oblique subduction and a gradual transition to a transform plate boundary during the latest Neoproterozoic and Cambrian (Murphy & Nance 1991). This was associated with the development of a crustal-scale, sinistral strike-slip shear zone/fault system and the transpressive amalgamation at 550–540 Ma of the Saint-Malo and Mancellian terranes with the previously amalgamated Trégor–La Hague and Saint-Brieuc terranes (Strachan *et al.* 1996). Cumulative lateral displacements were likely substantial (at least tens of kilometres; Treloar & Strachan 1990). The cause of the *c.* 550–540 Ma intracrustal magmatism in the Saint-Malo and Mancellian terranes remains enigmatic and the relative contributions of transpressional thickening, magmatic underplating and slab break-off require further evaluation. A key question, currently unresolved, that has major implications for regional tectonic models relates to the age and affinities of the unexposed basement that underlies the Saint-Malo and Mancellian terranes.

Concluding remarks

The Cadomian Orogenic Belt in North Armorica comprises a number of shear zone and fault-bounded microcontinental, volcanic arc and marginal basin terranes that developed along an active margin of Gondwana in the late Neoproterozoic (750–540 Ma). Although these terranes are characterized by geological histories that differ in detail, there is no evidence that any are far-travelled or truly 'exotic' to one another. In fact, sufficient points of commonality exist to support the view that they may have been 'proximal' terranes that evolved along the same plate margin (e.g. Strachan *et al.* 1996; Chantraine *et al.* 2001). Many published tectonic models are speculative because of a lack of unambiguous evidence for the location and direction of dip of associated subduction zones and the palaeogeographic setting of constituent terranes. Models that stress the potential importance and magnitude of strike-slip displacements during accretion may be highly relevant to our understanding of the belt, with the implication that none of the terranes originated in their present relative positions. Both tectonic studies (e.g. Strachan *et al.* 1989; Treloar & Strachan 1990; Egal *et al.* 1996) and detrital mineral studies (e.g. Fernadez-Suarez *et al.* 2002) highlight the importance of strike-slip, margin-parallel terrane transport in the development of the Cadomian Belt. It must be recognized that not only have the various parts of the observed belt been displaced to their current configuration, but also that significant parts of the orogen are likely to have been dispersed elsewhere.

The Bohemian Massif (U. L., K. D., A. G., T. J. & R. L. R.)

Introduction

The Bohemian Massif represents the most prominent inlier of basement rocks in Central Europe. This inlier records a long geological history that includes the formation of cratonic basement at *c.* 2 Ga, protracted Cadomian orogenic processes that culminated at *c.* 570 Ma, and variable reworking during the Variscan Orogeny at *c.* 330 Ma. The latter resulted from the collision of the supercontinents Laurussia and Gondwana. In places, the massif underwent the highest known Variscan metamorphic overprint, ultra-high pressure conditions being documented by metamorphic diamond in the Erzgebirge (Saxo-Thuringian Zone; Massonne 1999). In contrast, other units of the massif show only a very low grade tectonometamorphic overprint and well preserved remnants of the Cadomian basement and its Early Palaeozoic overstep sequence. The Bohemian Massif is bounded to the north and NE by the Mid-German Crystalline Zone, which is the best candidate for the suture of the Rheic Ocean. To the south and SE the Bohemian Massif is bordered by the Meso-Cenozoic orocline of the Alps and the Carpathians. Steep faults in the SW such as the Franconian Line and the Danube Fault juxtapose the basement rocks against Mesozoic platform sediments and the molasse of the Alps (Fig. 3.14).

The principal division of the Bohemian Massif into the Saxo-Thuringian and Moldanubian zones was by Kossmat (1927). The Teplá–Barrandian Unit and the Moravo-Silesian Zone (Fig. 3.14) were introduced later into the literature by various authors and under different names. The oldest geological component of the Bohemian Massif is represented by remnants of Palaeo- to Mesoproterozoic cratonic basement slivers such as the Dobra gneiss (1.38 Ga) and the Svetlik gneiss (2.05–2.1 Ga) (Fig. 3.14).

In most papers concerning palaeogeographic aspects, the Bohemian Massif is interpreted to be a part of the 'classical' Armorica microplate as defined by van der Voo (1979). Finger *et al.* (2000*a*) could demonstrate that a part of the massif shows strong affinities with Avalonia. These rock units are preserved in the Brunovistulian Block that represents the Neoproterozoic basement of the Moravo-Silesian Zone.

Whether or not Armorica actually rifted away from Gondwana to exist as an independent microplate or terrane in Ordovician to Devonian times has been discussed in recent years. The interpretation of palaeomagnetic data by Tait *et al.* (1997, 2000) suggests that Armorica did separate from Gondwana and drifted across the Rheic Ocean during Ordovician to Devonian time, and that a large distance was maintained between Armorica and the Gondwana mainland during at least the Silurian. In contrast, the interpretation of palaeobiogeographical, geochronological geochemical and sedimentological data has led to the conclusion that Armorica was never separated from the Gondwana mainland in pre-Pangaean time (Robardet 2002; Linnemann *et al.* 2004). Other authors in their plate-tectonic reconstructions argue for a

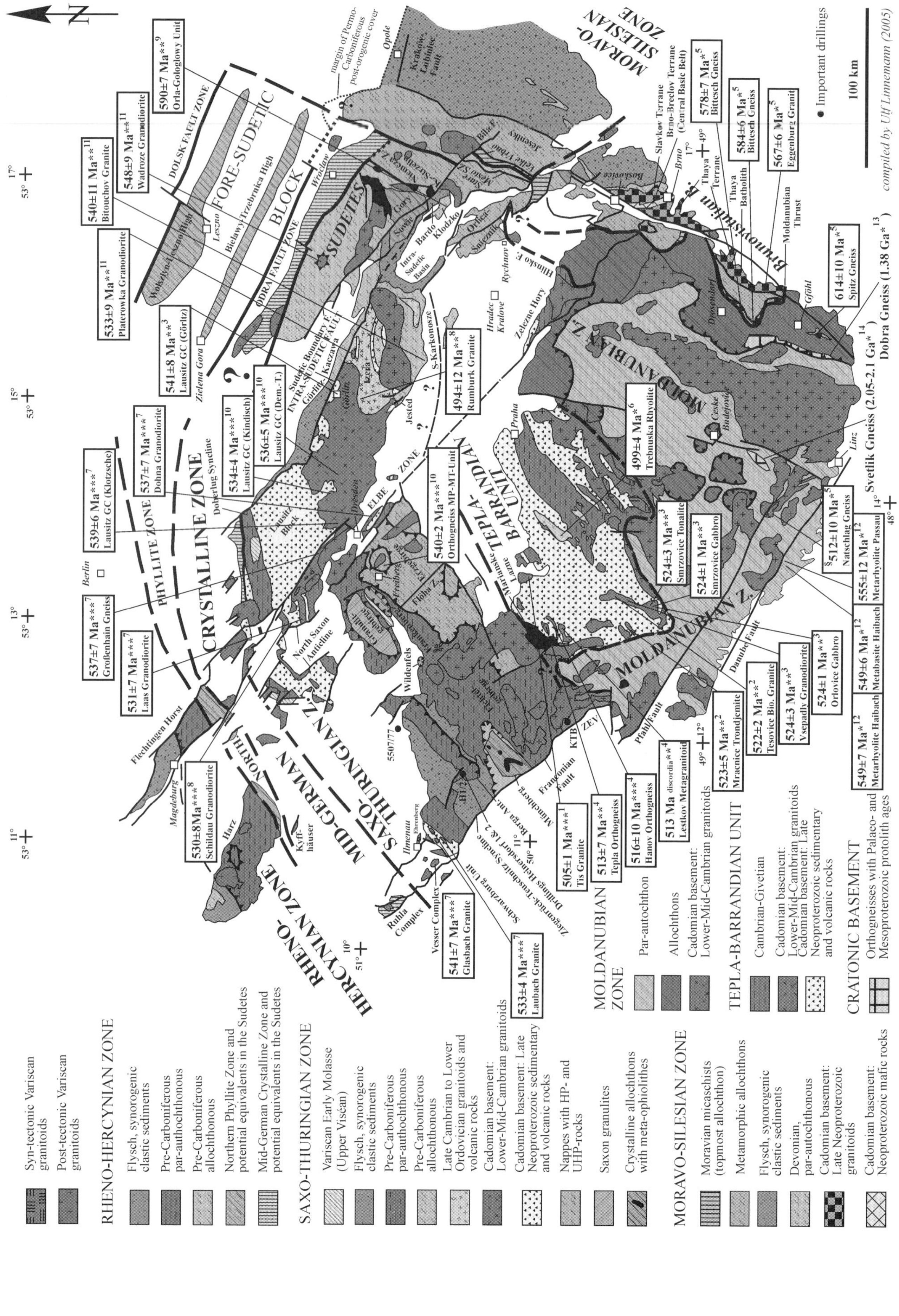
Syn-tectonic Variscan granitoids
Post-tectonic Variscan granitoids
RHENO-HERCYNIAN ZONE
Flysch, synorogenic clastic sediments
Pre-Carboniferous par-authochthonous
Pre-Carboniferous allochthonous
Northern Phyllite Zone and potential equivalents in the Sudetes
Mid-German Crystalline Zone and potential equivalents in the Sudetes
SAXO-THURINGIAN ZONE
Variscan Early Molasse (Upper Viséan)
Flysch, synorogenic clastic sediments
Pre-Carboniferous par-authochthonous
Pre-Carboniferous allochthonous
Late Cambrian to Lower Ordovician granitoids and volcanic rocks
Cadomian basement: Lower-Mid-Cambrian granitoids
Cadomian basement: Late Neoproterozoic sedimentary and volcanic rocks
Nappes with HP- and UHP-rocks
Saxon granulites
Crystalline allochthons with meta-ophiolithes
MORAVO-SILESIAN ZONE
Moravian micaschists (topmost allochthon)
Metamorphic allochthons
Flysch, synorogenic clastic sediments
Devonian, par-autochthonous
Cadomian basement: Late Neoproterozoic granitoids
Cadomian basement: Neoproterozoic mafic rocks
MOLDANUBIAN ZONE
Par-autochthon
Allochthons
Cadomian basement: Lower-Mid-Cambrian granitoids
TEPLA-BARRANDIAN UNIT
Cambrian-Givetian
Cadomian basement: Lower-Mid-Cambrian granitoids
Cadomian basement: Late Neoproterozoic sedimentary and volcanic rocks
CRATONIC BASEMENT
Orthogneisses with Palaeo- and Mesoproterozoic protolith ages
• Important drillings
100 km
compiled by Ulf Linnemann (2005)
RHENO-HERCYNIAN ZONE
MID-GERMAN CRYSTALLINE ZONE
NORTHERN PHYLLITE ZONE
SAXO-THURINGIAN Z.
TEPLA-BARRANDIAN UNIT
MOLDANUBIAN Z.
MORAVO-SILESIAN ZONE
FORE-SUDETIC BLOCK
SUDETES
ELBE ZONE
DOLSK FAULT ZONE
ODRA FAULT ZONE
INTRA-SUDETIC FAULT
Sudetic Boundary F.
Brunovistulian T.
Moldanubian Thrust
Danube Fault
Pfahl Fault
Franconian Fault
Berlin
Magdeburg
Dresden
Praha
Wroclaw
Opole
Brno
Linz
Görlitz
Zielona Gora
Leszno
Hradec Kralove
Ceske Budejovice
Harz
Kyffhäuser
Flechtingen Horst
North Saxon Anticline
Ruhla Complex
Vesser Complex
Ilmenau
Wildenfels
KTB
ZEV
Lausitz Block
Doberlug Syncline
Intra-Sudetic Basin
Bardo
Klodzko
Jested
Jesenky
Góry Sowie
Zelezne Hory
Drosendorf
Gföhl
Thaya Terrane
Thaya Batholith
Slavkov Terrane
Brno-Brecloy Terrane (Central Basic Belt)
Kraków-Lubliniec Fault
margin of Permo-Carboniferous post-orogenic cover
Schwarzburg Unit
Ziegenrück-Teuschnitz-Syncline
Drillings Hohendorf 1&2
Berga Ant.
Münchberg
Fichtelgebirge
Erzgebirge
Granulitgebirge
Frankenberg
Flöha Z.
Freiberg
Marianske Lazne
5507/77
53° 11°
53° 13°
53° 15°
53° 17°
51° 10°
50° 11°
49° 12°
48° 14°
49° 17°
N
541±7 Ma***[7] Glasbach Granite
533±4 Ma***[7] Laubach Granite
530±8Ma***[8] Schildau Granodiorite
531±7 Ma***[7] Laas Granodiorite
537±7 Ma***[7] Großenhain Gneiss
539±6 Ma***[7] Lausitz GC (Klotzsche)
537±7 Ma***[7] Dohna Granodiorite
534±4 Ma***[10] Lausitz GC (Kindisch)
536±5 Ma***[10] Lausitz GC (Dem.-T.)
541±8 Ma**[3] Lausitz GC (Görlitz)
533±9 Ma**[11] Platerowka Granodiorite
540±11 Ma**[11] Bitouchov Granite
548±9 Ma**[11] Wadroze Granodiorite
590±7 Ma**[9] Orla-Gogolowy Unit
494±12 Ma**[8] Rumburk Granite
540±2 Ma***[10] Orthogneiss MP-MT-Unit
499±4 Ma*[6] Trebnuska Rhyolite
524±3 Ma**[3] Smrzovice Tonalite
524±1 Ma**[3] Smrzovice Gabbro
512±10 Ma*[5] Natschlag Gneiss
555±12 Ma*[12] Metarhyolite Passau
549±6 Ma*[12] Metabasite Haibach
549±7 Ma*[12] Metarhyolite Haibach
524±1 Ma**[3] Orlovice Gabbro
524±3 Ma**[3] Vsepadly Granodiorite
522±2 Ma**[2] Tesovice Bio. Granite
523±5 Ma**[2] Mracnice Trondjemite
513 Ma discordia**[4] Lestkov Metagranitoid
516±10 Ma***[4] Hanov Orthogneiss
513±7 Ma**[4] Tepla Orthogneiss
505±1 Ma***[1] Tis Granite
614±10 Ma*[5] Spitz Gneiss
567±6 Ma*[5] Eggenburg Granit
584±6 Ma*[5] Bittesch Gneiss
578±7 Ma*[5] Bittesch Gneiss
Svetlik Gneiss (2.05-2.1 Ga*[14])
Dobra Gneiss (1.38 Ga*[13])

short distance between Armorica and Gondwana and the development of a narrow ocean (McKerrow *et al.* 2000). Finally, another group of geologists suggests the separation of a Hun Superterrane from Gondwana by the opening of the Neotethys in the late Ordovician (von Raumer *et al.* 2003); in their palaeogeographic reconstruction, the Hun Superterrane incorporates the Armorican microplate.

For the late Neoproterozoic to Early Cambrian timespan that is relevant for Cadomian orogenic processes, we prefer to use the term 'Cadomia' *sensu* Nance & Murphy (1996) for the non-Avalonian part of the Bohemian Massif and for the Armorican Massif to prevent confusion. The terms 'Armorica' or 'Armorican Terrane Assemblage' are both overloaded with different interpretations and meanings. 'Cadomia' is composed of the Cadomian basement and older cratonic blocks of the Armorican Massif, the French Massif Central and the Bohemian Massif without the Brunovistulian Block in the Moravo-Silesian Zone (Fig. 3.3). The Cadomian basement in the Bohemian Massif is divided into Avalonia-type and Cadomia-type terranes, units, zones or complexes. Such a classification was proposed by Murphy *et al.* (2004) for peri-Gondwana. Avalonia-type crustal units are characterized by detrital zircons showing Mesoproterozic ages and recycled *c.* 1.0–1.3 Ga juvenile crust. These terranes were accreted to the Gondwanan margin by *c.* 650 Ma. Cadomia-type rock complexes formed along the West African margin by the recycling of ancient (2–3 Ga) West African crust.

Rock complexes with Cadomian basement occur in all zones and units of the Bohemian Massif overprinted by the Variscan Orogeny to different metamorphic degrees under varying geotectonic regimes and styles. In this chapter it is not useful to divide the Bohemian Massif into various terranes because it appears that the entire basement of the massif belonged to one plate from the Neoproterozoic to the Lower Devonian, with the exception of the Brunovistulian Block in the Moravo-Silesian Zone (Linnemann *et al.* 2004). According to Finger *et al.* (2000*a*), the Brunovistulian and most proably the cratonic basement imprint of the 1.38 Ga Dobra gneiss were parts of the Avalonia microcontinent and remnants of Avalonia-type terranes, respectively. In contrast, all other late Neoproterozoic to Devonian parts of the Bohemian Massif belonged to Armorica or alternatively to the Cadomia-type part of peri-Gondwana. This situation suggests a bending of the suture of the Rheic Ocean in the northeastern part of the Bohemian Massif from a NW to a NNE trending direction (Figs 3.3 & 3.14). The oceanic suture between Cadomia-type and Avalonia-type rock complexes is the only terrane boundary *sensu stricto* in the Bohemian Massif. This structure crops out in the exposed units of the Mid-German Crystalline Zone and must be hidden and incorporated in the Variscan fault-and-thrust system between the Moravo-Silesian and the Moldanubian zones (Fig. 3.14).

In this general description we prefer the structural divisions of the Bohemian Massif based on Kossmat (1927) and other authors into zones and units reflecting predominantly the different degree and style of tectonometamorphic overprint during the Variscan Orogeny. Additionally we take into account that the Brunovistulian Block in the Moravo-Silesian Zone most probably is an Avalonia-type terrane (Finger *et al.* 2000*a*). During the Variscan Orogeny in the Bohemian Massif, the Cadomian basement was fragmented and overprinted by various processes and is preserved today in different Variscan tectonometamorphic zones or units. None of these includes the complete geological history of the Cadomian basement. In the following sections the Cadomian basement from different parts of the Bohemian Massif is first described. This is followed by a discussion of palaeogeographical constraints. Geochronological data of relevant igneous rocks are shown in Table 3.3. Table 3.4 presents a summary of the main orogenic processes related to Cadomian and earlier events.

The Cadomia-type part

The Saxo-Thuringian Zone

The Saxo-Thuringian Zone was first described by Kossmat (1927). The Precambrian and related post-kinematic intrusive rocks were first described in terms of Cadomian basement by Linnemann & Buschmann (1995*a*, *b*). A number of very low grade metamorphic rock complexes of Cadomian basement and the Palaeozoic overstep sequence are characteristic features of the Saxo-Thuringian Zone (Figs 3.14–3.16). In the earlier literature, Precambrian orogenic processes were referred to as the 'Assyntian' or 'Baikalian' orogenies (Stille 1948). Whether or not the Saxo-Thuringian Zone was affected by an orogenic event between the Precambrian and Early Palaeozoic successions has been debated since the fundamental fieldwork and documentation of H.R. von Gaertner (1944) in the Schwarzburg Unit (Fig. 3.17). Due to the lack of robust geochronological data and the fact that no clear unconformity between Precambrian and Early Palaeozoic rocks was observable in surface outcrops, the debate of an orogenic event in that time interval continued until the mid-1990s (Deubel 1959; Falk 1974; Bankwitz & Bankwitz 1995). The first clear evidence of the Cadomian unconformity in the Saxo-Thuringian Zone resulted from drillings, especially from the drill core 5507/77 near Gera (Linnemann & Buschmann 1995*a*) and later from the drillings of the Torgau-Doberlug Syncline (Buschmann 1995; Buschmann *et al.* 2006).

In the Saxo-Thuringian Zone the Cadomian basement consists of volcanosedimentary complexes with maximum ages of sedimentation around *c.* 570 Ma. These complexes were intruded by post-kinematic plutons at *c.* 540–530 Ma (Linnemann *et al.* 2000; Tichomirowa 2003; Gehmlich 2003). All sections are rootless and the nature of the underlying basement is unknown. An overview of stratigraphy, geochronology and model ages is given in Figures 3.16 and 3.28.

The Saxo-Thuringian Zone forms the northeastern part of the Bohemian Massif (Figs 3.14 & 3.15). It consists of Cadomian basement units, which are overlain by a Palaeozoic overstep sequence (Figs 3.16 & 3.28). The parautochthonous and weakly metamorphosed part of the Saxo-Thuringian Zone forms a Variscan NE- to east-trending fold-and-thrust belt which consists of the Schwarzburg Unit, the North Saxon Antiform, the Berga Antiform and the Lausitz Block, and the Torgau-Doberlug and Ziegenrück-Teuschnitz synclines (Figs. 3.15 & 3.16). In addition,

Fig. 3.14. Cadomian orogenic imprints in the Variscan basement of the Bohemian Massif and ages of plutons and volcanic rock related to Cadomian geotectonic processes (map compiled after Franke & Żelaźniewicz 2002; Kodzdrói *et al.* 2001; Linnemann & Schauer 1999). Data sources: [1]Venera *et al.* (2000); [2]Zulauf *et al.* (1997); [3]Dörr *et al.* (2002); [4]Dörr *et al.* (1998); [5]Friedl *et al.* (2004); [6]Drost *et al.* (2004); [7]Linnemann *et al.* (2000); [8]Hammer *et al.* (1999); [9]Mazur *et al.* (2004); [10]Tichomirowa (2003); [11]Żelaźniewicz *et al.* (2004); [12]Teipel *et al.* (2004); [13]Gebauer & Friedl (1994); [14]Wendt *et al.* (1993). *U–Pb SHRIMP; **U–Pb TIMS; ***Pb/Pb TIMS; §age of the youngest concordant zircon of the Natschlag gneiss. Lausitz GC, Lausitz Granitoid Complex.

Table 3.3. *Summary of isotopic data derived from meta-igneous units in the Bohemian Massif.*

Unit	Age (± 2σ error)	Number of analyses (N), comments	Reference
Moldanubian Zone			
Protolith ages/Proterozoic basement (overprinted by Variscan high-grade metamorphism)			
Rajov metatonalite (South Bohemia)	2048 ± 24 Ma MSWD = 0.4	N = 4, unabr zircon UI age	Wendt *et al.* (1993)
Harazim metagranodiorite (South Bohemia)	2060 ± 24 Ma MSWD = 5.7	N = 5, unabr zircon UI age	Wendt *et al.* (1993)
Harazim metagranodiorite (South Bohemia)	2061 ± 12 Ma	N = 3, unabr zircon evaporation, mean date from 3 crystals	Wendt *et al.* (1993)
Světlík metaquartzdiorite (South Bohemia)	2109 ± 60 Ma MSWD = 1.7	N = 3, unabr zircon UI age	Wendt *et al.* (1993)
Dobra metagranodiorite *(most probably to Bruno-vistulian Block)*	1336 ± 18 Ma (95% conf.), MSWD = 228	N = 5, unabr zircon UI age	Friedl *et al.* (2004)
Dobra metagranodiorite (*metamorphic age*)	*c.* 600 Ma	zircon, SHRIMP	Friedl *et al.* (2004) Gebauer & Friedl (1993)
Late Neoproterozoic protoliths (overprinted by Variscan high-grade metamorphism)			
Spitz gneiss (Austria)	614 ± 10 Ma (95% conf.)	N = 7, zircon SHRIMP	Friedl *et al.* (2004)
Metarhyolite (*subduction signature*)	555 ± 12 Ma (95% conf.)	N = 5, zircon SHRIMP	Teipel *et al.* (2004)
Metarhyolite (*subduction signature*)	549 ± 7 Ma (95% conf.)	N = 8, zircon SHRIMP	Teipel *et al.* (2004)
Metabasite (*enriched MORB*)	549 ± 7 Ma (95% conf.)	N = 7, zircon SHRIMP	Teipel *et al.* (2004)
Teplá-Barrandian Unit			
Domažlice crystalline complex (*metamorphic age*)	*c.* 50–540 Ma	3 paragneisses from 3 locations; monazite U–Th–Pb model ages	Zulauf *et al.* (1999)
Lower Cambrian intrusions in the Teplá-Barrandian Unit			
Všepadly granodiorite	524 ± 3 Ma	N = 1, unabr zircon	Dörr *et al.* (2002)
Všepadly granodiorite (*cooling age*)	513.2 ± 1.2 Ma MSWD = 1.14	$^{40}Ar/^{39}Ar$, hbl isotope corr date	Dallmeyer & Urban (1998)
Orlovice gabbro	524 ± 0.8 Ma	N = 6, unabr zircon	Dörr *et al.* (2002)
Smržovice tonalite	523 ± 3 Ma	N = 4, unabr zircon UI age	Dörr *et al.* (2002)
Smržovice gabbro	523 ± 1 Ma	N = 1, unabr zircon	Dörr *et al.* (2002)
Mračnice trondjemite	523 + 4/−5 Ma	N = 4, unabr zircon LI age	Zulauf *et al.* (1997)
Těšovice biotite granite	522 + 3.4/−2.7 Ma	N = 4, abr zircon UI age	Zulauf *et al.* (1997)
Těšovice biotite granite	522 ± 1.5 Ma	N = 1, abr zircon	Zulauf *et al.* (1997)
Lestkov metagranitoid	*c.* 513 Ma	N = 4, zircon, UI age	Dörr *et al.* (1998)
Teplá orthogneiss	513 + 6/−7 Ma	N = 4, zircon, UI age	Dörr *et al.* (1998)
Brunovistulian Block			
Slavkov Unit			
Ludvíkov Gneiss	684.5 ± 0.9 Ma	N = 3, unabr zircon evaporation, mean date from 3 crystals	Kröner *et al.* (2000)
Paragneisses (metam. age)	660–610 Ma	K–Ar, mica	Dudek & Melková (1975)
Filipovice orthogneiss	612.7 ± 1.0 Ma	N = 3, unabr zircon evaporation, mean date from 3 crystals	Kröner *et al.* (2000)
Paragneisses from bore-holes Gottwaldov, Kojetin, Kozlovice, Krasna (metamorphic age)	608-576 Ma (549 ± 19 Ma)	U–Th–total Pb, monazite	Finger *et al.* (1999)
Brno Pluton	*c.* 585 Ma	N = 3, unabr zircon UI age	van Breemen *et al.* (1982)
Diorites (metam.? cooling age)	555–605 Ma	K–Ar, mica/hbl	Dudek & Melková (1975)
Central Basic Belt			
Metarhyolite (quarry at Opalenka near Kuřim)	725 ± 15 Ma	N = 4, unabr zircon evaporation, mean date from 3 fractions	Finger *et al.* (2000*a*)
Thaya Unit			
Granodiorite - Gumping	595 ± 1 Ma	zircon, evaporation	Kröner (unpublished data) from Finger *et al.* (2000*b*)
Keprník Gneiss	584 ± 8 Ma MSWD = 0.06) 583.8 ± 0.9 Ma	N = 6, unabr zircon UI age N = 3, unabr zircon evaporation, mean date from 3 crystals	Kröner *et al.* (2000)
Bittesch Orthogneiss	584 ± 6 Ma (95% conf.) 578 ± 6 Ma (95% conf.)	N = 20, zircon SHRIMP N = 11, zircon SHRIMP	Friedl *et al.* (2004)
Eggenburg Metagranite	567 ± 6 Ma (95% conf.)	N = 13, zircon SHRIMP	Friedl *et al.* (2004)
Branná Unit	553 ± 7 Ma MSWD = 0.44 554.6 ± 1.2 Ma	N = 4, unabr zircon UI age N = 3, zircon evaporation, mean date from 3 crystals	Kröner *et al.* (2000)

(*continued*)

Table 3. (*continued*)

Unit	Age (± 2σ error)	Number of analyses (N), comments	Reference
Sudetes			
Kłodzko Metamorphic Unit			
Plagioclase gneiss within metagabbro	590.1 ± 7.2 Ma MSWD = 0.31	N = 4, unabr zircon UI age	Mazur *et al.* (2004)
Fore Sudetic Block			
Wądroże Wielkie gneiss	548 ± 9 Ma MSWD = 0.83	N = 5, unabr/abr zirc LI age	Żelaźniewicz *et al.* (2004)
South Krkonoše Unit			
Bitouchov granodiorite	540 +11/-10 Ma MSWD = 1.37	N = 7, unabr/abr zirc UI age	Żelaźniewicz *et al.* (2004)
Saxo-Thuringian Zone			
Lausitz Block			
Wüsteberg tuffite	565 ± 3 Ma	N = 4, zircon evaporation, mean date from 4 crystals	Linnemann *et al.* (2000)
Two-mica granodiorite	550 ± 10 Ma to 569 ± 25 Ma	mean ages, single zircon evaporation	Tikhomirowa (2002)
Muscovite-bearing biotite granodiorite	543 ± 5 Ma	mean age, single zircon evaporation	Tikhomirowa (2002)
Ostlausitz granodiorite (biotite granodiorite)	540 ± 7 Ma		Korytowski *et al.* (1993)
Westlausitz granodiorite (biotite granodiorite)	539 ± 6 Ma	N = 7, zircon evaporation, mean date from 7 crystals	Linnemann *et al.* (2000)
Granodiorite: Kindisch	539 ± 5 Ma	N = 10	Tichomirowa *et al.* (2001)
Demitz	533 ± 5 Ma	N = 14	
Steinberg (biotite granodiorite)	534 ± 6 Ma	N = 10, zircon evaporation, mean date from N crystals	
Platerówka granodiorite (biotite granodiorite)	533 ± 9 Ma MSWD = 0.47	N = 4, unabr/abr zirc LI age	Żelaźniewicz *et al.* (2004)
Elbezone			
Dohna granodiorite	537 ± 7 Ma	N = 11, zircon evaporation, mean date from 11 crystals	Linnemann *et al.* (2000)
Großenhain orthogneiss	537 ± 7 Ma	N = 5, zircon evaporation, mean date from 5 crystals	Linnemann *et al.* (2000)
North Saxon Anticline			
Laas granodiorite	531 ± 7 Ma	N = 5, zircon evaporation, mean date from 5 crystals	Linnemann *et al.* (2000)
Schildau Massif	530 ± 8 Ma	no details given	Hammer *et al.* (1999)
Schwarzburg Unit			
Glasbach granite	541 ± 7 Ma	N = 4, zircon evaporation, mean date from 4 crystals	Linnemann *et al.* (2000)
Laubach granite	533 ± 10 Ma	N = 7, zircon evaporation, mean date from 7 crystals	Linnemann *et al.* (2000)
Erzgebirge Block (overprinted by Variscan high-grade metamorphism)			
Augengneiss – Bärenstein (lower shear zone after Sebastian 1995)	575 ± 4 Ma	N = 2, SHRIMP (possibly inherited)	Mingram *et al.* (2004)
Augengneiss – Wolkenstein (lower shear zone after Sebastian 1995)	567 ± 7 Ma	N = 2, SHRIMP (possibly inherited)	Mingram *et al.* (2004)
Inner Grey Gneiss	541 ± 4 Ma	mean from U–Pb and evaporation ages	Tichomirowa *et al.* (2001), Tikhomirowa (2002)
Inner Grey Gneiss (MP-MT unit after Sebastian 1995)	528 ± 6 Ma	N = 8, SHRIMP zircon rims	Tichomirowa *et al.* (2001)
Augengneiss – Mědĕnec (MP-MT unit after Sebastian 1995)	524 ± 10 Ma	N = 9, concordia age LA-ICP-MS	Košler *et al.* (2004)

the Saxo-Thuringian Zone is transected by the NW–SE trending Elbe Zone (Fig. 3.15).

In the Saxo-Thuringian Zone a number of units and zones exist which include occurrences of Cadomian basement with different tectonostratigraphic histories (Linnemann & Romer 2002; Buschmann *et al.* 2006; Linnemann *et al.* 2007). Recent results show strong evidence for the existence of (i) backarc basins, (ii) retro-arc basins and a related remnant basin, and (iii) a shelf basin with passive margin deposits. The Cadomian basin types and the important Kernzone Complex of the Schwarzburg Unit are characterized as follows.

(i) *Cadomian backarc basin.* A first basin type includes the Cadomian volcanosedimentary units of the Altenfeld Formation (*c.* 1000 m) of the Katzhütte group in the NW part of the Schwarzburg Unit (Fig. 3.17) and of the Rothstein Formation (*c.* 1200 m) in the Torgau-Doberlug Syncline (Figs. 3.18 & 3.19). The latter, with the exception of the Rothstein Rock exposure, is known only from drillholes. Both formations are characterized by lithological units containing thick siliciclastic sucessions with volcanics and layers of massive and bedded black chert (Fig. 3.20D). Available data suggest an origin in a backarc setting for the Rothstein Formation (Buschmann 1995;

Table 3.4. *Tectonomagmatic events in the most important parts of the Bohemian Massif*

	Cadomia-type units			Avalonia-type units
Approx. age	Saxo-Thuringian Zone	Teplá-Barrandian Unit	Moldanubian Zone	Brunovistulian Block
2 Ga			Formation of Palaeoproterozoic protoliths (Svetlík gneiss) (c. 2.0 Ga)	
1.4 Ga				Formation of Mesoproterozoic protoliths (Dobra gneiss) (*c*. 1.38 Ga)
725 Ma				Rifting of Slavkov Terrane from Gondwana (Northern South America?) – Central Basic Belt
680 Ma				Granitoid magmatism (c. 685 Ma)
620 Ma				
620–580 Ma		Accumulation of volcanosedimentary sequence, subduction-related and within-plate volcanism ?	I-type granitoid magmatism (*c*. 614 Ma)	**Regional deformation** (medium pressure amphibolite facies grade) (I-type) Granitoid magmatism (*c*. 613–585 Ma)
570–550 Ma	Accumulation of volcanosedimentary succession in (i) backarc basin (ii) retro-arc basin	Accumulation of detrital sequence		(I/A-type) Granitoid magmatism (*c*. 570–555 Ma)
550–540 Ma	Trans-/compression and regional deformation	Regional deformation (LP-HT)	Subduction-related and within-plate volcanism (*c*. 550 Ma)	
540-520 Ma	Emplacement of plutons	Emplacement of plutons in transtensive shear-zones		
	Onset Cambrian deposition	Onset Cambrian deposition		Onset Cambrian deposition

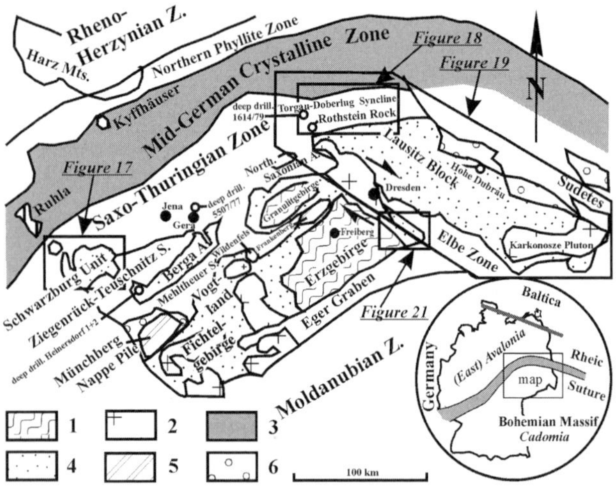

Fig. 3.15. Location and tectonostratigraphic units of the Saxo-Thuringian Zone and adjoining areas in the Mid-European Variscides. Position of relevant deep drillings (circles), locations and figures referred to in this paper are indicated. Key: 1, high grade rocks (Variscan overprint); 2, Variscan plutons (*c*. 330 Ma); 3, suture of the closure of the Rheic Ocean; 4, low grade Ediacaran (Cadomian) to Devonian rocks; 5, Variscan nappes; 6, Variscan wild flysch. Inset: simplified map of Germany showing the large suture zones between Avalonia, Cadomia and Baltica.

Linnemann *et al.* 2000, 2007). These backarc sediments are dominated by dark grey to black distal turbidites composed of intercalated greywacke and mudstone beds. In addition, mafic volcanics and pyroclastics occur.

The geochemistry of the volcanic rocks is diverse, comprising a suite of E-type MOR metabasalts to meta-andesites, calc-alcaline metabasalts, and subordinate alkaline metabasalts. Effusive alkaline metabasalts form mini pillow sequences, E-type MOR metabasalts large pillow sequences, calc-alkaline metabasalts and meta-andesites sills and dykes. The eruption characteristics, including the formation of thick pillow sequences and sill emplacement into unconsolidated sediments, point to seafloor spreading processes in thinned continental crust (Buschmann 1995). The presence of seafloor spreading processes is further indicated by relics of hydrothermal activity in volcanic and sediment-hosted submarine environments. The most spectacular hydrothermal products are bedded black cherts (Fig. 3.20A) that reflect discharges of silica-rich hydrothermal fluids. The formation of the basin in which the Rothstein Formation accumulated is attributed to pull-apart processes in a strike-slip dominated active continental margin environment. Strike-slip motions resulted in the resedimentation of unconsolidated sediments. Seafloor spreading and hydrothermal activity suggest the development of an oceanic substratum.

All sedimentological and geochemical data point to an origin for the Rothstein Formation in the central part of a backarc basin that developed on thinned continental crust (Buschmann 1995; Buschmann *et al.* 2006). Owing to its similar spatial position in the Saxo-Thuringian Zone and its similarity in lithology and geochemistry, the Altenfeld Formation is assigned to the same plate tectonic setting. Zircon data from a tuff layer suggest deposition of the Rothstein Formation that occurred at *c*. 566 Ma (Buschmann *et al.* 2001). For the Altenfeld Formation a maximum age of deposition is given by the age of the youngest population of detrital zircon grains (563 ± 6 Ma; U. Linnemann & T. Jeffries, unpublished data; Fig. 3.17). Geochemical data of

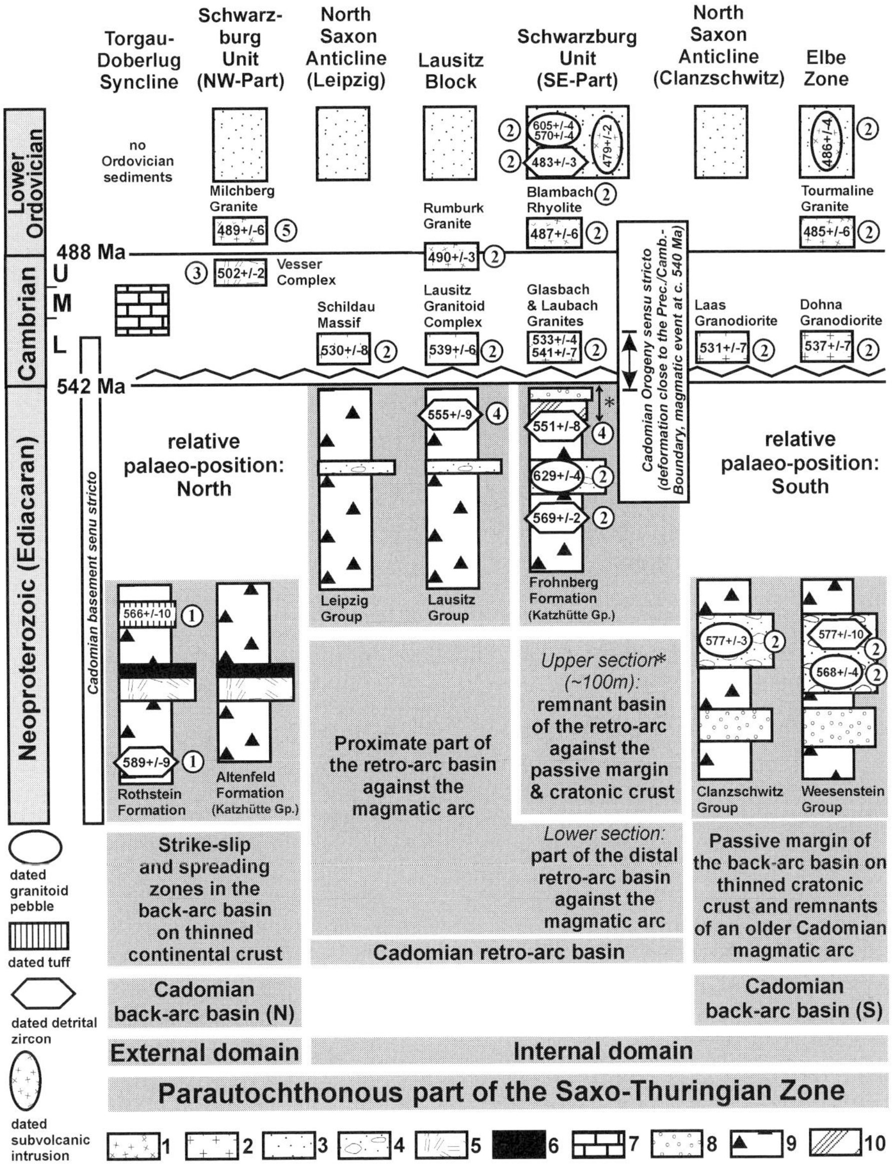

Fig. 3.16. Generalized lithostratigraphic profiles of parautochthonous units of the Saxo-Thuringian Zone with published geochronological data of the Cadomian basement and its Cambro-Ordovician overstep sequence (from Linnemann *et al.* 2007). Key: 1, Cambro-Ordovician rift-related igneous rocks; 2, Cadomian granitoids of the *c.* 540 Ma magmatic event; 3, Lower Ordovician siliciclastic sediments; 4, Late Neoproterozoic debris flows and glaciomarine tillites; 5, igneous rocks and metasediments of the Upper Cambrian Vesser Complex (predominantly mafic rocks); 6, Neoproterozoic hydrothermal black cherts; 7, Lower to Middle Cambrian sediments; 8, conglomerates, quartzites and quartzitic shales of the Purpurberg Quartzite (Weesenstein Group) and its equivalent in the Clanzschwitz Group; 9, greywackes and mudstones; 10, predominantly mudstones. Sources of geochronological data (circled numbers): 1, SHRIMP U–Pb (Buschmann *et al.* 2001); 2, TIMS Pb/Pb (Linnemann *et al.* 2000); 3, TIMS U–Pb (Kemnitz *et al.* 2002); 4, SHRIMP U–Pb (Linnemann *et al.* 2004); 5, LA-ICP-MS U–Pb (U. Linnemann & Gerdes A. unpublished data).

sedimentary and volcanic rocks of the Altenfeld and Frohnberg formations suggest an active margin setting (Buschmann 1995; Linnemann & Romer 2002). The Rothstein Formation is overlain by Lower to Middle Cambrian sediments (Figs. 3.16, 3.18 & 3.19), whereas the biotite-grade overprinted Altenfeld Formation is covered by Lower Ordovician siliciclastics of the Langer Berg Formation (Figs. 3.16 & 3.17), the contact being fault-bounded due to the lithological contrast. There is no evidence for a strong displacement. An angular unconformity between the Ediacaran and the Ordovician is assumed (Deubel 1959). The Rothstein Formation and its Cambrian cover sequence show contrasting degrees of metamorphism. In the case of the Rothstein Formation, the metamorphic grade is anchizone whereas the Cambrian sediments show only diagenetic features. Therefore it would appear that the Rothstein Formation was affected by Cadomian deformation and metamorphism around the Precambrian–Cambrian boundary (Buschmann 1995; Buschmann *et al.* 2006). No examples of 540–530 Ma plutons intrude either the Rothstein or Altenfeld formations. In contrast, early Cambrian plutons are present in all other areas (Fig. 3.16). Only a small granitoid body (the Milchberg Granite) crops out in the NW of the Schwarzburg Unit where it intrudes the Altenfeld Formation. Recent U–Pb dating characterize is the granite as an Ordovician intrusion (489 ± 6 Ma; U. Linnemann & A. Gerdes, unpublished data; Figs. 3.16 & 3.17).

(ii) *Kernzone Complex*. An important and very complicated

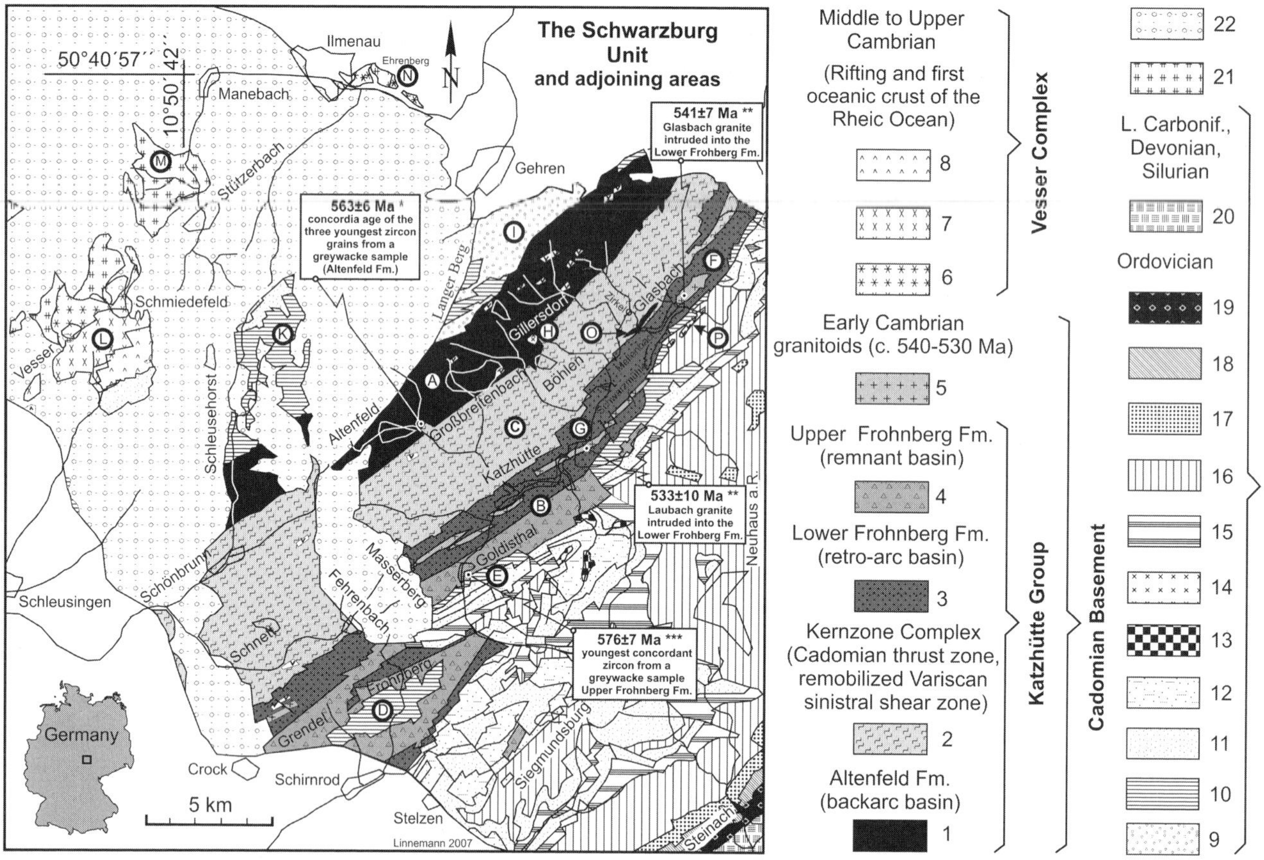

Fig. 3.17. Geological map of the Cadomian basement of the Schwarzburg Unit (latest Ediacaran to early Cambrian, Saxo-Thuringian Zone), the Vesser Complex (Middle to Upper Cambrian), and geological units of adjoining areas (complied by U. Linnemann based on von Gaertner (1944), Söllig (1953), Bankwitz *et al.* (1992), Wiefel (1996) and Linnemann (2003). Geochronological ages are taken from: *U. Linnemann & Jeffries (unpublished data, U–Pb on single zircons by LA-ICP-MS); **Linnemann *et al.* (2000; Pb/Pb evaporation age on single zircons); ***Linnemann *et al.* (2004; SHRIMP U–Pb). Key: 1, Altenfeld Formation (Ediacaran, younger than *c.* 562 Ma) with sedimentary deposits, black cherts, and volcanic rocks from a Cadomian backarc basin; 2, Kernzone Complex: very complex Cadomian thrust and shear zone reactivated during the Variscan orogeny as a sinistral shear zone; 3, Lower Frohnberg Formation consisting of monotonous greywacke-schist turbidites with fragments of black chert formed in the setting of a retro-arc or foreland basin, respectively; at the top of the profile are conglomerates and debris flows (latest Ediacaran to earliest Cambrian); 4, Upper Frohnberg Formation consisting mainly of shale and sandstones with fragments of black cherts, deposited in a remnant basin (latest Ediacaran to earliest Cambrian); 5, Early Cambrian intrusions of granites (*c.* 540 to 530 Ma); 6, Lower Rollkopf Formation dominated by schist, pyroclastic rocks and bimodal volcanic rocks (Middle to Upper Cambrian); 7, Upper Rollkopf Formation dominated by mafic magmatic rocks; 8, Neuwerk Formation dominated by mafic to intermediate lava flows; 9, Very mature coarse-grained sandstones and conglomerates of the Langer Berg Formation at the base of the Ordovician (Tremadoc); 10, low grade schists and thinly bedded sandy tempestites of the Goldisthal Formation (Tremadoc); 11, Very mature quartzites and sandstones of the Frauenbach Group (Tremadoc); 12, shists and thinly bedded sandy tempestites of the Frauenbach Wechsellagerung formation (Frauenbach group, Tremadoc); 13, porphyroids (Lower Ordovician, 479 ± 2 Ma; Linnemann *et al.* 2000); 14, Lower Ordovician intrusion of the Milchberg granite (489 ± 6 Ma; U. Linnemann & A. Gerdes, unpublished data); 15, Dachschiefer Formation composed of shales that have been locally mined for roofing tiles (base of the Phycodes Group, Tremadoc); 16, middle part of the Phycodes Group composed of shale, silt, and temestitic sandstones (Tremadoc); 17, Phycodenquarzit Formation, a quartzite at the top of the Phycodes Group (Tremadoc); 18, Griffelschiefer and Schmiedefeld formations consisting of dark shale and sedimentary iron ores (Arenig–Llanvirn); 19, Lederschiefer Formation composed of the Upper Ordovician glaciomarine tillite and a limestone horizon at its base (Ashgill); 20, Silurian, Devonian and Lower Carboniferous sedimentary rocks and volcanosedimentary complexes (not divided); 21, Variscan granitoid intrusions (*c.* 330–300 Ma); 22, Variscan Molasse (Upper Carboniferous and Rotliegend). Circled letters: A, Cadomian Basement at the NW flank of the Schwarzburg Unit (fragment of a backarc basin); B, Cadomian Basement at the SE flank of the Schwarzburg Unit (fragment of a retro-arc basin and overlying remnant basin); C, Kernzone Complex, characterized by phyllite tectonics under high greenschist to low amphibolite facies conditions; lithologies are shists, metagreywackes, quartzites, and more rare metavolcanic rocks and metagranites (the bulk of affected rock is derived from Ediacaran sediments; during Variscan reactivation Ordovician rocks became involved); D, Grendel Syncline; E, section at the roadcut of Wurzelberg Mountain. showing the contact of the Frohnberg Formation (latest Ediacaran–early Cambrian) and the overlying Goldisthal Formation (Tremadoc); F, Glasbach Granite (541 ± 7 Ma; Linnemann *et al.* 2000); G, Laubach Granite (533 ± 10 Ma; Linnemann *et al.* 2000); H, Milchberg Granite (489 ± 6 Ma; Linnemann & A. Gerdes, unpublished data); I, Langer Berg Mountain; K, Schleuse Horst; L, Vesser Complex; M, Variscan granitoids of the 'Thüringer Hauptgranit' exposed near Stützerbach; N, exposure of a part of the Vesser Complex at Ehrenberg Mountain near Ilmenau; O, gigantic Ediacaran chert body as a tectonic slice in the Kernzone Complex; P, Ordovician mafic intrusion at Mellenbach ('Diabas von Mellenbach').

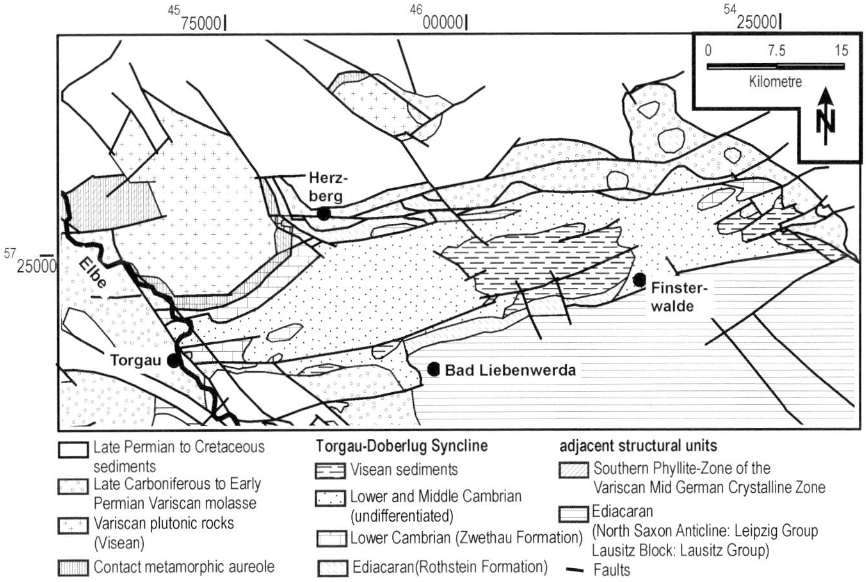

Fig. 3.18. Geological map of the Torgau-Doberlug Syncline showing distribution of rock complexes at the pre-Cenozoic surface deduced from geophysical mapping and drilling. The extent of the Torgau-Doberlug Syncline is defined by the presence of the Ediacaran Rothstein Formation, Cambrian units, and Viséan sediments (from Buschmann *et al.* 2001).

central element of the Schwarzburg Unit is the Kernzone Complex including the Blumenau Shear Zone. This structure divides the Schwarzburg Unit into NW and SE parts (Fig. 3.17). Rocks of the Kernzone Complex were metamorphosed under upper greenschist to low amphibolite facies conditions. Common lithologies include the schists, metagreywackes, quartzites and, more rarely, metavolcanic rocks and metagranites. The bulk of metamorphosed and deformed rocks were derived from Ediacaran sediments. More rare are early Cambrian metagranitoids and Ordovician magmatic rocks (Linnemann *et al.* 1999). The Kernzone Complex is deformed by a series of shear zones of which the Blumenau Shear Zone at the SE border of the complex is the most important. During the Variscan Orogeny, the Blumenau Shear Zone was active as a sinistral shear zone (Heuse *et al.* 2001). It is very likely that the Kernzone Complex formed during the late Neoproterozoic to early Cambrian because it divides two very different Cadomian basin types formed in different geotectonic settings. To the NW of the complex, a fragment of a backarc basin crops out (Altenfeld Formation), whereas at the SE flank the remnants of a retro-arc and remnant basin occur (Frohnberg Formation; Fig. 3.17). In this context the Kernzone Complex most probably represents part of a fold-and-thrust belt that developed during the formation of the Cadomian retro-arc basin (see later in this chapter and Fig. 3.27; Linnemann *et al.* 2007).

(iii) *Cadomian retro-arc basin*. The second group of Cadomian sedimentary units is represented by the Lausitz Group (*c.* 2000 m, Lausitz Block; Figs. 3.16 & 3.19), the Leipzig Group (*c.* 2000 m, North Saxon Anticline; Fig. 3.16), and the Frohnberg Formation (*c.* 1500 m) SE part of the Schwarzburg Unit; Figs. 3.16 & 3.17). Due to their striking similarities and regional linkage the Lausitz and Leipzig groups are grouped with the so-called Lausitz–Leizig Greywacke Complex (Fig. 3.19). All three units are characterized by monotonous, deep-marine-like sections of proximal to distal dark-grey to black turbidites composed of greywacke and mudstone couplets (Fig. 3.20E). Seismites indicate an active tectonic setting during basin formation. Intercalations of conglomerates contain remarkable fragments of black cherts (Fig. 3.20D) and other debris from older Ediacaran sediments and igneous arc-derived rocks (Fig. 3.20F). The pervasive occurrence of fragments of black chert in both the greywackes and conglomerates suggests that deposits from the Cadomian backarc basin became eroded, recycled and redeposited in the Cadomian retro-arc basin, remnants of which are presented by the Lausitz and Leipzig groups and the Frohnberg Formation (Linnemann *et al.* 2007).

SHRIMP U–Pb dating of detrital zircon grains from the Lausitz Group and the Frohnberg Formation indicates that they are younger than 555 ± 9 Ma and 551 ± 8 Ma, respectively (Linnemann *et al.* 2004). Recent data have shown that, at least in part, the sedimentary deposits of the Lausitz Group are even younger. A maximum age of deposition of 543 ± 4 Ma, obtained from the youngest zircon population from a chert-fragment-bearing conglomerate of the Lausitz Group from the Wetterberg quarry near Ebersbach, indicates a latest Ediacaran or earliest Cambrian maximum age of deposition for that conglomerate and a large part of the Lausitz Group (Fig. 3.19; Linnemann *et al.* 2007). These relatively young ages suggest a much younger age for the retro-arc basin than the *c.* 566 Ma Cadomian backarc basin. Because of the lithological analogies and the young maximum ages, the Lausitz and the Leipzig groups and Frohnberg Formation are classified as relicts of a Cadomian retro-arc basin (Linnemann *et al.* 2007). Geochemistry of the retro-arc sediments supports an active margin setting (Linnemann & Romer 2002). The Lausitz and Leipzig groups became deformed during Cadomian orognic processes (Fig. 3.20E) and are unconformably overlain by Lower Ordovician quartzites and mature conglomerates with pebbles of quartz, chert and hornfels. The Cadomian unconformity between the Lausitz and Leipzig groups and the overstepping Ordovican sediments is well-documented in drillhole 5507/77 (Fig. 3.15) and in the Hohe Dubrau area (Fig. 3.19) (Linnemann & Buschmann 1995*a*, *b*). In contrast to the lower part of the Frohnberg Formation (*c.* 1200 m, Fig. 3.15), which is composed mainly of thick-bedded greywackes similar

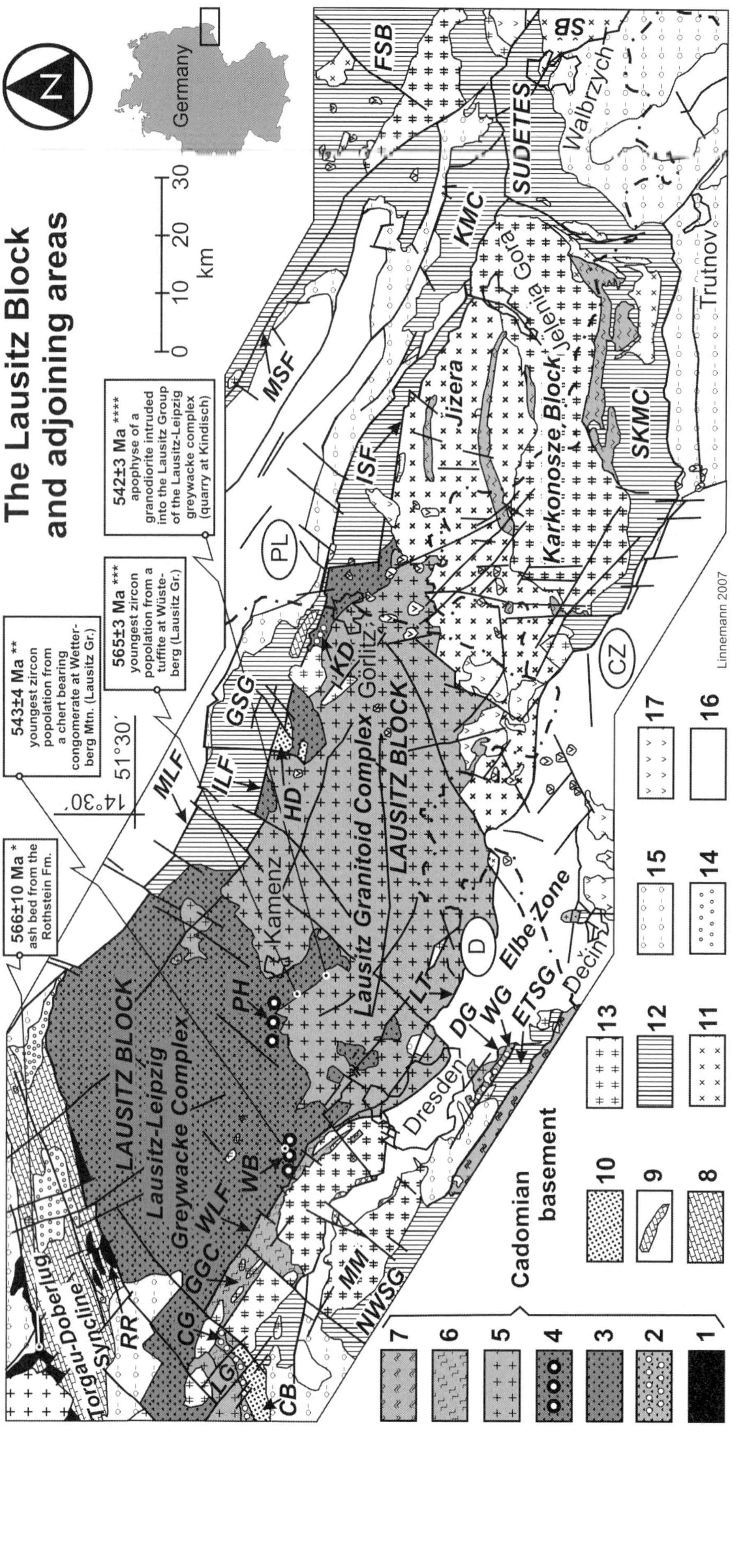
The Lausitz Block
and adjoining areas
N
Germany
0 10 20 30
km
566±10 Ma *
ash bed from the Rothstein Fm.
543±4 Ma **
youngest zircon population from a chert bearing congomerate at Wetterberg Mtn. (Lausitz Gr.)
565±3 Ma ***
youngest zircon popolation from a tuffite at Wüsteberg (Lausitz Gr.)
542±3 Ma ****
apophyse of a granodiorite intruded into the Lausitz Group of the Lausitz-Leipzig greywacke complex (quarry at Kindisch)
14°30´
51°30´
LAUSITZ BLOCK
Lausitz-Leipzig
Greywacke Complex
Lausitz Granitoid Complex
LAUSITZ BLOCK
Torgau-Doberlug
Syncline
Doberlug
RR
CG
GGC
WLF
WB
LG
CB
MM
NWSG
PH
Kamenz
HD
MLF
ILF
GSG
KD
Görlitz
PL
MSF
ISF
Jizera
KMC
FSB
SUDETES
SB
Walbrzych
Jelenia Góra
Karkonosze Block
SKMC
Trutnov
CZ
D
LT
DG
WG
ETSG
Elbe Zone
Dresden
Děčín
Linnemann 2007
Cadomian basement
1 2 3 4 5 6 7 8 9 10 11 12 13 14 15 16 17

to the Lausitz and Leipzig groups, the *c.* 300 m thick uppermost part was probably deposited in a shallower marine environment (Linnemann 2003). This suggests termination of the retro-arc basin regime and the onset of deposition in a remnant basin located in front of the outer retro-arc deposits.

The very topmost quartzite bed of the section, known as the 'Basisquarzit' (= basal quartzite), is traditionally interpreted as the base of the Palaeozoic overstep sequence (von Gaertner 1944). It is, however, more likely that this quartzite represents the final bed of a continuous upward-thickening succession of the Neoproterozoic remnant basin. There is no angular unconformity between the Upper Frohnberg Formation and the overstepping Ordovician strata of the Goldisthal Formation and no remarkable difference in metamorphic degree is observable, although there is a gap in sedimentation between the Cadomian basement and Ordovician overstep sequence that comprises the whole of the Cambrian (Linnemann *et al.* 2007). Therefore it is concluded that the final stage of Ediacaran sedimentation in the Schwarzburg Unit was a remnant basin that escaped Cadomian deformation.

All Ediacaran sedimentary units of the Cadomian retro-arc basin were intruded by voluminous granitoids at *c.* 540 Ma (Figs. 3.17 & 3.19) (Linnemann *et al.* 2000; Tichomirowa 2003; Gehmlich 2003). The largest exposed body is the Lausitz Granitoid Complex, which intruded into the Lausitz Group at 542 ± 3 Ma and covers an area of about 100 × 50 km (Fig. 3.19). The Frohnberg Formation in the Schwarzburg Unit was intruded by the Glasbach and Laubach Granites (541 ± 7 Ma, 533 ± 10 Ma; Fig. 3.17). The plutonic rocks of the Schildau Massif were intruded into the Leipzig Group at 530 ± 8 Ma (Fig. 3.16). Most granitoids were probably derived from melting of the Late Ediacaran greywackes or similar sediments, since they contain large numbers of inherited zircons with age spectra comparable to those in the sediments (Linnemann *et al.* 2000, 2004; Tichomirowa 2003; Gehmlich 2003). This interpretation is supported by geochemical data (Hammer 1996) and greywacke xenoliths in the granitoids (Fig. 3.20G). These *c.* 540 Ma granitoids record a relatively short-lived regime of high heat-flow, which can be attributed to slab break-off of the subducted oceanic plate.

(iv) *Passive margin shelf basin.* The third group of Ediacaran sedimentary units comprises passive margin successions characterized by highly mature quartzites, sandstones and quartz-rich shales deposited in a shallow marine environment (Linnemann 1991; Linnemann *et al.* 2007). The most prominent deposit of this type is represented by the Purpurberg Quartzite in the lower part of the Weesenstein Group (Figs. 3.20B & 3.21). Facies analysis of the Purpurberg Quartzite has shown that its deposition was caused by an extreme drop of sea level that is interpreted to be of glacio-eustatic origin (Linnemann 1991). The stratigraphic equivalent of the Purpurberg quartzite also occurs in the Clanzschwitz Group (North Saxon Anticline; Fig. 3.19). Other parts of the Weesenstein and Clanzschwitz Groups (*c.* 2000 m each) are likewise passive margin deposits but comprise quartz-rich mud- and siltstones. In the upper part of the two groups diamictites and layers with isolated pebbles (Fig. 3.20C) may be glaciomarine in origin (Linnemann 1991; Linnemann & Romer 2002). These passive margin deposits are situated in the North Saxon Anticline and the Elbe Zone (Fig. 3.2). Based on the spatial arrangement of the passive margin deposits in the Saxo-Thuringian Zone we assign these units to the passive margin of the same Cadomian backarc basin in which the Rothstein and Altenfeld formations were deposited. Detrital zircons and dated pebbles point to a minimum age of sedimentation of *c.* 570 Ma for the passive margin units (Linnemann *et al.* 2000, 2007). It is believed that the Weesenstein and

Fig. 3.19. Geological map of the Cadomian basement of the Lausitz Block (Saxo-Thuringian Zone) including deposits of the Lausitz–Leipzig Greywacke Complex (latest Ediacaran to early Cambrian), Early Cambrian intrusions of the Lausitz Granitoid Complex, and geological units of adjoining areas (complied by U. Linnemann based on Kodzdrói *et al.* (2001), Linnemann & Schauer (1999) and Linnemann & Romer (2002). Geochronological ages are taken from: *Buschmann *et al.* (2001; SHRIMP U–Pb); **Linnemann *et al.* (2007; U–Pb on single zircons by LA-ICP-MS); ***Linnemann *et al.* (2000; Pb/Pb evaporation age on single zircons); ****Linnemann *et al.* (unpublished data; SHRIMP U–Pb). Key: 1, Rothstein Formation (Ediacaran, *c.* 566 Ma) with sedimentary deposits, black cherts, and volcanic rocks from a Cadomian backarc basin in the Torgau-Doberlug Syncline; 2, diamictites intercalated with Ediacaran passive margin deposits (greywackes, quartzites quartz shists, shales) of the Weesenstein Group (Elbe Zone) and the Clanzschwitz Group (North Saxon Anticline); at Kunnersdorf near Görlitz, there is a small diamictite occurrence in the Lausitz Block; 3, Lausitz Group, consisting of monotonous greywacke-schist turbidites that formed in the setting of a retro-arc or foreland basin, respectively (latest Ediacaran to earliest Cambrian); 4, microconglomerate intercalated into the Lausitz Group and mainly composed by fragments of black chert; 5, Early Cambrian intrusions of granodiorites, granites, diorites and tonalites of the Lausitz Granitoid Complex and adjoining areas (*c.* 540 to 530 Ma); 6, Late Ediacaran sedimentary rocks and Early Cambian granitoid intrusions often strongly sheared and metamorphosed under upper greenshist to amphibolite facies conditions; 7, Late Ediacaran sedimentary rocks strongly sheared and metamorphosed under upper greenshist to amphibolite facies conditions (units of the Erzgebirge Block bordering on the Elbe Zone); 8, Lower to Middle Cambrian deposits; 9, Lower Cambrian as a few gigantic blocks in a Lower Carboniferous wild flysch deposit; 10, Lower Ordovician, very mature conglomerates, sandstones and shales (Tremadoc); 11, Lower Ordovician granitoids (*c.* 490 to 480 Ma); 12, Cambro-Ordovician, Silurian and Lower Carboniferous sedimentary rocks and volcanosedimentary complexes affected by the Variscan Orogeny (not subdivided); 13, Variscan granitoid intrusions (*c.* 335–300 Ma); 14, Variscan early molasse in the (Late Viséan) Torgau-Doberlug Syncline; 15, molasse deposits and volcanic rocks in Upper Carboniferous and Lower Permian basins; 16, Mesozoic and Cenozoic cover rocks; 17, Tertiary basalts. Abbreviations: CB, Collmberg Mountain (type area of the Collmberg Formation, Tremadoc); CG, Clanzschwitz Group (Ediacaran); DG, Dohna Granodiorite (Early Cambrian); ETSG, Elbtalschiefergebirge (Variscan schist belt, part of the complex dextral shear zone system of the Elbe Zone); FSB, Fore Sudetic block; GGC, Großenhain Gneiss Complex originated from Cadomian Basement rocks containing Ediacaran pararocks and Early Cambrian orthogneises; GSG, Görlitz-Schiefergebirge (Variscan wildflysch complex bordering to the north of the Lausitz Block); HD, area of the Hohe Dubrau (type area of the Dubrau Formation, Tremadoc); ILF, Intra Lausitz Fault; ISF, Intra Sudetic Fault; KD, roadcut near the village of Kunnersdorf (occurrence of a diamictite, Lausitz Group, Ediacaran); KMC, Kaczawa Metamorphic Complex; LG, Laas Granodiorite (Early Cambrian); LT, Lausitz Thrust; MLF, Main Lausitz Fault; MM, Meissen Massif (Variscan complex of granitoids); MSF, Marginal Sudetic Fault; NWSG, Nossen-Wilsdruff-Schiefergebirge (Variscan schist belt, part of the complex dextral shear zone system of the Elbe Zone); PH, village Petershain (occurrence of a microconglomerate containing black cherts, Lausitz Group, Ediacaran); RR, Rothstein Rock (type area of the Rothstein Formation, Ediacaran); SB, Sowie Mts. Block; SKMC, South Karkonosze Metamorphic Complex; WB, Wetterberg Mtn. with a quarry near the village Ebersbach (occurrence of a microconglomerate containing black cherts, Lausitz Group, Ediacaran); WG, Weesenstein Group (Ediacaran); WLF, West Lausitz Fault.

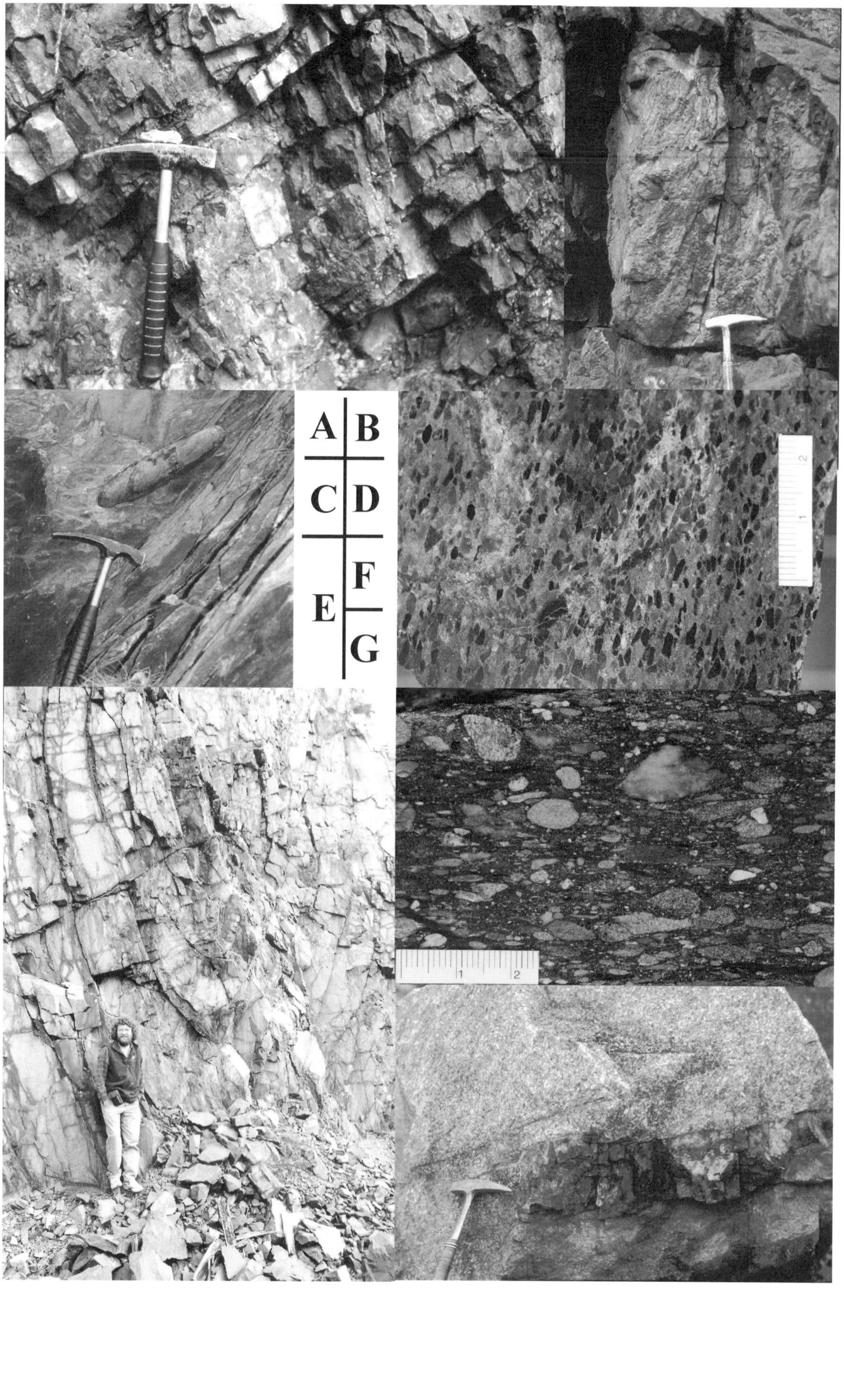
A B
C D
E F G
1 2
1 2

Clanzschwitz groups originally belonged to the same unit in pre-Variscan times that was subsequently disrupted by dextral Variscan strike-slip movement along the Elbe Zone (Linnemann 2003). A remnant between both units is represented by the Großenhain Gneiss Complex within the Elbe Zone (Fig. 3.19). Both the Wessenstein and Clanzschwitz groups were metamorphosed to biotite grade during Cadomian deformation. The Clanzschwitz Group is overlain by the transgressive Lower Ordovician Collmberg Quartzite (Fig. 3.19), whereas in the case of the Weesenstein Group the Ordovician overstep sequence is not preserved (Fig. 3.21). The Weesenstein and Clanzschwitz groups were intruded by the Dohna and Laas granodiorites at 537 ± 2 Ma (Fig. 3.21) and 531 ± 7 Ma, respectively (Linnemann *et al.* 2000).

In general, all Neoproterozoic sections within the Cadomian basement of the Saxo-Thuringian Zone would appear to be rootless due to Variscan stacking of the crust. An underlying cratonic basement is not known. However, NdT_{DM} model ages for the Late Neoproterozoic sediments range between 1.3 and 1.84 Ga (Linnemann & Romer 2002), clearly indicating that the source area of the Neoproterozoic sediments was dominated by old cratonic crust (Fig. 3.22). Furthermore, the sediments from the Cadomian backarc basins show much younger model ages (1.3–1.55 Ga) than the retro-arc deposits (1.2–1.84 Ga; Fig. 3.22). This is explained by the input of more juvenile material into the backarc sediments due to the magmatic activity caused by backarc speading. Additionally the model ages reflect a mixed age of old cratonic crust and Cadomian juvenile components (Linnemann & Romer 2002). εNd(T) values of Ediacaran sediments from Cadomian basins (−11 to −2) and *c.* 540 Ma Cadomian granitoids (−10 to −3) are plotted against the two known cratonic inliers in the Bohemian Massif (Fig. 3.23). The diagram shows that (i) the *c.* 540 Ma Cadomian granitoids of the Saxo-Thuringian Zone were largely derived from recycled deposits of Cadomian marginal basins, and (ii) no melt derived only from recycled cratonic basement exists (Linnemann & Romer 2002).

The U–Pb age spectra of detrital zircons of Ediacaran sediments from Cadomian basin deposits indicate that they display a common characteristic of the source area (Figs. 3.24 & 3.25). Investigated samples predominantly contain Late Neoproterozoic (*c.* 550–690 Ma) and Palaeoproterozoic (*c.* 1.8–2.2 Ga) grains. Samples also contain a small fraction of *c.* 700–950 Ma and a population of Palaeoarchaean grains (Figs 3.24 & 3.25; Linnemann *et al.* 2007). A common feature of all samples is an ‘age gap’ between 1.75 and 1.0 Ga, which is typical of a West African provenance and is diagnostic in distinguishing them from East Avalonia and Baltica (e.g. Nance & Murphy 1994). This is in agreement with the characteristic clusters of Palaeoproterozoic ages in the interval of *c.* 1.78–2.17 Ga. Such ages are typical of the western part of the Gondwana Supercontinent, which was affected by abundant magmatic intrusions (*c.* 1.8–2.2 Ga) during the Eburnean Orogeny (West African Craton). Furthermore, Neo- to Mesoarchaean zircon ages and additional Palaeoarchaean grains point to recycling of magmatic rocks formed during the Liberian and the Leonian orogenies, respectively. Both of these orogenies affected the West African Craton during the Archaean (Figs. 3.24 & 3.25; e.g. Rocci *et al.* 1991).

All features of the Cadomian basement in the Saxo-Thuringian Zone are summarized in a model shown in Figures 3.26 and 3.27 (Linnemann *et al.* 2007). The model explains the co-existence of the passive margin (Weesenstein and Clanzschwitz groups), the magmatic arc and in between a backarc basin (Rothstein and Altenfeld formations) in a strike-slip regime showing internal spreading zones (Fig. 3.26). The magmatic arc and the related marginal basins developed on an extremely stretched and thinned continental crust (Fig. 3.26). The recent Western Pacific region shows a similar situation (see inset Fig. 3.26). The magmatic arc is not preserved in the Saxo-Thuringian Zone. Its former existence is documented by the pyroclastic greywackes in the Rothstein and Altenfeld formations that show geochemical signatures of siliciclastic sediments derived from an evolved magmatic arc (Linnemann & Romer 2002). The *c.* 565–575 Ma population of detrital zircon dominates the age pattern in the pyroclastic greywacke turbidites (Fig. 3.25). Zircons of that population are typically needle-shaped and ideomorphic or only subrounded. Hence, they are interpreted to be derived directly from the magmatic arc source. These zircon grains reflect the main activity of the arc at *c.* 570 Ma and constrain the maximum age of deposition of the pyroclastic greywackes close to the age of sedimentation (Linnemann *et al.* 2007). This is supported by a dated ash bed from the Rothstein Formation at 566 ± 10 Ma (Buschmann *et al.* 2001). The maximum age of deposition for the Altenfeld Formation at 563 ± 6 Ma is in the same range (U. Linnemann & T. Jeffries, unpublished data; Fig. 3.17).

In the model presented here, the later stage of Cadomian basin development was characterized by the collision of the arc with the continent and consequent closure of the backarc basin to form a foreland fold-and-thrust belt and a related retro-arc basin (Fig. 3.27; Linnemann *et al.* 2007). The retro-arc basin is significantly younger than the backarc basin. New dating of detrital zircon grains has led to the discovery of a young zircon population showing a concordia age of 543 ± 4 Ma in the Lausitz Group (Linnemann *et al.* 2007; Fig. 3.19). This maximum age of deposition overlaps, within error, the Precambrian–Cambrian boundary and suggests a latest Ediacaran to earliest Cambrian age for the Cadomian retro-arc basin (Linnemann *et al.* 2007). Sediments are characterized by monotonous deep-marine deposits of the Lausitz and Leipzig groups and the Lower Frohnberg Formation (Fig. 3.27). The main lithology is repre-

Fig. 3.20. Photographs of rocks and outcrops of the Cadomian Basement (Saxo-Thuringian Zone, North Bohemian Massif). (**A**) Black bedded cherts of the Cadomian backarc basin cropping out at the Rothstein Rock (Ediacaran, *c.* 566 Ma, Rothstein Formation, Torgau-Doberlug Syncline). (**B**) Steeply dipping bed of the Purpurberg Quartzite with cross-bedding and wave ripples. The quartzite occurs in the lower part of the Weesenstein Group, which represents the passive margin of the Cadomian backarc basin (Ediacaran, *c.* 560–570 Ma, Weesenstein Group; Purpurberg Mountain near Niederseidewitz, Elbe Zone). (**C**) Stretched granitoid pebble in a mudstone matrix in the upper part of the Weesenstein Group. The diamicts of the Weesenstein Group are interpreted to be in part glaciomarine (Ediacaran, *c.* 560–570 Ma, Weesenstein Group; 100 m to the east of the railway station Weesenstein, Elbe Zone). (**D**) Microconglomerate of clasts of cherts and felsic rocks from the Lausitz Group demonstrating the redeposition of black cherts during the formation of the Cadomian retro-arc basin (Ediacaran, *c.* 540–545 Ma, Lausitz Group; Petershain near Kamenz, Lausitz Block). (**E**) Isoclinally folded greywacke turbidites of the Cadomian retro-arc basin (Ediacaran, Lausitz Group; quarry Butterberg near Bernsdorf, Lausitz Block). (**F**) Conglomerate from the Frohnberg Formation of the Katzhütte Group with pebbles of granitoids, black cherts, intraformational sedimentary clasts, and quartzites derived from the thrust belt feeding the retro-arc basin (Ediacaran, Katzhütte Group; Grendel Mountain near Schirnrod, Schwarzburg Unit). (**G**) Xenolite of a greywacke fragment in a 542 Ma granodiorite of the Lausitz Granitoid Complex of the Lausitz Block (Lausitz Granitoid Complex, 542 ± 3 Ma, latest Ediacaran to earliest Cambrian; quarry Kindisch near Elstra, Lausitz Block).

sented by thick-bedded greywacke mud turbidites. Features of rapid deposition such as water-escape structures, mud chips, seismites and slumps are common (Linnemann 2003). In addition, these sediments contain abundant fragments of black bedded chert derived from the Cadomian backarc basin (Fig. 3.20B) and conglomerates with pebbles of igneous rocks derived from the magmatic arc (Fig. 3.20F). Black cherts and greywacke deposits were reworked by fold-and-thrust belt activity from the

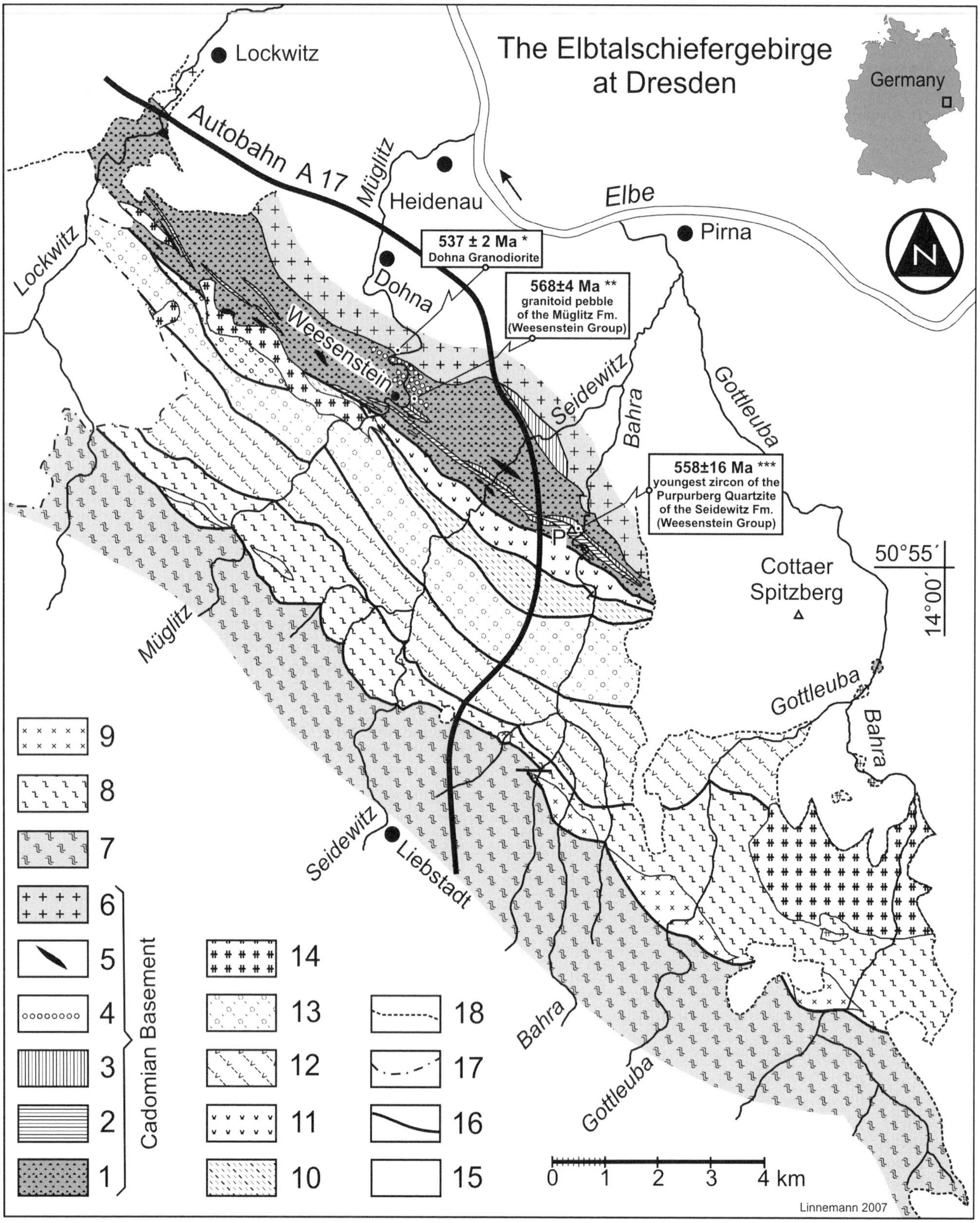

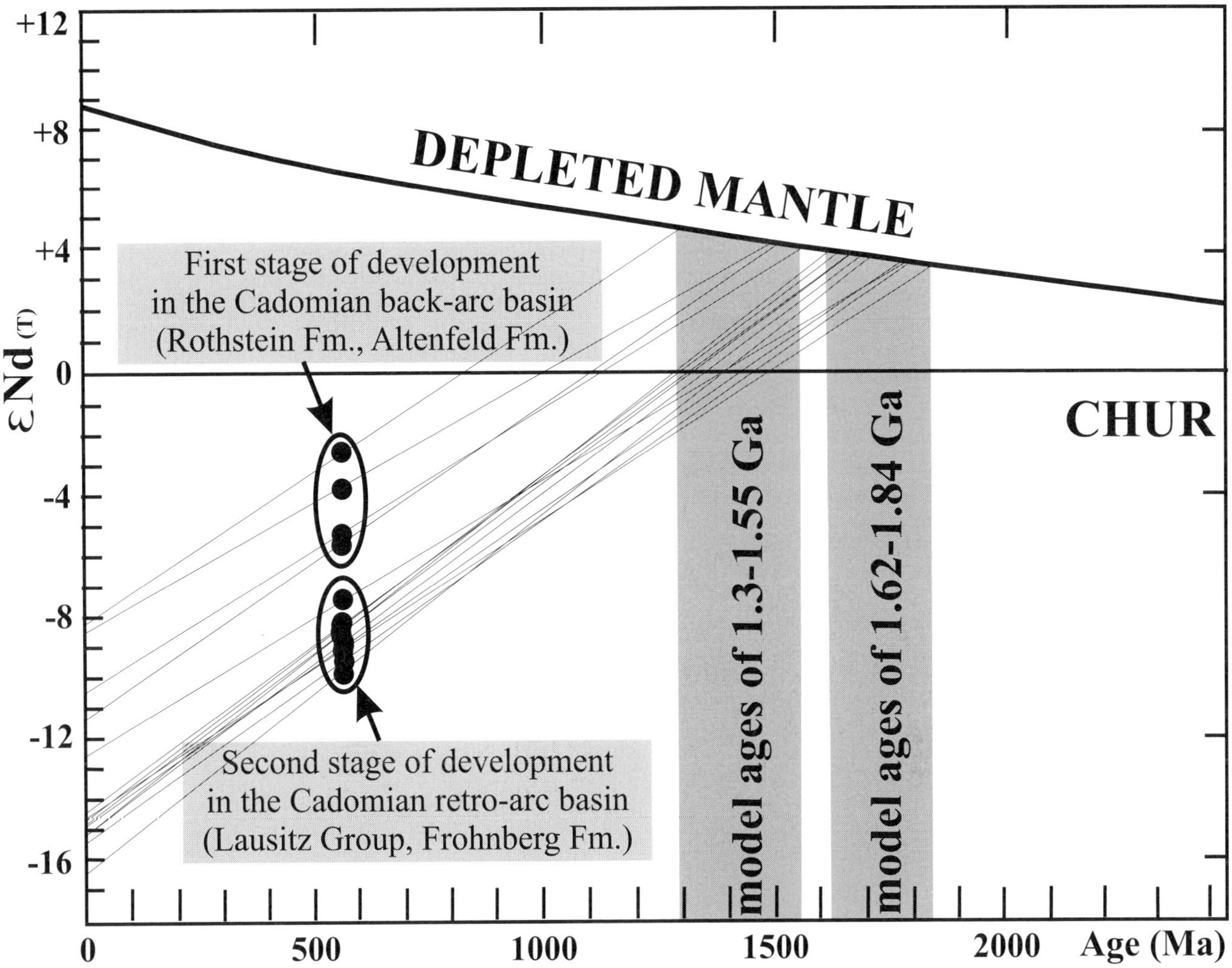

Fig. 3.22. T_{DM} Nd model ages (DePaolo 1981) of sedimentary rocks from the Cadomian basement of the Saxo-Thuringian Zone (data source: Linnemann & Romer 2002). CHUR, chondritic uniform reservoir.

Cadomian backarc basin and were redeposited in the retro-arc basin as conglomerates and other siliciclastics (Fig. 3.27).

In the outer part of the foreland basin close to the craton, a remnant basin dominates the depositional regime represented by the Upper Frohnberg Formation in the Schwarzburg Unit (Fig. 3.27; Linnemann *et al.* 2007). Sedimentary deposits are characterized by the increasing input of quartzites and quartz-rich shales of greater maturity than the greywacke deposits derived

Fig. 3.21. Geological map of the Cadomian passive margin deposits of the Weesenstein Group (Ediacaran) and the Early Cambrian Dohna Granodiorite in the Elbtalschiefergebirge at Dresden, which is a schist belt within the Variscan dextral shear zone system of the Elbe Zone (complied by U. Linnemann based on Pietzsch (1917), Linnemann (1991) and Kurze *et al.* (1992). Ages are taken from: *Linnemann & Jeffries (unpublished data; U–Pb on single zircons by LA-ICP-MS); **Linnemann *et al.* (2000; Pb/Pb evaporation age on single zircons); ***Linnemann *et al.* (2007; U–Pb on single zircons by LA-ICP-MS). Key: 1, Weesenstein Group containing the older Seidewitz Formation including the Purpurberg Quartzite and the younger Müglitz Formation (Ediacaran); 2, Purpurberg Quartzite and its equivalents (Seidewitz Formation, Weesenstein Group, Ediacaran); 3, quartz schists often under- and overlying the Purpurberg Quartzite (Seidewitz Formation, Weesenstein Group, Ediacaran); 4, isolated pebbles and conglomerates most probably glaciomarine in origin (Müglitz Formation, Weesenstein Group, Ediacaran); 5, metabasalts (Weesenstein Group, Ediacaran); 6, Dohna Granodiorite (Early Cambrian); 7, Cadomian metasediments of the Erzgebirge Block overprinted under upper greenschist to amphibolite facies conditions during Variscan orogenic processes; 8, Ordovician, Silurian and Devonian metasediments and volcanic rocks of the Elbtalschiefergebirge overprinted under greenschist facies conditions during Variscan orogenic processes (so-called 'Phyllitic unit' of the Elbe Zone or Mühlbach-Nossen Group); 9, Ordovician granite (Tourmaline Granite, 485 ± 6 Ma; Linnemann *et al.* 2000); 10, Ordovician metasediments of the Elbtalschiefergebirge overprinted under greenschist facies conditions during Variscan orogenic processes; 11, Devonian volcanosedimentary complex of the Elbtalschiefergebirge overprinted under greenschist facies conditions during Variscan orogenic processes; 12, Devonian 'diabase-limestone series' of the Elbtalschiefergebirge overprinted under greenschist facies conditions during Variscan orogenic processes; 13, Lower Carboniferous siliciclastic flysch deposits of the Elbtalschiefergebirge overprinted under greenschist facies conditions during Variscan orogenic processes; 14, Variscan (Lower Carboniferous) granitoids of Meissen and Markersbach; 15, Permian and younger sedimentary cover; 16, major faults and dextral shear zones of the Elbtalschiefergebirge that originated during Variscan orogenic processes; 17, limit of Permian (Rotliegend) sedimentary rocks; 18, limit of Cretaceous (Cenomanian–Turonian) sedimentary rocks. P, Top of Purpurberg Mountain, type locality for the Cadomian Purpurberg Quartzite (Linnemann 1991).

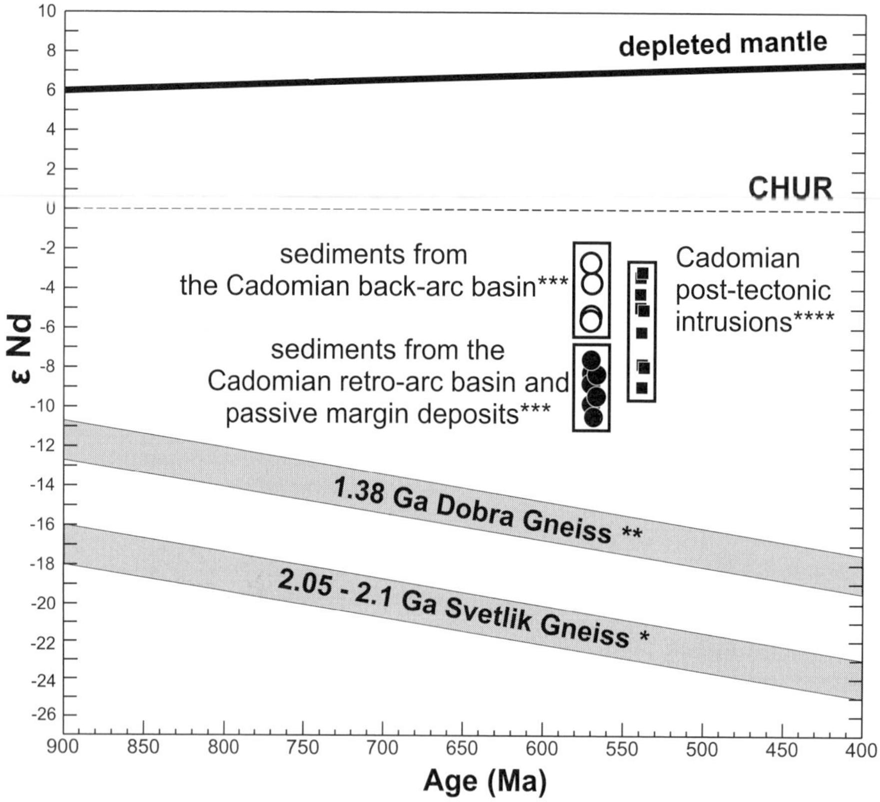

Fig. 3.23. Age versus initial εNd diagram for groups of Cadomian sediments and granitoids within the Saxo-Thuringian Zone (Bohemian Massif) and from the two known cratonic basement slivers within the Moldanubian Zone of the Bohemian Massif. Data are taken from: *Liew & Hoffmann (1988; Nd) and Wendt *et al.* (1993; protolith age); **Liew & Hoffmann (1988; Nd) and Gebauer & Friedl (1993; protolith age); ***Linnemann & Romer (2002); ****Linnemann & Romer (2002) and Hammer (1996). Note that the retro-arc and passive margin deposits have less evolved Nd isotopic compositions than the backarc basin deposits. The Cadomian granitoids show a similar Nd isotopic composition to the sediments.

from the retro-arc (Linnemann 2003). Fragments of very mature sandstones deposited in a passive margin setting dominate the composition of conglomerates in this part of the foreland basin (Linnemann 2003). The suggested model of late-stage basin development also explains why an angular Cadomian unconformity is developed at some localities but not at others (Fig. 3.27). In localities proximal to the foreland fold-and-thrust belt, sediments became deformed (Figs. 3.20E & 3.27). More distal deposits could easily escape deformation because the thick crust of the underlying craton protected this part of the section from the movement of the retro-arc forced by the arc–continent collision (Fig. 3.27). The remnant basin occurs only in the SE part of the Schwarzburg Unit and crops out best at the Wurzelberg section near Goldisthal (Fig. 3.17).

The latest stage of Cadomian orogenic processes in the Saxo-Thuringian Zone was accompanied by the intrusion at *c.* 540 to 530 Ma of plutons (Figs. 3.17, 3.19 & 3.21) into Cadomian sediments (Linnemann *et al.* 2000; Gehmlich 2003; Tichomirowa 2003). Major crustal thickening and high pressure–high temperature metamorphism are not observed in the Cadomian basement of the Saxo-Thuringian Zone. Zulauf *et al.* (1999) recognized a high temperature–low pressure metamorphism. Greywackes and other sedimentary rocks from the Cadomian marginal basins underwent a widespread anatectic event (Fig. 3.27) leading to the origin of large plutons such as the Lausitz Granitoid Complex which is the largest geological unit of the entire Saxo-Thuringian Zone (Fig. 3.19). All Cadomian plutons investigated thus far in the Saxo-Thuringian Zone contain a remarkable amount of inherited zircons (Linnemann *et al.* 2000, 2004; Gehmlich 2003; Tichomirowa 2003). Therefore, and also because of Nd-isotopic signatures similar to the Ediacaran sediments (Fig. 3.23), it is quite likely that during the magmatic event at *c.* 540–530 Ma Cadomian basins were recycled. The fact that this was possible without major crustal thickening makes the late Cadomian anatectic event in the Saxo-Thuringian Zone a remarkable phenomenon.

Delamination of the crust to produce the heat flow necessary to induce melting is excluded because there is no stacking and major thickening of the crust. Slab break-off of the underlying subducted plate, however, might have resulted in the extreme heat flow that would have been necessary to generate widespread anatexis (Fig. 3.27). Slab break-off was also used by Zulauf *et al.* (1999) and by Samson *et al.* (1997) to explain magmatic episodes of the Cadomian Orogeny older than 550–570 Ma observed in the Teplá–Barrandian Unit and in the North Armorican Massif, respectively. The onset of Cambrian sedimentation is recorded from *c.* 530–520 Ma onwards in the Zwethau, Tröbitz and Delitzsch formations of the Torgau-Doberlug Syncline (Lower to Middle Cambrian; Fig. 3.19). The gap in the sedimentary record between the Cadomian basement and Cambrian sediments probably represents *c.* 10 to 20 Ma (Linnemann 2003; Buschmann *et al.* 2006). Fossil fauna have been correlated with assemblages in Morocco (Buschmann *et al.* 2006 and references therein). The Lower Cambrian of the Görlitz

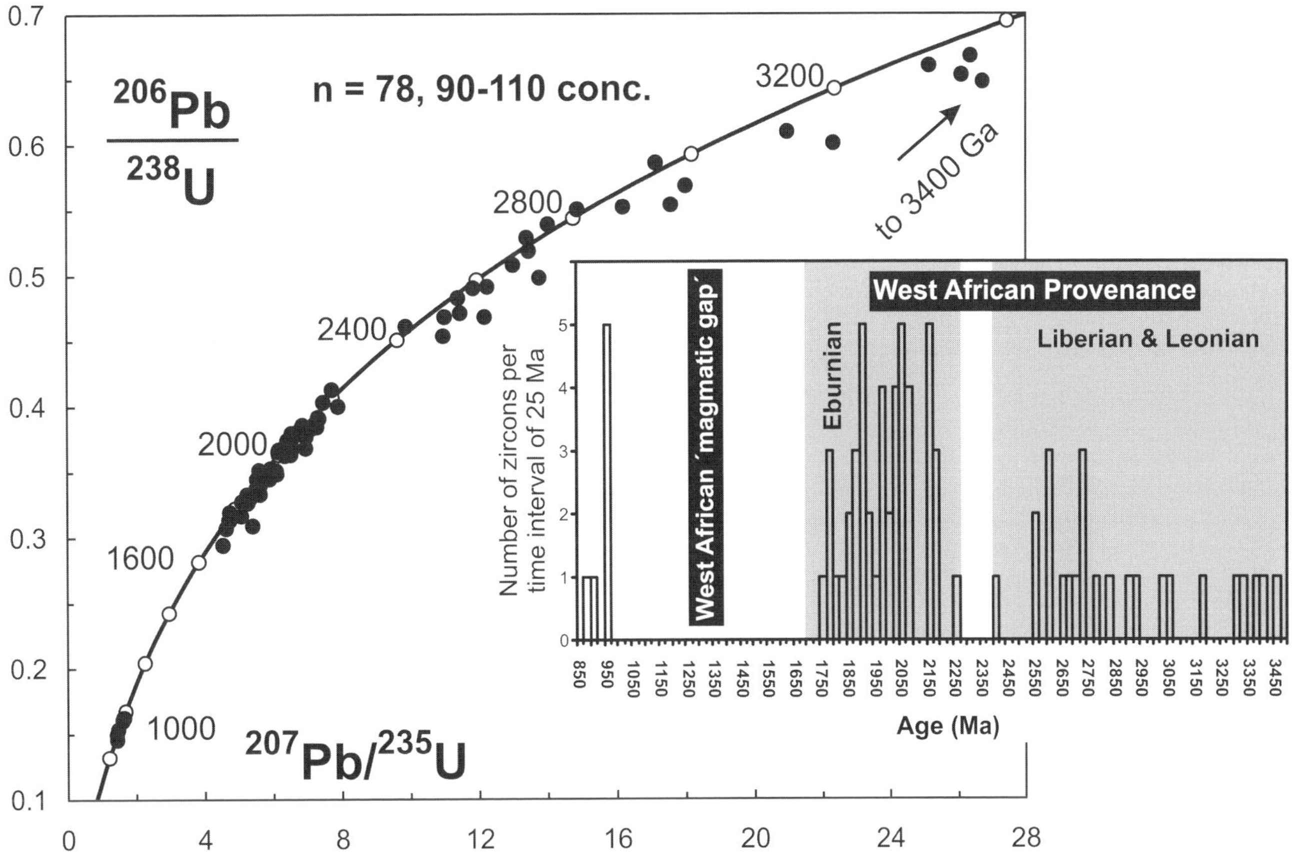

Fig. 3.24. Compilation of ages from detrital zircon older than 800 Ma from three Ediacaran sandstones from the Saxo-Thuringian Zone (samples from the Rothstein Formation and the Weesenstein and Lausitz Groups; data from Linnemann *et al.* 2007). Inset shows histogram, using 25 Ma increments, for the combined data sets. The histogram emphasizes the good agreement between detrital zircon populations and zircon ages from the West African Craton, including the 'magmatic gap'.

Schiefergebirge is an assemblage of blocks in a Variscan mélange matrix (Fig. 3.19; Linnemann & Schauer 1999). The same is the case for blocks of Middle Cambrian sediments in the Franconian Forest. The Heinersdorf 1 and 2 boreholes in the Berga Anticline contain Middle Cambrian sedimentary rocks but did not reach the contact with the underlying Cadomian basement (Wucher 1967).

Rock complexes composed of remnants from the Cadomian basement were incorporated in the Variscan nappe pile of the Erzgebirge Block (Fig. 3.28). Late Neoproterozoic metasediments and Early Cambrian igneous rock complexes, according to their geochronological ages and geochemical characteristics, are equivalent to all other units of the Cadomian basement in the Saxo-Thuringian Zone (Tichomirowa *et al.* 2001; Tichomirowa 2003). Late Neoproterozoic sediments that underwent a Variscan metamorphic overprint mostly occur in the HP-HT unit and in the 'Lower Shear Zone' of the Erzgebirge Block. Metamorphosed equivalents of the Early Cambrian plutons with protolith ages of *c.* 540–530 Ma are present in MP-MT units of the Variscan nappe pile. It cannot be excluded that the other tectonometamorphic complexes of the Erzgebirge Block, such as the HP-LT, MP-LT and the LP-LT units, also contain remnants of the Cadomian basement.

Sudetes

The Sudetes are a very complex part of the Bohemian Massif that became fragmented and deformed during the Variscan Orogeny (Fig. 3.14). The Western Sudetes are part of the Saxo-Thuringian Zone. These units belong to the eastern part of the Lausitz Block, the Görlitz Schiefergebirge, and the Izera-Karkonoše Block (Fig. 3.19). Other more eastern parts with relicts of Cadomian basement are represented by the Fore Sudetic Block, the Góry Sowie Massif, the Kłodzko Metamorphic Complex, the Orlica-Śnieżnik Massif and the Strzelin Massif (Figs. 3.19 & 3.29). Due to strong Variscan structuring and fragmentation the reconstruction of Cadomian orogenic processes in the Sudetes is very difficult. However, the main tectonostratigraphic and magmatic events are comparable to those in the Saxo-Thuringian Zone.

From the Izera-Karkonoše Block, two Cadomian granitoids were reported by Żelaźniewicz *et al.* (2004). The Bitouchov Granite gave an age of 540 ± 11 Ma. The Platerówka Granodiorite has been dated at 533 ± 9 Ma. From the Fore Sudetic Block the same authors published an age of 548 ± 9 Ma for the Wądroże Granodiorite. These data demonstrate the possible extension of remnants of the Cadomian basement from the Saxo-Thuringian Zone eastwards into the Sudetes. Although the field relations are uncertain due to the lack of outcrops, it has been suggested that all three granitoids intruded metasediments that were interpreted as Cadomian basement based on the minimum ages given by the intrusion of the intrusive granitoid plutons (Żelaźniewicz *et al.* 2004). Metasedimentary rocks from the Izera-Karkonoše Block may be present in the Velka Upa Group and the Czarnów Formation.

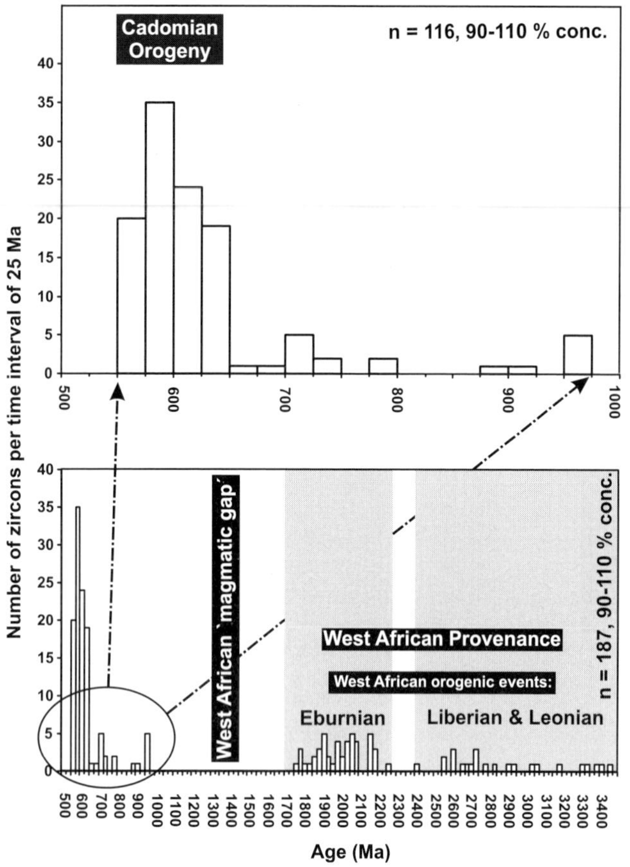

Fig. 3.25. Age compilation for detrital zircon from three Ediacaran sandstones from the Saxo-Thuringian Zone (samples from the Rothstein Formation and the Weesenstein and Lausitz Groups; data from Linnemann *et al.* 2007). Ages older than 1.7 Ga clearly indicate the West African provenance of Ediacaran sediments from the Saxo-Thuringian Zone.

The Kłodzko Metamorphic Unit is part of the Central Sudetes and consists of a number of separate tectonic units interpreted as thrust sheets. A plagioclase gneiss from the Orla-Gołogłowy Unit yields an upper intercept age of 590 ± 7 Ma that is interpreted to be the igneous crystallization age (Mazur *et al.* 2004). From the topmost Kłodzko Fortress Unit, zircon grains from a metatuffite were analysed by Pb/Pb dating. The youngest zircon population clusters around 590–600 Ma and is interpreted to reflect the maximum age of deposition of the metasediment (Mazur *et al.* 2004). The εNd values from volcanic and igneous rocks from both these units range between +3.2 and +8.9 reflecting a strong input of juvenile material indicating a suprasubduction origin (Kryza *et al.* 2003).

No Cadomian-age plutons are known from the Góry Sowie Massif and the Orlica-Śnieżnik Massif and the existence of (meta) sedimentary rocks related to the Cadomian basement is uncertain. The Stronie Formation in the Orlica-Śnieżnik Massif may represent a late Neoproterozoic metasequence (Fig. 3.29).

From the Strezelin gneiss in the easternmost part of the Fore-Sudetic Block, SHRIMP U–Pb inherited zircon cores of Palaeo- to Mesoproterozoic ages between 2.0 and 1.24 Ga are reported. These zircons show zoned rims of 600 ± 7 and 568 ± 7 Ma and are interpreted in terms of two different thermal events inducing zoned zircon overgrowth (Oberc-Dziedzic *et al.* 2003). Amphibolites, mica schists, calc-silicate rocks and marbles occur very close to the Strezelin gneiss but their contact is strongly mylonitized. These rocks were interpreted to be older than the Strezelin gneiss. The amphibolites are of within-plate tholeiitic character (Oberc-Dziedzic *et al.* 2003).

The Teplá–Barrandian Unit

The Teplá–Barrandian Unit, also known as the 'Bohemicum', was originally defined by Máška & Zoubek (1961). It is the classic area of the Cadomian basement in the Bohemian Massif and consists of the Teplá Block and the Barrandian (Fig. 3.14). According to Kossmat (1927) the Teplá–Barrandian Unit is part of the Moldanubian Zone. Due to its uniqueness it is described here separately. The largest part of the Teplá Block is composed of late Neoproterozoic rock units and Early to Mid-Cambrian and Variscan-age intrusions. Large massifs composed predominantly of Cadomian and Variscan mafic rocks, for example the Mariánzé Lazně Complex at the westernmost edge and the Neukirchen-Kdyně Massif at the Hoher Bogen in the southwesternmost corner of the Teplá–Barrandian Unit also occur. The metamorphic overprint of the Teplá Block was caused both by the Cadomian and the Variscan orogenic processes and this over-printing is generally much higher than in the tectonically detached Barrandian.

The Barrandian is a very weakly deformed and metamorphosed succession of sedimentary and volcanic rocks, which are Neoproterozoic to Mid-Devonian in age. This fragment of peri-Gondwanan rocks escaped completely the high-grade metamorphism of the Varsican Orogeny and is named in honour of the French engineer, geologist and palaeontologist Joachim Barrande (1799–1883). The easternmost part of the Teplá–Barrandian Unit is represented by the elezné Hory (Eisengebirge). The southernmost rocks of the Teplá–Barrandian Unit occur in the metamorphosed rocks of the so-called 'Islet Zone'. The Teplá–Barrandian Unit is bounded by the North, West and Central Bohemian shear zones. To the NE, the Teplá–Barrandian Unit is covered by Permian and Mesozoic to Cenozoic sediments.

Co-operation between Czech and French geologists during the Cold War allowed Precambrian sediments of the Barrandian to be compared with the Brioverian of the Armorican Massif, and their relation to Cadomian orogenic processes became established much earlier in the Teplá–Barrandian Unit (e.g. Chaloupský 1977; Chauvel & Rabu 1988; Zoubek in Zoubek *et al.* 1988; Křibek *et al.* 2000) than in other parts of the Bohemian Massif. The initial division of Precambrian sediments in the Teplá–Barrandian Unit into the classic Pre-Spilitic, Spilitic and Post-Spilitic stages was first proposed by Kettner (1917) (the term 'spilites' is used in the sense of metabasalts). The Precambrian was referred to as 'Algoncian' by the same author. The modern division of Late Neoproterozoic sediments is based on Röhlich (1965), Mašek & Zoubek (1980) and Mašek (2000). Alternative subdivisions have been proposed by Holubec (1966) and Röhlich (2000).

According to Mašek (2000; Fig. 3.30), the lowermost lithostratigraphic unit of the Neoproterozoic rocks in the Teplá–Barrandian Unit is represented by the Kralupy-Zbraslav Group which is subdivided into the Blovice Formation (*c.* 7000 m) at the base and the Davle Formation (*c.* 2000 m) in the upper part (Fig. 3.30). In the topmost part of the latter formation (Lečice Shale) a 200 m thick unit of dark shale occurs. Above the Kralupy-Zbraslav Group the *c.* 5.000 m thick Štěchovice Group represents the upper part of the late Neoproterozoic succession. Similar to the Saxo-Thuringian Zone, the section is rootless and

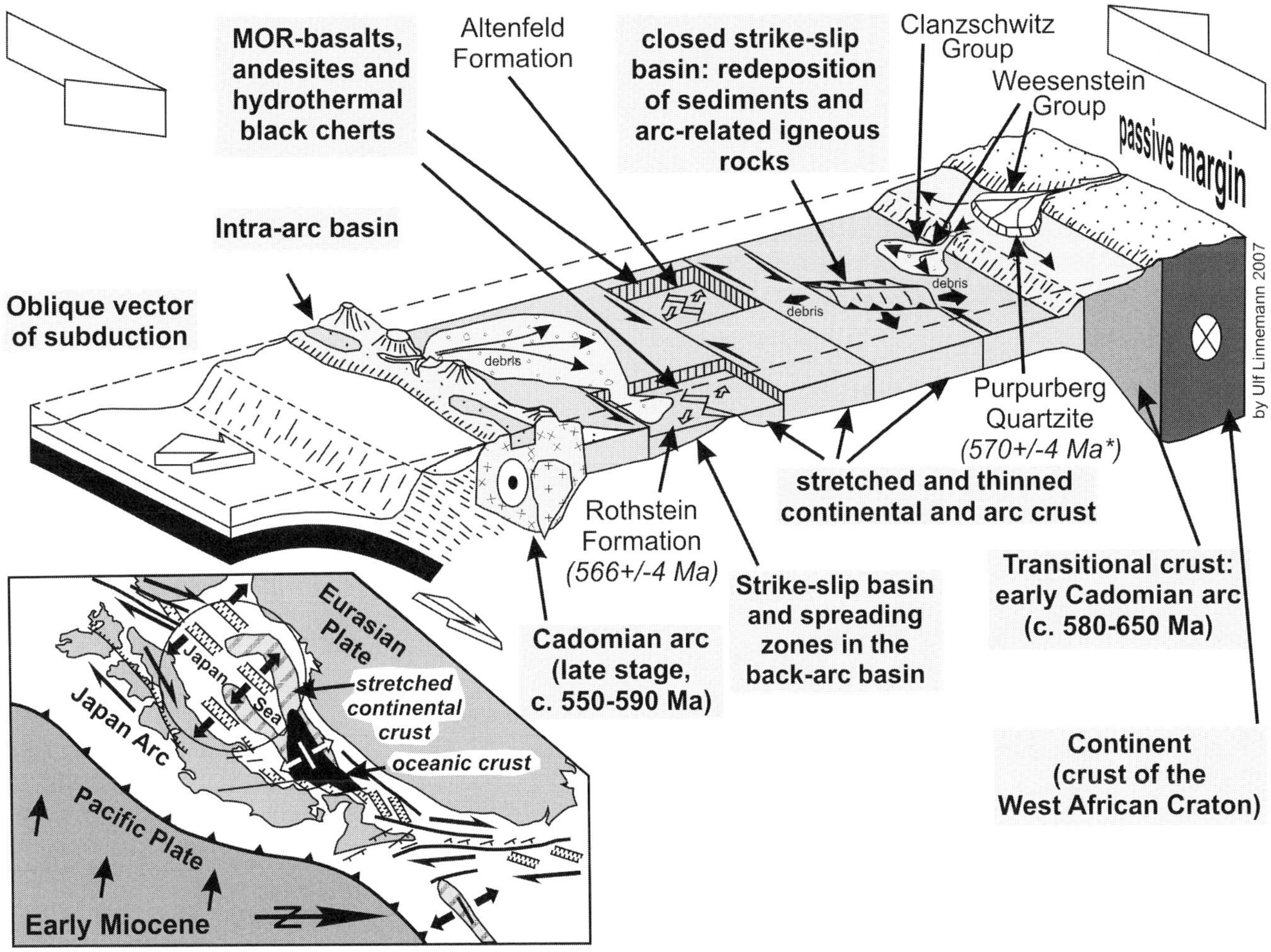

Fig. 3.26. Model for the plate tectonic development of the Cadomian backarc basin at *c.* 580 and 560 Ma based on data from the Saxo-Thuringian Zone (Bohemian Massif) (from Linnemann *et al.* 2007). Backarc basin consists of a passive margin towards the continent, represented by the Weesenstein and Clanzschwitz Groups and a margin characterized by more strongly stretched continental crust and the accumulation of predominantly arc-derived debris towards the arc. The backarc is documented by MOR-related rocks and hydrothermal black cherts recorded in the Altenfeld and the Rothstein Formations (*maximum age of deposition). Inset: sketch of analogous plate tectonic configuration represented by the opening of the Japan Sea in the Western Pacific region during the Early Miocene (after Jolivet *et al.* 1992). The backarc basin of the Japan Sea is largely floored by stretched continental crust.

the nature of the underlying basement is not known. The profile is dominated by greywacke and shale turbidites deposited in submarine fan settings. These sediments show characteristics both of proximal and distal turbidites. Also present are black cherts, conglomerates, volcanic rocks and carbonates. The Domažlice and Teplá crystalline complexes are considered to be metamorphosed equivalents of the Blovice Formation (e.g. Vejnar *et al.* 1984).

The Blovice Formation is the most extensive Late Neoproterozoic unit in the Teplá–Barrandian Unit and consists of weakly metamorphosed sedimentary and volcanic rocks such as greywacke and mudstone turbidites, black cherts, black shales, limestones and conglomerates, as well as some metabasalts and other volcanics. Volcanic rocks of the Blovice Formation show positive $\varepsilon Nd_{(t)}$ values ranging between +3.8 and +9.3 and demonstrate a strong juvenile magma source (Pin & Waldhausrova 2007; Fig. 3.30). The $\varepsilon Nd_{(t)}$ values of sedimentary rocks are measured in an interval of −0.1 to −12 (Pin & Waldhausrova 2007; Drost *et al.* 2007). The NdT_{DM} model ages were 1.2 to 2.2 Ga and reflect the strong input of old cratonic material into the sediments (Fig. 3.30).

The Davle Formation crops out probably only in anticlinal structures in the SE flank of the Barrandian and is characterized by acidic and intermediate volcanics with substantial amounts of pyroclastics, accompanied by greywacke and mudstone turbidites. Silicified black shales (Lečice Shale) comprise the youngest member of the Davle Formation.

The Štěchovice Group, lying concordantly on the Davle Formation, is developed and preserved exclusively in the SE part of the Barrandian. These siliciclastic sediments are represented by greywacke and mudstone turbidites that show characteristics of rapid sedimentation. No volcanics or silica-rich rocks are present. Distal thin-bedded turbidites dominate the lowermost part of the unit. More proximal and thick-bedded turbidites and intercalated conglomerates characterize the upper part of the Štěchovice Group. Diverse conglomerates are termed the Dobris Conglomerate, formerly supposed to be the base of the Post-Spilitic Stage. Intercalated conglomerates do not represent a single homogeneous horizon. Instead, these occur at different levels in the upper part of the Štěchovice Group. U–Pb dating of pebbles and detrital zircon gave a maximum age of sedimentation of 564 ± 16 Ma and 568 ± 3 Ma (Dörr *et al.* 2002; Drost *et*

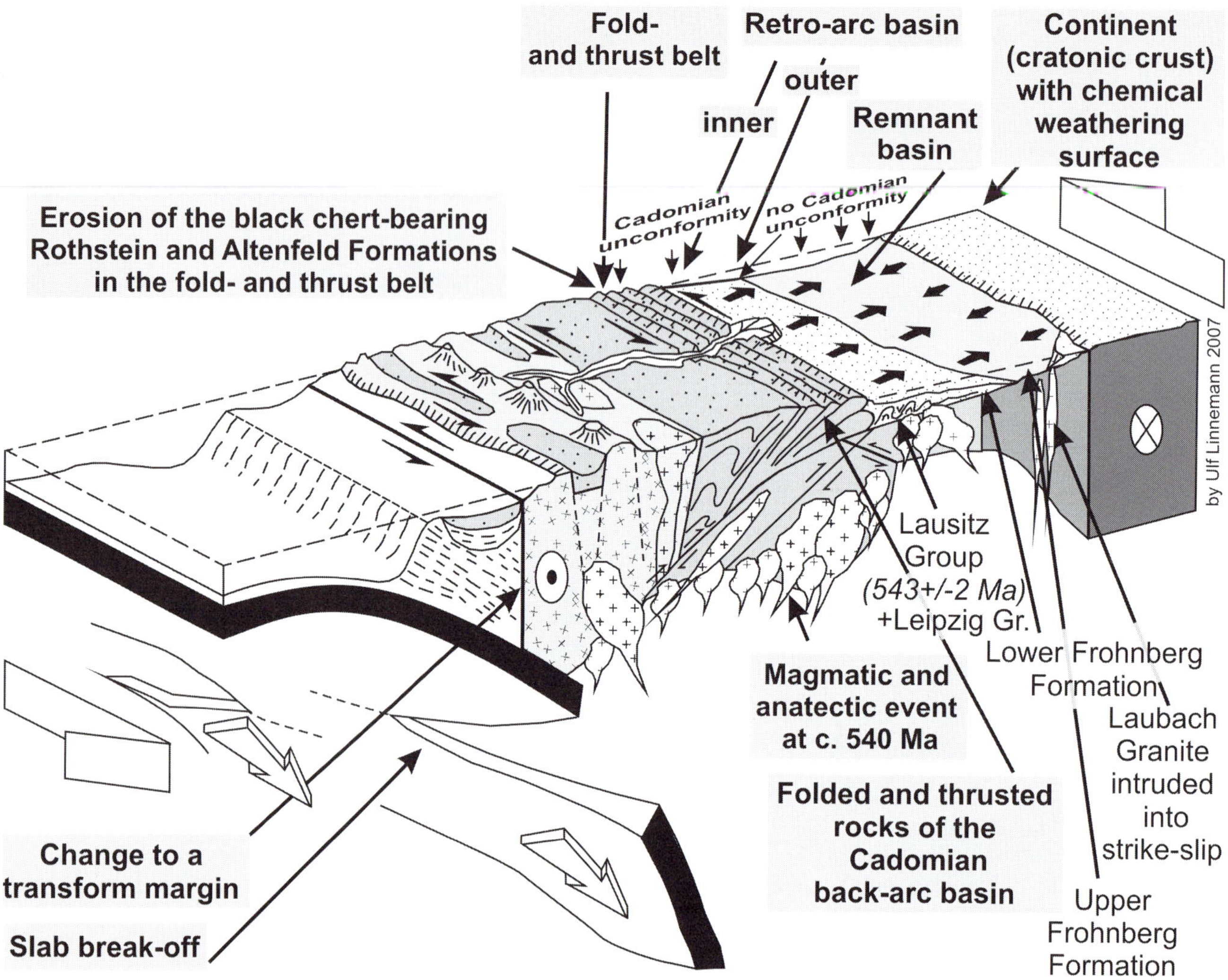

Fig. 3.27. Model for the plate tectonic evolution of the Cadomian retro-arc basin between *c.* 545 and 540 Ma based on data from the Saxo-Thuringian Zone (from Linnemann *et al.* 2007). There is no Cadomian angular unconformity on the outer margin of the retro-arc basin towards the continent because Late Neoproterozoic sediments were unaffected by deformation (e.g. upper section of Frohnberg Formation in SE part of Schwarzburg Unit). In contrast, closer to the fold-and-thrust belt in the inner part of the retro-arc basin, the sediments are deformed and an angular unconformity developed between Neoproterozoic retro-arc sediments and overstepping Cambro-Ordovician strata (e.g. Cadomian unconformities at top of the Lausitz and Leipzig Groups).

al. 2004). The $\varepsilon Nd_{(t)}$ values of sedimentary rocks range from −5.5 to −7.2 (Drost *et al.* 2007). The NdT_{DM} model ages range from 1.6–1.7 Ga and demonstrate the continuing strong input of old cratonic material into the basin (Fig. 3.30).

In their excellent review concerning the geological history of the Cadomian basement of the Teplá–Barrandian Unit, Kříbek *et al.* (2000) distinguish different volcanic belts. These are represented from NW to SE by the Stříbro-Plasy, Central and

Fig. 3.28. Late Neoproterozoic–earliest Paleozoic rock complexes of the Cadomian basement in the Saxo-Thuringian Zone (Bohemian Massif). Symbols: ***geological units that are present as Variscan nappes and thrust sheet complexes and shear zones containing low to high grade metamorphic rocks; The present-day tectonic assemblages do not reflect the original stratigraphic and spatial relationships; +detrital zircon; *tuff; #zircon from pebble. Sources of bio-, litho- and tectonostratigraphic data: low grade overprinted rocks of the Saxo-Thuringian Zone, Linnemann & Romer (2002); Erzgebirge in the Saxo-Thuringian Zone, Sebastian (1995); Sudetes, modified from Aleksandrowski & Mazur (2002); Barrandian, modified from Chlupáč (1993); Brunovistulian, mainly after Finger *et al.* (2000*a*), Jachowicz & Přichystal (1997), Fatka & Vavrdová (1998); Moldanubian nappe units, Fuchs (1976) and Thiele (1984). Sources of isotopic data (in Figs 3.29 and 3.30 also) [1]Linnemann & Romer (2002); [2]Linnemann *et al.* (2004); [3]Tichomirowa *et al.* (2001); [4]Kröner *et al.* (1995); [5]Hammer *et al.* (1999); [6]Tichomirowa (2003); [7]Dörr *et al.* (2002); [8]Drost *et al.* (2004); [9]Pin & Waldhausrova (2007); [10]Drost *et al.* (2007); [11]Zulauf *et al.* (1999); [12]Zulauf *et al.* (1997); [13]Dörr *et al.* (1998); [14]Venera *et al.* (2000); [15]Finger *et al.* (2000*a*); [16]Finger *et al.* (2000*b*); [17]Friedl *et al.* (2004); [18]Friedl *et al.* (1998); [19]Finger & Pin (1997); [20]Liew & Hofmann (1988); [21]van Breemen *et al.* (1982); [22]Friedl *et al.* (2000); [23]Hegner & Kröner (2000); [24]Kröner *et al.* (2000); [25]Pin *et al.* (2007); [26]Mingram *et al.* (2004); [27]Wendt *et al.* (1993); [28]Furnes *et al.* (1994); [29]Teipel *et al.* (2004); [30]Gebauer *et al.* (1989); [31]von Quadt & Gebauer (1993); [32]Mazur *et al.* (2004); [33]Żelaźniewicz *et al.* (2004); [34]Kröner *et al.* (2001); [35]Borkowska *et al.* (1980); [36]Crowley *et al.* (2002); [37]Oliver *et al.* (1993); [38]Kröner & Hegner (1998); [39]Turniak *et al.* (2000); [40]Kryza *et al.* (2003); [41]Oberc-Dziedzic *et al.* (2005); [42]Oberc-Dziedzic *et al.* (2003); [43]Košler *et al.* (2004); [44]Linnemann *et al.* (2000); [45]Buschmann *et al.* (2001).

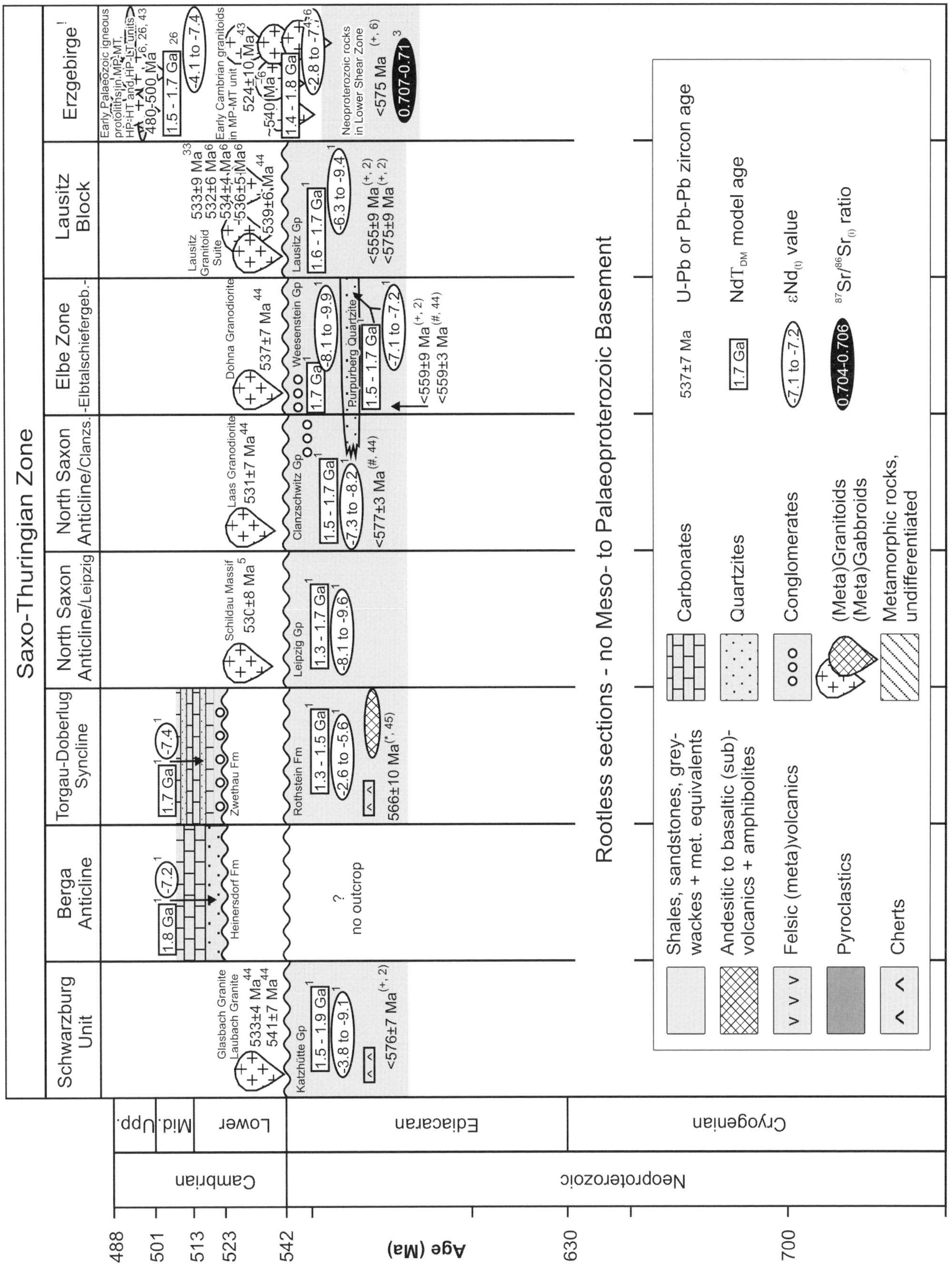
Saxo-Thuringian Zone
Schwarzburg Unit
Berga Anticline
Torgau-Doberlug Syncline
North Saxon Anticline/Leipzig
North Saxon Anticline/Clanzs.
Elbe Zone -Elbtalschiefergeb.-
Lausitz Block
Erzgebirge
Rootless sections - no Meso- to Palaeoproterozoic Basement
Glasbach Granite
Laubach Granite
533±4 Ma
541±7 Ma
Katzhütte Gp
1.5 - 1.9 Ga
-3.8 to -9.1
<576±7 Ma
Heinersdorf Fm
1.8 Ga
-7.2
no outcrop
Zwethau Fm
1.7 Ga
-7.4
Rothstein Fm
1.3 - 1.5 Ga
-2.6 to -5.6
566±10 Ma
Schildau Massif
530±8 Ma
Leipzig Gp
1.3 - 1.7 Ga
-8.1 to -9.6
Laas Granodiorite
531±7 Ma
Clanzschwitz Gp
1.5 - 1.7 Ga
-7.3 to -8.2
<577±3 Ma
Dohna Granodiorite
537±7 Ma
Weesenstein Gp
1.7 Ga
-8.1 to -9.9
Purpurberg Quartzite
1.5 - 1.7 Ga
-7.1 to -7.2
<559±9 Ma
<559±3 Ma
Lausitz Granitoid Suite
533±9 Ma
532±6 Ma
534±4 Ma
536±5 Ma
539±6 Ma
Lausitz Gp
1.6 - 1.7 Ga
-6.3 to -9.4
<555±9 Ma
<575±9 Ma
Early Palaeozoic igneous protoliths in MP-MT, HP-HT and HP-LT units
480-500 Ma
1.5 - 1.7 Ga
-4.1 to -7.4
Early Cambrian granitoids in MP-MT unit
524±10 Ma
~540 Ma
1.4 - 1.8 Ga
-2.8 to -7.1
Neoproterozoic rocks in Lower Shear Zone
<575 Ma
0.707-0.71
Cambrian
Lower
Mid.
Upp.
Ediacaran
Cryogenian
Neoproterozoic
Age (Ma)
488
501
513
523
542
630
700
Shales, sandstones, greywackes + met. equivalents
Andesitic to basaltic (sub)-volcanics + amphibolites
Felsic (meta)volcanics
Pyroclastics
Cherts
Carbonates
Quartzites
Conglomerates
(Meta)Granitoids
(Meta)Gabbroids
Metamorphic rocks, undifferentiated
537±7 Ma
U-Pb or Pb-Pb zircon age
1.7 Ga
NdT DM model age
-7.1 to -7.2
εNd(t) value
0.704-0.706
87Sr/86Sr(i) ratio

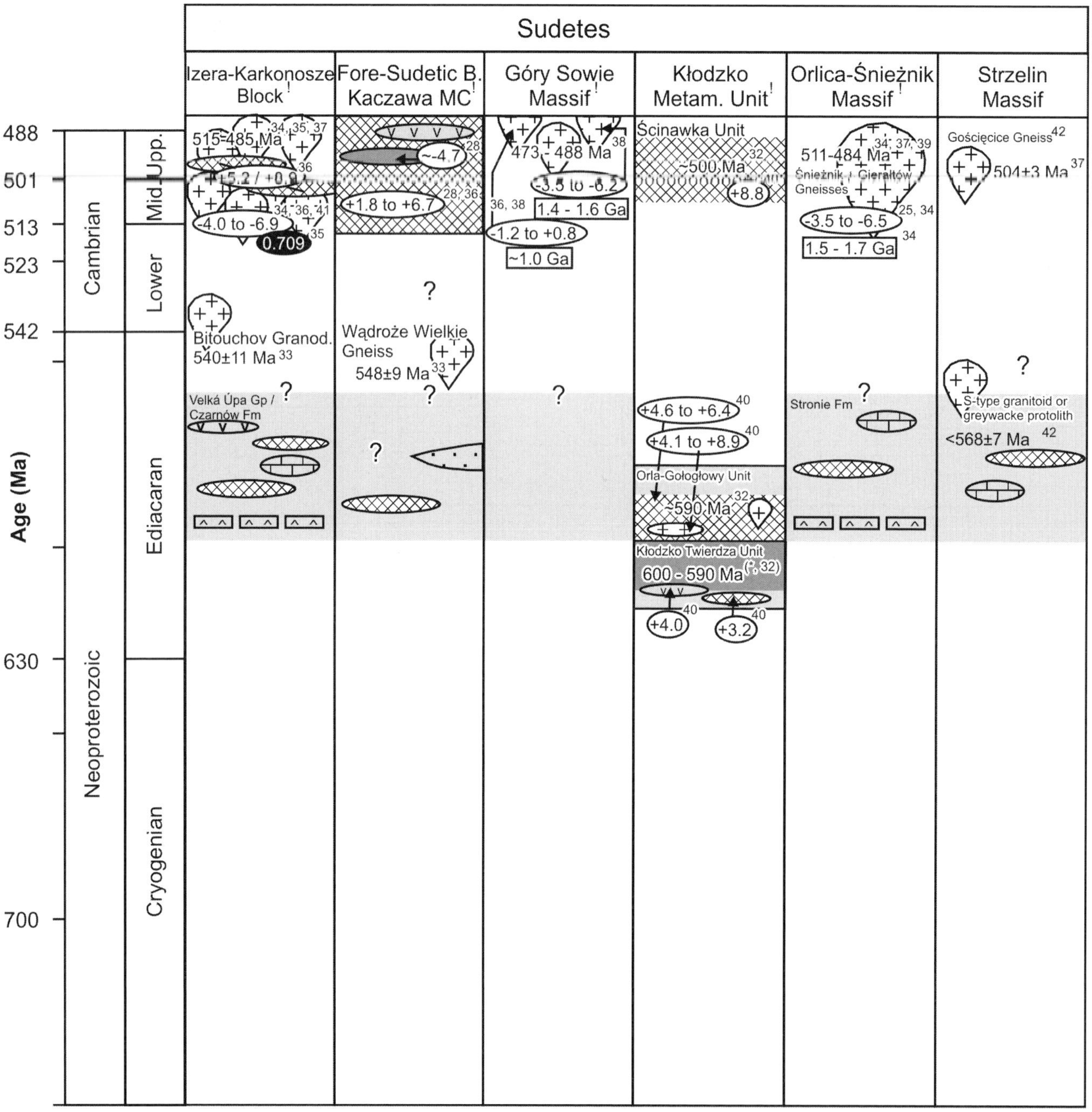

Fig. 3.29. Late Neoproterozoic to Earliest Paleozoic complexes of the Cadomian basement in the Sudetes (Bohemian Massif) (for explanation and data sources see Fig. 3.28 caption).

Nepomuk-Přibram volcanic belts containing late Neoproterozoic tholeiitic basalts. In the Jílové Zone to the SE a belt of late Neoproterozoic calc-alkaline igneous rocks including boninites occurs. The general interpretation of geochemical data, field observations and geophysical data led to the conclusion that the Teplá–Barrandian Unit, during its early stage of development, can be characterized as a basin with oceanic crust and operating subduction zones (e.g. Jakeš 1973). Subsequently, the magmas evolved from primitive to transitional and calc-alkaline with a higher degree of maturity, and the basin became increasingly shallow-water (summarized by Kříbek *et al.* 2000). These trends suggest a backarc setting in the later stage of the evolution of the Neoproterozoic of the Teplá–Barrandian Unit.

The age of the Cadomian deformation of the Teplá–Barrandian Unit is suggested to be *c.* 550 to 540 Ma based on a LP-HT metamorphic event determined from weighted average Th–U–Pb model ages on monazite of 540 ± 16 Ma, 542 ± Ma and 551 ± 19 Ma (Zulauf *et al.* 1999). The same authors concluded that the collision of a microterrane (island arc?) and Gondwana resulted in the LP-HT metamorphism that was associated with a slab break-off. Crustal thinning and exhumation of high-grade metamorphic rocks, accommodated by dextral transtension and

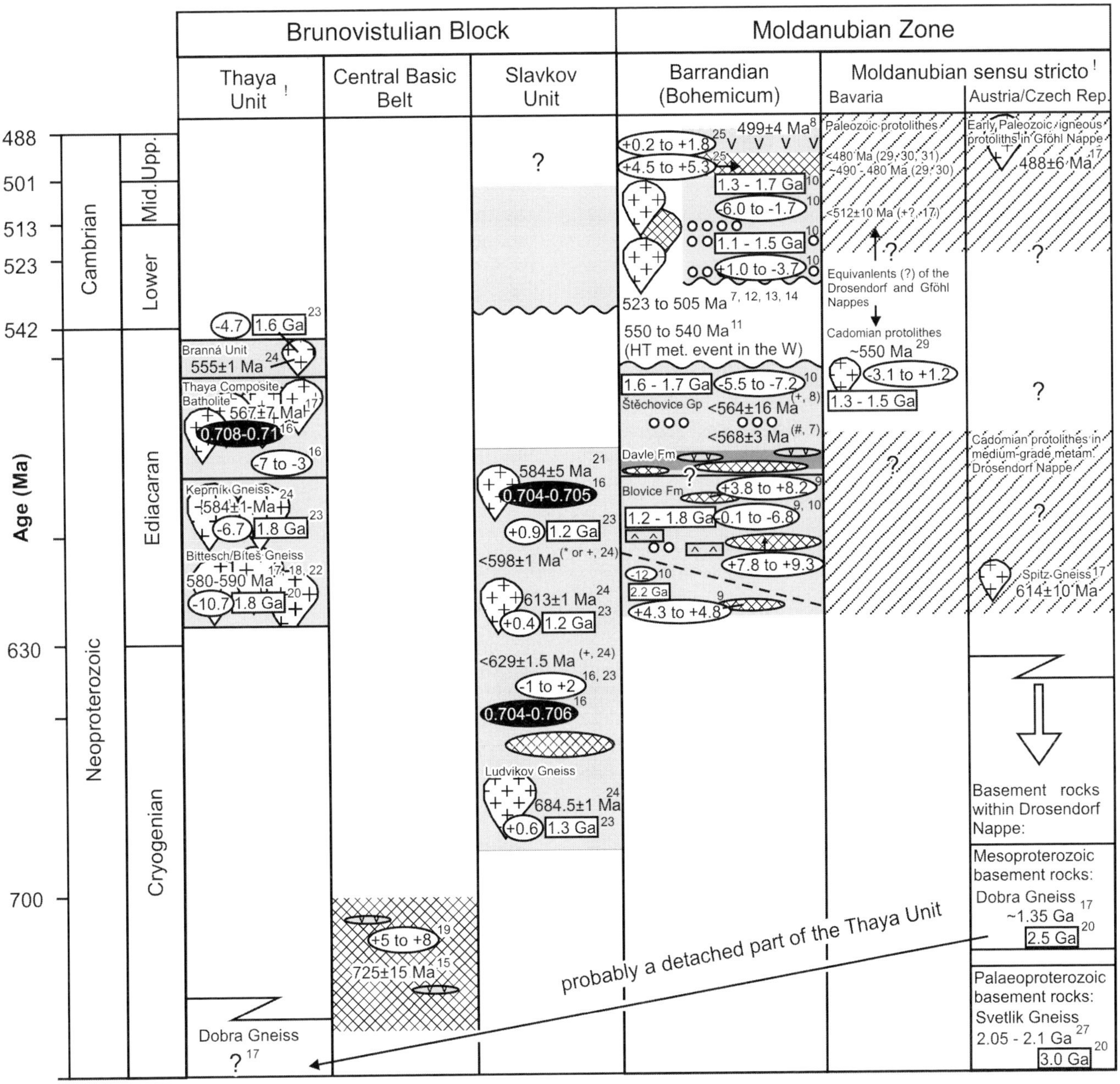

Fig. 3.30. Late Neoproterozoic to Earliest Paleozoic rock complexes of the Cadomian basement in the Moldanubian Zone and the Moravo-Silesian Unit (Bohemian Massif) (for explanation and data sources see Fig. 3.28 caption).

erosion of the orogenic wedge at *c.* 540–510 Ma and major crustal tilting of the Domažlice crystalline complex, pre-date the intrusion of synkinematic Cambrian granitoids. The generally calc-alkaline characteristics of these granitoids do not necessarily indicate prolonged Cadomian subduction during the Cambrian. The geochemical characteristics can also be well explained in terms of remelting of arc-type crust in a post-collisional high-heatflow regime (Zulauf *et al.* 1999). Intrusion ages of Cambrian plutons related to Cadomian orogenic processes range between 523 and 505 Ma (Zulauf *et al.* 1997; Dörr *et al.* 1998, 2002; Venera *et al.* 2000) (Fig. 3.30).

The western part of the Teplá–Barrandian Unit contains two remarkable complexes of mafic rocks. To the NW, the Mariánzé Lázně Complex occupies a large area. According to Timmermann *et al.* (2004) on the basis of U–Pb dating and geochemical data, this complex is a mixture of *c.* 540 Ma oceanic rocks juxtaposed at depth with lower crustal rocks of the structurally overlying Teplá–Barrandian Unit. The Neukirchen-Kdyně Massif in the area of the Hoher Bogen contains various acidic, intermediate, mafic and ultramafic rocks. In the SW of the massif, the Všeruby and Neukirchen Gabbro and the Hoher-Bogen Metagabbro yield Cambrian ages (Bues *et al.* 2002).

A clear angular Cadomian unconformity can be observed at various localities in the Teplá–Barrandian Unit. One of the most impressive outcrops is located north of Týrovice at the roadside in the valley of the Berounka river. Greywackes and mudstones of the Kralupy-Zbraslav Group were transgressed by Middle Cambrian sediments. The onset of Cambrian sedimentation in the Barrandian was clearly rift-related (Drost *et al.* 2004) and commenced in the Přibram-Jince area with the deposition of a series of lower Cambrian conglomerates and siliciclastic sediments with a thickness of *c.* 2500 m followed by the marine and fossiliferous Jince Formation (400 m). The latter is composed of mudstones and sandstones. The topmost part of the marine succession is represented by the conglomerates and sandstones of the Ohrazeníce Formation. In the Skryje area, Lower to Middle Cambrian sediments are only *c.* 200 m thick. In the upper Cambrian an extensive magmatic rift-related volcanic event occurred and is represented by the Strásice and Křivoklat-Rokycany complexes (Waldhausrová 1971). An Upper Cambrian rhyolite sample yields a SHRIMP U–Pb age of 499 ± Ma (Drost *et al.* 2004). The sedimentary part of the Cambrian section has been studied in detail by Havlicek (1971) and Kukal (1971).

The Moldanubian Zone

The Moldanubian Zone was first established by Kossmat (1927) and distinguished from the rest of the Bohemian Massif due to the dominance of crystalline units such as high-grade metamorphic rocks and voluminous granitoid plutons. The structure of the Moldanubian Zone is very complex. Litho- and tectonostratigraphic divisions differ from author to author and from country to country. Detailed discussion is beyond the scope of this chapter and we refer to the overview given by Fiala *et al.* (1995). In Figure 3.14, the Moldanubian Zone is divided only into relatively para-autochthonous complexes and allochthonous nappe structures. An important feature of the Moldanubian Zone is the occurrence of inliers of old cratonic basement such as the Mesoproterozoic Dobra Gneiss (1.38 Ga; Gebauer & Friedl 1993, 1994) and the Palaeoproterozoic Svetlik Gneiss (2.05–2.1 Ga; Wendt *et al.* 1993; Fig. 3.30). The Dobra Gneiss is commonly correlated with rock units of the Brunovistulian Block or the Moravo-Silesian Zone, respectively, and may represent a cratonic basement sliver that became detached by Variscan tectonics (Fig. 3.30).

A few isolated bodies related to Cadomian orogenic processes have been identified by new geochronological data (Friedl *et al.* 2004; Teipel *et al.* 2004). From the southeastern part of the Moldanubian Zone, Friedl *et al.* (2004) published a U–Pb age of the Spitz Gneiss of 614 ± 10 Ma. The youngest concordant grain from the Natschlag Gneiss gave a SHRIMP U–Pb age of 512 ± 10 Ma (Friedl *et al.* 2004). This rock may be correlated with the Cambrian plutonism related to the final stages of the Cadomian Orogeny. On the basis of U–Pb ages of inherited zircons from rocks of the Austrian part of the Bohemian Massif, Friedl *et al.* (2004) and concluded that rocks from North Africa were recycled in the Moldanubian Zone.

From a metarhyolite near Passau in the southern part of the Bohemian Massif, Teipel *et al.* (2004) published a SHRIMP U–Pb age of 555 ± 12 Ma. From a metarhyolite and a metabasite near Haibach the same authors have reported SHRIMP U–Pb ages of 549 ± 7 Ma and 549 ± 6 Ma, respectively. These ages are clearly related to the origin of the Cadomian basement. The cluster of SHRIMP U–Pb ages of inherited zircon is interpreted by Teipel *et al.* (2004) in terms of a West African provenance for late Neoproterozoic rocks of the Moldanubian Zone in the Bayerischer Wald.

The Avalonia-type part

The Brunovistulian Block

The easternmost part of the Bohemian Massif is occupied by the Moravo-Silesian Zone that contains a Cadomian basement unit called the Brunovistulian or Brunovistulian Block (Dudek 1980; Fig. 3.14). Cadomian basement and possible inliers of older cratonic rock units such as the 1.38 Ga Dobra Gneiss (Friedl *et al.* 2004) situated in the west of the Brunovistulian Block were overthrusted along major fault zones by the Moldanubian Gföhl Unit during Variscan collisional processes (Finger & Steyrer 1995). To the NE and east the Cadomian rocks and Palaeozoic overstep sequences are covered by thick Upper Carboniferous and Mesozoic and Cenozoic deposits.

Finger *et al.* (2000*b*) subdivided the Cadomian basement of the Moravo-Silesian Zone into three units: the western Thaya Terrane and the eastern Slavkov Terrane, which are separated by the Central Basic Belt. The narrow fault-bounded Central Basic Belt is mainly composed of metagabbros, metadiorites and metabasalts, with MORB/E-MORB geochemical signatures and εNd_{600} values of +5 to +8 (Finger & Pin 1997). Subordinate metarhyolites yielded a zircon evaporation age of 725 ± 15 Ma (Finger *et al.* 2000*a*). The Central Basic Belt is interpreted in terms of a remnant of a Neoproterozoic backarc basin that was closed during Cadomian arc–continent collision. In contrast, the Slavkov Terrane to the east (present directions) represents a part of the island arc, and the Thaya Terrane to the west appear to be part of the Neoproterozoic Gondwana margin (Finger & Pin 1997; Finger *et al.* 2000*a*, *b*).

The Slavkov Terrane comprises metamorphosed volcanosedimentary successions (today schists, phyllites, metabasites) in the SW separated by the Haná Fault Zone from metamorphosed clastic succession (today various types of schists and gneisses) in the NE (Dudek 1995). Zircon evaporation ages from a plagioclase-biotite schist from the Desná Dome gave an age of 598 ± 1 Ma (Kröner *et al.* 2000), which is interpreted as the maximum sedimentation age of the volcaniclastic precursor. Geochemical and isotope data of the metasedimentary rocks of the Slavkov Terrane suggest derivation from a young, poorly evolved arc-type source (Finger & Pin 1997; Finger *et al.* 2000*b*). I-type granitoids intruding these complexes at *c.* 613–580 Ma (van Breemen *et al.* 1982; Fritz *et al.* 1996; Kröner *et al.* 2000) have similar geochemical and isotopic characteristics (Fig. 3.17) and suggest that the rocks formed by melting of young calc-alkaline crust or contain significant contributions from a mantle source (Finger *et al.* 2000*b*). The $^{40}Ar/^{39}Ar$ hornblende age of 575.6 ± 2.2 Ma obtained from amphibolites of the Deblin Group (roof of the Brno Batholith) dates final cooling after the intrusion of the Brno Batholith.

In contrast, the Ediacaran rocks of the Thaya Terrane are characterized by crustal sources including cratonic components (Finger *et al.* 2000*b*). The Keprník and Bittesch gneisses yield zircon ages of 580–590 Ma for their I-type igneous protoliths, and contain abundant older Proterozoic xenocrysts (Kröner *et al.* 2000; Friedl *et al.* 2000, 2004). Nd isotope data ($\varepsilon Nd_{580} = -10.7$ to -6.7; $NdT_{DM} = 2.1$ to 1.8 Ga) also indicate incorporation of a cratonic source (Liew & Hofmann 1988; Hegner & Kröner 2000). The younger I/A-type granitoids of the Thaya composite batholith gave a U–Pb age of 567 ± 6 Ma (Friedl *et al.* 2004) and are characterized by Nd and Sr isotope signatures, which are somewhat more radiogenic than those of the Keprník and Bittesch gneisses but likewise correspond to involvement of old crustal rocks (Finger & Pin 1997).

Inherited zircon cores from the Bittesch Gneiss show ages corresponding to geological events at *c.* 1.2 Ga, 1.5 Ga and 1.65–1.8 Ga, which suggests linkage of the Thaya Terrane with the Gondwanan Amazonia Craton and the Avalonian terranes (Friedl *et al.* 2000, 2004). In contrast, zircon ages from most other units of the Bohemian Massif reflect the known geochronological record from the West African Craton.

Arc–continent collision is interpreted to have taken place at *c.*

600 Ma (Finger *et al.* 2000*b*); this is supported by the field observations of Leichmann (1996) and Finger *et al.* (2000*b*), who reported granitoids related to the Thaya Terrane intruding the western margin of the Central Basic Belt. Slavkov Terrane-related granitoids were emplaced in the eastern margin of the Central Basic Belt. The timing of the Cadomian regional metamorphism is not precisely constrained, but the intrusion of the Brno Pluton post-dates the deformation of its country-rock (Dudek 1995; Fritz *et al.* 1996). Total-Pb monazite ages (Finger *et al.* 1999), K–Ar hornblende and mica ages (Dudek & Melková 1975), and $^{40}Ar/^{39}Ar$ hornblende and mica ages (Fritz *et al.* 1996) (see Table 3.4) show a broad time interval (*c.* 660 to *c.* 550 Ma), but suggest regional Cadomian metamorphism between 620 and 580 Ma.

Ediacaran-age rocks of the Brunovistulian Block are unconformably overlain by Palaeozoic sedimentary rocks. Lower and Middle Cambrian siliciclastic sediments are known from boreholes. Their age was established by fossils (Bula *et al.* 1997 and references therein; Jachowicz & Přichystal 1997; Fatka & Vavrdová 1998). Lower Cambrian deposits attain a thickness of up to 3000 m. About 280 m of Middle Cambrian sediments were recognized in one borehole (Bula *et al.* 1997). A hiatus in sedimentation during the Upper Cambrian, has also been recognized from the Barrandian and the Saxo-Thuringian Zone (Linnemann & Romer 2002; Drost *et al.* 2004).

Palaeogeographical constraints in the Late Neoproterozoic

Palaeomagnetic data are not available for the Precambrian of the Bohemian Massif. All attempts to obtain results have failed. Fossil assemblages are very rare and represented in most cases by microfossils only. One exception is the recognition of hemispherical bodies on the surface of a mudstone of the Lausitz Group bed interpreted to be of organic origin (Linnemann 2007). However, to obtain palaeobiogeographical information from these fossils is almost impossible. The provenance of the Cadomian basement in the Bohemian Massif, however, may be constrained from age clusters of inherited and detrital zircons and from NdT_{DM} values in sediments and magmatic rocks. The latter provide information about the relative amounts of juvenile or continental crust which was involved during rock-forming processes.

The first successful attempt to use zircon ages in combinantion with NdT_{DM} values for the reconstruction of peri-Gondwana was published by Nance & Murphy (1994). The occurrence of Mesoproterozoic 'Grenvillian' zircon grains (*c.* 1.0 Ga) and the involvement of juvenile crust in a terrane was used by these authors to define its provenance from Amazonia. In contrast, a provenance from the West African Craton is suggested by an absence of Mesoproterozoic zircon grains in the inheritance of igneous rocks or in the age spectra of detrital zircon, and by NdT_{DM} values that reflect a recycled old cratonic crust (>2 Ga).

In recent years an increasing number of studies have demonstrated the recycling of juvenile Mesoproterozoic crust during the formation of Neoproterozoic rock complexes in Avalonia and related terranes (Murphy *et al.* 2004). For such Avalonia-type terrane units an Amazonian (South American) provenance is often suggested. For Cadomia a provenance from the West African Craton is assumed by most authors. According to Murphy *et al.* (2004) terranes with a clear West African provenance can be classified as Cadomia-type terranes.

Present data suggests that an Avalonia-type unit exists in the Brunovistulian Block of the Moravo-Silesian Zone. In contrast, only Cadomia-type units exist in the rest of the Bohemian Massif. Ediacaran rocks from the Brunovistulian Block demonstrate the clear occurrence of Mesoproterozoic zircon grains in the spectra of inherited zircon (Friedl *et al.* 2004). The timing of magmatic and tectonometamorphic events during the Ediacaran is comparable with similar processes in parts of Avalonia (Newfoundland) and does not conform with the geological evolution of the rest of the Bohemian Massif (Finger *et al.* 2000*a*). In addition, the Mesoproterozoic Dobra Gneiss (1.38 Ga) is located very close to the Moravo-Silesian Zone and could be a detached basement sliver from that unit that may have underlain the Neoproterozoic of the Moravo-Silesian Zone and became detached from its original position (Fig. 3.30).

For the Saxo-Thuringian Zone, a West African provenance based on SHRIMP U–Pb data of detrital and inherited zircons was first proposed by Linnemann *et al.* (2004). A recent study of detrital zircon grains from different sedimentary units of the Cadomian basement in the Saxo-Thuringian Zone shows similar results (Linnemann *et al.* 2007). The SHRIMP U–Pb data of zircons from the underlying greywacke beds of the 'Dobris Congomerate' in the Teplá–Barrandian Unit likewise support the idea of a West African provenance (Drost *et al.* 2004). Additional inherited zircon ages from the Moldanubian Zone show clearly that this part of the Bohemian Massif was derived from the periphery of the West African Craton (Friedl *et al.* 2004; Teipel *et al.* 2004). The occurrence of the cratonic inlier of the *c.* 2 Ga old Svetlik Gneiss in the Moldanubian Zone (Fig. 3.14) is another indication of the recycling of the margin of the West African Craton during the Cadomian Orogeny. The Eburnean Basement of the West African Craton formed at *c.* 2 Ga and is a typical and extensive part of the craton.

SHRIMP U–Pb ages of old zircon grain cores from the Strzelin Gneiss in the easternmost part of the Fore-Sudetic Block range between 1.24 and 2.0 Ga (Oberc-Dziedzic *et al.* 2003). Because of these rare Mesoproterozoic ages we cannot exclude that this rock complex was also part of an Avalonian-type terrane.

It seems that the largest part of the Bohemian Massif belongs to Cadomia-type units or terranes derived from the West African Craton and adjoining areas. Only in the case of the Brunovistulian Block of the Moravo-Silesian Zone does the available data set suggest an Avalonia-type unit that might have been derived from Amazonia. Due to their different geotectonic settings during the Palaeozoic and the drift of Avalonia, these crustal domains are divided from each other by the suture of the Rheic Ocean.

Concluding remarks

In the Bohemian Massif, Avalonia- and Cadomia-type units exist. Both were formed by diverse Cadomian orogenic processes at the Avalonian–Cadomian Active Margin during late Neoproterozoic and Early Palaeozoic times. In addition, in the Moldanubian Zone relicts of old cratonic basement exist, such as the Dobra Gneiss (1.38 Ga) and the Svetlik Gneiss (2.05–2.1 Ga). The Cadomian basement of the Avalonian part is restricted to the Brunovistulian in the Moravo-Silesian Zone. It is possible that the 1.38 Ga Dobra Gneiss was originally a part of the Avalonian segment that became detached and incorporated into the Moldanubian Zone. For the Avalonia-type unit (Brunovistulian Block) a provenance from Amazonia is suggested.

The largest part of the Bohemian Massif is composed of Cadomia-type terranes showing a distinct provenance from the West African Craton. The best preserved units crop out in the

Saxo-Thuringian Zone and in the Teplá–Barrandian Unit. No single zone in the Bohemian Massif records the complete geological history of the Cadomian Orogeny. Therefore, the reconstruction of the complex geotectonic and magmatic processes relies upon the combination of data from different regions.

Cadomian orogenic processes in the Bohemian Massif are characterized by a magmatic arc setting with marginal basins developed on stretched and thinned continental crust. The entire system developed within a strike-slip regime. Geochronological data indicate that the main activity of the continental magmatic arc culminated at *c.* 570 Ma. For that time, the initial stage of Cadomian basin development in a backarc setting can be reconstructed for the Saxo-Thuringian Zone. Sedimentary depositional systems are represented by (i) shallow marine passive margin sequences and (ii) arc-derived greywacke and mudstone turbidites showing an active margin provenance. Spreading zones with volcanic and hydothermal activity in the backarc basin were caused by internal strike-slip movements.

The later stage of Cadomian basin development (*c.* 543 Ma) is characterized by arc–continent collision and the change from a backarc basin to a retro-arc basin, a foreland fold-and-thrust belt and a remnant basin. Material from the continental island arc, the backarc basin and the underlying thinned continental crust were recycled and redeposited into the retro-arc basin. The later stage of Cadomian basin development was also characterized by a strike-slip regime.

During the latest stage of arc–continent collision, slab break-off in the subducted oceanic plate may have produced the enormous heat flow responsible for the anatectic event at 540–530 Ma culminating with the intrusion of voluminous granitoid plutons. All magmatic and geotectonic events of the Cadomian Orogeny took place without any major crustal thickening. High pressure–low temperature metamorphism is absent.

The strike-slip regime persisted into Early Cambrian times and was responsible for the change from an active margin setting to a transform margin. It is very likely that all reconstructed Cadomian orogenic processes were diachronous along the Avalonian–Cadomian Active Margin.

The Cadomian Orogeny: a synthesis

None of the preserved components of the Cadomian Orogen contains its complete story. Nevertheless, the type regions for the Cadomian Orogeny in the Armorican and Bohemian Massifs show similar features suggesting analogous Cadomian orogenic processes. The North Armorican Massif and the Saxo-Thuringian Zone show similar tectonostratigraphic units and orogenic zoning (Fig. 3.31). It would appear that in the North Armorican Massif only the outer zones of the orogen are preserved, whereas in the Saxo-Thuringian Zone unique occurrences of its inner part, such as highly mature passive margin deposits, exist towards the Gondwanan hinterland (Fig. 3.31). Although strike-slip related (sub-) horizontal long-distance displacements along major shear zones played an important role during Cadomian orogenic processes and Variscan reactivation, one can identify a clear orogenic zoning from the present NW to SE (Fig. 3.31).

In the outer regions of the orogenic belt, there exists a unit with remnants of magmatic arcs and related basins such as intra-arc basins. Adjacent is a zone containing *c.* 2 Ga slivers of cratonic basement (Icartian) which may have rifted off from the cratonic mainland before the Cadomian Orogeny and were later involved in the formation of Cadomian magmatic arcs. To the SE of the zone of magmatic arcs are volcanosedimentary basins, some with bedded black chert, that are related by most authors to backarc or inter-arc settings. Further to the SE, basins with clastic flysch-like sediments contain fragments of black cherts and older backarc sediments. These basins could be retro-arc basins containing the recycled debris derived from the backarc basins. Closure, erosion and deposition of the rocks from the backarc basins occurred during arc–continent collision. A belt of retro-arc basins was intruded by voluminous syn- to post-kinematic plutons during the Early Cambrian at *c.* 540–530 Ma. The part of the belt closest to the hinterland of Gondwana is

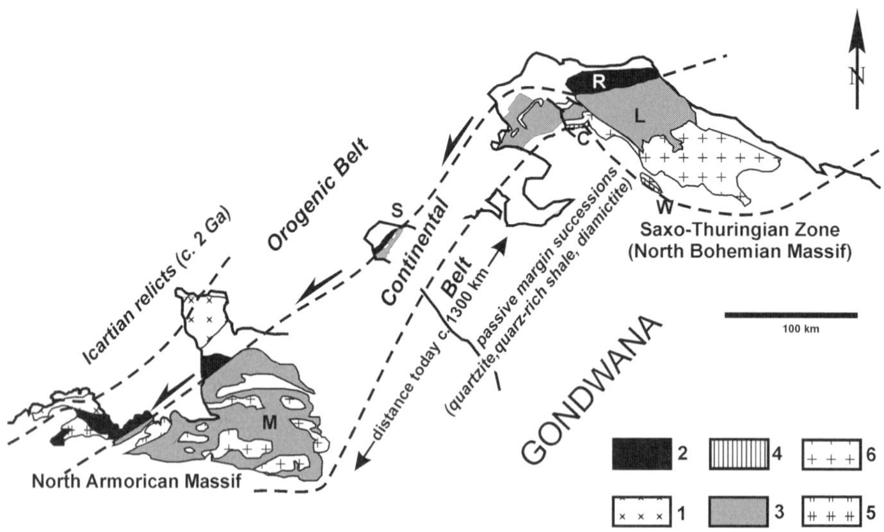

Fig. 3.31. Palinspastic reconstruction of the Cadomian orogenic zoning of the remnants of the Cadomian basement in the North Armorican Massif and in the North Bohemian Massif at the margin of the West African Craton as a part of the Gondwana supercontinent. Key: 1, magmatic arc and related basins; 2, Brioverian and equivalents in the orogenic belt; 3, Brioverian and equivalents in the continental belt; 4, passive margin deposits with quartzites, quartz schists, greywacke and diamictites; 5, migmatites; 6, post-tectonic granitoids, mostly anatectic (*c.* 540 Ma). Abbreviations: C, passive margin deposits of the Clanzschwitz Group; L, Lausitz Block; M, Mancellian Terrane; R, Rothstein Formation in the Torgau-Doberlug Sycline; S, Schwarzburg Unit; W, passive margin deposits of the Weesenstein Group.

represented by passive margin deposits (quartzites, quartz-rich shales, diamictites) that crop out in the Saxo-Thurinian Zone (Fig. 3.31).

Although our descriptions are dissimilar due to the variability of the different data sets and accessibility of parts of the orogenic belt, the common features of the Cadomian Orogeny can be summarized as follows.

- The Cadomian Orogen formed in an active margin setting at the periphery of Gondwana (peri-Gondwana) in the period *c.* 750 to *c.* 540–530 Ma.
- Most parts of the Cadomian Orogen were developed adjacent to the West African Craton as indicated by the distinctive provenance and basement inliers such as the *c.* 2 Ga Icartian and Svetlik gneisses.
- The Cadomian Orogen represents a cordillera-type orogen formed in an active plate margin setting, similar to the present-day regions of the western and eastern Pacific Ocean (e.g. Japan Sea, Baja California).
- The absence of Cadomian high-grade and/or high-pressure regional metamorphic rocks such as granulite and eclogite suggests that orogenic activity took place without major thickening and subduction of continental crust.
- Formation of magmatic arcs and related marginal basins was accompanied by major strike-slip related displacements of different geotectonic units and characterized by long-distance transcurrent slivering along the orogenic belt.
- *Juvenile island arc.* Cadomian-style subduction recorded from *c.* 750 Ma onwards resulted in the formation of a number of juvenile felsic island arcs at *c.* 750–650 Ma.
- *Continental magmatic arc.* A possible collision of early juvenile island arcs resulted in significant crustal thickening at *c.* 620 Ma. From *c.* 620 Ma onwards, there is evidence for continued subduction and intrusion of numerous calc-alkaline plutons. Melts are mantle-derived, but significantly contaminated by *c.* 2 Ga cratonic basement.
- *Early marginal basins.* By *c.* 610 Ma, mafic volcanism and formation of marginal basins and accumulation of sediments commenced. Deposition continued until *c.* 585 Ma.
- *Tectonomagmatic pulse at 580–570 Ma.* Early basins became inverted, deformed and metamorphosed under a sinistral oblique regime and were intruded by plutons at *c.* 580–570 Ma. The magmatic pulse at *c.* 580–570 Ma could have resulted from slab break-off following collision processes with the outboard arc systems or terranes.
- *Late marginal basins.* Cessation of calc-alkaline magmatism shortly afterwards (570–560 Ma) is thought to reflect increasingly oblique subduction and a gradual transition to a transform margin. Outboard magmatic arc activity continued on evolved, but stretched and thinned, continental crust. Under a crustal-scale, transpressive strike-slip shear zone/fault system several backarc basins originated. Passive margin sequences reflect the continental margin location of some of these basins. The diverse backarc basins closed during the final collision with the magmatic arc and Gondwana or the margin of the West African Craton. Collisional processes resulted in the formation of a fold-and-thrust belt and the development of retro-arc basins in which the debris of the magmatic arc and the backarc basins accumulated.
- *Final tectonomagmatic pulse at c. (550–) 540 (–530) Ma.* Transpressive amalgamation of the magmatic arc and possibly other terranes led to transpressional thickening, magmatic underplating and slab break-off to generate the thermal energy to generate widespread anatexis and voluminous intracrustal plutonism during latest Ediacaran and earliest Cambrian time.
- In the parts of the retro-arc basins proximal to the fold-and-thrust belt, sediments became deformed. More distal to the belt and much closer to the cratonic hinterland, sediments of the retro-arc basins or related remnant basin, escaped final Cadomian deformation. Therefore in some places the Cadomian unconformity is visible, in others it is not.
- Erosion and denudation lasted for *c.* 10 Ma and the onset of rift-related Cambrian sedimentation followed at *c.* 530 Ma.

The authors are very grateful to J. D. Inglis (Syracuse, USA), T. Brewer (Leicester, UK) and T. McCann (Bonn, Germany) for their critical reviews. We would like to thank C. Quesada (Madrid, Spain), F. Pereira (Evora, Portugal), O. Fatka (Prague, Czech Republic), P. Kraft (Prague, Czech Republic), J. Marek (Prague, Czech Republic), V. Kachlik (Prague, Czech Republic), M.-P. Dabard (Rennes, France), M. Robardet (Rennes, France), J.-P. Lefort (Rennes, France), D. Nance (Athens, USA), B. Murphy (Antigonish, Canada), S. Mazur (Wroclaw, Poland), P. Alexandowski (Wroclaw, Poland), R. Kryza (Wroclaw, Poland), G. Zulauf (Frankfurt/M, Germany), B. Buschmann (Freiberg/Sa, Germany), E. Bozkurt (Ankara, Turkey), J. Fernández-Suárez (Madrid, Spain), G. Gutiérrez-Alonso (Madrid, Spain) and C. Pin (Clermont-Ferrand, France) for many fruitful and interesting discussions and inspirations concerning the Cadomian Orogeny. B. Walker (Dresden) is thanked for the compilation of the reference list.

References

ALEKSANDROWSKI, P. & MAZUR, S. 2002. Collage tectonics in the northeasternmost part of the Variscan Belt: the Sudetes, Bohemian Massif. *In*: WINCHESTER, J. A., PHARAOH, T. C. & VERNIERS, J. (eds) *Palaeozoic Amalgamation of Central Europe*. Geological Society, London, Special Publications, **201**, 237–277.

ANDRAMAFOROHATRA, J. & DE LA BOSSE, H. 1998. Mise en evidence de témoins granitiques tardicadomiens à 540 dans le region de Belle-Isle-en-Terre, Massif Armoricain. *Bulletin de la Société Géologique de la France*, **8**, 279–287.

AUVRAY, B. 1979. Genèse et évolution de la croûte continentale dans le Nord du Massif armoricain. PhD thesis, University of Rennes.

AUVRAY, B., CHARLOT, R. & VIDAL, P. 1980. Données nouvelles sur le Protérozoïque inférieur du domaine nord-amoricain (France): âge et signification. *Canadian Journal of Earth Sciences*, **17**, 532–538.

AVIGAD, D., KOLODNER, K., MCWILLIAMS, M., PERSING, H. & WEISSBROD, T. 2003. Origin of northern Gondwana Cambrian sandstone revealed by detrital zircon SHRIMP dating. *Geology*, **31**, 227–230.

BALÉ, P. & BRUN, J. P. 1983. Les chevauchements cadomiens de la baie de Saint-Brieuc (Massif Armoricain). *Comptes Rendus de l'Académie des Sciences, Paris*, **297**, 359–362.

BALÉ, P. & BRUN, J. P. 1989. Late Precambrian thrust and wrench zones in northern Brittany (France). *Journal of Structural Geology*, **11**, 391–405.

BALLÈVRE, M., LE GOFF, E. & HÉBERT, R. 2001. The tectonothermal evolution of the Cadomian belt of northern Brittany, France: a Neoproterozoic volcanic arc. *Tectonophysics*, **331**, 19–43.

BANKWITZ, P. & BANKWITZ, E. 1995. Proterozoikum/Schwarburger Antiklinorium. *In*: SEIDEL, G. (ed.) *Geologie von Thüringen*. E. Schweitzerbart'sche Verlagsbuchhandlung, Stuttgart, 46–77.

BANKWITZ, P., BANKWITZ, E., KRAMER, W. & PIN, C. 1992. Early Paleozoic bimodal volcanism in the Vesser area, Thuringian Forest, east Germany. *Zentralblatt für Geologie und Paläontologie, Teil 1*, **9/10**, 1113–1132.

BARROIS, C. 1899. *Sketch of the Geology of Brittany.* Proceedings of the Geologists' Association, **16**.

BERTRAND, L. 1921. *Historié de la formation du sous-sol de la France. Les anciennes mers de la France et leurs dépots.* Flammarion éditeur, Paris.

BOHER, M., ABOUCHAMI, W., MICHARD, A. N., ALBAREDE, F. & ARNDT, N. 1992. Crustal growth in West Africa at 2.1 Ga. *Journal of Geophysical Research*, **97**, 345–369.

BORKOWSKA, M., HAMEURT, J. & VIDAL, P. 1980. Origin and age of Izera gneisses and Rumburk granites in the Western Sudetes. *Acta Geologica Polonica*, **30**, 121–145.

BOUOUGRI, E. H. & SAQUAQUE, A. 2004. Lithostratigraphic framework and correlation of the Neoproterozoic northern West African Craton passive margin sequence (Siroua-Zenaga-Bouazzer Elgraara Inliers, Central Anti-Atlas, Morocco): an integrated approach. *Journal of African Earth Sciences*, **39**, 227–238.

BROWN, M. 1974. *The Petrogenesis of the St Malo Migmatite Belt, North-Eastern Brittany, France*. PhD thesis, University of Keele.

BROWN. M. 1978. The tectonic evolution of the Precambrian rocks of the St-Malo region, Armorican Massif, France. *Precambrian Research*, **6**, 1–21.

BROWN, M. 1979. The petrogenesis of the St Malo Migmatite Belt, with particular reference to the diatexites. *Neues Jahrbuch für Abhandlungen*, **135**, 48–74.

BROWN, M. 1995. The late-Precambrian geodynamic evolution of the Armorican segment of the Cadomian belt (France): distortion of an active continental margin during south-west directed convergence and subduction of a bathymetric high. *Géologie de la France,* **3**, 3–22.

BROWN, M. & D'LEMOS, R. S. 1991. The Cadomian granites of Mancellia, northeast Armorican Massif of France: relationship to the St. Malo migmatite belt, petrogenesis and tectonic setting. *Precambrian Research*, **51**, 393–427.

BROWN, M., POWER, G. M., TOPLEY, C. G. & D'LEMOS, R. S. 1990. Cadomian magmatism in the North Armorican Massif. *In*: D'LEMOS, R. S., STRACHAN, R. A. & TOPLEY, C. G. (eds) *The Cadomian Orogeny.* Geological Society, London, Special Publications, **51**, 181–213.

BRUN, J.-P. 1977. La zonation structurale des dômes gneissiques. Un example: le massif de Saint-Malo (Massif armoricain France). *Canadian Journal of Earth Sciences*, **14**, 1697–1707.

BRUN, J.-P. 1992. Discussion on the timing and kinematics of Cadomian deformation (Northern Brittany). *Journal of the Geological Society, London*, **149**, 151–152.

BRUN, J.-P. & BALÉ, P. 1990. Cadomian tectonics in northern Brittany. *In*: D'LEMOS, R. S., STRACHAN, R. A. & TOPLEY, C. G. (eds) *The Cadomian Orogeny.* Geological Society, London, Special Publications, **51**, 95–114.

BRUN, J.-P. & MARTIN, H. 1978. Relations métamorphisme-déformation au cours de l'évolution d'un dôme migmatique: le massif de Saint-Malo (France). *Bulletin de la Société Géologique de France*, **20**, 91–101.

BRUN, J.-P., GUENNOC, P., TRUFFERT, C., VAIRON, J. & THE ARMOR WORKING GROUP OF THE GEOFRANCE 3-D PROGRAM 2001. Cadomian tectonics in northern Brittany: a contribution of 3-D crustal-scale modelling. *Tectonophysics*, **331**, 229–246.

BUES, C., DÖRR, W., FIALA, J., VEJNAR, Z. & ZULAUF, G. 2002. Emplacement depths and radiometric ages of Paleozoic plutons of the Neukirchen-Kdyne massif: differential uplift and exhumation of Cadomian basement due to Carboniferous orogenic collapse (Bohemian Massif). *Tectonophysics*, **352**, 225–243.

BUŁA, Z., JACHOWICZ, M. & ZABA, J. 1997. Principal characteristics of the Upper Silesian Block and Malopolska Block border zone (southern Poland). *Geological Magazine*, **134**, 669–677.

BUNEL, H. 1835. Observations sur les terrains intermédiaires du Calvados. *Mémoires de la Société Linnéenne de Normandie*, **5**, 99–100.

BUSCHMANN, B. 1995. Geotectonic facies analysis of the Rothstein Formation (Neoproterozoic, Saxothuringian Zone, Germany). Bergakademie Freiberg, Freiberg.

BUSCHMANN, B., NASDALA, L., JONAS, P., LINNEMANN, U. & GEHMLICH, M. 2001. SHRIMP U–Pb dating of tuff-derived and detrital zircons from Cadomian marginal basin fragments (Neoproterozoic) in the northeastern Saxothuringian Zone (Germany). *Neues Jahrbuch für Geologie und Paläontologie, Monatshefte,* **2001**(6), 321–342.

BUSCHMANN, B., ELICKI, O. & JONAS, P. 2006. The earliest unconformity in the Bohemian Massif: palaeogeographic affinities of Ediacaran (terminal Neoproterozoic) and Cambrian strata from the Torgau-Doberlug Syncline (Saxo-Thuringian Zone, Germany). *Precambrian Research*, 147, 387–403.

CABANIS, B., CHANTRAINE, J. & RABU, D. 1987. Geochemical study of Brioverian (Late Proterozoic) volcanic rocks in the northern Armorican Massif (France). *In:* PHAROAH, T. C., BECKINSALE, R. D. & RICKARD, D. (eds) *Geochemistry and Mineralization of Proterozoic Volcanic Suites*. Geological Society, London, Special Publications, **33**, 525–539.

CALVEZ, J. Y. & VIDAL, PH. 1978. Two billion years old relicts in the Hercynian belt of Western Europe. *Contributions to Mineralogy and Petrology*, **65**, 395–399.

CHANTRAINE, J., EGAL, E., THIÉBLEMONT, D., GOFF, E. L., GUERROT, C., BALLÈVRE, M. & GUENNOC, P. 2001. The Cadomian active margin (North Armorican Massif, France): a segment of the North Atlantic Panafrican belt. *Tectonophysics*, **331**, 1–18.

CHAUVEL, J. J. & RABU, D. 1988. Brioverian in Central Brittany. *In:* ZOUBEK, V., CONGE, D., KOZOUKHAROV, H. & KRAUTNER, G. (eds) *Precambrian in Younger Fold Belts.* Wiley, Chichester, 462–470.

CHLUPÁČ, I. 1993. *Geology of the Barrandian – A Field Trip Guide.* Senckenberg-Buch **69**, Verlag W. Kramer, Frankfurt.

COCHERIE, A., CHANTRAINE, J., FANNING. C. M., DABARD, M. - P., PARIS, F., LE HÉRISSE, A. & EGAL, E. 2001. Datation U/Pb: âge Briovériaen de la série d'Erquy (Massif armoricain, France). *Comptes rendus de l'Académie des Sciences, Paris*, **333**, 427–434.

COGNÉ, J. 1959. Données nouvelles sur l'antécambrien dans l'ouest de la France: Pentévrien et Briovérien en baie de St Brieuc (Côtes-du-Nord). *Bulletin de la Société Géologique de France*, **I**(1), 112–118.

COGNÉ, J. 1962. Le Briovérien: esquisse des caractères stratigraphiques, métamorphiques, structuraux et paléogéographiques de l'Antécambrien récent dans le Massif armoricain. *Bulletin de la Société Géologique de France*, **IV**(3), 413–430.

COGNÉ, J. 1972. Le Massif Armoricain et sa place dans la structure des socles ouest-européens. L'arc hercynien Ibero-Armoricain. *Histoire structurale du Golfe du Gascogne*. Editions, Technip, Paris, **1**, 1–23.

CONEY, P. J., JONES, D. L. & MONGER, J. W. H. 1980. Cordilleran suspect terranes. *Nature,* **288**, 329–333.

CROWLEY, Q. G., TIMMERMANN, H., NOBLE, S. R. & HOLLAND, J. G. 2002. Palaeozoic terrane amalgamation in Central Europe: a REE and Sm-Nd isotope study of the pre-Variscan basement, NE Bohemian Massif. *In:* WINCHESTER, J. A., PHARAOH, T. C. & VERNIERS, J. (eds) *Palaeozoic Amalgamation of Central Europe.* Geological Society, London, Special Publications, **201**, 157–176.

DABARD, M. P. 1997. *Les formations à cherts carbonés (phtanites) de la châine cadomienne. Genèse et signification géodynamique – exemple du segment armoricain*. PhD thesis, University of Rennes.

DALLMEYER, R. D., & URBAN, M. 1998. Variscan vs. Cadomian tectonothermal activity in northwestern sectors of the Teplá-Barrandian Zone, Czech Republic: constraints from $^{40}Ar/^{39}Ar$ ages: Geologisches Rundschau, **87**, 691–702.

DALLMEYER, R. D., D'LEMOS, R. S., STRACHAN, R. A. & MUELLER, P. A. 1991*a*. Tectonothermal chronology of early Cadomian arc development in Guernsey and Sark, Channel Islands. *Journal of the Geological Society, London*, **148**, 691–702.

DALLMEYER, R. D., STRACHAN, R. A. & D'LEMOS R. S. 1991*b*. Chronology of Cadomian tectonothermal activity in the baie de Saint-Brieuc (north Brittany), France: evidence from $^{40}Ar/^{39}Ar$ mineral ages. *Canadian Journal of Earth Sciences*, **28**, 762–773.

DALLMEYER, R. D., D'LEMOS, R. S. & STRACHAN, R. A. 1992. Timing of post-tectonic Cadomian magmatism on Guernsey, Channel Islands : evidence from $^{40}Ar/^{39}Ar$ mineral ages. *Journal of the Geological Society, London*, **149**, 139–147.

DALLMEYER, R. D., D'LEMOS, R. S. & STRACHAN, R. A. 1994. Timing of Cadomian and Variscan tectonothermal activity, La Hague and Alderney, North Armorican Massif; evidence from $^{40}Ar/^{39}Ar$ mineral ages. *Geological Journal*, **29**, 29–44.

DARLET, P. A., DABARD, M. P. & CHAUVEL, J. - J. 1990. Les metasediments briovériens de la Rance (Bretagne du Nord) dans l'évolution cadomienne du Massif Armoricain. *Comptes Rendus de l'Académie des Sciences, Paris*, **30**, 779–786.

DENIS, E. & DABARD, M. P. 1988. Sandstone petrography and geochemistry of Late Proterozoic sediments of the Armorican Massif (France) – a key to basin development during the Cadomian Orogeny. *Precambrian Research*, **42**, 189–206.

DEPAOLO, D.J. 1981. Neodymium isotopes in the Colorado Front Range and crust-mantle evolution in the Proterozoic. *Nature*, **291**, 193–197.

DEUBEL, F. 1959. Über die Grenze Vorpaläozoikum/Paläozoikum in Böhmen und Thüringen. *Geologie*, **8**, 818–827.

DISSLER, E., DORÉ, F., DUPRET, L., GRESSELIN, F. & LE GALL, J. 1988. L'evolution geodynamique cadomienne du Nord-Est du Massif Armoricain. *Bulletin de la Société Géologique de France*, **5**, 801–814.

D'LEMOS, R. S. & BROWN, M. 1993. Sm-Nd isotope characteristics of late Cadomian granite magmatism in northern France and the Channel Islands. *Geological Magazine*, **130**, 797–804.

D'LEMOS, R. S., DALLMEYER, R. D. & STRACHAN, R. A. 1992*a*. $^{40}Ar/^{39}Ar$ dating of plutonic rocks from Jersey, Channel Islands. *Proceedings of the Ussher Society*, **8**, 50–53.

D'LEMOS, R. S., BROWN, M. & STRACHAN R. A. 1992*b*. Granite magma generation, ascent and emplacement within a transpressional orogen. *Journal of the Geological Society, London*, **149**, 487–490.

D'LEMOS, R. S., MILLER, B.V. & SAMSON, S. D. 2001. Precise U–Pb zircon ages from Alderney, Channel Islands: growing evidence for discrete Neoproterozoic magmatic episodes in northern Cadomia. *Geological Magazine*, **138**, 719–726.

DORÉ, F. 1994. Cambrian of the Armorican Massif. *In*: KEPPIE, J. D. (ed.) *Pre-Mesozoic Geology of France*. Springer-Verlag, Berlin, 136–141.

DORÉ, F., DUPRET, L. & LE GALL, J. 1985. Tillites et tilloides du Massif armoricain. *Palaeogeography, Palaeoclimatology, Palaeoecology*, **51**, 85–96.

DÖRR, W., FIALA, J., VEJNAR, Z. & ZULAUF, G. 1998. U–Pb zircon ages and structural development of metagranitoids of the Teplá crystalline complex: evidence for pervasive Cambrian plutonism within the Bohemian massif (Czech Republic). *Geologische Rundschau*, **87**, 135–149.

DÖRR, W., ZULAUF, G., FIALA, J., FRANKE, W. & VEJNAR, Z. 2002. Neoproterozoic to Early Cambrian history of an active plate margin in the Teplá–Barrandian unit – a correlation of U–Pb isotopic-dilution-TIMS ages (Bohemia, Czech Republic). *Tectonophysics*, **352**, 65–85.

DROST, K., LINNEMANN, U. *et al.* 2004. New data on the Neoproterozoic – Cambrian geotectonic setting of the Teplá–Barrandian volcano-sedimentary successions: geochemistry, U–Pb zircon ages, and provenance (Bohemian Massif, Czech Republic). *International Journal of Earth Sciences* (*Geologische Rundschau*), **93**, 742–757.

DROST, K., ROMER, R. L., LINNEMANN, U., FATKA, O. KRAFT, P. & MAREK, J. 2007. Nd–Sr–Pb isotopic signatures of Neoproterozoic – Early Paleozoic siliciclastic rocks in response to changing geotectonic regimes: A case study from the Barrandian area (Bohemian Massif, Czech Republic). *In*: LINNEMANN, U., NANCE, R. D., KRAFT, P & ZULAUF, G. (eds). *The Evolution of the Rheic Ocean: From Avalonian-Cadomian Active Margin to Alleghenian-Variscan Collision*. Geological Society of America, Boulder, Colorado, Special Papers, **423**, 191–208.

DUDEK, A. 1980. The crystalline basement block of the Outer Carpathians in Moravia: Bruno-Vistulicum. *Rozpravy Československé Akademie Ved, Rada Matematickych a Prirodnich Ved*, **90**(8), 3–85.

DUDEK, A. 1995. VIII.B.4 Metamorphic Evolution. *In*: DALLMEYER, D. R., FRANKE, W. & WEBER, K. (eds) *Pre-Permian Geology of Central and Eastern Europe*. Springer-Verlag, Berlin, 508–511.

DUDEK, A. & MELKOVA, J. 1975. Radiometric age determination in the crystalline basement of the Carpathian Foredeep and of the Moravian Flysch. *Věstník Ústředního řstavu geologického*, **50**, 257–264.

DUFRENOY, P. A. 1838. Mémoire sur láge et la composition des terrains de transition de lOuest de la France. *Annales des Mines*, **3**, 91–100.

DUPRET, L., CABANIS, B., LE GALL, J. & DORÉ, F. 1985. Les tholéiites d'arc briovériennes de Montsurvent (Manche): leur place dans l'évolution orogénique cadomienne du nord de l'Armorique. *Hercynica*, **1**, 39–46.

DUPRET, L., DISSLER, E., DORÉ, F., GRESSELIN, F. & LE GALL, J. 1990. Cadomian geodynamic evolution of the northeastern Armorican Massif (Normandy and Maine). *In*: D'LEMOS, R. S., STRACHAN, R. A. & TOPLEY, C. G. (eds) *The Cadomian Orogeny*. Geological Society, London, Special Publications, **51**, 115–131.

EGAL, E., GUERROT, C., LE GOFF, E., THIÉBLEMONT, D. & CHANTRAINE, J. 1996. The Cadomian orogeny revisited in northern Brittany (France). *In:* NANCE, R. D. & THOMPSON, M.D. (eds) *Avalonian and Related Peri-Gondwanan Terranes of the Circum-North Atlantic*, Geological Society of America, Boulder, Colorado, Special Publications, **304**, 281–318.

FALK, F. 1974. Jungproterozoikum. *In*: HOPPE, W. & SEIDEL, G. (eds) *Geologie von Thüringen*. VEB Hermann Haak, Gotha/Leipzig, 119–142.

FATKA, O. & VAVRDOVÁ, M. 1998. Early Cambrian Acritarcha from sediments underlying the Devonian in Moravia (Minín 1 borehole, Czech Republic). *Vìstník Èeského geologického ústavu*, **73**, 55–60.

FERNÁNDEZ-SUÁREZ, J., GUTIÉRREZ-ALONSO, G. & JEFFRIES, T. E. 2002. The importance of along-margin terrane transport in northern Gondwana: insights from detrital zircon parentage in Neoproterozoic rocks from Iberia and Brittany. *Earth and Planetary Science Letters*, **204**, 75–88.

FIALA, J., FUCHS, G. & WENDT, J. I. 1995. Stratigraphy. In: DALLMEYER FRANKE, W. & WEBER, K. (eds) *Pre-Permian Geology of Central and Eastern Europe*, Springer, Berlin, 417–428.

FINGER, F. & PIN, C. 1997. Arc-type crustal zoning in the Bruno-Vistulicum, Eastern Czech Republic: a trace of the late Proterozoic Euro-Gondwana margin. *Journal of the Czech Geological Society*, **42**, 53.

FINGER, F. & STEYRER, H. P. 1995. A tectonic model for the eastern Variscides: Indications from a chemical study of amphibolites in the south-eastern Bohemian Massif. *Geologica Carpathica*, **46**, 137–150.

FINGER, F., SCHITTER, F., RIEGLER, G. & KRENN, E. 1999. The history of the Brunovistulian: total-Pb monazite ages from the metamorphic complex. *GeoLines*, **8**, 21–23.

FINGER, F., HANZL, P., PIN, C., VON QUADT, A. & STEYRER, H. P. 2000*a*. The Brunovistulian: Avalonian Precambrian sequence at the eastern end of the Central European Variscides. *In*: FRANKE, W., HAAK, U., ONCKEN, O. & TANNER, D. (eds) *Orogenic Processes: Quantification and Modelling in the Variscan Belt*. Geological Society, London, Special Publications, **179**, 103–112.

FINGER, F., TICHOMIROWA, M., PIN, C. HANŽL, P. 2000*b*. Relics of an early-Panafrican metabasite-metarhyolite formation in the Brno Massif, Moravia, Czech Republic. *International Journal of Earth Sciences* (*Geologische Rundschau*), **89**, 328–335.

FRANKE, W. & ŻELAŹNIEWICZ, A. 2002. Structure and evolution of the Bohemian arc. *In*: WINCHESTER, J. A., PHARAOH, T. C. & VERNIERS, J. (eds) *Palaeozoic Amalgamation of Central Europe*. Geological Society, London, Special Publications, **201**, 279–293.

FRIEDL, G., MCNAUGHTON, N., FLETCHER, I. R. & FINGER, F. 1998. New SHRIMP-zircon ages for orthogneisses from the south-eastern part of the Bohemian Massif (Lower Austria). *Acta Universitatis Carolinae - Geologica*, **42**, 251–252.

FRIEDL, G., FINGER, F., MCNAUGHTON, N. J. & FLETCHER, I. R. 2000. Deducing the ancestry of terranes: SHRIMP evidence for South America-derived Gondwana fragments in Central Europe. *Geology*, **28**, 1035–1038.

FRIEDL, G., FINGER, F., PAQUETTE, J. L., VON QUADT, A., MCNAUGHTON, N. J. & FLETCHER, I. R. 2004. Pre-Variscan geological events in the Austrian part of the Bohemian Massif deduced from U–Pb zircon ages. *International Journal of Earth Sciences* (*Geologische Rundschau*), **93**, 802–823.

FRITZ, H., DALLMEYER, D. R. & NEUBAUER, F. 1996. Thick-skinned versus thin-skinned thrusting: Rheology controlled thrust propagation in the Variscan collisional belt (The southeastern Bohemian Massif, Czech Republic – Austria). *Tectonics*, **15**, 1389–1413.

FUCHS, G. 1976. Zur Entwicklung der Böhmischen Masse. *Jahrbuch der Geologischen Bundesanstalt Wien*, **119**, 45–61.

FURNES, H., KRYZA, R., MUSZYNSKI, A., PIN, C. & GARMANN, L. B. 1994. Geochemical evidence for progressive, rift-related early Palaeozoic volcanism in the western Sudetes. *Journal of the Geological Society, London*, **151**, 91–109.

GAPAIS, D. & BALÉ, P. 1990. Shear zone pattern and granite emplacement within a Cadomian sinistral wrench zone at St. Cast, N. Brittany. *In*: D'LEMOS, R. S., STRACHAN, R. A. & TOPLEY, C. G. (eds) *The Cadomian Orogeny*. Geological Society, London, Special Publications, **51**, 169–180.

GAPAIS, D. & LE CORRE, C. 1980. Is the Hercynian belt of Brittany a major shear zone? *Nature*, **288**, 574–576.

GARLAN, T. 1985. Sédimentologie du Briovérien supérieur de Normandie et du Maine. PhD thesis, University of Caen.

GEBAUER, D. & FRIEDL, G. 1993. 1.38 Ga protolith age for the Dobra orthogneiss (Moldanubian Zone of the Southern Bohemian Massif,

NE-Austria: Evidence from ion-microprobe (SHRIMP) dating of zircon. *European Journal of Mineralogy, Beihefte*, **5**, 115.

GEBAUER, D. & FRIEDL, G. 1994. A 1.38 Ga protolith age for the Dobra orthogneiss (Moldanubian Zone of the Southern Bohemian Massif, NE-Austria): Evidence from ion-microprobe (SHRIMP) dating of zircon. Abstract for: Evolution of Variscan (Hercynian) and comparable Palaeozoic orogenic belts. Prague, February 17–19. *Journal of the Czech Geological Society*, **39**(1), 34–35.

GEBAUER, D., WILLIAMS, I. S., COMPSTON, W. & GRÜNENFELDER, M. 1989. The development of the Central European continental crust since the Early Archaean based on conventional and ion-microprobe dating of up to 3.84 b.y. old detrital zircons. *Tectonophysics*, **157**, 81–96.

GEHMLICH, M. 2003. Die Cadomiden und Varisziden des Saxothuringischen Terranes - Geochronologie magmatischer Ereignisse. Freiberger Forschungshefte, **C 500**.

GIBBONS, W. & POWER, G. M. 1975. The structure and age of the gneisses of Sark, Channel Isles. *Procedings of the Ussher Society*, **3**, 244–251.

GRAINDOR, M. J. 1957. Le Briovérien dans le Nord-Est du Massif armoricain. *Memoires pour servir a l'explication de la carte Géologique detailée de la France.* Imprimerie Nationale, Paris.

GRAVIOU, P. 1992. Reconaissance d'une suture majeure au sein de la châine cadomienne. *Comptes Rendus de l'Académie des Sciences, Paris*, **315**, 1799–1802.

GRAVIOU, P. & AUVRAY, B. 1985. Caractérisation pétrographique et géochimique des granitoïdes cadomiens du domaine nord-armoricain: implications géodynamiques. *Comptes Rendus de l'Académie des Sciences, Paris*, **303**, 315–318.

GRAVIOU, P. & AUVRAY, B. 1990. Late Precambrian M-type granitoid genesis in the Cadomian belt of NW France. *In*: D'LEMOS, R.S., STRACHAN, R.A. & TOPLEY, C.G. (eds) *The Cadomian Orogeny.* Geological Society, London, Special Publications, **51**, 231–244.

GRAVIOU, P., PEUCAT, J. J., AUVRAY, B. & VIDAL, PH. 1988. The Cadomian Orogeny in the northern Armorican Massif: petrological and geochronological constraints on a geodynamic model. *Hercynica*, **4**, 1–13.

GUERROT, C. 1989. *Archéen et Protérozoïque dans la chaîne hercynienne ouest-européenne: géochimie, isotopique (Sr-Nd-Pb) et géochronologie U–Pb sur zircons.* PhD thesis, University of Rennes.

GUERROT, C. & PEUCAT, J. J. 1990. U–Pb geochonology of the Upper Proterozoic Cadomian orogeny in the northern Armorican Massif, France. *In*: D'LEMOS, R.S., STRACHAN, R.A. & TOPLEY, C.G. (eds) *The Cadomian Orogeny.* Geological Society, London, Special Publications, **51**, 13–26.

HAMMER, J. 1996. Geochemie und Petrogenese der cadomischen und spätvariszischen Granitoide der Lausitz. Freiberger Forschungshefte, **C 463**.

HAMMER, J., EIDAM, J., RÖBER, B. & EHLING, B. - C. 1999. Prävariscischer and variscischer granitoider Magmatismus am NE-Rand des Böhmischen Maasivs - Geochemie und Petrogenese. *Zeitschrift für Geologische Wissenschaften*, **27**, 401–415.

HAVLÍCEK, V. 1971. Stratigraphy of the Cambrian of Central Bohemia. *Sborník geologických ved*, **20**, 7–50.

HÉBERT, R. 1993. Evolution tectonométamorphique d'un arc insulaire au Protérozoïque supérieur le domaine de saint-Brieuc (Massif armoricain). Document du BRGM, **228**.

HÉBERT, R. & BALLÈVRE, M. 1993. Petrology of staurolite-bearing metapelites from the Cadomian belt, northern Brittany (France): Constraints on low-pressure metamorphism: *Bulletin de la Société Géologique de France*, **164**, 215–228.

HÉBERT, R., HALLOT, E., GUERROT, C. & CHANTRAINE, J. 1993. New structural, petrological and radiometric constraints within the Cadomian belt: chronology of events in the Baie de Saint-Brieuc, northern Armorican Massif (France). *Comptes Rendus de l'Académie des Sciences de Paris*, **316**, 395–401.

HEGNER, E. & KRÖNER, A. 2000. Review of Nd isotopic data and xenocrystic and detrital zircon ages from the pre-Variscan basement in the eastern Bohemian Massif: speculations on palinspastic reconstructions. *In*: FRANKE, W., HAAK, U., ONCKEN, O. & TANNER, D. (eds) *Orogenic Processes: Quantification and Modelling in the Variscan Belt.* Geological Society, London, Special Publication, **179**, 113–129.

HEUSE, T., KRONER, U. & BARTL, S. 2001. Tektono- und lithostratigraphische Kartierungskriterien für den Zentralteil des Schwarzburger Sattels. *Geowissenschaftliche Mitteilungen von Thüringen*, **9**, 107–124.

HOLUBEC, J. 1966. Stratigraphy of the Upper Proterozoic in the core of the Bohemian Massif (the Teplá–Barrandian Region). *Rozpr Cesk Akad Ved Rada mat prir Ved Praha*, **76**(S4), 1–62.

INGLIS, J. D., SAMSON, S. D., D'LEMOS, R. S. & HAMILTON, M. 2004. U–Pb geochronological constraints on the tectonothermal evolution of the Paleoproterozoic basement of Cadomia, La Hague, NW France. *Precambrian Research*, **134**, 293–315.

INGLIS, J. D., SAMSON, S. D., D'LEMOS, R. S. & MILLER, B. V. 2005. Timing of Cadomian deformation and magmatism within La Hague, NW France. *Journal of the Geological Society, London*, **162**, 389–400.

JACHOWICZ, M. & PŘICHYSTAL, A, 1997. Lower Cambrian sediments in deep boreholes in south Moravia. *Vìstník Èeského geologického ústavu*, **72**, 329–332.

JAKEŠ, P. 1973. Ophiolites derived tectonically from the ocean floor or derived non-tectonically from island arcs? *Vestnik Ceského geologického ústavu*, **48**, 359–364.

JÉGOUZO, P. 1980. The South Armorican Shear Zone. *Journal of Structural Geology*, **2**, 39–47.

JOLIVET, L. M., FOURNIER, P., HUCHNON, V. S., ROZHDESTVENSKIY, K. F. S. & OSCORBIN, L. S. 1992. Cenozoic intracontinental dextral motion in the Okhotsk-Japan Sea region. *Tectonics*, **11**, 968–977.

JONIN, M. 1981, Un batholite fini-précambrien: le batholite Mancellian (Massif armoricain, France). Etude pétrographique et géochemique. PhD thesis, University of Brest.

KEMNITZ, H., ROMER, R. L. & ONCKEN, O. 2002. Gondwana breakkup and the northern margin of the Saxothuringian belt (Variscides of Central Europe). *Geologische Rundschau*, **91**, 246–259.

KERFORNE, F. 1901. Discordance du Cambrien sur le Precamrien, près de Rennes. *Bulletin de la Société Géologique de France*, **4**(I), 2587–2590.

KETTNER, R. 1917. Versuch einer stratigraphischen Einteilung des böhmischen Algonkiums. *Geologische Rundschau*, **8**, 169–188.

KODZDRÓI, W., KRENTZ, O. & OLPETAL, M. (eds). 2001. *Geological map Lausitz-Jizera-Karkonosze* (*without Cenozoic sediments*). Scale 1:100 000, Warsaw.

KORYTOWSKI, A., DÖRR, W. & ŻELAŹNIEWICZ, A. 1993. U–Pb dating of (meta)granitoids in the NW Sudetes (Poland) and their bearing on tectono-stratigraphic correlation. *Terra Abstracts*, **5**, 331–332.

KOŠLER, J., BOWES, D. R., KONOPÁSEK, J. & MÍKOVÁ, J. 2004. Laser ablation ICPMS dating of zircons in Erzgebirge orthogneisses: evidence for Early Cambrian and Early Ordovician granitic plutonism in the western Bohemian Massif. *European Journal of Mineralogy*, **16**, 15–22.

KOSSMAT, F. 1927. Gliederung des varistischen Gebirgsbaues. *Abhandlungen des Sächsischen Geologischen Landesamtes*, **1**, 1–39.

KŘÍBEK, B., POUBA, Z., SKOCEK, V. & WALDHAUSROVA, J. 2000. Neoproterozoic of the Teplá–Barrandian Unit as a part of the Cadomian orogenic belt: A reveiw and correlation aspects. *Vestnik Ceskeho geologickeho ustavu*, **75**(3), 175–196.

KRÖNER, A. & HEGNER, E. 1998. Geochemistry, single zircon ages and Sm–Nd systematics of granitoid rocks from the Góry Sowie (Owl Mts), Polish West Sudetes: evidence for early Palaeozoic arc-related plutonism. *Journal of the Geological Society, London*, **155**, 711–724.

KRÖNER, A., WILLNER, A. P., HEGNER, E., FRISCHBUTTER, A., HOFMANN, J. & BERGNER, R. 1995. Latest Precambrian (Cadomian) zircon ages, Nd isotopic systematics and P-T evolution of the granitoid orthogneisses of the Erzgebirge, Saxony and Czech Republic. *Geologische Rundschau*, **84**, 437–456.

KRÖNER, A., ŠTÍPSKÁ, P., SCHULMANN, K. & JAECKEL, P. 2000. Chronological constraints on the pre-Variscan evolution of the northeastern margin of the Bohemian Massif, Czech Republic. *In:* FRANKE, W., HAAK, U., ONCKEN, O. & TANNER, D. (eds) *Orogenic Processes: Quantification and Modelling in the Variscan Belt.* Geological Society, London, Special Publications, **179**, 175–197.

KRÖNER, A., JAECKEL, P., HEGNER, E. & OPLETAL, M. 2001. Single zircon ages and whole-rock Nd isotopic systematics of early Palaeozoic granitoid gneisses from the Czech and Polish Sudetes (Jizerské hory, Krkonose Mountains and Orlice-Snezník Complex). *International Journal of Earth Sciences* (*Geologische Rundschau*),

90, 304–324.

Kryza, R., Mazur, S. & Pin, C. 2003. Subduction- and non-subduction-related igneous rocks in the Central EuropeanVariscides: geochemical and Nd isotope evidence from the Klodzko Metamorphic Complex, Polish Sudetes. *Geodinamica Acta*, **16**, 39–57.

Kukal, Z. 1971. Sedimentology of Cambrian deposits of the Barrandian area (Central Bohemia). *Sborník geologických ved*, **20**, 53–100.

Kurze, M., Linnemann, U. & Tröger, K.-A. 1992. Weesensteiner Gruppe und Altpaläozoikum in der Elbtalzone (Sachsen). *Geotektonische Forschungen*, **77**, 101–167.

Lees, G. J. & Roach, R. A. 1993. A preliminary volcanological and petrochemical survey of the late Precambrian calc-alkaline volcanic succession of eastern Jersey, Channel Islands. *Proceedings of the Ussher Society*, **8**, 145–150.

Lees, G. J., Roach, R. A., Shufflebotham, M. & Griffiths, N. J. 1987. Upper Proterozoic volcanism in the Northern Massif Armoricain, France. *In:* Pharoah, T. C., Beckinsale, R. D. & Rickard, D. (eds) *Geochemistry and Mineralization of Proterozoic Volcanic Suites*. Geological Society, London, Special Publications, **33**, 503–523.

Le Gall, J. 1993. *Pétrogenèse des magmas andésitiques et ignimbritiques et leur signification dans l'évolution géodynamique cadomienne, Réconstitution des dynamisms éruptifs d'une province paléovolcanique: l'exemple du graben cambrien du Maine (est du Massif armoricain)*. Memoir de Géosciences, Université de Rennes, **52**.

Le Gall, J. & Barrat, J.-A. 1987. Pétrologie du magmatisme taricadomien du domaine nord-armoricain: l'exemple des complexes basique d'Ernée et de Trégomar. *Geology de la France*, **1**, 3–22.

Le Gall, J. & Mary, G. 1983. Place et signification du complexes basique de Brée et d'autres venues gabbroïques et doléritiques dans l'histoire cadomo-varisque de l'est du Massif armoricain. *Bulletin de la Société Géologique et Minéralogie de Bretagne*, **15**, 169–180.

Le Goff, E., Egal, E. & Chantraine, J. 1994. Structures et champ de déformation dans la chaîne cadomienne à la Ouest de la Baie de Saint-Breuc (Nord du Massif Armoricain, France). *Comptes Rendus de l'Académie des Sciences, Paris*, **319**, 231–238.

Leichmann, J. 1996. *Geologie und Petrologie des Brünner Massifs*. PhD Thesis, University of Salzburg.

Liew, T. C. & Hofmann, A. W. 1988. Precambrian crustal components, plutonic associations, plate environment of the Hercynian Fold Belt of central Europe: indications from a Nd and Sr isotopic study. *Contributions to Mineralogy and Petrology*, **98**, 129–138.

Linnemann, U. 1991. Glazioeustatisch kontrollierte Sedimentationsprozesse im Oberen Proterozoikum der Elbezone (Weesensteiner Gruppe/Sachsen), *Zentralblatt für Geologie und Paläontologie*, **I**(12), 2907–2934.

Linnemann, U. 2003. Sedimentation und geotektonischer Rahmen der Beckenentwicklung im Saxothuringikum (Neoproterozoikum-Unterkarbon), *In:* Linnemann, U. (ed.) Das Saxothuringikum-Abriss der präkambrischen und paläozoischen Geologie von Sachsen und Thüringen, *Geologica Saxonica*, **48/49**, 71–110.

Linnemann, U. 2007. Ediacaran rocks from the Cadomian Basement of the Saxo-Thuringian Zone (NE Bohemian Massif, Germany): Age constraints, geotectonic setting and basin development). *In:* Vickers-Rich, P. & Komarower, P. (eds) *The Rise and Fall of the Ediacaran Biota*. Geological Society, London, Special Publications, **286**, 35–51.

Linnemann, U. & Buschmann, B. 1995*a*. Der Nachweis der cadomischen Diskordanz in einer Tiefenbohrung bei Gera und deren Bedeutung für das proterozoisch-paläozoische Standardprofil im Schwarzburger Antiklinorium. *Geowissenschaftliche Mitteilungen von Thüringen*, **3**, 1–11.

Linnemann, U. & Buschmann, B. 1995*b*. Die cadomische Diskordanz im Saxothuringikum (oberkambrisch-tremadocische overlap-Sequenzen). *Zeitschrift für Geologische Wissenschaften*, **23**, 707–727.

Linnemann, U. & Romer, R. L. 2002. The Cadomian orogeny in Saxo-Thuringia, Germany: geochemical and Nd-Sr-Pb isotopic characterization of marginal basins with constraints to geotectonic setting and provenance. *Tectonophysics*, **352**, 33–64.

Linnemann, U. & Schauer, M. 1999. Die Entstehung der Elbezone vor dem Hintergrund der cadomischen und variszischen Geschichte des saxothuringischen Terranes. *Zeitschrift für Geologische Wissenschaften*, **27**, 529–561.

Linnemann, U., Gehmlich, M., Heuse, Th., Schauer, M. 1999. Die Cadomiden und Varisziden im Thüringisch-Vogtländischen Schiefergebirge (Saxothuringisches Terrane). *Beiträge zur Geologie von Thüringen*, **6**, 7–39.

Linnemann, U., Gehmlich, M. *et al.* 2000. From Cadomian subduction to Early Paleozoic rifting: the evolution of Saxo-Thuringia at the margin of Gondwana in the light of single zircon geochronology and basin development (Central European Variscides, Germany). *In*: Franke, W., Haak, U., Oncken, O. & Tanner, D. (eds) *Orogenic Processes: Quantification and Modelling in the Variscan Belt*. Geological Society, London, Special Publications, **179**, 131–153.

Linnemann, U., McNaughton, N. J., Romer, R. L., Gehmlich, M., Drost, K. & Tonk, C. 2004. West African provenance for Saxo-Thuringia (Bohemian Massif): Did Armorica ever leave pre-Pangean Gondwana? – U/Pb-SHRIMP zircon evidence and the Nd-isotopic record. *International Journal of Earth Sciences,*, **93**, 683–705.

Linnemann, U., Gerdes, A., Drost, K. & Buschmann, B. 2007. The continuum between Cadomian Orogenesis and opening of the Rheic Ocean: Constraints from LA-ICP-MS U–Pb zircon dating and analysis of plate-tectonic setting (Saxo-Thuringian zone, NE Bohemian massif, Germany), *In:* Linnemann, U., Nance, R.D., Kraft, P. & Zulauf, G. (eds) *The Evolution of the Rheic Ocean: From Avalonian-Cadomian Active Margin to Alleghenian-Variscan Collision*. Geological Society of America, Boulder, Colorado, Special Papers, **423**, 61–96.

McKerrow, W. S., Mac Niocail, C., Ahlberg, P. E., Clayton, G., Cleal, C. L. & Eagar, M. C. 2000. The Late Palaeozoic relations between Gondwana and Laurussia. *In*: Franke, W., Haak, U., Oncken, O. & Tanner, D. (eds) *Orogenic Processes: Quantification and Modelling in the Variscan Belt*. Geological Society, London, Special Publications, **179**, 2–20.

Mašek, J. 2000 Stratigraphy of the Proterozoic of the Barrandian area. *Vestnik Ceskeho geologickeho ustavu*, **75**, 197–199.

Mašek, J. & Zoubek, J. 1980. Proposal of new terms for main lithostratigraphic units of the Barrandian Proterozoic. *Vestnik Ceskeho geologickeho ustavu*, **55**, 121–123.

Máška, M. & Zoubek, V. 1961. Area of the intramontane Block. *In*: Zoubek *et al.* (eds) *Tectonic Development of Czechoslovakia*. Nakl. CSAV, Praha, 25–50.

Massonne, H.-J. 1999. A new occurrence of microdiamonds in quartzofeldspathic rocks of the Saxonian Erzgebirge, Germany, and their metamorphic evolution. *Seventh International Kimberlite Conference*, Cape Town, South Africa, April 1999. 533–539.

Mazur, S., Turniak, K. & Bröcker, M. 2004. Neoproterozoic and Cambro-Ordovician magmatism in the Variscan Klodzko Metamorphic Complex (West Sudetes, Poland): new insights from U/Pb zircon dating. *International Journal of Earth Sciences (Geologische Rundschau)*, **93**, 758–772.

Miller, B. V., Samson, S. D. & D'Lemos, R. S. 1999. Time span of plutonism, fabric development, and cooling in a Neoproterozoic magmatic arc segment: U–Pb age constraints from syn-tectonic plutons, Sark, Channel Islands, UK. *Tectonophysics*, **312**, 79–95.

Miller, B. V., Samson, S. D. & D'Lemos, R. S. 2001. U–Pb geochronological constraints on the timing of plutonism, volcanism, and sedimentation, Jersey, Channel Islands, UK. *Journal of the Geological Society, London*, **158**, 243–252.

Mingram, B., Kröner, A., Hegner, E. & Krentz, O. 2004. Zircon ages, geochemistry, and Nd isotopic systematics of pre-Variscan orthogneisses from the Erzgebirge, Saxony (Germany), and geodynamic interpretation. *International Journal of Earth Sciences (Geologische Rundschau)*, **93**, 706–727.

Murphy, J. B. & Nance, R. D. 1991. Supercontinent model for the contrasting character of Late Proterozoic orogenic belts. *Geology*, **19**, 469–472.

Murphy, J. B., Strachan, R. A., Nance, R. D., Parker, K. D. & Fowler, M. B. 2000. Proto-Avalonia: A 1.2–1.0 Ga tectonothermal event and constraints for the evolution of Rodinia. *Geology*, **28**, 1071–1074.

Murphy, J. B., Pisarevsky, S. A., Nance, R. D. & Keppie, J. D. 2004. Neoproterozoic Early Paleozoic evolution of peri-Gondwanan terranes: implications for Laurntia-Gondwana connections. *International Journal of Earth Sciences*, *(Geologische Rundschau)*, **93**, 659–682.

Nagy, E. A., Samson, S. D. & D'Lemos, R. S. 2002. U–Pb

geochronological constraints on the timing of Brioverian sedimentation and regional deformation in the St. Brieuc region of the Neoproterozoic Cadomian orogen, northern France. *Precambrian Research*, **116**, 1–17.

NANCE, R. D. & MURPHY, J. B. 1994. Contrasting basement isotopic signatures and the palinspastic restoration of peripheral orogens: Example from the Neoproterozoic Avalonian-Cadomian belt. *Geology*, **22**, 617–620.

NANCE, R. D. & MURPHY, J. B. 1996. Basement isotopic signatures and Neoproterozoic paleogeography of Avalonian – Cadomian and related terranes in the circum-North Atlantic. *In*: NANCE, R. D. & THOMPSON, M. D. (eds) *Avalonian and Related Peri-Gondwanan Terranes of the Circum-North Atlantic*. Geological Society of America, Boulder, Colorado, Special Papers, **304**, 333–346

NANCE, R. D., MURPHY, J. B. & KEPPIE, J. D. 2002. A Cordilleran model for the evolution of Avalonia. *Tectonophysics*, **352**, 11–31.

OBERC-DZIEDZIC, T., KLIMAS, K., KRYZA, R. & FANNING, C. M. 2003. SHRIMP U–Pb zircon geochronology of the Strzelin gneiss, SW Poland: evidence for a Neoproterozoic thermal event in the Fore-Sudetic Block, Central European Variscides. *International Journal of Earth Sciences*, **92**, 701–711.

OBERC-DZIEDZIC, T., PIN, C. & KRYZA, R. 2005. Early Palaeozoic crustal melting in an extensional setting: petrological and Sm–Nd evidence from the Izera granite-gneisses, Polish Sudetes. *International Journal of Earth Sciences*, **94**, 354–368.

OLIVER, G. J. H., CORFU, F. & KROGH, T. E. 1993. U–Pb ages from SW Poland: evidence for a Caledonian suture zone between Baltica and Gondwana. *Journal of the Geological Society, London*, **150**, 355–369.

PASTEELS, P. & DORÉ, F. 1982. Age of the Vire-Carolles Granite. *In:* ODIN, G. S. (ed.) *Numerical Dating in Stratigraphy Part II*. Wiley, New York, 784–790.

PEUCAT, J. - J. 1986. Behaviour of Rb-Sr whole-rock and U–Pb zircon systems during partial melting as shown in migmatitic gneisses from the St, Malo Massif, NE Brittany, France. *Journal of the Geological Society, London*, **143**, 875–886.

PEUCAT, J. J., HIRBEC, Y., AUVRAY, B., CONGÉ, J. & CORNICHET, J. 1981. Late Proterozoic zircon age from a basic-ultrabasic complex: a possible Cadomian orogenic complex in the Hercynian belt of western Europe. *Geology*, **9**, 169–173.

PIETZSCH, K. 1917. Das Elbtalschiefergebiet südwestlich von Pirna. *Zeitschrift der Deutschen Geologischen Gesellschaft*, **69**, 177–286.

PIN, C. & WALDHAUSROVÁ, J. 2007. Sm-Nd isotope and trace element study of the Late Proterozoic metabasalts ('spilites') from the Central Barrandian Domain (Bohemian Massif, Czech Republic). *In:* LINNEMANN, U., NANCE, R. D., ZULAUF, G. & KRAFT, P. (eds) *The Geology of Peri-Gondwana: The Avalonian-Cadomian Belt, adjoining Cratons and the Rheic Ocean (IGCP 497)*. Geological Society of America, Boulder, Colorado, Special Paper **423**, 231–247.

PIN, C., KRYZA, R., OBERC-DZIEDZIC, T., MAZUR, S., TURNIAK, K. & WALDHAUSROVÁ, J. 2007. The diversity and geodynamic significance of Late Cambrian (ca. 500 Ma) felsic anorogenic magmatism in the northern part of the Bohemian Massif: a review based on Sm-Nd isotope and geochemical data. *In:* LINNEMANN, U., NANCE, R.D., KRAFT, P. & ZULAUF., G. (eds) *The Evolution of the Rheic Ocean: From Avalonian-Cadomian Active Margin to Alleghenian-Variscan Collision*. Geological Society of America. Boulder, Colorado, Special Papers, **423**, 209–229.

PITON, P. 1985. *Etude géochemichimique et géochronologicique de la croûte continentale protérozoïque (1.8–2.2) au Nord du Massif Armoricain*. Rapport de Diplôme d'Etudes Appronfodies, University of Rennes, France.

POWER, G. M. & GIBBONS, W. 1994. Precambrian rocks in the Channel Islands. *In:* GIBBONS, W. & HARRIS, A. L. (eds) *A Revised Correlation of Precambrian Rocks in the British Isles*. Geological Society, London, Special Reports, **22**, 98–104.

POWER, G. M., BREWER, T. S., BROWN, M. & GIBBONS, W. 1990*a*. Late Precambrian foliated plutonic complexes of the Channel Islands and La Hague: early Cadomian plutonism. *In*: D'LEMOS, R. S., STRACHAN, R. A. & TOPLEY, C. G. (eds) *The Cadomian Orogeny*. Geological Society, London, Special Publications, **51**, 215–229.

POWER, G. M., BREWER, T. S. & D'LEMOS, R. S. 1990*b*. The post-tectonic Cadomian plutonic complex of La Hague, Manche, N. France. *In*: D'LEMOS, R. S., STRACHAN, R. A. & TOPLEY, C. G. (eds) *The Cadomian Orogeny*. Geological Society, London, Special Publications, **51**, 261–272.

RABU, D., CHAUVEL, J. J. & CHANTRAINE, J. 1983. Nouvelles propositions pour la lithostratigraphie du Briovérian (Protérozoique supérieur) et pour l'évolution géodynamique cadomienne en bai de Saint-Brieuc (Massif Armoricain). *Bulletin de la Société Géologique de France*, **25**, 615–621.

RABU, D., CHANTRAINE, J., CHAUVEL, J. J., DENIS, E., BALE, P. & BARDY, P. 1990. The Brioverian (Upper Proterozoic) and the Cadomian orogeny in the Armorican Massif. *In*: D'LEMOS, R. S., STRACHAN, R. A. & TOPLEY, C. G. (eds) *The Cadomian Orogeny*. Geological Society, London, Special Publications, **51**, 81–94.

RAINBIRD, R. H., HEAMAN, L. M. & YOUNG, G. M. 1992. Sampling Laurentia: Detrital zircon geochronology offers evidence for an extensive Neoproterozoic river system originating form the Grenville orogen. *Geology*, **20**, 351–354.

RAINBIRD, R. H., MCNICOLL, V. J., THERIAULT, R. J., HEAMAN, L. M., ABBOT, J. G., LONG, D. G. F. & THORKELSON, D. J. 1997. Pan-continental river system draining Grenville orogen recorded by U–Pb and Sm-Nd geochronology of Neoproterozoic quartzarenites and mudrcoks, northwestern Canada. *Journal of Geology*, **105**, 1–17.

ROACH, R. A. 1966. Outline and guide to the geology of Guernsey. *Transactions of the Societe Guernesiaise*, **18**, 751–776.

ROACH, R. A., LEES, G. J. & SHUFFLEBOTHAM, M. M. 1990. Brioverian volcanism and Cadomian tectonics, Baie de St Brieuc, Brittany: stages in the evolution of a late Precambrian ensialic basin. *In*: D'LEMOS, R. S., STRACHAN, R. A. & TOPLEY, C. G. (eds) *The Cadomian Orogeny*. Geological Society, London, Special Publications, **51**, 41–68.

ROBARDET, M. 2002. Alternative approach to the Variscan Belt in southwestern Europe: Preorogenic paleobiogeographical constraints. *In*: MARTINEZ CATALAN, J. R., HATCHER, R. D. J., ARENAS, R. & DIAZ GARCIA, F. (eds) *Variscan-Appalachian Dynamics: The building of the Late Paleozoic Basement*. Geological Society of America, Boulder, Colorado, Special Papers, **364**, 1–15.

ROBERTS, D. & SIEDLECKA, A. 2002. Timanian orogenic deformation along the northwestern margin of Baltica, Northwest Russia and Northeast Norway, and Avalonian-Cadomian connections. *Tectonophysics*, **352**, 169–184.

ROCCI, G., BRONNER, G. & DESCHAMPS, M. 1991. Crystalline Basement of the West African Craton. *In*: DALLMEYER, R. D. & LECORCHE, J. P. (eds). *The West African Orogens and Circum-Atlantic Correlatives*. Spinger, Berlin, 31–61.

RÖHLICH, P. 1965. Geologische Probleme des mittelböhmischen Algonkiums. *Geologie*, **14**(4), 373–403.

RÖHLICH, P. 2000. Some stratigraphic problems of the Barrandian Neoproterozoic. *Vestnik Ceskeho geologickeho ustavu*, **75**(3), 201–204.

RYAN, P. D. 1973. The solid geology of the area between Binic and Brehec, Côtes-du-Nord, France. PhD thesis, University of Keele.

SAMSON, S. D. & D'LEMOS, R. S. 1998. U–Pb geochronology and Sm-Nd isotopic composition of Proterozoic gneisses, Channel Islands, UK. *Journal of the Geological Society, London*, **155**, 609–618.

SAMSON, S. D. & D'LEMOS, R. S. 1999. A precise late Neoproterozoic U–Pb zircon age for the syntectonic Perelle quartz diorite, Guernsey, Channel Islands, UK. *Journal of the Geological Society, London*, **156**, 47–54.

SAMSON, S. D., MILLER, B. V. & D'LEMOS, R. 1997. U–Pb geochronological investigation of exposed basement within the Cadomia terrane, Channel Islands: a test of indirect methods of basement characterization. 7th *International Goldschmidt Conference*, Tucson, Arizona.

SAMSON, S. D., D'LEMOS, R. S., BLICHERT-TOFT, J. & VERVOORT, J. D. 2003. U–Pb geochronology and Hf-Nd isotope compositions of the oldest Neoproterozoic crust within the Cadomian Orogen: new evidence for a unique juvenile terrane. *Earth and Planetary Science Letters*, **208**, 165–180.

SAMSON, S. D., D'LEMOS, R. S., MILLER, B. V. & HAMILTON, M. 2005. Neoproterozoic paleogeography of the Cadomia and Avalon terranes: constraints from detrital zircon U–Pb ages. *Journal of the Geological Society, London*, **162**, 65–71.

SCHNEIDER SANTOS, J. O., HARTMANN, L. A., GAUDETTE, H. E., GROVES, D., MCNAUGHTON, N. J. & FLETCHER, I. R. 2000. A new understanding of the provinces of the Amazon craton based on integration of field mapping and U–Pb and Sm-Nd geochronology. *Gondwana Research*, **3**, 453–488.

SEBASTIAN, U. 1995. Die Strukturentwicklung des spätorogenen Erzgebirgsaufstiegs in der Flöhazone – Ein weiterer Beitrag zur postkollisionalen Extension am Nordrand der Böhmischen Masse. *Freiberger Forschungshefte*, **C461**, 114.

SHUFFLEBOTHAM, M. M. 1990. The geology of the Penthievre crystalline massif: a reappraisal of the type-Pentevrian area, northern Brittany. *In*: D'LEMOS, R. S., STRACHAN, R. A. & TOPLEY, C. G. (eds) *The Cadomian Orogeny*. Geological Society, London, Special Publications, **51**, 27–39.

SÖLLIG, A. 1953. *Ergebnisse einer Teilkartierung im Altpaläozoikum und Algonkium des Schwarzburger Sattels (Blatt Eisfeld)*. Diploma paper, Friedrich-Schiller-Universität, Jena.

SQUIRE, A. D. 1974. Brioverian sedimentology and structure of Jersey and adjacent areas. PhD thesis, University of London.

STILLE, H. 1948. Die assyntische Ära und der vor-, mittel und nachassyntische Magmatismus. *Zeitschrift der deutschen geologischen Gesellschaft*, **98**, 152–165.

STRACHAN, R. A. & ROACH, R. A. 1990. Tectonic evolution of the Cadomian belt in north Brittany. *In*: D'LEMOS, R. S., STRACHAN, R. A. & TOPLEY, C. G. (eds) *The Cadomian Orogeny*. Geological Society, London, Special Publications, **51**, 133–150.

STRACHAN, R. A., TRELOAR, P. J., BROWN, M. & D'LEMOS, R. S. 1989. Cadomian terrane tectonics and magmatism in the Armorican Massif. *Journal of the Geological Society, London*, **146**, 423–426.

STRACHAN, R. A., BROWN, M., D'LEMOS, R. S., ROACH, R. A. & TRELOAR, P. J. 1992. Reply to Discussion on the timing and kinematics of Cadomian deformation (Northern Brittany). *Journal of the Geological Society, London*, **149**, 152–155.

STRACHAN, R. A., D'LEMOS, R. S. & DALLMEYER, R. D. 1996. Neoproterozoic evolution of an active plate margin: North Armorican Massif, France. *In:* NANCE, R. D. & THOMPSON, M. D. (eds) *Avalonian and Related Peri-Gondwanan Terranes of the Circum-North Atlantic*, Geological Society of America, Boulder, Colorado, Special Publications, **304**, 319–332.

SUTTON, J. & WATSON, J. V. 1957. The structure of Sark, Channel Islands. *Proceedings of the Geologists' Association*, **68**, 179–203.

TAIT, J. A., BACHTADSE, V., FRANKE, W. & SOFFEL, H. C. 1997. Geodynamic evolution of the European Variscan fold belt: paleomagnetic and geological constraints. *Geologische Rundschau*, **86**, 585–598.

TAIT, J., SCHÄTZ, M., BACHTADSE, V. & SOFFEL, H. 2000. Palaeomagnetism and Palaeozoic palaeogeography of Gondwana and European terranes. *In*: FRANKE, W., HAAK, V., ONCKEN, O. & TANNER, D. (eds) *Orogenic Processes: Quantification and Modelling in the Variscan Belt*. The Geological Society, London, Special Publications, **179**, 21–34.

TASSINARI, C. C. G. & MACAMBIRA, M. J. B. 1999. Geochronological provinces of the Amazonian Craton. *Episodes*, **22**, 174–182.

TEIPEL, U., EICHHORN, R., LOTH, G., ROHRMÜLLER, J., HÖLL, R. & KENNEDY, A. 2004. U–Pb SHRIMP and Nd isotopic data from the western Bohemian Massif (Bayerischer Wald, Germany): Implications for Upper Vendian and Lower Ordovician magmatism. *International Journal of Earth Sciences* (*Geologische Rundschau*), **93**, 782–801.

THIÉBLEMONT, D., EGAL, E., GUERROT, C. & CHANTRAINE, J. 1999. Témoins d'une subduction éocadomienne (665–655 Ma) en Bretagne nord: arguments géochimiques. *Géologie France*, **1**, 3–11.

THIELE, O. 1984. Zum Deckenbau und Achsenplans des Moldanubikums der südlichen Böhmischen Masse (Österreich). *Jahrbuch der Geologischen Bundesanstalt Wien*, **126**, 513–523.

TICHOMIROWA, M. 2001. *Die Gneise des Erzgebirges - hochmetamorphe Äquivalente von neoproterozoisch-frühpaläozoischen Grauwacken und Granitoiden der Cadomiden*. PhD thesis, TU Bergakademie Freiberg.

TICHOMIROWA, M. 2003. *Die Gneise des Erzgebirges - hochmetamorphe Äquivalente von neoproterozoisch-frühpaläozoischen Grauwacken und Granitoiden der Cadomiden*. Freiberger Forschungshefte, **C 495**.

TICHOMIROWA, M., BERGER, H.-J. *et al*. 2001. Zircon ages of high-grade gneisses in the Eastern Erzgebirge (Central European Variscides) – constraints on origin of the rocks and Precambrian to Ordovician magmatic events in the Variscan foldbelt. *Lithos*, **56**, 303–332.

TIKHOMIROVA, M. 2002. Zircon inheritance in diatexite granodiorites and its consequence on geochronology – a case study in Lusatia and the Erzgebirge (Saxo-Thuringia, Eastern Germany). *Chemical Geology*, **191**, 209–224.

TIMMERMANN, H., STEDRÁ, V., GERDES, A., NOBLE, S. R., PARRISH, R. R. & DÖRR, W. 2004. The problem of dating high-pressure metamorphism: a U–Pb isotope and geochemical study on eclogites and related rocks of the Mariánské Lázne Complex, Czech Republic. *Journal of Petrology*, **45**, 1311–1338.

TOPLEY, C. G., BROWN, M., D'LEMOS, R. S., POWER, G. M. & ROACH, R. A. 1990. The Northern Igneous Complex of Guernsey, Channel Islands. *In*: D'LEMOS, R. S., STRACHAN, R. A. & TOPLEY, C. G. (eds) *The Cadomian Orogeny*. Geological Society, London, Special Publications, **51**, 245–260.

TRELOAR, P. J. & STRACHAN, R. A. 1990. Cadomian strike-slip tectonics in NE Brittany. *In*: D'LEMOS, R. S., STRACHAN, R. A. & TOPLEY, C. G. (eds) *The Cadomian Orogeny*. Geological Society, London, Special Publications, **51**, 151–168.

TRIBE, I. 1994. Deformation of granitoid rocks, channel Islands UK: Implications for the structural evolution of a late Precambrian arc. PhD thesis, Oxford Brookes University.

TRIBE, I. R. & D'LEMOS, R. S. 1996. Significance of a hiatus in down-temperature fabric development within syn-tectonic quartz diorite complexes, Channel Islands, UK. *Journal of the Geological Society, London*, **153**, 127–138.

TRIBE, I. R., STRACHAN, R. A. & D'LEMOS, R. S. 1996. Neoproterozoic shear zone tectonics within the Icartian basement of Guernsey and Sark, Channel Islands. *Geological Magazine*, **133**, 177–192.

TURNIAK, K., MAZUR, S. & WYSOCZANSKI, R. 2000. SHRIMP zircon geochronology and geochemistry of the Orlica-Śnieżnik gneisses (Variscan belt of Central Europe) and their tectonic implications. *Geodinamica Acta*, **13**, 293–312.

UNRUG, R. 1996. *Geodynamic Map of Gondwana Supercontinent Assembly – 1:10.000.000*. Council for Geosciences, Bureau de Recherches Géologiques et Minières, Orléans.

VAN BREEMEN, O., AFTALION, M., BOWES, D. R., DUDEK, A., MÍSAR, Z., POVONDRA, P. & VRÁNA, S. 1982. Geochronological studies of the Bohemian massif, Czechoslovakia, and their significance in the evolution of central Europe. *Transactions of the Royal Society of Edinburgh, Earth Sciences*, **73**, 89–108.

VAN DER VOO, R. 1979. Paleozoic assembly of Pangea: a new plate tectonic model for the Taconic, Caledonian and Hercynian orogenies. *EOS Transactions of the American Geophysical Union*, **60**, 241.

VAN DER VOO, R. 1988. Paleozoic paleogeography of North America, Gondwana, and intervening displaced terranes: Comparisons of paleomagnetism with paleoclimatology and biogeographical patterns. *Geological Society of America Bulletin*, **100**, 311–324.

VEJNAR, Z. 1984. *Geology of the Domazlice region*. Věstnik Ústředniho řstavu geolgického.

VENERA, Z., SCHULMANN, K. & KRÖNER, A. 2000. Intrusion within a transitional tectonic domain: the Cistá granodiorite (Bohemian Massif) – structure and rheological modelling. *Journal of Structural Geology*, **22**, 1437–1454.

VON GAERTNER, H. R. 1944. Die Schichtgliederung der Phyllitgebiete in Thüringen und Nordbayern und ihre Einordnung in das stratigraphische Schema. *Jahrbuch der Reichsanstalt für Bodenforschung*, **62**, 54–80.

VON QUADT, A. & GEBAUER, D. 1993. Sm-Nd and U–Pb dating of eclogites and granulites from the Oberpfalz, NE Bavaria, Germany. *Chemical Geology*, **109**, 317–339.

VON RAUMER, J. F., STAMPFLI, G. M. & BUSSY, F. 2003. Gondwana-derived microcontinents – the constituents of the Variscan and Alpine collisional orogens. *Tectonophysics*, **365**, 7–22.

WALDHAUSROVÁ, J. 1971. The chemistry of the Cambrian volcanics in the Barrandian area. *Krystallinikum*, **8**, 45–75.

WATTS, M. J. & WILLIAMS, G. D. 1979. Fault rocks as indicators of progressive shear deformation in the Guingamp region, Brittany. *Journal of Structural Geology*, **1**, 323–332.

WENDT, J. I., KRÖNER, A., FIALA, J. & TODT, W. 1993. Evidence from zircon dating for existence of approximately 2.1 Ga old crystalline basement in southern Bohemia, Czech Republic. *Geologische Rundschau*, **82**, 42–50.

WENT, D. & ANDREWS, M. 1990. Post-Cadomian erosion, deposition and basin development in the Channel Islands and northern Brittany. *In*: D'LEMOS, R. S., STRACHAN, R. A. & TOPLEY, C. G. (eds) *The Cadomian Orogeny*. Geological Society, London, Special Publications, **51**, 293–304.

WIEFEL, H. 1996. *Geologische Karte des Thüringer Waldes*. Scale 1:100 000. Thüringer Geologische Landesanstalt.

WINDLEY, B. F. 1995. *The Evolving Continents*, 3rd Ed. Wiley, Chichester.

WINTERER, E. L. 1964. Late Precambrian pebbly mudstone in Normandy, France: tillite or tilloid? *In*: NAIRN, A. E. (ed.) *Problems in Paleoclimatology*. Wiley, London, 159–178.

WUCHER, K. 1967. Ergebnisse der Kartierungsbohrung Geinersdorf 1/60 und 2/62 (Thüringisches Schiefergebirge). *Jahrbuch für Geologie*, **1**, 297–323.

ZEH, A., BRÄTZ, H., MILLAR, I. L. & WILLIAMS, I. S. 2001. A combined zircon SHRIMP and Sm-Nd isotope study of high grade paragneisses from the Mid-German Crystalline Rise: evidence for northern Gondwanan and Grenvillian provenance. *Journal of the Geological Society, London*, **158**, 983–994.

ŻELAŹNIEWICZ, A., DÖRR, W. *et al.* 2004. The eastern continuation of the Cadomian orogen: U–Pb zircon evidence from Saxo-Thuringian granitoids in south-western Poland and the northern Czech Republic. *International Journal of Earth Sciences (Geologische Rundschau)*, **93**, 773–781.

ZOUBEK, V., COGNÉ, J., KOZOUKCHAROV, D. & KRÄUTNER, M. G. 1988. *Precambrian in Younger Fold Belts*. Wiley, Chichester.

ZULAUF, G., DÖRR, W., FIALA, J. & VEJNAR, Z. 1997. Late Cadomian crustal tilting and Cambrian transtension in the Teplá–Barrandian unit (Bohemian Massif, Central European Variscides). *Geologische Rundschau*, **86**, 571–584.

ZULAUF, G., SCHITTER, F., RIEGLER, G., FINGER, F., FIALA, J. & VEJNAR, Z. 1999. Age constraints on the Cadomian evolution of the Teplá Barrandian unit (Bohemian Massif) through electron microprobe dating of metamorphic monazite. *Zeitschrift der deutschen geologischen Gesellschaft*, **150**(4), 627–639.

4 Cambrian

GERD GEYER (co-ordinator), OLAF ELICKI, OLDŘICH FATKA & ANNA ŻYLIŃSKA

The Cambrian (*c.* 545–488 Ma) is probably the most poorly studied and least documented of all Phanerozoic systems in Central Europe. Cambrian deposits in Central Europe are generally of limited extent, often largely covered by vegetation and slightly to strongly metamorphosed so that data on depositional environments and palaeogeographic history are very limited. Regional differences in the tectonic and resulting sedimentary history as well as faunal characteristics indicate a melange of plates and terranes in a configuration that differs extremely from their original spatial distribution.

Despite considerable interest in the Cambrian on a global scale following recognition of the 'Cambrian Explosion', and various areas with peculiar regional and faunal aspects (e.g. Burgess Shale, Chengjiang, Kaili), there has been a lack of detailed and general research on the Cambrian of Central Europe during the last two decades. Relevant studies have concentrated on a few areas such as Lusatia, the Holy Cross Mountains, or the Barrandian area in Bohemia.

The most relevant surface exposures are found in Bohemia, the Franconian Forest area in Bavaria, in western Thuringia, in the Lusatia area in Saxony, the Holy Cross Mountains of southern Poland, and in the Brabant Massif of Belgium (Fig. 4.1). These outcrops are of relatively limited extent, but some yield important fossil assemblages. In addition, Cambrian strata are known from a number of drillholes such as in the Delitzsch-Torgau-Doberlug Syncline of NW Saxony, Upper Silesia, and a large area in north and east Poland, which is part of the East European Platform. Our knowledge of the Cambrian is supplemented by radiometric data that indicate a Cambrian age for numerous slightly to heavily metamorphosed igneous rock complexes in such areas as the surroundings of the Bohemian Massif, the Mid-German Crystalline High, and the southern and central Alps.

Stratigraphic concepts (G.G.)

The Cambrian System is currently the only Phanerozoic system without a complete formally agreed internal subdivision. Up until now, only the base of the Cambrian (Landing 1994) and one internal global stratotype section and point (GSSP) (defining the base of the Paibian Stage and the Furongian Series which replaced the traditional Upper/Late Cambrian) (Peng *et al.* 2004) have been established. This lack of formal international stages and series partly reflects the scarcity of suitable biostratigraphic markers for intercontinental correlation as well as the pronounced faunal provincialism. For the western Gondwanan and Avalonian areas (which constitute most of the crustal fragments discussed below), Siberian and Iberian local stages have been in use, although they are unsuitable in many cases. A formal unified chronostratigraphic scale for West Gondwana has been introduced recently, based on the well-developed Early/Lower and Middle Cambrian of southern Morocco, Spain, and southern France (Geyer & Landing 2004). The base of the Cambrian is not recognizable in Central Europe with any certainty because of the lack of continuous sections that traverse the Ediacaran–Cambrian boundary interval and include index fossils.

The trilobite genus *Paradoxides* (*sensu lato*) is the traditional index fossil of the Middle Cambrian in the Early Palaeozoic continent Baltica (= East European Platform) but is also known from Bohemia and several regions in Germany and southern Europe so that these regions were termed the 'Acadobaltic faunal province'. The occurrence of *Paradoxides* (*sensu lato*) is a convenient biostratigraphic (and chronostratigraphic) marker. However, the first appearance date (FAD) of this genus has been shown to be diachronous (e.g. Geyer 1998, 2005; Geyer & Palmer 1995; Geyer & Landing 2004).

The Cambrian–Ordovician boundary still poses a major problem in Central Europe, basically because it was traditionally recognized by means of different fossil groups in different regions, such as graptolites in Europe, trilobites in North America, and conodonts in Australia and eastern Asia. This results in notable inconsistencies in the regional boundary levels. The first appearance of the planktonic graptolite *Rhabdinopora flabelliformis*, traditionally used as the base of the Ordovician in Europe, has been disregarded as defining the base of the Ordovician on a global scale because of its limited occurrence. Acritarchs, which can be used biostratigraphically in Central Europe for some of the Cambrian rocks, have not yet attained a level of international investigation that provides a base for the unequivocal recognition of the Cambrian–Ordovician boundary, although efforts have been made and single studies show a refined zonal stratigraphy (Vecoli 1996).

Palaeogeography and tectonic evolution (G.G. & O.F.)

For the Cambrian period, regional differences in the tectonic and sedimentary history as well as faunal characters clearly indicate a melange of plates and terranes of different palaeogeographical origin in present-day Central Europe. Most of the Variscan basement areas trace their evolution back to an origin in the Neoproterozoic Avalonian–Cadomian belt at the margin of Gondwana (Nance & Murphy 1994; Torsvik *et al.* 1996). It is generally agreed that during the Neoproterozoic the area at the NW margin of Gondwana was a Pacific/Andean-type orogen. The so-called Tornquist Ocean was subducted beneath this Neoproterozoic continental block, and the Cadomian basement of western and central Europe is generally derived from this zone (Cogne 1990). Central Europe south of the Iapetus Suture and west of the Trans-European Suture Zone was part of the Gondwanan margin and thus displays vestiges of the Cadomian orogenic event with resulting latest Neoproterozoic and early Cambrian magmatism. With the beginning of the Cambrian, the northern West Gondwanan margin, which became the southern and central parts of future Europe, included extended areas of similar sedimentary facies with reduced detrital input and carbonate platforms (Courjault-Radé *et al.* 1992). Investigation

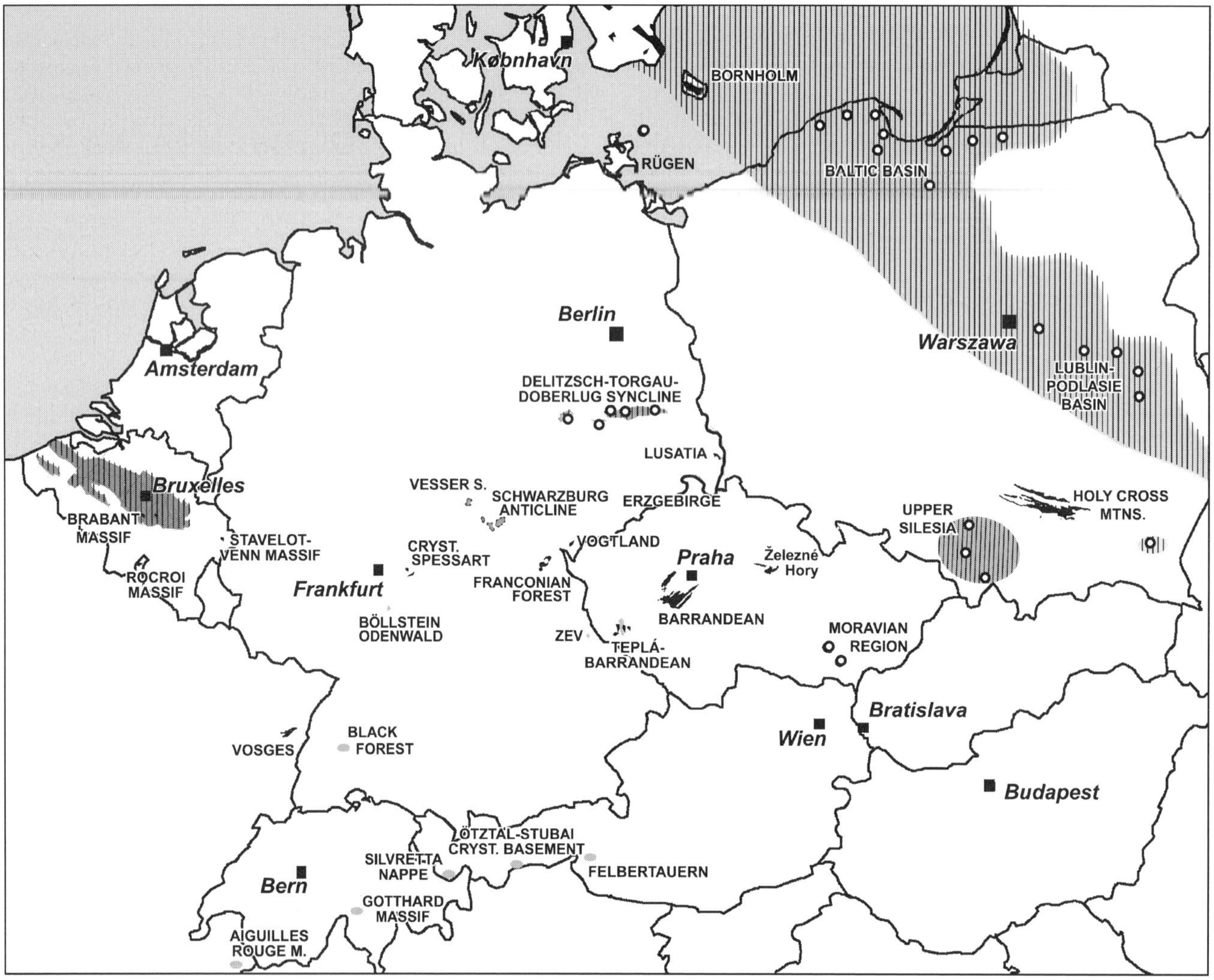

Fig. 4.1. Distribution of Cambrian rocks in present-day Central Europe. Black areas, outcrops of Cambrian rocks; grey areas, outcrops of rocks with Cambrian portions or suspect Cambrian rocks; grey with vertical lines, subsurface Cambrian rocks or subsurface rocks with Cambrian portions. Major structural areas indicated.

on magmatic and metamorphic rocks allows several different events to be identified. A subduction–collision scenario appears to have continued into the Cambrian and served for the formation and accretion of island-arc systems in the central part of Europe. Central European Neoproterozoic amphibolites of MOR and island-arc composition have been identified (Schaltegger & Gebauer 1999), which formed part of the Avalonian–Cadomian belt. These vestiges of a latest Proterozoic oceanic crust and Cambrian island arcs (Schaltegger *et al.* 1997; Bankwitz & Bankwitz 1997) demonstrate the existence of different tectonic regimes along the Gondwanan margin, which led to the development of a magmatic arc recognizable over a distance of more than 800 km from eastern France to western Poland along the Mid-German Crystalline High. This scenario is well illustrated for the Moldanubian domain and particularly the Barrandian and Teplá-Barrandian where the Proterozoic–Cambrian boundary is present within an angular unconformity. The Cambrian succession commences with a thick sequence of probable continental clastic deposits in graben structures. Synsedimentary faults and volcanic rocks at various stratigraphic levels indicate ongoing tectonic activity during the early Cambrian so that a dextral

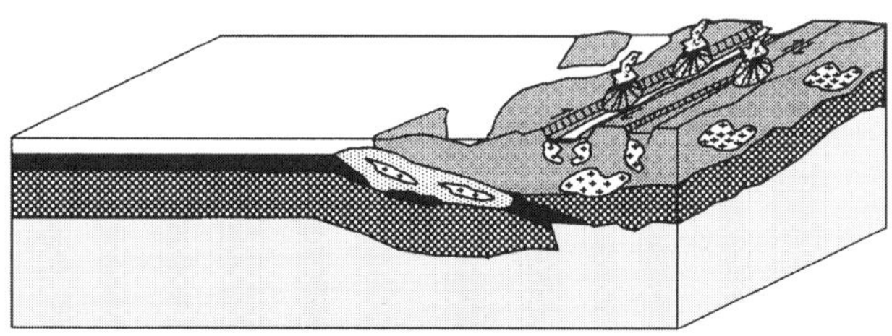

Fig. 4.2. Block diagram illustrating the structural situation in the early Cambrian for the Teplá–Barrandian/Moldanubian Zone. Slightly modified after Dörr *et al.* (2002, fig. 13), based on Zulauf *et al.* (1999).

transtensional regime (Fig. 4.2) can be assumed to have followed the Ediacaran Cadomian subduction and microplate collisional stages (Zulauf *et al.* 1999). From Early Cambrian times onward, Central European basement areas showed a rifting tendency which culminated in a breakup into peri-Gondwanan microcontinents as documented by the presence of ultramafic, mafic and granitoid rocks from the basement areas (Linnemann *et al.* 2000).

The majority of rocks in Central Europe identified as Cambrian in age belongs to the Saxothuringian domain ('Saxo-Thuringian Terrane' *sensu* Linnemann & Schauer 1999). This part of the Variscan Orogen represents part of a so-called 'Armorican Terrane Collage' which (from a more stratigraphic viewpoint) is otherwise called Cadomia (Linnemann *et al.* 1999, 2000). Palaeogeographically, this assemblage of comparatively small crustal fragments shows, in part, a strong relationship to the Ediacaran–Early Palaeozoic continent Gondwana, although the microplates in the western part apparently belong to the synchronous crustal segment termed Avalonia.

The question of whether the single Ediacaran–Cambrian crustal fragments that are now amalgamated in Central Europe have a Gondwanan or Avalonian (*sensu stricto*) signature is usually difficult to answer. Where the original depositional history can be reconstructed (as for the Franconian Forest or Lausitz area in Germany or the Barrandian sector in Bohemia), the lithological succession suggests a Gondwanan provenance whereas geochemical data are often interpreted as having Avalonian signatures. The reason for this is that the identification as Avalonian in such cases is understood in a different way from a Cambro-Ordovician stratigraphic and palaeogeographic viewpoint (compare Nance *et al.* 1991; Landing 1996; Dörr *et al.* 2004). From a traditional viewpoint, both the stratigraphy of the depositional environments, and faunal aspects of the Cambrian of the Saxothuringian domain all indicate signatures similar to those of the typical West Gondwanan succession, whereas Cambrian Avalonian-type rock successions and faunas are unknown from Central Europe except for the NW part suggesting that the model of Avalonia as an insular continent during the Cambrian (Landing 1996) is confirmed.

A particularly instructive example with alternative interpretations is the Brunovistulian Terrane. Palaeogeographical aspects of this area were discussed by Fatka & Vavrdová (1998), Moczydłowska (1997, 1998), Belka *et al.* (2000, 2002), Cocks (2002), Vavrdová (2004) and Nawrocki *et al.* (2004*a*, *b*). Moczydłowska (1997, 1998) and Vavrdová (2004) recognized the Brunovistulian Terrane as the distal segment of what is termed 'East Avalonia', while Belka *et al.* (2000, 2002) preferred its peri-Baltic origin. Fatka & Vavrdová (1998) suggested that the Early Cambrian position of the Brunovistulian Terrane was in the area between the recent southern to eastern margins of Baltica, or to the west of Avalonia. Winchester *et al.* (2002) reconstructed the Brunovistulian Terrane as occupying a position between Baltica and Gondwana during the Proterozoic to Early Cambrian, subsequently detaching from Gondwana but still attached to Baltica, and finally moving along the southern margin of Baltica. Cocks (2002) critically summarized palaeontological data and concluded that the faunal evidence is questionable in terms of whether the latter formed part of Baltica or Gondwana, or was a separate and independent terrane. Nawrocki *et al.* (2004*b*) combined palaeontological and palaeomagnetic data and concluded that the Brunovistulian Terrane was separated from peri-Gondwana and transported to its present position near the SW edge of the East European Platform by the Middle Ordovician.

The presence of major volcanic suites with oceanic signatures in most of the Central European Cambrian areas have led to the assumption that these crustal fragments were formed in a sector off the continent Gondwana. This region is usually termed peri-Gondwana. The precise palaeogeography of this sector during the Cambrian is far from being well understood. Africa and the microplates recognized in southern and Central Europe as well as in the Near East, are dealt with in palaeogeographic reconstructions as parts of a jigsaw puzzle. However, this approach does not consider the fact that Cadomian and Cambrian tectonometamorphic and tectonomagmatic events in Central Europe led to a further complication.

Precise palaeogeographic situations for Central Europe during the Cambrian cannot be reconstructed at the present time. However, a distinct line of evolution from the Neoproterozoic to the Ordovician ended with the Ordovician breakup of Gondwana. This development is overprinted by a complex regional evolution that finally led to the Variscan collision and orogeny in western and central Europe as a result of the docking to the enlarged Laurussian continental block. The Cadomian orogenic belt was active during the latest Neoproterozoic to earliest Cambrian at *c.* 540 Ma. Orogenic activity led to deformation and the final consolidation of the African basement and terranes of a Neoproterozoic island arc belt (e.g. Linnemann & Buschmann 1996).

During the Lower and Middle Cambrian, the formation and closure of local, small sedimentary basins on the Cadomian basement took place in the central and southern European areas. In these basins shelf sediments were of such thickness that parts of these successions survived the subsequent Cambro-Ordovician erosion. Strike-slip activity is considered to be responsible for local differences in sedimentological evolution (Linnemann 1996; Linnemann *et al.* 1998). However, other possibilities based on factors such as the progressive migration of Cambrian faunas, suggest that shelf areas were extensive with variable subsidence and resultant differences in sediment accumulation. Uplift and associated chemical weathering during the Late Cambrian resulted in for terrestrial conditions in Central Europe with the formation of a chemically weathered crust, which formed under humid, subtropical conditions (Linnemann *et al.* 1998).

At the Cambrian–Ordovician turnover, the European microplates were located, together with the North African part of western Gondwana, in high southern latitudes. The early Ordovician probably saw the maximum continental dispersal after the breakup of the Neoproterozoic continent of Rodinia, which began at *c.* 750 to 725 Ma.

The Cambrian continent of Baltica is represented by the East European Platform in north, east, and SE Poland and the adjacent Baltic Sea. Although no outcrops exist, the development of depositional environments and temporal and spatial facies distribution is fairly well known due to the numerous boreholes. Two sedimentary basins were active during the Cambrian: the Baltic Basin in the north, and the Lublin-Podlasie Basin in the south, separated by the Mazurian High formed by crystalline basement.

The widely discussed southwestern rim of Baltica is fairly well constrained for the Cambrian. Subsidence and resulting sediment accumulation show significant differences between the sedimentary basin of Upper Silesia and the peri-Tornquist basins in southeastern and northern Poland (Poprawa *et al.* 1999, 2005; Buła 2000; Poprawa & Pacześna 2002). Consequently, the SW margin of Baltica and the Brunovistulicum record different tectonic settings: the Brunovistulian Terrane was controlled by Ediacaran to earliest Cambrian orogenic convergence whereas the SW margin of Baltica records divergent plate margins interpreted as a result of the breakup of Baltica–Pannotia (Poprawa *et al.* 2005).

The Holy Cross Mountains (Góry Świętokrzyskie) in southern Poland appear to represent the suture between Baltica and peri-Gondwana. The facies development and tectonic evolution suggest a subdivision of the Holy Cross Mountains into a (southern)

Kielce and a (northern) Łysogóry region, separated by the Holy Cross Fault (Fig. 4.3). Magnetostratigraphic and biogeographic data from the Furongian suggest a close association with Baltica for the Łysogóry Block (e.g. Nawrocki 2004; Żylińska 2002) although provenance analysis suggest a peri-Gondwanan origin (Valverde-Vaquero *et al.* 2000). In contrast, the depositional history and faunas indicate a peri-Gondwanan character for the Kielce Block (e.g. Żelaźniewicz 1998; Żylińska & Masiak, in press).

West-Central Europe (G.G.)

Cambrian rocks in NW Central Europe are rare and confined to the Subhercynian and Rhenohercynian Zone and the Brabant Massif. All of the Cambrian rocks underwent pronounced metamorphism during Caledonian deformation so that structural complications do not permit the reconstruction of a detailed stratigraphy. The Lower Cambrian generally consists of mainly psammitic rocks, while the Middle and Upper Cambrian (Furongian) are primarily pelitic units with turbiditic intervals, suggesting deposition on an open shelf with a considerable deepening at the beginning of the Middle Cambrian. The lithostratigraphic signature suggests that during the Cambrian this part of Central Europe belonged to the Avalonian microcontinent (Fig. 4.3).

Ardennes

Lower Palaeozoic rocks are found in the Ardennes with Cambrian strata included in four inliers (traditionally termed 'massifs'). All of these underwent moderate Caledonian deformation and are now covered unconformably by Devonian rocks.

Two extended outcrop areas in the Ardennes anticline, the Rocroi Massif near Revin in the western Ardennes and the Stavelot-Venn Massif near Malmédy (Fig. 4.1), probably include a relatively complete sequence of Cambrian rocks. In contrast, only Middle Cambrian to Furongian strata appear to be represented in the small anticlines of Serpont and Givonne (Waterlot 1974; Walther *et al.* 1985) (Fig. 4.3).

The Lower Cambrian consists of the Deville Group, 200 to 400 m of quartz arenites with progressively larger amounts of phyllitic schists towards the top. Acritarchs from the upper Deville Group provide biostratigraphic control (Martin 1975). The Revin Group (Revinien) includes up to 2000 m of black shales and quartz arenites. The psammitic units allow the succession to be subdivided into five lithostratigraphic units. Acritarchs suggest Middle Cambrian and Furongian ages for the Revin Group (Martin 1975; Vanguestaine 1974, 1978, 1992; Ribecai & Vanguestaine 1993; Verniers *et al.* 2001).

All of the Cambrian strata were originally deposited on an open shelf. The basin increased in depth at the beginning of the Middle Cambrian, and the sediments became increasingly pelitic, with fine-grained clastics being regularly interrupted by turbiditic deposits.

Rocroi Massif

In the Rocroi Massif (Figs 4.1, 4.3), the Deville Group is subdivided into the La Longue-Haie and the Les Quatre Fils Aymon formations (Beugnies 1960). The former (Assise de la Longue-Haie) is dominated by cross-bedded quartzites without any known fossils. The Les Quatre Fils Aymon Formation (Assise de Quatre-Fils-Aymon, also 'Les 4 Fils Aymon') is a variable sequence of quartzites with greenish or reddish slates. The arenites often form lenticular bodies and may grade into conglomerates. The thickness varies from 95 to 270 m. The ichnofossil *Oldhamia* was discovered in the lower part of the formation and appears to suggest an early to earliest Cambrian age. However, acritarchs from the upper part of the formation, above the *Oldhamia*-bearing strata, appear to indicate a mid-Early to early Middle Cambrian age (Vanguestaine 1992).

The Revin Group is subdivided into four formations (Beugnies 1960). The 40–120 m thick Rocher de l'Uf Formation (Assise de Rocher de l'Uf) corresponds to the former unit Rv1. It comprises alternating quartzites and slates in the lower part, grading into more shale-dominated strata in the upper part. Acritarchs from the upper part have been dated as Early or Middle Cambrian in age (Vanguestaine 1992). The Roches à Sept Heures Formation (Assise de Roches-à-Sept-Heure, = Rv2) consists of up to 400 m of quartzites with minor slate intercalations. Fine-grained clastics gradually become more prominent up-section. Acritarchs have been found in several layers and indicate a late Middle Cambrian age (Vanguestaine 1992). The Anchamps Formation (Assise de Anchamps, = Rv3) is composed of micaceous black slates at the base, black pyritic quartzites in the middle, and an alternation of black psammitic slates and pyritic quartzites in the upper part. The formation varies between 200 and 600 m in thickness. Acritarchs are found in many layers and indicate a late Middle to Late Cambrian age (Vanguestaine 1992). The Petite Commune Formation (Assise de Petite Commune, formerly Rv4 and Rv52) is made up of *c.* 500 m of black quartzites alternating with micaceous and pyritic slates. Acritarchs from the lowest part of the formation were dated as Middle to Late Cambrian (Vanguestaine 1978), whereas those from the upper part of the formation indicate a Late Cambrian age (Roche *et al.* 1986; Ribecai & Vanguestaine 1993).

Stavelot-Venn Massif

In the Stavelot-Venn Massif, the Deville Group is subdivided into the Hour (up to slightly more than 150 m) and the Bellevaux (150 m in thickness) formations (Geukens 1999). This subdivision reflects the differences between the quartzite-dominated (lower) and shale-dominated (upper) development of the Lower Cambrian. The Belleveaux Formation includes the ichnofossils *Oldhamia antiqua* and *O. radiata* which suggest an early to earliest Cambrian age. However, acritarchs from the upper part of the formation above the *Oldhamia*-bearing strata appear to indicate a mid-Early to early Middle Cambrian age (Vanguestaine 1974, 1992). Therefore, it is believed that the lithologically uniform formation represents two different stratigraphic intervals (Fig. 4.3).

For the Stavelot-Venn Massif, the Revin Group is subdivided into the Wanne (Rv1 and Rv2 of the former nomenclature), La Venne (Rv3 and Rv4) and La Gleize formations (Rv5) (Geukens 1999). The Wanne Formation consists of dark greenish slates with greyish quartzites and rare conglomerates. Acritarchs from the formation are dated as Early? to Middle Cambrian (Vanguestaine 1974, 1992). In addition, inarticulate brachiopods found in the lowermost part of the Revin Group of the Stavelot Massif (Vanguestaine & Rushton 1979) appear to confirm an early Middle Cambrian age for these strata. The La Venne Formation consists of variable clastic rocks of Middle to Late Cambrian age (Ribecai & Vanguestaine 1993). It includes sills and dykes of rhyolitic to rhyodacitic volcanics, the age of which remains unclear. The La Gleize Formation comprises black and silty slates (termed 'quartzophyllade') with occasional calcareous nodules towards the top. A Late Cambrian age for the formation is based on acritarchs (Vanguestaine 1974).

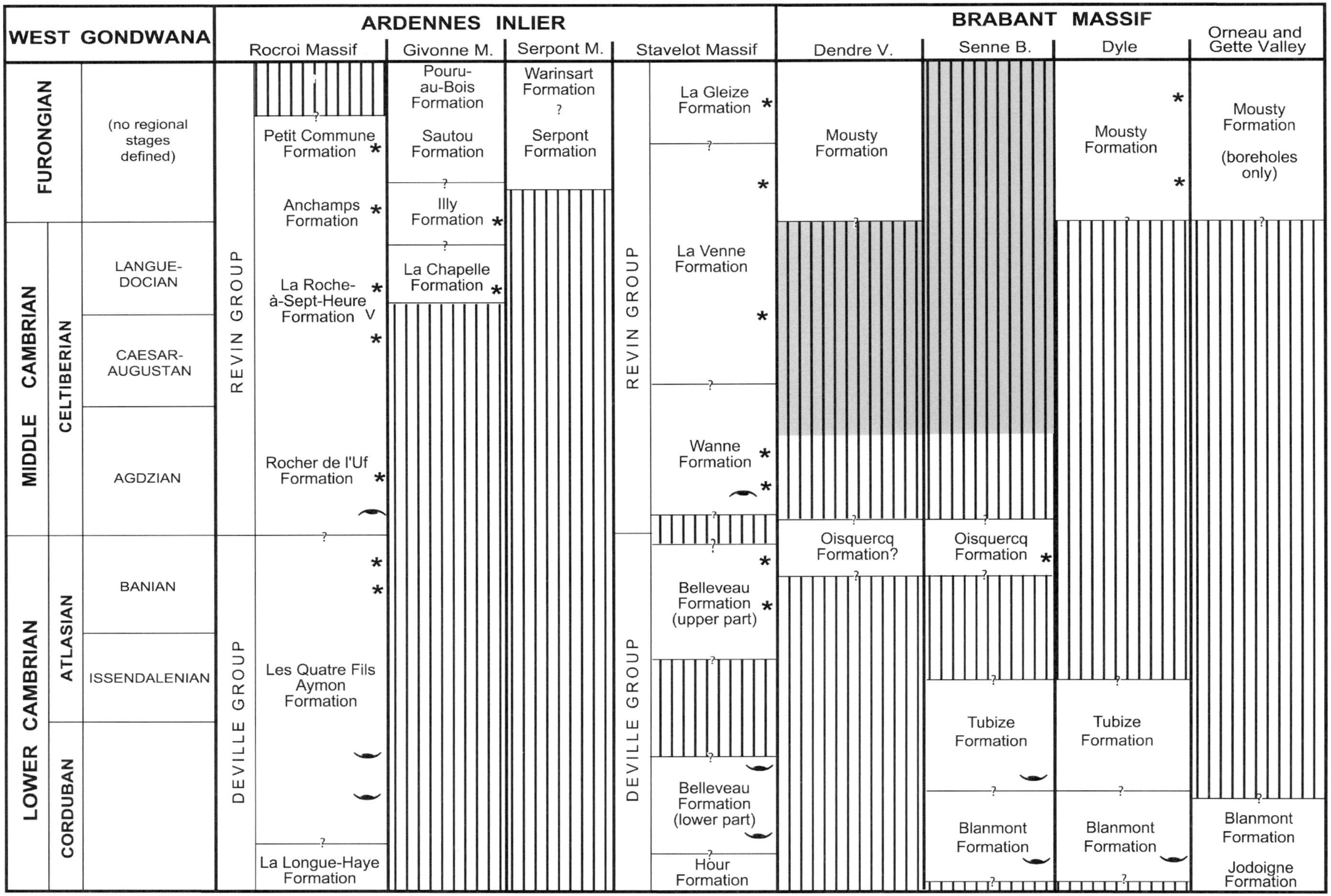

Fig. 4.3. Correlation table with Cambrian lithostratigraphic units of the Ardennes and the Brabant Massif. Vertical hatching, hiatus or missing interval; grey portions, stratigraphic gaps, possibly created by tectonic fault contacts. Asterisks, datable organic microfossils; convex-up symbol, macrofauna; convex-down symbol, trace fossils; V, levels with volcanism. Geochronology according to the West Gondwanan Standard (Geyer & Landing 2004). Modified from Verniers *et al.* (2002).

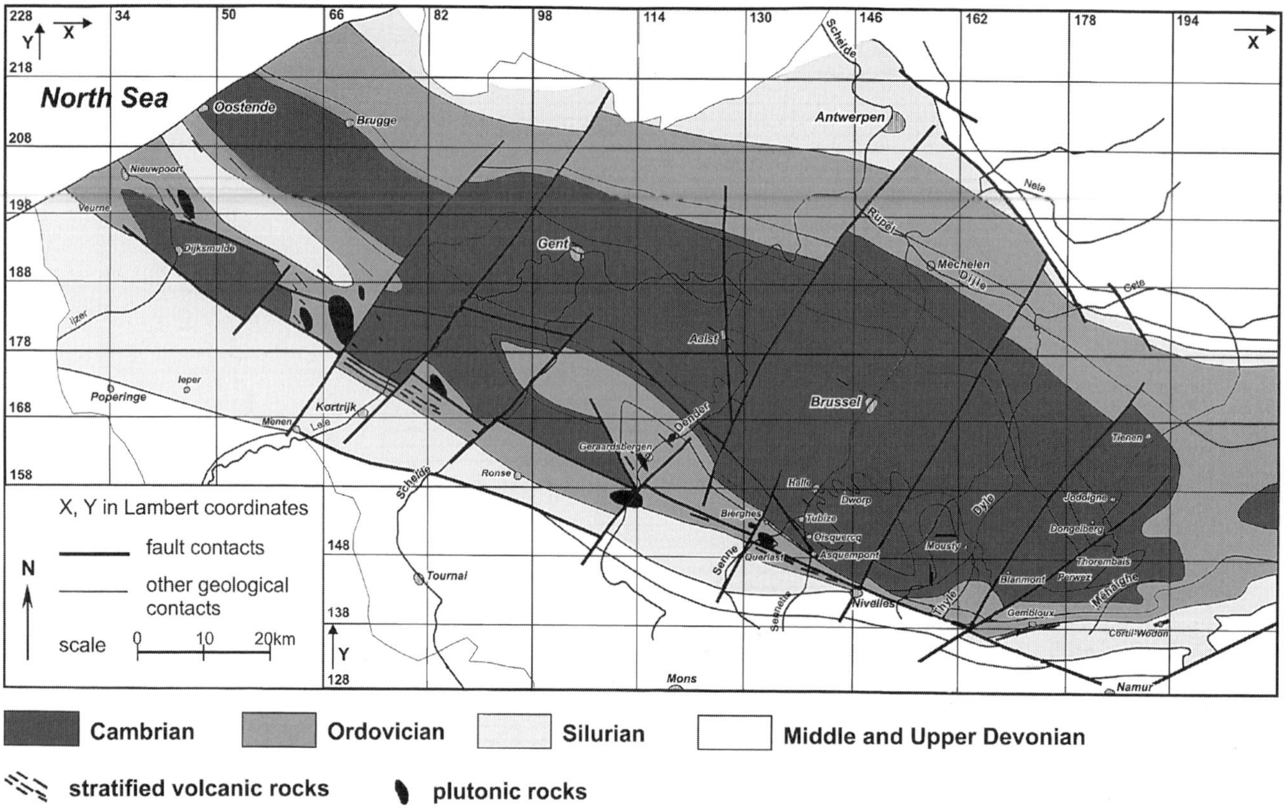

Fig. 4.4. Geological sketch map of the Brabant Massif basement rocks showing the vast area of Cambrian rocks covered by thin late Cenozoic deposits. Slightly modified after De Vos *et al.* (1993) and Herbosch *et al.* (2002, fig. 1).

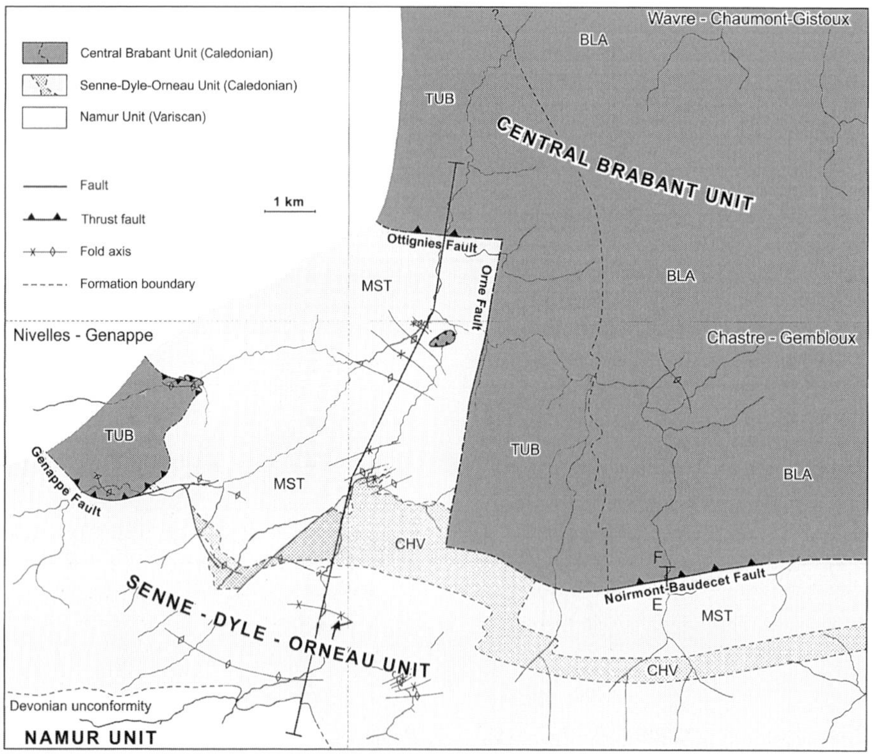

Fig. 4.5. Structural map of the Dyle Valley in the Brabant Massif (compare Fig. 4.4) illustrating the distribution of Cambrian and Ordovician strata. Abbreviations: BLA, Blanmont Formation; CHV, Chevlipont Formation ; MST, Mousty Formation; TUB, Tubize Formation. Asterisk indicates outcrop of the Tubize Formation at Mont-Saint-Guibert (see Fig. 4.6). Slightly modified from Herbosch *et al.* (2002, fig. 10).

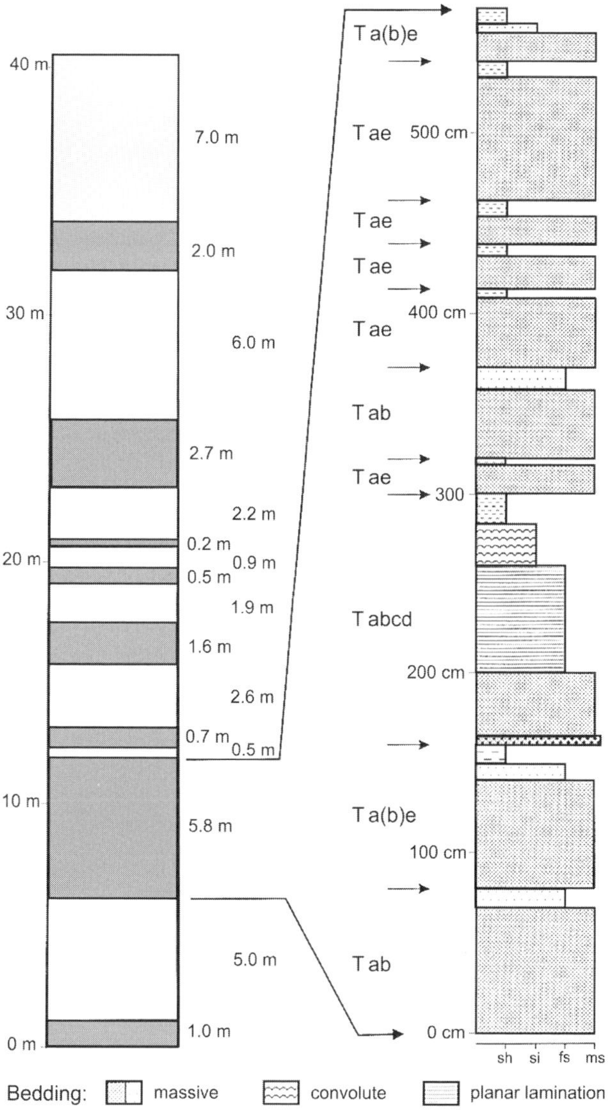

Fig. 4.6. Generalized stratigraphic section of the lower part of the Tubize Formation south of the church of Mont-Saint-Guibert, Brabant Massif (see Fig. 4.5). Left column: grey areas, rocks present in the outcrop; white areas, covered intervals. Right column: detailed lithological log from the section, which exemplifies different types of turbiditic Bouma sequences. Grain-size scale: sh, shale; si, siltstone, fs, fine-grained sandstone; ms, medium-grained sandstone. Redrawn from Herbosch *et al.* (2002, fig. 13).

Brabant Massif

The Brabant Massif between the Ardennes and the West Netherlands Basin is a large NW–SE directed anticline consisting of Palaeozoic basement rocks. Rocks with a high probability of Cambrian age are unknown from outcrops but are recognized from drillholes and underlie the Mesozoic and Cenozoic cover (Figs 4.1, 4.4, 4.5).

The stratigraphy of the area has recently been revised (Verniers *et al.* 2001; Herbosch *et al.* 2002) leading to a slight change in nomenclature. The Lower Cambrian is believed to consist of locally coarse quartz arenites and feldspathic sandstones with a total thicknes of more than 1000 m. The lowermost Cambrian is probably represented by the Blanmont Formation, a unit of variegated quartzitic sandstones with intercalations of slates. The strata contain the ichnofossil *Oldhamia*, which is believed to indicate an earliest Cambrian age. The Assise de Dongelberg (Legrand 1968) is a synonym for the Blanmont Formation.

The overlying Tubize Formation (Assise de Tubize) is a thick (more than 1500 m?) unit of mudstones and siltstones with intercalations of sandstones, some of which show turbiditic Bouma-type sequences (Piessens *et al.* 2004; Fig 4.6). The trace fossil *Oldhamia* is again found in some layers. The stratigraphic assignment as earliest Cambrian is based on similarities with the rocks of the Deville Group of the Ardennes and on the frequent occurrence of *Oldhamia*. Provenance analysis indicates that the rocks were derived from a nearby crystalline area with Cadomian and pre-Cadomian characteristics (Waterlot 1974; Bless *et al.* 1980; Walther 1980).

The Oisquercq Formation (Assise d'Oisquerq) generally consists of variable greenish and blackish slates. Acritarchs in its upper Asquempont Member indicate a latest Early to early Middle Cambrian age (Vanguestaine 1992). The > 500 m of bluish or dark grey shales or slates of the Mousty Formation (Assise de Mousty) are correlated with the Middle Cambrian to Furongian Revin Group of the Ardennes. Acritarchs from the top of the formation indicate a Middle or Late Cambrian age (Martin 1975; Vanguestaine 1992), but the associated graptolite *Rhabdinopora* indicates an earliest Tremadocian age.

The rock succession from the Brabant Massif area also indicates a development from shallower to deeper shelf environments during the Cambrian. Such a pattern is known from other regions of Avalonia (Walther 1980). However, the age and spatial relation to the Ardennes region are, at least for the upper half of the Cambrian, consistent with the acritarch biostratigraphy and lateral variations in facies (Vanguestaine 1991; Herbosch *et al.* 1991).

Mid-German Crystalline High (G.G.)

From Ediacaran–earliest Cambrian to Ordovician times a magmatic arc developed in Central Europe south of the Mid-German Crystalline High (Mitteldeutsche Kristallinschwelle). This arc, now recognizable over more than 800 km from present-day eastern France to western Poland, involved Cambro-Ordovician bimodal volcanism and related basic intrusions dated at 500 Ma (Bankwitz *et al.* 1994). Furongian-age volcanic complexes are known from the northern Vosges, the northern Thuringian Forest mountains (Vesser and Ilmenau areas), the Delitzsch areas (north of Leipzig) and from the Bober-Katzbach mountains (Góry Kaczawskie) (Bankwitz *et al.* 1994; Bankwitz & Bankwitz 1997). This zone is a metallogenic belt created by exhalative volcanic activity. The belt comprises volcanics, tuffs, pyroclastics, volcanic conglomerates and sediments with occasional magnetite ores, and marks a plate boundary (Bankwitz & Plotnikov 1994), probably representing the rim of the peri-Gondwanan microplate swarm. The rocks are now characterized by a greenschist metamorphic facies with intensive faulting and often blastomylonitic deformation.

Apart from the igneous rocks, areas of the Mid-German Crystalline High exclusively include moderately to heavily metamorphosed Cambrian rocks, the precise age of which is largely unknown. Major areas with supposed Cambrian rock units include the Odenwald region, the Spessart, and the Ruhla Crystalline Complex.

In the **Böllstein Odenwald**, the Cambrian is thought to be represented within a variable suite of mica schists with intercalations of calc-silicates, graphitic quartzites and amphibolites. The metamorphosed rocks were primarily shales and greywackes with minor volcanics, and range in age from latest Proterozoic to Silurian.

The crystalline basement of the **Spessart** includes various, easily distinguishable metamorphic and magmatic units deformed during the Variscan Orogeny. The present interpretation of the complex suggests a (?Early) Cambrian age for the Alzenau and the Elterhof formations (Hirschmann & Okrusch 1988). The Alzenau Formation (Amphibolit-Paragneis Serie) consists mainly of amphibolites and biotite and hornblende gneisses with minor graphitic schists and graphitic quartzites. The protoliths of the formation were volcanic and sedimentary rocks. The amphibolites were derived from metamorphosed basalts and tuffs of apparent oceanic origin (either ocean floor or island arcs tholeiitic basalts; Okrusch *et al.* 1985). The Elterhof Formation (Gestreifte Paragneisserie mit Marmor- und Amphiboliteinschaltugen or körnig-streifige Paragneisserie) is petrographically variable. However, it includes stratiform dolomitic marbles forming distinct bands up to 20 m in thickness, although alternating amphibolites, pyroxene-hornblende gneisses, acidic gneisses, and graphitic paragneisses constitute most of this thick formation. The chemical signature of the amphibolites indicates volcanic rocks of island arc origin. The age interpretation as Early Cambrian is based mainly on the comparison with similarly variable units in eastern Bavaria, Bohemia and the Sudetic Mountains. It is thus highly hypothetical but fits best with our present understanding of the geodynamic scenario during the latest Neoproterozoic–Ordovician periods.

In the **Ruhla Crystalline Complex**, the extensive Ruhla Formation is probably, in part, of Cambrian age. The relevant portions consist of mica schists with stratiform metabasites, some quartzites, and graphitic schists. Deformation and amphibolite-facies metamorphism occurred during early Devonian times.

Vesser Zone

The Vesser area in NW Thuringia is now interpreted as a marginal element of the Saxothuringian Zone. It includes a metamorphic, multiply deformed, and possibly rift-related (Bankwitz *et al.* 1994) succession of bimodal extrusives, siliciclastics and minor carbonates (Bankwitz & Bankwitz 1997). The thickness of this unit is at least 1400 m. Fossils are unknown. Kemnitz *et al.* (2002) published radiometric U—Pb zircon ages (501–517 Ma) from the rift-related magmatic rocks. These ages suggest that the rocks are an equivalent to the Late Cambrian–Early Ordovician overlap sequence on the Cadomian basement as identified in other regions of the Saxothuringian Zone.

The geochemical and petrological characteristics of the Vesser Zone units are unique for the Saxothuringian Zone and suggest that this area is a relict of the Rheic Ocean (Schäfer *et al.* 1998) related to the separation of Avalonia from Gondwana (Kemnitz *et al.* 1998). The Vesser Zone is thus considered to represent an Armorican fragment (Schäfer *et al.* 1998).

Saxothuringian Zone

Lusatia and NW Saxony (O.E.)

Lusatia and NW Saxony in eastern Germany represent structural elements of the Saxothuringian Zone (Saxo-Thuringian Terrane *sensu* Linnemann & Schauer 1999) (Fig. 4.7). This area, in particular, well illustrates the tectonostratigraphic history of the domain. The Cadomian Orogeny along the northern margin of Gondwana led to deformation just prior to the onset of the Cambrian (*c.* 550 Ma; Linnemann & Buschmann 1995*a*, *b*). The collision of continental and oceanic microplates created a number of crustal slices, or terranes. Lower and Middle Cambrian sediments were deposited on extended areas of the Cadomian basement (rather than within local basins). Other areas, such as the Cadomian basement of the Schwarzburg Anticline, were subject to subaerial exposure and chemical weathering (Linnemann & Buschmann 1995*b*). Weathered rocks of this crustal segment were locally reworked by fluvial processes suggesting a relative sea-level lowstand (Linnemann 1996). A general transgression, leading to submergence of the entire Cadomian basement, commenced in the Tremadocian (Linnemann & Buschmann 1995*b*).

Two Cambrian-bearing synclines flank the Lusatian Block to the east (the Görlitz Syncline) and NW (the Delitzsch-Torgau-Doberlug Syncline). Both of these areas represent parts of the same depositional region, now separated by younger tectonic elements. Cambrian deposits are known from the Görlitz and the Delitzsch-Torgau-Doberlug synclines from (1) rare surface outcrops within a small and fault-bounded area near the German-Polish border east of Dresden (Görlitz Syncline) and from (2) subsurface occurrences in the vicinity of Leipzig (Delitzsch-Torgau-Doberlug Syncline) (Fig. 4.8). Sedimentary rocks of possible Late Cambrian age are known from the northern Saxonian Slate Mountains (Sächsisches Schiefergebirge; Elicki 1997, 2003).

The structural relationship between the Cadomian-consolidated basement (the Lusatian Block) and the sedimentary successions within the marginal synclines is not fully understood. The Görlitz Syncline contains fragmentary and isolated relicts of stratigraphically narrow intervals (Göthel 2001), whereas the Delitzsch-Torgau-Doberlug Syncline includes a well-preserved early Palaeozoic tectonostratigraphic suite (Jonas *et al.* 2000). Interpretation of these sedimentary bodies as olistoliths, which are now dislocated within a Variscan flysch matrix, have been widely discussed and accepted over the last decades (e.g. Franke 1984; Linnemann & Buschmann 1995*b*; Linnemann & Schauer 1999). Nevertheless, for the Cambrian deposits transgressive sedimentation (overlap sequences) following the Cadomian unconformity (including a period of late Neoproterozoic–earliest Cambrian denudation) is an alternative interpretation (Buschmann *et al.* 1995; Göthel 2001). For the Cambrian, the general characteristics of the Saxothuringian Zone are: (1) the absence of earliest Cambrian strata; (2) the deposition of later Early to Middle Cambrian sediments, possibly on a morphologically differentiated western Gondwanan shelf area or in connected pull-apart basins; and (3) an intensive uplift-related denudation during the Furongian (Linnemann *et al.* 2000; Elicki & Wotte 2003).

Görlitz Syncline

The Cambrian succession of the Görlitz Syncline (Fig. 4.9) is incomplete and represents only a short stratigraphic episode, which is lithostratigraphically known as the Charlottenhof Formation. The thickness of the entire Cambrian succession is *c.* 250 to 300 m. Due to tectonic complications, the overlying and underlying units are unknown. The Lower Cambrian sequence is at least partly overthrusted on the Upper Carboniferous (Westphalian B) of the Northern Sudetic Basin.

The Cambrian suite is characterized by shallow-marine massive dolostones, overlying bedded (and nodular near the top) limestones (Ludwigsdorf Member) and siliciclastics at the top (*Lusatiops* Member). Indications of basic magmatism (diabases

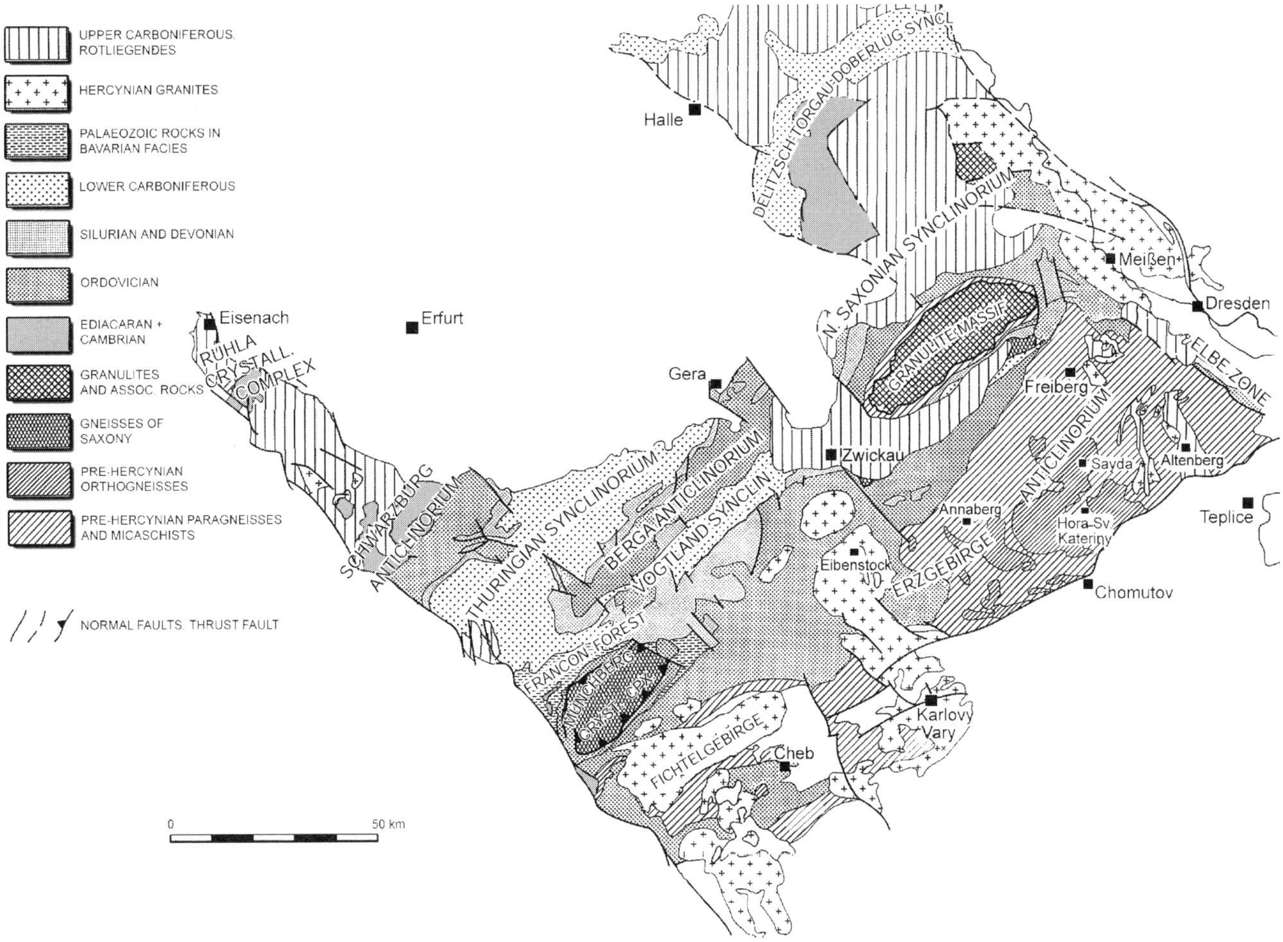

Fig. 4.7. Geological sketch map of the Saxothuringian and NE Bavarian–West Bohemian basement. The relatively limited areas with Cambrian rocks are marked in grey (areas in the Franconian Forest and the Schwarzburg Synclinorium slightly exaggerated). Modified after Walther (1992, fig. 89).

and diabase tuffs) were reported from drillcores by Hirschmann (1967), but rejected as equivalents for the siliciclastics and for the carbonates by Göthel & Elicki (1997).

The deepest part of the Cambrian succession (lower Ludwigsdorf Member) is characterized by massive and coarse dolostones with a maximum thickness of about 100 m. Because of a strong diagenetic overprint the coarse-grained dolostones do not show any biosedimentary patterns or biotic remains. Former reports and interpretations of single problematic remains (as archaeocyaths by Schwarzbach (1934) or 'stromatolites' by N. Hermsdorf, pers. comm.) are unsustainable after critical reevaluation (Elicki 1997).

This lower Ludwigsdorf Member is overlain by bedded and partly nodular limestones with (thin) claystone intercalations (upper Ludwigsdorf Member, up to 80 m). Sedimentary facies patterns allow a disctinction between: (1) well-bedded and more or less siliciclastic-influenced, bioclastic wackestone, packstone and floatstone (nodular near the top of the member) with phosphatic black pebbles, load casts, wave ripples, cross-bedding, and small channels (Elicki & Schneider 1992), which indicate a shallow marine environment. Some horizons of this facies are rich in shelly fossils such as poriferid spicules, hyoliths, echinoderms, chancelloriids, bivalves, gastropods, brachiopods, trilobites, cyanobacteria and other calcimicrobes, and shelly fossils of problematic systematic position (Rozanov *et al.* 1969; Elicki 1994, 1996, 1998; Elicki & Schneider 1992; Geyer & Elicki 1995). These organisms indicate a late Early Cambrian age (higher Banian (Atlasian) *sensu* Geyer & Landing 2004). The second facies type consists of regular alternations of 'Zebra limestone' and dolomitic limestone. The 'Zebra limestone' includes slumping structures and reworking as well as indications of tepee structures, and contains sulphates. The only fossil remains are rare and poorly preserved (allochthonous) calcimicrobial colonies as well as unidentifiable shelly remains reported from a single horizon at the base (Elicki 1992). Deposition took place in a restricted lagoonal environment (Elicki & Schneider 1992).

The overlying *Lusatiops* Member (90 to 120 m) is represented by claystones and siltstones with minor sandstone intercalations. The fossil content consists of trilobites (e.g. *Lusatiops, Serrodiscus*), brachiopods and hyoliths (Richter & Richter 1924; Freyer 1981*a*; Geyer & Elicki 1995), which indicates an insignificantly younger age than that of the upper Ludwigsdorf Member. Taphonomic considerations suggest a quieter and deeper depositional environment (Geyer & Elicki 1995).

Delitzsch-Torgau-Doberlug Syncline

The Cambrian rocks of the Delitzsch-Torgau-Doberlug Syncline (Fig. 4.2) do not represent a continuous succession (Fig. 4.10). Structural complications and facies changes suggest an estimated thickness of *c.* 1500 to 2000 m (Brause 1969), with the Lower and Middle Cambrian units being of similar thickness.

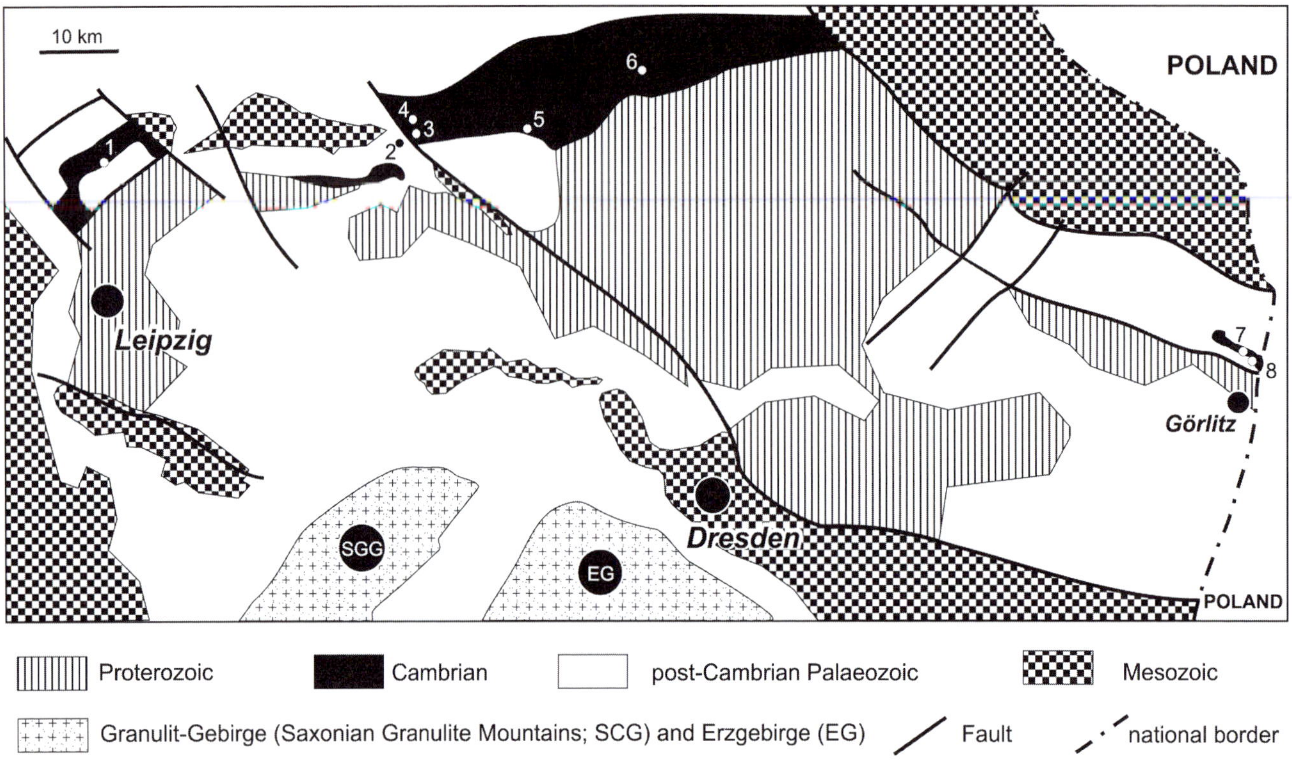

Fig. 4.8. Geological sketch map of the eastern part of the Saxothuringian Zone illustrating Cambrian outcrops and subsurface occurrences in Saxonia and adjacent areas. Important Cambrian localities: 1, Delitzsch; 2, Torgau; 3, Zwethau; 4, Rosenfeld; 5, Tröbitz; 6, Doberlug-Kirchhain; 7, Charlottenhof; 8, Ludwigsdorf.

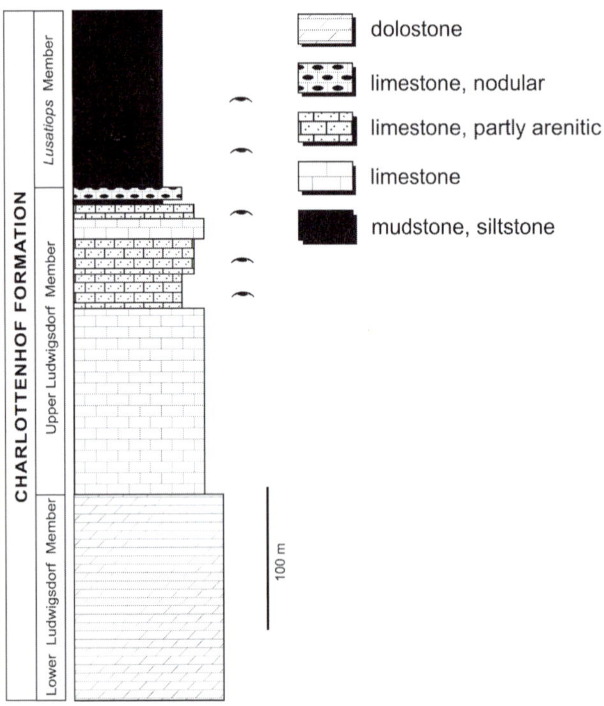

Fig. 4.9. Generalized lithological and lithostratigraphical section of the Lower Cambrian rocks in the Ludwigsdorf area, Lusatia. Lenticular symbols indicate stratigraphic position of trilobite occurrences.

Cambrian sedimentation commenced following a stratigraphic and structural gap (i.e. Cadomian unconformity) with localized conglomerates (debris flows), followed by Lower Cambrian shallow marine carbonates and minor siliciclastics (Zwethau Formation, Falkenberg Group) (Freyer & Suhr 1987; Elicki 1992, 1999*a*, *b*; Buschmann *et al.* 1995).

The Lower Cambrian Zwethau Formation consists of shallow-marine carbonates and siliciclastics (Fig. 4.11). The lower part of the succession (Torgau Member) predominantly consists of carbonates, whereas the upper part (Rosenfeld Member) comprises alternations of siliciclastic–carbonate rocks and pure siliciclastics. Intrusive volcanics (diabases) rarely occur. The thickness of the Zwethau Formation is estimated to be more than 800 m. The Torgau Member consists of fossiliferous, and partly oolitic or intraclastic, limestones and dolostones. Calcimicrobial biogenetic carbonates are common. Widespread fossils include archaeocyaths, cyanobacteria and other calcimicrobes (Freyer & Suhr 1987, 1992; Elicki & Debrenne 1993; Elicki 1999*a*), although skeletal fossils (the trilobite *Dolerolichia*, chancelloriids, and small shelly fossils) are also found (Sdzuy 1962; Elicki 1994). Remarkable, too, is the occurrence of calcimicrobial-archaeocyathan build-ups (Wotte 2004). Small-scale ripples, cross-bedding, bioturbation, mudcracks and local evaporites indicate a shallow subtidal to intertidal ramp environment with shoals and lagoons (Elicki 1992, 1999*a*). The thickness of the Torgau Member is up to >500 m. Its age, indicated by archaeocyaths, is mid-Early Cambrian (middle Issendalenian (Atlasian) *sensu* Geyer & Landing 2004).

The Rosenfeld Member comprises alternating limestones,

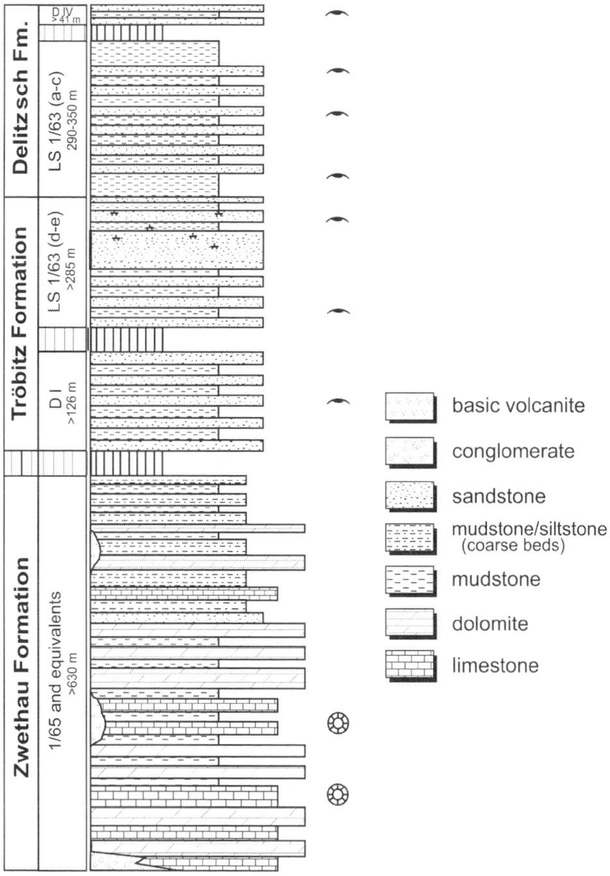

Fig. 4.10. Generalized composite lithological and lithostratigraphical section of the Lower and Middle Cambrian in the Delitzsch-Torgau-Doberlug Syncline. Lenticular symbol, stratigraphic position of trilobite; divided circle symbol, archaeocyath occurrence.

dolostones and siliciclastics with some diabases in the upper part. Freyer & Suhr (1987) indicated that, in contrast to the underlying Torgau Member, the amount of siliciclastics is much higher and redeposition patterns can often be observed. The thickness of the member is up to 280 m. Fossils of this member include redeposited acritarchs, cyanobacteria and shelly remains.

The Middle Cambrian (Arenzhain Group, with Tröbitz and Delitzsch formations) is represented by siliciclastics with rare carbonate intercalations (Brause 1969, 1970; Elicki 1997). Unfortunately, the transition from Lower to Middle Cambrian is unknown, leading to the assumption that a significant stratigraphic gap exists between the Lower and Middle Cambrian formations in this area.

The Tröbitz Formation consists of more than 400 m of quartzitic sandstones alternating with micaceous claystones. Locally, thin calcareous layers were observed at the transition to the overlying Delitzsch Formation. The depositional environment is interpreted as a low-energy siliciclastic shelf. The age is indicated by trilobites (*Paradoxides, Condylopyge, Ornamentaspis*) as early Middle Cambrian (*'Paradoxides' insularis* biozone) (Sdzuy 1957*a*, *b*, 1958, 1970, 1972), corresponding to the middle to upper Agdzian (Celtiberian) *sensu* Geyer & Landing (2004). Additional fossils include hyoliths, brachiopods and trace fossils. The latter are concentrated in the upper part of the formation.

These classical assemblages are now supplemented by a recently identified assemblage with trilobites such as *Protolenus (Hupeolenus)* aff. *termierelloides*, *Cambrunicornia* n. sp., as well as inarticulate brachiopods (*Trematobolus* sp., unidentified acrotretoid; B. Buschmann, O. Elicki, and G. Geyer, unpubl. data, 2006). This fauna features the lowest Middle Cambrian and represents the oldest known Middle Cambrian from the Lusatian/Saxonian segment of the Saxothuringian. The strata, identified from a drillcore west of Doberlug-Kirchhain, are in a facies somewhat different from that of the typical Tröbitz Formation.

The Delitzsch Formation consists of more than 350 m of mainly quartzitic sandstones alternating with micaceous claystones. In contrast to the Tröbitz Formation, the claystone is mostly greenish and has a higher mica content. Cross-bedding and abundant trace fossils were observed in the upper part of small-scale (up to 10 cm) sedimentary cycles within the formation. Some thin calcareous horizons occur in the middle part of the formation. Deposition is again interpreted to have taken place on a quiet siliciclastic shelf, but now with occasional periods of higher energy. According to the trilobites (e.g. *Paradoxides, Condylopyge, Ellipsocephalus, Bailiella, Solenopleura, Badulesia*) the age is early to mid-Middle Cambrian (*'Paradoxides' insularis* to lowest *Paradoxides paradoxissimus* biozone) *sensu* Sdzuy (1957*a*, *b*, 1958, 1970) (higher Agdzian to early Caesaraugustan (Celtiberian) *sensu* Geyer & Landing 2004). Additional fossils from the formation include hyoliths, brachiopods, echinoderms and helcionellids. The Cambrian strata of the Delitzsch-Torgau-Doberlug Syncline are unconformably overlain by Mesozoic or Cenozoic rocks.

Northern Saxonian Slate Mountains

The Northern Saxonian Slate Mountains (Nordsächsisches Schiefergebirge) are situated *c.* 30 km south of Leipzig (Fig. 4.7). In the vicinity of Borna-Hainichen, a unit comprising alternating quartzitic to conglomeratic greywackes and quartzitic sandstones with minor claystone and siltstone intercalations is termed the Collmberg Formation (Pietzsch 1910). The thickness of this formation is hard to estimate because of the complex regional deformation. However, it is assumed to be at least 1000 m (Berger & Docekal 1997: 341–342). From some horizons poorly preserved inarticulate brachiopods and some ichnofossils were recovered. Based on the trace fossil *Cruziana semiplicata*, a Late Cambrian to Early Ordovician age has been assumed for the possibly shallow marine siliciclastic shelf deposits (Freyer 1981b; Elicki 1997). The lower as well as the upper boundary of the formation are unknown.

Franconian-Thuringian Slate Mountains

The Thuringian-Franconian Slate Mountains (Thüringisch-Fränkisches Schiefergebirge) are situated in the east-central part of Germany and mainly consist of rocks of upper Proterozoic to Dinantian age within incomplete successions. Cambrian and probable Cambrian rocks are known from subunits in the eastern (Berga Anticline) and western Schwarzburg Anticline regions of the Thuringian sector and in NE Bavaria (Franconian Forest, or Frankenwald). In contrast, relict deposits of Cambro-Ordovician age have been identified in outcrops south of Leipzig.

Berga Anticline (O.E.)

Two old drillings from the core of the Berga Anticline ('Bergaer Sattel'; Fig. 4.7) near the village of Heinersdorf penetrated Ordovician rocks (local equivalents of the Tremadocian Frauenbach Formation) and reached a 'Limestone Member' and an underlying, non-penetrated 'Quartzitic Shale Member' (Wucher

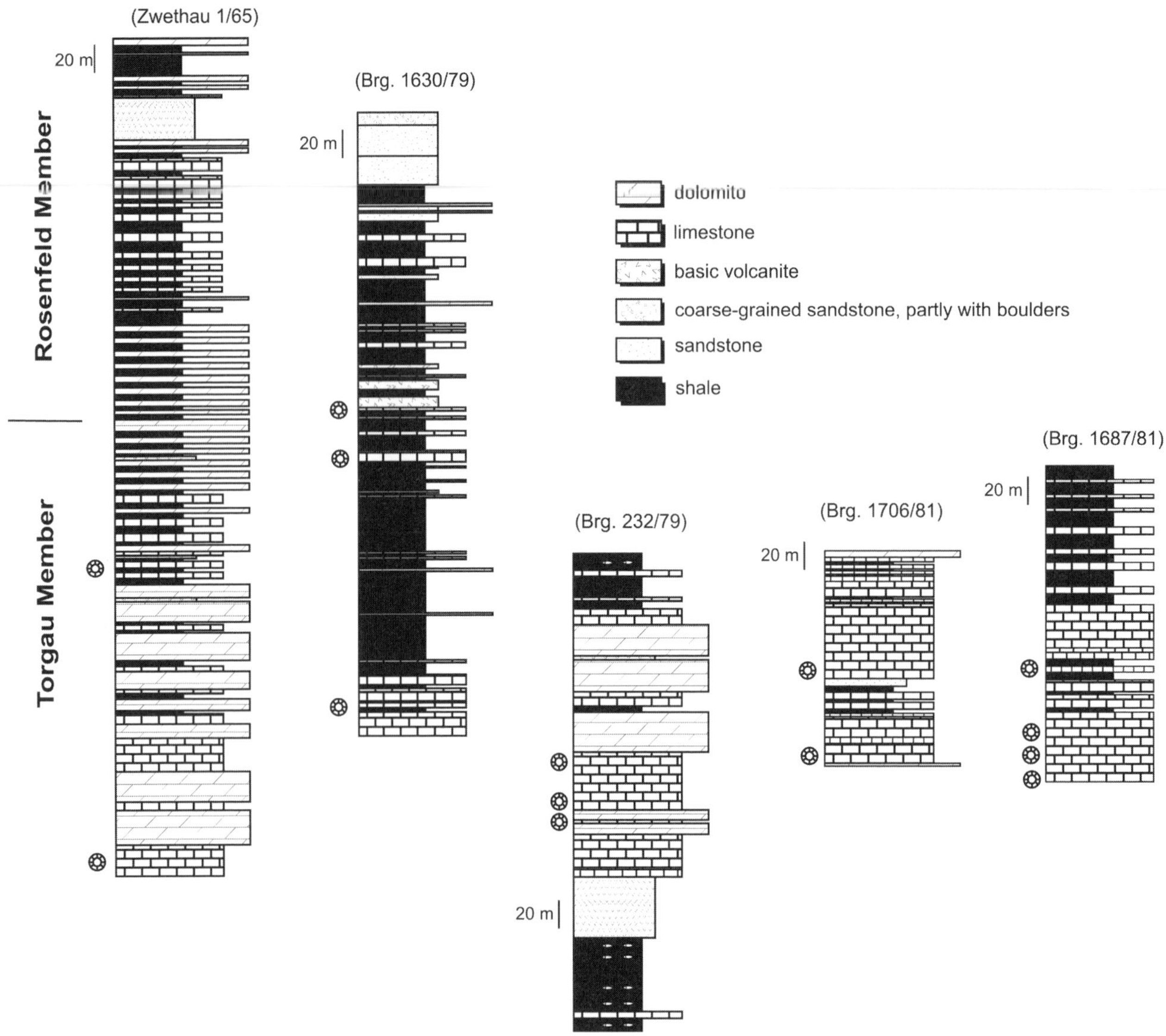

Fig. 4.11. Lithological sections of Lower and Middle Cambrian rocks from borings in the Delitzsch-Torgau-Doberlug Syncline and their tentative correlation. Members of the Zwethau Formation are only clearly recognizable in the cores of the Zwethau 1/65 borehole.

1965; Blumenstengel 1980). Both low-grade metamorphic members are unified into the Heinersdorf Group.

The Quartzitic Shale Member is at least 70 m thick and consists of alternating siltstones, shales, arkoses, and feldspar-bearing sandstones with fining-upward cycles. The age can only be estimated by the lithostratigraphic context. The so-called 'Neumühle Member' (a greenschist-grade metamorphic unit with a comparable petrographic signature and a thickness of more than 200 m) identified from about 50 km to the NE may be a lateral equivalent of the Quartzitic Shale Member.

The overlying Limestone Member (350–400 m thick) shows a gradual transition from the Quartzitic Shale Member below. It is developed as an alternation of claystones and carbonate beds. The carbonate intercalations include calcareous shales, flaser-bedded carbonates, dolostones and partly sapropelitic rocks. The thickest carbonate beds are 10–15 cm. From these layers, Blumenstengel (1980) described poorly preserved Cambrian chancelloriids, hyoliths and stromatolites. Regional similarities with well-dated deposits suggest a late Early to early Middle Cambrian age (Elicki 1997). A deeper or outer shelf environment is likely.

Schwarzburg Anticline (O.E.)

The Goldisthal Formation in the Schwarzburg Anticline (Fig. 4.7) is traditionally assumed to be of Cambrian age. It is a siliciclastic succession *c.* 200 to 300 m thick (locally probably up to 700 m). Petrographically, the unit consists of claystone/siltstone slates with some intercalated effusive volcanics (tuffs, pyroclastics) (Bankwitz & Bankwitz 1975; Bankwitz *et al.* 1997). Locally, quartzites and conglomerates occur at its base (Bankwitz 1970). Particularly in the upper portion, fining-upward cycles and hummocky cross-stratification are widely found. Several strata are rich in simple trace fossils. The lithological boundaries with the underlying Neoproterozoic greywackes of the Upper Katzhütte Formation (or the Frohnberg Group) and the overlying Tremadocian Frauenbach Formation (quartzites) are abrupt.

A Cambrian age was (and, in part, still is) assumed based on lithostratigraphic data. The suite of siliciclastic rocks has been interpreted as a continuous succession from the Neoproterozoic into the Lower Ordovician (Bankwitz 1970). However, new investigations on acritarchs (Estrada *et al.* 1994; Heuse *et al.* 1994) and on regional aspects (Linnemann & Buschmann 1995*b*) indicate that the formation includes either entirely Lower Ordovician strata or may, at best, extend into the Furongian with its lowest beds. Deposition took place on a siliciclastic shelf below storm wave base (for the lower part of the succession) and at or above the storm wave base (for the upper portion) (Lützner 1991; Bankwitz *et al.* 1997: 42–43).

Franconian Forest (Frankenwald) (G.G.)

The Franconian Forest area (Frankenwald; Figs. 4.7 & 4.12) contains a few Cambrian outcrops of very limited extent. Almost all of the Cambrian localities are fault-bounded and usually do not allow details of the rock successions to be recognized. However, the outcrops clearly represent a number of different formations, which are usually fossiliferous and present a relatively complete series through the Middle Cambrian succession (Geyer & Wiefel 1997; Fig. 4.13). The Lower Cambrian is not recognized with any certainty, although its upper part is supposedly represented by the Tiefenbach Formation. Ages of older rock units in the area have not yet been determined. Furongian (Late/Upper Cambrian) rocks are assumed to be absent due to a phase of both non-deposition and erosion at the end of the Cambrian. Upper Middle Cambrian rocks are hence unconformably overlain by the Tremadocian Leimitz Formation.

For most of the Palaeozoic, two distinctly different facies developments are recognizable in the area, termed the Bavarian and the Thuringian facies series. For Cambrian rocks, such a distinction cannot be recognized although the Cambrian localities are almost entirely confined to an area that includes nearly exclusively rocks of the Bavarian Facies series.

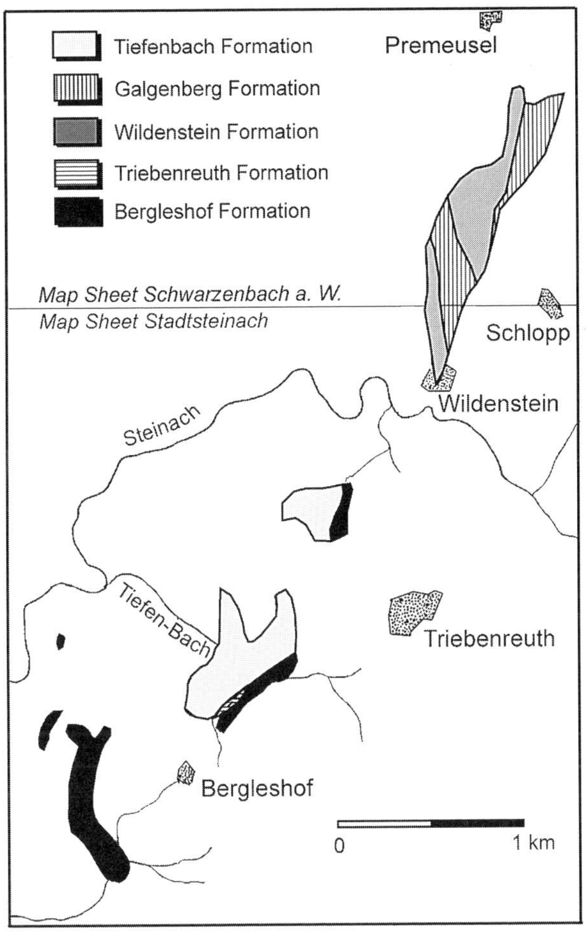

Fig. 4.12. Sketch map of the distribution of Cambrian rocks in the Wildenstein area, Franconian Forest, representing the vast majority of Cambrian outcrops in the Franconian Forest. Redrawn and modified from the geological map sheets 1:25 000 Stadtsteinach and Schwarzenbach am Wald and Sdzuy (1964, fig. 1).

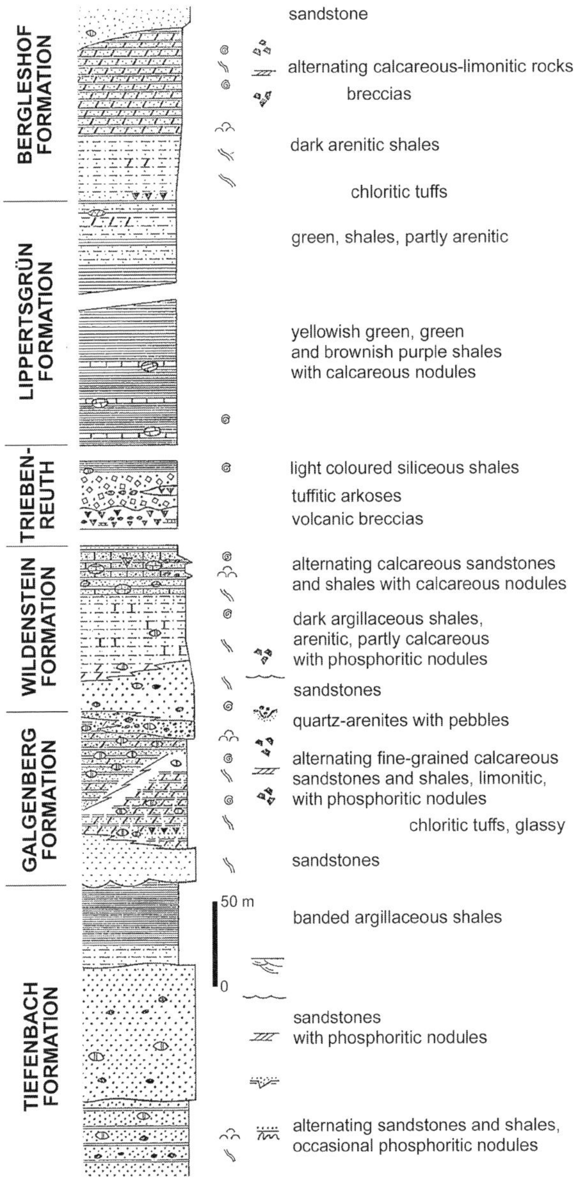

Fig. 4.13. Composite section of the Cambrian rocks in the Franconian Forest. Left-hand column of symbols indicates occurrence of body fossils, shell fragments and ichnofossils; right-hand column of symbols indicates sedimentary features such as brecciated clasts, erosional surfaces, cross-bedding and soft deformation. Modified from Ludwig (1969, fig. 1).

The Cambrian formations overlie, apparently disconformably, different lithostratigraphic units, which are attributed, without any strong arguments, to the Proterozoic. The Lower Cambrian (?) to lowermost Middle Cambrian Tiefenbach Formation is generally an alternation of quartz arenites and shales. The variable lithologies range from coarse- or medium-grained grey quartz arenites and greywackes; dark grey to blackish quartzitic siltstones; and dark grey to black or greyish green shales, often banded or flaser-bedded and micaceous, and with occasional bioturbation. Due to tectonic deformation, the thickness of the formation is speculative. Echinoderm debris, trilobite and brachiopod shell hash are reported from some hardgrounds of the lower part of the formation (Ludwig 1969). The same member may be extensively bioturbated, and vertical burrows of the *Skolithos*-type are frequent.

The Galgenberg Formation (Wurm 1924, 1925*a*, *b*; Sdzuy 1964) is dominated by dark grey to blackish grey medium-grained and medium grey, calcareous fine-grained sandstone. Subordinate are yellowish, greywacke-type calcareous sandstones, which form lenses at various levels. Synsedimentary breccias include phosphoritic nodules and grade into layers with intraclasts. Bioturbation and shell hash layers, as well as numerous sedimentary structures (with flaser-bedding, small-scale cross-bedding, scour surfaces, channel fills, and synsedimentary slumping) occur. The thickness of the Galgenberg Formation is unknown, but presumably exceeds 100 m. The early Middle Cambrian (= early Celtiberian) age is proved by a fauna in the fine-grained, calcareous sandstones, which is rich in individuals but relatively poorly diverse. Trilobites dominate this fauna and include *Ornamentaspis*, *Kingaspidoides*, *Latikingaspis*, *Parasolenopleura*, *Bailiella*, *Paradoxides*, *Acanthomicmacca* and *Wurmaspis*. These genera indicate an age of *Cephalopyge* and lower *Ornamentaspis frequens* zones (Agdzian Stage of the West Gondwanan standard). Echinoderms are frequent. Additional fossils include brachiopods and hyoliths. Trace fossils are relatively frequent.

The lower Middle Cambrian (= early Celtiberian) Wildenstein Formation (Wurm 1924, 1925*a*, *b*; Sdzuy 1964) is dominated by light grey to greenish grey greywackes and fine- to medium-grained sandstones which are frequently calcareous, and partly calcareous shales. Intercalated are calcareous nodules up to decimetre-size and, more rarely, quartz arenitic horizons. The thickness is probably in the order of 100 to 200 m. Fairly rich trilobite assemblages indicate an age of the upper *Ornamentaspis frequens* to lower *Kymataspis* zones (Agdzian Stage). These trilobites include *Paradoxides* s.l., *Condylopyge*, *Bailiella*, *Parasolenopleura*, *Acanthomicmacca*, *Ornamentaspis*, *Kingaspidoides* and *Dawsonia*. In addition, brachiopods, molluscs, sponge spicules, echinoderm debris, and the problematic *Oxyprymna schloppensis* are also found. Trace fossils are present throughout.

The Triebenreuth Formation (Gaertner *et al.* 1968; Ludwig 1969; Geyer & Wiefel 1997) is known only from two localities and is of unknown thickness. This formation consists of a volcanogenic to siliciclastic succession, and presumably includes a range of volcanogenic arkoses, shales and volcanic breccias. The succession is dominated by grey to greenish shales in the upper part and coarser-grained sandstones with subordinate poorly sorted quartz arenites. The geological situation suggests that this succession is dovetailed with diabase-keratophyre breccias. The tuffitic arkoses are coarse-grained quartzo-feldspathic sandstones, which were formed from reworked porphyritic crystal tuffs (Ludwig 1969). The base of the formation is apparently formed by volcanogenic conglomerates with shales. Clasts include spilites, granites, diabases and lydites up to 8 cm in diameter. The formation is interpreted as representing a phase of explosive volcanism that led to the expulsion of acidic as well as basic rocks (Ludwig 1969).

Fossil remains were found in siliceous shales with volcanogenic components. The majority includes trilobites (with *Conocoryphe* and Eodiscina, Corynexochida and Ptychopariida), which prove a middle Celtiberian age. Additional fossils include brachiopods and sponge spicules. Algal structures are reported from the volcanogenic conglomerates (Ludwig 1969).

The middle Middle Cambrian (late Caesaraugustan) Lippertsgrün Formation (Wurm 1928) generally consists of relatively monotonous shales with some calcareous nodules. The normal lithology includes light greenish to light grey, rarely reddish, thin, resistant slaty shales with minor amounts of mica, and subordinate slaty quartz arenite beds. Small carbonate nodules occur in several horizons. Local erosional surfaces are covered by shell-bearing, cross-bedded shales. The formation most probably grades into rocks assigned to the Bergleshof Formation (Sdzuy 1966). The most complete section (at Lippertsgrün) is *c.* 80 m thick.

Fossils are generally rare and only known from the calcareous layers and nodules. They include the trilobite genera *Solenopleuropsis*, *Sao*, *Paradoxides*, *Conocoryphe*, *Condylopyge*, *Parabailiella*, *Bailiella*, *Hypagnostus*, *Peronopsis*, *Phalagnostus*, *Ctenocephalus* and *Acontheus*. Carpoids (Cincta) and probable eocrinoids were also found.

The upper Middle Cambrian (Languedocian) Bergleshof Formation (Horstig 1954; Sdzuy 1966) consists of a variable succession of fine- and medium grained sandstones and siltstones. The typical succession includes an alternation of medium grey to dark and bluish grey, slaty siltstones and grey, greenish or grey brown, slightly calcareous, poorly sorted sandstones. The sandstones are generally lenticular or form flaser-shaped bodies in the siltstones. The top of the formation consists of grey, relatively coarse quartz arenites. Its thickness was estimated to be less than 100 m (Sdzuy 1966).

Layers with accumulated billingsellid brachiopod shells and echinoderm ossicles occur, as do hyoliths, echinoderms, trilobites, and rare inarticulate brachiopods. The trilobites include *Proampyx*, *Parasolenopleura*, *Holocephalina*, *Bailiella?*, *Jincella*, *Peronopsis* and *Leiagnostus?*. Reitz (1991) studied acritarchs from this formation and assigned them (obviously erroneously) to the Lower Tremadocian.

Northern margin of the Bohemian Massif

The northern rim of the Bohemian Massif consists of a number of metamorphic blocks that appear to include Cambrian rocks. Nevertheless, direct evidence of a Cambrian age is lacking in most cases, the age being derived either from radiometric dating (which is usually poorly constrained) or from lithostratigraphic sequence stratigraphic correlation with other areas. In addition, thicknesses produced for these metamorphic suites are only suggested, and further information is required for confirmation.

NE Bavaria and Vogtland (G.G.)

Metamorphosed Cambrian rocks exist in the Fichtelgebirge area (Fig. 4.7) and the adjacent northern Oberpfälzer Wald. Underlying the otherwise datable Lower Ordovician Frauenbach Group is an assemblage of quartzites, mica schists, phyllites, epigneisses, and rhyolithic metavolcanites known as the Warmensteinach Group. Local petrographic differences facilitate subdivision into different formations (Stettner 1988, 1997*a*). In the Waldsassen area (southeastern Fichtelgebirge) and the adjacent

Oberpfälzer Wald the Stiftsberge Group appears to replace the Warmensteinach Group. It similarly consists of quartzites, mica schists, and metarhyolithic rocks (Stettner 1988, 1997*a*). The lower Cambrian and/or uppermost Neoproterozoic is represented in the central Fichtelgebirge by the Arzberg Group, which is divided into the Wunsiedel Formation (marbles, calc-silicates, and graphitic schists, up to *c.* 1000 m thick) and the Alexandersbad Formation (phyllitic metapelites, quartzites, meta-arkoses, leucocrate and basaltic metavolcanites, up to *c.* 2000 m thick) (Stettner 1988, 1992).

The ages and structural history of the rocks of the **Münchberg Crystalline Complex** (Münchberger Gneismasse; Fig. 4.7) pose a major problem that strongly affects the tectonostratigraphic and structural interpretation of the area. Traditional interpretations suggest a late Neoproterozoic to Cambrian age for the Marienweiher Group (formerly termed the 'Liegendserie') which consists essentially of fine-grained gneisses interpreted as metahornfels, based on a loose framework of radiometric datings and regional comparisons (Stettner 1997*a*). The Stammbach Group (formerly known as the 'Hangendserie') consists primarily of basic metavolcanics such as eclogites, garnet amphibolites, chlorite amphibolites, and some intermediate and even acidic metavolcanics (hornblende gneisses etc.) (Werner 1993; Stettner 1997*a*). U–Pb datings of zircons from eclogites indicate an age of 525 ± 40 Ma (Gebauer & Grünenfelder 1979), but other analyses (e.g. Müller-Sohnius *et al.* 1987) suggest a somewhat younger age. On the other hand, supposed protoliths indicate an oceanic-setting volcanic suite that would favour a Cambro-Ordovician age (Stettner 1988, 1997*b*).

In the **Vogtland** area, developed as a syncline (Fig. 4.7), the Cambrian appears to be represented by the Brambach, Raun and Klingenthal groups (Hoth & Berger 1980; Berger & Hoth 1997). The Brambach Group consists of mica paragneisses and mica schists with minor intercalations of quartzites and metagreywackes. Its interpretation as relatively early Cambrian in age is based purely on lithostratigraphic correlations with similar units in the Fichtelgebirge and Erzgebirge areas. The Raun Group (dominantly mica schists with frequent intercalations of quartzites) is even less clearly correlatable. The Klingenthal Group consists of phyllites with considerable amounts of heteroclastic quartzites and quartzitic schists and locally metalydites. Some poorly preserved trace fossils were found in the upper part and possible organic microfossil remains were recovered from the lower Kraslice Formation (Oruzinsky & Kribek 1981). The interpretation of the group as late Cambrian or Cambro-Ordovician in age is based on reasonable lithostratigraphic correlations.

Erzgebirge (O.E.)

The Erzgebirge in SE Germany (Fig. 4.7) represents a tectonically complex unit with a complicated metamorphic overprint. It is, in general, composed of high- to medium-pressure para- and orthogneisses, mica schists and phyllites with intercalations of metabasalts, metarhyolites, metagranites and marbles. To what degree the metamorphosed Cambrian rocks are present is still a matter of discussion. The present stratigraphic interpretation is largely based on lithostratigraphic correlations (Leonhardt *et al.* 1997).

The Klinovec Group, a unit with greenschist- to amphibolitic-grade metamorphic facies, consists of mica schists, quartzites, gneisses, metagreywackes and up to 260 m of dolomite marbles (in its lower part). The succession has a mean thickness of *c.* 1300 m. Galenite Pb/Pb ages range from 522 to 555 Ma (Legler 1985) and thus suggest an Early Cambrian age.

This unit is considered to be overlain by about 900–1100 m of a greenschist-metamorphic, folded succession of mica schists, calcite marbles (in lower parts), quartzites and basic volcanics, termed the Jachimov Group. According to regional lithostratigraphic correlation it is assumed to have a probable Middle Cambrian age. A radiometric single zircon dating of muscovite gneiss from the group revealed an age of 555 ± 5 Ma (Kröner *et al.* 1995).

The youngest unit is believed to be the Thum Group, a folded, greenschist-metamorphic complex of phyllites, quartzites, basic metavolcanics, and in the lower part up to 80 m of calcite marbles. This suite is interpreted as late Cambrian to early Ordovician in age from a comparison with the neighbouring Vogtland area. The estimated mean total thickness is *c.* 1300 m. Chitinozoans have been identified from the carbonates (Herold Formation; Behr *et al.* 1965), but these identifications were subsequently challenged by Eisenack (1966). A Rb–Sr age from a phyllite has been determined as 497 Ma (Legler 1985).

Based purely on geochemical analyses and radiometric dating, the basement in the Erzgebirge region is interpreted as consisting of medium pressure–medium temperature orthogneisses of latest Proterozoic protolith ages (560 to 540 Ma; Kröner *et al.* 1995). High pressure–high temperature and high pressure–low temperature orthogneisses/metarhyolites of a gneiss-eclogite unit show a protolith age of *c.* 490 to 480 Ma (Kröner *et al.* 1995; Quadt & Gebauer 1998; Mingram *et al.* 1998) and suggest formation close to the Cambrian–Ordovician boundary. Chemical signatures of these rocks and the association with metarhyolites and eclogites (derived from mafic to ultramafic magmatic protoliths), marbles and clastic metasediments suggest that the origin of this association was in an extensional regime. The high-grade metamorphic rocks support a model of transition from a continental rift setting with bimodal magmatism to a passive margin with highly mature sediments (Mingram & Rötzler 1998). It is supposed that the Cambro-Ordovician magmatism was initiated by crustal thinning associated with rifting. The high maturity of the sediments suggests a sea-level lowstand accompanied by chemical weathering prior to the rift-related subsidence (Mingram & Rötzler 1998).

Saxonian Granulite Massif (O.E.)

The Saxonian Granulite Massif (Sächsisches Granulitgebirge; Fig. 4.7) in eastern Germany includes metamorphosed rocks of probable Cambrian age within its slaty margin. This suite, termed the Rabenstein Formation (Freyer *et al.* 1982), is represented by approximately 250 to *c.* 500 m of metamorphic and strongly folded rocks in the southeastern surroundings of the massif. It consists of alternating basic volcanics and phyllites with metagreywackes occurring locally in its upper part. Metamorphosed carbonates (25 to 35 m of calcite marbles) and amphibole schists are also present. The lower and upper boundaries of the formation are unknown.

Freyer *et al.* (1982) described problematic organic remains as crinozoa (originally interpreted as archaeocyaths, radiolarians and algae), hyolithelminths, sponge spicules, stromatolites, acritarchs and sporomorphs. Kurze (1962) reported some foraminifera from probably correlative horizons in the area. However, these assignments remain problematic. The Cambrian age of this formation is based mainly on regional lithostratigraphical arguments. Deposition probably took place on a deep shelf below storm wave base.

Elbe Zone (O.E.)

The Elbe Zone (Fig. 4.7) is part of an important crustal segment, which can be traced over lateral distances of several hundreds of kilometres (Kurze *et al.* 1997). Probable Cambrian rocks form a succession of slates and recrystallized limestones within a small area near Dresden (at the Wackerbarths Ruhe castle). The foliated and (during the Variscan Orogeny) contact-metamorphosed carbonates are lithostratigraphically comparable to other regions of the Saxothuringian Zone. Biostratigraphic or radiometric data are unknown from these rocks.

Brunovistulian Terrane (O.F.)

The Brunovistulian Terrane (Brunovistulicum or Bruno-Vistulicum) comprises the Upper Silesian Block and the Moravian Block (Dudek 1980) to the south in the Moravo-Silesian region. Recent interpretations by Franke & Żelaźniewicz (2002) suggest that these areas formed part of a so-called 'Silesian Terrane Assemblage' together with the Małopolska and Łysogóry terranes, whereas most authors believe that the Łysogóry Block was never part of, or connected with, the Brunovistulian Terrane during the early Palaeozoic.

The basement of the Brunovistulian Terrane is composed of Cadomian plutonic and metamorphic rocks (Dudek 1980; Finger *et al.* 2000*a*, *b*), in some areas with greenschist-grade Late Proterozoic metasediments (Żelaźniewicz *et al.* 2001). The signature of these basement rocks provides clues for the interpretation of its palaeogeographic and tectonic evolution (see Linnemann *et al.* 2008).

Knowledge about the subsurface development of the Lower Palaeozoic in this area has been based on about 450 boreholes penetrating Precambrian and Lower Palaeozoic successions, which were drilled between 1959 and 1992. The core material reveals the complex distribution pattern of the predominantly Lower Cambrian and, more rarely, Middle Cambrian and Furongian, Ordovician and Silurian rocks. All Cambrian rocks are flat-lying, unmetamorphosed sediments which were formed in shallow marine shelf environments in normal stratigraphic order.

Ediacaran and Cambrian units unconformably overlie the Cadomian basement (Kotas 1973, 1982; Moczydłowska 1997, 1998; Buła 2000; Fig. 4.14). The basal post-orogenic sediments are post-Cadomian terrestrial units (alluvial fan and braided-river deposits) of probably latest Ediacaran age (Potrójna Formation of the Potrójna IG-1 borehole and the lowest part of the Goczałkowice Formation of the Goczałkowice IG-1 borehole; Moczydłowska 1997). These continental deposits are transgressively overlain by coastal deposits as shown by the succession in the Goczałkowice IG-1 borehole. The base of the Cambrian is located at about this change in facies.

Detailed sedimentological and palaeontological studies have succeeded in distinguishing five Cambrian lithostratigraphical units, termed the Borzęta, Goczałkowice, Sosnowiec, Potrójna and Jaszczurowa formations (Kowalczewski 1990; Buła 2000). The siliciclastic succession increases in thickness to the east (Buła & Jachowicz 1996). The formations represent the Lower and Middle Cambrian and Furongian. However, only an Early Cambrian age can be proven by macrofossils (trilobites in the Pszczyna Member of the Goczałkowice Formation). Other levels have been dated by means of trace fossils and well-preserved and diverse assemblages of organic-walled microfossils, primarily acritarchs. Thus, in some cases there are differing views on the age of the formations. As an example, the Potrójna Formation is considered to represent the Lower Cambrian (Kowalczewski 1990) or Vendian (= Ediacaran) (Moczydłowska 1997; Buła 2000).

The **Borzęta Formation** has been recognized in the Borzęta IG-1 borehole (south of Kraków), and equivalents are known to be present in several other boreholes. The thickness of the formation may be up to *c.* 670 m (Buła & Jachowicz 1996). The Borzęta Formation has been subdivided into the Myślenice, Osieczany and Rajbrot members. The Myślenice Member is composed of claystones intercalated with siltstones, whereas sandy siltstones and fine-grained sandstones dominate the Osieczany Member. The youngest Rajbrot Member includes fine- to medium-grained sandstones. The Borzęta Formation was established in the marginal eastern part of the Brunovistulian Terrane (Buła & Jachowicz 1996; Buła 2000). Taxa of organic-walled microfossils (prasinophytes such as *Leiosphaeridia* and *Tasmanites*; acritarchs such as *Leiovalia* and *Granomarginata*; filamentous algae and animal rests such as *Ceratophyton*) are interpreted to indicate its sub-Holmian age (Buła *et al.* 1997).

The *Holmia* Zone age is generally accepted for sediments of the **Goczałkowice Formation** distinguished by Kotas (1982) and formalized by Buła (2000). In its type section (the Goczałkowice IG-1 borehole) it is *c.* 370 m thick and sandwiched between intrusive igneous rocks that directly overlie a metamorphic sequence and the regional unconformity at the base of the Old Red Sandstone. Known thickness ranges between *c.* 70 m and *c.* 2000 m (Buła 2000). The gently dipping (5–16°) Goczałkowice Formation has been subdivided into the Mogilany, Głogoczów and Pszczyna members. The Mogilany Member (also termed the 'Scolithos (*Skolithos*) Sandstones') consists of basal conglomerates and a fining-upward sandy unit, passing gradually into sandstones with intercalations of siltstones and conglomerates. Fine-grained to medium-grained sandstones with siltstones are assigned to the overlying Głogoczów Member (also termed the Bioturbated Sandstones). The overlying siltstones and fine-grained sandstones belong to the Pszczyna Member (also termed 'Siltstones with trilobites'). According to Kotas (1982), the lower part of the formation was deposited in a tidal to beach zone, the central part of the formation in a tidal zone, and the highest parts below wave base and influenced by turbidity currents.

Orłowski (1975*b*) described trilobites from the Goczałkowice IG-1 borehole. Their taxonomic position has been recently discussed in Nawrocki *et al.* (2004*b*), who confirmed the presence of the Baltic endemic genus *Schmidtiellus* associated with palaeogeographically more widely distributed genera (*Ornamentaspis* and *Strenuaeva*). Based on detailed analyses of acritarch distribution, Moczydłowska (1998) extended the stratigraphic correlation of the Goczałkowice Formation and suggested that equivalents of the *Schmidtiellus mickwitzi* and *Holmia kjerulfi* biozones are represented. In addition, Moczydłowska (1998) supposed that the *Protolenus* Zone and *Acadoparadoxides oelandicus* Superzone could be represented in the youngest parts of the succession.

The **Sosnowiec Formation** (Kowalczewski 1990; Buła 2000) is composed of alternating fine- and medium-grained sandstones and horizontally laminated sandy siltstones of *c.* 195 m thick, and the formation is cut by a gabbrodiorite and diabase sill with a thickness of up to 92 m (Moczydłowska 1998). The basal part of the Sosnowiec Formation was referred to the Radocha Member by Kowalczewski (1990), and consists of conglomeratic and greywacke sandstones to polymodal, poorly sorted quartz sandstones, passing into fine-grained quartzitic and muddy sandstones with slighly bioturbated mudstone intercalations. The Sosnowiec Formation has been interpreted as having been formed in a shallow-marine environment, both in the tidal zone and on a shelf below storm wave base (Moczydłowska 1997, 1998).

Inarticulate brachiopods (lingulids and acrotretids) occur in the lower part of the formation (Moczydłowska 1998; Biernat &

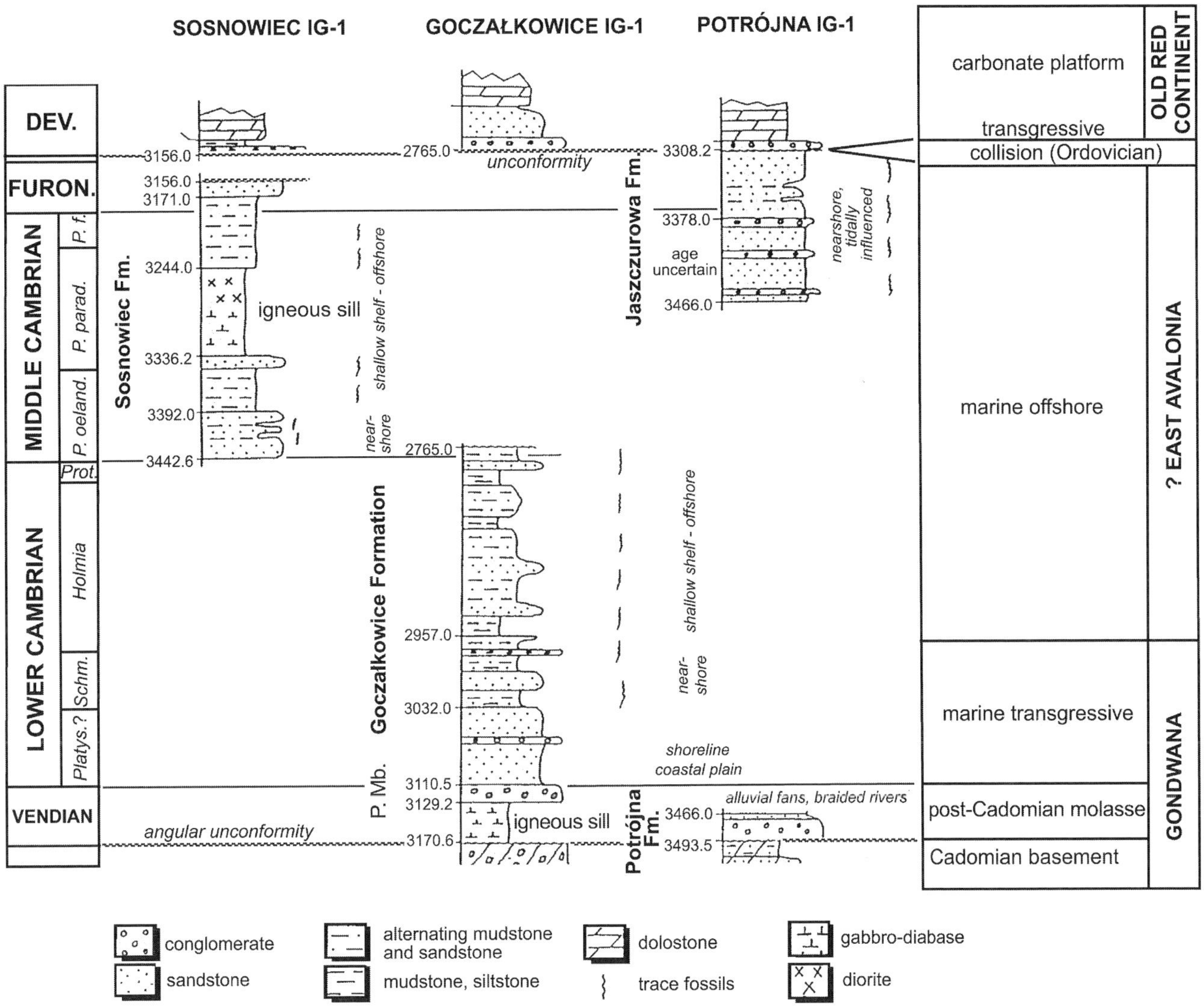

Fig. 4.14. Stratigraphy and depositional environments of Cambrian rocks from boreholes of the Upper Silesian Block. Lithostratigraphy according to Kotas (1973), Ślączka (1985) and Kowalczewski (1990), biostratigraphy according to Moczydłowska (1991, 1995*a*, 1997). Interpretation of geotectonic history modified after Moczydłowska (1997). Abbreviations: *Platys.*, *Platysolenites* biozone; *Schm.*, *Schmidtiellus mickwitzi* biozone; *Prot.*, *Protolenus* biozone; *P. oeland.*, *Paradoxides (Acadoparadoxides) oelandicus* biozone; *P. parad.*, *Paradoxides (P.) paradoxissimus* biozone; *P. f.*, *Paradoxides (P.) forchhammeri* biozone; FURON., Furongium; P. Mb., Pszczyna Member. Modified after Moczydłowska (1997, fig. 2).

Baliński 1973) and simple trace fossils (vertical and oblique burrows) have been identified from just above the Radocha Member. Rich acritarch assemblages have been studied by Buła & Jachowicz (1996) and Moczydłowska (1998) and contain more than 60 species including such stratigraphically important genera as *Heliosphaeridium*, *Alliumella*, *Skiagia*, *Cristallinium*, *Eliasum*, *Timofeevia* and *Adara*. Based on these acritarchs the age is interpreted as Middle Cambrian (Buła *et al.* 1997) and, more precisely, refers to the lower and middle Middle Cambrian *Acadoparadoxides oelandicus* and *Paradoxides paradoxissimus* Superzones (Moczydłowska 1998). However, the presence of equivalents of the *Paradoxides forchhammeri* Superzone (uppermost Middle Cambrian) and lowermost Furongian cannot be excluded for the highest levels.

The **Potrójna** and **Jaszczurowa formations** were established on the basis of data from the Potrójna IG-1 borehole (Kowalczewski 1990). The *c.* 25 m thick Potrójna Formation did not provide any fossils, and there are differing views on its age (Kolwaczewski 1990; Moczydłowska 1997, 1998; Buła 2000). However, acritarchs have been found in the overlying Jaszczurowa Formation. The **Jaszczurowa Formation** (*c.* 160 m thick) consists of quartz sandstones with mudstone intercalations and conglomeratic interbeds, with its lower part designated the Mucharz Member (Kowalczewski 1990). The formation is interpreted to have been deposited under nearshore, tidally influenced conditions (Ślączka 1985). Acritarch species such as *Timofeevia lancarae* (Cramer & Diez), *T. pentagonalis* (Vanguestaine), and *T. phosphoritica* Vanguestaine suggest a Furongian age (Moczydłowska 1997, 1998).

Tectonostratigraphic position

Although the stratigraphic succession can only be reconstructed from an incomplete record displayed by the few boreholes, the suggested depositional history is crucial for the interpretation of the tectonostratigraphic position of the Brunovistulian Terrane and Central Europe in general. The absence of any tectonic deformation other than normal faulting and of any distinct metamorphism of Cambrian rocks suggests that Upper Silesia was part of a stable crustal block limited by deep faults (Moczydłowska 1997). The coastal marine earliest Cambrian deposits overlying a post-Cadomian continental succession of probably latest Proterozoic/Ediacaran age can be interpreted as Gondwanan- or peri-Gondwanan-type rocks deposited along the Gondwanan continental margin (Fig. 4.14). The majority of Cambrian rocks, including the Goczałkowice, Sosnowiec and Jaszczurowa formations, are shallow marine to offshore siliciclastics which reflect transgressional–regressional cycles with brief shallowing episodes and resulting breaks in deposition (Kotas 1982; Moczydłowska 1997). The incompleteness of the Cambrian sequences and the presence of regional angular unconformities can be interpreted as being related to deposition in a pull-apart basin. This interpretation is in accordance with the suggestion that the Upper Silesian Block was part of East Avalonia and records the break-off from Gondwana and drift toward Baltica (Moczydłowska 1995*a*, *b*, 1997). However, the trilobites found in the Goczałkowice Formation are unknown from typical Avalonian formations and suggest a closer relationship with the Kielce Block of the Holy Cross Mountains (see next section).

The development of the southern segment of the Brunovistulian Terrane can be reconstructed from successions in **southern Moravia**. A distinctly different stratigraphic subdivision has been established although the general difference results mainly from a larger hiatus developed in Moravia. An Early Cambrian age has been assumed by Roth (1981) for a thick, basal clastic unit (underlying the biostratigraphically dated Devonian). In turn, lithological, palynological and ichnological data proving the Cambrian age have been published only during the last decade based on data from three boreholes (Měnín 1, Němčičky 3 and Němčičky 6). Jachowicz & Přichystal (1997), Vavrdová (1997) and Fatka & Vavrdová (1998) have documented the presence of well-diversified and preserved Lower Cambrian acritarchs and prasinophytes. The list of species and stratigraphical correlation has been further elaborated by Vavrdová & Bek (2001). Several ichnofossil taxa have been documented by Mikuláš & Nehyba (2001) and Vavrdová *et al.* (2003). A detailed correlation with the lithostratigraphic units of the Polish sector of the Brunovistulian Terrane has not yet been established.

Holy Cross Mountains (A.Ż.)

The Holy Cross Mountains (Góry Świętokrzyskie) represent a *c.* 100 km long and 30 km wide area, which formed on the foreland of the Variscan Orogen and is located close to the margin of the East European Platform. Differences in facies development, stratigraphy, and tectonic evolution have resulted in a traditional subdivision of the Holy Cross Mountains into two regions, a (southern) Kielce and a (northern) Łysogóry region, separated by the WNW–ESE trending Holy Cross Fault (Fig. 4.15). The distribution of Cambrian strata differs in the two regions (Figs. 4.16 & 4.28). Lower and Middle Cambrian strata with an uncomformably lying blanket of upper Furongian strata crop out in the Kielce region, whereas in the Łysogóry region only Middle Cambrian to Furongian strata can be observed. Non-conventional techniques (such as magnetostratigraphy) performed in the Łysogóry Block suggest a close association of this region with the Cambrian continent of Baltica (i.e. the East European Platform; Nawrocki 2004). This is partly supported by biogeographic data based on Furongian trilobites (Żylińska 2002) from the upper part of the succession. In contrast, provenance analysis suggests that Cambrian detrital material was derived from a Gondwanan source (Valverde-Vaquero *et al.* 2000), which can also be confimed by early Furongian trilobites (Żylińska 2002). In turn, the depositional history and faunas (particularly trilobites from the Lower–Middle Cambrian boundary interval and the

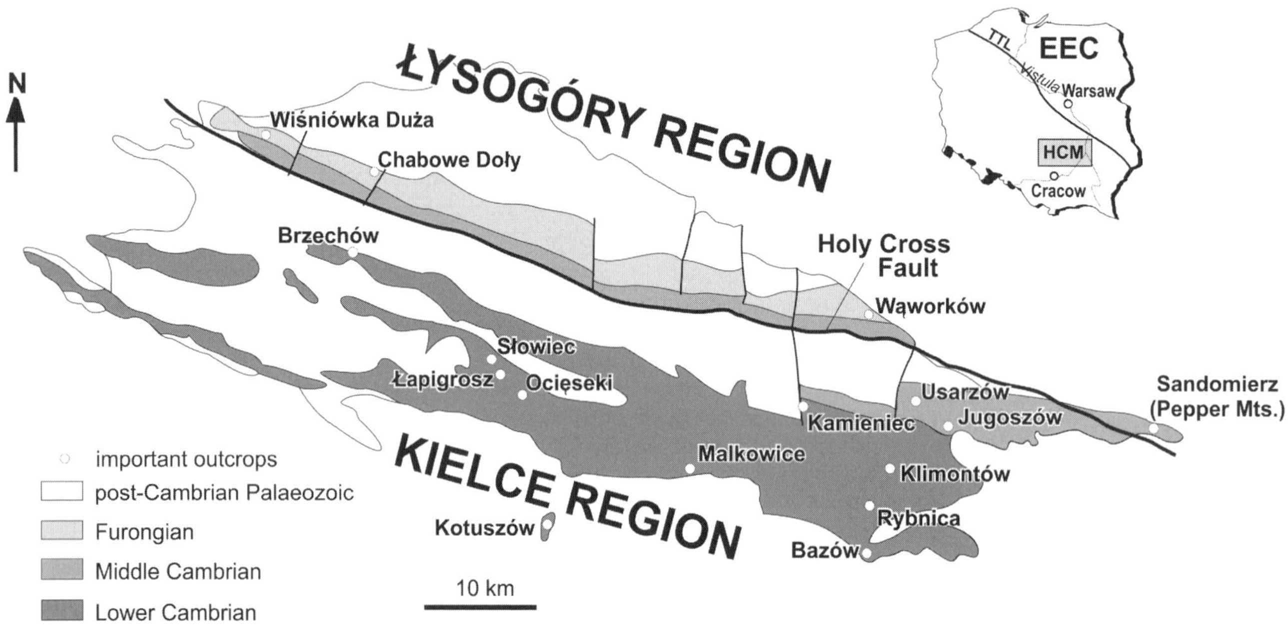

Fig. 4.15. Geological sketch map of the Holy Cross Mountains (Góry Świętokrzyskie) with its two facies regions with important outcrops. Insert shows position in southern Poland and with respect to the East European Platform (EEP) and the Tornquist-Teisseyre lineament (TTL).

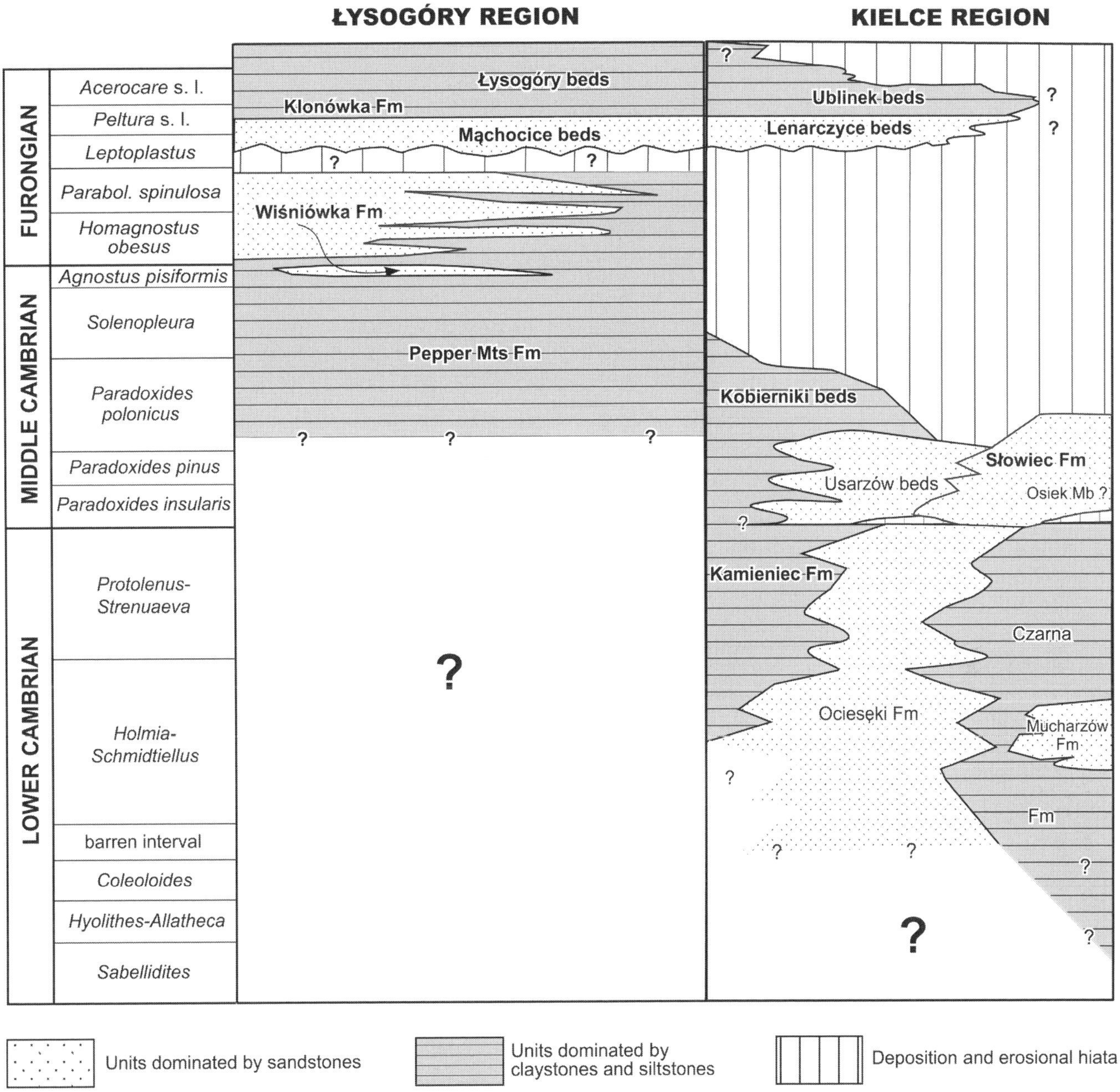

Fig. 4.16. Stratigraphic table of the Cambrian units of the Holy Cross Mountains illustrating the differences for the northern Łysogóry and southern Kielce regions. After Kowalczewski *et al.* (2006).

early Middle Cambrian) indicate a peri-Gondwanan character for the Kielce Block (e.g. Żelaźniewicz 1998; Żylińska & Masiak in press).

The Cambrian rocks in the Holy Cross Mountains (HCM) are developed as a thick siliciclastic unit (Fig. 4.16). The fossil content is dominated by trilobites, although their distribution in the succession is patchy. Accompanying fauna include brachiopods, hyoliths, bradoriids, eocrinoids and primitive echinoderms (*Velumbrella* and *Rotadiscus*). Biostratigraphic conclusions are supplemented by acritarchs. Analysis of the biotic assemblages can be supplemented by the analysis of ichnofossils. The entire succession, considered to be up to 3500 m thick (Orłowski 1992*b*), is dominated by sandstones in the western, and mudstones and shales in the eastern part of the area. Thus far, the Precambrian–Cambrian boundary has not been recognized in the area. Lithostratigraphic subdivisions were determined by Orłowski (1975*a*, 1988, 1992*b*; Kowalczewski 1990, 1995; Szczepanik *et al.* 2004*a*, *b*).

The lowest, 'sub-trilobitic' part of the Cambrian sequence comprises clay- and siltstone shales with intercalations of fine-grained siltstones and sandstones of the Czarna Formation (*c.* 800 m), present in cores and exposures in the southernmost part of the area. Palaeontologic data are restricted here to rare hyolithids, bradoriids, calcimicrobes and trace fossils (Kowalski

1983; Orłowski & Waksmundzki 1986). Acritarch data (with *Skiagia*) also suggest the younger, *Holmia-Schmidtiellus* Zone for the formation. Sandstone intercalations in the Czarna Formation increase towards its top. Laterally, the formation passes into the Ociesęki Formation, which can be traced in the central part of the area. The thickness of the Ociesęki Formation is up to 1100 m. This formation includes fine-grained, thin- to medium-bedded sandstones with mudstone intercalations, and is extremely rich in trace fossils in some intervals (Orłowski 1989; Orłowski & Żylińska 2002). It was deposited within a shallower area on a siliciclastic shelf, which could have been of tectonic origin (Kowalczewski 1990). Generally the fauna is dominated by trilobites (Orłowski 1985*a*). The lower part of the Ociesęki Formation contains endemic holmiids (*Holmia*, *Schmidtiellus* and *Kjerulfia*), accompanied by *Strenuella* and ellipsocephalines such as *Ornamentaspis* (*Holmia-Schmidtiellus* Zone). The succeeding trilobite assemblage is characterized by *Kingaspidoides sanctacrucensis* and *Issafeniella orlowinensis* (*Protolenus-Strenuaeva* Zone).

In the eastern part of the HCM the Ociesęki Formation passes into the finer siliciclastics of the Kamieniec Formation (*c.* 600 m). This is represented by clay and clay-siltstone shales with intercalations of fine-grained sandstones. Here, body and trace fossils are rare. The lower part of the formation is characterized by the presence of holmiids (*Holmia-Schmidtiellus* Zone), whereas the upper part is characterized by antatlasiinae and protoleniinae (*Protolenus-Strenuaeva* Zone) (Bednarczyk *et al.* 1965; Orłowski 1985*a*). The strata were deposited in low-energy, deep-water environments, possibly the outer shelf (Mizerski *et al.* 1991). Most probably the upper part of the *Protolenus-Strenuaeva* Zone can be attributed to the Celtiberian (Agdzian) (Geyer & Landing 2004), although the scarcity of outcrops with the relevant trilobites both in the Ociesęki and Kamieniec formations does not allow detailed correlation.

The coarsest siliciclastics in the HCM successions occur in the lower part of the Middle Cambrian and comprise the Słowiec Formation (*c.* 100 m) in the central, and the Usarzów Formation (*c.* 400 m) in the eastern part of the area. The strata are developed as medium-grained, bedded sandstones and coarse-grained thick-bedded sandstones. The fauna is dominated by paradoxidids and ellipsocephalines (representatives of *Ornamentaspis*; Bednarczyk 1970; Orłowski 1965, 1985*b*), and the lower part of the Słowiec Formation appears to be coeval with the *Cephalopyge notabilis* Zone in Iberia and Morocco (lower Celtiberian (Agdzian); Geyer & Landing 2004; Żylińska & Masiak 2007). The Osiek sandstones, a 30-m unit from the southernmost part of the area and considered earliest Cambrian in age by Kowalski (1983), also possibly belong to this formation (Kowalczewski 1990). Apart of the Usarzów and Słowiec formations, the Middle Cambrian probably also includes clay- and siltstone facies from the Lenarczyce IG-1 borehole (Szczepanik *et al.* 2004*a*), which contains the acritarchs *Eliasum laniscum*, *Cristallinum cambriense* and *Adara alea*, and lacks representatives of *Timofeevia* and *Skiaga*. These strata are informally included with the Kobierniki beds (Szczepanik *et al.* 2004*b*).

Middle Cambrian sedimentation terminates in the Kielce region in the *Paradoxides polonicus* Zone. From this interval the oldest known deposits from the Łysogóry region, the Góry Pieprzowe Formation (Pepper Mountains Shales, 400 m), crop out in the picturesque Pepper Mountains (Góry Pieprzowe) cut by the Vistula River just north of the historical town of Sandomierz (easternmost part of the area). The formation is represented by clay- and siltstone shales with thin sandstone intercalations, suggesting an outer shelf environment (Kowalczewski 1990) with local storm deposits (Przewłocki 2000). The trilobite data from this formation are scanty, and include *Solenopleura*, *Solenopleurina* and poorly preserved agnostids, apparently indicating the uppermost Middle Cambrian *Solenopleura canaliculata* Zone (Orłowski 1985*b*). Interesting stratigraphic conclusions, however, can be drawn from the analysis of acritarchs, which also point to the possible Late Cambrian age of the formation (Szczepanik 2001). Shales of the Pieprzowe Formation are partly overlain by fine-grained, well-sorted quartzitic sandstones with intercalations of mudstones and silty shales of the Wiśniówka Formation (thickness disputed; from 80–200 m when strong folding is assumed (Kowalczewski *et al.* 1986; Kowalczewski 1995) to 400–1400 m as a simple structure (Orłowski 1975*a*; Orłowski & Mizerski 1995)). The formation is known from its particularly rich trilobite ichnocoenosis, dominated by *Cruziana semiplicata* and *Rusophycus polonicus* (e.g. Radwański & Roniewicz 1963; Orłowski *et al.* 1970; Orłowski 1992*a*; Orłowski & Żylińska 1996; Żylińska 1999). These sediments were deposited in both inner and outer shelf environments (Studencki 1994). Sandstones of this formation form the highest elevations in the HCM area, termed the Main Range. The trilobite assemblages from this formation are restricted to the easternmost (Wąworków) and westernmost part of the area (Wiśniówka) (Orłowski 1968*b*) and are dominated by *Aphelaspis rara* accompanied by rare olenids (Żylińska 2001). The uppermost Cambrian formation in the HCM area is represented by the Klonówka Shales (400 m; Orłowski 1975*a*), documented from small exposures and boreholes northwards of the Main Range (Orłowski 1968*a*; Tomczykowa 1968; Tomczykowa & Tomczyk 2000; Żylińska 2002). Clayey and silty shales with mudstone and sandstone intercalations suggest deposition on the outer shelf. The sandstone intercalations are more common in the lower part of this unit (Mąchocice beds *sensu* Tomczykowa 1968), whereas the upper part, dominated by shales, reflects deepening of the basin (Łysogóry beds *sensu* Tomczykowa 1968). The rich and diverse trilobite assemblages are represented by the Olenidae, dominated by species of *Leptoplastides* as well as *Parabolina* (*Neoparabolina*) *frequens* (Tomczykowa 1968; Żylińska 2001), and show a distinct biogeographic evolution, from assemblages dominated by endemic faunas from the *Peltura minor* Zone to faunas dominated by Baltic faunas in the *Acerocare sensu lato* Zone, as known from the margins of the Baltica palaeocontinent (Nikolaisen & Henningsmoen 1985). Strata of this formation apparently pass into the Lower Ordovician as documented by graptolites (Tomczykowa 1968) (Fig. 4.17).

Recently, Furongian strata (*c.* 95 m) have also been discovered in the Kielce region, in the Lenarczyce PIG 1 borehole (Szczepanik *et al.* 2004*a*, *b*). They unconformably overlie the Kobierniki beds, displaying a different tectonic style in relation the latter. Similar to the Klonówka Formation from the Łysogóry Region, they are also bipartite, with the lower part dominated by sandstones with intercalated claystone–sandstone–siltstone beds and conglomerates with phosphatic nodules (Lenarczyce beds of Szczepanik *et al.* 2004*b*). The upper part is dominated by dark grey siltstones with thin intercalations of fine sandstones (Ublinek beds of Szczepanik *et al.* 2004*b*). Both Furongian-age units are documented by rich acritarch assemblages and the trilobite *Peltura* sp. has been found in the upper unit, and they probably range from the *Protopeltura praecursor* Zone (Lenarczyce beds) to the *Acerocare sensu lato* Zone (Ublinek beds). The Furongian strata are unconformably overlain by Lower Ordovician conglomerates, characteristic of the transgressive Ordovician succession in the Kielce region.

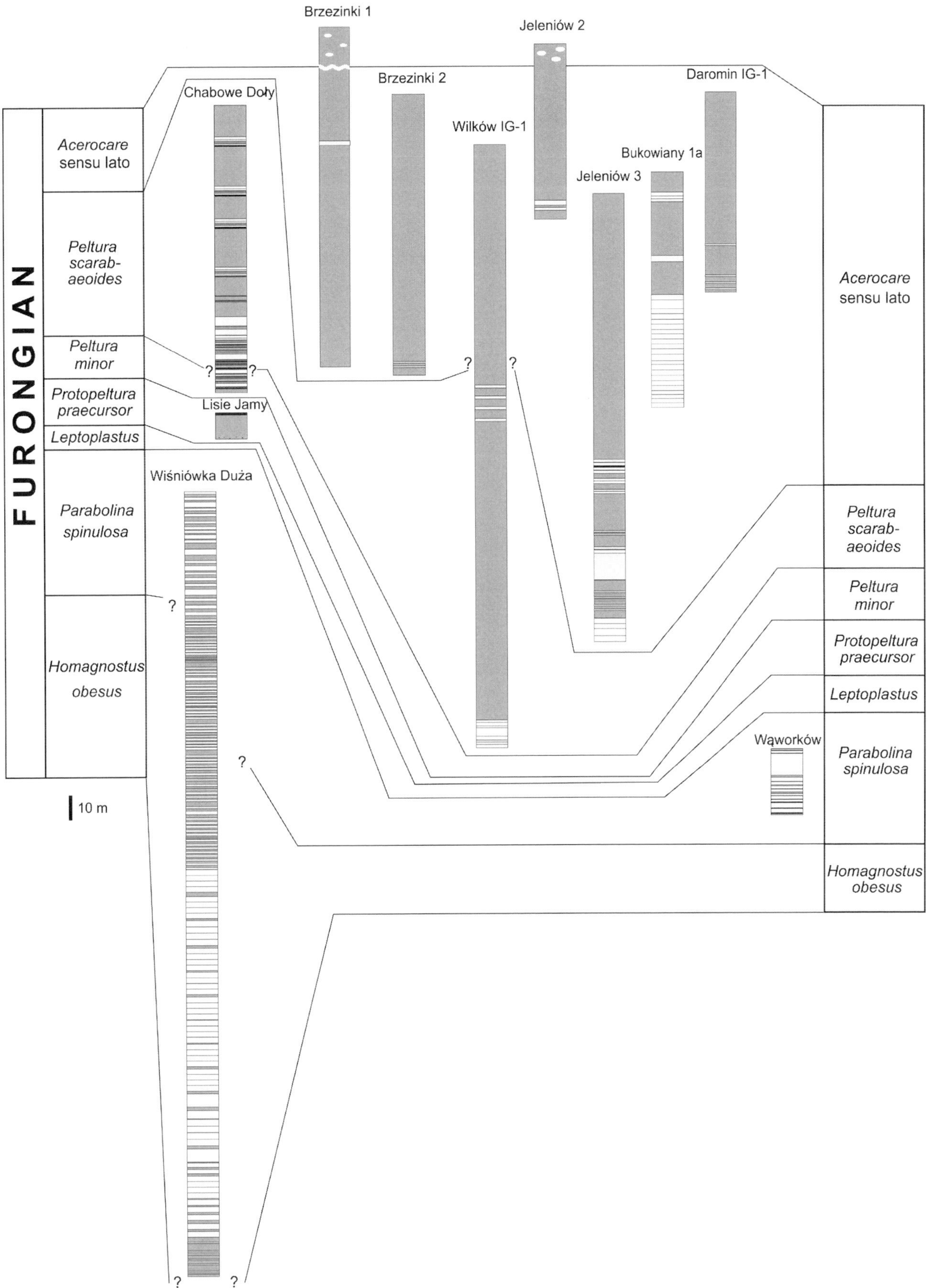

Fig. 4.17. Furongian sections from the Holy Cross Mountains, and their correlation and biostratigraphy. Slightly modified after Żylińska (2001); Daromin IG-1 log after Tomczykowa & Tomczyk (2000).

Bohemian Massif

Bohemicum (O.F.)

The Bohemian Massif includes four areas with Cambrian rocks, the age of which has been proved by palaeontological data. Three of them are situated within the Bohemicum (Fig. 4.18). The fourth belongs to the Moravo-Silesian region (Brunovistulian Terrane). Slightly metamorphosed Cambrian sediments are supposedly present in several other units of the Bohemian Massif, for example in the Islet Zone of the Central Bohemian Pluton (Havlíček *et al.* 1958; Kachlík 1992; Kachlík & Janoušek 2001) and/or in the Rožmitál Islet (in the SW continuation of the Příbram-Jince Basin). However, palaeontological data indicating a Cambrian age for these units have not yet been found.

The Bohemicum, located in the central part of the Bohemian Massif, contains unmetamorphosed to slightly metamorphosed Proterozoic and Lower Palaeozoic rocks (Cambrian to Devonian), which represent three tectonosedimentary cycles (Proterozoic, Cambrian, and Ordovician through Devonian). Cambrian through Devonian sedimentary rocks are, in general, richly fossiliferous and usually associated with subordinate volcanics.

The Cambrian sedimentary rocks represent the second tectonosedimentary cycle, and are preserved in three separate areas (in the Příbram–Jince and Skryje–Týřovice basins, and in the Železné Hory area; Fig. 4.18). Coeval plutonic rocks have been documented from several surface granitic bodies distributed in the western segment of the Teplá-Barrandian region (Zulauf *et al.* 1997; Dörr *et al.* 2002; see below).

The first stratigraphic subdivision of the Barrandian area was published by Barrande (1846, 1852), who distinguished three main faunas. The oldest fauna, designated by him as '*faune primordiale*', corresponds to the Middle Cambrian. The stratigraphic subdivision of the Cambrian in the area was subsequently studied and improved by Krejčí (1877), Lipold & Krejčí (1860), Kettner & Kodym (1919), Kettner & Bouček (1937), Havlíček & Šnajdr (1951*a*, 1957) and Havlíček (1950, 1954, 1967, 1968, 1970, 1971, 1998). A modern stratigraphic subdivision has been proposed by Havlíček (1971). The most detailed study of the lithology was published by Kukal (1971), whereas data on the volcanic rocks were summarized by Waldhausrová (1971) and Štorch (1998).

Příbram–Jince Basin

Proterozoic sedimentary rocks in the Příbram–Jince Basin are unconformably overlain by a *c.* 2500 m thick Cambrian succession of generally coarse-grained clastics (dominated by terrigenous conglomerates and sandstones with subordinate greywackes) passing upwards into finer-grained sediments (greywackes and shales) and finally into volcanics (Figs. 4.19 & 4.28). The stratigraphical subdivision of the Příbram–Jince Basin is based on an apparent repetition of comparatively fine- and coarse-grained lithotypes. More than a century of research on the stratigraphical subdivision finally resulted in the distinction of 13 lithostratigraphical units grouped into seven formations (Havlíček 1971).

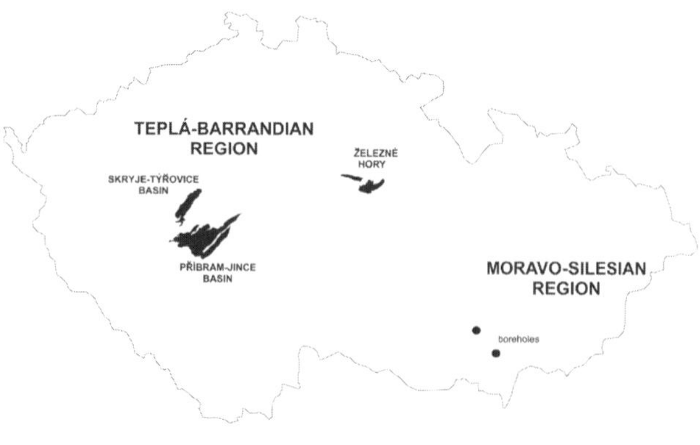

Fig. 4.18. Occurrences of Cambrian rocks in the three outcrop areas and location of boreholes with datable Cambrian rocks in the Czech Republic.

The **Žitec-Hluboš Formation** (Havlíček 1971; 100–600 m in thickness) rests with a prominent angular disconformity on the Neoproterozoic Štěchovice Group in the southern segment, or on the Kralupy-Zbraslav Group in the Brdy segment of the Příbram–Jince Basin. It consists of two types of conglomerates differing from each other mainly in colour. The older Žitec Conglomerate (Pošepný 1888) is petromictic, poorly sorted to unsorted, and composed of angular pebbles varying from several millimetres to a few tens of centimetres in diameter. The matrix is formed of coarse-grained greywackes and claystones, generally rich in unstable components. The second member, the Hluboš Conglomerate (Kettner 1915), is polymictic, red-brown to pink in colour (due to finely dispersed hematite in its cement) and contains a higher proportion of stable components. According to Kukal (1962, 1971), mudflow sedimentation on alluvial fans dominated during the time in which the member was deposited. This formation is overlain by subarkoses, subgreywackes and greywackes of the **Sádek Formation** (Kettner 1915; 250–1200 m) with a clayey matrix and rare intercalations of shales and conglomerates. Kukal (1963) interpreted the Sádek Formation as a continental complex, which originated in environments represented by alluvial plains, alluvial lakes, floodplains, and stream channels passing into deltaic and prodeltaic sedimentation. Mud-cracks have been documented at several localities (Kettner 1925; Kukal 1963). The overlying **Holšiny-Hořice Formation** (Havlíček 1971; 300–2000 m) comprises three members. The older Holšiny Conglomerate (Havlíček 1950) is well sorted and monomictic with dominant white and reddish quartz pebbles and siliceous cement. An erosional contact almost always separates the Sádek Formation from the overlying Holšiny Conglomerate.

The Hořice Sandstone (Havlíček 1950) is usually developed as a sequence of sandstones intercalated with claystone laminae. An intercalation of thinly laminated clayey shales (Paseky Shales; Havlíček 1950), several metres thick, occurs at various levels within the lower part of the Holšiny-Hořice Formation. The Paseky Shales contain the oldest macroscopic fossil assemblage within the Barrandian area (see 'Palaeontology and biogeography' section for details). The first fossil discovered in the Paseky Shales – the enigmatic crustacean-type *Kodymirus vagans* – was described by Chlupáč & Havlíček (1965) from Kočka. At that time, this locality was supposed to represent a marginal facies of the Middle Cambrian Jince Formation. The proper stratigraphical assignment of this locality to the Paseky Shale Member (about 800 to 1500 m below the Jince Formation) was later recognized by Havlíček (1968). Additional endemic arthropods (*Kockurus grandis* and *Vladicaris subtilis*) have been subsequently described. The absence of many groups which are typical for coeval Cambrian marine assemblages, such as trilobites, echinoderms and brachiopods, combined with the presence of spectacular arthropods and the scarcity of process-bearing acritarchs, most probably reflects a very peculiar environment, possibly characterized by restricted marine conditions. Kukal (1971, 1995) assumed sedimentation on alluvial plains with transition

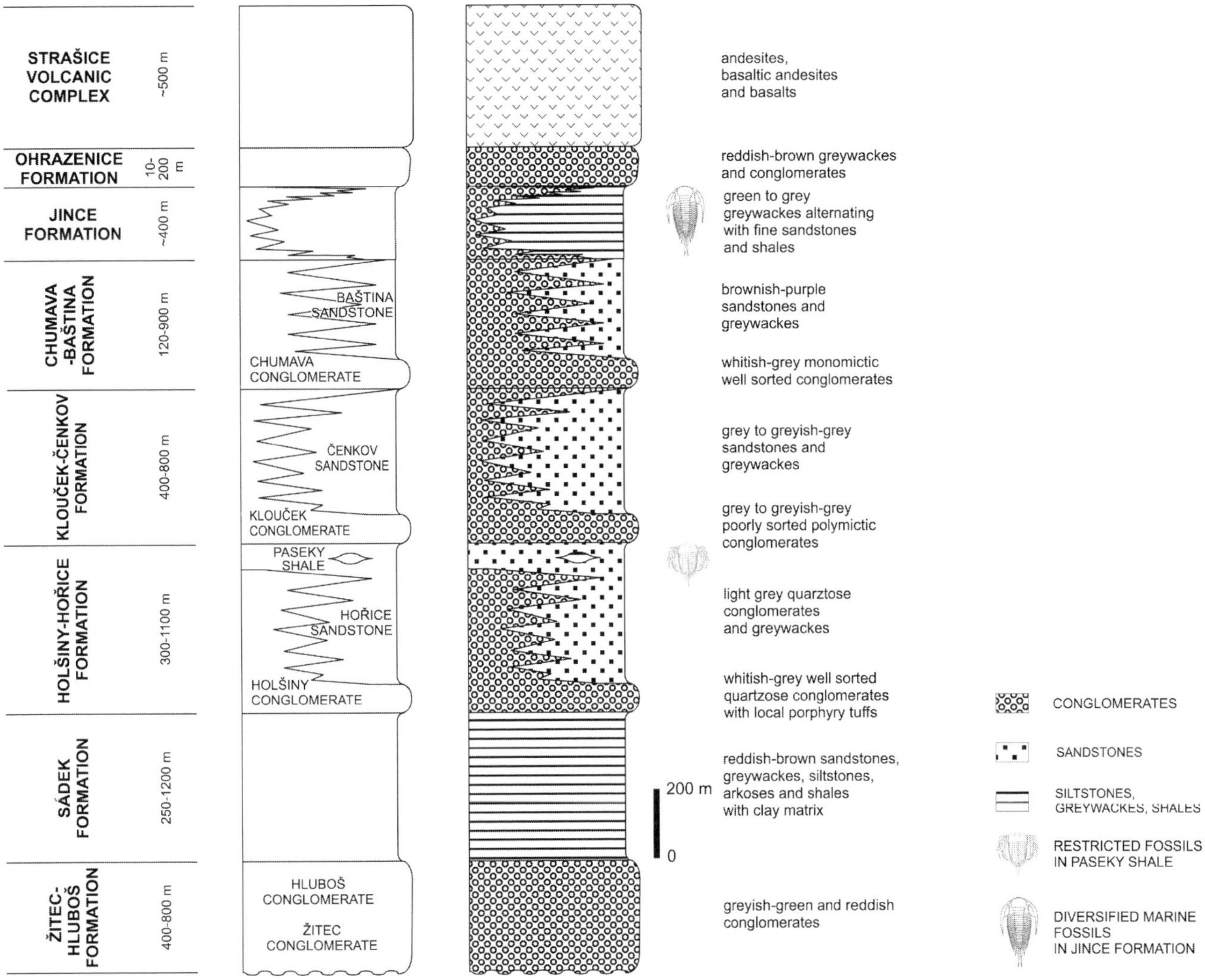

Fig. 4.19. Stratigraphy of the Cambrian rocks in the Příbram–Jince Basin (after Havlíček 1971).

into marine bays. Based on the ichnoassemblage, Mikuláš (1995) suggested a brackish or non-marine habitat, which would be the oldest known to date. However, current data do not present sufficient evidence for a non-marine depositional environment.

The **Klouček-Čenkov Formation** (Havlíček 1971; 400–800 m) covers a smaller area than the underlying formation. Its base is marked by the sudden appearance of the polymictic, poorly sorted Klouček Conglomerate (Havlíček 1950) with quartz and siliceous pebbles associated with clasts of tuff, andesite and volcanic glass. The conglomerates gradually pass into the Čenkov Sandstone (Havlíček 1950). The petrographic composition of these sandstones is identical with the Klouček Conglomerate, differing only in terms of grain size. In some areas, weathered pyroclastic intercalations are very common. The Klouček-Čenkov Formation represents a typical continental succession created by alluvial fans, streams, channels and lakes within an intermontane depression (Kukal 1971). The overlying **Chumava-Baština Formation** (Havlíček 1971; 100–900 m) commences with the sudden appearance of the monomictic well-sorted Chumava Conglomerate (Havlíček 1950) composed of white quartz and quartzite pebbles. These conglomerates pass into a sequence of fine-grained greywackes and arkoses of the Baština Sandstone (Havlíček 1967). Kukal (1971) interpreted these rocks as deposits of alluvial fans passing into fluvial sediments and redeposited gravels.

From the onset of Cambrian sedimentation, the area of maximum subsidence was situated in the western, so-called Brdy segment of the Příbram–Jince Basin. The Middle Cambrian marine ingression, represented by the **Jince Formation** (established by Lipold & Krejčí 1860 as the '*Ginetzer Schichten*'; 100–450 m), is, however, associated with a marked change of basin geometry (Havlíček 1971). During the Middle Cambrian, the depocentre shifted eastwards to the Litavka river valley area. Both the sedimentary facies and the distribution of fossil assemblages at this time indicate a pronounced bathymetric zonation. Bathymetric analyses of the Middle Cambrian biota indicate shallow marine assemblages in coarser-grained sediments in the WNW part of the basin, whereas deeper marine faunal elements are known from the areas to the east. Thus, the Příbram–Jince Basin is interpreted as deepening and opening eastwards. The Jince Formation includes comparatively rich assemblages of skeletal fossils such as trilobites and echinoderms, many of which are frequent collectors' items (see 'Palaeontology and biogeography' section for details).

The Jince Formation is sharply overlain by the well-sorted quartzose conglomerates and coarse sandstones of the **Ohraze-**

nice Formation (Havlíček 1950; up to 250 m thick, with rare intercalations of greywackes or subarkoses). According to Kukal (1963, 1971) the conglomerates represent fluvial channel deposits and reworked alluvial fans. During the final period of sedimentation the depocentre shifted once more to the west where the conglomerates of the **Pavlovsko Formation** (Havlíček 1954; up to 250 m) rest partly on the older strata of the Příbram–Jince Basin and partly on the Upper Proterozoic Kralupy-Zbraslav Group. According to Kukal (1963, 1971), these sediments were predominantly deposited by mudflows and more rarely in fluvial environments.

Marine regression associated with the filling of the Příbram–Jince Basin during late Middle and early Furongian times terminates the second tectonosedimentary cycle.

Skryje–Týřovice Basin

In the Skryje–Týřovice Basin (Figs. 4.20 & 4.28), an approximately 200 m thick succession of the Middle Cambrian Jince formations unconformably overlies the slightly metamorphosed Neoproterozoic Kralupy-Zbraslav Group. The basal sediments of the monomictic, white to grey Mileč Conglomerate and Sandstone (Kettner 1915; up to 10 m thick) are restricted to the area of Týřovice and Skryje. Fossils are preserved only as fragments due to deposition under high-energy, shallow-marine environments. The fauna includes common articulate brachiopods (*Jamesella* and *Pompeckium*; Havlíček 1970), locally common helcionelloid molluscs (*Helcionella*; Smetana 1918), and polymeroid trilobites (*Perneraspis*, *Germaropyge*, *Ptychoparioides*, and rarely *Paradoxides*; Růžička 1939; Šnajdr 1958). Kukal (1971) interpreted the dominant part of the succession as having formed on beaches, coastal barriers and bars. In some places, a particular facies of light sandstones (termed the Orthis Sandstone) with very common brachiopods can be distinguished (e.g. Kettner 1923). In the NE part of the basin such light coloured lithotypes are usually overlain by darker polymictic conglomerates, greyish green shales and greywackes of the Týřovice Greywacke and Conglomerate (Kettner & Kodym 1919). The Týřovice Conglomerates commonly contain pebbles of Proterozoic shales and greywackes, and are developed only in the NE part of the basin. These may represent the basal Cambrian strata in places where the Mileč Conglomerates and Sandstones are missing. More or less fragmentary shells of locally common articulate brachiopods (*Pompeckium* and *Jamesella*), helcionelloid molluscs (*Helcionella*) and within some levels spectacular graptoloids (Bouček 1938) and the lightly sclerotized *Wiwaxia* cf. *corrugata* (Matthew) are found (Fatka *et al.* in press). In the SE part of the basin, the oldest sediments are represented by greywackes commonly containing the brachiopod *Pompeckium*

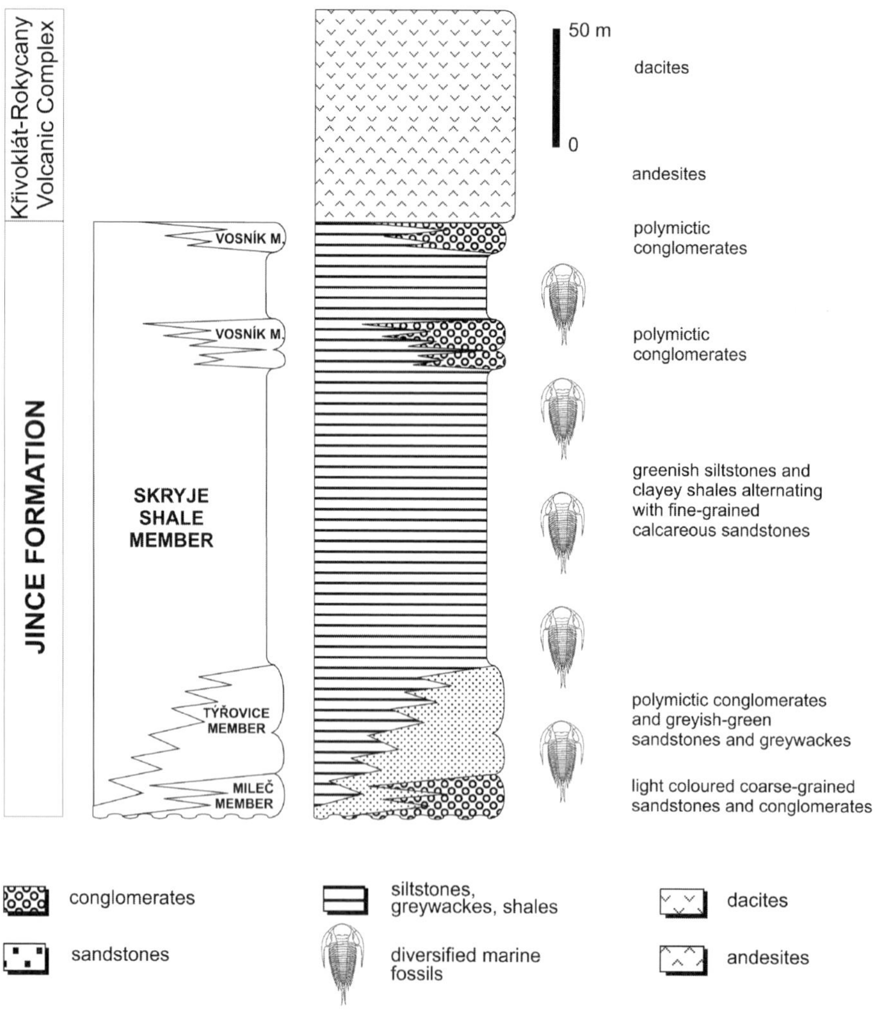

Fig. 4.20. Stratigraphy of the Cambrian rocks in the Skrye–Týřovice Basin (after Havlíček 1971)

and locally the trilobite *Ellipsocephalus*. However, the predominant and most widespread lithofacies is represented by the clayey shales and greywackes of the Skryje Shale (Jahn 1895; up to 200 m thick) with silty and sandy intercalations. This contains a well-diversified fauna (Fig. 4.21) including comparatively rare inarticulate brachiopods as well as the abundant articulate brachiopod *Bohemiella romingeri*, polymeroid trilobites (e.g. *Conocoryphe*, *Ctenocephalus*, *Eccaparadoxides*, *Ptychoparia*, *Sao*, *Agraulos*, *Skrejaspis*, *Jincella*), miomeroid trilobites (e.g. *Condylopyge*, *Pleuroctenium*) at higher levels, occasionally common cinctan and stylophoran echinoderms (*Trochocystites*, *Ceratocystis*), rare eocrinoids and edrioasteroids, molluscs, hyoliths, and moderately preserved and diversified acritarchs.

In the NE part of the basin the cessation of Middle Cambrian deposition more or less coincided with the deposition of several, metre-thick, tongue-like bodies of the polymictic Vosník Conglomerate (Kettner & Kodym 1919). These conglomerates appear in the higher parts of the Skryje Shale and in some places represent the uppermost part of the preserved succession immediately below the Křivoklát-Rokycany Volcanic Complex.

Kukal (1971) interpreted the deposition of the Týřovice Greywacke and Conglomerate, the Skryje Shale and the Vosník Conglomerate as being related to turbidity currents on steep slopes within the sedimentary basin.

The Middle Cambrian sedimentary rocks are covered by extrusives (dacites, andesites and rhyolites) of the Furongian Křivoklát-Rokycany Volcanic Complex.

Železné hory area

Havlíček (1949) published a preliminary report on the Middle Cambrian fauna discovered near the town of Heřmanův Městec in greywackes, sandstones and conglomerates of the Senice Shales. The fossil record was complemented by Havlíček & Šnajdr (1951*b*) and Šnajdr (1958). Tectonically deformed trilobite fauna from the Senice Shales incorporates five species known from the both the Příbram–Jince and Skryje–Týřovice basins.

Cambrian volcanic activity

The Furongian in the Barrandian area is only represented by volcanic complexes. Volcanic activity was concentrated in two separate areas.

The **Strašice Complex** shows a marked increase in calc-alkaline volcanism during the Late Cambrian (no radiometric dates available), which may correlate with intrusives of some West Bohemian plutons (Stod, Hanov, Lestkov massiv, Kdyně massiv).

The **Křivoklát-Rokycany Complex** is represented by tholeiitic and calc-alkaline subaerial basalts, andesites and rhyolites demonstrating several geochemical characteristics of an intraplate setting. These volcanic rocks have been dated at 494 Ma

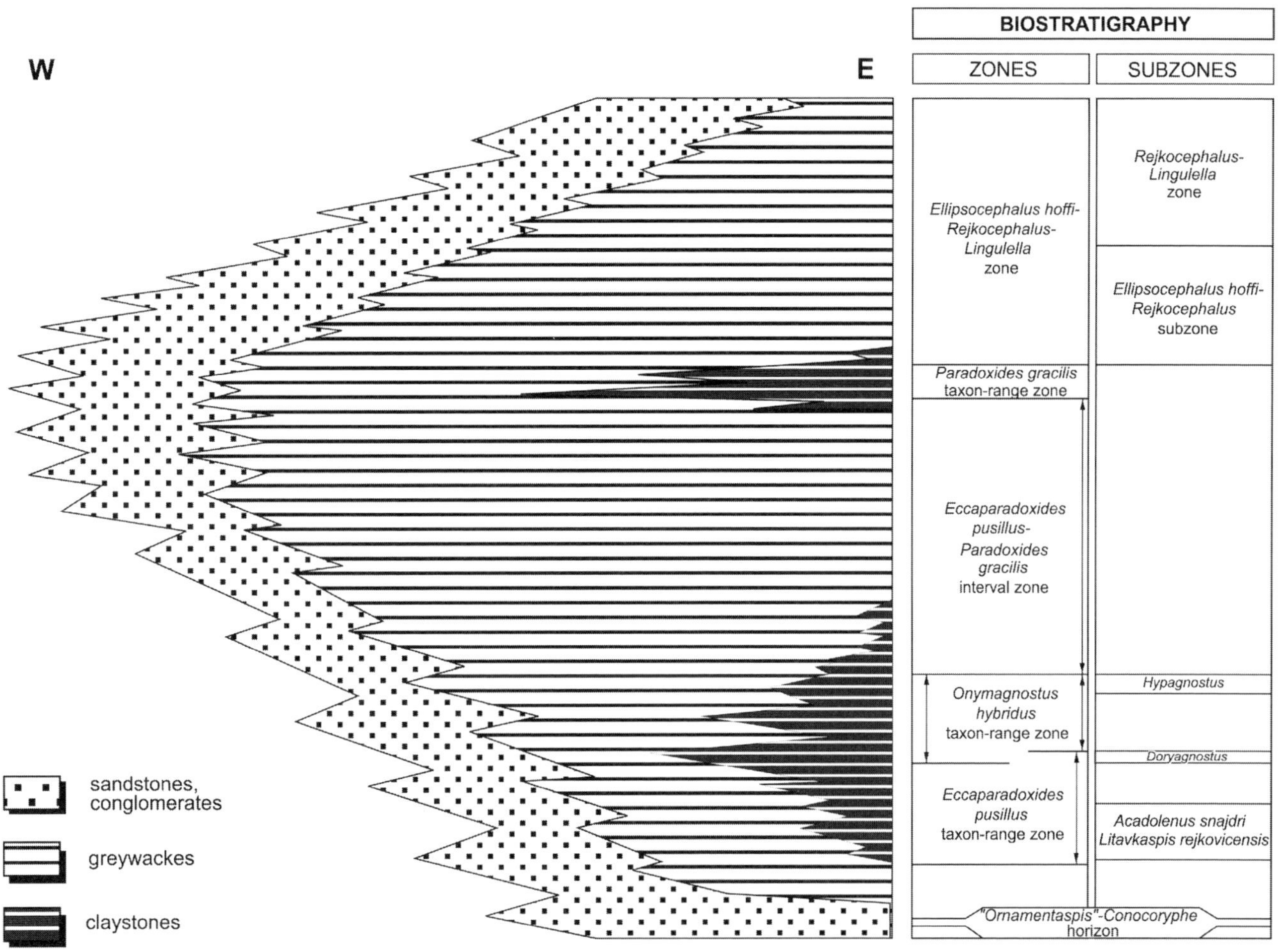

Fig. 4.21. Lithostratigraphy and temporal and spatial facies development of the Jince Formation in Příbram-Jince basin. Biostratigraphy after Fatka (2006).

with $(^{87}Sr/^{86}Sr)_i = 0.703$ (recalculated data from Vidal *et al.* 1975) and recently the clastic zircons provided a U–Pb SHRIMP age of 499 ± 4 Ma (Drost *et al.* 2004). The chemistry of the Křivoklát-Rokycany Complex apparently reflects a calc-alkaline precursor of the melt rather than an active continental marginal or island arc setting (as previously suggested by Patočka *et al.* 1993). Additionally, the structural criteria do not indicate any onset of compressional tectonics such as would be associated with subduction processes.

The Late Cambrian to Early Ordovician calc-alkaline basaltic to rhyolitic volcanism (Fediuk & Röhlich 1960; Fiala 1971) terminates the Cadomian tectonometamorphic megacycle. Older fissures, which provided access to the surface for the Furongian volcanics of the Strašice Complex, were reactivated during the Ordovician (Komárov Fault) and thus have a close relationship to both of these volcanic episodes. Mantle-derived alkaline and tholeitic magmas of Ordovician through Devonian ages reflect the ongoing and increasingly rapid extension of the Cadomian crust during the onset of the subsequent sedimentary cycle.

Western margin of the Bohemian Massif (G.G.)

The so-called **Erbendorf-Vohenstrauss Zone** (ZEV) differs both lithologically and structurally from the adjacent areas suggesting a unique tectonometamorphic history. Data from Cambrian rocks on the NW margin of the Bohemian Massif in Bavaria, however, remain somewhat speculative. The Cambrian appears to be represented by the Neustadt Group, which is considered to be Cambro-Ordovician in age (Hirschmann *et al.* 1994; Hirschmann & Stettner 1997). Geochemical characteristics of those metabasites, metagranites and metasediments were combined with radiometric, petrologic and structural data, but interpretations are still equivocal. The paragneisses of the southern part of the Erbendorf-Vohenstrauss Zone were dated as Lower Cambrian on the basis of Rb–Sr isotopic data (530 Ma; Weger *et al.* 1998). Metasedimentary and associated metavolcanic rocks suggest that clastic sedimentation occurred not far from a Gondwanan source area and close to a volcanic island arc. The paragneisses of the SE part of the Erbendorf-Vohenstrauss Zone were apparently formed in a similar environment, but even closer to the Gondwanan margin (Weger *et al.* 1998).

The region of the **Bayerischer Wald** and the **Oberpfälzer Wald** as part of the Moldanubicum *sensu stricto* is even more problematic in terms of age assignment (compare Stettner *et al.* 1997). The southern Bavarian Forest region (Bayerischer Wald) contains evidence of acidic and basic magmatism at *c.* 550 Ma. Leucocratic metamagmatites and metabasites from the Passau area yielded zircons with SHRIMP ages of 555 ± 12 Ma, 549 ± 7 Ma and 549 ± 6 Ma, respectively (Teipel *et al.* 2002). In the northern to middle part of the Bavarian Forest area, zircons from leucocratic metamagmatites indicate minimum SHRIMP ages of 486 ± 7 Ma and 480 ± 6 Ma, indicative of acidic magmatism close to the Cambrian–Ordovician boundary (Teipel *et al.* 2002).

The **Teplá-Barrandian** in western Bohemia and easternmost Bavaria (Fig. 4.22) is a supracrustal complex which, other than the adjacent Moldanubicum *sensu stricto* and the adjacent Saxothuringian Zone, was not affected by high temperature–low pressure metamorphism during the Carboniferous, and grades into the regular sedimentary and magmatic rocks of the Barrandian. In contrast to the Moldanubicum *sensu stricto* the thickened Cadomian crust collapsed close to the Proterozoic–Cambrian boundary allowing deposition of a thick sequence of probably continental clastic deposits in graben structures. Data from the Barrandian and the Teplá-Barrandian suggest that synsedimentary faults and volcanic rocks at various stratigraphic levels indicate ongoing tectonic activity during the early Cambrian, so that a dextral transtensional regime can be assumed to have followed the Ediacaran Cadomian subduction and microplate collisional stages (Zulauf *et al.* 1999; Dörr *et al.* 2002; Dallmeyer & Urban, 1998; Fig. 4.2). Cadomian crustal stacking and regional metamorphism is indicated by Barrovian-type metamorphic isograds in the Domažlice crystalline complex, the formation of which was dated from metamorphic monazite at 551 Ma to 540 Ma (Zulauf *et al.* 1999; Timmermann *et al.* 2002). Faulting led to the exhumation of amphibolite-facies rocks of the Domažlice crystalline complex, associated with eastward tilting of the crustal slice responsible for the present pattern of metamorphic isograds (Zulauf 1997). Mylonitized amphibolites of the Neukirchen-Kdyne igneous complex are interpreted as metamorphic basalts derived from N-type MOR basalts (Propach & Pfeiffer 1998). The erosion of upper crustal levels of this region is documented by Lower Cambrian continental sediments in the Barrandian Basin. During late Early Cambrian, the Mracnice and Stod plutons, and probably the Pobezovice Pluton, intruded the crust (Zulauf 1997). The geomorphology created during Cadomian orogenic deformation was largely removed by the early Middle Cambrian as suggested by the presence of marine Middle Cambrian deposits. However, magmatic activity continued, and the intrusion of associated magmatic dykes continued until early Ordovician times (Zulauf 1997). These plutons and dykes intruded into roughly NE–SW trending transtensive shear zones. According to the interpretation of Zulauf (1997), this Cambrian transtension heralded the large-scale Ordovician rifting that led to the separation of the Teplá-Barrandian, as a part of Armorica, from Gondwana.

Black Forest and Vosges area (G.G.)

The basement of the Moldanubian Zone in western Central Europe is exposed in the Black Forest mountains (Schwarzwald) in SW Germany and the Vosges area in neighbouring NE France. Both areas represent a coherent region, which is now separated by the Rhine Valley Graben. However, the vestiges of Cambrian history differ considerably between the two areas.

The Cambrian in the **Black Forest** is known only from some magmatic rocks, the radiometric signature of which indicates intrusion during the Cambrian (and the Ordovician) and also suggests that anatexis continued after the Cadomian event (Maass 1974, 1981; Fluck *et al.* 1980; Maass *et al.* 1980). The best known of these rocks are the Schapbach Gneisses. These are metaplutonic orthogneisses from the central part of the Black Forest, which have granitic, granodioritic, trondhjemitic and tonalitic compositions. An intrusion age of roughly 520 Ma was determined for their protoliths.

Isotopic and chemical compositions of the Cambrian and Ordovician granitoids have been studied primarily in the Black Forest and have triggered a debate as to whether their origin is arc- or rift-related. Recent investigations on the zircon ages constrained the emplacement of tonalitic, trondhjemitic and granodioritic orthogneisses in the Black Forest between 510 and 500 Ma (Chen *et al.* 2000). Chemical data (Nd values, REE patterns, K_2O/Na_2O ratios, etc.) suggest evidence for melting of amphibolites and contributions from an evolved crust. In addition, magmatic emplacement was followed by a high temperature event at *c.* 480 Ma, which was interpreted by Chen *et al.* (2000) as resulting from lithospheric thinning in a marginal basin behind a Cambrian magmatic arc.

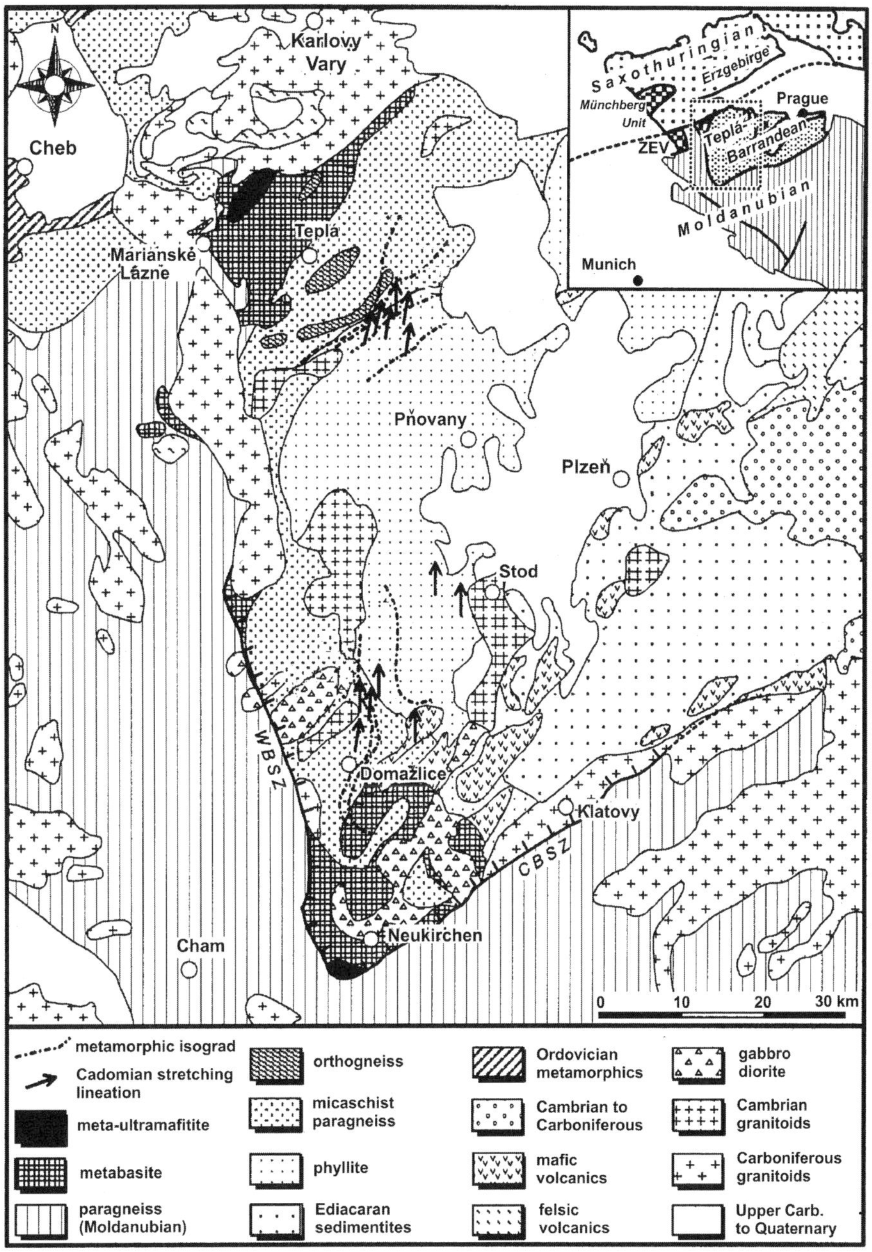

Fig. 4.22. Simplified geological map of the Teplá-Barrandian Unit and adjacent areas of the Moldanubicum (*sensu stricto*) and Saxothuringian Zone. CBSZ, Central Bohemian shear zone; WBSZ, West Bohemian shear zone; ZEV, Zone of Erbendorf-Vohenstrauß. Arrows indicate sense of D_2 shearing of the hanging wall. Slightly modified after Dörr *et al.* (2002, fig. 3).

In the northern **Vosges area** phyllitic schists, with intercalations of porphyroides, metasandstones and conglomerates, are termed Villé Schists (Weiler Schiefer or Schistes de Villé) (Fluck *et al.* 1984; Fig. 4.23). This extensive and heterogenous unit was originally thought to be Proterozoic in age but now is believed to include rocks of Cambrian and Ordovician age and appears to represent a primarily sedimentary unit. The central Vosges contains a highly metamorphic unit termed the St. Marie-aux-Mînes Gneisses, which consist of a lower amphibolite complex with probably younger garnet- and sillimanite-bearing paragneisses and kinzigite gneisses. As with the Villé Schists, this unit represents a primarily sedimentary unit with volcanogenic portions reflecting evolution during the lower part of the Cambrian, whereas the upper part of the Cambrian appears to have been composed of greywackes with some basaltic and rhyolitic episodes.

The margin of Baltica (G.G. & A.Ż.)

Northern, central and eastern Poland and the adjacent north-eastern tip of Germany were parts of the continent of Baltica during the Cambrian. The Cambrian strata in the subsurface of Poland are located in a zone in which a system of NW–SE directed peri-Tornquist basins such as the Baltic Basin and the Lublin-Podlasie Basin interfere with NE–SW directed structures.

Commencing in the mid-twentieth century, numerous deep

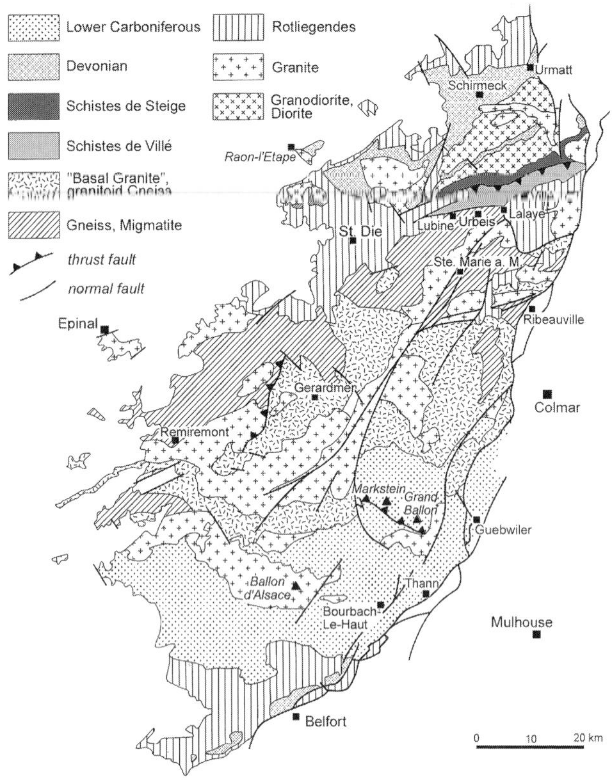

Fig. 4.23. Sketch map of the Palaeozoic rocks in the Vosges area with Cambrian units (Schistes de Steige, Schistes de Villé) in grey. Redrawn and slightly modified after Walther (1992, fig. 85).

borings penetrated the sedimentary cover in central and northern Poland and reached Cambrian subsurface deposits of the East European Platform. Additional data, primarily collected in the 1980s, came from around the Rügen area, northern Germany. The data from the cores revealed the thickness and lithological characteristics and enabled comparison of the stratigraphic sequences with previously known sections from outcrop areas such as Bornholm, southern Sweden, and Estonia.

Data from early borings in the Polish area resulted in the identification of a latest Proterozoic (so-called Vendian) stratigraphic complex (e.g. Sokolov 1952, 1953), which indicated the Cambrian lower boundary interval. Detailed reviews on the Cambrian strata and fossils followed in the 1960s (see Lendzion 1976, 1983*a*, *b*; Areń & Lendzion 1978; Mens *et al.* 1990).

The Baltic Basin in Poland

The northwestern and northeastern part of Poland with the northeastern tip of Germany represent the western flank of a *c.* 600 km wide depositional basin termed the 'Baltic Basin' or 'Baltic Syneclise'. This basin developed in the Ediacaran after the breakup of Rodinia and formed a passive continental margin, as subsidence decreased in latest Proterozoic times, on the western slope of the palaeocontinent of Baltica. The Ediacaran through Lower Palaeozoic successions can be identified as four depositional cycles, generally developed as consecutive transgressive–regressive cycles. The first of these depositional cycles includes the Ediacaran through Middle Cambrian deposits and thus the majority of the Cambrian rocks (Jaworowski 2001).

Distinct differences exist in the depositional history and, as a result, in the stratigraphy of the Cambrian in the Baltica sector in Poland. A western area shows a relatively complete but thin stratigraphic sequence of Cambrian rocks with a transition from the Ediacaran and with some hiatuses in the Middle Cambrian and Furongian (Fig. 4.24). In NE Poland Cambrian rocks are represented only by an upper Lower to lower Middle Cambrian clastic sequence bound by major unconformities and resting on Precambrian crystalline basement rocks, and by a stratigraphically restricted Furongian/Upper Cambrian alum shale unit (Fig. 4.25). In SE Poland, in contrast, there is a continuous transition from the Ediacarian to the Middle Cambrian and a hiatus in the Furongian (Fig. 4.26). The differences between the areas are due to a western high-gradient, *c.* 150 km wide sub-basin adjacent to the Tornquist Ocean, which was affected by thermal and sedimentary loading with considerable subsidence. A general gradual widening of the Baltic Basin is explained by extensional forces in the Tornquist rift zone towards the end of the Early Cambrian. In the mid-Middle Cambrian, apparent compression of a rift-to-drift stage created an upwarping of the central part (Poprawa *et al.* 1999; Greiling *et al.* 1999; Šliaupa 2001, 2002)

Northwestern Poland

The stratigraphic succession in NW Poland (Fig. 4.24) is characterized by typical Baltica-type thin, clastic formations in the Lower and Middle Cambrian with a tendency to successively finer-grained material and thin, alum shale deposits in the Furongian. Nine formations have been distinguished.

The oldest rocks are summarized as the **Żarnowiec Formation** (Lendzion 1970; *c.* 150 m in thickness), named after the Żarnowiec IG-1 borehole (corresponding to the Smołdzino Formation of Bednarczyk & Turnau-Morawska 1975). The formation is composed of sandstones varying grain sizes, horizontal laminated or cross-bedded, with local intercalations of siltstones. Fossils are absent, and it is believed that the Ediacaran–Cambrian boundary is located within this formation. The sediments represent alluvial and braided fans, as well as fan and braided deltas, with a gradual change from terrestrial to marine conditions (Jaworowski 1997). Analysis of the sandstones confirms the regional origin of the material. The crystalline basement in the Kościerzyna IG-1 region mainly yielded enderbitic detritus while that in the Gdańsk IG-1 region provided migmatitic detritus, which was transported from the craton interior (Sikorska 2000). These results are consistent with the view that these strata on the Pomeranian Caledonides foreland were formed on the passive margin of Baltica. The overlying **Kluki Formation** (Bednarczyk & Turnau-Morawska 1975; thickness up to 200 m) is a variably coloured clastic unit dominated by quartzose siltstones. Hyoliths, the index fossil *Mobergella*, acritarchs and other fossil remains place it into the traditional Baltic *Mobergella* biozone. This formation is partly overlain by the **Nowęcin Formation** (Bednarczyk & Turnau-Morawska 1975; maximum thickness *c.* 13 m), comprising sandstones with varying grain sizes, with conglomerates in its bottom and with mudstone intercalations. It most probably corresponds to the *Schmitdiellus mickwitzi* biozone.

The overlying **Łebsko Formation** (Bednarczyk & Turnau-Morawska 1975; maximum thickness 185 m) is generally composed of quartzose and glauconitic sandstones with thin siltstone interbeds and typical phosphatic nodules. Although the fossil assemblages found in this formation have a low diversity (with hyoliths, inarticulate brachiopods, bradoriids and trilobites), the youngest trilobites (e.g. *Paradoxides insularis*, *Bailiella emarginata*; Bednarczyk 1984) indicate that the formation extends into the lower part of the Middle Cambrian. The transitional lower

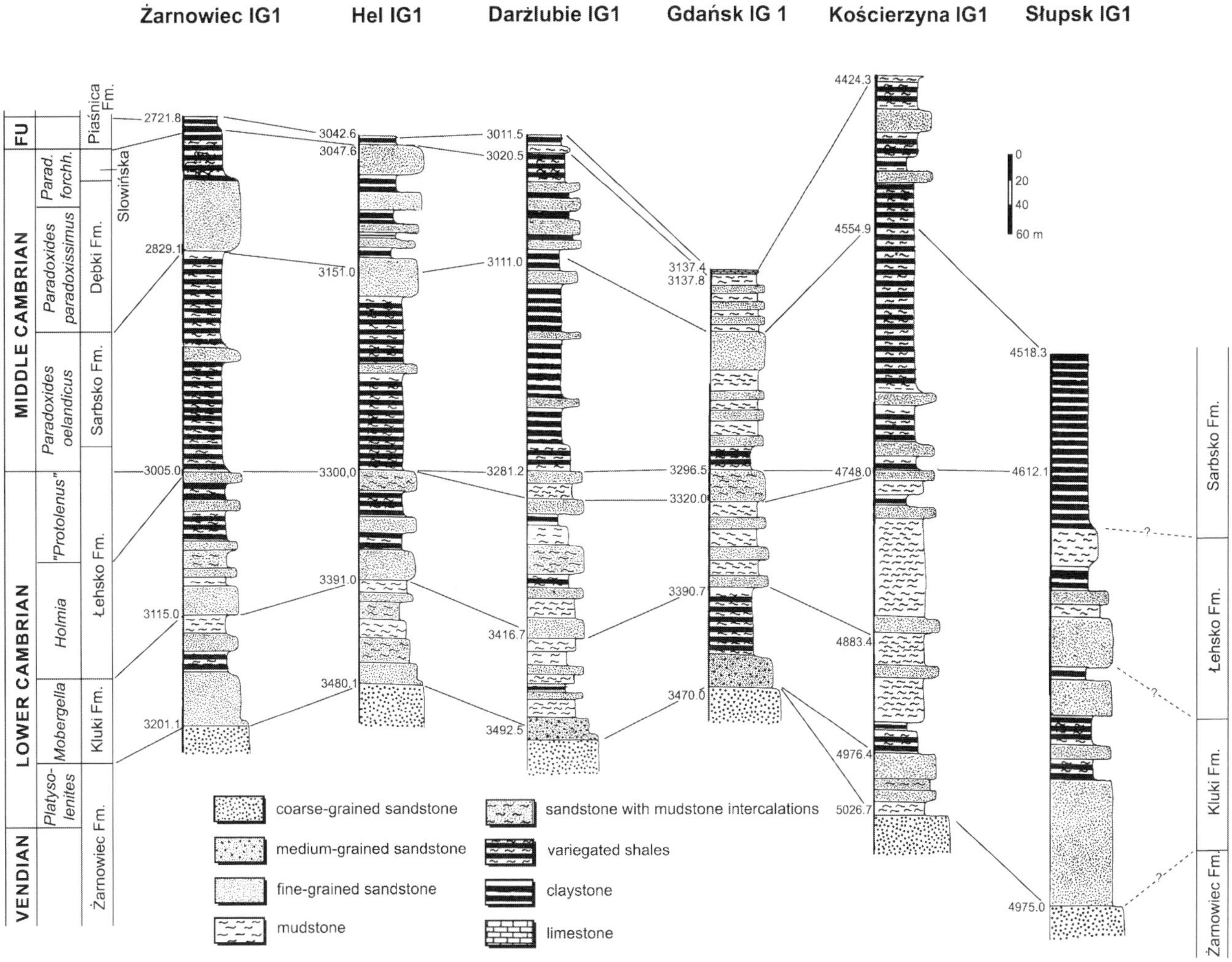

Fig. 4.24. The Cambrian in the Baltic Basin in NW Poland, illustrated by cores from important boreholes. Biostratigraphy according to Lendzion (1976, 1983*a*, *b*). Redrawn and supplemented with lithostratigraphic data after Pacześna (1996, fig. 2).

boundary and lithological similarities with the formations in southern Sweden suggest that the Łebsko Formation ranges from the *Holmia inusitata* to the *Paradoxides insularis* zones.

The **Sarbsko Formation** (Bednarczyk & Turnau-Morawska 1975; maximum thickness 210 m) consists of alternating grey quartzose sandstones or siltstones and dark argillitic shales. Very fossiliferous strata include a diverse fauna, mainly composed of trilobites and inarticulate brachiopods (Bednarczyk 1984). Index trilobites permit a precise biostratigraphical placement from the *Paradoxides insularis* through the *Ptychagnostus praecurrens* and (due to an apparent diachronism of the upper boundary) locally into the *Triplagnostus gibbus* zones. The **Dębki Formation** (Bednarczyk & Turnau-Morawska 1975; maximum thickness 80 m), similar to the Sarbsko Formation, is composed of alternating quartzose sandstones and dark argillitic shales. Its fauna of trilobites and inarticulate brachiopods includes such characteristic species as *Paradoxides paradoxissimus*, *Ellipsocephalus lejostracus* and *Triplagnostus gibbus* (Bednarczyk 1984) and represents the *T. gibbus* and *P. atavus* zones of the Middle Cambrian. The overlying **Osieki Formation** (Bednarczyk & Turnau-Morawska 1975; maximum known thickness 31 m) is an areally restricted unit of quartzose, calcareous siltstones with calcareous lenses and argillaceous interbeds. Fossils are unknown, and its stratigraphic assignment to the upper Middle Cambrian *P. punctuosus* Zone is based on the conformable contact with the Dębki Formation and an unconformity at the top.

The **Białogóra Formation** (Bednarczyk & Turnau-Morawska 1975) is a thin cover of quartzose, glauconitic sandstones with phosphoritic and argillaceous nodules (up to 4 m thick). Trilobites and brachiopods (such as *Lejopyge laevigata* and *Billingsella exporrecta*) allow precise correlation and assignment to the *Erratojincella brachymetopa* and *Lejopyge laevigata* zones of the uppermost Middle Cambrian.

The **Słowińska Formation** (Bednarczyk & Turnau-Morawska 1975) is a typical alum shale unit rich in organic matter (kerogen) and with organic-rich limestone beds and lenses (stinkstone). Although the maximum thickness does not exceed 12 m, a rich fauna with typical trilobites (e.g. *Agnostus pisiformis*, *Homagnostus obesus*, *Olenus wahlenbergi*, *O. truncatus*, *Parabolina spinulosa*) and conodonts (e.g. *Westergaardodina tricuspidata*) occurs, indicating the *Agnostus pisiformis* through *Parabolina spinulosa* zones of the early Furongian.

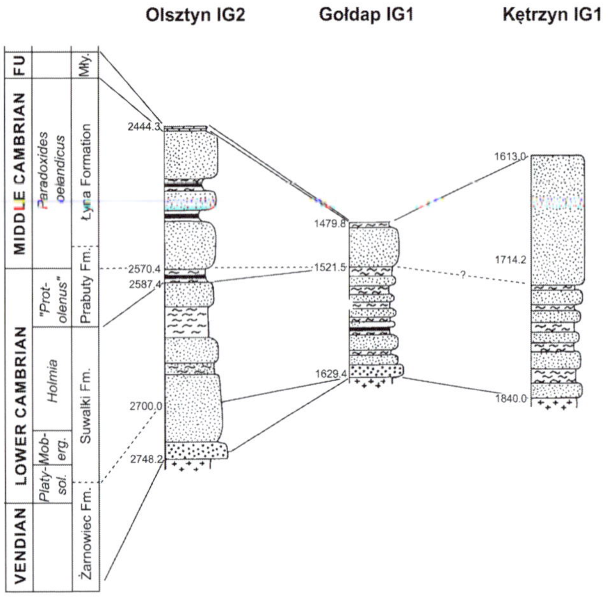

Fig. 4.25. The Cambrian in the Baltic Basin in NE Poland, illustrated by cores from important boreholes. Biostratigraphy according to Lendzion (1976, 1983*a*, *b*). Redrawn and supplemented with lithostratigraphic data after Paczeŝna (1996, fig. 3).

The **Piaśnica Formation** (Bednarczyk & Turnau-Morawska 1975) is similar to the Słowińska Formation, consisting of black, organic-rich argillites with interbeds and lenses of coarsely crystalline and often biodetritic limestones. The thickness reaches only 6 m, and the formation rests on the underlying Słowińska Formation with an erosional hiatus. The faunas consist of typical trilobites (e.g. *Peltura scarabaeoides*, *Sphaerophthalmus alatus*, *Parabolina lobata*) and conodonts, which suggest an age corresponding to the late Furongian *Peltura* and *Acerocare* zones. A continuation into the earliest Ordovician has been suggested.

Northern and northeastern Poland

The subsurface Cambrian rocks in north and NE Poland present incomplete successions of clastic Lower and Middle Cambrian rocks, overlain by a thin cover of alum shale (Fig. 4.25).

The basal unit is known as the **Suwałki Formation**. Resting disconformably on the crystalline basement, the Suwałki Formation (established as a group; Znosko 1961; up to 80 m) consists of sandstones and siltstones with intercalated iron-rich oolites and sideritic nodules. The strata yielded different fossil assemblages with trilobites such as *Holmia kjerulfi* and *Strenuaeva primaeva*, brachiopods and *Hyolithellus*. In addition, diverse ichnofossil and acritarch assemblages have been described. The trilobites clearly indicate a *Holmia kjerulfi* Zone age, in contrast to an earlier assumed late Proterozoic age (Znosko 1961).

The overlying **Prabuty Formation** (Lendzion 1986; maximum thickness 80 m) is composed of quartzose and glauconitic sandstones alternating with siltstones and argillites and containing phosphatic nodules. Macrofaunas include various brachiopods, hyoliths, and a trilobite assigned to *Ellipsocephalus*. Various trace fossils are known from the sandstones, and acritarchs were also studied. Although the biostratigraphic data are not unequivocal and prove only a lower Middle Cambrian age, it is assumed that the formation ranges from the *Holmia kjerulfi* into the *P. insularis* zone and has a strongly diachronous upper boundary. The **Łyna Formation** (Lendzion 1986; maximum thickness 140 m) develops from the Prabuty Formation and in places appears to be a lateral equivalent of the latter. The formation is dominated by usually weakly cemented sandstones with occasional calcareous nodules and some siltstone intercalations. The rare and poorly preserved fossils include the trilobite *Ellipsocephalus* cf. *polytomus*, inarticulate brachiopods, and a few trace fossils. The age of the formation is lower Middle Cambrian.

The Łyna Formation is disconformably overlain by the **Młynary Formation** (Lendzion 1986). The formation includes such trilobites as *Agnostus pisiformis*, *Olenus* cf. *truncatus*, and *Homagnostus obesus*, as well as the brachiopod *Orusia lenticularis*, indicative of an age corresponding to the *A. pisiformis* through *Parabolina spinulosa* zones. It is therefore an equivalent of the Słowińska Formation identified from NW Poland. However, the Młynary Formation is composed of fine-grained calcareous sandstones and finely crystalline limestones rather than alum shale. It is disconformably overlain by lower Ordovician rocks.

The Lublin-Podlasie Basin

The Lublin-Podlasie Basin in SE Poland (Fig. 4.26) formed as a result of the NW–SE directed peri-Tornquist depression crossed by the NE–SW directed so-called Orsha-Volhyn Aulacogen. As with the Baltic Basin, this basin developed in the Ediacarian following the breakup of Rodinia on the western slope of the palaeocontinent of Baltica. It consists of two structural subbasins: the Podlasie depression and the Lublin slope. Both subbasins record a nearly continuous succession of unmetamorphosed sedimentary rocks from the Ediacarian to the mid-Middle Cambrian, which (in contrast to Neoproterozoic rocks) constitute a relatively uniform cover. This similarity of the Cambrian sedimentary cover is a result of the development of a single depositional system extending from the late Ediacaran into the Middle Cambrian with deposition in open marine, shoreface and offshore sedimentary environments (Poprawa & Paczeŝna 2002; Paczeŝna & Poprawa 2004) (Fig. 4.27). Comparison with the Baltic Basin suggests that the deposits of late Middle and early Late Cambrian age were subsequently eroded during the latest Cambrian and earliest Ordovician (Paczeŝna & Poprawa 2004).

The base of the traditional Cambrian lies in the **Włodawa Formation**, a roughly 100 m thick unit of shales grading into fine- and medium-grained sandstones. The traditional Proterozoic–Cambrian boundary on the East European Platform is indicated by the base of the *Platysolenites antiquissimus* Zone in the upper part of the formation. The overlying **Mazowsze Formation** (Lendzion 1984; maximum thickness 175 m) consists of quartz sandstones with glauconite, phosphoritic concretions, and siltstone interbeds. The strata may also rest either on the Lublin Formation referred to the Ediacaran (upper Vendian) or on the crystalline basement. According to acritarch biostratigraphic data, the global Ediacarian–Cambrian boundary is located in the lowermost part of this formation. The formation also contains evidence of a remarkable radiation of metazoans as indicated by a diverse trace fossil assemblage whcih inhabited the open, shallow-marine environment (Paczeŝna 1986, 1996, 2001). This assemblage in the Mazowsze Formation appears to feature the *Trichophycus pedum* assemblage zone, the base of which indicates the Global Stratotype Section and Point in SE Newfoundland. The overlying **Kaplonosy Formation** (Lendzion 1984; maximum thickness 200 m) is represented by sandstones

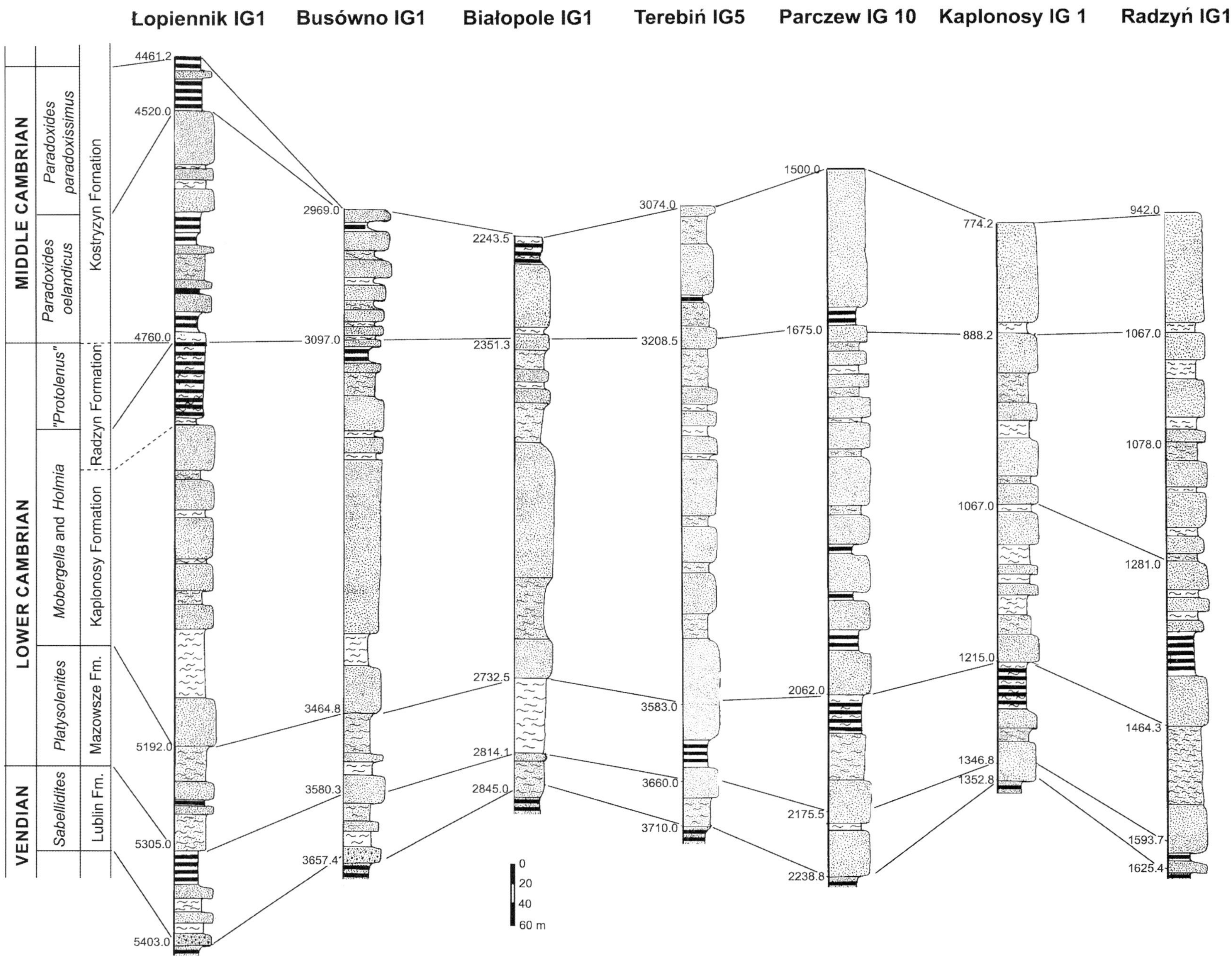

Fig. 4.26. The Cambrian in the Lublin-Podlasie Basin in SE Poland, illustrated by cores from important boreholes. Biostratigraphy according to Lendzion (1976, 1983*a*, *b*). Redrawn and supplemented with lithostratigraphic data after Pacześna (1996, fig. 4).

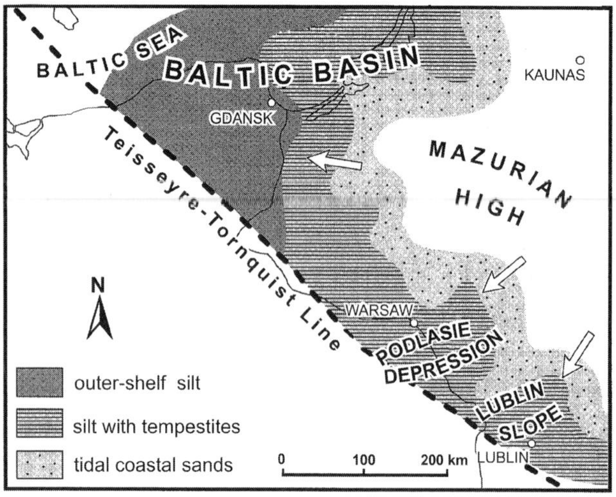

Fig. 4.27. Simplified palaeogeographic reconstruction of the southwestern rim of the East European Platform illustrating Middle Cambrian sedimentary facies. Modified after Jaworowski (1997) and Valverde-Vaquero *et al.* (2000, fig. 3B).

(fine-grained or varying in grain size) with few dark siltstone interbeds. The beds are often strongly bioturbated. Acritarch data are indicative of the *Holmia* zone. It is overlain by the **Radzyń Formation** (Lendzion, 1984; maximum thickness about 30 m), which comprises bioturbated siltstones with claystone and sandstone interbeds. The formation includes numerous trilobites (with *Schmidtiellus* sp., *Holmia* sp. and *Strenuaeva primaeva* in its lower part, and *Kingaspis* (*K.*) *borealis* and *Ellipsocephalus* cf. *hoffii* in its upper part) and is an equivalent of the *Holmia kjerulfi* and '*Ornamentaspis*' *linnarssoni* zones. The overlying unit is termed the **Kostrzyń Formation** (Lendzion 1986; maximum thickness less than 115 m), which consists of fine-grained sandstones, with episodic coarse quartz arenite and siltstone interbeds. The fauna is sparse, but the presence of *Ellipsocephalus polytomus* and *Paradoxides sjoegreni* indicates a lowermost Middle Cambrian *Eccaparadoxides oelandicus* superzone. The Kostrzyń Formation is unconformably overlain by Lower Ordovician (Tremadocian) strata.

Rügen area

Three major NW–SE striking fault zones affect the northeasternmost part of Germany in the area of the island of Rügen. They include the Trans-European Fault (see also discussion in McCann & Krawczyk 2001), the Caledonian Deformation Front north of Rügen, and the Tornquist Zone with the Sorgenfrei-Tornquist Zone and the Teisseyre-Tornquist Zone *sensu stricto*. Among the boreholes in this area, the G14–1/86 well (located in the Baltic Sea *c.* 35 km NE of Rügen) penetrated a Cambro-Ordovician succession on the southern margin of Baltica (Franke *et al.* 1994; Beier & Katzung 1999).

The Cambrian rock succession at this location includes 316 m of clastic sedimentary rocks of probable Early Cambrian age. The succession begins on top of an erosional unconformity with predominantly coarse sandstones and conglomerates of subtidal origin, grading into finer-grained, well-washed quartz arenitic sandstones with occasional pebble horizons. The upper half of the supposed Lower Cambrian rocks consists of laminated, often calcareous sandstones with occasional shale interbeds. Siltstones are frequently bioturbated, with *Diplocraterion*-type trace fossils. The uppermost Lower Cambrian strata are glauconite-bearing fine-grained sandstones with flaser-bedded siltstones and with an increasing amount of chloritized argillaceous clasts that suggest a very shallow marine origin. A coarse sandstone cap 1 m thick is possibly an equivalent of the Rispebjerg Formation which is present in Bornholm and Scania (Franke *et al.* 1994; Beier & Katzung 1999).

The Middle Cambrian succession begins with grey, calcareous shales with pyrite and phosphoritic nodules, grading into pyritic and phosphoritic argillites with some limestone beds and typical Scandinavian *P. paradoxissimus* and *P. forchhammeri* stage faunal elements. This *c.* 7 m thick Middle Cambrian sequence is capped by a 1.6 m thick limestone with phosphoritic concretions and anthraconitic layers, which appears to represent a condensed unit of Exsulans and Andrarum limestones. Furongian strata are primarily monotonous alum shales of *c.* 22 m in thickness, with a higher amount of pyrite (Beier & Katzung 1999).

Petrographic evidence suggests that the Cambrian sedimentary successions from the G14-1/86 well were derived from a stable cratonic area. Rare volcanic fragments suggest coeval erosion of subordinate volcanics (Franke *et al.* 1989; McCann 1998).

Palaeogeography and geotectonic history

Baltica was an independent continent separated from other crustal segments of present-day Central Europe through most of the Cambrian period. Recent investigations (Nawrocki *et al.* 2004; Nawrocki 2004*b*) suggest that the Baltic plate moved during the Ediacaran and Cambrian from moderate southern latitudes to a more equatorial position in southern latitudes during which it rotated anticlockwise. The European passive margin of Baltica was separated from an active margin of Gondwana/Avalonia.

Subsidence analyses of the basins in Poland through the Neoproterozoic and Cambrian indicates that a synrift phase in the Ediacarian is followed by a transition from a synrift to post-rift phase in latest Ediacarian and earliest Cambrian times, as indicated by coarse-grained continental conglomerates and arkoses with a lateral facies shift to sandstones and even mudstones, overlain by various shallow-marine clastic rocks. The remaining Early Cambrian through Middle Ordovician was a period characterized by post-rift thermal subsidence of a passive continental margin, as indicated by the presence of marine sedimentary units with a clear tendency towards finer-grained rocks plus a lateral expansion of the basins (Poprawa & Paczesna 2004; Poprawa *et al.* 2005) (Fig. 4.27).

The nature of the European margin, however, has been the subject of extended discussion. Two rival hypotheses have been suggested for the geotectonic position of the Pomerania-Rügen Caledonides. Analyses of gravity and magnetic data appeared to favour the presence of the so-called Circum-Fennosarmatian Caledonides. An alternative explanation claimed that the craton of Baltica (or the European Platform) extended to the west and reached northern Germany and Pomerania. Data from boreholes and seismic studies led to additional hypotheses such as the concept of an aulacogen and a major strike-slip fault, now proved to be incorrect. However, even today two different explanations exist. A number of scientists assume that Baltica extends to the Elbe-Odra Line, whereas others believe that the presence of deformation in Rügen and Pomerania are a result of terrane tectonics (see also discussion in Krawczyk *et al.* 2008). Recent data from analyses of differences in the crustal structure in northern Germany and western Poland, and the concept of a counterclockwise rotation of Baltica during the Ordovician, proved by palaeomagnetic data, seems to favour the latter model of proximal terranes detached from the cratonic margin farther to the SE and then re-accreted (Dadlez 2000; Nawrocki *et al.* 2004*a*).

Alpine realm (G.G.)

The presence of Cambrian sedimentary rocks in the Alps has not been proven directly. However, there is little doubt that more-or-less strongly metamorphosed suites such as in the Upper Austro-Alpine nappes ('Oberostalpin'), but also little metamorphosed units such as in the South-Alpine Karawanken and Carnic Alps and also many basement blocks which compose the Variscan areas in the Alps, include Cambrian rocks (Fig. 4.1).

Pre-Carboniferous basement areas in the Alps contain relicts of different structural areas including oceanic crust, magmatic arcs, subduction zones and granitoid belts. Oceanic plagiogranites and island-arc type granitoids of the Silvretta Nappe have been dated at 523 ± 30 Ma (Müller *et al.* 1996), and island-arc type tonalites and gabbros from the Silvretta Nappe have revealed an almost identical age of *c.* 523 Ma (Schaltegger *et al.* 1997), which would bring them into the late Early Cambrian. Rocks of Cambro-Ordovician age in this region usually feature a complex association of different types of metasediments and amphibolite units, as well as ultramafic rocks. The tectonic mixing is indicative of a complex Cambrian and post-Cambrian subduction scenario. Cambrian rocks were consumed during the closure of a backarc basin (*c.* 525 Ma; Schaltegger *et al.* 1997), whereas the later Cambrian featured an oceanic stage of Palaeotethys with active margin magmatism during the Furongian (Müller *et al.* 1995, 1996; Schaltegger *et al.* 1997). The northern Helvetic basement in the Central Alps includes rocks that formed during Ordovician tectonic events, which commenced immediately after the end of the Cambrian (Schaltegger *et al.* 2003) and ended with granitoid magmatism during Ordovician and Silurian continental collision. The Gotthard (Sergeev & Steiger 1995) and the Aiguilles Rouges massifs (Raumer *et al.* 2003) have similar tectonostratigraphic histories. A similar Palaeozoic evolution is assumed for Penninic basement blocks. However, the Penninic basement records magmatism along an active continental margin at 520 Ma (Schaltegger *et al.* 2002) and intracontinental rifting at 500 Ma (Bussy *et al.* 1996), whereas the backarc volcanism in the (Helvetic) Aar and Gotthard massifs appears to have a maximum age of *c.* 480 Ma (Schaltegger *et al.* 2003). All of these Alpine basement blocks may have been derived from Gondwana and are interpreted, for the Ordovician, as parts of the Hun Superterrane (European Hunic Terrane; Stampfli 2000; Stampfli *et al.* 2002; Raumer *et al.* 2002, 2003).

Many of the basement areas in the central and western Alps may correspond to the 'Noric Composite Terrane' (Frisch & Neubauer 1989) with elements of 'Speik' (Cambrian ophiolite) and 'Celtic' (Precambrian–Cambrian magmatic arc) terranes, and an early Palaeozoic cover (Raumer 1998). The similarity of the pre-Variscan evolution in present-day Alpine and 'classical' Variscan realms is striking. The evolution of the 'Noric Composite Terrane' parallels that of the 'Saxothuringian Composite Terrane' (Linnemann & Buschmann 1995*a*), which contained a Late Proterozoic–Cambrian arc, on which a Cambrian rift system developed, and an Ordovician overstep sequence. In addition, Raumer & Chiaradia (1992) have compared a supposed Proterozoic–Early Palaeozoic scheelite–diopside–skarn-bearing series from the External Massifs with the fossil-bearing Proterozoic–Cambrian sequences from the Delitzsch-Torgau-Doberlug Syncline.

U–Pb SHRIMP analyses of zircons from ore bodies in the well-known Felbertal tungsten deposit in the Tauern Window of the central Alps in Austria indicate that basalt emplacement occurred at the latest Neoproterozoic–Cambrian boundary interval (547 ± 27 Ma; Eichhorn *et al.* 2000) and included gabbroic and pyroxenic melts as well as normal 'I-type' granitoids in the Early Cambrian (529 ± 18 Ma; Eichhorn *et al.* 2000). The change from a volcanic to a magmatic event suggests a shift from a magmatic arc to a (ensialic) backarc setting in the early Cambrian in this area. The metamorphic Austro-Alpine basement to the south of the Tauern Window includes Upper Ordovician pelitic and volcanosedimentary units with elements of N-MORB volcanism and calc-alkaline volcanic arc basalts, the protolith ages of these ranging from about 590 to 540 Ma (Schulz 1997; Schulz *et al.* 1999; Schulz & Bombach 2002).

The geochemistry of granitoid rocks of the central Alps, namely the Ötztal-Stubai Crystalline Basement and the Silvretta Nappe, suggests the formation of a (now tectonically compressed) island-arc and backarc basin during the early Cambrian (Schaltegger *et al.* 1997; Klötzli-Chowanetz et al. 1997; Schermaier *et al.* 1998), probably distal to the NW margin of the Gondwana continent. This configuration was followed by an arc–continent collision during the latest Cambrian to early Ordovician, which created large-scale crustal anatexis in the region (Schermaier *et al.* 1998). Subduction in the region is also inferred for the Penninic Habach Formation (Steyrer & Höck 1985), the Cambrian evolution of which is similar to that of the Ötztal-Stubai Crystalline Basement.

In summary, the Late Proterozoic–Cambrian succession of the present-day western and central Alps shows evidence of tectonic activity, which involved oceanic domains after the Cadomian evolution along the Gondwanan margin. A pre-existing rift pattern was followed by sedimentary basins, which were filled by detritus from different cratonic regions. Evidence of Cambrian–Ordovician arc magmatism is provided by metavolcanics. The similarities in terms of magnetic evolution and the development of sedimentary basins suggest that the basement areas of the western and central Alps can be interpreted either as a lateral prolongation of the Armorican sector of western peri-Gondwana or as a microcontinent located between the eastern Alps and Armorica (Raumer 1998).

Palaeontology and biogeography (O.E., O.F., G.G. & A.Ż.)

Cambrian fossils in Central Europe are relatively rare. This is a result of (i) the fairly widespread Variscan metamorphism; (ii) the comparatively small outcrop size of relatively undeformed Cambrian rocks; and (iii) the unfavourable outcrop situation. The best studied faunas are derived from several complexes: (1) Lower Cambrian carbonates such as in the Lusatia area; (2) Lower Cambrian shallow marine clastic deposits such as in the Holy Cross Mountains; (3) Lower to Middle Cambrian shallow marine clastic deposits such as in Bohemia, the Franconian Forest and the Holy Cross Mountains; (4) Middle Cambrian deeper marine deposits such as in the Delitzsch-Torgau-Doberlug Syncline; and (5) Furongian open shelf deposits such as in the Holy Cross Mountains. These will be described in more detail below (Fig. 4.28). In addition, a fairly good fossil record is known from the boreholes of north and east Poland, which are part of the East European Platform and thus of the palaeocontinent Baltica.

Lower Cambrian carbonates of the Delitzsch-Torgau-Doberlug Syncline

Locally these include an association of archaeocyaths and calcimicrobes. Particularly noteworthy are associations documented in the Zwethau Formation of the Delitzsch-Torgau-Doberlug Syncline with *Dictyocyathus, Protopharetra, Erismacoscinus* and other archaeocyath genera associated with *Epiphyton, Renalcis, Girvanella, Proaulopora, Kordephyton, Botomaella, Subtifloria* and other calcimicrobes as well as some skeletal metazoans (Freyer & Suhr 1987, 1992; Elicki & Debrenne 1993; Elicki 1999*a*). The archaeocyath–calcimicrobe assemblages form carpets and small reef-like mounds typical of shallow-marine low-latitude environments of that time. These archaeocyath assemblages have been compared with associations from Spain and Morocco and a tentative stratigraphical context has been suggested (Elicki & Debrenne 1993). The associations appear to be coeval with the (mid-Lower Cambrian) archaeocyathan zones I to III of Moreno-Eiris *et al.* (1995), corresponding to the middle Issendalenian Stage of the Atlasian Series of West Gondwana (Geyer & Landing 2004). As with other archaeocyath assemblages, the key to their stratigraphic significance is based on the comparison with archaeocyaths of the Siberian platform. However, their precise age remains a matter of discussion and somewhat unclear since the biostratigraphic significance of the West Gondwanan archaeocyath faunas is not yet fully understood and generic assignments continue to be the subject of systematic concepts.

Lower Cambrian carbonates of the Görlitz Synclinorium

Some horizons of the Lower Cambrian limestones known from the Görlitz Syncline include a rich association of shelly fossils such as poriferid spicules (e.g. *Dodecaactinella, Eiffelia*), hyoliths (e.g. *Microcornus, Conotheca, Lenalituus*), echinoderm ossicles, chancelloriids (e.g. *Allonnia, Archiasterella, Chancelloria*), bivalves (*Pojetaia, Fordilla*), and other molluscs (e.g. *Yuwenia, Beshtashella, Pelagiella, Anabarella*), brachiopods, trilobites (*Lusatiops, Ferralsia, Calodiscus*), cyanobacteria and other calcimicrobes (*Epiphyton, Obruchevella, Endoconchia*) and shelly fossils of problematic systematic position (e.g. *Rhombocorniculum, Coleoloides, Microcoryne, Cambroclavus, Halkieria*) (Elicki 1994, 1996, 1998, 2000; Elicki & Schneider 1992; Geyer & Elicki 1995).

Coeval faunas from shales or mixed deposits include paucispecific typical Mediterranean-type trilobite assemblages such as the faunas with *'Holmia' zimmermanni*, *Serrodiscus silesius*, *Lusatiops lusaticus*, and *'Acanthomicmacca' schwarzbachi* from the Charlottenhof Formation of Lusatia (Geyer & Elicki 1995). These trilobites of the Görlitz Syncline indicate a higher Banian age.

Lower Cambrian Ociesęki Formation

These shallow-marine siliciclastics from the Holy Cross Mountains (HCM) contain an assemblage of trilobites accompanied by numerous trace fossils. Particularly fossiliferous are strata observed in the Ociesęki Range in the central part of the HCM area. The lower part of this unit (*Holmia-Schmidtiellus* Zone) contains endemic holmiids such as *Holmia, Schmidtiellus* and *Kjerulfia* (Orłowski 1985*a*) indicating a strong relationship to faunas known from coeval strata in Scandinavia. These holmiids, however, are accompanied by endemic species with a West Gondwanan (*Berabichia, Issafeniella, Strenuella* and *Ornamentaspis*), as well as a Laurentian aspect (*Atops*) (Orłowski 1985*a*, *c*). The upper part of this unit with *Kingaspidoides sanctacrucensis* and *Issafeniella orlowinensis* corresponds to the Protolenus-*Strenuaeva* Zone of the Lower Cambrian (Orłowski 1985*a*) and can be correlated with the uppermost Banian (Geyer & Landing 2004; Żylińska & Masiak 2007). The shallow-marine environment has been determined on the basis of ichnofossil assemblages of trilobite as well as non-trilobite origin (Orłowski 1989, 1992*a*; Orłowski & Żylińska 2002).

Lower Cambrian Paseky Shales

The Paseky Shales yielded the enigmatic crustacean-type arthropod *Kodymirus vagans* (Chlupáč & Havlíček 1965). In addition to the type locality (Kočka), five fossiliferous Paseky Shale localities have been re-investigated (Chlupáč *et al.* 1996). These studies have included an integrated study of primary producers of organic matter (organic-walled microfossils (Fatka & Konzalová 1996) and macroscopic algae (Steiner & Fatka 1996), consumers (such as the aglaspid merostome *Kodymirus vagans* Chlupáč & Havlíček, 1965 associated with two other non-trilobite arthropods; Chlupáč 1996), trace fossils (Mikuláš 1995) and sedimentology (Kukal 1995). The stratigraphic ranges of all of the taxa recognized at the Kočka locality have been recently summarized by Fatka *et al.* (2004). The absence of many groups, typical for coeval Cambrian marine assemblages, such as trilobites, echinoderms and brachiopods, combined with the presence of spectacular arthropods and the scarcity of process-bearing acritarchs, most probably reflects a very peculiar environment, possibly characterized by restricted marine conditions.

Middle Cambrian shallow faunas in Bohemia and the Franconian Forest

The Middle Cambrian faunas from shallow-marine sandstones and siltstones in Bohemia and the Franconian Forest are dominated by moderately diverse trilobite associations. The oldest assemblages consist of Mediterranean-type genera and species that clearly indicate a West Gondwanan affinity (according to the depositional environments). During the Middle Cambrian the affinities to trilobite faunas known from Baltica appear to become successively stronger. Particularly noteworthy are the trilobites from the Bergleshof Formation of the Franconian Forest, which contain 'Scandinavian' genera and species although it represents an extremely shallow depositional environment.

The Jince Formation of Bohemia provides the best preserved and diversified Cambrian faunas in Central Europe preserved in environmentally controlled assemblages (Fig. 4.29). For the Litavka river valley three bathymetrically dependent invertebrate assemblages have been established and briefly characterized (Fatka 2000). The oldest and youngest levels of the Jince Formation are characterized by a shallow marine *Lingulella*-dominated assemblage containing rare ellipsocephalid (*Germaropyge*) and conocoryphid (*Conocoryphe*) trilobites associated with rare paradoxidids (*Acadoparadoxides, Rejkocephalus*). A deeper-marine assemblage is dominated by polymeroid trilobites (*Ellipsocephalus*, *Conocoryphe*, complemented by paradoxidids such as *Hydrocephalus*, *Acadoparadoxides*, *Eccaparadoxides*, *Rejkocephalus*, ptychoparioids such as *Ptychoparia*, and solenopleurids such as *Jincella*, *Solenopleurina*), which are usually associated with common miomeroid trilobites such as *Peronopsis*, *Phalagnostus* and *Phalacroma*, rare articulate brachiopods, bradoriids (*Konicekion*), other bivalved crustaceans, and hyolithids

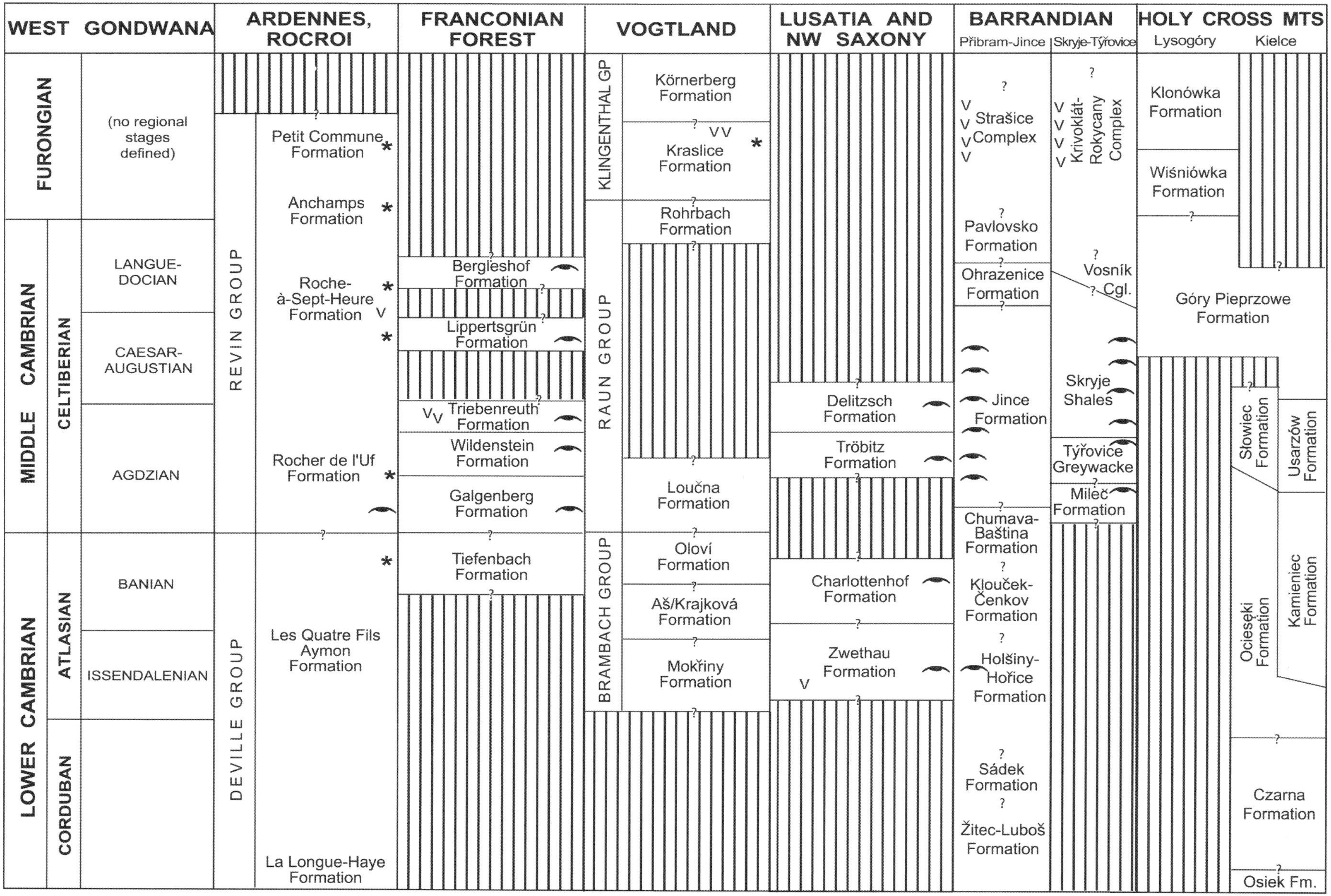

Fig. 4.28. Correlation table with Cambrian lithostratigraphic units of important and relevant areas in Central Europe. Vertical hatching, hiatus. Asterisks, datable organic microfossils; convex-up symbol, macrofauna; V, levels with volcanism. Geochronology according to the West Gondwanan Standard (Geyer & Landing, 2004).

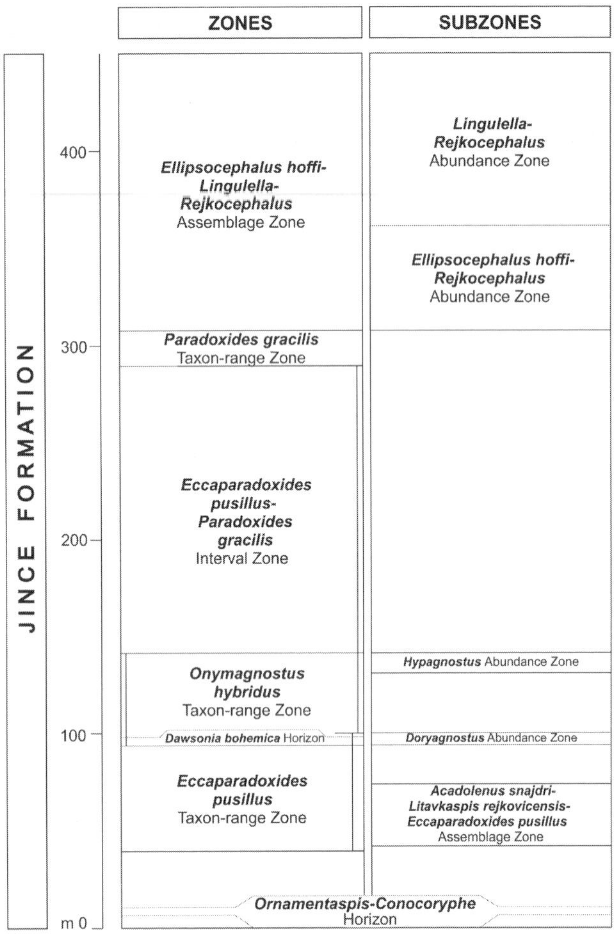

Fig. 4.29. Revised biostratigraphic concept for the Jince Formation, Middle Cambrian, Bohemia (according to Fatka 2005).

(Šnajdr 1975; Havlíček 1970; Chlupáč & Kordule 2002; Valent 2006). Fine-grained sedimentary rocks characterize the deepest environments and are dominated by agnostoids (*Onymagnostus*, *Tomagnostus*, *Hypagnostus*, *Doryagnostus*) and associated rare polymeroids (paradoxidids, conocoryphids and *Luhops*), foraminifers and paragastropod molluscs (Horný 1963, 1964; Bubík & Fatka 2001). Several layers at different stratigraphical levels within the succession include exceptionally preserved echinoderms. Relatively frequent are cinctans (*Asturicystis*), in the trilobite-dominated assemblage of the *Eccaparadoxides pusillus* Biozone (Fatka & Kordule 2001), whereas articulated eocrinoids (*Akadocrinus*, *Acanthocystites*, *Vyscytis*) usually occur with the ctenocystid *Etoctenocystis* and the edrioasteroid *Stromatocystites* in the *Paradoxides gracilis* Biozone (Fatka & Kordule 1985, 1990, 1991; Parsley & Prokop 2004). Traditional body fossil assemblages are complemented by well-diversified trace fossils (Mikuláš 2000) and problematica.

The stratigraphic ranges of all taxa recognized in the Jince Formation of the Příbram–Jince Basin have been recently summarized by Fatka *et al.* (2004). The common benthic assemblages provide a good basis for biostratigraphical subdivision of the formation. Its biostratigraphic concepts have recently been summarized (Fatka 2006). The units of the revised subdivision of the Jince Formation are shown in Figure 4.29.

The Middle Cambrian fauna in the Skryje–Týřovice Basin is fairly distinct from that of the Příbram–Jince Basin. In particular, the Skryje Shale includes a diverse fauna: rare inarticulate brachiopods (such as *Acrothele*, *Luhotreta*; Mergl & Šlehoferová 1990); the abundant articulate brachiopod *Bohemiella romingeri*; diversified polymeroid trilobites (e.g. *Conocoryphe*, *Ctenocephalus*, *Eccaparadoxides*, *Luhops*, *Ptychoparia*, *Solenopleurina*, *Sao*, *Agraulos*, *Skrejaspis*, *Jincella*; Šnajdr 1958; Álvaro *et al.* 2004); in higher stratigraphical levels also relatively common miomeroid trilobites (e.g. *Condylopyge*, *Pleuroctenium*, *Skryjagnostus*, *Phalagnostus*, *Peronopsis*; Šnajdr 1958; Fatka *et al.* 2004); in some levels very common cinctan and stylophoran echinoderms (*Trochocystites*, *Ceratocystis*; Jaekel 1901; Friedrich 1993; Lefebvre & Fatka 2003); rare eocrinoids (*Luhocrinus*, *Lichenoides*; Prokop & Fatka 1985; Parsley & Prokop 2004); edrioasteroids (*Stromatocystites*; Pompeckj 1896); rare molluscs (*Cambretina*, *Costipelagiella*; Horný 1963, 1964); hyoliths (*Maxilites*; Marek 1972, 1981); and moderately preserved and diversified acritarchs (e.g. *Timofeevia*, *Cristallinium*, *Eliasum*; Vavrdová 1982; Chlupáč & Fatka 1999). This fauna shows a closer relationship to faunas from other parts of Central Europe (e.g. the Lippertsgrün Formation of the Franconian Forest) and SW Europe (e.g. the Ossa Morena area of Spain).

The younger of the Middle Cambrian faunas from the Barrandian region in Bohemia shares genera with West Gondwana and Baltica but does not show a perfect fit with associations from either of these two areas. Thus, the fossils suggest that the Bohemian region may have been an area distant from both of the Cambrian continents. It should be mentioned that these faunal data are in accordance with palaeomagnetic data, which suggest that the Barrandian region was separated from Gondwana in the Ordovician (Tait *et al.* 2000).

Lower to Middle Cambrian faunas from the Holy Cross Mountains

The trilobite faunas from the Lower–Middle Cambrian boundary interval of the Kielce region of the southern Holy Cross Mountains show a strong affinity to the West Gondwanan regions in Iberia and the Moroccan Atlas regions. They share several genera (e.g. *Cobboldites*, *Palaeolenus*, *Myopsolenites*, *Issafeniella*, *Hamatolenus*, *Protolenus* (*Hupeolenus*) and a possible *Strettonia*; Geyer & Landing 2004; Żylińska & Masiak 2007). The similarities in terms of depositional environments also suggest that there is a significant relationship between this region and West Gondwana. As in the Franconian Forest and the Barrandian regions, the faunas would appear to become successively more Baltic during the Middle Cambrian.

The Middle Cambrian in the Delitzsch-Torgau-Doberlug Syncline

The Middle Cambrian succession in the Delitzsch-Torgau-Doberlug Syncline is dominated by fine-grained clastic deposits. Shales of a relatively shallow marine environment grade into dark shales of a more quiet-water facies with a typical middle shelf fossil assemblage that brackets West Gondwanan and Baltic associations (Fig. 4.30). The 'classical' trilobite fauna includes such common genera as *Condylopyge*, *Peronopsis*, *Acadoparadoxides*, *Eccaparadoxides*, *Conocoryphe*, *Solenopleura* and *Badulesia* (Sdzuy 1957*a*, *b*, 1958, 1970). Additional fossils include hyoliths, brachiopods and occasional helcionellids. Recently discovered, unpublished trilobites include *Protolenus* (*Hupeolenus*) and other protolenid trilobites in association with inarticulate bra-

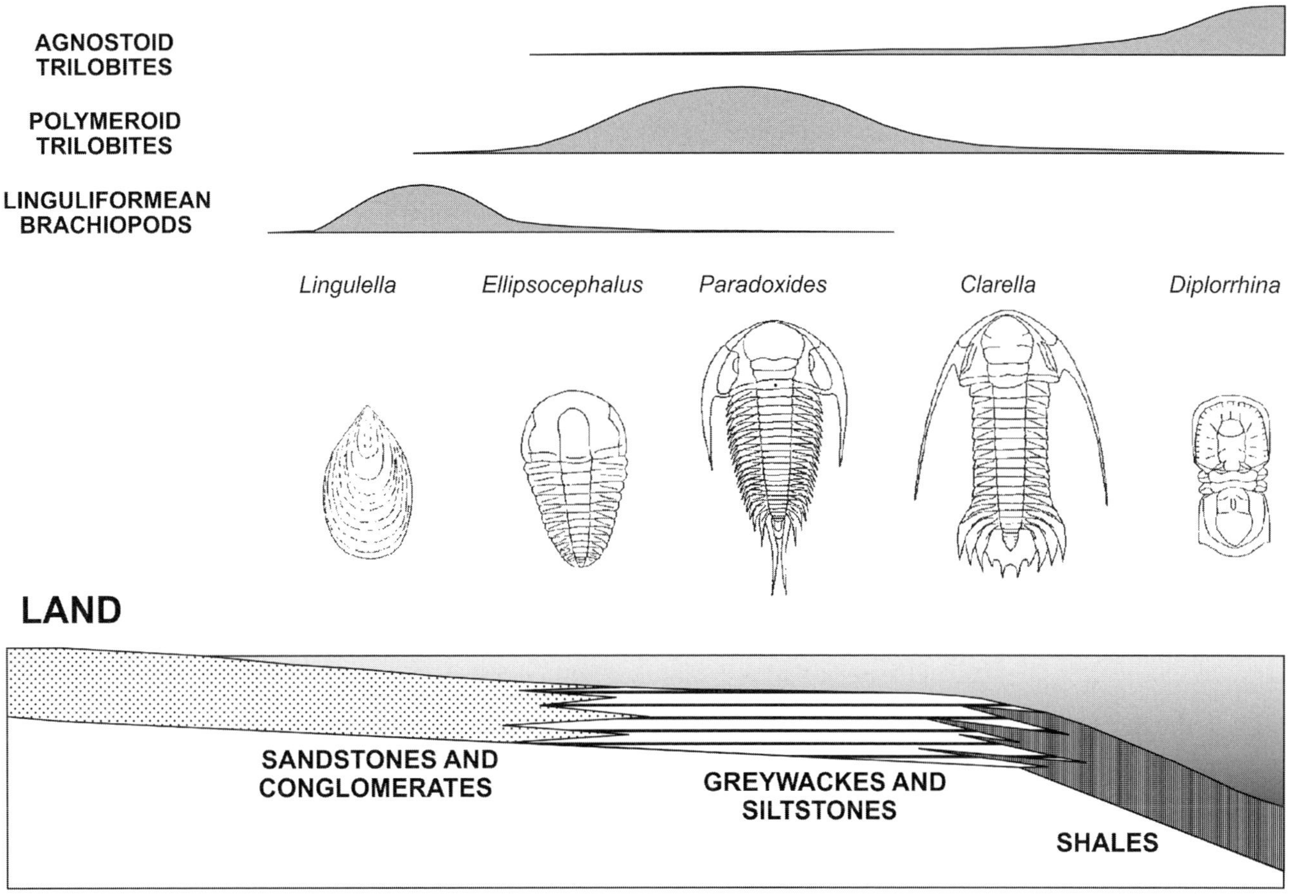

Fig. 4.30. Spatial distribution of lithofacies and biofacies in the Middle Cambrian with relative frequence of linguliformean brachiopods, polymeroid and agnostoid trilobites.

chiopods (B. Buschmann, O. Elicki & G. Geyer, unpubl. data, 2006), all with strong affinities to West Gondwana (southern Morocco, Cantabrian Mountains of northern Spain). They represent the lowermost Middle Cambrian upper *Hupeolenus* and/or lowest *Cephalopyge* zones (lower Agdzian) in the West Gondwanan stratigraphy.

Furongian faunas from the Holy Cross Mountains

Furongian faunas are known from open shelf deposits in the Łysogóry region of the Holy Cross Mountains (Żylińska 2001, 2002). Remarkably, the trilobite faunas are fairly similar to olenids known from the classical Furongian of Scandinavia, particularly from the *Peltura* Zones upwards (Fig. 4.17). The Baltica elements in the faunal assemblages increase distinctly towards the top of the succession and this increase is paralleled by a decrease in the endemic components. In the topmost part of the sequence the influx of olenids of Gondwanan (Argentina, Bolivia) and Avalonian aspect (various species of *Leptoplastides* and *Parabolina* (*Neoparabolina*) *frequens*) can be observed, suggesting comparisons with the trilobite assemblages observed at the margins of Baltica (Nikolaisen & Henningsmoen 1985) rather than those of the restricted-environment Alum Sea faunas. The lower part of the sequence (e.g. Wiśniówka Formation) is characterized by a low-diversity, inner-shelf assemblage sharing taxa with Avalonian exposures in Great Britain. Due to the widespread erosional and depositional(?) hiatus for the Furongian of peri-Gondwana, a comparison with the few known (and usually shallow marine) West Gondwanan trilobite faunas does not provide reliable clues for palaeobiogeographic connections and relationships, although similarities can be noted with the Late Cambrian trilobites from the Iberian Mountains (Shergold & Sdzuy 1991). It is possible, however, that, in contrast to the Kielce region, the Łysogóry region may represent an area with peri-Baltica rather than peri-Gondwanan affinities as indicated by detrital input in Cambrian rocks of the Łysogóry region with signatures of the East European Platform (Belka *et al.* 2002). Recent studies show that the Furongian strata from the Kielce region, particularly their upper part (Szczepanik *et al.* 2004*a*, *b*), reflect a very similar open shelf environment, with rare fauna identical to that from Łysogóry, which would suggest close proximity of this area to Baltica by the end of the Furongian.

Organic-walled microfossils

Organic-walled microfossils, particularly acritarchs, have also helped to establish a biochronology in otherwise sparsely fossiliferous strata. Taxonomically diverse acritarch associations allow the recognition of zones in the Lower and Middle Cambrian and Furongian series in some areas, such as in Upper Silesia (Moczydłowska 1991, 1998). In this area, four extended zones are recognizable for the Early Cambrian, termed the

Asteridium tornatum–Comasphaeridium veltetum, the *Skiagia ornata–Fimbriaglomerella membranacea*, the *Heliosphaeridium dissimilare–Skiagia ciliosa*, and the *Volkovia dentifera–Liepaina plana* zones, which are also recognizable elsewhere (Moczydłowska 1991, 1998; Moczydłowska & Vidal 1986). The acritarch associations clearly indicate that this suspect terrane is a distal portion of peri-Gondwana (or East Avalonia *sensu* Moczydłowska 1998) rather than of Baltica. The correlation between datings based on trilobites and acritrachs has been attempted for Furongian strata from the Holy Cross Mountains (Żylińska & Szczepanik 2002).

The fossil record of the Baltic and Lublin-Podlasie basins

Lower Cambrian through Furongian successions of the Baltic and Lublin-Podlasie basins in north and east Poland, known only from numerous boreholes, have body and trace fossils typical of Baltica. Lower Cambrian faunas include a fairly paucispecific amount of trilobites and remains of other arthropods, but a number of additional shelly faunas, such as the index fossil *Mobergella* (an operculum of uncertain systematic affinity) or hyoliths (Keller & Rozanov 1979; Lendzion 1983*a*). Rich ichnofossil assemblages are known from most of the Lower Cambrian clastic rocks throughout the area that allowed comprehensive palaeoecological analyses. Paczešna (1996) distinguished 19 ichnocoenoses, which depict changes in ecology and depositional environment. In addition, acritarch assemblages also permit stratigraphic assignments into the traditional Lower Cambrian biozones of Baltica (Jankauskas & Lendzion 1992, 1994).

Middle Cambrian strata of the Baltic and Lublin-Podlasie basins yield fairly rich fossil assemblages with typical Baltica-type trilobites (e.g. *Ellipsocephalus lejostracus*, *Paradoxides insularis*, *Paradoxides paradoxissimus*, *Bailiella emarginata*, *Triplagnostus gibbus* and *Lejopyge laevigata*), inarticulate and articulate brachiopods (e.g. *Billingsella exporrecta*). The Furongian is developed in a typical thin alum-shale facies with its characteristic trilobite (*Agnostus pisiformis*, *Homagnostus obesus*, *Olenus wahlenbergi*, *O. truncatus*, *Parabolina spinulosa*, *P. lobata*, *Peltura scarabaeoides*, *Sphaerophthalmus alatus*) and brachiopod (e.g. *Orusia lenticularis*) faunas.

Geochronology (O.E. & G.G.)

Radiometric age determinations for the Cambrian have suffered from severe methodological problems. Datings of pre-Middle-Ordovician rocks up to *c.* 1990 and Rb–Sr ages, in particular, are generally not consistent and often distinctly too high. Recent datings appear to identify, more reliably, Cambrian fractions within metamorphic suites. Table 4.1 records selected data with the most valuable information for distinct regions under discussion. Most of the datings has been based on zircons from boulders and tuffs and often refers to a maximum age.

Summary

Cambrian deposits in Central Europe are generally of limited extent, often largely covered by vegetation and slightly to strongly metamorphosed. Regional differences in the tectonic and resulting sedimentary history as well as faunal characters indicate a melange of plates and terranes in a configuration that differs extremely from their original spatial distribution.

The most relevant surface exposures are found in Bohemia, the Franconian Forest area in NW Bavaria, west Thuringia, the Lusatia area in east Saxony, in the Holy Cross Mountains of south Poland, and in the Brabant Massif of south Belgium and its vicinity. These outcrops are of relatively limited extent, but some yield important fossil assemblages. In addition, Cambrian strata are known from a number of drillholes such as in the Delitzsch-Torgau-Doberlug Syncline of NW Saxony, Upper Silesia, and north and east Poland. Our knowledge of the Cambrian is supplemented by radiometric data that indicate a Cambrian age for numerous slightly to heavily metamorphosed igneous rock complexes in such areas as the surroundings of the Bohemian Massif, the Mid-German Crystalline High, or in the southern and central Alps.

Although the outcrops and subsurface occurrences of Cambrian rocks belong to different geotectonic units, most of them show a signature of stratigraphy, sedimentary history and, in part, faunal characters which indicate that they formed as parts of the palaeocontinent Gondwana (or peri-Gondwana). Although tectonic activity was relatively weak in the traditional West Gondwanan areas (Morocco, Iberia, Sardinia), Central Europe south of the Iapetus Suture and west of the Trans-European Suture Zone was part of an Andean-type margin during Ediacaran and early Cambrian times, characterizing the Cadomian orogenic event. In contrast, the Cambrian subsurface outcrops in north, NE and SE Poland as well as those offshore the island of Rügen in the Baltic Sea are parts of Baltica (or the East European Platform). Only the Cambrian rocks known from Brabant and the Ardennes regions show a lithostratigraphic signature that matches those known from Great Britain and the Avalonian part of North America, so that they appear to have been part of an insular continent of Avalonia during the Cambrian.

The major part of the Cambrian deposits in Central Europe is overlain by younger strata and usually poorly studied. Exact and/or total thicknesses are unknown from most of the areas. The lithologic variability reflects greatly differing facies, which are a response to highly different environmental conditions that, in turn, reflect a fine-scaled palaeogeography. A clear trend shows a variety of different Lower Cambrian deposits that vary laterally as well as vertically from shales to sandstones, limestones and dolostones of different microfacies. In contrast, Middle Cambrian deposits are dominated by fine-grained sandstones with a tendency to increased grain size towards the top. The Furongian (Upper Cambrian) is partly represented by sandstones, often quartz arenites, more often by hiatuses due to both non-deposition and subsequent erosion. Depending on regional differences, sedimentary rocks are often replaced, or at least interrupted, by volcanic successions.

The fossil content is rather diverse, but best represented by late Early Cambrian through middle Middle Cambrian trilobites. Archaeocyaths apparently had one of their major occurrences in this area, and they show surprisingly good preservation. In addition, numerous Early Cambrian small shelly fossils are known.

References

Álvaro, J. J., Vizcaino, D., Kordule, V., Fatka, O. & Pillola, G.L. 2004. Some solenopleurine trilobites from the Languedocian (Late Mid Cambrian) of Western Europe. *Géobios*, **37**, 135–147.

Areń, B. & Lendzion, K. 1978. Charakterystyka stratygraficzno-litologiczna wendu i kambru dolnego. *Prace Instytutu Geologicznego*, **40**, 7–49.

Bankwitz, P. 1970. Über die Basisfolge des Kambriums (Goldisthaler Schichter) und ihre Grenze zum liegenden Proterozoikum (Katzhüt-

Table 4.1. *Summary of radiometric age determinations*

Age	Method	Rock type, location	Reference
Lusatia and Northwest Saxony			
539 ± 6 Ma (average age)	$^{207}Pb/^{206}Pb$, single zircon evaporation	Granodiorite near Dresden-Klotzsch	Linnemann *et al.* (2000)
562 ± 4 Ma	$^{207}Pb/^{206}Pb$, single zircon evaporation	Tuff intercalation in greywackes	Tichomirowa *et al.* (1997)
Schwarzburg Anticline			
533 ± 4 Ma (average age)	$^{207}Pb/^{206}Pb$, single zircon evaporation	Granite near Katzhütte	Linnemann *et al.* (2000)
541 ± 7 Ma (average age)	$^{207}Pb/^{206}Pb$, single zircon evaporation	Granite near Mellenbach	Linnemann *et al.* (2000)
504 ± 7 Ma (inherited/detritic zircons)	$^{207}Pb/^{206}Pb$, single zircon evaporation	Rhyolitic boulders from basal cgl. of Goldisthal Group (Ord.), Mellenbach	Linnemann *et al.* (2000)
507 ± 4 Ma (inherited/detritic zircons)	$^{207}Pb/^{206}Pb$, single zircon evaporation	Rhyolitic boulders from basal cgl. of Goldisthal Group (Ord.), Mellenbach	Linnemann *et al.* (2000)
511 ± 2 Ma (inherited/detritic zircons)	$^{207}Pb/^{206}Pb$, single zircon evaporation	Rhyolitic boulders from basal cgl. of Goldisthal Group (Ord.), Mellenbach	Linnemann *et al.* (2000)
530 ± 9 Ma (inherited/detritic zircons)	$^{207}Pb/^{206}Pb$, single zircon evaporation	Rhyolitic boulders from basal cgl. of Goldisthal Group (Ord.), Mellenbach	Linnemann *et al.* (2000)
543 ± 6 Ma (inherited/detritic zircons)	$^{207}Pb/^{206}Pb$, single zircon evaporation	Rhyolitic boulders from basal cgl. of Goldisthal Group (Ord.), Mellenbach,	Linnemann *et al.* (2000)
492 ± 1 Ma (inherited/detritic zircons)	$^{207}Pb/^{206}Pb$, single zircon evaporation	Congl. arkose from basal cgl. of Goldisthal Gp (Ord.), Viehberg near Katzhütte	Linnemann *et al.* (2000)
493 ± 4 Ma (inherited/detritic zircons)	$^{207}Pb/^{206}Pb$, single zircon evaporation	Congl. arkose from basal cgl. of Goldisthal Gp (Ord.), Viehberg near Katzhütte	Linnemann *et al.* (2000)
527 ± 17 Ma (inherited/detr. zircons)	$^{207}Pb/^{206}Pb$, single zircon evaporation	Congl. arkose from basal cgl. of Goldisthal Gp (Ord.), Viehberg near Katzhütte	Linnemann *et al.* (2000)
527 ± 9 Ma (inherited/detritic zircons)	$^{207}Pb/^{206}Pb$, single zircon evaporation	Congl. arkose from basal cgl. of Goldisthal Gp (Ord.), Viehberg near Katzhütte	Linnemann *et al.* (2000)
Northern Saxonian Anticline			
531 ± 7 Ma (average age)	$^{207}Pb/^{206}Pb$, single zircon evaporation	Granodiorite near Dürrenberg/Laas	Linnemann *et al.* (2000)
Erzgebirge			
555–522 Ma	$^{207}Pb/^{206}Pb$	Galenite, Klinovec Group	Legler (1985)
555 ± 5 Ma	$^{207}Pb/^{206}Pb$, single zircon evaporation	Muscovite gneiss, Jachimov Group	Kröner *et al.* (1995)
Elbe Zone			
537 ± 7 Ma (average age)	$^{207}Pb/^{206}Pb$, single zircon evaporation	Granodiorite near Dresden-Torna,	Linnemann *et al.* (2000)
537 ± 4 Ma (average age)	$^{207}Pb/^{206}Pb$, single zircon evaporation	Orthogneiss near Großenhain	Linnemann *et al.* (2000)
Münchberg Crystalline Complex			
525 ± 40 Ma	U–Pb datings of zircons	Eclogites	Gebauer & Grünenfelder (1979)
Bavarian Forest and Teplá-Barrandian			
555 ± 12 Ma	Zircons SHRIMP ages	Leucocratic metamagmatites and metabasites from the Passau area	Teipel *et al.* (2002)
549 ± 7 Ma	Zircons SHRIMP ages	Leucocratic metamagmatites and metabasites from the Passau area	Teipel *et al.* (2002)
549 ± 6 Ma	Zircons SHRIMP ages	Leucocratic metamagmatites and metabasites from the Passau area	Teipel *et al.* (2002)
480 ± 6 Ma (minimum ages)	Zircons SHRIMP ages	Leucocratic metamagmatites	Teipel *et al.* (2002)
486 ± 7 Ma (minimum ages)	Zircons SHRIMP ages	Leucocratic metamagmatites	Teipel *et al.* (2002)
551 Ma to 540 Ma	Metamorphic monazite	Domažlice crystalline complex	Zulauf *et al.* (1999), Timmermann *et al.* (2002)
541 + 7/−8 Ma (intrusion age)	U–Pb isotopic dilution TIMS age	Zgorzelec granodiorite	Dörr *et al.* (2002)
524 ± 7 Ma (intrusion age)	U–Pb isotopic dilution TIMS age	Všepadly granodiorite	Dörr *et al.* (2002)
515.9 ± 1.3 Ma (post-magm. cooling)	$^{40}Ar/^{39}Ar$ hornblende age	Všepadly granodiorite	Dallmeyer & Urban (1998)
522 ± 2 (intrusion age)	U–Pb isotopic dilution TIMS age	Tešovice granite	Zulauf *et al.* (1997), Dörr *et al.* (2002)
522 ± 6 Ma (mimimum age of intrusion)	U–Pb isotopic dilution TIMS age	Smržovice tonalite	Dörr *et al.* (2002)
523 ± 3 Ma (mimimum age of intrusion)	U–Pb isotopic dilution TIMS age	Smržovice tonalite	Dörr *et al.* (2002)
523 ± 1 Ma (mimimum age of intrusion)	U–Pb isotopic dilution TIMS age	Smržovice gabbro	Dörr *et al.* (2002)
523 + 4/−5 (intrusion age)	U–Pb isotopic dilution TIMS age	Mračnice trondhjemite	Zulauf *et al.* (1997)
524 ± 0.8 (age of emplacement)	U–Pb isotopic dilution TIMS age	Orlovice gabbro	Dörr *et al.* (2002)

ter Schichten) an der SE-Flanke des Schwarzburger Sattels. *Geologie*, **19**(9), 1023–1047.

BANKWITZ, E. & BANKWITZ, P. 1975. Zur Sedimentation proterozoischer und kambrischer Gesteine im Schwarzburger Antiklinorium. *Zeitschrift für geologische Wissenschaften*, **3**, 1279–1305.

BANKWITZ, E. & BANKWITZ, P. 1997. Ein kambro-ordovizischer magmatic belt [sic] in Mitteleuropa (Süddeutschland bis Westsudeten/Polen). *In*: BÜCHEL, G. & LÜTZNER, H. (eds) *Regionale Geologie von Mitteleuropa.* Geodynamische Prozesse zwischen Alpen und Nordatlantik, **149**. Hauptversammlung der Deuten Geologischen Gesellschaft und Jahreshauptversammlung der Fachsektion Geoinformatik, Jena, 27.09.–3.10.1997. Kurzfassungen der Vorträge und Poster, 25.

BANKWITZ, P. & PLOTNIKOV, L. M. 1994. Suturen in der mitteleuropäischen Erdkruste und die Position von KTB (Eine Diskussionsbemerkung). *KTB Report* **94–3**, 9–17.

BANKWITZ, P., BANKWITZ, E., KRAMER, W. & PIN, C. 1994. Early Paleozoic bimodal volcanism in the Vesser area, Thuringian Forest, eastern Germany. *Zentralblatt für Geologie und Paläontologie, Teil I*, **1992**(9/10), 1113–1132.

BANKWITZ, E., BANKWITZ, P., FALK, F., LÜTZNER, H. & MANN, M. 1997. Schwarzburger Antiklinorium. *In*: Stratigraphie von Deutschland II. Teil I: Thüringen, Sachsen, Ostbayern. *Courier Forschungsinstitut Senckenberg*, **200**, 19–54.

BARRANDE, J. 1846. *Notice préliminaire sur le système silurien et les Trilobites de Bohême.* Leipzig.

BARRANDE, J. 1852. *Système Silurien du centre de la Bohême. Vol. I. Recherches paléontologiques I: Trilobites*. Prague, Paris.

BEDNARCZYK, W. 1970. Trilobite fauna [sic] of the Lower *Paradoxides oelandicus* Stage from the Brzechów area in the western part of the Świętokrzyskie Mts. *Bulletin de l'Académie Polonaise des Sciences, Série des Sciences Géologique et Géographiques*, **18**, 29–35.

BEDNARCZYK, W. 1984. Biostratigraphy of the Cambrian deposits in the Łeba area. *Acta Geologica Polonica*, **34**, 95–109.

BEDNARCZYK, W. & TURNAU-MORAWSKA, M. 1975. Litostratygrafia osadów kambru i wendu w rejonie Łeby. *Acta Geologica Polonica*, **25**, 537–566.

BEDNARCZYK, W., JURKIEWICZ, H. & ORŁOWSKI, S. 1965. Lower Cambrian and its fauna from the boring of Zaręby near Łagów (Holy Cross Mts.). *Bulletin de l'Académie Polonaise des Sciences, Série des Sciences Géologique et Géographiques*, **13**, 231–236.

BEIER, H. & KATZUNG, G. 1999. Lithologie und Strukturgeologie des Altpaläozoikums in der Offshore-Bohrung G 14–1/86 (südliche Ostsee). *Greifswalder Geowissenschaftliche Beiträge*, **6**, 327–345.

BEHR, H.-J., JORDAN, H. & WEBER, W. 1965. Ein paläontologischer Beleg für das Alter der Vergneisung im Erzgebirge – Chitonozoen in den Phyllitarealen von Hermsdorf-Rehefeld. *Monatsberichte der Deutschen Akademie der Wissenschaften*, **7**, 408–415.

BELKA, Z., AHRENDT, H., FRANKE, W. & WEMMER, K. 2000. The Baltica-Gondwana suture in central Europe: evidence from K–Ar ages of detrital muscovite and biogeographical data. *In*: FRANKE, W., HAAK, W., ONCKEN, O. & TANNER, D. (eds) *Orogenic Processes: Quantification and Modelling in the Variscan Belt of Central Europe*. Geological Society, London, Special Publications, **179**, 87–102.

BELKA, Z., VALVERDE-VAQUERO, P., DÖRR, W., AHRENDT, H., WEMMER, K., FRANKE, W. & SCHÄFER, J. 2002. Accretion of first Gondwana-derived terranes at the margin of Baltica. *In*: Winchester, J. A., Pharaoh, T. C. & Verniers, J. (eds) *Palaeozoic Amalgamation of Central Europe*. Geological Society, London, Special Publications, **201**, 19–36.

BERGER, H.-J. & DOCEKAL, A. 1997. Nordsächsisches Schiefergebirge. *In*: Stratigraphie von Deutschland II. Teil I: Thüringen, Sachsen, Ostbayern. *Courier Forschungsinstitut Senckenberg*, 200, 335–342.

BERGER, H.-J. & HOTH, K. 1997. Vogtland. *In*: Stratigraphie von Deutschland II. Teil I: Thüringen, Sachsen, Ostbayern. *Courier Forschungsinstitut Senckenberg*, 200, 147–174.

BEUGNIES, A. 1960. Les grands traits de la géologie du Massif cambrien de Rocroi. *Bulletin de la Société Belge*, 69, 83–91.

BIERNAT, G. & BALIŃSKI, E. 1973. Fauna z otworów wiertniczych Sosnowiec IG-1 i Goczałkowice IG-1 (Stromatoporoidea, Tabulata, Brachiopoda i Trilobita). *Geological Quarterly*, **17**, 629–630.

BLESS, M. J. M., BOUCKAERT, J. & PAPROTH, E. (eds) 1980. Prepermian around the Brabant Massif in Belgium, the Netherlands and the Federal Republic of Germany. *Mededelingen van Rijks geologische Dienst*, **32**(14), 1–179.

BLUMENSTENGEL, H. 1980. Zur Paläontologie und Biostratigraphie der Heinersdorfer Serie (Kambrium) im Thüringer Schiefergebirge. *Freiberger Forschungshefte*, **C 348**, 61–68.

BOUČEK, B. 1938. Nález dendroidů v českém kambriu (Discovery of dendroids in Bohemain Cambrian). *Věda přírodní*, **19**(1), 18.

BRAUSE, H. 1969. Das verdeckte Altpaläozoikum der Lausitz und seine regionale Stellung. *Abhandlungen der Deutschen Akademie der Wissenschaften, Klasse Bergbau, Hüttenwesen und Montangeologie*, **1968**, 1–143.

BRAUSE, H. 1970. Ein neuer wichtiger Aufschluß im Kambrium von Doberlug-Kirchhain. *Geologie*, **19**, 1048–1065.

BUBIK, M. & FATKA, O. 2001. Vinice near Jince. *In:* M. BUBÍK, M. (ed.) *Fossil Record of Agglutinated Rhizopods (Foraminifers, Testaceans) on the Territory of Czech Republic*. Czech Geological Survey, Prague, 18–20.

BUŁA, Z. 2000. The Lower Palaeozoic of Upper Silesia and West Małopolska. *Prace Państwowego Instytutu Geologicznego*, **171**, 1–89.

BUŁA, Z. & JACHOWICZ, M. 1996. The Lower Palaeozoic sediments in the Upper Silesian Block. *Geological Quarterly*, **40**, 299–336.

BUŁA, Z., JACHOWICZ, M. & ŻABA, J. 1997. Principal characteristics of the Upper Silesian Block and Małopolska Block border zone (southern Poland). *Geological Magazine*, **134**, 669–677.

BUSCHMANN, B., LINNEMANN, U., SCHNEIDER, J. & SÜSS, T. 1995. Die cadomische Entwicklung im Untergrund der Torgau-Doberluger Synklinale. *Zeitschrift für geologische Wissenschaften*, **23**, 729–749.

BUSSY, F., SARTORI, M. & THÉLIN, P. 1996. U-Pb zircon dating in the middle Penninic basement of the Western Alps (Valais, Switzerland). *Schweizerische Mineralogische und Petrographische Mitteilungen*, **76**, 81–84.

CHEN, F., HEGNER, E. & TODT, W. 2000. Zircon ages and Nd isotopic and chemical compositions of orthogneisses from the Black Forest, Germany: evidence for a Cambrian magmatic arc. *International Journal of Earth Sciences*, **88**, 791–802.

CHLUPÁČ, I. 1996. Lower Cambrian arthropods from the Paseky Shale (Barrandian area, Czech Republic). *Časopis pro Mineralogii a geologii*, **40**, 9–36.

CHLUPÁČ, I. & FATKA, O. 1999. Day 1, stop 3: Luh near Skryje. *In:* KRAFT, J., KRAFT, P. & $FATKA, O. (eds) *Excursion Guide Barrandian*. VIIIth ISOS. Západočeské muzeum v Plzni, 13–14.

CHLUPÁČ, I. & HAVLÍČEK, V. 1965. *Kodymirus* n. g., a new aglaspid merostome of the Cambrian of Bohemia. *Sborník geologických Věd, Paleontologie*, **6**, 7–20.

CHLUPÁČ, I. & KORULE, V. 2002. Arthropods of Burgess Shale type from the Middle Cambrian of Bohemia (Czech Republic). *Věstník Českého geologického ústavu* **77**, 167–182.

CHLUPÁČ, I., KRAFT, J. & KRAFT P. 1996. Geology of fossil sites with the oldest Bohemian fauna (Lower Cambrian, Barrandian area). *Časopis pro Mineralogii a geologii* **40**, 1–8.

COCKS, L. R. M. 2002. Key Lower Palaeozoic faunas from near the Trans-European Suture Zone. *In*: WINCHESTER, J. A., PHARAOH, T. C. & VERNIERS, J. (eds) *Palaeozoic Amalgamation of Central Europe*. Geological Society, London, Special Publications, **201**, 37–46.

COGNE, J. P. 1990. The Cadomian Orogeny and its influence on the Variscan evolution of Western Europe. *In:* D'LEMOS, R. S., STRACHAN, R. A. & TOPLEY, C. G. (eds) *Cadomian Orogeny.* Geological Society, London, Special Publications, **51**, 305–311.

COURJAULT-RADÉ, P., DEBRENNE, F. & GANDIN, A. 1992. Palaeogeographic and geodynamic evolution of the Gondwana continental margins during the Cambrian. *Terra Nova*, **4**, 657–667.

DADLEZ, R. 2000. Pomeranian Caledonides (NW Poland), fifty years of controversies: a review and a new concept. *Geological Quarterly*, **44**, 221–236.

DALLMEYER, R. D. & URBAN, M. 1998.Variscan vs. Cadomian tectonothermal activity in northwestern sectors of the Teplá-Barrandian zone, Czech Republic: constraints from $^{40}Ar/^{39}Ar$ ages. *Geologische Rundschau*, **87**, 94–106.

DE VOS, W., VERNIERS, J., HERBOSCH, A. & VANGUESTAINE, M. 1993. A new geological map of the Brabant Massif. *Geological Magazine*, **130**, 605–611.

DÖRR, W., ZULAUF, G., FIALA, J., FRANKE, W. & VEJNAR, Z. 2002.

Neoproterozoic to Early Cambrian history of an active plate margin in the Teplá-Barrandian unit – a correlation of U-Pb isotopic-dilution-TIMS ages (Bohemia, Czech Republic). *Tectonophysics*, **352**, 65–85.

Dörr, W., Finger, F., Linnemann, U. & Zulauf, G. 2004. The Avalonian-Cadomian belt and related peri-Gondwana terranes. *International Journal of Earth Sciences*, **93**, 657–944.

Drost, K., Linnemann, U. *et al.* 2004. New data on the Neoproterozoic-Cambrian geotectonic setting of the Teplá-Barrandian volcano-sedimentary successions: geochemistry, U-Pb zircon ages, and provenance (Bohemian Massif, Czech Republic). *International Journal of Earth Sciences*, **93**, 742–757

Dudek, A. 1980. The crystalline basement block of the Outer Carpathians in Moravia: Bruno-Vistulicum. *Rozpravy Československé Akademie Věd*, **90**, 81–85.

Eichhorn, R., Höll, R., Loth, G. & Kennedy, A. 2000. Implications of U-Pb SHRIMP zircon data on the age and evolution of the Felbertal tungsten deposit (Tauern Window, Austria). *International Journal of Earth Sciences*, **88**, 496–512.

Eisenack, A. 1966. Besprechung zu BEHR, JORDAN & WEBER: Ein paläontologischer Beitrag für das Alter der Vergneisung im Erzgebirge – Chitinozoen in den Phyllitarealen von Hermsdorf-Rehefeld. *Zentralblatt für Geologie und Paläontologie, Teil II*, 1966, 62.

Elicki, O. 1992. *Faziesanalyse der unterkambrischen Karbonate Deutschlands*. PhD thesis, Freiberg University.

Elicki, O. 1994. Lower Cambrian carbonates from eastern Germany: palaeontology, stratigraphy and palaeogeography. *Neues Jahrbuch für Geologie und Paläontologie, Abhandlungen*, **191**, 69–93.

Elicki, O. 1996. Die Gastropoden und Monoplacophoren der unterkambrischen Görlitz Fauna. *Paläontologie, Stratigraphie, Fazies, 3. Freiberger Forschungshefte, C*, **464**, 145–173.

Elicki, O. 1997. Biostratigraphic data of the German Cambrian - present state of knowledge. *Paläontologie, Stratigraphie, Fazies, 4. Freiberger Forschungshefte, C*, **466**, 155–165.

Elicki, O. 1998. First report of *Halkieria* and enigmatic globular fossils from the Central European Marianian (Lower Cambrian, Görlitz Syncline, Germany). *Revista Española de Paleontología, No. extr. Homenaje al Prof. Gonzalo Vidal*, 51–64.

Elicki, O. 1999*a*. Palaeoecological significance of calcimicrobial communities during ramp evolution: an example from the Lower Cambrian of Germany. *Facies*, **41**, 27–40.

Elicki, O. 1999*b*. Beitrag zur Lithofazies und zur Lithostratigraphie im Unterkambrium von Doberlug-Torgau. *Paläontologie, Stratigraphie, Fazies, 7. Freiberger Forschungshefte, C*, **481**, 107–119.

Elicki, O. 2000. Die kambrische 'Görlitz-Fauna': Charakteristik und Bedeutung für die stratigraphische und paläogeographische Rekonstruktion Mitteleuropas. *Zeitschrift für geologische Wissenschaften*, **28**, 11–32.

Elicki, O. 2003. Das Kambrium Sachsens. *Veröffentlichungen des Museums für Naturkunde Chemnitz*, **26**, 41–62.

Elicki, O. & Debrenne, F. 1993. The Archaeocyatha of Germany. *Paläontologie, Stratigraphie, Fazies, 1. Freiberger Forschungshefte, C*, **450**, 3–40.

Elicki, O. & Schneider, J. 1992. Lower Cambrian (Atdabanian/Botomian) shallow-marine carbonates of the Görlitz Synclinorium (Saxony/Germany). *Facies*, **26**, 55–66.

Elicki, O. & Wotte, T. 2003. Cambroclaves from the Cambrian of Sardinia (Italy) and Germany: constraints for the architecture of western Gondwana and the palaeogeographical and paleoecological potential of cambroclaves. *Palaeogeography, Palaeoclimatology, Palaeoecology*, **195**, 55–71.

Estrada, S., Heuse, T. & Schulz, E. 1994. Zur regionalgeologischen Interpretation kambro-ordovizischer Schichten an der Nordwest-Flanke des Schwarzburger Antiklinorium, Thüringen. *Zeitschrift für geologische Wissenschaften*, **22**, 537–553.

Fatka, O. 2000. Das Mittlere Kambrium bei Jince, Tschechische Republik. *In:* Pinna, G. & Meischner, D. (eds) *Europäische Fossillagerstätten*. Springer-Verlag, Heidelberg, 21–23, 49–50, 244.

Fatka, O. 2006. Biostratigraphy of the Jince Formation (Middle Cambrian) in the Příbram-Jince Basin: Historical review. *Acta Universitatis Carolinea Geologica*, **47**(1–4), 53–61.

Fatka, O. & Konzalová, M. 1996. Microfossils of the Paseky Shales (Lower Cambrian, Czech Republic). *Journal of the Czech Geological Society*, **40**, 55–66.

Fatka, O. & Kordule, V. 1985. *Etoctenocystis bohemica* gen. et sp. nov. - new Ctenocystoid from Czechoslovakia (Echinodermata, Middle Cambrian). *Věstník Ústředního ústavu geologického*, **60**, 225–229.

Fatka, O. & Kordule, V. 1990. *Vyscystis ubaghsi* gen. et sp. nov., Lepidocystid eocrinoid (Middle Cambrian, Czechoslovakia). *Věstník Ústředního ústavu geologického*, **65**, 315–319.

Fatka, O. & Kordule, V. 1991. *Akadocrinus knizeki* sp. nov., gogiid eocrinoid from Czechoslovakia (Echinodermata, Middle Cambrian). *Věstník Ústředního ústavu geologického*, **66**, 239–243.

Fatka, O. & Kordule, V. 2001. *Asturicystis havliceki* sp. nov. (Echinodermata, Homostelea) from the Middle Cambrian of Bohemia. *Journal of the Czech Geological Society*, **46**, 189–193.

Fatka, O. & Vavrdová, M. 1998. Early Cambrian Acritarcha from sediments underlying the Devonian in Moravia (Měnín 1 borehole, Czech Republic). *Bulletin of the Czech Geological Survey*, **73**, 55–60.

Fatka, O., Kordule, V. & Szabad, M. 2004. Stratigraphic distribution of Cambrian fossils in the Příbram-Jince Basin (Barrandian area, Czech Republic). *Senckenbergiana lethaea*, **84**, 369–384.

Fatka, O., Kraft, P. & Szabad, M. (in press) *Wiwaxia* Walcott, 1911 from the 'Middle' Cambrian of the Barrandian area (Czech Republic). *Neues Jahrbuch für Geologie und Palaeontologie, Abhandlungen*.

Fediuk, F. & Röhlich, P. 1960. Die Basalschichten des Ordovizium in Prag-Troja (Nordwestflügel des Barrandiums). *Acta Universitatis Carolinea, Geologica*, **1960**, 75–93.

Fiala, F. 1971. The Upper Proterozoic and Lower Palaezoic geosynclinal volcanism of the Barrandian. *Sborník geologických Věd, Geologie*, **19**, 7–97.

Finger, F., Hanžl, P., Pin, C., Von Quadt, A. & Steyrer, H. P. 2000*a*. The Brunovistulian: Avalonian Precambrian sequence at the eastern end of the Central European Variscides? *In*: Franke, W., Haak, V., Oncken, O. & Tanner, D. (eds) *Orogenic Processes: Quantification and Modelling in the Variscan Belt*. Geological Society, London, Special Publications, **179**, 103–112.

Finger, F., Tichomirowa, M., Pin, C. & Hanžl, P. 2000*b*. Relics of an Early Panafrican metabasite metarhyolite formation in the Brno Massif, Moravia, Czech Republic. *International Journal of Earth Sciences*, **89**, 328–335.

Fluck, P. Maass, R. & Raumer, J. F. von 1980. The Variscan units east and west of the Rhine Graben. *In*: Cogné, J. & Slansky, M. (eds) *Géologie de l'Europe du précambrien aux basins sédimentaires post-Hercyniens*. Mémoir BRGM, **108**, 112–131.

Fluck, P. Maass, R. & Raumer, J. F. von 1984. Neuere Ergebnisse zur Entwicklung des Varistikums in den Vogesen. *Fortschritte der Mineralogie*, **62**, 37–47.

Franke, W. 1984. Variszischer Deckenbau im Raume der Münchberger Gneismasse – abgeleitet aus der Fazies, Deformation und Metamorphose im umgebenden Paläozoikum. *Geotektonische Forschungen*, **68**, 1–253.

Franke, W. & Żelaźniewicz, A. 2002. Structure and evolution of the Bohemian Arc. *In*: Winchester, J. A., Pharaoh, T. C. & Verniers, J. (eds) *Palaeozoic Amalgamation of Central Europe*. Geological Society, London, Special Publications, **201**, 279–293.

Franke, D., Hoffmann, N. & Kamps, H. J. 1989. Alter und Struktur des Grundgebirges im Nordteil der DDR. *Zeitschrift für Angewandte Geologie*, **35**, 289–296.

Franke, D., Gründel, J., Lindert, W., Meissner, B., Schulz, E., Zagora, I. & Zagora, K. 1994. Die Ostseebohrung G 14 – eine Profilübersicht. *Zeitschrift für geologische Wissenschaften*, **22**, 235–240.

Freyer, G. 1981*a*. Die unterkambrische Brachiopodenfauna des Görlitzer Schiefergebirges. *Abhandlungen und Berichte des Naturkundemuseums Görlitz*, **54**, 1–20.

Freyer, G. 1981*b*. Die Cruzianen von Hainichen bei Borna und das Alter ihrer Fundschichten. *Freiberger Forschungshefte*, C **363**, 51–55.

Freyer, G. & Suhr, P. 1987. Über Archaeocyathinen-Funde und den lithologischen Aufbau des Unterkambriums im Gebiet von Torgau. *Zeitschrift für geologische Wissenschaften*, **15**, 665–680.

Freyer, G. & Suhr, P. 1992. Über Algengemeinschaften aus der unterkambrischen Zwethauer Folge des Gebietes von Torgau. *Freiberger Forschungshefte, C*, **445**: 49–65.

Freyer, G., Geissler, E., Hoth, K. & Nhuan, T. T. 1982. Die

Altersstellung des Karbonatgesteinshorizonts Rabenstein-Auerswalde und ihre Bedeutung für die Geologie des süd-östlichen Schiefermantels des sächsischen Granulitgebirges. *Zeitschrift geologische Wissenschaften*, **10**(11), 1403–1424.

FRIEDRICH, W-P. 1993. Systematik und Funktionsmorphologie mittelkambrischer Cincta (Carpoidea, Echinodermata). *Beringeria*, **7**, 3–190.

FRISCH, W. & NEUBAUER, F. 1989. Pre-Alpine terranes and tectonic zoning in the eastern Alps. *In:* DALLMEYER, R. D. (ed.) *Terranes in the Circum-Atlantic Paleozoic Orogens*. Geological Society of America, Special Papers, **230**, 91–100.

GAERTNER, H. R. *et al.* 1968. *Saxothuringikum in Bavaria*. International Geological Congress, 23th Session, Guide to excursion C 34.

GEBAUER, D. & GRÜNENFELDER, M. 1979. U-Pb zircon and Rb-Sr mineral dating of eclogites and their country rocks; example, Münchberg Gneiss Massif, Northeast Bavaria. *Earth and Planetary Science Letters*, **42**, 35–44.

GEUKENS, F. 1999. Notes accompagnant une révision de la carte structurale du Massif de Stavelot. *Aardkundige Mededelingen*, **9**, 183–190.

GEYER, G. 1998. Intercontinental, trilobite-based correlation of the Moroccan early Middle Cambrian. *Canadian Journal of Earth Sciences*, **35**, 374–401.

GEYER, G. 2005. The base of a Middle Cambrian series: are suitable concepts in reach? *Geosciences Journal*, **9**, 81–99.

GEYER, G. & ELICKI, O. 1995. Lower Cambrian trilobites from the Görlitz Synclinorium (Germany) – review and new results. *Paläontologische Zeitschrift*, **69**, 87–119.

GEYER, G. & LANDING, E. 2004. A unified Lower–Middle Cambrian chronostratigraphy for West Gondwana. *Acta Geologica Polonica*, **54**, 179–218.

GEYER, G. & PALMER, A. R. 1995. Neltneriidae and Holmiidae (Trilobita) from Morocco and the problem of Early Cambrian intercontinental correlation. *Journal of Paleontology*, **69**, 459–474.

GEYER, G. & WIEFEL, H. 1997. Fränkisch-Thüringisches Schiefergebirge. *In*: Stratigraphie von Deutschland II. Teil I: Thüringen, Sachsen, Ostbayern. *Courier Forschungsinstitut Senckenberg*, **200**, 57–102.

GÖTHEL, M. 2001. Das autochthone und allochthone Paläozoikum des 'Görlitzer Schiefergebirges' (Mitteleuropäische Varisziden, Deutschland). *Zeitschrift für geologische Wissenschaften*, **29**, 55–73.

GÖTHEL, M. & ELICKI, O. 1997. Das Görlitzer Unterkambrium – eine variszische Duplexstruktur im Görlitzer Synklinorium? (Saxothuringikum, Sudetische Terrane, Sachsen). *Natur und Landschaft in der Niederlausitz*, **17**, 99–116.

GREILING, R.O., JENSEN, S. & SMITH, A. G. 1999. Vendian–Cambrian subsidence of the passive margin of western Baltica – application of new stratigraphic data from the Scandinavian Caledonian margin. *Norsk Geologisk Tidsskrift*, **79**, 133–144.

HAVLÍČEK, V. 1949. Palaeontological proof of Cambrian in the Železné Hory area. *Věstník Státního geologického Ústavu Československé Republiky*, **25**, 331–333 [in Czech].

HAVLÍČEK, V. 1950. Zpráva o geologickém mapování na Jinecku. *Věstník Státního geologického Ústavu Československé. Republiky*, **25**, 98–103.

HAVLÍČEK, V. 1954. Kambrium a ordovik na listě Žd'ár mapy 1:10 000. *Zprávy o geologických výzkumech v roce 1953*, 33–35.

HAVLÍČEK, V. 1967. Doplňky ke stratigrafii brdského kambria. (*Zprávy o geologických výzkumech v roce 1966*, 104–106.

HAVLÍČEK, V. 1968. Nejstarší makrofaunistický obzor v ČSSR. *Časopis pro Mineralogii a geologii*, **13**, 211–212.

HAVLÍČEK, V. 1970. *Pompeckium* and *Jamesella* (Orthacea, Brachiopoda) in the Middle Cambrian of Bohemia. *Věstník Ústředního ústavu geologického*, **45**, 289–290.

HAVLÍČEK, V. 1971. Stratigraphy of the Cambrian of Central Bohemia. *Sborník geologických Věd, Geologie*, **20**, 7–52.

HAVLÍČEK, V. 1998. Cambrian. *In*: CHLUPÁČ, I. , HAVLÍČEK, V., KŘÍŽ, J., KUKAL, Z. & ŠTORCH, P. (eds) *Palaeozoic of the Barrandian*. Český geologický ústav, Prague, 19–38.

HAVLÍČEK, V. & ŠNAJDR, M. 1951*a*. Cambrian and Ordovician at Jince and at the Brdy Hills. *Sborník Ústředního Ústavu geologického, odd. geologický*, **18**, 145–276 [in Czech].

HAVLÍČEK, V. & ŠNAJDR, M. 1951*b*. Report on new data about the Železné hory Lower Palaeozoic. *Věstník Ústředního ústavu geologického*, **26**, 28–29 [in Czech].

HAVLÍČEK, V. & ŠNAJDR, M. 1957. Stratigraphy and facies at the Brdy Cambrian. *Sborník Ústředního Ústavu geologického, odd. geologický*, **23**, 487–522 [in Czech].

HAVLÍČEK, V., HORNÝ, R., CHLUPÁČ, I. & ŠNAJDR, M. 1958. *Průvodce ke geologickým exkurzím do Barrandienu (Führer zu den geologischen Exkursionen in das Barrandium)*. Ústřední ústav geologický, Prague [in Czech and German].

HERBOSCH, A., VANGUESTAINE, M., DEGARDIN, J. M., DEJONGHE, L., FAGEL, N. & SERVAIS, T. 1991. Etude lithostratigraphique et sédimentologique du sondage de Lessines (bord meridional du Massif de Brabant, Belgique). *Annales de la Société Géologique de Belgique*, **114**, 195–212.

HERBOSCH, A., VERNIERS, J., DEBACKER, T., BILLIAERT, B., DE SCHEPPER, S. & BELMANS, M. 2002. The lower Palaeozoic stratigraphy and sedimentology of the Brabant Massif in the Dyle and Orneau Valleys and of the Condroz Inlier at Fosses; an excursion guidebook. *Geologica Belgica*, **5**, 71–143.

HEUSE, T., ERDTMANN, B.-D. & KRAFT, P. 1994. Upper Ordovician microfossils (acritarchs, chitinozoans) and graptolites from the Schwarzburg Anticline, Thuringia (Germany). *Veröffentlichungen des Naturhistorischen Museums Schleusingen*, **9**, 41–68.

HIRSCHMANN, G. 1967. Assyntische und variszische Baueinheiten im Grundgebirge der Oberlausitz (unter besonderer Berücksichtigung des östlichen Görlitzer Schiefergebirges). *Freiberger Forschungshefte, C*, **212**, 1–146.

HIRSCHMANN, G. & OKRUSCH, M. 1988. Spessart-Kristallin als Bestandteil der Mitteldeutschen Kristallinzone – ein Vergleich. *Neues Jahrbuch für Geologie und Paläontologie, Abhandlungen*, **177**, 1–39.

HIRSCHMANN, G. & STETTNER, G. 1997. Zone Erbendorf-Vohenstrauß (ZEV) bzw. Neustädter Scholle. *In*: Stratigraphie von Deutschland II. Teil I: Thüringen, Sachsen, Ostbayern. *Courier Forschungsinstitut Senckenberg*, **200**, 364–371.

HIRSCHMANN, G., STETTNER, G. & ROHRMÜLLER, J. 1994. *Das lithologische Profil der nördlichen ZEV.* KTB Report **94–2, B1**.

HORNÝ, R. 1963. Lower Paleozoic Bellerophonthina (Gastropoda) of Bohemia. *Sborník geologických Věd Paleontologie*, **2**, 57–164.

HORNÝ, R. 1964. The Middle Cambrian Pelagiellacea of Bohemia (Mollusca). *Sborník Národního muzea řada B*, **20**, 133–140.

HORSTIG, G. VON 1954. Das Mittelkambrium von Stadtsteinach im Frankenwald. *Neues Jahrbuch für Geologie und Paläontologie, Monatshefte*, **1954**, 448–456.

HOTH, K. & BERGER, H.-J. 1980. *Stratigraphie undn lithologisch-fazielle Charakteristik der Parametamorphite und Sedimente im Erzgebirge/Vogtland*. Report, VEB Geologische Forschung und Erkundung Freiberg.

JACHOWICZ, M. & PŘICHYSTAL, A. 1997. Lower Cambrian sediments in deep boreholes in south Moravia. *Věstník Českého geologického Ústavu*, **72**, 329–332.

JAEKEL, O. 1901. Über Carpoiden, eine neue Klasse von Pelmatozoen. *Zeitschrift der deutschen geologischen Gesellschaft*, 666–677.

JAHN, J.J. 1895. Ueber die geologischen Verhältnisse des Cambrium von Tejřovic und Skrej in Böhmen, *Jahrbuch der k. k. geologischen Reichsanstalt, Wien*, **45**, 641–790.

JANKAUSKAS, T. & LENDZION, K. 1992. Lower and Middle Cambrian acritarch-based biozonation of the Baltic Syneclise and adjacent areas (East European Platform). *Przegląd Geologiczny*, **40**, 519–524.

JANKAUSKAS, T. & LENDZION, K. 1994. Biostratigraphic correlation of Lower and Middle Cambrian sections in the Baltic Syneclise and adjacent areas. *Przegląd Geologiczny*, **42**, 365–370.

JAWOROWSKI, K. 1997. Depositional environments of the Lower and Middle Cambrian sandstone bodies. Polish Part of the East European Platform. *Biuletyn Państwowego Instytutu Geologicznego*, **377**, 1–112.

JAWOROWSKI, K. 2001. Upper Vendian–Middle Cambrian depositional sequence in the Polish part of the Baltic Basin and the occurrence of hydrocarbon-bearing sandstones. *1st Nordic-Baltic Cambrian Geo-Energy Seminar*, 15–16 September 2001, Jægersborg-Gentofte, Denmark, Abstracts.

JONAS, P., BUSCHMANN, B. & GAITZSCH, B. 2000. Unterkambrischer und unterkarboner Vulkanismus der Torgau-Doberluger Synklinale (NE Saxothuringische Zone). *Zeitschrift für geologische Wissenschaften*, **28**, 157–175.

Kachlík, V. 1992. Representation and relationship of the Proterozoic and Paleozoic units of the Central Bohemian Pluton's mantle and possibilities of their correlation. *In*: Kukal, Z. (ed.) *Proceedings of the 1st International Conference on the Bohemian Massif.* Czech Geological Survey, Prague, 144–149.

Kachlík, V. & Janoušek, V. 2001. Cambrian in the Netvořice-Neveklov Metamorphic Islet (roof of the Central Bohemian Pluton). *Geolines*, **13**, 73–74.

Keller, B. M. & Rozanov, A. Yu. (eds) 1979. *Stratigrafiya verkhnedokembriyskikh otlozheniy zapada Vostochno-Evropeyskoy platformy.* Nauka, Moscow.

Kemnitz, H., Romer, R. L., Ahrendt, H., Wemmer, K., Rosner, M. & Oncken, O. 1998. Post-Cadomian separation of Armorica from Gondwana: The Vesser Zone – a Saxothuringian witness. *Terra Nostra*, **98**, 87–88.

Kemnitz, H., Romer, R.L. & Oncken, O. 2002. Gondwana break-up and the northern margin of the Saxothuringian belt (Variscides of Central Europe). *International Journal of Earth Sciences*, **91**, 246–259.

Kettner, R. 1915. On the Žitec Conglomerate – lowermost horizon in the Bohemian Cambrian. *Rozpravy II. třídy České akademie věd a umění*, **24**(34), 1–64 [in Czech].

Kettner, R. 1923. The Skryje-Týřovice Cambrian and its surrouding. *Sborník Státního geologického ústavu ČSR*, **3**, 5–63 [in Czech].

Kettner, R. 1925. Geology of Příbram in the light of recent research. *Sborník Státního geologického ústavu ČSR*, **5**, 1–52 [in Czech].

Kettner, R. & Bouček, B. 1937. *Tableaux synoptiques des formations du Barrandien.* Travaux de l'Institut de Géologie et de Paléontologie de l'Université Charles a Praha.

Kettner, R. & Kodym, O. 1919. New stratigraphy of the Barrandian area. *Časopis Musea Království Českého*, **93**, 47–55 [in Czech].

Klötzli-Chowanetz, E., Klötzli, U. & Koller, F. 1997. Lower Ordovician migmatisation in the Oetztal crystalline basement (Eastern Alps, Austria): linking U-Pb and Pb-Pb dating with zircon morphology. *Schweizerische Mineralogische und Petrographische Mitteilungen*, **77**, 315–324.

Kotas, A. 1973. Profil utworów paleozoicznych w otworach wiertniczych Sosnowiec IG-1 i Goczałkowice IG-1. *Kwartalnik Geologiczny*, **17**, 626–627.

Kotas, A. 1982. Zarys budowy geologicznej Górnośląskiego Zagłębia Węglowego. *In*: Różkowski, A. & Ślósarz, J. (eds) *Przewodnik LIV Zjazdu Polskiego Towarzystwa Geologicznego, Sosnowiec*, 23–25 września 1982. Wydawnictwa Geologiczne, 45–72.

Kowalczewski, Z. 1990. Grubookruchowe skały kambru na środkowym południu Polski (litostratygrafia, tektonika, paleogeografia). *Prace Państwowego Instytutu Geologicznego*, **131**, 1–82.

Kowalczewski, Z. 1995. Fundamental stratigraphic problem of the Cambrian in the Holy Cross Mts. *Geological Quarterly*, **39**, 449–470.

Kowalczewski, Z., Kuleta, M., Lisik, R. & Moczydłowska, M. 1986. New data on Cambrian and Lower Ordovician rocks in the vicinities of Wiśniówka in the Góry Świętokrzyskie Mts. *Geological Quarterly*, **30**, 201–228 [in Polish].

Kowalczewski, Z., Żylińska, A. & Szczepanik, Z. 2006. Cambrian in the Holy Cross Mts.. *In*: Skompski, S. & Żylińska, A. (eds) *LXXVII Zjazd Naukowy PolskiegoTowarzystwa Geologicznego, Ameliówka k. Kielc, 28–30 czerwca 2006, Materiały konferencyjne.* Panstwowy Instytut Geologiczny, Warszawa, 15–28 [in Polish].

Kowalski, W. R. 1983. Stratigraphy of the Upper Precambrian and lowest Cambrian strata in southern Poland. *Acta Geologica Polonica*, **33**, 183–218.

Krawczyk, C. M., McCann, T., Cocks, L. R. M., England, R., McBride, J. & Wybraniec, S. 2008. Caledonian tectonics. *In*: McCann, T. (ed.) *The Geology of Central Europe. Volume 1: Precambrian and Palaeozoic.* Geological Society, London, 303–381.

Krejčí, J. 1877. *Geologie čili nauka o útvarech zemských se zvláštním ohledem na krajiny československé.* Praha.

Kröner, A., Willner, A. P., Hegner, E., Frischbutter, A., Hofmann, J. & Bergner, R. 1995. Latest Precambrian (Cadomian) zircon ages, Nd isotopic systematics and P-T evolution of granitoid orthogneisses of the Erzgebirge, Saxony and Czech Republic. *Geologische Rundschau*, **84**, 437–456.

Kukal, Z. 1962. Petrological study of Hluboš congomerates of the Bohemian Cambrian. *Zprávy o geologických výzkumech v roce 1961*, 79–80 [in Czech].

Kukal, Z. 1963. Report on petrological research of the Barrandian Cambrian. *Zprávy o geologických výzkumech v roce 1962*, 99–100 [in Czech].

Kukal, Z. 1971. Sedimentology of Cambrian deposits of the Barrandian area (Central Bohemia). *Sborník geologicky Věd, řada Geologie,* **20**, 53–100.

Kukal, Z. 1995. The Lower Cambrian Paseky Shale: sedimentology. *Journal of the Czech Geological Society*, **40**, 67–78.

Kurze, M. 1962. Mitteilung über zwei Fossilfunde (Foraminiferen) aus den Kalken der Arnsdorfer Schuppenzone. *Monatsberichte der Deutschen Akademie der Wissenschaften*, **4**, 134–138.

Kurze, M., Frischbutter, A., Linnemann, U. & Tröger, K.-A. 1997. Elbezone. *In*: Stratigraphie von Deutschland II. Teil I: Thüringen, Sachsen, Ostbayern. *Courier Forschungsinstitut Senckenberg*, **200**, 266–296.

Landing, E. 1994. Precambrian–Cambrian boundary global stratotype ratified and a new perspective of Cambrian time. *Geology*, **22**, 179–184.

Landing, E. 1996. Avalon: Insular continent by the latest Precambrian. *In*: Nance, R. D. & Thompson, M. D. (eds) *Avalon and Related Peri-Gondwanan Terranes of the Circum-North Atlantic.* Geological Society of America, Special Papers, **304**, 29–63.

Lefebvre, B. & Fatka, O. 2003. Palaeogeographical and palaeoecological aspects of the Cambrian-Ordovician radiation of echinoderms in Gondwanan Africa and peri-Gondwanan Europe. *Palaeogeography, Palaeoclimatology, Palaeoecology*, **195**, 73–97.

Legler, C. 1985. Die schichtgebundenen Mineralisationen des Erzgebirges. Probleme der Paragenese. *Freiberger Forschungshefte, C*, **401**.

Legrand, R. 1968. *Le Massif du Brabant.* Mémoires pour servir à l'explication des Cartes géologiques et minières de la Belgiques, **9**, 1–148.

Lendzion, K. 1970. Eocambrian and Cambrian formations in borehole Żarnowiec IG-1. *Przegląd Geologiczny*, **18**, 303–344 [in Polish].

Lendzion, K. 1976. Stratygrafia kambru zachodniej części syneklizy perybałtyckiej. *Biuletyn Instytutu Geologicznego*, **270**, 213–219.

Lendzion, K. 1983*a*. Rozwój kambru osadów platformowych Polski. *Prace Instytutu Geologicznego*, **105**.

Lendzion, K. 1983*b*. Biostratygrafia osadów kambru w polskiej części platformy wschodnioeuropejskiej. *Kwartalnik Geologiczny*, **27**, 669–694.

Lendzion, K. 1984. Basement rocks of the productive Carboniferous – Cambrian. *In:* Harasimiuk, M. (ed.) *Przewodnik LVI Zjazdu Polskiego Towarzystwa Geologicznego, Lublin,* 127–130.

Lendzion, K. 1986. Sedimentation of the Vendian-Cambrian marine sequence, Poland. *Geological Magazine*, **123**, 361–365.

Leonhardt, D., Hoth, K., Berger, H.-J. & Koch, E. A. 1997. Erzgebirge. *In*: *Stratigraphie von Deutschland II. Teil I: Thüringen, Sachsen, Ostbayern.* Courier Forschungsinstitut Senckenberg, **200**, 176–227.

Linnemann, U. 1996. Die Lagerungsverhältnisse und die sedimentäre Entwicklung des basalen Altpaläozoikums vom Langen Berg bei Gehren (Schwarzburger Anticlinorium, Saxothuringicum). *Beiträge zur Geologie von Thüringen, Neue Folge*, **1996**, 73–84.

Linnemann, U. & Buschmann, B. 1995*a*. Der Nachweis der cadomischen Diskordanz in einer Tiefbohrung bei Gera und deren Bedeutung für das proterozoisch-paläozoische Standardprofil im Schwarzburger Antiklinorium. *Geowissenschaftliche Mitteilungen von Thüringen*, **3**, 1–11.

Linnemann, U. & Buschmann, B. 1995*b*. Die cadomische Diskordanz im Saxothuringikum (oberkambrisch-tremadocische overlap-Sequenzen). *Zeitschrift für geologische Wissenschaften*, **23**(5/6), 707–727.

Linnemann, U. & Buschmann, B. 1996. Die cadomische Diskordanz im Saxothuringikum. *In*: Feldmann, L. & Quade, H. (eds) *Der Untergrund der Norddeutschen Tiefebene. Materialien zur 5. Jahrestagung, Clausthal-Zellerfeld, 8.-12.05.1996.* Exkursionsführer und Veröffentlichungen, **197**, 49–50.

Linnemann, U. & Schauer, M. 1999. Die Entstehung der Elbezone vor dem Hintergrund der cadomischen und variszischen Geschichte des Saxothuringischen Terranes – Konsequenzen aus einer abgedeckten geologischen Karte. *Zeitschrift für geologische Wissenschaften*, **27**, 529–561.

Linnemann, U., Gehmlich, M., Tichomirova, M. & Buschmann, B.

1998. Die Präkambrium-Paläozoikum-Grenze im Saxothuringischen Terrane der zentraleuropäischen Varisziden. *Terra Nostra*, **98**, V210-V211.

LINNEMANN, U., GEHMLICH, M., HEUSE, T. & SCHAUER, M. 1999. Die Cadomiden und Varisziden im Thüringisch-Vogtlandischen Schiefergebirge (Saxothuringisches Terrane). *Beiträge zur Geologie von Thüringen, N.F.*, **6**, 7–39.

LINNEMANN, U., GEHMLICH, M., *et al.* 2000. From Cadomian subduction to Early Palaeozoic rifting: the evolution of Saxo-Thuringia at the margin of Gondwana in the light of single zircon geochronology and basin development (Central European Variscides, Germany). *In*: FRANKE, W., HAAK, V., ONCKEN, O. & TANNER, D. (eds) *Orogenic Processes: Quantification and Modelling in the Variscan Belt*. Geological Society, London, Special Publications, **179**, 131–153.

LINNEMANN, U., D'LEMOS, R. *et al.* 2008. Cadomian tectonics. *In:* MCCANN, T. (ed.) *The Geology of Central Europe. Volume 1: Precambrian and Palaeozoic*. Geological Society, London, 103–154.

LIPOLD, M.V. & KREJČÍ, J. 1860. *Verhandlungen der k. k. Geologischen Reichsanstalt*, Wien, Sitzungsbericht vom 24. April, 88–91.

LUDWIG, V. 1969. Lithologische Untersuchung des Kambriums im Frankenwald (Bayern). *Geologisches Jahrbuch*, **87**, 89–156.

LÜTZNER, H. 1991. Schichtungsgefüge der Goldisthal- und Frauenbach-Folge (Kambro-Ordovizium) im Thüringischen Schiefergebirge. *Wissenschaftliche Beiträge der Ernst-Moritz-Arndt-Universität Greifswald, Fortschritte der Geologie*, 27–31.

MCCANN, T. 1998. Lower Palaeozoic evolution of the northeast German Basin/Baltica borderland. *Geological Magazine*, **135**, 129–142.

MCCANN, T. & KRAWCZYK, C. M. 2001. The Trans-European Fault: a critical reassessment. *Geological Magazine*, **135**, 129–142.

MAASS, R. 1974. Ein strukturgeologischer Beitrag zum Paläozoikum des Südschwarzwaldes. *Berichte der Naturforschenden Gesellschaft Freiburg i. Br.*, **64**, 25–38.

MAASS, R. 1981. The Variscan Black Forest. *Geologie en Mijnbouw*, **70**, 137–143.

MAASS, R., GRIMM, B., VOGT, C., WEINZIEHR, R. & WICKERT, 1980. Das Variszikum der südlichen Teile von Schwarzwald und Vogesen. *Nachrichten der deutschen geologischen Gesellschaft*, **23**, 10–13.

MAREK, L. 1972. Middle Cambrian Hyolithes: *Maxlites* gen. n. *Časopis Národního Muzea*, **141**, 69–72.

MAREK, L. 1981. Middle Cambrian hyolithid family Parentilitidae fam. nov. *Časopis Národního Muzea*, **150**, 163–168.

MARTIN, F. 1975. Acritarches du Cambro-Ordovicien du Massif du Brabant, Belgique. *Bulletin de l'Institut Royal des Sciences naturelles de Belgique, Sciences de la Terre*, **51**, 1–33.

MENS, K., BERGSTRÖM, J. & LENDZION, K. 1990. *The Cambrian on the East European Platform. Correlation chart and explanatory notes*. International Union of Geological Sciences, Publication No. **25**.

MERGL, M. & ŠLEHOFEROVÁ, P. 1990. Middle Cambrian inarticulate brachiopods from Central Bohemia. *Sborník geologických Věd, Paleontologie*, **31**, 67–104.

MIKULÁŠ, R. 1995. Trace fossils from the Paseky Shales (Early Cambrian, Czech Republic). *Journal of the Czech Geological Society*, **40**, 37–54.

MIKULÁŠ, R. 2000. Trace fossils from the Middle Cambrian of the Barrandian area (Central Bohemia, Czech Republic). *Czech Geological Survey, Special Papers*, **12**, 1–29.

MIKULÁŠ, R. & NEHYBA, S. 2001. Trace fossils of presumed Lower Cambrian age in Borehole Měnín – 1 in South Moravia. *Geologické výzkumy na Moravě a ve Slazsku v roce*, **2000**, 47–50.

MINGRAM, B. & RÖTZLER, K. 1998. Cambro-Ordovician rifting and passive margin sediments conserved in Variscan HP thrust units of the Erzgebirge. *Acta Universitatis Carolinae, Geologica*, **42**, 306–307.

MINGRAM, B., RÖTZLER, K., TICHOMIROWA, M., HAASE, G. & TEUFEL, S. 1998. Early Palaeozoic rift-related magmatism and passive margion sediments preserved in HP units of the Erzgebirge. *Terra Nostra*, **1998**, 105–107.

MIZERSKI, W., ORŁOWSKI, S. & WAKSMUNDZKI, B. 1991. New data on geology of the Kamieniec Shale Formation (Lower Cambrian, Holy Cross Mts). *Geological Quarterly*, **35**, 149–162.

MOCZYDŁOWSKA, M. 1991. Acritarch biostratigraphy of the Lower Cambrian and the Precambrian-Cambrian boundary in southeast Poland. *Fossils and Strata*, **29**.

MOCZYDŁOWSKA, M. 1995*a*. Neoproterozoic and Cambrian successions deposited on the East European Platform and Cadomian basement area in Poland. *Geophysica et Geodaetica*, **39**, 276–283.

MOCZYDŁOWSKA, M. 1995*b*. Cambrian microplankton distribution in Iberia and Baltica, and possible palaeogeographic relationships. *In*: RODRIGUEZ ALONSO, M. D. & GONZALO CORRAL, J. C. (eds) *XIII Reunion de Geología del Oeste Peninsula, Annual IGCP Project 319–320 Meeting,* Communicaciones, Signo S.I., Salamanca, 117–120.

MOCZYDŁOWSKA, M. 1997. Proterozoic and Cambrian successions in Upper Silesia: an Avalonian terrane in southern Poland. *Geological Magazine*, **134**, 679–689.

MOCZYDŁOWSKA, M. 1998. Cambrian acritarchs from Upper Silesia, Poland – biochronology and tectonic implications. *Fossils and Strata*, **48**.

MOCZYDŁOWSKA, M. & VIDAL, G. 1986. Lower Cambrian acritarch zonation in southern Scandinavia and southeastern Poland. *Geologiska Föreningen i Förhandlingar*, **108**, 201–223.

MORENO-EIRIS, E., PEREJÓN, A., RODRÍGUEZ, S. & FALCES, S. 1995. Palaeozoic Cnidaria and Porifera from Sierra Morena. *In*: PEREJÓN, A. (ed.) *VII International Symposium on Fossil Cnidaria and Porifera, Field trip D*, 1–68.

MÜLLER-SOHNIUS, D., VON DRACH, V., HORN, P. & KÖHLER, H. 1987. Alterbestimmungen an der Münchberger Gneismasse, Nordost-Bayern, *Neues Jarbuch für Mineralogie, Abhandlungen*, **156**, 175–206.

MÜLLER, B., KLÖTZLI, U. & FLISCH, M. 1995. U-Pb and Pb-Pb dating of the older orthogneiss suite in the Silvretta nappe, eastern Alps: Cadomian magmatism in the upper Austro-Alpine realm. *Geologische Rundschau*, **84**, 457–465.

MÜLLER, B., KLÖTZLI, U., SCHALTEGGER, U. & FLISCH, M. 1996. Early Cambrian oceanic plagiogranite in the Silvretta Nappe, Eastern Alps; geochemical, zircon U-Pb and Rb-Sr data from garnet-hornblende-plagioclase gneisses. *Geologische Rundschau*, **85**, 822–831.

NANCE, R. D. & MURPHY, J. B. 1994. Contrasting basement isotopic signatures and the palinspastic restoration of peripheral orogens; example from the Neoproterozoic Avalonian-Cadomian Belt. *Geology*, **22**, 617–620.

NANCE, R. D., MURPHY, J. B., STRACHAN, R. A., D'LEMOS, R. S. & TAYLOR, G. K. 1991. Late Proterozoic tectonostratigraphic evolution of the Avalonian and Cadomian terranes. *Precambrian Research*, **53**, 41–78.

NAWROCKI, J. 2004. New Vendian-Cambrian paleogeography of Baltica. *In:* CHOI, D. K. (ed.), *Ninth International Conference of the Cambrian Stage Subdivision Working Group,* Taebaek, Korea, Abstracts with Program. Seoul National University, Seoul, 20–23.

NAWROCKI, J., BOGUCKIJ, A. & KATINAS, V. 2004*a*. New Late Vendian palaeogeography of Baltica and the TESZ. *Geological Quarterly*, **48**, 309–316.

NAWROCKI, J., ŻYLIŃSKA, A., BUŁA, Z., GRABOWSKI, J., KRZYWIEC, P. & POPRAWA, P. 2004*b*. Early Cambrian location and affinities of the Brunovistulian terrane (Central Europe) in the light of palaeomagnetic data. *Journal of the Geological Society, London*, **161**, 513–522.

NIKOLAISEN, F. & HENNINGSMOEN, G. 1985. Upper Cambrian and lower Tremadoc olenid trilobites from the Digermul Peninsula, Finnmark, northern Norway. *Norges Geologiske Undersøkelse, Bulletin,* **400**, 1–49.

OKRUSCH, M., MÜLLER, R. & EL SHAZLY, S. 1985. Die Amphibolite, Kalksilikatgesteine und Hornblendegneise der Alzenauer Gneis-Serie am Nordwest-Spessart. *Geologica Bavarica*, **87**, 5–37.

ORŁOWSKI, S. 1965. A revision of the Middle Cambrian fauna from the Słowiec Hill (Holy Cross Mountains). *Biuletyn Geologiczny Uniwersytetu Warszawskiego*, **6**, 134–146 [in Polish with English summary].

ORŁOWSKI, S. 1968*a*. Cambrian of the Łysogóry Anticline in the Holy Cross Mountains. *Biuletyn Geologiczny Wydziału Geologii*, **10**, 153–222 [in Polish with English summary].

ORŁOWSKI, S. 1968*b*. Upper Cambrian fauna of the Holy Cross Mts. *Acta Geologica Polonica,* **18**, 257–291.

ORŁOWSKI, S. 1975*a*. Cambrian and Upper Precambrian lithostratigraphic units in the Holy Cross Mts. *Acta Geologica Polonica*, **25**, 431–448 [in Polish with English summary].

ORŁOWSKI, S. 1975*b*. Lower Cambrian trilobites from Upper Silesia

(Goczałkowice borehole). *Acta Geologica Polonica*, **25**, 377–383.

Orłowski, S. 1985*a*. Lower Cambrian and its trilobites in the Holy Cross Mts. *Acta Geologica Polonica*, **35**, 231–250.

Orłowski, S. 1985*b*. New data on the Middle Cambrian trilobites and stratigraphy in the Holy Cross Mts. *Acta Geologica Polonica*, **35**, 251–263.

Orłowski, S. 1985*c*. A trilobite with North American affinity in the Lower Cambrian of Poland. *Journal of Paleontology*, **59**, 975–978.

Orłowski, S. 1988. Stratigraphy of the Cambrian system in the Holy Cross Mts. *Geological Quarterly*, **32**, 525–532.

Orłowski, S. 1989. Trace fossils in the Lower Cambrian sequence in Świętokrzyskie Mountains, Central Poland. *Acta Palaeontologica Polonica*, **34**, 211–231.

Orłowski, S. 1992*a*. Trilobite trace fossils and their stratigraphic significance in the Cambrian sequence of the Holy Cross Mountains, Poland. *Geological Journal*, **27**, 15–34.

Orłowski, S. 1992*b*. Cambrian stratigraphy and stage subdivision in the Holy Cross Mountains, Poland. *Geological Magazine*, **129**, 471–474.

Orłowski, S. & Mizerski, W. 1995. Jeszcze raz o budowie geologicznej Góry Wiśniówki (Góry Świętokrzyskie). *Przegląd Geologiczny*, **43**, 11–14.

Orłowski, S. & Waksmundzki, B. 1986. The oldest Hyolitha in the Lower Cambrian of the Holy Cross Mountains. *Acta Geologica Polonica*, **36**, 225–232.

Orłowski, S. & Żylińska, A. 1996. Non-arthropod burrows from the Middle and Late Cambrian of the Holy Cross Mountains, Poland. *Acta Palaeontologica Polonica*, **41**, 385–409.

Orłowski, S. & Żylińska, A. 2002. Lower Cambrian trace fossils from the Holy Cross Mountains, Poland. *Geological Quarterly*, **46**, 135–146.

Orłowski, S., Radwański, A. & Roniewicz, P. 1970. The trilobite ichnocoenosis in the Cambrian sequence of the Holy Cross Mountains. *In*: Crimes, T. P. & Harper, J. C. (eds) *Trace Fossils. Geological Journal, Special Issue*, **3**, 345–360.

Oruzinsky, V. & Kribek, B. 1981. Extractable organic compounds associated with the metamorphosed stratiform Cu-deposit of Tisova (Czechoslovakia). *Mineralium Deposita*, **16**, 437–446.

Pacześna, J. 1986. Upper Vendian and Lower Cambrian ichnocoenoses of Lublin region. *Biuletyn Instytutu Geologicznego*, **355**, 31–47.

Pacześna, J. 1996. The Vendian and Cambrian ichnocoenoses from the Polish part of the East-European Platform. *Prace Państwowego Instytutu Geologicznego*, **152**.

Pacześna, J. 2001. Zastosowania skamieniałości śladowych w analizie facjalnej i wysokorozdzielczej stratygrafii sekwencji – przykład z kambru polskiej części kratonu wschodnioeuropejskiego. *Przegląd Geologiczny*, **49**(12), 1137–1146.

Pacześna, J. & Poprawa, P. 2004. Eustatic versus tectonic control on development of the Neoproterozoic and Cambrian successions at the Lublin-Podlasie Basin (SW margin of Baltica) – sequence stratigraphy and subsidence analysis combined. *In:* Choi, D. K. (ed.), *Ninth International Conference of the Cambrian Stage Subdivision Working Group,* Taebaek, Korea, Abstracts with Program. Seoul National University, Seoul, 47–55.

Parsley, R. & Prokop, R. J. 2004. Functional morphology and paleoecology of some sessile Middle Cambrian echinoderms from the Barrandian region of Bohemia. *Bulletin of Geosciences (Praha)*, **79**(3), 147–156.

Patočka, F., Vlašímský P. & Blechová K. 1993. Geochemistry of Early Palaeozoic Volcanics of the Barrandian Basin (Bohemian Massif, Czech Republic): Implications for paleotectonic reconstructions. *Jahrbuch der Geologischen Bundesanstalt*, **136**, 873–896.

Peng, S., Babcock, L. E., Robison, R. A., Lin, H., Rees, M. N. & Saltzman, M. R. 2004. Global Standard Stratotype Section and Point (GSSP) of the Furongian Series and Paibian Stage (Cambrian). *Lethaia*, **37**, 351–374.

Piessens, K., de Vos, W., Herbosch, A., Debacker, T. & Verniers, J. 2004. *Lithostratigraphy and geological structure of the Cambrian rocks at Halle-Lembeek (Zenne valley, Belgium)*. Geological Survey of Belgium, Professional Papers, **2004/1-300**.

Pietzsch, K. 1910. Cruzianen aus dem Untersilur des Leipziger Kreises. *Zeitschrift der Deutschen Geologischen Gesellschaft*, **62**, 571–582.

Pompeckj, J. F. 1896. Die Fauna des Cambrium von Tejřovic und Skrej in Böhnen. *Jahrbuch der kaiserlich-königlichen Reichsanstalt*, **45**, 2/3, 495–615.

Poprawa, P. & Pacześna, J. 2002. Rozwój ryftu w późnym neoproterozoiku-wczesnym paleozoiku na lubelsko-podlaskim skłonie kratonu wschodnioeuropejskiego – analiza subsydencji i zapisu facjalnego. *Przegląd Geologiczny*, **50**, 49–61.

Poprawa, P. & Pacześna, J. 2004. Late Neoproterozoic to Early Cambrian break-up of Rodinia/Pannotia recorded by development of sedimentary basins at the SW margin of Baltica. *In:* Choi, D. K. (ed.), *Ninth International Conference of the Cambrian Stage Subdivision Working Group,* Taebaek, Korea, Abstracts with Program. Seoul National University, Seoul, 56–61.

Poprawa, P., Šliaupa, S., Stephenson, R. & Lazauskiene, J. 1999. Vendian–Early Palaeozoic subsidence history of the Baltic Basin: geodynamic implications. *Tectonophysics*, **314**, 219–239.

Poprawa, P., Pacześna, J. & Buła, Z. 2005. Comparative analysis of the late-most Neoproterozoic to Cambrian tectonic evolution of sedimentary basins at the western Baltica and Brunovistulian Terraine [sic] (Central Europe). *Acta Micropalaeontologica Sinica*, **22**, 152–153.

Pošepný, F. 1888. Ueber die Adinolen von Příbram in Böhmen. *Tschemark's mineralogische und petrologische Mitteilungen*, **10**, 175–202.

Prokop, R.J. & Fatka, O. 1985. *Luhocrinus monicae* gen. et sp. nov. (Eocrinoidea) from the Middle Cambrian of Bohemia. *Věstník Ústředního ústavu geologického*, **60**(4), 231–234.

Propach, G. & Pfeiffer, T. 1998. Ocean floor basalt, not continental gabbro: a reinterpretation of the Hoher Bogen amphibolites, Teplá-Barrandian, Bohemian massif. *Geologische Rundschau*, **87**, 303–313.

Przewłocki, Z. 2000. *Sedimentary environment of the Cambrian Pepper Mountains Shale Formation, Holy Cross Mountains*. PhD Thesis, Faculty of Geology, University of Warsaw [in Polish].

Quadt, A. von & Gebauer, D. 1998. Evolution of eclogitic rocks in the Erzgebirge: A conventional and SHRIMP U-Pb zircon and Sm-Nd study. *Acta Universitatis Carolinae, Geologica*, **42**, 306–307.

Radwański, A. & Roniewicz, P. 1963. Upper Cambrian trilobite ichnocoenosis from Wielka Wiśniówka (Holy Cross Mountains, Poland). *Acta Palaeontologica Polonica*, **8**, 259–280 [in Polish].

Raumer, J. F. von 1998. The Palaeozoic evolution in the Alps: from Gondwana to Pangea. *Geologische Rundschau*, **87**, 407–435.

Raumer, J. F. von & Chiaradia, M. 1992. Bedeutung vor- und frühpaläozoischer Lithologien für Wolfram-Mineralisationen. *Zentralblatt für Geologie und Paläontologie, Teil I*, **1992**(1): 87–91.

Raumer, J. F. von, Stampfli, G. M., Borel, G. & Bussy, F. 2002. Organization of pre-Variscan basement areas at the north Gondwanan margin. *International Journal of Earth Sciences*, **91**, 35–52.

Raumer, J. F. von, Stampfli, G. M. & Bussy, F. 2003. Gondwana-derived microcontinents – the constituents of the Variscan and Alpine collisional orogins. *Tectonophysics*, **365**, 5–22.

Reitz, E. 1991. Acritarchen des Unter-Tremadoc aus dem westlichen Frankenwald, NE-Bayern. Lower Tremadocian acritarchs from the western Franconian Forest, northeastern Bavaria. *Neues Jahrbuch für Geologie und Palaeontologie, Monatshefte*, **1991**(2), 97–104.

Ribecai, C. & Vanguestaine, M. 1993. Latest Middle-Late Cambrian acritarchs from Belgium and northern France. *Special Papers in Palaeontology*, **48**, 45–55.

Richter, R. & Richter, E. 1924. Eine cambrische Fauna im Niederschlesischen Schiefergebirge. *Centralblatt für Mineralogie, Geologie und Paläontologie; Beilagenband* (1923), **23**, 730–735.

Roche, M., Sabir, M., Steemans, P. & Vanguestaine, M. 1986. Palynologie du sondage et de la région de Willerzie. *Aardkundige Mededelingen*, **3**, 149–190.

Roth, Z. 1981. Spodní kambrium na Moravě? *Časopis pro Mineralogii a geologii*, **26**, 1–6.

Rozanov, A. Yu., Missarzhevskij, V. V., *et al. The Tommotian Stage and the Problem of the Lower Boundary of the Cambrian.* Akademiya Nauk SSSR, Trudy geologiceskago instituta, **206**, Nauka, Moscow [in Russian].

Růžička, R. 1939. Trilobiti nejstarší české kambrické fauny od Týřovic z Kamenné hůrky. *Rozpravy II. třídy České Akademie*, **49**(30), 1–12 [in Czech and German].

Schäfer, F., Kemnitz, H., Romer, R. L. & Oncken, O. 1998. The NW-Thuringian Vesser zone – a Variscan retro-wedge. *Acta Universitatis Carolinae, Geologica*, **42**, 327–328.

SCHALTEGGER, U. & GEBAUER, D. 1999. Pre-Alpine geochronology of the Central, Western and Southern Alps. *Schweizerische Mineralogische und Petrographische Mitteilungen*, **79**, 79–87.

SCHALTEGGER, U., NÄGLER, T. N., CORFU, F., MAGGETTI, M., GALETTI, G. & STOSCH, H. G. 1997. A Cambrian island arc in the Silvretta nappe: constraints from geochemistry and geochronology. *Schweizerische Mineralogische und Petrographische Mitteilungen*, **77**, 337–350.

SCHALTEGGER, U., GEBAUER, D. & QUANDT, A. VON. 2002. Mafic-ultramafic rock association at Loderico-Biasca (Lower Penninic nappes, Ticino, Switzerland): Cambrian oceanic magmatism and its bearing on early Paleozoic paleogeography. *Chemical Geology*, **186**, 269–283.

SCHALTEGGER, U., ABRECHT, J. & CORFU, F. 2003. The Ordovician orogeny in the Alpine basement: constraints from geochronology and geochemistry in the Aar Massif (Central Alps). *Schweizerische Mineralogische und Petrographische Mitteilungen*, **83**, 183–195.

SCHERMAIER, A., SCHINDLMAYR, A., FINGER, F., ROBERTS, M. & PIN, C. 1998. Evidence for Early Paleozoic subduction and arc-continent collision in the Ötztal-Silvretta Crystalline Basement (western Eastern Alps). *Acta Universitatis Carolinae, Geologica*, **42**, 330–331.

SCHULZ, B. 1997. Pre-Alpine tectonometamorphic evolution in the Austroalpine basement to the south of the central Tauern Window. *Schweizerische Mineralogische und Petrographische Mitteilungen*, **77**, 281–297.

SCHULZ, B. & BOMBACH, K. 2002. Neoproterozoic to Early Palaeozoic evolution from active to passive margin setting in a north-Gondwanan domain (Austroalpine basement to the south of the Tauern Window, Eastern Alps). *In*: *Geo 2002. Planet Erde: Vergangenheit, Entwicklung, Zukunft*, 1.-5. Oktober 2002 in Würzburg. Programm und Kurzfassungen, 305.

SCHULZ, B., SCHÖNHOFER, R. & DINKELMEYER, R. 1999. Geochemical signatures of pre-Variscan active and passive margin magmatism in the Austroalpine basement to the south of the Tauern Window, Eastern Alps. *European Journal of Mineralogy Supplements*, **11**, 207.

SCHWARZBACH, M. 1934. Das Cambrium der Oberlausitz. *Abhandlungen der Naturforschenden Gesellschaft Görlitz*, **32**, 7–54.

SDZUY, K. 1957*a*. Revision der mittelkambrischen Trilobiten von Doberlug. *Senckenbergiana lethaea*, **38**, 7–28.

SDZUY, K. 1957*b*. Alter und tiergeographische Stellung des Mittelkambriums von Doberlug. *Geologie*, **6**, 465–475.

SDZUY, K. 1958. Tiergeographie und Paläogeographie im europäischen Mittelkambrium. *Geologische Rundschau*, **1**, 450–461.

SDZUY, K. 1962. Neue kambrische Fossilien aus Bohrungen im Gebiet von Doberlug. *Geologie*, **11**, 1087–1101.

SDZUY, K. 1964. Das Kambrium des Frankenwaldes. 1: Erforschungsgeschichte, Vorkommen und Stratigraphie. *Senckenbergiana lethaea*, **45**(1–4), 201–221.

SDZUY, K. 1966. Das Kambrium des Frankenwaldes. 2: Die Bergleshof-Schichten und ihre Trilobiten-Fauna. *Senckenbergiana lethaea*, **47**, 57–86.

SDZUY, K. 1970. Mittelkambrische Fossilien aus neuen Bohrungen bei Doberlug und bei Delitzsch. *Geologie*, **19**, 1066–1085.

SDZUY, K. 1972. Das Kambrium der acadobaltischen Faunenprovinz. Gegenwärtiger Kenntnisstand und Probleme. *Zentralblatt für Geologie und Paläontologie, Teil II*, **1972**(1/2), 1–91.

SERGEEV, S. & STEIGER, R. H. 1995. Improving the resolution of single grain U/Pb dating by use of zircon extracted from feldspar: application to the Variscan magmatic cycle in the Central Alps. *Earth and Planetary Science Letters*, **134**, 37–51.

SHERGOLD, J. H. & SDZUY, K. 1991. Late Cambrian trilobites from the Iberian Mountains, Zaragoza Province, Spain. *Beringeria*, **4**, 193–235.

SIKORSKA, M. 2000. Provenance petrological study of the Upper Vendian and Cambrian clastic material; foreland of the Pomeranian Caledonides (northern Poland). *Geological Quarterly*, **44**(3), 237–248.

ŚLĄCZKA, A. 1985. Kambr. Eokambr. *In*: ŚLĄCZKA, A. (ed.) *Profile głębokich otworów wiertnicznych Instytutu Geologicznego, Potrójna IG 1*, **59**, 38–41.

ŠLIAUPA, S. 2001. Genesis and geodynamic evolution of the Baltic Cambrian basin. *1st Nordic-Baltic Cambrian Geo-Energy Seminar,* 15–16 September 2001, Jægersborg-Gentofte, Denmark, Abstracts.

ŠLIAUPA, S. 2002. Baltijos kambro baseino genesė ir geodinaminė evoliucija. *Geologija*, **37**, 31–43.

SMETANA, V. 1918. On some gastropods from the Cambrian of Skryje-Týřovice. *Rozpravy České Akademie Věd a Umění*, tř. II, **27**, 1–13 [in Czech].

ŠNAJDR, M. 1958. Trilobiti českého středního kambria. *Rozpravy Ústředního Ústavu geologického*, **24**, 1–280 [in Czech with English summary].

ŠNAJDR, M. 1975. *Konicekion* nov. gen. from the Middle Cambrian of Bohemia (Ostracoda). *Věstník Ústředního ústavu geologického*, **50**, 153–156.

SOKOLOV, B. S. 1952. On the age of the old sedimentary cover of the Russian Platform. *Izvestiya Akademii Nauk SSSR, Seriya Geologicheskaya*, **1952**(5), 21–31 [in Russian].

SOKOLOV, B. S. 1953. Stratigraphical scheme of the Lower Palaeozoic (pre-Devonian) deposits of the northwestern part of the Russian Platform. *In*: *Devon Russkoy platformy*. Gostoptekhizdat, Moscow, 16–38 [in Russian].

STAMPFLI, G. M. 2000. Tethyan oceans. *In*: BOZKURT, E., WINCHESTER, J. A. & PIPER, J. D. A. (eds) *Tectonics and Magmatism in Turkey and the Surrounding Areas*. Geological Society, London, Special Publications, **173**, 1–23.

STAMPFLI, G. M., RAUMER, J. VON & BOREL, G. D. 2002. Paleozoic evolution of pre-Variscan terranes: from Gondwana to the Variscan collision. *In*: MARTINEZ CATALAN, J., HATCHER, R. D., ARENAS, R. & DIAZ GARCÍA, F. (eds) *Variscan-Appalachian Dynamics: The Building of the Late Paleozoic Basement*. Geological Society of America, Special Papers, **364**, 263–280.

STEINER, M. & FATKA, O. 1996. Lower Cambrian tubular micro- to macrofossils from the Paseky Shales of the Barrandian area (Czech Republic). *Paläontologische Zeitschrift*, **70**, 275–299.

STETTNER, G. 1988. The Moldanubian Region in the Bavarian segments of the Bohemian Massif – Federal Republic of Germany. *In*: ZOUBEK, V. with COGNÉ, J., KOZHOUKHAROV, D. & KRAUTNER, H. G. (eds) *Precambrian in Younger Fold Belts*. Wiley, Chichester, 252–267.

STETTNER, G. 1992. *Geologie im Umfeld der Kontinentalen Tiefbohrung Oberpfalz. Einführung und Exkursionen*. Bayerisches Geologisches Landesamt, München.

STETTNER, G. 1997*a*. Münchberger Kristallinkomplex. *In*: Stratigraphie von Deutschland II. Teil I: Thüringen, Sachsen, Ostbayern. *Courier Forschungsinstitut Senckenberg*, **200**, 104–113.

STETTNER, G. 1997*b*. Fichtelgebirge und nördlicher Oberpfälzer Wald. *In*: Stratigraphie von Deutschland II. Teil I: Thüringen, Sachsen, Ostbayern. *Courier Forschungsinstitut Senckenberg*, **200**, 114–144.

STETTNER, G., ROHRMÜLLER, J. & HIRSCHMANN, G. 1997. Bayerischer Wald/Oberpfälzer Wald. *In*: Stratigraphie von Deutschland II. Teil I: Thüringen, Sachsen, Ostbayern. *Courier Forschungsinstitut Senckenberg*, **200**, 373–399.

STEYRER, H. P. & HÖCK, V. 1985. Geochemistry of the metabasites in the Habach Formation (Salzburg/Hohe Tauern/Austria). *Ofioliti*, **10**, 441–456.

ŠTORCH, P. 1998. Volcanism. *In*: CHLUPÁČ, I., HAVLÍČEK, V., KŘÍŽ, J., KUKAL, Z. & ŠTORCH, P. (eds.) *Palaeozoic of the Barrandian*. Český geologický ústav, Prague, 149–164.

STUDENCKI, M. 1994. Wiśniówka Duża quarry. *In*: KOWALCZEWSKI, Z, SZULCZEWSKI, M, MIGASZEWSKI, Z. & JARECKA-STRYCZ, K. (eds) *Europrobe Trans-European Suture Zone Workshop. Excursion Guide-book. The Holy Cross Mts.*, 51–57.

SZCZEPANIK, Z. 2001. Acritarchs from Cambrian deposits of the southern part of the Łysogóry unit in the Holy Cross Mountains, Poland. *Geological Quarterly*, **45**, 117–130.

SZCZEPANIK, Z., TRELA, W. & SALWA, S. 2004*a*. Upper Cambrian in the Kielce Region of the Holy Cross Mts – preliminary report. *Przegląd Geologiczny*, **52**, 895–898 [in Polish].

SZCZEPANIK, Z., TRELA, W., ŻYLIŃSKA, A. & SALWA, S. 2004*b*. Nowe kambryjskie i ordowickie jednostki litostratygraficzne w regionie kieleckim Gór Świętokrzyskich. *Posiedzenia Naukowe Państwowego Instytutu Geologicznego*, **61**, 57–58.

TAIT, J., SCHÄTZ, M., BACHTADSE, V. & SOFFEL, H. 2000. Palaeomagnetism and Palaeozoic palaeogeography of Gondwana and European terranes. *In*: FRANKE, W., HAAK, W., ONCKEN, O. & TANNER, D. (eds) *Orogenic Processes: Quantification and Modelling in the Variscan Belt of Central Europe*. Geological Society, London,

Special Publications, **179**, 21–34.

TEIPEL, U., EICHHORN, R., HÖLL, R. & KENNEDY, A. 2002. Late Neoproterozoic and Lower Ordovician magmatic events in the western Bohemian Massif (Bayerischer Wald, Germany) – preliminary results from SHRIMP dating. *In*: NIEBUHR, B. (ed.) *Geo 2002. Planet Erde: Vergangenheit, Entwicklung, Zukunft*, 1.-5. Oktober 2002 in Würzburg. Programm und Kurzfassungen, 327–328.

TICHOMIROWA, M. LINNEMANN, U. & GEHMLICH, M. 1997. Zircon ages as magmatic time marks – comparison of the crustal evolution in different units of the Saxothuringian Zone. *Terra Nostra*, **97**, 137–141.

TIMMERMANN, H., KRENN, E., DÖRR, W., FINGER, F. & ZULAUF, G. 2002. Cadomian and Variscan metamorphism in the Teplá Barrandian unit, northwestern Bohemian Massif: implications for the resetting of the U-Pb systematics in monazites. *In*: *Geo 2002. Planet Erde: Vergangenheit, Entwicklung, Zukunft*, 1.-5. Oktober 2002 in Würzburg. Programm und Kurzfassungen, 331–332.

TOMCZYKOWA, E. 1968. Stratigraphy of the uppermost Cambrian deposits in the Świętokrzyskie Mountains. *Prace Instytutu Geologicznego*, **54**, 5–85 [in Polish with English summary].

TOMCZYKOWA, E. & TOMCZYK, H. 2000. Starszy paleozoik z otworu Daromin IG-1 – potwierdzenie budowy terranowej bloku łysogórskiego i małopolskiego (Góry Świętokrzyskie). *Biuletyn Państwowego Instytutu Geologicznego*, **393**, 167–203.

TORSVIK, T. H., SMETHURST, M. A. *et al.* 1996. Continental break-up and collision in the Neoproterozoic and Palaeozoic – a tale of Baltica and Laurentia. *Earth-Science Reviews*, **40**, 229–258.

VALENT, M. 2006. Stratigraphic distribution of the class Hyolitha (Mollusca) in the Barrandian area (Czech Republic). *Acta Universitatis Carolinea Geologica*, **47**, 183–188.

VALVERDE-VAQUERO, P., DÖRR, W., BELKA, Z., FRANKE, W., WISZNIEWSKA, J. & SCHASTOK, J. 2000. U–Pb single-grain dating of detrital zircon in the Cambrian of central Poland: implications for Gondwana versus Baltica provenance studies. *Earth and Planetary Science Letters*, **184**, 225–240.

VANGUESTAINE, M. 1974. Espèces zonales d'acritarches du Cambro-Trémadocien de Belgique et de l'Ardenne française. *Revue of Palaeobotany and Palynology*, **18**, 63–82.

VANGUESTAINE, M. 1978. Critères palynostratigraphiques conduisant à la reconnaissance d'un pli couché Revinien dans le sondage de Grand-Halleux. *Annales de la Société Géologique Belgique*, **100**, 249–276.

VANGUESTAINE, M. 1991. Datation par Acritarches des couches cambro-tremadociennes les plus profondes du sondage de Lessines (bord meridional du Massif du Brabant, Belgique). Acritarch biostratigraphy of Cambrian-Tremadocian horizons from the Lessines borehole, southern edge of the Brabant Massif, Belgium. *Annales de la Société Géologique de Belgique*, **114**, 213–231.

VANGUESTAINE, M. 1992. Biostratigraphie par acritarches du Cambro-Ordovicien de Belgique et des régions limitrophes: synthèse et perspectives d'avenir. *Annales de la Société Géologique Belgique*, **115**, 1–18.

VANGUESTAINE, M. & RUSHTON, A. W. A. 1979. Decouverté d'un brachiopode inarticulé, *Acrothele* cf. *bergeroni* Walcott, dans le Revinien inférieur de Trois-Ponts, Cambrien du Massif de Stavelot, Belgique. *Annales de la Société Géologique Belgique*, **102**, 295–301.

VAVRDOVÁ, M. 1982. Phytoplankton communities of Cambrian and Ordovician age of Central Bohemia. *Věstník Ústředního Ústavu geologického*, **57**, 145–155.

VAVRDOVÁ, M. 1997. Akritarcha kambrického stáří z bazálních klastik v podloží moravského devonu. *Zemní plyn, Nafta*, **42**, 31–32.

VAVRDOVÁ, M. 2004. The Brunovistulicum: assumptions and data. *Zeitschrift der deutschen geologischen Gesellschaft*, **155**, 1–9.

VAVRDOVÁ, M. & BEK, J. 2001. Further palynomorphs of Early Cambrian age from clastic sediments underlaying the Moravian Devonian. *Bulletin of the Czech Geological Survey*, **76**, 113–126.

VAVRDOVÁ, M., MIKULÁŠ, R. & NEHYBA, S. 2003. Lower Cambrian siliciclastic sediments in Southern Moravia (Czech Republic) and their paleogeographic constraints. *Geologica Carpatica*, **54**, 67–79.

VECOLI, M. 1996. Stratigraphic significance of acritarchs in Cambro-Ordovician boundary strata, Hassi-Rmel area, Algerian Sahara. *Bolletino della Società Paleontologica Italiana*, **35**, 3–58.

VERNIERS, J., HERBOSCH, A. *et al.* 2001. Cambrian-Ordovician-Silurian lithostratigraphic units (Belgium). *In*: BULTYNCK, P. & DEJONGHE, L. (eds) *Guide to a Revised Stratigraphic Scale of Belgium. Geologica Belgica*, **4**, 5–38.

VIDAL, P., AUVRAY, B., CHARLOT, R., FEDIUK, F., HAMEURT, J. & WALDHAUSROVÁ, J. 1975. Radiometric age of volcanics of the Cambrian "Křivoklát-Rokycany" Complex (Bohemian Massif). *Geologische Rundschau*, **64**, 563–570.

WALDHAUSROVÁ, J. 1971. The chemistry of the Cambrian volcanics in the Barrandian area. *Krystalinikum*, **8**, 45–75.

WALTHER, R. 1980. Lower Paleozoic paleogeography of the Brabant Massif and its southern surroundings adjoining area. *Mededelingen van Rijks geologische Dienst*, **32**, 14–25.

WALTHER, R. 1992. *Geologie von Mitteleuropa*, 5th edn. E. Schweizerbarth'sche Verlagsbuchhandlung (Nägele u. Obermiller), Stuttgart.

WALTHER, R., SPAETH, G. & KASIG, W. 1985. An outline of the geological structure of the northeastern Hohes Venn area and of its northern foreland. *Neues Jahrbuch für Geologie und Paläontologie, Abhandlungen*, **171**, 207–216.

WATERLOT, G. 1974. Paléozoique du Nord de la France et de Belgique (Ardennes et Boulonnais). *In*: DÉBELMAS, J. (ed.) *Géologie de la France, I, Vieux massifs et grands bassins sédimentaires*. Doin, Paris, 42–62.

WEGER, M., LOTH, G., HÖLL, R., MASCH, L., KÖHLER, H. & ROHRMÜLLER, J. 1998. The early tectonometamorphic evolution of the southern zone of Erbendorf-Vohenstrauss (ZV, NW-Bohemian Massif, Bavaria) inferred from geochemical, radiometric, petrologic, and structural characteristics. *Acta Universitatis Carolinae, Geologica*, **42**, 357.

WERNER, C.-D. 1993. Hangend-Serie und Prasinit-Serie von Frankenberg und Wildenfels. *Zeitschrift für Geologische Wissenschaften*, **21**, 247–259.

WINCHESTER, J.A., PHARAOH, T.C. & VERNIERS, J. 2002. Palaeozoic amalgamation of Central Europe: an introduction and synthesis of new results from recent geological and geophysical investigations. *In*: WINCHESTER, J. A., PHARAOH, T. C. & VERNIERS, J. (eds) *Palaeozoic Amalgamation of Central Europe*. Geological Society, London, Special Publications, **201**, 1–18.

WOTTE, T. 2004. Nachweis und fazielle Position von Reef Mounds im Unterkambrium von Doberlug-Torgau. *Neues Jahrbuch für Geologie und Paläontologie, Abhandlungen*, **233**, 397–422.

WUCHER, K. 1965. Ergebnisse der Kartierungsbohrungen Heinersdorf 1/60 und 2/62, Thüringisches Schiefergebirge. *Geologisches Jahrbuch*, **1** (for 1965), 297–323.

WURM, A. 1924. Über ein Vorkommen von Mittelcambrium (Paradoxides-Schichten) im bayrischen Frankenwald. *Zeitschrift der deutschen geologischen Gesellschaft*, **75** (Monatsberichte für 1923), 135–136.

WURM, A. 1925*a*. Geologie von Bayern. Teil 1: Nordbayern, Fichtelgebirge und Frankenwald. *In*: *Handbuch der Geologie und Bodenschätze Deutschlands. II. Abt.: Regionale Geologie Deutschlands*. Borntraeger, Berlin.

WURM, A. 1925*b*. Über ein Vorkommen von Mittelcambrium (Paradoxidesschichten) im bayrischen Frankenwald bei Wildenstein südlich Presseck. *Neues Jahrbuch für Mineralogie, Beilagen-Band*, **52**, Abt. B, 71–93.

WURM, A. 1928. Ueber eine neue mittelcambrische Fauna aus dem bayerischen Frakenwald und ihre Bedeutung für die Stratigraphie des älteren Paläozoikums (Conocoryphe-Schichten von Lippertsgrün). *Neues Jahrbuch für Mineralogie, Geologie und Paläontologie Beilagen-Band*, **59**, Abt. B, 31–47.

ŻELAŹNIEWICZ, A. 1998. Rodinian-Baltican link of the Neoproterozoic orogen in southern Poland. *Acta Universitatis Carolinae, Geologica*, **42**, 509–515.

ŻELAŹNIEWICZ, A., SEGHEDI, A., JACHOWICZ, M., BOHIŃSKI, W., BUŁA, Z. & CWOJDZIŃSKI, S. 2001. U-Pb SHRIMP data confirm the presence of a Vendian foreland flysh basin next to the East European Craton. *In*: PHARAOH, T. C. & GEE, D. (eds) *ESF EUROPROBE Meeting "Neoproterozoic–Early Palaeozoic time-slice Symposium: Orogeny and cratonic response on the margins of Baltica"*, Ankara, Abstracts, 98–100.

ZNOSKO, J. 1961. W sprawie pozycji stratygraficznej eokambryjskich sparagmitów i niektórych młodoprekambryjskich formacji. *Kwartalnik Geologiczny*, **5**, 737–774.

ZULAUF, G. 1997. Von der Anchizone bis zur Eklogitfazies: Angekippte Krustenprofile als Folge der cadomischen und variscischen Orogenese im Teplá-Barrandium (Böhmische Masse). *Geotektonische Forschungen*, **89**.

ZULAUF, G., DÖRR, W., FIALA, J. & VEJNAR, Z. 1997. Late Cadomian crustal tilting and Cambrian transtension in the Teplá-Barrandian unit (Bohemian Massif, Central European Variscides). *Geologische Rundschau*, **86**, 571–584.

ZULAUF, G., SCHITTER, F., RIEGLER, G., FINGER, F., FIALA, J. & VEJNAR, Z. 1999. Age constraints on the Cadomian evolution of the Teplá Barrandian unit (Bohemian Massif) through electron microprobe dating of metamorphic monazite. *Zeitschrift der deutschen geologischen Gesellschaft*, **150,** 627–640.

ŻYLIŃSKA, A. 1999. Trace maker of Cruziana semiplicata in the Upper Cambrian of the Holy Cross Mts. (Poland). *In*: PALMER, A. R. (ed.) *Abstracts, Laurentia 99, V Field Conference of the Cambrian Stage Subdivision Working Group*, Utah, Nevada California, U.S.A., 62–63.

ŻYLIŃSKA, A. 2001. Late Cambrian trilobites from the Holy Cross Mountains, central Poland. *Acta Geologica Polonica*, **51**, 333–383.

ŻYLIŃSKA, A. 2002. Stratigraphic and biogeographic significance of Late Cambrian trilobites from Łysogóry (Holy Cross Mountains, central Poland). *Acta Geologica Polonica*, **52**, 217–238.

ŻYLIŃSKA, A. & MASIAK, M. 2007. Cambrian trilobites from Brzechów, Holy Cross Mountains (Poland) and their significance in stratigraphic correlation and biogeographic reconstructions. *Geological Magazine,* **144**, 661–686.

ŻYLIŃSKA, A. & SZCZEPANIK, Z. 2002. The relationship between acritarch and trilobite biozones from the Upper Cambrian of the Holy Cross Mountains, Poland – a preliminary report. *In*: ALVARO, J.J. & CLAUSEN, S. (eds), *VIII Conference of the Cambrian Stage Subdivision Working Group, ISCS,* Caunes-Minervois, Languedoc, September 12–14 2002; Programme and Abstracts, 41.

5 Ordovician

THOMAS SERVAIS (co-ordinator), JERZY DZIK, OLDRICH FATKA, THOMAS HEUSE, MARCO VECOLI & JACQUES VERNIERS

The Ordovician outcrops of central Europe belong to various areas with, in general, a very complex tectonic evolution. In this chapter, we review the localities that were attributed to peri-Gondwanan terranes. We do not therefore include a detailed description of areas attributed to the Baltica palaeocontinent (Denmark, southern Sweden, Baltic States, northeastern Poland).

Whereas the northwestern part of Central Europe belonged, during the Ordovician, to the eastern part of the microcontinent of Avalonia (Belgium, western and northern Germany, possibly northwestern Poland), the outcrop areas of the Rhenohercynian, Saxothuringian and Moldanubian zones have mostly been assigned to Gondwana-derived terranes (such as, in palaeogeographical terms, Armorica or the Armorican Terrane Assemblage (ATA) and Perunica) or to sedimentary basins in the vicinity of the Gondwanan supercontinent. Our review includes the Avalonian sequences of Belgium in the northwestern part of the investigated area of Central Europe, continues into western, northern and eastern Germany and extends into northwestern and southern Poland. The review of the Ordovician of the Saxothuringian and the Moldanubian zones includes the outcrop areas of southeastern Germany, the Czech Republic and southwestern Poland. The Ordovician from the pre-Variscan parts of the Alpine mountain chains of Switzerland, Austria and northern Italy are also briefly discussed.

In terms of stratigraphy, the British Ordovician series and stages were generally used as a standard in most parts of western and central Europe. However, in the last two decades, significant progress has been made in Ordovician series and stage boundary definitions on a global scale. A tripartite division of the Ordovician System into Lower, Middle and Upper Ordovician series has been accepted. Seven global stages and their bases have recently been defined. Because of poor exposure and generally incomplete sequences, none of the global stratotypes (Global Stratotype Section and Point: GSSP) has been defined in Central Europe, but instead, these have been established in successions in China, North America and Scandinavia. Because of the peri-Gondwanan affinities of the Ordovician sequences in Central Europe, we follow here in general the classic British Ordovician, but also introduce the new global stratigraphical schemes where possible.

The new Ordovician timescale

For historical reasons and because of biostratigraphic difficulties, the chronostratigraphy of the Ordovician System is difficult to apply internationally. The Ordovician was a time of high faunal provincialism due to a wide dispersion of the main continental plates, ranging from high latitudes to equatorial locations. Consequently, different regional series and/or stages have been erected for most of these palaeoplates. The British Ordovician series and stages acted more or less as a standard, especially in western, southern and central Europe (while the Baltic stratigraphy was used in northern and northeastern Europe). However, the definition of these historical subdivisions of the Ordovician System in the British Isles does not fulfil the recommendation of the International Commission on Stratigraphy (ICS) and of the International Union of the Geological Sciences (IUGS) for the definition of global stratotypes.

In the 1990s, significant progress was made in Ordovician series and stage boundary definitions by the Ordovician Subcommission of Stratigraphy (ISOS). Figure 5.1 summarizes the current status of the stratigraphical subdivisions of the Ordovician. A tripartite division of the Ordovician System into Lower, Middle and Upper Ordovician series has been accepted and it was recently decided that seven global stages should be defined. The boundaries of these stages are defined on the basis of the occurrences of either graptolite or conodont species. The first stage that was approved by the IUGS and the ICS in early 1997 was the second global stage (stage 4) of the Middle Ordovician Series, the Darriwilian, with its GSSP near Huangnitang, Changshan, SE China. All other stage boundaries have now been defined: the first stage of the Lower Ordovician Series (stage 1), the Tremadocian, has its base (Cambrian–Ordovician boundary stratotype) at Green Point, Newfoundland. The GSSP of the second stage (stage 2, Floian) of the Lower Ordovician Series is located at the Diabasbrottet section at Mount Hunneberg, Sweden, and the GSSP of the first stage (stage 5, Sandbian) of the Upper Ordovician Series is at Fågelsång, Scånia, Sweden. The two last GSSP of the first stage of the Middle Ordovician (stage 3, Dapingian) and the third stage of the Upper Ordovician Series (stage 7, Hirnantian) have recently been selected in China.

From the seven global stages, the following names have been ratified: Tremadocian for stage 1, Floian for stage 2, Dapingian for stage 3, Darriwilian for stage 4, Sandbian for stage 5, Katian for stage 6 and Hirnantian for stage 7.

In addition to the global series and stages and the regional series and stages, Webby *et al.* (2004) also introduced the concept of 'time-slices' that allow a more precise international correlation. With the last decision of the ISOS to distinguish three global stages in the Upper Ordovician Series (instead of two) the time-slices in the Upper Ordovician should be renumbered, as indicated in Figure 5.1.

The classic British series names (Tremadoc, Arenig, Llanvirn, Caradoc, Ashgill) will not be used as Global Stages, and will thus be abandoned for global correlation, with the exception of the Tremadocian, which will be used globally. However, the historical British names will probably continue to be used as regional series in central and western Europe. For a revision of the current state of Ordovician stratigraphy, see Cooper & Sadler (2004).

The Cambrian–Ordovician boundary

The Cambrian–Ordovician boundary has only recently been approved by the ICS in the year 2000 after several decades of negotiations within the Cambrian–Ordovician Boundary Working Group. The boundary is defined by the GSSP in the Green Point

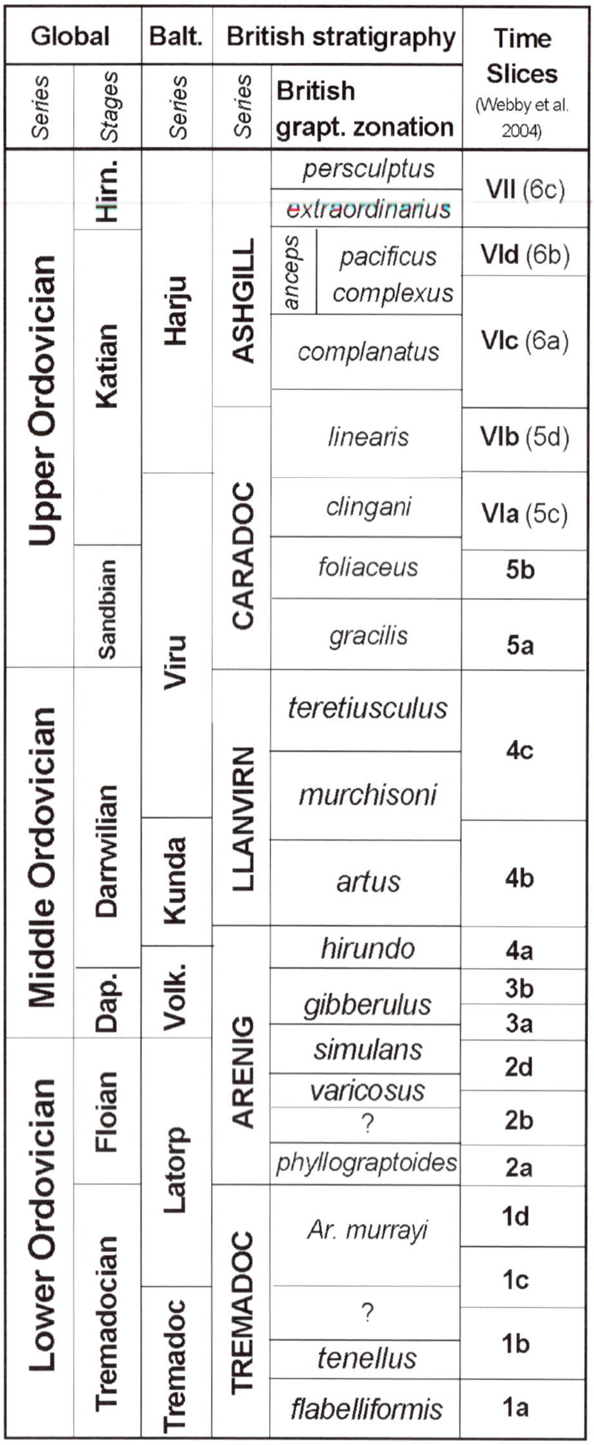

Fig. 5.1. Standard Ordovician chronostratigraphy (global series and stages) and correlation with Baltic regional series, British regional series and graptolite zonation, and Ordovician time-slices, after Webby *et al.* (2004). For explanation, see text.

section of western Newfoundland in Bed 23 of the measured section in the lower Broom Point Member, Green Point Formation (Cooper *et al.* 2001). The Ordovician begins with the first appearance of the conodont *Iapetognathus fluctivagus*, 4.8 m below the appearance of planktonic graptolites.

The Cambrian–Ordovician boundary can only be observed in a few localities in Central Europe and cannot be located with precision, because carbonate strata with conodonts are usually absent. However, the presence of planktonic dendroid graptolites in some sections allows a correlation with the boundary. In eastern Belgium, for example, Erdtmann (1986) revised the graptolites of the Solwaster (Stavelot Inlier) section and concluded that the fauna is identical to those from the Cambrian–Ordovician boundary sections near Oslo (Norway), Digermulen (northeastern Norway), Green Point (Canada, stratotype section) and Dayangcha (NE China). Similar graptolite faunas are also found in the Brabant Massif in Belgium, indicating the presence of the lower Tremadocian, and possibly of the Cambrian–Ordovician boundary.

Acritarchs also allow the Cambrian–Ordovician boundary to be roughly located, although precise index taxa have not yet been identified. Acritarchs from the Cambrian–Ordovician boundary interval have been identified in different outcrop areas of Belgium (Servais *et al.* 1993; Verniers *et al.* 2001). The acritarchs from the Traischbach Serie in the Black Forest (Schwarzwald), southwestern Germany, also correspond to a late Cambrian–early Ordovician interval, possibly including the boundary (Montenari & Servais 2000). The Cambrian–Ordovician boundary cannot be located precisely in any of the other areas discussed in this chapter.

The Cambrian–Ordovician boundary at Green Point lies at the peak of the largest positive excursion in the $\delta^{13}C$ curve through the boundary interval (Cooper *et al.* 2001) and during the global marine transgression that followed the *Acerocare* regressive event. In the absence of fossils, isotope studies could possibly help in the future to locate the Cambrian–Ordovician boundary by means of chemostratigraphy.

Ordovician palaeoclimate and sea-level changes

Models for Ordovician palaeoclimatology and palaeoceanography are rare and remain much more speculative than palaeogeographical reconstructions (Barnes 2004). Golonka *et al.* (1994) published palaeotemperature and palaeoclimate maps, constructed using a palaeoclimate modelling program developed by Parrish (1982). There is broad agreement that the latest-Ordovician glaciation was due to the formation of an ice cap, following the movement of Gondwana towards the South Pole, which was located in present-day North Africa (e.g. Ghienne 2003).

There is less agreement, however, on models of oceanic circulation, which vary greatly from one author to another. Bergström (1990), for example, presented an oceanic current circulation model for the Early Ordovician, which included a current in the southern hemisphere that originated in South America, drifted eastwards along the margin of European Gondwana (passing through the Tornquist Sea), to reach areas such as Saudi Arabia and South China. Finney & Chen (1990) presented another circulation pattern model for the same time interval, with a current flowing in exactly the opposite direction, originating around Arabia, drifting westwards on the northern margin of European Gondwana (passing through the Tornquist Sea), to reach Gondwanan Argentina. A more recent model published by Christiansen & Stouge (1999) also discusses the Arenig. These authors suggested that the temperate low-pressure zones were located at 50° latitude, and the subtropical high-pressure zones at 25° latitude. As the different models are contradictory, we avoid drawing oceanic circulation patterns on the palaeogeographical map presented here.

During the Early Ordovician the areas discussed in this paper were all located at high latitudes of the southern hemisphere, and, at least during the earliest Ordovician, were all parts of the Gondwanan margin. Some elements drifted from the supercontinent of Gondwana and moved northwards into lower latitudes to enter warmer waters. In several of the investigated areas, the water temperatures were sufficiently high to allow the deposition of limestones during the latest Ordovician (Ashgill, but pre-Hirnantian) Boda Event, a warming event recorded largely in central and western European successions (Fortey & Cocks 2005). This warming event was partly recorded by the deposition of carbonate rocks (mostly carbonaceous shales, but also limestones), such as in the Condroz Inlier in Belgium, or in the Schwarzburg Anticline in Germany, but also in the Czech and Polish successions. The Hirnantian cooling event was also recorded in several sections, as for example in the Lederschiefer Formation in east Germany and the Kosov Formation in the Czech Republic (presence of diamictites).

Isotope analyses are rare in the Ordovician of Central Europe. In the Holy Cross Mountains (Poland) the $\delta^{18}O$ record in the Mójcza Limestone and Zalesie Formation (Trela 2000; K. Małkowski unpublished data) suggests a decrease in the mean water temperature during the Middle Ordovician and then an increase during the Late Ordovician of about 8°C. This increase may be connected with the drift of the Małopolska Massif (with the Holy Cross Mountains at its northern margin) towards the equator (Lewandowski 1993). The warming tendency seems to terminate near the end of the limestone sedimentation in the middle Ashgill.

Another important issue with Ordovician oceanography is the exent of sea-level oscillations. During the Early Palaeozoic the variation in eustasy was considered to have resulted in the greatest extent of cratonic submergence in the Phanerozoic (Hallam 1992), and the high sea level was perhaps the highest in the entire Phanerozoic. Little attention, however, has been devoted to the reconstruction of Ordovician sea-level fluctuations. A North American sea-level curve for the Ordovician was published by Ross & Ross (1992, 1995). More recently, Nielsen (2004) published a reconstruction of the sea level of Baltoscandia. In the later study, the Ordovician was subdivided into three highstand and three lowstand 'intervals', with some correlation at second-order level with the Laurentian curve. Videt *et al.* (2006) presented a first sea-level curve for North African peri-Gondwana. Sea-level reconstructions for central and western European areas have not yet been presented and much further work needs to be done.

Tectonic setting and palaeomagnetism

It is beyond the scope of this chapter to review the tectonic setting and the palaeomagnetic data of the Lower Palaeozoic of Central Europe. Multidisciplinary studies have been undertaken in the 1990s and early 2000s within an EU-funded research network named PACE (Palaeozoic Amalgamation of Central Europe), of which the most significant new results have been published by Winchester *et al.* (2002*a*). According to these results, a simplified model can explain the accretion scenario of Central European terranes with Baltica in three stages: an end-Cambrian accretion of the Bruno-Silesian, Lysogóry and Małopolska terranes of Poland and the Czech Republic (Belka *et al.* 2002); a late Ordovician accretion of Avalonia (Verniers *et al.* 2002); and an early Carboniferous accretion of the Armorican Terrane Assemblage (Winchester *et al.* 2002*b*), which had coalesced during the Late Devonian–Early Carboniferous.

The westernmost parts of the regions analysed here represent parts of Avalonia. While there is no doubt of the attribution of the Belgian and north German succession to Avalonia, the Ordovician succession of Hessen (western Germany) also belongs to this microcontinent. These areas were displaced by thrusting during the Variscan Orogeny, since they contain typical Avalonian faunas and therefore lie north of the Rheic Suture (Cocks *et al.* 1997). This suture lies between the Avalonian margin of Laurussia and the Armorican part of the Saxothuringian Zone and other possible components of the Armorican Terrane Assembly (Tait *et al.* 1997).

During the opening phase of the Rheic Ocean in Arenig times, Avalonia rifted away from Gondwana. By late Silurian the ocean was closing rapidly, with a magmatic arc generated by southward subduction in the Mid-German Crystalline High (Dallmeyer *et al.* 1995). The northern margin of the ocean briefly experienced backarc extension with initiation of the Rhenohercynian Basin, and formation of the Lizard-Giessen ophiolites (Franke 1995) in a marginal basin at *c.* 397 Ma. (Clark *et al.* 1998). These basins were inverted/obducted during overthrusting by the Normannian Terrane *c.* 20 Ma. later (Sandeman *et al.* 2000), during the Ligerian Phase. Subsequently the Rhenohercynian Basin was transformed to a foreland basin and received sediments from the eroded, uplifted Saxothuringian terranes (Franke 1998; Franke *et al.* 1999); deformation continued until the late Carboniferous (Oncken *et al.* 1999).

The achievements of the important PACE project included basin analysis and the history of formation to deformation of the sedimentary basins, palaeomagnetic and geophysical studies of the deep crust to determine the exact boundaries of the terranes such as the SW limit of Baltica and the boundaries between Far Eastern Avalonia and Avalonia, and the location of other Palaeozoic deformation belts in Central Europe (EUROPROBE and PACE projects; Verniers *et al.* 2002; Winchester *et al.* 2002*a*, *b*, 2006; Pharaoh *et al.* 2006). A German multidisciplinary study on the Variscides produced new models for the Ordovician protoliths within the different Variscan structural units and on a smaller scale for the Rügen Caledonides (Katzung 2001). The history of plate movements in the Silurian was further elucidated with evidence for the presence and movement of the separate terrane of Far Eastern Avalonia (Verniers *et al.* 2002) and the palaeogeographical position of the two parts of the Holy Cross Mountains during the Ordovician, attached to Baltica but possibly not fixed.

Palaeogeographical setting

Progress made during the last 25 years in understanding Ordovician palaeogeography has resulted from the integration of tectonic, stratigraphical, palaeoclimatological, palaeomagnetic and palaeontological evidence. On the Ordovician maps of Scotese & McKerrow (1990) and similar subsequent reconstructions, the main continents are Laurentia (North America) and Siberia, both found in equatorial positions, Baltica, located at intermediate latitudes in the southern hemisphere, and the supercontinent Gondwana, by far the largest plate, which included South America, Africa, Antarctica, Australia, India and other marginal (micro)plates. This large continent extended from the South Pole to the Equator but remained separated from Laurentia throughout the Ordovician. As indicated by Cocks & Torsvik (2004), the present-day North Atlantic area, which in the Ordovician included Laurentia, Baltica and West Gondwana and their adjacent terranes and island arcs, is today palaeogeographically well constrained. The different basins of central Europe were located in the vicinity of the supercontinent of Gondwana (Cocks & Torsvik 2004).

Figure 5.2 illustrates the position of the different areas discussed in this paper at a global level. During the Ordovician, the northwestern parts of Central Europe belonged to the microcontinent Avalonia that included the eastern seaboard of North America from Cape Cod, Massachusetts, through eastern Newfoundland, Canada, the southern part of the British Isles, Belgium, and the central European parts south of the Trans-European Suture Zone, that crosses Poland in a NW–SE direction. While Cocks *et al.* (1997) placed the northeastern margin of Avalonia (the Avalonia–Baltica boundary) close to the Elbe line in central-northern Germany, it is now widely accepted that Belgium, the western and northern parts of Germany and possibly the northwestern part of Poland (western Pomerania, Koszalin-Chojnice Zone) belong to the easternmost extension of the microcontinent (Servais & Fatka 1997; Winchester *et al.* 2002*b*), but recent studies put the Polish part on the southernmost edge of the Baltica plate (Poprawa 2006). It is also widely accepted that this microcontinent separated from Gondwana during the Early Ordovician (probably in the Arenig), and rapidly drifted northwards during the Ordovician, closing the Iapetus Ocean, but leaving behind a widening Rheic Ocean, up to the docking with Baltica during the latest Ordovician (Ashill, in terms of British series; see also Krawczyk *et al.* 2008).

The exact palaeogeography of the other parts of Central Europe is still the subject of much debate. Although there is no question that the Rhenohercynian and Saxothuringian areas, but also the other Variscan Massifs encountered in the Vosges Mountains (eastern France), the Black Forest (southeastern Germany), the Bohemian Massif, and in the Alps (Grauwackenzone, Karnische Alpen and Karawanken), were all located on the margin of Gondwana, it is not clear if they belonged to independent microplates or microcontinents.

For example, Havlíček *et al.* (1994) defined a Perunica microcontinent, mostly on the basis of faunal (brachiopod and trilobite genera) palaeobiogeographical evidence, involving major parts of the Bohemian Massif. They assumed that this microplate was located between the Gondwana and Baltica palaeocontinents. Tait *et al.* (1997) provided palaeomagnetic data supporting this area as an independent palaeogeographical unit, but this was not accepted by Lewandowski (2003). Perunica was possibly not a separate continent, but only a rift basin structure. Future research is needed to provide additional information in order to test whether the Bohemian Massif was indeed a separate microcontinent, an independant terrane, or if it constituted a sedimentary basin located in the vicinity of the Gondwanan supercontinent during the Ordovician.

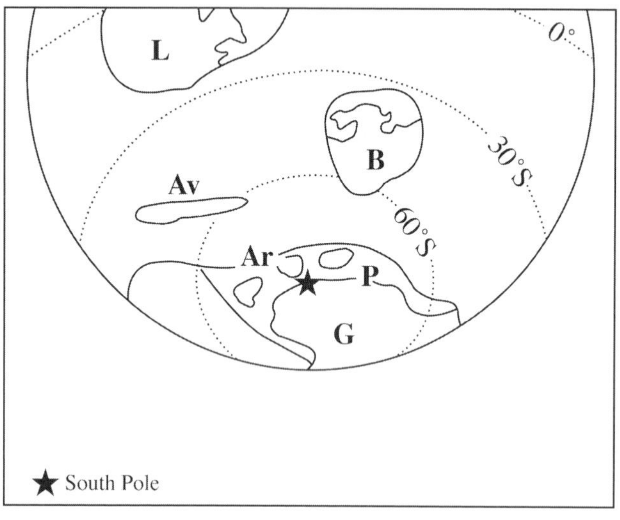

Fig. 5.2. Tilted Middle Ordovician (Llanvirn) palaeogeographical reconstruction, after Robardet (2003), placing Armorica (Ar) and Perunica (P) on the periphery of Gondwana. Abbreviations: Av, Avalonia; B, Baltica; G, Gondwana; L, Laurentia. The star represents the South Pole (located in northwestern Africa).

On the other hand, palaeomagnetic data from the Bohemian Massif, the Armorican Massif (Brittany), the eastern Pyrenees (northeastern Spain and southern France), the Catalan Coastal Ranges of northeastern Spain), the Grauwackenzone and the Carnic Alps of Austria were used by several authors to develop the definition of a so-called Armorican Terrane Assemblage (e.g. Tait *et al.* 1997). A major tectonic discontinuity between Bohemia and the more westerly units was noted. These Palaeozoic elements were considered to have rifted from the margin of Gondwana at some time in the (middle) Ordovician and subsequently drifted northwards. The exact position of these elements is discussed in several papers by the different 'schools' (e.g. Tait *et al.* 1997; Raumer *et al.* 2003). However, the concept of an independent 'microplate' of Armorica (or Armorican Terrane Assemblage) has been questioned by Robardet (2003), who considered that although repeatedly discussed for more than 20 years, the concept of an Armorica microplate may be fiction.

In the Polish parts, the palaeomagnetic evidence (Lewandowski 1993) indicates that the Małopolska Massif followed Gondwana in its drift across the South Pole during the Cambrian and Early Ordovician. In the Arenig it was located at the latitude of about 60°S, reaching the latitude of 10°S in the Late Silurian. This was the position near the western margin of the continent of Baltica. The palaeogeographical position of the Holy Cross Mountains has also been debated for many years. While some of the Ordovician faunas are generally of Baltic affinity, there were also influences of the Celtic province faunas of Wales and the Armorican Massif, as well as of the cold-water Gondwana faunas of Thuringia and the Moldanubicum (Dzik 1990, 1999*b*). Even more profound is the difference in facies and faunas between the Baltic-like Ordovician of the Małopolska Massif and the Sudetes (Dzik 1990). These were possibly terranes or microcontinents located in different climatic zones, separated by an ocean, and they were unified not earlier than the Variscan Orogeny. It has been suggested that the southern Polish massifs represent a separate pre-Caledonian Małopolska microcontinent (e.g. Dzik 1978, 1983; Lewandowski 1993).

Belgium

Outcrops of Lower Palaeozoic formations in Belgium are present in six areas: the Brabant Massif (Fig. 5.3, 1) in the north and NW, the Condroz Inlier, also called the 'Bande condrusienne' or 'Bande de Sambre-et-Meuse' in the centre (Fig. 5.3, 2), and, in the southeastern part, four inliers are located in the Ardennes (Fig. 5.3, 3–6). Traditionally these inliers were called 'massifs' in the regional literature. Two inliers are larger (Stavelot and Rocroi inliers, Fig. 5.3, 3 & 4) and two smaller (Givonne and Serpont inliers, Fig. 5.3, 5 & 6). Some of the Belgian outcrop areas extend into surrounding countries: the Brabant Massif into the Netherlands and France, the Stavelot Massif into Germany where it is called (Stavelot-) Venn Massif (Fig. 5.3, 3), and the Rocroi Massif (Fig. 5.3, 4) and the Givonne inlier (Fig. 5.3, 5) into France. All these massifs or inliers are composed of

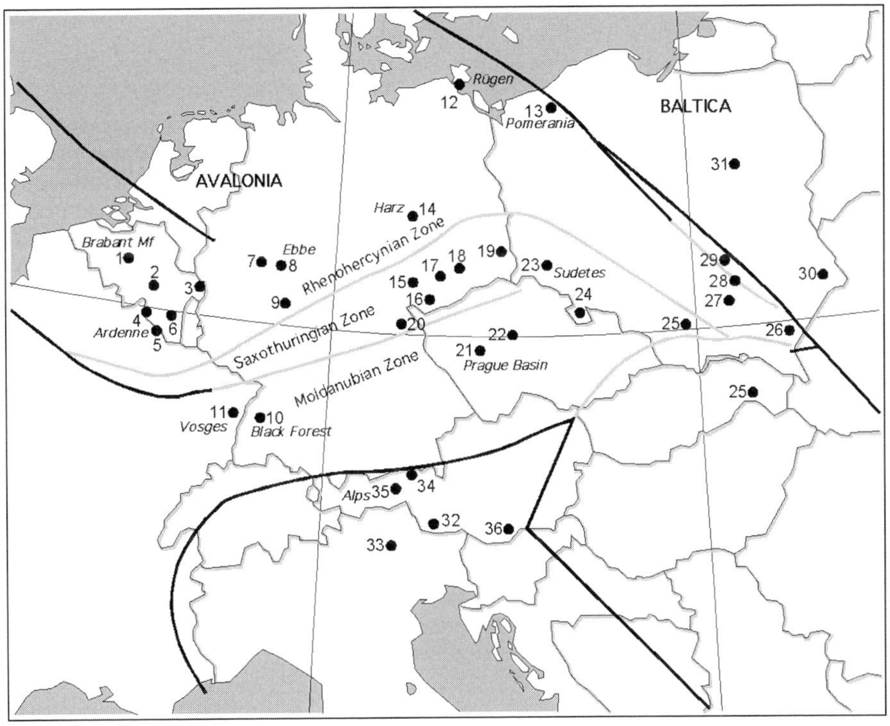

Fig. 5.3. Sketch map of Central Europe showing the distributional areas of Ordovician rocks (black dots) discussed in the text. 1, Brabant Massif, Belgium; 2, Condroz Inlier, Belgium; 3, Stavelot Inlier, Belgium, Germany; 4, Rocroi Inlier, Belgium, France; 5, Givonne Inlier, Belgium, France; 6, Serpont Inlier, Belgium; 7, Remscheid-Altena Anticline, Germany; 8, Ebbe Anticline, Germany; 9, Hessen area, Germany; 10, Black Forest (Schwarzwald), Germany; 11, Vosges Mountains, France; 12, Rügen Island, Baltic Sea, Germany; 13, Koszalin-Chojnice region, Pomerania, Poland; 14, Harz Mountains, Germany; 15, Schwarzburg Anticline, Germany; 16, Vogtland area, Germany; 17, Berga Anticline; 18, Frankenberg-Wildenfels areas; 19, Lusatia (Lausitz) area; 20, Münchberg area, Franconia, Germany; 21, Prague Basin, Czech Republic; 22, Železné Hory area, Czech Republic; 23, Kaczawa Hills, Sudetes, Poland; 24, Bardo Hills, Sudetes, Poland; 25, Myszków area, Małopolska Massif, Poland; 26, eastern part of the Małopolska Massif, Poland; 27, Łagów region, Holy Cross Mountains, Poland; 28, Kielce region, Holy Cross Mountains, Poland; 29, Łysogóry region, Poland; 30, marginal zone of the East European Platform, Poland; 31, numerous boreholes all over the East European Platform, Poland; 32, Carnic Alps, Austria, Italy; 33, Agordo-Cereda phyllitic area, Italy; 34, Northern Greywacke Zone outcrops, Austria; 35, Innsbruck Quarzphyllit Group outcrops, Austria; 36, Carinthian outcrops, Austria.

Cambrian to Silurian rocks and they are all unconformably covered by Devonian rocks. The unconformity has been called Caledonian in the literature and resulted from orogenic deformations caused by the Avalonia–Baltica–Laurentia collisions during the Late Ordovician to Early Devonian.

In the last decades, these different outcrop areas have been stratigraphically restudied, with most investigations concentrating on the Cambrian to Silurian successions of the Brabant Massif. The stratigraphy of the Condroz inlier, that includes sediments of Ordovician and Silurian age, has partly been reviewed and is currently still under revision. The Ardenne inliers are mostly completely devoid of macrofossils. Most biostratigraphical information is from palynomorphs which allowed the attribution of these successions to a succession ranging from the Lower Cambrian to the Middle Ordovician.

The present review of the Ordovician in Belgium thus concerns the three major outcrop areas, from NW to SE: the Brabant Massif, the Condroz inlier and the Ardenne inliers. A complete review of all Ordovician lithostratigraphical units is given by Verniers *et al.* (2001). The thickness of most units is generally estimated from geometry because only rarely is the entire column of the unit observable. These thicknesses are not corrected for compaction nor for tectonic thickening and hence should be regarded with caution.

Figure 5.5 correlates the different units of Belgium, from NW to SE, the Brabant Massif, the Condroz Inlier and the Ardenne Inliers. All of these successions are currently assigned to the eastern extension of the microcontinent Avalonia. Figure 5.5 shows the correlation of the Ordovician successions of Belgium with the GSSP.

Brabant Massif

The Brabant Massif (Figs 5.3, 1) contains a thick siliciclastic, often turbiditic, and rather complete Lower Palaeozoic sequence, extending from the lowest Cambrian to the uppermost Silurian. Its continuation below the Devonian cover can be traced by boreholes and geophysical data to the north beneath the Campine Basin as far as the Roermond Graben, to the south at least beneath the northern half of the Namur 'synclinorium' and to the west under the North Sea into the concealed Caledonides of East Anglia. The entire fold belt is termed the Anglo-Brabant fold belt (Pharaoh *et al.* 1993) and because there was more deformation than folding, it has been renamed as the Anglo-Brabant Deformation Belt (Winchester *et al.* 2002*a*, *b*). Syntheses on the geology of the Brabant Massif can be found in Legrand (1968) and De Vos *et al.* (1993) with greater detail about the deformation in Debacker *et al.* (2005). The Ordovician succession was first reviewed in Servais *et al.* (1993), but subsequently additional formations have been described and further biostratigraphical data on palynomorphs and graptolites have become available. The stratigraphical scheme summarized here is based on that of Verniers *et al.* (2001).

The different Ordovician outcrop or subsurface areas in the Brabant Massif include, from west to east: the Western Flanders borehole area, the Dendre river valley, the Senne river basin, the Dyle river basin, the Gette river valley, and the Orneau river valley. A great number of formation names have been used in the Belgian literature. In the 1990s, this stratigraphical nomenclature was reviewed and rationalized, and most formations have now been correlated between the different outcrop areas. The Ordovician succession in the Brabant Massif can thus be divided from the base to the top into 11 formations, the oldest formation being of late Cambrian/early Ordovian age (Mousty Formation) and the youngest formation extending into the Silurian (Brûtia Formation; Fig. 5.5).

Mousty Formation

This formation is probably mostly of Late Cambrian age, with only its uppermost parts extending into the Ordovician. It largely comprises shales or slates, sometimes mudstones, and locally centimetre- to decimetre-thick fining-upward sandstone or siltstone bands, that are interpreted as distal turbidites. The middle part of the formation is clearly more silty with pyritic shale gradually passing downwards into a pyritic siltstone and sometimes sandstone (Court-St-Etienne borehole). The Tangissart Member at the top of the formation is characterized by an increasingly recurrent black shale or slate with abundant millimetre-scale light-coloured silty laminae. The disappearance of the last black shale interval marks the boundary with the overlying Chevlipont Formation.

The Mousty Formation is observed only in exposures in the Dyle river valley and from boreholes in the Dendre and Orneau valleys. The upper member (Tangissart Member) is defined along the Ottignies–Charleroi railway section, north of the Laroche railway station. Graptolites (*Rhabdinopora* sp.) and acritarchs in the Tangissart Member prove the early Tremadocian age of the upper part of the formation (Lecompte 1949; Martin 1969*a*, *b*; André *et al.* 1991). The thickness is difficult to estimate. Verniers *et al.* (2001) tentatively suggested a thickness of more than 500 m.

Chevlipont Formation

The age of this formation has been established since the late 1940s, when graptolites indicated an earliest Ordovician age. An early Tremadocian age is provided for the lower part of the formation by the presence of the dendroid graptolites *Rhabdinopora flabelliformis* ssp. *socialis* and *R. typica* (Lecompte 1948, 1949), and based on acritarchs (Martin 1969*a*, *b*, 1976; Vanguestaine in André *et al.* 1991). The formation consists of siltstones (called 'quartzophyllade' in the older literature) with characteristic wavy bedding consisting of millimetre-scale alternations of siltstones and clayey siltstones and mudstones. Silty laminae occur characteristically in small lenses a few centimetres long and a few millimetres thick with oblique lamination. Small slumps are common. This facies has been interpreted as low density turbidites. It is sometimes replaced by decimetre-thick beds showing graded bedding (typical Tcde Bouma sequences) (Herbosch *et al.* 1991). The formation, which is probably up to 150–200 m thick, is present throughout the entire Brabant Massif; from the Dendre Valley (Lessines borehole), the Marke, Senne and Sennette valleys and Dyle Basin, and in many boreholes in the southern or northern part of the massif (see De Vos *et al.* 1993). In the Condroz Inlier, the same formation (previously named the Wépion Formation) occurs (see below).

Abbaye de Villers Formation

This formation, with a thickness estimated to be between 100 and 150 m, yields no macrofossils, but was dated with palynomorphs. Chitinozoans in the lower third of the formation in the Dyle Basin contain *Eremochitina brevis* (Samuelsson & Verniers 2000) within an assemblage similar to that occurring in the Grès Armoricain Formation (Fm) in Brittany, indicating the middle Arenig, Whitlandian (*pro parte*) or possibly upper Arenig (Paris 1981). Acritarchs in the middle and upper part of the formation indicate the upper Arenig or post-Arenig according to Martin (1976) and Vanguestaine (in André *et al.* 1991), which is corroborated by the chitinozoans (Samuelsson & Verniers 2000). An important time gap is thus present between the Chevlipont and the Abbaye de Villers formations. No macrofossils are found in the unnamed member. Lithologically, the Abbaye de Villers Fm is composed of siltstone to mudstone, with a distinct lenticular lamination; there is frequent and abundant bioturbation often along bedding planes. Oblique stratification occurs at a metre scale. The contact with the underlying Chevlipont Fm was not observed anywhere, but it marks a very distinct sedimentological break. The formation was first described in the Thyle Valley, just north of the old abbey at Villers, in the railway section between Court-St-Etienne and Fleurus, but it is widespread in the Senne and Dyle basins.

Tribotte Formation

This formation, originally described from the Tribotte locality in the Dyle valley, is present in the Senne and Dyle basins and in the Orneau valley. In the Dyle Basin the lower third contains clayey sandstone and siltstone with coarse laminations strongly bioturbated; some beds have metre-scale oblique stratification. The upper two-thirds comprises sandstone and siltstone, clearly with more clay than below. Bioturbation is intense with oblique to vertical burrows predominant ('Fucoides' in the older literature). The upper part shows typical structures (e.g. vertical bioturbation, flaser bedding) indicating an intertidal depositional environment. The thickness is estimated to be 200 to 300 m in the Dyle Basin and 150 to 200 m in the Senne Basin. Neither macrofossils nor acritarchs have been found. A poorly preserved chitinozoan assemblage containing *Euconochitina vulgaris* indicates a middle Arenig to early Llanvirn age (Verniers *et al.* 1999; Samuelsson & Verniers 2000).

Rigenée Formation

A formation with this name was first described from a section north of Rigenée in the Vallon des Goutailles in the Dyle Basin. This formation is composed of slates or mudstones, indistinctly coarsely laminated or massive, locally containing pyrite. The lower boundary of the unit marks a rapid and sharp upward change from light clayey siltstone to dark mudstone, probably corresponding to a rapid increase in water depth. Similar lithologies are also present in the Orneau Valley and in the Senne Basin, but these units were historically attributed to different formations. It is now evident that the Rigenée Fm is present in most parts of the Brabant Massif. The age is most probably Llanvirn; graptolites in the lower or middle part of the formation in the Sennette Valley belong to *D. artus* and the *D. murchisoni* biozones, based on the restudy of the fauna by Maletz & Servais (1998), thus corresponding to the entire Abereiddian. Acritarchs from the Dyle Basin indicated an (early) Llanvirn age (Martin

1969*a*). Acritarchs studied by Servais (1991*a*, 1993*a*) indicate a late Arenig or younger age for the base of the formation. A poor assemblage of chitinozoans with *Lagenochitina obeligis* and *Cyathochitina calix* indicates the same broad age bracket (Verniers *et al.* 1999; Samuelsson & Verniers, 1999, 2000). The formation is 150 to 200 m thick in the Dyle valley, and about 150 m in the Senne valley, and probably about 150 to 200 m in the Orneau valley.

Ittre Formation

This formation is characterized by an alternation of distinct laminae or beds, with sharp bedding planes of fine sandstone, siltstone and mudstone. The sandstone beds are commonly over 5 cm and up to several decimetres thick and show parallel, oblique and convolute lamination, fining-upward graded bedding, with load casts and other current marks at their base. The siltstones shows parallel lamination and the mudstones are massive. They were deposited as turbidites of the Bouma type (Servais 1991*b*). The occurrence of a megaslump within the Ittre and the Bornival formations has been observed in the Sennette valley (Servais 1991*b*; Debacker *et al.* 2001). Thin volcanic tuffs or volcanosedimentary beds have been described in the basal part of the formation in the Sennette valley (Corin 1964; Martin & Rickards 1979). The formation was first described in the Sennette valley, along the large canal section south of Asquempont, but was later also identified in the Lessines borehole (Dendre), and in the Senne and Dyle basins and the Orneau valley. The thickness is estimated to be over 180 m in the Sennette valley. The graptolites were redescribed by Maletz & Servais (1998), who indicated a time interval ranging from the *gracilis* to the *multidens* biozones (Aurelucian or Burrellian), with a preference for the latter biozone. The chitinozoans are moderately well preserved and belong to the *Spinachitina cervicornis* biozone. They indicate a probable Burellian age but a Chenyan age cannot be excluded (Vanmeirhaeghe *et al.* 2005; Vanmeirhaeghe 2006).

Bornival Formation

While most formations have been known for almost a century, the upper Ordovician Bornival, Huet and Madot formations were not described until the 1990s.

Van Grootel *et al.* (1997) first described the Bornival Fm in the Sennette valley. This unit comprises centimetre-scale alternations of dark grey micaceous siltstone and dark grey to blackish mudstone with abundant silt grains distributed throughout. In the Sennette valley the formation can be subdivided into (at least) three members. Because of faulted contacts, neither the lower boundary, supposedly with the Ittre Fm, nor the upper boundary, supposedly with the Huet Fm, is observed. Also the boundaries between the three members are not observed because of faulted contacts. The thickness in the Sennette valley is estimated at a minimum of 244 m. No macrofossils have been found and only rare and longer ranging chitinozoans have been recovered, defining an interzone with a likely Cehneyan age (Vanmeirhaeghe *et al.* 2005; Vanmeirhaeghe 2006).

Huet Formation

Van Grootel *et al.* (1997) described the Huet Fm near the village of Fauquez in the Sennette valley, where the Bornival Fm was also described. The fossiliferous unit is composed of greenish to grey mudstones, siltstones and poorly sorted fine-grained sandstones, with characteristic orange-yellow alveoli of decalcified fossil fragments. Some levels are rich in (often) decalcified macrofossils (bryozoans, brachiopods, crinoids, cystoids and trilobites). At the base of the beds the macrofossils are often fragmented. All of the features mentioned suggest that this unit was deposited as a tempestite. The base of the formation is not observed due to the presence of faults; the top is a rapid transition into the Fauquez Fm. The formation is also present in the Orneau valley and possibly in boreholes of Western Flanders (Legrand 1968). It has a minimum thickness of 60 m. The brachiopods suggest a Caradoc or Ashgill age for the formation (Maillieux 1926). While the cystoids indicate a (late) Caradoc age (Regnéll 1951), the trilobites point to a (mid-) Caradoc age (Maillieux 1926). A subsequent collection of trilobites, studied by Richter & Richter (1951), indicates an (early) Ashgill age. The location of this collection is unclear and it cannot be excluded that it was sampled from the overlying Madot Fm. The chitinozoan assemblages belong to the *Fungochitina spinifera* biozone occurring in the uppermost Onnian to lower Cautleyan. Combined with the macrofauna, a latest Caradoc age can be deduced, corresponding to a late Oandu to early Vormsi age in Baltoscandian terms (Vanmeirhaeghe *et al.* 2005; Vanmeirhaeghe 2006).

Fauquez Formation

Several lithological units have been named after the village of Fauquez in the Sennette valley. The classic 'assise de Fauquez', however, is the well known graptolitic shale unit that can still be found at several outcrops near the village. This formation is composed of an alternation of black slates and dark grey silty slates with abundant graptolites that are found from about 4 m above the base. The rhythmicity, enhanced by the numerous pyritic beds, and the presence of rich graptolite levels suggests deposition of low-density turbiditic currents (Herbosch *et al.* 1991). The dark colour indicates anoxic conditions. The lower boundary is a rapid transition, with a distinct change in the grain size, colour and bed thickness from decimetre-scale in the Huet Fm to centimetre- or millimetre-scale in the Fauquez Fm. The upper boundary of the Fauquez Fm with the Madot Fm is unobserved but supposed to be gradual. The formation is also present in the Dendre valley (described as 'Unité I', in the Lessines borehole by Herbosch *et al.* 1991). The thickness of the formation is estimated to exceed 35 m in the Sennette valley and 58 m in the Lessines borehole. The graptolites indicate a *clingani* or *linearis* biozone. Maletz & Servais (1998) reinvestigated the old collections and analysed new material, but did not discover the nominal species of the biozones. They concluded that the assemblage possibly belongs to the *Pleurograptus linearis* and/or upper part of the *Dicranograptus clingani* Biozone (corresponding to the upper Caradoc and lowermost Ashgill of the British chronostratigraphy). The area has been remapped in detail by Verniers *et al.* (2005). The chitinozoans of the Fauquez Fm belong to the *Fungochitina spinifera* Zone, indicating a Rakvere to early Vormsi age in terms of Baltoscandian stages and a late Onnian to earliest Cautleyan age in terms of British (sub) stages. Graptolites and chitinozoans combined indicate a late Onnian to Pusgillian or earliest Cautleyan age (latest Caradoc to earliest Ashgill) (Vanmeirhaeghe *et al.* 2005; Vanmeirhaeghe 2006). Acritarchs are very poorly preserved and do not provide any additional information.

Madot Formation

During their reinvestigation of the Upper Ordovician stratigraphy of the Fauquez area, Van Grootel *et al.* (1997) and Verniers *et al.* (2005) described this formation from the eastern side of the Sennette valley and from the Bois des Rocs on the western side. The formation contains many volcanic rocks (volcanic breccias containing shale fragments and volcanosedimentary rocks), interstratified with heterogeneous coarse siltstones and mudstones rich in macrofossils. Also interbedded are shales and fine siltstones which commonly contain dispersed single crystals of volcanic origin. The upper boundary of the formation is limited by a fault in the Sennette valley; the lower boundary with the Fauquez Fm is probably gradual but has not been observed so far. The formation has been subdivided in the Sennette valley into seven members. Originally described from the Senne and Sennette valleys, the Madot Formation, or some of its members, also occur in the Nivelles area and in the Orneau valley, as well as in many boreholes in the central part of West Flanders and southern East Flanders. The Lichtervelde Fm in the Lichtervelde borehole (−291 to 415,10 m) is also equivalent to parts of the Madot formation (Legrand 1964, 1966; Martin 1969*a*). The estimation of the thickness of the Madot formation is *c.* 100 m on the east side of the Sennette valley and more than 220 m on the west side of the valley, due to a marked thickness increase of the volcanics of Member 6 (only 10 m in the east and estimated at over 130 m in the west in the Bois des Rocs area). The macrofauna include crinoids, bryozoans, brachiopods, trilobites, corals and pelmatozoans. The trilobites determined by Richter & Richter (1951), possibly collected from this formation in the Orneau valley, indicate an early Ashgill age. Graptolites are absent in the outcrop area but were found in several levels in the Lichtervelde borehole and were restudied by Maletz & Servais (1998). They belong, for the middle part of the formation in the borehole, to the *D. complanatus* biozone, and for the top of the formation in the borehole to the *D. anceps* biozone (coarsely middle part of the Ashgill). Chitinozoans are the only (micro)-fossil group providing an age for the outcrop area. The lower part of the formation belongs to the *Fungochitina spinifera* Zone indicating a Rakvere to early Vormsi age, and the upper part can be attributed to the *Tanuchitina bergstroemi* Zone indicating a middle Vormsi to lower Pirgu age in terms of the Baltoscandian stages. Combined, a Pusgillian to Cautleyan age in terms of British (sub) stages is proposed for the Madot Formation. The early Ashgill age for the peak of extrusive volcanism in the Brabant Massif can also be confirmed (Vanmeirhaeghe *et al.* 2005; Vanmeirhaeghe 2006).

Brûtia Formation

This formation, described on the basis of a recent mapping campaign, comprises the transition from the Ordovician to the Silurian. Many of the palaeontological data are very old and need to be revised. Trilobites in the lower/middle part of the formation were described by Malaise (1903) without giving an age. The rare chitinozoans from the mottled mudstone member are dominated by a species resembling *Belonechitina* cf *gamachiana*, indicating a mid- or late Ashgill age, or possibly a Hirnantian age (Samuelsson & Verniers 1999, 2000). The upper parts of the formation are Silurian. Graptolites described from different valleys indicate a Rhuddanian (Early Llandovery) age for the top of the unit. A mid-Ashgill to Rhuddanian age is proposed for the formation. While the lower unit contains mudstone and slate, a mottled grey mudstone member occurs, a few metres thick, consisting of dark grey lenses (about 1–2 mm wide and 2–5 mm long) in a medium grey mudstone, interpreted as a bioturbated horizon (*Chondrites* sp.). A fine-grained, very hard quartzitic tuff represents the topmost part of the formation; it is transformed by weathering into kaolinite. The type area of this formation is in the Orneau valley, near Gembloux, in the hamlet of Grand-Manil, between the localities Try-al-Vigne and Brûtia. The formation is also present in the Thisnes valley (Monstreux-Nivelles), but absent in the Sennette valley, due to the presence of faults. In the Orneau valley its thickness is between 80 and 100 m, with about 40 m for the 'eurite' of Grand-Manil in the Orneau valley.

Condroz Inlier

The Condroz Inlier (Figs 5.3 & 5.4) is composed of at least four tectonic wedges transported along the Midi Overthrust in the Variscan Deformation Front. The most northerly wedge is situated at Ombret, with a major central part and two smaller southern parts, the Oxhe Inlier in the east and the Puagne area in the SW. The Condroz Inlier contains Ordovician and Silurian (no Cambrian) siliciclastic sedimentary rocks deposited in a deep shelf environnment, that is generally not turbiditic, with the exception of the succession at Ombret.

The Ordovician succession is, as in the Brabant Massif, incomplete. Historically, the formation names were different from those of the Brabant Massif, although some similarities, both in lithology and biostratigraphy, are evident. The Ordovician lithostratigraphy of the Condroz Inlier was reviewed by Martin (1969*a*) with no subsequent detailed studies for decades.

A total of nine formations have commonly been used in published literature from the Condroz Ordovician. While the ages of the oldest (Chevlipont, early Tremadocian, and Huy, early Llanvirn) and the youngest (Fosses and Génicot, Ashgill) formations are well constrained, the ages and the relations between the Sart-Bernard, Ombret, Vitrival-Bruyère, Oxhe and Basse-aux-Cannes formations remained unclear until recent chitinozoan data clarified the succession of these formations. These latter formations show a mixture of lithologies, but they are probably all of similar age (Caradoc). They have been described from different areas in the Condroz Inlier, and none is completely known, due to poor outcrop conditions. A renewed field campaign allowed revision of the stratigraphical succession and accurate dating with chitinozoans. New formations were introduced and others reduced to member level (Vanmeirhaeghe & Verniers 2004; Vanmeirhaeghe 2006; Owens & Servais 2007).

Chevlipont Formation

This formation, comprising laminated shales, siltstones and sandstones ('quartzophyllades' in the older literature), is recognized only from the Wépion borehole (at the depths 216 to 423,90 m), where it has a thickness of over 140 m. Graulich (1961) suggested that the formation is of early Tremadocian *Rhabdinopora [Dictyonema] flabelliforme* aff. *norvegica* graptolite Biozone age. Subsequently, Martin (1969*a*) described a rich acritarch assemblage typical for the Tremadocian. Verniers *et al.* (2001) used the name of the formation described from the Brabant Massif for the same unit: Chevlipont Fm (= 'Assise de Wépion' introduced by Martin (1969*a*), but not 'Grès de Wépion', an Emsian formation described by Graulich (1961).

Huy Formation

This formation is the oldest formation cropping out in the Condroz area. It is composed of black, silky, finely micaceous shale, with cone-in-cone structures and white-veined blackish quartzite beds. It was first described from the sections near the railway station of the town of Huy in the Meuse valley. The thickness is probably over 150 m, and exceeds 120 m in the Wépion borehole. The age of the formation has been reviewed by Servais & Maletz (1992). The restudy and recollection of graptolites proved beyond doubt the presence of pendent didymograptids, present only in the Abereiddian part of the Llanvirn; other species indicate the *D. artus* Biozone, lower Abereiddian,

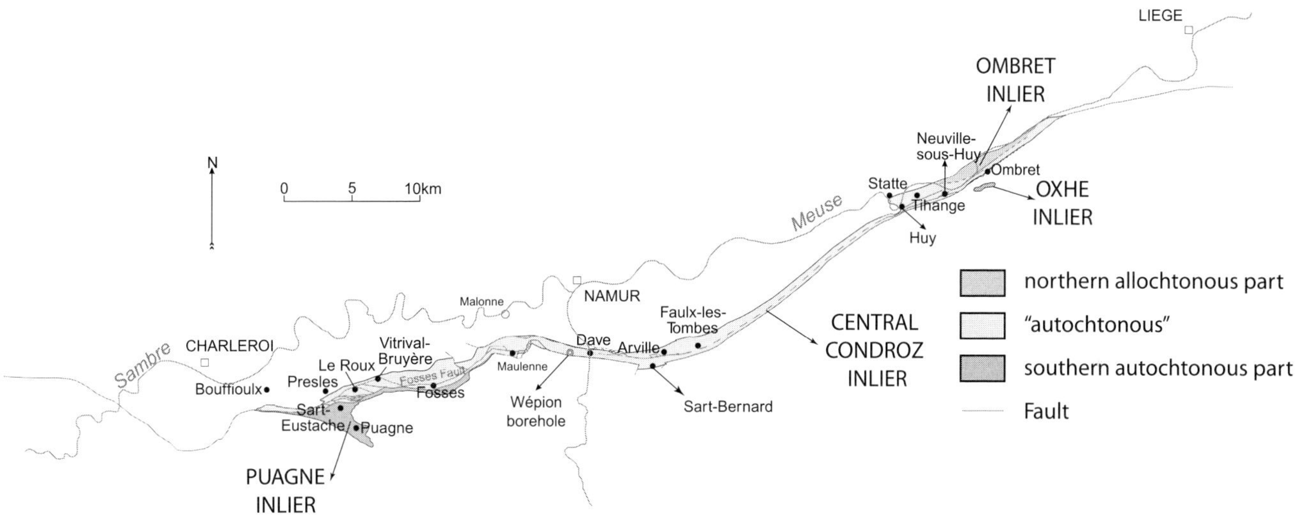

Fig. 5.4. Map of the outcrop area of the Condroz Inlier, Belgium (after Vanmeirhaeghe 2006).

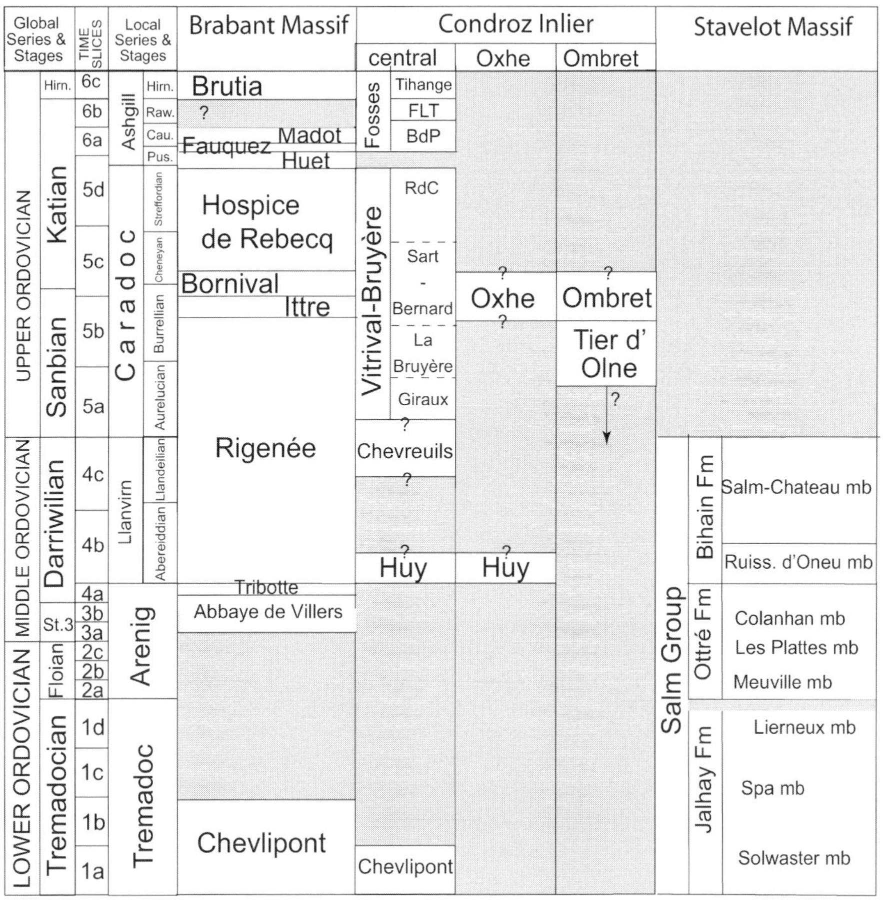

Fig. 5.5. Ordovician successions of Belgium and correlation with the Global Series and Stages, Time-slices and local series and stages from the British Isles (adapted from Verniers *et al.* 2001; Vanmeirhaeghe 2006; Owens & Servais 2007).

lower Llanvirn (Servais & Maletz 1992). The characteristic acritarch assemblage indicates an interval of late Arenig to early Llanvirn age (Servais & Maletz 1992). Trilobites are rare in the formation, and have recently been redescribed (Owens & Servais 2007). Other parts of the poor macrofauna are phyllocarids (Vannier *et al.* 2003) and the ichnofossil *Tomaculum* (Servais & Maletz 1992). Recent studies on chitinozoans indicate the presence of the *Siphochitina formosa* Zone in the upper part of the formation and confirm the early Llanvirn age (Vanmeirhaeghe 2006).

Sart-Bernard Formation
Verniers *et al.* (2001) considered the Sart-Bernard Fm to be equivalent to the Huy Fm. However, Servais & Maletz (1992) indicated, on the basis of literature data, that the formation must be younger than the early Llanvirn. The trilobites, reinvestigated by Owens & Servais (2007), suggest an early Caradoc age. A late Llanvirn to middle Caradoc age is also confirmed by graptolites and chitinozoans (Owens & Servais 2007). The formation is mainly composed of micaceous and sandy shales, and is thus also lithologically rather different from the Huy Fm. The formation was first described from the railway section NW of the village of Sart-Bernard (Servais & Maletz 1992). The thickness has not yet been estimated. The relation to the other formation of similar (Caradoc) age was considered unclear (see below). Vanmeirhaeghe (2006) lowers this unit to the rank of a member within the Vitrival-Bruyère Formation. Using chitinozoans, he dated the lower part as Middle Burrellian to Cheneyan and the upper part as Cheneyan.

Vitrival-Bruyère Formation
This formation of fine micaceous shales with quartzite intercalations and, in its higher part, clayey quartzitic sandstones in thick beds was described from outcrops near of the hamlet of La Bruyère, east of the village of Vitrival. Graptolites in a locality at Vitrival-Bruyère in the upper part of the unit were described in the early 1920s. Graptolites were originally identified as belonging to the *Climacograptus peltifer* Biozone, which was confirmed by Bulman (1950). The levels should thus belong to an interval equivalent to the *foliaceus* Biozone, uppermost Aurelucian and most of the Burrellian stages of the Caradoc (Fortey *et al.* 1995). Vanmeirhaeghe (2006) divided the formation into four members from bottom to top: the Giraux, La Bruyère, Sart Bernard and Rue de Courrière members. They range in age throughout the Caradoc from the Aurelucian to the Streffordian. Verniers *et al.* (2001) noted that the Oxhe Fm is possibly a lateral equivalent or even a synonym of the unit in the Oxhe inlier. Vanmeirhaeghe (2006) correlates the latter more accurately with the Sart-Bernard Member. The thickness of the Vitrival-Bruyère Fm with its four members is estimated by Vanmeirhaeghe (2006) at 380 to 480 m.

Basse-aux-Canes Formation
Delcambre & Pingot (2004) described this unit of silty shales and siltstones in the westernmost part of the Condroz Inlier, in the Puagne area, SE of Sart-Eustache. According to the geological map of Delcambre & Pingot (2004), the formation overlies the Sart-Bernard and Vitrival-Bruyère formations, and its thickness is estimated to be between 100 and 150 m. No macrofossils have been described from this unit, but its stratigraphical position implies a Caradoc or possibly early Ashgill age, as the formation underlies the Fosses Fm. The Basse-aux-Canes Fm slightly resembles the Vitrival-Bruyère Fm, but it does not have the characteristic sandstone beds of the latter. Verniers *et al.* (2001) noted that the Basse-aux-Canes Fm is possibly a lateral facies equivalent of (a part of) the latter formation, a result not contradicted by Vanmeirhaeghe (2006), who indicated a latest Caradoc to early Ashgill (mid-Katian) age for the (sampled part of) the formation.

Ombret Formation and Tier d'Olne Formation
A sandstone–slate unit was described as the Ombret Fm by Martin *et al.* (1970) from a large exposure along the road, NW of the church of Ombret, in the eastern part of the Condroz Inlier. Lithologically, the upper sandstone–slate member has similarities with the Ittre Fm in the Brabant Massif. The thickness of the upper sandstone–slate member has been estimated to be about 300 m, and the lower black shale member over 50 m. The Ombret Fm was restricted to the sandstone unit by Verniers *et al.* (2001) and the lower shale unit was termed the Tier d'Olne Fm. Based on acritarchs and chitinozoans, a definite Late Ordovician age was proposed, probably Caradoc (Martin *et al.* 1970). The biostratigraphy has been restudied in more detail with chitinozoans, which indicate a late Soudleyan to Cheneyan (mid-Caradoc) age for the unit (Vanmeirhaeghe 2006)

Oxhe Formation
This formation, described from the small inlier around the locality Fond d'Ohe, is characterized by micaceous mudstones with intercalated quartzitic sandstones and siltstone beds (Verniers *et al.* 2001). This formation was investigated by Dean (1991), who estimated the thickness to be 200 m; they studied the trilobites and attributed the formation to the Longvillian (upper substage of the Burrellian, Caradoc), an age confirmed by chitinozoans (De Geest 1998 cited in Vanmeirhaeghe 2006). The relation with the above-described formations remains unclear. Similar in lithology, the Oxhe formation has been included as a part of the Vitrival-Bruyère Fm by Michot (1954), but is kept as a separate formation by Vanmeirhaeghe (2006).

Fosses Formation
The Fosses Fm is one of the classic formations in the Condroz Inlier. It was recognized many years ago, on the basis of its abundant fossils. The formation was long thought to be composed of two members. The lower Bois de Presles Member is composed of limestones and calcareous shales with numerous brachiopods, trilobites, bryozoans, corals, crinoids, cystoids, and other echinoderm debris, molluscs and algae (Tourneur *et al.* 1993). The upper Faulx-les-Tombes Member is composed of sandy shales with a macrofauna, described as bioturbated calcareous shales by Tourneur *et al.* (1993). This formation has been described east of Fosses. Its thickness was estimated at about 100 m (Tourneur *et al.* 1993). The study of the brachiopods and trilobites indicated a pre-Hirnantian Ashgill age (Sheehan 1987; Lespérance & Sheehan 1987). Vanmeirhaeghe & Verniers (2004) recently confirmed this age (upper Pushgillian to Rawtheyan) using chitinozoans. They could also prove that a coarse crinoidal limestone above a basal conglomerate (called the Cocriamont conglomerate) was not situated at the base of the formation but in the middle of the overlying Génicot Fm of Silurian age (Aeronian). Vanmeirhagehe (2006) introduced a third, sandy laminated member in the top of the formation: the Tihange Member.

Génicot Formation
The Génicot Formation recently defined as a mappable unit by Delcambre & Pingot (2004), in the western part of the Condroz Inlier, is composed of black shales, dark grey laminated siltstones and frequent thin beds of sandstones. It was thought at one stage to represent the Hirnantian glaciation. However, brachiopods and graptolites proved the middle Llandovery age of the formation (Vanmeirhaeghe 2006) thus defining the presence of a large hiatus between the top of the Fosses Formation and the base of the Génicot Formation.

Ardenne inliers

The two main Ardenne inliers (Stavelot and Rocroi; Fig. 5.3, 3 & 4) are similar in composition. Both contain a thick siliciclastic, often turbiditic succession which extends in age from the earliest Cambrian to the Middle Ordovician. The two smaller inliers (Givonne and Serpont; Fig. 5.3, 5 & 6) display only parts of this sequence. These four inliers are all situated in the Variscan Ardennes Allochthon, which was thrust from the south and SE for about 10–100 km into their present position. Syntheses of the Stavelot Massif have been published by Geukens (1986, 1999), and of the Rocroi Massif by Beugnies (1963). Macrofossils are usually absent, except for some lower Tremadocian graptolites. The dating of the Ardenne Massifs was possible using acritarch assemblages; the major results were summarized in Vanguestaine (1992).

Whereas many similarities can be found between the Ordovician successions of the Brabant Massif and the Condroz Inlier, the stratigraphical succession in the Ardennes is rather different. Classically, the Ardenne succession has been subdivided into three major parts: the Deville, Revin and Salm groups. The latter group is Ordovician, covering an interval from the Tremadocian to the Llanvirn. The Ordovician sequence of the Salm Group is mostly developed in the Stavelot Massif, where three formations have been described, from base to top: the Jalhay, Ottré and Bihain formations. These were previously named Salmian 1, 2 and 3, but Geukens (1999) renamed these units following the recommendations of the National Commission of Stratigraphy of Belgium. Vanguestaine (1992) summarized the sedimentological analyses of the sequence. According to the different studies, the Salmian 1 (now Jalhay Fm) can be considered as a bathyal deposit, mostly turbiditic, passing into slope deposits of the Salmian 2 (Ottré Fm) and shelf deposits with sand bars of the Salmian 3 (Bihain Fm).

A single lithostratigraphical unit in the Rocroi Massif, the Vieux-Moulins de Thilay Fm, can be attributed to the Middle Ordovician. It is possible that in the two smaller inliers (Givonne, Serpont) the youngest units extend into the lowermost Ordovician, but palaeontological data are absent.

Jalhay Formation

This formation is composed of slates, sandstones and silty slates (called 'quartzophyllade' in the older literature) subdivided into three members. The Solwaster Member contains dendroid graptolites known since the end of the nineteen century. The graptolitic shales with *Rhabdinopora flabelliformis* were for a long time the only known macrofossil occurrence in the Ardenne 'Massifs'. The Spa Member comprises dark bluish grey and greenish grey sandy so-called 'quartzophyllades'. Many sandstone beds have sedimentary structures (convoluted bedding and cross-stratification). The Lierneux Member displays olive green clayey and greyish 'quartzophyllades', green sandy slates with reddish patches, and magnetite-bearing slates. White sandstone beds are found in the Lierneux-Vielsalm-Bihain area, as well as traces of bioturbation. The thickness of the Jalhay Fm has been estimated to be over 400 m (Laloux *et al.* 1996) with the Lierneux Member being about 80 m thick, and the Solwaster Member about 60–150 m thick. The thickness of the Spa Member has not yet been estimated. An early Tremadocian age is shown by the presence of the dendroid graptolite *Rhabdinopora flabelliformis* in the lower two members of the formation. Vanguestaine (1974, 1992) confirmed this age on the basis of acritarchs. The uppermost part of the formation, the Lierneux Member, lacks macrofossils but was dated using palynomorphs with the typical *messaoudensis–trifidum* acritarch assemblage indicating a late Tremadocian age for this part of the formation (Vanguestaine & Servais 2002).

Ottré Formation

The Ottré Fm comprises a succession of *c.* 150 m of slates and silty slates (so-called 'quartzophyllades') subdivided into three members. The Meuville Member is *c.* 70 m thick and is composed of slates with calcareous nodules, that are locally garnet-bearing. The *c.* 30 m thick Les Plattes Member is constituted by purplish slates and can be easily recognized because of its calcareous beds which contain iron and manganese nodules. In the southern part of the Stavelot Inlier the slates contain beds with intercalated coticule, andalusite, garnet and volcanic rocks. The base is formed by a slaty sandstone containing coarse ottrelite crystals. The third member, the Colanhan Member, comprises *c.* 50 m of purple slates with rare levels of green slates in the north and, in the southern part, purple and green compact layered slates, containing the mineral ottrelite. For many years, neither micro- nor macrofossils were observed in the members of the Ottré Fm. Stratigraphically, the formation is situated between the late Tremadocian (mid-Early Ordovician) Lierneux Member and the mid-Middle to Late Ordovician Salm-Château Member. Very recently, Vanguestaine *et al.* (2004) provided preliminary information on a conodont assemblage from the base of the Ottré Fm, which suggested an uppermost Tremadocian age. It might be possible that an important stratigraphical gap, covering at least the early and middle parts of the Arenig, is present in the Stavelot Massif, which would be similar to the hiatus observed in the Brabant Massif and the Condroz Inlier.

Bihain Formation

This formation is divided into two members. The Ruisseau d'Oneu Member is composed of sandy slates with thin sandstone beds with convolute lamination and slumping. Black silty slates (so-called 'quartzophyllades') and sandstones characterize the Salm-Château Member with cleaved 'quartzophyllade', thick bands of slates, often rich in pyrite, with much bioturbation. A stratotype has not yet been defined, the type area of the formation being in the Bihain Syncline. This formation is found only in the southern part of the Stavelot Massif. In the metamorphic zones the slate contains small crystals of a chloritoid (ottrelite). The thickness is estimated as follows: Salm-Château Member: over 150 m; Ruisseau d' Oneu Member: over 30 m; the age of most parts of this formation are unknown. Mid- and/or Late Ordovician acritarchs (Vanguestaine 1974) and chitinozoans (Vanguestaine unpublished work 1973) have been found in the Salm-Château Member.

Petites Tailles Formation

Geukens (1999) described this formation which is characterized by quartzitic sandstones, breccias, and chloritic slates and conglomeratic sandstones which contain greenish shale pebbles. This unit overlies the Bihain Fm and is present only in the southern part of the Stavelot Inlier, where it has a thickness of over 75 m. A stratotype has not yet been defined. Without macrofossils and only long-ranging acritarchs present (Vanguestaine unpublished work 1973), the precise stratigraphical position is unclear. However, it is probably of Middle or Late Ordovician age. Geukens (1999) accepts for at least part of the formation a lithological equivalence with the Colanhan Member of the Ottré Fm.

Vieux-Moulins de Thilay Formation

A single lithological unit from the Rocroi Inlier has been attributed to the Ordovician. This unit, consisting of grey laminated siltstones and slates ('quartzophyllades zonaires' from the literature) is found north of the hamlet of Vieux-Moulins de Thilay. Its stratigraphical position is unknown, as is its thickness. Some outcrops west of Vieux-Moulins de Thilay provided samples that are dated with acritarchs as Arenig to Llanvirn (Early to Middle Ordovician) by Roche *et al.* (1986). New evidence on the range of the genus *Frankea* and the species *F. sartbernardensis* (Servais 1993*b*) restrict the age of the unit to the late Arenig to Llanvirn (Middle Ordovician).

Western Germany

Ordovician sediments have been recorded from several localities in western Germany, including the Stavelot-Venn-Anticline, some locations in Hessen, and a few localities in the Vosges Mountains (eastern France) and the Black Forest (southwestern Germany). The successions from western and northeastern Germany (Rügen) that provided biostratigraphical information are correlated with the Belgian sequences that belong to Eastern Avalonia (Fig. 5.6).

Stavelot-Venn Anticline

While the Rocroi Inlier extends to northern France, where the same names are used for the lithological units as in Belgium, the Stavelot Inlier extends into western Germany (Fig. 5.3, 3), where different formation names were used. German stratigraphers adopted some of the names of the units attributed by the Belgian colleagues and mostly followed their stratigraphical scheme. Only the uppermost Cambrian (Revin Group) and the lowermost part of the Ordovician (lower part of the Salm Group) are found in outcrops in western Germany. Equivalents of the Jalhay Fm and the base of the Ottré Fm are recognized in western Germany. Following the recommendations of the German Stratigraphical Commission, Ribbert *et al.* (2001) defined two lithostratigraphical units, from base to top: the Wehebach-Jalhay Formation and the Thönbach-Ottré Formation. The first has exposures with graptolitic levels in Germany, particularly in the Wehebach valley to the south of Schevenhütte. The Thönbach-Ottré Fm is found to the NE of the Wehebach dam, on the northern side of the Thönbach river valley. As in Belgium, this unit (equivalent of the lower part of the Ottré Fm) yields no fossils and remains undated.

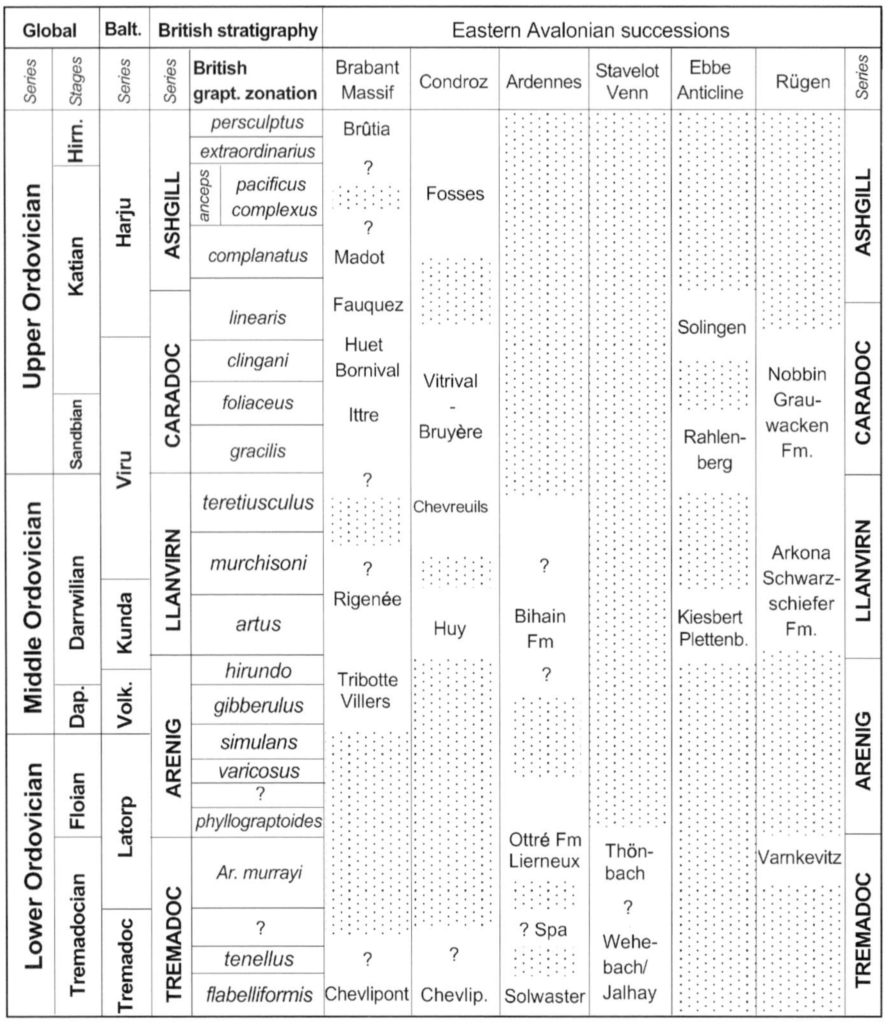

Fig. 5.6. Ordovician successions in the eastern part of the microcontinent Avalonia and correlation with the British regional series and graptolite zonation. Correlations based on: Verniers *et al.* (2001) for the Brabant Massif; Verniers *et al.* (2001), Vanmeirhaeghe (2006) and Owens & Servais (2007) for the Condroz Inlier; Verniers *et al.* (2001) for the Ardenne massifs; Ribbert *et al.* (2001) for the Stavelot Venn area; Eiserhardt *et al.* (2001) and Samuelsson *et al.* (2002*b*) for the Ebbe Anticline; Beier *et al.* (2001) and Servais *et al.* (2001) for the subsurface of Rügen.

Ebbe and Remscheid anticlines

Ordovician and Silurian rocks, structurally surrounded by Lower Devonian (Lochkovian) rocks, occur in the Solingen-Remscheid-Altena (Fig. 5.3, 7) and the Ebbe anticlines (Fig. 5.3, 8) in the Sauerland, east of Cologne, Germany. The Ordovician succession is surprisingly similar, both in terms of litho- and biostratigraphy, with the Ordovician sequence in the Condroz Inlier in Belgium. In addition, the successions of Rügen and western Pomerania (see below) show many similarities, so that it appears that the Ordovician of the Sauerland might have been deposited in the same basin and that this basin was located between the Condroz and Rügen depositional areas.

Following previous authors, Eiserhardt *et al.* (2001) named the entire succession 'Herscheider Schichten'. They reviewed the lithologies in detail and included complete fossil lists in their study. Maletz & Servais (1993) and Samuelsson *et al.* (2002*b*) provided additional stratigraphical information.

Four formations have been described. These have a total thickness of 800 m and comprise a monotonous siliciclastic succession of mudstones and siltstones. The lithological monotony of the succession makes it difficult to distinguish between the four lithological units. In terms of sedimentological interpretations, no detailed investigations have been published so far.

Plettenberger Bänderschiefer

The Plettenberger Bänderschiefer, up to 65 m thick, is a dark compact mudstone with abundant thin pyrite-rich siltstones (Eiserhardt *et al.* 1981). This formation has a similar fossil assemblage to the Huy Fm in the Condroz Inlier in Belgium and their graptolites and acritarchs are almost identical (Maletz & Servais 1993). The latter fossil groups allows the Plettenberger Bänderschiefer to be placed in the early Llanvirn *D. artus* graptolite Biozone (Maletz & Servais 1993). Subsequently Maletz (*in* Eiserhardt *et al.* 2001) recognized the *Holmograptus lentus* graptolite Biozone (lower part of the *artus* zone). Other fossil groups have been recorded, including trilobites, foraminiferan and trace fossils (for a complete review, see Eiserhardt *et al.* 2001). The trilobites of the Huy Fm, recently redescribed by Owens & Servais (2007), are almost identical to those described by Koch (1999, and references therein). Owens & Servais (2007), interpret the trilobite assemblage of the Plettenberger Bänderschiefer as representing the outer shelf/upper slope biofacies, typical of Eastern Avalonian assemblages, that are frequently recorded from the British Isles. Samuelsson *et al.* (2002*b*) confirmed the age of the Plettenberger Bänderschiefer with chitinozoa that also indicate an early Abereiddian (earliest Llanvirn) age.

Unterer (Kiesberter) Tonschiefer

The overlying 150–200 m thick Unterer (Kiesberter) Tonschiefer is a dark mudstone with rare sandstone beds. It has almost the same fossil content as the underlying formation, with many fossil groups present, including acritarchs and graptolites (Maletz & Servais 1993), trilobites (Koch 1999 and references therein), ostracods (Schallreuter & Koch 1999), phyllocarids (Koch & Brauckmann 1998), conularids and brachiopods. The age of this formation is slightly younger than that of the Plettenberger Bänderschiefer. Maletz & Servais (1993) included the formation in the *D. artus* graptolite Biozone. Additional studies allowed the Unterer (Kiesberter) Tonschiefer to be placed in the *Nicholsonograptus fasciculatus* Biozone (Maletz in Eiserhardt *et al.* 2001), which corresponds to the upper part of the *D. artus* Biozone. This age has been corroborated by chitinozoan data (Samuelsson *et al.* 2002*b*).

The (Rahlenberger) Grauwackenschiefer

The (Rahlenberger) Grauwackenschiefer is *c.* 300 m thick, and contains true thin-bedded greywackes, and siliceous concretions in the upper part. Eisenack (1939) described acritarchs and chitinozoans that are presently of low stratigraphical significance. Macrofossil groups are very rare but include conulariids, trilobites, malacostraceans and graptolites. These, however, do not provide a more precise biostratigraphy. More recent investigations of chitinozoans by Samuelsson *et al.* (2002*b*) have allowed the (Rahlenberger) Grauwackenschiefer to be placed in the Aurelucian (lowest Caradoc).

Oberer (Solinger) Tonschiefer

The youngest Ordovician unit present in the Ebbe and Remscheid anticlines is the Oberer (Solinger) Tonschiefer which is up to 200 m thick. This formation contains significantly fewer fossils, with only rare trilobites, very few graptolites and phyllocarids, and rare trace fossils present. The age attribution was therefore problematical, and only a post-Llanvirn age was suggested until the recent chitinozoan investigations indicated attribution to the upper Caradoc (Samuelsson *et al.* 2002*b*).

The Ordovician sequence of the Ebbe and Remscheid anticlines ranges, therefore, from the lower Llanvirn to the upper Caradoc. The Tremadocian, Arenig and Ashgill have not yet been recorded. The uppermost Ordovician Herscheider Schichten rocks are separated by a hiatus from the overlying uppermost Silurian rocks of the Köbbinghäuser Schichten (Timm *et al.* 1981). Maletz & Servais (1993) and Samuelsson *et al.* (2002*b*) supported the suggestion of Jaeger (1967) that the Herscheider Schichten were deposited in the same basin as the Belgian Condroz Inlier and the Rügen successions. Nd-isotope studies of Samuelsson *et al.* (2002*b*) show a remarkable similarity between the Ebbe Inlier, the Stavelot-Venn Inlier, the Brabant Massif, the English Lake District and the Welsh Basin (Gerdes *et al.* 2001). This conclusion supports the idea of a continuation of Eastern Avalonia east of the Brabant Massif into northern Germany (e.g. Maletz & Servais 1993; Pharaoh 1999).

Hessen, western Germany

Several areas include Ordovician rocks, such as the Rhenohercynian Zone found in central Germany as far south as the Northern Phyllite Belt (Fig. 5.3, 9). Although the precise age of the sediments is generally not well known, some units, such as the 'Giessener Grauwacke', for example, were traditionally placed in the Ordovician.

The exact age and the palaeogeographical attribution of most Ordovician units of Hessen remain problematical. Struve (1975) summarized the few palaeontological data. Some remnants (?olistoliths) of Ordovician rocks have been described from several places. The 'Andreasteich-Quarzit' (=Giessener Grauwacke) provided some trilobites and brachiopods indicating a Caradoc age. In the area around Münzenberg, some fossiliferous erratics have been named '*henningsmoeni*-Quarzit' and '*grandior*-Quarzit', following the identification of the trilobite species *Ogygiocaris henningsmoeni* and *Neseuretus grandior*, respectively. These erratics of unknown origin are possibly of Arenig to Llanvirn age (Struve 1975 and references therein).

More recently Reitz *et al.* (1995) reported the first recording of Lower Palaeozoic acritarchs from Hessen. The attribution of the assemblage of the 'Bierstatt-Phyllit' to the Arenig remains problematical, since the illustrated specimens (Reitz *et al.* 1995) do not justify an early Ordovician age.

In the Stratigraphical Table of Germany (Menning & Hendrich 2005) a series of formations (Mömbris, Hörstein-Huckelheim and Geiselbach formations) from the Spessart area were placed in the Late Cambrian to Silurian interval. None of these formations has been precisely dated and their stratigraphical position thus remains highly speculative.

Black Forest (Schwarzwald), southwestern Germany

The Black Forest (Figs 5.3, 10, & 5.7) is a part of the Variscan orogenic belt, which extends from the Iberian Peninsula to the Bohemian Massif. Being located in the south of the Rhenohercynian Zone, the area has usually been attributed to the Saxothuringian Zone (northernmost part of the Black Forest) or the Moldanubian Zone (southern part) (Kossmat 1927). Most of the sedimentary rocks of the Black Forest are metamorphozed and many of these metasedimentary units have been considered, until recently, to be devoid of fossils. The Vosges Mountains in Alsace, eastern France, display the same complex tectonometamorphic evolution as the Black Forest. Age information is, so far, extremely rare, and remains sometimes problematical.

The northernmost zone of the Schwarzwald is named the Baden-Baden–Gaggenau Zone. It contains crustal fragments of low to medium metamorphic grade, such as a greenschist facies metasedimentary rock unit and higher grade kyanite–garnet mica schists, gneisses and amphibolites (Wickert *et al.* 1990).

Within the metasedimentary units, Sittig (1965) distinguished four units based on lithological differences, of which some samples in the uppermost one, the 'Traischbach-Serie', comprise a pelitic succession with intercalated carbonate olistoliths, and have recently been dated as late Cambrian–early Ordovician. Montenari *et al.* (2000) and Montenari & Servais (2000) provided palynological data from three of 133 palynological samples. The acritarchs show taxa of the galeate plexus, as well as diacromorph acritarchs, which suggests that these rocks are of late Cambrian, or most probably early Ordovician age. These results allow correlation with the Villé schists of the northern Vosges Mountains (Fig. 5.3, 11 & 5.6) where Reitz & Wickert (1989) identified a similar acritarch assemblage.

To the south, most parts of the Moldanubian Black Forest provide no biostratigraphical data, except for the low grade metasediments of the Badenweiler–Lenzkirch Zone, where two chitinozoan genera indicate a Wenlock to Pridolí age (Middle to Upper Silurian) for the deposition of the phyllite protoliths in the Badenweiler–Lenzkirch Zone (Montenari *et al.* 2000).

Palynological data have recently been confirmed and extended by Vaida *et al.* (2004). Many of the age data in the Stratigraphical Table of Germany (Menning & Hendrich 2005) for the Black Forest remain speculative. Only the position of the Traischbachschiefer Formation is justified; the position of the Nordrach-Leptinit Formation in the Darriwilian is based on radiometric data.

Northern Germany, NW Poland

Rügen

Since the early 1960s numerous boreholes have been drilled by the East German oil and gas industry in NE Germany (Fig. 5.8, 2).

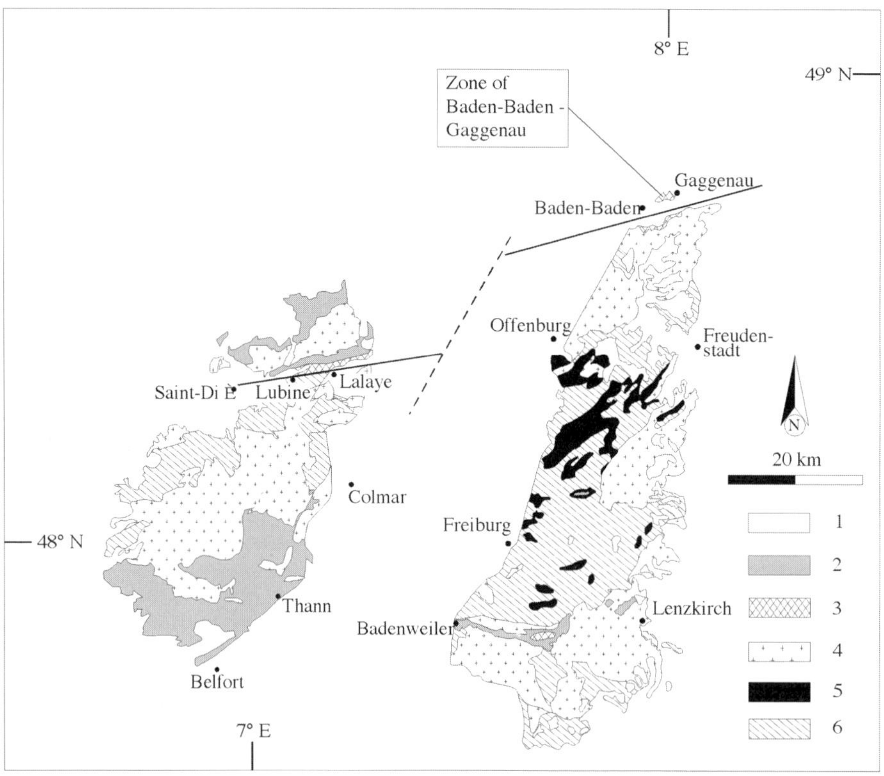

Fig. 5.7. Map of the outcrop areas of the Vosges Mountains (eastern France) and the Black Forest (southwestern Germany) after Montenari & Servais (2000). Key: 1, post-Variscan; 2, Devonian–Carboniferous; 3, Early Palaeozoic; 4, granites; 5, orthogneiss; 6, paragneiss.

Lower Palaeozoic rocks have been encountered in several boreholes on the island of Rügen (Rügen 5, Rügen 3h, Arkona 101, Lohme 2/70, Binz 1/73), on the German mainland (Loissin 1), and, in the late 1980s, in offshore boreholes of the 'Petrobaltic' project (H2–1/1990, K5–1/1988, G14–1/1986) (Figs 5.3, 12 & 5.8). The borehole G14 encountered rocks which are comparable to the Baltic succession known from Scania and Bornholm Island, and which are completely different from those of the other eight boreholes discussed in the present study. Only a small portion of the results of the original geological, petrographical and palaeontological investigations on the Rügen Ordovician has been published (e.g. Jaeger 1967, Schmidt & Franke 1977).

In the 1990s, several research groups started the reinvestigation of the boreholes. Publications included new palaeontological (Servais & Katzung 1993; Servais 1994), sedimentological and petrographical (Giese *et al.* 1994; Kurze *et al.* 1995; McCann 1998), and tectonic results (Katzung *et al.* 1993), as well as the first data from the 'Petrobaltic' boreholes (e.g. Rempel 1992; Hoth *et al.* 1993; Piske *et al.* 1994; Franke *et al.* 1994).

Subsequent studies concerned tectonic (Franke *et al.* 1996; Giese *et al.* 1997), palaeontological (Servais & Molyneux 1997; Maletz 1998; Samuelsson *et al.* 2001; Samuelsson & Servais 2001; Servais *et al.* 2001) and geochemical analyses (McCann 1998), as well as isotope studies, such as U–Pb data on detrital zircons (Tschernoster *et al.* 1997) and $^{40}Ar/^{39}Ar$ data on detrital muscovites (Dallmeyer *et al.* 1999).

According to these studies, the thickness of the Ordovician succession of Rügen is significant. For instance the Rügen 5 borehole encountered more than 3000 m of Ordovician rocks strongly tectonized and dislocated (Franke & Illers 1994; Beier & Katzung 1999). The sedimentation of the pelitic–clastic sequence probably commenced in the latest Tremadocian (Servais & Katzung 1993) and ended in the early Caradoc (Jaeger 1967; Maletz 1998; Samuelsson *et al.* 2000, 2001). In the easternmost part it possibly extended into the late Caradoc to early Ashgill (Vecoli & Samuelsson 2001). However, although numerous new palaeontological data are now available, some parts of the succession remain undated and older or younger sediments may be present. The succession indicates a strongly active continental margin tectonic setting (Giese *et al.* 1994; McCann 1998), consisting perhaps of an allochthonous Caledonian nappe complex thrust onto the Baltic shield (Franke *et al.* 1996; Giese *et al.* 1997; Dallmeyer *et al.* 1999). Overviews of the biostratigraphy, based on macrofossils (mostly graptolites) and organic-walled microfossils (acritarchs and chitinozoans) have been published by Maletz (2001), Servais *et al.* (2001), Samuelsson *et al.* (2001) and Beier *et al.* (2001).

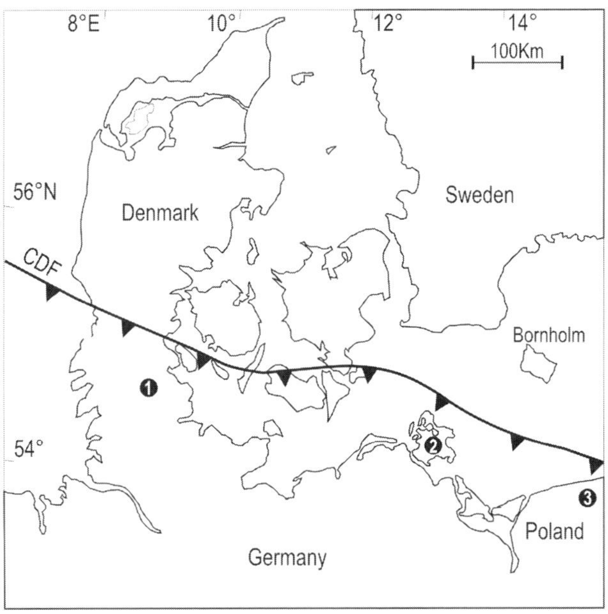

Fig. 5.8. Map of the subsurface localities with Ordovician rocks in northeastern Germany and northwestern Poland, including: 1, Schleswig-Holstein (subsurface sediments attributed to the Ordovician without biostratigraphical control); 2, the well dated borehole sections of the island of Rügen (Germany); 3, the Upper Ordovician of the boreholes of the Koszalin-Chojnice area (Poland).

Graptolites were found in most of the boreholes of the Rügen Ordovician (Rügen 5, Rügen 3h, Arkona 101, Binz 1/73, H2–1/1990, K5–1/1988). In some cores, the graptolite remains were unsufficient to provide precise biostratigraphical data. In many parts of the cores, graptolites were not present, and these parts of the sections remained undated. The graptolites were originally investigated by Jaeger (1967) who attributed them to his biozones 5 to 9. They were reviewed in detail by Maletz (1998), who attributed the different levels to the following graptolite biozones: *Holmograptus lentus* (early Abereiddian, early Llanvirn), *Pterograptus elegans* and *Pseudoamplexograptus distichus* (*D. murchisoni* Biozone, upper Abereiddian, middle Llanvirn), *Hustedograptus teretiusculus* (Llandeilian, late Llanvirn) and *Nemagraptus gracilis* (early Caradoc).

With the aim of better understanding the biostratigraphy of the Rügen boreholes, including the parts without graptolites, investigations on acritarchs and chitinozoa have been undertaken based on samples from all of the boreholes. Except for the sediments of the Loissin 1 borehole, which remain undated but have tentatively been placed in the Precambrian and named the Pommern Gruppe (Schwarbe-Buntschiefer Formation and Lubmin-Sandstein Formation), all boreholes contain Ordovician successions of black shales or greywackes ranging in age from early Llanvirn to early Caradoc (for a detailed review see Servais *et al.* 2001).

More recently three formal formations have been described from the Ordovician of Rügen: from base to top, the Varnkevitz-Sandstein Formation, the Arkona Schwarzschiefer Formation and the Nobbin-Grauwacken Formation, all of which belong to the Wittow Gruppe (Beier *et al.* 2001).

Varnkevitz-Sandstein Formation

The Rügen 5 borehole provides not only the thickest Ordovician succession (3172 m of tectonically dislocated rocks), but also the most complete. The lowermost part of the borehole (cored intervals between 3891.9 and 3869.9 m) remains undated. The oldest dated rocks are black shales recovered from five cored intervals which yielded organic-walled microfossils in samples between depths of 3836.4 and 3615.8 m. The acritarchs from this interval can be correlated with a succession observed in the English Lake District (Molyneux & Rushton 1988) and they indicate a late Tremadocian age (Servais & Molyneux 1997). The same samples also provided chitinozoa typical of the oldest chitinozoa biozone established in North Gondwana, suggesting a late (but not latest) Tremadocian age (Samuelsson 1999; Samuelsson *et al.* 2001). This sandstone–shale interval of late Tremadocian age has been found only in the Rügen 5 borehole and is now named the Varnkevitz-Sandstein Formation (Beier *et al.* 2001).

Arkona Schwarzschiefer Formation
The black shale unit with graptolites of Llanvirn age, originally named 'Oberere Didymograptusschiefer' (e.g. Jaeger 1967), has recently been defined as the Arkona Schwarzschiefer Formation. Tectonically disturbed at the base and top, the formation is found in the Arkona 101, Rügen 5, Lohme 2, Binz 1, H2 and K5 boreholes. Its thickness is estimated to be over 1000 m. The age ranges from the Llanvirn possibly into the Caradoc (Maletz 1998; Samuelsson *et al.* 2000; Samuelsson & Servais 2001; Vecoli & Samuelsson 2001).

Nobbin-Grauwacken Formation
The greywacke–shale sequence, previously named 'untere Dicellograptusschiefer' (e.g. Jaeger 1967), is now defined as the Nobbin-Grauwacken Formation (Beier *et al.* 2001). This succession was recovered only from the Rügen 3h and Rügen 5 boreholes. Some graptolite levels are found in both boreholes and indicate an age ranging from the middle Llanvirn into the early Caradoc (Maletz 1998). This age is confirmed by acritarch and chitinozoan studies (Servais *et al.* 2001).

Schleswig-Holstein, northern Germany

Three boreholes in Schleswig-Holstein (two drillings between the cities of Schleswig and Kiel and one drilling on the island of Westerland) provided slightly metamorphozed silty shales that have been attributed to the Cambro-Ordovician (Fig. 5.8, 1). Two lithological units were named 'Westerland-Einheit' and 'Flensburg-Einheit' and have been placed into the Ordovician (Schenck 2002). However, there is no biostratigraphical evidence for this assignation. T. Heuse (pers. comm. cited in Schenck 2002) mentioned organic debris. Other palynological investigations did not provide any biostratigraphical result (T. Servais & G. Katzung, unpublished data).

Pomerania, NW Poland

The Ordovician sedimentary rocks of West Pomerania, NW Poland (Fig. 5.3, 13; 5.8, 3), are closely comparable with those of coeval strata present in the Rügen subsurface, and show clear similarities in terms of their stratigraphical and petrographical composition (e.g. Franke 1994). They have, therefore, been assigned by several authors to the Rügen-type Ordovician (see Samuelsson *et al.* 2002*a*). These rocks are well-known from several boreholes in the area (Bednarczyk 1974; Samuelsson *et al.* 2002*a*) and show evidence of folding and faulting. The entire Ordovician succession, consisting of siltstones with rare calcareous intercalations and concretions, was attributed to the Ordovician following graptolite investigations (Bednarczyk 1974). Samuelsson *et al.* (2002*a*) dated the successions from several boreholes with chitinozoans, and attributed them to the upper Darriwillian (Llandeilian, Llanvirn) to middle Caradoc (*teretiusculus* to *clingani* graptolite biozones). The Ordovician succession is, like the Rügen Ordovician, overlain by thick Devonian, Carboniferous and Permian successions.

Correlation with the succession of the subsurface of Rügen is straightforward; quantitative assessment of microfossil assemblages of Pomerania and adjacent areas, interpreted as belonging to Avalonia, demonstrates the temporal and spatial association of Pomerania (and the Helgoland-Pomerania Deformation Belt) with the easternmost extension of Avalonia. A number of studies have demonstrated that the onset of foreland basin sedimentation began in the mid-Caradoc to late Ashgill in the Rügen area (e.g. Vecoli & Samuelsson 2001). In Pomerania, clastic sedimentation was active at least up to the middle Caradoc, further constraining the lower time limit for the start of the deformation of the Helgoland-Pomerania Deformation Belt (Samuelsson *et al.* 2002*a*).

Harz – Nördliche Phyllitzone, Germany

In the NW of the Mid-German Crystalline High, or Mid-German Crystalline Rise (= Mitteldeutsche Kristallinzone, MKZ), the Northern Phyllite Zone (Nördliche Phyllitzone) occurs between Eschwege in Hessen and Frankurt/Oder at the German–Polish border (Figs 5.3, 14 & 5.9). This zone has a length of about 300 km and a width of 10 to 20 km. The geotectonic evolution of this zone, however, remains unclear (e.g. Burmann *et al.* 2001). All of the units within this zone have been considered to be Devonian in age (e.g. Schwab 1976). However, more recently some parts were interpreted as Ordovician and initial datings with acritarchs provided evidence for a definitive Ordovician age of some parts of the succession (summary in Burmann *et al.* 2001). The stratigraphically best known area is the Wippraer Zone, east of the Harz Mountains, where some acritarchs allowed a clear attribution to the Ordovician (Burmann *et al.* 2001). Within the Wippra Gruppe, the Rammelburg-Phyllit-Quarzit Formation provided acritarchs attributed to the Arenig–Llanvirn interval (Burmann 1973*a*, *b*; Sehnert 1991) that can now be interpreted as latest Arenig to Llanvirn in age. Further east, near Paschleben, the Edderitz-Köthen-Tonschiefer Complex remains mostly undated, except for some beds within the Plömnitz-Tonschiefer Formation that include acritarch assemblages similar to those from the Rammelburg-Phyllit-Quarzit Formation (Burmann 1973*b*). A further acritarch assemblage, that is most probably Arenig in age, has been recently been described from the Natho Formation of the Pakendorf Group (discussion in Burmann *et al.* 2001).

Saxothuringia, Germany

The Ordovician of Saxothuringia (Figs 5.3, 15–20 & 5.9) is by far the most extensively studied succession of Germany, with the stratigraphical sections of the Schwarzburg Anticline having served as the standard (German 'Richtprofil') for many years.

A historical overview of the development of the litho- and biostratigraphy of Germany is given in the Stratigraphische Kommission Deutschlands (1997, 2001*a*, *b*). Selected stratigraphic schemes for the German Ordovician were compiled in the Stratigraphic Table of Germany (Hoth *et al.* 2002; Heuse & Maletz 2005). Large areas of Saxothuringia consist of metamorphic complexes without, or with only non-significant, index fossil records. These mainly tectonostratigraphic units have also been revised in the above-mentioned compilations.

As noted above, the Rhenohercynian areas and the Rügen Ordovician are considered to belong to the microcontinent of Avalonia. The Saxothuringian and Moldanubian zones show a comparable tectonostratigraphic history during the Early Palaeozoic with significant affinities in provenance and tectonic setting to Armorica. Avalonia was apparently separated from Armorica, and the Saxothuringian area probably drifted off from Gondwana later (Tait *et al.* 1997; Torsvik 1998). It is still a matter of discussion whether Saxothuringia was a separate part of the Armorican Terrane Assemblage during the early Palaeozoic (e.g. Tait *et al.* 1997; Schätz *et al.* 2002) or if they belonged to the north Gondwanan shelf throughout the Early Palaeozoic (Carls 2001; Linnemann *et al.* 2004). During the Late Ordovician, Saxothuringia was still located at high latitudes and was affected

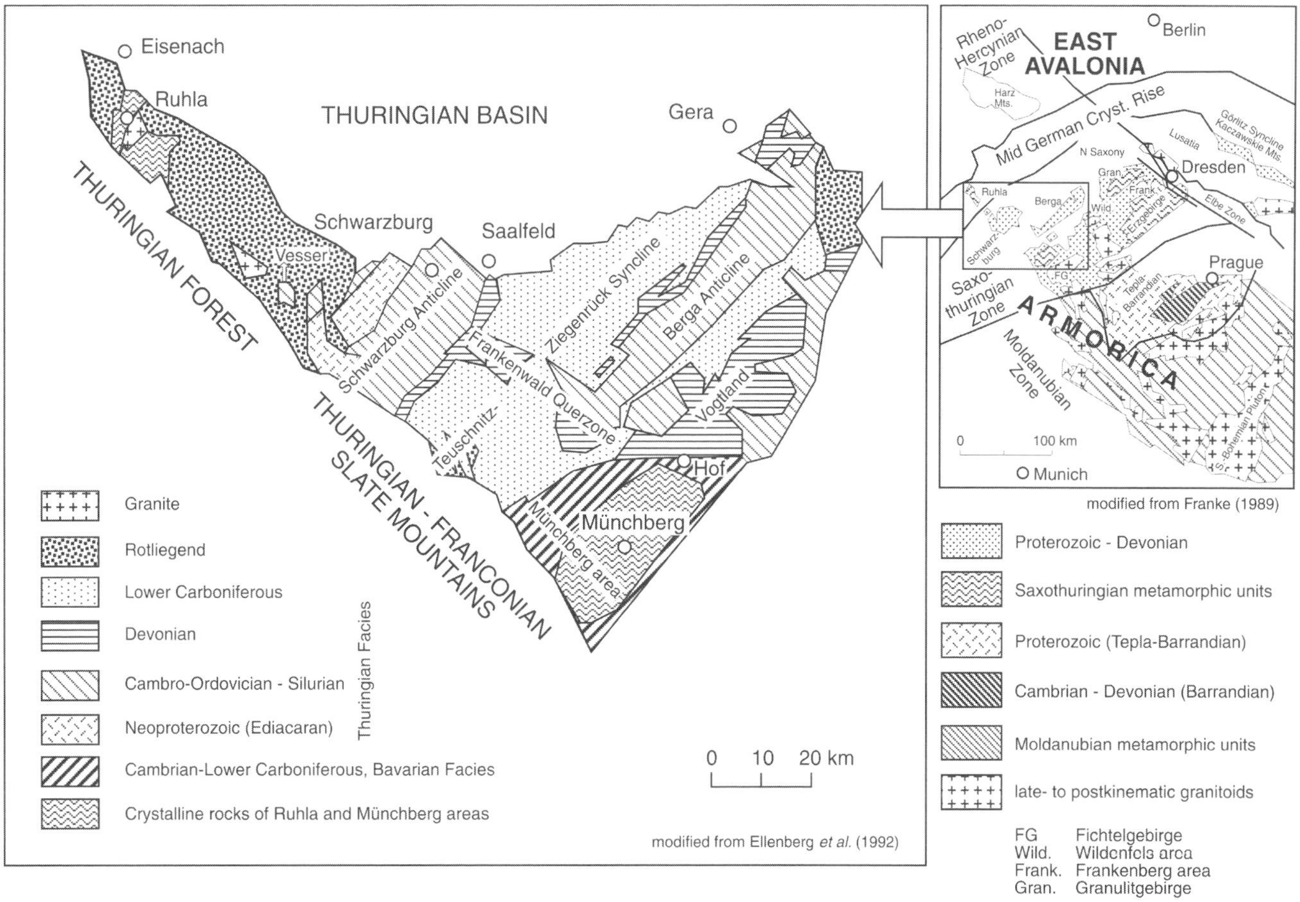

Fig. 5.9. Basement map of Saxothuringia (after Heuse 2000).

by the Hirnantian glaciation (e.g. diamictite in the Lederschiefer Fm). Schätz *et al.* (2002) published palaeomagnetic data from a 478 ± 1.8 Ma pyroclastic flow from Franconia which suggested a position of 63°S (+21.7°, −17.3°) and a weaker signal from the Upper Ordovician Döbra sandstone which gave a palaeolatitude of 38°S (±18.5°).

The Ordovician sedimentary succession on the Cadomian basement of the Saxothuringian Zone is characterized by a passive margin signature (e.g. Linnemann *et al.* 2000; Linnemann & Romer 2002). According to SHRIMP geochronology on detrital and inherited zircon grains as well as Nd-isotopic data, the sediments were derived from a West African craton (Linnemann *et al.* 2004), indicating a unique source area for the sediment supply to the Saxothuringian basin throughout the Ordovician.

High rates of siliciclastic sedimentation existed in the Early Ordovician (Lützner *et al.* 1986; Ellenberg *et al.* 1992), and were interpreted as indicating a rift setting corroborated by Cambro-Ordovician bimodal magmatism in Saxothuringia (e.g. Behr *et al.* 1984; Franke 1989; Erdtmann 1991; Bankwitz & Bankwitz 2000, 2002; Linnemann & Heuse 2000). Acid to basic magmatites and pyroclastic rocks were reported from anchi- to high grade metamorphic regions of Vesser (Bankwitz *et al.* 1994; Kemnitz *et al.* 1999, 2002), the Schwarzburg Anticline (Gehmlich *et al.* 1997; Gehmlich 2003; Linnemann *et al.* 2000, 2004), the Vogtland (e.g. Werner *et al.* 2005), the Franconian Forest (e.g. Schätz *et al.* 2002), the Münchberg Massif (e.g. Müller-Sohnius *et al.* 1987; Gebauer 1993), the Fichtelgebirge (e.g. Mielke *et al.* 1996; Siebel *et al.* 1997; Rohrmüller *et al.* 2000), the Frankenberg area (Rötzler *et al.* 1999), the Erzgebirge (Kröner *et al.* 1996; Krenz *et al.* 1996; Mingram 1998; Tichomirowa *et al.* 2001; Tichomirowa 2002), the Granulitgebirge (e.g. Kröner *et al.* 1998; Romer & Rötzler 2001), the Elbe Zone (Gehmlich *et al.* 1997; Linnemann *et al.* 2000; Mohnicke *et al.* 2001; Gehmlich 2003) and Lusatia (Hammer *et al.* 1999).

Mingram (1996, 1998) and Mingram & Rötzler (1999) published geochemical data from the metamorphic Erzgebirge, indicating a nappe pile of at least five metamorphic units. Geochemical signatures show significant correspondence, independent of high to low grade metamorphic overprinting. The data allow a correlation of the protoliths of these metamorphic units with the anchimetamorphosed sedimentary sequences of the Early Palaeozoic in the Schwarzburg Anticline.

Additionally, Meinhold *et al.* (2005) reported geochemical and palynological data from low grade metamorphic phyllites of the Greiz area (Vogtland). A correlation of the protolith with the Early Ordovician Goldisthal, Frauenbach and *Phycodes* units was inferred.

Palynological data, as well as discoveries and taxonomic revisions of graptolites, brachiopods and conodonts, have led to the construction of a more detailed biostratigraphic standard scale for Saxothuringia. Palaeogeographically, the graptolite, brachiopod and conodont faunas as well as the palynomorphs indicate a position of the Saxothuringian Zone in high to mid-latitudes during the entire Ordovician (e.g. Heuse 2000 and references therein). A review of biostratigraphically significant macro- and microfossils is given by Heuse (2000) and updated herein.

Two different sedimentary and magmatic regimes in the Palaeozoic (Ordovician–Devonian) of Saxothuringia are recognizable and are attributed to different facies (e.g. Wurm 1925; Behr *et al.* 1984; Franke 1984). In the Ordovician, the autochthonous Thuringian Facies has been interpreted as a widespread homogenous basin, with dominant siliciclastic sedimentation, whereas the allochthonous Bavarian Facies has been regarded as more differentiated, comprising siliciclastics as well as magmatic rocks. The Thuringian Facies is established in the Schwarzburg and Berga anticlines, including parts of the Franconian Forest, the Vogtland and northern Saxony. The Bavarian Facies is typically developed in the areas of Münchberg (Franconia), Wildenfels and Frankenberg as well as in the Elbe Zone (Saxony). Because it is still a point of discussion whether or not the different facies settings can be recognized east of the Elbe river, the facies affiliation of Lusatia remains doubtful.

Local palynofacies differences can also be observed between these two 'facies' areas of the Saxothuringian Zone. Palynologically, a 'Thuringian development' may be attributed to a more proximal depositional setting to the west (Schwarzburg Anticline) and deepening to the SE (Vogtland). In contrast, the Bavarian Facies perhaps reflects deeper water conditions (e.g. Heuse 2000).

The Schwarzburg Anticline (Figs 5.3, 15 & 5.9) constitutes the type area of the Ordovician of Saxothuringia. Siliciclastic sedimentation commenced during the late Tremadocian and continued up into the Ashgill Lederschiefer Fm. The succession also includes intervals of very condensed sedimentation or non-deposition (e.g. Tierberg and Schmiedefeld ironstone beds). The neighbouring Berga Anticline, Vogtland and Frankenwald 'Querzone' (NE Franconia) contain only sparse index fossils (Fig. 5.9). Generally, these areas are lithostratigraphically correlated with the Schwarzburg Anticline. The fossiliferous sections are discussed separately below.

Figure 5.10 is a stratigraphical correlation table for the different units of the Schwarzburg Anticline, the type area of the German Ordovician, and the adjacent sucessions with the British and Baltoscandian graptolite and conodont standard scales. The graptolite zonation for Germany has recently been revised by Maletz (Heuse & Maletz 2005).

Schwarzburg Anticline (Thuringia)

The stratigraphical succession of the Schwarzburg Anticline (Schwarzburg Antiklinorium) (Figs 5.3, 15, 5.9 & 5.11) is the most complete Ordovician succession in Germany. A number of formations have been described since the last century (e.g. Gaertner 1934, 1944, 1964; summarized in Bankwitz *et al.* 1997). These are briefly discussed below, from the base to the top of the Ordovician.

The Ordovician sequences consist of more than 3500 m of mainly siliciclastic sediments with minor intercalations of acid and basic magmatites, and tuffs and tuffites in the Early Ordovician. Chamositic ironstone horizons in the Middle and Late Ordovician correspond well with coeval occurrences in the peri-Gondwanan realm (e.g. Young 1992; Petránek & van Houten 1997). The rift-controlled sedimentation of *c.* 3000 m of late Tremadoc Goldisthal, Frauenbach and *Phycodes* units is accompanied by bimodal magmatism. The Middle to Late Ordovician Gräfenthal Group is characterized by condensed sedimentation (Lützner *et al.* 1986). Linnemann & Heuse (2000) and Winsemann & Jonen (2000) published a sequence stratigraphical study of the Ordovician of the Schwarzburg Anticline, and correlated third-order sequences with global sequences reported by Ross & Ross (1992). However, the correlation of the Saxothuringian sequences remains problematic, because the biostratigraphic zonation of most units is deficient.

Goldisthal Formation

The Goldisthal Formation, *c.* 300 m thick, consists of shales, minor intercalations of acid tuffs and tuffites. Falk (1964), Falk & Lützner (1991) and Falk *et al.* (2000) described sedimentological, bathymetrical, geochemical and stratigraphical aspects of the succession in detail. The formation possibly spans the Cambro-Ordovician transition. A poorly preserved Upper Cambrian to Lower Ordovician acritarch association from the NW flank of the Schwarzburg Anticline was described by Heuse in Estrada *et al.* (1994). The acritarchs are identifiable only to genus level due to subsequent thermometamorphic alteration by Variscan-age granites. Single zircon evaporation data (Pb/Pb) from rhyolitic pebbles from the base of the Goldisthal Fm (493 ± 5 Ma; Gehmlich *et al.* 1997) and from tuffs from the upper part (488 ± 15 Ma; M. Gehmlich in Falk *et al.* 2000) confirm the Tremadocian age.

Frauenbach Group

The Frauenbach Group, *c.* 650 m thick, consists of an alternation of quartzitic sandstones and shales with minor intercalations of subvolcanic porphyroid magmatites (Lützner *et al.* 1986; Falk & Lützner 1991; Falk *et al.* 2000). The group is subdivided into three formations: Lower Frauenbachquarzit Fm, the Frauenbach-Wechsellagerung Fm and Upper Frauenbachquarzit Fm. Single zircon data (Pb/Pb) from the Bärentiegel-Porphyroid (Upper Frauenbachquarzit Fm, 479 ± 5 Ma; Gehmlich *et al.* 1997) confirm a late Tremadocian age. Freyer & Wiefel (1991) figured a poorly preserved obolid from the Frauenbach-Wechsellagerung Fm as *Leptembolon* sp. Re-examination of the specimen revealed

Fig. 5.10. Correlation of the Ordovician of Saxothuringia with the British and Baltoscandian standard schemes (modified from Webby *et al.* 2004). The graptolite zonation for Germany was made by J. Maletz (Heuse & Maletz 2005), and is based on records from various regions within Germany and does not represent a purely Saxothuringian succession. Sources: (1) Heuse in: Estrada *et al.* (1994); (2) Loretz (1880), Koliha (1926), Gaertner (1944), Falk (1956), Müller (1956), I. Puura (unpubl.), Heuse & Puura (2000); (3) Volk (1938, 1966), Schmidt (1939), Heuse *et al.* (1994); (4) Dahlgrün (1930), Heuse (1989), Heuse *et al.* (1994), Marek *et al.* (2003); (5) Knüpfer (1967), Fuchs (1989, 1990); (6) Blumenstengel & Schallreuter (1998), Schallreuter (2005); (7) Knüpfer (1967, 1968), Fuchs (1989, 1990), Dzik (1990), Ferretti & Barnes (1997), Schallreuter & Hinz-Schallreuter (1998), Schallreuter (2005); (8) Reitz & Heuse (1994); (9) Burmann (1968, 1970, 1973c, 1976), Burmann *et al.* (1994), Walter & Berger (1998); (10) Burmann & Waskowiak (1968), Burmann (1976); (11) Heuse (1994), Reitz & Heuse (1994); (12) Sdzuy (1970), Peng (1990); (13) Barrande (1868), Pompeckj (1896), Kolíha (1937), Sdzuy (1955*a*, *b*, 1961, 1964, 1971, 1979), Peng (1990), Reitz (1991); (14) Sdzuy (1964, 1971); Sdzuy *et al.* (2001); (15) Gümbel (1864), Wurm (1961); (16) Sannemann (1953, 1955), Rein (1986); (17) Pietzsch (1910), Kolíha (1926, 1937), Bartsch (1956), Seilacher (1970), Freyer (1981), Fortey & Seilacher (1997), I. Puura (unpublished data), Heuse & Puura (2000); (18) Geinitz (1873), Schwarzbach (1934, 1936), Kolíha (1937), Lee (1938), Pietzsch (1962), Freyer (1967), Freyer & Wiefel (1991), I. Puura (unpublished data), Heuse & Puura (2000); (19) Burmann (1965, quoted in Brause *et al.* 1997), Paalits & Erdtmann (1993); (20) Burmann (2001).

Intern. Stages	British Series	British Zones	Baltoscandian Zones		German Zones	Saxothuringia				
		graptolites/trilobites	graptolites/trilobites	conodonts	graptolites	Schwarzburg Anticline	Berga Anticline Vogtland	Frankenberg/ Wildenfels areas	Münchberg area	Lusatia N Saxonia
HIRN.	ASHGILL	N. persculptus	N. persculptus	?	N. persculptus	Lederschiefer Fm. acrit., cryptospores (8)	Lederschiefer Fm.	Silurberg Fm. acritarchs (11)	Döbra Fm.	Eichberg Fm. acritarchs (19)
		N. extraordinarius	N. extraordinarius	A. ordovicicus	N. extraordinarius	Hangende Kalkbank conodonts, trilobites, ostracodes (7)				
KATIAN		D. anceps	?		no graptolites					
	CARADOC	D. complanatus	D. complanatus							
		P. linearis	P. linearis	A. superbus		Oberes Lager ostracodes (6)				
		D. clingani	D. clingani							
SANDB.		D. foliaceus	D. foliaceus	A. tvaerensis	D. multidens	Lagerquarzit	Hauptquarzit Fm.	Hauptquarzit Fm.		
		N. gracilis	N. gracilis	P. anserinus	N. gracilis	Schmiedefeld Fm.			Gösmes Formation (Randschiefer)	
DARRIWILIAN	LLANVIRN	H. teretiusculus	H. teretiusculus	P. serra	H. teretiusculus	Gräfenthal Group				
		D. murchisoni	D. murchisoni	E. suecicus	P. distichus Pterogr. elegans				conodonts (16)	
		D. artus	D. artus	E. variabilis	N. fasciculatus H. lentus	Unteres Lager conodonts (5)	chitinozoans (20)			Klitten Fm. acritarchs (19)
3	ARENIG	E. hirundo	E. hirundo	B. norrlandicus P. originalis B. navis/triangularis	E. hirundo	Griffelschiefer Fm. grapt., trilob., acrit., chitinozoans (4)	Griffelschiefer Fm., acritarchs, chitinozoans (9, 20)	Griffelschiefer Fm., acritarchs, (10)	Griffelschiefer Fm., acrit., chitinoz. (20)	
FLOIAN		I. gibberulus								
		D. simulans	P. elongatus P. densus	O. evae	P. densus	Tierberg Beds			Leuchtholz Beds brachiop. (15)	
		D. varicosus ?	D. balticus	P. elegans	D. balticus					
		P. phyllograptoides	P. phyllograptoides		P. phyllograptoides	? Phycodes Group graptolites, acritarchs (3), brachiopods (2)	Phycodes Group	Phycodes Group	Vogtendorf Fm. brachiop., trilob. (14)	
TREMADOCIAN	TREMADOC	A. murrayi	H. copiosus A. murrayi	P. proteus	A. murrayi					
		no graptolite zone defined A. sedgwickii C. salopiensis	(Ceratopyge Limestone) (Ceratopyge Shale)	P. deltifer	no graptolites	Frauenbach Group Goldisthal Fm. acritarchs (1)	Weißelster Group		?	Dubrau Fm. brachiopods (18)
		A. tenellus	B. kjerulfi A. tenellus	C. angulatus		?		Kühberg unit trilob. (12)	Leimitz Fm. trilobites, acrit. brachiop. (13)	?
		R. flabelliformis	R. flabelliformis R. parabola	I. fluctivagus	R. flabelliformis					
			Acerocare	C. lindstromi C. intermedius C. proavus	no graptolites					Collmberg Fm. trace fossils, brachiop. (17)

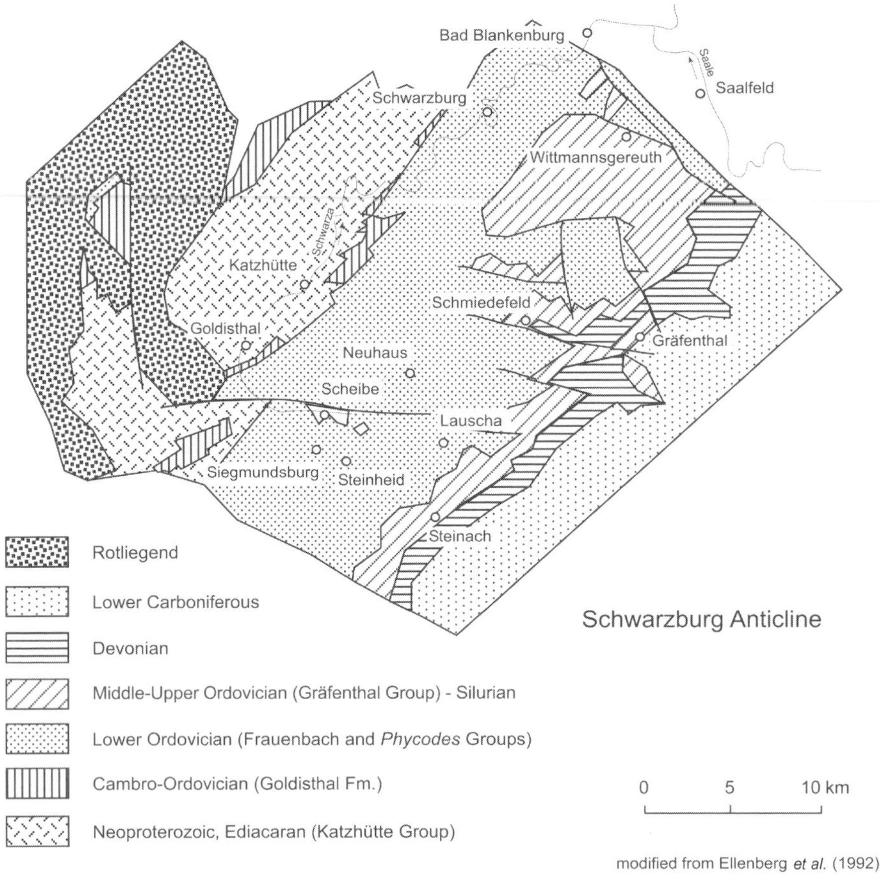

Fig. 5.11. Map of the outcrop areas of the Schwarzburg Anticline (after Heuse 2000).

a lack of distinct features allowing any conclusive generic assignment and thus the report of the occurrence of *Leptembolon* in Germany should be disregarded (I. Puura unpublished data; Heuse & Puura 2000).

Phycodes Group

The *Phycodes* Group consists of *c.* 3000 m of uniform siliciclastic rocks (sandstones, siltstones), lithostratigraphically subdivided into three formations: the Phycodendachschiefer Fm the Phycodenschiefer Fm, the Phycodenquarzit Fm and additional subunits (Gaertner 1934, 1944; Lützner *et al.* 1986). Sedimentological features and aspects of basin development were reported by Lützner *et al.* (1986), Ellenberg *et al.* (1992) and Mann (1996). Loretz (1880), Kolíha (1926), Gaertner (1944), Falk (1956) and Müller (1956) described 'inarticulate' brachiopods from sandy siltstones near Siegmundsburg. The locality was considered to belong lithostratigraphically to the top of the underlying Frauenbach Group. New data based on mapping indicate that the strata correlate with the base of the *Phycodes* Group (Biewald & Voigt 2004). The brachiopod specimens were identified by Puura (in Heuse & Puura 2000) as *Hyperobolus feistmanteli*. In the Prague Basin, this species occurs in the upper part of the Třenice Formation (Havlíček 1982; Havlíček & Fatka 1992), perhaps stratigraphically corresponding to the *Ceratopyge* Shale or *Ceratopyge* Limestone of Baltoscandia.

Acritarch assemblages and a taxonomic revision of the graptolite *Dictyonema* cf. *uralensis* (Volk 1966), now (?) *Araneograptus* cf. *murrayi* (Heuse *et al.* 1994), correlate the entire Phycodenschiefer Fm (lower part of the Phycodes Group) with the late Tremadocian *A. murrayi–H. copiosus* graptolite Zone or the base of the *P. proteus* conodont Zone. Whether or not the overlying Phycodenquarzit Fm extends into the lower Arenig as stated by Volk (1938) and Schmidt (1939) remains doubtful because of the very poor preservation of the graptolites (Heuse *et al.* 1994). Trace fossils (e.g. the name-giving *Phycodes circinatum*) occur frequently (revised by Lützner & Mann 1988; Mann 1996).

Gräfenthal Group

This group is subdivided into several formations. These are described below from base to top.

The Griffelschiefer Formation consists of 80–140 m of dark to black shales with only rare intercalations of siltstones. The shale was mined until the 1960s for production of pencil slate (Griffelschiefer). Locally, a clastic ironstone horizon up to 2 m thick (Tierberg Bed) is developed at the base of the formation (Hetzer 1958; Lützner *et al.* 1986; Ellenberg *et al.* 1992).

The trilobite fauna so far reported from different localities in the Griffelschiefer Fm was revised by Marek *et al.* (2003). From specimens of *Expansograptus* cf. *hirundo* and organic-walled microfossils from the Großer Tierberg hill (NE Steinach), the Griffelschiefer Fm can be correlated with the upper Arenig

E. hirundo Zone (Heuse *et al.* 1994). The stratigraphic position of the lower boundary of the Griffelschiefer Fm is uncertain, as is the age of the underlying Tierberg Bed (Tierberg Erzhorizont), which probably includes the entire *D. extensus* Zone (either due to condensed sedimentation or to non-deposition). The acritarch assemblages recovered also correspond to this age interval (e.g. Heuse 1989). Lithostratigraphically the Tierberg Bed may correspond to the middle Arenig ironstone horizons, and the superadjacent lower bed of the Schmiedefeld Erzhorizont may thus be correlated with the lower Llanvirn ironstones, which are very widespread in the peri-Gondwana shelf region (e.g. Young 1992).

The Schmiedefeld Formation consists of two 8–20 m thick clastic chamositic ironstone horizons (Lower and Upper Ore Beds) with an intercalated sandstone (Lagerquarzit, 0–30 m). They are overlain by a limestone layer up to 1 m thick (Kalkbank). Despite more than 60 years of exploitation, the genetic aspects and the origin of the oolitic and detrital iron ores of Thuringia remain obscure. Deposition in a shallow marine lagoonal environment was favoured by Hetzer (1958) and Reh (1964), whereas Fuchs (1990) postulated deeper shelf conditions with reworking processes, summarized by Reh (1964) and lastly by Trappe & Ellenberg (1994) and Ellenberg (2000). The latter assumed a sequence stratigraphical approach for the formation of the iron ores and genetically associated phosphoritic layers at the base and top, superimposed by reworking, concentration and winnowing processes. Such interlinking with a transgression stage was mentioned in the 1940s, as summarized by Petránek (1964). In contrast Linnemann & Heuse (2000) correlated the deposition of the iron ores with a third-order lowstand systems tract. Due to the large time interval, including the entire Llanvirn and Caradoc (at least about 20 Ma), up to four or more third-order sequences might be included (but not preserved) in the 50 m thickness of the Schmiedefeld Fm. In general, a terrestrial hinterland as source area for the iron ores is supposed and also a magmatic component has been discussed, but the genetic and transport processes are poorly understood.

Comprehensive descriptions of the conodont, trilobite and ostracode faunas of the Schmiedefeld Fm from samples from the former mines at Gebersdorf and Schmiedefeld as well as from boreholes were published by Knüpfer (1967, 1968). The lower bed of the Schmiedefeld Erzhorizont was correlated with the former Llanvirn/Llandeilo boundary and the Hangende Kalkbank (top of the upper bed of the Schmiedefeld Erzhorizont) with the Caradoc/Ashgill transition. Additional recordings in the mine at Wittmannsgereuth and revisions of Knüpfer's conodont assemblages by Fuchs (1989, 1990) confirmed the stratigraphic position of the Hangende Kalkbank as uppermost Caradoc/lowest Ashgill (*Amorphognathus superbus* Zone) or probably the succeeding *A. ordovicicus* Zone (Dzik 1990). The lower bed of the Schmiedefeld Erzhorizont was dated as the *Eoplacognathus variabilis* Zone (Arenig/Llanvirn transition). Investigations of the conodont fauna from a Kalkbank sample near Schmiedefeld by Ferretti & Barnes (1997) revised its stratigraphical position to the *A. ordovicicus* Zone (Ashgill). Consequently, the Schmiedefeld Fm currently includes the Llanvirn, Caradoc and lower Ashgill - in terms of the revised British standard series - as a result of condensed sedimentation or depositional gaps. From drillings NE of Rudolstadt Blumenstengel & Schallreuter (1998) reported ostracode faunas from the upper bed of the Schmiedefeld Erzhorizont.

This unique limestone layer (Kalkbank) within the siliciclastic Ordovician succession of Saxothuringia correlates well with the early Ashgill (pre-Hirnantian) carbonate successions in central and western Europe, considered to be indicative of a pre-glacial warming event (Boda Event; Fortey & Cocks 2005).

The Lederschiefer Formation is about 250 m thick and is a glaciomarine diamictite (Katzung 1961, 1999; Steiner & Falk 1981). Its shaly matrix contains dropstones and striated pebbles. The pebble content is composed of quartzites, silicified carbonates, shales and oolites (Katzung 1961, 1999).

The fauna obtained from pebbles of the diamictite yielded echinoderm remains, brachiopods, trilobites, corals, bryozoans and others (compiled by Schallreuter 2005). The trilobites probably indicate a Caradoc age (Struve 1959, 1962). A limestone pebble derived from mine tailings near Saalfeld, Thuringia, included ostracodes of Caradoc age, described by Blumenstengel (1965) and revised by Schallreuter & Hinz-Schallreuter (1998) and Schallreuter (2005). Although most of the taxa are endemic, palaeogeographical relations to Baltica, but also to regions in lower latitudes were discussed. According to Schallreuter *et al.* (2006), the ostracodes from the limestone pebbles as well as from the Upper Ore Bed and the Kalkbank of the underlying Schmiedefeld Formation have species in common with the central Iran Anarak fauna. An acritarch and cryptospore association has been reported by Reitz & Heuse (1994). The palynomorphs from the Gräfenthal Horst (north Franconia) confirm a Late Ordovician age (Ashgill). Weyer (1967) reported the warm-water coral *Tetradium* sp. from a quarzitic pebble of the Lederschiefer, attributed to the upper Caradoc, but its organic origin is currently under discussion (D. Weyer pers. comm. 2004).

Typical elements of the Hirnantian fauna are missing, and the age of the formation is constrained only by the under- and overlying units (early Ashgill Kalkbank/early Llandovery Lower Graptolithenschiefer Fm); however, the deposition of the Lederschiefer diamictite corresponds well with the Late Ordovician glaciation event.

Berga Anticline, Vogtland

In the Berga Anticline and the Vogtland (Figs 5.3, 16, 17 & 5.9) several formations have been described that can be correlated with the standard succession of the Schwarzburg Anticline (summarized in Geyer & Wiefel 1997). The Weißelster Group was lithologically correlated with the Frauenbach Group. The Phycodes Group, which is also found in the Vogtland, is overlain by the Griffelschiefer Fm. Burmann (1968, 1970, 1976) and Burmann *et al.* (1994) described chitinozoans and a highly diversified acritarch association from the Griffelschiefer Fm of the locality Unterhermsgrün. No stratigraphically significant graptolites have been detected thus far, but the palynomorphs are comparable with the association from the upper Arenig *E. hirundo* Zone of the Schwarzburg Anticline. Walter & Berger (1998) described a poorly preserved acritarch assemblage from the northwestern margin of the Erzgebirge from the Zwickau borehole (Saxony). According to the fossil content the sediment is considered to be a time equivalent of the Griffelschiefer Fm.

Between the Griffelschiefer Fm and the Lederschiefer Fm, a further quartzitic unit up to 150 m thick, the Hauptquarzit Fm, occurs. Ellenberg (1998) interpreted the sedimentological features as delta fan deposition. Burmann (2001) mentioned findings of *Dicranochitina* sp. (= ?*Sagenachitina* sp.) in the Hauptquarzit from near Gräfenwarth and correlated the sequence with the Bohemian Šárka Formation (early Llanvirn). The age assignment of the formation is based on this single taxon. However, the occurrence of the genus *Dicranochitina* Burmann 1973c (= ?*Sagenachitina*) is regionally restricted to late Arenig/early Llanvirn strata of Saxothuringia and Bohemia (Burmann *et al.* 1994) so that the upper range of the taxon remains unknown. Accordingly, a younger age for the formation is possible.

Wildenfels and Frankenberg areas, Elbe Zone

Shaly successions with alternating siltstones and sandstones and intercalations of mostly acid to intermediate magmatites show similarities to Thuringian Facies as well as Bavarian Facies sequences, and occur further east in the Wildenfels (Fig. 5.3, 17) and Frankenberg (Fig. 5.3, 18) areas, and in the Elbe Zone (Fig. 5.9; e.g. Pietzsch 1962; Kurze 1966, 1968, 1993; Schwan 1974, 2002; Kupetz 1987, 2000; Kurze & Tröger 1990; Kurze *et al.* 1992; Schreiber 1992; Linnemann 1994; Linnemann & Schauer 1999). The competitive scenarios concerning the tectonostratigraphic character of the anchi- to low-grade metamorphic successions (e.g. authochthonous versus allochthonous models) are not discussed herein.

Biostratigraphical data are also rare in these areas. Burmann & Waskowiak (1968) and Burmann (1976) described an acritarch flora from the Griffelschiefer Fm of the Loh valley corresponding to the Griffelschiefer Fm associations from the Vogtland and the Schwarzburg Anticline. The holotype of a single graptolite, *Didymograptus saxonicus*, from the same locality has subsequently disappeared. In the Frankenberg area, slates from the Silurberg Fm near Obermühlbach yielded an Upper Ordovician (Caradoc/Ashgill) acritarch association (Heuse 1994; Reitz & Heuse 1994).

Northern Saxony and Lusatia

Berger & Docekal (1997) summarized the stratigraphical and structural aspects of the Nordsächsisches Schiefergebirge (northern Saxony; Fig. 5.9), which is barren of index fossils. Several authors have mentioned brachiopods from silty sandstones of the area of Otterwich-Hainichen (northern Saxony) since the nineteenth century (summarized in Bartzsch 1956). However, the available specimens are either no longer available or too poorly preserved for any specific or generic assignment and cannot be used for any biostratigraphic conclusions (I. Puura unpublished data; Heuse & Puura 2000). Pietzsch (1910) described the trace fossil *Cruziana furcifera* (now *C. semiplicata*) from the Collmberg Fm. According to Seilacher (1970), Freyer (1981) and Fortey & Seilacher (1997), *C. semiplicata* is stratigraphically restricted to the Upper Cambrian. Fortey & Seilacher (1997) attributed *C. semiplicata* to the Upper Cambrian trilobite genus *Maladioidella*. Nevertheless, the authors mentioned that *C. semiplicata* occurs also in Tremadocian sediments and, thus, the trace fossil may have been produced by different animals.

In the Lausitz area (= Lusatia; Figs 5.3, 19 & 5.9), close to the German–Polish border, a series of formations have been described (summarized in Brause *et al.* 1997). The biostratigraphical information remains limited, and only approximate ages can be provided for this succession (Fig. 5.10).

Dubrau Formation

The Dubrau Formation consists of more than 100 m of conglomerates, quartzites and silty shales. Initial records of 'inarticulate' brachiopods by Geinitz (1873) were considered as unidentifiable by Kolíha (quoted in Schwarzbach 1934). 'Inarticulate' brachiopods from the Dubrau Quartzite near Groß-Radisch have been identified by I. Puura (pers. comm. 1998, 2003) as *Libecoviella arachne*. This species has also been described from the Bohemian Třenice Fm. Early reports of this species from the Dubrau quartzite by Schwarzbach (1934, 1936), Kolíha (1937), Lee (1938) and Pietzsch (1962) are corroborated by the revision of I. Puura (pers. comm., 1998, 2003). In Bohemia, *L. arachne* occurs in the upper part of the Třenice Fm perhaps corresponding with the '*Ceratopyge* Shale' or '*Ceratopyge* Limestone' of Baltoscandia. Consequently, the lingulate coquina in the Dubrau Quartzite and in the base of the *Phycodes* Group are considered to be coeval and can be correlated with the upper part of the Třenice Fm in the Prague Basin too, as already suggested by Kolíha (1937), who correlated the Dubrau Quartzite with the Libecov Horizon.

Klitten Formation

Silty shales of the subsurface Klitten Formation were penetrated by several borholes. The 454/71 well yielded an acritarch association from a horizon below the Eichberg Fm, stratigraphically assigned to the Arenig and/or Llanvirn (Paalits & Erdtmann 1993). Burmann (1970) reported the acritarch *Lusatia dendroidea* from well Reichwalde (upper Cambrian?/Tremadocian).

Eichberg Formation

The subsurface Eichberg Formation consists of sandstones with intercalated silty shales. Paalits & Erdtmann (1993) described an Upper Ordovician acritarch association (Caradoc/Ashgill) from well 454/71, comparable with the palynoflora from the Frankenberg area (Heuse 1994).

Franconia (Münchberg area, Fichtelgebirge)

Ordovician rocks are found at several locations in Franconia, NE Bavaria: in the Franconian Forest (Frankenwald) and in the Fichtelgebirge (Figs 5.3, 20 & 5.9). They are exposed as olistoliths in a number of isolated klippen around the Münchberg Massif (= Münchberger Gneis-Masse, assigned to the Thuringian as well as to the Bavarian Facies), and in the Fichtelgebirge (Thuringian Facies; summarized in Geyer & Wiefel 1997; Stettner 1997; Gandl 1998). Today, the Bavarian Facies of the Frankenwald is interpreted as a Variscan nappe thrust onto the autochthonous Thuringian Facies. As with the successions discussed above, many parts of the sequences in Franconia remain undated and fossils are extremely rare. From base to top, the following formations have been described.

Kühberg unit

An isolated block of shales occurs near Stadtsteinach (west of the Münchberg Massif). Sdzuy (1970) correlated the Kühberg schists lithostratigraphically with the Frauenbach Group of Thuringia (upper Tremadocian) whereas he compared the trilobite fauna to the Leimitz Fm (lower Tremadocian).

Leimitz Formation

Isolated occurrences of the Leimitz Formation (*c.* 100 m thick) are composed of shaly siltstones with minor intercalations of sandstones (e.g. Horstig 1957). The fauna of the formation was first described by Barrande (1868) and Pompeckj (1896). The trilobites, brachiopods and cystoids were revised by Sdzuy (1955*a*, *b*, 1961, 1964, 1971, 1979) and correlated with the lower Tremadocian. Peng (1990) corroborated an early Tremadocian age for the trilobite fauna of the Leimitz Fm. Reitz (1991) reported an acritarch association from slates near Hof. This fossiliferous horizon is barren of macrofossils and considered to be a lithological equivalent of the Leimitz Fm.

Vogtendorf Formation

The Vogtendorf Formation consists mainly of intermediate to basic tuffs and tuffites with minor intercalations of shales and siltstones. The brachiopod and trilobite faunas of the Vogtendorf Beds are probably late Tremadocian in age (Sdzuy 1964, 1971; Sdzuy *et al.* 2001).

Gösmes Formation (former 'Randschiefer')
The Gösmes Formation consists of several hundred metres of shales, sandstones, and intermediate to basic magmatites, tuffs and tuffites (e.g. Wurm 1925, 1961; Horstig & Stettner 1970, 1976; Sdzuy 1971). Sannemann (1953, 1955) documented conodont faunas from three different localities of the Randschiefer Fm near Helmbrechts (NW part of the Münchberg area) that he assigned to the Arenig, the Arenig–Caradoc and Llanvirn–Caradoc intervals. Additional conodont findings by Rein (1986) were considered to be indicative of probably Arenig age. Ignimbrites from the Vogtendorf Fm have been dated with an age of 478.2 ± 1.8 Ma by single zircon analysis (Schätz *et al.* 2002) corresponding to an Arenig age. The Lower Erzhorizont near Leuchtholz (Frankenwald Querzone) yielded a brachiopod fauna first described by Gümbel (1864) and now assigned to the Arenig (K. Sdzuy unpublished data, quoted in Wurm 1961).

Döbra Formation
The succession consists of about 40 m of fine-grained sandstones with shaly siltstones and black shales towards the top of the formation (Greiling 1957). Poorly preserved conodonts, brachiopods and graptolites are not indicative of a precise age assignment (e.g. Sannemann 1955; Greiling 1961, 1966). However, the stratigraphic position is largely determined by the ages of the underlying and overlying units.

Czech Republic

The Bohemian Massif includes three areas with Ordovician rocks, the ages of which are based on palaeontological data. All these areas are situated within the Teplá-Barrandian area (= Bohemicum) (Figs 5.3, 21, 22 & 5.12). Slightly metamorphosed Ordovician sediments are presumed in several other units of the Bohemian Massif, e.g. in the Islet Zone of the Central Bohemian Pluton (Havlíček *et al.* 1958). There are no palaeontological data to prove an Ordovician age for these rocks. The onset of the Ordovician was characterized by an extensive transgression which can be traced across part of the Bohemian Massif. However, the most complete and important sections are concentrated in the Teplá-Barrandian area.

The Teplá-Barrandian area, located in the central part of the Bohemian Massif, contains unmetamorphosed to slightly metamorphosed Proterozoic and Lower Palaeozoic rocks (Cambrian to Devonian). Ordovician sedimentary rocks represent the older portion of the third tectonosedimentary cycle, being preserved in two separate areas: the Prague Basin (Figs 5.3, 21 & 5.12) and the Železné Hory area (Figs 5.3, 22 & 5.12).

The Ordovician, Silurian and Devonian sediments were considered by Havlíček (1981) to have been deposited in the Prague Basin, which is a tectonically controlled, longitudinal depression striking SW–NE between Praha and Plzeň (Fig. 5.12). The successions are predominantly clastic, with subordinate volcanics and volcaniclastic rocks. The succession reaches a thickness of more than 2000 m in the axial part of the basin. The Ordovician rocks transgressively overlie either Cambrian successions of the Příbram-Jince and Skryje-Týřovice basins or folded Neoproterozoic rocks.

The first stratigraphical subdivision of the Teplá-Barrandian area was published by Barrande (1846, 1852), who distinguished three main faunas (see Chapter 3). The second of these faunas was established in his *étage des quartzites D* and corresponds to the Ordovician. The subdivision of the Ordovician was subsequently studied and improved by a great number of authors. The most detailed studies of the lithology were published by Kukal (1956, 1961, 1962, 1963*a*, *b*), while volcanic rocks were studied by Fiala (1971*a*, *b*), Patočka & Vlašímský (1986) and Patočka *et al.* (1993). The data have been summarized recently by Štorch (1998).

More than a century of research on the stratigraphical subdivision of the region finally resulted in the distinction of the main lithostratigraphical units grouped into 12 formations (Havlíček & Vaněk 1966). Havlíček & Marek (1973) defined four regional chronostratigraphical units, namely the Dobrotivá, Beroun, Králodvor and Kosov series in the Middle and Upper Ordovician. The applicability of these units for stratigraphical correlation was subsequently discussed by Havlíček and Fatka (1992) and Fatka

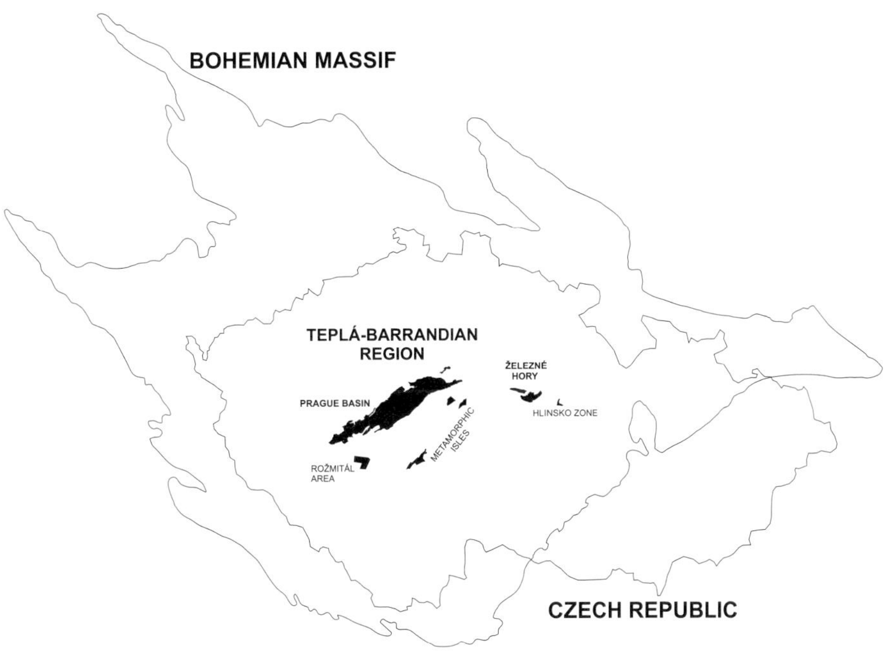

Fig. 5.12. Map of the outcrop areas of Bohemia, Czech Republic. The classic Ordovician sections are all located in the Prague Basin.

et al. (1995), who prefered their usage as stages. These chronostratigraphical units (or stages) are, however, only used at a regional (Bohemian) level. The classic correlation of the chronostratigraphical and lithostratigraphical units used in the Barrandian area is presented in Figure 5.13, and tentatively correlated with the international standard.

Třenice Formation

It is generally supposed that the Early Ordovician sea penetrated into central Bohemia from the east and formed in its western segment a shallow and narrow bay (Havlíček 1998) which subsequently widened and deepened. The basal Třenice Formation (established by Kettner (1916); thickness usually up to 30 m, exceptionally to 70 m) overlies with a prominent angular disconformity either the Neoproterozoic Štěchovice or Kralupy-Zbraslav groups or Cambrian rocks of the Příbram-Jince or Skryje-Týřovice basins. It consists of medium- to coarse-grained sandstones, greywackes and arkoses with intercalated conglomerates. In several parts of the basin Upper Cambrian clasts of volcanics can comprise up to 50% of all grains. Kukal (1963*a*) interpreted the origin of these sediments as due to rapid erosion of the high relief of Proterozoic and Cambrian rocks surrounding the Prague Basin. The fauna is dominated by large, most probably infaunal linguliformean brachiopods associated with trilobites, gastropods (Horný 1999) and in some layers also by echinoderms, rare graptoloids (summary in Růžička 1926, 1927; Havlíček & Vaněk 1966; Mergl 1994) and trace fossils (Mikuláš 1993, 1998). In the analysis of benthic assemblages four main faunas were established by Havlíček (1982) that are all confined to shallow-water sediments in the western part of the basin.

Global Series	Global Stages	Balt. Series	British Series	British grapt. zonation	Bohemian stages	Bohemian formations
Upper Ordovician	Hirn.	Harju	ASHGILL	persculptus	Kralodvorian /Kosovian	Kosov Fm.
				extraordinarius		
	Katian			anceps: pacificus		Kralodvor Fm.
				anceps: complexus		
				complanatus		
			CARADOC	linearis	Berounian	Bohdalec Fm.
		Viru		clingani		Zaho. Fm.
	Sandbian			foliaceus		Letna Fm.
				gracilis		Liben Fm.
Middle Ordovician	Darriwilian		LLANVIRN	teretiusculus	Oretanian/Dobrotivian	Dobrotiva Fm.
				murchisoni		Sarka Fm.
		Kunda		artus		
			ARENIG	hirundo		
	Dap.	Volk.		gibberulus	Arenigian	Klabava Fm.
Lower Ordovician	Floian	Latorp		simulans		
				varicosus		
				?		
				phyllograptoides		
	Tremadocian		TREMADOC	Ar. murrayi		
		Tremadoc		?	Tremadocian	Milina Fm.
				tenellus		Trenice Fm.
				flabelliformis		

Fig. 5.13. Tentative correlation of the local Bohemian stages and formations with the global series and stages and the regional Baltic and British series.

Mílina Formation

The overlying Mílina Formation, up to 30 m thick, has been interpreted as a regressive succession, although based on the presence of volcanic glass fragments within the greywackes it has also been suggested that the formation is merely a facies equivalent of the Třenice Formation. It consists of cherts intercalated with greywackes and siltstones. Chert beds usually contain 90–95% SiO_2 and *c.* 40% of the total formation thickness. Havlíček & Fatka (1992) have correlated the Mílina Formation with the *Ceratopyge* Regressive Event. The faunas of the Milina Fm are dominated by 'inarticulate' brachiopods, associated with 'articulate' brachiopods, sponges, diverse trilobites (e.g. Havlíček & Vaněk 1966; Mergl 1984) and trace fossils (Mikuláš 1993, 1998). Kolíha (1937) proposed six fossil assemblages. Subsequently, Havlíček & Vaněk (1966) criticized the absence of any section where such zonation could be documented and explained the assemblages in terms of their relation to lithofacies. These latter authors distinguished three horizons based on the brachiopod genus *Jivinella* for the Třenice and Mílina formations. Havlíček (1982) established two faunas dominated by inarticulate brachiopods: the weakly diversified *Leptembolon* Community with small 'inarticulates' and 'articulate' brachiopods, sponges, and fragmentary trilobites; and the *Thysanotos* community with large 'inarticulate' brachiopods and trace fossils. More recently, Mergl (2002) ascribed the low diversity *Leptembolon* Community to the *Acrotreta* Community Group.

Klabava Formation

The Klabava Formation is up to 300 m thick. Detailed sedimentological analyses were published, distinguishing three different members (Kraft & Kraft 2003), namely: the Mýto Shale Member, the Olešná Member, and the Ejpovice Member. The Mýto Shale Member represents a clayey shale facies with subordinate tuff beds. It represents the most richly fossiliferous lithotype with graptolites and 'inarticulate' brachiopods as the dominant groups. Greywackes, siltstones and silty shales were assigned to the Olešná Member. Redeposited tuffs and tuffaceous rare iron ore lenses represent the Ejpovice Member. The tuffaceous beds contain brachiopods (e.g. Mergl 2002), bryozoans, sponge spicules and locally conodonts. The long development of the biostratigraphic subdivision of the Klabava Formation has been recently summarized by Kraft & Kraft (1999, 2004) who defined five graptolite biozones. Four acritarch assemblages have also been distinguished (Vavrdová 1993), while Paris & Mergl (1984) and Fatka (1993) recognized a total of four chitinozoan biozones.

Šárka Formation

The ongoing marine transgression resulted in a rapid change of facies characterized by the sudden appearance of black shales and oolithic iron ores belonging to the Sárka Formation of which the thickness ranges from several metres to 350 m. Clayey shales dominate in the subsiding central part of the basin, while iron ore deposits are characteristic of the more stable marginal areas. Black shales contain between 0.3 and 1.5% of C_{org} and the silt and sand content increases to the NE. According to Kukal (1962), the source of the iron was related to lateritic weathering of shales and basaltic tuffs. Other possibilities, however, were discussed by Petránek (1991) and Petránek & van Houten (1997). In these studies the importance of submarine exhalation of fluids and/or the combination of microbial and other diagenetic processes affecting clayey sediments with the mobilization of iron, especially coupled with upwelling, could play a decisive role.

The richly diversified macrofauna includes conulariids, brachiopods, hyolithids, gastropods, bivalves, agnostoid and polymeroid trilobites, phyllocarid crustaceans, ostracodes, echinoderms, graptoloids, cephalopods, ichnofossils and different types of *incertae sedis* and problematica. Palynomorphs are represented by acritarchs and chitinozoans.

Earlier biostratigraphic subdivisions of the Šárka Formation based on graptolites have been critically evaluated by Kraft & Kraft (1999), who distinguish only two biozones: *Corymbograptus retroflexus* and the *Didymograptus clavulus*. Paris & Mergl (1984) and Fatka *et al.* (1996) documented the *Cyathochitina campanulaeformis*, *Belonechitina micracantha typica* and *Armoricochitina armoricana* chitinozoan biozones.

Havlíček (1982) recognized the trilobite-dominated *Euorthisina* Community inhabiting a subtidal, soft-bottom environment. The deeper-water black shales with dominant pelagic graptolites were described as a ‘pelagic fauna dominated by graptolites’ (Havlíček & Vaněk 1990; Havlíček 1998). Associated are common and well diversified palynomorphs represented by acritarchs and chitinozoans. The suprabenthic to epiplanktic *Paterula* Community Group (with *Rafanoglossa* Community; Mergl 2002) is associated with phyllocarid crustaceans (*Caryocaris*) and graptolites, conularids, nautiloids, trilobites and common microplankton (acritarchs and chitinozoans).

Dobrotivá Formation

The Dobrotivá Formation ranges in thickness from several metres to 350 m. The progressive deepening of the central segment of the basin with dominant sedimentation of black shales was acompanied by the onset of deposition of sand on the shallower margins of the basin. The sand supply decreased during deposition of this formation, but two main lithofacies can be recognized: the deeper Dobrotivá black shales and the shallower Skalka Quartzite. In his sedimentological analyses, Kukal (1963*b*) interpreted the black, clayey shales with muscovite as sediments of an anaerobic environment deposited at a depth of several hundred metres, while the Skalka Quartzites represent nearshore, shallow-water facies. In some areas a delta setting was supposed by Kukal (1956). Well diversified macrofauna include conulariids, brachiopods, hyolithids, cephalopods, bivalves, gastropods, agnostoid and polymeroid trilobites, phyllocarid crustaceans, ostracodes, echinoderms, graptoloids, cephalopods and ichnofossils. Microfossils are represented by poorly diversified acritarchs and chitinozoans. Following the study of Bouček (1973) two graptolite biozones are defined for the black shale facies of the Dobrotivá Formation: the *Hustedograptus teretiuscullus* Zone and the *Cryptograptus* aff. *tricornis* Zone (Kraft & Kraft 1999).

Havlíček (1982) distinguished the shallow-water ‘*Tigillites*’ Community in marginal parts of the basin with quartz-rich sedimentation containing the monotonous assemblage of *Skolithos*, while the black shale lithofacies in the central segment of the basin was ascribed to the *Paterula circina* Community.

Libeň Formation

The general development of the Libeň Formation (thickness 50 to 250 m) is very similar to that of the older Dobrotivá Formation with three basic lithotypes: black shales, pyroclastics and quartzites. Mature quartzose sandstones of the Řevnice Quartzites represent the whole sequence of the Libeň Formation in the western sector of the basin and along its northwestern and southeastern margins, while in the other parts the Řevnice Quartzite is developed mainly in the lowermost quarter of the formation. The sediments of the Řevnice Quartzite contain fine- to medium-grained quartz (usually 0.1 mm in diameter) and have been interpreted as very shallow-water environment sediments from a prograding sandy shoreline (sand bars and barriers) by Kukal (1956) and Chlupáč & Kukal (1988). In the central part of the basin, black, clayey shales represent sediments deposited in poorly oxic to anoxic environments. A narrow belt of pyroclastics of the Komárov Volcanic Complex has been established in the northwestern margin of the basin. ‘Articulate’ brachiopods, trilobites, bivalves, cephalopods, ostracods and echinoderms, rare gastropods, hyoliths and trace fossils are present. Havlíček (1982) and Havlíček & Vaněk (1990) distinguished three fossil assemblages: the shallow water ‘*Tigillites*’, the high energy *Drabovia dux* and the deeper water *Paterula* community.

Letná Formation

The Letná Formation ranges in thickness from 40 to 650 m.The base of the formation is established by the rapid appearance of rhythmic sedimentation with fine-grained sandstones (commonly with calcitic cement), alternating with sandy to clayey siltstones and sandstones. Dark shales to siltstones dominate in the central part of the basin, while siltstones and sandy shales prevail in the western segment. The fauna incorporates common brachiopods, trilobites, ostracodes, gastropods, cephalopods, problematical arthropods, conulariids, echinoderms and trace fossils (Mikuláš 1993, 1998) and have been ascribed the *Drabovia redux* and *Bicuspina* communities (e.g. Havlíček & Vaněk 1990). The *Drabovia redux* Community is related to the sorted quartzose sandstones with local clay fragments, while the relatively deeper *Bicuspina* Community is confined to fine, non-sorted greywackes in higher levels of the formation.

Vinice Formation

Increased deepening of the basin resulted into the dominance of black shale sedimentation as represented by the Vinice Formation (20–450 m in thickness). Kukal (1961) intepreted the black clayey shales, which are rich in organic matter, as having been deposited in a wide bay. In the NW flank of the basin, the base of the formation is marked by a level that contains iron ooids. Petrological and mineralogical studies were published by Petránek *et al.* (1988). The formation contains brachiopods, gastropods, conulariids, rare cephalopods, bivalves, trilobites, echinoderms, ostracodes and hyolithids associated with graptolites and ichnofossils. The deepening of the basin resulted in the establishment of a fauna dominated by the uniform *Paterula bohemica* Community. Havlíček & Vaněk (1990) also distinguished the *Deanaspis–Aegiromena* Community, characteristic of the poorly oxygenated soft-bottom and deeper-water environment containing a poor sessile benthos. A different assemblage, in general resembling the older *Bicuspina* Community, has been discovered in the Zdice-Nučice iron ore horizon with common brachiopods and gastropods, echinoderms, trilobites, and orthocone nautiloids. Planktonic forms are represented by comparatively rare graptoloids, conulariids and cyclopygid and odontopleurid trilobites. Horný (1999) established the gastropod *Nonorios* Communitiy.

Zahořany Formation

The Zahořany Formation (70–400 m thick) comprises mainly siltstones with intercalations of sandstones (commonly with carbonate cement) and silty shales representing a shallowing of the basin. Typical storm sediments with hummocky cross-stratification and wave-ripples are present in the shallowest parts of the basin (Havlíček & Štorch 1990). Trilobites, ostracodes, common and diversified gastropods, conulariids, cephalopods, echinoderms and ichnofossils are present. Havlíček (1982,1998) and Havlíček & Vaněk (1990) have established two fossil assemblages, namely the *Drabovia latior* (= *Drabovia-Aegiromena* fauna) and *Paterula* communities. The *Drabovia latior* Community is widely distributed, while the *Paterula* Community is typical of the black shale lithofacies in the highest levels of formation.

Bohdalec Formation

The base of this formation (20–200 m in thickness) is usually marked by the Karlík oolithic iron ore horizon. In the NW segment, synsedimentary faulting produced two parallel zones. The sediments are thus attributed to two basic lithotypes: black claystones and alternations of shales, siltstones and fine sandstones. Where present, the *'Polyteichus'* facies is developed in the lower and middle parts of the formation while the dark clayey shales consitute the upper levels of the formation. A richly diversified fauna of conulariids, bryozoans, brachiopods, gastropods, rare cephalopods, bivalves, hyoliths, ostracodes, trilobites and ichnofossils is present. Havlíček (1982, 1998) and Havlíček & Vaněk (1990) assigned the faunas to five main assemblages: *Hirnantia plateana*, *Onniella michlensis*, *Onnia abducta* (all three within the *'Polyteichus'* facies), *Svobodaina ellipsoides* and *Paterula* communities. The *Hirnantia plateana* Community laterally grades into the *Onniella michlensis* Community related to the black silty shales with carbonate cement. The deeper-water *Onnia abducta* Community was distinguished by Havlíček & Vaněk (1990) in claystones. Dark claystones and silty shales contain the *Svobodaina ellipsoides* Community with dominant brachiopods, associated with bryozoans, bivalves and trilobites. The widely distributed *Paterula* Community is characterized by small epiplanktonic brachiopods and sparse ichnofossils.

Králodvor Formation

The base of this formation (25–200 m in thickness) is marked by a discontinuous Podolí iron ore horizon overlain by claystones with siltstone intercalations. The upper parts of the formation are represented by clayey carbonates followed by silty shales. The faunal assemblages contain common gastropods, cephalopods, ostracodes, trilobites, rare conulariids, hyoliths and acritarchs. Havlíček (1982), Havlíček & Vaněk (1990) and Štorch & Mergl (1989) distinguished five main communities. The claystone sequence is generally poor in benthic elements and the *Rafanoglossella* Community is named after the infaunal lingulid brachiopod associated with trilobites, hyolithids, bivalves and gastropods. A temporary shallowing of the environment resulted in the development of the *Dedzetina* Community with dominant brachiopods associated with trilobites and graptolites. The epiplanktonic *Chonetoidea radiulata* Community contains minute brachiopods and ostracodes associated with miomeroid and polymeroid trilobites. Higher layers of the formation (represented by a discontinuous layer of carbonates) contain the deep-water *Proboscisambon* Community. The uppermost part of the Královdvor Formation formed by greenish silty shales contains the *Mucronaspis* Community. Havlíček (1998) distinguished four successive trilobite faunas within the Královdvor Formation: *Nankinolithus granulatus* Horizon, *Tretaspis seticornis seticornis* Horizon, *Marekolithus kosoviensis* Horizon and the *Mucronaspis grandis* Community.

Kosov Formation

The lithology of this formation (40–200 m thick) reflects both the glacio-eustatic and tectono-eustatic changes characterizing the end of the Ordovician. Kukal (1963*b*) interpreted the presence of large slumps as induced by strong tectonic activity on the sea bottom. Detailed analyses of the sedimentology and the associated fauna of the Králodvor–Kosov boundary interval were published by Brenchley & Štorch (1989) and Štorch & Mergl (1989), respectively. The base of the Kosov Formation coincides with the deposition of a 0.1 m thick diamictite bed conformably overlying the shales of the Královdvor Formation. Štorch (1990) documented pebbles and dropstones, indicating the presence of floating ice. The *c.* 1.5–8 m thick succession of silty and sandy shales separates the second diamictite layer; the overlying sequence consists of shales and sandy shales, greywackes and quartzose sandstones. The upper part of the formation consists of bioturbated claystones (locally containing carbonate concretions) separated at some localities from the overlying Silurian graptolitic shales by a level of bioturbated mudstones. The generally poor fauna includes brachiopods, gastropods, rare conulariids, trilobites, acritarchs, cryptospores and chitinozoans, while other faunal groups are extremely rare and/or even absent. The lowermost levels of this formation are characterized by the low-diversity *Mucronaspis* Community, while higher levels of the formation are characterized by an assemblage dominated by infaunal bivalves of the genus *Modiolopsis*, followed by the brachiopod-dominated *Hirnantia sagittifera* Community.

Volcanic activity

The period of Late Cambrian acid to intermediate calc-alkaline volcanism was followed by intensive Early Ordovician basaltic volcanism. These latter volcanics comprise tholeiitic and alkaline basalts, trachybasalts, trachyandesites and tephrites. Tuffs and tuffaceous shales also occur, beginning in the Tremadocian. Extensive volcanic activity since the latest Tremadocian is confined to the area of the Komárov Volcanic Centre, striking *c.* 20° NNE–SSW. The most common products are tuffs, agglomerates and hyaloclastites, which represent time equivalents of the shaly and sandy lithofacies of the Klabava, Šárka and Dobrotivá formations (late Early to Middle Ordovician), being mostly developed in the northern limb of the basin. The stratigraphical range and the thickness of the volcanic sequences decrease in the southern part of the basin, where rare basalts occur. Aphanitic and amygdaloidal basalts, hyaloclastites, tuffs and tuffaceous sediments represent the Late Ordovician volcanic activity of several eruption centres.

Poland

The Ordovician rocks in Poland occur in several terranes with different geological histories, bordering the eastern branch of the Variscan orogenic belt (e.g. Żelaźniewicz 1997; Aleksandrowski *et al.* 2000; Winchester 2002). In the SW, exposures of strata representing mostly deep-water sediments are known from two tectonically distinct regions of the Sudetes. In the NE, a much more complete geological record of the period is available; these Ordovician occurrences belong to the Małopolska Massif and the marginal part of the East European Platform (Fig. 5.14).

The Tornquist-Teisseyre Lineament (also referred as the Trans-European Suture Zone) separates the East European Platform from the Małopolska Massif and the core of the Variscides. The Ordovician deposits of northern Poland are present at great depths: in the Toruń 1 borehole, in northern Poland, they are at *c.* 5 km depth. They are probably metamorphosed in the axial area of the Variscan orogenic belt (where the base of the Permian is at a depth of more than 8 km; Pożaryski *et al.* 1992). The facies gradients in marginal parts of the surrounding regions strongly suggest that these were sediments deposited in a deep-water Ordovician sea or ocean, that was probably an eastern branch of the Iapetus, usually referred to as the Tornquist Sea (e.g. Żelaźniewicz 1997).

The southern part of Poland comprises a series of massifs which were consolidated in pre-Ordovician times. The region is subdivided by the Kraków-Lubliniec Fault Zone into two units: the larger Małopolska Massif to the east and the westerly Upper Silesian Block. All of the Ordovician strata were probably removed from the surface of the latter by pre-Mid-Devonian erosion (e.g. Belka *et al.* 2002), so it is not possible to establish its biogeographical affinities. In contrast, the Małopolska Massif contains an excellent palaeontological record on its northern margin, in the area of the Holy Cross Mountains, where the Ordovician is exposed in a series of narrow belts.

In the central belt of the Holy Cross Mountains, the Ordovician was first documented with fossils. The research was initiated by Roemer (1866) who introduced the allegedly Devonian brachiopod species *Orthis kielcensis*. This age misidentification was corrected by Hempel (1867) and the evidence for the Ordovician age was first explicitly presented by Michalski (1883), who found *O. kielcensis* and several other brachiopods indicative of an Ordovician age for the formation. The overlying organodetrital Mójcza limestone was first noted at Międzygórz by Zeuschner (1869) but its Ordovician age was determined first by Gürich (1901) in the type locality.

Before the *in situ* Ordovician was identified, fossils of that age were widely known and studied in northern and central Poland

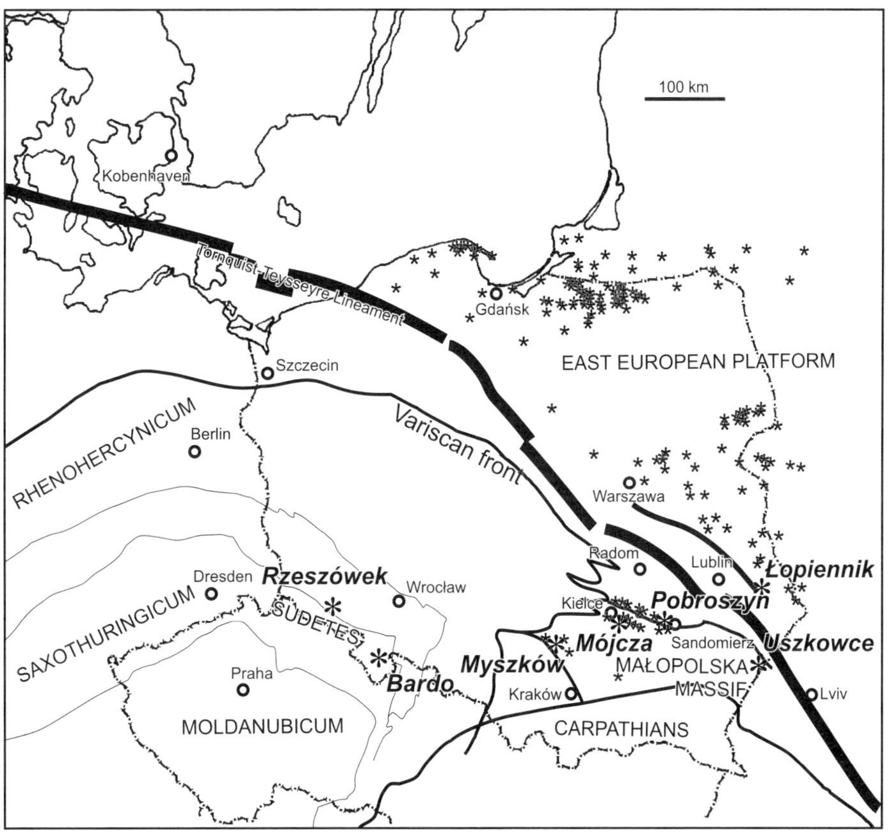

Fig. 5.14. Location of exposures of the Ordovician in the Sudetes and boreholes (asterisks) that reached Ordovician strata on the Małopolska Massif and the East European Platform in Poland (background map modified after Pożaryski *et al.* 1992). Larger asterisks indicate the most important localities discussed in the text. Moldanubicum = Moldanubian Zone; Rhenohercynicum = Rhenohercynian Zone; Saxothuringicum = Saxothuringian Zone; Tornquist-Teysseyre = Tornquist-Teisseyre.

within glacial erratic boulders occurring in the Quaternary till. These came from the East European Platform (craton), including Scandinavia and the bottom of the Baltic Sea. Some of the strata represented in boulders are unknown from exposures and they are of special importance to palaeontological research owing to their exquisite preservation. Numerous taxa of macrofossils were identified in this material in the early stages of scientific research. The most important are probably those of the Late Ordovician 'Sadewitzer Kalk' described by Roemer (1885) who published a general review of fossils from erratic boulders of Baltic origin. In more recent times Ordovician organic-walled, phosphatic and calcite fossils extracted from limestone matrix with acid were studied extensively by Kozłowski (e.g. Kozłowski 1963, 1968) and his pupils. Among their works, the monographic description of polychaete jaws by Kielan-Jaworowska (1966) and the identification of organic walls in calcareous algae by Kozłowski & Kaźmierczak (1968) are of most general interest.

The Ordovician exposures represented in each of the tectonic units of the territory of Poland are reviewed, from the Sudetes in the SW, crossing the Variscan Belt and the Tornquist-Teysseyre Lineament to reach the marginal part of the East European Platform (Fig. 5.14). The Ordovician strata on the East European Platform proper do not differ significantly from those exposed in the Baltic area and Scandinavia, so only the fact of their presence in deep boreholes is acknowledged here.

The Sudetes Mountains

The Sudetes Mountains (Figs 5.3, 23, 24, 5.14 & 5.15) are extremely complex in terms of their tectonic evolution and comprise mainly crystalline and metamorphosed rocks (see also Kroner *et al.* 2008; McCann *et al.* 2008). The age of metamorphism is Palaeozoic (Smulikowski 1995; Winchester *et al.* 2002*b*) and the oldest reliably dated sedimentary strata are Ordovician. The numerous tectonic units in the area probably represent different belts, but their correspondence to those of the German Variscides is not clear. Ordovician strata are documented by biostratigraphical data in one area only and inferred from their stratigraphical position relative to dated lowermost Silurian strata in other localities. These occurrences are, however, of much palaeobiogeographical significance.

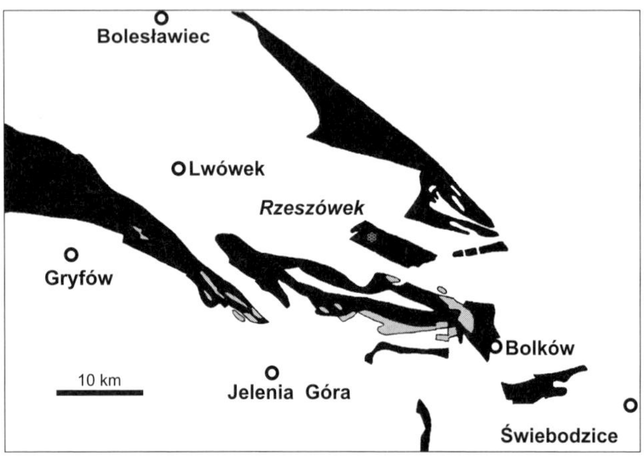

Fig. 5.15. Extent of outcrops of variously metamorphosed lower Palaeozoic sedimentary clastic rocks (black) and the Wojcieszów (Kauffung) Limestone in the Sudetes (simplified after Urbanek *et al.* 1975). Ordovician age is determined by conodonts from the Radzimowice (Altenberg) Slate at Rzeszówek (asterisk; Urbanek & Baranowski 1986) and suggested for the Wojcieszów Limestone (Skowronek & Steffahn 2000).

As a result of changed political boundaries after World War II and also the resettlement of population many local topographic names were changed, and this has influenced the geological nomenclature. Old names of lithostratigraphical units are given here in parentheses.

Bardo Hills
The Karkonosze Mountains, forming the crystalline and metamorphic core of the Sudetes, probably belong to the northern margin of the Bohemian Massif. Unmetamorphosed Early Palaeozoic deposits in the Bardo Hills of the Kłodzko area may represent the same belt (Fig. 5.3, 24). In trenches dug to access Ordovician–Silurian boundary strata, Teller (1962) identified from a number of localities a unit of conglomerate with quartz mudstone pebbles separating a black quartz sandstone series from the overlying shales, of palaeontologically-dated Silurian age. The Ordovician age of the sandstone unit is only inferred from its position below the Llandovery-age rocks. However, similar lithologies are found in Ordovician successions in the Prague Basin of Bohemia (Kosov Formation). The tectonic affinity of this part of the Sudetes to the Czech succession is, thus, at least not contradicted by the stratigraphic evidence.

Kaczawa Hills
This is a tectonically very complex part of the northern Sudetes (Figs 5.3, 23 & 5.14) with an unclear relationship to other parts of the Variscan belt. The oldest palaeontologically dated stratigraphic unit in the area is the Radzimowice (Altenberg) Slate, a sericite–quartz–albite phyllite that has yielded some taxonomically undeterminable conodonts. However, their morphology suggests that the unit cannot be older than early Ordovician (Urbanek & Baranowski 1986).

The Radzimowice Slate is comformably overlain by the Wojcieszów (Kauffung) Limestone. Except for the presence of the possible phyllocarid crustacean *Silesicaris*, hexactinellid sponge spicules, and agglutinating foraminifera, no convincing fossils are known from these strata. Their age must therefore be Ordovician or younger (Skowronek & Steffahn 2000).

A much more diverse fossil record is contained within the overlying strata of an unnamed metasedimentary formation, where the sediments appear to be mainly deep marine (Baranowski 1975), and crop out near Rzeszówek (Figs 5.14, 5.15). In one locality sideritic intercalations within the phyllitic slate yielded extremely abundant conodont assemblages (Baranowski & Urbanek 1972). These conodonts indicate an interval of a few zones of the upper Arenig. The oldest assemblage includes several species typical of the temperate-water Baltic faunas (including the stratigraphically important *Paroistodus parallelus*). Most surprisingly, the dominant species of the assemblage belongs to the genus *Plectodina*, unknown from the Baltic faunas of this age and suggesting Avalonian biogeographic affinities. Similarly hybrid is the composition of the youngest of the identified assemblages, containing the Baltic-type *Lenodus* and primitive *Icriodella*, foreign to Baltic assemblages but known from somewhat younger strata of the Armorican Massif and Thuringia (Dzik 1990).

The weakly palaeontologically-documented upper Ordovician (possible *Amorphognathus* conodonts) occurs in this area near Lubań (Chorowska *et al.* 1981), and is represented by sericitic shales with lydites in its topmost part, located below the Llandovery strata dated using graptolites.

Northward and westward of the Sudetes the Ordovician strata are hidden beneath Variscan and later sedimentary cover.

The Małopolska Massif

The most southwesterly located occurrence of Ordovician limestone with conodonts is within the Kraków-Lubliniec Fault Zone, along the margin of the Małopolska Massif (Fig. 5.3, 25; 5.14). The composition of the fossil assemblages is the same as in the Holy Cross Mountains on the opposite side of the Massif. The Małopolska Massif would, therefore, appear to have been an area of rather uniform sedimentation, at least in the Middle Ordovician.

The Ordovician strata are widely distributed across the Małopolska Massif but most of the published information is derived from the area of the Holy Cross Mountains. The Ordovician rocks in this area are generally poorly exposed and there are only a few localities where a significant part of the Ordovician rock column can be studied without special excavations. During the Early Ordovician, coarse clastics were deposited, while the Middle and Upper Ordovician strata are extremely condensed stratigraphically, being represented by limestones with phosphate ooids and coated grains, and marly shales, respectively. The data on the three facies belts of the Holy Cross Mountains are relatively well known from exposures and are supplemented by the borehole evidence across the main Małopolska Massif area and the marginal Myszków area to the SW. Figure 5.16 illustrates a simplified representation of the Ordovician facies distribution across the Małopolska Massif.

Myszków area

Fossiliferous Ordovician strata have been recovered by deep boreholes near Myszków (Figs 5.3, 25 & 5.14) and Zawiercie (Piekarski & Siewniak-Witruk 1974; Nehring-Lefeld & Szymański 1998). According to Nehring-Lefeld *et al.* (1992) the Ordovician succession commences with a conglomerate including claystone pebbles. Unlike the rest of the Małopolska Massif, there are no Lower Ordovician sandstones. Immediately above the conglomerate, more or less metamorphically altered limestone occurs (conodont CAI index 8 indicates heating to about 600°C). The conodont assemblages indicate that most of the Middle Ordovician is represented in 20 to 40 m of the limestone (Dzik & Pisera 1994). The upper Ordovician part of the succession is also in a limestone facies but locally the rock is finely crystalline.

In borehole BM 152, located *c.* 20 km to the west, the Ordovician is represented by at least 100 m of fine-grained clastics with rare sandstone and limestone intercalations, overlain by up to 60 m of carbonates and 30 m of sandstones of unknown age. Conodonts are reported from a limestone bed in the middle of the shaly part of the section which, if correctly determined, are Middle Ordovician in age.

Interior of the Małopolska Massif

Close to the northeastern tip of the massif, in boreholes Uszkowce 1 and 4 (Figs 5.3, 26, 5.14 & 5.16), Lower Ordovician sandstones (20–30 m thick) occur above a Cambrian shale. One of the shale intercalations contains dendrograptid rhabdosomes (Tomczyk 1962) and these have been used to support the Tremadocian age for the strata. A discontinuity with pebbles separates the sandstone succession from a *c.* 10 m thick unit of limestones, marls and mudstones with ferruginous ooids. A graptolite shale represents the Upper Ordovician, and is topped by a mudstone with gravel intercalations (Tomczyk 1962).

A similar Ordovician succession is reported from the southern margin of the massif, at Mędrzechów, as well as in its centre, in the Książ Wielki IG 1 borehole (Jurkiewicz 1975; Moryc & Nehring-Lefeld 1997; Modliński & Szymański 2001*a*, *b*). A conglomerate bed is followed by *c.* 14 m of sandstones (with *Isograptus* in the middle), that change gradually into limestones with phosphorite concretions, which may indicate stratigraphic condensation. A dolomitized limestone in the Książ Wielki IG 1 borehole is up to 13 m thick and contains the conodont *Scabbardella* in its lower half (W. Bednarczyk in Jurkiewicz 1975), which suggests a Late Ordovician age.

In all of the boreholes from the Małopolska Massif that are complete enough to ensure a reliable record, the succession of the Ordovician strata and the thicknesses of particular units are closely similar to those in the central facies belt of the Holy Cross Mountains. Invariably there is a stratigraphically condensed limestone succession overlying a thicker sandstone and underlying a graptolite shale or marl. The entire Małopolska Massif was apparently an area of carbonate sedimentation during the Middle Ordovician (at least during the Llanvirn, and in central parts also during the Caradoc). Despite the high degree of stratigraphic condensation, this limestone unit is surprisingly lithologically uniform across the entire area. Based on conodont studies, the succession appears to contain much of the Ordovician sequence apart from a large gap in the lower part. Such a situation could hardly be possible if any part of the Małopolska Massif was elevated and exposed to erosion during the sedimentation of the limestone. It would, therefore, appear that the entire surface of the Małopolska Massif was already peneplained in the Early Ordovician and then submerged.

Holy Cross Mountains

In the Holy Cross Mountains (Figs 5.3, 28, 5.16 & 5.17), four distinct facies regions, arranged roughly from the south to north, have been recognized: the Brzeziny-Zbrza, Łagów, Kielce, and Łysogóry regions.

Brzeziny-Zbrza area

The deepest-water facies of the Ordovician in the area is represented at Brzeziny, near the western end of the Holy Cross Mountains (Fig. 5.17). Graptolite shales predominate (Tomczyk 1962; Tomczyk & Turnau-Morawska 1964). A glauconitic sandstone occurs in the lowermost part of the section and passes upwards into mudstones, claystones and graptolite shales, with various species of *Didymograptus* and *Phyllograptus* indicative of an early Llanvirn age. These shales continue throughout the Llanvirn (Bednarczyk 1966*a*, *b*), with abundant chamosite ooids and rare intercalations of limestone at the top. In the graptolite shales of the overlying Morawica Beds a complete Caradoc graptolite succession is represented (Modliński & Szymański 2001*a*, *b*). The Brzeziny-Zbrza Ordovician deposits may be a continuation of the same facies belt as that located about 40 km to the east, in the Łagów region (see below).

Łagów Region

The Łagów facies region of Czarnocki (1928) represents an area of sedimentation in shallower water than that near Brzeziny, but definitely deeper than most of the Małopolska Massif, since it would appear to be distal to the source area for coarse clastics. The best exposure of the Ordovician in the area is at Zalesie Nowe near Łagów (Figs 5.3, 27 & 5.17), where the Ordovician commences with a thin layer of pebbles, a few centimetres thick, overlain by *c.* 3 m of shale, and then by *c.* 4.5 m of tuffaceous

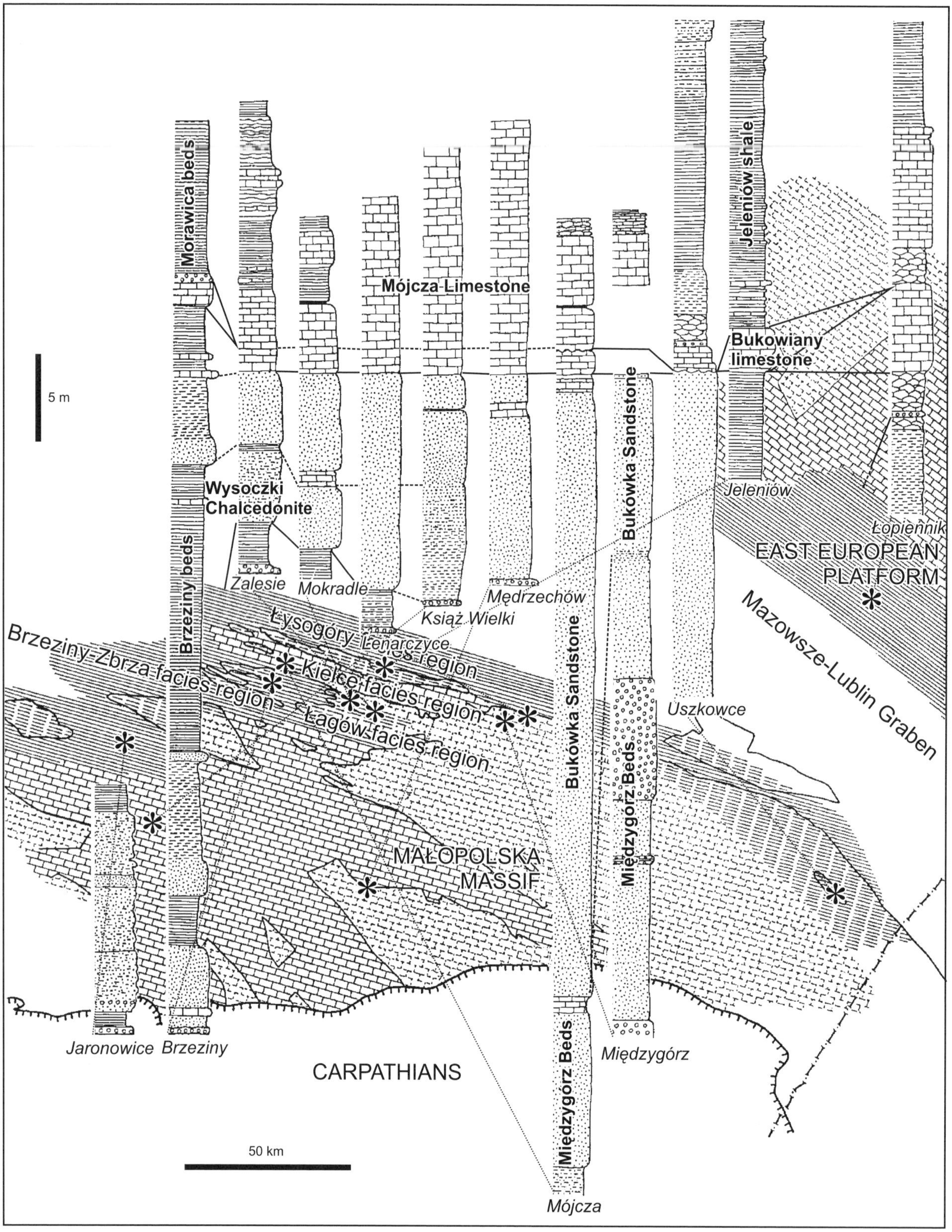

Fig. 5.16. A simplified presentation of the Ordovician facies distribution over the Małopolska Massif (modified from Dzik & Pisera 1994). Rock columns of representative boreholes and exposures are arranged according to the gradient from the deep-water Brzeziny area and the Łagów facies region (Zalesie) on the left, to the shallow-water Kielce facies region in the centre (Mójcza, Międzygorz), and to the Łysogóry facies region on the right (Jeleniów). Also shown is the most proximal locality on the East European Platform. The brick pattern shows the regions without any Caradoc shales within the limestone succession; the oblique lines show those regions where the shales are represented (broken lines are used for the areas where the Ordovician rocks have been removed by erosion but are presumed to occur originally).

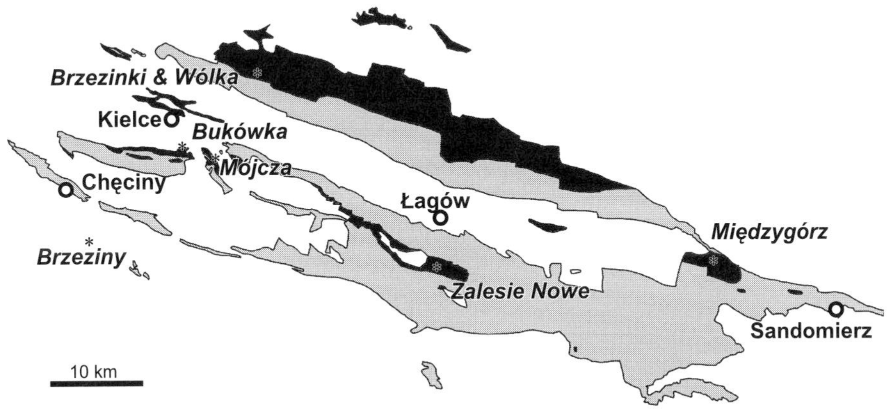

Fig 5.17. Extent of outcrops of the Cambrian (grey) and Ordovician to Silurian (black) strata in the Holy Cross Mountains (the Ordovician is of too small a thickness to be mapped at this scale). Locations of exposures discussed in the text are indicated by asterisks).

mudstones with cherty intercalations (= Wysoczki Chalcedonite Formation) (Chlebowski 1981; J. Samsonowicz *in* Kozłowski 1948). The chert is rich in detritus of carbonized rhabdosomes of benthic graptolites, well-preserved acritarchs, phosphatic conodont elements, and brachiopods. The conodonts indicate a late Tremadocian age (*Paltodus deltifer* Zone), corroborating the age indicated by acritarchs.

In the same localities, a coarser-grained glauconitic sandstone (Międzygórz Formation) occurs above the cherty unit, followed by the brachiopod-rich Bukówka Sandstone (Czarnocki 1928). The Mokradle borehole in this area revealed that these latter two formations are separated by a carbonate intercalation. Above the Bukówka Sandstone the phosphate ooid-rich Mójcza Limestone Formation occurs. Both at Zalesie and in the nearby Mokradle borehole, the transition from the Mójcza Limestone carbonates to siliciclastic sedimentation took place immediately after the deposition of a K-bentonite layer. The facies change would appear coeval with the worldwide *Nemagraptus gracilis* transgression (Dzik 1994). Six K-bentonite intercalations in total occur within the shale member of the Mójcza Limestone Formation (Chlebowski 1971), that ends with a purple-red shale with small ferruginous concretions.

Immediately above the Mójcza Limestone Formation the only undolomitized limestone bed in the succession is present, with numerous worm burrows and phosphatic *Sphenothallus* tubes (Dzik & Pisera 1994), separated by a thin intercalation of claystone from the dolomite and limestone beds of the overlying Zalesie Formation. They contain the zonal index species *Amorphognathus ordovicicus*. A bentonite horizon marks the top of the dolomitic unit.

The overlying shale member of the Zalesie Formation is 7 m thick and comprises shales with marly parts and thin sandy intercalations. Rare specimens of *Mucronaspis* and *Hirnantia* have been found immediately above this bentonite in a marly shale. These strata at nearby Bardo Stawy contain a diverse assemblage of trilobites (Kielan 1960) and brachiopods (Temple 1965).

Sedimentation continued in the area across the Silurian–Ordovician boundary. The terminal Ordovician strata in the Łagów facies region are represented by a black graptolite shale which passes up into a radiolarite within the basal Bardo beds. Only *Normalograptus* occurs (Dzik 1999*a*), for example, from a sandy claystone topping the Ordovician succession near Bardo Stawy (Masiak *et al.* 2003). The earliest Silurian graptolite *Parakidograptus acuminatus* occurs in the claystones immediately above (Kielan 1960; Masiak *et al.* 2003).

Kielce Region

The Kielce region extends from Kielce to Sandomierz to the ESE (Figs 5.3, 28 & 5.17). This region is the best exposed part of the Ordovician in the Holy Cross Mountains (e.g. Bednarczyk 1971). The main difference with respect to the Łagów region is the presence of a coarsely clastic and relatively thick lower succession, a completely organodetrital limestone middle part, and a thin marly cover on the top. In both regions the lower clastic unit comprises two parts: a basal section composed of generally darker, glauconite-rich sandstones (Międzygórz Formation), and an upper part composed of the light, originally calcareous Bukówka Sandstone Formation.

Near Kielce, the Ordovician succession commences with a thin (up to 0.6 m) layer of conglomerate with Cambrian sandstone pebbles (e.g. Bednarczyk 1964). An overlying sandstone and mudstone succession includes clay intercalations with a 15 cm thick bentonite (Chlebowski 1981). A glauconitic sandstone with cherty intercalations occurs higher in the section. At Międzygórz near Sandomierz a much thicker succession of conglomerates and dark glauconitic sandstones from the Międzygórz Formation occurs (Tomczyk 1954). Claystone intercalations and pebbles within the conglomerate contain probable debris of *Dictyonema* (Znosko & Chlebowski 1976) and a chitinozoan assemblage of the *Cyathochitina primitiva* Zone (Chlebowski & Szaniawski 1974), which ranges throughout most of the Arenig in Baltica (Nõlvak & Grahn 1993). The Międzygórz Formation is thus not older than the Arenig, which is consistent with the dating of the base of the Ordovician sequence in the Łagów facies region.

Apparently a pre-Arenig erosional event resulted in the peneplaining of the Sandomirian (late Cambrian) Orogen on the Małopolska Massif before the Early Ordovician clays were deposited. The Międzygórz conglomerate beds may have been deposited at the beginning of a worldwide early Arenig regression, and were most probably the result of a local debris flow (Dzik & Pisera 1994). The quartz pebbles in the conglomerate indicate the presence of a yet unrecognized crystalline source for the Małopolska Massif (Turnau-Morawska 1960).

Above the conglomerate in the Międzygórz succession a dark glauconitic sandstone with the obolid brachiopod *Thysanotos siluricus* occurs, and this is restricted to the late Tremadocian–early Arenig (Bednarczyk 1964, 1971). The contact with the overlying strata is tectonic at Międzygórz, but in the Bukówka quarry a marly reddish limestone bed (Czarnocki 1928), and elsewhere perhaps even shale (Tomczyk & Turnau-Morawska 1964), separates the dark-coloured Międzygórz Formation from the light-coloured orthid-rich Bukówka Sandstone Formation.

The Bukówka Sandstone Formation is very fossiliferous, with a spongy appearance resulting from numerous empty spaces left by the dissolved shells of brachiopods, including *Orthis kielcensis* (Dzik & Pisera 1994). These fossils are clearly of late Arenig age but do not allow more precise dating. Conodonts from the limestone intercalations suggest a late Volkhovian age (Dzik 1990). The occurrence at Międzygórz of the Llanvirn age nautiloid *Cyclolituites* (Dzik 1984*a*) and the enigmatic monoplacophoran(?) mollusc *Pilina* (known also from the Mójcza section; Dzik & Pisera 1994) suggests that carbonate sedimentation continued slightly longer near Sandomierz than near Kielce.

There is a continuous transition from the Bukówka Sandstone to the 8 m thick Mójcza Limestone in its type locality. In the lowermost part of the section the rock changes from limey sandstones to a sandy limestone (strongly recrystallized biosparite) with rare glauconite grains. Quartz grains in these beds are well sorted and poorly rounded, similar to those in the Bukówka Sandstone Formation, and thus probably from the same source. Phosphatic 'inarticulate' brachiopods occur in great numbers in these transitional strata (Bednarczyk & Biernat 1978). About 1.5 m from the base of the section, a horizon includes phosphorite pebbles (Ordovician or Cambrian extraclasts), burrows and perhaps erosional pockets representing a sedimentary discontinuity (perhaps with some submarine erosion) (Dzik 1990).

The typical feature of the Mójcza Limestone is the abundance of grains coated with phosphatic envelopes, and ferruginous ooids (Turnau-Morawska 1961; Łącka 1990). Macrofossils are abundantly only in the lower part of the Mójcza Limestone in its type locality, where two bedding planes, probably of tempestitic origin, are crowded with orthid shells and trilobite carapaces (Czarnocki & Samsonowicz 1913; Bednarczyk 1964, 1966*b*; Dzik & Pisera 1994), and close to the K-bentonite layer (the same as in Zalesie Nowe) 4.5 m above the base (Ryka & Tomczyk 1959), where rare endoceratid nautiloid phragmocones occur. Above the bentonite, macrofossils are extremely rare. Lithologically the Mójcza Limestone is monotonous, comprising mainly strongly recrystallized biosparite, biomicrosparite and sometimes poorly washed biosparite (packstone to grainstone).

About 3.2 m above the bentonite at Mójcza, the organodetrital limestone is replaced by an argillaceous limestone and marls up to 2.8 m thick. Ooids disappear and only rare cauliflower structures can be found. This marks the base of the lower carbonate member of the Zalesie Formation, better represented in the Łagów region. Poorly preserved orthoceratid phragmocones occur in these strata in the Międzygórz quarry, where they are dolomitic and less thick than at Mojcza, possibly due to tectonic reduction. Some light-coloured shale may represent remains of the terminal shale unit of the Zalesie Formation. The similarity to Mójcza is further strengthened by the abundance of agglutinating foraminifera (indicative of a supply of fine quartz grains to the area, earlier missing), which at Mójcza are restricted to the uppermost part of the carbonate succession (Dzik & Pisera 1994).

The Mójcza Limestone is an extreme case of stratigraphic condensation (Fig. 5.18). Abundant conodonts (Spasov & Teller 1963; Bergström 1971; Bednarczyk 1971; Dzik 1976, 1978, 1990, 1994) allow precise subdivision and time correlation of the Mójcza Limestone in all of the sections where they have been studied. Its base is dated as the top of the Arenig (early Kundan, *Lenodus variabilis* Zone), its top as the latest Caradoc (*A. superbus* Zone). The carbonate member of the Zalesie Formation represents much of the Ashgill. Altogether, at least 30 million years are represented by a dozen metres of rock succession, with only one biostratigraphically identifiable gap. The formation is widely distributed across the Kielce and Łagów facies regions in the Holy Cross Mountains, as well as in the subsurface Małopolska Massif.

The Ordovician strata are overlain at Mójcza by a soft sandy marl, 10 cm thick, containing Ordovician and Silurian conodonts (well preserved but fragmented), and a thick *K-bentonite* bed. A 1–2 cm thick limestone intercalation yielded the latest Llandovery *Pterospathodus celloni* Zone assemblage, which indicates that the topmost Ashgill and most of the Llandovery strata are missing. At Międzygórz the oldest Silurian black graptolite shale represents the middle Llandovery *Monograptus gregarius* Zone (Tomczyk 1954).

Łysogóry Region

Strata of the Łysogóry Region are exposed along the northern slope of the Łysogóry mountain range (Figs 5.3, 29, 5.16 & 5.17). This region differs markedly from the Kielce and Łagów regions in that coarse clastics are absent in the Early Ordovician. This makes the region comparable to sections of the East European Platform. These lithological similarities are also supported by palaeomagnetic and geochronological data (Lewandowski 1993; Belka *et al.* 2002). The most complete section of the Ordovician of this region is documented in the Jeleniów IG 2 and several other boreholes (Tomczykowa 1968).

The position of the Cambrian–Ordovician boundary within the Łysogóry succession is still a controversial issue. Based on the occurrence of the acritarchs *Peteinosphaeridium* and *Vulcanisphaera* in claystone intercalations, Kowalczewski *et al.* (1984, 1986) suggested that the base of the Tremadocian is in the Klonówka Shale Formation. However, if the presence of *Peteinosphaeridium* is confirmed, these rocks would be of latest Tremadocian or younger age. On the other hand, the overlying Łysogóry beds yielded a late Cambrian trilobite assemblage with *Parabolina* (Modliński & Szymański 2001*a*), while a small rhabdosome fragment, possibly representing the earliest Tremadocian index pelagic graptolite *Rhabdinopora*, has been encountered in the upper part of this unit (Tomczykowa 1968).

The most complete exposure of the Ordovician in the Łysogóry Region is at Pobroszyn near Opatów. The Klonówka Shale Formation from this locality yielded acritarchs that suggested late Cambrian to Tremadocian ages (Trela *et al.* 2001). The limestone and phosphorite intercalation in the overlying shales above may correspond to the subsurface Bukowiany limestone formation. In the lowermost limestone horizon (middle of the section), conodonts of the late Arenig *Baltoniodus navis* Zone occur. The palaeobiogeographical affinities of the conodonts are unusual, given the presence of typical Baltic elements, but the dominant species is related to '*Baltoniodus*' *bohemicus* from the Klabava Formation of Bohemia (Dzik 1984*b*), while some rare forms are similar to species from the North American Midcontinent. About 3 m upsection, rare conodonts in a laminated limestone bed suggest latest Volkhovian or Kundan age.

The overlying Jeleniów Shale Formation, found in boreholes (Samsonowicz 1934; Tomczykowa 1968), may reach a thickness of *c.* 150 m. The lower part of the formation contains a 1.5 m thick limestone intercalated with shales that include *Hustedograptus*-like biserial graptolites, which indicate a Llanvirn age (Tomczykowa 1968). Above this limestone a thick succession of mostly graptolite shale (about 150 m) with thin limestone inter-

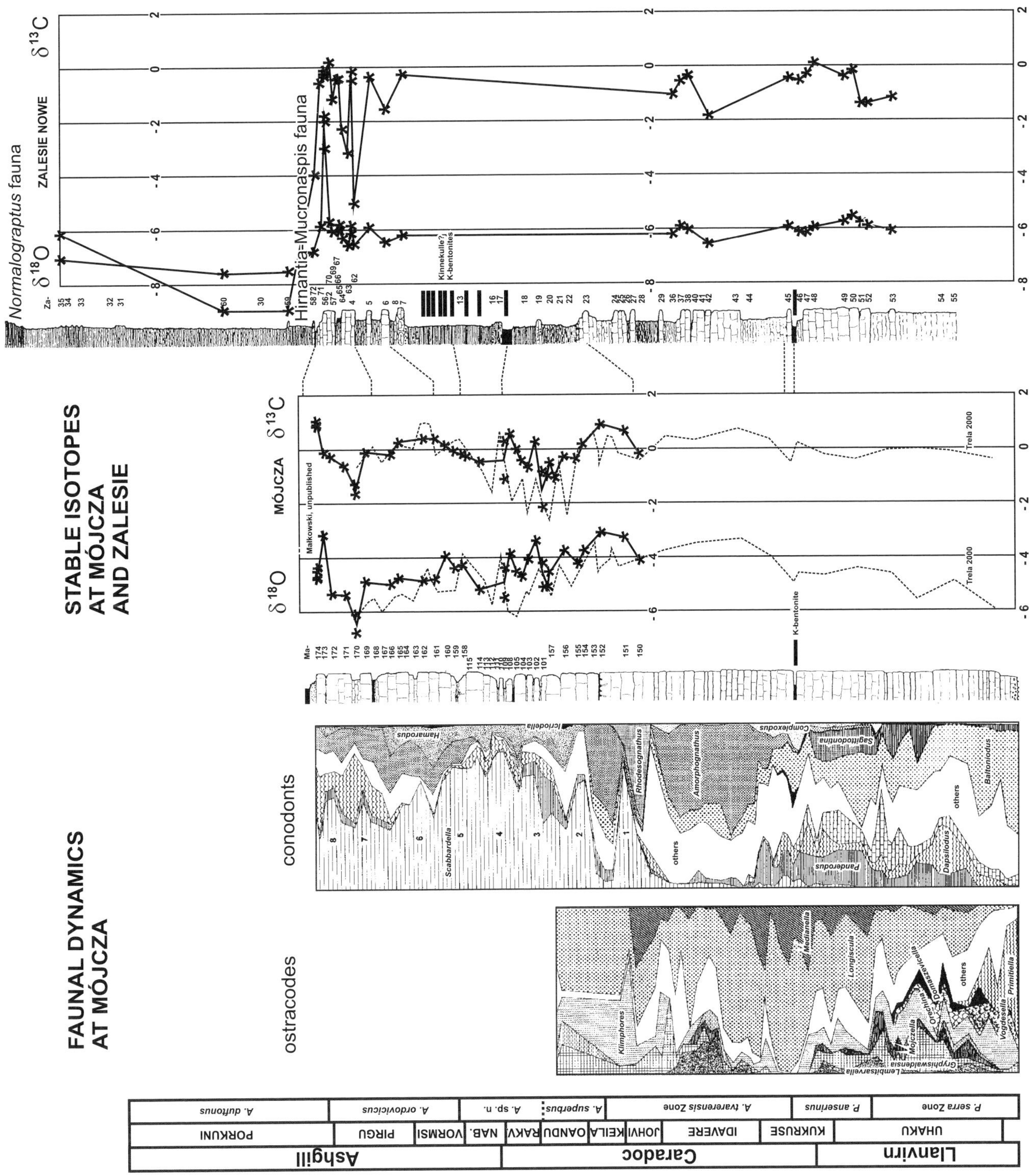

Fig. 5.18. Logs of percentages contribution of ostracod and conodont genera to samples taken from the post-Arenig (8.5 m thick) part of the section at Mójcza and the nearby Zalesie Nowe section (upper 21.0 m), and stable oxygen and carbon isotope measurements for the same sections in the Holy Cross Mountains, Poland (Małkowski, unpublished; curve of Trela (2000) is shown by broken line; the Zalesie Nowe record may be diagenetically altered in some parts of the section). Age correlation is based on a succession of zonally diagnostic conodont species and bentonite horizons; note occurrences of the *Hirnantia–Mucronaspis* and *Normalograptus faunas* at Zalesie Nowe. Numbers represent episodes of increased contribution of the conodont *Scabbardella*, of Gondwanan affinities and with extremely deep basal cavity of elements suggestive of cold- or deep-water preferences; note profound rebuilding of the assemblage after a period of disturbance near the end of sedimentation of organodetrital limestone with phosphate-coated grains near the end of the *Amorphognathus superbus* Zone.

calations is present (Tomczyk 1962). *Nemagraptus* is present in the lower part of the shale, giving a Caradocian age.

The topmost limestone intercalation in shales at Pobroszyn has yielded specimens of conodonts indicative of the late Caradoc or Ashgill, and possibly corresponding to the Wólka beds. The informal Wólka Siltstone Formation, of possibly 100 m thickness, yielded a diverse trilobite and brachiopod *Foliomena* assemblage of early Ashgill (Rawtheyan) age (Kielan 1960; Cocks & Rong 1988; Mergl 1990).

The sediments of the Łysogóry region were deposited in an open-marine environment that extended northward. It is a matter of controversy whether the region was located close to the Kielce region in the Ordovician or whether the dislocation along the northern margin of the Łysogóry range marks a large-scale migration of terranes. In this case the Łysogóry region would be related to the East European Platform (e.g. Lewandowski 1993). The late Llanvirn to Caradoc palaeopole of 11°N, 46.8°E recorded in the Mójcza Limestone Formation at Mójcza fits the polar wander path for Baltica, making any post-Ordovician rotation of the area relative to Baltica unlikely according to Schätz *et al.* (2006).

The East European Platform Margin in northern Poland (Baltica)

In northern Poland, Ordovician sediments are recorded only from boreholes (Figs 5.3, 31 & 5.14). In NW Poland, the highly folded Ordovician succession in the marginal part of the Tornquist-Teysseyre Zone was recovered in the Koszalin-Chojnice region of Pomerania (see above). This can be compared and correlated with the Ordovician of the Island of Rügen, north eastern Germany, and thus be attributed to Avalonia (see above).

In NE Poland, northward of the Łysogóry Region, Ordovician strata are only found at depths of up to 5 km (Pożaryski & Tomczyk 1993). In this areas at the margin of the East European Platform, the thickness of both the Ordovician and its younger strata cover increase distally.

The margin of the East European Platform is formed by the Mazowsze-Lublin Graben (established in the late Devonian), where the presence of Ordovician sediments at up to 8 km depth has been interpreted based on seismic data (Pożaryski & Tomczyk 1993). On the NE side of the Tornquist-Teisseyre Lineament relatively deep-water Ordovician sediments are present. The borehole closest to the Małopolska Massif on the East European Platform with recordings of Ordovician strata is the Łopiennik IG 1 borehole (Figs 5.3, 30 & 5.14).

The major carbonate platform of the Baltic region is known from hundreds of deep boreholes in NE Poland (Fig. 5.3, 31; Modliński 1975, 1982, 1984). The Ordovician was cored in this area for the first time in 1958 in the Żebrak IG 1 borehole near Siedlce in the Podlasie area.

The Ordovician rocks were of importance in seismic research since they provide the only reflective horizon within the lower Palaeozoic strata. Successions in this region are similar to those in the Baltic outcrops. They differ from those of the Małopolska Massif in the absence of any thick post-Tremadocian, Early Ordovician clastic sequences. In the Łopiennik IG 1 borehole a thin conglomerate bed occurs above a 48 m thick series of mudstones of Tremadocian age (Modliński 1984). The time equivalents of the Międzygórz conglomerates and the Bukówka Sandstone are represented in this borehole by argillaceous limestones. The Caradoc comprises a graptolite-rich shale, but in other boreholes, towards the centre of the platform, it is replaced by an argillaceous limestone. In the NW a black bituminous Tremadocian shale with *Rhabdinopora* occurs, whereas conglomerates and sandstones occur near the northeastern boundary of Poland (Modliński & Szymański 1997; Bednarczyk 1998). Generally, shales dominate in the western part of the area, except for the expansion of limestone facies near the Arenig–Llanvirn boundary. The graptolite shale covered almost the entire Polish part of the East European Platform in the middle Caradoc. Modliński & Szymański (1997) introduced local names for the lithological units of the subsurface Ordovician, that extend (under different names) to other Baltic regions.

The Alpine Chain: Austria, northern Italy

More than 50% of the Alps expose fragments of the Palaeozoic basement (Fig. 5.19). The Alpine mountain chain comprises different pre-Variscan 'building blocks' which were derived from the Gondwana margin (e.g. Raumer *et al.* 2003). The tectonic and metamorphic history of the region is complex and most of the basement pieces of the Alps do not represent the direct southern continuation of the Variscan elements observed in the Vosges-Black Forest or the Bohemian Massif (Raumer 1998). Biostratigraphical evidence remains rare and most parts of the thick successions of phyllitic rocks remain undated, although the Ordovician age of several units has been advanced in different tectonic models (e.g. Raumer *et al.* 2003).

Only the Upper Ordovician succession of the Carnic Alps (Figs 5.3, 32 & 5.19) is well known, precisely dated and studied in detail. The Caradoc–Ashgill succession is known from the Cellon section in the Carnic Alps near the Austrian–Italian border and from other outcrops in this area. The Uggwaschiefer Formation is the oldest unit that was precisely dated with chitinozoans (Priewalder 1997). In its upper part, these siliciclastics contain Rawtheyian-age shelly faunas. This formation is overlain by the Uggwakalk Formation, also of Ashgill age, deposited in a deep shelf and representing carbonate debris flow accumulations (Berry 2003). The Wolayerkalk Formation is coeval and was deposited in an inner shelf environment. Conodonts, chitinozoans and trilobites precisely date the succession, while many other fossil groups are also present, including crinozoan–bryozoan reef-building organisms. Correlations based on faunas present in strata below and above the reefs indicate a *Normalograptus extraordinarius* graptolite zone age, which correlates with the latest-Ordovician glaciation on Gondwana. *Hirnantia* faunas overlie the carbonates, with the graptolite *Persculptograptus persculptus* occurring in interbedded silty

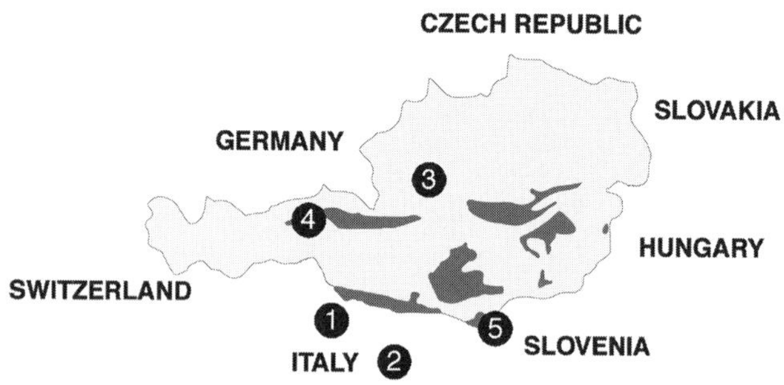

Fig. 5.19. Map of Austria indicating outcrop areas in the Alpine Chain where Ordovician (micro)fossils allowed biostratigraphical dating: 1, Agordo-Cereda phyllitic area (northern Italy); 2, Carnic Alps (Austria-Italy); 3, Northern Greywacke Zone (Eastern Austrian Alps); 4, Innsbruck Quarzphyllit Group (Austria); 5, Carinthia. See text for explanation.

shales within a sandstone–siltstone succession (Berry 2003). The Val Visdende schists underlying the Uggwaschiefer Formation are usually attributed to the Ordovician, but biostratigraphical evidence has never been reported.

Most of the older Palaeozoic elements of the Alps remain undated, and precise biostratigraphical evidence for pre-Late Ordovician rocks have not yet been provided. Acritarchs indicate Cambro-Ordovician ages for sediments from different parts of the Alpine chain, but most of these age assignments remain speculative.

In the Eastern Italian Alps (Figs 5.3, 33 & 5.19), initial biostratigraphical evidence for an Early Palaeozoic age was provided by acritarchs. The South-Alpine Metamorphic Basement in this area consists of thick phyllitic sequences metamorphosed under greenschist facies which has been attributed to Variscan activities (Sassi *et al.* 1974*a*, *b*; Sassi & Spiess 1993). The depositional age of the protoliths of these phyllites has been the subject of controversy (Sassi *et al.* 1984 and references therein). Because fossil evidence is scant, for a long time the only chronostratigraphical constraints were indirectly derived from the geological interpretation of radiometric dates (ages of the metamorphic phases and of rhyolithic rocks interbedded within the phyllitic sequence; Sassi *et al.* 1974*a*, *b*, 1984). These data show that the age of the protoliths of the phyllites should range between the pre-late Caradocian and the late Silurian/early Devonian (Sassi *et al.* 1974*a*, *b*, 1984). The first reliable biostratigraphical datings of the South-Alpine phyllites were published by Sassi *et al.* (1984) and Kalvacheva *et al.* (1986) and concerned a moderately to poorly preserved organic-walled microfossil assemblage, comprising acritarchs found in a single level of the Col di Foglia Formation in the Agordo-Cereda phyllitic area (Figs 5.3, 33 & 5.19). The acritarch assemblage was interpreted as indicative of an age range between the late Cambrian and the Tremadocian (Kalvacheva *et al.* 1986). However, the taxonomical assignments of many acritarch forms in Kalvacheva *et al.* (1986) are questionable and certainly require revision in the light of the progress of knowledge in the field of Palaeozoic palynology over the last 20 years. Nonetheless, a re-examination of the original palynological material by Kalvacheva from the Col di Foglia Formation, and the finding in this material of some acritarch morphotypes not previously recorded (Vecoli *et al.* 2008) confirms a latest Cambrian to early Tremadocian age of the assemblage. A new sampling campaign in the Agordo area (2003–04) failed to find the palyniferous horizon in the Col di Foglia Formation which yielded the original assemblage described by Kalvacheva *et al.* (1986) (M. Vecoli *et al.* 2008). On the other hand, a recent investigation of the metasediments of the Agordo area by Dieni *et al.* (2005) reported the finding of graptolites of Aeronian (Middle Llandovery, Early Silurian) age and of corals indicative of a probable Devonian age. These findings corroborate the so-called 'pre-Carboniferous hypothesis' for the age of the South Alpine Metamorphic Basement of the Eastern Alps, proposed by Sassi *et al.* (1984). According to this hypothesis, the South-Alpine Basement in Italy can be subdivided into a 'lower phyllitic complex' of pre-Caradoc age and an 'upper phyllitic complex' of Silurian to early Devonian age.

Other biostratigraphical information based on acritarchs was provided from different locations in Austria by E. Reitz in the late 1980s and early 1990s. An initial publication (Reitz & Höll 1989) concerned the Northern Greywacke Zone in the Eastern Austrian Alps (Figs 5.3, 34 & 5.19). Four out of 72 samples provided acritarchs attributed to the Tremadocian. However, the poorly to moderately well preserved specimens can, in the light of currently available biostratigraphical information, be attributed to a Late Cambrian to Early Ordovician interval, while the presence of the genus *Veryhachium*, if confirmed, should restrict the assemblage to a Tremadocian of younger age. Subsequently, Reitz & Höll (1990) illustrated some other poorly preserved acritarchs from the Innsbruck Quarzphyllit Group (Figs 5.3, 35 & 5.19) that they also attributed to the Early Ordovician. However, the illustrated specimens could also be Late Cambrian in age. In a further publication, Reitz & Höll (1991) illustrated specimens from the Unterer Wildschönauer Schiefer that they attributed to the Arenig. Some of the specimens illustrated in Reitz & Höll (1989, 1991) were republished by Reitz & Höll (1992) to confirm the Tremadocian and Arenig age, although they changed the taxonomic attribution of many specimens at the specific and even generic level. Reitz (2004) subsequently illustrated a similar assemblage from Carinthia (Figs 5.3, 36 & 5.19).

As for the Italian succession dated by acritarchs in the 1980s, new sampling campaigns were also conducted in the late 1990s and early 2000s in Austria, in order to palynologically investigate the Innsbrucker Quartzphyllit Group, the Val Visdende Formation and the Unterer Wildschönauer Schiefer. However, the palynological studies were unsuccessful, since palyniferous horizons were not discovered (T. Servais & W. Mette, M. Vecoli & H. Heinisch, unpublished data).

Summary

The Ordovician type sections for global correlation, that have been defined in the last decade, are situated either in Sweden (Baltica continent), North America (Laurentia) or the Yangtze Platform (South China continent). The historical sections in the British Isles have been abandoned and will serve only as regional standards in the future. The Ordovician outcrops and subsurface sections of Central Europe are, compared to the new standards, of limited value for international correlation and for the understanding of the Ordovician world.

The Ordovician of Central Europe belongs to various areas with, in general, a very complex tectonic evolution. The localities described here (from Belgium, Germany, the Czech Republic, Poland and the Alpine region) are part of a vast region affected by the Variscan Orogeny and, in palaeogeographical tems, all of these areas belonged to peri-Gondanan terranes, with the exception of the northeastern part of Poland that belonged to the Baltica palaeocontinent. It is today widely accepted that, during the Ordovician, the eastern part of the microcontinent of Avalonia included Belgium, western and northern Germany, and possibly northwestern Poland. It is less clear to what entity the outcrop areas of the Rhenohercynian, Saxothuringian and Moldanubian zones belonged. It is evident that they must be attributed to Gondwana-derived terranes (such as, in palaeogeographical terms, Armorica or the Armorican Terrane Assemblage and Perunica) or to sedimentary basins in the vicinity of the Gondwanan supercontinent. However, it is still not clear whether the different areas were separate microcontinents or simply tectonically separated units (terranes) or different sedimentary basins. Ongoing and future research will possibly provide answers to these questions.

Our review includes the Avalonian sequences of Belgium in the northwestern part of the investigated area of Central Europe, continues into western, northern and eastern Germany, and extends into northwestern and southern Poland. The review of the Ordovician of the Saxothuringian and the Moldanubian zones includes the outcrop areas of southeastern Germany, the Czech

Republic and southwestern Poland. The Ordovician from the pre-Variscan parts of the Alpine mountain chains of Switzerland, Austria and northern Italy are also briefly discussed. For each individual area, we present the stratigraphical succession, based on the most recent results, and correlate the successions with the modern standard of Ordovician stratigraphy, including the time-slice subdivision of Webby *et al.* (2004). The most complete successions are those from Belgium (Brabant Massif and Condroz Inlier), from Saxothuringia (Schwarzburg Anticline) and from Bohemia (Prague Basin). Most of the other areas present generally isolated outcrops in a complex tectonic context that cannot easily be integrated in a complete stratigraphical succession. Such areas include the Ardennes (Belgium and western Germany), the Black Forest (Schwarzwald, southwestern Germany), the Ebbe Mountains and isolated outcrops in Hessen (western Germany), the subsurface Ordovician of northern Germany, many of the areas in eastern Germany (such as the Harz Mountains or the Lausitz region, for example), and most areas in Poland, but also the basement of the Alps, which is very poorly understood but includes some Ordovician fragments.

Due to the complex tectonic history, most rocks are poorly preserved, and fossils are often absent in the sedimentary successions, which makes a perfect understanding of the stratigraphical succession difficult in many areas. Furthermore the development of the different sedimentary basins is mostly unknown thus far. Sedimentary analyses are absent from many areas, and the precise relationship between the different successions presented in this review remains preliminary.

Fossils are absent in many sedimentary units, and the mostly siliciclastic successions, that are now attributed to cold-water environments of high latitudes in the southern hemisphere, provide, compared to the palaeocontinents at low latitudes, only few fossils. An exception is the Prague Basin, an area that is famous for its excellent palaeontological content. The development of carbonate rocks in the latest Ordovician was possibly due to a global warming event. In summary, a high number of micro- and macrofossils have been described from the various successions of the Ordovician of Central Europe, allowing not only international correlation and palaeogeographical attributions, but also a first interpretation of the palaeoenvironment.

This review was only possible thanks to the detailed field work of many colleagues from the different areas investigated who are here gratefully acknowledged. We particularly thank D.A.T. Harper (Copenhagen) and A.W. Owen (Glasgow), both titular members of the Ordovician Subcommission (IUGS-UNESCO), and the editor, for their careful and constructive reviews of the text. This paper is a contribution to the IGCP projects 499 'The Rheic Ocean' and 503 'Ordovician Palaeogeography and Palaeoclimate'.

References

ALEKSANDROWSKI, P., KRYZA, R., MAZUR, S., PIN, C. & ZALASIEWICZ, J. A. 2000. The Polish Sudetes; Caledonian or Variscan? *Transactions of the Royal Society of Edinburgh, Earth Sciences*, **90**, 127–146.

ANDRÉ, L., HERBOSCH, A., VANGUESTAINE, M., SERVAIS, T., VAN GROOTEL, G., LOUWYE, S. & VERNIERS, J., 1991. Guidebook of the excursion on the stratigraphy and magmatic rocks of the Brabant Massif, Belgium. *In*: ANDRÉ L., HERBOSCH A., VANGUESTAINE M. & VERNIERS J., *Proceedings of the International Meeting on the Caledonides of the Midlands and the Brabant Massif*, Brussels, 20–23 September 1989. *Annales de la Société géologique de Belgique*, **114** (2), 283–323.

BANKWITZ, P. & BANKWITZ, E. 2000. Terrane fragment belt along the northern margin of the Saxothuringian microplate. *Acta Universitatis. Carolinae, Geologica*, **42**, 423–431.

BANKWITZ, P. & BANKWITZ, E. 2002. Kritische Bestandsaufnahme der Tektonik des Schwarzburg-Antiklinoriums (Proterozoikum) und der Vesser-Zone (Kambrium). *Zeitschrift für Geologische Wissenschaften*, **30**(4/5), 235–262.

BANKWITZ, P., BANKWITZ, E., KRAMER, W. & PIN, C. 1994. Early Paleozoic bimodal volcanism in the Vesser area, Thuringian Forest, eastern Germany. *Zentralblatt für Geologie und Paläontologie*, **I** (1992)9/10, 1113–1132.

BANKWITZ, E., BANKWITZ, P., FALK, F., LÜTZNER, H. & MANN, M. 1997. Schwarzburg Antiklinorium. *In*: *Stratigraphie von Deutschland II. Teil I: Thüringen, Sachsen, Ostbayern*. Stratigraphische Kommission Deutschlands, Stuttgart, Courier Forschungsinstitut Senckenberg, **200**, 19–54.

BARANOWSKI, Z. 1975. Flysch facies in the epimetamorphic of the northern part of the Kaczawa Mountains (SW Poland). *Rocznik Polskiego Towarzystwa Geologicznego*, **45**, 21–40.

BARANOWSKI, Z. & URBANEK, Z. 1972. Ordovician conodonts from the epimetamorphic complex from Rzeszówek in the Kaczawa Mts (western Sudetes). *Bulletin de l'Academie Polonaise des Sciences, Séries de Sciences de la Terre*, **20**, 211–216.

BARNES, C. 2004. Ordovician Oceans and Climate. *In*: WEBBY, B.D., PARIS, F., DROSER, M.L. & PERCIVAL, I. (eds) *The Great Ordovician Biodiversification Event*. Columbia Press, New York, 72–76.

BARRANDE, J. 1846. *Notice préliminaire sur le système silurien et les trilobites de Bohème*. Leipzig.

BARRANDE, J. 1852. *Système silurien du centre de la Bohème*. Prague.

BARRANDE, J. 1868. Silurische Fauna aus der Umgebung von Hof in Bayern. *Neues Jahrbuch für Mineralogie*, 641–696.

BARTSCH, L. 1956. *Geologische Untersuchungen im Nordsächsischen Sattel unter Berücksichtigung neuer Fossilfunde aus dem anstehenden Gestein*. Diplomarbeit, Math.- Naturwiss. Fakultät, Univ. Greifswald.

BEDNARCZYK, W. 1964. Stratygrafia i fauna tremadoku i arenigu (oelandianu) regionu kieleckiego Gór Świętokrzyskich. *Biuletyn Geologiczny Uniwersytetu Warszawskiego*, **4**, 3–127, 188–197.

BEDNARCZYK, W. 1966*a*. Litologia dolnego ordowiku (elandu) w regionie kieleckim Gór Świętokrzyskich. *Biuletyn Geologiczny Uniwersytetu Warszawskiego* **8**, 4–53.

BEDNARCZYK, W. 1966*b*. Stratygrafia wapieni z Mójczy pod Kielcami, Góry Świętokrzyskie. *Acta Geologica Polonica*, **16**, 107–123.

BEDNARCZYK, W. 1971. Stratigraphy and palaeogeography of the Ordovician in the Holy Cross Mts. *Acta Geologica Polonica*, **21**, 574–616.

BEDNARCZYK, W. 1974. The Ordovician in the Koszalin-Chojnice region (western Pomerania). *Acta Geologica Polonica*, **24**, 581–600.

BEDNARCZYK, W. 1998. Stratigraphical position of the "Obolus Sandstones" (Upper Cambrian-?lowermost Ordovician) in Poland. *Bulletin of the Polish Academy of Sciences, Earth Sciences*, **46**, 43–54.

BEDNARCZYK, W. & BIERNAT, G. 1978. Inarticulate brachiopods from the Lower Ordovician of the Holy Cross Mountains, Poland. *Acta Palaeontologica Polonica*, **23**, 293–310.

BEDNARCZYK, W., CHLEBOWSKI, R. & KOWALCZEWSKI Z. 1966. Utwory ordowiku w synklinie bardziańskiej. *Kwartalnik Geologiczny*, **10**, 705–723.

BEHR, H.-J., ENGEL, W., FRANKE, W., GIESE, P. & WEBER, K. 1984. The Variscan Belt in Central Europe; main structures, geodynamic implications, open questions. *Tectonophysics*, **109**, 15–40.

BEIER, H. & KATZUNG, G. 1999 Strain-Analyse an schwach deformiertem Altpaläozoikum der Tiefbohrungen Rügen 5/66 und G14/1–86 (NE-Deutschland). *Zentralblatt für Geologie und Paläontologie*, **1998**(1/2), 59–80.

BEIER, H., KATZUNG, G., MALETZ, J. & SCHALLREUTER, R. 2001. 4.31.Vorpommern und angrenzende Ostsee (Rügen-Kaledoniden). *In*: *Stratigraphie von Deutschland II, Ordovizium, Kambrium, Vendium, Riphäikum. Teil III.* Stratigraphische Kommission Deutschlands. Courier Forschungsinstitut Senckenberg, **235**, 120–134.

BELKA, Z., VALVERDE-VAQUERO, P., DÖRR, W., AHRENDT, H., WEMMER, K., FRANKE, W. & SCHÄFER, J. 2002. Accretion of first Gondwanan-derived terranes at the margin of Baltica. *In*: WINCHESTER, J.A., PHARAOH, T.C. & VERNIERS, J. (eds) *Palaeozoic Amalgamation of Central Europe*. Geological Society, London, Special Publications, **201**, 19–36.

BERGER, H.-J. & DOCEKAL, A. 1997. Nordsächsisches Schiefergebirge.

In: *Stratigraphie von Deutschland II. Teil I: Thüringen, Sachsen, Ostbayern.* Stratigraphische Kommission Deutschlands, Courier Forschungsinstitut Senckenberg, **200**, 335–342.

BERGSTRÖM, S.M. 1971. Conodont biostratigraphy of the Middle to Upper Ordovician of Europe and eastern North America. *Geological Society of America, Memoir*, **127**, 83–157.

BERGSTRÖM, S.M. 1990. Relations between conodont provincialism and the changing palaeogeography during the Early Palaeozoic. In: MCKERROW, W.S. & SCOTESE, C.R. (eds) *Palaeozoic Palaeogeography and Biogeography.* Geological Society, London, Memoirs, **12**, 105–122.

BERRY, W. B.N., 2003. Late Ordovician environmental changes in Carnic Alps and central Nevada: a comparative study. *Bulletin de la Société Géologique de France*, **174,** 211–216.

BEUGNIES, A. 1963. Le Massif de Rocroi. *Bulletin du Service de la Carte Géologique de France*, **270**, 59.

BIEWALD, W. & VOIGT, T. 2004. *Geologische Karte von Thüringen 1:25 000, Blatt 5532 Steinach.* Thüringer Landesanstalt für Umwelt und Geologie, Jena.

BLUMENSTENGEL, H. 1965. Zur Ostracodenfauna eines Kalkgerölls aus dem Thüringer Lederschiefer (Ordovizium). *Freiberger Forschungshefte*, C, **182**, 63–78.

BLUMENSTENGEL, H. & SCHALLREUTER, R. 1998. Ordovizische Ostrakodenfaunen Thüringens. *Beiträge zur Geologie von Thüringen, Neue Folge*, **5**, 13–22.

BOUČEK, B. 1973. *Lower Ordovician graptolites of Bohemia.* Academia, Prague, 1–185.

BRAUSE, H., ELICKI, O. & KEMNITZ, H. 1997. Lausitzer Masse (9). *In*: *Stratigraphie von Deutschland II, Ordovizium, Kambrium, Vendium, Riphäikum; Teil I: Thüringen, Sachsen, Ostbayern.* Stratigraphische Kommission Deutschlands, Courier Forschungsinstitut Senckenberg, **200**, 299–334.

BRENCHLEY, P.J. & ŠTORCH, P. 1989. Environmental changes in the Hirnantian (Upper Ordovician) of the Prague Basin, Czechoslovakia. *Geological Journal*, **24**, 165–181.

BULMAN, O.M.B. 1950. On some Ordovician graptolite assemblages of Belgium. *Bulletin de l'Institut Royal des Sciences Naturelles de Belgique*, **26**(5), 1–8.

BURMANN, G. 1968. Diacrodien aus dem unteren Ordovizium. *Paläontologische Abhandlungen, B*, **2**(4), 637–652.

BURMANN, G. 1970. Weitere organische Mikrofossilien aus dem unteren Ordovizium. *Paläontologische Abhandlungen, B*, **3**(3/4), 289–332.

BURMANN, G. 1973*a*. Das Ordovizium der nördlichen Phyllitzone. Teil 1 : Paschlebener Vorsprung. *Zeitschrift geologischer Wissenschaften*, **1.7**, 757–787.

BURMANN, G. 1973*b*. Das Ordovizium der nördlichen Phyllitzone. Teil 2 : Wippraer Zone. *Zeitschrift geologischer Wissenschaften, Themenhefte*, **1**, 9–43.

BURMANN, G. 1973*c*. Chitinozoen aus dem Arenig. *Abhandlungen des Zentral Geologischen Instituts Berlin*, **18**, 129–160.

BURMANN, G. 1976. Übersicht über das ordovizische Mikroplankton im Südteil der DDR (Vogtland, Wildenfelser Zwischengebirge). *Jahrbuch für Geologie*, **7/8**(1971/72), 47–62.

BURMANN, G. 2001. Acritarchen von Stadtsteinach/Frankenwald und die biostratigraphische Stellung des Griffelschiefers (Arenig) im Saxothuringikum nebst Bemerkungen zum Hauptquarzit und zur Dicranochitina-Provinz. *Neues Jahrbuch für Geologie und Paläontologie, Abhandlungen*, **220**, 25–62.

BURMANN, G. & WASKOWIAK, R. 1968. Zur Einstufung des Griffelschiefers auf Blatt Kirchberg-Wildenfels. *Geologie*, **17**(9), 1116.

BURMANN, G., ERDTMANN, B.-D., FATKA, O., HEUSE, T., KRAFT, P. & WALTER, H. 1994. The Ordovician chitinozoan genus Dicranochitina Burmann 1973. Poster session, Commission Internationale de Microflore du Paleozoique (CIMP), Subcommission Chitinozoans, Abstracts, Sheffield, September 1994.

BURMANN, G. , EHLING, B.-C., FRANZKE, H.J., HOTH, K., KOPP, J. & WUNDERLICH, J. 2001. Nördliche Phyllitzone (an der MKZ). *In*: *Stratigraphie von Deutschland II, Ordovizium, Kambrium, Vendium, Riphäikum. Teil III.* Stratigraphische Kommission Deutschlands, Courier Forschungsinstitut Senckenberg, Frankfurt, **235**, 11–67.

CARLS, P. 2001. Kritik der Plattenkinematik um das Rhenohercynicum bis zum frühen Devon. *Braunschweiger geowissenschaftliche Arbeiten*, **24**, 27–108.

CHLEBOWSKI, R. 1971. Petrografia utworów ordowiku rejonu synkliny bardziańskiej w południowej części Gór Świętokrzyskich. *Archiwum Mineralogiczne*, **29**, 193–304.

CHLEBOWSKI, R. 1981. Rozwoj stosunków litofacjalnych w ordowiku Gór Świętokrzyskich. *Przegląd Geologiczny*, **29**, 346–350.

CHLEBOWSKI, R. & SZANIAWSKI, H. 1974. Chitinozoa from the Ordovician conglomerate at Międzygórz in the Holy Cross Mts. *Acta Geologica Polonica*, **24**, 221–230.

CHLUPÁČ, I. & KUKAL, Z. 1988. Possible global events and the stratigraphy of the Paleozoic of the Barrandian (Cambrian-Middle Devonian, Czechoslovakia). *Sborník geologických Věd, Geology*, **43**, 83–146.

CHOROWSKA, M., RADLICZ, K. & TOMCZYK, H. 1981. Utwory ordowiku, syluru i dewonu w Lubaniu (Sudety, Góry Kaczawskie). *Kwartalnik Geologiczny*, **25**, 243–263.

CHRISTIANSEN, J.L. & STOUGE, S., 1999. Oceanic circulation as an element in palaeogeographical reconstructions: the Arenig (early Ordovician) as an example. *Terra Nova*, **11**, 73–78.

CLARK, A.H., SCOTT, D.J., SANDEMAN, H.A., BROWLEY, A.V. & FARRAR, E. 1998. Siegenian generation of the Lizard ophiolite: U-Pb zircon age data for plagiogranite, Porthkerris, Cornwall. *Journal of the Geological Society, London*, **155**, 595–598.

COCKS, L.R.M. & RONG, J.-Y. 1988. A review of the Lower Ordovician Foliomena brachiopod fauna with a new data from China, Wales, and Poland. *Palaeontology*, **31**, 53–67.

COCKS, L.R.M. & TORSVIK, T.H. 2004. Major Terranes in the Ordovician. *In*: WEBBY, B.D., PARIS, F., DROSER, M.L. & PERCIVAL, I. (eds) *The Great Ordovician Biodiversification Event.* Columbia Press, New York, 61–67.

COCKS, L.R.M., MCKERROW, W.S. & VAN STAAL, C.R. 1997. The margins of Avalonia. *Geological Magazine*, **134**, 627–636.

COOPER, R.A. & SADLER, P.M., 2004. The Ordovician Period. *In*: GRADSTEIN, F.M., OGG, J.G. & SMITH, A.G. (eds) *A Geologic Time Scale 2004*. Cambridge University Press, 165–187.

COOPER, R.A., NOWLAN, G.S. & WILLIAMS, S.H., 2001. Global stratotype and point for base of the Ordovician System. *Episodes*, **24**, 19–28.

CORIN, F. 1964. Sur les roches éruptives de la tranchée d'Hasquempont, canal de Charleroi. *Bulletin de la Société belge de Géologie*, **72**(1963), 94–98.

CZARNOCKI, J. 1928. Profil dolnego i górnego ordowiku w Zalesiu pod Łagowem w porównaniu z ordowikiem innych miejscowości środkowej części Gór Świętokrzyskich. *Sprawozdania Państwowego Instytutu Geologicznego*, **4**, 555–568.

CZARNOCKI, J. & SAMSONOWICZ, J. 1913. Drobny przyczynek do znajomości syluru Gór Kielecko-Sandomierskich. *Sprawozdania z posiedzeń Towarzystwa Naukowego Warszawskiego, Wydział* III, **6**, 487–491.

DAHLGRÜN, F. 1930. Die stratigraphische Bedeutung der neuen Graptolithenfunde im Thüringer Untersilur. *Beiträge zur Geologie Thüringen*, **2**, 23–28.

DALLMEYER, R.D., FRANKE, W. & WEBER, K. (eds) 1995. *The Pre-Permian geology of Central and Eastern Europe.* Springer, Berlin.

DALLMEYER, R.D., GIESE, U., GLASMACHER, U. & PICKEL, W. 1999. First 40Ar/39Ar age constraints for the Caledonian evolution of the Trans-European Suture Zone in NE-Germany. *Journal of the Geological Society, London*, **156**, 279–290.

DEAN, W., 1991. Ordovician trilobites from the inlier at Le Petit Fond d'Oxhe, Belgium. *Bulletin de l'Institut Royal des Sciences Naturelles de Belgique, Sciences de la Terre*, **61**, 135–155.

DEBACKER, T., SINTUBIN, M. & VERNIERS, J. 2001. Large-scale slumping deduced from structural relations and sedimentary features: a study in the Lower Palaeozoic Anglo-Brabant fold belt, Belgium. *Journal of the Geological Society, London*, **158**, 341–352.

DEBACKER T., DEWAELE S., SINTUBIN M., VERNIERS J., MUCHEZ PH. & BOVEN A. 2005. Timing and duration of the progressive deformation of the Brabant Massif, Belgium. *Geologica Belgica* **8**, 20–34.

DELCAMBRE, B. & PINGOT, J.L. 2004. *Carte géologique de Wallonie n°46 - 7/8 et Notice explicative de la carte Fontaine-l'Evêque Charleroi.* Ministère de la Région Wallonne, Namur, 1–108.

DE VOS, W., VERNIERS, J., HERBOSCH, A. & VANGUESTAINE, M. 1993. A new geological map of the Brabant Massif. *In*: PHARAOH, T.C., MOLYNEUX, S.G., MERRIMAN, R.J., LEE, M.K. & VERNIERS, J. (eds) Special Issue on the Caledonides of the Anglo-Brabant Massif. *Geological Magazine*, **130**, 605–611.

DIENI, I., GIORDANO, D., LOYDELL, D.K. & SASSI, F.P. 2005. Discovery of Llandovery (Silurian) graptolites and probable Devonian corals in the Southalpine Metamorphic Basement of the Eastern Alps (Agordo, NE Italy). *Geological Magazine*, **142**, 1–5.

Dzik, J. 1976. Remarks on the evolution of Ordovician conodonts. *Acta Palaeontologica Polonica* **21**, 395–455.

Dzik, J. 1978. Conodont biostratigraphy and paleogeographical relations of the Mójcza Limestone (Holy Cross Mts, Poland). *Acta Palaeontologica Polonica*, **23**, 51–72.

Dzik, J. 1983. Relationships between Ordovician Baltic and North American Midcontinent conodont faunas. *Fossils and Strata*, **15**, 59–83.

Dzik, J. 1984*a*. Phylogeny of the Nautiloidea. *Palaeontologia Polonica*, **45**, 1–219.

Dzik, J. 1984*b*. Early Ordovician conodonts from the Barrandian and Bohemian-Baltic faunal relationships. *Acta Palaeontologica Polonica*, **28**, 327–368.

Dzik, J. 1990. Conodont evolution in high latitudes of the Ordovician. *Courier Forschungsinstitut Senckenberg*, **117**, 1–28.

Dzik, J. 1994. Conodonts of the Mójcza Limestone. *Palaeontologia Polonica*, **53**, 43–128.

Dzik, J. 1999*a*. The Ordovician of the Holy Cross Mountains. Excursion 1, day 1, stop 1: Mójcza. Excursion 1, day 1, stop 2: Zalesie Nowe. Excursion 1, day 2, stop 4: Międzygórz. *In*: Dzik J., Lindemann U., & Heuse T. (eds) *International Symposium on the Ordovician System*, ISOS Prague 1999, Pre-Conference Fieldtrip. Excursion Guide Poland and Germany 3–13, 18–23.

Dzik, J. 1999*b*. Evolution of the Late Ordovician high-latitude conodonts and dating of Gondwana glaciations. *Bolletino della Societa Paleontologica Italiana*, **37**, 237–253.

Dzik, J. & Pisera, A. 1994. The Mójcza Limestone and its sedimentation. *Palaeontologia Polonica*, **53**, 5–41.

Eisenack, A. 1939. Chitinozoen und Hystrichosphaeriden im Ordovicium des Rheinischen Schiefergebirges. *Senckenbergiana Lethaia*, **21**, 135–152.

Eiserhardt, K.H., Heyckendorf, K. & Thombansen, E. 1981. Zur Stratigraphie und Tektonik des nördlichen Ebbe-Teilsattels (Sauerland, Rheinisches Schiefergebirge). *Mitteilungen aus dem Geologisch-Paläontologischen Insitut der Universität Hamburg*, **50**, 199–238.

Eiserhardt, K.H., Koch, L. & Maletz, J. 2001. Nördliches Rheinisches Schiefergebirge. *In*: *Stratigraphie von Deutschland II, Ordovizium, Kambrium, Vendium, Riphäikum. Teil III.* Stratigraphische Kommission Deutschlands, Courier Forschungsinstitut Senckenberg, Frankfurt, **235**, 90–108.

Ellenberg, J. 1998. Der ordovizische Hauptquarzit in Thüringen - eine Fan-Delta-Bildung? *Geowissenscaftliche Mitteilungen von Thüringen*, **6**, 7–20.

Ellenberg, J. 2000. Die Bildung oolithischer Eisenerze im thüringischen Ordovizium. *Geowissenschaftliche Mitteilungen von Thüringen, Beiheft*, **9**, 57–82.

Ellenberg, J., Falk, F. & Lützner, H. 1992. Siliciclastic shelf sedimentation of Early Palaeozoic deposits in the Thüringer Schiefergebirge (Thuringian Slate Mountains). *Excursion Guide-Book, 13th IAS Regional Meeting Sedimentology*, Excursion. B1, Jena 1992, 133–158.

Erdtmann, B.D. 1986. Comments on some earliest Ordovician (Salmian) graptolites from Solwaster, Massif de Stavelot, Belgian Ardennes. *Aardkundige Mededelingen*, **3**, 75–86.

Erdtmann, B.D. 1991. The post-Cadomian Early Palaeozoic tectonostratigraphic of Germany (Attempt at an analytical review). *Annnales de la Société géologique de Belgique*, **114**, 19–43.

Estrada, S., Heuse, T. & Schulz, E. 1994. Zur regionalgeologischen Interpretation kambro-ordovizischer Schichten an der Nordwest-Flanke des Schwarzburger Antiklinoriums, Thüringen. *Zeitschrift für Geologische Wissenschaften*, **22**, 537–553.

Falk, F. 1956 Ergebnisse einer Teilkartierung im Algonkium und Altpaläozoikum und der Südost-Flanke des Schwarzburgers Sattels (Blatt Steinach). *Berichte der Geologischen Gesellschaft der DDR*, **2**, 38–41.

Falk, F. 1964 Die konglomeratische Arkose der Goldisthaler Schichten. Ein grobklastischer Horizont im fraglichen Kambrium des Schwarzburger Sattels in Thüringen. *Abhandlungen der Deutschen Akademie der Wissenschaften zu Berlin, Klasse für Bergbau, Hüttenwesen und Montangeologie*, **2**, 3–26.

Falk, F. & Lützner, H. 1991. Schichtungsgefüge der Goldisthal- und Frauenbach-Folge (Kambro-Ordovizium) im Thüringischen Schiefergebirge. *Wissenschaften Beiträge Universität Greifswald, Fortschrift der Geologie*, 27–31.

Falk, F., Hahne, K. & Lützner, H. mit einem Beitrag von Gehmlich, M. 2000. Der Beginn siliziklastischer Sedimentation im Altpaläozoikum des Saxothuringikums: Sedimentologie und Geochemie des Richtprofils Goldisthal (Thüringisches Schiefergebirge). *Zeitschrift der Deutschen geologischen Gesellschaft*, **151**(4), 365–414.

Fatka, O. 1993. Chitinozoans and acritarchs in latest Tremadoc-early Arenig sediments of the Prague Basin, Czech Republic. *In*: Molyneux S.G. & Dorning K.J. (eds) *Contributions to Acritarch and Chitinozoan Research.* London, Special Papers in Palaeontology, **48**, 29–36 .

Fatka, O., Kraft, J., Kraft, P., Mergl, M., Mikuláš, R. & Štorch, P. 1995. Ordovician of the Prague Basin: Stratigraphy and development. *In*: Cooper, J.D., Droser, M.L. & Finney, S.C. (eds) *Ordovician Odyssey: Short Papers for the 7th International Symposium on the Ordovician System*, Pacific Section Society for Sedimentary Geology (SEPM), book **77**, 241–244. Fullerton, California.

Fatka, O., Kraft, J. & Kraft, P. 1996. Paleontological stratigraphical relations on the Arenig - Llanvirn boundary in the Prague Basin (Ordovician, Bohemia). *In*: Baldis, B. & Acenolaza, F.G. (eds) *El Paleozoico inferior en el nortoeste del Gondwana*, 263–264, INSUGEO, Tucuman, Argentina.

Ferretti, A. & Barnes, C.R. 1997. Upper Ordovician conodonts from the Kalkbank limestone of Thuringia, Germany. *Palaeontology*, **40**, 15–42.

Fiala, F. 1971*a*. Ordovician diabase volcanism and biotites lamprophyres of the Barrandian. *Sborník geologických Věd, Geologie*, **19**, 7–97.

Fiala, F. 1971*b*. The Upper Proterozoic and Lower Paleotoic geosynclinal volcanism of the Barrandian area and the Železné hory Mts. *Krystaliniukum*, **8**, 7–25.

Finney, S.C. & Chen Xu, 1990. The relationship of Ordovician graptolite provincialism to palaeogeography. *In*: McKerrow, W.S. & Scotese, C.R. (eds) *Palaeozoic Palaeogeography and Biogeography.* Geological Society, London, Memoirs, **12**, 123–128.

Fortey, R.A. & Cocks, L.R.M. 2005. Late Ordovician global warming – The Boda event. *Geology*, **33**(5), 405–408.

Fortey, R.A. & Seilacher, A.1997. The trace fossil Cruziana semiplicata and the trilobite that made it. *Lethaia*, **30**, 105–112.

Fortey, R.A., Harper, D.A.T., Ingham, J.K., Owen, A.W. & Rushton, A.W.A. 1995. A revision of Ordovician series and stages from the historical type area. *Geological Magazine*, **132**(1), 15–30.

Franke, D. 1994. The deformational history of the Caledonian terranes at Baltica's southwest margin. *Zeitschrift für Geologische Wissenschaften*, **22**, 67–80.

Franke, D. & Illers, K. 1994. The Caledonian deformational fabric of the Ordovician of Rügen island (N.E. Germany). *Zeischrift für Geologische Wissenschaften*, **22**, 359–370.

Franke, D., Gründel, J., Lindert, W., Meissner, B., Schulz, E., Zagora, I. & Zagora, K. 1994. Die Ostseebohrung G14 – eine Profilübersicht. *Zeitschrift für Geologische Wissenschaften*, **22**, 235–240.

Franke, D., Hoffmann, N. & Lindert, W. 1996. The Variscan Deformation Front in East Germany. Part 2 : Tectonic Interpretation. *Zeitschrift für angewandte Geologie*, **42**, 44–56.

Franke, W. 1984. Variszischer Deckenbau im Raume der Münchberger Gneismasse - abgeleitet aus Fazies, Deformation und Metamorphose im umgebenden Paläozoikum. *Geotektonische Forschungen*, **68**, 1–253.

Franke, W. 1989. Tectonostratigraphic units in the Variscan Belt of Central Europe. Geological Society of America, Special Papers, **230**, 67–90.

Franke, W. 1995. Rhenohercynian foldbelt : authochton and nonmetamorphic nappe units – stratigraphy. *In*: Dallmeyer, R.D., Franke, W. & Weber, K. (eds) *The Pre-Permian Geology of Central and Eastern Europe*. Springer, Berlin, 33–49.

Franke, W. 1998. Tectonic and plate tectonic units at the North Gondwanan margin : evidence from the Central European Variscides. *Schriften des Staatlichen Museums für Mineralogie und Geologie zu Dresden*, **9**, 132–134.

Franke, W., Altherr, R., Haak, V. & Oncken, O. 1999. Orogenic processes – quantification and modelling in the Variscides: summary of summaries. Geologische Vereinigung. 89. Jahrestagung, Freiberg, Sachsen. *Terra Nostra*, **99**, 7–10.

Freyer, G. 1967. Schichtung und Biostratinomie im unteren Teil des Dubrauquarzit-Komplexes. *Abhandlungen und Berichte des Naturkundemuseums Görlitz*, **42**, 1–11.

Freyer, G. 1981. Die Cruzianen von Hainichen bei Borna und das Alter ihrer Fundschichten. *Freiberger Forschungshefte, C*, **363**, 51–55.

Freyer, G. & Wiefel, H. 1991. Ein Brachiopodenfund aus der Frauenbach-Wechsellagerung des Schwarzburger Sattels. *Veröffentlichungen Naturhistorischen Museums Schleusingen*, **6**, 60–62.

Fuchs, A. 1989. Zur Stratigraphie der Gräfenthaler Schichten (Ordovizium) in Thüringen auf der Grundlage von Conodontenuntersuchungen. *Veröffentlichungen Naturhistorischen Museums Schleusingen*, **4**, 78–82.

Fuchs, A. 1990. Bemerkungen zur Stratigraphie und Charakter der ordovizischen Conodontenfauna Thüringens. *Neues Jahrbüch für Geologie und Paläontologie, Monatshefte*, **4**, 193–214.

Gaertner, H.-R.v. 1934. Schichtenfolge und Tektonik im mittleren Teil des Schwarzburger Sattels. *Jahrbuch der preussichen Geologischen Landesanstalt Berlin*, **54**(1933), 1–36.

Gaertner, H.-R.v. 1944. Die Schichtgliederung der Phyllitgebiete in Thüringen und Nordbayern und ihre Einordnung in das stratigraphische Schema. *Jahrbuch der Reichsstelle für Bodenforschung*, **62**(1941), 54–80.

Gaertner, H.-R.v. 1964. Stratigraphie und Tektonik am Südostabfall des Schwarzburger Sattels (Thüringen). *Geologie*, **13**(6–7), 682–700.

Gebauer, D. 1993. Geochronologische Übersicht. *In*: Bauberger, W. (ed.), *Geologische Karte von Bayern 1:25 000*, Erläuterungen zum Blatt Nr. 6439 Tännesberg. *Bayerisches Geologisches Landesamt*, 104 S., München.

Gehmlich, M. 2003. Die Cadomiden und Varisziden des Saxothuringischen Terranes: Geochronologie magmatischer Ereignisse. *Freiberger Forschungshefte, C*, **500**, 1–129.

Gehmlich, M., Linnemann, U., Tichomirowa, M., Lützner, M. & Bombach, K. 1997. Datierung und Korrelation neoproterozoisch-frühpaläozoischer Profile des Schwarzburger Antiklinoriums und der Elbezone auf der Basis der Geochronologie von Einzelzirkonen. *Zeitschrift für Geologische Wissenschaften*, **25**(1/2): 191–201.

Geinitz, H.B. 1873. Paläontologische Mittheilungen aus dem Mineralogischen Museum in Dresden. *Sitzungsberichte der naturwissenschaftlichen Gesellschaft Isis in Dresden*, **1872**, 125–131.

Gerdes, A., Pharaoh, T., Timmermann, H., Verniers, J., Vecoli, M., Noble, S. & Parrish, R. 2001. Nd-isotope record of clastic sediments in response to orogenic processes: from Gondwana break-up to Palaeozoic amalgamation of Central Europe. *Abstracts Geologische Vereinigung – Deutsche Geologische Gesellschaft 2001 Margins Meeting*, 2–6 October Kiel, Germany.

Geukens, F. 1986. Commentaire à la carte géologique du Massif de Stavelot. *Aardkundige Mededelingen*, **3**, 15–29.

Geukens, F. 1999. Notes accompagnant une révision de la carte structurale du Massif de Stavelot. *Aardkundige Mededelingen*, **9**, 183–190.

Geyer, G. & Wiefel, H. 1997. Fränkisch-Thüringisches Schiefergebirge. *In*: *Stratigraphie von Deutschland II. Teil I: Thüringen, Sachsen, Ostbayern*. Stratigraphische Kommission Deutschlands Courier Forschungsinstitut Senckenberg, **200**, 57–102.

Ghienne, J.F. 2003. Late Ordovician sedimentary environments, glacial cycles, and post-glacial transgression in the Taoudeni Basin, West Africa. *Palaeogeography, Palaeoclimatology, Palaeoecology*, **189**, 117–145.

Giese, U., Katzung, G. & Walter, R. 1994. Detrital composition of Ordovician sandstones from the Rügen boreholes: implications for the evolution of the Tornquist Ocean. *Geologische Rundschau*, **83**, 293–308.

Giese, U., Katzung, G., Walter, R. & Weber, J. 1997. The Caledonian deformation of the Brabant Massif and the Early Palaeozoic in northeast Germany : compared. *Geological Magazine*, **134**, 637–652.

Golonka, J., Ross, M.I. & Scotese, C.R. 1994. Phanerozoic paleogeographic and paleoclimatic modeling maps. *In*: *Pangea: Global Environments and Resources*. Canadian Society of Petroleum Geologists, Memoirs, **17**, 1–47.

Graulich, J.M. 1961. *Le sondage de Wépion*. Mémoires pour servir à l'explication des Cartes géologiques et minières de la Belgique, **2**, 1–102.

Greiling, L. 1957. Das Gotlandium des Frankenwaldes (Bayerische Entwicklung). *Geologisches Jahrbuch*, **73**, 301–356.

Greiling, L. 1961. Die stratigraphische Stellung des Döbrasandsteins im Frankenwald – Stellungnahme zur gleichnamigen Arbeit von Stein 1960. *Neues Jarhbuch für Geologie und Paläontologie Monatsheft*, 47–53.

Greiling, L. 1966. Sedimentation und Tektonik im Paläozoikum des Frankenwaldes. *Erlanger Geologische Abhandlungen*, **63**, 1–60.

Gümbel, C.W. 1864. Orthis vom Leuchtholz. *Neues Jahrbuch für Mineralogie, Geologie und Paläontologie*, **460**.

Gürich, G. 1901. Nachträge zum Paleozoicum in polnischen Mittelgebirge. *Neues Jahrbuch für Geologie und Paläontologie, Beilage Band*, **13**, 331–387.

Hallam, A. 1992. *Phanerozoic Sea-Level Changes*. Columbia University Press, New York.

Hammer, J., Eidam, J., Röber, B. & Ehling, B.-C. 1999. Prävariszischer und variszischer granitoider Magmatismus am NE-Rand des Böhmischen Massivs - Geochemie und Petrogenese. *Zeitschrift für Geologische Wissenschaften*, **27**, 401–415.

Havlíček, V. 1981. Development of a linear sedimentary depression examplified by the Prague Basin (Ordovician – Middle Devonian, Barrandian area – central Bohemia). *Sborník geologických Věd, Geologie*, **35**, 7–48.

Havlíček, V. 1982. Ordovician in Bohemia: development of the Prague Basin and its benthic communities. *Sborník geologických Věd, Geologie*, **37**, 103–136.

Havlíček, V. 1998. Ordovician. *In*: Chlupáč, I., Havlíček, V., Kříž, J., Kukal, Z. & Štorch, P. (eds) *Palaeozoic of the Barrandian (Cambrian to Devonian)*. Czech Geological Survey, 39–79.

Havlíček, V. & Fatka, O. 1992. Ordovician of the Prague Basin (Barrandian area, Czechoslovakia), *In*: Webby, B. & Laurie, J. (eds) *Global Perspectives on Ordovician Geology*. Balkema, Rotterdam, 461–472.

Havlíček, V. & Marek, L. 1973. Bohemian Ordovician and its international correlation. *Časopis pro mineralogii a geologii*, **18**, 225–232.

Havlíček, V. & Štorch, P. 1990. Storm sandstones and benthic fauna in the Zahořany Formation in Prague (Upper Ordovician, Czechoslovakia). *Věstník Ústředního Ústavu geologického*, **65**, 167–172.

Havlíček, V. & Vaněk, J. 1966. The biostratigraphy of the Ordovician of Bohemia. *Sborník Ústředního ústavu geologického, Paleontologie*, **8**, 7–69.

Havlíček, V. & Vaněk, J. 1990. Ordovician invertebrate communities in black-shale lithofacies (Prague basin, Czechoslovakia). *Věstník Ústředního Ústavu geologického*, **65**, 223–236.

Havlíček, V., Horný, R., Chlupáč, I. & Šnajdr, M. 1958. *Průvodce ke geologickým exkursím do Barrandienu*. Praha, 1–180.

Havlíček, V., Vaněk, J. & Fatka, O. 1994. Perunica microcontinent in the Ordovician (its position within the Mediterranean Province, series division, benthic and pelagic associations). *Sborník geologických věd, Geologie*, **46**, 23–56.

Hempel, J. 1867. Description geologique des environs de Kielce, de Chenciny et de Malogoszcz. *Annales des Mines Serie III*, **12**, 141–183.

Herbosch, A., Vanguestaine, M., Degardin, J.M., Dejonghe, L., Fagel, N. & Servais, T. 1991. Etude lithostratigraphique, biostratigraphique et sédimentologique du sondage de Lessines (Massif du Brabant). *Annales de la Société géologique de Belgique*, **114**(1), 195–212.

Hetzer, H. 1958. Feinstratigraphie, Sedimentationsverhältnisse und Paläogeographie des höheren Ordoviziums am SE-Rand des Schwarzberger Sattels. *Geologie*, **23**, 1–96.

Heuse, T. 1989. Acritarchen aus dem Griffelschiefer (Ordovizium) des Schwarzburger Antiklinoriums. *Veröffentlichungen Naturhistorischen Museums Schleusingen*, **4**, 69–77.

Heuse, T. 1994. Upper Ordovician acritarchs from the Frankenberger Zwischengebirge (Saxony), Eastern-Germany. *Neues Jarhbuch für Geologie und Paläontologie, Abhandlungen*, **3**, 357–391.

Heuse, T. 2000. Biostratigraphy and biofacies of the Ordovician of Saxo-Thuringia: a review. *Acta Universitatis Carolinae, Geologica*, **42**(1998), 423–431.

Heuse, T. & Maletz, J. 2005. Ordovizium in Deutschland. *In*: Menning, M. & Hendrich, A. (eds), *Erläuterungen zur Stratigraphischen Tabelle von Deutschland 2005 (ESTD 2005), Newsletter on Stratigraphy*, **41**(1–3), 32–37.

Heuse, T. & Puura, I. 2000. Biostratigraphical aspects of the Cadomian unconformity in Saxo-Thuringia. *In*: *An International Symposium in Honour of Hanns Bruno Geinitz, Abstracts and Excursion Guid*, Lange, J. M. Linnemann, U. Thalheim, K. Kunzmann, L. Schneider, J. Voigt, T. (eds) *Schriften des Staatlichen Museums für Mineralogie und Geologie*, Dresden, **11**, 75–79.

Heuse, T., Erdtmann, B.-D. & Kraft, P. 1994. Early Ordovician microfossils (acritarchs, chitinozoans) and graptolites from the Schwarzburg Anticline, Thuringia (Germany). *Veröffentlichungen Naturhistorischen Museums Schleusingen*, **9**, 41–68.

Horný, R. 1999. Gastropod, paragastropod and tergomyan biostratigraphy of the Ordovician of Bohemia. *In*: Kraft, P. & Fatka, O. (eds) *Quo vadis Ordovician? Acta Universitatis Carolinae, Geologica*, **43**(1–2), 397–400.

Horstig, G. von 1957. *Erläuterungen zur Geologischen Karte von Bayern 1:25.000*. Blatt 5637 Hof. Bayerisches Geologisches Landesamt, 111 S., München.

Horstig, G. von & Stettner, G. 1970. *Erläuterungen zur Geologischen Karte von Bayern 1:25.000*. Blatt 5736 Helmbrechts. Bayerische Geologische Landesamt, 176 S., München.

Horstig, G. von & Stettner, G. 1976. *Erläuterungen zur Geologischen Karte von Bayern 1:25.000*. Blatt 5735 Schwarzenbach am Wald. Bayerische Geologische Landesamt, 178 S., München.

Hoth, K., Zagora, I. & Zagora, K. 1993. Die überregionale Bedeutung der offshore-Bohrungen im Seegebiet zwischen Bornholm und Rügen für die Kenntnis der paläozoischen Entwicklung am Südwestrand der Osteuropäischen Plattform. *In*: Scholle, T. & Krauss, M. (eds) *Exkursionsführer. Gesellschaft für Geologische Wissenschaften*. 05.–10.10.1993, Prora/Rügen und Rønne/Bornholm, Berlin, 34–42.

Hoth, K., Heuse, T., Leonhardt, D. & Subkommission Riphäikum-Silur 2002. Ordovician. *In*: *Stratigraphic Table of Germany 2002*. German Stratigraphic Commission, Stuttgart.

Jaeger, H. 1967. Ordoviz auf Rügen. Datierung und Vergleich mit anderen Gebieten (Vorläufige Mitteilung). *Beriche der Deutschen Gesellschaft geologischer Wissenschaften. A. Geologie und Paläontologie*, **12**, 156–176.

Jurkiewicz, H. 1975. Budowa geologiczna podłoża mezozoiku centralnej części niecki miechowskiej. *Biuletyn Instytutu Geologicznego*, **283**, 5–150.

Kalvacheva, R., Sassi, F.P. & Zanferrari, A. 1986. Acritarch evidence for the Cambrian age of phyllites in the Agordo area (South-Alpine basement of eastern Alps, Italy). *Review of Palaeobotany and Palynology*, **48**, 311–326.

Katzung, G. 1961. Die Geröllführung des Lederschiefers (Ordovizium) an der SE-Flanke des Schwarzburger Sattels (Thüringen). *Geologie*, **10**(7), 778–802.

Katzung, G. 1999. Records of the Late Ordovician glaciation from Thuringia, Germany. *Zeitschrift der Deutsche geologische Gesellschaft*, **150**(3), 595–617.

Katzung, G. 2001. The Caledonides at the southern margin of the East European Craton. *Neues Jahrbuch für Geologie und Paläontologie, Abhandlungen*, **222**(1/2), 3–54.

Katzung, G., Giese, U., Walter, R. & Winterfeld, C. 1993. The Rügen Caledonides, northeast Germany. *Geological Magazine*, **130**, 725–730.

Kemnitz, H., Schäfer, F., Romer, R.L. & Rosner, M. 1999. Die geodynamische Geschichte eines variszischen Retrokeils im NW-Teil des Saxothuringikums unter besonderer Berücksichtigung des Vesser-Gebietes. *Beiträge zur Geologie von Thüringen Neue Folge*, **6**, 41–91, Jena.

Kemnitz, H., Romer, R.L. & Oncken, O. 2002. Gondwana break-up and the northern margin of the Saxothuringian belt (Variscides of Central Europe). *International Journal of Earth Sciences (Geologische Rundschau)*, **91**, 246–259.

Kettner, R. 1916. Příspěvek k petrografii vrstev krušnohorských (d1a), část 1 a 2. *Rozpravy II, třídy České akademie věd a umění*, **25**, 1–24.

Kielan, Z. 1960. Upper Ordovician trilobites from Poland and some related forms from Bohemia and Scandinavia. *Palaeontologia Polonica*, **11**, 1–198.

Kielan-Jaworowska, Z. 1966. Polychaete jaw apparatuses from the Ordovician and Silurian of Poland and a comparison with modern forms. *Palaeontologia Polonica*, **16**, 1–152.

Knüpfer, J. 1967. Zur Fauna und Biostratigraphie des Ordoviziums (Gräfenthaler Schichten) in Thüringen. *Freiberger Forschungshefte, C*, **220**, 1–119.

Knüpfer, J. 1968. Ostracoden aus dem Oberen Ordovizium Thüringens. *Freiberger Forschungshefte, C*, **234**, 5–29.

Koch, L. 1999. Die Familie Cyclopygidae (Trilobita) im Ordovizium des Ebbe-Sattels und Remscheider Sattels (Rheinisches Schiefergebirge, Deutschland). *Neues Jahrbuch für Geologie und Paläontologie, Abhandlungen*, **213**, 375–431.

Koch, L. & Brauckmann, C. 1998. Phyllocariden (Crustacea) aus dem Unteren Llanvirn (Ordovizium des Ebbe-Sattels (Rheinisches Schiefergebirge). *Neues Jahrbuch für Geologie und Paläontologie, Monatshefte*, **1998**, 55–64.

Kolíha, J. 1926. Facies baltico-polonais de'l Ordovicien inférieur en Boheme. *Věstník Ústředního Ústavu geologického*, **2**, 317–328.

Kolíha, J. 1937. Sur le Trémadocien et sur l'Arénigien inférieur en Boheme. *Bulletin de la Société Géologique de France*, **7**(5), 477–495.

Kossmat F. 1927. Gliederung des varistischen Gebirgsbaues. *Abhandlungen Sächsiches Geologischen Landesamt*, **1** (1927) 1–39.

Kowalczewski, Z., Kuleta, M., Lisik, R. & Moczydłowska, M. 1984. Nowe dane o skałach kambru z okolic Wiśniówki w Górach Świętokrzyskich. *Kwartalnik Geologiczny*, **28**, 448–449.

Kowalczewski, Z., Kuleta, M., Lisik, R., & Moczydłowska, M. 1986. New data on the Cambrian and Lower Ordovician in the vicinities of Wiśniówka, Holy Cross Mts. *Bulletin of the Polish Academy of Sciences, Earth Sciences*, **34**, 353–362.

Kozłowski, R. 1948. Les graptolites et quelques nouveaux groupes d'animaux de Tremadoc de la Pologne. *Palaeontologia Polonica*, **3**, 1–235.

Kozłowski, R. 1963. Sur la nature de Chitinozoaires. *Acta Palaeontologica Polonica*, **8**, 425–449.

Kozłowski, R. 1968. Nouvelles observations sur les Conulaires. *Acta Palaeontologica Polonica*, **13**, 497–535.

Kozłowski, R. & Kaźmierczak, J. 1968. On two Ordovician calcareous algae. *Acta Palaeontologica Polonica*, **13**, 325–343.

Kraft, J. & Kraft, P. 1999. Graptolite biozones of the Bohemian Lower and Middle Ordovician and their historical development. *Journal of the Czech Geological Society*, **44**, 53–62.

Kraft, P. & Kraft, J. 2003. Facies of the Klabava Formation (?Tremadoc – Arenig) and their fossil content (Barrandian area, Czech Republic). *In*: Albanesi, G., Beresi, M. & Peralta, S.H. (eds) *Ordovician from the Andes*. INSUGEO, Tucuman, Series Correlación Geológica, **17**, 321–325.

Kraft, P. & Kraft, J. 2004. Boundary interval of graptolite biozones in the Klabava Formation (Ordovician, Prague Basin). *Zprávy o geologických výzkumech v roce 2003*, 25–27 [in Czech].

Krawczyk, C.M., McCann, T., Cocks, L.R.M., England, R., McBride, J. & Wybraniec, S. 2008. Caledonian tectonics. *In*: McCann, T. (ed.) *The Geology of Central Europe. Volume 1: Precambrian and Palaeozoic*. Geological Society, London, 303–381.

Krenz, O., Kröner, A., Jaeckel, P. & Leonhardt, D. 1996. Neue Zirkondatierungen an Metamorphiten des Westerzgebirges. *Terra Nostra*, **96**(2), 127–130.

Kröner, A., Willner, A.P. & Teufel, S. 1996. Time of formation, peak HP-HT metamorphism and rapid cooling of quartz-feldspar rocks in the Central Erzgebirge, Saxony, Germany. *Terra Nostra*, **96**, 133–135.

Kröner, A., Jaeckel, P., Reischmann, T. & Kroner, U. 1998. Further evidence for an early Carboniferous ($\sim$340 Ma) age of the high-grade metamorphism in the Saxonian granulite complex. *Geologische Rundschau*, **86**, 751–766.

Kroner, U., Mansy, J.-L. *et al.* 2008. Variscan tectonics. *In*: McCann, T. (ed.) *The Geology of Central Europe. Volume 1: Precambrian and Palaeozoic*. Geological Society, London, 599–664.

Kukal, Z. 1956. A lithological investigation of the Skalka and Drabov Beds of the Ordovician of the Barrandian. *Sborník Ústředního Ústavu geologického, Oddíl geologický*, **23**, 215–295 [in Czech with English summary].

KUKAL, Z. 1961. Petrographical investigation of the Černín Beds of the Ordovician in the Barrandian area. *Věstník Ústředního Ústavu geologického*, **36**, 35–47 [in Czech with English summary].

KUKAL, Z. 1962. Petrographical investigation of the Ordovician Šárka Beds in the Barrandian area. *Sborník Ústředního Ústavu geologického, Oddíl geologický*, **27**, 175–214 [in Czech with English summary].

KUKAL, Z. 1963*a*. Composition and origin of the Ordovician sediments of the Třenice and Mílina Beds. *Sborník Ústředního Ústavu geologického, Oddíl geologický*, **28**, 265–307 [in Czech with English summary].

KUKAL, Z. 1963*b*. The results of sedimentological investigation of the Ordovician in the Barrandian area. *Sborník Ústředního Ústavu geologického, Oddíl geologický*, **28**, 103–137 [in Czech with English summary].

KUPETZ, M. 1987. Stratigraphie und regionale Stellung des variszisch gefalteten Paläozoikums im Nossen-Wilsdruffer Schiefergebirge. *Freiberger Forschungshefte, C*, **415**, 5–41.

KUPETZ, M. 2000. Stratigraphie und Deckentektonik im variszischen Paläozoikum des Nossen-Wilsdruffer Schiefergebirges. *Zeitschrift für Geologische Wissenschaften*, **28**(1/2), 247–272.

KURZE, M. 1966. Die tektonisch-fazielle Entwicklung im Nordostteil des zentralsächsischen Lineaments. *Freiberger Forschungshefte C*, **201**, 1–89.

KURZE, M. 1968. Ergebnisse einer Spezialkartierung im Altpaläozoikum südlich Frankenberg /Sachsen. *Freiberger Forschungshefte, C*, **241**, 45–56.

KURZE, M. 1993. Das mittelsächsische Altpaläozoikum und seine Beziehung zu benachbarten Gebieten. *Zeitschrift für Geologische Wissenschaften*, **21**(1/2), 153–162.

KURZE, M & TRÖGER 1990. Die Entwicklung des Paläozoikums in der Elbtalzone (Saxothuringikum). *Neues Jahrbuch für Geologie und Paläontologie Monatshefte*, **3**, 43–53.

KURZE, M., LINNEMANN, U., LOBST, R. & LORENZ, W. 1992. Sedimentological features and depositional models in low and medium grade metamorphic sequences in Saxony. *Excursion Guide-Book, 13th IAS Regional Meeting on Sedimentology*, Jena, 189–214.

KURZE, M., GERISCH, R., STARKE, R., KLEEBERG, R. & KLEMM, W. 1995. Das Ordovizium vom Rügener Typ. Neue Erkenntnisse sedimentologischer Untersuchungen. *Zentralblatt für Geologie und Paläontologie I*, **1–2**, 205–215.

ŁĄCKA, B. 1990. Mineralogy of the Ordovician ferruginous coated grains (the Holy Cross Mountains and Northern Poland). *Archiwum Mineralogiczne*, **45**, 79–92.

LALOUX, M., DEJONGHE, L., GEUKENS, F., GHYSEL, P. & HANCE, L. 1996. *Carte Géologique de Wallonie, 1:25.000, Notice Explicative*. Limbourg-Eupen 43/5–6.

LECOMPTE, M. 1948. Existence du trémadocien dans le massif du Brabant. *Bulletin de l'Académie Royale des Sciences, Lettres et Beaux Arts de Belgique, Classe des Sciences, 5è Série*, **34**, 677–687.

LECOMPTE, M. 1949. Découverte de nouveaux gites à Dictyonema dans le Tremadocien du massif du Brabant. *Bulletin de l'Institut royal des Sciences Naturelles de Belgique*, **25**(45), 1–8.

LEE, C. 1938. Schichtenfolge und Bau des Oberlausitzer Schiefergebirges. *Geotektonische Forschungen*, **2**, 1–55.

LEGRAND, R. 1964. Le forage de Lichtervelde (1935–1939); la coupure faunique entre le Silurien et l'Ordovicien par rapport à l'unité géologique de l'Ashgillien. *Bulletin de la Société belge de Géologie*, **73**, 53–64.

LEGRAND, R. 1966. *Sondage à Deerlijk*. Professional Paper of the Belgian Geological Survey, **4**, 1–17.

LEGRAND, R. 1968. *Le Massif du Brabant*. Mémoires pour servir à l'explication des Cartes géologiques et minières de la Belgique, **9**, 1–148.

LESPÉRANCE, P.J. & SHEEHAN, P.M. 1987. Trilobites et Brachiopodes ashgilliens (Ordovicien supérieur) de l'"Assise" de Fosse, Bande de Sambre-Meuse (Belgique). *Bulletin de l'Institut royal des Sciences Naturelles de Belgique, Sciences de la Terre*, **57**, 91–123.

LEWANDOWSKI, M. 1993. *Paleomagnetism of the Paleozoic rocks of the Holy Cross Mts (central Poland) and the origin of the Variscan Orogen*. Institute of Geophysics of the Polish Academy of Sciences, **A-23**, 3–84.

LEWANDOWSKI, M. 2003. Assembly of Pangea : combined paleomagnetic and paleoclimatic approach. *Advances in Geophysics*, **46**, 199–236.

LINNEMANN, U.-G. 1994. Geologischer Bau und Strukturentwicklung der südlichen Elbezone. *Abhandlungen des Staatlichen Museums für Mineralogie und Geologie, Dresden*, **40**, 7–36.

LINNEMANN, U. & HEUSE, T. 2000. The Ordovician of the Schwarzburg Anticline: Geotectonic setting, biostratigraphy and sequence stratigraphy (Saxo-Thuringian Terrane, Germany). *Zeitschrift der Deutschen geologischen Gesellschaft*, **151**(4), 471–491.

LINNEMANN, U. & ROMER, R.L. 2002. The Cadomian orogeny in Saxo-Thuringia, Germany: geochemical and Nd-Sr-Pb isotopic characterization of marginal basins with constraints to geotectonic setting and provenance. *Tectonophysics*, **352**, 33–64.

LINNEMANN, U. & SCHAUER, M. 1999. Die Entstehung der Elbezone vor dem Hintergrund der cadomischen und variszischen Geschichte des Saxothuringischen Terranes – Konsequenzen aus einer abgedeckten geologischen Karte. *Zeitschrift für Geologische Wissenschaften*, **27**(5/6), 529–561.

LINNEMANN, U., GEHMLICH, M., *et al.* 2000. From Cadomian subduction to Early Palaeozoic rifting: the evolution of Saxo-Thuringia at the margin of Gondwana in the light of single zircon geochronology and basin development (Central European Variscides, Germany). *In*: FRANKE, W. *et al.* (eds) *Orogenic Processes: Quantification and Modelling in the Variscan Belt*. Geological Society, London Special Publications, **179**, 131–153.

LINNEMANN, U., MCNAUGHTON, N.J., ROMER, R.L., GEHMLICH, M., DROST, K. & TONK, C. 2004. West African provenance for Saxo-Thuringia (Bohemian Massif): Did Armorica ever leave pre-Pangean Gondwana? - U/Pb-SHRIMP zircon evidence and the Nd-isotopic record. *International Journal of Earth Sciences (Geologische Rundschau)*, **93**, 683–705.

LORETZ, H. 1880. Petrefactenfunde im Thüringer Schiefergebirge. *Zeitschrift der Deutschen geologischen Gesellschaft*, **32**, 632–636.

LÜTZNER, H. & MANN, M. 1988. Arthropodenfährten aus der Phycoden-Folge (Ordovizium) des Schwarzburger Antiklinoriums. *Zeitschrift für Geologische Wissenschaften*, **16**(6), 493–501.

LÜTZNER, H., ELLENBERG, J. & FALK, F. 1986. Entwicklung der Sedimentationsrate und der Ablagerungsprozesse im Altpaläozoikum Thüringens. *Zeitschrift für Geologische Wissenschaften*, **1**(14), 83–93.

MCCANN, T. 1998. Lower Palaeozoic evolution of the northeast German Basin/Baltica borderland. *Geological Magazine*, **135**, 129–142.

MCCANN, T., SKOMPSKI, S., *et al.* 2008. Carboniferous. *In*: MCCANN, T. (ed.) *The Geology of Central Europe. Volume 1: Precambrian and Palaeozoic*. Geological Society, London, 411–529.

MAILLIEUX, E. 1926. Remarques sur l'Ordovicien de la Belgique. *Bulletin de la Société belge de Géologie, de Paléontologie et d'Hydrologie*, **36**, 67–85.

MALAISE, C. 1903. Compte rendu de la session extraordinaire de la Société Géologique de Belgique tenue à Namur, le 19, 20 septembre 1903. *Annales de la Société géologique de Belgique*, **30**, 132–134.

MALETZ, J. 1998. Die Graptolithen des Ordoviziums von Rügen (Norddeutschland, Vorpommern). *Paläontologische Zeitschrift*, **72**, 351–372.

MALETZ, J. 2001. Ordovician graptolite biostratigraphy of the Rügen wells, NE-Germany. *Neues Jahrbuch für Geologie und Paläontologie, Abhandlungen*, **222**, 55–72.

MALETZ, J. & SERVAIS, T. 1993. Acritarchs and graptolites from the Early Llanvirn (Ordovician) of the Herscheider Schichten (Rheinisches Schiefergebirge, Germany). *Neues Jahrbuch für Geologie und Paläontologie, Abhandlungen*, **190**, 125–142.

MALETZ, J. & SERVAIS, T. 1998. Upper Ordovician graptolites from the Brabant Massif, Belgium. *Geobios*, **31**(1), 21–37.

MANN, M. 1996. Paläobathymetrie und Beckenentwicklung im Altpaläozoikum des Schwarzburger Antiklinoriums (Thüringen). *Beiträge zur Geologie von Thüringen, Neue Folge*, **3**, 33–56.

MAREK, J., BARTZSCH, K., DROST, K., FATKA, O., KRAFT, P. & LINNEMANN, U. 2003. Revision of trilobites of the Griffelschiefer Formation (Ordovician, Schwarzburg anticline, Germany): preliminary results. *In*: ALBANESI, G.L., BERESI, M.S. & PERALTA, S.H. (eds) *Ordovician from the Andes*. INSUGEO, Tucuman, Serie Correlatión Geológica, **17**, 321–325.

MARTIN, F. 1969*a*. Les acritarches de l'Ordovicien et du Silurien belge. Détermination et valeur stratigraphique. *Mémoires de l'Institut Royal des Sciences naturelles de Belgique*, **160**, 1–176.

Martin, F. 1969*b*. Ordovicien et Silurien belge: données nouvelles apportées par l'étude des acritarches. *Bulletin de la Société belge de Géologie*, **77**, 175–181.

Martin, F. 1976. Acritarches du Cambro-Ordovicien du Massif du Brabant, Belgique. *Bulletin de l'Institut royal des Sciences naturelles de Belgique*, **51**(1975), 1–33.

Martin, F. & Rickards, B. 1979. Acritarches, chitinozoaires et graptolithes ordoviciens et siluriens de la vallée de la Sennette (Massif du Brabant, Belgique). *Annales de la Société géologique de Belgique*, **102**, 181–197.

Martin, F., Michot, P. & Vanguestaine, M. 1970. Le flysch caradocien d'Ombret. *Annales de la Société géologique de Belgique*, **93**, 337–362.

Masiak, M., Podhalańska, T. & Stempień-Sałek, M. 2003. Ordovician–Silurian boundary in the Bardo Syncline, Holy Cross Mountains, Poland – new data on fossil assemblages and sedimentary succession. *Geological Quarterly*, **47**, 311–329.

Meinhold, G., Hahn, T., Heuse, T. & Mingram, B. 2005. Geochemistry and palynology of metasediments from the phyllitic Greiz area (Saxo-Thuringia, Germany). *Neues Jahrbuch für Geologie und Paläontologie, Abhandlungen*, **237**(3), 423–452.

Menning, M. & Hendrich, A. (eds) 2005. Erläuterungen zur Stratigraphischen Tabelle von Deutschland 2005 (ESDT 2005). *Newsletters on Stratigraphy*, **41**, 1–405.

Mergl, M. 1984. Fauna of the upper Tremadocian of Central Bohemia. *Sborník geologických Věd, Paleotologie*, **26**, 9–46.

Mergl, M. 1990. Late Ordovician Foliomena brachiopod fauna from the Holy Cross Mountains (Poland). *Časopis pro Mineralogii a Geologii*, **35**, 147–154.

Mergl, M. 1994. Trilobite fauna from the Třenice Formation (Tremadoc) in Central Bohemia. *Folia Musei Rerum naturalium Bohemiae occidentalis, Geologica*, **39**, 1–31.

Mergl, M. 2002. Linguliformean and craniiformean brachiopods of the Ordovician (Třenice to Dobrotivá formations) of the Barrandian, Bohemia. *Acta Musei Nationalis Pragae, Series B, Natural History*, **58**(1–2), 1–82.

Michalski, A. 1883. *Krótkie sprawozdanie z badań geologicznych dokonanych latem 18882 r. w guberni kieleckiej*. Pamiętnik Fizjograficzny 3.

Michot, P. 1954. Le Silurien. *In*: *Prodrôme d'une description géologique de la Belgique*. Imprimerie Vaillant Carmanne, Liège, 39–82.

Mielke, H., Rohrmüller, J. & Gebauer, D. 1996. Ein metalateritisches Denudations-Niveau als lithologisch und zeitlich korrelierbarer Bezugshorizont in Phylliten, Glimmerschiefern und Gneisen des ostbayerischen Grundgebirges. *Geologie Bavarica*, **101**, 139–166.

Mikuláš, R. 1993. New information on trace fossils of the Early Ordovician of Prague Basin (Barrandian area, Czech Republic). *Časopis pro Mineralogii a Geologii*, **38**, 171–182.

Mikuláš, R. 1998. Ordovician of the Barrandian area: Reconstruction of the sedimentary basin, its benthic communities and ichnofacies. *Journal of the Czech Geological Society*, **43**, 143–159.

Mingram, B. 1996. *Geochemische Signaturen der Metasedimente des erzgebirgischen Krustenstapels*. Scientific Technical Report GFZ Potsdam, **96/04**.

Mingram, B. 1998. The Erzgebirge, Germany, a subducted part of northern Gondwana: geochemical evidence for repetition of Early Palaeozoic metasedimentary sequences in metamorphic thrust units. *Geological Magazine*, **135**(6), 785–801.

Mingram, B. & Rötzler, K. 1999. Geochemische, petrologische und geochronologische Untersuchungen im Erzgebirgskristallin – Rekonstruktion eines Krustenstapels. *Schriftenreihe Forschung Geowissenschaften*, **9**, 1–80.

Modliński, Z. 1975. Podstawowe zmiany facjalne i miąższościowe ordowiku platformowego północno-wschodniej Polski. *Kwartalnik Geologiczny*, **19**, 503–513.

Modliński, Z. 1982. Rozwój litofacjalny i paleotektoniczny ordowiku na obszarze platformy prekambryjskiej w Polsce. *Prace Instytutu Geologicznego*, **102**, 1–66.

Modliński, Z. 1984. Stratygrafia potremadockich osadów ordowiku Lubelszczyzny. *Kwartalnik Geologiczny*, **28**, 1–15.

Modliński, Z. & Szymański, B. 1997. The Ordovician lithostratigraphy of the Peribaltic Depression (NE Poland). *Kwartalnik Geologiczny*, **41**, 273–288.

Modliński, Z. & Szymański, B. 2001*a*. Stratygrafia i litologia ordowiku Suwalszczyzny (północno-wschodnia Polska). *Biuletyn Państwowego Instytutu Geologicznego*, **394**, 55–71.

Modliński, Z. & Szymański, B. 2001*b*. The Ordovician stratigraphy and palaeogeography of the Nida-Holy Cross Mts. area, Poland: a review. *Kwartalnik Geologiczny*, **45**, 417–433.

Mohnicke, M., Kurze, M. & Tichomirowa, M. 2001. Petrogenese und Altersstellung der Gesteine des "Coswiger Komplexes" (Elbezone). *Zeitschrift für Geologische Wissenschaften*, **29**(5/6), 505–519.

Molyneux, S. G. & Rushton, A. W. A. 1988. The age of the Watch Hill Grits (Ordovician), English Lake District: structural and palaeogeographical implications. *Transactions of the Royal Society Edinburgh, Earth Sciences*, **79**, 43–69.

Montenari, M., & Servais, T. 2000. Lower Palaeozoic (Late Cambrian – Early Ordovician) acritarchs from the metasedimentary Baden-Baden-Zone (Schwarzwald, SW-Germany). *Review of Palaeobotany and Palynology*, **113**, 73–85.

Montenari, M., Servais, T. & Paris, F. 2000. Palynological dating (acritarchs and chitinozoans) of Lower Paleozoic phyllites from the Black Forest/southwestern Germany. *Comptes Rendue de l'Académie de Sciences Paris, Sciences de la Terre et des planètes*, **330**, 493–499.

Moryc, W. & Nehring-Lefeld, M. 1997. Ordovician between Pilzno and Busko in the Carpathian Foreland (southern Poland). *Kwartalnik Geologiczny*, **41**, 139–150.

Müller, A.H. 1956. Die Brachiopodenreste aus der Frauenbach-Serie (Tremadoc) von Siegmundsburg bei Steinheid (Thüringen). *Berichte der Geologischen Gesellschaft der DDR*, **2**, 51–56.

Müller-Sohnius, D., Drach, V. von, Horn, P. & Köhler, H. 1987. Altersbestimmungen an der Münchberger Gneismasse, Nordost-Bayern. *Neues Jahrbuch für Mineralogie, Abhandlungen*, **156**, 175–206.

Nehring-Lefeld, M. & Szymański, B. 1998. Ordovician stratigraphy in the Żarki-Myszków area (NE margin of the Upper Silesian coal basin). *Kwartalnik Geologiczny*, **42**, 29–40.

Nehring-Lefeld, M., Modliński, Z., & Siewniak-Madej, A. 1992. Biostratigraphy of the old Paleozoic carbonates in the Zawiercie area (NE margin of the Upper Silesian coal basin). *Kwartalnik Geologiczny*, **36**, 171–198.

Nielsen, A.T. 2004. Ordovician sea-level changes: A Baltoscandian Perspective. *In*: Webby, B.D., Paris, F., Droser, M.L. & Percival, I. (eds) *The Great Ordovician Biodiversification Event*. Columbia Press, New York, 84–93.

Nolvak, J. & Grahn, Y. 1993. Ordovician chitinozoan zones from Baltoscandia. *Review of Palaeobotany and Palynology*, **79**, 249–269.

Oncken, O., Winterfeld, von, C. & Dittmar, U. 1999. Accretion of a rifted passive margin : The Late Paleozoic Rhenohercynian fold and thrust belt (Middle European Variscides). *Tectonics*, **18**, 75–91.

Owens, R.M. & Servais, T. 2007. The Ordovician of the Condroz Ridge, Belgium : trilobites from the south-eastern margin of Avalonia. *Palaeogeography, Palaeoclimatology, Palaeoecology*, **245**, 272–294.

Paalits, I. & Erdtmann, B.-D. 1993. The first record of Ordovician acritarchs from the Görlitz Synclinorium, Saxony, 63. *Abstrakte der 63. Jahrestagung der Paläontologischen Gesellschaft, Prag*, **18**.

Paris, F. 1981. *Les Chitinozoaires dans le Paléozoïque du Sud-Ouest de l'Europe*. Mémoire de la Société Géologique et Minéralogique de Bretagne, **26**, 1–412.

Paris, F. & Mergl, M. 1984. Arenigian chitinozoans from the Klabava Formation, Bohemia. *Review of Palaeobotany and Palynology*, **43**, 33–65.

Parrish, J.T. 1982. Upwelling and petroleum source beds, with reference to Paleozoic. *AAPG. Bulletin*, **66**(6), 750–774.

Patočka, F. & Vlašímský, P. 1986. Geochemistry of volcanic rocks of the early Palaeozoic of the Barrandian. *Zprávy o geologických výzkumech v roce* 1984, 159–160, [in Czech].

Patočka, F., Vlašímský, P. & Blechová, K. 1993. Geochemistry of Early Palaeozoic volcanics of the Barrandian Basin (Bohemian Massif, Czech Republic): Implications for paleotectonic reconstructions. *Jahrbuch der Geologischen Bundesanstalt*, **136**, 837–896.

Peng, S. 1990. Tremadoc stratigraphy and trilobite faunas of northwestern Hunan. *Beringeria*, **2**, 3–171.

Petránek, J. 1964. Gemeinsame Merkmale der Eisenerzlager im böhmischen und thüringischen Ordovicium. *Abhandlungen der*

Deutschen Akademie der Wissenschaften zu Berlin, Klasse für Bergbau, Hüttenwesen und Montangeologie, **2**, 79–95.

PETRÁNEK, J. 1991. Ordovician oolitic ironstones and their source of iron. *Věstník Ústředního Ústavu geologického*, **66**, 321–327.

PETRÁNEK, J. & VAN HOUTEN, F.B. 1997. Phanerozoic ooidal ironstones. *Czech Geological Survey Special Papers*, **7**, 1–71.

PETRÁNEK, J., DUREMBERG, D. & MELKA, K. 1988. Oolitic iron ore deposits at Chrustenice (Ordovician, Bohemia). *Sborník geologických Věd, Ložisková Geologie*, **28**, 9–55.

PHARAOH, T.C. 1999. Palaeozoic terranes and their lithospheric boundaries within the Trans-European Suture Zone (TESZ) : a review. *Tectonophysics*, **314**, 17–41.

PHARAOH, T.C., MOLYNEUX, S.G., MERRIMAN, R.J., LEE, M.K. & VERNIERS, J. 1993. The Caledonides of the Anglo-Brabant Massif reviewed. *In*: PHARAOH, T.C., MOLYNEUX, S.G., MERRIMAN, R.J., LEE, M.K. & VERNIERS, J. (eds) Special Issue on the Caledonides of the Anglo-Brabant Massif. *Geological Magazine*, **130**, 561–562.

PHARAOH T. C., WINCHESTER J. A., VERNIERS J., LASSEN A., SEGHEDI A. 2006. The Western Accretionary Margin of the East European Craton: an overview. *In*: GEE D. G. & STEPHENSON R. A. (eds) *European Lithosphere Dynamics*. Geological Society, London, Memoirs, **32**, 323–332.

PIEKARSKI, K. & SIEWNIAK-WITRUK, A. 1978. O występowaniu ordowiku w okolicy Mrzygłodu. *Przegląd Geologiczny*, **26**, 647–648.

PIETZSCH, K. 1910. Cruzianen aus dem Untersilur des Leipziger Kreises. *Zeitschrift der Deutschen geologischen Gesellschaft*, **62**, 571–582.

PIETZSCH, K. 1962. Das Schiefergebirge der Lausitz. *In*: PIETZSCH, K. (ed.) *Geologie von Sachsen*. Deutscher Verlag der Wissenschaften, Berlin, 168–191.

PISKE, J., RASCH, J., NEUMANN, E. & ZAGORA, K. 1994. Geologischer Bau und Entwicklung des Präperms der Insel Rügen und des angrenzenden Seegebietes. *Zeitschrift für Geologische Wissenschaften*, **22**, 11–226.

POMPECKJ, J.F. 1896. Ein neuentdecktes Vorkommen von Tremadoc-Fossilien bei Hof. *Berichte des nordoberfränkischen Vereinigung Natur, Geschichte, Landeskunde*, 89–101.

POPRAWA P. 2006. Development of the Caledonian collision zone along the western margin of Baltica and its relation to the foreland basin. *Prace Państwowego Instytutu Geologicznego*, **186**, 189–214 [in Polish with extended abstract in English].

POŻARYSKI, W. & TOMCZYK, H. 1993. Przekrój geologiczny przez Polskę południowo-wschodnią. *Przegląd Geologiczny*, **31**, 687–695.

POŻARYSKI, W., GROCHOLSKI, A., TOMCZYK, H., KARNKOWSKI, P., & MORYC, W. 1992. Mapa tektoniczna Polski w epoce waryscyjskiej. *Przegląd Geologiczny*, **30**, 643–650.

PRIEWALDER, H. 1997. The distribution of Chitinozoans in the Cellon Section (Hirnantian – Lower Lochkovian). A preliminary report. *Berichte der Geologischen Bundesanstalt*, **40**, 74–85.

RAUMER, J. VON 1998. The Palaeozoic evolution of the Alps from Gondwana to Pangea. *Geologische Rundschau*, **87**, 407–435.

RAUMER, J. VON, STAMPFLI, G.M. & BUSSY, F. 2003. Gondwana-derived microcontinents – the constituents of the Variscan and Alpine collisional orogens. *Tectonophysics*, **365**, 7–22.

REGNÉLL, G. 1951. Revision of the Caradocian–Ashgillian cystoid fauna of Belgium with notes on isolated pelmatozoan stem fragments. *Mémoires de l'Institut Royal des Sciences Naturelles de Belgique*, **120**, 1–47.

REH, H. 1964. Zur Entstehung der oolithischen Eisenerze im Ordovizium Thüringens. *Abhandlungen der Deutschen Akademie der Wissenschaften zu Berlin, Klasse für Bergbau, Hüttenwesen und Montangeologie*, **2**, 97–133.

REIN, B. 1986. Erläuterungen zur geologischen Spezialkartierung im Paläozoikum des Frankenwaldes auf Blatt 5735 Schwarzenbach am Wald, Blatt 5736 Helmbrechts. unveröff. Diplomarbeit, Paläontologisches Institut Universität Würzburg.

REITZ, E. 1991. Acritarchen des Unter-Tremadoc aus dem westlichen Frankenwald, NE-Bayern. *Neues Jahrbuch für Geologie und Paläontologie, Monatshefte*, **2**, 97–104.

REITZ, E. 2004. Unterdordovizische Acritarchen aus der Magdalensberg-Gruppe in Ostkärnten und aus der Salzburger Grauwackenzone (Österreich). *Jahrbuch der Geologischen Bundesanstalt*, **137**, 661–668.

REITZ, E. & HEUSE, T. 1994. Palynofazies im Oberordovizium des Saxothuringikums. *Neues Jahrbuch für Geologie und Paläontologie, Monatshefte*, **6**, 348–360.

REITZ, E. & HÖLL, R. 1989. Unterordovizische Acritarchen aus der Nördlichen Grauwackenzone (Ostalpen). *Jahrbuch der Geologischen Bundesanstalt*, **132**, 761–774.

REITZ, E. & HÖLL, R. 1990. Biostratigraphischer Nachweis von Unterordovizium in der Innsbrucker Quarzphyllitserie (Ostalpen). *Jahrbuch der Geologischen Bundesanstalt*, **133**, 603–610.

REITZ, E. & HÖLL, R. 1991. Biostratigraphischer Nachweis von Arenig in der Nördlichen Grauwackenzone (Ostalpen). *Jahrbuch der Geologischen Bundesanstalt*, **134**, 329–344.

REITZ, E. & HÖLL, R. 1992. Palynological evidence for Lower Ordovician rocks (Tremadoc and Arenig) in the Northern Greywacke Zone (Eastern Alps). *Terra Nova*, **4**, 198–207.

REITZ, E. & WICKERT, F. 1989. Late Cambrian to Early Ordovician acritarchs from the Ville-Unit, northern Vosges Mountains (France). *Neues Jahrbuch für Geologie und Paläontologie, Monatshefte*, **6**, 375–385.

REITZ, E., ANDERLE, H.J. & WINKELMANN, M. 1995. Ein erster Nachweis von Unterordovizium (Arenig) am Südrand des Rheinischen Schiefergebirges im Vordertaunus: der Bierstadt-Phyllit (Bl. 5915 Wiesbaden). *Geologisches Jahrbuch Hessen*, **123**, 25–38.

REMPEL, H. 1992. Erdölgeologische Bewertung der Arbeiten der gemeinsamen Organisation « Petrobaltic » im deutschen Schelfbereich. *Geologisches Jahrbuch, D*, **99**, 3–33.

RIBBERT, K.H., SERVAIS, T. & VANGUESTAINE, M. 2001. Stavelot-Venn-Antiklinale (28). *In*: *Stratigraphie von Deutschland II, Ordovizium, Kambrium, Vendium, Riphäikum. Teil III*. Stratigraphische Kommission Deutschlands. Courier Forschungsinstitut Senckenberg, **235**, 68–89.

RICHTER, E. & RICHTER, R. 1951. Trilobiten von Ashgill-Alter aus dem Massif du Brabant (Grand-Manil). *Bulletin de l'Institut Royal des Sciences Naturelles de Belgique*, **27**(16), 1–6.

ROBARDET, M. 2003. The Armorica 'microplate': fact or fiction? Critical review of the concept and contradictory palaeobiogeographical data. *Palaeogeography, Palaeoclimatology, Palaeoecology*, **195**, 125–148.

ROCHE, M., SABIR, M., STEEMANS, P. & VANGUESTAINE, M. 1986. Palynologie du sondage et de la région de Willerzie. *Aardkundige Mededelingen*, **3**, 149–190.

ROEMER, C. F. 1885. Lethaea erratica oder Aufzählung und Beschreibung der in der norddeutschen Ebene vorkommenden Diluvial-Geschiebe nordischer Sedimentär-Gesteine. *In*: DAMES, W. & KAYSER, E. (eds) *Palaeontologische Abhandlungen*, **2**(5), 1–173.

ROEMER, F. 1866. Geognostische Beobachtungen in polnischen Mittelgebirge. *Zeitschrift der Deutschen geologischen Gesellschaft*, **18**, 667–690.

ROHRMÜLLER, J., GEBAUER, D. & MIELKE, H. 2000. Die Altersstellung des ostbayerischen Grundgebirges. *Geologia Bavarica*, **105**, 73–84.

ROMER, R.L. & RÖTZLER, J. 2001. P-T-t evolution of ultrahigh-temperature granulites from the Saxon Granulite Massif, Germany. Part II: Geochronology. *Journal of Petrology*, **42**, 2015–2032.

ROSS, C.A. & ROSS, J.R.P. 1995. North American depositional sequences and correlations. *In*: COOPER, J.D., DROSER, M.L. & FINNEY, S.C. (eds) *Ordovician Odyssey – Short Papers for the 7th International Symposium on the Ordovician System*. Pacific Section Society for Sedimentary Geology (SEPM), book **77**, 309–313, Fullerton, California.

ROSS, J.R.P. & ROSS, C.A. 1992. Ordovician sea-level fluctuations. *In*: WEBBY, B.D. & LAURIE, J.R. (eds) *Global Perspectives on Ordovician Geology*. Balkema, Rotterdam, 327–335.

RÖTZLER, J., CARSWELL, D.A., GERSTENBERGER, H. & HAASE, G. 1999. Transitional blueschist-epidote amphibolite facies metamorphism in the Frankenberg massiv, Germany, and geotectonic implications. *Journal of Metamorphic Geology*, **16**, 109–125.

RŮŽIČKA, R. 1926. Fauna vrstev Eulomových rudního ložiska u Holoubkova (v Ouzkém). Část I. (Trilobiti). *Rozpravy II třídy České Akademie*, **35**(39), 1–26.

RŮŽIČKA, R. 1927. Fauna vrstev Eulomových rudního ložiska u Holoubkova (v Ouzkém). Část II. *Rozpravy II třídy České Akademie*, **36**(60), 1–21.

RYKA, W. & TOMCZYK, H. 1959. Bentonity w osadach staropaleozoicznych Gór Świętokrzyskich. *Kwartalnik Geologiczny*, **3**, 689–711.

SAMSONOWICZ, J. 1934. *Objaśnienie arkusza Opatów*. PIG, Warszawa.

SAMUELSSON, J. 1999. Ordovician Chitinozoa from Rügen, North-East Germany. *Acta Universitatis Carolinae Geologica*, **43**, 295–297.

SAMUELSSON, J. & VERNIERS, J. 1999. Middle to Late Ordovician Chitinozoan Biozonation of the Sennette, Dyle-Thyle and Orneau Valleys, Brabant Massif, Belgium. *Acta Universitatis Carolinae-Geologica*, **43**, 299–302.

SAMUELSSON, J. & VERNIERS, J. 2000. Ordovician chitinozoan biozonation of the Brabant Massif, Belgium. *Review of Palaeobotany and Palynology*, **113**, 105–129.

SAMUELSSON, J. & SERVAIS, T. 2001. Chitinozoa biostratigraphy of subsurface Ordovician sediments from the Lohme 2/70 well, Island of Rügen (NE-Germany). *Neues Jahrbuch für Geologie und Paläontologie, Abhandlungen*, **222**, 73–90.

SAMUELSSON, J., VERNIERS, J. & VECOLI, M. 2000. Chitinozoan faunas from the Rügen Ordovician (Rügen 5/66 and Binz 1/73 wells), NE-Germany. *Reviews of Palaeobotany and Palynology*, **113**, 131–144.

SAMUELSSON, J., VECOLI, M. & BEIER, H. 2001. Palynostratigraphy (Chitinozoa and acritarchs) of the G-14/1–86 borehole, southern Baltic Sea. *Neues Jahrbuch für Geologie und Paläontologie, Abhandlungen*, **222**, 91–122.

SAMUELSSON, J., VECOLI, M., BEDNARCZYK, W.S. & VERNIERS, J. 2002*a*. Timing of the Avalonia-Baltica plate convergence as inferred from palaeogeographic and stratigraphic data of chitinozoan assemblages in west Pomerania, northern Poland. *In*: WINCHESTER, J.A., PHARAOH, T.C. & VERNIERS, J. (eds) *Palaeozoic Amalgamation of Central Europe*. Geological Society, London, Special Publications, **201**, 95–114.

SAMUELSSON, J., GERDES, A., KOCH, L., SERVAIS, T. & VERNIERS, J. 2002*b*. Chitinozoa and Nd isotope stratigraphy of the Ordovician rocks in the Ebbe Anticline, NW Germany. *In*: WINCHESTER, J.A., PHARAOH, T.C. & VERNIERS, J. (eds) *Palaeozoic Amalgamation of Central Europe*. Geological Society, London, Special Publications, **201**, 115–131.

SANDEMANN, H.A.I., CLARK, A.H., SCOTT, D.J. & MALPAS, J.G. 2000. The Kennack Geniss of the Lizard Peninsula, Cornwall, SW England: commingling and mixing of mafic and felsic magmas accompanying Givetian continental incorporation of the Lizard ophiolite. *Journal of the Geological Society, London*, **157**, 1127–1242.

SANNEMANN, D. 1953. Neue stratigraphische Ergebnisse im Paläozoikum des Frankenwaldes auf Grund von Conodontenfunden. *Neues Jahrbuch für Geologie und Paläontologie, Monatshefte*, **11**, 480–482.

SANNEMANN, D. 1955. Ordovicium und Oberdevon der bayerischen Fazies des Frankenwaldes nach Conodontenfunden. *Neues Jahrbuch für Geologie und Paläontologie, Abhandlungen*, **102**, 1–36.

SASSI, F.P. & SPIESS, R. 1993. The South-Alpine metamorphic basement in the Eastern Alps. *In*: VON RAUMER, J.F., NEUBAUER, F. (eds) *Pre-Mesozoic Geology of the Alps*. Springer, Berlin, 599–607.

SASSI, F.P., ZANFERRARI, A. & ZIRPOLI, G. 1974*a*. Some considerations on the South Alpine basement of the Eastern Alps. *Neues Jahrbuch für Geologie und Paläontologie, Monatshefte*, **10**, 609–624.

SASSI, F.P., ZANFERRARI, A. & ZIRPOLI, G. 1974*b*. Nuovi dati sulla stratigrafia e I caratteri chimico-petrografici nei porfiroidi del Comelico (Alpi Orientali). *Memorie della Società Geologica Italiana*, **20**, 425–434.

SASSI, F.P., KALVACHEVA, R. & ZANFERRARI, A. 1984. New data on the age of deposition of the South-Alpine phyllitic basement in the Eastern Alps. *Neues Jahrbuch für Geologie und Paläontologie, Monatshefte*, **12**, 741–751.

SCHALLREUTER, R. 2005. Die Fauna der Klasten des Lederschiefers (Ordoviz, Thüringen). *Zeitschrift für Geologische Wissenschaften*, **33**(4/5), 283–343.

SCHALLREUTER, R. & HINZ-SCHALLREUTER, I. 1998. Ein Geschiebe aus Armorica im Thüringer Lederschiefer. *Archiv für Geschiebekunde*, **2**(5), 323–360.

SCHALLREUTER, R. & KOCH, L. 1999. Ostrakoden aus dem Unteren Llanvirn (Ordoviz) von Kiesbert (Ebbe-Sattel, Rheinisches Schiefergebirge. *Neues Jahrbuch für Geologie und Paläontologie, Monatshefte*, **1999**, 477–489.

SCHALLREUTER, R., HINZ-SCHALLREUTER, I., BALINI, M. & FERRETTI, A. 2006. Late Ordovician ostracoda from Iran and their significance for palaeogeographical reconstructions. *Zeitschrift für geologische Wissenschaften*, **34**(5), 293–345.

SCHÄTZ, M., REISCHMANN, T., TAIT, J., BACHTADSE, V., BAHLBURG, H. & MARTIN, U. 2002. The Early Palaeozoic break-up of northern Gondwana, new palaeomagnetic and geochronological data from the Saxothuringian basin, Germany. *International Journal of Earth Sciences (Geololische Rundschau)*, **91**, 838–849.

SCHÄTZ, M., ZWING, Z., TAIT, J., BELKA, Z., SOFFEL, H.C. & BACHTADSE, V. 2006. Paleomagnetism of Ordovician carbonate rocks from Malopolska Massif, Holy Cross Mountains, SE Poland — Magnetostratigraphic and geotectonic implications. *Earth and Planetary Science Letters*, **244**, 349–360

SCHENCK, P.F. 2002. Untergrund Schleswig. *In*: *Stratigraphie von Deutschland II, Ordovizium, Kambrium, Vendium, Riphäikum. Teil III*. Stratigraphische Kommission Deutschlands, Courier Forschungsinstitut Senckenberg, **235**, 158–162.

SCHMIDT, K. & FRANKE, D. 1977. Zur lithologisch-faziellen Entwicklung des Präperms im Nordteil der DDR. *Zeitschrift für angewandte Geologie*, **23**, 541–547.

SCHMIDT, W. 1939. Graptolithen aus dem Phycodenquarzit Thüringens. *Zeitschrift der Deutschen geologischen Gesellschaft*, **91**, 177–187.

SCHREIBER, A. 1992. Das Paläozoikum in der Zone Münchberg-Frankenberg und die Herkunft der dortigen Kristallinschollen - ein Beitrag zur Deckenfrage im Saxothuringikum. *Geologisches Jahrbuch A*, **135**, 3–135.

SCHWAB, M. 1976. Beiträge zur Tektonik der Rhenoherzynischen Zone im Unterharz. *Jahrbuch für Geologie*, **5/6**, 9–117.

SCHWAN, W. 1974. Die Sächsischen Zwischengebirge und Vergleiche mit der Münchberger Gneismasse und anderen analogen Kristallinvorkommen im Saxothuringikum. *Erlanger geologische Abhandlungen*, **99**, 1–189.

SCHWAN, W. 2002. Under-forward thrust, a compressional normal fault in orogenic belts: Its structural type and kinematic process. *Zeitschrift der Deutschen geologischen Gesellschaft*, **153**(1), 51–76.

SCHWARZBACH, M. 1934. Beiträge zur Geologie des Bober-Katzbach-Gebirges I. *Centralblatt für Mineralogie, Geologie und Paläontologie, B*, 401–413.

SCHWARZBACH, M. 1936. Oberlausitzer Schiefergebirge und Boberkatzbachgebirge - ein stratigraphisch-tektonischer Vergleich. *Abhandlungen der Naturforschenden Gesellschaft, Görlitz*, **32**, 31–63.

SCOTESE, C.R. & MCKERROW, W.S. 1990. Revised world maps and introduction. *In*: MCKERROW, W.S. & SCOTESE, C.R. (eds) *Palaeozoic Palaeogeography and Biogeography*. Geological Society, London, Memoirs, **12**, 1–21.

SDZUY, K. 1955*a*. Die Fauna der Leimitz-Schiefer (Tremadoc). *Abhandlungen der Senkenbergischen Naturforschenden Gesellschaft*, **492**, 1–74.

SDZUY, K. 1955*b*. Cystoideen aus dem Leimitz-Schiefer (Tremadoc). *Senckenbergiana lethaia*, **35**, 269–276.

SDZUY, K. 1961. Neue Funde aus den Leimitz-Schiefern (Tremadoc). *Senckenbergiana lethaia*, **42**, 227–243.

SDZUY, K. 1964. Das Kambrium des Frankenwaldes. 1: Erforschungsgeschichte, Vorkommen und Stratigraphie. *Senckenbergiana lethaia*, **45**, 201–221.

SDZUY, K. 1970. Die Kühberg-Schichten des westlichen Frankenwaldes sind "thüringisches" Unter-Tremadoc. *Neues Jahrbuch für Geologie und Paläontologie Monatshefte*, **11**, 674–682.

SDZUY, K. 1971. The Ordovician in Bavaria. *Mémoires du Bureau de Recherches Géologiques et Minières*, **73**, 379–390.

SDZUY, K. 1979. Two rare trilobites from the Tremadoc Leimitz Shales, Germany. *Alcheringa*, **3**, 63–72.

SDZUY, K., HAMMANN, W. & VILLAS, E. 2001. The Upper Tremadoc fauna from Vogtendorf and the Bavarian Ordovician of the Frankenwald (Germany). *Senckenbergiana lethaea*, **81**(1), 207–261.

SEHNERT, M. 1991. Zur Biostratigraphie epizonaler Metamorphite der Wippraer Einheit/Unterharz. *Zentralblatt für Geologie und Paläontologie, Teil 1*, **5**, 1366.

SEILACHER, A. 1970. Cruziana stratigraphy of "non-fossiliferous" Palaeozoic sandstones. *In*: CRIMES, T.P. & HARPER, J.W. (eds) *Trace Fossils. Geological Journal, Special Issues*, **3**, 447–476.

SERVAIS, T. 1991*a*. Contribution to the stratigraphy of the Ordovician Rigenée Formation (Brabant Massif, Belgium) with a preliminary study on acritarchs. *Annales de la Société Géologique de Belgique*, **114**, 233–245.

SERVAIS, T. 1991*b*. Discovery of turbiditical levels in the Late Ordovician of the Sennette valley (Brabant Massif, Belgium). *Annales de la Société Géologique de Belgique*, **114**, 147–251.

SERVAIS, T. 1993*a*. *A critical review of some Ordovician acritarch taxa and their stratigraphical implications in Belgium and Germany.*

PhD thesis, Université de Liège.

SERVAIS, T. 1993*b*. The Ordovician acritarch Frankea. *In*: MOLYNEUX, S.G. & DORNING, K.J. (eds) *Contributions to Acritarch and Chitinozoan Research*. Special Papers in Palaeontology, **48**, 79–95.

SERVAIS, T. 1994. The Ordovician acritarchs from Rügen (NE-Germany) palaeobiogeographical evidence for the attribution to Eastern Avalonia. *Neues Jahrbuch für Geologie und Paläontologie, Monatshefte*, **1994**, 566–580.

SERVAIS, T. & FATKA, O. 1997. The recognition of the Trans-European-Suture Zone by the palaeogeographic distribution pattern of Early to Middle Ordovician acritarchs. *Geological Magazine*, **134**, 617–625.

SERVAIS, T. & KATZUNG, G. 1993. Acritarch dating of Ordovician sediments of the Island of Rügen (NE-Germany). *Neues Jahrbuch für Geologie und Paläontologie Monatshefte*, **1993**(12), 713–723.

SERVAIS, T. & MALETZ, J. 1992. Lower Llanvirn (Ordovician) graptolites and acritarchs from the «assise de Huy», Bande de Sambre-et-Meuse, Belgium. *Annales de la Sociéte géologique de Belgique*, **115**, 265–285.

SERVAIS, T. & MOLYNEUX, S.G. 1997. The messaoudensis-trifidum acritarch assemblage (Ordovician: late Tremadoc – early Arenig) from the subsurface of Rügen (Baltic Sea, NE-Germany). *Palaeontographica Italica*, **84**, 113–161.

SERVAIS, T., HERBOSCH, A. & VANGUESTAINE, M. 1993. Review of the stratigraphy of the Ordovician in the Brabant Massif, Belgium. *Geological Magazine*, **130**, 699–710.

SERVAIS, T., SAMUELSSON, J., SEHNERT, M., VECOLI, M., GIESE, U. & VERNIERS, J. 2001. Ordovician palynomorphs from the subsurface of Rügen (NE-Germany) : review and perspectives. *Neues Jahrbuch für Geologie und Paläontologie. Abhandlungen*, **222**, 241–258.

SHEEHAN, P.M. 1987. Late Ordovician (Ashgillian) Brachiopods from the region of the Sambre and Meuse rivers, Belgium. *Bulletin de l'Institut Royal des Sciences Naturelles de Belgique, Sciences de la Terre*, **57**, 5–81.

SIEBEL, W., RASCHKA, H., IRBER, W., KREUZER, H., LENZ, K.-L., HÖHNDORF, A. & WENDT, I. 1997. Early Paleozoic acid magmatism in the Saxothuringian belt: New insights from a geochemical and isotopic study of orthogneisses and metavolcanic rocks from the Fichtelgebirge, SE Germany. *Journal of Petrology*, **38**, 203–230.

SITTIG E. 1965. Der geologische Bau des variszischen Sockels nordöstlich von Baden-Baden (Nordschwarzwald). *Oberrheinische geologische Abhandlungen*, **14**, 167–297.

SKOWRONEK, A. & STEFFAHN, J. 2000. The age of the Kauffung Limestone (W Sudetes, Poland) - a revision due to new discovery of microfossils. *Neues Jahrbuch für Geologie und Paläontologie, Monatshefte*, **2000**, 65–82.

SMULIKOWSKI, W. 1995. Evidence of glaucophane-schist facies metamorphism in the East Karkonosze complex, West Sudetes, Poland. *Geologische Rundschau*, **84**, 720–737.

SPASOV, C. & TELLER, L. 1963. Konodonty ot ordovikskite varovicy pri s. Mójcza v Górach Świętokrzyskich, Polša. *Trudove vrkhu geologskie na Bulgaria*, **5**, 75–83.

STEINER, J. & FALK, F. 1981. The Ordovician Lederschiefer of Thüringia. *In*: HAMBREY, M.J. & HARLAND, W.B. (eds) *Earth's Pre-Pleistocene Glacial Record*. Cambridge University Press, Cambridge, 579–581.

STETTNER, G. 1997. Fichtelgebirge und nördlicher Oberpfälzer Wald. *In*: *Stratigraphie von Deutschland II. Teil I: Thüringen, Sachsen, Ostbayern*. Stratigraphische Kommission Deutschlands, Courier Forschungsinstitut Senckenberg, **200**, 115–144.

ŠTORCH, P. 1990. Upper Ordovician – lower Silurian sequences of the Bohemian Massif, central Europe. *Geological Magazine*, **127**, 225–239.

ŠTORCH, P. 1998. Volcanism. *In*: CHLUPÁČ, I., HAVLÍČEK, V., KŘÍŽ, J., KUKAL, Z. & ŠTORCH, P. (eds) *Palaeozoic of the Barrandian (Cambrian to Devonian)*. Czech Geological Survey, 149–164.

ŠTORCH, P. & MERGL, M. 1989. Králodvor-Kosov boundary and the late Ordovician environmental changes in the Prague Basin (Barrandian area, Bohemia). *Sborník geologických Věd, Geologie*, **44**, 117–153.

STRATIGRAPHISCHE KOMMISSION DEUTSCHLANDS. 1997. Stratigraphie von Deutschland II. Ordovizium, Kambrium, Vendium, Riphäikum. Teil I: Thüringen, Sachsen, Ostbayern. *Courier Forschungsinstitut Senckenberg*, **200**, 1–437.

STRATIGRAPHISCHE KOMMISSION DEUTSCHLANDS. 2001*a*. Stratigraphie von Deutschland II: Ordovizium, Kambrium, Vendium, Riphäikum. Teil II: Baden-Württemberg, Bayern, Hessen, Rheinland-Pfalz, Nordthüringen, Sachsen-Anhalt, Brandenberg. *Courier Forschungsinstitut Senckenberg*, **234**, 1–236.

STRATIGRAPHISCHE KOMMISSION DEUTSCHLANDS. 2001*b*. Stratigraphie von Deutschland II. Ordovizium, Kambrium, Vendium, Riphäikum. Teil III: Nordthüringen, Sachsen-Anhalt, Brandenburg, Nordrhein-Westfalen, Niedersachsen, Mecklenburg-Vorpommern, deutscher Anteil Ostsee, Schleswig-Holstein, deutscher Anteil Nordsee. *Courier Forschungsinstitut Senckenberg*, **235**, 1–186.

STRUVE, W. 1959. Volkops volki n. g., n. sp., ein Phacopinae aus dem deutschen Ordovizium. *Senckbergiana lethaia*, **40**(1/2), 29–45.

STRUVE, W. 1962. Einige Trilobiten aus dem Ordovizium von Hessen und Thüringen. *Senckbergiana lethaia*, **43**(2), 151–180.

STRUVE, W. 1975. Die ältesten Fossilien Hessens. *Natur und Museum*, **105**, 262–282.

TAIT, J.A., BACHTADSE, V., FRANKE, W. & SOFFEL, H.C. 1997. Geodynamic evolution of the European Variscan fold belt: palaeomagnetic and geological constraints. *Geologische Rundschau*, **86**, 585–598.

TELLER, L. 1962. Zagadnienie granicy ordowik-sylur w Górach Bardzkich. *In*: PASSENDORFER, E. (ed.) *Księga pamiątkowa ku czci profesora Jana Samsonowicza*. Wydawnictwa Geologiczne, Warszawa, 171–186.

TEMPLE, J.T. 1965. Upper Ordovician brachiopods from Poland and Britain. *Acta Palaeontologica Polonica*, **10**, 379–450.

TICHOMIROWA, M. 2002. Die Gneise des Erzgebirges - hochmetamorphe Äquivalente von neoproterozoisch-frühpaläozoischen Grauwacken und Granitoiden. *Freiberger Forschungshefte, C*, **495**, 1–222.

TICHOMIROWA, M., BERGER, H.-J., KOCH, E. A., BELYATSKI, B. V., GÖTZE, J., KEMPE, U., NASDALA, L. & SCHALTEGGER, U. 2001. Zircon ages of highgrade gneisses in the Eastern Erzgebirge (Central European Variscides) - constrains on origin of the rocks and Precambrian to Ordovician magmatic events in the Variscan foldbelt. *Lithos*, **56**, 303–332.

TIMM, J., DEGENS, E.T. & WIESNER, M.G. 1981. *Erläuterungen zur Geologischen Karte des zentralen Ebbe-Antiklinoriums 1 :25.000*. Mitteilungen aus dem Geologisch-Paläontologischen Insitut der Universität Hamburg, **50**, 59–75.

TOMCZYK, H. 1954. Stratygrafia gotlandu niecki międzygórskiej w Górach Świętokrzyskich na podstawie fauny z łupków graptolitowych. *Biuletyn Instytutu Geologicznego*, **93**, 1–66.

TOMCZYK, H. 1962. Problem stratygrafii ordowiku i syluru w Polsce w świetle ostatnich badań. *Prace Instytutu Geologicznego*, **35**, 1–134.

TOMCZYK, H. & TURNAU-MORAWSKA, M. 1964. Stratygrafia i petrografia ordowiku Brzezin koło Morawicy w Górach Świętokrzyskich. *Acta Geologica Polonica*, **14**, 501–546.

TOMCZYKOWA, E. 1968. Stratygrafia osadów najwyższego kambru w Górach Świętokrzyskich. *Prace Instytutu Geologicznego*, **54**, 1–85.

TORSVIK, T.H. 1998. Palaeozoic palaeogeography: a North Atlantic viewpoint. *GFF*, **120**, 109–118.

TOURNEUR, F., VANGUESTAINE, M., BUTTLER, C., MAMET, B., MOURAVIEFF, N., POYT, E. & PRÉAT, A. 1993. A preliminary study of Ashgill carbonate beds from the lower part of the Fosses Formation (Condroz, Belgium). *Geological Magazine*, **130**, 673–679.

TRAPPE, J. & ELLENBERG, J. 1994. The phosphorites from the Ordovician iron ore deposits in Thuringia, Germany: Transgressive and regressive depositional systems-shift facies. *Zentralblatt für Geologie und Paläontologie Teil I*, **1992**(11/12), 1387–1402.

TRELA, W. 2000. Zapis izotopowy (18O i 13C) w wapieniach górnego ordowiku z Mójczy w Górach Świętokrzyskich. *Przegląd Geologiczny*, **48**, 736–740.

TRELA, W., SALWA, S. & SZCZEPANIK, Z. 2001. The Ordovician rocks of Pobroszyn in the Łysogóry region of the Holy Cross Mountains, Poland. *Kwartalnik Geologiczny*, **45**, 143–154.

TSCHERNOSTER, R., KRAMM, U., GIESE, U. & GLODNY, J. 1997. The evolution of the Baltica/Gondwana suture along the TESZ during Lower Palaeozoic times – Implications from detritus analysis and isotope studies. *Terra Nostra*, **97**, 148–152.

TURNAU-MORAWSKA, M. 1960. Zlepieniec glaukonitowy dolnego ordowiku okolic Międzygórza w Górach Świętokrzyskich. *Acta Geologica Polonica*, **10**, 123–148.

TURNAU-MORAWSKA, M. 1961. Charakterystyka petrograficzna ordowiku w Mójczy koło Kielc w porównaniu z równowiekowymi osadami innych obszarów Polski. *Kwartalnik geologiczny*, **5**, 775–787.

URBANEK, Z. & BARANOWSKI, Z. 1986. Revision of age of the

Radzimowice Slates from Góry Kaczawskie Mts (western Sudetes, Poland) based on conodonts. *Annales Societatis Geologorum Poloniae*, **56**, 399–408.

URBANEK, Z., BARANOWSKI, Z. & HAYDUKIEWICZ, A. 1975. Geologiczne konsekwencje występowania dewońskich konodontow w metamorfiku północnej części Gór Kaczawskich. *Geologia Sudetica*, **10**, 155–169.

VAIDA, M., HANN, H.P., SAWATZKI, G. & FRISCH, W. 2004. Ordovician and Silurian protolith ages of metamorphosed clastic sedimentary rocks from the southern Schwarzwald, SW Germany: a palynological study an dits bearing on the Early Palaeozoic geotectonic evolution. *Geological Magazine*, **141**, 629–643.

VAN GROOTEL, G., VERNIERS, J., GEERKENS, B., LADURON, D., VERHAEREN, M., HERTOGEN, J. & DE VOS, W. 1997. Timing of magmatism, foreland basin development, metamorphism and inversion in the Anglo-Brabant fold belt. *Geological Magazine*, **134**, 607–616.

VANGUESTAINE, M. 1974. Espèces zonales d' Acritarches du Cambro-Trémadocien de Belgique et de l'Ardenne Française. *Review of Palaeobotany and Palynology*, **18**, 63–82.

VANGUESTAINE, M. 1992. Biostratigraphie par acritarches du Cambro-Ordovicien de Belgique et des régions limitrophes: synthèse et perspectives d'avenir. *Annales de la Société géologique de Belgique*, **115**, 1–18.

VANGUESTAINE, M. & SERVAIS, T. 2002. Early Ordovician acritarchs of the Lierneux Member (Stavelot Inlier, Belglium) : stratigraphy and palaeobiogeography. *Bulletin de la Société géologique de France*, **173**, 561–568.

VANGUESTAINE, M., BREUER, P. & LEHNERT, O. 2004. Discovery of an Early Ordovician conodont fauna in the Salm Group of the Stavelot Inlier, Belgium. *Bulletin de l'Institut royal des Sciences naturelles de Belgique, Sciences de la Terre*, **74**, 39–48.

VANMEIRHAEGHE, J. 2006. *The evolution of the Condroz-Brabant Basin from Middle Ordovician to Llandovery: lithostratigraphical and chitinozoan biostratigraphical approach*. PhD thesis, Ghent University.

VANMEIRHAEGHE, J. & VERNIERS, J. 2004. Chitinozoans bio- and lithostratigraphical study of the Ashgill Fosses and Génicot Formations (Condroz Inlier, Belgium). *Review of Palaeobotany and Palynology*, **130**, 241–267.

VANMEIRHAEGHE, J., STORME, A., VAN NOTEN, K., VAN GROOTEL, G. & VERNIERS, J. 2005. Chitinozoan biozonation and new lithostratigraphical data in the Upper Ordovician of the Fauquez and Asquempont areas (Brabant Massif, Belgium) *Geologica Belgica*, **8**, 145–159.

VANNIER, J., RACHEBOEUF, P.R., BRUSSA, A., WILLIAMS, M., RUSHTON, A.W.A., SERVAIS, T. & SIVETER, D.J. 2003. Cosmopolitan arthropod zooplankton in the Ordovician seas. *Palaeogeography, Palaeoclimatology, Palaeoecology*, **195**, 173–191.

VAVRDOVÁ, M. 1993. Acritarch assemblages in the Arenig Series of the Prague Basin, Czech Republic. *Special Papers in Palaeontology*, **48**, 125–139.

VECOLI, M. & SAMUELSSON, J. 2001. Quantitative evaluation of microplankton palaeobiogeography in the Ordovician - Early Silurian of the northern Trans European Suture Zone: implications for the timing of the Avalonia-Baltica collision. *Review of Palaeobotany and Palynology*, **115**, 43–68.

VECOLI, M., DIENI, I., SASSI, F.P. & SERVAIS, T. 2008. *The Cambrian acritarchs of Col di Foglia (South-Alpine basement of the Eastern Alps, Agordo Italy)*. Rendiconti Accademia dei Lincei, Scienze Fisiche e Naturali. Springer-Verlag.

VERNIERS, J., SAMUELSSON, J., VAN GROOTEL, G., DE GEEST, P. & HERBOSCH, A. 1999. The Ordovician in Belgium: new litho- and biostratigraphical data with Chitinozoa from the Brabant Massif and the Condroz Inlier (Belgium). *Acta Universitatis Carolinae Geologica*, **43**, 93–96.

VERNIERS, J., HERBOSCH, A., *et al.* 2001. The Lower Palaeozoic formations of Belgium. *Geologica Belgica*, **4**, 1–28.

VERNIERS, J., PHARAOH, T., *et al.* 2002. The Cambrian to mid Devonian basin development and deformation history of eastern Avalonia, east of the Midlands Microcraton: new data and a review. In: WINCHESTER, J. A., PHARAOH T. C. & VERNIERS, J. (eds) *Palaeozoic Amalgamation of Central Europe*. Geological Society, London, Special Publications, **201**, 47–93.

VERNIERS, J., VAN GROOTEL, G. & DEBACKER, T. 2005. Upper Ordovician lithostratigraphy and structural architecture of the Fauquez area (Brabant Massif, Belgium). *Geologica Belgica*, **8**, 160–175.

VIDET, B., PARIS, F. & RUBINO, J.L. 2006. Ordovician sea-level changes: a northern Gondwana example. *In*: SENNIKOV, N.V., KANYGIN, A.V., OBUT, O.T. & KIPRIYANOVA, T.P. (eds) *Palaeogeography and global correlation of Ordovician Events*. Russian Academy of Sciences, Siberian Branch, Novosibirsk, 71–73.

VOLK, M. 1938. Die Zone des Phycodenquarzits am Thierberg bei Steinach. *Zentralblatt für Mineralogie, Geologie und Paläontologie, B*, **12**, 450–454.

VOLK, M. 1966. Dictyonema cf. uralensis Obut aus den Phycodenschiefern bei Steinach. *Geologische Blätter von Nord-Ost Bayern*, **16**(2/3), 162–163.

WALTER, H. & BERGER, H.-J. 1998. Paläozoische Mikrofossilien vom Nordrand des Erzgebirges (Sachsen). *Neues Jahrbuch für Geologie und Paläontologie. Abhandlungen*, **209**(1), 1–32.

WEBBY, B.D., COOPER, R.A., BERGSTRÖM, S.M. & PARIS, F. 2004. Stratigraphic framework and time-slices. *In*: WEBBY, B.D., PARIS, F., DROSER, M.L. & PERCIVAL, I. (eds) *The Great Ordovician Biodiversification Event*. Columbia Press, New York, 41–47.

WERNER, C.-D., LITTMANN, S. & GEHMLICH-PALEN, R.. 2005. Zur Geochemie und Geochronologie der cadomisch-frühvariszischen Granitoide im sächsisch-thüringischen Vogtland. *Zeitschrift für geologische Wissenschaften*, **33**(2–3), 81–108, Berlin.

WEYER, D. 1967. Ein Korallenfund aus dem Ordoviz von Thüringen. *Geologie*, **16**, 922–927.

WICKERT F., ALTHERR R. & DEUTSCH M. 1990. Polyphase Variscan tectonics and metamorphism along a segment of the Saxothuringian-Moldanubian boundary: The Baden-Baden Zone, northern Schwarzwald (F.R.G.), *Geologische Rundschau*, **79**, 627–647.

WINCHESTER, J.A. 2002. Palaeozoic amalgamation of Central Europe; new results from recent geological and geophysical investigations. *Tectonophysics*, **360**, 5–21.

WINCHESTER, J.A., PHARAOH, T.C. & VERNIERS, J. (eds) 2002*a*. *Palaeozoic Amalgamation of Central Europe*. Geological Society, London, Special Publications, **201**.

WINCHESTER, J.A., PHARAOH, T.C. & VERNIERS, J. 2002*b*. Palaeozoic amalgamation of Central Europe : an introduction and synthesis of new results from recent geological and geophysical investigations. *In*: WINCHESTER, J.A., PHAROAH, T.C. & VERNIERS, J. (eds) *Palaeozoic Amalgamation of Central Europe*. Geological Society, London, Special Publications, **201**, 1–18.

WINCHESTER, J.A., PHARAOH, T.C., VERNIERS, J., IOANE, D. & SEGHEDI, A. 2006. Palaeozoic accretion of Gondwana-derived terranes to the East European Craton: recognition of detached terrane fragments dispersed after collision with promontories. *In*: GEE, D.G. & STEPHENSON, R.A. (eds) *European Lithosphere Dynamics*. Geological Society, London, Memoirs, **32**, 323–332.

WINSEMANN, J. & JONEN, A. 2000. Stapelungsmuster eines sturmdominierten Schelfsystems: Die unterordovizische Goldisthaler Folge und Frauenbach-Gruppe des Thüringer Schiefergebirges. *Zeitschrift der Deutschen geologischen Gesellschaft*, **151**(3), 287–307.

WURM, A. 1925. *Geologie von Bayern. Teil 1: Nordbayern, Fichtelgebirge und Frankenwald*. Handbuch der Geologie und Bodenschätze Deutschlands, **373 S**, Bornträger, Berlin.

WURM, A. 1961. *Geologie von Bayern I, Frankenwald, Münchberger Gneismasse, Fichtelgebirge*. Nördlicher Oberpfälzer Wald. **555 S**, Borntraeger, Berlin.

YOUNG, T.P. 1992. Ooidal ironstones from Ordovician Gondwana: a review. *Palaeogeography, Palaeoclimatology, Paleaoecology*, **99**, 321–347.

ZEUSCHNER, L. 1869. Uber die neuentdeckte Silurformation von Kleczanow. *Zeitschrift der Deutschen geologischen Gesellschaft*, **21**, 257–262.

ŻELAŹNIEWICZ, A. 1997. The Sudetes as a Palaeozoic orogen in central Europe. *Geological Magazine*, **134**, 691–702.

ZNOSKO, J. & CHLEBOWSKI, R. 1976. Rewizja stratygrafii dolnego ordowiku w Górach Świętokrzyskich. *Rocznik Polskiego Towarzystwa Geologicznego*, **46**, 135–157

6 Silurian

JACQUES VERNIERS (co-ordinator), JÖRG MALETZ, JÍŘÍ KŘÍŽ, ŽIVILĖ ŽIGAITĖ, FLORENTIN PARIS, HANS PETER SCHÖNLAUB & RYSZARD WRONA

In an overview of the Silurian of Central Europe, it is important to realize that during this period the study area was spread more widely over the globe than nowadays because at least two oceans were present in the area which have since disappeared. Several palaeocontinents such as Baltica or Gondwana, smaller palaeoplates such as Avalonia and Far Eastern Avalonia, and Peri-Gondwana terranes such as Perunica, were separated by the Tornquist Sea and the Rheic Ocean. These palaeocontinents were brought together in the present-day configuration by closing of the oceans and the subsequent orogenic collisions, respectively termed the Caledonian and Variscan orogenies. Plate movements before and during the Alpine orogeny also brought pieces of northern Gondwana into the study area. These Proto-Alps are now included in the basement of the Alps and are observable in several tectonic windows (e.g. Carnic Alps).

The Ordovician–Silurian boundary

The Global Stratotype Section and Point (GSSP) of the base of the Silurian System, equivalent to the base of the Llandovery Series and the Rhuddanian Stage, is situated outside the study area in Dob's Linn, near Moffat in the Southern Uplands of Scotland (Bassett 1985; Cocks 1985; Holland 1985; Williams & Ingham 1989). The base was put at the lowest occurrence level of the graptolite *Parakidograptus acuminatus* s.l. After a systematic revision of the fauna another species was selected as index fossil. The lowest occurrence level of *Akidograptus ascensus* was then chosen to indicate the base of the Silurian (Melchin 2004). The Silurian System consists of four series, from bottom to top: Llandovery, Wenlock, Ludlow and Přídolí. The lower three are further subdivided into seven stages (Fig. 6.1).

Palaeoclimate

During the Silurian the climate was recovering globally from the Hirnantian glaciation (latest Ordovician), during which an icecap was centred on a large part of Gondwana. At the transition from the Ordovician to the Silurian, the Hirnantian icehouse climate changed rapidly to a greenhouse climate (Hallam & Wignall 1997; Brenchley *et al.* 2001). Hence, reconstructions all show a temperature increase through the Silurian (Frakes *et al.* 1992; Azmy *et al.* 1998). On the other hand, many parameters presumably connected with biotic productivity were declining during the Silurian such as low atmospheric oxygen, low organic carbon in the oceans (see below) and depletion of δC^{13} (Wilde *et al.* 1991; see below).

Budyko *et al.* (1987) reconstructed a global inventory of the masses of major lithofacies for the Silurian and proposed a global average temperature of 20°C, which is warmer than the modern global average of 15°C. The temperature was somewhat lower than during the even warmer Ordovician and Devonian, explained by a combination of increased land area in the northern hemisphere and evaporative cooling in the shallow tropical embayments (Wilde *et al.* 1991). The Silurian atmospheric carbon dioxide has been calculated as *c.* 900 ppm (three times modern values) which represents a clear decline from the 1200 ppm values in the Upper Ordovician (Budyko *et al.* 1987). According to the model of these authors, atmospheric oxygen declined from 65% of the present atmospheric level (PAL) at the base of the Silurian to *c.* 35% PAL in the late Silurian, following a general trend from the Early Ordovician to the Mid-Devonian. It resulted in anoxic water below about 100 m and in most bottom waters of the middle and deeper shelf during highstands. The atmospheric oxygen started to rise again in the Devonian due to the rapid spreading of plants over the continents. Berner (1989) also obtained a decline of atmospheric oxygen during the Silurian, but listed higher values (from 75% to 60% PAL).

According to Wilde *et al.* (1991) the low atmospheric oxygen levels, together with a warmer climate than today, imply that the oceanic surface mixed water contained less oxygen during the Silurian than today. This resulted in a thick pycnocline throughout the Silurian. For a tropical position they calculated a pycnocline with oxygen levels near zero between depths of 100 m and 3000 m, with some oxygen above and below. Due to this low oxygen level in the Silurian oceans, far less organic matter was required to trigger anoxic conditions in the seas than in other periods.

The area of continental shelf-seas decreased during the Silurian from about 40% to 30% (Budyko *et al.* 1987). This decrease reduced the area of surface water in contact with the atmosphere and also the volume of seawater saturated with oxygen in the upper part of the water column. This factor also contributed to an increase of the anoxic zone in deeper waters during the Silurian. The result is that worldwide Silurian sediments were often deposited in anoxic conditions and are dark in colour due to a certain amount of organic matter. On the other hand, the combination of high carbon dioxide levels and higher temperature allowed deposition of calcium carbonate in tropical shelf-seas, where a relatively alkaline pH of *c.* 8 prevailed (Wilde *et al.* 1991).

Sea-level changes

Silurian shelf sequences were well exposed on large areas of North America (Laurentia), Norway, Gotland, Estonia (Baltica), England and Wales (Avalonia), Siberia, South China and Australia (Gondwana). In the stratigraphical succession on stable platforms, shifts are repeatedly observed from shallow coral–stromatoporoid communities, through brachiopod-dominated communities to deeper high-diversity graptolite assemblages and back to the shallow communities. This variation has allowed

CHRONOSTRATIGRAPHY Global Standard Stratigraphy				LOCATION OF BASAL BOUNDARY STRATOTYPE (GSSP)
System	Series		Stages	
SILURIAN		Přídolí		Požáry Section Barrandian
SILURIAN	UPPER	Ludlow	Ludfordian	Sunnyhill Quarry Ludlow District
SILURIAN	UPPER	Ludlow	Gorstian	Pitch Coppice Ludlow District
SILURIAN	LOWER	Wenlock	Homerian	Whitwell Coppice Wenlock District
SILURIAN	LOWER	Wenlock	Sheinwoodian	Hughley Brook Wenlock District
SILURIAN	LOWER	Llandovery	Telychian	Cefn Cerig Section Llandovery District
SILURIAN	LOWER	Llandovery	Aeronian	Trefawr section,Cwm-coed-Aeron Farm Llandovery District
SILURIAN	LOWER	Llandovery	Rhuddanian	Dob's Linn Southern Uplands of Scotland

Fig. 6.1. The standard Silurian chronostratigraphy (after Holland 1989).

several authors to reconstruct eustatic sea-level curves for the Silurian (McKerrow 1979; Lenz 1982; Johnson *et al.* 1991, 1998; Ross & Ross 1996; Loydell 1998; Johnson 2006) (Fig. 6.2). The eustatic curves of the latter three studies often show the same trend, though around several levels the trends are opposite (Fig. 6.2). Further studies are needed to explain the discrepancies, which are possibly due to less refined dating of the levels. Loydell (1998) had the advantage to study the graptolitic facies of the deeper shelf and he could easily link all the bathymetric excursions to the graptolite biozonations (Kaljo *et al.* 2003; Melchin *et al.* 2004). According to Loydell (1998) four highstands occurred in the Llandovery following the important sea-level low in the lower and middle part of the Hirnantian (uppermost Ordovician) with a postulated drop of 40 m (Long 1993), 100 m (Brenchley 1989) or much more (Moreau *et al.* 2004). These are the Rhuddanian/Aeronian transition (between the *Coronograptus cyphus* and the *Coronograptus gregarius* zones), the mid-Aeronian (lowest *Stimulograptus sedgwickii* Zone), the lowest Telychian (*Spirograptus turriculatus* Zone) and the upper Telychian (*Monoclimacis crenulata* Zone and higher Telychian biozones). Community replacements in the Llandovery are symmetrical from bottom to top, suggesting gradual sea-level changes. Several authors explain the Llandovery eustatic changes as being related to the melting of remaining ice sheets, proven by the presence of several diamictite horizons in the early Aeronian, in the early and in the late Telychian in South America (Brenchley & Newall 1984; Grahn & Caputo 1992; Caputo 1998; Kaljo *et al.* 2003).

At its base (*Cyrtograptus centrifugus* and *Cyrtograptus murchisoni* zones) and at its top (*Pristiograptus/Colonograptus ludensis* Zone), the Wenlock is characterized by a low sea level according to the curve of Loydell (1998). A high sea level can be observed within the upper part of the Sheinwoodian (from the *Monograptus riccartonensis* to the *Cyrtograptus ellesae* zones). The Ludlow, according to the same curve, shows a high sea level in the mid-Gorstian (*Lobograptus scanicus* Zone), a low sea level in

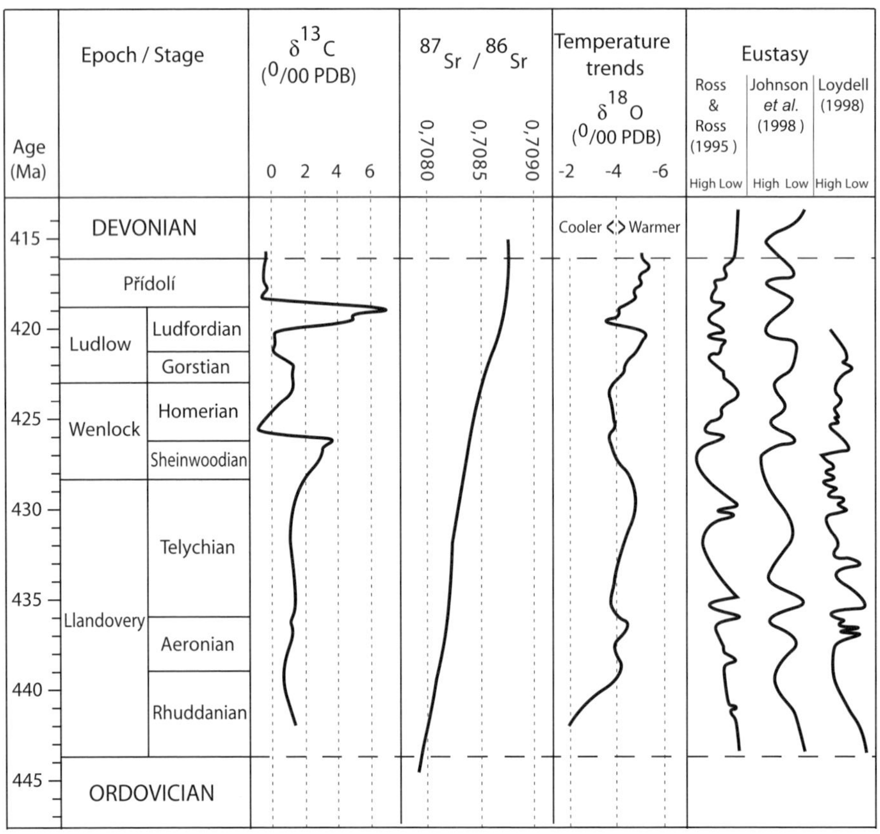

Fig. 6.2. Silurian sea-level curves and strontium, carbon and oxygen isotope curves (after Melchin *et al.* 2004).

the basal Ludfordian (*Saetograptus leintwardinensis* Zone) and a high sea level at the top of the Ludfordian. For the Přídolí there is no clear eustatic signature (Fig. 6.2).

Global and regional events

As Calner & Eriksson (2006) put it: 'the view of the Silurian as a time of stable greenhouse conditions has been successively challenged during the last decade. The increased use of carbon and oxygen stable isotopes in stratigraphy has resulted in one of the most important advances in Silurian research for decades'. According to them, the geochemical data repeatedly showed that the Silurian was characterized by recurrent anomalies in the global carbon cycle (GCC). The size of these anomalies is comparable or larger than those associated with Phanerozoic mass extinctions. Studies integrating these data with graptolite, conodont and chitinozoan biostratigraphy indicated that the Silurian GCC anomalies are coupled with moderate to substantial diversity changes and/or extinctions among the contemporary marine faunas. The GCC anomalies are also closely related to sea-level changes that in many cases appear to be global. Much work still needs to be done before the total rate of extinction and origination associated with the GCC anomalies can be estimated.

Silurian isotope curves have been produced for carbon, oxygen and strontium (Melchin *et al.* 2004). The $\delta^{18}O$ curves published by Samtleben *et al.* (1996), Wenzel & Joachimski (1996), Bickert *et al.* (1997), Azmy *et al.* (1998) and Heath *et al.* (1998) show similar trends with a reduction in values of all three isotopes, possibly indicative of an overall global warming. Detailed $\delta^{13}C$ curves for the Llandovery were published by Melchin & Holmden (2006). Smaller-scale events visible in the $\delta^{13}C$ and $\delta^{18}O$ curves in the Llandovery and early Wenlock in Estonia were thought to coincide with glacial events as described by Caputo (1998) in South America (Kaljo *et al.* 2003). However the youngest glaciation in South America was shown to be poorly dated and was proven not to be early Wenlock but Llandovery in age (Caputo 2003). Several events in the late Wenlock and Ludlow correspond in two of the three curves of Figure 6.2 with eustatic sea-level drops. Apparent asynchroneities between the three curves may be the result of correlative inaccuracies rather than real offsets in the timing of the events (Melchin *et al.* 2004). The carbon isotope curves show the same general trend as the oxygen isotope curves. Smaller-scale positive excursions coincide with eustatic lowstands (Melchin *et al.* 2004). Strontium isotope curves by Ruppel *et al.* (1998) and Azmy *et al.* (1999) indicate that the $^{87}Sr/^{86}Sr$ ratio increased steadily upwards through the Silurian, and this trend has been attributed to higher input of radiogenic Sr by rivers induced by climatic warming (Melchin *et al.* 2004).

The Silurian Period does not witness major extinction events, such as the latest Ordovician extinction event linked to the Hirnantian glaciation, nor is there any evidence of other important events such as major sea-level drops, meteorite impacts, extensive plateau basalt effusions, or falls in primary production. However there are still some 15 medium-order bioevents recorded during the Silurian (Fig. 6.3), of which two are particularly important: the severe extinction event of conodonts at the base of the Wenlock (Ireviken Event; Jeppsson 1987), where a turnover in the acritarch flora is observed with more originations than extinctions (Gelsthorpe 2004), and a major crisis amongst the graptolites, called the Great Crisis or the *lundgreni* Event (Jaeger 1991; Koren' 1987; Porębska *et al.* 2004). Three further events are observed: an early Ludfordian event has been recognized affecting the graptolites, ostracoderms, acritarchs and conodonts; a mid-Ludfordian event, the Lau event (Jeppsson 1987) with profound changes in the vertebrates, graptolites, corals and conodonts; and a mid-Přídolí event with extinction amongst corals, conodonts, graptolites and a diversity rise in the chitinozoans (Kaljo *et al.* 1991) (Fig. 6.3). Štorch (1995) discussed eight larger and smaller extinction events in graptolite evolution for the Silurian (Fig. 6.3), based mainly on research in Bohemia, but incorporating information from other regions. He gave an accurate estimation of the graptolite diversity changes in central Europe during the Silurian. The changes in diversity were compared with worldwide sea-level changes, showing that the adaptive radiations of Silurian graptolites occurred during sea-level rises when anoxic environments with black shale deposits were brought onto the shelves.

According to Melchin *et al.* (2004) some of the extinction events in the Llandovery and Wenlock correlate well with the $\delta^{18}O$ and $\delta^{13}C$ positive excursions linked with sea-level lowstands, some of which may be related to glacial episodes described by Caputo (1998, 2003) in South America (see above). The models proposed by Jeppsson (1990, 1998), modified and restricted to low latitudes by Bickert *et al.* (1997), attempted to relate all bioevents, especially those affecting conodonts, to changes in climate, oceanic circulation and terrestrial input. In these models humid episodes with high bioproductivity and mainly estuarine circulation on the shelf, alternated with more arid episodes with low bioproductivity and mainly anti-estuarine circulation on the shelf (see Calner *et al.* 2004; Johnson, 2006). However, the model ought to take into account more lithological, geochemical and palaeontological parameters in order to be more useful for the delimitation of the climatic–oceanic episodes (Kaljo *et al.* 2003)

Central European Silurian stratigraphy

The term Silurian was introduced by Murchison in 1837, named after the Celtic people living in Wales whom the Romans called the *Silures*. The history of subdivision of the Silurian System has been summarized in Holland (1989). Recent advances in the chronostratigraphy and biostratigraphy of the Silurian are summarized in Melchin *et al.* (2004). The Silurian is mainly dated using biozonations based on three fossil groups: graptolites used to define the bases of the series, stages and substages, conodonts used in shallow-marine carbonate facies, and the chitinozoans used in both shallow-marine and basinal facies. Three other fossil groups are also used for biozonation: the sporomorph groups, acritarchs and spores, and the vertebrates (Fig. 6.3).

As mentioned above, the Silurian System is divided into four series. The Llandovery Series is divided into three stages, from bottom to top: Rhuddanian, Aeronian and Telychian. The GSSPs of the bases of the latter two stages are situated in southern Wales near the town of Llandovery. The base of the Aeronian is located in the Trefawr forestry road section, 500 m north of Cwm-coed-Aeron Farm, just below the lowest occurrence level of the graptolite *Monograptus austerus sequens* representing the *Demirastrites triangularis* Zone (Cocks 1989). The GSSP of the base of the Telychian is situated in a small abandoned quarry on the west side of the Cefn Cerig road, approximating as closely as possible to the base of the *Spirograptus turriculatus* Zone (Cocks 1989). Diagnostic graptolites are not abundant in the GSSP area and the stratotype for the base of the Telychian is taken at a point just above the highest record of the brachiopod *Eocoelia intermedia* and below the earliest record of the succeeding species *Eocoelia curtisi* (Bassett 1985; Cocks 1985, 1989). The base of the Telychian is now usually shown at the base of the

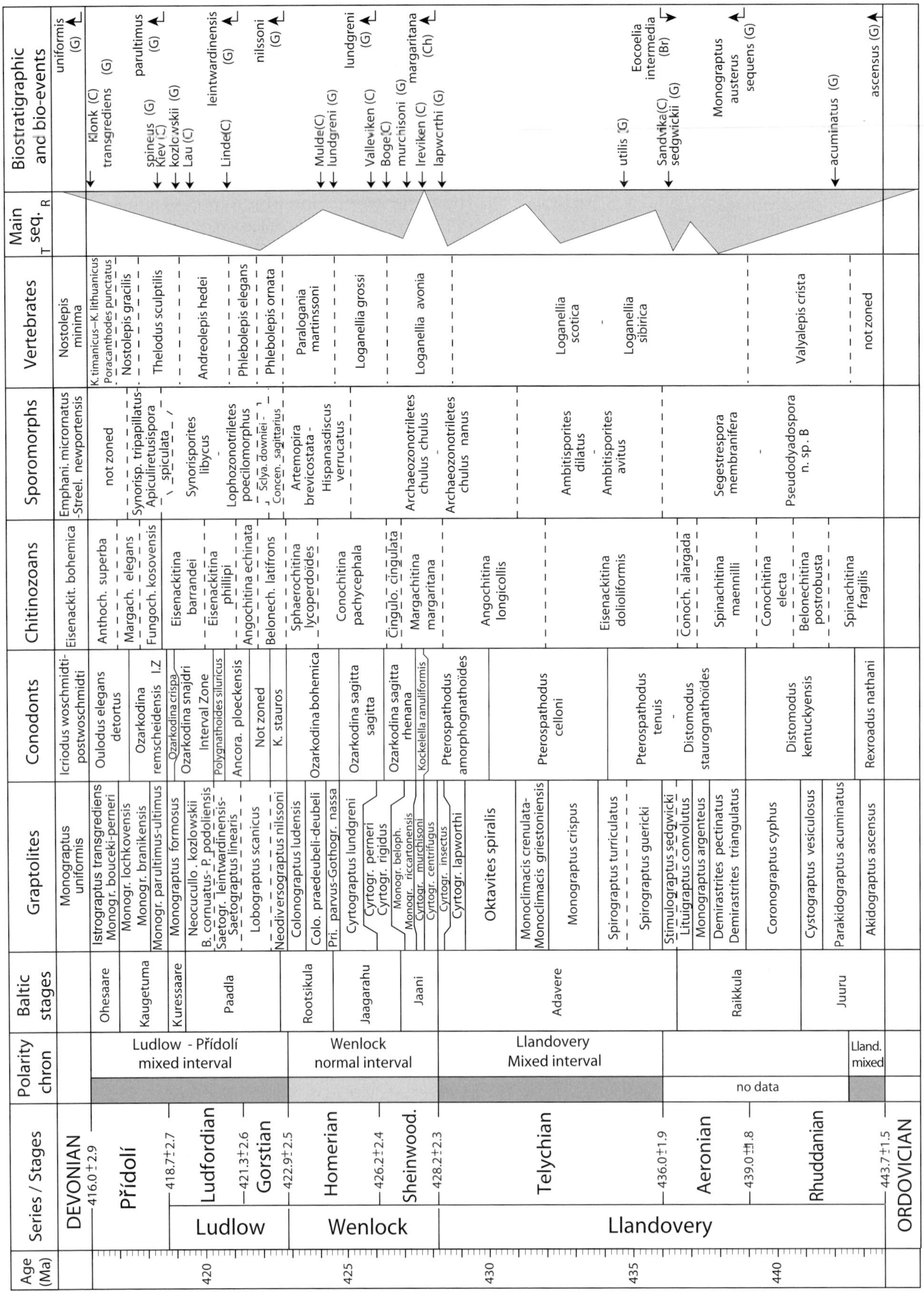

Fig. 6.3. Silurian biozonations and chronostratigraphy (graptolites, conodonts, chitinozoans, acritarchs, spores, vertebrates) (after Melchin *et al.* 2004). Right column: horizontal arrows indicate events with their names; (G): events in graptolites; (c) events in conodonts. Arrows pointing towards top: lowest occurrence datums of graptolite (G) and chitinozoan (Ch) species; arrows pointing down: highest occurrence datum of brachiopod species (Br).

Spirograptus guerichi Zone (see Koren' *et al.* 1995, 1996). This is based on the study by Loydell *et al.* (1993), who described *Spirograptus guerichi* as a new species, restricted to the lower part of the former *Spirograptus turriculatus* Zone and effectively subdividing the *Spirograptus turriculatus* Zone into a lower *Spirograptus guerichi* Zone and an upper *Spirograptus turriculatus* Zone. Following refined correlation studies, it remains unclear what the precise correlation is of the GSSP with the graptolite biozonation, but it appears to be within the upper part of the *Stimulograptus sedgwickii* Zone (Melchin *et al.* 2004).

The base of the Wenlock Series is defined in Shropshire (Welsh Borderland, UK) in Ape Dale, NW of the prominent ridge of Wenlock Edge. The section is a small outcrop along Hughley Brook, near Leasows Farm, close to the village of Hughley. The chosen level at the base of unit G in Bassett *et al.* (1975, p. 13 and fig. 5) contains no graptolites but was presumed to correlate with the base of the *Cyrtograptus centrifugus* Zone (Bassett 1985; Holland 1985; see history in Holland 1989 and details in Bassett 1989). Later studies in the GSSP on conodonts and chitinozoans indicated that the GSSP level might correspond with the base of the *Cyrtograptus insectus* Zone (Jeppsson 1997; Mullins 2000) or the base of the *Cyrtograptus murchisoni* Zone (Loydell *et al.* 2003; Mullins & Aldridge 2004). A working group of the International Subcommission of the Silurian Stratigraphy (ISSS) is currently addressing this problem. The Wenlock Series is divided into two stages: Sheinwoodian below and Homerian above. The base of the Homerian is situated in a stream section in Whitwell Coppice, north of Homer and Much Wenlock (Shropshire) at the first occurrence level of the graptolite *Cyrtograptus lundgreni* (Bassett 1985, 1989; Holland 1985).

The base of the Ludlow Series is also defined in Shropshire (Welsh Borderland, UK) close to the town of Ludlow, in a quarry at Pitch Coppice. White (1981) found few poorly preserved and not firmly determined graptolites in the GSSP and more graptolites 375 m and 30 km away. Lawson & White (1989) indicate that these records provide a strong indication that the lowest beds of the Ludlow series in the type area should be assigned to the *Neodiversograptus nilssoni* Zone. The base of the latter biozone is taken as the base of the Ludlow Series. The Ludlow Series is divided into two stages: the Gorstian below and the Ludfordian above. The base of the Ludfordian is situated in the Sunnyhill Quarry close to Ludlow at a horizon considered to correlate with the first occurrence of the graptolite *Saetograptus leintwardinensis* (Holland *et al.* 1980; Bassett 1985; Holland 1985; Lawson & White 1989).

The base of the Přídolí Series is defined in the Prague basin at the GSSP in the Požáry Quarries near Praha-Řeporyje (see below), where it coincides with the base of the *Monograptus parultimus* Zone within bed no. 96 of the Požáry Formation (Bassett 1985; Holland 1985; Kříž *et al.* 1986; Kříž 1989) (see below Fig. 6.19 below). The Přídolí Series is not subdivided into stages.

The base of the Devonian, the first GSSP of a system to be decided upon (in 1968), is also defined in the Prague basin (Chlupáč *et al.* 1972) at the first appearance level of the graptolite index species *Monograptus uniformis* in the Klonk section near Suchomasty (see Belka & Narkiewicz 2008).

Several absolute timescales have been proposed for the Silurian although few radiometric age determinations are available. Melchin *et al.* (2004) have evaluated all available data from radiometric age dating, the presumed evolutionary rates of graptolites and conodonts, and the observed stratigraphical thicknesses. After statistical analyses of the data they proposed the following radiometric ages for the bases of the series: Llandovery, 443.7 Ma; Wenlock, 428.2 Ma; Ludlow, 422.9 Ma; and Přídolí, 418.7 Ma. These are also the ages used in the international timescale of Gradstein *et al.* (2004).

Tectonic setting

The Silurian sediments in Central Europe were deposited on several palaeocontinents: (1) on the large Baltica palaeocontinent; (2) on the edge of the very large Gondwana palaeocontinent, a part often referred to as northern Gondwana; (3) on one of the many smaller palaeoplates that rifted off the Gondwana palaeocontinent during Palaeozoic times, such as Avalonia and Far Eastern Avalonia; and (4) the many other peri-Gondwana terranes. A distinction is made between plates and terranes. Plates were moving in previous times and could include a continent, i.e. an area of continental crust. When the plates lost their identity by subsequent plate tectonic movements they are referred to as palaeoplates. Terranes are usually smaller and are at present bordered by fault contacts. They represent the collided or docked parts of previously moving (micro)plates.

Silurian palaeocontinents

The following palaeocontinents or terranes are recognized to have existed in the Silurian within the present-day study area. In most cases they have become amalgamated and have moved considerably in relation to their Silurian position. Several of them were later emplaced in different tectonic regimes of the Variscan and Alpine orogenies and represent various structural levels in foldbelts and nappe-stacks (Fig. 6.4):

Baltica. Although the area is large, it represents only a smaller part of the huge palaeocontinent Baltica. In the north and east of the study area much Silurian is present in the subsurface, but outcrops are only present in SE Bornholm and in both the Łysogory and the northern margin of Małopolska parts (Kielce region) of the Holy Cross Mountains. There is agreement that the two terranes Łysogory and Małopolska were attached to Baltica in the Silurian.

Avalonia. Silurian strata from this palaeomicrocontinent are present in outcrops and the subsurface of the Anglo-Brabant Deformation Belt, in the Condroz Inlier (a thrust wedge in the Variscan front), in the northern Rhenish Massif (= Rheinisches Schiefergebirge) and in the Northern Phyllite Zone; they belong to the eastern part of Avalonia. Parts of Avalonia, now attached to North America from Newfoundland to Long Island, are called western Avalonia, while other parts of Avalonia now found in southern Ireland, southern Britain, and NW Europe, are called eastern Avalonia. Cocks *et al.* (1997) demonstrated with faunal similarities that in the Ordovician and Silurian, both parts formed only one palaeocontinent, Avalonia, which was split into two by the Mesozoic opening of the Atlantic Ocean.

Far Eastern Avalonia. This is a completely subsurface terrane, introduced by Verniers *et al.* (2002), where no Silurian sediments were proven until now (Silurian sediments from boreholes in the Koszalin-Chojnice Zone in NW Poland were estimated to belong to Far Eastern Avalonia, but are now thought to have been deposited on the edge of Baltica). The terrane has a boundary in the north and NE with Baltica along the Thor Suture, a remnant of the Tornquist Sea, and along the Heligoland-Pomeranian Deformation Belt, the orogenic belt resulting from the collision

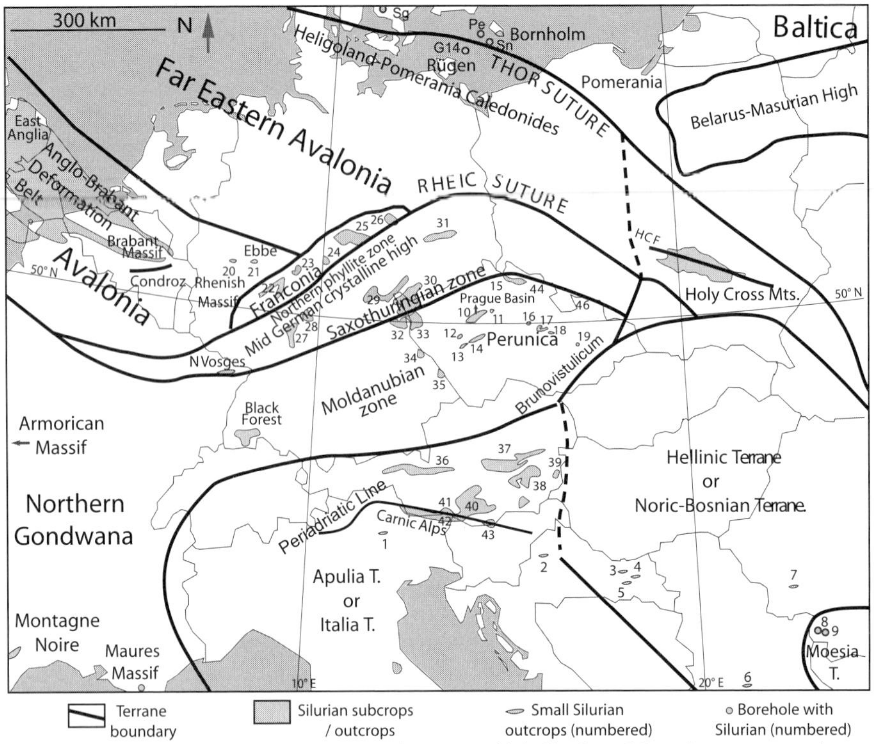

Fig. 6.4. Outcrop areas and subsurface presence of the Silurian in Central Europe, with indication of the palaeoterranes and sutures from the Silurian period and the names of post-Silurian tectonic units and faults. HCF, Holy Cross Mountains Fault; P.L., Periadriatic Line. Locations: 1, Agordo; 2, Medvednica Mountains; 3, Psunj Mountains; 4, Papuk Mountains; 5, Krndija Mountains; 6, Kacanik-Veles/Raska; 7, Suprabetic Nappes; 8, Girla Mare borehole; 9, Oprisor borehole; 10, Prague Basin; 11, NE Prague; 12, Rožmital; 13, Mirovice; 14, "Islet" zone; 15, Lužice Region and Ještěd mountains; 16, Vápenný Podol; 17, Hlinsko; 18, Polička; 19, Stínava; 20, Remscheid inlier; 21, Ebbe inlier; 22, Dill & Lahn Synclines; 23, Kellerwald; 24, Werra; 25, Upper and Lower Harz Mountains; 26, Flechtingen Horst; 27, Odenwald; 28, Spessart; 29, Thüringer Wald; 30, Erzgebirge; 31, Lausitz; 32, Oberfranken; 33, Vogtland; 34, Ober-Pfalz; 35, Bavarian Forest; 36, western Grauwacke Zone; 37, eastern Grauwacke Zone; 38, Graz; 39, South Burgenland; 40, Gurktal Nappes; 41, Nötsch; 42, Carnic Alps; 43, Karawanken Mountains; 44, Kaczawa Mountains; 45, Bardo Mountains; Sg., Slagerse-1 borehole; G14 borehole; Pe: Pernille-1 borehole; Sn: Stina-1 borehole.

of Baltica and Far Eastern Avalonia. In the NW it has a boundary with Laurentia along the Iapetus Suture, the remnant of the Iapetus Ocean, and along the Scottish-Scandinavian Caledonian Orogenic Belt. In the SE it has a boundary with peri-Gondwana terranes along the Rheic Suture, and in the SW with Avalonia along a fault zone north of the Anglo-Brabant Deformation Belt. The Ordovician fauna and microflora in the Rügen boreholes have an Avalonian affinity (Servais & Katzung 1993; Servais 1994; Katzung, 2001) and therefore it was previously included in Avalonia. Due to the presence of a subduction-related calc-alkaline magmatic arc running from East Anglia to the southern Brabant Massif, an oceanic crust had to be postulated NE of the Anglo-Brabant Deformation Belt and hence to the NE of it the presence of another Peri-Gondwana Terrane: Far Eastern Avalonia. This microcontinent is supposed to have moved away from Gondwana close to Avalonia (Verniers *et al.* 2002).

Peri-Gondwana and northern Gondwana

Northern Gondwana was part of the large Gondwana palaeocontinent. The peri-Gondwana terranes are a number of units of various size located along the southern margin of the Rheic Ocean and north of northern Gondwana. Some of these separated in a northward drift possibly as early as the Early Ordovician, but others separated during the Devonian (Paris & Robardet 1990). These units are present in many outcrop or subcrop areas in the southern part of Central Europe. They are sometimes referred to as Northern Gondwana Terranes. They are now incorporated in tectonic units of a later deformation age as the Variscan Moldanubian or Saxothuringian zones, or as the Alpine Hellenic or Apulia Terranes.

Peri-Gondwana terranes

An extensive overview of the complex terrane history of the terranes 4 to 7 in the Variscides is given by Franke (2000), McKerrow *et al.* (2000) and Tait *et al.* (2000).

Franconia. This hypothetical suspect terrane (Franke 2000), is present in two structures: (a) the upper plate of the Mid-German Crystalline High and (b) the Giessen-Werra-Harz Nappes, cropping out in the Dill Syncline near Marburg/Giessen, in Kellerwald, in the Upper Harz and in the Lower Harz. The root of this nappe is located between the Northern Phyllite Zone and the Mid-German Crystalline High; this plate might be linked or positioned very close to the following plate.

Saxothuringia (Saxo-Thuringia in Franke 2000). This is present in the Saxothuringian zone of Kossmat (1927). A well-studied outcrop area with Silurian rocks occurs in the Thuringia Forest (= Thüringer Wald), Oberfranken and Vogtland (Germany) (Heuse *et al.* 2006) and the northern part of the Vosges.

Perunica. This terrane contains the classic Prague Basin in the southeastern Barrandian Region of the Bohemian Massif. It is situated in the middle of the highly metamorphosed and deformed Moldanubian zone of the Variscan Belt. However, it escaped most of this deformation and displays one of the most complete Silurian outcrop areas in the world. Comparable to other areas of peri-Gondwana or northern Gondwana, it drifted from high southern to low northern latitudes during the Palaeozoic. Havlíček *et al.* (1994) argued the existence of this independent palaeoplate on the basis of brachiopod and trilobite assemblages. This was challenged by others (Servais & Lehnert 2006) who claimed that Perunica was only a part of northern Gondwana.

Moldanubia. This is a terrane where the Silurian crops out in the Bavarian Forest, the Oberpfalz Forest and in the southern Black Forest.

Brunovistulicum. This is a terrane which was strongly deformed during the Variscan Orogeny and consequently is of disputed origin, but because of its Neoproterozoic basement is considered as a part of peri-Gondwana.

Eastern Alps. Silurian sediments are present in **four tectonic units in the eastern Alps** (Austria and northern Italy). Three are situated north of an important Alpine structural divide, the Periadriatic Line, and one south of it: the Carnic Alps with the well-known Silurian outcrop area. They are situated in the present-day northern part of the Italia Terrane, also called the Apulia Terrane (Stampfli *et al.* 1998; Cocks & Torsvik 2002). The Silurian deposits with graptolites in the Medvednica Mountain, north of Zagreb, Croatia (Sremac & Mihajlović-Pavlović 1983) could also belong to this terrane.

Noric-Bosnian Terrane. (Raumer & Neubauer 1993). Also called the Hellenic Terrane (Stampfli *et al.* 1998; Cocks and Torsvik 2002), the Noric-Bosnian Terrane yields Silurian rocks in SE Serbia, in the Suva Planina Mountain, west of Kosovska–Mitrovica (Kříž & Veselinovič 1975) and in the South Carpathians (Iordan & Stanoiu 1993). The possible presence of Silurian rocks is mentioned in the Psunj, Papuk and Krndija mountains of northern Croatia (Jerenic *et al.* 1994) and also in SE Serbia, in Raška and East of Kosovska–Mitrovica (Kostić-Podgorska 1956).

Northern Gondwana terranes

Montagne Noire. This region contains Silurian sediments in tectonic thrust wedges and klippes on the southern edge of the Massif Central. They belong to the large northern Gondwana terrane, together with areas outside the study area in SW Europe such as the South Armorican Domain, Aquitaine and the Pyrenees.

Maures Massif. The Maures Massif (southern France) in the SW of the study area (Fig. 6.4) has one metamorphic Silurian formation containing rare fossiliferous lenses. The massif is not thought to belong to a separate palaeoplate. However, the fauna discovered is too limited and the facies too metamorphosed to link the Maures Massif either to the previous area or to the basement of the western Alps.

Peri-Gondwana or northern Gondwana terranes

Moesia Terrane. This terrane, located south of the Carpathians, with Silurian in the subsurface, was also a part of northern Gondwana or a derived peri-Gondwana palaeoplate. It is linked by several authors to the Bruno-Moravo-Silesia Terrane, north of the Carpathians. In the study area, Silurian rocks were demonstrated in two boreholes, Oprisor and Girla Mare (Iordan *et al.* 1985; Iordan 1984, 1999*a*, *b*; Yanev 2000; Vaida *et al.* 2005; Vaida & Verniers 2006).

Palaeomagnetism

The few palaeomagnetic studies over the last 15 years in Silurian strata of the study area have proposed a generally consistent position for Laurentia, Baltica and Avalonia. Laurentia, with its numerous warm-water carbonates and evaporites, was positioned around the equator and in the two subtropical belts through most of the Cambrian until the Devonian (MacNiocaill & Smethurst 1994; Torsvik *et al.* 1996). Baltica was situated in the southern subtropical belt, with Gotland at about 20°S in the Wenlock (Claesson 1979; Trench & Torsvik 1991). The Iapetus Ocean lay between Laurentia on one side and Baltica, Far Eastern Avalonia and Avalonia on the other side. The ocean closed in the early Wenlock in Ireland and in the early to mid-Wenlock in northern England (Hutton & Murphy 1987; Arthurton *et al.* 1988; Soper & Woodcock 1990; Kneller 1991; Kneller *et al.* 1993). In the early Wenlock, the latitude of the Welsh part of Avalonia (just west of Fig. 6.4) was estimated at about 13°S (Trench & Torsvik 1991).

The main divergences in published models relate to the very large palaeocontinent of Gondwana and especially northern Gondwana and the peri-Gondwana terranes. Two contrasting apparent polar wander paths for the Palaeozoic have been proposed for Gondwana: path X (Bachtadse & Briden 1991) and path Y (Schmidt & Embleton 1990) (see Van der Voo 1993; Tait *et al.* 1997). One model (Scotese & Barret 1990), also supported by lithofacies data, shows the South Pole in the latest Ordovician in West Africa moving to southern Brazil in the earliest Devonian. This implies that during the Silurian there was a northward shift of the areas of northern Gondwana and the peri-Gondwana palaeoplates from higher to lower latitudes, from cool to temperate waters.

Terranes 4 to 9 and 11 have been grouped by early palaeomagnetic studies as ‘Armorica’ (Van der Voo 1979, 1982) or later in the ‘Armorican Terrane Assemblage’ (Tait *et al.* 1997). Recent authors such as Fortey & Cocks (2003), restrict Armorica to the Armorican Massif, Montagne Noire and the Iberian Peninsula. Others (e.g. Robardet 2001, 2003; Linnemann *et al.* 2004) have considered this grouping as oversimplified and as neglecting the geological evidence from Iberia and France and reject the term altogether. These authors argue that several of these areas, mostly the western areas from Portugal up to Saxothuringia, remained attached to mainland northern Gondwana until at least the Devonian while other parts, such as Perunica, might have separated much earlier.

Data for the supposed position of Perunica (Bohemian Massif) have been summarized by Krs *et al.* (1986, 1987). Early Mid-Cambrian greywackes from the Příbram–Jince Basin, a few tens of kilometres west of the Prague Basin, show a palaeolatitude of about 39°S, and Late Cambrian andesites from this basin have a palaeolatitude of about 29°S. Early Ordovician chert and tuffaceous rock in the Prague Basin record a palaeolatitude of 28°S, and Lower Devonian micritic limestones in this basin show a latitude of 5–9°S. The independent development of Perunica is suggested because of its probable rotation, as shown by changes in palaeomagnetic directions from about 65° in the Mid Cambrian through 90° in the Late Cambrian, to 127–132° in the Early Ordovician (Krs *et al.* 1986). These data support the older data of Burrett (1983), who interpreted the apparent polar wander path and suggested that the Bohemian Massif moved independent from 'Armorica' during the Early Palaeozoic. Palaeomagnetic data from the Grauwacke Zone in the Upper Australo-Alpine Nappe Complex gave a palaeolatitude of 45–50°S for the late Silurian. The Proto-Alps in the Silurian, according to this study, were not part of northern Gondwana but a separate (peri-Gondwana) terrane (Schätz *et al.* 2002).

Palaeogeographic setting

The complex palaeoplate configuration of the study area reflects the involvement of very different tectonic settings during the Silurian. On larger stable palaeocontinents such as the interior parts of Baltica and northern Gondwana, the subsidence rates were rather low, with little or no tectonic deformation. At the edges of the large palaeocontinents passive margins developed with higher subsidence rates. On the active margins of the palaeocontinents high subsidence occurred with turbiditic sedimentation and tectonic deformation as in the Scandinavian Caledonides, the Heligoland-Pomerania Caledonides and the Anglo-Brabant Deformation Belt.

In SW **Baltica**, the stable platform on the East European Craton behaved mainly as a passive margin. Subsidence analysis of the Silurian strata indicates high subsidence in the Baltic Sea area, increasing in space towards its border with the Tornquist Sea (Oslo Graben, northern Poland and Lithuania) and increasing in time towards the upper Silurian (Poprawa *et al.* 1999; Lazauskienė *et al.* 2002). This can be explained by a combination of tectonic stresses induced by the mid-Silurian onset of the Laurentia–Baltica collision and later nappe formation and stacking in the Scandinavian Caledonides (i.e. the Scandian Orogeny) on one side, and the closure of the Tornquist Sea and soft docking and collision of Far Eastern Avalonia with Baltica, on the other. This closure commenced in the Ashgill and continued throughout the Silurian with the formation of the Heligoland-Pomerania-Deformation Belt (Winchester & the PACE TMR Network Team 2002; Winchester *et al.* 2002), formerly called the North German–Polish Caledonides, or the Pomeranian Caledonides (Dadlez 2000). It is sometimes differentiated from west to east into the Schleswig Caledonides, the Rügen Caledonides and the Pomeranian Caledonides (Katzung 2001). The Baltic Basin is thus interpreted as a flexural foreland basin formed during the oblique collision of Baltica and Far Eastern Avalonia (Poprawa *et al.* 1999) to which an additional tectonic load at the SW margin is required for the model to explain fully the shape of the Baltic Basin (Lazauskienė *et al.* 2002). There was no tectonic deformation after the Silurian or Lochkovian and hence the sediments are mostly subhorizontally stratified, and they underwent only very low heating by sediment load and burial metamorphism.

The docking of Far Eastern Avalonia and Baltica in the late Ordovician to Silurian is connected with the development of a foreland basin at the southern rim of Baltica. The sedimentation shows a distinct shift in the area of maximum deposition northwards over time. In the mid-Llandovery (*Spirograptus guerichi* Zone) more than 200 m of clastic, deep-water sediments were deposited in the southern Baltic Sea (G-14 borehole, Adlergrund, north of Rügen Island; Maletz 1997), after which the depocentre shifted northwards to the island of Bornholm and southern Sweden (Skåne/Scania), forming the *Colonus* Shale Basin in the Wenlock and Ludlow (Lindström 1960). The depocentre was migrating away from the approaching Caledonian Deformation Front (Vejbaek *et al.* 1994; McCann 1996; Katzung 2001). The final stage in the development of this basin can be recognized in the shallow-marine to terrestrial sediments of the Öved-Ramsasa Group in the Ludlow–Přídolí of Skåne/Scania in southern Sweden (Beier *et al.* 2000).

Avalonia and Far Eastern Avalonia detached from Gondwana in the Early to early Mid-Ordovician (for an overview see Prigmore *et al.* 1997). Prior to rifting, it was located in an area extending from what is now the northern coast of South America to somewhere between the northern coast of South America and Senegal in west Africa, where the terranes of Florida and Carolina were situated. This is based not only on the similarities of Proterozoic zircon ages but also on faunal evidence. The fauna from the British part of Avalonia is more similar to that of South America than to the fauna of north and west Africa (Villas *et al.* 2006).

The Rheic Ocean formed when northern Gondwana and Avalonia separated. Avalonia and Far Eastern Avalonia moved during the Mid- and Late Ordovician with a high drift rate towards Baltica, while the ocean in between, the Tornquist Sea, was being subducted. Far Eastern Avalonia collided with Baltica by early or mid-Ashgill times. This can be proven by the presence on the southern edge of Baltica of early or mid-Ashgill age sediments which contain reworked Lower Ordovician microfossils with a typical high latitude signature, the same as in Lower Ordovician sediments from high latitude parts of Gondwana. This can only be explained by accepting that by early or mid-Ashgill times both continents were in contact, with no ocean in between and that high latitude Lower Ordovician sediments were eroded and transported over the Tornquist Suture onto low latitude Baltica (Vecoli & Samuelsson 2001*a*, *b*). Following the onset of collision, the eastern part of Avalonia is postulated to have rotated 55° anticlockwise from the late Caradoc/Ashgill until the Mid-Devonian (Piper 1997). This resulted in short-lived subduction magmatism from Caradoc to Llandovery times and a slow long-lived deformation from the late Llandovery to early Eifelian in the Anglo-Brabant Deformation Belt (Debacker 2001; Verniers *et al.* 2002).

In the study area, Avalonia exhibits two different depositional settings in the Silurian: (1) a deep shelf with a mainly graptolitic shale facies in a sucession several hundred metres thick is present in the prolongation of the Midlands microcraton south of East Anglia, into the southern Brabant Massif and in the Condroz Inlier; and (2) a deeper environment in a postulated foreland basin setting, with at least 3 (possibly 7) km thick pile of Silurian turbidites deposited in the central and northern part of the Brabant Massif (Van Grootel *et al.* 1997; Verniers *et al.* 2002). Beneath the southern bight of the North Sea and under East Anglia, a Silurian succession several kilometres thick is suggested from geophysical modelling; it lies in the prolongation of the central and northern parts of the Brabant Massif (Lee *et al.* 1993). Deformation, under greenschist metamorphic conditions, transformed these regions into a slate belt, the Anglo-Brabant Deformation Belt. This belt occurs on the eastern end of Avalonia and close to the southern edge of Far Eastern Avalonia (Verniers *et al.* 2002).

To the south, are some areas that have been interpreted as belonging to Avalonia. In the intensely deformed Variscides of Central Europe several Silurian sedimentary or volcanic successions are present, sometimes highly metamorphosed and deformed by the Variscan Orogeny. However, this has not precluded a reconstruction of their Silurian palaeogeography. According to the hypothesis proposed by Franke (2000), a volcanic island arc formed at the southern margin of Avalonia during the late Silurian and early Devonian, as a result of the subduction to the north of the Rheic Ocean beneath Avalonia. Remnants of this arc are found in the lower plate of the Mid-

German Crystalline High and in the Northern Phyllite Zone present in the southern Taunus. These areas, if they belong to Avalonia, would represent a third depositional setting in Silurian times.

Peri-Gondwana and northern Gondwana

Although present over a wide area, and possibly on different palaeoplates, the Silurian sediments of peri-Gondwana and northern Gondwana are all rather similar to each other. Local variations include extensional basins with synsedimentary faults and half-graben structures with varying degrees of subsidence, or a volcanic archipelago with various water depths. If deformation occurred, it can mainly be attributed to later Variscan or Alpine tectonism.

In general, according to Kříž *et al.* (2003), there are two types of sedimentary successions on Silurian northern and peri-Gondwana: proximal and distal. The proximal type of succession with largely coarse-grained siliciclastic terrigenous sediments is found to the south and so is located outside of our study area, and close to the emergent land mass in north Africa. The second type of succession is distal with a low terrigenous influx and rather thin successions, sometimes not thicker than 50 m for the entire Silurian. This suggests low subsidence rates on a distal, more outer-shelf environment on a extensive passive margin. They consist typically of either black shales or calcareous shales and limestones (Berry & Boucot 1967). Within this second distal type of succession, five main facies developed: a deep shelf ‘Thuringian facies’, a possibly deeper shelf ‘Bavarian facies’, a ‘Shelly fauna facies’ on a moderately to shallow shelf, a very shallow ‘Prague Basin facies’ around the emerging volcanic archipelagos, and an unnamed facies proximal to local land areas with a high siliciclastic input sometimes of turbiditic nature.

The ‘Thuringian facies’ is the most common facies and consists of thin sequences of dark shales and cherts in the lower part of the Silurian. The absence of current indicators points to a calm, relatively deep environment, under dysaerobic or anoxic conditions. In the upper part of the Silurian typically thick aerobic limestone beds occur (Ockerkalke in Germany) with interbedded thin quartzite beds, possibly of turbiditic origin. These were deposited in Saxothuringia, in the Prague Basin on Perunica and in the Carnic Alp, where locally it is called the Bischofalm facies.

The Bavarian facies is quite similar to the Thuringian facies but consists of grey-green shales with interbedded graptolitic shales and alum shales; in addition, no limestones developed in the upper part of the Silurian. It is sometimes interpreted as a deeper shelf facies, too deep for limestone development. Alternatively, it has been interpreted as a shallower shelf facies (Kurze & Tröger 1990). It occurs in narrow bands on Saxothuringia.

The Shelly fauna facies mostly consists of thick calcareous shales and limestone nodules and lenses with shelly fauna, often dominated by bivalves (*Cardiola*), and sometimes by other shelly fossils as in the cephalopod limestones. It is supposed to have been deposited on an intermediate shelf, below wave base but within the reach of surface currents ventilating the seafloor. It occurs in Saxothuringia, the Prague Basin on Perunica, the Carnic Alps where it is subdivided in a shallower Wolayer Facies and a deeper Plöcken Facies, and also in the Montagne Noire. In the Carnic Alps there is a facies intermediate between this and the Thuringian facies called the Findenig facies.

The Prague Basin facies consists of thick basic volcanics and bioclastic carbonates, rich in brachiopods, bivalves, cephalopods, crinoids, corals and trilobites (Kříž 1991, 1998*b*). They were deposited in very shallow water on the gentle slopes of the Wenlock–Ludlow volcanic archipelago in the Prague Basin. The environment around the emerging volcanoes offered a unique environment favourable for the establishment of distal benthic communities (Kříž *et al.* 2003). The facies also developed in the volcanic centre of Graz in the Rannach Nappe and possibly in the Gurktal Nappe.

A fifth facies developed proximal to land areas and records a high input of siliciclastic material, sometimes turbiditic. It is situated in some outcrop areas of Austria north of the Periadriatic Line, such as the Glemmtal unit in the western Grauwacke Zone.

In addition to sedimentological similarities, close faunal relationships exist between all of the northern and peri-Gondwana areas in and outside the study area such as Morocco, Spain, the Pyrenees, Montagne Noire, the South Armorican domain, Sardinia and the Carnic Alps (Kříž & Serpagli 1993; Robardet *et al.* 1994; Kříž 1996, 1999*b*; Kříž *et al.* 2003; Robardet 2003). These relationships suggest that all of these areas were located in Silurian times on the margin of northern Gondwana. The latter regions, however, differ slightly in their Silurian sedimentological and faunal characteristics from those of the Algerian Sahara, southern Morocco, central Iberia and, more surprisingly, from those of the northern and central Armorican Domain. This suggests that the Montagne Noire and related areas were located northwards of the northern and central Armorican Domain during the Silurian (Robardet *et al.* 1994; Paris 2000). According to Robardet (2003) faunal and lithofacies evidence indicate that these microplates were attached to mainland northern Gondwana at least until the Devonian.

Franconia and Saxothuringia

The palaeoplates of Franconia and Saxothuringia were situated close to each other, and south of the Rheic Ocean, which was decreasing in width during the Silurian. The sediments in the Giessen-Werra-Harz Nappes are supposed to have originated between Franconia and Saxothuringia. They contain a tectonic melange of Silurian shales and limestones. Diamictites, recognized in the uppermost Ordovician strata, are used as an argument to locate Franconia as a peri-Gondwana terrane in high latitudes and hence close to northern Gondwana.

On Saxothuringia, Silurian shelf and slope sediments were deposited that now crop out in the Thuringia Forest (= Thüringer Wald). From the same location the uppermost Ordovician diamictites in the *Lederschiefer* unit have been used to attribute Saxothuringia to a peri-Gondwana terrane or to northern Gondwana. Not only does the fluxoturbiditic glaciomarine nature of the diamictites point to a high latitude, also the fossils in the clasts in the diamictites are not originally from the Thuringia Forest but are very similar to the northern Gondwana areas such as on the Armorican Massif, the Iberian Peninsula and north Africa. One of the latter areas is supposed to be the source area of the diamictites (Schallreuter & Hinz-Schallreuter 1998).

Between Saxothuringia and Perunica a short-lived Saxothuringian ocean is postulated that later subducted to the south below Bohemia (see Franke 2000). Evidence from zircons from Saxothuringia would indicate that this microcontinent remained attached to mainland northern Gondwana until Variscan times (Linnemann *et al.* 2004).

Another series of outcrops is probably related to a peri-Gondwana terrane, but without clear evidence. They comprise low-, medium- or high-grade metamorphic rocks of which radiometric ages and, in some cases, convincing fossil evidence indicate Silurian protoliths. They include part of the northern

Vosges (France), where the Steige Slate represents the less metamorphic Palaeozoic unit (Piqué *et al.* 1994). In this slate, Doubinger (1963) and Doubinger & von Eller (1963) discovered remains of microfossils they referred to as Silurian chitinozoans. The identifications of these possible microfossils were, however, subsequently questioned, due to their poor preservation (F. Paris in Kříž *et al.* 2003). As demonstrated in a similar context in the Black Forest in Germany (Montenari *et al.* 2000), scanning electron microscope investigations are required to ensure that these remains are chitinozoans before giving any firm age assignment to the Steige Slate.

Perunica

The Perunica Terrane (Havlíček *et al.* 1994), simlar to other peri-Gondwana terranes, drifted from a high southern latitude in the Cambrian to a low northern latitude in the latest Palaeozoic. It now lies within the Moldanubian Zone, a Variscan structural unit. The northerly position of the Perunica terrane in the late Silurian is reflected by the presence of carbonate platforms, characterized by very rich brachiopod-dominated communities and other benthic elements (e.g. corals, crinoids, trilobites and molluscs). The bivalve-dominated communities described from northern Gondwana in SW Europe by Kříž (1999*b*) are closely related to each other but are different from the communities in the Prague Basin on Perunica. These bivalve-dominated communities in Perunica show a generally higher diversity than those of northern Gondwana in SW Europe.

Early Silurian sedimentation on the Perunica Terrane was still influenced by the Late Ordovician glaciation. Seawater temperature was probably relatively low in the post-glacial period. In the early Silurian the sea was anoxic or strongly dysaerobic and black graptolitic shales were deposited in all of the basins of the Bohemian Massif. During the Telychian, the sea temperature slowly increased and intense water circulation commenced (Holland & Bassett 2002). In the Prague Basin, this is related to the deposition of calcareous shales. In the Wenlock, and especially during the Homerian, the limestone lithofacies in the Prague Basin is indicative of a further increase in temperature, reaching a maximum in the Přídolí (Kříž 1991). The development of this partly climatically influenced facies could correspond to a Silurian position for the Perunica Terrane of between 30° to 10°S palaeolatitude.

Moldanubia

Four areas, all in Germany, with medium- to high-grade metamorphism were shown to contain organic-walled microfossils such as acritarchs and chitinozoans, indicating in part a Silurian age. The Badenweiler-Lenzkirch Zone and the South Schwarzwald Granite and Gneiss Complex in the Black Forest belong to the Moldanubian Zone and provided an age for the protolith that was partly Silurian (Hann *et al.* 1995; Sawatski *et al.* 1997; Montenari *et al.* 2000; Vaida *et al.* 2000, 2004). The earlier two publications did not provide convincing photographs but the later two publications proved the presence of the organic-walled microfossils. In the Spessart and also in the northern Bavarian Forest, possible Silurian spores have been reported (Reitz 1987, 1992).

The Montagne Noire and northern Gondwana in SW Europe

Local Silurian palaeomagnetic data are absent or not convincing enough to provide any precise palaeogeographic location for the Montagne Noire during Silurian times. Sedimentological and faunal criteria are therefore used for the approximate palaeogeographic position of the area. The development of carbonates, especially in the Wenlock and Přídolí of the Montagne Noire, contrasts with their scarcity during the Ordovician on northern Gondwana. This indicates a northern drift of the area towards latitudes moderate enough to allow the deposition of these carbonates.

Lenses of crinoidal limestones and Silurian graptolite-bearing black shales have been reported (Gueirard *et al.* 1970) in a structurally complex and epizonal metamorphic context from the SW part of the Maures Massif, SE France (Crevola & Pupin 1994).

Peri-Gondwana in Brunovistulicum

The Silurian succession in the Moravo-Silesian region in the Drahany Highland, Stínava (see below and Fig. 6.4) most probably belongs to the Brunovistulicum microplate, which is considered to be a part of peri-Gondwana (Fig. 6.4). Remains of this microplate now form the southeastern Bohemian Massif (Suk 1979; Dudek 1980). For this reason, the Bohemian Massif is assumed to be a composite unit formed by elements from northern Gondwana, Perunica and other peri-Gondwana terranes. The presence of Baltica in Brunovistulicum was also suggested by Cambrian trilobites with a supposedly Baltican affinity (Havlíček *et al.* 1994), but the Neoproterozoic basement points to a Gondwana affinity. Perunica extends northwards to the Mid-European Rheic Suture that originated during the closure of the Rheic Ocean and now represents the boundary between the Saxothuringian and Rhenohercynian zones in the Variscan Orogen (Burrett & Griffiths 1977). The tectonic contact between Perunica and Brunovistulicum is associated with overthrusts and nappes with a pronounced eastern vergence, originally recognized by Suess (1912). Pre-Devonian sediments of unknown original extent and thickness were deposited in the basin between Perunica and Brunovistulicum, but were mostly destroyed during the collision of these blocks (Havlíček *et al.* 1994). South of Perunica, a Moldanubian Ocean has been postulated and on the ill-defined Moldanubia palaeoplate poorly dated and strongly metamorphic Silurian rocks have been recognized (Kříž *et al.* 2003).

Further south, the Proto-Alps Terrane, possibly belonging to the present-day Apulia Terrane, displays a northern Gondwana appearance. During the Silurian the Proto-Alps continued to shift from higher to lower latitudes. Based on the evidence presented below, it is estimated that the Austrian inliers were at a palaeolatitude of approximately 30–40°S. During the Silurian, close faunal relations existed with northern Europe (H. P. Schönlaub in Kříž *et al.* 2003).

Peri-Gondwana in the Proto-Alps

Within the Proto-Alps Terrane, the Silurian of the Carnic Alps, one of the numerous Austrian Silurian inliers, has been studied in much detail. Silurian faunas, generally typified as cosmopolitan, give little information of the palaeolatitudinal position of the individual areas (Schönlaub 1992). However, conodonts suggest a close affinity with coeval faunas from central, southern and southwestern Europe but differ from the more equatorial position of Avalonia and southern Baltica in terms of their more diverse assemblages (Bergström 1990; Aldridge & Schönlaub 1989). Silurian trilobites from the Carnic Alps are closely related to those from Bohemia, as well as other central European regions and Morocco (Alberti 1969, 1970; Santel 2001). Acritarch distributions suggest an intermediate position for the Austrian inliers between the high latitude *N. carminae* and the tropical *Domasia-Deunffia* biofacies (Priewalder 1987). Chitinozoans show a close relationship with Bohemia especially in the upper Ludlow to lower Lochkovian (Paris & Kříž 1984; Kříž *et al.* 1986; Dufka 1992; Kříž 1992). Chitinozoans in the Ashgill show a clear high latitude (peri-Gondwana) affinity (Priewalder 1997). Beginning in the late Llandovery, nautiloids became the predominant organisms in the carbonate facies of the Austrian Alps. The orthoceratids, abundant in the Wenlock and Ludlow, decreased during the Přídolí (Ristedt 1968, 1969; Schönlaub & Histon 1999) and are again abundant in the latest Přídolí (Histon 2002). Their diverse faunas appear to be closely related to those of Bohemia as well as to those of the Montagne Noire and Sardinia (Kříž & Serpagli 1993; Kříž 1996, 1998*b*, 1999*a*). The distribution of other molluscs, in particular bivalves, generally resembles that of the nautiloids. According to Kříž (1979), the Silurian cardiolid bivalves from the Carnic Alps and the western Grauwacke Zone inhabited a warm equatorial belt or were dispersed by surface currents.

In general, the corals were prominent constituents of a probable shallow-marine environment in the tropical belt. During the early Silurian, only weak indications of provincialism are seen among tabulate and rugose corals at the generic level. However, long-lived and far-floating (teleplanic) larvae might also have been transported by ocean currents over long distances (Kaljo & Klaamann 1973; Pickett 1975; McLean 1985; Pedder & Oliver 1990).

Rugose and tabulate corals occur in the upper Llandovery of central Carinthia and in the Ludlow near Graz, but are very rare in the shallow-marine and locally coated-grain-bearing limestones in the upper Llandovery of the western Grauwacke Zone (Schönlaub 1994*b* and references therein).

Ninety-seven K-bentonite levels have been recorded from the Upper Ordovician (Ashgill) to Lower Devonian (Lochkovian) successions of the Carnic Alps (sections Cellon, Oberbuchach, Nölblinggraben and Zollnersee Hütte). They occur in shallow- to deep-marine fossiliferous sediments which suggest a constant movement from a moderately cold climate of approximately 50° S in the Upper Ordovician, to the tropical belt in the Devonian. The distribution of the volcanic ash levels is consistent with current palaeogeographical reconstructions for the position of the Carnic Alps during the late Ordovician to early Devonian interval (Histon *et al.* 2007).

Sedimentary rock precursors in different basement areas of the Alps, such as quartz phyllites and amphibolite-grade metamorphic rocks in Austria, may also be Silurian in age (H. P. Schönlaub in Kríž 2003). In the Palaeozoic basement of the French Alps, greenschist occurs in the Belledonne Massif. A Silurian age was suggested without, however, firm fossil evidence (Ménot *et al.* 1994; F. Paris in Kříž *et al.* 2003). Recently at Agordo, NE Italy, middle Llandovery graptolites were discovered in the South Alpine greenschist metamorphic basement (Dieni *et al.* 2005).

Peri-Gondwana in SE Europe

In Croatia, Sremac & Mihajlović-Pavlović (1983) described Silurian sediments and graptolites in a deep-water marine facies. In eastern Serbia, Ludlow, Přídolí and Lochkovian bivalves have been described from the Suva Planina Mountains (Kříž & Veselinovič 1975). Another richly fossiliferous Silurian outcrop area occurs in the Godeanu Mountains (Romania) in the Danubian nappes of the Southern Carpathians. All three areas probably belong to the Noric-Bosnian Terrane.

Moesia, partly represented in the SE corner of the study area, is known only from the subsurface. In the Walachian Plain (Romania), several boreholes passed through Silurian strata dated with fossils, belonging to the characteristic fauna of northern Gondwana or peri-Gondwana. They are located on the western Moesia palaeoplate with a postulated northern Gondwana affinity (Iordan 1999*a*, *b*; Yanev 2000; Vaida *et al.* 2004; Vaida *et al.* 2005; Vaida & Verniers 2006).

Outside the study area, in Bulgaria, the Noric-Bosnian Terrane (Hellenic Terrane) has a typical Gondwana fauna in the Ordovician (Gutierrez-Marco *et al.* 2003).

Laurussia

To be exhaustive on the Silurian it should be mentioned that by the end of the Silurian a new plate tectonic configuration emerged following the joining of Baltica Avalonia and Laurentia to form the huge Laurussia palaeocontinent. The mostly Devonian–Lower Carboniferous sedimentation in the Rhenohercynian Basin locally commenced in the latest Silurian on the consolidated early Caledonian basement of the southern passive margin of this palaeocontinent. This is the case in northern France in the subsurface of the Liévin (Racheboeuf 1986), in the southern part of the Ardennes (Belgium and northern France; Godefroid 1995; Godefroid & Cravatte 1999), south of the Stavelot Massif (Petites Tailles Fm; Geukens 1965, 1999; Verniers *et al.* 2001), in the Rhenish Massif (= Rheinisches Schiefergebirge) of Sauerland (Köbbinghausen Schichten; Deutsche Stratigraphische Kommission 2002) and in the Hunsrück-Taunus area (Graue Phyllite Formation; Deutsche Stratigraphische Kommission 2002).

Silurian basins of Central Europe

In Central Europe the Silurian is only rarely present in outcrops. Even in the subsurface it is not present over large areas (Fig. 6.4). Well-studied successions with their lithofacies, biofacies and special features are discussed below for each palaeocontinent clockwise and starting in the north.

Baltica

The Silurian is remarkably undeformed and well-preserved on both sides of the southern half of the Baltic Sea, on the SW part of Baltica. This palaeocontinent is formed by the Baltic Shield, a Precambrian crystalline craton, called the East European Craton (EEC). Areas of this craton covered with Palaeozoic strata are often referred to as the East European Platform (EEP). Well-studied areas with Silurian outcrops include Gotland and Skåne/Scania in southern Sweden, the Oslo Graben, most of Estonia (all areas outside the study area) and Bornholm (Denmark). An overview of the stratigraphy and palaeogeography of Baltica, including the parts outside the study area, is given by Baarli *et al.* (2003). In the subsurface of Latvia (outside the study area), Lithuania, the Kaliningrad enclave, northern Poland and northern Germany (Adlergrund, Vorpommern) many boreholes have facilitated an accurate reconstruction of the Silurian basin on a passive margin and on what used to be called the (East) Baltic Syneclise or recently more simply the Baltic Basin. Silurian deposits in the Polish sector of the EEP are limited to the SW by the Trans-

European Suture Zone. In the SE, and on the margin of the Baltica palaeocontinent one can observe the Silurian outcrops of the Holy Cross Mountains (Poland), with well-studied and fossiliferous sections.

The Baltic Basin

A shelf-sea covered the southern Baltic Sea area. The seafloor sloped to the SW, towards the WNW–ESE edge of the Baltica continent forming a series of depth-related belts, each with a different facies. Jaanusson (1976) and Nielssen (1995) recognized three different facies belts, broadly orientated parallel to the Silurian coastline, and called confacies belts. Closest to the inferred palaeocoastline is the shallowest Lithuanian Confacies Belt, at an intermediate depth is the Central Baltoscandian Confacies Belt and in the deepest part of the shelf the (Oslo-) Scandian Confacies Belt (Fig. 6.5).

The shelf had a NE–SW directed axial part, the Latvian saddle or the Livonian tongue which formed the deepest part of the sea and was the area of maximum subsidence (Fig. 6.6). Gentle deformation, after the Silurian, altered the original depositional pattern, when parts were uplifted as the Baltic Shield, the smaller Łeba High, the important Belarus–Masurian High (in the literature its Polish part is referred to as the Belarus–Suwałki Uplift, or Anteclise; Paskevicius 1997) and the Sławatycze Horst. On these uplifted areas the Silurian was partly or completely eroded. Coeval subsidence in the Baltic Basin, the Podlasie Basin and the Bug Basin, resulted in the preservation of the Silurian strata. As a result the Silurian can now be observed in a series of tectonic structures from NW to SE: (1) the SE-dipping slope of the Baltic Shield; (2) (the axial part of) the Baltic Basin; and (3) the NW-dipping slope of the Belarus–Masurian High, the Podlasie Basin and the Bug Basin (Fig. 6.7).

The axial part of the Baltic Basin contains the thickest Silurian succession and stratigraphically the most complete geological section in Baltica, from the lowest Llandovery to the top of the Přídolí. There are only locally small stratigraphical breaks on the east and SE margins of the basin in the lower and upper parts of the section. Investigations of various faunal groups show that sedimentation was nearly continuous in the axial and deepest part of the basin. Short breaks in sedimentation occur and are indicated by pyritized rock surfaces and sudden changes in rock composition. The maximum thickness of the Silurian in the study area lies in Lithuania, in the Baltic Basin, and amounts to 836 m (Nida-1 borehole) (Fig. 6.7).

The Silurian in the Baltic Basin commences with a lower Llandovery limestone complex which is up to 40 m thick. Black mudstones, dark grey clays and marl deposits overlie this complex. The Přídolí strata comprise greenish grey carbonaceous clays, marls and limestones.

The thickness of the Silurian on the NW-dipping slope of the Belarus–Masurian High varies from 251 m in the central part of Lithuania (Prienai-3 borehole) (Fig. 6.6) to 140 m in the Vilnius area. From the Llandovery to the Ludlow there was an intermediate zone between the clay facies with a graptolite fauna in the Baltic Basin and a carbonaceous facies with a benthic fauna on the NW-dipping slope of the Belarus–Masurian High. This intermediate zone is important because it can be used for correlation in the Baltic area of stratigraphical units between the deep clay and the shallow carbonate facies, and also with other

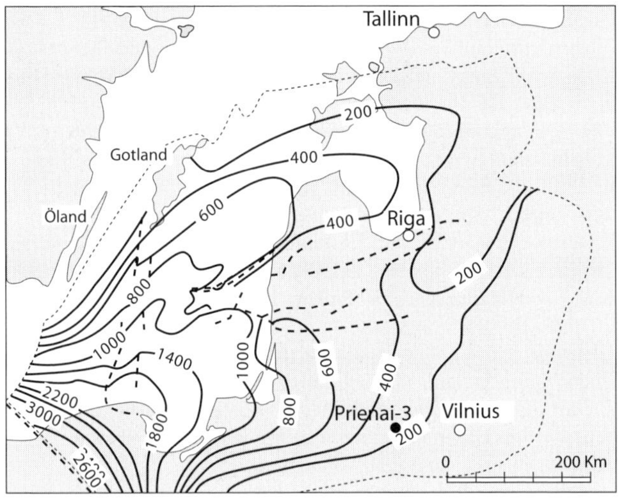

Fig. 6.6. Uncompacted thickness of the Silurian sediments in the Baltic Basin (after Lazauskienė *et al.* 2002).

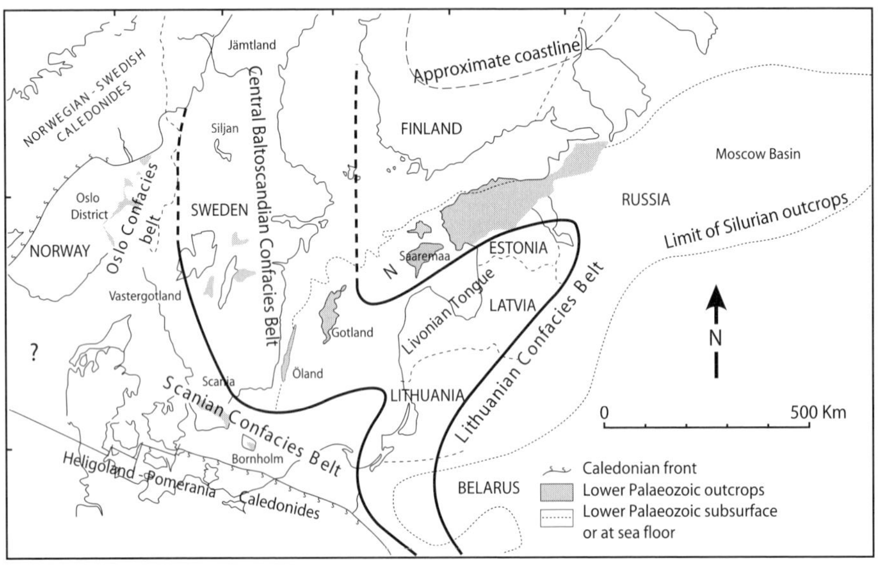

Fig. 6.5. Silurian outcrop and subcrop map of Baltic area with confacies belts (after Jaanuson 1976; Nielsen 1995).

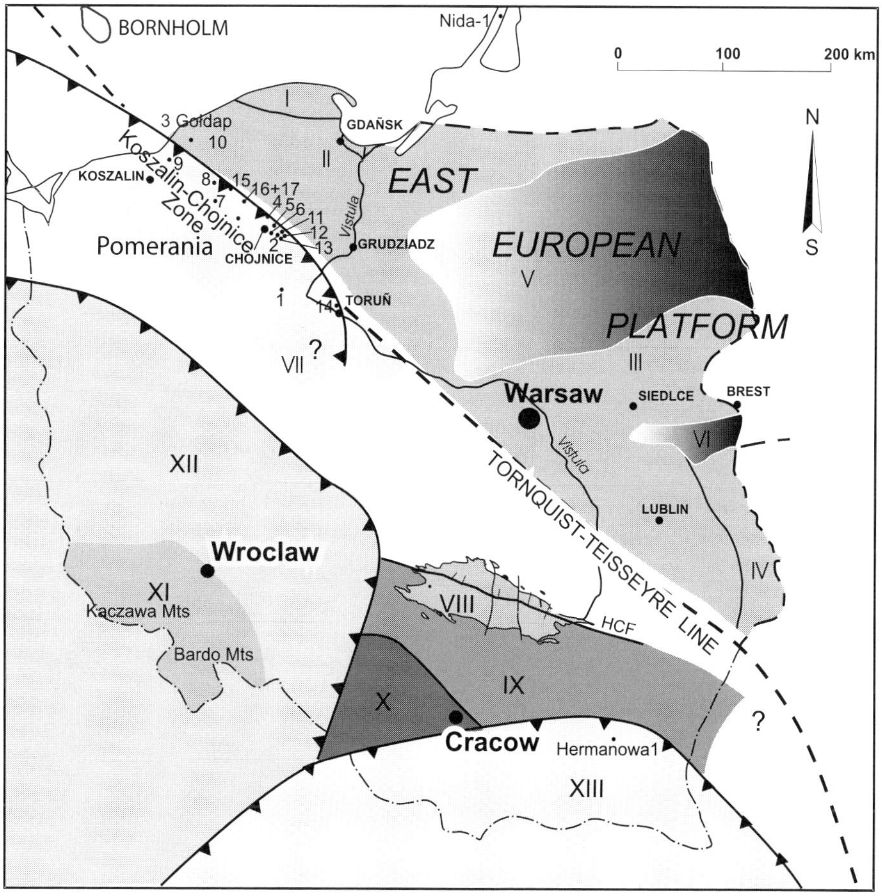

Fig. 6.7. Simplified structural map of Poland. Names of structures, I, Łeba Elevation; II, Baltic Basin; III, Podlasie Depression; IV, Bug Depression; V, Masurian-Suwałki Uplift; VI, Sławatycze Horst; VII, Pomerania, Koszalin-Chojnice Zone; VIII, Holy Cross Mountains; IX, Małopolska Massif; X, Upper Silesia Massif; XI, Sudetes; XII, Variscides; XIII, Carpathians. Boreholes described in text: 1, Bydgoszcz IG1; 2, Chojnice-3; 3, Darłowo-2; 4, Gołdap IG1; 5, Lutom-1; 6, Nicponie 1; 7, Okukino-1; 8, Polanów-1; 9, Skibno-1; 10, Słupsk IG1; 11, Stobno-1; 12, Stobno-2; 13, Stobno-3; 14, Toruń IG1; 15, Trzebelino; 16, Wierzchocina-1; 17, Wierzchocina-4.

areas throughout Europe. A carbonate clayey facies without graptolites in the Upper Ludlow and Přídolí represents a regression in the basin. The thickness of these Upper Ludlow and Přídolí strata ranges from 71 m in Jezioro Okrągłe (Poland) to 522 m in the Kunkojai-12 borehole (Lithuania). During Ludlow and Přídolí times the Silurian basin became significantly narrower, but the accumulation rate increased.

Baltica in Lithuania

Litho- and biostratigraphy

The lower Silurian (Llandovery and Wenlock) contains four rather thin formations. At the bottom the Apascia – Staciunai Formation contains nodular limestone deposits and is well exposed in the western part of Lithuania. It is dated as early Llandovery. The upper Llandovery and Wenlock are represented by the Barta Group, with three formations, from bottom to top, the Rasyte Formation (containing the *Coronograptus cyphus – Stimulograptus sedgwickii* and *Cyrtograptus lapworthi – Rastrites linnaei* zones), the Ragaine Formation (containing the *Monograptus riccartonensis – Cyrtograptus centrifugus*, *Cyrtograptus radians – Monograptus flexilis* and *Testograptus testis* zones) and the Siesartis Formation (containing the *Colonograptus ludensis – Gothograptus nassa* Zone). The formations are comprise black and dark grey mudstone with graptolite faunas in the western part of Lithuania, and marls with intercalated nodular limestones in the eastern part.

The upper Silurian contains four thick formations: two in the Ludlow, the Rusne Formation (*Neodiversograptus nilssoni*, *Lobograptus scanicus–Lobograptus progenitor* and *Pseudomonoclimacis tauragensis* graptolite zones) and the Pagegiai Formation (*Slovinograptus balticus* and *Monograptus formosus–Monograptus valleculosus* Zones); and two in the Přídolí, the Minija Formation (*Monograptus ultimus–Monograptus parultimus* and lower part of the *Neolobograptus lochkovensis* graptolite zone) and the Jura Formation (also *Neolobograptus lochkovensis* graptolite zone). The entire upper Silurian is represented by extensive calcareous clay deposits with nodular limestone units in the western part of Lithuania and by shallow marine basin and marginal facies with limestones, marls or dolomite, and detritic nodular limestones tracing a reef belt in the eastern part.

Lithofacies and palaeogeography

Four successive periods of basin development can be distinguished: mid Llandovery–early Wenlock (sequences S1 to S3), late Wenlock–early Ludlow (sequences S4 to S6), late Ludlow (sequences S7 and S8) and Přídolí (sequences S9 and S10) (Paskevicius 1997) (Fig. 6.8). These periods can be related to eustatic sea-level changes, tectonic development and local facies distribution.

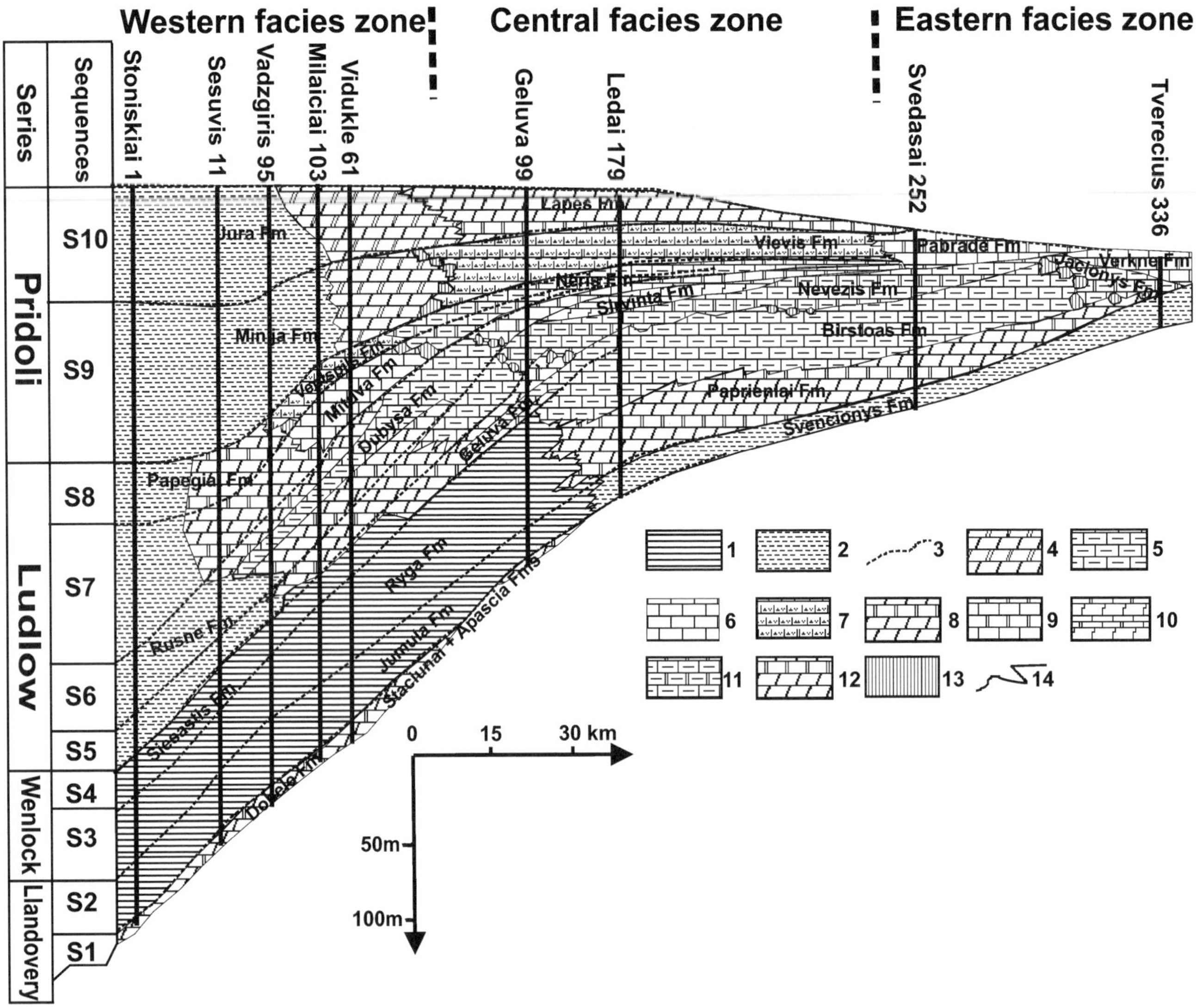

Fig. 6.8. Lithofacies cross-section throughout the central and eastern parts of the Silurian Baltic Basin, distribution of depositional sequences and main tectonosedimentary stages of foreland basin development. Key: 1, shales; 2, carbonaceous claystones; 3, boundaries of sequences; 4, carbonaceous marlstones; 5, clayey limestones; 6, limestones; 7, organoclastic limestones; 8, clayey limestones and marlstones; 9, dolomites; 10, dolomites and gypsum; 11, clayey dolomites; 12, dolomites interbedded with marlstones; 13, reefs; 14, boundaries of formations (modified after Lazauskienė *et al.* 2003).

At the end of the Ordovician the entire East Baltic area was covered by a very shallow marine and marginal basin. In SW Lithuania a stratigraphical hiatus is present. This was related to a large marine regression, presumably caused by the major eustatic sea-level fall due to the Hirnantian glaciation in the southern hemisphere.

In the early Llandovery a marine transgression took place in Lithuania from the west and NW over the low relief formed at the end of the Ordovician. Organic-material-rich clayey sedimentation was predominant across the area. In the late Llandovery and early Wenlock the sea further transgressed eastwards causing a corresponding distribution of facies and facies changes from lagoonal, through shallow-marine to shelf facies. Intensive volcanic activity occurred at that time in the zone of active Caledonian deformation in the west and south. Air-borne ashes were introduced into the Baltic Basin and bentonite layers formed, which were preserved in the deep-marine area. In the late Llandovery a shallower marine facies prevailed in the basin, explained as being the result of basin infill at the end of a highstand. In the east and SE of the Baltic Basin the presence of a shoreline and deltas has been proven. However, in the south the facies is devoid of terrigenous material with no indication of marginal environments, indicating a possible southward extension of the basin. The Wenlock–Ludlow transgression and increasing sedimentation rates are interpreted as the consequence of the regional tectonic activity at the SW margin of Baltica. In the early Wenlock the basin became shallower and a gradual replacement of the clayey facies by an organic-rich facies with dark graptolitic shales occurred. Due to the sea-level changes different facies overlap, and can repeat each other, forming sequences (see Fig. 6.8); the shallow carbonate shelf moved westwards. In SW Lithuania dolomitization of the carbonate deposits was common (Lazauskiene *et al.* 2003).

A regression commenced in the late early Wenlock in the Baltic Basin and continued through the remaining part of the Wenlock and into the early Ludlow. The area occupied by the basin decreased in size and regressive deposits were formed under conditions of rapid sedimentation. Metabentonite interlayers are less frequent due to either reduced volcanic activity in the Caledonian deformation zone, or the subsequent erosion and removal of the volcanic material. At the end of the Wenlock in central Lithuania a short transgression is noted. During the early Ludlow, on the NW-dipping slope of the Belarus–Masurian Uplift, dolomitic lagoonal deposits were formed containing nests of gypsum. During the late Ludlow the size of the Baltic Basin was generally decreasing, despite the occurrence of a short-term transgression. In the latest Ludlow the basin became shallower, ending with a transition from a marine basin to a lagoon with lower, or sometimes higher, water salinity. The late Ludlow maximum sea-level low is thought to represent the Lau Event (see Fig. 6.2) first observed in the Silurian of Gotland, north of the study area (Martma *et al.* 2005).

After the most prominent transgression at the beginning of the Přídolí, a marine regression occurred at the end of the Přídolí. The large thickness of Přídolí strata in SW Lithuania indicates rapid sedimentation within a basin that remained rather small. In the peripheral parts of the basin, Přídolí strata are less thick, and represented by shallow-marine and marginal red-coloured beach and lagoonal facies, usually rich in vertebrate fossils. The extensive terrigenous deposits indicate that the shoreline, at the end of the Silurian, extended in a SW–NE arc across central Lithuania. Three short-term marine transgressions took place during the late Přídolí, though a general regressive trend in the basin and a retreat of the shoreline to the west have been confirmed (Paskevicius 1997; Lazauskiene *et al.* 2003).

Baltica in Poland

In Poland the Silurian bio- and chronostratigraphy is mainly based on graptolites and can easily be correlated with the international standard stratigraphical framework (Fig. 6.9) (for summary see Tomczykowa & Tomczyk 1981; Tomczyk 1970; 1991; Urbanek & Teller 1997; Jaworowski 2000; Modliński & Szymański 2001; Podhalańska 2003; Szymański & Modliński 2003). Undisturbed Silurian strata have been encountered subsurface in hundreds of deep boreholes throughout the Palaeozoic sedimentary cover above the Polish portion of the Baltica. These deposits yield rich and varied fossil assemblages that have attracted the interest of many geologists and palaeontologists since the mid-1950s. The history of research and the results have been summarized by Teller (1997). The Palaeozoic sedimentary cover of the East European Platform was later affected by slight uplift or subsidence, resulting in the formation of two uplifted areas, the Masurian-Suwałki Uplift, part of the Belarus–Masurian High and the Sławatycze Horst, where the Palaeozoic strata were eroded, and three intervening structural depressions, with Palaeozoic strata still present and conventionally called, from NW to SE, the Baltic, Podlasie and Bug basins (Fig. 6.7).

The Koszalin-Chojnice Zone

The Silurian of Western Pomerania is known from 15 deep boreholes (Bydgoszcz IG1, Chojnice-3, Klosnowo IG1, Lutom-1, Nicponie 1, Okukina-1, Polanów-1, Skibno-1, Stobno-1, Stobno-2, Stobno-3, Toruń IG1, Trzebielino and Wierzchocina-1 and -4: Fig. 6.7) located in a narrow NW–SE zone in northwestern Poland, the so-called Koszalin-Chojnice Structural Zone (e.g. Dadlez 1978, 2000; Teller 1969, 1974, 1997; Teller & Korejwo 1968*a*, *b*, *c*; Tomczyk 1968, 1980; Katzung 2001; Podhalańska & Modliński 2006). This structure comprises strongly folded Lower Palaeozoic rocks and is situated to the SW of the Teisseyre-Tornquist Line (i.e. the western margin of East European Platform). The deformation at a high structural level, probably accompanied by thrusting, is indicated by the steeply dipping beds, predominantly more than 45° and some overturned bedding, changing over short distances within the boreholes, as well as the high frequency of slickenside and intense fracturing (Dadlez 1978). It is regarded as a continuation of the Danish–North German Caledonides (Pożaryski 1990), part of the Heligoland-Pomerania-Deformation Belt (Winchester & the PACE TMR Network 2002). The Silurian sequence in Western Pomerania is considered by Jaworowski (2000) as a Caledonian accretionary prism that supplied clastic material for the proximal Baltic Basin in Eastern Pomerania. The Silurian succession is composed of black and grey graptolitic shaly siltstones and mudstones, several hundred metres thick. In terms of age, the succession includes Llandovery, Wenlock and Ludlow units unconformably overlain by Devonian and younger deposits (Modliński 1968, Teller & Korejwo 1968*a*, *b*, *c*; Czermiński 1967; Dadlez 1967; Hajłasz 1967).

The Silurian in this area appears to be lithologically similar to the Ordovician, but differs in terms of the numerous siltstone intercalations which are found in the Llandovery and the upper Ludlow to Přídolí, and which are considered to be turbiditic in origin (Teller 1969, 1974; Tomczyk 1987). The Ludlow succession occasionally contains dolomites, whereas the Přídolí beds consist of dark grey siltstones with very thin intercalations of limestones, partly organo-detritic, and containing abundant faunas of bivalves, brachiopods, ostracods, sporadically trilobites, tentaculites and crinoid stems. Graptolites enable the Llandovery, Wenlock and Ludlow to be recognized, whereas ostracods are used for the Přídolí (Żbikowska 1974; Teller 1974). Intercalated limestones and dolomites also indicate shallow-water depositional conditions at this time, but sediment influx from the south at the beginning and the end of the Silurian is marked (Katzung 2001). Samuelsson *et al.* (2002) noted that the chitinozoans of the Upper Ordovician and Silurian in the boreholes are similar to those of Avalonia and are generally closer to Baltica than to northern Gondwana. Newer studies suggest that the area was situated on the edge of Baltica, close to the suture (Poprawa 2006).

Other Polish parts of the Baltic Basin

The Polish sector of the (Peri-)Baltic Basin extends from Koszalin in the west to Jezioro Okrągłe at the Lithuanian border in the east, and includes smaller structures, the Łeba Elevation (High) and Gdańsk Bay. The Silurian typically consists of monotonous, almost exclusively siliciclastic deposits represented by claystones and mudstones with thin sandstone interbeds and carbonate intercalations (for a summary see Jaworowski 2000; Szymański and Modliński 2003). The succession is *c.* 300 m thick (Gołdap IG1 borehole) in the eastern part of the region increasing to *c.* 3300 m (Słupsk IG1 borehole) in the western part (Teller 1969; Tomczyk 1968, 1970). The Llandovery deposits contain black claystones with graptolites and thin dolomitic limestones with a benthic fauna (brachiopods) or subordinate nodular limestone, and they unconformably overlie the Upper Ordovician. The Wenlock rocks are monotonous, locally calcareous claystones with abundant and well preserved graptolites, probably deposited during the Silurian transgression and substantial expansion of the basin (Teller 1997). The Ludlow rocks are similar but with a different graptolitic fauna (Urbanek 1966, 1970, 1997; Kozłowska-Dawidziuk 1995). The Přídolí rocks comprise claystones with frequent carbonate intercalations with an abundant benthic fauna deposited in a regressive setting. In the upper part of the succession these deposits are truncated by pre-Permian erosion.

The Silurian sedimentation of this region is comparable to the

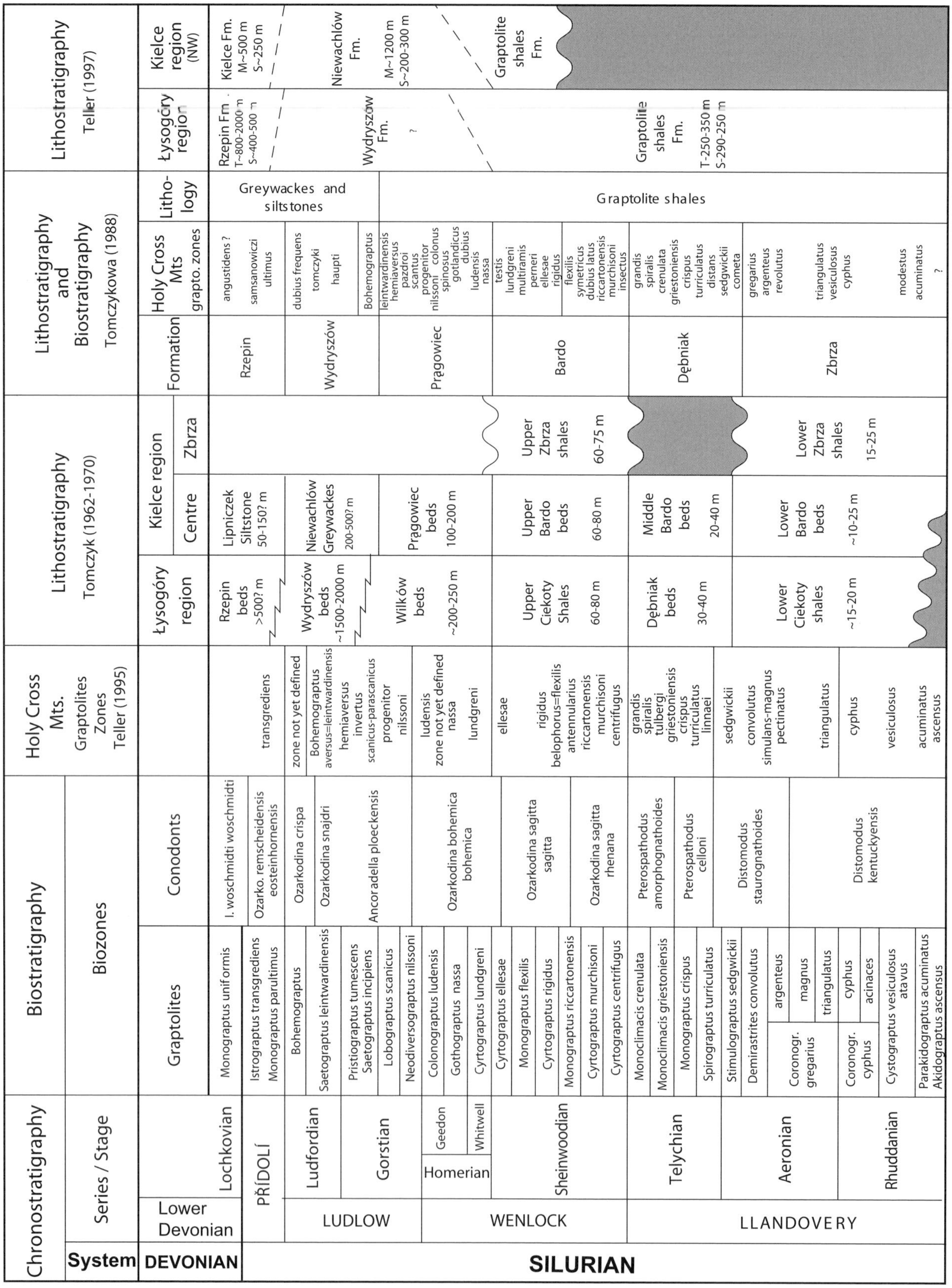

Fig. 6.9. Silurian litho- and biostratigraphy of the Holy Cross Mountains (after Modliński & Szymański 2001).

Silurian facies belts elsewhere in the Baltic Basin, suggesting that the calcareous claystones with carbonate intercalations were deposited in the relatively shallow conditions of an epicontinental sea (Einasto *et al.* 1986). In the western part of the Baltic Basin in northern Poland (Pomerania) the Silurian graptolitic clays and muds are located in the foreland of the Pomeranian Caledonides and are considered by Jaworowski (2000) to have been deposited in a hemipelagic environment with anoxic bottom waters. Their deposition under low-energy conditions during the Wenlock and Ludlow was repeatedly disturbed by silty debris flows and turbidity currents of clastic material derived from the Caledonian Orogen. The basin extended along the SW edge of the East European Craton coincident with the collision zone of Baltica and Far Eastern Avalonia (Jaworowski 2000).

Podlasie Basin

The Silurian deposits in this basic are mainly mudstones and shales with graptolites, ranging from 300 m to 1300 m in thickness, and unconformably overlying the Ordovician succession. They are unconformably overlain by Carboniferous, Permian or Jurassic strata. These deposits yield abundant and diverse graptolites and trilobites, used as the basis for biostratigraphy (Urbanek 1963, 1966, 1970, 1971; Tomczykowa 1988; Teller 1971). They are sparsely bioturbated and were deposited in relatively shallow-marine conditions in an epicontinental, outer-shelf environment.

Bug Basin

In this basin the Silurian-age deposits range from Wenlock through to the Devonian. The sediments are similar to those of the Podlasie Basin, namely graptolitic clayey and muddy calcareous shales with limestone nodules and intercalations, which increase in relative proportion upwards. The carbonate layers contain an abundant and diverse benthic fauna, including trilobites, bivalves, crinoids, brachiopods and tentaculites (Teller 1964; Korejwo & Teller 1964; Tomczykowa & Tomczyk 1979; Tomczykowa 1975). These undisturbed Silurian and Lower Devonian deposits extend southwards into the Ukrainian portion of the East European Platform. The basin shows signs of relatively shallow conditions on an epicontinental, outer-shelf environment, with a gradual shallowing and disappearing sea during the latest Silurian (Přídolí) (Teller 1997).

Holy Cross Mountains

There is an ongoing discussion as to the palaeoplate affinity of the Holy Cross Mountains during the Cambrian (either Baltica or Gondwana), but there is also consensus that during the Silurian it was attached to Baltica (see Belka *et al.* 2002; Cocks 2002; Cocks and Torsvik 2005; Elicki *et al.* 2008; Krawczyk *et al.* 2008). The Silurian of the Holy Cross Mountains has been studied since the mid-nineteenth century by several generations of prominent geologists and palaeontologists (Modliński & Szymański 2001). The Palaeozoic core of the Holy Cross Mountains is divided into two tectonostratigraphic, structurally different geological regions: the northern Łysogóry region and the southern Kielce region, separated by the Holy Cross Fault (Fig. 6.10).

The Silurian deposits of the Holy Cross Mountains are important for the understanding the amalgamation history of the Polish portion of the Trans-European Suture Zone (Belka *et al.* 2002). Fossil assemblages and sedimentary facies allow the succession to be subdivided into several local, poorly defined and thus informal biostratigraphical and lithostratigraphical units (Fig. 6.11 & 6.12) (for summaries see Tomczykowa & Tomczyk 1981; Modliński & Szymański 2001).

Łysogóry region

The Silurian of the Łysogóry region occurs in several outcrops and tens of boreholes (for details of location and review see Modliński & Szymański 2001). The Silurian succession consists of clastic marine sediments, without sedimentary breaks. These comprise black and grey shales with abundant graptolites (Ciekoty Beds) representing the Llandovery to Lower Ludlow deposits, and are approximately 100 to 140 m thick. They are overlain by grey and greenish clayey siltstones with carbonate interbeds and lenses (Wilków Beds), assigned to the upper

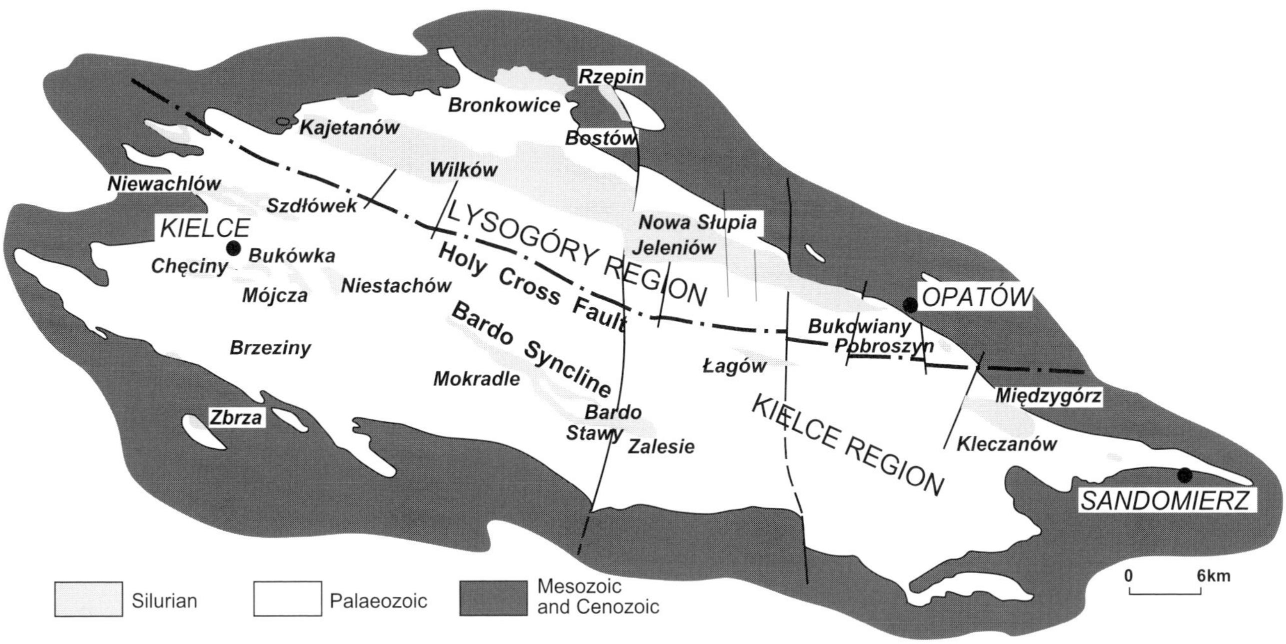

Fig. 6.10. Distribution of the Silurian deposits in the Holy Cross Mountains (modified after Modlinski & Szymanski 2001).

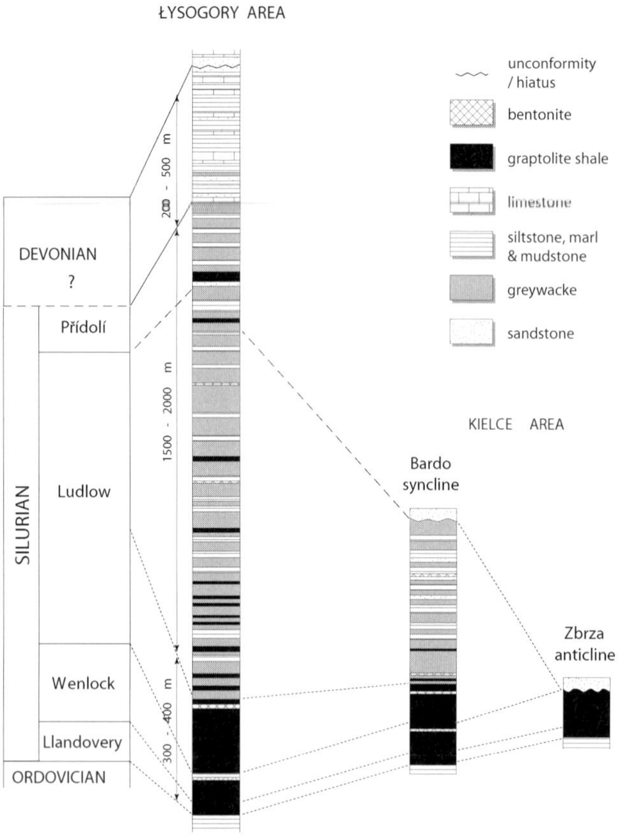

Fig. 6.11. Simplified lithostratigraphical profiles of the Silurian in the Góry Świetokrzyskie (Holy Cross Mountains) (modified after Tomczyk 1970; Modlinski & Szymański 2001).

Wenlock (Homerian) to lowermost Ludlow (Gorstian), and which are up to 250 m in thickness. These sediments are overlain by the Wydryszów Beds and Rzepin Beds with a poorly defined boundary in between. The latter are very thick, up to 2000–2500 m, and consist of grey-greenish shaly siltstones, mudstones and fine-grained sandy shales, sandstones and greywackes with numerous volcaniclastic interbeds and organoclastic limestones yielding bivalves, brachiopods, ostracods, trilobites, crinoids and sporadically graptolites (Tomczykowa & Tomczyk 1981). The fine-laminated mudstones and siltstones of the Rzepin Beds (at the Ludlow–Přídolí boundary) also contain abundant *Cooksonia*-like land plant fragments (Bodzioch *et al.* 2003). The succession is assigned to the Upper Ludlow to Přídolí and passes upward into the coarser-grained Devonian sediments (Tomczykowa & Tomczyk 1981; Tomczykowa 1988). The continuous sedimentary succession from the Silurian through the Devonian is similar to that on the foreland of the East European Platform, indicating a possible palaeogeographical connection (Stupnicka *et al.* 1991; Stupnicka 1992).

The graptolitic shales of the Llandovery to lower Ludlow were deposited in a deep marine basin (Porębska 2002), whereas the huge complex of clastic deposits (mainly greywackes) forming the Wydryszów and Rzepin beds were rapidly deposited (turbiditic character) in a shallow-water setting, which, at the Silurian–Devonian transition, gradually evolved to a shallow marine, nearshore and lagoonal, periodically brackish environment (Malec 2001; Kozłowski 2002, 2003; Bodzioch *et al.* 2003).

Kielce region

The Kielce region forms the northern margin of the Małopolska Massif to which it structurally belongs (Pożaryski & Tomczyk 1968). The Silurian of the region crops out in the central Kielce-Łagów Synclinal Zone, the Zbrza Anticline, the Bardo Syncline, the Łagów Syncline and the Międzygórz Syncline, as well as in more than a dozen boreholes (for detail of location and review see Tomczyk 1970; Modliński & Szymański 2001).

The Silurian succession of the central Kielce region consists of monotonous dark-coloured graptolitic shales with lydite interbeds. They are distinguished as the Bardo Beds and assigned to the Llandovery and lower Wenlock (Sheinwoodian). Their total thickness varies from 80 m to 150 m (Tomczykowa & Tomczyk 1981; Modliński & Szymański 2001). The Bardo Beds are overlain by the Prągowiec Beds of late Wenlock and early Ludlow (Gorstian) age. They comprise up to 200 m of calcareous grey and black shales and siltstones with graptolites and syngenetic calcareous concretions (Tomczykowa & Tomczyk 1981; Modliński & Szymański 2001). The overlying Niewachlów Beds represent a rapid change in sedimentation and are composed of sandstones and coarse-grained greywackes up to 500 m thick, which are assigned to the upper Ludlow (Malec 1991; Modliński & Szymański 2001). The Niewachlów Beds are overlain by the Lipniczek Beds, assigned to the uppermost Ludlow–Přídolí, and composed of siltstones and fine-grained sandstones, with marine fauna at the top (Tomczykowa & Tomczyk 1981; Modliński & Szymański 2001). These greywackes contain sedimentary structures indicating turbiditic transport. They were mainly derived from local volcanic sources (Przybyłowicz & Stupnicka 1991; Malec 1993). Lithologically, the entire succession resembles the Silurian of the Łysogóry region, albeit thinner and with numerous distinct sedimentary breaks, supposedly caused by pre-Variscan, synorogenic movements.

A somewhat different Silurian succession is present in the southwestern part of the Kielce Region, in the Zbrza Anticline (Fig. 6.10). Only Llandovery and Wenlock sediments up to 100 m thick are recognized (Lower and Upper Zbrza Shales). These sediments comprise black and grey bituminous clayey, siliceous and calcareous shales with rare limestone concretions (Deczkowski & Tomczyk 1969; Modliński & Szymański 2001).The erosionally truncated top of the Silurian succession in the Kielce region is unconformably overlain by Lower Devonian (Emsian) deposits; the break has been interpreted as a result of latest Silurian–pre-Emsian tectonic movements (Malec 1993; Przybyłowicz & Stupnicka 1989; Modliński & Szymański 2001).

Radom region

Silurian rocks are also encountered in three deep boreholes in the Radom region, adjacent and to the north of the Łysogóry region (for review see Modliński & Szymański 2001). The most representative section for this region is known from the entirely cored Ciepielów IG1 borehole. There, the Silurian is a monotonous sequence of grey laminated claystones and siltstones with graptolites (Tomczyk 1974) and chitinozoans (Wrona 1980) and with intercalated sandy and muddy siltstones, which may be calcareous. All of these deposits contain rare calcareous concretions and volcanogenic clasts. The biostratigraphy is based on a fairly diverse graptolite fauna, especially in the uppermost Ludlow and Přídolí (Tomczyk 1974; Tomczyk *et al.* 1977; Tomczyk & Tomczykowa 1983). The Silurian strata were not fully penetrated by the boreholes, but their thickness varies from *c.* 400 m to 650 m and they probably extend from the Upper Ludlow (Ludfordian), through the Přídolí, passing upwards with sedimentary continuity into Lower Devonian (Lochkovian) marine deposits (Żelichowski 1972; Tomczyk 1974; Tomczyk & Tomczykowa 1983; Teller 1997). The Silurian succession in the Radom region has so far not been subdivided into regional formal lithostratigraphic units (Modliński & Szymański 2001). The upper Ludlow beds in this region may be correlated with an informal regional stratigraphic unit, the lower Rzepin Beds of the Łysogóry region and with the uppermost part of the upper Siedlce Beds of the East European Platform (Tomczyk 1974). The Přídolí deposits may be considered as equivalents of another informal regional stratigraphic unit, the Upper Rzepin Beds of the Łysogory region, and can be correlated with the Podlasie Beds of the East European Platform (Tomczyk 1974).

Southern part of the Małopolska and Upper Silesia massifs

The Silurian of the southern part of the Małopolska Massif is known only from deep boreholes located to the south of the Holy Cross Mountains in the Nida region (Miechów Trough) and extending to the south to the Upper Silesia Massif (for details of location and review see Buła 2000; Modliński & Szymański 2001). There the Silurian strata comprise graptolitic shales with thin-bedded cherts (lydites in the Alpine region terminology), organoclastic limestones and dolomitic limestones, and fine-grained sandstone intercalations, unconformably overlying the Ordovician rocks. At the top of the section the mainly black or grey clayey shales are dominated by an unfossiliferous clastic unit comprising mainly siltstones, sandstones and conglomerates and comparable with the greywackes in the adjacent Kielce region (Jurkiewicz 1975). The total thickness can amount to several hundreds of metres. The sediments are erosionally truncated and unconformably overlain by Devonian or Triassic rocks (Jaworowski *et al.* 1967; Bednarczyk *et al.* 1968; Jurkiewicz 1975).

The Silurian succession of the Małopolska Massif has been compiled from discontinuous core profiles and comprises Llandovery, Wenlock and Ludlow deposits, which can be correlated with the coeval sequence exposed in the southern region of the Holy Cross Mountains (the Zbrza Anticline), and represents the same facies and graptolite succession (Bednarczyk *et al.* 1968). The Silurian succession of the Upper Silesia Massif has been subdivided into well defined lithostratigraphic units (Buła 2000): the Zawiercie (Ordovician–Lower Silurian), Mrzygłód (Wenlock–Ludlow) and Łapczyca formations (Přídolí).

Early Silurian, Llandovery and Wenlock sequences of the Małopolska and Upper Silesia massifs were deposited on a deep marine shelf that gradually evolved to a shallow marine environment. The coarse-grained, sandy and conglomerate deposits of Ludlow age from the Upper Silesia Massif characterize the general regressive pattern of the basin and this can be interpreted in terms of Caledonian tectonic events (summarized in Buła 2000; Modliński & Szymański 2001).

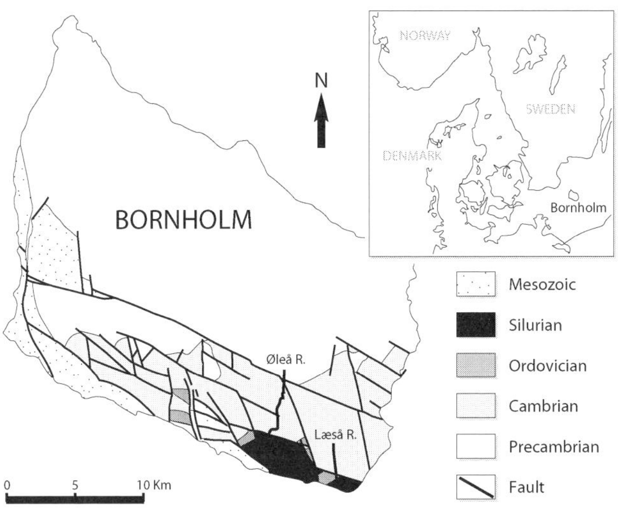

Fig. 6.12. Outcrop map of the Silurian in Bornholm (after Pedersen & Klitten 1990).

Baltica in Bornholm and surrounding area

Bornholm and the surrounding area formed an integral part on the southwesternmost edge of Baltica in the Silurian. It lies in the (Oslo-) Scanian Confacies Belt of Jaanusson (1976) and Nielssen (1995), the deepest facies belt of the Baltic Basin. Subsidence analysis of the Upper Ediacaran (ex-Vendian) and Lower Palaeozoic of the basin indicates that in the Silurian there was a marked increase in subsidence and sediment thickness, especially along its southwestern margin (Fig. 6.6), adjacent to the location of the Heligoland-Pomerania Caledonides (North German–Polish Caledonides).

The island of Bornholm is situated on a SW-dipping fault block with gently south dipping Silurian strata covering the thin Ordovician and Cambrian deposits. The Silurian crops out in a few block-faulted areas in the southeastern part of the island (Fig. 6.12). The outcrop area is a strip 12 km long and 3 km wide parallel to the coast. It extends from the southern tip of the island to the lower reaches of the Øleå River and also along the lower reaches of the Læså River. Exposures occur only along the rivers and along the coast where it is found in blocks on the beach or below sea level (up to 300–400 m off the coast and at depths of up to 6 m) (Bjerreskov 1975; Bjerreskov & Jørgensen 1983; Bjerreskov & Stouge 1985; Böhnke & Katzung 2001). Two formations have been defined and these contain predominantly grey to black marine shales with a total thickness of 160 m (Bjerreskov 1975) or nearly 200 m (Böhnke & Katzung 2001). There are also more than 20 bentonite beds (Bjerreskov 1975).

The Llandovery Rastrites Shales Formation consists of grey to dark grey silty shales with some pyrite and organic matter; in the middle part, calcareous interbeds up to 6 cm thick also occur. The contacts with the underlying Ashgill Tommarp Mudstone Formation and with the dark shales of the *Glyptograptus persculptus* Zone, a high γ-ray interval, are observed only in boreholes. The Rhuddanian has a weak fining-upward trend from silt-streaked shales and silty shales into grey mud shales. The shales are interpreted as density flows which occurred in an area of increasing depth (middle to outer shelf) and under anoxic to weakly oxic conditions (Pedersen 1989). The *Akidograptus ascensus, Parakidograptus acuminatus* and *Cystograptus vesiculosus* zones (Rhuddanian) are recognized in the Billegrav-1 and Bavnegård boreholes near the Øleå River (Bjerreskov 1975; Koren' & Bjerreskov 1997). The *Cystograptus vesiculosus, Monograptus revolutus, Coronograptus gregarius, Lituigraptus convolutus, Spirograptus turriculatus, Monograptus crispus, Monoclimacis griestoniensis, Oktavites spiralis* and *Cyrtograptus lapworthi* zones from the upper Rhuddanian to upper Telychian are also recognized (Bjerreskov 1975). The thickness is estimated at 85 m (Bjerreskov 1975) or at 125 m (Böhnke & Katzung 2001).

The Wenlock *Cyrtograptus* Shales Formation contains grey and greenish grey shales with limestone concretions and some clay beds. The formation crops out in the lower reaches of the Læså River (Bjerreskov 1975; Pedersen & Klitten 1990). The only lower Wenlock graptolite biozone recognized is the *Cyrtograptus centrifugus* Zone, with a thickness of more than 30 m. No other graptolite biozone has been identified on Bornholm with the exception of the 25 m thick *Cyrtograptus lundgreni* Zone (Homerian, Upper Wenlock) discovered in blocks on the beach east of the Øleå River. Whether hiatuses or a simple condensed section are present remains unsolved. The highest biozone recognized contains dark grey silty-sandy mudstones with intercalations of light grey tuffs and limestones. Six lithotypes are recognized including laminated siltstones, mudstones, microscale flaser to lenticular bedded silt-mudstones, limestones, graded sandstones and tuffs. These suggest an environment of deposition of between *c.* 300 and 1000 m depth (Bjerreskov & Jørgensen 1983) and is interpreted as deep-marine with anoxic conditions. The sediments are predominantly mud deposits or hemipelagites with intercalated silt and fine sand produced as a result of bottom currents and intermittent turbidite flows from the northern carbonate shelf platform (Böhnke & Katzung 2001), and air-fall water-deposited tuffs (Bjerreskov & Jørgensen 1983). In the tuffaceous sediments a $^{207}Pb/^{206}Pb$ analysis on single zircons gave a mean age of 430 ± 1.9 Ma. The subduction-related calc-alkaline magmatic arc volcanism giving rise to the tuffs is interpreted to be the result of subduction preceding the closure of the Tornquist Sea (Obst *et al.* 2002). The sediments of the Bornholm succession underwent early diagenesis often with pyrite formation in graptolites prior to compaction (Bjerreskov 1991) and later deep burial diagenesis exceeding 90°C due to the accumulation of *c.* 4 km of upper Silurian sediments (Vejbaek 1985; Pedersen 1989).

The Silurian succession in offshore boreholes in the area surrounding Bornholm exhibits a similar facies but is often thicker. Additional Silurian intervals, not found in Bornholm, are present in the boreholes and the succession is therefore considered to be more complete. Seismic stratigraphy indicates that the offshore Silurian succession has a thickness of more than 4 km (Vejbaek 1985). The Pernille-1 borehole (Fig. 6.4) contains at least 47 m of Silurian deposits dated with graptolites as the *Cyrtograptus ellesae* Zone (Upper Sheinwoodian), whereas the Stina-1 borehole has 399 m of Silurian (Katzung 2001). The Slagerse-1 borehole on Sjælland, with 150 m of grey shales and siltstones, belongs to the *Monograptus crispus* Zone (Poulsen 1974).

Baltica in Germany

The Ordovician and Silurian strata of the G-14 borehole (Adlergrund, north of Rügen island, offshore Germany), situated south of Bornholm, are closely comparable to the Skåne/Scania and Bornholm successions and belong to the Scanian Confacies belt (see above) deposited on Baltica. Beier *et al.* (2000) described the Silurian succession and suggested that it was deposited in a peripheral foreland basin during the collision of Far Eastern Avalonia and Baltica. The Silurian is underlain by a typical Baltica shelf succession of Ordovician age, followed by turbiditic Llandovery silt- and sandstones, interpreted as the northward prograding basin fill of this foreland basin.

The base of the Silurian is characterized by a change from the sandy-silty Jerrestad Formation of Ashgill age (Beier *et al.* 2001) to the mudstones of the lower part of the *Rastrites* Shale Formation (Katzung *et al.* 2004). The mudstones grade upward into graded silt- and sandstones with intercalated coarse sandstones with clay clasts and microconglomerates, containing coral and brachiopod fragments. McCann (1996) differentiated four facies types in the *Rastrites* Shale Formation. Layers of laminated mudstones (1), up to 4.5 mm thick, represent the hemipelagic background sedimentation. Graded mudstones (2) with rare *Planolites* sp. are regarded as low energy turbidites. Graded silt- and sandstones (3) in layers up to 24 cm thick, showing sharp bases and rippled surfaces, internally with cross-bedding, lamination or well developed Bouma sequences, rarely bioturbated, indicate low energy turbidites, at least in part produced by storm events. Microconglomerates (4) are represented by up to 20 cm thick layers, mostly graded, containing predominantly mudstone intraclasts and some bioclasts. Trace fossils are common indicating oxic conditions during deposition. Silurian graptolites are found *c.* 72 m above the lithologically defined base of the Silurian. A 240 m thick interval in the drillcore can be referred to the *'Monograptus' gemmatus* Subzone of the *Spirograptus guerichi* Zone (Maletz 1997). The *Rastrites* Shale Formation of Skåne/Scania and Bornholm are much more strongly condensed than the succession in the G-14 drillcore and an increase in sedimentation rate occurred during deposition of the Wenlock–Ludlow *Colonus* Shales. The Silurian succession of the G-14 drillcore was interpreted to represent the deep-water phase of a peripheral foreland basin, filled during the late Mid-Ordovician until the mid- to late Llandovery (Beier *et al.* 2000). At that marked time the foreland basin was dominated by subsidence and received an increasing amount of sediment. Due to the continuing obduction of the accretionary wedge onto the SW margin of Baltica, the depocentre of the basin migrated in a SW–NE direction (Beier *et al.* 1997; Maletz *et al.* 1997). The main depocentre is located in Skåne/Scania in the Late Wenlock–Early Ludlow, as is seen from the up to 600 m thick *Colonus* Shales, overlain by the Öved-Ramsåsa Group, a shallow marine carbonate-rich succession, representing the final filling of the basin, and overlain by terrestrial deposits.

Northern Gondwana and peri-Gondwana terranes

The Franconia Terrane in the Harz Mountains (Oberharz, Niederharz)

The Silurian rocks of the Harz Mountains are generally considered to be preserved as olistostromes in a Lower Carboniferous matrix. Identifiable successions are lacking and the Silurian of the Harz Mountains has to be pieced together from numerous, biostratigraphically datable, small outcrops (Alberti 1995; Burmann 2006). The thickness of the Silurian has been estimated by various authors to be between 50 and 200 m, but could be considerably higher. The lithology is dominated by dark grey to black, often finely laminated shales in the Llandovery to early Ludlow. Silicified shales or cherts are locally known from around the town of Benneckenstein in the central Harz region. Sandstones and greywackes are locally present. Limestones first appear in the upper Wenlock in the eastern part of the Harzgeröder Zone (in the eastern Harz) (see Fig. 6.13). In the Ludlow the lithologies are less monotonous. Limestones occur and sandstone and tuffs are common.

Graptolites are the most common faunal elements, known from numerous localities. They are exclusively present in the early Silurian. In the Upper Wenlock diverse shelly faunas with brachiopods, gastropods, nautiloids, crinoids and bivalves appear, but are most common from the *Neodiversograptus nilssoni* Zone onwards. Jaeger (1991) proved the presence of all Silurian stages based on graptolites and considered that there are no large

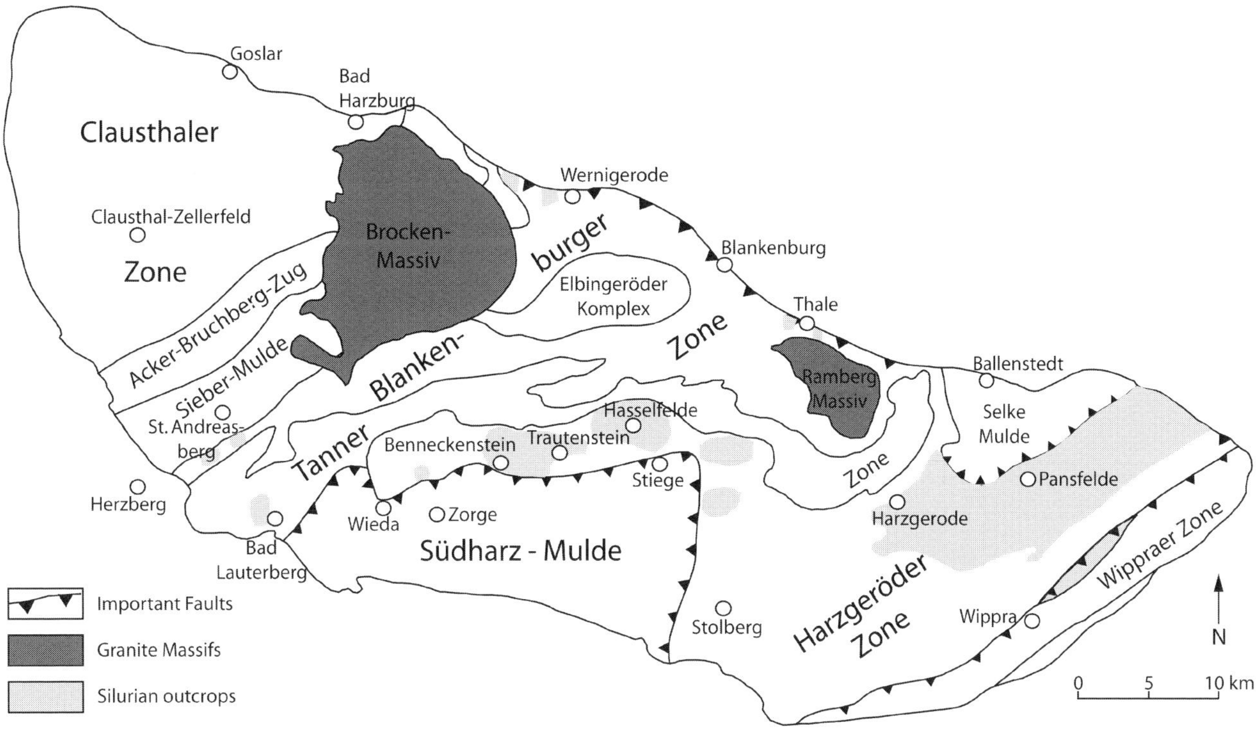

Fig. 6.13. Map of the Harz Mountains in Germany, with large tectonic units (zones, muldes (synclines), etc.) (after Burmann 2006).

biostratigraphical gaps, even though not all of the graptolite biozones have been recognized.

The faunas of the Silurian of the Harz Mountains are closely comparable to those of Saxothuringia. The *Scyphocrinus* horizon has been used, especially for correlation purposes, in the Upper Silurian of Germany. Interestingly, the lithofacies differs considerably from that of Saxothuringia. Basically the entire Silurian succession of the Harz Mountains consists of monotonous, so-called 'Graptolite Shales' (Reuter 1960; Jaeger 1991). It lacks, however, the typical cherts and alum shales of the Unterer Graptolithenschiefer (Lower Graptolite Shale) Formation of Saxothuringia. Lithostratigraphic differentiation, however, is almost impossible due to the strong tectonic overprint and the lack of exposure. Limestones are rare in this succession and are restricted to the upper Silurian, but some of the old records may have been misidentified and belong instead to the Lower Devonian. The limestones in the upper part of the Harz Silurian are less compact than the Ockerkalk in Saxothuringia. The Silurian–Devonian boundary is represented by fossil evidence without any lithological change in the western part of the Harzgeröder Zone (Maronde 1966, 1968). Jaeger (1991) considered the Silurian of the Harz Mountains to be closely comparable to the Unterer Steinhornschichten of the Kellerwald region.

The Franconia Terrane in the Dill Syncline, Marburg and Giessen

Little is known about the Silurian succession of this region since exposure is poor (Anderle 2006*a*, *b*). Despite this, Upper Silurian rocks have been identified as the oldest preserved in the area. The succession of the Lindener Mark region near Giessen covers only a minor part of the Silurian. The Ostrakodenkalk Formation (Ostracod Limestone Formation) consists of *c.* 5–10 m of thick, mostly massive limestones, grading upwards into bedded limestones with cherts. The limestones contain a poorly preserved fauna of ostracods, brachiopods, trilobites and graptolites. The age of the unit appears to be late Wenlock to Ludlow or Přídolí (Jaeger 1962). It is overlain by the Orthocerenkalk Formation (Orthoceras Limestone Formation) which comprises thin (*c.* 5 m) silty shales with carbonate concretions, rich in nautiloids, brachiopods, bivalves and graptolites, indicating a Přídolí to Lochkovian age. Schallreuter (2001*a*, *b*) suggested that there was a close relationship between the diverse ostracod fauna of the Lindener Mark area and faunas from Bohemia, Baltica and even North America.

The Taunus region of the SE Rhenish Massif (= Rheinisches Schiefergebirge) contains Silurian brachiopod faunas with *Dayia shirleyi*, indicating a Přídolí age in the Kellerskopf Formation (= Graue Phyllite), overlain by fossiliferous shales of Lochkovian, early Devonian age. However, older Silurian strata are unknown (Anderle 2006*a*).

The Franconia Terrane in the Kellerwald

Graptolites in the Kellerwald region (Kupfahl 1953, 1954) indicate the presence of Wenlock to Přídolí strata in a number of poorly exposed successions. The Silurian is exposed in small slivers, rarely more than 10 m thick, exposing only minor parts of the time interval. Jaeger (1962) revised the stratigraphy and suggested that the complete Silurian succession comprised graptolite shales with rare limestone intercalations. Anderle (2006*b*) differentiated the Altenteich unit of Llandovery to Wenlock age and the Lower and Upper Steinhorn units. The limestones in the Lower Steinhorn Unit contain graptolites (*Istrograptus transgrediens*), conodonts and crinoids (*Scyphocrinus* horizon) of Přídolí age, while the Upper Steinhorn Unit yielded graptolites of the *Monograptus hercynicus* Zone of early Devonian age (Jaeger 1962; Anderle 2006*b*).

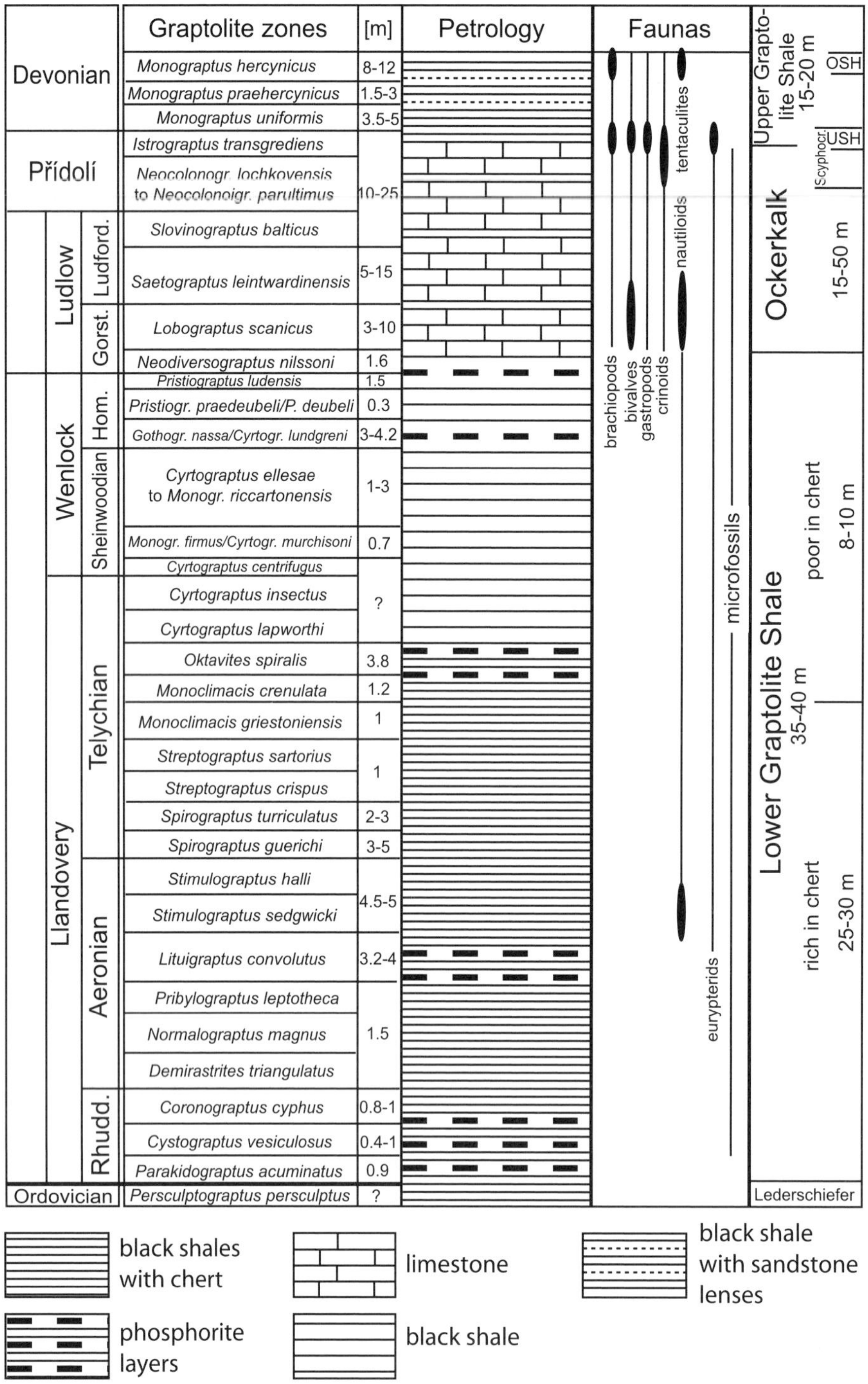

Fig. 6.14. Silurian to basal Devonian succession in Thuringia. OSH, Oberer Schalenbank Horizont (upper shelly horizon); USH, Unterer Schalenbank Horizont (lower shelly horizon). Some graptolite biozones are combined due to limited space (modified after Maletz & Katzung 2003).

The Saxothuringian Terrane in Thüringen, Oberfranken and Vogtland

This region includes one of the best studied Silurian areas in Germany, as noted in the detailed review on the Silurian of Germany published by the German Stratigraphical Commission (Deutsche Stratigraphische Kommission 2002, 2006). The succession in the Thuringia Forest (= Thüringer Wald) is the only complete succession with a well-established graptolite biozonation, but it has been condensed to less than 100 m (Maletz & Katzung 2003).

Silurian strata are widely distributed in Thuringia, Vogtland, Saxonia, Erzgebirge and the Lausitz regions (Fig. 6.4). In general, the successions are fairly similar, but poor preservation and limited exposures often make a detailed comparison difficult. The succession in the Erzgebirge differs considerably from that of the Thuringian succession due to the presence of tuffs which comprise up to 40% of the succession thickness (Geissler & Schauer 2006). Slightly different successions are also found in the Görlitzer and Frankenberger Schiefergebirge. The Silurian of the Görlitzer Schiefergebirge (Lausitz region), in general, matches the Saxothuringian succession. However, the Ockerkalk equivalent appears to be less massive (maximum thickness *c.* 10 m) and is dated at early to mid-Ludlow (Brause 2006). In the Frankenberger Zwischengebirge the Silurian succession is more condensed than in Saxothuringia, but the outcrop situation is extremely poor. The Ockerkalk Formation is represented by a few metres of greenish-grey shales (Jaeger 1977; Kurze 2006).

The Silurian in Thüringen, Oberfranken and Vogtland is represented by the Gräfenwarth Group, which extends up into the Lower Devonian (Emsian). It is dated using a rich graptolite fauna that has been studied for more than 150 years (Geinitz 1842; Maletz 2001). The group includes the Unterer Graptolithenschiefer Formation (Lower Graptolite Shale Formation), the Ockerkalk Formation and the Oberer Graptolithenschiefer Formation (= Upper Graptolite Shale Formation) (Fig. 6.14). The tripartite differentiation of the Silurian to Lower Devonian succession was originally documented by Gümbel (1879), based on initial mapping in Thuringia. The Silurian is widely distributed, but is poorly exposed in the region and strongly deformed by later Variscan activity. The Silurian part of the Gräfenwarth Group is less than 100 m thick (Maletz & Katzung 2003). An extensive overview of the geology of Saxothuringia was recently produced by Linnemann *et al.* (2003*a*, *b*) and Linnemann (2003*a*, *b*, *c*), outlining the stratigraphical successions of the region, the tectonic evolution and interpreting the plate tectonic history of the terrane and its relationship to the peri-Gondwana terrane assemblage.

The base of the Silurian coincides with a lithological change close to the base of the Unterer Graptolithenschiefer Formation. The Upper Ordovician glaciomarine Lederschiefer Formation (Katzung 1999) grades conformably into graptolitic shales with chert beds (= Kieselschiefer, also called lydites) (Schauer 1971). The base of the Silurian is lithologically considered to coincide with the base of the first chert horizon and biostratigraphically coincides with the first occurrence of the graptolite *Akidograptus ascensus* (Jaeger 1988).

The Unterer Graptolithenschiefer Formation is estimated to be *c.* 35–40 m thick, but complete sections through the unit do not exist. Alum shales (dark, organic-rich shales, containing iron pyrite, forming alum when weathering) with a large number of chert horizons characterize its lower part. The formation shows a typical alternation of chert beds, up to 10 cm thick, and thin alum shales beds. This alternation grades upwards into alum shales of the upper unit of the Unterer Graptolithenschiefer Formation. Phosphate concretions, often containing well-preserved fossils, are common and usually concentrated at certain levels.

The base of the Ockerkalk Formation coincides with the first distinct limestone bed. The Ockerkalk Formation consists of platy to thick-bedded limestones and intercalated thin shale layers. The formation is variable in thickness, partly due to tectonic deformation, and reaches 15 to 50 m. The amount of limestone in this formation varies between 60 and 80%. Diverse shelly faunas have been found in the limestones, whereas the intervening shale layers may contain poorly preserved graptolite faunas.

The Oberer Graptolithenschiefer Formation consists of black shales with rare beds and lenses of fine sandstones and siltstones. It is lithologically distinct from the underlying Unterer Graptolithenschiefer Formation. Only the lowermost 1 m belongs to the Silurian *Istrograptus transgrediens* Zone, whereas the major part of the formation is now considered to belong to the Devonian *Monograptus uniformis* to *Monograptus hercynicus* zones.

The Unterer and Oberer Graptolithenschiefer formations of the Gräfenwarth Group have been interpreted as deep-water sediments deposited under low energy conditions by most authors. Coarser-grained clastics and sedimentary structures are rare and restricted to a few levels, especially in the upper part of the succession, carrying specific names such as lower shelly horizon (USH), upper shelly horizon (OSH) and *Scyphocrinus* horizon. Hundt (1965) documented current activity in the Unterer Graptolithenschiefer Formation, based on the presence of current ripples and current-orientated graptolite accumulations. The sedimentation rate was extremely low so that individual graptolite biozones may be only a few centimetres thick. They are often difficult to detect in an outcrop due to the strong tectonic deformation, which may locally squeeze out some of the biozones (Jaeger 1991). The Ockerkalk Formation corresponds to a time of sea-level fall and more oxic conditions than the under- and overlying formations. A diverse benthic fauna appears, including trilobites, nautiloids, gastropods, brachiopods, bivalves and crinoids. The Oberer Graptolithenschiefer Formation corresponds to a return to anoxic conditions, as can be seen by the return to the deposition of black, organic-rich shales. However, an increase in sediment input led to a distinct thickness increase and the formation of decimetre-thick siltstone and sandstone beds.

Saxothuringian Zone in the Spessart Crystalline Complex

The metamorphic series of the Spessart Crystalline Complex is a large NE–SW orientated antiform in which two tectonic complexes can be differentiated (Kroner *et al.* 2008; McCann *et al.* 2008). The tectonically lower complex in the central Spessart is formed by the Mömbris, Hörstein-Huckelheim and Geiselbach formations. The upper complex is represented by the Alzenau Formation to the north and the Elterhof Formation to the south. Granitoid intrusions in the Mömbris Formation represent a Silurian age (418 Ma; Dombrowski *et al.* 1995). In addition, the Spessart Crystalline Complex yielded pteridophyte spores in garnet-bearing schists at *c.* 500 m above the base of the Geiselbach Formation in the lower complex. This is a quartzite unit formerly correlated with the early Ordovician Frauenbach and Phycoden groups of Thuringia. The unit is now referred to the Wenlock–Ludlow (Reitz 1987). Detailed information about the lithology and palaeontology of the unit was given by Hirschmann & Okrusch (2001).

Saxothuringian Zone in the Black Forest (= Schwarzwald)

Silurian rocks are found in the Badenweiler-Lenzkirch Zone, up to 5 km wide, in the southern Black Forest (Fig. 6.4). Greywackes of Late Devonian to Early Carboniferous age are well-known, but more recently acritarchs and chitinozoans suggest Early Ordovician to Silurian ages for some strata. The rocks are interpreted as a tectonic melange which includes units of Silurian age. The more highly metamorphosed rocks have yielded only rare and poorly preserved microfossils, of which the biostratigraphical value can be questioned. Similarly, the metamorphic Wetzldorf-Group and the Anglmühle Unit of the Bavarian Forest (= Bayerischer Wald) and Upper Pfalz Forest (=Oberpfälzer Wald) (Fig. 6.4) are referred to the Silurian on the basis of a rare fossil record. Stettner (2006) suggested a comparison of the phyllites of the Wetzldorf Group with the Lower Graptolite Shale Formation of Saxothuringia, partly based on the record of radiolarians and graptolites by Stürmer (1962).

The Moldanubian Zone with the Prague Basin on the Perunica Terrane (Czech Republic)

Most of the Silurian rocks of the Bohemian Massif were deposited on the Perunica Terrane which is regarded by some authors as a microplate that detached from Gondwana during Ordovician times (Havlíček *et al.* 1994) (Figs. 6.4 & 6.15). The existence of Perunica in the Silurian is supported by palaeomagnetic data (Krs *et al.* 1986) and by the distribution of Ordovician and Silurian fossils. It includes the Prague Basin (Fig. 6.15), the most fossiliferous Silurian outcrop area of Central Europe and a complete Silurian succession only slightly disturbed by later deformation (Fig. 6.16). The region was incorporated during the Variscan Orogen as a separate terrane between the Gondwana and Baltica palaeocontinents (Fig. 6.4).

Another Silurian outcrop area in the Bohemian Massif locally called the 'Islet Zone' (Fig. 6.17), is situated on the central Bohemian granite pluton and includes successions altered by contact and regional metamorphism. Other metamorphosed and biostratigraphically dated Silurian deposits are developed in the Rožmitál area, the Mirovice metamorphic area, the Železné Hory Mountains (Vápenný Podol), in the Hlinsko Lower Palaeozoic Region and in the Lužice Region (Fig. 6.17). Silurian rocks were also recorded from a borehole in the Bohemian Cretaceous Basin near Jablonné, Podještědí (Fig. 6.17) (Kříž *et al.* 2003). The Silurian sequence in the Moravo-Silesian area in the Drahany Highland, Stínava (Fig. 6.17) is similar to that of the Prague Basin (Kettner & Remeš 1935). It is considered to be part of the Brunovistulicum terrane (Havlíček 1980).

Facies and volcanism

Three of the distal Silurian facies in northern Gondwana occur in Bohemia, namely: the shelly fauna facies, the Prague Basin facies and the Thuringian facies (Kříž *et al.* 2003). They are sometimes called the 'Variscan' facies (Kříž *et al.* 2003), but are not to be confused with the tectonic events bearing the same name over large parts of Europe. The **shelly fauna facies** is characteristic for the upper Wenlock and the Ludlow of deeper parts of the Prague Basin. The shallow-marine **Prague Basin facies** (Kříž *et al.* 2003) differs from the Thuringian facies in its greater thickness (250–580 m).

In the Prague Basin, the earliest Silurian sediments were pelitic, black graptolitic shales. During the Telychian (Llandovery) and especially in the early Sheinwoodian (Wenlock), the presence of carbonates increased progressively. In the mid-Llandovery (Aeronian) and from the mid-Wenlock to the mid-Ludlow, sedimentation in the Prague Basin was influenced by the activity of a volcanic archipelago with the resultant deposition of volcaniclastics and related magma intrusions, tuffs and lava flows (Fig. 6.18). Several volcanic centres developed along synsedimentary growth faults and at their junctions (Fig. 6.18). Around the volcanic centres shallow-marine highs with slopes developed, favourable for the growth of the rich benthic faunas, mentioned above. During periods of minimal sediment supply isolated carbonate platforms developed. In the region of the Svatý Jan Volcanic Centre (Fig. 6.18), reworked limestone pebbles and fragments of algae, bryozoans, corals and stromatoporoids document intertidal or shallow subtidal environments during the late Wenlock (Kříž 1991, 1998b). This would suggest the existence of an emergent volcanic island at least at that time (see below). Temporary emergence of other volcanic centres is also indicated by the presence of volcaniclastic rocks and subaerial lava flows (Kříž 1992).

In the late Wenlock (Homerian) and the Ludlow eustatic movements were a primary cause of basin shallowing (Kříž 1991, 1998*b*). In the deeper parts of the basin but still in a shelf environment, eustasy had less influence and the facies and pelitic or tuffaceous-pelitic sedimentation of the shelly fauna facies was dominant. During periods of volcanic activity the tuffaceous admixture of volcaniclastics increased.

The activity of the Svatý Jan Volcanic Centre (Fig. 6.18) terminated during the latest Wenlock and early Ludlow (Gorstian) with repeated basaltic lava flows into the shallow-marine environment (Fiala 1982). A thick lava sheet (up to 60 m thick) covers earlier volcaniclastic deposits. The latest Přídolí brachiopod and crinoid limestones of the carbonate platform locally overlay the lava sheet. This indicates that during the late Homerian, Gorstian, Ludfordian, and most of the Přídolí, the top of the volcano was above sea level. A land flora with a highly diverse assemblage of trilete miospores and cryptospores of the *Artemypyra brevicosta–Hyspanaediscus verrucatus* Assemblage Zone was discovered in close proximity to the Svatý Jan Volcanic Centre in upper Wenlock tuffaceous shales and limestones (Dufka 1995*a*,*b*). These fossils support the hypothesis that the volcano formed, at least at that time, a subaerial volcanic island.

The Přídolí in the Prague Basin is characterized by laminated limestones and calcareous shales. The international boundary stratotype of the base of the Přídolí Series is defined in this basin (Fig. 6.19) (Kříž *et al.* 1986). The late Přídolí regression led to the deposition of shallow-marine bioclastic limestones with abundant cephalopods, bivalves, brachiopods and crinoids (similar to the Carnic Alps, eastern Serbia, and the Montagne Noire). In other parts of the world, this lowstand culminated in a transition from marine to non-marine facies (Kříž *et al.* 2003).

The most important feature for correlation with other Gondwana regions is the cephalopod limestone lithofacies which is found in the shelly fauna facies of the Prague Basin form at least 11 horizons (Kříž 1998*a*), notably across the Homerian–Gorstian boundary, in the late Gorstian, the early and late Ludfordian, and in the earliest and latest Přídolí. This lithofacies was deposited below storm wave base, but within the reach of surface currents ventilating the sea bottom (Ferretti & Kříž 1995). It is characterized by the presence of recurring communities dominated by bivalves (*Cardiola* Community Group; Kříž 1999*a*), and by abundant cephalopods and other molluscs whose larvae could be

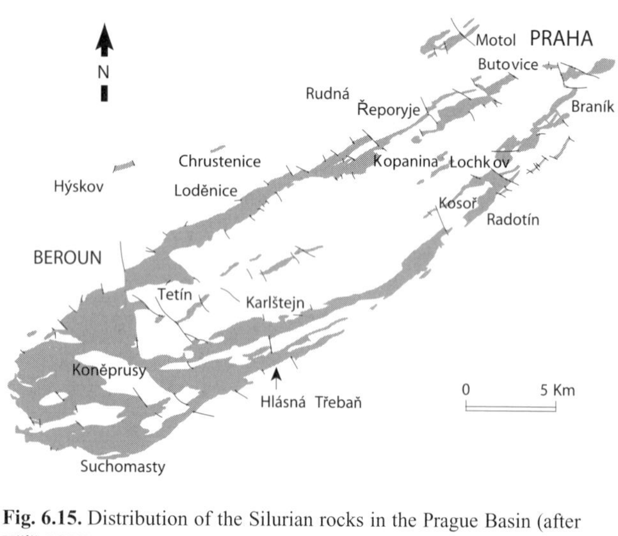

Fig. 6.15. Distribution of the Silurian rocks in the Prague Basin (after Kříž 1992).

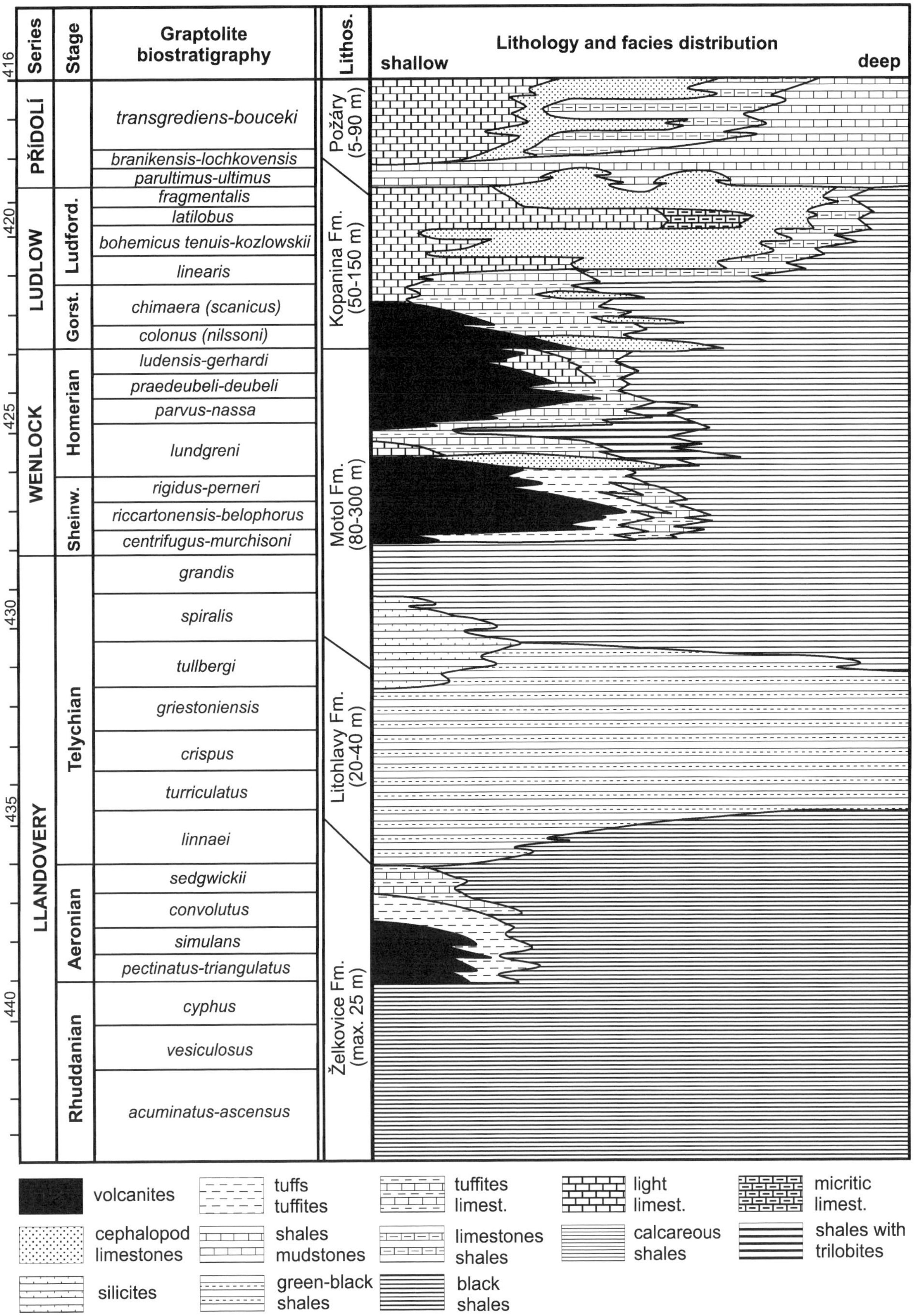

Fig. 6.16. Silurian biostratigraphy, lithostratigraphy and facies development of the Prague Basin (after Kříž 1992, 1999*b*).

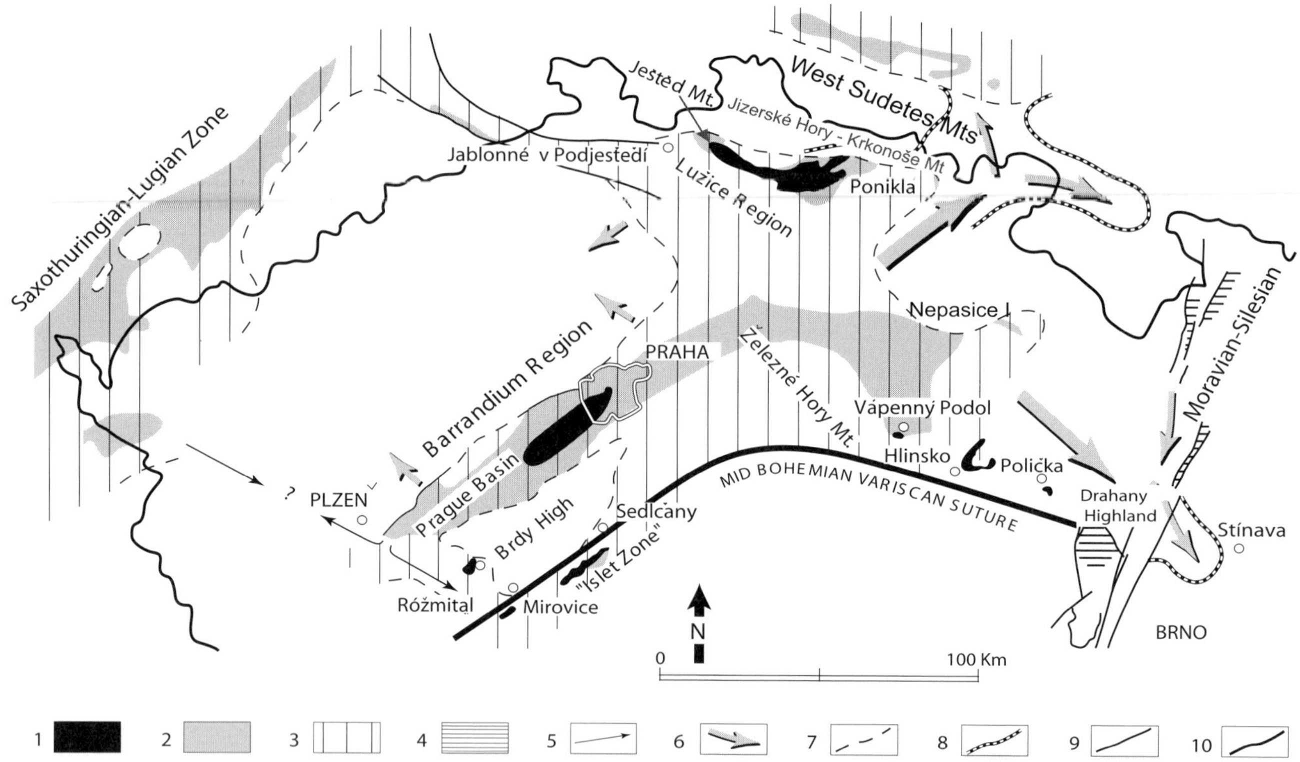

Fig. 6.17. Silurian in the Bohemian Massif. Key: 1, Silurian outcrops and subcrop; 2, Ordovician outcrops and subcrop; 3, presumed original extent of Ordovician; 4, crystalline rocks of possible Ordovician and Silurian age; 5, zone of possible communication between the middle Bohemian Region and the Saxothuringian–Lugian Zone; 6, Silurian transgression directions outside the Ordovician basins; 7, presumed Ordovician shoreline; 8, presumed Silurian shoreline where it differs from the Ordovician seashore; 9, younger faults responsible for the reduction of the original areal extent of Palaeozoic basin; 10, younger faults bordering limnic Permo-Carboniferous of the Boskovice Furrow (after Havlíček 1980).

transported by surface currents to other northern Gondwana areas (Kříž 1998*a*).

The Silurian near Stínava in the Drahany Highland (Fig. 6.17) is similar to that of the deeper parts of the Prague Basin. The lower part of the sequence is represented by black graptolitic shales (upper Telychian) and the upper part by a shelly fauna facies, represented by calcareous shales with limestone nodules (lower Gorstian) containing cephalopods, bivalves (*Cardiola, Patrocardia* and *Dualina*) and crinoids. The thickness is *c.* 100 m (Bouček 1935).

Outside the Prague Basin and the Moravo-Silesian region, the Silurian of the Bohemian Massif contains the **Thuringian facies** typically developed in the Thuringia Forest (see above) (Kříž *et al.* 2003). In central Bohemia in the 'Islet Zone' and the Železné Hory Mountains, the Silurian (Llandovery–Ludlow) contains black graptolitic shales and, in the upper part, black crinoidal limestones with cephalopods. In the 'Islet Zone' Silurian rocks have undergone contact metamorphism into chiastolite–cordieritite graphitic slates and hornfels, with graptolites still recognizable and hence allowing a general dating as the Silurian (Chlupáč 1989). Thin carbonate beds with remains of *Scyphocrinites* are developed at the Přídolí–Lochkovian boundary.

The Silurian in the Jizerské Hory–Krkonoše Mountains (west Sudetes) is developed as dark graphitic phyllites and laminated cherts (Chlupáč 1993). A limestone at the top of the Silurian near Křižany, in the Ještěd Mountains (Fig. 6.17), exhibits a lithology similar to that of the Ockerkalk Group known elsewhere in the Saxothuringian Zone. Black graptolitic shales with cherts constitute the thin Silurian succession of the Sedlčany-Krásná Hora metamorphic 'Islet Zone', the Rožmitál area and the Mirovice metamorphic area (Havlíček 1977; Štorch *et al.* 1984) and of the Hlinsko Region (Horný 1956). The entire Silurian in the Rožmitál area is only 40–50 m thick (Havlíček 1977).

The Thuringian facies is believed to be a deep-marine deposit. However, in the Silurian of the Bohemian Massif there are indications of possible subaerial emergence during Silurian times. In the Rožmitál area, Lower Devonian conglomerates contain boulders of Cambrian sandstones and of non-metamorphosed or slightly metamorphosed Proterozoic sedimentary rocks. The presence of volcaniclastic rocks supports the idea that exposed areas of Cambrian rocks in the Brdy High (also called Brdy Elevation) (Fig. 6.17) were not covered by the Ordovician and Silurian seas (Havlíček 1980). Additionally, in the Nepasice borehole (Fig. 6.17) near the town of Hradec Králové, Lower Devonian sediments have been documented directly overlying a Proterozoic basement (Havlíček *et al.* 1994).

Lithostratigraphy in the Prague Basin

The Želkovice, Litohlavy, and Motol formations span the early Silurian (Kříž 1975) (Fig. 6.16). The base of the Želkovice Formation correlates with the Ordovician–Silurian boundary and is identified by the dark shales of the *Akidograptus ascensus – Parakidograptus acuminatus* zones. The Litohlavy

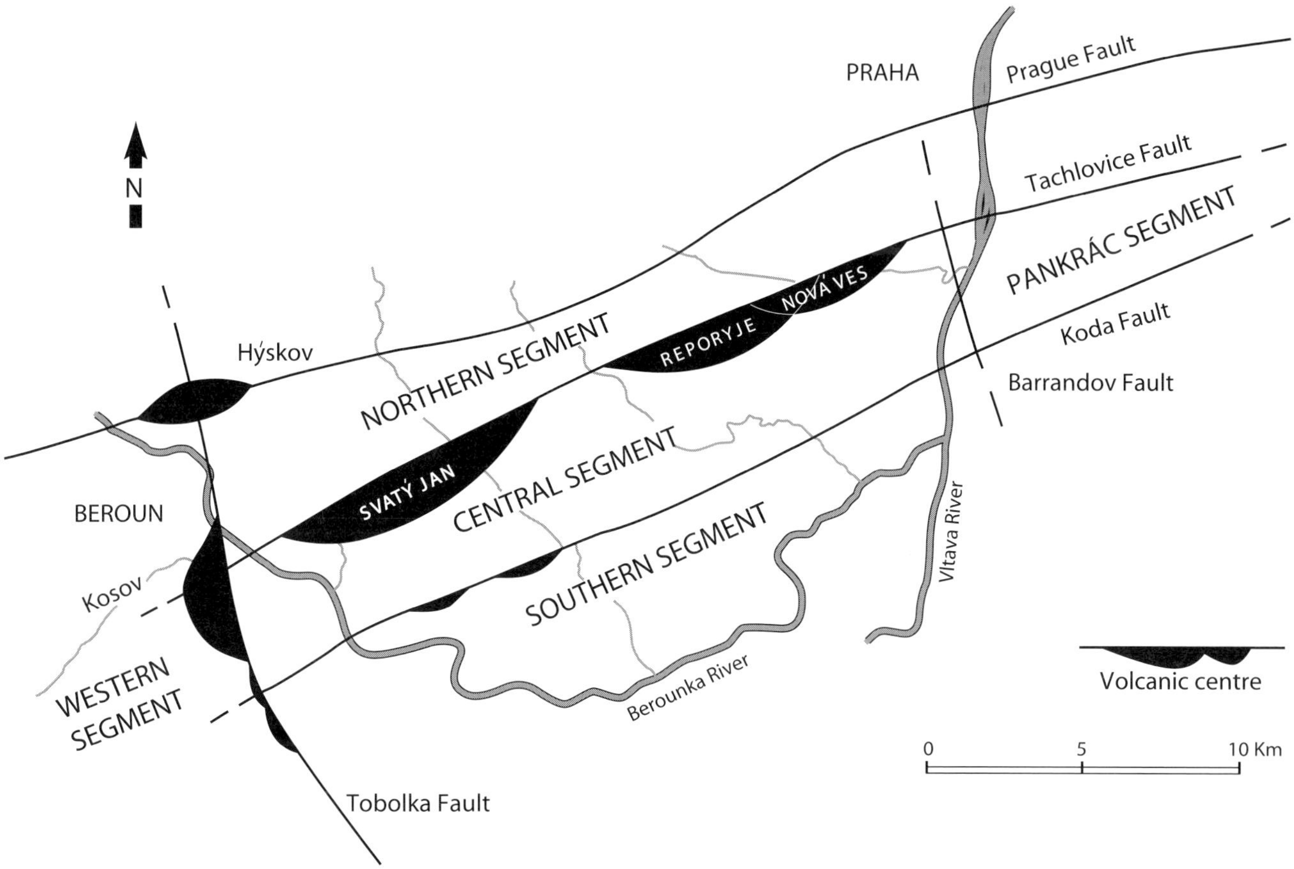

Fig. 6.18. Map with Silurian synsedimentary tectonics, volcanic centres and segments (sub-basins) in preserved parts of the Prague Basin (after Kříž 1991).

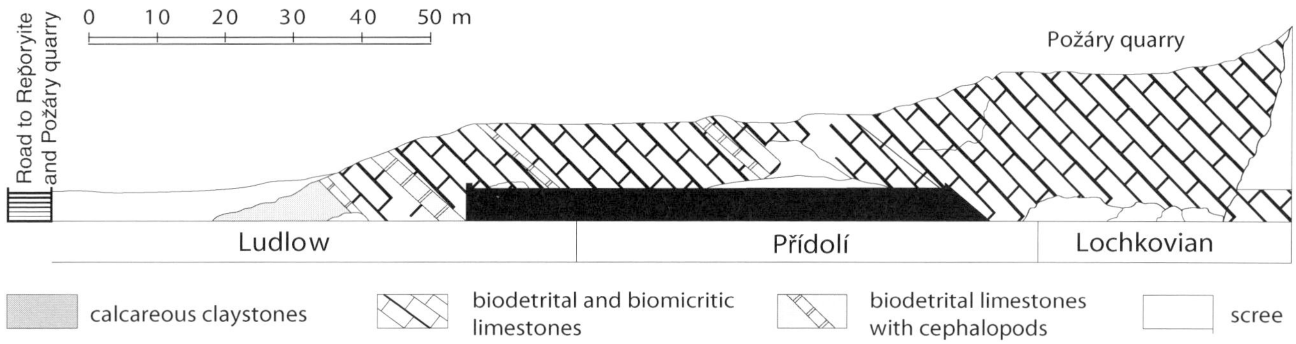

Fig. 6.19. Schematic representation of the GSSP international boundary stratotype in the Požáry quarry near Praha-Řeporyje (after Kříž *et al.* 1986).

Formation, developed as an alternation of black graptolite shales and greenish bioturbated calcareous claystones, overlies these dark shales. A thick band of green calcareous shales to clayey limestones defines the base of the Litohlavy Formation. Above this formation the Motol Formation is developed as calcareous dark shales.

The late Silurian (Ludlow and Přídolí) is represented by the Kopanina and the Požáry formations (Fig. 6.16). Considerable facies diversity has led to the definition of the lower boundary of the Kopanina Formation at the base of the *Colonograptus colonus* (= *Neodiversograptus nilssoni*) Zone. The lower boundary of the Požáry Formation at its stratotype in the Požáry Quarries near Praha-Řeporyje (Fig. 6.18) is defined just above the top of the massive cephalopod limestone (Kříž *et al.* 1986).

The thickness of the Silurian rocks in the Prague Basin (Fig. 6.16) is more than 580 m in those regions of igneous activity where thick volcaniclastic successions occur (Kosov and Svatý Jan volcanic centres). In the deep-water areas of the western and southern parts of the Prague Basin, the prevailing shelly fauna facies sediments have a maximum thickness of 450 m. The thickness is much less in the eastern part of the southern and northern segments of the basin.

Biostratigraphy of Perunica

The Silurian fossils of Bohemia have been studied for more than 230 years since the first reports by Zeno (1770). Joachim Barrande (1799–1883) travelled throughout the whole area of central Bohemia during 1840–46 (Barrande 1846) and surprised the scientific world in 1852 with the first volume of his monumental work 'Silurian System in Central Bohemia' with accurate drawings and detailed observations. Between 1872 and 1881 he published a further 20 volumes describing more than 2000 Silurian species of fossil crustaceans, cephalopods, brachiopods and bivalves.

In all of the Bohemian regions, the Silurian is dated mostly by graptolites (Štorch 1994). At Poniklá in the west Sudetes (Fig. 6.17) Horný (1964) described well preserved graptolites from phosphatic concretions embedded in graphitic phyllite, which indicated a Wenlock age. The upper part of the Ockerkalk Group limestone contains columnals and stems of the crinoid *Scyphocrinites*, the only fossil which could be identified. *Scyphocrinites* typically occurs in the uppermost Silurian and in the lowermost Devonian. Graptolites in shales of the Rožmitál area of the Bohemian 'Islet Zone' indicate the presence of the Telychian–Sheinwoodian boundary, the Homerian and the early Gorstian. In the Mirovice metamorphic 'islet', graptolites allowed the rocks to be dated to the Homerian. Two horizons in the Sedlčany-Krásná Hora metamorphic 'islet' are dated by graptolites as mid-Llandovery and as close to the Llandovery–Wenlock boundary. In the Železné Hory Mountains, the Silurian phyllitic black shales are dated using the presence of late Telychian, Sheinwoodian, Homerian and earliest Gorstian graptolites. The occurrence of *Scyphocrinites* in upper black limestones indicates a Přídolí age. In the Lower Palaeozoic Hlinsko Region several graptolite biozones from the Aeronian and Telychian have been recognized.

In the Prague Basin, the entire Silurian succession is palaeontologically well-dated (see Kříž 1991, 1992, 1998*b*) (Fig. 6.16). Štorch (1994) recognized 27 graptolite biozones in the Llandovery and Wenlock. Štorch (1995*a*) recognized eight graptolite biozones in the Ludlow, and Jaeger (in Kříž *et al.* 1986) six graptolite biozones in the Přídolí. Walliser (1964) identified 11 successive conodont biozones in the Wenlock to Přídolí carbonate facies. Schönlaub (in Kříž *et al.* 1993) studied the Wenlock–Ludlow boundary conodonts. A detailed (bed by bed) conodont biostratigraphy was established for the Ludlow–Přídolí boundary interval (Schönlaub in Kříž *et al.* 1986). Chitinozoan biostratigraphy for the Ludlow–Přídolí boundary interval was documented in detail by Paris (in Kříž *et al.* 1986). Dufka (1992, 1995*a*) and Dufka *et al.* (1995) recognized the Lower Silurian chitinozoan biozones from the global chitinozoan biozonation of the Silurian (Verniers *et al.* 1995). The Silurian near Stínava is correlated with the *Oktavites spiralis* and *Stomatograptus grandis* zones (late Telychian) and the *Cyrtograptus lundgreni* and *Colonograptus ludensis* zones (Homerian). Co-occurrences of graptolites and bivalves (*Cardiola, Patrocardia* and *Dualina*) allowed correlation of the uppermost part of the succession with the *Colonograptus colonus* Zone of the lower Ludlow (Bouček 1935). Bouček & Přibyl (1955) and Hansch (1993) developed a biostratigraphy based on ostracods, and Kříž (1999*c*) established a biostratigraphy of the upper Wenlock, Ludlow and Přídolí based on bivalve-dominated communities occurring in the shales and cephalopod limestone levels. He correlated these communities in the shelly fauna facies between Perunica and other areas of Gondwana such as eastern Serbia and Moesia or the East European Platform (Kříž *et al.* 2003). Each cephalopod limestone level in the Prague Basin has been well dated by graptolites and conodonts (Kříž 1998*a*).

Italia or Apulia Terrane (Austria)

In the Austrian part of the Alps well-studied, sometimes richly fossiliferous, Silurian outcrops occur in different nappes or tectonic windows. The most continuous sections are present in the Carnic and Karawanken Alps (northern margin of the Southern Alps), which show a facies very similar to northern Gondwana and are thought, by some authors, to belong to the same terrane. Further to the north, the Grauwacke Zone trends east–west from Styria via Salzburg to the Tyrol. In the Rannach Nappe, near Graz (Styria) and in the Gurktal Nappe, a facies similar to the Prague Basin facies is present.

Since the discovery of Silurian fossils in the Austrian Alps by Hauer (1847), knowledge of the Silurian has increased considerably. The Cellon section in the Carnic Alps became not only the stratotype section for the Silurian in the eastern and southern Alps (Fig. 6.20), but also the standard for the globally applicable conodont biozonation (Walliser 1964). An extensive summary of the Silurian of Austria is given in Schönlaub (1993, 1994*b*) and Kříž *et al.* (2003); the summary below is based on these works, as well as Schönlaub & Histon (1999) and Schönlaub *et al.* (2004).

The Silurian deposits of Austria vary from shallow-marine carbonates to graptolitic shales, with regionally similar thicknesses (normally not more than 60 m) and differences in facies patterns, rates of subsidence, source areas, fossil distributions, degrees of volcanic activity and the spatial and temporal relationships of climate-sensitive rocks such as coral and algal-bearing limestones, ooliths, coal beds and evaporites (Schönlaub 1993). The strata can be locally very fossiliferous with different distinct faunal assemblages containing varying amounts of nautiloids, trilobites, bivalves, brachiopods, graptolites, conodonts, foraminiferans, acritarchs, chitinozoans and ostracods. Complete but condensed successions occur in the carbonate-dominated facies and a continuous succession is present in the graptolite-bearing sequences, especially in the Carnic and Karawanken Alps. In other areas, due to poor preservation, the lack of fossils and metamorphic overprinting, stratigraphic continuity has yet to be demonstrated. The strata have mostly been dated by graptolites and conodonts, but other groups such as trilobites, bivalves, chitinozoans and acritarchs have also been used. Brachiopods, bivalves, and nautiloids are used for palaeoecological and palaeogeographical reconstructions.

The Periadriatic Line is a structurally important Alpine fault zone (Fig. 6.20). The Carnic and Karawanken Alps are located to the south of it whereas the Grauwacke Zone, the Rannach Nappe in the Graz area and the Gurktal Nappe are to the north. Although there are some similarities between these two areas, separated by the fault zone, there are also important differences. For example, siliciclastic rocks of Ordovician to Devonian age, as well as intense basic volcanism indicating crustal extension, dominate the area north of the fault, while thick carbonate deposits of Devonian age including reef limestones mainly occur in the Carnic Alps to the south of the Periadriatic Line. Interestingly, in the stratigraphic record corals make their first appearance north of the Periadriatic Line in the late Llandovery, i.e. some 20 Ma earlier than in the south where they first show up in strata of Ludlow age. The increased input of siliciclastic material of Ordovician to Silurian age suggests proximity to a continental source for the northern area, while the region to the

Fig. 6.20. Main regions of anchizonal to lower greenshist metamorphosed Palaeozoic strata in Austria. The Periadriatic Line separates the Carnic Alps and the Karawanken Mountains (Southern Alps) from other Alpine Palaeozoic remnants belonging to the Eastern Alps. The Wildseeloder and the Glemmtal units are too small to be located on the map: they are situated in the Kitzbühel Alps (after Hubmann *et al.* 2003).

south may represent an open marine setting far from any continental influence.

Carnic and Karawanken Alps

In the Carnic Alps, the Silurian transgression begins in the earliest Llandovery. At present, there is no record in the Carnic Alps of the *Akidograptus ascensus* Zone, the basal biozone of the Silurian, but the overlying *Parakidograptus acuminatus* Zone is present and situated very close to the Ordovician–Silurian boundary. Due to the unconformable contact separating the Ordovician and Silurian strata in both the Carnic and Karawanken Alps, a sedimentary rock unit is inferred to be missing. This gap may cover several Llandovery and Wenlock conodont biozones. In some locations where basal Lochkovian strata disconformably rest upon Upper Ordovician limestones, the Silurian may even be absent (Schönlaub 1971).

Four major depth- and energy-related (litho)facies have been distinguished for the Silurian of the Carnic and Karawanken Alps reflecting areas with different rates of subsidence. The first two are equivalent to the shelly fauna facies of northern Gondwana and the fourth to the Thuringian facies. The third facies is intermediate. The Wolayer facies was deposited in the shallowest, current-influenced environment. The Plöcken facies represents a moderately deep-marine, but still shelf environment and is exposed in the classic Cellon section. The Findenig facies is intermediate between the deeper shelf and the basinal environment and the graptolitic Bischofalm facies was deposited in a euxinic stagnant water basinal setting. Sediments from the Llandovery to the lower Ludlow indicate a steady trend of basin subsidence and an accompanying increase in water depth. This rise in sea level apparently decreased in the Upper Ludlow and possibly ceased during the Přídolí leading to uniform limestone deposition. Simultaneously, the black graptolitic shales in the Bischofalm facies were replaced by the green and grey shales of the Middle Bischofalm Shale. The appearance of the Bischofalm facies signals the reestablishment of a deep-water graptolitic environment at the top of the Silurian which subsequently persisted until the end of the Lochkovian (Fig. 6.21).

In the shallow environment of the **Wolayer facies** fossiliferous limestones with abundant orthocone nautiloids, numerous trilobites and bivalves, small brachiopods, gastropods, crinoids and rare corals were deposited (Gaertner 1931; Haas 1968; Ristedt 1968; Kříž 1979, 1999*b*; Schönlaub 1980; Histon 1999, 2002; Ferretti *et al.* 1999; Santel 2001). The Wolayer facies is named after the Upper Ordovician cystoid-bearing Wolayer Limestone. Due to the slow rate of subsidence and a hiatus at the base this facies is only 10 to 15 m thick. The type sections, including the Rauchkofel Boden section (Figs. 6.20 & 6.22), are located in the Lake Wolayer region of the central Carnic Alps (Gaertner 1931; Schönlaub 1971, 1980; Schönlaub & Histon 1999; Ferretti *et al.* 1999). This unit is overlain by the grey highly fossiliferous Orthoceras Limestone of latest Llandovery, Wenlock and Ludlow age (Ferretti 2005). Where this facies occurs the uppermost Ordovician to the lower Silurian strata are missing, in a sedimentary gap ascribed to the Hirnantian eustatic sea-level fall and its aftermath. In the overlying strata, conodonts of the *Ozarkodina sagitta* Zone (basal Homerian) are fairly abundant. The appearance of the conodont index species *Kockelella variabilis* suggests the base of the Ludlow as is the case in Bohemia (H.P. Schönlaub in Kříž *et al.* 1993). The overlying Cardiola Formation corresponds to the *Polygnathoides siluricus* conodont Zone of the same formation in the Cellon section. It is succeeded by pinkish and greyish limestones corresponding to the Alticola and Megaerella limestones of the Cellon section. However, diagnostic conodonts have not yet been found in these latter units at Rauchkofel Boden except from the uppermost limestone level. This highest conodont fauna contains common specimens of the *Ozarkodina remscheidensis eosteinhornensis*

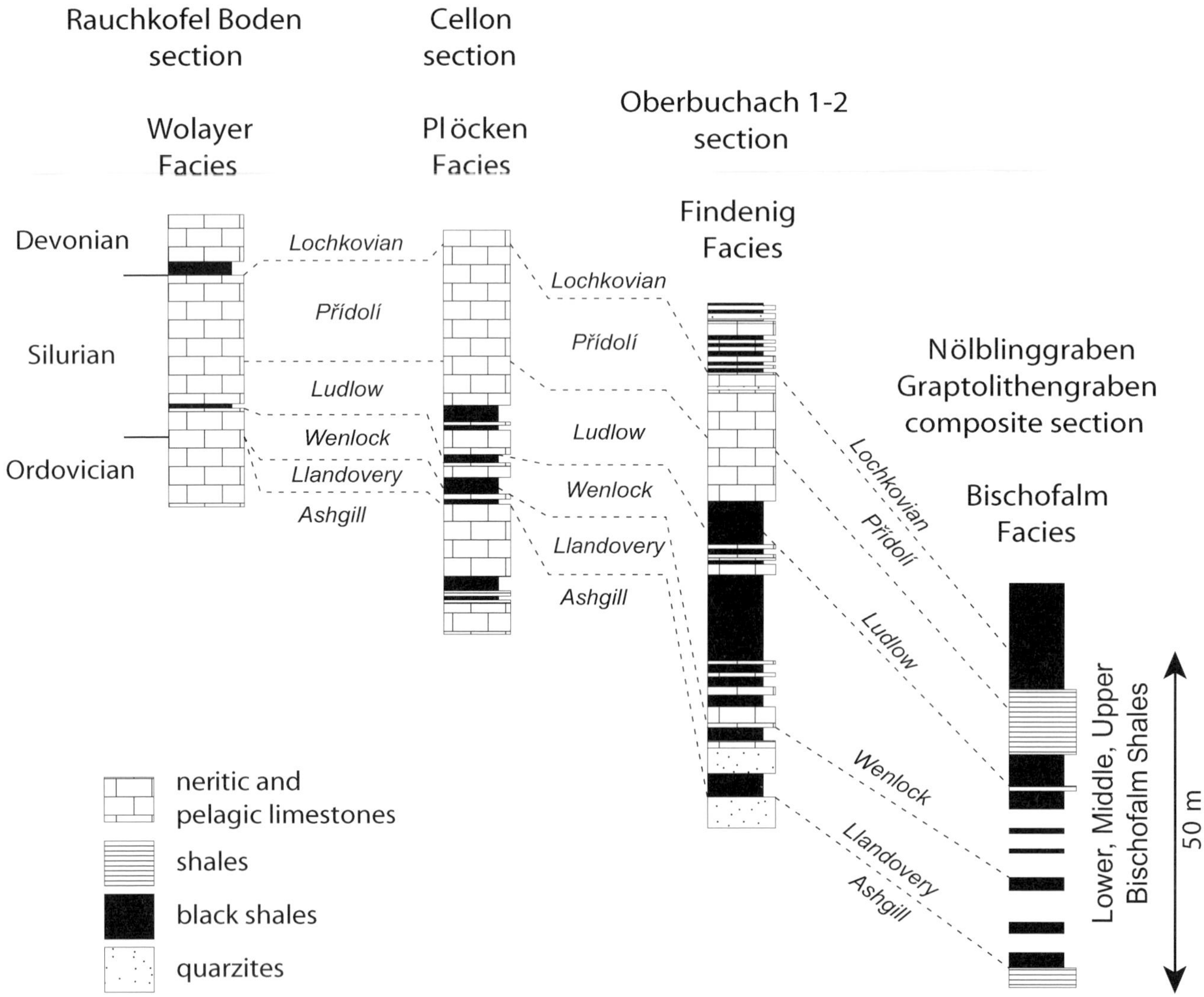

Fig. 6.21. Lithology and lithostratigraphical column of the Silurian of the four different lithofacies of the Carnic Alps. The black shales include C_{org}-rich graptolite-bearing shales and cherts and C_{org}-rich carbonates of the Wolayer facies; other shales are C_{org}-poor. In the Nölblinggraben-Graptolithengraben composite section Lower Silurian sediments are not continuously exposed (after Wenzel 1997).

Zone, indicating that the uppermost Přídolí is in bed number 331 (Schönlaub 1980; Schönlaub & Kreutzer 1994; Kříž 1999*b*) and the Silurian–Devonian boundary is situated just above this level and below the beds containing *Scyphocrinites*. Preliminary palaeoecological and palaeogeographical analyses from the Rauchkofel Boden section indicate a shallow-marine depositional environment with significant current activity. Orthocone cephalopod conchs are orientated from SW to NE in the Kok Limestone and change to a NNE to SSW direction in the overlying Lochkovian suggesting significant changes in the oceanic circulation systems around the Silurian–Devonian boundary (O. K. Bogolepova in Schönlaub & Kreutzer 1994).

The **Plöcken facies** contains, from bottom to top (Fig. 6.23), the Plöcken Formation (calcareous sandstone, Hirnantian, 4.8 m); the Kok Formation (pelagic ferruginous nautiloid limestones, with shale interbeds at the base, upper Llandovery to Wenlock, 13 m); the Cardiola Formation (alternating black limestones, marls and shales, Ludlow, 3.5 m); the Alticola Limestone (grey and reddish), nautiloid-bearing limestones, Ludlow to Přídolí, 20 m); and the Megaerella Limestone (grey, fossiliferous limestones, Přídolí, 8 m). It forms part of a continuous Ordovician to Devonian succession where several small hiatuses were recognized reflecting sea-level changes within an overall shallow-marine to moderately deep marine environment (Schönlaub & Histon 1999). The type section is the classic 60 m thick Cellon section, well-known for its globally applicable Silurian conodont biozonation established by Walliser (1964), and refined and partly revised in other areas. According to Schönlaub (1985, 1988) and Schönlaub & Sheehan (2003) the Ordovician–Silurian boundary can be drawn between the Plöcken and Kok Formations. From the basal Kok Formation at least six graptolite and two conodont biozones are missing in the lower Silurian. Deposition began in the late Llandovery, within the *Pterospathodus celloni* Zone. Based on graptolites and conodonts, the Llandovery–Wenlock boundary should be placed between samples 11 and 12 of Walliser (1964) (Fig. 6.23) and the thickness of the Llandovery does not exceed *c.* 3 m (Schönlaub 1997). The boundary between the Wenlock and Ludlow is drawn between

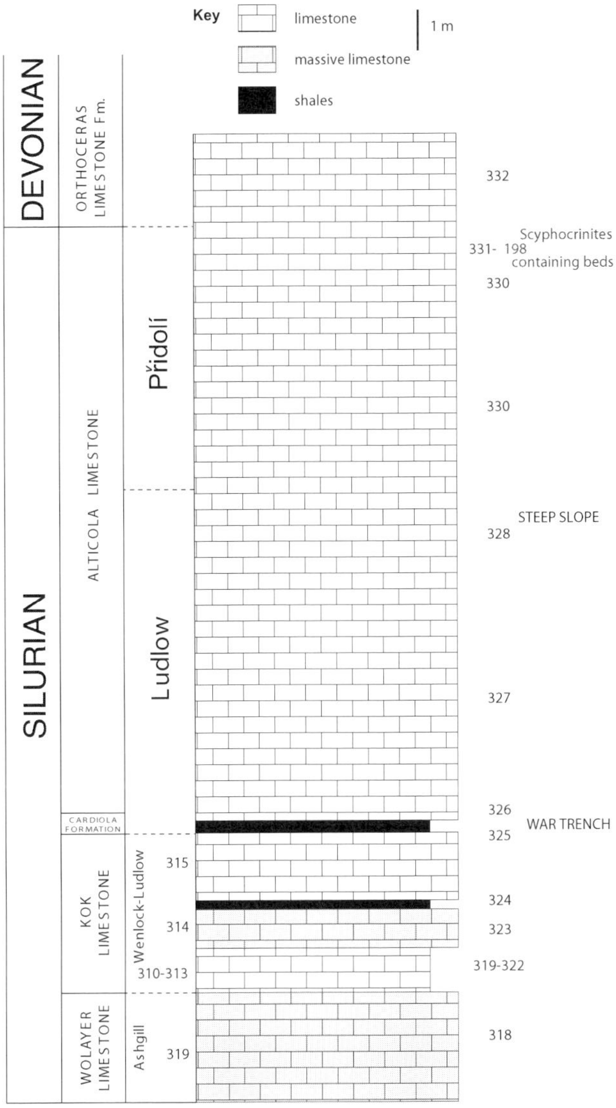

Fig. 6.22. Stratigraphic column of the Rauchkofel Boden section in the Wolayer facies. Numbers to the right of the column indicate sample numbers (after Ferretti *et al.* 1999).

samples 15B1 and 15B2 of Walliser (1964) through correlation with the stratotype in England. The thickness of the Wenlock in the Cellon section is not more than *c.* 5 m. Graptolites of the *Cyrtograptus rigidus* Zone in the shaly interbed between samples 12B and 12C clearly indicate a late Sheinwoodian age. Schönlaub (1994*b*) inferred that the lowest Homerian is missing.

The Ludlow–Přídolí boundary was placed, based on conodonts at 8 m above the base of the Alticola Limestone, a few centimetres above sample 32, by Walliser (1964) (see H.P. Schönlaub in Kříž *et al.* 1986) through comparison with the Bohemian sections and subsequently, confirmed by the occurrence, at that level of the basal Přídolí index graptolite *Monograptus parultimus* and of the basal Přídolí bivalve *Cardiolinka bohemica* community (Kříž 1999*b*). The thickness of the Ludlow is 16.45 m. The Silurian–Devonian boundary at Cellon is placed at the bedding plane between samples 47A and 47B of Walliser (1964). The Přídolí attains a thickness of *c.* 20 m at the Cellon section. The distribution of acritarchs, chitinozoans, brachiopods, bivalves, and taxonomically unrevised nautiloids and trilobites is summarized in Schönlaub & Kreutzer (1994).

A depositional gap around the base of the Silurian was caused by the latest Ordovician glacially induced sea-level fall, as in the Wolayer facies. Thereafter renewed sedimentation began in a moderately shallow-marine environment extending up into the earliest Wenlock. Bioturbated wackestones with algae and lumachelles indicate a very shallow-marine to intertidal environment at the base of sample 11 of Walliser (1964) (Fig. 6.23). Later in the Wenlock a progressive deepening occurred interrupted, however, at a sequence boundary by a hiatus at the Wenlock–Ludlow boundary. During the deposition of the Cardiola Formation, a pelagic offshore environment is indicated by the presence of radiolarian-bearing, black, marly interbeds and pelagic limestones with a diverse *Cardiola docens* community and *Cardiola pectinata* subcommunity (Kříž 1999*b*). The overlying Alticola Limestone reflects stable conditions in a pelagic setting that ended with a short regressive pulse with laminated grainstones and lumachelles recorded at a possible sequence boundary (Walliser 1964). A further deepening trend at the base of the Megaerella Limestone can be assumed (for details see Kreutzer in Schönlaub & Kreutzer 1994; Schönlaub & Histon 1999; Histon *et al.* 1999).

The **Findenig facies** (Figs. 6.20–6.24) is intermediate between the deeper shelf and the starved basinal environment and comprises interbedded, black, graptolitic shales, marls and blackish limestone beds. It is best developed in the Oberbuchach section where fossils, other than graptolites and conodonts, are very rare. The 10 m thick Hirnantian siliciclastic Plöcken Formation contains the graptolite ?*Normalograptus normalis* *c.* 1 m below its top; it is locally covered at its base by quartzose sandstones and higher up by a mixed argillaceous–calcareous lithology referred to as the Nölbling Formation (*c.* 50 m) of Llandovery to early Ludlow age. The lower siliciclastic part of the formation is overlain by an unnamed unit of interbedded laminated pyritic sandstones, bedded black cherts and black shales with a graptolite fauna of the *Coronograptus gregarius* and the *Demirastrites triangulatus* zones (early Aeronian). It is not yet clear whether the lower Llandovery is missing or simply unfossiliferous at this section. Higher up in the succession, a second horizon of graphitic sandstones with a late Llandovery age occurs. This is inferred from the presence of diagnostic conodonts of the *Pterospathodus celloni* Zone in limestones overlying this siliciclastic interval. The limestones are overlain by an alternating sequence of dark argillaceous limestones, black graptolite shales and cherts. Index conodonts occur only from the uppermost Llandovery to the Wenlock while graptolites occur over a larger range from the uppermost Llandovery to the Ludlow. In the overlying shales, graptolites occur at several levels indicating the *Monograptus riccartonensis* Zone (Sheinwoodian) up to the *Neodiversograptus nilssoni* Zone (lower Gorstian). At the Oberbuchach section the Wenlock–Ludlow boundary is placed some 40 m above the base of the graptolite-bearing sequence starting near the base of the Silurian. Strata containing the remaining part of the Ludlow and Přídolí are up to 20 m thick. This interval consists of lithologically distinct grey, and almost unfossiliferous pyritiferous limestones with a characteristic weathered surface (Schönlaub 1980).

The stagnant-water graptolitic **Bischofalm facies** is represented by 60–80 m of black siliceous shales, black cherty beds, and clayey alum shales containing abundant graptolites. The purely graptolitic facies is best exposed in the so-called 'Graptolithengraben' north of the Obere Bischofalm in the central Carnic Alps. Graptolites and a few conodonts on bedding planes

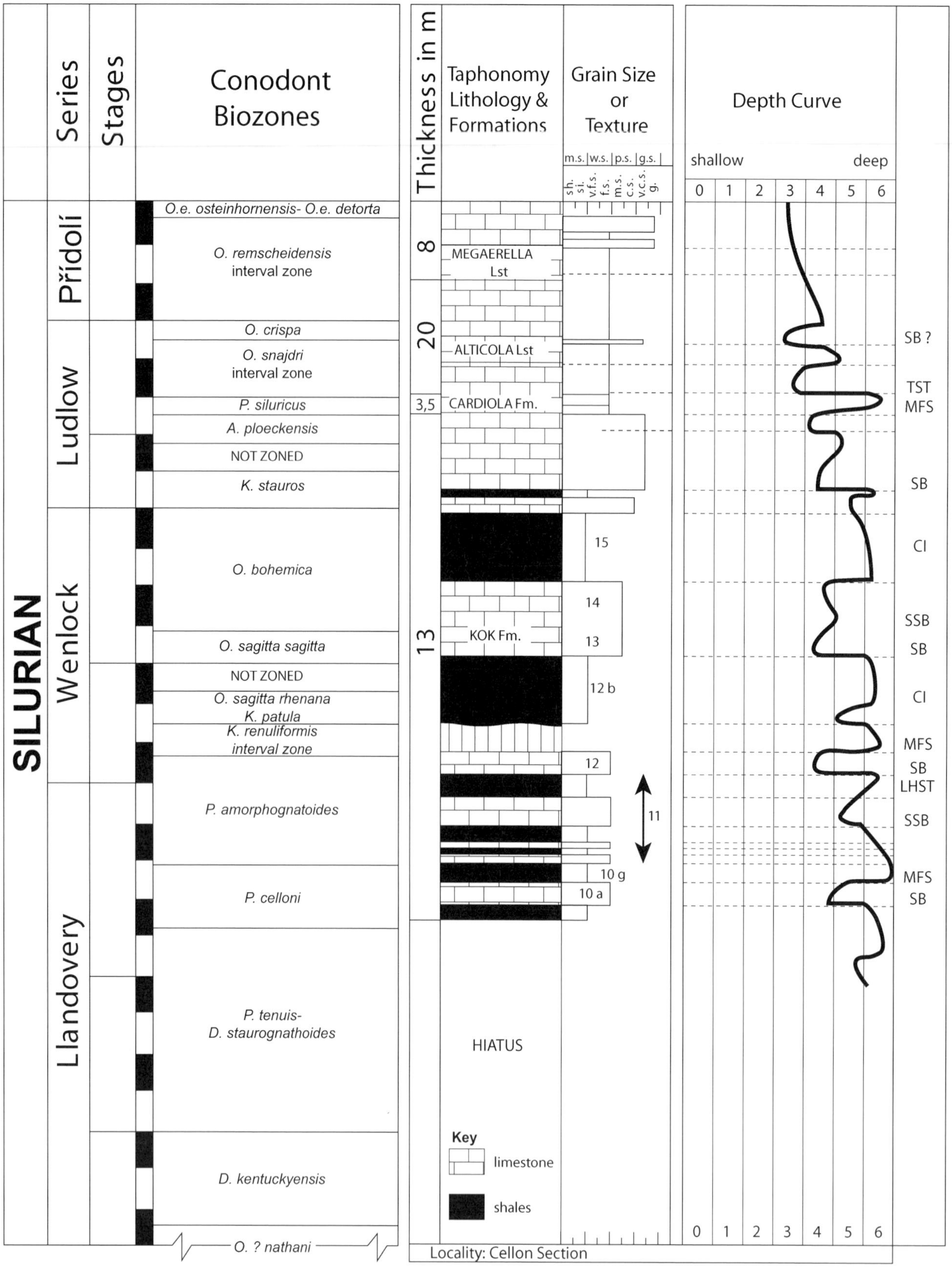

Fig. 6.23. Conodont stratigraphy, lithology, grain size, significant taphonomic features, depth curve and elements of the sequence stratigraphy of the Silurian of the Cellon section (modified from Schönlaub 1997). Abbreviations in grain size/texture column: m.s., mudstone; w.s., wackestone; p.s., packstone; g.s., grainstone; sh., shale; s.i., silt; v.f.s., very fine sand; f.s., fine sand; m.s., medium sand; c.s., coarse sand; v.c.s., very coarse sand; g., gravel; numbers represent Walliser samples. Abbreviations in depth curve column: CI, contourite interval; LHST, late highstand systems tract; MFS, maximum flooding surface; SB, sequence boundary; SSB, subsequence boundary; TST, transgressive systems tract (modified from Schönlaub 1997).

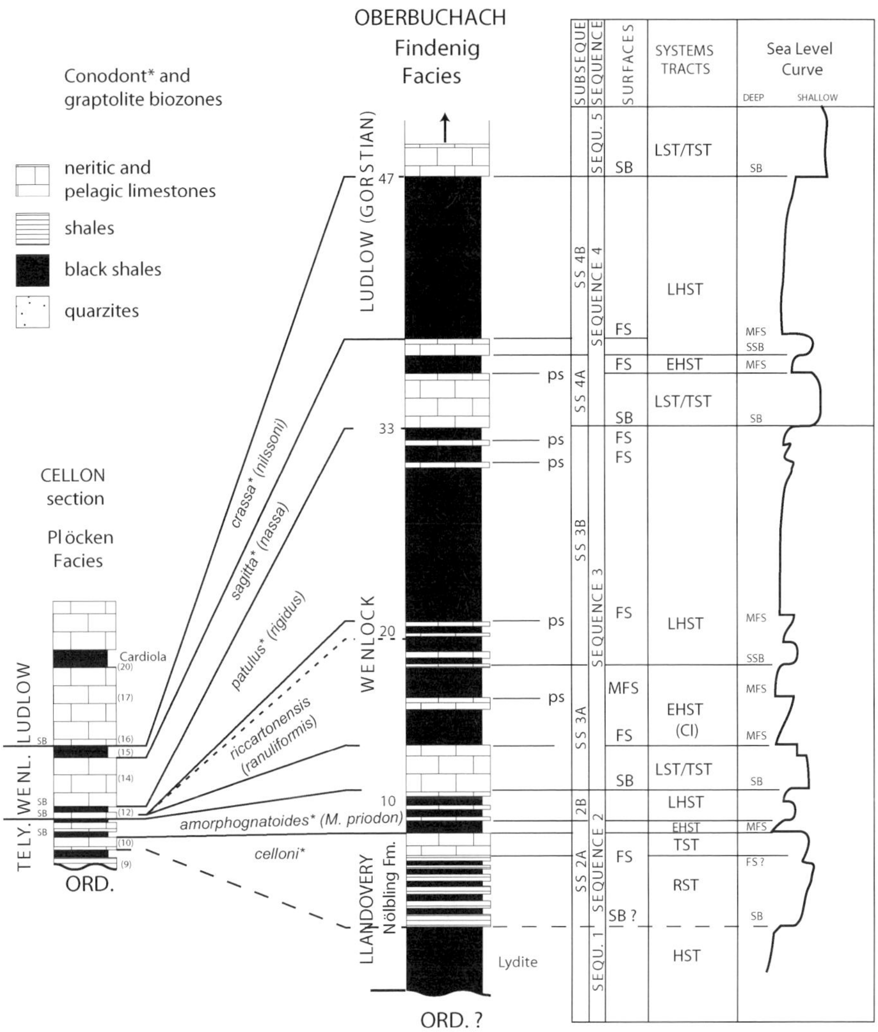

Fig. 6.24. Correlation and sequence stratigraphical interpretation of the Llandovery to Lower Ludlow of the Carnic Alps (after Brett & Schönlaub, 1998). Abbreviations: CI, contourite interval; EHST, early highstand systems tract; FS, flooding surface; HST, highstand systems tract; LHST; late highstand systems tract; LST/TST, lowstand systems tract and transgressive systems tract; MFS, maximum flooding surface; SB, sequence boundary; SSB, subsequence boundary; SS2A, subsequence 2A; TST, transgressive systems tract (after Brett & Schönlaub 1998).

are the only fossils found and have been known in the Alps since their discovery by Stache (1872). The graptolites are common in many layers both in the alum shales and in the siliceous shales. Some intervals, however, are almost barren of graptolites. The best exposed, and tectonically least disturbed, section is the main section (Hauptprofil), which has been studied in great detail since 1965 (e.g. Jaeger 1975; Flügel *et al.* 1977; Jaeger & Schönlaub 1977, 1980, 1994; Schönlaub 1985). Jaeger (1975) subdivided the Bischofalm facies into three units, the Lower, Middle and Upper Bischofalm Shales. They are included in a monotonous sequence of interbedded radiolarian-bearing cherts and alum shales. The cherts dominate the Llandovery and Wenlock while the shales prevail in the upper part of the succession. The green and grey shales of the Middle Bischofalm Shales yield only a few graptolites in very thin layers (H. Jaeger in Flügel *et al.* 1977). An almost 20 m thick undisturbed sequence is present in the main section and covers the interval from the *Pristiograptus ludensis* Zone of the uppermost Wenlock to the Lower Devonian *Monograptus hercynicus* Zone. The thickness of the composite section of graptolite-bearing Silurian to Lochkovian rocks ranges from 50 to 100 m. The very complete graptolite biozonal succession and the condensed nature of the succession are probably due to a very low, but nevertheless continuous, rate of deposition. The environmental conditions were anoxic or strongly dysaerobic except during the short interval corresponding to the deposition of the Middle Bischofalm Shale (Jaeger 1975; Flügel *et al.* 1977). The Silurian–Devonian boundary is within a homogenous black-shale facies of the Upper Bischofalm Shale, with no physical break at the boundary (H. Jaeger in Flügel *et al.* 1977). However, a distinct change in facies from green and grey shales to black shales preceded by one graptolite biozone marks the faunal change at the boundary. Additionally the Middle Bischofalm Shale occupies the same stratigraphic position as the non-graptolitic Ockerkalk of Thuringia.

The abundant presence of K-bentonite horizons in the Llandovery–middle Ludlow successions of the Carnic Alps is similar to that recorded in the British Isles, Sweden, Canada and North America, and documents widespread volcanism related to the closing of the Iapetus Ocean. Silurian K-bentonites of Přídolí age may be comparable with those described from Podolia

(Ukraine) for which a source area in the Rheic Ocean has been indicated (Bergström *et al.* 1998; Huff *et* al. 1998, 2004).

The Grauwacke Zone

The locations of the outcrop areas described below are given for their respective Austrian provinces (Länder), for example Styria or Salzburg Land. The Grauwacke Zone which contains a thick Lower Palaeozoic succession extends over a great distance in a series of different nappes; in the east (Steiermark) it comprises the eastern Grauwacke Zone and in the west (Salzburg Land and Tyro) the western Grauwacke Zone. The Silurian of the Grauwacke Zone in the Tyrol and Salzburg Land is fairly well known, but less so in Steiermark. The bio- and chronostratigraphy is based on conodonts and graptolites (Schönlaub & Kreutzer 1994).

In the eastern Grauwacke Zone, the 50 m thick Silurian succession shows two successive facies: (1) a crinoid- and nautiloid-bearing limestone with local intercalations of Llandovery basic volcanics, overlain by (2) black graptolite shales. These two facies pass laterally and vertically into interbedded limestones and shales, overlain by upper Ludlow and Přídolí limestones (Schönlaub 1979; Schönlaub & Heinisch 1993).

In the western Grauwacke Zone close together and more specifically in the Kitzbühel Alps two other lithofacies occur (Heinisch 1988). Both facies are incorporated in the two nappes: the Wildseeloder Unit in the north and the Glemmtal Unit in the south. Lithologically the Silurian comprises black shales, locally with graptolites, cherts, siliceous pelagic limestones to condensed cephalopod limestones, and dolomitic rocks (a variation of the shelly fauna facies).

In the Wildseeloder Unit the thick Upper Ordovician Blasseneck Quartz porphyry is overlain by several metres of Silurian deposits. Two important sections are representative for the two facies: the classic graptolite-bearing Lachtal-Grundalm section and the Spießnägel section. The first of these is located near the village of Fieberbrunn, east of Kitzbühel. There are no biostratigraphic data to locate precisely the Ordovician–Silurian boundary (Schönlaub & Kreutzer 1994). The less than 5 m thick middle and upper Llandovery 'Lydit-Kieselkalk-Komplex' comprises a mixed pelagic shale-limestone succession. The basal chert interval comprises black, massive cherts (lydites in the Alpine terminology) with radiolarian-bearing dolostones and reddish, cherty limestones that grade vertically into crinoidal limestones. The fauna consists of ostracods, foraminifers, brachiopods, radiolarians, conodonts and echinoderms. Additionally, bivalves, solitary corals, trilobites and orthocone nautiloids occur in the lower 1.40 m of the crinoidal limestone. The basal 2.10 m of the crinoidal limestone succession is assigned to the *Pterospathodus celloni* Zone (Telychian) and the upper part to the *Pterospathodus amorphognathoides* Zone (Llandovery–Wenlock transition). It is overlain by a 5 m thick Dolomit-Kieselschiefer-Komplex (Bedded Dolostone-Chert Formation) (Mostler 1966) a lithology resembling the Nölbling Formation of the Finderig facies in the Carnic Alps. The only identifiable graptolites occur in a horizon near the top of the Dolomit-Kieselschiefer-Komplex (Jaeger 1978); *Bohemograptus bohemicus* are characteristic of the *Neodiversograptus nilssoni* Zone (lower Gorstian).

The Spießnägel section, south of the village of Kirchberg and west of Kitzbühel, is one of the few sections with an exposed transition from presumably Upper Ordovician greywackes into the basal Silurian. According to Al-Hasani & Mostler (1969), the Silurian begins with 0.70 m of bioturbated mudstones with varying amounts of siliciclastic and tuffaceous material grading into 0.15 m of wackestones. Both units contain diagnostic conodonts of the *Pterospathodus celloni* Zone (late Llandovery). It is succeeded by 1.10 m of limestones with interbedded shales and thin limestone lenses, containing packstones with thin hash layers of bivalves, brachiopods, ostracods and echinoderms. The limestones are in turn overlain by greyish laminated dolostones assigned to the *Pterospathodus amorphognathoides* Zone and the lower Wenlock *Kockelella patula* Zone.

In the western part of the western Grauwacke Zone (Tyrol), the Dolomit-Kieselschiefer-Komplex is covered in turn by upper Silurian platform carbonates with dolomitic rocks and magnesite. According to Mostler (1966), the base of these carbonates can be assigned to the *Polygnathoides crassa* Zone near the Wenlock–Ludlow boundary or to the base of the overlying *Ancoradella ploeckensis* Zone near the Gorstian–Ludfordian boundary. No record of the Přídolí is yet available, although it may be present in the recrystallized dolostones.

The Glemmtal Unit comprises, in contrast to the Wildseeloder Unit, a mainly siliciclastic succession of more than 1000 m known as the Wildschönau Group with two formations. Locally, the Klingler Kar Formation comprises up to 50 m of intercalated condensed pelagic limestones, marls, interbedded cherts, siliceous shales and basalts. The lower part of Klingler Kar Formation is dated as upper Silurian by conodonts. The facies laterally grades into the turbiditic facies of the Löhnersbach Formation which is not yet dated.

Graz area

The best outcrops for the Palaeozoic of the Graz area are in the Rannach Nappe, the uppermost nappe of the Graz Thrust Complex. The Silurian is dominated by alkaline mafic lavas and volcaniclastics, suggesting an initial rift stage. The latter are overlain by Upper Silurian and Devonian carbonates. The Silurian of the Graz area is best displayed in the Eggenfeld section.

According to Fritz & Neubauer (1988) and Neubauer (1989), sedimentation in the Silurian Kehr Formation was controlled mainly by volcanism. During the early Ludlow, the eastern depositional area was characterized by a proximal shallow-marine setting with lavas and coarse lapilli tuffs, whereas the western area was characterized by a distal facies with intercalations of lapilli-rich beds, agglomerates, shales, and pelagic limestones. The Kehr Agglomerate consists of 1 to 3% quartzite, dolostones, chert and reworked limestones clasts. During late Silurian times, the volcanic centres were covered by fossiliferous carbonates that include a *c.* 4 m thick bedded dolostone with lenses of crinoid-, brachiopod-, trilobite- and nautiloid-bearing dolomitic limestones interbedded with tuffs and tuffitic shales. Based on conodonts, a late Ludlow (Ludfordian) to Přídolí age has been determined for the upper part of this succession (Ebner 1994).

Fossils in the Eggenfeld section allow dating of the Silurian volcanic activity in the eastern Alps. Even with poor outcrop conditions a well-constrained lithostratigraphy has been established (Ebner 1994). The succession commences with massive green basalts that interfinger with pinkish to greenish tuffs with graptolites, which are overlain by dark dolostones (unit D/1) containing crinoids, brachiopods, nautiloids and tabulate corals (*Favosites* sp.) and are succeeded by tuffs and tuffaceous shales. Further up-section a second unit (unit D/2) of dark dolostones contains lens-like accumulations of crinoids, brachiopods, trilobites, nautiloids and rare corals (e.g. *Syringaxon* sp.). The uppermost Silurian (unit D/3) consists of tuffs and tuffaceous shales again with intercalations of dark dolostones containing the same fossil groups as the underlying unit. The biostratigraphy is based on the fairly abundant conodonts and brachiopods present in all of the calcareous levels. Diagnostic conodonts in the lower dolostone (unit D/1) immediately above the basalts indicate a Ludfordian age for the end of basaltic volcanism. This age is confirmed by the graptolite *Bohemograptus bohemicus tenuis* in the basal volcaniclastic layer which interfingers with the green basalt (Hiden 1996). The index species of the *O. snajdri* Zone in association with *Ozarkodina remscheidensis eosteinhornensis* in the second carbonate level (D/2) also indicates a Ludfordian age.

Similar environmental conditions are suggested for the upper Silurian rocks of the other nappes in the Graz area, where the pelagic, nodular limestone development persisted from the late Silurian through the Devonian. Indeed both carbonate production and volcanic activity subsequently increased in Devonian times.

Burgenland

During the last 20 years new fossil assemblages have been discovered from weakly metamorphosed interbedded limestones, dolomites and shales in the south of Burgenland, eastern Austria (Fig. 6.20). Based on index conodonts, an Upper Silurian portion can be clearly identified underlying a Lower Devonian rugose- and tabulate-coral-bearing carbonate succession suggesting a similar development to that of the Graz area (Schönlaub 1984, 1994*a*). Of particular interest is the first record of serpulid tubes in the late Silurian carbonate beds near Kirchfidisch and the recognition of Silurian–Devonian boundary beds with conodont index species by Suttner & Luckender (2004) and Suttner *et al.* (2004).

The Gurktal Nappe

The Gurktal Nappe, situated in central Carinthia and southern Styria, is composed of several hundred metres of volcanic and siliciclastic rocks with intercalated limestones. The Lower Palaeozoic of the Gurktal Nappe system is characterized by volcanic rocks occurring at different times, and which are of varying intensity and of different geochemical character, reflecting different palaeotectonic settings (Loeschke & Heinisch 1993). The Silurian part of the succession comprises coral-bearing bioclastic limestone lenses at the Llandovery–Wenlock transition and locally 5–10 m thick limestones and dolostones of late Silurian age. Facies development suggests a subdivision into a carbonate-rich and a carbonate-poor facies. However, bad exposures and poor fossil control do not allow the reconstruction of a composite Silurian section (Buchroithner 1979; Ebner *et al.* 1990; Schönlaub & Heinisch 1993).

Montagne Noire, France

The Palaeozoic sediments of the Montagne Noire are widely exposed and richly fossiliferous (Feist in Robardet *et al.* 1994). Silurian strata, however, are less well represented since they occur only in restricted areas in the south of the Albigeois area, the region around the city of Albi, and in the Cabrières Klippes (Fig. 6.25). Elsewhere, for example, in the 'nappes' area, Silurian deposits are absent and lower Lochkovian nearshore strata rest unconformably on the Lower Ordovician (e.g. Mont Peyroux). Thus, in the area of the nappes, pre-Devonian erosion removed a large part of the Lower Palaeozoic succession (Feist & Echtler in Feist *et al.* 1994). No recent studies have been carried out on the black carbonates and shales cropping out south of the Albigeois age, which have yielded orthocone cephalopods and bivalves (Chaubet 1937), as well as conodonts (Centène 1977) of Silurian age.

The most complete Silurian sequence of the Montagne Noire is exposed in its SE part (Chaubet 1937), in the Falgairas and Laurens plateaus belonging to the Cabrières Klippes (Feist & Echtler in Echtler *et al.* 1994). This Silurian succession, including Llandovery to Přídolí strata, is interpreted as part of a huge olistolith preserved in a Visean matrix (Engel *et al.* 1982). The setting of this gravitationally transported Silurian material corresponds to the early activity of the Variscan Orogeny. The geographical origin of this allochthonous Silurian unit would have been situated north of the area from which the 'nappes' themselves were derived. The latter are presently devoid of any Silurian deposits but Silurian black shales are known from

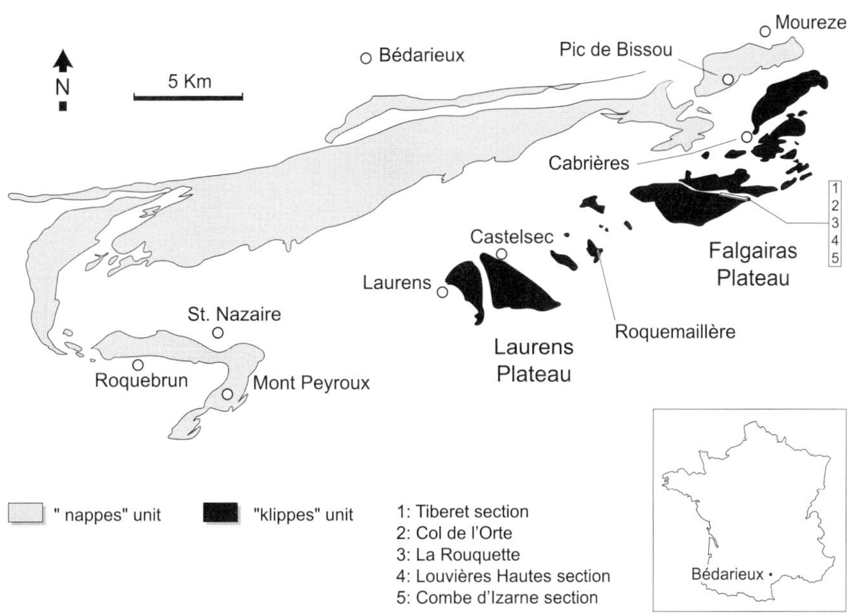

Fig. 6.25. Map of the Silurian of the Montagne Noire (after De Bock 1982).

localities further west in the subsurface of Aquitaine (Paris & Le Pochat 1994).

In the klippes of the Cabrières area, the Silurian succession commences with the Roquemaillère Black Shale (*c.* 50–70 m thick). These condensed anoxic deposits include limestone nodules and are fossiliferous. The Roquemaillère Black Shale ranges from mid-Llandovery to early Wenlock in age, with the oldest recorded graptolites in the area of late Aeronian age. The limestones and shales of the Falgairas Formation (50 to 60 m) overlie these fairly deep-water black shale deposits. The Falgairas Formation begins with platy to nodular limestone beds of Wenlock age including minor rare calcareous black shale interbeds. Ludlow strata are less fully developed. They are represented by rare metres of black calcareous mudstones and nodules with rare bedded or nodular limestones. Sedimentation was more active during the Přídolí (Feist 1977; De Bock 1982) as limestones and dolostones progressively replaced the black shales. The upper part of the Falgairas Formation records a shallowing trend, concomitant with the progressive input of detrital quartz (Feist 1977). Sandy limestones passing upward into calcareous sandstones replace the lower Přídolí limestones or dolomitic limestones. The local Silurian succession ends with the upper Přídolí Falgairas Sandstone (9 m) (Fig. 6.26). This nearshore sandy body is exposed on the Falgairas Plateau, in the Tiberet and Louvières Hautes sections (De Bock 1982) (Fig. 6.25).

Correlation

In addition to the numerous benthic fossils including trilobites, brachiopods, bivalves, ostracods and crinoids (Feist 1977), the Silurian strata of the Montagne Noire also yield a nectic and pelagic fauna (e.g. nautiloid orthocones, graptolites, conodonts, chitinozoans and acritarchs). Several Silurian graptolite biozones, i.e. *Stimulograptus sedgwickii*, *Neodiversograptus nilssoni*, *Lobograptus scanicus*, *Bohemograptus bohemicus* and *Monograptus ultimus*, in the Cabrières area (e.g. in the Roquemaillère and Combe d'Izarne sections) have been identified by Centène & Sentou (1975). Diagnostic conodont assemblages are also present all along the Silurian sequence in the Cabrières Klippes. Most of the conodont zones of the late Llandovery (i.e. *Pterospathodus celloni*, *Pterospathodus amorphognathoides)*, Wenlock (e.g. *Kockelella patula*, *Ozarkodina sagitta*), Ludlow (i.e. *Ancoradella ploeckensis*, *Polygnathoides siluricus*, *Ozarkodina snajdri*, *Ozarkodina crispa*) and Přídolí (*Ozarkodina remscheidensis eosteinhornensis*) have been documented by Feist & Schönlaub (1974), and by Centène & Sentou (1975) in the calcareous beds and nodules of the Roquemaillère and Falgairas strata (Fig. 6.25). Deflandre (1942, 1946) reported organic-walled microfossils from the Roquemaillère Limestone (calcareous nodules of Wenlock age and undifferentiated limestone samples from the Roquemaillère section, and from the Ribouryrel Brook). These pioneer studies of acritarchs and chitinozoans gave no accurate age assignment, but later investigations by De Bock (1982) on other sections in the Falgairas and Laurens plateaus recorded chitinozoan assemblages with *Urnochitina urna* and *Eisenackitina bohemica*. These two species are the chitinozoan index fossils for the Přídolí (Verniers *et al.* 1995) and for the early Lochkovian (Paris *et al.* 2000) respectively. Thus, they allow the Silurian–Devonian boundary to be fairly precisely located in the Cabrières Klippes succession (De Bock 1982) (Fig. 6.26).

Among the benthic fauna recorded in the Silurian of the Montagne Noire, special attention has been paid to the numerous and diverse bivalves recovered from Wenlock to Přídolí strata. These bivalves allowed accurate correlation at a specific level with contemporaneous assemblages of the Prague Basin (Kříž 1996). Moreover, five distinct bivalve-dominated communities (i.e. *Cardiola* community group) have been identified in the Silurian of the Cabrières Klippes. These are also recognized in the Prague Basin and the Carnic Alps as well as in Sardinia, and the Armorican Massif (Kříž & Paris 1982). This similarity would suggest that there were close palaeobiogeographical relationships between these different regions during the Silurian. The closest relation, however, was with Sardinia, which shared three bivalve communities with the Montagne Noire (Kříž & Serpagli 1993). The bivalve diversity was higher in the Prague Basin than in Sardinia or the Montogne Noir region which were typical northern Gondwana areas (Kříž 1999*a*, *c*).

In contrast to numerous other French Silurian localities, which are devoid of trilobites (Paris in Kříž *et al.* 2003), is the occurrence of a fairly diverse trilobite fauna in the Falgairas Formation limestones and shales (see Feist 1977). The trilobites are well represented in the Wenlock part of the formation, and to a lesser extent in the Ludlow and Přídolí beds.

Lithofacies and biofacies

No sequence stratigraphical analysis is available for the Silurian of the Cabrières Klippes and therefore the regional sea-level fluctuations can only be broadly evaluated based on the lithological evidence, and on the composition of the recorded faunal assemblages (Paris, fig. 18 in Kříž *et al.* 2003). Black shales are usually regarded as representing condensed sedimentation, under low energy conditions, and probably in fairly deep marine environments. This is consistent with the prevailing occurrences of the pelagic and epipelagic faunas in these black shales from the late Llandovery to the early Přídolí. However, limestones with bivalve-dominated communities (Kříž 1996) indicate episodic shallowing events, and/or the occurrence of temporary currents interrupting or ameliorating the anoxic to highly dysaerobic conditions prevailing on the local sea bottom. As in many other northern or peri-Gondwana regions, the late Silurian regression is recognized in the sedimentary record of the Montagne Noire. The progressive shallowing is recorded by the fauna with dominantly benthic assemblages, and for the conodonts by the replacement of the ozarkodinid species with the icriodontid species, which are regarded as representative of more shallow-marine environments (Feist 1977). Hydrodynamic energy increased significantly during the Přídolí, corresponding with increased input of clastic material; the acme was the deposition of the Falgairas Sandstone in a high-energy environment.

Avalonia

In Central Europe, the Silurian rocks of Avalonia only crop out in the Brabant Massif and in the Condroz Inlier of Belgium, in the Ebbe and Remscheid inliers of Germany and in one location in the Boulonnais Inlier, northern France. In the subsurface, the Silurian is more widely present, with many boreholes in the Brabant Massif (Legrand 1969; De Vos *et al.* 1993), in northernmost France in the Artois-Boulonnais structural high (CFP. *et al.* 1965, 1966; Legrand 1969), in the subsurface of East Anglia, UK (see Molyneux 1991) and in the Dutch sector of the Southern Bight of the North Sea (Van Adrichem, pers. comm. 1994; Cocks *et al.* 2003).

Belgium

The **Brabant Massif** is a mostly concealed basement unit with outcrops only in its southeastern part (Fig. 6.27). The Brabant Massif forms a part of the larger Anglo-Brabant foldbelt

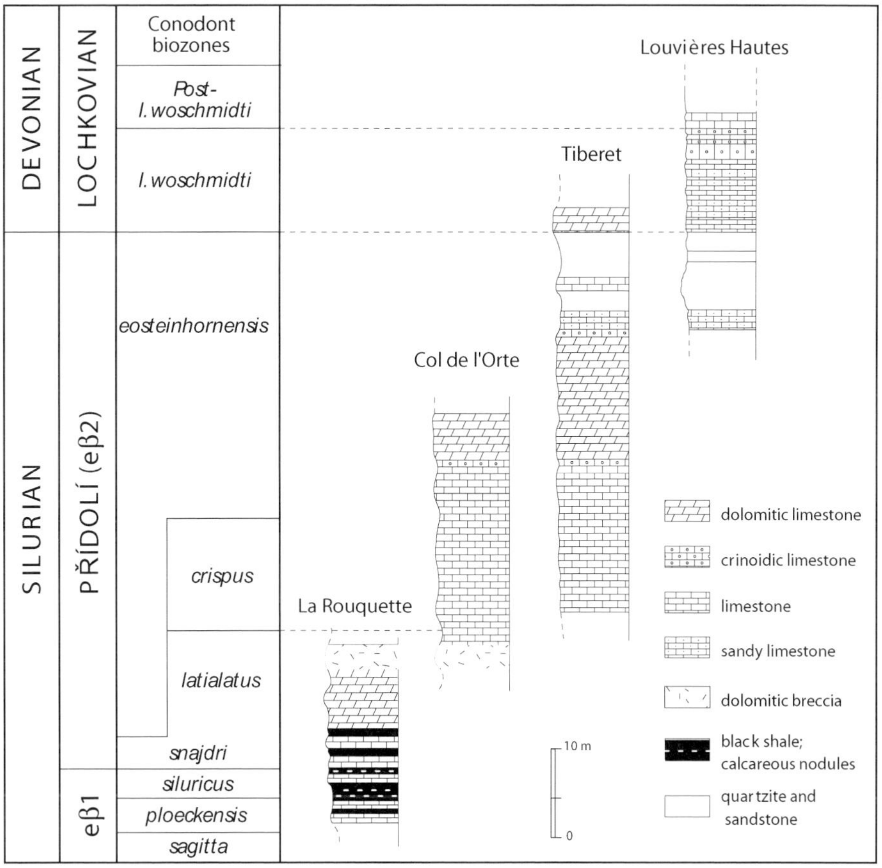

Fig. 6.26. Lithostratigraphy of the Silurian of the Montagne Noire (after Feist 1977).

(Pharaoh *et al.* 1993) later called the Anglo-Brabant Deformation Belt (Winchester *et al.* 2002) The rocks were folded, cleaved and faulted during the Silurian to early Eifelian Brabantian Orogeny (a late phase of the large Caledonian Orogeny) and metamorphosed to the zeolite or greenschist facies. One area escaped cleavage formation and metamorphism, namely the SW subcrop part. The orogeny resulted in the formation of an anticlinorial structure for the Brabant Massif (Van Grootel *et al.* 1997; Verniers *et al.* 2002). Silurian rocks are present along the southern rim: cropping out in the eastern part and in the west in numerous boreholes in southern Flanders (southern part of the provinces of East and West Flanders) and northern France. There is also poorly documented Silurian in a subcrop area with few boreholes in the NW, called the central rim, and in the north of the Brabant Massif (i.e. the northern rim). Further north, one borehole is reported in the Dutch sector of the North Sea and several extending below the upper Palaeozoic cover: Loenhout, Booischot and Kortgene (Legrand 1969; Verniers & Van Grootel 1991). The Silurian of the Brabant Massif was extensively reviewed by Verniers & Van Grootel (1991) and its lithostratigraphy formally revised in Verniers *et al.* (2001) (Fig. 6.27).

Graptolites are almost the only macrofauna present in the Llandovery to lower Ludfordian rocks. Most of the graptolite biozones as defined by Rickards (1976) have been recognized (Martin 1969). Organic microfossils, such as acritarchs and chitinozoans, have been used to date thick units when graptolites are missing. The following chitinozoan biozones have been demonstrated: *Belonechitina postrobusta*, *Spinachitina maennili*, *Eisenackitina dolioliformis*, *Angochitina longicollis*, *Margachitina margaritana*, *Cingulochitina cingulata*, *Sphaerochitina lycoperdoides* and *Angochitina elongata* (Martin 1969; Verniers 1983; Van Grootel 1990; Louwye *et al.* 1992; Verniers *et al.* 1995). These biozones were calibrated in relation to the graptolite biozonation by Verniers (1999).

Two different sedimentary environments are present in the Brabant Massif: (1) a deep shelf with graptolitic shales in the western part of the southern rim, and (2) a slope environment with proximal and distal turbidites in the outcrop area and along the central and northern rims. The Condroz Inlier (see below) is situated in the prolongation of the first area. Together they could have formed a large gently subsiding shelf basin with a highly subsiding slope basin filled with turbidites to the north (Verniers *et al.* 2002).

The Deerlijk and Lust formations are redorded in the subsurface of southern East and West Flanders. The Deerlijk Formation (more than 83 m thick) contains greenish and dark grey, homogenous or laminated shales or mudstones and clayey sandstones, interpreted as alternating distal turbidites and/or (laminated) hemipelagites. The sediments are often rich in graptolites with rare orthocone nautiloids, sponge spicules, foraminifers and benthic algae (Martin 1974). The Lust Formation (more than 54 m thick) is similar but contains up to 45% clayey fine-grained sandstones.

The Silurian (Rhuddanian to Gorstian) lithostratigraphy of the slope environment with proximal and distal turbidites is best studied in the outcrop area (see reviews in Verniers *et al.* 2001;

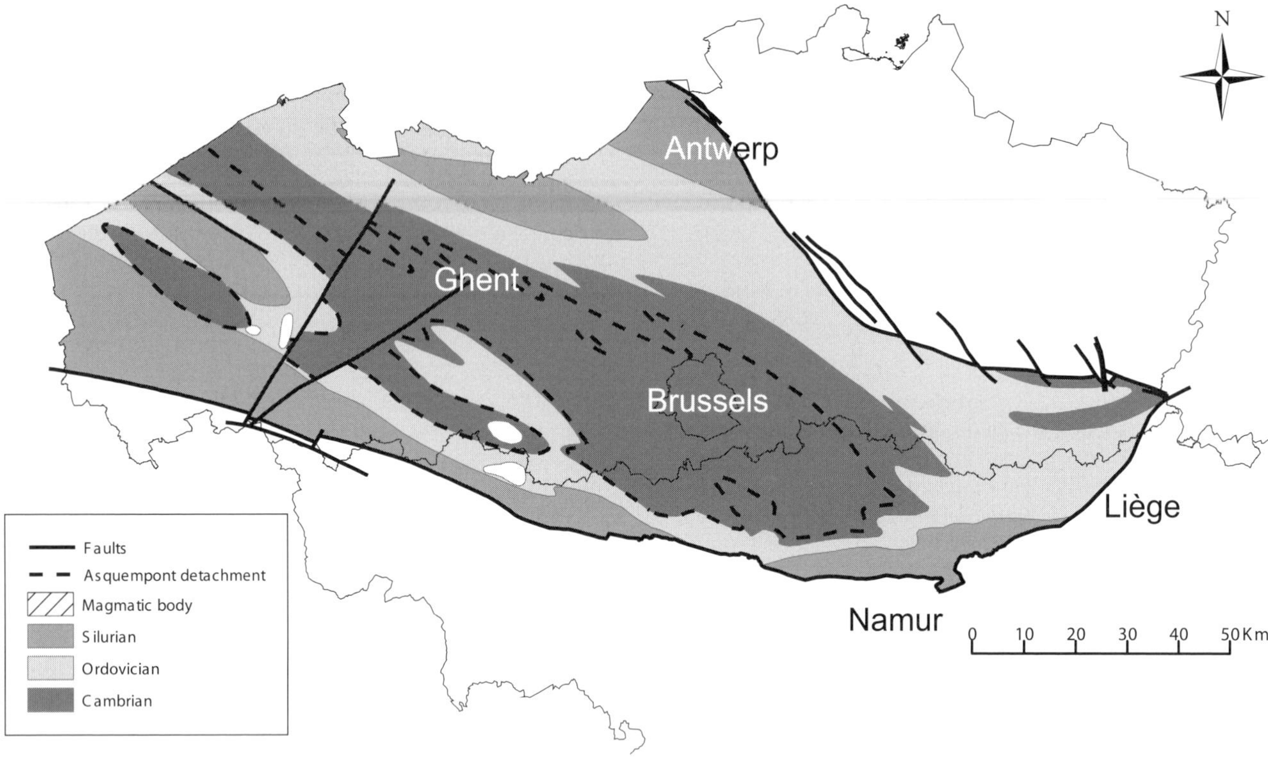

Fig. 6.27. Geological subcrop map of the Brabant Massif (Belgium) (modified after De Vos *et al.* 1993 and Piessens *et al.* 2005).

Herbosch *et al.* 2002). The lithology is mostly clayey to sandy siltstones or mudstones with rare sandstones. The colour is greenish and grey in the Llandovery and medium to dark grey in the Wenlock and Ludlow. Dark grey laminated hemipelagites, which may contain graptolites, are often intercalated. The Silurian comprises a non-turbiditic facies up to the mid-Telychian and above a Bouma-type turbiditic facies up to the Přídolí.

The Silurian succession commences with the greenish to dark grey mudstones of the Brûtia Formation (80 to 100 m thick), with some bioturbated levels; this is capped at the top by a white quartzitic tuff, locally weathered into kaolinite, with graptolites of the *Cystograptus vesiculosus* Zone in shales just above. These shales are finely bedded and together with laminated siltstones or fine sandstones form the Bois Grand-Père Formation (between 200 and 500 m thick), which is poorly dated to being mid-Rhuddanian to mid-Telychian in age. Bouma-type turbidites are first recorded in the mid-Telychian greenish grey to dark grey Latinne Formation (*c.* 180 m thick) and continue into the overlying medium to dark grey Hosdin Formation (*c.* 150 m thick) and the more than 626 m thick green Fallais Formation, all of which are late Telychian in age based on chitinozoans. The latter formation comprises several members, one which is a 26 m thick fining-upward volcaniclastic layer (Pitet Member). The overlying coarser-grained unit of the Corroy Formation is observed in many valleys, has a thickness of between 100 and 140 m and contains centimetre- to decimetre-scale sandstones, siltstones and shales (Herbosch *et al.* 2002). Graptolites from the *Cyrtograptus centrifugus*, *Cyrtograptus murchisoni, Monograptus riccartonensis* and *Cyrtograptus rigidus* zones have been reported (Legrand 1961; Verniers & Rickards 1978), together with rare *Conularia, Cardiola* sp., *Orthoceras* sp. and burrows (see Verniers & Van Grootel 1991). Thicker-bedded but lower-energy turbidites, always alternating with laminated hemipelagites, occur in the Les Vallées (more than 225 m), Vissoul (more than 30 m), Fumal (330 m) and Vichenet (more than 350 m) formations. These units are mostly composed of fine silty slate, clayey siltstones, and rare very fine-grained sandstones. The Vissoul Formation has a greater amount of siltstones or sandstones and the Vichenet Formation includes very thick, slightly calcareous slate units in between the laminated hemipelagites. These four formations are dated as mid-Wenlock to Homerian with chitinozoans (post-*Cingulochitina burdinalensis* interzone, *Eisenackitina lagena*, *Conochitina pachycephala*, *subcyatha* and *Sphaerochitina lycoperdoides* zones; Verniers 1999). Further to the west in the Senne valley the Petit Roeulx, Steenkerque and Froide Fontaine formations represent the lateral equivalents.

The lower Ludlow (Gorstian) is contained within the Ronquières Formation. It is more than 538 m thick and includes turbidites with intercalated laminated hemipelagites. Near the base of the formation 30 m of laminated hemipelagites with six metabentonite levels may represent the highest sea-level part of a megacycle. The top part of the Ronquières Formation is the only place in the Silurian of the Brabant Massif where complete Bouma sequences (Tabcde) occur. Graptolites and chitinozoans date the formation clearly as Gorstian, with the *Neodiversograptus nilssoni*, *Lobograptus scanicus* and possibly the *Pristiograptus tumescens* zones represented (Louwye *et al.* 1992).

One uppermost Gorstian to lower Ludfordian unit (the Bellegem Formation) and one unit (the Rekkem Formation) tentatively dated as Přídolí were described from a few boreholes in the SW part of the massif. They contain grey thin-bedded turbidites, often with burrowed horizons and fragments of the archaeostracan *Ceratiocaris* sp. (Van Grootel 1990).

To produce the observed pre-Givetian cleavage Michot (1978), Debacker (2001) and Verniers *et al.* (2002) suggested that sedimentation continued in the Brabant Massif after the Gorstian with deposition of another 4 to 6 km of upper Silurian sediments. Using reworked palynomorphs in the Lower Devonian south of the massif, Steemans (1989) postulated that sedimentation in the Brabant Massif could have continued until early in the Lochkovian.

Several parts of the lithostratigraphy in outcrops and boreholes are still insufficiently known in detail; this is especially true for much of the Llandovery, Ludfordian and Přídolí. The six boreholes along the northern rim contain distal turbidites alternating with laminated hemipelagites. These were dated as Wenlock and Ludlow using chitinozoans, but no formal lithostratigraphy has been applied.

The **Condroz Inlier** (Belgium), previously called the Condroz Ridge or the Sambre and Meuse Belt or Strip, is a 69 km long and 0.5 to 2 km wide belt situated south of the Sambre river from east of Charleroi to Namur, and south of the Meuse river from Namur to SW of Liège. The Variscan anticlinorial fold and fault structure, including the important Midi Thrust Fault between the Namur and Dinant synclinoria, brought discontinuous sections of Ordovician and Silurian strata in tectonic wedges to the present-day surface in this area. An overview of the older literature is given in Michot (1954) and Martin (1969). The Silurian part of the inlier was gently folded, without a clear cleavage development in Přídolí to early Lochkovian times, and later deformed during the Variscan Orogeny with faulting, folding and local cleavage development.

In contrast to the Brabant Massif, the Llandovery to Ludlow sedimentary rocks are much thinner, non-turbiditic, and richer in macrofauna (mostly graptolites with rare brachiopods, trilobites, crinoids, ostracods, conularids and *Orthoceras* sp.). The most recent graptolite study was by Maes *et al.* (1979). Most of the graptolite biozones, as defined by Rickards (1976), have been found, with the exception of the *Atavograptus atavus*, *Coronograptus cyphus*, *Gothograptus nassa* and the *Pristiograptus? ludensis* zones. However, a new study to establish recently defined biozones is necessary. Acritarch studies have elucidated a few of the previously undated sections (Martin 1969), and unpublished chitinozoan studies show the presence of the *Eisenackitina dolioliformis*, *Angochitina longicollis*, *Margachitina margaritana*, *Cingulochitina cingulata* and *Angochitina elongata* zones. A recent stratigraphical study by Vanmeirhaeghe and co-workers on the Llandovery indicated the presence of a hiatus from the Hirnantian to the lower Aeronian in the SW part of the Condroz Inlier (Puagne area and around Fosses), while in the central and NE part the succession appears to be continuous (Vanmeirhaeghe 2006).

The rocks are mostly siliciclastic and include shales, siltstones, some fine-grained sandstones and calcareous shales. They were deposited in an outer shelf environment (Verniers *et al.* 2001) (Fig. 6.28). A recently defined unit (Tihange) consists of greenish fine shales with Rhuddanian graptolites (Vanmeirhaeghe 2006). The dark grey or greenish Dave Formation (*c.* 280 m) contains silty shales with sandstones within the *Lituigraptus convolutus* and *Spirograptus turriculatus* zones. Several purple-red coloured intervals occur in the *Monoclimacis griestoniensis* and *Monoclimacis crenulata* zones and in the lower Sheinwoodian. The Criptia Group is a poorly studied possible lateral equivalent of the Dave Formation in the SW of the Condroz Inlier (Puagne area). This contains thick dark grey, greenish or yellowish fine and rather uniform shales. The Naninne Formation (more than *c.* 90 m) shows a rapid change to green or grey laminated, locally calcareous silty shales, fine sandstones, and shales (Maes *et al.* 1979). The Jonquoi Formation (*c.* 300 m) marks a return to homogenous or laminated greyish brown or green shales and mudstones with some levels of calcareous nodules (Maes *et al.* 1979). Lateral equivalents in the westernmost end of the inlier are the Longues Royes and Moncheret formations dated as Ludlow and/or Přídolí on the basis of trilete spores (Steemans in Delcambre & Pingot 2000). The Thimensart Formation (*c.* 100 m) contains grey laminated fine-grained sandstones and olive-green shales. The poorly studied, and supposedly thick, Colibeau Formation contains dark shales with rare sandstone beds. In the Telychian part of the Dave Formation and in the Naninne Formation, 11 acid volcanic or volcanosedimentary layers of approximately 1 m thick are present (Maes *et al.* 1979).

Northern part of Rhenish Massif (= Rheinisches Schiefergebirge)

The fairly restricted Silurian strata outcropping in the Ebbe and Remscheid anticlines, as well as in the Müsener Horst, are considered to be situated on the southern rim of eastern Avalonia (Verniers *et al.* 2002) (Fig. 6.4). A considerable gap exists in the Ebbe anticline between the Upper Ordovician (mid-Caradoc) of the Herscheid Group (Eiserhardt *et al.* 2001) and the Upper Silurian of the Köbbinghausen Formation. The uppermost unit of the Herscheid Group, the Solingen Shale Formation, contains chitinozoans of Caradoc age (Maletz 2000; Samuelsson *et al.* 2001). A stratigraphical gap, ranging from the late Caradoc (Upper Ordovician) to the Přídolí (late Silurian), has been noted, since fossils in the overlying Silurian Köbbinghausen Formation indicate a Přídolí age. The Köbbinghausen Formation is virtually identical in terms of its lithology to the underlying Solingen Shale Formation. Both are poor in fossils and it is possible that a considerable time interval may be present in the strongly condensed succession of the Köbbinghausen Formation, although without fossil evidence it is difficult to prove.

The Köbbinghausen Formation is estimated to be *c.* 300 m thick, including the Ocker Limestone. The lower part of the formation contains monotonous shales, grading upwards into more carbonaceous and fossil-rich layers, the *Dayia* beds (with *Dayia shirleyi,* formerly identified as *Dayia navicula* and with the trilobite *Acaste dayiana*) and the widely distributed *Scyphocrinus* horizon. The *Scyphocrinus* horizon contains *Scyphocrinites elegans* in the Ebbe anticline and can be correlated widely into the German Silurian. The units of the formations are, however, identified mostly by their fossil content and not in terms of their lithology. The upper part of the Köbbinghausen Formation consists of an alternating carbonate and carbonaceous shale succession, 200 m thick, formerly identified as the Ockrige Kalke (not to be confused with the Ockerkalk of the Thuringia Forest). The Silurian–Devonian boundary lies within this unit, as proven by trilobites and conodonts (see Timm 1981; Luppold 1995).

Far eastern Avalonia

No Silurian sediments have yet been recognized in boreholes from this completely subsurface terrane. Radiometric age dating of North Sea boreholes attests to intrusive, metamorphic or deformation events during the Silurian (Frost *et al.* 1981; Pharaoh 1999; Katzung 2001). A review of the Rügen Caledonides is also provided by Katzung *et al.* (2004) and Katzung & Feldrappe (2004).

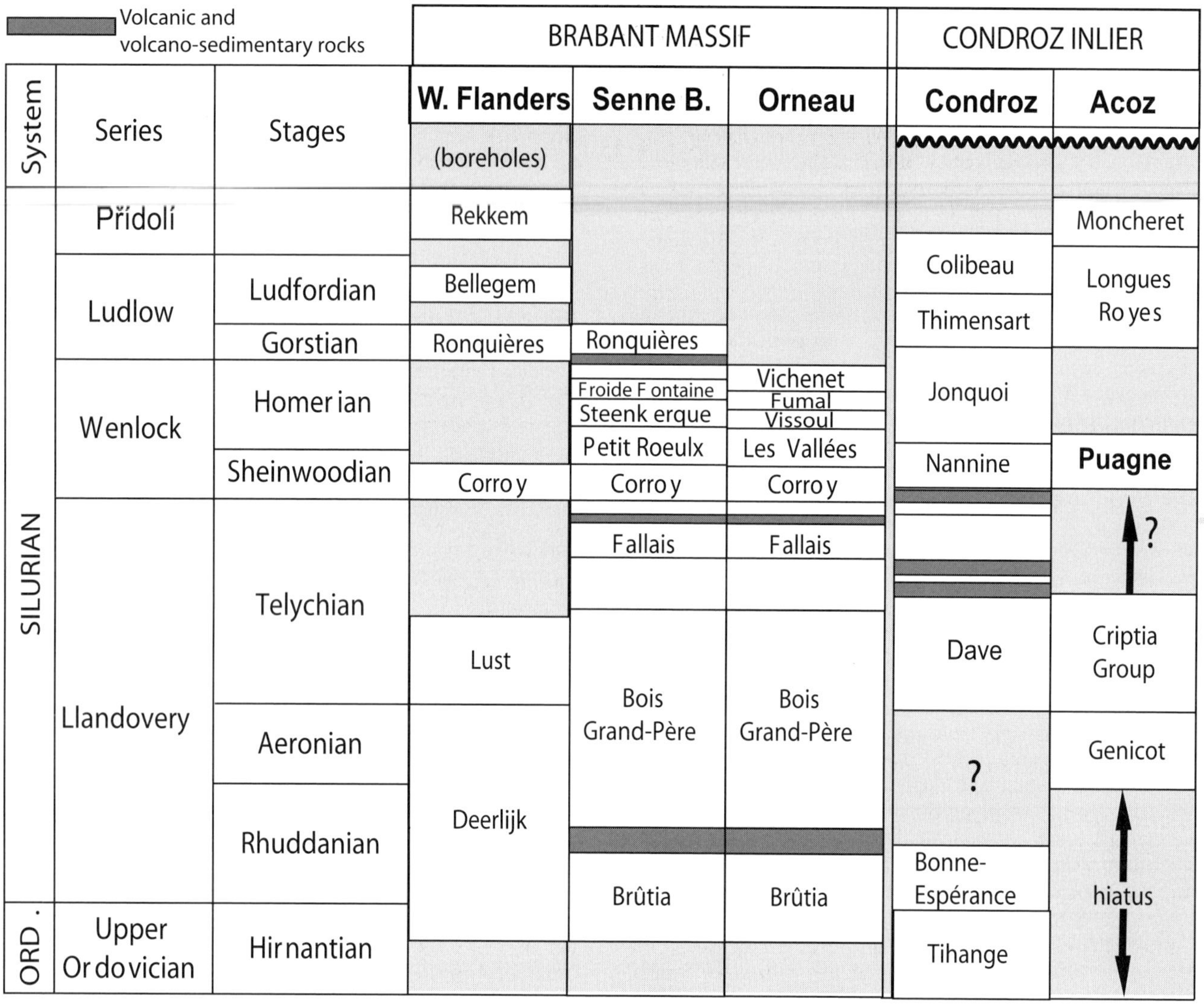

Fig. 6.28. Silurian lithostratigraphy of the Brabant Massif and the Condroz Inlier (Belgium) (modified after Verniers *et al.* 2001).

Areas of unclear palaeogeographical affinity

Carpathian Foreland

Dozens of deep boreholes have been drilled in the Silurian succession of the Carpathian Foreland (Tomczyk 1963, 1970; Moryc 1992, 1996; Maksym *et al.* 2003). The sediments are mainly monotonous graptolitic shales interbedded with thin radiolarian shales, assigned to the Llandovery, and marly mudstones of early Wenlock age which include the *rigidus* and *belophorus* graptolite zones (Maksym *et al.* 2003). The thickness of the Silurian deposits varies from a dozen metres in the northern Carpathian Foredeep to over 180 m in the southern part. The thickest, most complete and representative Silurian core was recovered from the Hermanowa-1 borehole (Maksym *et al.* 2003) (Fig. 6.7). The region possibly represents the easternmost part of the Małopolska Massif and indicates a similar but more variable depositional environment to that in the neighbouring central and western part of that massif (Nida Region). During the Llandovery and Wenlock, the graptolitic claystone facies was deposited in the pelagic, deep-water basin, with periodically oxygen-deficient conditions occurring at the bottom. The sediment accumulation rate was low and, thus, favourable for the deposition of a condensed radiolarian unit with sedimentary gaps (Maksym *et al.* 2003).

Sudetes (Polish part)

The Silurian in the Polish part of the Sudetic Mountains occurs in two areas: the Bardo Mountains (Góry Bardzkie) and the Kaczawa Mountains (Góry Kaczawskie) (Fig. 6.7). A historical review of Silurian research in the Sudetes has been published by Oberc (1970). All of the Palaeozoic successions in the Sudetes including the Silurian have been assigned to the Variscan Saxothuringian Zone, part of the structural mosaic of the European Variscides. The relationship the Saxothuringian Zone and the Barrandium Region as well as with other units of the Bohemian Massif has long been debated (Kryza *et al.* 1994; Żelaźniewicz 1997; Cymerman *et al.* 1997; Aleksandrowski 2003). There is agreement among Polish geologists that the Silurian in the Sudetes mountains (Bardo Mountains and Kaczawa Mountains) belong to Peri-Gondwana and Gondwana. Moreover, in the last year, archaeocyothids were found in the Cambrina of the Kaczawa Mountains that are very similar assemblages to those known from the neighbouring Goerlitz Synclinorium in Saxony.

The Silurian deposits of both regions in the Sudetes are developed in the Bavarian facies, one of the five northern or peri-Gondwana facies. As mentioned above it consists of greenish-grey shales interbedded with black graptolitic cherts and alum shales. The main difference with the Thuringian facies is the absence of carbonate sedimentation during the Ludlow and Přídolí. The sediments were deposited in the pelagic zone of a basin developed on oceanic-type crust. They probably represent fragments of an accretionary prism (Wajsprych 1986; Baranowski *et al.* 1987; Franke *et al.* 1993; Kryza & Muszyński 2003; Żelaźniewicz 2003), located in the southern part of the subtropical Rheic Ocean (Porębska & Sawłowicz 1997; Porębska & Koszowska 2001; Aleksandrowski *et al.* 2000). These accretionary tectonostratigraphic units finally coalesced during the Variscan Orogeny (Kryza & Muszyński 2003; Żelaźniewicz 2003).

The Silurian deposits in the Bardo Mountains are known from many small outcrops, but the most complete section is exposed at Żdanów (Fig. 6.29). The section commences with Ordovician shales intercalated with thin sand and silt turbidites overlain by Silurian shales which are increasingly interbedded up-section with light-coloured radiolarian cherts (Porębska 1982; Wyżga 1987; Porębska & Sawłowicz 1997; Porębska & Koszowska 2001). The Silurian part of the section is *c.* 48 m thick and is divided informally into the Lower and Upper Graptolitic Shales with the Green Shales in between (Fig. 6.29). The lithology of the units is characterized by bedded radiolarian cherts, siliceous and clayey shales with phosphate nodules and thin tuff layers. The shale units are finely laminated without any bioturbation, and yielding only planktonic fossils including abundant graptolites and acritarchs and rare conodonts and radiolarians. These deposits pass into Lower Devonian deposits represented by two highly condensed units of black graptolitic shales separated by grey-greenish shales.

The well-documented section with rich graptolite assemblages forms the basis for the biostratigraphy and chronostratigraphy of the Silurian–Lower Devonian interval in the Bardo Mountains (Malinowska 1955; Porębska 1982). Graptolite biozones from the Llandovery *Parakidograptus acuminatus* Zone up to the Pragian *craigensis* Zone can be recognized. The Silurian–Devonian boundary beds in the Bardo Mountains contain the uppermost Přídolí index graptolites *Monograptus transgrediens transgrediens* Perner and the first Lochkovian graptolites *Monograptus uniformis uniformis* Pribyl (Porębska 1982; Porębska & Sawłowicz 1997).

The Silurian rocks in the Kaczawa Mountains are deformed by folding or overthrusting (Baranowski *et al.* 1990, 1998). They are exposed as part of the Kaczawa Complex in small isolated outcrops, and it is often difficult to establish a continuous section. These variably thick sequences (up to several tens of metres) are mainly black siliceous and clayey slates with graptolites, cherts, quartzites and volcanogenic rocks, underlain by a thick complex of slates and phyllites, probably of Ordovician age. They are usually black in colour or rarely brown and grey. They were probably deposited in a pelagic setting. The rocks contain abundant graptolites with biozones assigned to the Llandovery, Wenlock and lower Ludlow (Jerzmański 1965; Kornaś 1977; Haydukiewicz 1987; Baranowski *et al.* 1987, 1990). Fossils diagnostic for the base of the Silurian, the *Coronograptus gregarius* Zone and the *Demirastrites triangulatus* Subzone (Llandovery) are recorded (Kornaś 1974), but there are none for the upper Ludlow and Přídolí. However, sedimentation was probably continuous from the Silurian into the Lower Devonian, since the latter is also represented by graptolite-bearing pelagic deposits (Baranowski *et al.* 1987). The lithology indicates the Bavarian facies. Ordovician low-grade metamorphic rocks of sedimentary and volcanic origin represent deposits developed during the initial rifting processes within a continental crust environment, while the Silurian–Devonian graptolitic shales and cherts (locally called lydites) with related T-MORB to N-MORB type lavas, represent oceanic-trench-type deposits of a mature basin developed on an oceanic-type crust (Furnes *et al.* 1994).

Summary

The Silurian is a relatively short period of approximately 28 Ma, but important in the history of life because it records the Llandovery recovery in biodiversity following the latest-Ordovician Hirnantian extinction event, the second most important one in the Phanerozoic. This recovery was followed by a peak in biodiversity for several groups in the Wenlock, Ludlow and Přídolí. The uppermost Wenlock and Ludlow witnessed the conquest of the continental realm with the first vascular land plants. Warm and increasing temperatures throughout the Silurian in a greenhouse climate, with some short icehouse intervals, went in tandem with low oxygen levels in the atmosphere and ocean surface waters. Low levels of organic matter were enough to cause anoxic conditions in the sea below 100 m depth, more than in any other period, leading to the characteristic anoxic and dark Silurian sediments. Strong anomalies in the global carbon cycle closely related to sea-level changes have recently been shown to influence the biodiversity of the marine faunas with radiations and extinctions.

During the Silurian Central Europe was more widely spread out over the globe than at present, with at least two intervening oceans which have since disappeared. Several large palaeocontinents such as Baltica in the NE and the northern part of Gondwana in the south were present, together with smaller palaeoplates such as Avalonia in the west, Far Eastern Avalonia in the north and NW and peri-Gondwana terranes such as Perunica in the centre. These were separated by the two oceans: the Tornquist Sea and the Rheic Ocean. These palaeocontinents were brought together by subduction of oceanic crust, closing of the oceans and the subsequent orogenic collisions, respectively termed the Caledonian and Variscan orogenies. The Alpine Orogeny also brought pieces of peri-Gondwana into the study area. They are included in the basement of the Alps in several tectonic windows (e.g. Carnic Alps) and are called the Proto-Alps.

Some palaeocontinents such as Baltica were situated in subtropical latitudes where locally small reefs could develop on shallow platforms. Other palaeocontinents such as northern Gondwana or peri-Gondwana were situated at high latitudes and characterized by the occurrence of glacial or glaciomarine diamictites in the Hirnantian, immediately preceding the Silurian. Avalonia and Far Eastern Avalonia were situated at temperate or low latitudes. The latter had collided with Baltica in the Ashgill and the former collided with Laurentia and Far Eastern Avalonia during the Silurian (see also Krawczyk *et al.* 2008).

Outcrop areas are rather scarce in Central Europe, but some subcrop areas have been drilled, thus providing additional information. The Silurian on Baltica is present in outcrops in the two parts of the Holy Cross Mountains, on Bornholm and in many boreholes of northern Poland NE of the Trans-European Suture Zone and in Lithuania. The Silurian of Far Eastern Avalonia is only recognized in boreholes from western Pomerania in the Koszalin-Chojnice Zone. Avalonia in the study area has outcrops in Belgium (Brabant Massif, Condroz Inlier) and Germany (Ebbe

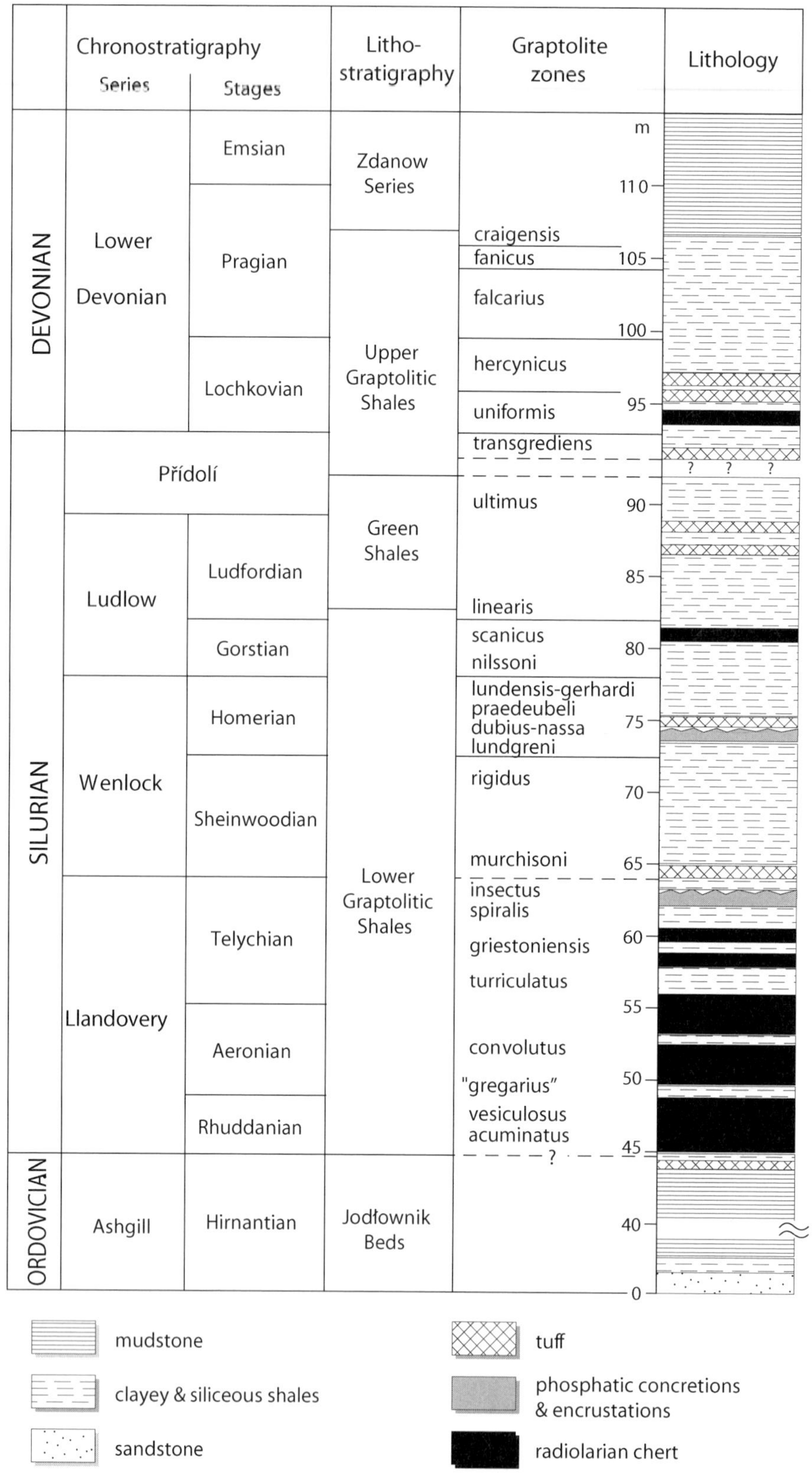

Fig. 6.29. Synthetic section of the Silurian strata in the Bardo Mountains with graptolite biostratigraphy after Malinowska (1955), Porębska (1982), Sawłowicz & Porębska (1998) and Porębska & Koszowska (2001).

anticline) and is recorded from boreholes in the Brabant Massif, North Sea and East Anglia. Peri-Gondwana terranes, within the Variscan structures of Franconia, the Mid-German Crystalline High, Saxothuringian Zone and the Moldanubian Zone, are found in many small outcrop areas of central Germany with complete successions in the Thüringer Wald, and in the Czech Republic particulary in the large, well-known outcrop area of the Prague Basin. Other peri-Gondwana terranes are found in Alpine tectonic units such as the Apulia/Italia Terrane and smaller outcrop areas, often deformed, north and south of the Periadriatic Line in Austria, northern Italy and Croatia. A well-studied outcrop area is present in the Carnic Alps succession. Another Alpine terrane, the Hellenic/Bosnian-Noric Terrane, contains small outcrop areas in Croatia and Serbia. Moesia is another peri-Gondwana terrane with Silurian present in several boreholes. In the study area, Silurian outcrops of northern Gondwana are only found in the Montagne Noire and the Maures Massif.

The depositional facies, often containing anoxic sediments such as black shales or calcareous shales and limestones, varies according to three factors: (1) local water depths are modulated by eustasy, and vary from lagoonal, shallow, middle or deep shelf environments and down the continental slope to deep marine units with turbidites or radiolarites; (2) the facies are also controlled by palaeolatitude, for example carbonates with occasional small reefs were only formed in shallow and middle shelf environments at low latitude, i.e. on Baltica and in the upper Silurian of Avalonia; (3) different tectonic settings, such as passive or active margins, result in important differences in sediment thickness. For example, Baltica (except along its SW margin) peri-Gondwana and northern Gondwana did not suffer crustal mobilization during the Silurian hence sedimentation occurred mainly on a passive margin, distal to the emergent land masses south of the study area. The low terrigenous influx and reduced subsidence resulted in successions of only a few tens to hundreds of metres in thickness. However, in faulted areas with active half-graben structures and sometimes intraplate volcanism, there were marked thickness variations as is the case in the Prague Basin and the Grauwacke Zone of Austria.

In the many peri-Gondwana outcrop areas, five main facies developed. The most common is the low energy, relatively deep shelf Thuringian facies with thin sequences of dysaerobic or anoxic dark shales and cherts in the lower Silurian with no current indicators, and typically thick aerobic limestones with interbedded thin quartzite beds in the upper Silurian (termed Ockerkalke in Germany). This facies type is found in Saxothuringia, parts of the Prague Basin and the Carnic Alps (where it is called the Bischofalm facies). The Bavarian facies is similar to the Thuringian facies and contains grey-green shales with interbedded graptolitic shales and alum shales. However there are no limestones in the upper Silurian. It is sometimes interpreted as having been deposited on a very deep shelf with radiolarites. It occurs in narrow bands in the Saxothuringian Zone.

The shelly fauna facies contains thick calcareous shales with limestone nodules and lenses rich in shelly fauna (cephalopod limestones) and was supposedly deposited on a moderately shallow to shallow shelf. It occurs in Saxothuringia, the Prague Basin, the Montagne Noire and the Carnic Alps, where it is subdivided into a shallower Wolayer facies and a less shallow Plöcken facies. There is also a facies intermediate between the shelly fauna facies and the Thuringian facies termed the Findenig facies. The Prague Basin facies formed in very shallow water on the gentle slopes of the Wenlock–Ludlow volcanic archipelago in the Prague Basin, and contains thick units of basic volcanics and macrofauna-rich bioclastic carbonates. The facies also developed in Austria. A fifth facies developed in Austria proximal to continental areas with a high input of siliciclastic material. This facies is sometimes turbiditic in nature.

The two Silurian active margins of Central Europe: the Heligoland-Pomerania Deformation Zone between Baltica and Far Eastern Avalonia, and the Anglo-Brabant Silurian Basin and Deformation Zone between Far Eastern Avalonia and Avalonia were areas of pronounced subsidence as recorded in the sometimes kilometre thick successions which were deposited. Due to subsequent Variscan or Alpine deformation and metamorphic overprinting with greenschist or amphibolite metamorphic facies in several areas, it is not always possible to deduce the original composition of the rocks.

References

Alberti, G. K. B. 1969. Trilobiten des jüngeren Siluriums sowie des Unter- und Mitteldevons. I. *Abhandlungen der senckenbergischen naturforschenden Gesellschaft*, **520**, 1–692.

Alberti, G. K. B. 1970. Trilobiten des jüngeren Siluriums sowie des Unter- und Mitteldevons. II. *Abhandlungen der senckenbergischen naturforschenden Gesellschaft*, **525**, 1–233.

Alberti, G. H. B. 1995. Zu Kenntnis und Problemen der Biostratigraphie und der faziellen Entwicklung des Silurs und Devons im Harz. *Nova Acta Leopoldina*, NF 71, Nr. 291, 23–39.

Aldridge, R. J. & Schönlaub, H. P. 1989. Conodonts. *In:* Holland, C. H. & Bassett, M. G. (eds) *A Global Standard for the Silurian System*. National Museum of Wales, Cardiff, Geological Series, **9**, 274–279.

Aleksandrowski, P. 2003. The Intra-Sudetic fault zone – example of a strike-slip terrane boundary. *In:* Ciężkowski, W., Wojewoda, J. & Żelaźniewicz, A. (eds) *Sudety zachodnie: od wendu do czwartorzędu*. WIND, Wrocław, 105–118 [in Polish with English summary].

Aleksandrowski, P., Kryza, R., Mazur, S., Pin, C. & Zalasiewicz, J. A. 2000. The Polish Sudetes: Caledonian or Variscan? *Transactions of the Royal Society of Edinburgh, Earth Sciences*, **90**, 127–146.

Al-Hasani, N. & Mostler, H. 1969. Zur Geologie der Spießnägel südlich Kirchberg (Nördliche Grauwackenzone, Tirol). *Veröffentlichung des Universität Innsbruck, 9, Alpenkundliche Studien*, **5**, 5–26.

Anderle, H.-J. 2006*a*. 5.5 Taunus. *In*: Heuse, T. & Leonhardt, D. (eds) *Deutsche Stratigraphische Kommission. Stratigraphie von Deutschland VII, Silur.* Schriftenreihe der Deutschen Gesellschaft für Geowissenschaften, **46**, 45–48.

Anderle, H.-J. 2006*b*. 5.6 Ostrand Rheinisches Schiefergebirge. *In*: Heuse, T. & Leonhardt, D. (eds) *Deutsche Stratigraphische Kommission. Stratigraphie von Deutschland VII, Silur.* Schriftenreihe der Deutschen Gesellschaft für Geowissenschaften, **46**, 49–56.

Arthurton, R. S., Johnson, E. H. & Mundy, D. J. C. 1988. *Geology of the country around Settle.* Memoir of the British Geological Survey, England & Wales, sheet **60**.

Azmy, K., Veizer, J., Bassett, M. G. & Copper, P. 1998. Oxygen and carbon isotopic composition of Silurian brachiopods: implications for coeval seawater and glaciations. *Geological Society of America Bulletin*, **110**, 1499–1512.

Azmy, K., Veizer, J., Wenzal, B., Bassett, M. G. & Copper, P. 1999. Silurian strontium isotope stratigraphy. *Geological Society of America Bulletin*, **111**, 475–483.

Baarli, B. G., Johnson, M. E. & Antoshkina, A.I. 2003. Silurian stratigraphy and palaeogeography of Baltica. *In:* Landing, E. & Johnson, M. E. (eds) *Silurian Lands and Seas, Paleogeography outside of Laurentia* (Proceedings of the James Hall Symposium, Rochester, NY, USA, August 1996). New York State Museum Bulletin, **493**, 3–34.

Bachtadse, V. & Briden, J. C. 1991. Palaeomagnetism of Devonian ring complexes from the Bayuda Desert, Sudan - new constraints on the apparent polar wander path for Gondwanaland. *Geophysical Journal International*, **104**, 635–646.

Baranowski, Z., Haydukiewicz, A., Kryza, R., Lorenc, S., Muszyński, A. & Urbanek, Z. 1987. Rozwój struktury wschodniej części Gór Kaczawskich na podstawie dotychczasowego rozpoznania stratygrafii, warunków sedymentacji i wulkanizmu. (Sudetes, Poland). *Przewodnik 58 Zjazdu PTG w Wałbrzychu*, 61–73.

Baranowski, Z., Haydukiewicz, A., Kryza, R., Lorenc, S., Muszyński, A., Solecki, A. & Urbanek, Z. 1990. Outline of the geology of the Góry Kaczawskie (Sudetes, Poland). *Neues Jahrbuch für Geologie und Paläontologie, Abhandlungen*, **179,** 223–257.

Baranowski, Z. Haydukiewicz, A., Kryza, R., Lorenc, S., Muszyński, A. & Urbanek, Z. 1998. The lithology and origin of the metasedimentary and metavolcanic rocks of the Chełmiec Unit (Góry Kaczawskie, Sudetes). *Geologia Sudetica*, **31**, 33–59 [in Polish with English summary].

Barrande, J. 1846. Notice préliminaire sur le système silurien et les trilobites de Bohême. Leipzig, 1–97.

Barrande, J. 1852. *Système silurien du centre de la Bohème*. Prague, Vol. I–VII.

Bassett, M. G. 1985. Towards a 'Common Language' in stratigraphy. *Episodes*, **8**, 87–92.

Bassett, M. G. 1989. The Wenlock Series in the Wenlock area. *In:* Holland, C. H. & Bassett, M. G. (eds) *A Global Standard for the Silurian System*. National Museum of Wales, Cardiff, Geological Series, 9, 51–73.

Bassett, M. G., Cocks, L. R. M., Holland, C. H., Rickards, R. B. & Warren, P. T. 1975. The type Wenlock Series. *Institute of Geological Sciences Report*, **75**(13), 19.

Bednarczyk, W., Korejwo, K., Łobanowski, H. & Teller, L. 1968. Stratigraphy of the Paleozoic sediments from borehole Strożyska 5 (Miechów Trough, S Poland). *Acta Geologica Polonica*, **18**, 679–689 [in Polish with English summary].

Beier, H., Maletz, J., & Katzung, G. 1997. Entwicklung eines kollisionsbedingten, altpaläozoischen Vorlandbeckens am SW-Rand der Osteuropäischen Plattform. *Schriftenreihe der Deutschen Geologischen Gesellschaft*, **2**, 26–27.

Beier, H., Maletz, J. & Böhnke, A. 2000. Development of an Early Palaeozoic foreland basin at the SW margin of Baltica. *Neues Jahrbuch für Geologie und Paläontologie, Abhandlungen*, **218**, 129–152.

Beier, H., Hinz-Schallreuter, I., Katzung, G., Maletz, J. & Schallreuter, R. 2001. 4.32 Adlergrand (Vorlord der Rügen-Kaledoniden) (32). *In:* Beier, H., Burmann, G., *et al.* (Stratigraphische Kommission Deutschland; Herausgeber) *Stratigraphie von Deutschland II. Ordovizium, Kambrium, Vendium, Riphäikum Teil III: Nordthüringen, Sachsen-Anhalt, Brandenburg, Nordrhein-Westfalen, Niedersachsen, Mecklenburg-Vorpommern, deutscher Anteil Ostsee, Schleswig-Holstein, deutscher Anteil Nordsee.* Courier Forschungsinstitut Senckenberg, **235,** 135–151.

Belka, Z. & Narkiewicz, M. 2008. Devonian. *In:* McCann, T. (ed.) *The Geology of Central Europe. Volume 1: Precambrian and Palaeozoic*. Geological Society, London, 383–410.

Belka, Z., Valverde-Vaquero, P., Dörr, W., Ahrendt, H., Wemmer, K., Franke, W., Schäfer, J., 2002. Accretion of first Gondwana derived terranes at the margin of Baltica. *In:* Winchester, J. A., Pharoah, T. C. & Verniers, J. (eds) *Palaeizoic Amalgamation of Central Europe.* Geological Society, London, Special Publications, **201**, 19–36.

Bergström, S. M. 1990. Relations between conodont provincialism and the changing palaeogeography during the Early Palaeozoic. *In*: McKerrow, W.S. & Scotese, C.R. (eds) *In Palaeozoic Palaeogeography and Biogeography.* Geological Society, London, Memoirs **12**, 105–121.

Bergström, S. M., Huff, W. D. & Kolata, D. R. 1998: Silurian K-Bentonites in North America and Europe. – *Temas Geológico-Mineros ITGE*, **23**, 54–56.

Berner, R. A. 1989. Biogeochemical cycles of carbon and sulphur and their effect on atmospheric oxygen over Phanerozoic times. *Palaeogeography, Palaeoclimatology, Palaeoecology*, **75**, 97–122.

Berry, W. B. N. & Boucot, A. J. 1967. Pelecypod-graptolite association in the Old World Silurian. *Geological Society of America Bulletin*, **78**, 1515–1521.

Bickert, T., Pätzold, J., Samtleben, C. & Munnecke, A. 1997. Paleoenvironmental changes in the Silurian indicated by stable isotopes in brachiopod shells from Gotland, Sweden. *Geochimica et Cosmochimica Acta*, **61**, 2717–2130.

Bjerreskov, M. 1975. Llandoverian and Wenlockian graptolites from Bornholm. *Fossils and Strata*, **8**, 1–93.

Bjerreskov, M. 1991. Pyrite in Silurian graptolites from Bornholm, Denmark. *Lethaia*, **24**, 351–361.

Bjerreskov, M. & Jørgensen, K. Å. 1983. Late Wenlock graptolite-bearing tuffaceous sandstone from Bornholm, Denmark. *Bulletin of the Geological Society of Denmark*, **31**, 129–149.

Bjerreskov, M. & Stouge, S. 1985. Field excursion guide Bornholm, Denmark. *In: Graptolite Working Group of the International Palaeontological Association, Third International Conference*. Institute of Historical Geology and Palaeontology, University of Copenhagen, Copenhagen, 1–23.

Bodzioch, A., Kozłowski, W. & Popławska, A. 2003. A *Cooksonia*-type flora from the Upper Silurian of the Holy Cross Mountains, Poland. *Acta Palaeontologica Polonica*, **48**, 653–656.

Böhnke, A. & Katzung, G. 2001. The Middle Silurian from Bornholm (Denmark) – sedimentology, petrology and age. *Neues Jahrbuch für Geologie und Paläontologie, Abhandlungen*, **222**, 161–191.

Bouček, B. 1935. O silurské fauně od Stínavy (západně od Plumlova) na Drahanské vysočině. *Časopis Vlasteneckého Spolku musejního v Olomouci*, **48**(3–4), 129–138.

Bouček, B. &, Přibyl, A. 1955. O silurských ostrakodech a stratigrafii vrstev budňanských (eß) z nejbližšího okolí Kosova a Koledníku u Berouna. *Sborník Ústředního Ústavu geologického, oddělení paleontologické*, **21**(1954), 577–662.

Brause, H. 2006. 5.18 Lausitz (Görlitzer Schiefergebirge). *In*: Heuse, T. & Leonhardt, D. (eds) *Deutsche Stratigraphische Kommission. Stratigraphie von Deutschland VII, Silur.* Schriftenreihe der Deutschen Gesellschaft für Geowissenschaften, **46**, 118–122.

Brenchley, P. J. 1989. The Late Ordovician extinction. *In:* Donovan, S. K. (ed.) *Mass Extinctions: Processes and Evidence*. Belhaven Press, London, 104–132.

Brenchley, P. J. & Newall, G. 1984. Late Ordovician environmental changes and their effect on faunas. *In:* Bruton, D. L. (ed.) *Aspects of the Ordovician System*. Geological Contributions of the University of Oslo, **295**, 65–79.

Brenchley, P. J., Marshall, J. D. & Underwood, C. J. 2001. Do all mass extinctions represent an ecological crisis? Evidence from the late Ordovician. *Geological Journal*, **36**, 329–340.

Brett, C. E. & Schönlaub, H. P. 1998. *In:* Ferretti, A. & Histon, K. (eds) *Cephalopod Limestone Biofacies, Carnic Alps, Austria* (SW Iberia Field Meeting of the Subcommission Silurian Stratigraphy, Madrid, 15–18 June), Abstract Volume.

Buchroithner, M. F. 1979. Biostratigraphische und fazielle Untersuchnungen im Paläozoicum von Mittelkärnten. *Carinthia* II, **169**(89), 71–95.

Budyko, M. I., Ronov, A. B. & Yanshin, A. L. 1987. *History of the Earth's Atmosphere* (translated from Russian by Lemeshko, S.F. & Yanuta, A.L.). Springer-Verlag, Berlin.

Buła, Z. 2000. The Lower Palaeozoic of Upper Silesia and West Małopolska. *Prace Państwowego Instytutu Geologicznego*, **171**, 1–71.

Burmann, G. 2006. 5.7 Harz. *In*: Heuse, T. & Leonhardt, D. (eds) *Deutsche Stratigraphische Kommission. Stratigraphie von Deutschland VII, Silur.* Schriftenreihe der Deutschen Gesellschaft für Geowissenschaften, **46**, 56–80.

Burrett, C. F. 1983. Palaeomagnetism and the Mid-European Ocean – an alternative interpretation of Lower Palaeozoic apparent polar wander. *Geophysical Journal of the Royal Astronomical Society*, **72**, 523–534.

Burrett, C. F. & Griffiths, J. 1977. A case for Mid-European Ocean. In: *La Chaine Varisque d'Europe Moyene et Occidentale* (Colloquium internationale CNRS), **243**, 523–534.

Calner, M. & Eriksson, M. E. 2006. Silurian research on the crossroads. *GFF*, 128, 73–74

Calner, M., Jeppsson, L. & Munnecke, A. 2004. The Silurian of Gotland – Part I: Review of the stratigraphic framework, event stratigraphy, and stable carbon and oxygen isotope development. *Erlanger geologische Abhandlungen, Sonderband*, **5**, 113–131.

Caputo, M. V. 1998. Ordovician-Silurian glaciations and global sea-level changes. *In:* Landing, E. & Johnson, M. E. (eds) *Silurian Cycles: Linkages of Dynamic Stratigraphy with Atmospheric, Oceanic and Tectonic Changes*. New York State Museum Bulletin, **491**, 15–25.

Caputo, M. V. 2003. Silurian glacial paleogeography in South America. *In:* Landing, E. & Johnson, M. E. (eds) *Silurian Lands and Seas, Paleogeography outside of Laurentia* (Proceedings of the James Hall Symposium, Rochester, NY, USA, August 1996). New York State Museum Bulletin, **493**, 57.

Centène, A. 1977. Nouvelles données biostratigraphiques (Conodontes) sur le Silurien des Monts de Lacaune (Montagne Noire, France). *In: 5ème Réunion annuelle des Sciences de la Terre, Rennes, Société Géologique de France*, 140.

Centène, A. & Sentou, G. 1975. *Graptolites et conodontes du Silurien du Midi méditerranéen. Essai de correlation entre les deux échelles.* Thesis, University of Montpellier.

CFP, Copesep, RAP & SNPA. 1965. Contribution à la connaissance des bassins paléozoiques du Nord de la France. *Annales de la Société géologique du Nord*, **85**, 273–281.

CFP, Copesep, RAP & SNPA. 1966. Contribution à la connaissance des bassins paléozoiques du Nord de la France (suite). Tableaux des Core-drills. *Annales de la Société géologique du Nord*, **86**, 115–128.

Chaubet, M. C. 1937. *Contribution à l'étude géologique du Gothlandien du versant méridional de la Montagne Noire.* Travaux du Laboratoire de Géologie de l'Université de Montpellier.

Chlupáč, I. 1989. Stratigraphy of the Sedlčany – Krásná Hora Metamorphic "Islet" in Bohemia (Proterozoic? to Devonian). *Časopis pro Mineralogii a Geologii*, **34**(1), 1–16.

Chlupáč, I. 1993. Geology of the Barrandian. A field trip guide. *Senckenberg-Buch*, **69**, 1–163.

Chlupáč, I., Jaeger, H. & Zikmundová, J. 1972. The Silurian–Devonian boundary in the Barrandian. *Bulletin Canadien Petrologie et Géologie*, **20**, 104–174.

Claesson, K. C. 1979. Early Palaeozoic geomagnetism of Gotland. *Geologiska Föreningens i Stockholm Förhandlingar*, **101**, 149–155.

Cocks, L. R. M. 1985. The Ordovician–Silurian boundary. *Episodes*, **8**, 98–100.

Cocks, L. R. M. 1989. The Llandovery Series in the Llandovery area. *In:* Holland, C. H. & Bassett, M. G. (eds) *A global standard for the Silurian System*. National Museum of Wales, Cardiff, Geological Series, **9**, 36–50.

Cocks, L. R. M. 2002. Key Lower Palaeozoic faunas from near the Trans-European Suture Zone. *In:* Winchester, J.A., Pharoah, T.C. & Verniers, J. (eds) *Palaeozoic Amalgamation of Central Europe.* Geological Society, London, Special Publications, **201**, 37–46.

Cocks, L. R. M. & Torsvik, T. H. 2002. Earth geography from 500 to 400 million years ago: a faunal and palaeomagnetic review. *Journal of the Geological Society, London*, **159**, 631–644.

Cocks, L. R. M. & Torsvik, T. H. 2005. Baltica from the late Precambrian to mid-Palaeozoic times: The gain and loss of a terrane identity. *Earth-Science Reviews*, **72**, 39–66.

Cocks, L. R. M., Mckerrow, W. S. & Van Staal, C. R. 1997. The margins of Avalonia. *Geological Magazine*, **134**, 627–636.

Cocks, L. R. M., McKerrow, W. S. & Verniers, J. 2003. The Silurian of Avalonia. *In:* Landing, E. & Johnson, M. E. (eds) *Silurian Lands and Seas, Paleogeography outside of Laurentia* (Proceedings of the James Hall Symposium, Rochester, NY, USA, August 1996). New York State Museum Bulletin, **493**, 35–53.

Crevola, G. & Pupin, J. P. 1994. Crystalline Provence: structure and Variscan evolution. *In:* Keppie, J. D. (ed.) *Pre-Mesozoic Geology in France and Related Areas*. Springer-Verlag, Berlin, 426–441.

Cymerman, Z., Piasecki, M. A. J. & Seston, R. 1997. Terranes and boundaries in the Sudetes, northeast Bohemian Massif. *Geological Magazine*, **134**, 717–725.

Czermiński, J. 1967. Metamorficzne podłoże Dewonu w Gościnnie k. Kołobrzegu. *Kwartalnik geologiczny*, **11**, 693–696.

Dadlez, R. 1967. New sections of the Zechstein substratum in North-West Poland. *Kwartalnik geologiczny*, **11**, 572–586 [in Polish with English summary].

Dadlez, R. 1978. Sub-Permian rock complexes in the Koszalin-Chojnice Zone. *Kwartalnik geologiczny*, **22**, 269–301 [in Polish with English summary].

Dadlez, R. 2000. Pomeranian Caledonides (NW Poland), fifty years of controversies: a review and a new concept. *Geological Quarterly*, **44**, 221–236.

Debacker, T. N. 2001. *Palaeozoic deformation of the Brabant Massif within eastern Avalonia, how, when and why?* PhD Thesis, University of Ghent.

De Bock, F. 1982. Présence de chitinozoaires dans le passage Silurien-Dévonien de la Montagne Noire sud-orientale (Hérault, France). *Geobios*, **15**, 845–871.

Deczkowski, Z. & Tomczyk, H. 1969. Older Palaeozoic in borehole Wilków (northern part of the Świętokrzyskie Mountains) (in Polish). *Kwartalnik Geologiczny*, **13**, 14–24.

Deflandre, G. 1942. Sur les microfossiles des calcaires siluriens de la Montagne Noire, les Chitinozoaires (Eisenack). *Comptes Rendus des Sciences de l'Académie des Sciences*, Paris, **215**, 286–288.

Deflandre, G. 1946. Microfossiles des calcaires siluriens de la Montagne Noire. *Annales de Paléontologie*, **31**(1944–1945), 41–75.

Delcambre, B. & Pingot, J. L. 2000. *Carte géologique de Wallonie n°46 - 7/8 and Notice explicative de la carte Fontaine-l'Evêque - Charleroi.* Ministère de la Région Wallonne, Namur.

Deutsche Stratigraphische Kommission 2002. Menning M. & Hendrich, A. (eds) *Stratigraphische Tabelle von Deutschland 2002.* E. Stein GmbH, Potsdam.

Deutsche Stratigraphische Kommission - Silur Subkommission. 2006. Silur Monographie. *Deutsche Geologische Gesellschaft.Schriftenreihe der deutschen Gesellschaft für Geowissenschaften.* **46**, 1–189.

De Vos, W., Verniers, J., Herbosch, A. & Vanguestaine, M. 1993. A new geological map of the Brabant Massif. *In:* Pharaoh, T. C., Molyneux, S. G., Merriman, R. J., Lee, M. K. & Verniers, J. (eds) Special Issue on the Caledonides of the Anglo-Brabant Massif. *Geological Magazine*, **130**, 605–611.

Dieni, I., Giordano, D., Loydell, D. K. & Sassi, F. P. 2005. *Discovery of Llandovery (Silurian) graptolites and probable Devonian corals in the Southalpine Metamorphic Basement of the Eastern Alps (Agordo, NE Italy). Geological Magazine*, **142**, 1–5.

Dombrowski A, Henjes-kunst F, Höhndorf A, Kröner, A., Okrusch, M. & Richter, P. 1995. Orthogneisses in the Spessart crystalline complex, north-west Bavaria - Silurian granitoid magmatism at an active continental-margin. *Geologische Rundschau* **84**, 399–411.

Doubinger, J. 1963. Chitinozoaires ordoviciens et siluriens des schistes de Steige dans les Vosges. *Bulletin du Service de la Carte géologique d'Alsace-Lorraine*, **16**, 125–136.

Doubinger, J. & von Eller, J. P. 1963. Présence de spongiaires dans les schistes précambriens métamorphiques des Vosges. *Bulletin du Service de la Carte géologique d'Alsace-Lorraine*, **16**, 111–123.

Dudek, A. 1980. The crystalline basement block of the outer Carpathians in Moravia: Bruno – Vistulicum. *Rozpravy České Akademie Věd, Řada matematicko- přírodovědná Věd*, **90**(8), 1–85.

Dufka, P. 1992. Lower Silurian chitinozoans of the Prague Basin (Barrandian, Czechoslovakia) – preliminary results. *Revue de Micropaléontologie*, **35**(2), 117–126.

Dufka, P. 1995*a*. Chitinozoans from the Sheinwoodian/Homerian boundary beds (Wenlock) in the Prague Basin (Barrandian, Bohemia). *Review of Palaeobotany and Palynology*, **86**, 135–145.

Dufka, P. 1995*b*. Upper Wenlock miospores and cryptospores derived from a Silurian volcanic island in the Prague Basin (Barrandian area, Bohemia). *Journal of Micropaleontology*, **14**, 67–79.

Dufka, P., Kříž, J. & Storch, P. 1995. Silurian graptolites and chitinozoans from the Uranium Industry boreholes drilled in 1968 - 1971 (Prague Basin, Bohemia). *Vestnik Ceského geologického ustavu*. **70**, 5–14.

Ebner, F. 1994. Section 1, Silurian/Devonian boundary section of Eggenfeld/Paleozoic of Graz. *In:* Schönlaub, H. P. & Kreutzer, L. H. (eds) *Field Meeting Eastern + Southern Alps, Austria 1994, Subcommission on Silurian Stratigraphy.* Berichte Geologische Bundesanstalt, Wien, **30**, 77–82.

Ebner, F., Fenninger, A., *et al.* 1990. Stratigraphic Correlation Forms of Paleozoic units in Austria. *In:* Sassi, F. P. & Zanferrari, A. (eds) Pre-Variscan and Variscan Events in the Alpine-Mediterranean Belts, Stratigraphic Correlation Forms. *Rendiconti Società Geologica Italiana*, **12**, 213–239.

Einasto, R. Z., Abushik, A. F., Kaljo, D. P., Koren', T. N., Modzelevskaya, T. L. & Nestor, H. Z. 1986. Silurian sedimentation and the fauna of the East Baltic and Podolian marginal basins: a comparison. *In*: Kaljo, D. & E. Klaamann, E. (eds) *Theory and Practice of Ecostratigraphy.* Institute of Geology, Academy of

Sciences of the Estonian SSR, Tallinn, 65–72.

Eiserhardt, K.-H., Koch, L. & Maletz, J. 2001. 4.29 Nördliches Rheinisches Schiefergebirge (29). *In:* Beier, H., Burmann, G., *et al.* (Stratigraphische Kommission Deutschland; Herausgeber) *Stratigraphie von Deutschland II. Ordovizium, Kambrium, Vendium, Riphäikum Teil III: Nordthüringen, Sachsen-Anhalt, Brandenburg, Nordrhein-Westfalen, Niedersachsen, Mecklenburg-Vorpommern, deutscher Anteil Ostsee, Schleswig-Holstein, deutscher Anteil Nordsee.* Courier Forschungsinstitut Senckenberg, **235**, 90–108.

Engel, W., Feist, R. & Franke, W. 1982. Le Carbonifère antéstéphanien de la Montagne Noire: rapports entre mise en place des nappes et sédimentation. *Bulletin du Bureau de Recherches Géologiques et Minières* (BRGM), **4**, 341–389.

Feist, R. 1977. *Le Siluro-Devonien du sud-est de la Montagne Noire (Hérault, France) et ses faunes de trilobites.* Doctoral thesis, University of Montpellier.

Feist, R. & Schönlaub, H. P. 1974. Zur Silur/Devon Grenze in der östlichen Montagne Noire Südfrankreichs. *Neues Jahrbuch für Geologie und Paläeontologie*, **4**, 200–219.

Feist, R., Echtler, H., Galtier, J. & Mouthier, B. 1994. Biostratigraphy and dynamics of the nonmetamorphic sedimentary record. *In:* Keppie, J. D. (ed.) *Pre-Mesozoic Geology in France and Related Areas.* Springer-Verlag, Berlin, 289–304.

Ferretti, A. 2005. Ooidal ironstones and laminated ferruginous deposits from the Silurian of the Carnic Alps, Austria. *Bollettino della Società Paleontologica Italiana*, **44**, 263–278.

Ferretti, A., Histon, K. & Schönlaub, H. P. 1999. The Silurian and early Devonian of the Rauchkofel Boden Section, Carnic Alps, Austria. *In:* Histon, K. (ed.) *V International Symposium on Cephalopods – Present and Past. Carnic Alps Excursion Guidebook.* Berichte Geologische Bundesanstalt, Wien, **47**, 55–62.

Ferretti, A., & Kříž, J. 1995. Cephalopod limestone biofacies in the Silurian of the Prague Basin, Bohemia. *Palaios*, **10**(3), 240–253.

Fiala, F. 1982. Basaltoid diabase from Hostim with indications of hematite mineralization. *Časopis pro Mineralogii a Geologii*, **27**(2), 173–185.

Flügel, H. W., Jaeger, H., Schönlaub, H. P. & Vai, G. B. 1977. Carnic Alps. *In:* Martinson, A. (ed.) *The Silurian–Devonian Boundary.* IUGS Series A, **5**, 126–142.

Fortey, R. A. & Cocks, L. R. M. 2003. Palaeontological evidence bearing on the global Ordovician–Silurian continental reconstructions. *Earth-Science Reviews*, **61**, 245–307.

Frakes, L. A., Francis, J. E. & Syktus, J. L. 1992. Climate Modes of the Phanerozoic: The History of the Earth's Climate over the Past 600 Million Years. Cambridge University Press, Cambridge.

Franke, W. 2000. The mid-European segment of the Variscides: tectonostratigraphic units, terrane boundaries and plate tectonic evolution. *In:* Franke, W., Haak, V., Oncken, O. & Tanner, D. (eds) *Orogenic Processes: Quantification and Modelling in the Variscan Belt.* Geological Society, London, Special Publications, **179**, 35–61.

Franke, W., Żelaźniewicz, A., Porębski, S. J. & Wajsprych, B. 1993. Saxothuringian Zone in German and Poland: differences and common features. *Geologische Rundschau*, **82**, 583–599.

Fritz, H. & Neubauer, F. 1988. Geodynamic aspects of the Silurian and Early Devonian sedimentation in the Paleozoic of Graz (Eastern Alps). *Schweizerische Mineralogische und Petrographische Mitteilungen*, **68**, 359–367.

Frost, R. T. C., Fitch, F. L. J. & Miller, J. A. 1981. The age and nature of the crystalline basement of the North Sea Bassin. *In: Petroleum Geology of the Continental Shelf of North-West Europe.* Institute of Petroleum, London, 43–47.

Furnes, H., Kryza, R., Muszyński, A., Pin, C. & Garmann, L. B. 1994. Geochemical evidence for progressive, rift-related early Palaeozoic volcanism in the western Sudetes. *Journal of the Geological Society, London*, **151**, 91–109.

Gaertner, H. R. von 1931. *Geologie der Zentralkarnischen Alpen. Denkschrifte der Österreichische Akademie der Wissenschaften, Wien, Mathematik-Naturwissenschaften Klasse*, **102**, 113–199.

Geinitz, H. B. 1842. Über Graptolithen. *Neues Jahrbuch für Mineralogie, Geognosie, Geologie und Petrefakten-Kunde, Stuttgart*, Jahrgang **1842**, 697–701.

Geissler, E. & Schauer, M. 2006. 5.15 Erzgebirge (Erzgebirgsnordrandzone). *In*: Heuse, T. & Leonhardt, D. (eds) *Deutsche Stratigraphische Kommission. Stratigraphie von Deutschland VII, Silur.* Schriftenreihe der Deutschen Gesellschaft für Geowissenschaften, **46**, 103–105.

Gelsthorpe, D. N. 2004. Microplankton changes through the early Silurian Ireviken extinction event on Gotland, Sweden. *Review of Palaeobotany and Palynology*, **130**, 89–103.

Geukens, F. 1965. Problème stratigraphique relatif aux planchettes Odeigne-Bihain. *Bulletin de la Société belge de Géologie*, **74**, 214–219.

Geukens, F. 1999. Notes accompagnant une révision de la carte structurale du Massif de Stavelot. *Aardkundige Mededelingen*, **9**, 183–190.

Geyer, G., Elicki, O., Fatka, O. & Żylińska, A. 2008. Cambrian *In:* McCann, T. (ed.) *The Geology of Central Europe. Volume 1: Precambrian and Palaeozoic.* Geological Society, London, 155–202.

Godefroid, J. 1995. *Dayia shirleyi* Alvarez & Racheboeuf, 1986, un brachiopode silurian dans les «Schistes de Mondrepuis» à Muno (sud de la Belgique). *Bulletin de l'Institut Royal des Sciences Naturelles de Belgique, Sciences de la Terre*, **65**, 269–272.

Godefroid, J. & Cravatte, T. 1999. Les brachiopodes et la limite Silurien/Dévonien à Muno (sud de la Belgique). *Bulletin de l'Institut Royal des Sciences Naturelles de Belgique, Sciences de la Terre*, **69**, 5–26.

Gradstein, F. M., Ogg, J. G. & Smith, A. G. 2004. *A Geologic Time Scale 2004.* Cambridge University Press, Cambridge.

Grahn, Y. & Caputo, M. V. 1992. Early Silurian glaciations in Brazil. *Palaeogeography, Palaeoclimatology, Palaeoecology*, **99**, 9–15.

Gueirard, S., Waterlot, G., Gherzi, A. & Samat, M. 1970. Sur l'âge llandovérien supérieur à taranonien inférieur des schistes à graptolites du Fenouillet, massif des Maures (Var). *Bulletin de la Société Géologique de France*, **12**(7), 195–199.

Gümbel, C. W. 1879. Geognostische Beschreibung des Fichtelgebirges mit dem Frankenwalde und dem westlichen Vorlande. Justus Perthes.

Guterriez-marco, J. C., Yanev, S., *et al.* 2003. New biostratigraphical data from the Ordovician of Bulgaria. *Serie Correlacion Geologica*, **17**, 79–85.

Haas, W. 1968. Trilobiten aus dem Silur der Karnischen Alpen. *Carinthia* II, **SH27**, 23.

Hajłasz, B. 1967. O znalezieniu *Tentaculites* sp. w otworze wiertniczym Gościnno IG-1. *Kwartalnik Geologiczny*, **11**, 697–698. [in Polish with English summary].

Hallam, A. & Wignall, P. B. 1997. *Mass Extinctions and their Aftermath.* Oxford University Press, Oxford.

Hann, H. P., Swatzki, G. & Vaida, M. 1995. Chitinozoen und Acritarchen des Ordoviciums aus metamorphen Grauwacken der Zone von Badenweiler-Lenzkirch, Schwarzwald, SW-Deutschland. *Neues Jahrbuch für Geologie und Paläontologie, Monatshefte*, **6**, 375–383.

Hansch, W. 1993. Stratigraphical, palaeoecological and palaeobiogeographical aspects of the Upper Silurian ostracod faunas of Baltoscandia and Central Europe. *In: Proceedings of the 11th International Symposium on Ostracoda*, Warrnambool, Victoria, Australia, 1991, 23–37.

Hauer, F. von 1847. Versteinerungen von Dienten, in Salzburg. *Haidingers Berichte*, **1**, 10.

Havlíček, V. 1977. The Paleozoic (Cambrian – Devonian) in the Rožmitál area. *Věstník Ústředního Ústavu geologického*, **52**, 81–92.

Havlíček, V. 1980. Vývoj paleozoických pánví v Českém masívu (kambrium - spodní karbon); Development of Paleozoic basins in the Bohemian Massif (Cambrian – Lower Carboniferous). *Sborník geologických Věd, Geologie*, **34**, 31–65.

Havlíček, V., Vaněk, J. & Fatka, O. 1994. Perunica microcontinent in the Ordovician (its position within the Mediterranean Province, series division, benthic and pelagic associations). *Sborník geologických Věd, Geologie*, **46**, 23–56.

Haydukiewicz, A. 1987. Sekwencja stratygraficzna w kompleksie kaczawskim. *Przewodnik 58 Zjazdu PTG w Wałbrzychu*, 95–102.

Heath, R. J., Brenchley, P. J. & Marshall, J. D. 1998. Early Silurian carbon and oxygen stable-isotope strtaigraphy of Estonia: implications for climate change. *In:* Landing, E. & Johnson, M. E. (eds) *Silurian Cycles: Linkages of Dynamic Stratigraphy with Atmospheric, Oceanic and Tectonic Changes.* New York State Museum Bulletin, **491**, 313–327.

HEINISCH, H. 1988. Hinweis auf die Existenz eines passiven Kontinentalrandes im Altpaläozoikum der Nördlichen Grauwackenzone – Ostalpen. *Schweizerische Mineralogische und Petrographische Mitteilungen*, **68**, 407–418.

HERBOSCH, A., VERNIERS, J., DEBACKER, T., BILLIAERT, B., DE SCHEPPER, S. & BELMANS, M. 2002. The Lower Palaeozoic stratigraphy and sedimentology of the Brabant Massif in the Dyle and Orneau valleys and of the Condroz Inlier at Fosses: an excursion guidebook. *Geologica Belgica*, **5**(3–4), 71–143.

HEUSE, T., SCHAUER, M., MALETZ, J. & MEISEL, S. 2006. Regionale Einheiten 5.11 Thüringisch-Fränkisches Schiefergebirge. *In:* STRATIGRAPHISCHE KOMMISSION DEUTSCHLANDS (eds). *Stratigraphie von Deutschland (Band 4): Silur-Monographie.* Schriftenreihe der Deutschen Geologischen Gesellschaft, **46**, 86–91.

HIDEN, H. R. 1996. Zur Fossilführung des Basalanteils der hohen Deckengruppe des Grazer Paläozoikums (Österreich). *Mitteilungen der Abteilung für Geologie und Paläontologie, Landesmuseum Joanneum*, **54**, 77–91.

HIRSCHMANN, G. & OKRUSCH, M. 2001. Spessart und Rhön (18) – Teil der MKZ. In: HOTH, K. & LEONHARDT, D. (eds) Stratigraphie von Deutschland II, Ordovizium, Kambrium, Vendium, Riphäikum. Teil II: Baden-Württemberg, Bayern, Hessen, Rheinland-Pfalz, Nordthüringen, Sachsen-Anhalt, Brandenburg. Courier Forschungsinstitut Senckenberg, **234**, 93–108.

HISTON, K. 1998. Die Nautiloideen-Fauna aus dem Silur der Karnischen Alpen. *Geologischen und Paläontogischen Mitteilungen*, **32**, 105–115.

HISTON, K. 1999. Revision of Silurian Nautiloid Cephalopods from the Carnic Alps (Austria) – The HERITSCH (1929) Collection in the Geological Survey of Austria. *Abhandlungen Geologische Bundesanstalt, Wien.*, **56**, 229–258.

HISTON, K. 2002. A nautiloid assemblage from the Upper Silurian (Pridoli) of the Carnic Alps, Austria. *In*: WYSE JACKSON, P. N., PARKES, M. A. & WOOD, R. (eds) *Studies in Palaeozoic Palaeontology and biostratigraphy in honour of Charles Hepworth Holland.* Special Papers in Palaeontology, **67**, 115–133.

HISTON, K. & SCHÖNLAUB, H. P. 1999. Taphonomy, paleoecology and bathymetric implications of the nautiloid fauna from the Silurian of the Cellon Section (Carnic Alps, Austria). *In: Proceedings of the First International Conference on North Gondwanan Mid-Palaeozoic Biodynamics (IGCP Project 421).* Abhandlungen Geologische Bundesanstalt, Wien, **54**, 259–274.

HISTON, K., FERRETTI, A. & SCHÖNLAUB, H. P. 1999. Silurian Cephalopod Limestone Sequence of the Cellon Section, Carnic Alps, Austria. *In:* HISTON, K. (ed.) *V International Symposium Cephalopods – Present and Past. Carnic Alps Excursion Guidebook.* Berichte Geologische Bundesanstalt, Wien, **47**, 46–54.

HISTON, K., KLEIN, P., SCHÖNLAUB, H. P. & HUFF, W. D. 2007. Lower Palaeozoic K-bentonites from the Carnic Alps, Austria. *Austrian Journal of Earth Sciences*, **100**, 26–42.

HOLLAND, C. H. 1985. Series and stages of the Silurian System. *Episodes*, **8**, 101–103.

HOLLAND C. H. 1985 Series and stages of the Silurian System, *Episodes*, **8**, 101–103

HOLLAND, C. H. 1989. History. *In:* HOLLAND, C. H. & BASSETT, M. G. (eds) *A global standard for the Silurian System.* National Museum of Wales, Cardiff, Geological Series, **9**, 8–22.

HOLLAND, C. H. & BASSETT, M. G. (eds). 2002. *Telychian rocks of the British Isles and China (Silurian, Llandovery Series) An Experiment to Test Precision in Stratigraphy.* National Museum of Wales, Geological Series, **21**, 1–210.

HOLLAND, C. H., LAWSON, J. D., WALMSLEY, V. G. & WHITE, D. E. 1980. Ludlow stages. *Lethaia*, **13**, 268.

HORNÝ, R. 1956. Nové poznatky o biostratigrafii skutečsko-hlinského siluru. *Věstník Ústředního Ústavu geologického*, **31**, 126–131.

HORNÝ, R. 1964. New graptolites from the metamorphosed Silurian in Bohemia. *Časopis Národního Muzea, oddíl přírodovědný*, **133**(4), 224.

HUBMANN, B., POHLER, S., SCHÖNLAUB, H. P. & MESSNER, F. 2003. Paleozoic Coral-Sponge Bearing Successions in Austria. *9th International Symposium on Fossil Cnidaria and Porifera*, Graz, Austria, 3–7 August 2003. Excursion A1. Berichte Geologische Bundesanstalt, Wien, **61**, 1–91.

HUFF, W. D., BERGSTRÖM, S. M., KOLATA, D. R. & SUN, H. 1998. The Lower Silurian Osmundsberg K-Bentonite. Part II: mineralogy, geochemistry, chemostratigraphy, and tectonomagmatic significance. *Geological Magazine*, **135**, 15–26.

HUFF, W. D., MÜFTÜOGLU, E., BERGSTRÖM, S. M. & KOLATA, D. R. 2004. Resolving questions of consanguinity between the Late Ordovician Deicke, Millbrig and Kinnekulle K-Bentonites in North America and Baltoscandia. *Geological Society of America Abstracts with Programs*, **36**, 246.

HUNDT, R. 1965. *Aus der Welt der Graptolithen.* Commercia Verlag, Berlin-Bonn.

HUTTON, D. H. W. & MURPHY, F. C. 1987. The Silurian of the Southern Uplands and Ireland as a successor basin to the end-Ordovician closure of Iapetus. *Journal of the Geological Society*, London, **144**, 765–772.

IORDAN, M. 1984. Biostratigraphy of the Silurian and Devonian in the Moldavian and Moesian Platforms (Romania). *Anuarul Institutului de Geologie si Geofizica, Stratigrafie-Paleontologie*, **LXIV**, 259–267.

IORDAN, M. 1999*a*. The Palaeozoic brachiopods of Romania. *Geo-Eco-Marina*, **4**(1999), 135–145.

IORDAN, M. 1999*b*. Biostratigraphy of the Paleozoic in the Moesian Platform. *In:* Excursion Guide of the Joint Meeting of EUROPROBE TESZ, PANCARDI and Georift Projects "Dobrogea – the interface between the Carpathians and the Trans-European Suture Zone", Tulcea, 1999. *Romanian Journal of Tectonics and Regional Geology*, **77**(suppl. 2), 18–23.

IORDAN, M. & STANOIU, I. 1993. On the Lower Silurian macrofauna (Llandoverian) from the Valea Izvorului Formation (South Carpathians). *Academia Română, Revue Roumaine de Géologie*, **37**, 63–76.

IORDAN, M., ILIESCU, V., VISARION, A., BALTREŞ, A., SĂNDULESCU, E. & SEIFERT, K. 1985. Litho- and biostratigraphy of the Palaeozoic sequences in the Oprişor and Gârla Mare boreholes (southwestern part of the Moesian Platform). *Dări de seamă ale şedinţelor, Stratigrafie*, **69**, 5–28.

JAANUSSON, V. 1976. Faunal dynamics in the Middle Ordovician (Viruan) of Baltoscandia. *In:* BASSETT, M. G. (ed.) *The Ordovician System. Proceedings of Palaeontogical Association Symposium*, Birmingham, 1974, 301–326.

JAEGER, H. 1962. Das Silur (Gotlandium) in Thüringen und am Ostrand des Rheinischen Schiefergebirges (Kellerwald, Marburg, Gießen). Symposium-Band d. 2. Internationalen Arbeitstagung über die Silur /Devon - Grenze und die Stratigraphie von Silur und Devon. Bonn-Bruxelles 1960. *Schweizerbart'sche Verlagsbuch*, Stuttgart, 108–135.

JAEGER, H. 1975. Die Graptolithenführung im Silur/Devon des Cellon-Profils (Karnische Alpen). *Carinthia* II, **165**(85), 111–126.

JAEGER, H. 1977. Das Silur/Lochkov-Profil im Frankenberger Zwischengebirge (Sachsen). *Freiberger Forschungshefte*. **C326**, 45–59.

JAEGER, H. 1978. Graptolithen aus dem Silur der Nördlichen Grauwackenzone (Ostalpen). *Mitteilungen. Österreichische Geologische Gesellschaft*, **69**(1976), 89–107.

JAEGER, H. 1988. Devonian Graptoloidea. *In:* MCMILLAN, N. J., EMBRY, A. F. & GLASS, D. J. (eds) *Devonian of the World.* Vol. 3. Proceedings of the Second International Symposium on the Devonian System Calgary, Alberta.

JAEGER, H. 1991. Neue Standard-Graptolithenzonenfolge nach der "Großen Krise" an der Wenlock/Ludlow-Grenze (Silur). *Neues Jahrbuch für Geologie und Paläontologie, Abhandlungen*, **182**, 303–354.

JAEGER, H. & SCHÖNLAUB, H. P. 1977. Das Ordoviz/Silur-Profil im Nölblinggraben (Karnische Alpen, Österreich). *Verhandlungen der Geologische Bundesanstalt, Wien*, **1977**, 349–359.

JAEGER, H. & SCHÖNLAUB, H. P. 1980. Silur und Devon nördlich der Gundersheimer Alm in den Karnischen Alpen (Österreich). *Carinthia* II, Klagenfurt, **170**(90), 403–444.

JAEGER, H. & SCHÖNLAUB, H. P. 1994. Section 4 „Graptolithengraben (graptolite gorge) north of Upper Bischofalm. *In:* SCHÖNLAUB, H. P. & KREUTZER, L. H. (eds) *Field Meeting Eastern + Southern Alps, Austria 1994.* Subcommission on Silurian Stratigraphy. Berichte Geologische Bundesanstalt, Wien, **30**, 97–100.

JAWOROWSKI, K. 2000. Facies analysis of the Silurian shale-siltstone succession in Pomerania (northern Poland). *Geological Quarterly*, **44**, 297–315.

JAWOROWSKI, K. JURKIEWICZ, H. & KOWALCZEWSKI, Z. 1967. Sinian and Palaeozoic in the borehole Jaronowice IG-1. *Kwartalnik Geologiczny*, **11**, 21–38 [in Polish with English summary].

JEPPSSON, L. 1987. Lithological and conodont distributional evidence for episodes of anomalous oceanic conditions during the Silurian. *In:* ALDRIDGE, R. J. (ed.) *Palaeobiology of Conodonts*. Ellis Horwood, Chichester, 129–145.

JEPPSSON, L. 1990. An oceanic model for lithological and faunal changes tested on the Silurian record source. *Journal of the Geological Society*, London, **147**, 663–674.

JEPPSSON, L. 1997. A new latest Telychian, Sheinwoodian and Early Homerian (Early Silurian) Standard Conodont Zonation. *Transactions of the Royal Society of Edinburgh, Earth Sciences*, **88**, 91–114.

JEPPSSON, L. 1998. Silurian oceanic events: a summary of general characteristics. *In:* LANDING, E. & JOHNSON, M. E. (eds) *Silurian Cycles: Linkages of Dynamic Stratigraphy with Atmospheric, Oceanic and Tectonic Changes*. New York State Museum Bulletin, **491**, 239–257.

JERENIC, G., PAMIC, J., SREMAC, J. & SPANIC, D. 1994. Palynological and organic-petrographic data on very low-and low-grade metamorphic rocks in the Slavonian Mountains (Northern Croatia). *Geologia Croatica*, **47**(2), 149–155.

JERZMAŃSKI, J. 1965. Geology of north-eastern part of the Kaczawa Mts. And their eastern extent. *Biuletyn Instytutu Geologicznego*, **185**, 109–195 [in Polish].

JOHNSON, M. E. 2006. Relationship of Silurian sea-level fluctuations to oceanic episodes and events. *GFF*, **128**, 115–121.

JOHNSON, M. E., KALJO, D. & RONG, J.-Y. 1991. Silurian eustasy. *In:* BASSETT, M. G., LANE, P. D. & EDWARDS, D. (eds) *The Murchison Symposium. Proceedings of an International Conference on The Silurian System*. Special Papers in Palaeontology, **44**, 145–163.

JOHNSON, M. E., RONG, J.-Y. & KERSHAW, S. 1998. Calibrating Silurian eustasy against the erosion and burial of coastal paleotopography. *In:* LANDING, E. & JOHNSON, M. E. (eds) *Silurian Cycles: Linkages of Dynamic Stratigraphy with Atmospheric, Oceanic and Tectonic Changes*. New York State Museum Bulletin, **491**, 3–13.

JURKIEWICZ, H. 1975. The geological structure of the basement of the Mesozoic in the central, part of the Miechów Trough. *In:* Geological Research in the Świetokrzyskie Region. *Biuletyn Instytutu Geologicznego*, **283**, 5–83 [in Polish with English summary].

KALJO, D. & KLAAMANN, E. 1973. Ordovician and Silurian Corals. *In:* HALLAM, A. (ed.) *Atlas of Paleobiogeography*. Elsevier, Amsterdam, 37–45.

KALJO, D. NESTOR, H. & EINASTO, R. 1991. Aspects of Silurian carbonate platform sedimentation. *In:* BASSETT, M. G., LANE, P. D. & EDWARDS, D. (eds) *The Murchison Symposium. Proceedings of an International Conference on The Silurian System*. Special Papers in Palaeontology, **44**, 205–224.

KALJO, D., MARTMA, T., MÄNNIK, P. & VIIRA, V. 2003. Implications of Gondwana glaciations in the Baltic late Ordovician and Silurian and a carbon isotopic test of environmental cyclicity. *Bulletin de la Société Géologique de France*, **174**, 59–66.

KATZUNG, G. 1999. Records of the late Ordovician glaciation from Thuringia, Germany. *Zeitschrift der deutschen geologischen Gesellschaft*, **150**, 595–617.

KATZUNG, G. 2001. The Caledonides at the southern margin of East European Craton. *Neues Jahrbuch für Geologie und Paläontologie, Abhandlungen*, **222**, 3–53.

KATZUNG, G. & FELDRAPPE, H. 2004. Kaledonisches Stockwerk. *In:* KATZUNG, G. (ed.) *Geologie von Mecklenburg-Vorpommern*. Schweizerbart'sche Verlagsbuchhandlung, 371–383.

KATZUNG, G., MALETZ, J. & FELDRAPPE, H. 2004. Altpaläozoikum. *In:* KATZUNG, G. (ed.) *Geologie von Mecklenburg-Vorpommern*. Schweizerbart'sche Verlagsbuchhandlung, 51–69.

KETTNER, R. & REMEŠ, M. 1935. *Sur la trouvaille des schistes siluriens a Graptolites en Moravia*. Věstník Královské České Společnosti Nauk, Třída II, 1–11.

KNELLER, B. C. 1991. A foreland basin on the southern margin of Iapetus. *Journal of the Geological Society, London,* **148**, 207–210.

KNELLER, B. C., KING, L. M. & BELL, A. M. 1993. Foreland basin development and tectonics on the northwest margin of Eastern Avalonia. *Geological Magazine*, **130**, 691–697.

KOREJWO, K. & TELLER, L. 1964. Upper Silurian non-graptolite fauna from Chełm borehole (Eastern Poland). *Acta Geologica Polonica*, **14**, 233–400 [in Polish with English summary].

KOREN', T. N. 1987. Graptolite dynamics in Silurian and Devonian time. *Bulletin of the Geological Society of Denmark*, **35**, 149–159.

KOREN', T. & BJERRESKOV, M. 1997. Early Llandovery monograptids from Bornholm and the southern Urals: taxonomy and evolution. *Bulletin of the Geological Society of Denmark*, **44**, 1–43.

KOREN', T. N., LENZ, A. C., LOYDELL, D. K., MELCHIN, M. J., ŠTORCH, P. & TELLER, L. 1995. Generalized graptolite zonal sequence defining Silurian time intervals for global paleogeographic studies. *Lethaia*, **28**, 137–138.

KOREN', T. N., LENZ, A. C., LOYDELL, D. K., MELCHIN, M. J., ŠTORCH, P. & TELLER, L. 1996. Generalized graptolite zonal sequence defining Silurian time intervals for global paleogeographic studies (corrections). *Lethaia*, **29**, 59–60.

KORNAŚ, I. 1974. Dolny sylur w Górach Kaczawskich na podstawie fauny graptolitowej. Sprawozdania z posiedzeń naukowych Instytutu Geologicznego. *Kwartalnik geologiczny*, **18**, 433–434 [in Polish].

KORNAŚ, I. 1977. Problemy litostratygraficzne dolnego syluru w Sudetach Zachodnich. Sprawozdania z posiedzeń naukowych Instytutu Geologicznego. *Kwartalnik Geologiczny*, **21**, 388–389.

KOSTIĆ-PODGORSKA, V. 1956. Kačanik-Veles Serie and Stantrg Series. *In:* PETKOVIĆ, K. V. (ed.) *Lexique Stratigraphique Internationale. Volume 1 Europe, Fascicule 12A: Yougoslavie*. CNRS., Paris, **16–17**, 36.

KOSSMAT, F. 1927. Gliederung des varistischen Gebirgsbaues. *Abhandlungen des Sächsischen Geologischen Landesamtes*, **1**, 1–39.

KOZŁOWSKI, W. 2002. Sedimentary environment of the Upper Silurian Carbonate beds from the Łysogóry Region of the Holy Cross Mts and its paleogeographical implications. *Przegląd Geologiczny*, **50**, 1206 [in Polish].

KOZŁOWSKI, W. 2003. A *Cooksonia*-type flora from the Upper Silurian of the Holy Cross Mountains, Poland. *Acta Palaeontologica Polonica*, **48**, 653–656.

KOZŁOWSKA-DAWIDZIUK, A. 1995. Silurian retiolitids of the East European Platform. *Acta Palaeontologica Polonica*, **40**, 261–326.

KRAWCZYK, C. M., MCCANN, T., COCKS, L. R. M., ENGLAND, R., MCBRIDE, J. & WYBRANIEC, S. 2008. Caledonian tectonics. *In*: MCCANN, T. (ed.) *The Geology of Central Europe. Volume 1: Precambrian and Palaeozoic*. Geological Society, London, 303–381.

KŘÍŽ, J. 1975. Revision of the Lower Silurian stratigraphy in central Bohemia. *Věstník Ústředního ústavu geologického*, **50**(5), 275–282.

KŘÍŽ, J. 1979. Silurian Cardiolidae (Bivalvia). *Sbornik Geologických Věd, Paleontologie*, **22**, 1–157.

KŘÍŽ, J. 1989. Přídolí Series in the Prague Basin (Barrandian area, Bohemia). *In:* HOLLAND, C. H. & BASSETT, M. G. (eds) *A Global Standard for the Silurian System*. National Museum of Wales, Cardiff, Geological Series, **9**, 90–100.

KŘÍŽ, J. 1991. The Silurian of the Prague Basin (Bohemia) – tectonic, eustatic and volcanic controls on facies and faunal development. *In:* BASSETT, M. G., LANE, P. D. & EDWARDS, D. (eds) *The Murchison Symposium. Proceedings of an International Conference on The Silurian System*. Special Papers in Palaeontology, **44**, 179–203.

KŘÍŽ, J. 1992. *Silurian field excursions: Prague Basin (Barrandian) Bohemia*. National Museum of Wales, Cardiff, Geological Series, **13**, 1–111.

KŘÍŽ, J. 1996. Silurian Bivalvia of Bohemian type from the Montagne Noire and Mouthoumet Massif, France. *Palaeontographica, Abteilung A*, **240**, 29–63.

KŘÍŽ, J. 1998*a*. Recurrent Silurian–lowest Devonian cephalopod limestones of Gondwanan Europe and Perunica. *In:* LANDING, E. & JOHNSON, M. E. (eds) *Silurian Cycles: Linkages of Dynamic Stratigraphy with Atmospheric, Oceanic, and Tectonic Changes*. New York State Museum Bulletin, **491**, 183–198.

KŘÍŽ, J. 1998*b*. Silurian. *In:* CHLUPÁČ, I., HAVLÍČEK, V., KŘÍŽ, J., KUKAL, Z. & ŠTORCH, P. (eds) *Paleozoic of the Barrandian (Cambrian to Devonian)*. Český geologický ústav, Praha, 79–101.

KŘÍŽ, J. 1999*a*. Bivalvia dominated communities of Bohemian type from the Silurian and Lower Devonian carbonate facies. *In:* BOUCOT A. J. & LAWSON, J. D. (eds) *Palaeocommunities: A Case Study from the Silurian and Lower Devonian*. Cambridge University Press, Cambridge, World and Regional Geology series **11**, 229–252.

KŘÍŽ, J. 1999*b*. Silurian and lowermost Devonian bivalves of Bohemian type from the Carnic Alps. *Abhandlungen der Geologischen Bundesanstalt*, **56**(1), 259–316.

Kříž, J. 1999*c*. Silurian Bivalvia – evolution, palaeocology, palaeogeography, importance for biostratigraphy and correlation. *Abhandlungen der Geologischen Bundesanstalt*, **54**, 377–384.

Kříž, J. & Paris, F. 1982. Ludlovian, Pridolian and Lochkovian in La Meignanne (Massif Armoricain): Biostratigraphy and correlations based on Bivalvia and Chitinozoa. *Geobios*, **15**, 391–421.

Kříž, J. & Serpagli, E. 1993. Upper Silurian and lowermost Devonian Bivalvia of Bohemian type from south-western Sardinia. *Bollettino della Societá Paleontologica Italiana*, **32**(3), 289–347.

Kříž, J. & Veselinovič, M. 1975. Ludlovian, Přídolian and Lochkovian bivalves from the Suva Planina Mountains. (Eastern Serbia, Yugoslavia). *Věstník Ústředního ústavu geologického*, Praha, **50**(6), 365–369.

Kříž, J., Jaeger, H., Paris, F. & Schönlaub, H. P. 1986. Přídolí – the fourth subdivision of the Silurian. *Jahrbuch der Geologischen Bundesanstalt, Wien*, **129**(2), 291–360.

Kříž, J., Dufka, P., Jaeger, H. & Schönlaub. H. P. 1993. The Wenlock/Ludlow boundary in the Prague Basin (Bohemia). *Jahrbuch der Geologische Bundesanstalt*, **136**(4), 809–839.

Kříž, J., Degardin, J.-M., *et al.* 2003. Silurian stratigraphy and paleogeography of north Gondwanan and Perunican Europe. *In:* Landing, E. & Johnson, M. E. (eds) *Silurian Lands and Seas, Paleogeography outside of Laurentia* (Proceedings of the James Hall Symposium, Rochester, NY, USA, August 1996). New York State Museum Bulletin, **493**, 105–178.

Kroner, U., Mansy, J.-L. *et al.* 2008. Variscan tectonics. *In*: McCann, T. (ed.) *The Geology of Central Europe. Volume 1: Precambrian and Palaeozoic*. Geological Society, London, 599–664.

Krs, M., Krsová, M., Pruner, P. & Havlíček, V. 1986. Paleomagnetism, paleogeography and multi-component analysis of magnetization of Ordovician rocks from the Barrandian area of the Bohemian Massif. *Sborník geologických Věd, užitá Geofyzika*, **20**, 9–45.

Krs, M., Krsová, M., Pruner, P., Chvojka, R. & Havlíček, V. 1987. Palaeomagnetism, palaeogeography and the multicomponent analysis of Middle and Upper Cambrian rocks of the Barrandian in the Bohemian Massif. *Tectonophysics*, **139**, 1–20.

Kryza, R. & Muszyński, A. 2003. The metamorphic Kaczawa Complex – fragment of Variscan accretionary prism. *In:* Ciężkowski, W., Wojewoda, J. & Żelaźniewicz, A. (eds) *Sudety zachodnie: od wendu do czwartorzędu*. WIND Wrocław, 95–104 [in Polish with English summary].

Kryza, R., Muszyński, A., Turniak, K. & Zalasiewicz, J. A. 1994. A Lower Palaeozoic shallow water sequence in the eastern European Variscides (SW Poland): provenance and depositional history. *Geologische Rundschau*, **83**, 5–19.

Kupfahl, H.-G. 1953. Untersuchungen im Gotlandium und Unterdevon des Kellerwaldes und bei Marburg. *Notizblatt Hessischen Landesamest für Bodenforschung zu Wiesbaden*, **81**, 96–128.

Kupfahl, H.-G. 1954. Graptolithen des Kellerwaldes. *Roemeriana*, **1**, 173–182.

Kurze, M. 2006. 5.16 Mittelsachsen. *In:* Deutsche Stratigraphische Kommission (ed.) *Silur Monographie.* Schriftenreihe der Deutschen Gesellschaft für Geowissenschaften, **46**, 106–110.

Kurze, M & Tröger, K.-A.1990. Die Entwicklung des Paläozoikums in der Elbtalzone (Saxothuringikum). *Neues Jahrbuch Geologie und Paläontologie, Monatshefte*, **1**, 43–53.

Lawson, J. D. & White, D. E. 1989. The Ludlow Series in the type area. *In:* Holland, C. H. & Bassett, M. G. (eds) *A Global Standard for the Silurian System*. National Museum of Wales, Cardiff, Geological Series, **9**, 73–90.

Lazauskienė, J., Stephenson, R., Sliaupa, S. & van Wees, J. D. 2002. 3-D flexural modelling of the Silurian Baltic Basin. *Tectonophysics*, **346**, 115–135.

Lazauskienė, J., Sliaupa, S., Brazauskas, A., Musteikis, P., 2003. Sequence stratigraphy of the Baltic Silurian succession: tectonic control on the foreland infill. *In:* McCann, T. & Saintot, A. (eds) *Tracing Tectonic Deformation Using the Sedimentary Record*. Geological Society, London, Special Publications, **208**, 95–115.

Lee, M., Pharaoh, T. C., Williamson, J. P., Green, C. A. & De Vos, W. 1993. Evidence on the deep structure of the Anglo-Brabant Massif from gravity and magnetic data. *Geological Magazine*, **130**(5), 575–582.

Legrand, R. 1961. Mise à jour de la légende stratigraphique du Gothlandien inférieur du massif du Brabant. *Bulletin de la Société belge de Géologie, Paléontologie et Hydrologie*, **70**, 186–195.

Legrand, R. 1969. *Le Massif du Brabant*. Mémoires pour servir à l'explication des Cartes géologiques et minières de la Belgique, **9**(1968).

Lenz, A.C. 1982 Ordovician to Devonian sea-level changes in western and northern Canada. *Canadian Journal of Earth Sciences*, **19**, 1919–1932.

Lindström, M. 1960. On some sedimentary and tectonic structures in the Ludlovian Colonus Shale of Scania. *GFF*, **82**, 319–341.

Linnemann, U. (ed.) 2003*a*. Das Saxothuringikum. Abriss der prakambrischen und paläozoischen Geologie von Sachsen und Thüringen. *Staatliches Museum fur Mineralogie und Geologie zu Dresden, Geologica Saxonica*, **48/49**, 1–159.

Linnemann, U. 2003*b*. Die Struktureinheiten des Saxothuringikums. *Geologica Saxonica*, **48/49**, 19–28.

Linnemann, U. 2003*c*. Sedimentation und geotektonischer Rahmen der Beckenentwicklung im Saxothuringikums (Neoproterozoikum-Unterkarbon). *Geologica Saxonica*, **48/49**, 71–110.

Linnemann, U., Elicki, O. & Gaitztsch, B. 2003*a*. Die Stratigraphie des Saxothuringikums. *Geologica Saxonica*, **48/49**, 29–70.

Linnemann, U., Romer, R. L., Gehmlich, M. & Drost, K. 2003*b*. Paläogeographie und Provenance des Saxothuringikums unter besonderer Beachtung der Geochronologie von prävaiszischen Zirkonen und der Nd-Isotopie von Sedimenten. *Geologica Saxonica*, **48/49**, 121–132.

Linnemann, U., McNaughton, N. J., Romer, R. L., Gehmlich, M., Drost, K. & Tonk, C. 2004. West African provenance for Saxothuringia (Bohemian Massif): Did Armorica ever leave pre-Pangean Gondwana? U/Pb-SHRIMP zircon evidence and the Nd-isotopic record. *International Journal Earth Sciences, Geologische Rundschau*, **93**, 683–705.

Loeschke, J. & Heinisch, H. 1993. Palaeozoic volcanism of the Eastern Alps and its palaeotectonic significance. *In*: Raumer, J. F. von & Neubauer, F. (eds) *The Pre-Mesozoic Geology in the Alps*. Springer, Berlin, 441–455.

Long, D. G. F. 1993. Oxygen and Carbon Isotopes and Event Stratigraphy near the Ordovician Silurian Boundary, Anticosti Island Quebec. *Palaeogeography, Palaeoclimatology, Palaeoecology*, **104**(1–4), 49–59.

Louwye, S., Van Grootel, G. &Verniers, J. 1992. The stratigraphy of the type locality of the ?late Wenlock/early Ludlow Mont Godart Formation and the early Ludlow Ronquières Formation, Brabant Massif, Belgium. *Annales de la Société géologique de Belgique*, **115**(1), 307–331.

Loydell, D. 1998. Early Silurian sea-level changes. *Geological Magazine*, **135**, 447–471.

Loydell, D. 2003. Late Telychian graptolites of the Rauchkofelboden section (Central Carnic Alps, Austria). *Jahrbuch der Geologische Bundesanstalt, Wien*, **143**, 57–61.

Loydell, D. K., Štorch, P. & Melchin, M. J. 1993. Taxonomy, evolution and biostratigraphical importance of the Llandovery graptolite *Spirograptus*. *Palaeontology*, **36**, 909–926.

Loydell, D. K., Männik, P. & Nestor, V. 2003. Integrated biostratigraphy of the lower Silurian of the Aizpute-41 core, Latvia. *Geological Magazine*, **140**, 205–229.

Luppold, F. W. 1995. Ein neues Vorkommen von Ober-Silur (Pridolium) – Conodonten auf Blatt Solingen (Rheinisches Schiefergebirge). *Courier Forschungsinstitut Senckenberg*, **188**, 1–9.

McCann, T. 1996. Silurian facies from the G-14 well, offshore Northern Germany. *Zeitschrift der Deutschen geologischen Gesellschaft*, **147**, 209–219.

McCann, T., Skompski, S., *et al.* 2008. Carboniferous. *In*: McCann, T. (ed.) *The Geology of Central Europe. Volume 1: Precambrian and Palaeozoic*. Geological Society, London, 411–529.

McKerrow, W. S. 1979. Ordovician and Silurian changes in sea level. *Journal of the Geological Society of London*, **136**, 137–45.

McKerrow, W. S., MacNiocaill, C., Ahlberg, P. E., Clayton, G., Cleal, C. J. & Eagar, R. M. C. 2000. The Late Palaeozoic relations between Gondwana and Laurussia. *In:* Franke, W., Haak, V., Oncken, O. & Tanner, D. (eds) *Orogenic Processes: Quantification and Modelling in the Variscan Belt*. Geological Society, London, Special Publications, **179**, 9–20.

McLean, R.A. 1985. New Early Silurian rugose corals from the Panuara area, central New South Wales. *Alcheringa*, **9**, 23–34.
MacNiocaill C. & Smethurst, M. A. 1994. Palaeozoic palaeogeography of Laurentia and its margins: a reassessment of palaeomagnetic data. *Geophysical Journal International*, **116**, 715–725.
Maes, G., Rickards, R. B., Rombouts, L. & Vandevelde, N. 1979. Silurian formations between Neuville-sous-Huy and Ombret: their correlation, age and structure. *Annales de la Société Géologique de Belgique*, **101**(for 1978), 31–36.
Maksym, A., Śmist, P., Pietrusiak, M. Staryszak, G. & Liszka, B. 2003. New data on development of the Lower Paleozoic sediments in the Sędziszów Małopolski-Rzeszów region baset on Hermanowa-1 borehole (SE Poland). *Przegląd Geologiczny*, **51**, 412–418 [in Polish with English summary].
Malec, J. 1991. "Szarogłazy niewachlowskie" w profilu górnego syluru regionu południowego Gór Świętokrzyskich. *Kwartalnik geologiczny*, **35**, 284–285 [in Polish with English summary].
Malec, J. 1993. Upper Silurian and Lower Devonian in the western Holy Cross Mts. *Kwartalnik geologiczny*, **37**, 501–536 [in Polish with English summary].
Malec, J. 2001. Sedimentology of deposits from around the Late Caledonian unconformity in the western Holy Cross Mts. *Geological Quarterly*, **45**, 397–415.
Maletz, J. 1997. Ordovician and Silurian strata of the G-14 well (Baltic sea): graptolite faunas and biostratigraphy. *Zeitschrift für Geologische Wissenschaften*, **25**, 29–39.
Maletz, J. 2000. Ordovician biostratigraphy of the Herscheid Schichten (Rheinisches Schiefergebirge, Germany). *Neues Jahrbuch für Geologie und Paläontologie, Abhandlungen*, **218**, 45–60.
Maletz, J. 2001. Graptolite research in Germany. *Geologica Saxonica, Abhandlungen des Staatlichen Museums für Mineralogie und Geologie Dresden*, **46/47**, 169–180.
Maletz, J. & Katzung, G. 2003. Silur. *In:* Seidel, G. (ed.). *Geologie von Thüringen*. Schweizerbart'sche Verlagsbuchhandlung, 130–140.
Maletz, J., Beier, H., Katzung, G. & Niedzwiedz, A. 1997. A Lower Palaeozoic (Ordovician – Silurian) foreland basin at the southwestern rim of Baltica. *Terra Nostra*, **97**(11), 81–84.
Malinowska, L. 1955. Stratygrafia gotlandu Gór Bardzkich. *Biuletyn Instytutu Geologicznego*, **95**, 5–81.
Maronde, H.-D. 1966. Der Grenzbereich Silur-Devon im Südwestharz. *Nachrichten der Akademie der Wissenschaften Göttingen, II. Mathematische Physikalische Klasse* **5**, 53–63.
Maronde, H.-D. 1968. Geologische Untersuchungen im Grenzbereich Silur-Devon des Südwest-Harzes. *Geologisches Jahrbuch*, **86**, 113–172.
Martin, F. 1969. Les acritarches de l'Ordovicien et du Silurien belge. Détermination et valeur stratigraphique. *Mémoires de l'Institut Royal des Sciences naturelles de Belgique*, **160**.
Martin, F. 1974. Ordovicien supérieur et Silurien inférieur à Deerlijk (Belgique). Palynofaciès et microfaciès. *Mémoires de l'Institut Royal des Sciences Naturelles de la Belgique*, **174**.
Martma T., Brazauskas A., Kaljo D., Kaminskas D. & Musteikis P. 2005. The Wenlock-Ludlow carbon isotope trend in the Vidukle core, Lithuania, and its relations with oceanic events. *Geological Quarterly*, **49**, 223–234.
Melchin, M. J. 2004. Restudying a global stratotype for the base of the Silurian: a report to the members of the Subcommission on Silurian Stratigraphy. *Silurian Times*, **12**, 12–13.
Melchin, M. J. & Holmden, C. 2006. Carbon isotope chemostratigraphy of the Llandovery in Arctic Canada: implications for global correlation and sea-level change. *GFF*, **128**, 173–180.
Melchin, M. J., Cooper, R. A. & Sadler, P. M. 2004. The Silurian Period. *In:* Gradstein, F.M., Ogg, J.G. & Smith, A.G. (eds) *A Geologic Time Scale*. Cambridge University Press, Cambridge, 188–201.
Ménot, R. P., von Raumer, J. F., Bogdanoff, S. & Vivier, G. 1994. Variscan Basement of the Western Alps: the External Cristaline Massifs. *In:* Keppie, J. D. (ed.) *Pre-Mesozoic Geology in France and Related Areas*. Springer-Verlag, Berlin, 458–466.
Michot, P. 1954. Le Silurien. *In:* Fourmarier, P. (ed.) *Podrôme d'une description géologique de la Belgique*. Imprimerie Vaillant Carmanne, Liège, 39–82.
Michot, P. 1978. La faille mosane et la phase hyporogénique bollandienne, d'âge emsien dans le rameau calédonien condruso-brabançon. *Annales de la Société géologique de Belgique*, **101**, 321–335.
Modliński, Z. 1968. Ordovician in West Pomerania. *Kwartalnik geologiczny*, **12**, 488–492 [in Polish with English summary].
Modliński, Z. & Szymański, B. 2001. The Silurian of the Nida, Holy Cross Mts. and Radom areas, Poland – a review. *Geological Quarterly*, **45**, 435–454.
Molyneux, S. G. 1991. The contribution of palaeontological data to an understanding of the Early Palaeozoic framework of eastern England. *Annales de la Société Géologique de Belgique*, **114**, 93–105.
Montenari, M., Servais, T. & Paris, F. 2000. Palynological dating (acritarchs and chitinozoans) of Lower Paleozoic phyllites from the Black Forest/southern Germany. *Comptes Rendus de l'Académie des Sciences, Série II (A), Sciences de la Terre et des Planètes*, **330**, 493–499.
Moreau, J., Ghienne, J.-F. Deynoux, M. & Rubino, J.-L. 2004. Ice-proximal sedimentary records of the Late Ordovician glacial cycles. *Erlanger geologische Abhandlungen, Sonderband*, **5**, 55.
Moryc, W. 1992. Budowa geologiczna utworów podłoza miocenu w rejonie Sędziszów Młp.-Rzeszów i ich perspektywność. *Nafta-Gaz*, **9/10**, 206–223.
Moryc, W. 1996. Budowa geologiczna podłoza miocenu w rejonie Pilzno-Dębica-Sędziszów Młp.-Rzeszów i ich perspektywność. *Nafta-Gaz*, **12**, 521–550.
Mostler, H. 1966. Zur Einstufung der „Kieselschiefer" von der Lachtal-Grundalm (Fieberbrunn, Tirol). *Verhandlungen der Geologische Bundesanstalt, Wien*, **1966**, 157–170.
Mullins, G. L. 2000. A chitinozoan morphological lineage and its importance in Lower Silurian stratigraphy. *Palaeontology*, **43**, 359–373.
Mullins, G. L. & Aldridge, R. J. 2004. Chitinozoan biostratigraphy of the basal Wenlock Series (Silurian) Global Stratotype Section and Point. *Palaeontology*, **47**, 745–773.
Neubauer, F. 1989. Lithostratigraphie und Strukturen an der Basis der Rannachdecke im zentralen Grazer Paläozoikum (Ostalpen). *Jahrbücher, Geologische Bundesanstalt, Wien*, **132**, 459–474.
Nielssen, A. T. 1995. Trilobite biostratigraphy, palaeoecology, and systematics of the Komstad Limestone and Huk Formations (lower Ordovician), southern Scandinavia. *Fossils & Strata*, **38**, 1–374.
Oberc, J. 1970. The Silurian: the Sudetes. *In*: Skołowski, S. (ed.) *Geology of Poland, Stratigraphy, Part 1, Pre-Cambrian and Palaeozoic*. Wydawnictwa Geologiczne, Warszawa, 278–283.
Obst, K., Böhnke, A., Katzung, G. & Maletz, J. 2002. Pb–Pb zircon dating of tuff horizons in the Cyrtograptus Shale (Wenlock, Silurian) of Bornholm. *Bulletin of the Geological Society of Denmark*, **49**, 1–8.
Paris, F. 2000. Early Palaeozoic paleobiogeography of northern Gondwana regions. *Acta Universitatis Carolinae Geologica*, **42**(1998), 473–483.
Paris, F. & Kříž, J. 1984. Nouvelles espèces de Chitinozoaires à la limite Ludlow- Přídolí en Tchécoslovaquie. *Review of Palaeobotany and Palynology*, **43**, 155–177.
Paris, F. & Le Pochat, G. 1994. The Aquitaine Basin. *In:* Keppie, J. D. (ed.) *Pre-Mesozoic Geology in France and Related Areas*. Springer-Verlag, Berlin, 405–415.
Paris, F. & Robardet, M. 1990. Early Palaeozoic palaeobiogeography of the Variscan regions. *Tectonophysics*, **177**, 193–213.
Paris, F., Winchester-Seeto, T., Boumendjel, K. & Grahn, Y. 2000. Toward a global biozonation of Devonian chitinozoans. *Courier Forschungsinsitut Senckenberg*, **220**, 39–55.
Paskevicius, J. 1997. *The Geology of the Baltic States*. Geological Survey of Lithuania, Vilnius.
Pedder, A. E. H. & Oliver, D. W. A. Jr. 1990. Rugose coral distribution as a test of Devonian palaeogeographic models. *In:* McKerrow, W. S. & Scotese, C.R. (eds) *Palaeozoic Palaeogeography and Biogeography*. Geological Society of London, Memoirs, **12**, 267–275.
Pedersen, G. K. 1989. The sedimentology of the Lower Palaeozoic black shales from the shallow wells Skelbro 1 and Billgrav 1, Bornholm, Denmark. *Bulletin of the Geological Society of Denmark*, **37**, 151–173.
Pedersen, G. K. & Klitten, K. 1990. Anvendelse af gamma-logs ved korrelation af marine skifre i vandforsyningsboringer på Bornholm. *Dansk Geologisk Forening, Årsskrift for* **1987–89**, 21–35.
Pharaoh, T. C. 1999. Palaeozoic terranes and their lithospheric

boundaries within the Trans-European Suture Zone (TESZ): a review. *Tectonophysics*, **314**, 17–41.

PHARAOH, T. C., MOLYNEUX, S. G., MERRIMAN, R. J., LEE, M. K. & VERNIERS, J. 1993. The Caledonides of the Anglo-Brabant Massif reviewed. *In:* PHARAOH, T. C., MOLYNEUX, S. G., MERRIMAN, R. J., LEE, M. K. & VERNIERS, J. (eds) Special Issue on the Caledonides of the Anglo-Brabant Massif. *Geological Magazine*, **130**(5), 561–562.

PICKETT, J.W. 1975. Continental reconstructions and the distribution of coral faunas during the Silurian. *Proceedings of Royal Society of New South Wales*, **108**, 147–156.

PIESSENS, K., DE VOS, W., BECKERS, R., *et al.* 2005. Opmaak van de pre-Krijt subcropkaart van het Massief van Brabant voor invoering in de Databank Ondergrond Vlaanderen. Final report to the 'Ministerie van de Vlaamse Gemeenschap, Afdeling Natuurlijke Rijkdommen en Energie'.

PIPER, J. D. A. 1997. Tectonic rotation within the British paratectonic Caledonides and Early Palaeozoic location of the orogen. *Journal of the Geological Society, London*, **154**, 9–13.

PIQUÉ, A., FLUCK, P., SCHNEIDER, J. L. & WHITECHURCH, H. 1994. The Vosges Massif. *In:* KEPPIE, J. D. (ed.) *Pre-Mesozoic Geology in France and Related Areas*. Springer-Verlag, Berlin, 416–425.

PODHALAŃSKA, T. 2003. The Ordovician/Silurian boundary – graptolitic biozonation and chronostratigraphy: problems and new ideas. *Przegląd Geologiczny*, **51**, 942–946 [in Polish with English summary].

PODHALAŃSKA, T. & MODLIŃSKI, Z. 2006. Stratigraphy and facies characteristics of the Ordovician and Silurian deposits of the Koszalin-Chojnice zone; similarities and differences to the western margin of the east European craton and Rügen area. *Prace Państwowego instytutu geologicznego*, **186**, 39–78 [in Polish with English summary].

POPRAWA, P., SLIAUPA, S., STEPHENSON, R. & LAZAUSKIENE, J. 1999. Late Vendian-Early Palaeozoic tectonic evolution of the Baltic Basin: regional tectonic implications from subsidence analysis. *Tectonophysics*, **314**, 219–239.

POPRAWA, P. 2006. Development of the Caledonian collision zone along the western margin of Baltica and its relation to the foreland basin. *Prace Państwowego Instytutu Geologicznego* (Warsaw) 186: 189–214. [in Polish with extended abstract in English]

PORĘBSKA, E. 1982. Latest Silurian and Early Devonian graptolites from Żdanów section, Bardo Mts. (Sudetes). *Annales Societatis Geologorum Poloniae*, **52**, 89–209.

PORĘBSKA, E. 2002. Mid-Silurian paleobiogeography of Małopolska and Łysogóry massifs based on graptolites. *Przegląd Geologiczny*, **50**, 1220 [in Polish].

PORĘBSKA, E. & KOSZOWSKA, E. 2001. Mazuelloids – apatitealgae from the Lower Paleozoic of Bardzkie Mts (Sudety Mts, SW Poland). *Przegląd Geologiczny*, **49**, 1050–1060 [in Polish with English summary].

PORĘBSKA, E. & SAWŁOWICZ, Z. 1997. Palaeoceanographic linkage of geochemical and graptolite events across the Silurian-Devonian boundary in Bardzkie Mountains (Southwest Poland). *Palaeogeography, Palaeoclimatology, Palaeoecology*, **132**, 343–354.

PORĘBSKA, E., KOZŁOWSKA-DAWIDZIUK, A. & MASIAK, M. 2004. The lundgreni event in the Silurian of the East European Platform, Poland. *Palaeogeography, Palaeoclimatology, Palaeoecology*, **213**, 271–294.

POULSEN, C. 1974. Further contributions to the knowledge of the Palaeozoic of Slagerse No. 1, western Sealand, *Danmarks Geologiske Undersøgelse, II ser.*, **101**, 1–72.

POŻARYSKI, W. 1990. The Middle Europe Caledonides – wrenching orogen composed of terranes. *Przeglad geologiczny*, **38**, 1–9 [in Polish, with English Summary].

POŻARYSKI, W. & TOMCZYK, H. 1968. Assyntian Orogeny in South-East Poland. *Biuletyn Instytutu Geologicznego*, **237**, 13–28 [in Polish with English summary].

PRIEWALDER, H. 1987. *Acritarchen aus dem Silur des Cellon-Profils, Karnische Alpen, Österreich.* Abhandlungen Geologische Bundesanstalt, Wien, **40**.

PRIEWALDER, H. 1997. The distribution of the Chitinozoans in the Cellon Section (Hirnantian – Lower Lochkovian). A Preliminary Report. *In:* SCHÖNLAUB, H. P. (ed.) *IGCP – 412 Inaugural Meeting Vienna, Guidebook*. Berichte Geologische Bundesanstalt, Wien, **40**, 74–85.

PRIGMORE, J. K., BUTLER, A. J. & WOODCOCK, N. H. 1997. Rifting during separation of Eastern Avalonia from Gondwana: Evidence from subsidence analysis. *Geology*, **25**, 203–207.

PRZYBYŁOWICZ, T. & STUPNICKA, E. 1989. Petrographic characteristics of Upper Silurian rocks from Niestachów (Świętokezyskie Mts.). *Archiwum Mineralogiczne*, **44**, 129–152.

PRZYBYŁOWICZ, T. & STUPNICKA, E. 1991. Manifestation of volcanism in Ordovician and Silurian of the southern part of Świętokrzyskie Mountains. *Archiwum Mineralogiczne*, **47**, 137–154.

RACHEBOEUF, P. R. (ed.) 1986. Le Groupe de Liévin. Pridoli – Lochkovien de l'Artois (N. France). *Biostratigraphie du Paléozoïque*, **3**, 1–215.

RAUMER, J. F. VON & NEUBAUER, F. (eds) 1993. *The Pre-Mesozoic Geology in the Alps*. Springer, Berlin.

REITZ, E. 1987. Palynologie in metamorphen Serien: Silurische Sporen aus einem granatführenden Glimmerschiefer des Vor-Spessart. *Neues Jahrbuch für Geologie und Paläontologie, Monatshefte*, **1987**, 699–704.

REITZ, E. 1992. Silurische Mikrosporen aus einem Biotit-Glimmerschiefer bei Riitsteig. Nördlicher Bayrischer Wald. *Neues Jahrbuch für Geologie und Paläontologie, Monatshefte*, **1992**, 351–358.

REUTER, F. 1960. Stratigraphische Probleme im Silur des Harzes. *In:* SVOBODA, J. U. A. (ed.) *Prager Arbeitstagung über die Stratigraphie des Silurs und des Devons* (1958). Nakladatelstvi Ceskoslovenske Akademie Ved, Prag. 303–312.

RICHTER, R. 1875. Aus dem Thüringischen Schiefergebirge 5. *Zeitschrift der Deutschen Geologischen Gesellschaft*, **27**, 261–273.

RICKARDS, R. B. 1976. The sequence of Silurian graptolite zones in the British Isles. *Geological Journal*, **11**, 153–188.

RISTEDT, H. 1968. Zur Revision der Orthoceratidae. Abhandlungen der Akademie der Wissenschaften in Wien, mathematische-naturwissenschaftliche Klass, **1968**, 213–287.

RISTEDT, H. 1969. Orthoceren als Leitfossilien des Silurs. *Carinthia II, Sonderheft*, **27**, 25–28.

ROBARDET, M. 2002. Alternative approach to the Variscan Belt in southwestern Europe: Preorogenic paleobiogeographical constraints. *In*: GARCIA, F. D., CATALAN, J. R. M., ARENAS, R. & HATCHER, R. D. JR. (eds) *Appalachian dynamics: The building of the late Paleozoic basement.* Geological Society of America, Special Paper, **364**, 1–15.

ROBARDET, M. 2003. The Armorica 'microplate': fact or fiction? Critical review of the concept and contradictory palaeobiogeographical data. *Palaeogeography, Palaeoclimatology, Palaeoecology*, **195**, 125–148.

ROBARDET, M., VERNIERS, J., FEIST, R. & PARIS, F. 1994. Le Paléozoïque anté-varisque de France, contexte paléogéographique et géodynamique. *Géologie de la France*, **3**, 3–31.

ROSS, C. A. & ROSS, J. R. P. 1996. Silurian sea-level fluctuations. *In:* WITZKE, B. J., LUDVIGSON, G. A. & DAY, J. (eds) *Paleozoic Sequence Stratigraphy: Views from the North American Craton.* Geological Society of America, Special Papers, **306**, 197–192.

RUPPEL, S. C., JAMES, E. W., BARRICK, J. E., NOWLAN, G. & UYENO, T. T. 1998. High-resolution $^{87}Sr/^{86}Sr$ record: evidence for eustatic control of seawater chemistry? *In:* LANDING, E. & JOHNSON, M. E. (eds) *Silurian Cycles: Linkages of Dynamic Stratigraphy with Atmospheric, Oceanic and Tectonic Changes*. New York State Museum Bulletin, **491**, 285–295.

SAMTLEBEN, C., MUNNECKE, A., BICKERT, T. & PÄTZOLD, J. 1996. The Silurian of Gotland (Sweden): facies interpretation based on stable isotopes in brachiopod shells. *Geologische Rundschau*, **85**, 278–292.

SAMUELSSON, J., VECOLI, M. & BEIER, H. 2001. Ordovician–Silurian palynostratigraphy (Chitinozoa and acritarchs) of the G14–1/86 borehole, southern Baltic Sea. *Neues Jahrbuch für Geologie und Paläontologie, Abhandlungen*, **222**, 91–122.

SAMUELSSON, J., VECOLI, M., BERNARDCZYK, W.S. & VERNIERS, J. 2002. Timing of the Avalonia–Baltica plate convergence as inferred from palaeogeographic and stratigraphic data of chitinozoan assemblages in West Pomerania, northern Poland. *In:* WINCHESTER, J. A., PHARAOH T. C. & VERNIERS, J. (eds) *Palaeozoic Amalgamation of Central Europe*. Geological Society, London, Special Publications, **201**, 95–113.

SANTEL, W. 2001. Trilobiten aus dem Silur der Karnischen Alpen/Österreich, Teil I. *Palaeontographica, Abt.*, **A262**, 87–191.

SAWATSKI, G., VAIDA, M. & HANN, H. P. 1997. Altpaläozoische Chitinozoen und Acritarchen in Gneisen des Südschwarzwalds, SW

Deutschland. *Neues Jahrbuch für Geologie und Paläontologie, Monatshefte*, **3**, 165–178.

Sawłowicz, Z. & Porębska, E. 1998. Geochemical study of the Ordovician/Silurian boundary section in Żdanów (Bardzkie Mts., Poland). *Mineralogical, Magazine*, **62**, 1326–1328.

Schallreuter, R. 2001*a*. Silurische Ostrakoden Deutschlands. *Neues Jahrbuch für Geologie und Paläontologie, Abhandlungen*, **218**, 23–43.

Schallreuter, R. 2001*b*. Ostrakoden aus dem Ostrakodenkalk (Silur) der Lindener Mark bei Gießen (Hessen). *Paläontologische Zeitschrift*, **74,** 517–531.

Schallreuter, R. & Hinz-Schallreuter, I. 1998. Ein Geschiebe aus Armorica im Thüringer Lederschiefer (Ordoviz). *Archiv für Geschiebekunde*, **2**, 332–360.

Schätz, M., Tait, J., Bachtadse, V., Heinisch, H. & Soffel, H. 2002. Palaeozoic geography of the Alpine realm, new palaeomagnetic data from the Northern Greywacke Zone, Eastern Alps. *International Journal of Earth Sciences*, **91**, 979–992.

Schauer, M. 1971. Biostratigraphie und Taxionomie der Graptolithen des tieferen Silurs unter besonderer Berücksichtigung der tektonischen Deformation. *Freiberger Forschungshefte, Paläontologie*, **C273**, 1–185.

Schmidt, P. W. & Embleton, B. J. J. 1990. The paleomagnetism of the Tumblagooda Sandstone, Western-Australia – Gondwana Paleozoic apparent polar wandering. *Physics of the Earth and Planetary Interiors*, **64**, 303–313.

Schönlaub, H. P. 1971. Palaeo-environmental studies at the Ordovician/Silurian boundary in the Carnic Alps. *In*: Babin, C. (ed.) *Colloque Ordovicien – Silurien*. Brest, September 1971. Mémoire du Bureau des Recherches Géologiques et Minières, **73**, 367–376.

Schönlaub, H. P. 1979. Die Nördliche Grauwackenzone. In: Das Paläozoikum von Österreich. *Abhandlungen Geologische Bundesanstalt, Wien*, **33**, 76–97.

Schönlaub, H. P. 1980. *Carnic Alps. Field Trip A. Guidebook Second European Conodont Symposium*. Abhandlungen Geologische Bundesanstalt, Wien, **35**, 5–57.

Schönlaub, H. P. 1984. Das Paläozoikum von Sulz bei Güssing im Südburgenland. *Jahrbuch Geologische Bundesanstalt, Wien*, **127**, 501–505.

Schönlaub, H. P. 1985. Das Paläozoikum der Karnische Alpen. *In: Arbeitstagung der Geologischen Bundesanstalt 1985*. Geologische Bundesanstalt, Wien, 34–52.

Schönlaub, H. P. 1988. The Ordovician–Silurian boundary in the Carnic Alps of Austria. *In*: Cocks, L. R. M. & Rickards, R. B. (eds) *A Global Analysis of the Ordovician–Silurian Boundary*. Bulletin of the British Museum (Natural History), Geology, **43**, 107–115.

Schönlaub, H. P. 1992. Stratigraphy, biogeography and paleoclimatology of the Alpine Paleozoic and its implications for plate movements. *Jahrbuch Geologische Bundesanstalt, Wien*, **135**, 381–418.

Schönlaub, H. P. 1993. Stratigraphy, biogeography and climatic relationships of the Alpine Palaeozoic. *In:* Raumer, J. F. von & Neubauer, F. (eds) *Pre-Mesozoic Geology in the Alps*. Springer, Berlin, 65–91.

Schönlaub, H. P. 1994*a*. Das Altpaläozoikum im Südburgenland. *In:* Lobitzer, H., Csaszar, G. & Daurer, A. (eds) *Jubiläumsschrift 20 Jahre Geologische Zusammenarbeit Österreich – Ungarn*, **2**, 365–377.

Schönlaub, H. P. 1994*b*. The faunal relationship of the Silurian of the Alps. In: Schönlaub, H. P. & Kreutzer, L. H. (eds) Field Meeting Eastern + Southern Alps, Austria 1994. Subcommission on Silurian Stratigraphy. Berichte Jahrbuch Geologische Bundesanstalt, Wien, **30**, 52–60.

Schönlaub, H. P. (ed.) 1997. *IGCP Project 421 North Gondwanan Mid-Palaeozoic Biodynamics, Inaugural Meeting Vienna, September 17–21, 1997, Guidebook*. Berichte Geologische Bundesanstalt, Wien, **40**.

Schönlaub, H. P. & Heinisch, H. 1993. The classic fossiliferous Palaeozoic units of the Eastern and Southern Alps. *In*: Raumer, J. F. von & Neubauer, F. (eds) *Pre-Mesozoic Geology in the Alps*. Springer, Berlin, 395–422.

Schönlaub, H. P. & Histon, K. 1999. The Palaeozoic of the Southern Alps. In: Histon, K. (ed.) *V International Symposium Cephalopds – Present and Past. Carnic Alps Excursion Guidebook*. Berichte Geologische Bundesanstalt, Wien, **47**, 6–30.

Schönlaub, H. P. & Kreutzer, L. H. (eds) 1994. *Field Meeting Eastern + Southern Alps, Austria 1994*. Subcommission on Silurian Stratigraphy. Berichte Geologische Bundesanstalt, Wien, **30.**

Schönlaub, H. P. & Seehan, P. M. 2003. Die Krise des Lebens am Ende des Ordoviziums. *In:* Hansch, W. (ed.) *Katastrophen in der Erdgeschichte – Wendezeiten des Lebens*. Museo, **19**, 82–98.

Schönlaub, H. P., Histon, K. & Pohler, S. 2004. The Paleozoic of the Carnic Alps. *In: Field Trip Carnic Alps 2004. University of Northern Colorado, Guidebook*. Geologische Bundesanstalt, Wien, 2–32.

Scotese, C. R. & Barrett, S. F. 1990. Gondwana's movement over the South Pole during the Paleozoic: evidence from lithologic indicators of climate. *In:* McKerrow, W. S. & Scotese, C. R. (eds) *Paleozoic Paleogeography and Biogeography*. Geological Society of London, Memoirs, **12**, 75–85.

Servais, T. 1994. The Ordovician acritarchs from Rügen, (NE-Germany) palaeobiogeographical evidence for the attribution to Eastern Avalonia. *Neues Jahrbuch für Geologie und Paläontologie, Abhandlungen*, **9**, 566–580.

Servais, T. & Katzung, G. 1993. Acritarch dating of Ordovician sediments of the island of Rügen (NE Germany). *Neues Jahrbuch für Geologie und Paläontologie, Monatshefte*, **1993**, 713–723.

Servais, T. & Lehnert, O. 2006. Good and bad terranes: Are some methods better than others for inferring palaeogeography? *In*: *Programme: Abstracts and Field Excursion Guides of the Annual Meeting of project IGCP 503, 'Changing palaeogeographical and palaeobiogeographical patterns in the Ordovician and Silurian'*, Glasgow, 28 August to 4 September 2006, 44.

Soper, N. J. & Woodcock, N. H. 1990. Silurian collision and sediment dispersal patterns in southern Britain. *Geological Magazine*, **127**, 527–542.

Sremac, J. & Mihajlović-Pavlović, M. 1983. *Graptoliti Medvednice (Zagrebačke gore)*. Rad JAZU, Zagreb.

Stache, G. 1872. Entdeckung von Graptolithenschiefer in den Südalpen. *Verhandlungen Geologische Bundesanstalt, Wien*, **1872**, 234–235.

Stampfli, G. M., Mosar, J., Marquer, D., Marchant, R., Baudin, T. & Borel, G. 1998. Subduction and obduction processes in the Swiss Alps. *Tectonophysics*, **296**, 159–204.

Steemans, P. 1989. Etude palynostratigraphique du Dévonien inférieur dans l'ouest de l'Europe. *Mémoire pour servir à l'Explication des Cartes Géologiques et Minières de Belgique*, **27.**

Stettner, G. 2006. Oberpfälzer Wald/Bayerischer Wald. *In;* Deutsche Stratigraphische Kommission (ed.) *Silur Monographie.* Schriftenreihe der Deutschen Gesellschaft für Geowissenschaften, **46**, 127–129.

Štorch, P. 1994. Graptolite biostratigraphy of the Lower Silurian (Llandovery and Wenlock) of Bohemia. *Geological Journal*, **29**, 137–165.

Štorch, P. 1995*a*. Biotic crises and post-crisis recoveries recorded by Silurian planktonic graptolite faunas of the Barrandian area (Czech Republic). *Geolines*, **3**, 59–70.

Štorch, P. 1995*b*. Upper Silurian (upper Ludlow) graptolites of the *N. inexpectatus* and *N. kozlowskii* biozones from Kosov Quarry near Beroun (Barrandian area, Bohemia). *Věstník Českého geologického Ústavu*, **70**(4), 65–89.

Štorch, P., Waldhauserová, J. & Žežulková, V. 1984. Starší paleozoikum rožmitálského vývoje v mirovickém metamorfovaném ostrově. *Věstník Ústředního Ústavu geologického*, **59**(2), 105–107.

Stupnicka, E. 1992. The significance of the Variscan orogeny in the Świętokrzyskie Mountains (Mid Polish Uplands). *Geologische Rundschau*, **81**, 561–570.

Stupnicka, E., Przybyłowicz, T. & Żbikowska, B. 1991. Age of the Niewachlów greywackes and shales from Widełki near Bardo (Holy Cross Mts.) *Przegląd Geologiczny*, **39**, 389–393 [in Polish with English summary].

Stürmer, W. 1962. Die ersten Fossilien aus Kieselschiefern der Waldsassener Schieferserie von Dobrigau (vorläufige Mitteilung). *Geologische Blätter NO-Bayern*, **12**, 129–133.

Suess, E. F. 1912. Die moravischen Fenster und ihre Beziehung zum Grundgebirge des Hohen Gesenkes. *Denkschriften Österreichische Akademie der Wissenschaften, Mathematisch-naturwissenschaftliche Klasse*, **88**, 54–629.

Suk, M. 1979. Hauptprobleme des tiefen Unterbaues der böhmischen

Masse. *Krystalinikum*, **14**, 109–118.

Suttner, T. & Lukeneder, A. 2004. Accumulations of Late Silurian serpulid tubes and their palaeoecological implications (Blumau-Formation; Burgenland; Austria). *Annalen des Naturhistorischen Museums in Wien, Wien*, **10A**, 175–187.

Suttner, T. J., Schönlaub, H. P. & Hubmann, B. 2004. New conodont data of the „Baron v. Kottwitz" section (Upper Silurian to Lower Devonian). *Berichte des Institutes für Erdwissenschaften Karl-Franzens-Universität Graz*, **9**, 395–397.

Szymański, B. & Modliński, Z. 2003. Revision of the Silurian stratigraphy in selected boreholes from the Polish sector of the Baltic Depression (Northern Poland). *Biuletyn Państwowego Instytutu Geologicznego*, **405**, 109–138 [in Polish with English summary].

Tait, J., Bachtadse, V., Franke, W. & Soffel, H. 1997. Geodynamic evolution of the European Variscan fold belt: palaeomagnetic and geological constraints. *Geologische Rundschau*, **86**, 585–598.

Tait, J., Schätz, M., Bachtadse, V. & Soffel, H. 2000. Palaeomagnetism and Palaeozoic palaeogeography of Gondwana and European terranes *In:* Franke, W., Haak, V., Oncken, O. & Tanner, D. (eds) *Orogenic Processes: Quantification and Modelling in the Variscan Belt*. Geological Society, London, Special Publications, **179**, 21–34.

Teller, L. 1964. Graptolite fauna and stratigraphy of the Ludlovian deposits of the Chełm borehole, eastern Poland. *Studia Geologica Polonica*, **13**, 1–88.

Teller, L. 1969. The Silurian biostratigraphy of Poland based on graptolites. *Acta Geologica Polonica*, **19**, 393–502.

Teller, L. 1971. The Upper Silurian of Poland in the light of recent investigations. *In*: Nalivkin, D.V. (ed.) *Silurian/Devonian Boundary and Silurian Biostratigraphy.* Trudy III Mezhdunarodnogo simpoziuma, Tom I, Leningrad 1968, 201–207.

Teller, L. 1974. The Silurian of the margin of the East European platform in the region of Miastko-Chojnice (NW Poland). *Acta Geologica Polonica*, **24**, 563–580.

Teller, L. 1997. The subsurface Silurian in the East European Platform. *In*: Urbanek, A. & Teller, L. (eds) Silurian graptolite faunas in the East European Platform: stratigraphy and evolution. *Palaeontologia Polonica*, **56**, 7–21.

Teller, L. & Korejwo, K. 1968*a*. Lower Silurian deposits from the borehole Lutom 1 (North-Eastern Poland). *Acta Geologica Polonica*, **18**, 293–301.

Teller, L. & Korejwo, K. 1968*b*. Early Paleozoic deposits in the deep substratum of north-western Poland. *Acta Geologica Polonica*, **18**, 613–619.

Teller, L. & Korejwo, K. 1968*c*. Stratigraphy of the Upper Silurian from the boreholes Chojnice 3 and Stobno 1, NW Poland. *Acta Geologica Polonica*, **18**, 750–759.

Timm, J. 1981. Zur Trilobitenstratigraphie des Silur?Devon-Grenzbereiches im Ebbe-Antiklinorium (Rheinisches Schiefergebirge. *Mitteilungen aus dem geologisch-paläontologischen Institut, Universität Hamburg*, **50**, 91–107.

Tomczyk, H. 1963. Ordovician and Silurian in the Basement of the Fore-Carpathian Depression. *Rocznik Polskiego Towarzystwa Geologicznego*, **33**, 289–320.

Tomczyk, H. 1968. Silurian stratigraphy in the Peribaltic areas of Poland, based on drilling data. *Kwartalnik Geologiczny*, **12**, 35–36.

Tomczyk, H. 1970. The Silurian. *In*: Skołowski, S. (ed.) *Geology of Poland, Vol. I, Stratigraphy, Part 1, Pre-Cambrian and Palaeozoic*. Wydawnictwa Geologiczne, Warszawa, 237–319.

Tomczyk, H. 1974. Sylur. *In*: Niemczycka, T. (ed.) Ciepielów IG-1. *Profile głębokich otworów wiertniczych Instytutu Geologicznego (Deep boreholes core profiles)*, **20**, 75–81. Wydawnictwa Geologiczne, Warszawa.

Tomczyk, H. 1980. Sylur w brzeżnej części platformy prekambryjskiej na tle wyników wiercenia Toruń 1. *Kwartalnik Geologiczny*, **24**, 421.

Tomczyk, H. 1987. Sylur. *In*: Raczyńska, A. (ed.) Budowa geologiczna wału pomorskiego i jego podłoża. *Prace Państwowego Instytutu Geologicznego*, **119**, 12–16.

Tomczyk, H. 1991. Silurian. *In:* Pajchlowa, M. (ed.) *Geology of Poland, Vol. III, Atlas of Guide and Characteristic Fossils, Part 1a, Palaeozoic (including Upper Proterozoic)*. Wydawnictwa Geologiczne, Warszawa, 224–233.

Tomczyk, H. & Tomczykowa, E. 1983. Marine Silurian/Devonian Junction beds in the area of the Middle Vistula south of Warsaw. *Biuletyn Instytutu Geologicznego*, **344**, 7–22 [in Polish with English summary].

Tomczyk, H., Pajchlowa, M. & Tomczykowa, E. 1977. Poland. *In*: Martinsson, A. (ed.) *The Silurian–Devonian boundary.* IUGS Series A, **5**, 65–83.

Tomczykowa, E. 1975. The trilobite subfamily Homalonotinae from the Upper Silurian and Lower Devonian of Poland. *Acta Palaeontologica Polonica*, **20**, 3–46.

Tomczykowa, E. 1988. Silurian and Lower Devonian biostratigraphy and palaeoecology in Poland. *Biuletyn Instytutu Geologicznego*, **359**, 21–41 [in Polish with English summary].

Tomczykowa, E. & Tomczyk, H. 1979. Middle Silurian in the Podlasie Depression (Eastern Poland). *Biuletyn Instytutu Geologicznego*, **318**, 59–103 [in Polish with English summary].

Tomczykowa, E. & Tomczyk, H. 1981. Rozwój badań syluru i najniszego dewonu w Górach Świętokrzyskich. *Przewodnik 53 Zjazdu PTG w Kielcach*, 42–57.

Torsvik, T. H., Smethurst, M. A., *et al.* 1996. Continental break-up and collision in the Neoproterozoic and Palaeozoic – a tale of Baltica and Laurentia. *Earth Science Reviews*, **40**, 229–258.

Trench, A. & Torsvik, T. H. 1991. A revised Paleozoic apparent wander path for south Britain (eastern Avalonia). *Geophysical Journal International*, **104**, 227–233.

Urbanek, A. 1963. On generation and regeneration of cladia in some Upper Silurian monograptids. *Acta Palaeontologica Polonica*, **8**, 135–254.

Urbanek, A. 1966. On the morphology and evolution of the Cucullograptinae (Monograptidae, Graptolithina). *Acta Palaeontologica Polonica*, **11**, 291–544.

Urbanek, A. 1970. Neocucullograptinae n. subfam. (Graptolithina) their evolutionary and stratigraphic bearing. *Acta Palaeontologica Polonica*, **15**, 163–388.

Urbanek, A. 1971. Zonal evolution of monograptids in the Siedlce Beds: a new chapter ingraptolite history. *In*: Nalivkin, D. V. (ed.) *Silurian/Devonian Boundary and Silurian Biostratigraphy*. Trudy III Mezhdunarodnogo simpoziuma, Tom I, Leningrad 1968, 222–228.

Urbanek, A. 1997. Late Ludfordian and early Přidoli monograptids from the Polish Lowland. *In*: Urbanek, A. & Teller, L. (eds) Silurian graptolite faunas in the East European Platform: stratigraphy and evolution. *Palaeontologia Polonica*, **56**, 87–232.

Urbanek, A. & Teller, L. (eds) 1997. Silurian graptolite faunas in the Eastern European Platform: stratigraphy and evolution. *Palaeontologia Polonica*, **56**, 1–272.

Vaida, M., & Verniers, J. 2006. Chitinozoan implications in the palaeogeography of the East Moesia, Romania. *Palaeogeography, Palaeoclimatology, Palaeoecology*, **241**, 561–571.

Vaida, M., Hann, H. P, Sawatzki, G., Frisch, W. & Loeschke, J. 2000. Sedimentation ages of paragneisses and lower grade metamorphic rocks from South Schwarzwald (Variscan Belt, Germany) and their significance for the recognition of the paleotectonic evolution. *Terra Nostra*, **1**, 128.

Vaida, M., Hann, H. P, Sawatzki, G. & Frisch, W. 2004. Ordovician and Silurian protolith ages of metamorphosed clastic sedimentary rocks from the southern Schwarzwald, SW Germany: a palynological study and its bearing on the Early Palaeozoic geotectonic evolution. *Geological Magazine*, **141**(5), 629–643.

Vaida, M., Seghedi, A., & Verniers, J. 2005. Northern Gondwanan affinity of the East Moesian Terrane based on chitinozoans. *Tectonophysics*, **410**, 379–387.

Van der Voo, R. 1979. Palaeozoic assembly of Pangea; a new plate tectonic model for the Taconic, Caledonian and Hercynian orogenies (abstract). *EOS Transactions AGU*, **60**, 241.

Van der Voo, R. 1982. Pre-Mesozoic paleomagnetism and plate tectonics. *Annual Review of Earth and Planetary Sciences*, **10**, 191–220.

Van der Voo, R. 1993. *Paleomagnetism of the Atlantic, Tethys and Iapetus Oceans*. Cambridge University Press, Cambridge.

Van Grootel, G. 1990. *Litho- en biostratigrafische studie met Chitinozoa in het westelijk deel van het Massief van Brabant.* PhD thesis, Rijksuniversiteit Gent.

Van Grootel, G., Verniers, J., Geerkens, B., Laduron, D., Verhaeren, M., Hertogen, J. & De Vos, W. 1997. Timing of subsidence-related magmatism, foreland basin development, metamorphism and inversion in the Anglo-Brabant fold belt. *Geological Magazine*, **134**(5), 607–616.

VANMEIRHAEGHE, J. 2006. *The evolution of the Condroz-Brabant Basin from Middle Ordovician to Llandovery: lithostratigraphical and chitinozoan biostratigraphical approach*. PhD Thesis, Ghent University.

VECOLI, M. & SAMUELSSON, J. 2001*a*. Reworked acritarchs as provenance indicators in the Lower Palaeozoic of Denmark. *Comptes Rendus de l'Académie des Sciences Paris, Sciences de la Terre et des Planètes*, **332**, 465–471.

VECOLI, M. & SAMUELSSON, J. 2001*b*. Quantitative evaluation of microplankton palaeobiogeography in the Ordovician-Early Silurian of the northern Trans European Suture Zone: implications for the timing of the Avalonia-Baltica collision. *Review of Palaeobotany and Palynology*, **115**, 43–68.

VEJBAEK, O. V. 1985. Seismic stratigraphy and tectonics of sedimentary basins around Bornholm, southern Baltic. *Danmarks Geologiske Undersøgelse, Ser.* **A8**, 1–30.

VEJBAEK, O. V., STOUGE, S. & POULSEN, K. D. 1994. Palaeozoic tectonic and sedimentary evolution and hydrocarbon prospectivity in the Bornholm area. *Danmarks Geologiske Undersøgelse, Ser.* **A34**, 1–23.

VERNIERS, J. 1983. *The Silurian of the Mehaigne area (Brabant Massif, Belgium), lithostratigraphy and features of the sedimentary basin.* Geological Survey of Belgium, Professional Papers **203.**

VERNIERS, J. 1999. Calibration of Wenlock Chitinozoa versus graptolite biozonation in the Wenlock of Builth Wells district (Wales, U. K.), compared with other areas in Avalonia and Baltica. *Bollettino della Società paleontologica italiana*, **38**(2–3), 359–380.

VERNIERS, J. & RICKARDS, B. 1978. Graptolithes et chitinozoaires Siluriens de la vallée de la Burdinale, Massif du Brabant, Belgique. *Annales de la Société Géologique de Belgique,* **101**, 149–161.

VERNIERS, J. & VAN GROOTEL, G. 1991. Review of the Silurian in the Brabant Massif, Belgium. *In:* ANDRÉ, L., HERBOSCH, A., VANGUESTAINE, M. & VERNIERS, J. (eds) Proceedings of the international meeting on the Caledonides of the Midlands and the Brabant Massif, Brussels, 20–23 September 1989. *Annales de la Société Géologique de Belgique*, **114**(1), 163–193.

VERNIERS, J., NESTOR, V., PARIS, F., DUFKA, P., SUTHERLAND, S. & VAN GROOTEL, G. 1995. A global Chitinozoa biozonation for the Silurian. *Geological Magazine*, **132**(6), 651–666.

VERNIERS, J., HERBOSCH, A., *et al.* 2001. Cambrian–Ordovician–Silurian lithostratigraphical units (Belgium). *In:* BULTYNCK, P. & DEJONGHE, L. (eds) Lithostratigraphical Scale of Belgium. *Geologica Belgica*, **4**(1–2), 5–38.

VERNIERS, J., PHARAOH, T., *et* al. 2002. The Cambrian to mid Devonian basin development and deformation history of eastern Avalonia, east of the Midlands Microcraton: new data and a review. *In:* WINCHESTER, J. A., PHARAOH T. C. & VERNIERS, J. (eds) *Palaeozoic Amalgamation of Central Europe*. Geological Society, London, Special Publications, **201**, 47–93.

VILLAS, E., GUTIÉRREZ-MARCO, J.C., CARLOTTO, V. & CÁRDENAS, J. 2006. Palaeogeographical significance of Arenig brachiopods with Celtic affinities from Peru. *In: Programme: Abstracts and Field Excursion Guides of the Annual Meeting of project IGCP 503, 'Changing palaeogeographical and palaeobiogeographical patterns in the Ordovician and Silurian'*, Glasgow, 29 August to 4 September 2006, 51.

WAJSPRYCH, B. 1986. Sedimentary record of tectonic activity on a Devonian-Carboniferous continental margin Sudetes. *In*: TEISSEYRE, A.K. (ed.) *IAS 7th Regional Meeting, 23–25 May 1986, Excursion Guidebook*. Ossolineum, Wrocław, 141–162.

WALLISER, O. H. 1964. Conodonten des Silurs. Abhandlungen des Hessischen Landesamtes für Bodenforschung, **41**, 1–106.

WENZEL, B. 1997. Isotopenstratigraphische Untersuchungen an silurischen Abfolgen und deren paläoozeanische Interpretation. *Erlanger geologische Abhandlungen, Sonderband,* **129**, 1–117.

WENZEL, B. & JOACHIMSKI, M. M. 1996. Carbon and oxygen isotopic composition of Silurian brachiopods (Gotland/Sweden): Palaeoceanographic implications. *Palaeogeography, Palaeoclimatology, Palaeoecology*, **122**(1–4), 143–166.

WHITE, D.E. 1981. The base of the Ludlow Series in the graptolitic facies. *Geological Magazine*, **118**, 566.

WILDE, P., BERRY, W. B. N. & QUINBY-HUNT, M. S. 1991. Silurian oceanic and atmospheric circulation and chemistry. *In:* BASSETT, M. G., LANE, P. D. & EDWARDS, D. (eds) *The Murchison Symposium. Proceedings of an International Conference on The Silurian System.* Special Papers in Palaeontology, **44**, 123–143.

WILLIAMS, S. H. & INGHAM, J. K. 1989. The Ordovician–Silurian boundary stratotype at Dob's Linn, southern Scotland. *In:* HOLLAND, C. H. & BASSETT, M. G. (eds) *A Global Standard for the Silurian System.* National Museum of Wales, Cardiff, Geological Series, **9**, 27–35.

WINCHESTER, J. A & THE PACE TMR NETWORK TEAM 2002. Palaeozoic amalgamation of Central Europe: new results from recent geological and geophysical investigations. *Tectonophysics*, **360**(1–4), 5–21.

WINCHESTER, J., PHARAOH, T. & VERNIERS, J. 2002. Palaeozoic amalgamation of Central Europe: an introduction and synthesis of new results from recent geological and geophysical investigations. *In:* WINCHESTER, J. A., PHARAOH T. C. & VERNIERS, J. (eds) *Palaeozoic Amalgamation of Central Europe*. Geological Society, London, Special Publications, **201**, 1–18.

WRONA, R. 1980. Upper Silurian–Lower Devonian Chitinozoa from subsurface of southeastern Poland. *Palaeontologia Polonica*, **41**, 103–165.

WYŻGA, B. 1987. Lower Palaeozoic of Bardo Mountains (Sudetes): a sequence of deep sea pelagic sediments. *Geologia Sudetica*, **22**, 119–145.

YANEV, S. 2000: Palaeozoic terranes of the Balkan Peninsula in the framework of Pangea assembly. *Palaeogeography, Palaeoclimatology, Palaeoecology*, **161**, 151–177.

ŻBIKOWSKA, B. 1974. Upper Silurian ostracod zones in the Chojnice area (NW Poland). *Bulletin de l'Académie Polonaise des Sciences. Série des Sciences de la Terre*, **22**, 45–48.

ŻELAŹNIEWICZ, A. 1997. The Sudetes as a Palaeozoic orogen in central Europe. *Geological Magazine*, **134**, 691–702.

ŻELAŹNIEWICZ, A. 2003. Developments in the geology of the crystalline basement of the West Sudetes in 1990–2003. *In*: CIĘŻKOWSKI, W., WOJEWODA, J. & ŻELAŹNIEWICZ, A. (eds), *Sudety zachodnie: od wendu do czwartorzędu.* WIND Wrocław, 7–15 [in Polish, with English summary].

ŻELICHOWSKI, A. 1972. Evolution of the geological structure of the Góry Świętokrzyskie and the River Bug. *Biuletyn Instytutu Geologicznego*, **263**, 7–97 [in Polish with English summary].

ZENO, F. 1770. Beschriebung des bei Prag von dem Wissehrader Tore gelegenen Kalksteinbruches, mit seinem Seeversteinerungen und anderen Fossilien. *Neue physicalische Belustigungen*, **2**, 362–420.

7 Caledonian tectonics

CHARLOTTE M. KRAWCZYK (co-ordinator), TOM McCANN, L. ROBIN M. COCKS, RICHARD W. ENGLAND, JOHN H. McBRIDE & STANISLAW WYBRANIEC

The Caledonian Orogeny lasted from the late Cambrian to the Devonian with the main collisional events occurring during Ordovician and Silurian times. Direct evidence of the extent of this orogenic event across central Europe is limited because of the lack of outcrops of this age. The Caledonian Orogeny, together with the subsequent Variscan and Alpine orogenies, is one of a succession of major tectonic events which have defined the geological evolution of Central Europe. Thus, the present configuration and condition of the lithosphere of central Europe is the result of superimposed periods of deformation (Fig. 7.1). Consequently, a wide range of investigative techniques needs to be employed to unravel these events in order to determine the properties of the various elements of the Caledonides and to elucidate the evolution of the Caledonian Orogeny. Additionally, evidence of the orogeny is deeply buried beneath thick successions of younger sediments, e.g. Dutch and North German Basin, or has been reworked extensively by later events, e.g. Belgium or to the SE of the Trans-European Suture Zone.

The word 'Caledonia', the Latin name for northern Scotland, was used by Eduard Suess (1885–1909) not only to describe a geographic region but also to indicate an orogen he termed 'Caledonisches Gebirge'. Furthermore, Suess was the first to put his definition into a tectonic context: 'Die in der Kaledonischen Faltungsära gebildeten Gebirge treten vor allem in Irland, Wales, Schottland und im Westteil Skandinaviens in Erscheinung'. [The mountains built during the Caledonian folding era appear particularly in Ireland, Wales, Scotland and in the western part of Scandinavia] (Suess 1885–1909, Vol. 3:182).

The Caledonian Orogeny had a complex history with several phases of deformation or magmatism occurring in time and space. Indeed, some early contractional events, for example the Grampian Orogeny on the Laurentian margin, were intense and widespread enough to have their own names (Woodcock & Strachan 2002*a*; Lambert & McKerrow 1976). The Caledonian Orogeny *sensu lato* includes 'all Cambrian to Devonian tectonic events associated with the development and the closure of those parts of the Iapetus Ocean, which were situated between Laurentia (to the NW), Baltica and Avalonia (to the SE and E)' (McKerrow *et al.* 2000: 1149). These events include the development and closure of the Iapetus Ocean and the Tornquist Sea (*c.* 570–360 Ma). The final collisional stage and suturing may be defined as the Caledonian Orogeny *sensu stricto* (from *c.* 440 Ma onwards). Thus, the European Caledonian Orogen encompasses the regions of Great Britain and Ireland, NW Scandinavia and parts of northern Germany and Poland (Fig. 7.1), as well as SE Greenland and Svalbard. Indeed, the Caledonian Orogeny was responsible for consolidating most of the British and Irish crust into the present pattern of fault-bounded blocks or terranes, each with a distinctive geological history. The internal geometry of the orogen is, therefore, dominated by the major block-bounding faults, mostly having the NE to SW 'Caledonoid' strike. Outside of Europe, the Northern Appalachians were also involved. The term 'Caledonian' should only be used in a geological context and never in terms of age definition (McKerrow *et al.* 2000).

As noted above, the Caledonian Orogeny comprises numerous different tectonic or orogenic phases. These resulted from several distinct events, e.g. earlier phases of ocean closure, arc–arc and arc–continent collisions, or later continent–continent collisions (Table 7.1). During their 210 Ma duration, the various orogenic phases were sometimes very localized, and were also often diachronous (McKerrow *et al.* 2000; Mac Niocaill *et al.* 1998; Torsvik *et al.* 1996, and references therein).

Western and Central Europe comprise a number of terranes derived from the three continental landmasses Laurentia, Baltica and Gondwana (with Avalonia as the northernmost terrane) which collided during Ordovician–Silurian times to form the basement of northern Europe. The existence of these terranes has been established or inferred from geological studies, although most of the boundaries are not exposed at the surface. However, deep seismic and potential field studies have provided some of the best constraints on the subsurface location of the structure and relationships between these terranes. This chapter presents a review of the current knowledge concerning the evolution of the various terranes, especially the boundaries and suture zones, documenting their amalgamation, and the internal tectonic structures and events of the Caledonian Orogeny in Central Europe. Of particular interest are the areas of the British, Belgian, southern Scandinavian and German Caledonides up to the Trans-European Suture Zone in Poland (Fig. 7.2).

There is an extensive literature on the Caledonian of western and Central Europe and key references include works by Blundell *et al.* (1992) on the European Geotraverse project, Dallmeyer *et al.* (1995), and more recently the results of the EUROPROBE Trans-European Suture Zone project (e.g. Gee & Zeyen 1996; Pharaoh *et al.* 1997; Thybo *et al.* 1999; Winchester *et al.* 2002*b*; Gee & Stephenson 2006). Recent data and reviews covering the areas of Britain (e.g. Woodcock & Strachan 2002*a*; Fortey & Cocks 1992; Cocks *et al.* 1997), Brabant, Condroz, and Stavelot-Venn (Verniers *et al.* 2002), Ebbe (Samuelsson *et al.* 2002*b*), Rügen (Vecoli & Samuelsson 2001; Servais *et al.* 2003) and Pomerania (Samuelsson *et al.* 2002*a*) have also been published. A variety of atlases of the NW and Central European region are also available. Seismic data have been published by DEKORP (Meissner & Bortfeld 1990) and by BIRPS (Klemperer & Hobbs 1992; Snyder & Hobbs 1999; Brewer *et al.* 1983) as well as in a range of individual papers (see below). These, together with the geological maps of Ziegler (1990), palaeogeographical reconstructions via the PALEOMAP project (Scotese

We would like to dedicate this chapter to the memory of Stuart McKerrow (1922–2004).

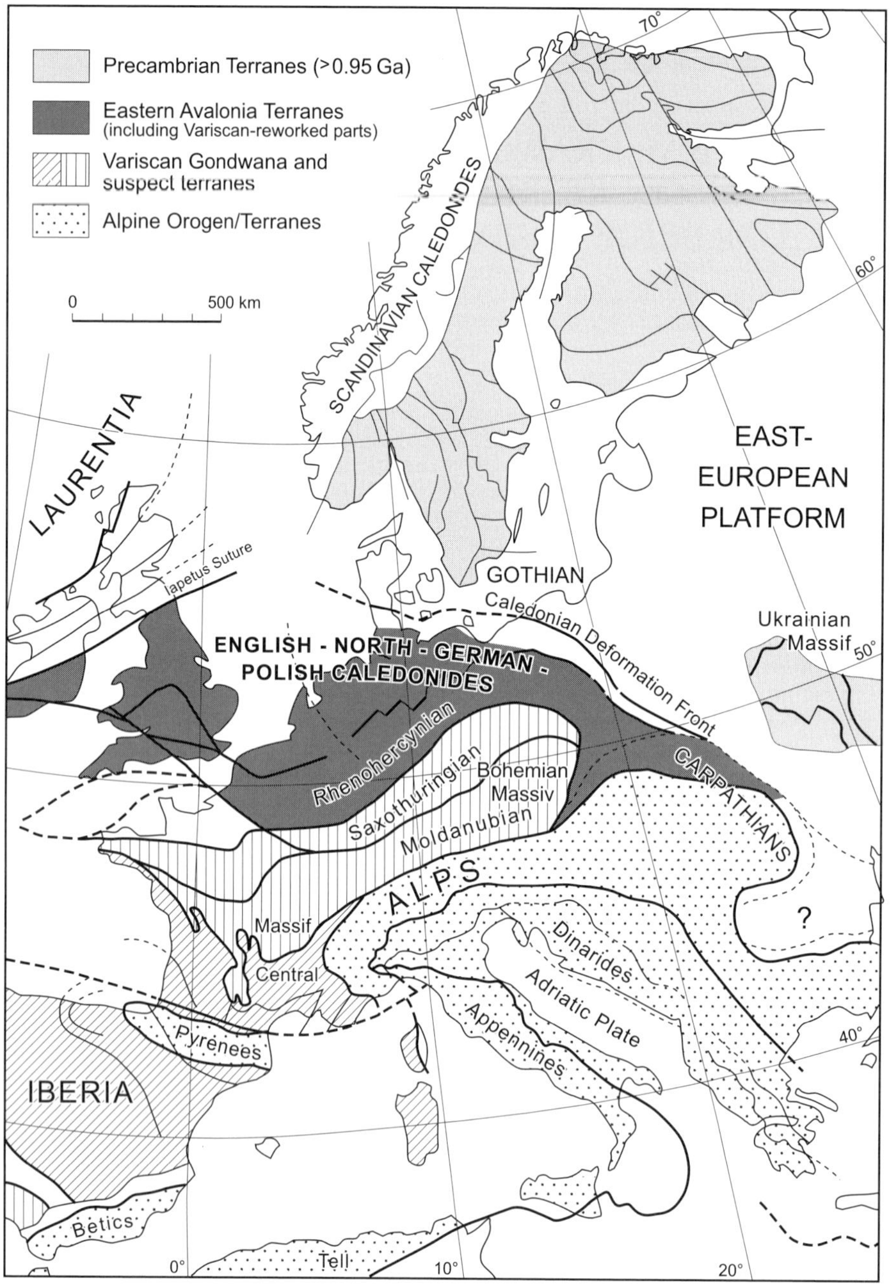

Fig. 7.1. Main tectonic units of Precambrian and Phanerozoic Europe (after Blundell *et al.* 1992).

1999), potential field data by the British Geological Survey (2007; Pedley *et al.* 1994), Banka *et al.* (2002) and Wybraniec *et al.* (1998), as well as the Geothermal Atlas of Europe by Hurtig *et al.* (1992) provide a very thorough database for the region.

Structural and palaeogeographic evolution of the Caledonides

The forelands of the North German–Polish Caledonides are formed by the Precambrian Fennoscandian Shield and its extension beneath the Baltic Basin (Depression) and the stable East European Platform (craton). The bulk of the Fennoscandian–Baltic craton was consolidated in pre-Grenvillian times (Ziegler 1990). The Caledonian orogenic cycle (late Cambrian to earliest Devonian) was governed by the sinistral oblique convergence and collision of the Laurentia–Greenland and Fennoscandian–Baltic cratons (Ziegler 1989). Orogenic activity involved the welding together of a series of microcontinental fragments/terranes (Fig. 7.3). A terrane is herein defined (after Oliver *et al.* 1993) as a fault-bounded geological entity (i.e. a tectonostratigraphical unit) of regional extent, characterized by a geological history which is different from that of adjacent terranes. Terrane boundaries may be transcurrent faults, thrust faults or a melange, being cryptic or polygenetic (Oliver *et al.* 1993).

Recognized terranes have been derived either by tectonic transport around the margin of Baltica (and are thus Baltica related), or from the Gondwana margin (Pharaoh 1999). These latter terranes were sequentially accreted to Baltica during the

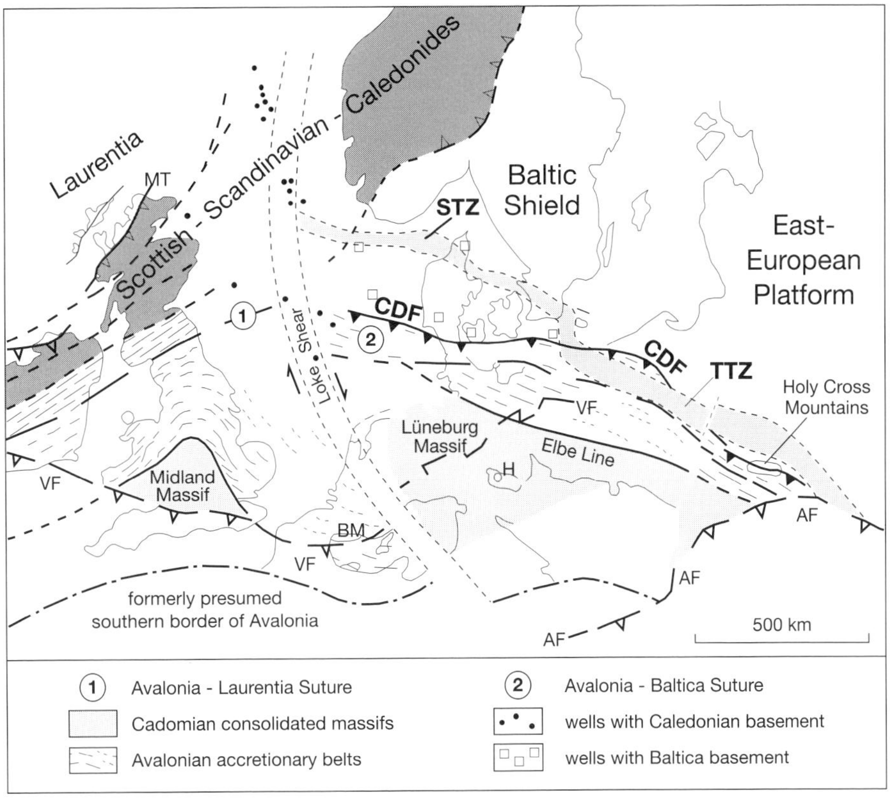

Fig. 7.2. Simplified overview map of the Caledonian triple-junction, showing the extent of the Caledonides, with main suture zones and tectonic boundaries as known at the beginning of the 1990s (after Blundell *et al.* 1992). Compare with younger, revised maps (e.g. Heeremanns *et al.* 2004; Fig. 7.26). Abbreviations: AF, Alpine Front; BM, Brabant Massif; CDF, Caledonian Deformation Front; H, Harz mountains; MT, Moine Thrust; STZ, Sorgenfrei-Tornquist Zone; TTZ, Tornquist-Teisseyre Zone; VF, Variscan Front.

late Ordovician (= Shelvian), late Silurian (= Scandian), and early Devonian (= Acadian) orogenic phases (Ziegler 1990). Orogenic activity and accretion continued into the late Carboniferous (see Chapter 9).

The most significant terrane boundaries present within the region of Central Europe are those associated with the destruction of oceanic lithosphere. These include: (a) the Caledonian Deformation Front, equivalent in parts to the Tornquist Suture (= Thor Suture; Berthelsen 1998) separating Avalonia and Baltica; (b) the Iapetus Suture (Harland & Gayer 1972) separating Avalonia and Baltica from the Laurentian terranes; and (c) the Rheic Suture (Cocks & Fortey 1982) separating Avalonia and other terranes (e.g. the Moravo-Silesian Terrane) which were accreted to Baltica in early Palaeozoic times, from the terranes of the Variscan internides (Saxothuringian and Moldanubian zones) which were accreted in late Palaeozoic times. Several other lineaments, for example, the Elbe Line, separate regions with contrasting geophysical characteristics within the major terranes outlined above, and may represent subterrane boundaries within composite terranes. The complexity of the situation in Central Europe, however, cannot be overstated (Fig. 7.3). For example, the Bohemian Massif alone contains a number of inferred oceanic sutures (Matte *et al.* 1990), many of which are 'cryptic' since they are largely unexposed and there is still considerable debate about their significance (Pharaoh 1999). Additionally, Oliver *et al.* (1993) have recognized a series of terranes within the Sudetes of SW Poland which have been interpreted as having formed part of the Tornquist Sea and which were later amalgamated during the collision of Baltica and Avalonia.

For the Caledonian Orogeny, the closure of the Iapetus Ocean and the Tornquist Sea are the most significant events. Both of these oceans, however, contained a complex array of secondary features, e.g. island arc, continental arcs, all of which must be integrated into the story (cf. Table 7.1). The Caledonian history of western and Central Europe is, thus, a complex one.

The following sections will outline the various large-scale elements of the Caledonian Orogeny *sensu lato* as well as the palaeontological, structural and other evidence from surface outcrops and borehole data. All of this provides the main framework for the orogeny, but it fails to define terranes and terrane boundaries which are buried beneath younger sediments. Furthermore, information concerning the subsurface nature of these structures as well as the deeper crust and upper mantle is missing.

Fossils and geological observations in the different Caledonide terranes

Although the most obvious features of the Caledonides are tectonic, the understanding of the histories of the various terranes and blocks which make up the Caledonides today have largely been put into perspective by the fossils within them. These are important in two separate and distinct ways: first, for dating the rocks in which they are found; and second, to indicate the relative affinities of the terranes at the time of deposition.

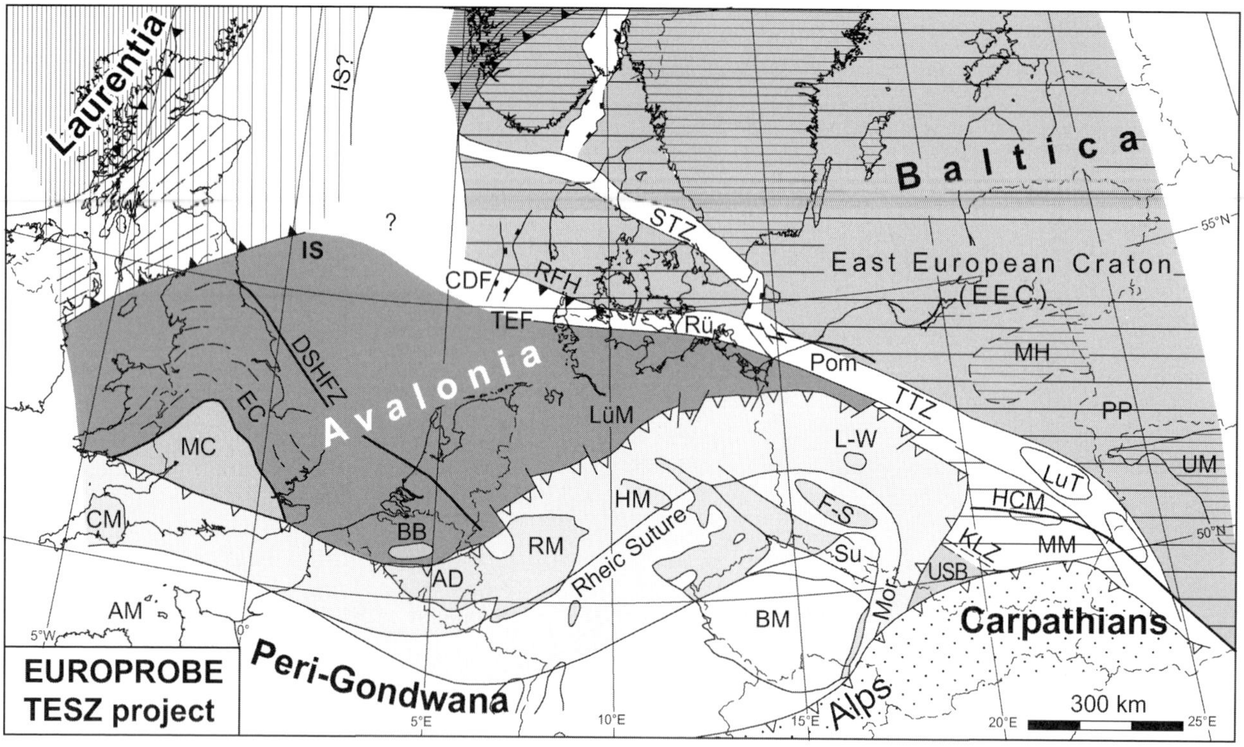

Fig. 7.3. Basement tectonic sketch map of the Trans-European Suture Zone and adjacent areas, slightly modified and simplified from Berthelsen (1992) and Pharaoh (1999). Compare with older map in Figure 7.1. Abbreviations: AD, Ardennes; AM, Armorican Massif; BB, Brabant Massif; BM, Bohemian Massif; C, Cadomia; CDF, Caledonian Deformation Front; CM, Cornubian Massif; DSHFZ, Dowsing-South Hewett Fault Zone; EC, Eastern English Caledonides; EEC, East European Craton; F-S, Fore-Sudetic Block; HM, Harz Mountains; HCM, Holy Cross Mountains; IS, Iapetus Suture (Avalonia–Laurentia); IS?, uncertain location of Laurentia–Baltica Suture; KLZ, Kraków-Lubliniec Zone; LuT, Lublin Trough; LüM, Lüneberg Massif; L-W, Leszno-Wolsztyn Basement High; MC, Midlands Microcraton; MH, Mazurska High; MM, Malopolska Massif; Mor, Moravia; Pom, Pomerania; PP, Pripyat Trough; RFH, Ringkøbing-Fyn High; RM, Rhenish Massif; Rü, Rügen Island; STZ, Sorgenfrei-Tornquist Zone; Su, Sudetes Mountains; TEF, Trans-European Fault Zone; TTZ, Tornquist-Teisseyre Zone; UM, Ukrainian Massif; USB, Upper Silesian Coal Basin.

Table 7.1. *Main phases of the Caledonian Orogeny, with type localities and key references*

Phase	Age (Ma)/Interval	Geodynamic setting	Geographic region	References
Penobscotian	490–485 Tremadoc	Obduction of Penobscot Arc onto Avalonian Iapetus margin	North Wales, SE Ireland, New England, Newfoundland	Brück *et al.* (1979), Woodcock (1990*a*), van Staal *et al.* (1998)
Grampian	478–468 Llanvirn	Impingement of mid-ocean ridge with northern margin of Iapetus. Collision of Arenig Ballantrae Arc with northern margin of Iapetus*	Scottish Highlands	Lambert & McKerrow (1976) Dewey & Mange (1999), Soper *et al.* (1999)
Humberian	480 Arenig	Arc–continent collision	Western Newfoundland	van Staal *et al.* (1998)
Taconian	455 Caradoc	Arc–continent collision	Mainland Maritimes, New England	Bird & Dewey (1970)
Finnmarkian	Early Ordovician	Arc–continent collision	Scandinavia	Ramsay *et al.* (1985), Stephens & Gee (1985), Gee (1987)
Shelveian	448–438 Ashgill	Avalonia–Baltica collision and closure of Tornquist Ocean	Eastern Wales	Toghill (1992); McKerrow *et al.* (1991), Cocks *et al.* (1997)
Scandian	428–387 (?) Wenlock–Emsian	Major nappe emplacement	Scandinavia; Scotland (south Midland Valley, southern Uplands)	Stephens & Gee (1985); Mykura (1991), McKerrow *et al.* (1991)
Acadian	394–380 Emsian–Eifelian	Impingement of Meguma terrane with Nova Scotia margin of Laurentia	Southern parts British Isles Brabant Massif /Belgium Northern Appalachians	McKerrow (1988), Woodcock (1990*a*), van Staal *et al.* (1998)

*Interpretation favoured today

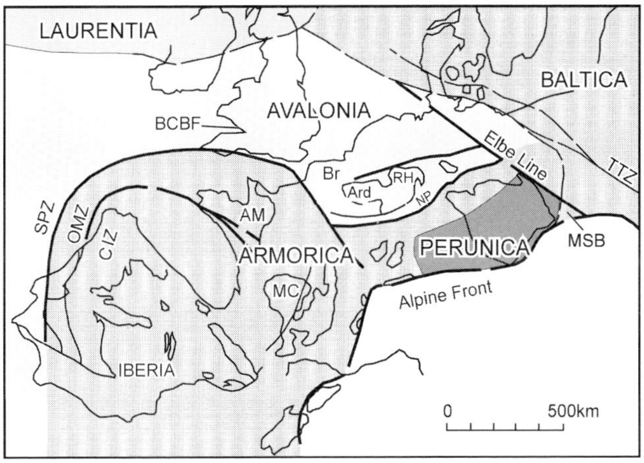

Fig. 7.4. Division of Gondwana into the terranes Avalonia, Ibero-Armorica and Perunica (Bohemia) in Central Europe. Abbreviations: AM, American Massif; Ard, Ardennes; BCBF, Bristol Channel-Bray Fault; Br, Brabant Massif; CIZ, Central Iberian Zone; MaCe, Massif Central; MSB, Moravo-Silesian Block; NP, Northern Phyllite Belt; OMZ, Ossa-Morena Zone; RH, Rhenoherzynian Zone; SPZ, South Portuguese Zone; TTZ, Tornquist-Teisseyre Zone.

The faunal data from all of the main Caledonian terranes have been extensively reviewed by Fortey & Cocks (2003), but are summarized below with respect to the separate terranes (Figs 7.4 & 7.5).

Fossils used for dating

The best fossils for dating are those with a long larval life and which thus spread very quickly as part of the plankton. On a global scale these are dependent on temperature (broadly latitude) but on a smaller scale this does not seem to have been so important. In both the Ordovician and Silurian, graptolites were by far the most rapidly evolving fossils of zonal value, and the 'standard' series of graptolite biozones was largely developed within the Caledonide terranes, particularly in Great Britain during the nineteenth century. Graptolites were at their most diverse and abundant in oceanic areas, and are less commonly found on the old continental shelf areas. Chitinozoa and acritarchs are also useful plankton for dating, although the duration of their zones was usually three or four times the length of the graptolite zones. The shelf areas had, however, some fossils useful for dating, particularly conodonts in limestones and also occasionally brachiopods (e.g. *Stricklandia* and *Eocoelia* in the Lower Silurian).

Fossils used to identify terrane affinities or separations

These are mostly the benthos that lived on the shelves around the terranes, particularly those whose larvae had a shorter life and which thus dispersed slowly. Chief amongst these are the trilobites (although not all trilobites were benthic), the articulated brachiopods and most of the ostracods. Cocks & Fortey (1982) and Fortey & Cocks (2003) discussed at length the principles involved in recognizing the individual terranes by faunal means. In brief: (a) the terrane affinities of fossils may only be sensibly gauged against others of similar age, and, equally, of comparable ecologies; (b) the faunal patterns of diverging terranes are different from those of terranes approaching each other – in the former, the common species dwindle with time, while in the latter they increase; and (c) appropriate allowances must be made for differences in both palaeotemperature and palaeolatitude. One of the strengths of fossils in their use for the differentiation of terranes is that the method is completely independent of the other principal method used, namely palaeomagnetism. Although the latter is more objective in determining palaeolatitudes and terrane rotations through time, it can tell us nothing about either palaeolongitude or terrane closeness, and, in addition, the original palaeomagnetic signature can sometimes be completely deleted by subsequent tectonic overprinting. In contrast, even though fossils are often badly deformed in tectonic belts, as long as they can be confidently identified then they continue to be of palaeogeographical use.

Gondwana

This was by far the largest-known landmass in the Lower Palaeozoic, and included a vast area of South America (Amazonia?), Africa, peninsular India, Antarctica, Australia and many peripheral areas, as well as Ibero-Armorica (most of Spain, Portugal and France) in the Ordovician. Gondwana encompassed the South Pole to 30°N and thus occupied climatic areas ranging from polar to equatorial. During the earliest Ordovician, the South Pole was occupied by what is located today in North Africa (probably Libya), and thus the area of today's southern Europe (termed Western Gondwana) was at high latitudes, with few carbonates and a relatively low diversity benthic fauna, termed the Mediterranean Province (e.g Havlíček *et al.* 1994). It can be seen from both faunal and palaeomagnetic studies that the various terranes rifted away from the margins of Gondwana at different times; relevant to Europe are Avalonia, which was probably derived from the north of South America in the earliest Ordovician, Perunica (Bohemia), which was probably part of North Africa and drifted from Gondwana during the late Llanvirn, and Ibero-Armorica (= Armorican Terrane Assemblage), about which there is controversy about the time of separation, but which probably did not leave Gondwana until the latest Silurian at the earliest.

Laurentia

This is the area occupying most of present-day North America and Greenland, but also includes Spitsbergen, Scotland, and NW Ireland. It was an independent landmass from late Proterozoic times. In the Early Ordovician, when the Iapetus Ocean separating Laurentia from the other terranes was at its widest, Laurentia straddled the palaeo-equator, and much of the craton was flooded by shallow seas. The faunas included particularly distinctive trilobites, termed the Bathyurid fauna (Cocks & Fortey 1982). Later in the Ordovician as Iapetus narrowed, the plankton of key benthic faunas were able to progressively cross the ocean, with a consequent steady overall reduction in endemicity, although many endemic elements still remained, for example the brachiopods in the late Ordovician Cincinnattian faunas of the mid-west of the USA.

Terranes accreted in early Palaeozoic time

Pharaoh (1999), Winchester *et al.* (2002*a*) and Cocks (2000) have provided a very detailed overview of the various continental landmasses and terranes which were involved in the complex accretionary processes in the Early Palaeozoic of Central and western Europe (Fig. 7.3). The main terranes are outlined below.

Baltica. Baltica forms more than half of modern Europe (Cocks 2000) and has an area of *c.* 8 million km^2. It comprises the exposed Baltic (Fennoscandian) and Ukrainian shields and the concealed East European Craton (EEC), extending as far east as

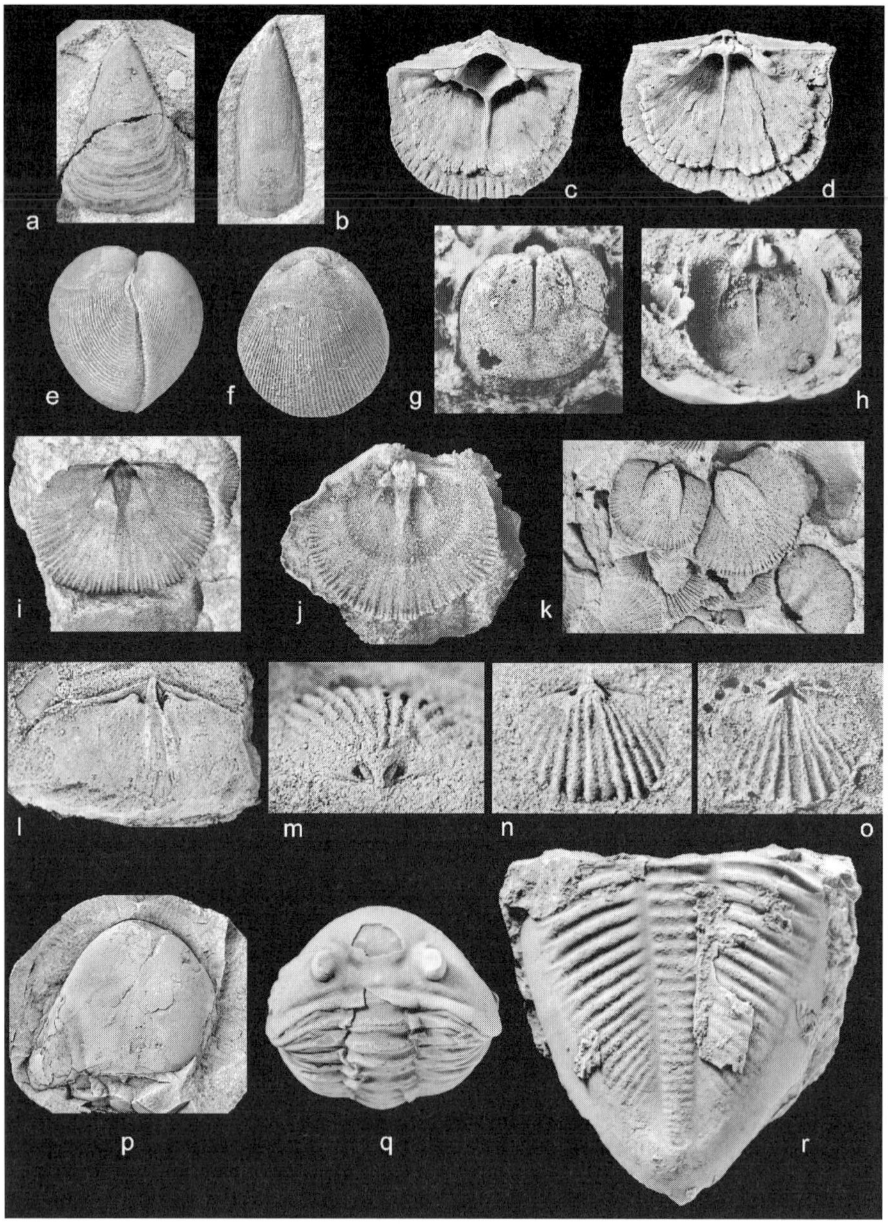

Fig. 7.5. Key fossils for terrane recognition. (**a**, **b**) Lower Ordovician Gondwanan brachiopods; (a) *Lingulepis crassipyxis Havlicek*, Gres de May (Arenig), France, B 5932, ×1; (b) *Ectenoglossa lesueuri (Rouault)*, Budleigh Salterton Pebble Bed (Arenig), Devon, England, B 14419, ×1. (**c**–**h**) Ordovician Baltic brachiopods; (c, d) *Estlandia marginata (Pahlen)*, Kukruse Formation (Caradoc), Estonia, BB 5177, ×1.8; (e–h) *Lycophoria nucella (Dalman)*; (e, f) Kunda Stage (Arenig), St Petersburg, Russia, B 987, ×1.5; (g, h) Upper Orthid Sandstone (Arenig), Holy Cross Mountains, Poland, BC 4883, ×2.8. (**i**–**k**) Middle Ordovician Gondwanan brachiopods, *Tafilaltia valpyana (Davidson)*; (i, j) Budleigh Salterton Pebble Bed (Llandeilo), Devon, England, B 15290, ×2.5; (k) Gres de petit May (Landeilo), Caen, France, B 85258, ×3. (**l**–**o**) Silurian cosmopolitan brachiopods; (l) *Stricklandia lens (J. de C. Sowerby)*, Wood Burn Formation (Llandovery), Girvan, Scotland, BB 32459, ×2.5; (m–o) *Eocoelia curtisi Ziegler*, Lower Camregan grits (Llandovery), Girvan, Scotland, (m, n) BB 32648, ×4, (o) BB 32583, ×4. (**p**) Upper Ordovician Perunican trilobite *Parabarrandia crassa (Barrande)*, Dobrotiva Formation (Caradoc), Bohemia, Czech Republic, ×0.8. (**q**, **r**) Lower Ordovician Baltic trilobites; (q) *Asaphus lepidurus Nieskowski*, Lepidurus Limestone (Arenig), St Petersburg, Russia, It 15430, ×1.5. (r) *Megistaspis (Megistaspidella) sp.*, Kunda Stage (Arenig), Oland, Sweden, BM 18382, ×0.75. All specimens except (p) in the Natural History Museum, London.

the Urals (Nikishin *et al.* 1996; Pharaoh 1999) and from the Trans-European Suture Zone (TESZ) to the SW into (and probably including) the Barents Sea, at least as far north as Franz Josef Land. The most southerly part of Baltica, located beneath the present-day North Sea, is marked by the triple-junction between Laurentia to the west, Avalonia to the south and Baltica to the east (Fig. 7.2). The complexities of the precise boundary definitions of Baltica have been recently discussed in detail by Cocks & Torsvik (2005).

Baltica evolved during several episodes of Precambrian crustal accretion and reworking (Gaál & Gorbatchev 1987), becoming an independent terrane in late Precambrian times following its separation from the rest of the Rodinian-Pannotia Supercontinent (Dalziel 1997). However, its integrity as a separate terrane has only been recognized for *c*. 30 years (Cocks & Torsvik 2005).

The basement rocks on Baltica are overlain by a thin, tectonically undisturbed Cambro-Ordovician shelf succession. Following collision with Avalonia, these were overlain by Silurian-age sediments deposited in a rapidly subsiding Silurian foredeep (Poprawa *et al.* 1999). Boreholes in the region contain up to 7 km of Silurian strata (Pharaoh 1999).

Because of its isolation, which was at its maximum in the latest Cambrian and earliest Ordovician, a high proportion of the benthic faunas on Baltica were endemic. Specific examples include distinctive asaphid trilobites, such as the megistaspids, and many brachiopods, for example, *Lycophoria*, the sole representative of its family and present in Baltica in rock-forming abundance in places. Like Avalonia, partly because it moved from higher to lower palaeolatitudes during the Ordovician, and partly because the surrounding Iapteus and Tornquist oceans became narrower as the Ordovician progressed, the benthic faunas became less endemic but more diverse with time, although even by latest Ordovician (mid-Ashgill) times there were still a large number of endemic strophomenoid brachiopods in, for example, the Boda Limestone bioherms of central Sweden (Cocks 2005).

In Central Europe the SW margin of Baltica is intimately bound up with the TESZ (Tornquist-Teisseyre Zone). This zone has been the subject of much recent research (for details see Winchester *et al.* 2002*a*). The TESZ forms the suture between Baltica (located to the NE of the TESZ, apart from the area of the Holy Cross Mountains; see below), and Avalonia (in its NW part) and Gondwana (in its SE part). However, the structures within the TESZ were formed as part of the complex series of late Palaeozoic events termed the Variscan Orogeny which affected various parts of the southern margin of the pre-existing Baltica Terrane in different ways, and thus the Upper Palaeozoic TESZ does not represent the Lower Palaeozoic margin of Baltica along all of its length.

The suture between Baltica and Laurentia can only be determined by unravelling the complexities of the Scandinavian Caledonides where deformation occurred both in the late Precambrian and the early Palaeozoic, culminating in the largely Silurian Scandian Orogeny (Cocks & Fortey 1998; Cocks & Torsvik 2005). The situation is complicated by the progressive accretion around the NW margin of Baltica of small terranes representing parts of island arcs, and sometimes carrying distinctive faunas. Within the Iapetus Ocean, offshore from both Laurentia and Baltica, there were various island arcs in the Ordovician from which many benthic faunas were preserved and described (see Table 7.1) (e.g. Harper *et al.* 1996; Neuman *et al.* 1997). A mix of endemic benthos, and a variety of immigrants from surrounding larger terranes, reflect their mid-oceanic original positions from the late Cambrian to the late Ordovician. Collision between these arcs and also their neighbouring palaeocontinents occurred at different times, mostly during the Ordovician and Silurian.

The island of Rügen lies to the south of the TESZ (Fig. 7.6) and its affinity has been in dispute. Deep-marine sediments in cores from Rügen lack terrane-diagnostic benthic macrofossils although they contain Lower and Middle Ordovician carbonate lithic fragments which were thought to be indicative of the warmer Baltica than the cooler and higher-latitude Gondwana/Avalonia (Cocks *et al.* 1997). However, more detailed micropalaeontological work (e.g. Servais & Fatka 1997; Vecoli & Samuelsson 2001) suggests that the sediments were deposited in a deep-marine basin within the Ran and Tornquist oceans, which lay between Baltica and Avalonia/Gondwana (Fig. 7.7), rather than representing facies marginal to Baltica itself (Cocks & Torsvik 2005) (for detailed discussion see 'Avalonia-related terranes' section below).

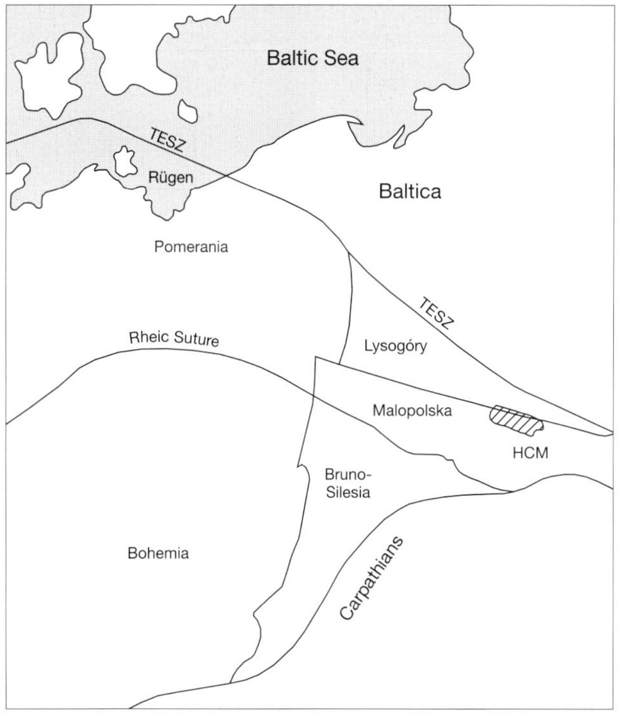

Fig. 7.6. Present-day situation of suture zones and landmasses (Cocks 2002). Abbreviations: HCM, Holy Cross Mountains; TESZ, Trans-European Suture Zone.

Baltica-related terranes. As noted above, the TESZ is a complex suture zone, comprising a tectonic collage of continental terranes which were accreted to the margin of the palaeocontinent of Baltica in early Palaeozoic time. In southern Poland, there is evidence for three fault-bounded crustal units (although geophysical evidence suggests that the situation is even more complex): the Lysogóry Unit, the Malopolska Massif and the Upper Silesian Massif (Figs 7.6 & 7.8). Most authors have concluded that these units were essentially adjacent to each other in the Lower Palaeozoic, or even formed a single amalgamated block (for details see Cocks 2002). Although the Holy Cross Mountains lie to the south of the TESZ today (Fig. 7.6), the Late Cambrian and Ordovician faunas from the area confirm the position of the area as an integral part within Baltica rather than as part of peri-Gondwana as suggested by some authors (e.g. Belka *et al.* 2002). Diagnostic Baltic faunas include trilobites, brachiopods and ostracods (for details see Cocks & Torsvik 2005).

The Holy Cross Mountains, Poland, provide excellent exposures of Palaeozoic rocks on the SW margin of the EEC. Data and fossils from the Holy Cross Mountains suggest that the region formed part of the Baltic craton before the Silurian (Nawrocki 2000), although this is not always clear (Pharaoh 1999). The NW-trending Holy Cross Fault separates rocks of the Lysogóry Unit from the Kielce Unit of the largely concealed

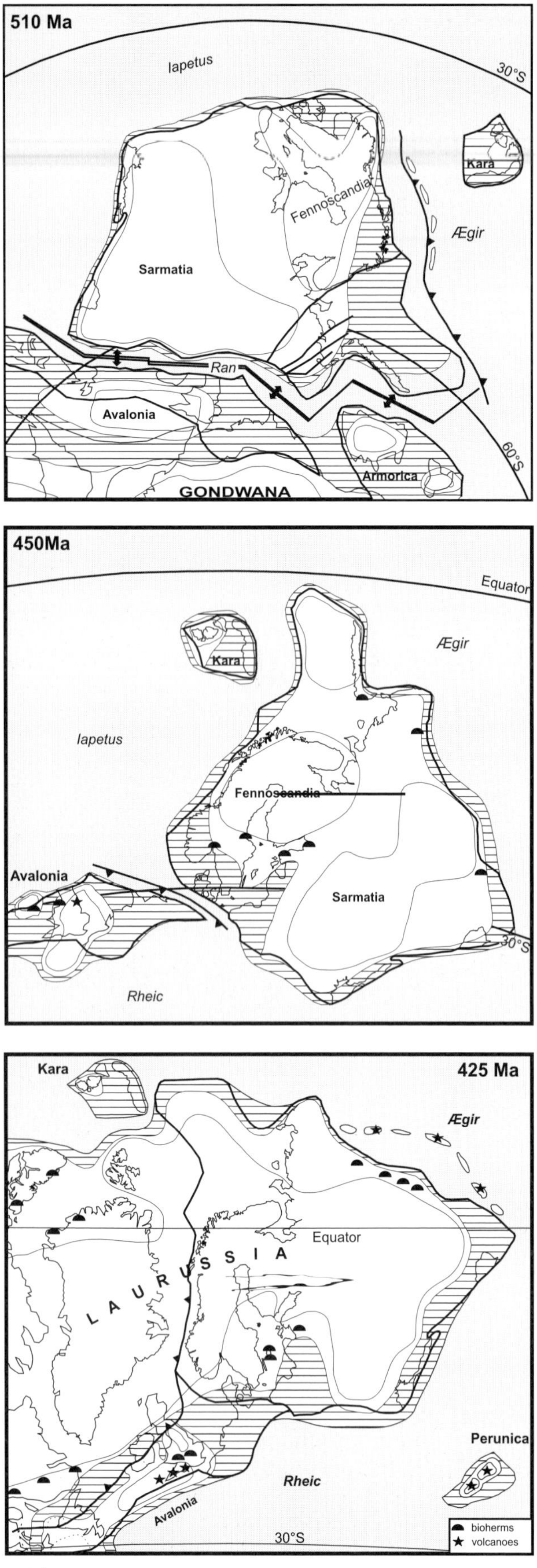

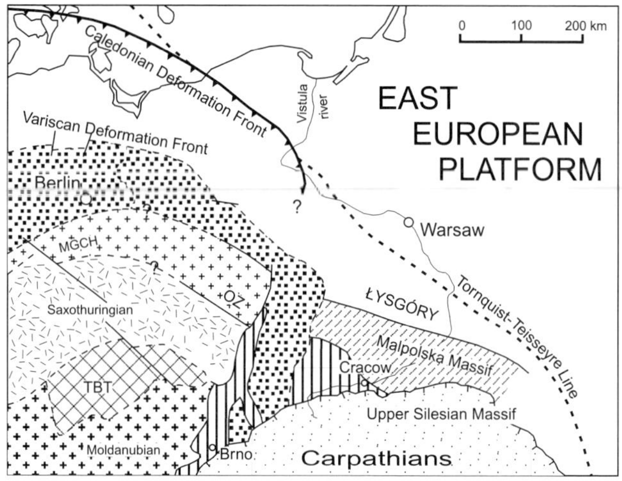

Fig. 7.8. Simplified structural map of crustal units in the Variscan foreland of southern Poland (after Belka *et al.* 2002). Abbreviations: MGCH, Mid-German Crystalline High; OZ, Odra Zone; TBT, Tepla-Barrandian Terrane.

Malopolska Massif to the south. As noted above three main areas can be recognized.

1. The **Lysogóry Unit** comprises a narrow band *c.* 30 km wide (based on geophysical evidence) which is composed of thick continental crust (Winchester *et al.* 2002*a*). The extent of the unit to the west, north and east is obscure (Fig. 7.8). Palaeozoic successions are exposed in the northern part of the Holy Cross Mountains. The oldest rocks are Lower Cambrian deformed shales (Fig. 7.9). An overlying sandstone succession (up to 1400 m thick) contains a rich assemblage of trilobite ichnofossils and has been interpreted as a shallow-marine offshore environment (Radwanski & Roniewicz 1960; Zylinska 2001, 2002). The overlying Silurian includes a prominent clastic unit comprising up to 1500 m of greywackes with interbedded volcaniclastics. The succession represents an upward evolution from deep- to shallow-marine deposition (Tomczyk 1970). Of note is the fact that there are no angular unconformities in the pre-Carboniferous stratigraphy.

2. The **Malopolska Massif** probably represents a composite terrane composed of the Malopolska and San blocks, which exhibited individual tectonic and sedimentary development during Early Palaeozoic times (Belka *et al.* 2000). The boundary between the blocks, however, cannot be precisely defined (Fig. 7.8), but differences in the crustal thicknesses suggest a NE–SW boundary close, and parallel to, the river Vistula (see Dadlez 2001). The western part (i.e. Malopolska Massif *sensu stricto*) is characterized by the pre-Ordovician deformation of Cambrian and Precambrian rocks (Fig. 7.9). Crustal thickness is 35–38 km, which is *c.* 10 km thinner than the San Block (for details see Belka *et al.* 2000).

The Malopolska Massif is bounded to the north and SW by the Holy Cross Fault and the Crakow Fault, respectively. Crystalline basement is unknown. Underlying the Palaeozoic is

Fig. 7.7. Palaeogeographic reconstructions based on palaeontological and palaeomagnetic data (after Cocks & Torsvik 2005). Baltica and adjacent terranes are shown at distinct times, from top to bottom: Lower to Middle Cambrian (510 Ma), Late Ordovician (Mid-Ashgill, 450 Ma), mid-Silurian (Wenlock, 425 Ma).

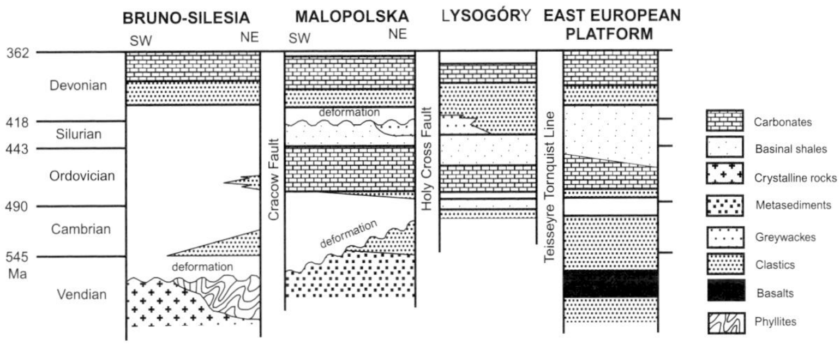

Fig. 7.9. Crustal units in the Variscan foreland in southern Poland and on the East European Platform near Warsaw (after Belka *et al.* 2002). Stratigraphic columns are separated by tectonic boundaries and show differences in occurrence of deformation events and stratigraphic gaps.

a thick (>3000 m) succession of weakly metamorphosed clastic rocks (Kowalczewski 1990) of probable Vendian age (based on acritarchs). Belka *et al.* (2000) have suggested that this basement was a late Precambrian forearc-trench system linked to the Avalonian/Cadomian active continental margin. The entire Precambrian to Middle Cambrian succession is folded and discordantly overlain by the Lower Ordovician. The Ordovician succession comprises offshore sandstones, open-marine carbonates and shales, and is interpreted as a transgressive succession. The overlying Silurian comprises a monotonous succession of graptolitic shales overlain by up to 1200 m of greywackes. This resembles the Silurian of the Lysogóry Unit, but is thinner.

3. The **Upper Silesian Massif/Bruno-Silesia** comprises the northern part of a larger block called the Brunovistulian (*sensu* Dudek 1980) or Bruno-Silesia as a palaeogeographical unit (Belka *et al.* 2000). However, there is uncertainty concerning the inferred outline and the position of terrane boundaries since much of the Brunovistulian is hidden beneath the Variscan overthrust (Moravo-Silesian Belt) and the Carpathians. It is separated from the Malopolska Massif by the Crakow Fault (Figs 7.8 & 7.9). The Brunovistulian comprises a thick autochthonous Palaeozoic sedimentary succession resting on Cadomian crystalline basement. The latter includes both intrusive rocks and their metamorphic cover (Dudek 1995). Basement is unconformably overlain by a thick Lower/Middle Cambrian succession deposited in a tide-dominated shallow offshore environment which grades upwards into a deeper offshore setting. Of note is a significant stratigraphic gap, comprising the Ordovician and Silurian (with the exception of some marine Ordovician preserved at the northern margin of the Upper Silesian Massif). Despite linkage of the Bruno-Silesian Block and the Malopolska Massif by some authors, Bula *et al.* (1997) have convincingly demonstrated that the two units have separate stratigraphical and tectonic developments. The evidence for any particular faunal geographic affinity for the Bruno-Silesian Block seems inconclusive and it could originally have been either peri-Gondwanan or peri-Baltic (Cocks 2002).

Winchester *et al.* (2002*a*, *b*) have noted the significant problems associated with trying to interpret the tectonic and sedimentary evolution of this complex region, as well as establishing the continental affinities of the various units. Further to the SE, and apparently sharing a similar Palaeozoic geological history, are the Central and Southern Dobrogea terranes and the Moesian Platform of southern Romania. However, further work is required to establish any clear relationship. Bruno-Silesia was considered by Moczydlowska (1997) to be a possible eastward extension of Avalonia. By contrast the Lysogóry Unit and Malopolska Massif were interpreted as being Baltica fragments (Dadlez *et al.* 1994; Pharaoh 1999). However, a link between the Bruno-Silesian Block and the Malopolska Massif is inferred because the Cambrian succession of the latter has been interpreted as an accretionary wedge to the Bruno-Silesian Block. If true, this would suggest that the two blocks have always been closely linked.

Cocks (2002) noted that there is no faunal evidence to suggest that these blocks were ever separated from Baltica. However, Belka *et al.* (2000) suggested that some Cambrian faunas have an affinity with Gondwana; he corroborated this with a provenance study on detrital muscovites. Cocks (2002), however, noted that since the discovery of late Neoproterozoic deformed basement to the Uralides in the east of Baltica (Glasmacher *et al.* 1999) the presence of pan-African-age detrital muscovites could not be used as proof of Gondwanan affinities. Despite these differences, both authors agree that since latest Cambrian times these various blocks were attached to the Baltica margin, thus precluding any possibility of them being part of Avalonia which was, in the early Ordovician, still attached to Gondwana.

The oldest rocks in the area crop out in the Malopolska Massif and comprise Vendian-age siltstones and volcaniclastic rocks (Bula *et al.* 1997). Lithologically, these rocks are similar to successions from the adjacent EEC (Moczydlowska 1995). The Cambrian strata in both the Lysogóry Unit and the Malopolska Massif contain a shelly fauna diagnostic of Baltica (Orlowski 1992), although those in the Lysogóry Unit are more similar to those of Armorica (Pharaoh 1999). Early Cambrian brachiopod faunas of the Malopolska Massif show mostly Gondwanan affinities. During the Middle Cambrian, progressive introduction of Baltica species together with sediments sourced from Baltica indicates the proximity to Baltica, with docking occurring in the Late Cambrian. In the Lysogóry Unit, however, Middle–Upper Cambrian rocks contain fossils which are absent from Baltica and instead show a more Gondwanan affinity. However, clastic material in the Cambrian succession was derived both from Baltic and Gondwanan sources suggesting that neither continent was distant (Belka *et al.* 2000). Ordovician faunas are well documented in the southern parts of the Holy Cross Mountains (Dzik *et al.* 1994) and show essentially Baltica affinities, confirming that the accretion of the Malopolska Massif to Baltica was complete by the end of the Cambrian.

Clastic sediment provenance, depositional history and palaeomagnetic data indicate that the amalgamation of the Malopolska Massif and the Lysogóry Unit was attained during the late Silurian. However, the presence of Devonian arc-related magmatism in the Jesiniky Mountains suggests that with SE-directed subduction on the NW margin of the Bruno-Silesian Block, its displacement along the TESZ margin of Baltica may have continued into late Palaeozoic time.

Actual docking of the Bruno-Silesian Block and the Malopolska Massif is recorded by the Sandomierz Phase of deformation in the late Cambrian (Belka *et al.* 2000). However, the absence of calc-alkaline volcanic rocks in the Cambrian successions of both units argues against them having had a tectonically independent existence.

The precise timing of the docking of the Lysogóry Unit with Baltica is still controversial, but appears to have been later than that of the Malopolska Massif. It has been suggested that these terranes rifted from the Gondwana margin in early Cambrian (Malopolska Massif) and earliest Ordovician (Lysogóry Unit) times and represent the earliest Gondwana-derived crust accreted to Baltica. In contrast, Moczydlowska (1995) argues that the faunal and lithological similarities indicate minimal separation from the EEC, while Dadlez *et al.* (1994) and Berthelsen (1998) suggest involvement of the Holy Cross Mountains in a marginal overthrust belt (as in Pomerania).

All of these blocks accreted to Baltica far earlier than any part of Avalonia, and indeed before Avalonia rifted away from Gondwana. The tectonic junction of these fragments with Avalonia has been termed the Moravian Line. Whether or not any substantial strike-slip occurred between the Holy Cross Mountains and the main EEC of Baltica along the Tornquist margin of Baltica remains uncertain. However, it is more likely to have been a few hundred kilometres rather than the thousands of kilometres postulated by some authors.

Many papers continue to suggest that Avalonian basement underlies parts of the Bohemian Massif (e.g. Finger *et al.* 2000). In this debate, establishing the late Ordovician timing of the accretion of Avalonia to Baltica (Vecoli & Samuelsson 2001) is a crucial piece of evidence that the basement of 'Far Eastern Avalonia' (see below) has Avalonian affinities, although rendered possibly suspect with respect to the main part of Avalonia by the Anglo-Brabant Deformation Belt, which is characterized by calc-alkaline magmatism.

Isotopic evidence from the Bohemian Massif for the timing of ophiolite generation and deformation clearly shows that accretion dates for these crustal blocks are much later than that of Avalonia, and that Devonian and Carboniferous subduction and collision of constituent blocks of the Armorican Terrane Assemblage predated accretion to the Laurentian supercontinent. However, as observed by Aleksandrowski & Mazur (2002) individual crustal blocks within the Armorican Terrane Assemblage (see below) appear to be continuous for long distances to the west, negating suggestions that separate Armorican and Perunican blocks existed (see also Verniers *et al.* 2008).

Avalonia. Avalonia is one of several exotic terranes within the Appalachian–Caledonide Orogen that are interpreted as having evolved along an active margin of Neoproterozoic Gondwana. Avalonia extends from New England and Atlantic Canada (sometimes referred to as West Avalonia) into southern Ireland and Britain and on through Belgium, the Netherlands and Germany into western Poland (sometimes referred to as East Avalonia) (Nance *et al.* 2002). The Brabant Massif of Belgium forms the eastward continuation of the East England Caledonides within the Anglo-Brabant Massif (Debacker *et al.* 2001) and has a similar early Palaeozoic history (Woodcock 1991). The Rhenohercynian nappes of Cornubia and central Germany are situated to the north of the Rheic Suture (Cocks *et al.* 1997) and, therefore, represent part of Avalonia detached during the Variscan Orogeny. Following opening and closing of the short-lived Rhenohercynian Ocean (Franke 1989) in Devonian times, this area was later incorporated into the Variscide foreland fold-and-thrust belt.

The Cambrian to Devonian palaeogeographical history of the Avalonia microcontinent and its relation to the palaeocontinents of Laurentia, Baltica and Gondwana have been established in recent years (Cocks 2000; Verniers *et al.* 2002; Winchester *et al.* 2002*b*). The margins of Avalonia have been discussed by Cocks *et al.* (1997) and Verniers *et al.* (2002), while the composition and history of the Avalonia microcontinent and related microplates are reviewed in Pharaoh (1999) and Verniers *et al.* (2002). The presence of Neoproterozoic basement has been proven in Eastern Avalonia (see Pharaoh 1999), and beneath the Harz Mountains (Breitkreuz & Kennedy 1999) and the south Hunsrück area (Molzahn *et al.* 1998) in Germany.

Avalonia is one of the largest of the peri-Gondwanan terranes, all of which are characterized by Neoproterozoic histories of continental margin subduction, and by the peri-Gondwanan faunas of their early Palaeozoic shallow-marine successions (e.g. Nance & Thompson 1996). The peri-Gondwanan evolution of Avalonia includes prolonged (*c.* 740–570 Ma) Neoproterozoic subduction and arc development, and its transition, without an intervening continental collision, to an early Cambrian platform environment. Subsequent rifting from Gondwana was followed by collision with Baltica/Laurentia during the Caledonide Orogeny (Nance *et al.* 2002).

The Precambrian basement rocks comprise metamorphosed arc-magmatic rocks, and marginal sedimentary basins accreted to the Rodinia–Pannotia (subsequently Gondwana) Supercontinent between 680 and 545 Ma (Pharaoh & Gibbons 1994). The heterogeneity of the terrane has been confirmed by geochemical and isotopic studies (for details see Pharaoh 1999) and these contrasts may reflect derivation of the Avalonian and Cadomian slivers from different parts of the Gondwana margin (O'Brian *et al.* 1993).

The opening of Iapetus is generally attributed to the rifting of Laurentia from Amazonia and Baltica (e.g. Dalziel 1992, 1997; Torsvik *et al.* 1996). The transition from a magmatic arc to a continental platform on Avalonia was broadly coeval with the rifting of Baltica (Nance *et al.* 2002). However, as Iapetus opened and Baltica separated from Amazonia, Gondwana rotated counterclockwise (on most palaeocontinental reconstructions), bringing its Avalonian margin into the Iapetus seaway. As a result, it is the Iapetus Ocean that closes as Avalonia separates from Gondwana, while the Rheic Ocean opens behind it (Nance *et al.* 2002).

The timing of the Early to Mid-Ordovician separation of Avalonia from West Africa is not well constrained by palaeontology (Cocks & Fortey 1982) or by palaeomagnetism (Trench *et al.* 1992). Avalonian faunas are recognized in southern Ireland, in Britain south of the Iapetus Suture and north of the Lizard Thrust, including the Rhenohercynian nappes of Cornubia, in the Brabant and Ardennes massifs of Belgium, and in the northern Phyllite Zone of Germany (Dallmeyer *et al.* 1995; Cocks *et al.* 1997). Cladistic analyses have shown that until the Arenig, shelf benthic faunas of Avalonia clustered with those of Gondwana (i.e. including distinctive Neseuretid–Calymenid trilobites and Mediterranean Province brachiopods) (Fortey & Cocks 1992) but by the Late Llanvirn, separation was wide enough (>1000 km) to reduce faunal interchange with Gondwana and allow interchange with Laurentia and Baltica (thus eradicating the Gondwanan signature). These changes were enhanced by the fact that the Ordovician Avalonian drift history was also from higher to lower latitudes within the southern hemisphere (Cocks & Torsvik

2002), which meant that the benthic faunas became progressively more diverse.

Palaeomagnetic (Trench *et al.* 1992) and Sm/Nd data (Thorogood 1990) suggest that separation of Avalonia from Gondwana had taken place in the Arenig, which is consistent with evidence of rift-related magmatism elsewhere in Avalonia (Barr *et al.* 1994). Minor bimodal rift volcanism that is predominantly of Middle Cambrian age may record the initial rifting from Gondwana (Nance *et al.* 2002). Additionally, widespread Arenig subsidence also suggests that rift–drift transition had been accomplished by the Early Ordovician. If true, the separation of Avalonia from Gondwana closely followed the unrelated initiation of subduction of the Iapetus Ocean at *c.* 510 Ma (e.g. van Staal *et al.* 1998), and might record a switch from transform motion to spreading along the Avalonian plate margin in response to Iapetus slab pull (Nance *et al.* 2002). Verniers *et al.* (2002) have proposed a new model which suggests that during early Cambrian times an extensional basin, or series of basins, developed which formed a rift along the Anglo-Brabant-Ardennes area (e.g. Crowley *et al.* 2000; Winchester *et al.* 2002*a*) soon after the Neoproterozoic Pan-African Orogeny or even during it (Murphy *et al.* 1999). Hence, a zone of crustal weakness was established which formed a template controlling structural evolution from the Cambrian to the Middle Devonian.

A late Ordovician calc-alkaline arc can be traced from northern England to Belgium and is inferred to result from the subduction of Iapetus/Tornquist oceanic lithosphere beneath Avalonia (Pharaoh *et al.* 1993). The progressive change in the onset of arc volcanism across Avalonia during Ordovician times testifies to counterclockwise rotation of Avalonia with respect to this subduction zone (Pharaoh *et al.* 1995). With the cessation of Iapetus Ocean subduction during Caradoc–Ashgill times, tectonic instability increased. The subsequent docking of Avalonia with Baltica (late Ashgill–early Silurian), associated with an anticlockwise rotation, led to the subduction of oceanic crust in the basin lying west of Eastern Avalonia, giving rise to calc-alkaline subduction magmatism.

The anticlockwise rotation of Wales and the Midlands Microcraton, relative to the Lake District in Caradoc to mid-Devonian times is considered to be the driving mechanism for the further evolution of the Anglo-Brabant Deformation Belt which forms a folded, faulted and weakly metamorphosed belt in the subcrop of East Anglia (eastern England) and continues under the North Sea into the Brabant Massif. The rotation implies a scissors-like closure for the Anglo-Brabant Deformation Belt leading to *c.* 500 km of shortening in its southern part (i.e. Brabant Massif). This movement explains the short-lived late Ordovician–earliest Silurian subduction of oceanic crust as revealed by subduction-related magmatism. Volcanism ceased in the Brabant Massif area in the early Llandovery, possibly because the oceanic crust had subducted completely and the continental crust of the Southern North Sea–Lüneburg Terrane had approached the SW limit of Baltica. Continuing rotation created the Silurian foreland basin of the central and northern Brabant Massif (van Grootel *et al.* 1997) and, possibly, the Anglian Basin in eastern England (Pharaoh *et al.* 1993). Rotation may also have led to the Wenlock–early Eifelian Brabantian phase of shortening and inversion of the central steep belt of the Brabant Massif. Some areas were shielded from deformation, including the SW Brabant Shelf and the Condroz Shelf, possibly due to their being underlain by Neoproterozoic basement or a more rigid, low density body at depth. The end of rotation by Eifelian time terminated the Brabantian phase of deformation.

Avalonia fragments. From Cambrian through to Tremadocian times, Avalonia was attached, or close to, northern Gondwana. However, by Llanvirn times it was moving away, leading to the opening of the Rheic Ocean (Verniers *et al.* 2002). A 45°S position is estimated for Avalonia at the end of the mid-Ordovician. By the late Ordovician a position of between 30° and 40°S has been deduced, and Avalonia was close to Baltica. By the Wenlock, Avalonia was located at 13°S, Baltica rotated substantially anticlockwise from the late Cambrian to the early Ordovician (*c.* 100°) and from late Ordovician to mid-Silurian (*c.* 50°) (for details see Verniers *et al.* 2002).

Verniers *et al.* (2002) have detailed the various fragments of Avalonia which are distributed across central and western Europe (Fig. 7.10). While some of these fragments are unequivocally considered to form part of Avalonia, others are more controversial, and are discussed in more detail below (see also Verniers *et al.* 2008). In contrast to the situation in England, Wales and southern Ireland (for an overview see Woodcock & Strachan 2002*a*), the record in central Europe is more fragmentary and difficult to reconstruct. Verniers *et al.* (2002) have summarized the main fragments present, including the following.

Brabant Massif. Vanguestaine (1992) subdivided the Lower Palaeozoic into three sedimentary megacycles (Lower Cambrian–Tremadocian, Arenig–Caradoc, Ashgill–Silurian) which are comparable with the megasequences identified by Woodcock (1990*b*) in the Welsh Basin. The base of megasequence 1, on a supposed Neoproterozoic craton, has not been observed. However, the presence of Proterozoic crystalline basement beneath, or close to, the Brabant Massif is inferred from zircons which were formed during a magmatotectonic event dated at latest Neoproterozoic, and are contained within metavolcanic rock fragments in the Lower Cambrian sediments, and as inclusions and xenoliths in the Upper Ordovician magmatic rocks (Verniers *et al.* 2002). The megasequence, with a minimum thickness of 3700 m, contains earliest Cambrian to early Tremadocian terrigenous and deep-marine sediments with greywackes, sandstones and thick units of pelagic to hemipelagic mudstones (Verniers *et al.* 2002). The depositional environment is deeper than shelf as evidenced by the turbiditic units. At the top of the megasequence is a substantial hiatus extending from the Lower Tremadocian through to the upper part of the Middle Arenig (*c.* 12 Ma; Verniers *et al.* 2001).

Megasequence 2 comprises terrigenous shallow shelf sediments changing at the top to a deeper-marine setting with turbidite development. The onset of deeper water conditions follows a long interval (hiatus or condensed sedimentation of *c.* 7 Ma; Fig. 7.10). They are thinner than the underlying megasequence (*c.* 850 m minimum). The stratigraphic contact with the overlying megasequence 3 is unknown because of the presence of faults. However, the transition from megasequence 2 to 3 is coincident with a dramatic change from deep water to shelf conditions. Megasequence 3 records shelf deposition in the upper Caradoc–middle Llandovery which evolved into a thick foreland basin unit in the upper Llandovery to Prídolí. The megasequence thickness is estimated at >3200 m in the area of the Brabant Massif.

Condroz Inlier. Only the upper part of megasequence 1, comprising low-density distal turbidites, is present. Megasquence 2 unconformably overlies 1, commencing with a basal conglomerate overlain by Llanvirn graptolitic mudstones. Following a hiatus or condensed section of *c.* 6 Ma, the mudstones are overlain by mid-Caradoc siltstones with rare quartzites. The depositional environment was probably a deep shelf.

Deposition of megasequence 3 began after a probable hiatus from the mid-Caradoc–mid-Ashgill (*c.* 9 Ma). The mega-

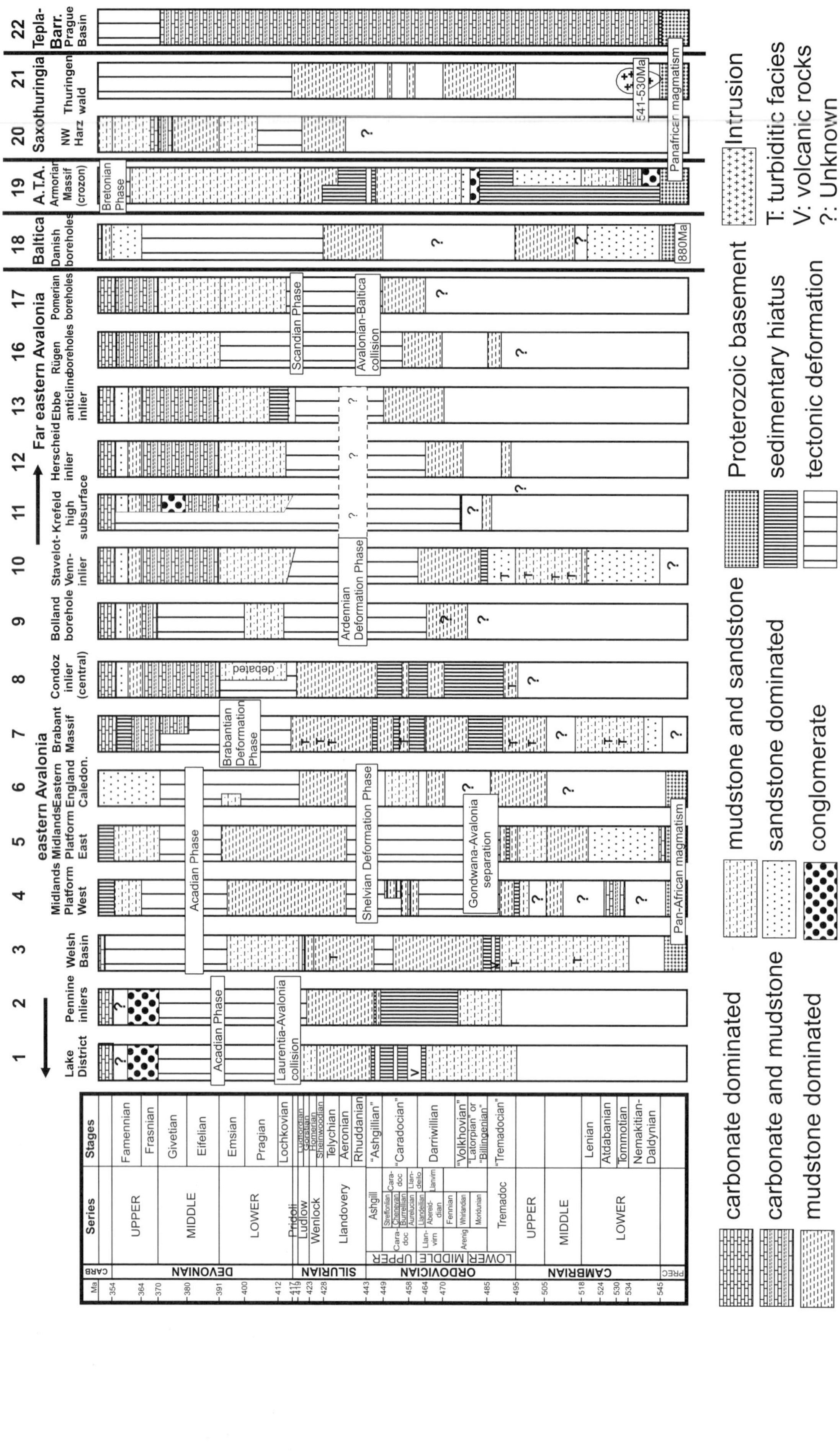

Fig. 7.10. Cambrian to Devonian successions in different parts of Avalonia with main tectonic phases (after Verniers *et al.* 2002).

sequence commences with a (calcareous) mudstone unit, at the top of which are two coarse sand or conglomeratic beds (tentatively dated as Hirnantian) whose deposition is related to the eustatic sea-level fall at the time. In the Silurian muddy deposition continued until late Ludlow times. No turbidites have been observed within the megasequence (Fig. 7.10) suggesting that deposition occurred in a deep shelf setting (Verniers *et al.* 2002).

Ardennes inliers & Bolland borehole. In the Ardennes region there are four inliers present. In the two largest (Stavelot-Venn, Rocroi) only megasequences 1 and 2 are present. The two smaller inliers (Serpont, Givonne) contain only parts of megasequence 1. Megasequence 3 is absent due to the deformation of the inliers during the Ardennian Phase in late Ordovician times (Michot 1980).

Megasequence 1 in the Stavelot-Venn Inlier commences above basement with an inferred Neoproterozoic age. The succession comprises sandstones and mudstones deposited in a nearshore to intertidal setting on a shallow shelf. Up-section the succession becomes progressively deeper (to deep shelf). The presence of a coarse turbiditic facies further up-section has been interpreted as indicating deposition on the lower slope (upper fan area of a submarine fan system) (von Hoegen *et al.* 1985; Verniers *et al.* 2001). The top of the Cambrian terminates with black mudstones which are overlain by Tremadocian turbidites which grade upwards into mudstones and sandstones. The Tremadocian succession has been interpreted as representing the filling of an epicontinental basin (Lamens 1985, 1986). The overall thickness of megasequence 1 is estimated to exceed 2400 m, with only 300–400 m for the Cambrian shelf deposits.

According to Vanguestaine (1992) there is a gap in sedimentation between megasequences 1 and 2, coincident with a dramatic change in the depositional setting from a shallow platform at the top of the former to a probable deep-water environment at the commencement of the lattter. Sediments are predominantly mudstones and siltstones with rare turbidites. The succession passes upwards into siliciclastic pelagic to turbiditic sediments of late Arenig or Llanvirn age deposited in a deep-marine setting (Lamens 1988). The thickness of megasequence 2 is estimated at >330 m.

While megasequence 1 in the Ardennes inliers is similar to that of the Brabant Massif, the Lower Cambrian part in the Ardennes was deposited on a shelf whereas the Brabant Massif succession was deposited in deep water. In Upper Cambrian and Tremadocian times, the depositional environments are similar.

The Bolland borehole is 3001 m deep and in its lowermost 126 m contains folded mudstones of Arenig–Llanvirn age (Michot 1979). However, the affinity of these sediments with those of either the Brabant Massif or the Ardennes inliers has not yet been established (Verniers *et al.* 2002).

For the Belgian data it is clear that megasequence 1 is similar in all of the three main areas – the Brabant Massif, the Condroz Inlier and the Ardennes inliers. Megasequence 2, while similar in the former two, differs substantially in the latter. The separation into three different areas is even more pronounced during the depositon of megasequence 3, where shelf deposits (Brabant Massif, Condroz Inlier) contrast with a turbiditic basin (north and central Brabant Basin) and uplift and deformation (Ardennes inliers).

Krefeld High. A 100 × 25 km, NNW–SSE-orientated subcrop high has been postulated beneath the city of Krefeld (Germany) and between the cities of Düsseldorf and Nijmegen. This high supposedly contains Proterozoic and Lower Palaeozoic rocks. The presence of thick Givetian-age conglomerates near Düsseldorf suggests a local source area. Boreholes from the area have encountered Eifelian shallow-marine sediments beneath these conglomerates. It has been suggested that this high was uplifted from the Givetian until the end of the Famennian (Verniers *et al.* 2002).

Ebbe & Remscheid inliers of the Rhenish Massif (= Rheinisches Schiefergebirge). Ordovician and Silurian rocks occur in the Ebbe and the Solingen-Remscheid-Altena anticlines in the Sauerland, east of Cologne (Germany). These sediments are usually considered to be part of Avalonia. The strongly deformed Herscheider Schichten form a monotonous 800 m thick clastic succession of mudstones and siltstones. Isotope work carried out by Samuelsson *et al.* (2002*b*) suggested that the Herscheider Schichten were deposited in the same sedimentary basin as the rocks of Rügen (see below), supporting the idea of a continuation of eastern Avalonia eastwards of the Brabant Massif into northern Germany. Although both areas have a similar deeper shelf depositonal setting as indicated by lithology, sedimentary structures and ichnofossils, they could also have been deposited on two different, but neighbouring, microplates of Avalonia (i.e. the Avalonian Terrane Assemblage of Verniers *et al.* 2002).

Southern North Sea–Lüneburg Terrane (= Far Eastern Avalonia). The crust beneath the Anglo-Dutch Basin is generally considered to be part of Eastern Avalonia or an Avalonia-related microplate. However, for 250 km to the NE of the Dowsing-South Hewett Fault Zone and north of the Rhenish Massif, pre-Devonian basement is not encountered by deep boreholes (Pharaoh 1999). Strong geophysical constrasts in the crust on either side of this lineament led Pharaoh *et al.* (1995) to suggest that it might be an Ordovician terrane boundary.

Boreholes have cored basement on the Mid-North Sea/ Ringkøbing Fyn High penetrating schists which underwent prograde greenschist-amphibolite metamorphism at 450–425 Ma and retrogression at 415–400 Ma (Frost *et al.* 1981). The former is considered to mark the docking of Avalonia to Baltica in the late Ordovician, coeval with the Shelveian Event in England and the Ardennian Phase in Belgium. Further evidence for the timing of docking comes from the sedimentary succession of the G14 borehole near the island of Rügen (north Germany; for location see Fig. 7.54) which began to receive reworked Gondwanan fossils of Llanvirn age and Neoproterozoic micas in early mid-Ashgill time (Samuelsson *et al.* 2002*a*, *b*). Detritus was fed across the suture zone from the North German–Polish Caledonides (= Heligoland-Pomerania Deformation Belt of Verniers *et al.* 2002). Further deformation occurred during the development of a foredeep on the margin of the East European Craton (EEC) throughout the Silurian culminating in the Scandian Phase.

Danish and Northern Germany subsurface, Rügen & Pomerania. In the Danish subsurface, the Lower Palaeozoic consists of relatively undeformed (?), predominantly pelitic rocks (Poulsen 1974; Frost *et al.* 1981; Michelsen & Nielsen 1991; Vejbaek *et al.* 1994). The Slagelse 1 borehole of western Sjaelland provides a good overview of the Cambrian to Lower Silurian (400 m) stratigraphy of the area. The Lower Cambrian comprises a basal quartzitic sandstone (6 m) overlain by a shale succession (22 m) rich in trilobites and graptolites. This is overlain by a Middle–Upper Cambrian black shale succession ('Alum Shales', 27 m). The overlying 106 m comprise undifferentiated Ordovician-lowermost Silurian. The Lower Silurian (Llandovery) comprises finely laminated silts and shales (174 m).

Similar, but incomplete, Lower Palaeozoic successions are encountered in several other wells drilled in the Norwegian–Danish Basin (e.g. Novling 1, Terne 1, Pernille 1) and in the G14 borehole located in the southernmost Baltic Sea (McCann 1998). These successions are closely correlatable, both lithologically and biostratigraphically, with the classic platform, Lower Palaeozoic successions of Scania (Sweden) and Bornholm (e.g. Michelsen & Nielsen 1991; Samuelsson *et al.* 2001). The Lower Palaeozoic sedimentary successions of the Danish subsurface were deposited on the SW passive margin of Baltica with the main detrital input being derived from the stable cratonic regions of the Baltic Shield (for details see Verniers *et al.* 2002). The Lower Palaeozoic sediments of the Norwegian–Danish Basin as well as those of the Bornholm area indicate a long period of tectonic stability between the early Cambrian and the latest Ordovician–earliest Silurian. This was followed by a period of foreland basin-type rapid sedimentation beginning in the Llandovery and continuing into Wenlock times (Vejbaek *et al.* 1994).

The Ordovician sediments of western Pomerania (NW Poland) are similar to the coeval strata present in the Rügen subsurface. The entire Ordovician succession consists of siltstones with rare carbonaceous intercalations and may be subdivided into five lithological units ranging in age from late Darriwillian (Llandeilian) to late Caradoc (Samuelsson *et al.* 2002a). The onset of foreland basin sedimentation began in the Rügen area in mid-Caradoc to late Ashgill times. In Pomerania, clastic sedimentation was active at least to the mid-Caradoc. According to Cocks (2002), acritarch and chitinozoan evidence from the Rügen and Pomerania areas suggests that sediments in both were deposited in an ocean basin within the Tornquist Sea rather than on either Baltica or Avalonia themselves.

Rhenohercynian zones of Avalonia in central Germany (Giessen nappes). The Rhenohercynian nappes of central Germany extend as far south as the Northern Phyllite Zone and to the north of the Rheic Suture (see Kroner *et al.* 2008; McCann *et al.* 2008). They represent a part of Avalonia displaced by thrusting during the Variscan Orogeny. Initial opening of the Rheic Ocean in Arenig times coincided with the rifting of Avalonia from Gondwana. By late Silurian times the ocean was closing rapidly, with the formation of a magmatic arc, generated by southward-directed subduction, in the area of the Mid-German Crystalline High (see Kroner *et al.* 2008; McCann *et al.* 2008). The northern margin of the ocean briefly experienced backarc basin extension with the initiation of the Rhenohercynian Basin and the formation of the Lizard-Giessen ophiolites (W. Franke 1995) in a marginal basin at *c.* 397 Ma (Clark *et al.* 1998).

There is ongoing discussion concerning which parts of the fragmentary Central European record comprise parts of Avalonia or are better considered as Avalonia-related fragments. Precambrian and early Palaeozoic basement exposed in central England, Belgium and western Germany is widely accepted as forming part of Avalonia (Verniers *et al.* 2002). Avalonian basement in central England typically consists of late Proterozoic intrusive, volcanic and sedimentary rocks and, as with the Bruno-Silesian Block, was also affected by latest Proterozoic to pre-Lower Cambrian deformation.

Boreholes in eastern England reveal that the Midlands Microcraton is bounded to the NE by a Caledonian deformation belt. Late Ordovician calc-alkaline volcanic rocks are present within this belt and extend from eastern England to Belgium (André 1991; Pharaoh *et al.* 1993). The southern end of this belt is exposed in the Brabant Massif of Belgium and hence it has been termed the Anglo-Brabant Deformation Belt (Winchester *et al.* 2002*a*). This belt is inferred to have developed in early Devonian (Acadian) time above a zone of crustal suturing inherited from the late Ordovician soft collision of Avalonia and Baltica. The presence of the Anglo-Brabant Deformation Belt calls into question whether the basement further east, i.e. NE of the Dowsing-South Hewett Fault Zone–Lower Rhine Lineament (Pharaoh 1999) also forms part of Avalonia. Pharaoh *et al.* (1993) suggested that this lineament may separate crusts with differing structures, juxtaposed by late Ordovician subduction, the inferred cause of the calc-alkaline volcanism noted above. In this area, the crystalline basement is generally not exposed. Rare exposures (e.g. Wartenstein Gneiss (Molzahn *et al.* 1998) and Ecker Gneiss (Baumann *et al.* 1991)) typically have calc-alkaline compositions and late Neoproterozoic ages which are broadly comparable to Avalonian basement in central England. Hence, despite the presence of the intervening Anglo-Brabant Deformation Belt, the basement of this area is generally linked with that of Avalonia. However, since so many crustal basement fragments both in Avalonia and the Variscides of Central Europe appear to record late Proterozoic Cadomian deformation, it is the timing of the docking of these individual crustal blocks with Baltica which is most likely to decide their affinities (Verniers *et al.* 2002).

Avalonia-related terranes and subterranes. The precise boundaries of Avalonia are not always clear, and thus for certain areas there is ongoing discussion about what precisely constitutes the core of Avalonia (Winchester *et al.* 2002*a*; Verniers *et al.* 2002). For example, the crust beneath the Anglo-Dutch Basin is generally considered to be part of Avalonia. However, the affinity of the region to the NE of the Elbe Line is less certain. It has been suggested that this region, along the northern margin of Avalonia, may comprise an early Palaeozoic accretionary collage (Pharaoh *et al.* 1995).

Boreholes on the Mid-North Sea High and the Ringkøbing-Fyn High have penetrated phyllitic schists of uncertain provenance. Ar/Ar plateau ages derived from these strata correspond to the Avalonia–Baltica collision in late Ordovician times (Pharaoh 1999). In addition, deep boreholes in Denmark (south Jylland) and northern Germany (Schleswig-Holstein) to the NE of the Elbe Line have encountered similar rocks. However, the question remains as to whether the crust beneath the Mid-North Sea High–Ringkøbing-Fyn High region and to the north of the Elbe Line is fundamentally different from that underlying the Anglo-Dutch Basin and the Lüneberg region (i.e. belonging to what D. Franke (1995) has called the East Elbe Suspect Terranes; see also discussion in Franke & Hoffmann 1999).

Arguments favouring the Elbe Line (for location see Figs 7.2 & 7.4) as the NE boundary of Avalonia include the evidence of continuity between Baltica and part of northern Germany (Tanner & Meissner 1996). Seismic evidence includes the increased crustal velocities in the area extending from the Baltic Shield to the Elbe Line (Rabbel *et al.* 1995) and the observed velocity changes across the Elbe Line between crust derived from Baltica and that derived from Eastern Avalonia (Abramowitz *et al.* 1999). Furthermore, xenoliths have been recorded from Permian basalts in northern Germany which show affinities to Proterozoic anorthosite massifs of the Baltic Shield (Kämpf *et al.* 1994). In this model, the area between the Caledonian Deformation Front and the Elbe Line is interpreted as part of an allochthonous sheet which was thrust onto the passive margin of Baltica (Tanner & Meissner 1996). The region between the Caledonian Deformation Front and the Elbe Line, therefore, is a complex one with affinities to both Baltica and Avalonia. Indeed, Abramowitz *et al.* (1997), Meissner & Krawczyk (1999) and Meissner *et al.* (2002) have suggested that the region may be a composite microcontinent made up of a collage of terranes accreted in front of the original microcontinent of Avalonia. While the palaeontological evidence would suggest that the area is clearly part of Eastern

Avalonia, the precise geological setting is unclear, and probably involves a degree of (?low-angle) thrusting.

Another problematic area is the region to the NE of the Lower Rhine Lineament. For 250 km to the NE of this lineament and its continuation, the Dowsing-South Hewett Fault Zone, the nature of the pre-Devonian basement is unknown. Strong geophysical contrasts in the crust on either side of this zone led Pharaoh *et al.* (1995) to infer the presence of an Ordovician terrane boundary.

Further to the east, near the island of Rügen, deep boreholes have cored thrusts and deformed sandstones of Ordovician age (Katzung *et al.* 1993). The Lower Palaeozoic succession from the G14 well is comparable to similar successions further north on Bornholm and Scania in terms of, for example, stratigraphic thickness, lithofacies, and faunal contents (Franke 1993; McCann 1996, 1998). Thus, the region around the G14 well can be assigned to Baltica, suggesting that no large-scale strike-slip movements can have occurred along the Sorgenfrei-Tornquist Zone in Phanerozoic times (Franke 1990*b*).

The precise position of the Rügen area on the Avalonian microcontinent, however, has not been fully determined. Some authors favour a position on the eastern margin of the Avalonia microcontinent (Torsvik *et al.* 1992, 1993) while others (e.g. Channell *et al.* 1993) appear to suggest that this part of NE Germany was part of a separate Western Avalonia microcontinent. Microfossil data, however, would suggest that the Rügen area was unequivocally part of Eastern Avalonia (Servais 1994; Servais & Fatka 1997), although Cocks *et al.* (1997) contest this interpretation since it is based upon poorly diagnostic deep-water faunas and the high proportion of carbonate grains in the Rügen rocks which are not recorded elsewhere from the higher-latitude parts of peri-Gondwana. Detrital muscovites within the Rügen succession yield an Ar/Ar age of *c.* 609 Ma indicating a 'Cadomian' age continental source (Dallmeyer *et al.* 1999), together with a significant contribution from an immature volcanic arc source (McCann 1998). The source arc is not exposed and may lie concealed within the Southern North Sea Terrane. Thus, the Rügen rocks may represent a lower erosion level of the same terrane or a contrasting thrust-controlled externide zone of the German Caledonides (Pharaoh 1999). However, in order to clarify this situation, detailed studies of the protolith and provenance of the Mid-North Sea High and the Ringkøbing-Fyn High are required.

Armorica. Situated to the south of Avalonia, Armorica covers much of modern France, including the Massif Central. The region also extends eastwards (i.e. Bohemian Massif) and the Saxothuringian Zone was apparently part of the Armorican Terrane. There is, however, much confusion as to what Armorica is (see summary in Robardet 2003). Recent research suggests that it may not have comprised a single block, and it has been suggested that it is better to refer to it as the Armorican Terrane Assemblage (Tait *et al.* 2000). The concept of a microplate or microcontinent of Armorica was originally based on palaeomagnetic data (Van der Voo 1979, 1982; Jones *et al.* 1982). For the Late Proterozoic and Cambrian, palaeomagnetic data for Armorica were similar to those obtained from Gondwana, suggesting that both were closely associated. Similar results were later obtained for the Ordovician (see references in Perroud *et al.* 1984). Conversely, the palaeomagnetic data for the Late Devonian indicated that Armorica and Gondwana were in different latitudes and that Armorica was, at that time, juxtaposed with – or very close to – the southern margin of Laurussia. Thus, between the Ordovician and Late Devonian, Armorica had moved independently in terms of latitude, justifying the status of a microplate.

However, the palaeogeographical evolution of Armorica during the Late Ordovician, Silurian and Devonian is much more enigmatic (Fig. 7.11). Both the scarcity of palaeomagnetic data as well as the ambiguity of some of the data have made it difficult to define the palaeogeographical evolution. Various scenarios for this period have been proposed (for summary see Robardet 2003), all of which suggest that Armorica and Gondwana were separated by a wide ocean at some period of the Palaeozoic. More recently, however, there have been critical reviews of the palaeomagnetic data, intimating that reliable data for the Silurian and Devonian of southern and central Europe (i.e. Armorica) are very rare (Robardet 2003).

The sedimentary history of North Africa and in the Variscan regions of southern and central Europe are very similar, suggesting that there was no separation of these areas at any time between the Late Ordovician and the Late Devonian (Robardet 2003; Fig. 7.11). The geographical distribution and evolution of lithofacies, the persistent affinities of planktonic organisms, as well as various marine invertebrate, vertebrate and botanical groups suggest that all of these regions remained closely linked during the entire Palaeozoic. Furthermore, there is no evidence for the proximity of Armorica and Baltica–Avalonia (or later Laurussia) between Late Ordovician and Early Devonian times. These two palaeogeographical units remained separated by the Rheic Ocean during this period, and closure of the oceanic faunal barrier did not begin before the Silurian (?Pragian). If we accept the existence of an Armorica microplate/terrane, then this would imply that there were two distinct major oceans present in the Palaeozoic, one separating Armorica from Baltica–Avalonia, and the second one between Armorica and Gondwana (Robardet 2003). Consequently, two distinct oceanic suture belts should be identifiable within the Variscan Belt, although the evidence suggests that there is only one (i.e. Rheic Ocean). However, if we accept that Armorica remained part of Gondwana, then the formation of the belt would have resulted in the closure of just one ocean (i.e. Rheic Ocean; Robardet 2003). Tait *et al.* (1997) have noted that the location of the early Palaeozoic Rheic suture zone between Avalonia and Armorica has been a matter of some debate, but is thought to be located at the southern margin of the Northern Phyllite Zone of Central Europe (see Kroner *et al.* 2008; McCann *et al.* 2008). Greenschist metamorphism characterizes this zone which is a tectonic collage containing remnants of both southern Avalonia and northern Armorica, including Ordovician quartzites, Ordovician/Silurian magmatic arc rocks and mid-upper Devonian clastic successions (Anderle *et al.* 1995; Franke & Oncken 1995).

Palaeobiogeographical analysis allows models for two oceans to be dismissed because they are inconsistent with the global distribution of faunas and floras during pre-Carboniferous times. A model for a single major Variscan ocean, proposed by several authors (e.g. Quesada 1991; Quesada *et al.* 1994) for the Iberian part of the Variscan belt also proves suitable for the French part of the belt. Therefore, the reality of an Armorica microplate detached from Gondwana, and having had an independent latitudinal evolution separate from that of Gondwana during part of the Palaeozoic, cannot be considered as established. Lithological indicators of palaeoclimate and palaeobiogeographical data concerning both marine vertebrates and invertebrates, and also land plants, are in total harmony and indicate that the southern European regions have remained permanently in close connection with Gondwana, of which they comprised the northern margin (Fig. 7.11). Thus, according to Robardet (2003), the concept of a separate microcontinent of Armorica can be considered a fiction. The true nature of Armorica/Gondwana, however, remains to be

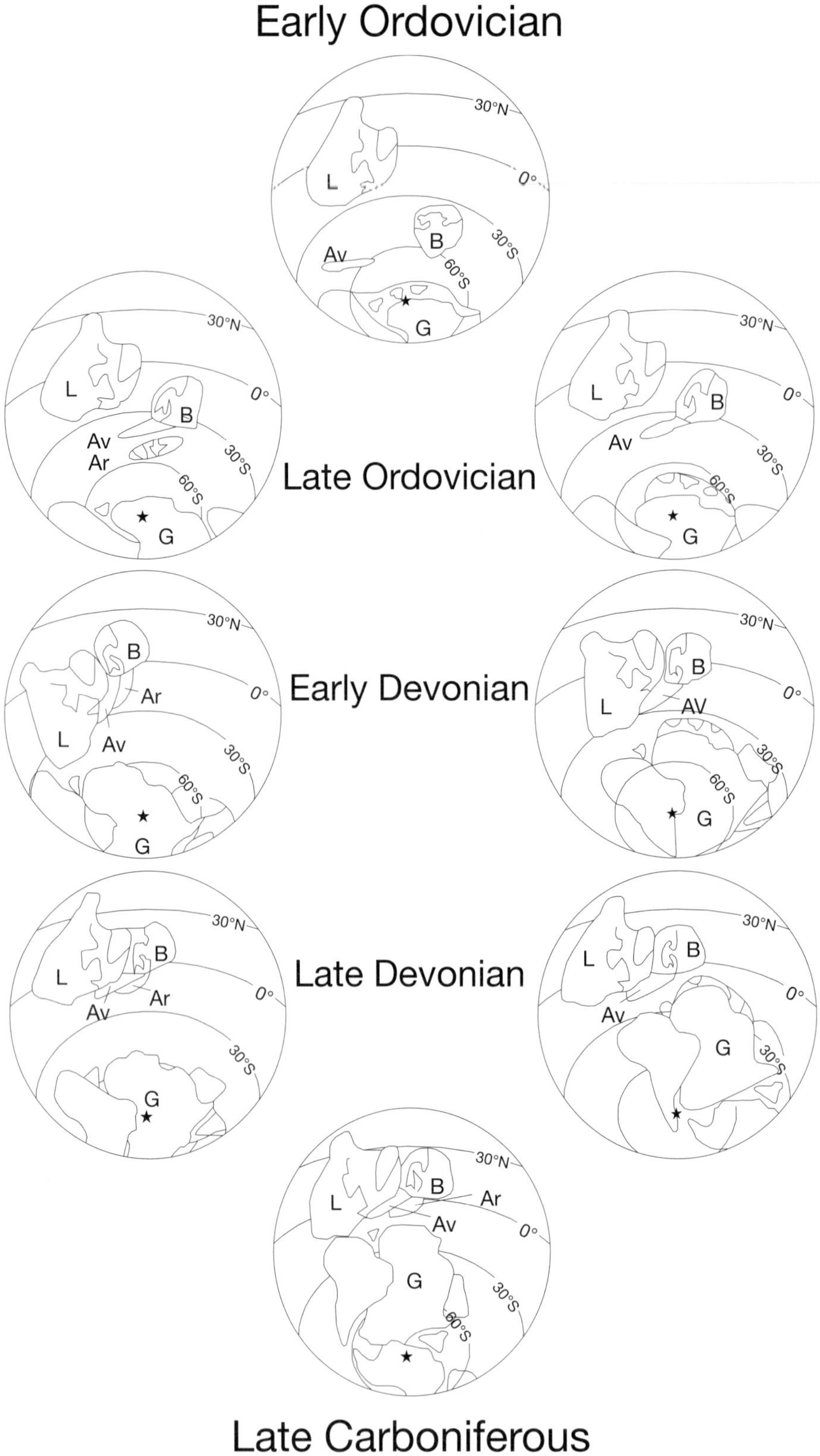

Fig. 7.11. Models of palaeogeographical evolution between Early Ordovician and Late Carboniferous (Robardet 2003). Abbreviations: Ar, Armorica; Av, Avalonia; B, Baltica; G, Gondwana; L, Laurentia.

established. The observed complexity (e.g. the differences between the western and eastern parts of 'Armorica') suggest, perhaps, that the region was located on a Gondwana plate (i.e. peri-Gondwanan fragment), but possibly in the form of an archipelago (e.g. Australia & Indonesia) and so distant enough to have some degree of independence in terms of collisional/indentor tectonics, while at the same time remaining close enough to maintain palaeomagnetic and biological cohesion with Gondwana.

Alpine fragments. Within the modern Alpine setting, particularly in Austria, there are numerous peri-Gondwanan fragments preserved in a variety of tectonic settings (e.g. Schönlaub 1997). A recent review has noted that many of the terranes, chiefly consist of Cambro-Ordovician arc-related metavolcanics with late Ordovician granitoids (von Raumer 1998). The oldest shelly faunas known from the area are late Caradoc brachiopods which clearly form part of the Mediterranean province (Havlíček *et al.* 1994). There is, therefore, no known faunal date which can be used to separate these fragments from Gondwana. Palaeomagnetic data, however, suggest some separation, at least by the late Silurian (Cocks 2000).

Caledonian granites

Laurentia, Baltica and northern Avalonia

An extensive suite of granitic plutons and associated dykes were intruded in the Caledonian Orogen during Silurian and Devonian times (Figs 7.12 & 7.13). These intrusions are known collectively as the Older and Newer Granites in Britain and Ireland (Woodcock & Strachan 2002*a*). Rb–Sr whole-rock geochronology indicates different intrusion ages in different parts of the orogen, but the mean age is approximately 400 Ma (Woodcock 2002*a*). The Older Granites are represented by S-type peraluminous granitic intrusions, which formed during the early stages of continent collision (Pitcher 1993). In the Caledonian Orogen these are associated with the high-grade Grampian regional metamorphism at *c.* 470 Ma (Dewey & Mange 1999) and are located within slate belts on both sides of the Iapetus Suture (Soper 1986). In contrast, the Newer Granites were emplaced around 40–50 Ma later and into the metamorphic part of the Caledonides following the cessation of the main Caledonian tectonic events. These granites display an I-type composition (Atherton & Ghani 2002). I-type granites are characteristic of subduction zones and post-collisional settings where the melting of the country-rocks usually results from decompression following crustal thickening (Pitcher 1993).

Contemporaneous lamprophyre and appinite intrusions associated with many of the late granites are thought to have been derived from partial melts of subcontinental mantle. In order to preserve their primitive chemistry, these melts must have risen rapidly through the crust, with ascent along crustal-scale strike-slip faults being the most likely mechanism (Woodcock & Strachan 2002*b*; Figs 7.13 & 7.14). There are marked geochemical differences between the granites across the Iapetus Suture. To the north of the suture, granites contain a lower proportion of

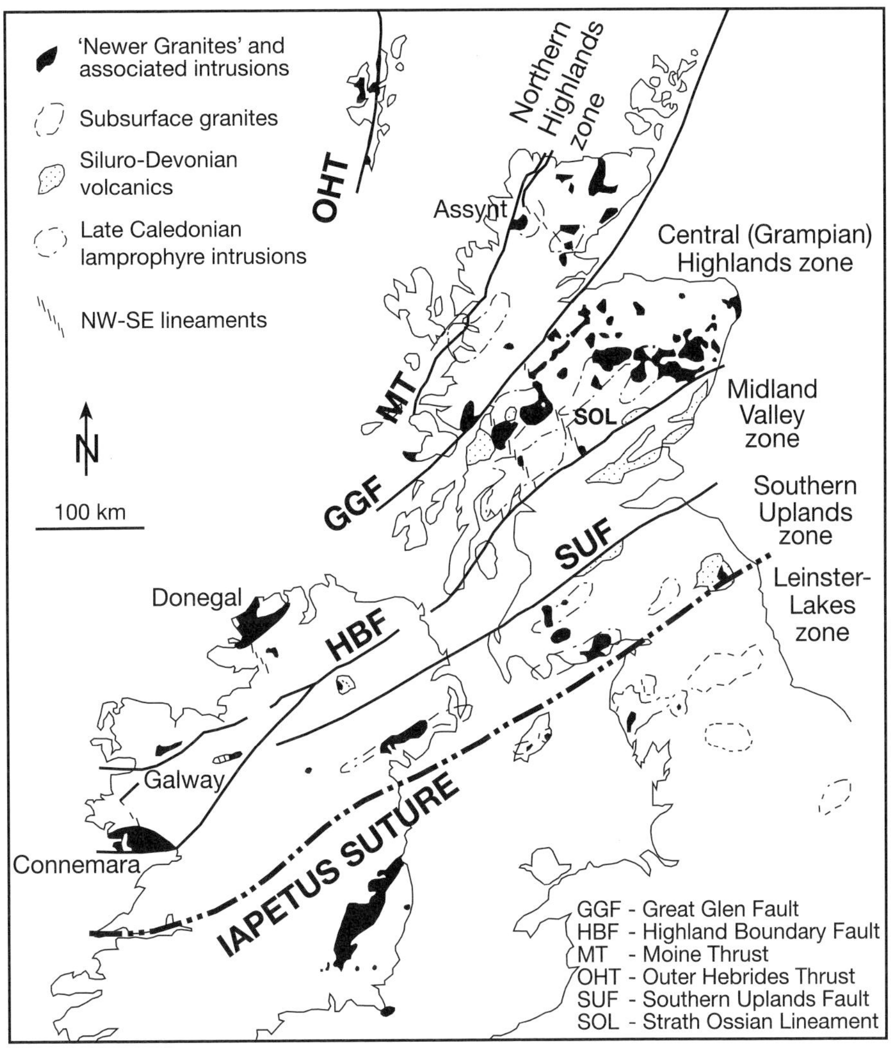

Fig. 7.12. Location of the Siluro-Devonian Newer Granites and major Caledonian fault systems (after Woodcock & Strachan 2002*b*; Jaques & Reavy 1994).

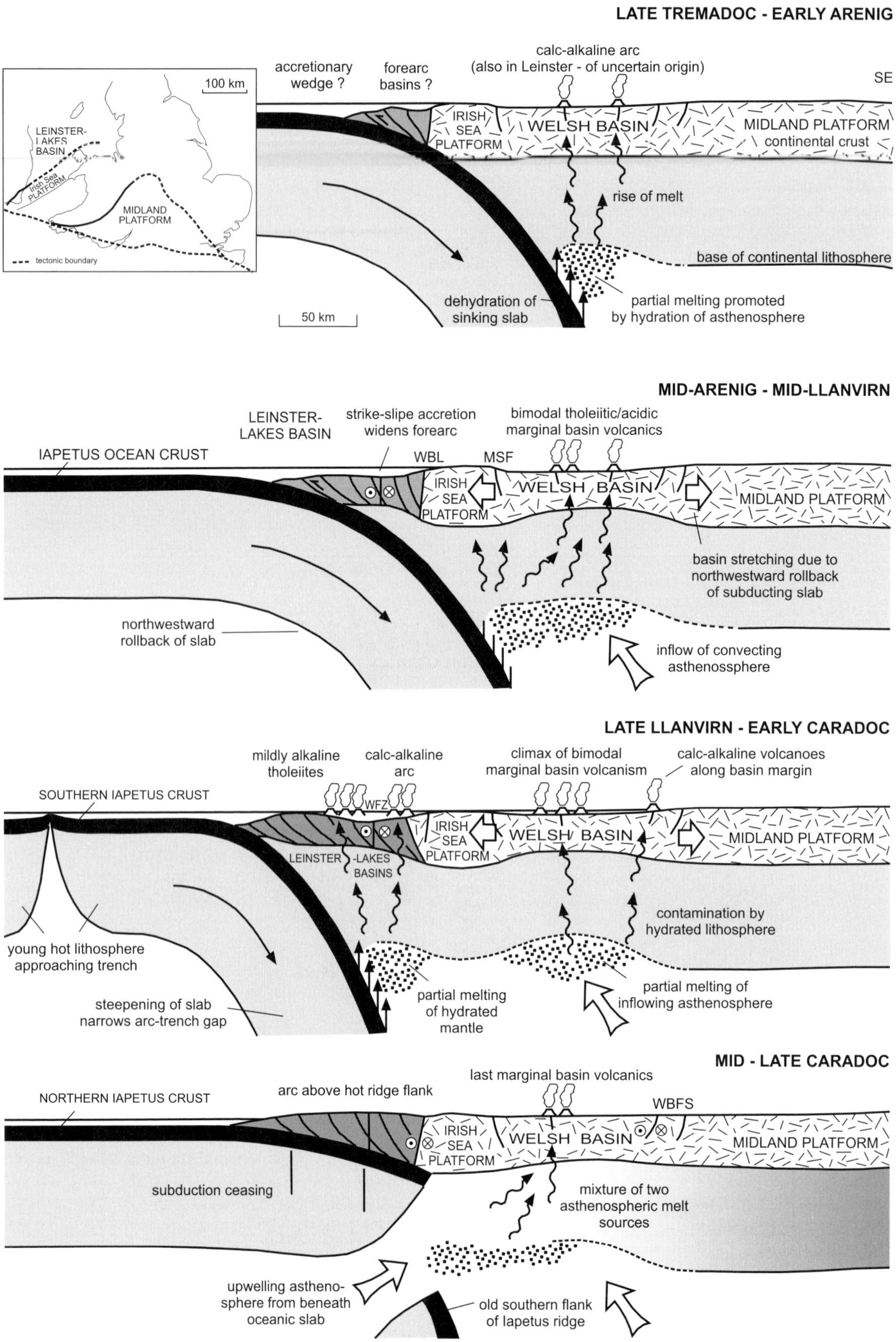

Fig. 7.13. Different modes of magma generation illustrated for the British Caledonides (after Woodcock 2002*a*). Abbreviations: MSF, Menai Strait Fault; WBFS, Welsh Borderland Fault System; WBL, Wexford Boundary Lineament; WFZ, Wicklow Fault Zone.

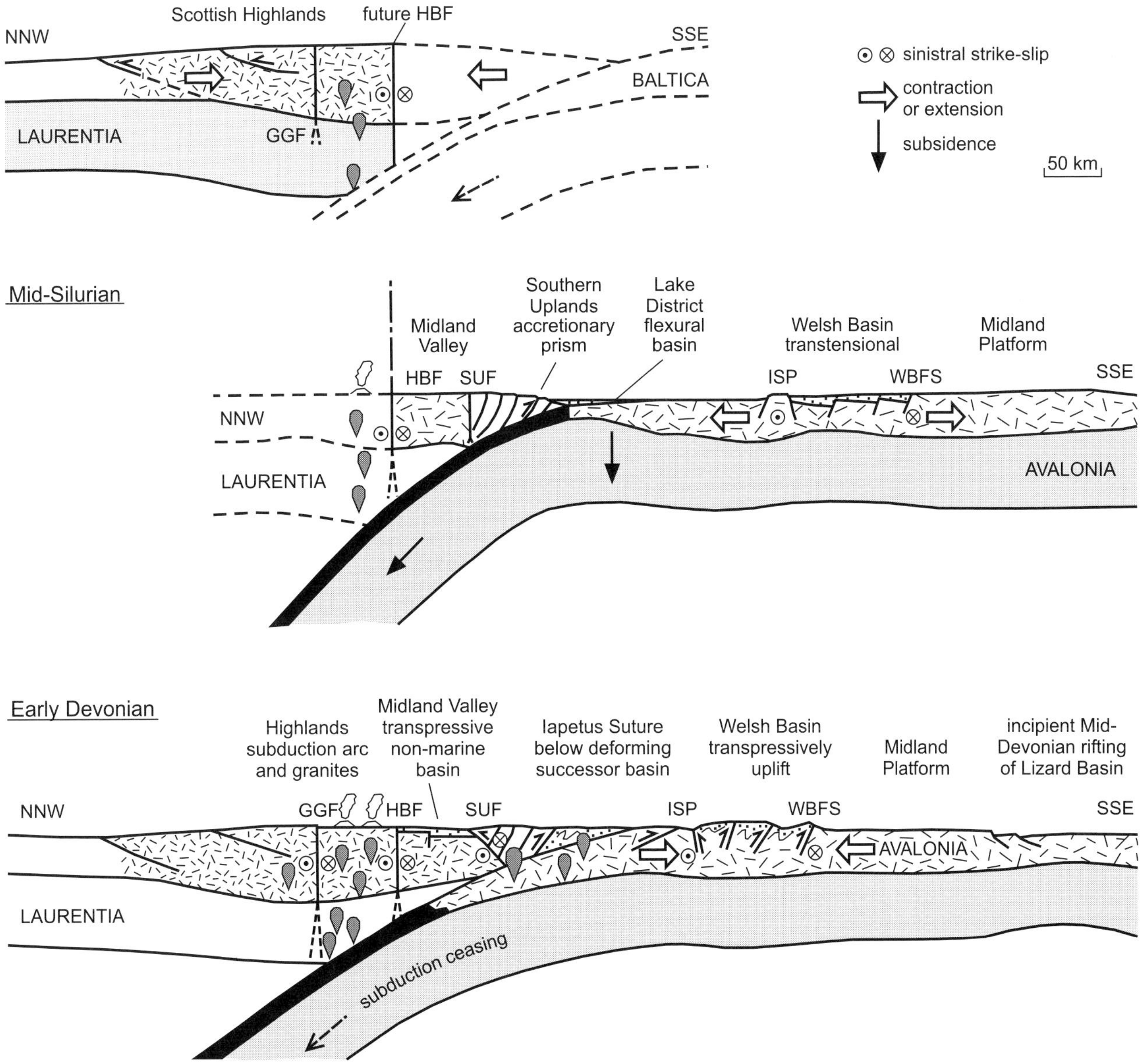

Fig. 7.14. The British Caledonides during Mid-Silurian and Early Devonian (after Woodcock 2002*a*). Abbreviations: GGF, Great Glen Fault; HBF, Highland Boundary Fault; ISP, Irish Sea platform; SUF, Southern Upland Fault; WBFS, Welsh Borderland Fault System.

radiogenic strontium (^{87}Sr) than those to the south, and a greater proportion of low ^{87}Rb–^{86}Sr values. These differences reflect contrasts in the magma chemistry across the suture and probably indicate a greater contribution from mantle rocks beneath the Southern Uplands (Scotland) (Woodcock & Strachan 2002*b*).

Elsewhere in the Caledonian region magmatic intrusions have also been recorded. For example, in eastern Svalbard–central Nordaustlandet magmatic anomalies appear to be part of a *c.* 300 km long linear belt of late Silurian or early Devonian post-tectonic plutonism that characterizes the Caledonian basement of the region. They appear to have been intruded at the end of the main period of Caledonian terrane assemblage of the north-western Barents Shelf (Gee *et al.* 1999).

The precise mechanisms by which granitic magmas are generated and subsequently ascend through the crust are geotectonic problems that have preoccupied geologists for many years (Woodcock & Strachan 2002*b*). Detailed field studies of granite plutons in the Caledonides have revealed that, in some areas, the deep-seated Caledonian strike-slip faults exerted a fundamental control on the siting and ascent of magmas derived from the upper mantle and lower crust (Figs 7.13 & 7.14). These magmas probably ascended the fault zones as narrow dykes that coalesced at mid- to high-crust levels to form the elongate to subrounded, steep-sided plutons characteristic of the Newer Granites. Contemporaneous lamprophyre and appinite intrusions associated with many of the Newer Granites are thought to derive from partial melts of subcontinental mantle.

There is a close spatial and temporal relationship between the late Caledonian granites in Scotland and Ireland (Fig. 7.15) and sinistral strike-slip faults and shear zones (Jacques & Reavy 1994). The tectonic environment at the time of plutonism (425–400 Ma) has been related (Soper & Hutton 1984; Hutton 1987) to the final oblique closure of the Iapetus Ocean. Hutton & Reavy (1992) have argued that the Caledonides during this period were

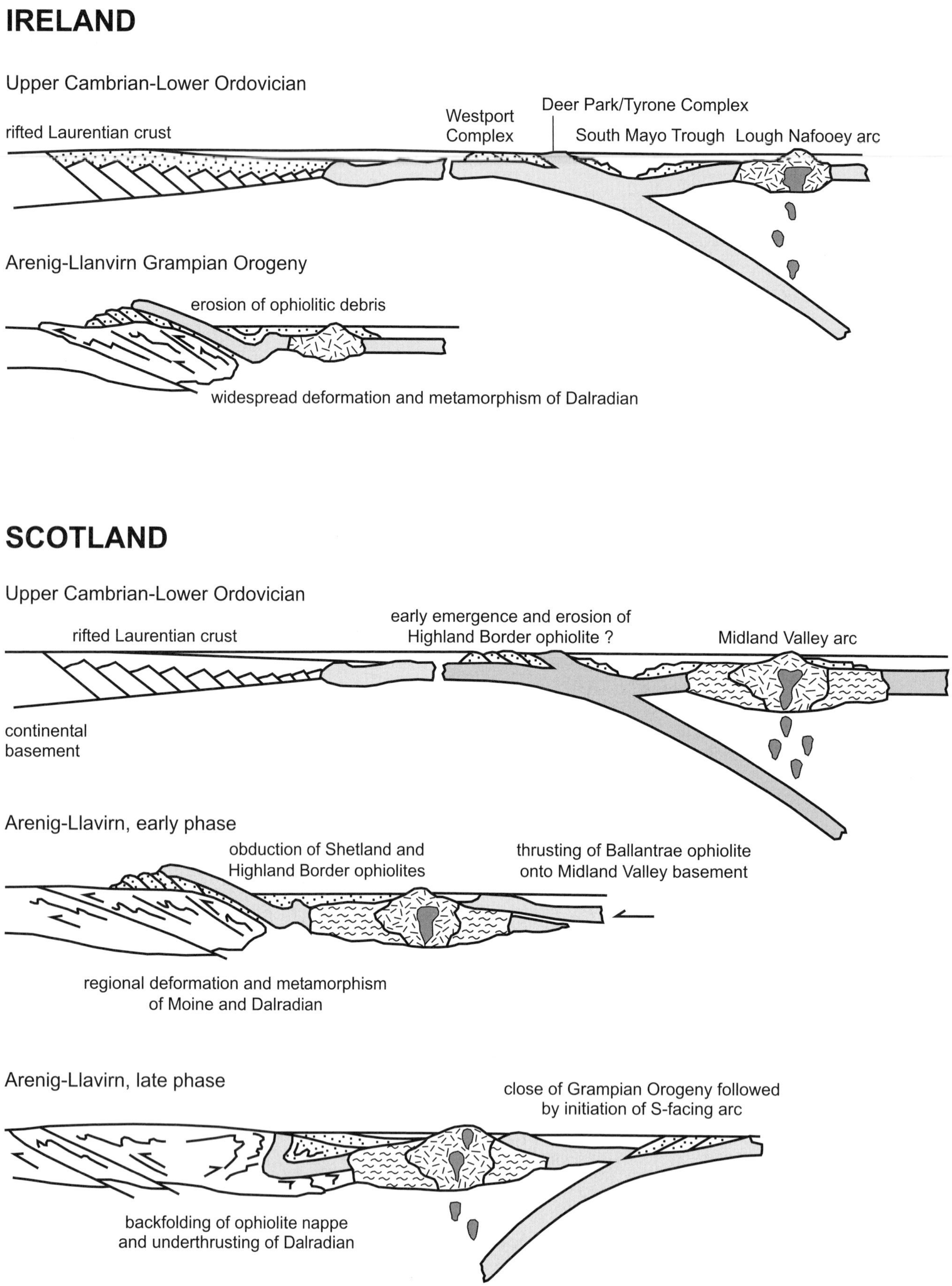

Fig. 7.15. Distinct phases of Grampian Orogeny in Ireland and Scotland (after Woodcock 2002*a*).

subjected to a gross sinistral transpressional deformation, and that NE–SW trending transpressional faults and shear zones are the typical expression of this event. Hutton & Reavy (1992) have further argued that such structures not only controlled the siting and ascent of magma and the emplacement of plutons, but may also have been responsible for granite genesis due to anatexis of thickened crust at the lower limits of transpressional crustal-scale shear zones detaching into the Moho.

Jacques & Reavy (1994) have analysed late Caledonian granites in Scotland. The analysed granites (Argyll suite) are intimately associated with synplutonic Caledonian NE–SW orientated shear zones and pre-Caledonian reactivated NW–SE lineaments (Fig. 7.16). Hutton & Reavy (1992) suggested that Caledonian magmatism was caused by anatexis of thickened crust by mantle-derived melts at the lower ends of NE–SW trending transpressional faults detaching into the Moho, within an overall setting of a thickened lithosphere consequent to the final transpressional Caledonian collision. Jacques & Reavy (1994) suggest that transpressional activity in some areas reactivated older lineaments. Movement resulted in the development of a series of mobile blocks bounded by ductile shear zones in two directions (Fig. 7.16). As the blocks interacted with the gross transpressional system, they may have been rotated to some extent. Transtensional voids would thus be created at certain intersections and crustal thickening accentuated along the Caledonian shear zones. This may have provided the mechanistic framework for the evolution of some late Caledonian granites. Thus, the main cause for the abundance of plutons within the Argyll suite may be the concentration of closely spaced lineament intersections within a section of the crust. The spatial distribution of the plutons is thus closely tied to the likely sites of extensional zones between the hypothetical crustal blocks (Fig. 7.16).

Despite the important role of strike-slip fauts in Newer Granite emplacement, other structures evidently controlled the final emplacement geometries of some plutons, e.g. the development of the Moine Thrust Zone and late Caledonian extension (Woodcock & Strachan 2002*b*). Volcanic successions of Siluro-Devonian age occur in some areas close to the Iapetus Suture and these are genetically related to the granites. The generally calc-alkaline geochemistry is most readily explained if it was generated in one or more subduction-related volcanic arcs, e.g. the Midland Valley and Grampian area where the arc was associated with the northward subduction of Iapetus Ocean crust beneath Laurentia. However, this model does not easily explain the plutons close to the suture and fails to account for those south of the suture (Woodcock & Strachan 2002*b*).

One suggestion is that these southern granites were intruded above a second subduction zone, dipping northward beneath Eastern Avalonia from the Rheic Ocean. However, three other processes of magma generation, not necessarily related to a subduction arc, have been invoked to explain the lower-crustal and uppermost-mantle melting required to source the Newer Granites (Woodcock & Strachan 2002*a*). These include: (1) melting by raised temperatures during tectonic thickening of

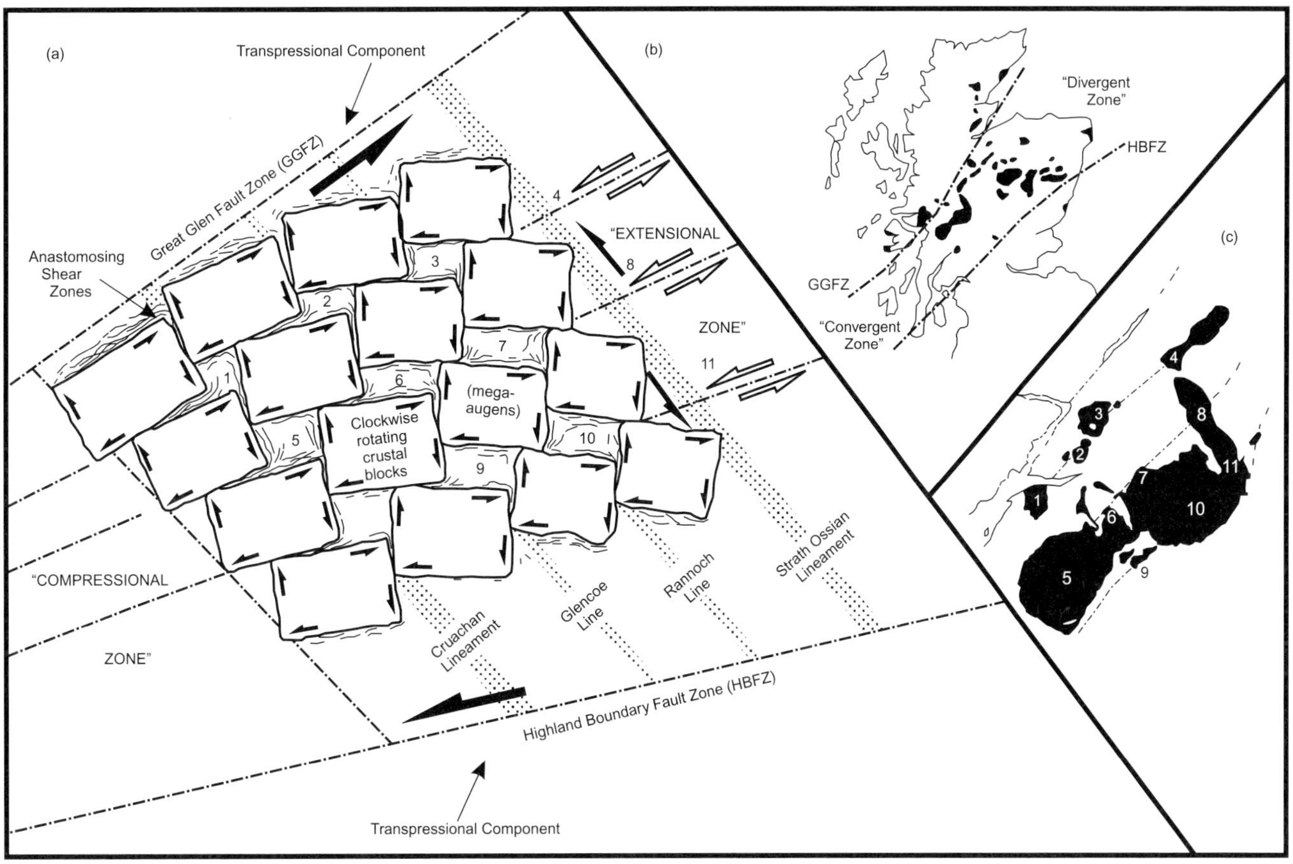

Fig. 7.16. Model of Grampian crustal structure (after Jaques & Reavy 1994). (**a**) Crustal block rotations at intersections of lineaments between the Great Glen Fault and the Highland Boundary Fault. (**b**) Location of faults and intrusive centres (see also Fig. 7.12). (**c**) Intrusive centres; numbers correspond to the numbers in (a).

the crust: evidence now suggests that the main thickening in the Scottish Highlands occurred 50 Ma earlier, during the Grampian event; (2) melting by lower pressures during rapid uplift and erosion: this hypothesis involves far more uplift than has occurred anywhere south of the Highland Boundary Fault, where low-grade metamorphic successions are still exposed at the surface rather than high-grade rocks exhumed from the lower crust; and (3) melting by delamination of the lithosphere, replacing old cold lithosphere by hotter rocks that could promote partial melting. The relative contribution of these processes to the formation of the Newer Granites, however, has yet to be assessed.

Avalonia magmatism. Igneous activity occurred across the Caledonian-influenced area, from Ireland and Scotland through to northern Germany and Poland. However, direct evidence of this activity is not always clear. The earliest evidence of magmatic activity associated with the Caledonian events is found in the Ordovician of Scotland and Ireland and associated with the Grampian Orogeny which occurred along the southern margin of the Laurentian microcontinent (see Strachan 2002*a* for details). Of particular interest are the ophiolite-related successions (e.g. Ballantrae Igneous Complex, the Shetland ophiolite, Clew Bay Complex) and the associated volcanic arc associations (e.g. South Mayo Trough, Midland Valley) (Strachan 2002*a*, *b*; Dewey & Mange 1999; Brown 1991; Friedrich *et al.* 1999). The subsequent mid-Ordovician–Silurian magmatic history of the Laurentian microcontinent was related to the development of an active plate margin following the onset of northerly directed subduction of Iapetus oceanic crust.

Throughout Cambrian and earliest Ordovician times the Gondwana margin was a passive one, in terms of the relative lack of volcanic activity and the fact that Iapetus Ocean crust was not subducting beneath it. However, it underwent sporadic localized crustal extension, maybe on a transform-faulted margin similar to that inherited from the Cadomian deformational phase. The earliest magmatic activity on Avalonia has been recorded from early Cambrian times with the presence of tuffs from the Welsh Basin. Later in the Cambrian there is evidence of volcanic activity in the Leinster Basin (Ireland) and arc volcanism in the Welsh Basin (Brasier *et al.* 1992; Prigmore *et al.* 1997; Woodcock 2002*a*). These episodes were related to subduction activity marking the onset of a volcanically dominated phase of Avalonian history. They are, however, minor when compared with the large volumes of volcanics which dominate the late Tremadoc to Caradoc successions during the subsequent Ordovician subduction of Iapetus Ocean crust (Woodcock 2002*a*). The earlier phases predate major volcanism and their plate tectonic driving forces are, thus, less clear. However, as noted by Woodcock (2002*a*), they may represent a continuation of the regional transtensional tectonics that characterized the late stages of the evolution of the Avalonian margin of Gondwana. These early rifting events may also have led to modest separation from Gondwana, distancing the two areas sufficiently to sever sedimentary links while maintaining faunal links (Woodcock 2002*a*).

The onset of a major volcanic episode in the Tremadoc (early Ordovician) was related to the beginning of arc volcanism and the related subduction of Iapetus Ocean lithosphere. The consensus view, however, is that a separate microcontinent of Eastern Avalonia detached itself by subduction-related rifting only during Arenig or Llanvirn times. The resulting continental fragment moved northward opening the Rheic Ocean in its wake. Ahead of it, the Iapetus Ocean closed progressively until by earliest Silurian time, Eastern Avalonia began to impinge on the Laurentian continent to the north. The closure of this segment of Iapetus involved oceanic subduction at both its Laurentian and Avalonian margins and abundant igneous rocks are the most characteristic geological signature of this subduction beneath Avalonia.

The igneous rocks that distinguish the Ordovician geological record of **England and Wales** are highly variable in terms of their composition, age and palaeogeographic position (Woodcock 2002*b*; Kokelaar 1988; Kokelaar *et al.* 1984; Pharaoh *et al.* 1993). The tectonic significance of this variation, however, is only partly understood. The volcanic and plutonic rocks can be grouped into a number of igneous provinces, each with some petrological coherence. Each province can also be tentatively assigned to a distinctive position within a subduction system. The provinces include the Leinster and Lake District basins (immature and mature volcanic arcs), eastern England (mature volcanic arc), the Welsh Basin (volcanic arc, backarc basin), and the Midland Platform (end of subduction) (for details see Woodcock 2002*b*). The subduction-related history can be subdivided into four stages (see also Fig. 7.13): (1) Late Tremadoc-early Arenig: onset of arc volcanism; (2) Mid-Arenig–mid-Llanvirn; development of a marginal basin in Wales; (3) Late Llanvirn–early Caradoc: arc volcanism climax; and (4) Mid-Cardoc: shutdown of arc volcanism and cessation of subduction beneath Eastern Avalonia. The abrupt shutdown of volcanic activity in mid-Caradoc times may have been related to the fact that at this time the Avalonian margin overran the active spreading ridge in the Iapetus Ocean, resulting also in a phase of strike-slip tectonics along the Avalonia margin. A modern analogue is provided by the Californian margin of North America (for details see Woodcock 2002*b*).

Following shutdown of the main volcanic centres in mid-Caradoc times, there was a marked reduction in magmatic activity. However, sporadic volcanic rocks and related minor intrusives provide evidence for both local and remote igneous activity throughout the uppermost Ordovician and through to the lower Devonian. Indeed, some recorded thin tuffs may even have been sourced on the southern rim of Laurentia. Late Ordovician volcanism occurred in NW England, an area that had already hosted Caradoc and earlier centres.

Silurian volcanic centres were distributed along an east–west orientated zone through southern Eastern Avalonia, close to or just within the later Variscan Belt and thus located in areas not previously associated with volcanic activity (Woodcock 2002*b*). The Silurian volcanics are geochemically different from the earlier Ordovician, suggesting a dominant source from within-plate rifting and only minor contributions from a relict or active subduction zone. This possible episode of extension or transtension along the southern belt of Eastern Avalonia is related to a major outstanding problem of Avalonian geology – the history of the southern edge of the microcontinent.

In the **Brabant Massif** there were three periods of volcanism between the Cambrian and the mid Devonian: two minor ones intercalated in the Lower Cambrian and in the lower Tremadocian of megasequence 1, and one major period at the top of megasequence 2 and extending into the base of megasequence 3 (see above for details of megasequences). The interstratified magmatic rocks recorded from the Cambrian succession have not yet been studied in detail. In recent boreholes in the Marcq Valley (25 km WSW of Brussels), altered porphyritic volcanic rocks have been found intercalated with Tremadocian turbiditic metasediments. These are described as acid volcanic rocks of ignimbritic origin or as a submarine mass-flow equivalent. They may indicate magmatic activity before or during the rifting episode related to initial Avalonia separation from Gondwana in the early Ordovician.

Late Ordovician to early Silurian magmatism is well documented in the Anglo-Brabant Deformation Belt. It occurs along the southern part of the main axis of the Brabant Massif. It is calc-alkaline and probably related to the subduction of oceanic lithosphere beneath the Avalonia microcontinent (André *et al.* 1986; Pharaoh *et al.* 1993; Van Grootel *et al.* 1997). It is mostly dacitic in composition with some rhyolites, rare andesites and basalts. In the Belgian part, the volcanic rocks are mostly ash-flow tuffs and volcanic breccias with some lava flows as well as some sills.

Some extrusive volcanism continued into the Llandovery in the eastern part of the Brabant Massif and into the lower Wenlock in the Condroz Inlier. Thus the period of extrusion is concentrated in the lower Ashgill (Verniers & Van Grootel 1991; Van Grootel *et al.* 1997). The decreasing age of magmatism towards the east of the belt was commented on by Pharaoh *et al.* (1995) and related to the oblique convergence of Avalonia with Baltica, involving significant anticlockwise rotation of the Avalonia microplate.

Central European terranes

All of the various tectonic units discussed within this section were originally located to the south of the Mid-German Crystalline High and thus the events described were Gondwanan events (e.g. all of the 490–480 Ma ages were related to Ordovician rifting along the Gondwanan margin). Nonetheless, we will present a brief overview of some of these events here, since they are related to the broader interpretation of Caledonian events, for example with respect to the rifting of the various continental fragments from Gondwana as well as their coming together to form Laurussia.

In parts of Gondwana (i.e. Saxothuringian Zone and adjacent areas), the petrogenesis of pre-Variscan granitic orthogneisses and acid metavolcanic rocks has been recently summarized by several authors (e.g. Kröner *et al.* 1994*a*, *b*; Sommermann *et al.* 1992). A comparative summary of the available ages (Fig. 7.17) from the northernmost element of the Saxothuringian Zone, the Mid-German Crystalline High and the Northern Phyllite Zone reveals that the intrusive ages span *c.* 200 Ma, although in different areas the age data fall into restricted ranges (Siebel *et al.* 1997).

The data reveal evidence of a Cadomian magmatic event (600–540 Ma) and a conspicuous lack of magmatic activity during the Cambrian (Fig. 7.17). Magmatic ages intimately associated with the Caledonian Orogeny (520–480 Ma) are dominant in granitoids of the western Sudetes (Borkowska *et al.* 1980; Korytowski *et al.* 1993; Oliver *et al.* 1993) and these granitoids are related to the closure of the Tornquist Ocean (Oliver *et al.* 1993; Kröner *et al.* 1994*a*, *b*). Ordovician and Silurian ages are recorded from the granites of the Fichtelgebirge (see below), Spessart and southern Taunus areas (Fig. 7.17).

The granitoids of the Mid-German Crystalline High (e.g. Spessart) have been related to a collision-zone tectonic setting by Okrusch & Richter (1986) and were redefined by Dombrowski *et al.* (1995) as forming part of a volcanic arc assemblage reflecting Silurian magmatism at *c.* 440–410 Ma along an ancient active margin between the Saxothuringian and Rhenohercynian zones. An active margin setting was also inferred for Silurian metavolcanic rocks (*c.* 440–420 Ma) of the Northern Phyllite Zone (i.e. Taunus area) by Meisl (1990). Both areas are distinct from the Saxothuringian *sensu stricto*, where no Silurian igneous activity has been recorded.

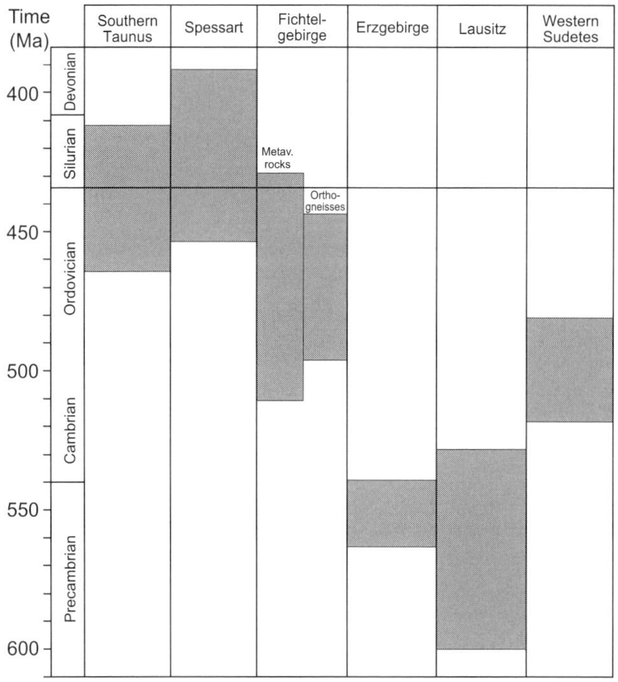

Fig. 7.17. Pre-Variscan magmatic activity in the Saxothuringian belt and adjacent areas (after Siebel *et al.* 1997).

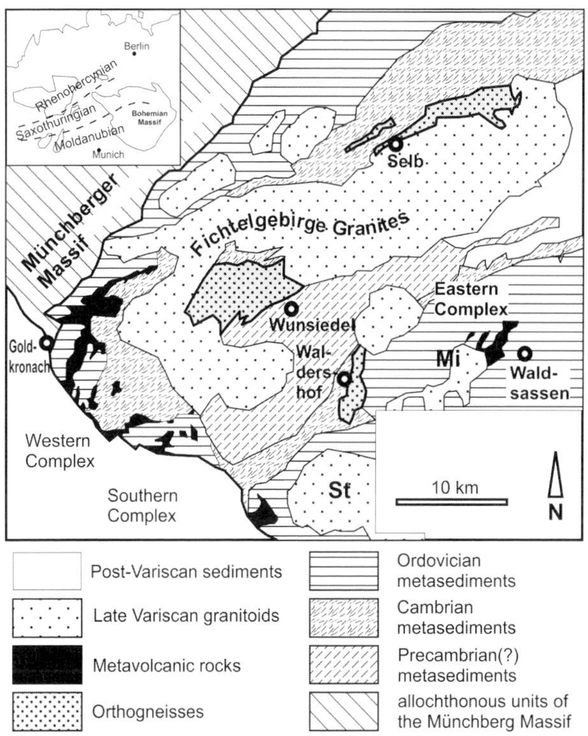

Fig. 7.18. Geological overview of the Fichtelgebirge (after Siebel *et al.* 1997). Mi, Mitterteich granite; St, Steinwald granite.

Early Palaeozoic acid and basic volcanogenic rocks have been documented in various parts of the Saxothuringian Zone. Cambro-Ordovician volcanic rocks of the Vesser area, on the northern margin of the Saxothuringian Zone, are thought to have been generated within a backarc setting involving ensialic crust (Bankwitz *et al.* 1992).

Siebel *et al.* (1997) note that different subunits of the Saxothuringian Zone are characterized by granitoids with distinct emplacement ages (e.g. Precambrian, Cambro-Ordovician, Ordovician and Silurian) (Figs 7.17 & 7.18). This fact excludes a common pre-Devonian magmatic evolution of the subunits.

Fichtelgebirge. The regional geology of the Fichtelgebirge has been extensively reviewed by other workers (for details see Siebel *et al.* 1997). The region belongs, together with the Erzgebirge, to the Saxothuringian Zone of the Variscan Orogen (see Kroner *et al.* 2008; McCann *et al.* 2008; Timmerman 2008). It is predominantly characterized by weakly metamorphosed sediments and several generations of igneous rocks. Within the Fichtelgebirge, metabasic rocks of Cambro-Ordovician age are rare and their geochemical characteristics show affinities to modern within-plate basalts (Okrusch *et al.* 1989). Extension of the Saxothuringian Zone during the Early Palaeozoic is documented in the sedimentary record. The Cambro-Ordovician sediments of the Fichtelgebirge are part of a neritic to hemipelagic succession (Franke *et al.* 1993, 1995) which might have been deposited in an intracratonic basin or a 'failed rift' (Franke & Oncken 1995). It is likely that the genesis of the orthogneisses and metavolcanic rocks is related to extensional forces. Lithospheric extension, thinning and heating owing to extra mantle input may have caused the heating of fusible crustal rocks followed by partial melting of these rocks (Siebel *et al.* 1997).

Metavolcanic rocks include ignimbritic units, lava flows, tuff deposits and associated volcaniclastic and sedimentary successions (Siebel *et al.* 1997). The principal types are interpreted as lava plains, air-fall tuffs and tuffites (Emmert & Stettner 1995). The age of the adjacent country-rocks (phyllites, quartzites, shales) is Late Cambrian–Early Ordovician (Emmert & Stettner 1995).

Early Palaeozoic acid magmatism in the Fichtelgebirge commenced with the intrusion of orthogneiss precursors during the Early and Middle Ordovician and ended with the eruption of the volcanic successions during the Late Ordovician (Fig. 7.18). K–Ar ages of the investigated rocks range from 316 to 298 Ma for muscovites and from 306 to 280 Ma for biotites, demonstrating thermal influences from Late Visean to Late Stephanian granite intrusions (Siebel *et al.* 1997).

Rb–Sr whole-rock isochron ages of the Wunsiedel gneiss of 480 ± 4 Ma are regarded as an estimate for its presumably Early Ordovician age of intrusion. Ordovician ages (composite age: 560 ± 45 Ma) have been confirmed by recent U–Pb and Rb–Sr analyses for orthogneiss bodies from the Fichtelgebirge (for details see Siebel *et al.* 1997; Fig. 7.18).

Lusatia (Lausitz). Lower Cambrian and Lower Ordovician ages have been reported from Lusatian granitoids and greywackes. The same ages and geochemical patterns were noted for other Saxothuringian units, suggesting that magmatic activity took place in several Saxothuringian units within the same time interval. This conclusion confirms that several Saxothuringian units may have undergone a similar tectonometamorphic evolution during Neoproterozoic–Ordovician time as part of the Avalonian Cadomian orogenic belt at peri-Gondwana (Tichomirowa *et al.* 2001; Rötzler *et al.* 1997).

Sudetes, Poland. The occurrence of Early Palaeozoic alkali basalts and tholeiites within the western Sudetes indicates emplacement in an extensional geodynamic setting (Furnes *et al.* 1989). Nonetheless, the geodynamic evolution of this area is highly complex and is still a matter of debate (see below, and McCann *et al.* 2008; Timmerman 2008).

The Góry Sowie Massif, situated at the NE margin of the Bohemian Massif (Fig. 7.19) is one of the oldest tectonostratigraphic units of the West Sudetes. It is bounded on all sides by predominantly brittle faults (Crowley *et al.* 2001) and is subdivided into two parts by the NW–SE trending Sudetic Marginal Fault (Fig. 7.20). The Góry Sowie Block has been variously interpreted as a klippe of the Moldanubian Nappe, an intramontane 'median massif', an 'orogenic diapir', a thrust sheet, a microcontinent, a terrane, or fragment of a terrane (for details see Kröner & Hegner 1998). The block comprises granitoid gneisses with a trondhjemite–granodiorite composition which are tectonically interlayered with pelitic to semi-pelitic as well as rare quartzo-feldspathic paragneisses, subordinate lenses of amphibolite, metagabbro, marble and graphitic schists. The metasediments initially formed a sandstone–mudstone–greywacke association of presumed late Proterozoic to early Cambrian age (for details see Kröner & Hegner 1998).

Geochemical and isotopic analysis by Kröner & Hegner (1998) has led to a variety of conclusions, which they stress may only be valid for the Góry Sowie Block and not the West Sudetes in general. The Góry Sowie terrane underwent its main magmatic, tectonic and metamorphic evolution in early Palaeozoic times, particularly in the Ordovician. Emplacement ages for pre-tectonic granitoids are between 487 and 482 Ma and thus define a rather short period of granitoid activitiy. These are comparable in age with the Śnieżnik granite gneiss (488 +4/−7 Ma) and the Kowary granite gneiss (490 ± 5 Ma) to the SE and west of the Góry Sowie Block, respectively (U–Pb zircon dates after Oliver *et al.* 1993), as well as with granitoid gneisses in the southern continuation of the Śnieżnik Complex in the Orlice Hory Mountains of the Czech Republic (480–510 Ma; Kröner *et al.* 1994*b*). Geochemically, the Góry Sowie granitoids are I-type granites with a calc-alkaline affinity and fall within the field of the volcanic arc granites (Kröner & Hegner 1998).

The most likely tectonic setting satisfying the chemical and isotopic data for these rocks is a Japan-type active margin, in line with the suggestion of Oliver *et al.* (1993) who proposed a period of subduction and arc magmatism in part of the West Sudetes in early Ordovician times. This interpretation also agrees with palaeomagnetic data suggesting that at least 2000 km of oceanic crust was consumed during the Ordovician when the Tornquist Ocean decreased rapidly in size from over 5000 km to less than 1000 km (Trench *et al.* 1992) and when the Avalonian superterrane, to which the Góry Sowie Block, according to Kröner & Hegner (1998), most probably belonged, approached Baltica (Tait *et al.* 1995; Torsvik *et al.* 1996). Convincing evidence for the subduction of oceanic crust beneath Eastern Avalonia was provided by Noble *et al.* (1993) and faunal data summarized by McKerrow & Scotese (1990) are also compatible with this model.

The proposed scenario of the Sudetes evolution assumes that the Tornquist Ocean was probably first subducted beneath the Gondwana-derived Avalonia terrane along short-lived subduction zones, so that Avalonia grew into a superterrane as it approached Baltica. In this scenario, the metasedimentary assemblage in the Góry Sowie Block, into which the granitoids were emplaced, probably represents an early Palaeozoic accretionary wedge whose sediments were derived from a mixture of older crust and detritus from juvenile arc material. Noble *et al.* (1993) suggested a similar evolution for the Caledonian domain of eastern England where they proposed that an

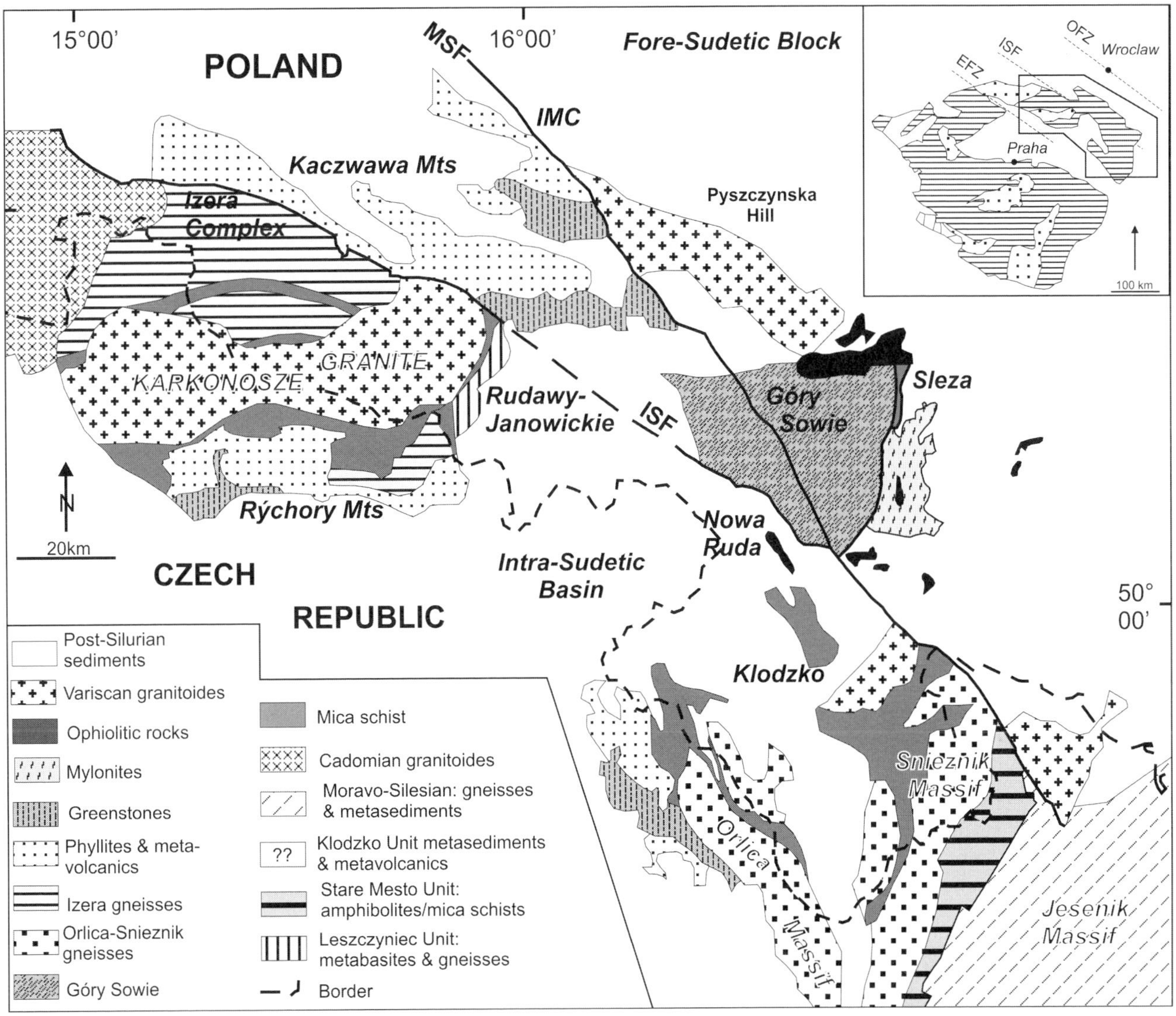

Fig. 7.19. Simplified geological map of the NE Bohemian Massif area with tectonic boundaries (after Crowley *et al.* 2000). Abbreviations: EFZ, Elbe Fault Zone; IMC, Imbramowice Metamorphic Complex; ISF, Intra-Sudetic Fault; MSF, Marginal Sudetic Fault; OFZ, Odra Fault Zone.

Ordovician (421–457 Ma) continental arc originated from subduction of the Tornquist Ocean below Avalonia. A zircon age of 499 +8/−3 Ma was obtained by Sandeman *et al.* (1997) from the Man of War Gneiss (Lizard ophiolite complex, SW England) has also been related to this subduction event. Oliver *et al.* (1993) speculated on a connection between Poland and the British Caledonides and proposed that closure of the Tornquist Ocean occurred initially in the Sudetes and involved the accretion of various terranes, probably in a sinistral transpressive regime. Additionally, Moczydlowska (1997) proposed that Upper Silesia formed part of Avalonia (based on faunal similaries) in Cambrian times.

These various conclusions, however, are at variance with tectonic scenarios that envisage Cambro-Ordovician rifting in the West Sudetes based on the presence of bimodal magmatism (e.g. Franke *et al.* 1993; Żelaźniewicz 1997). However, according to Kröner & Hegner (1998) such extension is unlikely for the Góry Sowie Block, at least in the Ordovician, since the 487–482 Ma old granitoid rocks were intensely deformed soon after their emplacement, a deformational event which is suggested to be related to subduction and terrane amalgamation.

A similar evolution has recently been reported from the German Deep Drilling Site (KTB) in Bavaria, southern Germany, where granitoid magmatism at *c.* 480 Ma was followed by high-grade metamorphism at *c.* 400 Ma and prolonged cooling to 370–380 Ma (O'Brian *et al.* 1997). These data support the idea that much of the metamorphic evolution in the Góry Sowie Block probably occurred in late Ordovician and early Silurian times, when the palaeolatitudes of Avalonia and Baltica were broadly similar, and the Tornquist Ocean was about to close (Torsvik & Trench 1991). Additionally, Brueckner *et al.* (1996) envisaged a scenario in the Góry Sowie Block area during closure of the Tornquist Ocean, in which the 402 Ma age reflects a rifting event, probably associated with a marginal basin-type spreading centre.

The Avalonia terrane developed as an active margin on its Baltica-facing side, along which accretionary wedges and sub-

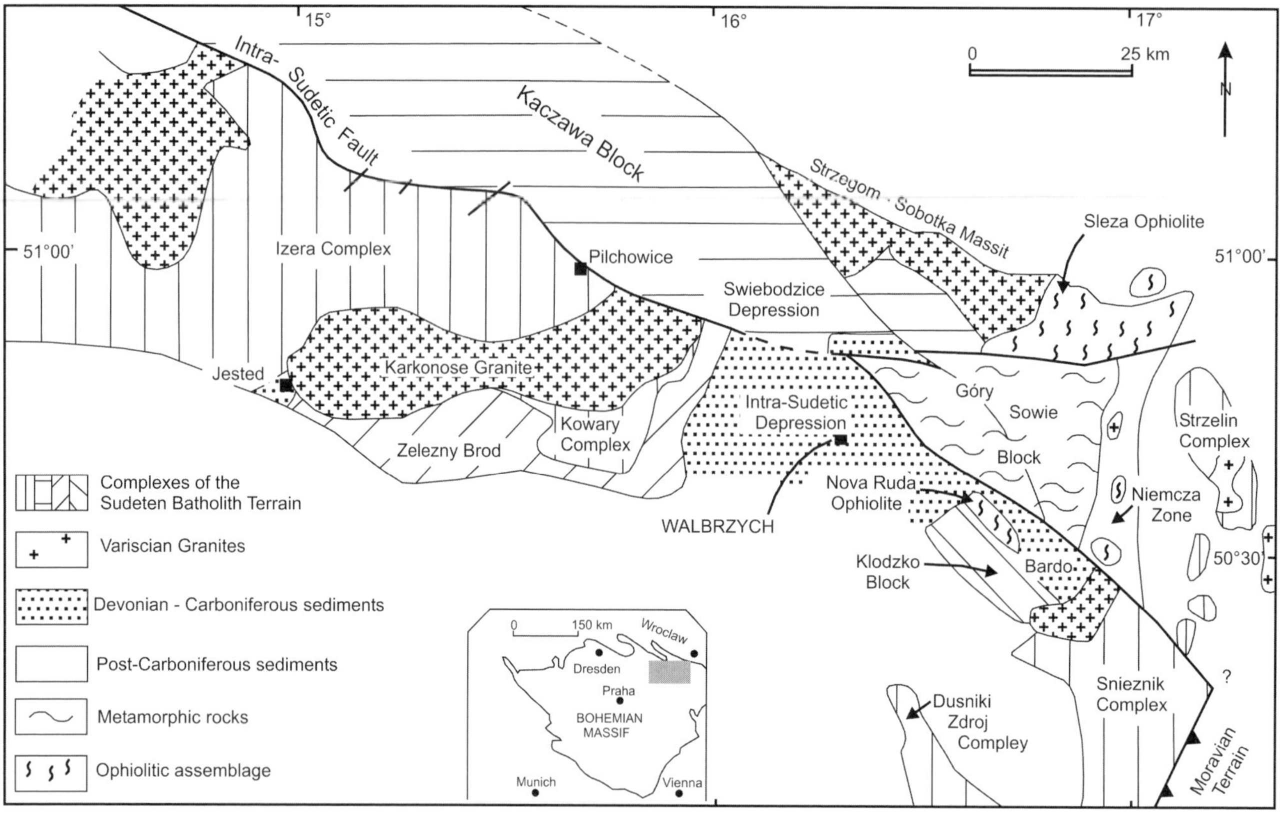

Fig. 7.20. West Sudetes with major rock types and terrains (after Kröner & Hegner 1998).

duction-related granitoids were formed and immediately deformed after their generation (Kröner & Hegner 1998). Final collision of Avalonia with Baltica, involving significant crustal thickening in the latest Silurian to early Devonian, was probably responsible for the local formation of high-pressure granulites in the Góry Sowie Block at *c.* 400 Ma which were then tectonically sandwiched between the ortho- and paragneisses during their ascent from lower crustal levels.

New zircon ages and geochemical data from the West Sudetes led Oliver *et al.* (1993) to propose a model whereby considerable narrowing of the Tornquist Ocean occurred during the Ordovician, followed by continuing ocean-floor and island-arc magmatic activity in the Silurian and final transpressive ocean closure by the mid-Devonian. This view was supported by Johnston *et al.* (1994) who concluded that the Tornquist Ocean suture – the collision zone between Baltica and a variety of suspect terranes originally derived from Gondwana – is located in the Sudetes Mountains, and that much of the observed orogenic activity can be ascribed to the Caledonian events of northern and east-central Europe. Kröner & Hegner (1998) reported an idea that Avalonia, on its way from Gondwana to Baltica, did not travel as a tectonically passive terrane, but that successively younger accretionary wedge terranes were added to its Baltica-facing active margin. This would suggest that Avalonia already possessed an internally complex structure when it docked with Baltica in late Silurian times (Berthelsen 1992). Franke *et al.* (1993) and Zelazniewicz & Franke (1994) strongly oppose this view and consider the early Palaeozoic magmatic rocks in the West Sudetes as resulting from a major rifting episode.

Geological significance of Ordovician magmatism. Protoliths of Erzgebirge eclogites (Fig. 7.21) with high εNd values and MORB composition have been identified as mantle-derived mafic to ultramafic magmatic rocks, perhaps basalts of an extensional regime, continental rift setting or a backarc spreading ridge (Schmädicke *et al.* 1995). The association of crustally generated high-silica rhyolites with tholeiitic to alkaline basalts in a thick sedimentary succession as suggested for the Erzgebirge could be interpreted in terms of an extensional setting (Mingram *et al.* 1998, 2004).

Mineralogical and geochemical features of schists from the Erzgebirge correspond to Ordovician sedimentary successions from the Saxothuringian Zone (e.g. the Schwarzburg Anticline west of the Erzgebirge, the Elbe Zone and the NE border of the Erzgebirge) (Mingram 1998) (Figs 7.22 & 7.23). The Ordovician sedimentary succession of the Schwarzburg Anticline consists of highly mature quartzites and shales overlain by sedimentary iron ores and glaciomarine diamictites, and has a thickness of *c.* 3500 m (Linnemann & Romer 2002). This enormous thickness of shelf sediments reflects extreme subsidence, probably due to rifting processes which culminated in the separation of the Saxothuringian terrane from NW Gondwana at the end of the early Ordovician (Linnemann *et al.* 2000). Single zircon ages for intercalated rhyolites at 487 ± 6 Ma, a porphyroid at 479 ± 5 Ma and a tuff at 484 ± 10 Ma reflect Tremadocian ages for the Schwarzburg succession (Gehmlich *et al.* 1997; Linnemann *et al.* 2000; Tichomirowa *et al.* 2001; Figs 7.24 & 7.25).

Similar ages of 489 ± 2 Ma (sericite gneiss) and 477 ± 5 Ma (tourmaline granite) were found in the Elbe Zone (Gehmlich *et al.* 1997; Linnemann *et al.* 2000; Tichomirowa *et al.* 2001; Figs

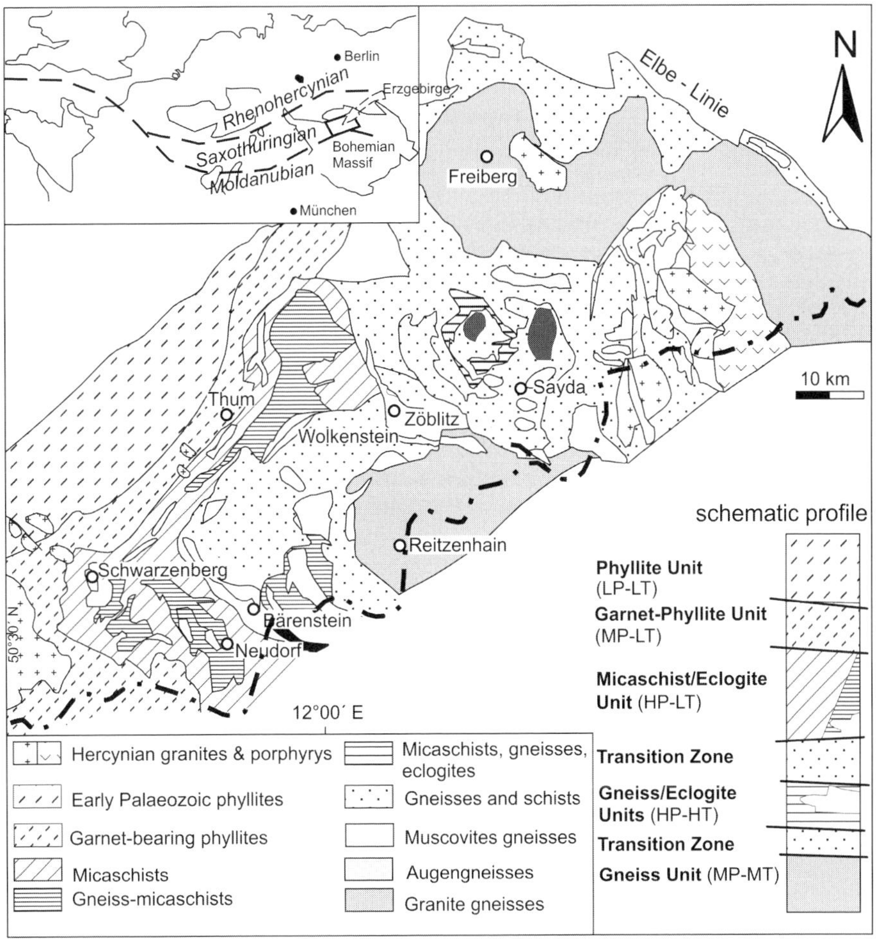

Fig. 7.21. Simplified geological map of the Erzgebirge area with tectonic boundaries (after Mingram *et al.* 2004).

7.24 & 7.25). These high-silica early Ordovician rhyolites are comparable to the protoliths of the muscovite gneisses in the high-pressure Gneiss/Eclogite and Micaschist/Eclogite units of the Erzgebirge (Mingram *et al.* 2004).

Bimodal volcanism as well as mafic to ultramafic intrusive complexes suggest that there were extensive rifting processes during the earliest Ordovician; the products were later incorporated into the Variscan Belt (e.g. Pin 1990; Kryza & Pin 2002; Kemnitz *et al.* 2002). In contrast to other early Ordovician bimodal suites preserved in the European Variscides (e.g. the Polish West Sudetes; Kryza *et al.* 1994), the magmatic suites of the Erzgebirge do not exhibit alkaline or peralkaline compositions. According to Pin (1990), most of the bimodal suites are associated with rhyolites which probably resulted from melting at the base of the crust, possibly above an ascending plume. The generation of crustally derived magmatic rocks in a rift-related setting implies mantle-induced thermal activity. The results may also indicate a change from a passive to active margin setting as suggested for early Ordovician successions of the English Lake District (Cooper *et al.* 1993) or the Central Iberian Zone (Valverde-Vaquero & Dunning 2000).

Geophysical evidence of the Caledonian Orogeny

Central Europe has been covered with a dense network of seismic profiles over the past 30 years, including a mix of normal-incidence deep reflection traverses and wide-angle reflection profiles. Together, these provide greater detail, on a continental scale, than anywhere else in the world. Normal-incidence reflection data image sharp, near-horizontal boundaries but processing techniques are not sufficiently advanced to image vertical boundaries directly. Additionally, the images captured are distortions because of the imperfections of migrating 2D sections of individual lines. Furthermore, many of the patterns observed at depth in the crust are interference composites. Wide-angle seismic data recorded on long profiles offer complementary information about the physical properties of the subsurface but lack the horizontal resolution of normal-incidence profiling. Interpretation of wide-angle data is accomplished via inverse modelling to produce cross-sections of the crust and lithosphere.

Geophysical evidence portrays the nature of the crust and upper mantle at the present day, including all of the post-Caledonian changes which have occurred to modify its characteristics. Thus anything attributed to the Caledonides must take subsequent changes into account. Such changes include events related to the Variscan Orogeny, Mesozoic basin evolution, Alpine deformation transmitted into the foreland and other Cenozoic extensional events (e.g. Rhine Graben).

Research into Caledonian geology is often hampered by concealment of large parts of the Caledonides beneath younger sedimentary successions. For that reason, large seismic cam-

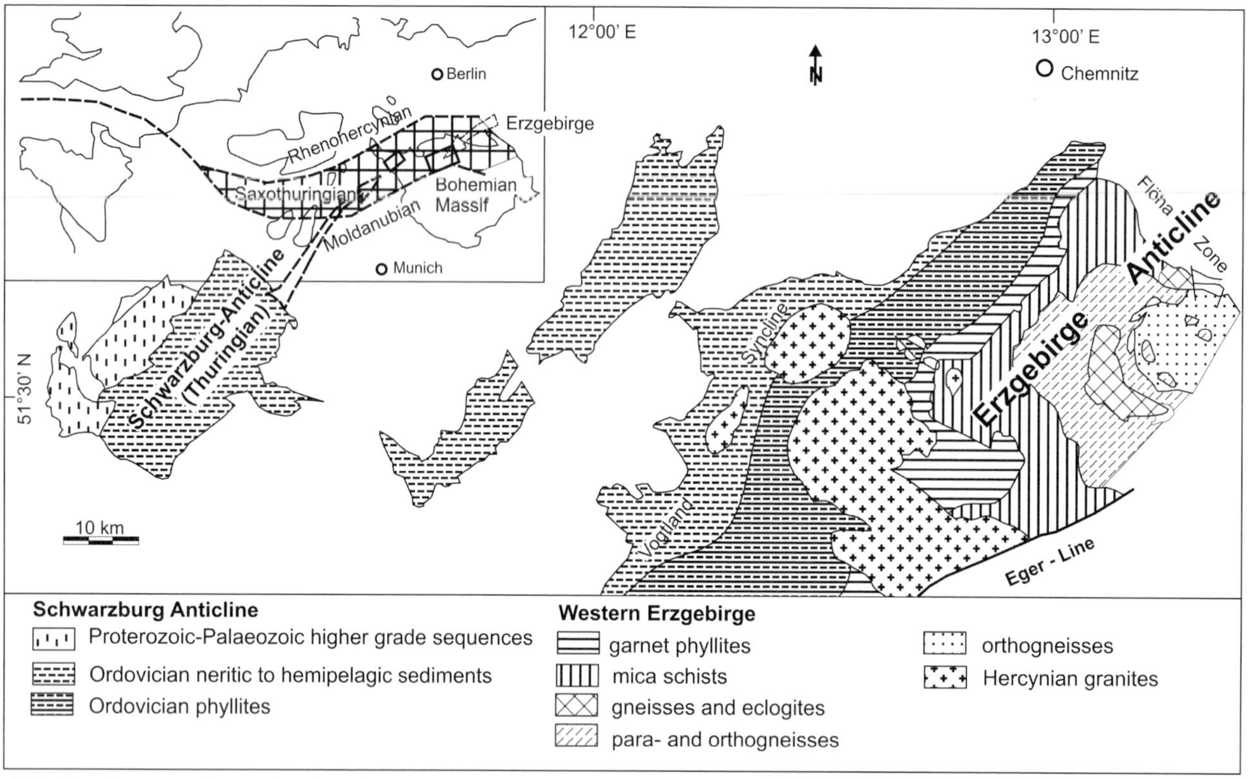

Fig. 7.22. Overview map of exposed Variscan basement in Thuringia and the Erzgebirge (after Mingram *et al.* 1998).

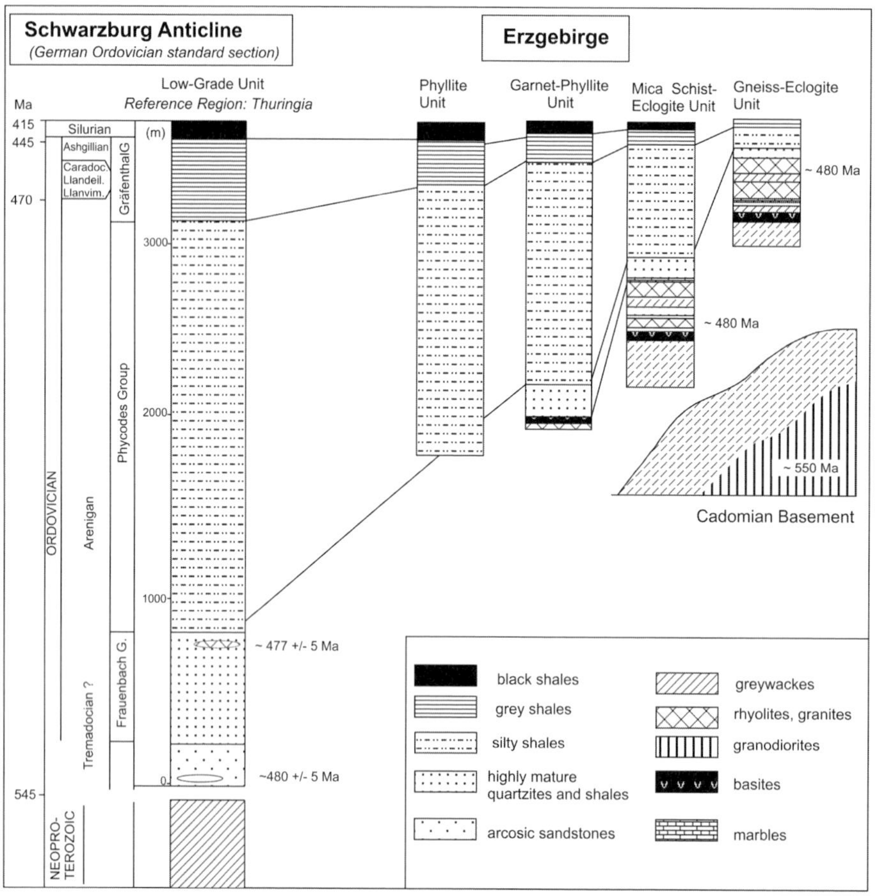

Fig. 7.23. Lithological profiles of the German Ordovician standard section in correlation with metamorphic units of the Erzgebirge (after Mingram *et al.* 1998).

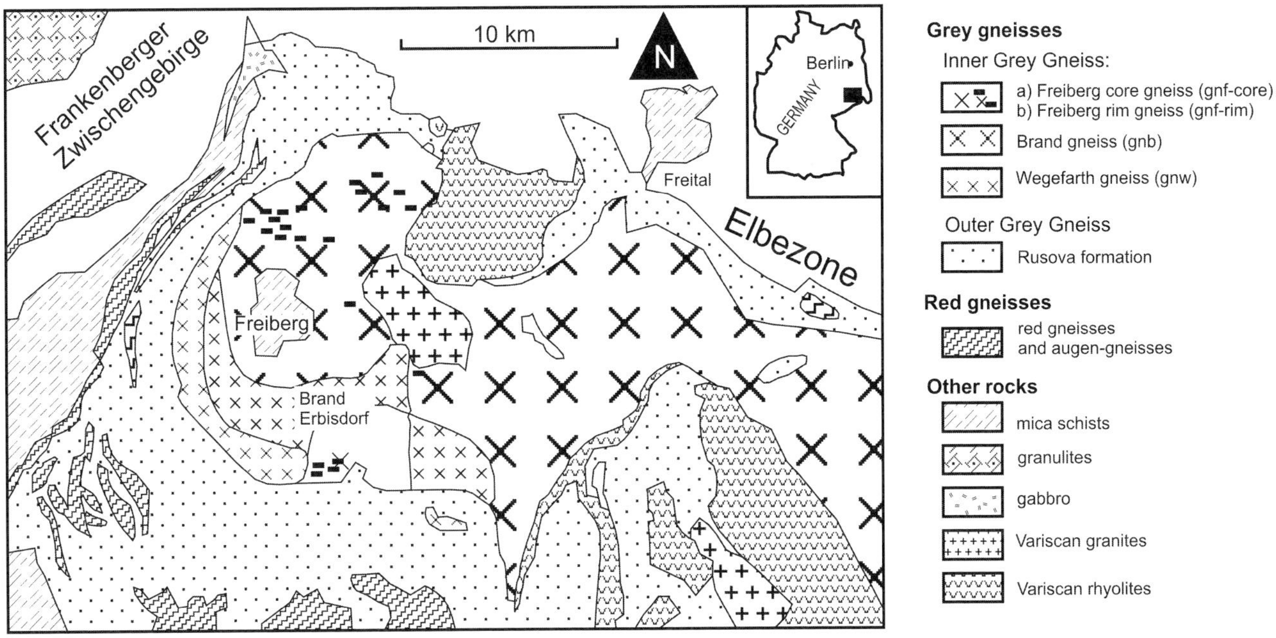

Fig. 7.24. The Eastern Erzgebirge area and adjacent regions with distribution of xenoliths (after Tichomirowa *et al.* 2001).

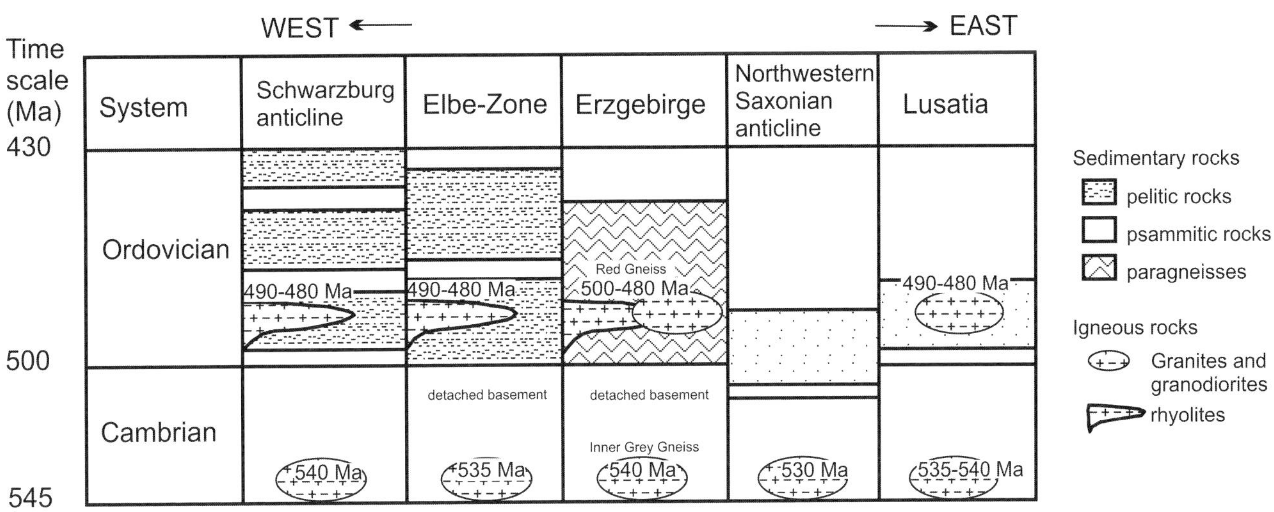

Fig. 7.25. Sedimentary and igneous rocks of Cambrian and Ordovician age in the Eastern Erzgebirge area and adjacent regions (after Tichomirowa *et al.* 2001).

paigns have been very important in investigating the lithospheric structures corresponding to the different terranes that collided following the closure of, for example, the Iapetus Ocean and the Tornquist Sea. National programmes such as BIRPS (British Institutions Reflection Profiling Syndicate) and DEKORP (DEutsches KOntinentales Reflexionsseismisches Programm), with both land-based and offshore activities, and later the Polish surveys (LT, POLONAISE, CELEBRATION) have been important initiatives in the last two decades (see overview of principal surveys in Table 7.2 & Fig. 7.26).

Within Central Europe, a series of tectonic zones and sutures have been identified (e.g. Franke *et al.* 1995). The key suture is the southern boundary of Avalonia which extends from Cornwall into France below the Paris Basin to emerge south of the Ardennes (Cocks *et al.* 1997). The Rhenohercynian Zone to the south and east of the Ardennes has intermittent Lower Palaeozoic outcrops ranging from the Rhenish Massif (= Rheinisches Schiefergebirge) through the Harz Mountains to the Magdeburg and Thuringia areas in eastern Germany. Most of the autochthonous parts of this region were probably part of Avalonia (Cocks 2000; see discussion above). To the immediate south of the Rhenohercynian Zone lay the Northern Phyllite Zone which is separated to the north and south by sutures as well as by different tectonic and metamorphic realms. This region probably also formed part of Avalonia, although there is no faunal evidence to support this (Cocks 2000; Kroner *et al.* 2008; McCann *et al.* 2008).

Table 7.2. *Summary of major seismic surveys in Central Europe crossing terranes affected by the Caledonian Orogeny*

Program /Survey	Profile	Geographic region	Regional/local tectonic setting	Regime	Seismic data type*	Reference†
BIRPS		British Isles				
	WIRE	West of ireland	Iapetus Suture	Offshore	Refl	Klemperer (1989)
	SWAT	East of Ireland	Iapetus Suture	Offshore	Refl	BIRPS & ECORS (1986)
	WINCH	East of Ireland	Iapetus Suture	Offshore	Refl	Brewer *et al.* (1983)
	NEC	North Sea	Iapetus Suture	Offshore	Refl	Freeman *et al.* (1988)
	NSDP, MOBIL	North Sea	Avalonia	Offshore	Refl	Klemperer (1988), Blundell *et al.* (1991, 1993)
	SALT	North Sea	Avalonia	Offshore	Refl	Barton (1986)
VARNET-96		SW Ireland	Iapetus Suture	Onshore	Refra	Landes *et al.* (2000)
BGS		England	Avalonia	Onshore	Refl	Chadwick *et al.* (1989)
MONA LISA		North Sea	Iapetus Suture	Offshore	Refl	MONA LISA Working Group (1997*a*, *b*)
CST	CCT	South Norway, SW Sweden	Transscand. Igneous Belt	Onshore	Refl, Refra	Hurich *et al.* (1989)
	WCT	SW Norway	Scandinavian Caledonides	Offshore	Refl	Hurich & Kristoffersen (1988)
FG		SW Norway	Scandinavian Caledonides	Onshore	Refra	Sellevoll & Warrick (1971)
Mobil Search		Baltic Sea - Skagerrak	Tornquist Zone	Offshore	Refl	Husebye *et al.* (1988)
FENNOLORA		Norway	Fennoscandian Shield	Onshore	Refra	Guggisberg *et al.* (1991)
EUGEMI (part of EGT)		From Germany to Italy	Central Europe	Onshore	Refra	EUGEMI Working Group (1990)
EUGENO-S (part of EGT)	E1	Denmark/Germany	Baltic Shield - Northgerman Basin	Onshore	Refra	EUGENO-S Working Group (1988); Thybo (1990)
NARS (part of EGT)		Southern Sweden to Alboran Sea	Avalonia and its sutures	Onshore	Refra	Nolet *et al.* (1986)
BELCORP	BELCORP-1	Belgium	Caledonian *and* Variscan influenced basement	Onshore	Refl	Bouckaert *et al.* (1988)
Industry lines	8601, 8606	Netherlands	Avalonia	Onshore	Refl	Remmelts & Duin (1990)
	unspecified	NW Germany	NW German Basin	Onshore	Refl	Dohr (1983); Trappe (1989)
EUGEMI		NW Germany	NW German Basin: Caledonian *and* Variscan influenced basement	Onshore	Refra	Aichroth *et al.* (1992)
ZIPE		NE Germany	NE German Basin	Onshore	Refra	Rabbel *et al.* (1995)
NGM		NE Germany	NE German Massif	Onshore	Refra	Rabbel *et al.* (1995)
NIZUSE		NE Germany	NE German Basin	Onshore	Refl	Horst *et al.* (1994)
KMgW		East German	Northgerman Massif	Onshore	Refra	Lange (1973); Hoffmann *et al.* (1989)
SEMLJA		East Germany	Northgerman Massif	Onshore	Refra	Lange (1978); Hoffmann *et al.* (1989)
Zentroseis		East Germany	Northgerman Massif	Onshore	Refra	Bormann *et al.* (1989)
DEKORP-BASIN 96	BASIN 9601	NE Germany: from Rügen Island to Harz mountains	NE German Basin: Caledonian *and* Variscan influenced basement	Onshore	Refl, Refra	DEKORP-BASIN Research Group (1999)
Petrobaltic		Southern Baltic Sea	Avalonia-Baltica suture	Offshore	Refl	Rempel (1992); Schlüter *et al.* (1997)
SASO		Southern Baltic Sea	Avalonia-Baltica suture	Offshore	Refl	Schlüter *et al.* (1997)
BABEL	BABEL A	Southern Baltic Sea	Avalonia-Baltica suture; Tornquist Zone	Offshore	Refl, Refra	BABEL Working Group (1991, 1993)
PQ2		Southern Baltic Sea	Avalonia-Baltica suture	Offshore	Refl	DEKORP-BASIN Research Group (1998)
				registered onshore	Refra	Bleibinhaus *et al.* (1999)
TOR		NE Germany, Denmark, Sweden	Precambrian, Caledonian *and* Variscan influenced basement	Onshore	RF	TOR Working Group (1999)
LT	LT-7	Poland	Trans-European Suture Zone	Onshore	Refra	Guterch *et al.* (1994)
POLONAISE		Poland	Trans-European Suture Zone	Onshore	Refra	Guterch *et al.* (1998; 1999)
CELEBRATION		Poland	Trans-European Suture Zone	Onshore	Refra	Guterch *et al.* (2003)

*Refl, near-vertical incidence reflection seismic data; Refra, wide-angle refraction/reflection seismic data; RF, receiver function data.
†Only the reference for the first publication of a data set is given; references of enhanced data processing and/or interpretations are given in the text.

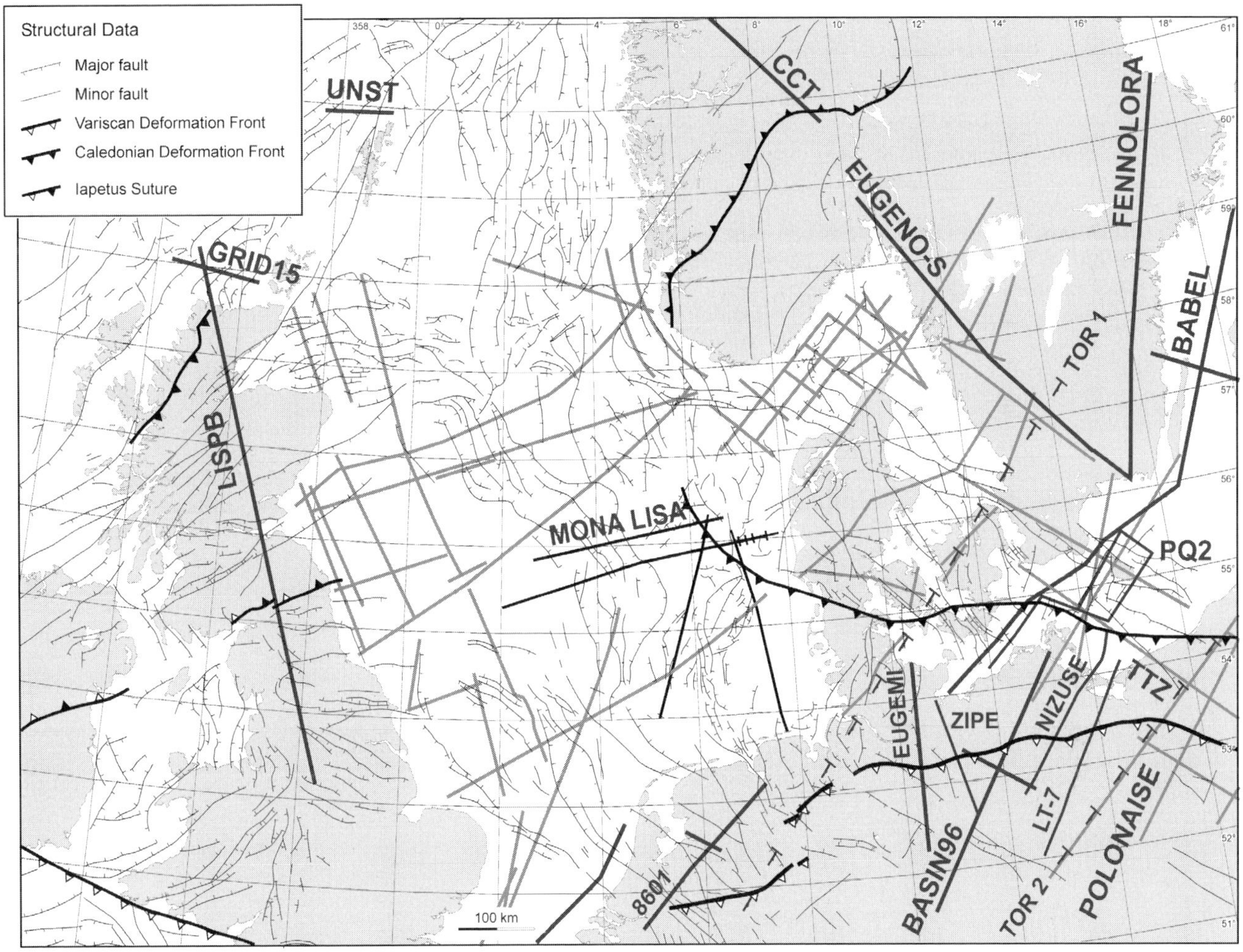

Fig. 7.26. Overview of principal seismic surveys crossing distinct terranes involved in the Caledonian Orogeny (see also Table 7.2). Base map refers to most recent structural compilations (after Heeremanns *et al.* 2004). Grey lines represent deep seismic reflection profiles; solid black lines are profiles referred to in the text.

Regional overview based on seismic data

Substantial deep structural information on the Central European Caledonides has been gathered by a series of geophysical surveys across different parts of the orogen (Fig. 7.26 & Table 7.2). These data are reviewed below in order to focus on important areas of the Caledonides.

The British Caledonides

The area of the British and Irish Caledonides is one of the classic regions for the study of Caledonian geology and includes exposure of the former Iapetus Suture. For this reason, any discussion of Caledonian tectonics and geodynamic evolution must consider these areas. The crustal structure of the region may be subdivided into two parts, the first is located to the north of the former Iapetus Suture and the second to the south of it (Fig. 7.12). Three different terranes have been recognized in Scotland, including the Outer Hebrides, the NW Highlands, and the Dalradian or Central Highlands zones. An additional terrane, the Outer Isles Terrane, may also exist. These terranes are separated by a series of major structures. The most northerly of these, the late-Caledonian Moine Thrust Zone, is roughly coincident with the Grenville Front (van Staal *et al.* 1998), and separates the Outer Isles from the NW Highland Zone to the south. In turn, the NW Highland Zone is bounded along its southern margin by the Great Glen Fault which separates it from the Dalradian or Central Highlands Zone. A thin sliver of oceanic crust was thrust up along the southern margin of the Dalradian zone. This probable remnant of the Iapetus Ocean, or a related backarc basin, is referred to as the Central Highlands zone. To the south of the Central Highlands Zone are the Midland Valley zone and the Southern Uplands Zone. This complex is recognized as a preserved example of an accretionary prism which formed along the northern margin of the Iapetus Ocean during the final stages of its closure during the Late Ordovician (Leggett *et al.* 1979).

To the south of the Southern Uplands Zone are a group of four terranes (Lake District, North Wales, Eastern England, Midland Microcraton) which are generally related to Gondwana (Avalonia) based on their contained faunas. They are thought to have originated by rifting of the northern margin of the Gondwana supercontinent during the early Palaeozoic and are separated from the terranes to the north by the Iapetus Suture, which is not exposed at the surface, although a zone of thrusting exposed on the northern coast of the Isle of Man suggests the suture must subcrop immediately offshore.

The Midland Microcraton is situated to the south of the Welsh Basin, Lake District and Eastern England terranes. Small exposures of this block crop out in central England (e.g. Charnwood and Nuneaton inliers; Noble *et al.* 1993) and these reveal that the block has a stratigraphy which is distinct from the terranes to the north. In particular, it lacks any record of subduction-related magmatism and there is no record of Silurian deep-marine sedimentation.

The various terranes described above are well established, although some of the tectonic relationships are a matter of debate and some terranes are regarded as suspect. Caledonian structures have previously been imaged by the BIRPS Group on several seismic normal-incidence reflection profiles around the UK and Ireland between the Atlantic Ocean and the North Sea (e.g Klemperer & Hobbs 1992 and references therein; data from the NEC, NSDP85 and MOBIL surveys for the western North Sea).

Moine Thrust and associated structures. The NW boundary of the Caledonian Orogen is usually taken to be along the Moine Thrust zone, where metamorphic rocks of the Scottish Highlands were thrust westward over the Hebridean Platform (Woodcock & Strachan 2002*b*). Its opposing boundary lies along the Welsh Borderland Fault System, where folded and cleaved rocks of the Welsh Basin are upfaulted against the weakly deformed Midland Platform. This deformation can be traced in the subsurface around a northern promontory of the platform and southeastwards beneath eastern England. The southern Caledonian deformation front is obscured at both its ends beneath the northern edge of the Variscan Belt. The extent and history of the Caledonian Orogen is uncertain south of the Variscan Front (Woodcock & Strachan 2002*b*).

The Moine Thrust zone has an onland length of *c.* 200 km but its total length may exceed 500 km. Displacement along the thrust zone is of the order of 150 km, and at present it dips at *c.* 8–12° to the east as a result of late Palaeozoic or Mesozoic tilting. The Moine Thrust can be traced to the north coast of Scotland where it strikes approximately NE–SW and dips at *c.* 10° to the SE (Figs 7.26 & 7.27). To the east, Moine rocks with a foliation dipping at about 15°, but steepening to near-vertical around inliers of Lewisian basement, are cut by normal faults which bound or cut through inliers of Devonian and/or Triassic sediments (the sediments are terrestrial and their exact ages are unknown). At the Faraid Head location a thrust, interpreted as

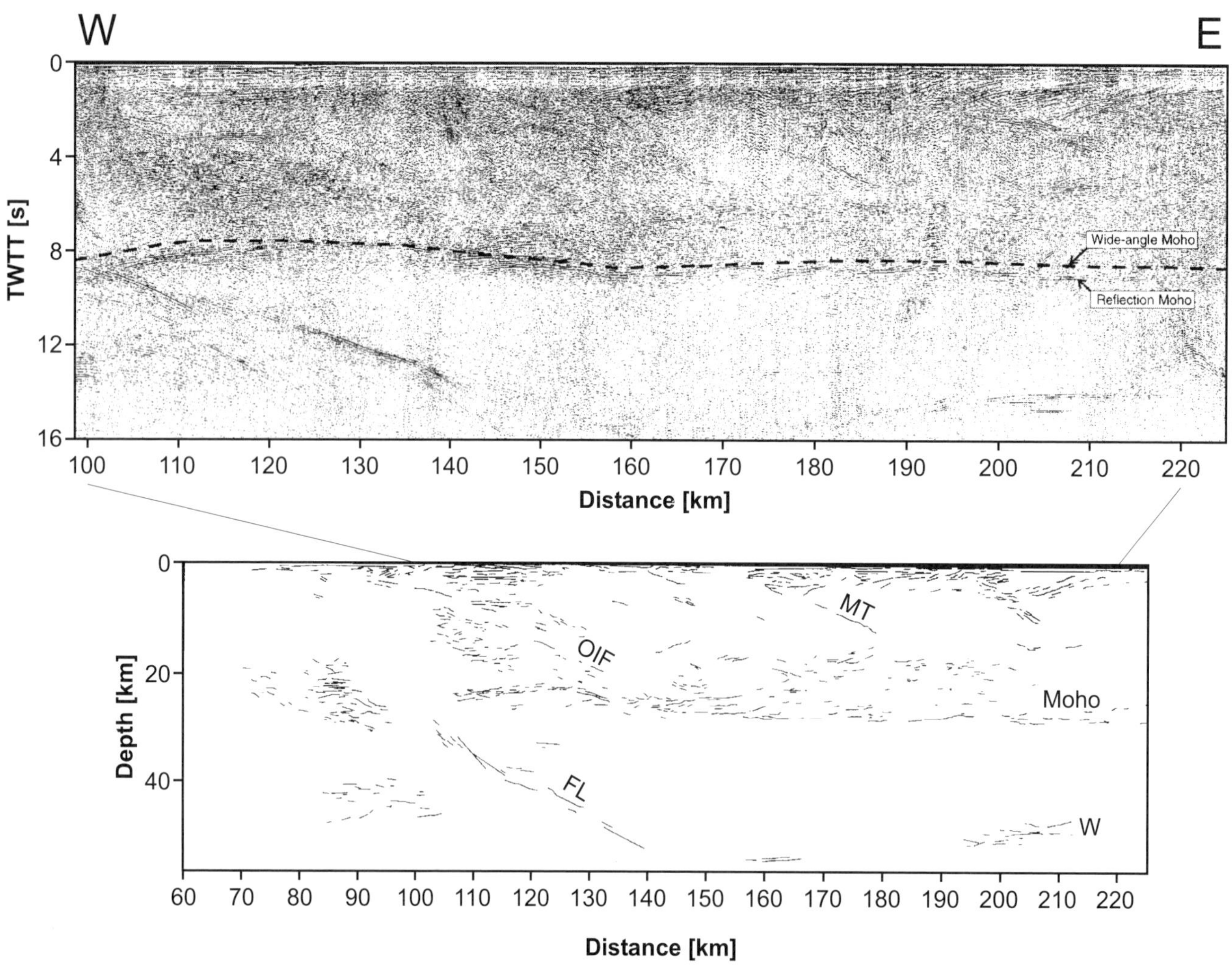

Fig. 7.27. The GRID-17 reflection seismic profile (top) north of Scotland and the depth-migrated line drawing (bottom) with interpretation after Flack *et al.* (1990). Abbreviations: FL, Flannan reflector; MT, Moine Thrust; OIF, Outer Isles fault; W, the W reflector.

the Moine Thrust, is offset sinistrally on a coast-parallel normal fault and downthrown to the north from the main fault zone which crops out inland further to the east. The Moine Thrust zone can be subdivided into two parts. The structurally higher part of the thrust zone comprises a thick belt (hundreds of metres) of mylonites formed from intense ductile shear and recrystallization of Moine metasediments and associated slices of Lewisian basement. The mylonites are characterized by the widespread recrystallization of quartz and the growth of new muscovite and chlorite within the greenschist facies. The structurally lower part of the thrust zone is a complex stack of thrust sheets that incorporates Lewisian basement and Cambro-Ordovician sedimentary rocks. The thrusts are sharp brittle structures lacking much mylonite (Woodcock & Strachan 2000*b*).

Between 1983 and 1985 BIRPS shot a grid of deep seismic profiles (some extending to 60 s TWT) across the West Orkney Basin and the Shetland Platform with the aim of tracing the Moine Thrust offshore toward the Shetland Isles. Immediately offshore from Scotland, the profiles image a series of well developed half grabens bounded by east-dipping normal faults which can be traced through relatively unreflective mid-crust to the top of a reflective lower crust. The top of the lower crust lies at approximately 7 s TWT. The base of the crust was defined as the base of the reflective crust at about 9 s TWT. This has been confirmed as the Moho following a more recent deep refraction profile (Morgan *et al.* 1994) and a reinterpretation of the LISPB profile (Barton 1992), which also show that the crustal velocities vary between *c.* 5 km/s near the surface to 6.8 km/s at the Moho (Fig. 7.28). The most spectacular result of the deep reflection profiling is the appearance on most of the sections of reflections from beneath the Moho (e.g. profile GRID-17, Fig. 7.27). There are two main features, the first of which is a series of discontinuous reflectors extending from the base of the crust, from a point just to the west of the downward projection of the Flannan Fault (which bounds a Triassic half-graben west of the Outer Isles within the Outer Isles Terrane), to a TWT of at least 30 s (*c.* 120 km depth) and dipping at about 30° to the east. This is generally referred to as the Flannan reflector. The second feature is a package of reflectors orientated subparallel to the Moho and imaged to the east of the Flannan reflector at about 15 s TWT or *c.* 45 km depth, termed the 'W reflector' (see below) (Fig. 7.27).

The observed crustal reflectivity is generally interpreted as imaging the Devonian West Orkney Basin (Coward & Enfield 1987). Snyder (1990) contrasted the geometry of the Moine Thrust onshore with the geometries of faults offshore and demonstrated that if one of the faults imaged offshore was the Moine Thrust reactivated as a normal fault, then the thrust would have to have been substantially rotated into a steeper dip, perhaps along the coast-parallel fault cropping out at Faraid Head. More recent work favours the faults offshore cutting the existing Caledonian structures, but with the possibility that they represent reactivated normal faults which accommodated Neoproterozoic Moine sedimentation (England 1999). To the north of the West

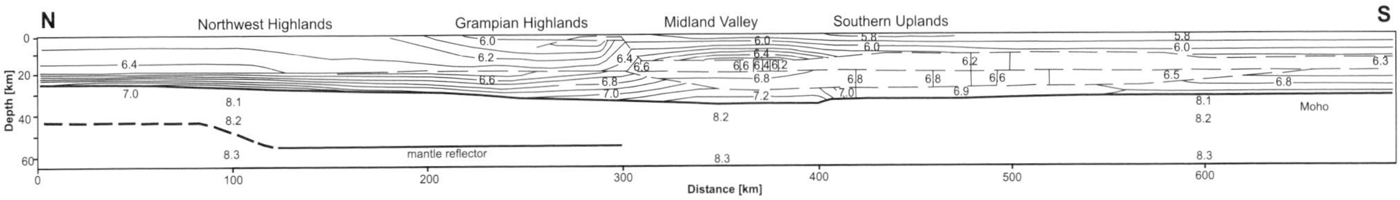

Fig. 7.28. Seismic P-wave velocity (in km/s) model of crust and uppermost mantle structure beneath northern UK based on the LISPB seismic refraction profile (modified from Barton (1992); for location see Fig. 7.26). Thin lines are velocity contours (interval is 0.1 km/s), dashed lines are first-order discontinuities.

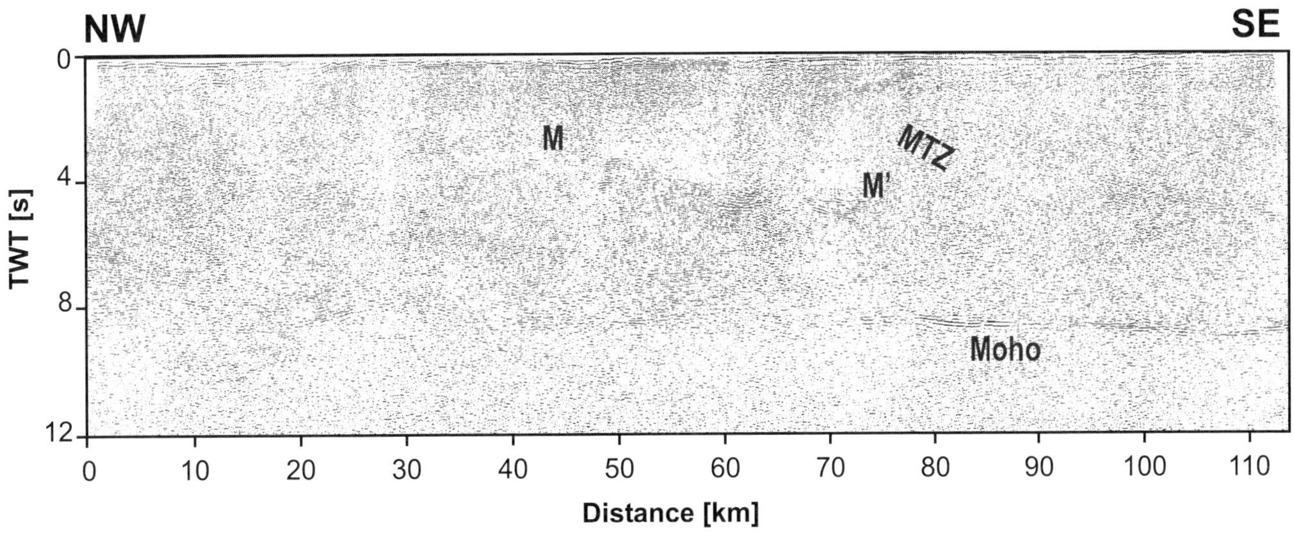

Fig. 7.29. Time migration of the BIRPS GRID-15 profile (modified from McBride & England 1995) from offshore north of the Scottish Highlands. MTZ, Moine Thrust Zone (Snyder 1990). M–M′ reflection is discussed in McBride & England (1995).

Orkney Basin a change occurs in the mid-crustal reflectivity and only a single zone of dipping reflectivity is seen (Fig. 7.29). This can be traced with the aid of commercial deep seismic data toward the Shetland Isles and projected to the Wester Keolka Shear Zone exposed on the coast of Shetland separating Lewisian from amphibolite grade (?)Moine rocks (McBride & England 1995, 1999). Whether this fault and associated reflectivity can be correlated directly with the Moine Thrust onshore in Scotland is not clear; its geometry must change significantly along strike to do so.

Snyder & Flack (1990) mapped the sub-Moho Flannan reflector in three dimensions beneath the Outer Isles and SW toward Ireland. Its subsurface geometry matches the strike swing of the Outer Isles Fault and the Caledonian structures of NW Britain and Ireland. An examination of the WINCH and WIRE reflection profiles north and west of Ireland indicates that the reflector is offset by the Great Glen Fault which can be traced offshore (Snyder & Flack 1990; Flack *et al.* 1990). This strongly suggests that the Flannan reflector predates the Great Glen Fault, leading to much debate about the origin of the reflector (cf. Morgan *et al.* 1994; Warner *et al.* 1996). Most recent theories have suggested that it represents a relict subduction zone on the basis of modelling the velocity structure using inverse techniques to match the observed reflectivity (Mosegaard *et al.* 1997; Brittan & Warner 1997). However, there is no direct surface geological evidence for pre-Caledonian subduction in NW Scotland consistent with the geometry of the reflector.

The reflectivity subparallel to the Moho, sometimes referred to as the W reflector (W being the inverse of M) has been the subject of a recent wide-angle/refraction experiment intended to determine its origin from a study of the seismic velocities either side of it (Morgan *et al.* 2004; Price & Morgan 2000). Like the Flannan reflector, there appears to be a decrease in velocity across it and the reflections originate from a layer of discrete thickness. Perhaps surprisingly, considering the region was once part of the Caledonian Orogen, the Moho is now relatively flat. This led Hynes & Snyder (1995) to suggest, on the basis of calculations of metamorphic conditions at the base of the crust during the Caledonian Orogeny, that the region between the Moho and the W reflector is formed from eclogitized lower crust. This would be consistent with the results of the wide-angle study but conflicts with McClay *et al.* (1986) who, based on their interpretation of the deep seismic data from the West Orkney Basin, suggest that the orogen underwent extensional collapse during the Devonian. Extensional structures in the Moine Rocks (Holdsworth 1989) also support the idea of orogenic collapse.

The W reflector is seen to extend as far north and west as its intersection with the Flannan reflector. There is no evidence in migrated sections that the two reflectors cross-cut. The W reflector can be traced to the south on the LISPB refraction profile (as interpreted by Barton 1992) and in the WINCH and WIRE profiles. These data indicate that it is also cut by the Great Glen Fault and, perhaps more importantly, that the reflector extends only as far south as the Highland Boundary Fault. It is not seen on the WINCH profile in the region of the North Channel (although these data are of poor quality in this region) and it is not seen on the better quality MOBIL3, NEC and SWABS profiles east of the UK, all of which begin to the south of the Highland Boundary Fault (Bois *et al.* 1990). Due to thick near-surface sediments causing seismic attenuation, there is very little sub-Moho reflectivity on the NSDP85–7 profile which does cross the Highland Boundary Fault, so determination of the precise end of the W reflector is difficult in this area. Given its observed spatial extent, it would appear that the W reflector is restricted to the former Laurentian lithosphere and, in the absence of other evidence, its relationship to the Flannan reflector suggests that both of these deep reflectors originate from structures or processes that predate the Caledonian Orogeny.

The Great Glen Fault. The Great Glen Fault can be traced south from the Shetland Islands, to the east of the Orkney Islands and across Scotland, and a splay of the fault has been mapped in NW Ireland. Its trace is crossed by a number of deep seismic profiles (Fig. 7.26).

To the north and south of the Shetland Isles the Great Glen Fault is imaged on the UNST, Fairisle and NSDP84–3 profiles by its continuation as the Walls Boundary Fault (McBride 1994) (Fig. 7.30). At the near surface it marks the edge of some small Devonian basins and so must have a near-vertical orientation. Projecting the fault down into the lower crust, it does not appear to cut lower crustal reflectivity, but at the Moho it is manifested on the UNST profile as a large diffractive feature. Initially, it was thought that this diffraction originated from a physical step at the base of the crust (McGeary 1989) but more recent work indicates that the diffraction collapses to a synformal structure interpreted as an inverted flower structure (McBride 1995). The formation of flower structures is associated with strike-slip deformation, which is consistent with the known history of displacements on the fault (Stewart *et al.* 1999, 2001). A similar picture is observed from the WINCH and WIRE profiles where they cross the fault in the North Channel west of Scotland. The Great Glen Fault cuts, and sinistrally displaces (by *c.* 60 km), the Flannan reflector (see above) so that it appears twice on the WIRE profiles (Snyder & Flack 1990). Before recent studies on the displacement history of the Great Glen Fault (Stewart *et al.* 1999, 2001) this was the best evidence of it having a history of strike-slip displacement, which had previously only been inferred from an apparent offset of metamorphic isograds in the Central Highlands (Winchester 1974). The width of the fault zone in the mantle, where it cuts the Flannan Fault, is no more than 10 km. It is not known how deep the fault penetrates into the mantle lithosphere but the intersection with the Flannan reflector is at *c.* 40 km beneath the present surface. The Great Glen Fault is not seen on the LISPB

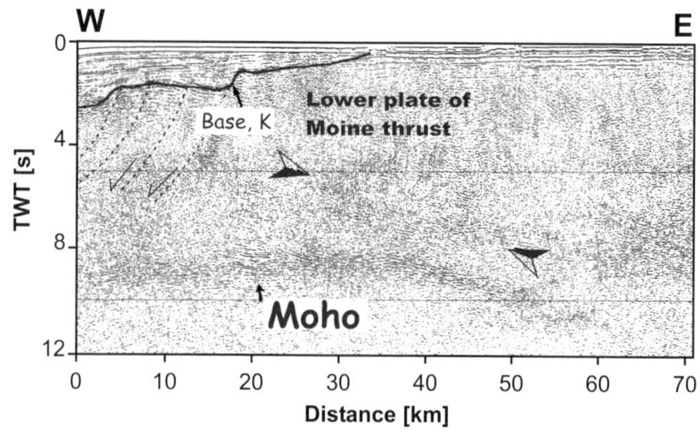

Fig. 7.30. Finite-difference migration of the BIRPS UNST profile (modified from McBride & England 1995) from offshore north of Shetland. K, Cretaceous. Arrows indicate package of bright reflections structurally beneath the Moine thrust. The location of geological structures is from Ritchie *et al.* (1987) and British Geological Survey (1989).

profile (Barton 1992) and there appears to be no major change in physical properties of the crust or mantle across it.

The Highland Boundary Fault. The Highland Boundary Fault may originally have defined the SE limit of rifted Laurentian crust during the Neoproterozoic development of Iapetus. It was later transformed into a collisional suture by accretion of the Midland Valley basement and arc during the Grampian Orogeny (Woodcock & Strachan 2000*a*). The Great Glen Fault and the Highland Boundary Fault both reach a depth of at least 40 km. These faults have a long history.

Strike-slip activity. There is some geological evidence for appreciable strike-slip motion across the major Caledonian boundaries north of the Iapetus suture, and also to the south (e.g. the relationship between volcanism within the Welsh Basin and Lake District areas and their reconstruction as part of an active plate margin). For example, Moinian sediments in the Northern Highlands lie unconformably on Lewisian basement and were influenced by Proterozoic magmatism (*c*. 1050 Ma) and folding almost contemporary with the Grenville mobile episode in the Laurentian Shield. South of the Great Glen Fault, Moinian sediments continue, apparently without any major unconformity, into the Dalradian succession which includes early Palaeozoic sediments (Esang & Piper 1984).

The Great Glen Fault is the largest of a set of NE–SW strike-slip faults with an important component of early Devonian movement. However, Late Silurian–Devonian movements along this fault were probably below the limits of palaeomagnetic detectability (Esang & Piper 1984). With the addition of the FAST profile to the north of Shetland it is possible to demonstrate that the structure of the Great Glen Fault at the Moho varies along strike, with the fault producing local thickening and thinning of the crust, as is frequently observed on active strike-slip fault systems at the surface. These observations are excellent evidence that major strike-slip faults can traverse the entire crust and penetrate into the upper mantle. Further support for this observation comes from a study of xenoliths found in Carboniferous volcanic pipes on either side of the Great Glen Fault in central Scotland (Canning *et al.* 1998). This demonstrated that the fault juxtaposed mantle with different records of metasomatism over a distance of no more than 10 km, indicating the deformation of the mantle that accommodated strike-slip motion at the surface was relatively discrete.

A crucial component of this palaeocontinental collage is the Highland Boundary Fault and associated structures. The large Caledonian sinistral strike-slip displacements imply that the Scottish Highlands were located further to the NE during early Silurian times, probably opposite Scandinavia. This palaeogeography makes it likely that Caledonian displacements on the Moine Thrust zone resulted from the collision of Laurentia and Baltica, rather than from contemporaneous Caledonian processes south of the Highland Boundary Fault (Woodcock & Strachan 2000*b*).

Iapetus Suture. The signatures and location of the Iapetus Suture across Ireland and surrounding seas have been previously reviewed by Landes *et al.* (2005). North–south trending refraction seismic lines crossing the area of the presumed Iapetus Suture from the COOLE '85 (Lowe & Jacob 1989) and VARNET '96 (Landes *et al.* 2000) surveys consistently show that this suture zone can be characterized by a thickening of the mid-crustal layer (Vp *c*. 6.4 km/s) found roughly between 11 and 20 km depth. This might be interpreted as tectonic stacking of different crustal units during collision of Laurentia to the north and Avalonia to the south (Landes *et al.* 2005). Abramovitz *et al.* (1999) suggested that the Iapetus Suture zone might have affected all crustal layers. Despite this, a marked difference in the velocity structure of Avalonia is noted by Abramovitz *et al.* (1999) comparing the results from the VARNET and MONA LISA data, with low lower-crustal velocities in a two-layered Avalonian crust on the VARNET and a high-velocity lower crust in a three-layered model on the MONA LISA model. However, comparing the Iapetus and Tornquist sutures across the North Sea, it can be seen that both suture zones separate Proterozoic shield-type crust from Palaeozoic Avalonia crust. Differences in sub-Moho velocities may be attributed to a change in subduction polarity between the Tornquist Sea and the Iapetus Ocean.

The Scandinavian Caledonides

The large-scale geology of SW Scandinavia is dominated by a fundamental structural transition from the Precambrian Baltic Shield, exposed in SW Sweden and southern Norway, to the Phanerozoic sedimentary basins in the Danish area. It is well documented that relatively thin crystalline Baltica crust underlies the entire Danish area, including the eastern North Sea, as well as parts of the North German Basin (Lassen & Thybo 2004). This structural transition is also clearly imaged in the Moho topography. Across the Baltic Shield, crustal thickness reaches 40–50 km, but in the Danish area it varies between 20 and 38 km (Fig. 7.31). However, this change in thickness is not caused by the presence of a Baltica margin, but is an intra-Baltica feature.

A variety of geophysical studies have been undertaken in the region in order to elucidate the deep structure of the area. One of the first studies was the FENNOLORA project which gathered deep seismic data from the entire Scandinavian peninsula and was aimed at the uppermost mantle, but also provided some crustal information (e.g. Stangl 1990; Guggisberg *et al.* 1991). Subsequently, the EUGENO-S and EUGEMI projects, which were part of the European Geotraverse where observations were made along five profiles, were designed to investigate the margin of the Baltic Shield where previous studies had observed that there were large differences in crustal structure across the edge of the shield (Fig. 7.32). More recent deep seismic projects have primarily focused on the acquisition of coincident normal-incident reflection and refraction/wide-angle reflection data; for example the MOBIL Search investigation of the Oslo-Skagerrak Graben system and the Norwegian–Danish Basin (Kinck *et al.* 1991; Lie & Husebye 1994); the BABEL project which aimed to image the Sorgenfrei-Tornquist Zone, Caledonian collision structures, and Proterozoic orogenic structures in the Baltic Sea (see below) (BABEL Working Group 1993); and the MONA LISA project which addresses structures related to Caledonian collision and deep mantle features in the North Sea (MONA LISA Working Group 1997*b*).

The 1979 Fennoscandian project (FENNOLORA) was aimed at the determination of the detailed structure in the Earth's mantle down to a depth of *c*. 400 km. Beneath the Fennoscandian Shield the crust is characterized by a smooth increase in P-wave velocity down to the Moho which lies at depths of *c*. 50 km in the southern part of the shield and *c*. 45 km further north. The mean crustal velocity is 6.6 km/s. At the base of the crust the velocity increases gradually from *c*. 7 km/s to 8.0–8.4 km/s in a 5–10 km thick crust–mantle transition zone. The relatively homogeneous nature of the Baltic Shield crust borders on areas with more differentiated crustal velocities, e.g. southern tip of Sweden (38 km) and beneath the northern Caledonides (43 km) (Guggisberg *et al.* 1991).

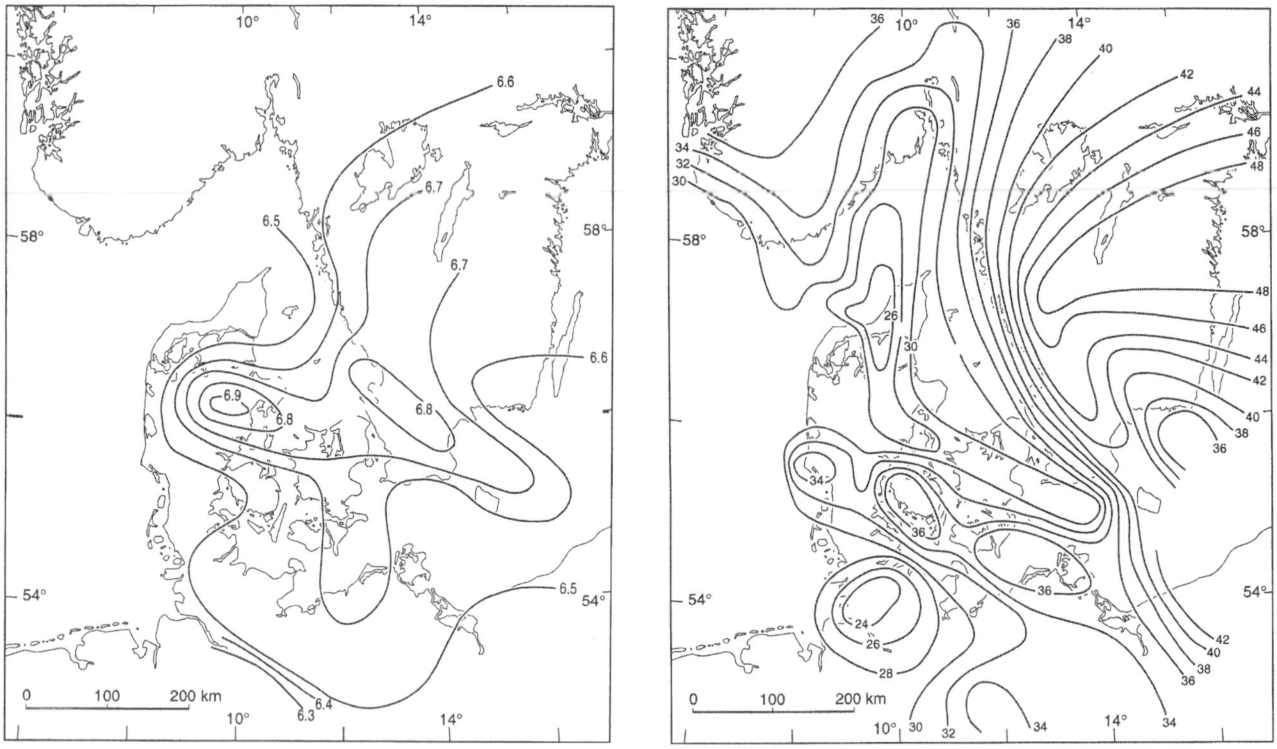

Fig. 7.31. (Left) Average crustal P-wave velocities through metamorphic and crystalline basement (in km/s). (Right) Depth to Moho (in km) for northern Europe (after EUGENO-S Working Group 1988; Thybo 2001, 1997).

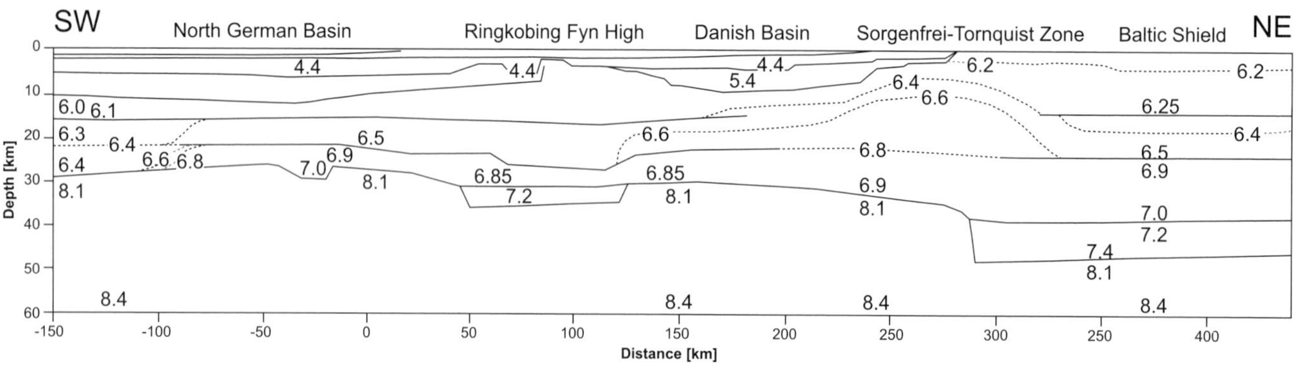

Fig. 7.32. P-wave velocity profile from south to north of the combined EUGENO-S 1 and EUGEMI profiles (after Thybo 2001).

As part of the Central Scandinavian Transect (Hurich & Kristoffersen 1988; Hurich *et al.* 1989), the Central Caledonian Transect acquired reflection seismic data in central Scandinavia between 1988 and 1992 (see Fig. 7.26 & Table 7.2). Strong reflectivity in the upper to middle crust down to 25 km depth is interpreted to have been caused by Caledonian tectonics (Juhojuntti *et al.* 2001). Thin, stacked sedimentary thrust sheets are seen down to 15 km depth, where a strong reflector marks the top Precambrian basement (Fig. 7.33). Below, the lower crust is almost transparent with some large-scale shear zones inferred to have developed in the Trans-Scandinavian Igneous Belt during Caledonian overprinting.

The MONA LISA project was designed to address deep structural and geodynamic problems in the SE North Sea area, with the main aim being to image plate collision structures and reflectivity characteristics of the crust and lithospheric mantle in order to map the Caledonian Deformation Front and possible remnants of the Tornquist and Iapetus oceanic lithospheric plates (MONA LISA Working Group 1997*a*, *b*). East of the MONA LISA survey area, integrated reflection and refraction deep seismic profiles across the Caledonian Deformation Front include the BABEL profile in the SW Baltic Sea (BABEL Working Group 1993) and the onshore LT-7 profile in NW Poland (Guterch *et al.* 1994; Dadlez 1997). Additionally, two refraction profiles across Caledonian structures are available in the vicinity of the MONA LISA study area: the LISPB profile across onshore Britain, designed to image crustal and upper mantle velocity structure (Bamford *et al.* 1976, 1978); and the European

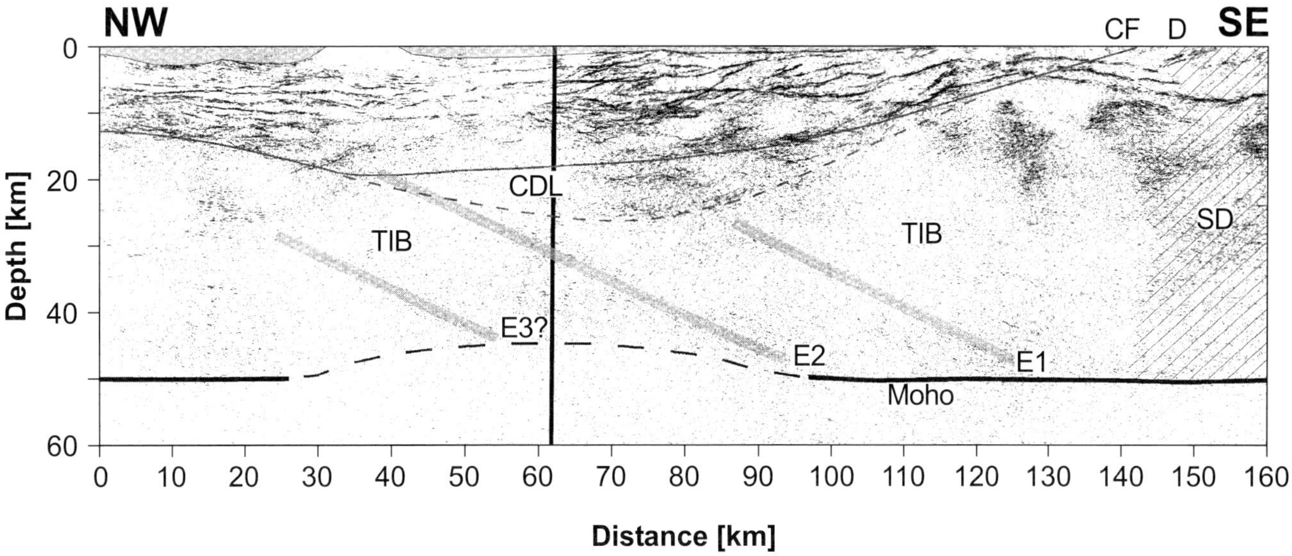

Fig. 7.33. Interpreted crustal structure along the reflection seismic profile CCT (Central Caledonian Transect) in southern Sweden (after Juhojuntti *et al.* 2001). Abbreviations: CDL, Caledonian deformation limit; CF, Caledonian front; D, dolerites; E, lower crustal shear zones; SD, Palaeozoic rocks; TIB, Trans-Scandinavian Igneous Belt.

Geotraverse (EGT) crustal refraction profiles from two sub-projects EUGENO-S (EUGENO-S Working Group 1988) and EUGEMI (EUGEMI Working Group 1990; Aichroth *et al.* 1992). The transition zone between the East European Platform and Phanerozoic Europe has been covered by many deep seismic profiles in Poland (e.g. Guterch *et al.* 1986).

The four MONA LISA seismic profiles show a variety of dipping and subhorizontal reflectors from the mantle and the crust that have been recorded up to 24 s TWT over much of the area (e.g. profile P2 in Fig. 7.34). Unusually low intracrustal reflectivity, in particular the poorly reflective lower crust, is marked when compared to the UK (Klemperer & Hobbs 1992), the German area to the SW (Meissner & Bortfeld 1990) and the Skagerrak area to the NE (Husebye *et al.* 1988). It is, however, similar to areas further to the east in northern Germany (Brink *et al.* 1992; Horst *et al.* 1994) and along the BABEL profile in the Baltic Sea (BABEL Working Group 1993). Subhorizontal reflectors were observed on two perpendicular profiles at *c.* 21 s TWT which is near the base of the lithosphere. These deep reflections are sharper than expected from a thermal–rheological transition, suggesting that they were generated at lithological or tectonic interfaces.

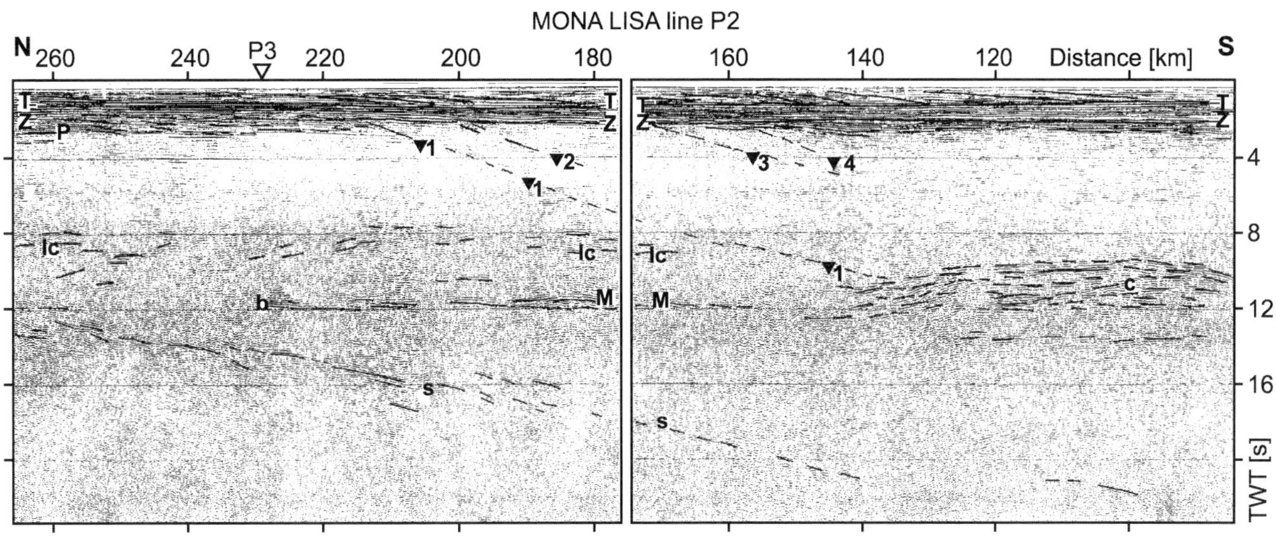

Fig. 7.34. Part of profile MONA LISA P2 (MONA LISA Working Group 1997*b*; Abramovitz & Thybo 2000), as unmigrated reflection seismic section overlain by interpreted line-drawing. Abbreviations: b, location of possible Moho offset; c, subhorizontal reflectivity in lower crust; lc, lower crustal reflections from Baltica; M, reflection Moho; s, south-dipping mantle event; P, Palaeozoic sediments; T, base Tertiary; Z, base Zechstein. Numbered arrows represent Caledonian thrusts.

The MONA LISA data reveal that there are three crustal types imaged in the southern North Sea. The first of these is a typical three-layered shield-type crust beneath the Ringkøbing-Fyn High, which contrasts with the second type comprising a two-layered crust of Caledonian origin to the south. Between them is a third type which is complex and transitional. In addition, sharp and strong normal-incidence and wide-angle reflections from the Moho were recorded to the south of the inferred Caledonian Deformation Front in contrast to the less distinct reflections further north (Abramovitz *et al.* 1998; Abramovitz & Thybo 2000).

Abramovitz *et al.* (1998) suggest that the closure of the Tornquist Sea occurred along a north-dipping subduction zone which was later overprinted by a late or post-Caledonian south-dipping shear zone. Sub-Moho velocities are 7.8–7.9 km/s beneath 35–34 km thick Baltica crust, and 8.1–8.3 km/s beneath 26–25 km thick Caledonian crust. The subhorizontal Moho across the collision zone implies late or post-Caledonian re-equilibration of the seismological Moho. Low-velocity upper mantle (7.8–8.1 km/s) to the north of the former Baltica lower crust is inferred to occur in eclogite facies following pressure-induced metamorphism as a result of lithospheric flexure during the Caledonian Orogeny (Abramovitz *et al.* 1998).

The Belgian and Dutch parts of the Caledonides

The Brabant Massif in Belgium represents the structural continuation of the Welsh Massif in Britain and forms the south-eastern extremity of the Anglo-Brabant Fold Belt. The Brabant Massif is that part of the Caledonides which is exposed in Belgium. The Belgian Geotraverse 1986 was gathered in the framework of BELCORP (Belgian Continental Reflection Seismic Program) that shot a series of near-vertical incidence reflection seismic profiles to fill the gap between related programmes in France and Germany. The 130 km long, north–south trending profile BELCORP '86 thereby covered, in its central part, the Caledonian Brabant Massif, a stable foreland block of the Variscides north of the Midi Fault, which is flanked on both sides by Upper Carboniferous and younger strata (Bouckaert *et al.* 1988). It is difficult to draw any far-reaching conclusion from the profile, since commercial data processing did not sufficiently consider the amplitude information.

The reflection seismic structure across the Brabant Massif is characterized by short reflectors (less than 1 km) which is a typical pattern for old shield areas. Summarizing these relatively incoherent features in terms of dip enabled them to be grouped into zones of 5–15 km thickness and tens of kilometres in length. These zones are interpreted to mark sedimentary wedges of Caledonian age with major décollement surfaces developed within the wedges (Bouckaert *et al.* 1988). This might appear as a rather unspecific interpretation compared to current standards of seismic quality, but this data set helped, at the time, to corroborate the concept of large-scale crustal shear zones (e.g. Reston & Blundell 1987). Furthermore, more detailed evidence for the trace of the Caledonian Deformation Front (CDF) in Belgium came much later from geological fieldwork which demonstrated the strong similarities between the different parts of the British and Belgian Caledonides (Sintubin 1999; Verniers *et al.* 2002; see below).

The available deep seismic Dutch profiles were shot parallel to the Belgian and Dutch coastline (Remmelts & Duin 1990; Duin *et al.* 1995). A highly reflective lower crust is observed between 7 and 11 s TWT in the Lower Saxony Basin (which extends eastwards into Germany), with an almost horizontal and undisturbed Moho at *c.* 30 km depth. A correlation with structures in the British Caledonides (Blundell *et al.* 1991) which cross onto the northern edge of the Brabant Massif is, however, not possible.

The North German Caledonides and Baltic Sea area

The North German Caledonides extend as sediment-buried basement due north of the Variscan deformed foreland, that comprises the sedimentary North German Basin. This is a Mesozoic successor basin, often divided into the NW or Lower Saxony Basin and the NE German Basin (Pratsch 1979; Brink *et al.* 1992). Thus, the extent of the Caledonides between the Netherlands and Poland is mainly evinced from deep seismic studies (e.g. Dohr 1983, 1989; Trappe 1989; Meissner & Bortfeld 1990) or inferred by potential field methods (see below).

Various possible boundaries have been proposed to mark the southern limit of Baltica exposed at surface, and the northeastern margin of Avalonia. The NW–SE trending and deep-reaching fault zone north of Rügen, the CDF, and further inland the Elbe Line are the features most often considered. New experiments to better image the complex region between the CDF and the Elbe Line, and thus to locate the trace of the CDF and the nature of Baltica and its southern limit, began in the mid-1990s. In order to outline the seismic signatures recorded in this area, the results of different types of seismic measurements including the EGT/Thumark, BASIN '96, BABEL, PQ2 and TOR surveys will be examined (see also Table 7.2 & 7.26).

Since structural imaging is often hampered by the masking effect of the Zechstein salt at the base of the sedimentary basin, onshore refraction seismic studies were used to reveal the velocity structure of crustal and sub-Moho levels. The Thumark profile in the southern half of the NE German Basin south of the Elbe Line shows basement velocities of only *c.* 6.2–6.4 km/s down to the Moho, where they jump to 8.1 km/s (Bankwitz 1985; Hoffmann *et al.* 1989; Bormann *et al.* 1989; Förste & Schulze 1991). This is also consistent with results from the EUGEMI profile (EUGEMI Working Group 1990; Aichroth *et al.* 1992). A Moho offset from 30 to 33 km depth is postulated to occur at the Gardelegen Fault between the Variscan foredeep and the North German Basin (Fig. 7.35).

North of the Elbe Line, the crustal velocity structure changes in the North German Basin (Rabbel *et al.* 1995; Bleibinhaus *et al.* 1999). The most striking feature is a *c.* 10 km thick high-velocity layer in the lower crust between *c.* 20 and 30 km depth (Fig. 7.36). Its P-wave velocity is 6.9–7.5 km/s, which is typical for shield crust. This high-velocity lower crust extends between Elbe Line and the CDF (Rabbel *et al.* 1995). Other surveys in the North German Basin support this finding (Aichroth *et al.* 1992; EUGENO-S Working Group 1988, Schulze & Lück 1992). While Hoffmann (1990) interpreted the area south of the Elbe Line as an additional microcontinent in the Caledonian collision, its interpretation as an important tectonic element marking the southernmost extent of Baltica evident in the NE German Basin was only later more precisely stated (DEKORP-BASIN Research Group 1999; Krawczyk *et al.* 1999; Bayer *et al.* 1999; Hoffmann & Franke 1997). High-velocity lower crust is also found further to the west in the North Sea on the MONA LISA profile 1 (Abramovitz *et al.* 1999), and to the east in Poland in the POLONAISE 97 data (Grad *et al.* 1999, 2002*a*; *b*, Janik *et al.* 2002).

A first complete structural image of both Variscan- and Caledonian-influenced areas of the North German Caledonides was provided by the BASIN '96 survey (DEKORP-BASIN Research Group 1999; Krawczyk *et al.* 1999), calibrated in the Mesozoic part by numerous wells (Hoth *et al.* 1993). The survey

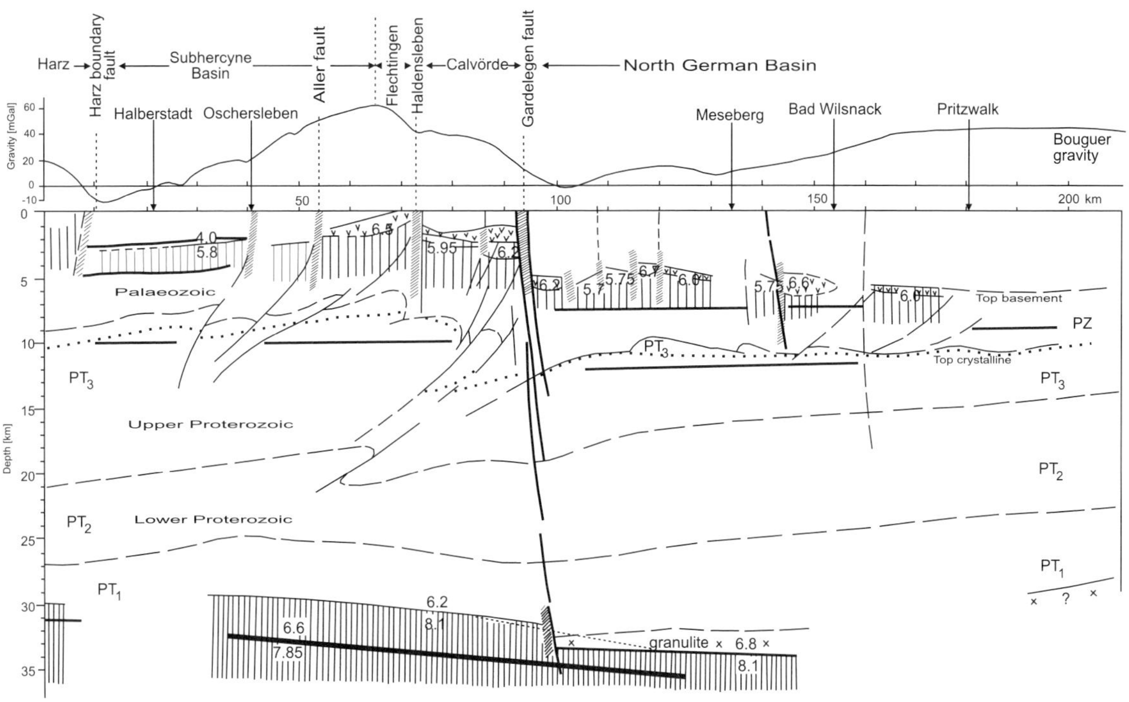

Fig. 7.35. Profile Thumark-Nord north of the Elbe Line with crustal velocities and refraction seismic interfaces (after Bankwitz 1985). PT_1–PT_3 label different Proterozoic units.

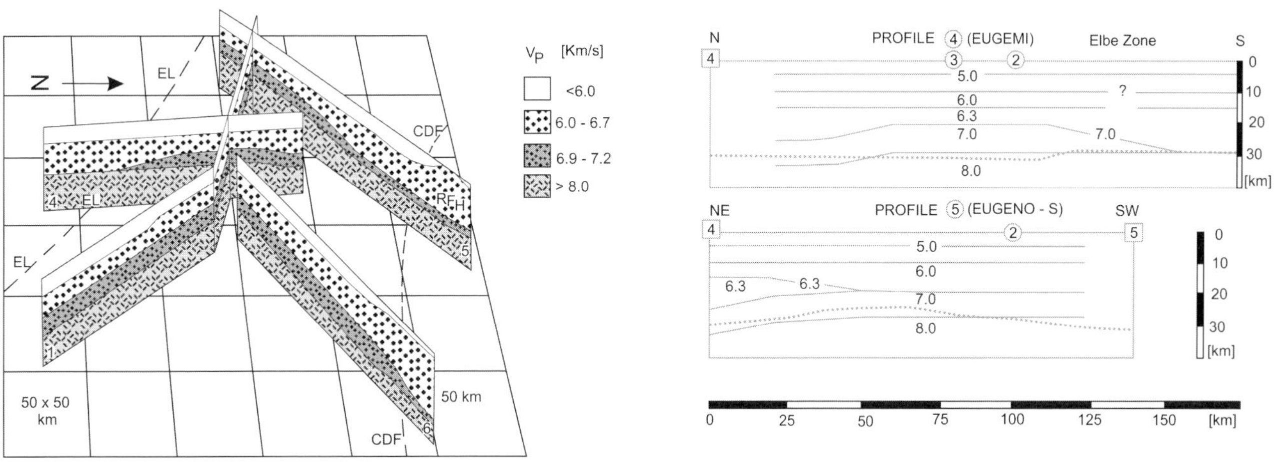

Fig. 7.36. (Left) Fence diagram of refraction seismic profiles in the NW German Basin. (Right) 2D P-wave velocity profiles (values in km/s) of selected lines close to the Elbe Line (after Rabbel *et al.* 1995). CDF, Caledonian Deformation Front; EL, Elbe Line.

revealed that the Moho is a continuous structure across the entire BASIN 9601 profile (Fig. 7.37) and is located at the base of a *c.* 2–4 km thick reflector band at 30 km depth beneath the basin (35 km depth at the margins). The Variscan structures along the southern basin margin show the basal Zechstein reflector with *c.* 5 km vertical offset at the Gardelegen Fault (Kossow *et al.* 2000). Offsets of 2.5 km are observed at the Haldensleben Fault, and at the Harz Boundary Fault, underlain by strong mid-crustal reflectors at *c.* 20 km depth. These offsets have been geometrically modelled to mark the mid-crustal expression of the fault zones observed at surface (Kossow & Krawczyk 2002). At the northern margin of the basin, the Zechstein reflectors shallow from 3 to 1.5 km depth and are truncated by the overlying Triassic. The landward extension to the north of profile BASIN 9601 across Rügen, however, could not resolve structures below the Mesozoic basin (Mayer *et al.* 2000). Despite this, moderate-amplitude SW-dipping reflectors are observed down to *c.* 20–25 km depth on BASIN 9601, beneath the northern third of the basin (Fig. 7.37), and this surface is interpreted as the Caledonian suture. The wedging form of this structure suggests that Baltica crust extended much farther southward below northern Germany (i.e. into Avalonia) than previously thought, possibly as far south as the basin centre or the Elbe Line (Krawczyk *et al.* 1999).

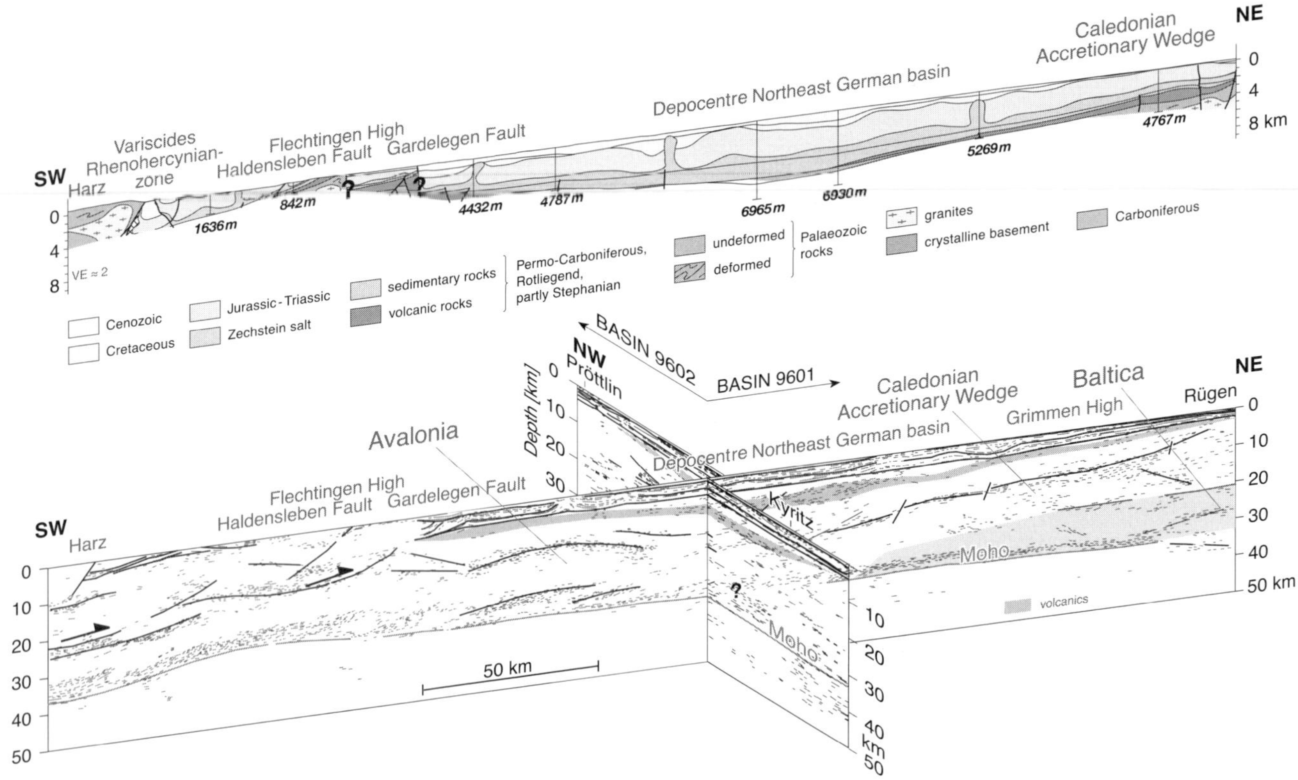

Fig. 7.37. (Top) Generalized geological cross-section running parallel to the BASIN 9601 profile. (Bottom) Interpreted line drawing of the BASIN 96 reflection seismic survey across the NE German Basin with main tectonic and stratigraphic features.

At the eastern border of the NE German Basin, eight deep seismic profiles have been combined into a consistent 110 km long NE–SW traverse (i.e. NIZUSE), recorded down to 12 s TWT (Hoffmann *et al.* 1989; Horst *et al.* 1994). This profile has yielded some mid-crustal reflectivity particularly in the area of Variscan influence (see also Hoffmann *et al.* 1996) and Moho reflectivity at *c.* 28 km depth (Fig. 7.38). Because of the limited recording length, the Moho is not observed at the northern rim of the NE German Basin on this profile. This would correlate with the observation that the Moho was found along BASIN 9601 at *c.* 32 km depth (see Fig. 7.37). There are also no direct indications for a deep-reaching fault such as the Trans-European Fault, as suggested some time ago by Berthelsen (1984) (see also McCann & Krawczyk 2001).

The North German Basin–Baltic Sea area comprises the transition from the Precambrian Fennoscandian Shield to the north and its extension beneath the North German mainlands where it is overlain by a thick sedimentary, and in parts volcanic, succession of Carboniferous to post-Permian age. Several regional reflection seismic surveys in the SW Baltic Sea have focused on the uppermost representation of sedimentary sequences in order to map the Mesozoic evolution of the area (e.g. Erlström *et al.* 1997; Deeks & Thomas 1995; Thomas *et al.* 1993; Piske & Neumann 1990) or to investigate the hydrocarbon prospectivity (e.g. Petrobaltic survey; Rempel 1992; JEBCO survey, unpublished) or both (SASO project; Schlüter *et al.* 1997; Piske *et al.* 1994).

The data from the PQ2 survey provides well-constrained images of upper crustal reflectivity and lower crustal/uppermost mantle reflections in the SW Baltic Sea (DEKORP-BASIN Research Group 1998; Meissner & Krawczyk 1999; Krawczyk *et al.* 2002). In the area of the Caledonian suture, a reflection pattern was observed with opposing dips in the upper crust and the uppermost mantle (Fig. 7.39). Detailed analysis of dipping reflections in the upper crust has provided evidence for two different sets of reflections, which are separated by the O-horizon (Fig. 7.40), which is the main decollement of the Caledonian deformation complex (see also Piske *et al.* 1994; Schlüter *et al.* 1997). South-dipping reflections above the O-horizon are interpreted as Caledonian thrust structures. Beneath the O-horizon, SW-dipping reflections in the upper crust are interpreted as ductile shear zones and crustal deformation features that evolved during the Sveconorwegian Orogeny (see Krawczyk *et al.* 2002).

The Caledonian deformation complex is subdivided into: (1) south-dipping foreland thrusts in the north (Avalonian thrusts onto Baltica); (2) the south-dipping suture itself that shows increased reflectivity; and (3) apparently NE-dipping down-faulted sedimentary horizons south of the Avalonia–Baltica suture, which may have been reactivated during Mesozoic normal faulting (Fig. 7.40). The Moho reflections at 28–35 km depth appear to truncate a north-dipping mantle structure, which may represent remnant structures from Tornquist Sea closure or late-collisional compressional shear planes in the upper mantle. A contour map of these mantle reflections indicates a consistent northward dip, which is steepest where there is strong bending of the Caledonian Deformation Front (Krawczyk *et al.* 2002). The thin-skinned character of the Caledonian deformation complex and the fact that north-dipping mantle reflections do not truncate the Moho would suggest that Baltica crust was not mechanically involved in the Caledonian collision and, therefore, escaped deformation in this area.

Combining onshore and offshore data into an interpreted line-

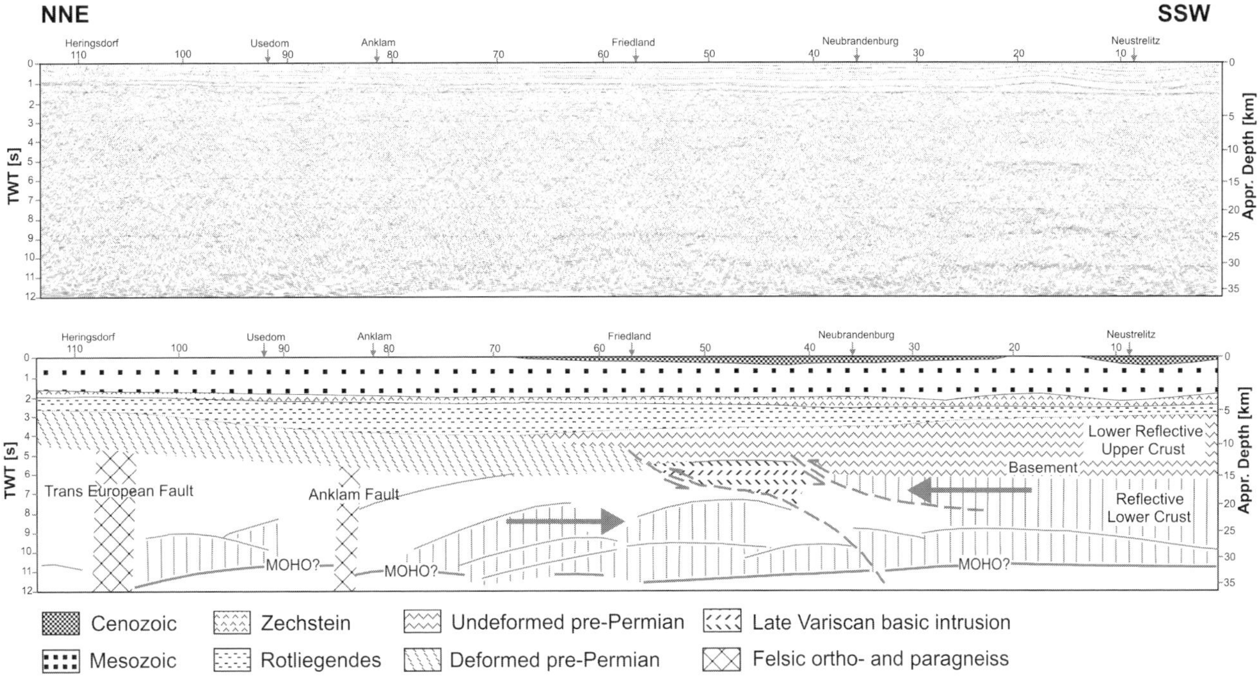

Fig. 7.38. (Top) Migrated reflection seismic profile NIZUSE between Usedom and Neustrelitz, NE Germany (see Fig. 7.26 for location). (Bottom) Interpretation (after Horst *et al.* 1994).

drawing together with a velocity model (Fig. 7.39), the extent of Baltica below the NE German mainland is clear, both structurally and also in terms of wide-angle velocity modelling. A high-velocity lower crust with P-wave values of 7.0 km/s indicates strong Baltica affinities (Bleibinhaus *et al.* 1999), even though the structural continuity is disturbed by a velocity decrease in area north of the CDF. This coincides with the area of sporadic reflections, not from a single continuously reflecting element, generated in the uppermost mantle (Fig. 7.39). These strong reflections apparently dip to the NE, and may mark the traces of Tornquist Sea subduction (Meissner & Krawczyk 1999). Such reflections are also observed on the BABEL profile A that was the first survey to recognize the Moho and sub-Moho levels structurally in the area (BABEL Working Group 1993) (Fig. 7.41).

Proterozoic terrane accretion has taken place north of the TESZ, as evidenced by wedging structures (e.g. Fig. 7.39; DEKORP BASIN Research Group 1998), and already proposed by the BABEL Working Group (1990). Revision of both structural and velocity data from BABEL led Abramovitz *et al.* (1997) to propose an intermediate terrane between a northern Svecofennian-deformed terrane and a southern terrane of Gothian age (Fig. 7.42). This intermediate terrane is characterized by a high crustal reflectivity which is not resolved in the wide-angle data but is instead deduced from different sub-Moho velocities.

The largest seismic antenna in Europe thus far was organized by the TOR project in 1996–97 (TOR Working Group 1999; Gregersen *et al.* 2002). In addition to Moho depth and upper lithospheric structures along the northern part of the BASIN'86 survey (Gossler *et al.* 1999), Gregersen *et al.* (2002) also detected deep lithospheric differences down to 300 km depth and defined lateral boundaries. At the northern rim of the Tornquist Zone and between the Elbe Line and the CDF, two marked lateral changes in velocity perturbation are interpreted to mark the transition from fast P-wave velocities beneath the Baltic Shield to slower values below the Variscan-influenced areas further to the south (Gregersen *et al.* 2002). A clear, horizontal lithosphere/asthenosphere boundary is defined at 125 km depth in the area extending between the Sorgenfrei-Tornquist Zone (STZ) and the Elbe Line (Cotte *et al.* 2002).

An almost vertical boundary between eastern and western Europe is also shown in the model of Schweitzer (1995) based on regional seismic-wave propagation studies. Based on these data he also proved a *c.* 200 km deep root of the Tornquist Zone. Further evidence for such a deep-reaching Baltica–Avalonia basement subdivision has been provided by the mapping of random lithospheric heterogeneities in northern and central Europe (Hock *et al.* 2004). These authors extended the TOR array further to the south and have characterized the lithosphere in terms of the scattering from seismic coda analyses. Scattering values are lowest in the North German Basin (Qs = *c.* 100), which cannot be explained by scattering in the crust only, while scattering is about four times more extensive in the Baltic Shield area (Hock *et al.* 2004).

The Trans-European Suture Zone

The structure and evolution of the contact between Precambrian Europe and the younger, Phanerozoic terranes in the SW is still one of the most interesting geological and tectonic problems in Europe. The TESZ represents the most prominent lithospheric boundary in Europe north of the Alpine–Carpathian orogenic front (Grad *et al.* 2002*a*, *b*). It is a broad and complex zone of terrane accretion separating ancient lithosphere of the Baltic Shield and the EEC from the younger lithosphere of western and southern Europe (Pharaoh 1999). This 2000 km long boundary zone between the two crustal domains is a deep-seated structure penetrating the entire lithosphere (e.g. Zielhuis & Nolet 1994; Schweitzer 1995). It transects the European continent from the North Sea in the NW to the Black Sea in the SE (Fig. 7.3). Over

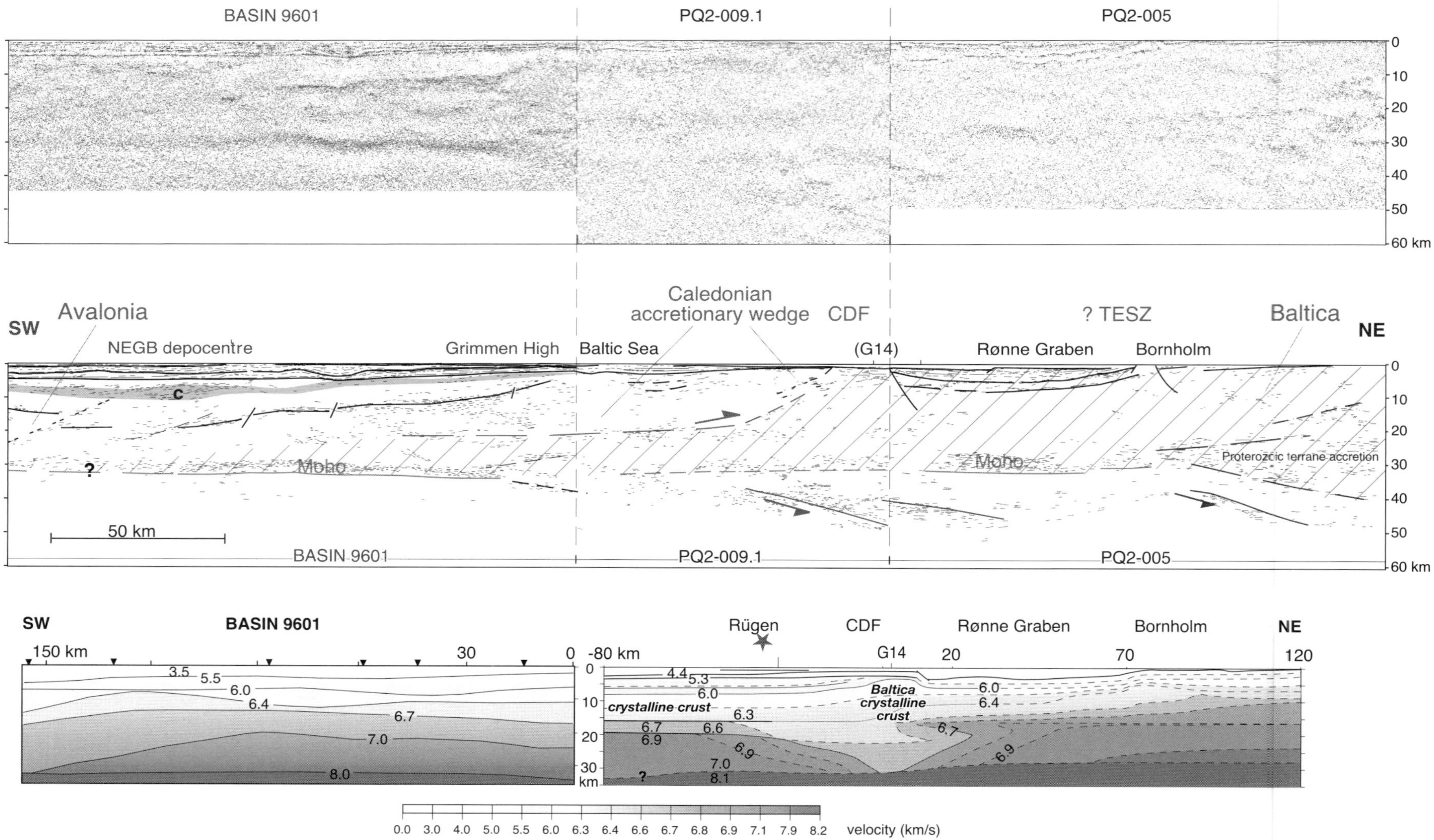

Fig. 7.39. Extent of Baltica between SW Baltic Sea and NE German Basin. (Top) Migrated reflection seismic profiles BASIN 9601, PQ2-9.1 and PQ2-5 (see Fig. 7.26 for locations). (Middle) Interpreted linedrawing and (Bottom) refraction seismic velocity model. Abbreviations: CDF, Caledonian Deformation Front; NEGB, NE German Basin; TESZ, Trans-European Suture Zone. G14 shows location of well in the Baltic Sea.

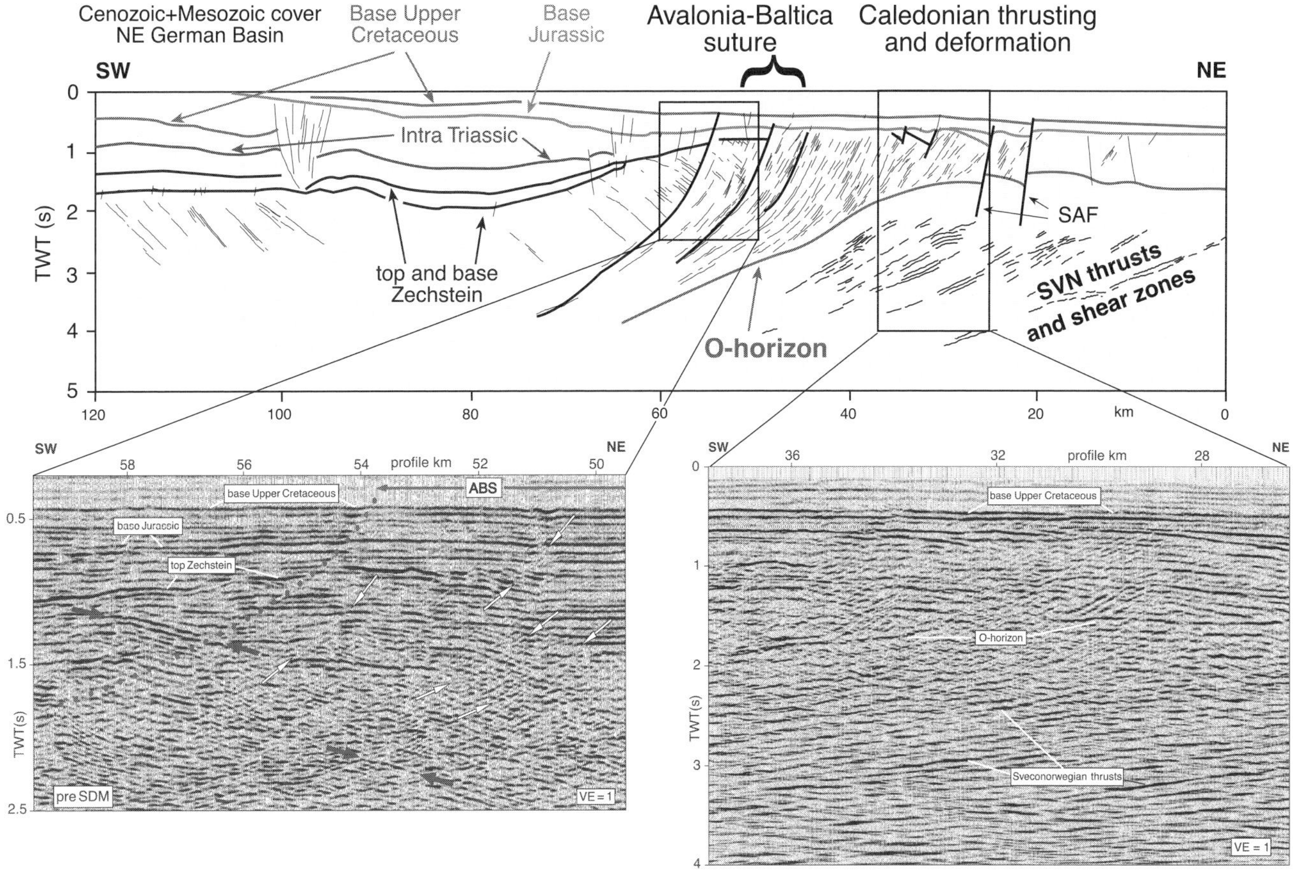

Fig. 7.40. (Top) Interpretation and (Bottom) prestack depth-migrated reflection seismic images of the northern rim of the NE German Basin in the southern Baltic Sea (after Krawczyk *et al.* 2002). The O-horizon, the main Caledonian decollement, is interpreted to separate Caledonian thrusts from Sveco-Norwegian thrust structures found below.

much of its length, the TESZ is concealed beneath younger basins.

The crystalline basement of the EEC adjacent to the TESZ comprises mainly granulites, migmatites, anorthosites and granite-gneisses (Ryka 1984). This is overlain by sedimentary successions, the oldest (Late Precambrian–Middle Ordovician) of which comprises thin shelf sediments which were deposited in a passive margin setting at the SW edge of Baltica (Poprawa *et al.* 1999). As a consequence of the subsequent collision with Avalonia, the shallow marginal basin was succeeded by a rapidly subsiding late Silurian foreland basin which developed on the SW Baltica margin (Poprawa *et al.* 1999). The total thickness of Silurian fine-grained sediments filling this foreland basin locally exceeds 4500 m.

The Early Palaeozoic platform of Western Pomerania includes a succession comprising mainly dark shales of Ordovician–Silurian age (e.g. Bednarczyk 1974) which were intensely deformed in late Silurian times (Dadlez 1978) presumably due to the collision of Baltica and Avalonia. They are roughly equivalent to the deformed Ordovician strata known from Rügen and interpreted as a fragment of a Caledonian fold-and-thrust belt (Giese *et al.* 1995). Both in Rügen and Pomerania, Lower Palaeozoic sediments may represent more proximal parts of the same Caledonian foreland basin whose distal part onlaps the SW margin of the EEC (Poprawa *et al.* 1999).

A series of deep seismic studies cross this part of Central Europe and comprise mainly seismic refraction experiments. These include the LT, TTZ and POLONAISE'97 (Guterch *et al.* 1994, 1999; Jensen *et al.* 1999; Grad *et al.* 1999, 2002*a*, *b*; Janik *et al.* 2002), CELEBRATION 2000 (Guterch *et al.* 1998, 2003), ALP 2002 (Brückl *et al.* 2003), and SUDETES 2003 (Grad *et al.* 2003) surveys. The latter surveys focus mainly on the younger tectonic units of Central Europe. The POLONAISE profiles, however, cross the TESZ, and presumably the coincident location of the CDF in Poland, and will be outlined in more detail below (see Figs 7.26 & 7.43).

POLONAISE'97 investigated the velocity structure across the TESZ between the EEC and the Palaeozoic Platform (PP). The PP is characterized by a crustal thickness of *c.* 30 km and mantle velocities of $Vp > 8.25$ km/s. The EEC shows a crustal thickness of 40–50 km and mantle velocities of Vp *c.* 8.1 km/s. The TESZ between these areas attains intermediate values with a crustal thickness of *c.* 40–45 km (Grad *et al.* 2002*a*). In general, the seismic reflectivity in the uppermost mantle down to 90 km depth is less pronounced in the EEC than in the PP and the TESZ (Grad *et al.* 2002*b*). A set of reflections, or a form of reflective boundary, is always found *c.* 10 km below the Moho. Since this feature is noted along all of the profiles, it cannot be attributed to a specific tectonic unit (Guterch *et al.* 1999). Additionally, these findings corroborate results from a profile constructed from 2D seismic velocity data between southern Sweden and northern Poland (Grad *et al.* 2002*b*). This latter profile images the fine detail of the uppermost mantle with alternating high- and low-velocity layers down to *c.* 120 km

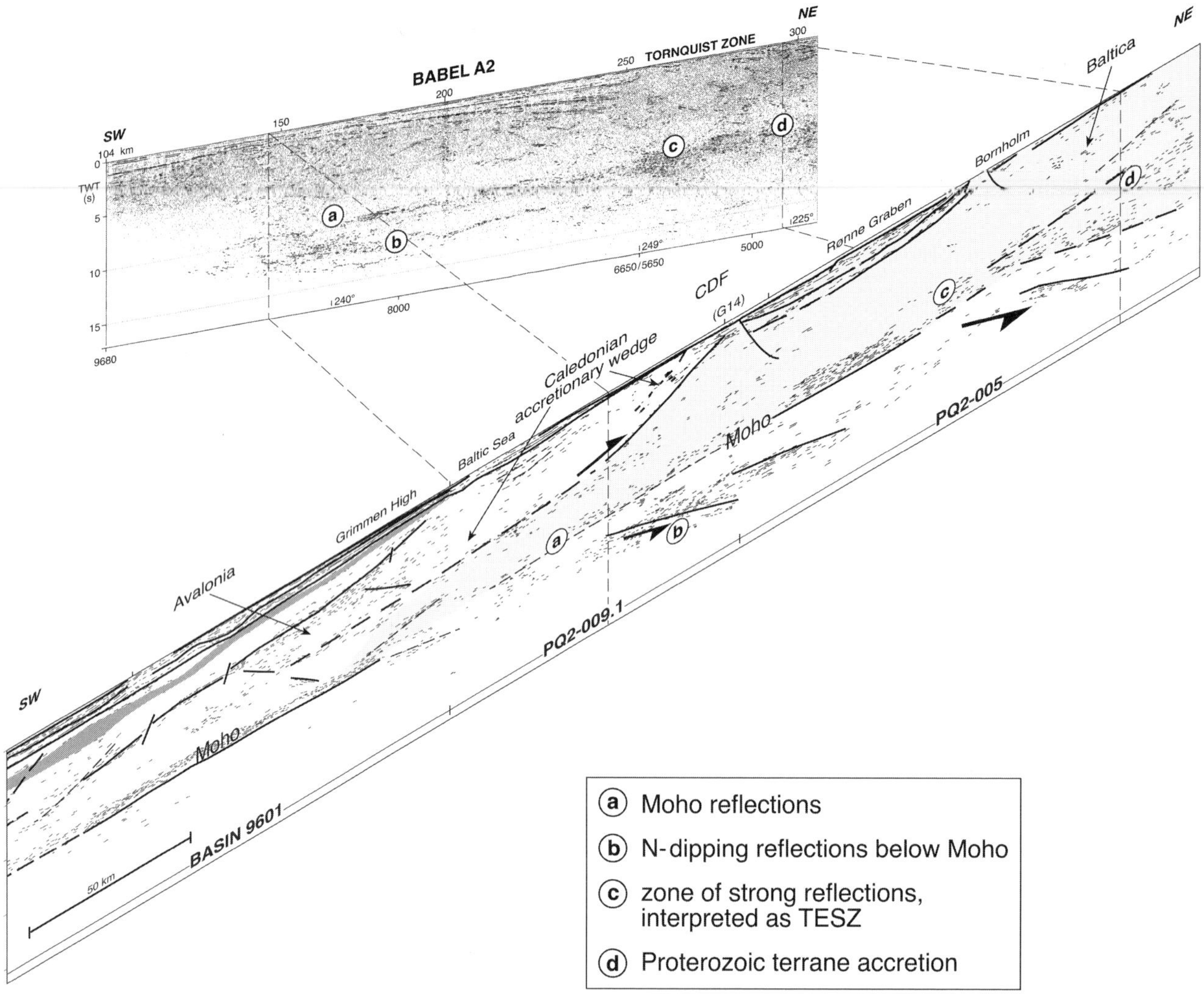

Fig. 7.41. Comparison of the reflection seismic features imaged on profiles BASIN 9601 and BABEL A in the southern Baltic Sea area (see Fig. 7.26 for locations). CDF, Caledonian Deformation Front; TESZ, Trans-European Suture Zone. G14 shows location of well in the Baltic Sea.

depth (Grad *et al.* 2002*b*), comparable with the EUGENO-S and FENNOLORA models (EUGENO-S Working Group 1988; Guggisberg *et al.* 1991).

The contrasting seismic structure of the crust, accompanied by a large Moho step, indicates the essential difference between the EEC and the basement of the PP juxtaposed along the TESZ; this is also observed off northern Denmark (e.g. Lie & Andersson 1998). The contact corresponds to a distinct vertical discontinuity penetrating the upper and middle crust along the boundary between the Tornquist-Teisseyre Zone (TTZ) and the SW margin of the EEC. Thus, at higher crustal levels the TTZ would appear to represent an important geological boundary between Baltica and terranes accreted to its SW margin during the Palaeozoic. However, in the lower crust this boundary is not evident and the EEC basal layer appears to continue further to the SW of the TTZ (Grad *et al.* 2002*a*, *b*). This could mean that the boundary is a low-angle interface such as that observed beneath the NE German Basin.

The Moho topography is strongly undulating in southern Scandinavia and NE Europe (Fig. 7.31). Jensen *et al.* (2002) have shown that there are similarities between the areas of the TTZ in Poland and the Fennoscandian Border Zone which is partly coincident with the STZ in Denmark. In both of these areas, the Moho dips steeply from a crustal thickness of *c.* 32 km in the young PP and basin areas to *c.* 45 km in the old Precambrian platform and Baltic Shield area (Jensen *et al.* 2002).

Potential field data

Potential field data are ideally suited for providing an overview of existing terranes and sutures in an area. The potential field maps of Europe document the most fundamental features of the entire crust, thereby allowing structures from the overlying sedimentary basins to be separated from the underlying Palaeozoic basement. It should, however, be noted that modelling subsurface bodies from potential field data is inherently ambiguous unless constrained by other information (e.g. seismic reflection or refraction data).

The most recent and consistent **gravity map of Europe** synthesizes the existing data from western Europe with new data gathered from the eastern European countries (Bachmann &

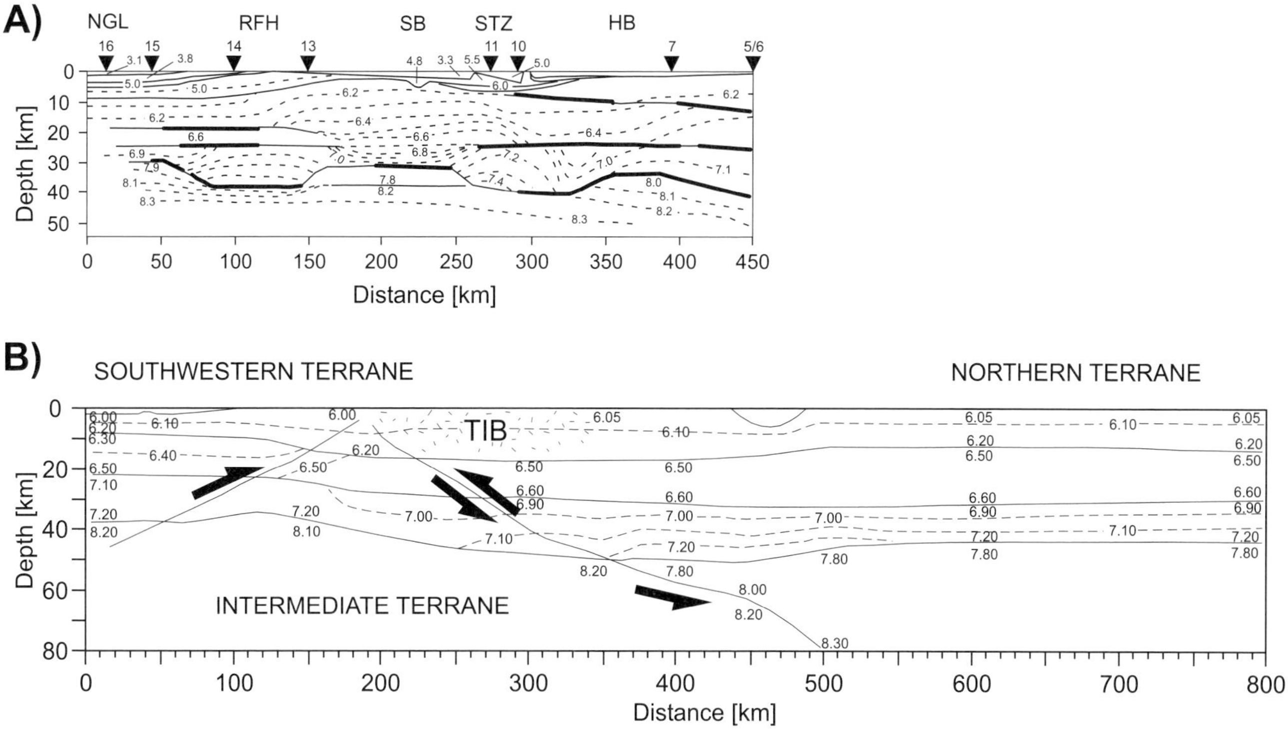

Fig. 7.42. Different refraction seismic velocity models along profile BABEL A in the Baltic Sea area (see Fig. 7.26 for location) according to (**A**) BABEL Research Group (1991) and (**B**) Abramovitz *et al.* (1997), suggesting different tectonic models or terrane assemblages. P-wave velocities are given in km/s. Abbreviations: HB, Hanø Basin; NGL, North German Basin; RFH, Ringkøbing-Fyn High; SB, Skurup Basin; STZ, Sorgenfrei-Tornquist Zone; TIB, Transscandinavian Igneous Belt.

Grosse 1989; Grosse *et al.* 1990, 1992; Lee *et al.* 1990; Blundell *et al.* 1992; British Geological Survey 1997; Wybraniec *et al.* 1998; Banka *et al.* 2002). With a resolution of 5 × 5 km, the gravity image shows Bouguer anomaly data on land and free-air anomaly data offshore (Fig. 7.44). All major large-scale tectonic features and suture zones can be recognized on the map. A pronounced gravity low is recorded across Fennoscandia and for the Alps. The CDF is generally associated with a NW–SE trending, sharp gravity lineament across northern Germany. A possible link to a similar element beneath Scotland may also be present. To the SE, a similarly orientated feature is evident below Poland (Fig. 7.44).

On a more regional scale, 2D and 3D gravity modelling has been performed in the NE German Basin. Along the DEKORP-BASIN '96 profile, the 3D gravity model proposed a high-density lower crust below the centre of the NE German Basin and an increased thickness of low-density upper crust below the margins to fit observed and modelled data (Barrio-Alvers *et al.* 1998; Scheck *et al.* 1999). Geodynamically, such a supply of high-density material could be related to the transfer of mantle material into the crust by magmatic underplating. Indications corroborating this hypothesis are also provided by seismic data (Rabbel *et al.* 1995; Krawczyk *et al.* 1999), studies on magmatic activity (Benek *et al.* 1996) and subsidence analysis (van Wees *et al.* 2000; Scheck & Bayer 1999).

The continuation of Baltica below the north German mainland has also been modelled by gravity data, which suggest the presence of an Avalonian accretionary wedge on top of Baltica crust that thins towards the NE German Basin depocentre (Kuder 2002; Fig. 7.45). An alternative interpretation is that there may be a low-density body below the northern margin of the Southern Permian Basin. This, however, is a less favourable model since reprocessing of seismic data in the region has indicated the presence of an accretionary wedge (see above; McCann & Krawczyk 2001; Lassen *et al.* 2001; Krawczyk *et al.* 2002).

In NW and central Poland, where Baltica basement is also largely concealed beneath the Mesozoic cover, a detailed gravity model of the crustal structure along the TESZ, based on the *Gravimetric Atlas of Poland* (Krolikowski & Petecki 1995) has been constructed (Krolikowski & Petecki 1997). This Bouguer anomaly map suggests the presence of anomalous masses in the Lower Palaeozoic successions. In order to interpret these anomalies, a zone of 35 km crustal thickness overlying a dense upper mantle has been suggested for the region along the TESZ (Krolikowski & Petecki 1997). This model is supported by the seismic data (Guterch *et al.* 1994, and references therein) as well as the most recent potential field maps of Poland (Petecki *et al.* 2003*a*, *b*). These maps show a clear separation between the Palaeozoic Platform and the East European Craton.

The **magnetic field of Europe** map has been compiled using a 2 × 2 km grid from various data sets (Lee *et al.* 1990; Wonik *et al.* 2001; British Geological Survey 1998; GeoFrance 3D 2005; Banka *et al.* 2002). A strong contrast can be observed between the highly magnetic crust of the EEC and the less magnetic Palaeozoic-accreted terranes of Central Europe. Additionaly, there is an irregular pattern across Baltica (Fig. 7.46). A SE–NW trending maximum occurs along the trace of the TESZ and terminates with a sharp gradient in the Polish Trough (Williamson *et al.* 2002). The CDF can be traced along this prominent pattern of magnetic anomalies from northern Poland to the North

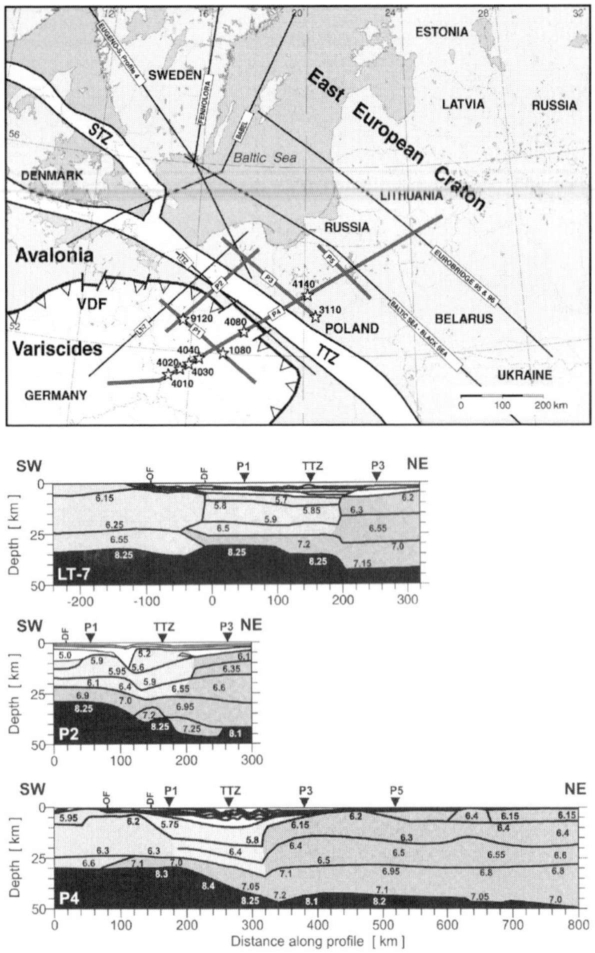

Fig. 7.43. Location map of the seismic campaign POLONAISE'97 and profiles LT-7 and TTZ in Poland. The classification of different crustal types identified across the Trans-European Suture Zone is given along three profiles (after Grad *et al.* 2002*a*, *b*).

Sea, where it either vanishes towards the west, or is offset northwards and continues as far as Scotland (Fig. 7.46).

Various attempts at 2D and 3D combined structural, magnetic and/or gravimetric modelling have been made with the aim of unravelling the possible evolution of observed structural features. Aeromagnetic and Bouguer anomaly data were investigated, together with structural field observations, at the southeastern extremity of the Lower Palaeozoic Anglo-Brabant Belt in Belgium. This area is north of the CDF, and is the only location there where it is not concealed (Sintubin 1999; Sintubin & Everaerts 2002). This integrated approach led to the production of a new conceptual tectonic map of the investigated area with several domains defined from the potential field data that reveal the tectonic lineaments separating those different domains (Fig. 7.47). The metamorphic contour line separating the diagenetic zone in the SW of the Brabant Massif, probably coincides with the CDF (Woodcock 1991). To the SW of this zone, no Caledonian deformation has been observed (van Grootel *et al.* 1997), and the presence of cratonic basement is assumed (Verniers *et al.* 2002; Fig. 7.48). A compressional wedge model has been proposed for the Brabant Massif, with Precambrian basement controlling its kinematics (Sintubin & Everaerts 2002). This is in contrast to the classic interpretation of an anticlinorial culmination (van Grootel *et al.* 1997).

In the southern North Sea, combined magnetic–gravimetric 2.5D modelling reveals anomalies related to a sharply defined Avalonia–Baltica Suture along the EUGENO-S and MONA LISA 1 and 2 profiles (Williamson *et al.* 2002). Here, the magnetic Baltica basement dips to the SW below the non-magnetic Avalonia basement and always penetrates the entire crust down to *c.* 30–35 km depth. The internal structure of Baltica is interpreted as being imbricated (Williamson *et al.* 2002). Low-amplitude magnetic anomalies to the north and south of the suture zone are interpreted to be caused by a mid-crustal magnetic body, which might represent the 'missing arc' related to an inferred southward subduction of the Tornquist Sea.

Images from deep seismic data between the North Sea and Poland reveal a variety of structural features in the crust. These are also indicated in the magnetic field and a pseudogravity map (Bayer *et al.* 2002). This type of map is a spatial integration across the magnetic anomalies based on the assumption that the magnetic signature correlates with the density. From these data, the CDF is interpreted to coincide with the continental slope of Baltica, which terminates at the lower Elbe-Odra Line (Bayer *et al.* 2002). This lineament is interpreted as forming the boundary between Baltica- and Avalonia-dominated areas, with a high lower crustal velocity always found to the north of it. The potential field data additionally suggest that Baltica may extend below the Southern Permian Basin to the south. This suggestion was previously proposed, but based only on seismic data (e.g. Rabbel *et al.* 1995; Krawczyk *et al.* 1999).

The **thermal structure** of an area can be used in conjunction with gravity and density data to explain phenomena such as convective flow in the upper crust, mineral dissolution/precipitation, and also the mechanical state of the deeper crust, which is crucial to the understanding of the geodynamic evolution of a region. With the commencement of geothermal investigations in Europe, 1D or 2D geothermal maps were constructed on the base of 1D extrapolations to depth of geothermal well data (e.g. Cermak & Hurtig 1979; Haenel & Staroste 1988; Hurtig *et al.* 1992). While the *Atlas of Geothermal Resources in the European Community, Austria and Switzerland* (Haenel & Staroste 1988) displays maps only for NW Europe, the *Geothermal Atlas of Europe* project (Hurtig *et al.* 1992) provided for the first time a compilation of geothermal data for all of Europe.

Low surface heat-flow values between 30 and 50 mW/m^2 are observed for the Fennoscandian area, while higher values, typically 60–80 mW/m^2, are noted from younger units located further south (Fig. 7.49; Majorowicz *et al.* 2003). Close to Norway the transition between these two domains, which are interpreted as being separated by the CDF, occurs over a relatively short distance of *c.* 20 km. Further to the west this zone of separation appears to broaden to at least 50 km width (Fig. 7.49). A data gap exists between Copenhagen and the northern margin of the NE German Basin, although this is partly filled in terms of temperature distribution at depth (Schellschmidt *et al.* 1999).

Low temperatures are characteristic for the Baltic Shield area, where there is also a stable temperature–depth distribution. This pattern changes, when moving southwards and crossing the area of the CDF to one where strong lateral variations are observed in the upper kilometres of the sedimentary successions (Fig. 7.49; Hurtig *et al.* 1992). Furthermore, the relationship between near-surface heat-flow density and depth of the Moho differs for the various tectonic units (Fig. 7.50; Hurtig 1995). Schellschmidt *et al.* (1999) corroborated this relationship in greater detail for the

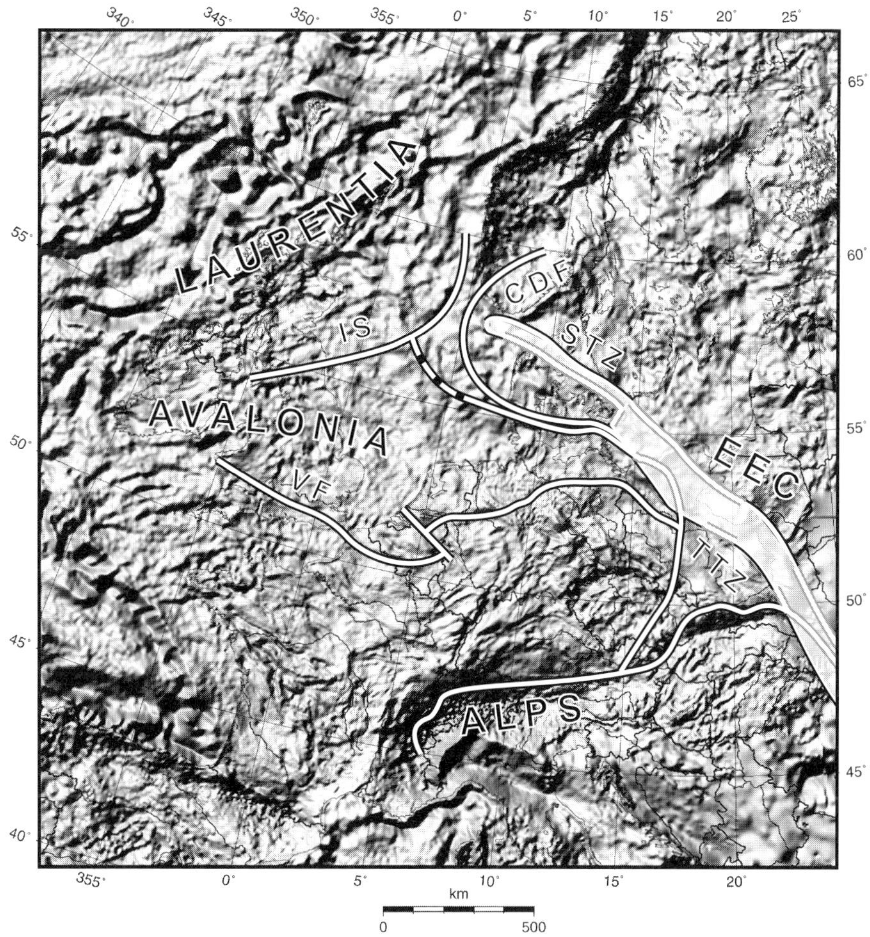

Fig. 7.44. Gravity map of Central Europe overlain with main tectonic features and fault zones (see Wybraniec *et al.* 1998; Banka *et al.* 2002; Fig. 7.3). The Caledonian Deformation Front (CDF) and the Trans-European Suture Zone can be recognized as generally NW–SE trending gravity lineaments. Abbreviations: EEC, East-European Craton; IS, Iapetus Suture; STZ, Sorgenfrei-Tornquist Zone; TTZ, Tornquist-Teisseyre Zone; VF, Variscan Deformation Front.

area between Rügen and Flechtingen in the NE German Basin. They note that temperatures down to 1.5 km depth are colder on Rügen (40–50°C) than in the basin (background temperature is *c.* 60°C, three isolated maxima show 70°C).

A variety of local- to regional-scale geothermal models, using 1D to 3D modelling approaches, have been proposed for those areas of Central Europe influenced by Caledonian tectonics (e.g. Cermak & Bodri 1986; Wolfgramm 2002; Vosteen *et al.* 2004). However, models investigating the sedimentary infill and using these results to draw conclusions on the extent of underlying Caledonian basement are rare since extrapolation of surface values to depth is problematic. Despite this, available thermal models consistently indicate a temperature gradient from the colder, Baltica-influenced regions towards the warmer, southerly located and Variscan-dominated areas.

Three-dimensional geothermal modelling for specific areas reveals the different character of underlying terranes. Extending the work of Galson & Rybach (1985), Cermak & Bodri worked with a 2D approach (1986) and calculated a 3D geothermal model along the length of the EGT (Cermak & Bodri 1995). This model assessed the crustal temperature field on a more continental scale (Fig. 7.51). While low crustal temperatures (400–500°C at the Moho) and low heat flow values (*c.* 30 mW/m^2) characterize the Baltic Shield and the North German Caledonides, higher values are typical of the Mediterranean area (50–60 mW/m^2). In addition, the CDF marks the northern boundary of the transitional zone from low to high values.

In the northern part of the NE German Basin, the depth-dependent 1D temperature distribution from drillhole data implies a moderately steep geothermal gradient within the Permo-Carboniferous volcanic rocks (Wolfgramm 2002). This suggests a stronger fluid circulation, and this may be facilitated by the intensely altered rocks close to the CDF. These results are consistent with the modelling of the thermal regime in the NE German Basin by 2D inversion. Here, a significant high thermal conductivity of the pre-Permian basement in the northern part of the NE German Basin is observed, coincident with the suggested location of the CDF (Vosteen *et al.* 2004).

Determination of the past heat-flow history of an area relies on indirect evidence, and thus the resolution of palaeothermal reconstructions with 1D vitrinite reflectance and temperature data is limited, as has been shown by Nielsen (1995) for the North Sea region. To overcome part of this problem, attempts are underway to derive a thermal–compositional tomography from seismic data in order to provide an independent constraint on estimates of the temperature structure of the crust. These studies

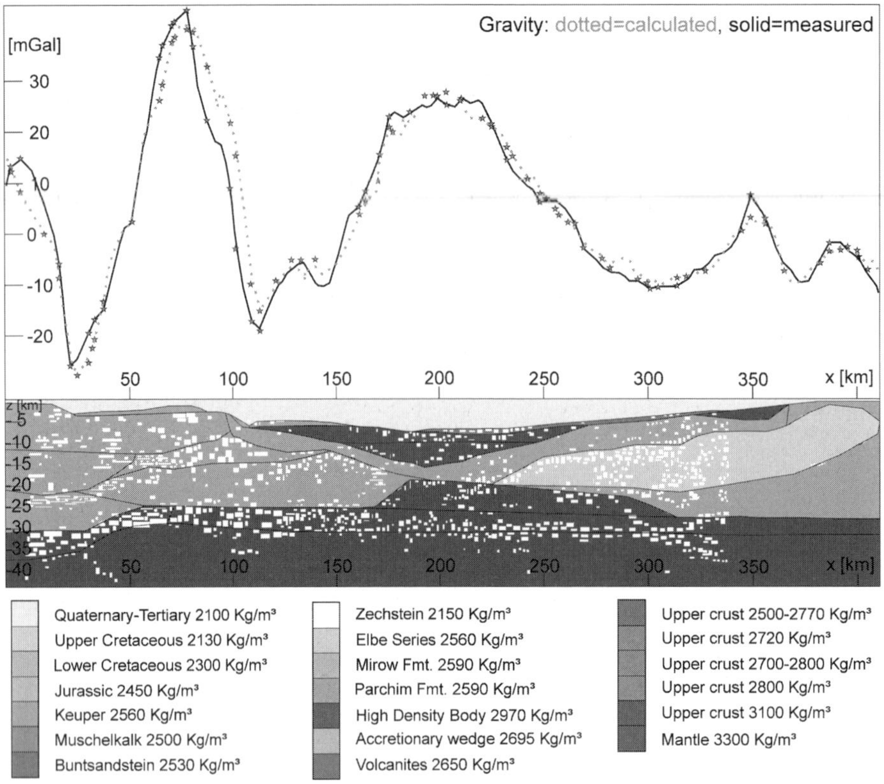

Fig. 7.45. 2D cross-section from 3D gravity modelling in the NE German Basin (from Kuder 2002). An Avalonian accretionary wedge (Akkretionskeil) on top of Baltica crust (right half of profile) and Variscan influenced areas (left part of profile) are evident.

are carried out in conjunction with work on surface heat-flow (Furlong *et al.* 1995).

A very comprehensive 2D numerical analysis of the thermal structure of the lithosphere across the Baltic Shield and northern Tornquist Zone exists along the FENNOLORA and EUGENO-S/EUGEMI profiles (Balling 1995; for location of profiles and seismic structure see above). Geographically, heat-flow densities were compiled for the Fennoscandian Shield, the Danish area and northern Germany along 2300 profile-kilometres. Upper mantle heat-flow values in southern shield areas (i.e. close to the CDF) are 30–40 mW/m^2 (Fig. 7.52), and are interpreted to be mainly related to heat flux from the upper mantle rather than upper crustal heat generation (Balling 1995). Petrological discussions suggest that the present deep crust is at lower temperature–pressure conditions than in the early stages of the Caledonian activity. However, the present-day deepest parts of the crust in the shield areas are still in the eclogite stability field (Balling 1995). Therefore, together with the absence of significant gravity anomalies in those regions (see above), it can be suggested that eclogites are in abundance in the shield crust.

Geological and geophysical synthesis

In describing the Caledonian Orogeny in Central Europe, a series of tectonic zones and sutures can be identified from the various types of data. These datasets and results are outlined in detail above. Below, a summary in terms of nature, location and significance of the main structures is provided.

Oceanic sutures

The most important Caledonian sutures are those representing the former oceans, Iapetus, Tornquist and Rheic. The Iapetus Ocean separated Baltica and Avalonia from Laurentia between early Cambrian and mid-Silurian times (110 Ma). That part of the ocean separating Avalonia from Baltica is usually referred to as the Tornquist Sea. This closed at about Ordovician–Silurian boundary times (443 Ma), whilst the main Iapetus Ocean did not close until the mid-Silurian.

Tornquist Suture

The Tornquist Suture forms the boundary between Baltica to the north and Avalonia to the south (Fig. 7.3). The subsurface location of the suture has been inferred primarily from the variations in the velocity structure and thickness of the Avalonia and Baltica plates as indicated by a number of seismic wide-angle profiles (MONA LISA Working Group 1997*a*, *b*). Thybo (1990) presented an interpretation of the combined EUGEMI and EUGENO-S profiles (Fig. 7.26) which showed that the Baltic Shield was characterized by thick (*c.* 45 km) crust with high (>7 km/s) lower-crustal velocities. The crust thins to 29–32 km beneath the North Danish Basin before increasing to 36 km beneath the Ringkøbing-Fyn High. South of the high, the crust beneath the North German Basin is between 26 and 30 km thick. Changes in velocity and velocity gradients at the southern edge of the Ringkøbing-Fyn High, within the Palaeozoic strata and crystalline basement, were interpreted as indicators of the suture (e.g. Thybo 1990; Krawczyk *et al.* 1999). Further to the south, at the Elbe Line, a pronounced change in lower-crustal velocity

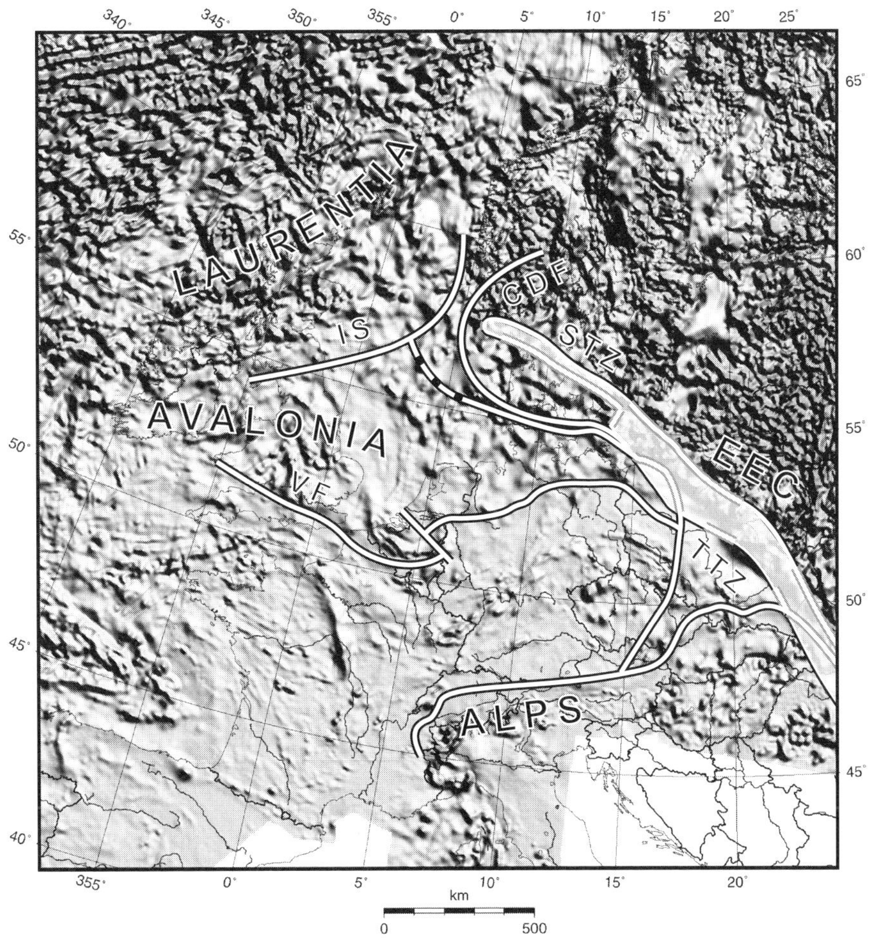

Fig. 7.46. Magnetic map of Central Europe overlain with main tectonic features and fault zones (see Wybraniec *et al.* 1998; Banka *et al.* 2002; and Fig. 7.3). The East European Craton (EEC) has a highly magnetic crust in contrast to the less magnetic Palaeozoic-accreted terranes of Central Europe. Abbreviations: CDF, Caledonian Deformation Front; IS, Iapetus Suture; STZ, Sorgenfrei-Tornquist Zone; TTZ, Tornquist-Teisseyre Zone; VF, Variscan Deformation Front.

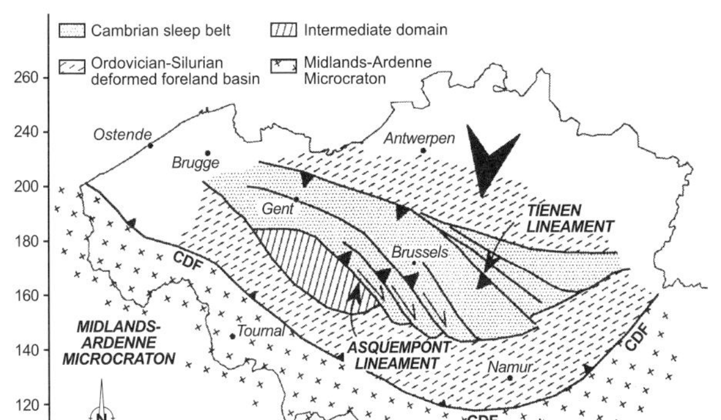

Fig. 7.47. Tectonic sketch map of the Brabant Massif (after Sintubin & Everaerts 2002). The Caledonian Deformation Front (CDF) separates the Midland-Ardenne Microcraton from foreland strata forming the southern boundary of a compressional wedge.

from *c.* 7.0 to 6.5 km/s at the Moho has been identified. This was also considered to represent the location of the suture.

Similar interpretations have been made from the BABEL data (BABEL Working Group 1993) where the crust beneath the Møn High (the eastern extension of the Ringkøbing-Fyn High) is modelled as having a thickness of 38 km and velocities of >7.0 km/s. This is in marked contrast to the values obtained to the south of the high, where the crustal thickness is *c.* 30 km while the velocities are <6.9 km/s. This change in velocity structure and crustal thickness has been found beneath and south of the inferred near-surface location of the Avalonia–Baltica suture.

Meissner *et al.* (1994) reviewed the available near-normal-incidence reflection data across the eastern part of Avalonia and concluded that the edge of this microcontinent could be defined on the basis of reflectivity patterns. The edge of Avalonia is thus defined by reflectivity broadly coincident with the changes in velocity structure recognized from wide-angle profiling.

The BABEL near-normal-incidence data (BABEL Working Group 1993) provides an image of the suture closest to the MONA LISA profiles. North of the Møn High, a group of weak, SW-dipping reflectors between 4 and 6 s TWT were interpreted as representing the Caledonian Deformation Front. NE of this feature, gently NE-dipping reflectors occur in the upper mantle at between 10 and 12.5 s TWT. This reflectivity pattern is interpreted as the leading edge of Baltica indenting the lower crust of Avalonia. Part of Avalonia was thrust over the southern edge of Baltica, and the lower Avalonia crust and upper mantle were thrust beneath Baltica (MONA LISA Working Group 1997*a*, *b*).

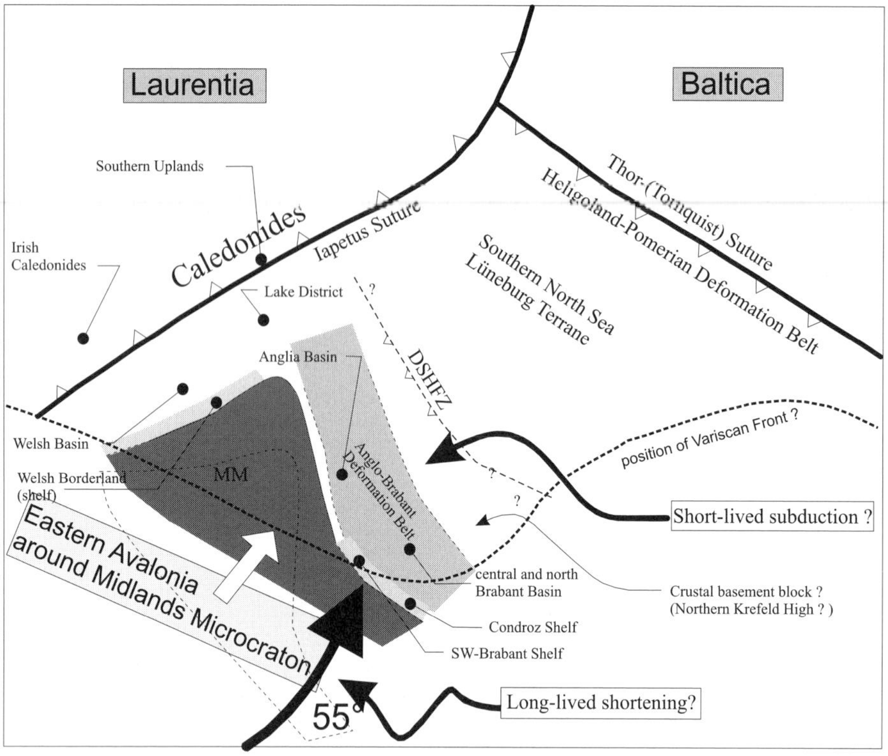

Fig. 7.48. Sketch of the Palaeozoic deformation of the Brabant Massif as part of Avalonia (after Debacker 2001). DSHFZ, Dowsing-South Hewett Fault Zone; MM, Midlands Microcraton.

The presence of an ocean (i.e. Tornquist Sea) between Baltica and Gondwana was first proposed on the basis of palaeobiogeographic studies (Cocks & Fortey 1982), and later supported by palaeomagnetic evidence that showed that the part of Avalonia beneath southern Britain has a different apparent polar wander path from that of Baltica between the Early and Late Ordovician (Torsvik *et al.* 1992; Fig. 7.53). Dating of core material from exploration wells in the North Sea, Denmark and northern Germany has revealed the presence of low-grade (greenschist facies) metamorphism that occurred at *c.* 440 Ma (Frost *et al.* 1981) and is interpreted as resulting from Caledonian deformation.

As noted above, two possible boundaries have been suggested for the NE margin of Avalonia, the Caledonian Deformation Front (CDF) and the Elbe Line (Tanner & Meissner 1996). The CDF forms the northern limit of the Caledonian margin thrust belt which developed following the oblique collision between Avalonia and Baltica, and comprises a series of south/SW-dipping seismic reflectors which are interpreted as having been formed by the thrusting of Caledonian metamorphosed shelf slope sediments over the Precambrian basement of Baltica (Franke 1990*a*; Thomas *et al.* 1993; Krawczyk *et al.* 2002). North of the CDF, the undeformed Lower Palaeozoic succession is found in downfaulted and slightly SW-dipping blocks (Thomas *et al.* 1993).

Magnetotelluric data in the NE German Basin between the island of Rügen and the river Elbe shows good conductors in the middle Cambrian–Lower Ordovician Scandinavian alum shales, which extend between the trace of the CDF at surface and from there further inland (Fig. 7.54; Hoffmann & Franke 1997; Hoffmann *et al.* 1998, 2001). This good conductor lies at 5 km depth below the Baltic Sea area and deepens towards the SW down to *c.* 12 km depth. Thus, it may mark the Caledonian overthrust zone on top of Baltica-derived basement. Alternative interpretations also exist: Hoffmann & Stiewe (1994) argued for Late Carboniferous mafic intrusions at the southern end of Baltica where it meets Avalonia, while Krauss (1994) suggested Caledonian oceanic or island arc ophiolites. This would imply an early Caledonian southward subduction of oceanic crust of the Tornquist Sea below Avalonian continental crust. According to the different findings of geophysical measurements, Hoffmann & Franke (1997) suggest a bipolar character of the Tornquist closure with southward subduction of Baltica during Ordovician, and northward subduction during Silurian and early Devonian.

The suture forming the SW margin of Baltica is mostly conflated within the later Sorgenfrei-Tornquist Zone which runs from the Laurentia–Avalonia–Baltica triple-junction, now hidden beneath the North Sea, and which merges with the northern Carpathian margin at the western Black Sea (Cocks 2000). The precise location of the SW edge of Baltica (i.e. that part of Baltica to the south of the Sorgenfrei-Tornquist Zone) is not known. This is largely as a result of masking by younger sediments (Tanner & Meissner 1996). Many workers have identified the strike-slip Trans-European Fault, which runs through parts of northern Germany and south of Rügen to join up with the Teisseyre-Tornquist Zone in Poland, as the southern boundary of Baltica (Berthelsen 1992; EUGENO-S Working Group 1988; Franke 1990*a*, 1993; Hoffmann 1991). More recent work, however, has questioned the existence and supposed importance of this structure (McCann & Krawczyk 2001).

A terrane accretionary model, rather than intraterrane fault movement, is now the favoured interpretation for the observed

Fig. 7.49. Heat-flow map of Central Europe (simplified and summed contours from map of Hurtig 1995).

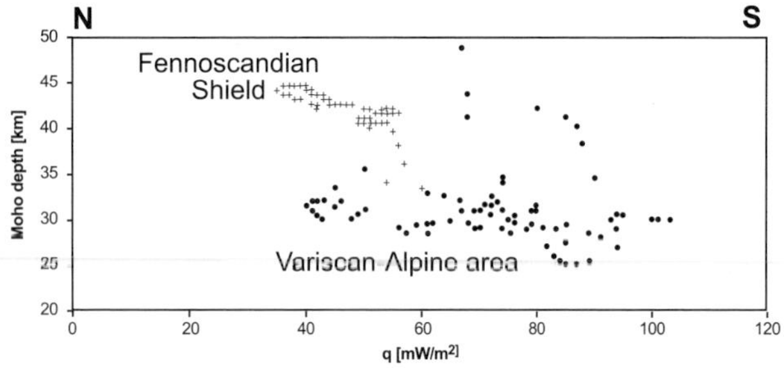

Fig. 7.50. Heat-flow density measured along a profile between Kola peninsula and Tunisia (after Hurtig 1995). Heat flow increases with decreasing crustal thickness for the Fennoscandian Shield, while there is almost no trend in the Variscan-Alpine influenced parts.

geology of the former Baltica–Avalonia accretion zone (Ziegler 1990; Berthelsen 1992; Torsvik *et al.* 1993; Meissner *et al.* 1994; Tanner & Meissner 1996; Krawczyk *et al.* 2002). In such a model, the northward convergence and accretion of Gondwana-derived continental fragments to the southern margin of the newly forming Laurussian supercontinent is envisaged (Ziegler 1989). These fragments would have included Avalonia, which drifted northwards and docked with Baltica by Ordovician–Silurian times (Cocks & Fortey 1990; Thomas *et al.* 1993; Torsvik *et al.* 1992).

Iapetus Suture

The suture between the former Laurentia and the eastern part of Avalonia can justifiably be regarded as the most important Caledonian structural feature of Britain and Ireland (Woodcock & Strachan 2000*b*). The suture runs from SW Ireland, in a northwesterly direction NW of the Isle of Man and into the Solway Firth along the England–Scotland border (Fig. 7.12). Despite this, the Iapetus Suture is the least well exposed of the region's Palaeozoic terrane boundaries and along much of its length is obscured by post-Caledonian strata.

The Iapetus Suture zone represents the former site of the Iapetus Ocean and separates Laurentia or Laurentia-derived crust to the north from Gondwanan terranes to the south (Fig. 7.1). Along the Scottish border the suture is constrained by faunal contrasts in Ordovician rocks to lie between the Lake District and the northern belt of the Southern Uplands. The accretionary model for the Silurian rocks of the Southern Uplands implies that they were scraped onto the Laurentian margin before collision and that the suture lies to the SE of them (Woodcock & Strachan 2002*a*). Within the Iapetus Suture zone in Canada, Britain and Ireland, there are also the accreted relicts of many smaller terranes derived from mid-Iapetus island arcs; these were outlined and reviewed for the European sector by Armstrong & Owen (2001). Deep seismic surveys in the Irish Sea and North Sea are an additional guide to the position of the suture. These show a prominent reflector separating southern crust containing reflecting surfaces from less reflective crust to the north. Contours on this reflector project up to the surface at about the so-called 'Solway Line'. The seismic data indicate that the Iapetus Suture is a northward-dipping plane, and that the crust of Eastern Avalonia has underthrust the Laurentian margin (Woodcock & Strachan 2002*a*).

The Iapetus Suture has been imaged by several seismic reflection and refraction surveys around the UK (e.g. Freeman *et*

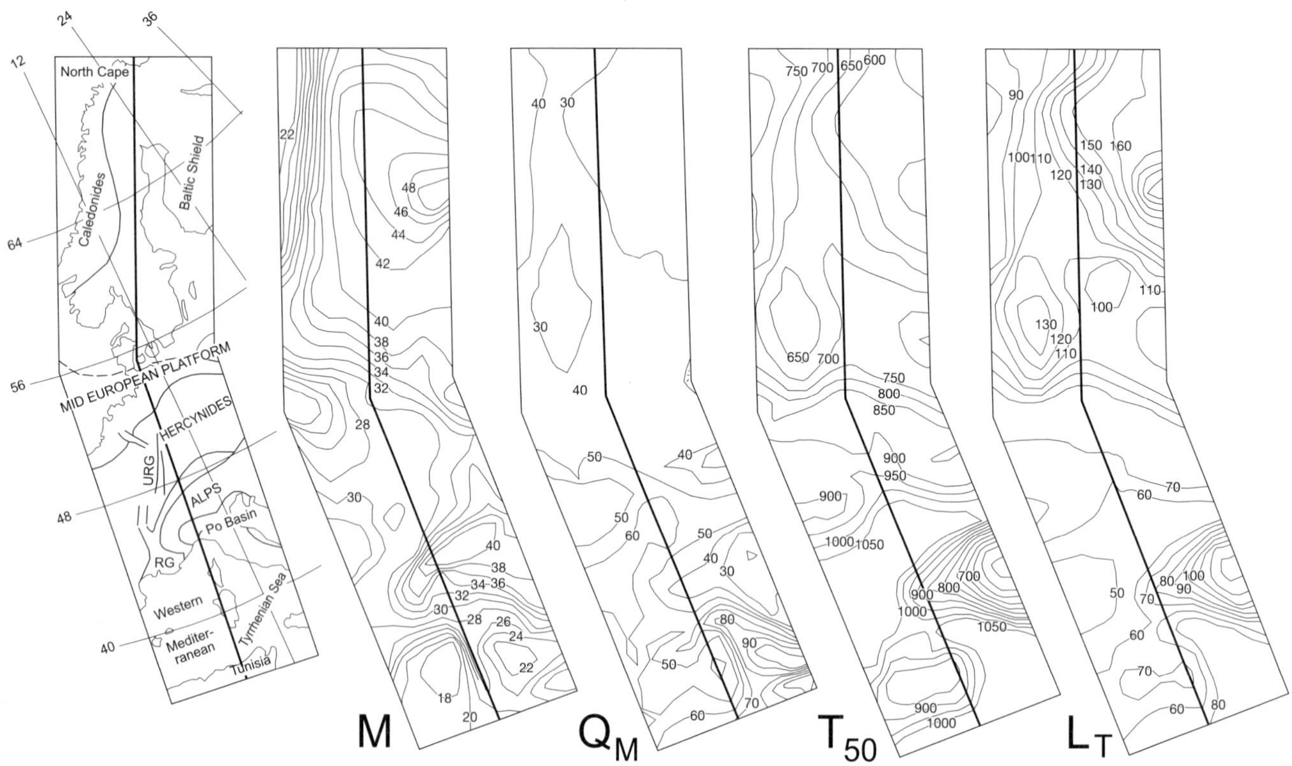

Fig. 7.51. Results from 3D temperature modelling along the European Geotraverse (after Cermak & Bodri 1995). From left: location of the EGT line and tectonic units; crustal thickness (km); calculated Moho heat flow (mW/m²); calculated temperature distribution at 50 km depth (°C); calculated thermal thickness of the lithosphere (km).

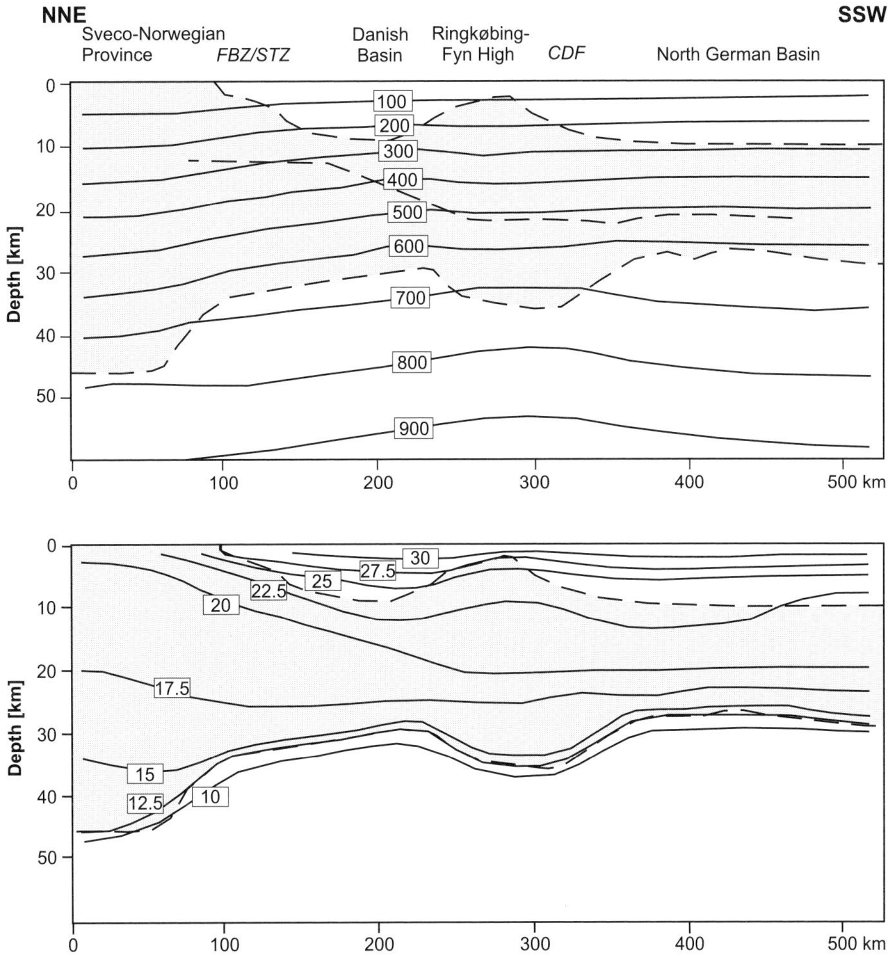

Fig. 7.52. Cross-section of the thermal structure along profile EUGENO-S 1 (see Fig. 7.26). Observed heat flow isotherms (top; in °C) and temperature gradients (bottom; in °C/km) are given (after Balling 1995). Dashed lines are seismic boundaries. CDF, Caledonian Deformation Front; FBZ/STZ, Fennoscandian Border Zone/Sorgenfrei-Tornquist Zone.

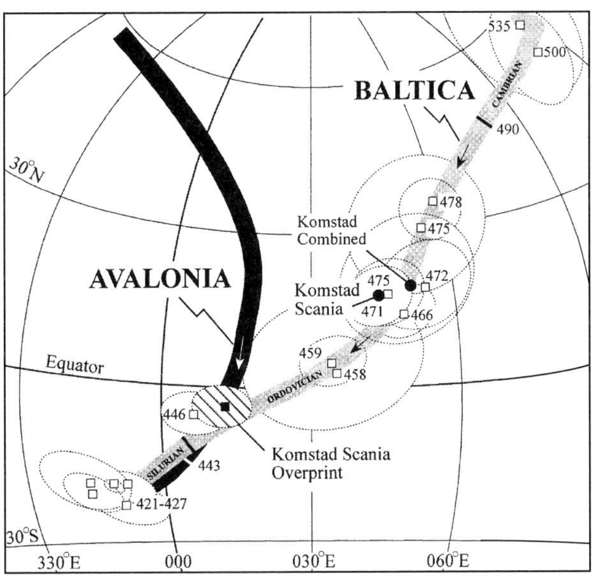

Fig. 7.53. Cambrian to Mid-Silurian palaeomagnetic poles from Baltica and apparent polar wander path for Avalonia (numbers indicate million years; ellipses mark 95% confidence; after Torsvik & Rehnström 2003).

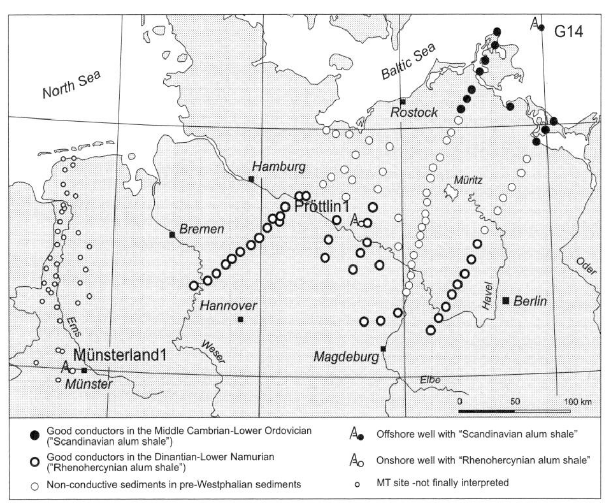

Fig. 7.54. Result of magnetotelluric surveys at the northern rim of the North German Basin (after Hoffmann *et al.* 2001).

al. 1988; Soper *et al.* 1992*a*; Landes *et al.* 2000, 2005). Geological studies suggest that before the ocean closed in the Silurian, subduction had occurred both to the south and the north of the ocean (van Staal *et al.* 1998). The northern margin of Iapetus is imaged, from east to west on the NEC, WINCH and WIRE1B profiles (Fig. 7.26). On all of these there is evidence for a boundary within the middle and lower crust dipping at about 30° to the north which projects upward to the subcrop of the suture at the surface, as inferred from surface geology. On the WIRE1B profile there is a discrete reflector apparently overthrust by Variscan-age structures (Klemperer *et al.* 1991). Beneath the North Channel and offshore NE England, where the suture is also imaged on the MOBIL profiles, there is a north-dipping boundary separating unreflective middle crust to the south from reflective middle crust to the north, which Soper *et al.* (1992*b*) interpreted as the suture, although Freeman *et al.* (1988) interpreted the same structure as the base of a remnant slab of oceanic crust. Immediately to the north of the suture the lower crust reflectivity is dominated by two parallel reflectors immediately above the Moho. These have been interpreted as remnant Iapetus oceanic crust (Freeman *et al.* 1988) or as resulting from magmatic underplating (Smith 1992). The SWABS profile (Singh *et al.* 1998) indicates that the base of the crust was formed from layered high-velocity rocks, which would appear to confirm the underplating hypothesis. South of the Iapetus Suture, Soper *et al.* (1992*b*) interpreted two north-dipping reflectors near the base of the crust as thrusts resulting from the deformation of the Lake District/Eastern England terranes during the northward movement of the Midland Microcraton. Further evidence for this so-called Acadian deformation event is found in the form of similar north-dipping reflectors in the middle and lower crust in commercial deep reflection profiles in the east Irish Sea (England & Soper 1997; Woodcock 1990*a*).

The Iapetus Suture cannot be traced to the west of Ireland owing to a lack of suitably orientated profiles in this region. It can, however, be traced into the North Sea on the MOBIL and NSDP85–7 profiles to the point where signal penetration to the deep crust diminishes with the increase in thickness of the Cretaceous and Tertiary sediments of the Central Graben (Husebye *et al.* 1988). The profiles indicate that the strike of the north-dipping boundary and the two Moho-parallel reflectors swing from a NE–SW orientation across the UK and Ireland into a more east–west orientation. This observation contradicts gravity data which show a lineament in the same position and orientation as the suture onshore and which crosses the northern part of the Central Graben and swings toward Norway. This change in strike mirrors the change in geometry of the Great Glen Fault and may be an indication of the original shape of the Laurentian margin and the geometry of the main suture between Laurentia and Baltica. Since it is probable that the main suture between Avalonia and Baltica was the last to form (Torsvik 1998; Soper *et al.* 1992*b*; MONA LISA Working Group 1997*a*; Winchester *et al.* 2002*a*), with a significant component of strike-slip deformation, the east–west trend of the Iapetus Suture in the deep seismic data may represent this suture, and links up with the Caledonian Deformation Front (Thor Suture) beneath Denmark, the North German Caledonides and the Baltic region.

Baltica-Laurentia structures. Approximately 10 000 km of deep (16 s TWT) seismic reflection data were gathered on a series of cruises in the Norwegian offshore area in 1987 as part of the MOBIL Search project. Results included the production of a revised Moho map for the region of south Scandinavia (Kinck *et al.* 1991) which demonstrated that the thinnest crust was to be found in the North Sea and Skagerrak areas (*c.* 20 km) whilst east-central Sweden is underlain by very thick crust (up to *c.* 50 km).

The crust beneath the Caledonides of western Norway is relatively thick (*c.* 40 km) and this is similar to the Caledonide crust on the British side of the Viking/Central Graben. Analysis of two geological–geophysical transects across the Norwegian Caledonides, the West Coast Transect (WCT) and the Central Caledonides Transect (CCT) (Fig. 7.26), together with pressure–temperature–time data, form the basis for the reconstruction of the kinematic evolution of the Scandinavian Caledonides collisional orogenic belt (Hurich 1996). The foreland and hinterland of the Caledonian orogen show distinctly different geophysical signatures. In the WCT, the upper and middle crust of the foreland is characterized by several discrete bands of north-dipping reflectors within a background of weak, sparse sub-horizontal events. The lower crust shows a zone of strong, spatially dense reflectors that abruptly terminate at the Moho (Deemer & Hurich 1991). In contrast, the hinterland is characterized by numerous upper and middle crustal reflections, a zone of sparse reflectivity in the lower crust and a distinct band of Moho reflectors (Hurich 1996). In the CCT the foreland is characterized by strong continuous zones of reflectivity in the upper and middle crust, sparse, weak lower crustal reflectivity and the lack of a distinct reflection Moho (Hurich 1996). In the hinterland, the Moho is also indistinct, but the upper and middle crustal reflectivity is strong, albeit discontinuous.

Reconstruction of the geodynamic evolution of the area suggests that the early stages of the collision between Baltica and Laurentia involved intracontinental subduction of a minimum of 150 km of Baltic continental crust to a depth of at least 100 km. The subduction juxtaposed strong, Laurentian upper mantle rocks in the upper plate against weak, Baltic rocks in the lower plate. The growth of high-pressure mineral assemblages resulted in partial to complete eclogitization and the growth of coesite in the subducted crust. Strain localization in eclogitic shear zones initiated large-scale thrust faults and a décollement, resulting in delamination. After decoupling, the partially eclogitized part of the subducted crust shortened and rose due to continued convergence between Laurentia and Baltica and the buoyant forces resulting from the density contrast between the crustal material and the upper mantle. This unroofing and shortening resulted in the development of a crustal-scale duplex and an increase in the thickness of the crystalline crust to *c.* 60 km (Hurich 1996). Thrust-related thickening was accompanied by the extrusion of crustal rocks from the subduction zone. Thrusting also resulted in the incorporation of fragments of the Laurentian upper mantle into the Baltic crust.

Rheic Suture

The Rheic Suture separates Avalonia or Avalonian-derived crust to the north from Armorican terranes (Fig. 7.4). To the south of Avalonia, the Rheic Suture marks the closure of the Rheic Ocean in early Devonian times. The BIRPS offshore profiles, SWAT 6/7, 8 and 9 image south-dipping sets of mid-crustal reflectors correlated with the Lizard Boundary Thrust underlying the Lizard Ophiolite (Leveridge *et al.* 1984), but they could also be related to Variscan deformation. In Germany, the DEKORP-2 profile images the suture as a series of SE-dipping reflectors in the mid-crust (Meissner & Bortfeld 1990).

With the exception of WINCH 4, which shows very little deep structure, there is a gap in deep reflection profiling to the south of the Iapetus Suture (Fig. 7.26). Barton (1992) describes a southward decrease in seismic P-wave velocity across the suture from the LISPB data, which probably reflects the change from Laurentia- to Gondwana-derived continental lithosphere. In the North Sea, the MOBIL4 and MOBIL5 profiles are rather short but image reflective lower crust at their SW ends. The SNST83 profile runs diagonally across the southern North Sea from offshore East Anglia. At its SW end this profile exhibits a thick layer of lower crustal reflectivity between 7 and 12 s TWT and from the base of this layer a package of southerly inclined reflectivity extends into the upper mantle. A similar pattern of reflectivity is observed on the adjacent MOBIL6 and MOBIL7 profiles (England 1995). The Dutch profile (Duin *et al.* 1995) shot parallel to the Belgian and Dutch coast along strike from the lower crustal reflectivity observed on the SNST line does not show this reflectivity, although the data are poor owing to the line having been shot in poor weather (Remmelts & Duin 1990). Immediately south of the mantle reflector on MOBIL6, a series of enigmatic (?)mid-crustal thrust structures is observed (Blundell *et al.* 1991) which cross onto the northern edge of the Anglo-Brabant Massif. England (1995) tentatively interpreted these structures as the remnants of the south-directed Ordovician subduction zone which was the source of the calc-alkaline intrusive and volcanic rocks of the concealed Caledonides of eastern England and the Brabant Massif in Belgium. If this interpretation is correct it raises a question concerning the origin of the basement of the southern North Sea, which is interpreted from the dating of borehole samples of the basement (see above) as part of Avalonia (Frost *et al.* 1981). This problem has been partly avoided through the introduction of the term Far Eastern Avalonia to describe the crustal fragment of Avalonia that collided with Baltica along the Tornquist Suture (Winchester *et al.* 2002*a*).

Other important tectonic structures

Possible terrane and subterrane boundaries

There are a number of possible terrane and subterrane boundaries across the region of Central Europe associated with the accretion processes of the various terranes. It is, however, not always easy to obtain evidence for these structures which clearly shows their defining character, and therefore much of the evidence for these structures is circumstantial. Thus, while important, these structures cannot be definitively recognized as terrane boundaries.

The most westerly of these possible structures is the Dowsing-South Hewett Fault Zone which extends from northern Britain into the Netherlands (Verniers *et al.* 2008). This fault zone controlled the margin of the Anglo-Dutch Basin during the development of the Carboniferous and younger basins and their subsequent inversion. Seismic profiles demonstrate contrasting reflectivity patterns on either side of the fault zone. This, together with the reactivation history of the zone, led Pharaoh *et al.* (1993, 1995) to suggest that the zone is a possible Ordovician terrane boundary.

Various lines of geophysical and geological evidence suggest that the Elbe Line is a significant crustal boundary (see above). The Elbe Line separates regions with distinct basement tectonic grains, and crustal velocities are seen to vary across it (e.g. Betz *et al.* 1987; Rabbel *et al.* 1995). These geophysical contrasts led Tanner & Meissner (1996) to favour the Elbe Line as the true NE boundary of Avalonia. However, as noted above, the lineament more likely represents an intra-Avalonian, subterrane boundary, or a more significant terrane boundary between the southern North Sea-Lüneberg Terrane and East Elbian Suspect Terranes proposed by D. Franke (1995).

The Krakow-Lubliniec Zone, the SE continuation of the Elbe Line, is a narrow belt of intense ductile–brittle deformation and magmatic intrusion separating crustal blocks with contrasting early Palaeozoic histories (Bula *et al.* 1997) and may therefore represent a terrane boundary (Dadlez *et al.* 1994). It separates the Malopolska Massif (with Baltican affinity) from the Moravo-Silesian Massif (with suspected Gondwanan affinity).

The seismic structure of the crust between the TTZ and the Krakow-Lubliniec Zone generally resembles that known from northern Germany and southern Denmark (Aichroth *et al.* 1992; Rabbel *et al.* 1995; Abramovitz *et al.* 1998) and is typical of the basement domain that extends between the CDF in the NE and the Elbe Line to the SW. Like the basement of central Poland, that domain shows low P-wave velocities down to a depth of *c.* 20 km and relatively high velocities below. Furthermore, its lower crust is characterized by a high vertical velocity gradient and strong reflectivity.

Major crustal tectonic zones that are not crustal boundaries

The Sorgenfrei-Tornquist Zone (STZ) lies at the southern limit of the exposed Fennoscandian Shield (Fig. 7.3). It is an important intraplate tectonic zone underlying a complex graben system of Carboniferous–Mesozoic age (Thybo 1997, 2001). The STZ is offset by the Rønne Graben and continues into northern Poland as the Tornquist-Teisseyre Zone (TTZ) where it is associated with a steep crustal discontinuity (Guterch *et al.* 1986). In both locations these structures are related to steps in the depth of the Moho. The nature of the tectonic linkage between the shallow and deep crustal expression of the TTZ is a critical question which was investigated by the POLONAISE project. Initial results from this project (Guterch *et al.* 1999) note the control exercised by this structure on the location and evolution of the Mid-Polish Trough, particularly in terms of its asymmetry.

The TTZ also lies to the north of the inferred location of the Caledonian Deformation Front (Thor Suture) at the Krakow-Lubliniec Zone. Berthelsen (1998) has suggested that the STZ and TTZ are the crustal expression of a 'boundary detachment' resulting from the contrasting mechanical response of the lithospheres of the EEC and 'Mobile Europe' to Variscan orogenic compression, isolating the crust to the north from post-Variscan extension and inversion.

The Elbe Fault Zone of the Bohemian Massif, frequently confused in the literature with the Elbe Line (Scheck *et al.* 2002), is a relatively insignificant late transverse Variscan structure and is not a terrane boundary. The Intra-Sudetic Fault was inferred to be the possible location of the Tornquist Suture by Johnston *et al.* (1994) but more recent studies confirm the correlation of Sudetic terranes across it and indicate that it is an entirely superficial late Variscan structure.

Trans-European Suture Zone. Seismic traverses (including normal incidence and wide-angle reflection seismics) in the North Sea (BIRPS, MONA LISA), Germany (DEKORP BASIN'96) and Poland (POLONAISE) all show that a high density basement, linked with that of the EEC, extends with progressive attenuation some 150 km to the SW of the TTZ (DEKORP-BASIN Research Group 1998; Guterch *et al.* 1999). This would suggest that the TESZ is not a steep feature, as originally inferred, but dips to the SW at a shallow angle, perhaps no more than 20°, so that Baltica crust extends as far south as the Elbe Line in Germany (see above).

Similarly, in Poland, while the sub-Permian location of the Baltica margin runs approximately parallel to the TTZ, seismic lines from the POLONAISE'97 profiles (Guterch *et al.* 1999) suggest that its attenuated edge at the base of the crust aligns with the Krakow-Lubliniec Zone, the SE continuation of the Elbe Line (= Kracow Fault) (Winchester *et al.* 2002*b*). This coincidence may have governed the location of, and later movements along, the Krakow-Lubliniec Zone. Further to the NW, the continuation of the TESZ bends westwards, passes south of Denmark and traversing the SE North Sea curves to the NW to meet the Iapetus Suture at a triple point junction 300 km east of the Scottish coast (Pharaoh 1999).

The TESZ area in northern Poland has been investigated extensively over the last decade by a number of seismic refraction profiles, including, LT-7, TTZ and P1–P5 within the POLONAISE'97 experiment (Guterch *et al.* 1994, 1999; Grad *et al.* 1999; Jensen *et al.* 1999, 2002; Sroda & POLONAISE Working Group 1999; Janik *et al.* 2002). Information from the LT-7 profile indicated that the crustal thickness in the area of the TESZ is intermediate between that of the EEC to the east (*c.* 42 km) and that of the area to the SW of the Variscan Front (*c.* 30 km) (Guterch *et al.* 1994).

The STZ is the NE border of the Tornquist Fan, which broadly divides Central Europe (Phanerozoic) from Fennoscandia (Proterozoic), whereas Central Europe and eastern Europe are separated by the TTZ. These two branches of the Tornquist Zone join between Germany and Sweden. Cotte *et al.* (2002) have shown that the STZ is a major tectonic feature within the entire lithosphere. The transition across the STZ is sharp and determined to be very steeply dipping to the SW.

The so-called Tornquist Fan has an ambiguous origin: on the one hand, the STZ marks a sharp transition in the geophysical properties of the crust from the Baltic Shield (35–45 km thick) to the Tornquist Fan (*c.* 32 km thick) (e.g. Tryggvason *et al.* 1998). On the other hand, the basement of the Tornquist Fan is seismically similar to that of the Baltic Shield. Additionally, the northern rim of the Tornquist Zone shows a marked difference in terms of crustal velocities (Gregersen *et al.* 2002) and there are generally large lateral differences of 4–6% in P-wave velocity. Compounding the ambiguity is the fact that lateral variations in the structure of the lithosphere are not as well constrained as crustal ones, since no dense passive seismic experiment has, thus far, been carried out in the area. Crustal models have been determined by reflection and refraction seismic studies such as FENNOLORA (Guggisberg *et al.* 1991), EUGENO-S (EUGENO-S Working Group 1988) and BABEL (e.g. BABEL Working Group 1991, 1993).

It is now well known that pronounced lateral variations exist in the lithosphere between the 'old' eastern Europe and the 'young' Central Europe. Several surface wave studies have focused on this structure. Snieder (1988) determined a very sharp transition in lithospheric structure from one side of the TTZ to the other at depths of between 100 and 200 km, while Zielhuis & Nolet (1994) determined the seismic structure down to 140 km, below which the contrast from one side to the other is smaller. Subsequently, Marquering & Snieder (1996) noted the existence of lateral variations in lithospheric structure in the vicinity of the TTZ at depths of 80 and 140 km, representing a sharp transition. North of the TTZ several studies have provided information about the thickness of the Baltic Shield and demonstrated how it varies from one side of the STZ to the other (e.g. Husebye & Hovland 1982; Sacks *et al.* 1979).

Fennoscandian Border Zone. The Fennoscandian Border Zone is a complex tectonic zone defining the border between the stable Fennoscandian-Baltic Shield and the metastable Danish Basin and Polish Trough. It forms the NW part of the Tornquist-Teisseyre Zone, one of the major European tectonic sutures (Liboriussen *et al.* 1987). Little is known about the Early Palaeozoic tectonic evolution of the Fennoscandian Border Zone mainly due to tectonic overprinting during the Late Palaeozoic and Mesozoic. However, it has been argued that this structure has existed since Precambrian times based on the presence of Precambrian-age sutures in the southern part of the Fennoscandian Shield bordering the Fennoscandian Border Zone (Strömberg 1976). In Denmark, Lower Palaeozoic strata are partly preserved beneath the Danish Basin and to a greater degree in half-grabens associated with the Fennoscandian Border Zone, while in Poland strata of similar age also increase in thickness towards the SE prolongation of the Fennoscandian Border Zone (Liboriussen *et al.* 1987). It is postulated that its evolution was dominated during the Early Palaeozoic by tensional tectonics and during the Late Palaeozoic by dextral wrench movements. The latter led to the development of curvilinear faults and the subsidence of pull-apart basins.

Caledonian geodynamics

The geodynamic evolution of the three main continental fragments within the Caledonian Orogeny are summarized and discussed below.

Late Neoproterozoic (<750 Ma) to Early Ordovician Laurentian margin of Iapetus

The Late Neoproterozoic was a period of major tectonic, climatic, biogeochemical and biological change. The collision of Laurentia, Baltica and Amazonia (*c.* 1.1–1.0 Ga) resulted in the formation of the supercontinent Rodinia. The notion of a late Proterozoic supercontinent was initially postulated in the 1970s when geologists noted the existence of a number of 1100–1000 Ma 'mobile belts' (Dewey & Burke 1973). The interval from *c.* 1100 to 900 Ma marked a geologically important period during the formation of the Neoproterozoic supercontinent (Meert & Torsvik 2003). The Grenvillian, Sveconorwegian and slightly older Kibaran orogenic belts are thought to mark the sutures between the various elements of the Rodinian supercontinent (see Dalziel 1997; McCann 2008; Meert & Powell 2001; Powell *et al.* 2001). The lack of palaeomagnetic data for this time period (Meert & Powell 2001) means that any reconstruction must be extremely fluid. Furthermore, any reconstruction must also fit the palaeogeographic constraints imposed by younger (pre-breakup) palaeomagnetic data and must also be geologically reasonable.

The Late Proterozoic breakup of Rodinia resulted from the separation of East Antarctica from western Laurentia and was followed by the amalgamation of the west and east Gondwana blocks to form the Vendian supercontinent (Strachan & Holdsworth 2002). The Late Neoproterozoic breakup of this supercontinent resulted in the separation of Laurentia, Baltica and Amazonia to form the Iapetus Ocean (Fig. 7.55).

Most recent palaeogeographic reconstructions assume that the proto-Andean margin of Amazonia formed the conjugate margin to Laurentia during Late Proterozoic times (*c.* 750 Ma), a period when Laurentia formed the core of the Rodinia supercontinent (e.g. Hoffmann 1991). The breakup of Rodinia supposedly took place in several stages (Dalziel 1992) which is consistent with the longevity of rift-related magmatism and the formation of several large Neoproterozoic rift basins in Laurentia (e.g. Moine & Dalradian, Scotland; van Staal *et al.* 1998). The final breakup, including the separation of Laurentia from Amazonia, occurred at some stage between 590 and 550 Ma. This separation, together with the break-off of Baltica, led to the opening of the Iapetus Ocean (van Staal *et al.* 1998; Fig. 7.55; see also McCann 2008).

Traditional Rodinia models argue that breakup of the supercontinent commenced with the 800–700 Ma opening of the Palaeo-Pacific Ocean between Laurentia and Australia-Antarctica (Fig. 7.56). This model has recently been challenged by a number of authors who argued that the East Gondwana landmass was not a coherent block in mid-Neoproterozoic time (e.g. Meert 2001, 2003; Boger *et al.* 2002; Foden et al. 2001; Wingate & Giddings 2000). Hartz & Torsvik (2002) suggest that the 'traditional' fits of Baltica and Laurentia should be abandoned since they are largely based on incorrect geological links between a small amount of 'Grenvillian' crystalline rocks in southern Baltica and Grenvillian belts in Laurentia. These include the linking of the 1.2–0.9 Ga Sveconorwegian (Scandinavia) and Grenville (eastern North America) orogenies which Johansson *et al.* (2001) have noted is characterized instead by large differences in the structural histories and diachroneity of events between rocks in the two areas. Furthermore, the Grenvillian events in east Greenland are defined by 1.0–0.93 Ga allochthonous granites (Watt & Thrane 2001) and no documented autochthonous Grenvillian–Sveconorwegian deformation exists along the northern margins of east Greenland and west Norway. The late Precambrian basin evolution in these two regions also varies considerably with respect to tectonic setting, depositional age and environment, basin fill thickness, lithologies, and igneous activity (Kumpulainen & Nystuen 1985) and the overall conclusion must be that the Precambrian geology of east Greenland and that of Scandinavia are not as compatible as implied in traditional late Precambrian reconstructions (Hartz & Torsvik 2002). Another problem is the apparent dischroneity and different styles of rifting within the Iapetus Ocean. Southern Iapetus, along eastern North America, probably opened from south to north during the latest Ediacaran (570–550 Ma) (Cawood *et al.* 1996). However, palaeomagnetic, palaeontological and geological evidence suggests that by 580 Ma, Baltica was inverted in the southern hemisphere, implying that the northern Iapetus Ocean opened earlier (*c.* 600 Ma) than the southern Iapetus (Meert *et al.* 1998; Torsvik & Rehnström 2001; Torsvik *et al.* 1996) (see below).

According to Hartz & Torsvik (2002), Baltica was geographically inverted throughout the Neoproterozoic (Fig. 7.56). McCann (2008) provides an extensive and critical discussion of this model. The breakup that led to the formation of the Iapetus Ocean was initiated at a junction between a rift (Laurentia–Gondwana), a right-lateral fault (Laurentia–Baltica) and a trench (inverted Baltica–Gondwana), thereby linking the late Precambrian Iapetus opening to the Timanian and Avalonian orogenies between Gondwana and Baltica.

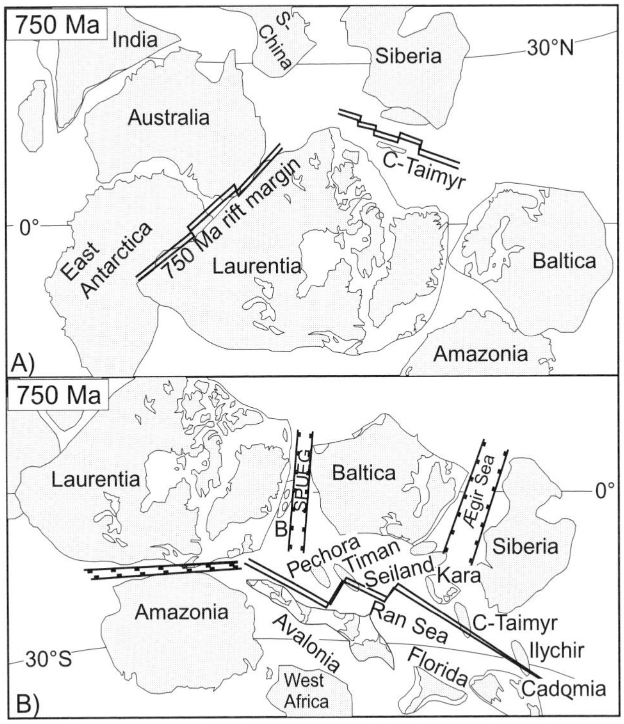

Fig. 7.56. Alternative reconstructions of Rodinia at *c.* 750 Ma (after Hartz & Torsvik 2002). (**A**) Traditional reconstruction; (**B**) revised reconstruction with southern peri-Urals–East Greenland (SPUEG) model (B, Barentia).

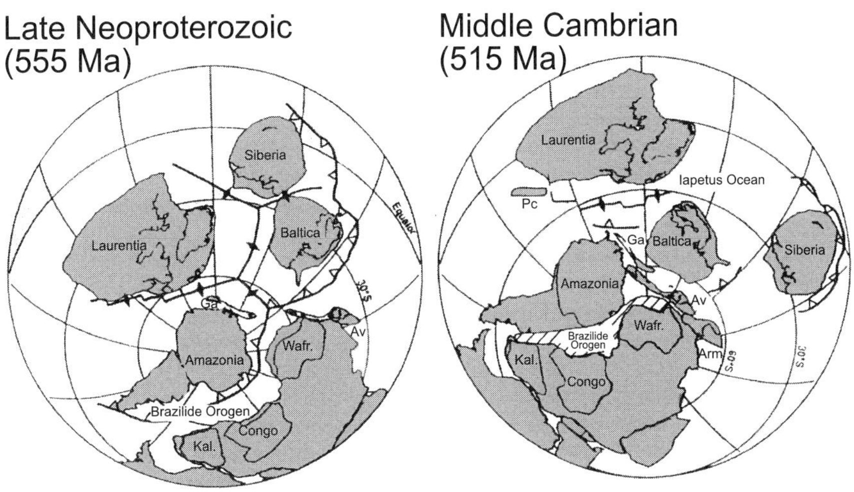

Fig. 7.55. Reconstructions of the separation of Laurentia and Baltica at distinct times with the opening of the Iapetus Ocean (after van Staal *et al.* 1998). Abbreviations: Arm, Armorica; Av, Avalonia; Ga, Ganderia; Kal, Kalahari craton; Pc, Perunica; Wafr, West Africa.

In contrast to the Neoproterozoic–Ediacaran geological non-correlation between western Baltica (Scandinavia) and east Greenland, there are remarkable correlations between east Greenland and Russian Baltica (southern peri-Urals). Late Mesoproterozoic to earliest Neoproterozoic metasedimentary rocks are intruded by early Neoproterozoic granites ranging in age from *c.* 1.0 to 0.93 Ga in both east Greenland (Watt & Thrane 2001) and the southern peri-Urals (Khain 1985; Maslov *et al.* 1997). In both areas, the intrusions are considered anorogenic and subsequent early Neoproterozoic basins are typical marine rifts, filled with mudstones, sandstones and limestones (Henriksen & Higgins 1976; Khain 1985).

The notion that Baltica and Laurentia were drifting independently is supported by the conclusions of Elming & Mattsson (2001) who suggested that rifting between the two continents began at *c.* 1.27 Ga. It should be noted that the southern peri-Urals–East Greenland fit between Laurentia and Baltica can be reconciled with palaeomagnetic data from the 1100 to 1000 Ma interval (Hartz & Torsvik 2002) but it is incompatible with *c.* 930–850 Ma poles (Meert & Torsvik 2003; Fig. 7.57). Another alternative argued by Hartz & Torsvik (2002) is that Siberia was not located near Laurentia at 750 Ma but was situated close to Baltica and separated from it by the Aegir Sea. Palaeoreconstructions (Fig. 7.57) based on the available poles show that Baltica may have had a close association with Siberia. Indeed, the reconstruction at 975 Ma (Fig. 7.57) is similar to that advocated by Hartz & Torsvik (2002).

The final breakup of Rodinia occurred between *c.* 590 and 550 Ma (Fig. 7.58) and supposedly took place in several stages (Dalziel 1992), an idea which is consistent with the longevity of rift-related magmatism and the formation of several large Neoproterozoic rift basins in Laurentia (van Staal *et al.* 1998). The timing of the final breakup is broadly synchronous with the assembly of Gondwana (e.g. Meert 2003; Powell & Pisarevsky 2002; Boger *et al.* 2002). Meert (2001) suggested, on the basis of palaeomagnetic data, that final Gondwana assembly did not occur until sometime in the 550–530 Ma interval. This is consistent with a number of new geological findings dividing east Gondwana into several different blocks (Fitzsimons 2000; Boger *et al.* 2002; Meert 2003). Figure 7.59 is adopted from Meert & Torsvik (2003) and shows the position of the major continental blocks at 580 Ma. Hartz & Torsvik (2002) suggest that there was a second rift event, marked by Vendian-age dykes younging northwards along Laurentia's eastern margin. This has been interpreted as being related to the northward opening of the Iapetus Ocean by 570 Ma with rifting continuing until 550 Ma. The opening of the Iapetus Ocean and the origin of the Laurentian margin are best constrained by the onset of thermal-subsidence-driven sedimentation of the continental margins (Bond & Kominz 1984). This occurred between 590 and 550 Ma in the northern Appalachians, which is consistent with the youngest known U–Pb zircon ages of synrift magmatism along the Laurentian margin (e.g. Kumarapeli *et al.* 1989; Cawood *et al.* 1996; McCausland *et al.* 1997). Van Staal *et al.* (1998) also

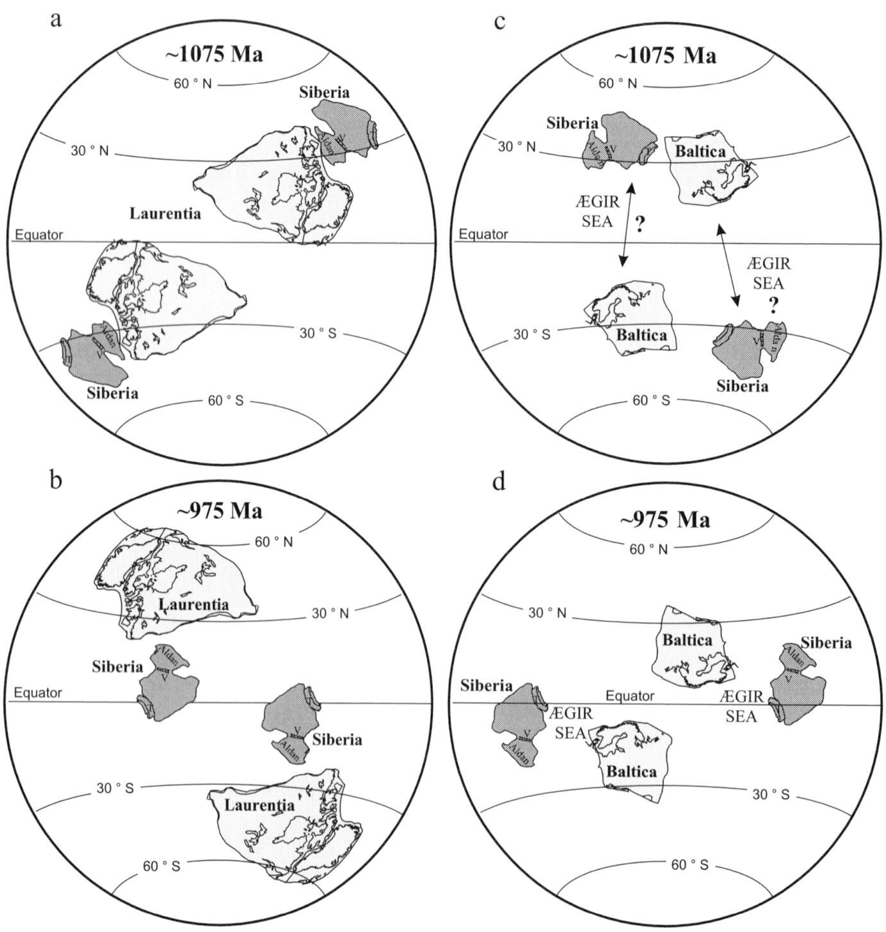

Fig. 7.57. Possible reconstructions at 1075 Ma (**a**, **c**) and at 975 Ma (**b**, **d**) using different poles for reconstruction (after Meert & Torsvik 2003).

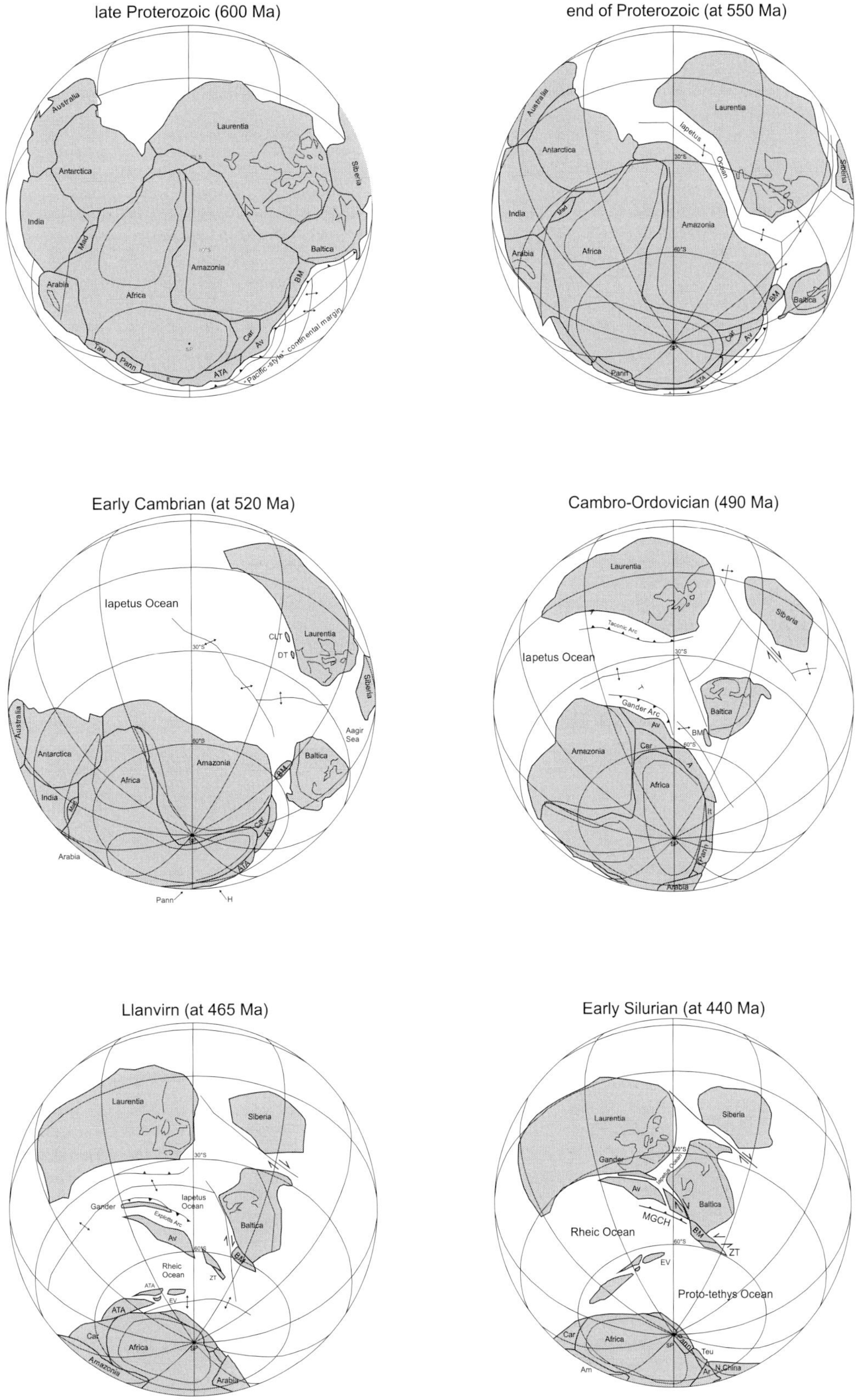

Fig. 7.58. Reconstructions showing continental distributions at different times (after Winchester *et al.* 2002*a*). Abbreviations: ATA, Armorican Terrane Assemblage; Av, Avalonia; BM, Bruno-Silesia-Moesia; Car, Carolina; CLT, Chain Lakes Terrane; DT, Dashwood Terrane; EV, Eastern Variscides; MGCH, Mid-German Crystalline High; Pann, Pannonia; Tau, Taurus; ZT, Zonguldak Terrane.

suggest that the rift–drift transition in Scotland occurred in the very latest Precambrian, and thus the final breakout of Laurentia took place within a relatively short time period of <40 Ma. However, as is common with modern rifted margins, rifting events leading to the formation of Iapetus were probably diachronous, both along and across the trace of the margin. For example, in parts of the Appalachians rifting may have occurred during two distinct episodes, an earlier unsuccessful one at *c.* 700–760 Ma and a second between 550–600 Ma leading to continental separation (Tull *et al.* 1998).

Once opening commenced, the Iapetus Ocean continued to widen throughout the Cambrian at *c.* 7.5 cm/a across the Amazonian-Laurentian sector. While Amazonia, Avalonia and Baltica moved to the south and Siberia moved east with respect to Laurentia, the latter continued migrating further northwards towards the equator (MacNiocaill & Smethurst 1994). The absence of any significant differential movement between Amazonia, Avalonia and Baltica during the Cambrian suggests that these were more or less fellow travellers. Spreading rates in the Tornquist Sea were, therefore, probably low or zero soon after the rift–drift transition. Baltica, however, was far enough from Avalonia and, by implication, Amazonia during the Early to Mid-Cambrian to have significantly more diverse trilobite faunas than Avalonia (McKerrow *et al.* 1992).

The Grampian Orogeny: Mid-Ordovician arc–continent collision

Intra-Iapetus subduction on the Gondwanan side must have commenced by at least 513 Ma (van Staal *et al.* 1998). Subduction led to the formation of a series of backarc or successor pull-apart basins with related magmatic activity (e.g. Tull *et al.* 1998). Subsequently, subduction commenced along the Laurentian margin.

During Lower–Middle Ordovician times there is extensive evidence in Scotland and Ireland for the obduction of ophiolites (Fig. 7.15). For example, the Shetland ophiolite comprises a series of nappe structures which exhibit a layered sequence with olivine-rich lherzolite–harzburgite at the base, overlain by (in ascending order) dunites, banded gabbros, wehrlite-clinopyroxenites and a sheeted dyke complex composed mainly of microgabbros (Flinn 1985). Similar ophiolitic units are also found in Ireland and elsewhere in Scotland and provide evidence for obduction onto the Laurentian margin during Early–Middle Ordovician times. However, whether the ophiolites originally formed oceanic crust *sensu stricto* or marginal basin crust is difficult to assess, partly due to the ambiguity of the geochemical studies (Strachan 2002*b*). In some areas (e.g. Balantrae Complex, Scotland), the close association of the ophiolite with volcaniclastic sediments implies a marginal basin setting. Volcanic arcs along the northern margin of the Iapetus Ocean are well documented from western Ireland (i.e. Mayo Trough). Here it can be demonstrated that Early Ordovician sedimentation occurred at the same time as erosion of an ophiolite and metamorphic complex to the north and eruption of a volcanic arc to the south (Dewey & Ryan 1990; Dewey & Mange 1999).

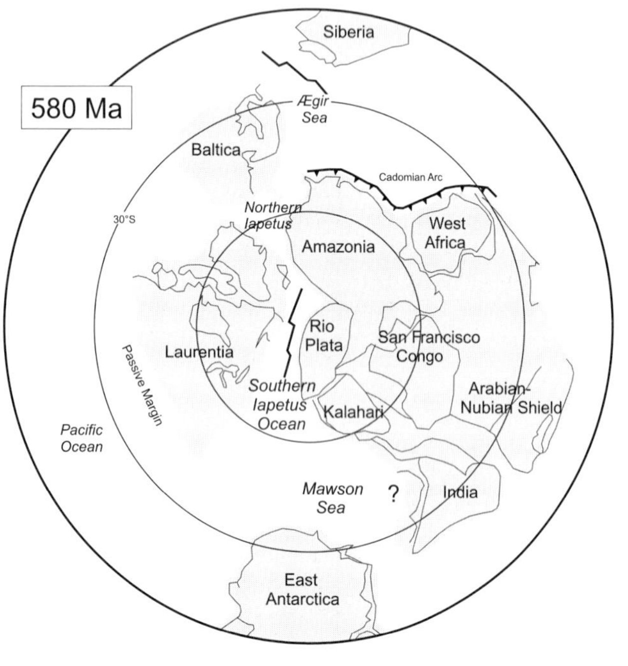

Fig. 7.59. Reconstruction at 580 Ma (after Meert & Torsvik 2003).

The geodynamic evolution of the Grampian Orogeny in Ireland and Scotland is similar to that proposed for the early Middle Ordovician Taconic Orogeny in North America (Fig. 7.60). This resulted from the emplacement of major ophiolite nappes onto the Laurentian margin during arc–continent collision. In North America (e.g. Newfoundland) the ophiolites are preserved as thrust klippes which overlie Cambro-Ordovician shelf carbonates equivalent to those exposed in NW Scotland (Strachan 2002*a*). Obduction of the Baie Verte Oceanic Tract and Penobscot Arc onto, respectively, the Laurentian and the Ganderian/Avalonia margins began nearly coevally between 490 and 485 Ma (Tremadoc) (van Staal *et al.* 1998).

Obduction on both microcontinent margins was followed by an arc polarity reversal which led to arc formation along parts of the Laurentian margin (e.g. Notre Dame and Popelogan-Victoria arcs). This arc polarity reversal was the result of the Grampian Orogeny (= Taconic Orogeny, North America) which marked the initial stages of narrowing of the Iapetus Ocean and probably resulted from the collision of the passive Laurentian margin with a volcanic arc developed above an intra-oceanic subduction zone (Clift *et al.* 2004; Strachan 2002*a*). It should, however, be noted that these orogenic events cannot be related exclusively to the interaction of one single volcanic/magmatic arc with the Laurentian margin during the Ordovician as noted by various geological and geophysical data (for details see van Staal *et al.* 1998).

Mid-Ordovician to Silurian tectonics on the northern Iapetus margin

During the late Arenig–early Caradoc (475–455 Ma) Iapetus was being subducted at a rate of *c.* 12.5 cm/a. This is not surprising considering that Iapetus was subducting beneath both the Laurentian and Avalonian margins (van Staal *et al.* 1998). After the Caradoc, convergence between Laurentia and Avalonia continued with NW-directed subduction. Coeval subduction of the Tornquist Sea beneath Avalonia is thought to have been responsible for the Caradoc phase of backarc volcanism in England, Wales and SE Ireland. This volcanism ceased in the Ashgill, presumably due to the onset of collision between Baltica and Avalonia.

Geological evidence indicates that Laurentia and Avalonia collided diachronously during the Silurian. In Britain and Ireland, the progressive underthrusting of Avalonia beneath

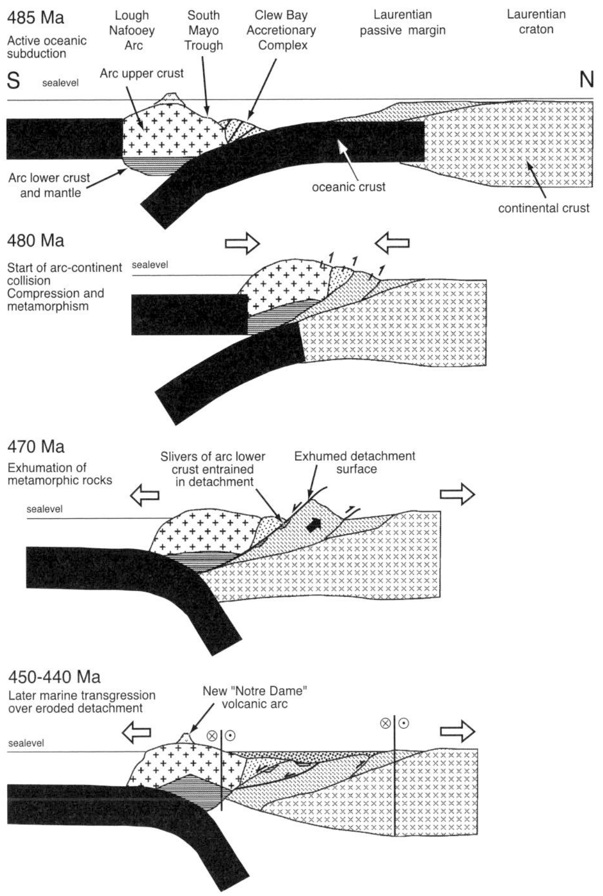

Fig. 7.60. Initial stages of narrowing of the Iapetus Ocean during the Ordovician, that probably resulted from collision of the passive Laurentian margin with a volcanic arc developed above an intra-oceanic subduction zone (after Clift *et al.* 2004).

Laurentia is best reflected in the distribution of the Wenlock and Ludlow-age foreland basin deposits in central Ireland and north England (Hutton & Murphy 1987; King 1994) and the uplift of the Midland Valley (van Staal *et al.* 1998). It is likely that narrow seaways persisted after the start of collision, such as the foreland basin between present-day Timor and Australia (cf. van Staal & de Roo 1995). Deformation of the foreland basin sediments during the Emsian in north England and Wales may represent the final stages of Avalonian–Laurentian collision.

The development of the eastern Gondwanan margin of the Iapetus Ocean in the northern Appalachians and the British and Irish Caledonides is poorly understood (van Staal *et al.* 1998). Closure of the Rheic Ocean probably involved a significant anticlockwise rotation of Gondwana to allow convergence between NW Africa and Laurentia. Thus the Laurentian margin became the locus of significant dextral transpression during the Devonian (Dewey 1982; van Staal & de Roo 1995). The polarity of subduction of the Rheic Ocean was probably also towards the NW. Late Ordovician–Silurian arc-like magmatism in Avalonia (Barr *et al.* 2001; Greenough *et al.* 1993) is most simply explained by such a setting because Laurentia, with its accreted terranes, represented the upper plate. Cessation of this volcanism coincided with later collision termed the Acadian Orogeny (Devonian).

Late Neoproterozoic to Ordovician history of Avalonia and Baltica

Baltica

The East European Craton forms the core of Baltica today. During the Precambrian Baltica was made up of three terranes: Fennoscandia, Samartia and Volgo-Uralia (Cocks & Torsvik 2005). Soon after 1.9 Ga, Fennoscandia was accreted to Volgo-Uralia and Samartia creating a Protobaltica Terrane (McCann 2008). However, current thinking supports the existence of a subsequent supercontinent – Rodinia – in which Protobaltica formed a unit, and which consolidated at *c.* 1100–1000 Ma, and most probably began to breakup at *c.* 0.75 Ga (see above). Within the Rodinian collage, the Baltica area was adjacent to, and probably welded to, Laurentia, a terrane which included most of North America, Greenland, Scotland and the northern parts of England and Ireland. Laurentia was, in turn, attached to the South American terranes of Rio Plata and Amazonia, and possibly also West Africa. A reconstruction at *c.* 750 Ma, shortly after initiation of the breakup of Rodinia (Torsvik 2003), shows a spreading ocean between the combined Laurentia–Protobaltica and other terranes to the east, and Australia, East Antarctica and further blocks to the west (Fig. 7.56).

Baltica became an independent terrane when it split off from Laurentia, leaving a widening Iapetus Ocean between the two. When this rifting actually commenced is a matter of some uncertainty. However, the southern part of Iapetus, between Laurentia and South America, appeared to have opened first soon after 580 Ma (Hartz & Torsvik 2002; Fig. 7.59) and the rifting migrated gradually northwards until Baltica finally separated from Laurentia between 570 and 550 Ma, relatively near the end of Precambrian times (Fig. 7.58).

During the mid- and late Ediacaran (*c.* 560–550 Ma) the current northern part of the NW margin of Baltica changed from an extensional tectonic regime to an active margin (i.e Timanide Orogeny). This was a period of active accretion in which various microcontinental blocks were united with Baltica to form a much expanded terrane area in Lower Palaeozoic times (Cocks & Torsvik 2005).

Evidence of the existence of a wide Iapetus Ocean separating Laurentia and Avalonia/Gondwana is found in Britain and Ireland. This includes: (a) major differences in the age and character of the Precambrian basement (up to 2900 Ma north of Iapetus, no older than 700 Ma to the south); (b) contrasts in sedimentary facies especially between the warm-water Cambro-Ordovician carbonate-rich successions to the north and the carbonate-poor sediments deposited up until Late Ordovician times, to the south; (c) faunal contrasts (see above); and (d) palaeomagnetic evidence (Woodcock 2002*b*).

In general, the Cambrian sediments of Baltica lie unconformably upon the metamorphosed Precambrian and thus represent a transgression, which had begun in the late Ediacaran (Martinsson 1974). Much of the area of Baltica was a shelf environment for most of the Cambrian and thus formed part of the global trend towards steadily rising eustatic sea-level. Artyushkov *et al.* (2000) have analysed Cambrian sea-level changes in detail in Estonia and the St Petersburg area of Russia, and concluded that sea-level changes were predominantly caused by tectonic factors. However, there are no known tectonic events which might have caused these changes, and a wider survey is required.

In the the NW part of Baltica, there were substantial Late Cambrian orogenic events, termed the Finnmarkian Orogeny (see review by Andreasson 1994) which peaked between 505 and 500 Ma, and which were probably associated with subduction within the Aegir Ocean between Baltica and Siberia (Fig. 7.58). Neighbouring terranes to Baltica at that time chiefly consisted of Gondwana to the south, and it would appear that Avalonia had not yet become detached from Gondwana at this time. Precisely when detachment occurred is poorly constrained, although an Early Ordovician date is most probable and is supported by faunal evidence (Fortey & Cocks 2003). To the east of Avalonia lay the Armorican Terrane Assemblage (= Armorica, although see discussion above) whose tectonically complex and varied parts make up most of France and the Iberian Peninsula today. Armorica did not become detached from the main part of Gondwana until the Early Devonian (see Torsvik & Cocks 2004).

Because of Baltica's rotation, which gathered pace in late Cambrian times, there must have been progressive and substantial strike-slip movement in the Ran Ocean between Baltica and Gondwana. The Ran Ocean initially developed as an arm of the Precambrian Iapetus Ocean (Hartz & Torsvik 2002). Subsequently, the Ran was united with the Rheic Ocean, but the latter name cannot be used prior to the Early Ordovician, since the Rheic only came into existence after Avalonia left Gondwana leaving a spreading centre and a widening Rheic Ocean between the two (Cocks & Torsvik 2005).

Baltica travelled steadily northwards towards the palaeo-equator throughout the Ordovician, and it is noticeable that the relatively thin-bedded Early Ordovician limestones in the east Baltic area are of cooler-water origin (Jaanusson 1973). The Iapetus Ocean was at its widest at about Cambrian–Ordovician boundary times (490 Ma) and the Ran Ocean can also be shown to have widened significantly by the Early Ordovician by comparison with the Cambrian, and thus Baltica was at its most isolated (Cocks & Torsvik 2005). As the Ordovician progressed, the surrounding oceans to the south and west of Baltica became steadily narrower.

There is no evidence of any oceanic islands which subsequently amalgamated with Baltica along its southern margins (TESZ and eastwards), but on its western margin, and originally within the Iapetus Ocean, there are a series of suspect terranes, some within the Scandinavian Caledonides. Faunas in these areas show little affinity with contemporary Ordovician faunas from Baltica (for details see Neuman & Harper 1992; Harper *et al.* 1996), though this is probably related to the fact that some of these developed on islands located on the shelf within an arc in the Iapetus Ocean near Laurentia and with good faunal contact with the latter. Furthermore, in the Early Ordovician, today's NW Baltica margin faced northwards, towards the northern Iapetus Ocean and on into the Panthalassic Ocean (Fig. 7.6). Some of these suspect terranes were emplaced onto Baltica in the Silurian only after the entire terrane had rotated by *c.* 90°, at a time when the NW Baltica margin faced Laurentia. Thus, these faunas (e.g. Holonda fauna) were transported eastwards by nappes of Silurian age onto the Baltica craton, and represent faunas which originally lived a great distance away from the autochthonous Ordovician faunas of the Baltica microcontinent.

Palaeomagnetic data indicate that Baltica was situated in subtropical latitudes by the end of the Ordovician. This is confirmed by the absence of glaciogenic sediments during the latest Ordovician global glacial inverval. The Hirnantia brachiopod fauna (Rong & Harper 1988) is well known from the Hirnantian of the Holy Cross Mountains (Poland), the Oslo region, and elsewhere on Baltica, although it is not necessarily indicative of cool water temperatures, but may reflect the greater depths that could briefly support benthic faunas due to increased oxygenation within the global oceans (Cocks & Torsvik 2005).

Fortey & Cocks (2005) have presented evidence for a mid-Ashgillian global warming event (their Boda Event) which preceded the Hirnantian mass extinction at the end of the Ordovician and the climatic deterioration and oceanic and/or atmospheric changes that accompanied it (Gibbs *et al.* 1997; Brenchley *et al.* 2003).

Baltica-related units

Although Belka *et al.* (2000) and Valverde-Vaquero *et al.* (2000) suggested that the Holy Cross Mountain terranes of Poland (the Lysogóry, Malopolska and Upper Silesian massifs) had Cadomian/Gondwanan links, those conclusions are controversial (Cocks 2002). Understanding the complex history of these blocks is difficult because of their shared history *vis-à-vis* Baltica (see discussion above). Moreover, their position with respect to each other and to Baltica during the Cambrian was fundamentally different from that of today. The three areas, although yielding faunal linkages to Baltica, are exotic terranes derived from the Gondwana margin rather than being displaced fragments of the Baltica crust. Their apparent faunal linkages to Baltica during the Cambrian are because they were situated closer to Baltica than to any other part of Gondwana at the time.

The Upper Silesian Massif and Avalonia share the same geotectonic derivation, although they have disparate drift histories. During the Early Cambrian the former was already separated from the Gondwana margin and lay close to Baltica. Except for a single record of Baltica conodonts in the Ordovician of the massif, there is no other evidence for the palaeogeographical association of the Upper Silesian Massif after post-mid-Cambrian and pre-early Devonian times. Belka *et al.* (2000) suggested an amalgamation of the Upper Silesian Massif and the Malopolska Massif during the Early Devonian.

Avalonia

In contrast to its Neoproterozoic history as an active plate margin, the Avalonian margin of Gondwana was mostly quiescent throughout Cambrian times. During the Tremadoc, a volcanic arc was re-established along the Gondwana margin signalling the onset of the subduction of Iapetus beneath the Avalonian segment of Gondwana. This continued through much of the Ordovician (Woodcock 2002*a*). The Early Ordovician was a period of high sea levels with the development of carbonate platforms and oceans with abundant black shales. It was also a time of widespread suprasubduction-zone ophiolites obducted shortly after generation, and abundant blueschists and arc magmatism. Substantial reorganization of plate boundaries also occurred, perhaps caused by the high rates of relative plate motion with widely dispersed continents and little continent collision (Dewey 1988).

From the palaeontological and palaeomagnetic evidence it is clear that the Cambrian to mid-Devonian evolution of Avalonia has to be divided into three or four periods, each with a different geodynamic context (Fig. 7.61). The first period (early Cambrian–Tremadoc) was when Avalonia was still attached to Gondwana at high latitude, close to the northern part of South America or west Africa (McKerrow *et al.* 1992; Cocks & Torsvik 2002). This was followed in Arenig–mid Ashgill times by a period when Avalonia was isolated as a separate microcontinent, drifting away from Gondwana towards Baltica. In late Ashgill–mid Devonian times, Avalonia docked with Baltica and together they collided with Laurentia by the Wenlock. Finally, in early–mid-Devonian times there was a phase of extensional rifting, forming the Rhenohercynian Basin, which was coeval with compression in the Anglo-Brabant Deformation Belt (Verniers *et al.* 2002). These periods can also be correlated with the three megasequences which were defined by Woodcock (1990*b*) in the Welsh Basin: (1) megasequence 1 (Lower Cambrian–Upper Tremadocian) can be equated with the period when Avalonia was attached to Gondwana; (2) megasequence 2 (Lower Arenig–middle Ashgill) represents the period when Avalonia detached from Gondwana creating the Rheic Ocean in its wake and the closure of the Iapetus Ocean and the Tornquist Sea in front of it; and (3) megasequence 3 (upper Ashgill–Lower Devonian) with the collision of Baltica and Laurentia when subduction magmatism of the Iapetus Ocean beneath Avalonia ceased and collision with Laurentia was initiated.

When Avalonia detached from Gondwana during the late Tremadoc to early Arenig, subduction underflow occurred beneath the Irish, Welsh and Lake District segments of the microcontinent. The Midlands Microcraton and the Anglo-Brabant area were rotated approximately 55° anticlockwise in relation to the Lake District (Piper 1997). The Anglo-Brabant area was, in this scenario, orientated parallel to the transport direction and remained a passive margin, with an oceanic basin to the east, separating it from the rest of Avalonia (Winchester *et al.* 2002*a*). Initiation of subduction beneath Avalonia in the Arenig approximately coincided with its breaking away from Gondwana and the opening of the Rheic Ocean.

The precise timing of Avalonia's accretion to Laurentia and Baltica is uncertain, but is likely to have been accomplished by the mid-Silurian. The Avalonia–Baltica convergence story and the destruction of the intervening Tornquist Sea were first inferred on faunal and palaeomagnetic grounds (e.g. Cocks & Fortey 1990; McKerrow *et al.* 1991; Torsvik *et al.* 1993; Cocks 2000; von Raumer *et al.* 2002), but with contributions and refinements from tectonic, sedimentary, petromagnetic, seismic and isotopic data (see reviews in Pharaoh 1999; McCann & Krawczyk 2001). Faunal evidence supports progressive links with Laurentia and Baltica by the Late Ordovician (Williams *et al.* 1995) while palaeomagnetic data suggest that any palaeolatitudinal separation between Avalonia and Laurentia had disappeared by the Early Silurian (e.g. Miller & Kent 1988; Trench & Torsvik 1992).

The Ordovician sediments of the Rügen area, located offshore of Avalonia, are markedly different from those of the G14 area to the north (for details see McCann 1998). The petrographic signature is that of an arc, with the geochemistry indicating a strong active continental margin signature (McCann 1998). Lithic fragments suggest the possibility of input from a subduction zone or rifted continental margin. The dominance of pyroxene derived from orogenic tholeiitic basalts, and chromite from alpine peridotites and ophiolitic sequences, in the Rügen 5/66 borehole confirms that basin igneous activity was also important (Giese *et al.* 1994). Large-scale volcanism in Eastern Avalonia commenced in the early Ordovician, initially in the Welsh Basin and later in the Leinster Basin and Lake District. Indeed, the period of Lake District subduction-related volcanism led to the eruption of significant amounts of tholeiitic basalts and andesites passing southwards into calc-alkaline andesites and rhyolites (Bevins *et al.* 1985). The extent of this volcanic activity can be gauged from the presence from early Ordovician times of volcanic ash beds, presumably derived from eruptions within the Iapetus region of the Caledonides, on the southern margins of Baltica (Huff *et al.* 1992). The major period of tectonic and associated volcanic activity may have been associated with the closure of the Iapetus Ocean between Eastern Avalonia and Laurentia, although this arc system may as easily have formed due to closure of the Tornquist Sea (Pickering *et al.* 1988). Volcanic activity, however, continued intermittently into the early (Bevins *et al.* 1985) and later Silurian (Teale & Spears 1986).

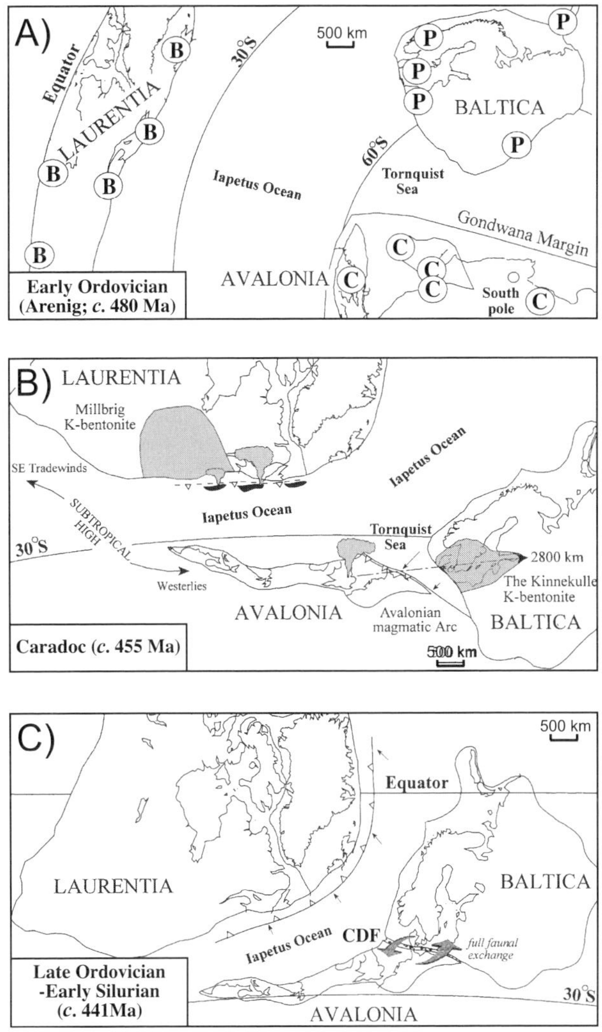

Fig. 7.61. Reconstructions of Laurentia, Baltica and NW Gondwana at distinct times (after Torsvik & Rehnström 2003). (**A**) Early Ordovician reconstruction and the distribution of Arenig–Llanvirn platform trilobites (Cocks & Fortey 1990): B, *Bathyurid*; P, *Ptychopygine/Megalaspid*; C, *Calymenacean*; D, *Dalmanitacean*. (**B**) Mid-Caradoc and (**C**) Late Ordovician–Early Silurian reconstruction. CDF, Caledonian Deformation Front.

Biostratigraphic and palaeomagnetic data suggest closure of the Tornquist Sea between Baltica and Avalonia by latest Ordovician times (Cocks & Fortey 1982, 1990; Tanner & Meissner 1996; Torsvik *et al.* 1990). Closure of the Tornquist Sea by the late Ordovician would suggest that there was a unified sedimentary system spanning the remnant Tornquist Ocean by this time (Ashgill). The actual situation, however, is a little more complex. The thick (hundreds of metres) Ordovician succession on Rügen largely comprises turbiditic sandstones and mudstones whereas sediments of similar age to the north on Bornholm and Scania are dominantly black shales and carbonates (<100 m thick). Also there is a distinct lack of Silurian sediments in the Rügen area, unlike the Llandovery shelf succession described from the G14 well (McCann 1996). Assuming a position along the eastern extension of Avalonia, the Cambro-Ordovician succession of Rügen would represent the link of both domains, implying that the closure of the Tornquist Sea had already begun by the middle Ordovician (Giese *et al.* 1994). Accepting the approximate Caradoc–Ashgill palaeogeographic position of Rügen as suggested by Torsvik *et al.* (1992) would imply the existence of strong links, both biological and sedimentological, by this time at the latest.

The lack of continuity has been interpreted as the result of later left-lateral displacement of the Rügen area relative to Baltica (Giese *et al.* 1994). Recent palaeomagnetic research, however, indicates that the mid-Silurian palaeopoles from Eastern Avalonia and Baltica differ (Torsvik *et al.* 1992, 1996), implying that the Tornquist Sea existed until late Ordovician times (Scotese 1984; Oliver *et al.* 1993; Torsvik *et al.* 1993).

Controversies and open questions about ocean closures

The relative and actual timing of closure of the oceans and collision of Laurentia, Baltica and Avalonia is still the subject of some debate. The East Greenland and Scandinavian Caledonides are of similar age and show similar deformation kinematics, changing from sinistral oblique to orthogonal closure during the Silurian. Soper & Woodcock (1990) argued that Avalonia collided with Laurentia by anticlockwise rotation during the Silurian. Closure of the Tornquist Sea is indicated by the Late Ordovician convergence of apparent polar wander paths for Baltica and Avalonia (Fig. 7.53) and faunal mixing (Torsvik & Rehnström 2003). Scotese *et al.* (1990) used the similarities in apparent polar wander paths and a breakdown in faunal provinciality to suggest that the closure of the Tornquist Sea occurred before the Ashgill (Late Ordovician). This timing is consistent with the clustering of the metamorphic ages along the suture at *c.* 440 Ma (Frost *et al.* 1981).

The existence of a Caledonian Orogen between Avalonia and Baltica has been demonstrated by different basement ages found in well cores sampled across the suture (Frost *et al.* 1981; Nielsen & Japsen 1990). To the NE of the suture, Precambrian basement ages are 880–415 Ma whereas to the SW Caledonian ages of 450–415 Ma have been measured in metamorphic basement rocks. In the southern North Sea, Precambrian crystalline basement has been dated as 844–781 Ma which would situate the area on Baltica (Scheck *et al.* 2002).

Avalonia and the European massifs were located at high southerly latitudes during the Early Ordovician and fringed the northern margin of Gondwana. Avalonia separated from Gondwana during the Arenig–Llanvirn and closure of the Tornquist Sea is indicated by the Late Ordovician convergence apparent polar wander paths for Baltica and Avalonia and faunal mixing. The early Ashgill Shelvian event (Woodcock 1990*a*; Toghill 1992) is broadly coeval with Taconic (455–442 Ma) deformation along the Laurentian margin (Robinson *et al.* 1998) and an Ashgill cooling event (446–449 Ma) was recorded in the western Baltica nappes (Andersen *et al.* 1998; Torsvik 1998; Eide *et al.* 1999). The tectonic setting for the latter is uncertain since we do not know whether these cooling ages relate to interaction between Baltica and Avalonia or Baltica and an Iapetus island arc or microcontinent.

The Tornquist Sea may have had a substantial east–west width, undetectable from palaeomagnetic data. Late Ordovician faunal exchange does not prove that the sea was fully closed, but suggests that it did not act as a significant barrier in Caradoc and Ashgill times. Volcanic ash remnants (K-bentonites; see also Fig. 7.61) are preserved as clay-rich layers in marine sediments and many Palaeozoic K-bentonites have been tied to Plinian eruptions associated with subduction-related, Andean-type magmatism in or near north England (Huff *et al.* 1992; Fortey *et al.* 1996). In Baltica, Ordovician K-bentonites are widespread. They vary in age from Llandeilo to Ashgill, but the thickest and most regionally important is the Mid-Caradoc (*D. multidens* graptolite Zone) Kinnekulle K-bentonite in Sweden (Bergström *et al.* 1995). This has a maximum thickness of 2 m and extends for more than 1000 km in an east–west pattern. The preserved ash volume is estimated at nearly 1000 km^3 (Huff *et al.* 1996). The layer has been dated to 457 $\pm$ 2 Ma (U–Pb zircon age cited in Tucker & McKerrow 1995) and 455 $\pm$ 2 Ma ($^{40}Ar/^{39}Ar$ biotite; Min *et al.* 2001) and was originally said to be coeval with the Milbrig K-bentonite in North America, although Haynes *et al.* (1995) have disproved this, and it is now known that the latter is at least 3 Ma younger. Both units show compositional affinity with volcanic arc granites and indicate generation along convergent continental plate margins (Huff *et al.* 1992). Huff *et al.* (1992) argued for a common volcanic centre within the Iapetus Ocean but Torsvik & Rehnström (2003) suggest that the Kinnekulle K-bentonite was associated with a Mid-Caradoc calc-alkaline magmatic event in Avalonia and that westerlies brought huge ash fallouts over Baltica. Volcanic vents on an arc or microplate, which was undergoing collision with Laurentia, were probably the source for the younger Millbrig K-bentonite.

Avalonia left Gondwana in the Early Ordovician. Rifting was followed by the progressive narrowing and eventual elimination of the western part of the Tornquist Sea, which separated Avalonia from Baltica for most of the Ordovician. The exact nature and timing of the Avalonia–Baltica collision is not easily observed in the geological record. Protuberant orogenic structures or high-pressure rock products are missing or buried, and the collision is therefore often described as a soft docking event, governed by strike-slip convergence and occurring at about the Ordovician–Silurian boundary (443 Ma; Torsvik & Rehnström 2003). In Avalonia, Late Ordovician folding, faulting and igneous activity is known from the Welsh Borderlands area and this may relate to Avalonia–Baltica docking (Torsvik & Rehnström 2003). There is much disagreement concerning the direction of the subduction following this docking. Poprawa *et al.* (1999), for example, suggest subduction to the north (Fig. 7.62), although the geological and geophysical evidence suggests that the converse is true. A substantial slab of Baltica today is beneath Avalonia to the south of the TESZ. The centre of Baltica was little affected by the Avalonia–Baltica collision.

Soper *et al.* (1992*a*), however, have noted that Baltica and Avalonia had very different deformational histories before the Early Silurian and so argued that it was unlikely that the two continents were coupled before this time.

By mid-Silurian times, faunal evidence suggests that Avalonia and Baltica were separated by <1000 km (McKerrow & Cocks

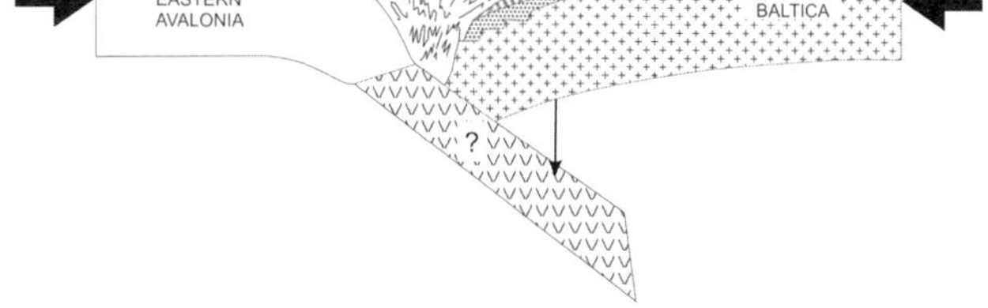

Fig. 7.62. Sketch of Late Silurian (Ludlow–Pridoli) plate convergence between Avalonia and Baltica including foredeep development (after Poprawa *et al.* 1999).

1986), while palaeomagnetic data suggest a distance of *c.* 1000–1500 km (Torsvik *et al.* 1993), largely longitudinal (MacNiocaill & Smethurst 1994). Thus, observed differences in the sedimentological successions of Baltica and Avalonia may be interpreted in terms of longitudinal variations, together with the variable tectonic environments that prevailed in the individual regions (McCann 1998).

Using kinematics of deformation to infer directions of continental convergence, Soper *et al.* (1992*a*) showed that for Avalonia to converge sinistrally with Laurentia and for Baltica to close orthogonally with Laurentia, the Tornquist Sea must first be subducted beneath Avalonia and Baltica. However, there is little direct geological evidence of subduction to the north or south of the Baltica–Avalonia suture with the exception of detritus of arc affinity in the Avalonia sucession in the Rügen area (e.g. Katzung *et al.* 1993; McCann 1998). This would imply that subduction of oceanic material occurred to the south, beneath Avalonia, during closure of the Tornquist Sea.

Geological and geophysical data from onshore and offshore northern Denmark indicate that Silurian rocks form the upper part of a conformable succession that commenced in the Lower Cambrian. The Upper Silurian section is considerably expanded relative to the lower part of the succession, which would imply an increase in subsidence, consistent with the succession being deposited in a foredeep as Avalonia was thrust over Baltica. The remains of this foreland basin imply a substantial sediment source close to the southern margin of Baltica beginning at the end of the Ordovician. Such a source could have been produced at the onset of the Avalonia–Baltica collision. However, there are insufficient data for the foreland basin to allow the direction of subduction to be determined.

Final closure, when it occurred, was complex. In one model, the overthrusting of Avalonia in a NE/ENE direction led to the formation of the seismic reflectors which are interpreted as the Caledonian Deformation Front (Tanner & Meissner 1996). Deeper seismic reflectors have been interpreted as remnants of a subduction zone, where oceanic material from the Tornquist Sea was subducted beneath an active continental margin, followed by the continent–continent collision which created the thrusts and nappes of the CDF (Tanner & Meissner 1996). A second model involves initial collision occurring in SW Poland (Oliver *et al.* 1993) leading to later transpressional closure along the Elbe Line combined with an anticlockwise rotation (Torsvik *et al.* 1993) of Avalonia and the main subduction of crust at the Iapetus Suture. Thus, the area between the CDF and the Elbe Line could have been incorporated as a microterrane. Given the complexity of the situation in SW Poland where six terranes were accreted along the Trans-European Suture Zone (Oliver *et al.* 1993), it is clearly possible that similar situations are to be recognized along the length of the Tornquist margin resulting from the complex amalgamation of Avalonia and Baltica.

Longevity and regional importance of orogenic structures

Direct and indirect evidence of the extent of the Caledonian orogenic event across Central Europe has been presented and discussed. Since persistent Caledonian-derived structures may also have exerted influence on the later Variscan and Alpine orogenies in the region (see Frotzheim *et al.* 2008; Kroner *et al.* 2008; Scheck-Wenderoth *et al.* 2008; Voigt *et al.* 2008) some selected points are discussed below, that are based on examples available in the NE German Basin (NEGB) and its extension to the north across the Caledonian Deformation Front (CDF). The modelling studies underline that the present configuration and condition of the lithosphere of Central Europe is the result of superimposed periods of deformation, which were presumably not independent through time.

Finite element-based **dynamic modelling** studies of the recent European stress field have been applied on various scales south of the CDF subcrop. On the larger scale of northern Europe, a viscous thin-sheet approach uses the crustal thickness defined by seismic data as an initial model condition to reproduce local strain and stress fields (Marotta *et al.* 2001; Fig. 7.63). These models predict a rather undeformable area in the northern part of the NEGB. Furthermore, deviations of the maximum-stress directions occur, shifting from NW–SE in the western parts to NE–SW in the eastern regions (Marotta *et al.* 2002).

The influence of different underlying terranes on the **neotectonic stress field** has also been considered. Sim *et al.* (1999) constructed the first map of the recent stress for the East European Platform, while a more detailed approach was performed by Kaiser *et al.* (2005). The western part of the East European Platform and the eastern part of the West European Plate underwent WNW–ESE regional compression, whereas the northern and central East European Platform underwent east–west compression (Sim *et al.* 1999). A thin-shell modelling approach based on GPS and structural data along the northernmost margin of the NEGB has been performed in order to understand the transition between the Southern Permian Basin and the Baltic Shield. Major deviations of the stresses occur along strong contrasts in the lithospheric structure and influence the stress pattern significantly (Kaiser *et al.* 2005).

The studies outlined above demonstrate that the key factors in controlling the compressive stresses in the NEGB are the Alpine collision and the local horizontal rheological heterogeneity caused by the underlying Baltica basement structure. Thus, a relatively strong lithosphere below the northern margin of the NEGB constituting the transition to the Baltic Shield may explain the fan-like stress field patterns observed today (Marotta *et al.* 2001).

On a more local scale in the NEGB, 3D finite-element modelling of the regional thermal field may also be interpreted as representing the influence of the Baltic Shield for the given temperature distribution at the top of the Rotliegend (e.g. Bayer *et al.* 1997; Ondrak *et al.* 1999). While these authors define the transitional character of the CDF towards the south, Balling (1995) discussed it with respect to its separation towards the north (see above).

Structural modelling and balancing studies have been widely used to understand the geometric, kinematic and dynamic evolution of all types of geological structures (see references in McClay 1996). Such forward models provide insights into the progressive structural evolution of a tectonic setting. Generally, this approach is applied to areas which are well studied in terms of age-dated sedimentary successions and tectonic transport directions. There are, however, almost no such models for the deeply buried Central European Caledonides, since they are generally only concerned with sedimentary basin infill. Those studies which are relevant provide some ideas concerning the stress and deformation states and these observations can be discussed in combination with other results.

Based on well data and depth maps, an initial digital 3D structural model was constructed for the NEGB and gravity-modelled in 2D. The model was successively expanded both with the addition of geophysical (mainly seismic) data and also in terms of width and depth (Scheck & Bayer 1999; Scheck-Wenderoth & Lamarche 2005). Combined volumetric and back-

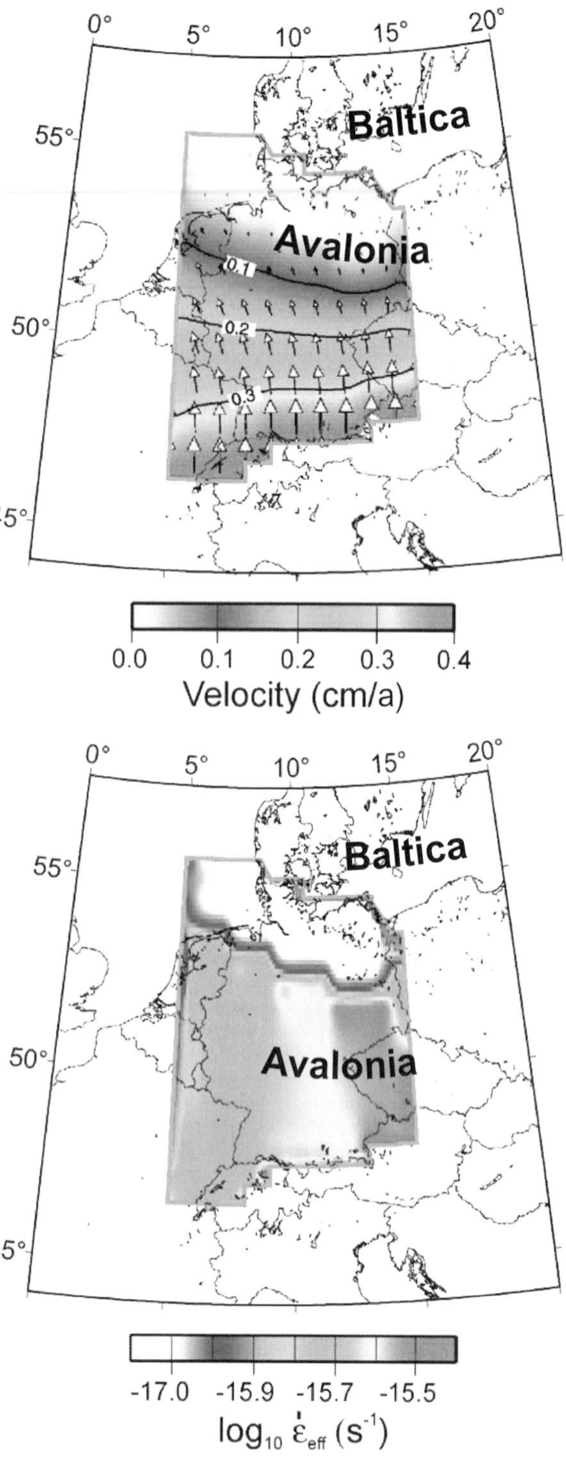

Fig. 7.63. Finite-element stress-field modelling (after Marotta *et al.* 2002) with deformation velocity field (top) and effective strain rate (bottom).

stripping methods indicate that about one-third of the basin's subsidence was caused by tectonics, including basal underplating, before basin evolution began. The underlying Baltica basement is marked by the presence of a high-density body in the lower crust.

In the NEGB, the only palinspastic reconstructions available are by Kossow & Krawczyk (2002), based on results from the BASIN'96 and commercial seismic surveys (Krawczyk *et al.* 1999; Kossow *et al.* 2000). A flexural cantilever model was also applied for forward modelling of the initial phase of NEGB formation in combination with detailed analysis of core material (Rieke *et al.* 2001). The formation of the NEGB was initiated during the Early Permian and its evolution was largely controlled by normal faulting related to deep-seated ductile shearing, with a gently dipping northern basin margin. The cantilever model predicts a stretching factor of $\beta = 1.2$ in the basin centre and 1.0 at the margins, which would have only a slight effect on the crustal structure. Interestingly, deformation intensity decreases from south to north, and most of the deformation was accommodated at distinct fault zones in the southern half of the NEGB (e.g. Fig. 7.64). This was most likely related to Alpine-induced intraplate deformation (Kossow & Krawczyk 2002). Despite the decoupling behaviour of the underlying salt, the fact that Baltica crust extends southwards to beneath the NEGB depocentre might also influence the degree of stress transfer through the basin at later times. However, there is no evidence for older crustal extension or Caledonian-age deformation that can be resolved.

Summary

The scenario of the Caledonian Orogeny is of a number of larger and smaller continental masses rifting from Gondwana and drifting north at various rates, sometimes rotating, eventually coming into contact with Laurentia and Baltica. The Gondwanan fragments formed an archipelago and moved rather like the fragments of Indonesia or the Philippines have been doing over the last 50 Ma. Furthermore, they did not all collide during the Caledonian Orogeny, but also in the subsequent Variscan Orogeny (see Kroner *et al.* 2008; McCann *et al.* 2008). A review of the present state of knowledge on the Caledonian Orogeny in Central Europe can be summarized in a number of salient points.

- The existence of the terranes of Laurentia, Baltica and Gondwana has been inferred from geological studies, deep seismic profiling and potential field studies.
- The main oceanic sutures are the Tornquist Suture/Caledonian Deformation Front (CDF) and the Iapetus Suture.
- There is still ongoing debate concerning subduction polarity during ocean closures, or even a possible flip in subduction polarity.
- A possible deep-reaching link between the CDF and the Iapetus Suture cannot be definitively proven.
- The Iapetus Suture is defined across Ireland and Great Britain.
- The CDF is defined in Scandinavia and between the North Sea and the Baltic Sea.
- The southernmost extent of Baltica is located between the Southern Permian Basin depocentre and the Elbe Line and involves thrusting of Avalonia onto Baltica.
- The thin-skinned nature of the Caledonian deformation complex is evident in the Baltic Sea area. Here, Baltica crust escaped deformation during Caledonian collision.
- The longevity and regional importance of orogenic structures in the central European region is evident. Pre-existing tectonic structures of presumably Caledonian origin were reactivated during subsequent tectonic events. The physical properties of inherited lithospheric structure thus determine both the physical effects and the areal extent of processes active up to the present day.

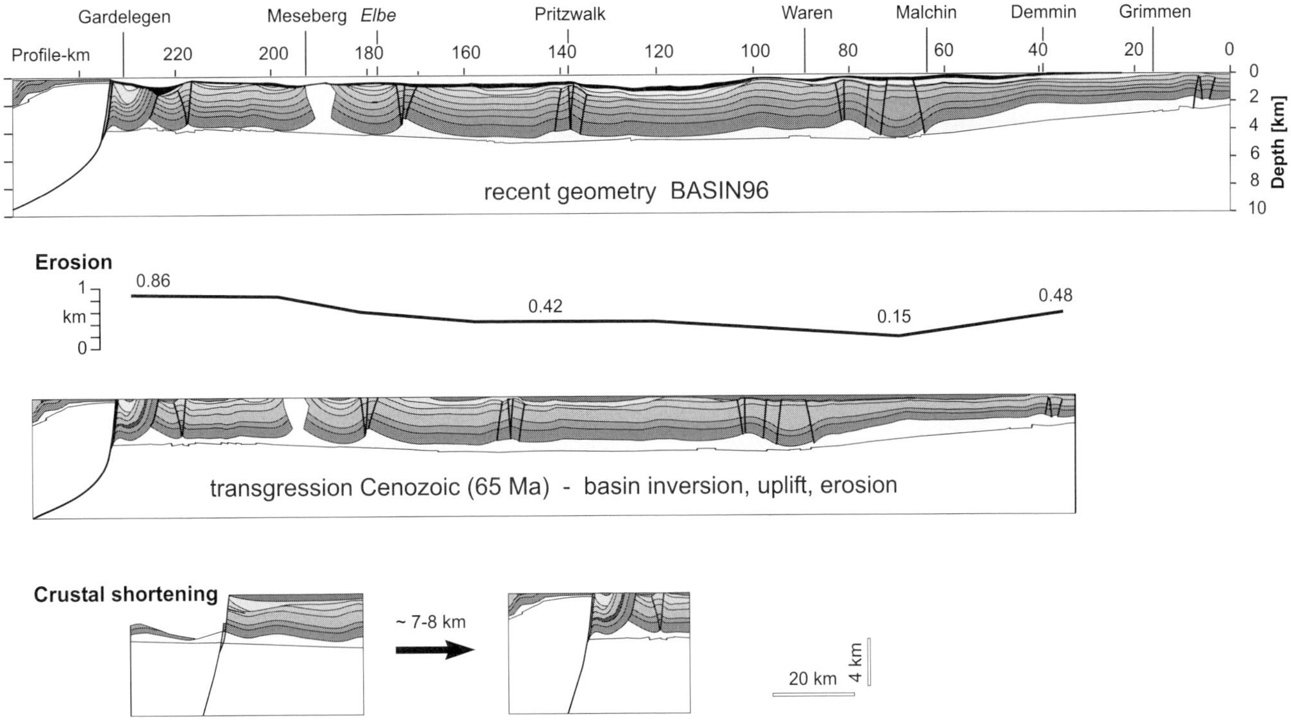

Fig. 7.64. (Top) Recent geometry of the basin infill. (Middle) Erosion during the Cretaceous–Tertiary inversion. (Bottom) Crustal shortening accommodated at the Gardelegen fault during that inversion phase quantified from restoration of profile BASIN 9601 in the NE German Basin. Deformation seems to be highest in the SW half of the profile in the Variscan-influenced domains, whereas the Baltica-influenced middle to lower crust in the NE half of the profile hampers deformation transfer through the crustal levels.

Some ongoing problems, however, still remain, including the following.

- Precisely when and how did the various terranes collide?
- Does surface geology indicate the exhumation of crustal roots not only in western Norway or north of Scotland, but also in other areas?
- Can exhumation rates be deduced precisely enough to reconstruct the destruction of the Caledonian mountain ranges?
- Was the southern Avalonia margin a passive margin formed by rifting from Gondwana in Early Ordovician time, or was it formed by later rifting? Did it face a major Rheic Ocean to the south, or a narrower mainly intracontinental basin?
- Did Avalonia ride up over Baltica, or was it a sliver of crust that was obducted, leaving lower crust and mantle lithosphere to subduct?
- Exactly when did Avalonia, the Armorican Terrane Assemblage and the other European peri-Gondwanan terranes leave Gondwana?
- Why are the Caledonides of northern Germany buried beneath a thick covering of Mesozoic sediments?
- How deep do the observed lineaments reach today? Do they exist below the lithosphere–asthenosphere boundary?
- How important were strike-slip movements?

Sincere thanks are given to D. Blundell, C. van Staal and B. Mingram. They all made very helpful suggestions to unravel the Caledonian mysteries, especially D. Blundell who put a lot of effort into shaping this chapter.

References

Abramovitz, T. & Thybo, H. 2000. Seismic images of Caledonian, lithosphere-scale collision structures in the southeastern North Sea along MONA LISA Profile 2. *Tectonophysics*, **317**, 27–54.

Abramovitz, T., Berthelsen, A., *et al.* 1997. MONA LISA – Deep seismic investigations of the lithosphere in the southeastern North Sea. *Tectonophysics*, **269**(1–2), 1–19.

Abramovitz, T., Thybo, H. & MONA LISA Working Group 1998. Seismic structure across the Caledonian Deformation Front along MONA LISA profile 1 in the southeastern North Sea. *Tectonophysics*, **288**, 153–176.

Abramovitz, T., Landes, M., Thybo, H., Jacob, A.W.B. & Prodehl, C. 1999. Crustal velocity structure across the the Tornquist and Iapetus Suture Zones – a comparison based on MONA LISA and VARNET data. *Tectonophysics*, **314**(1–3), 69–82.

Aichroth, B., Prodehl, C. & Thybo, H. 1992. Crustal structure along the central segment of the EGT. *Tectonophysics*, **207**, 43–64.

Aleksandrowski, P., Mazur, S. 2002. Collage tectonics in the northeasternmost part of the Variscan Belt: The Sudetes, Bohemian Massif. *In*: Winchester, J. A, Pharaoh, T. C. & Verniers, J. (eds) *Palaeozoic Amalgamation of Central Europe*. Geological Society, London, Special Publications, **201**, 237–277.

Anderle, H.J., Franke, W. & Schwab 1995. Rhenohercynian foldbelt – Metamorphic units (Northern Phyllite Zone). *In*: Dallmeyer, R.D., Franke, W. & Weber, K. (eds) *Pre-Permian Geology of Central and Eastern Europe*. Springer, Berlin, 99–109.

Andersen, T.B., Berry, H.N., Lux, D.R. & Andresen, A. 1998. The tectonic significance of pre-Scandian 40Ar/39Ar phengite cooling ages in the Caledonides of western Norway. *Journal of the Geological Society, London*, **155**, 297–309.

André, L. 1991. The concealed crystalline basement in Belgium and the 'Brabantia' microplate concept: constraints for the Caledonian magmatic and sedimentary rocks. *Annales de la Societé Geologique de Belgique*, **114**, 117–140.

André, L., Hertogen, J. & Deutsch, S. 1986. Ordovician–Silurian magmatic provinces in Belgium and the Caledonian orogeny in middle Europe. *Geology*, **14**, 879–882.

ANDRÉASSON, P.G. 1994. The Baltoscandian margin in Neoproterozoic–early Paleozoic times. Some constraints on terrane derivation and accretion in the Arctic Scandinavian Caledonides. *Tectonophysics*, **231**, 1–32.

ANDRÉASSON, P.G. & ALBRECHT, L. 1995. Derivation of 500 Ma eclogites from the passive margin of Baltica and a note on the tectonometamorphic heterogeneity of eclogite-bearing crust. *Geological Magazine*, **132**, 729–738.

ARMSTRONG, H.A. & OWEN, A.W. 2001. Terrane evolution of the paratectonic Caledonides of northern Britain. *Journal of the Geological Society, London*, **158**, 475–486.

ARTYUSHKOV, E.V., LINDSTRÖM, M. & POPOV, L.E. 2000. Relative sea-level changes in Baltoscandia in the Cambrian and Early Ordovician: the predominance of tectonic factors and the absence of large scale eustatic fluctuations. *Tectonophysics*, **320**, 375–407.

ATHERTON, M.P. & GHANI, A.A. 2002. Slab breakoff: a model for Caledonian, Late Granite syn-collisional magmatism in the orthotectonic (metamorphic) zone of Scotland and Donegal, Ireland. *Lithos*, **62**, 65–85.

BABEL WORKING GROUP 1990. Evidence for early Proterozoic plate tectonics from seismic reflection profiles in the Baltic shield. *Nature*, **348**, 34–38.

BABEL WORKING GROUP 1991. Deep seismic survey images crustal structure of Tornquist-Zone beneath southern Baltic Sea. *Geophysical Research Letters*, **18**, 1091–1094.

BABEL WORKING GROUP 1993. Deep seismic reflection/refraction interpretation of BABEL profiles A and B in the southern Baltic Sea. *Geophysical Journal International*, **112**, 325–343.

BACHMANN, G.H. & GROSSE, S. 1989. Struktur und Entwicklung des Norddeutschen Beckens - geologische und geophysikalische Interpretation einer verbesserten Bouguer-Schwerekarte. *Niedersächsische Akademie der Geowissenschaften*, **2**, 23–47.

BALLING, N. 1995. Heat flow and thermal structure of the lithosphere across the Baltic Shield and northern Tornquist Zone. *Tectonophysics*, **244**(1–3), 13–50.

BAMFORD, D., FABER, S., *et al.* 1976. A lithospheric seismic profile in Britain. I. Preliminary results. *Geophysical Journal of the Royal Astronomical Society*, **54**, 43–60.

BAMFORD, D., NUNN, K., PRODEHL, C. & JACOB, B. 1978. LISPB-IV. Crustal structure of Northern Britain. *Geophysical Journal of the Royal Astronomical Society*, **54**, 43–60.

BANKA, D., PHARAOH, T.C., WILLIAMSON, J.P. & TESZ PROJECT POTENTIAL FIELD CORE GROUP 2002. Potential field imaging of Palaeozoic orogenic structure in northern and central Europe. *Tectonophysics*, **360**, 23–45.

BANKWITZ, P. 1985. *Komplexinterpretation des tiefenseismischen Profils THUMARK - Bericht der SAG Tiefenerkundung*. Akademie der Wissenschaften, ZIPE, Potsdam.

BANKWITZ, P., BANKWITZ, E., KRAMER, W. & PIN, C. 1992. Early Paleozoic bimodal volcanism in the Vesser area, Thuringian Forest, eastern Germany. *Zentralblatt für Geologie und Paläontologie*, **Part 1**(9–10), 1113–1132.

BARR, S.M., BEVIER, M.L., WHITE, C.E. & DOIG, R. 1994. Magmatic history of the Avalon Terrane of Southern New Brunswick, Canada, based on U-Pb (zircon) geochronology. *Journal of Geology*, **102**, 399–409.

BARR, S.M., WHITE, C.E., CULSHAW, N.G. & KETCHUM, J.W.F. 2001. Geology and tectonic setting of Paleoproterozoic granitoid suites in the Island Harbour Bay area, Makkovik Province, Labrador. *Canadian Journal of Earth Sciences*, **38**(3), 441–463.

BARRIO-ALVERS, L., BAYER, U., GÖTZE, H.-J. & SCHECK, M. 1998. Density structure of the Northeast German Basin: 3D modelling along the DEKORP line BASIN96. *Annales Geophysicae*, **16**, C23.

BARTON, P.J. 1986. The relationship between seismic velocity and density in the continental crust - a useful constraint? *Geophysical Journal of the Royal Astronomical Society*, **87**(1), 195–208.

BARTON, P.J. 1992. LISPB revisited: A new look under the Caledonides of northern Britain. *Geophysical Journal International*, **110**, 371–391.

BAUMANN, A., GRAUERT, B., MECKLENBURG, S. & VINX, R. 1991. Isotopic age determinations of crystalline rocks of the upper Harz Mountains, Germany. *Geologische Rundschau*, **80**(3), 669–690.

BAYER, U., SCHECK, M. & KOEHLER, M. 1997. Modelling of the 3D thermal field in the northeast German basin. *Geologische Rundschau*, **86**, 241–251.

BAYER, U., SCHECK, M., *et al.* 1999. An integrated study of the NE German Basin. *Tectonophysics*, **314**(1–3), 269–283.

BAYER, U., GRAD, M., *et al.* 2002. The southern margin of the East European Craton: New results from seismic sounding and potential fields between the North Sea and Poland. *Tectonophysics*, **360**(1–4), 301–314.

BEDNARCZYK, W. 1974. The Ordovician in the Koszalin-Chojnice region (Western Pomerania). *Acta Geologica Polonica*, **24**(4), 581–600.

BELKA, Z., AHRENDT, H., FRANKE, W., SCHÄFER, J. & WEMNER, K. 2000. The Baltica/Gondwana suture in Central Europe: evidence from K/Ar ages of detrital muscovites and biogeographic data. *In*: FRANKE, W., ALTHERR, R., HAAK, V., ONCKEN, O. & TANNER, D. (eds) *Orogenic Processes: Quantification and Modelling in the Variscan Belt*. Geological Society, London, Special Publications, **179**, 87–102.

BELKA, Z., VALVERDE-VAQUERO, P., DÖRR, W., AHRENDT, H., WEMMER, K., FRANKE, W. & SCHÄFER, J. 2002. Accretion of first Gondwana-derived terranes at the margin of Baltica. *In*: WINCHESTER, J.A., PHARAOH, T.C. & VERNIERS, J. (eds) *Palaeozoic Amalgamation of Central Europe*. Geological Society, London, Special Publications, **201**, 19–36.

BENEK, R., KRAMER, W., *et al.* 1996. Permo-Carboniferous Magmatism and related subsidence of the NE German Basin. *Tectonophysics*, **266**, 379–404.

BERGSTRÖM, S.M., HUFF, W.D., KOLATA, D. & BAUERT, K. 1995. Nomenclature, stratigraphy, chemical fingerprinting, and areal distribution of some Middle Ordovician K-bentonites in Baltoscandia. *Geologiska Föreningens Forhaldligar*, **117**, 1–13.

BERTHELSEN, A. 1984. The early (800–300 Ma) crustal evolution of the off-shield regions of Europe. Mobile Europe. *In*: GALSON, D.A. & MUELLER, ST. (eds) *First EGT Workshop – The northern Segment*. ESF, Strasbourg, 125–142.

BERTHELSEN, A. 1992. Mobile Europe. *In*: BLUNDELL, R., FREEMAN, R. & MUELLER, ST. (eds) *A Continent Revealed - The European Geotraverse*. Cambridge University Press, Cambridge, 11–32.

BERTHELSEN, A. 1998. The Tornquist Zone northwest of the Carpathians: an intraplate pseudosuture. *Geologiska Föreningens Forhaldligar*, **120**, 223–230.

BETZ, D., FÜHRER, G., GREINER, G. & PLEIN, E. 1987. Evolution of the Lower Saxony Basin. *Tectonophysics*, **137**, 127–170.

BEVINS, R.E., STILLMANN, C.J. & FURNES, H. 1985. A review of Caledonian volcanicity in the British Isles and scandinavia. *In*: GAYER, R.A. (ed.) *The Tectonic Evolution of the Caledonide-Appalachian Orogen*. Vieweg & Sohn, Wiesbaden, 80–96.

BIRD, J.M. & DEWEY, J.F. 1970. Lithospheric plate-continental margin tectonics and the evolution of the Apalachian Orogen. *Geological Society of America Bulletin*, **81**, 1031–1060.

BIRPS & ECORS 1986. Deep seismic reflection profiling between England, Ireland and France. *Journal of the Geological Society, London*, **143**, 45–52.

BLEIBINHAUS, F., BEILECKE, T., BRAM, K. & GEBRANDE, H. 1999. A seismic velocity model for the SW Baltic Sea derived from BASIN'96 refraction seismic data. *Tectonophysics*, **314**, 269–284.

BLUNDELL, D.J., HOBBS, R.W., KLEMPERER, S.L., SCOTT, R.R., LONG, R.E., WEST, T.E. & DUIN, E. 1991. Crustal structure of the central and southern North Sea from BIRPS deep seismic reflection profiling. *Journal of the Geological Society of London*, **148**, 445–457.

BLUNDELL, D., FREEMANN, R. & MUELLER, S. 1992. *A Continent Revealed – The European Geotraverse*. Cambridge University Press, Cambridge.

BLUNDELL, D. J., HOBBS, R. W., KLEMPERER, S. L., SCOTT-ROBINSON, R., LONG, R. E., WEST, T. E. & DUIN, E. 1993. Crustal structure of the Central and Southern North Sea from BIRPS deep seismic profiling. *Journal of the Geological Society, London*, **148**, 445–457.

BOGER, S.D., WILSON, C.J.L. & FANNING, C.M. 2002. Early Paleozoic tectonism within the East Antarctic Craton; the final suture between East and West Gondwana ? *Geology*, **29**(5), 463–466.

BOIS, C., LEFORT, J.-P., LE GALL, B., SIBUET, J.-C., GARIEL, O., PINET, B. & CAZES, M. 1990. Superimposed Variscan, Caledonian and Proterozoic features inferred from deep seismic profiles recorded between southern Ireland, southwestern Britain and western France. *Tectonophysics*, **177**, 15–37.

Bond, G.C. & Kominz, M.A. 1984. Construction of tectonic subsidence curves for the early Paleozoic miogeocline, southern Canadian Rocky Mountains; implications for subsidence mechanisms, age of breakup, and crustal thinning. *Geological Society of America Bulletin*, **95**(2), 155–173.

Borkowska, M., Hamcurt, J. & Vidal, P. 1980. Origin and age of Izera gneisses and Rumburk granites in the Western Sudetes. *Acta Geologica Polonica*, **30**, 121–145.

Bormann, P., Bankwitz, P. & Schulze, A. 1989. Geophysikalische Ergebnisse und geologische Konsequenzen tiefenseismischer Untersuchungen in der DDR. *Freiberger Forschungshefte*, **C440**, 72–103.

Bouckaert, J., Fock, W. & Vandenberghe, N. 1988. First results of the Belgian geotraverse 1986 (BELCORP). *Annales de la Societe Geologique de Belgique*, **111**, 279–290.

Brasier, M.D., Ingham, J.K. & Rushton, A.W.A. 1992. Cambrian. *In*: Cope, J.C.W., Ingham, J.K. & Rawson, P.F. (eds) *Atlas of Palaeogeography and Lithofacies*. Geological Society, London, Memoirs, **13**, 13–18.

Breitkreuz, C. & Kennedy, A. 1999. Magmatic flare-up at the Carboniferous/Permian boundary in the NE German Basin revealed by SHRIMP zircon ages. *Tectonophysics*, **302**(3–4), 307–326.

Brenchley, P.J., Carden, G.A., *et al.* 2003. High-resolution stable isotope stratigraphy of Upper Ordovician sequences: Constraints on the timing of bio-events and environmental changes associated with mass extinction and glaciation. *Geological Society of America Bulletin*, **115**, 89–104.

Brewer, J.A., Matthews, D.H., Warner, M.R., Hall, J., Smythe, D.K. & Whittington, R.J. 1983. BIRPS deep seismic reflection studies of the British Caledonides. *Nature*, **305**(5931), 206–210.

Brink, H.J., Durschner, H. & Trappe, H. 1992. Some aspects of the late and post-Variscan development of the Northwestern German Basin. *Tectonophysics*, **207**, 65–95.

British Geological Survey 1989. *Miller Solid Geology Sheet, 61°N-02°W*, 1:250.000, HMSO, London.

British Geological Survey 1997. *Colour shaded relief gravity anomaly map of Britain, Ireland and adjacent areas*, 1:1.500.000. BGS, Keyworth, Nottingham.

British Geological Survey 1998. *Colour shaded relief magnetic anomaly map of Britain, Ireland and adjacent areas*, 1:1.500.000, BGS, Keyworth, Nottingham.

British Geological Survey 2007. World Wide Web address: http://www.bgs.ac.uk/geophysics/home.html.

Brittan, J. & Warner, M. 1997. Wide-angle seismic velocities in heterogeneous crust. *Geophysical Journal International*, **129**(2), 269–280.

Brown, P.E. 1991. Caledonian and earlier magmatism. *In*: Craig, G.Y. (ed.) *Geology of Scotland*. Geological Society, London, 229–295.

Brück, P.M., Colthurst, J.R.J., *et al.* 1979. Southeast Ireland: Lower Palaeozoic stratigraphy and depositional history. *In*: Harris, A.L., Holland, C.H. & Leake, B.E. (eds) *The Caledonides of the British Isles Reviewed*. Geological Society, London, Special Publications, **8**, 533–544.

Brückl, E., Bodoky, T., *et al.* 2003. ALP 2002 seismic experiment. *Studia Geophysica et Geodaetica*, **47**(3).

Brueckner, H.K., Blusztajn, J. & Bakun-Czubarow, N. 1996. Trace element and Sm-Nd ‚age' zoning in garnets from peridotites of the Caledonian and Variscan mountains and tectonic implications. *Journal of Metamorphic Geology*, **14**, 61–73.

Bula, Z., Jachowicz, M. & Zaba, J. 1997. Principal characteristics of the Upper Silesia Block and Malopolska Block border zone. *Geological Magazine*, **134**, 669–677.

Canning, J.C., Henny, P.J., Morrison, M.A., van Calsteren, P.W.C., Gaskarth, J.W. & Swarbrick, A. 1998. The Great Glen Fault: a major lithospheric boundary. *Journal of the Geological Society, London*, **155**, 425–428.

Cawood, P.A., Van Gool, J.A.M. & Dunning, G.R. 1996. Geological development of eastern Humber and western Dunnage zones: Corner Brook – Glover Island region, Newfoundland. *Canadian Journal of Earth Sciences*, **33**, 182–198.

Cermak, V. & Bodri, L. 1986. Two-dimensional temperature modelling along five East-European geotraverses. *Journal of Geodynamics*, **5**(1–3), 241–251.

Cermak, V. & Bodri, L. 1995. Three-dimensional deep temperature modelling along the European geotraverse. *Tectonophysics*, **244**(1–3), 1–11.

Cermak, V. & Hurtig, E. 1979. Heat flow map of Europe. *In*: Cermak, V. & Rybach, L. (eds) *Terrestrial Heat Flow in Europe (map enclosure)*. Springer, Berlin, 328.

Chadwick, R.A., Pharaoh, T.C. & Smith, N.J.P. 1989. Lower crustal heterogeneity beneath Britain from deep seismic reflection data. *Journal of the Geological Society, London*, **146**, 617–630.

Channell, J.E.T., McCabe, C. & Woodcock, N.H. 1993. Palaeomagnetic study of Llandovery (Lower Silurian) red beds in north-west England. *Geophysical Journal International*, **115**, 1085–1094.

Clark, A.H., Scott, D.J., Sandeman, H.A., Browley, A.V. & Farrar, E. 1998. Siegenian generation of the Lizard ophiolite: U-Pb zircon age data for plagiogranite, Porthkerris, Cornwall. *Journal of the Geological Society, London*, **155**, 595–598.

Clift, P.D., Dewey, J.F., Draut, A.E., Chew, D.M., Mange, M. & Ryan, P.D. 2004. Rapid tectonic exhumation, detachment faulting and orogenic collapse in the Caledonides of western Ireland. *Tectonophysics*, **384**, 91–113.

Cocks, L.R.M. 2000. The Early Palaeozoic geography of Europe. *Journal of the Geological Society, London*, **157**, 1–10.

Cocks, L.R.M. 2002. Key Lower Palaeozoic faunas from near the Trans-European Suture Zone. *In*: Winchester, J.A., Pharaoh, T.C. & Verniers, J. (eds) *Palaeozoic Amalgamation of Central Europe*. Geological Society, London, Special Publications, **201**, 37–46.

Cocks, L.R.M. 2005. Strophomenate brachiopods from the late Ordovician Boda Limestone of Sweden: their systematics and implications for palaeogeography. *Journal of Systematic Palaeontology*, **3**, 243–282.

Cocks, L.R.M. & Fortey, R.A. 1982. Faunal evidence for oceanic separations in the Palaeozoic of Britain. *Journal of the Geological Society, London*, **139**, 465–478.

Cocks, L.R.M. & Fortey, R.A. 1990. Biogeography of Ordovician and Silurian faunas. *In*: McKerrow, W.S. & Scotese, C.R. (eds) *Palaeozoic Palaeogeography and Biogeography*. Geological Society, London, Memoirs, **12**, 97–104.

Cocks, L.R.M. & Fortey, R.A. 1998. The Lower Palaeozoic margin of Baltica. *Geologiska Föreningens Forhaldligar*, **120**, 173–179.

Cocks, L.R.M. & Torsvik, T.H. 2002. Earth geography from 500 to 400 million years ago: a faunal and palaeomagnetic review. *Journal of the Geological Society, London*, **159**, 631–644.

Cocks, L.R.M. & Torsvik, T.H. 2005. Baltica from the late Precambrian to mid-Palaeozoic times: The gain and loss of a terrane's identity. *Earth-Science Reviews*, **72**, 39–66.

Cocks, L.R.M., McKerrow, W.S. & van Staal, C.R. 1997. The margins of Avalonia. *Geological Magazine*, **134**, 627–636.

Cooper, A.H., Millward, D., Johnson, E.W. & Soper, N.J. 1993. The early Palaeozoic evolution of northwest England. *Geological Magazine*, **130**, 711–724.

Cotte, N., Pedersen, H.A. & TOR Working Group 2002. Sharp contrast in lithospheric structure across the Sorgenfrei-Tornquist Zone as inferred by Rayleigh wave analysis of TOR1 project data. *Tectonophysics*, **360**, 75–88.

Coward, M.P. & Enfield, M.A. 1987. The structure of the West Orkney and ajacent basins. *Petroleum Geology of Northwest Europe. Proceedings of 3rd conference, London, 1986. Vol. 2*, 687–696.

Crowley, Q.G., Floyd, P.A., Winchester, J.A., Franke, W. & Holland, J.G. 2000. Early Palaeozoic rift-related magmatism in Variscan Europe: fragmentation of the Armorican terrane assemblage. *Terra Nova*, **12**(4), 171–180.

Dadlez, R. 1978. Sub-Permian rock complexes in the Koszalin-Chojnice Zone. *Kwartalnik Geologiczny*, **22**, 269–301 [in Polish, English summary].

Dadlez, R. 1997. Seismic profile LT-7 (northwest Poland): geological implications. *Geological Magazine*, **134**, 653–659.

Dadlez, R. 2001. Holy Cross Mts area – crustal structure, geophysical data and general geology. *Geological Quarterly*, **45**, 99–106.

Dadlez, R., Kowalczewski, Z. & Znosko, J. 1994. Some key problems of the pre-Permian tectonics of Poland. *Kwartalnik Geologiczny*, **38**, 169–189.

Dallmeyer, R.D., Franke, W. & Weber, K. 1995. *Pre-Permian Geology of Central and Eastern Europe*. Springer Verlag, Berlin.

Dallmeyer, R.D., Giese, U., Glasmacher, U. & Pickel, W. 1999. First

40Ar/39Ar age constraints for the Caledonian evolution of the Trans-European Suture Zone in NE Germany. *Journal of the Geological Society, London*, **156**, 279–290.

Dalziel, I. 1992. On the organisation of American plates in the Neoproterozoic and the breakout of Laurentia. *Geological Society of America Today*, **2**, 238–241.

Dalziel, I. 1997. Overview: Neoproterozoic–Palaeozoic geography and tectonics: review, hypotheses and environmental speculations. *Geological Society of America Bulletin*, **109**, 16–42.

Debacker, T.N. 2001. *Palaeozoic deformation of the Brabant Massif within eastern Avalonia: how, when and why?* PhD thesis, Ghent University, Belgium.

Debacker, T.N., Sintubin, M. & Verniers, J. 2001. Large-scale slumping deduced from structural relations and sedimentary features: a study in the Lower Palaeozoic Anglo-Brabant fold belt, Belgium. *Journal of the Geological Society, London*, **158**, 341–352.

Deeks, N.R. & Thomas, S.A. 1995. Basin inversion tectonics in a strike-slip regime: The Tornquist Zone, Southern Baltic Sea. *In*: Buchanan, J. G. & Buchanan, P. G. (eds) *Basin Inversion*. Geological Society, London, Special Publications, **88**, 319–338.

Deemer, S.J. & Hurich, C.A. 1991. Comparison of coincident high-resolution wide-aperture and CDP profiling along the southwest coast of Norway. *In*: Meissner, R., Brown, L., Dürbaum, H.J., Franke, W., Fuchs, K. & Seifert, F. (eds) *Continental Lithosphere: Deep Seismic Reflections*. Geodynamic Series, **22**, 435–442.

DEKORP-BASIN Research Group 1998. Survey provides seismic insights into an old suture zone. *Eos, Transactions*, **79**(12), 151,159.

DEKORP-BASIN Research Group 1999. The deep crustal structure of the Northeast German basin: New DEKORP-BASIN'96 deep-profiling results. *Geology*, **27**(1), 55–58.

Dewey, J.F. 1982. Plate tectonics and the evolution of the British Isles. *Journal of the Geological Society, London,* **139**, 317–414.

Dewey, J.F. 1988. Lithospheric stress, deformation and tectonic cycles: the disruption of Pangea and the closure of Tethys. *In*: Audley-Charles, M.G. & Hallam, A. (eds) *Gondwana and Tethys*. Geological Society, London, Special Publications, **37**, 23–40.

Dewey, J.F. & Burke, K.C. 1973. Tibetan, Variscan, and Precambrian basement reactivation: products of continental collision. *Journal of Geology*, **81**(6), 683–692.

Dewey, J.F. & Mange, M. 1999. Petrography of Ordovician and Silurian sediments in the western Irish Caledonides: tracers of a short-lived Ordovician continent-arc collision orogeny and the evolution of the Laurentian Appalachian-Caledonian margin. *In*: MacNiocaill, C. & Ryan, P.D. (eds) *Continental Tectonics*. Geological Society, London, Special Publications, **164**, 55–107.

Dewey, J.F. & Ryan, P.D. 1990. The Ordovician evolution of the South Mayo Trough, western Ireland. *Tectonics*, **9**, 887–901.

Dohr, G. 1983. Ergebnisse geophysikalischer Untersuchungen über den Bau des Nordwestdeutschen Beckens. *Erdöl Erdgas*, **99**, 252–267.

Dohr, G. 1989. Ergebnisse geophysikalischer Arbeiten zur Untersuchungen des tieferen Untergrundes in Norddeutschland. *Niedersächsische Akademie für Geowissenschaften*, **2**, 4–22.

Dombrowski, A., Henjes-Kunst, F., Höhndorf, A., Kröner, A., Okrusch, M. & Richter, P. 1995. Orthogneisses in the Spessart crystalline complex, north-west Bavaria, Silurian granitoid magmatism at an active continental margin. *Geologische Rundschau*, **83**, 399–411.

Dudek, A. 1980. The crystalline basement block of the Outer Carpathians in Moravia: Bruno-Vistulicum. *Rozprawy Ceskoslovenské Akademie Ved*, **90**, 1–85.

Dudek, A. 1995. Moravo-Silesian Zone – metamorphic evolution. *In*: Dallmeyer, R.D., Franke, W. & Weber, K. (eds) *Pre-Permian Geology of Central and Eastern Europe*. Springer-Verlag, Berlin, 508–511.

Duin, E., Rijkers, R. & Remmelts, G. 1995. Deep seismic reflections in the Netherlands: an overview. *Geologie en Mijnbouw*, **74**(3), 191–197.

Dzik, J., Olempska, E. & Pisera, A. 1994. Ordovician carbonate platform ecosystem of the Holy Cross Mountains. *Palaeontologia Polonica*, **53**, 1–315.

Eide, E.A., Torsvik, T.H., Andersen, T.B. & Arnaud, N. 1999. Early Carboniferous unroofing in Western Norway: the tale of alkali feldspar thermochronology. *Journal of Geology*, **107**, 353–374.

Elming, S.A. & Mattsson, H. 2001. Post Jotnian basic intrusions in the Fennoscandian Shield, and the break up of Baltica from Laurentia; a palaeomagnetic and AMS study. *Precambrian Research*, **108**(3–4), 215–236.

Emmert, U. & Stettner, G. 1995. *Erläuterungen zur Geologischen Karte von Bayern* 1:25000, Map sheet No. 6036 Weidenberg. Bayerisches Geologisches Landesamt, Munich, 1–239.

England, R. W. 1995. The seismic signatures of the Iapetus Ocean in the southern North Sea. *Studia Geophysica et Geodetica*, **39**, 235–240.

England, R. W. 1999. Deep structure of North West Europe from Deep Seismic Profiling: The link between basement tectonics and basin development. *In*: Mohriak, W. & Talwani, M. (eds) *Atlantic Rifts and Continental Margins*. AGU Geophysical Monograph, **115**, 57–83.

England, R. W. & Soper, N. J. 1997. Lower crustal structure of the East Irish Sea from deep seismic reflection data. *In*: Meadows, N. S., Trueblood, S. P., Hardman, M. & Cowan, G. (eds) *Petroleum Geology of the Irish Sea and Adjacent Areas*. Geological Society, London, Special Publications, **124**, 61–72.

Erlström, M., Thomas, S.A., Deeks, N. & Sivhed, U. 1997. Structure and tectonic evolution of the Torquist Zone and adjacent sedimentary basins in Scania and the southern Baltic Sea area. *Tectonophysics*, **271**, 191–215.

Esang, C.B. & Piper, J.D.A. 1984. Palaeomagnetism of Caledonian intrusive suites in the Northern Highlands of Scotland: Constraints to tectonic movements within the Caledonian Orogenic Belt. *Tectonophysics*, **104**, 1–34.

EUGEMI Working Group 1990. The European Geotraverse seismic refraction experiment of 1986 from Genova, Italy, to Kiel, Germany. *Tectonophysics*, **176**, 43–57.

EUGENO-S Working Group 1988. Crustal structure and tectonic evolution of the transition between the Baltic Shield and the North German Caledonides (the EUGENO-S Project). *Tectonophysics*, **150**, 253–348.

Finger, F., Hanzl, P., Pin, C., von Quadt, A. & Steyrer, H.P. 2000. The Brunovistulian; Avalonian Precambrian sequence at the eastern end of the Central European Variscides? *In*: Franke, W., Haak, V., Oncken, O. & Tanner, D. (eds) *Orogenic Processes; Quantification and Modelling in the Variscan Belt*. Geological Society, London, Special Publications, **179**, 103–112.

Fitzsimons, I.C.W. 2000. Grenville-age basement provinces in East Antarctica: evidence for three separate collisional orogens. *Geology*, **28**(10), 879–882.

Flack, C.A., Klemperer, S.L., McGeary, S., Snyder, D.B. & Warner, M.R. 1990. The reflective upper mantle of the UK. *Geology*, **18**, 528–532.

Flinn, D. 1985. The Caledonides of Shetland. *In*: Gee, D.G. & Sturt, B.A. (eds) *The Caledonide Orogen – Scandinavia and Related Areas*. John Wiley, Chichester, 1159–1171.

Foden, J., Barovich, K., Jane, M. & O'Halloran, G. 2001. Sr-isotopic evidence for Late Neoproterozoic rifting in the Adelaide geosyncline at 586 Ma: Implications for a Cu ore forming fluid flux. *Precambrian Research*, **106**(3–4), 291–308.

Förste, C. & Schulze, A. 1991. Recent DSS investigations in the North German Basin and the Variscan fold belt. *22nd assembly of the European Seismological Commission*, Barcelona.

Fortey, R.A. & Cocks, L.R.M. 1992. The Early Palaeozoic of the north Atlantic Region as a test case for the use of fossils in continental reconstruction. *Tectonophysics*, **206**, 147–158.

Fortey, R.A. & Cocks, L.R.M. 2003. Palaeontological evidence bearing on global Ordovician–Silurian continental reconstructions. *Earth-Science Reviews*, **61**, 245–307.

Fortey, R.A. & Cocks, L.R. M. 2005. Late Ordovician global warming – The Boda Event. *Geology*, **33**, 405–408.

Fortey, N.J., Merriman, R.J. & Huff, W.D. 1996. Silurian and late Ordovician K-bentonites as a record of late Caledonian volcanism in the British Isles. *Transactions of the Royal Society of Edinburgh*, **86**, 167–180.

Franke, D. 1990*a*. Der präpermische Untergrund der Mitteleuropäischen Senke - Fakten und Hypothesen. *Niedersächsische Akademie der Geowissenschaften Veröffentlichungen*, **4**, 19–75.

Franke, D. 1990*b*. The northwest part of the Tornquist-Teisseyre Zone – platform margin or intraplate structure? *Zeitschrift für angewandte Geologie*, **36**, 45–48.

Franke, D. 1993. The southern border of Baltica – a review of the present state of knowledge. *Precambrian Research*, **64**, 419–430.

Franke, D. 1995. The North Variscan Foreland. *In*: Dallmeyer, D., Franke, W. & Weber, K.W. (eds) *The Pre-Permian Geology of Central and Eastern Europe*. Springer, Berlin, 554–566.

Franke, D. & Hoffmann, N. 1999. Das Elbe-Lineament – bedeutende Geofraktur oder Phantomgebilde? – Teil 1: Die Referenzgebiete. *Zeitschrift für geologische Wissenschaften*, **27**(3/4), 279–318.

Franke, W. 1989. Variscan plate tectonics in Central Europe – current ideas and open questions. *Tectonophysics*, **169**, 221–228.

Franke, W. 1995. Rhenohercynian foldbelt: autochton and nonmetamorphic nappe units – stratigraphy. *In*: Dallmeyer, D., Franke, W. & Weber, K. (eds) *The Pre-Permian Geology of Central and Eastern Europe*. Springer-Verlag, Berlin, 33–49.

Franke, W. & Oncken, O. 1995. Zur prädevonischen Geschichte des Rhenohercynischen Beckens. *Nova Acta Leopoldina*, NF, **71**(291), 53–72.

Franke, W., Zelazniewicz, A., Porebski, S.J. & Wajspyrch, B. 1993. The Saxothuringian zone in Germany and Poland: differences and common features. *Geologische Rundschau*, **82**, 583–599.

Franke, W., Dallmeyer, R.D. & Weber, K. 1995. Geodynamic evolution. *In*: Dallmeyer, R.D., Franke, W. & Matte, P. (eds) *Pre-Permian Geology of Central and Eastern Europe*. Springer-Verlag, Berlin, 579–593.

Freeman, B., Klemperer, S. L. & Hobbs, R. W. 1988. The deep structure of northern England and the Iapetus Suture zone from BIRPS deep seismic reflection profiles. *Journal of the Geological Society, London*, **145**, 727–740.

Friedrich, A.M., Bowring, S.A., Martin, M.W. & Hodges, K.V. 1999. Short-lived continental magmatic arc at Connemara, western Irish Caledonides: implications for the age of the Grampian orogeny. *Geology*, **27**, 27–30.

Frost, R.T.C., Fitch, F.J. & Miller, J.A. 1981. The age and nature of the crystalline basement of the North Sea Basin. *In*: Illing, L.V. & Hobson, G.D. (eds) *Petroleum Geology of the Continental Shelf of North-West Europe*. Institute of Petroleum, London, 43–57.

Froitzheim, N., Plašienka, D. & Schuster, R. 2008. Alpine tectonics of the Alps and western Carpathians. *In*: McCann, T. (ed.) *The Geology of Central Europe. Volume 2: Mesozoic and Cenozoic*. Geological Society, London, 1141–1232.

Furlong, K.P., Spakman, W. & Wortel, R. 1995. Thermal structure of the continental lithosphere: constraints from seismic tomography. *Tectonophysics*, **244**(1–3), 107–117.

Furnes, H., Kryza, R. & Muszynski, A. 1989. Geology and geochemistry of Early Palaeozoic volcanics of the Swierzawa unit, kaczawa Mts. W. Sudetes, Poland. *Neues Jahrbuch für Geologie und Paläontologie*, **3**, 136–154.

Gaál, G. & Gorbatchev, R. 1987. An outline of the Precambrian evolution of the Baltic Shield. *Precambrian Research*, **35**, 15–52.

Galson, D.A. & Rybach, L. 1985. Recommendations for an integrated programme of geothermal studies along the EGT. *Terra Cognita*, **5**, 371–374.

Gee, D.G. 1987. The Scandinavian Alum Shales - Mid Cambrian to Tremadoc deposition in response to early Caledonian subduction. *Norsk geologisk Tidskrift*, **67**, 233–235.

Gee, D.G. & Stephenson, R.A. 2006. *European Lithosphere Dynamics*. Geological Society, London, Memoirs, **32**.

Gee, D.G. & Zeyen, H. 1996. *EUROPROBE 1996 – Lithosphere dynamics – Origin and Evolution of Continents*. EUROPROBE Secretariat, Uppsala.

Gee, D.G., Johansson, A., Larionov, A.N. & Tebenkov, A.M. 1999. A Caledonian granitoid pluton at Djupkilsodden, Central Nordaustlandet, Svalbard: Age, magmatic signature and tectonic significance. *Polarforschung*, **66**, 19–32.

Gehmlich, M., Tichomirowa, M., Linnemann, U., Lützner, H. & Bombach, K. 1997. Kalibrierung und Korrelation des saxothuringischen Ordoviziums durch die Einzelzirkon-Evaporisationsmethode. *Terra Nostra*, **97**, 43–46.

GeoFrance 3D 2005. World Wide Web address: http://www2.brgm.fr/geofrance3d/

Gibbs, M.T., Barron, E.J. & Kump, L.R. 1997. An atmospheric CO2 threshold for glaciation in the Late Ordovician. *Geology*, **25**, 447–450.

Giese, U., Katzung, G. & Walter, R. 1994. Detrital composition of Ordovician sandstones from the Rügen boreholes: implications for the evolution of the Tornquist Ocean. *Geologisches Rundschau*, **83**, 293–308.

Giese, U., Dallmeyer, R.D., Kramm, U. & Mingram, B. 1995. First isotope investigations (U/Pb, Ar/Ar) of Early Palaeozoic rock samples from the Rügen boreholes. *Nachrichten Deutsche Geologische Gesellschaft*, **54**, 71–72.

Glasmacher, U.A., Reynolds, P., Alekseyev, A.A., Puschkov, V.N., Taylor, K. Gorozhanin, V. & Walter, R. 1999. 40Ar/39Ar thermochronology west of the main Uralian fault, Southern Urals, Russia. *Geologische Rundschau*, **87**(4), 515–525.

Gossler, J., Kind, R., Sobolev, S.V., Kämpf, H., Wylegalla, K. & Stiller, M. 1999. Major crustal features between the Harz Mountains and the Baltic Shield derived from receiver functions. *Tectonophysics*, **314**(1–3), 321–334.

Grad, M., Janik, T., *et al.* 1999. Crustal structure of the mid-Polish trough beneath the Teisseyre-Tornquist Zone seismic profile. *Tectonophysics*, **314**(1–3), 145–160.

Grad, M., Guterch, A. & Mazur, S. 2002*a*. Seismic refraction evidence for crustal structure in the central part of the Trans-European Suture Zone in Poland. *In*: Winchester, J. A., Pharoah, T. C. & Veniers, J. (eds) *Palaeozoic Amalgamation of Central Europe*. Geological Society, London, Special Publications, **201**, 295–309.

Grad, M., Keller, G.R., Thybo, H., Guterch, A. & POLONAISE Working Group 2002*b*. Lower lithospheric structure beneath the Trans-European Suture Zone from POLONAISE'97 seismic profiles. *Tectonophysics*, **360**, 153–168.

Grad, M., Spicak, A., Keller, G.R., Guterch, A., Broz, M., Hegedus, E. & SUDETES 2003 Working Group 2003. SUDETES 2003 seismic experiment. *Studia Geophysica et Geodaetica*, **47**(3).

Greenough, J.D., Kamo, S.L. & Krogh, T.E. 1993. A Silurian U-Pb age for the Cape St Mary's sills, Avalon Peninsula, Newfoundland, Canada: Implications for Silurian orogenesis in the Avalon Zone. *Canadian Journal of Earth Sciences*, **30**, 1607–1612.

Gregersen, S., Voss, P. & TOR Working Group 2002. Summary of project TOR: delineation of a stepwise, sharp, deep lithosphere transition across Germany-Denmark-Sweden. *Tectonophysics*, **360**, 61–73.

Grosse, S., Conrad, W., Behr, H.J. & Heinrich, T. 1990. Major gravity axes and anomalies in central Europe. *In*: Freemann, R., Giese, P. & Mueller, S. (eds) *The European Geotraverse: Intergrative Studies*. European Science Foundation, Strasbourg, 135–146.

Grosse, S., Behr, H.J., Edel, J.B. & Heinrich, T. 1992. The gravity field along the central segment of the EGT. *Tectonophysics*, **207**, 97–121.

Guggisberg, B., Kaminski, W. & Prodehl, C. 1991. Crustal structure of the Fennoscandian Shield: A traveltime interpretation of the long-range FENNOLORA seismic refraction profile. *Tectonophysics*, **195**, 105–137.

Guterch, A., Grad, M., Materzok, R. & Perchuc, E. 1986. Deep structure of the earth's crust in the contact zone of the Palaeozoic and Precambrian platforms in Poland (Tornquist-Teisseyre Zone). *Tectonophysics*, **128**, 251–279.

Guterch, A., Grad, M., *et al.* 1994. Crustal structure of the transition zone between Precambrian and Variscan Europe from new seismic data along LT-7 profile (NW Poland and eastern Germany). *Compte Rendu Academie Sciences Paris*, **319**(2), 1489–1496.

Guterch, A., Grad, M., Thybo, H., Keller, G.R. & Miller, K.C. 1998. POLONAISE '97 – an international seismic experiment between Precambrian and Variscan Europe in Poland. *Eos, Transactions*, **79**(26), 302, 305.

Guterch, A., Grad, M.,Thybo, H., Keller, G.R. & The POLONAISE Working Group 1999. POLONAISE'97 – an international seismic experiment between Precambrian and Variscan Europe in Poland. *Tectonophysics*, **314**, 101–121.

Guterch, A., Grad, M., *et al.* 2003. Deep crustal structure along CELEBRATION 2000 seismic trans-Carpathian profiles extending from the East European Craton to the Pannonian basin. *Publications of the Institute of Geophysics, Polish Academy of Sciences, Series M*, **28**(363), 103–104.

Haenel, R. & Staroste, E. 1988. *Atlas of Geothermal Resources in the European Community, Austria and Switzerland*. Verlag Th. Schaefer,

Hannover.

Harland, W.B. & Gayer, R.A. 1972. The Arctic Caledonides and earlier Oceans. *Geological Magazine*, **109**, 289–384.

Harper, D.A.T., Mac Niocaill, C. & Williams, S.H. 1996. The palaeogeography of early Ordovician Iapetus terranes: an integration of faunal and palaeomagnetic constraints. *Palaeogeography, Palaeoclimatology, Palaeoecology*, **121**, 297–312.

Hartz, E.H. & Torsvik, T.H. 2002. Baltica upside down: a new plate tectonic model for Rodinia and the Iapetus Ocean. *Geology*, **30**, 255–258.

Havlíček, V., Vanek, J. & Fatka, O. 1994. Perunica microcontinent in the Ordovician (its position within the Mediterranean Province, series division, benthic and pelagic associations). *Sbornik geologickych véd Geologie*, **46**, 23–56.

Haynes, J.T., Melson, W.G. & Kunk, M.J. 1995. Composition of biotite phenocrysts in Ordovician tephras casts doubt on the proposed trans-Atlantic correlation of the Millbrig K-bentonite (United States) and the Kinnekulle K-bentonite (Sweden). *Geology*, **23**, 847–850.

Heeremans, M., Faleide, J.I. & Larsen, B.T. 2004. Late Carbonifeous – Permian of Central Europe: an introduction to a new regional map. *In*: Wilson, M., Neumann, E.-R., Davies, G.R., Timmermann, M.J., Heeremans, M. & Larsen, B.T. (eds) *Permo-Carboniferous Magmatism and Rifting in Europe*. Geological Society, London, Special Publications, **223**, 75–88.

Henriksen, N. & Higgins, A.K. 1976. *East Greenland Caledonian Fold Belt*. Groenlands Geol. Unders., Copenhagen, 182–246.

Hock, S., Korn, M., Ritter, J.R.R. & Rothert, E. 2004. Mapping random lithospheric heterogeneities in northern and central Europe. *Geophysical Journal International*, **157**, 251–264.

Hoffmann, H., Bölsche, J., Horst, W., Lange, W., Palesch, F., Pomeranceva, V. & Wruck, K. 1989. Tiefenseismische Untersuchungen in der DDR durch den VEB Kombinat Geophysik - Stand und Ergebnisse. *Zeitschrift für Angewandte Geologie*, **35**, 308–314.

Hoffmann, H., Stiewe, H. & Pasternak, G. 1996. Struktur und Genese der Mohorovicic-Diskontinuität (Moho) im Norddeutschen Becken - ein Ergebnis langzeitregistrierter Steilwinkelseismik. *Zeitschrift für Angewandte Geologie*, **42**, 138–148.

Hoffmann, N. 1990. Zur paläodynamischen Entwicklung des Präzechsteins in der Nordeutschen Senke. *Niedersächsische Akadademie Geowissenschaften Veröffentlichungen*, **4**, 5–18.

Hoffmann, N. & Franke, D. 1997. The Avalonia–Baltica Suture in NE Germany – new constraints and alternative interpretations. *Zeitschrift für geologische Wissenschaften*, **25**(1/2), 3–14.

Hoffmann, N. & Stiewe, H. 1994. Neuerkenntnisse zur geologisch-geophysikalischen Modellierung der Pritzwalker Anomalie im Bereich des Ostelbischen Massivs. *Zeitschrift für geologische Wissenschaften*, **22**(1/2), 161–171.

Hoffmann, N., Jödicke, H., Fluche, B., Jording, A. & Müller, W. 1998. Modellvorstellungen zur Verbreitung potentieller präwestfalischer Erdgas-Muttergesteine in Norddeutschland - Ergebnisse neuer magnetotellurischer Messungen. *Zeitschrift für Angewandte Geologie*, **44**, 140–158.

Hoffmann, N., Jödicke, H. & Gerling, P. 2001. The distribution of pre-Westphalian source rocks in the North German Basin; evidence from magnetotelluric and geochemical data. *Geologie en Mijnbouw. Netherlands Journal of Geosciences*, **80**(1), 71–84.

Hoffmann, P.F. 1991. Did the breakout of Laurentia turn Gondwanaland inside-out? *Science*, **252**, 1409–1412.

Holdsworth, R. E. 1989. Late brittle deformation in a Caledonian thrust wedge: new evidence for gravitational collapse in the Moine thrust sheet, Sutherland, Scotland. *Tectonophysics*, **170**, 17–28.

Horst, W., Flueh, E.R. & Börngen, M. 1994. NIZUSE: a deep seismic reflection line in north-eastern Germany. *Geologische Rundschau*, **83**, 161–169.

Hoth, K., Rusbült, J., Zagora, K., Beer, H. & Hartmann, O. 1993. Die tiefen Bohrungen im Zentralabschnitt der Mitteleuropäischen Senke - Dokumentation für den Zeitabschnitt 1962–1990. *Schriftenreihe für Geowissenschaften*, Heft **2**, 145 pp.

Huff, W.D., Bergström, S.M. & Kolata, D.R. 1992. Gigantic Ordovician volcanic ash fall in North America and Europe: biological, tectonomagmatic, and event-stratigraphic significance. *Geology*, **20**, 875–878.

Huff, W.D., Kolata, D.R., Bergström, S.M. & Zhang, Y.S. 1996. Large-magnitude Middle Ordovician volcanic ash falls in North America and Europe: dimensions, emplacement and post-emplacement characteristics. *Journal of Volcanology and Geothermal Research*, **73**, 285–301.

Hurich, C.A. 1996. Kinematic evolution of the lower plate during intracontinental subduction: An example from the Scandinavian Caledonides. *Tectonics*, **15**, 1248–1263.

Hurich, C.A. & Kristoffersen, Y. 1988. Deep structure of the Caledonide orogen in southwestern Norway: New evidence from marine seismic reflection profiling. *In*: Kristoffersen, Y. (ed.) *Progress in Studies of the Lithosphere in Norway*. Geological Survey of Norway, 66–69.

Hurich, C.A., Palm, H., Dyrelius, D. & Kristoffersen, Y. 1989. Deformation of Baltica continental crust during Caledonian intracontinental subduction: views from seismic reflection data. *Geology*, **17**, 423–425.

Hurtig, E. 1995. Temperature and heat-flow density along European transcontinental profiles. *Tectonophysics*, **244**(1–3), 75–83.

Hurtig, E., Cermal, V., Haenel, R. & Zui, V. 1992. *Geothermal Atlas of Europe*. Hermann Haak Verlagsgesellschaft mbH, Gotha.

Husebye, E.S. & Hovland, J. 1982. On upper mantle seismic heterogeneities beneath Fennoscandia. *The Structure of the Lithosphere-Asthenosphere in Europe and the North Atlantic*, **90**(1–2), 1–17.

Husebye, E.S., Ro, H.E., Kinck, J.J. & Larsson, F.R. 1988. Tectonic studies in the Skagerrak province: the 'MOBIL Search' cruise. *Norges Geologiske Undersøkelse Bulletin*, **3**, 14–20.

Hutton D.H.W. 1987. Strike-slip terranes and a model for the evolution of the British and Irish Caledonides. *Geological Magazine*, **124**, 405–425.

Hutton, D.H.W. & Murphy, F.C. 1987. The Silurian of the Southern Uplands and Ireland as a successor basin to the end-Ordovician closure of Iapetus. *Journal of the Geological Society, London*, **144**, 765–772.

Hutton, D.H.W. & Reavy, R.J. 1992. Strike-slip tectonics and granite petrogenesis. *Tectonics*, **11**, 960–967.

Hynes, A. & Snyder, D.B. 1995. Deep-crustal mineral assemblages and potential for crustal rocks below the Moho in the Scottish Caledonides. *Geophysical Journal International*, **123**(2), 323–339.

Jaanusson, V. 1973. Aspects of carbonate sedimentation in the Ordovician of Baltoscandia. *Lethaia*, **6**, 11–34.

Jacques, J.M. & Reavy, R.J. 1994. Caledonian plutonism and major lineaments in the SW Scottish Highlands. *Journal of the Geological Society, London*, **151**, 955–969.

Janik, T., Yliniemi, J., Grad, M., Thybo, H., Tiira, T. & POLONAISE Working Group 2002. Crustal structure across the TESZ along POLONAISE'97 seismic profile P2 in NW Poland. *Tectonophysics*, **360**, 129–152.

Jensen, S.L., Janik, T., Thybo, H. & POLONAISE PROFILE i Working Group 1999. Seismic structure of the Palaeozoic Platform along POLONAISE'97 profile 1 in northwestern Poland. *Tectonophysics*, **314**, 123–143.

Jensen, S.L., Thybo, H. & POLONAISE'97 Working Group 2002. Moho topography and lower crustal wide-angle reflectivity around the TESZ in southern Scandinavia and northeastern Europe. *Tectonophysics*, **360**, 187–213.

Johansson, A., Maluski, H. & Gee, D.G. 2001. Ar-Ar dating of Caledonian and Grenvillian rocks from northeasternmost Svalbard; evidence of two stages of Caledonian tectonothermal activity in the High Arctic?. *Norsk Geologisk Tidsskrift*, **81**(4), 263–281.

Johnston, J.D., Tait, J.A., Oliver, G.J.H. & Murphy, F.C. 1994. Evidence for a Caledonian orogeny in Poland. *Transactions of the Royal Society of Edinburgh, Earth Sciences*, **85**, 131–142.

Jones, D.L., Silberling, N.J. & Coney, P.J. 1982. Timing of major accretionary events in Alaska. *Eos, Transactions*, **63**(45), 913–914.

Juhojuntti, N., Juhlin, C. & Dyrelius, D. 2001. Crustal reflectivity underneath the Central Scandinavian Caledonides. *Tectonophysics*, **334**, 191–210.

Kaiser, A., Reicherter, K., Hübscher, C. & Gajewski, D. 2005. Variation of the present-day stress field within the North German Basin – insights from thin shell FE modeling based on residual GPS velocities. *Tectonophysics*, **397**, 55–72.

Kämpf, H., Korich, D. & Brause, H. 1994. Anorthosit-, Gabbro- und Ilmeniterz-Xenolithe in permischen Basalten der Tiefbohrung

Schwerin 1, Ein petrographisches Indiz für das Ostelbische Massif. *Zeitschrift für geologische Wissenschaften*, **22**, 115–128.

Katzung, G., Giese, U., Walter, R. & von Winterfeld, C. 1993. The Rügen Caledonides, northeast Germany. *Geological Magazine*, **130**, 725–730.

Kemnitz, H., Romer, R.L. & Oncken, O. 2002. Gondwana break-up and the northern margin of the Saxothuringian belt (Variscides of Central Europe). *International Journal of Earth Science (Geologische Rundschau)*, **91**, 246–259.

Khain, V.E. 1985. *Geology of the USSR: First part - Old cratons and Paleozoic fold belts*. Beitraege zur Regionalen Geologie der Erde, **17**.

Kinck, J.J., Husebye, E.S. & Lund, C.-E. 1991. The South Scandinavian crust: structural complexities from seismic reflection and refraction profiling. *Tectonophysics*, **189**, 117–133.

King, L.M. 1994. Subsidence analysis of eastern Avalonian sequences: implications for Iapetus closure. *Journal of the Geological Society, London*, **151**, 647–657.

Klemperer, S.L. 1988. Crustal thinning and nature of extension in the northern North Sea from deep seismic reflection profiling. *Tectonics*, **7**(4), 803–821.

Klemperer, S.L. 1989. Seismic reflection evidence for the location of the Iapetus suture west of Ireland. *Journal of the Geological Society, London*, **146**, 409–412.

Klemperer, S.L. & Hobbs, R.W. 1992. *The BIRPS Atlas: Deep Seismic Reflection Profiles around the British Isles*. Cambridge University Press, Cambridge.

Klemperer, S.L., Ryan, P.D. & Snyder, D.B. 1991. A deep seismic reflection transect across the Irish Caledonides. *Journal of the Geological Society, London*, **148**, 149–164.

Kokelaar, B.P. 1988. Tectonic controls of Ordovician arc and marginal basin volcanism in Wales. *Journal of the Geological Society, London*, **145**, 759–776.

Kokelaar, B.P., Howells, M.F., Bevins, R.E., Roach, R.A. & Dunkley, P.N. 1984. The Ordovician marginal basin of Wales. *In*: Kokelaar, B.P. & Howells, M.F. (eds) *Marginal Basin Geology*. Geological Society, London, Special Publication, **16**, 245–269.

Korytowski, A., Dorr, W. & Zelazniewicz, A. 1993. U-Pb dating of (meta)granitoids in the NW Sudetes (Poland) and their bearing on tectono-stratigraphic correlation. *Terra Nova*, **5** (Abstract Supplement 1), 331–332.

Kossow, D. & Krawczyk, C.M. 2002. Structure and quantification of factors controlling the evolution of the inverted NE German Basin. *Marine and Petroleum Geology*, **19**(5), 601–618.

Kossow, D., Krawczyk, C.M., McCann, T., Strecker, M. & Negendank, J.F.W. 2000. Style and evolution of salt pillows and related structures in the northern part of the Northeast German Basin. *International Journal of Earth Sciences*, **89**(3), 652–664.

Kowalczewski, Z. 1990. Grubookruchowe skaly kambru na srodkowym poludniu Polski. *Prace Panstwowego Instytutu Geologicznego*, **131**, 1–82.

Krauss, M. 1994. The tectonic structure below the Southern Baltic Sea and its evolution. *Zeitschrift für geologische Wissenschaften*, **22**(1/2), 19–32.

Krawczyk, C.M., Stiller, M. & DEKORP-BASIN Research Group 1999. Reflection seismic constraints on Paleozoic crustal structure and Moho beneath the NE German Basin. *Tectonophysics*, **314**(1–3), 241–253.

Krawczyk, C.M., Eilts, F., Lassen, A. & Thybo, H. 2002. Seismic evidence of Caledonian deformed crust and uppermost mantle structures in the northern part of the Trans-European Suture Zone (TESZ), SW Baltic Sea. *Tectonophysics*, **360**(1–4), 215–244.

Krolikowski, C. & Petecki, Z. 1995. *Gravimetric Atlas of Poland*. Polish Geological Institute, Warsaw.

Krolikowski, C. & Petecki, Z. 1997. Crustal structure at the Trans-European Suture Zone in northwest Poland based on gravity data. *Geological Magazine*, **134**(5), 661–667.

Kröner, A. & Hegner, E. 1998. Geochemistry, single zircon ages and Sm-Nd systematics of granitoid rocks from the Góry Sowie (Owl Mts), Polish West Sudetes: evidence for early Palaeozoic arc-related plutonism. *Journal of the Geological Society, London*, **155**, 711–724.

Kröner, A., Hegner, E. & Jaeckel, P. 1994*a*. Pb-Pb and U-Pb zircon ages and Nd isotopic systematics for metamorphic rocks from the Góry Sowie block, West Sudetes, Poland and geodynamic significance. *Journal of the Czech Geological Society*, **39**, 60.

Kröner, A., Jaeckel, P. & Opletal, M. 1994*b*. Pb-Pb and U-Pb zircon ages for orthogneisses from eastern Bohemia: Further evidence for a major Cambro-Ordovician magmatic event. *Journal of the Czech Geological Society*, **39**, 61.

Kroner, U., Mansy, J.-L. *et al.* 2008. Variscan tectonics. *In*: McCann, T. (ed.) *The Geology of Central Europe. Volume 1: Precambrian and Palaeozoic*. Geological Society, London, 599–664.

Kryza, R. & Pin, C. 2002. Mafic rocks in a deep-crustal segment of the Variszides (the Góry Sowie, SW Poland): evidence for crustal contamination in an extensional setting. *International Journal of Earth Science (Geologische Rundschau)*, **91**, 1017–1029.

Kryza, R., Muszynski, A., Turniak, K. & Zalasiewicz, J.A. 1994. A lower Palaeozoic shallow water sequence in the eastern European Variscides (SW Poland) – provenance and depositional history. *Geologische Rundschau*, **83**(1), 5–19.

Kuder, J. 2002. *3D Schwerefeldmodellierung zur Erfassung des tiefen Untergrundes im Nordost-Deutschen Becken*. PhD thesis, FU Berlin.

Kumarapeli, P.S., Dunning, G.R. Pintson, H. & Shaver, J. 1989. Geochemistry and U-Pb zircon age of comenditic metafelsites of the Tibbit Hill Formation, Quebec Appalachians. *Canadian Journal of Earth Sciences*, **26**, 1374–1383.

Kumpulainen, R. & Nystuen, J.P. 1985. Late Proterozoic basin evolution and sedimentation in the westernmost part of Baltoscandia. *In*: Gee, D. (ed.) *The Caledonide Orogen – Scandinavia and Related Areas*. John Wiley & Sons. Chichester, 213–232.

Lambert, R.S. & McKerrow, W.S. 1976. The Grampian Orogeny. *Scottish Journal of Geology*, **12**, 271–292.

Lamens, J. 1985. Transition from turbidite to shallow-water sedimentation in the lower Salmian (Tremadocian-lower Ordovician) of the Stavelot Massif, Belgium. *Sedimentary Geology*, **44**, 121–142.

Lamens, J. 1986. Depositional history of Salmian (Lower Ordovician sediments in Belgium). *Aardkundige Mededelingen*, **3**, 125–138.

Lamens, J. 1988. Sedimentology of the Cambro-Ordovician of the Caledonian Stavelot Massif. Excursion A-5. *In*: Herbosch, A. (ed.) *IAS, 9th European Regional Meeting, Excursion Guidebook Leuven-Belgium*. Belgian Geological Survey, 67–76.

Landes, M., Prodehl, C., Hauser, F., Jacob, A.W.B. & Vermeulen, N.J. 2000. VARNET-96: Influence of the Variscan and Caledonian orogenies on crustal structure in SW Ireland. *Geophysical Journal International*, **140**(3), 660–676.

Landes, M., Ritter, J.R.R., Readman, P.W. & O'Reilly, B.M. 2005. A review of the Irish crustal structure and signatures from the Caledonian and Variscan Orogenies. *Terra Nova*, **17**, 111–120. DOI: 10.1111/j.1365-3121.2004.00590.x.

Lange, W. 1973. *Beiträge zur regionalen tiefenseismischen Erkundung des Nordteils der DDR (KMgW- und SEMLJA-Programm 1963–1971)*. PhD thesis, Karl-Marx-Universität Leipzig (Dissertation A1973).

Lange, W. 1978. [Investigation of the structure of the Earth's Crust and the Upper Mantel in the DDR] *In*: Sollogub, V. B., Chekunov, A. V., Kaluzhnaya, L. T. *et al.* (eds) [*The Structure of the Earth's Crust and the Upper Mantle of Central and Eastern Europe*] (In Russian). Naukova, Dumka, Kiev, 52–71.

Lassen, A. & Thybo, H. 2004. Seismic evidence for Late Proterozoic orogenic structures below the Phanerozoic sedimentary cover in the Kattegat area, SW Scandinavia. *Tectonics*, **23**(2), 1–25.

Lassen, A., Thybo, H. & Berthelsen, A. 2001. Reflection seismic evidence for Caledonian deformed sediments above Sveconorwegian basement in the southwestern Baltic Sea. *Tectonics*, **20**, 268–276.

Lee, M. K., Pharaoh, T. C. & Soper, N. J. 1990. Structural trends in central Britain from images of gravity and aeromagnetic fields. *Journal of the Geological Society, London*, **147**, 241–258.

Leggett, J. K., McKerrow, W. S. & Eales, M. H. 1979. The southern Uplands of Scotland, a lower Palaeozoic accretionary prism. *Journal of the Geological Society, London*, **136**, 755–770.

Leveridge, B.E., Holder, M.T. & Day, G.A. 1984. Thrust nappe tectonics in the Devonian of south Cornwall and the western English Channel. *In*: Hutton, D.W. & Sanderson, D.J. (eds) *Variscan Tectonics of the North Atlantic Region*. Geological Society, London, Special Publications, **14**, 103–112.

Liboriussen, J., Ashton, P. & Thygesen, T. 1987. The evolution of the Fennoscandian border zone in Denmark. *Tectonophysics*, **137**, 21–29.

LIE, J.E. & ANDERSSON, M. 1998. The deep-seismic image of the crustal structure of the Tornquist Zone beneath the Skagerrak Sea, northwestern Europe. *Tectonophysics*, **287**(1–4), 139–155.

LIE, J.E. & HUSEBYE, E. 1994. Simple-shear deformation of the Skagerrak lithosphere during the formation of the Oslo Rift. *Tectonophysics*, **232**, 133–141.

LINNEMANN, U. & ROMER, R.L. 2002. The Cadomian orogeny in Saxo-Thuringia, Germany. Geochemical and Nd-Sr-Pb isotopic characterisation of marginal basins with constraints to geotectonic settings and provenance. *Tectonophysics*, **352**, 33–64.

LINNEMANN, U., GEHMLICH, M., *et al.* 2000. From Cadomian subduction to early Palaeozoic rifting: the evolution of Saxo-Thuringia at the margin of Gondwana in the light of single zircon geochronology and basin development (central European Variscides, Germany). *In*: FRANKE, W., HAAK, V., ONCKEN, O. & TANNER, D. (eds) *Orogenic Processes: Quantification and Modelling in the Variscan Belt of Central Europe*. Geological Society, London, Special Publications, **179**, 131–153.

LOWE, C. & JACOB, A.W.B. 1989. A north–south seismic profile across the Caledonian Suture zone in Ireland. *Tectonophysics*, **168**, 297–318.

MCBRIDE, J.H. 1994. Structure of a continental strike-slip fault from deep seismic reflection: Walls Boundary fault, northern British Caledonides. *Journal of Geophysical Research*, **99**, 23985–24005.

MCBRIDE, J. H. 1995. Does the great Glen fault really disrupt Moho and upper mantle structure? *Tectonics*, **14**, 422–434.

MCBRIDE, J. H. & ENGLAND, R. W. 1995. Problems interpreting deep crustal reflectors beneath the Moine thrust between Shetland and the Scottish Highlands. *Geological Magazine*, **132**, 465–472.

MCBRIDE, J.H. & ENGLAND, R.W. 1999. Window into the Caledonian orogen: Structure of the crust beneath the East Shetland platform, United Kingdom. *Geological Society of America* Bulletin, **111**, 1030–1041.

MCCANN, T. 1996. Silurian facies from the G-14 well, offshore Northern Germany. *Zeitschrift der Deutschen Geologischen Gesellschaft*, **147**, 209–217.

MCCANN, T. 1998. Lower Palaeozoic evolution of the northeast German Basin/Baltica borderland. *Geological Magazine*, **135**, 129–142.

MCCANN, T. 2008. Introduction and overview. *In*: MCCANN, T. (ed.) *The Geology of Central Europe. Volume 1: Precambrian and Palaeozoic.* Geological Society, London, 1–20.

MCCANN, T. & KRAWCZYK, C.M. 2001. The Trans-European Fault: a critical reassessment. *Geological Magazine*, **138**, 19–29.

MCCANN, T., SKOMPSKI, S., *et al.* 2008. Carboniferous. *In*: MCCANN, T. (ed.) *The Geology of Central Europe. Volume 1: Precambrian and Palaeozoic*. Geological Society, London, 411–529.

MCCAUSLAND, P.J.A., HODYCH, J.P. & DUNNING, G.R. 1997. Evidence from western Newfoundland for the final breakup of Rodinia? U-Pb age and palaeolatitude of the Skinner Cove volcanics. *Geological Society of Canada/Mineralogical Association of Canada Annual Meeting, Program with Abstracts*, Ottawa, Ontario, A-96.

MCCLAY, K.R. 1996. Recent advantages in analogue modelling: uses in section interpretation and validation. *In*: BUCHANAN, P.G. & NIEUWLAND, D.A. (eds) *Modern Developments in Structural Interpretation, Validation and Modelling.* Geological Society, London, Special Publications, **99**, 201–226.

MCCLAY, K.R., NORTON, M.G., CONEY, P. & DAVIS, G.H. 1986. Collapse of the Caledonian Orogen and the Old Red Sandstone. *Nature*, **323**(6084), 147–149.

MCGEARY, S. 1989. Reflection seismic evidence for a Moho offset beneath the Walls Boundary strike-slip fault. *Journal of the Geological Society, London*, **146**, 261–269.

MCKERROW, W.S. 1988. Wenlock to Givetian deformation in the British Isles and the Canadian Appalachians. *In*: HARRIS, A.L. & FETTES, D.J. (eds) *The Caledonian-Appalachian Orogen*. Geological Society, London, Special Publications, **38**, 437–448.

MCKERROW, W. S. & COCKS, L. R. M. 1986. Oceans, Island Arcs and Olistostromes: The Use of Fossils in Distinguishing Sutures, Terranes and Environments around the Iapetus Ocean. *Journal of the Geological Society, London*, **143**, 185–191.

MCKERROW, W.S. & SCOTESE, C.T. (eds) 1990. *Palaeozoic Palaeogeography and Biogeography.* Geological Society, London, Memoirs, **12**.

MCKERROW, W.S., DEWEY, J.F. & SCOTESE, C.F. 1991. The Ordovician and Silurian development of the Iapetus Ocean. *Special Papers in Palaentology*, **44**, 165–178.

MCKERROW, W.S., SCOTESE, C.R. & BRASIER, M.D. 1992. Early Cambrian reconstructions. *Journal of the Geological Society, London*, **149**, 599–606.

MCKERROW, W.S., MCNIOCAILL, C. & DEWEY, J.F. 2000. The Caledonian Orogeny redefined. *Journal of the Geological Society, London*, **157**, 1149–1154.

MACNIOCAILL, C. & SMETHURST, M.A. 1994. Palaeozoic palaeogeography of Laurentia and its margins: a reassessment of palaeomagnetic data. *Geophysical Journal International*, **116**, 715–725.

MACNIOCAILL, C., SMETHURST, M.A. & RYAN, P.D. 1998. Oroclinal bending in the Caledonides of western Ireland: a mid-Palaeozoic feature controlled by a pre-existing structural grain. *Tectonophysics*, **299**, 31–47.

MAJOROWICZ, J.A., CERMAK, V., SAFANDA, J., KRZYWIEC, P., WROBLEWSKA, M., GUTERCH, A. & GRAD, M. 2003. Heat flow models across the Trans-European Suture Zone in the area of the POLONAISE'97 seismic experiment. *Physics and Chemistry of the Earth*, **28**, 375–391.

MAROTTA, A.M., BAYER, U., SCHECK, M. & THYBO, H. 2001. The stress field below the NE German Basin: effects induced by the Alpine collision. *Geophysical Journal International*, **144**, F8–F12.

MAROTTA, A.M., BAYER, U., THYBO, H. & SCHECK, M. 2002. Origin of the regional stress in the North German basin: results from numerical modelling. *Tectonophysics*, **360**, 245–264.

MARQUERING, H. & SNIEDER, R. 1996. Shear-wave velocity structure beneath Europe, the northeastern Atlantic and western Asia from waveform inversions including surface-wave mode coupling. *Geophysical Journal International*, **127**(2), 283–304.

MARTINSSON, A. 1974. The Cambrian of Norden. *In*: HOLLAND, C.H. (ed.) *Cambrian of the British Isles, Norden and Spitsbergen*. John Wiley, New York, 185–283.

MASLOV, A.V., ERDTMANN, B.D., IVANOV, K.S., IVANOV, S.N., & KRUPENIN, M.T. 1997. The main tectonic events, depositional history, and the palaeogeography of the Southern Urals during the Riphean–early Palaeozoic. *Tectonophysics*, **276**(1–4), 313–335.

MATTE, P.H., MALUSKI, H., RAJLICH, P. & FRANKE, W. 1990. Terrane boundaries in the Bohemian Massif: result of large scale Variscan shearing. *Tectonophysics*, **177**, 151–170.

MAYER, P., KRAUSS, M. & VORMBAUM, M. 2000. Der Strukturbau des Vorpommern-Störungssystems im Bereich der NE-Fortsetzung des DEKORP-Profils BASIN 9601 (DFG-Projekt VPSS I). *Zeitschrift für geologische Wissenschaften*, **28**(3/4), 397–404.

MEERT, J.G. 2001. Growing Gondwana and rethinking Rodinia; a paleomagnetic perspective. *Gondwana Research*, **4**(3), 279–288.

MEERT, J.G. 2003. A note on the impact factor of Gondwana Research. *Gondwana Research*, **6**(4), 901.

MEERT, J.G. & POWELL, C. MCA. 2001. Assembly and breakup of Rodinia. *Precambrian Research*, **110**(1–4).

MEERT, J.G. & TORSVIK, T.H. 2003. The making and unmaking of a supercontinent: Rodinia revisited. *Tectonophysics*, **375**, 261–268.

MEERT, J.G., TORSVIK, T.H., EIDE, E.A. & DAHLGREN, S. 1998. Tectonic significance of the Fen Province, S. Norway – constraints from geochronology and paleomagnetism. *Journal of Geology*, **106**(5), 553–564.

MEISL, S. 1990. Metavolcanic rocks in the „Northern Phyllite Zone' at the southern magin of the Rhenohercynian belt. *In*: *Terranes in the Circum-Atlantic Paleozoic Orogens, Field Guide Rhenohercynian Belt and Mid-German Crystalline Rise*. International Conference on Paleozoic Orogens in Central Europe, Göttingen-Gießen, IGCP **223**, 25–42.

MEISSNER, R. & BORTFELD, R.K. (eds) 1990. *DEKORP-Atlas. Results of Deutsches Kontinentales Reflexionsseismisches Programm*. Springer-Verlag, Berlin.

MEISSNER, R. & KRAWCZYK, C.M. 1999. Caledonian and Proterozoic terrane accretion in the South-West Baltic Sea. *Tectonophysics*, **314**(1–3), 255–267.

MEISSNER, R., SADOWIAK, P. & THOMAS, S.A. 1994. Avalonia, the third partner in the Caledonian collisions: Evidence from deep seismic data. *Geologische Rundschau*, **83**, 186–196.

MEISSNER, R., THYBO, H. & ABRAMOVITZ, T. 2002. Interwedging and inversion structures around the trans-European suture zone in the Baltic Sea, a manifestation of compressive tectonic phases. *Tectono-*

physics, **360**, 265–280.

Michelsen, O. & Nielsen, L.H. 1991. Well records on the Phanerozoic stratigraphy in the Fennoscandian Border Zone, Denmark. *Danmarks Geologiske Undersolgelse, Series A*, **29**, 1–37.

Michot, P. 1979. La faille mosane et la phase hyporogénique bollandienne, d'age emsien dans le rameau calédonien condruso-brabacon. *Annales de la Société Géologique de Belgique*, **101**, 321–335.

Michot, P. 1980. Le segment tectogénique calédonien belge. *Mémoires de l'Académie Royale de Belgique, Classe des Sciences, Collection in – 8o, 2ème série*, **43**, 1–61.

Miller, J.D. & Kent, D.V. 1988. Regional trends in the timing of Alleghanian remagnetization in the Appalachians. *Geology*, **16**(7), 588–591.

Min, K., Renne, P.R. & Huff, W.D. 2001. 40Ar/39Ar dating of Ordovician K-bentonites in Laurentia and Baltoscandia. *Earth and Planetary Science Letters*, **185**, 121–134.

Mingram, B. 1998. The Erzgebirge, Germany – a subducted part of northern Gondwana: geochemical evidence for repetition of early Palaeozoic metasedimentary sequences in metamorphic thrust units. *Geological Magazine*, **135**, 785–801.

Mingram, B., Rötzler, K., Tichomirowa, M., Haase, G. & Teufel, S. 1998. Early Paleozoic rift-related magmatism and passive margin sediments preserved in HP units of the Erzgebirge. *Terra Nostra*, **98**, 105–107.

Mingram, B., Kröner, A., Hegner, E. & Krentz, O. 2004. Zircon ages, geochemistry, and Nd isotopic systematics of pre-Variscan orthogneisses from the Erzgebirge, Saxony (Germany) and geodynamic interpretation. *International Journal of Earth Sciences (Geologische Rundschau)*, **93**, 706–727.

Mocyzdlowska, M. 1995. Neoproterozoic and Cambrian successions deposited on the East European Platform and Cadomian basement area in Poland. *In*: Gee, D.G. & Beckholmen, M. (eds) *The Trans-European Suture Zone: EUROPROBE, Liblice 1993. Studies in Geophysics and Geodesy*, **39**, 276–285.

Moczydlowska, M. 1997. Proterozoic and Cambrian successions in Upper Silesia: an Avalonian terrane in southern Poland. *Geological Magazine*, **134**, 679–689.

Molzahn, M., Anthes, G. & Reischmann, T. 1998. Single zircon Pb/Pb age geochronology and isotope systematics of the Rhenohercynian basement. *Tera Nostra*, **98**(1), 67–68.

MONA LISA Working Group 1997*a*. MONA LISA – Deep seismic investigations of the lithosphere in the southeastern North Sea. *Tectonophysics*, **269**, 1–19.

MONA LISA Working Group 1997*b*. Closure of the Tornquist Sea: constraints from MONA LISA deep seismic reflection data. *Geology*, **25**, 1071–1074.

Morgan, J.V., Hadwin, M., Warner, M.R., Barton, P.J. & Morgan, R.P.Ll. 1994. The polarity of deep seismic reflections from the lithospheric mantle: evidence for a relict subduction zone. *Tectonophysics*, **232**, 319–328.

Mosegaard, K., Singh, S., Snyder, D. & Wagner, H. 1997. Monte Carlo analysis of seismic reflections from Moho and the W reflector. *Journal of Geophysical Research,* **B102**(2), 2969–2981.

Murphy, J.B., Keppie, J.D. & Nance, R.D. 1999. Fault reactivation within Avalonia: Plate margin to continental interior deformation. *Tectonophysics*, **305**(1–3), 183–204.

Mykura, W. 1991. Old Red Sandstone. *In*: Craig, G. Y. (ed) *Geology of Scotland*. Geological Society, London, 297–346.

Nance, R.D. & Thompson, M.D. (eds) 1996. *Avalonian and related Peri-Gondwanan terranes of the Circum-North Atlantic*. Geological Society of America, Special Paper, **304**, 390 pp.

Nance, R.D., Murphy, J.B. & Keppie, J.D. 2002. A Cordilleran model for the evolution of Avalonia. *Tectonophysics*, **352**, 11–31.

Nawrocki, J. 2000. Late Silurian paleomagnetic pole from the Holy Cross Mountains: constraints for the post-Caledonian tectonic activity of the Trans-European Suture Zone. *Earth and Planetary Science Letters*, **179**, 325–334.

Neuman, R.B. & Harper, D.A.T. 1992. Palaeogeographic significance of Arenig-Llanvirn Toquima Table Head and Celtic brachiopod assemblages. *In*: Webby, B.D. & Laurie, J.R. (eds) *Global Perspectives on Ordovician Geology*. Balkema, Rotterdam, 241–254.

Neuman, R.B., Bruton, D.L. & Pojeta, J. 1997. Fossils from the Ordovician 'Upper Hovin Group' (Caradoc-Ashgill), Trondheim region, Norway. *Norges Geologiske Undersøkelse Bulletin*, **432**, 25–58.

Nielsen, S.B. 1995. An upper limit to palaeo heat flow: theory and examples from the Danish Central Graben. *Tectonophysics*, **244** (1–3), 137–152.

Nielsen, L.H. & Japsen, P. 1991. *Deep wells in Denmark, 1935–1990*. Geological Survey of Denmark, **A31**.

Nikishin, A.M., Ziegler, P.A., *et al.* 1996. Late Precambrian to Triassic history of the East European craton: dynamics of sedimentary basin evolution. *Tectonophysics*, **268**, 23–63.

Noble, S.R., Tucker, R.D. & Pharaoh, T.C. 1993. Lower Palaeozoic and Precambrian igneous rocks from eastern England, and their bearing on late Ordovician closure of the Tornquist Sea: constraints from U-Pb and Nd isotopes. *Geological Magazine*, **130**, 835–846.

Noblet, C. & Lefort, J.P. 1990. Sedimentological evidence for a limited separation between Armorica and Gondwana during the early Ordovician. *Geology*, **18**, 303–306.

Nolet, G., Dost, B. & Paulssen, H. 1986. Intermediate wavelength seismology and the NARS experiment. *Annales Geophysicae*, **4**, 305–314.

O'Brian, P.J., Duyster, J., Grauert, B., Schreyer, W., Stöckhert, B. & Weber, K. 1997. Crustal evolution of the KTB drill site: From oldest relics to the late Hercynian granites. *Journal of Geophysical Research*, **102**, 18203–18220.

O'Brian, S.J., Wardle, R.J. & King, A.F. 1993. The Avalon Zone: a Pan-African terrane in the Appalachian Orogen of Canada. *Geological Journal*, **18**, 195–222.

Okrusch, M. & Richter, P. 1986. Orthogneisses of the Spessart crystalline complex, northwest Bavaria, indicators of the geotectonic environment? *Geologische Rundschau*, **75**, 555–568.

Okrusch, M., Seidel, E., Schüssler, U. & Richter, P. 1989. Geochemical characteristics of metabasites in different tectonic units of the northeast Bavarian crystalline basement. *In*: Emmermann, R. & Wohlenberg, J. (eds) *The German Deep Drilling Program (KTB)*. Springer-Verlag, Berlin, 67–81.

Oliver, G.J.H., Corfu, F. & Krogh, T.E. 1993. U-Pb ages from SW Poland: evidence for a Caledonian suture zone between Baltica and Gondwana. *Journal of the Geological Society, London*, **150**, 355–369.

Ondrak, R., Förster, A., Scheck, M. & Gerisch, R. 1999. The present-day temperature field of the Northeast German Basin – a comparison of temperature measurements and 3D-modeling. *In:* Thiry, M. (ed.) *Transport in Sedimentary Systems: From the Pore-to the Basin-Scale*. Université Louis Pasteur, Strasbourg, Mémoire **99**, 111–114.

Orlowski, S. 1992. Cambrian stratigraphy and stage subdivision in the Holy Cross Mountains, Poland. *Geological Magazine*, **129**, 471–474.

Pedley, R.C., Busby, J.P. & Dabek, Z.K. 1994. *GRAVMAG User Manual*. British Geological Survey, Keyworth, Nottingham.

Perroud, H., Van der Voo, R. & Bonhommet, N. 1984. Paleozoic evolution of the Armorica Plate on the basis of paleomagnetic data. *Geology*, **12**(10), 579–582.

Petecki, Z., Wybraniec, S., Polechonska, O. & Ciesla, E. 2003*a*. *Magnetic anomaly map of Poland*, 1:1.000.000, Geological Institute, Warszaw.

Petecki, Z., Wybraniec, S., Polechonska, O. & Ciesla, E. 2003*b*. *Magnetic anomaly map of Poland - Pseudogravimetric Map*, 1:1.000.000, Geological Institute, Warszaw.

Pharaoh, T.C. 1999. Palaeozoic terranes and their lithospheric boundaries within the Trans-European Suture Zone (TESZ): a review. *Tectonophysics*, **314**, 17–41.

Pharaoh, T.C. & Gibbons, W. 1994. Precambrian rocks in England and Wales south of the Menai Fault System. *In*: Gibbons, W. & Harris, A.L. (eds) *A Revised Correlation of Precambrian Rocks in the British Isles*. Geological Society, London.

Pharaoh, T.C., Brewer, T.S. & Webb, P.C. 1993. Subduction-related magmatism of late Ordovician age in eastern England. *Geological Magazine*, **130**, 647–656.

Pharaoh, T.C., England, R. & Lee, M.K. 1995. The concealed Caledonide basement of Eastern England and the southern North Sea – a review. *In*: Gee, D.G. & Beckholmen, M. (eds) *The Trans-European Suture Zone: EUROPROBE, Liblice 1993*. Studia Geophysica et Geodaetica, **39**, 330–346.

Pharaoh, T.C., England, R.W., Verniers, J. & Zelazniewicz, A.

1997. Introduction: geological and geophysical studies in the Trans-European Suture Zone. *Geological Magazine*, **134**(5), 585–590.

Pickering, K.T., Bassett, M.G. & Siveter, D.J. 1988. Late Ordovician–early Silurian destruction of the Iapetus Ocean: Newfoundland, British Isles and Scandinavia – a discussion. *Transactions of the Royal Society of Edinburgh, Earth Sciences*, **79**, 361–382.

Pin, C. 1990. Variscan oceans: Ages, origins and geodynamic implications inferred from geochemical and radiometric data. *Tectonophysics*, **177**, 215–227.

Piper, J.D.A. 1997. Tectonic rotation within the British paratectonic Caledonides and early Palaeozoic location of the orogen. *Journal of the Geological Society, London*, **154**(1), 9–13.

Piske, J. & Neumann, E. 1990. Tektonische Gliederung des prävariszischen Untergrundes in der südwestlichen Ostsee. *Zeitschrift für angewandte Geologie*, **36**, 342–346.

Piske, J., Rasch, H.-J., Neumann, E. & Zagora, K. 1994. Geologischer Bau und Entwicklung des Präperms der Insel Rügen und des angrenzenden Seegebietes. *Zeitschrift für geologische Wissenschaften*, **22**(1/2), 212- 226.

Pitcher, W.S. 1993. *The Nature and Origin of Granite*. Blackie, London.

Poprawa, P., Sliaupa, S., Stephenson, R. & Lazauskiene, J. 1999. Late Vendian–Early Palaeozoic tectonic evolution of the Baltic Basin: Regional tectonic implications from subsidence analysis. *Tectonophysics*, **314**(1–3), 219–239.

Poulsen, C. 1974. Further contributions to the knowledge of the Palaeozoic of Slagelse no. 1, Western Sealand. *Danmarks Geologiske Undersogelse*, Series **II**, 101.

Powell, C.McA. & Pisarevsky, S.A. 2002. Late Neoproterozoic assembly of East Gondwana. *Geology*, **30**(1), 3–6.

Powell, C.McA. , Jones, D.L., Pisarevsky, S.A. & Wingate, M.T.D. 2001. Palaeomagnetic constraints on the position of the Kalahari Craton in Rodinia. *Precambrian Research*, **110**(1–4), 33–46.

Pratsch, J.C. 1979. Regional structural elements in Northwest Germany. *Journal of Petroleum Geology*, **2**, 159–180.

Price, C. & Morgan, J. 2000. Lithospherice structure North of Scotland II, Poisson's ratios and waveform modelling. *Geophysical Journal International*, **142**, 737–734.

Prigmore, J.K., Butler, A.J. & Woodcock, N.H. 1997. Rifting during separation of Eastern Avalonia from Gondwana: evidence from subsidence analysis. *Geology*, **25**, 203–207.

Quesada, C. 1991. Geological constraints on the Paleozoic tectonic evolution of tectonostratigraphic terranes in the Iberian Massif. *Tectonophysics*, **185**(3–4), 225–245.

Quesada, C., Fonseca, P.E., Munha, J., Oliveira, J.T. & Ribeiro, A. 1994. The Beja-Acebuches Ophiolite (southern Iberia Variscan fold belt) – geological characterization and geodynamic significance. *Boletin Geologico y Minero*, **105**(1), 3–49.

Rabbel, W., Förste, K., Schulze, A., Bittner, R., Röhl, J. & Reichert, J.C. 1995. A high velocity layer in the lower crust of the North German Basin. *Terra Nova*, **7**, 327–337.

Radwanski, A. & Roniewicz, P. 1960. Ripple markes and other sedimentary structures of the Upper Cambrian at Wielka Wisniowka (Holy Cross Mountains). *Acta Geologica Polonica*, **10**, 371–399.

Ramsay, D.M., Sturt, B.A., Zwann, K.B. & Roberts, D. 1985. Caledonides of northern Norway. *In*: Gee, D.G. & Sturt, B.A. (eds) *The Caledonide Orogen – Scandinavia and Related Terranes*. Wiley, Chichester, 163–184.

Remmelts, G. & Duin, E.J.T. 1990. Results of a regional deep seismic survey in the Netherlands. *In: The Potential of Deep Seismic Profiling for Hydrocarbon Exploration, Proceedings of 5th IFP conference*, Arles, 1989, 335–343.

Rempel, H. 1992. Erdölgeologische Bewertung der Arbeiten der Gemeinsamen Organisation 'Petrobaltic' im deutschen Schelfbereich. *Geologisches Jahrbuch*, **D99**, 3–32.

Reston, T.J. & Blundell, D. 1987. Possible mid-crustal shears at the edge of the London Platform. *Geophysical Journal of the Royal Astronomical Society*, **89**, 251–158.

Rieke, H., Kossow, D., McCann, T. & Krawczyk, C.M. 2001. Tectono-sedimentary evolution of the northernmost margin of the NE German Basin between uppermost Carboniferous and Late Permian (Rotliegend). *Geological Journal*, **36**, 19–38.

Ritchie, J.D., Hitchen, K. & Mitchell, J.G. 1987. The offshore continuation of the Moine Thrust north of Shetland as deduced from basement isotope ages. *Scottish Journal of Geology*, **23**(2), 163–173.

Robardet, M. 2003. The Armorica 'microplate'; fact or fiction? Critical review of the concept and contradictory palaeobiogeographical data. *Palaeogeography, Palaeoclimatology, Palaeoecology*, **195**(1–2), 125–148.

Robinson, P., Tucker, R.D., Bradley, D., Berry, H.N., Berry, I.V. & Osberg, P.H. 1998. Palaeozoic orogens in New England, USA. *Geologiska Föreningens Forhaldligar*, **120**, 119–148.

Rong, J.Y. & Harper, D.A.T. 1988. A global synthesis of the latest Ordovician Hirnantian brachiopod fauna. *Transactions of the Royal Society of Edinburgh, Earth Sciences*, **79**, 383–402.

Rötzler, K., Mingram, B., Tichomirowa, M. & Teufel, S. 1997. Frühpaläozische Vulkanitalter metamorpher Einheiten am NW-Rand des Bömischen Massivs. *Terra Nostra*, **97**, 142–145.

Ryka, W. 1984. Precambrian evolution of the East-European Platform in Poland. *Kwartalnik Geologiczny*, **26**(2), 257–272.

Sacks, I.S., Snoke, J.A. & Husebye, E.S. 1979. Lithosphere thickness beneath the Baltic Shield. *Structure and Compositional Variations of the Lithosphere and Asthenosphere*, **56**(1–2), 101–110.

Samuelsson, J., Vecoli, M. & Beier, H. 2001. Ordovician-Silurian palynostratigraphy (Chitinozoa and acritarchs) of the G14–1/86 borehole, southern Baltic Sea. *Neues Jahrbuch für Geologie und Paläontologie*, **222**, 259–290.

Samuelsson, J., Vecoli, M., Bednarczyk, W.S. & Verniers, J. 2002*a*. Timing of the Avalonia–Baltica plate convergence as inferred from palaeogeographic and stratigraphic data of chtinozoan assemblages in west Pomerian, northern Poland. *In:* Winchester, J.A., Pharaoh, T.C. & Verniers, J. (eds) *Palaeozoic Amalgamation of Central Europe*. Geological Society, London, Special Publications, **201**, 95–114.

Samuelsson, J., Gerdes, A., Koch, L., Servais, T. & Verniers, J. 2002*b*. Chitinozoan and Nd isotope stratigraphy of the Ordovician rocks in the Ebbe Anticline, NW Germany. *In:* Winchester, J.A., Pharaoh, T.C. & Verniers, J. (eds) *Palaeozoic Amalgamation of Central Europe*. Geological Society, London, Special Publications, **201**, 115–132.

Sandeman, H.A., Clark, A.H., Styles, M.T., Scott, D.J., Malpas, J.G. & Farrar, E. 1997. Geochemistry and U-Pb and 40Ar-39Ar geochronology of the Man of War Gneiss, Lizard Complex, SW England: pre-Hercynian arc-type crust with a Sudetes-Iberian connection. *Journal of the Geological Society, London*, **154**, 403–417.

Scheck, M. & Bayer, U. 1999. Evolution of the Northeast German Basin – inferences from a 3-D structural model and subsidence analysis. *Tectonophysics*, **313**, 145–169.

Scheck, M., Barrio-Alvers, L., Bayer, U. & Götze, H.-J. 1999. Density structure of the Northeast German Basin – 3D modelling along the DEKORP line BASIN96. *Physics and Chemistry of the Earth*, **24**(3), 221–230.

Scheck, M., Thybo, H., Lassen, A., Abramovitz, T. & Laigle, M. 2002. Basement structure in the southern North Sea, offshore Denmark, based on seismic interpretation. *In:* Winchester, J.A., Pharaoh, T.C. & Verniers, J. (eds) *Palaeozoic Amalgamation of Central Europe*. Geological Society, London, Special Publications, **201**, 311–326.

Scheck-Wenderoth, M. & Lamarche, J. 2005. Crustal memory and basin evolution in the Central European Basin System – New insights from a 3D structural model. *Tectonophysics*, **397**(1–2), 143–165.

Scheck-Wenderoth, M., Krzywiec, P., Zühlke, R., Maystrenko, Y. & Froitzheim, N. 2008. Permian to Cretaceous tectonics. *In:* McCann, T. (ed.) *The Geology of Central Europe. Volume 2: Mesozoic and Cenozoic*. Geological Society, London, 999–1030.

Schellschmidt, R., Hägedorn, F. & Fesche, H.-W. 1999. Das Temperaturfeld in Nordostdeutschland - neue Messungen, neue Karten, neue Interpretationen. *GGA Report*, 1–23.

Schlüter, H.-U., Best, G., Jürgens, U. & Binot, F. 1997. Interpretation reflexionsseismischer Profile zwischen baltischer Kontinentalplatte und kaledonischem Becken in der südlichen Ostsee - erste Ergebnisse. *Zeitschrift der Deutschen Geologischen Gesellschaft*, **148**(1), 1–32.

Schmädicke, E., Mezger, K., Cosca, M.A. & Okrusch, M. 1995. Variscan Sm-Nd and Ar-Ar ages of eclogite-facies rocks from the

Erzgebirge, Bohemian Massif. *Journal of Metamorphic Geology*, **13**, 537–552.

Schönlaub, H.P. 1997. The Silurian of Austria. *Berichte Geologische Bundesanstalt*, **40**, 20–41.

Schulze, A. & Lück, E. 1992. Some results of deep seismic investigations in East Germany. *Physics of the Earth and Planetary Interiors*, **69**, 231–238.

Schweitzer, J. 1995. Blockage of regional seismic waves by the Teisseyre-Tornquist zone. *Geophysical Journal International*, **123**, 260–276.

Scotese, C.R. 1984. Paleozoic palaeomagnetism and the assembly of Pangaea. *In*: Van der Voo, R., Scotese, C.R. & Bonhommet, N. (eds) *Plate Reconstruction from Paleozoic Paleomagnetism*. AGU Geodynamics Series no. **12**, Washington DC, 1–10.

Scotese, C. 1999. PALEOMAP Project. World Wide Web address: http://www.scotese.com/pzanim.htm.

Scotese, C.R., Van der Voo, R., Mueller, R.D. & McKerrow, W.S. 1990. Phanerozoic plate tectonic reconstructions results of the PALEOMAP project. *Geological Society of America Bulletin*, **22**(7), 229–230.

Sellevoll, M.A. & Warrick, R.E. 1971. A refraction study of the crustal structure in southern Norway. *Seismological Society of America Bulletin*, **61**, 457–471.

Servais, T. 1994. Ordovician acritarchs from Rügen (NE-Germany): palaeobiogeographical evidence for the attribution to Eastern Avalonia. *Neues Jahrbuch für Geologie und Paläontologie, Monatshefte*, **9**, 566–580.

Servais, T. & Fatka, O. 1997. Recognition of the Trans-European Suture Zone (TESZ) by the palaeobiogeographical distribution pattern of early to middle Ordovician acritarchs. *Geological Magazine*, **134**, 617–625.

Servais, T., Li, J., Molyneux, S. & Raevskaya, E. 2003. Ordovician organic-walled microphytoplankton (acritarch) distribution: The global scenario. *Palaeogeography, Palaeoclimatology, Palaeoecology*, **195**(1–2), 149–172.

Siebel, W., Raschka, H., Irber, W., Kreuzer, H., Lenz, K.-L-, Höhndorf, A. & Wendt, I. 1997. Early Palaeozoic acid magmatism in the Saxothuringian Belt: New insights from a geochemical and isotopic study of orthogneisses and metavolcanic rocks from the Fichtelgebirge, SE Germany. *Journal of Petrology*, **38**, 203–230.

Sim, L.A., Korcemagin, V.A., Frischbutter, A. & Bankwitz, P. 1999. The neotectonic stress field pattern of the East European Platform. *Zeitschrift für geologische Wissenschaften*, **27**(3/4), 161–181.

Singh, S.C., Hague, P.J. & McCaughey, M. 1998. Study of the crystalline crust from two-ship normal incidence and wide-angle experiment. *Tectonophysics*, **286**, 79–92.

Sintubin, M. 1999. Arcuate fold and cleavage patterns in the southeastern part of the Anglo-Brabant fold belt (Belgium): tectonic implications. *Tectonophysics*, **309**, 81–97.

Sintubin, M. & Everaerts, M. 2002. A compressional wedge model for the Lower Palaeozoic Anglo-Brabant Belt (Belgium) based on potential field data. *In* Winchester, J.A., Pharaoh, T.C. & Verniers, J. (eds) *Palaeozoic Amalgamataion of Central Europe*. Geological Society, London, Special Publications, **201**, 327–344.

Smith, K. 1992. Carboniferous magmatism in the Iapetus convergence zone: evidence from deep seismic reflection profiles. *Journal of the Geological Society, London*, **149**, 907–914.

Snieder, R. 1988. Large-scale waveform inversions of surface waves for lateral heterogeneity; 2, Application to surface waves in Europe and the Mediterranean. *Journal of Geophysical Research*, **B93**(10), 12,067–12,080.

Snyder, D.B. 1990. The Moine thrust in the BIRPS data set. *Journal of the Geological Society, London*, **147**, 81–86.

Snyder, D.B. & Flack, C.A. 1990. A Caledonian age for reflectors within the mantle lithosphere north and west of Scotland. *Tectonics*, **9**(4), 903–922.

Snyder, D. & Hobbs, R. 1999. *The BIRPS Atlas II – A Second Decade of Deep Seismic Reflection Profiling*. Geological Society, London.

Stephens, M.B. & Gee, D.G. 1985. A tectonic model for the evolution of the eugeoclinal terranes in the central Scandinavian Caledonides. *In:* Gee, D.G. & Sturt, B.A. (eds) *The Caledonide Orogen – Scandinavia and Related Terranes*. Wiley, Chichester, 953–978.

Sommermann, A.E., Meisl, S. & Todt, W. 1992. Zirkonalter von drei verschiedenen Metavulkaniten aus dem Südtaunus. *Geologisches Jahrbuch Hessen*, **120**, 67–76.

Soper, N.J. 1986. The Newer Granite problem: a geotectonic view. *Geological Magazine*, **123**, 227–236.

Soper, N.J. & Hutton, D.H.W. 1984. Late Caledonian sinistral displacements in Britain: implications for a three plate collision model. *Tectonics*, **3**, 781–794.

Soper, N.J. & Woodcock, N.H. 1990. Silurian collision and sediment dispersal patterns in southern Britain. *Geological Magazine*, **127**(6), 527–542.

Soper, N. J., Strachan, R. A., Holdsworth, R. E., Gayer, R. A. & Greiling, R. O. 1992*a*. Sinistral transpression and the closure of Iapetus. *Journal of the Geological Society, London*, **149**, 871–880.

Soper, N. J., England, R. W., Snyder, D. B. & Ryan, P. D. 1992*b*. The Iapetus Suture zone in England, Scotland and Eastern Ireland: a reconcilliation of the geological and deep seismic data. *Journal of the Geological Society, London*, **149**, 697–700.

Soper, N.J., Ryan, P.D. & Dewey, J.F. 1999. Age of the Grampian orogeny in Scotland and Ireland. *Journal of the Geological Society*, **156**(6), 1231–1236.

Sroda, P. & POLONAISE profile P3 Working Group 1999. P- and S-wave velocity model of the southwestern margin of the Precambrian East European Craton: Polonaise '97 profile P3. *Tectonophysics*, **314**(1–3), 175–192.

Stangl, R. 1990. *Die Struktur der Lithosphäre in Schweden, abgeleitet aus einer gemeinsamen Interpretation der P- und S-Wellen Registrierungen auf dem FENNOLORA-Profil*. PhD thesis, University of Karlsruhe.

Stewart, M.A, Strachan, R.A & Holdsworth, R.E. 1999. The structure and early kinematic history of the Great Glen Fault Zone, Scotland. *Tectonics*, **18**, 326–342.

Stewart, M.A., Strachan, R.A, Martin, M.W. & Holdsworth, R.E. 2001. Constraints on early sinistral displacements along the Great Glen Fault Zone, Scotland: structural setting, U-Pb geochronology and emplacement of the syn-tectonic Clunes tonalite. *Journal of the Geological Society, London*, **158**, 821–830.

Strachan, R. 2002*a*. The Grampian Orogeny: Mid-Ordovician arc–continent collision along the Laurentian margin of Iapetus. *In*: Woodcock, N.H. & Strachan, R. (eds) *Geological History of Britain and Ireland*. Blackwell Science, Oxford, 88–106.

Strachan, R. 2002*b*. Mid-Ordovician to Silurian sedimentation and tectonics on the northern active margin of Iapetus. *In*: Woodcock, N.H. & Strachan, R. (eds) *Geological History of Britain and Ireland*. Blackwell Science, Oxford, 107–123.

Strachan, R. & Holdsworth, R.E. 2002. Late Neoproterozoic (<750 Ma) to Early Ordovician passive margin sedimentation along the Laurentian margin of Iapetus. *In*: Woodcock, N.H. & Strachan, R. (eds) *Geological History of Britain and Ireland*. Blackwell Science, Oxford, 73–87.

Strömberg, A.G.B. 1976. A pattern of tectonic zones in the western part of the East European Platform. *Geologiska Föreningens Forhaldligar*, **98**, 227–243.

Suess, E. 1885–1909. *Das Antlitz der Erde*. Tempsky & Freytag, Prag-Wien-Leipzig, 3 vols.

Tait, J., Bachtadse, V. & Soffel, H.C. 1995. Upper Ordovician palaeogeography of the Bohemian Massif: implications for Armorica. *Geophysical Journal International*, **122**, 211–218.

Tait, J.A., Bachtadse, V., Franke, W. & Soffel, H.C. 1997. Geodynamic evolution of the European Variscan fold belt: palaeomagnetic and geological constraints. *Geologische Rundschau*, **86**(3), 585–598.

Tait, J., Schätz, M., Bachtadse, V. & Soffel, H. 2000. Palaeomagnetic and Palaeozoic palaeogeography of Gondwana and European terranes. *In*: Franke, W., Altherr, R., Haak, V., Oncken, O. & Tanner, D. (eds) *Orogenic Processes: Quantification and Modelling in the Variscan Belt*. Geological Society, London, Special Publications, **179**, 21–34.

Tanner, B. & Meissner, R. 1996. Caledonian deformation upon southwest Baltica and its tectonic implications: Alternatives and consequences. *Tectonics*, **15**, 803–812.

Teale, C.T. & Spears, D.A. 1986. The mineralogy and origin of some Silurian bentonites, Welsh Borderland, U.K. *Sedimentology*, **33**(5), 757–765.

Thomas, S.A., Sivhed, U., Erlström, M. & Seifert, M. 1993.

Seismostratigraphy and structural framework of the SW Baltic Sea. *Terra Nova*, **5**, 364–374.

Thorogood, E. 1990. Provenance of the pre-Devonian sediments of England and Wales: Sm-Nd isotopic evidence. *Journal of the Geological Society, London*, **147**, 591–594.

Thybo, H. 1990. A seismic model along the EGT profile – from the North German Basin into the Baltic Shield. *In*: Freeman, R. & Mueller, S. (eds) *Proceedings of the 5th Study Centre on the European Geotraverse Project*. ESF, Strasbourg, 99–108.

Thybo, H. 1997. Geophysical characteristics of the Tornquist Fan area, northwest Trans-European Suture Zone: indication of late Carboniferous to early Permian dextral transtension. *Geological Magazine*, **134**, 597–606.

Thybo, H. 2001. Crustal structure along the EGT profile across the Tornquist Fan interpreted from seismic, gravity and magnetic data. *Tectonophysics*, **334**, 155–190.

Thybo, H., Pharaoh, T.C. & Guterch, A. 1999. Geophysical investigations of the Trans-European Suture Zone. *Tectonophysics*, **314**(1–3).

Tichomirowa, M., Berger, H.-J., *et al.* 2001. Zircon ages of high-grade gneisses in the Eastern Erzgebirge (Central European Variscides) – constraints on origin of the rocks and Precambrian to Ordovician magmatic events in the Variscan foldbelt. *Lithos*, **56**, 303–332.

Timmerman, M. J. 2008. Palaeozoic magmatism. *In*: McCann, T. (ed.) *The Geology of Central Europe. Volume 1: Precambrian and Palaeozoic*. Geological Society, London, 665–748.

Toghill, P. 1992. The Shelveian event, a Late Ordovician tectonic episode in Southern Britain (Eastern Avalonia). *Proceedings of the Geologists' Association*, **103**, 31–35.

Tomczyk, H. 1970. Silurian. *In*: Sololowski, S., Cieslinski, S. & Czerminski, J. (eds) *Geology of Poland, Vol. 1, Part 1: Precambrian and Paleozoic*. Wydawnictwa Geologiczne, Warszawa, 237–319.

TOR Working Group 1999. Important findigs expected from Europe's largest seismic array. *Eos, Transactions*, **80**(1), 1, 6.

Torsvik, T.H. 1998. Palaeozoic palaeogeography: a North Atlantic viewpoint. *Geologiska Föreningens Forhaldligar*, **120**, 109–118.

Torsvik, T.H. 2003. The Rodinia jigsaw puzzle. *Science*, **300**, 1379–1381.

Torsvik, T.H. & Cocks, L.R.M. 2004. Earth geography from 400 to 259 Ma: a palaeomagnetic, faunal and facies review. *Journal of the Geological Society, London*, **161**, 555–572.

Torsvik, T.H. & Rehnström, E.F. 2001. Cambrian palaeomagnetic data from Baltica: implications for true polar wander and Cambrian palaeogeography. *Journal of the Geological Society, London*, **158**, 321–329.

Torsvik, T.H. & Rehnström, E.F. 2003. The Tornquist Sea and Baltica-Avalonia docking. *Tectonophysics*, **362**, 67–82.

Torsvik, T.H. & Trench, A. 1991. The Ordovician history of the Iapetus ocean in Britain: new palaeomagnetic constraints. *Journal of the Geological Society, London*, **148**, 423–425.

Torsvik, T.H., Olesen, O., Ryan, P.D. & Trench, A. 1990. On the palaeogeography of Baltica during the Palaeozoic: new palaeomagnetic data from the Scandinavian Caledonides. *Geophysical Journal International*, **103**, 261–279.

Torsvik, T.H., Smethurst, M.A., van der Voo, R., Trench, A., Abrahamsen, N. & Halvorsen, E. 1992. Baltica: a synopsis of Vendian-Permian palaeomagnetic data and their palaeotectonic implications. *Earth Science Reviews*, **33**, 133–152.

Torsvik, T.H., Trench, A., Svensson, I. & Walderhaug, H.J. 1993. Palaeogeographic significance of mid-Silurian palaeomagnetic results from southern Britain – major revision of the apparent polar wander path for eastern Avalonia. *Geophysical Journal International*, **113**, 651–668.

Torsvik, T.H., Smethurst, M.A., *et al.* 1996. Continental break-up and collision in the Neoproterozoic and Palaeozoic – A tale of Baltica and Laurentia. *Earth Science Reviews*, **40**, 229–258.

Trappe, H. 1989. Deep seismic profiling in the North German Basin. *First Break*, **7**, 173–184.

Trench, A. & Torsvik, T.H. 1992. The closure of the Iapetus Ocean and Torniquist Sea: New Palaeomagnetic Constraints. *Journal of Geological Society, London*, **149**, 867–870.

Trench, A., Torsvik, T.H. & McKerrow, W.S. 1992. The palaeogeographic evolution of Southern Britain during early Palaeozoic times: a reconciliation of palaeomagnetic and biogeographic evidence. *Tectonophysics*, **201**, 75–82.

Tryggvason, A., Lund, C.-E. & Friberg, M. 1998. A two-dimensional seismic velocity model across the transition zone between the Baltic Shield and the North German Basin – the EUGENO-S profile 1 revisited. *Tectonophysics*, **290**, 47–58.

Tucker, R.D. & McKerrow, W.S. 1995. Early Palaeozoic chronology: a review in light of new U-Pb zircon ages from Newfoundlland and Britain. *Canadian Journal of Earth Sciences*, **32**, 368–379.

Tull, J.F., Kish, S.A., Groszos, M.S. & Campbell, S.K. 1998. Laurentian magmatism of the southern Appalachian Blue Ridge: Post-Iapetan rifting. *Geological Society of America* Bulletin, **110**, 1281–1303.

Valverde-Vaquero, P. & Dunning, G.R. 2000. New U-Pb ages for Early Ordovician magmatism in Central Spain. *Journal of the Geological Society, London*, **157**, 15–26.

Valverde-Vaquero, P., Dörr, W., Belka, Z., Franke, W., Wisniewska, J. & Shastok, J. 2000. U-Pb single-grain dating of detrital zircons in the Cambrian of central Poland: implications for Gondwana versus Baltica provenance studies. *Earth and Planetary Science Letters*, **184**, 225–240.

Van der Voo, R. 1979. Paleozoic assembly of Pangea: A new plate tectonic model for the Taconic, Caledonian and Hercynian orogenies. *Eos Transactions of the AGU*, **60**, 241.

Van der Voo, R. 1982. Pre-Mesozoic paleomagnetism and plate tectonics. *Annual Review of Earth and Planetary Sciences*, **10**, 191–220.

Van Grootel, G., Verniers, J., Geerkens, B., Laduron, D., Verhaeren, M., Hertogen, J. & De Vos, W. 1997. Timing of magmatism, foreland basin development, metamorphism and inversion in the Anglo-Brabant Fold Belt. *Geological Magazine*, **134**, 607–616.

Vanguestaine, M. 1992. Biostratigraphie par acritarches du Cambro-Ordovicien de Belgique et des regions limitrophes: synthèse et perspectives d'avenir. *Annales de la Société Géologique de Belgique*, **115**, 1–18.

Van Staal, C.R. & de Roo, J.A. 1995. Mid-Paleozoic tectonic evolution of the Appalachian Central Mobile Belt in Northern New Brunswick, Canada: Collision, extensional collapse and dextral transpression. *In*: Hibbard, J., Van Staal, C.R. & Cawood, P. (eds) *Current Perspectives in the Appalachian-Caledonian Orogen*. Geological Association of Canada, Special Paper, **41**, 95–114.

Van Staal, C.R., Dewey, J.F., Mac Niocaill, C. & McKerrow, W.S. 1998. The Cambrian–Silurian tectonic evolution of the Northern Appalachians and British Caledonides: history of a complex, west and southwest Pacific-type segment of Iapetus. *In*: Blundell, D. & Scott, A.C. (eds) *Lyell: The Past is the Key to the Present*. Geological Society, London, Special Publications, **143**, 199–242.

Van Wees, J.-D., Stephenson, R.A., *et al.* 2000. On the origin of the Southern Permian Basin, Central Europe. *Marine and Petroleum Geology*, **17**, 43–59.

Vecoli, M. & Samuelsson, J. 2001. Quantitative evaluation of microplankton palaeogeography in the Ordovician–Early Silurian of the northern Trans European Suture Zone: implications for the timing of the Avalonia-Baltica collision. *Review of Palaeobotany and Palynology*, **115**, 43–68.

Vejbaek, O.V., Stouge, S. & Poulsen, K.D. 1994. Palaeozoic tectonic and sedimentary evolution and hydrocarbon prospectivity in the Bornholm area. *Danmarks Geologiske Undersogelse, Serie A*, **34**.

Verniers, J. & Van Grootel, G. 1991. Review of the Silurian in the Brabant Massif, Belgium. *In*: André, L, Herbosch, A., Vanguestaine, M. & Verniers, J. (eds) *Proceedings of the International Meeting on the Caledonides of the Midlands and the Brabant Massif*, Brussels, 20–23 September 1989. Annales de la Société géologique de Belgique, **114**, 163–193.

Verniers, J., Herbosch, A., *et al.* 2001. Cambrian–Ordovician–Silurian lithostratigraphical units (Belgium). *In*: Bultynck, P. & Dejonghe, L. (eds) *Guide to a Revised Lithostratigraphic Scale of Belgium*. Geologica Belgica, **4**, 5–38.

Verniers, J., Pharaoh, T., *et al.* 2002. The Cambrian to mid Devonian basin development and deformation history of Eastern Avalonia, east of the Midlands Microcraton: new data and a review. *In*: Winchester, J.A., Pharaoh, T.C. & Verniers, J. (eds) *Palaeozoic Amalgamation of Central Europe*. Geological Society, London, Special Publications, **201**, 47–93.

VERNIERS, J., MALETZ, J., KŘÍŽ, J., ŽIGAITĖ, Ž., PARIS, F., SCHÖNLAUB, H. P. & WRONA, R. 2008. Silurian. *In*: MCCANN, T. (ed.) *The Geology of Central Europe. Volume 1: Precambrian and Palaeozoic*. Geological Society, London, 249–302.

VOIGT, S., WAGREICH, M., *et al.* 2008. Cretaceous. *In*: MCCANN, T. (ed.) *The Geology of Central Europe. Volume 2: Mesozoic and Cenozoic*. Geological Society, London, 923–997.

VON HOEGEN, J., LEMME, B., ZIELINSKI, J. & WALTER, R. 1985. Cambrian and Lower Ordovician in the Stavelot-Venn Massif. A model for depositional history. *Neues Jahrbuch für Geologie und Paläontologie*, **171**, 217–235.

VON RAUMER, J.F. 1998. The Palaeozoic evolution in the Alps: from Gondwana to Pangea. *Geologische Rundschau*, **87**, 407–435.

VON RAUMER, J.F., STAMPFLI, G.M., BOREL, G. & BUSSY, F. 2002. Organisation of pre-Variscan basement areas at the north-Gondwanan margin. *International Journal of Earth Sciences*, **91**, 35–52.

VOSTEEN, H.-D., RATH, V., SCHMIDT-MUMM, A. & CLAUSER, C. 2004. The thermal regime of the Northeastern-German Basin from 2-D inversion. *Tectonophysics*, **386**, 81–95.

WARNER, M., MORGAN, J., BARTON, P., MORGAN, P., PRICE C. & JONES, C. 1996. Seismic reflections from the mantle represent relict subduction zones within the continental lithosphere. *Geology*, **24**, 39–42.

WATT, G.R. & THRANE, K. 2001. Early Neoproterozoic events in East Greenland. *Precambrian Research*, **110**(1–4), 165–184.

WILLIAMS, S.H., HARPER, D.A.T., NEUMAN, R.B., BOYCE, W.D. & MAC NIOCAILL, C. 1995. Lower Paleozoic fossils from Newfoundland and their importance in understanding the history of the Iapetus Ocean. *In*: HIBBARD, J, VAN STAAL, C.R. & CAWOOD, P. (eds) *Current Perspectives in the Appalachian-Caledonian Orogen*. Geological Association of Canada, Special Papers, **41**, 115–126.

WILLIAMSON, J.P., PHARAOH, T.C., BANKA, D., THYBO, H., LAIGLE, M. & LEE, M.K. 2002. Potential field modelling of the Baltica-Avalonia (Thor-Tornquist) suture beneath the southern North Sea. *Tectonophysics*, **360**, 47–60.

WINCHESTER, J.A. 1974. The zonal pattern of regional metamorphism in the Scottish Caledonides. *Journal of the Geological Society, London*, **130**, 509–524.

WINCHESTER, J.A. & THE PACE TMR NETWORK TEAM 2002*a*. Palaeozoic amalgamation of Central Europe: new results from recent geological and geophysical investigations. *Tectonophysics*, **360**, 5–21.

WINCHESTER, J.A., PHARAOH, T.C. & VERNIERS, J. 2002*b*. *Palaeozoic Amalgamation of Central Europe*. Geological Society, London, Special Publications, **201**.

WINGATE, M.T.D. & GIDDINGS, J.W. 2000. Age and palaeomagnetism of the Mundine Well dyke swarm, Western Australia – implications for an Australia–Laurentia connection at 755 Ma. *Precambrian Research*, **100**(1–3), 335–357.

WOLFGRAMM, M. 2002. *Fluidentwicklung und Diagenese im Nordostdeutschen Becken - Petrographie, Mikrothermometrie und Geochemie stabiler Isotope*. PhD thesis, Martin-Luther-Universität Halle-Wittenberg.

WONIK, T. 1992. *Kompilation und Interpretation der magnetischen Anomalien der Totalintensität in Zentral- und Nordeuropa*. PhD thesis, University Göttingen.

WONIK, T., TRIPPLER, K., GEIPEL, H., GREINWALD, S. & PASHKEVITCH, I. 2001. Magnetic anomaly map for Northern, Western and Eastern Europe. *Terra Nova*, **13**, 203–213.

WOODCOCK, N.H. 1990*a*. Transpressive Acadian deformation across the Central Wales Lineament. *Journal of Structural Geology*, **12**(3), 329–337.

WOODCOCK, N.H. 1990*b*. Sequence stratigraphy of the Palaeozoic Welsh Basin. *Journal of the Geological Society, London*, **147**, 537–547.

WOODCOCK, N.H. 1991. The Welsh, Anglian and Belgian Caledonides compared. *Annales de la Société Géologique de Belgique*, **114**, 5–17.

WOODCOCK, N.H. 2002*a*. The Cambrian and earliest Ordovician quiescent margin of Gondwana. *In*: WOODCOCK, N.H. & STRACHAN, R. (eds) *Geological History of Britain and Ireland*. Blackwell Science, Oxford, 141–152.

WOODCOCK, N.H. 2002*b*. Ordovician volcanism and sedimentation on Eastern Avalonia. *In*: WOODCOCK, N.H. & STRACHAN, R. (eds) *Geological History of Britain and Ireland*. Blackwell Science, Oxford, 153–167.

WOODCOCK, N.H. & STRACHAN, R.A. 2002*a*. *Geological History of Britain and Ireland*. Blackwell Science, Oxford.

WOODCOCK, N.H. & STRACHAN, R.A. 2002*b*. The Caledonian Orogeny: a multiple plate collision. *In*: WOODCOCK, N.H. & STRACHAN, R.A. (eds) *Geological History of Britain and Ireland*. Blackwell Science, Oxford, 187–205.

WYBRANIEC, S., ZHOU, S., *et al.* 1998. New map compiled of Europe's gravity field. *Eos*, **79**(37), 437–442.

ZELAZNIEWICZ, A. 1997. The Sudetes as a Palaeozoic orogen in central Europe. *Geological Magazine*, **134**, 691–702.

ZELAZNIEWICZ, A. & FRANKE, W. 1994. Discussion on U-Pb ages from SW Poland: evidence for a Caledonian suture zone between Baltica and Gondwana. *Journal of the Geological Society, London*, **151**, 1050–1052.

ZIEGLER, P.A. 1989. *Evolution of Laurussia – A Study in Late Palaeozoic Plate Tectonics*. Kluwer, Dordrecht.

ZIEGLER, P.A. 1990. *Geological Atlas of Western and Central Europe*. Shell Internationale Petroleum Maatschappij B.V.

ZIELHUIS, A. & NOLET, G. 1994. Deep seismic expression of an ancient plate boundary in Europe. *Science*, **265**, 79–81.

ZYLINSKA, A. 2001. Late Cambrian trilobites from the Holy Cross Mountains, Central Poland. *Acta Geologica Polonica*, **51**, 333–383.

ZYLINSKA, A. 2002. Stratigraphic and biogeographic significance of Late Cambrian trilobites from the Holy Cross Mountains. *Acta Geologica Polonica*, **82**, 217–238.

8 Devonian

Z. BEŁKA & M. NARKIEWICZ

The term 'Devonian' was first used in the 1830s to describe a succession of rocks in Devonshire, a large county in SW England. The Devonian period is an interval of about 57 Ma in the middle part of the Palaeozoic with remarkable geological variety. It was influenced by active plate movements producing a wide range of environments and facies. The geosphere and biosphere were undergoing extraordinary changes during this period which can claim a remarkable number of biotic 'firsts', including the first appearance of vascular plants, ammonoids, insects and amphibians. A major radiation of fish took place and the Devonian has therefore been called 'the age of fish'. The development of the earliest land floras, forests and soils on land areas changed the biosphere by transforming the terrestrial environment and linking it more closely with the aquatic realm. This influenced the evolution of tetrapods and their colonization of the land. The global climate switched from the hot greenhouse phase of the earlier Palaeozoic to the cold icehouse phase of the later Palaeozoic. This change was associated with one of the most important extinction phases of the Earth's history (the late-Frasnian biodiversity crisis). The greenhouse climate favoured the development of reef ecosystems dominated by rugose corals and stromatoporoids. They reached their acme in the Middle Devonian and disappeared in the late Frasnian, probably due to the global expansion of black, anoxic sediments on the shelves. Recent data suggest that the Frasnian crisis was triggered in large part by pulses of global cooling which caused substantial regression phenomena (Joachimski & Buggisch 2002; Dopieralska 2003). A gradual cooling during the late Famennian resulted in continental glaciations in the southern hemisphere.

The frequent occurrence of conodonts and their wide geographical distribution in Devonian rocks has led to the use of these microfossils for worldwide correlation. Other fossil groups, especially ammonoids, have been used to construct biozonal schemes for the Devonian System, but the resolution of these schemes is always inferior to the conodont scale (Figs 8.1 & 8.2). The Devonian timescale is characterized by a high resolution to the biozone level and is calibrated with methodically consistent zircon and monazite isotopic ages (Kaufmann 2006). In the highly resolved part of the scale (mid-Eifelian to upper Famennian), zonal resolution averages 0.6 Ma. The age interpolations for the base and top of the Devonian system are 418.1 ± 1.1 Ma and 360.7 ± 0.7 Ma, respectively (Figs 8.1 and 2). The seven stages of the Devonian (Lochkovian, Pragian, Emsian, Eifelian, Givetian, Frasnian, Famennian) display remarkable differences in their duration. The longest stages are the Emsian and the Famennian with durations of about 18 Ma and 15 Ma, respectively. In contrast, the Pragian represents only about 3 Ma. The type section of the Silurian–Devonian boundary is located at Klonk in the Barrandian. The base of the Devonian System (also the base of the Lochkovian stage) is defined at the first appearance of the graptolite *Monograptus uniformis* in the upper part of Bed No. 20 (Chlupáč & Kukal 1977).

Palaeogeography and geotectonics

During Devonian times, the area of Central Europe was divided into three parts which belonged to different large geotectonic units. The Precambrian and early Palaeozoic basement, which extends from England to Poland and Ukraine and is located north of the Rheic Suture and, in the east, north of the Carpathian Front, formed the southern passive continental margin of Euramerica (Fig. 8.3). This palaeocontinent is also often called Laurussia or the Old Red Continent. The crust situated west of the Moldanubian Thrust and south of the Rheic Suture but north of the Alpine Front constitutes a complex tectonic collage of Gondwana-derived massifs and terranes, which has been termed by Tait *et al.* (1997) the Armorican Terrane Assemblage (ATA). In the mid-European segment of the ATA, three belts can be distinguished (Fig. 8.4): the Saxothuringian, the Tepla-Barrandian, and the Moldanubian (Franke 2000). Fragments of the Cadomian and Palaeozoic basement exposed within the Alpine-Carpathian Orogen belong to another principal geotectonic unit of the central European crust. There is ongoing debate about whether they represent different tectonostratigraphic terranes derived from the Gondwana margin or are parts of a single coherent plate (e.g. Tait *et al.* 1997; Stampfli & Borel 2002). In this paper we describe these fragments as the South European terranes (Fig. 8.3).

The palaeogeographic position of Laurussia (Euramerica) during the Devonian is well constrained. This large continent with extremely wide shelf areas was positioned in equatorial latitudes (Tait *et al.* 2000; Stampfli & Borel 2002; Lewandowski 2003) so that the equator ran across the central European shelf. To the south, Euramerica was bordered by an oceanic realm, the parts of which are usually termed the Rheic Ocean and the Variscan Sea. It separated the southern margin of Laurussia from the NW margin of Gondwana. The ATA and the South European terranes occurred within this realm.

Although there is general agreement that both megacontinents, Laurussia and Gondwana, and all terranes located in between drifted northwards during Devonian times, different scenarios of geotectonic evolution are postulated (for review, see Dopieralska 2003). This is primarily because the position of Gondwana is palaeomagnetically poorly constrained (Tait *et al.* 2000). As a consequence, the oceanic realm between Laurussia and Gondwana is presented, in contrasting reconstructions, as being either a quite narrow or a very wide ocean. The first scenario can be inferred from biogeographic data whereas the second is postulated on the basis of palaeomagnetic measurements (e.g. McKerrow *et al.* 2000; Tait *et al.* 2000). According to McKerrow *et al.* (2000), the African margin of Gondwana was already close to Laurussia in mid-Devonian time. In contrast, Tait *et al.* (2000), who used only palaeomagnetic data, favoured a separation of megacontinents by an ocean at least 3000 km wide during the Late Devonian. The concept of close proximity of the NW margin of Gondwana to the SW margin of Laurussia in the early Emsian was suggested by Lewandowski (2003), who postulated

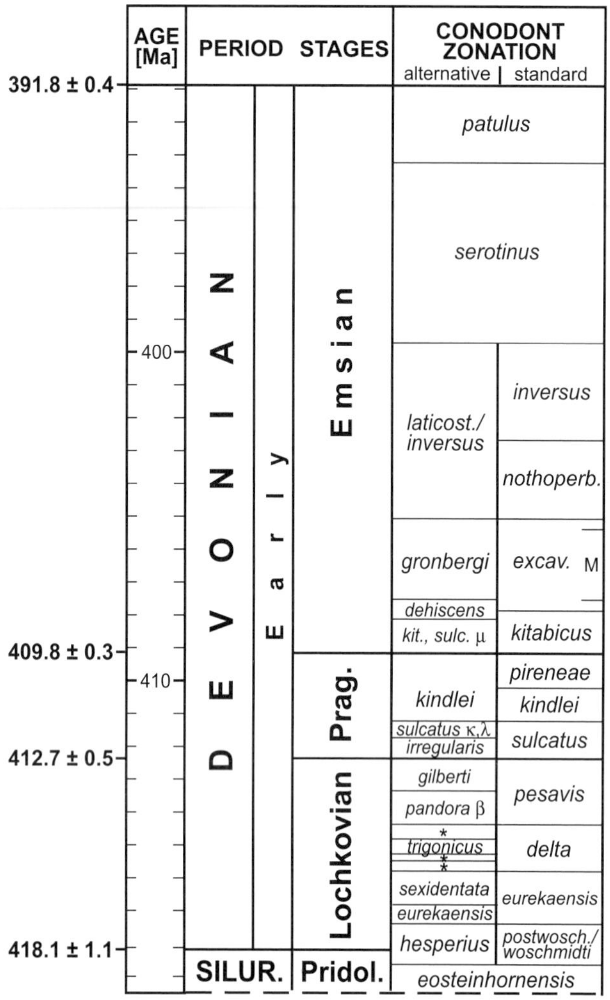

Fig. 8.1. Early Devonian timescale (from Kaufmann 2006). Conodont zones not labelled are (in ascending order): *omoalpha*, *eleanorae*, *alexii*.

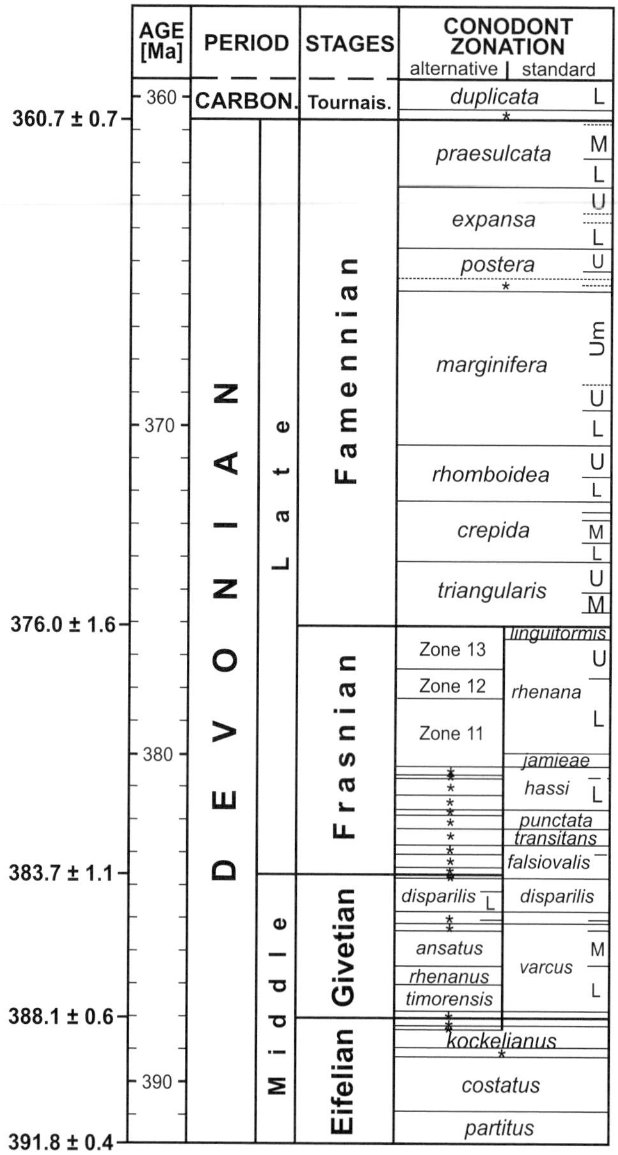

Fig. 8.2. Middle and Late Devonian timescale (from Kaufmann 2006). Conodont zones not labelled are (in ascending order): Eifelian: *australis*, *eiflius*, *ensensis*; Givetian: *hemiansatus*, *latifossatus*, *hermani*, *norrisi*; Frasnian: Zone 1–10; Famennian: *trachytera*; Tournaisian: *sulcata*.

only a narrow oceanic domain (*c.* 400 km) between both margins.

In the majority of palaeogeographical reconstructions, terranes forming the ATA are regarded as plates which moved northwards more or less individually. New extensive geological data from the Variscan orogen (for summary, see Franke 2000) revealed a very complex drift and accretionary scenario for the ATA (Fig. 8.4) The accretion of the Saxothuringian Terrane to the Avalonian margin of Euramerica resulted in the closure of the Rheic Ocean still in the Emsian. Subsequently, in the Eifelian, rifting within this terrane led to the formation of a new oceanic realm. This narrow space, usually termed the Rhenohercynian Ocean, was closed in the early Frasnian. The Tepla-Barrandian accreted to the Saxothuringian in the late Emsian. The Moldanubian and the South European terranes, however, remained isolated during the Devonian. Most probably, they were amalgamated not before the final collision of Gondwana and Euramerica in Late Carboniferous times.

Despite current uncertainty and open questions concerning Devonian palaeogeography, it is evident that most of the areas of Central Europe were located in the tropical realm. Although the position of the Moldanubian and the South European terranes is not directly proved by data, their location in the subtropics can be inferred from the warm-water character of the fauna and the occurrence of reef carbonates in the Carnic Alps (Kreutzer 1992).

Sea-level changes and seawater circulation

The commonly accepted sea-level curve for the Devonian has been constructed based on deepening events recognized on Euramerican platforms (Johnson *et al.* 1985; Johnson & Sandberg 1988; Sandberg *et al.* 1992). The overall sea-level rise during Early and Middle Devonian times was punctuated by several short regression phases. The maximum sea-level high-stand was attained during the 'semichatovae' transgression in the middle Frasnian. Following this most prominent Devonian transgression the sea level began to fall. There is no doubt that

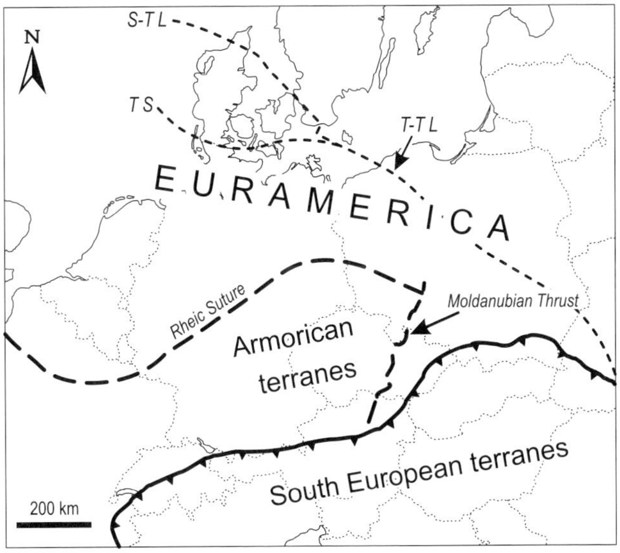

Fig. 8.3. Map of Central Europe showing the major palaeogeographical and geotectonic units during the Devonian Euramerica (Laurussia). Abbreviations: S-TL, Sorgenfrei-Tornquist Line; TS, Thor Suture; T-TL, Teisseyre-Tornquist Line.

the strong transgressive–regressive drops in the late Famennian were related to glacial phases in the southern hemisphere. Recent isotopic data, however, suggest that first glacial-induced regressions may have already occurred in late Frasnian times (Joachimski & Buggisch 2002) and thus confirm an earlier concept of Narkiewicz & Hoffman (1989).

Information on seawater circulation in Central Europe during the Devonian can be taken from models proposed by Heckel & Witzke (1979) and Oczlon (1990). Although both circulation patterns have been constructed based on different kinds of geological data, they show quite similar pictures with a general westward flow along the southern margin of Euramerica. This South Equatorial Current was responsible for the migration of coral larvae and conodont animals in a westerly direction (e.g. Belka 1998; Pedder 1999).

Depositional environments – inner shelf of Euramerica (Laurussia)

During the Devonian the northernmost part of Central Europe was characterized by a predominantly erosional regime connected with the internal, most elevated areas of Laurussia. Deposition was often intermittent and confined to narrow depocentres framed by faults and/or bordered by palaeoelevations of Caledonian or cratonic basement (Fig. 8.5). This is the case for the North Sea area influenced by the Mid-North Sea High (MNSH), the fault-bounded Rügen Depression, the Baltic Basin located between the Baltic Shield and the Mazury–Byelorussian High, and less clearly for the area of Pomerania. The palaeohighs supplied clastic material to nearshore marine systems. Marine deposition, including the development of shallow-water platform carbonate systems, was related mostly to eustatic highstands, particularly in late Emsian and late Givetian–Frasnian times. Typically, the Early Devonian is a period of either erosion or continental sedimentation whereas the Late Devonian, and in particular the Frasnian, displays maximum marine influence.

North Sea

In central and southern North Sea regions, south of the Orcadian Basin, the Devonian has been penetrated by tens of wells located in the eastern UK sector, scattered near the Auk, Argyll and Emble fields in the south (Fig. 8.5), and near the Buchan and Stirling fields in the north (Richards 1990; Cameron *et al.* 1992; Marshall *et al.* 1996; Marshall & Hewett 2003). The Devonian succession is generally developed in a typical continental Old Red Sandstone facies. Deposition of terrigenous sediments of extremely variable thickness took place within fault-bounded, mostly NE-striking depocentres such as the Midland Valley in central Scotland. These were separated by palaeoelevations consisting of the strongly deformed Lower Palaeozoic units intruded by granitic plutons of the Caledonian Orogen (Richards 1990). The most prominent of these, the east–west trending MNSH, with an axis more or less along the present 55°N, is interpreted as forming a positive topographic structure during most of the Devonian, with the Carboniferous onlapping its margins (Marshall *et al.* 1996).

The Old Red Sandstone deposition has been described in detail based mainly on onshore sections in the UK (for a recent review see Trewin & Thirlwall 2002) as well as on offshore occurrences in petroleum basins (see Marshall & Hewett 2003). The depositional systems range from proximal alluvial fans to aeolian, braided river and lacustrine systems, the latter commonly containing rich fish faunas. According to Cameron *et al.* (1992), the Devonian in the southern and central North Sea area is relatively thin (up to 1000–1400 m) and patchily distributed, with the Lower Devonian mostly missing.

In the central North Sea area, south of the Highland Boundary Fault, the Lower Old Red Group comprises uppermost Silurian to middle Emsian sandy deposits, with volcanics and shales in the Lochkovian (Marshall & Hewett 2003). Sediment thickness

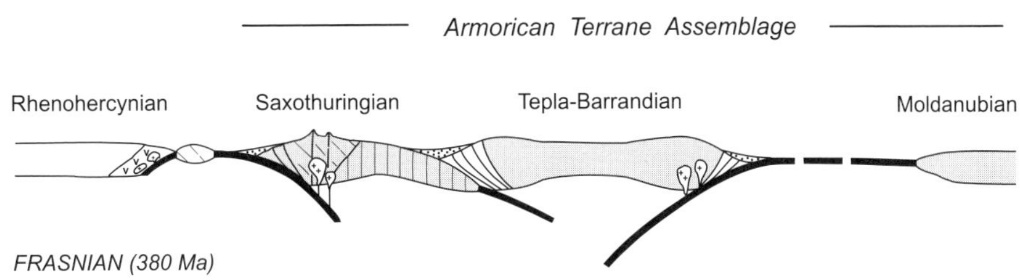

Fig. 8.4. Scheme showing the geotectonic reconstruction of the mid-European Variscides during Frasnian times (adapted from Franke 2000). The Rhenohercynian represented the marginal part of the Euramerica shelf.

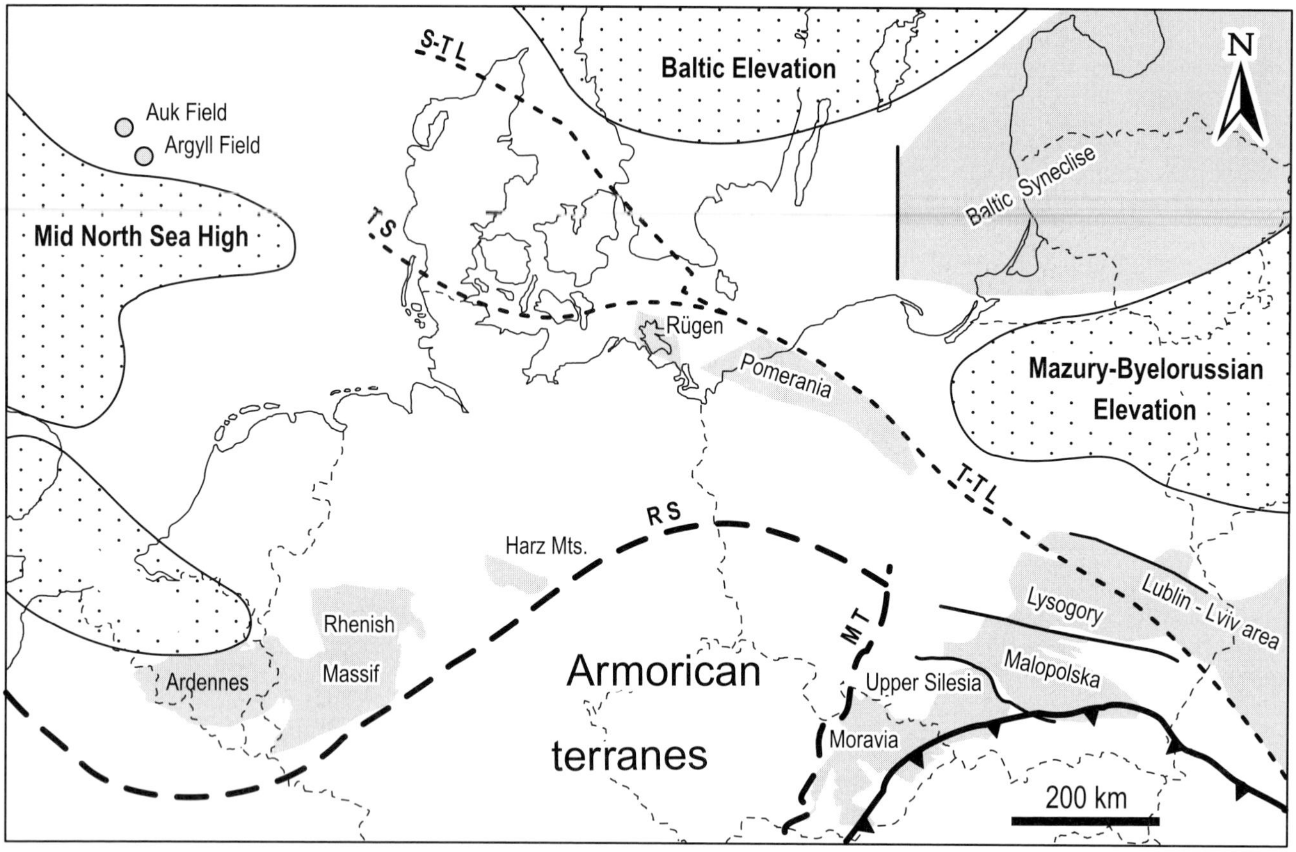

Fig. 8.5. Main Devonian outcrop and shallow subcrop areas (shaded) in Central Europe representing sediments deposited on the shelf of Euramerica. For the North Sea only the most important oil fields are shown, in which the Devonian rocks were encountered. The elevated land areas during the Devonian are stippled and the major faults are indicated. The map is based on various sources cited in the text. Abbreviations: MT, Moldanubian Thrust; RS, Rheic Suture; S-TL, Sorgenfrei-Tornquist Line; TS, Thor Suture; T-TL, Teisseyre-Tornquist Line.

varies from zero in the central North Sea and over the MNSH, to 600 m further northwards. The Middle Old Red Group (= Kyle Group) rests on an unconformity and corresponds to the Givetian (without the uppermost part). It may attain thicknesses ranging from 100–200 m in the Argyll Field area to 500 m in the Emble Field to the east, where it probably overlies Middle Devonian rhyolite. In complete sections, it is composed of a sandstone unit overlain by two marine calcareous/dolomitic horizons separated by a mudstone-to-sandstone unit. The limestones contain an open-marine fauna of brachiopods, corals and crinoids and may be traced as a prominent seismic horizon towards the east and north.

In the area of the Argyll and Auk fields north of the MNSH (Fig. 8.5) the marine succession rests directly upon the Caledonian basement. The age of the lower carbonate horizon is most likely Givetian which would enable these rocks to be correlated with marginal marine clastics described from the onshore and offshore Orkney area (Marshall *et al.* 1996). The carbonates are interpreted as evidence of two marine incursions related to eustatic highstands during the late Givetian–early Frasnian. The incursions followed either the structural trend of the 'proto-Central Graben' (Glennie 1990; Cameron *et al.* 1992) or the southwestern margin of the Fennoscandian High ('Tornquist Zone' of Marshall *et al.* 1996).

The Upper Old Red Group is composed of sandstones and conformably overlies the Kyle Group. It corresponds to the uppermost Givetian and Upper Devonian and grades upwards into Lower Carboniferous sandstones and mudstones.

NE Germany (Rügen area)

The Devonian depositional development of the Rügen area has been recently described by McCann (1999) based partly on earlier work by Zagora (summarized in 1995) and Piske *et al.* (1994). The succession is up to 3 km thick and occurs in a local NW-trending fault-bounded depocentre (Rügen Depression) subcropping between the Arkona (NE) and Stralsund (SW) uplifts (Fig. 8.5). It rests unconformably upon the Ordovician substrate, and is overlain either conformably by the Early Carboniferous strata or unconformably by the Upper Carboniferous or Permian. The base of the Devonian is, in places, developed as terrigenous coarse- to fine-grained deposits, up to 10 m thick, containing late Emsian spores and plant remains (Zagora 1995). Elsewhere, deposition commenced in the Eifelian. The Middle Devonian is biostratigraphically documented by palynomorphs and ostracods whereas the Upper Devonian contains marine fauna, including stratigraphically important brachiopods and ostracods.

The Eifelian attains a thickness of up to 800 m and is composed of mostly non-marine mudstones to quartz siltstones and sandstones with subordinate calcareous marine horizons. The transitional Eifelian–Givetian interval displays a lateral change from more sand-rich and thinner (*c.* 200 m) development in the NW to a fine-grained, more mud-rich, and thicker (up to 430 m) succession in the SE. The Givetian is up to 700 m thick and comprises mainly siliciclastic rocks ranging from quartz sandstones to mudstones. The upper part, attaining a thickness of *c.* 100 m (Zagora 1995), is composed of grey to dark grey mudstones, siltstones and sandstones with marly intercalations and marine fossils, including brachiopods indicative of a late Givetian age (Schmidt & Franke 1977). The total thickness of the Middle Devonian is up to 1500–1800 m.

The Frasnian deposits comprise mostly fossiliferous, commonly biostromal, limestones. These are locally dolomitized, with thicknesses decreasing from 900 m in the north to *c.* 200 m towards the south. Fossil assemblages may be diverse and include locally abundant corals, stromatoporoids, brachiopods, molluscs, crinoids and bryozoans. Terrigenous intercalations are rare and confined mainly to the northwestern part of the Rügen area. The interpretation of seismic data suggests synsedimentary fault activity, with sediments prograding basinwards from localized highs of footwall blocks. This is taken by McCann (1999) as evidence for a lithospheric stretching regime. The Famennian attains a thickness of nearly 200 m, and displays bipartite development. The lower part is lithologically similar to the Frasnian, overlain by a complex of mixed shallow-marine carbonates (oolites included) and terrigenous brown-coloured claystones to sandstones. The total maximum thickness of the Upper Devonian is *c.* 1100 m.

Pomerania

The Devonian of Pomerania (NW Poland) occurs as a wedge-shaped subcrop area striking NW along the Teisseyre-Tornquist Line (Figs 8.3 & 8.5). Unpublished data from petroleum exploration wells in the offshore area of Poland suggest that it forms almost continuous subcrops with the Devonian of Rügen. The NE limit of the Devonian subcrops is clearly erosional; towards the SW the Devonian dips progressively beneath the Permian–Mesozoic of the Mid-Polish Trough and occurs too deep to be penetrated by boreholes. Devonian rocks unconformably overlie the Lower Palaeozoic sediments of the Heligoland-Pomerania Deformation Belt (Winchester and the PACE TMR Network Team 2002). The Devonian succession of Pomerania has been encountered in tens of deep wells and the stratigraphic framework was established by Dadlez (1978).

Devonian deposition commenced after a period of erosion which lasted for much of the Early Devonian. The lowermost deposits are most likely early to middle Eifelian in age (Turnau & Matyja 2001). These are siliciclastic alluvial and alluvial/deltaic facies in the NE of the area (Jamno Formation (Fm)) grading towards the south and SW into a nearshore clastic/carbonate system (Studnica and Tuchola formations). In the Givetian, the marine nearshore clastic and clastic/carbonate systems of the Silno and Chojnice formations (SW) and the Miastko-Sianów formations (NE) predominate. The uppermost Middle Devonian is represented by regressive nearshore marine clastic deposits (Wyszebórz Fm) grading landwards (i.e. towards the NE and north) into alluvial facies. The overall thickness of the Middle Devonian ranges from *c.* 300 m in the SE to 800 m in the NW (Matyja 1998).

The Frasnian shows typical transgressive development (Matyja 1993). The nearshore clastic systems are replaced by mixed carbonate/siliciclastic alluvial/deltaic to open-marine (outer platform) cyclically arranged facies (Dadlez & Dadlez 1986) succeeded by a coral/stromatoporoid biostromal platform. The total thickness of these shallow-marine sediments (i.e. Koczała Fm) ranges from *c.* 150 to 500 m. Towards the SW, they grade into more open-marine, dark marly facies (Człuchów Fm). The lower part of this unit is composed of dark-coloured marly limestones with brachiopods and crinoids, and with silty/sandy intercalations in places. The lower Frasnian succession is overlain by up to 900 m of dark pyrite-rich laminated mudstones containing a pelagic fauna. Locally, these mudstones contain intercalations of platform-derived debris flows. During the middle Frasnian, carbonate platform development was terminated by a short-lived regression recorded by a sheet of up to 50 m of fine-grained sandstones with redeposited platform-derived skeletal material and intercalations of bioturbated mudstones. This unit is overlain by marly/carbonate fossiliferous deposits of an open subtidal shelf. In its upper Frasnian part it forms a lateral equivalent of the dark marly facies and starting from the early Famennian it stretches over the entire Pomerania area.

The Frasnian–Famennian boundary can be traced by changes in lithology and conodont biofacies (Matyja & Narkiewicz 1992). In the earliest Famennian, after a short black-shale depositional episode, the fossiliferous marls and marly limestones (*c.* 100 to 200 m thick) became widespread across the whole area. A global sea-level rise in the middle Famennian led to the development of stromatoporoid/coral mudmounds (60 to 220 m thick). The inter-mound facies is represented by nodular bioclastic limestones. At the beginning of the late Famennian, the northern part of the area became the site of peritidal siliciclastic/carbonate deposition with local evaporites (Kłanino Fm), 150 to almost 400 m thick. These deposits grade towards the south into more open-marine carbonates that are up to several hundred metres thick (Krojanty Fm). These are interpreted as representing a carbonate platform or a ramp. Shallow-marine development was terminated by the onlap of deeper-shelf marly deposits containing an open-marine fauna, including mainly bivalves, gastropods and brachiopods.

Eastern Baltic area

The Devonian of the eastern Baltic area (Fig. 8.5) is a part of the Main Devonian Field, the largest continuous area of Devonian shallow subcrops and outcrops in Europe. It stretches from northern Lithuania to Latvia, southern Estonia and neighbouring parts of the Baltic offshore in the west and Russia in the east. Its present southwestern erosional limit corresponds to the north-trending Rozewie Fault and the east–west trending system of Sambia faults (Kramarska 1999). Structurally, the area belongs to the Baltic Syneclise (Basin) whose axis/depocentre strikes from the Gdańsk Bay towards the NE. The basin is located between the elevated parts of the East European Craton (EEC; East European Platform), the Mazury-Byelorussian High in the south and the Baltic High in the north and NW. The following account is based on summaries published by Kleesment & Mark-Kurit (1997), Kurszs (1992), Paskevicius (1993) and Narbutas & Ungicius (2001).

The Devonian succession comprises part of the flat-lying Palaeozoic cover of the EEC. In the axial part of the basin there is a continuous transition from the Silurian into the Devonian within a sequence of shallow-marine deposits. The Devonian attains a maximum thickness of *c.* 1 km, for example in the offshore Kaliningrad region and in western Latvia. Towards the basin margins, the succession becomes thinner, less complete and unconformably overlies different Lower Palaeozoic units and, in some areas, the crystalline Precambrian basement. The upper part of the Devonian succession becomes progressively erosionally truncated away from the basin axis.

Overall, the sedimentary succession is mainly continental, including alluvial red-bed and lacustrine depositional systems and restricted marine siliciclastic or mixed clastic/carbonate/evaporite deposits with plant remains, fish fauna, occasional ostracods and lingulids. Open-marine carbonates and siliciclastics are limited to certain stratigraphic levels, mostly in the Middle and Upper Devonian. The succession includes characteristic erosional surfaces with associated sedimentary gaps, as well as depositional cyclicity of different orders. The biostratigraphy of the continental deposits is based mostly on spores and fish fauna. The marine horizons are partly correlated using brachiopods and conodonts.

The Lower Devonian consists exclusively of regionally discontinuous, mostly finer-grained siliciclastic units attaining an overall thickness of up to 300 m (Volkolakov *et al.* 1977). Alluvial to lacustrine systems prevail, with two levels displaying a marine influence, in the Pragian–lower Emsian and the upper Emsian, respectively. The younger marine band marks the onset of the major transgression which culminated in the deposition of the widespread middle Eifelian Narva Group. This is predominantly marine, attaining a thickness of 70–180 m, and comprises mudstone and marly/sandy carbonates with open-marine fossils including brachiopods, trilobites, tentaculitids, conodonts, crinoids, but also fish remains, conchostracans and ostracods, indicative of intermittently restricted conditions. The upper part of the Middle Devonian is siliciclastic and continental with some marginal marine influence in the presumed Givetian of Lithuania. The total thickness of the Middle Devonian varies between 50 and 460 m (Volkolakov *et al.* 1977).

The lower Frasnian is represented by terrigenous siliciclastics, mainly alluvial mudstones and sandstones that are partly red-coloured. They are overlain by a middle Frasnian carbonate complex which includes fossiliferous limestones and secondary dolostones with a marine fauna, as well as some beds of more restricted evaporitic facies containing dolomicrites and sulphates. The top of the carbonate succession is marked by an erosional surface which is overlain by continental reddish mudstones and sandstones ascribed to the lower part of the upper Frasnian. The upper Frasnian displays distinct lateral variability with continental red-bed deposits in Latvia grading northwards into open to restricted marine carbonates and clastics with sulphates.

The lower Famennian comprises mostly marine carbonates with an open-marine fauna of brachiopods, bivalves and conodonts, and rare evaporite/mudstone intercalations indicative of more restricted marine conditions. The upper Famennian is developed as alternating relatively thin siliciclastic horizons and more calcareous/dolomitic units, representing several transgressive–regressive cycles. Restricted marine and continental sandstones and mudstones are more important in the north, grading towards the south and west into carbonates with an open-marine fauna, including late Famennian spiriferids, conodonts and crinoids. Towards the north, in Estonia, the Famennian as well as the middle and upper Frasnian are missing. In Lithuania, the total thickness of the Famennian is *c.* 200 m (Paskevicius 1993), whereas the entire Upper Devonian is up to 500 m thick (Volkolakov *et al.* 1977).

Depositional environments – outer shelf of Laurussia (Euramerica) (Rhenohercynian Belt)

The outer shelf of Laurussia constitutes a complex system of carbonate platforms, intracratonic basins and intrabasinal highs. It extends from southern England through Belgium (Ardennes area), the central part of Germany to southern Poland and Moravia (Fig. 8.5). The best records of facies gradients, known from the Eastern Rhenish area and from Moravia, show a transition from an inner-shelf setting to a distal, outer-shelf environment. Restricted outcrops in the Harz Mountains represent a deep, volcanic-influenced intrashelf basin with Middle to Upper Devonian sediment infill. On the other hand, the outcrops and shallow subcrops of SE Poland formed a shelf area with a complex pattern of local depocentres and palaeohighs.

Sedimentation on the outer shelf of Laurussia was primarily controlled by: (1) pre-existing topography inherited from the pre-Devonian tectonic events; (2) the supply of clastics sourced from the eroded landmass of Euramerica, these being particularly important during the Early Devonian; (3) Emsian rifting processes that led to the development of the Rhenohercynian Ocean and were continuous throughout the Middle Devonian; and (4) eustatic events which controlled important depositional episodes particularly in the Late Devonian (e.g. reef development, clastic input).

Ardennes and neighbouring areas

The Devonian outcrops and subcrops in the Ardennes (Fig. 8.6) are distributed mainly in central and southern Belgium and in the cross-border areas with France (southern Ardennes) and Germany (Aachen area). They occur mainly to the south of the Variscan Front which is here represented by the major overthrust of the Faille du Midi (Midi Overthrust), and its eastern extension, the Aachen Thrust. This WSW–ENE striking structure displaced Carboniferous and underlying rocks up to *c.* 30 km towards the north. However, it did not obscure the general palaeogeographical pattern of the Devonian as deduced from the overthrust strata in the south (Ardennes and Vesdre nappes) and autochthonous successions in the north (Namur Synclinorium along the southern flank of the Brabant Massif). The basement of the Devonian belongs to the Avalonian Terrane (Verniers *et al.* 2002).

The main palaeogeographical feature was the elevated area of the Brabant Massif in the north (Fig. 8.5), which was characterized by an erosional or continental regime. To the south, the area passed into a more subsiding belt showing progressively greater sediment thicknesses and more marine influence in the coeval depositional units. Second-order structures partly controlling the sedimentation patterns, mostly in the Early Devonian and Eifelian, included elevated areas comprising Lower Palaeozoic rocks. These include the Rocroi and Givonne massifs south of the Dinant and Neufchateau synclinoriums, respectively, and the Stavelot Massif to the east (Fig. 8.6). Synsedimentary block-tectonics is thought to have influenced deposition, particularly during the Early and Late Devonian.

The palaeontological content of Devonian exposures in the Ardennes has been studied since the beginning of the nineteenth century and thus has considerably contributed to the conceptual development of Devonian stratigraphy. This is reflected in the names of the currently used Devonian stages: Givetian, Frasnian and Famennian.

Lower Devonian

In general, the succession consists of two complexes. The lower one, developed as thick siliciclastic units, comprises the Lochkovian to the lower part of the upper Emsian. Its biostratigraphy is based on brachiopods, spores and rare conodonts. The upper complex, however, is composed of mostly fossiliferous marly carbonates of late Emsian age. The Lower Devonian is absent in

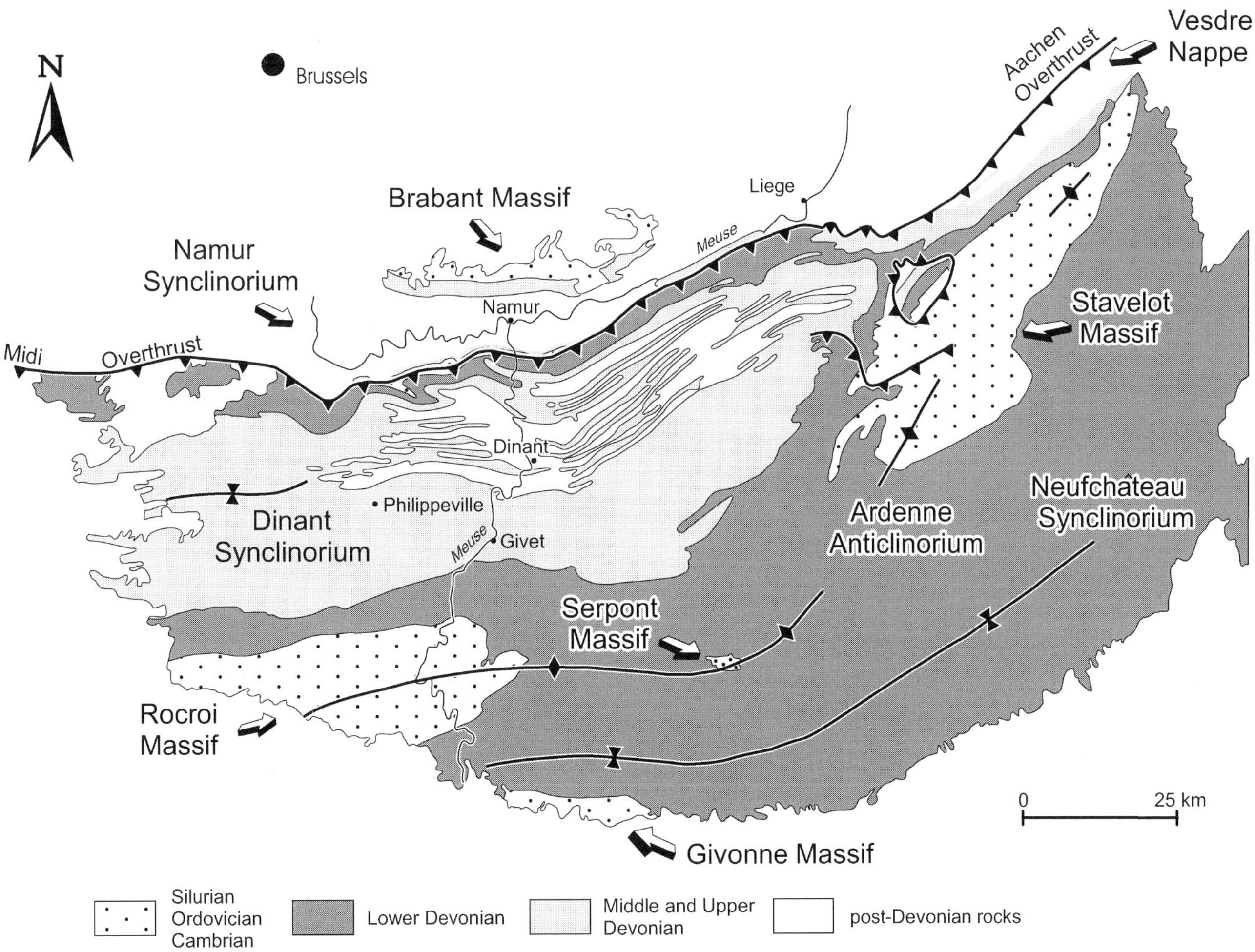

Fig. 8.6. Simplified geological map of the Ardennes (after Bultynck & Dejonghe 2001).

the area of the Brabant Massif due to the prevailing erosional regime.

The basal part of the Devonian is developed as transgressive clastic deposits, commonly with a basal conglomerate, unconformably overlying the Lower Palaeozoic basement and interpreted as alluvial to nearshore tidal marine systems (Godefroid *et al.* 1994; Bultynck & Dejonghe 2001). In the south (in the Dinant and the Neufchateau synclinoriums) sedimentation commenced during the Pridolian–early Lochkovian interval as evidenced by brachiopods, spores and conodonts. Towards the north the onset of deposition of the basal clastics occurred at a stratigraphically higher interval, i.e. in the late Lochkovian. This diachroneity is associated with a thickness decrease, from 4–5 km in the area of Neufchateau Synclinorium to 1–2 km in the northern part of the Dinant Synclinorium (Bultynck *et al.* 2000).

The southern part of the Ardennes is characterized by mixed continental to shallow-marine facies development. The facies range from shaly with diverse marine fauna and even some carbonate intercalations (Mondrepuis, Levrezy, and Ville formations) through nearshore restricted siliciclastics with a poor marine fauna (Saint-Hubert, Mirwart, La Roche, Pernelle and Vireux formations) to alluvial mottled sandstones and mudstones (Oignies and Chooz formations). In general, regressive episodes are noted in the late Lochkovian and in the mid-Emsian (Bultynck *et al.* 2000). The onset of the important transgressive period extending into the Middle Devonian took place during the late Emsian. It is developed as a thin (*c.* 20 m) basal sandstone unit grading upwards into thick (*c.* 300 m) fine-grained clastic deposits (Hierges Fm) with an abundant and diverse marine fauna.

The northern part of the Dinant Synclinorium is characterized by siliciclastics ranging from shales and mudstones to sandstones and conglomerates. In places these rocks contain plant remains and fish fauna, but notably lack marine fossils.

The upper, carbonate complex has been documented only in the southern flank of the Dinant Synclinorium. It is represented by a succession of alternating shales and limestones with abundant brachiopods (St-Joseph Fm) overlain by predominantly marly carbonates (Eau Noire Fm) containing rugose and tabulate corals. Both formations are *c.* 100–160 m thick and constitute the uppermost Emsian. They are interpreted as having been deposited on a carbonate/shaly ramp. The temporal equivalents of these facies in the north and NE are developed as siliciclastic successions attaining progressively more continental character (Hampteau, Wepion and Burnot formations).

Middle Devonian

In the southern part of the Dinant Synclinorium the Eifelian constitutes the major part of the transgressive–regressive cycle which commenced in the late Emsian (Bultynck *et al.* 1991). Its total thickness amounts to 650–700 m. The Eau Noire Fm is overlain by up to 380 m of thick biostromal (rugose/tabulate/stromatoporoid) limestones (Couvin Fm) roughly equivalent to the lower part of the Eifelian. The extent of the carbonate platform is controlled by the basement elevation of the Rocroi Massif. The Couvin Fm is overlain and replaced from the east by shales (Jemelle Fm) with silty/sandy intercalations in the lower part and small biohermal lenses in the upper part. It is gradually onlapped by up to 110 m of shales and sandstones (Lomme Fm). This latter formation represents a regressive phase which persisted into late Eifelian times.

The Lomme Fm is overlain by shales and marly nodular limestones (Hanonet Fm) with abundant brachiopods and rugose corals, and biostromal intercalations in the upper part. This unit reflects a transgressive regime constituting most of the Givetian but which had already commenced in the latest Eifelian. The Eifelian–Givetian boundary is in the lower part of the Hanonet Fm which is 30–70 m thick and may pass laterally into lenticular, partly crinoidal, limestone bodies *c.* 120 m thick named the X Formation (Bultynck *et al.* 2000). In the south, the Givetian is *c.* 450 m thick and forms a shallow-marine carbonate platform predominantly formed of coral/stromatoporoid biostromes with a slightly restricted back-reef facies (Trois-Fontanes, Mont d'Haurs and upper Fromelennes formations). The upper, mostly biostromal part of the Fromelennes Formation ranges upwards close to the Frasnian boundary (Fig. 8.7).

In the north, the Middle Devonian has reduced sediment thicknesses and more shallow-marine to continental development. Along the southern flank of the Namur Synclinorium, the Eifelian rests unconformably upon the Silurian. The lowermost part of the Eifelian is presumably represented by red sandstones and conglomerates with plant-bearing shales (Burnot Fm). This succession is overlain by mixed siliciclastic/carbonate deposits (Riviere Fm) constituting the main part of the Eifelian (50–100 m). The Givetian is 130–160 m thick and represented by shallow-marine carbonates (Nevremont Fm) overlain by a succession of shales, sandstones and various carbonates (Le Roux Fm).

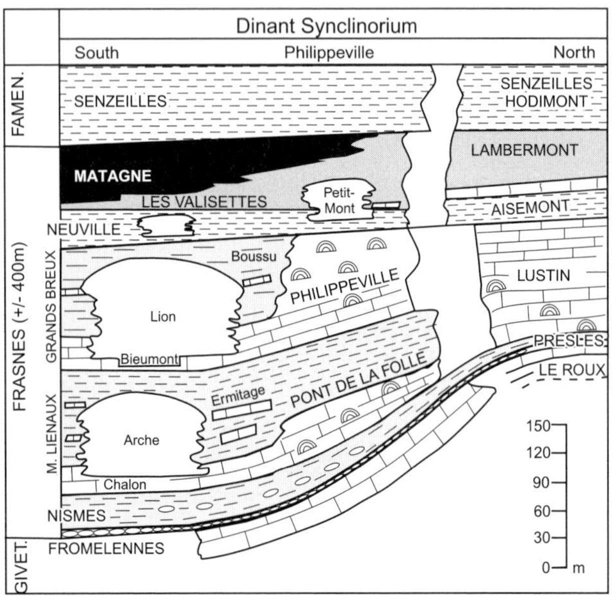

Fig. 8.7. Stratigraphy of the Frasnian in the Dinant Synclinorium (from Bultynck & Dejonghe 2001).

Further north, in the northern part of the Namur Synclinorium, the Givetian directly overlies the Silurian basement. The succession (Bois de Bordeaux Fm) is composed of basal siliciclastic red-beds overlain by various limestones, including oolites, bioclastic and biostromal units, and is capped by red sandy/silty shales with rare limestone intercalations (Bultynck & Dejonghe 2001).

Upper Devonian

The total thickness of the Frasnian is *c.* 400 m in the southern Dinant Synclinorium, gradually decreasing northwards to about half of this value (Boulvain *et al.* 1999). The succession commences with greenish shales containing subordinate nodular limestone horizons rich in brachiopods (Nismes Fm). It is interpreted as representing a global eustatic transgression and comprises a gradual onlap towards the north where it grades into Fe-oolitic carbonates (Presles Fm).

Each of the two succeeding formations (Moulin Lienaux and Grands Breux) represents a separate transgressive–regressive cycle (cf. Gouwy & Bultynck 2000). Both are composed of a basal marly calcareous ramp with localized development of large coral/stromatoporoid buildups west of the Meuse River. The buildups, referred to as the Arche and Lion members, display thicknesses of up to 120 m and 120–250 m, respectively. Their sedimentology, biota and palaeoecological aspects have been the subject of numerous studies (e.g. Lecompte 1970; Tsien 1979; Sandberg *et al.* 1992). The buildups are surrounded by shaly and marly sediments which partly represent post-buildup sediments with some calcareous/detrital intercalations sourced from the buildups. According to Sandberg *et al.* (1992), the Lion Member buildups display evidence of karst development as a result of a regressive episode and subsequent, or coeval, extensional tectonics and volcanism. Towards the north, in the Phillippeville area, the two cycles mentioned above are represented by two units of coral/stromatoporoid biostromal carbonates (Pont de la Folle Fm and the Phillippeville Fm) separated by a marly wedge thinning to the north (= upper Pont de la Folle Fm). The southern edge of the biostromal complexes shows pervasive dolomitization, and is probably controlled by a synsedimentary fault (da Silva & Boulvain 2002). To the north of the Dinant Synclinorium, both units merge into a single carbonate platform succession (Lustin Fm) which attains a thickness of *c.* 100 m (Fig. 8.7).

A transgressive event at the beginning of the late Frasnian led to the development of a third biohermal level, represented by numerous red-coloured stromatactis mounds (Petit-Mont Member). These are located within nodular marls (Neuville Fm) and are usually 20 to 80 m high; in places (Philippeville area) they protrude into the overlying shales (Valisettes Fm). According to Sandberg *et al.* (1992), the development of the stromatactis mounds was succeeded by a regression leading to deep karstification of the buildups.

The uppermost Frasnian is represented by black shales (Matagne Fm) with only rare limestone intercalations yielding mostly small pelagic fauna. In the south, this formation is 50 m thick, but northwards (in the Philippeville area) it thins progressively. Its equivalent, along the northern margin of the Dinant Synclinorium, consists of a 50 to 100 m thick succession of silty shales and limestone intercalations including a biostrome and brachiopod coquinas (Lambermont Fm). Da Silva & Boulvain (2002) suggested that the southern limit of this facies was controlled by synsedimentary faulting.

Much of the Famennian was characterized by the predominant erosional continental regime in the north (i.e. in the Namur area).

Only terrestrial red-coloured clastics and shallow-marine, littoral deposits (Samme Fm) of the uppermost Famennian are preserved (Fig. 8.8). The relatively simple pattern of increasing subsidence and marine influence to the south was complicated by differential tilting of basement blocks on the shelf (Dreesen *et al.* 1988). The Famennian is most completely developed in the Dinant Synclinorium whereas to the east, in the Vesdre Nappe, it is considerably reduced both in thickness and in terms of stratigraphy.

The Famennian displays a characteristic tripartite development, with basal (Famenne Group) and top (Strunian) units comprising predominantly fine-grained clastic and carbonate lithologies, whereas coarse siliciclastics, commonly arkosic sandstones, with subordinate carbonates (Condroz Group, Psammites du Condroz) form the middle part of the succession (Fig. 8.8). The basal part of the succession is uniformly developed as open-marine deposits mainly consisting of shales with siltstone intercalations which become more frequent upwards. Subordinate constituents include carbonate nodules, sandy/calcareous beds with crinoids and rhynchonellid brachiopods, and several Fe-oolitic horizons. The thickness increases from 50–60 m south of Namur to 260 m and more than 400 m towards the southern part of the Dinant Synclinorium.

The Condroz Group is up to 800 m thick in the Dinant Synclinorium and wedges out northwards. It begins with shaly deposits (Aye Fm) which pass northwards into more proximal sandy/silty facies (Esneux Fm) which represent a shallow subtidal environment. A eustatic transgressive event (Baelen Event; Dreesen *et al.* 1988) resulted in more open-marine carbonate deposition of nodular limestones with reworked skeletal material (Souverain-Pre Fm), and with localized development of crinoidal/stromatactoid mudmounds (20 to 150 m thick Baelen Member). In more proximal locations, the nodular limestones intercalate with sandstones and siltstones (Ciney Fm). The overlying, mostly sandy deposits (Montfort Fm) are interpreted as a sand-barrier complex with facies ranging from subtidal forebarrier to barrier and lagoonal/evaporitic backbarrier environments. The uppermost Condroz Group deposits represent a regressive interval composed of continental, clastic red-beds and lagoonal dolomites (Evieux Fm) grading laterally to marginal marine peritidal limestones (Beverire Fm).

The upper part of the Famennian is characterized by a transgressive regime that led to the deposition of marine shales with siltstones in the lower part, common sandy interbeds in the middle part, and crinoidal limestones in the upper part. These lithologies (Etroeungt Fm) are 120 to 200 m thick and pass laterally into 30–40 m thick coral/stromatoporoid biostromes (Dolhain Fm). The buildups are mostly known from the Vesdre area. Devonian sedimentation was terminated by the significant regression during the latest Famennian that was responsible for the deposition of extremely shallow-water carbonates and/or stratigraphic gaps in the Devonian–Carboniferous transition.

Rhenish Massif and Harz Mountains: a general structural background

The Rhenish Massif (Rheinisches Schiefergebirge) reveals the most complete palaeogeographical record of the Rhenohercynian Zone. Traditionally, it has been subdivided into a western part, including the Eifel Mountains, and an eastern one, with the boundary corresponding roughly to the Rhine River (Fig. 8.9). In fact, this subdivision also reflects a major difference in terms of the general development of the Devonian. The boundary may have been controlled by a deep crustal discontinuity (cf. Engel *et al.* 1983), for example, the elongated structural feature termed the Mid-Netherlands-Krefeld High by Dreesen *et al.* (1988) or a transform fault traceable as far as the North Sea (Franke 1995). In contrast to the relatively complex temporal and spatial relationships of the various Devonian units in the east, the succession of the Eifel area is simple, consisting mainly of marginal-marine to shallow-marine clastics and carbonates. Their development closely matches that of the Ardennes area (as shown by Weddige 1988), being a slightly more distal shelf equivalent to the deposits exposed in the southern Dinant Synclinorium.

The Devonian rocks of the Rhenish Massif and Harz Mountains can be subdivided into several structural/sedimentary belts striking SW–NE, parallel to the general strike of the Rhenohercynian Zone (Franke 2000). In the NW part the Devonian succession includes autochthonous deposits resting on Avalonian

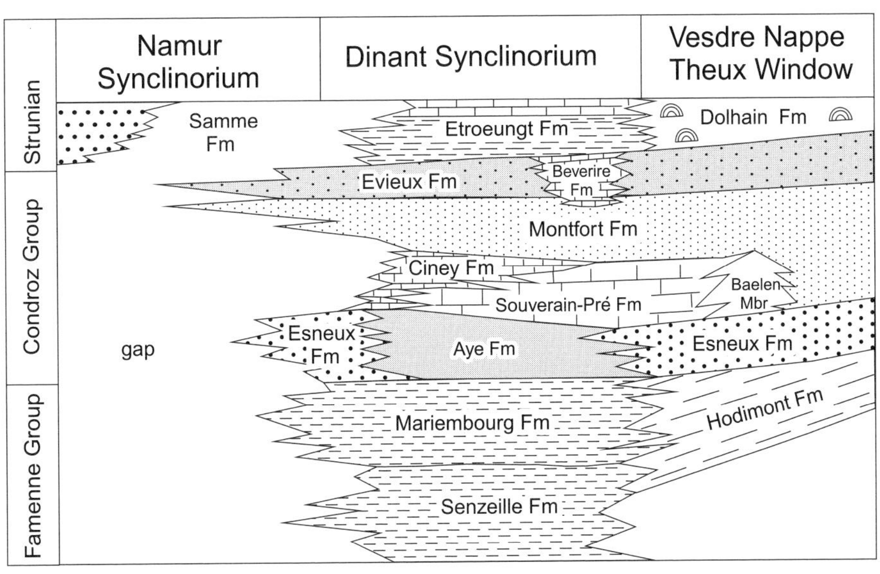

Fig. 8.8. Lithostratigraphy of the Famennian in Belgium (from Bultynck & Dejonghe 2001).

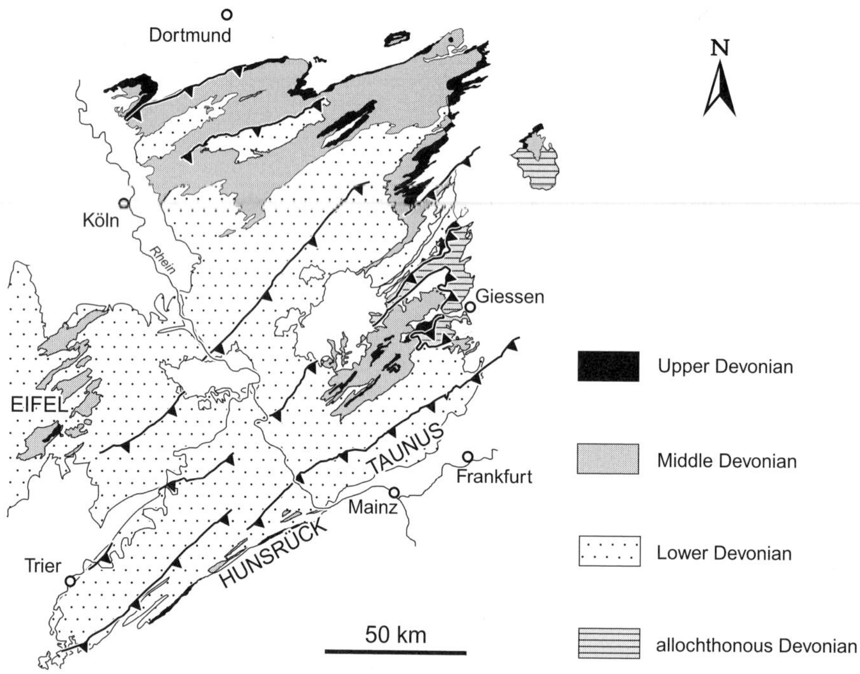

Fig. 8.9. Map showing distribution of the Devonian rocks in the Rhenish Massif (adapted from Oncken *et al.* 2000).

crust. These represent part of the complex southern margin of Euramerica, from clastic/carbonate shelf in the NW (stable shelf of Meischner 1971) through a deeper shaly volcanic basin in the centre (Dill and Lahn synclines) to the Taunus shelf margin in the SE (Franke 2000). Whereas the Taunus area seems to have its counterpart west of the Rhine, in the Hunsrück area, the deeper basin appears to be restricted to the eastern part of the Rhenish Massif. According to the palinspastic reconstruction by Franke (2000; see also Franke (1995) for a review, and Jansen *et al.* (2001) for alternative views), the allochthonous units recognized from the SE Rhenish Massif and from the Harz Mountains constitute fragments of the two southeasternmost belts of the Rhenohercynian Zone (Fig. 8.10). (1) Distal continental margin/slope with the rifted basement composed of Silurian–early Devonian magmatic arc and adjoining Armorican crust. The Devonian deposits, mostly ascribed to the so-called Hercynian Complex, comprise condensed pelagic limestones of Eifelian to late Famennian age, Famennian quartz arenites (Hoerre-Gommern Quartzite) intercalated with pelagic shales and radiolarites, Middle to Upper Devonian limestone turbidites and shales, overlain by greywackes, and early Devonian sandstones topped by Middle to Upper Devonian hemipelagic shales and limestones. (2) The oceanic Rhenohercynian crust overlain by condensed Emsian to late Devonian shales and radiolarian cherts, succeeded in turn by Frasnian greywackes with metamorphic material derived from the Mid-German Crystalline High. The latter deposits record the first stages of compressional deformation related to the development of the active SE margin of the Rhenohercynian Basin. The stratigraphic succession and depositional architecture of both belts, as restored from complex thrust-fold tectonics, can only be interpreted in a general way, and therefore will not be treated in detail here.

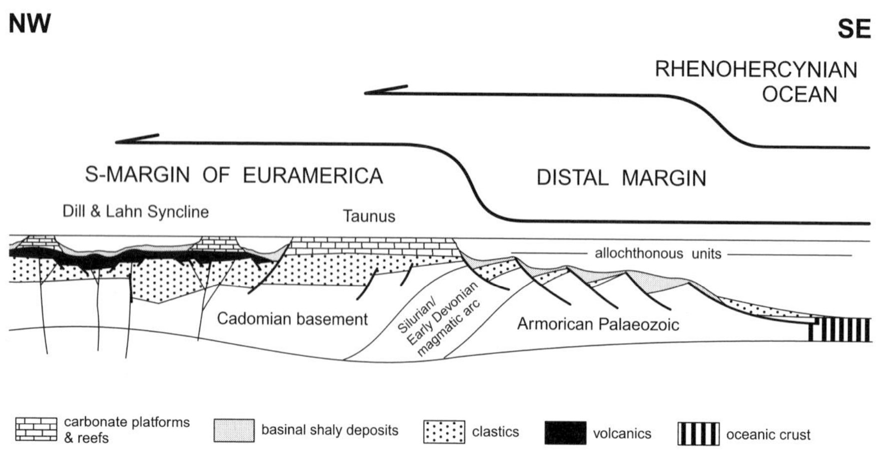

Fig. 8.10. Palinspastic restoration of allochthonous units in the Rhenish Massif (from Franke 2000).

Eifel Mountains

According to Franke (1995, 2000), the Devonian succession west of the Rhine differs from its eastern counterpart mainly in terms of its negligible volcanism and earlier development of a carbonate platform. The development of the latter began in the early Eifelian following a period of siliciclastic shelf sedimentation. The oldest known Devonian deposits are from the lower Emsian, although their basal contact is unclear. The lower Emsian can be exceptionally thick (*c.* 5 km in the Mosel Trough) in the region, and this appears to be typical for the entire Rhenish area (Jansen *et al.* 2001).

The Eifel Mountains are the type area for the Eifelian stage, for which the basal Global Stratotype Section and Point is located in the village of Schoenecken-Wetteldorf (Weddige *et al.* 1988). The base of the stage is established in the upper part of a mudstone and siltstone succession (Heisdorf Fm) which also contains some rare argillaceous limestone or calcareous sandstone beds. The Middle Devonian is developed as a succession of shallow-marine, mostly carbonate units, with marly horizons marking the relatively deeper-water intervals. According to Faber (1980), during the early Eifelian there was cyclic development of a carbonate, coral/stromatoporoid platform, whereas in the late Eifelian and early Givetian a flat shelf-lagoon system prevailed. The upper Givetian–lower Frasnian (Bolsdorf Schichten, Wallersheim Dolomit) is represented by a partly dolomitized biostromal platform. The whole Middle Devonian to lower Frasnian carbonate succession is up to 1000 m thick. During the middle Frasnian, variable subsidence created a platform-to-basin pattern in the shelf bathymetry (Krebs 1969). The upper Frasnian, however, again shows more uniform facies development; it is represented by monotonous shales that contain two anoxic carbonate levels known as the Kellwasserkalk. The Devonian succession is terminated by lower Famennian ostracod-bearing shales passing laterally into black marls.

Eastern Rhenish Massif (autochthonous part)

Lower Devonian

The lowermost Devonian clastics conformably overlie the Lower Palaeozoic sediments. It is generally assumed that clastic deposition, which developed and reached its maximum during the Early Devonian, was ultimately sourced by the Caledonian basement exposed to the NW. Only the southernmost clastic units, such as the Taunus Quartzite of Pragian–Emsian age, can be related to sources located within the Mid-German Crystalline High (Figs 8.4 & 8.9). It has been estimated that Lower–Middle Devonian shelf clastics may have attained a maximum thickness of 14 km (Oncken *et al.* 2000) with a depocentre located in the Siegen area. The biostratigraphy of these deposits is mostly based on brachiopods and trilobites, and to a lesser extent on bivalves and fish remains.

After a period of relatively uniform red-bed deposition in the Lochkovian, the depositional environment gradually began to differentiate into a still predominantly inner (proximal) shelf and an incipient outer shelf to the SE. Most of the area is characterized by thick successions of shallow-water marine clastics with impoverished fauna and locally abundant plant remains. More open-marine systems, partly with calcareous admixture and with a more diverse fauna occurred to the south. This fauna, including exceptionally well-preserved arthropods and crinoids, is particularly well-known from the lower Emsian Hunsrück Shale. This unit is composed of black shales with turbiditic silty and sandy intercalations and was deposited in a deeper basinal setting (Krebs 1979). The uppermost lower Emsian records regressive conditions with red-bed sedimentation, succeeded in the earliest late Emsian by a sandstone succession deposited in a tide-dominated setting (Ems Quartzite).

The late Emsian was a time of considerable retrogradation of proximal facies belt towards the NW, paralleled by the encroachment of more open-marine systems on continental and restricted marine facies. This was coincident with the culmination of felsic volcanism (Hauptkeratophyr) which had been developing since the Lochkovian mostly in the NW part of the eastern Rhenish Massif. Both the distribution of local depocentres and the volcanism were presumably related to synsedimentary block-faulting (Engel *et al.* 1983) resulting from the prevailing extensional regime and this was certainly connected with the onset of rifting along the southern margin of Avalonia where the Rhenohercynian Ocean had started to open (Franke 2000).

Middle Devonian

In the latest Emsian and the early Eifelian, three main depositional realms were established. In the NW area, the inner-shelf environments were polarized into more proximal nearshore marine facies with continental red-beds and rare coarse-grained intercalations, while more open-marine systems dominated to the SW (e.g. Lenneschiefer). The biota found includes fish and eurypterids, bivalves, ostracods, rare brachiopods and abundant plants. The outer-shelf environment comprises fine-grained sandstones with limestone intercalations, and a diverse open-marine fauna and common bioturbation (Goldring & Langenstrassen 1979). These facies graded to the SE into a much thinner black shale succession, known as the Wissenbach Shale with siltstone/sandstone turbiditic intercalations, and a mainly pelagic fauna of cephalopods and dacryoconarids, deposited in a deep marine setting. In the Lahn-Dill area typical trough–high (Schwellen-Becken) topography began to develop due to the development of volcanic basaltic highs. Condensed cephalopod limestones occur on the tops of the highs, which are surrounded by shale-dominated basins. In the late Eifelian the shelf edge was positioned along a line running from the SW (Siegen Anticline) to the NE (Ostsauerland Syncline), and this line probably corresponded to a basement fault.

The Eifelian–Givetian boundary is marked by a transgressive event recorded as a tongue of basinal shales (Odershausen Beds) within the coarse clastic shelf sequence. The lower Givetian is similar to the underlying succession comprising mainly siliciclastic shelf sediments. During the mid-Givetian the clastics sources were less effective as a result of progressive levelling of the Caledonian hinterland, and concomitant sea-level rise (Engel *et al.* 1983). This reduction in clastic input created suitable conditions for the establishment of carbonate platforms on the shelf, with the development of a widespread coral/stromatoporoid facies (Schwelm Limestone), which was succeeded by areally restricted reef complexes (Massenkalk) along the shelf margin and upon volcanic elevations in the Lahn-Dill area. The total thickness of the reef carbonates varies greatly, from 100 to 2000 m (Bender *et al.* 1977; Meischner, 1996). Between the buildups dark, marly well-bedded limestones and shales (Flinz facies), frequently with turbidites composed of reef-derived detrital material (allodapic limestones of Meischner, 1964), were deposited.

The Givetian was a time of intense submarine basaltic volcanism which developed along the fault-controlled shelf edge (Sauerland) where the volcanics are up to 250 thick (Bender *et al.* 1977), and in the Lahn-Dill area where they attain a thickness of hundreds of metres (Franke 1995). The volcanism, displaying geochemical signatures of a within-plate setting, contributed to the persistence of the earlier established high-to-basin topography, and to the development of characteristic deep marine volcaniclastics (Schalstein). In the deeper basinal areas ongoing deposition of black shales (Wissenbach Shale) led to a total accumulation of up to 1500 m of fine-grained siliciclastics. Volcanic ridges became sites of reef growth and, consequently, sources of bioclastic detritus transported into the adjoining basinal depressions (Fig. 8.11). Additionally, iron ores are associated with sites of submarine volcanism (e.g. the Meggen ore deposits in Sauerland).

Upper Devonian

The stage of reef growth and the corresponding pattern of depositional systems affected by synsedimentary block tectonics (Krebs 1979) persisted into mid-Frasnian times (Fig. 8.11). Reefs attained a high diversity, including tabulate corals and stromatoporoids as the main frame-builders, and contain a complex array of subenvironments including fully developed backreef and forereef areas with transitions to more off-reef areas (for review see Krebs 1974; Burchette 1981; Machel & Hunter 1994). In the mid-Frasnian, widespread reef growth ceased due to reasons which are not yet fully understood; probably a combination of global and regional factors was responsible (Eder & Franke 1982; Narkiewicz & Hoffman 1989). The termination of reef growth roughly coincides with the appearance of the lower Kellwasser horizon, a characteristic black, organic-rich limestone which is well-known from many localities in the Rhenish area and elsewhere. Following the cessation of widespread reef growth, the deposition of dark shales with calcareous nodules and cephalopod limestones prevailed in the formerly elevated areas, whereas in the deeper basins ostracod shales (Cypridinenschiefer) were deposited. However, the bottom relief was still largely controlled by inactive reef complexes and by volcanic/reef highs in the Lahn-Dill area. The elevations were sites of non-deposition and/or accumulation of condensed crinoidal and cephalopod limestones. During the latest Frasnian, anoxic conditions occurred again on the sea bottom and the upper Kellwasser horizon developed in many places on the shelf. This event, however, did not change the overall pattern of sedimentation which continued into the Famennian. In more elevated and relatively shallower-marine areas an earliest Famennian decrease in carbonate production can be observed (e.g. Buggisch 1972).

During the Famennian the shelf margin retreated to the NW corner of the Rhenish Massif (i.e. Ruhr area). Sedimentation was characterized by the intermittent influx of well-sorted sandy material which formed individual turbidites in shale successions, or predominantly sandy units (e.g. mid-Famennian Nehden Sandstone, or upper Famennian Dasberg Beds). The thickness of the Famennian decreases from *c.* 500 m in the NW to 150–300 m in the Lahn–Dill area. The Devonian–Carboniferous boundary occurs just above the base of the poorly fossiliferous Hangenberg Shale (Meischner 1996).

Harz Mountains

In the Harz Mountains structural equivalents of the central and southeastern part of the Rhenish Massif are exposed. Whereas the northwestern Harz area may be roughly considered to be an extension of the Dill Syncline, the remaining part is mostly composed of complex, thrusted and chaotic units comparable to the allochthonous units of the SE Rhenish Massif (Franke 2000). The boundary runs along the NW border of the Acker-Bruchberg Zone which comprises the Lower Carboniferous Hőrre-Gommern Quartzite. South of this line the autochthonous basement crops out in the Elbingerode Window.

The lowest part of the autochthonous Devonian succession is represented by sandstones which are interpreted as having been deposited in a clastic shelf environment (Franke 1995). In the latest Emsian, deposition of 600 m thick, deep basinal Wissenbach Shales began (Kulke 1997). The area was subsequently influenced by the development of basaltic volcanism which extended up into the early Givetian. Volcanic activity led to the development of a series of submarine complexes, more than 100 m thick, and composed of pillow-lavas and volcaniclastic sediments (Alberti & Walliser 1977). In the early Givetian, the volcanic highs became sites of reef growth in two areas, at Iberg (Franke 1973) and at Elbingerode. Reef development was terminated before the late Frasnian. In the western Harz Mountains, in a deeper setting, two Kellwasser limestone levels can be observed in the late Frasnian succession. In the Famennian, the former reefs as well as non-active volcanic highs were covered by condensed cephalopod limestones, whereas in the adjoining troughs shales with variable calcareous admixtures accumulated.

Southeastern Poland

The southern Polish shelf area can be subdivided into several regions which vary both in terms of their subsidence history and depositional development (Narkiewicz 2007). These areas largely correspond to sub-Devonian basement blocks with different early Palaeozoic accretion histories, and include, from west to east: (1) the Upper Silesian Block, (2) the Małopolska Block, (3) the Łysogóry-Radom Block, and (4) the margin of the EEC in the Lublin-Lviv area (Fig. 8.5). During the Devonian, these crustal units were roughly situated in their present-day positions (Belka

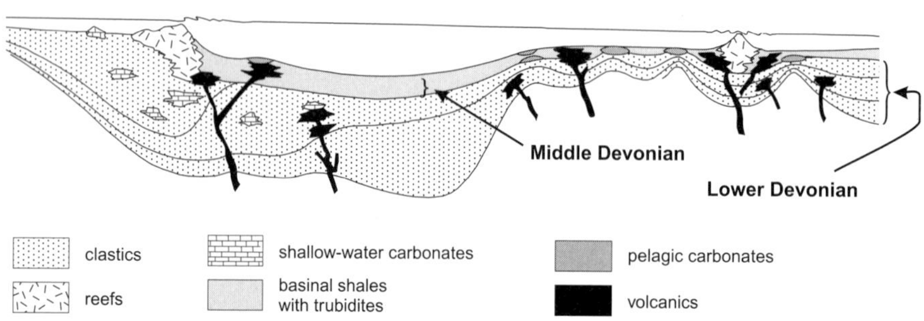

Fig. 8.11. Depositional model for the Frasnian of the eastern Rhenish Massif showing also thickness relationships during the Early and Middle Devonian (modified from Eder *et al.* 1977).

et al. 2000). Devonian rocks of the Małopolska and Łysogóry-Radom blocks are exposed in the southern and northern parts of the Holy Cross Mountains, respectively. The Devonian of the Upper Silesian Block can be directly studied only in small outcrops west of Cracow. In the remaining regions, the Devonian has been encountered in deep wells which are particularly numerous in southeastern Poland.

Lower Devonian

Continuous deposition across the lower system boundary is observed only in the Łysogóry-Radom Block and over a considerable part of the Lublin-Lviv area. In the former area, the Lochkovian is mainly represented by marine fine-grained clastics which vary in thickness from 170 to 700 m (Miłaczewski 1981). Further east, in the Lviv area, they correlate with a mudstone succession containing subordinate intercalations of fossiliferous limestones of probable tempestite origin (Drygant 1984; Pajchlowa & Miłaczewski 2003). In Łysogóry the lower Lochkovian is represented by mudstones (Bostów Beds) with trilobites and miospores (Racki & Turnau 2000). These deposits grade both upwards and laterally into alluvial and marginal marine clastics (Klonów Beds) of variable thickness (up to 200 m) (Szulczewski 1995; Kowalczewski *et al.* 1998). This latter unit is also found in the Małopolska Massif where it includes fine- to coarse-grained clastic rocks, mainly greywackes and lithic and quartz arenites. These were deposited in alluvial fan, braid-plain and braid-delta settings.

The Pragian and Emsian rocks are distributed across most of the Holy Cross Mountains area (Kowalczewski *et al.* 1998). The succession, usually several hundred metres thick, is mainly composed of sandy deposits interpreted as alluvial and marginal marine facies (Barcza Beds) overlain by sandstones with clayey/silty intercalations (Zagórze Fm) representing a shallow-marine, nearshore, storm-dominated depositional setting (Szulczewski 1995). North of the Holy Cross area, the equivalent of the Barcza Beds is up to 1300 m thick and represented by clastic deposits (Zwoleń Fm), mostly quartz sandstones and siltstones, with horizons of pedogenic dolomitic concretions, as well as plants, fish and Agnatha, eurypterids and lingulids. The depositional environment is interpreted as an alluvial system developed under arid climate conditions (Miłaczewski 1981).

In the remaining areas of southern Poland clastic sediments ascribed to different Lower Devonian levels unconformably overlie an uneven surface of the pre-Devonian basement. In the southern Holy Cross Mountains these deposits, previously named the 'placoderm sandstone' may locally attain a thickness of more than 200 m. They are subdivided into a lower mottled mudstone complex with tuffites (Haliszka Fm) assigned to the Pragian, and two sandy units separated by a mottled tuffite complex (Winna Fm), regarded as Emsian in age (Tarnowska 1976). The depositional environments are interpreted as being predominantly alluvial with some marginal marine influence along the northern margin of the area. In Upper Silesia, the Devonian clastics, developed as cyclically arranged conglomerates to mudstones, 0–130 m thick, are interpreted as having been deposited as part of an alluvial fan system (Narkiewicz 1996). These clastics rest unconformably on either the Cambrian marine clastics or Neoproterozoic crystalline basement. Their late Emsian age is confirmed by palynological data (Turnau 1974).

Contrary to the previously suggested provenance of clastic material from local reworked Lower Palaeozoic rocks (Miłaczewski 1981; Tarnowska 1988), new isotopic data (K–Ar cooling ages of detrital mica) revealed the presence of two mica populations widespread in the Lower Devonian of southern Poland (Belka *et al.* 2000). Both the older population, with *c.* 460 to 470 Ma ages, and the younger one, with *c.* 420 to 430 Ma ages, are interpreted as having been derived from a northerly direction, from sources dominated by Upper Silurian metamorphic rocks, most probably from the Scandinavian Caledonides.

Middle Devonian

Close to the end of Early Devonian times, the overall sedimentary regime changed from a siliciclastic to a predominantly shaly and carbonate one (Fig. 8.12). Only in basin-marginal settings (e.g. southern Upper Silesia, eastern part of the Lublin area) did clastic deposition continue into the Middle Devonian.

In the Łysogóry region, Emsian marine clastics grade upward into a predominantly marly uppermost Emsian–lower Eifelian sequence (Grzegorzowice Fm), which is fossiliferous and up to

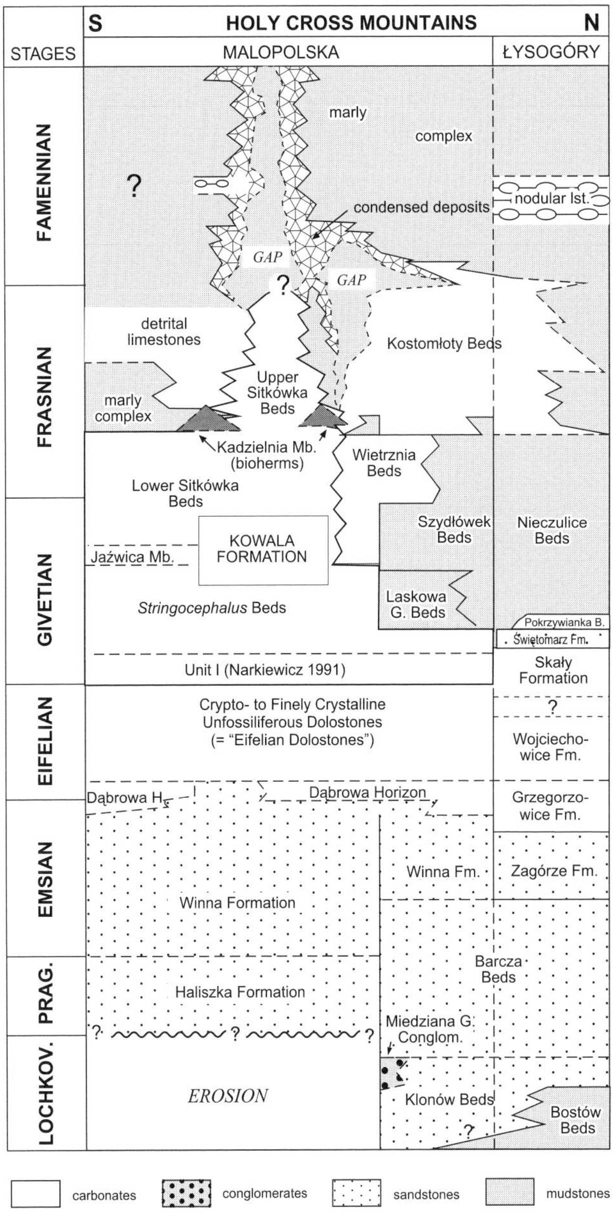

Fig. 8.12. Lithostratigraphy of the Devonian of the Holy Cross Mountains (adapted from Racki & Turnau (2000) and Racki & Narkiewicz 2000).

150 m thick. This is overlain by a thick (up to 550 m), mainly non-fossiliferous peritidal/lagoonal dolomites which pass upwards into peritidal cycles with *Amphipora* biostromes (Wojciechowice Fm). A deepening phase that began during the late Eifelian (Racki 1997) is manifested by a 260 m thick unit of fossiliferous shales (Skały Fm) which include an exceptionally well-preserved brachiopod fauna (Malec & Turnau 1997). Deepening was interrupted as a consequence of a sea-level drop during the middle Givetian and is marked by the deposition of shallow-marine clastics (Świętomarz Fm). This formation locally attains a thickness of up to 100 m. A renewed transgression in the middle Givetian led to localized growth of coral/stromatoporoid buildups followed by deep-water, marly facies (Nieczulice Fm, *c.* 350 m thick) with rare brachiopods and cephalopods (Malec & Turnau 1997).

North of Łysogóry, in the Radom area, the shelf exhibited the highest subsidence rate. Hence, the Middle Devonian is represented by predominantly shaly deposits and attains a thickness of more than 1300 m (Malec *et al.* 1996; Narkiewicz 2002). In this pelagic setting the sandy equivalents of the Świętomarz Fm, deposited during the short middle Givetian regressive phase are only 12 m thick. The overlying shale sequence is lithologically similar to the late Givetian equivalents in Łysogóry but it is less calcareous and, in contrast, contains more abundant ostracod assemblages as well as plant remains.

In the southern part of the Holy Cross Mountains, which constitutes a northern margin of the Małopolska Block, the basal part of the Middle Devonian is composed of open-marine fossiliferous carbonates and black shales (Dąbrowa unit), up to several tens of metres thick. This succession is considered to be latest Emsian–early Eifelian in age. It is succeeded by peritidal/lagoonal, usually laminated dolomicrites. Despite a lack of biostratigraphic evidence, it is generally assumed that at the time of the Eifelian–Givetian transition a shallow-marine, carbonate platform, dominated by coral/stromatoporoid facies (Kowala Fm), developed across the entire Małopolska area (Narkiewicz *et al.* 1990; Racki 1993). Its internal architecture and stratigraphy remain, in many places, unknown due to extensive secondary dolomitization (Narkiewicz 1991). The Givetian portion of the carbonate platform system is at least 500 m thick. Its basal part is composed of peritidal cycles with solution-collapse breccias marking primary supratidal evaporitic levels. These are overlain by more open-marine facies comprising rhythmically bedded coral/stromatoporoid biostromes and various peritidal limestones. The late Givetian transgression led to the development of marly and detrital limestones (Jaźwica Member) with a rich open-marine fauna, including brachiopods, corals, dacryoconarids and conodonts. These carbonates are covered by a thick coral/stromatoporoid platform (Sitkówka Beds) which ranges into the Frasnian.

In the central part of the Małopolska Block, the Middle Devonian succession is similar both in terms of lithologies and thicknesses to that of the southern Holy Cross Mountains (Jurkiewicz 1975). However, further south, in the basement of the Carpathian Foredeep, the early Eifelian fossiliferous Dąbrowa unit is lacking. Devonian clastics range up into the Eifelian and are directly covered by a thick coral/stromatoporoid complex (Kicuła & Żakowa 1972).

In the Lublin area, the Middle Devonian is characterized by a marked vertical variability related to cyclic changes in facies, ranging from open-marine shelf, through a carbonate platform, into lagoonal with evaporites, nearshore clastics and an alluvial system (Narkiewicz *et al.* 1998). Moreover, it displays a general trend of increasing thickness in a southeasterly direction, up to a maximum of 250 m. Three transgressive–regressive cycles can be distinguished, driven most probably by global eustatic transgressions (Narkiewicz *et al.* 1998; Narkiewicz & Narkiewicz 1992). Towards the NE, in the area with an elevated EEC basement, clastics constitute most of the Middle Devonian succession. These represent nearshore marine, and probably also alluvial, facies in their basal part. Carbonate deposition may have commenced as late as the late Givetian, or even the early Frasnian.

The Middle Devonian succession of the Upper Silesian Block varies strongly in thickness (locally >400 m) and thickens to the north. Unfortunately, it is stratigraphically very poorly constrained and the onset of carbonate deposition, for example, still remains undated (Matyja *et al.* 2001; Narkiewicz 2005). The carbonate succession overlies Lower Devonian clastics and commences, in the south, with marly, bioturbated dolomite mudstones which contain features characteristic of a subtidal to peritidal environment. The overlying thick carbonate succession is organized into several shallowing-upward cycles, each 3 to 10 m thick, and composed of alternations of coral/stromatoporoid biolithites, shallow-marine wackestones, and peritidal mudstones. This period of extreme shallow-water carbonate deposition persisted, most probably, until the late Givetian when it was replaced by coral/stromatoporoid facies. Toward the north, the Middle Devonian includes increasingly open-marine carbonates. These represent a carbonate platform system, with widespread, coral/stromatoporoid deposits, partly exposed to the west of Cracow (Narkiewicz & Racki 1984) and along the NE margin of the Upper Silesian Block (Racki 1993).

Upper Devonian

The Upper Devonian is rather poorly documented in the northern part of the Holy Cross Mountains although it attains a thickness of several hundred metres. The basal Frasnian is a monotonous marly sequence (Nieczulice Fm) deposited in a deep-water basinal environment, which also characterized the late Givetian sedimentation (Malec *et al.* 1995). The overlying succession of well-bedded carbonate mudstones and wackestones with cherts (Kostomłoty Beds) represents almost the entire Frasnian and ranges into the lowermost Famennian. During the early Famennian the carbonates were replaced by a thick, monotonous, shaly and marly succession which partly includes nodular limestone beds. Its lithological character suggests deposition in a deep basin.

In contrast to the Łysogóry region, the Upper Devonian in the southern part of the Holy Cross Mountains is well-documented and exhibits considerable variations in both facies and stratigraphy. The lower part of the Frasnian is represented by a coral/stromatoporoid facies of a shallow-water carbonate platform (Lower Sitkówka Beds), which was bordered by intracratonic basins to the south and north (Fig. 8.13). The internal architecture of the carbonate platform is biostratigraphically poorly constrained. As a result of a global sea level rise during the early Frasnian, the widespread carbonate platform shrank to isolated reef complexes (Narkiewicz 1988). The Dyminy Reef that developed close to Kielce, where it attains a thickness of 200–300 m, forms a typical example. Within all reef complexes a clear facies differentiation can easily be distinguished between the biostromal facies, similar in its lithological development to the Givetian counterparts, and the detrital forereef (slope) facies rimming the buildups. Early in the late Frasnian a strong transgressive pulse, preceded by forced regression, drowned the reef complexes and the shallow-water carbonate sedimentation survived only in the central area of the Dyminy Reef until the early Famennian

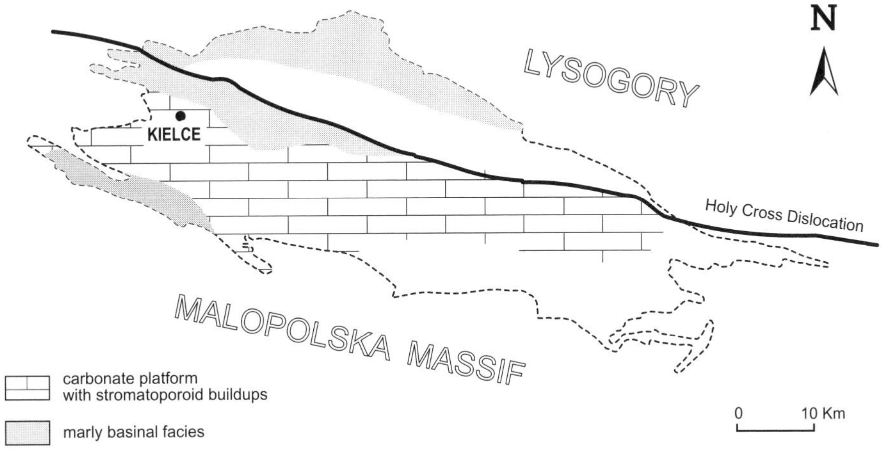

Fig. 8.13. Facies distribution in the Holy Cross Mountains during Frasnian time (modified from Racki 1993). Dashed line indicates the area of the Palaeozoic outcrops.

(Narkiewicz 1988; Narkiewicz & Hoffman 1989). The Frasnian–Famennian transition is generally associated with the presence of stratigraphic gaps (Szulczewski 1971) separating the Frasnian buildups from their Famennian pelagic cover. This cover is predominantly composed of condensed cephalopod wackestones with abundant nektonic (cephalopods, conodonts) and/or benthic (brachiopods, crinoids) fauna. On the flanks of the buildups or in proximal basinal settings, where the sedimentary record is more complete, there are intercalations of debris flows around the Frasnian–Famennian boundary. These reflect episodes of erosion and resedimentation related to a global drop in sea level during the latest Frasnian. Is some places, this boundary is marked by geochemical anomalies (Racki *et al.* 2002). Although the growth of the coral/stromatoporoid buildups was terminated during the late Frasnian, these structures were still responsible for differences in bathymetry on the shelf during the early Famennian. Regional extension during the Famennian caused the submarine highs to become submerged and a basinal depression developed in the southern Holy Cross Mountains. Consequently, the middle and the upper Famennian are represented mainly by marly sequences with numerous interbeds of nodular mudstones and wackestones. Only small fragments of the former carbonate platform displayed marked sedimentary and subsidence evolution involving the deposition of a condensed, pelagic cover, but not before late Famennian times (Szulczewski *et al.* 1996). North of the Holy Cross Mountains, in the Radom area, the base of the Upper Devonian coincides more or less with the development of a widespread carbonate platform extending far to the east (Malec *et al.* 1996). Unfortunately, post-Variscan erosion removed much of the Upper Devonian strata in the area.

In the Lublin region, the lower Frasnian succession (*c.* 100 m thick) is represented by a carbonate platform system overlain by a carbonate ramp 300 m thick, which locally contains stromatoporoid/algal buildups. The onset of ramp development coincided with increased tectonic subsidence related to the formation of the Lublin Graben (Narkiewicz *et al.* 1998; Narkiewicz 2003). Consequently, the Famennian attained, in places, a thickness of *c.* 2000 m. It commences with a poorly fossiliferous sequence composed of marl–limestone rhythmic beds, commonly disturbed by slumping (Bychawa Fm). These are overlain by nodular wackestones (Firlej Fm) which grade upwards into carbonate/siliciclastic deposits (Hulcza Fm) representing marginal marine and continental deposition along the northeastern active edge of the basin (Matyja & Żbikowska 1985). Towards the SW, the succession interfingers with open-marine marls (Niedrzwica Fm) containing cephalopod fauna and plant detritus.

In Upper Silesia, the Upper Devonian displays a persistent lateral facies trend from predominantly marly basinal facies in the north to more shallow-water carbonates in the south (Narkiewicz 1978, 2001, 2005). Nevertheless, its total thickness (*c.* 400–500 m) remains more or less constant throughout the area. In the north the lower Frasnian is developed as nodular limestones, *Styliolina* limestones and marly shales, whereas to the south the deposits grade into shallow-marine stromatoporoid biostromes associated with detrital carbonates. In the middle part of the Frasnian there is a characteristic regressive interval marked by a series of subaerial erosional discontinuities in the south, or as coarse-grained detrital limestones with a sandy admixture in the NE. The upper part of the Frasnian is developed as deep-water carbonate mudstones in the north and shallow-water mudstones to wackestones with oncoids and calcareous algae in the south. The Famennian in the proximal southern area is characterized by a considerable admixture of terrigenous quartz in the shallow-water shelf carbonates. In contrast, in the north, the Famennian is composed of basinal marly deposits with numerous intercalations of detrital packstones and nodular wackestones. The former include carbonate detritus transported from elevated areas on the shelf. West of Cracow, the upper Famennian includes very peculiar buildups containing stromatoporoid (labechiid) and coral fauna (Berkowski 2002), which did not have any similar counterparts in the world at that time. These were presumably located along sandy shoals fringing a shallow-water platform or upper ramp (Narkiewicz & Racki 1985). In the entire region the uppermost Famennian is missing, and a stratigraphic gap separates the Devonian from the Early Carboniferous rocks (Belka 1985). This discontinuity resulted from a latest Devonian global regression that affected most of the shelves worldwide.

Moravia

The Devonian occurs in a triangle-shaped area located east of the Moldanubian Thrust (Figs 8.5 & 8.14); its eastern limit is largely erosional and covered by Carpathian thrusts. To the north, the Devonian rocks can be traced across the Polish–Czech border where they connect with the southern part of the Upper Silesian Block and are similar to the Polish successions. The geological

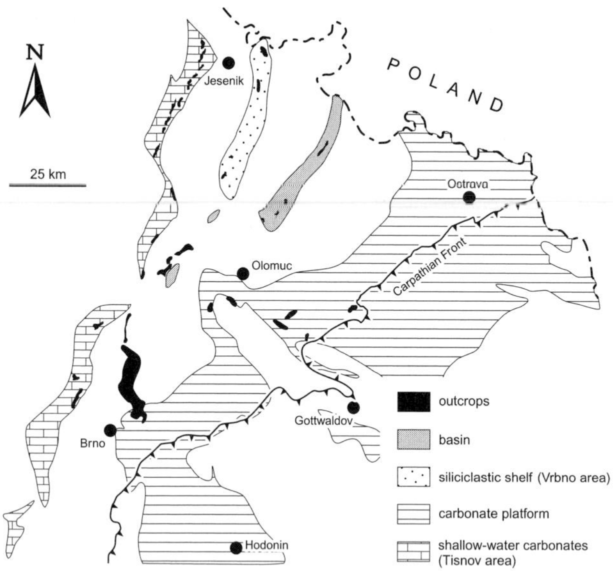

Fig. 8.14. Map of Moravia showing main depositional realms during the Devonian (after Hladil 1992). The distribution of Devonian rocks in the subsurface is based on borehole data.

information on the Devonian in Moravia is based on data from numerous boreholes and a limited number of outcrops. These document a range of depositional environments including a carbonate platform along the Carpathian front, a basinal realm north of Olomouc, a siliciclastic shelf in the Hruby Jesenik Mountains, and a shallow-water shelf facies along the Moldanubian Thrust in the west (Fig. 8.14). However, the Devonian rocks within both shelf environments are metamorphosed while basinal deposits are strongly deformed. This leads to serious difficulties when restoring the original depositional sequence and lateral facies relationships.

Lower Devonian

Within the carbonate platform, the Devonian succession commenced with clastics that overlie the Neoproterozoic basement or a thin Lower Palaeozoic cover (Fig. 8.15). These comprise conglomerates, quartz and arkosic sandstones and mudstones deposited in alluvial systems under arid or semiarid conditions (Dvořák 1990, 1998; Galle *et al.* 1988). Their thickness is highly variable, depending partly on pre-existing basement topography, ranging from tens of metres in the SE to *c.* 1600 m in the west. According to Kalvoda (2002), this thick clastic cover may be related to rifting processes. The clastics are mostly assigned to the Emsian although, in more proximal areas, they may have developed in Middle Devonian times (Chlupáč *et al.* 2000). Towards the NW, in the transitional zone between the carbonate platform and the basinal realm, the onset of clastic deposition, as represented by coarse-grained conglomerates, may have commenced in the Lochkovian (Hladil *et al.* 1999). These rocks are overlain by Emsian shales with basaltic tuffs, followed by carbonates (biostromes and calciturbidites) of late Emsian age.

In the basinal realm, sedimentation began in the Pragian as documented by dacryoconarid and trilobite fauna (Chlupáč *et al.* 2000). The basal deep-water shales attain a thickness of *c.* 200 m

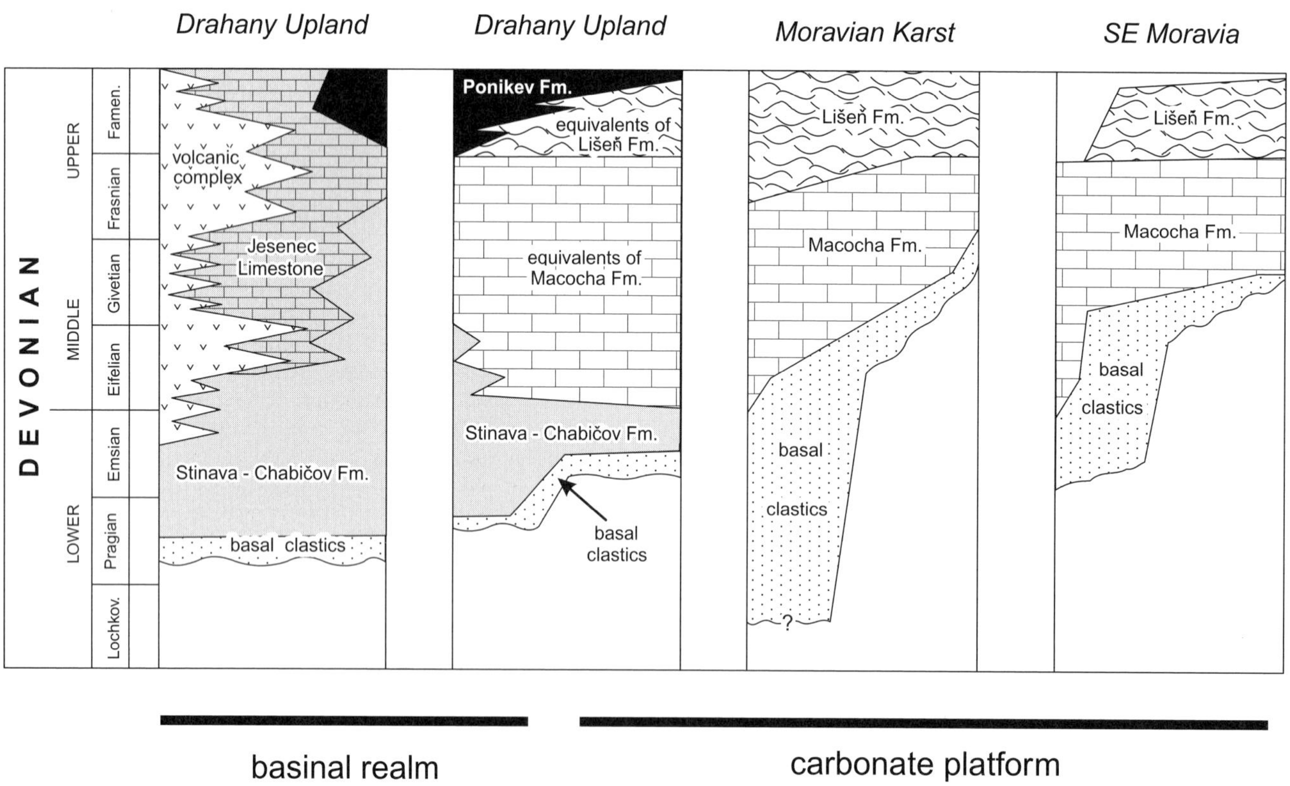

Fig. 8.15. Schematic correlation of the Devonian in Moravia (adapted from Kalvoda 2002). Columns show lithological variations across a facies transect from the carbonate platform in the SE into the basinal realm in the NW.

and contain much volcanic material (Kalvoda 2002). The input of volcanic and carbonate material into the basin was related to the formation of volcanic edifices which were later covered by carbonates (Hladil 1992).

The basal part of the succession of the siliciclastic shelf (Vrbno Group) is represented by the Lochkovian (?) to Pragian phyllites and quartzites. These clastic metasediments, *c.* 150 thick, are interpreted as representing a shallow-water facies which originally contained significant amounts of carbonate (Hladil *et al.* 1999). They are succeeded by a very thick (>1000 m) complex of metamorphosed sediments and volcanics which are considered to be Emsian to Famennian in age (Fig. 8.15). In the more proximal part of the shelf (Tisnov-type deposition) the Lower Devonian is represented by Pragian metasediments (quartzites and phyllites, *c.* 300 m thick) covered by the Emsian clastics which are very thick (*c.* 600 m) and contain brachiopod fauna. This succession is interpreted as having been deposited in estuarine, deltaic and neritic marine settings (Hladil 1994; Hladil *et al.* 1999).

Middle Devonian

In the carbonate platform area, the Middle Devonian is developed as a thick carbonate succession composed of several transgressive–regressive cycles (Hladil 1988) in which stromatoporoid biolithites constitute the predominant lithology. Each cycle is characterized by distinct stromatoporoid and rugose coral assemblages (Galle *et al.* 1988). In general, the lower part of the Middle Devonian succession is developed in a more restricted dolomitic facies, whereas later they display typical features of coral/stromatoporoid shallow-marine calcareous deposition (Dvořák 1990). On the carbonate platform margin to slope environment, coral-bearing limestones, mostly comprising packstones to mud-supported rudstones, were deposited (Hladil *et al.* 1999). In more distal (basinward) settings shales were deposited up to early Givetian times (Chlupáč *et al.* 2000).

In the basinal realm, shale deposition (Stinava-Chabicov Fm) was continuing during the Middle Devonian. This succession contains a significant proportion of submarine volcanics as well as calciturbidites comprising redeposited shallow-water skeletal material (Hladil 1992; Kalvoda 2002). The submarine volcanic rises formed by alkaline tholeitic basalts were sites of carbonate buildups. Reef fringes grade into steep slopes with slumps, debris flows and chaotic megabreccias whereas in adjoining basinal depressions calciturbidites pass laterally into condensed siliceous shales (Hladil 1992, 1994; Hladil *et al.* 1999). During the late Givetian some volcanic rises were exposed and subjected to weathering (Hladil 1994).

On the siliciclastic shelf (Vrbno area), alkaline trachytic volcanism prevailed in the Givetian and was related to the continental rifting (Hladil 1992; Hladil *et al.* 1999). The Middle Devonian sedimentary succession, originally represented by clastics, is now metamorphosed to phyllites, metaquartzites and greenschists. Limited and scattered carbonate occurrences include fossiliferous micrites, spiculites, small mudmounds and bioherms, iron-rich ooids, algal laminites, and calcareous/clastic sediments with plant detritus (Hladil 1994). In the tectonically disturbed and strongly metamorphosed Tisnov area, clastic deposition continued during the Eifelian and the sediments were overlain by a complex of carbonates (Vitosov and Heroltice limestones) which is mostly Givetian in age. These carbonates are *c.* 100 m thick (Chlupáč *et al.* 2000) and represent a carbonate platform system characterized by *Amphipora* biostromes intercalated with common algal laminites, coral/bryozoan bioherms, quartz sand and gravel beds, as well as relicts of evaporites indicative of a sabkha environment (Hladil 1992, 1994).

Upper Devonian

In the carbonate platform area, the Frasnian is represented by a complex of coarse-bedded stromatoporoid biolithites several hundred metres thick (Macocha Fm). Apart from abundant stromatoporoids, rugose and tabulate corals are also common in this reef facies. A mid-Frasnian event, equivalent to the Lower Kellwasser horizon, is marked by a general decline in carbonate production, karstification in proximal areas and sedimentation of dark carbonates in more offshore settings (Hladil 1994). Following this short event, the upper Frasnian displays maximum faunal diversity and a peak in coral/stromatoporoid platform development (Galle *et al.* 1988; Hladil 1988). The growth of coral/stromatoporoid buildups was terminated before the end of the Frasnian, but a few stromatoporoids and corals survived and became extinct during the early Famennian (Chlupáč 1988). Local karstification and lateritic weathering (Dvořák *et al.* 1986; Hladil 2003) point to sea-level fall during the latest Frasnian.

The Famennian was characterized by considerable facies variability (Kalvoda 2002), ranging from shallow-water limestones with abundant foraminifera and algae (Drazovice Limestone) to organodetrital calciturbidites intercalated by shales and cherts (Hady-Ricka Limestone), and marly (often nodular) hemipelagic wackestones with distal turbidites, rich in radiolarians and conodonts (Krtiny Limestone). Towards the NW, the pelagic carbonate sequences grade into deep-water radiolarian-bearing shales and radiolarites (Ponikev Fm).

Late Devonian deposition in the basinal realm was essentially a continuation of the volcanosedimentary deep-water, marine system that prevailed in the region during Middle Devonian times (Fig. 8.15). Monotonous shales rich in radiolarians and radiolarites constituted the dominant lithology. According to Hladil (1992, 1994), some isolated submarine volcanic rises capped by reefs were periodically emergent during the Frasnian, and subsequently drowned during the Famennian. Emergence and drowning phases are indicated by a sedimentary succession composed of red-coloured breccia layers overlain by pelagic sediments.

In the Vrbno area, the Frasnian is mostly represented by strongly metamorphosed carbonates. Beginning in late Frasnian times the amount of volcanic input decreased. Metamorphic rocks contain relicts of sedimentary structures suggesting the presence of reefs in more shallow-water settings and sandy deposition under terrestrial conditions (Hladil *et al.* 1999).

In the proximal part of the shelf (Tisnov area), only Frasnian rocks are present. These include shallow-water sediments, predominantly carbonates that, in places, are overlain by clastics. Within the lower part of the succession small scattered mudmounds are present (Hladil 1994). It is assumed that tectonic uplift resulted in the cessation of Late Devonian deposition, and subsequent erosion removed significant parts of the succession. The region was later subjected to intense tectonic deformation along the Moravian Shear Zone (Hladil *et al.* 1999).

Depositional environments – Armorican terranes

Tepla-Barrandian

Barrandian

The Barrandian area is world famous for its Devonian stratigraphy, containing sections with continuous marine sedimentation from the Silurian up into the Devonian. One of these, exposed at

Klonk (near Suchomasty), has been selected as the international type section of the Silurian–Devonian boundary. In addition, the Lochkovian and Pragian stages have been established here and defined by stratotypes (Chlupáč & Kukal 1986).

The Devonian rocks are well-exposed in the central part of the Barrandian, SW of Prague (Figs 8.16 & 8.17), where they form a 20 km long and 5 km wide syncline between the Litavka and Vltava rivers. The succession is represented by Early and Middle Devonian unmetamorphosed marine deposits, predominantly carbonates, which are locally extremely fossiliferous (Chlupáč *et al.* 1998). Most of the fauna shows a close relation to North African communities. The Devonian succession ranges from 450 to 800 m in thickness, with a general trend of increasing thickness from the NW to the SE. The sedimentary evolution reflects subsidence and extension in a passive-margin setting. Sedimentation was mainly in an offshore shelf environment. Facies development was generally characterized by the occurrence of two facies belts: the bioclastic and the micritic facies. The former represents a moderately shallow, high-energy environment that occupied the southwestern and western part of the Barrandian while the latter characterizes a much deeper pelagic environment that extended to the north and east. Bioclastic material, which was primarily deposited in the high-energy environment, was also periodically transported by turbidity currents into the deeper pelagic realm.

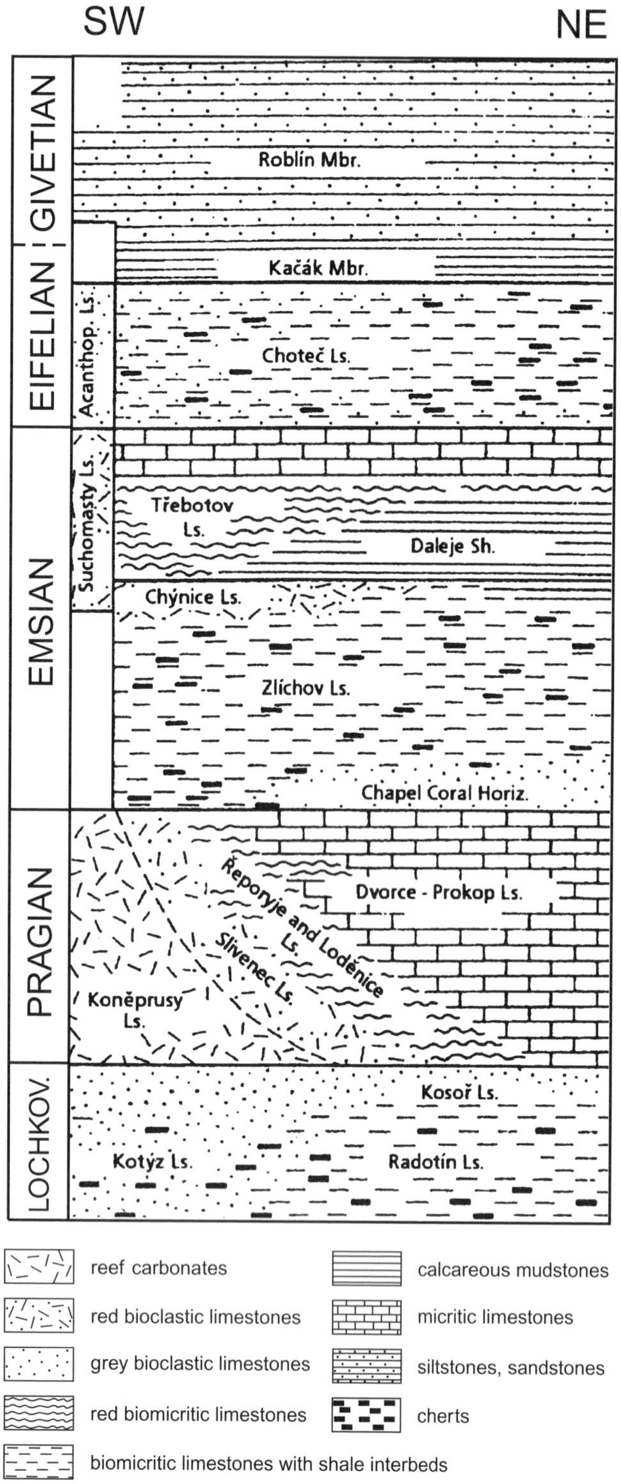

Fig. 8.16. Lithology and stratigraphy of the Devonian in the Barrandian area (after Chlupáč 1988).

During the Lochkovian, the bipartite facies pattern was clearly defined. The facies differentiation corresponds to the presence of volcanic highs that originated during Silurian times. In the NW, medium- to coarse-grained crinoidal packstones with thin shale intercalations (Kotys Limestone) accumulated (Fig. 8.16). Both fauna and lithology suggest deposition in a shallow subtidal environment (Chlupáč *et al.* 1998). Towards the south and east, the Kotys Limestone is laterally replaced by a succession composed of dark grey to black limestones alternating with black shales (Radotin Limestone). In contrast to the Kotys Limestone, the carbonate layers of this unit are mudstone and wackstone in texture and contain a fauna dominated by nektonic and planktonic organisms. Near Prague, the Radotin Limestone includes slump structures and debris flows. The lithology, sedimentary structures and the character of the fauna suggest deposition in a deep-water basinal environment and on the carbonate platform slope.

The Pragian shows marked facies differentiation that ranges from deep-water micritic carbonates to a shallow-water reef complex. The latter developed in the SW, in the Koneprusy area, and attains a thickness of *c.* 150 m. It mainly consists of coarse-grained crinoidal limestones and stromatoporoid biolithites. These carbonates host a large number of neptunian dykes and cavities filled by bioclastic material. Also typical are various fenestral and numerous stromatactis structures. The shallow-water character of the reef complex (Koneprusy Limestone) is documented by local erosion phenomena. The extraordinarily rich fauna is dominated by benthic organisms: brachiopods, crinoids, gastropods, bivalves, corals and trilobites. Carbonates of the reef complex pass basinwards into bedded crinoidal limestones (wackestones and/or packstones) with frequent stromatactis structures (Slivenec and Vinarice limestones). The detrital material is often sorted and its composition indicates derivation from the reef area. The fauna exhibits some similarity to that of the reef carbonates but it also shares many species that are known to occur more frequently in carbonate sediments deposited at greater depths. The deep-water, basinal facies of the Pragian is represented by micritic limestones (Dvorce-Prokop Limestone) that usually display a nodular appearance and are intercalated by thin shales. These mudstones and wackestones are often strongly bioturbated. Benthic organisms outnumber the nektonic ones and their skeletons are commonly intact, albeit typically small in size. Trilobites and brachiopods constitute the dominant components. Among the planktonic elements only dacryoconarid tentaculites are very abundant and also biostratigraphically important. In the deep-water, pelagic facies of the

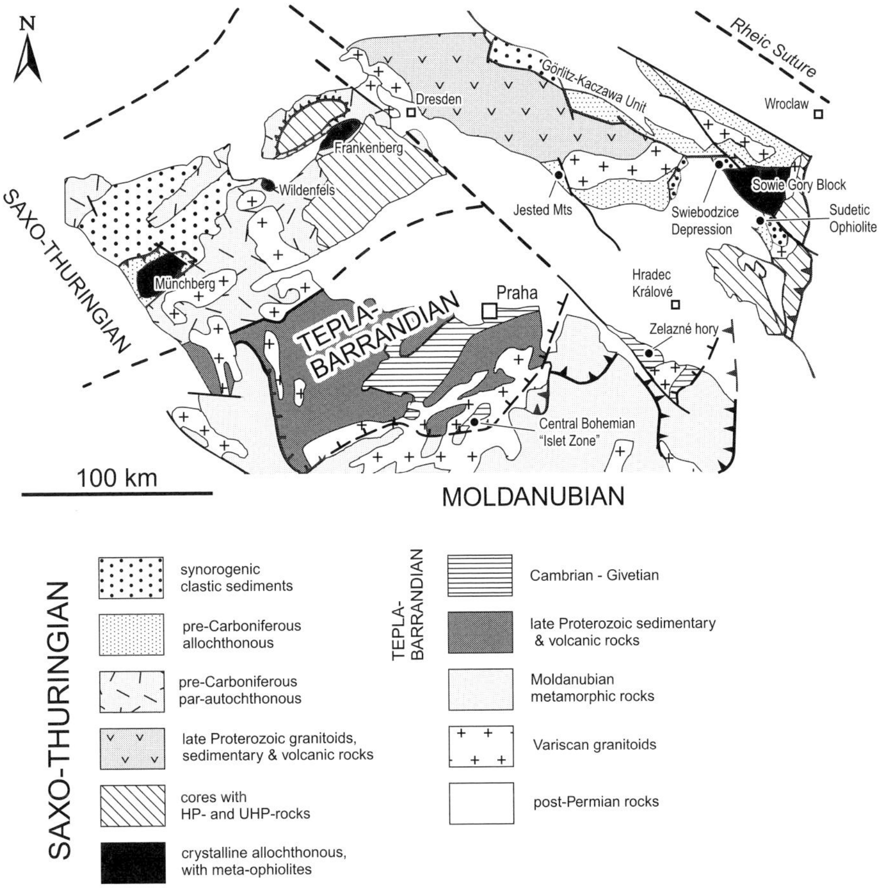

Fig. 8.17. Simplified geological map of the northern part of the Bohemian Massif (adapted from Franke & Zelazniewicz 2000). Tectonostratigraphic units containing Devonian rocks are labelled.

Prague area several mud mound structures have been recognized (Chlupáč *et al.* 1998).

In contrast to the Pragian, the Emsian is characterized by a uniform facies development (Fig. 8.16). During the early Emsian, a basinal facies represented by a monotonous sequence of limestones with shaly interbeds (Zlichov Limestone) was deposited across the entire area. The carbonates are predominantly mudstones and wackestones. To the east, at the base of the Emsian succession, there are debris flows several metres thick (Chapel Coral 'Horizon') composed of coarse reef-derived detritus. These units are usually associated with slumps in the overlying and underlying layers. Coarse-grained debris probably generated turbidity currents that distributed finer bioclastic material towards the SW. Thus, the distribution of bioclastic material reveals that its source area was located to the east or SE of the present-day occurrence of Devonian rocks in the Barrandian. The facies pattern of the upper Emsian resembles that of the Pragian and shows a transect from shallow-water carbonates (Suchomasty Limestone) in the SW through pelagic carbonates (Trebotov Limestone) in the central area to basinal shales (Daleje Shale) in the east and NE. The Suchomasty Limestone comprises crinoidal grainstones and packstones that rest disconformably on the Pragian reefs at Koneprusy or fill neptunian dykes in the reefs (Havlicek & Kukal 1991). Carbonates of the Trebotov Limestone are mainly nodular mudstones and wackestones. They also include submarine lava flows and tuff intercalations in the vicinity of Chynice and Chotec. According to Patocka *et al.* (1993), the geochemistry of the volcanic rocks corresponds to within-plate alkaline basalts. During the latest Emsian, the Trebotov Limestone attained its maximum lateral distribution extending across into the areas where previously the basinal Daleje Shale had been deposited (Fig. 8.16).

During the Eifelian, the facies differentiation exhibits a similar pattern to that of the lower Emsian. In the Koneprusy area, shallow-marine, coarse-grained crinoidal packstones and grainstones with rich benthic fauna (Acantopyge Limestone) accumulated, whereas in the remaining area, towards the NE, the succession is composed of fine-grained packstones and wackestones (Chotec Limestone) intercalated with shales or mudstones. This latter facies represents a deep environment with depths around the storm wave base. The bioclastic material and fauna in the coarser layers have a shallow-water character and were redeposited from a southwesterly direction. A different fauna occurs in the fine-grained lithologies; it is predominantly represented by cephalopods, small brachiopods and trilobites. In the Upper Eifelian a dramatic facies change can be observed. Black, organic-rich shales (Kacak Member) overlie the Chotec Limestone. The low-diversity fauna is dominated by planktonic and nektonic elements. The shales are overlain by 250 m of Givetian-age fine-grained clastics (Roblin Member), which constitute the youngest stratigraphical unit of the marine Palaeozoic in the Barrandian area. Sedimentary structures observed in siltstones and sandstones suggest significant influence of turbidity currents (Kukal & Jäger 1988).

Central Bohemian 'Islet Zone'
Small outcrops of Devonian rocks occur in the metamorphic roof of the Central Bohemian Pluton at the SE margin of the Tepla-Barrandian (Fig. 8.17). The lithologies and facies differ significantly from those of the Barrandian. In the Rozmitál area, about 90 km SW of Prague, the succession is composed of slightly metamorphosed fine-grained clastics with plant remains (Lochkovian to middle Emsian) overlain by sandstones with interbeds of polymictic conglomerates (middle Emsian to Eifelian). Dacryoconarid tentaculites and trilobites provide evidence for the Devonian age of the succession (Havlicek 1977). In the vicinity of Sedlcany, *c.* 80 km south of Prague, there is a unit of low-grade metamorphic rocks which comprise carbonates overlain by cordierite hornfels, quartzites and conglomerates. Although biostratigraphic data are not available, the sequence is regarded as ranging from Lochkovian to Emsian in age (Chlupáč 1988).

Zelezné Hory
The Palaeozoic rocks of the Zelezné Hory (Fig. 8.17) are metamorphosed. The succession exposed close to Vápenný Podol contains carbonates with the crinoid *Scyphocrinites* in the basal part. This crinoid genus is known to occur in the uppermost Silurian and in the Lochkovian. Although no fossils were found in the remaining part of this carbonate unit (Podol Limestone), it is considered to represent the Lochkovian and the Pragian, primarily because of lithological similarities to the Early Devonian sequence of the Barrandian (Chlupáč *et al.* 2000).

Basement of the North Bohemian Cretaceous Basin
In the area located directly north of the Zelezné Hory, Palaeozoic rocks are covered by thick Cretaceous cover. The geotectonic affinity of the basement is unclear. Franke & Zelazniewicz (2000) regarded this unit as part of the Bohemian Terrane (= Tepla-Barrandian). Devonian rocks have been pierced by a borehole near Hradec Králové (Fig. 8.17). The sequence commences with Frasnian–Famennian bioclastic and micritic limestones resting on volcanics of uncertain age (Chlupáč & Zikmundová 1976; Czech *et al.* 1989). The fossils (stromatoporoids, calcareous algae) present in the carbonates are characteristic of a shallow-water environment. They are overlain by pelagic wackestones which contain ammonoids and conodonts. This fauna is indicative of the late Famennian and early Carboniferous, and documents a continuous sedimentary transition from the Devonian into the Carboniferous.

Saxothuringian Belt

Sudetic Ophiolite
The Sudetic Ophiolite, a tectonostratigraphic unit bordering the Sowie Góry Block in the West Sudetes (Fig. 8.17), represents a unique place in the Variscan Belt where Devonian sediments deposited directly on the oceanic crust are preserved. In addition, these rocks have never been exposed to temperatures higher than *c.* 100°C (Dopieralska *et al.* 2006). The succession consists of carbonates resting on late Silurian/Early Devonian gabbros (Franke & Zelazniewicz 2000). It is *c.* 60 m thick and still easily accessible in the abandoned quarry at Dzikowiec (= Ebersdorf in the old German literature). The lower part of the section has a shallow-water character. It begins with a breccia composed of gabbro clasts overgrown by stromatoporoids and calcareous algae. Overlying are fine-grained, bioclastic wackestones to packstones (Main Limestone) bearing a unique fauna of colonial corals (Berkowski 2002). A foraminifer fauna provides evidence for the onset of shallow-water carbonate deposition during Famennian times (Lower *marginifera* Zone). Conodont fauna was only found in the topmost part of the Main Limestone and is indicative of the Lower *expansa* Zone. Thus, there is a stratigraphic gap between the Main Limestone and the overlying *Wocklumeria* Limestone (Dopieralska *et al.* 2006). The latter is condensed, *c.* 2.5 m thick, and composed of red cephalopod limestones, wackestones and mudstones, with a typical nodular appearance. The unit represents a phase of pelagic, deep-water deposition which persisted at least to the Lower *praesulcata* Zone. The palaeogeographic position of the Dzikowiec succession during the Late Devonian is unknown. Nd isotopic data obtained from conodonts suggest a position distant from any continental terrane (Dopieralska *et al.* 2006).

Świebodzice Depression
The Świebodzice Depression is a small synorogenic basin which developed adjacent to the Sowie Góry Block (Fig. 8.17) during the Late Devonian and underwent marked subsidence at that time. It contains a 2.5 km thick succession of Devonian coarse-grained, immature clastics ranging in age from late Frasnian to late Famennian. The succession represents sediments accumulated on the slope of a fan-delta and in a basin plain depositional system (Porebski 1981). An active fault zone separated the delta from the source area. Clastic material was stored in alluvial fans that prograded into the marine environment. In consequence, there is a lateral trend from terrestrial deposits in the south to marine deposits in the north. Conglomerates, coarse-grained sandstones and detrital limestones were deposited in the southern (proximal) part of the basin, whereas mudstones and siltstones predominated in the northern, more distal part. The coarse clastic material was supplied to the fan-delta slope by sediment gravity flows. Sedimentation in the deepest parts of the depositional lobes was controlled by low-density turbidity currents whereas in the interlobe areas it was dominated by mud suspension (Porebski 1990). The lithological spectrum of clastic material suggests a provenance from the Sowie Góry Block and from another massif located presumably beneath the Intra-Sudetic Basin, and now covered by Mesozoic sediments.

Görlitz-Kaczawa Unit
Fragments of the Devonian succession are found within various thrust sheets of the Görlitz-Kaczawa Unit (Fig. 8.17). They were all metamorphosed under greenschist facies conditions and now comprise monotonous siliceous shales with rare interbeds of cherts or polymictic clastics (Baranowski *et al.* 1990). This succession was deposited in a very deep basinal environment. It is unknown whether this basin was formed on continental or on oceanic crust. Baranowski *et al.* (1987) interpreted the Cambrian–Carboniferous sequence of the Görlitz-Kaczawa Unit as a fragment of an accretionary prism. The fauna collected from the metasediments includes conodonts, radiolarians and graptolites. Benthic fossils have not been found. Because of metamorphic overprinting datable microfossils are rare, and hence the Devonian sections are biostratigraphically poorly constrained. The Lochkovian has been documented in a sequence of graptolite shales exposed in the western part of the Görlitz-Kaczawa Unit (Jäger 1964). The most complete succession, *c.* 100 m in thickness, has been recognized in the central part. This comprises siliceous shales and cherts that range in age from Emsian to late Famennian.

Lausitz-Izera Unit
Devonian rocks are restricted to a small area in the Ještěd Mountains (Fig. 8.17), in the Bohemian part of the West Sudetes.

They are slightly metamorphosed and exposed in a section close to Jitrava. The succession commences with a thick complex of metabasalts and tuffs. The Devonian age is inferred from their stratigraphic position above Silurian limestones. The volcanics are overlain by phyllitic shales with intercalations of metaclastics. Overlying are up to 100 m of coral/stromatoporoid limestones, most probably Givetian and Frasnian in age (Chlupáč *et al.* 2000). The Famennian is represented by a condensed sequence of nodular limestones that contain cephalopods and a rich conodont fauna (Zikmundová 1964). Carbonate deposition terminated during latest Famennian times and was followed by the accumulation of shales and later, in the Early Carboniferous, by greywackes (Chlupáč 1993). The Devonian lithological succession of the Lausitz-Izera Unit reflects a progressive, stepwise lowering of the sediment surface through time from a carbonate platform to a deep-water basin.

East Thuringia and Vogtland

This area forms the largest outcrop of Devonian rocks within the Saxothuringian Belt and comprises a complete Devonian succession which belongs to the autochthon structural level of the belt (Fig. 8.17). The succession is up to 400 m thick and contains pelagic sediments and volcanics deposited in a deep-water, distal outer shelf within the Saxothuringian Terrane (Franke 2000). The oldest part of the succession consists of graptolite shales of Lochkovian age. The Pragian to Givetian interval includes pelagic carbonates (wackestones) and shales. Both lithotypes contain a very rich tentaculite fauna. In the Eifelian the tentaculite shales pass upwards into a monotonous black shale succession. The pelagic sedimentation, dominated by mud suspension, was interrupted in the early Frasnian by intense marine volcanism related to a pulse of crustal extension (Franke 1993). In consequence, metabasalts and tuffs covered the sea bottom and formed submarine highs. Their chemical composition corresponds to within-plate alkali basalts and tholeiites. The thickness of the Lower Frasnian volcanic complex is up to 150 m. Lateral to the volcanic highs, coarse-grained greywackes were deposited. On the highs, however, marly mudstones accumulated during the late Frasnian and the Famennian and these rocks typically contain small, early diagenetic carbonate nodules (Knotenkalk). In the upper Frasnian there are two horizons of organic-rich, black limestones (Kellwasserkalk) which are interpreted as representing two short regression events (Dopieralska *et al.* 2006). Between the highs the deep-water realm was characterized by the deposition of shales and marly mudstones. In the upper Famennian, the latter contain increasing amounts of small diagenetic carbonate nodules that usually form marker beds recognizable in many distant sections. The rocks contain a rich conodont, ostracod and cephalopod fauna, which is associated with deep-water solitary, rugose corals, small brachiopods and bivalves (e.g. Bartzsch & Weyer 1986). During the latest Famennian a series of turbidity current events occurred in eastern Thuringia and provided clastic material which is now interbedded within the marly mudstone succession. The mineral and clast composition of these turbidites reveals their derivation from the active margin of the Bohemian Terrane (= Tepla-Barrandian) and thus records the closure of the narrow Saxothuringian Ocean (Schäfer *et al.* 1997).

Münchberg Wildenfels, and Frankenberg nappes

The tectonic klippes of Münchberg, Wildenfels and Frankenberg represent the allochthon structural level of the Saxothuringian Belt (Fig. 8.17). They comprise lithological successions that were derived from the southern continental slope of the Saxothuringian Terrane (Franke 2000) and are usually termed the 'Bavarian facies'. The Palaeozoic fossiliferous sediments and volcanics are of very low metamorphic grade and occur predominantly in the basal parts of the nappes. Devonian rocks are represented by intraplate bimodal volcanics associated with pelites and sandstones (Wirth 1978). However, the most characteristic elements of the 'Bavarian facies' are radiolarian cherts which were deposited in the deepest parts of the slope, presumably below the calcite compensation depth (CCD). Nodular carbonate wackestones and mudstones occur as subordinate lithologies in the Münchberg Nappe. In the Wildenfels and Frankenberg units, however, carbonates constitute the more prominent fragments of the Devonian. In fact, their facies and lithological successions resemble to some extent those of the Saxothuringian autochthon (see above).

Elbe Valley

Exposures of Devonian rocks occur near Dresden and Pirna in the Elbe valley (Fig. 8.17). They contain mudstones usually rich in tentaculites in the Early and Middle Devonian part of the succession. During the early Frasnian there was a phase of volcanic activity which controlled the subsequent facies distribution in the area. The volcanic complex consists of intraplate basalts accompanied by tuffs. Local uplift supplied siliciclastic material to adjacent deeper parts on the slope. The Famennian includes basinal siliceous shales with cherts.

Depositional environments – South European terranes

Austro-Alpine Unit

Greywacke Zone

The Greywacke Zone, a long east–west trending belt of Palaeozoic rocks in Austria, constitutes the basement of the Northern Calcareous Alps. Devonian rocks occur both in the western and eastern segments of this unit (Fig. 8.18).

In the western part, which extends from Innsbruck in Tirol to Radstadt in the Salzburg area, two different facies realms can be recognized: the Wildseeloder and the Glemmtal units. The former includes Lower to Middle Devonian sediments deposited as part of a shallow-water carbonate platform (lagoonal and reef dolomites) which are covered by Frasnian pelagic limestones (Mostler 1968; Schönlaub 1979). In contrast, the Glemmtal Unit comprises deep-water basinal deposits represented by fine-grained clastics intercalated with cherts and siliceous shales. There are also packages of condensed carbonates that were deposited on seamounts. During the Pragian, and possibly also in the Middle and the Late Devonian, an intense phase of volcanic activity occurred in the western part of the Greywacke Zone. This produced several hundred metres of basic lavas with intercalations of volcaniclastic and clastic rocks (Kreutzer *et al.* 2000).

In the eastern part of the Greywacke Zone the Devonian is represented by a carbonate sequence 200 to 300 m thick. It includes Early Devonian pelagic, dacryoconarid-bearing wackestones with numerous bioclastic interbeds. These intercalations can attain a thickness of several metres and contain bioclasts of shallow-water fauna, predominantly of stromatoporoids, crinoids and brachiopods. They represent gravity flows of shallow-water detrital material which was transported into the basinal environment (Flajs & Schönlaub 1976). The Middle Devonian, which is also developed as pelagic wackestones, is biostratigraphically very poorly constrained and most probably not more than 50 m thick. The Upper Devonian has been recognized at only a few

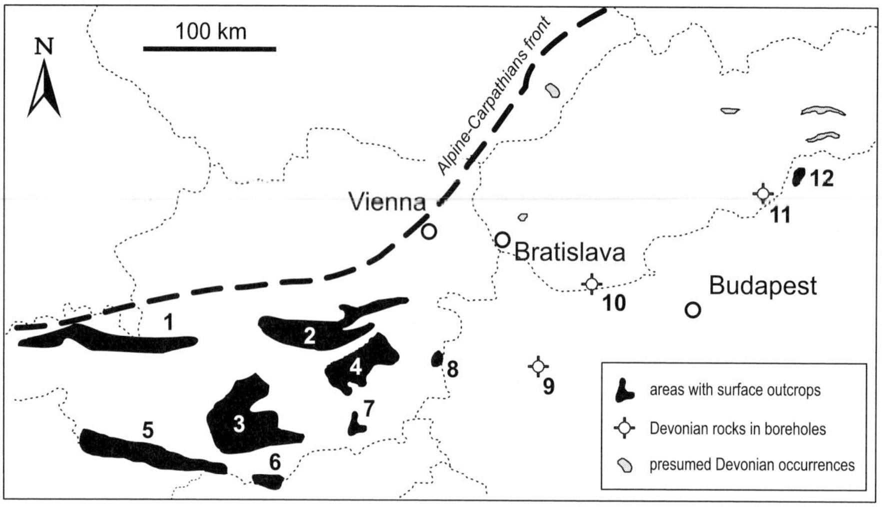

Fig. 8.18. Distribution of Devonian rocks in nappes of the Eastern Alps and the West Carpathians. Areas with extensive surface exposures are indicated in black. Key to locations: 1, western Greywacke Zone; 2, eastern Greywacke Zone; 3, western Gurktal Nappe; 4, Graz Palaeozoic; 5, Carnic Alps; 6, Karawanken Alps; 7, eastern Gurktal Nappe; 8, Palaeozoic of Burgenland; 9, Kékkút-4 borehole; 10, Zelezny Háj-1 borehole; 11, Blhovce borehole; 12, Szendröi Mountains. For more information on problematical Devonian occurrences in the West Carpathians see Chlupáč (1988).

localities. Conodont fauna provided evidence for pelagic carbonate deposition which lasted until early Famennian times (Kreutzer *et al.* 2000). Devonian rocks, especially in the Styrian segment of the eastern part of the Greywacke Zone (Fig. 8.18), are known to host rich iron-ore and magnesite deposits related to volcanic and metamorphic processes during the Carboniferous.

Graz Palaeozoic

In the middle part of Styria there is a relatively large area north and west of Graz where thick Palaeozoic rocks are exposed (Fig. 8.18). They form a complex pile of nappes with differing characteristic lithostratigraphic and metamorphic styles. According to Hubmann (in Kreutzer *et al.* 2000), three major nappes can be recognized.

The lower nappe is characterized by greenschist- to amphibolite-grade metamorphic rocks. Devonian rocks attain a thickness of at least 1200 m. Their stratigraphy is based on rare coral finds. The Lower Devonian is dominated by phyllites intercalated with volcaniclastics. These rocks grade into the overlying calcareous schists and marbles (Schöckelkalk) of the Middle Devonian. Because of metamorphic overprinting, it is difficult to reconstruct the depositional environment of these strata. Hubmann (1993) suggested deposition on an offshore shelf located on a passive continental margin.

The intermediate nappe includes more than 1000 m of Devonian rocks which underwent greenschist-grade metamorphism. Despite rare conodont and coral records, which provided the Early and Middle Devonian ages, the stratigraphy of the sequence is very poorly constrained. Hubmann (in Kreutzer *et al.* 2000) recognized two facies groups, both representing deposits of an offshore environment. The 'Laufnitzdorf sequence' is dominated by pelagic carbonates whereas the 'Kalkschieferfolge' is primarily composed of calcareous schists, sandstones and dolomitic marbles.

The upper nappe comprises a complete record of the Devonian and consists of predominantly nearshore carbonates which have been overprinted by greenschist-grade metamorphism. The succession is up to 1800 m thick and displays lateral differences in lithology and stratigraphic content (Schönlaub 1979). These lateral differences led to the identification of two facies zones: the Rannachfacies and the Hochlantschfacies. The former begins with Lochkovian-age shallow-water clastics and carbonates, intercalated with volcaniclastic rocks. However, much of the Lower Devonian is represented by very thick (*c.* 1000 m) peritidal sandstones, which grade into *Amphipora*-rich dolomites at the beginning of the Eifelian. Givetian deposition continued with the accumulation of fossiliferous wackestones, which mark the onset of gradual lowering of the sediment surface. Consequently, pelagic cephalopod wackestones and mudstones were deposited during the Late Devonian. The sequence of the Hochlantschfacies is several hundred metres thick and ranges in age from Pragian to Famennian. It differs from the Rannachfacies primarily in the occurrence of volcanic rocks in the Eifelian and the development of small coral buildups during the Givetian.

It is likely that the nappes of the Graz Palaeozoic continue eastwards in the subsurface. Early Devonian carbonates and shales, which show similarities to the coeval rocks of the upper nappe, were drilled beneath the Tertiary of the Styrian Basin and are also known from small exposures in the southern part of Burgenland and in the adjacent region of Hungary (Schönlaub in Kreutzer *et al.* 2000).

Gurktal Nappe

The Gurktal Nappe consists of low- to greenschist-grade metamorphic rocks of volcanic and sedimentary origin. The stratigraphic age of the sediments is poorly constrained. Isolated records of conodonts prove the Devonian age of most of the carbonate complexes and provide evidence of a significant facies differentiation within the Gurktal Nappe (Neubauer & Pistotnik 1984).

In the Murau Unit, which constitutes the lower part of the Gurktal Nappe, the Lower Devonian is represented by phyllites, calcareous phyllites and black schists. These rocks are interpreted to have been deposited in a basinal, anoxic realm. Overlying is a sequence of marbles, more than 300 m thick. The conodont fauna from the base of this complex, which was illustrated by Buchroithner (1978), is indicative of an Eifelian age. Its specific composition suggests deposition in an offshore environment, presumably on the carbonate platform.

In the Stolzalpe Unit, the upper part of the Gurktal Nappe, three different facies belts can be recognized in the Lochkovian and Pragian. These are the sandy 'Pranker facies', the fine-clastic 'Magdalensberg facies' and the carbonate-rich 'Auen facies'. The sandy 'Pranker facies' comprises several hundred metres of sandstones with rare dolomite lenses deposited presumably in a shallow-water environment. The 'Magdalensberg facies', also several hundred metres thick, represents basinal deposits. It is dominated by fine-grained metaclastics with interbeds of metagreywackes and also includes rare marble lenses (Buchroithner 1979). These carbonates are mainly recrystallized bioclastic rudstones and packstones of turbiditic origin. The allochthonous character of the bioclastic material is indicated by the occurrence of reworked Silurian conodonts. The 'Auen facies' is represented by several metres of strongly recrystallized limestones and dolomites. In the latest Pragian facies differentiation disappeared (Neubauer & Pistotnik 1984) and carbonates became widespread across the entire area. Sediments of the Emsian and Eifelian are primarily pelagic wackestones (Flaserkalk). A significant volcanic event took place during late Emsian times. It was responsible for the deposition of thick tuffs with volcaniclastic interbeds. The Givetian and the Upper Devonian are only locally preserved and developed as coarse-grained biodetrital carbonates within shales.

Pohorje Trough

The Pohorje Trough occupies a small area in northern Slovenia (Fig. 8.18), 50 km west of Maribor. It is located north of the Periadriatic Fault System and contains an early Palaeozoic, low-grade metamorphosed sedimentary succession that includes *c.* 100 m of marine Devonian rocks. The Devonian consists of argillaceous schists overlain by thick-bedded limestones, and clastic rocks (siltstones and greywackes) in its upper portion (Mioc 1977). The conodont fauna recovered from the carbonates suggests an Early Devonian age. The carbonates comprise mudstones alternating with crinoidal rudstones and biodetrital packstones. Unfortunately, their depositional environment remains unrecognized because of thermal overprinting and recrystallization. The general lithology would appear to indicate a deep-water setting for the entire Devonian sequence.

South Alpine Unit (Southern and Dinaric Alps)

Carnic Alps

The Variscan basement of the Carnic Alps consists of three major nappes: the Fleons Nappe (FN) in the western part, the Cellon-Kellewand Nappe (CKN) in the central segment, and the Hochwipfel Nappe (HN) in the central-eastern Carnic Alps. According to Läufer *et al.* (1999), the FN and CKN comprise sedimentary rocks deposited on a passive continental margin, while the HN is characterized by melange-like deposits.

The low-grade metamorphosed sedimentary sequence of the FN is stratigraphically poorly constrained but there is a clear subdivision into a northern pelagic and a southern shallow-marine belt. In the north, only the Lower Devonian is preserved and is represented by basinal shales and pelagic carbonates (Flaserkalk, Plattenkalk). In the south, the Devonian succession begins with pelagic carbonates followed by Middle Devonian massive limestones interpreted as fragments of a carbonate platform and reefs (Hinderer 1992). The Upper Devonian has not been recognized in the Fleons Nappe.

Devonian rocks of the Cellon-Kellewand Nappe display significant variations in thickness and facies. Tectonic, stratigraphical and facies reconstructions revealed a general facies pattern similar to that of the FN, with a shallow-water realm in the south and pelagic and basinal deposition in the north (Kreutzer 1992). The shallow-water deposits attain a thickness of more than 1100 m, whereas the pelagic sequence is only *c.* 200 m thick. The Lochkovian begins with crinoidal packstones and grainstones (neritic Rauchkofelkalk) in the south. Their offshore equivalents comprise mudstones with abundant radiolarians (pelagic Rauchkofelkalk) and basinal shales (Bischofalmschiefer). In Pragian times the facies differentiation became much more pronounced. The shallow-water realm yields a facies transect from intertidal deposits (algal laminites and amphiporoid packstones/floatstones) to thick detrital carbonates (bioclastic grainstones) with small coral/stromatoporoid buildups (Hohe Warte Kalk). Further offshore, in the 'transitional' facies (Kreutzer 1992), tentaculitid wackestones with interbeds of allodapic packstones and grainstones accumulated. These carbonates grade laterally into red nodular carbonates (wackestones and mudstones) which may contain shale interbeds (Bandel 1974). Both the lithology and the nektonic and planktonic fauna (cephalopods, conodonts, radiolarians and tentaculites) are indicative of deposition on a pelagic high. The most distal basinal deposits are represented by shales with radiolarites and sandstone interbeds (Zollner Formation). The facies differentiation created during the Pragian remained relatively stable until the end of Frasnian time. Coral/stromatoporoid buildups attained their climax during Givetian and early Frasnian times. This period was also one in which non-deposition locally affected the pelagic swell (Bandel 1974). In the Famennian, the entire area underwent a progressive lowering of the sediment surface and as a result uniform pelagic conditions with deposition of cephalopod mudstones/wackestones (Flaserkalk) were established.

In contrast to other nappes, the Hochwipfel Nappe includes Devonian rocks occurring only as huge olistoliths within the Carboniferous deep-marine deposits, and hence there is no continuous stratigraphic sequence (Läufer *et al.* 1999). The allochthonous blocks comprise a broad spectrum of carbonates, shales and radiolarites, which are Early to Late Devonian in age (Schönlaub 1979).

Karawanken Alps

The Devonian succession is exposed in a series of isolated outcrops on both sides of the Austrian–Slovenian border (i.e. Seeberger Aufbruch area, Jezersko area). It commences with black, pelagic limestones (mudstones, wackestones) containing cephalopods and conodonts of Lochkovian age. Pelagic sedimentation continued up into the late Emsian and resulted in the deposition of wackestones with conodonts and styliolinids (Kristic *et al.* 1988). In the late Emsian an extensive carbonate platform was established which persisted up to the late Frasnian. During that time a 500 m thick complex of very pure carbonates was deposited. This includes coral/stromatoporoid buildups that interfinger with forereef packstones and rudstones, and with backreef mudstones (Rantitsch 1992). The core facies of the buildups is dominated by tabulate corals, stromatoporoids and the green alga *Renalcis*. Fragments of these organisms constitute the dominant components of the forereef carbonates. The backreef facies is represented mostly by monotonous mudstones with algal laminites. The laminites contain sedimentary structures characteristic of intertidal and supratidal environments, such as 'bird's-eyes' and 'teepees' (Rantitsch 1992). Accumulations of volcaniclastic material are only locally recognized within the

Middle Devonian part of the reef complex (Loeschke & Rolser 1971). This led Rantitsch (1992) to compare the buildups with recent atoll reefs. The carbonate platform of the Karawanken region presumably extended in a southeasterly direction into the Central Slovenia, where Ramovs (1979) identified blocks of Emsian coral/stromatoporoid limestones within younger sediments. In late Frasnian early Famennian times, carbonate-platform deposition terminated as a result of emergence. This is revealed primarily by diagenetic overprinting of reef carbonates under meteoric/vadose conditions (Rantitsch 1992). The Famennian carbonate succession is punctuated with several stratigraphic gaps and units of intraformational conglomerates. These document a gradual drowning of the carbonate platform and its transformation into a more basinal setting.

Summary

The Devonian period, an interval of *c.* 57 Ma, occurred in the middle of the Palaeozoic and straddled the Caledonian and Variscan orogenies, the events which dominated the evolution of Central Europe through Palaeozoic times. The area of Central Europe was divided into three parts, namely Laurussia (Euramerica) to the north, with the Armorican and South European terranes situated to the south (Fig. 8.3). The continent of Laurussia (Euramerica), which formed as a result of the collision of Laurentia, Baltica and Avalonia (Caledonian Orogeny; see Krawczyk *et al.* 2008), lay to the north, whereas the peri-Gondwanan terranes (Armorican, South European) lay to the south (Fig. 8.3). These latter terranes originally formed the extended northern continental shelf of the Gondwana plate. Oblique NE-directed convergence between the southern Gondwana plate and the northern plate of Laurussia began in the Silurian, and this movement was associated with the subduction of the Rheic Ocean beneath the northern plate (see Kroner *et al.* 2008). Diachronous collision between the two plates, and various intervening blocks, extended from the Early Devonian through to the Early Carboniferous. These active tectonics resulted in the formation of a wide range of depositional environments; the relatively widespread deposition of continental sediments (particularly in the early Devonian), gave rise to the informal term 'Old Red Continent' for the continent of Laurussia (Euramerica).

The Devonian was the time of the greatest diversity of marine fauna in the Palaeozoic. On land, vascular plants and forests became established and, before the end of the period, land tetrapods appeared (House & Gradstein 2004). Extensive carbonate production at the time was associated with the dominant greenhouse climate which favoured the development of reef ecosystems characterized by rugose corals and stromatoporoids. These reached their acme in the Middle Devonian and disappeared in the late Frasnian, probably due to the global expansion of anoxic black sediments in the shelf regions. This crisis was probably related to the pulses of global cooling which caused substantial regressions and severely affected the shelfal areas. Cooling (which ultimately resulted in continental glaciations in the southern polar areas of Gondwana), was related to the switch from the hot greenhouse phase of the earlier Palaeozoic to the cold icehouse phase of the later Palaeozoic and also coincided with the latest-Frasnian biodiversity crisis, one of the most important extinction events which severely affected numerous marine groups including the brachiopods, trilobites, ammonites, conodonts and acritarchs as well as jawless fish and all placoderms.

The depositional environments along the southern shelf of Laurussia can be subdivided into two main areas. The inner shelf (North Sea, NE Germany, Pomerania, eastern Baltica area) is characterized by a predominantly erosional regime where deposition was often intermittent and confined to narrow depocentres. Early Devonian sedimentation was predominantly continental whereas the Late Devonian is the period of maximum marine influence, mainly related to eustatic highstands (e.g. late Emsian, late Givetian–Frasnian).

The outer shelf of Laurussia (Euramerica) comprises a complex system of carbonate platforms, intracratonic basins and intrabasinal highs extending from southern England through Belgium (Ardennes), central Germany (Rhenish Massif, Harz Mountains) to southern Poland and Moravia. The most complete sedimentary successions show a transition from an inner shelf setting to a distal, outer shelf environment. Sedimentation was predominantly controlled by the pre-existing topography, the supply of clastic material from the eroding Caledonian uplands, Emsian rifting (which led to the formation of the Rhenohercynian Ocean) and eustatic events.

The two southerly terranes are more fragmentary in style, and reconstruction of the precise sequence of events becomes increasingly difficult. The Armorican terranes comprise the Tepla-Barrandian (e.g. Barrandian area, Central Bohemia, Zelezné Hory) and the Saxothuringian (e.g. Sudetic Ophiolite, Görlitz-Kaczawa Unit, East Thuringia & Vogtland) areas contain mixed sedimentary and magmatic successions which vary markedly over time. The southernmost South European terranes (South Alpine Unit, Southern & Dinaric Alps are again fragmentary, although individual successions can be up to 1.8 km thick.

References

ALBERTI, H. & WALLISER, O. H. 1977. Überblick über die paläogeographische und tektonische Enwicklung des Westharzes. *In*: *Field-guide of the Geological Meeting "Geotagung '77", Göttingen*, part I, 172–178.

BANDEL, K. 1974. Deep-water limestones from the Devonian-Carboniferous of the Carnic Alps, Austria. *In:* HSÜ, K. J. & JENKYNS, H. C. (eds) *Pelagic Sediments: On Land and Under the Sea.* International Association of Sedimentologists, Special Publications, **1**, 93–115.

BARANOWSKI, Z., HAYDUKIEWICZ, A., KRYZA, R., LORENC, S., MUSZYNSKI, A. & URBANEK, Z. 1987. Rozwoj struktury wschodniej czesci Gor Kaczawskich na podstawie dotychczasowego rozpoznania stratygrafii, warunkow sedymentacji i wulkanizmu. *In*: *Field-guide of the 58 Annual Meeting of the Geological Society of Poland*, 61–73.

BARANOWSKI, Z., HAYDUKIEWICZ, A., KRYZA, R., LORENC, S., MUSZYNSKI, A., SOLECKI, A. & URBANEK, Z. 1990. Outline of the geology of the Gory Kaczawskie (Sudetes, Poland). *Neues Jahrbuch für Geologie und Paläontologie, Abhandlungen*, **179**, 223–257.

BARTZSCH, K. & WEYER, D. 1986. Biostratigraphie der Devon/Karbon-Grenze im Bohlen-Profil bei Saalfeld (Thüringen, DDR). *Zeitschrift für Geologische Wissenschaften*, **14**, 147–152.

BELKA, Z. 1985. Lower Carboniferous conodont biostratigraphy in the northeastern part of the Moravia – Silesia Basin. *Acta Geologica Polonica*, **35**, 33–60.

BELKA, Z. 1998. Taxonomy, phylogeny and biogeography of the late Famennian conodont genus *Mashkovia*. *Journal of Micropalaeontology*, **17**, 119–124.

BELKA, Z., AHRENDT, H., FRANKE, W. & WEMMER, K. 2000. The Baltica-Gondwana suture in central Europe: evidence from K/Ar ages of detrital muscovites. *In*: FRANKE, W., ALTHERR, R., HAAK, V. & ONCKEN, O. (eds) *Orogenic Processes – Quantification and Modelling in the Variscan Belt of Central Europe*, Geological Society, London, Special Publications, **179**, 87–102.

BENDER, P., EDER, W., ENGEL, W., FRANKE, W., LANGENSTRASSEN, F., WALLISER, O. H. & WITTEN, W. 1977. Paläogeographische Entwicklung des östlichen Rheinischen Schiefergebirges, demonstriert an einem Querschnitt. *In*: *Field-guide of the Geological Meeting 'Geotagung '77', Göttingen*, part I, 1–57.

Berkowski, B. 2002. Famennian Rugosa and Heterocorallia from southern Poland. *Palaeontologia Polonica*, **61**, 1–87.

Boulvain, F., Bultynck, P., *et al.* 1999. Les Formations du Frasniens de la Belgique. *Memoires of the Geological Survey of Belgium*, **44**, 1–126.

Buchroithner, M. F. 1978. Biostratigraphische Untersuchungen im Paläozoikum der Steiermark. *Mitteilungen des naturwissenschaftlichen Vereins, Steiermark*, **108**, 77–93.

Buchroithner, M. F. 1979. Biostratigraphische und fazielle Untersuchungen im Paläozoikum von Mittelkärnten. *Carinthia II*, **169**, 71–95.

Buggisch, W. 1972. Zur Geologie und Geochemie der Kellwasserkalke und ihren begleitenden Sedimente (Unteres Oberdevon). *Abhandlungen des Hessischen Landesamts für Bodenforschung*, **62**, 1–68.

Bultynck, P. & Dejonghe, L. 2001. Devonian lithostratigraphic units (Belgium). *In*: Bultynck, P. & Dejonghe, L. (eds) *Lithostratigraphic scale of Belgium. Geologica Belgica*, **4**(1–2), 39–69.

Bultynck, P., Coen-Aubert, M., *et al.* 1991. Les formations du Devonien Moyen de la Belgique. *Memoire pour servir a l'Explication des Cartes Geologiques et Minieres de la Belgique*, **30**, 1–106.

Bultynck, P., Coen-Aubert, M. & Godefroid, J. 2000. Summary of the state of correlation in the Devonian of the Ardennes (Belgium – NE France) resulting from the decisions of the SDS. *Courier Forschungsinstitut Senckenberg*, **225**, 91–114.

Burchette, T. P. 1981. European Devonian reefs: a review of current concepts and models. *In*: Toomey, D. F. (ed.) *European Fossil Reef Models*. SEPM Special Publications, **30**, 85–142.

Cameron, T. D. J., Crosby, A., Balson, P. S., Jeffery, D. H., Lott, G. K., Bulat, J. & Harrison, D. J. (eds) 1992. *The Geology of the Southern North Sea*. British Geological Survey, London.

Chlupáč, I. 1988. The Devonian of Czechoslovakia and its stratigraphic significance. *In*: McMillian, N. J., Embry, A. F. & Glass, D. J. (eds) *Devonian of the World*. Canadian Society of Petroleum Geologists, Memoirs, **14**(1), 481–497.

Chlupáč, I. 1993. Stratigraphic evaluation of some metamorphic units in the north part of the Bohemian Massif. *Neues Jahrbuch für Geologie und Paläontologie, Monatshefte*, **188**, 363–388.

Chlupáč, I. & Kukal, Z. 1977. The boundary stratotype at Klonk. The Silurian–Devonian Boundary. *IUGS Series A*, **5**, 96–109.

Chlupáč, I. & Kukal, Z. 1986. Reflexion of possible global Devonian events in the Barrandian area, C.S.S.R. *Lecture Notes Earth Sciences*, **8**, 171–179.

Chlupáč, I. & Zikmundová, J. 1976. The Devonian and Lower Carboniferous in the Nepasice bore in east Bohemia. *Vestnik Ústredniho ústavu geologického*, **51**, 269–278.

Chlupáč, I. , Havlicek, V., Kriz, J., Kukal, Z. & Storch, P. 1998. *Palaeozoic of the Barrandian (Cambrian to Devonian)*. Czech Geological Survey, Prague.

Chlupáč, I., Galle, A., Hladil, J. & Kalvoda, J. 2000. Series and stage boundaries in the Devonian of the Czech Republic. *Courier Forschungsinstitut Senckenberg*, **225**, 159–172.

Czech, S., Havlicek, V. & Zikmundová, J. 1989. The Upper Devonian and Lower Carboniferous in north-eastern Bohemia (based on boreholes in the Hradec Králové area). *Vestnik Ústredniho ústavu geologického*, **64**, 65–76.

Dadlez, J. & Dadlez, R. 1986. Late Devonian transgression in the Unisław sequence, central Poland. *Kwartalnik Geologiczny*, **30**, 445–472.

Dadlez, R. 1978. Sub-Permian rock complexes in the Koszalin-Chojnice zone. *Kwartalnik Geologiczny*, **22**, 270–301 [in Polish with English summary].

Da Silva, A.-Ch. & Boulvain, F. 2002. Sedimentology, magnetic susceptibility and isotopes of a middle Frasnian carbonate platform: Tailfer section, Belgium. *Facies*, **46**, 89–102.

Dopieralska, J. 2003. *Neodymium isotopic composition of conodonts as a palaeoceanographic proxy in the Variscan oceanic system*. PhD thesis, Justus-Liebig-University, Giessen.

Dopieralska, J., Belka, Z. & Haack, U. 2006. Geochemical decoupling of water masses in the Variscan oceanic system during Late Devonian times. *Palaeogeography, Palaeoclimatology, Palaeoecology*, **240**, 108–119.

Dreesen, R., Paproth, E. & Thorez, J. 1988. Events documented in Famennian sediments (Ardenne-Rhenish Massif, Late Devonian, NW Europe). *In:* McMillan, N. J., Embry, A. F. & Glass, D. J. (eds) *Devonian of the World*. Canadian Society of Petroleum Geologists, Memoirs, **14**(2), 295–308.

Drygant, D.M. 1984. *Correlation and Conodonts of the Silurian and Lower Devonian Deposits of Volhynia and Podolia*. Naukovaya Dumka, Kiev [in Russian].

Dvořák, J. 1990. Geology of Palaeozoic sediments of the deep borehole Jablunka 1 (Beskydy Mts., NE Moravia) – comparison with the deep borehole Munsterland 1. *Sbornik geologickich Ved, Geologie*, **45**, 65–90.

Dvořák, J. 1998. Lower Devonian Basal Clastics – Old Red Formation, Southern Moravia, Czech Republic. *Bulletin of the Czech Geological Survey*, **73**(4), 271–279.

Dvořák, J., Friáková, O., Hladil, J., Kalvoda, J. & Kukal, Z. 1987. Geology of Paleozoic in the vicinity of quarries of Mokrá Cement Factory (Moravian Karst). *Sbornik geologickich Ved, Geologie*, **42**, 41–88.

Eder, W. & Franke, W. 1982. Death of Devonian reefs. *Neues Jahrbuch für Geologie und Paläontologie, Abhandlungen*, **163**, 241–243.

Eder, W., Engel, W., Franke, W., Langenstrassen, F., Walliser, O. H. & Witten, W. 1977. Überblick über die paläogeographische Entwicklung des östlichen Rheinischen Schiefergebirges. *In*: *Fieldguide of the Geological Meeting "Geotagung '77", Göttingen*, part I, 1–11.

Engel, W., Franke W. & Langenstrassen, F. 1983. Palaeozoic sedimentation in the northern branch of the Mid-European Variscides – Essay of an interpretation. *In:* Martin, H. & Eder, F. W. (eds) *Intracontinental Fold Belts*. Springer-Verlag. Berlin, 9–41.

Faber, P. 1980. Fazies-Gliederung und -Entwicklung im Mittel-Devon der Eifel (Rheinisches Schiefergebirge). *Mainzer Geowissenschaftliche Mitteilungen*, **8**, 83–149.

Flajs, G. & Schönlaub, H.P. 1976. Die biostratigraphische Gliederung des Altpaläozoikums am Polster bei Eisenerz (Nördliche Grauwackenzone, Österreich). *Verhandlungen der Geologischen Bundesanstalt*, **1976**, 257–303.

Franke, W. 1973. Bau und Entwicklunsgeschichte des Iberger Riffes (Mitteldevon bis Unterkarbon III, NW-Harz, W-Deutschland). *Geologisches Jahrbuch, A*, **11**, 3–127.

Franke, W. 1993. The Saxonian Granulites – a metamorphic core complex? *Geologische Rundschau*, **82**, 505–515.

Franke, W. 1995. Rhenohercynian foldbelt: autochthon and nonmetamorphic nappe units-stratigraphy. *In:* Dallmeyer, D., Franke, W. & Weber, K. (eds*) Pre-Permian Geology of Central and Western Europe*. Springer, Berlin, 33–49.

Franke, W. 2000. The mid-European segment of the Variscides:tectonostratigraphic units, terrane boundaries and plate tectonic evolution. *In*: Franke, W., Altherr, R., Haak, V. & Oncken, O. (eds) *Orogenic Processes – Quantification and Modelling in the Variscan Belt of Central Europe*. Geological Society, London, Special Publications, **179**, 35–61.

Franke, W. & Zelazniewicz, A. 2000. The eastern termination of the Variscides: terrane correlation and kinematic evolution. *In*: Franke, W., Altherr, R., Haak, V. & Oncken, O. (eds) *Orogenic Processes – Quantification and Modelling in the Variscan Belt of Central Europe*. Geological Society, London, Special Publications, **179**, 87–102.

Galle, A., Hladil, J., Friáková, O., Kalvoda, J., Krejči, Z. & Zukalová V. 1988. Biostratigraphy of Middle and Upper Devonian carbonates of Moravia (Czechoslovakia). *In*: McMillian, N. J., Embry, A. F. & Glass, D. J. (eds) *Devonian of the World*. Canadian Society of Petroleum Geologists, Memoirs, **14**(3), 633–645.

Glennie, K. W. 1990. North Sea history and structural framework. *In:* Glennie, K.W. (ed.) *Introduction to the Petroleum Geology of the North Sea*, 3rd edn. Blackwell, Oxford

Godefroid, J., Blieck, A. *et al.* 1994. Les formations du Devonien inferieur du Massif de la Vesdre, de la Fenetre de Theaux et du Synclinorium de Dinant (Belgique, France). *Memoire pour servir a l'Explication des Cartes Geologiques et Minieres de la Belgique*, **38**, 1–144.

Goldring, R. & Langenstrassen, F. 1979. Open shelf and near-shore clastic facies in the Devonian. *In:* House, M. R., Scrutton, C. T. & Bassett, M. G. (eds) *The Devonian System*. Special Papers in Palaeontology, **23**, 81–97, The Palaeontological Association, Aberystwyth.

Gouwy, S. & Bultynck, P. 2000. Graphic correlation of Frasnian sections (Upper Devonian) in the Ardennes, Belgium. *Bulletin de l'Institut Royal des Sciences Naturelles de Belgique, Sciences de la Terre*, **70,** 25–52.

HAVLICEK, V. 1977. The Paleozoic (Cambrian–Devonian) in the Rozmital area. *Vestnik Ústredniho ústavu geologického*, **52**, 81–94.

HAVLICEK, V. & KUKAL, Z. 1991. Sedimentology, benthic communities and brachiopods in the Suchomasty (Dalejan) and *Acanthopyge* (Eifelian) Limestones of the Koneprusy area (Czechoslovakia). *Sbornik geologickich Ved, Paleontologie*, **31**, 105–205.

HECKEL, P. H. & WITZKE, B. J. 1979. Devonian world paleogeography determined from distribution of carbonates and related lithic paleoclimatic indicators. *In:* HOUSE, M. R., SCRUTTON, C. T. & BASSETT, M. G. (eds) *The Devonian System*. Special Papers in Palaeontology, **23**, 99–123.

HINDERER, M. 1992. Die vulkanoklastische Fleonsformation in den westlichen Karnischen Alpen - Sedimentologie, Petrographie und Geochemie. *Jahrbuch der Geologischen Bundesanstalt*, **135**, 335–379.

HLADIL, J. 1988. Structure and microfacies of the Middle Devonian and Upper Devonian carbonate buildups in Moravia, Czechoslovakia. *In*: MCMILLIAN, N. J., EMBRY, A. F. & GLASS, D. J. (eds) *Devonian of the World*. Canadian Society of Petroleum Geologists, Memoirs, **14**(2), 607–618.

HLADIL, J. 1992. Zonality in the Devonian carbonate sediments in Moravia (CSFR). *Proceedings of 1st International Conference on the Bohemian Massif*, Prague, Czechoslovakia, 26 September–3 October 1988, 121–126.

HLADIL, J. 1994. Moravian Middle and Late Devonian buildups: evolution in time and space with respect to Laurussian shelf. *Courier Forschungsinstitut Senckenberg*, **172,** 111–125.

HLADIL, J. 2002. Geophysical records of dispersed weathering products on the Frasnian carbonate platform and early Famennian ramps in Moravia, Czech Republic: proxies for eustasy and palaeoclimate. *Palaeogeography, Palaeoclimatology, Palaeoecology*, **181,** 213–250.

HLADIL, J., MELICHAR, L., *et al.* 1999. The Devonian in the easternmost Variscides, Moravia: a holistic analysis directed towards comprehension of the original context. *Abhandlungen der Geologischen Bundesanstalt*, **54,** 27–47.

HUBMANN, B. 1993. Ablagerungsraum, Mikrofazies und Paläoökologie der Barrandeikalk-Formation (Eifelium) des Grazer Paläozoikums. *Jahrbuch der Geologischen Bundesanstalt*, **136**, 393–461.

JÄGER, H. 1964. *Monograptus hercynicus* in den Westsudeten und das Alter der Westsudeten-Hauptfaltung. *Geologie*, **13**, 249–273.

JANSEN, U., KONIGSHOF, P., PLODOWSKI, G., SCHINDLER, E. & SCHINDLER, T. 2001. Pre-Conference field trip (V2): Rhein/Mosel area and Lahn/Dill Synclines, Rheinisches Schiefergebirge. *In:* JANSEN, U., KONIGSHOF, P., PLODOWSKI, G. & SCHINDLER, E. (eds) *Field Trips Guidebook, 15th International Senckenberg Confernece*, 11–21 May 2001, 47–54.

JOACHIMSKI, M. M. & BUGGISCH, W. 2002. Conodont apatite $\delta^{18}O$ signatures indicate climatic cooling as a trigger of the Late Devonian mass extinction. *Geology*, **30**, 711–714.

JOHNSON, J. G. & SANDBERG, C. A. 1988. Devonian eustatic events in the Western United States and their biostratigraphic responses. In: MCMILLAN, N. J., EMBRY, A. F. & GLASS, D. J. (eds) *Devonian of the World*. Canadian Society of Petroleum Geologists, Memoirs, **14**(3), 171–178.

JOHNSON, J. G., KLAPPER, G. & SANDBERG, C. *A.* 1985. Devonian eustatic fluctuations in Euramerica. *Geological Society of America Bulletin*, **96**, 567–587.

JURKIEWICZ, H. 1975. The geological structure of the basement of the Mesozoic in the central part of the Miechów Trough. *Biuletyn Instytutu Geologicznego*, **283**, 5–100 [in Polish with English summary].

KALVODA, J. 2002. Late Devonian – Early Carboniferous foraminiferal fauna: zonations, evolutionary events, paleobiogeography and tectonic implications. *Folia Facultatis Scientiarum Naturalium Universitatis Masarykianae Brunensis, Geologia*, **39**, 1–213.

KAUFMANN, B. 2006. Calibrating the Devonian Time Scale – a synthesis of U–Pb ID-TIMS ages and conodont stratigraphy. *Earth Science Review*, **76**, 175–190.

KICUŁA, J. & ŻAKOWA, H. 1972. Devonian and Carboniferous in the basement of the southern part of the Miechów syncline. *Rocznik Polskiego Towarzystwa Geologicznego*, **42**, 165–228 [in Polish with English summary].

KLEESMENT, A. & MARK-KURIK , E. 1997. Devonian. *In:* RAUKAS, A. & TEEDUMAE, A. (eds) *Geology and Mineral Resources of Estonia*, Tallin, 107–123. Estonian Academic Publishers.

KOWALCZEWSKI, Z., JAWOROWSKI, K. & KULETA, M. 1998. Klonów Beds (uppermost Silurian-?lowermost Devonian) and the problem of Caledonian deformations in the Holy Cross Mts. *Geological Quarterly*, **42,** 341–378.

KRAMARSKA, R. (ed.) 1999. *Geological Map of the Baltic Sea Bottom without Quaternary Deposits*. Polish Geological Institute, Gdańsk-Warszawa.

KREBS, W. 1969. Über Schwarzschiefer und bituminöse Kalke im mitteleuropaischen Variscikum. Teil 1. Vorkommen in tiefen Becken und in abgeschlossenen Teilbecken des offenen Meeres oder auf dem Schelf. *Erdöl und Kohle-Erdgas-Petrochemie*, **22**(1), 2–6.

KREBS, W. 1974. Devonian carbonate complexes of Central Europe. *In:* LAPORTE, L. F. (ed.) *Reefs in Time and Space*. SEPM, Special Publications, **18**, 115–208.

KREBS, W. 1979. Devonian basinal facies. *In:* HOUSE, M.R., SCRUTTON, C. T. & BASSETT, M. G. (eds) *The Devonian System*. Special Papers in Palaeontology, **23**, 125–139, The Palaeontological Association, Aberystwyth.

KREUTZER, L. H. 1992. Palinspastische Entzerrung und Neugliederung des Devons in den Zentralkarnischen Alpen aufgrund von neuen Untersuchungen. *Jahrbuch der Geologischen Bundesanstalt*, **135**, 261–272.

KREUTZER, L. H., SCHÖNLAUB, H. P. & HUBMANN, B. 2000. The Devonian of Austria. *Courier Forschungsinstitut Senckenberg*, **225**, 173–183.

KRISTIC, B., GRUBIC, A., RAMOVS, A. & FILIPOVIC, I. 1988. The Devonian of Yugoslavia. *In:* MCMILLAN, N. J., EMBRY, A. F. & GLASS, D. J. (eds) *Devonian of the World*. Canadian Society of Petroleum Geologists, Memoirs, **14**(1), 499–506.

KUKAL, Z. & JÄGER, O. 1988. Siliciclastic signal of the Variscan orogenesis: the Devonian Srbsko Formation of Central Bohemia. *Vestnik Ústredniho ústavu geologického*, **63**, 65–81.

KULKE, H. 1997. Der Harz (Norddeutschland): geologisch-lagerstattenkundlicher Überblick, historische Baumaterialien (Natursteine, Gipsmörtel, Schlackensteine, Blei). *Mitteilungen der Österreichischen Mineralogischen Gesellschaft*, **142,** 43–84.

KURSZS, V.M. 1992. *Devonian Terrigeneous Sedimentation in the Main Devonian Field*. Zinantne, Riga (in Russian).

LÄUFER, A. L., HUBICH, D. & LOESCHKE, J. 1999. Geodynamic evolution of the Variscan Fold-and-thrust belts of the Carnic Alps. *Terra Nostra*, **99**(1), 136.

LECOMPTE, M. 1970. Die Riffe im Devon der Ardennen und ihre Bildungsbedingungen. *Geologica et Palaeontologica*, **4**, 25–71.

LEWANDOWSKI, M. 2003. Assembly of Pangea: combined paleomagnetic and paleoclimatic approach. *Advances in Geophysics*, **46**, 199–236.

LOESCHKE, J. & ROLSER, J. 1971. Der altpaläozoische Vulkanismus in den Karawanken (Österreich). *Zeitschrift der deutschen geologischen Gesellschaft*, **122**, 145–156.

MCCANN, T. 1999. Middle to Late Devonian basin evolution in the Rügen area, NE Germany. *Geologie en Mijnbouw*, **78**, 57–71.

MCKERROW, W. S., MAC NIOCAILL, C., AHLBERG, P. E., CLAYTON, G., CLEAL, C. J. & EAGAR, R. M. C. 2000. The late Palaeozoic relations between Gondwana and Laurussia. *In*: FRANKE, W., ALTHERR, R., HAAK, V. & ONCKEN, O. (eds) *Orogenic Processes – Quantification and Modelling in the Variscan Belt of Central Europe*. Geological Society, London, Special Publications, **179**, 9–20.

MACHEL, H. G. & HUNTER, I. G. 1994. Facies models for Middle to Late Devonian shallow-marine carbonates, with comparisons to modern reefs: a guide for facies analysis. *Facies*, **30,** 155–176.

MALEC, J. & TURNAU, E. 1997. Middle Devonian conodont, ostracod and miospore stratigraphy of the Grzegorzowice-Skały section, Holy Cross Mountains, Poland. *Bulletin of the Polish Academy of Science, Earth Sciences*, **45**, 67–86.

MALEC, J., TARNOWSKA, M., LENARTOWICZ, L., FIJAŁKOWSKA, A., TURNAU, E., ROMANEK, A. & FILIPIAK, P. 1995. *Lithostratigraphical, Biostratigraphical and Geochemical Correlations of the Devonian Deposits in the Holy Cross Mts.* PGI report, 1–202, Warszawa [in Polish].

MALEC, J., MIŁACZEWSKI, L., NARKIEWICZ, K. & NARKIEWICZ, M. 1996. Stratigraphy of the Devonian in the Szwejki IG 3 deep well, Central

Poland. *Geological Quarterly*, **40**, 367–392.

Marshall, J. & Hewett, T. 2003. Devonian. *In:* Evans, D., Graham, C., Armour, A. & Bathurst, P. (eds) *The Millenium Atlas: Petroleum Geology of the Central and Northern North Sea*. Geological Society, London, 65–81.

Marshall, J., Rogers, D. A. & Whiteley, M. J. 1996. Devonian marine incursions into the Orcadian Basin, Scotland. *Journal of the Geological Society, London*, **153**, 451–466.

Matyja, H. 1993. Upper Devonian of Western Pomerania. *Acta Geologica Polonica*, **43**, 27–94.

Matyja, H. 1998. Depositional architecture of the Devonian basin in the Pomerania-Kujawy area. *Prace Państwowego Instytutu Geologicznego*, **165**, 73–88 [in Polish with English summary].

Matyja, H. & Narkiewicz, M. 1992. Conodont biofacies succession near the Frasnian/Famennian boundary: some Polish examples. *Courier Forschungsinstitut Senckenberg*, **154**, 125–147.

Matyja, H. & Żbikowska, B. 1985. Stratigraphy of Devonian carbonate sequence in borehole sections in Lublin area. *Przegląd geologiczny*, **33**, 259–263 [in Polish with English summary].

Matyja, H., Tomaś, A., Lipiec, M. & Turnau, E. 2001. Stratigraphy of the Devonian–Carboniferous carbonates (Palaeozoic basement of the Polish Outer Carpathians). *Prace Państwowego Instytutu Geologicznego*, **174**, 33–60 [in Polish with English summary].

Meischner, D. 1964. Allodapische Kalke, Turbidite in Riff-nahen Sedimentations-Becken. *In*: Bouma, A. H. & Brouwer, A. (eds) *Turbidites*. Developments in Sedimentology, **3**, 156–191.

Meischner, D. 1971. Clastic sedimentation in the Variscan Geosyncline east of the River Rhine. Sedimentology of parts of Central Europe. *In:* Müller, G. (ed.) Guidebook. VIII International Sedimentological Congress, Heidelberg, August 31–September 3, 1971, 9–42.

Meischner, D. 1996. Bausteine der Paläogeographie. Devonischer Schelf, nördliches Schiefergebirge. *In:* Franke, W., Meischner, D. & Oncken, O. (eds) *Geologie eines passiven Plattenrandes: Devon und Unterkarbon im Rechtsrheinischen Schiefergebirge*. Exkursionen der Geologischen Vereinigung, **3**, 9–13.

Miłaczewski, L. 1981. The Devonian of the south-eastern part of the Radom-Lublin area (eastern Poland). *Prace Instytutu Geologicznego*, **101** [in Polish with English summary].

Mioc, P. 1977. Geologic structure of the Drava Valley between Dravograd and Selnica. *Geologia*, **20**, 193–230 [in Slovenian with English summary].

Mostler, H. 1968. Das Silur im Westabschnitt der Nördlichen Grauwackenzone (Tirol und Salzburg). *Mitteilungen der Gesellschaft der Geologischen Bergbaustudenten Österreichs*, **18**, 89–150.

Narbutas, V. & Ungicius, A. 2001. *Middle Devonian Terrigeneous Complexes in Lithuania*. Akstis, Vilnius [in Lithuanian with English summary].

Narkiewicz, M. 1978. Stratigraphy and facies development of the Upper Devonian in the Olkusz-Zawiercie area, southern Poland. *Acta Geologica Polonica*, **28**, 415–470 [in Polish with English summary].

Narkiewicz, M. 1988. Turning points in sedimentary development in the Late Devonian in southern Poland. *In:* McMillan, N. J., Embry, A. F. & Glass, D. J. (eds) *Devonian of the World*. Canadian Society of Petroleum Geologists, Memoirs, **14**, 619–635.

Narkiewicz, M. 1991. Mesogenetic dolomitization processes: an example from the Givetian to Frasnian of the Holy Cross Mountains, Poland. *Prace Państwowego Instytutu Geologicznego*, **132**, 1–54 [in Polish with English summary].

Narkiewicz, M. 1996. Devonian stratigraphy and depositional environments in proximity of the Sub-Carpathian Arch: Lachowice 7 well, southern Poland. *Geological Quarterly*, **40**, 65–88.

Narkiewicz, M. 2001. Lithostratigraphy, depositional environments and diagenesis of the Devonian and Carboniferous carbonates in the Rajbrot-Tarnawa area (Sub-Carpathian Arch). *Prace Państwowego Instytutu Geologicznego*, **174**, 9–32 [in Polish with English summary].

Narkiewicz, M. 2002. Middle Devonian epicontinental basin development in SE Poland: a role of crustal discontinuities. *Proceedings International Symposium on Geology of the Devonian System*, Syktyvkar, 9–12 July 2002, 30–32.

Narkiewicz, M. 2003. Tectonic controls of the Lublin Graben (Late Devonian – Carboniferous). *Przegląd Geologiczny*, **51**, 771–776.

Narkiewicz, M. 2005. Devonian and Carboniferous carbonate complex in the southern part of the Upper Silesian Block. *Prace Państwowego Instytutu Geologicznego*, **182**, 1–46 [in Polish with English summary].

Narkiewicz, M. 2007. Development and inversion of Devonian and Carboniferous basins in the eastern part of the Variscan foreland (Poland). *Geological Quarterly*, **51**, 231–256.

Narkiewicz, M. & Hoffman, A. 1989. The Frasnian/Famennian transition: the sequence of events in southern Poland and its implications. *Acta Geologica Polonica*, **39**, 13–28.

Narkiewicz, M. & Narkiewicz, K. 1992. Transgressive pulse in the Upper Frasnian of the Janczyce I section (Holy Cross Mts.): sedimentology and conodont biofacies. *Geological Quarterly*, **36**, 281–304.

Narkiewicz, M. & Racki, G. 1984. Stratigraphy of the Devonian of the Dębnik Anticline. *Kwartalnik Geologiczny*, **28**, 513–546.

Narkiewicz, M. & Racki, G. 1985. Major features of the Late Devonian paleogeography in the near-shore shelf area of southern Poland. *Przegląd Geologiczny*, **33**, 271–274 [in Polish with English summary].

Narkiewicz, M., Racki, G. & Wrzołek, T. 1990. Lithostratigraphy of the Devonian stromatoporoid–coral carbonate sequence in the Holy Cross Mountains. *Kwartalnik Geologiczny*, **34**, 433–456 [in Polish with English summary].

Narkiewicz, M., Miłaczewski, L., Krzywiec, P. & Szewczyk, J. 1998. Outline of the Devonian depositional architecture in the Radom-Lublin area. *Prace Państwowego Instytutu Geologicznego*, **165**, 57–72 [in Polish with English summary].

Neubauer, F. & Pistotnik, J. 1984. Das Altpaläozoikum und Unterkarbon des Gurktaler Deckensystems (Ostalpen) und ihre paläogeographischen Beziehungen. *Geologische Rundschau*, **73**, 149–174.

Oczlon, M. S. 1990. Ocean currents and unconformities: The North Gondwana Middle Devonian. *Geology*, **18**, 509–512.

Oncken, O., Plesch, A., Weber, J., Ricken, W. & Schrader, S. 2000. Passive margin detachment during arc-continent collision (Central European Variscides). *In*: Franke, W., Altherr, R., Haak, V. & Oncken, O. (eds) *Orogenic Processes – Quantification and Modelling in the Variscan Belt of Central Europe*. Geological Society, London, Special Publications, **179**, 199–216.

Pajchlowa, M. & Miłaczewski, L. 2003. Korelacja biostratygraficzna dewonu Polski z innymi obszarami. *In:* Malinowska, L. (ed.) *Budowa geologiczna Polski t.III, Atlas skamienialości przewodnich i charakterystycznych*, część 1b, zesz. 1, 20–26.

Paskevicius, J. (ed.) 1993. *Catalogue of the Vendian-Devonian stratotypes of Lithuania*. Geological Society of Lithuania. Vilnius.

Patocka, F., Vlasimsky, P. & Blechova, K. 1993. Geochemistry of Early Paleozoic volcanics of the Barrandian Basin (Bohemian Massif, Czech Republik): implication for paleotectonic reconstructions. *Jahrbuch der Geologischen Bundesanstalt*, **136**, 873–896.

Pedder, A. E. H. 1999. Paleogeographic implications of a Devonian (Givetian, Lower *Varcus* Subzone) rugose coral fauna from the Mader Basin (Morocco)). *Abhandlungen der geologischen Bundesanstalt*, **54**, 385–434.

Piske, J., Rasch, H.-J., Neumann, E. & Zagora, K. 1994. [Geological structure and development of the pre-Permian of the Isle of Rügen and adjacent offshore.] *Zeitschrift der Geologischen Wissenschaften*, **22**, 211–226 [in German with English summary].

Porebski, S. J. 1981. Swiebodzice succession (Upper Devonian–Lower Carboniferous; Western Sudetes): a prograding, mass-flow dominated fan-delta complex. *Geologica Sudetica*, **16**, 101–192.

Porebski, S. J. 1990. Onset of coarse clastic sedimentation in the Variscan realm of the Sudetes (SW Poland): an example from the upper Devonian-lower Carboniferous Swiebodzice succession. *Neues Jahrbuch für Geologie und Paläontologie, Abhandlungen*, **179**, 259–274.

Racki, G. 1993. Evolution of the bank to reef complex in the Devonian of the Holy Cross Mountains. *Acta Palaeontologica Polonica*, **37**, 87–182.

Racki, G. 1997. Devonian eustatic fluctuations in Poland. *Courier Forschungsinstitut Senckenberg*, **199**, 1–12.

Racki, G. & Narkiewicz, M. 2000. Tectonic versus eustatic controls of sedimentary development of the Devonian in the Holy Cross Mts., Central Poland. *Przegląd Geologiczny*, **48**, 65–76 [in Polish with English summary].

Racki, G. & Turnau, E. 2000. Devonian series and stage boundaries in

Poland. *Courier Forschungsinstitut Senckenberg*, **225**, 145–158.

RACKI, G., RACKA, M., MATYJA, H. & DEVLEESCHOUWER, X. 2002. The Frasnian–Famennian boundary in the South Polish-Moravian shelf basins: integrated event-stratigraphical approach. *Palaeogeography, Palaeoclimatology, Palaeoecology*, **181**, 251–297.

RAMOVS, A. 1979. The Devonian of the Karawanke Mts. *Field Meeting on the Paleozoic in Slovenia, NW Yugoslavia*, Yugoslaw committee for IGCP, Project No. 5, 10–13.

RANTITSCH, G. 1992. Fazies und Diagenese devonischer Riffkalke des Seeberger Aufbruches (Kärnten, Österreich). *Jahrbuch der Geologischen Bundesanstalt*, **135**, 273–285.

RICHARDS, P. C. 1990. Devonian. *In:* GLENNIE, K. W. (ed.) *Introduction to the Petroleum Geology of the North Sea*. 3rd edn. Blackwell, Oxford.

SANDBERG, C. A., ZIEGLER, W., DREESEN, R. & BUTLER, J. L. 1992. Conodont biochronology, biofacies, taxonomy, and event stratigraphy around Middle Frasnian Lion mudmound (F2h), Frasnes, Belgium. *Courier Forschungsinstitut Senckenberg*, **150**, 1–87.

SCHÄFER, J., NEUROTH, H., AHRENDT, H., DÖRR, W. & FRANKE, W. 1997. Accretion and exhumation at a Variscan active margin, recorded in the Saxothuringian flysch. *Geologische Rundschau*, **86**, 599–611.

SCHMIDT, K. & FRANKE, D. 1977. Zur lithologisch-faziellen Entwicklung des Präperms im Nordteil der DDR. *Zeitschrift der Angewandten Geologie*, **23**, 541–548.

SCHÖNLAUB, H. P. 1979. Das Paläozoikum in Österreich. *Abhandlungen der Geologischen Bundesanstalt*, **33**, 1–124.

STAMPFLI, G. M. & BOREL, G. D. 2002. A plate tectonic model for the Paleozoic and Mesozoic constrained by dynamic plate boundaries and restored synthetic oceanic isochrones. *Earth and Planetary Science Letters*, **196**, 17–33.

SZULCZEWSKI, M. 1971. Upper Devonian conodonts, stratigraphy and facial development in the Holy Cross Mts. *Acta Geologica Polonica*, **21**, 1–129.

SZULCZEWSKI, M. 1995. Depositional evolution of the Holy Cross Mts. (Poland) in the Devonian and Carboniferous – a review. *Geological Quarterly*, **39**, 471–488.

SZULCZEWSKI, M., BELKA, Z. & SKOMPSKI, S. 1996. The drowning of a carbonate platform: an example from the Devonian–Carboniferous of the southwestern Holy Cross Mountains, Poland. *Sedimentary Geology*, **106**, 21–49.

TAIT, J. A., BACHTADSE, V., FRANKE, W. & SOFFEL, H. 1997. Geodynamic evolution of the European Variscan fold belt: paleomagnetic and geological constraints. *Geologische Rundschau*, **86**, 585–598.

TAIT, J., SCHÄTZ, M., BACHTADSE, V. & SOFFEL, H. 2000. Palaeomagnetism and Palaeozoic palaeogeography of Gondwana and European terranes. *In*: FRANKE, W., ALTHERR, R., HAAK, V. & ONCKEN, O. (eds) *Orogenic Processes – Quantification and Modelling in the Variscan Belt of Central Europe*. Geological Society, London, Special Publications, **179**, 21–34.

TARNOWSKA, M. 1976. Lithological correlation of the Lower Devonian in the eastern part of the Góry Świętokrzyskie. *Biuletyn Instytutu Geologicznego*, **296**, 75–128 [in Polish with English summary].

TARNOWSKA, M. 1988. Zarys historii sedymentacji osadow dewonu dolnego west poludniowej czesci Gor Swietokrzyskich. *Geological Quarterly*, **32**, 242–243.

TREWIN, N. H. & THIRLWALL, M. F. 2002. Old Red Sandstone. *In:* TREWIN, N. H. (ed.) *Geology of Scotland*, 4th edn, Geological Society of London, 213–249.

TSIEN, H. H. 1979. Paleoecology of algal-bearing facies in the Devonian (Couvinian to Frasnian) reef complexes of Belgium. *Palaeogeography, Palaeoclimatology, Palaeoecology*, **27**, 103–127.

TURNAU, E. 1974. Microflora from core samples of some palaeozoic sediments from beneath the Flysch Carpathians (Bielsko-Wadowice area, southern Poland). *Rocznik Polskiego Towarzystwa Geologicznego*, **44**, 143–169.

TURNAU, E. & MATYJA, H. 2001. Timing of the onset of Devonian sedimentation in northwestern Poland: palynological evidence. *Rocznik Polskiego Towarzystwa Geologicznego*, **71**, 67–74.

TURNAU, E. & RACKI, G. 1999. Givetian palynostratigraphy and palynofacies: new data from the Bodzentyn Syncline (Holy Cross Mountains, central Poland). *Review of Palaeobotany and Palynology*, **106,** 237–271.

VERNIERS, J., PHARAOH, T., *et al.* 2002. The Cambrian to mid Devonian basin development and deformation history of Eastern Avalonia, east of the Midlands Microcraton: new data and a review. *In:* WINCHESTER, J. A, PHARAOH, T. C & VERNIERS, J. (eds) *Palaeozoic amalgamation of Central Europe*. Geological Society, London, Special Publications, **201**, 47–93.

VOLKOLAKOV, F. K., POLIVKO, I. A., AGALTSOVA, E. N. & YAKOVLEVA, V. I. 1977. *Geological Structure and Petroleum Aspects of the Offshore Part of the Baltic Syneclise*. Zinantne, Riga [in Russian].

WEDDIGE, K. 1988. Eifel conodonts. *Courier Forschungsinstitut Senckenberg*, **102**, 103–110.

WEDDIGE, K., WERNER, R. & ZIEGLER, W. 1988. Stop A 14: Wetteldorfer Richtschnitt, S. Schönecken Lower/Middle Devonian boundary stratotype. *Courier Forschungsinstitut Senckenberg*, **102**, 111–118.

WINCHESTER, J.A. & the PACE TMR Network Team. 2002. Palaeozoic amalgamation of Central Europe: new results from recent geological and geophysical investigations. *Tectonophysics*, **360**, 5–21.

WIRTH, R. 1978. *Geochemie und Petrographie der paläozoischen Magmatite des Frankenwaldes. Diabase - Keratophyre - Pikryte*. PhD thesis, Universität Würzburg.

ZAGORA, K. 1995. Devonian of the Isle of Rügen. *Studia geophysica et geodaetica*, **39**, 298–301.

ZIKMUNDOVÁ, J. 1964. Conodontenfunde im Devon des Jested-Gebirges. *Vestnik Ústredniho ústavu geologického*, **39**, 455–457.

9 Carboniferous

TOM McCANN (co-ordinator), STANISLAW SKOMPSKI, EDOUARD POTY, MICHEL DUSAR, ANNA VOZÁROVÁ, JÖRG SCHNEIDER, ANDREAS WETZEL, KARL KRAINER, KRISTIJAN KORNPIHL ANDREAS SCHÄFER, MICHAEL KRINGS, STANISLAV OPLUSTIL & JENNY TAIT

The Carboniferous (359.2–299 Ma, Gradstein *et al.* 2004) succession of Central Europe records one of the most important time periods with respect to European geology, since it marks the final collision of Gondwana with the northern continent of Laurussia (i.e. Laurentia, Baltica and Avalonia). Oblique convergence resulted in collisional processes which created a mountain belt extending from Russia, through western Europe and into North America. The climax of the Variscan Orogeny was the formation of the supercontinent Pangaea leaving a relict Palaeotethys to the east (Scotese & Langford 1995) (Fig. 9.1). The Variscan belt is a broad (*c.* 1000 km) complex curvilinear feature extending across Europe and marking the zones of Variscan-age deformation (Figs 9.2 & 9.3). The final phase of Variscan activity was also a period of terrane mobility and tectonic instability in the Central European region with sinistral wrench faulting causing widespread rifting of the northern European crust (Pegrum 1984*a*, *b*; Ziegler 1990).

The Carboniferous succession in Central Europe is generally dominated by marine sediments (both clastic and carbonate) in the lower part of the succession.. The clastic sediments tend to be deeper-water shelf or turbiditic successions, although in some areas (e.g. Belgium, northern Germany) limestones are locally important or even dominant, particularly during the Tournaisian and Visean. In late Carboniferous times, successions are predominantly continental with some coal-bearing units being deposited (particularly in Westphalian times). An exception to the dominantly sedimentary record is provided by volcanic and volcaniclastic rocks which are abundant in the northern part of the area and are largely related to the onset of Stephanian–early Permian wrench tectonics (see McCann *et al.* 2008).

The dramatic economic expansion during the eighteenth and nineteenth centuries across Central and western Europe was largely based on the exploitation of the Westphalian-age coal deposits. This so-called 'Industrial Revolution' was the basis for the subsequent development of the major manufacturing centres of nineteenth century Europe. Other economically valuable materials found in Carboniferous-age strata, notably heavy ceramic and refractory clays, iron ores, oil shales and high-purity limestones, also played an important role on the location of heavy industry during the Industrial Revolution (Besley 1990). Much of our knowledge of the Carboniferous is founded on studies to exploit these natural resources (Guion *et al.* 2002). In the 1960s in the Netherlands and Germany, renewed exploration of the Carboniferous resulted in the discovery of a number of gas fields and reservoirs in Westphalian C and D sandstones (Besley 1990).

In recent times, there have been a number of volumes and papers published on the Carboniferous and related Variscan history of Europe (e.g. Franke 1990; Dallmeyer *et al.* 1995; Zdanowski & Zakowa 1995; Franke *et al.* 2000). Despite this, there is still considerable confusion concerning various aspects of the Carboniferous evolution in the region, in particular, the precise definition of the various units/terranes, their geological history and their evolution. This confusion is further compounded by the widespread use of local terminology and stratigraphy. The following sections, therefore, provide an initial introduction to the tectonic setting of the Variscan area, indicating some of the controlling factors influencing the origin and development of the Carboniferous basins in Central Europe (see Kroner *et al.* 2008) for more details on the tectonic history, and Timmermann 2008 for the magmatics). This is followed by a broad regional overview outlining the main tectonic and magmatic events across Central Europe.

The Devonian–Carboniferous boundary

The Devonian–Carboniferous boundary is of particular geological interest since it is the sixth largest extinction event within the Phanerozoic record (Fig. 9.4). Many organisms were affected including conodonts, ammonoids, echinoderms, trilobites, ancient agnatian and placodermian fishes, continental plant communities, ostracodes, foraminifera, stromatoporoids and corals (e.g. Kullmann 2000; Olempska 1997; Bojar & Neubauer 2003).

The base of the Carboniferous System is defined at the first appearance of the conodont *Siphonodella sulcata* within the evolutionary lineage from *Siphonodella praesulcata* to *Siphonodella sulcata* (Paproth *et al.* 1991). This entry follows the Hangenberg Event, responsible for the demise of Devonian fauna and reflecting a drastic sea-level drop (Paproth 1986). The Devonian–Carboniferous boundary is located at the base of Bed 89 in Trench E' at La Serre in southern France (Flajs & Feist 1988). Secondary markers also found in Bed 89 include the trilobites *Belgibone abruptirhachis*, *Archegonus (Phillibole)* and *Carbonocoryphe*. Menning *et al.* (2006) note that this section is located within an allochthonous turbiditic succession and since many of its critical conodont data are based on reworked faunas, it has been the subject of much criticism.

The Early Tournaisian of western Europe includes an anoxic black shale above the basal *Siphonodella crenulata* Zone, and this shale is developed both in basinal facies (Lower Alum Shale and Rußschiefer, Germany) and in shallow-water facies (Schistes du Pont d'Arcole, Belgium). Cephalopod limestones and ostracod shales below this semiglobal event are similar to the Frasnian/Fammenian dysphotic/aphotic facies development (Menning *et al.* 2006). In Central Europe, the trilobite *Belgibole abruptirhachis* occurs immediately above the Hangenberg Schiefer interval

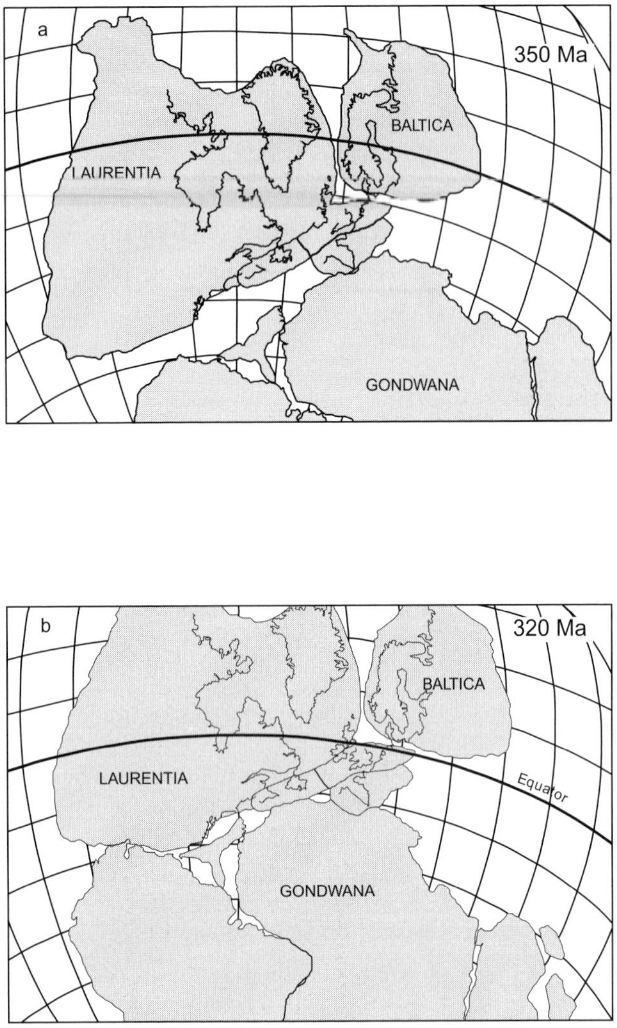

Fig. 9.1. Continental dispositions in (**a**) Early (Visean, 350 Ma) and (**b**) Late (Westphalian, 320 Ma) Carboniferous times (after McKerrow *et al.* 2000).

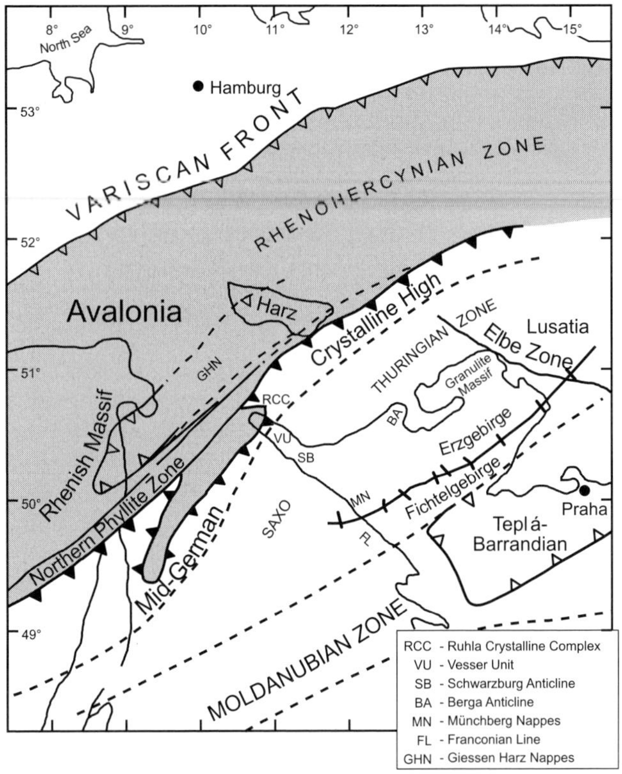

Fig. 9.2. Geological overview map of the Central European Variscides showing the major fault zones and structural units (after Schäfer *et al.* 2000).

in various cephalopod-bearing sections from the Rhenish Massif, Germany (= Rheinisches Schiefergebirge), the Holy Cross Mountains (Poland) and the Carnic Alps (Austria) (e.g. Becker 1988; Becker *et al.* 1984; Schönlaub *et al.* 1992; Higgs *et al.* 1993; Luppold *et al.* 1994; Korn *et al.* 1994; Clausen *et al.* 1994). In southern Belgium, this event is recorded by a metre-thick bed of rudstones and grainstones with lithoclasts, ooids, crinoids and a reworked Devonian fauna. Interestingly, the Hangenberg Schiefer is interpreted as representing an anoxic event, which was perhaps related to a short transgressive pulse (Walliser 1995). Worldwide, geochemical data have revealed evidence of weak iridium anomalies close to the bottom and the top of these shales. The origin of these anomalies is still under debate but they may be related to limestone–shale lithological transitions or even an asteroid impact (Claeys *et al.* 1992; Bojar & Neubauer 2003).

Palaeoclimate and sea-level changes

The Carboniferous was a time of major changes in the Earth's palaeogeography, climate and carbon cycle. During the Early Carboniferous, Gondwana moved northward and collided with Laurussia, then continued rotating clockwise and suturing with it through the rest of the Carboniferous (Wicander & Monroe 1993) (Fig. 9.1). By the end of the Carboniferous and Early Permian, major continents (Laurussia, Gondwana, Siberia, Kazakhstania and China) either collided or moved close to one another and the supercontinent Pangaea began taking shape (Mii *et al.* 2001). In terms of climate, the Carboniferous was marked by the drawdown of atmospheric CO_2 and significant cooling, as well as major glaciations and sharp sea-level fluctuations leading to the development of cyclic marine sequences. Increased continental weathering rates and aridity were a feature of the late Carboniferous.

Combined with the evolution of land plants and the Variscan Orogeny, the Carboniferous was the time of formation of the world's largest coal deposits (Buggisch *et al.* 2003; see below). The evolution of terrestrial plants and widespread coal formation created a new organic carbon reservoir and changed the global carbon cycle (Berner 1987, 1997). Atmospheric concentrations of carbon dioxide (CO_2) in the Early Carboniferous were approximately 1500 ppm (parts per million), but by the Middle Carboniferous had declined to about 350 ppm – comparable to average CO_2 concentrations today. Additionally, changes in the chemistry, circulation and temperature of Carboniferous oceans, and the waxing and waning of Carboniferous glaciers, is recorded in the isotopic compositions of marine carbonates (e.g. Veizer *et al.* 1986; Grossman 1994).

Several mechanisms have been proposed to explain Carboniferous glaciation. These include tectonic triggering (Veevers & Powell 1987; Powell & Veevers 1987), changes in the CO_2

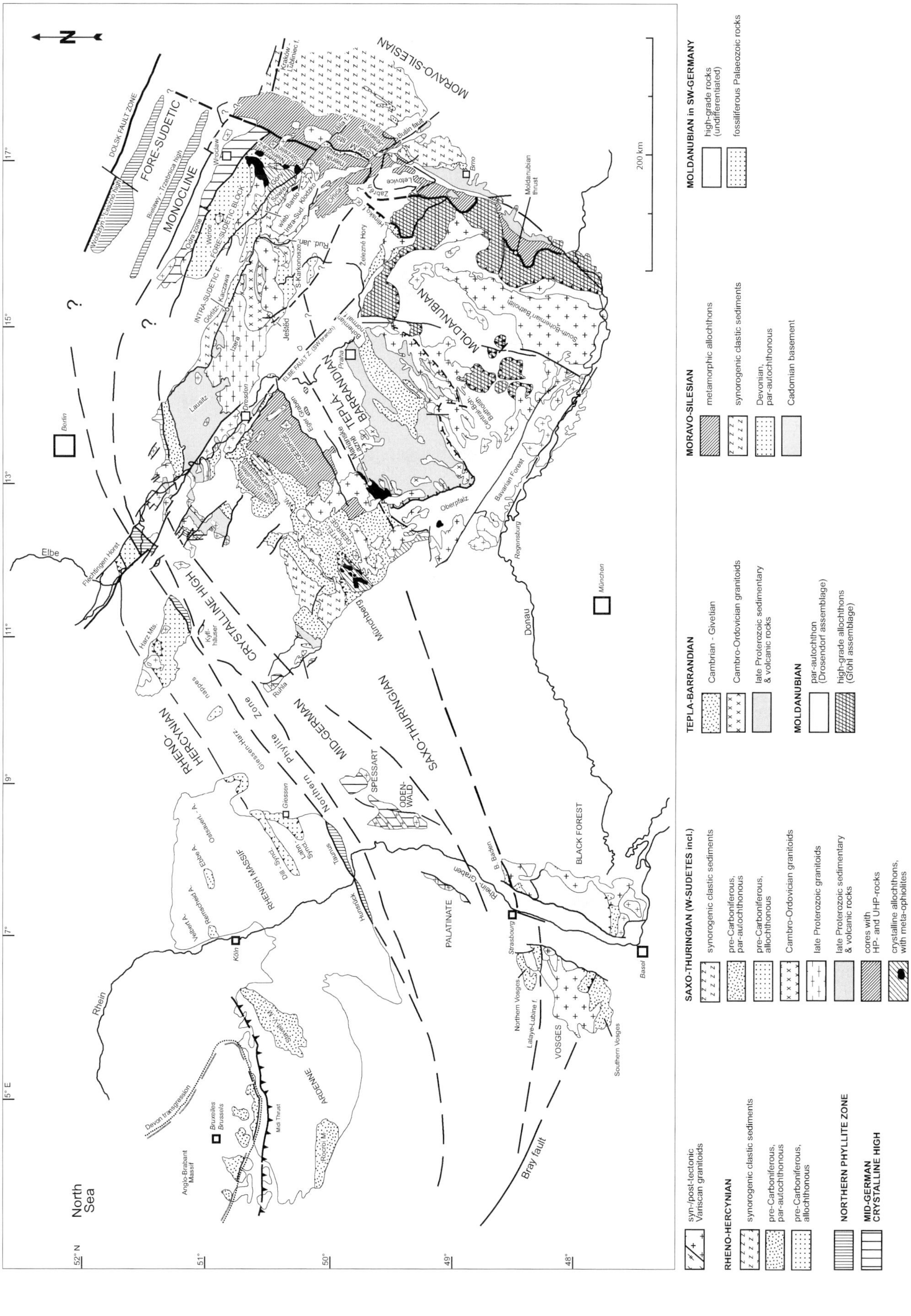

Fig. 9.3. Structural map of the Variscan Belt in Central Europe (after Franke 2000).

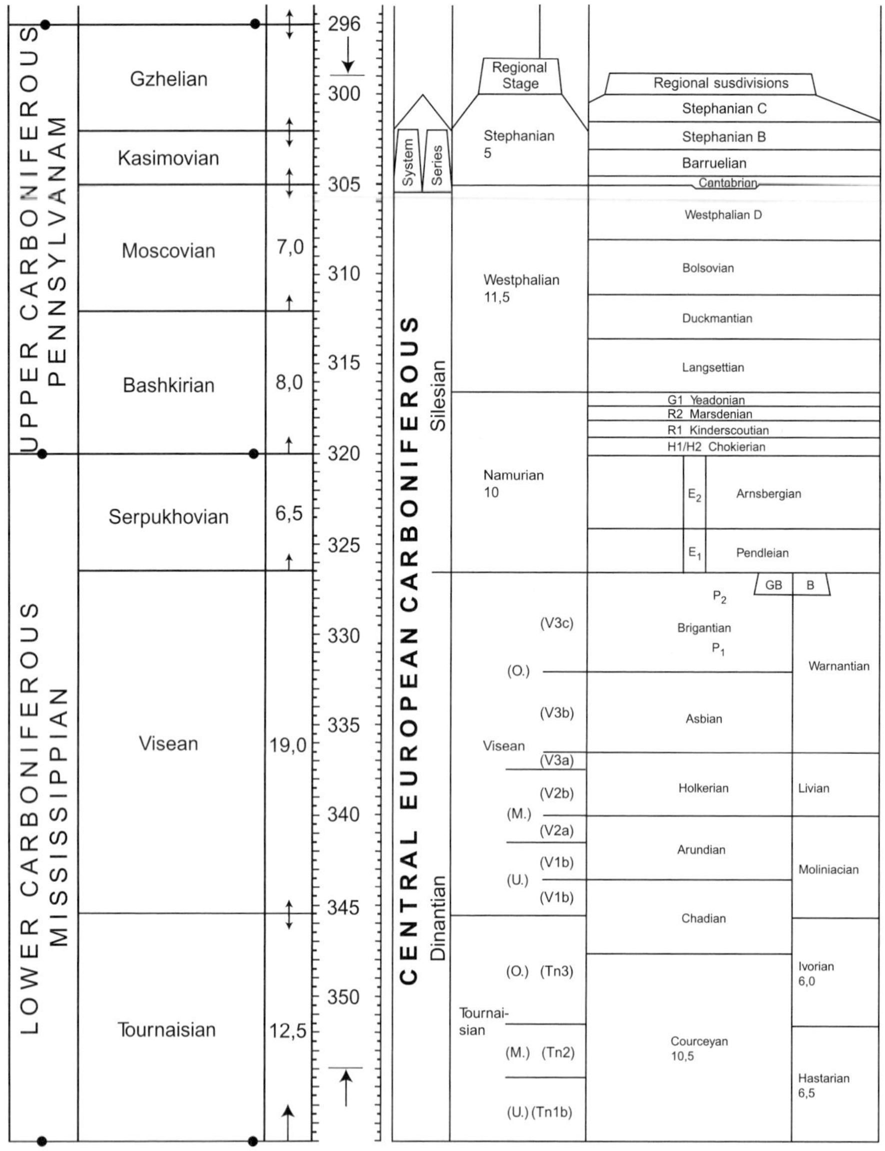

Fig. 9.4. Carboniferous stratigraphy of Central Europe (Deutsche Stratigraphische Kommission 2002).

content (Crowley & Baum 1991; Beerling 2002), the closure of the equatorial seaway between Laurussia and Gondwana (Rowley *et al.* 1985) with resultant changes in the oceanic circulation patterns and warming of higher latitude areas (Raymond *et al.* 1989), and possible increased precipitation over Gondwana (Smith & Read 2000). However, the relationships between insolation changes, ocean circulation and ice buildups are complex (Denton 2000). More recent work has concentrated on observed isotopic changes. Grossman *et al.* (2002) have noted that there were marked increases in both $\delta^{18}O$ and $\delta^{13}C$ in the mid-Namurian (i.e. Serpukhovian–Bashkirian boundary). Indeed, this boundary is coincident with a major Palaeozoic bioevent with major extinctions and originations of conodont and ammonoid fauna as well as turnover in crinoids, corals and foraminifers (e.g. House 1993; Walliser 1995). The isotopic evidence would suggest that there was a major transition at this time from a greenhouse to an icehouse climate mode.

Average global temperatures in the early Carboniferous were hot – approximately 22°C. Throughout the Tournaisian and Visean (i.e. Dinantian) the climate was seasonal, possibly monsoonal. Clay minerals in pre-early Visean palaeosols suggest that the climate was dry with seasonal wetting, whereas after this time the climate became humid with seasonal drying (Guion *et al.* 2002). Abundant meteoric cements in Late Dinantian shelf carbonates record humid conditions during emergence, whereas leached evaporites in late Visean peritidal carbonates in Derbyshire (England) imply a semiarid climate (Guion *et al.* 2002). Stable isotope analyses of brachiopods (Bruckschen & Veizer 1997) indicate two marked phases of glacial activity at this time, one in the late Tournaisian and a second in the mid-late Visean. Evidence for these intervals is based on the presence of early Visean glaciogenic sediments in South America and Africa (Veevers & Powell 1987) and late Visean strata from western Argentina (Gonzalez 1990) and Australia (Claoué-Long *et al.* 1995). These phases indicate the probable onset of glacio-eustacy in the Early Carboniferous with the coeval regular fluctuations in sea level which this implies (Fig. 9.5). These fluctuations led to a progressive increase in the amplitude of sea-level oscillations as indicated by the thickness and compositional changes in cyclothems during the late Visean (Wright & Vanstone 2001). This would imply that some threshold was crossed in Asbian (Visean) time that triggered a shift in the behaviour of the Gondwanan ice sheets. From this point onwards they were able to grow sufficiently large to trigger decimetre-scale sea-level oscillations, and also become sensitized to changes in solar insolation (Wright & Vanstone 2001).

The mid-Carboniferous was a time of great global environmental change. The Pangaean supercontinent was beginning to assemble, and this increased landmass led to the development of an increasingly continental climate (Crowley & North 1991). Additionally, the Earth's climate was entering an icehouse mode, with cooling and glaciation indicated by glacial sediments (e.g. Veevers & Powell 1987; Claoue-Long *et al.* 1995; Frakes *et al.* 1992), brachiopod migration patterns (Raymond *et al.* 1989), sea-level decline (Ross & Ross 1987, 1988) and $\delta^{18}O$ increases (Grossman *et al.* 2002; Mii *et al.* 2001; Bruckschen *et al.* 2001). However, uncertainty still remains over the precise timing of the onset of the glaciation, a reflection of the poor biostratigraphical control of the identified glacial deposits and the difficulties associated with the interpretation and correlation of the various proxy indicators used (Wright & Vanstone 2001). Significant glaciation probably began in the southern hemisphere (spreading later to the northern hemisphere) in the late Namurian A, or at the Namurian A–B boundary although northern high latitudes remained warm into the Namurian A and cooled at the transition from Namurian A to B (for details see Raymond & Metz 2004). Cooling during the Mid-Carboniferous reduced average global temperatures to about 12°C (comparable to the average global temperature on Earth today). Mii *et al* (2001) have suggested that there was a 5°C decline (from −25°C to −20°C) in low-latitude sea-surface temperature with the onset of the mid-Carboniferous glaciation. This is consistent with estimates of tropical sea-surface temperature change associated with the last glacial to interglacial transition (e.g. Guilderson *et al.* 1994; Stute *et al.* 1995).

The Namurian and Westphalian climates of Central Europe were tropical ever-wet to seasonally wet, punctuated by short-term fluctuations which were linked with the southern hemisphere glaciation. The effect of high-latitude glaciation on the tropical climate is controversial, but seasonally drier climates are most likely during periods of intense glaciation and lowered sea level. Melting of ice and the rise in sea level promoted wetter equatorial climates with peat formation, giving rise to the abundant coal deposits in parts of Central Europe (e.g. Ruhr

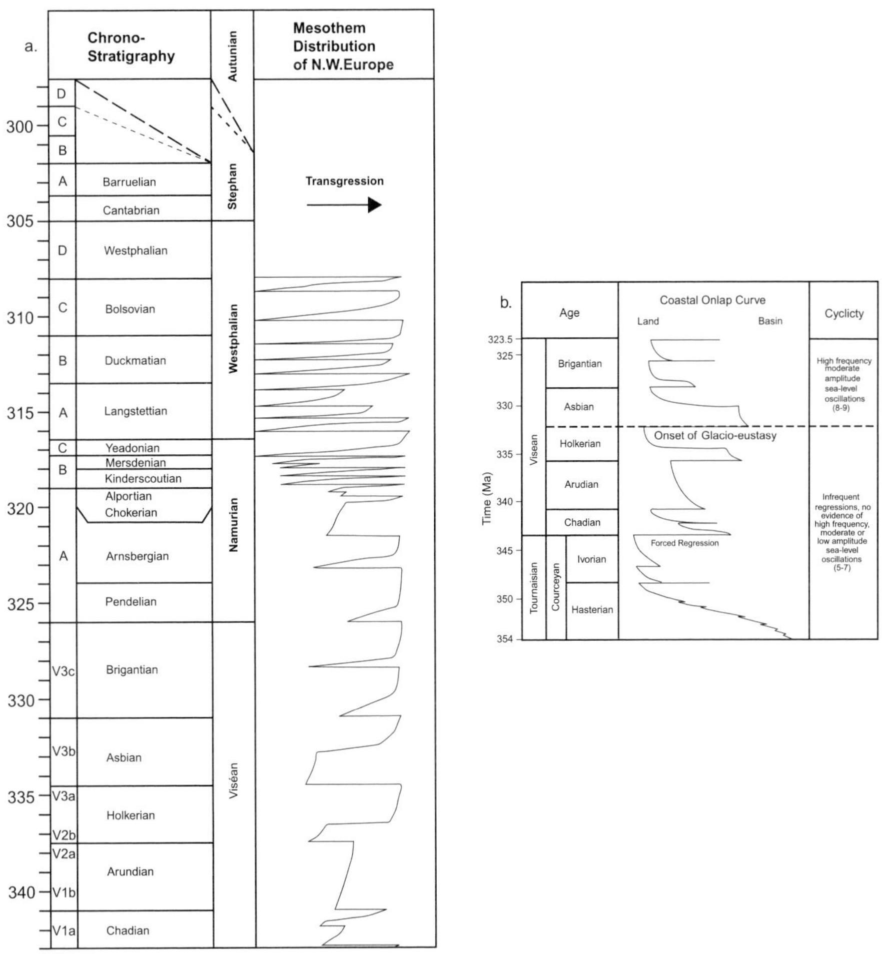

Fig. 9.5. Sea-level curves for the Carboniferous. (**a**) The transgressive–regressive (i.e. coastal onlap) depositional sequences of NW Europe which provides an approximation of eustatic sea-level changes (after Ross & Ross 1987). (**b**) Synthesis of the Lower Carboniferous sea-level and cyclicity record. Coastal onlap curve from Ross & Ross (1988) (after Wright & Vanstone 2001).

area, Germany; Silesia, Poland; Teichmuller 1973). Arid conditions in the late Westphalian and Stephanian are indicated by calcretes and red-beds (e.g. Saar-Nahe Basin, NE German Basin). During the late Stephanian the warm to temperate humid climate changed to a seasonally dry, warm climate corresponding to the late Palaeozoic climate scenario in the European Variscides (Scotese & McKerrow 1990). However, the regional climatic zones may have been locally modified by elevations within the Variscan mountain belt (e.g. Becq-Giraudon & van den Driessche 1994). The increasing aridity at the end of the Carboniferous was, perhaps, caused by the development of a rain shadow situated NW of the evolving Variscan mountains which gave shelter from the dominant trade winds blowing from the east (Guion *et al.* 2002). However, other models prefer the reorganization of major ocean currents for the increasing aridity of northern Pangaea given that the Variscan mountains had probably been eroded down to lower mountain ranges by the latest Westphalian and could not, therefore, have acted as moisture shelters (Schneider *et al.* 2005*a*; Roscher & Schneider 2006).

As noted above, the Carboniferous was a period of intense glacial activity with related sea-level changes. Glaciation lasted over 100 Ma (i.e. Permo-Carbonifeous), with the peak extent being about 60 Ma. Ice sheet growth continued until it reached as estimated area of 21×10^6 km^2 during the Carboniferous and Permian – similar to that of the Pleistocene ice sheet (23.5×10^6 km^2). The Visean–Namurian boundary coincided with a temporary eustatic lowstand in sea level. Sea levels rose sharply during the early Namurian, dropped sharply in the middle Namurian and rose again in the Westphalian (Ross & Ross 1987). Crowley & Baum (1991) have used ice sheet area–volume relations to estimate the magnitude of Carboniferous sea-level fluctuations with a best estimate of 60 ± 15 m. However, Buggisch *et al.* (2003) have suggested up to 200 m based on the $\delta^{13}C$ and $\delta^{18}O$ records (e.g. Grossman *et al.* 2002).

Central European Carboniferous stratigraphy

The name 'Carboniferous' is derived from the Italian *carbonarium* (charcoal producer) or Latin *carbo* (charcoal) and *ferous* (i.e. bearing) and both mark the episode of massive coal formation which took place at that time (Davydov *et al.* 2004). The term Carboniferous was first used in an adjectival sense by Kirwan (1799) and by Farey in 1811 as an informal term for the section heading to describe 'Coal-measures or Carboniferous strata' (Ramsbottom 1984).

Although the Carboniferous was one of the earliest geological periods to be established, it is one of the most complex and confusing in terms of global stratigraphic classification and correlation (Davydov *et al.* 2004). The reason for this is that the assembly of the supercontinent Pangaea led to the almost total separation of tropical and subtropical shelves. Additionally, the Gondwana glaciation and consequent climatic changes and sea-level fluctuations resulted in marked biogeographic differentiation (Davydov *et al.* 2004).

The basic European stratigraphy for the Carboniferous was established by the early twentieth century, although it is still being refined (e.g. Menning *et al.* 2001, 2006; Weyer *et al.* 2005; Wrede *et al.* 2005). The succession in Central Europe is divided into two major successions: the Lower Carboniferous or Dinantian (i.e. Tournaisian and Visean) and the Upper Carboniferous or Silesian (i.e. Namurian, Westphalian and Stephanian). These are only broadly concurrent with the global subdivision of Mississippian (Tournaisian, Visean, Serpukhovian) and Pennsylvanian (Bashkirian, Moscovian, Kasimovian, Gzelian) (Fig. 9.4). These are separated by the global Mid-Carboniferous Boundary, defined at the lowest occurrence of the conodont *Declinognathodus noduliferus* and locaed at the base of the Pennsylvanian (Menning *et al.* 2006). The subsystems of the western and central European Carboniferous – Dinantian and Silesian – are, like the related stages of Tournaisian through to Stephanian, defined as chronostratigraphic units which contain biostratigraphic zones (cf. Menning *et al.* 2000, 2006). In some areas (e.g. Belgium) these subdivisions correspond to lithological ones, i.e. the Dinantian is made up of carbonate rocks while the Silesian comprises siliciclastics with coal seams in its upper part. In other areas, this lithological subdivision is complicated by the presence of the mainly clastic Culm and the carbonate Kohlenkalk facies (see below).

The Carboniferous stratigraphy has recently been updated for Germany correlating the existing Carboniferous subdivisions with the global standards (Deutsche Stratigraphische Kommission 2002; Menning *et al.* 2006) (Fig. 9.4). The Central European divisions are the ones which will be used in this contribution. In the literature, however, local substage names are still widely used.

Biostratigraphy

During the Carboniferous period, sedimentation was coeval with the Variscan Orogeny and the strong tectonic control on Carboniferous sedimentation resulted in mobile, unstable basins, with sedimentary successions that show marked temporal and spatial variations both in lithofacies and thickness. This has promoted a proliferation of stratigraphic units of only local importance, making regional correlations difficult. In addition, the range of Carboniferous depositional environments, from continental to deep marine, has caused considerable correlation problems through the Central European region and beyond, since many index fossils are facies controlled (Guion *et al.* 2002). Successful correlation requires a multidisciplinary approach combining biostratigraphy, radiometric dating, volcanic horizon correlation, facies analysis, well-log interpretation, seismic stratigraphy and geochemistry (Guion *et al.* 2002).

The Lower Carboniferous was originally subdivided using corals and brachiopods for platform carbonates, and goniatites and bivalves for basinal successions (e.g. Brenckle & Lane 1987). More recently, a cyclostratigraphic approach has been used involving dated cycles based on foraminifera, conodonts and macrofauna. In some areas detailed biozonation syntheses have been published (e.g. Hance *et al.* 2001).

The Silesian succession is largely subdivided on the basis of fossiliferous marine bands containing goniatites, conodonts, ostracods, foraminifera and palynomorphs. Some 60 goniatite-bearing marine bands occur within the Namurian alone, with fewer in the Westphalian where marine incursions were rarer. In the absence of marine sediments, non-marine bivalve assemblages and plants (macroflora) may be useful. Additionally, spores are invaluable in Upper Carboniferous correlation, especially in the subsurface (e.g. Wagner 1984; Higgs *et al.* 1988; Peppers 1996). and the classical macro- and microflora biostratigraphic subdivisions have an increasingly eco-biostratigraphical character in the latest Carboniferous. Additionally, biozonations based on vertebrates (Zafic 2000); amphibians (Werneburg 1996; Werneburg & Schneider 2006), insects (Hmich *et al.* 2003; Schneider *et al.* 2004; Schneider & Werneburg 2006), and freshwater xenacanth shark teeth (Schneider & Zajíc 1994) are increasingly used to provide local to interregional correlations of continental and mixed continental/marine successions.

The detailed and high-resolution conodont–ammonoid zonation for the Carboniferous, with 35–40 zones, is in marked contrast to the 21 radiometric dates available (Menning *et al.* 2000; Davydov *et al.* 2004). Given that the duration of the Carboniferous is close to 60 Ma, the average resolution of a zone is, therefore, equivalent to 1.5–1.0 Ma whereas the radiometric resolution is much less.

Cyclicity

One of the most prominant aspects of the geology of the European Carboniferous is the pronounced cyclicity present in many siliciclastic and carbonate successions (Besley 1990) (Fig. 9.5). The origin and controls of this cyclicity have been much debated since its initial description in the 1930s. In the 1970s, Ramsbottom proposed that major breaks in sedimentation were associated with sea-level lowstands. He termed the successions between these breaks 'mesothems' which could be subdivided into a number of smaller-scale cycles, termed 'cyclothems' (Ramsbottom 1979). Ramsbottom's hypothesis marked the first attempt to formulate a consistent sequence stratigraphy for any part of the NW European Carboniferous and it has been very influential in terms of providing a framework for the worldwide correlation of the Carboniferous (Ross & Ross 1985), for regional stratigraphic subdivision (Ramsbottom *et al.* 1978) and as a framework for sedimentary facies models (Besley 1990).

Cyclothemic sedimentation patterns are characteristic of late Visean shallow-marine carbonate sequences. In terms of depositional style, these are comparable with successions produced during the late Pleistocene ice-house period (Wright & Vanstone 2001). Many late Visean shelf carbonate units comprise successive shallowing-upward cycles whose deposition may be related to eustatic sea-level variations estimated to be of the order of tens of metres (Süss *et al.* 2000, 2001, 2002). Such high-frequency, low-amplitude eustasy reflects the repeated growth and melting of small continental icecaps during the early stages of accretion of the major Gondwana ice sheet. Climate modelling suggests that the eustatic effects of glaciation were most important during the Namurian and early Westphalian. This is in agreement with the suggestion of Wright & Vanstone (2001) that glacio-eustasy commenced abruptly at *c.* 330 Ma (early Asbian), most likely with a 100 ka eccentricity signal with a progressive increase in amplitude of the 100 ka oscillation through the late Visean. The cyclic nature of Late Carboniferous sedimentary sequences is related to third-order glacio-eustatic sea-level fluctuations with amplitudes of *c.* 100–200 m (Süss *et al.* 2002). These can be related to ice sheet development in the southern polar regions of Gondwana (Hambrey & Harland 1981; Caputo & Crowell 1985).

Ramsbottom's ideas on cyclicity have prompted attempts to construct a global Carboniferous correlation framework based on sea-level changes (Guion *et al.* 2002). Early attempts were not initially accepted until work by Exxon geologists in the early 1980s introducted the concept of unconformity-bounded sequences that led to the application of sequence stratigraphy to Carboniferous sedimentation. The influence of glacio-eustatic fluctuations in sea level on facies architecture in Carboniferous successions is now widely accepted (e.g. Wright 1987; Walkden 1987; Wilson *et al.* 1990; Vanstone 1996). This knowledge has enabled sequence stratigraphy to be applied to successions within the Central European Carboniferous record, for example the recognition of nine third-order sequences in the Dinantian of Belgium and northern France (Hance *et al.* 2001, 2002). In Belgium, marker horizons such as marine bands and volcanic ash layers provide the framework for a subdivision of the Coal Measures Group, based on the cyclothem concept (Renier 1930; Van Leckwijck & Fiege 1969) as well as facilitating isochronous correlations between different basins (Van Leckwijck 1948; Delcambre 1987, 1996; Van Skenwinkel 1990). In the Ruhr Basin, eleven fourth-order sequences consisting of 49 parasequences and high-frequency sequences have been defined (Süss *et al.* 2007).

Magnetostratigraphy

The most significant feature of the Carboniferous magnetic field is a long late Carboniferous interval of reverse polarity. This Late Carboniferous–Early Permian Kiaman Superchron is the longest known period of predominantly reversed polarity (Palmer *et al.* 1985; Davydov *et al.* 2004; Weyer & Menning 2006).

Tectonic setting (T.M. & K.Ko.)

Introduction and tectonic zones

The European Variscan fold-and-thrust belt can be traced on the basis of geological and geophysical data as a broad arc extending from southern Ireland and the adjacent offshore areas across Central Europe into Poland. The reconstructed geometry of the Variscan belt can be subdivided into a number of distinct geotectonic zones that show a general continuity around the belt (Figs 9.2 & 9.3). Many of these zones are separated by steeply-dipping faults or shear zones, which record an oblique collisional event across the orogen (see also Kroner *et al.* 2008). A number of these boundaries are also considered to represent suture zones (see below), the sites of Palaeozoic oceans that separated fragments of continental crust in the form of microplates or smaller terranes. Variscan suture zones are marked by the occurrence of ophiolite assemblages and calc-alkaline continental arc volcanics of Devonian or Carboniferous age. Older ophiolite assemblages of Cambrian–Silurian age are preserved along some boundaries, and relate to the initial stages of continental fragmentation that occurred along the northern margin of Gondwana during the Early Palaeozoic. The high-grade core of the Variscan orogen runs through central and NW Spain, France, Germany and the Bohemian area of the Czech Republic (Figs 9.2 & 9.3).

Basin evolution and form within the evolving Variscan Orogen was determined by a variety of factors, in particular, the position relative to the orogenic front. Immediately adjacent to, and to the north of, the Variscan Deformation Front, narrow foreland basins, termed the external basins (Variscan externides), are interpreted to have formed mainly as a result of subsidence related to thrust loading of the crust. However, tectonic activity also played an important role, particularly at the onset of orogenic activity as a result of tectonic-loading-induced extension, and in the final phases of orogenic activity where compression and transpression occurred. The related internal basins (Variscan internides) comprise those basins formed within the orogenic belt, which were largely structurally controlled.

Variscan Europe was traditionally subdivided into four distinct zones with different sedimentary/magmatic and tectonomagmatic histories (Fig. 9.3). From north to south these are the Westphalian (originally the Subvarisches Saumsenke, i.e. Variscan foredeep), Rhenohercynian, Saxothuringian and Moldanubian zones which together comprise the western Variscan foldbelt between the rivers Maas and Elbe (Kossmat 1927). Subsequent modification of Kossmat's original scheme has included the identification of the Mid-German Crystalline High (MGCH) by Scholtz (1930), who distinguished it as the 'Spessart Schwelle', a zone of crystalline rocks located between two zones of Palaeozoic sedimentation, i.e. between Kossmat's Rhenohercynian and Saxothuringian zones. More recently, Zimmerle (2000) provided an overview of the history and evolution of the MGCH. The extension of Kossmat's divisions to the Elbe Lineament was described by Möbus (1966) and Brause (1979), while equivalents of Kossmat's zones east of the Elbe Lineament were identified by Dvorak & Paproth (1969). Kossmat's original classification has also been extended to the West European Variscides (Ellenberger & Tamain 1980). The South Portuguese Zone is presumed to be the equivalent of the Rhenohercynian Zone, while the Ossa-Morena Zone has been compared to the MGCH. The Northern and Central Armorican zones are counterparts of the Saxothuringian Zone while the West Asturian-Leonaise Zone is also, in parts, comparable to the latter. The prolongation of the Moldanubian Zone to the west can be found in the south Armorican Zone and in most parts of the Galician-Castilian Zone (Schöneberg & Neugebauer 1994; Dallmeyer & Martinez Garcia 1990).

Using Kossmat's classification as a base and taking modern plate tectonic theory into account, all of these zones (i.e. the original four of Kossmat, together with the MGCH and the Northern Phyllite Zone, see below) can be defined as palaeogeographically coherent units that broadly retained their material integrity throughout the Variscan orogenic event (Franke 1989; Franke *et al.* 1995*a*; Oncken 1997; Plesch & Oncken 1999). Franke (2000) provides a detailed overview of the main tectonostratigraphic units of the Variscan belt in Central Europe, including their palaeogeographic affinities and plate kinematic evolution (see Dallmeyer *et al.* 1995). Modern tectonic concepts agree in a bilateral symmetry with south-directed subduction on the northern flank (i.e. Rhenohercynian, Saxothuringian) and north-directed subduction on the southern flank (i.e. Moldanubian)

The **Rhenohercynian Zone** (RHZ) is exposed from Cornwall through to the Moravo-Silesian Belt and possibly into southern Portugal (for details see Franke 2000) (Figs 9.2 & 9.3). All segments of the RHZ have similar sedimentary records, which are dominated by Devonian clastic shelf sediments derived from Caledonian sources to the north and deposited along the southern, passive margin of Laurussia. The Giessen-Werra-Harz nappes also contain Devonian MORB-type metabasalts and pelagic sediments (Franke 2000). Sedimentary cover was shortened by *c.* 50% (equivalent to 180 km) (Oncken *et al.* 1999) during final collision between 330 Ma and 300 Ma.

The RHZ records two convergent episodes, the first being the closure of the Rheic Ocean between Avalonia and Armorica (or more correctly, the Armorican Terrane Assemblage (ATA); (for details see Tait *et al.* 1996; Tait 1999, Fernández-Suárez *et al.* 2002; for an alternative view see Robardet 2003). Following a period of extension in Early to Mid-Devonian times, a tremendous amount of siliciclastic material from the reworked Caledonian orogen in northern Europe (i.e. Old Red Continent) was transported into the RHZ area forming the Rhenohercynian Basin. This is excellently exposed in its central part along the Mid-Rhine Valley (Stets & Schäfer 2002). This basin was subsequently inverted and incorporated into the present-day Rheinisches Schiefergebirge (Meyer & Stets 1975, 1994, 1996). The second episode was the closure of the Rhenohercynian Ocean, which occurred in late Devonian–early Carboniferous times (Franke 2000). Incipient collision was heralded at the base Tournaisian of the SE Hörre-Gommern allochthon where northerly derived sandstone turbidites with Caledonian detritus are intercalated with turbidites derived from the MGCH (Ricken *et al.* 2000). The final closure of the Rhenohercynian Ocean is indicated by the presence of turbidites on the foreland, as documented in the Late Tournaisian succession of the NW Harz Mountains (Ribbert 1998).

The RHZ is presently interpreted as a foreland fold-and-thrust belt which developed from a Devonian and lower Carboniferous passive margin (Oncken & Weber 1995) during south-directed subduction, and partial underthrusting beneath the Saxothuringian Zone to the SE (Franke & Oncken 1990; Oncken 1997). Accretion and oblique collision from Visean to early Namurian times resulted in metamorphism (342–332 Ma) with amphibolite-facies peak conditions (Okrusch 1995; Todt *et al.* 1995; Reischmann & Anthes 1996).

Southward subduction and closure of the Rhenohercynian Ocean resulted in the formation of a late Devonian–mid-Carboniferous magmatic arc on the northern margin of the Armorican microplate (Franke 1989; Reischmann & Anthes 1996; Anthes & Reischmann 2001; Zeh & Brätz 2004). This composite unit comprises older basement, arc intrusions and material accreted from the Rhenohercynian plate (Oncken 1997) and has been referred to as the Mid-German Crystalline High, a 75–100 km wide unit that can be traced for a distance of *c.* 1000 km (see below).

The **Northern Phyllite Zone** (NPZ) is a narrow belt of rocks situated between the very low-grade parts of the Rhenish Massif and the Harz Mountains to the NW and the MGCH to the SE (Fig. 9.3). The NPZ contains a very complex association of protoliths (Anderle 1987; Anderle *et al.* 1995), including metamorphosed equivalents of the Rhenohercynian shelf succession (south Hunsrück), Late Silurian–Early Devonian metavolcanic, arc-related rocks (south Taunus), as well as Ordovician metapelites whose microflora is similar to ones from Armorica or Gondwana (for details see Franke 2000). In general, westerly parts of the NPZ contain rocks derived from the shelf of the Rhenohercynian Ocean while eastern parts contain portions of Armorica. Thus, the interpretation of the NPZ is that it contains a segment of the 'Rheic' Suture (Cocks & Fortey 1982) between Avalonia and the ATA. Evidence of this suture is contained within arc-derived metavolcanic rocks in the south Taunus area which record the consumption of the ocean, while greywackes and metabasalts in the SE part of the NPZ probably represent the suture zone (Franke 2000).

The **Mid German Crystalline High** (= Mid-German Crystalline Zone) is a complex zone located to the south of the RHZ, whose structure, metamorphic grade and composition change considerably along strike (Flöttmann & Onken 1992; Okrusch & Weber 1996; Oncken 1997; Franke 2000). The marked dischroneity is evidenced by the fact that while subsidence and sedimentation were occurring in the Saar region (Germany) in a forearc basin formed by subduction erosion (Oncken 1997), in the Odenwald and Palatinate areas (Germany) metamorphism and plutonism were active (see Krohe & Willner 1995; Altherr *et al.* 1999). Amphibolite-facies metamorphism has also been recorded (e.g. Spessart, Ruhla) as has the presence of high-pressure eclogites (Will & Schmädicke 2001) Protolith ages and palaeogeographic derivation of the metamorphic rocks in the MGCH are largely unknown (Franke 2000). As in the NPZ, the boundary faults (which probably represent late-Variscan strike-slip faults) trend ENE, while the internal subdivision of the zone is broadly NE–SW.

The MGCH represents the SE active margin of the Rhenohercynian Zone. Firstly, it formed the source area for Early and

Mid-Devonian shallow-marine siliciclastics which were supplied to the high-energy depositional environments along the southern margin of the Rhenohercynian Basin (Siegennian Taunus Quartzite in the Taunus and Hunsrück areas; Stets & Schäfer 2002). Secondly, Frasnian–Namurian deep-marine sediments were deposited within the Rhenohercynian Ocean (Fig. 9.3). The region is characterized by arc-related plutonism which was of particular importance in mid-Carboniferous times (see also Anthes 1998; Anthes & Reischmann 1996). Rocks derived from Silurian–early Devonian arc-related magmatism are also present.

The **Saxothuringian Zone** (STZ) comprises the northern parts of the Vosges and Black Forest areas, as well as the Thuringian-Saxonian Schiefergebirge (i.e. Thüringerwald), Fichtelgebirge and Erzgebirge areas. The parautochthon consists of Cadomian basement and a Cambro-Ordovician rift succession overlain by late Ordovician to Famennian pelagic sediments and Famennian to Visean deep-marine sediments derived from the active Bohemian margin to the SE (Fig. 9.3) (Franke 1984*a*, *b*; Franke & Stein 2002; Linnemann *et al.* 2002; Linnemann 2004). The sediments of the overlying allochthon (Münchberg, Wildenfels, Frankenberg) include deeper-water equivalents of the Ordovician–Devonian of the parautochthon as well as metamorphosed early Ordovician MORB-type mafic rocks (Franke 2000). The deepest structural level of the STZ is present in the Saxonian Granulite and Erzgebirge areas where rocks, probably derived from the orogenic root and later intruded, are exposed.

Both the sediments and the geochemistry of the magmatic rocks in the STZ indicate that early Palaeozoic crustal extension occurred. However, some authors suggest that the region was an active margin setting (e.g. Oliver *et al.* 1993; Kröner & Hegner 1998). Late Devonian–early Carboniferous metamorphic rocks have preserved mineral ages of *c.* 480 Ma, which have been interpreted as evidence for an early Palaeozoic convergent episode, although these ages may also be explained as products of magmatism and metamorphism in the lower crust of an early Palaeozoic rift (Franke 2000). Indeed, Cambro-Ordovician rift metamorphism has been documented in the region (e.g. Kröner *et al.* 2000).

Collision occurred in late Devonian times as indicated by the deposition of Famennian turbidites derived from the Teplá-Barrandian margin (see below) on the continental Saxothuringian foreland basin (Franke & Engel 1986; Schäfer *et al.* 1997). The suture south of the STZ has been inferred from allochthonous Lower Palaeozoic MORB-type rocks which are now present between the STZ and the Teplá-Barrandian of Bohemia (Linnemann *et al.* 2002), and from palaeomagnetic data showing that the STZ and the Bohemian Massif rotated independently during Silurian and Devonian times (Tait *et al.* 2002).

The **Moldanubian Zone** (MZ) comprises two parts, the northernmost Teplá-Barrandian (TB) and the more southerly Moldanubian Zone *sensu stricto* (Fig. 9.3). The TB and the high-grade metamorphic rocks adjacent to the SE were termed the Moldanubian Region by Kossmat (1927). The MZ (including the Teplá-Barrandian area) comprises most of the Vosges/Black Forest area and the Bohemian Massif.

The TB, which can be considered to be part of the ATA, is the best-preserved fragment of the Cadomian Orogen in Central Europe (Zulauf 1997; Chlúpác 1993) (see Chapters 2 & 3). Its NW and SE margins comprise eclogite facies and mantle rocks and probably represent sutures which were later overprinted by normal faulting (Zulauf 1997; Bues & Zulauf 2000). The Moldanubian Zone *sensu stricto* (MZss) is a complex association of rocks whose protolith ages are mostly unknown. The structure of the MZss is dominated by the large-scale, thrust-related inversion of metamorphic facies which has emplaced high-pressure rocks (Gföhl Assemblage) over the parautochthonous Drosendorf Assemblage (Petrakakis 1997). In the MZ, the onset of collision is less well constrained than in other Variscan zones, varying between the areas of the Bohemian Massif and the southern Black Forest and Vosges (Late Visean) (Franke 2000).

A characteristic feature of the NE termination of the European Variscides are changes in the orientation of the structural grain from NE–SW in Germany to NW–SE and then NNE–SSW in the southwesternmost part of Poland (Mazur *et al.* 2006). This oroclinal bending is a major feature in the eastern extremity of the Variscan foldbelt and may be related to indentor tectonics (Tait *et al.* 1997). This is partially explained by the fact that this area was the approximate site of the first collision between Eurasia and Gondwana, which subsequently continued their convergence mostly through the rotation of both continents, pivoting about a pole located in the centre of the future Pangaea (Perini *et al.* 2004). Geological evidence for this, however, is not so clear (for details see Tait *et al.* 2004).

Fault systems and suture zones

The Variscan Orogen was characterized by a particularly long period of intracontinental deformation, associated with the collision of Laurussia and Gondwana (which rotated clockwise). This rotation added a SW–NE stress direction to the northern one, resulting in the convoluted shape of the Variscan Orogen strike-slip zones (Franke *et al.* 1995*a*). As noted above, the general strike along the northern margin of the Variscan Front varies across western and Central Europe (e.g. Bachtadse & Van der Voo 1986; Bachtadse 1990), and these changes may reflect large-scale thrust rotations (e.g. in the area of the Rhenish Massif and the Ardennes), and/or pervasive strike-slip tectonics in the more internal parts of the orogen (e.g. Harz Mountains, Franconian Forest, Massif Central) (Badham 1982; Tait *et al.* 1997). Matte (1991) has noted that during the Variscan Orogeny the central part of the Variscan belt was the site of particularly complex deformation involving subduction, collision, indentation tectonics, obduction and block rotation (Matte 1991), leading to the development, in this region in particular, of a complex regional geometry.

Along the northern margin of Gondwana, rifting resumed during the Early Carboniferous reflecting an intraplate stress reorganization following the docking of the various rifted terranes (i.e. mainly ATA, partly equivalent to the composite Hun Terrane *sensu* Stampfli *et al.* 2001) to the Laurussian margin. From the late Early Carboniferous onwards, Laurussia was transected by the Arctic–North Atlantic Rift System which was partially superimposed on the Caledonian suture zone (Ziegler 1990). Indeed, Variscan exploitation of older Caledonian structures has been reported from other areas (e.g. offshore Ireland; for details see Shannon 1991; McCann 1996). It is, therefore, highly likely that within Central Europe older structures, both Caledonian and early Variscan, were reactivated by later Variscan tectonic activity (Butler *et al.* 1997).

The post-collisional evolution of Europe (i.e. the latest Carboniferous–Early Permian) was characterized by the formation of a series of rift- and wrench-induced basins across Central Europe, together with significant magmatic events. This was related to the change in convergence of Gondwana and Laurussia from an oblique collisional zone to dextral translation. At the same time, the Variscan Orogen was transected by conjugate wrench faults, partly terminating in pull-apart basins (Ziegler 1990; Matte 1991).

Palaeomagnetism (J.T.)

By latest Devonian times, Laurentia and Baltica had finally amalgamated forming the Laurussian continent which was situated in equatorial latitudes. Throughout early and mid-Palaeozoic times, various Gondwana-derived continental fragments were gradually accreted to the southern margin of Baltica and Laurentia, as evidenced by both palaeomagnetic and faunal evidence. Palaeomagnetic data indicate that the southern margin of the Baltic shield was at 20°S in the earliest Carboniferous. Similarly, data from the Anglo-Brabant Massif, Moravia, and the Mid-German Crystalline High indicate palaeolatitudes of 20–24°S, demonstrating the amalgamation of these terranes with Baltica. The apparent polar wander (APW) path for Gondwana is slightly more complex. Late Devonian data are strongly in favour of a South Pole situated in central Africa, thus requiring oceanic separation between northern Africa and southern Laurasia. Extrapolation between the Late Devonian, the mid-Carboniferous and the early Late Carboniferous data from Gondwana indicate gradual closure of this ocean at this time.

Laurussia remained more or less stationary during the Devonian; the equator crossed Hudson Bay and southern Greenland. APW paths for Gondwana indicate that the continent followed a complex drift pattern involving major latitudinal and probably also longitudinal displacements. Following collision with Laurussia there was joint clockwise rotation and northward drift during the Early Carboniferous. This northward movement ceased at the transition from the Early to Late Carboniferous, although clockwise rotation continued during the late Carboniferous–Early Permian (total of 54°).

The amount of crustal shortening during the Variscan Orogeny is difficult to ascertain. However, based purely on palaeomagnetic data, Jones *et al.* (1979) suggested that post-Devonian crustal shortening between the Armorican Massif and the northern Variscan foreland does not exceed the palaeomagnetic margin of error and thus cannot be more than 500 km. Moreover, the Carboniferous APW paths of Africa, Europe and North America are, at present, not sufficiently constrained, so there is no quantitative evaluation of the total amount of crustal shortening (Aïfa *et al.* 1990).

Palaeogeographic setting

The Variscan Orogeny of Europe is correlative to the final stages of the Appalachian orogenies along eastern North America, and marks the final amalgamation of Laurasia and Gondwana to form the supercontinent Pangaea (Fig. 9.1). The major oceans separating the various plates constituting Europe had closed by the Devonian, although final consolidation into the present-day configuration of Variscan Europe did not occur until some time later (Tait *et al.* 1997; von Raumer *et al.* 2003). This period was one of large-scale shearing and transpressional movement with the opening of small oceanic basins (Franke 1989; Matte 1991; Quesada 1991). Large-scale differential rotation and deformation occurred throughout the Variscan Belt (e.g. Tait *et al.* 1996).

Within Europe, basement terranes form a collage of blocks with both continental and oceanic affinities. Throughout the Palaeozoic, the northern (i.e. Baltica and Laurentia) and southern (Gondwana) continents interacted through a series of rifting and collisional events involving island arcs, rifted continental fragments, and the major continents. Various terranes of Gondwanan affinity drifted northward across the palaeo-oceans that existed between Gondwana and Laurussia, and collided with both Laurentia and Baltica beginning with the accretion of Avalonia during the Ordovician–Silurian. Although the existence of Pangaea is universally accepted, the identities and precise geometries of the final colliding blocks with southern Europe are still debated (e.g. Stampfli & Borel 2002; Muttoni *et al.* 2003).

As noted above, the Devonian to Early Carboniferous evolution of the Variscan orogenic system was governed by alternating tensional and compressional tectonic cycles, reflecting the development of an essentially intracontinental Pacific-type, back-arc system (Ziegler 1990). Four principal phases of deformation and exhumation may be recognized, each of which lasted between 20 to 39 Ma and was restricted in geographical extent. These phases probably resulted from the successive docking of continental lithospheric fragments along the southern margin of Laurussia. The first of these phases – the Ligerian (Late Silurian–Early Devonian) – was contemporaneous, but not cogenetic, with the Acadian phase of the Caledonian Orogeny (Cogné & Lefort 1985). The remaining three phases – Bretonian (Late Devonian–Early Carboniferous), Sudetian (Visean–early Namurian) and Asturian (Westphalian–Early Permian) – are assigned to the Variscan Orogeny (Warr 2000).

An increasing amount of palaeomagnetic data (e.g. Bachtadse *et al.* 1995; Tait *et al.* 1996) has resulted in a more complex view of the Variscan Orogeny, involving the microcontinents of Avalonia and Armorica (i.e. ATA). Avalonia comprises the autochthonous parts of the Rhenohercynian Zone, while Armorica consists of the northern and central Armorican zones in France, the Saxothuringian Zone and the Teplá-Barrandian (Franke *et al.* 1995). As deduced from palaeomagnetic data, the ocean separating these plates must have had a minimum width of *c.* 1000 km (Franke *et al.* 1995). The Variscan orogenic cycle commenced when the fragments comprising the ATA detached from Gondwana in Middle Ordovician times and drifted northwards. The Rheic Ocean was formed in middle/late Ordovician times between Avalonia/Baltica and the ATA (Armorica) (Cocks & Fortey 1982). Subsequent narrowing of this ocean occurred during the late Ordovician–earliest Devonian and was accompanied by the formation of an oceanic island arc between Avalonia and the ATA. The closure of the Rheic Ocean was completed by early Devonian times (Franke 1992; Franke *et al.* 1995). The Rheic Suture extends eastwards from SW England to the south of the Ardennes and to the south of the Rhenohercynian Zone (Franke *et al.* 1995; Cocks *et al.* 1997). There is, however, still some discussion about the precise position of the suture east of Dresden, and in the Czech Republic and southern Poland (Belka *et al.* 2002; Finger *et al.* 2002; for discussion concerning the precise location see McKerrow *et al.* 2002).

Extensional movements along the Rheic Suture began in late early to middle Devonian times leading to the formation of a broadly east–west trending oceanic or backarc basin of limited width. Basin formation led to the renewed separation of Avalonia from the ATA to the south. The resultant ocean, termed the Rhenohercynian Ocean (Oncken 1997; Oncken *et al.* 1999; Warr 2000), constituted a basin (i.e. Rhenohercynian Basin) which extended from southern England to an area east of the Harz Mountains (Germany) and was primarily filled with clastic sediments derived from the Caledonides located to the north (Walliser & Alberti 1983; Franke 2000; Oncken *et al.* 2000). Early Devonian volcanic activity (basalts with a tholeiitic affinity) confirms the extensional character of the Rhenohercynian Basin (Matte 1986*a*). With continued extension, oceanic crust was generated within the basin. The Rhenohercynian Ocean attained a maximum width of only *c.* 400 km and was, therefore, not wide enough to be verifiable by palaeomagnetic methods (Franke & Oncken 1990).

The onset of SE-directed subduction during the late Devonian–early Carboniferous led to the closure of the Rhenohercynian Basin. The occurrence of microdiamond-bearing ultrahigh-pressure rocks in the Erzgebirge region of Germany indicates that subduction of continental crust took place at depths in excess of 100 km (Massonne 2001). Franke (1992) has suggested that subduction of oceanic crust in the region was complete by the end of Devonian times, and thus the geodynamic evolution of the region from the Early Carboniferous onwards may be considered to be the collisional stage of the Variscan Orogeny (Henk 1997). Oncken (1998), however, has noted that subduction erosion occurred from Famennian through to Visean times, thus suggesting a slightly later Carboniferous collisional stage for the Variscan Orogeny. Subduction resulted in the evolution of a magmatic arc – the Mid-German Crystalline High – within the overriding Saxothuringian microplate (Oncken *et al.* 2000) which is considered to be the core of an orogenic wedge (Oncken 1997).

During the early stage of underthrusting of the Rhenohercynian continental crust beneath the Saxothuringian microplate (330–320 Ma) laterally heterogeneous accretion occurred leading to thickening of the crust and rapid uplift (Oncken 1997). From the Visean onwards the deformation front within the Rhenohercynian Zone propagated northwestwards until the end of Westphalian B times (312 Ma; Lorenz & Nicholls 1984). As a consequence of the collision most of the Saxothuringian Zone should be underlain by material from the Rhenohercynian Zone (Von Seckendorff *et al.* 2004*b*).

During the late Visean the Variscan orogenic system entered the Himalayan-type continent–continent collision stage (Fig. 9.6). Continued dextral-oblique convergence of Gondwana and Laurussia together with the progressive closure of the Proto-Atlantic and western Proto-Tethys oceans, was accompanied by the propagation of the collisional front into the Appalachian and central Mediterranean domains. At the same time, significant crustal shortening was achieved in the Variscan foldbelt as indicated by the occurrence of major nappe structures (in part involving basement) in the Moldanubian area, the southern Massif Central, and also the Variscan externides (Burg *et al.* 1984; Behr *et al.* 1982; Cazes *et al.* 1986; Behr & Heinrichs 1987). The northern Rhenohercynian Basin shelf was destabilized at the end of the Visean and subsided rapidly during the Namurian (e.g. Oncken *et al.* 2000, and references therein), in response to thrust loading and the development of the Variscan foreland basin (Fig. 9.6). This event coincided with the end of the main rifting phase of graben systems in Britain, the termination of lateral movements along the Arctic–North Atlantic megashear and the beginning of crustal extension in the Norwegian–Greenland Sea area (Ziegler 1990). Additionally, Bruguier *et al.* (2003) have suggested that extension was partly related to buoyancy control between the continental and oceanic part of the slab whereupon the continental crust detached from the oceanic crust and underplated Laurussia resulting in uplift and the formation of extensional volcanosedimentary basins (e.g. French Massif Central, Vosges, Black Forest, Alps) (Fig. 9.6)

The Lower Carboniferous successions of Central Europe are typically divided into two broad depositional areas, one of which comprises mainly carbonates (i.e. Kohlenkalk), the other mainly. clastics (i.e. Culm). The carbonate facies, predominantly fossil-rich shallow-marine limestones, developed along the southern margin of Laurussia and built up an extensive platform extending from Ireland via the Ardennes across into Poland. Deposition consisted predominantly of algal reefs, calcareous debris and dark bituminous limestones. Fossils contained within these sequences include bryozoans, corals, productids, goniatites and crinoids (e.g. Kasig 1980; Kasige & Wilder 1983; Amler 1987, 2004). The Kohlenkalk facies attained thicknesses of between 300 and 700 m. To the south of the Kohlenkalk facies was an interdigitated contact (marked by reef debris and calciturbidites) with the synorogenic clastic-dominated, shelf and deeper-marine Culm facies. Indeed, the boundary between the calcareous Kohlenkalk facies in the north and the clastic Culm facies to the south can only be traced with difficulty.

Variscan deformation continued into the Late Carboniferous, with widespread Variscan deformation consisting of thrust and wrench faulting, folding, post-tectonic granite emplacement and the accumulation of thick continental sediments in the developing foredeeps (Fig. 9.7). At this time, the marked differences in, for example, tectonosedimentary evolution, between the Variscan orogen and its foreland decreased. Lithologically, the Upper Carboniferous successions are relatively monotonous. Only the Namurian A still shows the orogen–foreland differentiation. During Late Carboniferous times, paralic coalfields extended across Laurussia from North America through Britain and on into the Rhenohercynian Zone of Germany and eastwards to Silesia and the Donets region (Ukraine). These provide the source for most of the gas found in the Rotliegend sandstones of the Southern Permian Basin . The succession in this region is also marked by numerous marine horizons and clearly most of the Westphalian deposition (largely non-marine) occurred close to sea level (Paproth 1991).

Following the Late Westphalian–Early Stephanian consolidation of the Variscan foldbelt, convergence between Gondwana and Laurussia apparently changed from an essentially north–south directed collision to an east–west orientation. Counterclockwise rotation of the southern African plate against the stable northern European craton resulted in wrenching along NW–SE trending faults (Ziegler 1990). Continental-scale dextral shears (e.g. Tornquist-Teisseyre Zone) were linked by secondary sinistral and dextral shear systems (Ziegler *et al.* 2006). The westward translation of Gondwana relative to Laurussia along a 3500 km long dextral megashear has been proposed, but recent work from northern Spain does not support the existence of such a major structure (Weil *et al.* 2001). It is, however, possible that a series of smaller shear structures were active.

The collision between Gondwana and Laurussia continued to develop until late in the Carboniferous/Early Permian, when the intercontinental collision began to affect the northwestern part of Africa (e.g. west African orogens have an age of Late Carboniferous to Early Permian; Lécorché *et al.* 1989). As a result of the extensive collision episode, a central Pangaean mountain range was formed extending from Mexico to Poland (Golanka & Ford 2000) and southwards into Morocco (Pique & Michard 1989). Late Carboniferous events were also marked in the Alps, Carpathians (Dallmeyer *et al.* 1996; Vozárová & Vozár 1988) and the Rhodope area (Yanev 1992).

Beginning in late Early Carboniferous times, compression within Central Europe was continuous for *c.* 45 Ma, with Variscan deformation advancing northward at a rate of *c.* 0.5 cm/year (Ahrendt *et al.* 1983). This is only slightly less than the value of 0.79 cm/year which characterized Alpine orogenic compression (Schmid *et al.* 1996). In western Germany, for example, a series of major north- and south-verging, basement-cored nappes were emplaced during Namurian and Westphalian times. The resultant thickening of continental crust by thrusting and stacking ultimately formed the Variscan Orogen (Henk 1997; Plesch & Oncken 1999). The widespread occurrence of low-pressure metamorphic rocks and late- to post-orogenic calc-alkaline intrusives in the internal parts of the Variscan Orogen

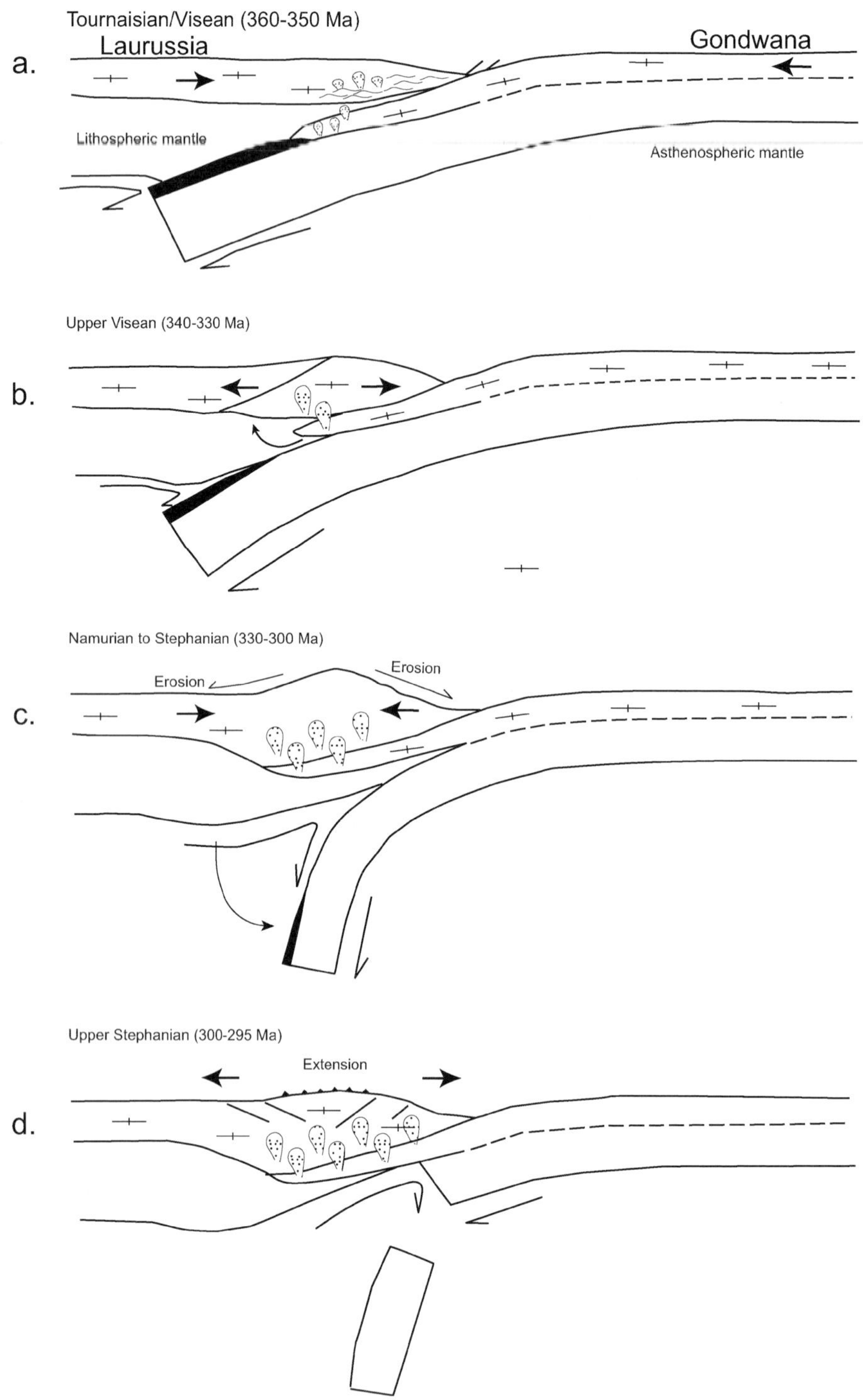

Fig. 9.6. Simplified 'accordian'-like model illustrating the proposed post-collision (340–290 Ma) evolution of the Variscan belt. Vertical scale enlarged (after Bruguier *et al.* 2003). (**a**) Tournaisian–Visean: this is a post-collisional situation. The subducted slab is under extension due to opposite internal forces resulting from buoyancy contrasts between the continental and the oceanic crust. (**b**) Upper Visean: continental crust detaches from oceanic crust and begins underplating beneath the Armorica/Laurussia block by upward rotation, while delamination of the Gondwana lithospheric mantle is initiated. Early lithospheric thinning and asthenospheric rise are responsible for rapid heating, resulting in the first magmatic pulse accompanied by high-pressure/high-temperature metamorphism in the lower crust and uplift and volcanic activity at the surface. Extension is localized along major crustal-scale faults. (**c**) Namurian–Stephanian: Delamination of the continental crust continues, resulting in both steepening of the sinking slab (as it loses its lighter part) and an increase in lithospheric thinning. The degree of melting increases, leading to magmatism in the deep and middle crust. Uplift results in the formation of topographic relief with accompanying erosion. (**d**) Upper Stephanian: the slab is under extension, and finally breaks off. The uplift of hot asthenospheric material triggers a second magmatic/volcanic flare-up. Heat advection allows the lower crust to flow leading to extension and reduced uplift.

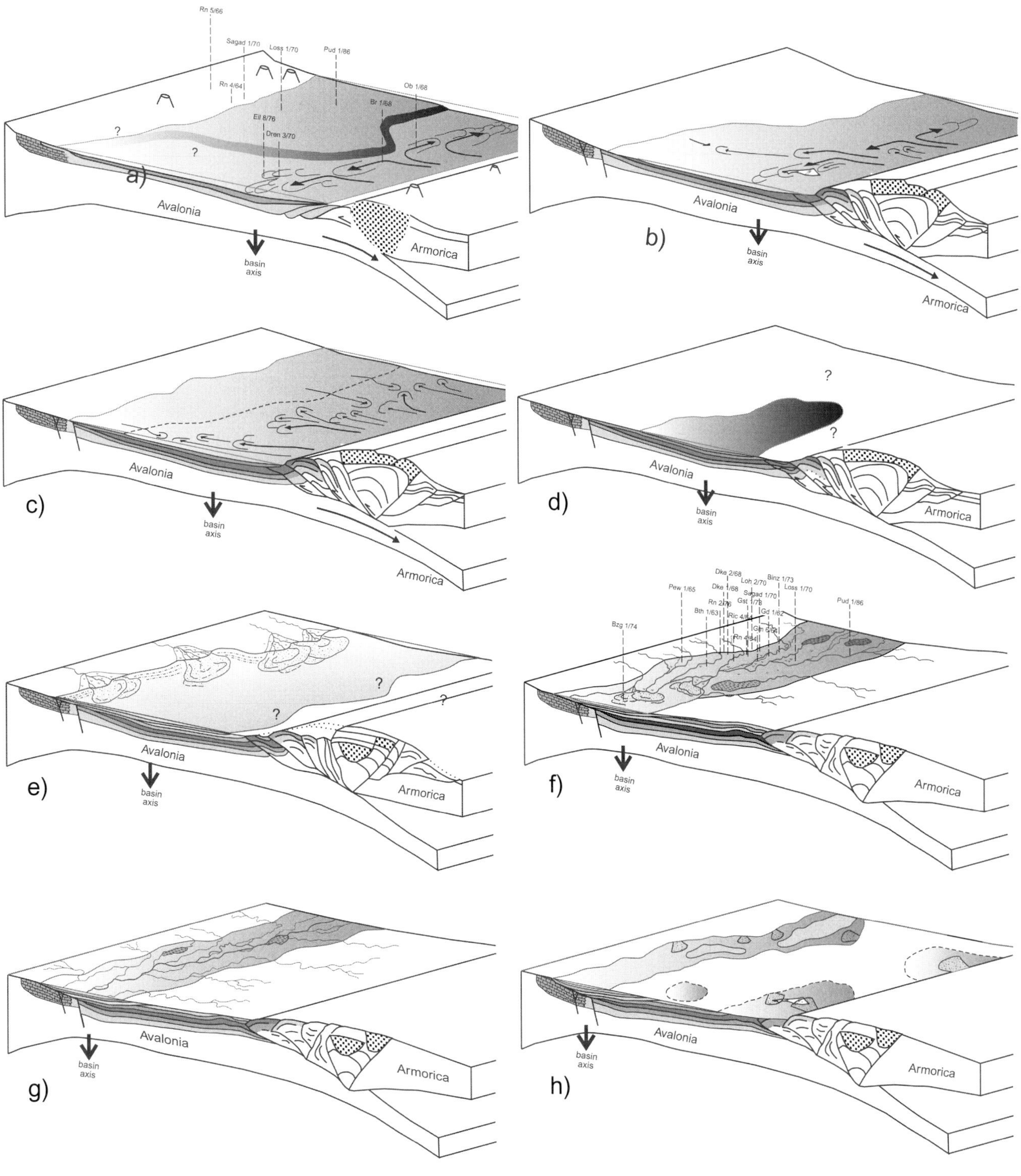

Fig. 9.7. Schematic block diagram showing the tectonosedimentary evolution of the Visean–Stephanian succession in the NE German Basin: (**a**) Upper Visean, (b) Namurian A, (**c**) Namurian B, (**d**) Namurian C, (**e**) Westphalian A, (**f**) Westphalian B, (**g**) Westphalian C/D, (**h**) Stephanian (after Kornpihl 2005).

suggest that significant amounts of crustal shortening ocurred, accompanied by crustal delamination, subduction and anatectic remobilization of lower crustal material and partial melting of upper mantle material (Ziegler 1984, 1986). Rough estimates of total crustal shortening yield a value of *c.* 150–200 km (Ziegler 1990), with an average of 14 km/Ma of shortening in the Rhenohercynian Zone (Plesch & Oncken 1999). This is less than half of the crustal shortening recorded from the Alpine Realm during the Tertiary which was estimated at *c.* 500 km (Dewey *et al.* 1989; Schmid *et al.* 1996). As noted above, orogenic activity peaked during the Namurian and early Westphalian. By latest Westphalian time, the Variscan foldbelt of western and Central Europe was consolidated and inactive (Ziegler 1990). However, there is evidence of localized tectonic activity, for example the presence of angular unconformities between the Westphalian and Permian deposits in the NE German Basin (McCann 1999). In Germany, post-orogenic uplift of the Rhenohercynian thrust belt, prior to deposition of the Permian succession, amounted to 2–3 km (along the northern margin), increasing southwards to 6 km (Littke *et al.* 2000) with values of up to 4 km in the area of the Saar-Nahe Basin (Schäfer & Korsch 1998).

Variscan magmatism

The late stages of the evolution of the Variscan continent–continent collisional belt in Carboniferous–Permian times were characterized by crustal extension, high heat flow, post-kinematic emplacement of granitoids and widespread late-orogenic volcanism (von Seckendorff *et al.* 2004*a*, *b*; see also Kroner *et al.* 2008; Timmermann 2008). This crustal extension also resulted in thinning of the ductile lower crust, and the formation of numerous late-orogenic intermontane basins in the brittle upper crust.

Variscan granitoids

The ascent of granitoid magmas and their emplacement in the middle and upper crust is a typical feature of orogenic events. These magmas may form from constrasting source regions and each source rock/source region is likely to produce fairly distinct granitoid types and granitoid suites (e.g. White & Chappell 1983; Pitcher 1983). The timing and localities of partial melt formation during an orogenic cycle depend on the crustal composition, as well as the thermal and tectonic state within the orogen (England & Thompson 1984; White & Chappell 1983). Thus, the granite inventory of an orogenic belt can be seen as an important reflection of the deep structures and processes that are commonly the least accessible and the worst exposed to direct observation (Finger *et al.* 1997).

The pre-collisional framework of the European Variscides comprised a ridge–trough system of continental microplates, intervening basins and partly oceanic rifts, which evolved in the early Palaeozoic along the fragmented margin of northern Gondwana (e.g. Pin 1990; Finger & Steyrer 1995). The similar tectonic style and the occurrence of remnants of oceanic crust and subduction-related volcanic rocks at both sides of the Varican orogenic belt, led Matte (1986*a*) to propose a bilateral suduction model which involved the subduction of two oceanic plates beneath a central Moldanubian complex. The two major pre-collisional subduction systems developed within the ridge–trough system and included a northern, south-dipping system (probably a tandem of two staggered subduction zones that led to the closure of the Saxothuringian and Rhenohercynian basins) and a southern, north-dipping system that led to the closure of a Massif Central–intra-Alpine ocean(s) (Finger *et al.* 1997). Accoring to Matte (1986*a*), both of these Variscan subduction systems acted only in an early orogenic stage (i.e. between Ordovician and Devonian times) and were superseded by collisional tectonics as Gondwana and Laurussia moved closer together. Other models suggest the closing of just one Variscan ocean (e.g. Ziegler 1986), or explain the Variscan tectonic events as predominantly ensialic (e.g. Neubauer 1988).

During the Variscan collision, the Precambrian basement of the ridges amalgamated with the sedimentary basin fill and some ocean floor of the early Palaeozoic rift basins. Thrusting, concomitant with crustal stacking, was most intense in the late Devonian and early Carboniferous and led to widespread metamorphism (especially in the internal Moldanubian Zone). Parts of the Moldanubian and Saxothuringian zones were additionally affected by (?subduction-zone related) late Devonian/early Carboniferous eclogite/granulite-facies metamorphism (O'Brian & Carswell 1993).

Nappe stacking ceased in the late Visean and gave way to transpressional and transtensional tectonics in response to the dextral wrenching of Gondwana relative to Laurussia (Schaltegger & Corfu 1995). During this late Visean phase, large parts of the Moldanubian Zone underwent intense high-T/low-P regional metamorphism and anatexis (Finger *et al.* 1997). In the late Carboniferous and Permian, the Variscan crust underwent post-orogenic uplift and extension.

The Variscan granitoid plutonism of central Europe is distinctly heterogeneous both in terms of age and the regional distribution of the various granite types (Fig. 9.8.; see below). Variscan granites represent a mixture from a number of sources such as (a) lithospheric mantle, (b) lower continental crust, (c) oceanic crust of an ancient detached slab, and (d) asthenospheric mantle (Schaltegger 1997) (Fig. 9.9). Some of these components may also be included within ancient polyphase, high-grade rocks of the lower crust and may reflect subduction-related geochemical characteristics that are inherited from pre-Carboniferous tectonic stages. When such a variety of source rocks became partially molten during a period of lithospheric tectonic activity, a highly diverse granitoid suite formed in one single pulse inheriting these geochemical characteristics. Compositional regional variations between plutons of the same magmatic pulse may, therefore, reflect inhomogeneities within the lower crustal source rocks, as well as variable influence from the lithospheric mantle (Schaltegger 1997).

A characteristic of the Variscan Orogen and its northern foreland is the large volume of syn- and post-tectonic granitoids of Carboniferous and early Permian age (Von Seckendorff *et al.* 2004*b*). An extensive suite of lamprophyre dykes, which may or may not be associated with the granites, occurs in the Black Forest (Wimmenauer 1973; Hegner *et al.* 1998), Vosges (Turpin *et al.* 1988), the Bohemian Massif (Nemec 1974), the Rhenohercynian Zone in SW England (Leat *et al.* 1987) and the Variscan basement of the Alps (Oberhänsli 1986) (Fig. 9.10). Lamprophyre dykes in the Saxothuringian Zone have ages of Visean-Namurian (334–323 Ma) and Stephanian–early Permian (297–295 Ma) (both $^{40}Ar/^{39}Ar$; von Seckendorff *et al.* 2004*b*).

Several unique features of the Variscan granitoids do not fit with a simple Himalayan-style continent–continent collsion model (Finger *et al.* 1997). Unlike the Himalayas, pre-collision I-type rocks, that might represent the former arc above a subduction zone, are not dominant. This suggests that subduction played a lesser role in the pre-collisional evolution of the Variscan crust, and supports the idea that any presumed (intra-Variscan) oceans (e.g. Rhenohercynian, Saxothuringian, Massif Central, Intra-Alpine) were small. Secondly, a major characteristic of the European Variscides is widespread post-collisional melting of the lower crust and the production of voluminous high-temperature S-type and high-K I-type granite magmas. Such conditions are special, requiring a large heat input, possibly through mantle-derived mafic magmas, into a lower crust apparently devoid of older, infertile granitic or granulitic material. Models that invoke delamination of a granulitic lower crust (orogenic root) could explain the high heat flow at this stage (Gayk & Kleinschrodt 2000) (Fig. 9.11). Finally, there are a

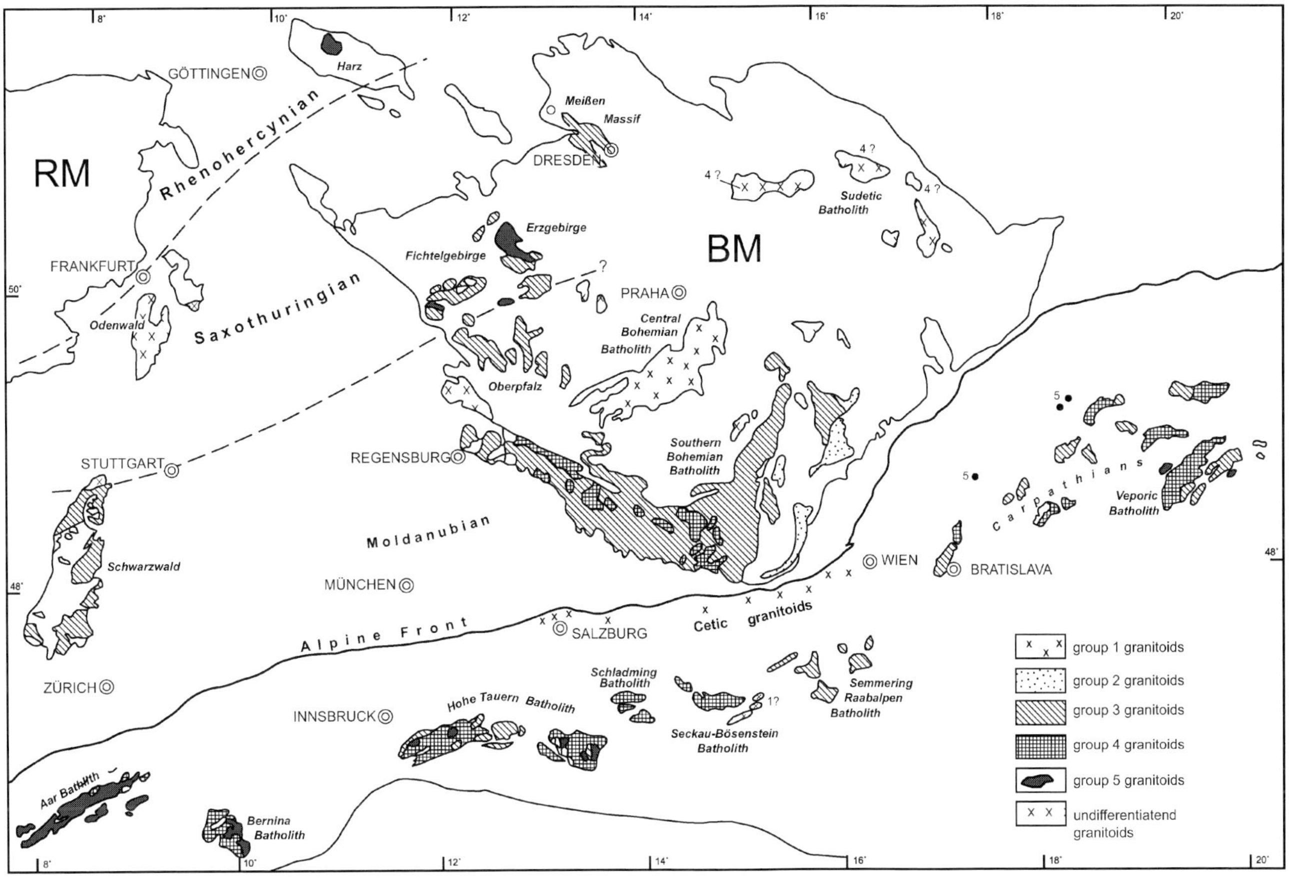

Fig. 9.8. Geological sketch map showing the position of Variscan granitoid plutons within Central Europe. BM, Bohemian Massif; RM, Rhenish Massif (based on Finger *et al.* 1997).

particularly large number of late-stage calc-alkaline, I-type plutons within Variscan Europe. According to Finger *et al.* (1997), this may point towards a protracted evolutionary history of transpressive and transtensive crustal dynamics with, or without, renewed subduction of ocean crust.

The spatial distribution of Variscan granitoids in central Europe is shown in Figure 9.8. Intrusion activity was concentrated along the central axis of the foldbelt (Moldanubian Zone) where granitoid rocks constitute more than 50% of the crystalline basement. The largest Moldanubian batholiths are the Southern Bohemian Batholith, the Central Bohemian Batholith, the Oberpfalz Batholith and the Black Forest batholithic area. Plutons are also present in the northern external zones of the orogen, for example, in the Saxothuringian (Odenwald, Fichtelgebirge/Erzgebirge, Sudetic Batholith) and Rhenohercynian (Harz) zones. Variscan-age plutons are also abundant (20–30%) in the intra-Alpine/Carpathian realm, an area which represents the former southern flank of the orogen.

According to Finger *et al.* (1997) five genetic groups can be recognized, where each magmatic pulse can be characterized by its own typical geochemical signature (Fig. 9.8).

1. Late Devonian–early Carboniferous 'Cordilleran' I-type granitoids (*c.* 370–340 Ma) comprising mainly tonalites and granodiorites which occur in association with diorites and gabbros. These rocks form plutonic massifs in the Saxothuringian Zone, Central Bohemia and the intra-Alpine Variscides and can be interpreted as volcanic arc granites related to the subduction of early Variscan oceans. The Odenwald granitoids are typical of this group and are of particular importance. They have a chemical signature indicative of a volcanic arc setting and include some gabbro bodies, diorites, tonalites and large masses of normal and high-K granodiorites (Henes-Klaiber 1989, 1992; Altherr *et al.* 1999). In a corresponding tectonic position, the granitoids and lower Carboniferous volcanics of the northern Vosges and the northern Black Forest are geochemically similar to the Odenwald rocks. This northern belt continues eastwards into the Bohemian Massif, while similar granitoids have also been recorded in the Central Bohemian Batholith.

An interesting intra-Alpine normal-K granitoid suite are the so-called 'Cetic granitoids', which are exotic boulders of Na-rich, hornblende-bearing tonalites to granodiorites with very primitive Sr-isotope characteristics embedded in the northern Alpine flysch. These rocks may represent pieces of the former 'Cetic Massif', now buried beneath the Alps (Frasl & Finger 1988). They also appear to be the only rocks which may have formed contemporaneously with early Variscan subduction processes and may, therefore, represent magmas derived from subduction, modified by assimilation–fractional crystallization-type processes during ascent and emplacement.

Within existing tectonic models (e.g. Franke 1989), the northern Variscan granitoid belt (e.g. Odenwald, northern Black Forest), may be related to the southward subduction of northern Variscan oceans. In the south of the Variscan foldbelt, the northward closure of the southern Variscan oceanic realm(s) may

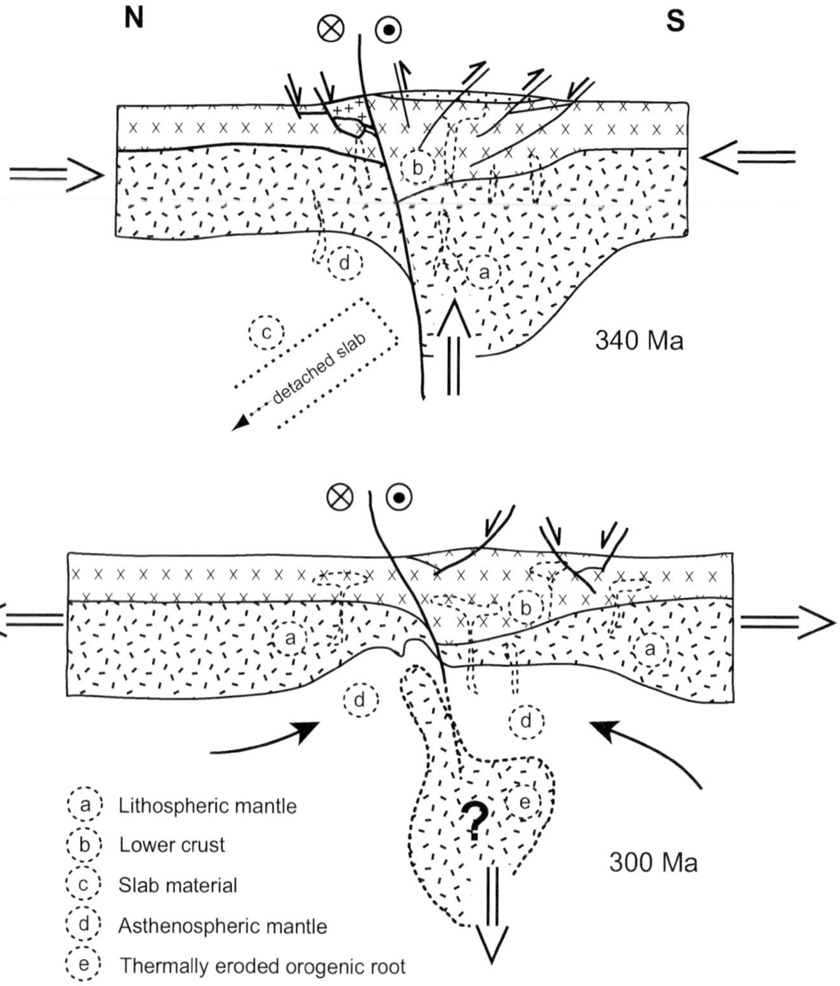

Fig. 9.9. Hypothetical cross-sections through the lithosphere in the central part of the Variscan belt at 340 and 300 Ma. Locations (a) to (e) denote possible source regions for late Variscan granitoid melts (after Rey *et al.* 1997).

have caused some early-Variscan subduction-related plutonism in the eastern and western Alps (Bonin *et al.* 1993). However, the precise locations of the appropriate subduction zones are still controversial (e.g. Frisch & Neubauer 1989; Schermaier *et al.* 1997).

2. Early Carboniferous, deformed S-type granite/migmatite associations (*c.* 340 Ma) which occur in the footwall of a major thrust in southern Bohemia (Gföhl Nappe) and appear to represent a phase of syncollisional crustal melting related to nappe stacking. Deformed granites are also present in the SE Bohemian Massif (e.g. Gföhl and Wolfshof gneisses).

3. Late Visean and early Namurian S-type and high-K, I-type granitoids (*c.* 340–310 Ma) which are mainly granitic in composition. These granitoids represent the largest volume of Variscan plutonism and are particularly abundant within the Moldanubian Zone, for example in the Southern Bohemian Batholith where they constitute >60% of the outcrop area. This zone experienced a high heat flow at this time, probably as a consequence of post-collisional extension and magmatic underplating.

The rocks within this group can be broadly divided into four subgroups. The first consists of moderately peraluminous, S-type granites with some muscovite or cordierite and sometimes garnet (e.g. Eisgarn granite, Southern Bohemian Batholith; Falkenberg and Flossenburg granites, NE Bavaria). Subgroup II comprise highly potassic, weakly peraluminous, sometimes metaluminous, K-feldspar megacrystic biotite granitoids (e.g. Weinsberg granite, Southern Bohemian Batholith; Kristallgranit I, Bavaria). Subgroup III have high K_2O_2 contents and mostly contain megacrystic K-feldspar (e.g. 'durbachite' plutons, Bohemia; Guiv metasyenite, Aar Massif). Subgroup IV are leucocratic, near-minimum peraluminous (S-type) granites (e.g. Altenberg granite, South Bohemian Batholith).

Crustal material appears to have been the major source of the magmas mainly from lower crust but some material may also have been derived from the middle crust.

4. Post-collisional, epizonal I-type granodiorites and tonalites (*c.* 310–290 Ma) are found throughout the Central European Variscides, although most of them occur in the Alps and Carpathians. These late I-type plutons may be related to renewed subduction along the southern foldbelt flank, and/or to extensional decompression melting near the crust–mantle boundary. Post-collisional mantle or slab melting may have occurred in connection with remnant subduction zones below the orogen undergoing thermal relaxation and dehydration.

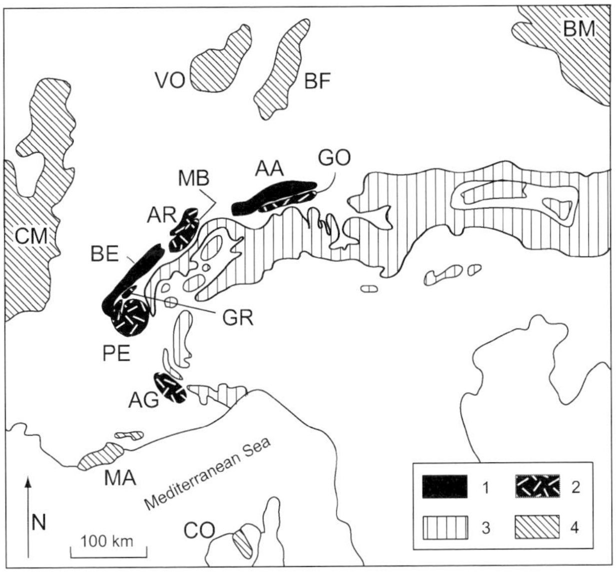

Fig. 9.10. Variscan basement in Central Europe. Key: 1 & 2, External Crystalline Massifs of the Alps (1, Belledonne, BE; 2, Aar, AA; Aiguilles-Rouges, AR; Argentera, AG; Gotthard, GO; Grandes Rousses, GR; Mont Blanc, MB; Pelvoux, PE); 3, Variscan basement of the Alpine nappes; 4, European Variscan framework (Black Forest, BR; Bohemian Massif, BM; Corsica, CO; Massif Central, CM; Maures-Tanneron, MA; Vosges, VO) (after Debon *et al.* 1998).

5. Late Carboniferous–Permian leucogranites (*c.* 300–250 Ma) many of which are similar to sub-alkaline A-type granites. These are mainly found in the Alps and Carpathians. Some of these late-stage leucogranites appear to be highly fractionated S-type granites (e.g. Fichtelgebirge, Erzgebirge), and may be related to late-stage, local crustal fusion events, related to thrust reactivation and further magmatic underplating as suggested by seismic data from the margins of the Rhenohercynian and Saxothuringian zones. However, most of the leucogranites (especially those in the Alps) are distinct high-Y magmas, which probably required a separate magma source. The precise sources of the granites, however, are not well constrained, and it would appear that they were derived from a variety of different sources, with some of the leucogranites representing highly fractionated S-type melts.

Carboniferous volcanism

The Carboniferous–Early Permian in NW Europe was a period of widespread basin formation that was associated in many areas with magmatic activity (see Timmermann 2008). Extensive volcanism, dyke and sill emplacement took place at various times across the deformation belt (Timmermann 2004). The settings in which this magmatic activity occurred are the northern foreland, broadly defined as the region to the north of the Variscan fold-and-thrust belt, and the internal parts of the Variscan orogen. The earliest phase of volcanic activity occurred in the Lower Carboniferous and was mainly restricted to the Lublin Basin (SE Poland, Ukraine) where extrusives included altered trachybasalt and tephrite lavas and tuffs. A second phase of extension in late Visean times followed a phase of major uplift and erosion and was associated with extrusive activity (Timmermann 2004).

Carboniferous volcanic activity was predominantly confined to the Stephanian–early Rotliegend period. At this time, a series of magmatic provinces developed within the collage of basement terranes of varying ages and lithospheric characteristics (including thicknesses) which had been brought together during the preceding Variscan Orogeny. The widespread nature of volcanic

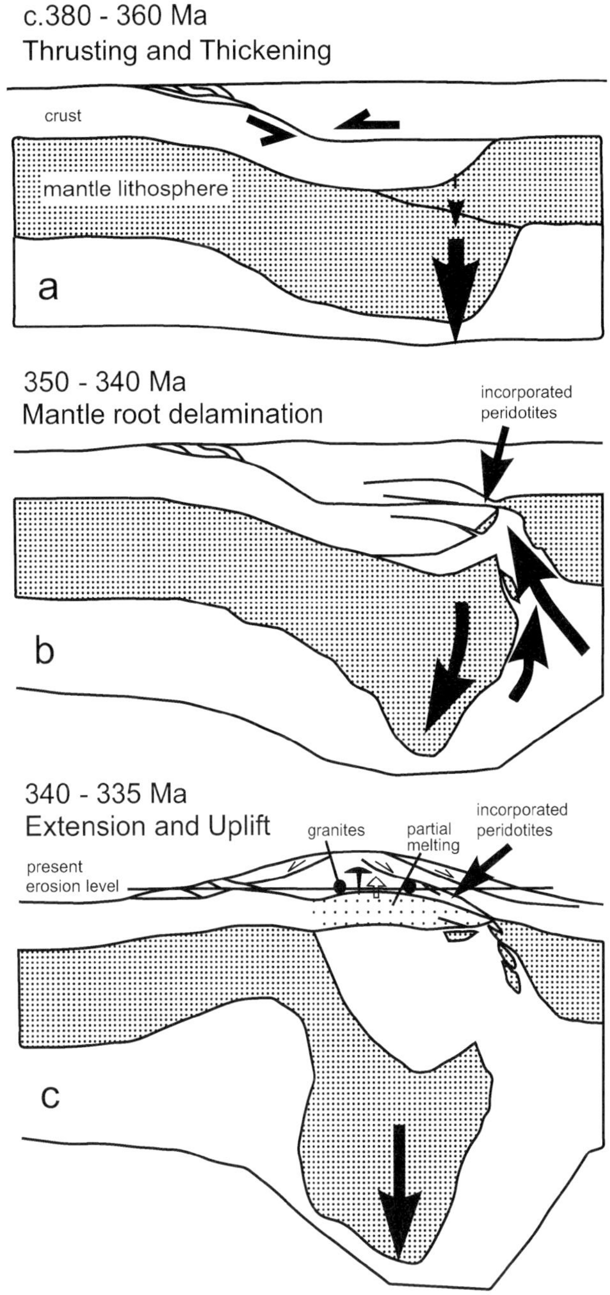

Fig. 9.11. Schematic sketch of a scenario to explain the observed features as outlined in the text. (**a**) Thickened cold lithosphere after thrusting and incorporation of high pressure units. (**b**) Delamination of the cold mantle root, affecting deeper parts of the crust. Peridotites were incorporated in marginal parts of the still thickened crust. (**c**) Rapid uplift and exhumation following delamination. Synchronous high heat flow by hot mantle at the base of the crust. Uplift and high heat flow cause significant partial melting and intrusion of granites in the middle to upper crustal levels (after Gayk & Kleinschrodt 2000).

activity at this time would suggest that there was, perhaps, a common tectonomagmatic event with a common causal mechanism (Timmermann 2004). Magmatism covered a wide range of rock types, both on a regional and local scale, ranging from highly alkaline to tholeiitic basalts to trachytes and rhyolites. The most intensive magmatism occurred in the Oslo Graben (*c.* 120 000 km^3; Neumann *et al.* 1992) and the NE German Basin (*c.* 48 000 km^3). In the former, the dominant mantle source component for the mildly alkali basalts to subalkaline magmatism is geochemically similar to the prevalent mantle (PREMA) component (Neumann *et al.* 2004). Rifting and magmatism in the area was, therefore, probably related to local decompression and thinning of highly asymmetric lithosphere in response to regional stretching north of the Variscan Front, implying that the PREMA source was located in the lithospheric mantle (Neumann *et al.* 2004). However, since PREMA sources are widely accepted to be plume-related, the possibility of a plume located beneath the area cannot be disregarded. Locally, there is also evidence of other sources. For example, in NE Germany, mantle lithosphere modified by subduction-related fluids from Variscan subduction systems has provided an important magma source (Neumann *et al.* 2004).

Although on a regional scale the Stephanian–early Permian magmatic event appears the most significant, locally it was predated by a Visean magmatic phase which may have had a different geodynamic setting (Wilson *et al.* 2004). This magmatism was coeval with the later stages of a 45 Ma period of Late Devonian–Early Carboniferous (375–330 Ma) tholeiitic flood basalt volcanism in the Maritimes Basin of eastern Canada which has been inferred to be plume-related (Dessureau *et al.* 2000). Dessereau *et al.* (2000) have suggested that plume-related magmatic activity may have gradually migrated eastwards during Late Devonian–Permian times, from the Maritimes Basin through Scotland to Norway and northern Germany. Plate tectonic reconstructions, however, suggest that there was rapid northward plate motion during the Carboniferous which, when combined with the latest-Carboniferous phase of wrench tectonics, makes it difficult to constrain potential hot-spot tracks. Consequently, such a hypothesis remains highly speculative (Wilson *et al.* 2004).

Within the internal Variscides volcanic activity has been recorded from several basins (e.g. Illfeld Basin, Thuringia Forest area, Saar-Nahe Basin, Alps, Massif Central). Carboniferous–Permian magmatism was unevenly distributed across the region. Despite this, peak magmatic activity was concentrated in a narrow timespan from *c.* 300 to 280 Ma (Neumann *et al.* 2004).

Sedimentary basins and magnetic activity

The following sections will outline the main zones of Carboniferous sedimentation and magmatic activity. The various zones correspond broadly to those initially defined above, with the exception of the region of the Sudetes, which because of the complexity of the region will be treated separately. The sedimentary basins which formed during the Carboniferous were intimately connected with the Variscan Orogen. Therefore, the basins will be described in three main groups: those occurring to the north of the Variscan Orogenic Front, i.e. to the north and along the northern margin of the Rhenohercynian Zone; those within the Rhenohercynian Zone; and those located further south, in the Saxothuringian and Moldanubian zones and beyond.

The North Variscan Foreland Basin

The most northerly basins discussed are the north German Basin and the Western Pomerania region of north Poland, which formed to the north of the Variscan Orogenic Front in northern Germany and Poland, respectively (Fig. 9.2). These basins, partly equivalent to the Westphalian of Kossmat (1927), contain a thick Carboniferous succession extending from Tournaisian through to Stephanian. The Lublin-Lvov Basin, situated on the margin of the East European Platform, is also described within this section. This basin developed far to the east of the Variscan Orogenic Front and its geotectonic position is similar to that of the abovementioned basins.

North German Basin (T.M. & K.Ko.)

The Devonian–Carboniferous transition, which was a global sea-level lowstand, is marked by a regional unconformity across northern Germany. Initial sedimentation in the north German Basin was of Tournaisian age and carbonate rich. These sediments were restricted to the Rügen Depression, a tectonically controlled NW–SE trending basin located on the island of Rügen (McCann 1999). By early Visean times carbonate deposition (Kohlenkalk) was established across the northern basin margin where the reefs formed part of an extensive carbonate platform which stretched in a broad band across western and Central Europe (Fig. 9.7). The middle Visean is largely missing (McCann 1999) and this fact, coupled with the presence of an intra-early Visean angular unconformity and some late Tournaisian tuffaceous intercalations, is indicative of localized tectonic activity (Hoffmann *et al.* 1975; Schmidt & Franke 1975). The late Visean saw the development of thick clastic units, comprising deeper-water sandstones and mudstones (Culm), south of Rügen. This may have been related to coeval crustal flexure due to the onset of the final phase of Variscan deformation, which led to a change in sedimentary style to the south. In the eastern part of the Rhenish Massif (= Rheinisches Schiefergebirge) area and in the Harz Mountains, both part of the Rhenohercynian Zone, late Visean and Namurian-age deep-marine successions are up to 3000 m thick (e.g. Eder *et al.* 1983).

In NE Germany the Visean–Namurian boundary is marked by a regional, partly erosional unconformity, probably underlined by the temporary eustatic sea-level lowstand at this time, and the transition from predominantly carbonate to clastic sedimentation (Figs, 9.8 & 9.12). In the Rügen area, the Namurian is absent, with an angular unconformity between the Visean and the Westphalian. This regional unconformity has been related to the uplift of a fault-bounded NW–SE trending anticlinal horst during late Visean–early Namurian times (Hoffmann *et al.* 1975). Contemporaneous block faulting in Pomerania, with uplifts of *c.* 1000 m, have also been noted (Racyznska 1987; Ziegler 1990). This phase of deformation is interpreted as reflecting the buildup of compressional stresses resulting from the collision of the orogenic wedge with the passive margin of the Rhenohercynian Basin. Published seismic data from Rügen (Piske *et al.* 1994; McCann 1999) clearly show the extent of post-Visean block faulting.

In outer shelf areas, the Visean–Namurian hiatus is generally absent, and Namurian deep-water shales transgress over the shallow-marine late Visean carbonates. This northward migration of the depocentres has been interpreted as reflecting the progradation of the orogenic front (Franke & Engel 1988) and may also be indicative of increased basin subsidence at the time. However, petrographical data suggests that Namurian-age sediments were

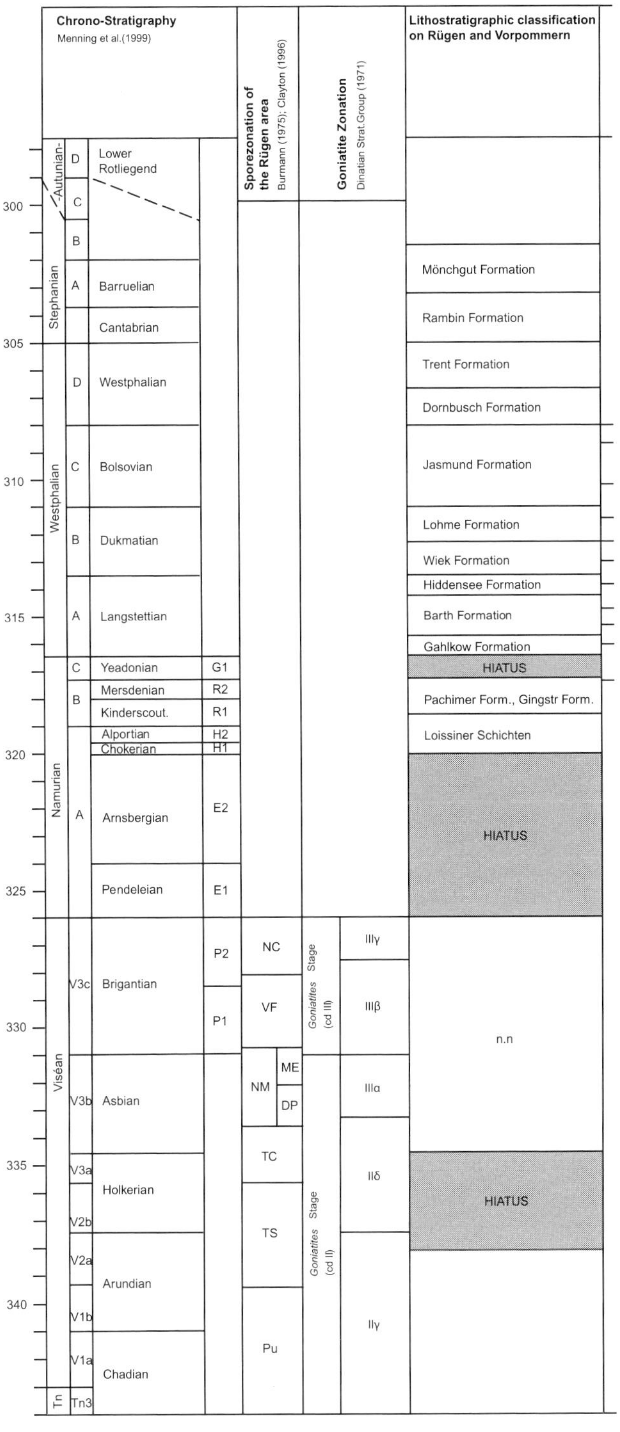

Fig. 9.12. Bio- and lithostratigraphy of the NE German Variscan foreland basin (after Kornpihl 2005).

prograding from north to south in the northern part of the basin (McCann 1999).

The early Namurian transgression across the northern shelf of the Rhenohercynian Basin was followed by a late Namurian regression which was coupled with an increase in clastic supply from the northern basin margin. The relatively high sedimentation rates led to the establishment of paralic conditions, initially along the southern basin margin, as clastic fans prograded northwards into the trough. With time, the northward-prograding fans coalesced with southward prograding deltas leading to the establishment of stable paralic conditions and the joining together of the Rhenish Massif and north German depositional systems. By early Westphalian times the Variscan foreland basin had shallowed to the extent that paralic conditions prevailed across the region (up to 3.5 km of coal measures were deposited). Westphalian sediments (predominantly continental with some evidence of brackish water conditions) are more widespread that those of the Namurian and overstep the Namurian basin margins in northern Germany and Pomerania (McCann 1999).

Short-term glacio-eustatic sea-level fluctuations in Westphalian times were responsible for cyclical, short-lived marine transgressions leading to the deposition of widespread, frequently carbonate-rich, marine bands. The transgressions originated from the south Portuguese and Cornwall segments of the Variscan foreland basin where deep-marine conditions were maintained in Westphalian A and B times (Ziegler 1989). There is, however, some evidence that transgressions entered from the Moscow Platform via the Baltic Depression (Basin) and possibly via the Dnepr-Donets Basin to the east (Allen & Dineley 1988). The presence of tuffs and rhyolites in the Westphalian C and D units, respectively, was possibly related to increased tectonic instability (McCann 1999).

Sedimentation in the Stephanian was mainly continental, with fluvial and lacustrine depositional environments predominating. Temporary marine incursions in Stephanian times, represented by calcareous intercalations with marine faunas, are related to transgressions. Localized volcanic activity is suggested by the presence of tuffs, particulary in the central part of the region (e.g. Berlin area). Tectonic instability increased with the onset of the early Permian and the development of a complex system of conjugate sinistral and dextral shears (Ziegler 1990; Benek *et al.* 1996).

Western Pomerania, NW Poland (S.S.)

In Western Pomerania Early Carboniferous deposition was associated with an evolving carbonate ramp and shelf system, intercalated with volcaniclastic units (Fig. 9.13). This was part of the vast carbonate platform which occupied the Rhenohercynian Shelf and extended via Pomerania in the NW through the Holy Cross Mountains into Upper Silesia and the Moravo-Silesian domain in the south (Zelichowski 1983a, *b*, 1995; see below). Intermittent backarc extension, punctuated by bimodal magmatic activity, resulted in modificaiton of this carbonate platform.

In comparison to the distinctly differentiated Lower Carboniferous sedimentary environments in southern Poland, the region of NW Poland would appear to be more unified (Zelichowski 1987; Lipiec & Matyja 1998). The Tournaisian and Visean strata, usually >1000 m thick, represent a continuous regressive cycle, commencing with offshore marly shales and sandy tuffaceous deposits. During the Tournaisian these were diachronously overlain by shallow-water oolitic and algal limestones, marls and sandstones with abundant admixtures of pyroclastic material (Fig. 9.13). Sedimentation terminated with the Upper Visean sandy/shaley units, deposited in deltaic, lagoonal and tidal flat environments. The Mid-Carboniferous break in sedimentation was relatively short and by Westphalian A times the deposition of a mainly sandy/shaley succession (with rare coal beds) had already begun (Boikowski & Dembowski 1972; Boikowski & Zelichowski 1980). These units are overlain by a mudstone-dominated succession with intercalated sandstones and conglomerates, the latter containing volcanic clasts. The precise

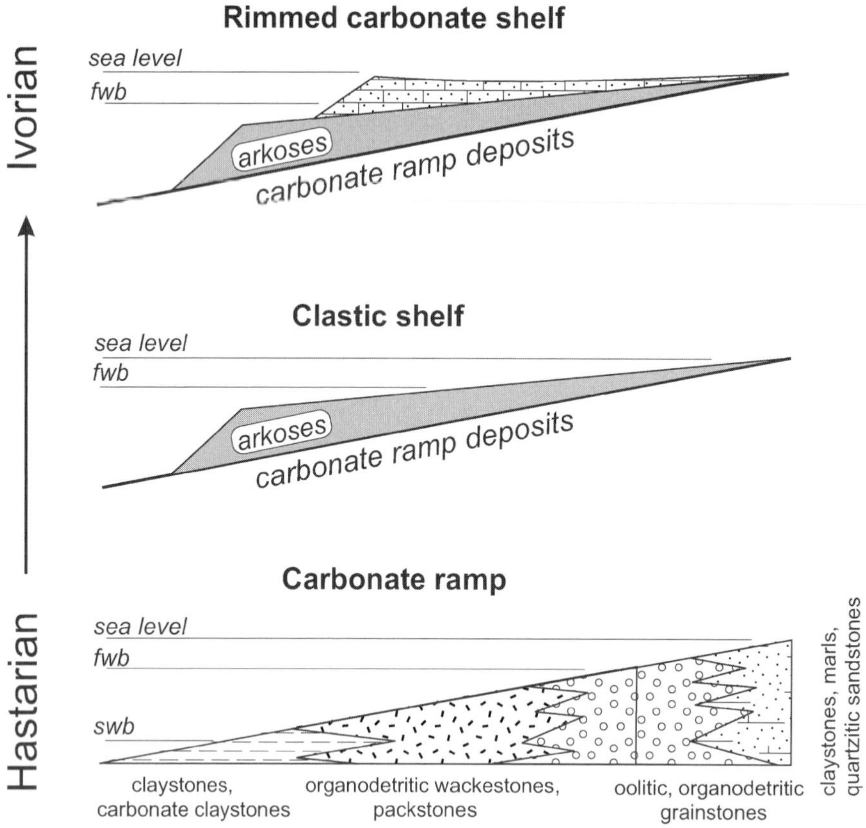

Fig. 9.13. Model of ramp to rimmed shelf evolution in the Tournaisian of the Western Pomerania (after Lipiec & Matyja 1998).

stratigraphic position (either Stephanian or Early Permian) is still under discussion. The west–east orientation of facies boundaries suggests that the Pomeranian succession extends far to the NE, although subsequent erosion has removed any evidence of this.

The Variscan foreland (Rhenohercynian Zone)

The Rhenohercynian Zone (RHZ) was defined by Kossmat (1927) based upon metamorphic grades, magmatic belts and sedimentary facies, and can be interpreted, within a plate tectonic context, in terms of pre-orogenic extensional basins and foreland basins (Franke 1995*a*) (Figs. 9.2 & 9.3). Towards the east, the RHZ can be traced from the Rhenish Massif to the Harz Mountains, the Flechtingen Block and on through the subsurface east of the Bohemian Massif into the Moravo-Silesian Belt of the Czech Republic and Slovakia (Franke & Zelazniewicz 2000). To the west of the Ardennes, the RHZ is covered by the Mesozoic succession of the Paris Basin. During the period of overall convergence and compression, there was a phase of extensional tectonics and basin development in the RHZ (Behr *et al.* 1984; Engel *et al.* 1983*a*; Floyd 1995) with associated volcanism. In Germany, lower Devonian acid volcanics are followed by mid-Devonian intraplate continental tholeiites, whereas in SW England basic and acid volcanics occur throughout the Devonian. These latter suites have been interpreted as a rifted passive continental margin which became truly oceanic towards the south (Floyd 1984; Thirwall 1988). This basin probably extended eastwards into the RHZ of Germany where thick accumulations of Devonian shelf and hemipelagic sediments are associated with bimodal volcanism. MORB-type mafic rocks clearly suggest extension and opening of the Rhenohercynian Basin along the suture zone of the former Rheic Ocean. This oceanic basin must have remained fairly limited, however, as it cannot be recognized in the palaeomagnetic record. The onset of convergence is documented by the late Devonian (Frasnian–Famennian) deep-marine sediments and foreland-basin sedimentation throughout the entire RHZ from Portugal eastwards into Moravia (Czech Republic) (Franke & Engel 1986).

The main part of the RHZ is occupied by Early–Middle Devonian shelf clastics and carbonates which represent the passive southern margin of Laurussia. These are overlain by hemipelagic shales and early Carboniferous deep-marine sediments. This succession is overlain by a nappe comprising early Devonian MORB-type metabasalts, condensed mud-rich successions, and Frasnian–early Carboniferous deep-marine clastics representing the oceanic part of the Rhenohercynian Basin (Franke 1995*b*).

The Northern Phyllite Zone (NPZ), consisting of sedimentary and volcanic rocks overprinted by Carboniferous deformation and greenschist-grade metamorphism (Massonne 1995), forms the SE margin of the RHZ. The palaeogeographic derivation of the protoliths within the NPZ is heterogeneous and not fully understood (Franke 1995*a*). To the SE the RHZ is bounded by the Mid-German Crystalline High (MGCH) (for background see Franke 1995*a*), which formed the active southern margin of the Rhenohercynian Basin, i.e. as a site of synorogenic magmatism and as the source of synorogenic clastics transported towards the NW into the Rhenohercynian Basin. The age and tectonic setting of the protoliths of the MGCH are largely enigmatic (Franke 1995*a*).

The Rhenohercynian Basin developed during the Devonian and Early Carboniferous under a predominantly extensional regime in the areas of the Mid-European and North German–Polish Caledonides and was an important site of deposition through to Stephanian times. This basin is essentially a series of related basins/sub-basins extending along strike for over 2500 km from southern Ireland through to the eastern Czech Republic (Moravia). As noted, the area is segmented into various sub-basins including, for example, the French Basin, the Rhenish Basin (including the Ruhr Basin) and the Silesian Basin. By late Carboniferous times some of these basins had coalesced, leading to the development of a broad area of continental sedimentation in Stephanian times, which continued up into the Permian (see McCann *et al.* 2008).

Volcanism in the RHZ occurred intermittently throughout the Devonian and into the Lower Carboniferous, with major activity culminating in the Upper Devonian–Dinantian. Submarine basalts are mainly associated with sedimentary basinal facies, although interpretation of their tectonic environment suggests two different eruptive regimes: (1) as the volcanic carapace of oceanic crust, and (2) local volcanism associated with basin margin rifts (Floyd 1995). In the Lahn-Dill region of the Rhenish Massif and the Harz Mountains there are Upper Devonian and Lower Carboniferous basinal sequences with tholeiitic and alkaline within-plate basalts (Platen 1991; Floyd 1995). These volcanic rocks display affinities to intracontinental low-potassic tholeiites (Werner & Roesler 1979; Wedepohl *et al.* 1983; Flick & Nesbor 1988). Walliser (1981) related this Early Carboniferous volcanic pulse to the reactivation of pre-existing tensional fault systems. In the eastern parts of the Rhenish Basin, rift-related volcanism reached a second peak at the transition from the Tournaisian to Visean, with the widespread extrusion of the submarine Deckdiabas. These dolerite spilites are commonly associated with ultrabasic sills and rare pyroclastics.

Although no arc basalts have been positively identified in the RHZ, evidence for the previous existence of an arc coeval with the ocean basin was considered to be preserved in Devonian sediments (Floyd & Leveridge 1987; Floyd *et al.* 1990). These sediments contain large quantities of arc-derived calc-alkaline acidic clasts, although model ages indicate that they were derived from older arc systems tectonically not connected to the Devonian ocean (Floyd *et al.* 1991). If southerly directed subduction was involved in the closure of the Devonian Rhenohercynian Ocean, then typical arc volcanism is missing, or the arc was not located at the southern margin of the zone, but elsewhere. Early Carboniferous I-type granites in the northern Vosges, Black Forest and Odenwald may represent the plutonic roots of arc magmatism related to such subduction (Liew & Hofmann 1988). Additionally, work on metavolcanic rocks, of possible late Silurian to Devonian age, from the NPZ indicates the presence of both mid-ocean ridge basalts and within-plate basalts as well as an extensive calc-alkaline meta-andesite to metarhyolite suite (South Taunus area), which may represent the missing arc (Meisl 1995). More recently, subduction-related eclogite facies metabasites of MORB affinity have been recorded from the east Odenwald (Will & Schmädicke 2001). These rocks have a minimum age of 360 Ma (Scherer *et al.* 2002).

Western Rhenoherycnian Zone (Belgium and France) (M.D. & E.P.)

The Devonian–Carbonifeous synorogenic sedimentary successions of Belgium and northern France were deposited along the southern edge of the Laurussian continent, more precisely its eastern Avalonian component (Fig. 9.14). They are separated by an angular unconformity from low-grade, mainly fine-grained, siliciclastic deposits, ranging from lowermost Cambrian to uppermost Silurian in age. The Caledonian deformation event, progressively affecting these basement rocks from south to north, ranges from Llandovery to late Early Devonian, and possibly extends into the Eifelian (see Chapter 7).

During the Dinantian, two major marine basins, separated by the Anglo-Brabant Massif, developed in Belgium and northern France: the Campine Basin to the north and the Namur-Dinant Basin to the south (Figs 9.14–9.17). During the Silesian, they were both part of the NW European paralic coal basin evolving in the north Variscan foreland. However, both areas had different tectonosedimentary histories. Whereas the Namur-Dinant Basin (NDB) is folded and outcropping, resulting in the many classical Palaeozoic type sections, the Carboniferous of the Campine Basin is only slightly faulted and tilted and is entirely concealed beneath younger strata (e.g. Langenaeker & Dusar 1992; Dreesen *et al.* 1995; Langenaekar 2000; Kosters & Donselaar 2003). The lithostratigraphy of the region was established by Paproth *et al.*

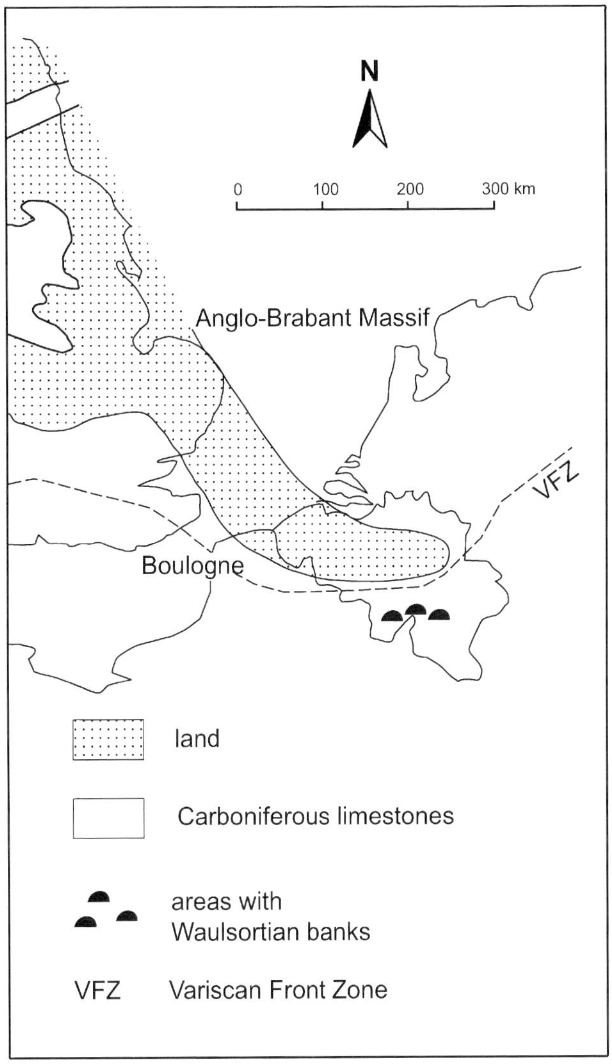

Fig. 9.14. General context of Dinantian sedimentation in NW Europe. Distribution of emergent areas and Waulsortian buildups at the end of the Tournaisian (after Hance *et al.* 2001).

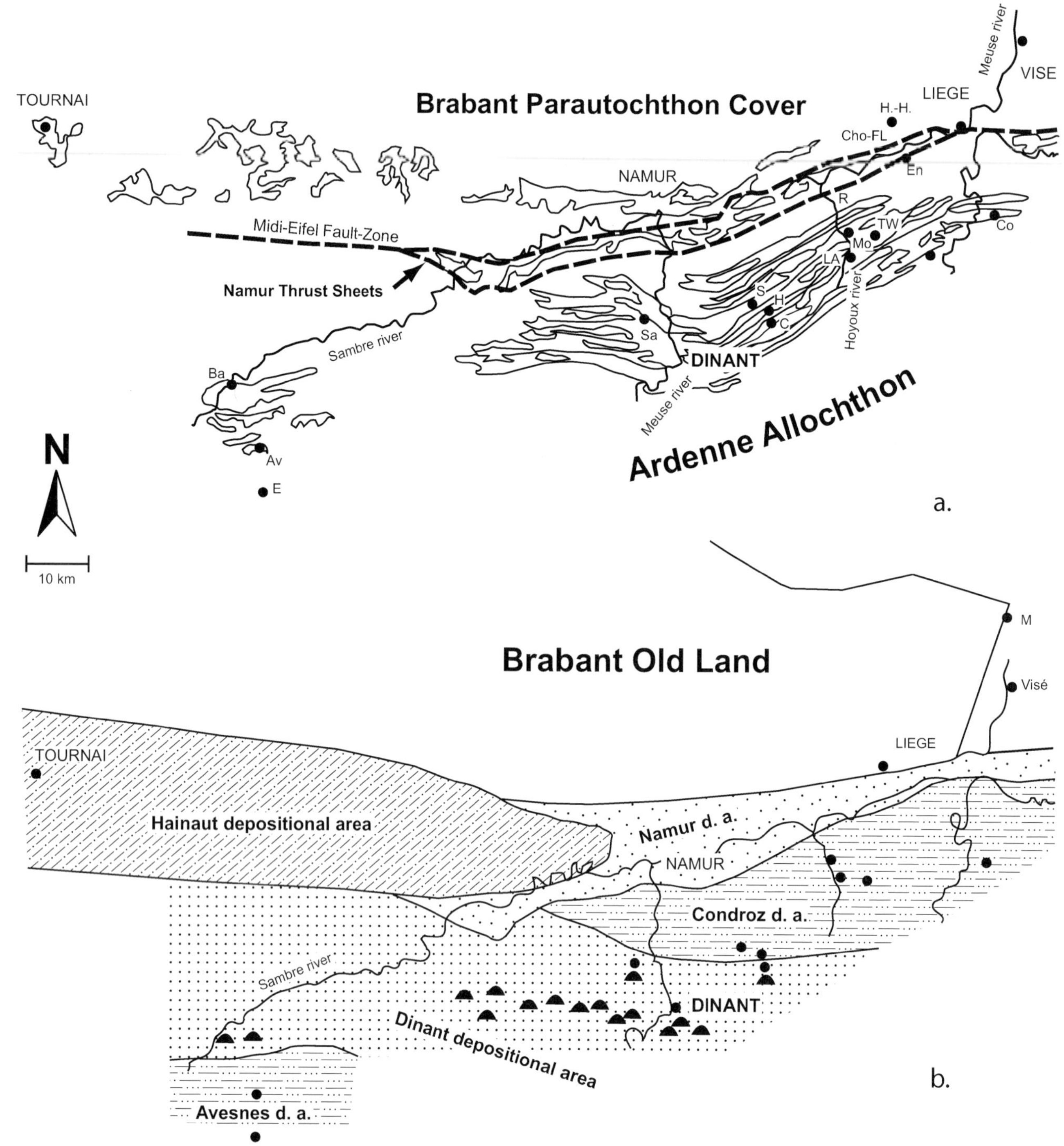

Fig. 9.15. (**a**) Distribution of Dinantian rocks in southern Belgium and northern France (Avesnois). Key sections: H.-H., Horion-Hozémont; Cho-FL, Chokier-Flémalle; En, Engihoul; Ba, Bachant; Av, Avesnes; E, Etroeungt; Sa, Salet; S, Sovet; H, Halloy; C, Ciney; R, Royseux; Mo, Modave; LA, Les Avins; TW, Terwagne; Co, Comblain. (**b**) Dinantian sedimentation areas in the Namur-Dinant Basin (not palinspastic). M, Maastricht (after Poty *et al.* 2001).

(1983*a*, *b*) and updated by Delmer *et al.* (2002) and Poty *et al.* (2002*b*) (Figs 9.18 & 9.19). Additionally, a synthesis of the biozonation has been published by Conil *et al.* (1991) and Hance *et al.* (2001).

The Campine Basin, the Anglo-Brabant Massif and the northern part of the NDB (i.e. Namur Synclinorium/sub-basin) belong to the structural Brabant Zone or the Brabant Parautochthon of Hance *et al.* (1999) and Mansy *et al.* (1999), as the southern wedge of the Variscan foreland. The remainder of the NDB (i.e. Dinant Synclinorium/sub-basin and the southern part of the Namur Synclinorium/sub-basin) belongs to the structural Rhenohercynian Zone affected by Variscan folding, and is part of the folded Ardenne Allochthon of the same authors. The Dinant Synclinorium/sub-basin is thrust over the Namur Synclinorium/

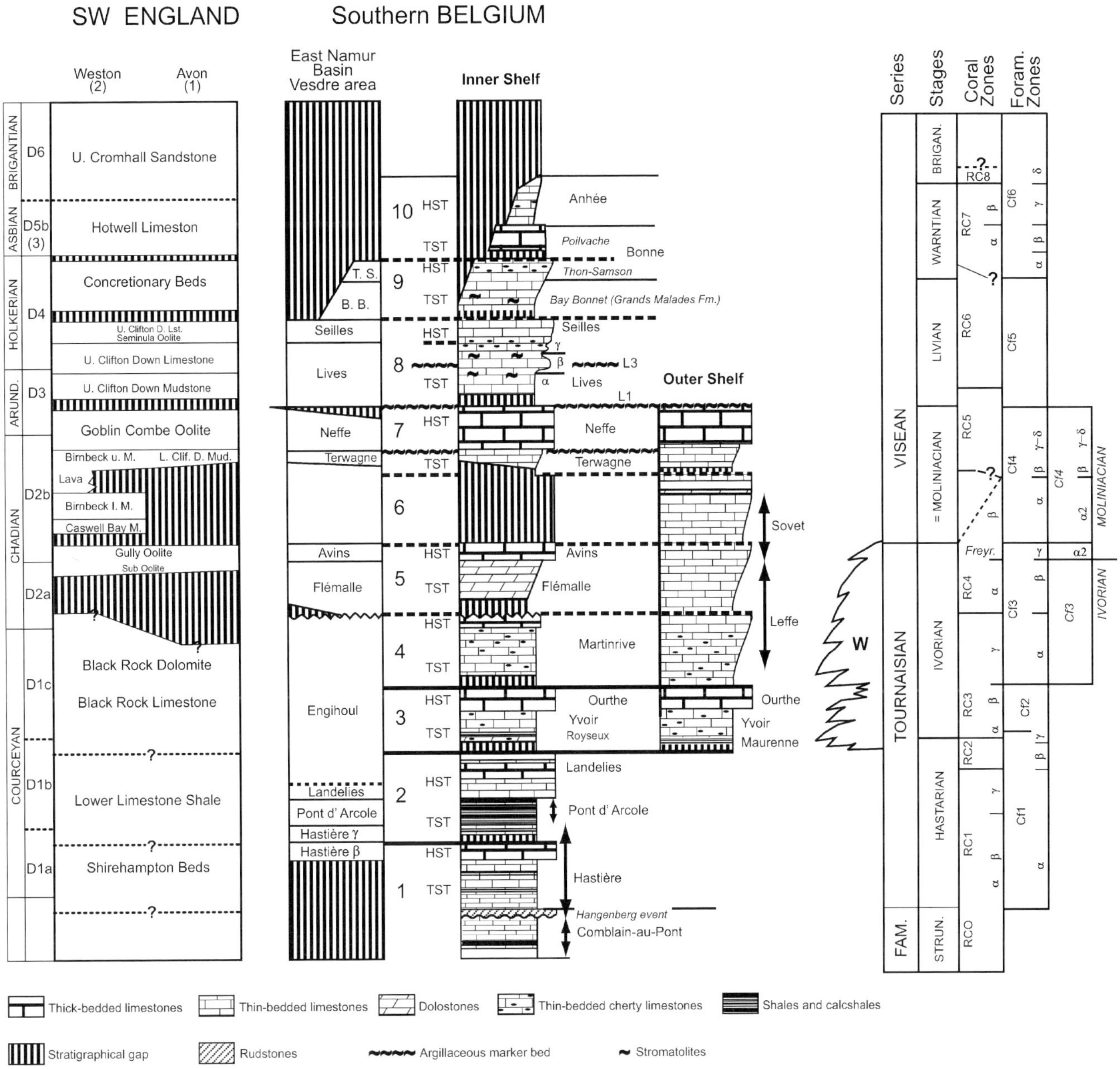

Fig. 9.16. Correlation of the Lower Carboniferous stratigraphy from southern Belgium westwards to England.

sub-basin along the line of the Midi-Eifel faults (Faille du Midi et eifélienne), a series of major thrusts along the southern boundary of the Anglo-Brabant Massif (Hollmann & Walter 1995).

Two major unconformities mark the tectonic uplift and erosion phases affecting the Carboniferous strata. The Asturian unconformity is linked to the northward progradation of the Variscan thrust front, which resulted in the deposition of a more complete Westphalian foreland succession in the Campine Basin. This terminated in mid-Westphalian D times. The Cimmerian unconformity at the top of a truncated Permian to Jurassic sequence was created by the uplift of the Anglo-Brabant Massif and the downwarp of the Roer Valley Graben (the western branch of the Lower Rhine Graben) to the north and the Paris Basin to the south. The Upper Cretaceous–Cenozoic succession has remained subhorizontal, with a slight northward dip towards the North Sea Basin or the Roer Valley Graben. The succession progressively covers all of the older strata, from the Jurassic in the Roer Valley Graben to the Cambrian at the core of the Anglo-Brabant Massif in central Belgium, as well as the Variscan and Caledonian massifs in the Ardennes to the south (Giese *et al.* 1997).

Between the twelfth and twentieth centuries, more than 2 billion tonnes of coal were mined in Belgium. The Coal Measures Group includes all of the Silesian coal-bearing siliciclastic sediments which show a rhythmic succession of coal–mudstone–sandstone sequences, totalling up to *c.* 4000 m in thickness, and with an age of Arnsbergian/Namurian A through to Westphalian D (e.g. Guian *et al.* 1995). The base of the Belgian Coal Measures Group coincides with the top of the Dinantian carbonate deposits. A hiatus, corresponding at least to the uppermost Visean and to the base of the Namurian (Pendleian), accompanied by a sharp lithological change, generally marks this

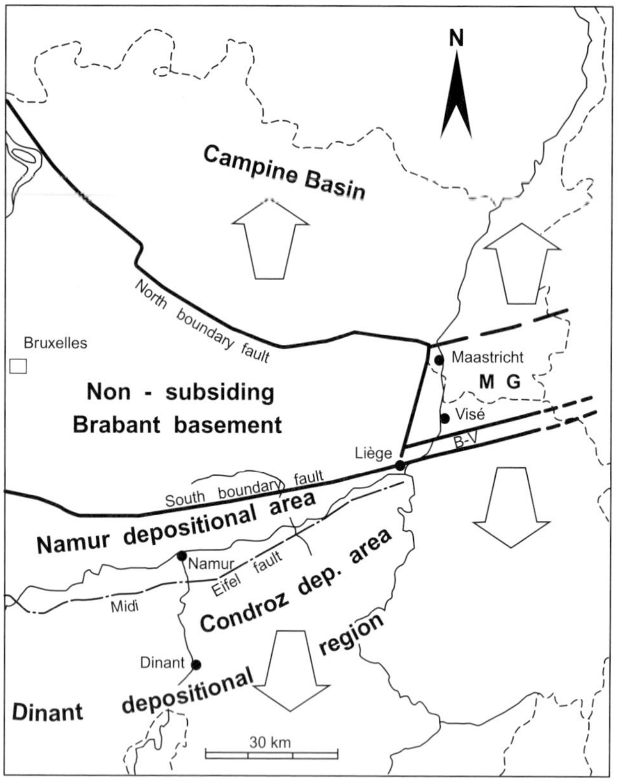

Fig. 9.17. Main tectonosedimentary units in northeastern Belgium (non-palinspastic map) during the Visean. The Variscan Midi-Eifel Fault is drawn because it hides the fault which originally separated the Namur sedimentation area from the Condroz sedimentation area. Abbreviations: MG, Visé-Maastricht graben; B-V, Booze-Le Val Dieu ridge separating the Visé-Maastricht sedimentation area from the Namur-Dinant Basin during the Visean (after Poty 1997).

boundary. The top of the Coal Measures Group is unconformably overlain by late Permian or younger deposits.

Paralic facies conditions prevailed during the deposition of the Belgian Coal Measures Group. This was related to basinal sag to the north of the Variscan Front, as well as the increasing supply of siliciclastic sediments in a closing intracontinental depression. Silesian sediments were derived partly from the Anglo-Brabant Massif, but the rising Variscan mountains to the south increasingly sourced the area. From base to top of this regressional succession there is a gradual transition from basinal mudstones and marine prodelta to lower/upper delta plain, through to lower/upper alluvial plain with a braided river system (Langenaeker & Dusar 1992; Paproth *et al.* 1996). The Namurian commences with the drowning of the karstified Visean carbonates initially covered by basinal mudstones with associated turbidites, and subsequently overlain by deltaic sequences from the Kinderscoutian onwards. However, in the southernmost coal basins (Dinant Synclinorium and the eastern part of the Namur Synclinorium) closer to the Variscan Deformation Front, coal deposition began in Arnsbergian (lower Namurian) times. In late Namurian times and extending up into the lowermost Westphalian A, these deltas gave way to sheet deltas. Deltaic sedimentation ceased with the depositon of a significant marine marker bed (i.e. Finefrau Nebenbank) which can be correlated across Central Europe.

Around the Westphalian A–B boundary, continuing uplift of the Variscan hinterland changed the hydrographic profile of the fluvial systems draining into the paralic basin, resulting in a shift from lower to upper alluvial plain conditions (Paproth *et al.* 1996). The Aegir Marine band, corresponding to a eustatic sea-level rise and marking the Westphalian B–C boundary, is the only widespread eustatic marine incursion. In the Campine Basin, Westphalian sedimentation terminated with the deposition of the Westphalian D Neeroeteren Sandstones. These are lithologically similar to the Tübbergen Sandstones in the Lower Saxony Basin between the Netherlands and Germany. These coarse-grained sandstones were sourced from the south and occur in stacked channel deposits (individual units are up to 30 m thick). The sandstones are highly porous (up to 20%) and permeable, but porosity has been compromised by subsequent kaolinization.

The Namur-Dinant Basin (NDB)

The NDB evolved in Devonian times as two distinct basins (i.e. the Namur Basin to the north, and the Dinant to the south). By early Carboniferous times, the basins were connected and extended across the region to the south of the Anglo-Brabant Massif (Fig. 9.16). Dinantian rocks, mainly carbonates, crop out in the Namur and Dinant synclines, and in their eastern (Vesdre area in eastern Belgium and Aachen area in western Germany), and western (Boulonnais and Avesnois in northern France) extensions (e.g. Bless *et al.* 1976, 1980, 1984). The depositional area was more than 250 km long with a width (after tectonic compression) of *c.* 50 km.. During the Dinantian, the NDB was part of a larger marine area extending to the west. It can be subdivided into six tectonosedimentary sub-basins (Poty 1997; Hance *et al.* 2001) (Fig. 9.15b), as outlined below.

The **Hainaut sedimentation area** (HSA) ('Auge Hennuyère' of Paproth *et al.* 1983*a*, *b*) is characterized by high subsidence rates, with the accumulation of up to 2500 m of Dinantian-age rocks. The deposits are mainly limestones (except for the shaly Pont d'Arcole Formation) with some secondary dolomitization. Lithostratigraphically, the carbonate successions differ from those of the other areas, and are only partly understood. In the Saint-Ghislain borehole (Groessens *et al.* 1982; Rouchy *et al.* 1993), the lower and middle Visean includes anhydrite units (up to 200 m thick), which are the only well preserved evaporite deposits known in the basin. However, the presence of solution collapse breccias and calcite pseudomorphs testifies to the presence of more widespread evaporite deposits. The classical Tournaisian type sections around the city of Tournai are located in the HSA.

The **Namur sedimentation area** (NSA) displays an incomplete stratigraphic succession characterized by a proximal carbonate facies. The total thickness increases locally from south to north due to inverse tectonic tilting (Poty 1997). The Dinantian succession in the Boulonnais area (northern France) is similar (Hoyez 1971; Poty 1994) and suggests that this area was connected to the NSA.

The **Condroz sedimentation area** (CSA) comprises exposures of proximal carbonate facies. The stratigraphic record, however, contains gaps in the NE part of the area, although to the SW the succession is more complete.

The **Dinant sedimentation area** (DSA) ('Auge dinantaise' of Paproth *et al.* 1983*a*) was strongly influenced from the Hastarian (lower Tournaisian) to the lowermost Visean by the development of thick Waulsortian buildups (Lees 1997; Lees *et al.* 1977). Relatively deep-water sedimentation occurred in the depression situated between the the inner shelf area to the north and the Waulsortian buildups to the south.

The **Visé-Maastricht sedimentation area** (VSA) ('Visé-Puth Basin' of Bless *et al.* 1976) is located along the eastern border of

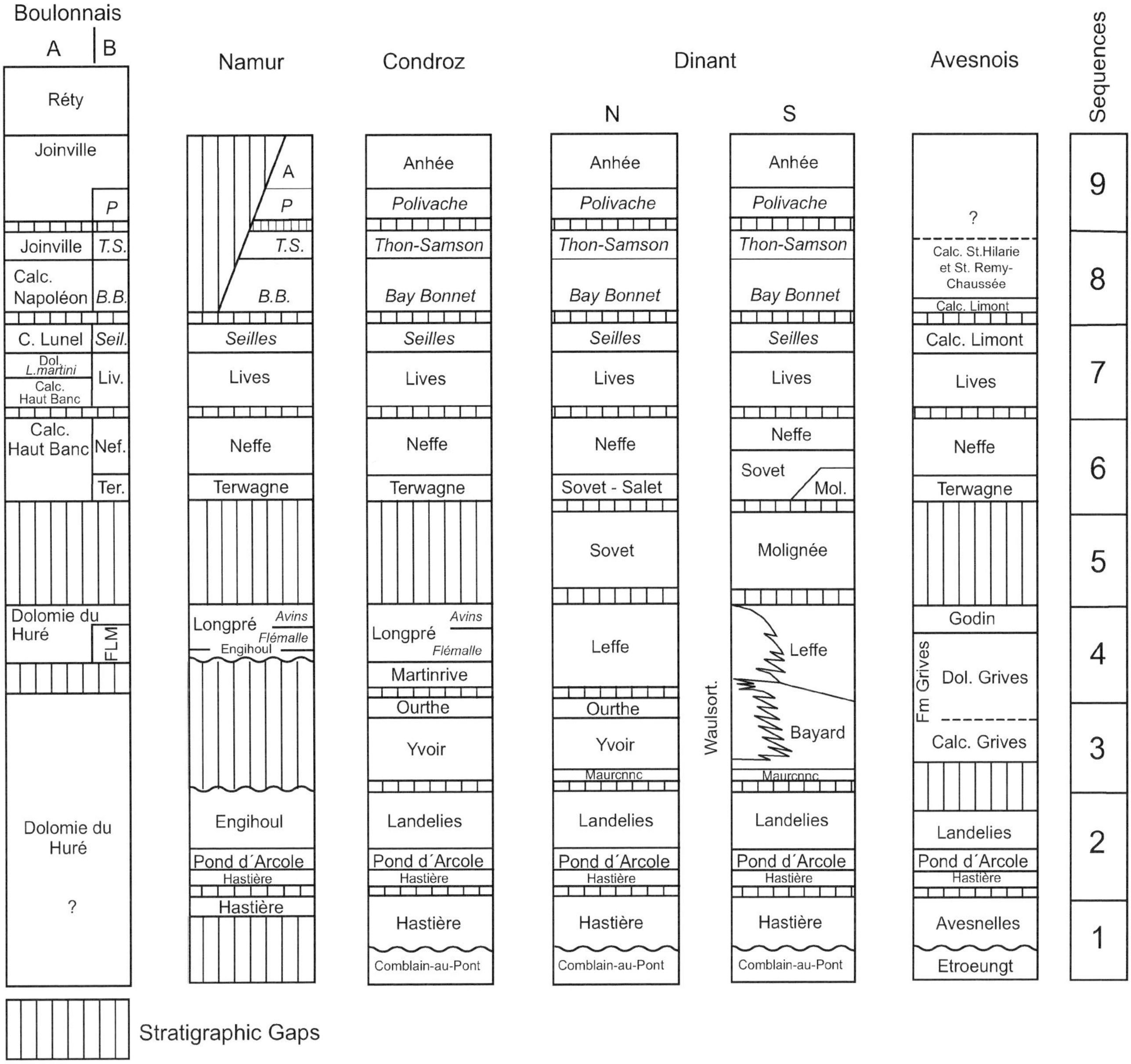

Fig. 9.18. Dinantian lithostratigraphy of the Namur-Dinant Basin in southern Belgium (except the Visé-Maastricht sedimentation area and the Hainaut sedimentation area) and northern France. Part B of the Boulonnais column indicates those formations of the Namur sedimentation area which have been recognized there. The lithostratigraphic units in italics are members, the others are formations (after Hance *et al.* 2001).

the Anglo-Brabant Massif and experienced block faulting during the Devonian–Carboniferous (Poty 1991) (Fig. 9.17). It was connected to the NSA during the Upper Devonian and Tournaisian and then evolved into a rapidly subsiding graben filled with deep-marine carbonate turbidites, with a connection to the Campine Basin. However, by Visean times it was separated from the NSA by the Booze-Val-Dieu Ridge (Poty 1997). The Dinantian of this area crops out only in the vicinity of Visé (site of the classic type sections) although the succession here includes numerous stratigraphic gaps. It is also known from boreholes in the vicinity of Maastricht (Netherlands).

The southern **Avesnois sedimentation area** (ASA) occurs in northern France, although its eastern extent is unknown. It displays local facies for the Tournaisian and, in Visean times, facies similar to those in the NSA and CSA (Mansy *et al.* 1989; Hance *et al.* 2001). The area of northern Avesnois, which contains Dinant-type facies, is included in the DSA.

The early Tournaisian succession of the NDB displays similar facies throughout the region, reflecting the homoclinal carbonate ramp setting (e.g. Fraaye & Werver 1990). Lithostratigraphic correlations across the region are well documented. Towards the north, however, the formations thin and the stratigraphic gaps increase. During the late Tournaisian, the depostional pattern was more varied. Inner ramp facies cover the Namur-Condroz and southern Avesnois areas, whereas outer ramp facies and an associated discontinuous barrier of Waulsortian buildups were

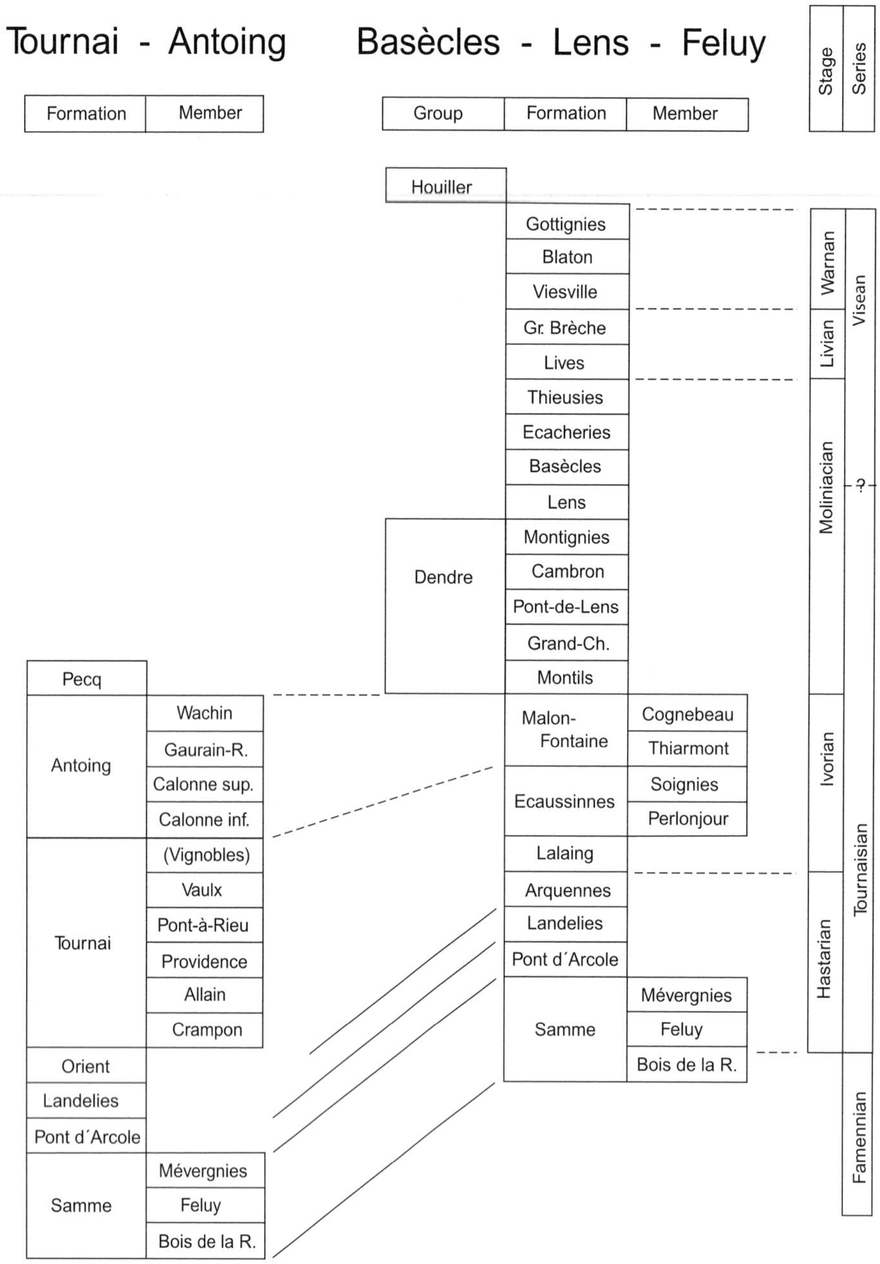

Fig. 9.19. Lithostratigraphy of the Hainaut sedimentation area (after Poty *et al.* 2001).

restricted to the Dinant area. During the latest Tournaisian the ramp evolved to a shelf with a break in slope separating the inner and outer shelf facies. A deep-water basin with Culm facies may have existed to the SE (as in SW England and SW Ireland) but this is not exposed and, moreover, the shallow platform situated to the SW (i.e. ASA) may possibly have extended to the east.

Submarine topographic irregularities produced during the Tournaisian as a result of variable sedimentation rates, and the growth of Waulsortian buildups, were smoothed out by early Visean times. Subsequently, middle and late Visean sedimentation across the area from western Germany to southern Ireland was governed by an aggrading shelf. During the mid-Visean, open-marine facies were restricted to the north, while evaporites developed in the south. This inversion of the normal pattern was probably related to an early phase of Variscan shortening. Late Visean deposits are locally absent. In the DSA this stratigraphic gap comprises the uppermost Visean and basal Namurian, whereas in the northern part of the NSA (i.e. that part surrounding the Anglo-Brabant Massif) it includes the uppermost mid-Visean and extends through to mid-Namurian times. Visean-age limestones are thus diachronously overlain by Namurian siliciclastics. The stratigraphic gap is, at least partly, due to the erosion and karstification of Visean carbonates following the global sea-level fall at the Visean–Namurian boundary.

Boulonnais (northern France)

The Boulonnais region of northern France comprises an Upper Palaeozoic massif which corresponds to the westernmost part of the Namur-Dinant Basin. In this region, Late Devonian clastics are thrust-faulted over Early Carboniferous limestones with minor tectonically intercalated coal slices (Mansy *et al.* 1999). The Dinantian of this region is mainly exposed in two of the three tectonic units recognized by Colbeaux & Leplat (1985), namely the central allochthonous Haut-Banc Unit and the northern autochthonous Ferques Unit. The southern allochthonous Hydrequent Unit comprises only small Carboniferous tectonic wedges. The Midi Fault in the Boulonnais is a complex rather than a singular fault zone and it does not represent the emergent front of the Variscan orogen which, based on the degree of deformation within the coal basin, suggests that the front lies several kilometres to the north (Mansy *et al.* 1999).

The lithostratigraphic units defined in the Dinantian of the Boulonnais (Colbeaux *et al.*, 1985) are similar to those of the NSA (Fig. 9.18) in Belgium (Colbeaux *et al.* 1985; Hoyez 1971; Poty 1994). The Huré Dolomite is equivalent to part of the Belgian Namur Dolomite Group. The lower and middle parts of the Haut-Banc Limestone are equivalent to the Avins Member of the Longpré Formation. The upper part of the Haut-Banc Limestone, the *Siphonodendron martini* Dolomite, and the Lunel and Napoleon limestones are equivalent to units recognized in the Lives and Grand-Malades formations. The lower and middle parts of the Joinville Limestone are equivalent to the Bonne River Formation; its upper part can be correlated with the Anhée Formation, but does not show exactly the same lithological characteristics. Some differences in the lithostratigraphy can occur between (and within?) the tectonic units of the Boulonnais (as in the NSA in Belgium). Thus, the Rety Dark Limestone can be correlated with the middle and upper parts of the Joinville Limestone (Colbeaux *et al.* 1985).

Campine Basin

The Campine Basin is a Variscan foreland basin filled with Devonian and Carboniferous sediments. The basin is located on the northern flank of the Anglo-Brabant Massif and has a length of 100 km in Belgium, extending northwards and eastwards into the Netherlands, the southern North Sea and Germany. An important transpressional fault zone of Palaeozoic age divides the basin into western and eastern sub-basins. During the Carboniferous, block faulting with syntectonic differences in palaeosubsidence resulted in marked differences in the burial history and sedimentological styles of these sub-basins, with frequent hiatuses, especially in the Tournaisian and uppermost Visean to lower Namurian.

The depositional succession is characterized by Kohlenkalk carbonates and comprises shallow-marine carbonates, dolomitized in the lower part, with reefal and peri-reefal units on local highs, commonly brecciated and karstified in the Visean. Karstified reefal limestones constitute good, but overmature, reservoirs and hence are used for gas storage. Towards the east, the Campine Basin grades into the VSA. Economically, the Campine Basin is the largest coal basin in Belgium and forms the continuation of the South Limburg coalfield of the Netherlands and the northern part of the Aachen and Ruhr coalfields in Germany.

The Devonian and Carboniferous succession of the Campine Basin is unconformably overlain by Permian to Cenozoic deposits and is thus concealed, with the exception of outcrops in the VSA. During the Givetian and Upper Devonian, sediments were only deposited in the Campine Basin during periods of sea-level highstands, except where thick, locally derived, partly conglomeratic units were deposited in small half-grabens (Muchez & Langenaeker 1993). During the Carboniferous basin subsidence increased considerably. However, despite this the Tournaisian is mostly absent. The Visean succession varies in thickness with respect to the general half-graben structure of the basin, but also locally as a result of block faulting. The maximum drilled thickness for the Visean varies between *c.* 760 m (eastern Campine Basin) and 314 m (western Campine Basin). Deposits are typically Kohlenkalk-type, shallow-marine limestones comprising microbial/cryptalgal buildups; the lower part (*c.* 300 m) of these units is intensely dolomitized. The east–west striking Hoogstraten Fault, situated more or less along the Belgium-Netherlands border, was originally considered to form the boundary between the Campine shelf and the deeper basinal environments to the north (i.e. in the Netherlands). It is, however, now interpreted as a listric fault affecting only Silesian strata. In the eastern part of the Campine Basin, towards the VSA, the Dinantian sequence is more complete and includes Tournaisian deposits.

No lithostratigraphic units have been formally defined in the concealed Campine Basin, but Muchez *et al.* (1987) recognized four Visean-age carbonate megasequences, namely earliest, early, mid- and early late Visean. The uppermost Visean is usually absent, due to both the global sea-level fall at the Visean–Namurian boundary and local tectonics. Emergence was accompanied by brecciation and karstification, responsible for the creation of a saline aquifer with good reservoir properties which, as noted above, is currently used for natural gas storage.

In the Campine Basin, Variscan deformation was generally limited to block faulting and tilting of the strata, partly along strike-slip faults during the Asturian phase of deformation (Late Carboniferous). Major NW–SE faults were reactivated during the Cimmerian (Mid-Jurassic) and Neogene extensional tectonic phases, associated with the development of the Roer Valley Graben, and inverted during the upper Cretaceous (Bouckaert & Dusar 1987). Cimmerian uplift of the Anglo-Brabant Massif resulted in downwarp of the Campine Basin towards the Roer Valley Graben in the NE. In this way, a more complete post-Carboniferous sedimentary record characterizes the Campine Basin from SW to NE.

In Upper Carboniferous times, the basin was broadly subdivided into two major parts. The first of these, the Antwerp Campine sub-basin to the west, contains only Namurian and Lower Westphalian deposits, with limited coal reserves but good access to the deep saline aquifer in the underlying Visean. To the east was the Limbourg Campine sub-basin, which contains a fully preserved coal-rich Westphalian succession. The boundary between both parts was probably transitional but largely coincides with the NNW–SSE trending Beringen-Rauw fault system, the outer tectonic boundary of the Roer Valley Graben boundary faults and the western limit of coal mining. The Limbourg Campine sub-basin can be further subdivided into an eastern and western coalfield by the north–south trending transpressive Donderslag Fault, which has markedly affected burial history, coal rank (Langenaeker 1992) and coal-seam distribution (Dusar 1989).

Wallonian Basin

The Wallonian Basin is located to the south of the Anglo-Brabant Massif and extends along the Sambre-Meuse axis from the French Nord-Pas-de-Calais coalfield in the west to the southern parts of the German Aachen and Ruhr coalfields in the east (Fig. 9.17). It is one of the most important coal-producing basins in Belgium and is subdivided into several mining districts, namely Borinage (or Mons), Centre, Charleroi and Basse-Sambre districts, all of which are located in the province of Hainaut and the Liège district.

The Wallonian Basin was located at the front of the Variscan fold-and-thrust belt and was thus strongly affected by the Asturian phase (Late Carboniferous) of Variscan folding and was squeezed against the rigid Anglo-Brabant Massif to the north. Folds in the basin are steep and asymmetrical. The Midi Thrust forms the southern limit to this basin: Lower Devonian rocks are thrust at a low angle over the coal basin whose real southern extension is yet unknown. In this way compressive tectonics dominated the southern part of the basin. Later deformation was possibly associated with the dissolution of the underlying carbonate/evaporite beds.

The Hainaut mining district comprises two distinctive tectonic units. The lower parautochthonous unit consists of an imbricate fan, which is separated by reverse faults. The upper allochthonous unit is composed of deformed overthrust sheets ('nappes' or massifs; = Great Superficial Massif of Delmer 2004) wherein a more complete section of the Coal Measures has been preserved. The boundary between these units is presumably faulted or thrusted, although a more recent interpretation suggests that it may represent a fossil peneplain, eroded during the Westphalian C before the emplacement of the Great Superficial Massif. This massif would originally have been deposited on the Anglo-Brabant Massif to the north, from where it became detached and slid, under force of gravity, towards the south into the Hainaut mining district, where it covered the parautochthonous unit. Gravity sliding was triggered by the dissolution of the evaporites intercalated with the Devono-Carboniferous carbonates of the HSA, underlying the Hainaut Basin, as demonstrated from the Saint-Ghislain borehole (Groessens *et al.* 1982). The Great Superficial Massif was subsequently partially overthrusted by the Midi Massif from south to north.

Rhenish Massif (Germany)

The Rhenish Massif (= Rheinische Schiefergebirge), which evolved from a rifted passive margin to an orogenic belt following collision with an upper plate continental fragment, the Mid-German Crystalline High, may be subdivided into two main regions, east and west of the Rhine (Fig. 9.20). The region to the west (= Linksrheinisches Schiefergebirge) comprises three main areas: the Eifel, Mosel and Hunsrück areas. Carboniferous sedimentation in this area was mainly restricted to the area around Aachen (extending through to the Anglo-Brabant Massif) (Wrede & Zeller 1991). The Rhenish Massif east of the Rhine (= Ruhr Basin) exhibits all of the features typical of a foreland basin generated by the tectonic load of the advancing active margin (Brix *et al.* 1988; Drozdzewski & Wrede 1994, 1997). Here, a post-rift sequence developed following the Early Devonian. Both a shallow-marine shelf and a deeper-marine basin evolved as a result of locally varying subsidence rates. On the shelf, clastic deposits and shallow-marine carbonates were deposited and accompanied by the development of a reef facies (Kohlenkalk). In the deeper parts of the basin turbiditic limestones and shales (Culm) predominate. The lower Lower Carboniferous pelagic shales are overlain by fine- to coarse-grained, mostly turbiditic clastic rocks deposited in a foreland-basin setting.

Basement in the Rhenish Massif consists of lower Palaeozoic sediments and medium-grade gneisses of Cadomian age. The early (continental) rifting event is not reliably documented. Peak crustal stretching occurred during the late Lower Devonian, and is recorded by strong subsidence, the formation of growth faults and widespread volcanism (Oncken & Weber 1995). During the Devonian and Lower Carboniferous a thick sedimentary succession (3–12 km) of shallow marine clastics and carbonates accumulated. The southernmost imbricates of the Rhenohercynian Zone probably represent the strongly telescoped slope and rise successions bordering a minor southern oceanic basin indicated by MOR-type metabasalts (Dittmar & Oncken 1992).

A widespread shale horizon occurs at the Devonian–Carboniferous boundary, and the overlying Tournaisian-age black shales and cherts and Early–Middle Visean cherts are correlatable across much of Europe (Engel *et al.* 1983a). In the **eastern Rhenish Massif**, northward migration of the deep-marine basin, which began in the south during Early Visean times, reflects the prograding front of the Variscan orogenic belt (Ricken *et al.* 2000) (Figs 9.21, 9.22). The basin represents the early stage of foreland basin sedimentation in the region, with basin fill being predominantly turbiditic. This deep-marine succession occurs in two main areas: the eastern Rhenish Massif and the Harz Mountains (Ricken *et al.* 2000). The central part of the basin, where clastic sediments were supplied from the Mid-German Crystalline High, was subsequently buried by younger sediments.

The turbiditic succession has a preserved thickness of >2 km (particularly in the Harz area), and was accumulated in two main stages. The first of these was related to trench–forearc sedimentation in a remnant oceanic setting during pre-collisional convergence. These turbidites were accreted into an accretionary wedge from Frasnian through to Tournaisian times (Engel & Franke 1983). Following convergence, turbidite sedimentation continued with deposition on the subsiding continental shelf, which was loaded by overthrusting of the prograding orogenic wedge (Franke *et al.* 2000) and followed by post-collisional turbidite sedimentation in Late Visean times.

Toward the north and west, the deep-marine deposits were gradually replaced during Namurian times by shallower marine to continental sediments, which persisted until Late Westphalian times (Littke *et al.* 2000). In Late Namurian times, subsidence was overcompensated by the clastic input, and sedimentation continued in the paralic environment of the Ruhr Basin at the NW margin of the Rhenish Massif.

In the **western Rhenish Massif** the sedimentary succession commences with an initial phase of neritic clastic and carbonate sedimentation, succeeded by hemipelagic environments, some deep-marine clastics and, finally, coal-bearing continental sedimentation (Walter *et al.* 1985; Franke 1995b). A carbonate platform, representing the southeasternmost part of the Carboniferous Limestone Platform, developed from latest Devonian times onwards. This platform can be traced eastwards towards Düsseldorf (Germany). Across much of the Rhenish Massif and also in boreholes in northern Germany, the platform facies is laterally replaced by dark shales with limestone turbidites (see above). Only far to the NE, on the southern shore of the Baltic Sea, have Early Carboniferous platform carbonates been encountered. In the northern part of the Rhenish Massif west of the Rhine, this carbonate platform is overlain by coal-bearing continental sequences of Late Carboniferous age.

During and after the NW propagation of the Variscan fold-and-thrust belt, a large system of foreland continental basins developed to the north of it. The largest, the **Ruhr Basin**, is situated at the NW margin of the Rhenohercynian Belt, to the east of the exposed Anglo-Brabant Massif (Fig. 9.20). The basin today has a length of *c.* 150 km in a SW–NE direction and a width of *c.* 80 km (e.g. Drozdzewski 1993). The Upper Carboniferous strata of the Ruhr Basin were conformably deposited on the older shelf succession. They reflect the later,

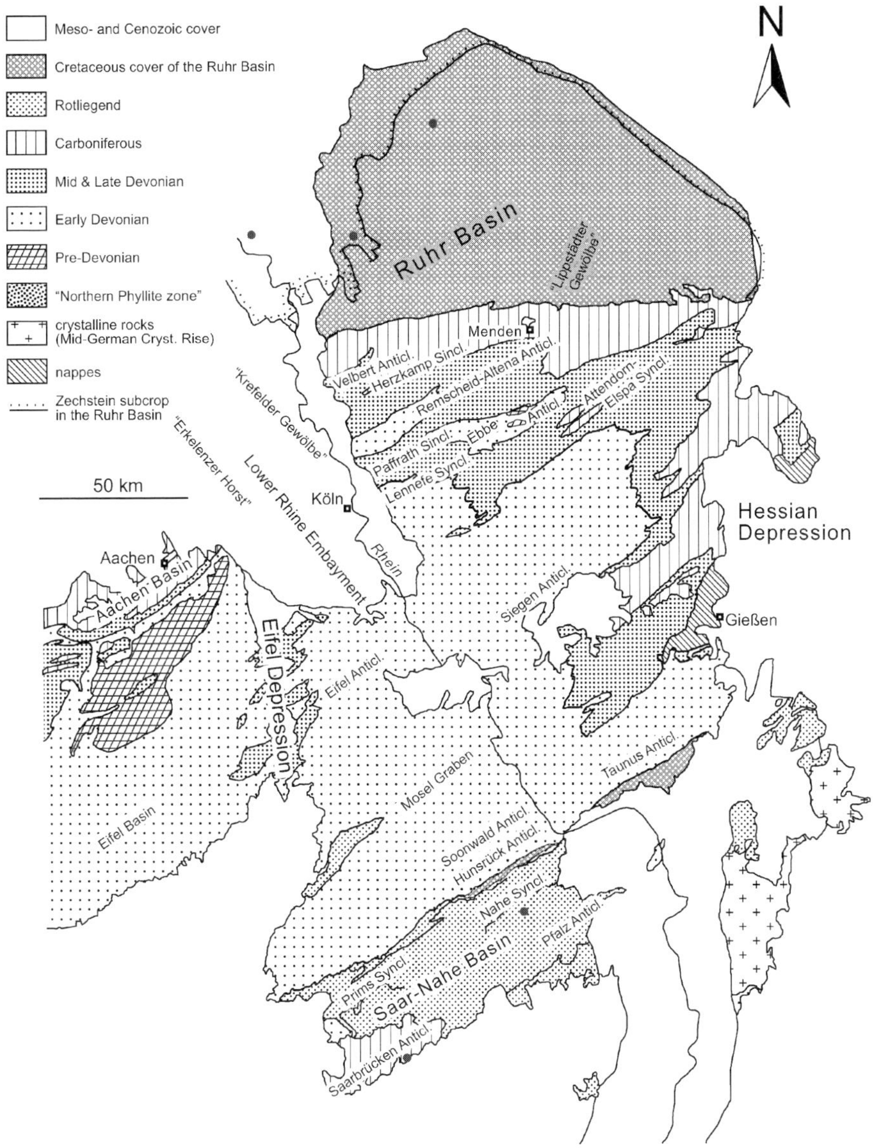

Fig. 9.20. Geological map of the Carboniferous basins in the Rhenish Massif area (after Littke *et al.* 2000).

basin-full stage of the Variscan Orogen in a typical peripheral foreland basin setting. The sedimentary environments developed from deep-marine to shallow-marine and finally to paralic (Drozdzewski 1993; Süss *et al.* 2002, 2006). Coal formation began in Namurian C times and continued at least until Westphalian C times. Regionally, the Namurian and Westphalian strata thin from *c.* 5 km in the south to <3000 m in the north.

In addition to the main basinal area of the Rhenish Massif there are a series of related allochthonous units (Figs 9.23 & 9.24). These units, however, occur in areas of low relief and the resulting scarcity of outcrops precludes detailed analysis. The tectonic nature of several units is, therefore, problematic. The **Giessen-Harz Nappe** (= Giessen Werra Südharz/Selke Nappe) is exposed at the SE margin of the Rhenish Massif. The base of this unit comprises a tectonic mélange of sedimentary rocks (including early Ordovician quartzites, Silurian shales and limestones, and early Devonian limestones, sandstones, conglomerates and debris flows) which contains an Armorican fauna (Franke & Oncken 1995; Plusquellec & Jahnke 1999). This is overlain by isolated tectonic slices of metabasalt with both a MOR and intraplate signature (Floyd 1995; Ganssloser *et al.* 1995). In some cases the basalts are overlain by condensed shales and bedded radiolarian cherts of Devonian age (Figs 9.23 & 9.25). The main part of the Giessen-Harz Nappe consists of turbidites and intercalated shales. The age of the turbidites varies across the area from Early Frasnian to Early Famennian (for details see Franke 1995*b*) extending up into the Early Carboniferous (Dörr 1990) reflecting the early orogenic activities of the Mid-German Crystalline High (Franke 2000). Both compositionally and lithologically, the succession is very similar to units from the Lizard Complex in Cornwall.

The **Hörre-Gommern Unit** is subdivided into two parts: a clastic-rich part intercalated with shales, bedded cherts and occasional limestones (in the SE), and sandstones and quartzites (in the NW). The latter is a belt of quartz arenites intercalated with pelagic shales and radiolarian cherts of Famennian–Visean

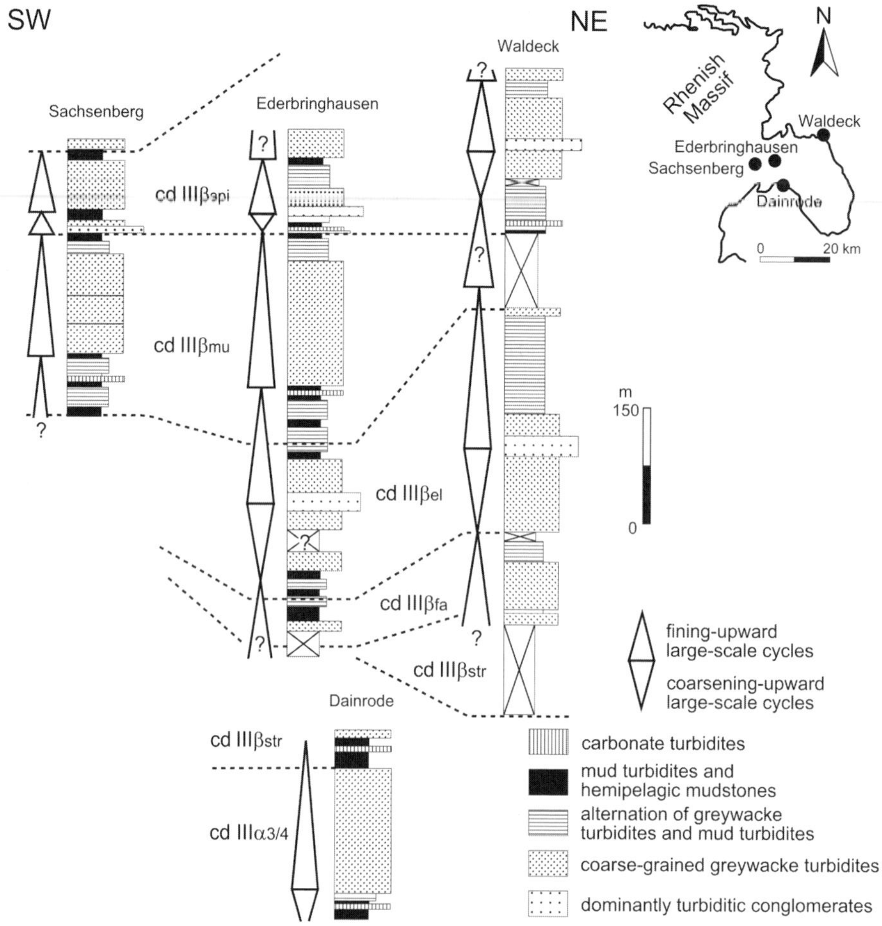

Fig. 9.21. Correlation of large-scale turbidite cycles that show the transition from coarse-grained clastics to mud turbidites with intercalated carbonate turbidites at the eastern Rhenish Massif. Large-scale cycles were interpreted to indicate dominantly accommodation cycles at the former shelf of the Mid-German Crystalline High (after Schrader 1999; Ricken *et al.* 2000).

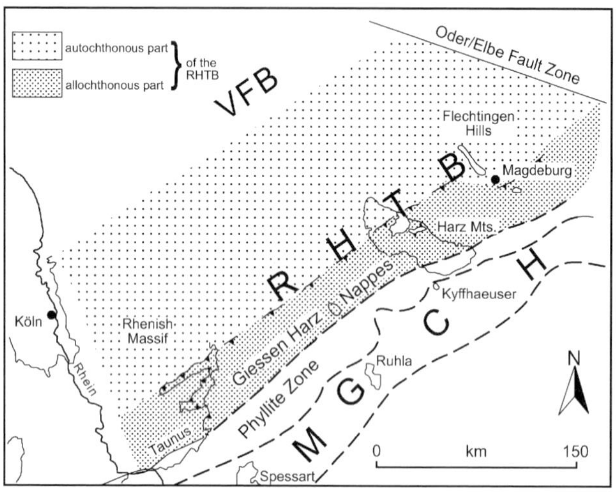

Fig. 9.22. The Rhenohercynian Turbidite Basin (RHTB) with the allochthonous (i.e. the Giessen-Harz Nappes) and the autochthonous portion (MGCH, Mid-German Crystalline High). To the NW the transition towards the Variscan Foreland Basin (VFB) is shown (after Ricken *et al.* 2000).

age, which extends *c.* 300 km from the Rhenish Massif into the Flechtingen Horst (Meischner 1991; Jäger & Gursky 1998). The source of the siliciclastic material was probably Pomerania (Poland) at the western margin of the Baltic Shield where coeval shallow-marine sandstones occur (Franke 2000). Tectonic lenses of the Hörre-Gommern Quartzite have also been found within the Hessen Shale Unit (= Hessische Schieferserie) and on top of the parautochthon of the Lahn Syncline (Wierich & Vogt 1997). The SE part of the Hörre-Gommern Unit comprises Middle Devonian–early Tournaisian limestone turbidites, intercalated with shales and rare sandstones, and overlain by Famennian–early Visean greywackes (Bender & Homrighausen 1979; Franke 2000).

The **Harz Mountains** are situated along strike to the NE of the Rhenish Massif and have a similar geological history (for details see Wachendorf 1986; Franke 1995*b*; Schwab 1979) (Fig. 9.24). Crystalline Cadomian basement (Ecker Gneiss) is overlain by late Early Devonian shelf sediments. Crustal extension in early Middle Devonian times led to a change in sedimentation to increasingly hemipelagic, with facies patterns and subsidence being controlled by extensional faults (Franke 1995*b*). In the NW Harz the pattern continues with late Middle Devonian volcanism, Givetian–Frasnian reef

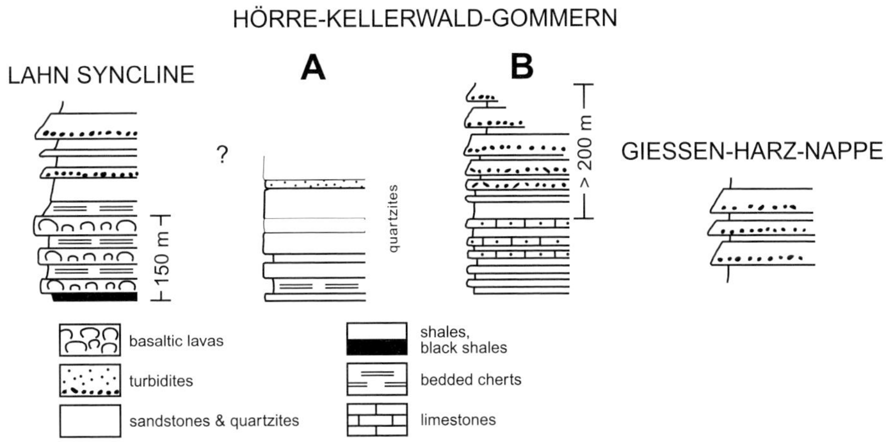

Fig. 9.23. Stratigraphic columns documenting facies and thickness for the main parts of the Rhenohercynian Zone east of the River Rhine (after Franke 1995*b*).

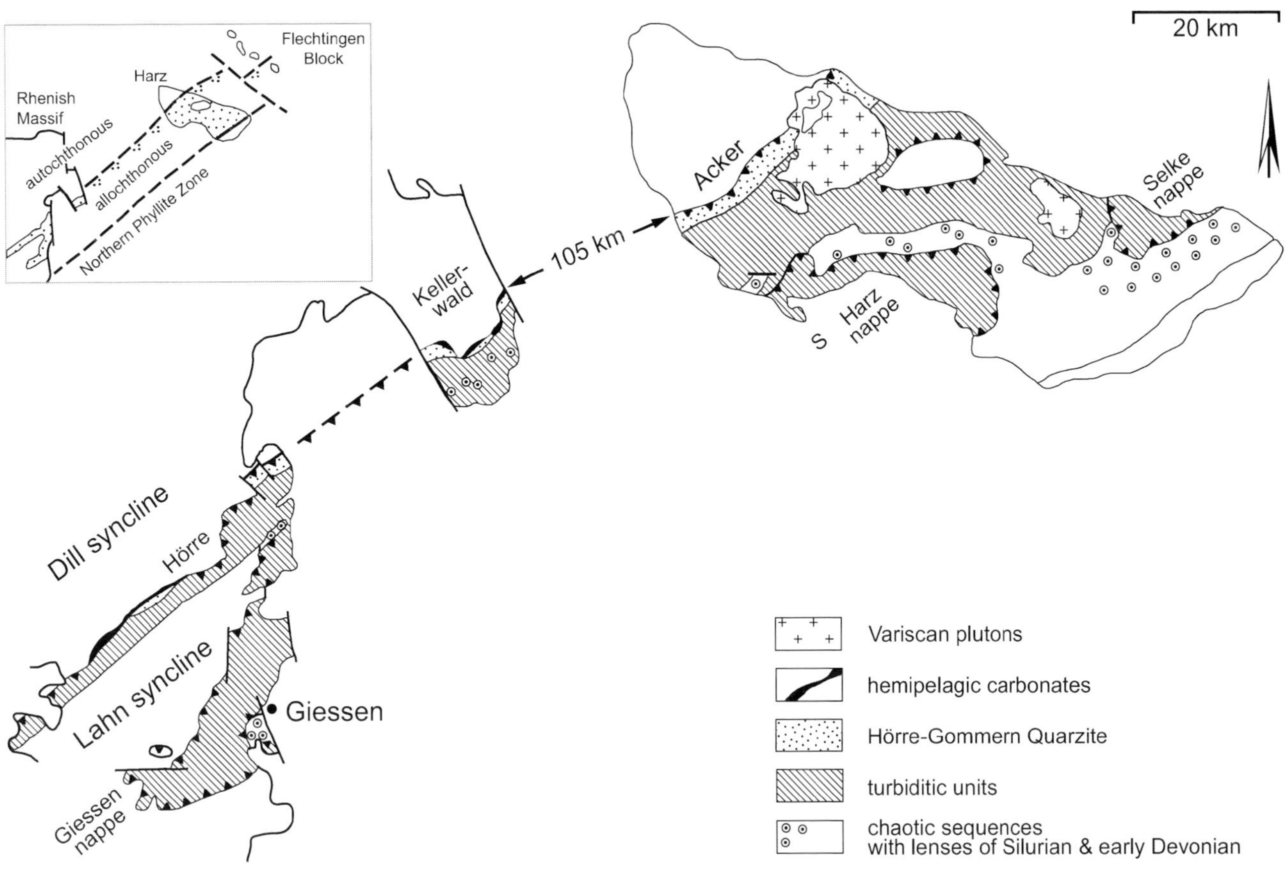

Fig. 9.24. Allochthonous units (proven and suspected) in the Rhenish Massif and Harz Mountains area (after Franke 1995*b*).

development, Famennian–Early Carboniferous hemipelagic carbonates, mudstones and cherts as well as Early Carboniferous basaltic volcanism (Franke 1995*b*). This succession is broadly similar to parts of the eastern Rhenish Massif (e.g. Dill Syncline).

Carboniferous deep-marine sediments are also present in the **Flechtingen Horst (Block)** area (Fig. 9.24), a narrow horst zone 50 km NE of the Harz Mountains. Small early Carboniferous quartzite outcrops represent the easternmost known extension of the quartz arenites of the NW Hörre-Gommern Unit (i.e. Hörre-Kellerwald-Acker quartzite).

Malopolska Region (S.S.)

In western Poland, the fold-and-thrust belt of the Variscan externides and their associated foreland basin are almost entirely buried beneath up to 8 km of Permian–Cenozoic-age sediments. The SW and central parts of this belt cover the area of southern

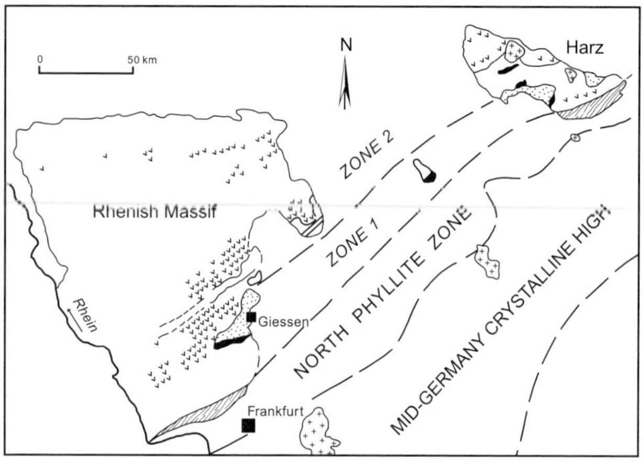

Fig. 9.25. Map of central Germany showing the distribution of Devonian and Lower Carboniferous basaltic rocks. Zone 1, Devonian MORB-like lavas at the base of the Giessen-Harz Nappe (black areas). Zone 2, Devonian and Lower Carboniferous lavas with witin-plate basalt affinities (vvv ornament) (after Floyd 1995).

Wielkopolska and involve a more than 2 km thick late Visean to Westphalian syn- to post-orogenic deep-marine succession (Mazur *et al.* 2006). In the northern Wielkopolska, Kuiavia, northern Malopolska and Upper Silesia regions of Poland, the Variscan foreland basin developed on top of a collage of probable Neoproterozoic terranes which were accreted to the East European Platform during Early Palaeozoic times.

Subsidence of the Variscan foreland basin in front of the evolving orogen commenced in the late Visean and Namurian A and was accompanied by a rapid north- and northeastward advance of late-orogenic clastic systems across Malopolska, Wielkopolska and Upper Silesia to reach the Lublin-Lvov Basin in the east and Pomerania in the north. Rapid subsidence of this foreland basin was locally interrupted in the late Namurian and earliest Westphalian, resuming in the Westphalian A/B. This hiatus is thought to be related to intraplate compressional deformation reflecting increasing collisional coupling between the Variscan orogenic wedge and its foreland (Ziegler *et al.* 2002; Mazur *et al.* 2006).

During the Early Carboniferous, the Malopolska Region, together with its Moravian extension (eastern Czech Republic), constituted part of the Laurussian shelf, which was bounded to the south by a Pre-Carpathian landmass (Fig. 9.26). Generally, the area was typified by shallow-marine carbonate platforms, where the sediment thickness was up to 1000 m (Paszkowski & Szulczewski 1995). The Devonian–Carboniferous boundary is usually located within peritidal facies with very poor biostratigraphic control, although it may also fall within a non-depositional gap. Shallow-water carbonate sedimentation persisted up to the end of Visean times. A well-defined deepening episode was first noted here by the end of Tournaisian times. The carbonate platforms were separated by deep, tectonically controlled troughs which were infilled by Culm facies, and partly intercalated with carbonate sediment gravity flows (Zakowa 1980; Zakowa & Migaszewski 1995). The Early Carboniferous basin of the Holy Cross Mountains is a particularly fine example of such an intrashelf trough succession. The Upper Visean Galezice Debrite Member, the only carbonate unit in this typical Culm succession, is interpreted here as a submarine fan deposit (Belka *et al.* 1996). By the end of the Visean/beginning of the Namurian, sedimentation in the region had terminated as a result of faulting and uplift (Kedzior 2001). Tectonic inversion of the region is interpreted to have been related to the Sudetic tectonic phase of the Variscan Orogen.

Lublin-Lvov Basin, (eastern Poland and Ukraine) (S.S.)

The Lublin Basin (eastern Poland) and its Ukrainian extension (Lvov-Volhynian Basin) developed in the Late Visean as a consequence of accelerated subsidence in the pericratonic areas and repeated incursions of a shallow sea (Fig. 9.27). The onset of deposition was preceded by a volcanic episode. During maximum flooding episodes (beginning in the Namurian), the Lublin Sea may have been connected with the Dnepr-Donets and Moscovian basins through local depressions of the East European Craton.

Marine influence continued in this area up to the end of the Westphalian A (although by this time it was only in the form of short ingressions). The extremely regular sedimentary cyclicity, related to the deltaic depositional environment, is the most characteristic feature of this paralic succession, especially in the Lower Namurian part (Skompski 1996). The coal seams from this part of the succession are numerous and laterally continuous, but economically insignificant since they are extremely thin. In the upper part of the succession, which was deposited in a limnic/fluvial environment and is dominated by fine-grained siliciclastic sediments, a typical coal-bearing association is developed. Within the economically important Lublin Formation, 25 coal beds have been recognized with thicknesses ranging from 0.8 to 3.8 m. However, only eight of these can be treated as stable seams, preserved over a broad area. Sedimentation in the region terminated in Westphalian C or D times.

The Northern Phyllite Zone

The Northern Phyllite Zone (NPZ) was initially defined by Scholtz (1934) as a narrow metamorphic zone separating the Rhenohercynian Zone (RHZ) to the north from the Mid-German Crystalline High (MGCH) to the south (Figs 9.3 & 9.28). The NPZ is characterized by greenschist-grade metamorphism which is intermediate between the low-grade rocks to the NW and the medium- to higher-grade rocks to the SE in the MGCH. The NPZ can be characterized as a heterogeneous tectonic assemblage of rocks containing probable remnants of the distal, SE part of the Rhenohercynian Basin passive margin, together with rocks derived from a Silurian magmatic arc, and exotic sedimentary successions possibly derived from the Armorican Terrane Assemblage (ATA) (Franke 1995*a*). During the Ordovician and Silurian it formed part of the incipient stage of the Rhenohercynian rift, but probably represents an Ordovician–Silurian magmatic arc. The original site of this arc is unknown, but from Early Devonian times onwards it was overlain by clastic sequences similar to those from the Rhenohercynian Zone further to the north (Anderle *et al.* 1995). In early Devonian times the NPZ was situated to the NW of the opening Rhenoherycnian Ocean, proximal to its passive northern margin (Anderle *et al.* 1995).

The NW margin of the NPZ is a thrust fault, while the SW margin is not exposed. The precise nature and the evolution of the contact between the NPZ, the MGCH and the Saxothuringian Zone is unclear because of a lack of exposures. However, results from the DEKORP 1 seismic line (DEKORP Research Group 1991) suggest large-scale thrusting of the Saxothuringian Zone over the NPZ (Oncken *et al.* 1995). Between the NPZ and the MGCH there is a narrow belt of unmetamorphosed greywackes

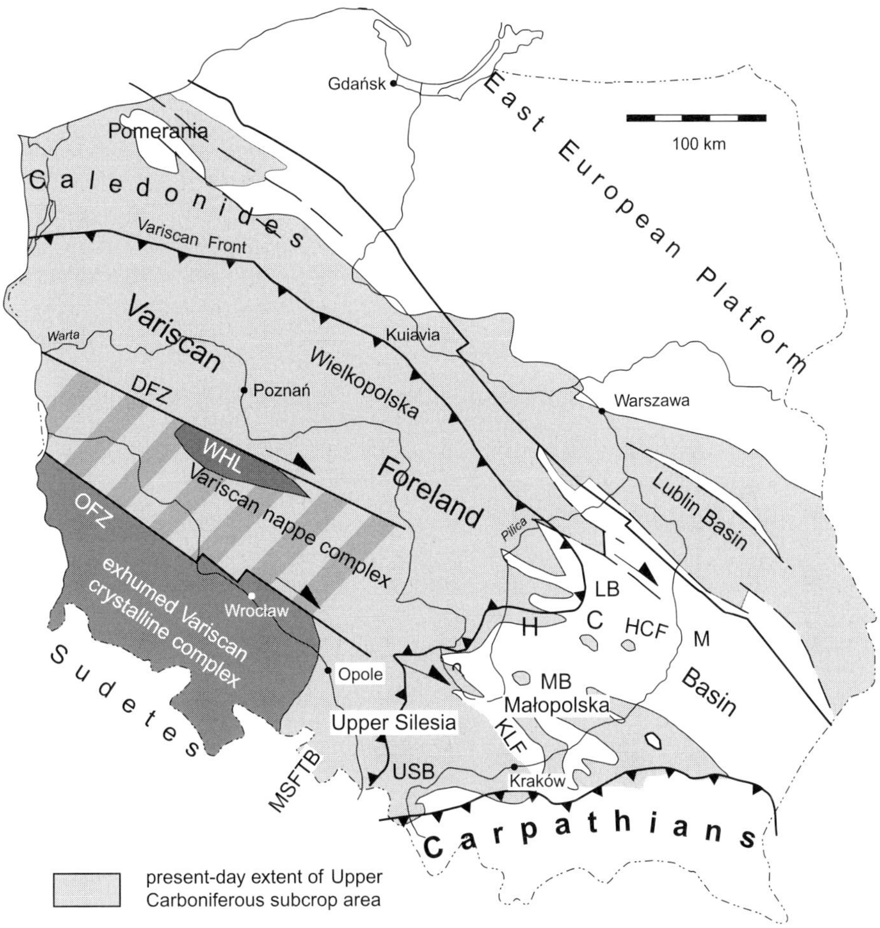

Fig. 9.26. Variscan external zone of Poland. The location of the Variscan Front according to Pozaryski *et al.* (1992) is indicated. Abbreviations: DFZ, Dolsk Fault Zone; HCF, Holy Cross Fault; HCM, Holy Cross Mountains; KLF, Kraków-Lubliniec Fault; LB, Lysogóry Block; MB, Malopolska Block; MSFTB, Moravo-Silesian Fold-and-Thrust Belt; OFZ, Odra Fault Zone; USB, Upper Silesian Block; WHL, Wolsztyn-Leszno High (after Mazur *et al.* 2006).

which probably represent part of the Giessen Nappe (i.e. part of the Rhenohercynian Basin ocean floor and its sedimentary cover) (Anderle *et al.* 1995).

The NPZ is exposed in a number of areas, including an isolated outcrop within the Saar-Nahe Basin, with more extensive outcrops along the southern margins of the Rhenish Massif and the Harz Mountains (Anderle *et al.* 1995) (Fig. 9.28). Along strike the NPZ may be subdivided into three areas – from west to east, the south Hunsrück, the south Taunus and the Wippra units – although there are significant differences both in terms of stratigraphy and lithology between these units as well as with the adjacent RHZ succession. The **South Hunsrück Unit** comprises mainly metapelites and metabasalts with intercalated lenses of quartzite, sandstone and limestone. To the NW there is a faulted contact with the RHZ, with tectonic lenses of gneiss (including the Wartenstein Gneiss) located on the RHZ side of the fault (Anderle *et al.* 1995). To the SE the NPZ is bounded by Early Permian continental sediments and volcanics. In its southwestern part the south Hunsrück succession is divided by a thrust fault (Wiesbachtal Mylonite). The succession between the Wartenstein Gneiss and the Wiesbachtal Mylonite is mainly Devonian in age (Berger *et al.* 1991).

The southeasternmost part of the NPZ in the region (i.e. SE of the Wiesbachtal Mylonite) comprises metapelites, metacherts and metabasalts with rare intercalated phyllites and iron ores (Vorsoonwald Series). These have been interpreted as equivalents of Early Carboniferous successions from the Lahn and Dill synclines of the unmetamorphosed Rhenish Massif (Berger *et al.* 1991) although Oncken *et al.* (1995) have suggested that they represent a metamorphosed part of the Giessen Nappe. Geochemical analysis of the basalts reveals their within-plate signature suggesting tectonic mixing of oceanic crust with rocks of the stretched continental margin.

The **South Taunus Unit** comprises mainly acid and intermediate volcanics with intercalated metasediments (NW part) as well as metasediments with rare metabasalts (SE part). The oldest rocks in the area are metavolcanics interpreted to have been derived from Ordovician–Silurian lavas. The metasediments are subdivided into lower (Eppstein Formation) and upper (Lorsbach Formation) units, of early Devonian age (Anderle *et al.* 1995). The former comprises phyllites with sandy laminae and, in places, greywackes and arkoses. The Lorsbach Formation is more variable, consisting of slates, quartzites and rare cherts. The Taunus Quartzite (of Lower Devonian, Siegen stage, in age) is a prominent orthoquartzite deposited in a high-energy tidal setting (Stets & Schäfer 2002)

The **Wippra Unit** is a belt of low-grade rocks located at the SE margin of the Harz Mountains. It comprises metapelites,

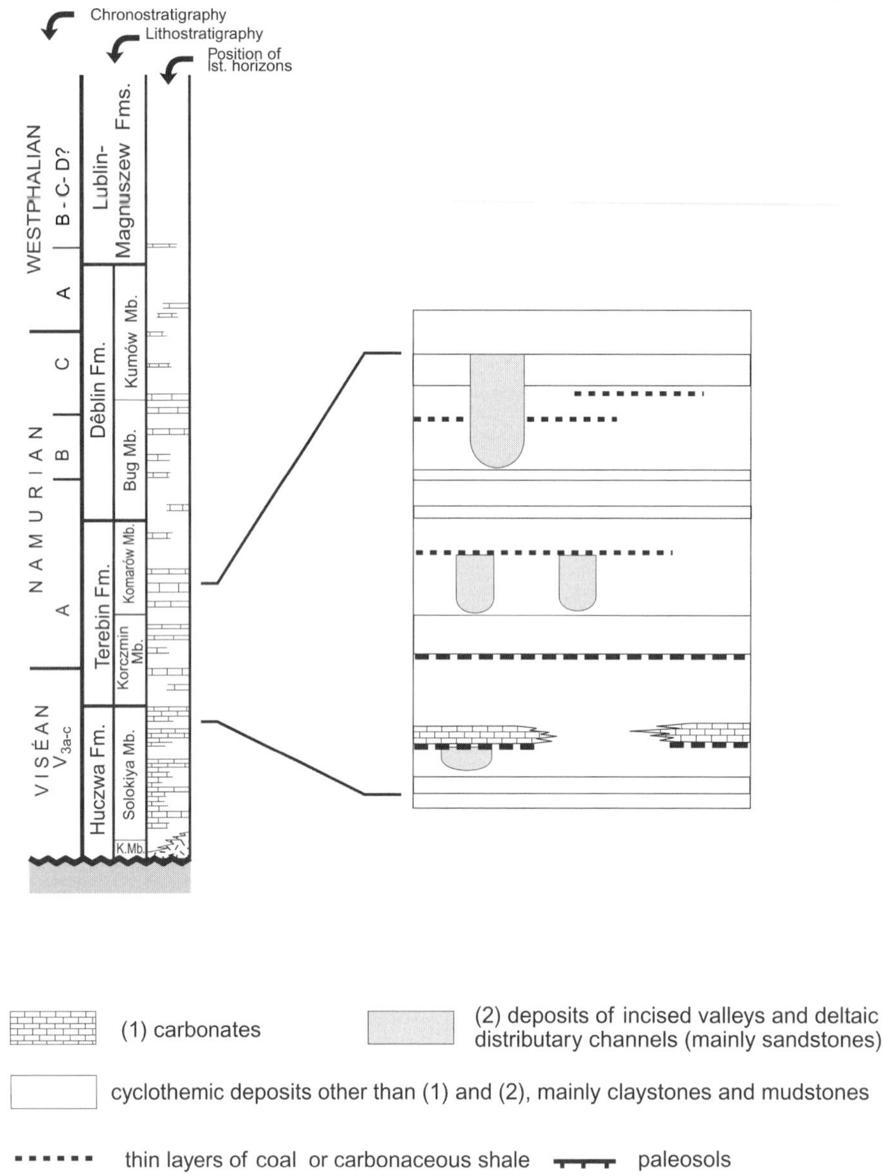

Fig. 9.27. Scheme of the Yoredale-type cyclothems in the Lower Carboniferous of the Lublin Carboniferous Basin (after Skompski 2003; stratigraphic columns after Porzycki & Żelichowski (in Porzycki 1979) and Porzycki & Zdanowski 1995).

quartz sandstones, slates, greywackes, volcanics (including metabasalts and comagmatic pyroclastics), rare limestones and siderites which have been subdivided into seven lithological units (for details see Anderle *et al.* 1995).

Volcanic successions occur in parts of the NPZ. In the South Hunsrück and Wippra units (Harz Mountains), there are frequent occurrences of ocean-floor basalts. In the South Hunsrück Unit bands of MORB alternate with bands of intraplate basalts probably representing a tectonic mix, possibly related to thinned continental and oceanic crust of Devonian age at the southern, passive margin of Laurussia (Meisl 1995). Devonian MORB-type rocks in the Wippra Unit could also be interpreted as part of the Giessen-Werra-Harz Nappe, but their assumed late Devonian age does not fit this concept, since the oceanic crust in the nappe is dated as pre-middle Devonian, and the late Devonian is already represented by turbidites which indicate plate convergence (Meisl 1995). The volcanics of the Wippra Unit differ, and suggest more of an affinity with Armorica (= ATA).

In the South Taunus Unit the rare metabasalts also show MORB signatures, possibly representing parts of the early Devonian ocean floor. The South Taunus Unit is dominated by metavolcanics which were derived from an island arc or active margin setting. The destruction of this arc was possibly related to the closure of the Rheic Ocean between Avalonia and Armorica (= ATA).

Ordovician–Silurian post-collisional granitoids and Silurian clastic sediments are also present in the Spessart region (see below) of the MGCH suggesting a magmatic and sedimentary development similar to that of parts of the NPZ. It is possible that at that time parts of the NPZ and MGCH were coherent before being separated by Devonian rifting which created the Rhenohercynian Ocean (Anderle *et al.* 1995). In early Devonian

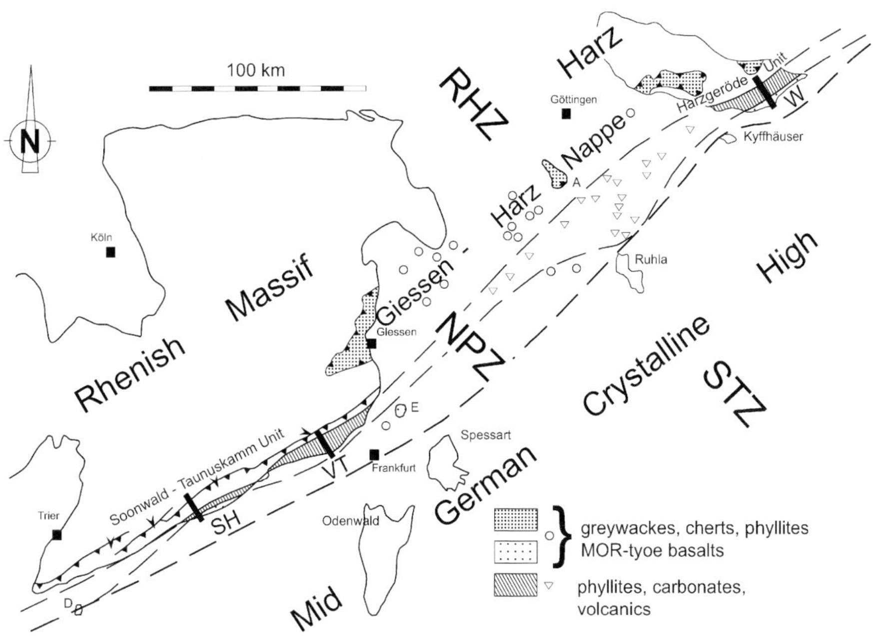

Fig. 9.28. Geological setting of the Northern Phyllite Zone in the Rhenish Massif and Harz Mountains. A, Albungen-Werna; D, Düppenweiler; E, Erbstadt; NPZ, Northern Phyllite Zone; RHZ, Rhenohercynian Zone; SH, South Hunsrück; STZ, Saxothuringian Zone; VT, Vordertaunus (South Taunus) (after Anderle *et al.* 1995).

times, the NPZ was situated to the NW of the opening Rhenohercynian Ocean, proximal to its passive margin.

The NPZ possibly represents the southern prolongation, i.e. slope of the Rhenohercynian passive continental margin (e.g. Dittmar & Oncken 1992) or possibly an exotic terrane, of Saxothuringian provenance, within the accretionary wedge formed in front of the approaching Saxothuringian active margin. However, since the westerly parts of the NPZ contain rocks derived from the Rhenohercynian Shelf and, further east, portions of the ATA, it has been suggested that the NPZ contains a segment of the Rheic Ocean Suture (Cocks & Fortey 1982) between Avalonia and the ATA (Fig. 9.29). Indeed, Franke (2000) has suggested that the Rheic Suture is marked by the arc-derived metavolcanic rocks of the South Taunus Unit, which

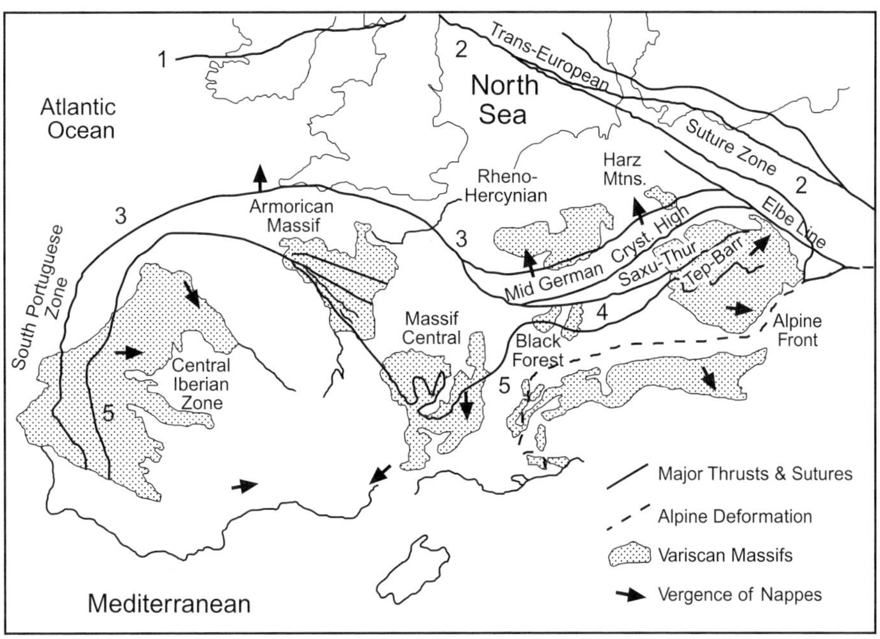

Fig. 9.29. Palaeozoic sutures in Europe (after Franke *et al.* 1995*a*; Matte *et al.* 1990; McKerrow *et al.* 2000). Key: 1, Iapetus Ocean separating Avalonia from Laurentia; 2, Tornquist Sea, separating Avalonia from Baltica; 3, Rheic Ocean, separating Avalonia from the Armorican Terrane Assemblage; 4, ophiolites south of the Saxothuringian Zone; 5, Ligerian-Massif Central-Moldanubian ophiolites.

record the consumption of that ocean. Additionally, the greywackes and metabasalts SE of the NPZ probably mark the Rhenohercynian Ocean Suture Zone.

The Saxothuringian Zone

The Saxothuringian Zone (STZ) of Kossmat (1927) is one of the most complex parts of the Variscan Belt in central Europe (Franke 1995*c*) (Fig. 9.3). It consists of a northwestern crystalline belt, the Mid-German Crystalline High and a southeastern part, dominated by relatively low-grade sedimentary and volcanic rocks. These successions were deposited in a Cambrian–Devonian rift basin which was closed in Devonian–Early Carboniferous times and is referred to as the Saxothuringian Basin (Franke 1995*c*). A detailed overview of the region has recently been published (see Linnemann 2004)

The primary relationship between the Saxothuringian Basin and the Mid-German Crystalline High (MGCH) is unknown (Franke 1995*c*). Today, the crystalline rocks are juxtaposed against the relatively low-grade material by a south-facing reverse fault (possibly a major strike-slip zone). There are no clastic sediments in the Saxothuringian Basin which can be unequivocally referred to sources in the MGCH. To the SE the Saxothuringian Basin was bounded, at least from early Famennian times onwards by the Cadomian Teplá-Barrandian (see below), which represents the upper plate and active margin during Variscan convergence.

Mid-German Crystalline High (= Mid-German Crystalline Zone)

The MGCH, which represents the northern margin of Armorica (= ATA), is a NE–SW striking belt of aligned basement complexes several hundred kilometres long, which is locally exposed in the Palatine Forest, the Odenwald, Spessart, Ruhla and Kyffhäuser crystalline complexes and extends into SE Poland (where it can only be recognized in boreholes) (Figs 9.3, 9.30) (see also Kroner *et al.* 2008; Timmermann 2008). It comprises a series of metamorphic and magmatic complexes, the former composed of metasedimentary and migmatic as well as metamagmatic rocks (orthogneisses) which vary in terms of the degree and type of metamorphism. The magmatic complexes are represented by plutonic bodies of variable composition (mafic, intermediate, acid), age of intrusion (pre-Variscan, Variscan) and degree of deformation (late kinematic, post-kinematic) (Hirschmann 1995). The evolution of the MGCH has recently been summarized by Oncken (1997), Okrusch & Weber (1996), with contributions in Dallmeyer *et al.* (1995) and Zimmerle (2000).

The origin of the MGCH at the northern margin of the Saxothuringian Zone is still debated. High-pressure metamorphic rocks (e.g. eclogites, blueschists) are often taken as evidence for subduction-related processes at convergent plate margins, and in the Variscan Fold Belt widespread occurrences of eclogites associated with gneisses provide ample evidence of high-pressure metamorphism during the Variscan collision. In Central Europe, eclogites are found in various crystalline complexes (e.g O'Brian 1997; O'Brian *et al.* 1990; Okrusch *et al.* 1991), for example the Black Forest, the Münchberg Gneiss Complex, the Oberpfalz, the Erzgebirge, the Sudetes, various parts of the Bohemian Massif and the eastern Odenwald (Will & Schmädicke 2001). Furthermore, plutonic rocks, many of which can be characterized as I-type granites with a distinct subuction signature, are abundant (Henes-Klaiber 1992).

Reischmann & Anthes (1996) postulated a three-phase geodynamic development of the MGCH involving oceanic crust subduction beneath the MGCH. Subduction commenced in the Silurian, with the south-directed subduction of the Rheic Ocean. This was followed by the south-directed late Devonian subduc-

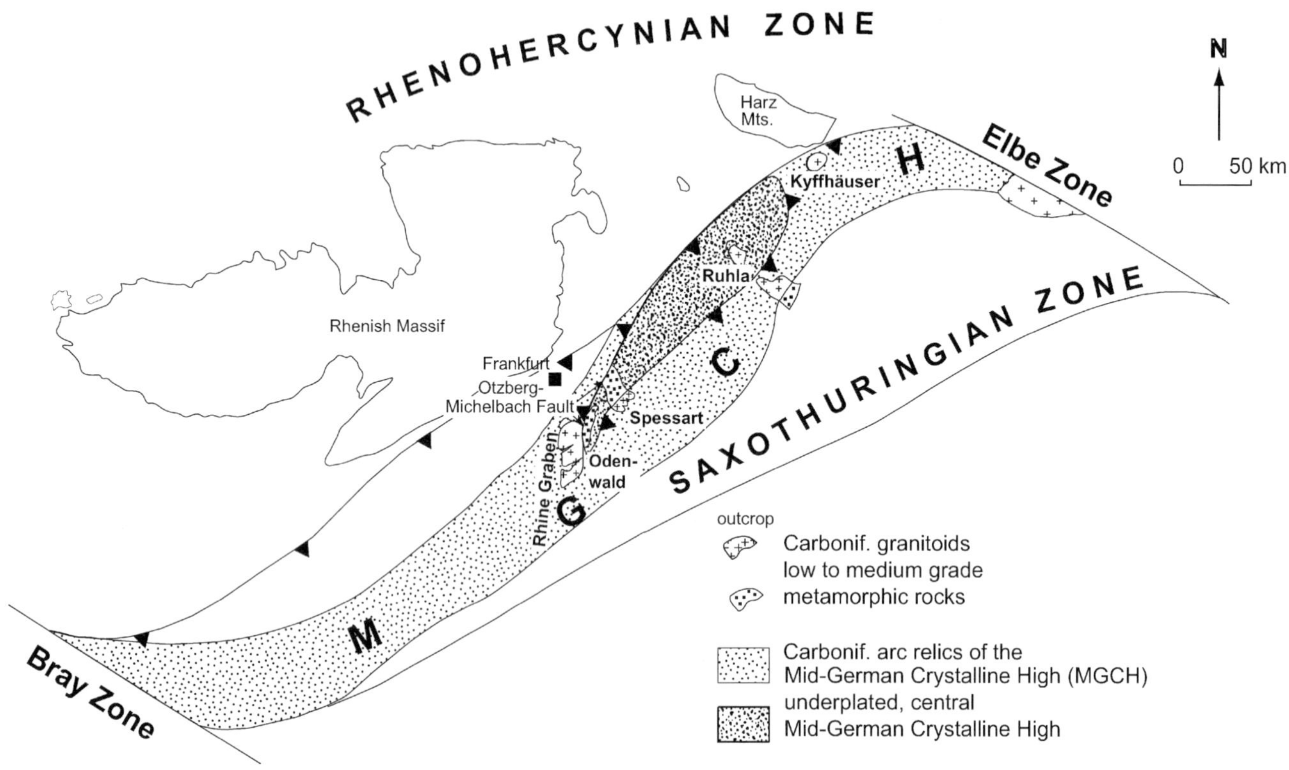

Fig. 9.30. Schematic map of the Mid-German Crystalline High (after Stein 2001).

tion of the Rhenohercynian Ocean and later by the north-directed lower Carboniferous subduction of the Saxothuringian Ocean. Since the Variscan Orogeny developed in a transpressional regime due to oblique collision, much of the pre-Variscan history has been erased (Stein 2001). However, palaeontological evidence suggests that at least part of the MGCH belonged to Avalonia in pre-Devonian times (Franke *et al.* 1995*a*). Additionally, rocks with a distinct island arc signature on both plate margins suggest the early closure of the Rheic Ocean in Silurian or early lower Devonian times. During the late lower Devonian development of the Rhenohercynian Ocean, and its final closure at about the time of the lower–upper Carboniferous boundary, these Avalonian parts were amalgamated into Armorica (Franke & Oncken 1995). This two-ocean scenario was postulated by Altenberger & Besch (1993) based on geochemical data from the eastern Böllstein Odenwald, although Robardet (2003) would question this scenario. Zeh (1996) notes that different parts of the MGCH have different orogenic developments and his model also involves a second south-dipping subduction zone at the southern border at *c.* 380–360 Ma.

Oncken (1997) introduced a new model for the MGCH which is based on the orogenic wedge theory of Platt (1988). This model envisages a combination of frontal and basal accretion, together with subduction erosion as being responsible for the MGCH as an orogenic wedge. According to Oncken (1997), the MGCH represents the site of a plate boundary between a northern lower and a southern upper plate, the Rhenohercynian (southern part of Avalonia) and Saxothuringian (assumed to be the northern rim of Armorica) belts, respectively. Oblique convergence and subduction led to the amalgamation of the two plate fragments in the early Carboniferous (Oncken 1997; Will 2001). Oncken (1997) thus interpreted the MGCH as an orogenic wedge and subdivided it into two distinct basement units: a lower plate unit of Rhenohercynian basement rocks exposed in the eastern Odenwald, the Spessart and Ruhla crystalline complexes, which tectonically underlies an upper plate unit of Carboniferous magmatic arc relicts. Rocks from this upper unit are exposed in the Kyffhäuser Crystalline Complex and the western part of Odenwald. However, Altherr *et al.* (1999) have challenged this interpretation, since chemical and isotopic signatures from granitoids beneath the western Odenwald do not show evidence for underplated Avalonian material beneath large parts of the MGCH.

Odenwald Crystalline Complex

The Odenwald Crystalline Complex is the largest exposure of the MGCH in Central Europe (Fig. 9.31). The complex can be subdivided into the Böllstein Odenwald and the Bergsträsser Odenwald, separated by an important strike-slip zone, the Otzberg Zone (Stein 2001). Other authors, however, favour a four-fold subdivision (e.g. Krohe 1991, 1992, 1994). More recently, Stein & Dietl (2001) and Stein (2000) suggested that the Odenwald can be subdivided into three main geotectonic units: the Böllstein Odenwald, the Frankenstein Complex, and the Bergsträsser Odenwald. All of these units may have had a common sedimentation and early tectono–metamorphic development related to a medium pressure metamorphic event at *c.* 400–375 Ma. Subsequently, they were separated and experienced different tectonomagmatic histories before being juxtaposed along two major strike-slip zones: one between the Frankenstein Complex and the Flasergranitoid Zone, and the second, the Otzberg Zone, between the Böllstein and the Bergsträsser Odenwald (Stein & Dietl 2001).

In Krohe's (1991) four-fold division of the Odenwald Crystalline Complex, the western part (Bergsträsser Odenwald, Units I–III) belongs to the Saxothuringian upper plate, and the eastern part (Böllsteiner Odenwald, Unit IV) to the subducted Rhenohercynian lower plate unit (*sensu* Oncken 1997). Thus, in this scheme the Otzberg Shear Zone separates the Böllsteiner Odenwald from the Bergsträsser Odenwald. This shear zone is a fundamental boundary of the Mid-German Crystalline High (e.g. Krohe 1991; Oncken 1997; Will 2001). Units I–III underwent low–medium pressure metamorphism at high temperatures (e.g. Okrusch 1995; Okrusch *et al.* 2000; Will & Schmädicke 2003), whereas recently discovered eclogite relicts in the orthogneiss core of Unit IV (Will & Schmädicke 2001) indicate a high pressure metamorphic event. The initial isotopic signature and chemical characteristics of the syncollisional Odenwald granitoids (Units II and III) are similar to those of contemporaneous granitoids from other parts of the STZ (e.g. northern Vosges, Black Forest, Central Bohemian Batholith).

The **Böllstein Odenwald** forms a NNE–SSW striking antiformal structure with an orthogneiss core and a metasedimentary schist mantle; the latter probably represents the oldest paragneiss unit of the Odenwald, and geochemically has been characterized as a volcaniclastic series with an island-arc signature (Altenberger & Besch 1993). Geochemical compositions of the Böllstein orthogneisses suggest S-type granitoids as magmatic protoliths (Altenberger *et al.* 1990) which are interpreted as synorogenic intrusions developed during a continent–island arc collision. Reischmann *et al.* (2001) regard the orthogneiss educts as hybrid magmas which assimilated crustal material but which cannot be clearly classified as I- or S-type granitoids. This is an important difference with the mainly I-type granitoids of the Bergsträsser Odenwald (Stein 2001).

Metabasic rocks have been reported from the Böllsteiner Odenwald and provide evidence of ecologite facies metamorphism (Will & Schmädicke 2001) – the first record of a high-pressure metamorphic event in the MGCH. The eclogites formed from a tholeiitic protolith that may have originated in a continental extension zone or a young oceanic ridge or backarc environment. Subduction and subsequent collision are likely to have led to their juxtaposition with the surrounding orthogneisses in the Böllsteiner Odenwald (Will & Schmädicke 2001).

The **Bergsträsser Odenwald** consists of *c.* 90% calc-alkaline intrusives and 10% metamorphic rocks with a variegated lithology. The magmatic rocks are separated by six narrow zones of metamorphic rocks (= Schieferzüge) which vary in width from several hundred metres to several kilometres (Fig. 9.31). The four most important ones are (from north to south): the Eberstadt-Rossdorf Schieferzug, the Auerbach-Grossbieberau Schieferzug, the Heppenheim Schieferzug (= Hauptschieferzug), and the Weinheim-Waldmichelbach Schieferzug (= Schöllenagglomerat). The lithologies are predominantly gneisses and schists with rare cherts and metaphosphorites, derived mainly from greywacke protoliths (Stein 2001). Marbles and calc-silicate rocks are intercalated into the clastic metasediments. These metasediments alternate with amphibolites, that are derived from basalts of tholeiitic composition (Stein 2001). Indeed, Poller *et al.* (2001) have suggested that all of the amphibolites in the succession were derived from island-arc tholeiites extruded in a backarc setting.

The three major units of the Bergsträsser Odenwald include the Frankenstein Complex, the Flasergranitoid Zone, and the southern Bergsträsser Odenwald (Willner *et al.* 1991). There are different intrusion depths and ages for the Frankenstein Gabbro and for the plutons of the Flasergranitoid Zone (Stein & Dietl

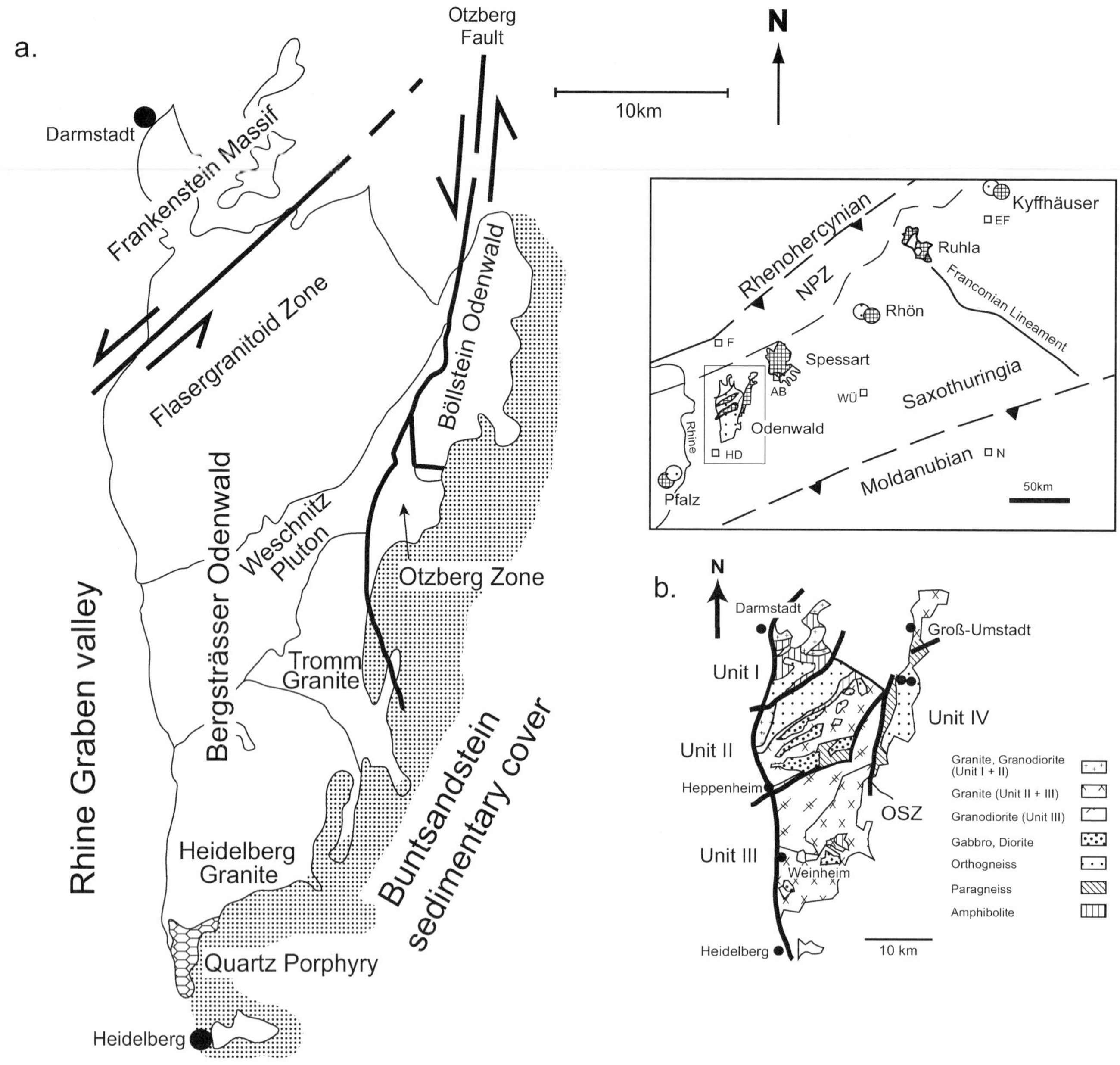

Fig. 9.31. (**a**) The Odenwald Crystalline Complex with its main structural units, the Böllstein Odenwald, the Otzberg Zone and the Bergsträsser Odenwald. (**b**) Geological sketch map of the Odenwald Crystalline Complex; Units I–III constitute the Bergsträsser Odenwald; Unit IV is the Böllsteiner Odenwald; OSZ, Otzberg Shear Zone. The boundaries between the respective units are after Krohe (1996) (after Stein 2001; Will & Schmädicke 2001).

2001). The former is derived from mantle melts intruded *c*. 360 Ma ago (Kirsch *et al.* 1988), which is *c*. 20 Ma before pluton emplacement in the Flasergranitoid Zone (Kreuzer & Harre 1975). Igneous activity commenced with the Frankenstein Complex at *c*. 365 Ma and continued in the central and southern parts after a hiatus of *c*. 25 Ma (i.e. at 340–325 Ma). The magmatic rocks, which are interpreted as having formed part of a magmatic arc, comprise gabbros, gabbrodiorites, diorites, granodiorites and granites which are calc-alkaline and have an I-type signature. They have a typical subduction-related geochemistry. The intrusions are either large, diapiric plutons (e.g. Frankenstein Complex, and the Weschnitz, Tromm & Heidelberg plutons) or are intimately associated with metamorphic rocks in the Flasergranitoid Zone (Stein 2001).

The igneous rocks of the Bergsträsser Odenwald show a general zonation from north (predominantly basic gabbros and diorites) to south (acidic granites) (Stein 2001). This zonation is reflected in the geochemistry which indicates a continuous shift from subduction-related island-arc granitoids in the northern part to an increasing influence of an active continental margin towards the south (Stein 2001). This zonation is in marked contrast to the complex lithological association of the Flasergranitoid Zone where gabbros and basic hornblende diorites (= Basischen Hauptgesteinszug) are concentrated in the south, while to the north the lithologies change to more acidic biotite diorites, granodiorites and granites. This internal zonation is completely different from that of the main series and may indicate, perhaps, a second north-dipping subduction zone (Reischmann & Anthes 1996).

The Bergsträsser Odenwald is characterized by a polymetamorphic history, with an earlier medium-pressure regional metamorphism, which must be older than 365 Ma, and a younger penetrative low-pressure/high-temperature imprint. Both events reached amphibolite facies conditions. Metamorphism can be correlated with granitoid intrusions (Stein 2001).

The **Otzberg Zone** is a vertical, sinistral shear zone, up to 1 km wide, consisting of NNE–SSW striking gneisses and amphibolites. It is interpreted as an extensional/transtensional strike-slip zone formed under brittle to ductile conditions during lower Carboniferous times (Krohe 1994). The westernmost part of the this zone, the Otzberg Fault, is regarded as the boundary between the Böllstein and Bergsträsser units of the Odenwald (Fig. 9.31).

Spessart

The Spessart Crystalline Complex comprises a suite of quartz diorite–granodiorite arranged in a series of formations or complexes (e.g. Okrusch & Richter 1986; Dombrowski *et al.* 1995; Hirschmann 1995; Weber 1995*b*) (Fig. 9.32). The central part of the Spessart Crystalline Complex comprises gneisses of granitic to granodioritic composition. These were deformed and metamorphosed under lower amphibolite facies conditions together with the pelitic, psammitic and subordinate calcareous sediments (Dombrowski *et al.* 1995). From south to north within the complex, the formations comprise: the Elterhof Formation (striated paragneisses with intercalations of amphibolites, marbles, calc-silicates rocks, hornblende gneisses and graphic quartzites); the Schweinheim Formation (biotite gneiss with interlayers of muscovite-biotite schists and some more quartzitic rocks); the Rotgneiss Complex (metagranitic muscovite-biotite gneiss with intercalated metabasites and ultramafics); the Mömbris Formation (metapelitic staurolite-bearing paragneisses with intercalated metapsammitic paragneisses and calc-silicate rocks); the Geiselbach Formation (garnet-bearing mica schists and quartz mica schists with intercalations of sericite quartzites); and the Alzenau Formation (alternating amphibolites, biotite gneisses, granitic gneisses and minor quartzites and calc-silicate rocks).

These formations may be grouped into at least two tectonic units. The lower unit comprises the Mömbris (Cambrian–Ordovician age; Bederke 1957) and Geiselbach (Upper Silurian age; Reitz 1987) formations and probably the Rotgneiss Complex (Hirschmann 1995). The Elterhof and Alzenau formations (Precambrian–Cambrian age; Fluck *et al.* 1980; Hirschmann & Okrusch 1988) constitute the upper unit. The position of the Schweinheim (?Haibach) Formation with respect to the upper and lower units is not clear (Hirschmann 1995). K–Ar ages from Spessart cluster between 324 and 318 Ma indicating rapid cooling in early Late Carboniferous times (Nasir *et al.* 1991). Similar ages (323–315 Ma) were obtained from the area east of Spessart (Schmidt *et al.* 1986).

Geochemical analysis of lamprophyres within the Spessart Crystalline Complex suggests that they were derived from a mantle source which was metasomatized during a previous (?Devonian) subduction event (von Seckendorff *et al.* 2004*b*). The generation of the parent melts was possibly triggered by partial melting of metasomatized mantle due to lithospheric detachment, removal and replacement of metasomatized lithospheric mantle by upwelling hot asthenospheric mantle. Lithospheric mantle detachment may have caused post-collisional Namurian uplift and cooling of the crust, and facilitated emplacement of lamprophyre dykes along fault zones at high crustal levels (von Seckendorff *et al.* 2004*b*).

Ruhla Crystalline Complex

The Ruhla Crystalline Complex (RCC) consists of four distinct tectonometamorphic units which were juxtaposed during Variscan convergence and late Variscan extension (Zeh 1996; Zeh *et al.* 2000) (Fig. 9.33). The Brotterode Formation and the Central

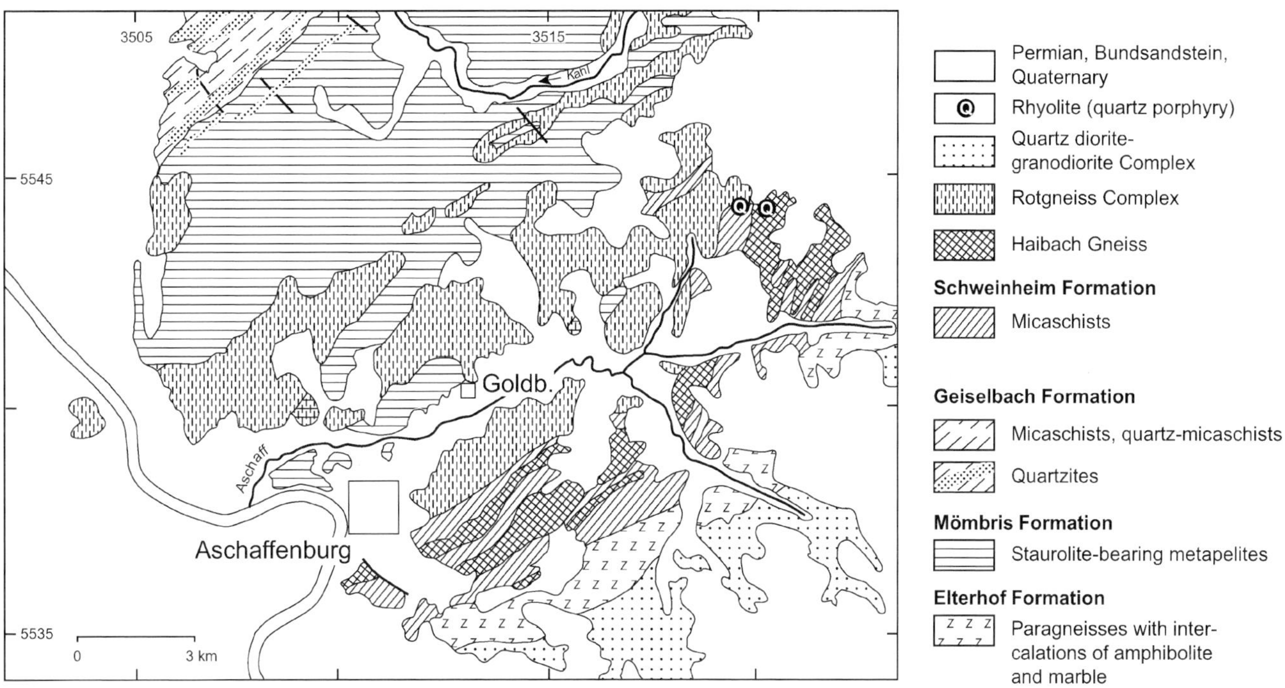

Fig. 9.32. Geological sketch map of the Spessart Crystalline Complex (after Okrusch 1983; Dombrowski *et al.* 1995).

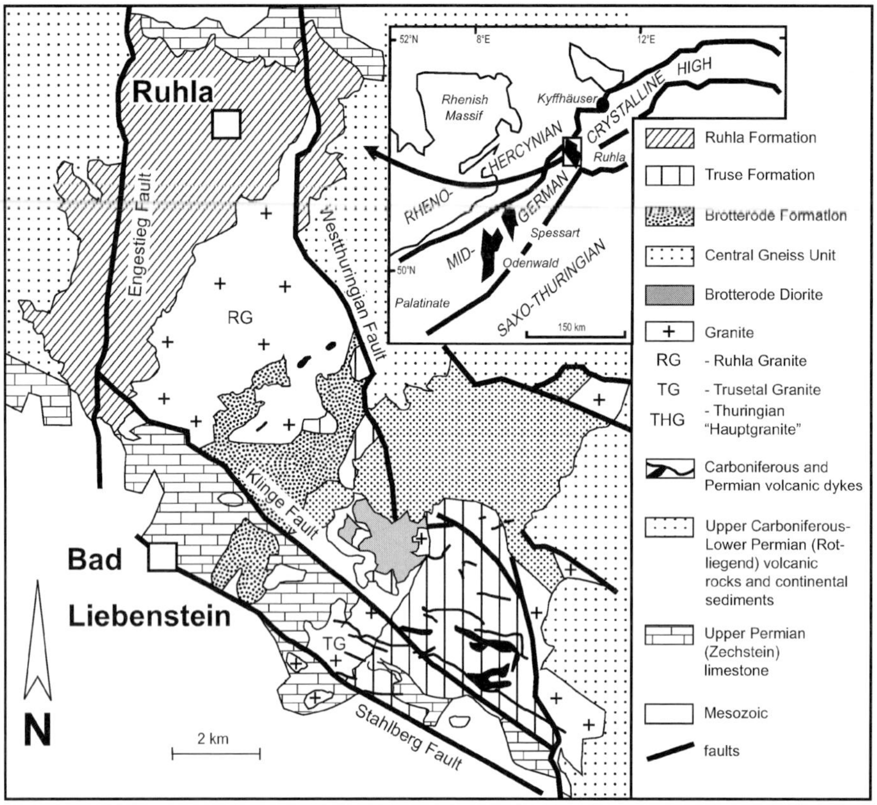

Fig. 9.33. Location of the Ruhla Crystalline Complex within Europe and a simplified geological map of the area (after Zeh *et al.* 2003).

Gneiss Unit consist predominantly of leucocratic migmatic gneisses, hornblende gneisses, amphibolites, metapelites and rare marbles. On either side are the Truse and Ruhla formations which predominantly consist of metapelties, quartzites and amphibolites. Bodies of granitic orthogneiss of variable size are intercalated within the metasediments of the Ruhla Formation (Zeh *et al.* 2000). These units are transected by the NW-trending Klinge and Stahlberg faults and the north–south trending Engestieg and West Thuringian faults. To the east of the West Thuringian Fault, rocks, including the Brotterode Formation and the Thuringian Main Granite, are directly overlain by a Stephanian-age continental succession comprising sediments and related volcanic rocks. To the west of the Engestieg Fault the Ruhla Formation is overlain by Stephanian volcanic rocks (Ilmenau Formation) and Lower Permian fanglomerates (Eisenach Formation) (Zeh *et al.* 2000) (see below).

The succession in the Ruhla area may be subdivided into western, central and eastern units (Fig. 9.33) which are separated from one another by NNE–SSW trending tectonic zones along which Variscan-age intrusions (e.g. Ruhla granite) occur. The western section (Thal-Ruhla area) comprises the Ruhla Group which consists of mica schists with intercalated metabasites, albite gneisses, quartzites, quartzitic mica schists, graphitic mica schists and rare calc-silicate rocks in its lower part (Cambrian age). The upper part (Ordovician–Silurian age) consists of mica schists and metapelitic gneisses, quartzitic mica schists and quartzites. The central section comprises the Liebenstein and the Schmalwasserstein-Rennweg areas which are separated from one another by a north–south trending fault zone which is partly marked by dioritic and granitic intrusions (Wunderlich 1989). The former is occupied by the Liebenstein Formation (biotite gneisses with intercalated quartz-feldspar schists and biotite schists in the lower part; migmatic biotite gneisses, hornblende gneisses and metatectic amphibolites in the upper part). Pre- to synkinematic granites are intimately connected with the Liebenstein Formation. The Schmalwasserstein-Rennweg area is characterized by metablastic or phylonitic gneisses. The eastern section comprises the Truse (biotite and muscovite-biotite gneisses with intercalated amphibolites, amphibolite-biotite schists, quartzites, graphitic quartzites, garnet-rich layers and calc-silicate rocks) and Hohleborn (feldspar gneisses alternating with muscovite-biotite gneisses with intercalated metabasalts) formations.

Both the Spessart and Ruhla areas expose, in the cores of antiformal structures, deeper structural levels containing higher grade equivalents of the Silurian–early Devonian magmatic arc rocks exposed in the Taunus segment of the NPZ (Franke 2000). According to Oncken (1997) these relationships indicate tectonic underplating of the Rhenohercynian crust beneath the nascent active margin of the MGCH (Fig. 9.34).

Between 425 and 395 Ma, the Truse Formation and the Central Gneiss Unit were intruded by numerous granites, which were subsequently overprinted during Variscan deformation (Brätz 2000). A second magmatic event is recorded by the Thuringian Hauptgranite which intruded the eastern margin of the RCC at *c.* 337 Ma (Zeh *et al.* 2000). During Late Carboniferous–early Permian times the RCC was extensively block faulted and intruded by the Trusetal Granite (*c.* 298 Ma), the Ruhla Granite (*c.* 295 Ma), the Brotterrode Diorite (*c.* 289 Ma) (Zeh *et al.* 2000), and numerous porphyry dykes.

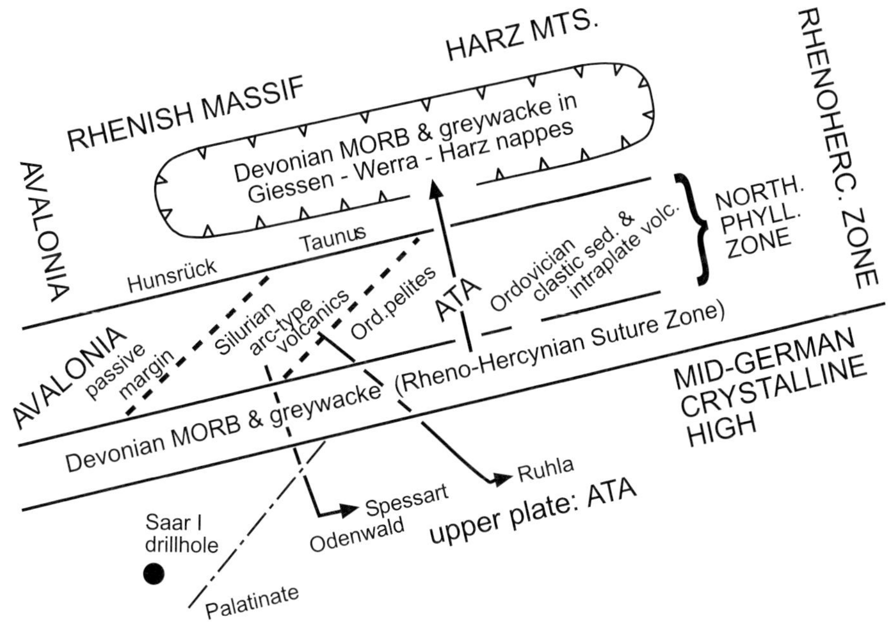

Fig. 9.34. Schematic representation of the tectonic relationships between the Rhenohercynian Zone, the Northern Phyllite Zone and the Mid-German Crystalline High. The Silurian arc-type volcanics illustrated between stippled lines mark the Rheic Suture Zone (ATA, Armorican Terrane Assemblage) (after Franke 2000).

Erzgebirge Crystalline Complex

The Erzgebirge forms part of the Saxothuringian Zone and is located at the northern margin of the Bohemian Massif, the largest outcrop of Variscan metamorphic basement in Central Europe (Figs 9.35 & 9.36). It has an exposed length of *c.* 120 km and a width of c. 45 km. To the SE, the Erzgebirge is bounded by the Eger Line, a NE–SW trending Cenozoic fault zone, whereas to the NE the border is defined by the Elbe Lineament, a strike-slip zone active since pre-Variscan times (see Scheck *et al.* 2002). It is composed of crystalline units, the Erzgebirge Crystalline Complex (ECC), in its main part and low-grade Palaeozoic successions at its western part and along the margins. The metamorphic gap between the high-grade crystalline rocks and the overlying low-grade successions is best explained by later tectonic juxtaposition, probably along an extensional detachment.

The lithological assemblage of the Erzgebirge represents a tectonic crustal stack originating from Variscan subduction and the collision of continental crust, followed by the exhumation of overthickened crust (Rötzler *et al.* 1998; Kroner & Hahn 2004). Detailed petrological studies have identified several metamorphic units with different pressure/temperature histories (e.g. Schmädicke *et al.* 1995; Schmädicke & Evans 1997; Willner *et al.* 1997; Rötzler *et al.* 1998). The Erzgebirge consists of at least five tectonometamorphic units, comprising, from base to top: (1) a medium pressure/medium temperature Gneiss Unit consisting of granitic orthogneisses, minor metagreywackes and small amphibolite bodies; (2) a high pressure/high temperature Gneiss/Eclogite Unit which includes granulite facies ortho- and paragneisses, eclogites, microdiamond-bearing gneisses, migmatitic gneisses, and kyanite-bearing micaschists; (3) a high pressure/low temperature Micaschist/Eclogite Unit containing mica schists, eclogites, granite orthogneisses and paragneisses; (4) a medium pressure/low temperature Garnet-Phyllite Unit with garnet-bearing phyllites, quartzites and minor amphibolites; and (5) a low pressure/low temperature Phyllite Unit comprising phyllites and quartz phyllites (Mingram *et al.* 2004). This latter unit represents an early Palaeozoic-age neritic to hemipelagic sedimentary succession belonging to the Thuringian facies (see Saxothuringian Basin section below). The high pressure/high temperature Gneiss/Eclogite Unit is separated from the medium-pressure/medium temperature Gneiss Unit and the Micaschist/Ecogite Unit by a complex zone of biotite- and two mica gneisses with intercalations of amphibolites, marbles and paragneisses. These zones with local mylonites and ultramylonites containing retrograded rocks from both units were named 'transition zones' by Sebastian (1995) and Rötzler *et al.* (1999). They are interpreted as large-scale shear zones.

The Erzgebirge was probably part of the Avalonian microcontinent derived from the NW margin of Gondwana. There are conflicting interpretations for the evolution of the Erzgebirge (for details see Mingram *et al.* 2004; Kroner & Hahn 2004). The schists and quartzites of the Phyllite Unit, the Garnet-Phyllite Unit, the Micaschist/Eclogite Unit and the Gneiss/Eclogite Unit of the Erzgebirge are lithologically and geochemically equivalent to Tremadoc and Arenig Thuringian facies which were probably deposited on a passive shelf margin of Gondwana (Mingram 1998). Ordovician protolith ages for the muscovite gneisses documents a phase of early Palaeozoic magmatism. The association of crustally generated high-silica rhyolites with tholeiitic to alkaline basalts in a thick (up to 3500 m) sedimentary succession, as suggested for the Erzgebirge, could be interpreted in terms of an extensional setting. The enormous thickness of shelf sediments reflects extreme subsidence, probably due to rifting, which culminated in the separation of the Saxothuringian Zone from NW Gondwana at the end of the early Ordovician (Linnemann *et al.* 2000). Bimodal volcanism as well as mafic to ultramafic intrusive complexes indicate extensive rifting processes in the Variscan Belt during the earliest Ordovician. In contrast to other early Ordovician bimodal suites from the

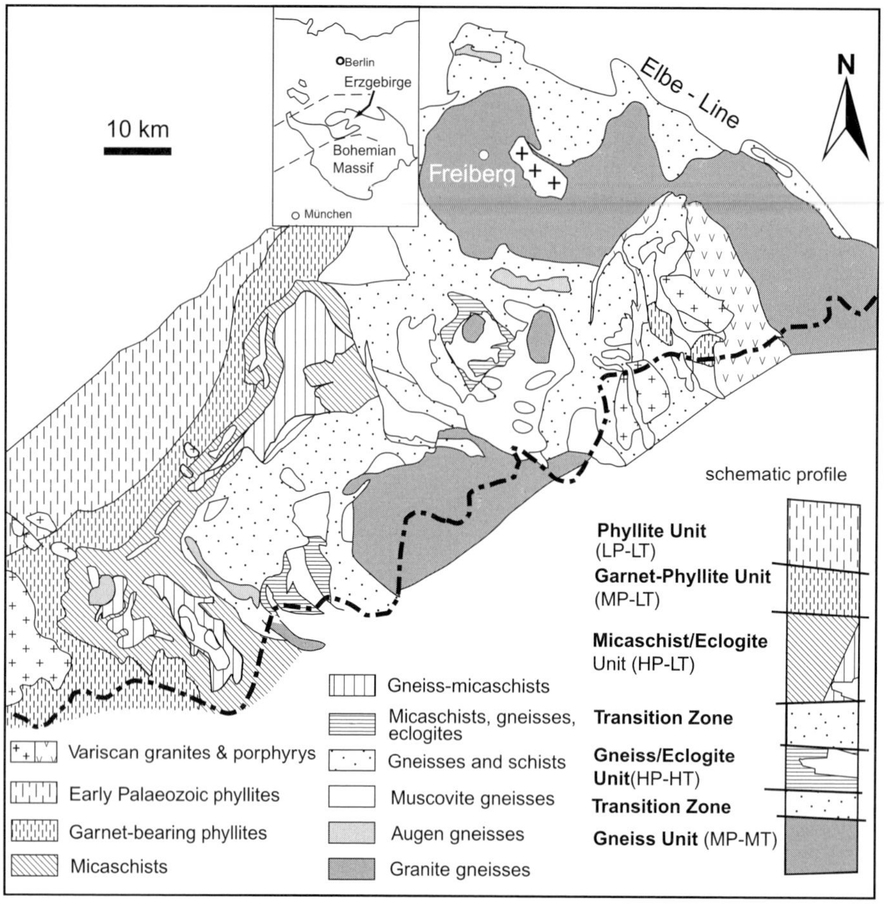

Fig. 9.35. Simplified geological map of the Erzgebirge (after Pietzsch 1962). The schematic profile depicts the different thrust units of the Erzgebirge as proposed by Willner *et al.* (1997) and Rötzler *et al.* (1998). The black lines indicate the tectonic boundaries (after Mingram *et al.* 2004).

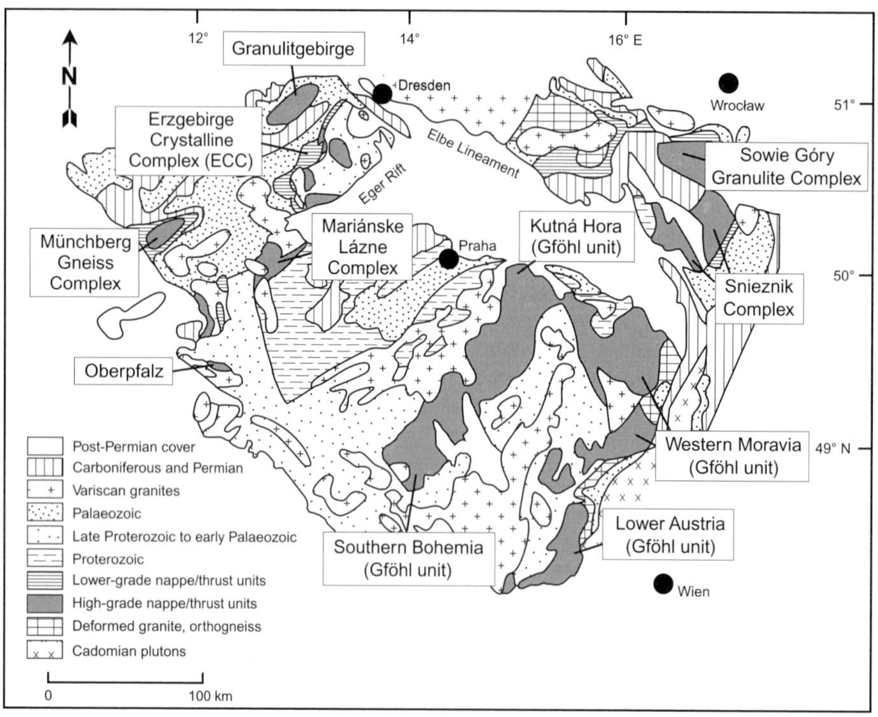

Fig. 9.36. Simplified geological map of the Bohemian Massif (after O'Brian & Carswell 1993; Matte *et al.* 1990).

European Variscides (e.g. Polish West Sudetes; Kryza *et al.* 1994), the magmatic suites of the Erzgebirge do not exhibit alkaline or peralkaline compositions. The protoliths represent crustal melts. According to Pin (1990), most of the bimodal suites are associated with rhyolites which probably resulted from melting at the base of the crust, possibly above an ascending plume. The generation of crustally derived magmatic rocks in a rift-related setting implies mantle-induced thermal activity. They could also indicate a change from passive to active margin setting as suggested for early Ordovician successions of the English Lake District (Cooper *et al.* 1993). Indeed, similar volcanic rocks are known from other, low-grade regions of the Saxothuringian Zone where they form part of a thick passive-margin succession reflecting a rift setting.

The early Palaeozoic succession underwent ultra-high-pressure metamorphism during late Palaeozoic subduction, followed by exhumation and stacking. This was followed by a major metamorphic and deformation overprint in the early Carboniferous, with a peak age of high pressure metamorphism at *c.* 342 Ma (Von Quadt & Gebauer 1998). Massonne (1999) discovered microdiamonds in quartz-muscovite gneisses from the Erzgebirge in Saxony providing evidence of ultra-high-pressure conditions (Nasdala & Massonne 2000).

Evidence from garnet-bearing ultramafic rocks suggests that mantle slices came into contact with crustal rocks either prior to, or during, the Variscan high-pressure compressional stage. An extensional phase preceding the Variscan compression has been invoked to explain a number of other features (e.g. Carswell 1991), and was certainly responsible for generating the magmatic protoliths of the eclogites. On the other hand, mantle and crustal rocks might have been brought into contact by compression during the Variscan orogenesis. A possible explanation is that the downgoing (continental) plate could have 'sampled' slices from the subcontinental mantle in the upper plate as a result of tectonic erosion during the early stages of crustal thickening.

The eastern Erzgebirge is the easternmost part of the Fichtelgebirge-Erzgebirge anticline, where late- to post-orogenic Variscan uplift and exhumation processes accompanied intense felsic (rhyolitic and granitic) magmatism that was controlled by brittle-fracture tectonics and formed the eastern Erzgebirge volcanoplutonic complex (Stemprok *et al.* 2003) (Fig. 9.37). The orogenic collapse of the Saxothuringian Zone as part of the Variscan Orogenic Belt contributed to the formation of volcano-tectonic intramontane depressions and strike-slip-related pull-apart basins in the marginal and central parts of the Erzgebirge. Caldera complexes were also formed (Müller *et al.* 2005). The eastern Erzgebirge volcanoplutonic complex displays textural and chemical evidence for multiple interactions of mafic and silicic magmas (Müller *et al.* 2005).

All metamorphic ages recorded at the eastern margin of the Saxothuringian Zone point to an important Early Carboniferous tectonothermal event (Konopásek & Schulmann 2005). However, other tectonic units which are exposed further to the west record Devonian metamorphic and cooling ages. This discrepancy may suggest that the Devonian metamorphism and early exhumation of certain units occurred during subduction of the Saxothuringian Ocean below the active continental margin of the Moldanubian-Teplá-Barrandian root zone. However, their final emplacement must have taken place during the Early Carboniferous collision. Franke (2000) proposed that southeastward subduction of the Saxothuringian Ocean commenced at least 395 Ma ago. However, the contemporaneous exhumation of high-pressure granulites and of underlying low-temperature eclogites during the Early Carboniferous (e.g. Schmädicke 1994) indicates the preservation of an active and long-lasting subduction-zone-thickened continental root system which terminated by thrust bounding of the Saxothuringian-Teplá-Barrandian and Moldanubian terranes at *c.* 340 Ma (Pitra *et al.* 1999; O'Brian 2000). Konopásek & Schulmann have presented a new model for the Carboniferous collision in the region, proposing that the juxtaposition of rocks

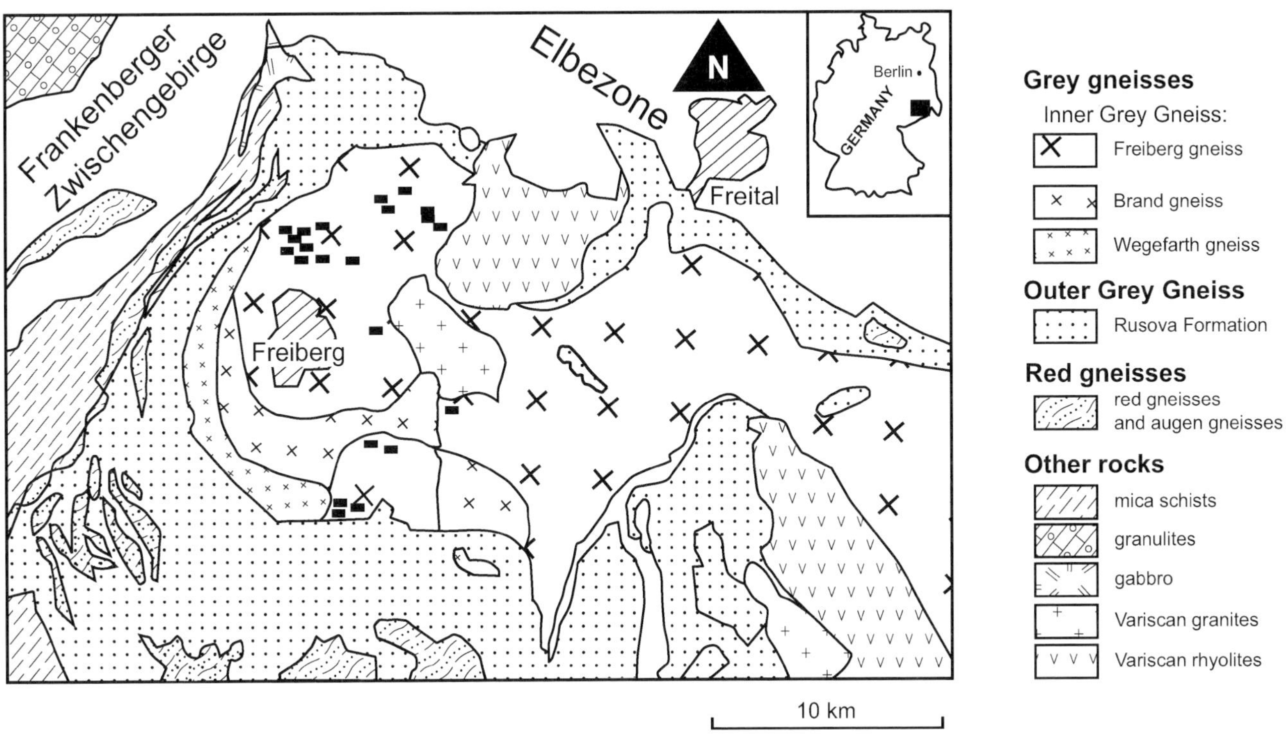

Fig. 9.37. Generalized map of the Freiberg anticline (Eastern Erzgebirge, Germany) (after Tichomirowa *et al.* 2001).

with contrasting thermal histories is the result of the subduction of the Saxothuringian crust and its later collision with a thickened continental orogenic root (Fig. 9.38). The short time-span between the formation ages of high-pressure rocks and the cooling ages of the host lithologies suggests that the exhumation of the coupled subduction zone–orogenic root system was extremely rapid (Konopásek & Schulmann 2005).

Krušné hory Mountains (Erzgebirge) area

The Saxothuringian domain of the Krusné hory (Erzgebirge) Mountains in the northern Bohemian Massif consists of metamorphic rocks exhibiting a polyphase tectonometamorphic evolution. Large amounts of eclogites have been noted (Klápová 1990; Hurych & Brueckner 1995). According to Schmädicke *et al.* (1995), the Saxothuringian eclogites of the Erzgebirge occur in three distinct high-pressure units which differ in peak P-T conditions and show a systematic decrease in the intensity of metamorphic conditions to the SW. The age of eclogite facies metamorphism is estimated at 360–333 Ma. Four major tectonometamorphic events have been recognized in the Czech part of the succession (Klápová *et al* 1998).

The structurally deepest Upper Proterozoic complex consists of grey gneisses and schists, metagreywackes, mica schists and metaconglomerates (cf. Fig. 9.37). The pre-Variscan metamorphic grade of basement rocks reached conditions of partial anatexis in the vicinity of red gneiss intrusions, and is dated at *c.* 573 Ma (Krentz 1985). In the hanging wall of this metamorphic succession, a supposed Lower Palaeozoic (Cambrian) succession occurs. This differs from the underlying rocks by the frequent occurrence of quartzites, quartz metaconglomerates and dolomite marble (Hoth *et al.* 1983). In the highest structural position, significant portions of allochthonous eclogitic rocks occur. These are surrounded by orthogneisses and a varied group of rocks including metapelites, marbles and magnetite skarns (Lorenz 1989).

Erzgebirge granitoids

Granitic rocks constitute a large proportion of the exposed rocks in the Erzgebirge and exhibit a great variety of compositional and textural types (Fig. 9.39). Indeed, the region contains almost the complete suite of silicic igneous rocks that characterize collisional settings, and these were emplaced within a small volume of the upper crust. The emplacement ages of the three groups of granites in the German Erzgebirge overlap between *c.* 325 and 318 Ma and emplacement occurred at shallow crustal levels. It post-dates the peak of collisional orogenic activity by *c.* 20 Ma and was associated with extension during post-orogenic collapse (Förster *et al.* 1999).

The plutons in the Erzgebirge region are composite, with up to four cogenetic subintrusions. Furthermore, the granites may be classified into three major groups, including low-F biotite granites, low-F two-mica granites, and high-F, high P_2O_3 Li-mica granites (Förster *et al.* 1999). The Erzgebirge late-collisional granites share many of the mineralogical and compositional features of granites from elsewhere in the German and European Variscides, with one important difference: highly evolved Li-mica and topaz-bearing granites are compartively abundant in the Erzgebirge. A possible explanation for this is, in part, the specific characteristics of the source rocks (Förster *et al.* 1999).

One of the key features of the Variscan Erzgebirge, as noted above, is the coeval emplacement of diverse granite types in a closely confined area shortly after the end of collisional orogeny. A commonly favoured hypothesis for the origin of the Variscan S-type granites involves crustal melting in response to radioactive heating within a thickened (>60 km) crust. The span of time between the culmination of thickening and the emplacement of the late-collisional granites is at least 20 Ma which is sufficient to allow for the buildup of radioactive heat to values

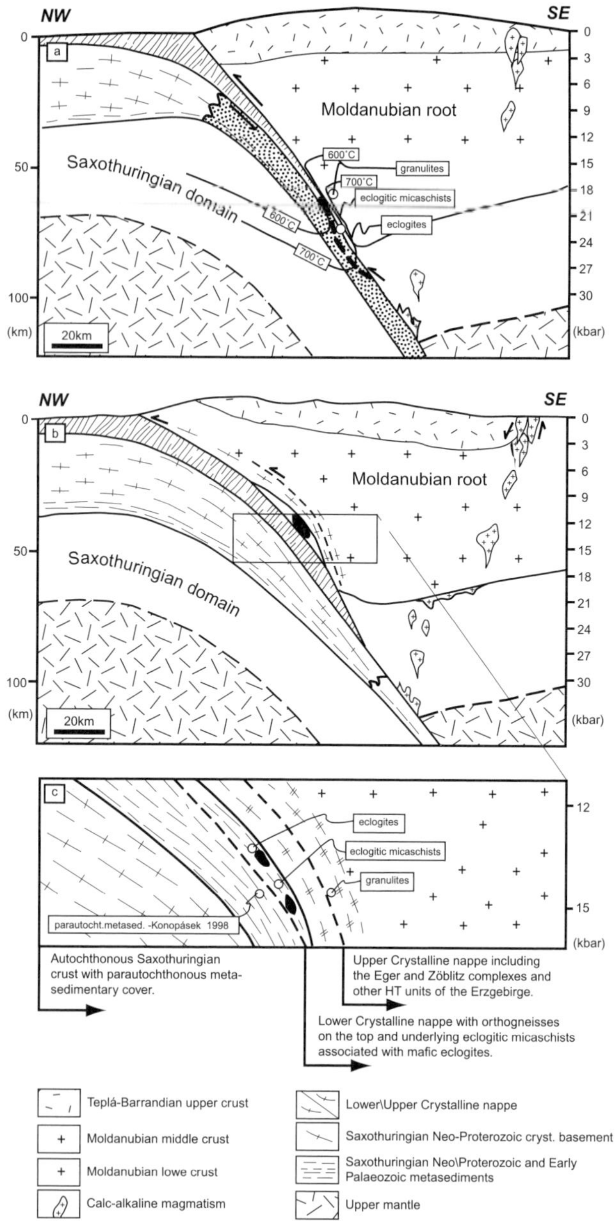

Fig. 9.38. Position of key lithologies indicated in a geological model of the western margin of the Bohemian Massif. (**a**) Situation corresponding to the time when all allochthonous lithologies (o) reached their maximium pressure (e.g. at 342 Ma). The bold dashed line indicates the detachment of a part of the eclogitized mafic Saxothuringian crust, together with overlying eclogitized metasediments, and their juxtaposition to the upper plate. (**b**) During the continuing collision, a new intraroot thrust was activated bringing granulitized lower crust (the Upper Crystalline nappe) on top of the mid-crustal rocks. At this stage, parautochthonous metasediments reach their maximum pressure. (**c**) Detailed Fig. 9.showing final relationships between parautochthonous unit and both Lower and Upper crystalline nappes prior to their final exhumation. After this stage, the whole lithological sequence was probably mechanically coupled, since the cooling ages from both allochthonous and parautochthonous rocks are identical (after Konopákek & Schulmann 2005).

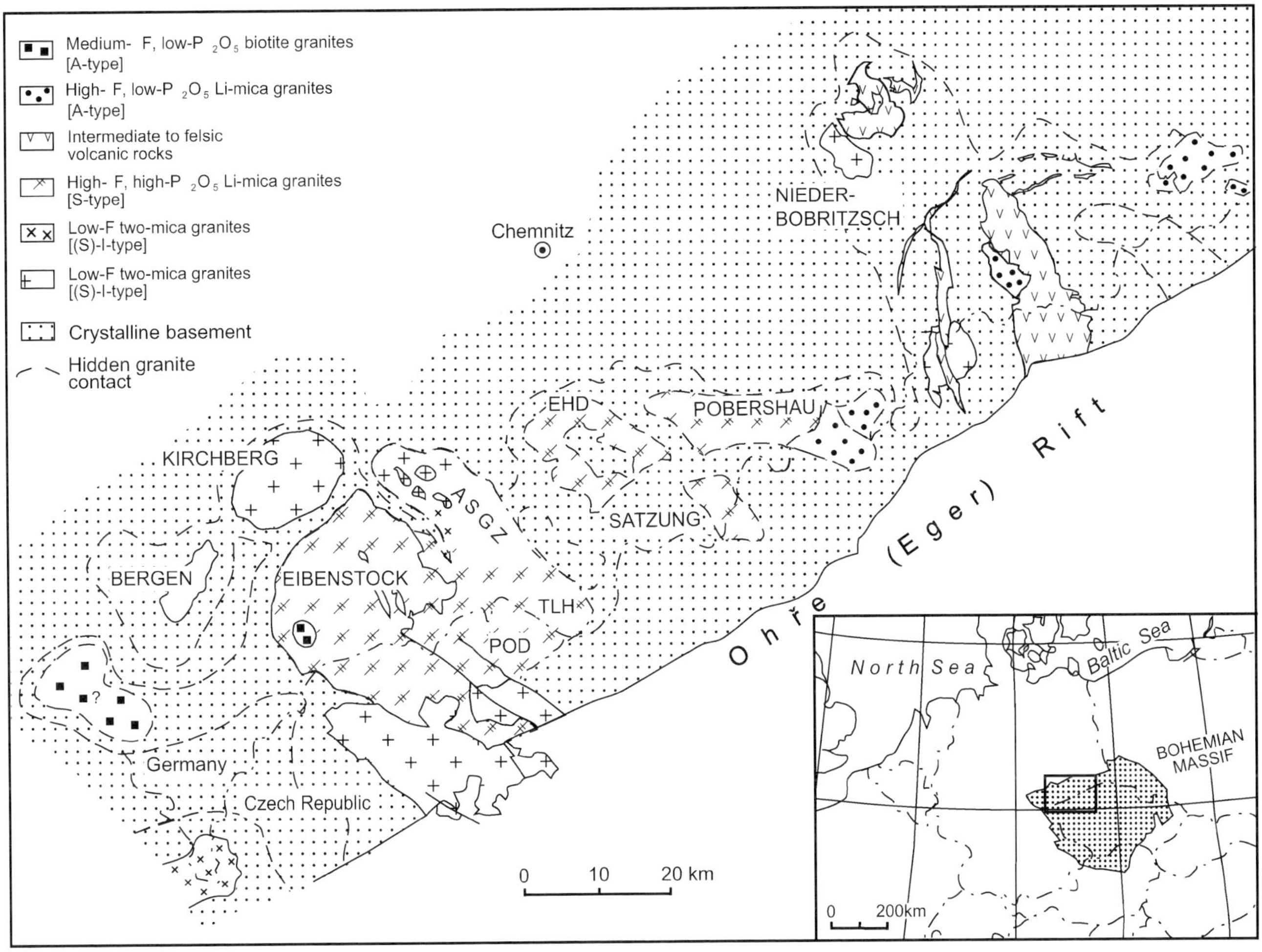

Fig. 9.39. Generalized geological map of the Erzgebirge showing the distribution of the different groups of Variscan granites and rhyolites. Abbreviations: ASGZ, Aue Schwarzenberg Granite Zone; EHD, Ehrenfriedersdorf; POD, Podlesi; TLH, Tellerhaüser (after Förster *et al.* 1999).

high enough for anatexis (Zen 1992). An alternative or additional heat source could have been the intrusion or underplating of the crust by hot mafic magma derived from the mantle (Förster *et al.* 1999). Delamination of the lithospheric mantle after crustal thickening is a possible mechanism to increase the heat input to the base of the crust and may have occurred in the Variscides (Zulauf 1997). Thus, the current evidence favours an intracrustal origin for the Erzgebirge granites with heat provided by radioactive decay in a thickened continental crust. Additional heat from mafic magmas cannot be ruled out, but does not seem to be required to explain the granites and is not supported by any tangible observations.

Northern Oberpfalz

Granitoids are represented in the northern Oberpfalz by a number of small- to medium-sized (*c.* 10–150 km^2) intrusive bodies (Fig. 9.40). The granites vary from coarse-grained, strongly porphyritic, mildly peraluminous, biotite and biotite-muscovite granites to weakly porphyritic or medium- to fine-grained, strongly peraluminous, biotite-muscovite and muscovite granites (Siebel *et al.* 2003). The granites are exposed in different areas (including the Moldanubian and Saxothuringian zones). Geochronological data suggest that metamorphic conditions in the Variscan crust produced different granite types at different times. The data support a model involving an early period of mantle-related magmatism which post-dates the final convergence stage of the Variscides. This magmatic activity was coeval with the thermal peak of regional metamorphism and was followed by a late period of crustal anatexis which was probably related to post-collisional extension of the Variscan crust (Siebel *et al.* 2003).

Fichtelgebirge Crystalline Complex

Together with the Erzgebirge, the Fichtelgebirge forms an anticlinal zone situated in the NW part of the Bohemian Massif (Fig. 9.41). It is predominantly characterized by weakly metamorphosed sediments and several generations of igneous rocks. Metasediments mainly comprise Cambrian–Ordovician successions and belong to the Thuringian facies which is typified by a neritic (Cambrian, Early Ordovician) and hemipelagic (Late Ordovician and younger) succession (Franke *et al.* 1995*a*) (see Saxothuringian Basin section, below). However, the age of the pre-Ordovician succession is controversial. Anticlinal structures originated in the Late Palaeozoic during a phase of late Variscan deformation under greenschist to lower amphibolite facies conditions (Stein 1988).

Magmatic activity in the Fichtelgebirge was marked by polyphase plutonic and volcanic cycles. The predominant

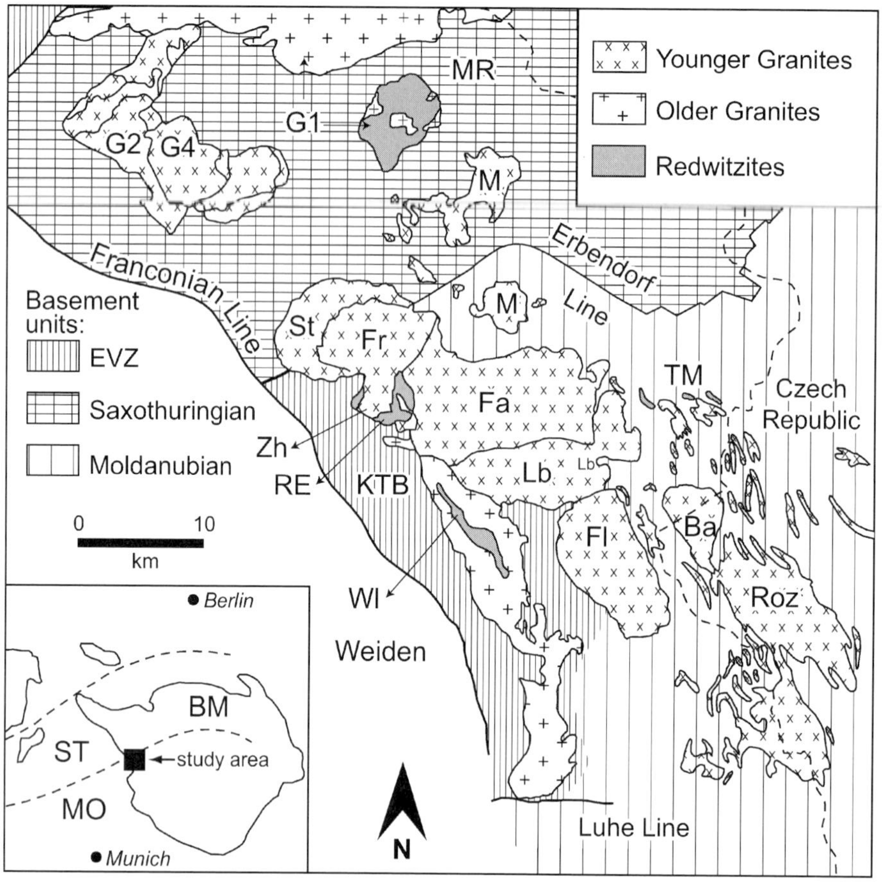

Fig. 9.40. Simplified geological map of the crystalline basement of NE Bavaria showing granitoid distribution. Abbreviations: Bä, Bärnau; EVZ, Erbendorf-Vohenstrauß Zone. Fa, Falkenberg; Fl, Flossenbürg; Fr, Friedenfels; Gl, Weißenstadt-Marktleuthen; G2-G4, younger granite series from the Fichtelgebirge; KTB, German deep-drilling project location; L, Leuchtenberg; Lb, Liebenstein; M, Mitterteich; MR, Marktredwitz; RE, Reuth-Erbendorf; Roz, Rozvadov granitoids; St, Steinwald; WI, Wurz-Ilsenbach; Zh, Zainhammer; (after Siebel *et al.* 2003).

magmatic sequences, both pre- and late Variscan, consist of large volumes of felsic rocks, while metabasites, characterized by amphibolitic and doleritic rocks of alkaline basaltic character, are subordinate (Okrusch *et al.* 1989). The pre-Variscan felsic igneous rocks have variously been referred to as gneisses, porhphyry gneisses and porphyry mylonites. Texturally, they can be subdivided into granitic orthogneisses and acid metavolcanic rocks. The orthogneisses form three main plutonic bodies, the largest of which, the Wunsiedel Gneiss (*c.* 40 km^2), is also the most variable in terms of its petrography and geochemistry. It was intruded into marbles and calc-silicate rocks and was later intruded by late Variscan granites. The Selb Granitic Gneiss straddles the border between Saxony, Bavaria and Bohemia, and extends up to 15 km in length and 1–2 km in width. It is in contact with both a quartz-mica schist unit and a late Variscan granite. The Waldershof Gneiss forms an elongate body 5 km long and 1–2 km wide. Contact metamorphism was due to intrusion of the Variscan Mitterteich Granite (Siebel 1995). Recent studies have examined similar, pre-tectonic gneisses and volcanic rocks from adjacent areas (e.g. southern Taunus, Spessart, Thuringia Forest, Erzgebirge, western Sudetes; see Siebel *et al.* 1997) integrating discussions of the compositional and temporal relationships of the Fichtelgebirge Early Palaeozoic magmatic rocks into a wider regional context (Fig. 9.42).

Saxonian Granulite Massif

The Saxonian Granulite Massif (SGM) is an oval dome structure extending over 45 km along strike within the Saxothuringian Zone in eastern Germany (Fig. 9.43). The NE margin of the SGM is faulted against weakly metamorphosed rocks of the Elbe Fault Zone, while the NW and SW margins are unconformably overlain by post-orogenic rocks. The structurally lower part of the SGM comprises high-pressure granulites (Granulite Complex) and is separated from the overlying low-pressure rocks (Schist Cover) by a detachment zone that contains up to a few hundred metres of isolated bodies of cordierite gneiss and orthogneiss, as well as a meta-ophiolite unit (Rötzler 1992; Reinhardt & Kleemann 1994; Rötzler *et al.* 2004).

The Granulite Complex is dominated by felsic granulites and retrograde products thereof, with rare interlayers of mafic granulite up to a few metres thick. The compositional layering is parallel to the main mylonitic foliation of the granulites. Also locally present are larger bodies of mafic granulite and serpentinized garnet peridotite. Geochemical and isotopic evidence indicates that the vast majority of the granulites have igneous protoliths and may have been derived from tholeiitic and calc-alkaline volcanic suites (Werner 1987).

Rötzler *et al.* (2004) have interpreted the rapid exhumation of the Saxonian granulites as due to the delamination of the subducted lithospheric slab and subsequent buoyant ascent.

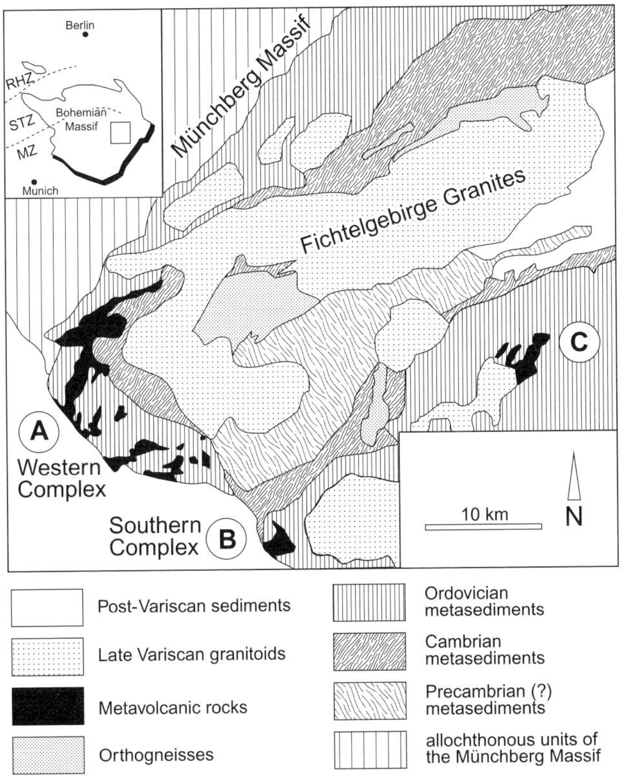

Fig. 9.41. Geological sketch map of the Fichtelgebirge. Inset map shows location of the area (box) in relation to the Rhenohercynian Zone (RHZ), Saxothuringian Zone (STZ) and Moldanubian Zone (MZ) (after Siebel *et al.* 1997).

Indeed, there are striking parallels between the SGM and the Erzgebirge to the SE, such as the age of metamorphism, the rapid ascent of high-pressure rocks to a mid-crustal level, the stress field evolution, and the cooling history. These similarities imply that the high-pressure rocks in these areas were jointly buried in the southern collision zone of the Saxothuringian Plate and then exhumed by the same mechanism (Willner *et al.* 2000). Thus, the spatial relationship between the SGM and the Erzgebirge may result from propagating orogenic growth and the ascent of high-pressure bodies during continued plate collision (Matte 1998).

Northern Vosges

The Vosges Massif is divided into two parts by a large shear zone (Lalaye-Lubine Shear Zone) which separates the Saxothuringian Zone from the Moldanubian Zone in the central and southern Vosges Mountains (Fig. 9.44). The northern Vosges, to the north of the shear zone, consists of non- and weakly metamorphosed Devonian–Lower Carboniferous active continental margin successions intruded by a calc-alkaline magmatic series ranging from diorites to granites with Late Visean ages (Hess *et al.* 1995). The central and southern parts of the Vosges Massif to the south of the shear zone belong to the Moldanubian Zone (see below).

The northern Vosges shows a decrease in metamorphic grade and stratigraphic age from south to north. In the southern part of the area, phyllites of Cambro-Ordovician to Silurian age are reported (Reitz & Wickert 1989). Further north there are Devonian metabasalts, pyroclastics, conglomerates and sandstones overlain by Frasnian black shales and Famennian–early Carboniferous greywackes, conglomerates and mudstones (Braun & Maass 1992). These early Carboniferous deep-marine sediments are up to 1000 m thick and were presumably derived from southern sources. Sedimentation was accompanied by repeated submarine volcanic activity that terminated with a pulse of subaerial, explosive volcanism (Wickert & Eisbacher 1988). Volcanic activity was coeval with the intrusion of I-type granitoids. Syndepositional deformation of the Carboniferous succession by wrench faulting and north-verging thrusting occurred around 350–330 Ma (Eisbacher *et al.* 1989) while post-tectonic I-type granitoids were intruded in the NW between 336 and 330 Ma (Altherr *et al.* 2000).

Exotic metamorphic nappes

As previously noted, the parautochthonous Thuringian facies of the Saxothuringian Basin was overthrust from the SE by nappes which have survived as a series of tectonic klippen (i.e. Münchberg, Wildenfels, Frankenberg; Fig. 9.3). These klippen represent erosional remnants of one of several nappe piles which were originally more extensive, and possibly laterally continuous with one another (Franke *et al.* 1995*b*). Within each of the klippen, the tectonic sequence is characterized by inversion of the stratigraphy and the metamorphic grades. The lower nappes comprise, from bottom to top, proximal Early Carboniferous deep-marine sediments, Silurian–Devonian bedded chert, and a volcanosedimentary succession of largely Ordovician age (Franke *et al.* 1995*b*).

The **Münchberg Nappe Pile** consists of four metamorphic nappes. The Upper Sequence (= Hangend-Serie) consists mainly of banded hornblende gneisses and amphibolites, as well as paragneisses with some calc-silicate rocks and marbles. The Lower Sequence (= Liegend-Serie) comprises pelitic and quartzofeldspathic paragneisses with intercalations of an extensive sheet of felsic orthogneisses and some smaller metagabbro and metagranodiorite bodies. The Marginal Amphibolites (= Rand-Amphibolit-Serie) comprise mainly massive, banded and schistose amphibolites. The upper and lower parts of this unit have undergone pervasive mylonitization. The final unit, the Prasinite-Phyllite Sequence (= Prasinit-Phyllit-Serie) consists of metamorphosed basic to intermediate lavas and pyroclastics. The metavolcanics are subalkaline with a calc-alkaline tendency and are compatible with an island arc or backarc setting (Oppermann 1985). Overall, the lithologies comprise sediments and mafic volcanic rocks in amphibolite facies (= ?Cadomian basement of the Teplá-Barrandian margin), eclogites and serpentinites (= Saxothuringian oceanic or thinned continental crust), amphibolite-grade metasediments intruded by felsic and mafic plutons, metabasalts and marble lenses, a prasinite–phyllite sequence.

The **Wildenfels Complex** comprises mainly banded hornblende gneisses and garnet amphibolites in the upper part. There are also some felsic intrusive bodies (now orthogneiss). Moving down-sequence is a unit of mica schists and metaquartzites which is underlain by unmetamorphosed Palaeozoic successions of the Thuringian facies (Franke *et al.* 1995*b*).

The **Frankenberg Complex** consists of three units which are very similar to the Münchberg succession. The top unit is the Hornblende-Muscovite Series which comprises hornblende-bearing schists, metapelites, some orthogneisses, and frequent intercalations of mafic rocks. The middle unit, the Frankenberger Gneiss, consists of orthogneisses with some intercalated metasediments. The bottom unit, the Prasinit Series, is composed of

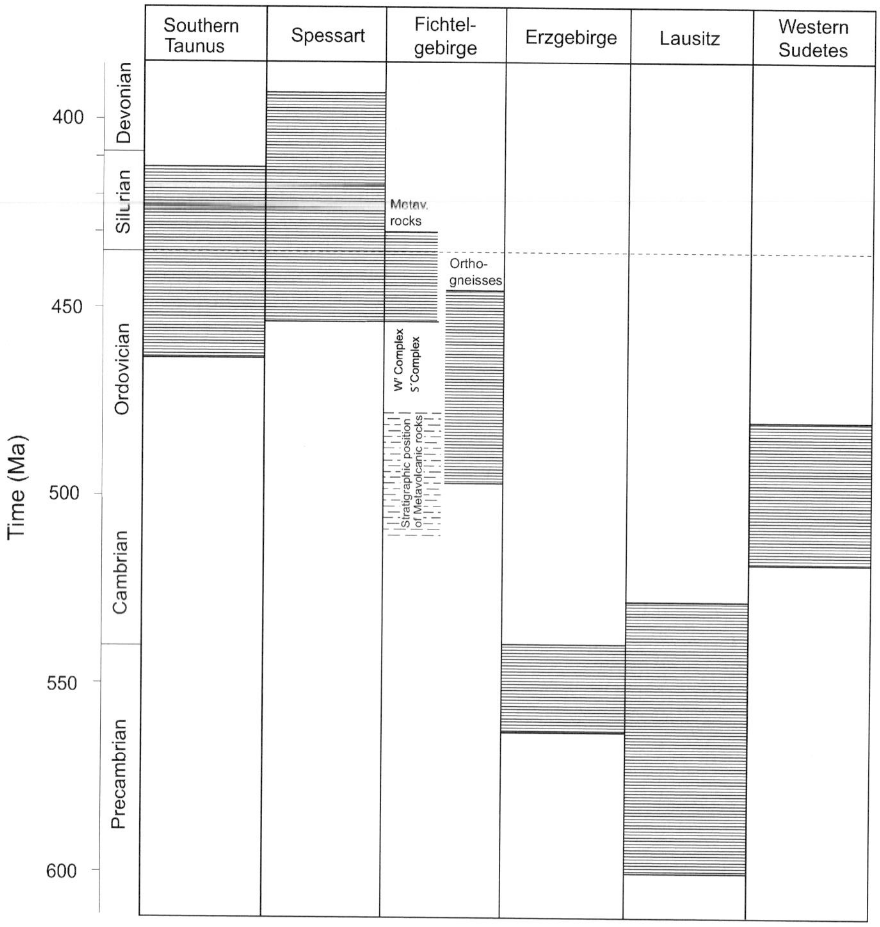

Fig. 9.42. Time chart showing pre-Variscan magmatic activity within the Saxothurianian Zone and adjacent areas. References: Southern Taunus (Sommermann *et al.* 1992); Spessart (Lippolt 1986; Dombrowski *et al.* 1995); Fichtelgebirge (Teufel 1988; Wiegand 1996; Mielke *et al.* 1996; Siebel *et al.* 1997); Erzgebirge (Kröner *et al.* 1995); Lausitz (Korytowski *et al.* 1993; Kröner *et al.* 1994); Western Sudetes (Borkowska *et al.* 1980; Korytowski *et al.* 1993; Oliver *et al.* 1993). Time scale based on Odin (1994) (after Siebel *et al.* 1997).

various mafic lavas, rare rhyolites, pyroclastic rocks, metapelites, cherts, marbles and some greywackes.

Saar-Nahe Basin

This Permo-Carboniferous basin extends from SW Germany into France and is filled with exclusively continental sediments (Fig. 9.45). The basin has a half-graben morphology, being bordered to the north by the south-dipping Hunsrück Boundary Fault (HBF) which, according to Henk (1993), is a detachment soling out at mid-crustal levels (at a depth of *c.* 16 km). Alternatively, the South Hunsrück Fault is interpreted to be a steep dextral strike-slip fault with a considerable transtensional component of movement (Korsch & Schäfer 1991, 1995). The basin has a SW–NE lateral extent of *c.* 300 km and is *c.* 100 km wide. The evolution of the Saar-Nahe Basin (SNB) is closely related to the complex kinematics of the HBF which is part of a prominent suture zone separating the Rhenohercynian and the Saxothuringian zones. Subsidence along this fault probably began in the Namurian and definitely occurred during the Westphalian with an increasing subsidence rate after Westphalian C–D times (von Seckendorff *et al.* 2004*a*). The sedimentary development of the SNB was characterized by a continuous shift of the depocentre nearly parallel to the fault. During the Namurian and Westphalian the depocentre was situated in the southwesternmost part of the basin, and moved towards the ENE during the Stephanian and Rotliegend. The two most important Carboniferous units are the Westphalian Saarbrücken Group and the overlying Stephanian Ottweiler Group (Schäfer 2006).

In the Saar 1 borehole, the crystalline basement of the SNB is represented by a weakly deformed aplitic albite granite which is covered by a non-metamorphosed shelf succession (*c.* 1000 m thick) of Middle–Upper Devonian marine limestones with intercalated clastics (Hering *et al.* 1976; Kneuper 1976) (Fig. 9.46). Upper Devonian sandstones contain much volcanic debris suggesting an island arc as the source area (Weber 1995*b*). Marine sedimentation, comprising black alum shales with intercalated cherts and volcanic-rich turbidites, continued into the Lower Carboniferous. At the end of the Lower Carboniferous there was a marked phase of thrusting within the Saxothuringian Zone (Sudetian phase). Indeed, evidence of thrusting and crustal stacking can be seen in the DEKORP 1C profile (see Weber 1995, fig. 2). The timing of this event correlates with the switch from subduction to collision in the Eastern Rhenish Massif (Weber 1995*b*).

Basin initiation was in latest Namurian or possibly earliest Westphalian times and is marked by the deposition of a basal conglomerate (Weber 1995*b*). There is an angular unconformity between the Westphalian D and Stephanian A, indicative of a reorganization of fault kinematics in the area. In the area of the

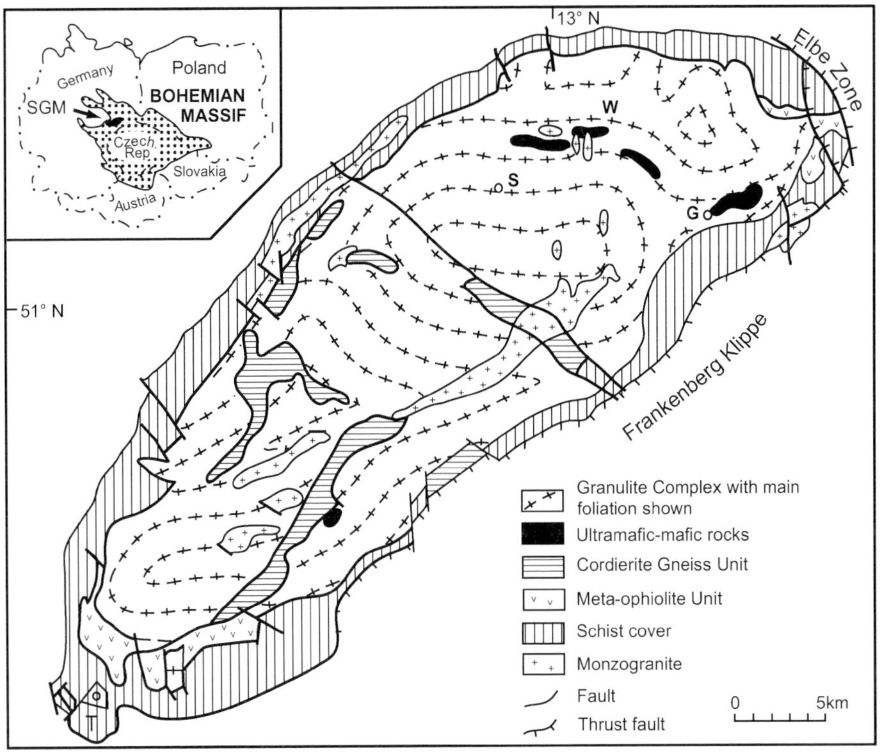

Fig. 9.43. Geoloigcal map of the Saxonian Granulite Massif (SGM). Locations: G, Greifendorf; S, Schweikershain; T, Tirschheim; W, Waldheim (after Rötzler *et al.* 2004).

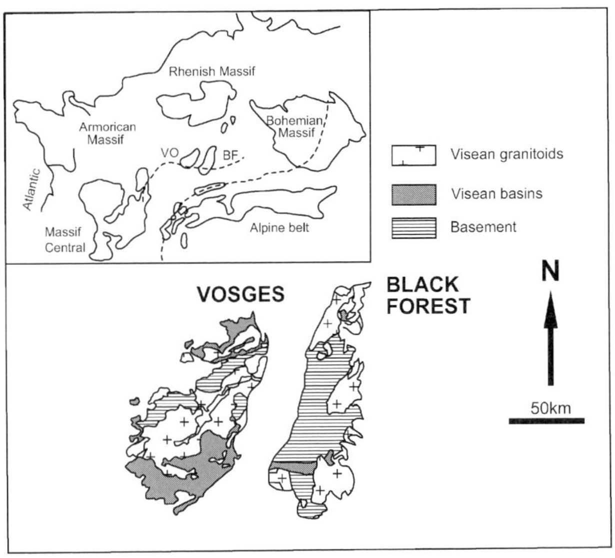

Fig. 9.44. Geological sketch map of the Vosges and Black Forest area (after Schaltegger 1997).

Saarbrücken Anticline, this resulted in the removal of up to 1.5 km of sediment (Engel 1985). Major uplift in the SE part of the basin elevated the basement to provide a new sediment source (Korsch & Schäfer 1995). The sedimentary infill of the basin is dominated by continental clastics, predominantly fluvial, lacustrine and alluvial fan deposits, which accumulated from latest Namurian C–Westphalian A to Permian times (Weingart 1976; Schäfer 1989, 2005; Schäfer & Korsch 1998).

Within the *c*. 6500 m thick continental sedimentary fill of the basin, a large variety of igneous rocks were emplaced over a period of *c*. 4 Ma from 296 to 293 Ma, as high-level intrusions and lava flows, extrusive domes, diatremes and pyroclastic deposits, ranging in composition from basalt and basaltic andesite to rhyodacite, rhyolite and trachyte (von Seckendorff *et al.* 2004*a*). Composite intrusive–extrusive complexes consist of andesite, rhyodacite and alkali feldspar trachyte with up to 10 wt% K_2O. The geochemical characteristics of the most primitive mafic magmas indicate a slightly enriched upper mantle source modified by subduction-related fluids.

The geodynamic evolution of the SNB is coupled with the development of the Mid-German Crystalline High (von Seckendorff *et al.* 2004*a*). According to the models of Franke & Oncken (1990) and Oncken (1997), the closure of the ocean between the Rhenohercynian and Saxothuringian microplates began in the Late Devonian (Frasnian, *c*. 375 Ma). The oblique southward subduction of oceanic crust beneath the Saxothuringian microplate ended in the Tournaisian (*c*. 355 Ma; Franke & Oncken 1990). To the NW of the MGCH, the 'Devonian Saar precursor basin' (*sensu* Oncken 1997) can be interpreted as a forearc basin on the edge of an evolving continental margin (Oncken 1997). In the 'precursor Saar Basin', flat-lying undeformed Middle Devonian–Lower Carboniferous marine sediments overlie a weakly metamorphosed granite of Ordovician age (Sommermann 1993).

The west–east extensional stress regime that dominated during the period of post-Westphalian basin formation (Stollhofen 1998) was oblique to the pre-existing fault pattern, the common slip direction was maintained by transtensional strike-slip movements

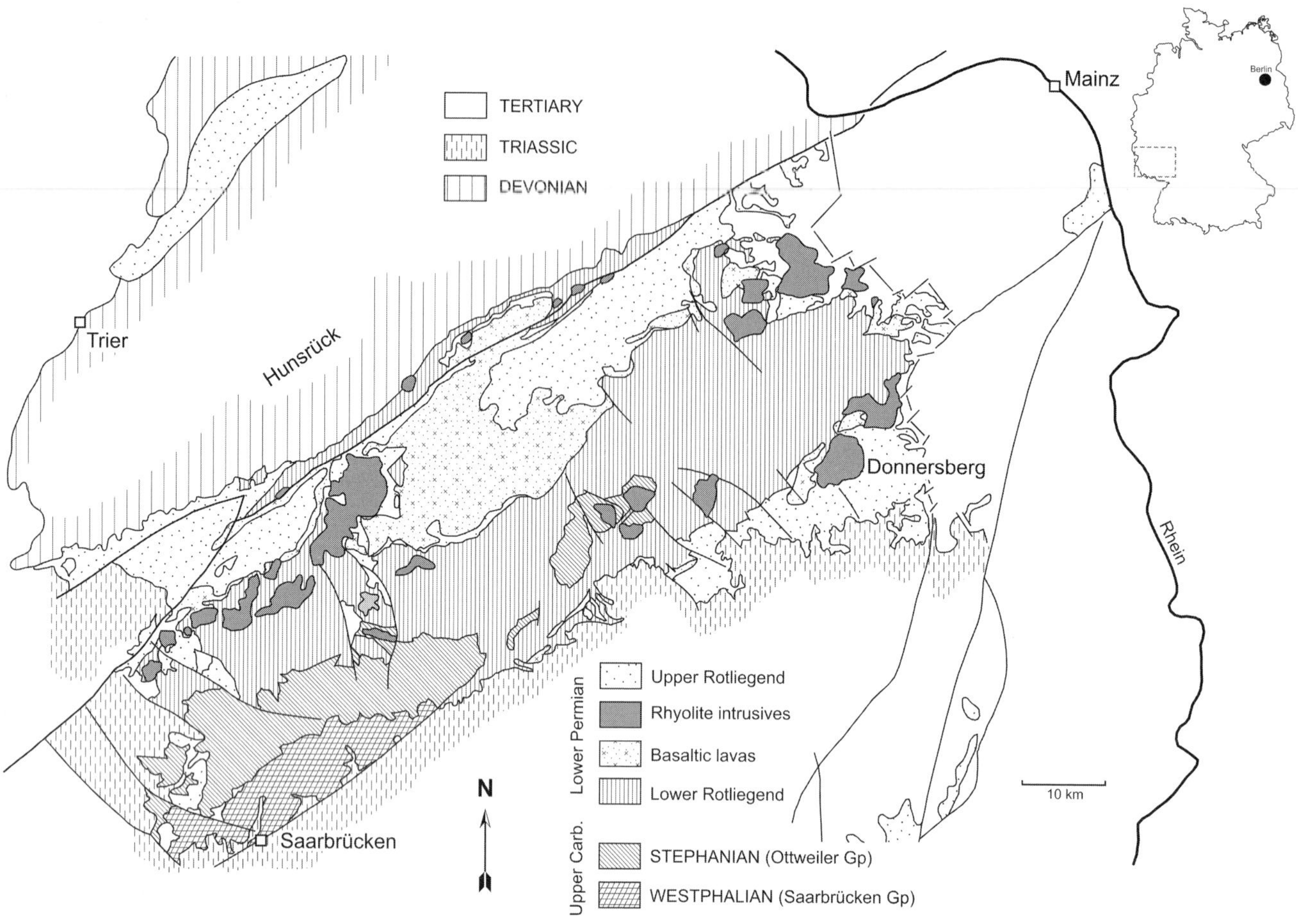

Fig. 9.45. Geological sketch map of the Saar-Nahe Basin, SW Germany.

on the transfer faults and oblique-slip motion on the normal faults (Stollhofen *et al.* 1999). An angular unconformity, indicative of a reorganization of fault kinematics in the area, underlies the synrift megasequence. The Stephanian–Lower Rotliegend pre-volcanic synrift sequence has a thickness of 3.8–4.7 km and was deposited over a period of *c.* 14 Ma (see also McCann *et al.* 2008). It comprises lacustrine, fluviodeltaic and fluvial sediments with minor limestones, coals and pyroclastic fallout deposits (Stollhofen *et al.* 1999). The fallout deposits are predominantly rhyolitic to ryhodacitic and were probably derived from sources in the Black Forest and Vosges areas (Königer & Lorenz 2002; Königer *et al.* 2002). The pre-rift succession is overlain by the Upper Rotliegend volcanic synrift sequence (1.1 km thick, deposited over *c.* 4 Ma) where widespread bimodal calc-alkaline magmatism occurred with subvolcanic intrusion of rhyolitic/dacitic domes and basaltic to andesitic sills and dykes. Magma generation can be attributed to underplating and the intrusion of pulses of mantle-derived melts into the crust inducing partial melting. The extrusive rocks are interbedded with fluvial sediments and minor lacustrine units (Stollhofen 1994).

Saale Basin/Thuringia Forest Basin (J.S.)

The Saale Basin *sensu stricto* is a continental basin, 150 km long and 90 km wide, extending from the Thuringian Forest in the SW across the present-day Thuringian Basin as far as the Elbe Line to the NE where it joins with the Variscan foreland basin (Fig. 9.47). It originated during latest Carboniferous times as a half-graben structure located along the suture between the Rhenohercynian and the Saxothuringian zones and above the MGCH (Schneider *et al.* 2005*b*). The older Carboniferous sediments were eroded and deformed to differing degrees. Subsidence commenced in early Stephanian times and was related to Variscan post-collisional extension (Asturian event). Older Carboniferous sediments, of Visean to Westphalian age, have been reported from boreholes (Steinbach 1990; Gaitzsch *et al.* 1999). These rest unconformably on deformed Precambrian–Cambrian sediments.

The oldest formation, the Klitzschmar Formation (Visean, with a maximum thickness of 1400 m) comprises alluvial fan, debris flow and fluvial channel coarse clastics as well as floodplain, lacustrine and palustrine deposits with local impure coals and carbonaceous shales. In the upper part of the formation dacitic to rhyolitic pyroclastics (up to 20%) may be present (Fig. 9.48). The unconformably overlying Namurian A Sandersdorf Formation (>400 m) comprises braided river deposits from a floodplain environment with transitions to coastal plain and shallow-marine/brackish deposits. The latter are indicated by the occurrence of the inarticulate brachiopod Orbiculoidea. Additionally, the oldest known flying insect, *Delitzschala bitterfeldensis* (Brauchman & Schneider 1996), was found in these deposits. Deposition occurred in a wide valley which to the north extended into a

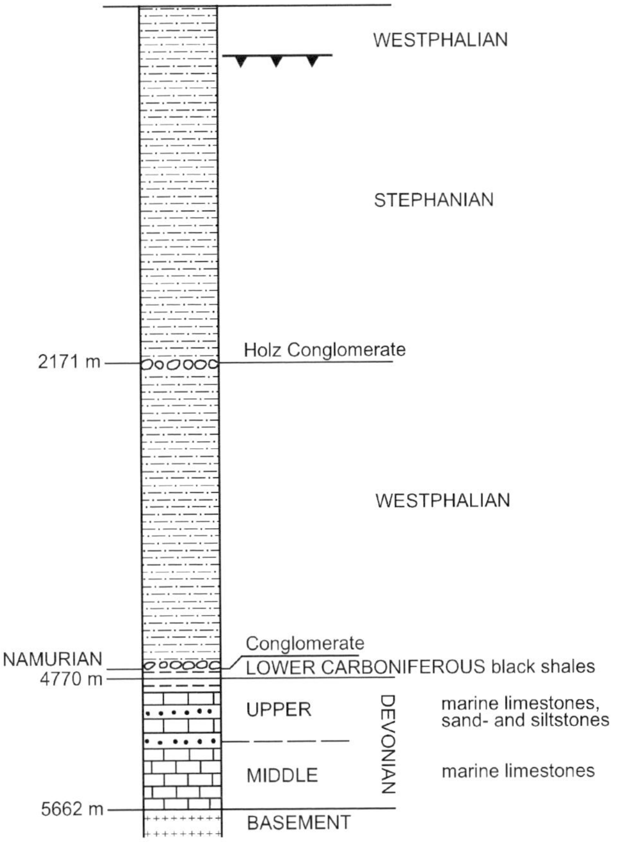

Fig. 9.46. Stratigraphy of the Saar 1 borehole, from the Saar-Nahe Basin, SW Germany.

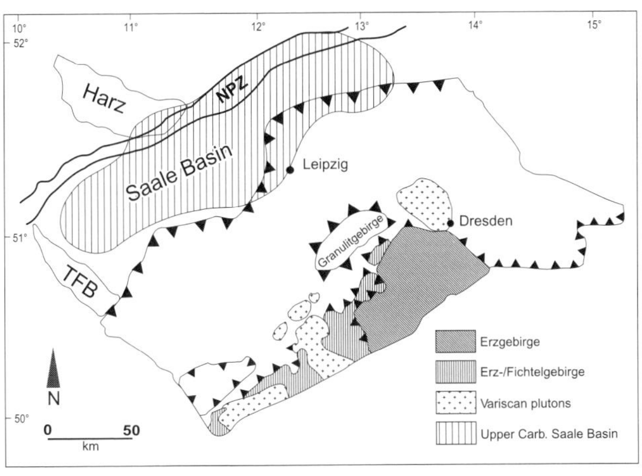

Fig. 9.47. The Saale Basin in Upper Carboniferous times. Abbreviations: NPZ, Northern Phyllite Zone; TFB, Thuringia Forest Basin (after Linnemann *et al.* 2003).

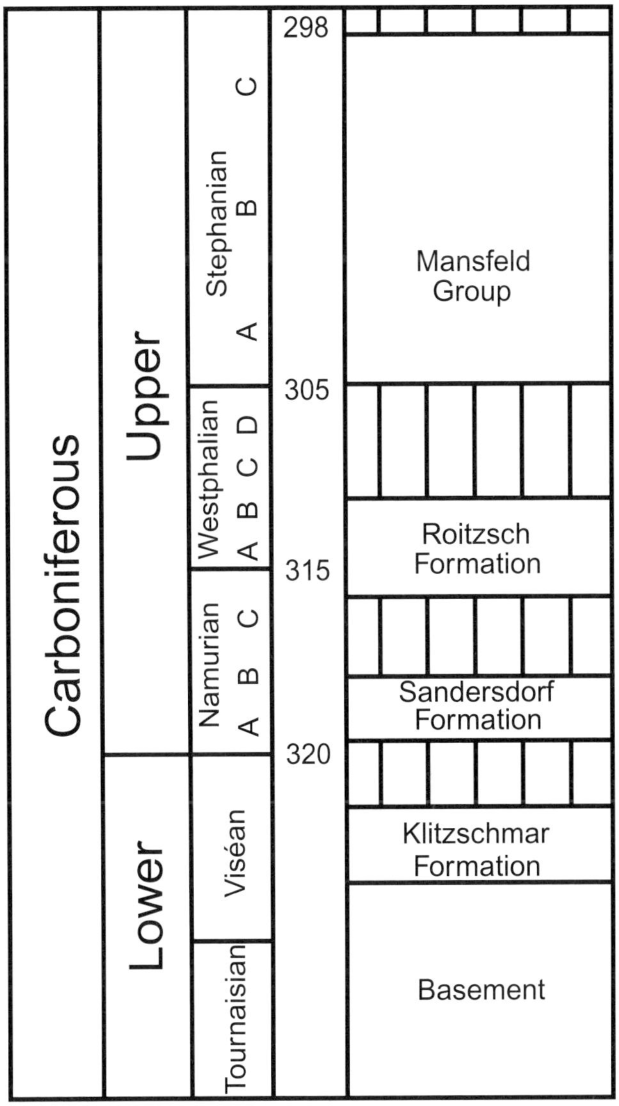

Fig. 9.48. Schematic profile of the Carboniferous in the Saale Basin (after Rappsilber 2003).

shallow-marine bay. This opened into the Variscan foreland basin to the west of Berlin (Gaitzsch *et al.* 1999). These deposits thus form an important link between the earliest continental successions of the Variscan internides and the marine turbidites deposited in the foreland basin to the north.

Isolated wells in the area of Leipzig have cored Westphalian-age continental sediments of the Roitzsch Formation which unconformably overlie the Namurian strata. This succession attains a minimum thickness of 450 m, and commences with a basal unit (150 m) of coarse clastics (Westphalian A) overlain by conglomerate/rooted soil cycles (Westphalian B/C) and containing coal seams (several decimetres to 1.0 m thick) and siltstones rich in plant detritus. The Westphalian deposits are related to the southern border of the Variscan foredeep as a valley fill which formed part of the drainage system dewatering the Central Bohemian Basin and the Westphalian-age basins of the Erzgebirge (Gaitzsch *et al.* 1999: Oplustil & Pesek 1998). In Westphalian C times, there is evidence for cyclicity in upper delta plain sediments in the region of the Boizenburg 1/74 well (for well location see Hoth *et al.* 1993).

The unconformably overlying Stephanian succession commenced with the deposition of the red-beds of the Mansfeld Group (up to 1500 m) which were deposited in a proximal to distal alluvial fan setting. The Mansfeld Group can be subdivided into three upward-fining megacycles: (1) the Gorenzen Formation (60–100 m) comprising an alluvial fan/debris flow association with related fluvial channels and lacustrine deposits with impure coal seams and carbonaceous shales; (2) the Rothenburg Formation (*c.* 600 m) consisting of an alluvial fan/braided river to alluvial plain and lacustrine succession with minor carbonaceous shales; and (3) the Siebigerode Formation (200–800 m) which comprises an alluvial fan/braided river system with associated floodplains and lacustrine to palustrine successions and coal seams. A characteristic feature of the Siebigerode Formation is the deposition of arkosic sandstones rich in kaolinized feldspar which were derived from palaeohighs of the MGCH to the south and SE of the basin. The uppermost part of all three formations comprises lacustrine black shales and coal seams. The latter have been worked in the Plötz-Wettin region (north of Halle). Macroflora, insects and amphibians have provided a Stephanian C age for the facies (Schneider *et al.* 2005*a*, *b*). U–Pb-SHRIMP ages are 297 ± 3 to 301 ± 3 Ma (Breitkreuz & Kennedy 1999).

The onset of the Rotliegend is heralded by the volcanotectonic reorganization of the basins (Schneider *et al.* 2000). This event is indicated by the rapid deposition of the Kieselschiefer-Quarzit-Konglomerat (chert-quartzite conglomerate) at the base of the Lower Rotliegend Halle Formation.

The Thuringian Forest Basin (Lützner *et al.* 2003; Anderas *et al.* 2005; Schneider 1996) is filled with up to 6000 m of sediments and volcanics of Carboniferous and Permian age (Fig. 9.47 & 9.49). The basin is *c.* 40–60 km wide, orientated NW–SE and formed on top of the deeply eroded Visean-age Thuringian Main Granite and the Ruhla Crystalline Complex (see above). Basin formation was related to tectonic activity, including NW–SE orientated block faulting/uplift (i.e. perpendicular to the trend of the Saale and Kraichgau basins). Additionally, the Thuringian Forest Basin was at least partially connected to the Saale Basin to the NNE via extensive fluvial systems.

Sedimentation commenced in the uppermost Stephanian with the deposition of up to 100 m of conglomerates and coarse arkosic sandstones. These were derived from the underlying weathered granites. Overlying sediments include fluvial to lacustrine fine-grained clastics with fossiliferous lake horizons and thin coal seams. Deformation and related magmatic activity led to the deposition of up to 1000 m of intermediate to acidic pyroclastics and lavas. The Stephanian C age of the sediment/volcanic succession of the possibly coeval Möhrenbach (in the SE) and Georgenthal (in the NW) formations has been confirmed by biostratigraphy and radiometric data (Schneider 2001; Lützner 2005). Deposition of the basal Rotliegend Illmenau Formation followed an extensive erosional hiatus of regional extent.

In the Thuringian Forest region, evidence for Late Carboniferous/Early Permian mantle upwelling is provided by intrusions of I-type granites and diorites, dykes of tholeiitic within-plate basalts, lamprophyres, granites and diorite-porphyry as well as widespread rhyolites and andesitic volcanism (Zeh *et al.* 2000). It is also possible that the combined effects of transtensional movements and mantle upwelling played an important role in the evolution of the prominent NE-trending intramontane continental basins formed by the inversion of the MGCH.

Saxothuringian Basin

The geological evolution of the main part of the Saxothuringian Basin (STB) has been summarized by Franke (1984*a*, *b*), Franke *et al.* (1995*b*), Franke & Stein (2000), Linnemann *et al.* (2000) and Kroner & Hahn (2004) (Fig. 9.50). The basin represents a Cambro-Ordovician rift basin which probably developed on Cadomian crust. The present-day northern boundary of the STB is a south-facing reverse fault which separates the basin from the

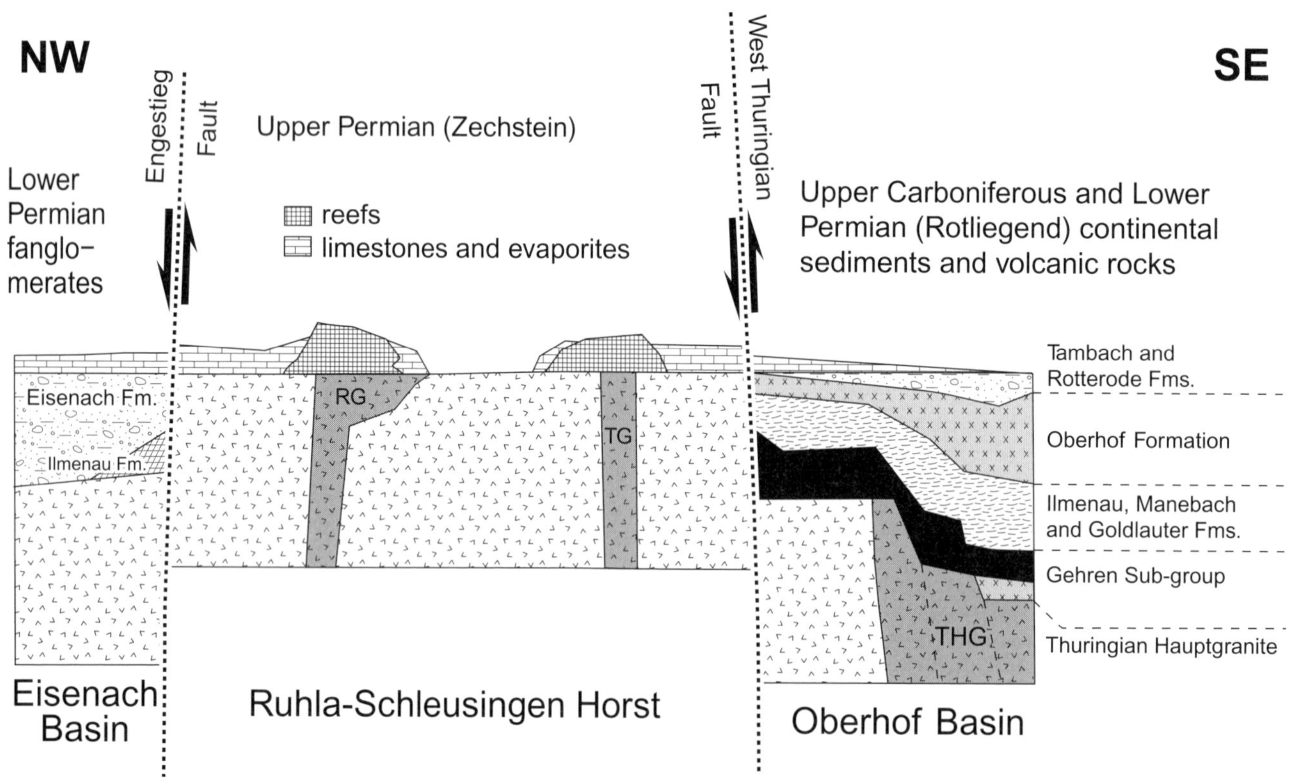

Fig. 9.49. Schematic geological profile through the Ruhla Crystalline Complex and its surroundings (after Zeh & Brätz 2004).

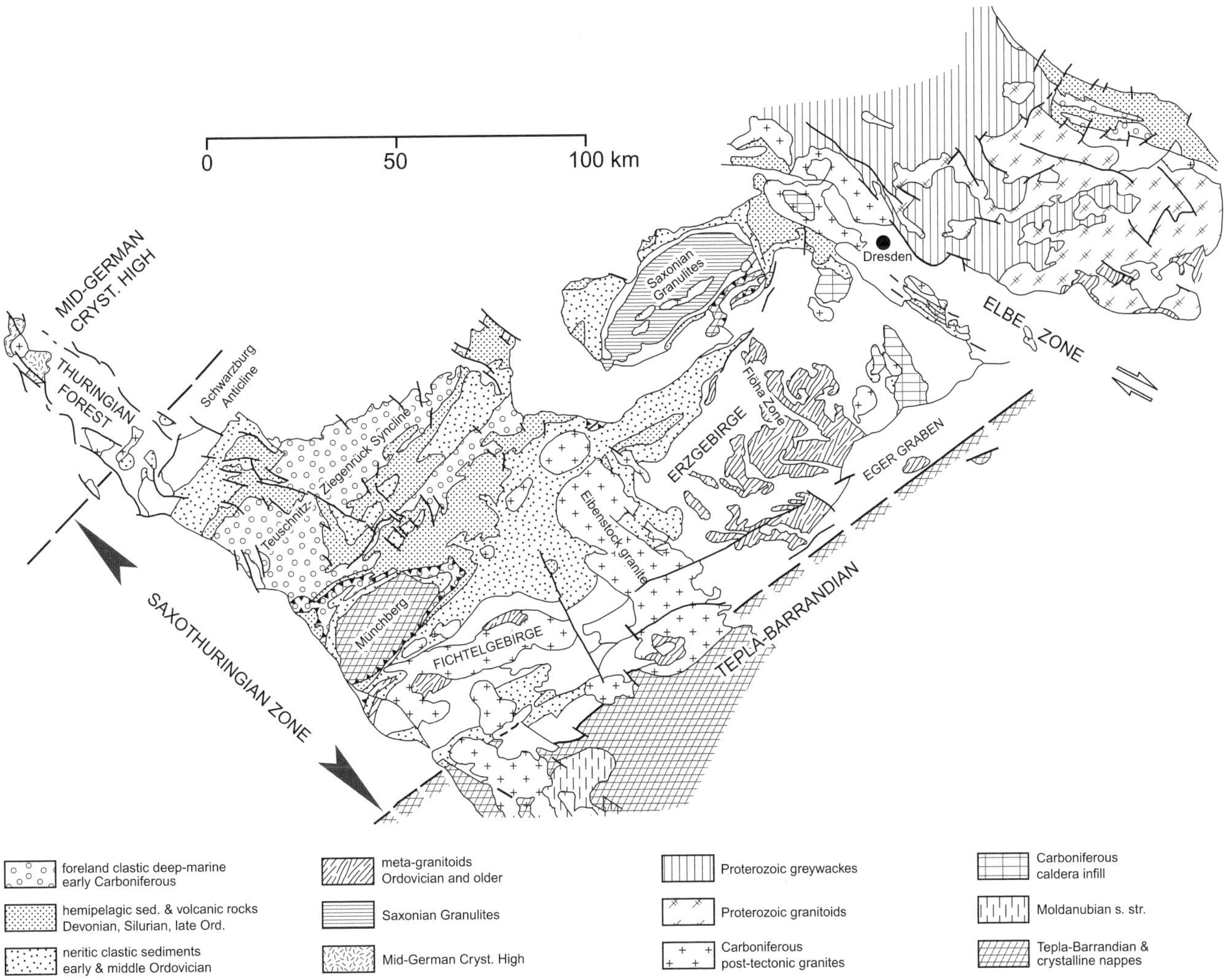

Fig. 9.50. Geological sketch map of the Saxothuringian Zone of SW Germany (after Falk *et al.* 1995).

Mid-German Crystalline High (MGCH), although it is possible that this boundary is a strike-slip zone (Franke *et al.* 1995*b*). Although the MGCH was subjected to uplift from Frasnian times onwards, there are no clastic sediments in the STB which can unequivocally be referred to sources within the high. To the SE, the STB was bounded (from early Famennian times onwards) by the Cadomian Teplá-Barrandian, which represents the upper plate and active margin during the Variscan convergence (Franke 1995*c*).

Crustal extension is documented by the presence of thick neritic clastic successions and bimodal volcanics (Franke *et al.* 1995*a*, *b*). The rift sequence was subsequently overlain by late Ordovician–Famennian pelagic sediments and Famennian–Visean deep-marine sediments derived from the active Bohemian margin to the SE (Franke 2000). The overlying allochthon (Münchberg, Wildenfels, Frankenberg tectonic klippen) contains, in its lower part, deeper-water equivalents of the Ordovician–Devonian parautochthonous succession and proximal deep-marine sediments. The upper allochthon comprises early Ordovician MORB-type mafic rocks metamorphosed to eclogite facies. The allochthon is derived from a root zone located between the Fichtelgebirge/Erzgebirge and the Teplá-Barrandian Unit (Franke 2000).

The internal organization of the STB is complex since it contains high-grade metamorphic rocks and fossiliferous sequences in close proximity. The main part of the STB comprises Cambro-Ordovician clastic sediments and volcanic rocks which represent the passive margin of a continental block to the NW (Franke 1995*c*). The shelf succession is overlain by late Ordovician to Early Carboniferous hemipelagic sediments and intraplate volcanic rocks followed by distal deep-marine sediments (i.e. Thuringian facies). Deeper-marine environments are contained in the Münchberg, Wildenfels and Frankenberg tectonic klippen which were derived from areas to the SE (Franke 1995*c* (see above)). The klippen may be interpreted as erosional remnants of a formerly coherent nappe pile. The nappes were derived from a position to the SE of the present-day Fichtelgebirge, now marked by the Saxothuringian/Teplá-Barrandian boundary (Behr *et al.* 1982; Franke 1984*a*, *b*). An oceanic part of the STB is represented by the eclogites contained within the crystalline nappes of the Münchberg Klippe.

The STB was fed and eventually overthrust by the NW active margin of the Teplá-Barrandian (Schäfer *et al.* 1997). The earliest preserved clastic sediments within the basin are of early Famennian age (*c.* 370 Ma) and include clasts derived from Early Palaeozoic continental slope sediments, originally deposited at the NW margin of the STB as well as ultramafic (?ophiolite) material. The presence of this material indicates that the narrow Saxothuringian Ocean had closed by this time. In late Devonian through to early upper Visean times, synorogenic clastic sediments were deposited on Saxothuringian thinned continental crust. The deeper part of the basin migrated northwestwards with time, followed by the front of tectonic deformation (Franke 1984*a*, *b*; Franke & Engel 1986; Falk *et al.* 1995).

The deepest parts of the STB are exposed in the Saxonian Granulite area and the Erzgebirge (see above). These areas expose high-pressure/medium-temperature and high-pressure/high-temperature rocks which were probably derived from the orogenic root to the SE and were subsequently intruded into the lower crust of the northwestern foreland (Henk *et al.* 2000; Krawczyk *et al.* 2000). Although the sedimentary record and the geochemical character of the mafic magmatic rocks indicate early Palaeozoic extension, some observations have been interpreted as suggesting the presence of an active margin setting in the STZ and its equivalents in the Sudetes (e.g. Oliver *et al.* 1993; Kröner & Hegner 1998). Geochemically, many Cambro-Ordovician felsic magmatic rocks have active margin signatures. However, there is no sedimentary or tectonic evidence of an early Palaeozoic convergent setting. Thus the signature may be a relict Cadomian one or related to the extraction of melts from a Cadomian arc (Floyd *et al.* 2000). Late Devonian–early Carboniferous metamorphic rocks have preserved mineral ages at *c*. 480 Ma, suggesting that this event is related to an early Palaeozoic convergent episode (Franke 2000). However, the main metamorphic minerals in these rocks have been dated as late Devonian and early Carboniferous suggesting that the early Palaeozoic mineral ages are relicts of metamorphic and magmatic processes whose P-T environment is unknown. With respect to the information derived from the sedimentary record and the geochemical affinities of the mafic magmatic rocks, it would appear that these dates are best explained as products of magmatism and metamorphism in the hot lower crust of an early Palaeozoic (presumably Cambro-Ordovician) rift (Franke 2000).

Two facies realms can be identified within the STB (Fig. 9.51). The northernmost realm (Thuringian) apparently represents the passive margin of a continental unit (not yet identified) to the NW which was the source of the Cambro-Ordovician clastics. However, it cannot be excluded that strike-slip movements have displaced the STB rocks and that their present juxtaposition with the MGCH is coincidental (Falk *et al.* 1995). Such an interpretation is supported by the observation that the MGCH did not shed clastic material into the STB in early Carboniferous times, although it was an important source area for the deep-marine sediments of the Rhenohercynian Zone to the north.

The **Thuringian facies** commences with Proterozoic-age rocks which are overlain by a Cambrian succession. The transition from the greywacke-dominated Precambrian into the Cambrian is well exposed in some areas (Falk *et al.* 1995). Cambrian lithologies vary from area to area and include (1) silty shales with intercalated quartzites and felsic volcanic rocks, (2) bimodal volcanics, and (3) carbonates. The basal Ordovician comprises shallow-marine quartzites and shales, locally with conglomerates (up to 2000 m thick). In many areas there are intercalated felsic and mafic volcanic rocks (both effusives and sills). The uppermost Ordovician is thinner (hundreds of metres) and comprises shales with intercalated quartzites, oolitic ironstones and limestones. At the top of the Ordovician there is a shaley succession dominated by glaciomarine diamictites (Falk *et al.* 1995). The Silurian succession is dominated by black shales and cherts whose deposition was probably related to the post-glacial sea-level rise (Wignall 1991). These pass upwards into Ludlow-age limestones with local intercalations of basic lavas and tuffs. The oldest Devonian rocks are black shales with rare intercalations of pelagic limestone of Gedinnian–Givetian age, which are overlain by turbidites, probably derived from the NW. The Frasnian succession comprises hundreds of metres of sedimentary and volcanic rocks whose origin was related to rifting activity. The remainder of the Frasnian and Famennian are composed of shales with nodular limestones.

Over much of the Thuringian facies realm, synorogenic

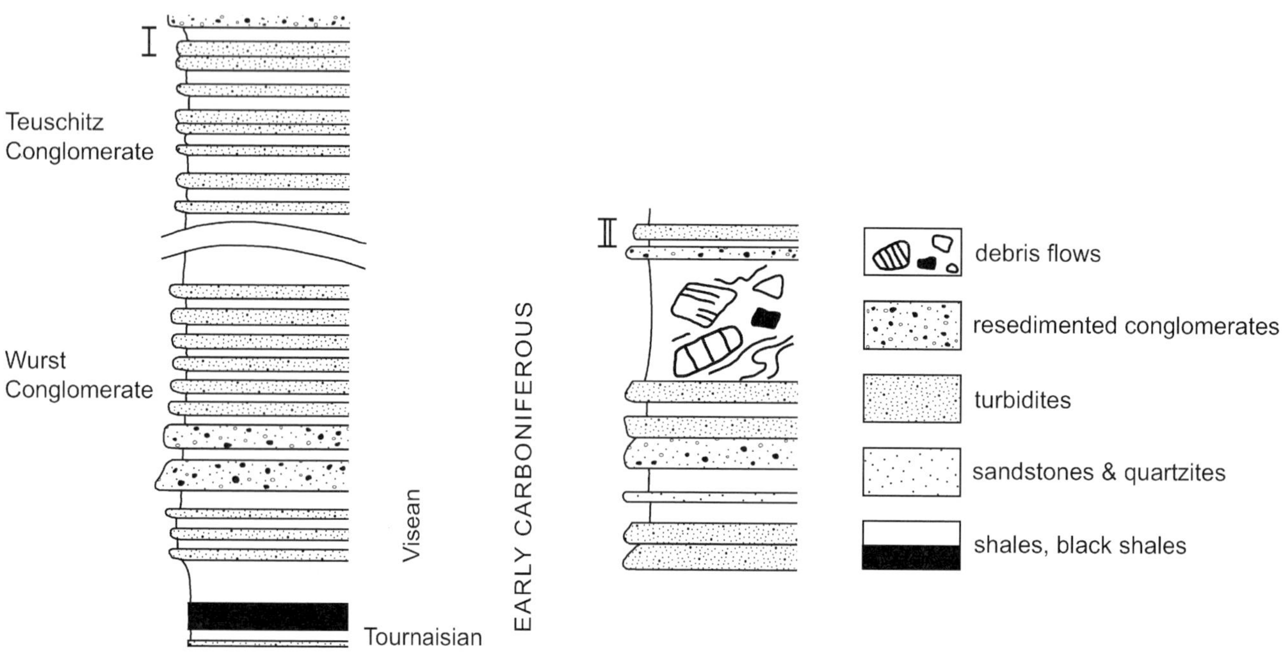

Fig. 9.51. Lithostratigraphic colums for the Thuringian facies autochthon (I) and the Bavarian facies allochthons (II) in the western part of the Saxothuringian Zone (Münchberg nappes and surrounding areas) (after Falk *et al.* 1995).

siliciclastic sedimentation did not begin before the basal Visean, with notable exceptions (e.g. Fichtelgebirge where it began in early Famennian times). However, the occurrence of the earliest deep-marine sediments along the southern margin of the STB is consistent with the concept of a northward-prograding orogenic front (Falk *et al.* 1995). Widespread deep-marine sedimentation commenced at about the Tournaisian–Visean boundary and extends into the late Visean. The Early Carboniferous succession comprises *c.* 3 km of classical turbidites with some conglomeratic intervals and rare felsic tuffs. Clastics were shed from the rising thrust mass onto the adjacent Thuringian facies province (i.e. to the north).

The **Bavarian facies** is restricted to certain tectonic units, at least part of which represent nappes. The Cambrian of the Bavarian facies has survived only in the form of large olistoliths or gravitational nappes (up to 2 km^2 in size) and comprising greywackes, shales and rare tuffs. Some other olistoliths contain shales with early Ordovician faunas, with the main Ordovician unit being contained in a nappe and comprising bimodal volcanics overlain by shales and sandstone turbidites (Falk *et al.* 1995). The Silurian succession is predominantly black shales and bedded cherts with rare felsic and mafic tuffs. Limestones occur in the late Silurian. The Devonian comprises radiolarian cherts, greywackes, limestones, shales and tuffs. With the exception of one occurrence of bedded chert, the Early Carboniferous consists of coarse-grained sandstones, conglomerates, greywackes, slumps, debris flows and intercalated mudstones. Clasts of limestone within the debris flows and conglomerates have been dated as Tournaisian and Visean in age and were derived from areas on a growing accretionary prism (Behr *et al.* 1982).

The Early Palaeozoic Bavarian facies successions tend to represent deeper-marine conditions than their Thuringian equivalents, and were originally deposited further to the SE (later displaced as a result of nappe formation). Additionally, it cannot be excluded that some of the Bavarian pelagic successions were originally underlain by oceanic crust, as suggested by the MORB-type chemistry of eclogites overriding the Bavarian facies nappes (Tischendorf *et al.* 1995). The southernmost passive margin of the STB has not, thus far, been identified, and it may be hidden among the metamorphic rocks of the Teplá-Barrandian (Falk *et al.* 1995).

Erzgebirge Basin (J.S.)

The Erzgebirge Basin, situated in SE Germany, is 70 km long and 30 km wide (Fig. 9.52). It comprises a series of sub-basins with differing sedimentation histories and tectonic development (Schneider *et al.* 2005*c*). The area forms an important link between the inner Variscides of the Bohemian Massif and the outer Variscides of the Rhenohercynian Zone and the Variscan foreland basin in northern Germany and Poland.

To the NE of the Erzgebirge Basin, tectonically bounded turbiditic successions, of probable Early Carboniferous age, and fan delta units (Striegis Formation, 650–1000 m) are exposed. Overlying the Striegis Formation are the delta fan/alluvial fan sequences of the Visean-age Hainichen Subgroup (>1000 m). The early to post-orogenic Visean sediments of the Hainichen Subgroup were deposited at the top and/or in front of the gravitational collapsing pile of Variscan nappes probably in a piggyback basin located between the metamorphic complexes of the Erzegebirge and Saxonian Granulite massifs.

The Upper Carboniferous succession is restricted to the Zwickau and Oelsnitz subbasins (Fig. 9.52). These cover an area

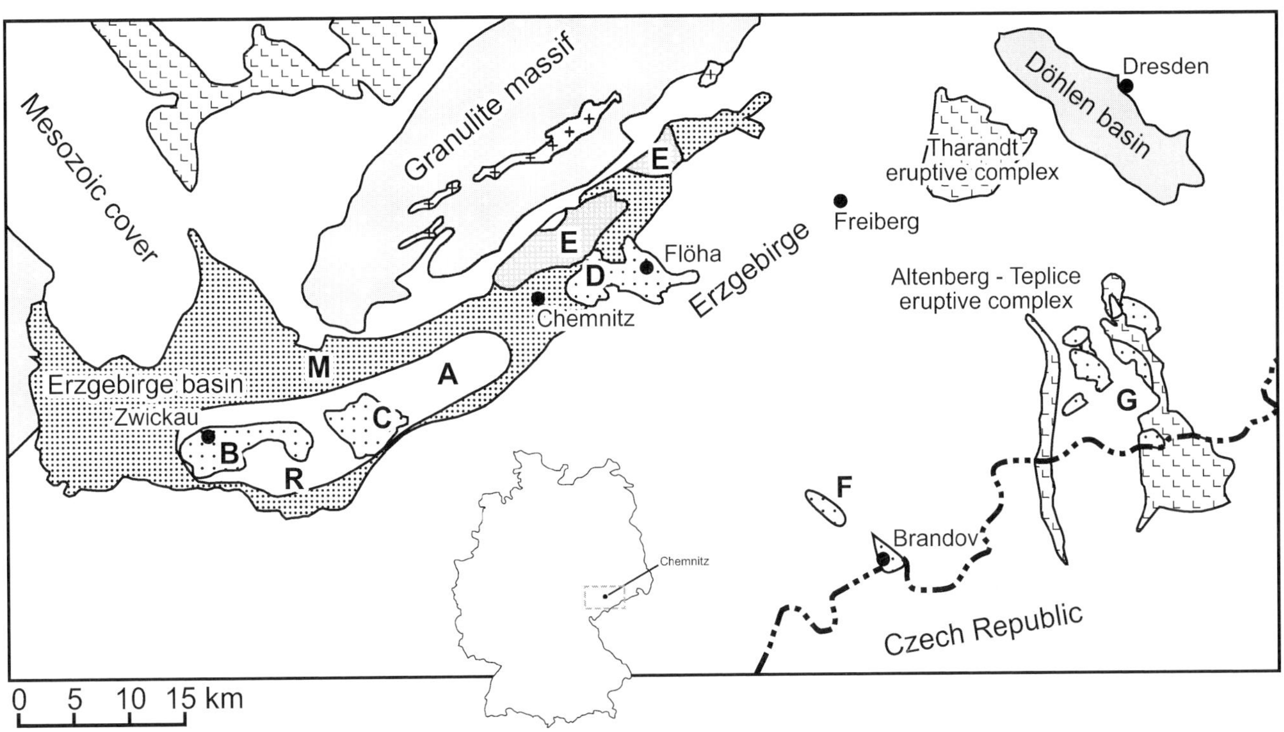

Fig. 9.52. Simplified map of the Erzgebirge Basin and related sub-basins in SE Germany: A, distribution of Westphalian-age sediments in the Zwickau-Oelsnitz area; B, Zwickau sub-basin; C, Oelsnitz sub-basin; D, Flöha sub-basin; E, remnants of the Hainichen sub-basin; F, Olbernhau-Brandov sub-basin; G, eruptive complex; M, Mülsen fan, Zwickau sub-basin; R, Reinsdorf fan, Zwickau sub-basin (after Schneider 2005*d*).

25 km long and 7 km wide and contains a Westphalian succession which is unconformably overlain by Rotliegend sediments. Basement comprises mainly isoclinal to imbricate folded, unmetamorphosed Upper Devonian and Silurian rocks. The origin of these subbasins was related to increased tectonic activity during the Westphalian D (Schneider *et al.* 2006). The Zwickau Formation (c. 350 m) comprises alluvial fan units interbedded with alluvial plain/floodplain and swamp deposits. Rare volcanic ash horizons are present. Coal seams of decimetres to 10 m thickness were exploited from 1348 to 1978. A Westphalian D to lowermost Stephanian age is confirmed by macro- and microfloras.

As noted above, granitoid intrusion also occurred in the area, and associated extrusives include rhyolites, rhyodacites, andesites and, locally, basalts and lamprophyres. These occur in spatial association with the granitoid series, but are always stratigraphically younger. The precise genetic connection between the intrusive and extrusive activity is not always clear (Tischendorf *et al.* 1995).

While Carboniferous sedimentation and volcanic activity is mainly concentrated in the Erzgebirge Basin, there is some evidence of sedimentation around the towns of Brandov and Mikulov in the Czech Republic (Fig. 9.52). The sedimentary fill of the Brandov area is divided into two parts. The lower succession, up to 250 m thick, consists of coal-bearing fluvial strata dominated by coarse-grained arkosic sandstone to conglomerates. The rich flora indicates a middle Westphalian age corresponding to the Radnice Member in central and western Bohemia. Overlying is a coal-barren unit, *c.* 50 m thick, which is separated from the lower unit by an erosional contact. The age of this unit is unclear (due to the absence of fossils), but is probably Stephanian C or early Permian (Pešek *et al.* 1998).

Carboniferous rocks between the towns of Mikulov and Teplice are represented by a rhyolite body up to 2000 m thick, with intercalated sediments indicating the polyphase history of this volcanosedimentary complex. Breiter (1997) distinguished four volcanic phases; the first two are of middle and upper Westphalian age, the third Stephanian and the final one Permian.

Torgau-Doberlug Basin, Germany (J.S.)

The Torgau-Doberlug Basin represents the eastern branch of the Delitzsch-Torgau-Doberlug Synclinal Zone (DTDS), a narrow ENE-striking Variscan structural belt at the NE margin of the Saxothuringian Zone (Fig. 9.53). The DTDS is bordered by the Neoproterozoic Lusatian-Leipzig Greywacke complexes to the south and by part of the MGCH to the north. The preserved depositional area is *c.* 22 × 8 km, and sediment thickness is up to *c.* 700 m.

The Upper Visean of this area records the transition from deep-marine to continental deposition. In areas of contemporaneous deposition, deep-marine or continental facies were deposited, reflecting the interplay of orogenic collapse and related extensional basin formation as well as high-frequency sea-level fluctuations. These early continental basins include both terrestrial areas close to the Bohemian Massif and shallow marine-paralic-fluviatile dominated areas distal to it. The onset of early Variscan continental deposition in the DTDS is related to the *crenistria*-event sea-level rise (Herbig 1998). Cyclic fining-upward sequences are interpreted in terms of the interplay of extensional phases and glacio-eustatic sea-level fluctuations. Contemporaneous intrusion of andesitic to rhyodacitic sills occurred and was probably linked to the Pretzsch-Prettin intrusion complex (*c.* 330 Ma).

Four units are recognized in the basin: the Finsterwalde, Doberlug, Kirchhain and Werenzhain formations (Nöldeke 1976). The succession commences with clays, greywackes and shallow marine carbonates at the base followed by several fining-upward cycles of conglomerates, greywackes and shales with frequently intercalated coal seams.

Teplá-Barrandian (Bohemicum, Moldanubian Zone)

The Moldanubian Zone is a complex association of lithologies whose protolith ages are mostly unknown. Originally the zone comprised the Teplá-Barrandian (TB) and the adjacent high-grade metamorphic rocks to the SE (Kossmat 1927) (Figs 9.3, 9.54 & 9.55). It has, however, become customary to differentiate

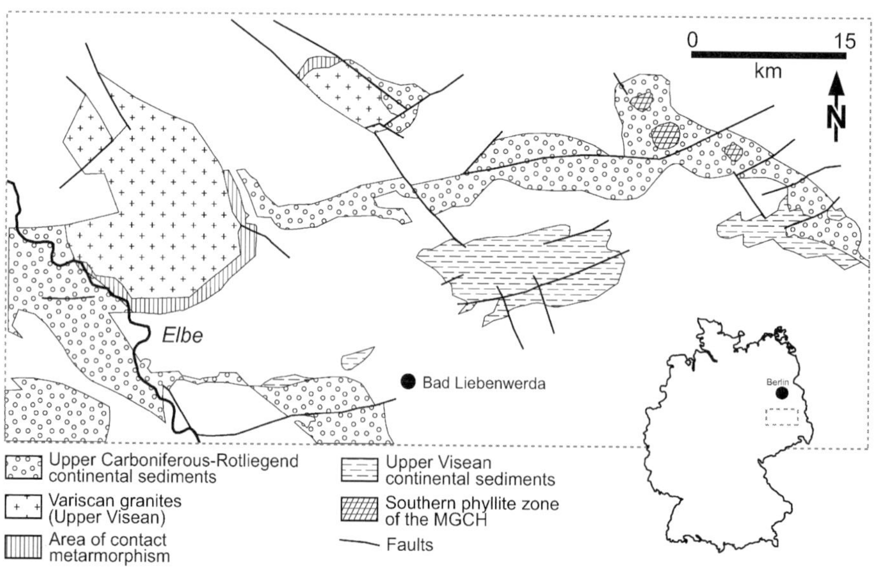

Fig. 9.53. Geological map of the Torgau-Doberlug Basin showing the distribution of Carboniferous and Lower Permian sediments.

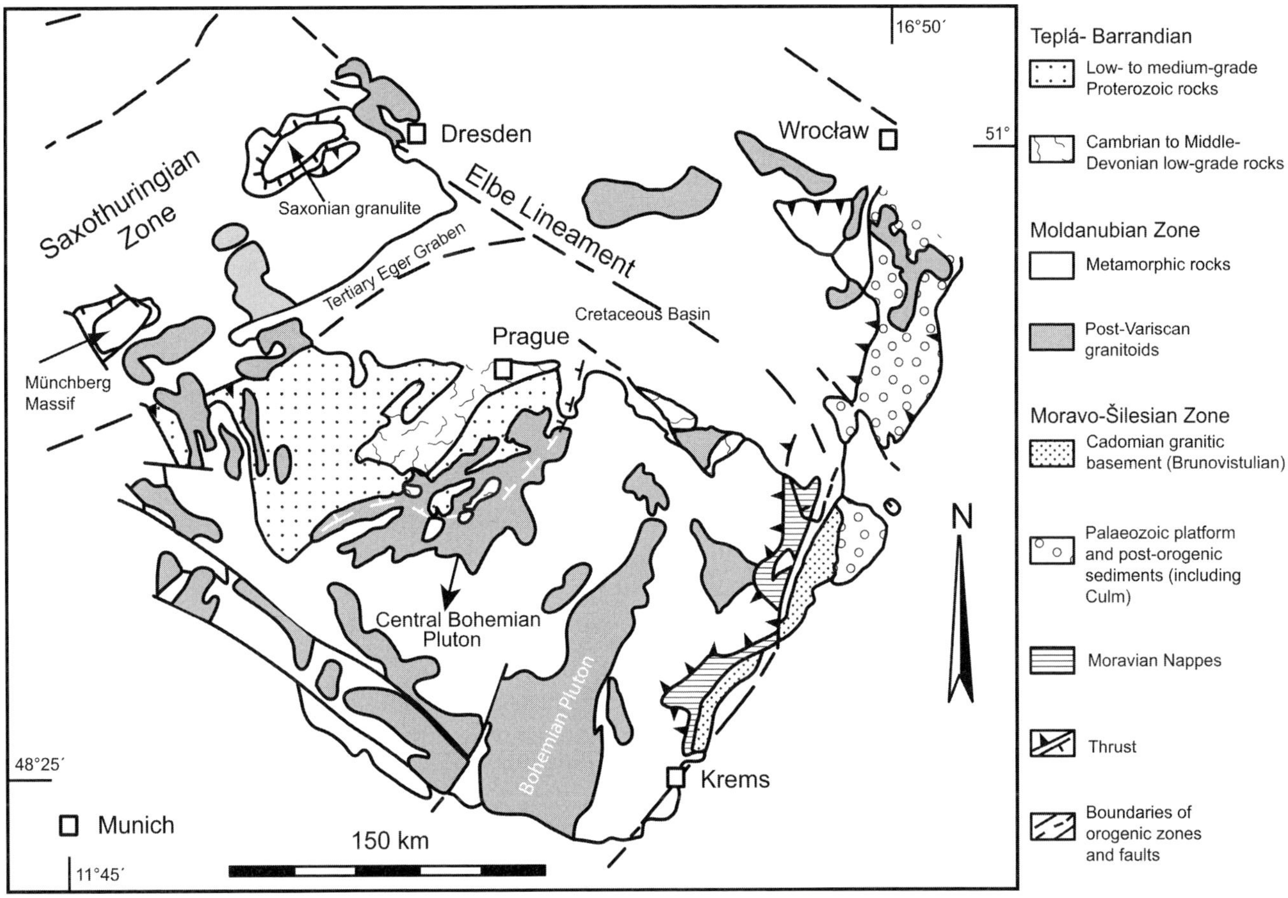

Fig. 9.54. Simplified geological map of the SE part of the Variscan orogen in Central Europe (after Petrakakis 1997).

the high-grade, SE part of the Bohemian Massif as the Moldanubian *sensu stricto* (see below), while the term Teplá-Barrandian is reserved for the part to the NW. The TB is considered to be part of the Armorican Terrane Assemblage as documented by evidence of the late Ordovician glaciation as well as palaeomagnetics and biogeography (e.g. Franke 2000: McKerrow *et al.* 2000; Tait *et al.* 2000; see also Robardet 2003, and Krawczyk *et al.* 2008). The TB was thrust over Moldanubian rocks as a complete crustal block during the Variscan convergence in Upper Devonian times (Zulauf 1997).

In the TB is where non-metamorphic Cambrian to mid-Devonian sediments and volcanics rest unconformably on an Upper Proterozoic basement. The latter shows a higher degree of deformation and metamorphism than the Palaeozoic cover (Zulauf *et al.* 1997). This situation is in marked contrast to the surroundings of the TB. Within the Moldanubian *sensu stricto* located to the west and SW of the TB, as well as the Saxothuringian Zone to the north, penetrative Variscan deformation associated with late-stage high-temperature/low-pressure metamorphism has largely destroyed the earlier structures and fabrics (e.g. Carswell & Jamtreit 1990). Thus, the Teplá-Barrandian (= Bohemicum) is the best preserved fragment of the Cadomian Orogen in central Europe (see Linnemann *et al.* 2008*a*, *b*).

In the eastern part of the Bohemian Massif, the crustal block of the TB unit is separated from the surrounding Moldanubian and Saxothuringian units by large-scale shear zones (Bues & Zulauf 2000). Some of these shear zones (e.g. Hoher Bogen Shear Zone, West Bohemian Shear Zone) were active at the same time, primarily during Lower Carboniferous crustal extension, and can thus be combined and referred to as the Bohemian Shear Zone (Zulauf *et al.* 1998; Scheuvens & Zulauf 2000). Structural and kinematic investigations of these shear zones indicate ductile normal movements and subsidence of the TB with respect to the surrounding Saxothuringian and Moldanubian units, due to elevator-style tectonics related to orogenic collapse in the Variscan internides (e.g. Zulauf 1994; Behrmann & Tanner 1997; Zulauf *et al.* 1998; Bues *et al.* 1998).

Late Proterozoic sediments and arc-related volcanic rocks constitute the basement (Zulauf 1997; Zulauf *et al.* 1999), and are overlain by Cambrian–middle Devonian sedimentary and volcanic rocks. The geochemistry of the volcanic rocks suggests that they were formed in a subduction setting (e.g. Patocka *et al.* 1993). However, a transtensional setting is suggested by the intrusion history of Cambrian-age plutons (Zulauf *et al.* 1999) while Ordovician–middle Devonian rocks indicate both subsidence and extension in an intracontinental/passive-margin setting (Franke 2000). Subsequent uplift is indicated by the Middle Devonian sediments while Late Devonian plutons probably record subduction beneath the SE margin of Bohemia.

Basins of central and western Bohemia (S.O.)

The Late Palaeozoic continental strata in central and western Bohemia are formally subdivided into the Pilsen, Manětín, Žihle,

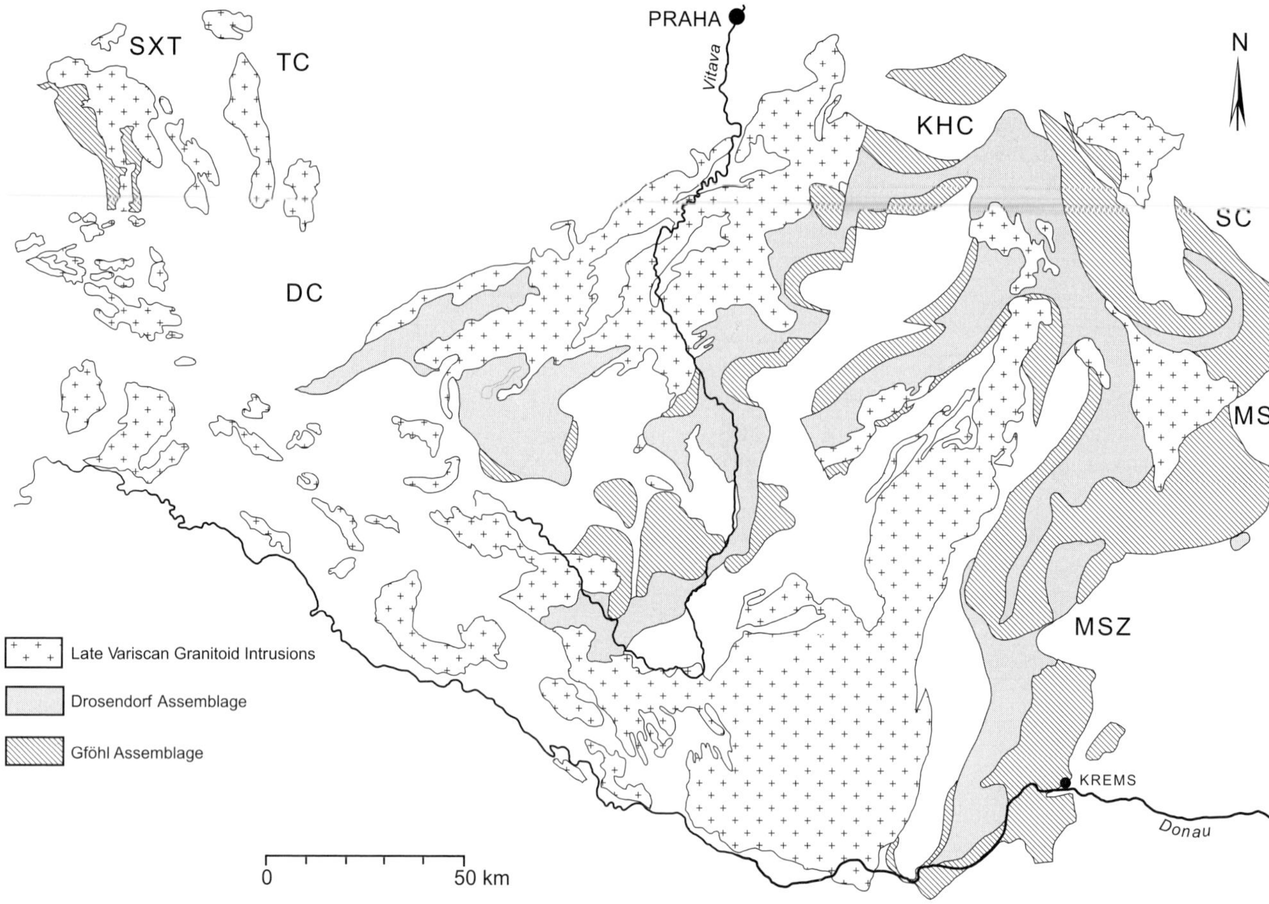

Fig. 9.55. Terranes within the Moldanubian Zone. Abbreviations: DC, Domažlice Crystalline Complex; KHC, Kutná Hora Crystalline Complex; MZS, Moravo-Silesian Zone; SC, Svratta Crystalline Complex; SXT, Saxothuringian Basin; TC, Teplá Crystalline Complex (after Fiala *et al.* 1995).

Radnice, Kladno-Rakovník and Mšeno-Roudnice basins (Fig. 9.56) which originally formed a single basin on the Cadomian crust of the Teplá-Barrandian. Deposition in this area commenced around the Westphalian A–B boundary and continued, with several interruptions, until the end of the Carboniferous (and possibly into the lowermost Permian). Basin-wide alternation of thick red and grey-coloured units has been used to subdivide the basin fill into the Lower Grey (Kladno), Lower Red (Týnec), Upper Grey (Slaný) and Upper Red (Líně) formations (Weithofer 1896, 1902). The Kladno Formation consists of the Radnice (Westphalian C) and Nýřany (upper Westphalian D–Cantabrian) members separated by a basin-wide hiatus. The deposits of the Radnice Member are interpreted as fluvial systems, possibly tectonically controlled (Opluštil 2003). The NNE-SSW striking valleys, within which the fluvial sediments were deposited, are interpreted as pull-apart basins (Pašek & Urban 1990) within which thickness varies between 0 and 280 m. Reconstruction of the valley system indicates that the eastern part of the area was drained in a NW direction to the North Variscan Foreland Basin (Opluštil *et al.* 1999).

The Radnice Member of the Kladno Formation consists of fluvial, lacustrine to lacustrine delta and colluvial sediments. Coal seams are concentrated in the Pilsen, Radnice and Lubná groups. The most important of these, the Radnice Group, consists of the Lower and Upper Radnice coals separated by the volcaniclastics of the Whetstone Horizon dated by the Ar^{40}/Ar^{39} method as 309.3 ± 4.6 Ma (Lippolt *et al.* 1986). Locally more than 10 m thick, the Upper Radnice Coal is economically the most important seam of the continental basins of the Bohemian Massif. Widespread volcaniclastic horizons in the northern part of the Kladno-Rakovník Basin are accompanied by rhyolite effusions with a cumulative thickness of over 100 m.

Deposition of the overlying Nýřany Member followed the basin-wide hiatus related to local tectonics (i.e. Leonian tectonic phase) (Oplustil & Pešek 1998) and quickly extended across much of the present-day basin. This 300–500 m thick unit is of fluvial origin and is characterized by the predominance of coarse- to medium-grained pebbly arkoses intercalated with fine-grained sandstones and mudstones. Sediments are arranged into 10–20 m thick fining-upward cycles which may overlie thin coal seams. The overlying Týnec (Lower Red) Formation (Barruelian) is similar both in terms of facies and architecture to the Nýřany Member. The relative absence of coal seams in the Týnec Formation is interpreted as a response to climatic change from humid to drier or (?) seasonal conditions (Pešek 1994; Oplustil & Pešek 1998). The average thickness of this formation varies between 150 and 200 m. Sediments of the Nýřany Member and the Týnec Formation represent deposits of a large braidplain system supplied with clastics mainly from the source area located to the south in the Central Bohemian Pluton (Pešek 1994).

The overlying Slaný (Upper Grey) Formation (Stephanian B)

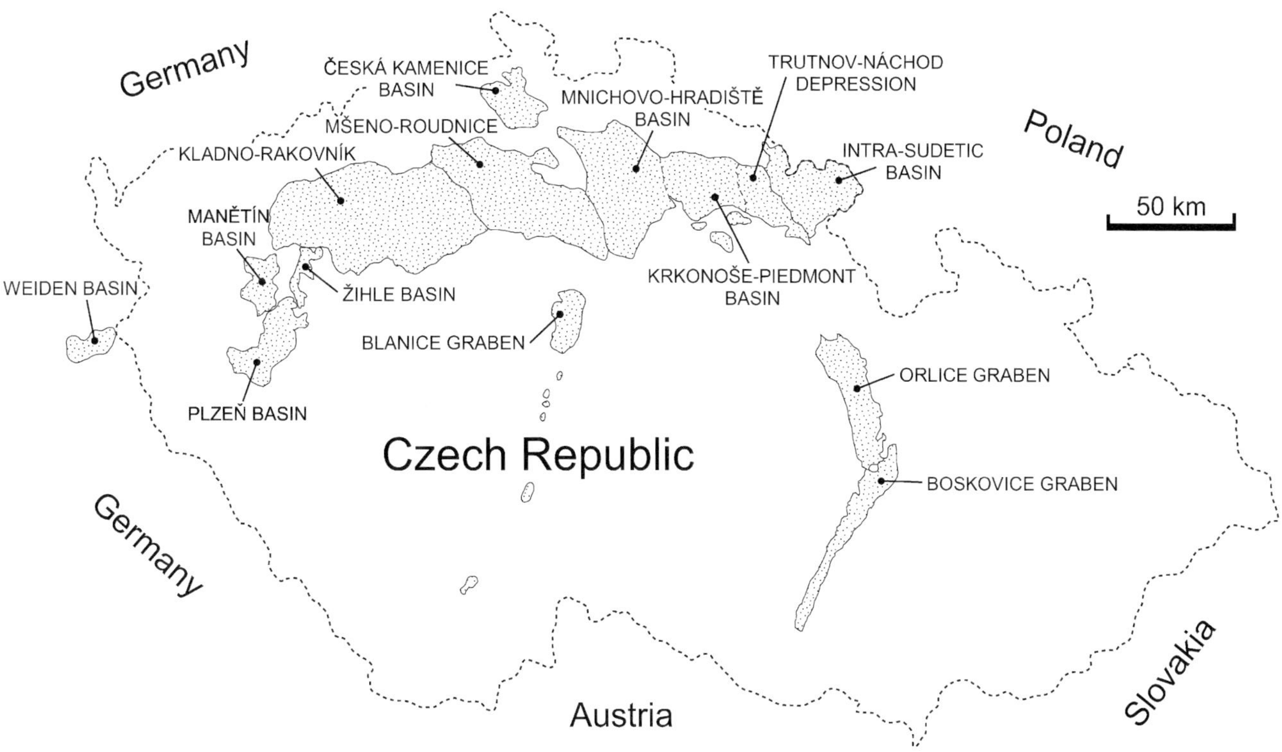

Fig. 9.56. Location map of the main Carboniferous and Permian basins in the Czech Republic.

consists of *c*. 200–250 m of fluvial, lacustrine and deltaic sediments. The lower part of the formation is represented by sediments deposited in a fluvial system similar to that of the Týnec Formation but under more humid climatic conditions (e.g. Mělník group of coals). These basal, and locally coal-bearing, fluvial strata are overlain by more than 100 m of upward-coarsening lacustrine to lacustrine delta sediments (Skoček 1990). The lacustrine succession commences with a horizon of dark laminated claystone, traceable across the basin (Skoček 1990). The laminated units are overlain by prodeltaic mudstones and fine- to medium-grained sandstones which pass upward into ripple-bedded sandstones and finally to medium- to coarse-grained arkosic fluvial sandstones with rare pebbly sandstones. The thickness and extent of the lacustrine to deltaic succession suggests that a lake occupied the entire basin. However, the precise mode of formation of this lake has not yet been adequately explained. The upper part of the Slaný Formation consists of fluvial strata with laterally persistent lacustrine horizons.

The youngest sediments in central and western Bohemia are those of the Líně (Upper Red) Formation (Stephanian C–?lower Rotliegend). These are separated from the underlying Slaný Formation by a basin-wide erosional surface with palaeorelief of up to 200 m (Skopec *et al.* 2000). Palaeobotanical data indicate a hiatus between these two formations (Šetlík 1977). Deposition of the Líně Formation occurred under generally drier climatic conditions compared to the previous formation, as indicated by the frequent calcrete horizons. The presence of rare horizons, some tens of metres thick, of grey or mottled mudstones, locally with cherts, volcaniclastics, lacustrine carbonates or thin coal seams, may indicate temporary periods with a more humid climate. The thickness of the Líně Formation is locally up to *c*. 1000 m. Rare biostratigraphic data suggest a Stephanian C age, although, based on lithological correlation with the adjacent Sudetic area, the upper part of the formation may be of lower Rotliegend age (Pešek 1994).

Moldanubian Zone *sensu stricto* (Moldanubian Zone)

As noted above, the Teplá-Barrandian and the high-grade metamorphic rocks adjacent to the SE have been summarized by Kossmat (1927) as the 'Moldanubian Region'. It has since become customary to consider the high-grade area of the Bohemian Massif as the Moldanubian *sensu stricto*. The Moldanubian Zone is characterized by an abundance of Late Variscan plutonic rocks of predominantly granitoid composition. These rocks are considered to be parts of one of the principal magmatic complexes, the Moldanubian (South Bohemian) Pluton and the Central Bohemian Pluton (Holub *et al.* 1995, 1997).

Bohemian Massif

The Bohemian Massif, exposed in adjacent parts of Poland, the Czech Republic, Austria and Germany, is an important part of the Variscides of Central Europe (Suk 1984) (Figs 9.3 & 9.36). It comprises major structural zones which from NW to SE are the Saxothuringian, Teplá-Barrandian, Moldanubian and Moravo-Silesian (Seston *et al.* 2000). The Moldanubian Zone of the Bohemian Massif represents a major crystalline segment within the eastern part of the mid-Palaeozoic Variscan fold-and-thrust belt (Fiala *et al.* 1995) (Fig. 9.55). Variscan granitoids were concentrated along the central axis of the fold belt (Moldanubian Zone) where they constitute more than 50% of the crystalline basement (Fig. 9.6). The largest Moldanubian batholiths are the Southern Bohemian Batholith, the Central Bohemian Batholith, the Oberpfalz Batholith and the Black Forest batholithic terrain.

The Moldanubian Zone experienced high heat flow in late Visean–early Namurian times, probably as a consequence of post-collisional extension and magmatic underplating (Finger *et al.* 1997). The lower crust is, therefore, considered to be the major source for the magmas, although some material may have been derived from the middle crust (Finger *et al.* 1997).

The Moldanubian granitoids can be broadly divided into four subgroups (Finger *et al.* 1997). The first consists of moderately peraluminous, S-type granites with some muscovite or cordierite and sometimes garnet (e.g. Eisgarn granite, Southern Bohemian Batholith; Falkenberg and Flossenburg granites, NE Bavaria). Subgroup II comprises highly potassic, weakly peraluminous, sometimes metaluminous, K-feldspar megacrystic biotite granitoids (e.g. Weinsberg granite, Southern Bohemian Batholith; Kristallgranit I, Bavaria). Subgroup III has high K_2O_2 contents and mostly contains megacrystic K-feldspar (e.g. 'durbachite' plutons, Bohemia; Guiv metasyenite, Aar Massif). Subgroup IV contains leucocratic, near-minimum peraluminous (S-type) granites (e.g. Altenberg granite, South Bohemian Batholith).

The geology of the Moldanubian is complex and poorly understood. In particular, it is difficult to correlate the various regions in terms of their geological evolution. For that reason, Fiala *et al.* (1995) have proposed that the region be subdivided into four areas ('terranes' according to Fiala *et al.* 1995), which have experienced different geological histories and are now separated by major thrusts or strike-slip faults. However, Franke (2000) has modified this based on subsequent work (e.g. Petrakakis 1997) to include just two, what he terms, assemblages (see below). In this scheme, the structure of the Moldanubian is dominated by the large-scale, thrust-related inversion of metamorphic facies which has emplaced high-pressure rocks (Gföhl Assemblage) over the parautochthonous Drosendorf Assemblage. Thrusting of the Gföhl rocks was to the north and east (Büttner & Kruhl 1997) over a distance of at least 150 km (Franke 2000).

The **Drosendorf Assemblage** consists of a basal part (Ostrong Unit or Monotonous Series) comprising metasedimentary and felsic meta-igneous rocks with rare ecologite lenses (Fig. 9.36). This unit is structurally overlain by the Variegated Series (Bunte Serie) with a basal sheet of orthogneiss and a variety of metamorphosed sedimentary and igneous rocks. Amphibolites within the Variegated Series were derived from within-plate and ocean-floor basalts (Patocka 1991).

The second unit, the **Gföhl Assemblage**, commences with the amphibolites of the Raabs-Meisling Unit. These are overlain by the Gföhl Orthogneiss, which is derived from an Ordovician granitoid. The upper part of the Gföhl Assemblage comprises felsic granulites, partly retrogressed into amphibolite facies (Carswell & O'Brian 1992). Isotopic data from the assemblage are complicated, but suggest that the peak metamorphic pressures were achieved before 340 Ma (Early Carboniferous).

The best palaeogeographic suggestion, albeit with a still insufficient factual basis, is that there was opening of a Massif Central/Moldanubian Ocean in Ordovician times (represented by ultramafic and magmatic rocks in the Gföhl Assemblage) and a southerly continental microplate (contained in the Drosendorf Assemblage). As noted by Franke (2000), one of the many remaining questions concerns the palaeogeographic position of the Drosendorf Terrane, which could either represent the northern margin of mainland Gondwana, or else another member of the Armorican Terrane Assemblage (see Krawczyk *et al.* 2008).

Apart from the late Devonian metagranitoids at the SE margin of the Teplá-Barrandian, it is possible to distinguish two age clusters of granitoids: *c.* 340 Ma and 325–305 Ma (see Franke 2000). Both periods post-date the tectonic emplacement of the Gföhl over the Drosendorf assemblages and the younger generation also post-date later metamorphic events (Büttner & Kruhl 1997; Finger & Clemens 1995). The older age group is mainly represented by the Central Bohemian Batholith, which may be referred to a volcanic arc. The younger group are related to either post-collisional extension in contact with mantle or mantle-derived melts, or renewed subduction (Finger *et al.* 1997).

South and Central Vosges and Black Forest

The Central Vosges area consists of high-grade metamorphic successions that were intruded by numerous granitoid plutons (granodiorites, anatectic granites and late-stage leucogranites) (Fig. 9.57). The metamorphic rocks are subdivided into an upper high-grade unit containing granulites with garnet peridotites and kinzigitic metapelites and an underlying lower-grade metasedimentary unit, containing mainly migmatitic gneisses and anatectic granites. Both underwent high-temperature/low-pressure metamorphism and ductile deformation during exhumation from the middle crust in Visean times (Schaltegger *et al.* 1996). In the central parts of the Vosges and Black Forest areas, the presence of this mantle material and high-pressure metamorphic rocks is interpreted as being related to the presence of a subduction zone.

The rocks of these areas have been involved in SE displacement and, as mentioned elsewhere, they probably correlate with the Moldanubian rocks of the Bohemian Massif. This comparison is supported by the similar isotopic records from both areas. Ultramafic rocks and eclogites, as noted above, are usually interpreted as remnants of a Massif Central/Moldanubian Ocean which separated the Bohemian and the Moldanubian areas (Matte *et al.* 1990). However, the Variscan basement of SW Germany does not contain rocks comparable with the Teplá-Barrandian; instead equivalents of the Saxothuringian (i.e. northern Black Forest and Vosges) are directly juxtaposed against the Moldanubian.

The **Southern Vosges Basin** represents a local extensional domain situated between large dextral strike-slip faults which accommodated the dextral slip component of oblique compression (Fig. 9.58). The extension was restricted in time and took place within a short period of *c.* 3 Ma. The basin is part of a larger zone, probably consisting of a number of such basins, extending from the northernmost Massif Central (Morvan-Beaujolais) through the Southern Vosges into the Black Forest (Badenweiler-Lenzkirch Zone), suggesting formation by an orogen-wide process (e.g. Rey *et al.* 1992). The southern margin of this basin is formed by terrestrial and fluvial deposits of Permian age. Three lithologically and stratigraphically different units can be distinguished in this southern part (Maass 1988). The Lower Unit forms the base of the sequence with monotonous Upper Devonian (Famennian) marine shales (the so-called 'Treh-shales') and greywackes of turbiditic origin (Maass & Stoppel 1982). The sediments are associated with bimodal volcanic rocks such as low-K rhyolites and basalts (Schaltegger *et al.* 1996) and were probably deposited during a phase of active extension. The Middle Unit succession, which also comprises marine turbiditic shales and greywackes, yields fossils of Early to Late Visean age (Eisele *et al.* 2002). Synsedimentary volcanic activity of andesitic to rhyolitic composition has been documented by the occurrence of 'pepperites' in places where lavas intruded wet sediments (Maass 1992). Olistostromes and slump structures document tectonic instability and a high submarine relief during Late Visean times (Eisele *et al.* 2002). The Upper Unit constitutes the final Variscan sedimentation event. It contains predominantly terrigenous sequences of ?Namurian to Westpha-

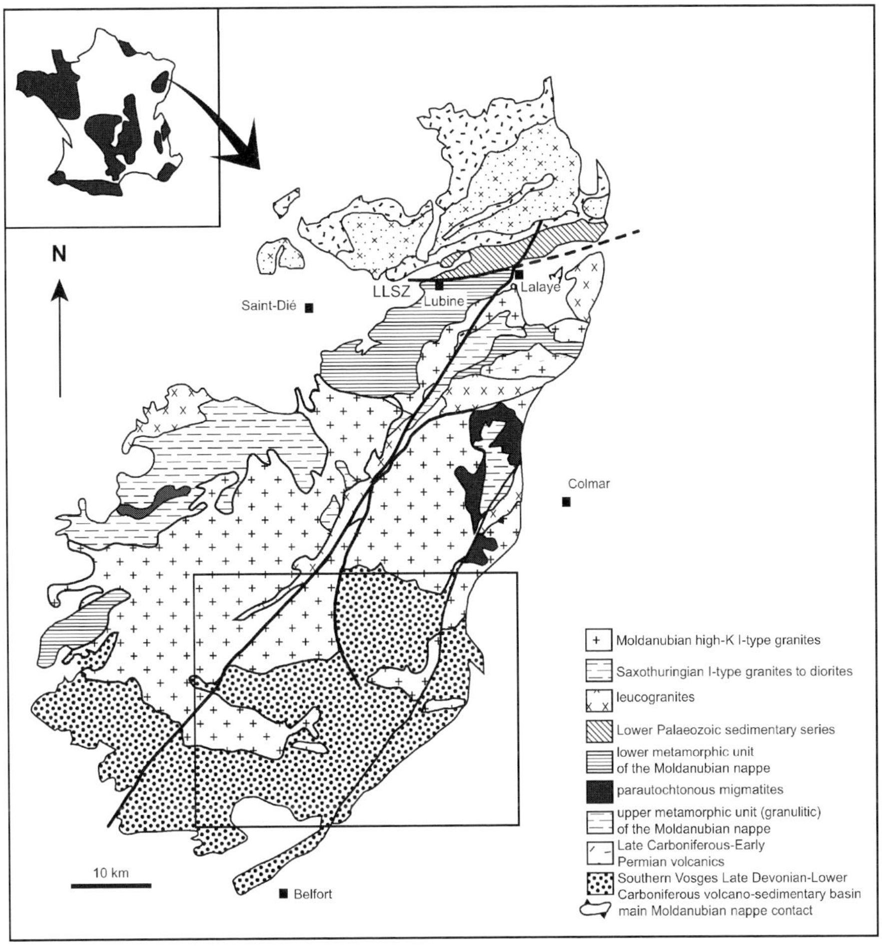

Fig. 9.57. Structural map of the Vosges, eastern France. LLSZ, Lalaye-Lubine shear zone. Rectangle indicates area of Figure 9.58 (after Schaltegger *et al.* 1996).

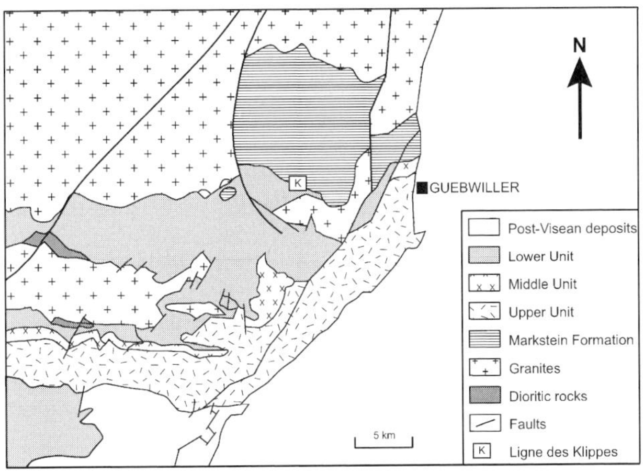

Fig. 9.58. Geological sketch map of the volcanosedimentary formations of the Southern Vosges (after Schaltegger *et al.* 1996). Location indicated on Figure 9.57.

lian age. The base of this succession is composed of a possibly brackish, coal- and plant-bearing greywacke–pelite sequence (Schneider 1990) with intercalated pyroclastic rocks of ignimbritic origin. These deposits are overlain by sediments indicating a rapid change from a brackish to a continuous terrestrial environment, documented by several facies associations, such as river channel fills, fluvial sandbars, and floodplain sediments (Maass & Schneider 1995) and lake deposits (Eisele *et al.* 2002). These are interrupted by more or less structureless high-energy deposits of arkosic arenites and primary pyroclastic and volcaniclastic deposits.

The northern part of the basin contains up to 4000 m of thick marine clastics (Famennian–Late Visean) with only rare volcanics. The succession is considered to be the distal equivalent of the southern part of the basin. Here the extensive volcanism is represented by tuffs and reworked volcanics. A layer containing lenses of high-grade gneisses, serpentinites and mafic rocks with oceanic affinities (Schneider *et al.* 1990), of unknown age and origin, separates the two parts of the basin, and has been interpreted as an olistostrome (Maass & Stoppel 1982) although it has also been considered as a succession of tectonic lenses marking a south-vergent thrust fault (Schaltegger *et al.* 1996). These olistostromic rocks constitute the oldest strata in the basin and are possibly contemporaneous with basin opening. The sediments at the base of the Middle Unit show evidence of extensive reworking, synsedimentary deformation and block-tilting and have been interpreted in terms of a later Visean extensional event. Subsequent sedimentation evolved from a marine to a continental setting. Deformation of the sedimentary pile may be interpreted in terms of a third tectonic phase slightly before, or during, the emplacement of the magmatic rocks.

Geochemical and petrographic analysis has revealed that the timing of subduction-related volcanism and sedimentation in the Southern Vosges Basin coincided, with arc magmatism in the Mid-German Crystalline High and the closure of the Rhenohercynian Ocean. Eisele *et al.* (2002) favour an extensional backarc setting on thinned continental crust for the Southern Vosges in Late Devonian to Early Carboniferous times, related to the development of a magmatic arc to the north of the basin. The onset of convergence and continental collision in Early to Mid-Carboniferous time could have transformed the basin into a retro-arc foreland basin with sedimentation, synchronous igneous activity and deep-marine units grading into continental sediments in the Upper Unit (Eisele *et al.* 2002).

The **Black Forest** (= Schwarzwald) is a bivergent structure with a high-grade core (Zentralschwarzwälder Gneiskomplex) which was emplaced by dextral transpression over low-grade units on either side of the Baden-Baden Zone to the north and the Badenweiler-Lenkirch Zone to the south (Maass 1981; Eisbacher *et al.* 1989; Krohe & Eisbacher 1988) (Fig. 9.59). The high-grade core is a complex tectonic assemblage comprising Ordovician-age metasediments (Hanel *et al.* 1998) and metagranitoids (Chen *et al.* 2000). Ultramafic rocks and eclogites occur as small isolated bodies (Kalt & Altherr 1996), the latter probably being derived from Ordovician magmatic rocks (Kalt *et al.* 1990).

It is assumed that an oceanic suture is situated in the northern Black Forest separating the Moldanubian and Saxothuringian zones (e.g. Kossmat 1927; Franke 1989). Loeschke *et al.* (1998)

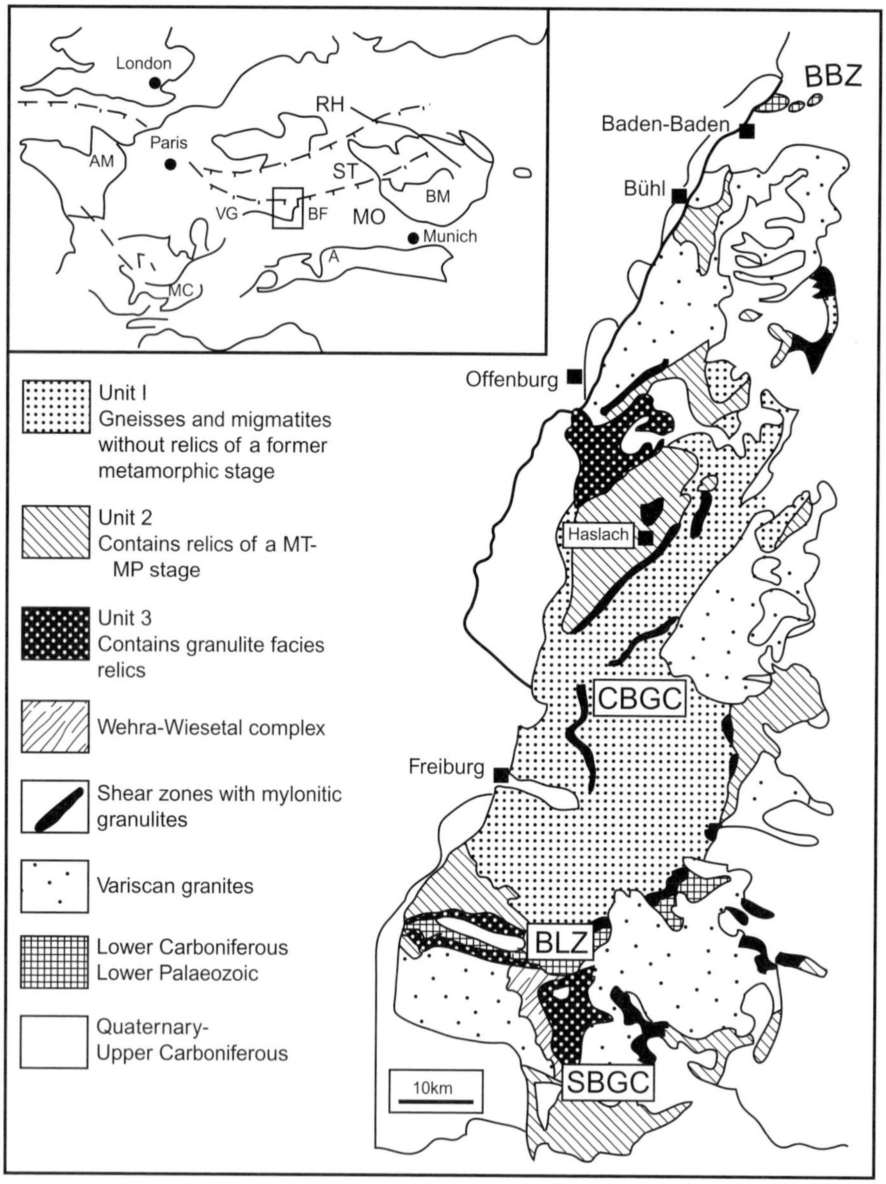

Fig. 9.59. Simplified geological map of the Black Forest. The metamorphic basement is subdivided into three tectonic units according to Hanel *et al.* (1999). The inset shows the Variscan massifs in Central Europe and their subdivision into three zones (RH, Rhenohercynian; ST, Saxothuringian; MO, Moldanubian) according to Kossmat (1927). Abbreviations: A, Alps; AM, Armorican Massif; BBZ, Baden-Baden Zone; BF, Black Forest; BLZ, Badenweiler-Lenzkirch Zone; BM, Bohemian Massif; CBGC, Central Schwarzwald Gneiss Complex; MC, Massif Central; SBGC, Southern Black Forest Gneiss Complex; VG, Vosges (after Marschall *et al.* 2003).

suggested that there was another suture zone in the southern Black Forest area, represented by the Badenweiler-Lenzkirch Zone (BLZ; Fig. 9.59) which consists of Ordovician to Lower Carboniferous metasedimentary and partly weakly metamorphosed volcanosedimentary successions (e.g. Altherr & Maass 1977; Hann *et al.* 1995; Güldenpfennig 1997, 1998; Montenari *et al.* 2000; Vaida *et al.* 2000, 2003). This zone separates the Central Black Forest Gneiss Complex from the Southern Black Forest Gneiss Complex. Both of these complexes are considered to form part of the Moldanubian Zone, but they appear to represent different crustal blocks. Both basement blocks are subdivided into internal nappes, which reflect the Variscan collisional event. The nappe boundaries are cut by large granitic bodies and granite porphyry dykes with late Visean intrusion ages (334–332 Ma; Brewer & Lippolt 1972; Schaltegger 2000). The Rand Granite situated at the southern margin of the Central Black Forest Gneiss Complex comprises metagranites, metatrondhjemites and biotite augen gneisses derived from monzogranites to granodiorites (Hann *et al.* 2003). Large parts of the granite originated from anatexis of continental crust. Hann *et al.* (2003) suggest that the Rand Granite developed along an active continental margin and therefore represents a possible root of a Variscan magmatic arc.

Detailed geochemical investigations have been carried out for the Lower Carboniferous sedimentary sequences of the BLZ (Güldenpfennig & Loeschke 1991; Güldenpfennig 1997), which led to the development of a plate tectonic model for this area (Loeschke *et al.* 1998). The zone contains Ordovician–Silurian clastic metasediments which are overlain by Frasnian shales and cherts, late Devonian–Tournaisian greywacke turbidites and Visean felsic volcanic rocks intercalated with continental sediments (Loeschke *et al.* 1998; Güldenpfennig 1998). This succession is interpreted as representing an early Palaeozoic extensional basin which later evolved into a Visean foreland basin adjacent to a north-dipping subduction zone and overprinted by transtension (Echtler & Chauvet 1992; Franke 2000).

The southernmost portion of the Black Forest is occupied by the Südschwarzwälder Gneiskomplex,. comprising anorthosites, amphibolites and anatectic gneisses (Franke 2000). Tectonometamorphic deformation in the region was at least partly coeval with clastic sedimentation in the BLZ to the north. However, since crustal stacking and marine sedimentation mutually preclude each other, the crystalline unit must have been juxtaposed against the sedimentary one at a later time by transpressional displacement (Franke 2000).

Basins in the Czech Republic (S.O.)

The territory of the Czech Republic is dominated by the Bohemian Massif, the largest Variscan unit of central Europe (Figs 9.3, 9.36). It was formed by the amalgamation of individual Armorican terranes (Saxothuringian, Teplá-Barrandian and Moldanubian) and their final collision with Avalonia and the Brunovistulian. Most of these collisional processes took place between the Devonian and the Lower Carboniferous. From the beginning of the Upper Carboniferous, the intensity of deformation gradually decreased and the Westphalian is characterized by the transition from the orogenic to the post-orogenic stage of development. As a consequence, crustal thickening related to compressional phases of the orogeny was replaced by extension and gravitational collapse.

The distribution, stratigraphic range and tectonosedimentary history of the Carboniferous deposits in the Czech Republic reflect the tectonic processes related to the Variscan Orogeny. Several types of sedimentary basins related to the particular geotectonic positions, developed. A foreland basin (the Moravo-Silesian Culm Basin and its successor, the Upper Silesian Coal Basin) was formed along the eastern margin of the amalgamated Armorican terranes (Moldanubian and Saxothuringian mainly) as a response to their collision with the western margin of the Brunovistulian (e.g. Rajlich 1980; Kumpera & Martinec 1995; Bábek *et al.* 2004). Initial deposition (Devonian) in this basin was marine, becoming more continental over time until Westphalian (and possibly Stephanian) times when it was exclusively continental. Another sedimentary basin existed from the end of the Middle Devonian through to the lower Visean along the contact of the Teplá-Barrandian and Saxothuringian Zone, probably following the NW–SE trending Elbe Fault Zone. The precise extent of this basin, however, is poorly constrained.

During the Westphalian and Stephanian, continental basins were formed in the interior of the Bohemian Massif. The formation of these basins was related to the final phase of Variscan orogenic activity, or even, post-orogenic relaxation. Volcanic activity accompanied deposition in most of these basins.

The youngest Late Palaeozoic basins in the Czech Republic are asymmetric, NNE–SSW striking grabens formed in the Stephanian C during a period of wrench faulting. The most completely preserved of these is the Boskovice Graben (Basin) located along the eastern margin of the Moravian Zone and interpreted as a transpressional strike-slip basin (Pešek *et al.* 2001) (Figs 9.60 & 9.61). It is about 3–10 km wide and nearly 100 km long filled with up to 3000 m of Late Stephanian–Permian age sediments. The basin consists of two main depocentres with deposition beginning in the Stephanian C in the southern depocentre and extending to the north in early Permian times. The Stephanian-age sediments are mainly fluvial, passing, along the eastern tectonically bounded margin, into the alluvial fan deposits of the Rokytná conglomerates. In the central and western parts of the basin, the basal conglomerates pass upward into sandstones and mudstones with three coal seams (Rosice-Oslavany Group).

A second important basin, the Blanice Graben (Basin), in the central part of the Moldanubian Zone consists of several isolated relicts of Stephanian C and early Permian sediments (Figs 9.61 & 9.62). Sediment thickness varies between a few hundred metres and *c.* 1000 m. Stephanian C strata are composed of sandstones and conglomerates with intercalated mudstones and one or two locally workable coal seams in the upper part of the unit.

North and east Bohemia (S.O.)

The Lower Carboniferous strata in north and east Bohemia crop out only in the Ještěd Hills (Fig. 9.36). The upward-coarsening succession is tens of metres thick and comprises shales and mudstones at the base and mudstones and sandstones with intercalated greywackes and conglomerates in its upper part. The age of the succession, based on a trilobite fauna, is Tournaisian.

SE of this area, a *c.* 300 m thick succession of non-metamorphosed Lower Carboniferous strata was encountered in deep boreholes beneath the Mesozoic cover near Hradec Králové (to the NE of Prague). The Tournaisian part of the succession is dominated by micritic limestones overlain by bioclastic crinoid limestones. During the Visean, carbonate deposition gave way to deep-marine (Culm) sedimentation. Further to the east, a mixed carbonate/siliciclastic succession is found in the Tournaisian part of the succession.

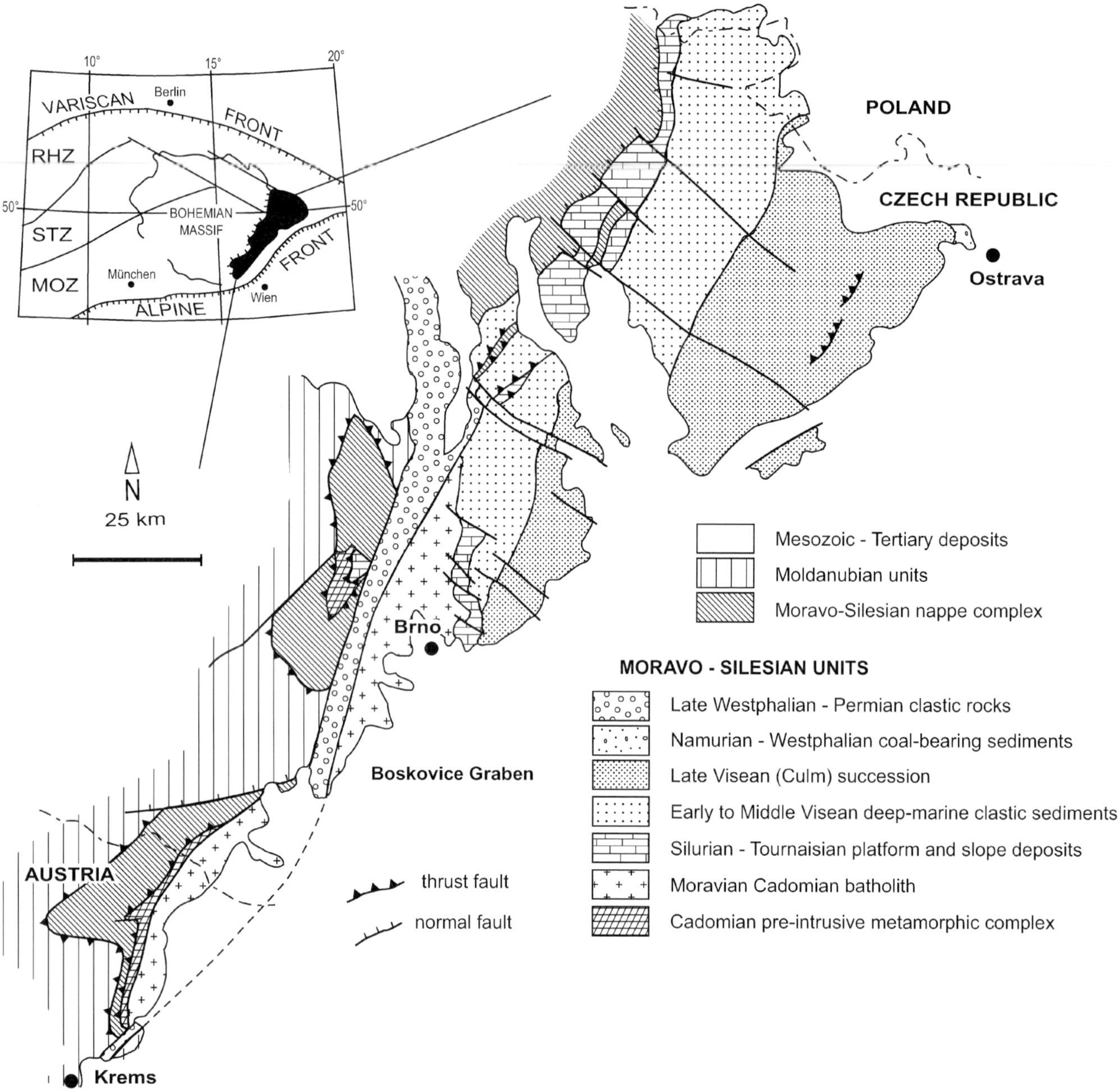

Fig. 9.60. Simplified tectonic map of the Moravo-Silesian unit. Inset shows location in relation to the Moldanubian Zone (MOZ), Rhenohercynian Zone (RHZ) and Saxothuringian Zone (STZ) (after Fritz & Neubauer 1995).

Carboniferous of the Eastern and Southern Alps (K.K.)

In the Eastern Alps Carboniferous sedimentary rocks are found in different tectonic units of the Upper Austro-Alpine nappe system, including the Veitsch Nappe of the Greywacke Zone, the Palaeozoic of Graz, the Gurktal Nappe, the Drau Range (Carboniferous of Nötsch), and the Steinach Nappe (Fig. 9.63). In the Southern Alps, Carboniferous rocks are restricted to the Carnic Alps and the Karawanken Mountains. Sedimentation was strongly influenced by Variscan deformation as well as climatic changes. As is the case for the Alpine area of northern Switzerland (see below), the sedimentary rocks have been overprinted by Alpine deformation.

The Carboniferous succession may be divided into synorogenic sediments deposited during the Variscan Orogeny (Tournaisian to Westphalian) and into Late- and Post-Variscan sedimentary rocks (Westphalian to upper Stephanian). Sediments were deposited in different environments, including non-marine siliciclastic, shallow- and deep-marine siliciclastic and carbonate-dominated settings.

According to Frisch & Neubauer (1989) the Rheic Ocean closed during the Devonian and Early Carboniferous as a result of the NW movement of a microplate termed the Noric Terrane, which was part of Africa (Gondwana). This closure resulted in the formation of a series of synorogenic basins related to extensional and compressional strike-slip movements between the microplate(s) and Gondwana (Vai & Cocozza 1986). In these synorogenic basins, shallow- to deep-marine sediments were

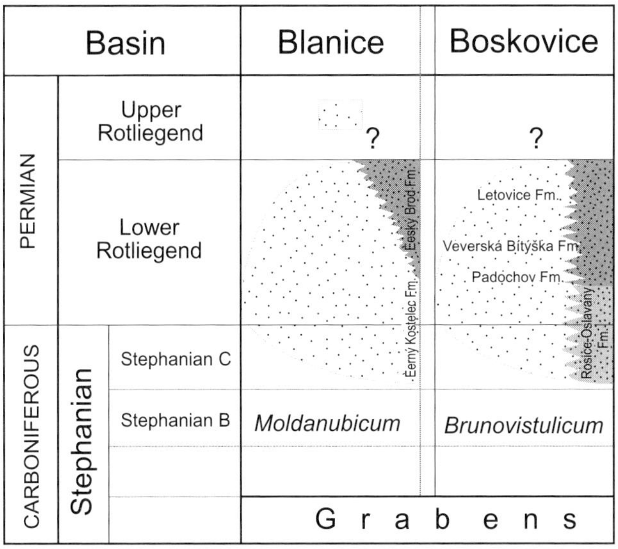

Fig. 9.61. Simplified stratigraphy of the Blanice and Boskovice grabens.

deposited during Tournaisian to mid-Westphalian times. Sedimentary rocks of this phase are exposed in the Veitsch Nappe, the Nötsch area, the Gurktal Nappe, the Palaeozoic of Graz and in the South Alpine Carnic Alps and Karawanken Mountains. During a compressional phase ('Carnic Phase' of Vai 1975; Castellarin & Vai 1981) the synorogenic sediments of the Carnic Alps and Karawanken Mountains were deformed. In the Eastern and Southern Alps the deformed Variscan basement (including the synorogenic sedimentary rocks) is unconformably overlain by the late- to post-Variscan continental sediments.

Veitsch and Noric nappes (Greywacke Zone)

The Carboniferous succession of the Veitsch Nappe is *c.* 650 m thick and, according to Ratschbacher (1984, 1987), may be subdivided into the Steilbachgraben (Early Visean), Triebenstein (Visean–Namurian) and Sunk (?Westphalian) formations (Fig. 9.64). Depositional environments range from mixed siliciclastic/carbonate shallow marine (Steilbachgraben Formation) to massive carbonates containing scattered bioherms (Triebenstein Formation) and shallow-marine siliciclastic/carbonate deposits, grading upward into coarse-grained fluvial deposits. The lower part of the succession contains marine fossils such as brachiopods, echinoderms, corals, bryozoans, trilobites and ostracods indicating a Visean age (Briegleb 1971; Felser 1977; Hahn & Hahn 1977; Krainer 1992). Plant fossils occur in the upper part of the succession and are dated as Westphalian A–C (van Amerom & Boersma 1974; Tenchov 1980).

In the Noric Nappe low-grade metamorphic sedimentary rocks of Carboniferous age are exposed in the Eisenerz area. A thin limestone breccia (up to 10 m) containing a mixed conodont fauna of Devonian to Visean age unconformably overlies Devonian limestones. According to Schönlaub (1982) this breccia was formed during a karstification event in the Visean. Above this breccia 100–150 m of slates of the Eisenerz Formation are developed with intercalated carbonate lenses containing Visean-age conodonts (Schönlaub 1979, 1982).

Carboniferous of Nötsch

The tectonically isolated Carboniferous succession of Nötsch, exposed at the southern margin of the Upper Austro-Alpine Drau Range in southern Austria, is a >1000 m thick, dominantly siliciclastic sedimentary sequence deposited within a deep-marine environment (Fig. 9.64). The sequence is divided (from base to top) into the Erlachgraben, Badstub and Nötsch formations (Krainer 1992, 1993).

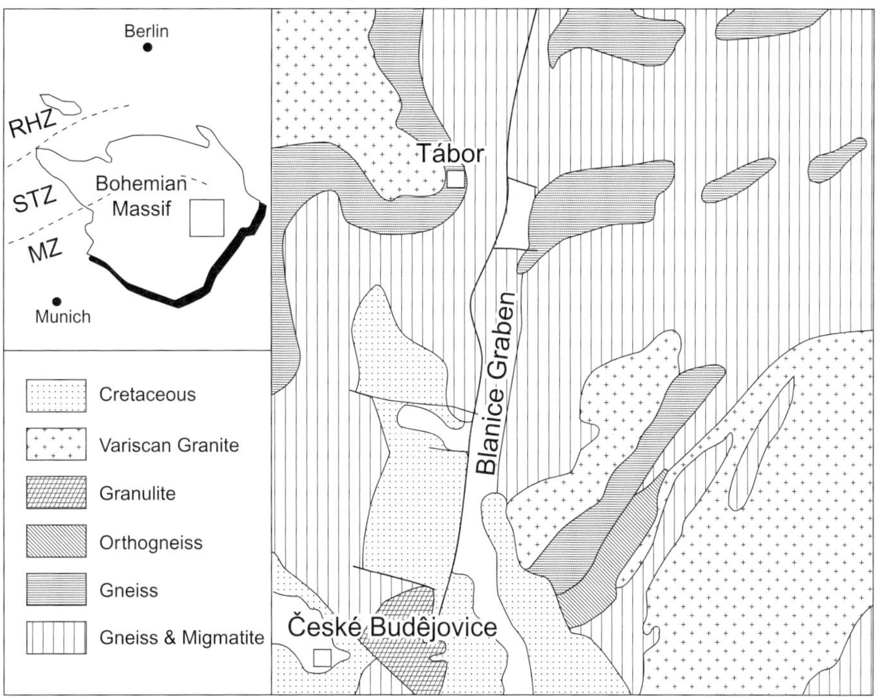

Fig. 9.62. Simplified geological map of the central part of the Bohemian Massif showing the location of the Blanice Graben. Inset shows location in relation to the Rhenohercynian Zone (RHZ), Saxothuringian Zone (STZ) and Moldanubian Zone (MZ) (after Walter 1995).

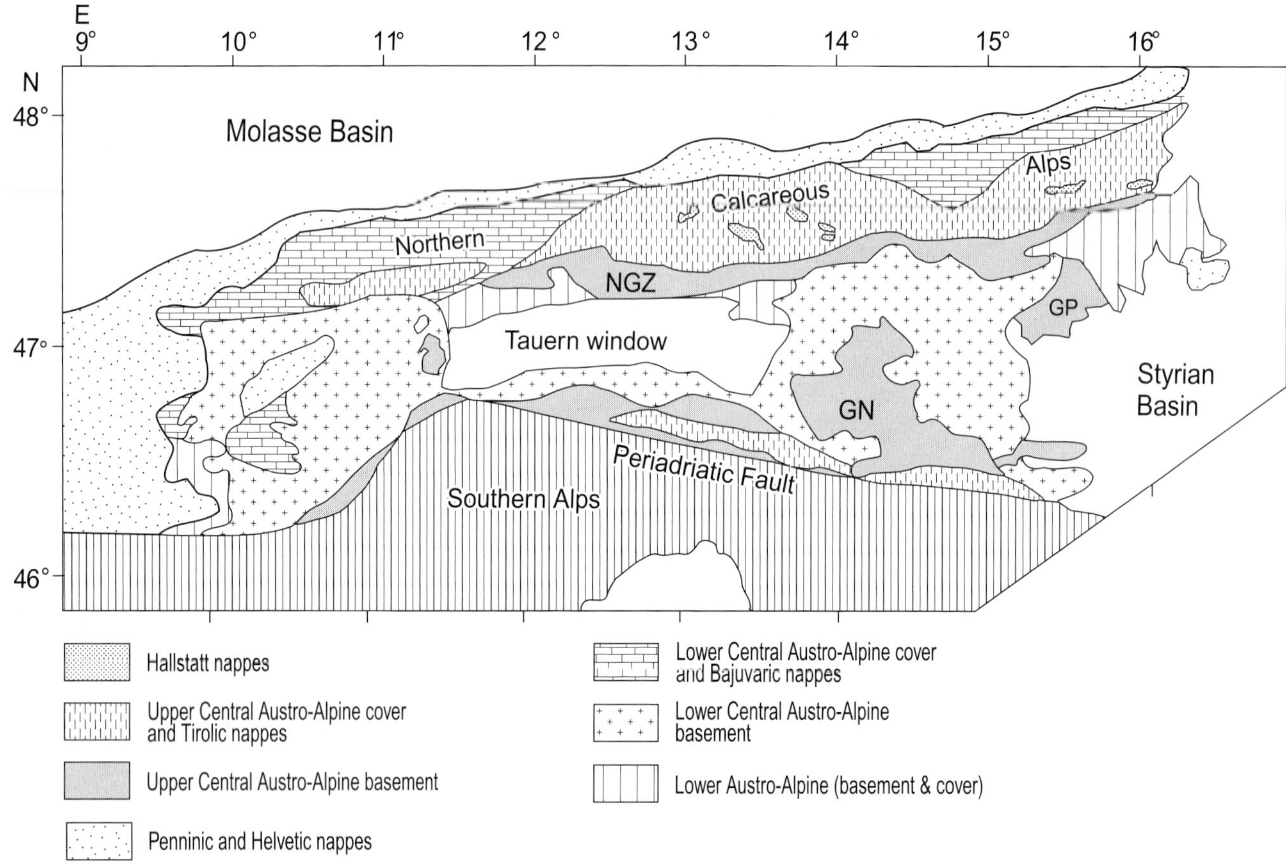

Fig. 9.63. Tectonic map of the Eastern Alps. Abbreviations: GN, Gurktal Nappe; GP, Graz Palaeozoic; NGZ, Northern Greywacke Zone (after Neubauer & Höck 2000; Janák *et al.* 2006).

The Erlachgraben Formation is several hundred metres thick and in the lower part is composed of siliciclastic sediments forming pronounced fining-upward sequences up to several metres thick. These sediments were probably deposited on the lower continental slope by gravity flows. The upper part consists mainly of fossiliferous siltstones and mudstones with intercalated fossiliferous marly and silty limestones composed of bioclastic wackestones and algal wackestone/packstones. These limestones contain a diverse fossil assemblage of foraminifers, algae and pseudo-algae, indicating mid-Namurian (Arnsbergian) age (Krainer & Vachard 2002). Well-preserved plant fossils from this sequence also suggest the same age (van Amerom & Kabon 1999).

The overlying Badstub Formation is approximately 400 m thick and consists of breccias, conglomerates, sandstones and fossiliferous siltstones which are extremely rich in amphibolite clasts. These deposits are interpreted as submarine gravity flows deposited in a proximal fan or slope environment along an active fault zone (Krainer & Mogessie 1991; Krainer 1992).

The Nötsch Formation is several hundred metres thick and consists, in the lower part, of fossiliferous shales and siltstones with intercalated sandstones and conglomerates, brachiopod coquina layers and thin fossiliferous silty limestones. The succession contains a rich fauna including brachiopods, echinoderms, bivalves, bryozoans, trilobites, gastropods, ostracods (Schönlaub 1985*a*; Schraut 1996, 1999; Hahn & Hahn 1987) and flora (van Amerom & Kabon 2000). The upper part of the Nötsch Formation is composed of polymictic conglomerates and sandstones similar to the lower part of the Erlachgraben Formation The sediments are interpreted to have been deposited in a continental slope environment (Krainer 1992).

Palaeozoic of Graz

In the Rannach Facies of the Upper Austro-Alpine Nappe System of the Palaeozoic of Graz, the Lower Carboniferous is represented by the Sanzenkogel Formation, a 30–35 m thick succession of condensed, greyish micritic limestones with intercalations of sandy shales, thin phosphorite and chert (lydite) beds in the lower part, unconformably overlying Middle and Upper Devonian limestones (Fig. 9.64). The Sanzenkogel Formation contains cephalopds and a mixed conodont fauna indicating an age ranging from Tournaisian to Namurian A. Stratigraphic gaps, represented by erosional horizons, and fissure fillings are interpreted to have been caused by synsedimentary tectonic movements, local subaerial exposure and karstification during the latest Tournaisian.

The Sanzenkogel Formation is disconformably overlain by the dominantly clastic Dult Formation which locally exceeds 70 m in thickness. The lower 10–15 m are represented by shallow-marine limestones containing conodonts ascribed to the Namurian B. The limestones are overlain by a shaly facies. Erosional surfaces, hematite crusts and desiccation cracks indicate periods of subaerial exposure (Ebner 1976, 1977; Ebner *et al.* 1987). In the Hochlantsch Facies the Devonian Hochlantsch Limestone is overlain by up to 100 m of pelagic limestones ('Mixnitz

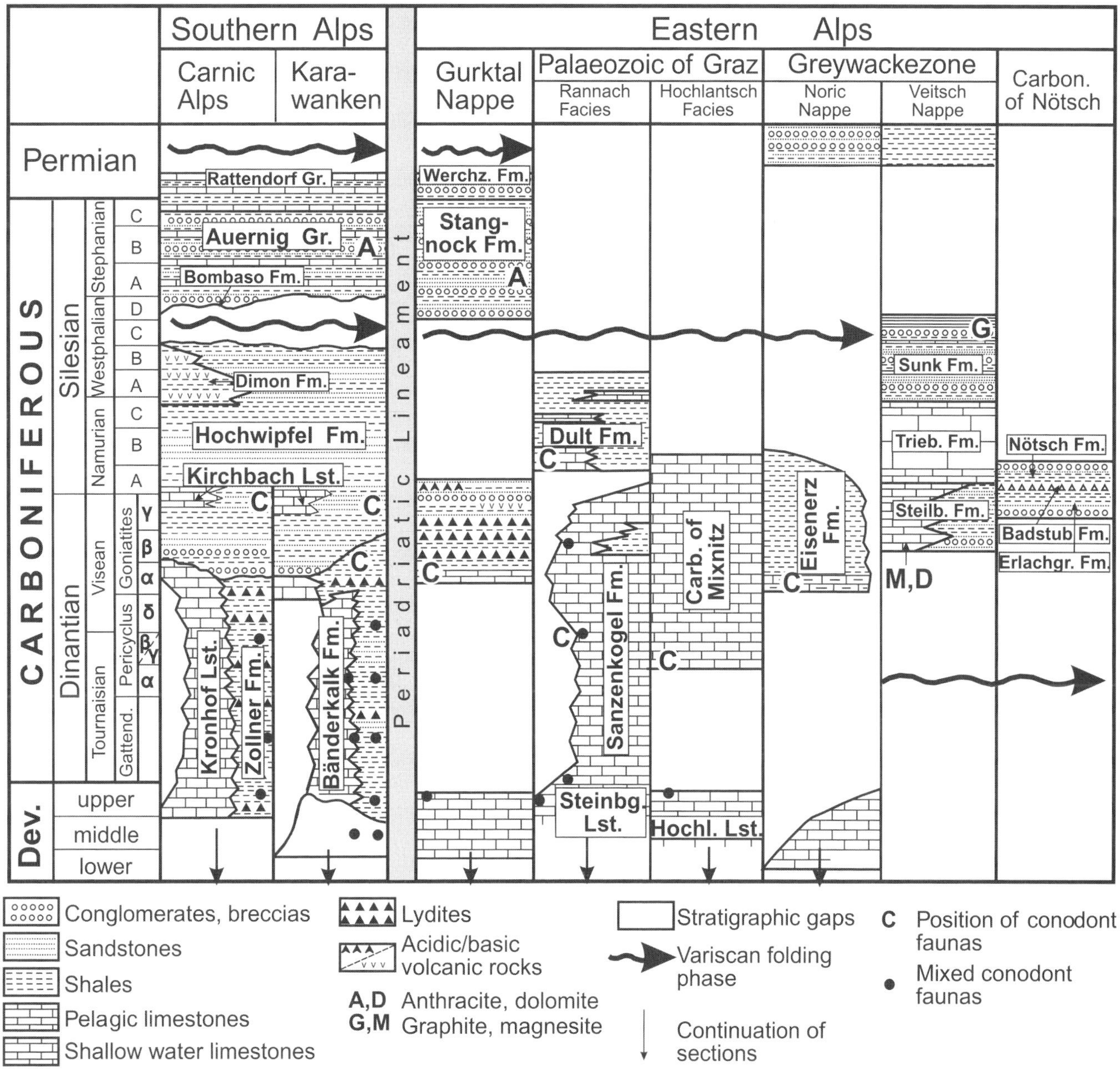

Fig. 9.64. Carboniferous lithology and stratigraphy of the Eastern and Southern Alps (after Ebner *et al.* 1987, 1991).

Carboniferous'), ranging in age from Late Tournaisian to Namurian B (Ebner 1978; Ebner *et al.* 1987).

Gurktal Nappe

The Variscan basement in the region of the Gurktal Nappe is composed of deformed and weakly metamorphosed sedimentary and volcanic rocks of Ordovician to Early Carboniferous age. The youngest rocks of the Variscan basement are represented by >20 m of shales and cherts resting on Upper Devonian-age limestones (Fig. 9.64). Based on conodonts, these rocks are assigned to the Late Tournaisian or Tournaisian–Visean boundary (Neubauer & Herzog 1985). Basement is unconformably overlain by a >400 m succession of non-marine clastics (Stangnock Formation) (Figs 9.64 & 9.65). The sediments are well exposed along the NW margin of the Gurktal Nappe in the Stangnock-Königstuhl area. The lowermost part is composed of polymictic conglomerates rich in gneiss clasts, and intercalated, immature, coarse-grained sandstones of a proximal fluvial environment. The main part of the succession consists of several poorly developed fining-upward megasequences containing sediments deposited within a gravelly braided river system at the base, grading upward into a gravel–sandstone facies with characteristic features of a meandering river system. At the top of these megasequences, several metres of thick dark shales containing abundant well-preserved plant fossils are developed. The shales are interpreted as overbank fines, deposited on floodplains and in oxbow lakes. Locally, the shales are overlain by thin anthracite seams. Conglomerates contain abundant quartz clasts. Sandstones are classified as moderately sorted, subangular lithic arenites and sublitharenites, with subordinate lithic wackes, all of which contain large amounts of polycrystalline quartz. In the uppermost part of the Stangnock Formation the sandstones

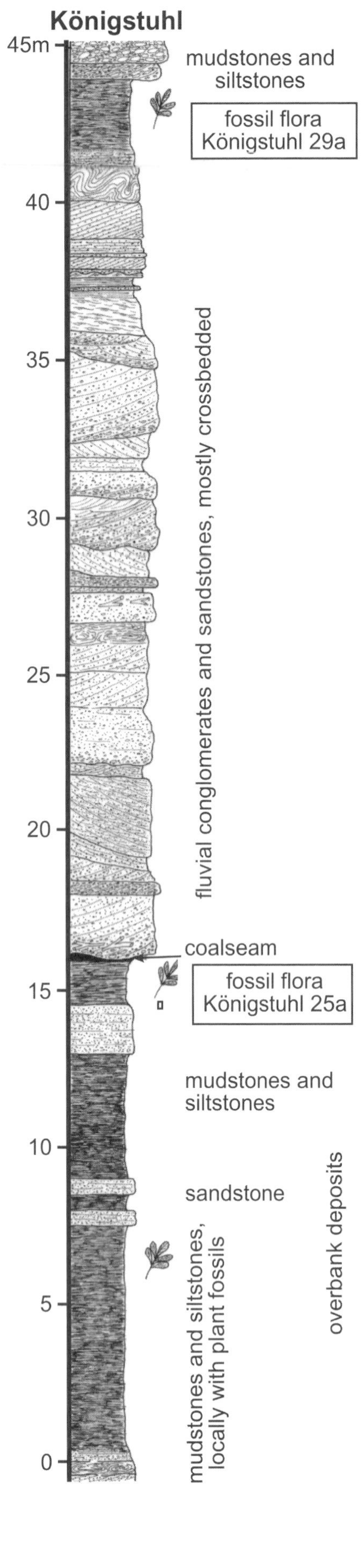

contain small amounts of volcanic debris indicative of the onset of volcanic activity during the latest Carboniferous.

Current directions show a distinct eastward trend indicating that the sediments of the Stangnock Formation were deposited in an approximately east–west trending intermontane basin. The sharp erosional contact at the base of the megasequence is interpreted as resulting from synsedimentary block faulting (Krainer 1989*a*, *b*, 1992, 1993). More than 100 species of plant fossils have been described from the interbedded shales and suggest a Stephanian age (Fritz *et al.* 1990; Fritz & Krainer 1997).

Steinach Nappe

Non-marine clastic sediments of Stephanian age unconformably overlie the Variscan basement of the Upper Austro-Alpine Steinach Nappe. The poorly exposed succession is composed of grey-coloured channel, bar and overbank sediments of a fluvial environment. Locally thin anthracite seams occur. Overbank shales contain well-preserved plant fossils. The fossil assemblage contains c. 30 taxa and is lower Stephanian in age. The facies and mineralogical composition of the sediments are very similar to the Stangnock Formation of the Gurktal Nappe (Fig. 9.64); the sediments were probably deposited in the same intermontane basin (Krainer 1990).

Carnic Alps and Karawanken Mountains (Southern Alps)

In the Carnic Alps and Karawanken Mountains, deposition of condensed pelagic goniatite and clymeniid limestones extended from the Frasnian–Famennian boundary up to the Late Tournaisian (Fig. 9.66). The Carboniferous part, which is generally 1–2 m thick although locally it can be up to 10 m thick, contains thin beds of carbonate breccia and a mixed conodont fauna. The radiolarian chert facies (i.e. Zollner Formation) also extends from the Pragian up to the Late Tournaisian (Schönlaub 1985*b*; Schönlaub & Heinisch 1993). These pelagic limestones and cherts are overlain by a >1000 m thick succession of deep-marine clastics termed the Hochwipfel Formation. Extensive palaeokarst features at the top of the Tournaisian limestone unit are indicative of a drop in sea level, followed by a rapid rise in sea level during the Late Tournaisian. Sedimentation of the clastic Hochwipfel Formation began near the Tournaisian–Visean boundary. These deep-sea sediments were deposited in a relatively narrow oceanic basin, which was formed during an extensional phase ('transform rifting phase' according to Vai & Cocozza 1986). The main lithotypes are pelitic sediments with intercalations of pebbly mudstones, chert and limestone breccias, conglomerates, and sandstones, interpreted to represent different types of gravity flows. Locally, well developed Bouma sequences occur. Ichnofossils as well as plant remains are present, while thin limestone intercalations yielded conodonts assigned to the Visean–Namurian boundary. Upwards the Hochwipfel Formation grades into volcaniclastic and volcanic rocks of the Dimon Formation (Tessensohn 1971, 1974*a*, *b*; van Amerom *et al.* 1984; Schönlaub 1985*b*; Spaletta & Venturini 1988; summary in Schönlaub & Heinisch 1993).

In the region the final phase of the Variscan Orogeny reached its climax between the Late Namurian and the Late Westphalian, resulting in extensive folding and overthrusting (Venturini

Fig. 9.65. Stratigraphic column through the Stangnock Formation (upper part) at Königstuhl showing the typical lithology of fluvial conglomerates and sandstones and thick overbank deposits containing plant fossils.

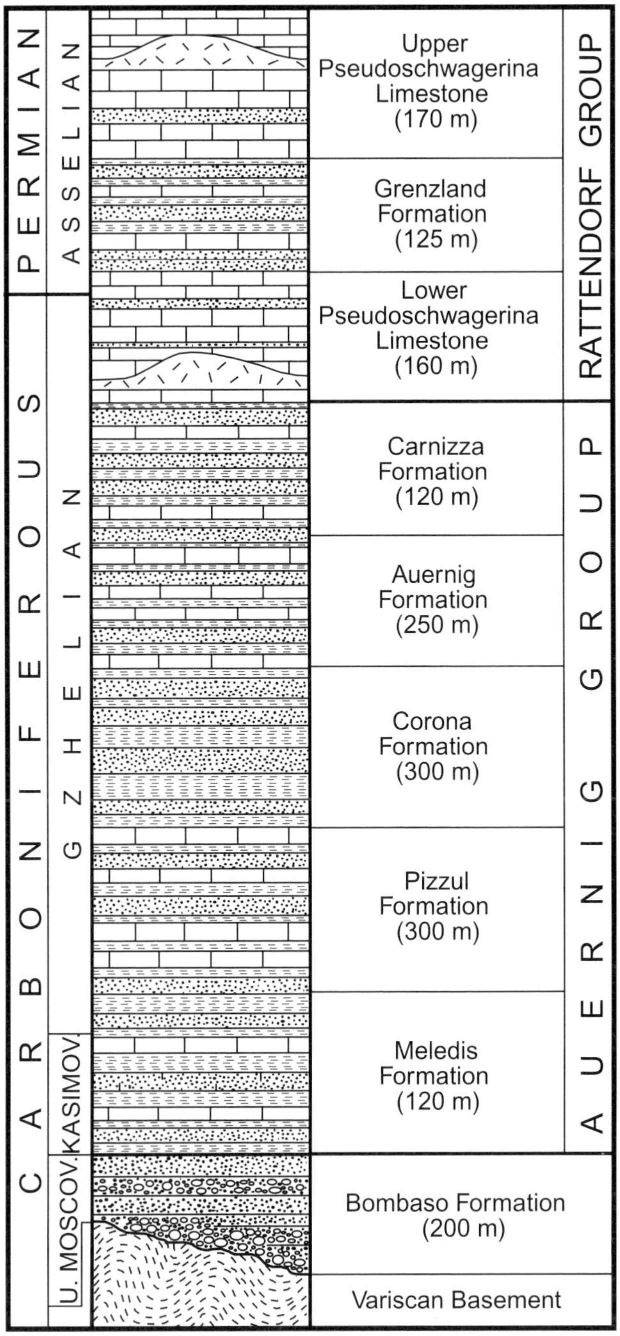

Fig. 9.66. Lithology and stratigraphy of the Late Carboniferous and Early Permian sedimentary sequence of the Carnic Alps.

1990a, b). Here, the deformed Variscan basement is unconformably overlain by a thick succession of Late Carboniferous and Early Permian deltaic and shallow-marine siliciclastic and carbonate sediments which were deposited in discrete basins formed by block-and-wrench faulting subsequent to the final phase of Variscan activity (Venturini 1982, 1990*a*, *b*). The late- to post-Variscan succession can be divided into two cycles which are separated by a major unconformity (see Massari 1986). The lowermost cycle is composed of Late Carboniferous to Early Permian deltaic to shallow-marine deposits, while the uppermost commences with the continental to shallow-marine clastics of the Gröden Formation.

The Carboniferous succession, which is well exposed in the Carnic Alps, is represented by the Bombaso Formation, the Auernig Group and the lower part of the Rattendorf Group (Fig. 9.66). The Bombaso Formation ranges in thickness from a few metres up to *c*. 200 m and consists of poorly sorted immature breccias and conglomerates which are predominantly composed of either radiolarian chert and volcanic clasts (Pramollo Member) or Silurian–Devonian carbonate clasts (Malinfier Horizon), and are derived from the Variscan basement. A fining-upward trend is observed within the succession. Rare skeletal grains (e.g. brachiopods, crinoids and fusulinids) are found indicating deposition in a marine environment. The sediments of the Bombaso Formation are interpreted as mass gravity flow deposits on fan deltas (Venturini 1990*a*, *b*; Krainer 1992). The fusulinid fauna indicates a late Westphalian age (Krainer & Davydov 1998; Davydov & Krainer 1999). The plant fossil assemblage contains *Neuropteris scheuchzeri* (Fritz & Krainer 1995) which according to Wagner (1984) is a key fossil for the Cantabrian (Early Stephanian).

The Bombaso Formation is overlain by the Auernig Group which is up to 1200 m thick and was subdivided into the Meledis, Pizzul, Corona, Auernig and Carnizza formations by Selli (1963). The Auernig Group is a cyclic succession of shallow-marine clastic and carbonate rocks. The Meledis, Corona and Carnizza formations are predominantly composed of clastic sediments. In contrast, the Pizzul and Auernig formations are mixed siliciclastic/carbonate units with several, intercalated fossiliferous limestone horizons up to 13 m thick. The main lithofacies are quartz-rich conglomerates deposited in a near-shore environment, trough cross-bedded and hummocky cross-bedded sandstones deposited on the shoreface, and bioturbated and locally fossiliferous siltstones and shales from an offshore shelf environment, as well as bedded and massive fossiliferous limestones. Massive limestones represent well preserved algal mounds (*Anthracoporella* mounds; Krainer 1995). In the Meledis Formation small mounds formed of auloporid corals are developed (Flügel & Krainer 1992).

In the upper part of the Auernig Group (i.e. Corona, Auernig and Carnizza formations) the above-mentioned lithotypes are arranged to form well developed transgressive–regressive cycles ('Auernig cyclothems') with thicknesses of 10–40 m (Fig. 9.67). The conglomerates of these cycles represent relative sea-level lowstands, while fossiliferous limestones are indicative of sea-level highstands. The formation of these cycles was related to eustatic sea-level changes caused by the Gondwana glaciation (Massari & Venturini 1990; Krainer 1992, 1995). Based on the rich fusulinid fauna the Auernig Group is dated as Stephanian–lowermost Rotliegend (Kahler 1983, 1985, 1986, 1989; Krainer & Davydov 1998). Plant fossils, which are known from many localities and different stratigraphic levels throughout the succession, indicate a Stephanian age (summary in Fritz *et al.* 1990; Fritz & Krainer 1993, 1994, 1995). The Auernig Group is conformably overlain by the Rattendorf Group, which is up to *c*. 450 m thick and divided into the Lower Pseudoschwagerina Limestone (LPL), the Grenzland Formation and the Upper Pseudoschwagerina Limestone.

The LPL is a cyclic succession up to 160 m thick, composed of shallow-marine limestones and thin intervals of clastic sediments forming three well-developed cycles. The clastic intervals form the base of the cycles and were deposited during relative sea-level lowstands in a nearshore environment. During transgression, well-bedded fossiliferous limestones and massive algal mounds accumulated (Samankassou 1997, 1998; Krainer *et al* 2003). During relative sea-level highstands, with water depths

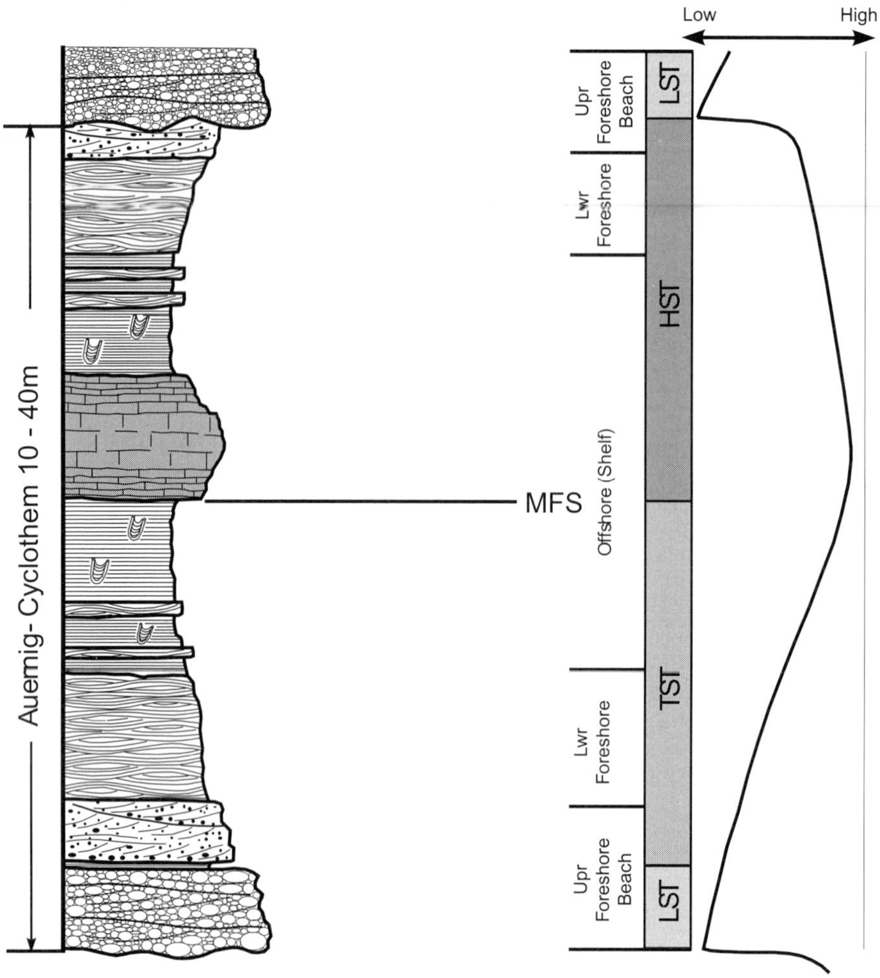

Fig. 9.67. Idealized Auernig Cyclothem from the Late Carboniferous Auernig Formation, Carnic Alps.

of some tens of metres, dark grey, bedded cherty limestones with dark marl intercalations were deposited (Homann 1969; Flügel 1974, 1977; Buggisch *et al.* 1976; Flügel *et al.* 1997; Forke *et al.* 1998; Forke & Samankassou 2000; Forke 2002). Based on fusulinids, much of the LPL is of latest Carboniferous age.

Tauern Window (T.M.)

The crystalline Moldanubian Zone in Austria comprises three main lithological units that generally strike NNE–SSW and dip ESE. These are the Monotone and Bunte series and the Gföhl Unit (see above). The Monotone Series is intruded in the west by the granitic Southern Bohemian Pluton but contact phenomena are missing. It is composed predominantly of stromatitic to nebulitic cordierite gneiss, cordierite-free paragneiss, quartzite, leucocratic siliminite gneiss, subordinate calc-silicate and rare ecologitic rocks (Fig. 9.68). It is separated from the overlying Bunte Series by a thin thrust boundary comprising blastomylonitic gneiss and amphibolite (South Bohemian Main Thrust) (Petrakakis 1997). The base of the Bunte Series comprises I-type Dobra granodioritic orthogneiss intercalated with amphibolite. Additional lithologies include marble, calc-silicate rocks, orthogneiss, graphite-bearing paragneiss and schists as well as quartzites and amphibolites (Petrakakis 1997). The Gföhl Unit comprises amphibolites, metasyenites, gneiss and granulites. Lenticular bodies of mantle-derived garnet lherzolite and garnet pyroxenite commonly occur at the margin or base of the gneiss and granulite.

There were three distinct pulses of magmatism in Early Carboniferous (Visean), Late Carboniferous (Stephanian) and Early Permian times, each of which involved granitoid intrusions and the coeval development of volcanosedimentary basins. A similar relationship has been reported for the Carboniferous parts of the basement of the Alps further to the west. Following the intrusion of subduction-related volcanic-arc granitoids (374 ± 10 Ma), collisional intrusive granitic, anatectic, and extrusive rhyolitic, dacitic rocks were produced over a short interval at *c.* 340 Ma. This Early Carboniferous magmatism, which produced relatively small amounts of melt, can be attributed to the amalgamation of the Gondwana-derived 'Tauern Window' terrane with Laurussia–Avalonia. Magmas are mainly of the high-K I-type and may have been generated during a short phase of decompressional melting of lithospheric mantle and lower crustal sources. In the Late Carboniferous, a second pulse of magmatism occurred producing batholiths of calc-alkaline I-type granitoids and minor coeval bodies of felsic and intermediate volcanics. Prior to this phase of magmatism, several kilometres of upper crust must have been eroded, because volcanosedimentary successions in the central Tauern Window area (Austria) rest unconformably on 340 Ma old granitoids.

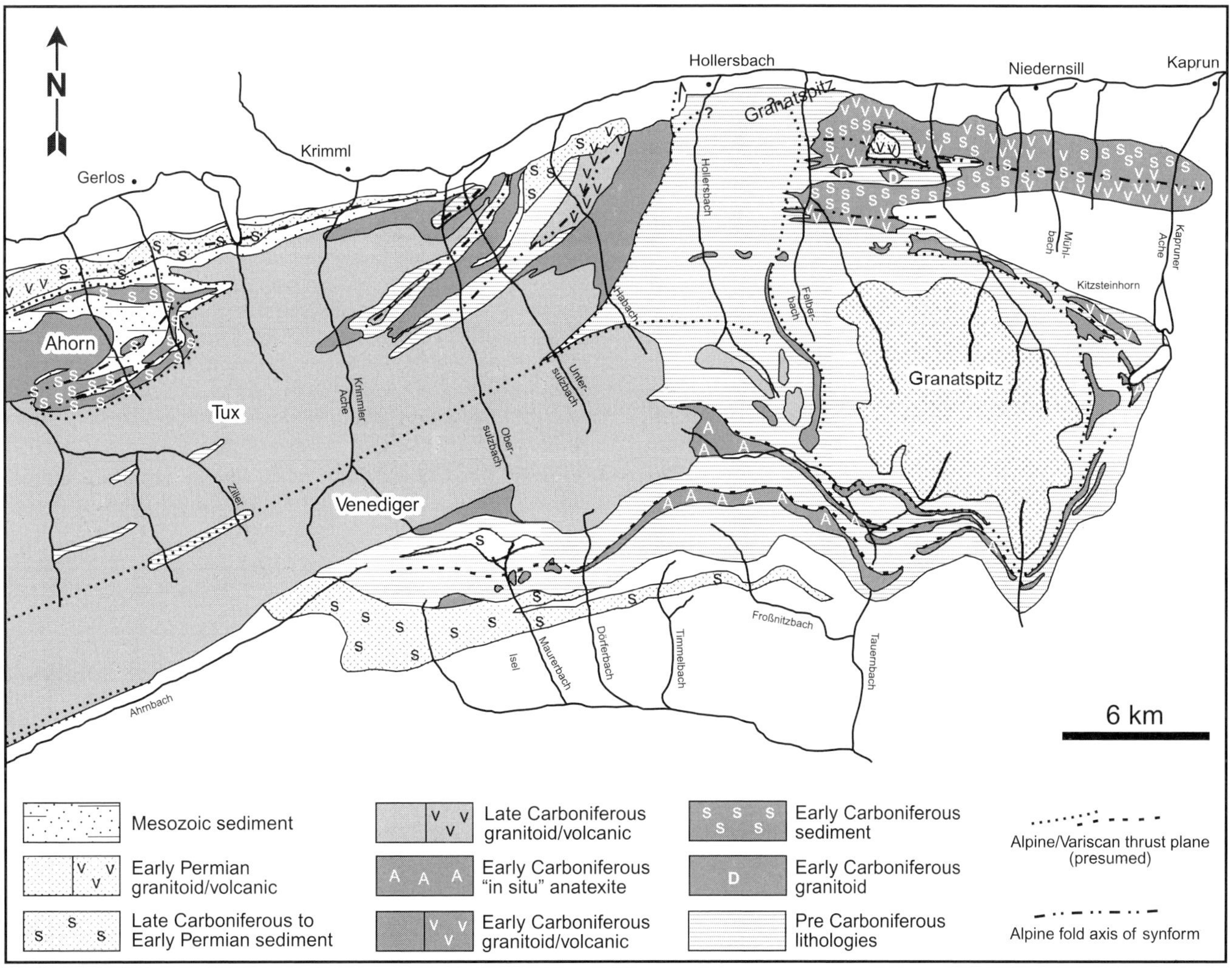

Fig. 9.68. Simplified geological map of the western and central Tauern Window (after Schmidegg 1961; Raith *et al.* 1980; Lammerer 1986; Finger *et al.* 1993; Kupferschmied 1993, 1994; Oehlke *et al.* 1993; Schermaier 1993; Kupferschmied & Höll 1994; Kupferschmied *et al.* 1994; Lammerer & Weger 1998; Weger 1998; Eichorn *et al.* 2000).

The **central Tauern Window** consists of pre-Variscan basement, which was intruded by Variscan granitoids and locally covered by Variscan volcanics and sediments. Basement consists of (metamorphosed) Cambro-Ordovician magmatites and sediments of an active continental margin (Habach Group: arc and forearc lithologies) and ensialic backarc lithologies (Stubach Group and equivalents) (Eichhorn *et al.* 1999).

Variscan volcanics and sediments have been identified in outcrops of the **western Tauern Window** which survived post-Variscan uplift, Alpine compressional tectonics and erosion in synforms. The Tux Nappe, for example, includes a 'Post-Variscan Transgression Series' which, in places, concordantly overlies a pre-Carboniferous basement. The unit comprises conglomerates and a thin rhyolite of unknown age. A similar sedimentary succession is present in the Southern and Northern Schönach Syncline. The 'Porphyrmaterialschiefer Series', preserved in the Tux Nappe at the northern margin of the Tauern Window, resembles the sediments of the two Schönach synclines (Eichhorn *et al.* 2000). In the Central Tauern Window, the existence of similar relicts of volcanosedimentary basins has been locally assumed, though is still unproven. The Venediger Nappe contains a 'Micaschist Unit' at the southern Tauern Window margin, and a leaf fragment from within the unit is probably of Late Carboniferous–Early Permian age (Franz *et al.* 1991). Early Carboniferous ages have been reported from the Habach area of the Tux Nappe and from the Amer and Hollersback areas of the Granatspitz Nappe (see Eichhorn *et al.* 2000).

Variscan granitoids in the western and central Tauern Window are mainly represented by Central Gneiss bodies. In addition to the more voluminous Central Gneiss plutons, thin elongate gneiss lamellae and minor intrusive gneiss bodies also occur (Lammerer & Weger 1998) (Fig. 9.68).

Amalgamation of the north-drifting northern portions of the 'Intra-Alpine' terrane hosting the Tauern Window area with Laurussia–Avalonia probably did not occur prior to the Late Devonian, as suggested by the primitive volcanic-arc chemistry of the Zwölferkogel granodioritic gneiss. This rock appears to be another representative of the Early Variscan, potentially subduction-related I-type granitoid group as defined by Finger *et al.* (1997). It is likely that the collision between the Tauern Window basement and Laurussia–Avalonia occurred between the Late Devonian and the Early Carboniferous. These events were accompanied by volcanic activity in a syncollisional extensional setting. Intramontane transtensional basins containing detrital sediments and calc-alkaline volcanics (or plutons and dykes at somewhat deeper levels) evolved during the Visean. The magmatic evolution of the Tauern Window area at that time is similar to the evolution reported from the 'External Domain' (Switzerland) where intramontane basins containing subaerial clastics and pyroclastic tuffs were probably formed between 345 and 335 Ma and were filled in an extensional setting (Schaltegger & Corfu 1992).

Bearing in mind the Late Carboniferous dextral westward wrenching of Gondwana relative to Laurussia (Arthaud & Matte 1977), dextral mega-strike-slip movements may have shifted the 'Intra-Alpine' terrane to the south of Armorica. Prior to that the 'Intra-Alpine' terrane may have been located in a SE extension of Armorica (Fig. 9.69). Accordingly, the Devonian Zwölferkogel gneiss and the Cetic granitoids may be related to a southward subduction of the Rhenohercynian Ocean. Since subduction probably ceased in the Late Devonian (Franke *et al.* 1995*a*), an active Andean-type margin setting is an unlikely explanation for the Early Carboniferous magma pulse. Schaltegger (1997) proposed that these magmas may have been generated during a short phase of decompressional melting of lithospheric mantle and lower crustal sources. This melting stage may have been triggered by detachment of the subducted oceanic slab or simply by thermal erosion of the Variscan orogenic root through a convecting mantle. An extensional regime for the emplacement of the granitoids is also suggested by their partly shallow intrusion into coeval volcanosedimentary basin successions (Eichhorn *et al.* 2000). The Variscan Carboniferous magmatic evolution of the Tauern Window area is similar to the evolution reported from the more westerly Aar, Aiguilles Rouges and Mont Blanc massifs of the External Domain in Switzerland. This fact supports the palaeogeographic interpretation of the Tauern Window as an eastward extension of the External Domain (Lammerer 1988; von Raumer 1998).

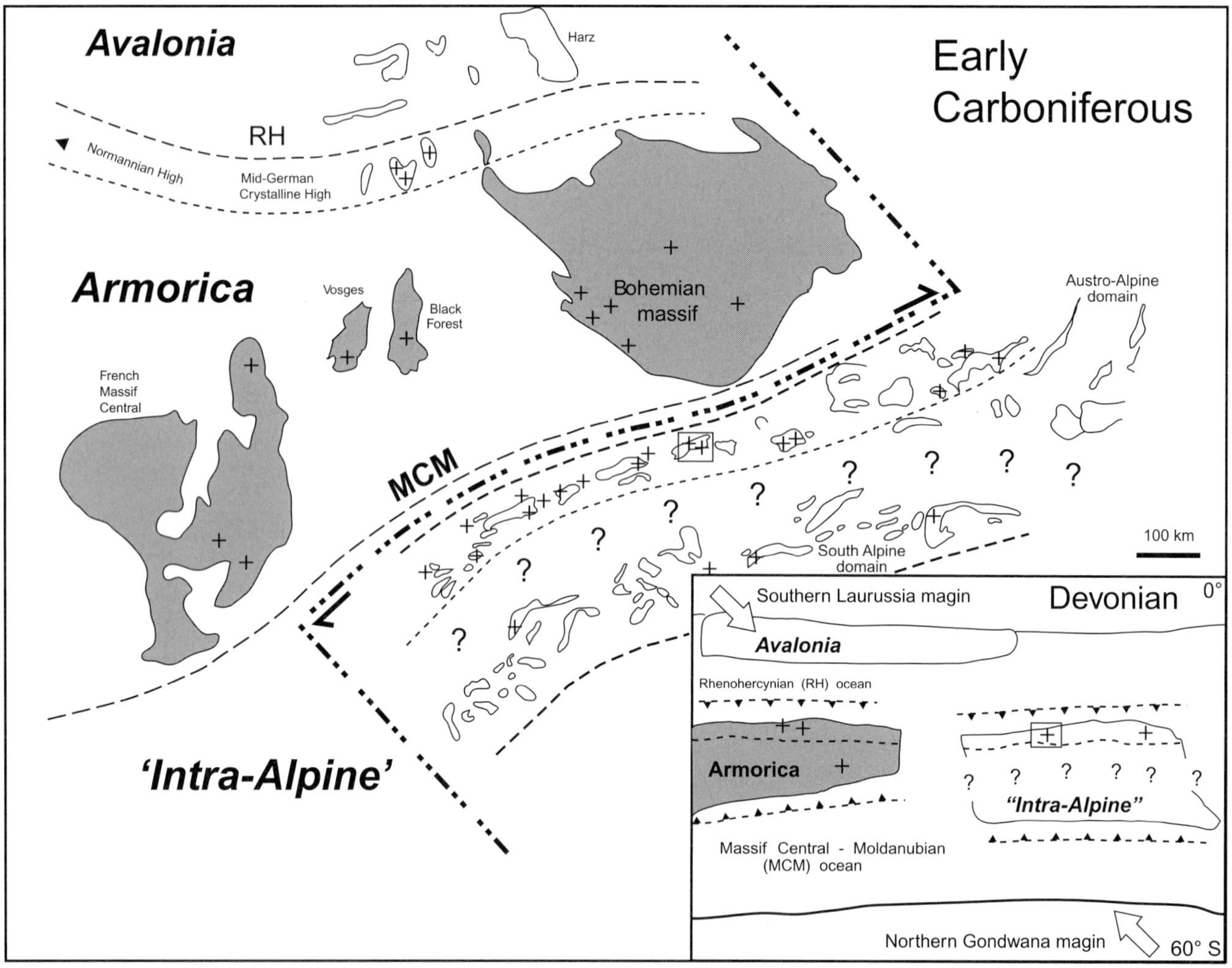

Fig. 9.69. Tentative palinspastic reconstruction of Central Europe in the Late Carboniferous (modified after Franke *et al.* 1995*a*; Finger *et al.* 1997; Tait *et al.* 1997; von Raumer 1998; Eichorn *et al.* 2000). Large-scale dextral strike-slip movements are assumed to have moved the 'Intra-Alpine' terrane from an easterly position to a southerly one relative to Armorica. Thus, the Devonian and Early Carboniferous granitoid intrusions (marked by crosses) in the northernmost parts of the Armorican and 'Intra-Alpine' terrane may have resulted from the same processes, i.e. the southward subduction of a Rhenohercynian ocean and their subsequent collision with Laurussia. The small square in the main diagram marks the approximate position of the Tauern Window (Eichorn *et al.* 2000).

Variscan Internides

The succession in the following regions were deposited within the Variscan fold-and-thrust belt and so have a tectonosedimentary evolution which is distinct from that of the basins and areas to the north. A selection of these areas is presented here in order to provide an idea of the range of environments present.

Massif Central (T.M.)

The French Massif Central is one of the most important exposures of the Variscan Internides (Fig. 9.70). It is one of the two main segments of the French Palaeozoic Orogen and covers an area of *c.* 80 000 km^2 in the central part of France. It is mainly composed of high-grade metamorphic rocks and widespread granitic plutons (Santallier 1994). The tectonometamorphic history of the region is a polyphase one, displaying all of the characteristics of a collisional belt (see references in Santallier 1994). Rifting commenced in the Cambro-Ordovician and was characterized by the widespread emplacement of basic magmas. This was followed by high-pressure metamorphism related to Silurian subduction. In Devonian and lower Carboniferous times collision led to polyphase metamorphism and local extension. The later Carboniferous was associated with strike-slip tectonics, characterized by major shear zones, and late thermal evolution associated with crustal thinning which resulted in low-pressure metamorphism and associated granitoid intrusion.

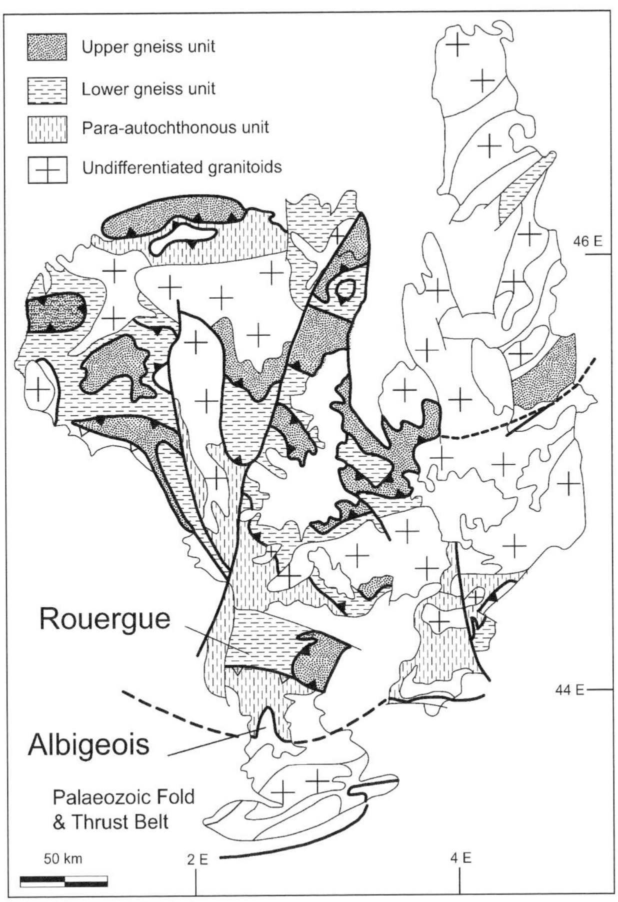

Fig. 9.70. Structural map of the French Massif Central (after Duguet & Faure 2004).

The Variscan is characterized in the Massif Central by a series of events occurring between the Late Silurian and Early Permian (e.g. Ledru *et al.* 1989, 1994, 2001; Faure *et al.* 2004). The Late Silurian–Middle Devonian, the northward continental subduction of the Gondwanan margin beneath the Armorica microplate was followed by the exhumation of high- and ultra-high pressure rocks (Lardeaux *et al.* 2001). A nappe stacking, coeval with the exhumation of high-pressure rocks, is evidenced by NE–SW trending lineations. During Late Devonian times, the occurrence of calc-alkaline magmatic rocks in the NE part of the Massif Central is interpreted as signalling a new subduction event and rifting period. The true collision between Laurussia and Gondwana occurred in the Late Devonian–Early Carboniferous. In the northern part of the Massif Central, compression ended in the Early Carboniferous, whereas it extended through to the Middle Carboniferous in more southerly parts (Talbot *et al.* 2005). This Middle Carboniferous compressional event is characterized by south-directed ductile shearing (Burg & Matte 1978). The entire Massif Central underwent extension due to late-orogenic collapse during Middle–Late Carboniferous times. This rifting event was diachronous, commencing earlier in the north than in the south (Talbot *et al.* 2005). A second extensional event, coeval with the formation of coal-bearing half-grabens, occurred during the Late Carboniferous–Early Permian. According to previous studies (e.g. Matte 1986*a*) nappe emplacement was accomodated by southward-directed thrusting. Six main units can be recognized (Ledru *et al.* 1994).

The first of these comprises the three **Upper Epizonal Units** located within the Upper Allochthon. The first two of these, the Thiviers-Payzac and the ophiolitic Génis units, occur in the SW part of the Limousin, although their precise tectonic position is not clear (due to deformation of the boundaries). The Génis Unit is a fragment of overturned oceanic crust which has Late Ordovician–Pridolian shales and limestones at its base, although the precise age is not known (Ledru *et al.* 1994). It has, however, been proposed that the unit is an Early Devonian olistostrome. The third unit (Brevenne Unit) occurs on the eastern margin of the Massif Central and corresponds to an extensional rift filled by MOR-type basalts, gabbros, and some rhyolites and trondhjemites. It was obducted during the Early Visean.

The **Upper Gneiss Unit** consists of leptynite–amphibolite complexes, which are composed of quartzo-feldspathic gneisses, amphibolites and amphibolite gneisses. The gneisses were derived from granites, acid volcanics, greywackes and acid tuffs while the amphibolites were derived from ultrabasic and basin plutons and volcanic or tuffaceous rocks (Ledru *et al.* 1994). The protoliths were emplaced in Late Cambrian–Early Ordovician times (Duthou *et al.* 1981; Pin & Lancelot 1982). High-pressure relicts (eclogite and granulite) occur in boudins. Eclogitization is dated as Late Silurian. The high-pressure rocks are related to continental subduction of the North Gondwana margin beneath Armorica. Migmatization and retrogression of eclogites occurred during exhumation processes at the end of the collisional stage. The migmatites are dated as Early Devonian.

The **Lower Gneiss Unit** is essentially composed of metagreywackes and metapelites metamorphosed under medium-pressure/medium-temperature conditions and intruded by Early Palaeozoic alkaline granites which were transformed into augen gneiss. Some of these gneisses belong to the pre-orogenic magmatic stage assumed to be of Cambrian–Ordovician age (Duguet & Faure 2004).

The **Presumed Autochthonous Complexes** are thought to be autochthonous since they occupy the lowest lithostratigraphic position and do not show any evidence of high-pressure metamorphism. They comprise metapelites and greywackes and correspond to platform and basin deposits of Late Precambrian–Early Palaeozoic age.

The **Southern Palaeozoic Nappe Complexes** contain pelitic, calcareous and volcanic rocks, with ages ranging from Cambrian through to Visean.

The **Post-Collision Deposits** rest unconformably on metamorphic basement, and comprise Middle Visean clastics and a thick late Visean volcanic unit. During the Middle Visean marine deposits were deposited over a broad area.

Devonian-Lower Carboniferous deposits are found exclusively in the Southern Montagne Noire, where they unconformably overlie Lower Palaeozoic strata (Feist *et al.* 1994). The succession comprises a nearly continuous carbonate succession with depositional environments ranging from nearshore and tidal facies through to neritic and pelagic facies related to the development of a carbonate platform. The Visean contains shelf and slope deposits including limestone turbidites. The uppermost Visean and lower Namurian are characterized by microbialite buildups locally rich in corals (Aretz 2002; Poty *et al.* 2002*a*, *b*) which may be found as olistoliths within the deeper-marine Carboniferous siliclastic deposits (Engel *et al.* 1981).

Deep-marine deposits became more significant in the later Carboniferous as a result of the approach of the orogenic front in early Visean times. Commencing with a thick debris-flow unit and associated olistoliths in the Upper Visean, a >4000 m thick succession was deposited. Environments included turbidite channels and overbank deposits, fans, slides and slumps. The thickest and best preserved deep-marine succession is in the SE Montagne Noire.

Variscan basins

Late Carboniferous sedimentation in the Massif Central region is characterized by the formation of continental basins in isolated intramontane troughs. From Stephanian–early Permian times, clastic-dominated coal-rich successions were deposited unconformably on exposed crystalline basement (Courel 1987). These basins document the advanced stages of intense crustal uplift and extension which dominated the late-orogenic evolution subsequent to major Variscan crustal shortening. Basin locations are frequently linked to large-scale late Variscan wrench-fault zones (Arthaud & Matte 1977), while basin formation was controlled by active fault zones (Bellot *et al.* 2005) (see also McCann *et al.* 2008).

The basins of the Massif Central are broadly similar in terms of their basin fill (thick clastic successions, widespread volcanic units) and their close association with faults (Bruguier *et al.* 1998). A typical example is the **Bosmoreau Basin**, a small graben filled with continental Carboniferous-age sediments and located in the NW part of the Massif Central. It is bounded by normal faults and developed in the hanging wall of the Argentat Fault (Fig. 9.71). Basin fill is estimated at *c.* 600 m and consists mainly of fluviolacustrine to palustrine sediments including coal seams (Bruguier *et al.* 2003). The succession can be subdivided into four sequences. The lowermost sequence consists of *c.* 30 m of fine-grained sediments and includes a 60 cm thick layer of volcanic ash (dated at Late Visean) (Bruguier *et al.* 2003). The three overlying sequences are upper Stephanian in age and unconformably overlie the earlier succession. Their deposition coincides with fault reactivation in the Bosmoreau Basin.

Variscan granitoids

During Middle and Late Carboniferous times, numerous plutons were emplaced in the Massif Central attesting to a significant crustal melting event resulting from the Early–Middle Carboniferous nappe stacking (Downes *et al.* 1997). While some of this was syncompressional (e.g. Duguet & Faure 2004), much of the granitic magmatism occurred during the last stages of the compressional event or during the period of late-orogenic extensional collapse (Duthou *et al.* 1984; Faure 1995).

Carboniferous of the Northern Alpine Region (A.W.)

Carboniferous rocks are exposed in Switzerland in outcrops north of the Penninic Front within the External Massifs and the Helvetic realm (e.g. Trümpy 1980; Fig. 9.72), all of which were subsequently affected by Alpine tectonic deformation. In addition, several Late Palaeozoic basins have been detected in the subsurface further to the north (Fig. 9.72). All of these basins have a multiphase history. Following initial basin formation during the late Carboniferous, they continued to fill during the Permian, and were often affected by later wrench tectonics (see Kroner *et al.* 2008; McCann *et al.* 2008). The Late Palaeozoic basins in northern Switzerland formed after the main Variscan Orogeny (dated in Switzerland as pre-late Westphalian), but in the French Alps, post-orogenic sedimentation commenced in the Upper Namurian (Trümpy 1980). Because all of these basins formed within the same timespan and in close proximity to one another, they exhibit striking similarities, including: (1) small, elongate mainly SW–NE trending graben or graben-like structure; (2) Carboniferous deposits are restricted to a narrow trough that is overlain by more extensive Permian deposits (see McCann *et al.* 2008); (3) basins are entrenched into crystalline basement belonging to the internal Variscides, hence representing intramontane basins (e.g. Trümpy 1980: Ziegler 1990); and (4) basins are filled with continental clastics, mainly eroded from the crystalline basement, and with volcanic/volcaniclastic material.

Two periods of late Variscan extension or transtension in the Alpine realm were suggested by Schaltegger & Corfu (1995), both of which were combined with magmatic activity. The short-lived Westphalian–Stephanian episode of extension, volcanism and plutonism (303–295 Ma) was followed by late orogenic extension lasting for 50 Ma in this part of the Variscides. This led to thinning of the crust and upwelling of hot mantle that caused high heat flow and the intrusion of mantle melts. In contrast, deformational structures and subduction-related calc-alkaline volcanics led Oberhänsli *et al.* (1988) to postulate that the Aar Massif was affected by very late Variscan nappe tectonics. While the intramontane volcaniclastic sediments accumulated, deeper parts of the same basin had already been wedged into the basement where they were intruded by granites.

Within the Carboniferous sedimentary succession, five principal lithofacies associations can be distinguished (e.g. Trümpy 1980; Capuzzo & Wetzel 2004). The first of these comprises mainly volcanics (rhyolites, rhyodacites, andesites and some sediments). The second consists of fluvial sediments with associated flood-plains (these often being rich in late Westphalian–Stephanian flora; e.g. Jongmans 1960), while the third is a lacustrine facies containing laminated mudstones with freshwater ostracodes, intercalated marly intervals and thin sands. Association four is an alluvial fan with typical fanglomerates and sheet flood deposits. Rare ignimbritic rhyolites may also occur. The final lithofacies association consists of a playa with fine-grained clastics and subordinate sandstones and freshwater limestones interfingering with fanglomerates. Volcanics are frequent in this association.

North-Swiss Permo-Carboniferous Basin

The North-Swiss Permo-Carboniferous Basin (NSB) in the subsurface of northern Switzerland was discovered early in

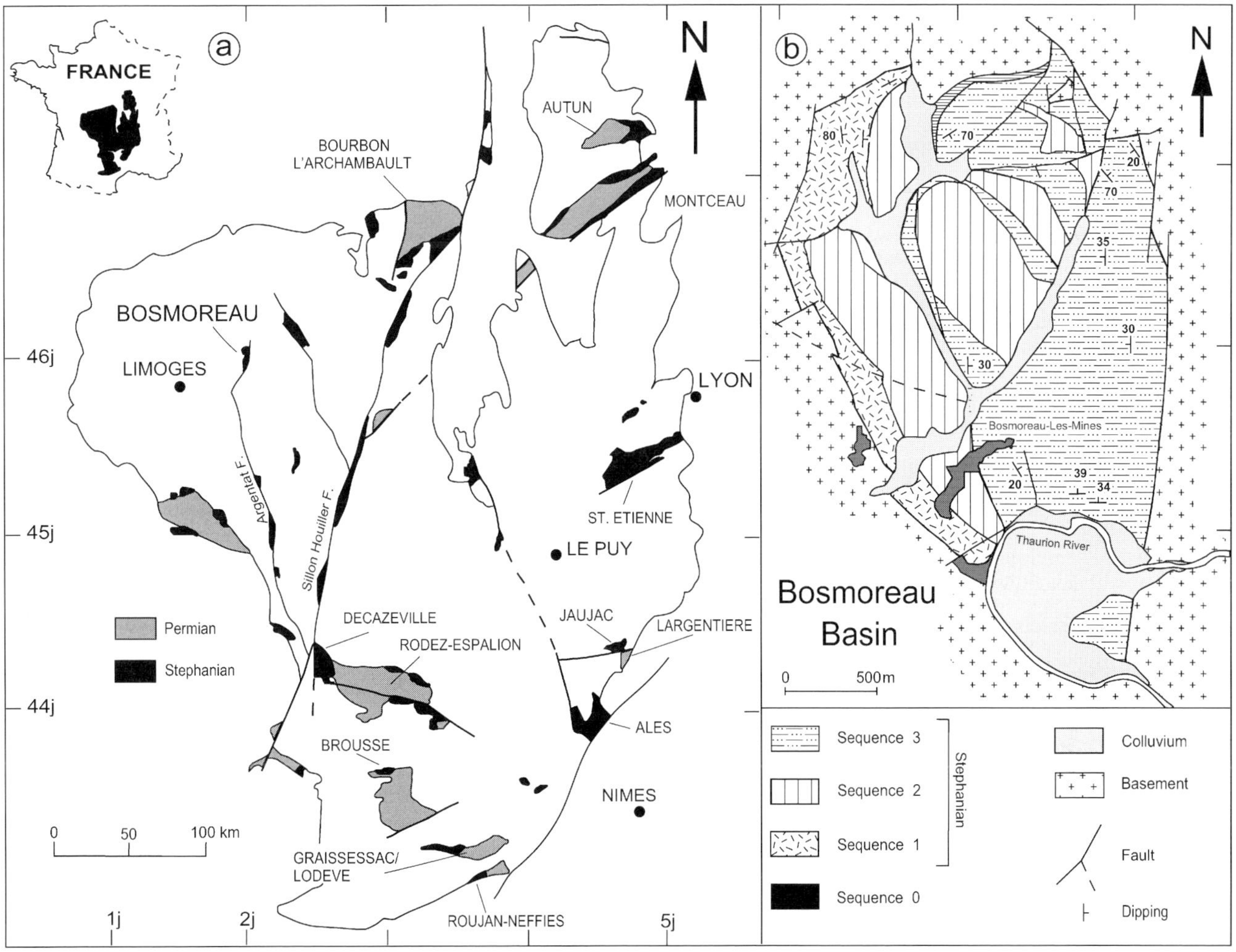

Fig. 9.71. (**a**) Location of the main Stephanian–Early Permian basins of the French Massif Central. (**b**) Simplified geological map of the Bosmoreau Basin (after Becq-Giraudon 1985; Bruguier *et al.* 2003).

the 1980s (e.g. Matter *et al.* 1987). Initially, it was called the Constance-Frick Trough (= Konstanz-Frick Trog; e.g. Laubscher 1987) and later when its continuation further to the west was recognized, the North-Swiss Permo-Carboniferous Basin (= Nordschweizer Permokarbon Trog; e.g. Thury *et al.* 1994). The basin is 10–12 km wide, situated within crystalline basement, and filled with >1500 m of locally derived continental clastics (Matter 1987; Blüm 1989). The NSB is disrupted by a series of NW–SE trending faults (Fig. 9.73). Carboniferous deposits have only been drilled at Weiach (min. 572 m) and appear to be restricted to a narrow, graben-like structure. Microfloral remains suggest a Stephanian age (Hochuli 1985) which was confirmed by the radiometric dating of zircons from ash layers (303 Ma in the middle of the drilled Carboniferous deposits and 298 Ma at the top; Schaltegger 1997). Extraformational clasts are similar to the crystalline basement now exposed in the Black Forest area.

The section at Weiach is subdivided into six units represented by mudstone- and sandstone-dominated intervals (see Matter 1987). These were interpreted as fluvial and floodplain deposits suggestive of an anastomosing river system (Matter 1987). A *c.* 2 m thick coal seam occurs within one of the sandstone units.

The palaeogeographic position within the Variscides, the variable sediment thickness, and the dominance of crystalline basement clasts derived from local sources suggest a pull-apart origin for the NSB (e.g. Matter 1987; Blüm 1989). The basin is interpreted as one of a series of en-echelon pull-apart basins that formed as a result of late Variscan wrench tectonics. Recently, Ziegler *et al.* (2006) have suggested that its formation was related to late Westphalian–Stephanian wrenching. During the Stephanian to Early Permian, strike-slip activity continued with the subduction of detached lithospheric slabs causing upwelling of hot mantle and a thermal pulse (see McCann *et al.* 2008). for further discussion of the NSB).

Entlebuch Basin

This basin was first recognized in the subsurface when an exploration well was drilled in 1980. Due to commercial interest, no seismic records or additional data have been published, and thus the width, internal structure and the continuation along strike are only tentatively known (Fig. 9.72; Bachmann *et al.* 1987). Palaeozoic sediments are present, but the basement was not reached. The Palaeozoic deposits are covered by 661 m of Mesozoic strata (Buntsandstein to Tithonian) and 4341 m of Tertiary-age continental sediments (Upper Freshwater Molasse). No detailed lithological descriptions have been published. The drilled Palaeozoic interval comprises sandstones with intercalated

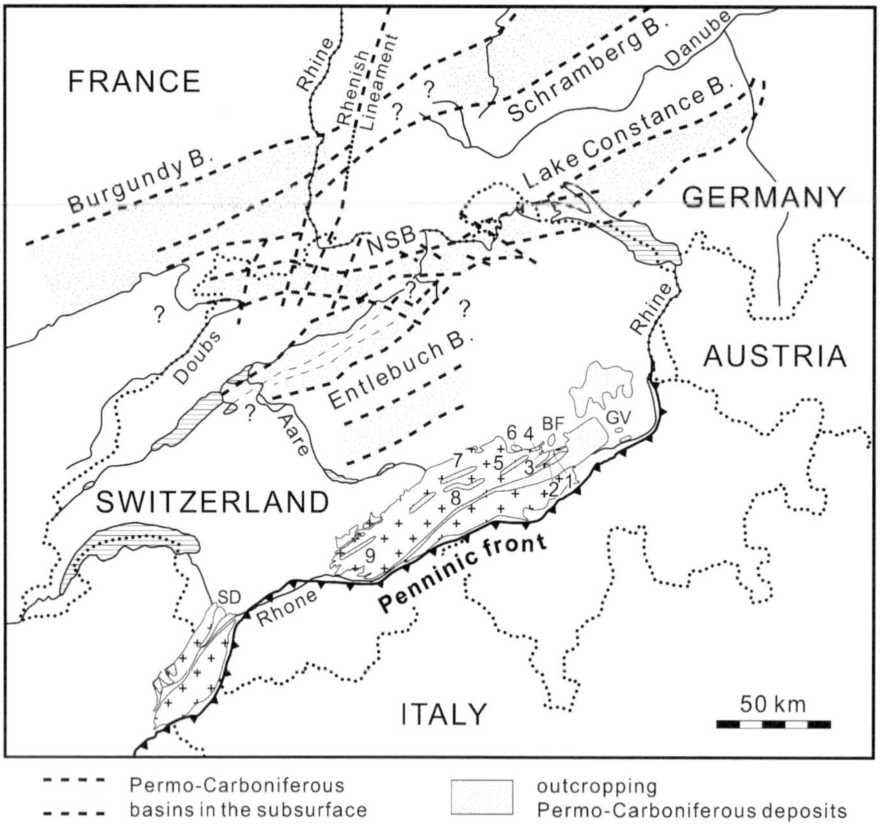

Fig. 9.72. Location of Permo-Carboniferous basins in Switzerland north of the Penninic front. Basins close to the Penninic front (Bifertengrätli (BF), 'Glarner Verrucano Basin' (GV), Salvan-Dorénaz Basin (SD)) are exposed (after Trümpy & Dössenegger 1972; Schweizerische Geologische Kommission 1980), including remnants (numbered; see Wetzel & Allia 2003 for details). Basins in the subsurface (North-Swiss Permocarboniferous Basin (NSB)) after Boigk & Schöneich (1974), Bachmann *et al.* (1987), Meier (1994), Wetzel & Allia (2003).

mudstones and mudstone intervals with rare coal seams. Microfloral remains have been dated as Westphalian B/C.

Salvan-Dorénaz Basin

The Salvan-Dorénaz Basin (SDB) forms a narrow, fault-bounded complex 'syncline' within the crystalline basement of the Aiguilles-Rouges and Arpille massifs (Figs 9.72 & 9.74). The basement represents updomed or thrusted units of the internal Variscides (von Raumer 1988). During the Carboniferous, bimodal magmatism occurred in the Aiguilles-Rouges Massif at deeper crustal to subvolcanic levels (Bussy *et al.* 2000). The Vallorcine (Brändlein *et al.* 1994) and Montenvers granites and the Fully granodiorite intruded syntectonically along SSW–NNE trending, dextral transtensional shear zones (Fig. 9.74).

The modern structure of the basin resulted from late Alpine, two-phase folding (Pilloud 1991). A low-angle regional unconformity between the Permo-Carboniferous and overlying Triassic rocks implies moderate inversion at that time (Pilloud 1991). Another phase of inversion and erosion occurred during Alpine deformation (Badertscher & Burkhard 1998). Therefore, the original SDB fill is only partially preserved. After restoration of the Alpine deformation, the SDB appears as an asymmetric graben elongated in a NE–SW direction and bounded by faults (Fig. 9.75). The fill is up to 1.7 km thick along the NW side of the basin and thins to the SE. It is separated from basement by a steep, SE-dipping mylonitic zone (MZ in Fig. 9.74). This fault was active during the Late Carboniferous as a right-lateral strike-slip, transtensional fault (Brändlein *et al.* 1994). The age of the basin fill was first determined from palaeoflora (Westphalian C–D at the base and Stephanian in the middle; Jongmans 1960). Later U–Pb radiometric age dating of zircons provided 308 ± 3 Ma for basal dacitic flows, and 295 ± 3 Ma for a tuff layer from an upper level extrapolated to 292 ± 3 Ma at the top of the basin fill (Capuzzo & Bussy 2000; Fig. 9.75). Conservative decompaction of the sediments would result in an initial thickness of 3500 m.

Four lithostratigraphical units have been recognized in the SDB (for details see Capuzzo & Wetzel 2004). Unit I (wet alluvial braided fan system) is up to 350 m thick adjacent to the western boundary fault and thins to the SE. It is composed of coarse-grained conglomerates (mainly metamorphic and igneous clasts). Debris flows dominate the proximal areas close to the footwall slope, whereas the distal parts are characterized by braided distributary channels. Unit II (anastomosed river system) comprises up to 650–700 m of dark mudstones deposited on a floodplain with some palaeosols, fine ash layers and rare lacustrine limestones, within which channelized sandstones, up to 5 m thick, occur. Unit III (meandering river system) crops out in the NE part of the basin and is at least 300–350 m thick. The boundary with Unit II is gradual, the amount of sand increases and the sandstone beds become more laterally continuous. These latter are fining upward and form multistorey sandstone bodies which alternate with metre-thick dark mudstones; they are interpreted as the deposits of a highly sinuous, sand-rich, mean-

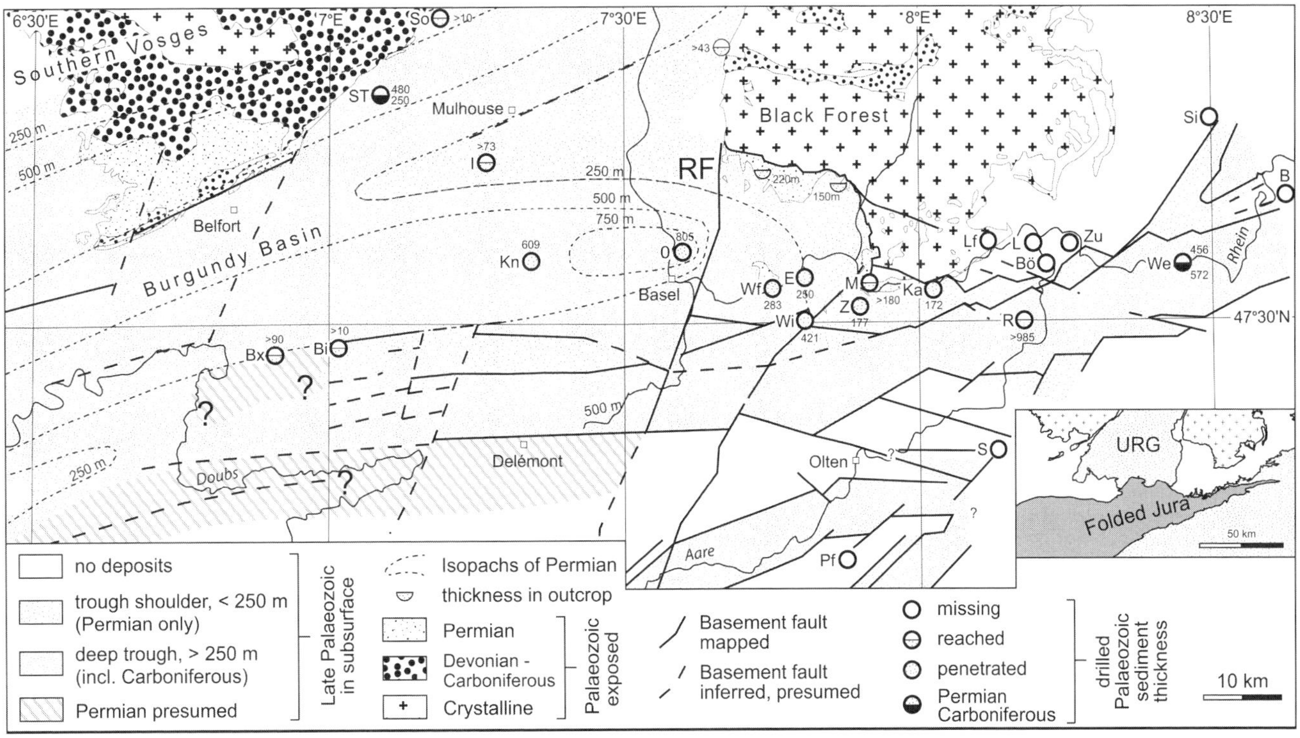

Fig. 9.73. North-Swiss Permocarboniferous Basin in map view. Compiled from data by Boigk & Schöneich (1974), Meier (1994), Wetzel & Allia (2003) and Ustaszewski *et al.* (2005). In addition, data from boreholes have been used: B, Benken (Nagra 2001); Bö, Böttstein (Nagra 1985); Bi, Bisel (unpublished); Bx, Buix (Schmidt *et al.* 1924); E, Engerfelden (Ryf 1984); I, Illfurth (unpublished); K, Kaisten (Nagra 1991); Kn, Knorrinque (unpublished); L, Leuggern (Peters *et al.* 1989); Lf, Laufenburg (Heusser 1926); M, Mumpf (Schmassmann & Bayramgil, 1946); O, Otterbach (Häring, 2002); Pf, Pfaffnau (Büchi *et al.* 1965); R, Riniken (Matter *et al.* 1987); S, Schafisheim (Matter *et al.* 1988*b*); Si, Siblingen (Nagra 1993); So, Soultz (unpublished); ST, south of Thann (Fluck *et al.* 1980); We, Weiach (Matter *et al.* 1988*a*); Wf, Weiherfeld (Schmassmann & Bayramgil 1946); Wi, Wintersingen (Schmassmann & Bayramgil 1946); Z, Zuzgen (Schmassmann & Bayramgil 1946); Zu, Zurzach (Cadish 1956).

dering river (Capuzzo & Wetzel 2004). Palaeoflow was to the SW (i.e. the opposite of that within Unit II). Unit IV ('dry' alluvial fan system) sediments thin from *c.* 700 m along the NW side of the basin to the SE and interfinger at different levels with Units II and III. The lower boundary of this unit is marked by a change in the colour of the sediment (from dark grey to reddish) which may reflect the onset of a climatic change to a warmer, at least seasonally dry climate. Unit IV exhibits all the characteristics of a prograding alluvial fan system.

Episodic tectonic movements occurred along transtensional NE–SW to NNE–SSW trending faults bordering the NW margin of the SDB. These faults acted as conduits for dacitic subaerial flows, (308 ± 3 Ma) (Capuzzo & Bussy 2000). At the same time granites and subvolcanic dykes intruded into the basement at *c.* 5 km depth. The conglomerates at the base of the basin fill succession contain boulders eroded from these intrusions, and thus basin formation and exhumation of basement rocks occurred almost synchronously, presumably due to wrench tectonic activity (Capuzzo & Bussy 2000). The coarse clastics of Unit I indicate a steep relief within the catchment (Fig. 9.76A). After the initial phase of basin formation, subsidence continued at a lower rate and tributaries incised backwards, as indicated by the widening of the detritus age spectrum to older ages upsection (i.e. 327 Ma). However, the tributaries did not incise deeper as the age of the youngest detritus remains constant, suggesting a marked decrease in exhumation and/or denudation rates at that time (Capuzzo *et al.* 2003; Capuzzo & Wetzel 2004). The development of an anastomosing river system (Unit II) implies increasing accommodation space (Fig. 9.76B). Wrench tectonics displaced the alluvial fan system (Unit IV) dextrally at the NW border fault. Shallow lacustrine and peat-swamp deposits restricted to the NW side of the basin indicate asymmetric differential subsidence (Stollhofen 1998). In the upper part of Unit II laterally persistent, thick, caliche-bearing palaeosols may reflect a period of tectonic quiescence. The anastomosing river deposits gradually change to a meandering system (Fig. 9.76C). The reversal in drainage direction from the NE (Unit II) to the SW (Unit III) probably reflects large-scale topographic reorganization in this part of the Variscides.

Glarner Verrucano Basin

The evolution of this basin has been greatly affected by the development of the Helvetic nappes (Fig. 9.72). The deposits, which are normally *c.* 1000 m thick (up to a maximum of 1600 m), accumulated in a SW–NE trending basin. After restoration for Alpine deformation, the basin is *c.* 25–30 km wide and located between the Aar and Gotthard massifs (Fig. 9.77). The basin can be traced for about 60 km along strike. Tectonic deformation, rare fossils, the scarcity of marker beds, and the common lateral facies changes make it difficult to establish a stratigraphy for the basin. Nonetheless, an age younger than Westphalian to early Stephanian is inferred. Two lithofacies associations occur: an alluvial fan facies along the NW margin grades into a silty playa lake facies, containing slumped horizons, tuffaceous layers, freshwater limestones and caliche horizons in the basin centre.

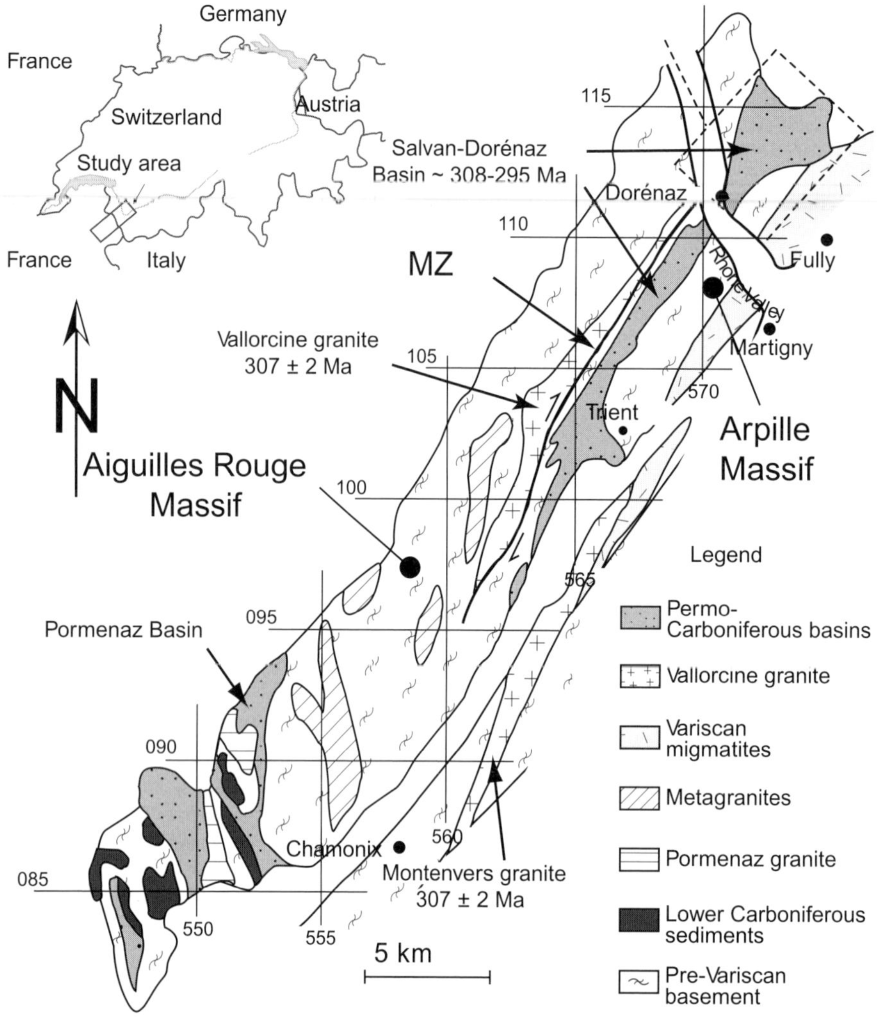

Fig. 9.74. Location map of the Salvan-Dorénaz Basin and geological map of the Aiguilles-Rouges Massif (modified after Brändlein *et al.* 1994). U–Pb ages are after Bussy *et al.* (2000) and Capuzzo & Bussy (2000); MZ refers to the northwestern bounding fault zone delimiting the basin. Geographical co-ordinates refer to Swiss map co-ordinates.

Aar Massif

Upper Carboniferous sediments are sporadically exposed along the entire length of the Aar Massif, especially along its northern border (Fig. 9.72). Very late Variscan (Oberhänsli *et al.* 1988) and Alpine (Franks 1968*a*) tectonics resulted in marked deformation, and thus, at the present time, the best exposures are those of the Bifferten inlier at the eastern side of the massif (for further details see Franks 1966, 1968*a*, *b*; Oberhänsli *et al.* 1988; Schaltegger & Corfu 1995). The Bifertengrätli Formation rests unconformably on the Tödi Granite and on parts of the central Aar Granite (Fig. 9.78), and is unconformably overlain by Triassic deposits. Three members have been distinguished within the formation: the Volcanic, Estuarine and Lacustrine members (for further details see Franks 1968*a*).

Western Carpathians, Slovakia (A.V.)

The evolution of the Western Carpathian Carboniferous basins is related to the Variscan collisional event. However, reconstruction of the area is complicated by the later destruction of large parts of the Late Variscan crust by subsequent Alpine compression. Like most of the other collisional foldbelts, the Western Carpathians have been traditionally divided into external and internal structural zones (for summary see Biely *et al.* 1996), based largely on the relative ages of the Alpine events and the intensity of their deformation and metamorphic effects. The two zones are: (1) the internal zone, related to the Late Jurassic high-pressure/low-temperature subduction event and the Early/Middle Cretaceous collision, followed by nappe stacking; and (2) the external zone, related to the Late Cretaceous/Early Paleocene to the Oligocene/Early Miocene subduction/accretion and collisional events. The fragments of the original Carboniferous sedimentary basin-fill successions as well as their substrate are preserved only within the internal structural zone (Fig. 9.79), as a part of the principal Alpine crustal-scale superunits (i.e. Veporicum, Zemplinicum and Gemericum), and two cover nappe systems (Hronic, Turnaic). The Carboniferous successions of the Western Carpathians are identified on the basis of litho- and biofacies data from a wide range of sedimentary environments, including deep- and shallow-marine through to paralic and continental settings.

Variscan convergence, from Latest Devonian/Lower Carboni-

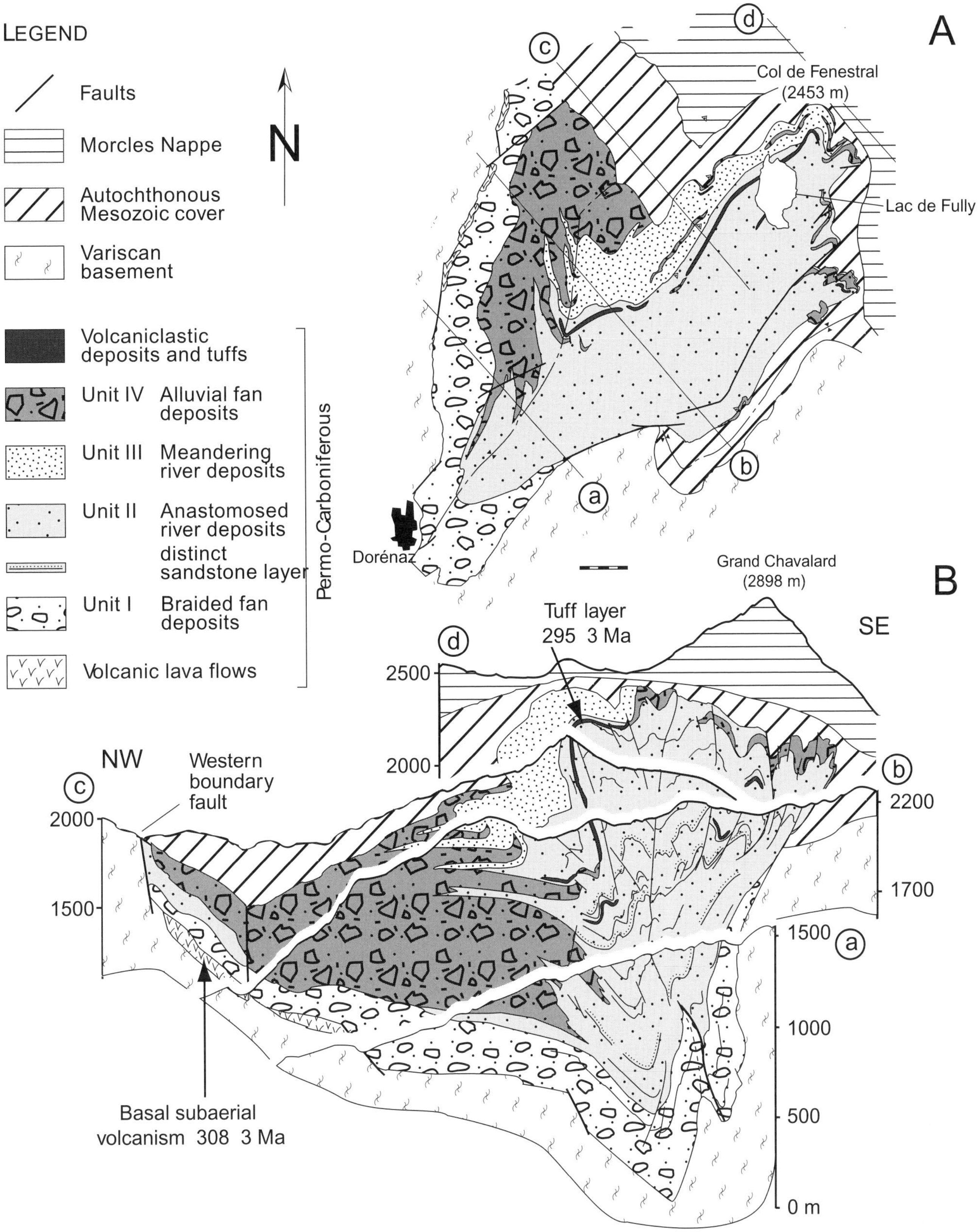

Fig. 9.75. Geological map (**A**) and multiple cross-sections (**B**) of the northern part of the Salvan-Dorénaz Basin (modified after Pilloud 1991) displaying the four lithostratigraphical units constituting the basin fill. The geometry of the Unit IV deposits in map view suggests right-lateral displacement during their accumulation (see text). The complex syncline structure formed during Alpine deformation. U–Pb ages of synsedimentary volcanic deposits are after Capuzzo & Bussy (2000).

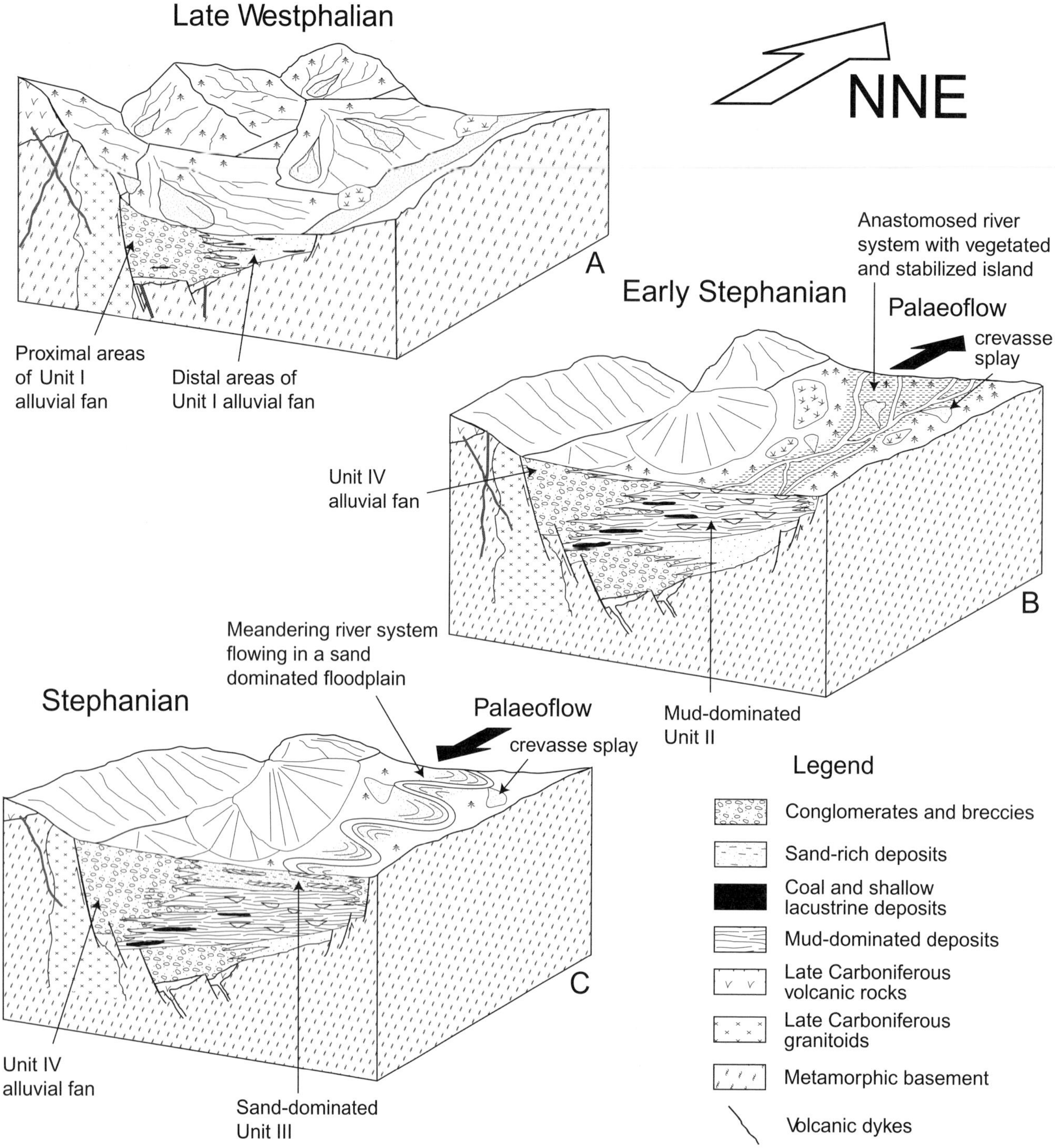

Fig. 9.76. Schematic block diagrams representing evolution of the Salvan-Dorénaz Basin. (**A**) Formation of the basin induced by asymmetric subsidence along northwestern bounding faults, which also favoured the emplacement of basal dacitic flows and controlled the deposition of Unit I alluvial fans derived from western source areas. (**B**) Deposition of the mud-dominated Unit II and establishment of an anastomosed river system with axial drainage to the NE; this change was possibly induced by increased differential tectonic subsidence within the basin. (**C**) Deposition of the sand-dominated Unit II by a meandering river system formed in a relatively steep fluvial valley presenting a reversal of its axial drainage towards the SW. In (b) and (c) prograding and retreating cycles of alluvial fans (Unit IV) are illustrated schematically, and their right-lateral displacement through time.

ferous times, led to the suturing of the Western Carpathian units and the closure of the SW remnants of the Rheic Ocean. The amount of convergence was gradually modified by the activity of large, mostly dextral transfer faults and the development of the Upper Carboniferous shallow-marine and continental basins. Relicts of the colliding microplates are preserved as fragments of crystalline basement rocks within the main crustal-scale units of the Alpine Western Carpathians internides (Fig. 9.79).

The **Central Western Carpathian Crystalline Zone** (CWCZ) comprises mainly medium- to high-grade mica schist/gneiss

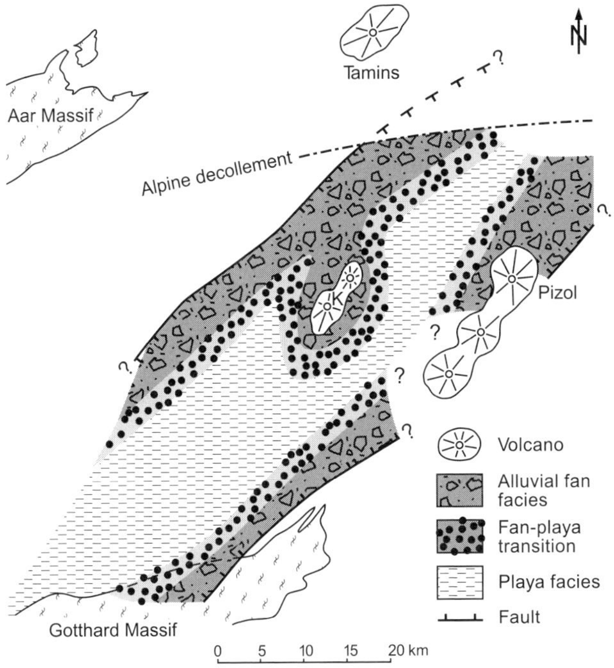

Fig. 9.77. Map of the Glarus Permo-Carboniferous Basin restored for Alpine tectonic deformation and thrusting (redrawn from Fisch & Ryf 1966).

lithological complexes with intercalations of basic and acid orthogneisses. Unique associations include leptinite–amphibolite complexes and granulitic paragneisses comprising symplectitic amphibolites and eclogites (Vozárová 1993*a*; Janák *et al.* 1996; Ludhová & Janák 1999). The character of Variscan metamorphism was polyphase. The older high-pressure phase is documented by eclogite relicts which were strongly re-equilibrated at high-pressure granulite and amphibolite facies pressure/temperature conditions during their exhumation. Rare U–Pb zircon data gave an age of 400–410 Ma (Tatra Mountains; Poller *et al.* 2000). High-pressure metamorphism may be related to subduction processes and the initial collision of the CWCZ continental blocks which drifted from the northern margin of Gondwana in Late Silurian/Early Devonian times. Collision of this peri-Gondwanan microplate (relict of the Tatro-Veporic Terrane which included the Tatra, Kohút, Byšta Terranes and the Ipoltica Suspect Terrane; Vozárová & Vozár 1996) with Laurussia during Late Devonian/Early Carboniferous times resulted in the second phase of metamorphism (low pressure/high temperature), dehydratation melting, and emplacement of granitoids resulting in crustal thickening. Magmatic activity was restricted to two periods, 360–340 Ma and 320–300 Ma (Rb–Sr, U–Pb data: Cambel & Král' 1989; Cambel *et al.* 1990). Fragments of this thick Variscan crust are inferred to be an integral part of the subsequent Alpine Western Carpathian Units, i.e. Tatricum, Northern and Southern Veporicum and Zemplinicum. It has been suggested (based on sedimentary provenance data) that the Upper Carboniferous basin of the Hronic Unit (rootless nappe) was located on the same microplate (i.e. the Ipoltica Suspect Terrane).

The **Northen Gemeric Zone** (NGZ) comprises relicts of Tournaisian–Visean deep-marine sediments and thrust wedges of pre-Carboniferous terranes (e.g. Spiš Composite terrane comprising the Klátov and Rakovec crystalline complexes; Vozárová & Vozár 1996). The higher-grade Klátov Complex consists of mainly amphibolites and meta-ultramafic rocks, with rare gneisses and calc-silicate metasediments, and is considered to be

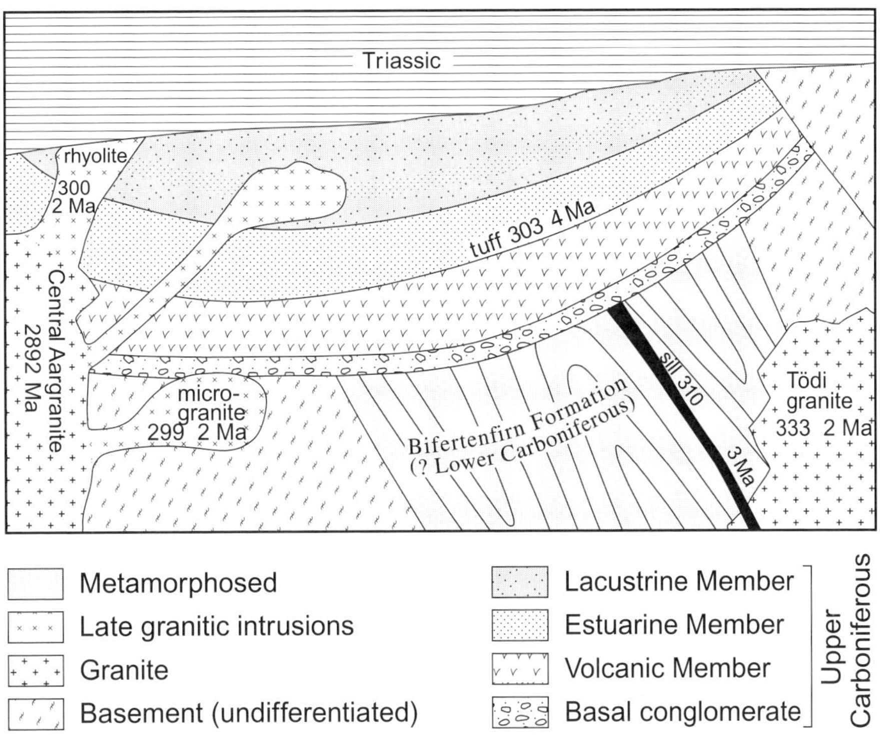

Fig. 9.78. Restored cross-section of a Permo-Carboniferous Basin within the Aar Massif (redrawn from Schaltegger & Corfu 1995); it has been affected heavily by Alpine tectonic deformation and represents the Biferten inlier (Franks 1968*a*). Radiometric age data after Schaltegger & Corfu (1995).

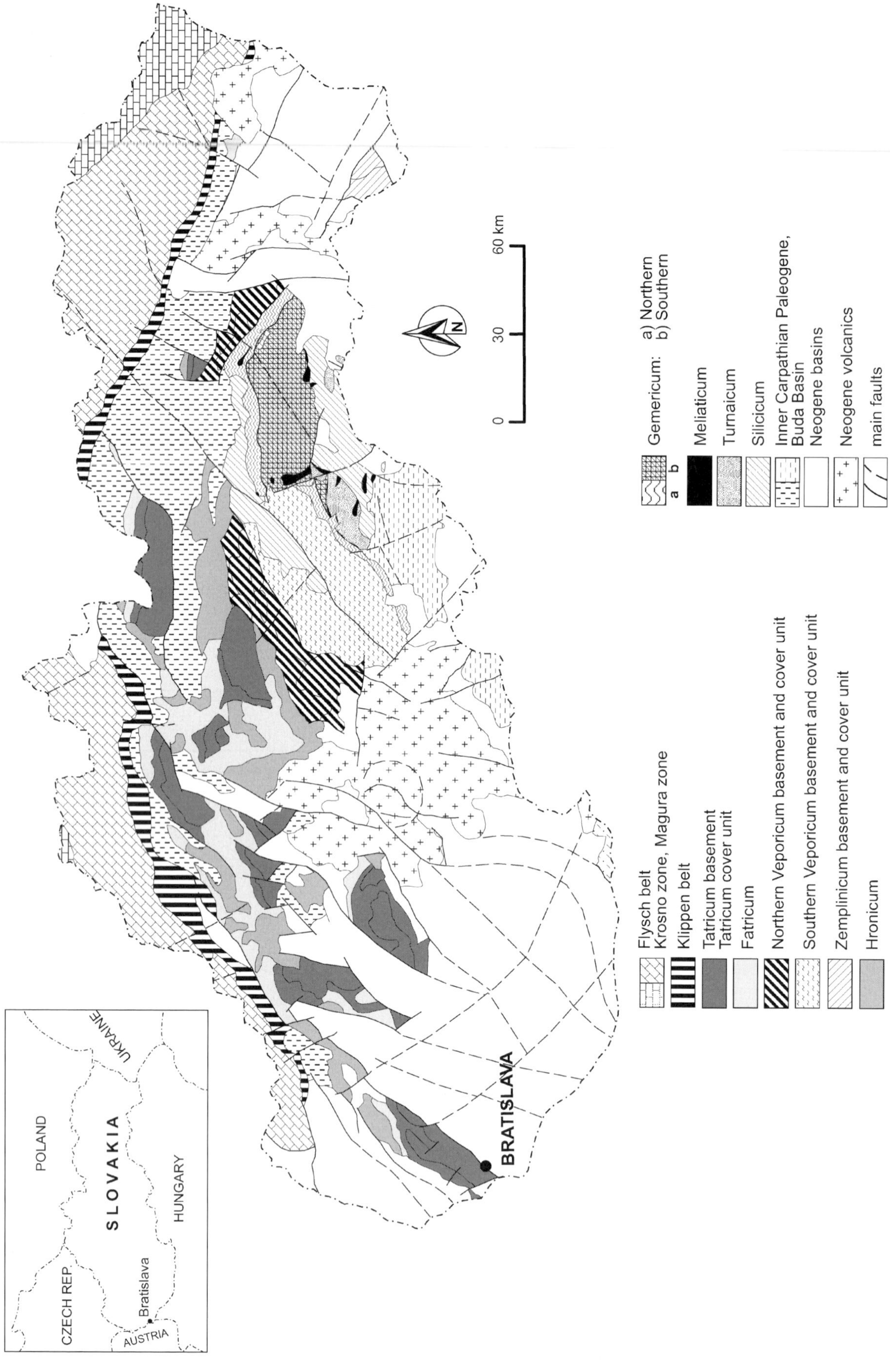

Fig. 9.79. Tectonic sketch of the Slovak part of Western Carpathians (modified after Biely *et al.* 1996).

part of an incomplete ophiolite suite (Spišiak *et al.* 1985). The low-grade Rakovec Complex is predominantly composed of tholeiitic metabasalts and their metavolcaniclastics associated with sandy/pelitic and Fe-rich metasediments and small bodies of gabbrodiorites as well as metakeratophyres. The magmatic rocks show geochemical characteristics close to E-MORB/OIT and island-arc basalts (Ivan 1994). This zone represents a relict of the Variscan collisional suture documenting the amalgamation of two microplates, represented by the CWCZ crystalline blocks of the Tatra-Veporic terrane on the one side, and the Spiš Composite terrane with the Rakovec and Klátov complexes from the NGZ on the other. Deformational and metamorphic events of both pre-Carboniferous terranes are dated as Late Devonian–Early Carboniferous.

The Tournaisian–Visean turbidite clastic wedges were derived from both sides of the supposed suture and cover the remains of the intrasuture remnant basin. Serpentinite fragments (i.e. outliers of oceanic crust) are an integral part of this Early Carboniferous deep-marine succession. The Early Carboniferous turbidites grade upwards into the uppermost Visean–lower Namurian shallow-water clastic carbonate units. This entire succession was extensively deformed (greenschist facies) prior to late Namurian–early Westphalian times. A new sedimentary cycle commenced in Westphalian–Stephanian times represented by the deposition of delta-fan/shallow-marine to paralic sediments.

The **Inner Western Carpathian Crystalline Zone** (IWCZ) is mostly composed of thick Early Palaeozoic volcanogenic sequences (e.g. Gelnica terrane; Vozárová & Vozár 1996). In addition to large quantities of redeposited acid to intermediate volcanic and volcaniclastic material derived from a coeval magmatic arc (Vozárová & Ivanička 1996), mature detritus derived from continental margin as well as basic and ultramafic debris is also present. Based on lithofacies and petrofacies analysis, the Gelnica Flysch has been interpreted as representing the infill of part of a forearc basin located along an active continental margin (Vozárová 1993*b*). The Ordovician–early Silurian stratigraphic range of of these metasediments (greenschist facies) was proven by the presence of agglutinated foraminifera (Vozárová *et al.* 1998; Soták *et al.* 2000).

U–Pb isotope data of zircon from associated metavolcanic rocks confirm the wide range of ages present, including: ^{206}Pb–^{238}U – from 391 to 497; Ma, ^{207}Pb–^{235}U, from 392 to 573 Ma (Cambel *et al.* 1977; Ščerbak *et al.* 1988). A Proterozoic age for the source area is indicated by U–Pb detrital zircon data of metasediments which range from 651 to 940 Ma (Cambel *et al.* 1977).

Relicts of the Western Carpathians Carboniferous sedimentary basin fill are preserved in the two Alpine Veporicum and Gemericum crustal-scale superunits as well as within the two rootless Hronicum and Turnaicum nappe sedimentary complexes (Fig. 9.80). The distribution of Carboniferous basins in time and space, as well as the lithologies of their infill, is indicative of a reverse polarity (i.e. southern) in comparison to the Alpine Orogeny. Variscan orogenic events resulted in unevenly consolidated continental crust which, after a short period of stability (from south to north in the Upper Permian/Lower Triassic to Middle/Upper Triassic) was incorporated again in the new orogenic Alpine cycle.

On the basis of Carboniferous basin-fill successions in the Western Carpathians, two different zones of continental collision were identified: (1) the Internal Zone, where collision stopped occurred during the Bretonian–Sudetian deformational phases, resulting in syncollision Lower Carboniferous sedimentation and a marine Upper Carboniferous perisutural basin (i.e. NGZ) as well as continental, backarc transpression basins (i.e. CWCZ); and (2) the External Zone, where collision terminated during the

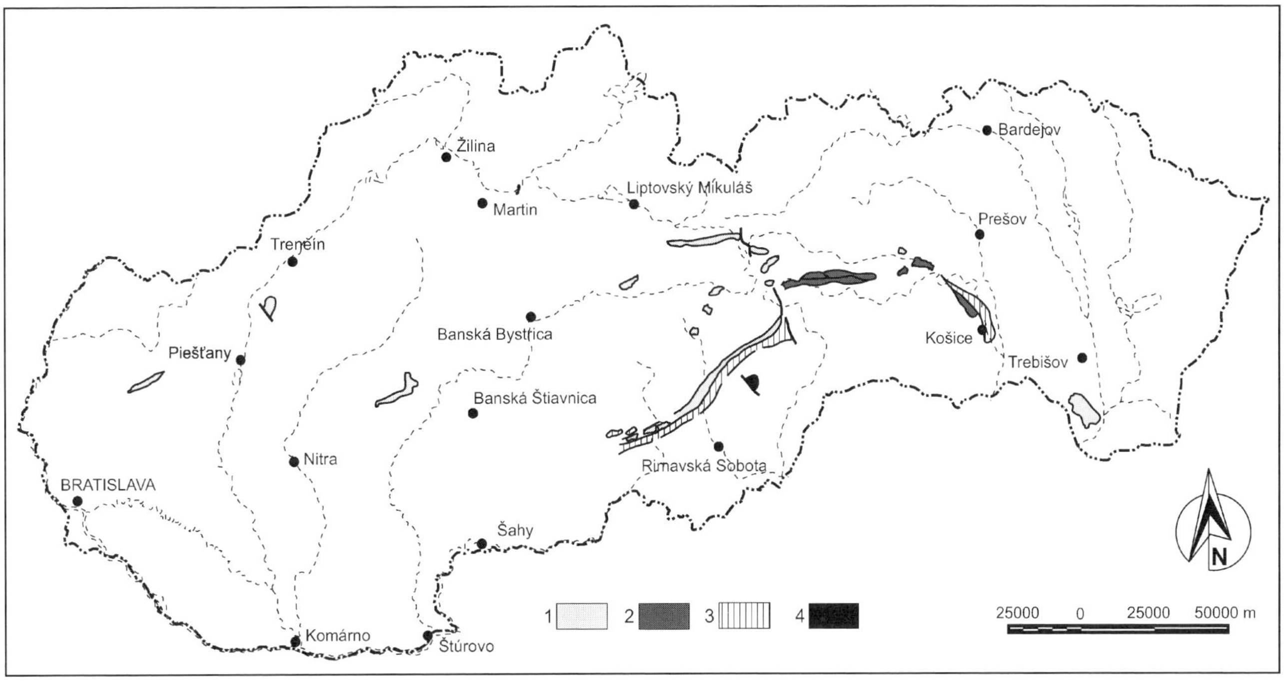

Fig. 9.80. Map of the Carboniferous surface occurrences in Slovakia. Key: 1, Continental Upper Carboniferous sedimentary rock complexes of the Central Western Carpathians Zone (CWCZ); 2, shallow-water to paralic Upper Carboniferous sedimentary rock complexes of the Northern Gemeric Zone; 3, Lower Carboniferous turbidites of the Northern Gemeric Zone; 4, Bashkirian turbidites of the Turnaicum Unit of the Inner Western Carpathian Zone.

Asturian phase (i.e. IWCZ). This zone is represented by upper Namurian–lower Westphalian deep-marine sediments (Turnaic Unit, Slovakia) and the Upper Westphalian–Stephanian marine basin (relicts preserved in northern Hungary).

Northern Gemeric Unit

This unit is the only one within the Carpathian succession which covers the entire Carboniferous. The basal formations, of Tournaisian–early Namurian age, are best preserved in tectonically–restricted relicts along the WSW and ESE boundaries of the Northern Gemeric Unit. These suffered significant shortening as a result of Alpine thrusting on the underlying Southern Veporic Unit. Despite the tectonic and metamorphic deformation, the present thickness of the Ochtiná Group (Vozárová 1996) is estimated at >1000 m. The Hrádok and Črmel' formations (the former in the WSW, the latter in the ESE) comprise siliciclastic turbidites composed mainly of black to grey sandy/shaly metasediments (Fig. 9.81). Intercalations of paraconglomerates and other coarse-grained conglomeratic mass flow complexes may occur at the formation base. Rarer lithofacies occurring only as thin layers and lenses include lydites, siliceous metapelites, laminated and massive limestones. The Hrádok Formation has been dated as Upper Tournaisian–Visean (Bajanik & Planderová 1985), while the Črmel' Formation has an age of Tournaisian–Visean (Snopková in Bajaník *et al.* 1984).

Both formations also contain volcanic rocks, namely metabasalts and related metavolcaniclastics and metadolerites with a tholeiitic magmatic trend, and a distinct N-MORB affinity. Slabs of ultramafic rocks (oceanic crust fragments), indicated by antigorite serpentinites and talc-tremolite schists, have also been described from the Hrádok Formation. $^{40}Ar/^{39}Ar$ ages from detrital white mica in this formation suggest a cooling age of 465 ± 2.5 Ma for the source area (Vozárová *et al.* 2005). Such an age suggests the existence of intra-Ordovician crustal blocks during the initial stage of Variscan collision-related suturing.

The Lubeník Formation (uppermost Visean–lower Namurian, based on conodont fauna; Kozur *et al.* 1976), which conformably overlies the Hrádok Formation, was deposited in a shallow-marine neritic and littoral environment. It comprises black metapelites, dolomitic limestones and dolomites, some of which

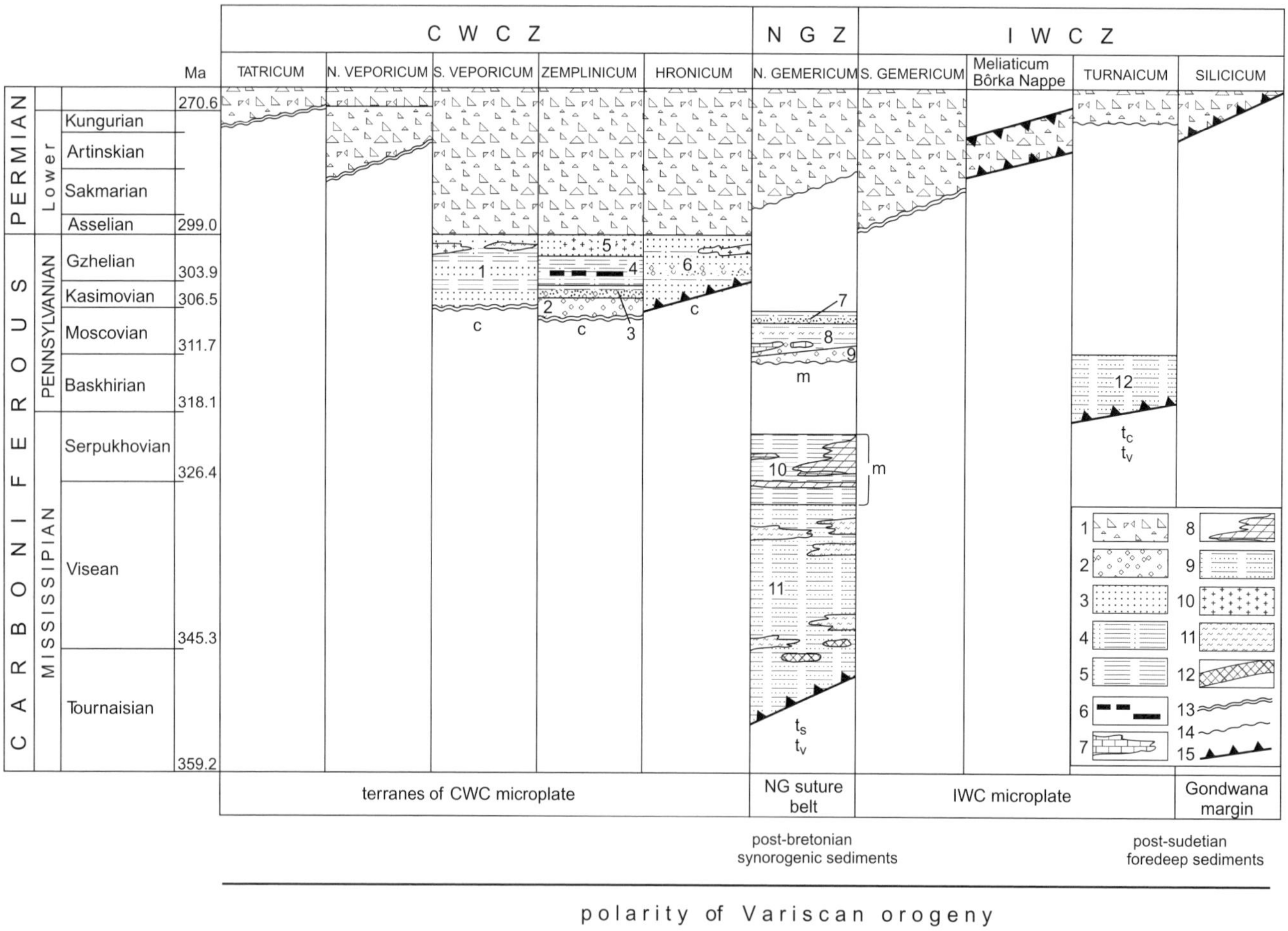

Fig. 9.81. Lithostratigraphical scheme of the Western Carpathian Carboniferous sequence. Key: 1, continental Lower Permian deposits in general; 2, polymict conglomerates; 3, arkosic and lithic sandstones; 4, alternating sandstones and shales; 5, shales, siltstones, mudstones; 6, coal seam; 8, limestone, dolomitic limestone; 9, dolomite, magnesite; 10, turbiditic deep-water slope deposits; olistostromes; 11, rhyolites-dacites and associated volcaniclastic; 12, basalts and associated volcaniclastics; 13, serpentinized ultramafic rocks; 13, unconformity; 14, disconformity; 15, thrust plane. Symbols: c, continental, alluvial, fluvio-lacustrine, lacustrine and swamp environment; m, shallow-water to paralic environment; t_s, siliciclastic turbidites; t_v, volcaniclastics turbidites; t_c, carbonate olistostrome. The lithostratigraphic units are indicated by the numerals 1–12 (in bold) on the stratigraphic columns. Key: *Southern Veporicum*: 1, Slatviná Fm; *Zemplinicum*: 2, Čerhov Fm; 3, Luhyňa Fm; 4, Tŕňa Fm; 5, Kašov Fm; *Hronicum*: 6, Nižná Boca Fm; *Northern Gemericum*: 7, Hámor Fm; 8, Zlatník Fm; 9, Rudňany Fm; 10, Lubeník Fm; 11, Hrádok and Črmel' fms; *Turnaicum*: 12, Turiec Fm.

show traces of a relict pel-intra-biosparite texture comprising pellets, microsparitic intraclasts and bioclasts. These latter include foraminifera, echinoderms, lamellibranchiata, ostracoda, bryozoa and algal fragments. The age of the formation is Late Visean (Mamet & Mišík 2003; Plašienka & Soták 2001).

Overlying the Lower Carboniferous succession is the Dobšiná Group (Westphalian A–D) whose strata are preserved within the Alpine North Gemeric Unit (Fig. 9.81). The tectonic setting for this group was coincident with ongoing continental collision. Sedimentation commenced with coarse-grained fan-delta deposition, including boulder conglomerates. Material was derived from the underlying pre-Upper Carboniferous complexes. Due to strong Alpine reworking, the direct contact between these deltaic-marine overstep sequences and basement is preserved only in some places. Basal conglomerates of the Rudňany Formation unconformably overstep the Klátov and Rakovec complexes, and the Lower Carboniferous deep-marine succession of the Črmel' Formation (see Vozárová & Vozár 1996; Vozárová 1996). Black shales (including a Westphalian A–B microflora) and sandstone intercalations are typical of the fining-upward Rudňany Formation. The results of $^{40}Ar/^{39}Ar$ multigrain analyses of detrital mica from the Rudňany Formation sandstone and the orthogneiss pebble muscovite from the associated conglomerate provide ages of 385 ± 3 Ma and 372 ± 2.5 Ma, which reflect a uniform Variscan source (Vozárová *et al.* 2005).

Overlying clastic/carbonate facies correspond to the basal part of the Zlatnik Formation (Westphalian) which is rich in trilobites and conodonts (Rakusz 1932; Kozur & Mock 1977). The upper part of the formation comprises fine-grained clastic metasediments associated with fine basaltic volcaniclastics and rare high-K tholeiitic basalts. Magmatic activity reflects the deepening of the Westphalian sedimentary basin. The end of sedimentation within the mid-Westphalian peripheral basin is reflected by the paralic sequence of the Hámor Formation, which is characterized by distinct cyclical, coarsening-upward shaley/sandy/conglomeratic sediments, the absence of synsedimentary volcanism, and the local occurrence of a ribbed coal seam. The microfloral assemblages provides an age of Westphalian D.

Turnaic Unit

The Turiec Formation (upper Namurian–lower Westphalian) is exposed as part of a tectonic underlier of the Silica Nappe in the internal part of the Western Carpathians. The BRU-1 borehole (Vozárová & Vozár 1992) enables the thrust-related position on the Meliatic Upper Jurassic olistostrome complex to be determined. The formation comprises distal turbidites (mainly $T_{d\text{-}e}$), including black phyllites, metasiltstones and metasandstones with intercalated mass flow deposits (e.g. gravitational slides, carbonate olistostrome) (Fig. 9.81). These latter deposits contain redeposited acid volcaniclastics and conglomerates. The volcaniclastics are related to the rhyolite–dacite group of the subalkaline magmatic series. The olistostrome consists of carbonate olistoliths ranging from decimetres to tens of metres in size and containing a Namurian B–Westphalian A fauna and microflora (Ebner *et al.* 1990).

Zemplinic Unit

The crystalline rocks of the Zemplinic Unit (Fig. 9.81), together with their Late Palaeozoic and Mesozoic envelope sequence, are made up of a tectonic horst striking NW–SE, uplifted from the basement of the Tertiary filling of the East Slovakian (Transcarpathian) Basin. The character of the crystalline rocks and the mainly Mesozoic evolution of the strata, suggest a possible correlation with the other units of the CWCZ (Vozárová 1989; Faryad 1995). The Zemplinic succession consists of four partial lithostratigraphic units (Čerhov, Luhyňa, Tŕňa, Kašov formations; Grecula & Együd 1982; Vozárová 1986). The Westphalian D–Stephanian stratigraphic range was established based on macro- and microfloral assemblages (Planderová *et al.* 1981).

Clast-supported, well-rounded, polymict conglomerates are the dominant lithofacies of the Čerhov Formation (Westphalian D–Stephanian A). These are interpreted mainly as braided-river deposits. The lithostratigraphic profile consists of repeated fining-upward cycles. Rare black shale and siltstone intercalations occur in the upper part of the sequence. The overlying Luhyňa Formation (Stephanian A–B) consists of fine-grained lacustrine sediments, including sandstones, mudstones and shales, with intercalated distal-fan stream deposits.

Cyclothems with thin coal seams represent the Tŕňa Formation (Lower Stephanian) which can be subdivided into two large cycles (several hundreds of metres thick). The lower cycle contains seven limnic–fluvial cyclothems with coal seams of variable thickness (from several centimetres up to 160 cm). Generally, the sediments are rich in mica and plant debris (fragments of tree trunk, bark). Distinct fining-upward cycles, comprising black shales with thin coal seams and occasionally dark clay lenses and limestone nodules, suggest a lacustrine–fluvial and swamp environment. The second cycle is characterized by alluvial stream-channel lithofacies, with sandstones dominant and an absence of the coal-bearing association. Units of rhyolite/dacite calc-alkaline volcaniclastics typify this part of the succession. Thick units of rhyolite/dacite volcaniclastics (including ignimbrites) and alluvial, stream-channel and flood-plain sediments are the predominant lithologies in the overlying the Kašov Formation (Stephanian B–C).

Southern Veporic Unit

The Southern Veporic basement was significantly deformed by later Alpine events. The main part of this basement consists of several types of low- to medium-grade metamorphic complexes, including occurrences of Al-Fe-rich metasediments (Korikovskij *et al.* 1989; Kováčik 1991; Meres & Hovorka 1991) and orthogneisses of magmatic arc provenance. In addition, relicts of migmatitized and strongly retrograde metamorphosed high-grade crystalline complexes have been identified (?Proterozoic in age; Bezák 1991). The Carboniferous cover is represented by an upward-coarsening succession of Stephanian-age sediments (Slatviná Formation). However, the precise relationship of the formation with the basement is difficult to determine due to either subsequent Alpine thrusting or the contact-thermal effects of younger granitoids (Vozárová & Vozár 1988; Vozárová 1990).

The Slatviná Formation (Stephanian C) comprises metasandstones and metapelites arranged in two large coarsening-upward regressive cycles (Fig. 9.81). These cycles are interpreted in terms of progradation from a deltaic to a fluvial setting. Anoxic conditions may occasionally develop resulting in black shale deposition. Abundant plant detritus is indicative of the proximity of vegetation-covered land.

Hronic Nappe Unit

This has been defined as a rootless megastructural Alpine unit consisting of several partial nappes: namely the Šturec and Choč nappes (Andrusov *et al.* 1973) (Fig. 9.81). Due to their internal structure and mutual relations as well as lithological characteristics, the partial nappes have been distinguished as mainly Triassic complexes. The Hronic Nappe units contain Late Palaeozoic volcanosedimentary formations, variably preserved as a consequence of tectonic compression during nappe thrusting.

There is no evidence of either the underlying pre-Stephanian sediments or of the immediate crystalline basement. Tectonic slices of granitoid blastomylonites found in the basal part of the Šturec Nappe may be partly indicative of its composition (Andrusov 1936; Vozárová & Vozár 1979). Petrographic data suggest an affinity to a magmatic arc source area (the hypothetical Ipoltica Terrane; Vozárová & Vozár 1996).

The rift-related Nižná Boca Formation (Westphalian–Stephanian) forms the oldest unit and is generally a regressive clastic succession with a clear coarsening-upward trend. It was deposited in a fluvial-lacustrine delta setting. Cogenetic, mostly subaerial, dacite volcanism is represented by abundant redeposited volcanogenic material mixed with non-volcanic detritus, as well as thin layers of dacitic tuffs and rare lava flows.

Dating of detrital white mica by $^{40}Ar/^{39}Ar$ from the Nižná Boca Formation yielded an age of 309 ± 3 Ma (basal sample) with ages of 318 ± 2 Ma and 329 ± 2 Ma, respectively, from the middle and upper parts of the formation (Vozárová *et al.* 2005). The time gap between the average cooling in the source region and the age of sedimentation at the basal part of the Nižná Boca Formation may not exceed 5 Ma, whereas at the top of the same formation this time difference reaches *c.* 20 Ma. These cooling ages of the source area reflect the development of the Hronicum rift and the heterogeneity of the source area.

Sudetes (T.M.)

The Sudetes, which straddle the boundary between Poland and the Czech Republic, represent the northeasternmost exposed part of the Variscan crystalline basement in Europe and constitute the NE margin of the Bohemian Massif (Zelazniewicz 1995; Mazur *et al.* 2006) (Figs 9.26 & 9.82). The succession contains variously metamorphosed volcanosedimentary successions and igneous suites of pre-Carboniferous age. The metamorphic rocks are onlapped by sediments deposited within Late Devonian–Carboniferous basins and are intruded by voluminous late- to post-orogenic Carboniferous granites (Cwojdzinski & Grocholski 1995; Mazur *et al.* 2006, 2007). The geology of the poorly exposed, and highly deformed and metamorphosed rocks of the Sudetes of SW Poland has long been controversial, with early interpretations ranging from that of Suess (1888) who considered the region to form part of the Variscan Belt, to that of Bederke (1924) who interpreted the region as Caledonian and reworked by later Variscan deformation (Collins *et al.* 2000; Mazur *et al.* 2006). The former theory presumes that the various Variscan zones as defined in western and Central Europe can be extended eastwards into Poland, while the latter introduces the idea of the presence of exotic terranes in the Sudetes with little or no affinity to the major Variscan tectonostratigraphic units. Additional problems (Alexandrowski 1995; Alexandrowski *et al.* 1997, 2000; Franke & Zelazniewicz 2000; Mazur *et al.* 2006, 2007) include whether the tectonic setting of igneous rocks and thermal events recently dated as Lower Palaeozoic (Oliver *et al.* 1993; Kröner & Hegner 1998) were extensional or compressional, the presence or absence of a widespread intra-Devonian unconformity, and the location and character of major tectonic boundaries (Matte *et al.* 1990; Franke *et al.* 1995*a*; Mazur *et al.* 2005, 2006; Alexandrowski *et al.* 1997).

The Sudetic region extends between the WNW–ESE trending Middle Odra Fault Zone to the NE, and the parallel Elbe Fault Zone to the SW (Fig. 9.26). It is usually subdivided into the West Sudetic domain (= Lugicum) and a smaller, East Sudetic domain (= Silesicum). The former comprises a series of distinct, fault-bounded Palaeozoic tectonic units with usually unclear mutual geometrical and genetic interrelationships. It is bisected by a prominent WNW–ESE to NW–SE fault zone (Intra Sudetic Fault), which parallels the Middle Odra and Elbe fault zones. Alexandrowski & Mazur (2002) follow other workers by considering the eastern part of this domain as the Central Sudetes.

Geologically, the Sudetes area forms the NE segment of the Variscides. As noted above, it is a complex area comprising a mosaic of geologically distinct, fault-bounded pre-Permian units affected by contractional, strike-slip and extensional tectonics and characterized by abrupt changes in the dominant structural trends (Aleksandrowski & Mazur 2002). To the NE, across the Middle Odra Fault Zone, these complexes adjoin Carboniferous unmetamorphosed deep-marine and continental successions of the Variscan externides, buried below the thick Permo-Mesozoic succession of the German–Polish Basin.

The apparently independent geological evolution of most geological units in the Sudetes, combined with the occurrence of ophiolite bodies along some of their boundaries, or of igneous rocks with MORB-type geochemical signature as well as high-pressure metamorphic rocks (blueschists, eclogites and granulites) strongly suggests that the area comprises a number of distinct tectono-stratigraphic terranes, separated by tectonic sutures and major faults/shear zones (Mazur *et al.* 2006). Aleksandrowski & Mazur (2002) favour the model of Matte *et al.* (1990), albeit with some variations, which suggests that the main Variscan terranes/tectonostratigraphic zones from the western and central Bohemian Massif as well as from Germany and France continue eastwards into the Sudetes.

The Sudetic segment of the Variscan Orogenic Belt formed by accretion of the following terranes: Moldanubian (Gföhl), Góry-Sowie-Kłodzko and NE Teplá Barrandian making up the Central Sudetes and the Lusatia-Izera and Brunovistulian terranes located in the West and Central Sudetes, respectively (Fig. 9.36). Between these larger units are a variety of smaller terranes compressed along their boundaries, including the Moravian Terrane and the SE Karkonosze and Kaczawa terranes (see below for details). Amalgamation of the Sudetic segment of the Variscan continued from Late Silurian through to Early Carboniferous times. It was not a continuous process of terrane convergence, but rather a sequence of distinct orogenic events separated by stages of extension and rifting. At the Devonian–Carboniferous boundary, a collision commenced between the Saxothuringian and the already amalgamated Teplá-Barrandian and Góry Sowie-Kłodzko terranes which led to the formation of the Saxothuringian suture zone. Accretion of the Sudetic area was completed by early/mid-Carboniferous times when collision between the Moldanubian Terrane and the Moravian and Brunovistulian terranes occurred. The evolution of the Sudetes was coeval with the closure of at least two marine basins floored by oceanic crust. Indeed, the vestiges of these basins are the allochthonous volcanosedimentary complexes of the Kaczawa Unit and of the southern and eastern Izera-Karkonosze Massif (see below). According to Mazur *et al.* (2006) the closure of these basins involved three collisional events that took place (1) at the turn from the Middle to the Late Devonian, resulting in the amalgamation of the Central Sudetes, (2) at the turn from the Devonian to the Carboniferous due to collision of the Lusatia-Izera Terrane with the Teplá-Barrandian and Góry-Sowie-Kłodzko terranes in the West Sudetes, and (3) during the Early Carboniferous collision of the East Sudetic orogenic wedge with the continental Brunovistulian Terrane.

As noted above, the Sudetes region represents the NE margin of the Bohemian Massif, and is mainly composed of upper Proterozoic and Palaeozoic, mostly S-type granitoids (Fig. 9.83).

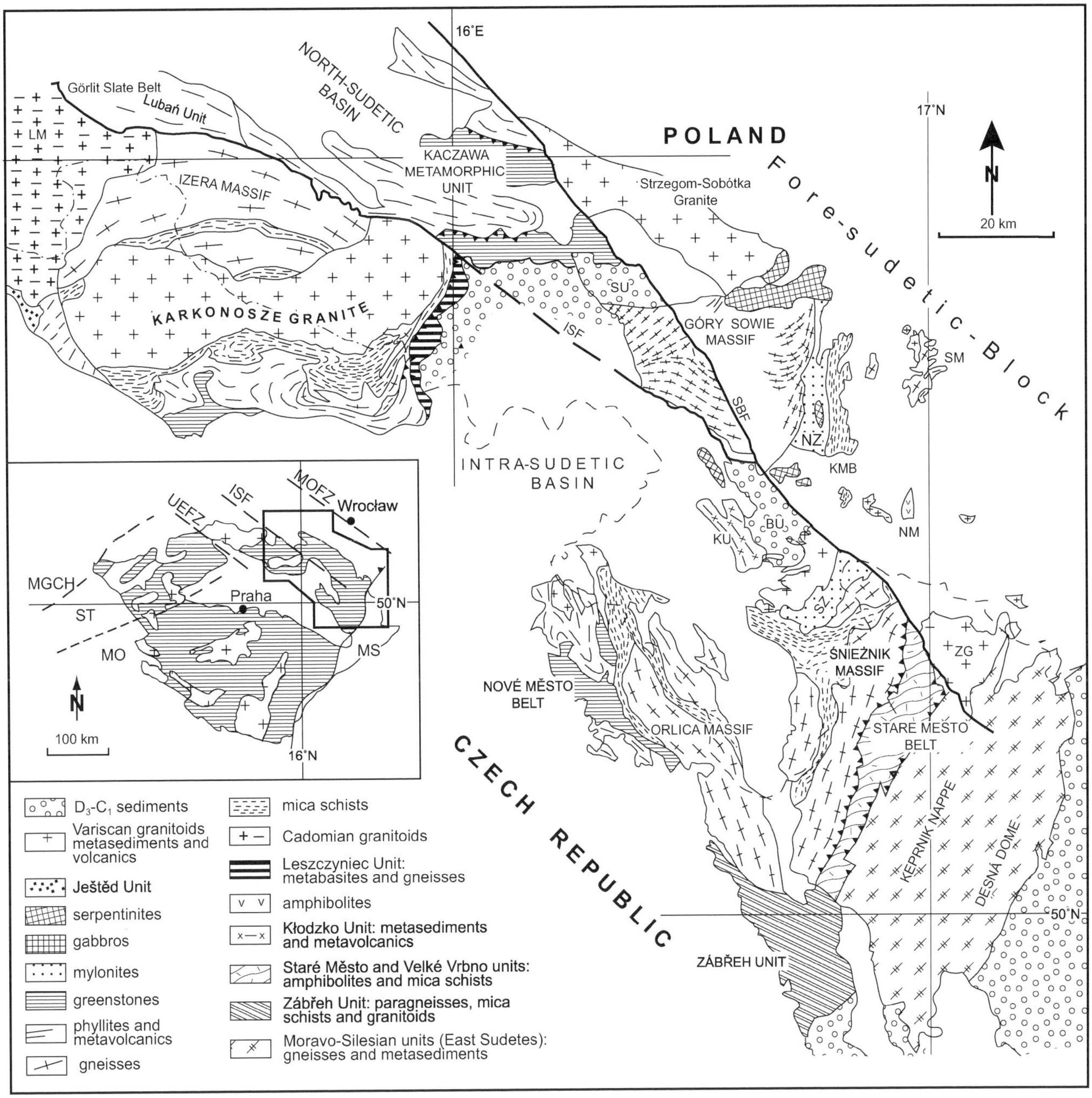

Fig. 9.82. Geological sketch map of the Sudetes. Abbreviations: BB, Bardo Basin: EFZ, Elbe Fault Zone; ISF, Intra-Sudetic Fault; KMB, Kamieniec Metamorphic Belt; KU, Kłodzko Metamorphic Unit; KZG, klodzko, Zloty Stok Granite; LG, Lustian Granitoid Massif; MGCH, Mid-German Crystalline High; MO, Moldanubian Zone; MS, Moravo-Silesian Zone; NM, Niedzwiedz Massif; NZ, Niemcza Shear Zone; OFZ, Odra Fault Zone; RT, Ramzová Thrust; SB, Swiebodzice Basin; SBF, Sudetic Boundary Fault; SCM, Strzelin Crystalline Massif; ST, Saxothuringian Zone; SZ, Skrzynka Shear Zone; ZG, Zulová Granite (after Aleksandrowski & Mazur 2002; Mazur *et al.* 2006).

It also includes a number of small, structural-stratigraphic domains (sometimes <50 km^2) which are generally bounded by tectonic breaks (Fig. 9.82).

Subsequent granitoid intrusion has increased the complexity of the region by obscuring fault boundaries. For example, the Main Intra-Sudetic Fault has been interpreted as a major structure separating Caledonian and Variscan elements in the Lugicum (Don *et al.* 1990), but in spite of marked lithological, stratigraphical and deformational differences across this fault, details of its character and kinematic evolution are unclear. Oliver *et al.* (1993) proposed that the fault was one of many faults between a collage of suspect terranes in the Sudetes marking a suture dividing Baltica from either Gondwana itself or from Gondwana derivatives. Cymerman & Piasecki (1994) and Cymerman *et al.* (1997) placed little importance on the Intra-Sudetic Fault. Their model proposed a ‘Central Sudetic Terrane’ formed of basinal/oceanic and ophiolitic rocks at the centre of the Polish Sudetes. Also unclear is the nature of the border between the West and East Sudetes.

Sedimentation commenced in the West Sudetes in late Proter-

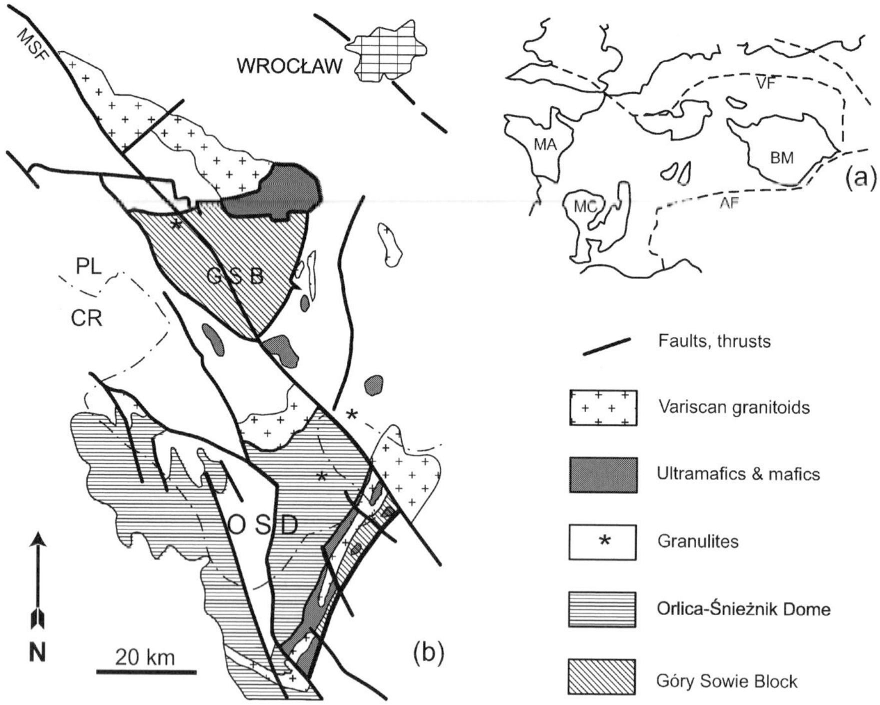

Fig. 9.83. (**a**) Location of the Sudetes granitoids in Variscan Europe. Abbreviations: AF, Alpine Front; BM, Bohemian Massif; MA, Massif Armoricain; MC, Massif Central; VF, Variscan Front. (**b**) Geological sketch map of the central Sudetes. Abbreviations: CR, Czech Republic; GSB, Góry Sowie Block; MSF, Marginal Sudetic Fault; OSD, Orlica-Snieznik Dome; PL, Poland (after Kryza *et al.* 1996).

ozoic times and extended up into the Palaeozoic (locally up to the early Late Carbonifeous) (Urbanek *et al.* 1995). In some areas early Carboniferous deep-marine successions and mélanges were deposited. For example, there is a 4000 m thick siliciclastic succession (sandstones, conglomerates, rare limestones) preserved within the Świebodzice Basin. Deposition occurred within a fan-delta system generated by alluvial fans prograding directly onto a marine, tectonically controlled slope (Porebski 1981). This synorogenic Late Frasnian–Early Tournaisian succession is in turn unconformably overlain by coal-bearing deposits of Late Visean–Early Namurian age (Teisseyre 1968).

Variscan magmatism is represented in the region by widespread granitic plutonism and contemporaneous Carboniferous and Early Permian bimodal volcanism (Kryza 1995*a*). Five main granitoid bodies are recognized: (1) the Karkonosze Granite; (2) the Strzegom-Sobótka Massif; (3) the Strzelin-Zulova Massif; (4) the Kłodzko-Zloty Stok Massif; and (5) the Niemcza granitoids. Magmatism probably commenced at *c.* 340–350 Ma and continued until around 260 Ma with at least two periods of pronounced activity at *c.* 325–330 Ma and 280 Ma (Kryza 1995*a*). A period of late Variscan metamorphism was related to granitic plutonism (Kryza 1995*b*)

West Sudetes

The area of the West Sudetes comprises three main units (Fig. 9.82), all of which were deformed between the Late Devonian and the Late Carboniferous (Mazur *et al.* 2006).

The **Lusatian Massif and Görlitz Slate Belt**: the Lusatian Massif comprises Cadomian, Neoproterozoic to Early Cambrian granitoids within non- to low-grade metamorphic Neoproterozoic greywackes (e.g. Kröner *et al.* 1994; Linnemann *et al.* 1998; Zelazniewicz *et al.* 2004). The Görlitz Slate Belt contains fragments of dismembered Lower Cambrian to Lower Carboniferous successions partly embedded in Lower Carboniferous deep-marine sediments (Mazur *et al.* 2006). This succession is a complex one, including turbidites, chert, limestones, conglomerates and mélange with olistoliths (Urbanek *et al.* 1995; Thomas 1990; Linnemann *et al.* 1998).

The **Kaczawa Unit** comprises several folded thrust sheets composed of numerous thrust slices and mélange bodies. The lower part of the succession (Ordovician) probably formed in an intracontinental rift (Kryza 1993; Seston *et al.* 2000) while the overlying E-MORB to N-MORB basalts and shales (Silurian), and slates and cherts (Devonian) accumulated in an oceanic rift setting (Furnes *et al.* 1994) The upper part of the succession (latest Devonian–Early Carboniferous) comprises low-grade metamorphosed siliciclastics, volcaniclastics, carbonates, basic and acid volcanics, pelagic clay and siliceous shales and deep-marine sediments, and has been interpreted as an accretionary prism complex (Baranowski *et al.* 1987, 1990; Collins *et al.* 2000; Seston *et al.* 2000; Mazur *et al.* 2006).

The **Karkonosze-Izera Massif** includes the Karkonosze Granite Pluton (dated at 329 ± 17 Ma, Rb–Sr whole rock isochron; Duthou *et al.* 1991) and its metamorphic envelope. The envelope comprises four different structural units, from base to top: (1) the Izera Kowary Unit, mainly comprising the Izera (Rumburk) Granite (Late Cambrian–Early Ordovician) which was transformed into the Izera/Kowary Gneiss by later deformation (this extends into the Czech Republic as the Krkonoše Gneiss); (2) the Jested Unit, comprising Middle to Late Devonian shallow-marine to hemipelagic sediments and minor volcanics and passing up into Tournaisian–lower Visean deep-marine clastics; (3) the South Karkonosze Unit, which probably represents a tectonically dismembered succession deposited in a large marine basin and consisting of metamorphosed Ordovician to Devonian sediments including bimodal volcanics; and (4) the Leszczyniec Unit, comprising both mafic and felsic intrusive and extrusive rocks (see Mazur & Aleksandrowski 2001; Aleksandrowski & Mazur 2002; Mazur *et al.* 2006).

Central Sudetes

The Central Sudetes comprises 12 distinct units (Fig. 9.84). These are briefly outlined below (for details see Alekandrowski & Mazur 2002; Mazur *et al.* 2006).

The **Góry Sowie Massif** is composed of amphibolite facies paragneisses thought to have been derived from uppermost Proterozoic protoliths accompanied by migmatites and orthogneisses and small ultramafic and granulitic bodies (Kryza 1981; Zelazniewicz 1990). A high-grade metamorphic event at *c.* 400 Ma (U–Pb zircon method; O'Brian *et al.* 1997) was followed by relatively rapid uplift of the massif associated with contemporaneous decompression under amphibolite facies conditions. Final exhumation of the massif is recorded by the presence of gneissic pebbles in uppermost Devonian–lowermost Carboniferous conglomerates in the Swiebodzice Sedimentary Unit (Porebski 1990). The massif is considered to be part of the Neoproterozoic crust which was subducted to mantle depths during the Late Silurian/Early Devonian and exhumed during the Late Devonian (Mazur *et al.* 2006).

The **Świebodzice Basin** comprises a succession *c.* 4000 m thick of Late Devonian–lowermost Carboniferous predominantly polymict conglomerates (Porebski 1991). In the lower part of the succession the conglomerates are interbedded with thick fossiliferous mudstones containing sandstone turbidites and rare limestone lenses. The basin fill is interpreted as a slope-type fan-delta complex. This small synorogenic basin straddles a fault system which separates the Kaczawa basement to the north from the Góry Sowie Block to the south.

The sedimentary succession of the **Bardo Basin** is predominantly marine (Chorowska & Wajsprych 1995). Upper Devonian limestones and Early Carboniferous deep-marine deposits are overlain by mélange which contains Lower Palaeozoic olistoliths and Devonian deep-marine sediments. Sedimentation in the region ceased with the onset of the Sudetic deformation phase (end Visean–early Namurian A). The deep-marine sedimentation pattern is also characteristic of the eastern Sudetes (e.g. Glubczyce Zone).

The **Central Sudetic Ophiolite** comprises several mafic/ultramafic bodies that crop out along the northern (Sleza Ophiolite), eastern (Szklary and Braszowice Serpentinite massifs) and southwestern (Nowa Ruda Gabbor-Diabase Massif) margins of the Góry Sowie Massif. The spatial retationship between this ophiolite and the Góry Sowie Massif, however, is unclear due to lack of exposure of the contact (see Mazur *et al.* 2006).

The **Kłodzko Metamorphic Unit** comprises six tectonic subunits, including (from base to top) the Maly Bozkow Subunit (Middle Devonian shelf succession), the Laczna Subunit (mélange, of unknown age), the Bierkowice Subunit (?Palaeozoic mafic volcanics), the Scinawka Subunit (MORB-type gabbros), the Orla-Gologlowy Subunit (Neoproterozoic MORB-type gabbros and mafic volcanics associated with some felsic subvolcanic rocks), and the Kłodzko Fortress Subunit (Neoproterozoic–Early Cambrian deep-marine sediments, basaltic lavas, pyroclastic sandstones, dacitic/andesitic tuffs). The tectonic nature of the contacts between the various parts of the Kłodzko Metamorphic Unit has been inferred from their contrasting metamorphic paths and differences in their lithostratigraphy and palaeoenvironmental affinities (Mazur 2003).

The **Intra-Sudetic Basin** a syn- to post-orogenic, relatively large (60 $\times$ 25 km) intramontane basin, which was initiated in mid-Visean times (Turnau *et al.* 2002) and remained active throughout the Carbonifeous and Permian accumulating *c.* 10 km of clastic sediments (Nemec *et al.* 1982; Dziedzic & Teisseyre 1990; Bossowski & Nowak 1995). Early Carboniferous sedimentation (alluvial fan conglomerates) reflected the rapid orogenic uplift of the basin's surroundings, accompanied by extensional collapse and exhumation of the deformed Sudetic crystalline complexes. This coarse-grained succession is overlain by latest Visean deltaic deposits that grade up into shallow-marine sediments. In the Late Carboniferous the most distinctive units are the coal-bearing formations (Walbrzych Formation, Namurian A–B; Zacler Formation, Westphalian A–C), which have been the basis of exploitation in the Lower Silesian Coal Basin. In the eastern parts of the Intra-Sudetic Depression, the Lower Namurian Walbrzych Formation lies discordantly on the older basement (related to the Sudetic phase of deformation); in the western part, however, a continuous transition is more probable.

In the Czech part of the basin, deposition began around the Visean–Namurian boundary and extended, with frequent hiatuses, into the lower Triassic. The Czech succession is subdivided into eight formations. The names of some of these formations are identical with those used in Poland. However, their stratigraphical ranges differ across the national boundaries (Bossowski & Nowak 1995). The Westphalian part of the succession is dominated by up to 1200 m of thick coal-bearing fluvial strata of the Žacléř Formation. Local hiatuses may be present between particular members. The succession comprises upward-fining cycles and contains more than 60 coal seams (each *c.* 1 m thick), concentrated mainly in the lower (upper Namurian–lower Westphalian B) and middle (Westphalian B) units. The final phase of deposition (Žacléř Formation) was accompanied by volcanic activity that produced a laterally extensive volcaniclastic horizon up to 100 m thick (Křenov rhyolitic tuff). The plateau age of this horizon was dated by Ar^{40}/Ar^{39} as 309.0 $\pm$ 3.7 Ma (Lippolt *et al.* 1986). Thin rhyolitic tuffs or effusive melaphyr layers (up to tens of metres thick) occasionally appear in the Westphalian B and C parts of the formation.

The overlying Odolov Formation (Westphalian D–Stephanian B; up to 1500 m thick) is separated from the underlying unit by a basin-wide unconformity. The formation comprises two upward-fining units – the Svatoňovice and the younger Jívka members – which contain red-coloured clastics with rare coal at the top. The arkosic parts of the Jívka Member contain silicified wood remains.

In the Westphalian D, the depocentre of the Intra-Sudetic Basin extended westward into the Krkonoše-Piedmont and Mnichovo Hradiště basins. Equivalents of the Odolov Formation in these basins include the Kumburk (Westphalian D–Barruelian) and Syřenov (Stephanian B) formations. The Kumburk Formation, which is up to 600 m thick, is characterized by alternating mudstones and arkosic sandstones. The Syřenov Formation is only *c.* 100 m thick and comprises fluvial, locally coal-bearing sediments.

Deposition in the basins of the Sudetic area was interrupted at the end of Stephanian B/early Stephanian C times. This basin-wide hiatus corresponds to a similar break in deposition between the Slaný and Líně formations in central and western Bohemia.

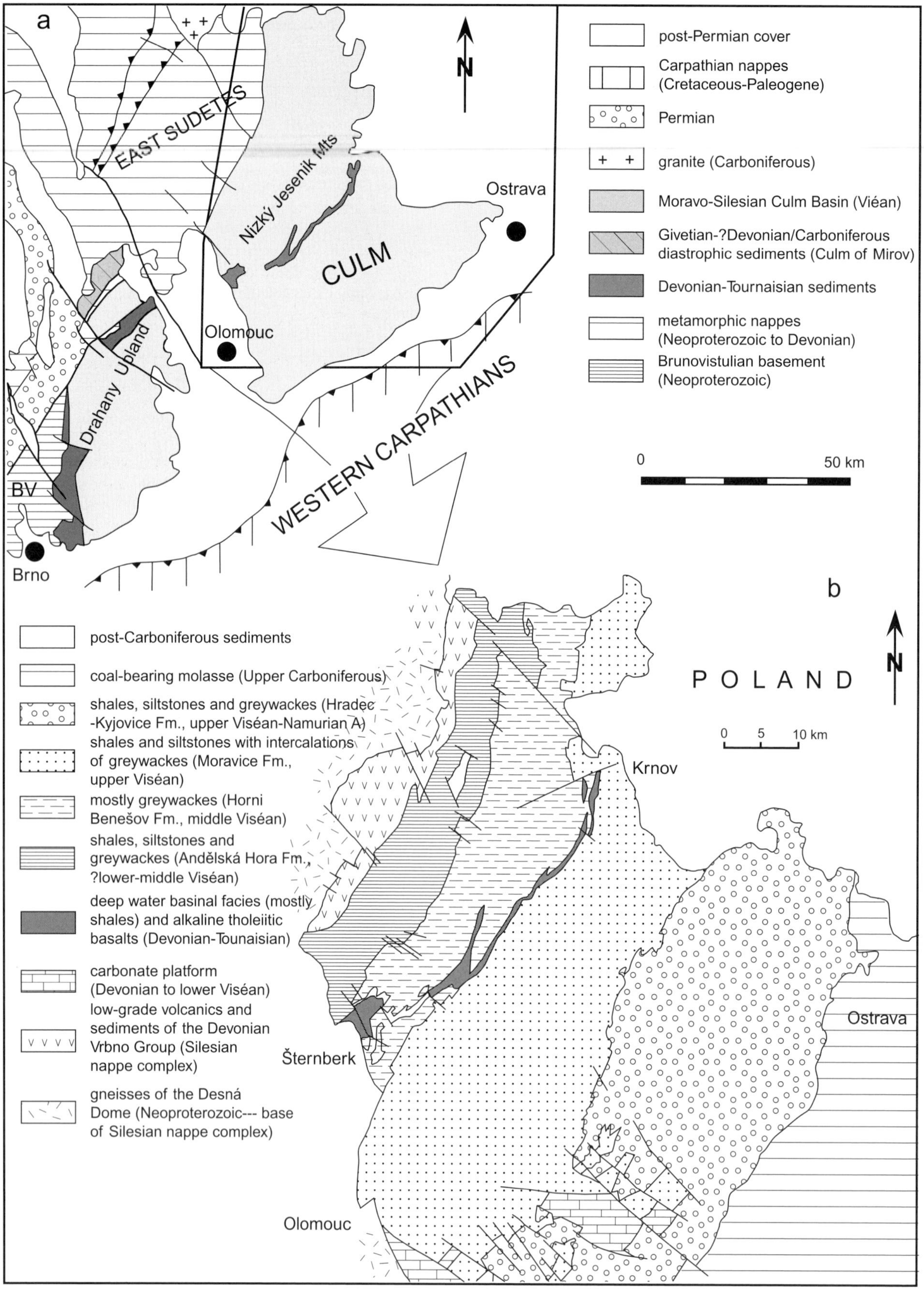

Fig. 9.84. (**a**) Moravo-Silesian fold-and-thrust belt on the eastern margin of the Bohemian Massif (BV, Neoproterozoic crystalline basement of the Brno Massif). (**b**) Simplified geological map of the northern part of the Moravo-Silesian fold-and-thrust belt (after Dvorák 1992; Mazur *et al.* 2006).

The duration of the hiatus increases to the east, such that in the Intra-Sudetic Basin it takes in much of the Stephanian C whereas in the Česká Kamenice, Mnichovo Hradiště and Krkonoše-Piedmont basins it correlates with the deposition of the coarse-grained braided river deposits of the Semily Formation. The widespread Ploužnice Horizon, comprising shales, bituminous shales and thin lacustrine carbonate and volcaniclastic beds, appears in the middle part of the succession. Bodies of andesitic basalt up to 100 m thick occur in this unit in the Mnichovo Hradiště Basin.

The **Orlica-Śnieżnik Massif** is composed mainly of amphibolite-grade orthogneisses and subordinate staurolite-grade variegated schists of presumably Neoproterozoic protoliths with eclogite and granulite lenses. Emplacement of the magmatic precursor to all textural varieties of orthogneisses was dated at *c.* 500 Ma (various methods; Oliver *et al.* 1993; Turniak *et al.* 2000; Kröner *et al.* 2001). The massif probably represents a gneissic dome, although other evidence suggests that it might represent a number of folded thrust sheets (see Mazur *et al.* 2006).

The **Nové Město Belt** is located along the SW margin of the Orlica-Snieznik Massif, and comprises mainly phyllites, greenstones and amphibolites whose ages remain unknown. They are, however, considered to be Late Proterozoic (Chaloupsky *et al.* 1995). MORB affinities for the amphibolites have been noted (Floyd *et al.* 1996). Two late-tectonic granitoid intrusions occur along the contact with the Orlica-Śnieżnik Massif.

The **Niemcza and Skrznka shear zones**: the former is a 5 km wide left-lateral strike-slip ductile shear belt comprising mylonitized gneisses (including lenses of non-mylonitic gneisses, amphibolites and quartzo-graphitic schists as well as granitoids and syenites/diorites) which extends along the eastern edge of the Góry Sowie Massif; the latter is a left-lateral, ductile strike-slip shear approximately 12 km long and 4 km wide, which exposes various blastomylonites, mylonites, cataclasites, gneisses and schists. This latter zone forms the boundary between the Bardo Basin and the Kłodzko-Zloty Stok Granitoid Massif to the NW and the Snieznik Masif to the SE. It is probable that both shear zones were linked in a single kinematic system and that the present offset between them is a result of late displacements along the Intra-Sudetic Fault (Mazur *et al.* 2006).

The **Kamieniec Metamorphic Belt** crops out as a north–south orientated schist belt to the east of the Niemcza Shear Zone. It comprises mica schists with intercalations of quartz-feldspar schists and marbles as well as subordinate lenses of quartz-graphite and amphibolite schists and eclogites.

The **Staré Město Belt** comprises a stack of thrust sheets adjacent to the Orlica-Snieznik Massif and separates the Central from the East Sudetes (i.e. the Lugian from the Moravo-Silesian areas). The belt is *c.* 40 km long and 4–5 km wide and comprises high-grade metasediments, banded felsic/amphibolite rocks with lenses of spinel peridotites at their base, sheared gabbros and a syntectonic tonalite intrusion (Schulmann & Gayer 2000). These are interpreted as products of a Cambro-Ordovician initial rift based on the igneous assemblage, and show features typical of a suture zone. The juxtaposition of dissimilar crustal domains of the East and Central Sudetes along the Staré Město thrust belt may have been triggered by the exhumation of plate fragments that, earlier, had been subducted westwards below the active margin of the West Sudetes during Early Carboniferous times (Schulmann & Gayer 2000). The timing and polarity of subduction of the oceanic domain which separated the East and Central Sudetes are still poorly constrained (Mazur *et al.* 2006).

The **Niedźwiedź Amphibolite Massif** is a 1.5 km thick succession of mainly MORB-type amphibolites and metagabbros (Awdankiewicz 2004).

East Sudetes

The East Sudetes form part of a collision-related deformational and metamorphic belt, nearly 50 km wide and 300 km long, that occupies the eastern margin of the Bohemian Massif extending from Lower Austria through Moravia to Silesia (Mazur *et al.* 2006). The belt comprises nappe piles that crop out from below the upper plate of the collision zone. The collision zone is represented in the Sudetic segment by the East Sudetic pile of nappes comprising, from top to bottom, the Velké Vbrno Nappe, the detached Branná Unit, and the lowermost Keprník Nappe (Fig. 9.82). This latter nappe rests on the parautochthonous gneisses of the Desná Dome (Schulmann & Gayer 2000). The entire eastern Bohemian collisional belt is commonly referred to as the Moravo-Silesian Zone (= 'Moravo-Silesicum' of Suess 1926) (e.g. Dallmeyer *et al.* 1995; Franke & Zelazniewicz 2000) or the Moravian Terrane (e.g. Matte 1991). There are five main units.

The **Velké Vbrno Nappe** is located east of the Staré Město Belt and comprises orthogneisses and metasediments. The gneisses have been dated at *c.* 574 Ma (Kröner *et al.* 2000). The protolith of the supracrustal series may have been a volcanosedimentary succession of possible Early Palaeozoic age (Mazur *et al.* 2007).

The **Branná Unit** and the **Keprník Nappe**: the former, which is thrust over the latter, comprises highly sheared Devonian metasediments including shallow-water conglomerates, quartzites, limestones, sericite and sericite-graphite phyllites, porphyroids and calcsilicate schists. The Keprník Nappe comprises a large granitic orthogneiss body (*c.* 546 Ma, U–Pb zircon dated; van Breemen *et al.* 1982) with subordinate staurolite-bearing metapelites, calc-silicate rocks and quartzites (Schulmann & Gayer 2000).

The **Desná Dome** comprises a metasedimentary Devonian-age succession (Vrbno Group) which overlies gneisses and mylonites (570–650 Ma, U–Pb zircon ages; Schulmann & Gayer 2000). The sediments, deposited in late Devonian times, are predominantly deep marine, up to 3000 m thick, and are associated with abundant metavolcanics. These are interpreted as having been deposited in an extensional basin and there is some geochemical evidence of an arc/backarc signature.

The **Strzelin Massif** can be subdivided into the structurally lower Strzelin Complex and the Stachów Complex. The former comprises Neoproterozoic gneisses (including the Strzelin and Nowolesie gneisses) accompanied by minor mica schists and amphibolites with a within-plate geochemical signature (Szczepanski & Obere-Dziedzic 1998). The outer envelope (Jeglowa Beds) consists of quartzites, quartz-sericite schists and metaconglomerates whose protoliths were deposited on a continental margin during Early- to Mid-Devonian times (Patocka & Szczepanski 1997). The Stachów Complex comprises two varieties of orthogneisses, both of Late Cambrian/Early Ordovician age (Oliver *et al.* 1993; Kröner & Mazur 2003).

The East Sudetic nappe stack is flanked to the SE by the external **Moravo-Silesian Fold-Thrust Belt** whch contains up to 7.5 km of a synorogenic Early Carboniferous succession and minor pre-orogenic Devonian-age sediments of the Moravo-Silesian Basin representing a fragment of the Variscan foreland basin system (Mazur *et al.* 2006) (Fig. 9.84). To the north and NE the deformed fold-thrust belt grades into the less deformed, large foreland basin of SW Poland and to the SE it is buried beneath the Neogene Carpathian foredeep and nappes. The fold-thrust belt is exposed in two areas, the Drahany Uplands and the Nizky Jesenik Mountains.

Recent palaeomagnetic studies have revealed that the Moravo-Silesian has been rotated clockwise through *c.* 90° (Tait *et al.* 2000) or even 180° (Franke & Zelazniewicz 2000), although the mechanics of this clockwise rotation have not been resolved. Rotation possibly commenced when central parts of the present-day Bohemian Massif were forced to intrude to the NE as an indenter (Lorenz & Nicholls 1984). Such an indenter might have been confined between the Moldanubian Thrust on the SE and shear zones on either side of the Mid-German Crystalline High (Schäfer *et al.* 2000). Once a north–south orientation of the Moravo-Silesian Zone was attained, further rotation could have been achieved when basement units to the SE (now concealed by the Carpathians and the Pannonian Basin) were displaced to the NW by strike-slip activity along the Tornquist-Teisseyre Zone, thus tightening the Bohemian Arc (Franke & Zelazniewicz 2000).

Drahany Basin

The Drahany Basin is one of two sub-basins which occur in the Moravo-Silesian area (Figs 9.85 & 9.86). It is located in the central part of the area and consists of several transverse blocks separated by deep-seated faults. Lithological and thickness differences in both the sedimentary and volcanic successions on the blocks reflect the differential mobility of the basement. The oldest sediments are Silurian shales which are overlain by late Early to Upper Devonian partly transgressive shales and related volcanics, as well as a thick Mid–early Upper Devonian reef limestone complex up to 1000 m thick (Dvorák 1995). Platform subsidence commenced in Upper Devonian times as evidenced by the deposition of shallow-marine limestones with intercalated dolomites which were overlain in places by black siliceous shales (Famennian–Tournaisian). These shales (Ponikev Formation) were deposited in the central and northern parts of the Drahany Upland. The overlying Lower and Middle Visean are characterized by trilobite-bearing shales (Brezina Formation).

The onset of deep-marine sedimentation is diachronous, commencing initially in the NW around the Devonian–Carboniferous boundary. It gradually expanded on the western block of the upland during Tournaisian times when more than 2000 m of turbidites were deposited (Protivanov Formation). The coeval strata on the Moravian Karst Block to the south have a total thickness of only some tens of metres (Dvorák 1995).

During the Middle Visean the zone of maximum subsidence shifted eastwards and included the main part of the Drahany Upland. Deposition of the Rozstání Formation (*c.* 1500 m of clastics) followed, as the depocentre shifted to the south during uppermost Visean times. The continental sediments of the Upper Visean Myslejovice Formation (up to 3000 m thick) constitute the final unit of this sedimentary succession. In Namurian times sediments (>1000 m of continental coal-bearing deposits) were laid down in the southern Menín Block.

Jeseníky Basin

Sedimentation in this basin commenced in Devonian times and was characterized by a period of intense volcanic activity (Figs 9.85 & 9.86). Volcanism was concentrated along three deep-seated fault zones running NNE–SSW across the area, and these narrow zones are separated by gneiss domes. Volcanic activity peaked during late Early–Middle Devonian in each of the three fault zones and tailed off through into the Lower Visean.

Devonian sediments are overlain by shales of the Ponikev

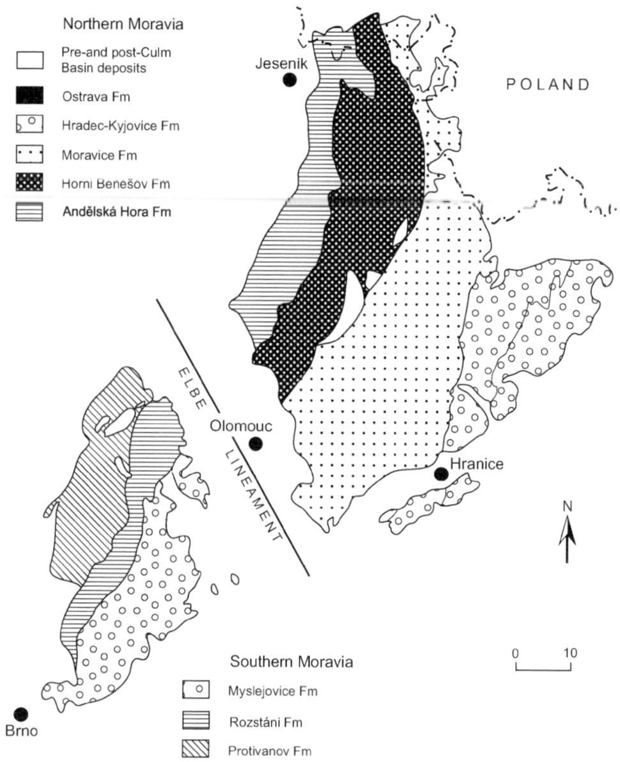

Fig. 9.85. Simplified geological map of formations within the Culm Basin (based on Czech Geological Survey maps; after Hartley & Otava 2001).

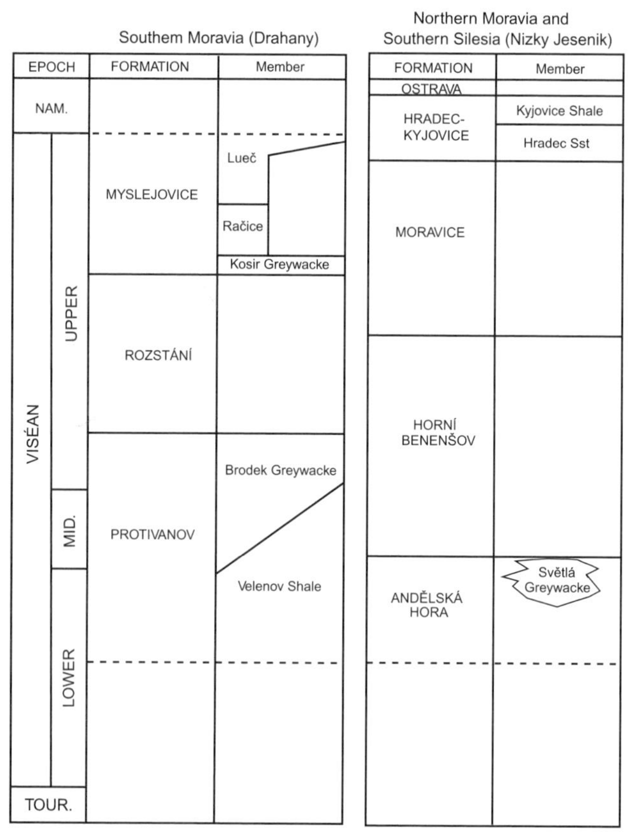

Fig. 9.86. Carboniferous stratigraphy of the Moravo-Silesian area (after Hartley & Otava 2001).

Formation (Upper Frasnian–Lower Visean). A period of uplift followed, as suggested by the presence of limestone breccias, quartz conglomerates and coarse-grained sandstones in narrow north–south trending depresssions between the volcanic ridges. Sedimentation commenced in the eastern part of the Jeseníky Basin in Middle Devonian times. The succession here is overlain by lenticular limestone breccias with phosphate nodules (Upper Famennian–Upper Visean). Clastic material was derived from the rising central massif to the west. Maximum sediment thickness in the region varies between 1500 and 2000 m.

Upper Silesian Basin, south Poland (S.S.)
The Upper Silesian Basin (USB) is situated directly on the foreland of the Moravo-Silesian branch of the Variscides (i.e. East Sudetes). During Early Carboniferous times the area of maximum subsidence within the foreland basin, which evolved at the Variscan Deformation Front, had shifted gradually to the east. This shifting resulted in accelerated subsidence of the area occupied by the starved basin in the west and the carbonate platforms in the east (Fig. 9.87). In Tournaisian and Visean times, progressive backstepping of the carbonate platform edge was paralleled by the eastward advance of a deep-water clastic fan system. This reflects the gradual elevation of the Moravo-Silesian-Sudetic mountain system which became sharply accentuated during the late Visean (Ziegler 1990). However, extensional (?transtensional) tectonics, controlling platform sedimentation, appear to have persisted into the Visean (Suk 1984). In this context, the occurrence of MORB-type ophiolites in the Sudetic Mountains of Poland yielding ages of 353 ± 21 Ma and 351 ± 16 Ma (Pin 1990) probably reflect the Early Carboniferous opening of small pull-apart basins in response to transtensional movement between the Moravo-Silesian Mountains and their foreland (Ziegler 1990).

The top of the Lower Carboniferous carbonate succession within the USB is covered by the marine Malinowice/Zalas siliciclastic series, with the Visean–Namurian boundary located somewhere in the middle of this unit (Fig. 9.88). The upper part of the section, with a thickness of >8000 m, consists of the coal-bearing Paralic Series, which is concordantly covered by three continental coal-bearing units (Kotas 1995). The Paralic Series was deposited on coastal and deltaic plains and attains a thickness of 3800 m in the western part of the USB. It contains numerous marine and tuff horizons, which are useful for correlation. Stratigraphically, this series represents the Namurian A.

In late Visean and early Namurian times, the clastic influx into the southern parts of the Rhenoherycnian Basin from the rising Mid-German Crystalline High increased sharply. A contemporaneous increase of clastic influx is observed into the Lower and Upper Silesian basins from the Sudetic and Moravo-Silesian ranges (Schmidt & Franke 1975; Franke *et al.* 1978; Franke & Engel 1988). The Upper Silesian Sandstone Series (Namurian A–Westphalian A) is dominated by coarse-grained sandstones and conglomerates (Kotas & Malczyk 1972) and is characterized by the presence of thick coal seams (up to 9% of the succession thickness). This series represents deposition on alluvial plains of braided river systems, which explains the numerous stratigraphic gaps within the complex.

The Mudstone Series (Westphalian A–Westphalian B; maximum thickness 2000 m) was deposited on the alluvial plains of a meandering river system, and is dominated by fine-grained deposits (Porzycki 1972). The coal seams are thin but numerous and constitute 5–7% of the total profile thickness. The sediments

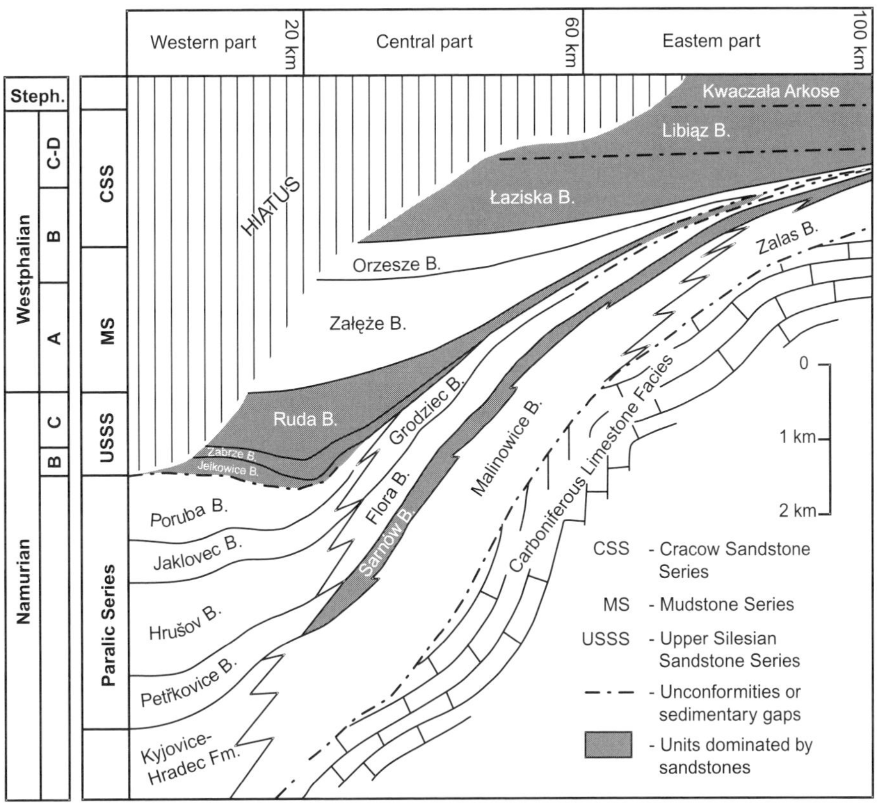

Fig. 9.87. Generalized cross-section of the Moravia-Silesia Basin in the early Visean (after Bełka 1987).

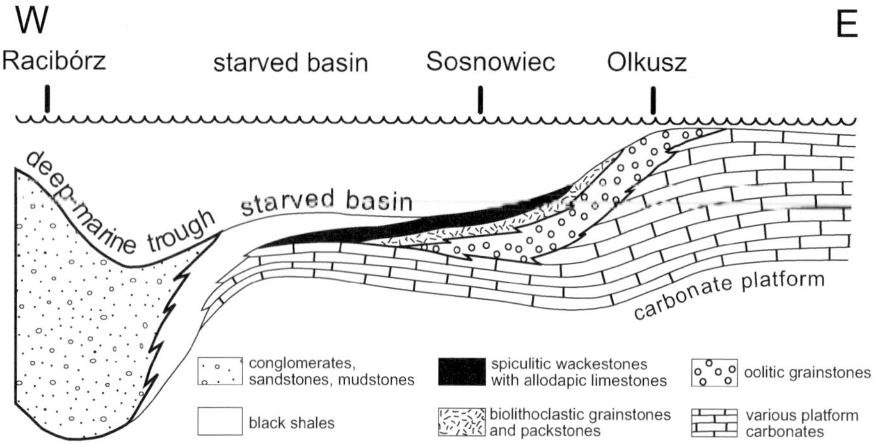

Fig. 9.88. Lithostratigraphic scheme of the Upper Carboniferous succession in the Upper Silesia Coal Basin (after Kotas 1995).

of the Cracow Sandstone Series (Westphalian B–Westphalian D, maximum thickness of 1600 m) were deposited in braided river systems on alluvial plains (Dembowski 1972; Gradziski *et al.* 1995). Coal seams are not numerous, but can be significant in thickness (up to 5 m).

Generally, in the coal-bearing succession the number of coal seams of economic value exceeds 200, with the total thickness of coal about being 340 m, and average coal potential about 4% (Jureczka & Kotas 1995). The succession was folded during the Asturian deformation phase and is unconformably covered by the Kwaczala Arkose, a red-coloured unit, representing the Stephanian. It comprises coarse-grained sandstones and conglomerates with characteristic silicified tree trunks.

Palaeontology

Plant fossils (M.K.)

In the Early Carboniferous plant distribution was dominated by climate, but some floras including lycophytes (e.g. *Lepidodendron, Lepidodendropsis*) and some seed plants (*Triphyllopteris, Sphenopteridium*) are characteristic of the European palaeoarea floras. Such floras are well known through southern Laurentia and Avalonia. Records of these floras in northern and eastern Germany and Silesia were reviewed by Meyen in Vakhrameev *et al.* (1978). Early Carboniferous (Visean) miospores show strong latitudinal variation (Clayton 1996). For example, the *Monilospora* microflora occurs in the northern parts of Baltica (as well as NW Canada and Svalbard), while the *Grandispora* microflora is well documented from northern Germany (Rügen), northern Poland (Pomerania) and Romania (Moesian Platform) (McKerrow *et al.* 2002). A third microflora (*Aratrisporites saharaensis*) is known from North Africa extending across into Syria and Saudi Arabia. There are no transitional associations between the first two microfloras and the most southerly one, suggesting that during Early Carboniferous times, the northern margin of Africa was apparently a barrier to plant migration (McKerrow *et al.* 2002).

In the Devonian, most plant life was confined to wet and humid habitats, whereas during the Carboniferous plants developed a wider range of ecological strategies, and began to colonize the drier areas (Taylor & Taylor 1993). In general, terrestrial plant life in the Devonian was characterized by the appearance and initial radiation of most major lineages of vascular plants, whereas plant life in the Carboniferous was characterized by increasing structural complexity and architectural and ecological diversification within these plant lineages (Kerp 2000; Krings *et al.* 2003).

The Early Carboniferous (Tournaisian and Visean) of Europe is relatively poorly understood with regard to plant life, due to the fact that only a small number of localities yield well-preserved floras (e.g. Scott *et al.* 1984). The plant megafossil record demonstrates that lycophytes, sphenophytes, coenopteridalean and filicalean ferns, and pteridosperms (seed ferns) were the major components of the Early Carboniferous vegetation, and had partitioned the land surface along distinct ecological lines (DiMichele *et al.* 1992). Arborescent lycophytes, which first appeared in the Middle to Late Devonian, were the prominent elements of the peat-forming swamp floras. Later, during the late Early Carboniferous, larger lycophyte trees, including *Lepidodendron* and *Lepidophloios*, appeared; these forms would eventually become the dominant constituents of most Late Carboniferous coal-swamp forests.

Late Carboniferous (Namurian B–Stephanian) plant macrofossils are known from innumerable localities throughout Europe. However, virtually all floras described from the Namurian to the Westphalian represent basinal, coal-forming facies, whereas relatively little is known of the vegetation that existed in the drier 'hinterland' environments (Kerp 1996, 2000). The fossil record indicates that more or less uniformly developed and ecologically stable swamp forest ecosystems were present in the vast wetlands at the continental margins. Smaller areas with swamp forest vegetation, the so-called intramontane basins, existed on the continents. Within these extensive coal-swamp ecosystems, the spatial distribution of the different plant groups again occurred along distinct ecological lines, similar to those described above for the Early Carboniferous (DiMichele *et al.* 1992).

In the Late Namurian and Westphalian peat-forming swamp forests, 70 to 80% of the biomass was produced by arborescent lycophytes. Three major types of Late Carboniferous arborescent lycophytes can be distinguished, *Lepidodendron, Lepidophloios* and *Sigillaria* (e.g. Remy & Remy 1977; Niklas 1997). They were large trees up to 40 m tall, with trunks of up to 2 m in diameter and crowns of dichotomously branched, leafy twigs. In addition, arborescent and herbaceous sphenophytes and coenopteridalean and filicalean ferns, and two groups of seed plants, i.e.

cordaites and pteridosperms, occurred in the Late Carboniferous coal-swamp forests. The sphenophytes were represented by two major groups, the calamites and sphenophylls.

During the Westphalian D–Stephanian transition, the vast wetland areas at the continental margins diminished and eventually vanished, and a dramatic turnover in the floral composition of the swamp forests occurred. The arborescent lycophytes disappeared almost completely, and subsequently, in the Stephanian, only *c*. 5% of the biomass was produced by lycophytes (dominantly of the *Sigillaria* type) (Kerp 1996, 2000). The typical Stephanian swamp forests were more dynamic ecosystems that covered only small areas. They were still dominated by swamp forest elements, but, for the first time, arborescent filicalean ferns (*Psaronius* species) formed a relatively closed canopy. It is interesting to note that, during the Stephanian, the proportional abundance of scrambling and climbing forms in the pteridosperms increased considerably (Kerp & Krings 1998).

The transitional period between the uppermost Stephanian and early Permian yields the first significant macrofossil record of 'hinterland' elements (Kerp 2000). In these environments, conifers were prominent components of the vegetation. In addition, specialized ferns and pteridosperms occurred.

Animal fossils

The Carboniferous was a period of marked changes in ocean circulation which resulted in biogeographic differentiation, particularly that of the terrestrial biota (Davydov *et al.* 2004). The period marked the appearence of reptiles (wtih amniotic egg reproduction) as well as the extinction, or decreasing role, of early Palaeozoic biota such as stromatoporoids, tabulate corals, trilobites, ostracods, and heavily armoured marine fish. In contrast, other organisms first appeared, or underwent rapid diversification (e.g. foraminifera, ammonoids, freshwater pelecypods, gastropods, sharks, ray-finned fishes, winged insects) (Davydov *et al.* 2004; Kullmann 2000).

Sea-level lowstands, represented in the sedimentary record as sequence boundaries, coincide with major evolutionary extinctions of shallow-water fauna. Indeed, the origination of species is closely associated with the subsequent transgressions. A major Palaeozoic bioevent corresponds to the Mid-Carboniferous boundary (Sepkoski 1995), with major extinctions and originations of conodont and ammonoid faunas, as well as turnover in crinoids, corals and foraminifera (House 1993; Walliser 1995).

The extensive development of damp forests as well as the increasing seasonality of climate and large-scale cycles of wet and dry phases triggered the evolution of non-marine vertebrates and invertebrates. The Carboniferous was the main phase of early tetrapod evolution. Around 40 different families arose in the Carboniferous. In addition to the amphibians there were the temnospondyls and anthracosaurs. This latter group were aquatic-adapted carnivores, mostly fish hunters. The first reptiles also appeared in the Late Carboniferous. These were initially small insect-eaters and the true radiation of the reptiles commenced later during the increasingly dry climate of the Permian period.

The Carboniferous was mainly the time of the giant arthropods and large winged insects such as *Meganeura*, terrestrial *Arthropleura* up to 2.5 m long, or semi-aquatic eurypterids up to 1.0 m long such as *Hibbertopterus*. The flourishing of these giant arthropods and their sudden disappearence at the Carboniferous–Permian boundary has been discussed as being related to the effect of the extremely high oxygen content of the atmosphere in the middle Carboniferous (see above). A more realistic reason for the flourishing of the arthropods was the absence of predators which could overcome these giants. Their extinction was linked to the changing of their environments – the disappearence of tropical forests and wet floodplains due to climatic changes during the Late Carboniferous and Early Permian. Non-marine fish faunas were dominated by ganoid fishes (palaeoniscids) and freshwater sharks (e.g. *Xenacanthus*, up to 1.0 m long, and *Orthacanthus*, up to 2.5 m long).

Summary

The Carboniferous Period is closely linked with the Variscan Orogeny which dominated many aspects of Carbonifeous basin development. In addition, tectonic, magmatic and metamorphic processes were closely linked with the ongoing orogenic activity (Figs 9.89 & 9.90). Compared with other orogens, the Variscan Orogeny had an unusally complex evolution and many aspects are the subject of active debate (Franke 2000). However, its development and architecture have become much clearer since the identification and correlation of plate tectonic elements (terranes, sutures, ophiolites, magmatic arcs, foreland fold-and-thrust belts and foreland basins), and the integration of palaeomagnetic and geophysical data with high-precision dating (Timmerman 2004). In general terms, the Variscan Orogeny may be considered the result of Devonian–early Carboniferous accretion onto the southern margin of Laurussia of various Gondwana-derived terranes or microplates (often composite) of predominantly Neoproterozoic (Cadomian/pan-African) crust, together with their Neoproterozoic–Silurian passive margin successions and accreted Ordovician–Devonian island arcs (Franke 1989; Ziegler 1990; Matte 1991).

A number of distinct zones have been recognized across Variscan Europe, each of which shows a different sedimentary, magmatic and tectonomagmatic history (Figs 9.89 & 9.90). These zones can be defined as palaeogeographically coherent units that broadly retained their material integrity throughout the Variscan Orogeny.

Devonian and Early Carboniferous (Dinantian)

Following the Silurian–early Devonian Caledonian Orogeny (see Krawczyk *et al.* 2008), northern Europe formed part of Laurussia (= Old Red Continent) and was covered by Silurian–Devonian, mainly continental sediments. Devonian rifting and spreading at the southern edge of this continent resulted in the formation of a broadly east–west trending oceanic or backarc basin of limited width. This basin (re-)separated Avalonia from the Armorican microplate to the south and is known as the Rhenohercynian Ocean. Dating of the Lizard Ophiolite (SW England) indicates that the opening and spreading was underway by 397 ± 2 Ma (Clark *et al.* 1998; Sandeman *et al.* 2000). Devonian–early Carboniferous sediments accumulated on the thinned and stretched crust of the northern passive margin, attaining thicknesses of >10 km in the Rhenish Massif area.

During the Early Carboniferous, a major complex of extensional basins was present to the north of the Anglo-Brabant Massif, extending across the region from Belgium through to Poland. Extension, basin formation and volcanism occurred in the foreland, while subduction and closure of the Rhenohercynian Ocean resulted in calc-alkaline, I-type arc magmatism on continental crust that probably belonged to the Armorican microplate (see discussion above) (Franke 1989; Oncken 1997). To the north, deposition was dominated by siliciclastic sediments, whereas carbonate sedimentation was more important in the south. Sedimentation was influenced by repeated extensional

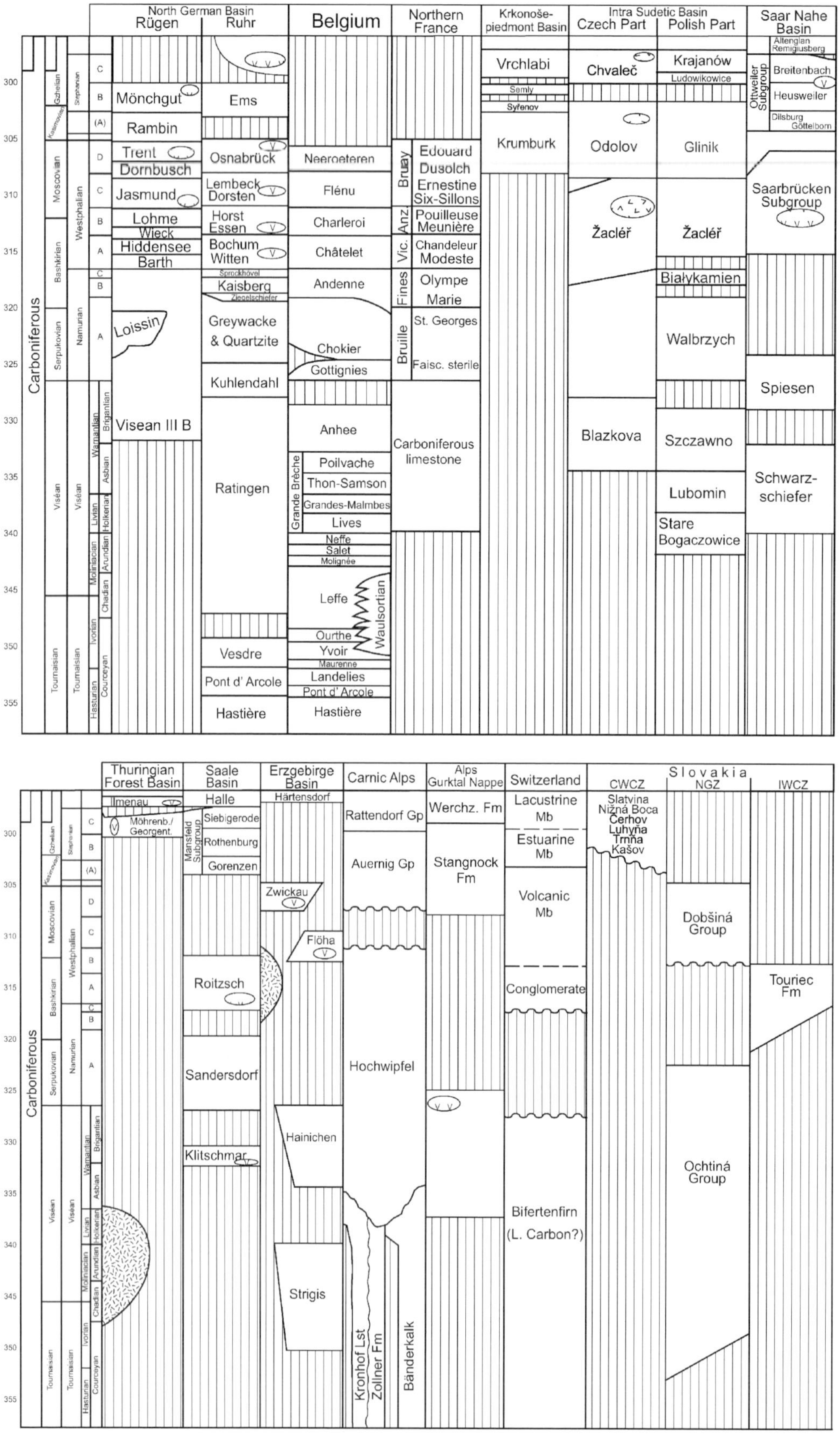

Fig. 9.89. Summary diagram of the Carboniferous stratigraphy across Central Europe (after Schneider & Roscher 2002; Schneider *et al.* 2005*d*; present study).

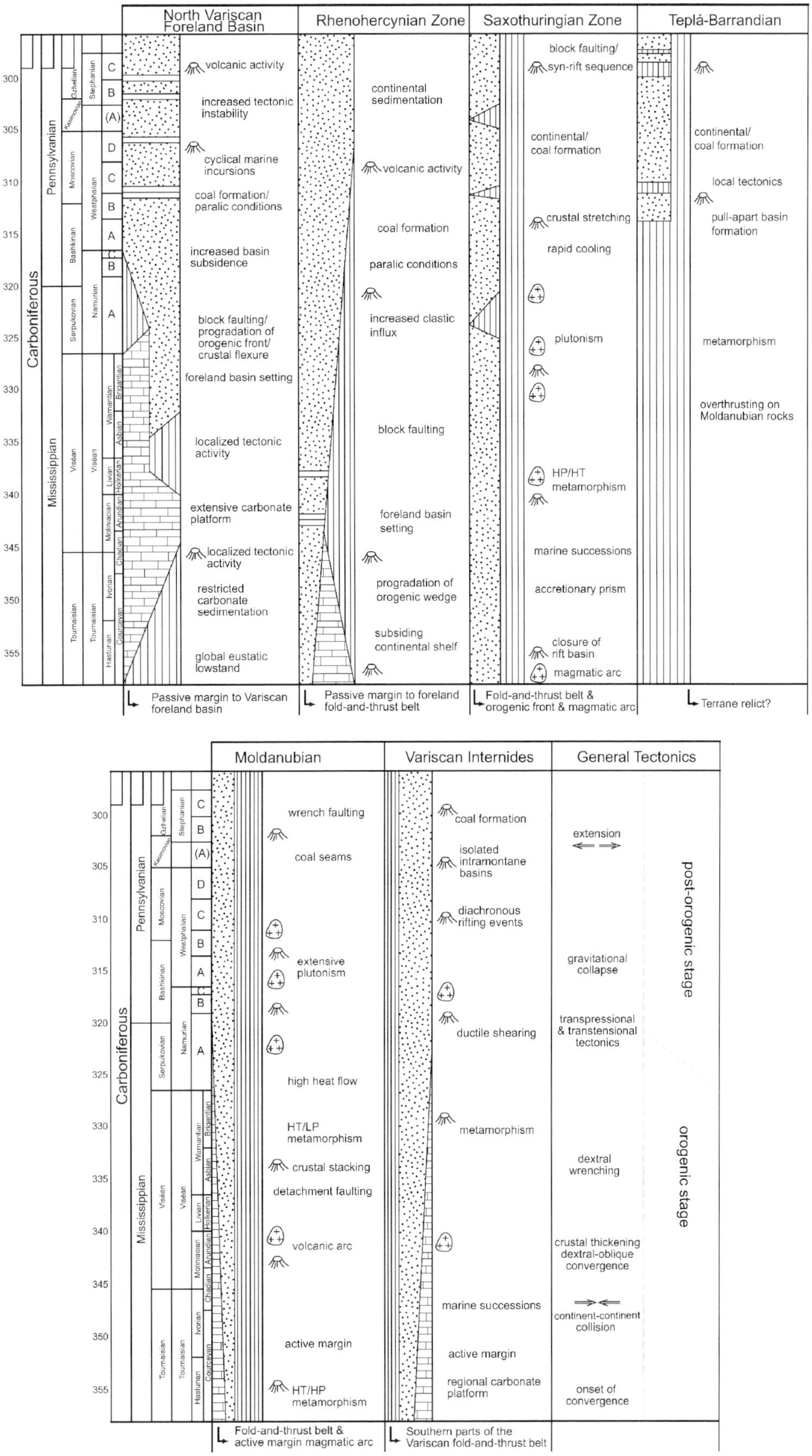

Fig. 9.90. Tentative chronology of events across Central Europe during Carboniferous times.

phases that led to the reactivation of siliciclastic sources and the uplift and erosion of carbonate platforms. Intervening periods of thermal subsidence were characterized by the nucleation and growth of carbonate platforms around basement highs and the shedding of carbonates into basins. Siliciclastic and carbonate sediments were deposited in shallow- and marginal-marine, deep-marine and pelagic settings in the Culm Basin and were eventually displaced towards the north during Variscan thrusting.

Magmatic activity across the region is predominantly of Visean age (325–335 Ma) and the resultant arc formed part of a composite unit, 75–100 km wide, known as the Mid-German Crystalline High. Accretion of the magmatic arc and oblique collision occurred in Visean–early Namurian times resulting in metamorphism (332–342 Ma) that reached amphibolite- to granulite-facies conditions in the Mid-German Crystalline High (Okrusch 1995; Reischmann & Anthes 1996; Will & Schmädicke 2003), and tectonic emplacement of slivers of eclogites in the eastern Odenwald (Will & Schmädicke 2001), although some parts escaped reworking (e.g. the northern Odenwald; Schubert *et al.* 2001). In the eastern extension of the Saxothuringian Zone, metamorphism reached ultra-high-pressure and high-temperature conditions at *c.* 341 Ma, as is clear from the presence of coesite and diamond in felsic (crustal) gneisses and the presence of intercalated garnet peridotites. This early Visean event reflects continental collision and the deep subduction of continental crust that was uplifted to mid-crustal levels within a few million years after the peak of metamorphism (Kröner & Willner 1998; O'Brian 2000; Hwang *et al.* 2001).

Namurian–Westphalian

Carbonate deposition continued into the lowermost Namurian across the region, although it was less extensive than the underlying Dinantian. As a result of lithospheric flexure, due to thrust-wedge loading, a series of foreland basins developed in front of the deformation zone in which subsidence (sometimes quite rapid) kept pace with infilling by fluviatile/lacustrine clastic sediments and thick coal measures, punctuated by several marine incursions.

The Westphalian was generally dominated by non-marine fluvial and lacustrine sedimentation. Several marine incursions occurred in the early Westphalian but they gradually became less frequent. Extensive peat deposits accumulated under humid climate conditions on a low-lying plain that extended across much of western and Central Europe in the Westphalian A and B. By the Westphalian C and D, thrust sheets had propagated northwards, forming a mountain chain across the region. These rising Variscan mountains shed immature lithic sediment northwards into the foreland basins to the north of the orogen. From late Westphalian times onwards the climate became increasingly arid leading to a decrease in peat development and its replacement by calcretes. This ariditiy would be characteristic of the succeeding Permian period (see McCann *et al.* 2008).

To the south of the Rhenohercynian Zone and within the Variscan Orogen (i.e. Mid-German Crystalline High), a series of basins (e.g. the NE-trending Saar-Nahe Basin) opened due to major strike-slip faulting and oblique extension. In the case of the Saar-Nahe Basin, development occurred along the Hunsrück Fault. The absence of older, pre-Namurian sediments suggests that this was an elevated area from the late Visean to the late Namurian. This period of non-deposition reflects Variscan compression, uplift, exhumation and cooling.

Continuing convergence may have led to lateral escape tectonics and the onset of orogenic collapse in the core of the orogen, while northward-migrating folding and thrusting of passive-margin sequences resulted in the formation of the Rhenohercynian foreland fold-and-thrust belt. From Westphalian time onwards, a regional-scale, approximately east–west orientated stress field was superimposed on a collapsing Variscan Orogen and its foreland.

Stephanian–early Rotliegend

The Stephanian is poorly preserved across much of the region since uplift and erosion accompanied Variscan compression in the latest Westphalian and Stephanian. Some significant strike-slip movement occurred during this late phase. Stephanian–Early Permian times a large number of fault-controlled basins were formed, accompanied by widespread volcanic and plutonic activity (Heeremans *et al.* 2004). The cause of this was related to a combination of outward-propagating orogenic collapse, detachment of dense mantle material and the effects of the regional extensional stress field.

We would like to thank T. Beilfuß, B. Krumbiegel and A. Siemes for the excellent diagrams. The manuscript was greatly improved by the comments of A. Hartley and T. Will.

References

AHRENDT, H., CLAUER, N., HUNZICKER, J. C. & WEBER, K. 1983. Migration and folding and metamorphism in the Rheinische Schiefergebirge deduced from K-Ar and Rb-Sr age determinations. *In:* MARTIN, H. & EDER, F. W. (eds) *Intracontinental Fold Belts – Case Studies in the Variscan Belt of Europe and Demara Belt in Namibia*. Springer-Verlag, Berlin, 323–388.

AÏFA, T., FEINBERG, H. & POZZI, J.-P. 1990. Devonian-Carboniferous palaeopoles for Africa: consequences for Hercynian geodynamics. *Tectonophysics*, **179**, 287–304.

ALEKSANDROWSKI, P. 1995. The significance of major strike-slip displacements in the development of Variscan structures of the Sudetes (SW Poland). *Przeglad Geologiczny*, **43**, 745–754.

ALEKSANDROWSKI, P. & MAZUR, S. 2002. Collage tectonics in the northeasternmost part of the Variscan Belt: the Sudetes, Bohemian Massif. *In:* WINCHESTER, J. A., PHARAOH, T. C. & VERNIERS, J. (eds) *Palaeozoic Amalgamation of Central Europe*. Geological Society, London, Special Publications, **201**, 237–277.

ALEKSANDROWSKI, P., KRYZA, R., MAZUR, R. & ZABA, J. 1997. Kinematic data on major Variscan strike-slip faults and shear zones in the Polish Sudetes, northeast Bohemian Massif. *Geological Magazine*, **134**, 727–739.

ALEKSANDROWSKI, P., KRYZA, R., MAZUR, S., PIN, C. & ZALASIEWICZ, J.A. 2000. The Polish Sudetes: Caledonian or Variscan? *Transactions of the Royal Society, Edinburgh*, **90**, 127–146.

ALLEN, K. C. & DINELEY, D. L. 1988. Mid-Devonian to mid-Permian floral and faunal regions and provinces. *In:* HARRIS, A. L. & FETTES, D J. (eds) *The Caledonian-Appalachian Orogen*. Geological Society, London, Special Publications **38**, 531–548.

ALTENBERGER, U. & BESCH, T. 1993. The Böllstein Odenwald – evidence for pre- to Early Variscan plate convergence in the Central European Vsriscides. *Geologisches Rundschau*, **82**, 475–488.

ALTENBERGER, U., BESCH, T., MOCEK, B., ZAIPENG, Y. & YONG, S. 1990. Geochemie und Geodynamik des Böllsteiner Odenwaldes. *Mainzer Geowissenschaftliche Mitteilungen*, **19**, 183–200.

ALTHERR, R. & MAASS, R. 1977. Metamorphite am Südrand der Zentralschwarzwälder Gneisanatexitmasse zwischen Geschwend und Bernau. *Neues Jahrbuch für Geologie und Paläontologie, Abhandlungen*, **154**, 129–154.

ALTHERR, R., HENES-KLAIBER, U., HEGNER, E., SATIR, M. & LANGER, C. 1999. Plutonism in the Variscan Odenwald (Germany): from subduction to collision. *International Journal of Earth Sciences*, **88**, 422–443.

ALTHERR, R., HOLL, A., HEGNER, E., LANGER, C. & KREUZER, H. 2000. High-potassium, calc-alkaline I-type plutonism in the European Variscides: Northern Vosges (France) and northern Schwarzwald (Germany). *Lithos*, **50**, 51–73.

AMLER, M. R. W. 1987. Fauna, Paläogeographie und Alter der

Kohlenkalk-Vorkommen im Kulm des östlichen Rheinischen Schiefergebirges (Dinantium). *Geologische Abhandlungen Hessen*, **88**, 339p.

AMLER, M. R. W. 2004. Bivalve biostratigraphy of the Kulm Facies (Early Carboniferous, Mississippian) in central Europe. *Newsletters on Stratigraphy*, **40**, 183–207.

ANDERLE, H.-J. 1987. The evolution of the South Hunsrück and Taunus Borderzone. *Tectonophysics*, **137**, 101–114.

ANDERLE, H.-J., FRANKE, W. & SCHWAB, M. 1995. Stratigraphy (Northern Phyllite Zone). *In:* DALLMEYER, R. D., FRANKE, F. & WEBER, K. (eds) *Pre-Permian Geology of Central and Eastern Europe*. Springer Verlag, Berlin, 99–107.

ANDREAS, D., LÜTZNER, H. & SCHNEIDER, J. W. 2005. Karbon des Thüringer Waldes - Kraichgau-Senke und Thüringisch-Fränkische Grabenzone. *In:* WREDE, V. (ed.) Stratigraphie von Deutschland. Oberkarbon. *Courier Forschungsinstitut Senckenberg*, **254**, 403–418.

ANDRUSOV, D. 1936. Sub-tatric nappes in Western Carpathians. *Carpathica*, **1**, 3–50 [in Czech].

ANDRUSOV, D., BYSTRICKÝ, J. & FUSAN, O. 1973: Outline of the Structure of the Western Carpathians. *Guidebook for Geological Excursion X Congress of CBGA*. D. Štúr Institute of Geology, Bratislava, 1–44.

ANTHES, G. 1998. *Geodynamische Entwicklung der Mitteldeutschen Kristallinschwelle: Geochronologie und Isotopengeochemie*. Thesis, Johannes Gutenberg Universität, Mainz.

ANTHES, G. & REISCHMANN, T. 1996. Geochronologie und Isotopengeochemie der NE Mitteldeutschen Kristallinschwelle. *Terra Nostra*, **96**(2), 9–10.

ANTHES, G. & REISCHMANN, T. 2001. Timing of granitoid magmatism in the eastern Mid German Crystalline Rise. *Journal of Geodynamics*, **31**, 119–143.

ARETZ, M. 2002. Rugose corals and associated carbonate microfossils from the Brigantian (Mississippian) of Castelsec (Montagne Noire, southern France). *Geobios*, **35**, 187–200.

ARTHAUD, F. & MATTE, P. 1977. Late Paleozoic strike-slip faulting in southern Europe and northern Africa. Result of a right lateral shear zone between the Appalachians and the Urals. *Geological Society of America, Bulletin*, **88**, 1305–1320.

AWDANKIEWICY, H. 2004. Sedimentation, volcanism and subvolcanic intrusions in a late Palaeozoic intramontane trough (the Intra-Sudetic Basin, SW Poland). *In:* BREITKREUZ, C. & PETFORD, N. (eds) *Physical Geology of High-Level Magmatic Systems*. Geological Society, London, Special Publications, **234**, 5–11.

BÁBEK, O., MIKULAS, R., ZAPLETAL, J. & LEHOTSKY, T. 2004. Combined tectonic-sediment supply-driven cycles in a Lower Carboniferous deep-marine foreland basin, Moravice Formation, Czech Republic. *International Journal of Earth Sciences*, **93**, 241–261.

BACHMANN, G. H., MÜLLER, M. & WEGGEN, K. 1987. Evolution of the Molasse Basin (Germany, Switzerland). *Tectonophysics*, **137**, 77–92.

BACHTADSE, V. 1990. Hercynian palaeomagnetism of Europe: arguments for large-scale complex thin-skinned rotations. *Cuadernos de Geologica Iberica*, **12**, 147–162.

BACHTADSE, V. & VAN DER VOO, R. 1986. Paleomagnetic evidence for crustal and thin-skinned rotations in the European Hercynides. *Geophysical Research Letters*, **13**, 161–164.

BACHTADSE, V., TORSVIK, T. H., TAIT, J. A. & SOFFEL, H. C. 1995. Paleomagnetic constraints on the paleogeographic evolution of Europe during the Paleozoic. *In:* DALLMEYER, R. D., FRANKE, W. & WEBER, K. (eds) *Pre-Permian Geology of Central and Eastern Europe*. Springer, Berlin, 567–578.

BADERTSCHER, N. & BURKHARD, M. 1998. Inversion alpine du graben Permo-Carbonifère de Salvan-Dorénaz et sa relation avec le chevauchement de la nappe de Morcles sus-jacente. *Eclogae Geologicae Helvetiae*, **91**, 359–373.

BADHAM, J. P. N. 1982. Strike-slip orogens – an explanation for the Hercynides. *Journal of the Geological Society, London*, **139**, 493–504.

BAJANÍK, Š. & PLANDEROVÁ, E. 1985. Stratigraphic position of the lower part of the Ochtiná Formation (between Magnezitovce and Magura). *Geologicke Práce, Správy*, **82**, 67–76 [in Slovak].

BAJANÍK, Š., IVANIČKA, J., MELLO, J., PRISTAŠ, J., REICHWALDER, P., VOZÁR, J. & VOZÁROVÁ, A. 1984: *Geological map of the Slovenské rudohorie Mts. – eastern part*, 1:50.000. D. Štúr Institute of Geology, Bratislava.

BARANOWSKI, Z., HAYDUKIEWICZ, A., KRYZA, R., LORENC, S., MUSZYNSKI, A. & URBANEK, Z. 1987. The development of the eastern part of the Góry Kaczawskie structure. *In:* BARANOWSKI, Z., GROCHOLSKI, A., MALINOWSKI, J., OBERC, J. & POREBSKI, S. (eds) *Field Guide of 58th Meeting of the Polish Geological Society, Walbrzych*. Institute of Geology & University of Wroclaw, Kraków, 61–114 [in Polish].

BARANOWSKI, Z., HAYDUKIEWICZ, A., KRYZA, R., LORENC, S., MUSZYNSKI, A., SOLECKI, A. & URBANEK, Z. 1990. Outline of the geology of the Góry Kaczawskie (Sudetes, Poland). *Neues Jahrbuch für Geologie und Paläontologie, Abhandlungen*, **179**, 223–257.

BECKER, R. T. 1988. Ammonoids from the Devonian-Carboniferous boundary in the Hasselback Valley (Northern Rhenish Slate Mountains). *Courier Forschungs-Institut Senckenberg*, **100**, 193–213.

BECKER, R. T., BLESS, J. J. M., *et al.* 1984. Hasselbachtal, the section best displaying the Devonian-Carboniferous boundary beds in the Rhenish Massif (Rheinisches Schiefergebirge). *Courier Forschungs-Institut Senckenberg*, **67**, 181–191.

BECQ-GIRAUDON, J. F. 1985. Synthèse des nouvelles données sur le Stéphanien de Bosmoreau-les-Mines (Creuse) et ses relations avec les témoins de meme age au sud de Bourgneuf. *Géologie France*, **4**, 363–380.

BECQ-GIRAUDON, J. F. & VAN DEN DRIESSCHE, J. 1994. Dépôts périglaciaires dans le Stéphano-Autunien du Massif Central; temoin de lèffondrement gravitaire d'un haut plateau hercynian. *Comptes Rendus de l'Academie des Sciences, Serie 2*, **318**, 675–682.

BEDERKE, E. 1924. Das Devon in Schlesien und das Alter der Sudetenfaulting. *Fortschritte der Geologie und Paläontologie*, **7**, 1–55.

BEDERKE, E. 1957. Alter und Metamorphose des kristallinen Grundgebirges im Spessart. *Abhandlungen des Hessischen Landesamt, Bodenforschung*, **18**, 7–19.

BEERLING, D. J. 2002. Low atmospheric CO2 levels during the Permo-Carboniferous glaciation inferred from fossil lycopods. *PNAS*, **99**, 12567–12571.

BEHR, H. J. & HEINRICHS, T. 1987. Geological interpretation of DEKORP 2–5: a deep seismic reflection profile across the Saxothuringian and possible implications for the late Variscan structural evolution of Central Europe. *Tectonophysics*, **142**, 173–202.

BEHR, H. J., ENGEL, W. & FRANKE, W. 1982. Variscan wildflysch and nappe tectonics in the Saxothuringian zone (Northeast Bavaria, West Germany). *American Journal of Science*, **282**, 1438–1470.

BEHR, H. J., ENGEL, W., FRANKE, W., GIESE, P. & WEBER, K. 1984. The Variscan Belt in Central Europe. Main structures, geodynamic implications, open questions. *Tectonophysics*, **109**, 15–40.

BEHRMANN, J. H. & TANNER, D. C. 1997. Carboniferous tectonics of the VAriscan basement collage in eastern Bavaria and western Bohemia. *Geologisches Rundschau*, **86**, 15–27.

BELKA, Z. 1987. The development and decline of a Dinantian carbonate platform: an example from the Moravia-Silesia Basin, Poland. *In:* MILLER, J., ADAMS, A. E. & WRIGHT, P. (eds) European Dinantian Environments. *Geological Journal*, **12**, 177–188.

BELKA, Z., SKOMPSKI, S. & SOBON-PODGÓRSKA, J. 1996. Reconstruction of a lost carbonate platform on the shelf of Fennosarmatia: evidence from Visean polymictic debrites, Holy Cross Mountains, Poland. *In:* STROGEN, P., SOMERVILLE, I. D. & JONES, G. L. (eds) *Recent Advances in Carboniferous Geology*. Geological Society, London, Special Publications, **107**, 315–329.

BELKA, Z., AHRENDT, H., FRANKE, W. & WEMMER, K. 2002. The Baltica-Gondwana suture in central Europe: evidence from K-Ar ages of detrital muscovites and biogeographical data. *In:* FRANKE, W., HAAK, V., ONCKEN, O. & TANNER, D. (eds) *Orogenic Processes: Quantification and Modelling in the Variscan Belt*. Geological Society, London, Special Publications, **179**, 87–102.

BELLOT, J. P., ROIG, J. Y. & GENNA, A. 2005. The Hospital coal basin (Massif Central, France): relay on the left-lateral strike-slip Argentat fault in relation to the Variscan postorogenic extension. *Bulletin Societé Géologique France*, **176**, 151–159.

BENDER, P. & HOMRIGHAUSEN, R. 1979. Die Hörre-Zone, eine Neudefinition auf lithostratigraphischer Grundlage. *Geologica et Palaeontologica*, **13**, 257–260.

BENEK, R., KRAMER, W., *et al.* 1996. Permo-Carboniferous magmatism and related subsidence of the NE German Basin. *Tectonophysics*, **266**, 379–404.

Berger, E., Torres, P. & Wefers, J. 1991. Zur Stratigraphie der Metamorphen Zone des Hunsrücksüdrandes im Rheinischen Schiefergebirge. *Neues Jahrbuch für Geologie und Paläontologie, Monatshefte*, **1991**, 737–746.

Berner, R. A. 1987. Models for carbon and sulfur cycles and atmospheric oxygen: application to Paleozoic geologic history. *American Journal of Science*, **287**, 177–196.

Berner, R. A. 1997. The rise of plants and their effect on weathering and atmospheric CO2. *Science*, **276**, 544–546.

Besley, B. M. 1990. Carboniferous. *In:* Glennie, K. W. (ed.) *Introduction to the Petroleum Geology of the North Sea.* Blackwell, Oxford, 104–136.

Bezák, V. 1991. Metamorphic conditions of the Veporic unit in the Western Carpathians. *Geologica Carpathica*, **42**, 219–222.

Biely, A., Bezák, V., *et al.* 1996. *Explanation to Geological Map of Slovakia*, 1:500 000. Geological Survey of Slovakia, Bratislava, 1–76.

Bless, M. J. M., Bouckaert, J., *et al.* 1976. Dinantian rocks in the subsurface north of the Brabant and Ardenno-Rhenish Massif in Belgium, The Netherlands and the Federal Republic of Germany. *Mededelingen Rijks Geologische Dienst*, **27**, 81–195.

Bless, M. J. M., Conil, R., Defourny, P., Groessens, E., Hance, L. & Hennebert, M. 1980. Stratigraphy and thickness variations of the Struno-Dinantian deposits around the Brabant Massif. *In:* Bless, M. J. M., Bouckaert, J. & Paproth, E. (eds) Pre-Permian around the Brabant Massif in Belgium, The Netherlands and Germany. *Mededelingen Rijks Geologische Dienst*, **32**, 56–65.

Bless, M. J. M., Bouckaert, J. & Paproth, E. 1984. Migration of facies belts as a response to Continental Drift during the late Devonian and Carboniferous. *Bulletin de la Société belge de Géologie*, **93**, 189–195.

Blüm, W. 1989. Faciesanalyse im Rotliegenden des Nordschweizer Permokarbon-Trogs (Hochrhein-Region zwischen Basel und Laufenburg). *Eclogae Geologicae Helvetiae*, **82**, 455–489.

Boigk, H. & Schöneich, H., 1974. Perm, Trias und älterer Jura im Bereich der Mittelmeer-Mjösen-Zone und des Rheingrabens. *In:* Illies, J. H.& Fuchs, K. (eds) *Approaches to Taphrogenesis*. Inter-Union Commission on Geodynamics, Scientific Report. Schweizerbart, Stuttgart, pp. 60–71.

Bojar, A.-V. & Neubauer, F. 2003. Geochemical anomaly across the Devonian Carboniferous extinction boundary, Eastern Alps, Austria. *Geophysical Research Abstracts*, **5**, 14913.

Borkowska, M., Hameurt, J. & Vidal, P. 1980. Origin and age of Izera gneisses and Rumburk granites in the Western Sudetes. *Acta Geologica Polonica,* **30**, 121–145.

Bojkowski, K. & Dembowski, Z. 1972. Die Palaeogeographie und Lithofazies des Karbon in Polen. *C.R. 7e Congress International Stratigraphy Geology Carboniferous*, Krefeld, **2**, 195–209.

Bojkowski, K. & Zelichowski, A. M. 1980. An outline of palaeogeography of the Namurian B-C and Westphalian in Poland. *Biuletyn Panstwowego Instytutu Geologicznego,* **328**, 37–59.

Bonin, B., Brändlein, P., *et al.* 1993. Late Variscan magmatic evolution of the Alpine basement. *In:* von Raumer, J. F. & Neubauer, F. (eds) *The Pre-Mesozoic Geology in the Alps*. Springer, Berlin, 171–202.

Bossowski, A. & Nowak, G. 1995. Comparison of the Upper Carboniferous in the Polish and Czech parts of the Intra-Sudetic Depression. *In:* Zdanowski, A. & Zakowa, H. (eds) The Carboniferous system in Poland. *Prace Panstwowego Instytutu Geologicznego,* **148**, 157–159.

Bossowski, A., Ihnatowicz, A., Mastalerz, K., Kurowski, L. & Nowak, G. J. 1995. Lithostratigraphy and sedimentologic-paleogeographic development; Intra-Sudetic Depression. *In:* Zdanowski, A. & Zakowa, H. (eds) *Prace Panstwowego Instytutu Geologicznego*, **148**, 142–147.

Bouckaert, J. & Dusar, M. 1987. Arguments géophysiques pour une tectonique cassante en Campine (Belgique), active au Paléozoique Supérieur et reactivée depuis le Jurassique Supérieur. *Annales de la Société Géologique du Nord*, **151**, 201–208.

Brändlein, P., Nollau, G., Sharp, Z. & Von Raumer, J. 1994. Petrography and geochemistry of the Vallorcine granite (Aiguilles-Rouges massif, Western Alps). *Schweizerische Mineralogische und Petrographische Mitteilungen*, **74**, 227–243.

Brätz, H. 2000. Radiometrische Altersdatierungen und geochemische Untersuchungen von Orthogneises, Graniten und Granitporphyren aus dem Ruhlaer Kristallin, Mitteldeutsche Kristallinzone. Dissertation, Doctor rerum naturalium, University of Würzburg, Germany.

Brauchmann, C. & Schneider, J. W. 1996. Ein unter-karbonisches Insekt aus dem Raum Bitterfeld/Delitzsch (Pterygota, Arnsbergium, Deutschland). *Neues Jahrbuch Geologie und Paläontologie, Monatshefte,* **1996**(1), 17–30.

Braun, A. & Maass, R. 1992. Oberdevonische Radiolarien aus dem Brueschtal (Nord-Vogesen, Elsaß) und ihr regionaler und stratigraphischer Zusammenhang. *Neues Jahrbuch für Geologie und Paläontologie, Abhandlung*, **185**, 161–178.

Brause, H. 1979. Mobilistische Aspekte zur Zonengliederung des mitteleuropäischen variszischen Tektogens. *Zeitschrift für geologische Wissenschaften*, **7**, 1113–1127.

Breiter, K. 1997. The Teplice rhyolite (Krušné hory Mts., Czech Republic) chemical evidence of a multiply exhausted stratigraphic magma chamber. *Vestnik Ceskeho Geologickeho Ustavu*, **72**(2), 205–213.

Breitkreuz, C. & Kennedy, A. 1999. Magmatic flare-up at the Carbonifeous/Permian boundary in the NE German Basin revealed by SHRIMP zircon ages. *Tectonophysics*, **302**, 307–326.

Brenckle, P. L. & Lane, R. H. 1987. *Selected Studies in Carboniferous Paleontology and Biostratigraphy.* Courier Forschungsinstitut Senckenberg, **98.**

Brewer, M. S. & Lippolt, H. J. 1972. Isotopische Altersbestimmungen an Schwarzwald-Gesteinen. *Fortschritte der Mineralogie*, **50**, 41–50.

Briegleb, D. 1971. Geologie der Magnesitlagerstätte am Sattlerkogel in der Veitsch *(Steiermark). Berg- und Hüttenmännische Monatshefte*, **116**, 359–375.

Brix, M. R., Drozdzewski, G., Greiling, R. O., Wolf, R. & Wrede, V. 1988. The N Variscan margin of the Ruhr coal district (Western Germany): structural style of a buried thrust front? *Geologische Rundschau*, **77**, 115–126.

Bruckschen, P. & Veizer, J. 1997. Oxygen and carbon composition of Dinantian brachiopods: Palaeoenvironmental implication for the Lower Carboniferous of Western Europe. *Palaeogeography, Palaeoclimatology, Palaeoecology*, **132**, 243–264.

Bruckschen, P., Veizer, J., Schwark, L. & Leythaeuser, D. 2001. Isotope stratigraphy for the transition from the late Palaeozoic greenhouse in the Permo-Carboniferous icehouse – new results. *Terra Nostra*, **4**, 7–11.

Bruguier, O., Becq-Giraudon, J. F., Bosch, D. & Lancelot, J. R. 1998. Late Visean hidden basins in the internal zones of the Variscan Belt: U-Pb zircon evidence from the French Massif Central. *Geology*, **26**, 627–630.

Bruguier, O., Becq-Giraudon, J. F., Clauer, N. & Maluski, H. 2003. From late Visean to Stephanian: pinpointing a two-stage basinal evolution in the Variscan belt. A case study from the Bosmoreau basin (French Massif Central) and its geodynamic implications. *International Journal of Earth Sciences*, **92**, 338–347.

Büchi, U. P., Lemcke, K., Wiener, G. & Zimdars, J. 1965. Geologische Ergebnisse der Erdölexplorationsbohrungen aus das Mesozoikum im Untergrund des schweizerischen Molassebeckens. *Bulletin der Vereinigung schweizerischer Petroleum-Geologen und -Ingenieure*, **32,** 7–38.

Bues, C. & Zulauf, G. 2000. Microstructural evolution and geologic significance of garnet pyriclasites in the Hoher Bogen shear zone (Bohemian Massif, Germany). *International Journal of Earth Sciences*, **88**, 803–813.

Bues, C., Behrmann, J., Franke, W., Pauli, C., Seibert, J. & Zulauf, G. 1998. Strukturentwicklung und Kinematik der Metabasite und Metapelite im Grenzbereich Teplá-Barrandium - Moldanubikum im Raum Hoher Bogen-Rittsteig (NE Bayern). *Zeitschrift der Deutsche Geologische Gesellschaft*, **149**, 329–343.

Buggisch, W., Flügel; E., Leitz, F. & Tietz, G.-F. 1976. Die fazielle und paläogeographische Entwicklung im Perm der Karnischen Alpen und in den Randgebieten. *Geologische Rundschau*, **65**, 649–690.

Buggisch, W., Alekseev, A. S., Davydov, V. I. & Joachimski, M. M. 2003. Stable oxygen and carbon isotopes – the problem of Late Carboniferous and Early Permian data. *Geophysical Research Abstracts*, **5**, 12888.

BURG, J. P. & MATTE, P. 1978. A cross section through the French Massif Central and the scope of its Variscan geodynamic evolution. *Zeitschrift der Deutschen Geologischen Gesellschaft*, **129**, 429–460.

BURG, J.-P., LEYRELOUP, A., MARCHAND, I. & MATTE, P. 1984. Inverted metamorphic zonation and large-scale thrusting in the Variscan Belt: an example in the French Massif Central. *In:* HUTTON, D. H. W. & SANDERSON, D. J. (eds) *Variscan Tectonics of the North Atlantic Region*. Geological Society, London, Special Publications, **14**, 47–61.

BUSSY, F., HERNANDEZ, J. & VON RAUMER, J. 2000. Bimodal magmatism as a consequence of the post-collisional readjustment of the thickened Variscan lithosphere (Aiguilles-Rouges and Mont-Blanc massifs, Western Alps). *Transactions of the Royal Society of Edinburgh: Earth Sciences*, **91**, 221–233.

BUTLER, R. W. H., HOLDSWORTH, R. E. & LLOYD, G. E. 1997. The role of basement reactivation in continental deformation. *Journal of the Geological Society, London*, **154**, 69–71.

BÜTTNER, S. & KRUHL, J. H. 1997. The evolution of a late-Variscan high-T/low-P region: the southeastern margin of the Bohemian Massif. *Geologische Rundschau*, **86**, 22–38.

CADISH, J., 1956. Über die Wiedererbohrung der Therme von Zurzach (Kt. Aargau). *Eclogae Geologicae Helvetiae*, **49**, 313–316.

CAMBEL, B. & KRÁL', J. 1989. Isotopic geochronology of the Western Carpathian crystalline complex: the present state. *Geologický Zborník, Geologica Carpathica*, **40**, 387–410.

CAMBEL, B., ŠČERBAK, N. P., KAMENICKÝ, L., BARTNICKIJ, J. N., & VESEL'SKÝ, J. 1977. Nekotoryje svedenija po geochronológie krystalinikuma Zapadnych Karpat na osnove dannych U-Th-Pb metoda. *Geologický Zborník, Geologica Carpathica* **28**, 243–259.

CAMBEL, B., KRÁL', J. & BURCHART, J. 1990. *Isotope Geochronology of the Western Carpathian Crystalline*. Monography, VEDA, Bratislava, 1–183.

CAPUTO, M. V. & CROWELL, J. C. 1985. Migration of glacial centers across Gondwana during Paleozoic Era. *Geological Society of America, Bulletin*, **96**, 1020–1036.

CAPUZZO, N. & BUSSY, F. 2000. High-precision dating and origin of synsedimentary volcanism in the Late Carboniferous Salvan-Dorénaz basin (Aiguilles-Rouges Massif, Western Alps). *Schweitzerische Mineralogische Petrographische Mitteilungen*, **80**, 147–167.

CAPUZZO, N. & WETZEL, A. 2004. Facies and basin architecture of the Late Carboniferous Salvan-Dorénaz continental basin (Western Alps, Switzerland/France). *Sedimentology*, **51**, 675–697.

CAPUZZO, N., HÄNDLER, R., NEUBAUER, F. & WETZEL, A. 2003. Post-collisional rapid exhumation and erosion during continental sedimentation: the example of the late Variscan Salvan-Dorénaz basin (Western Alps). *International Journal of Earth Sciences*, **92**, 364–379.

CARSWELL, D. A. 1991. Variscan high P-T metamorphism and uplift history in the Moldanubian Zone of the Bohemian Massif in Lower Austria. *European Journal of Mineralogy*, **3**, 323–342.

CARSWELL, D. A. & JAMTVEIT, B. 1990. Variscan Sm-Nd ages for the high pressure metamorphism in the Moldanubian Zone of the Bohemian Massif, Lower Austria. *Neues Jahrbuch für Mineralogie Abhandlungen*, **162**, 69–78.

CARSWELL, D. A. & O'BRIAN, P. 1992. High pressure Quartzo-Feldspathic Garnet Granulites in the Moldanubian Zone, Bohemian Massif in Lower Austria: Their P-T conditions for formation, uplift history and geotectonic significance. *In: Abstracts, International Workshop, High Pressure Granulites-Lower Crustal Metamorphism, Prague*, 4–5.

CASTELLARIN, A. & VAI, G. B. 1981. Importance of Hercynian tectonic within the framework of the Southern Alps. *Journal of Structural Geology*, **3**, 477–486.

CAZES, M., MASCLE, A., *et al.* 1986. Large Variscan overthrusts beneath the Paris Basin. *Nature*, **323**, 144–147.

CHALOUPSKY, J., CHLUPÁC, I., MASEK, J., WALDHAUSROVÁ, J. & CHÁB, J. 1995. VII Moldanubian Region, B. Teplá-Barrandian Zone, 1. Stratigraphy. *In:* DALLMEYER, R. D., FRANKE, W. & WEBER, K. (eds) *Pre-Permian Geology of Central and Eastern Europe*. Springer, Berlin, 379–391.

CHEN, F., HEGNER, E. & TODT, W. 2000. Zircon ages and Nd isotopic and chemical compositions of orthogneisses from the Black Forest, Germany: evidence for a Cambrian magmatic arc. *International Journal of Earth Sciences*, **88**, 791–802.

CHLUPÁC, I. 1993. *Geology of the Barrandian: A Field Guide*. Waldemar Kramer Verlag, Frankfurt.

CHOROWSKA, M. & WAJSPRYCH, B. 1995. Lithostratigraphy and sedimentologic-paleogeographic development; Góry Bardzkie Mts. *In:* Zdanowski, A. & ZAKOWA, H. (eds) *Prace Panstwowego Instytutu Geologicznego*, **148**, 137–140.

CLAEYS, P., CASIER, J. G. & MARGOLIS, S. V. 1992. Microtektites and mass extinctions: Evidence for a Late Devonian asteroid impact. *Science*, **257**, 1102–1104.

CLAOUE-LONG, J. C., COMPSTON, W., ROBERTS, J. & FANNING, C. M. 1995. Two Carboniferous ages: a comparison of SHRIMP zircon dating with conventional zircon ages and $^{40}Ar/^{39}Ar$ analysis. *In:* BERGGREN, W. A. *et al.* (eds) *Geochronology, Time Scales and Global Stratigraphic Correlation.* Society of Economic Paleontologists and Mineralogists, Special Publications, **54**, 1–27.

CLARK, A. H., SCOTT, D. J., SANDEMAN, H. A. BROMLEY, A. V. & FARRAR, E. 1998. Siegenian generation of the Lizard ophiolite: U-Pb zircon age data for plagiogranite, Porthkerris, Cornwall. *Journal of the Geological Society, London*, **155**, 595–598.

CLAUSEN, C.-D., KORN, D., *et al.* 1994. Die Devon-Karbon-Grenze bei Stockum (Rheinisches Schiefergebirge). *Geologie und Paläontologie in Westfalen*, **29**, 97–147.

CLAYTON, G. 1996. Mississipian miospores. *In:* JANSONIUS, J. & MCGREGOR, D.C. (eds) *Palynology, Principles and Applications.* American Association of Stratigraphic Palynologists Foundation, **2**, 589–596.

COCKS, L. R. M. & FORTEY, R. A. 1982. Faunal evidence for oceanic separation in the Palaeozoic of Britain. *Journal of the Geological Society, London*, **139**, 465–478.

COCKS, L. R. M., MCKERROW, W. S. & VAN STAAL, C. R. 1997. The margins of Avalonia. *Geological Magazine*, **134**, 627–636.

COGNÉ, J. & LEFORT, J. P. 1985. The Ligerian Orogeny: A proto-Variscan event, related to the Siluro-Devonian evolution of the Tethys I ocean. *In:* GEE, D. G. & STURT, B. A. (eds) *The Caledonide Orogen-Scandinavia and Related Areas*. John Wiley, Chichester, 1185–1195.

COLBEAUX, J.-P. & LEPLAT, J. 1985. Le Massif Paléozoïque du Bas-Boulonnais. *In:* COLBEAUX, J.-P. (ed.) *Géologie du Boulonnais.* Espace Naturel Régional, Sciences et Natures, **3**, 119–125.

COLBEAUX, J.-P., CONIL, R., COQUEL, R. & LEPLAT, J. 1985. Le Carbonifère. *In:* COLBEAUX, J.-P. (ed.) *Géologie du Boulonnais.* Espace Naturel Régional, Sciences et Natures, **3** 32–41.

COLLINS, A. S., KRYZA, R. & ZALASIEWICZ, J. 2000. Macrofabric fingerprints of Late Devonian–Early Carbonifeous subduction in the Polish Variscides, the Kaczawa complex, Sudetes. *Journal of the Geological Society, London*, **157**, 283–288.

CONIL, R., GROESSENS, E., LALOUX, M., POTY, E. & TOURNEUR, F. 1991. Carboniferous guide foraminifers, corals and conodonts in the Franco-Belgian and Campine Basins. *Courier Forschungs-Institut Senckenberg*, **130**, 15–30.

COOPER, A. H., MILLWARD, D., JOHNSON, E. W. & SOPER, N. J. 1993. The early Palaeozoic evolution of northwest England. *Geological Magazine*, **130**, 711–724.

COUREL, L. 1987. Les bassins houillers du Paléozoique supérieur continental du Massif Central français: dynamique sédimentaire. *Annales de la Société Géologique du Nord*, **106**, 229–237.

CROWLEY, T. J. & BAUM, S. K. 1991. Estimating Carboniferous sea-level fluctuations from Gondwanan ice extent. *Geology*, **19**, 975–977.

CROWLEY, T. J. & NORTH, G. R. 1991. *Palaeoclimatology.* Oxford University Press, Oxford Monographs on Geology and Geophysics, **18**.

CWOJDZINSKI, S. & GROCHOLSKI, A. 1995. Occurrence of Carboniferous rocks and paleotectonic conditions of sedimentation in the area of Poland: Sudetes and adjacent regions. *In:* ZDANOWSKI, A. & ZAKOWA, H. (eds) *Prace Panstwowego Instytutu Geologicznego*, **148**, 19–23.

CYMERMAN, Z. & PIASECKI, M. A. J. 1994. The terrane concept in the Sudetes, Bohemian Massif. *Geological Quarterly*, **38**, 191–210.

CYMERMAN, Z., PIASECKI, M. A. J. & SESTON, R. 1997. Terranes and terrane boundaries in the Sudetes, northeast Bohemian Massif. *Geological Magazine*, **134**, 717–725.

DALLMEYER, R. D. & MARTINEZ-GARCIA, E. 1990. *Pre-Mesozoic Geology of Iberia (IGCP-Project 233)*. Springer-Verlag.

DALLMEYER, R. D., FRANKE, W. & WEBER, K. (eds) 1995. *Pre-Permian*

Geology of Central and Eastern Europe. Springer-Verlag, Berlin.

Dallmeyer, R. D., Neubauer, F., Handler, R., Fritz, H., Mueller, W., Pana, D. & Putis, M. 1996. Tectonothermal evolution of the internal Alps and Carpathians: evidence from $^{40}Ar/^{39}Ar$ mineral and whole rock data. *Eclogae Geologicae Helvetiae*, **89**, 203–227.

Davydov, V. & Krainer, K. 1999. Fusulinid Assemblages and Facies of the Bombaso Fm. and basal Meledis Fm. (Moscovian - Kasimovian) in the Central Carnic Alps (Austria/Italy). *Facies*, **40**, 157–196.

Davydov, V., Wardlaw, B. R. & Gradstein, F. M. 2004. The Carboniferous Period. *In:* Gradstein, F. M., Ogg, J. G. & Smith, A. G. (eds) *A Geologic Time Scale 2004*. Cambridge University Press, 222–248.

Debon, F., Guerrot, C., Ménot, R.-P., Vivier, G. & Cocherie, A. 1998. Late Variscan granites of the Belledonne massif (French Western Alps): an Early Visean magnesian plutonism. *Schweizerische Mineralogische und Petrographische Mitteilungen* 78, 67–85.

DEKORP Research Group 1991. First results and preliminary interpretation of deep reflection seismic recordings along profile DEKORP 1. *Geophysical Journal International*, **106**, 203–227.

Delcambre, B. 1987. Application de la typologie du zircon à la téphostratigraphie du Westphalien C de la Belgique et des régions limitrophes. *Bulletin de la Société belge de Géologie*, 96: 129–136.

Delcambre, B. 1996. *Application de la typologie du zircon à la téphostratigraphie du Carbonifère de la Belgique et des régions limitrophes*. Doctoral thesis, Université Catholique de Louvain.

Delmer, A. 2004. Tectonique du front varisque en Hainaut et dans le Namurois. *Memoirs of the Geological Survey of Belgium*, **50**, 1–62.

Delmer, A., Dusar, M. & Delcambre, B. 2002. Upper Carboniferous lithostratigraphic units (Belgium). *In:* Bultynck & Dejonghe, (eds) Guide to a Revised Lithostratigraphic Scale of Belgium. *Geologica Belgica*, **4**(1–2), 95–103.

Dembowski, Z. 1972. The Cracow Sandstone Series of the Upper Silesian Coal Basin. *Prace Panstwowego Instytutu Geologicznego*, **61**, 535–537.

Denton, G. H. 2000. Does an asymmetric thermohaline-ice-sheet oscillator drive 100,000-yr glacial cycles? *Journal of Quaternary Science*, **15**, 301–318.

Deutsche Stratigraphische Kommission 2002. *Stratigraphische Tabelle von Deutschland 2002*. GeoForschungsZentrum, Potsdam; Forschungsinstitut Senckenberg, Frankfurt am Main..

Dessureau, G., Piper, D.J.W. & Pe-Piper, G. 2000. Geochemical evolution of earliest Carboniferous continental tholeiitic basalts along a crustal-scale shear zone, southwestern Maritimes basin, eastern Canada. *Lithos*, **50**, 27–50.

Dewey, J. F., Helman, M. L., Turco, E., Hutton, D. H. W. & Knott, S. D. 1989. Kinematics of the western Mediterranean. *In:* Coward, M. P., Dietrich, D. & Park, R. G. (eds) *Alpine Tectonics*. Geological Society, London, Special Publications, **45**, 265–283.

DiMichele, W. A., Hook, R. W., *et al.* 1992. Paleozoic terrestrial ecosystems. *In:* Behrensmeyer, A. K., Damuth, J. D., DiMichele, W. A., Potts, R., Sues, H. D. & Wing, S. L. (eds) *Terrestrial Ecosystems Through Time. Evolutionary Paleoecology of Terrestrial Plants and Animals*. The University of Chicago Press, 205–325.

Dittmar, U. & Oncken, O. 1992. Anatomie und Kinematik eines passiven varistischen Kontinentalrandes - zum Strukturbau des südwestlichen Rheinischen Schiefergebirges. *Frankfurter Geowissenschafliche Arbeiten*, **A12**, 34–37.

Dombrowski, A., Henjes-Kunst, F., Höhndorf, A., Kröner, A., Okrusch, M. & Richter, P. 1995. Orthogneisses in the Spessart crystalline complex, north-west Bavaria, Silurian magmatism at an active continental margin. *Geologische Rundschau*, **83**, 399–411.

Don, J. 1990. The differences in Palaeozoic facies-structural evolution of the West Sudetes. *Neues Jahrbuch für Geologie und Paläontologie Abhandlungen*, **178**, 307–328.

Don, J., Dumicz, M., Wojciechoska, I. & Zelazniewicz, A. 1990. Lithologiy and tectonics of the Orlica-Snieznik Dome, Sudetes - recent state of knowledge. *Neues jahrbuch für Geologie und Paläontologie Abhandlungen*, **179**, 139–188.

Dörr, W. 1990. Stratigraphie, Stoffbestand und Fazies der Gießener Grauwacke (östliches Rheinisches Schiefergebirge). *Geologische Abhandlungen Hessen*, Wiesbaden, **91.**

Downes, H., Shaw, A., Williamson, B. J. & Thirlwall, M. F. 1997. Sr, Nd and Pb isotopic evidence for the lower crustal origin of Hercynian granodiorites and monzogranites, Massif Central, France. *Chemical Geology*, **136**, 99–122.

Dreesen, R., Bossiroy, D., Dusar, M., Flores, R. M. & Verkaeren, P. 1995. Overview of the influence of syn-sedimentary tectonics and palaeo-fluvial systems on coal seam and sand body characteristics in the Westphalian C strata, Campine Basin, Belgium. *In:* Whateley, M. K. G. (ed.) *European Coal Geology*. Geological Society, London, Special Publications, **82**, 215–232.

Drozdzewski, G. 1993. The Ruhr coal basin (Germany): structural evolution of an autochthonous foreland basin. *International Journal of Coal Geology*, **23**, 231–250.

Drozdzewski, G. & Wrede, V. 1994. Faltung und Bruchtektonik - Analyse im Subvariscikum. *Fortschritte in der Geologie von Rheinland und Westfalen*, **38**, 7–187.

Drozdzewski, G. & Wrede, V. 1997. Die Variszische Faltungsfront in Nordostdeutschland - Hypothese und Fakten. *Zeitschrift der angewandten Geologie*, **43**, 104–111.

Duguet, M. & Faure, M. 2004. Granitoid emplacement during a thrusting event: structural analysis, microstructure and quartz c-axis patterns. An example from Hercynian plutons in the French Massif Central. *Journal of Structural Geology*, **26**, 927–945.

Duthou, J.-L., Couturie, J.-P., Mierzejewski, M. P. & Pin, C. 1991. Age determination of the Karkonosze granite using isochronone Rb-Sr whole rock method. *Przeglad Geologiczny*, **39**, 75–79 [Polish with English summary].

Dusar, M. 1989. The Westphalian C in the Campine Basin: coal content influenced by tectonics. *Annales de la Société géologique de Belgique*, **112**, 249–250.

Duthou, J. L., Piboule, M., Gay, M. & Dufour, E. 1981. Datations radio-métriques Rb-Sr sur les orthogranulites des Monts du Lyonnais (Massif Central francais). *Comptes Rendus de l'Academie des Sciences Paris* II, **292**, 749–752.

Duthou, J. L., Cantagrel, J. M., Didier, J. & Vialette, Y. 1984. Paleozoic granitoids from the French Massif Central: age and origin studied by 87Rb/87Sr system. *Physics of the Earth and Planetary Interiors*, **35**, 131–144.

Dvorák, J. 1992. Mutual relation of pre-flysch and flysch facies in the Moravian Palaeozoic. *Vestnik Ceskeho Geologickeho Ustavu*, **67**, 363–369.

Dvorák, J. 1995. Moravo-Silesian Zone: autochthon: stratigraphy. *In:* Dallmeyer, D., Franke, W. & Weber, K. (eds) *Pre-Permian Geology of Central and Western Europe*. Springer, Berlin, 475–489.

Dvorák, J. & Paproth, E. 1969. Über die Position und die Tektogenese des Rhenohercynikums und des Sudetikums in den Mitteleuropäischen Varisziden. *Neues Jahrbuch für Geologie und Paläontologie, Monatshefte*, **2**, 112–120.

Dziedzic, K. & Teisseyre, A. K. 1990. The Hercynian molasse and younger deposits in the Intra-Sudetic Depression, SW Poland. *Neues Jahrbuch für Geologie und Paläontologie, Abhandlungen*, **179**, 285–305.

Ebner, F. 1976. Die Schichtfolge an der Wende Unterkarbon/Oberkarbon in der Rannachfazies des Grazer Paläozoikums. *Verhandlungen der geologischen Bundesanstalt*, **1976**, 65–93.

Ebner, F. 1977. Die Gliederung des Karbons von Graz mit Conodonten. *Jahrbuch der geologischen Bundesanstalt*, **120**, 449–493.

Ebner, F. 1978. Stratigraphie des Karbons der Rannachfazies im Paläozoikum von Graz, Österreich. *Mitteilungen der Österreichischen Geologischen Gesellschaft*, **69**, 163–196.

Ebner, F., Fenninger, A. & Ratschbacher, L. 1987. Stratigraphy; tectonics, and paleogeography of low-grade Carboniferous strata in Austria (geotraverse B). *In:* Sassi, F.P. & Bourrouilh, R. (eds) *Correlation of Prevariscan and Variscan Events of the Alpine-Mediterranean Mountain Belt*. IGCP Project No. 5, Newsletter, **7**, 139–142.

Ebner, F., Vozárová, A., Straka, P. & Vozár, J. 1990. Carboniferous conodonts from Brusník Anticline (Southern Slovakia). *In:* Minaříková, D. & Lobitzer, H. (eds) *Thirty Years of Geological Cooperation between Austria and Czechoslovakia*. ÚÚG - GBA, Prague-Vienna, 249–252.

Ebner, F., Kovács, S. & Schönlaub, H. P. 1991. Das klassische Karbon in Österreich und Ungarn – ein Vergleich der sedimentären fossilführenden Vorkommen. *Jubiläumsschrift 20 Jahre Geologische Zusammenarbeit Österreich-Ungarn*, **1**, 263–294.

Echtler, H. P. & Chauvet, A. 1992. Carboniferous convergence and subsequent crustal extension in the southern Schwarzwald (SW Germany). *Geodinamica Acta*, **5**, 37–49.

Eder, F. W., Engel, W., Franke, W. & Sadler, P. M. 1983. Devonian and Carboniferous limestone turbidites of the Rheinisches Schiefergebirge and their tectonic significance. *In:* Martin, H. & Eder, F. W. (eds) *Intracontinental Fold Belts – Case studies in the Variscan Belt of Europe and the Demarara Belt of Namibia*. Springer Verlag, Berlin, 93–124.

Eichhorn, R., Höll, R., Loth, G. & Kennedy, A. 1999. Implications of U-Pb SHRIMP zircon data on the age and evolution of the Felbertal tungsten deposit (Tauern Window, Austria). *International Journal of Earth Sciences*, **88**, 496–512.

Eichhorn, R., Loth, G., Höll, R., Finger, F., Schermaier, A. & Kennedy, A. 2000. Multistage Variscan magmatism in the central Tauern Window (Austria) unveiled by U/Pb SHRIMP zircon data. *Contributions to Mineralogy and Petrology*, **139**, 418–435.

Eisbacher, G. H., Lüschen, E. & Wickert, F. 1989. Crustal-scale thrusting and extension in the Hercynian Schwarzwald and Vosges, Central Europe. *Tectonics*, **8**, 1–21.

Eisele, J., Gertisser, R. & Montenari, M. 2000. Geochemistry and provenance of Devono-Carboniferous volcano-sedimentary sequences from the Southern Vosges Basin and the geodynamic implications for the western Moldanubian Zone. *In:* Franke, W., Haak, V, Oncken, O. & Tanner, D. (eds) *Orogenic Processes: Quantification and Modelling in the Variscan Belt.* Geological Society, London, Special Publications, **179**, 433–444.

Eisele, J., Gertisser, R. & Montenari, M. 2002. Geochemistry and provenance of Devono-Carboniferous volcano-sedimentary sedimentary sequences from the Southern Vosges Basin and the geodynamic implications for the western Moldanubian Zone. *In:* Franke, W., Haak, V., Oncken, O. & Tanner, D. (eds) *Orogenic Processes: Quantification and Modelling in the Variscan Belt.* Geological Society, London, Special Publications, **179**, 433–444.

Ellenberger, F. & Tamain, G. 1980. Hercynian Europe. *Episodes*, **1980**(1), 22–27.

Engel, H. 1985. Zur Tektogenese des Saarbrücker Hauptsatttels und der Südlichen Randüberschiebung. *In:* Drozdzewski, G., Engel, H., Wolf, R. & Wrede, V. (eds) *Beiträge zur Tiefentektonik westdeutscher Steinkohlenlagerstätten.* Geologisches Landesamt Nordrhein-Westfalen, Krefeld, Fortschritte in der Geologie von Nordrheinland und Westfalen, 217–235.

Engel, W. & Franke, W. 1983. Flysch sedimentation: its relations to tectonism in the European Variscides. *In:* Martin, H. & Eder, F.W. (eds) *Intracontinental Fold Belts – Case Studies in the Variscan Belt of Europe and the Demarara Belt of Namibia.* Springer Verlag, Berlin, 289–321.

Engel, W., Feist, R. & Franke, W. 1981. Le Carbonifère antéstéphanien de la Montagne noire: rapports entre mise en place des nappes et sédimentation. *Buletin BRGM*, **4**, 341–389.

Engel, W., Franke, W. & Langenstrassen, F. 1983*a*. Paleozoic sedimentation in the northern branch of the Mid-European Variscides: Essay of an interpretation. *In:* Martin, H. & Eder, F. W. (eds) *Intracontinental Fold Belts – Case studies in the Variscan Belt of Europe and the Demarara Belt of Namibia*. Springer Verlag, Berlin, 9–41.

Engel, W., Flehmig, W. & Franke, W. 1983*b*. The mineral composition of Rhenohercynian flysch sediments and its tectonic significance. *In:* Martin, H. & Eder, F. W. (eds) *Intracontinental Fold Belts – Case Studies in the Variscan Belt of Europe and the Demarara Belt of Namibia*. Springer Verlag, Berlin 171–184.

England, P. C. & Thompson, A. B. 1984. Pressure-temperature-time paths of regional metamorphism of the continental crust. I. Heat transfer during the evolution of the regions of thickened continental crust. *Journal of Petrology*, **25**, 894–928.

Falk, F., Franke, W. & Kurze, M. 1995. V. B. 1 Stratigraphy. *In:* Dallmeyer, R. D., Franke, W. & Weber, K. (eds) *Pre-Permian of Central and Eastern Europe*. Springer, Berlin, 221–234.

Farey, J. 1811. A General View of the Agriculture and Minerals of Derbyshire. Vols I-III. McMillan, London.

Faryad, S. W. 1995: Geothermometry of metamorphic rocks from the Zemplinicum (Western Carpathians, Slovakia). *Geologica Carpathica*, **46**, 113–123.

Faure, M. 1995. Late orogenic Carbonifeous extensions in the Variscan French Massif Central. *Tectonics*, **14**, 132–153.

Faure, M., Ledru, P., Lardeaux, J. M. & Matte, P. 2004. Paleozoic orogenies in the French Massif Central. A cross section from Béziers to Lyon. *Pre-congress Excursion B22, 24th International Geological Congress,* Florence, 15–19 August.

Feist, R., Echtler, H, Galtier, J. & Mouthier, B. 1994. Biostratigraphy and dynamics of the nonmetamorphic sedimentary record. *In:* Keppie, J. D. (ed.) *Pre-Mesozoic Geology in France and Related Areas*. Springer-Verlag, Berlin, 289–304.

Felser, K. O. 1977. Die stratigraphische Stellung der Magnesitvorkommen in der östlichen Grauwackenzone (Steiermark, Österreich). *Berg- und Hüttenmännische Monatshefte*, **122**, 17–23.

Fernández-Suárez, J., Gutiérrez Alonso, G., Cox, R. & Jenner, G. A. 2002. Assembly of the Armorica microplate: A strike-slip terrane delivery? Evidence from U-Pb ages of detrital zircons. *Journal of Geology*, **110**, 619–626.

Fiala, J., Fuchs, G. & Wendt, J. I. 1995. Stratigraphy (Moldanubian Zone). *In:* Dallmeyer, R. D., Franke, W. & Weber, K. (eds) *Pre-Permian of Central and Eastern Europe.* Springer, Berlin, 417–428.

Fiege, K. & Van Lekwijck, W. 1969. Cyclicité dans le Namurien du synclinal de Namur (Belgique). *Service géologique de Belgique Professional Paper*, **7**, 91p.

Finger, F. & Clemens, J.D. 1995. Migmatization and "secondary" granitic magmas: effects of emplacement and crystallization of "primary" granitoids in Southern Bohemia, Austria. *Contributions to Mineralogy and Petrology*, **120**, 311–326.

Finger, F. & Steyrer, H. P. 1995. A tectonic model for the eastern Variscides: indications from a chemical study of amphibolites in the south-eastern Bohemian Massif. *Geologica Carpathica* **46**, 137–150.

Finger, F., Roberts, M. P., Haunschmid, B., Schermaier, A. & Steyrer, H. P. 1997. Variscan granitoids of central Europe: their typology, potential sources and tectonothermal relations. *Mineralogy and Petrology*, **61**, 67–96.

Finger, F., Hanzl, P., Pin, C., Von Quadt, A. & Steyrer, H. P. 2002. The Brunovistulian: Avalonian Precambrian sequence at the eastern end of the Central European Variscides? *In:* Franke, W., Haak, V., Oncken, O. & Tanner, D. (eds) *Orogenic Processes: Quantification and Modelling in the Variscan Belt.* Geological Society, London, Special Publications, **179**, 103–112.

Finger, F., Frasl, G., Haunschmid, B., Lettner, H., von Quadt, A., Schermaier, A., Schindlmayr, A. O. & Steyrer, H. P. 1993. The Zentralgneise of the Tauern Window (Eastern Alps): insight into an Intra-Alpine Variscan batholith. *In:* von Raumer, J. F. & Neubauer, F. (eds) *Pre-Mesozoic Geology in the Alps*. Springer, Berlin, 375–391.

Fisch, W. & Ryf, W. 1966. Der Verrucano in den Glarner Alpen, Atti del Symposium sul Verrucano. Societá Toscana di Scienze Naturali, Pisa, pp. 233–244.

Flajs, G. & Feist, R. 1988. Index conodonts, trilobites and environment of the Devonian-Carboniferous Boundary beds at La Serre (Montagne Noire, France). *Courier Forschungs-Institut Senckenberg*, **100**, 53–107.

Flick, H. & Nesbor, H. D. 1988. Der Vulkanismus in der Lahnmulde. *Jahresbericht und Mitteilungen des Oberrheinischen Geologischen Vereines*, **70**, 411–475.

Flöttmann, T. & Oncken, O. 1992. Constraints on the evolution of the Mid German Crystalline Rise - a study of outcrops west of the river Rhine. *Geologisches Rundschau*, **81**, 515–543.

Floyd, P. A. 1984. Geochemical characteristics and comparison of the basic rocks of the Lizard Complex and the basaltic lavas within the Hercynian troughs of SW England. *Journal of the Geological Society, London*, **141**, 61–70.

Floyd, P. A. 1995. Igneous activity. *In:* Dallmeyer, R. D., Franke, F. & Weber, K. (eds) *Pre-Permian Geology of Central and Eastern Europe*. Springer Verlag, Berlin, 59–81.

Floyd, P. A. & Leveridge, B. E. 1987. Tectonic environment of the Devonian Gramscatho basin, South Cornwall: framework mode and geochemical evidence from turbiditic sandstones. *Journal of the Geological Society, London,* **144**, 531–542.

Floyd, P. A., Leveridge, B. E., Franke, W., Shail, R. & Dorr, W. 1990. Provenance and depositional environment of Rhenohercynian synorogenic greywackes from the Giessen Nappe, Germany. *Geologische Rundschau*, **79**, 611–626.

Floyd, P. A., Shail, R., Leveridge, B. E. & Franke, W. 1991.

Geochemistry and provenance of Rhenohercynian synorogenic sandstones: implications for tectonic environment discrimination. *In:* MORTON, A. C., TODD, S. P. & HAUGHTON, P. D. W. (eds) *Developments in Sedimentary Provenance Studies*. Geological Society, London, Special Publications, **57**, 173–188.

FLOYD, P. A., WINCHESTER, J. A., CIESIELCZUK, J., LEWANDOWSKA, A., SZCZEPANSKI, J. & TURNIAK, K. 1996. Geochemistry of early Palaeozoic amphibolites from the Orlica-Snieznik dome, Bohemian massif: petrogenesis and palaeotectonic aspects. *Geologische Rundschau*, **85**, 225–238.

FLOYD, P. A., WINCHESTER, J., SESTON, R., KRYZA, R. & CROWLEY, Q. 2000. Review of geochemical variation in Lower Palaeozoic metabasites from the NE Bohemian Massif: intracratonic rifting and plume-ridge interactions. *In:* FRANKE, W., HAAK, V., ONCKEN, O. & TANNER, D. (eds) *Orogenic processes: Quantification and modelling in the Variscan Belt*. Geological Society, London, Special Publications, **179**, 155–174.

FLUCK, P., MAASS, R. & VON RAUMER, J. F. 1980. The Variscan units east and west of the Rhine Graben. *In:* COGNÉ, J. & SLANSKY, M. (eds) *Géologie de l'Europe du Précambrien aux bassins sédimentaires post-hercyniens*. Mémoires du Bureau de Recherches Géologiques et Miniéres, **108**, 112–131.

FLÜGEL, E. 1974. Fazies-Interpretation der unterpermischen Sedimente in den Karnischen Alpen. *Carinthia II*, **164**(84), 43–62.

FLÜGEL, E. 1977. Environmental models for Upper Paleozoic benthic calcareous algal communities. *In:* FLÜGEL, E. (ed.) *Fossil Algae*. Springer, Berlin, 314–343.

FLÜGEL, E. & KRAINER, K. 1992. Allogenic and autogenic controls of reef mound formation: Late Carboniferous auloporid coral buildups from the Carnic Alps, Italy. *Neues Jahrbuch für Geologie und Paläontologie, Monatshefte*, **185**, 39–62

FLÜGEL, E., FOHRER, B., FORKE, H., KRAINER, K. & SAMANKASSOU, E. 1997. Exkursion B 1, Cyclic sediments and algal mounds in the Upper Paleozoic of the Carnic Alps. 18th European Regional Meeting of Sedimentology, Heidelberg, 2–4 September. *Gaea heidelbergensis*, **4**, 79–100.

FORKE, H. C. 2002. Biostratigraphic Subdivision and Correlation of Uppermost Carboniferous/Lower Permian Sediments in the Southern Alps: Fusulinoidean and Conodont Faunas from the Carnic Alps (Austria/Italy), Karavanke Mountains (Slovenia), and Southern Urals (Russia). *Facies*, **47**, 201–276.

FORKE, H. C. & SAMANKASSOU, E. 2000. Biostratigraphical Correlation of Late Carboniferous (Kasimovian) Sections in the Carnic Alps (Austria/Italy): Integrated Paleontological Data, Facies, and Discussion. *Facies*, **42**, 177–210.

FORKE, H. C., KAHLER, F. & KRAINER, K. 1998. Sedimentology, microfacies and stratigraphic distribution of foraminifers of the Lower "*Pseudoschwagerina*" Limestone (Rattendorf Group, Late Carboniferous), Carnic Alps (Austria/Italy). *Senckenbergiana lethaea*, **78**, 1–39.

FÖRSTER, H.-J., TISCHENDORF, G., TRUMBULL, R. B. & GOTTESMANN, B. 1999. Late-collisional granites in the Variscan Erzgebirge, Germany. *Journal of Petrology*, **40**, 1613–1645.

FRAAYE, R. H. B. & WERVER, O. P. 1990. Trace fossils and their environmental significance in Dinantian carbonates of Belgium. *Palaeontologische Zeitschrift*, **64**, 367–377.

FRAKES, L. A., FRANCIS, J. E. & SYKTUS, J. I. 1992. *Climate Modes of the Phanerozoic: The History of the Earth's Climate over the Past 600 Million Years*. Cambridge University Press, Cambridge.

FRANKE, D. 1990. Der präpermische Untergrund der Mitteleuropäischen Senke - Fakten und Hypothesen. *In:* Geologie und Kohlenwasserstofferkundung in Präzechstein der DDR - Nordostdeutsche Senke. *Niedersächsische Akademie der Geowissenschaften, Veröffentlichungen*, Hannover, **4**, 19–75.

FRANKE, W. 1984*a*. Late events in the tectonic history of the Saxothuringian zone. *In:* HUTTON, D. H. W. & SANDERSON, D. J. (eds) *Variscan Tectonics of the North Atlantic Region*. Geological Society of London, Special Publications, **14**, 33–45.

FRANKE, W. 1984*b*. Variszischer Deckenbau im Raume der Münchberger Gneismasse, abgeleitet aus der Fazies, Deformation und Metamorphose im ungebenden Paläozoikum. *Geotektonische Forschungen*, **68**, 1–253.

FRANKE, W. 1989. Tectonostratigraphic units in the Variscan belt of Central Europe. *In:* DALLMEYER, R. D. (ed.) *Terranes in the Circum-Atlantic Paleozoic Orogens*. Geological Society of America, Special Papers, **230**, 67–90.

FRANKE, W. 1992. Phanerozoic structures and events in Central Europe. *In:* BLUNDELL, D., FREEMANN, R. & MUELLER, ST. (eds) *A Continent Revealed: the European Geotraverse*. Cambridge University Press, Cambridge, 164–180.

FRANKE, W. 1995*a*. Introduction (Rhenohercynian Foldbelt). *In:* DALLMEYER, R. D., FRANKE, W. & WEBER, K. (eds) *Pre-Permian Geology of Eastern and Central Europe*. Springer-Verlag, Berlin, 29–30.

FRANKE, W. 1995*b*. Stratigraphy (Autochthon and Nonmetamorphic Nappe Units). *In:* DALLMEYER, R. D., FRANKE, W. & WEBER, K. (eds) *Pre-Permian Geology of Eastern and Central Europe*. Springer-Verlag, Berlin, 33–49.

FRANKE, W. 1995*c*. Introduction (Saxothuringian Basin). *In:* DALLMEYER, R. D., FRANKE, W. & WEBER, K. (eds) *Pre-Permian Geology of Eastern and Central Europe*. Springer-Verlag, Berlin, 217–218.

FRANKE, W. 2000. The mid-European segment of the Variscides: Tectono-stratigraphic units, terrane boundaries and plate tectonic evolution. *In:* FRANKE, W., HAAK, V, ONCKEN, O. & TANNER, D. (eds) *Orogenic Processes: Quantification and Modelling in the Variscan Belt*. Geological Society, London, Special Publications, **179**, 35–62.

FRANKE, W. & ENGEL, W. 1986. Synorogenic sedimentation in the Variscan Belt of Europe. *Bulletin de la Société Géologique de France*, **1**, 25–33.

FRANKE, W. & ENGEL, W. 1988. Tectonic setting of synorogenic sedimentation in the Variscan Belt of Europe. *In:* BESLY, B. M. & KELLING, G. (eds) *Sedimentation in a Synorogenic Basin Complex. The Upper Carboniferous of North-West Europe*. Blackie, London, 8–17.

FRANKE, W. & ONCKEN, O. 1990. Geodynamic evolution of the North-Central Variscides: a comic strip. *In:* FREEMANN, R., GIESE, P. & MUELLER, ST. (eds) *The European Geotraverse: Integrative Studies*. European Science Foundation, Strasbourg, 187–194.

FRANKE, W. & ONCKEN, O. 1995. Zur prädevonischen Geschichte des Rhenohercynischen Beckens. *Nova Acta Leopoldina, Neue Folge*, **71**, 53–72.

FRANKE, W. & STEIN, E. 2000. Exhumation of high-pressure rocks in the Saxo-Thuringian belt: geological constraints and alternative models. *In:* FRANKE, W., HAAK, V, ONCKEN, O. & TANNER, D. (eds) *Orogenic Processes: Quantification and Modelling in the Variscan Belt*. Geological Society, London, Special Publications, **179**, 337–354.

FRANKE, W. & ŻELAŹNIEWICZ, A. 2000. The eastern termination of the Variscides: terrane correlation and kinematic evolution. *In:* FRANKE, W., HAAK, V., ONCKEN, O. & TANNER, D. (eds) *Orogenic Processes: Quantification and Modelling in the Variscan Belt*. Geological Society, London, Special Publications, **179**, 63–86.

FRANKE, W., EDER, W., ENGEL, W. & LANGENSTRASSEN, R. 1978. Main aspects of geosynclinal sedimentation in the Rhenohercynian zone. *Zeitschrift der Deutschen Geologischen Gesellschaft*, **129**, 201–216.

FRANKE, W., DALLMEYER, R. D. & WEBER, K. 1995*a*. Geodynamic evolution. *In:* DALLMEYER, R. D., FRANKE, W. & WEBER, K. (eds) *Pre-Permian Geology of Central and Eastern Europe*. Springer Verlag, Berlin, 579–593.

FRANKE, W., KREUZER, H., OKRUSCH, M., SCHÜSSLER, U. & SEIDEL, E. 1995*b*. Saxothuringian Basin: Exotic metamorphic nappes: stratigraphy, structure, and igneous activity. *In:* DALLMEYER, D., FRANKE, W. & WEBER, K. (eds) *Pre-Permian Geology of Central and Western Europe*. Springer, Berlin, 277–294.

FRANKE, W., HAAK, V., ONCKEN, O. & TANNER, D. (eds) 2000. *Orogenic Processes: Quantification and Modelling in the Variscan Belt*. Geological Society, London, Special Publications, **179.**

FRANKS, G. D. 1966. The development of the limnic Upper Carboniferous of the eastern Aar Massif. *Eclogae Geologicae Helvetiae*, **59**, 943–950.

FRANKS, G. D. 1968*a*. A study of Upper Paleozoic sediments and volcanics in the northern part of the eastern Aar Massif. *Eclogae Geologicae Helvetiae*, **61**, 49–140.

FRANKS, G. D. 1968*b*. The pre-Westphalian (Hercynian) metamorphism and structures of the Tödi area (Aar Massif). *Schweizerische Mineralogische und Petrographische Mitteilungen*, **48**, 667–694.

FRASL, G. & FINGER, F. 1988. The "Cetic Massif" below the Eastern Alps – characterized by its granitoids. *Schweizerische Mineralogische und Petrographische Mitteilungen*, **68**, 433–439.

FRANZ, G., MENGE, R. & MOSBRUGGER, V. 1991. Callipteridium sp. a plant fossil from the amphibolite facies basement/Tauern, Austria. *Terra Nova*, **3**, 137–141.

FRISCH, W. & NEUBAUER, F. 1989. Pre-Alpine terranes and tectonic zoning in the Eastern Alps. *In:* DALLMEYER, R. D. (ed.) *Terranes in the Circum-Atlantic Palaeozic Orogens*. Geological Society of America, Special Papers, **230**, 91–100.

FRITZ, A. & KRAINER, K. 1993. Eine neue Megaflora aus dem Stefan der Kronalpe. *Carinthia II*, **183**(103), 485–517.

FRITZ, A. & KRAINER, K. 1994. Die Megaflora "Garnitzenberg-Südost alpha" aus der Auernig-Schichtgruppe der Karnischen Alpen. *Carinthia II*, **184**(104), 441–465.

FRITZ, A. & KRAINER, K. 1995. Die Megafloren Tomritsch-3 und Tomritsch-6 aus den oberkarbonen Auernigschicthen der Karnischen Alpen. *Carinthia II*, **185**(105), 553–583.

FRITZ, A. & KRAINER, K. 1997. Eine oberkarbone Megaflora von der Nordseite des Stangnocks, Gurktaler Alpen. *Carinthia II*, **187**(107), 325–356.

FRITZ, A., BOERSMA, M. & KRAINER, K. 1990. Steinkohlenzeitliche Pflanzenfossilien aus Kärnten. *Carinthia II, Sonderheft*, **49**, 1–189.

FRITZ, H. & NEUBAUER, F. 1995. Moravo-Silesian Zone - Structure. *In:* DALLMEYER, R. D., FRANKE, W. & WEBER, K. (eds) *Pre-Permian Geology of Central and Eastern Europe*. Springer-Verlag, Berlin, 490–494.

FURNES, H., KRYZA, R., MUSZYNSKI, A., PIN, C. & GARMANN, L. B. 1994. Geochemical evidence for progressive, rift-related early Palaeozoic volcanism in teh western Sudetes. *Journal of the Geological Society, London*, **151**, 91–109.

GAITZSCH, B., RÖSSLER, R., SCHNEIDER, J. W. & Schretzenmayr, S. 1999. Neue Ergebnisse zur Verbreitung potentieller Muttergesteine im Karbon von Nord- und Mitteldeutschland. *Geologisches Jahrbuch*, **A 149**, 25–58.

GANSSLOSER, M., VIBRANS, E. & WACHSENDORF, H. 1995. Die Metabasalte des Harzes. *Zentralblatt für Geologie und Paläontologie*, **Teil 1**, 1103–1115.

GAYK, T. & KLEINSCHRODT, R. 2000. Hot contacts of garnet peridotites in middle/upper crustal levels: nw constraints on the nature of the late Variscan high-T/low-P event in the Moldanubian (central Vosges/NE France). *Journal of Metamorphic Geology*, **18**, 293–305.

GIESE, U., KATZUNG, G., WALTER, R. & WEBER, J. 1997. The Caledonian deformation of the Brabant Massif and the Early Palaeozoic in northeast Germany: compared. *Geological Magazine*, **134**, 637–652.

GOLANKA, J. & FORD, D. 2000. Pangaen (Late Carboniferous-Middle Jurassic) paleoenvironments and lithofacies. *Palaeogeography, Palaeogeography, Palaeoclimatology*, **161**, 1–34.

GONZALEZ, C. R. 1990. Development of late Palaeozoic glaciations of the South American Gondwana in western Argentina. *Palaeogeography, Palaeoclimatology, Palaeoecology*, **79**, 275–287.

GRADSTEIN, F. M., OGG, J. G. & SMITH, A. G. 2004. *A Geologic Time Scale 2004*. Cambridge University Press.

GRADZIŃSKI, R., DOKTOR, M. & SŁOMKA, T. 1995. Depositional environments of the coal-bearing Cracow Sandstone Series (upper Westphalian), Upper Silesia, Poland. *Studia Geologica Polonica*, **108**, 149–170.

GRECULA, P. & EGYÜD, K. 1982. Lithostratigraphy of Upper Paleozoic and Lower Triassic strata from the Zemplinske vrchy Mts. (SE Slovakia). *Mineralia Slovaca*, **14**, 3, 221–239 (in Slovak).

GROESSENS, E., CONIL, R. & HENNEBERT, M. 1982. *Le Dinantien du sondage de Saint-Ghislain. Stratigraphie et Paléontologie*. Mémoires explicatifs des cartes géologiques et minéralogiques de Belgique, Service Géologique de Belgique, **22**.

GROSSMAN, E. L. 1994. The carbon and oxygen isotope record during the evolution of Pangea: Carboniferous to Triassic. *In:* KLEIN, G. D. (ed) *Pangea: Paleoclimate, Tectonics, and Sedimentation During Accretion, Zenith, and Breakup of a Supercontinent*. Geological Society of America, Special Papers, **288**, 207–228.

GROSSMAN, E. L., BRUCKSCHEN, P., MII, H.-S., CHUVASHOV, B. I., YANCEY, T. E. & VEIZER, J. 2002. Carboniferous paleoclimate and global change: Isotopic evidence from the Russian Platform. *In: Carboniferous Stratigraphy and Paleogeography in Eurasia*. Institute of Geology and Geochemistry, Russian Academy of Sciences, Urals Branch, Ekaterinburg, 61–71.

GUILDERSON, T. P., FAIRBANKS, R. G. & RUBENSTONE, J. L. 1994. Tropical temperature variations since 20,000 years ago: modulating interhemispheric climate change. *Science*, **263**, 663–665.

GUION, P. D., FULTON, I. M. & JONES, N. S. 1995. Sedimentary facies of the coal-bearing Westphalian A and B north of the Wales-Brabant High. *In:* WHATELEY, M. K. G. & SPEARS, D. A. (eds) *European Coal Geology*. Geological Society, London, Special Publications, **82**, 45–78.

GUION, P. D., GUTTERIDGE, P. & DAVIES, S. J. 2002. Carboniferous sedimentation and volcanism on the Laurussian margin. *In:* WOODCOCK, N. & STRACHAN, R. (eds) *Geological History of Britain and Ireland*. Blackwell, Oxford, 227–270.

GÜLDENPFENNIG, M. 1997. Geologische Neuaufnahme der Zone von Badenweiler-Lenzkirch (Südschwarzwald) unter besonderer Berücksichtigung unterkarbonischer Vulkanite und Grauwacken. *Tübinger Geowissenschaftliche Arbeiten*, **A32**, 1–120.

GÜLDENPFENNIG, M. 1998. Zur tektonischen Stellung unterkarbonischer Grauwacken und Vulkanite der Zone von Badenweiler-Lenzkirch (Südschwarzwald). *Zeitschrift der deutschen geologischen Gesellschaft*, **149**, 213–232.

GÜLDENPFENNIG, M. & LOESCHKE, J. 1991. Petrographie und Geochemie der Zone von Badenweiler-Lenzkirch in der Umgebung von Präg (Südschwarzwald). *Jahreshefte des Geologischen Landesamts Baden-Württemberg*, **33**, 5–32.

HAHN, G. & HAHN, R. 1977. Trilobiten aus dem Unter-Karbon der Veitsch (Steiermark, Österreich). *Neues Jahrbuch für Geologie und Paläontologie, Monatshefte*, **3**, 137–143

HAHN, G. & HAHN, R. 1987. Trilobiten aus dem Karbon von Nötsch und aus den Karnischen Alpen Österreichs. *Jahrbuch der Geologischen Bundesanstalt*, **129**, 567–619

HAMBREY, M. J. & HARLAND, W. B. (eds) 1981. *Earth's Pre-Pleistocene Glacial Record*. Cambridge University Press, Cambridge.

HANCE, L., DEJONGHE, L., GHYSEL, P., LALOUX, M. & MANSY, J. L. 1999. Influence of heterogeneous lithostructural layering on orogenic deformation in the Variscan Front Zone (eastern Belgium). *Tectonophysics*, **309**, 161–177.

HANCE, L., POTY, E. & DEVUYST, F.-X. 2001. Stratigraphie séquentielle du Dinantien type (Belgique) et corrélation avec le Nord de la France (Boulonnais, Avesnois). *Bulletin de la Société Géologique de France*, **172**, 411–426.

HANCE, L., DEVUYST, F.-X. & POTY, E. 2002. Sequence stratigraphy of the Belgian Lower Carboniferous – Tentative correlation with the British Isles. *In:* HILLS, L. V., HENDERSON, C. M. & . BAMBER, E. W. (eds) *Carboniferous and Permian of the World*. Canadian Society of Petroleum Geologists, Memoirs, **19**, 41–51.

HANN, H. P., CHEN, F., ZEDLER, H., FRISCH, W. & LOESCHKE, J. 2003. The Rand Granite in the southern Schwarzwlad and its geodynamic significance in the Variscan belt of SW Germany. *International Journal of Earth Sciences (Geologische Rundschau)*, **92**, 821–842.

HANEL, M., KALT, A., MONTENARI, M. & WIMMENAUER, W. 1998. Bestimmung von Sedimentationsaltern hochmetamorpher Paragneise des Schwarzwaldes. *Terra Nostra*, **98/2**.

HANEL, M., MONTENARI, M. & KALT, A. 1999. Determining sedimentation ages of high-grade metamorphic gneisses by their palynological record: a case study in the northern Schwarzwald (Variscan Belt, Germany). *International Journal of Earth Sciences*, **88**, 49–59.

HANN, H. P., SAWATZKI, G. & VAIDA, M. 1995. Acritarchen und Chitinozoen des Ordoviziums aus metamorphen Grauwacken der Zone von Badenweiler-Lenkirch, Schwarzwald, SW-Deutschland. *Neues Jahrbuch für Geologie und Paläontologie, Monatshefte*, **1995**, 375–383.

HANN, H. P., CHEN, F., ZEDLER, H., FRISCH, W. & LOESCHKE, J. 2003. The Rand Granite in the southern Schwarzwald and its significance in the Variscan belt of SW Germany. *International Journal of Earth Sciences*, **92**, 821–842.

HÄRING, M. O., 2002. Sondierbohrung Otterbach, Basel: Der erste Schritt zur Entwicklung eines geothermischen Heiz-Kraftwerks nach dem Hot-Dry-Rock Verfahren. *Bulletin für Angewandte Geologie*, **7**, 19–30.

HARTLEY, A. J. & OTAVA, J. 2001. Sediment provenance and dispersal in a deep marine foreland bas*In:* the Lower Carboniferous Culm Basin, Czech Republic. *Journal of the Geological Society, London*, **158**,

137–150.

HEGNER, E., KÖLBL-EBERT, M. & LOESCHKE, J. 1998. The origin of post-collisional Variscan lamprophyres from the Black Forest, Germany; $^{40}Ar/^{39}Ar$ phlogopite ages, compositional, and isotope characteristics. *Lithos*, **45**, 395–411.

HENES-KLAIBER, U. 1989. Der Odenwald: Hinweise auf einen variszischen Arc-Magmatismus? *Beiheft zum European Journal of Mineralogy*, **1/1**, 66.

HENES-KLAIBER, U. 1992. *Zur Geochemie der variszischen Granitoide des Bergstraßer Odenwaldes*. Ph.D. Thesis, University of Karlsruhe.

HENK, A. 1993. Late orogenic evolution in the Variscan Internides: the Saar-Nahe Basin, Southwest Germany. *Tectonophysics*, **223**, 273–290.

HENK, A. 1997. Gravitational orogenic collapse vs plate-boundary stresses: a numerical modelling approach to the Permo-Carboniferous evolution of Central Europe. *Geologische Rundschau*, **86**, 39–55.

HENK, A., VON BLANCKENBURG, F., FINGER, F., SCHALTEGGER, U. & ZULAUF, G. 2000. Syn-convergent high-temperature metamorphism and magmatism in the Variscides: a discussion of potential heat sources. *In:* FRANKE, W., HAAK, V., ONCKEN, O. & TANNER, D. (eds) *Orogenic Processes: Quantification and Modelling in the Variscan Belt.* Geological Society, London, Special Publications, **179**, 35–61.

HERBIG, H.-G. 1998. The late Asbian transgression in the Central European Culm basins (Late Visean, cd III alpha). *Zeitschrift des Deutschen Geologischen Gesellschaft*, **149**(1), 39–58t.

HERING, O., ZIMMERLE, W., *et al.* 1976. Die Tiefbohrung Saar 1 – Petrographische Beschreibung und Deutung der erbohrten Schichten. *Geologisches Jahrbuch*, **A27**, 91–305.

HESS, J. C., LIPPOLT, H. J. & KOBER, B. 1995. The age of the Kagenfels granite (northern Vosges) and its bearing on the intrusion scheme of late Variscan granitoids. *Geologische Rundschau*, **84**, 568–577.

HEUSSER, H., 1926. Beiträge zur Geologie des Rheintales zwischen Waldshutund Basel. Beiträge zur Geologischen Karte der Schweiz, Neue Folge, 57, II. Abteilung. Geologische Kommission der Schweizerischen Naturforschenden Gesellschaft, Bern.

HIGGS, K., CLAYTON, G. & KEEGAN, J. B. 1988. Stratigraphy and systematic palynology of the Tournaisian rocks of Ireland. *Special Papers of the Geological Survey of Ireland*, **7**, 1–93.

HIGGS, K. T., STREEL, M., KORN, D. & PAPROTH, E. 1993. Palynological data from the Devonian–Carboniferous boundary beds in the new Stockum trench 2 and the Hasselbach borehole, northern Rhenish Massif. *Annales de la Société Géologique de Belgique*, **115**, 551–557.

HIRSCHMANN, G. 1995. Lithological Characteristics (Mid-German Crystalline Rise). *In:* DALLMEYER, R. D., FRANKE, F. & WEBER, K. (eds) *Pre-Permian Geology of Central and Eastern Europe.* Springer Verlag, Berlin, 155–163.

HIRSCHMANN, G. & OKRUSCH, M. 1988. Spessart-Kristallin und Ruhlaer Kristallin als Bestandteile der Mitteldeutschen Kristallinzone – ein Vergleich. *Neues Jahrbuch für Geologie und Paläontologie Abhandlungen*, **177**, 1–39.

HMICH, D., SCHNEIDER, J. W., SABER, H., & EL WARTITI, M. 2003. First Permo-Carboniferous insects (blattids) from North Africa (Morocco) – implications on palaeobiogeography and palaeoclimatology. *Freiberger Forschungshefte*, **C499**, 117–134.

HOCHULI, P. A. 1985. Palynostratigraphische Gliederung und Korrelation des Permo-Karbon der Nordostschweiz. *In:* MATTER, A., PETERS, T., ISENSCHMID, C., BLÄSI, H.-R., & ZIEGLER, H.-J. (eds) Sondierbohrung Weiach. Geologie. Landeshydrologie und -geologie, *Geologischer Bericht*, **6**, 185–194.

HOFFMANN, N., LINDERT, W., WEYER, D. & ILLERS, R.-H. 1975. Zum Unterkarbon-Vorkommen auf den Inseln Rügen und Hiddensee. *Zeitschrift für geologische Wissenschaften*, **3**, 851–873.

HOLLMANN, G. & WALTER, R. 1995. The Variscan deformation front between Stavelot-Venn Anticline and Brabant Massif - a balanced geological cross section along the Liège-Theux traverse. *Neues Jahrbuch für Geologie und Paläontologie, Monatshefte*, **2**, 92–104.

HOLUB, F. V., KLECKA, M. & MATEJKA, D. 1995. Igneous activity (Moldanubian Zone). *In:* DALLMEYER, R. D., FRANKE, W. & WEBER, K. (eds) *Pre-Permian Geology of Eastern and Central Europe*. Springer-Verlag, Berlin, 444–452.

HOLUB, F. V., COCHERIE, A. & ROSSI, P. 1997. Radiometric dating of granitic rocks from the Central Bohemian Plutonic Complex (Czech Republic): constraints on the chronology of thermal and tectonic events along the Moldanubian-Barrandian boundary. *Comptes Rendus de l'Academie des Sciences, Série II A*, **325**, 19–26.

HOMANN, W. 1969. Fazielle Gliederung der Unteren Pseudoschwagerinenkalke (Unter-Perm) der Karnischen Alpen. *Neues Jahrbuch für Geologie und Paläontologie, Monatshefte*, **1969**, 265–280.

HOTH, K., RUSBUELDT, J., ZAGORA, K., BEER, H. & HARTMANN, O. 1993. *Die tiefen Bohrungen im Zentralabschnitt der Mitteleuropäischen Senke - Dokumentation für den Zeitabschnitt 1962–1990.* Gesellschaft für Geowissenschaften e.V. Berlin.

HOTH, K., LORENZ, W. & BERGER, H. J. 1983. Die Lithostratigraphie des Proterozoikums im Erzgebirge. *Zeitschrift für Angewandte Geologie*, **29**, 413–418.

HOUSE, M. R. 1993. Fluctuations in ammonoid evolution and possibble environmental controls. *In:* HOUSE, M. R. (ed.) *The Ammonoidea: Environment, Ecology and Evolutionary Change*. Systematics Association, Special Volume **47**, Clarendon Press, Oxford, 13–34.

HOYEZ, B. 1971. Le Viséen du Boulonnais: analyse et corrélations séquentielles. *Annales de la Société Géologique du Nord*, **41**, 113–128.

HURYCH, V. & BRUECKNER, H. K. 1995. Eclogites from the Meluzina and Medenec localities at the Krusnè hory Mountains, western part of the Bohemian Massif, Czech Republic. *Journal of the Czech Geological Society*, **40**(3), A-17.

HWANG, S.L., SHEN, P., CHU, H.-T., YUI, T.F. & LIN, C.-C. 2001. Genesis of microdiamonds from melt and associated multiphase inclusions in garnet of ultrahigh-pressure gneiss from Erzgebirge, Germany. *Earth and Planetary Science Letters*, **188**, 9–15.

IVAN, P. 1994. Early Paleozoic of the Gemeric Unit (Inner Western Carpathians): Geodynamic setting as inferred from metabasalts geochemistry data. *Mitteilungeb der Österreichischen Geologischen Gesellschaft*, **86**, 23–31.

JÄGER, H. & GURSKY, H.-J. 1998. Quartzite im Unterkarbon der Hörre-Gommern-Zone: ein paläogeographisches Problematicum im Rhenoherzynikum. *In: Alfred-Wegener-Stiftung, GEO-BERLIN'98.* Gemeinsame Jahrestagung, DGG, DMG, GGW, Pal. Ges. Berlin, **98/3**, 148.

JANÁK, M., O'BRIEN, P. J., HURAI, V. & REUTEL, C. 1996. Metamorphic evolution and fluid composition of garnet-clinopyroxene amphibolites from the Tatra Mts., Western Carpathians. *Lithos*, **39**, 57–79.

JANÁK, M., FROITZHEIM, N., VRABEC, M., KROGH RAVNA, E. J. & DE HOOG, J. C. M. 2006. Ultrahigh-pressure metamorphism and exhumation of garnet peridotite in Pohorje, Eastern Alps. *Journal of Metamorphic Geology*, **24**, 19–31.

JONES, M., VAN DER VOO, R. & BONHOMMET, N. 1979. Late Devonian to early Carboniferous palaeomagnetic poles from the Armorican Massif, France. *Geophysical Journal of the Royal Astronomical Society*, **58**, 287–308.

JONGMANS, W. J. 1960. *Die Karbonflora der Schweiz (mit Beitrag von Ritter, E. Geologische Übersicht über die Karbonvorkommen der Schweiz).* Nr. **108**, Kummerly & Frty, Bern.

JURECZKA, J. & KOTAS, A. 1995. Coal deposits; Upper Silesian Coal Basin. *In:* ZDANOWSKI, A. & ZAKOWA, H. (eds) *Prace Panstwowego Instytutu Geologicznego*, **148**, 164–173.

KAHLER, F. 1983. Fusuliniden aus Karbon und Perm der Karnischen Alpen und der Karawanken. *Carinthia II, Sonderheft*, **41**, 1–107.

KAHLER, F. 1985. Oberkarbon und Unterperm der Karnischen Alpen. Ihre Biostratigraphie mit Hilfe der Fusuliniden. *Carinthia II, Sonderheft*, **42**, 1–93.

KAHLER, F. 1986. Ein Normalprofil der Fusuliniden-Stratigraphie im Oberkarbon und Unterperm der Karnischen Alpen. *Carinthia II*, **176**(96), 1–17.

KAHLER, F. 1989. Die Fusuliniden. *In:* EBNER, F. & KAHLER, F. (eds) *Catalogus Fossilium Austriae, Heft II/b/1: Foraminifera Palaeozoica.* Östereichische Akademie der Wissenschaften, Wien, 87–295.

KALT, A. & ALTHERR, R. 1996. Metamorphic evolution of garnet-spinel peridotites from the Variscan Schwarzwald (Germany). *Geologische Rundschau*, **85**, 211–224.

KALT, A., SCHLEICHER, H. & KRAMM, U. 1990. Karbonische Eklogitbildung im Schwarzwald. *Beiheft zum European Journal of Mineralogy*, **2**, 120.

KASIG, W. 1980. Dinantian carbonates in the Aachen region, F.R.G. *In:* BLESS, M. J. M., BOUCKAERT, J. & PAPROTH, E. (eds) *Pre-Permian around the Brabant Massif in Belgium, The Netherlands and Germany*. Mededelingen Rijks Geologische Dienst, **32**, 44–52.

KASIG, W. & WILDER, H. 1983. The sedimentary development of the western Rhenisches Schiefergebirge and the Ardennes (Germany/Belgium). *In:* MARTIN, H. & EDER, F. W. (eds) *Intracontinental Fold Belts – Case Studies in the Variscan Belt of Europe and the Demarara Belt of Namibia*. Springer Verlag, Berlin, 185–209.

KEDZIOR, A. 2001. Identification of the fluvial-channel tracts based on thickness: Zabrze Beds (Namurian B) in the Main Anticline and Bytom-Dabrowa trough of Upper Silesia Coal Basin. *Annales Societatis Geologorum Poloniae*, **71**, 21–34.

KERP, H. 1996. Der Wandel der Wälder im Laufe des Erdaltertums. *Natur und Museum*, **126**, 421–430.

KERP, H. 2000. The modernization of landscapes during the late Paleozoic-early Mesozoic. *In:* GASTALDO, R.A. & DIMICHELE, W.A. (eds) *Phanerozoic Terrestrial Ecosystems*. The Paleontological Society Papers, vol. **6**. The Paleontological Society, Lawrence, Kansas, 79–113.

KERP, H. & KRINGS, M. 1998. Climbing and scrambling growth habits: common life strategies among Late Carboniferous seed ferns. *Comptes Rendus des séances de l'Académie des Sciences de Paris, Série IIa*, **326**, 583–588.

KIRWAN, R. 1799. Additional Observations on the Proportion of Real Acid in the Three Ancient Known Mineral Acids and on the Ingredients in Various Neutral Salts and Other Compounds. George Bonham, Dublin.

KLÁPOVÁ, H. 1990. *Eclogites of the Bohemian part of the Saxothuringian*. Rozpravy. CSAV, 100, 5, 1–86.

KLÁPOVÁ, H., KONOPÁSEK, J. & SCHULMANN, K. 1998. Eclogites from the Czech part of the Erzgebirge: multi-stage metamorphic and structural evolution. *Journal of the Geological Society, London*, **155**, 567–583.

KNEUPER, G. 1976. Regional-geologische Folgerungen aus der Bohrung Saar 1. *Geologisches Jahrbuch*, **A27**, 499–510.

KÖNIGER, S. & LORENZ, V. 2002. Geochemistry, tectonomagmatic origin and chemical correlation of altered Carboniferous-Permian fallout ash tuffs in southwestern Germany. *Geological Magazine*, **139**, 541–558.

KÖNIGER, S., LORENZ, V., STOLLHOFEN, H. & ARMSTRONG, R. A. 2002. Origin, age and stratigraphic significance of distal fallout ash tuffs from the Carboniferous-Permian continental Saar-Nahe Basin (SW Germany). *International Journal of Earth Sciences (Geologische Rundschau)*, **91**, 341–356.

KONOPÁSEK, J. & SCHULMANN, K. 2005. Contrasting Early Carboniferous field geotherms: evidence for accretion of a thickened orogenic root and subducted Saxothuringian crust (Central European Variscides). *Journal of the Geological Society, London*, **162**, 463–470.

KORN, D., CLAUSEN, C.-D., BELKA, Z., LEUTERITZ, K., LUPPOLD, F. W., FEIST, R. & WEYER, D. 1994. Die Devon/Karbon-Grenze bei Drewer (Rheinisches Schiefergebirge). *Geologie und Paläontologie in Westfalen*, **29**, 71–95.

KORNPIHL, K. 2005. *Tectono-sedimentary evolution of the NE German Variscan foreland basin*. Unpublished Ph.D. thesis. University of Bonn, Germany.

KORIKOVSKIJ, S. P., KRIST, E. & BORONIKHIN, V. A. 1989. Staurolite-chloritoid schists from the Klenovec region: prograde metamorphism of high aluminia rocks of the Kohút zone – Veporides. *Geologica Carpathica*, **39**, 187–200.

KORSCH, R. J. & SCHÄFER, A. 1991. Interpretation of DEKORP deep-seismic reflection profiles IC and 9N acress the Variscan Saar-Nahe Basin, southwestern Germany. *Tectonophysics*, **191**, 127–146

KORSCH, R. J. & SCHÄFER, A. 1995. The Permo-Carboniferous Saar-Nahe Basin, south-west Germany and north-east France: basin formation and deformation in a strike-slip regime. *Geologische Rundschau*, **84**, 293–318.

KORYTOWSKI, A., DÖRR, W. & ZELAZNIEWICZ, A. 1993. U-Pb dating of (meta)granitoids in the NW Sudetes (Poland) and their bearing on tectono-stratigraphic correlation. *Terra Nova* **5**, 331–332.

KOSSMAT, F. 1927. Gliederung des varistischen Gebirgsbaues. *Abhandlungen des Sächsisches geologisches Landesamt*, **1**.

KOSTERS, M. & DONSELAAR, M. E. 2003. Syn-sedimentary faulting and the formation of localised reservoir sands: Carboniferous examples from the Campine Basin, Belgium and the Dutch offshore. *Petroleum Geoscience*, **9**, 309–319.

KOTAS, A. 1995. Lithostratigraphy and sedimentologic-paleogeographic development; Upper Silesian Coal Basin. *In:* ZDANOWSKI, A. & ZAKOWA, H. (eds) *Prace Panstwowego Instytutu Geologicznego*, **148**, 124–134.

KOTAS, A. & MALCZYK, W. 1972. The Upper Silesian Sandstone Series of the Upper Namurian Stage of the Upper Sielsian Coal Basin. *Prace Panstwowego Instytutu Geologicznego*, **61**, 462–466.

KOVÁČIK, M. 1991. High Al and Fe micaschists from the southeastern Veporicum: protolith and regional metamorphism. *Mineralia Slovaca*, **23**, 23–32 [in Slovak].

KOZUR, H. & MOCK, R. 1977. Erster Nachweis von Conodonten im Paleozoikum (Karbon) der Westkarpaten. *Časopsis pro Mineralogii a. Geologii*, **22**, 299–305.

KOZUR, H., MOCK, R. & MOSTLER, H. 1976. Stratigraphische Neuienstufung der Karbonatgesteine der unteren Schichtenfolgen von Ochtiná (Slovakei) in das oberste Vise-Serpukhovian (Namur A). *Geologische und Palaeontologische. Mitteilungen*, **6**, 1–29.

KRAINER, K. 1989*a*. Molassesedimentation im Oberkarbon der Ostalpen am Beispiel der Stangnock-Formation am NW- Rand der Gurktaler Decke (Österreich). *Zentralblatt für Geologie und Paläontologie*, **Teil I (1988)**, 807–820.

KRAINER, K. 1989*b*. Die fazielle Entwicklung der Oberkarbonsedimente (Stangnock-Formation) am NW-Rand der Gurktaler Decke. *Carinthia II*, **179(99)**, 563–601.

KRAINER, K. 1990. Ein Beitrag zum Oberkarbon der Steinacher Decke ("Karbon des Nößlacher Joches". Tirol). *Mitteilungen der Gesellschaft für Geologie und Bergbaustudenten Österreich*, **36**, 87–99.

KRAINER, K. 1992. Fazies, Sedimentationsprozesse und Paläogeographie im Karbon der Ost- und Südalpen. *Jahrbuch der Geologischen Bundesanstalt*, **135**, 99–193.

KRAINER, K. 1993. Late- and Post-Variscan Sediments of the Eastern and Southern Alps. *In:* VON RAUMER, J. F. & NEUBAUER, F. (eds) *Pre-Mesozoic Geology in the Alps*. Springer, Berlin, 537–564.

KRAINER, K. 1995. *Anthracoporella* Mounds in the Late Carboniferous Auernig Group, Carnic Alps (Austria). *Facies*, **32**, 195–214.

KRAINER, K. & DAVYDOV, V. 1998. Facies and biostratigraphy of the Late Carboniferous/Early Permian sedimentary sequence in the Carnic Alps (Austria/Italy). *Geodiversitas*, **20**(4), 643–662.

KRAINER, K. & MOGESSIE, A. 1991. Composition and significance of resedimented amphibolite breccias and conglomerates (Badstub Formation) in the Carboniferous of Nötsch (Eastern Alps, Austria). *Jahrbuch der Geologischen Bundesanstalt*, **134**, 65–81.

KRAINER, K. & VACHARD, D. 2002. Late Serpukhovian (Namurian A) microfacies and carbonate microfossils from the Carboniferous of Nötsch (Austria). *Facies*, **46**, 1–26.

KRAINER, K., FLÜGEL, E., VACHARD, D. & JOACHINSKI, M. M. 2003. A close look at Late Carboniferous algal mounds: Schulterkofel, Carnic Alps, Austria. *Facies*, **49**, 325–350.

KRAWCZYK, C.M., STEIN, E., CHOI, S., *et al.* 2000. Constraints on distribution and exhumation mechanisms of high-pressure rocks from geophysical studies - the Saxothuringian case between the Bary Fault and Elbe Line. *In:* FRANKE, W., HAAK, V., ONCKEN, O. & TANNER, D. (eds) *Orogenic Processes: Quantification and Modelling in the Variscan Belt*. Geological Society, Special Publications, **179**, 303–322.

KRAWCZYK, C. M., MCCANN, T., COCKS, L. R. M., ENGLAND, R., MCBRIDE, J. & WYBRANIEC, S. 2008. Caledonian tectonics. *In*: MCCANN, T. (ed.) *The Geology of Central Europe. Volume 1: Precambrian and Palaeozoic*. Geological Society, London, 303–381.

KRENTZ, O. 1985. Temperaturregime und Altersstellung der regionalen Metamorphose im mittleren Erzgebirge. *Freiberger Forschungshefte*, **C390**, 12–28.

KREUZER, H. & HARRE, W. 1975. K/Ar-Altersbestimmungen an Hornblenden und Biotiten des Kristallinen Odenwalds. *In:* AMSTUTZ, G. C., MEISL, S. & NICKEL, E. (eds) *Mineralien und Gesteine im Odenwald*. Aufschluß Sbd, **27**, 71–77-

KRINGS, M., KERP, H., TAYLOR, T. N. & TAYLOR, E. L. 2003. How Paleozoic vines and lianas got off the ground: on scrambling and climbing Carboniferous-Early Permian pteridosperms. *The Botanical Review*, **69**, 204–224.

KROHE, A. 1991. Emplacement of synkinematic plutons in the Variscan Odenwald (Germany) controlled by transtensional tectonics. *Geolo-*

gische Rundschau, **80**, 391–409.

KROHE, A. 1992. Structural evolution of intermediate-crustal rocks in a strike-slip and extensional setting (Variscan Odenwald, SW Germany): differential upward transport of metamorphic complexes and changing deformation mechanisms. *Tectonophysics*, **206**, 357–386.

KROHE, A. 1994. Verformungsgeschichte in der mittleren Kruste eines magmatischen Bogens. *Geotektonische Forschung*, **80**, 1–147.

KROHE, A. 1996. Variscan tectonics of central Europe: post accretionary intraplate deformation of weak continental lithosphere. *Tectonics*, **15**, 1364–1388.

KROHE, A. & EISBACHER, G. H. 1988. Oblique crustal detachment in the Variscan Schwarzwald, southwestern Germany. *Geologische Rundschau*, **77**, 25–43.

KROHE, A. & WILLNER, A. P. 1995. The Odenwald Crystalline Complex. *In:* DALLMEYER, R. D., FRANKE, W. & WEBER, K. (eds) *Pre-Permian Geology of Central and Eastern Europe*. Springer, Berlin, 174–181.

KRÖNER, A. & HEGNER, E. 1998. Geochemistry, single grain zircon ages and Sm-Nd systematics of granitoid rocks from the Góry Sowie (Owl Mts), Polish West Sudetes: evidence for early Palaeozoic arc-related plutonism. *Journal of the Geological Society, London*, **155**, 711–724.

KRÖNER, A. & MAZUR, S. 2003. Proterozoic and Palaeozoic crustal components across the East/Central Sudetes boundary at the eastern margin of the Bohemian Massif: new U/Pb and Pb/Pb single zircon ages from the eastern Fore-Sudetic block (SW Poland). *Journal of the Czech Geological Society*, **48**, 83–84.

KRÖNER, A. & WILLNER, A. P. 1998. Time of formation and peak of Variscan HP-HT metamorphism of quartz-feldspar rocks in the central Erzgebirge, Saxony, Germany. *Contributions to Mineralogy and Petrology*, **132**, 1–20.

KRÖNER, A., HEGNER, E., HAMMER, J., HAASE, G., BIELICKI, K. H., KRAUSS, M. & EIDAM, J. 1994. Geochronology and Nd-Sr systematics of Lusatian granitoids, significance for the evolution of the Variscan orogen in east-central Europe. *Geologische Rundschau*, **83**, 357–376.

KRÖNER, A., WILLNER, A. P., HEGNER, E., FRISCHBUTTER, A., HOMFANN, J. & BERGNER, R. 1995. Latest Precambrian (Cadomian)zircon ages, Nd isotope systematics and P-T evolution of granitoid orthogneisses of the Erzgebirge, Saxony and Czech Republich. *Geologische Rundschau*, **84**, 437–456.

KRÖNER, A., JAECKEL, P., HEGNER, E. & OPLETAL, M. 2001. Single zircon ages and whole-rock Nd isotopic systematics of early Palaeozoic granitoid gneisses from the Czech and Polish Sudetes (Jizerské hory, Krkonose and Orlice-Sneznik Complex). *International Journal of Earth Sciences*, **90**, 304–324.

KRÖNER, A., STIPSKÁ, P., SCHULMANN, K. & JAECKEL, P. 2000. Chronological constraints on the pre-Variscan evolution of the northeastern margin of the Bohemian Massif, Czech Republic. *In:* FRANKE, W., HAAK, V., ONCKEN, O. & TANNER, D. (eds) *Orogenic processes: Quantification and modelling in the Variscan Belt*. Geological Society, London, Special Publications, **179**, 175–198.

KRONER, U. & HAHN, T. 2004. Sedimentation, Deformation und Metamorphose im Saxothuringikum während der variszischen Orogenese: Die komplexe Entwicklung von Nord-Gondwana während kontinentaler Subduktion und schiefer Kollision. *In:* LINNEMANN, U. (ed.) Das Saxothuringikum. *Geologica Saxonica* **48/49**, 133–146.

KRONER, U., MANSY, J.-L. *et al.* 2008. Variscan tectonics. *In*: MCCANN, T. (ed.) *The Geology of Central Europe. Volume 1: Precambrian and Palaeozoic*. Geological Society, London, 599–664.

KRYZA, R. 1981. Migmatization in gneisses of the northern part of the Sowie Góry, Sudetes. *Geologica Sudetica*, **14**, 7–100. [Polish with English summary].

KRYZA, R. 1993. Basic metavolcanic rocks of the central Kaczawa Mts (Sudetes): a petrological study. *Acta Universitatis Wratislavaiensis, Prace Geoliczne-Minerabgiczne* **39** [in Polish with English summary].

KRYZA, R. 1995*a*. West Sudetes: Igneous Activity. *In:* Dallmeyer, R. D., FRANKE, W. & WEBER, K. (eds) *Pre-Permian Geology of Eastern and Central Europe*. Springer-Verlag, Berlin, 341–350.

KRYZA, R. 1995*b*. West Sudetes: metamorphic evolution. *In:* DALLMEYER, R. D., FRANKE, W. & WEBER, K. (eds) *Pre-Permian Geology of Eastern and Central Europe*. Springer-Verlag, Berlin, 351–359.

KRYZA, R., MUSZYNSKI, A., TURNIAK, K. & ZALASIEWICZ, J. A. 1994. A lower Palaeozoic shallow-water sequence in the eastern European Variscides (SW Poland): provenance and depositional history. *Geologische Rundschau*, **83**, 5–19.

KRYZA, R., PIN, C. & VIELZEUF, D. 1996. High-pressure granulites from the Sudetes (south-west Poland): evidence of crustal subduction and collisional thickening in the Variscan Belt. *Journal of Metamorphic Geology*, **14**, 531–546.

KULLMANN, J. 2000. Ammonoid turnover at the Devonian-Carboniferous boundary. *Revue de Paléobiology*, **8**, 169–180.

KUMPERA, O. & MARTINEC, P. 1995. The development of the Carboniferous accretionary wedge in the Moravian-Silesian Paleozoic basin. *Journal of the Czech Geological Society*, **40**, 47–64.

KUPFERSCHMIED, M. P. 1993. Structural studies in the Western Habach Group (Tauern Window, Salzburg, Austria). *Abhandlungen der Geologischen Bundesanst Wien*, **49**, 67–78.

KUPFERSCHMIED, M. P. 1994. Geologische Untersuchungen im Tauernfenster zwischen Hollersbachtal und Krimmler Achental. *Münchener Geologische Hefte*, **12**, 1–160.

KUPFERSCHMIED, M. P. & HÖLL, R. 1994. Die geologische Neuaufnahme der Habachmulde und ihre genetischen Implikationen für die Habachgruppe (Tauernfenster/Ostalpen). *Jahrbuch der Geologischen Bundesanstalt Wien*, **137**, 139–153.

KUPFERSCHMIED, M. P., HÖLL, R. & MILLER, H. 1994. Lithologische und strukturgeologische Untersuchungen in der Krimmler Gneiswalze (Tauernfenster/Ostalpen) und ihrem Umfeld. *Jahrbuch der Geologische Bundesanstalt Wien*, **137**, 155–170.

LANGENAEKER, V. 1992. *Coalification maps for the Westphalian of the Campine Coal Basin*. Belgian Geological Survey, Professional Papers, **256.**

LAMMERER, B. 1986. Das Authochthon im westlichen Tauernfenster. *Jahrbuch der Geologischen Bundesanstalt Wien*, **129**, 51–67.

LAMMERER, B. 1988. Thrust-regime and transpression-regime tectonics in the Tauern Window (Eastern Alps). *Geologische Rundschau*, **77**, 143–156.

LAMMERER, B. & WEGER, M. 1998. Footwall uplift in an orogenic wedge: the Tauern Window in the Eastern Alps of Europe. *Tectonophysics*, **285**, 213–230.

LANGENAEKER, V. 2000. The Campine Basin. Stratigraphy, structural geology, coalification and hydrocarbon potential for the Devonian to the Jurassic. *Aardkundige Mededelingen*, **10**, 1–142.

LANGENAEKER, V. & DUSAR, M. 1992. Subsurface facies analysis of the Namurian and earliest Westphalian in the western part of the Campine Basin (N Belgium). *Geologie en Mijnbouw*, **71**, 161–171.

LARDEAUX, J. M., LEDRU, P., DANIEL, I. & DUCHENE, S. 2001. The Variscan French Massif Central – a new addition to the ultra-high pressure metamorphic "club": exhumation processes and geodynamic consequences. *Tectonophysics*, **332**, 143–167.

LAUBSCHER, H. P. 1987. Die tektonische Entwicklung der Nordschweiz. *Eclogae Geologicae Helvetiae*, **80**, 287–303.

LEAT, P. T., THOMPSON, R. N., MORRISON, M. A., HENDRY, G. L. & TRAYHORN, S. C. 1987. Geodynamic significance of post-Variscan intrusive and extrusive potassic magmatism in southwest England. *Transactions of the Royal Society of Edinburgh: Earth Sciences*, **77**, 349–360.

LÉCORCHÉ, J. P., DALLMEYER, R. D. & VILLENEUVE, M. 1989. Definition of tectonostratigraphic terranes in the Mauritanide, Bassaride, and Rokelide orogens, West Africa. *In:* DALLMEYER, R. D. (ed.) *Terranes in the Circum-Atlantic Paleozoic orogens*. Geological Society of America, Special Papers, **230**, 131–144.

LEDRU, P., LARDEAUX, J. M., *et al.* 1989. Où sont les nappes dans le Massif central francais? *Bulletin de la Société Géologique de France*, **5**, 605–618.

LEDRU, P., AUTRAN, A. & SANTALLIER, D. 1994. Lithostratigraphy of Variscan terranes in the French Massif Central: A basis for paleogeographical reconstruction. *In:* KEPPIE, J. D. (ed.) *Pre-Mesozoic Geology in France and Related Areas*. Springer-Verlag, Berlin, 276–288..

LEDRU, P., COURRIOUX, G., DALLAIN, C., LARDEAUX, J. M., MONTEL, J. M., VANDERHAEGHE, O. & VITEL, G. 2001. The Velay dome (French Massif Central): melt generation and granite emplacement during orogenic evolution. *Tectonophysics*, **342**, 207–237.

LEES, A. 1997. Biostratigraphy, sedimentology and palaeobathymetry of

Waulsortian buildups and peri-Waulsortian rocks during the late Tournaisian regression, Dinant area, Belgium. *Geological Journal*, **32**, 1–36.

LEES, A., NOEL, B. & BOUW, P. 1977. The Waulsortian "reefs" of Belgium. A progress report. *Mémoires de l'Institut géologique de l'Université Catholique de Louvain*, **29**, 289–315.

LIEW, T. C. & HOFMANN, A. W. 1988. Precambrian crustal components, plutonic associations, plate environment of the Herycnian fold belt of central Europe: indications from a Nd and Sr isotopic study. *Contributions to Mineralogy and Petrology*, **98**, 129–138.

LINNEMANN, U. 2003. Die Struktureinheiten des Saxothuringikums. *In:* LINNEMANN, U. (ed.) Das Saxothuringikum. *Geologica Saxonica*, **48/49**, 19–28.

LINNEMANN, U. (ed.) 2004. Das Saxothuringikum. *Geologica Saxonica* **48/49**.

LINNEMANN, U., GEHMLICH, M., TICHOMIROVA, M. & BUSCHMANN, B. 1998. The Peri-Gondwanan basement of the Saxothuringian Composite Terrane. *In:* LINNEMANN, U., HEUSE, T., FATKA, O., KRAFT, P., BROCKE, R. & ERDTMANN, B.-D. (eds) *Pre-Variscan Terrane Analysis of "Gondwanan Europe"*. International Geological Conference Excursion Guide and Abstracts, Schriften des Staatlichen Museum Minerologie und Geologie, Dresden, **9**, 7–13.

LINNEMANN, U., GEHMLICH, M., *et al.* 2002. From Cadomian subduction to Early Palaeozoic rifting: the evolution of Saxo-Thuringia at the margin of Gondwana in the light of single zircon geochronology and basin development (central European Variscides, Germany). *In:* FRANKE, W., HAAK, V., ONCKEN, O. & TANNER, D. (eds) *Orogenic Processes: Quantification and Modelling in the Variscan Belt.* Geological Society, London, Special Publications, **179**, 131–153.

LINNEMANN, U., ROMER, R. L. *et al.* 2008*a*. Precambrian. *In*: MCCANN, T. (ed.) *The Geology of Central Europe. Volume 1: Precambrian and Palaeozoic*. Geological Society, London, 21–101.

LINNEMANN, U., D'LEMOS, R. *et al.* 2008*b*. Cadomian tectonics. *In*: MCCANN, T. (ed.) *The Geology of Central Europe. Volume 1: Precambrian and Palaeozoic*. Geological Society, London, 103–154.

LIPIEC, M. & MATYJA, H. 1998. Depositional architecture of the Lower Carboniferous sedimentary basin in Pomerania. *Prace Państwowego Instytutu Geologicznego*, **165**, 101–112.

LIPPOLT, H. J., HESS, J. C., HOLUB, V. M. & PESEK, J. 1986. Correlation of Upper Carboniferous deposits in the Bohemian Massif (Czechoslovakia) and in the Ruhr District (FR Germany): evidence from Ar-40/Ar-39 ages of tuff layers. *Zeitschrift der Deutschen Geologischen Gesellschaft*, **137**, 447–464.

LITTKE, R., BÜKER, C., HERTLE, M., KARG, H., STROETMANN-HEINEN, V. & ONCKEN, O. 2000. Heat flow evolution, subsidence and erosion in the Rheno-Hercynian orogenic wedge of central Europe. *In:* FRANKE, W., HAAK, V., ONCKEN, O. & TANNER, D. (eds) *Orogenic Processes: Quantification and Modelling in the Variscan Belt.* Geological Society, London, Special Publications, **179**, 231–255.

LOESCHKE, J., GÜLDENPFENNIG, M., HANN, H. P. & SAWATZKI, G. 1998. Die Zone von Badenweiler-Lenzkirch (Schwarzwald): eine variskiche Suturzone. *Zeitschrift der Deutschen Geologischen Gesellschaft*, **149**, 197–212.

LORENZ, V. & NICHOLLS, J. A. 1984. Plate and intraplate processes of Hercynian Europe during the late Paleozoic. *Tectonophysics*, **107**, 25–56.

LORENZ, W. 1989. Geological outline of the Erzgebirge Anticlinorium. *In:* TISCHENDORF, G. (ed.) *Silicic Magmatism and Metallogenesis of the Erzgebirge.* Veröffentlichungen des Zentralinstitut für Physik der Erde, Potsdam, **107**, 6–35.

LUDHOVÁ, L. & JANÁK, M. 1999: Phase relations and P-T path of cordierite-bearing migmatites, Western Tatra Mts., Western Carpathians. *Geologica Carpathica*, **50**, 283–293.

LUPPOLD, F. W., CLAUSEN, C.-D., KORN, D. & STOPPEL, D. 1994. Devon/Karbon-Grenzprofile im Bereich von Remscheid-Altenaer Sattel, Warsteiner Sattel, Briloner Sattel und Attendorn-Elsper Doppelmulde (Rheinisches Schiefergebirge). *Geologie und Paläontologie in Westfalen*, **29**, 7–69.

LÜTZNER, H. 2005. Thüringer Wald. *In:* MENNING, M., BENEK, R., *et al.* Das Rotliegend in der Stratigraphischen Tabelle von Deutschland 2002 - "Paternoster-Stratigraphie" auf dem Rückzug. *Newsletters on Stratigraphy*, **41**, 91–122.

LÜTZNER, H., LITTMANN, S., MÄDLER, J., ROMER, R. L. & SCHNEIDER, J. W. 2003. Stratigraphic and radiometric age data for the continental Permo-Carboniferous reference section Thüringer Wald, Germany. *Proceedings of the XVth International Congress on Carboniferous and Permian Stratigraphy*, Utrecht, 2003.

MCCANN, T. 1996. The North Celtic Sea Reflector – a possible Caledonian basement structure, offshore southern Ireland. *Tectonophysics*, **266**, 361–377.

MCCANN, T. 1999. The tectonosedimentary evolution of the northern margin of the Carboniferous foreland basin of NE Germany. *Tectonophysics*, **313**, 119–144.

MCCANN, T., KIERSNOWSKI, H., *et al.* 2008. Permian. *In*: MCCANN, T. (ed.) *The Geology of Central Europe. Volume 1: Precambrian and Palaeozoic*. Geological Society, London, 531–597.

MCKERROW, W. S., MACNIOCAILL, C., AHLBERG, P. E., CLAYTON, G., CLEAL, C. J. & EAGAR, R. M. C. 2000. The Late Palaeozoic relations between Gondwana and Laurussia. *In:* FRANKE, W., HAAK, V., ONCKEN, O. & TANNER, D. (eds) *Orogenic Processes: Quantification and Modelling in the Variscan Belt.* Geological Society, London, Special Publications, **179**, 9–20.

MCKERROW, W. S., MACNIOCAILL, C., AHLBERG, P. E., CLAYTON, G., CLEAL, C. J. & Eagar, R. M. C. 2002. The Late Palaeozoic relations between Gondwana and Laurentia. *In:* FRANKE, W., HAAK, V., ONCKEN, O. & TANNER, D. (eds) *Orogenic Processes: Quantification and Modelling in the Variscan Belt.* Geological Society, London, Special Publications, **179**, 9–20.

MAASS, R. 1988. Die Südvogesen in variszischer Zeit. *Neues Jahrbuch für Geologie und Paläontologie, Monatshefte*, **10**, 611–638.

MAASS, R. & SCHNEIDER, J. L. 1995. Die südlichen Vogesen. *Jahresbericht und Mitteilungen des Oberrheinischen Geologischen Vereines*, **77**, 139–153.

MAASS, R. & STOPPEL, D. 1982. Nachweis von Oberdevon bei Markstein (Bl. Munster, Südvogesen). *Zeitschrift der deutschen geologischen Gesellschaft*, **133**, 403–408.

MAMET, B. & MIŠÍK, M. 2003: Marine Carboniferous algae from metacarbonates of the Ochtiná Formation (Gemeric Unit, Western Carpathians). *Geologica Carpathica*, **54**, 3–8.

MANSY, J. L., CONIL, R., *et al.* 1989. Nouvelles données stratigraphiques et structurales sur le Dinantien dans l'Avesnois. *Annales de la Société géologique du Nord*, **108**, 125–142.

MANSY, J.-L., EVERAERTS, M. & DE VOS, W. 1999. Structural analysis of the adjacent Acadian and Variscan fold belts in Belgium and northern France from geophysical and geological evidence. *Tectonophysics*, **309**, 99–116.

MARSCHALL, H. R., KALT, A. & HANEL, M. 2003. P-T evolution of a Variscan lower-crustal segment: a study of granulites from the Schwarzwald, Germany. *Journal of Petrology*, **44**, 227–253.

MASSARI, F. 1986. Some thoughts on the Permo-Triassic evolution of the South-Alpine area (Italy). *Memorie della Società Geologica Italiana*, **34**, 179–188.

MASSARI, F. & VENTURINI, C. 1990. The significance of the Auernig Group cyclicity. *In:* VENTURINI, C. (ed.) *Field Workshop on Carboniferous to Permian Sequence of the Pramollo-Nassfeld basin (Carnic Alps), Guidebook*. Arti Grafiche Friulani, Udine, 81–86.

MASSONNE, H.-J. 2001. First find of coesite in the ultrahigh-pressure metamorphic region of the Central Erzgebirge, Germany. *European Journal of Mineralogy*, **13**, 565–570.

MASSONNE, H. J. 1995. Metamorphic evolution. *In:* DALLMEYER, R. D., FRANKE, W. & WEBER, K. (eds) *Pre-Permian Geology of Central and Eastern Europe*. IGCP **233**, Springer, Berlin, 132–137.

MASSONNE, H. J. 1999. A new occurrence of microdiamonds in quartzofeldspathic rocks of the Saxonian Erzgebirge, Germany, and their metamorphic evolution. *Proceedings of the 7th International Kimberlite Conference*, Capetown, 1998, 533–539.

MATTE, P. 1986*a*. Tectonics and plate tectonic models for the Variscan belt of Europe. *Tectonophysics*, **126**, 329–374.

MATTE, P. 1986*b*. La chaine varisque parmi les chaines paléozoique péri-atlantiques, modèle d'évolution et position des grands blocs continentaux au Permo-Carbonifère. *Bulletin de la Société Géologique de France,* **(8)II**, 9–24.

MATTE, P. 1991. Accretionary history and crustal evolution of the Variscan belt in Western Europe. *Tectonophysics*, **196**, 309–337.

MATTE, P. 1998. Continental subduction and exhumation of HP rocks in Paleozoic orogenic belts: Uralides and Variscides. *GFF*, **120**, 209–

222.
Matte, P., Maluski, H., Rajlich, R. & Franke, W. 1990. Terrane boundaries in the Bohemian Massif: results of large-scale Variscan thrusting. *Tectonophysics*, **177**, 151–170.
Matte, P. H. 1991. Accretionary history and crustal evolution of the Variscan belt in Europe. *Tectonophysics*, **196**, 309–337.
Matter, A. 1987. Faciesanalyse und Ablagerungsmilieus des Permokarbons im Nordschweizer Trog. *Eclogae Geologicae Helvetiae*, **80**, 345–367.
Matter, A., Peters, T. J., Bläsi, H.-R. & Ziegler, H.-J. 1987. *Sondierbohrung Riniken*. NAGRA Technischer Bericht, **86–02**, Wettingen.
Matter, A., Peters, T. J., Bläsi, H.-R., Meyer, J., Ischni, H., & Meyer, C. 1988*a*. *Sondierbohrung Weiach*. NAGRA Technischer Bericht, 86–01, Wettingen.
Matter, A., Peters, T. J., Bläsi, H.-R., Schenker, F. & Weiss, H.-P. 1988*b*. *Sondierbohrung Schafisheim*. Nagra Technischer Bericht. NAGRA, Baden.
Mazur, S. 2003. Structural evolution of the Kłodzko Metamorphic Complex and the implications for the Variscan tectonics of the Sudetes. *Acta Universitatis Wratislaviensis, Prace. Geololiczno-Mineralogiczne* **74**, 1–197.
Mazur, S. & Aleksandrowski, P. 2001. The Teplá(?)/Saxothuringian suture in the Karkonosze-Izera massif, Western Sudetes, Central European Variscides. *International Journal of Earth Sciences*, **90**, 341–360.
Mazur, S., Aleksandrowski, P., Kryza, R. & Oberc-Dziedzic, T. 2006. The Variscan Orogen in Poland. *Geological Quarterly*, **50**, 89–118.
Mazur, S., Aleksandrowski, P., Turniak, K. & Awdankiewicz, M. 2007. Geology, tectonic evolution and Palaeozoic magmatism of Sudetes - an overview. *In:* Kozlowski, A. & Wiszniewska, J. (eds) Granitoids in Poland. *Archivum Mineralogiae Monograph*, **1**, 59–87.
Mazur, S., Aleksandrowski, P. & Szczepanski, J. 2005. The presumed Teplá-Barrandian/Moldanubian terrane boundary in the Orlica Mountains (Sudetes, Bohemian Massif): structural and petrological characteristics. *Lithos*, **82**, 85–112.
Meier, B. P. 1994. *Untere Süsswassermolasse des zentralen und östlichen Mittellandes*. NAGRA Interner Bericht, 94–27. NAGRA, Wettingen.
Meischner, D. 1991. Kleine Geologie des Kellerwaldes (Exkursion F). *Jahresberichte und Mitteilungen des Oberrheinischen Geologischen Vereins*, **73**, 115–142.
Meisl, S. 1995. Igneous activity (Northern Phyllite Zone). *In:* Dallmeyer, R. D., Franke, W. & Weber, K. (eds) *Pre-Permian Geology of Eastern and Central Europe*. Springer-Verlag, Berlin, 118–131.
Menning, M., Weyer, D., Dorzdzewski, G., Van Amerom, H. W. J. & Wendt, I. 2000. A Carboniferous time scale 2000: Discussion and use of geological parameters as time indicators from Central and Western Europe. *Geologisches Jahrbuch*, **A156**, 3–44.
Menning, M., Belka, Z., *et al.* 2001. The optimal number of Carboniferous series and stages. *Newsletter Stratigraphy*, **38**, 201–207.
Menning, M., Alekseev, A. S., *et al.* 2006. Global time scale and regional stratigraphic reference scales of Central and West Europe, East Europe, Tethys, South China, and North America as used in the Devonian–Carboniferous–Permian Correlation Chart 2003 (DCP 2003), *Palaeogeography, Palaeoclimatology, Palaeoecology*, **240**, 318–372.
Méres, Š. & Hovorka, D. 1991: Geochemistry and metamorphic evolution of the Kohút crystalline complex mica schists (the Western Carpathians). *Acta Geologica et Geographica Universitatis Comenianae*, **47/1**, Bratislava, 15–66.
Meyer, W. & Stets, J. 1975. Das Rheinprofil zwischen Bonn und Bingen. *Zeitschrift des Deutschen Geologischen Gesellschaft*, **126**, 15–29.
Meyer, W. & Stets, J. 1994. Geologie des Ardennisch-Rheinischen Schiefergebirges. *In:* von Koenigswald, W. & Meyer, W. (eds) Pfeil München. *Erdgeschichte im Rheinland. Fossilien und Gesteine aus 400 Millionen Jahren*: 13–34.
Meyer, W. & Stets, J. 1996. *Das Rheintal zwischen Bingen und Bonn*. Sammlung geologischer Führer, **89**.
Mielke, H., Rohrmüller, J. & Gebauer, D. 1996. Ein metalateritisches Denudations-Niveau als lithologisch undzeitlich korrelierbarer Bezugshorizont in Phylliten, Glimmerschiefern und Gneisen des ostbayerischen Grundgebirges. *Geologica Bavarica*, **101**, 139–166.
Mii, H.-S., Grossman, E. L., Yancey, T. E., Chuvashov, B. & Egorov, A. 2001. Isotopic records of brachiopod shells from the Russian Platform – evidence for the onset of mid-Carboniferous glaciation. *Chemical Geology*, **175**, 133–147.
Mingram, B. 1998. The Erzgebirge, Germany – a subducted part of northern Gondwana: geochemical evidence for repetition of early Palaeozoic metasedimentary sequences in metamorphic thrust units. *Geological Magazine*, **135**, 785–801.
Mingram, B., Kröner, A., Hegner, E. & Krentz, O. 2004. Zircon ages, geochemistry, and Nd isotopic systematics of pre-Variscan orthogneisses from the Erzgebirge, Saxony (Germany), and geodynamic interpretation. *International Journal of Earth Science*, **93**, 706–727.
Möbus, G. 1966. Die variszische Tektogenese in der Elbtalzone. *Geologie*, **15**, H1.
Montenari, M., Servais, T. & Paris, F. 2000. Palynological dating (acritarchs and chitinozoans) of Lower Paleozoic phyllites from the Black Forest/southwestern Germany. *Comptes Rendus de l'Academie de Sciences, Paris, Earth and Planetary Sciences*, **330**, 493–499.
Muchez, P. & Langenaeker, V. 1993. Middle Devonian to Dinantian sedimentation in the Campine Basin (northern Belgium): its relation to Variscan tectonism. *In:* Frostick, L. & Steel, R. (eds) *Tectonic Controls and Signatures in Sedimentary Successions*. International Association of Sedimentologists, Special Publications **20**, 171–181.
Muchez, P., Viaene, W., Wolf, M. & Boukaert, J. 1987. Sedimentology, coalification pattern and paleogeography of the Campine-Brabant Basin during the Visean. *Geologie en Mijnbouw*, **66**, 313–326.
Müller, A., Breiter, K., Seltmann, R. & Pécskay, Z. 2005. Quartz and feldspar zoning in the eastern Erzgebirge volcano-plutonic complex (Germany, Czech Republic): evidence of multiple magma mixing. *Lithos*, **80**, 201–227.
Muttoni, G., Kent, D. V., Garzanti, E., Brack, P., Abrahamsen, N. & Gaetani, M. 2003. Early Permian Pangea "B" to Late Permian Pangea "A". *Earth and Planetary Science Letters*, **215**, 379–394.
Nagra, 1985. *Sondierbohrung Böttstein Untersuchungsbericht*. Beiträge zur Geologie der Schweiz – Geotechnische Serie, 67. Schweizerische Geotechnische Kommission, Bern.
Nagra, 1991. *Sondierbohrung Kaisten Untersuchungsbericht*. Beiträge zur Geologie der Schweiz – Geotechnische Serie, 83. Schweizerische Geotechnische Kommission, Bern.
Nagra, 1993. Sondierbohrung Siblingen Untersuchungsbericht. Beiträge zur Geologie der Schweiz – Geotechnische Serie, 86. Schweizerische Geotechnische Kommission, Bern.
Nagra, 2001. Sondierbohrung Benken. NAGRA, Technischer Bericht, 00–01. NAGRA, Wettingen.
Nasdala L. & Massonne, H. J. 2000. Microdiamonds from the Saxonian Erzgebirge, Germany: in situ micro-Raman characterisation. *European Journal of Mineralogy*, **12**, 495–498.
Nemec, D. 1974. Petrochemistry of the dyke rocks of the Central Bohemian Pluton. *Neues Jahrbuch für Mineralogie Monatshefte*, **1974**, 193–209.
Nemec, W., Porebski, S. J. & Teisseyre, A. K. 1982. Exploratory notes to the lithotectonic molasse profile of the Intra-Sudetic Basin, Polish part (Sudety Mts, Carboniferous-Permian). *Veröffentlingungen des Zentralinstituts für Physik der Erde, Akadamie der Wissenschaften der DDR*, **66**, 267–278.
Nasir, S., Okrusch, M., Kreuzer, H., Lenz, H. & Höhndorf, A. 1991. Geochronology of the Spessart Crystalline Complex, Mid-German Crystalline Rise. *Mineralogy and Petrology*, **44**, 39–55.
Neubauer, F. 1988. The Variscan orogeny in the Austroalpine and Southalpine domains of the Eastem Alps. *Schweizerische Mineralogische und Petrographische Mitteilungen*, **68**, 339–349.
Neubauer, F. & Herzog, U. 1985. Das Karbon der Stolzalpendecke Mittelkärntens – Implikationen für die variszische Paläogeographie und Orogenese im Ostalpin. *Anzeiger der Österreichischen Akademie der Wissenschaften, mathematisch-naturwissenschaftliche Klasse*, 1985.
Neubauer, F. & Höck, V. 2000. Aspects of geology in Austria and adjoining areas: Introduction. *Mitteilungen der Österreichischen Geologischen Gesellschaft*, **92**, 9–14.
Neumann, E.-R., Olsen, K. H., Baldridge, W. S. & Sundvoll, B.

1992. The Oslo Rift: a review. *Tectonophysics*, **208**, 1–18.

Neumann, E.-R., Wilson, M., Herremans, M., Spencer, E. A., Obst, K., Timmerman, M. & Kirstein, L. 2004. Carboniferous–Permian rifting and magmatism in southern Scandinavia, the North Sea and northern Germany: a review. *In:* Wilson, M., Neumann, E.-R., Davies, G. R., Timmerman, M. J., Herremans, M., & Larsen, B. T. (eds) *Permo-Carboniferous Magmatism and Rifting in Europe*. Geological Society, London, Special Publications, **223**, 11–40.

Niklas, K. J. 1997. *The Evolutionary Biology of Plants*. The University of Chicago Press.

Nöldeke, W. 1976. Das Obervisé von Doberlug-Kirchhain. *Jahrbuch für Geologie (für 1969/70)*, **5/6**, 589–706.

Oberhänsli, R. 1986. Geochemistry of meta-lamprophyres from the Central Swiss Alps. *Schweizerische Mineralogische und Petrographische Mitteilungen*, **66**, 315–342.

Oberhänsli, R., Schenker, F. & Mercolli, L. 1988. Indications of Variscan nappe tectonics in the Aar Massif. *Schweizerische Mineralogische und Petrographische Mitteilungen*, **68**, 509–520.

O'Brian, P. J. 1997. Garnet zoning and reaction textures in overprinted eclogites, Bohemian Massif, European Variscides: a record of their thermal history during exhumation. *Lithos*, **41**, 119–133.

O'Brian, P. J. 2000. The fundamental Variscan problem: high-temperature metamorphism at different depths and high-pressure metamorphism at different temperatures. *In:* Franke, W., Haak, V., Oncken, O. & Tanner, D. (eds) *Orogenic Processes: Quantification and Modelling in the Variscan Belt*. Geological Society, London, Special Publications, **179**, 369–386.

O'Brian, P. J. & Carswell, D. A. 1993. Tectonometamorphic evolution of the Bohemian massif: evidence from high pressure metamorphic rocks. *Geologische Rundschau*, **82**, 531–555.

O'Brian, P. J., Carswell, D. A. & Gebauer, D. 1990. Eclogite formation and distribution in the European Variscides. *In:* Carswell, D. A. (ed.) *Eclogite Facies Rocks*. Blackie, Glasgow, 204–224.

O'Brian, P. J., Kröner, A., Jaekel, P., Hegner, E., Zelazniewicz, A. & Kryza, R. 1997. Petrological and isotopic studies on Palaeozoic high pressure granulites, Góry Sowie (Owl) Mts., Polish Sudetes. *Journal of Petrology*, **38**, 433–456.

Odin, G. S. 1994. Geological time scale 1994. *Comptes Rendus Hebdomadaires des Séances de l'Académie des Sciences*, série II, **318**, 59–71.

Oehlke, M., Weger, M. & Lammerer, B. 1993. The "Hochfeiler Duplex" - imbrication tectonics in the SW Tauern Window. *Abhandlungen der Geologischen Bundesanstalt Wien*, **49**, 107–124.

Okrusch, M. 1983. The Spessart crystalline complex, northwest Bavaria. *Fortschritte der Mineralogie*, **61**, 135–169.

Okrusch, M. 1995. Mid-German Crystalline High - Metamorphic evolution. *In:* Dallmeyer, R. D., Franke, W. & Weber, K. (eds) *Pre-Permian Geology of Central and Eastern Europe*. Springer-Verlag, Berlin, 201–213.

Okrusch, M. & Richter, P. 1986. Orthogneisses of the Spessart crystalline complex, northwest Bavaria, indicators of the geotectonic environment? *Geologische Rundschau*, **75**, 555–568.

Okrusch, M. & Weber, K. 1996. Der Kristallinkomplex des Vorspessart. *Zeitschrifte für Geologische Wissenschaften*, **24**, 141–174.

Okrusch, M., Schubert, W. & Stähle, V. 2000. The Odenwald, Germany: Variscan metamorphic evolution and igneous events. *European Journal of Mineralogy, Beihefte*, **12**, 45–89.

Okrusch, M., Seidl, E., Schüssler, U. & Richter, P. 1989. Geochemical characteristics of metabasites in different tectonic units of the northeast Bavarian crystalline basement. *In:* Emmermann, R. & Wohlenberg, J. (eds) *The German Continental Deep Drilling Program (KTB)*. Springer, Berlin, 67–81.

Okrusch, M., Matthes, S., Klemd, R., O'Brian, P. J. & Schmidt, K. 1991. Eclogites at the northwestern margin of the Bohemian Massif: a review. *European Journal of Mineralogy*, **3**, 707–730.

Okrusch, M., Schubert, W. & Nasir, S. 1995. Igneous activity (Pre- to Early Variscan magmatism). *In:* Dallmeyer, R. D., Franke, F. & Weber, K. (eds) *Pre-Permian Geology of Central and Eastern Europe*. Springer Verlag, Berlin, 190–200.

Olempska, E. 1997. Changes in benthic ostracod assemblages across the Devonian-Carboniferous boundary in the Holy Cross Mountains, Poland. *Acta Palaeontologica Polonica*, **42**, 291–332.

Oliver, G. J. H., Corfu, F. & Krogh, T. E. 1993. U-Pb ages from SW Poland: evidence for a Caledonian suture zone between Baltica and Gondwana. *Journal of the Geological Society, London*, **150**, 355–369.

Oncken, O. 1997. Transformation of a magmatic arc and an orogenic root during oblique collision and its consequence for the evolution of the European Variscides (Mid-German Crystalline Rise). *Geologisches Rundschau*, **86**, 2–20.

Oncken, O. 1998. Evidence for subduction-erosion in a fossil collisional belt – Mid-European Variscides (Germany). *Geology*, **26**, 1075–1078.

Oncken, O. & Weber, K. 1995. Structure. *In:* Dallmeyer, R. D., Franke, F. & Weber, K. (eds) *Pre-Permian Geology of Central and Eastern Europe*. Springer Verlag, Berlin, 50–58.

Oncken, O., Franzke, H. J., Dittmar, U. & Klügel, T. 1995. Structure (Northern Phyllite Zone). *In:* Dallmeyer, R. D., Franke, F. & Weber, K. (eds) *Pre-Permian Geology of Central and Eastern Europe*. Springer Verlag, Berlin, 108–117.

Oncken, O., Von Winterfeld, C. & Dittmar, U. 1999. Accretion of a rifted passive margin the late Paleozoic Rhenohercynian fold and thrust belt (Middle European Variscides). *Tectonics*, **18**, 75–91.

Oncken, O., Plesch, A., Weber, J., Ricken, W. & Schrader, S. 2000. Passive margin detachment during arc-continent collision (Central European Variscides). *In:* Franke, W., Haak, V., Oncken, O. & Tanner, D. (eds) *Orogenic Processes: Quantification and Modelling in the Variscan Belt*. Geological Society, London, Special Publications, **179**, 199–216.

Opustil, S. 2005. Evolution of the Middle Westphalian river valley drainage system in central Bohemia (Czech Republic), and its palaeogeographic implication. *Palaeogeography, Palaeoclimatology, Palaeoecology*, **222**, 223–258.

Oplustil, S. & Pešek, J. 1998. Stratigraphy, palaeoclimatology and palaeogeography of the Late Palaeozoic continental deposits in the Czech Republic. *Geodiversitas*, **20**, 547–620.

Opluštil, S., Sýkorová, I. & Bek, J. 1999. Sedimentology, coal petrology and palynology of the Radnice Member in the S-E part of the Kladno-Rakovník Basin, Central Bohemia (Bolsovian). *Acta Universitatis Carolinae, Geologica*, 599–623.

Palmer, J. A., Perry, S. P. G. & Tarling, D. H. 1985. Carboniferous magnetostratigraphy. *Journal of the Geological Society, London*, **142**, 945–955.

Paproth, E. 1986. An introduction to a field trip to the late Devonian outcrops in the northern Rheinisches Schiefergebirge (Federal Republic of Germany). *Annales de la Société géologique de Belgique*, **109**(1), 275–284.

Paproth, E. 1991. Carboniferous palaeogeographic development in Central Europe. *Comptes Rendus Onzième Congrès International de Stratigraphie et de la Géologie Carbonifère* (Bejing 1987), **1**, 177–186.

Paproth, E., Conil, R., *et al.* 1983*a*. Bio- and Lithostratigraphic subdivisions of the Dinantian in Belgium, a review. *Annales de la Société géologique de Belgique*, **106**, 185–239.

Paproth, E., Dusar, M., *et al.* 1983*b*. Bio- and lithostratigraphic subdivisions of the Silesian in Belgium, a review. *Annales de la Société géologique de Belgique*, **106**, 241–283.

Paproth, E., Feist, R. & Flajs, G. 1991. Decision on the Devonian-Carboniferous boundary stratotype. *Episodes*, **14**, 331–336.

Paproth, E., Dusar, M., Verkaeren, P. & Bless, M. J. M. 1996. Stratigraphy and cyclic nature of Lower Westphalian deposits in the boreholes KB174 and KB206 in the Belgian Campine. *Annales de la Société géologique de Belgique*, **117**, 169–189.

Pašek, J. & Urban, M. 1990. The tectonic evolution of the Plzeň basin (Upper Carboniferous, West Bohemia): a review and reinterpretation. *Folia Museum Rerum naturalium Bohemiae occidentalis, Geologica*, Plzeň, **32**.

Paszkowski, M. & Szulczewski, M. 1995. Late Paleozoic carbonate platforms in Polish part of the Moravia-Małopolska shelf. *XIII International Congress Carbon.-Permian, Kraków*, Guide B-4, 5–12.

Patocka, F. 1991. Geochemistry and primary tectonic environment of the amphibolites from the Cesky Krumlov Varied Group (Bohemian Massif, Moldanubicum). *Jahrbuch für Geologie*, **134**, 117–133.

Patocka, F. & Szczepanski, J. 1997. Geochemistry of quartzites from the eastern margin of the Bohemian Massif (the Hruby Jeseník Mts. Devonian and the Strzelin Crystalline Massif). *Mineralogical*

Society of Poland, Special Papers, **9**, 151–154.

PATOCKA, F., VLASIMSKY, P. & BLECHOVÁ, K. 1993. Geochemistry of early Paleozoic volcanics of the Barrandian Basin (Bohemian Massif, Czech Republic): implications for paleotectonic reconstructions. *Jahrbuch der Geologischen Bundesanstalt*, **136/4**, 873–896.

PEGRUM, R. M. 1984*a*. Structural development of the southwestern margin of the Russian-Fennoscandian platform. *In:* SPENCER, A. M. *et al.* (eds) *Petroluem Geology of the North European Margin*. Norwegian Petroleum Society, Graham & Trotman, London, 359–369.

PEGRUM, R. M. 1984*b*. The extension of the Tornquist Zone in the Norwegian North Sea. *Norsk Geologisk Tidsskrift*, **64**, 39–68.

PEPPERS, R. A. 1996. Palynological correlation of major Pennsylvanian (Middle and Upper Carboniferous) chronostratigraphic boudaries in the Illinois and other coal basins. *Geological Society of America Memoir*, **188**, 1–111.

PERINI, G., CEBRIA, J. M., LOPEZ-RUIZ, J. & DOBLAS, M. 2004. Carbonifeorus-Permian mafic magmatism in the Variscan belt of Spain and France: implications for mantle sources. *In:* WILSON, M., NEUMANN, E.-R., DAVIES, G. R., TIMMERMAN, M. J., HERREMANS, M., & LARSEN, B. T. (eds) *Permo-Carboniferous Magmatism and Rifting in Europe*. Geological Society, London, Special Publications, **223**, 415–438.

PEŠEK, J. 1994. Carboniferous of Central and Western Bohemia (Czech Republic). Czech Geological Survey, Prague.

PEŠEK, J., HOLUB, V., JAROS, J., MALY, L., MARTÍNEK, K., PROUZA, V., SPUDIL, J. & TÁSLER, R. 2001. *Geologie a loziska svrchnopaleozoickych limnickych pánví v Ceské republice*. Cesky geologicky ústav, Praha.

PEŠEK, J., OPLUŠTIL, S., KUMPERA, O., HOLUB, V. & SKOČEK, V. 1998. *Paleogeographic Atlas, Late Paleozoic and Triassic Formations, Czech Republic*. Czech Geological Survey, Appendix 1–41, Prague, 1–53.

PETERS, T. MATTER, A., BLÄSI, H. R., ISENSCHMID, C., KLEBOTH, P., MEYER, C. & MEYER, J. 1989. *Sondierbohrung Leuggern*. Geologische Berichte, 9. Landeshydrologie und -geologie, Bern.

PETRAKAKIS, K. 1997. Evolution of Moldanubian rocks in Austria: review and synthesis. *Journal of Metamorphic Geology*, **15**, 203–222.

PIETZSCH, K. 1962. *Die Geologie von Sachsen*. VEB Deutscher Verlag der Wissenschaften, Berlin.

PILLOUD, C. 1991. Structures de déformation alpine dans le synclinal de permo-carbonifère de Salvan-Dorenaz (massif des Aiguilles Rouges). *Mémoire de Géologie (Lausanne)*, **9**, 1–100.

PIN, C. 1990. Variscan oceans: ages, origins and geodynamic implications inferred from geochemical and radiometric data. *Tectonophysics*, **177**, 215–227.

PIN, C. & LANCELOT, J. R. 1982. U-Pb dating of an early Palaeozoic bimodal magmatism in the French Massif Central and of its further metamorphic evolution. *Contributions to Mineralogy and Petrology*, **79**, 1–12.

PIQUE, A. & MICHARD, A. 1989. Moroccan Hercynides; a synopsis; the Paleozoic sedimentary and tectonic evolution at the northern margin of West Africa. *American Journal of Science*, **289**, 286–330.

PISKE, J., RASCH, H.-J., NEUMANN, E. & ZAGORA, K. 1994. Geologischer Bau und Entwicklung des Präperms der Insel Rügen und des angrenzenden Seegebietes. *Zeitschrift des Geologischen Wissenschaften*, **22**, 211–226.

PITCHER, W. S. 1983. Granite type and tectonic environment. *In:* HSU, K.J. (ed.) *Mountain Building Processes*. Academic Press, London, 19–40.

PITRA, P., BURG, J. P. & GUIRAUD, M. 1999. Late Variscan strike-slip tectonics between the Teplá-Barrandian and Moldanubian terranes (Czech Bohemian Massif): petrostructural evidence. *Journal of the Geological Society of London*, **156**, 1003–1020.

PLANDEROVÁ, E., SITÁR, V., GRECULA, P. & EGYÜD, K. 1981. Biostratigraphical evaluation of graphite shales of Zemplínske vrchy Hills. *Mineralia Slovaca*, **13**, 97–128 [in Slovak].

PLAŠIENKA, D. & SOTÁK, J. 2001: Stratigraphic and tectonic position of Carboniferous sediments in the Furmanec (Muráň Plateau, Western Carpathians). *Mineralia Slovaca*, **33**, 29–44.

PLATEN, K. M. 1991. *Geochemische Untersuchungen an Metabasalten aus dem Rhenohercynikum*. MSc thesis, Justus-Liebig University, Gießen.

PLATT, J. 1988. The mechanics of frontal imbrication: a first-order analysis. *Geologische Rundschau*, **77**, 577–590.

PLESCH, A. & ONCKEN, O. 1999. Orogenic wedge growth during collision – constraints on mechanics of a fossil wedge from its kinematic record (Rheno-"Hercynian fold-and-thust belt, Central Europe). *Tectonophysics*, **304**, 117–139.

PLUSQUELLEC, Y. & JAHNKE, H. 1999. Les tabulés de l'Erbslochgrauwacke (Emsien inférieur du Kellerwald) et le problème des affinités paléogéographiques de l'allochtone "Giessen-Harz". *Abhandlungen der Geologischen Bundesanstalt*, **54**, 435–451.

POLLER, U., ALTENBERGER, U. & SCHUBERT, W. 2001. Geochemical investigations of the Bergstrasser Odenwald amphibolites - implications for back-arc magmatism. *Mineralogy and Petrology*, **72**, 63–76.

POLLER, U., JANÁK, M., KOHÚT, M. & TODT, W. 2000. Early Variscan magmatism in the Western Carpathians: U-Pb zircon data of granitoids from the High Tatra Mts. (Slovakia). *International Journal of Earth Science*, **89**, 336–349.

POREBSKI, S. J. 1981. Swiebodzice succession (Upper Devonian-lowermost Carboniferous): a prograding, mass flow dominated fan-delta complex. *Geologia Sudetica*, **16**, 101–192.

POREBSKI, S. J. 1990. Onset of coarse clastic sedimentation in the Variscan realm of the Sudetes (SW Poland): an example from the Upper Devonian-Lower Carboniferous Swiebodzice succession. *Neues Jahrbuch für Geologie und Paläontologie, Abhandlungen*, **179**, 259–274.

POREBSKI, S. J. 1991. Onset of coarse clastic sedimentation in the Variscan realm of Sudetes (SW Poland): an example from the Upper Devonian–Lower Carboniferous Swiebodzice succession. *Neues Jahrbuch für Geologie und Paläontologie, Abhandlungen*, **179**, 259–274.

PORZYCKI, J. 1972. The Siltstone Series of the Lower Westphalian stage of the Upper Silesian Coal Basin. *Prace Instytutu Geologicznego*, **61**, 505–508.

PORZYCKI, J. 1979. Litostratygrafia osadów karbonu Lubelskiego Zagłębia Węglowego. *In:* MIGIER, T. (ed.) *Stratygrafia weglonosnej formacji karbonskiej w Polsce*. II Symposjum, Sosnowiec, 4–5 May 1977, 19–28 [in Polish].

PORZYCKI, J. & ZDANOWSKI, A. 1995. Lithostratigraphy and sedimentologic-paleogeographic development: southeastern Poland (Lublin Carboniferous Basin). *In:* ZDANOWSKI, A. & ZAKOWA, H. (eds) *Prace Panstwowego Instytutu Geologicznego*, **148**, 102–109.

POTY, E. 1991. Tectonique de blocs dans le prolongement oriental du massif du Brabant. *Annales de la Société géologique de Belgique*, **114**, 265–275.

POTY, E. 1994. Nouvelles précisions sur les corrélations stratigraphiques du Dinantien du Boulonnais et de la Belgique: application de la biozonation corallienne. *Comptes Rendus de l'Académie des Sciences, Paris*, **319**, 467–473.

POTY, E. 1997. Devonian and Carboniferous tectonics in the eastern and southeastern part of the Brabant Massif (Belgium). Aardkundige Mededelingen, *Leuven University Press*, **8**, 143–144.

POTY, E., HANCE, L., LEES, A. & HENNEBERT, M. 2001. Dinantian lithostratigraphic units (Belgium). *Geologica Belgica*, **4**, 69–94.

POTY, E., ARETZ, M. & BARCHY, L. 2002. Stratigraphie et sedimentologie des Calcaires Productus du Carbonifere inferieur de la Montagne Noire (Massif Central, France). *Comptes Rendus de l'Academie des Sciences, Geosciences*, **334**, 843–848.

POTY, E., ARETZ, M. & BARCHY, L. 2002*a*. Stratigraphie et sédimentologie des "calcaires à *Productus*" du Carbonifère inférieur de la Montagne noire (Massif central, France). *Comptes Rendus Geoscience*, **334**, 843–848.

POTY, E., HANCE, L., LEES, A. & HENNEBERT, M. 2002*b*. Dinantian lithostratigraphic units (Belgium). *Geologica Belgica*, **4**, 69–94.

POWELL, C. MCA. & VEEVERS, J. J. 1987. Namurian uplift in Australia and South America triggered the main Gondwanan glaciation. *Nature*, **326**, 177–179.

POZARYSKI, W., GROCHOLSKI, A., TOMCZYK, H., KARNKOWSKI, P. & MORYC, W. 1992. Mapa tektoniczna Polski w epoce waryscyjskiej. *Przeglad Geologiczny*, **11**, 643–651 [in Polish with English summary].

QUESADA, C. 1991. Geological constraints on the Paleozoic tectonic evolution of tectonostratigraphic terranes in the Iberian Massif. *Tectonophysics*, **185**, 225–245.

RACYZNSKA, A. (ed.) 1987. *Geological Structure of the Pomeranian Swell and its Basement*. Prace Institut Geologiczny Wydawnictwa Geologiczne Warszawa, **CYIX** [in Polish].

Raith, M., Mehrens, C. & Thöle, W. 1980. Gliederung, tektonischer Bau und metamorphe Entwicklung der penninischen Serien im südlichen Venediger-Gebiet, Osttirol. *Jahrbuch der Geologischen Bundesanstalt Wien*, **123**, 1–37.

Rajlich, P. 1980. Strain and tectonic styles related to Variscan transpression and transtension in the Moravo-Silesian Culmian Basin, Bohemian Massif, Czechoslovakia. *Tectonophysics*, **174**, 351–367.

Rakusz, Gy. 1932. Die oberkarbonischen Fosilien von Dobšiná und Nagyvisnyo. *Geologica Hungarica, Series Palaeontologica*, **8**, 1–219.

Ramsbottom, W. H. C. 1979. Rates of transgression and regression in the Carboniferous of NW Europe. *Journal of the Geological Society, London*, **136**, 147–154.

Ramsbottom, W. H. C. 1984. The founding of the Carboniferous System. *In:* Mackenzie, G. (ed.) *Compte Rendu 9ème Congrès International de Stratigraphie et de Géologie du Carbonifère 1979, Volume 1*. South Illinois University Press, Carbondale, 109–112.

Ramsbottom, W. H. C., Calver, M. A., Eagar, R. M. C., Hodson, F., Holliday, D. W., Stubblefield, C. J. & Wilson, R. B. 1978. *A Correlation of Silesian Rocks in the British Isles*. Geological Society, London, Special Reports, **10**.

Rappsilber, I. 2003. *Struktur und Entwicklung des nördlichen Saale Beckens (Sachsen-Anhalt): Geophysik und Geologie*. PhD thesis, Martin-Luther-Universität Halle-Wittenberg.

Ratschbacher, L. 1984. Beitrag zur Neugliederung der Veitscher Decke (Grauwackenzone) in ihrem Westabschnitt (Obersteiermark, Österreich). *Jahrbuch der Geologischen Bundesanstalt*, **127**, 423–453.

Ratschbacher, L. 1987. Stratigraphy, tectonics, and paleogeography of the Veitsch nappe (Greywacke zone, Eastern Alps, Austria): a rearrangement. *In:* Flügel, H. W., Sassi, F. P. & Grecula, P. (eds) *Pre-Variscan and Variscan Events in the Alpine-Mediterranean Mountain Belts*. Mineralia Slovaca-Monography, Alfa Publishers, Bratislava, 407–414.

Raymond, A. & Metz, C. 2004. Ice and its consequences: Glaciation in the Late Ordovician, Late Devonian, Pennsylvanian-Permian, and Cenozoic compared. *Journal of Geology*, **112**, 655–670.

Raymond, A., Kelley, P. H. & Lutken, C. B. 1989. Polar glaciers and life at the equator: The history of Dinantian and Namurian (Carboniferous) climate. *Geology*, **17**, 408–411.

Reinhardt, J. & Kleemann, U. 1994. Extensional unroofing of granulitic lower crust and related low-pressure, high-temperature metamorphism in the Saxonian Granulite Massif, Germany. *Tectonophysics*, **238**, 71–94.

Reischmann, T. & Anthes, G. 1996. Geochronology of the mid-German crystalline rise west of the River Rhine. *Geologisches Rundschau*, **85**, 761–774.

Reischmann, T., Anthes, G., Jaeckel, P. & Altenberger, C. 2001. Age and origin of the Böllsteiner Odenwald. *Mineralogy and Petrology*, **72**, 29–44.

Reitz, E. 1987. Palynologie in metamorphen Serien: I. Silurische Sporen aus einem granatführenden Glimmerschiefer des Vor-Spessart. *Neues Jahrbuch für Geologie und Paläontologie, Monatshefte*, **11**, 699–704.

Reitz, E. & Wickert, F. 1989. Late Cambrian to Early Ordovician acritarchs from the Villé unit, Northern Vosges Mountains (France). *Neues Jahrbuch für Geologie und Paläontologie, Monatshefte*, **6**, 375–384.

Remy, W. & Remy, R. 1977. *Die Floren des Erdaltertums. Einführung in die Morphologie, Anatomie, Geobotanik und Biostratigraphie der Pflanzen des Paläophytikums*. Essen (Glückauf Verlag), 468 pp.

Renier, A., 1930. Considerations sur la stratigraphie du terrain houiller de la Belgique. *Memoires du Musee royal d'histoire naturelle de Belgique*, **44**, 1–101.

Rey, P., Burg, J. & Caron, J. M. 1992. Middle and Late Carboniferous extension in the Variscan belt: structural and petrological evidences from the Vosges massif (eastern France). *Geodinamica Acta*, **5**, 17–36.

Rey, P., Burg, J.P. & Casey, M. 1997. The Scandinavian Caledonides and their relationship to the Variscan Belt. *In:* Burg, J.P. & Ford, M. (eds) *Orogeny Through Time*. Geological Society, London, Special Publications, **121**, 179–200.

Ribbert, K.-H. 1998. Vulkanopsammite in der pelitischen Facies des Unterkarbon II und III im Westharz. *Zeitschrift für Geologische Wissenschaften*, **26**, 445–468.

Ricken, W., Schrader, S., Oncken, O. & Plesch, A. 2000. Turbidite basin and mass dynamics related to orogenic wedge growth; the Rheno-Hercynian case. *In:* Franke, W., Haak, V., Oncken, O. & Tanner, D. (eds) *Orogenic Processes: Quantification and Modelling in the Variscan Belt*. Geological Society, London, Special Publications, **179**, 257–280.

Robardet, M. 2003. The Armorica "microplate": fact or fiction? Critical review of the concept and contradictory palaeobiogeographical data. *Palaeogeography, Palaeoclimatology, Palaeoecology*, **195**, 125–148.

Roscher, M. & Schneider, J. W. 2006. Early Pennsylvanian to Late Permian climatic development of central Europe in a regional and global context. *In:* Lucas, S. G., Cassinis, G. & Schneider, J. W. (eds) *Nonmarine Permian Chronology and Correlation*. Geological Society, London, Special Publications, **265**, 95–136.

Ross, C. A. & Ross, R. P. 1987. Late Paleozoic sea levels and depositional sequences. *In:* Ross C. A. & Haman, D. (eds) *Timing and Depositional History of Eustatic Sequences: Constraints on Seismic Stratigraphy*. Cushman Foundation for Foraminiferal Research, Special Publication, **24**, 137–149.

Ross, C. A. & Ross, J. R. P. 1988. Late Paleozoic transgressive-regressive deposition. *In:* Wilgus, C. K., Hasitngs, B. S., Kendall, C. G. StC, *et al.* (eds) *Seal-Level Changes An Integrated Approach*. Society of Economic Paleontologists and Mineralogists, Special Publication, **42**, 227–247.

Rötzler, J. 1992. Zur Petrogenese im Sächsischen Granulitgebirge. Die pyroxenfreien Granulite und die Metapelite. *Geotektonische Forschungen*, **11**, 1–100.

Rötzler, J. & Romer, R.L. 1995. P-T-t evolution of ultrahigh-temperature granulites from the Saxon Granulite Massif, Germany. Part I: Petrology. *Journal of Petrology*, **42**, 1995–2013.

Rötzler, J., Schumacher, R., Maresch, V. W. & Wilner, A. P. 1998. Contrasts in the metamorphic evolution and exhumation history of different metapelite units in the Western Erzgebirge (Saxony, Germany). *European Journal of Mineralogy*, **10**, 261–280.

Rötzler, J., Carswell, A. A., Gerstenberger, U. & Haase, G. 1999. Transitional blueschist-epidote-amphibolite facies metamorphism in the Frankenberg massif, Germany, and tectonic implications. *Journal of Metamorphic Geology*, **17**, 109–125.

Rötzler, J., Romer, R. L., Budzinski, H. & Oberhänsli, R. 2004. Ultrahigh-temperature high-pressure granulites from Tirschheim, Saxon Granulite Massif, Germany: P-T-t path and geotectonic implications. *European Journal of Mineralogy*, **16**, 917–937.

Rouchy, J. M., Groessens, E. & Laumondais, A. 1993. Dislocation des formations évaporitiques par la tectonique et la dissolution : le modèle des évaporites dinantiennes du domaine varisque franco-belge. *Bulletin de la Société Géologique de France*, **164**, 38–50.

Rowley, D. B., Raymond, A., Totman Parrish, J., Lottes, A. L., Scotese, C. R. & Ziegler, A. M. 1985. Carboniferous palaeogeographic, phytogeographic and palaeoclimatic reconstructions. *International Journal of Coal Geology*, **5**, 7–42.

Ryf, W. 1984. Thermalbohrung 1983 Engerfeld, Rheinfelden. *Rheinfeldener Neujahrsblätter*, 1984, 35–52.

Samankassou, E. 1997. Palaeontological response to sea-level change: distribution of fauna and flora in cyclothems from the Lower *Pseudoschwagerina* Limestone (latest Carboniferous, Carnic Alps, Austria). *Geobios*, **30**(6), 785–796.

Samankassou, E. 1998. Skeletal Framework Mounds of Dasycladacean Alga *Anthracoporella*, Upper Paleozoic, Carnic Alps, Austria. *Palaios*, **13**, 297–300.

Sandeman, H. A., Clark, A. H., Scott, D. J. & Malpas, J. G. 2000. The Kennack Gneiss of the Lizard Peninsula, Cornwall, SW England: comingling and mixing of mafic and felsic magmas accompanying Givetian continental incorporation of the Lizard ophiolite. *Journal of the Geological Society, London*, **157**, 1227–1242.

Santallier, D. S. 1994. Introduction. The Massif Central. *In:* Keppie, J. D. (ed.) *Pre-Mesozoic Geology in France and Related Areas*. Springer-Verlag, Berlin, 267–275.

Ščerbak, N. P., Bartnickij, J. N., Mickevič, N. J., Stepanjuk, L. M., Cambel, B. & Grecula, P. 1988. U-Pb radiometric data of the Modra granodiorite from the Malé Karpaty Mts. and the Early

Paleozoic porphyroide of the Spišsko-gemerské rudohoria Mts. (Western Carpathians). *Geologica Zborník, Geologica Carpathica*, **39**, 427–436 [in Russian, English abstract].

SCHÄFER, A. 1989. Variscan molasse in the Saar-Nahe Basin (W-Germany). Upper Carboniferous and Lower Permian. *Geologisches Rundschau*, **78**, 499–524.

SCHÄFER, A. 2005. Sedimentologisch-numerisch begründeter stratigraphischer Standard für das Permo-Karbon des Saar-Nahe-Beckens. *In:* WREDE, V. (ed.) Deutsche Stratigraphische Kommission, Stratigraphie von Deutschland, V. Das Oberkarbon (Pennsylvanium) in Deutschland. *Courier Forschungsinstitut Senckenberg*, **254**, 369–394.

SCHÄFER, A. & KORSCH, R. J. 1998. Formation and sediment fill of the Saar-Nahe Basin (Permo-Carboniferous, Germany). *Zeitschrift der Deutschen Geologischen Gesellschaft*, **149**, 233–269.

SCHÄFER, F., ONCKEN, O., KEMNITZ, H. & ROMER, R. 2000. Upper-plate deformation during collisional orogeny: a case study from the German Variscides (Saxo-Thuringian Zone). *In:* FRANKE, W., HAAK, V., ONCKEN, O. & TANNER, D. (eds) *Orogenic processes: Quantification and modelling in the Variscan Belt*. Geological Society, London, Special Publications, **179**, 281–302.

SCHÄFER, J., NEUROTH, H., AHRENDT, H., DÖRR, W. & FRANKE, W. 1997. Accretion and exhumation at a Variscan active margin, recorded in the Saxothuringian flysch. *Geologische Rundschau*, **86**, 599–611.

SCHALTEGGER, U. 1997. Magma pulses in the Central Variscan Belt: episodic melt generation and emplacement during lithospheric thinning. *Terra Nova*, **9**, 242–245.

SCHALTEGGER, U. 2000. U-Pb geochronology of the southern Black Forest Batholith (Central Variscan Belt): timing of exhumation and granite emplacement. *International Journal of Earth Sciences*, **88**, 814–828.

SCHALTEGGER, U. & CORFU, F. 1995. Late Variscan "Basin and Range" magmatism and tectonics in the Central Alps: evidence from U-Pb geochronology. *Geodinamica Acta*, **8**, 82–98.

SCHALTEGGER, U., SCHNEIDER, J. L., MAURIN, J. C. & CORFU, F. 1996. Precise U-Pb chronometry of 345–340 Ma old magmatism related to syn-convergence extension in the Southern Vosges (Central Variscan Belt). *Earth and Planetary Science Letters*, **144**, 403–419.

SCHECK, M., BAYER, U., OTTO, V., LAMARCHE, J., BANKA, D. & PHARAOH, T. 2002. The Elbe Fault System in North Central Europe – a basement controlled zone of crustal weakness. *Tectonophysics*, **360**, 281–299.

SCHERER, E. E., MEZGER, K. & MÜNKER, C. 2002. Lu-Hf ages of high pressure metamorphism in the Variscan fold belt of southern Germany. Goldschmidt Conference Abstract 2002. *Geochimica et Cosmochimica Acta,* Supplement, **66**, A677.

SCHERMAIER, A. 1993. *Gliederung der Zentralgneise im mittleren und westlichen Tauernfenster: Geologie - Petrographie - Zirkontypologie - Geochemie*. Ph.D. thesis, University of Salzburg, Austria.

SCHERMAIER, A., HAUSCHMID, B. & FINGER, F. 1997. Distribution of Variscan I- and S-type granites in the Eastern Alps: a possible clue to unravel pre-Alpine basement structures. *Tectonophysics*, **272**, 315–333.

SCHEUVENS, D. & ZULAUF, G. 2000. Exhumation, strainlocation, and emplacement of granitoids along the western part of the Central Bohemian shear zone (Bohemian Massif). *International Journal of Earth Sciences*, **89**(3), 617–630.

SCHOLTZ, R. 1930. Das Variszische Bewegungsbild entwickelt aus der inneren Tektonik eines Profiles von der Böhmischen Masse bis zum Massiv von Brabant. *Fortschritte der Geologie und Paläontologie*, **8**, 235–316.

SCHMÄDICKE, E. 1994. Die Eklogite des Erzgebirges. *Freiberg Forschungshefte*, **C456**.

SCHMÄDICKE, E. & EVANS, B. W. 1997. Garnet-bearing ultramafic rocks from the Erzgebirge, and their relation to other settings in the Bohemian Massif. *Contributions to Mineralogy and Petrology*, **127**, 57–74.

SCHMÄDICKE, E., MEZGER, K., COSCA, M. A. & OKRUSCH, M. 1995. Variscan Sm-Nd and Ar-Ar ages of eclogite-facies rocks from the Erzgebirge and their relation to other settings in the Bohemian Massif. *Journal of Metamorphic Geology*, **13**, 537–552.

SCHMASSMANN, H. & BAYRAMGIL, O. 1946. Stratigraphie, Petrographie und Paläogeographie der Perm-Formation im schweizerischen Tafeljura und die Steinkohlenfrage der Nordschweiz. *Tätigkeitsberichte der Naturforschenden Gesellschaft Baselland*, **15**, 8–117.

SCHMID, S. M., PFIFFNER, O. A., FROITZHEIM, N., SCHÖNBORN, G. & KISSLING, E. 1996. Geological-geophysical transect and tectonic evolution of the Swiss-Italian Alps. *Tectonics*, **15**, 1036–1064.

SCHMIDEGG, O. 1961. Geologische Übersicht der Venedigergruppe. *Verhandlungen der Geologischen Bundesanstalt Wien*, **11961**, 34–55.

SCHMIDT, C., BRAUN, L., PALTZER, G., MÜHLBERG, M., CHRIST, P. & JACOB, F. 1924. *Bohrungen von Buix bei Pruntrut und Allschwil bei Basel*. Beiträge zur Geologie der Schweiz – Geotechnische Serie, 10. Aschmann and Scheller, Zürich.

SCHMIDT, K. & FRANKE, D. 1975. Stand und Probleme der Karbonforschung in der Deutschen Demokratischen Republik. *Zeitschrift für Geologische Wissenschaften*, **3**, 819–849.

SCHMIDT, F.-P., GEBREYOHANNES, Y. & SCHLIESTEDT, M. 1986. Das Grundgebirge der Rhön. *Zeitschrift der Deutsche Geologische Gesellschaft*, **137**, 287–300.

SCHNEIDER, J. L. 1990. *Enregistrement de la dynamique varisque dans les Bassins volcano-sédimentaires dévono-dinantiens: exemple des Vosges du Sud (zone moldanubienne)*. PhD thesis, Strasbourg University, France.

SCHNEIDER, J. L., HASSENFORDER, B. & PAICHELER, J. C. 1990. Une ou plusieurs "Lignes des Klippes" dans les Vosges du Sud (France)? Nouvelles données sur la nature des "klippes" et leur signification dans la dynamique varisque. *Comptes Rendues de la Academie Sci Paris*, **311**(II), 1221–1226.

SCHNEIDER, J. W. 1996. Biostratigraphie des kontinentalen Oberkarbon und Perm im Thüringer Wald, SW-Saale-Senke - Stand und Probleme. *Beiträge zur Geologie von Thüringen, Neue Folge*, **3**, 121–151.

SCHNEIDER, J. W. 2001. Rotliegendstratigraphie – Prinzipien und Probleme. *Beiträge zur Geologie von Thüringen, Neue Folge*, **8**, 7–42.

SCHNEIDER, J. W. & WERNEBURG, R. 2006. Insect biostratigraphy of the European late Carboniferous and early Permian. *In:* LUCAS, S. G., CASSINIS, G. & SCHNEIDER, J. W. (eds) *Non-marine Permian Biostratigraphy and Biochronology*. Geological Society, London, Special Publications, **265**, 325–336.

SCHNEIDER, J. W. & ZAJÍC, J. 1994. Xenacanthodier (Pisces, Chondrichthyes) des mitteleuropäischen Oberkarbon und Perm - Revision der Originale zu Goldfuss 1847 und Fritsch 1879–1890. *Freiberger Forschungshefte*, **C452**, 1–151.

SCHNEIDER, J. W., HAMPE, O. & SOLER-GIJÓN, R. 2000. The Late Carboniferous and Permian: Aquatic vertebrate zonation in Southern Spain and German basins. *IGCP 328, Final Report*. Courier Forschungsinstitut Senckenberg, **223**, 543–561.

SCHNEIDER, J. W., LUCAS, S. G. & ROLAND, J. M. 2004. The blattida (insecta) fauna of Carrizo Arroyo, New Mexico - biostratigraphic link between marine and non-marine Pennsylvanian/Permian boundary profiles. *Bulletin of the New Mexico Museum of Natural History and Science*, **25**, 247–261.

SCHNEIDER, J. W., GORETZKI, J. & RÖSSLER, R. 2005*a*. Biostratigraphisch relevante nicht-marine Tiergruppen im Karbon der variscischen Vorsenke und der Innensenken. *In:* WREDE, V. (ed.) *Stratigraphie von Deutschland. Oberkarbon*. Courier Forschungsinstitut Senckenberg, **254**, 103–118.

SCHNEIDER, J. W., RÖSSLER, R., GAITZSCH, B. G., GEBHARDT, U. & KAMPE, A. 2005*b*. Saale-Senke. *In:* WREDE, V. (ed.) *Stratigraphie von Deutschland. Oberkarbon*. Courier Forschungsinstitut Senckenberg, **254**, 419–440.

SCHNEIDER, J. W., RÖSSLER, R., *et al.* 2005*c*. Vorerzgebirge-Senke und Erzgebirge. *In:* WREDE, V. (ed.) *Stratigraphie von Deutschland. Oberkarbon*. Courier Forschungsinstitut Senckenberg, **254**, 447–460.

SCHNEIDER, J. W., HOTH, K., GAITZSCH, B. G., BERGER, H. J., STEINBORN, H., WALTER, H. & ZEIDLER, M. 2005*d*. Carboniferous stratigraphy and development of the Erzgebirge Basin, East Germany. *Zeitschrift der Deutsche Gesellschaft für Geowissenschaften*, **156**, 431–466.

SCHNEIDER, J. W., KÖRNER, F., ROSCHER, M. & KRONER, U. 2006. Permian climate development in the northern peri-Tethys area - the Lodève Basin, French Massif Central compared in a European and global context. *Palaeogeography, Palaeoclimatology, Palaeoecology*,

240, 161–183

SCHOLTZ, H. 1930. Das varistische Bewegungsbild, entwickelt aus der inneren Tektonik eines Profils von der Böhmischen Masse bis zum Massiv von Brabant. *Fortschritte der Geologie und Paläontologie*, **8**, 235–316.

SCHOLTZ, H. 1934. Die Tektonik des Steinkohlenbeckens im Saar-Nahe-Gebiet und die Entstehungsweise der Saar-Saale-Senke. *Zeitschrift des Deutsche Geologische Gesellschaft*, **85**, 316–382.

SCHÖNEBERG, R. & NEUGEBAUER, J. 1994. *Einführung in die Geologie Europas*. Romback Verlag, Freiburg.

SCHÖNLAUB, H. P. 1979. Das Paläozoikum in Österreich. *Abhandlungen der Geologischen Bundesanstalt*, **33**, 1–124.

SCHÖNLAUB, H. P. 1982. Die Grauwackenzone in den Eisenerzer Alpen (Österreich). *Jahrbuch der geologischen Bundesanstalt*, **124**, 361–423.

SCHÖNLAUB, H. P. 1985*a*. Das Karbon von Nötsch und sein Rahmen. *Jahrbuch der geologischen Bundesanstalt*, **127**, 673–692.

SCHÖNLAUB, H. P. 1985*b*. Das Paläozoikum der Karnischen Alpen. *Arbeitstagung der Geologischen Bundesanstalt*. Kötschach-Mauthen, Gailtal, Kärnten, 34–52.

SCHÖNLAUB, H. P. & HEINISCH, H. 1993. The classic fossiliferous Paleozoic units of the Eastern and Southern Alps. *In:* VON RAUMER, J. F. & NEUBAUER, F. (eds) *Pre-Mesozoic Geology in the Alps*. Springer, Berlin, 395–422.

SCHÖNLAUB, H. P., ATTREP, M., *et al.* 1992. The Devonian/Carboniferous boundary in the Carnic Alps (Austria) – A multidisciplinary approach. *Jahrbuch der Geologischen Bundesanstalt*, **135**, 57–98.

SCHRADER, S. 1999. *Das Rhenohercynische Turbiditbecken (Spätes Viseum) - Faziesentwicklung und Beckenanalyse*. Ph.D. thesis, University of Cologne, Germany.

SCHRAUT, G. 1996. Die Arthropoden aus dem Unterkarbon von Nötsch (Kärnten/Österreich). *Abhandlungen der Geologischen Bundesanstalt*, **51**, 1–193.

SCHRAUT, G. 1999. Paläofaunistische Untersuchungen aus dem Unterkarbon von Nötsch (Kärnten/Österreich), Teil 2: Cephalopoda (Nautiloidea, Ammonoidea), Crustacea (Phyllocarida), Echinoidea. *Jahrbuch der Geologischen Bundesanstalt*, **141**, 503–517.

SCHUBERT, W., LIPPOLT, H. J. & SCHWARZ, W. 2001. Early- to Middle-Carboniferous hornblende $^{40}Ar/^{39}Ar$ ages of Bergsträsser Odenwald amphibolites and gabbros. *Mineralogy and Petrology*, **72**, 113–132.

SCHULMANN, K. & GAYER, R. 2000. A model for a continental accretionary wedge developed by oblique collision: the NE Bohemian Massif. *Journal of the Geological Society, London*, **157**, 401–416.

SCHWAB, M. 1979. Zum Deckenbau der Variszíden (Harz-Rheniden-Südwestural). *Zeitschrift für Geologische Wissenschaften*, **7**, 1131–1155.

SCHWEIZERISCHE GEOLOGISCHE KOMMISSION 1980. *Geologische Karte der Schweiz* 1:500000. Bern.

SCOTESE, C. R. & MCKERROW, W. S. 1990. Revised world maps and introduction. *In:* MCKERROW, W. S. & SCOTESE, C. R. (eds) *Palaeozoic Palaeogeography and Biogeography*. Geological Society, London, Memoirs, **12**, 1–21.

SCOTESE, C. R. & LANGFORD, R. P. 1995. Pangea and the paleogeography of the Permian. *In:* SCHOLLE, P. A., PERYT, T. M. & ULMER-SCHOLLE, D. S. (eds) *The Permian of Northern Pangea. Paleogeography, Paleoclimates, Stratigraphy. Vol. 1*. Springer, Berlin, 3–19.

SCOTT, A. C., GALTIER, J. & CLAYTON, G. 1984. Distribution of anatomically preserved floras in the Lower Carboniferous in Western Europe. *Transactions of the Royal Society Edinburgh*, **75**, 311–340.

SEBASTIAN, U. 1995. Die Strukturentwicklung des spätorogenen Erzgebirgsaufstieg in der Flöhazone - Ein weiterer Beitrag zur postkollisionalen Extension am Nordrand der Böhmischen Masse. *Freiberger Forschungshefte*, **C461**.

SELLI, R. 1963. Schema geologico delle Alpi Carniche e Giulie occidentali. *Giornale di Geologia*, **12**(30), 1–136.

SEPKOSKI, J. J. Jr. 1995. Patterns of Phanerozoic extinction: a perspective from global data bases. *In:* WALLISER, O. H. (ed.) *Global Events and Event Stratigraphy*. Springer-Verlag, Berlin, 35–51.

SESTON, R., WINCHESTER, J. A., PIASECKI, M. A. J., CROWLEY, Q. G. & FLOYD, P. A. 2000. A structural model for the western-central Sudetes: a deformed stack of Variscan thrust sheets. *Journal of the Geological Society, London*, **157**, 1155–1167.

ŠETLÍK, J. 1977. *Results on recent investigations on the Carboniferous flora of Bohemia*. Symposium on Carboniferous Stratigraphy, Czech Association of Geology, 315, 340.

SHAIL, R. K. & ALEXANDER, A. C. 1997. Late Carboniferous to Triassic reactivation of Variscan basement in the western English Channel: evidence from onshore exposures in south Cornwall. *Journal of the Geological Society, London*, **154**, 163–168.

SHANNON, P. M. 1991. The development of Irish offshore sedimentary basins. *Journal of the Geological Society, London*, **148**, 181–189.

SIEBEL, W. 1995. Constraints on Variscan granite emplacement in north-east Bavaria, Germany: further clues from petrogenetic studz of the Mitterteich granite. *Geologische Rundschau*, **84**, 384–398.

SIEBEL, W., RASCHKA, H., IRBER, W., KREUZER, H., LENZ, K. L., HÖHNDORF, A. & WENDT, I. 1997. Early Palaeozoic acid magmatism in the Saxothuringian Belt: new insights from a geochemical and isotopic study of orthogneisses and metavolcanic rocks from the Fichtelgebirge, SE Germany. *Journal of Petrology*, **38**, 203–230.

SIEBEL, W., CHEN, F. & SATIR, M. 2003. Late-Variscan magmatism revisited: new implications from Pb-evaporation zircon ages on the emplacement of redwitzites and granites in NE Bavaria. *Geologische Rundschau*, **92**, 36–53.

SKOČEK, V. 1990 Stephanian lacustrine delta sequence in central and northeastern Bohemia. *Sbornik geologiceskih, Geologie*, **45**, 91–122 [In Czech with English summary].

SKOMPSKI, S. 1996. Stratigraphic position and facies significance of the limestone bands in the subsurface Carboniferous succession of the Lublin Upland. *Acta Geologica Polonica*, **46**(3/4), 171–268.

SKOPEC, J., PEŠEK, J. & KOBR, M. 2000. Fosilní říční sít' na svrchu slánského souvrství v mšensko-roudnické pánvi. *Uhlí rudy geologický průzkum*, **5**, 3–11.

SMITH, L. B. & READ, J. F. 2000. Rapid onset of late Palaeozoic glaciation on Gondwana: evidence from Upper Mississippian strata of the Mid-continent United States. *Geology*, **28**, 279–282.

SOMMERMANN, A. 1993. Zirkonalter aus dem Granit der Bohrung Saar 1. *Beihefte zum European Journal of Mineralogy*, **5**, 145.

SOMMERMANN, A. E., MEISL, S. & TODT, W. 1992. Zirkonalter von drei verschiedenen Metavulkaniten aus dem Südtaunus. *Geologisches Jahrbuch Hessen*, **120**, 67–76.

SOTÁK, J., VOZÁROVÁ, A. & IVANIČKA, J. 2000. New microfossils from the Early Paleozoic formations of the Gemericum (Foraminiferida). *Geologica Carpathica*, Special Issue, **50**, 72–74.

SPALLETTA, C. & VENTURINI, C. 1988. Conglomeratic sequences in the Hochwipfel-Formation: a new paleogeographic hypothesis on the Hercynian flysch stage of the Carnic Alps. *Jahrbuch der Geologischen Bundesanstalt*, **131**, 637–647.

SPIŠIAK, J., HOVORKA, D. & IVAN, P. 1985. Klátov Group, the representative of the Paleozoic amphibolite facies metamorphites on the Inner Western Carpathians. *Geologické. Práce, Správy*, **82**, 205–220 [in Slovak].

STAMPFLI, G. M. & BOREL, G. D. 2002. A plate tectonic model for the Paleozoic and Mesozoic constrained by dynamic plate boundaries and restored synthetic oceanic isochrons. *Earth and Planetary Science Letters*, **196**, 17–33.

STAMPFLI, G. M., MOSAR, J., FAVRE, P., PILLEVUIT, A. & VANNAY, J.-C. 2001. Permo-Mesozoic evolution of the western Tethyan realm: the Neotethys/East- Mediterranean connection. *In:* ZIEGLER, P. A., CAVAZZA, W., ROBERTSON, A. H. F. & CRASQUIN-SOLEAU, S. (eds) *PeriTethys Memoir 6: Peritethyan rift/wrench basins and passive margins*. IGCP 369, Mémoires du Muséum national d'Histoire naturelle, Paris, 51–108.

STEIN, E. 1988. Die strukturgeologische Entwicklung im Übergangsbereich Saxothuringikum/Moldanubikum in NE-Bayern. *Geologica Bavarica*, **92**, 5–131.

STEIN, E. 2001. The geology of the Odenwald Crystalline Complex. *Mineralogy and Petrology*, **72**, 7–28.

STEIN, E. & DIETL, C. 2001. Hornblende thermobarometry of granitoids from the Central Odenwald (Germany) and their implications for the geotectonic development of the Odenwald. *Mineralogy and Petrology*, **72**, 185–207.

STEINBACH, V. 1990. Struktur und Entwicklung der karbonischen Molassen im Raum Delitzsch-Bitterfeld. *Freiberger Forschungshefte*, **C439**, 69–108.

STEMPROK, M., HOLUB, F. V. & NOVÁK, J. K. 2003. Multiple magmatic pulses of the eastern volcano-plutonic complex, Krusné Hory/Erzgebirge batholith, and their phosphorus contents. *Bulletin of*

Geosciences, **78**, 277–296.

STETS, J. & SCHÄFER, A. 2002. Depositional environments in the Lower Devonian siliciclastics of the Rhenohercynian Basin (Rheinisches Schiefergebirge, W-Germany) - Case studies and a model. *Contributions to Sedimentary Geology*, **22**.

STEWART, W. N. & ROTHWELL, G. W. 1993. *Paleobotany and the Evolution of Plants*. 2nd edn. Cambridge University Press, Cambridge.

STOLLHOFEN, H. 1994. Synvulkanische Sedimentation in einem fluviatilen Ablagerungsraum: Das basale 'Oberrotliegend' im permokarbonen Saar-Nahe Becken. *Zeitschrift der Deutschen Geologischen Gesellschaft*, **145**, 343–378.

STOLLHOFEN, H. 1998. Facies architecture variations and seismogenic structures in the Carboniferous-Permian Saar-Nahe basin (SW Germany): evidence for extension-related transfer fault activity. *Sedimentary Geology*, **119**, 47–83.

STOLLHOFEN, H., FROMMHERZ, B. & STANISTREET, I. G. 1999. Volcanic rocks as discriminants in evaluating tectonic versus climatic control on depositional sequences, Permo-Carboniferous continental Saar-Nahe Basin. *Journal of the Geological Society, London*, **156**, 801–808.

STUTE, M., FORSTER, M., *et al.* 1995. Cooling of tropical Brazil (50C) during the last glacial maximum. *Science*, **269**, 379–382.

SUK, M. (ed.) 1984. *Geological History of the Territory of the Czech Socialist Republic*. Academia Publishing House, Czech Academy of Science, Geological Survey, Prague.

SUESS, E. 1888. *Das Antlitz der Erde, II*. Tempsky, Wien.

SUESS, F. E. 1926. *Intrusionstektonik und Wandertektonik im variszischen Grundgebirge*. Borntraeger, Berlin.

SÜSS, M. P., DROZDZEWSKI, G. & SCHÄFER, A. 2000. Sequenzstratigraphie des kohleführenden Oberkarbon im Ruhr-Becken. *Geologisches Jahrbuch*, **A-156**, 45–106.

SÜSS, M. P., SCHÄFER, A. & DROZDZEWSKI, G. 2001. A sequence stratigraphic model for the Lower Coal Measures (Upper Carboniferous) of the Ruhr district, north-west Germany – Discussion. *Sedimentology*, **48**, 1171–1179.

SÜSS, M. P., DROZDZEWSKI, G. & SCHÄFER, A. 2002. The Ruhr Basin and the Aachen Basins – sedimentary environments, sequence stratigraphic model, and synsedimentary tectonics of Variscan foreland basins (Namurian B/C to Westphalian C; W. Germany). *In:* HILLS, L. V., HENDERSON, C. M. & BAMBER, E. W. (ed.) *Carboniferous and Permian of the World*. Canadian Society of Petroleum Geologists, Memoirs, **19**, 208–227.

SÜSS, M. P., DROZDZEWSKI, G. & SCHÄFER, A. 2006. Sedimentary environment dynamics and the formation of coal in the Pennsylvanian Variscan foreland in the Ruhr Basin (Germany, Western Europe). *International Journal of Coal Geology*, **69**, 267–287.

SZCZEPANSKI, J. & OBERC-DZIEDZIC, T. 1998. Geochemistry of amphibolites from the Strzelin crystalline massif, Fore-Sudetic Block, SW Poland. *Neues Jahrbuch für Geologie und Paläontologie, Abhandlungen*, **173**, 23–40.

TAIT, J. A. 1999. New Early Devonian paleomagnetic data from NW France: Paleogeography and implications for the Armorican microplate hypothesis. *Journal of Geophysical Research*, **104**(B2), 2831–2840.

TAIT, J. A., BACHTADSE, V. & SOFFEL, H. 1996. Eastern Variscan fold belt; paleomagnetic evidence for oroclinal bending. *Geology*, **24**, 871–874.

TAIT, J. A., BACHTADSE, V., FRANKE, W. & SOFFEL, H. C. 1997. Geodynamic evolution of the European Variscan Fold belt: palaeomagnetic and geological constraints. *Geologische Rundschau*, **86**, 585–598.

TAIT, J., SCHÄTZ, M., BACHTADSE, V. & SOFFEL, H. 2000. Palaeomagnetism and Palaeozoic palaeogeography of Gondwana and European terranes. *In:* FRANKE, W., HAAK, V., ONCKEN, O. & TANNER, D. (eds) *Orogenic Processes: Quantification and Modelling in the Variscan Belt*. Geological Society, London, Special Publications, **179**, 21–34.

TAIT, J. A., SCHÄTZ, M., BACHTADSE, V. & SOFFEL, H. 2002. Palaeomagnetism and Palaeozoic palaeogeography of Gondwana and European terranes. *In:* FRANKE, W., HAAK, V., ONCKEN, O. & TANNER, D. (eds) *Orogenic Processes: Quantification and Modelling in the Variscan Belt*. Geological Society, London, Special Publications, **179**, 21–34.

TAIT, J. BACHTADSE, V., FRANKE, W. & SOFFEL, H. C. 2004. Geodynamic evolution of the European fold belt: palaeomagnetic and geological constraints. *Geologische Rundschau*, **86**, 585–598.

TALBOT, J. Y., FAURE, M., CHEN, Y. & MARTELET, G. 2005. Pull-apart emplacement of the Margeride granitic complex (French Massif Central). Implications for the late evolution of the Variscan orogen. *Journal of Structural Geology*, **27**, 1610–1629.

TAYLOR, T. N. & TAYLOR, E. L. 1993. *The Biology and Evolution of Fossil Plants*. Prentice Hall, Englewood Cliffs.

TEICHMÜLLER, R. 1973. Die paläogeographische-fazielle und tektonische Entwicklung eines Kohlenbeckens am Beispiel des Ruhrkarbons. *Zeitschrift der Deutschen Geologischen Gesellschaft*, **124**, 149–165.

TEISSEYRE, H. 1968. Stratigraphy and tectonics of the wiebodzice depression. *Biuletyn Instytutu Geologicznego*, **222**, 71–106.

TENCHOV, Y. 1980. Die paläozoische Megaflora von Österreich. Eine Übersicht. *Verhandlungen der Geologischen Bundesanstalt*, **1980**, 161–174.

TESSENSOHN, F. 1971. Der Flysch-Trog und seine Randbereiche im Karbon der Karawanken. *Neues Jahrbuch für Geologie und Paläontologie, Abhandlungen*, **138**, 169–220.

TESSENSOHN, F. 1974*a*. Zur Fazies paläozoischer Kalke in den Karawanken (Karawankenkalke II). *Verhandlungen der Geologischen Bundesanstalt*, **1974**, 89–130.

TESSENSOHN, F. 1974*b*. Schichtlücken und Mischfaunen in paläozoischen Kalken der Karawanken. *Carinthia II*, **164**(84), 137–160.

TEUFEL, S. 1988. Vergleichende U-Pb- und Rb-Sr-Altersbestimmungen an Gesteinen der Übergangsbereiches Saxothuringikum/Moldanubikum, NE-Bayern. Göttinger Arbeiten für Geologie und Paläontologie, **49**, 1–87.

THIRWALL, M. F. 1988. Wenlock to mid-Devonian volcanism of the Caledonian-Appalachian orogen. *In:* HARRIS, A. L. & FETTES, D. J. (eds) *The Caledonian-Appalachian Orogen*. Geological Society, London, Special Publications, **38**, 415–428.

THOMAS, U. 1990. Unterkarbone Wildflysch-Ablagerungen im Südteil der DDR. *Zeitschrift für Angewandte Geologie*, **36**, 182–184.

THURY, M., GAUTSCHI, A., MAZUREK, M., *et al.* 1994. *Geology and hydrology of the crystalline basement of Northern Switzerland*. Nagra Technischer Bericht, NTB 93–01, Baden.

TICHOMIROWA, M., BERGER, H.-J., KOCH, E. A., *et al.* 2001. Zircon ages of high-grade gneisses in the Eastern Erzgebirge (Central European Variscides) - constraints on origin of the rocks and Precambrian to Ordovician magmatic events in the Variscan foldbelt. *Lithos*, **56**, 303–332.

TIMMERMAN, M. J. 2004. Timing, geodynamic setting and character of Permo-Carboniferous magmatism in the foreland of the Variscan Orogen, NW Europe. *In:* WILSON, M., NEUMANN, E.-R., DAVIES, G. R., TIMMERMAN, M. J., HERREMANS, M. & LARSEN, B. T. (eds) *Permo-Carboniferous Magmatism and Rifting in Europe*. Geological Society, London, Special Publications, **223**, 41–74.

TIMMERMAN, M. J. 2008. Palaeozoic magmatism. *In*: MCCANN, T. (ed.) *The Geology of Central Europe. Volume 1: Precambrian and Palaeozoic*. Geological Society, London, 665–748.

TISCHENDORF, G., FÖRSTER, H.-J., FRISCHBUTTER, A., KRAMER, W., SCHMIDT, W. & WERNER, C. D. 1995. Igneous activity (Autochthon and Nonmetamorphic Nappe Units). *In:* DALLMEYER, R. D., FRANKE, W. & WEBER, K. (eds) *Pre-Permian Geology of Eastern and Central Europe*. Springer-Verlag, Berlin, 249–259.

TODT, W. A., ALTENBERGER, U. & VON RAUMER, J. F. 1995. U-Pb data on zircons for the thermal peak of metamorphism in the Variscan Odenwald, Germany. *Geologische Rundschau*, **84**, 466–472.

TRÜMPY, R. 1980. *Geology of Switzerland – a Guide Book*. Part A: An Outline of the Geology of Switzerland. Wepf, Basel.

TRÜMPY, R. & DÖSSENEGGER, R. 1972. Permian of Switzerland. *In:* FAHLKE, H. (ed) *Rotliegend, Essays of European Lower Permian*. Brill, Leiden, 189–215.

TURNAU, E., ZELAZNIEWICZ, A. & FRANKE, W. 2002. Middle to early late Visean onset of late orogenic sedimentation in the Intra-Sudetic Basin, West Sudetes: miospore evidence and tectonics. *Geologia Sudetica*, **34**, 9–16.

TURNIAK, K., MAZUR, S. & WYSOCZANSKI, R. 2000. SHRIMP zircon geochronology and geochemistry of the Orlica-Snieznik gneisses (Variscan belt of Central Europe) and their tectonic implications. *Geodinamica Acta*, **13**, 1–20.

TURPIN, L., VELDE, D. & PINTE, G. 1988. Geochemical comparisons between minettes and kersantites from the Western European Hercynian orogen: trace element and Pb-Sr-Nd isotope constraints on their origin. *Earth and Planetary Science Letters*, **87**, 73–86.

Urbanek, Z., Zelazniewicz, A., Kemnitz, H., Hermsdorf, N. & Linnemann, U. 1995. Western Sudetes; Stratigraphy. *In:* Dallmeyer, R. D., Franke, W. & Weber, K. (eds) *Pre-Permian Geology of Eastern and Central Europe*. Springer-Verlag, Berlin, 315–327.

Ustaszewski, K., Schuhmacher, M. E. & Schmid, S. M., 2005. Simultaneous normal faulting and extensional flexuring during rifting: an example from the southernmost Upper Rhine Graben. *International Journal of Earth Sciences*, **94**, 680–696.

Vai, G. B. 1975. Hercynian basin evolution of the Southern Alps. *In:* Coy Squyres (ed.) *Geology of Italy, Vol 1*. Earth Sciences Society of the Libyen Arab Republic, Tripoli, 293–298.

Vai, G. B. & Cocozza, T. 1986. Tentative schematic zonation of the Hercynian chain in Italy. *Bulletin de la Société Géologique de France*, **8 II**(1), 95–114.

Vaida, M., Hann, H.P., Sawatzki, G., Frisch, W. & Loeschke, J. 2000. Sedimentation ages of paragneisses and lower grade metamorphic rocks from South Schwarzwald (Variscan belt, Germany) and their significance for the reconstruction of the paleotectonic evolution. *Terra Nostra*, **2000/1**, 128.

Vaida, M., Hann, H.P., Sawatzki, G., Frisch, W. & Loeschke, J. 2003. Ordovician and Silurian protolith ages of metamorphic rocks from the South Schwarzwald, SW Germany: a palynological study and its bearing on the Early Paleozoic geotectonic evolution. *Geological Magazine*, **141**, 629–643.

Vakhrameev, V. A., Dobruskina, I. A., Myen, S. V. & Zaklinsaja, E. D. 1978. *Paläozische und mesozoische Floren Eurasiens und Phytogeographie dieser Zeit*. Fischer, Jena.

Van Amerom, H. W. J. & Boersma, M. 1974. Vorläufige Untersuchungsergebnisse an älteren und neu aufgesammelten jungpaläozoischen Pflanzenfunden der Ostalpen (Österreich). *Carinthia II*, **164**, 9–15.

Van Amerom, H. W. J. & Kabon, H. 1999. Neue fossile Floren aus dem Nötscher Karbon (1. Teil). *Carinthia II*, **189**, 637–672.

Van Amerom, H. W. J. & Kabon, H. 2000. Neue fossile Floren aus dem Nötscher Karbon (2. Teil). *Carinthia II*, **190**, 483–516.

Van Amerom, H. W. J., Flajs, G. & Hunger, G. 1984. Die "Flora der Marinelli-Hütte" (Mittleres Visé) aus dem Hochwipfelflysch der Karnischen Alpen (Italien). *Mededelingen Rijks Geologische Dienst*, **37**, 21–41.

Van Breemen, O., Aftalion, M., Bowes, D. R., Dudek, A., Misar, Z., Povondra, P. & Vrana, S. 1982. Geochronological studies of the Bohemian massif, Czechoslovakia, and their significance in the evolution of Central Europe. *Transactions of the Royal Society of Edinburgh, Earth Sciences*, **73**, 89–108.

Van Leckwijck, W. 1948. Quelques observations sur les variations verticales des caractères lithologiques et fauniques de divers horizons marins du terrain houiller de Belgique. *Annales de la Société géologique de Belgique*, **72**, 439–468.

Van Steenwinkel, M. 1990. Sequence stratigraphy from "spot" outcrops – example from a carbonate-dominated setting: Devonian-Carboniferous transition, Dinant synclinorium (Belgium). *Sedimentary Geology*, **69**, 259–280.

Vanstone, S. D. 1996. The influence of climate change on exposure surface development: A case study from the Late Dinantian of England and Wales. *In:* Strogen, P., Somerville, I. D. & Jones, G. L. (eds) *Recent Advances in Carbonate Geology*. Geological Society, London, Special Publications, **107**, 281–301.

Veevers, J. J. & Powell, C. McA. 1987. Late Paleozoic glacial episodes in Gondwanaland reflected in transgressive-regressive depositional sequences in Euramerica. *Geological Society of America, Bulletin*, **98**, 475–487.

Veizer, J., Fritz, P. & Jones, B. 1986. Geochemistry of brachiopods: oxygen and carbon isotopic records of Paleozoic oceans. *Geochimica and Cosmochemica Acta*, **50**, 1679–1696.

Venturini, C. 1982. Il Bacino tardoercinico di Pramollo (Alpi Carniche): un evoluzione regolata dalla tettonica sinsedimentaria. *Memorie della Società Geologica Italiana*, **24,** 23–42.

Venturini, C. (ed.) 1990*a*. *Field workshop on Carboniferous to Permian Sequence of the Pramollo-Nassfeld Basin (Carnic Alps)*. Guidebook Arti Grafiche Friulani, Udine.

Venturini, C. 1990*b*. *Geologia delle Alpi Carniche centro orientali*. Edizioni dei Museo Friulano di Storia Naturale, Udine, **36**.

von Quadt, A. & Gebauer, D. 1998. Evolution of eclogitic rocks in the Erzgebirge: A conventional and SHRIMP U-Pb zircon and Sm-Nd study. *Acta Universitatis Carolinae-Geologica*, **42**, 324–325.

von Raumer, J. F. 1988. The Hercynian basement in the Helvetic realm Western and Central Alps. *Memorie della Società Geologica Italiana*, **29**, 57–69.

von Raumer, J. F. 1998. The Palaeozoic evolution in the Alps from Gondwana to Pangea. *Geologische Rundschau*, **87**, 407–435.

von Raumer, J. F., Stampfli, G. M. & Bussy, F. 2003. Gondwana-derived microcontinents - the constituents of the Variscan and Alpine collisional orogens. *Tectonophysics*, **365**, 7–22.

von Seckendorff, V., Arz, C. & Lorenz, V. 2004*a*. Magmatism of the late Variscan intermontane Saar-Nahe Basin (Germany): a review. *In:* Wilson, M., Neumann, E.-R., Davies, G. R., Timmerman, M. J., Herremans, M., & Larsen, B. T. (eds) *Permo-Carboniferous Magmatism and Rifting in Europe*. Geological Society, London, Special Publications, **223**, 361–391.

von Seckendorff, V., Timmerman, M. J., Kramer, W. & Wrobel, P. 2004*b*. New 40Ar/39Ar ages and geochemistry of late Carboniferous-early Permian lamprophyres and related volcanic rocks in the Saxothuringian Zone of the Variscan Orogen (Germany). *In:* Wilson, M., Neumann, E.-R., Davies, G. R., Timmerman, M. J., Herremans, M., & Larsen, B. T. (eds) *Permo-Carboniferous Magmatism and Rifting in Europe*. Geological Society, London, Special Publications, **223**, 335–359.

Vozárová, A. 1986. Problem of lithostratigraphical division of Permo-Carboniferous of the Zemplínske vrchy Hills and definition of the Luhyňa Formation. *Regionálna geologia Západných Karpát*, **21**, Dionýz Štúr Institute of Geology, Bratislava, 39–45 [in Slovak].

Vozárová, A. 1989. Petrology of crystalline rocks of Zemplinicum. *In:* Papanikolaou, D. & Sassi, F. (eds) *IGCP 276*. Geological Society of Greece, Athens, Newsletter, No. **1**, 97–104.

Vozárová, A. 1990. Development of metamorphism in the Gemeric/Veporic contact zone (Western Carpathians). *Geologicky Zbornik, Geologica Carpathica*, **41**, 475–502.

Vozárová, A. 1990. Development of metamorphism in the Gemeric/Veporic contact zone (Western Carpathians). *Geologicky Zborik, Geologica Carpathica*, **41**, 475–502.

Vozárová, A. 1993*a*. Pressure-temperature conditions of metamorphism in the northern part of the Branisko crystalline complex. *Geologica Carpathica*, **44**, 219–232.

Vozárová, A. 1993*b*. Provenance of the Gelnica Group metasandstones and relationship to paleotectonics of the sedimentary basin. *Západné Karpaty, seria Mineralogia, Petrgrafia, Geochemia, Metalogeneza*, **16**, 7–54 [in Slovak, with English summary].

Vozárová, A. 1996. Tectono-sedimentary evolution of Late Paleozoic sedimentary basins based on interpretation of lithostratigraphic data (Western Carpathians, Slovakia). *Slovak Geological Magazine*, **3–4**, 251–271.

Vozárová, A. & Ivanička, J. 1996. Geodynamic setting of Gelnica Group acid volcanism. *Slovak Geological Magazine*, **3–4**, D. Štúr Publ., Bratislava, 245–250.

Vozárová, A. & Vozár, J. 1979. Crystalline rocks in basal part of the Choč Nappe. *Geologické Práce, Správy*, **72**, 195–198 [in Slovak].

Vozárová, A. & Vozár, J. 1988. *Late Paleozoic in West Carpathians*. Monography. D. Štúr Institute of Geology Bratislava.

Vozárová, A. & Vozár, J. 1992. Tornaicum and Meliaticum in borehole Brusník, BRU-1, Southern Slovakia. *Acta Geologica Hungaricae*, **35**, 97–116.

Vozárová, A. & Vozár, J. 1996. Terranes of West Carpathian-North Panonian Domain. *Slovak Geological Magazine*, **1**, 61–83.

Vozárová, A., Soták, J. & Ivanička, J. 1998. Cambro-Ordovician fossils (conodonts, foraminifers, chitinous schields) from the metamorphic series of the Gemericum (Western Carpathians). *In: Tenth Meeting of European Union of Geosciences, Abstracts*, Vol. **4**. Cambridge University Press, New York, 266.

Vozárová, A., Frank, W., Král', J. & Vozár, J. 2005. $^{40}Ar/^{39}Ar$ dating of detrital mica from the Late Paleozoic sandstones in the Western Carpathians, Slovak Republic. *Geologica Carpathica*, **56,** 463–472.

Wachendorf, H. 1986. Der Harz - variszischer Bau und geodynamische Entwicklung. *Geologisches Jahrbuch*, **A 91**, 3–67.

Wagner, R. H. 1984. Megafloral zones of the Carboniferous. *IX*

International Congress of Stratigraphy and Geology of the Carboniferous, Compte Rendu **2**, 109–134.

Walkden, G. M. 1987. Sedimentary and diagenetic styles in late Dinantian carbonates of Britain. *In:* Miller, J., Adams, A. E. & Wright, Q. P. (eds) *European Dinantian Environments*. John Wiley, Chichester, 131–155.

Walliser, O. H. 1981. The geosynclinal development of the Rhenische Schiefergebirge (Rhenohercynian zone of the Variscides, Germany). *Geologie en Mijnbouw*, **60**, 91–96.

Walliser, O. H. 1995. Global events in teh Devonian and Carboniferous. *In:* Walliser, O. H. (ed.) *Global Events and Event Stratigraphy.* Springer-Verlag, Berlin, 225–250.

Walliser, O. H. & Alberti, H. 1983. Flysch, Olistostromes and nappes in the Harz Mountains. *In:* Martin, H. & Eder, F. W. (eds) *Intracontinental Fold Belts – Case studies in the Variscan Belt of Europe and the Damara Belt of Namibia.* Springer Verlag, Berlin, 144–169.

Walter, R. 1995. *Geologie von Mitteleuropa*. E. Schweizerbart'sche Verlagsbuchhandlung (Nägele u. Obermiller), Stuttgart.

Walter, R., Spaeth, G., & Kasig, W. 1985. An outline of the geological structure of the northeastern Hohes Venn area and its northern foreland. *Neues Jahrbuch für Geologie und Paläontologie Abhandlungen*, **171**, 107–216.

Warr, L. N. 2000. The Variscan Orogeny: the welding of Pangaea. *In:* Woodcock, N. & Strachan, R. (eds) *Geological History of Britain and Ireland.* Blackwell, Oxford, 271–294.

Weber, K. 1995*a*. Introduction (Mid-German Crystalline Rise). *In:* Dallmeyer, R. D., Franke, F. & Weber, K. (eds) *Pre-Permian Geology of Central and Eastern Europe.* Springer Verlag, Berlin, 153–154.

Weber, K. 1995*b*. The Spessart Crystalline Complex. *In:* Dallmeyer, R. D., Franke, F. & Weber, K. (eds) *Pre-Permian Geology of Central and Eastern Europe.* Springer Verlag, Berlin, 167–173.

Weber, K. 1995*c*. The Saar-Nahe Basin. *In:* Dallmeyer, R. D., Franke, F. & Weber, K. (eds) *Pre-Permian Geology of Central and Eastern Europe.* Springer Verlag, Berlin, 182–185.

Wedepohl, K. H., Meyer, K. & Muecke, G. K. 1983. Chemical composition and genetic relations of Meta-Volcanic rocks from the Rhenohercynian Belt of northwest Germany. *In:* Martin, H. & Eder, F. W. (eds) *Intracontinental Fold Belts – Case Studies in the Variscan Belt of Europe and the Damara Belt of Namibia.* Springer Verlag, Berlin, 231–256.

Weger, M. 1998. Duktile Kinematik kontinentaler Kruste am Beispiel der Zentralgneise des westlichen Tauernfensters (Ostalpen, Österreich und Italien). *Münchener Geologische Hefte*, **A22**, 1–268.

Weil, A. B., Van der Voo, R. & Van der Pluijm, B. A. 2001. Oroclinal bending and evidence against the Pangea megashear: The Cantabria-Asturias arc (northern Spain). *Geology*, **29**, 991–994.

Weingart, H. W. 1976. Das Oberkarbon in der Tiefbohrung Saar 1. *Geologisches Jahrbuch*, **A27**, 399–408.

Weithofer, K. A. 1896. Die geologischen Verhltnisse des Bayer-Schachtes und benachbarten teiles der Pilsner Kohlenmulde. *Osterreichische Zeitschrift für. Bergbau und Httenwesen*, **44**, 317–321, 331–335, 345–349, 355–357.

Weithofer, K. A. 1902. Geologische Skizze des Kladno - Rakonitzer Kohlenbeckens. *Verhandlungen der Kaiserlich-Kniglichen geologischen Reichsanstalt*, Wien, 399–420.

Werneburg, R. 1996. Temnospondyle amphibien aus dem Karbon Mitteldeutschlands. *Veröffentlichung des Naturhistorisches Museum, Schloß Bertholdsburg*, **11**, 23–64.

Werneburg, R. & Schneider, J. W. 2006. Amphibian biostratigraphy of the European Permo-Carboniferous. *In:* Lucas, S. G., Cassinis, G. & Schneider, J. W. (eds) *Non-marine Permian Biostratigraphy and Biochronology.* Geological Society, London, Special Publications, **265**, 201–215.

Werner, C.-D. 1987. Saxonian granulites: a contribution to the geochemical diagnosis of original rocks in high-metamorphic complexes. *Gerlands Beiträge zur Geophysik*, **96**, 271–290.

Werner, C.-D. & Rössler, H. J. 1979. Aussagenmöglichkeiten der initialen Magmatite für die Klärung struktureller Verhältnisse am Beispiel des Mitteleuropäischen Variszikum. *Zeitschrift für geologische Wissenschaften*, **7**, 353–366.

Wetzel, A. & Allia, V., 2003. Der Opalinuston in der Nordschweiz: Lithologie und Ablagerungsgeschichte. *Eclogae Geologicae Helvetiae*, **96**, 451–469.

Weyer, D. & Menning, M. 2006. Geologische Zeitskala, stratigraphische Nomenklatur und Magnetostratigraphie. *In:* Amler, M. R. W. & Stoppel, D. (eds) *Deutsche Stratigraphische Kommission: Stratigraphie von Deutschland VI - Unterkarbon (Mississippium).* Schriftenreihe der Deutsche Gesellschaft für Geowissenschaft, **41**, 27–50.

Weyer, D., Amler, M. R. W., Korn, D., Ribbert, K. H., Stoppel, D. & Weller, H. 2005. Das Unterkarbon (Mississipium) in der Stratigraphischen Tabelle von Deutschland 2002. *In:* Menning, M. & Hendrich, A. (eds) Erläuterungen zur Stratigraphischen Tabelle von Deutschland 2005. *Newsletters on Stratigraphy*, **41**, 71–76.

White, A. J. R. & Chappell, B. W. 1983. Granitoid belts and their distribution in the Lachlan Fold Belt, southeastern Australia. *Geological Society of America Memoir*, **159**, 21–34.

Wicander, R. & Monroe, J. S. 1993. *Historical Geology: Evolution of the Earth and Life through Time.* West Publishing, New York.

Wickert, F. & Eisbacher, G. H. 1988. Two-sided Variscan thrust tectonics in the Vosges Mountains, northeastern France. *Geodynamica Acta*, **2**, 101–120.

Wiegand, B. 1996. Isotopengeologische und geochemische Untersuchungen zur prävariscischen magmatischen und sedimentären Entwicklung im saxothuringisch-moldanubischen Übergangsbereich (Grenzgebiet BRD/CR). Unpublished Ph.D. thesis, University of Göttingen.

Wierich, F. & Vogt, W. 1997. Zur Verbreitung, Biostratigraphie und Petrographie unterkarbonischer Sandsteine des Hörre-Gommern-Zuges im östlichen Rhenoherzynikum. *Geologica et Palaeontologica*, **31**, 97–142.

Wignall, P.B. 1991. Model for transgressive black shales? *Geology*, **19**, 167–170.

Will, T. M. 2001. Paleostress-tensor analysis of late deformation events in the Odenwald Crystalline Complex and comparison with other units of the Mid-German Crystalline Rise, Germany. *Mineralogy and Petrology*, **72**, 229–247.

Will, T. M. & Schmädicke, E. 2001. A first report of retrogressed eclogites in the Odenwald Crystalline Complex: evidence for high-pressure metamorphism in the Mid-German Crystalline Rise, Germany. *Lithos*, **59**, 109–125.

Will, T. M. & Schmädicke, E. 2003. Isobaric cooling and anticlockwise P-T paths in the Variscan Odenwald Crystalline Complex, Germany. *Metamorphic Geology*, **21**, 469–480.

Willner, A. P., Massonne, H. J. & Krohe, A. 1991. Tectono-thermal evolution of a part of a Variscan magmatic arc: the Odenwald in the Mid-German Crystalline Rise. *Geologische Rundschau*, **80**, 369–389.

Willner, A. P. Rötzler, K. & Maresch, V. W. 1997. Pressure-temperature and fluid evolution of quartzo-feldspathic metamorphic rocks with a relic high-pressure, granulite-facies history from the central Erzgebirge (Saxony, Germany). *Journal of Petrology*, **38**, 307–336.

Willner, A. P., Krohe, A. & Maresch, W. V. 2000. Interrelated P-T-t-d paths in the Variscan Erzgebirge Dome (Saxony, Germany): constraints on the rapid exhumation of high-pressure rocks from the root zone of a collisional orogen. *International Geological Review*, **42**, 64–85.

Wilson, D., Davies, J. R., Fletcher, C. J. N. & Smith, M. 1990. Geology of the South Wales Coalfield, Part VI, the Country around Bridgend. HMSO, London.

Wilson, M., Neumann, E.-R., Davies, G. R., Timmerman, M. J., Heeremans, M. & Larsen, B. T. 2004. Permo-Carboniferous magmatism and rifting in Europe: introduction. *In:* Wilson, M., Neumann, E.-R., Davies, G. R., Timmerman, M. J., Herremans, M., & Larsen, B.T. (eds) *Permo-Carboniferous Magmatism and Rifting in Europe.* Geological Society, London, Special Publications, **223**, 1–10.

Wimmenauer, W. 1973. Lamprophyre, Semilamprophyre and anchibasaltische Ganggesteine. *Fortschritte der Mineralogie*, **51**, 3–67.

Wrede, V. & Zeller, M. 1991. Die Stratigraphische Einstufung der Bohrung Frenzer Staffel 1 (1985) (Inde-Mulde, Aachener Steinkohlenrevier). *Geologisches Jahrbuch*, **A 116**, 73–86.

Wrede, V., Kahlert, E., Kampe, A. & Thum, H. 2005. Das Oberkarbon (Silesium) in der Stratigraphischen Tabelle von Deutschland 2002. *In:* Menning, M. & Hendrich, A. (eds)

Erläuterungen zur Stratigraphischen Tabelle von Deutschland 2005. *Newsletters on Stratigraphy*, **41**, 77–89.

WRIGHT, V P. 1987. The evolution of the early Carbonifeous limestone province in SW Britain. *Geological Magazine*, **124**, 477–480.

WRIGHT, V. P. & VANSTONE, S. D. 2001. Onset of late Palaeozoic glacio-eustasy and the evolving climates of low latitude areas: a synthesis of current understanding. *Journal of the Geological Society of London*, **158**, 579–582.

WUNDERLICH, J. 1989. Neue Ergebnisse zur Geologie des südöstlichen Ruhlaer Kristallins. *Freiberger Forschungshefte*, **C429**, 7–32.

YANEV, Y. 1992. Configuration of the deep geophysical discontinuities beneath the territory of Bulgaria: discussion. *Geologica Balcanica*, **22**, 95–96.

ZAJÍC, J. 2000. Vertebrate zonation of the non-marine Upper Carboniferous-Lower Permian basins of the Czech Republic. *Courier Forschungs-Institut Senckenberg*, **223**, 563–575.

ZAKOWA, H. 1980. Main features of the Dinantian stratigraphy and development in the Holy Cross Mts. *In: Lithology and Tectonics of the Carboniferous Deposits. Comptes Rendus du 8e Congrès International de Stratigraphie et de Geologie Carbonifère*, **6**, 76–85 (Moscow 1975).

ZAKOWA, H. & MIGASZEWSKI, Z. 1995. Lithostratigraphy and sedimentologic-paleogeographic development; Góry Swietokrzyskie Mts. *In:* ZDANOWSKI, A. & ŻAKOWA, H. (eds) *Prace Panstwowego Instytutu Geologicznego*, **148**, 109–115.

ZDANOWSKI, A. & ZAKOWA, H. (eds) 1995. The Carboniferous system in Poland. *Prace Panstwowego Instytutu Geologicznego*, **148**, 1–215.

ZEH, A. 1996. Die Druck-Temperatur-Deformations-Entwicklung des Ruhlaer kristallins (Mitteldeutsche kristallinzone). *Geotek Forschungen*, **86**.

ZEH, A. & BRÄTZ, H. 2004. Timing of Upper Carboniferous-Permian horst-basin formation and magmatism in the NW Thuringian Forest, central Germany: a review. *In:* WILSON, M., NEUMANN, E.-R., DAVIES, G. R., TIMMERMAN, M. J., HERREMANS, M., & LARSEN, B. T. (eds) *Permo-Carboniferous Magmatism and Rifting in Europe*. Geological Society, London, Special Publications, **223**, 319–334.

ZEH, A., COSCA, M., BRÄTZ, H., OKRUSCH, M. & TICHOMIROWA, M. 2000. Simulatneous horst-basin formation and magmatism during Late VAriscan transtension: evidence from 40Ar/39Ar and 207Pb/206Pb geochronology in the Ruhla Crystalline Complex. *International Journal of Earth Science*, **89**, 52–71.

ZEH, A., WILLIAMS, I. S., BRÄTZ, H. & MILLAR, I. L. 2003. Different age response of zircon and monazite during the tectono-metamorphic evolution of a high grade paragneiss from the Ruhla Crystalline Complex, central Germany. *Contributions to Mineralogy and Petrology*, **143**, 691–706.

ŻELAŹNIEWICZ, A. 1990. Deformation and metamorphism in the Góry Sowie gneiss complex, Sudetes, SW Poland. *Neues Jahrbuch für Geologie und Paläontologie Abhandlungen*, **179**, 129–157.

ŻELAŹNIEWICZ, A. 1995. Western Sudetes: Introduction. *In:* DALLMEYER, R. D., FRANKE, W. & WEBER, K. (eds) *Pre-Permian Geology of Eastern and Central Europe*. Springer-Verlag, Berlin, 311–314.

ZELICHOWSKI, A. M. 1987. Development of the Carboniferous of the SW margin of the East-European Platform in Poland. *Przeglad Geologiczny*, **5**, 230–237.

ZELICHOWSKI, A. M. 1983*a*. Lithological associations in the Carboniferous in Poland. *Przeglad Geologiczny*, **XXXI**(362), 342–350.

ZELICHOWSKI, A. M. 1983*b*. The Carboniferous in western Pomerania. *Przeglad Geologiczny*, **XXXI**(362), 356–364.

ZELICHOWSKI, A. M. 1995. Lithostratigraphy and sedimentologic-paleogeographic development; Western Pomerania. *In:* ZDANOWSKI, A. & ZAKOWA, H. (eds) *Prace Panstwowego Instytutu Geologicznego*, **148**, 97–100.

ZEN, E. A. 1992. Using granite to image the thermal state of the source terrane. *Transactions of the Royal Society of Edinburgh, Earth Sciences*, **83**, 107–114.

ZIEGLER, P. A. 1984. Caledonian and Hercynian crustal consolidation of Western and Central Europe – a working hypothesis. *Geologie en Mijnbouw*, **63**, 93–108.

ZIEGLER, P. A. 1986. Geodynamic model for the Palaeozoic crustal consolidation of Western and Central Europe. *Tectonophysics*, **126**, 303–328.

ZIEGLER, P. A. 1989. *Evolution of Laurussia – A Study in Late Palaeozoic Plate Tectonics*. Kluwer Academic Publishers, Dordrecht.

ZIEGLER, P. A. 1990. *Geological Atlas of Western and Central Europe*. Shell Internationale Petroleum Maatschappij B.V.

ZIEGLER, P. A., BERTOTTI, S. & CLOETINGH, S. 2002. Dynamic processes controlling foreland development – the role of mechanical (de)coupling of orogenic wedges and forelands. *EGU Stephan Mueller Special Publication Series*, **1**, 17–56.

ZIEGLER, P. A., SCHUMACHER, M. E., DÈZES, P., VAN WEES, J.-D. & CLOETINGH, S. 2006. Post-Variscan evolution of the lithosphere in the area of the European Cenozoic Rift System. *In:* GEE. D.G. & STEPHENSON, R.A. (eds) *European Lithosphere Dynamics*. Geological Society, London, Memoirs, **32**, 97–112.

ZIMMERLE, W. 2000. The "Mitteldeutsche Schwelle" – Facts and Theories. *Zentralblatt für Geologie und Paläontologie Teil I*, **1999**, 163–184t.

ZULAUF, G. 1994. Ductile normal faulting along the West Bohemian Shear Zone (Moldanubian/Teplá-Barrandian boundary): evidence for late Variscan extensional collapse in the Variscan internides. *Geologische Rundschau*, **83**, 276–292.

ZULAUF, G. 1997. Von der Anchizone bis zur Eklogitfazies: Angekippte Krustenprofile als Folge der cadomischen und variscischen Orogenese im Teplá Barrandium (Böhmische Masse). *Geotektonische Forschungen*, **89.**

ZULAUF, G., DÖRR, W., FIALA, J. & VEJNAR, Z 1997. Late Cadomian crustal tilting and Cambrian transtension in the Teplá-Barrandian unit (Bohemian Massif, Central European Variscides). *Geologisches Rundschau*, **86**, 571–584.

ZULAUF, G., FIALA, J., KOTKOVÁ, J & MALUSKI, H. 1998. Rheologischer Kollaps eines unterkarbonischen Hochplateaus in der Böhmischen Masse: Die Rolle der Nordböhmischen Scherzone. *Freiburger Forschungshefte*, **C471**, 261–263.

ZULAUF, G., SCHITTER, F., RIEGLER, G., FINGER, F., FIALA, J. & VEJNAR, Z. 1999. Age constraints on the Cadomian evolution of the Teplá-Barrandian unit (Bohemian Massif) through electron microprobe dating of metamorphic monazite. *Zeitschrift der Deutschen Geologischen Gesellschaft*, **150**, 627–639.

10 Permian

TOM McCANN, HUBERT KIERSNOWSKI (co-ordinators), KARL KRAINER, ANNA VOZÁROVÁ, TADEUSZ M. PERYT, STANISLAV OPLUSTIL, HARALD STOLLHOFEN, JÖRG SCHNEIDER, ANDREAS WETZEL, FRÉDÉRIC BOULVAIN, MICHEL DUSAR, ÁKOS TÖRÖK, JÁNOS HAAS, JENNY TAIT & FRANK KÖRNER

The Permian (299–251 Ma; Wardlaw *et al.* 2004) succession of Central Europe records the change from a Pangaea configuration and compressive tectonic regime inherited from the Variscan Orogeny, to the development of the broad thermal subsidence-controlled Southern Permian Basin and its inundation by the Zechstein Sea. During latest Carboniferous–Early Permian times, the final phase of Variscan orogenic extension produced a series of small strike-slip and extensional continental basins across central and western Europe. Within these basins Stephanian and Lower Rotliegend continental successions were deposited. Subsequent thermal subsidence led to the gradual coalescence of these isolated basins to form the large Southern Permian Basin which extended across much of central and western Europe (Fig. 10.1).

Early Permian sedimentation was predominantly fluvial and lacustrine, changing later to aeolian. This change was due either to a significant climate change, or the result of a decline in relief of the surrounding uplands. By the end of the Early Permian extensive dunefields occupied the basin margins with saline lakes (playas) in the basin depocentres (Verdier 1996). A regional, possibly glacio-eustatic, rise in sea level later in Permian (Zechstein) times resulted in the rapid flooding (from the north) of the Southern Permian Basin. The Zechstein succession comprises a series of evaporitic cycles, and associated carbonates and muds, reflecting progressively greater evaporation and the shallowing either of the whole basin or the margins of the basin.

There has been a considerable amount of interest in the Permian in recent years, with a number of valuable books and papers published (e.g. Gebhardt *et al.* 1991; Plein 1993, 1995; Scholle *et al.* 1995*a*, *b*; Glennie 1998, 2005; Benek *et al.* 1996; Glennie *et al.* 2003; Schneider 2001; Menning *et al.* 2005; McCann *et al.* 2006; Wilson *et al.* 2004). This chapter provides an overview of the tectonic setting, indicating some of the controlling factors influencing the origin and development of the Permian basins in Central Europe. This is followed by an outline of the Permian stratigraphy within the main basins.

The Carboniferous–Permian boundary

The base of the Permian System is defined as the first occurrence of the isolated-nodular morphotype of the conodont *Streptognathodus isolatus* within the *S. 'wabaunsensis'* conodont chronocline, 27 m above the base of Bed 19 in the Aidaralash Creek section located in the Southern Ural Mountains of northern Kazakhstan (Wardlaw *et al.* 2004). Secondary markers include the termination of the *Prouddenites–Uddenites* ammonoid lineage at the top of Bed 19 and the introduction of *Svetlanoceras primore* and *Prostacheoceras principale* in Bed 20. Six metres higher up-section is the lowest fusilinid foram, *Sphaeroschwangerina*. Elsewhere, the first appearance of the 'isolated-nodular' morphotype of the *Streptognathodus 'wabaunsensis'* chronocline can be recognized in the basinal shales in Usolka in the lower part of Bed 16 (Chuvashov *et al.* 1990).

In Central Europe the boundary may be defined by fusilinids, for example, within the uppermost part of the Lower Pseudoschwagerina Limestone of the Rattendorf Group (see McCann *et al.* 2008). However, there are areas where the change is marked by lithological variations. For example, the Late Carboniferous–Early Permian of the Western Carpathians is characterized by the continental, coarse-grained siliciclastic formations associated with calc-alkaline acid/intermediate to basic and locally intermediate to basic tholeiitic volcanism. The Permian sedimentary succession oversteps either crystalline rock complexes or lithologically dissimilar Late Carboniferous sedimentary formations. Two forms of Late Carboniferous–Permian boundary have been recognized. In the first, there is a gradational change upward from the terrestrial Late Carboniferous strata to the coarsening-upward continental Permian deposits, without any indication of a stratigraphic hiatus. The second comprises an erosional contact between the Permian and the underlying Late Carboniferous sedimentary formations (see relevant sections for details).

Palaeoclimate and sea-level changes

The Pangaea supercontinent migrated *c.* 20° northward through the Permian and Triassic. A number of microcontinents (e.g. North and South China) traversed low latitudes during the Permian, but most of these collided with the main Pangaean landmass by the Late Triassic. During the Permian, continental size and aggradation were at a maximum; a considerable landmass was exposed and the supercontinent stretched almost from pole to pole. A single world ocean, Panthalassa, with a semi-enclosed Tethyan Sea, dominated the marine environment. The increase in landmass led to a number of significant climatic changes, including extreme seasonality and high aridity in continental interiors, and summer monsoons along the Tethyan margins (Gibbs *et al.* 2002). Indeed, the extensive marginal marine and continental environments of the latest Carboniferous and early Permian provide one of the best examples in the stratigraphic record where palaeoclimatic change can be demonstrated. The widespread Westphalian coal measures required high humidity and elevated water tables for vegetation to grow and be preserved. The transition to Stephanian, and then Permian, redbeds (still essentially non-marine) suggests that the increasing

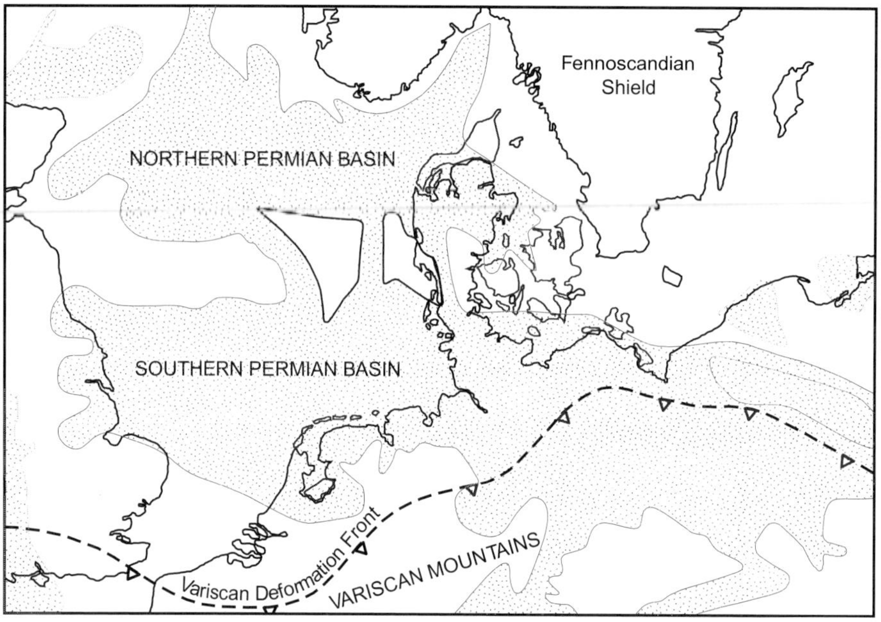

Fig. 10.1. Sketch map of the Northern and Southern Permian basins of NW Europe (after Stemmerik *et al.* 2000).

aridity was a gradual development and that this was coincident with the continued formation of Pangaea and the slow northward drift of the continents from a tropical to an arid climate belt. From the Sakmarian to the Wordian (Fig. 10.2), there was a distinct weakening of northern hemisphere summer monsoon precipitation corresponding with a reduction in coal deposition in this area, as well as major floral changes (Rees *et al.* 2002). Indeed, the Permian marine and terrestrial floras and faunas are indicative of increased provinciality and a highly differentiated biosphere (A.M. Ziegler 1990). The southern hemisphere summer monsoon was strengthened in the Wordian, relative to the Sakmarian. This change was due to an increase in southern land area in the subtropics (Gibbs *et al.* 2002).

The Permian Period contains the most recent transition from a major glaciation to a generally ice-free state (Rees *et al.* 2002). This change from icehouse to greenhouse conditions was probably caused by a rise in atmospheric CO_2 to around five times present levels (e.g. Berner & Kothavala 2001; Raymond & Metz 2004). Glaciations occurred during the Permo-Carboniferous from about 330 Ma to 240 Ma (i.e. extending into the Triassic) (Crowell 1983, 1995), with peak glaciation during the Westphalian (Parrish *et al.* 1986). Recent work has established the first numerical dates for the Carboniferous–Permian glaciation in Namibia and South Africa. These dates, based on U–Pb SHRIMP dating of zircon separates from fallout ash layers intercalated in Dwyka glacial deposits reveal that the end of the glaciation (*c.* 288 Ma) coincided, more or less, with the Carboniferous–Permian boundary (Bangert *et al.* 1999, 2000; Stollhofen *et al.* 2000). Thus, the idea that there was significant glacio-eustatic control on sedimentation and sea level during the Permian is questionable. However, other Gondwanan glaciations were slightly younger and there is evidence, for example, of glaciation in the northern hemisphere during both the Early (Sakmarian) and Late Permian (Kazanian) (for details see Raymond & Metz 2004). A glacial event in Oman (Al Khlata Formation) ended at *c.* 280 Ma (Osterloff *et al.* 2004) suggesting that both the Zechstein/Early Triassic (and earlier) depositional cyclicity in Europe, and similar cyclicity in Arabia, may have been controlled by continuing glacial eustacy.

The southern coast of Pangaea moved across the South Pole in the Early Permian and this roughly coincided with the cessation of significant continental glaciation (Veevers & Powell 1987). At the same time extensive evaporites and aeolian sandstones were deposited in the northern hemisphere, including Europe. In low latitudes, monsoonal climates were widespread (Parrish 1995). Deglaciation was apparently relatively rapid, being confined mainly to the Early Permian Sakmarian Stage (Veevers & Powell 1987). A thriving *Glossopteris* forest cover replaced the icesheets throughout much of southern Gondwana, and the classic and ubiquitous Gondwanan sequence, from tillites to coal-swamp deposits, indicates a major climate warming (Ziegler *et al.* 1997). Relatively minor upland glacial deposits persisted into the Artinskian in South Africa (Visser 1997). Apart from some evidence of ice activity in Australia and Siberia during the Capitanian Stage (Chumakov 1994; Barron & Fawcett 1995), generally ice-free conditions persisted through the Mesozoic and into the early Cenozoic (Gibbs *et al.* 2002).

The long-term global (eustatic) sea-level trend for the Carboniferous–Permian (and on into the Triassic) is a gradual fall. The eustatic fall in the Permian–Triassic is considered to have led to the lowest stand of global sea level experienced in the entire Phanerozoic and was coincident with the formation of Pangaea, or more specifically, the giant ocean of Panthalassa with its old and dense seafloor. The Permian System contains a great number and diversity of depositional sequences which illustrate sedimentary responses to a series of sea-level fluctuations (Ross & Ross 1995; Kiersnowski *et al.* 1995). These sea-level fluctuations (some of which were related to continued glacio-eustatic fluctuations; Martin 1981) had many different amplitudes and durations, and were accompanied by a wide spectrum of rates of deposition. In Central Europe the Lower Permian strata are mainly non-marine red-beds; the Upper Permian strata (Zechstein) have only a few marginal, restricted 'marine beds' in their lower part becoming predominantly cycles of evaporites and red-beds in their upper part. Marine fossils are found in only a few beds in the lower part of the Zechstein where they are part of ecologically stressed communities that typically have relatively few species and genera.

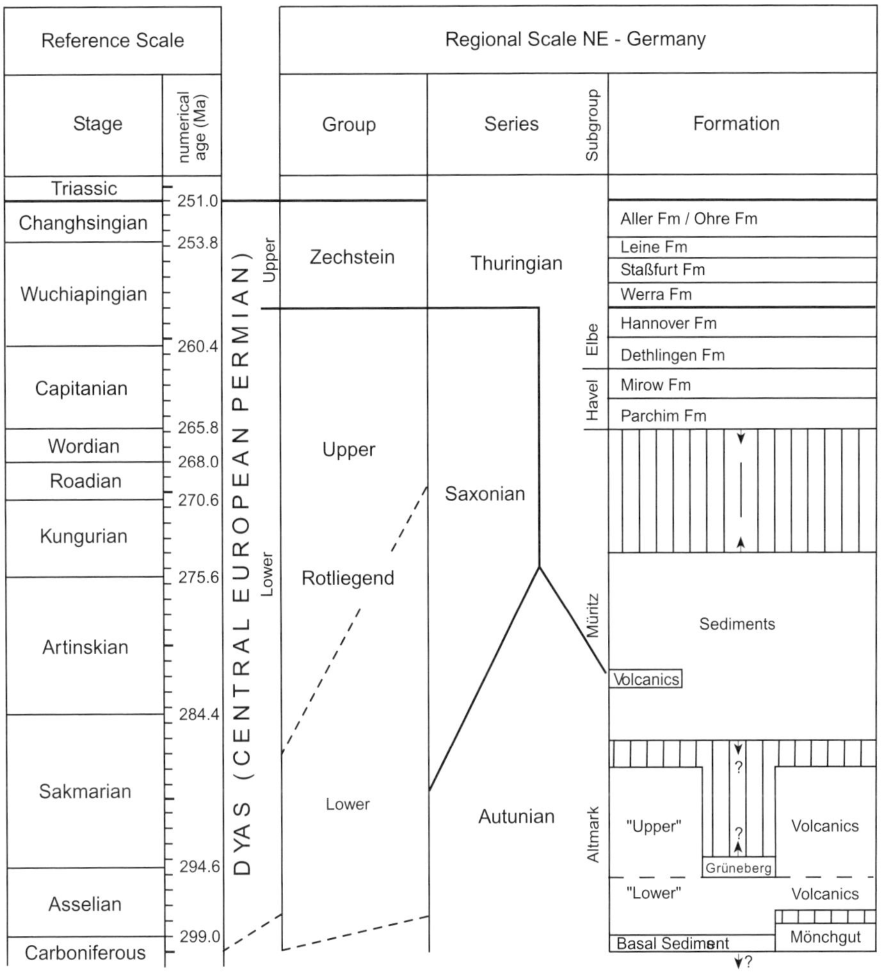

Fig. 10.2. Permian timescale (after Menning 1995; Gradstein *et al.* 2005; Deutsche Stratigraphische Kommission 2004).

Throughout Central Europe the Carboniferous–Permian succession, initially comprising siliciclastic carbonate sediments, followed by later deltaic, swampy and often non-marine red-bed sediments, reflects a falling trend in sea level. In Early Permian times, a eustatic sea-level lowstand, related to the Carboniferous–Permian glaciations on Gondwana, was accentuated by Variscan uplift; the end result of this combined tectono-eustatic event was the almost complete lack of marine influence on earliest Permian sedimentation in many areas. Later in the Permian, rising sea levels, coupled with incipient rifting in the North Atlantic or North Sea areas and erosion of the Variscan foreland uplifts, led to marine transgressions including, ultimately, the formation of the Zechstein Sea. The distribution of facies and the sedimentary history of the Zechstein Sea would appear to have been controlled by a combination of basin dynamics, climate changes and sea-level fluctuations.

Eustatic variations during the deposition of the Zechstein Limestone led to periodic subaerial exposure of the carbonate platforms and intense meteoric diagenesis of the exposed sediments (Peryt 1984). Sea-level fluctuations have also been inferred during the formation of the Z1 evaporites of the southern North Sea and elsewhere. The subsequent regression in the latest Permian was coincident with a global sea-level lowstand, the final coalescence of Pangaea and the onset of more humid climates. The influence of basin morphology on sedimentation was of lesser importance, and generally restricted, except in those situations where the pre-existing Variscan basement highs continued to be active. One such example is that of the Brandenburg-Wolsztyn-Pogorzela High in Poland which influenced sedimentation into Zechstein times (Kiersnowski *et al.* 1995).

Stratigraphy

The stratigraphy of the Permian is complicated by the fact that it is overwhelmingly a continental red-bed succession and lacks good biostratigraphic control (see below for exceptions). Thus the timespan of deposition of the Rotliegend, in particular, seems to change with each new publication, and dates for the start and end of the Permian period have varied even in recent years (Glennie 2005). Menning (2001) has provided an integrated Permian timescale based on absolute dates (U–Pb, Ar/Ar, Rb–Sr) and field indicators for the duration of Permian stages (Fig. 10.2). This work reveals that the duration of the Lower Permian Series is comparable with the duration of the Middle plus Upper Permian series. U–Pb zircon SHRIMP ages from the southern Ural Mountains indicate an age of 292 Ma for the Carboniferous–Permian (Gzelian–Lower Permian) boundary. In contrast, Ar/Ar sanidine ages from Central Europe provide an age of 296 Ma for the same boundary. More recently, Wardlaw *et al.* (2004) set the Carboniferous–Permian boundary at 299 ± 0.8 Ma. Additional chronostratigraphic data for the Permian is limited to isotope dating of late Tatarian (Zechstein) salts (Lippolt *et al.* 1993) and Rotliegend volcanic rocks (e.g. Lippolt *et al.* 1982; Lippolt & Hess 1989; Breitkreuz & Kennedy 1999).

The most recent stratigraphic table (Deutsche Stratigraphische Kommission 2002) includes the North German Basin (Fig. 10.3) and adjacent sub-basins and provides a useful frame of reference for other stratigraphic schemes for the Rotliegend of Central Europe. In particular, Upper Rotliegend lithostratigraphic units can be correlated across the entire Southern Permian Basin (e.g. Van Wijhe *et al.* 1980; Schöder *et al.* 1995; Karnkowski 1981; Pokorski 1976, 1981, 1988, 1989, 1990; Wagner 1994). A scarcity of well-defined chronostratigrahic boundaries has resulted in particular stratigraphic units having uncertain stratigraphic ranges as well as assumed hiatuses having uncertain timespans (Menning 1995; Menning *et al.* 2006; Deutsche Stratigraphische Kommission 2002).

The internationally accepted subdivisions of the Permian (Fig. 10.2) are frequently replaced in Central Europe (and, indeed, in other parts of Europe; for examples see McCann *et al.* 2006), by more localized terms. Of particular importance here are the subdivisions of the Early Permian which is broadly the Rotliegend (technically the Rotliegend overlaps into the uppermost Stephanian). The Rotliegend may be subdivided into the Lower and Upper Rotliegend (generally separated by the Saalian Unconformity Jackson 1997) and these subdivisions are broadly

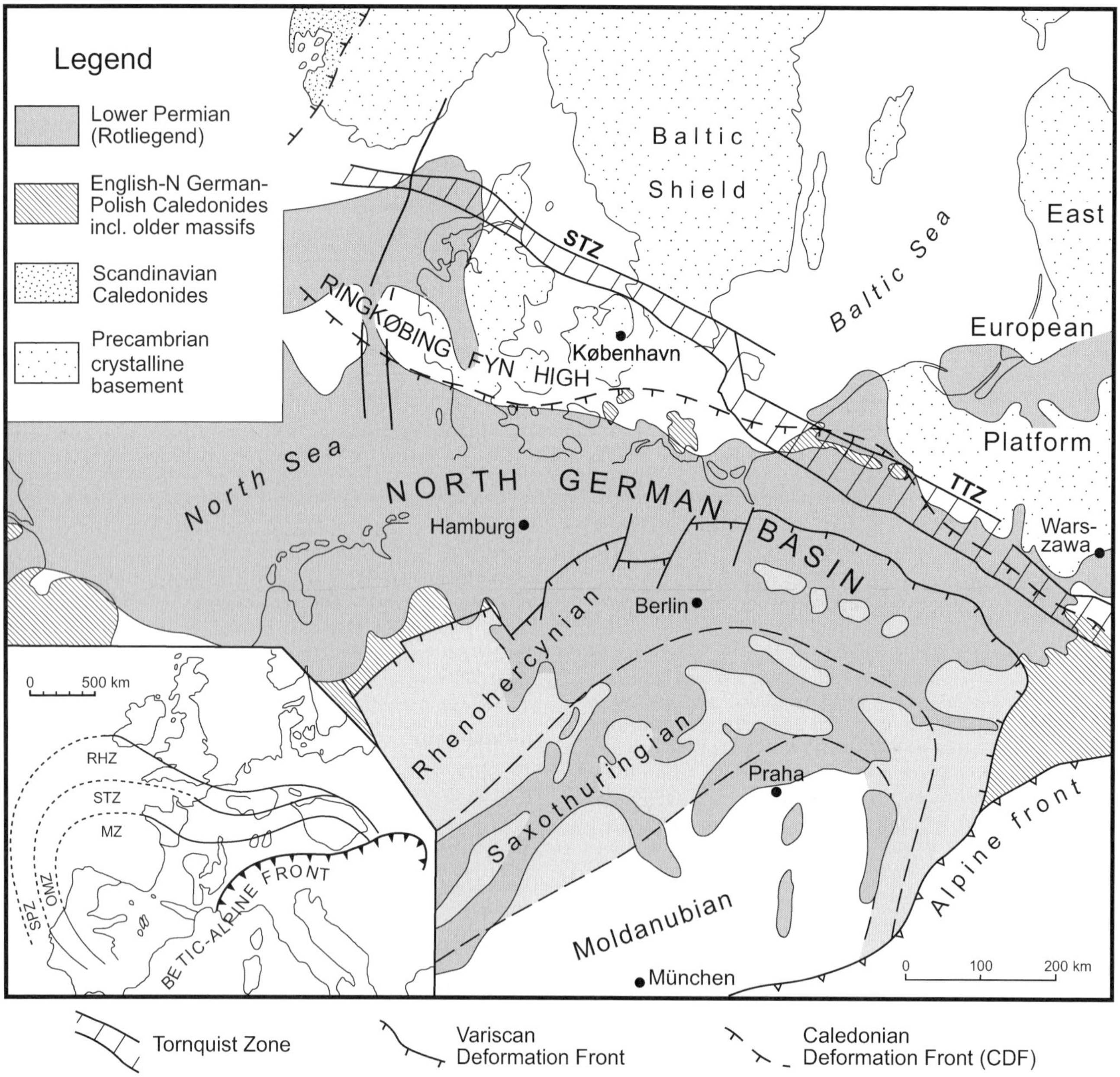

Fig. 10.3. Structural sketch map of the Lower Permian (Rotliegend) and the main tectonic elements of the Central European basement (modified from P. A. Ziegler 1990; Berthelsen 1992; Plein 1993). Key (main diagram): STZ, Sorgenfrei-Tornquist Zone; TTZ, Teisseyre-Tornquist Zone. Inset: MZ, Moldanubian Zone; OMZ, Ossa-Morena Zone; RHZ, Rhenohercynian Zone; SPZ, South Portuguese Zone; STZ, Saxothuringian Zone.

comparable to the largely discontinued terms Autunian and Saxonian. In addition, some authors (e.g. Hoffmann *et al.* 1989; Schneider & Gebhardt 1993) have subdivided the Upper Rotliegend into two major stratigraphical units: Upper Rotliegend 1 (UR 1) and Upper Rotliegend 2 (UR 2). UR 1 crops out in some basins in central Germany, and in these areas constitutes part of the continental succession that commenced deposition in the Late Carboniferous (e.g. Saar-Nahe Basin; Schneider 1996*a*). However, over much of the Southern Permian Basin, UR 1 is represented by the Saalian Unconformity, which resulted from the uplift and erosion associated with the northward development of the Variscan Orogeny (Glennie 2005). UR 2, overlying this unconformity, is an economically important gas-bearing unit which occupied much of the Southern Permian Basin prior to the latest Permian deposition of the marine Zechstein (Late Permian, also referred to as the Thuringian) (Fig. 10.2).

There are a number of correlation problems associated with the Permian succession. The first of these is the correlation of the Central European stratigraphy with the international marine standard (Schneider & Roscher 2002). There are also problems correlating the Southern Permian Basin succession with the smaller basins located within the Variscan orogenic belt. However, problems of internal correlation within the Southern Permian Basin also exist. While much correlation is litho- or biostratigraphic, magnetostratigraphy and isotope dating (see Menning 1995) are increasingly used for stratigraphic purposes.

Lithostratigraphy

The stratigraphy of the Permian succession in Central Europe is based, for the most part, on lithostratigrapic or allostratigraphic principles (e.g. Gralla 1988; Gast 1991, 1993; Plein 1990, 1995), largely due to the lack of proper regionally important guide fossils. Additional lithostratigraphical techniques used in reconstructing basin stratigraphy in the region include the correlation of palaeosols and lacustrine black shales (e.g. Holub & Tasler 1982; Nemec *et al.* 1982; Rößler 1997). However, there are many problems associated with a purely lithostratigraphical approach. Within the many intramontane basins, originally established in early Permian times, the succession extends from the Lower through to the Upper Rotliegend. In these basins, sedimentation was only temporarily interrupted at the end of the Lower Rotliegend (i.e. Saalian Unconformity) and resumed with the accumulation of coarse fluviatile clastics under increasingly arid conditions. In many places, however, there is no clear distinction between Lower and Upper Rotliegend sediments (Hoffmann *et al.* 1989), due particularly to the lack of fossils and the poorly defined chronostratigraphic/biostratigraphic Lower Rotliegend–Upper Rotliegend boundary (289–290 Ma).

Biostratigraphy

Microflora and macroflora are the best index fossils for Permian biostratigraphic use, although both are strongly facies linked. However, an ongoing problem is that certain flora specimens, including palynoflora characteristic for Late Carboniferous–Early Permian times, have such a wide stratigraphic range and timespan (e.g. coexistence of the 'Stephanian' and 'Autunian' floras; Broutin *et al.* 1990; see also Kozur 1978, 1980, 1988) that it is almost impossible to use them for sensitive stratigraphic correlation (Izart *et al.* 1998). This level of biostratigraphic uncertainty thus leads to distinct problems concerning the Carboniferous–Permian boundary (Schneider *et al.* 2000) with the result that in many basins, fluvial and alluvial red-beds of the 'Rotliegend lithofacies' may represent either Upper Carboniferous or Lower Permian deposits.

In some Early Permian basins within the Variscan domain, which are rich in floral and faunal remains, the local biostratigraphy may be based on ichnofossils (Boy & Martens 1991) more specifically, terapod and arthropod ichnia, freshwater fish (fish teeth and scales), insects, conchostracans (Schneider 1989; Zajic 1997), ostracods and hydromedusae (Gaupp *et al.* 2000). More advanced biostratigraphy using amphibian or xenacanth teeth makes correlation between separate basins possible (e.g. Schneider 1996*a*, 2001; Schneider & Zajíc 1994; Werneburg & Kiersnowski 1996; Zajíc 2000).

Correlation of Zechstein sequences with global standard stages using biostratigraphy is difficult because of the lack of index fossils in the Zechstein. However, palynomorphs and conodonts offer some prospect of biostratigraphic correlation in the future (Kiersnowski *et al.* 1995).

Sequence stratigraphy (cyclostratigraphy)

As noted above, the relatively barren nature of much of the Permian successions in Central Europe (particularly the Lower Permian) makes it difficult to achieve high-resolution correlation across the region. In addition, the relative isolation of many of the basins in the region means that each has its own individual stratigraphy (Schneider 2001). Correlating these separate basin stratigraphies into a unified Southern Permian Basin stratigraphy (including the southern North Sea Rotliegend Basin, North German Basin and the Polish basins) is only possible for the Upper Rotliegend 2 and the Zechstein (Kiersnowski *et al.* 1995). However, since these deposits are mainly barren (with the general exception of stratigraphically unimportant ichnofossils), it is difficult to correlate the timing of stratigraphically important depositional events, as well as to make genetic and stratigraphic connections between facies, across the region. Some authors have adapted methods of sequence stratigraphy, such as those used in lacustrine basins, in order to circumvent some of the inherent correlation problems. For example, the presence of a perennial and extensive saline lake in the Rotliegend of the Southern Permian Basin has facilitated the establishment of a series of sequences within the basin (e.g. Gast 1991, 1993; George & Berry 1993; Yang & Nio 1994; Gaupp *et al.* 2000). A similar approach has been used for the playa-lake succession of the Polish Rotliegend basin (Kiersnowski 1997).

The basin-wide extent of evaporitic and clastic sedimentation in the Zechstein of the Southern Permian Basin has proved much more amenable to sequence stratigraphic techniques. For example, Strohmenger *et al.* (1996*a*) divided the Main Dolomite into both sequences and parasequences and, subsequently, proposed a new subdivision of the upper part of the Zechstein into stratigraphic sequences (Strohmenger *et al.* 1996*b*). Wagner & Peryt (1997) proposed a sequence stratigraphic subdivision of the Polish Zechstein and identified four depositional sequences ranging in age from uppermost Rotliegend through to the lower part of the Buntsandstein. This work provided many new insights including, for example, the fact that the upper part of the Zechstein succession in the Polish Basin is the most complete and best known sequence representing the decline of Zechstein deposition (Peryt & Wagner 1998).

Other stratigraphic methods involve the identification of distinct cycles in the sedimentary record. These include climatostratigraphy (Yang & Baumfalk 1994; Gaupp *et al.* 1997), magnetostratigraphy (Menning *et al.* 1988) inorganic geochemistry (Hiete *et al.* 2006) or genetically related correlation (Hug &

Gaupp 2006) all of which have been applied locally in the Southern Permian Basin.

Tectonic setting (Late Carboniferous–Early Permian)

The Variscan Belt represents a mosaic of microplates and geological terranes that collided within an oblique (dextral) collisional regime from Late Devonian through to Carboniferous times (see Belka & Narkiewicz 2008; Krawczyk *et al*, 2008; McCann *et al.* 2008 for details). The final phase of collision, involving the southern margin of the Iberian–Armorica microplates and Gondwana, led to widespread deformation across the orogenic zone during the Late Carboniferous (Warr 2002). The resulting regional compression resulted in northward-directed thrust propagation and the development of an asymmetric foreland basin, with maximum subsidence occurring near the advancing thrust front. Crustal shortening in the Variscan fold-and-thrust belt largely terminated at the end of the Westphalian.

The Permo-Carboniferous collision of Gondwana and Laurussia, together with other continental fragments, led to the formation of the late Palaeozoic supercontinent of Pangaea. Surrounding this landmass was a large single ocean (Panthalassa), with a smaller sea, Tethys, covering much of what is now southern and central Europe. The sheer size of Pangaea meant that it was already unstable from the outset and the post-Variscan period was one of intense crustal re-equilibration and reorganization. The development of a complex system of conjugate sinistral and dextral shears has been recognized across Europe (Arthaud & Matte 1975; Mattern 2001; Ziegler 1982). Activity along this fault system, which extends far into the Variscan foreland, controlled the subsidence of the various Early Permian transtensional and pull-apart basins, and resulted in transpressional deformation which led to the formation of tectonically silled basins (e.g. Kiersnowski & Buniak 2006). This extensional activity was dominantly east–west orientated and was associated with block faulting (locally termed Franconian, i.e. Late Carboniferous/Early Rotliegend, and Saalian, i.e. Early/Late Rotliegend) (Figs 10.4 & 10.5). A series of roughly north–south striking grabens were formed (e.g. Gast 1988), and these acted as a feeder rift system for sediment transport from the Variscan hinterland to the subsiding foredeep in the north. The asymmetrical development of these grabens and the basinward differentiation in sedimentary facies within them, suggests that there were differences in stretching rates and amounts.

The deep-crustal fracturing across the region triggered a widespread phase of intrusive and extrusive magmatism which was characterized by a highly variable chemistry (e.g. Francis 1988; Neumann *et al.* 2004; see Timmermann 2008). This would suggest that the regional stress pattern that had affected western and central Europe during the Variscan Orogeny, changed fundamentally at the transition from the Westphalian to the Stephanian and early Permian. Indeed, a systematic rotation of the principal horizontal compressional stress-axis from an initial northerly direction to an east–west direction can be observed. This change in the compressional stress axis may have been related to changes in the convergence direction between Gondwana and Laurussia, presumably a consequence of lithospheric thickening in the Variscan fold-and-thrust belt having reached a critical level at which additional orthogonal strain could no longer be achieved. The ensuing locking of the Variscan subduction system and the subsequent disintegration of the Variscan fold-and-thrust belt was probably a result of the combined effects of dextral shear, gravitational collapse of the overthickened crust (probably aided by a far-field dextral extensional stress field; Torsvik & Van der Voo 2002), and possibly backarc extension related to post-orogenic steepening and decay of the north-dipping Tethys subduction zone (Jowett & Jarvis 1984). The combined effect of crustal re-equilibration and tectonic activity controlled the kinematic patterns and subsidence of the approximately 70 intermontane basins, all of which are characterized by a major strike-slip component in their deformational history. Western and central Europe was already thermally weakened by the preceding orogenies and many of the Carboniferous–Permian basins parallel long-lived Variscan or Caledonian lineaments (Henk 1993). Cooling of the lithosphere, following the initial period of block faulting, graben development and associated

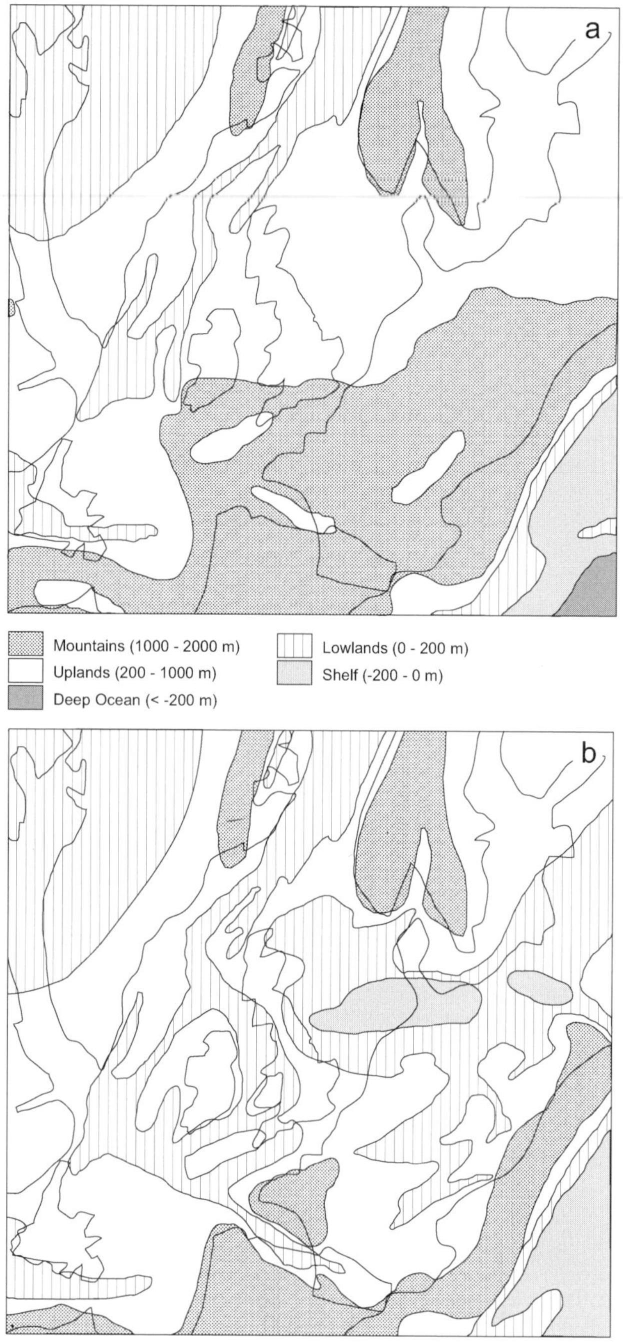

Fig. 10.4. Lower Permian palaeogeography: (**a**) Sakmarian and (**b**) Artinskian (after Scotese & Langford 1995; Scotese 2004).

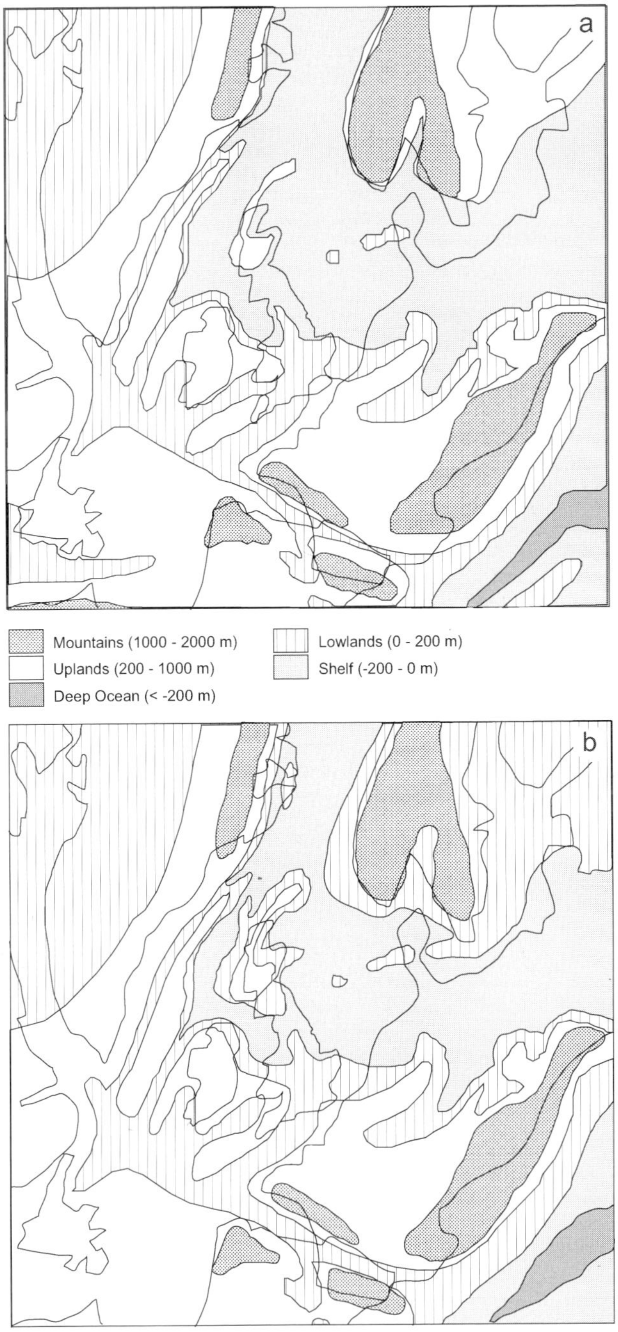

Fig. 10.5. Upper Permian palaeogeography: (**a**) Wordian and (**b**) Tartarian (after Scotese & Langford 1995; Scotese 2004)

volcanic activity, resulted in the development of a continuously expanding regional depression with little or no faulting (e.g. McCann 1998*b*, *c*).

A number of major post-Variscan tectonic structures have been recognized in Central Europe (Fig. 10.3). The Tornquist Zone, comprising a fan-shaped series of fault and fault-zone splays, was activated in the western Baltic area and Danish Scania (Berthelsen 1992). The Tornquist-Teisseyre Line forms the boundary between the stable Fennoscandian-East European Craton and the fragmented platform of western Europe. This much-debated lineament came into evidence during the Permo-Carboniferous. It is, however, not represented by a major single fracture, but rather corresponds to a broad zone of essentially NW–SE striking faults (Pozaryski & Dembowski 1984) with dextral offsets of up to 400 km suggested for the zone (Lewandowski 1993). In SE Poland the Tornquist-Teisseyre Fault Zone is superimposed on the Variscan Deformation Front and in northern Poland on the Caledonian Deformation Front. The faults within the Tornquist-Teisseyre Zone were of great important for the evolution of the Polish Rotliegend and Zechstein basins in that they controlled the location of the areas of maximum basin subsidence.

According to P.A. Ziegler (1990) a major post-Variscan wrench fault was also active along the present Elbe Fault System (northern Germany) as part of a complex megashear zone which developed in north central Europe between the Appalachians and Uralides as a result of dextral translation of the Armorican-European plate with respect to the African plate (Arthaud & Matte 1977). Along the SE end of this wrench fault, geological evidence for late Carboniferous dextral ductile shearing has been found in outcrops of the Elbe Zone (Mattern 1996). Scheck *et al.* (2002) have noted that maximum deformation occurred along the Elbe Fault System during late Carboniferous wrenching and volcanism (and subsequent to these events). Late Palaeozoic dextral transtensional movements with estimated displacements of 60–120 km have also been postulated.

Basin modelling

Despite the fact that there is no unified model for the Permian stretching/wrench event, recent studies agree on two fundamental and relevant points: (1) Permian rifting was widespread in Europe with progressively propagated development; and (2) its signature strongly influenced the evolution of the European lithosphere during Mesozoic and Cenozoic times (Sørensen 1986). It may not, however, be possible to provide more detailed models for the evolution of the region (see McCann *et al.* 2006). Numerical modelling of lithospheric rifting, for example, requires numerous parameters, amongst them the structure of the pre-rift crust and mantle lithosphere. Because the structure of the pre-Permian lithosphere has been obscured by repetitive tectonic phases in most parts of Europe, lithosphere-scale modelling of the Permian rifting remains difficult and modelling results need to be treated with a high degree of circumspection. Subsidence models across the Southern Permian Basin, however, indicate that: (1) latest Carboniferous–Early Permian rifting (290–305 Ma) was a widespread and dramatic event in Europe; and (2) the thermal signature of the Permian rifting was a significant control on the subsequent Mesozoic and Cenozoic evolution of the European lithosphere. Ziegler *et al.* (2006) have noted that subsidence modelling suggests that there was a period of Permo-Carboniferous 'stretching' extending from 300 to 280 Ma. This involved decoupled crustal extension and mantle/lithosphere attenuation. Such an assumption is compatible with the concept that during the Permo-Carboniferous re-equilibration of the crust–mantle boundary and crustal extension played a significant, but local, role (Van Wees *et al.* 2000). This can be concluded from the fact that some important Permo-Carboniferous troughs (e.g. Massif Central, Bohemian Massif) do not coincide with major Late Permian depocentres, while depocentres such as the Southern Permian Basin are not underlain by major Permo-Carboniferous basins (P.A. Ziegler 1990; Ziegler *et al.* 2006). This would suggest that during the Permo-Carboniferous tectono-magmatic event, uniform and/or depth-dependent lithospheric extension was, at a regional scale, only a contributing factor and not the dominant mechanism of crustal and mantle/lithosphere thinning. Mechanical stretching of the lithosphere, therefore, played a subordinate role whilst thermal thinning of the mantle/lithosphere and magmatic and erosional thinning of the crust predominated, providing the principal driving mechanism for the Late Permian and later subsidence of the Central European intracratonic basins.

Examination of a variety of different basins across Europe has allowed the successions within the basins to be compared and contrasted, as well as providing information on other aspects including basin form, the controls on basin formation, and the magmatic, tectonic and infill histories of the individual basins. The observed contrasts across the region suggest that the underlying processes that controlled the post-Variscan evolution of Europe were very different in the areas located in the former Variscan foreland basin, from those areas within the fold-and-thrust belt. Modelling of the post-Variscan Permian basins suggests that the controlling mechanisms on basin formation vary between the different areas. For example, Permian rifting in the former Variscan hinterland would appear to have been strongly controlled by the collapse of the mountain chain (Brunet & Le Pichon 1982; Prijac *et al.* 2000) with a possible far-field extensional component also being plausible (Henk 1999). Furthermore, this process may have been modified by the slow decay of the associated thermal anomaly (e.g. Paris Basin). In contrast, rifting in the former Variscan foreland would appear to have been dominated by late Variscan wrench tectonics (van Wees *et al.* 2000), particularly along the boundary between Precambrian and Phanerozoic Europe (Dadlez *et al.* 1995). Numerical modelling highlights the role of such lithospheric discontinuities in controlling rifting (Pascal & Cloetingh 2002; Pascal *et al.* 2003).

Palaeomagnetism (J.T.)

One of the more controversial problems of Permian palaeogeography concerns the initial configuration of Pangaea and its evolution through Triassic times. While it is clear that the classic Wegenerian Pangaea reconstruction (termed Pangaea A; Torsvik & Cocks 2004) was valid prior to opening of the Atlantic in Jurassic times, this configuration results in unacceptable continental overlap between Laurasia and Gondwana when Permian palaeomagnetic data from Gondwana are taken into consideration. Thus an alternative early Permian Pangaea B reconstruction was proposed where Gondwana was positioned further to the east (along lines of latitude) with respect to Laurasia. Clearly, however, this reconstruction had significant tectonic implications and some 3000 km of dextral shear between Gondwana and the northern continents would be required to transform the Permian Pangaea B into the Triassic–Jurassic Pangaea A configuration. As a result, this model has not been widely accepted, despite the (albeit limited) geological evidence for dextral shear and the supposition that the preservation of large-scale strike-slip faults is unlikely (Umhoefer 2000; Torsvik & Cocks 2004). A third variation, Pangaea C, resolves the overlap problems of Pangaea A and B, but would involve, among other things, a dextral strike-slip east–west movement of *c.* 8000 km between the northern and southern halves of Pangaea during the Triassic and Early Jurassic (Torsvik & Cocks 2004).

Over the years, several different arguments, such as inclination shallowing, age uncertainties and non-dipolar fields, have been proposed to discredit palaeomagnetic data that argue against a Permian Pangaea A (Rochette & Vandamme, 2001; Van der Voo & Torsvik, 2001). More recently, however, high quality Permian palaeomagnetic data have been obtained from different regions of Gondwana (Bachtadse *et al.* 2002; Torcq *et al.* 1997). These new observations sufficiently counter the various arguments put forward that the Early Permian Pangaea B model remains pertinent to latest Palaeozoic palaeogeography. This being the case, the Pangaea B to A transformation probably occurred in latest Permian–early Triassic times, being driven, at least in part, by subduction beneath the western margin of Gondwana (Muttoni *et al.* 2003). Nevertheless, until more definitive geological evidence for Permo-Triassic palaeogeography is available, the initial configuration of Pangaea will remain a matter of debate and discussion.

Palaeogeographic setting

By Early Permian times, most of the world's continents had collided to form the supercontinent of Pangaea. Scotese & Langford (1995) provide a detailed discussion of the history of Pangaea (e.g. in terms of research) extending from the initial naming of the supercontinent by Wegener (1912) to the present. The Permian world can be subdivided into four distinct plate-tectonic regions, namely Panthalassa, the Pangaean Rim, Central Pangaea and Tethys (Scotese & Langford 1995). Although it made up almost two-thirds of the Earth's surface, virtually nothing is known about Panthalassa (Scotese 1987). The only remnants are preserved in obducted ophiolites and volcanic seamounts which were emplaced along the rim of Pangaea, although many of these may have originated in marginal basins along the edge of Panthalassa. The Pangaean Rim, extending for almost 24 000 km around the outer edge of Pangaea, comprised a ring of subduction zones which were active throughout the Middle and Late Palaeozoic. Evidence of subduction-related orogenic activity and island arc complexes have been preserved within the rim area (Scotese & Langford 1995). Central Pangaea comprised the core of continents which were sutured together during the Variscan collisions. Along these sutures rose up two great mountain ranges: the Central Pangaean Mountain Range (Appalachian-Mauretanide-Variscan orogenic belts) and the Western Asian Mountain Range (Uralian & Altay-Sayan orogenic belts). The Central Pangaean Mountain Range was formed as a result of the diachronous collision between Laurussia and Gondwana. This collision took place earlier along the eastern part of the range (the Late Devonian Bretonic phase in Europe) and culminated during the latest Carboniferous–early Permian (Sakmarian). The European extension of these mountain ranges (Variscan mountains) separated northern Europe from Mediterranean Europe and North Africa (P.A. Ziegler 1989). The Permian in this southerly region (France, Italy, Spain) consists primarily of continental deposits, volcanics and igneous intrusions (e.g. Cassinis *et al.* 1995; McCann *et al.* 2006).

Cooling of the lithosphere, following the initial period of graben development and associated volcanic activity, resulted in the development of a continuously expanding regional depression with little or no faulting (e.g. McCann 1998*b*, *c*). Continued subsidence led to the development of a series of connected basins (i.e. Southern Permian Basin) extending from eastern England through to Poland (Fig. 10.1). Within these basins, sediments were predominantly clastic along the margins, with playa lakes developing in the basin centres reflecting the aridity of the climate. The flooding of the Southern Permian Basin in Zechstein times resulted in the formation of a major inland sea. The arid climate combined with repeated restriction of the seawater flux into the basin gave rise to the production of extensive carbonates, sulphates and halites (see below).

Stephanian–Lower Rotliegend magmatic activity

The rare occurrence of Lower Rotliegend sediments within the former Variscan foreland of Europe suggests that much of the area later occupied by the Southern Permian Basin was regionally uplifted and subjected to extensive erosion. This uplift, probably induced by a combination of wrench-related lithospheric deformation, magmatic inflation of the lithosphere and thermal erosion of the mantle lithosphere, was coupled with periods of significant magmatic activity (see also Timmermann 2008). Melt generation was probably related to localized divergent wrench-induced decompressional partial melting of the uppermost asthenosphere and the lithospheric thermal boundary layer, possibly combined with the impingement of a not-very-active mantle plume on the base of the lithosphere (Ziegler 1996). While the most voluminous and evolved magmatism occurred in the North German Basin (Benek *et al.* 1996), dated at 297–302 ± 3 Ma (Breitkreuz & Kennedy 1999), the Oslo Rift contains the most extensive and best preserved sequences of basaltic lavas associated with this event (Neumann *et al.* 1992). Volcanic activity, however, was widespread throughout Europe Neumann *et al.* 2004, with thick sequences being deposited elsewhere, for example, in Spain (e.g. Iberian region; Lago & Pocovi 1984) the Italian Alps (e.g. Collio Basin; Breitkreuz *et al.* 2001), Liguria and Sardinia (e.g. Cortesogno *et al.* 1998).

In northern Germany the centre of the Lower Rotliegend volcanism was in the Lower Elbe region (>2000 m thick) and to the north of Berlin (>2500 m thick) (Benek *et al.* 1996). The northern Germany volcanic centre had significant influence towards the east. The various volcanic rocks covering western and northwestern Poland (Ryka 1989; Jackowicz 1994) range from 200–400 m to over 1500 m in thickness. Small occurrences of volcanics interbedded with sedimentary rocks are also found on the southern margin of the Mid-North Sea and Ringkøbing-Fyn highs (Glennie 1990).

Intrusive activity also occurred across the region of Central Europe, although this was much more localized than in Carboniferous times. In the NW Thuringian Forest, for example, there was a magmatic event between 301 and 294 Ma during which granite bodies and granite porphyry dykes were intruded (Zeh & Brätz 2004). This event was coeval with the volcanism of the Möhrenbach Formation (SE Thuringian Forest), and was also synchronous with the intrusion of the Henneberg Granite in the Frankenwald area and intrusions in the Harz Mountains (Brocken Granite, Oker Granite, Ramberg Granite, Harzburg gabbronorite). Volcanic rocks exposed in the Ilfeld Basin, south of the Harz Mountains, belong to the same volcanic suite (Zeh & Brätz 2004). Late Carboniferous–Early Permian granite emplacement has also been reported from the Delitzsch Massif and the Flechtingen High (both in Germany), the Aar Massif of the Alps (e.g. Central Aar Granite), Black Forest (granite porphyry dykes, ignimbrites, tuffs), northern Vosges (ignimbrites, tuffs), Saar-Nahe Basin (granite porphyries, tuffs), the Oslo Rift, Cornwall (i.e. Cornish granites) and the Central Iberian Zone. In the Tauern Window area, subalkaline to alkaline granitoids and rhyolites are found in the Penninic, Austro-Alpine and South Alpine domains; these late-stage Variscan S- and A-type granites with Permian intrusion ages are associated with local extension along continental wrench zones that were running through the Variscan crust (Schaltegger 1997), or with a Permian rifting event heralding the onset of Alpine activity (von Quadt *et al.* 1999; Schaltegger 1997).

In summary, the Permo-Carboniferous magmatic event was widespread across Europe, and is assumed to result either from extension-induced decompression melting of lithospheric mantle or from mantle plume activity (Breitkreuz & Kennedy 1999; Timmermann 2004). Magmatic activity declined markedly after 290 Ma, with only restricted occurrrences reported (see Timmermann 2008).

Early–Middle Permian basins (Asselian–Capitanian)

Following the end of Variscan contraction in latest Carboniferous times, subsequent Stephanian–early Permian magmatic activity and basin formation in both the internal Variscides (i.e. the area within the Variscan fold-and-thrust belt and to the south of it) and the external Variscides (i.e. the area of the northern foreland basin, to the north of the Variscan Orogenic Belt) took place within a relatively short timespan (*c.* 290–300 Ma). In the internal parts of the orogen the basins tend to be small and isolated, but deep. Basin infill was relatively rapid with the accumulation of thick successions of Rotliegend-age aeolian and fluviatile/lacustrine sediments. Tectonic activity coincided with the overall change in climate (i.e. to semiarid) in the Stephanian–Early Permian. As noted above, many of the basins contain volcanics and the presence of these magmatic successions aids in distinguishing the Permian successions from the underlying Stephanian (see Plein 1995). The following sections examine a number of basins/regions across Central Europe in order to provide an overview both of the main basin types which formed at this time, and to outline the differences between the basins which formed in front of, and within, the Variscan Orogen. These differences were largely based on the fact that the Southern Permian Basin was drifting northward into the global desert belt (10–30° N) during the Permian, accentuated by the fact that the area of the Variscan Orogen created a possible rainshadow.

During the Rotliegend, two sag basins evolved: the Southern and the Northern Permian basins. These are separated by the Mid-North Sea and Ringkøbing-Fyn highs. The Southern Permian Basin is best known from hydrocarbon exploration, and is described in many publications (e.g. Glennie 1998; Plein 1995; P.A. Ziegler 1990). The Northern Permian Basin is less well explored and, as a consequence, not so well understood (Glennie *et al.* 2003).

Basins within the Variscan foreland

Southern Permian Basin (T.M.)

The Southern Permian Basin (SPB) comprises a series of connected basins extending from eastern England into Poland, a distance of some 1700 km. Its width ranges from 300 to 600 km (Van Wees *et al.* 2000). The basin developed between the northern foreland of the Variscan mountain belt and the Ringkøbing-Fyn High and is superimposed on the Late Carboniferous Variscan foredeep (Gaupp *et al.* 2000). The SPB had several narrow connections with adjacent basins (for details see Sørensen & Martinsen 1987) and possible temporary connections with the Tethys domain via the Polish-Dobrogea trough along a rift zone (Ziegler 1989) and with small basins in the Inner Variscan domain (Kiersnowski *et al.* 1995). The SPB is of great economic importance, containing a number of significant hydrocarbon accumulations, for example at Groningen and Salzwedel-Peckensen (e.g. McCann 1998*b*). The SPB is subdivided into the Southern North Sea–North German and the Polish sub-basins (see below) (Fig. 10.1).

In the area of the SPB, up to 800 m of Stephanian-age continental red-beds, containing correlative marine bands, were deposited in a broad successor basin to the Namurian–Westphalian Variscan foreland basin. During the late Stephanian–Early Permian, this basin was disrupted by predominantly transpressive wrench tectonics, as evidenced by the deep truncation of the Late Carboniferous series and the conspicuous absence of significant Early Permian basins (P.A. Ziegler 1990; McCann 1999). The phase of wrench tectonic activity was associated with the development of extensive volcanic fields (see above). Ziegler *et al.* (2006) suggested that crustal thinning within the North German part of the SPB could be interpreted in terms of a 'stretching' factor of 1.45. This thinning can be attributed to late Stephanian–Early Permian magmatic destabilization of the crust–mantle boundary which was accompanied by major thermal attenuation of the mantle lithosphere (Van Wees *et al.* 2000).

As noted earlier, the destruction of the subducted oceanic crust of the Rhenohercynian Ocean led to lithospheric stretching and associated calc-alkaline magmatic activity (e.g. Gast 1991; Benek *et al.* 1996; Stemmerik *et al.* 2000; Kockel 2002; Gast & Gundlach 2006) resulting in the formation of a series of north–south striking grabens (Gast 1988; Gast & Gundlach 2006) (Fig. 10.6). These grabens, which channelled sediment from the Variscan hinterland to the subsiding foredeep to the north, varied in their form and infill, suggesting that there were differences in both the rates and amounts of stretching. Clastics, derived from the Variscan highlands, were deposited along the southern margins of the SPB in a number of more or less discrete alluvial fans and cones (Fig. 10.7). These deposits grade both laterally, and northward, into wind-blown sands, forming a broad dune belt, that gives way to sandy and silty shales and cyclical halites in the basin axis. These latter were laid down as sabkha and desert lake deposits. The SPB mudflat deposits have their maximum lateral and vertical distribution in northern Germany (Gaupp *et al.* 2000) (Fig. 10.8). The clastic supply to the SPB from northern sources was rather limited in the area of the North Sea, Denmark and Germany (McCann 1998*c*), but more important in Poland where an eastern clastic source is also evident (Kiersnowski 1997).

Basin subsidence, which occurred mainly during the Rotliegend (e.g. McCann *et al.* 2000; van Wees *et al.* 2000), was coupled with the cyclical advance and retreat of dune and sabkha sands, respectively, towards the basin axis. Thermal subsidence was concentrated in those areas which had been the sites of maximum volcanic activity. However, it was also influenced by post-Variscan downwarping of the former foreland region (McCann *et al.* 2000). Several minor rift pulses characterized the Middle and Late Permian (Gast 1988; Plein 1995; Glennie 1998). These pulses occurred prior to and during deposition of the Upper Rotliegend Group (the Saalian and Altmark I to III pulses; Plein 1995) and the Zechstein (see below), although they are not recognized everywhere within the basin. By the end of Rotliegend times, the surface of the SPB had apparently subsided below the level of the world oceans, as is evident from the catastrophic transgression of the Zechstein Sea (Glennie 1990).

The following sections examine a series of representative sub-basins within the Southern Permian Basin. It should be pointed out, however, that the actual number of sub-basins is much larger (e.g. Gast 1988; Lützner 1988), and as such is beyond the scope of this overview.

Polish Basin (H.K.)

The Polish Basin, which formed in the Late Palaeozoic in front of the evolving Variscan orogen (Dadlez 1997), is one of a series of sedimentary basins which developed around the margins of the East European Craton (Nikishin *et al.* 1996). The extent of the Variscan Orogeny in Poland is still uncertain (Pozaryski *et al.* 1992; Dadlez *et al.* 1994) mainly due to the lack of reliable data documenting the regional extent and styles of orogenic deformation. However, recent studies of well data have provided new clues concerning the styles of deformation within the external Variscides as well as identifying the possible location of the front (and the possibly significant role of strike-slip movements within the frontal part of the orogenic zone) and the source area for the basin infill (Aleksandrowski *et al.* 2003; Jaworowski 2002). Extension, subsidence and basin development, subsequent to Variscan orogenic activity, were all parallel to the SW margin of the East European Craton and the Tornquist-Teisseyre Zone (Figs 10.9 & 10.10).

The Rotliegend basins in Poland may be subdivided into two groups: those which overlie Precambrian basement (e.g. Słupsk, Warmia, Cracow-Upper Silesia basins) and those, such as the Polish Basin and related sub-basins (e.g. Intra-Sudetic, North-Sudetic, Lower Silesian and Pomerania basins), which overlie Caledonian and/or Variscan basement (Karnkowski 1999) (Fig. 10.11). The main depocentre of the Polish Basin, the Mid-Polish Trough (MPT), was active during the basin's evolution, commencing in the early Permian and extending through to the Mesozoic. Throughout this period it remained a zone of maximum subsidence with almost uninterrupted sedimentation (Dadlez 1997; Kutek 2002). The present base of the Permian in the MPT is at a depth of 3–8 km (Marek 1988; Marek & Pajchlowa 1997; Dadlez 1998; Znosko 1999).

The development of the MPT depocentre began at the end of Lower Rotliegend times following volcanic activity (within the NW Mid-Polish Trough and the NE German Basin; see above and also Timmermann 2008) associated with regional wrenching and generalized crustal destabilization. The development of particular depositional systems was controlled both by the structural evolution of the extensional basin and coeval palaeoclimatic fluctuations. Additional controlling factors included changes in both source areas and sediment transfer paths over time. During the early stages of basin development, the main sediment source areas were located to the west and north, changing over time to NE and south. At later stages, supply from eastern and southern areas became dominant.

The Rotliegend succession in the Polish Basin may be subdivided into a pre- and post-volcanic period (Fig. 10.12). The former (i.e. Silesian Subgroup) commenced with the deposition of the Late Carboniferous–early Permian Dolsk Formation comprising fluviolacustrine sediments. The succession is a fragmentary one, but was probably originally more extensive. The Dolsk Formation is overlain by volcanics with a rhyolitic/basaltic composition (Wyrzeka Formation) which were associated with the Saalian event (Kiersnowski *et al.* 1995). Volcanic activity was limited to the western part of the basin. The thickness of this volcanic unit varies between 100 and 200 m (rarely up to *c.* 1000 m), which is considerably thinner than the coeval volcanics from the NE German Basin.

Cessation of volcanic activity was coeval with the initial development of the Polish Basin *sensu stricto*. The upper part of the Lower Rotliegend succession comprises mainly pyroclastic and fluvial and lacustrine siliciclastic deposits (Maliszewska *et al.* 2003). These sediments (e.g. Polwica Conglomerate Member) were derived from the erosion of the basement blocks displaced by the Saalian tectonic movements. These units were subsequently overlain by the Wielkopolska Subgroup, comprising three formations whose lithologies are related to their palaeogeographic positions (Fig. 10.12). The Ksiaż Wielkopolski Formation, distributed along the basin margins, comprises conglomerates and

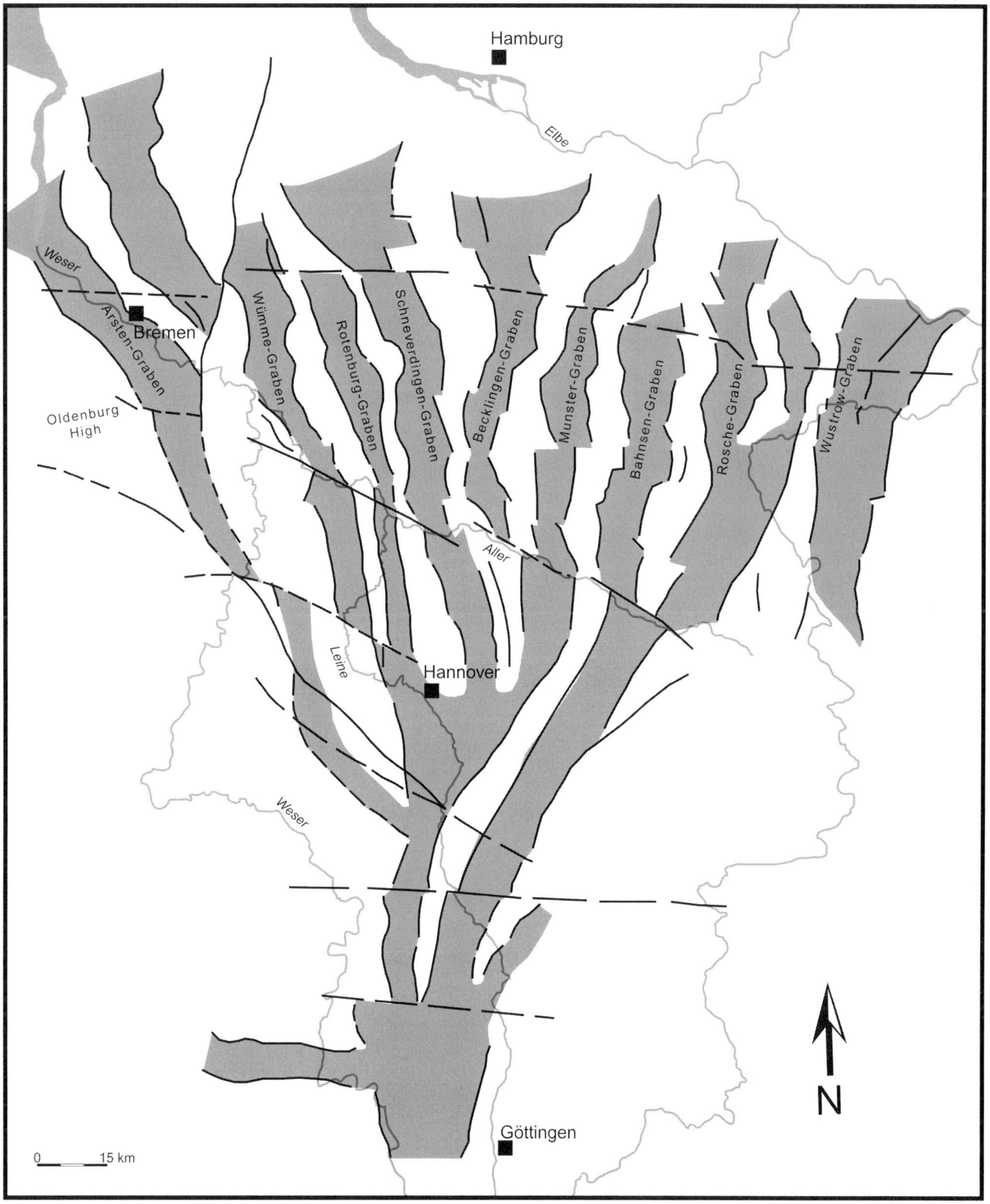

Fig. 10.6. Map of the Lower Saxony rift system, Germany, showing the most important border faults and strike-slip elements (after Gast 1988).

breccias with frequent thick sandstone interbeds and rarer intercalated claystones and silstones. These sediments represent an alluvial fan environment. The more basinward, but coeval Siekierki Formation comprises mainly sandstones with rare conglomerate intervals, while in the basin centre, the fine-grained Piła Formation is interpreted as continental sabkha deposits (Karnkowski 1999).

The palaeogeography and thickness distribution (with a maxi-

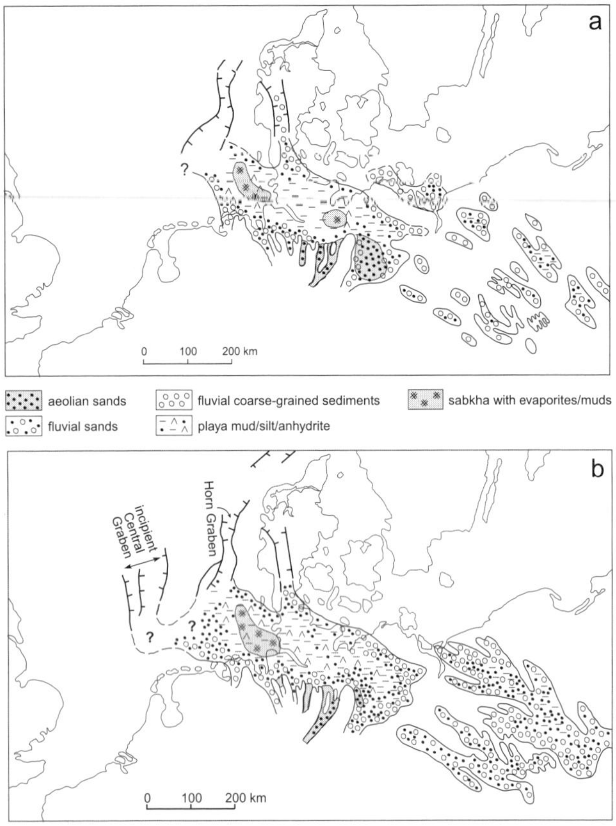

Fig. 10.7. Depositional environments of (**a**) the Rotliegend Parchim Formation and (**b**) the Mirow Formation in Central Europe (after Plein 1995; McCann 1998*b*).

mum of >1500 m) of the Upper Rotliegend deposits within the MPT reflects the activity of the main tectonic phases (of varying degrees of importance) which occurred during basin initiation and development (Kiersnowski 1998). Throughout the Permian the most distinctive structural elements in terms of basin palaeomorphology were the Brandenburg–Wolsztyn–Pogorzela (BWP) series of palaeohighs (Fig. 10.13). These highs, comprising of an extreme northward extension of the Variscan Exter-

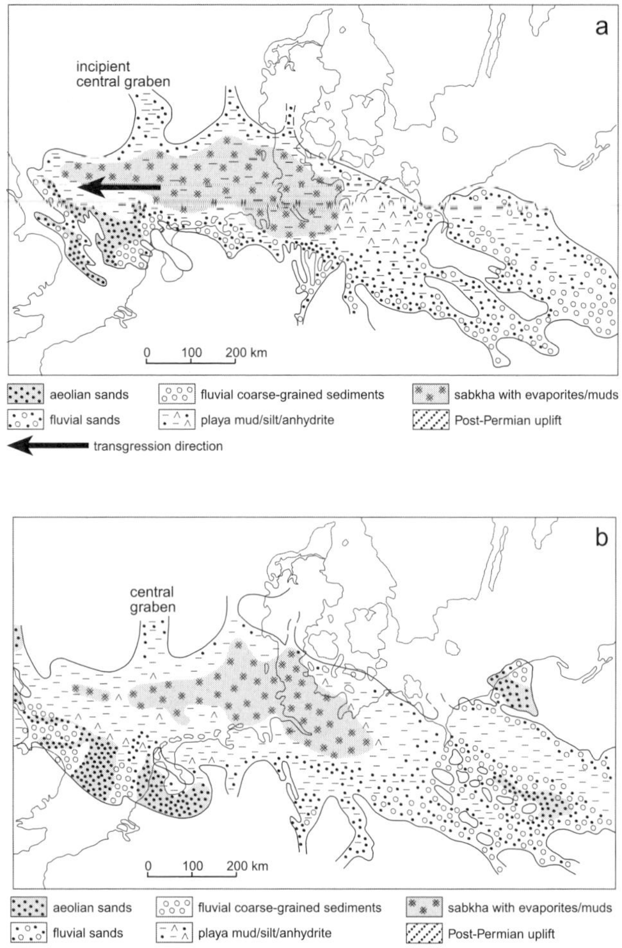

Fig. 10.8. Depositional environments of (**a**) the Rotliegend Dethlingen Formation and (**b**) the Hannover Formation in Central Europe (after Plein 1995; McCann 1998*b*).

nides, were formed mainly by Carboniferous–Permian volcanic rocks and subordinate Lower Carboniferous (and Devonian?) sedimentary rocks. The BWP highs are separated from the North German Basin to the west by the Liebenwalde-Torgelow and

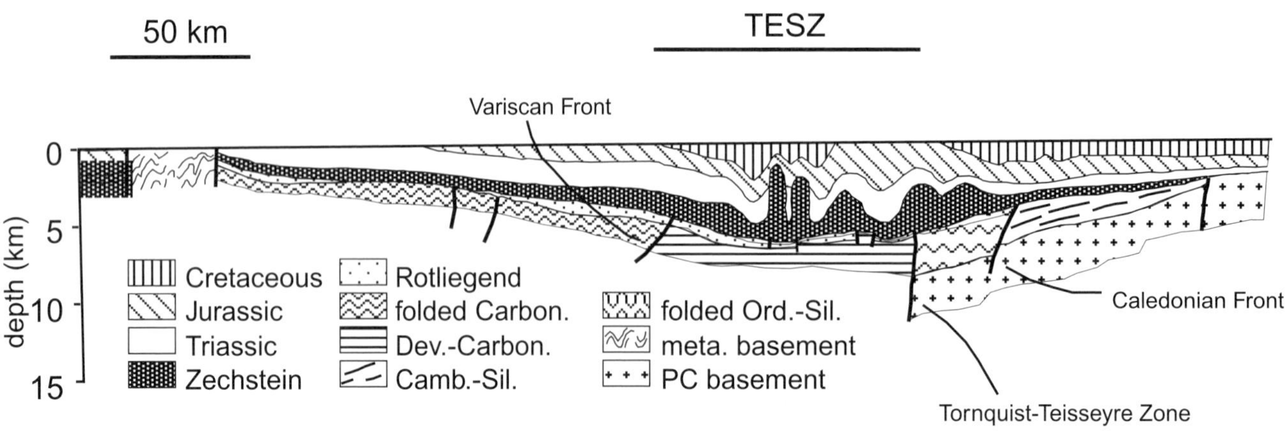

Fig. 10.9. Geological cross-section based on well and seismic data through the Polish Trough. The approximate location of the Trans-European Suture Zone (TESZ) (after Dadlez 1993) is indicated (after Stephenson *et al.* 2003). See Figure 10.10 for location.

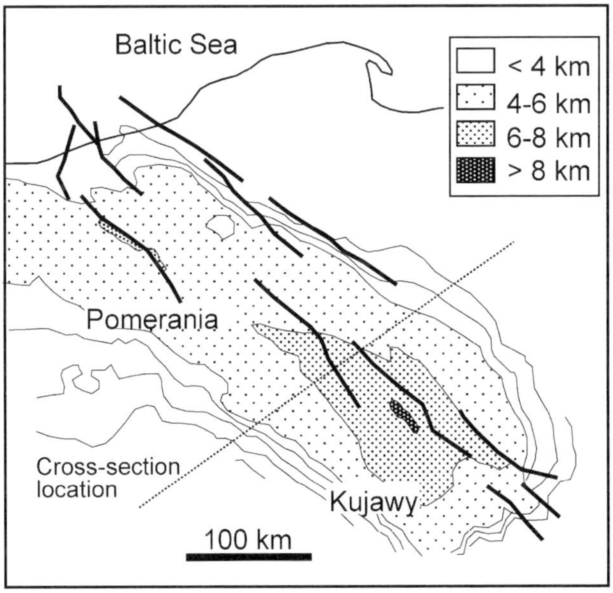

Fig. 10.10. Structure on the base of the Upper Rotliegend (depths below sea level) with the geographical segments of the Polish Basin. Contour intervals 500 m. The position of the cross-section across the basin (Fig. 10.9) is indicated (after Stephenson *et al.* 2003).

Lower Odra fault zones, and are bounded to the SW by the Dolsk Fault Zone and faults related to eastern Variscan tectonic structures. The palaeomorphology of the BWP highs was altered by several phases of volcanic activity which resulted in the deposition of thick units of lava or pyroclastic rocks (more than 400–500 m thick). Simultaneously, weathering processes modified the palaeomorphology allowing the removal of unconsolidated deposits.

Basin fill accumulation and distribution was controlled by differential subsidence related to synsedimentary tectonics and the locations of block-faulted structures. While the basin is polygenetic in origin, it generally displays the typical features of a half-graben with maximum subsidence along the NE margin (= Tornquist-Teisseyre Zone). This morphology is reflected in the lithofacies distribution in the basin, which is asymmetrical and displays an overall onlapping trend. The axial zone of the basin is dominated by fine-grained clastic lithofacies. The persistent occurrence of the playa environment together with the asymmetrical development of aeolian sediments along the SW margin, allows the Rotliegend Mid-Polish Trough to be categorized as a lacustrine half-graben. Aeolian sediments are found mainly in areas of lower subsidence, their distribution accentuating the basin asymmetry and clearly defining its less active SW margin. The development of the aeolian succession was controlled mainly by palaeoclimatic conditions involving at least three significant climatic shifts.

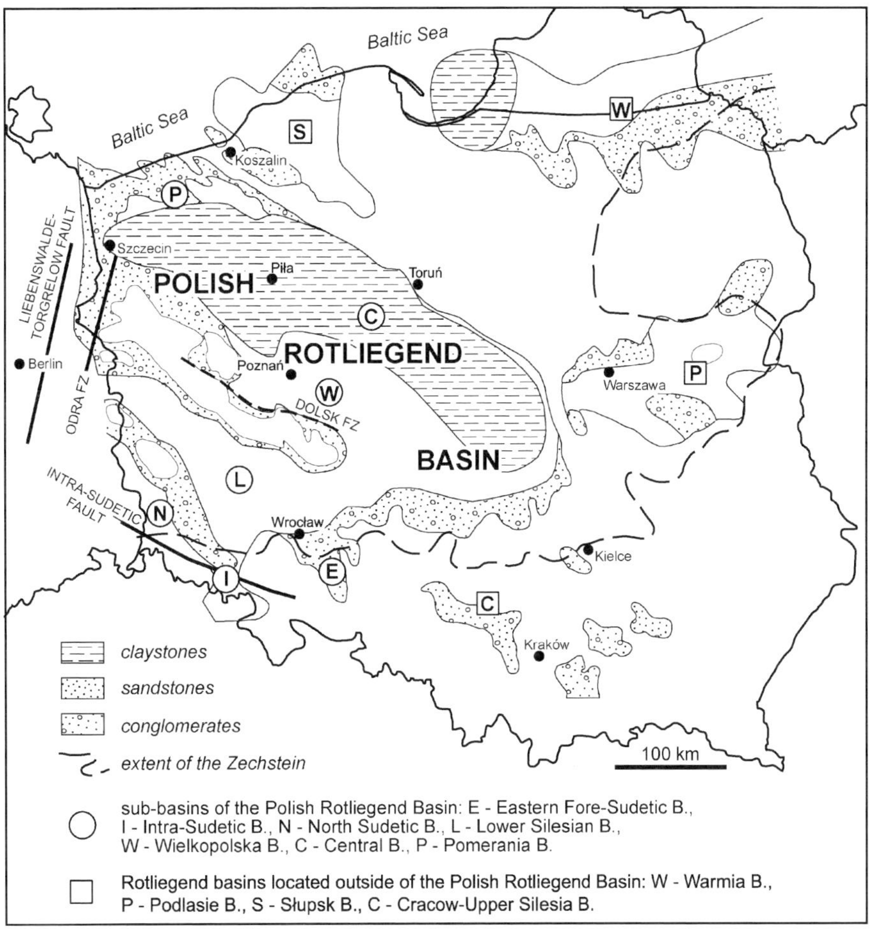

Fig. 10.11. Rotliegend basins of Poland (after Karnkowski 1999).

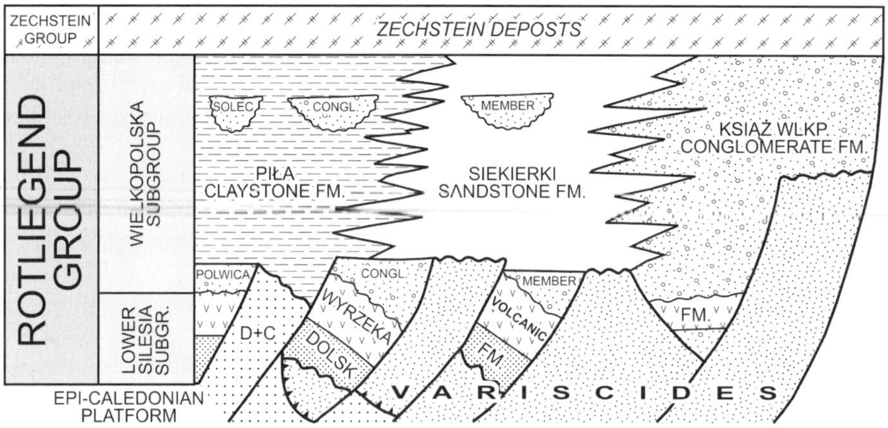

Fig. 10.12. Formal Rotliegend stratigraphy in the Polish Basin (after Karnkowski 1999).

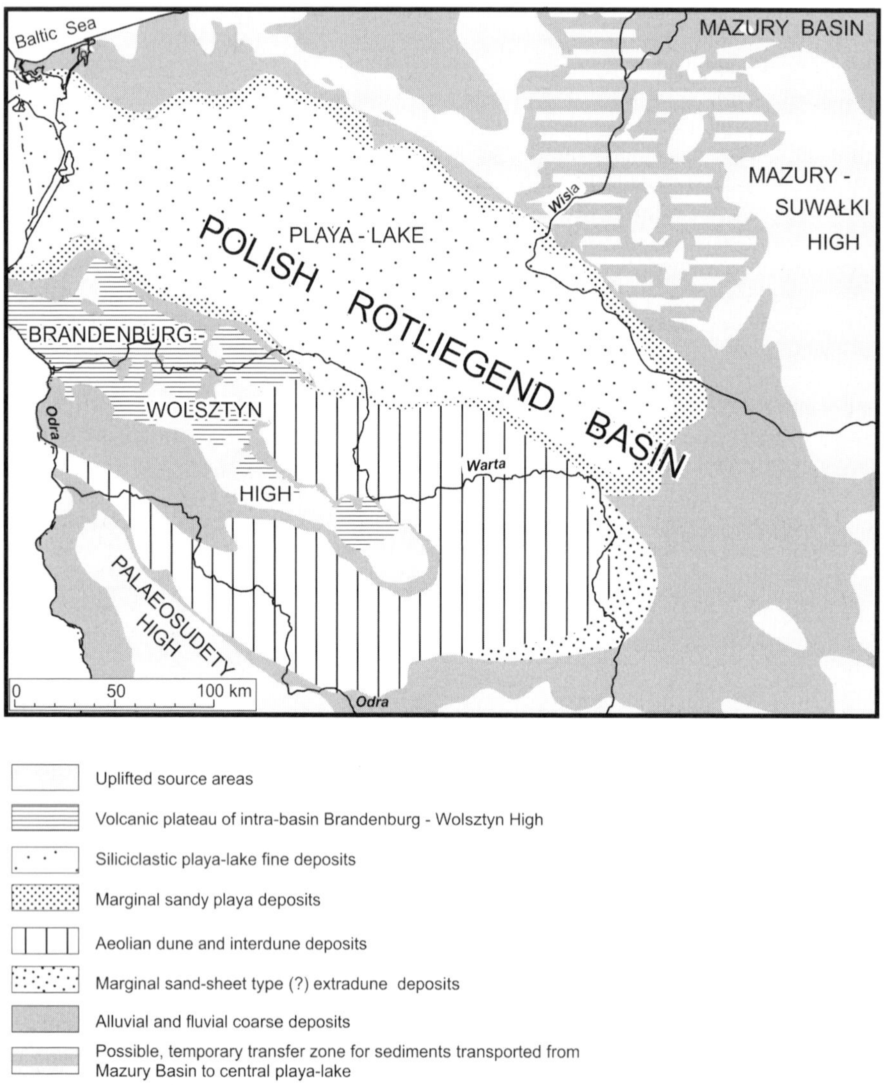

Fig. 10.13. Palaeogeography of the topmost Upper Rotliegend Basin in Poland (modified after Pokorski 1988: Kiersnowski 1997; Kiersnowski & Buniak 2006).

Northeast German Basin (T.M.)

The North German region, which later formed part of the SPB, comprises crystalline basement of varying Precambrian ages (for details see McCann 1998*a*) which is covered by a sedimentary succession, in excess of 12 km, of lower and upper Palaeozoic, Mesozoic and Cenozoic strata (Hoth *et al.* 1993). By the early 1990s almost 180 wells had been drilled into the Rotliegend of western Germany while in eastern Germany more than 1500 wells were drilled (McCann *et al.* 2000). The most important hydrocarbon reservoir unit in the NE German Basin (NEGB) is the Lower Permian Rotliegend succession of continental clastics, sourced from the underlying Westphalian coal measures and capped by Zechstein evaporites. Hydrocarbon quality, however, in the NEGB is poorer than in western Germany due to the high N_2 content (McCann 1998*b*).

The Rotliegend succession of the NEGB can be divided into four subgroups (Plein 1995). The oldest, the Altmark Subgroup, consists mainly of massive extrusive volcanic rocks and brecciated volcanic agglomerates with tuffaceous layers. The overlying Müritz Subgroup was deposited in local basins and comprises mainly grey, black and reddish-brown laminated siltstones and mudstones with intercalated fine-grained sandstones and occasional conglomeratic, pyroclastic and carbonate layers. At the onset of the uppermost Rotliegend, there was a clear change in basin geometry (e.g. Rieke *et al.* 2001, 2003). The subsequent Havel Subgroup comprises mainly fluvial and aeolian sandstones interbedded with mudflat sediments and rare lacustrine deposits in the basin centre. This subgroup may be divided into the Parchim (*c.* 700 m maximum thickness) and Mirow (*c.* 450 m maximum thickness) formations. The distribution of the Parchim Formation is largely restricted to the Havel-Müritz Basin in the south (Figs 10.7 & 10.14) and the West Mecklenburg Basin in the north. The sediments are predominantly fluviatile with rare aeolian sandstones and fine-grained playa-lake deposits. The deposition of the overlying Mirow Formation was coincident with basin enlargement as indicated by the wider areal distribution of the formation, when compared to the underlying Parchim Formation which was largely confined to graben structures (Figs 10.7 & 10.14). There is also less variation in the sedimentary facies, which are predominantly sandstones and fine-grained cyclic playa sediments. Aeolian sandstones and fluvial deposits were, however, locally important.

The overlying Elbe Subgroup comprises the Dethlingen (*c.* 650 m maximum thickness) and Hannover (*c.* 700 m maximum thickness) formations. Deposition of the Dethlingen Formation coincided with a period of thermal subsidence and pronounced basin enlargement (Figs 10.8 & 10.15) Sediment input from the southern margin was predominantly fluvial with cyclic sedimentation in the basin centre (Schneider & Gebhardt 1993; Gast 1991). Basin subsidence continued into Hannover Formation times. During this time interval variations in water-table levels led to changes in both halite and clastic distribution within the NEGB resulting in the development of thick clastic units along the southern margin of the basin, but significantly thinner units to the north. The central parts of the basin comprised playa-lake sediments which grade westwards into thick evaporites.

The NEGB shows no evidence of significant synsedimentary tectonism (e.g. Kossow *et al.* 2000), as described for the NW German Basin, the Dutch Basin (e.g. Verdier 1996, Geluk 2005) and the English Basin (e.g. George & Berry 1997). This absence presumably reflects the trend of gradually increasing tectonic activity in these latter areas, which may be related to the initiation of the breakup of Pangaea in the late Permian.

Hessian Basin (T.M.)

The Hessian Basin (= Hessischer Trog) divides the Rhenish Massif from the Harz and Thuringia Forest areas and connects the Saar-Nahe Basin with the Lower Saxony rift system (Fig. 10.15). Between the Saar-Nahe Basin and the Hessian Basin is a local high, the Sprendlinger Horst, which occasionally acted as a barrier to sediment transport between the two areas. The Lower Saxony rift system is the northward continuation of the Hessian Basin and comprises more than ten major rift basins (both grabens and half-grabens) (Fig. 10.6). The evolution of this series of north–south trending rifts was first described by Gast (1988). The precise onset of rift activity is not clear, but was probably subsequent to the earliest phase of Permian volcanicity, and related transtensional activity. The rifts are arranged in a fan shape extending northwards from the Hessian Basin. Sediment infill (up to 400 m) is Lower and Upper Rotliegend in age and comprises mainly dune sands and fanglomerates, although those basins to the east are filled mainly with volcanics. Transtensional movement led to the segmentation of some of the rifts in an east–west direction which affected the subsidence and infill histories (Gast 1988). Similar graben systems have also been reported from the area to the north of Berlin (Baltrusch & Klarner 1993). The Permian succession within the basin has been reported by Kowalczyk (1983) and Kulick *et al.* (1984), while Hug (2004) and Hug & Gaupp (2005) have outlined the Zechstein history.

The northern part of the Hessian Basin is pre-Permian in age. Basement is formed by part of the Variscan Rhenohercynian Zone, which can be correlated with the successions of the adjacent Harz and Rhenish Massif areas. Rotliegend volcanics have been reported from the basin, some of which have been reworked into rhyolite conglomerates (Kulick *et al.* 1984). Five formations have been recognized within the Hessian Basin, some of which can be correlated across into the Saar-Nahe Basin (Müller 1996). The lowermost unit (Lindheim Formation) is *c.* 220 m thick and comprises a series of fining-upward cycles (conglomerate to sand and silt), with intercalated calcareous nodules, coals and abundant plant material. Rare tuffs are also found. The overlying Altenstadt Formation (*c.* 180 m minimum) is finer-grained comprising cyclically arranged sands and silts, with rare conglomerates and acid pyroclastics.

A basalt unit frequently marks the base of the Düdelsheim Formation (100–200 m). This formation is predominantly clastic, comprising finer-grained sandstones and siltstones as well as rare conglomerates and carbonates. The sediments were deposited in a fluviolacustrine setting (Kowalczyk 1983). The overlying Schöneck Formation (150–250 m) also commences with a basalt unit, and comprises coarse- to fine-grained clastic rocks. Within the formation there are numerous tuff beds, providing evidence of coeval volcanic activity.

The uppermost unit (Bleichenbach Formation; minimum 300 m) comprises sandy siltstones and siltstones with some intercalated coarser-grained sandstones. In more central parts of the basin, the formation becomes increasingly muddy in character. The formation was deposited under arid conditions in a fluviolacustrine setting (Kowalczyk 1983).

Kraichgau-Saale Basin, Thuringia Forest area (T.M., J.S.)

The intermontane basins of the Thuringia Forest belong to a NE-trending system of basins, which include the Saar-Nahe-Werra and the Saale basins (Fig. 10.15) (Zeh & Brätz 2004). These basins result from the inversion of the NE-trending Mid-German Crystalline High (MGCH) which was established as a result of Variscan subduction and collision (360–320 Ma) in a NW-

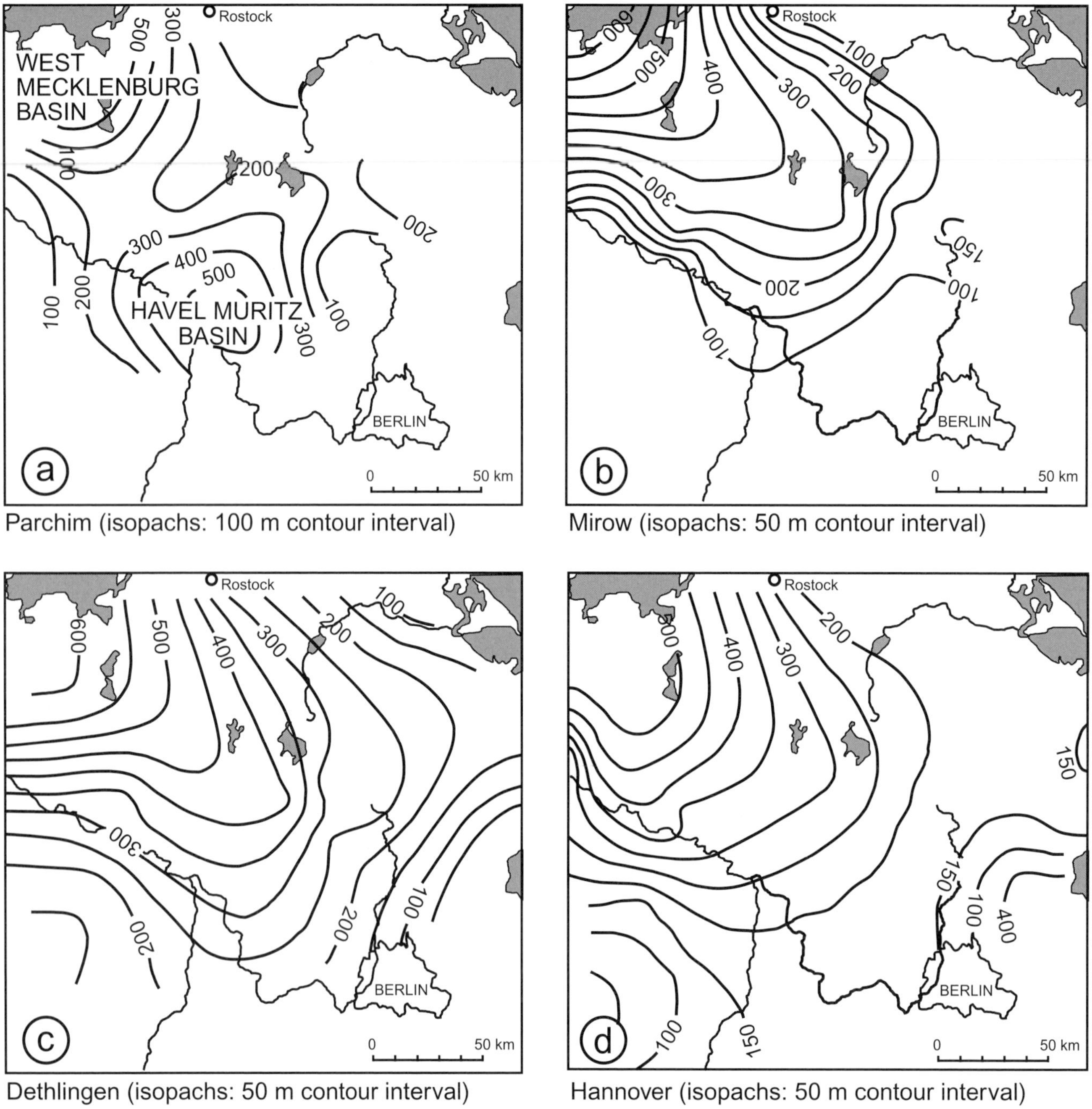

Fig. 10.14. Isopach maps of the Rotliegend (**a**) Parchim, (**b**) Mirow, (**c**) Dethlingen and (**d**) Hannover formations, NE German Basin (after McCann *et al.* 2000).

directed compressional stress field (e.g. Oncken 1988). Inversion of the MGCH in the NE began in Visean times whereas to the SW it began somewhat later (i.e. Westphalian).

The **Saale Basin** is located to the SE of the Harz Mountains. It has a length of *c.* 200 km and a width of 30–50 km and is the type locality of the Saalian unconformity which marks the boundary between the Lower and Upper Rotliegend in much of Central Europe (Fig. 10.15). The basin is orientated subparallel to the SW-striking grain of the Variscan basement and evolved under a tectonically mobile regime as reflected by the presence of syndepositional faulting, repeated changes in the direction of clastic influx and intrabasinal dispersion patterns, and by pronounced volcanic activity. The Late Carboniferous and Early Permian sediments contained in the basin were partly truncated prior to the deposition of the Rotliegend clastic succession. This, together with the fact that activity along the fault systems controlling basin subsidence also varied with time, suggests that the regional stress pattern changed during the Lower Rotliegend

Sedimentation, which was initially areally restricted, commenced in the Saale Basin during the early Westphalian. In the late Westphalian and Stephanian the area of the basin became enlarged, with coeval coal-bearing clastics locally attaining

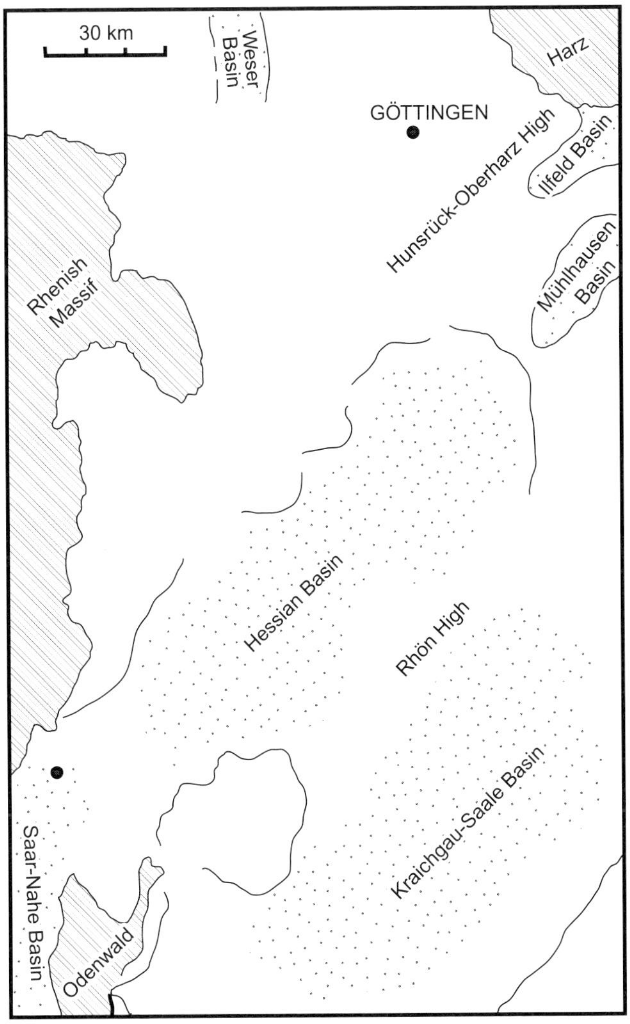

Fig. 10.15. Location map of the Hessian Basin, Kraichgau-Saale Basin and the Saar-Nahe Basin, Germany (after Kulick *et al.* 1984). The Palaeozoic crystalline massifs (Harz, Rhenish Massif, Odenwald) are also shown.

thicknesses of up to 1 km. These are disconformably overlain by early Lower Rotliegend fluvial clastics and volcanics. At the transition from early to late Lower Rotliegend the generally NE-trending basin was subdivided by a system of NW-striking faults into individual sub-basins. Syndepositional faulting, accompanied by widespread effusive volcanic activity, persisted until the end of the Lower Rotliegend and died out during the early Upper Rotliegend. The total thickness of volcanics and sediments is >2 km. Upper Rotliegend sands and conglomerates unconformably overlie the mildly deformed Lower Rotliegend succession.

As noted above, the basin is subdivided into a number of sub-basins, which contain distinct facies and/or formation names (for details see Schneider & Gebhardt 1993). The following description is, therefore, a generalized one and is based on the succession deposited in the NW part of the basin. The basal Hornburg Formation comprises a quartz conglomerate deposited in an alluvial fan setting with related fluvial sediments (Schneider & Gebhardt 1993). Associated lacustrine deposits were deposited in a playa setting. This is overlain by the Eisleben Formation which commences with a basal conglomerate. Subsequent sediments were deposited in sabkha/playa-lake settings with associated caliche development.

The basement of the NW part of the **Thuringia Forest** area forms part of the MGCH and is here exposed as the Ruhla Crystalline Complex (see also McCann *et al.* 2008) and transected by the Franconian Fault System (Fig. 10.16). Movements along this fault system caused intense block faulting that led to the formation of a series of deep sub-basins (e.g. Oberhof, Eisenach, Meinigen), which were subsequently filled with Permo-Carboniferous continental sediments and volcanics. These sub-basins occur close to horsts of metamorphic basement truncated by several plutons (Zeh & Brätz 2004). During the late Variscan, the Ruhla Crystalline Complex comprised three segments that each experienced a different post-orogenic exhumation history (Zeh *et al.* 2000). These segments are separated by the north–south trending Engestieg and West Thuringian faults, with segmentation resulting in localized differences in the sedimentary succession across the area.

The general Thuringia Forest succession commenced with basal conglomerates grading upwards into fluvial and lacustrine deposits with local coal-swamp development (Lützner *et al.* 2003). The lowermost unit is the Gehren Subgroup (up to 3100 m) comprising three volcanic sequences (Schneider *et al.* 1996*b*). The succession is dominanted by intermediate volcanics and pyroclastics with acid tuffs and ignimbrites cropping out to the SE. The overlying Ilmenau Formation (Lower Rotliegend) is *c.* 600 m thick and is separated from the older formations by an unconformity. The formation is characterized by bimodal (basalt–rhyolite) volcanism and intercalated sediments and volcaniclastics. The overlying Manebach and Goldlauter formations are predominantly clastic, the former also containing coal horizons. Deposition was in an alluvial setting, including river and lakes. Volcanic rocks in these formations are restricted to rare ash layers.

The Oberhof Formation (up to 1200 m) records the second phase of volcanic activity. It begins with a mixed association of trachyandesites, basalts and rhyolites evolving upward into extensive rhyolite domes, flows and subvolcanic intrusions. Interbedded are clastic sediments deposited in fluvial and lacustrine environments (Schneider *et al.* 1996*b*). This is overlain by fluvial clastics of the Rotterrode Formation deposited in a dominantly fluvial setting (Lützner 1981), followed by the alluvial sediments of the Tambach Formation (Schneider *et al.* 1996*b*).

The volcanosedimentary record of the Eisenach and Oberhof sub-basins suggests that basin formation was asynchronous and that sedimentation was heterogeneous (e.g. Lützner 1988; Lützner *et al.* 1995). According to Lützner (1988), the Oberhof Basin was, at least temporarily, activated as a pull-apart basin filled by Upper Carboniferous and Lower Permian (Rotliegend) volcanic rocks and sediments (up to 4 km), belonging to the Gehren Subgroup and the Manebach, Ilmenau, Goldlauter, Oberhof, Rotterode and Tambach formations.

A magmatic event with the intrusion of granite bodies and granite porphyry dykes occurred in the NW Thuringian Forest between 301 and 294 Ma (Zeh & Brätz 2004). This event was coeval with the volcanism of the Möhrenbach Formation (SE Thuringian Forest) as well as with a series of intrusions in the region (e.g. Frankenwald, Harz; see Zeh & Brätz 2004).

Ilfeld Basin (T.M.)

The Ilfeld Basin is a small NE–SW orientated basin located along strike from the larger Hessian Basin, and to the south of the Harz Mountains (Fig. 10.17). Büthe & Wachendorf (1997)

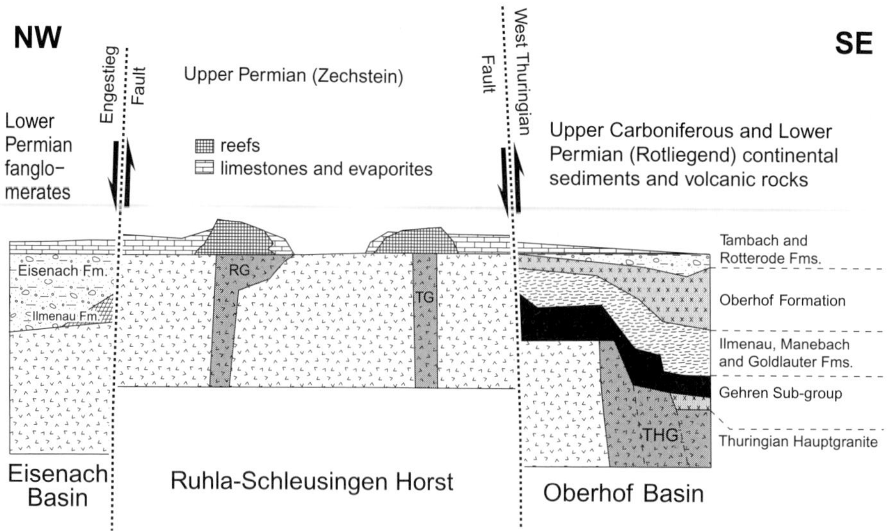

Fig. 10.16. Upper Carboniferous–Permian basins in the NW Thuringian forest area, central Germany (after Zeh & Brätz 2004).

have suggested that the basin formed as a pull-apart basin during dextral transtension along older, Variscan lineaments. The Permo-Carboniferous succession within the basin consists mainly of coarse and fine-grained siliciclastics with intercalated volcaniclastic sediments and lava flows (Paul *et al.* 1997; Paul 1999). The succession is *c.* 800 m thick, with tuffs, ignimbrites and latitic, trachytic and rhyolitic flows making up *c.* 400 m of the entire thickness. The sediments are predominantly continental, and thus it is difficult to distinguish between the Stephanian and Lower Rotliegend parts of the succession.

Post-Variscan sedimentation commenced in the Upper Carboniferous (Stephanian). The basal Fuchsburg Formation, consisting of alluvial fan deposits, was deposited above reddened and weathered Devonian greywackes and cherts. The overlying Netzkater Formation is characterized by fine-grained grey palustrial deposits. The presence of abundant plant fossils and related coals makes it possible to differentiate between a submerged algal flora and one developed in dry locations. The Neustadt Formation consists of reddish conglomerates, sandstones and clays deposited in alluvial fans. Latitic and trachytic lava flows reveal the initial impact of volcanic activity.

The fine-grained Baumgarten Formation was deposited in a playa lake environment. During humid periods a lake formed in the centre of the basin as revealed by the presence of laminated microbial carbonates with ostracods, *Xenacanthodian* teeth, and *Branchiosaurids* jaws. In arid periods the lacustrine environment was replaced by pedogenic carbonates forming soils and caliche nodules and root horizons.

Lava flows of rhyolitic composition covered large areas of the basin at the Permian–Carboniferous boundary. They have domal

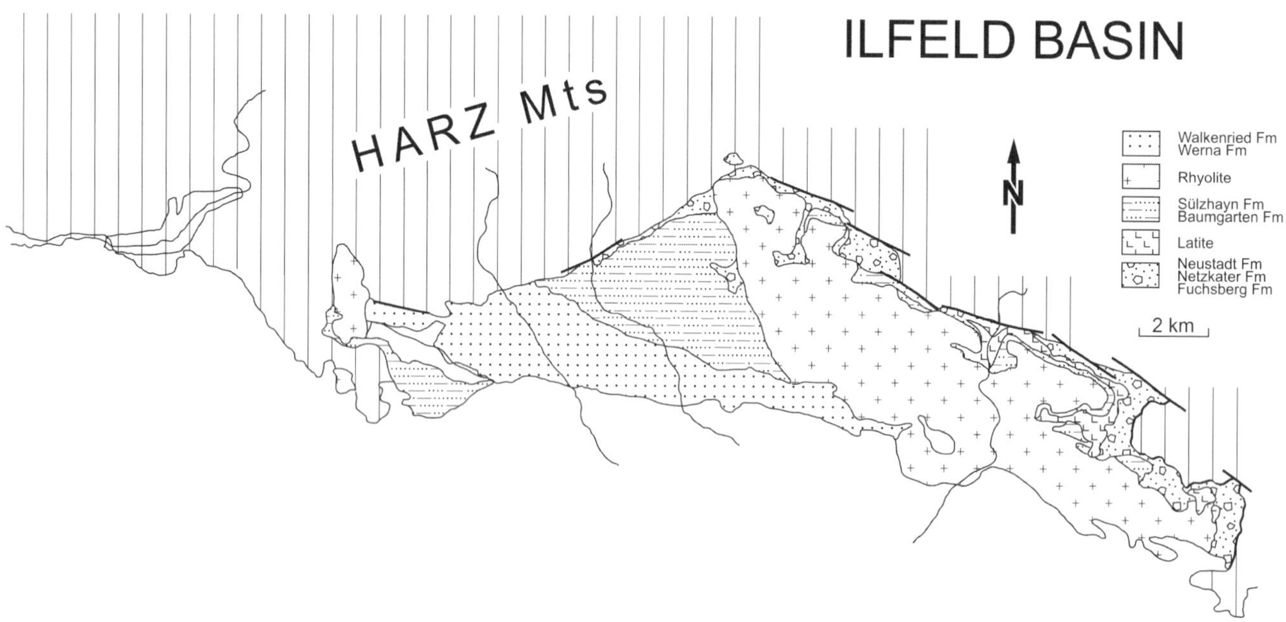

Fig. 10.17. Sketch map of the Ilfeld Basin, margin of the southern Harz area (after Paul *et al.* 1997).

structures and also penetrate into the playa deposits of the Baumgarten Formation (Paul *et al.* 1997). The lava flows were accompanied by the pyroclastic breccia, lapilli and ash tuffs forming the Sülzhayn Formation. This loosely aggregated material was easily reworked by fluvial processes. The overlying Werna Formation consists of coarse-grained sediments – frequently reworked volcanics – deposited as a series of alluvial fans. The sedimentary succession is terminated by the aeolian Walkenried Formation. The entire terrestrial Rotliegend succession is overlain by the marine Upper Permian Zechstein Group. However, the presence of an angular unconformity between both of these units suggests that there was a time gap, the duration of which is unknown.

Belgium (F.B., M.D.)

The sedimentary successions of southern Belgium (the Permian age of which, while likely, has not yet been proven) are dominated by continental deposits comprising alluvial fan, fluvial and fluviolittoral sediments, while in northern Belgium the successions mostly belong to the Zechstein Group. The formations of southern Belgium, however, are of relatively minor importance in terms of their geographical extent and thickness. Three depositional areas are considered (Fig. 10.18): the Malmédy Graben in the Eastern Ardennes, the northern part of Belgian Lorraine (NE part of the Paris Basin) and the subsurface of the Campine Basin (in NE Belgium). The latter forms the southernmost margin of the Southern Permian Basin.

The Malmédy Graben. The Malmédy Graben is *c.* 22 km long and varies in width from 0.5 to 2.5 km. The basin extends SW–NE across the Caledonian Stavelot Venn Massif in the eastern part of the Ardennes and to the NW of the Saar-Nahe Basin. Within the graben, Permian deposits overlie the folded and faulted Cambrian and Ordovician basement. The basin infill, comprising the Malmédy Formation, is considered to have been deposited in an alluvial fan and fluvial setting (Bultynck *et al.* 2001), and consists of: (1) a 15–60 m thick lower member with a basal conglomerate containing angular quartzitic clasts, overlain by alternating microconglomeratic sandstones and conglomerates with argillaceous matrix and rounded sandstone or shale clasts; (2) a 35–150 m thick conglomeratic member comprising predominantly limestone clasts in a calcareous/argillaceous matrix, with intercalated sandstone/siltstone/mudstone lenses; the limestone clasts are dated as Emsian to Lower Carboniferous on the basis of conodonts, corals and brachiopods, and are presumed to have been derived from the nearby Eifel region in Germany; and (3) a 30 m thick upper member consisting of a conglomerate rich in angular shale, sandstone and quartz clasts and rare limestones in an argillaceous matrix.

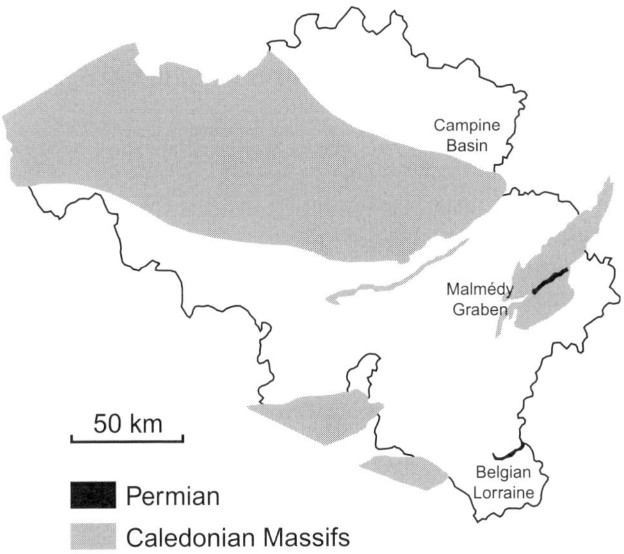

Fig. 10.18. Distribution of Permian successions in Belgium.

The Belgian Lorraine. The geometrically complex formations of the Belgian Lorraine constitute the NE part of the Paris Basin, and form the so-called Gulf of Luxemburg (Fig. 10.18). The nature of the sedimentary bodies in this area suggests deposition in a littoral environment characterized by mixed sedimentation with siliciclastics predominating.

The lowermost unit, the Habay Formation, is markedly continental and rests unconformably on the Lower Devonian basement. Its thickness is variable, with a maximum of 30 m close to the border between Belgium and Luxemburg. The precise age of the Habay Formation is unclear, with both Permian and Lower Triassic ages being assigned. The succession comprises alternating beds of conglomerate with shale and sandstone clasts, sand, argillaceous sand, conglomeratic dolomite and clay. Pedogenetic features are ubiquitous.

The Campine Basin. Permian to Lower Jurassic formations subcrop in the NE part of the Campine Basin and the Roer Valley Graben (NE Belgium and SE Netherlands) at the margin of the Southern Permian Basin (Fig. 10.18). The discovery of these concealed deposits resulted from salt exploration of the Campine Basin at the beginning of the twentieth century. During the 1980s renewed coal exploration resulted in extensive drilling of the area.

The Permian deposits display typical continental/marginal marine facies in continuity with similar formations in the southern Netherlands. Seismic surveys have revealed that the Permian rocks unconformably overlie eroded Carboniferous Coal Measures and gradually wedge out southwards, towards the Anglo-Brabant Massif, with Triassic deposits slightly overstepping the Permian strata. The timing of this disconformity is not precisely known, but it may post-date the Lower Triassic Buntsandstein. The transition between the Permian and the Triassic is conformable; the boundary is arbitrarily chosen, in conformity with current practice in the Netherlands (van Adrichem Boegaert & Kouwe 1993). The sedimentary evolution within this succession corresponds to a transition from a shallow marine to a floodplain environment.

The Permian Helchteren Formation (Dusar *et al.* 2001) is up to 45 m thick and comprises five members. The first member consists of a mineralized basal conglomerate which contains quartzite and limestone clasts (De Craen & Swennen 1992). It is overlain by grey silty carbonates alternating with marly mudstones and siltstones, constituting the second member, which is the thickest and most widespread and represents the fringe deposits of a marine carbonate ramp. The third member comprises grey to red silty mudstones with pedogenetic carbonate nodules and lagoonal fauna, passing upwards to massive sandstone beds, and locally to oolitic grainstones (fourth member), representing a beach barrier-shoreface. The fifth, and uppermost, member comprises massive red claystones with siltstone and sandstone intercalations. The second to fifth members represent a condensed Zechstein section.

The basal conglomeratic member bears a striking compositional resemblance to the Malmédy Formation and is, thus, tentatively assigned to the Rotliegend. The conglomerates appear to be derived from a similar source to that of the Malmédy Conglomerate, which could be the Eifel region in Germany, since the adjacent Anglo-Brabant Massif was probably a low-relief area at the time. It is, however, only locally preserved. In other parts of the Campine Basin it is reworked at the base of the overlying second member.

The Netherlands (T.M.)

Permian rocks occur over a large part of the subsurface of the Netherlands and form the lower part of the post-Variscan sedimentary succession (for details see Geluk 2005). The succession may be subdivided into three units: (1) the Lower Rotliegend consisting of a succession of volcanic and clastic rocks with a geographically limited distribution; (2) the Upper Rotliegend comprising fine-grained clastics and evaporites in the northern offshore and predominantly sandstones in the onshore and western offshore area; and (3) the Zechstein comprising a series of evaporites, carbonates and subordinate clastics.

The Lower Rotliegend occurs in distinct areas, namely the Ems Low, the Dutch Central Graben and the Horn Graben (offshore Denmark) (Fig. 10.19). In the Ems Low (maximum thickness 80 m), volcaniclastics represent the westernmost part of the Polish–German volcanic succession. In addition to volcanics (spilitic basalts) they comprise mudstones and tephra layers. In the Dutch Central Graben (maximum thickness *c.* 150 m), the succession comprises tephras and thick basaltic flows interbedded with claystones and sandstones. These correlate with the volcanics recorded in the Horn Graben by Stemmerik *et al.* (2000) (see below).

The Upper Rotliegend succession, with a maximum thickness of >700 m, consists of two formations: the Slochteren Formation comprising mainly sandstones and conglomerates, and its lateral equivalent, the Silverpit Formation composed of claystones, siltstones and evaporites. There is strong evidence for syndepositional fault movements within the Upper Rotliegend succession based on well data (Geluk 2005). In the western parts of the Central Netherlands Basin the thickness varies across major faults. Depositional environments were predominantly fluvial, aeolian and lacustrine/playa.

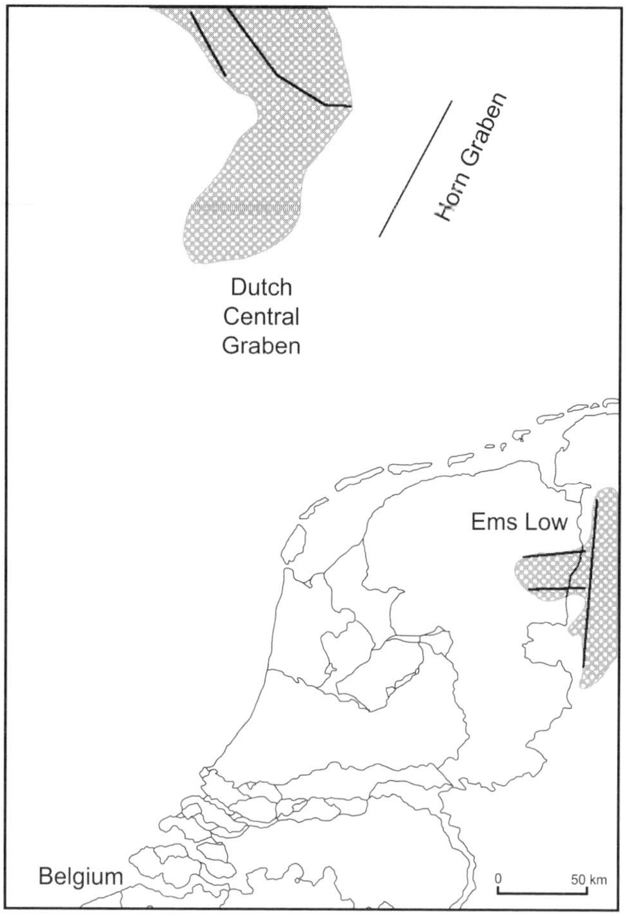

Fig. 10.19. Distribution of Lower Rotliegend successions in the Netherlands (after Geluk 2005).

Northern Permian Basin

As noted above, the sediments of the Northern Permian Basin are not as well understood as those of the Southern Permian Basin (SPB) (Glennie *et al.* 2003; Stemmerik *et al.* 2000) (Fig. 10.1). The sediments were deposited in a smaller basin, located to the north of the Ringkøbing-Fyn High, which was connected to the SPB via the Bamble and Glückstädt troughs and the Central and Horn grabens. The Rotliegend succession of the northern basin is less well known than that of the southern basin due to a lack of outcrops and limited drilling (Stemmerik *et al.* 2000). The Rotliegend sediments of Denmark were deposited in the Danish Central Graben and the Danish–Norwegian Basin, both of which are located north of the fragmented Mid-North Sea–Ringkøbing-Fyn High, and they comprise part of the succession deposited within the Northern Permian Basin (Fig. 10.20).

In the Danish North Sea the Rotliegend is subdivided into two formations. The lowermost Karl Formation is areally widespread and is defined to include the syntectonic volcanic, volcaniclastic and sedimentary fill of the Permian half-grabens in the Danish area. The succession is dominated by volcanics of alkaline basaltic composition and volcaniclastic sediments. With the exception of some isolated rhyolite flows (e.g. Horn Graben Rhyolite Member), the succession differs in composition from the more acid volcanics that characterize the Lower Rotliegend of the SPB (Aghabawa 1993). The rhyolitic flows were extruded as silicic lava which solidified as glass and were subsequently devitrified (Aghabawa 1993). The majority of the volcanics are basic in composition, and described as alkaline basalts, hawaiites and mugearites (Aghabawa 1993). Associated sediments are mainly volcaniclastic sandstones and conglomerates, probably deposited in alluvial and fluvial settings as well as siltstones and claystones deposited in lacustrine and sabkha environments (Stemmerik *et al.* 2000). The overlying Auk Formation represents the post-rift succession and comprises sandstones, pebbly sandstones and conglomerates with thin interbeds of silty mudstone, which were probably deposited in aeolian, fluvial and sabkha/lacustrine settings

Basins within the Variscan Orogen (intramontane basins)

Eastern and Southern Alps (K.K.)

Permian sedimentary rocks occur in all major tectonic units of the Eastern Alps (Penninic unit, Lower, Middle and Upper Austro-Alpine unit) and are widespread in the Southern Alps

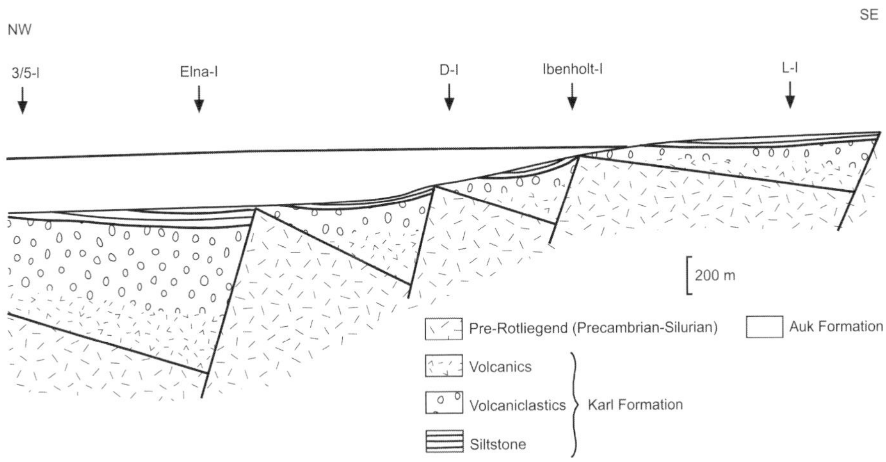

Fig. 10.20. Simplified SE–NW cross-section from the East North Sea-Ringkøbing-Fyn High to the Danish-Norwegian Basin showing the distribution of Rotliegend sediments. Note the onlap of the Auk Formation towards the Ringkøbing-Fyn High (after Stemmerik *et al.* 2000).

(Fig. 10.21). Permian sediments of the Penninic, Lower and Middle Austro-Alpine units were all deformed and overprinted by Alpine metamorphism (greenschist facies) to phengite and sericite schists, arkose schists, arkose gneisses quartzites and porphyroids.

In the Eastern Alps red-beds, mainly non-marine, were deposited under semiarid and arid climatic conditions during the transition from the Variscan to the Alpine tectonic cycle. Indeed, a change in facies and sediment colour is observed near the Carboniferous–Permian boundary indicating a climatic change from humid to arid conditions. In the eastern part of the Southern Alps (Karawanken Mountains, Carnic Alps) shallow-marine carbonate and siliciclastic sediments were the predominant deposits. In the western part of the region (i.e. Dolomites) the lower part of the Permian succession is represented by non-marine red-beds and thick volcanics, while in the upper part shallow-marine evaporitic and carbonate sediments interfinger with non-marine deposits (see summary in Krainer 1993).

Deposition of the late- to post-Variscan sediments commenced at different times: locally as early as the Late Carboniferous (Cantabrian/Late Moscovian, e.g. Carnic Alps, parts of Gurktal Nappe, Steinach Nappe) and as late as the Late Permian–Early Triassic (e.g. Middle Austro-Alpine Unit). However, the main phase of sedimentation was during the Early Permian (Northern Calcareous Alps, Drau Range, parts of Gurktal Nappe, Dolomites).

Sedimentation was influenced by fracture tectonics, particularly during the Lower Permian, and locally by volcanic activity. This most important tectonic event in the region (i.e. the Saalian event) separates the lower (Variscan) cycle (Upper Carboniferous–Lower Permian) from the upper (Alpine) cycle (Upper Permian–Lower Triassic). This boundary between the lower and upper cycles is marked by a stratigraphic gap and a significant change in facies. Synsedimentary block faulting resulted in the uplift and reworking of individual crustal blocks. In the Carnic Alps, for example, the basal breccia of the upper cycle (Tarvis Breccia) rests locally upon Devonian limestones. In the Eastern Alps Late Permian sediments of the upper cycle (Gröden Formation) contain large amounts of reworked acid volcanic rocks derived from the lower cycle, documenting the extensive nature of the block faulting and the rejuvenation of the relief caused by the Saalian event.

The Permian sediments of the Eastern Alps contain almost no fossils. However, rare plant fossils are present in the lowermost Permian of the western part of the Northern Calcareous Alps, the Gurktal Nappe and the Drau Range (van Amerom *et al.* 1982; Fritz & Boersma 1987*a*, *b*; Fritz *et al.* 1990).

Penninic, Lower and Middle Austro-Alpine tectonic units

In these three tectonic units (Tauern Window of the Penninic unit, Radstätter Tauern, Tarntaler Berge, Semmering and Wechsel, Stangalm and Brenner Mesozoic) the late- to post-Variscan clastic succession is up to 300 m thick (mostly <100 m) and metamorphically overprinted up to greenschist facies. Based on lithological correlation with the late- to post-Variscan sequence of the Upper Austro-Alpine units, the succession can be subdivided into a coarse-grained lower part termed the 'Alpine Verrucano' and ascribed to the Permian, and a finer-grained upper part termed the Lantschfeldquarzit, Semmeringquarzit or 'Skythquarzit' and ascribed to the Scythian (Frank 1972; Krainer 1984; Krois & Stingl 1989; Schünemann *et al.* 1982).

The age and depositional environment of these rocks are unclear due to later Alpine deformation and metamorphic overprinting. Locally, as in the Stangalm and Brenner Mesozoics, the entire succession is probably of Scythian age (Krainer 1984; Krois & Stingl 1989) (Fig. 10.22).

Upper Austro-Alpine tectonic units

The Permian succession of the Upper Austro-Alpine tectonic units (Northern Calcareous Alps, Gurktal Nappe and Drau Range) unconformably overlies the Variscan basement; this latter comprises schists and gneisses ('Gailtalkristallin') in the Drau Range and weakly metamorphosed sediments and volcanics in the Northern Calcareous Alps and Gurktal Nappe. Along the NW margin of the Gurktal Nappe, localized sedimentation commenced during the Late Carboniferous, and early Permian red-beds overlie the Late Carboniferous Stangnock Formation (Fig. 10.22).

The continental Permian succession is up to *c.* 150 m thick and divided into two lithostratigraphic units, separated by a major tectonic hiatus. Proximal to distal alluvial-fan sediments comprising red-coloured breccias, conglomerates and immature sandstones grade into fine-grained sandflat/playa complexes with caliche crusts and rare thin algal layers. These comprise the

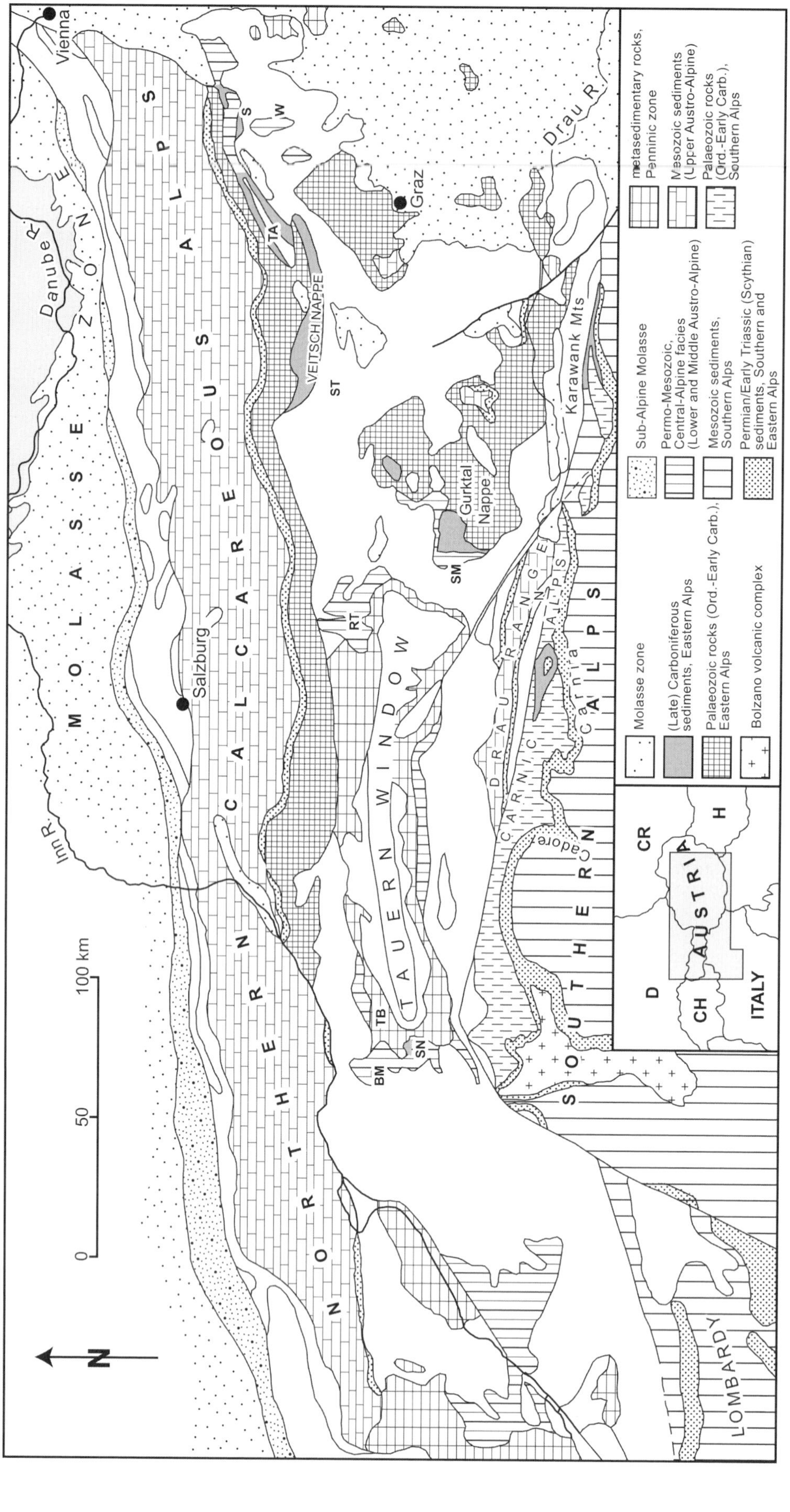

Fig. 10.21. Geological sketch map of the Eastern and Southern Alps (after Krainer 1993).

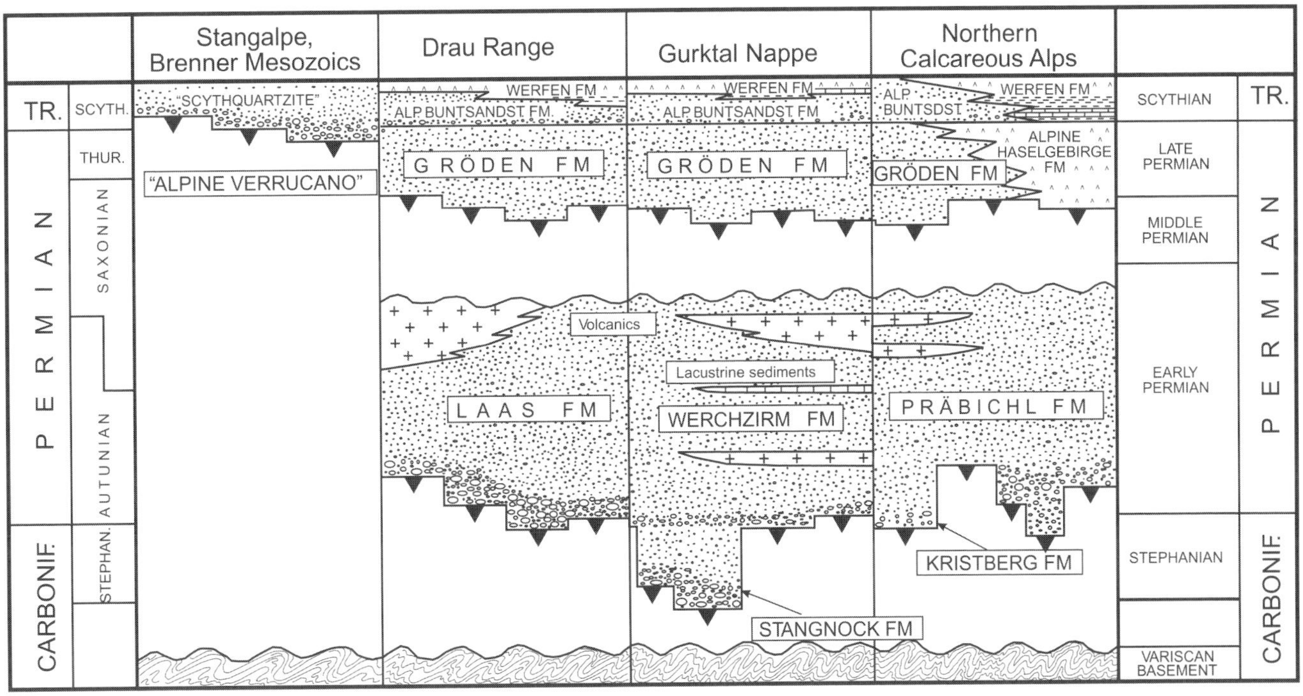

Fig. 10.22. Stratigraphic table of the Permian successions in the Penninic, Lower and Middle Austro-Alpine tectonic units.

lower unit, which includes the Kristberg Formation, the basal breccia and the Präbichl Formation in the Northern Calcareous Alps, as well as the Werchzirm (Gurktal Nappe) and Laas (Drau Range) formations (Fig. 10.22).

Lower Permian sediments are typically polymictic breccias and conglomerates containing abundant clasts derived from the local basement, as well as lithic arenites and greywackes (Krainer & Stingl 1986*a*, *b*; Krainer 1987*a*, 1989, 1990). Plant fossils from the lower part of this succession indicate an Early Permian age (van Amerom *et al.* 1982; Fritz & Boersma 1987*a*, *b*, 1988; Fritz *et al.* 1990). A sequence comprising up to *c*. 100 m of rhyolites and related rocks (including ignimbrites and pyroclastic flows) is widespread at the top of the early Permian succession. These volcanics can be correlated to the Bolzano Volcanic Complex of the Southern Alps.

The Early Permian succession is overlain by the Gröden Formation red-beds deposited in ephemeral braided rivers and playas (Fig. 10.23). The boundary between the two is marked by a sharp lithological change, possibly tectonically related, and correlated to the boundary between the Bolzano Volcanic Complex and the Gröden Formation in the Dolomites and between the Trogkofel Limestone and the Tarvis Breccia in the Carnic Alps. The Gröden Formation is up to *c*. 350 m thick in the Drau Range and Gurktal Nappe, and up to 800 m in the Northern Calcareous Alps.

Southern Alps: Carnic Alps and Karawanken Mountains

In the Carnic Alps and the Karawanken Mountains shallow-marine depositional environments prevailed throughout most of the Permian; non-marine sediments are restricted to the Tarvis Breccia and the Gröden Formation. The Permian succession here is 1000–2000 m thick and divided into the Rattendorf Group, the Trogkofel Group, the Tarvis Breccia, and the Gröden and Bellerophon formations (Fig. 10.24).

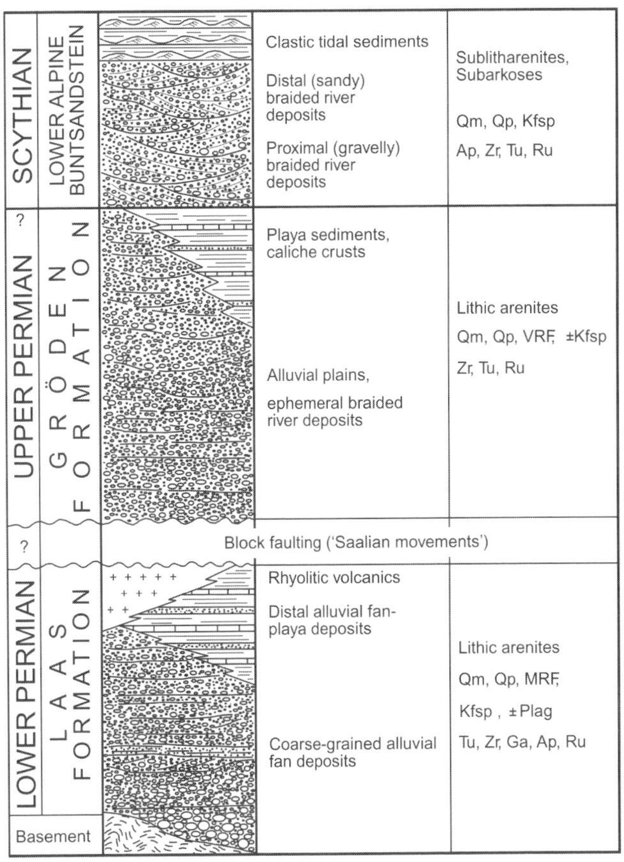

Fig. 10.23. Stratigraphic successions of the Upper Austro-Alpine tectonic units.

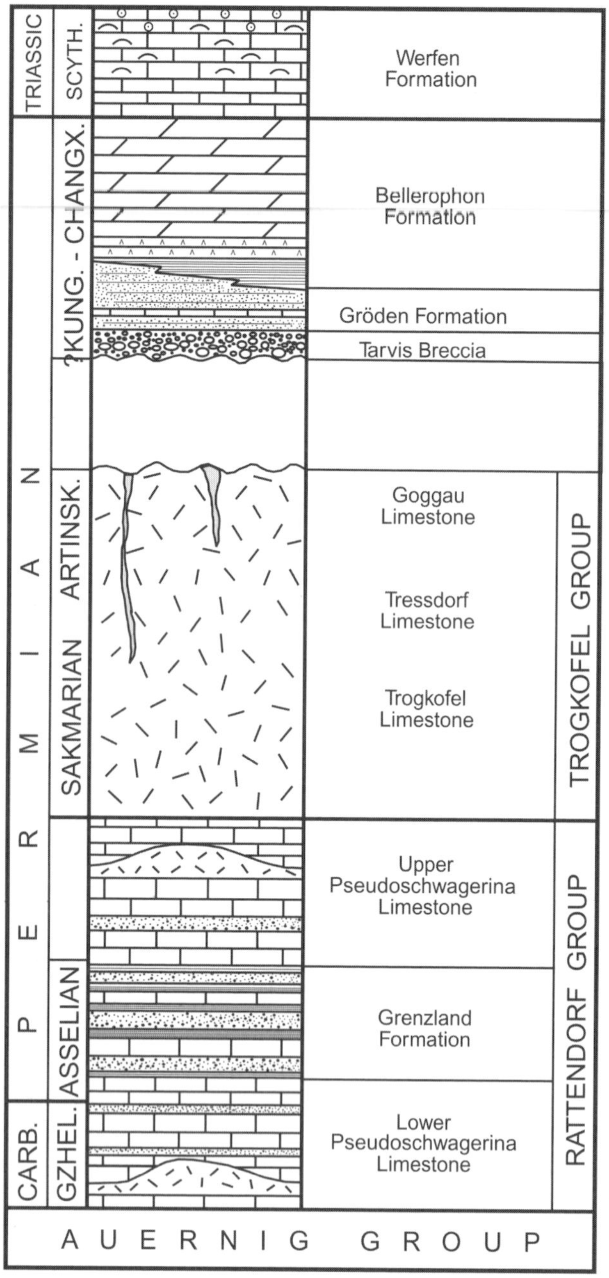

Fig. 10.24. Permian successions of the Rattendorf and Trogkoffel groups.

The Rattendorf Group comprises the Lower Pseudoschwagerina Limestone, the Grenzland Formation and the Upper Pseudoschwagerina Limestone. The facies of the Rattendorf Group is similar to that of the underlying Auernig Group, with sediments being deposited in nearshore, inner shelf and outer shelf environments. The Grenzland Formation is a cyclic succession of predominantly shallow-marine siliciclastic sediments including quartz-rich conglomerates, sandstones and siltstones intercalated with thin-bedded, fossiliferous limestones (Buttersack & Boeckelmann 1984; Boeckelmann 1985). A thin caliche horizon indicates subaerial exposure. Locally plant fossils occur within thin shale intervals (Fritz & Boersma 1984; Fritz & Krainer 2004). Limestones contain fusulinids indicating a middle-late Lower Rotliegend age (Krainer & Davydov 1998).

The Upper Pseudoschwagerina Limestone is predominantly composed of dark-grey, thin-bedded fossiliferous limestones and intercalated thin intervals of siltstone and sandstone, and rarely, well-rounded and well-sorted quartz-rich conglomerates. The limestones contain abundant fossils, particularly calcareous algae (Homann 1972), smaller foraminifera (Vachard & Krainer 2001), fusulinids, corals, bryozoans, brachiopods, echinoderm fragments and others. The limestone microfacies have been studied by a number of authors (e.g. Flügel 1968, 1974; Buttersack & Boeckelmann 1984). The cyclical pattern of sedimentation within this unit is indicative of repeated shifting from nearshore (siliciclastic sediments) to offshore environments (limestones) in an open marine shelf with normal water circulation (Flügel 1981; for summary see also Krainer 1993; Flügel *et al.* 1997).

The Trogkofel Group is mainly composed of up to *c.* 400 m of massive limestones (Trogkofel, Tressdorf and Goggau limestones). At Trogkofel the massive limestone facies of the Trogkofel Group consists mostly of boundstones forming large mounds. According to Flügel (1980, 1981) these mounds represent 'stratigraphic reefs' which probably formed in a shelf-margin environment adjacent to a shallow-water carbonate platform. The fusulinid assemblage indicates an early Sakmarian to late Artinskian age (Kahler & Kahler 1980; Kahler 1986; see also Krainer & Davydov 1998).

Locally, the bedded limestone facies of the Upper Pseudoschwagerina Limestone is erosively overlain by the coarse-grained, poorly sorted carbonate breccias of the Trogkofel Group. Tectonic pulses caused uplift and subaerial exposure of individual blocks resulting in erosion and reworking of previously deposited sediments, particularly *Tubiphytes/Archaeolithoporella* mounds, and the deposition of carbonate breccias and marly shales in an adjacent, relatively shallow basinal environment. These breccias indicate that in the Carnic Alps Early Permian deformation, commencing in the early Sakmarian, occurred.

In the southern Karawanken Mountains the 'Clastic Trogkofel beds' (Ljubljana beds), first recognized by Ramovs (1963), appear to be the stratigraphic equivalent of the Trogkofel Limestone (Ramovs 1972, 1974, 1983; Buser 1974). This sequence is 500–2000 m thick and is composed of red and grey, locally fossiliferous shales, siltstones, sandstones and quartz conglomerates. Thin limestone beds containing conodonts are intercalated. Large blocks of shallow-water limestones, probably representing olistoliths, also occur.

Limestones of the Trogkofel Group are unconformably overlain by the rocks of the Tarvis Breccia. The boundary between the two represents a hiatus caused by significant block faulting (i.e. Saalian event). The Tarvis Breccia is up to *c.* 200 m thick and composed mainly of limestone clasts derived from the Trogkofel Formation. The breccia is interpreted as a footwall fan deposit with associated proximal debris-flow deposits. Intercalated red siltstones and claystones with carbonate concretions are assumed to have formed in a coastal sabkha environment.

Dolomites

In the Dolomites (Alto Adige/South Tyrol and Trentino; northern Italy) the Permian succession is composed of two tectonosedimentary cycles. The lower cycle is of Early Permian age and includes the Ponte Gardena Conglomerate and the Bolzano Volcanic Complex (Atesino Platform Volcanics). The upper cycle consists of the ?Middle–Late Permian Gröden and Bellerophon formations (Fig. 10.25).

Locally, the coarse-grained red-beds of the Ponte Gardena Conglomerate unconformably overlie the Variscan basement which is mainly composed of quartz phyllites. The red-bed units are up to 200 m thick and were deposited in small intermontane basins formed by block faulting in the lowermost Permian (Krainer 1989). The Ponte Gardena Conglomerate is overlain by the volcanic rocks of the Bolzano Volcanic Complex (BVC), while locally the BVC rests directly on the Variscan basement. The BVC reaches a maximum thickness of *c.* 3000 m and covers an area of 4000 km^2. The succession consists mainly of ignimbrites and pyroclastics, with subordinate rhyodacitic lava flows and andesitic rocks. The BVC can roughly be divided into a lower part composed of andesitic to dacitic/rhyodacitic volcanics and an upper part of rhyodacitic to rhyolitic volcanic rocks (D'Amico & Segatta 1979; Brandner & Mostler 1982). Radiometric determinations indicate a mean age for the BVC of *c.* 276 Ma (i.e. Artinskian/Rotliegend) (Barth & Mohr 1994). Recently, Bargossi *et al.* (2004) reported zircon ages from the Bolzano Volcanic Complex ranging from 279.6 ± 1.1 Ma (Artinskian) to 274.1 ± 1.6 Ma (early Kungurian).

Alluvial and lacustrine sedimentary rocks are intercalated within the BVC at several stratigraphic levels, indicating periods of volcanic quiescence during sedimentation. These sedimentary rocks are derived from the underlying volcanics. Chert layers are interpreted as abiogenic precipitates analogous to modern volcanic lake settings. The chert layers yielded a rich, diverse and well-preserved miospore assemblage indicating semiarid to arid climatic conditions and an Artinskian to Kungurian age (Krainer & Spötl 1998; Hartkopf-Fröder *et al.* 2001).

In the Tregiovo Basin the uppermost part of the BVC contains a distinctive sedimentary unit with conglomerates and sandstones at the base, grading into lacustrine mudstones with intercalated stromatolitic carbonates and chert layers. This unit is termed the Tregiovo beds (Cassinis & Neri 1992; Barth & Mohr 1994). The BVC is unconformably overlain by the Gröden Formation, while locally the Gröden Formation rests directly on the Variscan basement.

Massif Central (T.M.)

Late Palaeozoic sedimentation in the Massif Central is characterized by the formation of continental basins in isolated intramontane troughs (Feist *et al.* 1994). The many Stephanian–Lower Rotliegend basins contain mainly alluvial, fluviatile and lacustrine clastic sediments as well as coal measures deposited in an active tectonic setting. The sediments were deposited on a variety of substrates which had been deformed and metamorphosed during the Variscan Orogeny (Châteauneuf & Farjanel 1989). Subsequent to Stephanian compression at the end of the Variscan Orogeny, a period of deformation was initiated during which new intracontinental basins formed, the thickness and extent of whose deposits differed from those which had accumulated during the Stephanian (Fig. 10.26). The Stephanian compressive basins of the Massif Central are long and narrow with varying thicknesses of sediment (Blés *et al.* 1989), whereas the extensional Permian basins contain successions up to several thousands of metres thick and are much more extensive, occupying the sites of the future large Mesozoic basins.

Basins in the Massif Central area that are located along approximately east-trending faults (e.g. the Saint-Etienne Basin) tend to have pull-apart geometries related to strike-slip movements along the basin margin faults. However, those located along approximately north-trending faults tend to be half-grabens, some of which are strongly asymmetric (Faure 1995; Mattauer & Matte 1998). Extensive soft-sediment deformation features reported from the Saint-Affrique and Cévennes basins testify to the syndepositional, extensional to transtensional

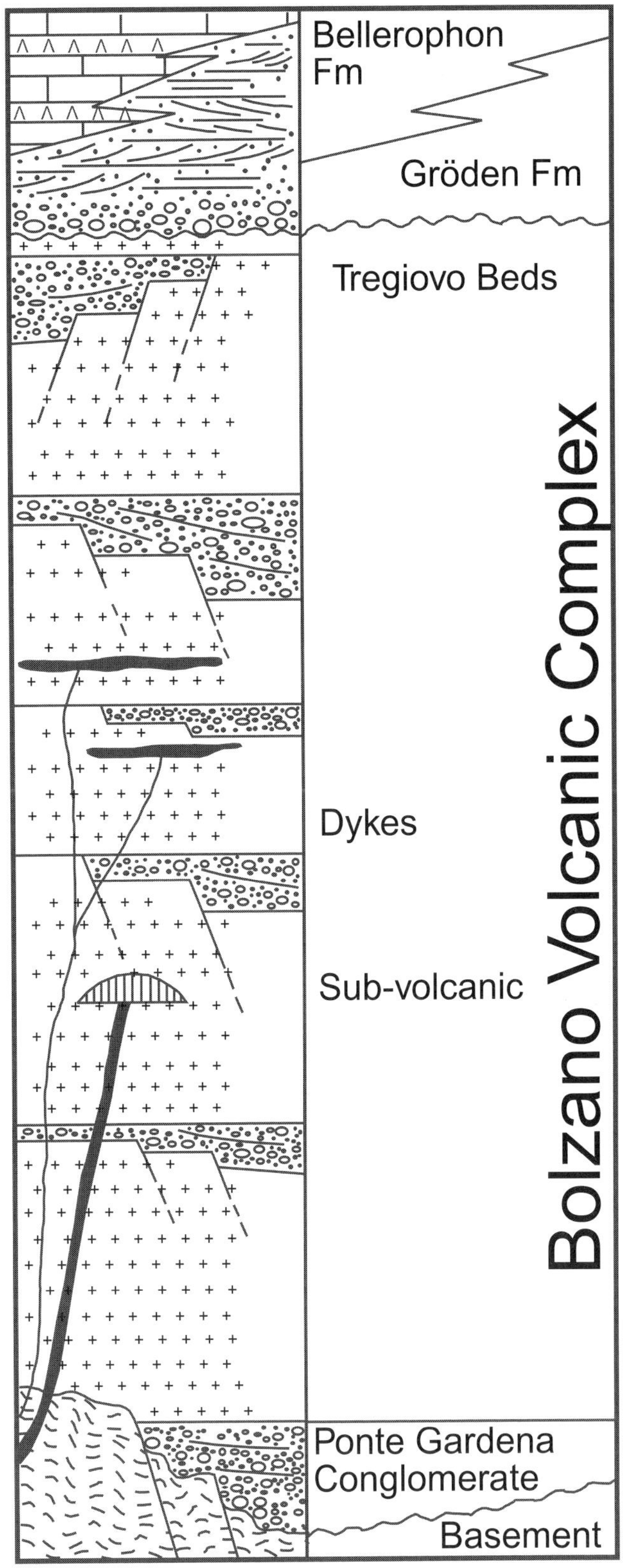

Fig. 10.25. Generalized lithostratigraphic log through the Bolzano Volcanic Complex.

character of the basin-margin faults (Legrand *et al.* 1994; Djarar *et al.* 1996).

Sedimentation within the Permian basins of the Massif Central was controlled by synsedimentary basin-margin faults, and

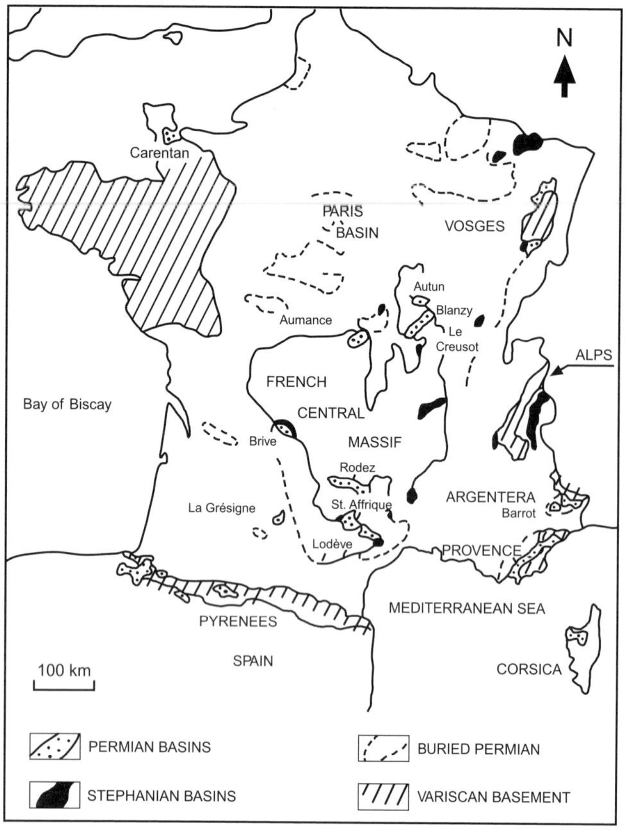

Fig. 10.26. Permian basins of France (after Cassinis *et al.* 1995).

activity along these faults resulted in considerable palaeorelief as well as the formation of basement horsts which were important sediment sources (e.g. alluvial cones). Some of the larger basins in the region were originally just smaller sub-basins with independent drainage systems that amalgamated during periods of accelerated subsidence (e.g. the Cévennes Basin; Djarar *et al.* 1996). The basins of the Massif Central contain only small volumes of magmatic rocks. These are limited to centimetre to decimetre-thick layers of strongly altered airfall tuffs and minor amounts of felsic-intermediate volcanic rocks. In the Rodez Basin, for example, *c.* 5 m of andesite lavas and *c.* 30 m of trachytic tuffs occur at the base of the Lower Rotliegend sedimentary succession and are probably of early Rotliegend age.

From Stephanian to Permian times, coarse-grained fluviolacustrine successions with intercalated coals and rare volcanic ash were deposited unconformably on the exhumed crystalline basement (Courel 1987; Cassinis *et al.* 1995). In France, the Permian succession is generally subdivided into Autunian and Saxonian strata (both Lower Permian) and into Thuringian deposits (considered to be more or less equivalent to the Upper Permian) (Cassinis *et al.* 1995). However, each basin developed in a general tensional context and in many areas there is no appreciable change in the sediment type from Upper Rotliegend through to Zechstein times. This lack of data leads many authors to prefer to classify the successions as Saxono-Thuringian in age.

The typical Lower Rotliegend (= Autunian) was defined in the Autun Basin located to the north of the Massif Central although similar and coeval deposits are also found in other areas, for example NW France (e.g. Carentan Basin; Chateauneuf & Farejanel 1989), the Paris Basin (Bouas 1987), the northern part of Pyrenees (Bixel & Lucas 1983) and SE France (Chateauneuf & Farjanel 1989). Upper Rotliegend (= Saxonian) deposts include those that overlie the Autunian, and are generally separated by an unconformity attributed to a period of intra-Permian (i.e. Saalian) deformation. The expression of this unconformity varies from changes in lithology in some basins to the presence of an actual angular unconformity in others (Cassinis *et al.* 1995).

Blanzy-Le Creusot Basin

This NE–SW trending basin is located in the northern part of the Massif Central (Figs 10.26 & 10.27). Basin initiation was in the Stephanian, when a thick (*c.* 1200 m) succession of clastic sediments was deposited. These include up to 600 m of coal deposits. The overlying Early Permian (Autunian) succession is composed of more than 1000 m of alternating sandstones, conglomerates and bituminous shales (Feist *et al.* 1994).

Autun Basin

The Autun Basin is an east–west orientated half-graben bounded by Devonian–Lower Carboniferous strata to the north and faulted against a batholith to the south (Feist *et al.* 1994). The fault is Early Permian in age and defines a wide ductile shear zone. The basin infill comprises mainly fluviolacustrine deposits with associated fossiliferous shales and rare volcanic ashes. Four units

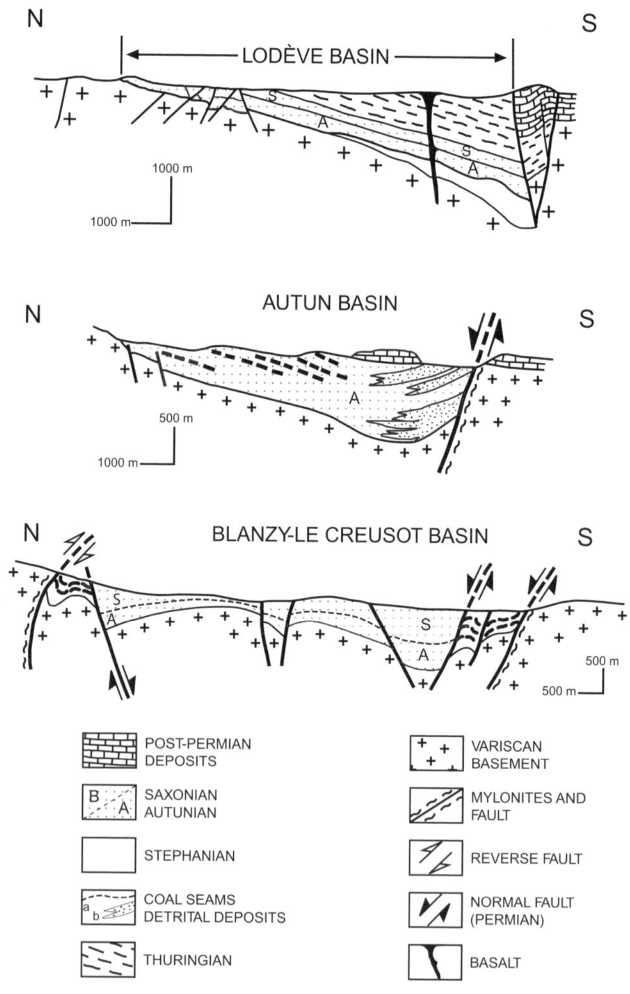

Fig. 10.27. Cross-sections through Permian basins in central France: Lodève Basin, south of the Massif Central (after Cassinis *et al.* 1995), Autun Basin, and Blanzy-Le Creusot Basin (after Blès *et al.* 1989). Locations are shown on Fig. 10.26.

can be distinguished, commencing with *c.* 250 m of coal measures intercalated with muddy sandstones and rare conglomerates. This is overlain a 400 m thick, lithologically similar unit, the main difference being the fewer coal-bearing intervals. Both of these intervals are Early Permian (Lower Rotliegend) in age. The next unit comprises 250 m of bituminous mudstones which are overlain by the 250 m thick Boghead Group comprising muddy and carbonaceous laminites with some sandier units at the top. While no Upper Rotliegend sediments are recorded, they are present in other basins in the area (e.g. Lodève Basin).

Lodève Basin (F.K.)

The basin is located on the southern rim of the Massif Central forming a half-graben structure with a supposed master fault to the south (Broutin *et al.* 1992; Deroin *et al.* 2001). There is an erosional contact between the underlying Stephanian – cropping out in the neighbouring Graissessac Basin – and the Permian deposits. Permian sedimentation commenced in the Asselian–Sakmarian (Lower Rotliegend) with the deposition of the grey facies comprising fluviatile clastics and lacustrine black shales. These contain abundant plant fossils (mainly conifers) as well as rare aquatic amphibians and fish remains. Tuff beds, of a calc-alkaline composition, are frequently intercalated. In the middle Sakmarian there is a transition from the lacustrine grey facies into the floodplain red facies. This is indicated by the appearance of red-brown mudstones and siltstones with intercalated calcisoils as well as braided river and sheetflood sediments. These Early Permian successions attain thicknesses of *c.* 700 to 800 m (Odin 1986).

Following a hiatus, the second sedimentary cycle began in Artinskian (Upper Rotliegend) times with the deposition of tectonically induced basal fanglomerates. These are overlain by *c.* 150 m of red-brown floodplain deposits comprising claystones/siltstones with intercalated calcisoils as well as braided river and sheetflood sediments. The succession continues with a transition into 1800 m of playa deposits with rare intercalated ash-fall tuff beds of an alkaline composition (Körner *et al.* 2003). Fossils of a seasonal fauna such as triopsids, conchostrachans as well as insects are found mainly in the channel units. The uppermost part of the Permian comprises *c.* 200 m of floodplain sediments, mainly braided river and sheetflood deposits but also meandering river sandstones, siltstones and mudstones. Root horizons, pedogenic calcisols, *Scoyenia* burrows as well as a rich tetrapod fauna are all characteristic of this unit; the unit extends into the Wuchiapingian (lowermost Zechstein; Schneider *et al.* 2006).

Saar-Nahe Basin (H.S.)

This Carboniferous–Permian basin extends from SW Germany into France and is filled with exclusively continental sediments with a maximum cumulative thickness of up to 10 km (Figs 10.28 & 10.29). During its post-Westphalian history, the basin had a half-graben geometry, being bordered to the north by the south-dipping Hunsrück Boundary Fault (HBF) which, according to Henk (1993), is a detachment soling out at mid-crustal levels (at a depth of *c.* 16 km). The evolution of the Saar-Nahe Basin is closely related to the complex kinematics of the HBF which is part of a prominent suture zone separating two of the main tectonostratigraphic units of the Variscan foldbelt: the Rhenohercynian and the Saxothuringian zones.

Basin initiation occurred in the latest Namurian or possibly earliest Westphalian; there is an absence of older Namurian sediments and the nature of the pre-rift rocks underlying the Carboniferous–Permian basin fill is essentially undeformed and unmetamorphosed (Veit 1976; Zimmerle 1976) suggesting that this was an elevated area from the late Visean to the late

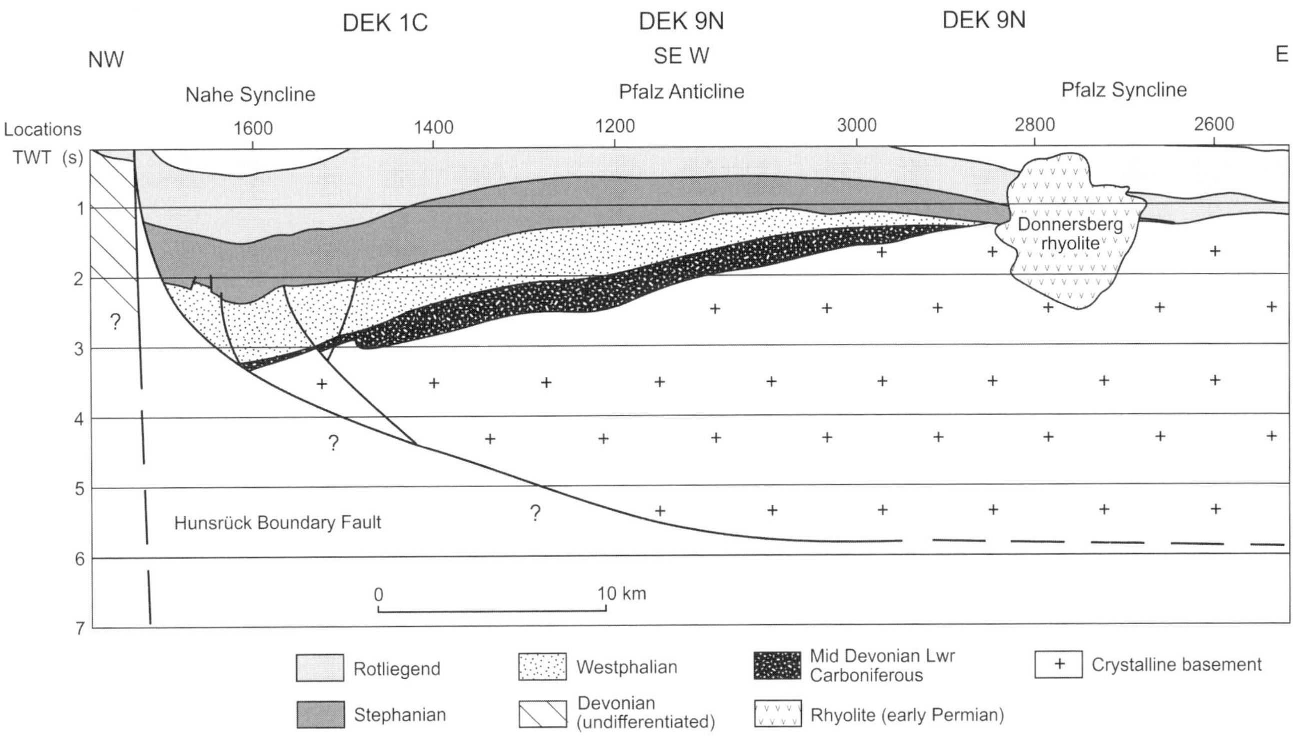

Fig. 10.28. Geological interpretation of DEKORP 1C and 9N showing the half-graben form of the Saar-Nahe Basin, Germany (after Henk 1993).

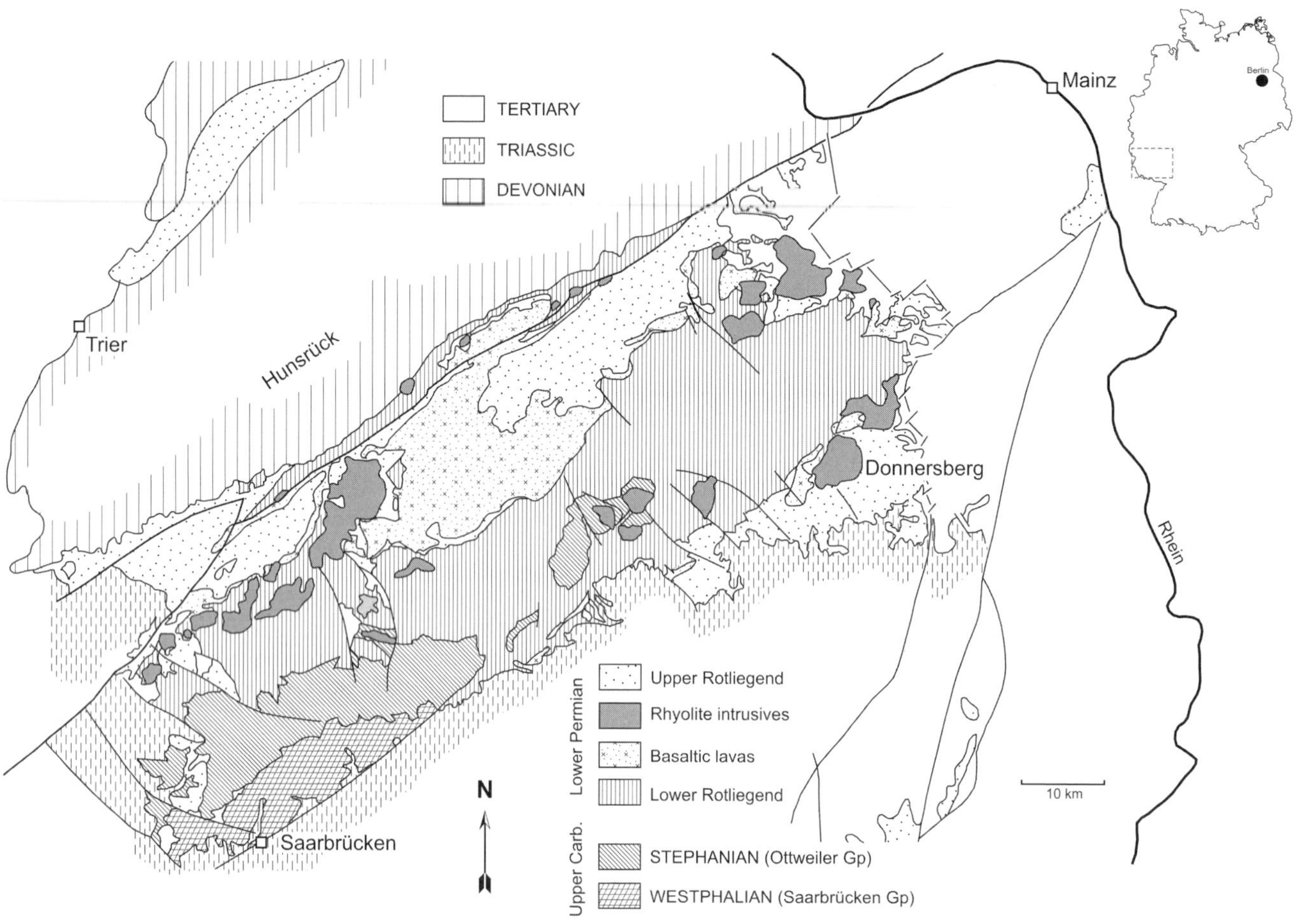

Fig. 10.29. Stratigraphic overview of the Saar-Nahe basin fill (after Schäfer 2005).

Namurian. This period of non-deposition reflects Variscan compression, uplift, exhumation and cooling, as indicated by 320–335 Ma $^{40}Ar/^{39}Ar$ cooling ages from the Mid-German Crystalline Rise (MGCR) basement rocks to the south (i.e. Odenwald and Erzgebirge). There is an angular unconformity between the Westphalian B and C (Veit 1976) and an erosional unconformity between the Westphalian D and the Stephanian A (Engel 1988), indicative of a major period of reorganization of fault kinematics in the area. Major uplift in the southeastern part of the basin elevated the MGCR basement to provide a new sediment source (Korsch & Schäfer 1995). The cause of this was probably a major change in plate boundary forces – the transition from intracontinental compression to dextral translation between Laurussia and Gondwana, which was about to collide with eastern to southeastern North America. Rotation and translation led to the reactivation of old lineaments (such as the northern boundary of the Saar-Nahe Basin, the HBF) and the establishment of large-scale, new fault systems in Europe (e.g. Arthaud & Matte 1977; P. A. Ziegler 1990). The west–east extensional stress regime that dominated during the post-Westphalian basin formation (Stollhofen 1998) was oblique to the pre-existing fault pattern; the common slip direction was maintained by transtensional strike-slip movements on the NW–SE trending transfer faults and oblique-slip motion on the SW–NE trending normal faults, such as the HBF (Schwab 1987; Stollhofen *et al.* 1999).

The basin fill subdivides naturally into four unconformity-bounded megasequences: (1) an initial Westphalian proto-rift sequence; (2) a Stephanian pre-volcanic synrift sequence; (3) a subsequent Lower Permian (Rotliegend) volcanic synrift sequence; and (4) a final late Lower to Upper Permian post-rift sequence (Stollhofen 1999). The Stephanian pre-volcanic synrift sequence has a thickness of 3.8–4.7 km and was deposited over a period of *c.* 14 Ma. It comprises lacustrine, fluviodeltaic and fluvial sediments with minor limestones, coals and pyroclastic fallout deposits (these latter derived from sources outside of the basin) (Stollhofen *et al.* 1999). A detailed stratigraphy of the basin has recently been published by Schäfer (2005).

The Stephanian sediments are overlain by the Rotliegend volcanic synrift sequence (1.1 km thick, deposited over *c.* 4 Ma, 290–286 Ma) which is composed mainly of pebbly arkosic sandstones and siltstones, interpreted as meandering fluvial stream deposits (Stollhofen 1994). Thin coals and biogenic limestones are rare within the succession and appear to be restricted to fault-controlled fluvial floodplain settings and small lakes. There is convincing evidence for synsedimentary tectonic control of intrabasinal sediment thickness, facies and drainage patterns (Stollhofen 1998). Thick basaltic to andesitic lava flows and, in particular, pyroclastic beds of rhyolitic to dacitic composition form laterally extensive marker horizons. The tuffs have been classified as RT (Rhyolitic Tuffs) I to IV by Haneke *et al.* (1979). Intrusions were generally emplaced at shallow depths and occur predominantly as rhyolitic to dacitic domes and basaltic to andesitic sills and dykes. Most of the widespread rhyolitic pyroclastics of the Rotliegend were erupted from tuff rings capping acidic intrusive/extrusive volcanic domes, whereas the more basinal counterparts were derived from lava fissure eruptions or explosive maar volcanoes (Stollhofen & Stanistreet 1994). Traditionally, magmatism has been related to the 'Upper Rotliegend' lithostratigraphy, but radiometric dating (293–288 Ma; Lippolt & Hess 1989) suggests that the onset of intrabasinal volcanism marked the beginning of the Permian (according to the numerical timescale of Gradstein & Ogg 1996). Consequently, underlying strata which have formerly been termed 'Lower Rotliegend' should be considered to be Upper Carboniferous in age (Schäfer & Korsch 1998). This is further confirmed by the radiometric dating of extrabasinally derived tuff layers interbedded with strata that have formerly been considered as Saar-Nahe 'Lower Rotliegend': $^{40}Ar/^{39}Ar$ ages of fallout tuff biotites of the Kusel Group and U/Pb SHRIMP dating of zircon separates of fallout tuffs from the Lebach Group reveal ages of 300 Ma and 297.0 ± 3.2 Ma (Königer *et al.* 2002), respectively, confirming their pre-Permian age. More recent discussion of the Permian–Carboniferous boundary, however, suggests that the correct age is 299 ± 0.8 Ma (Gradstein *et al.* 2004), thus suggesting that these tuffs straddle the boundary. The absence of contemporaneous volcanics within the Saar-Nahe Basin suggests an extrabasinal derivation for the volcanic ash. Given that the majority of the tuffs are rhyolitic to rhyodacitic and indicate petrographic and geochemical affinities to Moldanubian S-type granitoids, Königer & Lorenz (2002) have suggested that they were derived from the general region of the central and northern Black Forest (SW Germany) and the northern Vosges (eastern France), 100–150 km south of the Saar-Nahe Basin.

The generation of the basic to intermediate Rotliegend volcanics can be attributed to underplating and intrusion of pulses of mantle-derived melts into the crust, inducing partial melting (Arz 1996). Magma mixing, fractionation and wall rock contamination then produced the intermediate to acidic melts, some of which cooled slowly and rested for more than 25 000 years before they erupted (Arz 1996). Two magmatic cycles are obvious in the basin, each commencing with bimodal effusives and developing into voluminous acidic to intermediate intrusives. Faults located the positions of acidic volcanic domes, basaltic sills and dykes and many of the diatremes in the Saar-Nahe Basin; they also controlled the internal flow architecture of the intrusives (Haneke 1987) and the development of some of the ore mineralization. Most likely, these faults provided conduits to tap the melts and episodic tectonic events had a triggering effect on both intrusive and extrusive magmatic activity (Stollhofen 1997).

The post-rift unit involves a time span of about 36 Ma (286–250 Ma) recorded by reddish sandflat and playa deposits of which >2000 m are still preserved. These fine-grained and occasionally gypsum-bearing strata interfinger with coarse alluvial fan deposits, particularly at the tectonically active northern basin margin close to the HBF (Stapf 1982). Within the lithostratigraphic record distal to the master fault, the onset of post-rift development is marked by the deposition of a coarse-grained fluvial unit, termed the 'Quartzite-Conglomerate'. As revealed by field observations, section balancing and subsidence analysis (Henk 1993), at this stage depositional patterns were no longer controlled by active intrabasinal magmatism and tectonism but rather by the action of thermal subsidence, sediment load and compaction. Post-rift strata onlap the previously important tectonic boundaries, and both palaeocurrent data and sediment source characteristics indicate a reorientation of the drainage system towards the SW. During the Upper Permian, up to 3200 m of sediments were eroded during a phase of basin inversion which took place prior (Hertle & Littke 2001) to the deposition of the 'Stauf conglomerate and sandstone' (Henk 1992) which is considered to be equivalent to the Zechstein of the Southern Permian Basin (Dittrich 1996).

Bohemian Massif and related areas (Central Bohemian Basin, Boskovice Graben) (H.K., S.O.)

The western margin of the Bohemian Massif is marked by a system of NW-striking faults which delimit the Thuringian, Bohemian and Bavarian Forest blocks. Associated with these faults are a number of half-grabens that contain a thick Stephanian–early Permian clastic succession that is characterized by the presence of minor coal seams and occasionally lacustrine, kerogenous shales. The geometry of these basins is indicative of development in a divergent wrench setting (Schröder 1988) as shown on the reflection seismic profile across the Weiden Basin (associated with the Franconian Fault) which images the half-graben morphology and the *c.* 4 km of Permo-Carboniferous clastics (Dill & Hartkopf-Fröder 1990).

Biostratigraphically proven Permian deposits of continental origin are known only from the basins of the Sudetic area and from the Blanice and Boskovice grabens (Figs 10.30 & 10.31). The sedimentary record present within these basins reflects not only the tectonic controls on deposition but also the overall climatic trend of increasing aridity that culminated in the Upper Permian/Lower Triassic. There is no biostratigraphic evidence of Permian deposits either in central and western Bohemia or in the Krušné hory (Erzgebirge) Mountains area.

Sudetic area

In the Sudetic area, the Carboniferous–Permian boundary is usually marked by a hiatus extending either from the uppermost Stephanian C (e.g. Krkonoše-Piedmont and Mnichovo Hradiště basins) or comprising the entire Stephanian C (e.g. Intra-Sudetic Basin) (Figs 10.30 & 10.31). In terms of the early Permian succession, the Sudetic area can be subdivided into western (Mnichovo Hradiště, Krkonoše-Piedmont and Česká Kamenice basins) and eastern (Intra-Sudetic Basin; see below) parts.

The Permian succession of the western part of the Sudetic area, commences with the Vrchlabí Formation (early Permian). In the northern part of the basin (i.e. zone of maximum subsidence) c. 500 m of mudstones with rare sandstone intercalations are present. The lacustrine Rudnick Horizon (up to 60 m) comprises black shales and carbonates, grey mudstones and sandstones with plant remains and a rich vertebrate fauna in the lower part of the unit. To the south, the formation comprises mainly fluvial sandstones and the thickness decreases to *c.* 300 m. The observed lithological variation is interpreted as representing a transition from fluvial to lacustrine sedimentary systems. Volcanic rocks, including andesites, rhyolitic ignimbrites and tuffs, are intercalated with the sediments. In the northern part of the Mnichovo Hradi Basin their cumulative thickness reaches 500 m.

The Prosečné Formation (up to 400 m, Early Permian) overlies the Vrchlabí Formation and is characterized by red mudstones and claystones with rare sandstones. The stratigraphically important Kalná Horizon in the upper part of the formation is a lacustrine complex comprising bituminous shales, carbonates and grey mudstones. Basalts and rhyolite tuff beds are common throughout this unit.

Deposition of the overlying Chotěvice Formation (up to 400 m, Early Permian) commenced following a short hiatus related to the onset of the Saalian deformational event. Sedi-

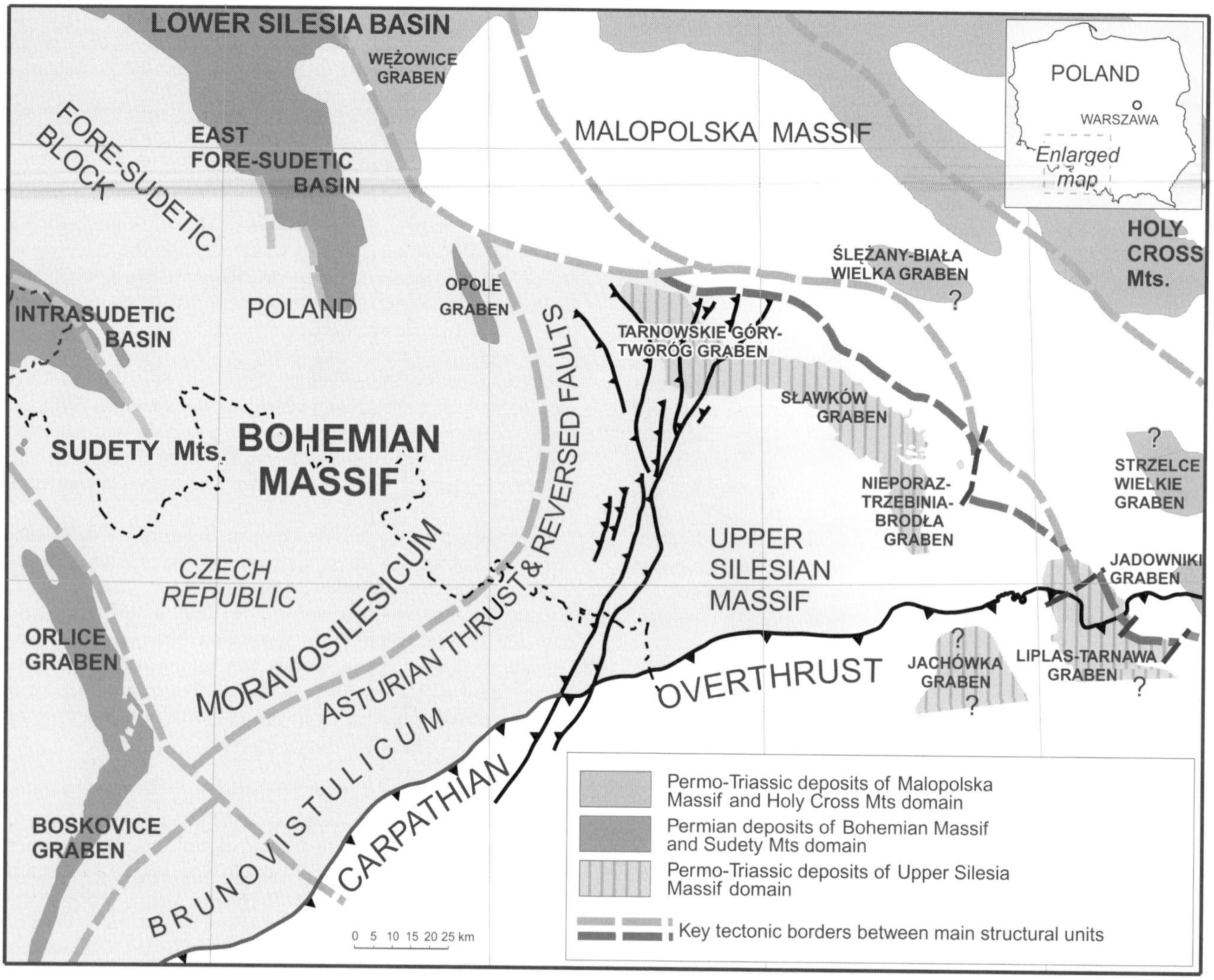

Fig. 10.30. Permian basins of Poland, the Czech Republic and the Slovak borderland (after Kiersnowski 2001).

ments include mudstones, sandstones and conglomerates arranged into cycles. Thin carbonate horizons and significant thicknesses of evaporites in the upper part of the formation suggest deposition in a semiarid to arid climate. Deposition of the formation was terminated by a prominent hiatus related to Saalian deformation.

Post-Saalian deposition in the Sudetic area continued following a hiatus in the NW–SE trending Trutnov-Náchod Depression, a sub-basin of the Krkonoše Piedmont Basin, and in the adjacent part of the Intra-Sudetic Basin. The Trutnov-Náchod sub-basin was most probably formed as a dextral pull-apart structure superimposed on older basin fill (Uličný 2004). Deposition in the sub-basin began in the Upper Rotliegend and comprised conglomerates and coarse-grained sandstones with red mudstone intercalations, up to 600 m thick (Trutnov Formation) deposited as part of an alluvial fan sequence. Part of the succession is interpreted as aeolian (Valín 1972).

Intra-Sudetic Basin

The Intra-Sudetic Basin is the largest of a group of intramontane basins which originated within the Variscan orogenic belt (Fig. 10.30). The basin is located on the northern margin of the Bohemian Massif and closely corresponds to the Eastern Fore-Sudetic Basin. It is *c.* 60 km long and 25 km wide and the infill is predominantly (*c.* 90%) Carboniferous in age (Karnkowski 1999).

The Intra-Sudetic Basin has a long geological history beginning in the Early Carboniferous as an intramontane basin. The Late Carboniferous sedimentary succession consists of conglomerates, sandstones, mudstones and coal seams with a thickness of up to 3000 m (Holub & Tasler 1982). The Late Carboniferous–Early Permian succession represents a transitional unit of mixed grey-black and red-coloured sediments, mainly sandstones, mudstones and conglomerates with thin intercalations of volcanic rocks. Also included are plant assemblages comprising a hydrophile flora and some precursors of the coming drier-climate flora (Schneider 1989). The position of the Carboniferous–Permian boundary within the Intra-Sudetic Basin has still to be defined, although the onset of Permian sedimentation is considered to correspond with the Glinik Beds which represent a transitional member between the typical coal-bearing unts of Namurian–Westphalian age and the Stephanian–early Permian red-beds (Karnkowski 1999).

The Early Permian (Lower Rotliegend) deposits are mostly

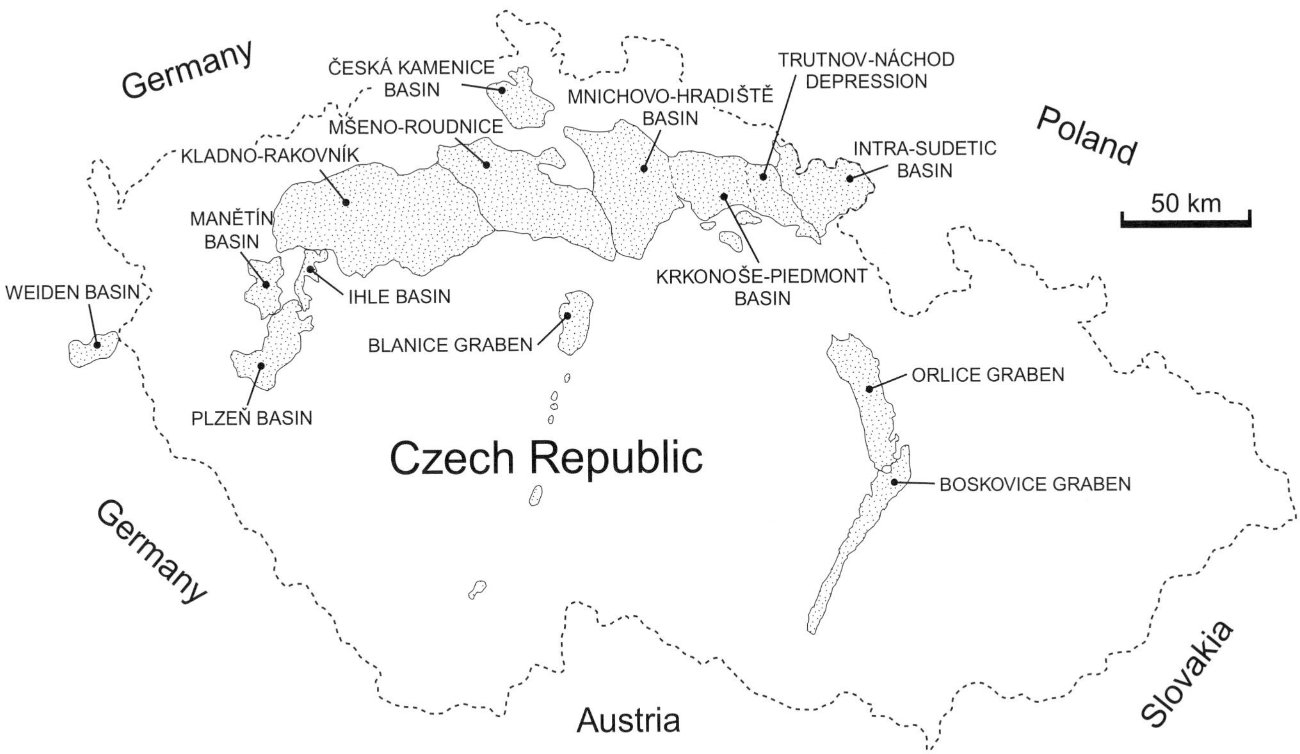

Fig. 10.31. Permian basins of the Czech Republic and SE Germany.

sandstones and mudstones with intercalations of black shales (Anthracosia and Walchia shales) and limestones rich in fossil material (both plants and animal) (Raczynski & Mastalerz 1996). The sediments are arranged into three megacycles with thicknesses of *c*. 250–300 m. The megacycle bases are erosional, and sedimentation commences with alluvial fan deposits (along the basin margins) overlain by fluviolacustrine deposits (Mastalerz *et al.* 1993; Karnkowski 1999). Within the Intra-Sudetic Basin the Upper Rotliegend (Saxonian) deposits are relatively minor when compared to the older units. They are *c*. 400–500 m thick, and comprise conglomerates, and sandstones with calcareous (caliche) interbeds which were formed under arid or semiarid climate conditions.

Important elements within the basin fill include the volcanic rocks (mostly trachybasalts and rhyolites) which comprise several lava or tuff layers (Nowakowski 1968) and which formed in three major effusive/eruptive cycles (Nemec *et al.* 1982). The origin of the volcanic cover, coeval with the synsedimentary tectonic movements which affected the basin margins and caused differential basin subsidence, led to abrupt changes in basin palaeomorphology and the development of a fluviolacustrine depositional system (Wojewoda & Mastalerz 1989). The total thickness of the Lower Rotliegend deposits exceeds 2000 m.

In addition to the Intra-Sudetic Basin there are several related intramontane or foredeep basins (grabens) within the internal and external Variscides which are coeval in age (Raczynski & Mastalerz 1996; Karnkowski 1999). These basins are also infilled with Upper Carboniferous–Lower Permian deposits.

In the Czech part of the Intra-Sudetic Basin the deposition of the Chvaleč Formation was continuous from the Upper Stephanian C through to the earliest Permian. The Lower Rotliegend is represented by the Broumov Formation. The lithology and sedimentary environments of these units are similar to their stratigraphic equivalents in the Krkonoše Piedmont Basin (see above) and include both lacustrine and volcanic horizons. The thickness of the Early Permian strata in the centre of the Czech Intra-Sudetic Basin is more than 1500 m.

North Sudetic Basin

The North Sudetic Basin is a fault-bounded, NE–SE oriented synclinal structure, which includes several grabens and horsts in its eastern and southern parts. Volcanism in the Sudetes region peaked in the early Permian and represented a distinctive, post-collisional, extension related magmatic episode (Awdankiewicz 1999*a*, *b*, 2006). The North Sudetic Basin formed much later than the Intra-Sudetic Basin, but despite this the features of Stephanian and Lower Permian basin infill are very similar in both (Mastalerz *et al.* 1993). The basin infill of the North Sudetic Basin has a thickness of over 1000 m, and an age range of Late Carboniferous–Late Permian. The Permo-Carboniferous stratigraphy of the North Sudetic Basin is informal. In general, the profile is subdivided into Lower Rotliegend–Autunian and Upper Rotliegend–Saxonian. The Lower Rotliegend stratigraphy has been determined by flora (including palynological data) and fauna, while the Upper Rotliegend stratigraphy, due to the scarcity of guide fossils, is predominantly based on lithostratigraphy (Mastalerz & Wojewoda 1988).

The lower part of the succession contains rocks of Early Carboniferous age comprising sandstones, mudstones (with Anthracosia Shales) and conglomerates of fluviolacustrine and subordinate alluvial fan origin arranged into two fining-upward megacycles (Karnkowski 1999). These are overlain by two units of volcanic rocks (Wielislawka Formation) which appear in the lower part of the Rotliegend succession (Mroczkowski & Skowronek 1980). The volcanic units contain rare intercalated sediments. The post-volcanic succession (Boleslawiec Formation) is composed of fluvial conglomerates and sandstones with numerous caliche horizons (Milewicz 1985), and locally aeolian sandstones, at the top of the formation.

Eastern Fore-Sudetic Basin

The Eastern Fore-Sudetic Basin is located within the Moravo-Silesian Variscides and is bounded to the north by the Odra Fault Zone. In terms of its tectonosedimentary evolution, the basin is very similar to both the Intra-Sudetic and North Sudetic basins. The Eastern Fore-Sudetic Basin is a narrow and deep, north–south orientated tectonic graben, containing up to 1100 m of coarse clastic continental deposits. The Permian deposits overlie Precambrian-age folded and faulted metamorphic rocks (Mastalerz & Raczyński 1993).

The Permian clastic sediments are predominantly coarse-grained alluvial-fan and finer-grained alluvial-plain deposits. In the lowermost part of the succession lacustrine and pedogenic carbonates and thin tuffs and coals are intercalated (Figs 10.30 & 10.32) The succession can be subdivided into three depositional megasequences, each separated by erosional phases (Kiersnowski 1995). In the latest stages of sedimentation, the alluvial fan deposits spread into the adjoining Lower Silesian sub-basin.

Poznan Basin

The Poznan Basin and several related tectonic grabens within and around the Brandenburg–Wolsztyn–Pogorzela highs are infilled with Lower Rotliegend sedimentary rocks which were deposited under warm and humid climatic conditions within alluvial and lacustrine environments. The basin infill often contains some pyroclastics. In several cases, volcanic rocks (lava or pyroclastics) occur beneath the sedimentary succession. The main sediment sources were volcanic rocks and clastic Carboniferous-age rocks derived from the Brandenburg–Wolsztyn–Pogorzela highs (Maliszewska *et al.* 2003). The maximum thickness of the Lower Rotliegend sedimentary sequence exceeds 500 m.

Following a period of erosion (with a concomitent gap in the sedimentary record), the sequence continues with Upper Rotliegend deposits which overlap the older strata. These form a succession of sandstones and conglomerates up to 1000 m thick which were deposited in aeolian and alluvial sedimentary environments (Kiersnowski 1998; Weihe 1997).

Blanice, Boskovice and Orlice grabens

Deposition in the Blanice and Boskovice grabens commenced in the uppermost Carboniferous and continued without interruption through to the Lower and on into the Upper Rotliegend (Figs 10.31 & 10.32). In the Blanice Graben the oldest Permian strata belong to the Lhotice Member (early Permian) represented by a 50–175 m thick complex of mostly fine-grained sandstones and mudstones with rare coal seams at the base. These are overlain by arkosic sandstones, mudstones, bituminous carbonates and shales. The succession is interpreted as a fluvial/lacustrine to lacustrine/delta succession. The lacustrine horizon, comprising carbonates and shales, contains a rich invertebrate and vertebrate fauna as well as plant remains. The uppermost succession in the basin is the Chýnov Member (Early Permian). This is characterized by alternating grey to red-grey arkosic sandstones with occasional conglomerate and mudstone intercalations, and rare thin bituminous carbonate beds. The lateral equivalent of this unit is a coarse-grained breccia deposited along the eastern, tectonically active, margin. The total thickness of the Permian succession in the Blanice Graben is *c.* 700 m.

The coeval equivalent of all of the above formations along the eastern, tectonically bounded basin margin is the Rokytná Conglomerate. This unit represents sediments deposited on

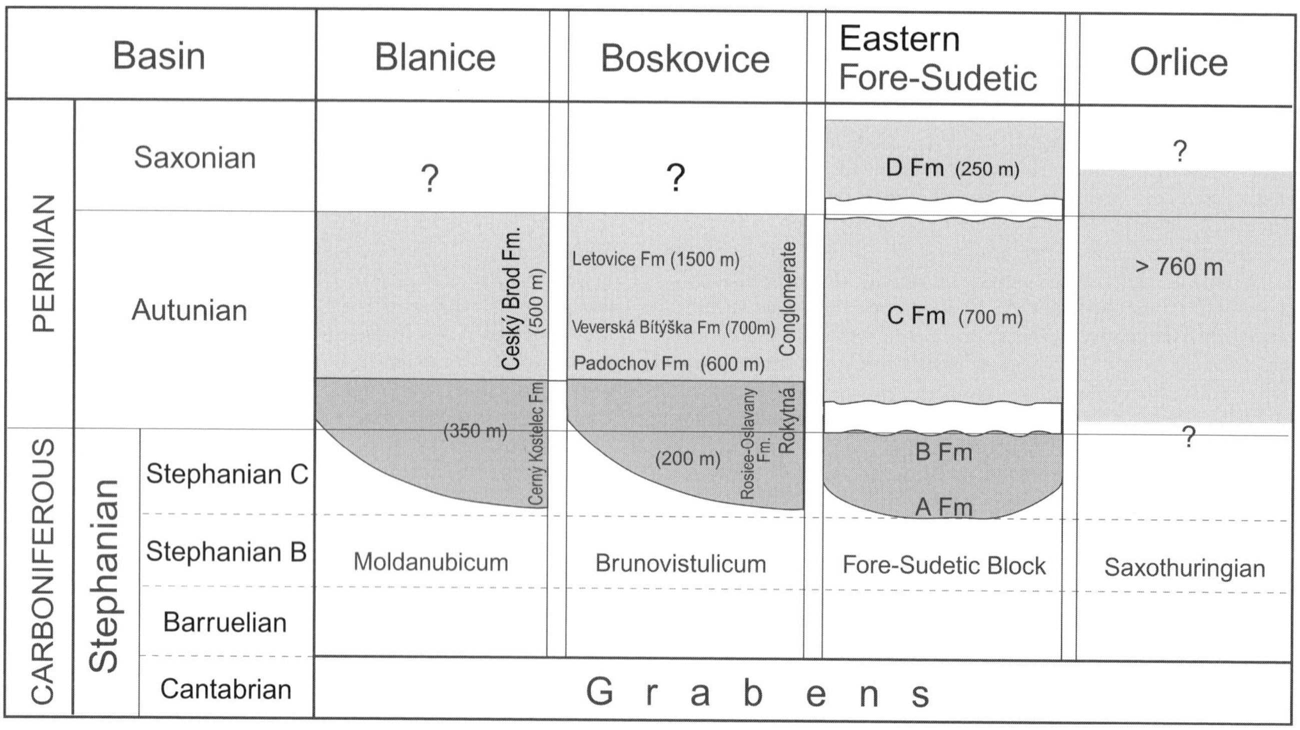

Fig. 10.32. Generalized stratigraphy of selected basins from Poland and the Czech Republic.

alluvial fans which pass westward into fluvial/lacustrine and lacustrine/delta environments.

The thickness of the Permian strata in the Boskovice Graben can reach up to 3000 m. It commences with up to 1000 m of red to grey mudstones, arkosic sandstones and conglomerates of the Padochov Formation (early Permian). Two lacustrine horizons of grey mudstones and carbonates in the upper part of the formation provided very rich fossil flora and fauna. In contrast, no fossils have been found in the overlying, up to 1500 m thick, Veverská Bítiška Formation. This is dominated by alternating mudstones and sandstones. The youngest part of the succession is the Letovice Formation (late Early Permian) which is best developed in the northern depocentre where it attains a thickness of up to 3000 m. This unit consists of red to locally grey sandstones and mudstones with several carbonate horizons. The youngest two horizons have yielded a rich flora and fauna, particularly vertebrates and insects (Pelek 2004).

The Orlice Graben is located to the SSE of the Intra-Sudetic Basin, in the continuation of the axis of the Trutnov-Náchod Depression. It is a NNW–SSE striking, *c.* 40 km long and 10 km wide half-graben structure with the main fault located along the eastern basin margin. To the south the Orlice Basin is connected with the Boskovice Graben. Its basin infill is only poorly known but is estimated to exceed 1000 m. It consists predominantly of red arkosic sandstones, locally with mudstones. While fossils are absent from the succession, the lithologies present are comparable with those of the Intra-Sudetic Basin, thus suggesting a Rotliegend age for the basin infill.

Southeast Variscides and their foredeep (H.K.)

Along the north and NE periphery of the Upper Silesian Coal Basin there are several tectonic grabens which are infilled with Permian deposits (Siedlecka 1964) (Fig. 10.30). The Cracow-Upper Silesia Basin is one of these, and is located in the marginal zone of the Upper Silesian Coal Basin. The basin is elongate in shape, and its tectonosedimentary evolution is related to the deformation of the adjacent Variscan foreland basin.

Basin infill comprises volcanic rocks (tuffs and lavas), alluvial fan conglomerates, limestones derived from spring travertines assigned to the Lower Rotliegend (Lipiarski 1971), and alluvial fan conglomerates and clayey/sandy/carbonaceous deposits of playa lakes, assigned to the Upper Permian (Upper Rotliegend and Zechstein age) (Kiersnowski & Maliszewska 1985; Kiersnowski 1991). The caliche horizons are interpreted as representing pronounced breaks in sedimentation. The total thickness of the various stratigraphic units is estimated at 1000 m. The stratigraphy used within the region is an informal one.

Liplas-Tarnawa Graben

In southern Poland, beneath and to the north of the Carpathian nappes, a series of tectonic grabens, filled with Permo-Triassic deposits, is recognized (Fig. 10.30). These deposits are only preserved in the areas of the axis of earlier Palaeozoic synclines, formed by Lower and partly Upper Carboniferous-age rocks, although it is believed that they constitute the remnants of more widespread units which were subsequently removed by erosion. During Permian times, these synclines were strongly deformed by tectonic block movements resulting in the origin of several tectonic grabens with characteristic subsidence patterns.

The Liplas-Tarnawa Graben contains Permo-Triassic deposits which are over 1400 m thick and comprise a series of sedimentary units (Kiersnowski 2001). The lowermost unit (Early Permian) comprises conglomerates, sandstones and siltstones deposited in a fluvial/floodplain setting. Coals are intercalated within this unit (Buła 2001). The overlying unit (Middle–Late Permian) contains conglomerates (braided river environment) and fine-grained sandstones, siltstones and claystones with interbedded anhydrites and gypsum, deposited within an alluvial plain to sabkha environment. These latter sediments are stratigraphically coeval with the Permian basins of the Northern Gemericum (Vozárová & Vozar 1988). The uppermost sedimentary complex is composed of claystones, siltstones, fine-grained sandstones and gypsum deposited within an alluvial plain/fluvial sedimentary environment (Zechstein–lowermost Triassic).

Döhlen Basin, Germany (T.M., J.S.)

The Döhlen Basin is an intramontane Rotliegend basin (Fig. 10.33). It is 22 km long and 6 km wide and formed at the end of the Carbonifeous as a result of rifting along the Elbe Fault Zone (= Elbtalzone = Elbe Lineament). The basin is of particular interest since it contains Permian coals (up to seven seams) which, in addition to their economic value, have yielded a rich fauna and flora (Barthel 1976; Schneider & Barthel 1997). Basin development and facies patterns were strongly controlled by synsedimentary tectonic activity (Hoffmann 2000). Around 50% of the basin fill comprises pyroclastics, in stark contrast to other Rotliegend basins where clastic (e.g. Saar-Nahe Basin) or volcanic (e.g. Saale Basin) material constitutes the fill. The source area for this volcanic material was local volcanic complexes. To the NW of the Döhlen Basin there are outcrops of rhyodacite/trachyandesitic lavas.

The sedimentary sequence (maximum 800 m thick) can be subdivided into four formations. The lowermost two formations (Unkersdorf, Döhlen) are of probable Late Carboniferous age. The Unkersdorf Formation is very variable in terms of its lithology. To the NE the base comprises a 75 m tuff member while to the south are breccias and conglomerates (which are overlain by the tuff member). Overlying this are conglomerates containing clasts of rhyolite, monzonite, quartz, gneiss, slates and other metasediments, which grade into sandstones to the NE. Volcanic units (mainly andesites) are also found within the Unkersdorf Formation. The overlying Döhlen Formation can be subdivided into two fining-upward sequences, each *c.* 50 m thick. Both of these can be further subdivided into thinner fining-upward cycles, each of which is capped with a coal bed (*c.* 1–6 m thick). Within this formation up to seven coal seams are developed indicative of the development of wetland environments in a fluviolacustrine setting. Volcanic products (e.g. crystal and ash tuffs) are also present.

The Niederhäslich Formation (?Asselian) commences with intercalated conglomerates, containing pebbles of gneiss and rarer rhyolites, and sandstones. This is overlain by sands and silts with locally intercalated tuffs. These are predominantly fluvio-lacustrine (including amphibian-like fossils; Witzmann & Schoch 2006) with evidence of soil formation. Also included are two lacustrine carbonate beds (Gebhardt & Schneider 1993) and rarer coal horizons. The total thickness is *c.* 320 m.

The uppermost Bannewitz Formation (more than 390 m thick and of ?Sakmarian age) commences with rhyolite-rich fanglomerates overlain by volcaniclastics with intercalated pyroclastics and volcaniclastics (including lahars). The upper part of the formation comprises volcaniclastics deposited in an alluvial setting with evidence of fluvial activity (Hoffmann 2000).

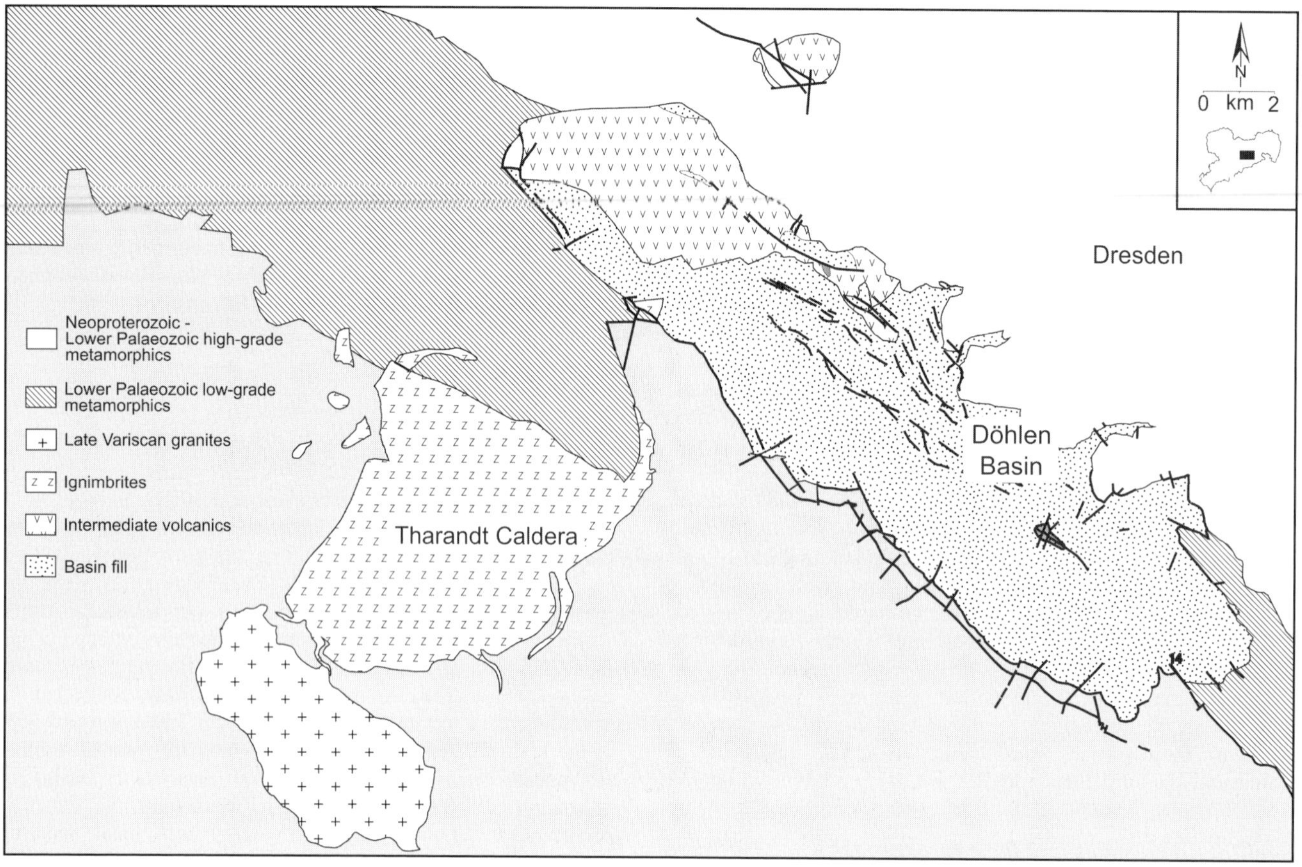

Fig. 10.33. Map of the Dohlen Basin in SW Germany, showing the location of the Permo-Carboniferous rocks in the area (after U. Hoffmann, unpublished work).

Erzgebirge Basin (= Chemnitz Basin), Germany (T.M., J.S.)

The Erzgebirge basin is 70 km long with a width of 30 km (Fig. 10.34). Basin initiation occurred in the Carboniferous, and the basin was intermittently infilled from Late Visean times onwards. The Rotliegend phase of basin evolution followed regional tectonic and volcanic activity and related basin reorganization (Kröner 1995)

The lowermost Härtensdorf Formation (*c.* 280 m) rests unconformably on Westphalian-age strata (Fischer 1991), and comprises two thick sediment cycles, up to *c.* 175 m thick. Sedimentation occurred under semiarid conditions and commenced with the deposition of a conglomerate with some locally developed coals grading up into fine-grained clastics with lacustrine carbonates. The depositional environment was alluvial fan/fluvial/lacustrine. The upper part of the formation also contains alluvial sediments with intercalated pyroclastics (Schneider *et al.* 2004). The overlying Planitz Formation (up to 170 m) commences with a tuff horizon and is characterized by volcanics, pyroclastics and clastic sediments deposited in an alluvial fan/lacustrine environment.

The base of the Leukersdorf Formation (up to 700 m) comprises erosive basal conglomerates. These are overlain by alluvial red-beds and rare carbonates deposited in a lacustrine/fluvial setting. Rare intercalated pyroclastic beds are related to a caldera eruption in the western part of the basin. This eruption formed the spectacular Petrified Forest of Chemnitz which resulted from a directed blast of the base surge eruption (Rößler 1995, 2001; Rößler & Barthel 1998). The Upper Rotliegend Mülsen/Gera Formation (up to 360 m) comprises fanglomerates and intercalated fine-grained clastics deposited in an alluvial-fan setting.

Switzerland (A.W.)

The pre-Mesozoic (Cambrian–Permian) basement of the Alps which is exposed in the external massifs and in tectonic windows below the Alpine nappes, contains late Carboniferous–early Permian granitoid intrusions, volcanic and sedimentary rocks. Late Carboniferous–early Permian basins contain alluvial, fluvial and lacustrine clastic sediments, coal measures, andesitic/rhyolitic lavas and rhyolitic tuffs (Schaltegger 1997; Cortesogno *et al.* 1998). The basins are narrow pull-apart structures which formed in an intracontinental, transtensional–transpressional setting. Locally preserved unconformities between a substrate of deformed and metamorphosed Visean rocks and the basin infill testify to the appreciable uplift (several kilometres) and erosion prior to *c.* 300 Ma (Schaltegger & Corfu 1992). The 290–310 Ma plutonic rocks are predominantly granitoids of calc-alkaline I-type, but the younger Permian magmatic rocks are generally more alkaline (Bussy & Cadoppi 1996). Magmatic activity continued into mid-Permian times, especially in the southern and SW Alps, Provence, northern Italy, Corsica and Sardinia (see also Timmermann 2008).

Permian rocks are exposed in Switzerland in outcrops north of

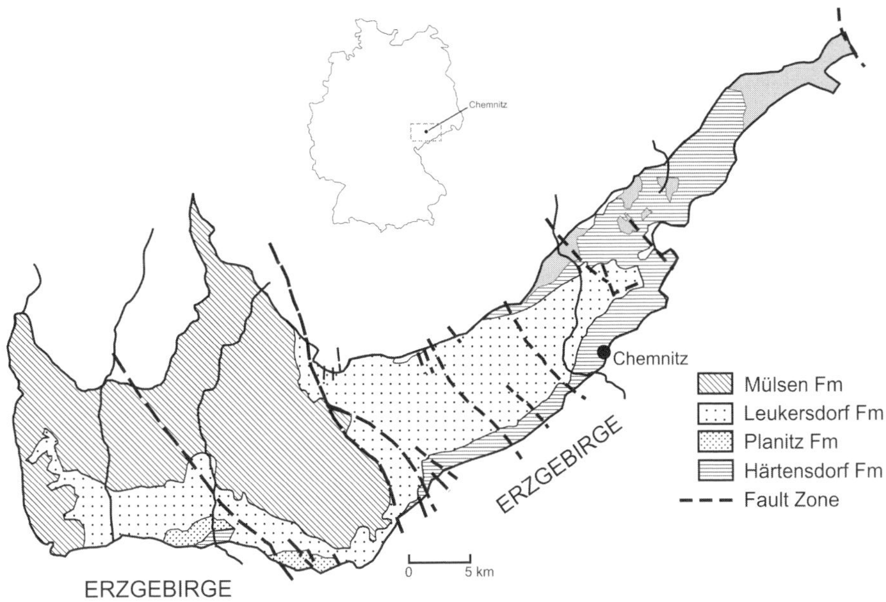

Fig. 10.34. Geological map of the Erzegebirge Basin/Chemnitz Basin

the Penninic front within the External Massifs and the Helvetic Realm (e.g. Trümpy 1980; Schweizerische Geologische Kommission 1980, Fig. 10.35). However, the basins containing these successions have been heavily affected by Alpine tectonic deformation.

North Swiss Permo-Carboniferous Basin

An overview of the Permian deposits within this basin has been published by Blüm (1989). The Permian sediments occupy a significantly wider area than the Carboniferous deposits (Fig. 10.36, see also McCann *et al.* 2008). It would appear that only Lower Permian deposits accumulated, but biostratigraphical dating using palynomorphs (Hochuli 1985) may be subject to some uncertainty due to palaeotopographic effects (e.g. Becq-Giraudon 1993). Except for Weiach, the Permian deposits rest on crystalline basement. At Weiach the lower boundary has been biostratigraphically dated as Stephanian (Hochuli 1985). The upper boundary of the Permian is marked by characteristic conglomerates ascribed to the Germanic Buntsandstein (e.g. Müller *et al.* 1984).

Permian deposits have mainly been correlated according to their lithology (e.g. Blüm 1989). Four lithological units can be identified including, from base to top, the Lacustrine Unit, the Lower Alluvial Fan Unit, the Playa Unit and the Upper Alluvial Fan Unit. The Lacustrine Unit is composed of several cycles, up to 20 m thick, which commence with dark laminated mudstones (lake), and are overlain by stromatolitic shallow-water sands and greyish muddy sandstones (lake margin) and fining-upward sandstone bodies (lake tributaries). The cyclic sedimentation suggests several episodes of lacustrine development rather than a stationary lake. At the top of the succession there is a coarsening-upward trend into the overlying unit. Within this lowermost succession there is an apparent record of climate change from wet to seasonally dry (Hochuli 1985; Matter 1987). At Weiach, the lacustrine unit is *c.* 140 m thick, while at Wintersingen (40 km west of Weiach) it is 76 m. At this latter location, the lake sediments were deposited later and rest on crystalline basement implying a degree of differential synsedimentary subsidence in the region (Matter 1987; Blüm 1989).

The Lower Alluvial Fan Unit comprises two progradational phases of an alluvial fan system which are documented by coarsening-upward successions, several tens of metres thick, with prominent conglomerates/breccias at the top. Unit thickness varies between 6 m and 150 m and thickness variations tend to occur across faults. The Playa Unit, comprising up to 250 m of playa mudstones, overlies the fan complex. Because of their fine-grained nature, these sediments may represent a period of relative tectonic quiescence (Blüm 1989). As in the previous unit, the thickness of the playa sediments also varies across faults. The final unit, the Upper Alluvial Fan Unit, overlies the playa deposits. The sediments coarsen upwards into breccias and conglomerates which were deposited on a prograding alluvial fan. Their thickness is relatively uniform. Extensive calcrete formation suggests intermittent sedimentation prior to the overstep of the Buntsandstein (Blüm 1989).

Tectonosedimentary development

The palaeogeographic position within the Variscides, the varying sediment thickness (especially across faults) and the dominance of crystalline basement clasts derived from local sources suggest a pull-apart origin for the North-Swiss Permo-Carboniferous Basin (e.g. Matter 1987; Blüm 1989). The overall basin structure is interpreted as an en-echelon series of pull-apart basins (e.g. Diebold *et al.* 1992) that formed as a result of late Variscan wrench tectonics (e.g. Arthaud & Matte 1977; Diebold *et al.* 1992). At Weiach, an unconformity between the Lacustrine Unit and the Lower Alluvial Fan Unit has been noted, where the vitrinite reflectance abruptly increases across the boundary (1252 m depth). Kempter (1987) calculated an erosional gap of *c.* 100–1200 m, a value later confirmed by Schegg & Leu (1998). In addition, transpressional structures (Fig. 10.37) were identified when interpreting seismic records (Laubscher 1987; Diebold *et al.* 1992) and basin inversion was deduced. Permian transpression affected the basin in the western part at the northern side and in the eastern part at its southern side (e.g.

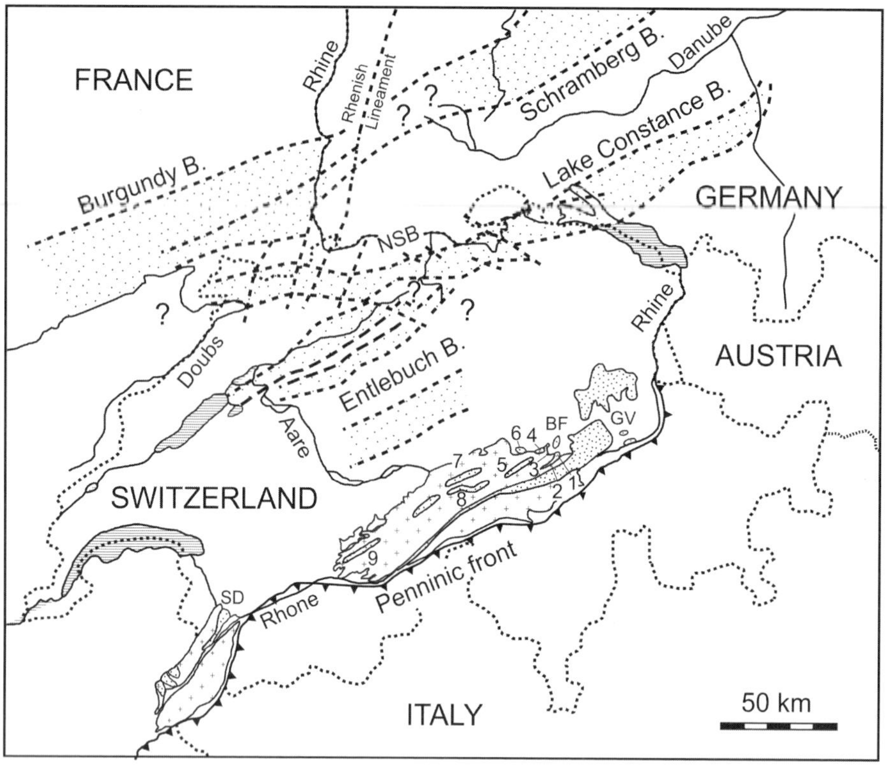

Fig. 10.35. Permo-Carboniferous basins in Switzerland north of the Penninic front. Basins close to the Penninic front (BF, Bifertengrätli; GV, 'Glarner Verrucano Basin'; SD, Salval-Dorénez Basin) are exposed (after Trümpy & Dössenegger 1972). Basins in the subsurface (NSB) after Boigk & Schöneich (1974), Bachmann *et al.* (1987), Meier (1994), Thury *et al.* (1994), Wetzel & Allia (2003). Remnants of Late Palaeozoic basins within the Aar Massif: Southeast: 1, Val Gliems Fm; 2, Sandpass Fm; East: 3, Biefertenfirn Fm; 4, Klein Tödi Fm; 5, Tscharren Fm; 6, Windgällen Fm; Central: 7, Trift Fm; 8, Dietschengletscher Fm; West: 9, Lötschenpass Fm.

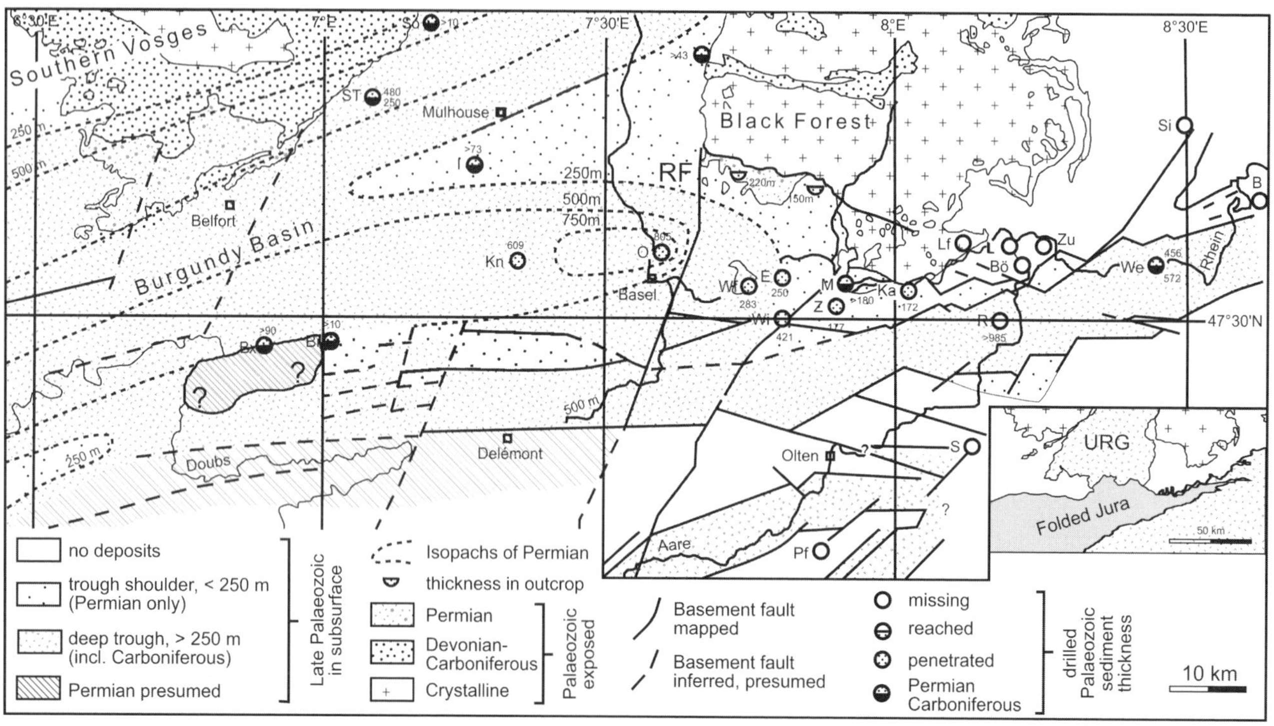

Fig. 10.36. North Swiss Permo-Carboniferous Basin in map view. Compiled from data by Boigk & Schöneich (1974), Diebold *et al.* (1992), Meier (1994), Thury *et al.* (1994), and Wetzel & Allia (2003). In addition, data from boreholes have been used (see McCann *et al.* 2006, for details).

Laubscher 1987; Diebold *et al.* 1992). These authors emphasized the structural similarity with Permo-Carboniferous basins in the French Central Massif (see compilation by Chateauneuf & Farjanel 1989). The transpressional episode has been related to the Saalian phase of tectonic deformation (Laubscher 1987) although Schaltegger (1997) has suggested on the basis of U–Pb dates that deformation was earlier (i.e. latest Stephanian).

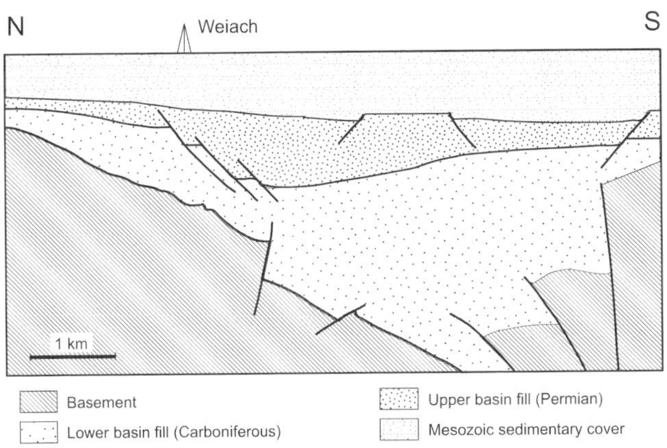

Fig. 10.37. North Swiss Permo-Carboniferous Basin cross-section close to the well at Weiach (see Fig. 10.28), and based on interpreted seismic records (Diebold *et al.* 1992; after Schaltegger 1997). The boundary between the lower and upper basin-fill units roughly corresponds to the Carboniferous–Permian boundary.

Slovakia and eastern Central Europe (A.V.)

Upper Carboniferous–Permian rifting and strike-slip displacements largely affected the western Carpathian Variscan orogenic belt in terms of the generation of a complex system of collisional retroarc and peripheral foreland basins (Fig. 10.38). The deposition of terrestrial overstep sequences commenced within the internal part of the Western Carpathians Variscan orogenic belt during the Late Carboniferous and continued throughout the Permian. In the external part of the Western Carpathians Variscan orogenic system, the deposition of continental sedimentary systems commenced at the beginning of the Permian. Generally, the Permian successions of the Western Carpathians are dominated in their lowermost part by continental, mainly coarse-grained red-beds, while the uppermost part is more variable, including continental to sabkha/lagoonal evaporitic units.

Due to subsequent pronounced Alpine deformation, the Permian deposits of this region are restricted in outcrop to one of three Variscan crustal zones in the Western Carpathians, namely the Central Western Carpathian Crystalline Zone (CWCZ), the Northern Gemeric Zone (NGZ) and the Inner Western Carpathian Zone (IWCZ). The main differences between these zones are in terms of their Variscan geodynamic evolution as well as in terms of the chronological and spatial evolution of the Late Palaeozoic basins contained within them. The lithostratigraphy of these various zones is outlined below.

Central Western Carpathian Crystalline Zone

The origin of the sedimentary basins within this zone was as a result of crustal thinning and transtensional–transpressional tectonics. The Early Permian sedimentary successions are preserved in one of five crustal-scale Alpine Units, namely: Tatricum, Northern and Southern Veporicum, Zemplinicum and Hronicum (Figs 10.38 & 10.39). The Permian sediments of the Tatricum, Veporicum and Zemplinicum units have all been deformed and overprinted by Alpine metamorphism (subgreenschist to greenschist facies) to phengite and sericite schists (Vozárová & Vozár 1988; Lupták *et al.* 2000, 2003).

The Carboniferous–Permian boundary in the CWCZ is either unconformable, where the Permian units overlie crystalline basement, or conformable, where the Stephanian grades into the early Permian but with a sharp change in sediment colour related to increasing aridity. The coarse-grained continental sediments of the Tatricum and Northern Veporicum units have an unconformable base (Fig. 10.40). The Tatricum sediments are not very thick, and are preserved only in isolated areas (e.g. High Tatra Vel'ká Fatra and Malé Karpaty mountains), with more important occurrences in, for example, the Považský Inovec Tribeč and Malá Fatra mountains. Common features of these successions include: the overstepping of Variscan crystalline basement, the presence of continental, proximal to distal braided/alluvial sediments (conglomerates, coarse-grained sandstones) and intercalated rhyolites showing a calc-alkaline trend. The sediments tend to lack biostratigraphical material. U–Pb isotope data from the rhyolite horizon gave a date of 280 Ma (Archangelskij in Rojkovič 1997).

In the Northern Veporicum (Fig. 10.39), the Early Permian succession is represented by the lower part of the L'ubietová Group, comprising the Brusno Formation (Fig. 10.32). The Brusno Formation consists of fluvial sediments deposited in low-sinuosity rivers. Coeval rift-related volcanic activity resulted in calc-alkaline dacite effusions associated with pyroclastic flows (ignimbrites) and epiclastic deposits. Rare andesite/basalts and their volcaniclastics show affinity to a tholeittic magmatic trend. The Early Permian age of the Brusno Formation is only provided by poor assemblages of monosaccate spores (Planderová in Planderová & Vozárová 1982).

Representatives of those Permian successions which are conformable on the Stephanian include the Early Permian deposits in the Southern Veporicum, Zemplinicum and Hronicum units (Fig. 10.40). The strongly deformed and metamorphosed metasediments of the Southern Veporicum Unit consist of coarse-grained sandstones and rare conglomerates (both metamorphosed) with abundant granitoid detritus (e.g. the Rimava Formation; Vozárová & Vozár 1982). Occasional units of rhyolites and rhyolitic tuffs are also present.

The Early Permian sediments of the Zemplinicum Unit (Cejkov Formation) were deposited in an alluvial fan setting alternating with floodplain or ephemeral lake deposits and calcrete horizons, all showing the typical features of semiarid/arid climatic conditions. Several layers of rhyolite tuffs are also present within the formation. The Lower Permian age of the Cejkov Formation is based on the abundance of species from the genera *Potonieisporites* and *Vittatina,* while the upper part of the same formation gave an age of Upper Permian (Planderová *et al.* 1981).

The Early Permian succession of the Hronicum Unit (the Malužiná Formation; Vozárová & Vozár 1988) comprises a thick unit of alternating conglomerates, sandstones and shales. Lenses of dolomite, gypsum and calcrete/caliche horizons occur locally. The sediments of the Malužiná Formation were deposited in fluvial and fluviolacustrine environments under semiarid/arid conditions. Fining-upward cycles of the order of several metres, as well as three regional megacycles, have all been recognized. The basal parts of each megacycle comprise channel-lag and point-bar deposits, laterally associated with floodplain and levee

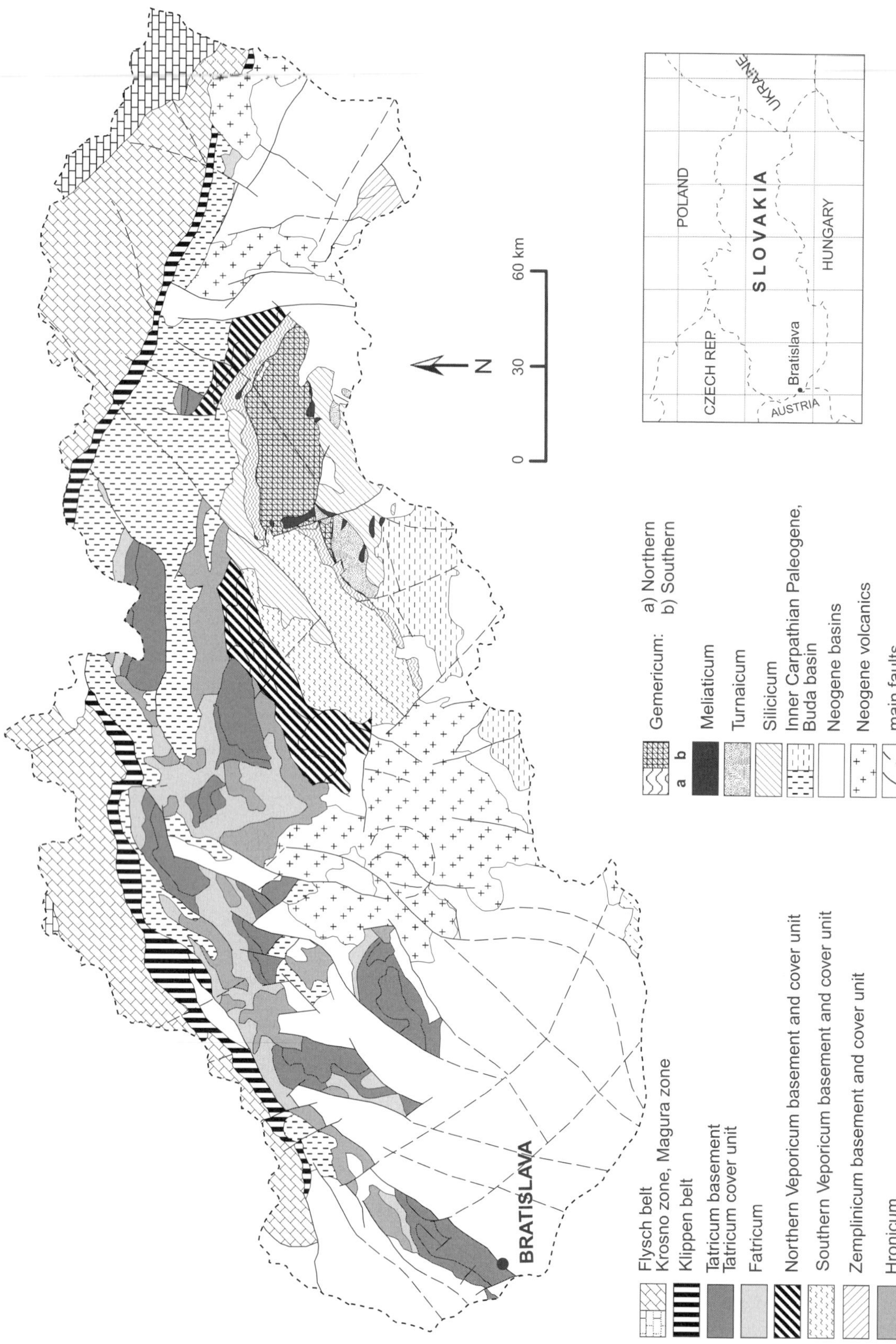

Fig. 10.38. Tectonic sketch of the Slovak part of Western Carpathians (modified after Biely *et al.* 1996).

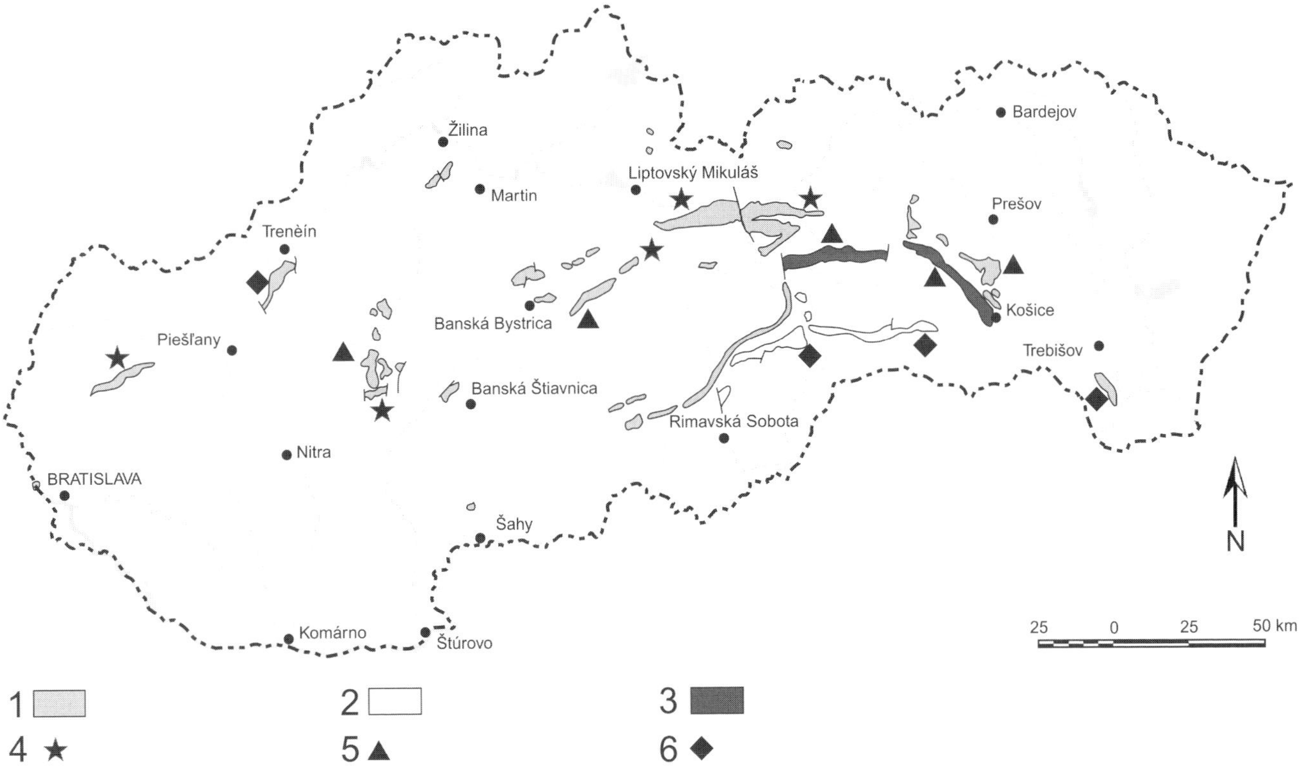

Fig. 10.39. Map of the Permian surface occurrences in the Slovak part of the Western Carpathians. Key: 1, Permian rock complexes of the Central Western Carpathians Zone; 2, Permian rock complexes of the Northern Gemeric Zone; 3, Permian rock complexes of the Inner Western Carpathian Zone; 4, andesite–basalt continental tholeiites and their volcaniclastics; 5, bimodal rhyolite–basalts and their volcaniclastics; 6, rhyolite–dacite calc-alkaline to alkaline volcanites and their volcaniclastics.

sequences. The upper part of each megacycle is characterized by playa, and rare sabkha and ephemeral lake deposits. An important phenomenon is the polyphase synsedimentary rift-related andesite/basalt volcanism with a continental tholeiitic magmatic trend (Vozár 1997; Dostál et al. 2003).

Microflora prove the Early Permian age of the first megacycle of the Malužiná Formation (Planderová & Vozárová 1982). The Lower–Upper Rotliegend microfloral assemblages correspond approximately to the first and second megacycles. This assumption is supported by ^{206}Pb–^{238}U and ^{207}Pb–^{235}U dating of 263 and 274 Ma from uranium-bearing layers of the upper part of the second megacycle (Legierski in Rojkovič 1997).

Northern Gemeric Zone

Within the NGZ continental Permian successions slightly overlap the deformed remnants of the late Carboniferous peripheral basin infill as well as all pre-Westphalian complexes of the zone itself (Figs 10.39 & 10.40). Coarse clastic sediments derived from the collisional belt predominate, and these are associated with bimodal andesite/basalt–rhyolite volcanism. The origin of the Early Permian basin here was related to post-Asturian transpression–transtension.

The basal part of the succession (Knola Formation) contains mostly poorly sorted conglomerates and breccias of variable thickness. Deposition was as mudflows, partially eroded by alluvial, mainly stream channel, deposits. The age of the sediments is not clear, due to the lack of fossils.

The overlying Petrova Hora Formation comprises clastic sediments arranged into fining-upward alluvial cycles. Depositional environments include fluvial and floodplain settings alternating with playa-lake settings at the tops of the megacycles. Bimodal volcanics and volcaniclastics are a significant feature of the Petrova Hora Formation. Analysis of the volcanic units indicates that activity was polyphase. The microflora found in the upper part of the Petrová Hora Formation are of Upper Rotliegend age (Planderová 1979).

Inner Western Carpathian Crystalline Zone

Within the IWCCZ (Figs 10.38 & 10.39), post-Variscan successions are developed only in the Southern Gemeric Unit and in the Bôrka Nappe. Here the Early Permian continental Rožňava Formation, a relict basin fill related to the initial stage of post-Variscan rifting, was deposited. This formation unconformably overlies the low-grade Early Palaeozoic volcanosedimentary complex of the Gelnica Group (Vozárová & Vozár 1993). The Rožňava Formation is subdivided into two large cycles, with conglomerate horizons at the base of each and a sandstone–mudstone member in their upper part. Sediments were deposited as stream-channel and sheetflood deposits. Both of the conglomeratic horizons interfinger with rhyolite–dacite subaerial volcaniclastics and rarer lava flows. The chemical composition of these volcanics is calc-alkaline. The Lower Permian age of the Rožňava Formation is based on the presence of microflora (Planderová 1980).

A high pressure/low temperature rock complex along the southern margin of the Southern Gemeric Unit is assigned to a single tectonic unit termed the Bôrka Nappe. This comprises a variable, discontinuous and intensely tectonically segmented

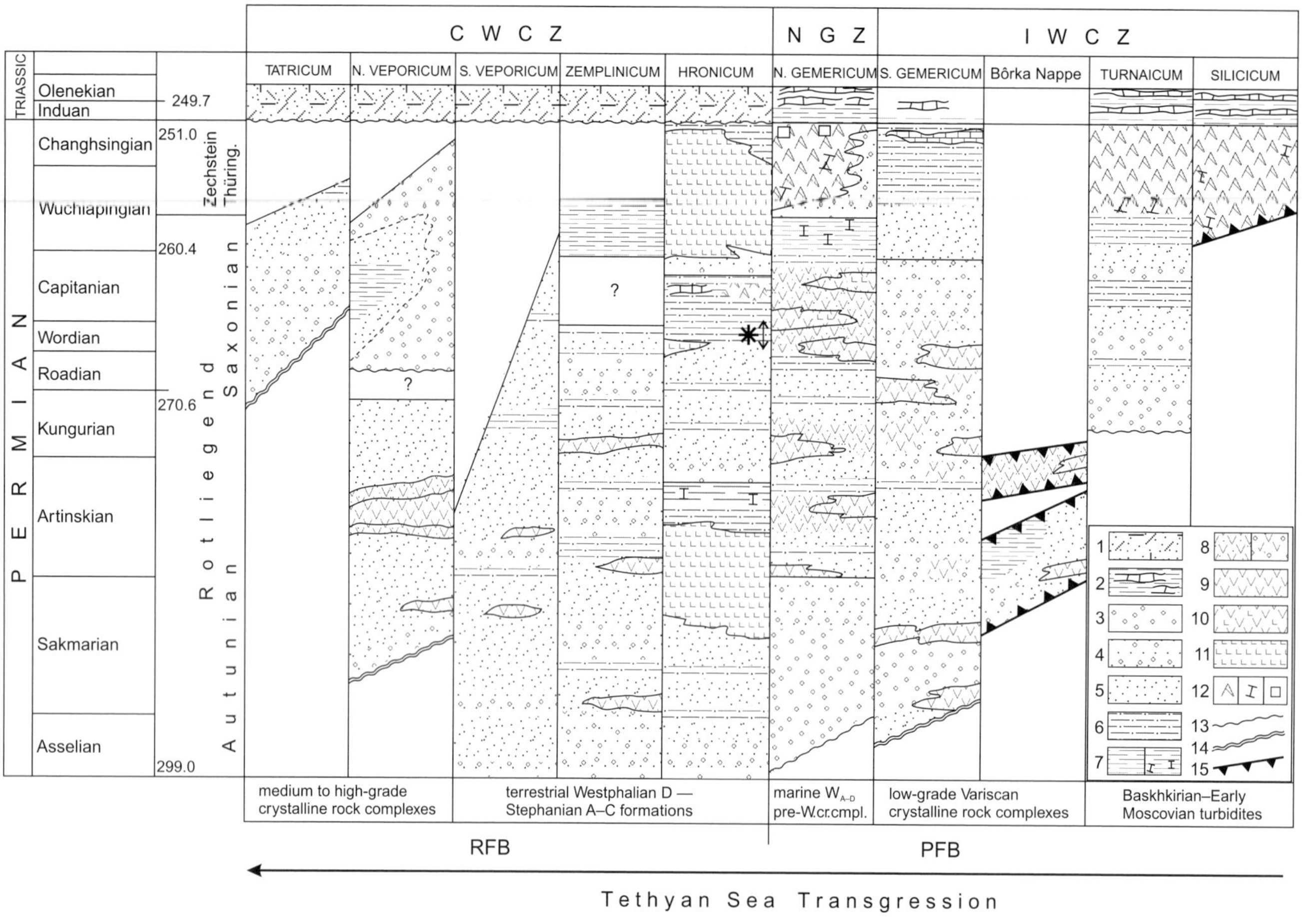

Fig. 10.40. Lithostratigraphical scheme of the Western Carpathian Permian sequence. Key: Lower Triassic: 1, quartzsose sandstones and fine-grained quartz conglomerates (Buntsandstein facies); 2, shallow-water shales, marls, carbonates. Permian continental: 3, polymict conglomerates; 4, sandy conglomerates with intercalations of sandstones; 5, arkosic and lithic sandstones; 6, alternating sandstones and shales; 7, shales, siltstones, mudstones; 8, fine-grained sediments with calcareous nodules and lenses; 8, acid volcaniclastics mixed with sandstones (left), conglomerates (right); 9, rhyolites–dacites and their volcaniclastics; 10, bimodal volcanic suite and their volcaniclastics; 11, andesite–basalt tholeiites and their volcaniclastics. Permian sabkha-lagoonal: 12, evaporites: sulphate (left), salt breccia (middle), dolomite (right). General explanations: 13, disconformity; 14, unconformity; 15, overthrust plane. Abbreviations: CWCZ, NGZ, IWCZ see text. RFB, Retroarc Foreland Basins; PFB, Peripheral Foreland Basins.

package comprising Permian and Mesozoic formations. Two tectonic outliers of the Early Permian metasediments have been recognized: the Bučina Formation, composed of felsitic rhyolite–dacites and their volcaniclastics, interdigitated with some non-volcanic quartzose detritus; and the Jasov Formation, composed predominantly of siliciclastic quartzose metasediments, with rare acid volcanics and volcaniclastics at the base. Both formations are lithologically very similar to the Early Permian of the Southern Gemeric Unit. They are interpreted as representing remnants of the rift-related basin fill.

Hungary (A.T., J.H.)

The Permian of Hungary may be subdivided into three major megatectonic units (Figs 10.41 & 10.42): the Transdanubian Central Range of central western Hungary, the Bükk Unit in northern Hungary and the Tisza Unit in southern Hungary. Each of these units displays a very different tectonosedimentary evolution.

Transdanubian Range

At the end of the Carboniferous and in the earliest Permian (280–300 Ma) S-type alkali granodioritic magma was intruded into the Variscan metamorphic complex leading to the formation of granite, granodiorite and quartzdiorite intrusions along the Balaton Lineament (a continuation of the Periadriatic Lineament). Uplift and denudation due to latest Carboniferous–Early Permian orogenic movements were followed by regional subsidence in the Middle Permian. At this stage, a remarkably thick terrestrial series began to accumulate in the SW part of the Transdanubian Range (Majoros 1983). In the Late Permian its NE part was subjected to marine inundation and alluvial, coastal plain, peritidal and subtidal lagoon facies developed. This pattern is very similar to that which developed in the Southern Alps at the same time (Val Gardena Formation–Bellerophon Formation).

In the Balaton Highland area, continental red-beds represent the Middle–Upper Permian (Balatonfelvidék Sandstone Formation, an equivalent of the Val Gardena Sandstone, Southern Alps). In this area, its thickness may reach 500–800 m. To the NE there is a significant reduction in formation thickness to only

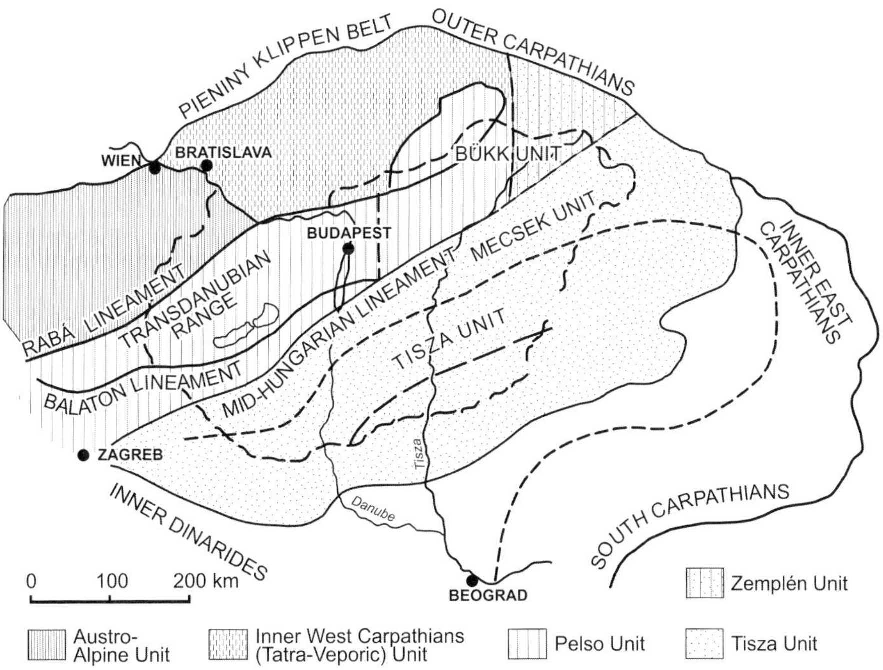

Fig. 10.41. Tectonic zones of the pre-Tertiary basement of the Pannonian Basin area (after Haas *et al.* 1995).

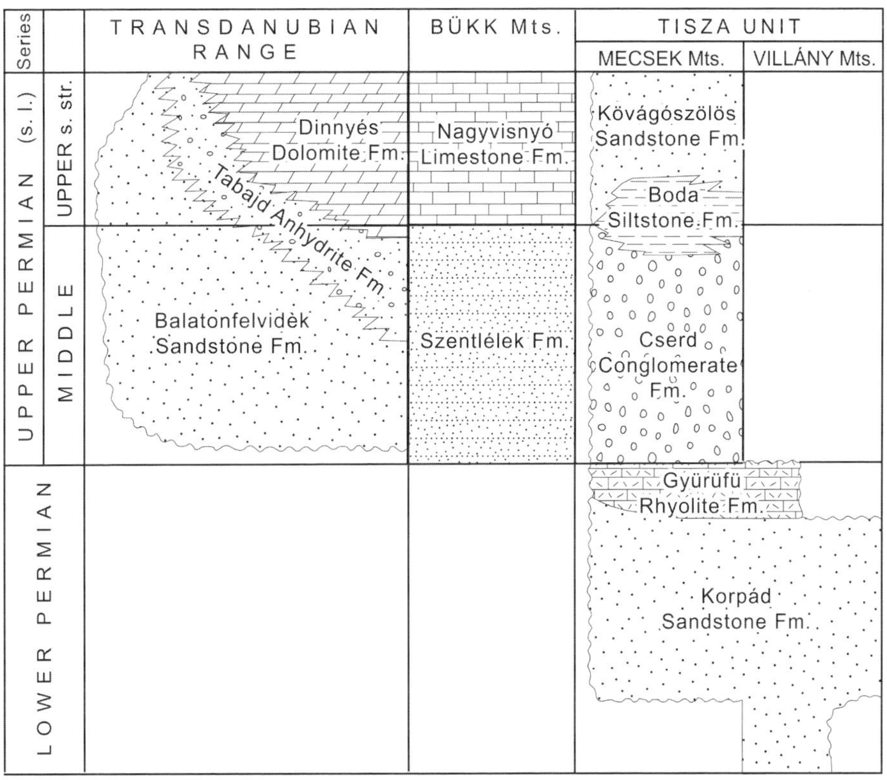

Fig. 10.42. Schematic lithostratigraphy of the Permian in Hungary.

150 m. In some parts of the Balaton Highlands coarse breccias occur at the base of the formation and these are interpreted as fanglomerates deposited under semiarid climatic conditions.

Generally, the Balatonfelvidék Sandstone Formation commences with a coarse clastic member which is made up of conglomerate–sandstone–siltstone cycles bounded by unconformities. Matrix-supported conglomerates in the lower part of the member belong to a proximal, upper alluvial fan facies. The clast-supported sandy conglomerates are of middle-fan affinity, whereas cycles with sandstone and siltstone beds (which are characteristic of the upper part of the formation) represent a distal fan environment as well as alluvial and coastal plains. The commonly occurring cross-bedded sandstone beds are channel deposits (point bar, channel bar, channel fill) while the siltstone layers represent floodplain sediments (Majoros 1998).

The formation is poor in fossils, but coalified plants, imprints of leaves and stems, and silicified trunks occasionally occur. Reptile footprints (*Korynichnium sphaerodactylum*) have been found in one locality.

Bükk Mountains
In the Late Permian the Bükk Parautochthonous Unit may have been located in the inner part of the large Palaeo-Tethys Gulf (Protić *et al.* 2000; Filipović *et al.* 2003). The Permian formations overlie the eroded surface of a shallow-marine Upper Carboniferous unit, although the contact between them is mostly disturbed by Alpine-age faults.

The Middle Permian succession begins with a 170–250 m thick sandstone and siltstone unit (Szentlélek Formation). Light-coloured sandstones characterize the basal part of the lower member of this formation. Sandstones constitute the middle part of the sequence, while the upper part consists of siltstones and fine-grained sandstones. Fossils are completely absent from the formation. Deposition probably occurred within a continental environment (i.e. alluvial and/or coastal plain). The upper member of the Szentlélek Formation is very similar to the Fiamazza facies of the Bellerophon Formation in the Southern Alps, which is Zechstein in age (see below).

Tisza Unit
The surface occurrence of the Permian formations is restricted to the western part of the Mecsek Mountains, southern Hungary. In the subsurface, Permian successions are found between the Mecsek and Villány zones, in the Dráva Basin and in parts of the Great Hungarian Plain. The thickness of the Permian sediments varies between 400 and 4000 m with an average of 2500 m according to core drilling in the Mecsek and Villány area (Barabás & Barabás-Stuhl 1998). In addition *c.* 800 m of volcanics were also drilled. Thus, the entire Permian succession is up to 5 km in the subsurface of southern Hungary.

Permian sediments overlie either Late Carboniferous continental sandstones (Téseny Sandstone) or the crystalline to granitic basement of the Tisza Unit. The oldest proven Permian sediments of the Tisza Unit belongs to the Korpád Sandstone Formation. The cyclic, variegated, but dominantly reddish quartz sandstones also contain conglomerates and frequent siltstone intercalations (dated as Early Permian) (Barabás-Stuhl 1981). An acid volcanic series overlies the red sandstones. This consists of rhyolite lavas and tuffs and less frequently pyroclastics or subvolcanic bodies (50–80 m and rarely up to 800 m; Gyürüfü Rhyolite Formation). The volcanic series, which often includes ignimbrites, is capped by coarse siliciclastic sediments of the Cserd Conglomerate Formation.

The Cserd Conglomerate Formation comprises conglomerates, poorly sorted coarse sandstones and sandy siltstones interpreted as fan and cyclic fluvial sediments (Jámbor 1964). The maximum thickness of this formation can reach 1000 m along the basin margins, with an average thickness of 300–400 m. The very thick occurrences are interpreted as basin marginal alluvial fans (Barabás & Barabás-Stuhl 1998). Pebbly sandstone beds form the basal unit of the overlying Boda Siltstone Formation. This 100 m thick transitional member passes into very thick red siltstones (Jámbor 1964). The monotonous parallel laminated red siltstones and clayey siltstones are interrupted by greenish-red fine sandstone and by thin clayey dolomites. The maximum thickness of 1000 m was documented in the main depositional areas, but red siltstone beds wedging out at the margins have a thickness of only a few tens of metres. The Boda Siltstone Formation has been dated as Late Permian (Barabás-Stuhl 1981).

Late Permian (Wuchiapingian–Changhsingian/zechstein) sedimentary basins (T.M., T.P.)

Zechstein Basin

During the Permian, the Arctic Sea occupied the Barents Shelf, the Sverdrup Basin, and the area of the Alaska North Slope and Chukotka. Its Late Permian southward transgression into the area of the Norwegian–Greenland Sea was preceded by an important intra-Kungurian pulse of rifting (Surlyk *et al.* 1986; P.A. Ziegler 1988, 1989). The progressive subsidence of the Norwegian–Greenland Sea Rift, combined with a glacio-eustatic sea-level rise due to deglaciation in Gondwana, resulted in the breakdown of the barriers that separated this seaway from the Northern and Southern Permian basins. This occurred at the transition from the Kungurian to the Ufimian. With the flooding of these intracontinental topographic depressions, the huge Zechstein inland sea was established. This had a length of *c.* 1700 km and a width of around 600 km. Rapid flooding of the palaeotopographic lows corresponding to the Southern Permian Basin (SPB) probably had a loading effect on the lithosphere causing further regional subsidence and the overstepping of the Rotliegend basin margins.

The Zechstein Basin extended from northern Britain, SE across the North Sea through the Netherlands, Belgium, Denmark, Germany and Poland to the edge of the Variscan Massifs (e.g. Harz area, Bohemian Massif) and further east to Lithuania. An arid climate combined with repeated restriction of the seawater influx into the SPB, which can be related to glacio-eustatic sea-level oscillations and possibly also to tectonic instability in the Arctic–North Atlantic domain, gave rise to the deposition of the cyclical Zechstein carbonate, sulphate and halite series. The primary thickness of the Zechstein series in the basinal areas is of the order of 1500–2000 m (Olsen 1987). The present thickness distribution is variable due to the effects of salt diapirism, subsurface leaching, and erosion at Mesozoic unconformities (Jenyon & Cresswell 1987).

The Zechstein succession was deposited in a huge drawdown seaway. Traditionally, four evaporitic cycles have been distinguished: Z1 Werra Series, Z2 Stassfurt Series, Z3 Leine Series, and Z4 Aller Series (Richter-Bernburg 1955). Subsequently, younger cycles – Ohre (Z5), Friesland (Z6) and Mölln (Z7) – were recognized in the axial parts of the North German Basin (Kulick & Paul 1987; Best 1989). In Poland, four cyclothems (PZ1–PZ4) are recognized, with the PZ4 cyclothem being subdivided into five subcyclothems (from PZ4a to PZ4e) reflecting climatic changes (Wagner 1994). The subcyclothem PZ4a in Poland corresponds to cyclothem Z4 in Germany, and cyclothem Z7 (Mölln) in Germany most probably corresponds to subcyclothem PZ4d in the Polish Basin (Wagner & Peryt 1997). Thus subcyclothem PZ4e in Poland is regarded as the youngest Zechstein deposit in the entire basin.

An ideal classic Zechstein cycle commences with a transgressive non-evaporitic 'shale' followed by carbonates and culminating in thick evaporites (mostly halite/anhydrite). The most complete classic cycles are found in the Z1 and Z2 cycles where lithologies have been related to higher-order glacio-eustatic sea-level fluctuations (Strohmenger *et al.* 1996*a*, *b*). Each successive cycle becomes progressively more evaporitic and within each cycle the facies also become progressively more evaporitic upwards. The cycle closes with a partial or full restriction of the basin, which is reflected by the deposition of sulphates (Z1) or halites (Z2 to Z7).

Sedimentation rates in the Zechstein were variable. In general, during periods of carbonate and anhydrite deposition on the shelves, slow sedimentation rates characterized the sediment-starved basinal areas. Conversely, sedimentation rates were probably rapid during the infilling of basinal areas with halites. The purity of these halites suggests that the waters entering the basins of Central Europe had already been depleted of their carbonate and sulphate content.

During the deposition of the Z1–Z3 Zechstein cycles there was a progressive overstepping of the basin margins, particularly in the North Danish embayment. A cyclical fall in sea level during the Zechstein Z4–Z7 times is reflected by a restriction of the area of sedimentation (Wagner *et al.* 1981). This restriction peaked at the transition from the evaporite-dominated Zechstein to the clastic-dominated Buntsandstein depositional regime, and coincided with a regional regression of the Boreal Sea (cf. Scotese & Langford 1995).

Two potash cycles in the Z1 are exploited in the Werra-Fulda Basin SE of Kassel (Germany) while two potash beds in the Z2 and Z3 are mined near Hannover, Germany. Potash is also extracted from the Aller Series (Z4) in the Netherlands (Warren 1999). Throughout Europe, and in the offshore southern Zechstein Basin in the North Sea, the Weissliegend is a significant gas reservoir. In Poland, carbonates of the first two cycles are important gas and oil reservoirs.

In Zechstein times, considerable differential fault movement occurred on the anhydrite platforms at the margins of the Zechstein Basin during the deposition of the Z1 (Werra) Formation (Geluk 1999). Loading of rapidly deposited anhydrite on a differentiated substrate is thought to have amplified these movements (Geluk 1999). Subsequent movements occurring during the deposition of the upper Zechstein are, however, more difficult to identify, although in places there is evidence of an unconformable contact. These movements, which show evidence of compressional activity, may possibly be linked to the Uralian Orogeny. As a result, source areas to the south of the basin were uplifted and sands shed into the basin in the western offshore area and the UK sector (Geluk 2005). Contemporaneously, in the central parts of the main basin, the youngest series of Zechstein halites were precipitated (Z6–Z7; Best 1989).

Zechstein transgresssion

Zechstein sedimentation commenced when the Rotliegend SPB, which lay below sea level, was flooded through a combination of rifting and eustatic sea-level rise (Glennie & Buller 1983). The trangression was geologically rapid as indicated by the regional distribution of the basal Zechstein Kuperschiefer which corresponds to a sharp chronostratigraphic marker. The Kupferschiefer is a black, highly organic, laminated shale which was deposited below wave base in anoxic waters. Reducing conditions were established shortly after the Zechstein transgression, whereby the influx of nutrient-rich fresh water from continental areas promoted a high organic productivity in surface waters and, together with the high evaporation rates, the development of permanent stagnant bottom-water conditions. After the initial transgression, water depths in the axial parts of the Southern Permian Basin may have been in the range of 200–300 m (P.A. Ziegler 1990).

During the initial transgression the topmost parts of the Rotliegend dune sands were reworked and form the so-called Weissliegend. These 10–50 m thick grey to white sands are structurally featureless and contain evidence of soft-sediment deformation. They are interpreted as representing aeolian sands which were redistributed by flash floods and wave action. In the area of the former Rotliegend playa lake, the uppermost red-beds often contain thin intervals which are characterized by the presence of marine shells. These marine bands indicate the occurrence of short-lived pre-Zechstein ingressions of the Boreal Sea into the Permian basins of Central Europe (Hoffmann *et al.* 1989). Two marine ingressions have been recorded by Legler (2005)

In basinal areas, the Kupferschiefer is generally *c.* 30 cm thick, and increases to a metre or more before onlapping against pre-existing topographic highs along the basin margin. On intrabasinal swells, the Kupferschiefer is developed in a calcareous facies and contains normal marine faunas that were deposited in well-aerated waters above the chemocline. The slope facies of such highs is characterized by laminated black shales containing calcareous tempestites (Paul 1982*a*, *b*). In basin margin areas, the Kupferschiefer contains land plants, reptiles and marine faunas (Jowett *et al.* 1987). Hirst & Dunham (1963) suggest that the rate of deposition of the Kupferschiefer was 30–40 cm in 17 ka. Additionally, Brongersma-Sanders (1971) has suggested that much of its organic matter was derived from organic blooms generated by trade-wind upwelling of nutrient-rich waters.

In the Campine Basin (NE Belgium), the Zechstein Limestone is underlain by a mineralized conglomerate. Although the mineralization is reminiscent of the Kupferschiefer, it results from several late diagenetic phases (successively, ankeritization, marcassite and pyrite formation) of cement replacement in both matrix and clasts (De Craen & Swennen 1992).

The Kupferschiefer was abruptly overlain by the Zechstein Limestone (Zechsteinkalk) which was deposited in oxygenated waters. In basin margin areas, and on intrabasinal highs, the Zechstein Limestone attains a thickness of 50–100 m and is developed in a shallow-water facies containing a rich fauna that displays a close affinity with the Arctic faunal province (Hollingworth & Pettigrew 1988). In Poland, however, Zechstein microfaunas also contain an admixture of Tethys-related species (Peryt & Peryt 1977), suggesting that the SPB was at least temporarily connected to the SE with the Tethys Ocean. Shelf carbonate banks, consisting of micrites, oolitic grainstones and patch reefs, were fringed by a reef belt formed mainly by bryozoans and stromatolites. Basinward these were offset by kerogenous micrites. Facies patterns were strongly influenced by the pre-existing relief over which the Zechstein series transgressed.

Sequence stratigraphy

Tucker (1991) published an alternative to the classic Zechstein stratigraphy based on what he interpreted as third-order sequences (ZS1–ZS7). In this scheme, Tucker correlates evaporites with sea-level lowstands and carbonates with sea-level highstands. This approach (i.e. correlating lowstands with evaporite deposition) is contrary to the classic approach but more realistic in terms of evaporite hydrology. Tucker's classification was criticized by Goodall *et al.* (1993) because it assumed that sea level exerted a one-to-one control on evaporite mineralogy and position, and subsequently by Strohmenger *et al.* (1996*a*) since the classification did not respect exposure surfaces and stratal geometries. Strohmenger *et al.* (1996*b*) provided a refined sequence stratigraphy for the lowermost Zechstein units (Z1 and Z2) using facies analysis geometric relationships and stacking patterns as seen in seismic data from the subsurface of Germany (Figs 10.43 & 10.44). Their analysis indicates that reworked sandstones of the Weissliegend and/or the Zechstein conglomerate record the initial transgression of the Zechstein Sea and overlie the Zechstein sequence boundary. Transgressive systems tract deposits of the first Zechstein sequence (ZS1) are represented by the deeper marine carbonates of the Mutterflöz. The overlying Kupferschiefer is interpreted as a condensed section indicating maximum flooding of the first Zechstein sequence. The bulk of the shallow-marine Zechstein limestone is interpreted as a highstand systems tract. This unit is separated from the thin uppermost part of the Zechstein Limestone by a karst horizon corresponding to an erosive sequence boundary. Correlatable shallow-water carbonates (oncolites and stromatolites) indicate lowstand systems tracts of the second Zechstein sequence (ZS2). The uppermost part of the Zechstein Limestone represents the transgressive systems tract of this sequence.

According to Strohmenger *et al.* (1996*a*, *b*) there is no indication of a sequence boundary at the top of the Zechstein Limestone (Ca-1) platform carbonates (in contrast with Tucker 1991). They argue that the Zechstein Limestone carbonates were in a basinal position and show a deepening-upward tendency towards the superimposed deposits of the Werra Anhydrite (A1) sulphates. They place the maximum flooding surface of the second Zechstein sequence at the top of the Zechstein Limestone platform carbonates. The overlying anhydrites of the A-1 sulphate platform are interpreted as predominantly highstand system tract deposits displaying an erosive sequence boundary (ZSB3) at its top. This contrasts markedly with the Tucker (1991) scheme. According to Strohmenger *et al.* (1996*a*, *b*) it is only in the slope and basinal environments that the A-1 also represents lowstand

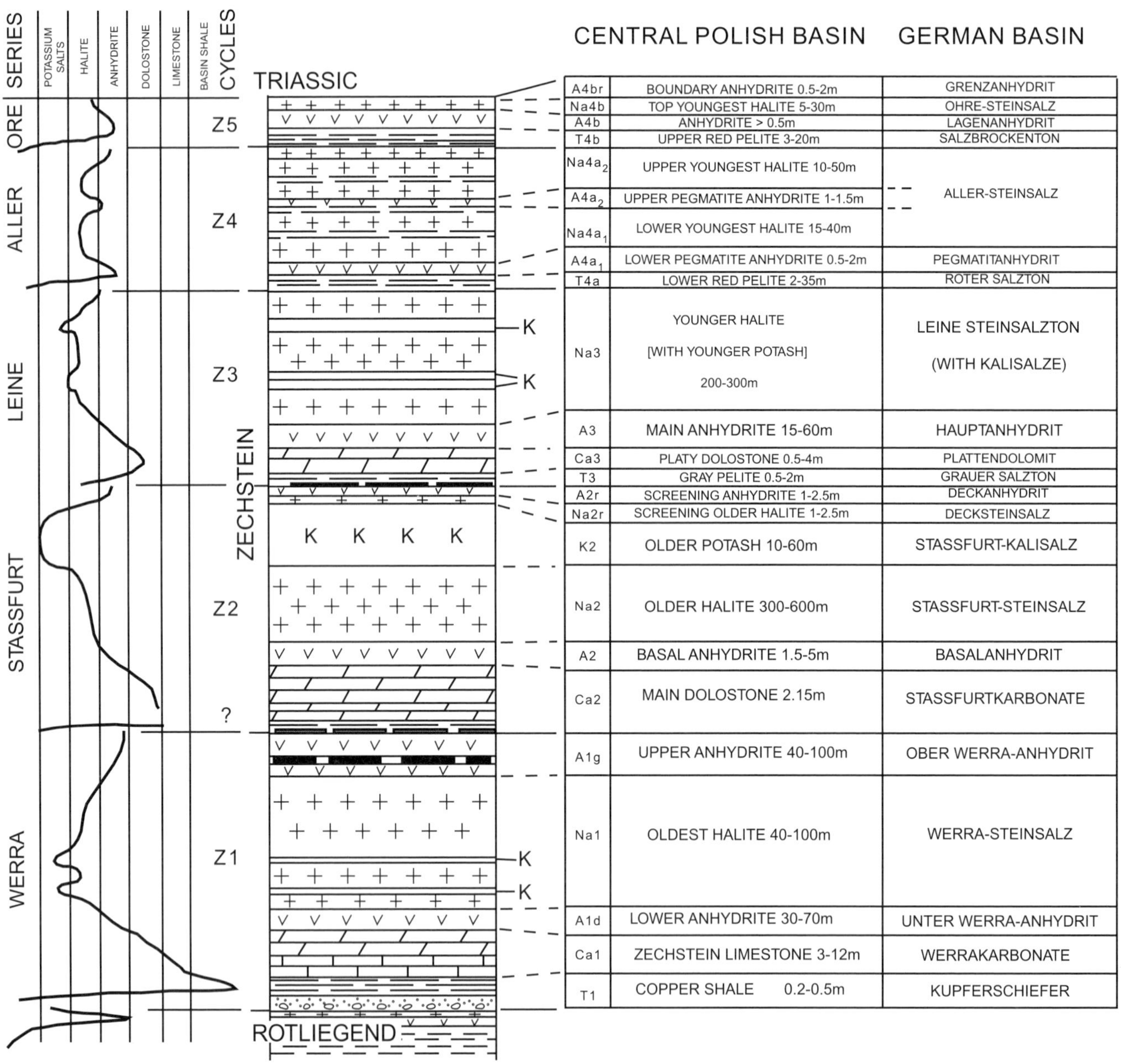

Fig. 10.43. Zechstein cyclicity in the Central Polish and German basins showing five intervals of shale, carbonate and evaporite cyclicity. The lower two cycles contain marginal-marine fossils which show some resemblance to those from the lower Kazanian of the Russian Platform. However, there is no precise biostratigraphic correlation between the two areas. Legend: K, potash salts; V, anhydrite & halite (after Ross & Ross 1995).

NORTH-WEST GERMANY		HESSIAN BASIN	CENTRAL POLAND	
BRÖKELSCHIEFER CYCLE	BRÖKELSCHIEFER	BRÖKELSCHIEFER - TON BRÖKELSCHIEFER - SANDSTEIN		REWAL FORMATION
MÖLLN CYCLE Z 7	MÖLLN-HALIT (Na7) MÖLNN-TON (T7)	MÖLNN-TON (T7)	PZ 4e PZ 4d	
FRIESLAND CYCLE Z 6	FRIESLAND-HALIT (Na6) FRIESLAND-ANHYDRIT (A6) FRIESLAND-TON (T6)	U. FRIESLAND-TON (T6r) FRIESLAND-SANDSTEIN (S6)	PZ 4c	
OHRE CYCLE Z 5	OHRE-HALIT (Na5) OHRE-ANHYDRIT (A5) OHRE-TON (T5)	U. OHRE-TON (T5r) OHRE-SANDSTEIN (S5)	PZ 4b	TOP YOUNGEST HALITE ($Na4b_2$) UPPER RED PELITE-UPPER PART ($T4b_2$) INTRASTRATAL HALITE ($Na4b_1$) UPPER RED PELITE-LOWER PART ($T4b_1$)
ALLER CYCLE Z 4	U. ALLER-ANHYDRIT (A4r) ALLER-HALIT (Na4) PEGMATITANHYDRIT (A4) ROTER SALZTON (T4)	U. ALLER-TON (T4r) ALLER-SANDSTEIN (S4)	PZ 4a	UPPER YOUNGEST CLAY HALITE ($Na4a_{2}t$) UPPER YOUNGEST HALITE ($Na4a_2$) UPPER PEGMATITE ANHYDRITE ($A4a_2$) LOWER YOUNGEST HALITE ($Na4a_1$) LOWER PEGMATITE ANHYDRITE ($A4a_1$) UNDERLYING HALITE ($Na4a_0$) LOWER RED PELITE (T4a)
LEINE CYCLE Z 3	U. LEINE-HALIT (Na3o) KALIFLÖZ (K3) L. LEINE-HALIT (Na3u) HAUPTANHYDRIT (A3) GRAUER SALZTON (T3)	U. LEINE-TON (T3r) A3 PLATTENDOLOMIT (Ca3) GRAUER SALZTON (T3)	PZ 3	YOUNGER CLAY HALITE (Na3t) YOUNGER HALITE (A3) YOUNGER POTASH (K3) MAIN ANHYDRITE (A3) PLATY DOLOMITE (Ca3) GREY PELITE (T3)
STASSFURT CYCLE Z 2	DECKANHYDRIT (A2r) DECKSTEINSALZ (Na2r) KALIFLÖZ STASSFURT (K2) STASSFURT-HALIT (Na2) BASALANHYDRIT (A2) STINKSCHIEFER (Ca2)	U. STASSFURT-TON (T2r) A2 HAUPTDOLOMIT (Ca2)	PZ 2	SCREENING ANHYDRITE (A2r) SCREENING OLDER HALITE (Na2r) OLDER POTASH (K2) OLDER HALITE (Na2) BASAL ANHYDRITE (A2) MAIN DOLOMITE (Ca2)
WERRA CYCLE Z 1	WERRA-ANHYDRIT (A1) ZECHSTEINKALK (Ca1) KUPFERSCHIEFER (T1)	U. WERRA-ANHYDRIT (A1r) U. WERRA-TON (T1r) U. WERRA-ANHYDRIT (A1o) U. WERRA-HALIT (Na1o) KALIFLÖZE (K1Th, K1H) L. WERRA-HALIT (Na1u) L. WERRA-ANHYDRIT (A1u) ZECHSTEINKALK (Ca1) KUPFERSCHIEFER (T1)	PZ 1	UPPER ANHYDRITE (A1g) OLDEST HALITE (Na1) LOWER ANHYDRITE (A1d) ZECHSTEIN LIMESTONE (Ca1) COPPER SHALE (T1)

Fig. 10.44. Classification and correlation of the Zechstein in NW Germany, the Hessian Basin and Central Poland (after Kulick & Paul 1987; Wagner 1991; Kiersnowski *et al.* 1995).

wedge deposits of the third Zechstein sequence (ZS3). Warren (1999) has noted that such contradictions underlie the difficulties in sequence stratigraphic analysis of evaporite deposits derived from geometries of marine systems. The terms lowstand and highstand system tract are largely defined by whether the shelf area of the basin is covered by water. Furthermore, it is assumed that there was a surface connection to the ocean, and so there were direct eustatic responses in brine base level within the evaporitic basin.

Zechstein cycles

Zechstein 1. The Weissliegend comprises fine- to medium-grained sandstones which are variously interpreted as aeolian, marine or fluvial/estuarine in origin (Fig. 10.44). Opinions on the age of the Weissliegend also differ, with suggested ages ranging from Early Permian to Late Permian and also straddling the boundary between the two (Kiersnowski *et al.* 1995). In some areas, the Weissliegend exhibits vertical subdivisions (Nemec & Porebski 1981) where the lower part is interpreted as aeolian while the upper part comprises massive sandstones with evidence of soft-sediment deformation. Localized bioturbation and the presence of a marine fauna indicates deposition in a shallow-marine setting (Kiersnowski *et al.* 1995). In Germany and Poland the Weissliegend grades into limestone (termed the Mutterflöz or Border Limestone) although it is only present in those areas which were shallow marine (Peryt 1989). Overlying this is the Kupferschiefer which is a laminated, bituminous marly shale containing a benthic fauna (Kiersnowski *et al.* 1995). This is interpreted as the true beginning of the Zechstein 1 cycle, which commenced with the Kupferschiefer transgression and was followed by the deposition of the open-marine Zechstein Limestone (Zechsteinkalk). Locally, however, the Weissliegend has an age coincident with the Z1 transgression (see below).

Deposition of the Z1 (Werra) Formation commenced with the Kupferschiefer, a 0.5 m thick finely laminated claystone, which in Germany has a total organic content of up to 5% (Frisch & Kockel 2003). The Z1 Formation was also deposited over much of the Netherlands, with the exception of the southern offshore area (Geluk 2005). The thickest occurrence, up to 500 m, is on the anhydrite platform to the south of the main basin. The formation onlaps the Anglo-Brabant Massif.

The Kupferschiefer is missing over parts of the Mid-North Sea High and on other highs where Rotliegend clastics are absent. The Zechstein Limestone (Zechsteinkalk) and its equivalents comprise mainly dolostones in marginal areas but pass into bituminous limestones in the lower part of the marginal slopes and across much of the basin (Kiersnowski *et al.* 1995). During the deposition of the Z1 carbonates the Mid-North Sea High became overstepped with algal mounds and banks developing on it. In much of northern Denmark, the Kupferschiefer and Z1 carbonates are missing by onlap. The maximum flooding surface of the first sequence occurs in the lower part of the Zechstein Limestone.

Carbonate platforms fringed the basin both to the north and the south. Reefs have been identified in the southern basin margin in Germany and the Netherlands (Johnson *et al.* 1994; Geluk 2000). In Poland (e.g. Peryt 1984; Sliwinski 1988) and Germany (Pöhlig 1986; Paul 1991) along the margin of the carbonate platform and in the shoals in the basin centre, two (or more; Sliwinski 1988) sets of peritidal layers appear within deposits representing a deeper depositional environment. These record lower-order sea-level changes.

Along the southern margin of the Southern Permian Basin, the Zechstein Limestone developed as a platform facies, covering an erosion surface developed on deformed Palaeozoic strata of the Campine Basin (Belgium) and the Anglo-Brabant Massif area, leaving only some remnants of earlier Permian clastic deposits (Geluk 2000). The sand content of the Zechstein Limestone is high, regularly exceeding 50%, but there are few signs of reworking of the onlapped strata. Following a fall in sea level, the overlying grey to red silty mudstones were deposited in a lagoon in which there is also evidence of soil development. The second to fifth members of the Helchteren Formation represent a condensed Zechstein section. Their impoverished nearshore to lagoonal fauna and land flora are already indicative of the latest Permian (Demaret-Fairon *et al.* 1985). The sandy carbonates of the second member are correlated with the Werra Formation (Z1 cycle), while the lagoonal mudstone of the third member is correlated with the Grey Salt Clay Member of the Leine Formation; the beach sandstone of the fourth member is also correlated to the Leine Formation (Z3 cycle) while the top claystone member is correlated to the Zechstein Upper Claystone Formation (Z4 cycle) (see below).

Carbonate buildups occur in the upper part of the Z1 cycle (Peryt 1978; Paul 1980, 1996). These consist of bryozoans, encrusting foraminifers and algae (stromatolites), with abundant aragonitic cementation. Zechstein reefs (usually 50 m thick) show a remarkable variety of geometries (Paul 1996). They are generally present along the basin margin (Paul 1995), forming barriers at the shelf edge (Peryt 1978), or occur above highs in the pre-Zechstein topography (Paul 1980; Dyjaczynski *et al.* 2001). Fringing reefs have extensive flats and very steep (sometimes vertical) forereef slopes; patch reefs formed archipelagos with nearly enclosed lagoons, and conical pinnacle reefs. 3D seismic studies in western Poland show a very irregular shape for the reef bodies (Dyjaczynski *et al.* 2001) and a sharp decrease in thickness, corresponding to facies changes at the margins of the reefs.

A rapid restriction of the Z1 basin, probably induced by a relative fall in sea level, resulted in the termination of the Zechstein Limestone and the development of the basinward-prograding banks of the Werra Anhydrite. In much of the Zechstein Basin the Werra Anhydrite (up to 600 m thick; Clark 1986) is divided into two main units, Lower and Upper, although these have not been recognized in NE England. The Lower Werra Anhydrite is transgressive and several cycles have been recognized (Richter-Bernburg 1955). The lowermost parts of these cycles contain nodular anhydrites which may have formed in subaerial to shallow subaqueous settings. These are succeeded by bedded anhydrites, which were also probably formed in shallow subaqueous settings. During the succeeding phases, there is a distinct palaeogeographical zonation, with varied development of the upper part of the cycle. In thicker sequences, massive anhydrites and local breccias occur, and in thinner sequences laminated anhydrites are common (Kiersnowski *et al.* 1995). The complex depositional pattern of the Lower Werra Anhydrite resulted from the interaction of a number of controlling factors, including the rates of sulphate deposition, subsidence and eustatic fluctuations. The thickest sequences of Zechstein 1 anhydrites usually occur in the broad marginal part of the sedimentary basin (Richter-Bernburg 1955; Peryt & Antonowicz 1990).

In southern Germany there are several sub-basins (e.g. Werra-Fulda Basin, Kraichgau-Saale Basin, parts of the Saar-Nahe Basin) which were separated by sills from the main Zechstein Basin. These sub-basins were strongly influenced by the proximity of their hinterlands, trapping the terrigeneous mud which was thereby prevented from reaching the main basin (Kiersnowski *et al.* 1995). A large lagoonal belt encloses the rims of these southern sub-basins (Fig. 10.44).

On the Zechstein 1 shelves the top of the Werra Anhydrite corresponds to a hiatus that was caused by a temporary lowstand in sea level. In basinal areas the Werra Anhydrite consists of finely laminated anhydrites and carbonates (20–30 m thick) that apparently were deposited in deeper waters. Thus, they testify to continued sediment starvation of the basin where deposition was continuous across the cycle boundary between Zechstein 1 and 2 (Wagner *et al.* 1981; Taylor 1986).

A system of sulphate platforms and adjoining chloride basins developed in the marginal part of the Z1 basin adjacent to the Werra carbonate platform. These show a wide spectrum of mainly subaqueous facies (Czapowski 1987; Peryt 1994). The sequences of the Lower Werra Anhydrite in basinal and slope areas show deepening-upward trends (Peryt 1994), although clear cyclicity in the Lower Werra Anhydrite and the Oldest Halite is lacking (cf. Peryt *et al.* 1993; Czapowski *et al.* 1993; Peryt 1994). However, in individual halite basins clear cycles, well marked by changes of bromine content (cycles of increasing-upward salinity), occur (Peryt *et al.* 1992).

In the Netherlands and Germany, Zechstein 1 facies and thickness patterns along the southern margin of the Southern Permian Basin were controlled by syndepositional extensional tectonics. This is evident from the accumulation of up to 300 m of halite, equivalent to the Werra Anhydrite, in differentially subsiding troughs that were controlled by reactivated Stephanian–Autunian fault systems. Such salt basins occur, for example, in the Emsland-Krefeld embayment which forms a re-entrant of the Zechstein 1 basin between the London–Brabant and Rhenish massifs, and in the northern parts of the Hessian Basin (Rost 1975; Weber 1979; M.A. Ziegler 1989). Similar patterns are also evident in northern Poland and on the Fore Sudetic monocline in SW Poland (Wagner *et al.* 1981).

During evaporite deposition in the central part of the basin, much of the Zechstein Limestone platform was exposed. Thus, the upper part of the Oldest Halite (i.e. Werra Halite, Na1) and the sulphate platforms were subjected to local dissolution of the previously precipitated evaporites. The thickness and character of the Oldest Halite (Werra Halite) are both related to its palaeogeographic position. In Germany, the halite is normally restricted to the peripheral sub-basins located south of the main basin (Fig. 10.45). In the former basins of the Lower Werra Anhydrite, halite sequences are thick and may have originated in a deep-water setting, whereas in the former shoal areas, shallow-water deposits were formed (Czapowski 1987).

In the peripheral part of the basin, the distinctive Anhydrite Breccia overlies the Zechstein Limestone deposits. This breccia represents a transgressive systems tract (TST) while the overlying massive anhydrites in the peripheral parts of the basin are a highstand systems tract (HST). The maximum flooding surface corresponds to deposition of the massive anhydrites. The lower portion of the Upper Anhydrite in the peripheral part of the basin is a TST while the middle and upper portions are HSTs (Peryt *et al.* 1996*a*). In the transitional zone between the basin centre and its margins, the part of the Upper Anhydrite below the Anhydrite Breccia is a lowstand systems tract (LST) precipitate which is overlain by a TST deposit and subsequently by HST deposits. In the basin centre exclusively TST deposits have been recognized.

At the close of the Zechstein 1 cycle, the deposition of the Upper Werra Anhydrite corresponded with a fall in sea level of several tens of metres (Peryt 1992) and the evaporite platform of PZ1 was once again subaerially exposed leading to karstification of the Upper Anhydrite (Strohmenger *et al.* 1996*a*). The Upper Werra Anhydrite is usually a few tens of metres thick and comprises mudstones (with anhydrite nodules) at the base, overlain by bedded and laminated anhydrite. In the peripheral parts of the basin interlayers of nodular anhydrite are common and dolomites have also been noted (Kiersnowski *et al.* 1995). The Upper Werra Anhydrite deposits close to the basin margin probably originated in shallow, subaqueous and subaerial depositional environments, while those in the basin centre formed under stable subaqueous conditions.

Zechstein 2. The Z2 (Stassfurt) Formation comprises a basal carbonate unit followed by the basal anhydrite and salt. Three depositional areas can be identified, namely platform, slope and basin. The platform facies comprises a complex of oolitic, peloidal, bioclastic and pisolitic packstones, grainstones and wackestones (Strohmenger *et al.* 1996*a*). Depositional environments range from sabkha, tidal flats, algal shoals, oolite shoals, lagoons and ooid bars (Strohmenger *et al.* 1996*a*). In the vicinity of the platform, slumps and turbidites of displaced shelf deposits have been identified (Van de Sande *et al.* 1996). The slope facies consists of calcitic and dolomitic mudstones, as well as redeposited platform sediments (Geluk 2005).

The contact between the Zechstein 1 and Zechstein 2 units is gradational in the central parts of the basin, but sharp in the marginal areas where a break in deposition is expressed by the presence of clastic deposits and brecciation and erosion of the uppermost parts of the Z1 units (Kiersnowski *et al.* 1995) (Fig. 10.45). The onset of the Z2 cycle was marked by a rise in sea level, a reduction in water salinities and the return to a carbonate depositional regime. This is reflected by the accumulation of the Z2 carbonates on the basin foreslopes and overlying banks of Werra Anhydrite. These carbonates are referred to as the Main Dolomite (i.e Hauptdolomit, Ca2) which has a maximum thickness of >200 m, thinning to a few tens of metres in the platform areas. In the carbonate platform areas the Main Dolomite exhibits a shallowing-upward sequence, with mudstones in the lower part and ooidal, peloidal and ooidal/peloidal grainstones in the upper part (Kiersnowski *et al.* 1995). Cyclicity has been recorded in the carbonate platform of the Main Dolomite in several parts of the basin, which may be related to eustatic sea-level variations and periods of subaerial exposure.

Near the basin margins, Main Dolomite carbonates pass landwards into a sabkha complex containing sulphate and siliciclastic deposits (Peryt & Kasprzyk 1992; Fijakowska & Peryt 1995). At the basin margins the banks of the Main Dolomite consist of micrites, algal laminites and, particularly along the outer bank margins, oolitic bars. A high rate of carbonate production resulted in continued basinward progradation of these banks involving the shedding of oolitic material over the bank edges and its transport by density currents to more basinal areas. These slope facies are dominated by slump and slide deposits, debris flows and turbidites (Peryt 1992; Kiersnowski *et al.* 1995). In slope areas, the Main Dolomite grades into the thinly bedded Stinkdolomit which gives way basinward to the *c.* 10 m thick Stinkschiefer (Ca2) (Wagner *et al.* 1981).

The width of the combined Werra Anhydrite (A1) and Main Dolomite (Ca2) banks is variable and depended on the gradient of the basin slope and also on syndepositional faulting controlling, in part, the bank geometries. However, syndepositional activity along faults appears to have decreased gradually during the period of accumulation of the Main Dolomite (M.A. Ziegler 1989). The carbonate banks are relatively narrow along the Danish margin of the Southern Permian Basin and also along the NE margin of the North Permian Basin. Broad platforms developed on the Mid-North Sea High, in the Emsland-Krefeld embayment in the Hessian Basin and in Poland. Water depths in basin areas were in excess of 300–400 m at the end of the period of Main Dolomite deposition.

In sequence stratigraphic terms, most of the Main Dolomite deposits are HST deposits, but subsequently sea level fell (by some 150–200 m) and a younger carbonate platform, attached to the previously deposited slope sediments, was formed (Zdanowski 2004). Thus, the upper portion of the Main Dolomite in the zone of the carbonate platform slope can be treated as the beginning of LST deposition of the third Zechstein sequence. Renewed restriction of the basin led to the termination of the Main Dolomite regime, the deposition of the generally thin Basal Anhydrite which blankets the carbonate banks, and ultimately the infilling of basinal areas by the massive Older Halite (i.e. Stassfurt-Halit, Na2) and K-Mg salts. The Basal Anhydrite is a transgressive sequence and comprises massive anhydrite formed in shallow subaqueous settings which passes upward into bedded and finely laminated anhydrite.

In basinal areas the deposits of the Older Halite attain a thickness of some 1500 m and more in the North Sea, over 400 m in Poland and some 200 m on the marginal platforms. Basin infilling by salt had a loading effect on the lithosphere and resulted in further basin subsidence. This increased subsidence was marked by the overstepping of the Stassfurt salts onto the Main Dolomite carbonate banks. Basin restriction reached a peak towards the end of the Zechstein 2 cycle when thick K-Mg salts were precipitated. Bromine contents suggest that at the beginning of deposition of the Older Halite the water depths were up to

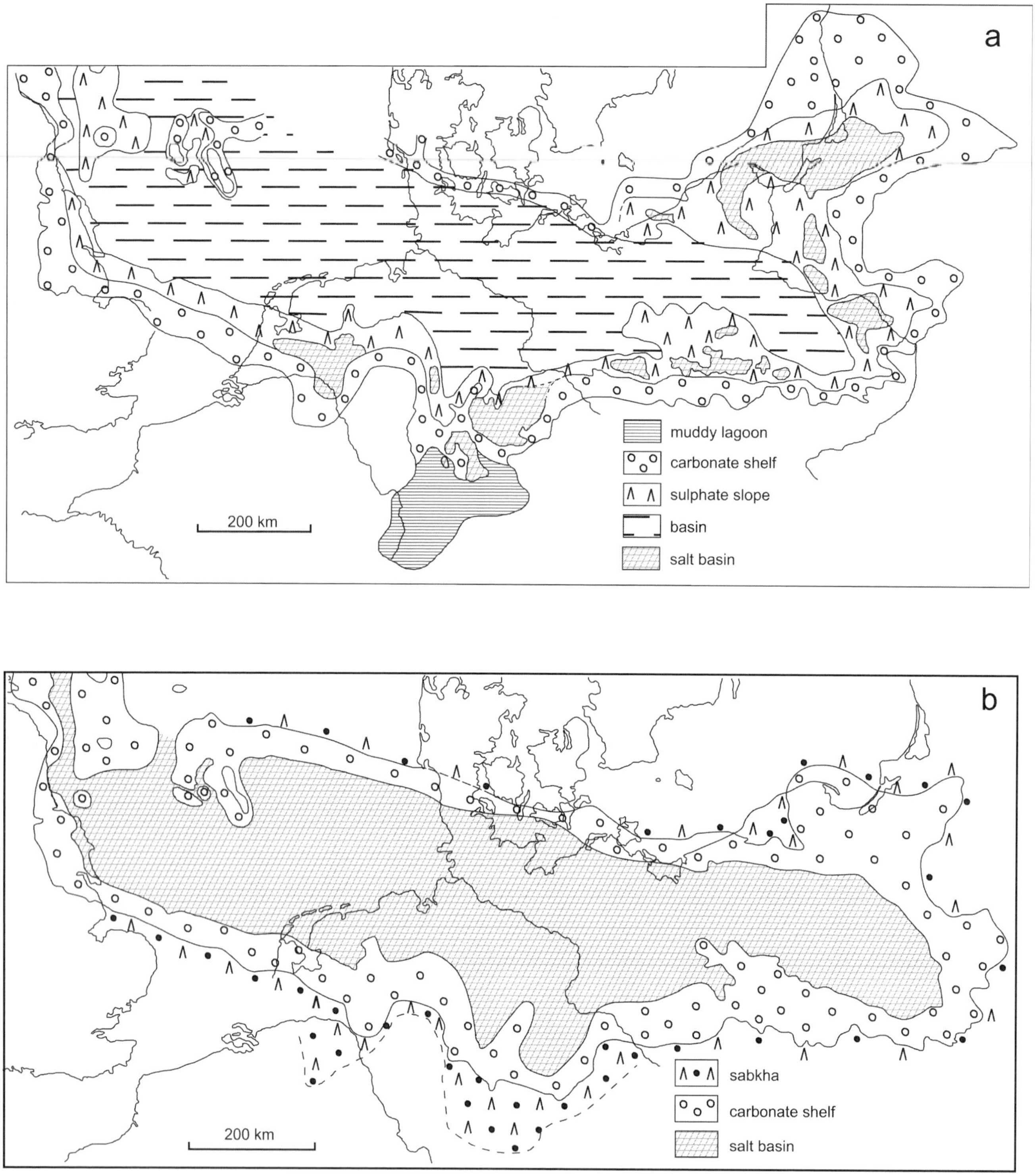

Fig. 10.45. (**a**) Zechstein 1 and (**b**) Zechstein 2 basins in Central Europe (after Kiersnowski *et al.* 1995).

140 m in the peripheral parts of the basin, and probably deeper in the basin centre.

The Older Halite (Na2) sedimentary basin evolved from a relatively deep open salt basin, with free brine exchange, to shallow closed salinas fringed by coastal salt pans along the basin borders. At the end of the period of Na2 deposition water depths were at a minimum (several centimetres to decimetres) in these salina/salt pan environments (Czapowski *et al.* 1990). This transition from a deep basin through to a shallow one was due to the high rate of basin fill. In addition to the infilling, the salt also onlapped the slopes of the former Ca2 platform extending as far as the top of the platform.

Continued subsidence of the Southern Permian Basin was coupled with a progressive overstepping of the basin margins. This is particularly evident in southern Denmark where shorelines advanced through Jutland into area of Sjaelland and in the Hessian Basin through which the Zechstein Sea ingressed (Boigk & Schöneich 1974).

The Draping Older Halite (i.e. Decksteinsalz, Na2r), Draping Anhydrite (i.e. Deckanhydrit, A2r) as well as the Grey Pelite (i.e. Grauer Salzton, T3), to which the maximum flooding surface is related, are TST deposits (Peryt *et al.* 1996*a*, *b*; Strohmenger *et al.* 1996b). The Grey Pelite deposits represent a condensed section.

Zechstein 3. The Z3 (Leine) Formation comprises the grey salt clay, carbonate, main anhydrite and salt. A decrease in basin restriction, coupled with a marine transgression, resulted in the deposition of the Platy Dolomite (i.e. Plattendolomite, Ca3) (Füchtbauer & Peryt 1981; Wagner *et al.* 1981; Clark 1986) (Fig. 10.46). This carbonate unit forms a broad, basinward-prograding platform attaining a maximum thickness of *c*. 90 m and thinning toward the basinal areas where it passes into shales. A common feature of the Platy Dolomite in many marginal areas of the Zechstein Basin is the recognition of only one shallowing-upward cycle and the presence of oolite shoals and lagoonal deposits in the area between those shoals and land areas (Kiersnowski *et al.* 1995).

In platform areas, the Platy Dolomite consists of cyclically alternating dolomitized micrites, algal laminites and nodular anhydrites. The bank-edge facies comprises bioclastic and oolitic grainstones (or sandstones in the Campine Basin, NE Belgium) that give way basinwards to thinly bedded carbonate mudstones and shales. Indeed, the extension of the Z3 sequence in the

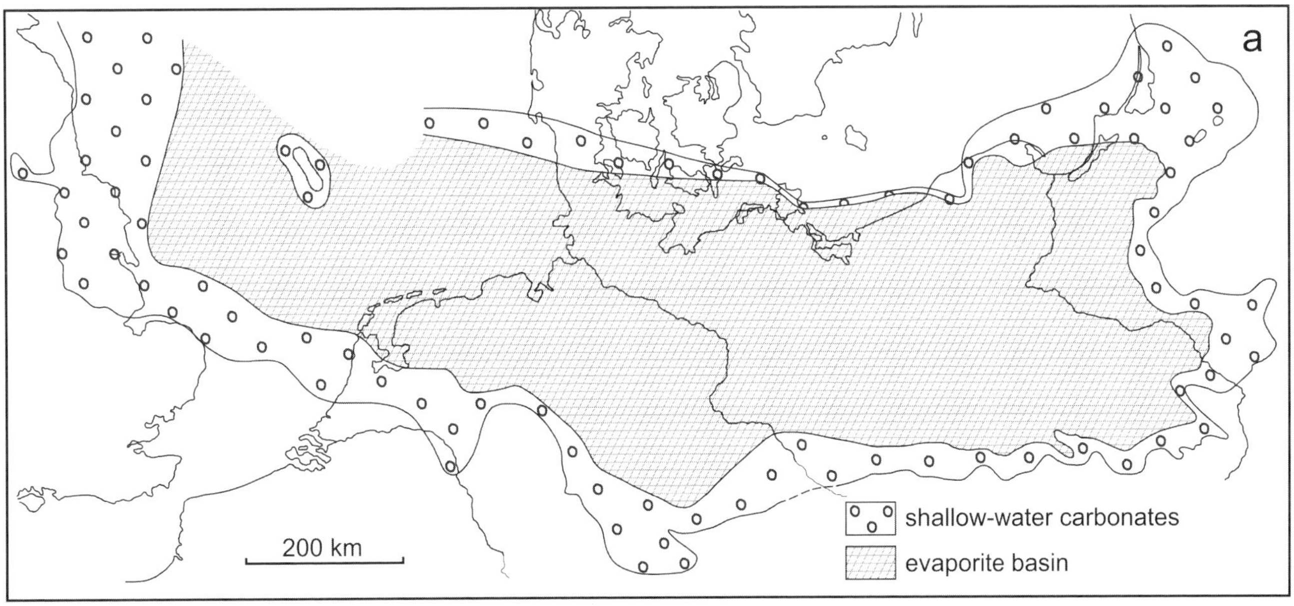

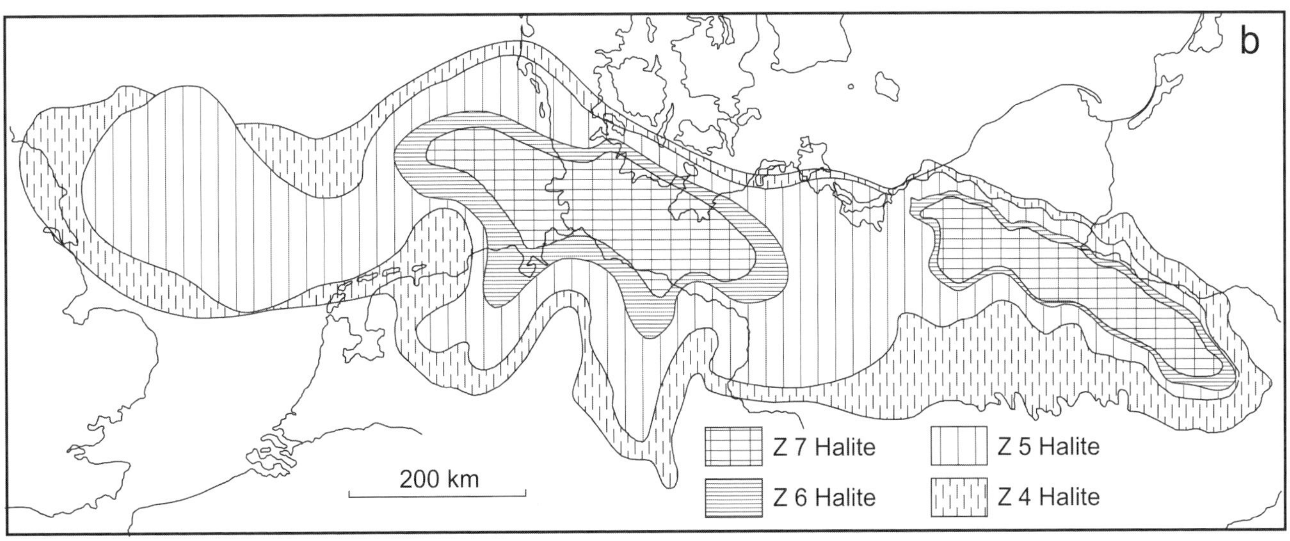

Fig. 10.46. (**a**) Zechstein 3 and (**b**) Zechstein 4–7 basins in Central Europe (after Kiersnowski *et al.* 1995).

Campine Basin (southern margin of the Southern Permian Basin) is marked by either massive coarsening-upward beach/barrier sandstone or oolite shoals. These are overlain by massive red claystones, reminiscent of the Bröckelschiefer, considered to be the transition to the Triassic Buntsandstein sequences. In the western offshore area of the Netherlands, the carbonates grades southwards into fluvial sandstones (Geluk *et al.* 1996).

As noted above, the Platy Dolomite is transgressive in nature (Gasiewicz 1990) and the transition between the Platy Dolomite and the Main Anhydrite occurred without any change in the sedimentary regime (Gasiewicz & Peryt 1994). It is, thus, possible that the sequence boundary occurs in the lowermost part of the Main Anhydrite at the boundary between the laminated anhydrite (= HA3 according to some authors) and the streaky anhydrite (HA4).

Renewed restriction of the basin resulted in the deposition of the Main Anhydrite (A3), which shows a slightly deepening-upward succession and is transgressive in Germany and Poland. The cycle commences with nodular anhydrite overlain by massive anhydrite and, rarely, laminated anhydrite. The entire Younger Halite (Na3, and the associated K-Mg salts in the basin centre) originated in shallow subaqueous environments (Czapowski 1993). The Younger Halite may be coeval with the Main Anhydrite (A3) of the basin margin (Kiersnowski *et al.* 1995).

By the time of deposition of the Younger Clay Halite (Na3) (particularly in the uppermost parts of this unit) sedimentation in the Zechstein depositional area had changed substantially. Within the halites, terrigeneous beds of considerable lithostratigraphic importance were deposited across the Zechstein Basin. The deposition of these interstratified terrigeneous beds resulted from marked changes in climate. The stable arid climate in the carbonate–terrigeneous cyclothems was replaced by alternating humid and dry periods with a general trend toward more humid climates. In humid periods, terrigenous or terrigenous/saline deposits were accumulated, and in dry periods evaporites were precipitated.

A distinctive feature of the terrigenous–evaporite cyclothems was the areal shrinkage of the evaporite basins and a simultaneous progradation of terrigenous deposits towards the basin centre. In the axial parts of the evaporitic basin in the Mid-Polish Trough and in northern Germany, the specific clayey-salt (zuber) facies occurs (Wagner 1994; Best 1989). In humid periods, red claystone with subordinate siltstone, deposited from waning bottom currents, alternated with halite intercalated with suspension-current-related terrigenous pelite. In dry periods, evaporite precipitation predominated.

Zechstein 4. The upper part of the Zechstein succession consists of halites, anhydrites and mudstones. The bromine content of the halite indicates that the lower part of the Zechstein 4 cycle in Poland originated as a shrinking marine basin whereas the upper part was formed in a continental setting (Kiersnowski *et al.* 1995) (Fig. 10.46). Halite of the lower part originated in shallow salt lagoons and later in salt pans within a sabkha depositional setting.

The Z4 (Aller) Formation is composed of a basal claystone, the red salt clay followed by the pegmatite anhydrite and salt. Along the southern basin margin, the formation comprises anhydritic claystones deposited in sahkhas, and medium- to coarse-grained fluvial sandstones in the western offshore area of the Netherlands (Geluk *et al.* 1996).

Zechstein 5. A brief interruption of evaporitic conditions marked the base of Zechstein 5 during which the generally <20 m thick Aller Carbonate and Anhydrite was deposited (Fig. 10.46). Halites and other salts were precipitated during the subsequent restricted phase of this cycle. The Zechstein 5 cycle commenced with an anhydrite layer and terminated with the accumulation of halites.

The Z5 (Ohre) Formation is limited to the NE part of the Netherlands and to the NW in the Dutch offshore. It comprises a basal claystone overlain by halite.

Zechstein 6 and 7. The Zechstein 6 and 7 units consist, in basinal areas, of cyclical salts and clays and, along the basin margins, of red shales and sands (Fig. 10.46). The Zechstein evaporitic system terminated with the deposition of the red Bröckelschiefer shales and sands which form the conventional base of the Buntsandstein. The transition to this depositional regime dominated the Early Triassic evolution of the Zechstein Basin and occurred within the upper part of the Tatarian, and thus preceded the onset of the Triassic (Fuglewicz 1987; Kulick & Paul 1987). These younger salts, reported from NW Germany (Best 1989), are absent from the Netherlands (Geluk 2005).

Zechstein successions within the Variscan Orogen (intramontane basins)

Eastern and Southern Alps: Upper Austro-Alpine tectonic units (K.K.)

At the top of the sedimentary succession within the Upper Austro-Alpine tectonic units there is a thick Zechstein-age unit. The sediments of the Gröden Formation are characterized by large amounts of reworked volcanic fragments derived from the rhyolitic volcanic rocks at the top of the Lower Permian succession (Fig. 10.22). The only fossils known from the Gröden Formation are rare silicified stem fragments in the Drau Range. Stratigraphically significant fossils, however, are lacking (Stingl 1983; Krainer 1987*a*, 1989, 1990; Spötl & Burns 1994).

In the central and eastern part of the Northern Calcareous Alps, a few hundred metres of halite-bearing marine evaporites, termed the 'Alpine Haselgebirge', were deposited in an approximately east–west trending Late Permian rift. This rift is represented today by a strongly deformed décollement horizon at the base of the Northern Calcareous Alps. In the central part of this rift (with its westernmost termination near Innsbruck), a central salt pan developed during the most arid stage. The salt pan is surrounded by sabkhas and saline mudflats. These mudflats probably represent the distal facies of large alluvial fan and floodplain systems on which fine-grained sediments were transported from the north, west and south into the subsiding rift basin. Analysis of the sulphur-bearing isotopes of sulphate minerals, bromide content in halites and rare marine bivalves indicates the marine origin of these evaporites (Spötl 1987, 1988, 1989, 1992).

The boundary between the Upper Permian Gröden Formation and the overlying Lower Triassic (Scythian) Alpine Buntsandstein Formation is marked by a sharp lithological change caused by a climatic shift from arid conditions during the Late Permian to more humid conditions during the Early Triassic (Figs 10.22 & 10.23). This boundary most probably corresponds to the Permian–Triassic boundary (Krainer 1987*b*). The Alpine Haselgebirge, in the central and eastern part of the Northern Calcareous Alps, is overlain by shallow-marine clastic and carbonate sediments of the Werfen Formation.

Southern Alps: Carnic Alps and Karawanken Mountains (K.K.)
The succession in this part of the Austrian Alps is predominantly shallow marine and calcareous. The succession culminates with the Tarvis Breccia which represents the transition to the Zechstein. The boundary between the Tarvis Breccia and overlying Gröden Formation is indicated by the first appearance of rhyolitic volcanic clasts. As a result of strong block faulting, the Gröden Formation overlies rocks of different age (Tarvis Breccia, Auernig Group, Hochwipfel Formation and Devonian reef limestones) and varies in thickness from 0 to 800 m. The Gröden Formation predominantly consists of red shales and siltstones with intercalated sandstones and dolomites, deposited on distal alluvial fans and alluvial plains, locally in a shallow-marine environment (Buggisch *et al.* 1976; Buggisch 1978; Ori & Venturini 1981; Farabegoli *et al.* 1986).

The Gröden Formation is conformably overlain by the Bellerophon Formation, which is *c.* 200 m thick and composed of an evaporitic facies in the lower part and well-bedded dolomites (wackestones, mudstones and grainstones) in the middle and upper parts, indicative of deposition in a restricted shelf environment. The overlying Werfen Formation consists of shallow-marine sediments which display similar facies and transgressive–regressive trends to the Dolomites (Buggisch 1974; Buggisch & Noe 1986; Boeckelmann 1988, 1990). The uppermost part of both the Bellerophon and Werfen formations and the Permian–Triassic boundary have been intensively studied in outcrop and in a 331 m deep borehole at Reppwand/Nassfeld (see Holser & Schönlaub 1991).

The Dolomites (K.K.)
Within the succession in the Dolomites, the Gröden Formation shows considerable lateral variations in thickness from 0 to 500m. The succession is subdivided into three terrigenous intervals separated by the carbonate and evaporitic sediments of the Bellerophon Formation and interpreted as transgressive–regressive events. According to Farabegoli & Viel (1982) these events were tectonically induced. The observed facies range from alluvial fan, braided stream and meandering stream deposits, to sandy lobes of terminal fan settings and coastal sabkha deposits. Sandstones are classified as feldspathic litharenites and lithic arkoses. Calcrete horizons, thin coal beds, plant fossils and tetrapod footprints are present, indicating deposition under semi-arid climatic conditions (Bosellini & Dal Cin 1968; Buggisch *et al.* 1976; Massari *et al.* 1988). The Bletterbach/Butterloch section contains one of the richest tetrapod footprint assemblages of the European Permian (e.g. Conti *et al.* 1977). Plant fossils are reported from several localities (e.g. Fritz & Krainer 1999). The middle part of the Bletterbach/Butterloch section contains a nautiloid-dominated marine fauna (Conti *et al.* 1977).

The Bellerophon Formation overlies, and in its lower part interfingers with, the Gröden Formation. This formation is up to 500 m thick and is subdivided into large-scale cycles with thicknesses of 50 to 200 m. The depositional environment ranges from coastal sabkha sediments composed of dolomites, marls, mudstones, nodular chicken-wire and laminated gypsum, grading into open-lagoon fossiliferous dolomitic limestones. Fine-grained red-bed sediments mark a regressive event and grade upward into dolomites of a restricted lagoon environment and finally into dark grey, fossiliferous shallow-marine limestones of the Badiota facies (Bosellini & Hardie 1973). The uppermost part of the Bellerophon Formation consists of 0.5–2 m thick dark marls and limestones containing a characteristic brachiopod and fusulinid fauna indicating a Late Permian (Dorashamian = Changhsingian) age (Neri & Posenato 1985).

The overlying Werfen formation (Servino Formation, Lombardy) is a marine succession of shallow-shelf clastic and carbonate sediments deposited in subtidal, tidal flat and supratidal mudflat environments. The Permian–Triassic boundary lies within the basal oolitic unit termed the Tesero horizon, which is developed west of the Cadore-Comelico area (for a detailed discussion of the Permian–Triassic boundary see Neri & Pasini 1985; Brandner 1988; Buggisch & Noe 1986).

Intra-Sudetic Basin (H.K.)
The Late Permian (Zechstein) part of the succession within the Czech part of the Intra-Sudetic Basin is represented by a 30–150 m thick complex of arkosic sandstones with conglomerate/breccia and mudstone intercalations (Bohuslavice Formation). The presence of dolomitic cement or concretions is suggestive of deposition in an arid climate (playa-lake deposits) although it has also been interpreted as a sabkha embayment of the Zechstein Sea.

The Intra-Sudetic and Krkonoše Piedmont basin succession is terminated by a complex of probable Triassic strata (Bohdašín Formation) up to 70 m thick, which unconformably overlies the Permian deposits. Biostratigraphic data are absent except for footprints of theropod dinosaurs indicating a middle Triassic to middle Jurassic age.

Western Carpathians (A.V.)
In the Western Carpathians, Late Permian sediments are found in all of the major tectonic units (Tatricum, Northern and Southern Veporicum, Zemplinicum, Hronicum, Northern and Southern Gemericum, Meliaticum, Turnaicum and Silicicum; Figs 10.38 & 10.39). Sedimentation patterns were more uniform and distinct lithofacies developed within each of the Variscan structural zones. At this time, sedimentation was influenced by the onset of transgression by Tethys from the SE to the NW.

Central Western Carpathian Crystalline Zone. In the Tatricum Unit there are no biostratigraphical data and thus it is impossible to clearly separate the Late and Early Permian strata. Conglomerates, sandstones and rare shales extend from the base to the upper part of the entire Permian sequence. There is a characteristic stratigraphic hiatus near the Permian–Lower Triassic boundary related to the discordant position of the siliciclastic Lower Triassic sediments, which overlie either the Permian sequence or overstep directly onto the Variscan crystalline basement. In some places, below the Lower Triassic, there is evidence of weathering of the crystalline basement.

In the Northern Veporicum Unit, the Late Permian Predajná Formation disconformably overlies the Brusno Formation. Polymict clastic sediments were deposited in an alluvial fan and floodplain sedimentary setting with isolated ephemeral lakes also present. Two regionally developed megacycles, containing thick horizons of conglomerates, are related to synsedimentary tectonic activity. The thickness of the second of these megacycles was partially reduced during pre-Triassic erosion. The Late Permian age of the Predajná Formation is based on the microflora present (Planderová & Vozárová 1982).

In the Permian of the Southern Veporicum Unit, the identification of the Late Permian succession is difficult due to the absence of biostratigraphical data and lithological differentiation, and also to the effects of Alpine reworking. The succession consists of monotonous coarse-grained arkosic sediments, which were overlain by the Lower Triassic quartz-rich sediments.

In the Zemplinicum, a complex of monotonous mudstones represents the youngest Permian lithostratigraphic unit (Černochov Formation). This relatively thin unit (*c.* 50 m) represents a playa association. The thin-bedded mudstones are intercalated with several thin layers of redeposited acid volcaniclastics and are unconformably overlain by Lower Triassic quartzose conglomeratic sandstones. The poorly preserved microflora provide a likely age of Late Permian.

The Late Permian sediments of the Hronicum were deposited in a complex rift system, which was related to extensional processes initiated during the Late Carboniferous–Early Permian. Sediments, including red-beds (the upper part of the Malužiná Formation), were deposited in fluvial or fluviolacustrine and ephemeral-lake environments under arid/semiarid climatic conditions. The upper part of the Malužiná Formation is conformably overlain by Lower Triassic siliciclastic sediments. The Late Permian age is proved by microflora (Planderová 1973). However, disparate age indications have been indicated from magnetostratigraphic investigations (Vozárová & Túnyi 2003). An important unit within the Hronicum Late Permian sediments is the product of the second eruption phase of mafic rift-related continental tholeiites and their volcaniclastics (Dostal *et al.* 2003).

Northern Gemeric Zone. The Rotliegend terrigenous and terrigenous/volcanogenic successions are overlain by a relatively mature sandy/conglomeratic horizon, containing pebbly material derived from the underlying Lower Permian. This evidence of erosion may have been a consequence of the break in sedimentation at the end of the Rotliegend, but biostratigraphic evidence to support this assumption is missing. Deposition was predominantly as alluvial, stream-channel sediments which grade upwards to salt pan, nearshore sabkha/lagoonal facies and gypsum shelf, with anhydrite/gypsum and salt breccia horizons and small accumulations of sylvite (in the Novoveská Huta Formation) (Kántor 1972 in Vozárová 1997). There is a gradual transition from the Late Permian siliciclastic–evaporitic sequence up to the Lower Triassic siliciclastic–carbonate sediments.

Inner Western Carpathian Zone. The gradually prograding Štítnik Formation (Southern Gemeric Unit) is a monotonous complex of cyclically alternating sandstones, siltstones and shales. Lenses of calcareous sandstones and dolomitic limestones with intercalations of shales occur only in its upper part. Thin lenses of phosphatic sandstones are rare. Phosphatic sandstones contain intraclasts of microphosphorites as well as apatite crystals within the matrix. The sediments contain a relatively high amount of rhyolite/dacite detritus (?synsedimentary or redeposited from the Rožňava Formation).

The sedimentary environment is interpreted as alluvial/lacustrine and lacustrine, with ephemeral high-alkaline lakes present in some areas, grading into nearshore, lagoonal/sabkha facies. The phosphatic sandstones originated in ephemeral eutrophic lakes as a result of the concentration of phosphorus due to iron redox cycling and the associated microbial effects (Vozárová & Rojkovič 2000). Age determinations are known only from the uppermost part of the Štítnik Formation. The Upper Permian was dated on the basis of plant remains (Šuf 1963).

Within the Inner Western Carpathian Zone (i.e. the Turnaic Unit), the continental red-beds unconformably overlying the middle Carboniferous deep-marine strata are most probably Late Permian in age (the Brusník Formation contains no biostratigraphical material and remains undated). The mostly coarse-grained siliciclastic sediments within the Turnaic Unit gradually grade up into evaporites. These Upper Permian–Lower Scythian evaporites are an integral part of both the Turnaic and Silicic units. They have been referred to the Perkupa Formation, with their main occurrence being in the Slovak Karst, Slovakia (Mello *et al.* 1997) and the Aggtelek Karst in northern Hungary (Kovács *et al.* 1989). The primary facies of the Perkupa Formation evaporites include anhydrites and anhydrite arenites, with either a massive or horizontally laminated structure. Clastic grains within the anhydrite arenites comprise anhydrite, dolomite and dolomitic shales. Bedding results from fine laminae of greenish-grey clays and/or yellow and yellowish-grey and black dolomites, alternating with anhydrite and anhydrite arenites. Evaporites of both units are overlain by Wesfen Formation sediments.

Hungary (A.T., J.H.)

Transdanubian Range. To the NE of the Balaton Highlands an evaporitic unit, the Tabajd Evaporite Formation, consisting of siltstone, dolomite, anhydrite and gypsum, overlies the Balatonfelvidék Sandstone Formation and partly interfingers with it. Within the evaporitic formation, dolomite and anhydrite concretions, nodules, laminae and thin beds occur intercalated with red or greenish-grey siltstones. The sedimentary environment of the Tabajd Evaporite Formation was a coastal sabkha setting where sulphate precipitation and dolomitization took place in the groundwater fluctuation zone under arid conditions. Sporomorphs, present in some layers, are essentially the same as the assemblage from the upper segment of the Balatonfelvidék Sandstone Formation, suggesting that the uppermost part of that formation is Upper Rotliegend–early Zechstein in age (Barabás-Stuhl 1975).

In the NE part of the Transdanubian Range the upper segment of the Upper Permian is represented by a cyclic lagoonal facies consisting predominantly of dolomite (Dinnyés Dolomite). This is underlain by an evaporitic siltstone unit and interfingers with it. The thickness of the formation is 200–300 m. It consists of grey and dark grey bituminous dolomite with interlayers of nodular anhydrite or gypsum.

The evaporite nodules indicate sabkha facies in a periodically desiccated lagoon. The laminated or locally fenestrally laminated bituminous dolomites represent intertidal–supratidal facies. The peloidal, calcareous algal, foraminiferal or ostracodal wackestone microfacies indicate a subtidal lagoon environment, whereas the oolitic, bioclastic grainstone suggests ooid shoals. The lagoonal dolomite is rich in Upper Permian calcareous algae, foraminifera and ostracods.

Bükk Mountains. The upper member of the Szentlélek Formation comprises alternating greenish-grey claystones, dolomites, gypsum and anhydrite. Evaporites predominate in the lower third of the member. Following the first appearence of dolomite, the unit becomes increasingly dolomitic. The dolomitic beds contain foraminifera and ostracodes; the small number of species and large number of specimens suggest a high-salinity environment, although the ostracode assemblage is Middle Permian in age (Fülöp 1994). The lithofacies and fossils suggest that there was an alternation of the sabkha and subtidal lagoonal environments. The features of this member are very similar to those of the coeval Fiamazza facies of the Bellerophon Formation in the Southern Alps (which is Late Permian (Zechstein) in age).

The evaporitic dolomite series of the Szentlélek Formation passes gradually upward into dark grey bituminous limestone, 170–260 m thick (Nagyvisnyó Limestone). In the lower part of the formation limestone and dolomite layers alternate, showing an upward-decreasing trend of dolomitization. Hermatypic corals (*Waagenophyllum*) and calcareous sponges were found in the topmost bed of the lower member. Above this bed, medium-bedded, dark grey to black limestones are intercalated with thin black shales. In the upper part of the formation limestone and marl layers alternate, and marl with calcareous nodules is typical. In some layers brachiopods and molluscs can be found in large quantities, while trilobites are rare.

The formation is very rich in microfossils. Calcareous algae are common, the quantity of benthic foraminifera is remarkable and the ostracode assemblage is also extremely rich in species. Based on the fossils present the Nagyvisnyó Limestone can be assigned to the Upper Permian (Fülöp 1994). The large amount of dasycladacean algae in the Nagyvisnyó Limestone clearly indicates a euphotic, subtidal, low-energy inner-shelf depositional environment. The faunal assemblage suggests normal-salinity marine conditions. The formation shows similarities with the Badiota facies of the Bellerophon Formation in the Southern Alps, the Slovenian Žažar Formation and the Dinaridic (west Serbian) Jadar Formation (Pešić *et al.* 1988).

Tisza Unit. The Late Permian Cserd Conglomerate Formation is overlain by the Kővágószölös Sandstone Formation (latest Permian), which is 150 to 1400 m thick (Barabás & Barabás-Stuhl 1998). This is subdivided into four units, each of them representing one megacycle. The oldest member consists of variegated coarse-grained sandstones and siltstones with carbonate concretions. In the overlying unit grey, coarse to fine sandstones and siltstones predominate. Silicified wood, organic carbon-rich sandstones and greenish sandstones also occur. The two upper members include sandstones and gravel-rich sandstones. The topmost beds of the Kővágószölös Sandstone Formation were recently considered to be Triassic based on sporomorphs (Barabás-Stuhl 1993).

Palaeontology

The greatest of all mass extinctions, with the eclipse of about 95% of all plant and animal species and 50% of all genera, occurred around the Permian–Triassic boundary (251 Ma) (Erwin 1995), although there is no consensus regarding the cause. Studies on a variety of Permian fossils including bryozoa (Ross 1995), sponges (Rigby & Senowbari-Daryan 1995), ammonoids (Nassichuk 1995) and pollen (Utting & Piasecki 1995) confirm these changes. Recent studies have suggested an extra-terrestrial meteor impact and worldwide catastrophic die-off of vegetation at the Permian–Triassic boundary (e.g. Ward *et al.* 2000; Becker *et al.* 2001). However, the evidence for a meteor impact is equivocal (see Farley *et al.* 2001), as is that for a global loss of vegetation. Rees (2002) analysed global Permian and Triassic plant data in a palaeogeographic context and revealed that the scale and timing of effects varied markedly between regions. The patterns are best explained by differences in geography, climate and fossil preservation rather than catastrophic events. In addition, although many plant species and genera became extinct, the larger groups of plants mostly persisted. Many genera of conifers that were dominant in the Late Permian disappeared, but were replaced by others, while in the Triassic conifers were also dominant. This is in marked contrast to the fauna; here several groups of animals, including trilobites, became completely extinct.

Permian terrestrial climate zones (biomes) have been determined from fossil floras and climate-sensitive sedimentary rocks (e.g. A.M. Ziegler 1990; Gibbs *et al.* 2002; Rees *et al.* 2002). Climates were well differentiated from the equator to the poles, although extensive continental icesheets were absent after the Early Permian and did not redevelop until the late Cenozoic (e.g. Markwick & Rowley 1998), indicating a latitudinal temperature gradient less severe than the present one. The Permian marine and terrestrial floras and faunas are indicative of increased provinciality and a highly differentiated biosphere (A.M. Ziegler 1990). At least a dozen marine biostratigraphic units characterized the Late Permian, and A.M. Ziegler (1990) has described eight major terrestrial floras ranging from the tropical Euramerican floras to the polar floras of the Far Eastern and Siberian regions. Widespread desert regions characterize the Late Permian, as evidenced by desert sediments and subtropical desert and mid-latitude desert floras (Barron & Fawcett 1995). A complete spectrum of temperate biomes and subtropical deserts existed in both hemispheres, but equatorial coals and tropical rainforests were only well developed throughout the Permian on the Chinese microcontinents (Rees 2002). Farther west in Euramerica, climate conditions changed gradually during the Early Permian such that the main coal-forming plants were replaced by ones adapted to drier conditions.

Apart from the more dramatic changes which occurred at the end of the Permian, the onset of the Permian was also marked by other significant changes. In Late Carboniferous times (Westphalian and ?Stephanian) uplift occurred across Central Europe and the earliest Permian was a time of subaerial erosion. Reduction in relief at this time would have contributed to reduced rainfall (Utting & Piasecki 1995). The general floral changes which occurred at the end of the Carboniferous and beginning of the Permian are associated with increasing aridity. In the Early Permian there was relatively low diversity of vegetation, whereas in the Late Permian the climate was maritime and inland tropical desert (Utting & Piasecki 1995).

Summary

The Permian succession of Central Europe records the change from a Pangaea configuration and compressive tectonic regime inherited from the Variscan Orogeny, to the development of the broad thermal subsidence-controlled Southern Permian Basin and its subsequent flooding by the Zechstein Sea (Figs 10.47 & 10.48). During the Permian, continent size and aggradation were at a maximum, and this continent, Pangaea, was surrounded by a single world ocean, Panthalassa, with a semi-enclosed Tethyan Sea. This singular continental configuration led to the development of extreme climatic conditions across the region.

Early Permian (Rotliegend)

During latest Carboniferous–Early Permian times, the final phase of Variscan orogenic extension resulted in the formation of a series of small strike-slip and extensional continental basins across central and western Europe. Rifting in the former Variscan foreland appears to have been dominated by late Variscan wrench tectonics (van Wees *et al.* 2000), particularly along the boundary between Precambrian and Phanerozoic Europe (Dadlez *et al.* 1995). Numerical modelling highlights the role of such lithospheric discontinuities in controlling rifting (Pascal & Cloetingh 2002; Pascal *et al.* 2003). In contrast, Permian rifting in the former Variscan hinterland seems to have been strongly controlled by the collapse of the mountain chain (Brunet & Le Pichon 1982; Prijac *et al.* 2000) with a possible far-field extension component also being plausible (Henk 1999). This process may also have been modified by the slow decay of the associated thermal anomaly (e.g. Paris Basin).

The extensive rift system which developed in late Carboniferous–early Permian times was associated with widespread mag-

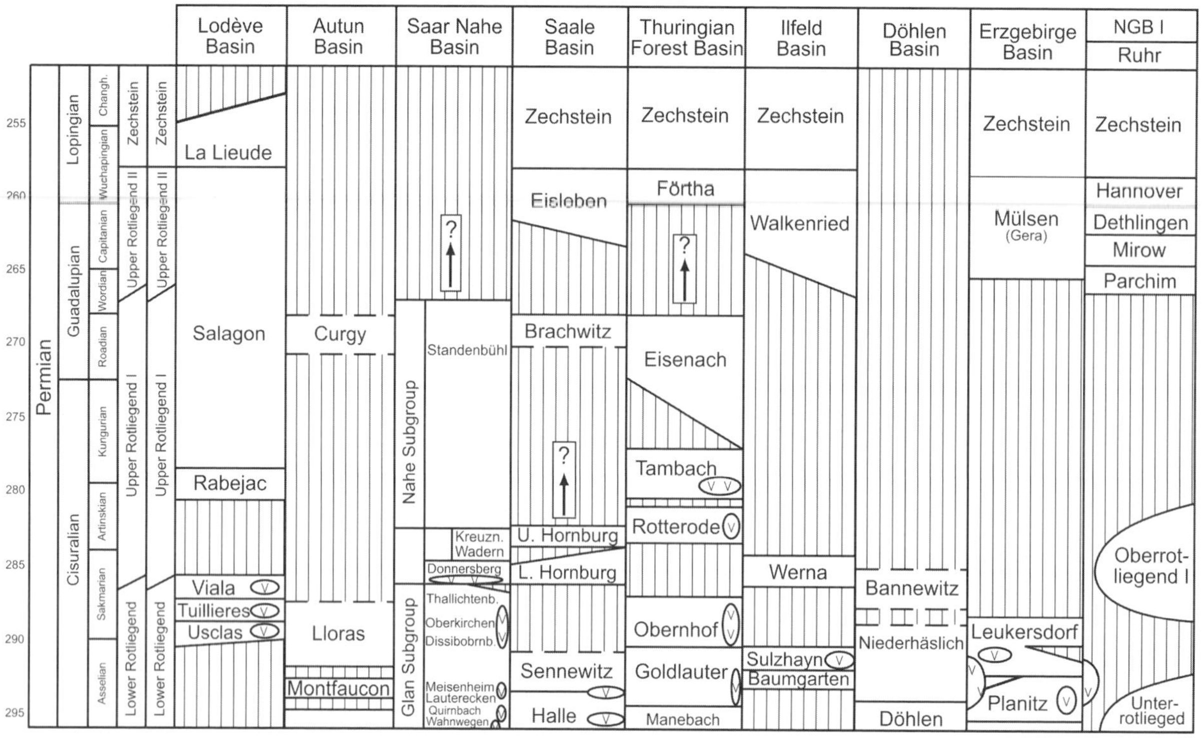

Fig. 10.47. Summary diagram of the Permian stratigraphy across Central Europe (after Schneider & Roscher 2002; Schneider *et al.* 2005; present study).

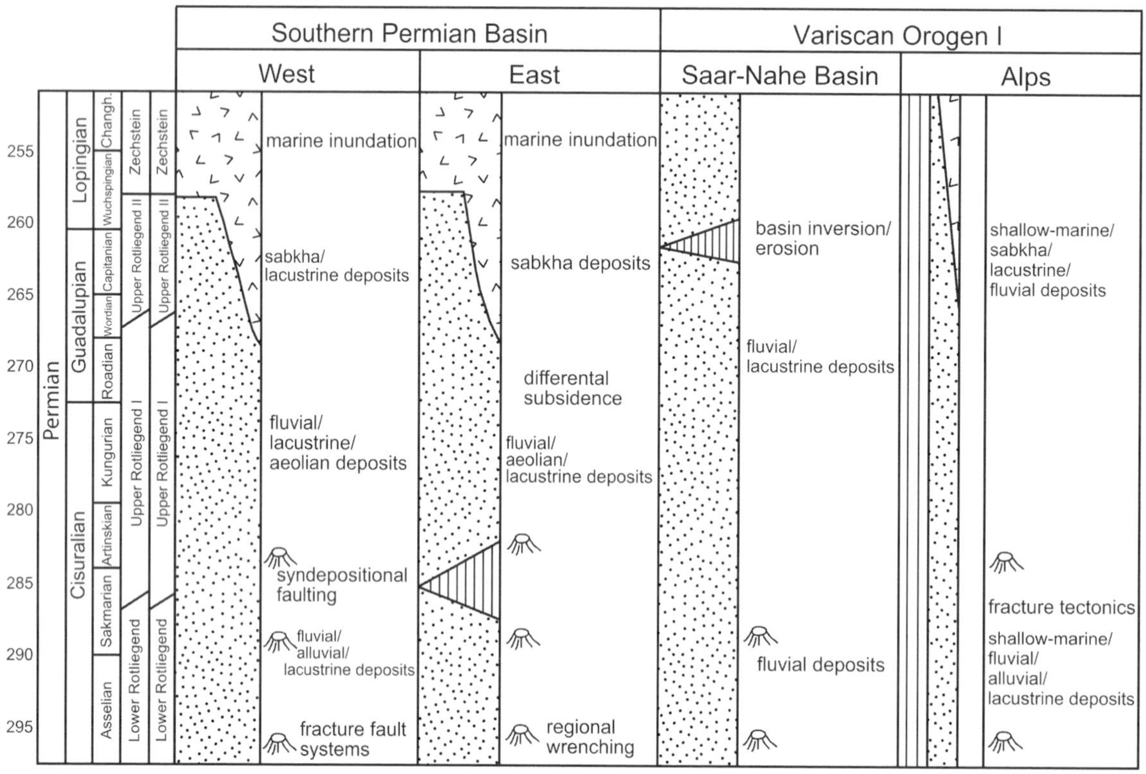

Fig. 10.48. Tentative chronology of events across Central Europe during Permian times.

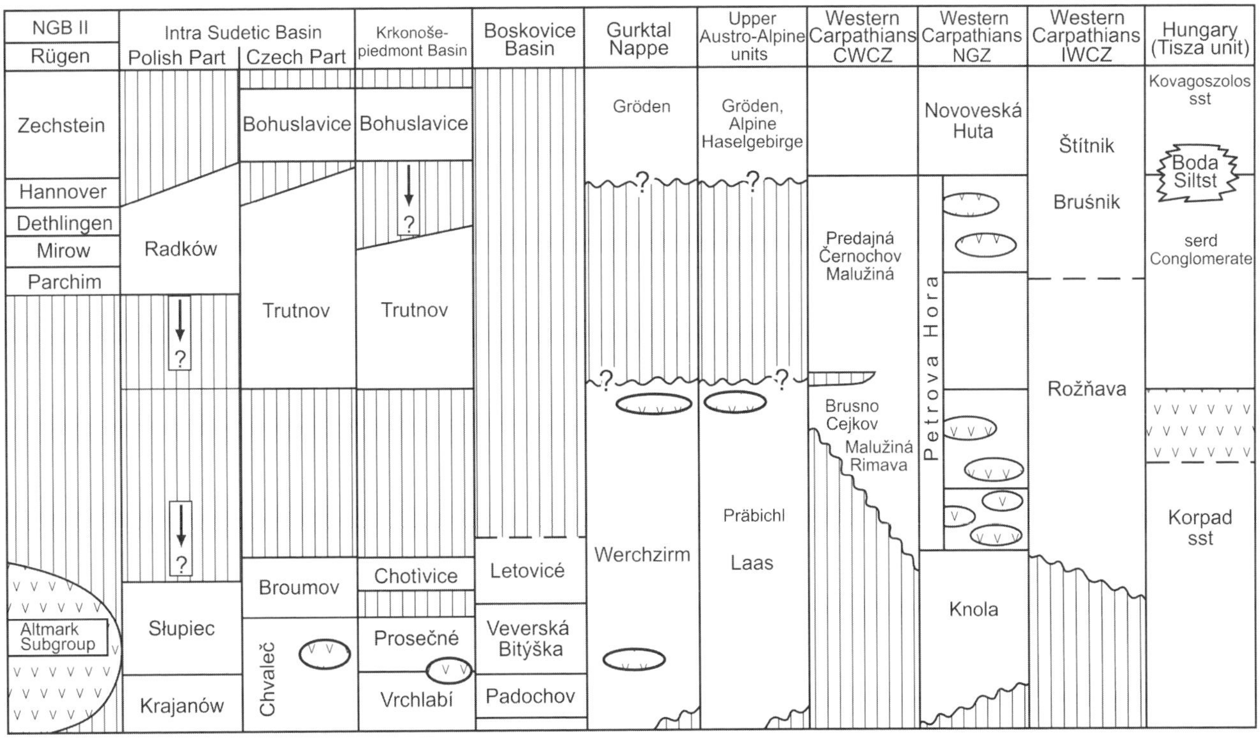

Fig. 10.47. (*continued*)

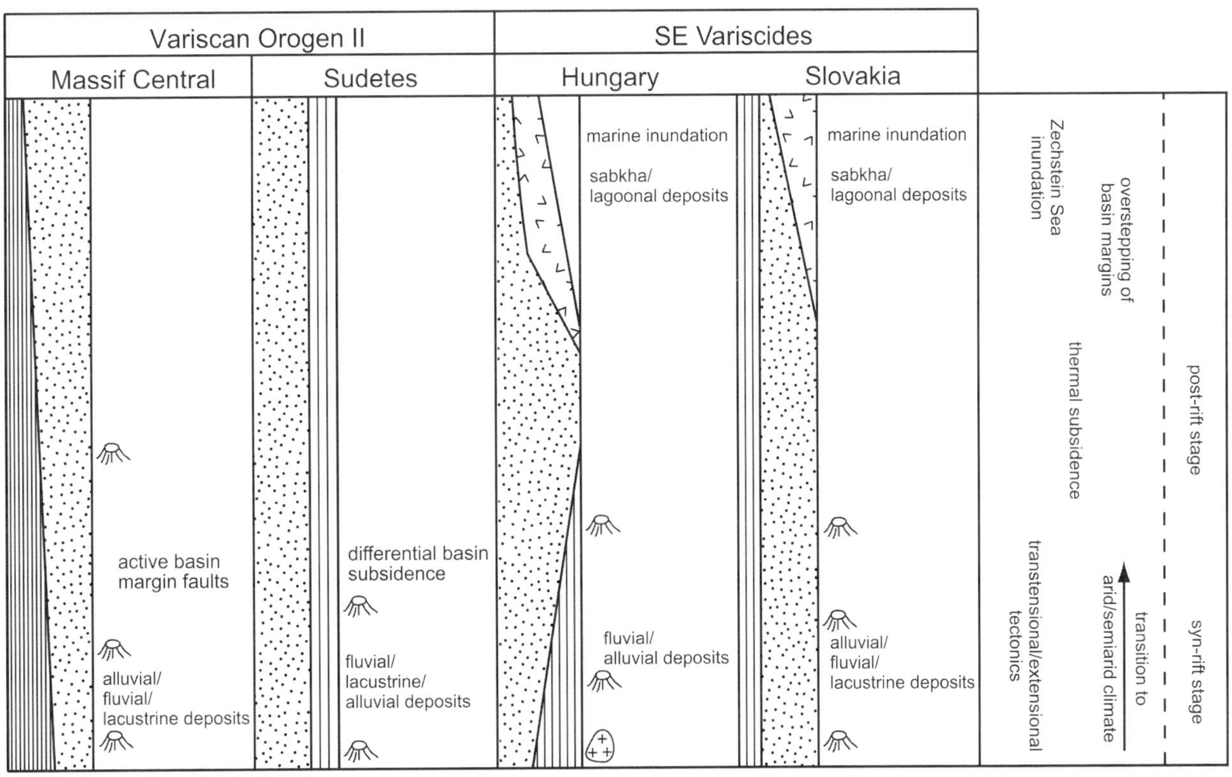

Fig. 10.48. (*continued*)

matism, developed within the northern foreland of the Variscan orogenic belt. Cooling of the lithosphere, following the initial period of block faulting, graben development and associated volcanic activity, resulted in a continuously expanding regional depression with little or no faulting. This resulted in the gradual coalescence of these isolated basins to form the large Southern Permian Basin which extended across much of central and western Europe. Both this, and the associated Northern Permian Basin, were initiated as broad crustal downwarps with little associated faulting. The two basins developed within a mosaic of basement terranes of different ages and geodynamic histories and their geometries show little apparent evidence of control by pre-existing structures (van Wees *et al.* 2000).

Early Permian sedimentation was predominantly fluvial, changing later to aeolian. This change was due either to a significant climate change or a decline in the relief of the surrounding uplands. By the end of the Early Permian extensive dunefields occupied the basin margins with saline lakes (playas) in the basin depocentres (Verdier 1996).

Late Permian (Zechstein)

Zechstein sedimentation commenced when the Rotliegend Southern Permian Basin, which lay below sea level, was flooded through a combination of rifting and eustatic sea-level rise (Glennie & Buller 1983). The trangression was geologically rapid as indicated by the regional distribution of the basal Zechstein Kuperschiefer, an organic-rich laminated shale, which corresponds to a sharp chronostratigraphic marker. Reducing conditions were established shortly after the Zechstein transgression resulting in the development of permanent stagnant bottom-water conditions. Following the initial transgression, water depths in the axial parts of the Southern Permian Basin may have been in the range of 200–300 m (P.A. Ziegler 1990). The Kupferschiefer was abruptly overlain by the Zechstein Limestone (Zechsteinkalk) and followed by the Werra Anhydrite. This cyclicity established the pattern for subsequent Zechstein deposition, with a total of between five and seven cycles recorded across the region. The conventional view of these cycles suggests that they correspond to phases of progressively greater evaporation and the shallowing of either the entire basin or its margins. They can, however, also be interpreted in sequence stratigraphic terms (Ruffell & Shelton 2002) where the thick gypsum deposits accumulated directly above the sequence boundary as a lowstand wedge passing upward into lowstand halites and potash deposits. The shallow-marine carbonate platforms represent the transgressive and highstand systems tracts. Care, however, should be used in such an approach given that there are problems with the use of sequence stratigraphic analysis in evaporitic successions, particularly in the assumption that there was a surface connection to the ocean and, therefore, a direct eustatic response in brine base level within the evaporitic basin (Warren 1999).

We would like to thank T. Beilfuß for the excellent diagrams. The manuscript was greatly improved by the comments of R. Gaupp, R. Gast and K. Glennie.

References

AGHABAWA, M. A. 1993. *Petrology and geochemistry of the Rotliegendes volcanic rocks in Denmark and their tectonic implications.* Dynamisk/stratigrafisk analyse af Palaeozoikum i Danmark. Danmarks Geologiske Undersøgelse Kunderapport 1993/35. Geological Survey of Denmark, Copenhagen.

ALEKSANDROWSKI, P., KIEŁT, M., KUROWSKI, L., MASTALERZ, K., MAZUR, S., NOWAK, J. & PASEK, P. 2003. Preliminary results of applying dipmeter data to structural study of the Carboniferous fold-thrust belt underlying the Fore-Sudetic Monocline, SW Poland. *Proceedings of the 8th Meeting of the Czech Tectonic Studies Group/1st Meeting of the Central European Tectonics Group*, 24–27 April, Hruba Skala. Geolines - Papers in Earth Sciences, **16**, 13.

ARTHAUD, F. & MATTE, P. 1975. Les décrochements tardi-Hercyniens du sud-ouest de l'Europe: géometrie et essais de réconstruction des conditions de la déformation. *Tectonophysics*, **25**, 139–170.

ARTHAUD, F. & MATTE, P. 1977. Late Paleozoic strike-slip faulting in southern Europe and northern Africa. Result of a right lateral shear zone between the Appalachians and the Urals. *Geological Society of America, Bulletin*, **88,** 1305–1320.

ARZ, C. 1996. Origin and petrogenesis of igneous rocks from the Saar-Nahe-Basin (SW-Germany): Isotope, trace element and mineral chemistry. PhD thesis, Geoscience Faculty, University of Würzburg.

AWDANKIEWICZ, M. 1999*a*. Volcanism in a late Variscan intramontane trough: Carboniferous and Permian volcanic centres of the Intra-Sudetic Basin, SW Poland. *Geologia Sudetica*, **32**, 13–47.

AWDANKIEWICZ, M. 1999*b*. Volcanism in a late Variscan intramontane trough: the petrology and geochemistry of the Carboniferous and Permian volcanic rocks of the Intra-Sudetic Basin, SW Poland. *Geologia Sudetica*, **32**, 83–111.

AWDANKIEWICZ, M. 2006. Fractional crystallisation, mafic replenishment and assimilation in crustal magma chambers: geochemical constraints from the Permian post-collisional intermediate-composition volcanic suite of the North-Sudetic Basin (SW Poland). *Geologia Sudetica*, **38**, 39–61.

BACHMANN, G. H., MÜLLER, M. & WEGGEN, K. 1987. Evolution of the Molasse Basin (Germany, Switzerland). *Tectonophysics*, **137**, 77–92.

BACHTADSE, V., R. ZÄNGLEIN, J. TAIT, AND H. C. SOFFEL, 2002. Palaeomagnetism of the Permo-Carboniferous (280Ma) Jebel Nehoud ring Complex, Central Sudan. *Journal of African Earth Sciences*, **35**, 89–97.

BALTRUSCH, S. & KLARNER, S. 1993. Rotliegend-Gräben in NE-Brandenburg. *Zeitschrift der deutschen geologischen Gesellschaft*, **144**, 173–186.

BANGERT, B., STOLLHOFEN, H., LORENZ, V. & ARMSTRONG, R. 1999. The geochronology and significance of ash-fallout tuffs in the glaciogenic, Carboniferous-Permian Dwyka Group of Namibia and South Africa. *Journal of African Earth Sciences*, **29**, 33–49.

BANGERT, B., STOLLHOFEN, H., GEIGER, M. & LORENZ, V. 2000. High-resolution tephrostratigraphy, fossil record and age of Carboniferous-Permian glaciomarine mudstones within the Dwyka Group of southern Namibia. *Communications of the Geological Survey of Namibia*, **12**, 235–245.

BARABÁS, A. & BARABÁS-STUHL, Á. 1998. Permian stratigraphy of Mecsek Mountains and its surroundings. *In:* BÉRCZI, I. & JÁMBOR, Á. (eds) *Magyarország geológiai képződményeinek rétegtana*. MOL Rt., Magyar Állami Földtani Intézet, Budapest, 187–215 [in Hungarian].

BARABÁS-STUHL, Á. 1975. Contribution to the biostratigraphy of the Upper Paleozoic in Transdanubia. *Földtani Közlöny*, **105**(3), 320–340 [in Hungarian].

BARABÁS-STUHL, Á. 1981. Microflora of the Permian and Lower Triassic sediments of the Mecsek Mountains (South Hungary). *Acta Geologica Hungarica*, **24**(1), 49–97.

BARABÁS-STUHL, Á. 1993. Palynological revaluation of Lower Triassic and Lower Anisian formations of Southeast Transdanubia. *Acta Geologica Hungarica*, **36**(4), 405–458.

BARGOSSI, G. M., KLÖTZLI, U. S., MAIR, V., MAROCCHI, M. & MORELLI, C. 2004. The Lower Permian Athesian volcanic Group (AVG) in the Adige Valley between Merano and Bolzano: A stratigraphic, petrographic and geochronological outline. Poster (abstract).

BARRON, E. J. & FAWCETT, P. J. 1995. The climate of Pangaea: a review of climate model simulations of the Permian. *In:* SCHOLLE, P. A., PERYT, T. M. & ULMER-SCHOLLE, D. S. (eds) *The Permian of Northern Pangea. Paleogeography, Paleoclimates, Stratigraphy*, Vol. 1. Springer, Berlin, 37–52.

BARTH, S. & MOHR, B. A. R. 1994. Palynostratigraphically determined age of the Tregiovo sedimentary complex in relation to radiometric emplacement ages of the Atesina Volcanic Complex (Permian, Southern Alps, N. Italy). *Neues Jahrbuch für Geologie und Paläontologie, Abhandlungen*, **192**, 273–292.

BARTHEL, M. 1976. Die Rotliegendflora Sachsens. *Abhandlungen des Staatlichen Museums für Mineralogie und Geologie Dresden*, **24**, 1–190.

BECKER, L., POREDA, R. J., HUNT, A. G., BUNCH, T. E. & RAMPINO, M. 2001. Impact event at the Permian-Triassic boundary: Evidence from extraterrestrial noble gases in fullerenes. *Science,* **291**, 1530–1533.

BECQ-GIRAUDON, J. F. 1993. Problèmes de la biostratigraphie dans le Paléozoique supérieur continental (Stéphanien-Autunien) du Massif Central. *Geodinamica Acta*, **6**, 219–224.

BENEK, R., KRAMER, W., *et al.* 1996. Permo-Carboniferous magmatism of the Northeast German Basin. *Tectonophysics*, **266**, 379–404.

BERNER, R. A. & KOTHAVALA, Z. 2001. GEOCARB III: A revised model of atmospheric CO2 over Phanerozoic time. *American Journal of Science*, **301**, 182–204.

BERTHELSEN, A. 1992. Mobile Europe. *In:* BLUNDELL, D., FREEMANN, R. & MUELLER, ST. (eds) *A Continent Revealed – The European Geotraverse*. Cambridge University Press, Cambridge.

BEST, G. 1989. Die Grenze Zechstein/Buntsandstein in Nordwest-Deutschland nach Borhmessungen. *Zeitschrift der deutschen geologischen Gesellschaft*, **140**, 73–85.

BIELY, A. (ED.), BEZÁK, V., ELEČKO, M. *et al.* 1996. Geological map of Slovakia, 1:500 000. *Ministry of the Environment of Slovak Republic, Geological Survey of Slovak Republic*, Bratislava.

BIXEL, F. & LUCAS, C. 1983. Magmatisme, tectonique et sédimentation dans les fossés stéfano-permiens des Pyrénées occidentales. *Revue de Géologie Dynamique et Géographie Physique*, **24**, 329–342.

BLÈS, J. L., BONIJOLY, D., CASTAING, C. & GROS, Y. 1989. Successive post-Variscan stress fields in the European plate (French Massif Central and its borders): comparison with geodynamic data. *Tectonophysics*, **169**, 79–111.

BLÜM, W. 1989. Faciesanalyse im Rotliegenden des Nordschweizer Permokarbon-Trogs (Hochrhein-Region zwischen Basel und Laufenburg). *Eclogae Geologicae Helvetiae*, **82**, 455–489.

BOECKELMANN, K. 1985. Mikrofazies der Auernig-Schichten und Grenzland-Bänke westlich des Rudnig-Sattels (Karbon- Perm; Kamische Alpen). *Facies*, **13**, 155–174.

BOECKELMANN, K. 1988. Die Werfener Schicthen in den Karnischen Alpen und westlichen Karawanken (Österreich und Italien). PhD Thesis, RWTH Aachen.

BOECKELMANN, K. 1990. The uppermost Permian and the Permian–Triassic boundary in the Gartnerkofel 1-Core. *In:* VENTURINI, C. & KRAINER, K. (eds) *Field workshop on Carboniferous to Permian sequence of the Pramollo-Nassfeld basin (Carnic Alps)*. Proceedings (preprint) Arti Grafiche Friulani, Udine, 74–79.

BOIGK, H. & SCHÖNEICH, H. 1974. Perm und Trias und älterer Jura im Bereich des südlichen Mittelmeer-Miöse-Zone und des Rheingrabens. *In:* ILLIES, J.H. & FUCHS, K. (eds) *Approaches to Tephrogenesis.* Inter-Union Commission on Geodynamics, Scientific Report No. 8. Schweizerbart, Stuttgart, 60–71.

BOSELLINI, A. & DAL CIN, R. 1968. Sedimentologia delle Arenarie di Val Gardena. Sezione di S. Martino in Badia (Bolzano). *Bollettino della Societa Geologica Italiana*, **187**, 401–423.

BOSELLINI, A. & HARDIE, L. A. 1973. Depositional theme of a marginal marine evaporite. *Sedimentology,* **20**, 5–27.

BOUAS, S. 1987. *Les bassins permo-carbonifères sous la couverture mésozoique du Sud du Bassin de Paris*. Rapport de stage de DEA, Université de Paris Sud, Institut Français du Pètrole, réf. **35 595**.

BOY, J. A. & MARTENS, T. 1991. Zur Problematik chronostratigraphischer Korrelationen im mitteleuropäischen Rotliegend. *Newsletter Stratigraphy,* **25**, 163–192.

BRANDNER, R. 1988. The Permian–Triassic boundary in the Dolomites (Southern Alps, Italy), San Antonio section. Rare Events in Geology (IGCP 199). *Bericht Geologische Bundesanstalt*, **15**, 49–56.

BRANDNER, R. & MOSTLER, H. 1982. Der geologische Aufbau des Schlerngebietes und seiner weiteren Umgebung. *In:* MOSTLER, H. (ed.) *Exkursionsführer zur Jahrestagung der Österreichischen Geologischen Gesellschaft.* Seis am Schiern, Österreichische Geologische Gesellschaft, Innsbruck, 1–42.

BREITKREUZ, C. & KENNEDY, A. 1999. Magmatic flare-up at the Carboniferous–Permian boundary in the NE German Basin revealed by SHRIMP zircon ages. *Tectonophysics*, **302**, 307–326.

BREITKREUZ, C., CORTESOGNO, L. & GAGGERO, L. 2001. Crystal-rich mass flow deposits related to the eruption of a sublacustrine silicic cryptodome (Early Permian Collio Basin, Italian Alps). *Journal of Volcanology and Geothermal Research,* **114**, 373–390.

BRONGERSMA-SANDERS, M. 1971. Origin of major cyclicity of evaporites and bituminous rocks: an actualistic model. *Marine Geology,* **11**, 123–144.

BROUTIN, J., DOUBINGER, J., *et al.* 1990. Le renouvellement des flores au passage carbonifere-permien: approches, stratigraphique, biologique, sédimentologique. *Comptes Rendues de la Academie de la Science*, Paris, Sér. II, **311**, 1563–1569.

BROUTIN, J., CHATEAUNEUF, J. J. & MATHIS, V. 1992. The Lodève basin. *Cahiers de Micropaléontologie*, **7**, 107–139.

BRUNET, M.-F. & LE PICHON, X. 1982. Subsidence of the Paris Basin. *Journal of Geophysical Research*, **87**, 8547–8560.

BUGGISCH, W. 1974. Die Bellerophonschichthen der Reppwand (Gartnerkofel) (Oberperm, Karnische Alpen). Untersuchungen zur Fazies und Geochemie. *Carinthia II,* **164**, 17–26.

BUGGISCH, W. 1978. Die Grödener Schicthen (Perm, Südalpen). Sedimentologische und geochemische Untersuchungen zur Unterscheidung mariner und kontinentaler Sedimente. *Geologische Rundschau*, **67**, 149–180.

BUGGISCH, W. & NOE, S. 1986. Upper Permian and Permian- Triassic boundary of the Carnia (Bellerophon Formation, Tesero horizon, northern Italy). *Memorie Società Geologica Italiana*, **34**, 91–106.

BUGGISCH, W., FLÜGEL, E., LEITZ, F. & TIETZ, G.-F. 1976. Die fazielle und paläogeographische Entwicklung im Perm der Karnischen Alpen und in den Randgebieten. *Geologische Rundschau*, **65**, 649–690.

BULTYNCK, P., GEUKENS, F. & SMOLDEREN, A. 2001. Permian lithostratigraphic units, Malmédy Graben (Belgium). *In:* BULTYNCK, P. & DEJONGHE (eds) Guide to a revised lithostratigraphic scale of Belgium. *Geologica Belgica*, **4**, 105–106.

BUŁA, Z. 2001. Lithology and stratigraphy of Carboniferous and Lower Permian clastics from the Tarnawa 1 borehole section. (In Polish with English summary). *In:* MATYJA, H. (ed.) *Paleozoic basement of the central segment of the Polish Outer Carpathians.* Prace Panstwowego Instytutu Geologicznego, **CLXXIV**, 61–65.

BUSER, S. 1974. Neue Feststellungen im Perm der westlichen Karawanken. *Carinthia II*, **164**, 27–37.

BUSSY, F. R. & CADOPPI, P. 1996. U-Pb zircon dating of granitoids from the Dora-Maira massif (western Italian Alps). *Schweizerische Mineralogische und Petrographische Mitteilungen*, **76**, 217–233.

BÜTHE, F. & WACHENDORF, H. 1997. Die Rotliegend-Entwicklung des Illfelder Beckens und des Kyffhäusers: Pull-Apart-Becken und Rhomb-Horst. *Zeitschrift für Geologische Wissenschaften*, **25**, 291–306.

BUTTERSACK, E. & BOECKELMANN, K. 1984. Palaeoenvironmental evolution during the Upper Carboniferous and the Permian in the Schulter -Trogkofel area (Carnic Alps, northern Italy). *Jahrbuch der geologischen Bundesanstalt*, **126**, 349–358.

CASSINIS, G. & NERI, C. 1992. Sedimentary and paleotectonic evolution of some Permian continental basins in the central Southern Alps, Italy. *Cuadernos de Geologia Ibérica*, **16**, 145–176.

CASSINIS, G., TOUTIN-MORIN, N. & VIRGILI, C. 1995. A general outline of the Permian continental basins in southwestern Europe. *In:* SCHOLLE, P. A., PERYT, T. M. & ULMER-SCHOLLE, D. S. (eds) *The Permian of Northern Pangea 2. Sedimentary Basins and Economic Resources*. Springer Verlag, Berlin, 137–157.

CHATEAUNEUF, J.-J. & FARJANEL, G. (eds) 1989. *Synthèse géologique des bassins permiens français*. Mémoires de Bureau de recherches géologiques et minières, **128**.

CHUMAKOV, N. M. 1994. Evidence of Late Permian glaciation in the Kolyma River Basin: a repercussion of the Gondwanan glaciations in northeast Asia? *Stratigraphy and Geological Correlation*, **2**, 130–150.

CHUVASHOV, B. I., DUPINA, I. G. V. MIZENS, G. A. & CHERNYKH, V. V. 1990. *Key sections of the Upper Carboniferous and Lower Permian of the western slope of the Urals and Preurals*. Russian Academy of Sciences, Ural Branch, Sverdlovsk, pp 3–367 [in Russian].

CLARK, D. N. 1986. The distribution of porosity in Zechstein carbonates. *In:* BROOKS, J., GOFF, J. C. & VAN HOORN, B. (eds) *Habitat of Palaeozoic gas in NW Europe.* Geological Society, London, Special Publications, **23**, 121–149.

CONTI, M. A., LEONARDI, G., MARIOTTI, N. & NICOSIA, U. 1977. Tetrapod footprints of the "Val Gardena Sandstone" (north Italy).

Their paleontological, stratigraphic and paleoenvironmental meaning. *Paleontologia Italiana*, **170**, 1–91.

Cortesogno, L., Cassinis, G., *et al.* 1998. The Variscan post-collisional volcanism in Late Carboniferous-Permian sequences of Ligurian Alps, Southern Alps and Sardinia (Italy). *Lithos*, **45**, 305–328.

Courel, L. 1987. Les bassins houillers du Paléozoique supérieur continental du Massif Central francais; dynamique sédimentaire. *Annales de la Societe Geologique Nord*, **106**, 229–237.

Crowell, J. C. 1983. Ice ages recorded on Gondwanan continents. *Transactions of the Geological Society of South Africa*, **86**, 237–262.

Crowell, J. C. 1995. The ending of the Late Paleozoic Ice Age during the Permian Period. *In:* Scholle, P. A., Peryt, T. M. & Ulmer-Scholle, D. S. (eds) *The Permian of Northern Pangea. Paleogeography, Paleoclimates, Stratigraphy*, Vol. 1. Springer, Berlin, 62–74.

Czapowski, G. 1987. Sedimentary facies in the Oldest Rock Salt (Na1) of the Leba elevation (northern Poland). *Lecture Notes in Earth Science*, **10**, 207–224.

Czapowski, G. 1993. Facies characteristics and distribution of the Zechstein (Upper Permian) salt deposits of PZ3 (Leine) cycle in Poland. *Bulletin of the Polish Academy of Sciences, Earth Science*, **41**, 229–237.

Czapowski, G., Antonowicz, L. & Peryt, T. M. 1990. Facies and palaeogeography of the Zechstein (Upper Permian) Older Halite (Na2) in Poland. *Bulletin of the Polish Academy of Sciences, Earth Science*, **39**, 45–55.

Czapowski, G., Peryt, T. M. & Antonowicz, L. 1993. Facies and palaeogeography of the Zechstein (Upper Permian) Oldest Halite (Na1) in Poland. *Bulletin of the Polish Academy of Sciences, Earth Science*, **41**, 217–227.

Dadlez, R. 1993. *Teisseyre-Tornquist tectonic zone in Poland*. Publications of the Institute of Geophysics, Polish Academy of Sciences, A20, 49–51.

Dadlez, R. 1997. Epicontinental basins in Poland: Devonian to Cretaceous – relationship between the crystalline basement and sedimentary infill. *Geological Quarterly*, **41**, 419–432.

Dadlez, R. 1998. *Map of the Zechstein-Mesozoic in Polish Lowlands*. Warszawa, Panstwowny Institut Geologiczny.

Dadlez, R., Kowalczewski, Z. & Znosko, J. 1994. Some key problems of the pre-Permian tectonics of Poland. *Geological Quarterly*, **38**, 169–190.

Dadlez, R., Narkiewicz, M., Stephenson, R. A., Visser, M. T. M. & Van Wees, J. D. 1995. Tectonic evolution of the Mid-Polish Trough: modelling implications and significance for central European geology. *Tectonophysics*, **252**, 179–196.

D'Amico, C. & Segatta, G. 1979. Le lava afiriche riolitiche-trachiriolitiche dei Lagorai sud-occidentale (Trentino) nel sistema vulcanico atesino dei Perrniano, gruppo inferiore. *Mineralogica Petrographica Acta*, **23**, 239–255.

De Craen, M. & Swennen, R. 1992. Sedimentology and diagenesis of the ankeritized basal Zechstein conglomerate in the Campine Basin (Bree borehole, NE Belgium). *Geologie en Mijnbouw*, **71**, 145–160.

Demaret-Fairon, M., Dusar, M. & Streel, M. 1985. Permian stratigraphy in the Campine Basin. *Annales de la Société géologique de Belgique*, **108**, 411–412.

Deroin, J. P., Bonin, B., *et al.* 2001. The Permian of Southern France: An overview. *Natura Bresciana, Annales della Museo Civico di Scienze Naturali*, 189–202.

Deutsche Stratigraphische Kommission 2002. *Stratigraphische Tabelle von Deutschland 2002*. GeoForschungsZentrum, Potsdam; Forschungsinstitut Senckenberg, Frankfurt am Main.

Diebold, P., Naef, H. & Ammann, M. 1992. *Zur Tektonik der zentralen Nordschweiz*. Geologische Berichte, **14**, Landesgeologie und-hydrologie, Bern.

Dill, H. & Hartkopf-Fröder, C. 1990. Stratigraphie und Lithologie des Permokarbons im Weidener Becken (NO-Bayern, B.R. Deutschland). *Zeitschrift der deutschen geologischen Gesellschaft*, **141**, 31–47.

Dittrich, D. 1996. Unterer Buntsandstein und die Randfazies des Zechsteins in der nördlichen Pfälzer Mulde (Exkursionen C1 am 11. und C2 am 12. April 1996). *Jahresbericht und Mitteilungen des Oberrheinischen Geologischen Vereines*, N.F., **78**, 71–94.

Djarar, L., Wang, H., Guiraud, M., Clermonte, J., Courel, L., Dumain, M. & Laversanne, J. 1996. Le bassin stéphanien des Cévennes (Massif Central): un exemple de relation entre sédimentation et tectonique extensive tardi-orogénique dans la chaine varisque. *Geodinamica Acta*, **9**, 193–221.

Dostal, J., Vozár, J., Keppie, J. D. & Hovorka, D. 2003. Permian volcanism in the Central Western Carpathians (Slovakia): Basin-and-Range type rifting in the southern Laurussian margin. *International Journal of Earth Sciences (Geologische Rundschau)*, **92**, 27–35.

Dusar, M., Langenaeker, V. & Wouters, L. 2001. Permian-Triassic-Jurassic lithostratigraphic units in the Campine Basin and the Roer Valley Graben (NE Belgium). *In:* Bultynck & Dejonghe (eds) Guide to a revised lithostratigraphic scale of Belgium. *Geologica Belgica*, **4**, 107–112.

Dyjaczynski, K., Górski, M., Mamczur, S. & Peryt, T. M. 2001. Reefs in the basinal facies of the Zechstein Limestone (Upper Permian) of Western Poland. *Journal of Petroleum Geology*, **24**, 265–285.

Engel, H. 1988. Neue Erkenntnisse zum geologischen Werdegang des Saarkarbons. *Documentae Naturae*, **44**, 8–22.

Erwin, D. H. 1995. The End-Permian mass extinction. *In:* Scholle, P. A., Peryt, T. M. & Ulmer-Scholle, D. S. (eds) *The Permian of Northern Pangea*, Vol. 1. Springer-Verlag, Berlin, 20–34.

Farabegoli, E. & Viel, G. 1982. *Il Permo-Scitico delle Alpi Meridionali*. Relazione finale Contr CREST 092.79.7 MPPI. Sallim Pertulosa.

Farabegoli, E., Flügel, E., Levanti, D., Noe, S., Pasini, M., Perri, M. C. & Venturini, C. 1986. The Tarvisio Breccia and Val Gardena Sandstone: Lithostratigraphic and Paleoenvironments. The Permian–Triassic Boundary – Carnia and Comelico. In: *Field Conference on Permian and Permian–Triassic Boundary in the South-Alpine Segment of the Western Tethys, Excursion Guidebook*. Societa di Geologica Italiana, Commerciale Pavese, Pavia, 55–72.

Farley, K. A., Mukhopadhyay, S., Isozaki, Y., Becker, L. & Poreda, R. J. 2001. An extraterrestrial impact at the Permian-Triassic boundary? *Science*, **293**, 2343.

Faure, M. 1995. Late orogenic Carboniferous extensions in the Variscan French Massif Central. *Tectonics*, **14**, 132–153.

Feist, R., Echtler, H., Galtier, J. & Mouthier, B. 1994. Biostratigraphy and dynamics of the nonmetamorphic sedimentary record. *In:* Keppie, J. D. (ed.) *Pre-Mesozoic Geology in France and Related Areas*. Springer-Verlag, Berlin, 289–304.

Fijalkowska, A. & Peryt, T. M. 1995. The PZ2 cycle in the Zechstein of the North Sudetic Trough (SW Poland). *Przeglad Geologiczny*, **43**, 31–34 [in Polish].

Filipović, I., Jovanović, D., Sudar, M., Pelikán, P., Kovács, S., Less, G. & Hips, K. 2003. Comparison of the Variscan–Early Alpine evolution of the Jadar Block (NW Serbia) and "Bükkium" (NE Hungary) terranes; some paleogeographic implications. *Slovak Geological Magazine*, **9**, 3–21.

Fischer, F. 1991. *Das Rotliegend des ostthüringsch-westsächsischen Raumes (Vorerzgebirgs-Senke, Nordwestsächsischer Vulkanitkomplex, Geraer Becken)*. PhD thesis, TU Bergakademie Freiberg.

Flügel, E. 1968. Bericht über fazielle und stratigraphische Untersuchungen im Perm der Karnischen Alpen. *Carinthia II*, **158**(78), 38–65.

Flügel, E. 1974. Fazies-Interpretation der unterpermischen Sedimente in den Karnischen Alpen. *Carinthia II*, **164**(84), 43–62.

Flügel, E. 1980. Die Mikrofazies der Kalke in den Trogkofel- Schichten der Karnischen Alpen. *In:* Flügel, E. (ed) Die Trogkofel-Stufe im Unterperm der Karnischen Alpen. *Carinthia II, Sonderheft* **36**, 51–99.

Flügel, E. 1981. Lower Permian Tubiphytes/Archaeolithoporella buildups in the Southem Alps (Austria and Italy). *In:* Toomey, D. F. (ed.) *European Fossil Reef Models*. Society of Economic Paleontologists and Mineralogists, Special Publications, **30**, 143–160.

Flügel, E., Fohrer, B., Forke, H., Krainer, K. and Samankassou E. 1997. Cyclic sediments and algal mounds in the Upper Paleozoic of the Carnic Alps (Excursion B 1 of the 18th IAS Regional European Meeting of Sedimentology, Heidelberg). *Gaea heidelbergensis*, **4**, 79–100.

Francis, E.H. 1988. Mid-Devonian to early Permian volcanism in Old World. *In:* Harris, A. L. & Fettes, D. J. (eds) *The Caledonian-Appalachian Orogen*. Geological Society, London, Special Publications, **38**, 573–584.

Frank, W. 1972. Permoskyth im Penninikum der Hohen Tauern. *Verhandlungen der geologischen Bundesanstalt*, **1972**, 151–153.

FRISCH, U. & KOCKEL, F. 2003. *Der Bremen-Knoten im Strukturnetz Nordwestdeutschlands.* Berichte aus dem Fachbereich der Geowissenschaften der Universität Bremen, **223**.

FRITZ, A. & BOERSMA, M. 1984. Fundberichte über Pflanzenfossilien aus Kärnten 1984. Beitrag 8: Grenzlandbänke (Rudnigsattel, Unterperm), Karnische Alpen. *Carinthia II,* **174**(94), 59–69.

FRITZ, A. & BOERSMA M. 1987*a*. Fundberichte über Pflanzenfossilien aus Kärnten 1987, Beitrag 15: Wunderstätten (Unterperm), St. Pauler Berge. *Carinthia II*, **177**(97), 381–394.

FRITZ, A. & BOERSMA M. 1987*b*. Fundberichte über Pflanzenfossilien aus Kärnten 1987, Beitrag 16: Kötschach, Gailtaler Alpen (Unterperm). *Carinthia II*, **177**(97), 395–407.

FRITZ, A. & BOERSMA M. 1988. Fundberichte über Pflanzenfossilien aus Kärnten 1988, Beitrag 17: Ulrichsberg (Unterperm), Kärnten. *Carinthia II*, **178**(98), 437–446.

FRITZ, A. & KRAINER, K. 1999. Eine Rotliegend-Flora aus dem Grödener Sandstein bei Mölten, Südtirol. *Der Schlern*, **73**(10), 637–650.

FRITZ, A. & KRAINER, K. 2004. Pflanzenfossilien aus dn Grenzlandbänken der Rattendorfer Alm (Kärnten, Karnische Alpen). *Carinthia II*, **194**(114), 445–454.

FRITZ, A., BOERSMA, M. & KRAINER, K. 1990. Steinkohlenzeitliche Pflanzenfossilien aus Kärnten. *Carinthia II,* **49**, 1–189.

FÜCHTBAUER, H. & PERYT, T. M. (eds) 1981. *Zechstein Basin with Emphasis on Carbonate Sequences*. Contributions to Sedimentology, **9**, Schweizerbart, Stuttgart.

FUGLEWICZ, J. 1987. Upper Permian assemblages of Gondwana miospores in sediments of lower Buntsandstein. *Przeglad Geologiczny*, **11**, 583–586.

FÜLÖP, J. 1994. *Geology of Hungary, Paleozoic II*. Akadémiai Kiadó, Budapest, 222–229 [in Hungarian].

GASIEWICZ, A. 1990. The Zechstein Platy Dolomite (Ca3) deposition in the Puck Bay region (northern Poland). *Przeglad Geologiczny*, **38**, 187–195 [in Polish with English summary].

GASIEWICZ, A. & PERYT, T. M. 1994. Biolaminites at the Zechstein (Upper Permian) Platy Dolomite (Ca3)-Main Anhydrite (A3) boundary: implications for evolution of an evaporite basin. *Beiträge zur Paläontologie,* **19**, 91–101.

GAST, R. 1988. Rifting im Rotliegenden Niedersachsens. *Die Geowissenschaften*, **6**, 115–122.

GAST, R. 1991. The perennial Rotliegend Saline Lake in NW Germany. *Geologisches Jahrbuch*, **A119**, 25–59.

GAST, R. 1993. Sequenzanalyse von äolischen Abfolgen im Rotliegenden und deren Verzahnung mit Küstensedimenten. *Geologisches Jahrbuch,* **131A**, 117–139.

GAST, R. & GUNDLACH, T. 2006. Permian strike-slip and extensional tectonics in Lower Saxony, Germany. *Zeitschrift der deutschen Gesellschaft für Geowissenschaften*, **157**, 41–56.

GAUPP, R., FORSTER, C. & BITZER F. 1997. *Die Elbe-Subgruppe (Oberrotliegend) im NW-deutschen Becken – fazielle Entwicklung und Zyklostratigraphie*. Technische Universität Bergakademie Freiberg. XLVIII. Berg- und Hüttenmännischer Tag, Kolloquium 1. Stratigraphie, Sedimentation und Beckenentwicklung im Karbon und Permian Abstracts, 16–17.

GAUPP, R., GAST, R. & FORSTER, C. 2000. Late Permian playa lake deposits of the Southern Permian Basin (Central Europe). *In:* GIERLOWSKI-KORDESCH, E. H. & KELTS, K. R. (eds) *Lake Basins Through Space and Time.* AAPG, Studies in Geology, **46**, 75–86. Tulsa, Oklahoma.

GEBHARDT, U. & SCHNEIDER, J. 1993. Paläoökologie und Paläobiogeographie "mariner" Kalkalgen im intrakontinental-lakustrische Niederhäslich-Kalk des intramontanen Döhlen Beckens (Unterrotliegend, Assel, Elbe-Zone). *Freiberger Forschungsheft*, **C450**.

GEBHARDT, U. SCHNEIDER, J. & HOFFMANN, N. 1991. Modelle zur Stratigraphie und Becnenentwicklung im Rotliegenden der Norddeutschen Senke. *Geologische Jahrbuch*, **127**, 405–427.

GELUK, M. C. 1999. Late Permian (Zechstein) rifting in the Netherlands: models and implications for petroleum geology. *Petroleum Geoscience*, **5**, 189–199.

GELUK, M. C. 2000. Late Permian (Zechstein) carbonate facies maps, the Netherlands. *Netherlands Journal of Geosciences/Geologie en Mijnbouw*, **79**, 17–27.

GELUK, M. C. 2007. Permian. *In:* WONG, TH. E., BATJES, D. A. J., DE JAGER, J. (eds) *Geology of the Netherlands*. Royal Dutch Academy of Science and Arts, Amsterdam.

GELUK, M. C., PLOMP, A. & VAN DOOM, T. H. M. 1996. Development of the Permo-Triassic succession in the basin fringe area, southern Netherlands. *In:* RONDEEL, H. E., BATJES, D. A. J., NIEUWENHUIJS, W. H. (eds) *Geology of Gas and Oil Under the Netherlands*. Kluwer, Dordrecht, 57–78.

GEORGE, G. T. & BERRY, J. K. 1993. A new lithostratigraphy and depositional model for the Upper Rotliegend of the UK Sector of the Southern North Sea. *In:* NORTH, C. P. & PROSSER, D. J. (eds) *Characterization of Fluvial and Aeolian Reservoirs*. Geological Society, London, Special Publications, **73**, 291–319.

GEORGE, G. T. & BERRY, J. K. 1997. Permian (Upper Rotliegend) synsedimentary tectonics, basin development and palaeogeography of the Southern North Sea. *In:* ZIEGLER, K., TURNER, P. & DAINES, S. R. (eds) *Petroleum Geology of the Southern North Sea: Future Potential*. Geological Society, London, Special Publications, **123**, 31–61.

GIBBS, M. T., REES, P. M., KUTZBACH, J. E., ZIEGLER, A. M., BEHLING, P. J. & ROWLEY, D. B. 2002. Simulations of Permian climate and comparison with climate-sensitive sediments. *Journal of Geology*, **110**, 33–55.

GLENNIE, K. W. 1990. Outline of North Sea history and structural framework. *In:* GLENNIE, K. W. (ed.) *Introduction to the Petroleum Geology of the North Sea*, 3rd edn. Blackwell Science, London, 34–77.

GLENNIE, K. W. 1998. Lower Permian-Rotliegend. *In:* GLENNIE, K. W. (ed.) *Petroleum Geology of the North Sea*. Blackwell Science, Oxford, 137–174.

GLENNIE, K. W. 2005. Regional tectonics in relation to Permo-Carboniferous hydrocarbon potential, Southern North Sea Basin. *In:* COLLINSON, J. D., EVANS, D. J., HOLLIDAY, D. W. & JONES, N. S. (eds) *Carboniferous Hydrocarbon Geology: The Southern North Sea and Surrounding Onshore Areas.* Occasional Publications Series, **7**, Yorkshire Geological Society, 1–12.

GLENNIE, K. W. & BULLER, A. T. 1983. The Permian Weissliegend of NW Europe: the partial deformation of eolian sands caused by the Zechstein transgression. *Sedimentary Geology*, **35**, 43–81.

GLENNIE, K. W., HIGHAM, J. & STEMMERIK, L. 2003. Permian. *In:* EVANS, D., GRAHAN, C., ARMOUR, A. & BATHURST, P. (eds) *The Millenium Atlas: Petroleum Geology of the Central and Northern North Sea*. Geological Society, London, 91–103.

GOODALL, I. G., HARWOOD, G. M., KENDALL, A. C., MCKIE, T. & TUCKER, M. E. 1993. Discussion on sequence stratigraphy of carbonate-evaporite basins: models and application to the Upper Permian (Zechstein) of Northeast England and adjoining North Sea. *Journal of the Geological Society, London,* **149**, 1050–1054.

GRADSTEIN, F. & OGG, J. 1996. A Phanerozoic time scale. *Episodes*, **19**, 3–5.

GRADSTEIN, F., OGG, J. & SMITH, A. G. 2004. *A Geologic Time Scale*. Cambridge University Press, Cambridge.

GRALLA, P. 1988. Das Oberrotliegende in NW-Deutschland. Litostratigraphie und Faziesanalyze. *Geologisches Jahrbuch*, **106A**, 3–59.

HAAS, J., KOVÁCS, S., KRYSTYN, L. & LEIN, R. 1995. Significance of Late Permian-Triassic facies zones in terrane reconstructions in the alpine-North Pannonian domain. *Tectonophysics*, **242**, 19–40.

HANEKE, J. 1987. *Zur Genese und stratigraphisch-tektonischen Stellung eines permokarbonen Rhyolith-Domes im Saar-Nahe-Gebiet (SW-Deutschland).* Pollichia Buch **10**, Pfalzmuseum für Naturkunde, Bad Dürkheim.

HANEKE, J., GÄDE, C.W. & LORENZ, V. 1979. Zur stratigraphischen Stellung der rhyolitischen Tuffe im Oberrotliegenden des Saar-Nahe-Gebietes und der Urangehalt des Kohlen-Tuff-Horizontes an der Kornkiste bei Schallodenbach/Pfalz. *Zeitschrift der deutschen geologischen Gesellschaft*, **130**, 535–560.

HARTKOPF-FRÖDER, CH., WOOD, G. D. & KRAINER, K. 2001. Palynology of the Permian Bolzano Volcanic Complex, Southern Alps, Italy, Part 1: Miospore preservation, quantitative spore color and quantitative fluorescence microscopy. *In:* GOODMAN, D. K. AND CLARKE, R. T. (eds) *Proceedings of the IX International Palynological Congress*, Houston, Texas, U.S.A., 1996. American Association of Stratigraphic Palynologists Foundation, 79–97.

HENK, A. 1992. Mächtigkeit und Alter der erodierten Sedimente im Saar-Nahe-Becken (SW-Deutschland). *Geologische Rundschau*, **81**, 323–331.

HENK, A. 1993. Late orogenic basin evolution in the Variscan Internides:

the Saar-Nahe Basin, southwest Germany. *Tectonophysics*, **223**, 273–290.

HENK, A. 1999. Did the Variscides collapse or were they torn apart? A quantitative evaluation of the driving forces for postconvergent extension in central Europe. *Tectonics*, **18**, 774–792.

HERTLE, M. & LITTKE, R. 2001. Coalification pattern and thermal modelling of the Permo-Carboniferous Saar-Nahe Basin (SW-Germany). *International Journal of Coal Geology*, **42**, 273–296.

HIETE, M., BERNER, U., HEUNISCH, C. & RÖHLING, H.-G. 2006. A high-resolution inorganic geochemical profile across the Zechstein-Buntsandstein boundary in the North Germany Basin. *Zeitschrift der deutschen Gesellschaft für Geowissenschaften*, **157**, 77–106.

HIRST, D.M. & DUNHAM, K.C. 1963. Chemistry and petrography of the Marl Slate of S.E. Durham, England. *Economic Geology*, **58**, 912–940.

HOCHULI, P. A. 1985. Palynostratigraphische Gliederung und Korrelation des Permo-Karbon der Nordschweiz. *Eclogae Geologicae Helvetiae*, **78**, 719–831.

HOFFMANN, U. 2000. *Proklastite und Silicite im Rotliegend des Döhlen-Becken - Stratigraphie, Genese und Paläontologie*. MSc thesis, TU Bergakademie Freiberg.

HOFFMANN, N., KAMPS, H.-J. & SCHNEIDER, J. 1989. Neuerkentnisse zur Biostratigraphie und Paläodynamik des Perms in der Nordostdeutschen Senke - ein Diskussionsbeitrag. *Zeitschrift für angewandte Geologie*, **35**, 189–207.

HOLLINGWORTH, N. & PETTIGREW, T. 1988. *Zechstein Reef Fossils and their Palaeoecology*. Palaeontological Association, University Printing House, Oxford.

HOLSER, W. T. & SCHÖNLAUB, H. P. (eds) 1991. *Permian–Triassic boundary in the Carnic Alps of Austria (Gartnerkofel region)*. Abhandlungen der Geologischen Bundesanstalt, **45**.

HOLUB, V. M. & TASLER, R. 1982. Explanatory notes to the lithotectonic molasse profile of the Intra-Sudetic (lower Silesian) Basin, Bohemian Part (CSSR, Bohemian Massif and Sudety Mts.; Dinantian – Lower Triassic) (Comment to Annex 24). *In:* LÜTZNER, H. & SCHWAB, G. (eds) *Tectonic Regime of Molasse Epochs*. Veröffentlichung der Zentralinsitut für Physik der Erde, Potsdam, **66**, 279–300.

HOMANN, W. 1972. Unter- und tief-mittelpermische Kalkalgen aus den Rattendorfer Schichten, dem Trogkofel-Kalk und dem Treßdorfer Kalk der Karnischen Alpen (Österreich). *Senckenbergiana Lethaea*, **53**, 135–313.

HOTH, K., RUSBÜLT, J., ZAGORA, K., BEER, H. & HARTMANN, O. 1993. Die tiefen Bohrungen im Zentralabschnitt der Mitteleuropäischen Senke - Dokumentation für den Zeitabschnitt 1962–1990. *Schriftenreihe Geowissenschaft*, **2**.

HUG, N. 2004. Sedimentgenese und Paläogeographie des höheren Zechstein bis zur Basis des Buntsandstein in der Hessischen Senke. *Geologische Abhandlungen Hessen*, **106**.

HUG, N. & GAUPP, R. 2006. Palaeogeographic reconstruction in unbedded and unfossiliferous red bed sediments by means of genetically-related correlation: Results from the upper part of the German Zechstein (Upper Permian). *Zeitschrift der deutschen Gesellschaft für Geowissenschaften*, **157**, 107–120.

IZART, A., VASLET, D., *et al.* 1998. Stratigraphic correlations between the continental and marine Tethyan and Peri-Tethyan basins during the Late Carboniferous and the Early Permian. *In:* CRASQUIN-SOLEAU, S., IZART, A., VASLET, D. & DE WEVER, P. (eds) Peri-Tethys: stratigraphic correlations 2. *Geodiversitas*, **20**(4), 521–595.

JACKOWICZ, E. 1994. Permian volcanic rocks from the northern part of the foresudetic monocline. *Prace Panstwowego Instytutu Geologicznego*, **145**.

JACKSON, J. A. 1997. *Glossary of Geology*. American Geological Institute, Alexandria, Virginia, USA.

JÁMBOR, Á. 1964. Carboniferous formations in the zone between Mecsek and Villány Mountains. *Magyar Állami Földtani Intézet Évi Jelentése 1967-ről*, 215–221 [in Hungarian].

JAWOROWSKI, K. 2002. Geotectonic significance of Carboniferous deposits NW of the Holy Cross Mts (central Poland). *Geological Quarterly*, **46**, 267–280.

JENYON, M. K. & CRESSWELL, P. M. 1987. The southern Zechstein salt basin of the British North Sea, as observed in regional seismic traverses. *In:* BROOKS, J. & GLENNIE, K. W. (eds) *Petroleum Geology of NW Europe*. Graham & Trotman, London, 277–292.

JOHNSON, H., WARRINGTON, G. & STOKER, S. J. 1994. Permian and Triassic of the Southern North Sea. *In:* KNOX, R. W. & CORDEY, W. G. (eds) *Lithostratigraphic Nomenclature of the North Sea*. British Geological Survey, Nottingham.

JOWETT, E. C. & JARVIS, G. T. 1984. Formation of foreland rifts. *Sedimentary Geology*, **41**, 51–72.

JOWETT, E. C., RYDZEWSKI, A. & JOWETT, R. J. 1987. The Kupferschiefer Cu-Ag ore deposits in Poland: a reappraisal of the evidence of their origin and presentation of a new genetic model. *Canadian Journal of Earth Sciences*, **24**, 2016–2037.

KAHLER, F. 1986. Ein Normalprofil der Fusuliniden-Stratigraphie im Oberkarbon und Unterperm der Karnischen Alpen. *Carinthia II*, **176**(96), 1–17.

KAHLER, F. & KAHLER, G. 1980. Fusuliniden aus den Kalken der Trogkofel-Schicthen der Karnischen Alpen. *In:* FLÜGEL, E. (ed.) Die Trogkofel-Stufe im Unterperm der Karnischen Alpen. *Carinthia II*, **36**, 183–254.

KARNKOWSKI, P. H. 1981. The current lithostratigraphy subdivision of the Rotliegendes in Poland and proposition of its formalization. *Kwartalnik Geologiczny*, **25**, 59–66 [in Polish with English abstract].

KARNKOWSKI, P. H. 1999. *Origin and evolution of the Polish Rotliegend Basin*. Polish Geological Institute, Special Papers, **3**, 1–93.

KEMPTER, E. H. K. 1987. Fossile Maturität, Paläothermogradienten und Schicktlücken in der Bohrung Weiach im Licthe von Modellberechnungen der thermischen Maturität. *Eclogae Geologicae Helvetiae*, **80**, 543–552.

KIERSNOWSKI, H. 1991. Litostratygrafia permu północno-wschodniego obrzeżenia Górnośląskiego Zagłębia Węglowego - nowa propozycja. *Przeglad Geologiczny*, **4**, 198–203.

KIERSNOWSKI, H. 1995. Origin and development of the Late Palaeozoic, Eastern Fore-Sudetic Basin. *Materialy Sesji*, **64**, Zjazdu PTG, 19–35.

KIERSNOWSKI, H. 1997. Depositional development of the Polish Upper Rotliegend Basin and evolution of its sediment source areas. *Geological Quarterly*, **41**, 433–456.

KIERSNOWSKI, H. 1998. Architektura depozycyjna basenu czerwonego spągowca. *In:* NARKIEWICZ, M. (ed.) *Analiza basenów sedymentacyjnych Niżu Polskiego*. Prace Panstwowego Instytutu Geologicznego, **CLXV**, 113–128.

KIERSNOWSKI, H. 2001. Permian-Triassic deposits of the Liplas-Tarnawa Basin. *In:* MATYJA, H. (ed.) *Paleozoic Basement of the Central Segment of the Polish Outer Carpathians*. Prace Panstwowego Instytutu Geologicznego, **CLXXIV**, 87–100 [in Polish with English summary].

KIERSNOWSKI, H. & BUNIAK, A. 2006. Evolution of the Rotliegend Basin of northwestern Poland. *Geological Quarterly*, **50**, 119–138.

KIERSNOWSKI, H. & MALISZEWSKA, A. 1985. Grubookruchowe osady czerwonego spągowca w rejonie siewiersko-olkuskim w świetle nowych badań. *Przegląd Geologiczny*, **4**, 181–192.

KIERSNOWSKI, H., PAUL, J., PERYT, T. M. & SMITH, D. 1995. Facies, palaeogeography and sedimentary history of the Southern Permian Basin in Europe. *In:* SCHOLLE, P., PERYT, T. M. & ULMER-SCHOLLE, D. S. (eds) *The Permian of Northern Pangea 2*, 119–136. Springer-Verlag, Berlin.

KOCKEL, F. 2002. Rifting processes in NW Germany and the German North Sea sector. *Netherlands Journal of Geosciences/Geologie en Mijnbouw*, **81**, 149–158.

KÖNINGER, S. & LORENZ, V. 2002. Geochemistry, tectonomagmatic origin, and chemical correlation of altered Carboniferous-Permian fallout ash tuffs in SW-Germany. *Geological Magazine*, **139**, 541–558.

KÖNIGER, S., LORENZ, V., STOLLHOFEN, H. & ARMSTRONG, R. A. 2002. Origin, age and stratigraphic significance of distal fallout ash tuffs from the Carboniferous-Permian continental Saar-Nahe Basin (SW-Germany). *International Journal of Earth Sciences (Geologische Rundschau)*, **91**, 341–356.

KÖRNER, F., SCHNEIDER, J. W., HOERNES, S., GAND, G. & KLEEBERG, R. 2003. Climate and continental sedimentation in the Permian of the Lodève Basin (Southern France). *Boletin Societa Geologica Italiana*, **2**, 185–191.

KORSCH, R. J. & SCHÄFER, A. 1995. The Permo-Carboniferous Saar-Nahe Basin, south-west Germany and north-east France: basin formation and deformation in a strike-slip regime. *Geologische Rundschau*, **84**, 293–318.

Kossow, D., Krawczyk, C. M., McCann, T., Strecker, M. & Negendank, J. F. M. 2000. Style and evolution of salt pillows and related structures in the northern part of the Northeast German Basin. *International Journal of Earth Sciences*, **89**, 652–664.

Kovács, S., Less, Gy., Piros, O., Reti, Zs. & Roth, L. 1989. Triassic Formation of the Aggtelek-Rudabánya Mts. (Northeastern Hungary). *Acta Geologica*, **31**, 31–63.

Kowalczyk, G. 1983. *Das Rotliegende zwischen Taunus und Spessart.* Geologische Abhandlungen Hessen, **84**.

Kozur, H. 1978. Beiträge zur Stratigraphie des Perm. Teil III (1). Zur Korrelation der überwiegend kontinental Ablagerungen des obersten Karbons und Perms von Mittel- und Westeuropa. *Freiberger Forschungshefte*, **C 342**, 117–142.

Kozur, H. 1980. Beiträge zur Stratigraphie des Perm. Teil III (2). Zur Korrelation der überwiegend kontinental Ablagerungen des obersten Karbons und Perms von Mittel- und Westeuropa. *Freiberger Forschungshefte*, **C 348**, 69–172.

Kozur, H. 1988. The age of the Central European Rotliegendes. *Zeitschrift für geologische Wissenschaften*, **16**, 907–915.

Krainer, K. 1984. Sedimentologische Untersuchungen an permischen und untertriadischen Sedimenten des Stangalm-Mesozoikums (Kärnten/Österreich). *Jahrbuch der geologischen Bundesanstalt*, **127**, 159–179.

Krainer, K. 1987*a*. Das Perm der Gurktaler Decke: eine sedimentologische Analyse. *Carinthia II*, **177**(97), 49–92.

Krainer, K. 1987*b*. Zusammensetzung und fazielle Entwicklung des Alpinen Buntsandsteins und der Werfener Schicthen im westlichen Drauzug (Kärnten/Osttirol). *Jahrbuch der geologischen Bundesanstalt,*, **130,** 61–91.

Krainer, K. 1989. Composition and evolution of Lower Permian molasse sediments (Ponte Gardena Conglomerate) at the base of the Bolzano Volcanic Complex, Southern Alps (N Italy). *Neues Jahrbuch für Geologie und Paläontologie Monatshefte*, **1989**, 400–424.

Krainer, K. 1990. Fazielle und sedimentpetrographische Untersuchungen im Perm des Drauzuges. *Mitteilungen der Östereichischen Geologischen Gesellschaft*, **82**, 49–78.

Krainer, K. 1993. Late- and Post-Variscan Sediments of the Eastern and Southern Alps. *In:* Von Raumer, J. F. & Neubauer, F. (eds) *Pre-Mesozoic Geology in the Alps*. Springer-Verlag, Berlin, 537–564.

Krainer, K. & Davydov, V. 1998. Facies and biostratigraphy of the Late Carboniferous/Early Permian sedimentary sequence in the Carnic Alps (Austria/Italy). *Geodiversitas*, **20,** 643–662.

Krainer, K. & Spötl, C. 1998. Abiogenic silica layers within a fluvio-lacustrine succession, Bolzano Volcanic Complex, northern Italy: a Permian analogue for Magadi-type cherts? *Sedimentology*, **45**, 489–505.

Krainer, K. & Stingl, V. 1986*a*. Alluviale Schuttfächersedimente im Ostalpinen Perm am Beispiel der Präbichlschicthen an der Typuslokalität bei Eisenerz/Steiermark (Österreich). *Mitteilungen der Östereichischen Mineralogischen Gesellschaft*, **78**, 231–249.

Krainer, K. & Stingl, V. 1986*b*. Perm, Unter- und Mitteltrias im Bereich von Wörgl bis Saalfelden O (Exkursion E am 3. und 4. April 1986). *Jahresbericht und Mitteilungen des Oberrheinischen Geologischen Vereines*, **68**, 93–103.

Kröner, U. 1995. Postkollisionale Extension – Ein Modell für die Entwicklung am Nordrand der Böhmischen Masse. *Freiberger Forschungshefte*, **C457**.

Krois, P. & Stingl, V. 1989. Die stratigraphische Neueinstufung der basalen klastischen Gesteine im Brennermesozoikum (Stubaier Alpen/Tirol). *Zentralblatt für Geologie und Paläontologie, Teil I*, **9**(19), 1455–1466.

Kulick, J. & Paul, J. 1987. *Zur Stratigraphie und Nomenklatur des Zechsteins*. Glossar Internationale Symposium, Zechstein 1987, Exkursionsführer, Niedersächsisches Landesamt für Bodenforschung, Hannover, II, 8–18.

Kulick, J., Leifeld, D., *et al.* 1984. Petrofazielle und chemische Erkundung des Kupferschiefers der Hessischen Senke und des Harz-Westrands. *Geologisches Jahrbuch*, **68D**, 1–226.

Kutek, J. 2002. The Polish Permo-Mesozoic Rift Basin. *In:* Ziegler, P. A., Cavazza, W., Robertson, A. H. F. & Crasquin-Soleau, S. (eds) *Peri-Tethys Memoir 6: Peri-Tethyan Rift/Wrench Basins and Passive Margins*. Mémoires du Muséum national d'Histoire naturelle, **186**, 213–236.

Lago, M. & Pocovi, A. 1984. Le vulcanisme calc-alcaline d'age Stephanien-Permien dans la Chaine Iberique (Est de l'Espagne). *Supplement Bulletin de Mineralogie, Paris*, **110,** 42–44.

Laubscher, H. P. 1987. Die tektonische Entwicklung der Nordschweiz. *Eclogae Geologicae Helvetiae*, **80**, 287–303.

Legler, B. 2005. *Marine Prä-Zechstein-Ingressionen in das Südliche Permbecken – Ein multidiziplinärer Ansatz.* DGMK/ÖGEW-Frühjahrstagung des Fachbereichs Aufsuchung und Gewinnung. Deutsche Wissenschaftliche Gesellschaft für Erdöl Erdgas und Köhle e.V., Celle, 127–135.

Legrand, X., Soula, J. C. & Rolando, J. P. 1994. The Saint-Affrique Permian Basin (southern France) – an example of a rollover controlled alluvial sedimentation during regional extensional tectonics. *Geodinamica Acta*, **7**, 103–120.

Lewandowski, M. 1993. Palaeomagnetism of the lower Palaeozoic rocks of Holy Cross Mountains: palaeogeographic and geotectonic implications. *Publications of the Institute of Geophysics, Series A: Physics of the Earth Interior*, **20**, 111–118.

Lipiarski, I. 1971. La flore du Permien inférieur apparaissant dans le travertin de Karniowice aux environs de Cracovie. *Prace Panstwowego Instytutu Geologicznego*, **58**, 5–112 [in Polish with French summary].

Lippolt, H. J., Hess, J. C. & Burger, K. 1984. Isotopische Alter von pyroklastischen Sanidinen aus Kaolin-Kohlentonsteinen als Korrelationsmarken für das mitteleuropäische Oberkarbon. *Fortschritte in der Geologie von Rheinland und Westfalen*, **32**, 119–150.

Lippolt, H. J. & Hess, J. C. 1989. Isotopic evidence for the stratigraphic position of the Saar-Nahe Rotliegend volcanism III. Synthesis of results and geological implications. *Neues Jahrbuch für Geologie und Paläontologie, Abhandlungen*, **9**, 539–552.

Lippolt, H. J., Raczek, J. & Schneider, H. 1982. Isotopenalter eines Unteren Rotliegend-Biotits aus der Bohrung Wrzesnia/Polen. *Aufschluß*, **33**, 13–25.

Lippolt, H. J., Hartmann, S. & Pilot, J. 1993. 40Ar/39Ar-dating of Zechstein potash salts: new constraints on the numerical age of the latest Permian and the P-Tr boundary. *Terra Abstracts*, **1,** 591.

Lupták, B., Janák, M., Plašienka, D., Schmidt, S. Th. & Frey, M. 2000. Chloritoid-kyanite schists from the Veporic Unit, Western Carpathians, Slovakia: implications for Alpine (Cretaceous) metamorphism. *Schweizerische Mineralogische und Petrographische Mitteilungen*, **80**, 213–223.

Lupták, B., Janák, M., Plašienka, D. & Schmidt, S. Th. 2003. Alpine low-grade metamorphism of the Permian-Triassic sedimentary rocks from the Veporic Superunit, Western Carpathians: phyllosilicate composition and "crystallinity" data. *Geologica Carpathica*, **54**(6), 367–375.

Lützner, H. 1981. Sedimentation der variszischen Molasse im Thüringer Wald. *Schriftenreihe geologische Wissenschaften*, **17**, 1–217.

Lützner, H. 1988. Sedimentology and basin development of intramontane Rotliegend basins in Central Europe. *Zeitschrift für geologische Wissenschaften*, **16**, 845–863.

Lützner, H., Andreas, D., Mädler, J., *et al.* 1995. Siles und Rotliegend. *In:* Seidel, G. (ed.) *Geologie von Thüringen*. Schweizerbart, Stuttgart, 188–277.

Lützner, H., Andreas, D., Mädler, J., *et al.* 2003. Siles und Rotliegend. *In:* Seidel, G. (ed.) *Geologie von Thüringen*. Schweizerbart, Stuttgart, 2nd edn, 215–301.

McCann, T. 1998*a*. Lower Palaeozoic evolution of the NE German Basin/Baltica Borderland. *Geological Magazine*, **135**, 129–142.

McCann, T. 1998*b*. Rotliegend prospectivity in the NE German Basin. *Petroleum Geoscience*, **4**, 17–27.

McCann, T. 1998*c*. Sandstone petrography and provenance of the Rotliegend of the NE German Basin. *Sedimentary Geology*, **116**, 177–198.

McCann, T. 1999. The tectonosedimentary evolution of the northern margin of the Carboniferous foreland basin of NE Germany. *Tectonophysics*, **313**, 119–144.

McCann, T., Krawczyk, C. M. & Rieke, H. 2000. Integrated basin analysis – an example from the Upper Rotliegend of the NE German Basin. *Erdöl Erdgas Kohle*, **116**, 261–266.

McCann, T., Pascal, C., *et al.* 2006. Post-Variscan (end Carboniferous-Early Permian) basin evolution in Western and Central Europe. *In:* Gee, D. G. & Stephenson, R. A. (eds) *European Lithosphere Dynamics*, Geological Society, London, Memoirs, **32**, 355–388.

Majoros, G. 1983. Lithostratigraphy of the Permian formations of the

Transdanubian Central Mountains. *Acta Geologica Hungarica*, **26**(1–2), 7–20.

Majoros, G. 1998. A Dunántúli-Középhegység úipaleozoós képz dményeinek rétegtana. (Lithostratigraphy of Late Palaeozoic formations of the Transnubian Central Range). *In:* Bérci, I. & Jámbor, Á. (eds) *Magyarország geológiai képz dményeinek rétegtana.* MOL Rt., MÁFi, Budapest, 119–147 [in Hungarian with English summary].

Maliszewska, A., Kiersnowski, H. & Jackowicz, E. 2003. Lower Rotliegend volcaniclastic rocks at Wielkopolska (Western Poland). *Prace Panstwowego Instytutu Geologicznego*, **CLXXIX**, 1–59 [in Polish with English summary].

Marek, S. (ed.) 1988. Paleothickness, facies and paleotectonic maps of the epicontinental Permian and Mesozoic in Poland. *Kwartalnik Geologiczny*, **32**.

Marek, S. & Pajchlowa, M. (eds) 1997. *Epikontynentalny perm i mezozoik w Polsce.* Prace Panstwowego Instytutu Geologicznego, **CLIII**.

Markwick, P. J. & Rowley, D. B. 1998. The geological evidence for Triassic to Pleistocene glaciations: Implications for eustasy. *In:* Pindell, J.L. & Drake, C.L. (eds) *Paleographic Evolution and Non-glacial Eustasy, Northern South America.* SEPM Special Publications, **58**, 17–43.

Martin, H. 1981. The Late Palaeozoic Gondwana glaciation. *Geologisches Rundschau*, **70**, 480–496.

Massari, F., Conti, M. A., *et al.* 1988. The Val Gardena Sandstone and Bellerophon Formation in the Bletterbach Gorge (Alto Adige, Italy): biostratigraphy and sedimentology. *Memoria Scienza Geologica*, **40**, 229–273.

Mastalerz, K. & Raczyński, P. 1993. Litostratygrafia i ewolucja basenu północnosudeckie go w karbonie i permie. *In:* Mastalerz, K. (ed.) *Baseny Sedymentacyjne: Procesy, Osady, Architektura*, Wrocław 1993, Przewodnik 2 Krajowego Spotkania Sedymentologów, 90–97.

Mastalerz, K. & Wojewoda, J. 1988. Rotliegendes sedimentary basins of the Sudetes, Central Europe. *In:* Kiersnowski, H. (ed.) *Rotliegendes Lacustrine Basins - Workshop.* Książ Castle 1988, Guidebook, 1–9.

Mastalerz, K., Kurowski, L. & Wojewoda, J. 1993. Litostratygrafia i ewolucja basenu śródsudeckiego w karbonie i permie. *In:* Mastalerz, K. (ed.) *Baseny Sedymentacyjne: Procesy, Osady, Architektura*, Wrocław 1993, Przewodnik 2 Krajowego Spotkania Sedymentologów, 65–73.

Mattauer, M. & Matte, P. 1998. Le bassin stéphanien de St-Etienne ne résulte pas d'une extension tardi-hercienne généralisée: c'est un bassin pull-apart en relation avec un décrochement dextre. *Geodinamica Acta*, **11**, 23–31.

Matter, A. 1987. Faciesanalyse und Ablagerungsmilieus des Permokarbons im Nordschweizer Trog. *Eclogae Geologicae Helvetiae*, **80**, 345–367.

Mattern, F. 1996. The Elbe Zone at Dresden – a Late Palaeozoic pull-apart intruded shear zone. *Zeitschrift der deutschen geologischen Gesellschaft*, **147**, 57–80.

Mattern, F. 2001. Permo-Silesian movements between Baltica and Western Europe: tectonics and "basin families". *Terra Nova*, **13**, 368–375.

McCann, T., Skompski, S. *et al.* 2008. Carboniferous. *In:* McCann, T. (ed.) *The Geology of Central Europe, Volume 1: Precambrian and Palaeozoic.* Geological Society, London, 411–529.

Meier, B. P. 1994. *Untere Süsswassermolasse des zentralen und östlichen Mittelandes.* NAGRA Interner Bericht, 94-27. NAGRA, Wettingen.

Mello, J. (ed.) *et al.* 1997. *Explanations to geological map of the Slovak Karst*, 1:50 000. Geological Survey of the Slovak Republic, Bratislava, 1–225 [in Slovak].

Menning, M. 1995. A numerical time scale for the Permian and Triassic periods. *In:* Scholle, P. A., Peryt, T. M. & Ulmer-Scholle, D. S. (eds) *The Permian of Northern Pangea. Paleogeography, Paleoclimates, Stratigraphy*, Vol. 1. Springer, Berlin, 77–97.

Menning, M. 2001. A Permian time scale 2000 and correlation of marine and continental sequences using the Illawarra reversal (265 Ma). *Natura Bresciana, Monografia*, **25**, 355–362.

Menning, M., Katzung, G. & Lützner, H. 1988. Magnetostratigraphic investigations in the Rotliegendes (300–252 Ma) of Central Europe. *Zeitschrift für geologische Wissenschaften*, **16**, 1045–1063.

Menning, M., Benek, R., *et al.* 2005. Das Rotliegend in der Stratigraphischen Tabelle von Deutschland 2002 - Paternoster-Stratigraphie auf dem Rückzug. *In:* Menning, M. & Hendrich, A. (eds) Erläuterungen zur Stratigraphischen Tabelle von Deutschland 2005. *Newsletters on Stratigraphy*, **41**, 93–122

Menning, M., Alekseev, A. S., Chuvashov, B. I., *et al.* 2006. Global time scale and regional stratigraphic reference scales of Central and West Europe, East Europe, Tethys, South China and North America as used in the Devonian-Carboniferous-Permian Correlation Chart 2003 (DCP 2003). *Palaeogeography, Palaeoclimatology, Palaeoecology*, **240**, 318–372.

Mroczkowski, J. & Skowronek, C. 1980. Osady czerwonego spągowca w Iwinaach (niecka północnosudecka). *Geologia Sudetica*, **XV**(1), 125–141.

Müller, H. 1996. Das Permokarbon im nördlichen Oberrheingraben. Paläogeographische und strukturelle Entwicklung des permokarbonen Saar-Nahe-Beckens im nördlichen Oberrheingraben. *Geologische Abhandlungen Hessen*, **99**.

Müller, W. H., Huber, M., Isler, A. & Kleboth, P. 1984. *Geologische Karte der zentralen Nordwestschweiz.* NAGRA Technische Berichte, **84–25**, NAGRA, Baden.

Muttoni, P. E., Kent, D. V., Garzanti, E., Brack, P., Abrahamsen, N. & Gaetani, M. 2003. the mid-Permian revolution from Pangea "B" to Pangea "A". *Earth and Planetary Science Letters*, **215**, 379–394.

Nassichuk, W. W. 1995. Permian ammonoids in the Arctic regions of the World. *In:* Scholle, P. A., Peryt, T. M. & Ulmer-Scholle, D. S. (eds) *The Permian of Northern Pangea. Paleogeography, Paleoclimates, Stratigraphy*, Vol. 1. Springer, Berlin, 210–235.

Nemec, W. & Porebski, S. J. 1981. Sedimentary environment of the Weissliegendes sandstones in Fore-Sudetic Monocline. Proceedings of the International Symposium on the Central European Permian, Geological Institute Warsaw. *National Geological Institute Press, Warszawa*, 281–293.

Nemec, W., Porębski, S. J. & Teisseyre, A. K. 1982. Explanatory notes to the lithotectonic molasse profile of the Intra-Sudetic Basin, Polish Part (Sudety Mts., Carboniferous – Permian (Comment to Annex 23). *In:* Lützner, H. & Schwab, G. (eds) *Tectonic Regime of Molasse Epochs.* Veröffentlichung des Zentralinstituts für Physik der Erde, **66**, 267–278.

Neri, C. & Pasini, M. 1985. A mixed fauna at the Permian–Triassic boundary, Tesero section, westem Dolomites (Italy). *Bollettino della Societa Palaeontologica Italiana*, **123**, 113–117.

Neri, C. & Posenato, R. 1985. New biostratigraphical data on uppermost Werfen Formation of westem Dolomites (Trento, Italy). *Geologisch-Paläontologische Mitteilungen Innsbruck*, **14**, 83–107.

Neumann, E.-R., Olsen, K. H., Baldridge, W. S. & Sundvoll, B. 1992. The Oslo Rift: a review. *Tectonophysics*, **208**, 1–18.

Neumann, E.-R., Wilson, M., Heeremans, M., Dunworth, E. A., Obst, K., Timmermann, M. J. & Kirstein, L. 2004. Permo-Carboniferous rifting and magmatism in northern Europe: a review. *In:* Wilson, M. *et al.* (eds) *Permo-Carboniferous Rifting and Magmatism in Europe.* Geological Society, London, Special Publications, **223**, 11–40.

Nikishin, A. M., Ziegler, P. A., *et al.* 1996. Late Permian to Triassic history of the East European Craton: dynamics of sedimentary basin evolution. *Tectonophysics*, **268**, 23–63.

Nowakowski, A. 1968. Wulkanity permskie Gór Suchych w niecce Śródsudeckiej. *Geologia Sudetica*, **4**, 400–408.

Odin, B. 1986. *Les formations permiennes, Autunien supérieur à Thuringien, du "bassin" de Lodève (Hérault, France): stratigraphie, minéralogie, paléoenvironments, corrélations.* Ph.D. thesis, Aix-Marseille University, France.

Olsen, J. C. 1987. Tectonic evolution of the North Sea region. *In:* Brooks, J. & Glennie, K. W. (eds) *Petroleum Geology of Northwest Europe*, Vol. 1. Graham & Trotman, London, 389–403.

Oncken, O. 1988. Aspects of the reconstruction of the stress history of a fold and thrust belt (Rhenish Massif, Federal Republic of Germany). *Tectonophysics*, **152**, 19–40.

Ori, G. G. & Venturini, C. 1981. Facies continentali permiane nelle Alpi meridionali. *Rendiconti della Societa' Geologica Italiana*, **3**, 21–24.

Osterloff, P., Penney, R., Aitken, J., Clark, N. & Al-Husseini, M. 2004. Depositional sequences of the Al Khlata Formation, subsurface interior Oman. *In*: Al-Hulleini, M. (ed.) *Carboniferous-Early*

Triassic Arabian Plate Stratigraphy. GeoArabia Special Publication, **3**, 61–81.

Parrish, J. M., Parrish, J. T. & Ziegler, A. M. 1986. Permian-Triassic paleogeography and paleoclimatology and implications for therapsid distribution. *In:* Hotton, N. H., MacLean, P. O., Roth, J. J. & Roth, E. C. (eds) *The Ecology and Biology of Mammal-like Reptiles.* Smithsonian Institute Press, Washington DC, 109–131.

Parrish, J. T. 1995. Geologic evidence of Permian Climate. *In:* Scholle, P. A., Peryt, T. M. & Ulmer-Scholle, D. S. (eds) *The Permian of Northern Pangea. Paleogeography, Paleoclimates, Stratigraphy,* Vol. 1. Springer, Berlin, 53–61.

Pascal, C. & Cloetingh, S. A. P. L. 2002. Rifting in heterogeneous lithosphere: inferences from numerical modeling of the northern North Sea and the Oslo Graben. *Tectonics,* **21**. DOI: 10.1029/2001TC901044.

Pascal, C., Cloetingh, S. A. P. L. & Davies, G. R. 2003. Asymmetric lithosphere as the cause of rifting and magmatism in the Permo-Carboniferous Oslo Graben. *In:* Wilson, M. (ed.) *Permo-Carboniferous Rifting and Magmatism in Europe.* Geological Society, London, Special Publications, **223**, 139–156.

Paul, J. 1980. Upper Permian algal stromatolitic reefs. *Contributions to Sedimentology,* **9**, 253–268.

Paul, J. 1982*a*. Zur Rand- und Schwellen-Fazies des Kupferschiefers. *Zeitschrift der deutschen geologischen Gesellschaft,* **133**, 571–605.

Paul, J. 1982*b*. Types of stratification in the Kupferschiefer. *In:* Einsele, G. & Seilacher, A. (eds) *Cyclic and Event Stratigraphy.* Springer-Verlag, Berlin, 476–481.

Paul, J. 1991. Zechstein carbonates — marine episodes of a hyperhaline sea. *Zentralblatt für Geologie und Paläontologie, Teil I,* 1029–1045.

Paul, J. 1995. Stromatolite reefs of the Upper Permian Zechstein Basin (Central Europe). *Facies,* **32**, 28–31.

Paul, J. 1996. Stromatolite reefs of the Upper Permian Zechstein Basin. *Göttinger Arbeiten zur Geologie und Paläontologie,* **Sb2**, 325–328.

Paul, J. 1999. Evolution of a Permo-Carboniferous Basin: The Ilfeld Basin and its relationship to adjoining Permo-Carboniferous structures in Central Germany. *Neues Jahrbuch Geologische Paläontologische Abhandlungen,* **214**, 211–236.

Paul, J., Wagner, K. & Weseling, C. 1997. Lithostratigraphie und Fazies des Ilfelder Beckens (Permokarbon, Harz). *Freiberger Forschungshefte* C466, **4**, 109–153.

Pelek, J. 2004. Late Paleozoic limnic basins and coal deposits of the Czech Republic. *Folia Musei Rerum Naturalium Bohemiae Occidentalis, Geologica,* Special Volume **1**, 1–188.

Peryt, T. M. 1978. Sedimentology and paleoecology of the Zechstein Limestone (Upper Permian) in the Fore-Sudetic area (western Poland). *Sedimentary Geology,* **20**, 217–243.

Peryt, T. M. 1984. Sedymentacja i wczesna diageneza utworów wapienia cechsztynskiego w Polsce zachodniej. *Prace Panstwowego Instytutu Geologicznego,* **109**.

Peryt, T. M. 1989. Basal Zechstein in southwestern Poland: sedimentation, diagenesis, and gas accumulations. *In:* Boyle, R. W., Brown, A. C., Jefferson, C. W., Jowett, E. C. & Kirkham, R. V. (eds) *Sediment-Hosted Stratiform Copper Deposits.* Geological Society of Canada, Special Papers, **36**, 601–625.

Peryt, T. M. 1992. Debris-flow deposits in the Zechstein (Upper Permian) Main Dolomite of Poland: significance for the evolution of the basin. *Neues Jahrbuch für Geologie und Paläontologie, Abhandlungen,* **185**, 1–19.

Peryt, T. M. 1994. The anatomy of a sulphate platform and adjacent basin system in the Łeba sub-basin of the Lower Werra Anhydrite (Zechstein, Upper Permian), northern Poland. *Sedimentology,* **41**, 83–113.

Peryt, T. M. & Antonowicz, L. 1990. Facies and palaeogeography of the Zechstein Lower Anhydrite (A1d) in Poland. *Przeglad Geologiczny,* **38**, 173–180 [in Polish with English summary].

Peryt, T. M. & Kasprzyk, A. 1992. Stratigraphy and sedimentary history of the Zechstein (Upper Permian) of the North Sudetic Trough (SW Poland). *Przeglad Geologiczny,* **40**, 457–467 [in Polish with English summary].

Peryt, T. M. & Peryt, D. 1977. Zechstein foraminifera from the Fore-Sudetic monocline areas (West Poland) and their palaeoecology. *Annales de la Société Géologique de Pologne,* **XLVII**, 301–326.

Peryt, T. M. & Wagner, R. 1998. Zechstein evaporite deposition in the Central European Basin: cycles and stratigraphic sequences. *Journal of Seismic Exploration,* **7**, 201–218.

Peryt, T. M., Czapowski, G. & Gasiewicz, A. 1992. Facies and paleogeography of the Zechstein of western part of the Peri-Baltic Syneclise (North Poland). *Przeglad Geologiczny,* **40**, 223–233 [in Polish with English summary].

Peryt, T. M., Orti, F. & Rosell, L. 1993. Sulfate platform-basin transition of the Lower Werra Anhydrite (Zechstein, Upper Permian), SW Poland: facies and petrography. *Journal of Sedimentary Petrology,* **63**, 646–658.

Peryt, T. M., Kasprzyk, A. & Antonowicz, L. 1996*a*. Upper Werra Anhydrite (Zechstein, Upper Permian) in Poland. *Bulletin of the Polish Academy of Science, Earth Science,* **44**, 121–130.

Peryt, T. M., Kasprzyk, A. & Czapowski, G. 1996*b*. Basal Anhydrite and Screening Anhydrite (Zechstein, Upper Permian) in Poland. *Bulletin of the Polish Academy of Science, Earth Science,* **44**, 131–140.

Pešić, L., Ramovš, A., Sremac, J., Pantić-Prodanović, S., Filipović, I., Kovács, S. & Pelikán, P. 1988. Upper Permian deposits of the Jadar region and their position within the Western Paleotethys. *Memorie della Società Geologica Italiana,* **34**, 211–219.

Planderová, E. 1973. Palinological research in the melaphyre serie of the Choč Unit in the NE part of the Nízke Tatry Mts. between Spišský Štiavnik and Vikartovce. *Geologické Práce, Správy,* **60**, Bratislava, 143–168 [in Slovak].

Planderová, E. 1979. Biostratigraphical valuation of the Carboniferous of the Choč Nappe, based on palynology. *Geologické Práce, Správy,* **72**, 31–60 [in Slovak].

Planderová, E. 1980. New results about age of "Rožňava-Železník Group". *Geologické Práce, Správy,* **74**, 113–128 [in Slovak].

Planderová, E. & Vozárová, A. 1982. Biostratigraphical correlation of the Late Paleozoic formations in the West Carpathians. *In:* Sassi, F. P. (ed.) *IGCP Project No. 5, Newsletter,* **4**, 67–71.

Planderová, E., Sitár, V., Grecula, P. & Együd, K. 1981. Biostratigraphical evaluation of graphite shales of Zemplín Island. *Mineralia slovaca,* **13**, 97–128.

Plein, E. 1990. The Southern Permian basin and its palaeogeography. *In:* Heiling, D, Rothe, P., Förstner, U. & Stoffers, P. (eds) *Sediments and Environmental Geochemistry.* Springer-Verlag, Heidelberg, 124–133.

Plein, E. 1993. Bemerkungen zum Ablauf der paläogeographischen Entwicklung im Stephan und Rotliegend des Norddeutschen Beckens. *Geologisches Jahrbuch,* **A131**, 99–116.

Plein, E. (ed.) 1995. *Norddeutsches Rotliegendbecken Rotliegend-Monographie Teil II.* Senckenbergische Naturforschende Gesellschaft, Frankfurt-am-Main.

Pöhlig, Ch. 1986. Sedimentologie des Zechsteinkalks und des Werra-Anhydrits (Zechstein 1) in Südost-Niedersachsen. *Göttinger Arbeiten zur Geologie und Paläontologie,* **30**.

Pokorski, J. 1976. Rotliegendes of the Polish Lowlands. *Przeglad Geologiczny,* **XXIV 278**, 318–323.

Pokorski, J. 1981. Paleogeography of the Upper Rotliegendes in the Polish Lowland. *Proceedings of International Symposium on Central European Permian,* Jabłonna, 1978. *National Geological Institute Press, Warszawa,* 56–68.

Pokorski, J. 1988. Rotliegendes lithostratigraphy in north-western Poland. *Bulletin of the Polish Academy of Sciences,* **36**, 99–108.

Pokorski, J. 1989. Evolution of the Rotliegendes Basin in Poland. *Bulletin of the Polish Academy of Sciences, Earth Sciences,* **37**, 49–55.

Pokorski, J. 1990. Czerwony spągowiec pobrzeża Pomorza Zachodniego i przyległego akwenu Bałtyku. Rotliegendes in the nortwesternmost Pomerania and the adjacent Baltic Basin. *Kwartalnik Geologiczny,* **34**, 79–92.

Pożaryski, W. & Dembowski, Z. (eds) 1984. *Geological map of Poland and adjoining countries without Cenozoic, Mesozoic and Permian formations,* scale 1:1.000.000. Wydawnictwa Geologiczny Warszawa.

Pożaryski, W., Grocholski, A., Tomczyk, H., Karnkowski, P. & Moryc, W. 1992. Mapa tektoniczna Polski w epoce waryscyjskiej. *Przeglad Geologiczny,* **11**, 643–650 [in Polish].

Prijac, C., Doin, M.P., Gaulier, J.-M., & Guillocheau, F. 2000. Subsidence of the Paris Basin and its bearing on the late Variscan lithosphere evolution; a comparison between PLATE and CHABLIS models. *Tectonophysics,* **323**, 1–38.

Protić, L., Filipović, I., *et al.* 2000. Correlation of the Carboniferous,

Permian and Triassic sequences of the Jadar Block, Sana-Una and "Bükkium" Terranes. *In:* Karamata, S. & Janković, S. (eds) *Proceedings of the International Symposium Geology and Metallogeny of the Dinarides and Vardar Zone.* Serbian Academy of Sciences and Arts, Collection and Monographs, Department of Natural Sciences, Mathemathics and Technical Sciences, **1**, 61–69.

Raczyński, P. & Mastalerz, K. 1996. New localities and findings of fossil flora and fauna from the Rotliegend of the Polish part of the Sudetes. Z badań karbonu i permu w Sudetach. *Acta Universitatis Wratislaviensis, Prace Geologiczno-Mineralogiczne LII*, **1795**, 107–116 [in Polish with English summary].

Ramovs, A.1963. Biostratigraphie der Trogkofel-Stufe in Jugoslawien. *Neues Jahrbuch für Geologie und Paläontologie Monatshefte*, **1963**, 382–388.

Ramovs, A. 1972. Mittelpermische Klastite und deren marine Altersäquivalente in Slowenien, NW Jugoslawien. *Verhandlungen der Geologischen Bundesanstalt*, **1972**, 35–45.

Ramovs, A. 1974. Fortschritte in der Untersuchung des Unter- und Mittel-Perms in Slowenien, NW Jugoslawien. *Carinthia II*, **164**(94), 99–103.

Ramovs, A. 1983. Unterperm-Conodonten aus den Karawanken/Slowenien, NW Jugoslawien. *Neues Jahrbuch der Geologie und Paläontologie, Abhandlungen*, **164**, 414–427.

Raymond, A. & Metz, C. 2004. Ice and its consequences: Glaciation in the Late Ordovician, Late Devonian, Pennsylvanian-Permian, and Cenozoic compared. *Journal of Geology*, **112**, 655–670.

Rees, P. M. 2002. Land-plant diversity and the end-Permian mass extinction. *Geology*, **30**, 827–830.

Rees, P. M., Ziegler, A. M., Gibbs, M. T., Kutzbach, J. E., Behling, P. J. & Rowley, D. B. 2002. Permian phytogeographic patterns and climate data/model comparisons. *Journal of Geology*, **110**, 1–31.

Richter-Bernburg, G. 1955. Stratigraphische Gliederung des deutschen Zechsteins. *Zeitschrift des deutsche geologische Gesellschaft*, **105**, 843–854.

Rieke, H., Kossow, D., McCann, T. & Krawczyk, C. M. 2001. Tectono-sedimentary evolution of the northernmost margin of the NE German Basin between uppermost Carboniferous and Late Permian (Rotliegend). *Geological Journal*, **36**, 19–38.

Rieke, H., McCann, T., Krawczyk, C. M. & Negendank, J. F. M. 2003. Evaluation of controlling factors on facies distribution and evolution in an arid continental environment: an example from the Rotliegend of the NE German Basin. *In:* McCann, T. & Saintot, A. (eds) *Tracing Tectonic Deformation Using the Sedimentary Record.* Geological Society, London, Special Publications, **208**, 71–94.

Rigby, J. K. & Senowbari-Daryan, B. 1995. Permian sponge biogeography and biostratigraphy. *In:* Scholle, P. A., Peryt, T. M. & Ulmer-Scholle, D. S. (eds) *The Permian of Northern Pangea. Paleogeography, Paleoclimates, Stratigraphy*, Vol. 1. Springer, Berlin, 153–166.

Rochette, P. & Vandamme, D. 2001. Pangea B: an artefact of incorrect paleomagnetic assumptions? *Annali di Geofisica*, **44**, 649–658.

Rojkovič, I. 1997. Uranium mineralization in Slovakia. *Acta Geologica Universitatis Comenianae, Bratislava*, 1–117.

Ross, J. R. P. 1995. Permian bryozoa. *In:* Scholle, P. A., Peryt, T. M. & Ulmer-Scholle, D. S. (eds) *The Permian of Northern Pangea.* Vol. 1, *Paleogeography, Paleoclimates, Stratigraphy.* Springer, Berlin, 196–209.

Ross, C. A. & Ross, J. R. P. 1995. Permian Sequence Stratigraphy. *In:* Scholle, P. A., Peryt, T. M. & Ulmer-Scholle, D. S. (eds) *The Permian of Northern Pangea.* Vol. 1, *Paleogeography, Paleoclimates, Stratigraphy.* Springer, Berlin, 98–123.

Rössler, R. 1995. Zur Entstehung des Versteinerten Waldes von Chemnitz. *Veröffentlichungen des Museums für Naturkunde Chemnitz*, **18**, 35–46.

Rössler, R. 1997. Permo-Carboniferous paleosols – timelines and environment indicators. *TU Bergakademie Freiberg. Kolloquium 1. Stratigraphie, Sedimentation und Beckenentwicklung im Karbon und Perm.* Abstracts der Vorträge, 37–38.

Rössler, R. 2001. *Der versteinerte Wald von Chemnitz.* Museum für Naturkunde Chemnitz, Chemnitz [in German with English summaries].

Rössler, R. & Barthel, M. 1998. Rotliegend taphocoenoses preservation favoured by rhyolitic explosive volcanism. *Freiberger Forschungshefte*, **C474**, 59–101.

Rost, U. 1975. Beziehungen zwischen den tektonischen und paläographischen Bedingunen und den lithofaziellen Verhältnissen im Werra- und Stassfurtkarbonat Thüringens. Zeitschrift für angewandte Geologie, **21**, 462–467.

Ruffell, A.H. & Shelton, R.G. 2002. Permian to Late Triassic post-orogenic collapse, early Atlantic rifting, deserts, evaporating seas and mass extinctions. *In:* Woodcock, N. & Strachan, R. (eds) *Geological History of Britain and Ireland.* Blackwell, Oxford, 297–313.

Ryka, W. 1989. Rotliegendes volcanics, sediment lithologies and paleoenvironments, and Polish basin history: an overview. *In:* Boyle, R. W., Brown, A. C., Jefferson, C. W., Jowett, E. C. & Kirkham, R. V. (eds) *Sediment-hosted Stratiform Copper Deposits.* Geological Association of Canada, Special Papers, **36**, 627–633.

Schäfer, A. 2005. Sedimentologisch-numerisch begründeter Stratigraphischer Standard für das Permo-Karbon des Saar-Nahe-Beckens. *Courier Forschungs Institut Senckenberg*, **254**, 369–394.

Schäfer, A. & Korsch, R. J. 1998. Formation and sediment fill of the Saar-Nahe Basin (Permo-Carboniferous, Germany). *Zeitschrift der deutschen geologischen Gesellschaft*, **149**, 233–269.

Schaltegger, U. 1997. Magma pulses in the Central Variscan Belt: episodic melt generation and emplacement during lithospheric thinning. *Terra Nova*, **9**, 242–245.

Schaltegger, U. & Corfu, F. 1992. The age and source for late Hercynian magmatism in the Central Alps: Evidence from precise U-Pb ages and initial Hf isotopes. *Contributions to Mineralogy and Petrology*, **111**, 329–344.

Scheck, M., Bayer, U., Otto, V., Lamarche, J., Banka, D. & Pharaoh, T. 2002. The Elbe Fault System in North Central Europe - a basement controlled zone of crustal weakness. *Tectonophysics*, **360**, 281–299.

Schegg, R. & Leu, W. 1998. Analysis of erosion events and palaeogeothermal gradients in the North Alpine Foreland Basin of Switzerland. *In:* Düppenbecker, S. & Iliffe, J. E. (eds) *Basin Modelling: Practice and Progress.* Geological Society, London, Special Publications, **141**, 137–155.

Schneider, J. 1989. Basic Problems of Biogeography and Biostratigraphy of the Upper Carboniferous and Rotliegendes. *Acta Musei Reginaehradecensis S. A.: Scientiae Naturales*, Hradec Kralove, **23**, 1–44.

Schneider, J. 1996*a*. Xenacanth teeth – a key for taxonomy and biostratigraphy. *Modern Geology*, **20**, 321–340.

Schneider, J. W. 1996*b*. Biostratigraphie des kontinentalen Oberkarbon und Perm im Thüringer Wald, SW-Saale-Senke - Stand und Probleme. *Beiträge zur Geologie von Thüringen, Neue Folge*, **3**, 121–151.

Schneider, J. 2001. Rotliegendstratigraphie – Prinzipien und Probleme. *Beiträge zur Geologie von Thüringen. N. F.*, **8**, 7–42.

Schneider, J. & Gebhardt, U. 1993. Litho- und Biofaziesmuster in intra- und extramontanen Senken des Rotliegend (Perm, Nord- und Ostdeutschland). *Geologisches Jahrbuch*, **A131**, 57–98.

Schneider, J. W. & Zajic, J. 1994. Xenacanthiden (Pices, Chondrichthyes) des mitteleuropäischen Oberkarbon und Perm - Revision der Originale zu Goldfuss 1847, beyrich 1848, Kner 1867 und Fritsch 1879-1890. *Freiberger Forschungshefte*, **C452**, 101–151.

Schneider, J. W. & Barthel, M. 1997. Eine Taphocoenose mit Arthropleura (Arthropoda) aus dem Rotliegend (Unterperm) des Döhlen-Becken (Elbe-Zone, Sachsen). *Freiberger Forschungshefte*, **C466**, 183–223.

Schneider, J. & Roscher, M. 2002. Charts for the non-marine/marine stratigraphic correlation of the Carboniferous and Pemian. *Workshop: Oberkarbon-Untertrias in Zentraleuropa: Prozesse und ihr Timing.* Website: www.geo.tu-freiberg.de/palaeo/index2.html

Schneider, J. W., Hampe, O. & Soler-Gijón, R. 2000. The Late Carboniferous and Permian Aquatic vertebrate zonation in Southern Spain and German basins. IGCP 328 Final Report. *Courier Forschungs-Institut Senckenberg*, **223**, 543–561.

Schneider, J. W., Gaitzsch, B., Legler, B., Rössler, R., Fischer, F. & Roscher, M. 2004. Permocarboniferous of the Erzgebirge Basin. Excursion Guide. Late Westphalian terrestrial biotas and palaeoenvironments of the Variscan foreland and adjacent intramontane basins. *2004 Central European Meeting of IGCP 469*, Freiberg, Germany.

SCHNEIDER, J. W., HOTH, K., GAITZSCH, B. G., BERGER, H. J., STEINBORN, H., WALTER, H. & ZEIDLER, M. K. 2005. Carboniferous stratigraphy and development of the Erzgebirge Basin, East Germany. *Zeitschrift der deutsche Gesellschaft für Geowissenschaftent*, **156**, 431–466.

SCHNEIDER, J. W., KÖRNER, F., ROSCHER, M. & KRONER, U. 2006. Permian climate development in the northern peri-Tethys area – the Lodève basin, French Massif Central. *Palaeogeography, Palaeoclimatology, Palaeoecology*, **240**, 161–183.

SCHÖDER, L., PLEIN, E., *et al.* 1995. Stratigraphische Neugliederung des Rotliegend im Norddeutschen Becken. *Geologisches Jahrbuch, Reihe A*, **148,** 3–21.

SCHOLLE, P. A., PERYT, T. M. & ULMER-SCHOLLE, D. S. (eds) 1995*a*. *The Permian of Northern Pangea. Volume 1 – Paleogeography, Paleoclimates, Stratigraphy.* Springer-Verlag, Berlin.

SCHOLLE, P. A., PERYT, T. M. & ULMER-SCHOLLE, D. S. (eds) 1995*b*. *The Permian of Northern Pangea. Volume 2 – Sedimentary Basins and Economic Resources.* Springer-Verlag, Berlin.

SCHRÖDER, B. 1988. Outline of the Permo-Carboniferous basins at the western margin of the Bohemian Massif. *Zeitschrift für geologische Wissenschaften*, **16**, 993–1001.

SCHÜNEMANN, M., VON GOSEN, W. & THIEDIG, F. 1982. Die Viktringer Permotrias und ihre Beziehung zur Gurktaler Decke (Kärnten - Österreich). *Mitteilungen des Geologisch-Paläontologischen Institutes der Universität Hamburg*, **53**, 191–206.

SCHWAB, K. 1987. Compression and right-lateral strike-slip movement at the Southern Hunsrück Borderfault (Southwest Germany). *Tectonophysics*, **137**, 115–126.

SCHWEIZERISCHE GEOLOGISCHE KOMMISSION. 1980. *Geologische Karte der Schweiz 1:500000*. Bern.

SCOTESE, C. R. 1987. Development of the Circum-Pacific Panthalassic Ocean during the Early Palaeozoic. *In:* MONGER, J. W. H. & FRANCETEAU, J. (eds) *Circum-Pacific Orogenic Belts and the Evolution of the Pacific Ocean Basin*. American Geophysical Union, Geodynamics Series, **18**, 49–57.

SCOTESE, C. R. 2004. A continental drift flipbook. *Journal of Geology*, **112**, 729–741.

SCOTESE, C. R. & LANGFORD, R. P. 1995. Pangea and the paleogeography of the Permian. *In:* SCHOLLE, P.A., PERYT, T.M. & ULMER-SCHOLLE, D.S. (eds) *The Permian of Northern Pangea.* Vol. 1, *Paleogeography, Paleoclimates, Stratigraphy.* Springer, Berlin, 3–19.

SIEDLECKA, A. 1964. Permian in the north-eastern border of the Upper Silesia Coal Basin. *Annales de la Société Géologique de Pologne*, **34**, 309–394 [in Polish with English summary].

ŚLIWINSKI, W. 1988. *Brzeżne facje cechsztynu w niecce północnosudeckiej*. W: Wybrane zagadnienia geologii złóż Polski Zachodniej. Materiały konferencji naukowej. Uniwersytet Wrocławski.

SØRENSEN, K. 1986. Danish Basin subsidence by Triassic rifting on a lithosphere cooling background. *Nature*, **319**, 660–663.

SØRENSEN, S. & MARTINSEN, B. B. 1987. A paleogeographic reconstruction of the Rotliegendes deposist in the northeastern Permian Basin. *In*: BROOKS, J. & GLENNIE, K. (eds) *Petroleum Geology of North West Europe*. Graham & Trotman, London, 497–508.

SPÖTL, C. 1987. Eine klastisch-evaporitische Oberperm-Entwicklung im Hallstätter Salzberg (Salzkammergut, Österreich). *Mitteilungen der Östereichischen Geologischen Gesellschaft*, **80**, 115–142.

SPÖTL, C. 1988. Schwefelisotopendatierungen und fazielle Entwicklung permoskythischer Anhydrite in den Salzberg- bauen von Dürrnberg/Hallein und Hallstatt (Österreich). *Mitteilungen der Gesellschaft für Geologie und Bergbaustudenten Österreich*, **34/35**, 209–229.

SPÖTL, C. 1989. The Alpine Haselgebirge Formation, northern Calcareous Alps (Austria): Permo-Scythian evaporites in an alpine thrust system. *Sedimentary Geology*, **65**, 113–125.

SPÖTL, C. 1992. Carbonates in Upper Permian evaporates of the Northern Calcareous Alps, Austria. *Geologische Rundschau*, **81**, 309–321.

SPÖTL, C. & BURNS, S. J. 1994. Magnesite diagenesis in red beds: a case study from the Permian of the Northern Calcareous Alps (Tyrol, Austria). *Sedimentology*, **41**, 543–565.

STAPF, K. R. G. 1982. Schwemmfächer- und Playa-Sedimente im Ober-Rotliegenden des Saar-Nahe-Beckens (Permokarbon, SW-Deutschland). Ein Überblick über Faziesanalyse und Faziesmodell. *Mitteilungen der Pollichia*, **70**, 7–64.

STEMMERIK, L., INESON, J. R. & MITCHELL, J. G. 2000. Stratigraphy of the Rotliegend Group in the Danish part of the Northern Permian Basin, North Sea. *Journal of the Geological Society, London*, **157**, 1127–1136.

STEPHENSON, R. A., NARKIEWICZ, M., DADLEZ, R., VAN WEES, J.-D. & ANDRIESSEN, P. 2003. Tectonic subsidence modelling of the Polish Basin in the light of new data on crustal structure and magnitude of inversion. *Sedimentary Geology*, **156**, 59–70.

STINGL, V. 1983. Ein Beitrag zur Fazies der Prebichlschichten zwischen St. Johann i. T. und Leogang (Tirol/Salzburg). *Geologisch-Paläontologische Mitteilungen, Innsbruck*, **12**, 207–233.

STOLLHOFEN, H. 1994. Synvulkanische Sedimentation in einem fluviatilen Ablagerungsraum: Das basale 'Oberrotliegend' im permokarbonen Saar-Nahe-Becken. *Zeitschrift der deutschen geologischen Gesellschaft*, **145,** 343–378.

STOLLHOFEN, H. 1997. The interaction of strike-slip deformation, sedimentation and magmatism: The Permo-Carboniferous Saar-Nahe-Basin (SW-Germany). 18th IAS Regional Meeting 1997, Heidelberg, Germany, Excursion A 1 Field guide. *Gaea heidelbergensis*, **4**, 1–13.

STOLLHOFEN, H. 1998. Facies architecture variations and seismogenic structures in the Carboniferous-Permian Saar-Nahe Basin (SW Germany): evidence for extension-related transfer fault activity. *Sedimentary Geology*, **119**, 47–83.

STOLLHOFEN, H. 1999. Sequence stratigraphy in continental settings - examples from the Permo-Carboniferous Saar-Nahe Basin, SW Germany. *Zentralblatt für Geologie und Paläontologie, Teil I*, **3–4**, 233–260.

STOLLHOFEN, H. & STANISTREET, I. G. 1994. Interaction between bimodal volcanism, fluvial sedimentation and basin development in the Permo-Carboniferous Saar-Nahe Basin (SW-Germany). *Basin Research*, **6**, 245–267.

STOLLHOFEN, H., FROMMHERZ, B. & STANISTREET, I. G. 1999. Volcanic rocks as discriminants in evaluating tectonic versus climatic control on depositional sequences, Permo-Carboniferous continental Saar-Nahe Basin. *Journal of the Geological Society, London*, **156,** 801–808.

STOLLHOFEN, H., STANISTREET, I. G., BANGERT, B. & GRILL, H. 2000. Tuffs, tectonism and glacially related sea-level changes, Carboniferous-Permian, Southern Namibia. *Palaeogeography, Palaeoclimatology, Palaeoecology*, **161**, 127–150.

STROHMENGER, C., VOIGT, E. & ZIMDARS, J. 1996*a*. Sequence stratigraphy and cyclic development of Basal Zechstein carbonate-evaporite deposits with emphasis on Zechstein 2 off-platform carbonates (Upper Permian, Northeast Germany). *Sedimentary Geology*, **102**, 33–54.

STROHMENGER, C., ANTONINI, M., JAGER, G., ROCKENBAUCH, K. & STRAUSS, C. 1996*b*. Carbonate reservoir facies distribution in relation to Zechstein sequence stratigraphy (upper Permian, northwest Germany) – an integrated approach. *Bulletin des Centres de Recherches Exploration-Production Elf Aquitaine*, **20**, 1–35.

ŠUF, J. 1963. Report of geological research in the Štítnik area. Geologické Práce, Správz, **27**, 63–68 (in Czech with German summary).

SURLYK, F., HURST, J. M., PIASEKI, S., ROLLE, F., SCHOLLE, P. A., STEMMERIK, L. & THOMSEN, E. 1986. The Permian of the western margin of the Greenland Sea – a future exploration target. *In:* HALBOUTY, M.T. (ed.) *Future Petroleum Provinces of the World.* American Association of Petroleum Geologists, Geological Memoir, **40**, 629–659.

TAYLOR, J. C. M. 1986. Late Permian Zechstein. *In:* GLENNIE, K. W. (ed.) *Introduction to the Petroleum Geology of the North Sea.* Blackwell, Oxford, 87–111.

THURY, M., GAUTSCHI, A., MAZUREK, M., *et al.* 1994. *Geology and Hydrogeology of the Crystalline Basement of Northern Switzerland.* NAGRA Technischer Bericht, 93-01. NAGRA, Wettingen.

TIMMERMAN, M. J. 2004. Timing, geodynamic setting and character of Permo-Carboniferous magmatism in the foreland of the Variscan Orogen, NW Europe. *In*: WILSON, M., NEUMANN, E.-R., DAVIES, G. R., TIMMERMAN, M. J., HEEREMANS, M. & LARSEN, B. T. (eds) *Permo-Carboniferous Magmatism and Rifting in Europe*. Geological Society, London, Special Publications, **223**, 41–74.

TIMMERMANN, M. J. 2008. Palaeozoic magmatism. *In:* MCCANN, T. (ed.) *The Geology of Central Europe, Volume 1: Precambrian and Palaeozoic*. Geological Society, London, 665–748.

Torcq, F., Besse, J., Vaslet, D., Marcoux, J., Ricou, L. E., Halawani, M., & Basahel, M. 1997. Paleomagnetic results from Saudi Arabia and the Permo-Triassic Pangea configuration. *Earth and Planetary Science Letters*, **148**, 553–567.

Torsvik, T. H. & Cocks, R. M. 2004. Earth geography from 400 to 250 Ma: a palaeomagnetic, faunal and facies review. *Journal of the Geological Society, London*, **161**, 555–572.

Torsvik, T. H. & Van der Voo, R. 2002. Refining Gondwana and Pangea palaeogeography: estimates of Phanerozoic non-dipole (octupole) fields. *Geophysical Journal International*, **151**, 771–794.

Trümpy, R. 1980. *Geology of Switzerland - a Guide-Book. Part A: An Outline of the Geology of Switzerland*. Wepf, Basel.

Trümpy, R. & Dössegger, R. 1972. Permian of Switzerland. *In*: Falke, H. (ed.) *Rotliegend, Essays on European Lower Permian*. Brill, Leiden, 189–215.

Tucker, M. E. 1991. Sequence stratigraphy of carbonate-evaporite basins: models and application to the Upper Permian (Zechstein) of Northeast England and adjoining North Sea. *Journal of the Geological Society, London*, **148**, 1019–1036.

Uličný, D. 2004. A drying-upward aeolian system of the Bohdasin Formation (early Triassic), Sudetes of NE Czech Republic: record of seasonality and long-term palaeoclimate change. *Sedimentary Geology*, **167**, 17-39.

Umhoefer, P. J. 2000. Where are the missing Faults in Translated Terranes? *Tectonophysics*, **326**, 23–35.

Utting, J. & Piasecki, S. 1995. Palynology of the Permian of Northern Continents: A review. *In:* Scholle, P. A., Peryt, T. M. & Ulmer-Scholle, D. S. (eds) *The Permian of Northern Pangea*. Vol. 1, *Paleogeography, Paleoclimates, Stratigraphy*. Springer, Berlin, 237–261.

Vachard, D. & Krainer, K. 2001. Smaller foraminifers, characteristic algae and pseudo-algae of the latest Carboniferous/Early Permian Rattendorf Group, Carnic Alps (Austria/Italy). *Rivista Italiana di Paleontologia e Stratigrafia*, **107**(2), 169–195.

Van Adrichem Boogaert, H. A. & Kouwe, W. F. P. (compilers) 1993. Stratigraphic nomenclature of the Netherlands, revision and update by RGD and NOGEPA. *Mededelingen Rijks Geologische Dienst*, **50**.

Van Amerom, H. W. J., Angerer, H. & Mostler, H. 1982. Über eine Autuno-Stephanische Flora aus den Kristbergschicthen im Montafon, Vorarlberg (Österreich). *Jahrbuch der geologischen Bundesanstalt*, **124**, 283–323.

Van de Sande, J. M. M., Reijers, T. J. A. & Casson, N. 1996. Multidisciplinary exploration strategy in the northeast Netherlands Zechstein 2 Carbonate play, guided by 3D seismic. *In:* Rondeel, H. E., Batjes, D. A. J. & Nieuwenhuijs, W. H. (eds) *Geology of Oil and Gas under the Netherlands*. Kluwer, Dordrecht, 125–142.

Van der Voo R. & Torsvik T. 2001. Evidence for late Palaeozoic and Mesozoic non-dipole Fields Provides and Explanation for the Pangaea Reconstruction Problem. *Earth and Planetary Science Letters*, **187**, 71–81.

Van Wees, J.-D., Stephenson, R. A., *et al.* 2000. On the origin of the Southern Permian Basin, Central Europe. *Marine and Petroleum Geology*, **17**, 43–59.

Van Wijhe, D. H., Lutz, M. & Kaasschieter, J. P. H. 1980. The Rotliegend in the Netherlands and its gas accumulations. *Geologie en Mijnbouw*, **59**, 3–24.

Veevers, J. J. & Powell, C. M. 1987. Late Paleozoic glacial episodes in Gondwanaland reflected in transgressive-regressive depositional sequences in Euramerica. *Society of America, Bulletin*, **98**, 475–487.

Veit, E. 1976. Geophysik und Bau des Untergrundes des Saarbrücker Hauptsattels. *Geologisches Jahrbuch*, **A 27**, 409–428.

Verdier, J. P. 1996. The Rotliegend sedimentation history of the Southern North Sea and adjacent countries. *In:* Rondeel, H.E., Batjes, D.A.J. & Nieuwenhuijs, W.H. (eds) *Geology of Gas and Oil under the Netherlands*. The Royal Geological and Mining Society of the Netherlands – KNGMG, 45–56.

Visser, J. N. J. 1997. A review of the Permo-Carboniferous glaciation in Africa. *In:* Martini, I. P. (ed.) *Late Glacial and Postglacial Environmental Changes: Pleistocene, Carboniferous-Permian, and Proterozoic*. Oxford University Press, Oxford, 169–191.

von Quadt, A., Haunschmid, B. & Finger, F. 1999. Mid Permian A-type plutonism in the eastern Tauern Window. *European Journal of Mineralogy*, **11**, 185.

Vozár, J. 1997. Rift-related volcanism in the Permian of the Permian of the Western Carpathians. *In:* Grecula, P., Hovorka, D. & Putiš, M. (eds) *Geological Evolution of the Western Carpathians*. Monograph, Mineralia slovaca, Bratislava, 225–234.

Vozárová, A. 1997. Upper Permian- Lower Triassic evaporites in the Western Carpathians (Slovakia). *Slovak Geological Magazine*, **3**, 223–230.

Vozárová, A. & Vozár, J. 1982. New lithostratigraphic units in the southern part of Veporicum. *Geologické práce, Správy*, 79, Bratislava, 169–194 [in Slovak].

Vozárová, A. & Vozár, J. 1993. Pre-Alpine Terranes in the Western Carpathians: reconstruction of collision and post-collision stages of the Variscan Orogeny. *In*: Rakús, M. & Vozár, J. (eds):*Geodynamický vývoj a hĺbinná stavba Západných Karpát*, Dionýz Štúr Institute of Geology, Bratislava, 129–143 [in Slovak].

Vozárová, A. & Rojkovič, I. 2000. Permian lacustrine phosphatic sandstone in the Southern Gemeric unit, Western Carpathians, Slovakia. *Geologica Carpathica*, **51**(4), 265–278.

Vozárová, A. & Túnyi, I. 2003. Evidence of the Illawara Reversal in the Permian sequence of the Hronic Nappe (Western Carpathians, Slovakia). *Geologica Carpathica*, **54**(4), 229–236.

Vozárová, A. & Vozár, J. 1988. *Late Paleozoic of West Carpathians*. Monograph., Dionýz Štúr Institute of Geology, Bratislava.

Wagner, R. 1991. Stratigraphie des höchsten Zechsteins im Polnischen Zentralbecken, *Zentralblatt für Geologie und Paläontologie*, **1**, 883–892.

Wagner, R. 1994. Stratigraphy and evolution of the Zechstein basin in the Polish Lowlands. *Prace Panstwowego Instytutu Geologicznego*, **146** [in Polish with English summary].

Wagner, R. & Peryt, T. M. 1997. Possibility of sequence stratigraphic subdivision of the Zechstein in the Polish Basin. *Geological Quarterly*, **41**, 457–474.

Wagner, R., Peryt, T. M. & Piatkowski, P. S. 1981. The evolution of the Zechstein sedimentary basin in Poland. *Proceedings of International Symposium on Central European Permian. National Geological Institute Press*, Warszawa, 69–83.

Ward, P. D., Montgomery, D. R. & Smith, R. 2000. Altered river morphology in South Africa related to the Permian–Triassic extinction. *Science*, **289**, 1740–1743.

Wardlaw, B. R., Davydov, V. & Gradstein, F. M. 2004. *In:* Gradstein, F. M., Ogg, J. G. & Smith, A. G. (eds) *A Geologic Time Scale 2004*. Cambridge University Press, 249–270.

Warr, L. N. 2002. The Variscan Orogeny: the welding of Pangaea. *In:* Woodcock, N. & Strachan, R. (eds) *Geological History of Britain and Ireland*. Blackwell, Oxford, 271–294.

Warren, J. K. 1999. *Evaporites – Their Evolution and Economics*. Blackwell, Oxford.

Weber, W. 1979. Zur synsedimentären Tektonik des Oberperms in Mitteleuropa. *Zeitschrift für geologische Wissenschaften*, **7**, 871–878.

Wegener, A. 1912. Die Entstehung der Kontinente. *Geologische Rundschau*, **3**, 276–292.

Weihe, T. 1997. *Sedimentologie und Fazies des Rotliegenden (unteres Perm) nördlich des Wolsztyn-Hochs, Westpolen, im Spiegel der Beckenentwicklung*. Clausthaler Geowissenschaften, Dissertation **51**, Clausthal, Germany.

Werneburg, R. & Kiersnowski, H. 1996. A discosauriscid amphibian from the Rotliegend of the Intrasudetic Basin in Poland and its biostratigraphic importance. *In:* Mastalerz, K. (ed.) *Z badań karbonu i permu w Sudetach*. Acta Universitatis Wratislaviensis No. **1795**. Prace geologiczno – mineralogiczne, **LII**, Wrocław, 117–124.

Wetzel, A. & Allia, V. 2003. Der Opalinuston in der Nordschweiz: Lithologie und Ablaagerungsgeschichte. *Eclogae Geologicae Helvetiae*, **96**, 451–469.

Wilson, M. Neumann, E.-R., Davies, G. R., Timmerman, M. J., Heeremans, M. & Larsen, B. T. (eds) 2004. *Permo-Carboniferous Magmatism and Rifting in Europe*. Geological Society, London, Special Publications, **223**.

Witzmann, F. & Schoch, R. R. 2006. Skeletal development of the temnospondyl *Acanthostomatops vorax* from the Lower Permian Döhlen Basin of Saxony. *Transactions of the Royal Society of Edinburgh, Earth Sciences*, **96**, 365–385.

Wojewoda, J. & Mastalerz, K. 1989. Ewolucja klimatu oraz

allocykliczność i autocykliczność sedymentacji na przykładzie osadów kontynentalnych górnego karbonu i permu w Sudetach. *Przeglad Geologiczny*, **37**, 173–180.

Yang, C. S. & Baumfalk, Y. A. 1994. Milankovitch cyclicity in the Upper Rotliegend Group of the Netherlands offshore. *In:* De Boer, P. L. & Smith, D. G. (eds) *Orbital Forcing and Cyclic Sequences*. International Association of Sedimentologists, Special Publications, **19**, 47–61.

Yang, C. S. & Nio, S. D. 1994. Applications of high-resolution sequence stratigraphy to the Upper Rotliegend in the Netherlands offshore. *In:* Weimer, P. & Posamentier, H. (eds) *Siliclastic Sequence Stratigraphy: Recent Advances and Applications*. American Association of Petroleum Geologists, Memoirs, **58**, 285–316.

Zajíc, J. 1997. Vertebrate biostratigraphy of the Bohemian Permo-Carboniferous limnic basins – recent results of investigation. *TU Bergakademie Freiberg. Kolloquium 1, Stratigraphie, Sedimentation und Beckenentwicklung im Karbon und Perm*. Abstracts der Vorträge, 47–49.

Zajíc, J. 2000. Vertebrate zonation of the non-marine Upper Carboniferous - Lower Permian basins of the Czech Republic. *Courier Forschungs-Institut Senckenberg*, **223**, 563–575.

Zdanowski, P. 2004. Wide restricted lagoons (salinas) of the main Dolomite as a final deposition of carbonate lowstand system tracts in the Gorzów Wielkopolski region. *IAS 23rd Meeting Coimbra-Portugal*, Abstracts Book, 291.

Zeh, A. & Brätz, H. 2004. Timing of Upper Carboniferous/Permian horst-basin formation and magmatism in the NW Thuringian Forest, Central Germany, a review. *In*: Wilson, M., Neumann, E.-R., Davies, G. R., Timmerman, M. J., Heeremans, M. & Larsen, B. T. (eds) *Permo-Carboniferous Magmatism and Rifting in Europe*. Geological Society, London, Special Publications, **223**, 319–334.

Zeh, A., Cosca, M. A., Brätz, H., Okrusch, M. & Tichomirowa, M. 2000. Simultaneous horst-basin formation and magmatism during Late Variscan transtension: evidence from $^{40}Ar/^{39}Ar$ and $^{207}Pb/^{206}Pb$ geochronology in the Ruhla Crystalline Complex. *International Journal of Earth Sciences*, **89**, 52–71.

Ziegler, A. M. 1990. Phytogeographic patterns and continental configurations during the Permian Period. *In:* McKerrow, W. S. & Scotese, C. R. (eds) *Palaeozoic Palaeogeography and Biogeography*. Geological Society, London, Memoirs, **12**, 263–379.

Ziegler, A. M., Hulver, M. L. & Rowley, D. B. 1997. Permian world topography and climate. *In:* Martini, I. P. (ed.) *Late Glacial and Postglacial Environmental Changes: Pleistocene, Carboniferous-Permian, and Proterozoic*. Oxford University Press, Oxford, 111–146.

Ziegler, M. A. 1989. North German Zechstein facies patterns with regard to their substrate. *Geologische Rundschau*, **78**, 105–127.

Ziegler, P. A. 1982. *Geological Atlas of Western and Central Europe*. Shell Internationale Petroleum Maatschappij B.V.

Ziegler, P. A. 1988. Laurussia – the Old Red Continent. *In:* McMillan, N. J., Imbry, A. F. & Glass, D. J. (eds) *Devonian of the World*. Canadian Society of Petroleum Geologists, Memoirs, **14**(1), 15–48.

Ziegler, P. A. 1989. *Evolution of Laurussia – A Study in Late Palaeozoic Plate Tectonics*. Kluwer Academic, Dordrecht.

Ziegler, P. A. 1990. *Geological Atlas of Western and Central Europe*, 2nd edn. Shell Internationale Petroleum Maatschappij B.V.

Ziegler, P. A. 1996. Geodynamic processes govern development of rifted basins. *In:* Roure, F., Ellouz, N., Shein, V. S. & Skvortsov, L. (eds) *Geodynamic Evolution of Sedimentary Basins*. Technip, Paris, 19–67.

Ziegler, P. A., Schumacher, M. E., Dèzes, P., Van Wees, J.-D. & Cloetingh, S. 2006. Post-Variscan evolution of the lithosphere in the area of the European Cenozoic Rift System. *In:* Gee, D. G. & Stephenson, R. A. (eds) *European Lithosphere Dynamics*. Geological Society, London, Memoirs, **32**, 97–112.

Zimmerle, W. 1976. Petrographische Beschreibung und Deutung der erbohrten Schichten. *Geologisches Jahrbuch*, **A 27**, 91–305.

Znosko, J. 1999. *Tectonic Atlas of Poland*. Panystwowy Instytut Geologiczny, Warszawa.

11 Variscan tectonics

UWE KRONER (co-ordinator), JEAN-LOUIS MANSY*, STANISŁAW MAZUR, PAWEŁ ALEKSANDROWSKI, HORST PETER HANN, HERMANN HUCKRIEDE, FRÉDÉRIC LACQUEMENT, JULIETTE LAMARCHE, PATRICK LEDRU, TIMOTHY C. PHARAOH, HUBERT ZEDLER, ARMIN ZEH & GERNOLD ZULAUF

The Variscan Orogeny is the major Middle to Late Palaeozoic tectonometamorphic event in Central Europe representing the final collision of Gondwana with the northern continent of Laurussia. Thus, large areas of the pre-Permian basement consist of continental crust that achieved its final form during this event. The Variscan Orogeny represents the European version of the evolution of the supercontinent of Pangaea at the end of the Palaeozoic. Western Pangaea, including the Variscan Orogen, formed as a result of the continuous closing of the oceanic domains between Gondwana and Laurussia (Old Red Continent: North American Craton + East European Craton + Avalonia). Coeval accretion of large volumes of oceanic crust along the Eastern Uralides and Altaids (Sengör *et al.* 1993) as well as Precambrian continental crust (e.g. Siberia and Kazakhstan) along the eastern edge of Laurussia represents the formation of eastern Pangaea. Following the Permian termination of collisional tectonics along western Pangaea there was ongoing convergence in the Asian part until the Early Mesozoic, as demonstrated by the evidence of Triassic continental subduction within the Qinling–Dabie–Sulu Belt between the northern Sino-Korean Craton and the southern Yangtze Craton (Ernst 2001, and references therein).

Despite the occurrence of both pre- and synorogenic subduction processes within the area of the Variscan Orogen, the accretion of juvenile crust plays a relatively minor role in terms of the crustal evolution of the region. Recycling of basement formed during the Neoproterozoic–Early Cambrian Cadomian Orogeny and its Early Palaeozoic cover can be considered to be one of the most striking features of the Variscides. The most prominent characteristic in terms of the collision between Gondwana and Laurussia is the Rheic suture. If we consider the evolution of the Rheic Ocean as being the result of the Cambro-Ordovician northward rifting of the peri-Gondwanan Avalonian terrane (Murphy *et al.* 2006, and references therein; see also Krawczyk *et al.* 2008) than we can trace the Rheic suture between continental crustal fragments accreted to Laurussia by the Caledonian Orogeny and Gondwana-type crust present south of the Rheic Ocean (Fig. 11.1). Laurussian crust in Central Europe is characterized by the Late Caledonian formation of the terrestrial Devonian and related marine sediments along the adjacent shelf (see Belka & Narkiewicz 2008). Detrital mica with a Caledonian age is exclusively found within these stratigraphic units (Huckriede *et al.* 2004) and represents the most distinctive feature for continental crust belonging to those areas to the north of the Rheic suture.

The Upper Ordovician glaciation of Gondwana (Sahara glaciation) is documented by the widespread occurrence of glaciomarine tillites in northern Africa and Europe (Ghienne 2003, and references therein). However, a contemporaneous tillite is not known from Avalonia, which must have drifted away from Gondwana earlier. Thus, the glaciomarine tillite represents an ideal tracer to assign individual crustal fragments to the southern passive margin of the Rheic Ocean, i.e. to Gondwana during the Upper Ordovician (Linnemann & Heuse 2000). The trace of the Rheic suture is shown in Figures 11.1 & 11.2.

Timing of Palaeozoic post-Caledonian orogenies between Gondwana and Laurussia

Post-Caledonian closing of the Rheic Ocean took place between the Silurian and the Permian. However, a simple correlation of orogenic events along the suture zone is not possible due to the lack of preserved evidence along its trace. Generally three differing, partly overlapping, regions have been distinguished along the Rheic suture zone affected by the Acadian, Variscan and Alleghanian orogenies (Fig. 11.2).

The Acadian Orogeny

The Acadian Orogeny was named after the original Acadian settlements along the Atlantic coast between Eastern Maine and Nova Scotia. In addition to the Northern Appalachians, this term is now also used for coeval tectonometamorphic and/or tectonomagmatic events in the Southern Appalachians and the Anglo-Brabant foldbelt of Ireland, Wales, England and Belgium. Originally, the term was used for post-Silurian, pre-Late Devonian deformation partly accompanied and followed by magmatism (e.g. Mansy *et al.* 1999). Indeed, new geochronological data (e.g. Sherlock *et al.* 2003) has evidenced a short-lived, compressional event at *c.* 400 Ma for the Anglo-Brabant foldbelt. However, in the Appalachians tectonometamorphic processes continued until the Lower Carboniferous (*c.* 345 Ma). This disparity in terms of the longevity of events across the region led to the view that the Acadian Orogeny was a prolonged event (Spear *et al.* 2002) and also to the subdivision of orogenic activity into Acadian and Neo-Acadian phases (e.g. Merschat *et al.* 2005). Despite the recent support of initial plate tectonic models which related the Acadian Orogeny to a Late Caledonian compressional event between Laurentia and Avalonia, there are stratigraphic, geochronological and tectonic arguments for a time

* This chapter is dedicated to the memory of Jean-Louis Mansy (deceased).

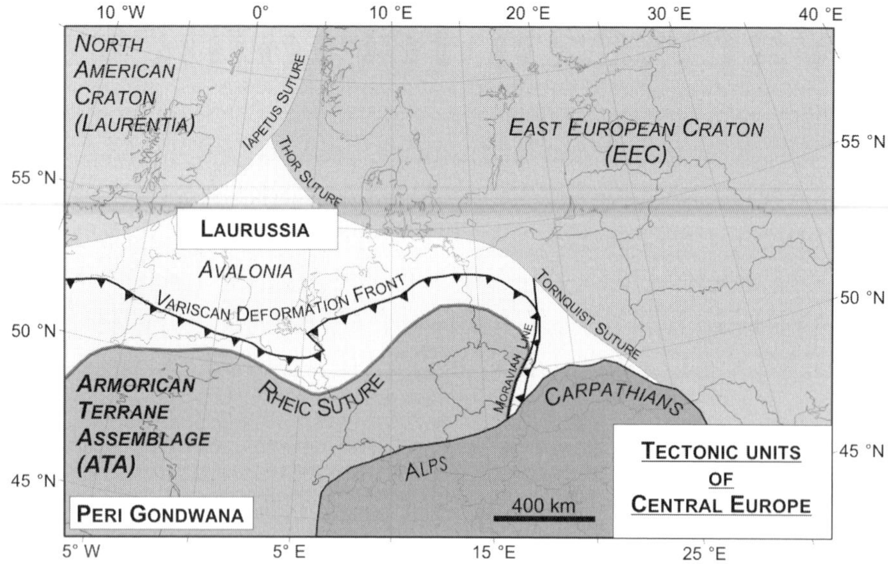

Fig. 11.1. Basement tectonic sketch map of Central Europe.

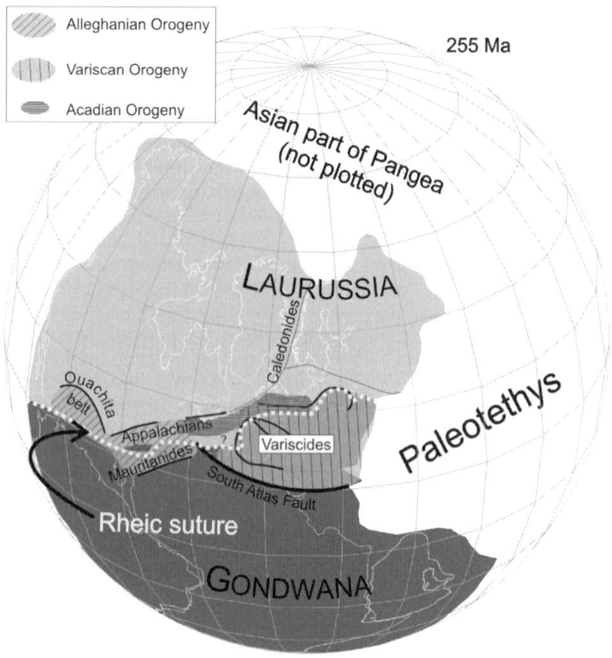

Fig. 11.2. Spatial distribution of the Palaeozoic orogenies related to the formation of the Rheic suture of Pangaea (Silurian–Late Permian) during ongoing convergence between Gondwana and Laurussia (orthographic projection).

gap of *c.* 20 Ma between the inferred end of Iapetus convergence (*c.* 420 Ma) and the apparent onset of Acadian deformation (for details see Soper & Woodcock 2003). Thus, the Acadian Orogeny, which affected continental crust north of the Rheic suture, is possibly an Early Variscan expression of initial collisional contacts with continental crust south of the Rheic suture as proposed by Soper & Woodcock (2003) and Meere & Mulchrone (2006), or may reflect Andean-type upper plate deformation during ongoing subduction of the Rheic Ocean (Murphy *et al.* 2004).

The Variscan Orogeny

This long-lasting orogeny (*c.* 120 Ma) is restricted to central, western and southeastern Europe. The *locus typicus* is situated in the German town of Hof (*Curia Variscorum sensu* Suess 1885). Subduction processes persisted from Late Silurian through to Early Carboniferous times as demonstrated by the age of high pressure metamorphism as well as the coeval existence of magmatic arcs (for details see below). Compressional tectonics lasted until the earliest Permian as demonstrated by the final deformation of the Cantabria-Asturias Arc in northern Spain (Weil *et al.* 2001), the syn- to post-kinematic intrusion of voluminous granitoids in Iberia (Fernández-Suárez *et al.* 2000) as well as the termination of ductile and brittle deformation along the northern external fold-and-thrust belt (Quinn *et al.* 2005). The remnants of the oldest subduction phase are referred to as the Early Variscan stage lasting from the Silurian to the Devonian (Matte 2001). During the Mid-Variscan stage (*c.* Early Carboniferous) continental crust to the south of the Rheic suture was affected by a large-scale tectonometamorphic overprint including high pressure–high temperature (HP/HT) metamorphism, and this was followed by a large-scale high temperature–low pressure (LP/HT) event. The final geometry of the orogen characterized by vast fold-and-thrust belts affecting voluminous foreland sediments is the result of the Late Variscan stage (Late Carboniferous–Early Permian).

The Alleghanian Orogeny

This orogeny was named after the Allegheny Mountains in the Southern Appalachians and lasted from 320 to 260 Ma (Hatcher 2002). This thrust-dominated event with an internal dextral strike-slip domain was coeval with the deformation of the Mauritanides in western Africa (Dallmeyer & Lécorché 1989) and the final structuring of the Ouachita Marathon Belt along the southern edge of the North American Craton (Poole *et al.* 2005). It thus represents the expression of the final closing of the Rheic Ocean. Post-Alleghanian cooling persisted until the Early Triassic (Wintsch *et al.* 2003). It is noteworthy that Permian compressional deformation along this part of the Rheic Suture Zone was contemporaneous with extensive rift-related volcanism in Central Europe (see McCann *et al.* 2008*b* & Timmerman 2008). Dextral shearing of Gondwana along the South Atlas Fault (Arthaud & Matte 1977) at this time could be an appropriate explanation for this phenomenon and demonstrates the diachronism during the formation of this part of Pangaea.

Architecture of the Variscan Orogen

The Variscan basement is overlain by Permo-Carboniferous volcanosedimentary successions of a post-Variscan extensional stage followed by Meso-Cenozoic platform sediments. Late Cretaceous to Tertiary extensional and compressional tectonics, related to the opening of the Atlantic Ocean and the Alpine Orogeny, respectively, led to a reactivation of the continental crust. Therefore, most of the exposures of Variscan complexes outside of the Alpine Orogen represent late Mesozoic to Cenozoic basement uplifts juxtaposed with the younger platform sediments. These outcrops are representative for the adjacent basement areas covered by younger sediments, as demonstrated by geophysical investigations (see below) as well as numerous drillholes. This region may be considered to be the type area of the Variscan Orogeny (Fig. 11.3). Since Alpine collisional tectonics affected continental crust with a Variscan basement similar to the Variscan crust of stable Central Europe (von Raumer & Neubauer 1993), the Variscan Orogen formed an area much larger in size. It is, however, beyond the scope of this chapter to describe the entire orogen, and therefore we will concentrate on the key areas of Central Europe which were undeformed, or only slightly deformed by subsequent Alpine tectonics.

The subdivision by Kossmat

The classical subdivision of the Variscan area of Europe into a number of distinct geotectonic zones was made by Kossmat in 1927 (see McCann *et al.* 2008*a* for details). From north to south these are the Westphalian (originally the Subvariszische Saumsenke, i.e. Variscan Foredeep), the Rhenohercynian, the Saxothuringian and Moldanubian zones (Fig. 11.3) which together constitute the western Variscan foldbelt (Kossmat 1927). Subsequent modifications of Kossmat's original scheme have included the identification of the Mid-German Crystalline High (MGCH) by Scholtz (1930) as a zone of crystalline rocks located between Kossmat's Rhenohercynian and Saxothuringian zones. The extension of Kossmat's divisions to the Elbe Lineament was described by Möbus (1966), while equivalents of Kossmat's zones east of the Elbe Lineament were identified by Dvorak & Paproth (1969). Kossmat's original classification has also been extended to the West European Variscides (Ellenberger & Tamain 1980). The South Portuguese Zone is presumed to be the equivalent of the Rhenohercynian Zone, while the Ossa-Morena Zone has been compared to the MGCH. The Northern and Central Armorican zones are counterparts of the Saxothuringian Zone while the West Asturian-Leonaise Zone is also, in parts, comparable to the latter. The prolongation of the Moldanubian Zone to the west can be found in the south Armorican Zone and in most parts of the Galician-Castilian Zone (Schönenberg & Neugebauer 1997; Dallmeyer & Martinez Garcia 1990).

The **Rhenohercynian Zone** (RHZ) extends from Cornwall through to the Moravo-Silesian Belt and possibly into southern Portugal, and comprises mainly Devonian-age clastics (for details see Franke 2000). The Giessen-Werra-Harz nappes also contain Devonian MORB-type metabasalts and pelagic sediments (Franke 2000). The **Northern Phyllite Zone** (NPZ) is a narrow belt of rocks situated between the very low-grade parts of the Rhenish Massif and the Harz Mountains of western Germany to the NW, and the MGCH to the SE. The NPZ contains a very complex association of protoliths with westerly parts generally comprising rocks derived from the shelf of the Rhenohercynian Zone, while eastern parts contain portions of Armorica. Thus, it has been suggested that the NPZ represents part of the 'Rheic' Suture Zone (Cocks & Fortey 1982).

The **Mid-German Crystalline High** (MGCH) is a complex zone located to the south of the RHZ, whose structure, metamorphic grade and composition change considerably along strike (Franke 2000; Okrusch & Weber 1996; Oncken 1997) (for details see below).

The **Saxothuringian Zone** (STZ) comprises the northern parts of the Vosges and Black Forest (Schwarzwald) areas, as well as the Thuringian-Saxonian Schiefergebirge (i.e. Thüringer Wald), Fichtelgebirge and Erzgebirge areas. The deepest structural level of the STZ is present in the high-grade metamorphic rocks of the Saxon Granulite Complex and the Erzgebirge.

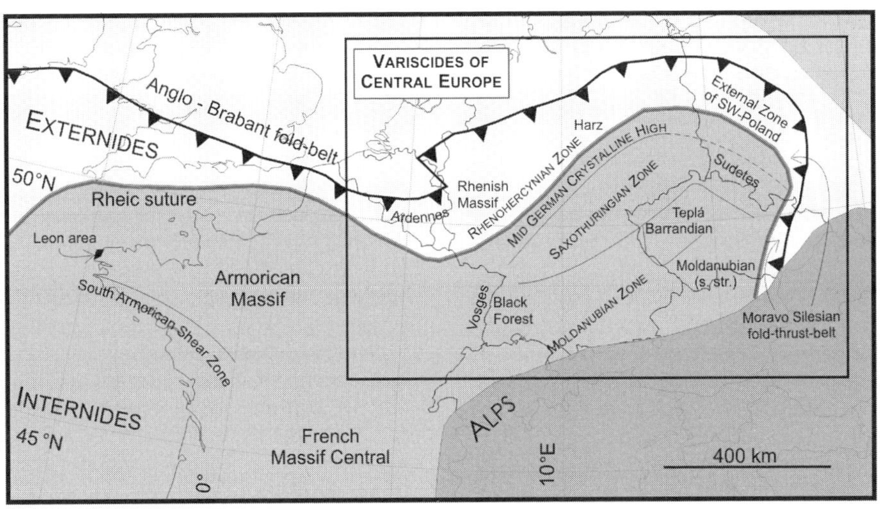

Fig. 11.3. Schematic sketch map of the Central European Variscides.

The **Moldanubian Zone** (MZ) comprises two parts, the northernmost Teplá-Barrandian (TB) and the more southerly Moldanubian Zone *sensu stricto*. The TB and the high-grade metamorphic rocks adjacent to the SE were termed the Moldanubian Region by Kossmat (1927). The MZ (including the TB area) comprises most of the Vosges/Black Forest area and the Bohemian Massif.

Internides and Externides of the Variscan Orogen in Central Europe

In general, the central European part of the Variscan Orogen can be subdivided into an internal and an external zone. The boundary between both zones constitutes the Rheic Suture Zone (Fig. 11.3). Thus, the entire external zone belongs to Laurussia whereas the Internides are built up either from the so-called Armorican Terrane Assemblage, which comprises peri-Gondwana crust initially situated south of the Rheic Ocean (Tait *et al.* 1997), or rock complexes that formed along the active margin of the Rheic Suture Zone. The external zone is considered as a (par-) autochthon. Taking this view, the narrow NE-SW striking allochthonous belt along the SE part of the RHZ of Germany (Northern Phyllite Zone, Giessen Harz Nappe) is different from the Externides of the Variscan orogen

The External Zone of the Variscan Orogen represents an extensive fold-and-thrust belt affecting low grade to non-metamorphic Palaeozoic sediments. The (par)autochthonous lithologies of the Ardennes–Brabant region, the Rhenish Massif, the Harz Mountains and the Moravo-Silesian Zone (Fig. 11.3) belong to this zone. The marine sedimentation which took place during the Devonian and the Early Carboniferous across large areas of the Externides was controlled by syn- to post-Acadian extension with localized zones of strong subsidence, submarine volcanism and marine platform sedimentation. These sedimentary successions were overlain by a Lower to Upper Carboniferous clastic wedge. This foreland basin became overfilled as a result of the high rates of sediment supply from the internal parts of the Variscides (Franke 2000, and references therein; see also McCann *et al.* 2008*a*). Paralic sedimentation continued until the Upper Carboniferous and led to the deposition of extensive coal deposits. In the Ruhr Basin, for example, coal formation commenced in the Namurian C, reached its maximum during the Westphalian A and B and ended during Westphalian D times (Drozdzewski 1993). These coal-rich successions are the youngest dated strata which were carried within the thrust sheets of the Variscan Externides (Ziegler 1990). Post-orogenic granites (293–295 Ma) of the Harz area (Baumann *et al.* 1991) define the uppermost time constraint for the Variscan Orogeny in Central Europe.

In contrast to the Variscan Externides, the internal zone is characterized by individual lithological complexes which preserve regional low- to high-grade metamorphism as well as synorogenic sedimentation and volcanism, both accompanied, and followed by, magmatism dominated by granites. Autochthonous Cadomian-age blocks are juxtaposed with allochthonous domains. The Armorican Massif of NW France, the Teplá-Barrandian of the Bohemian Massif and the Lusatian Block of the autochthonous domain of the Saxothuringian Zone are representative areas of Cadomian crust which were little (or not at all) affected by Variscan pervasive deformation (see Linnemann *et al.* 2008 & Timmerman 2008). In contrast with these zones of low strain, some units of the allochthonous domains contain information concerning Variscan subduction-related metamorphism (Fig. 11.3). For example, high pressure units occur: (1) in the Leon area of the Armorican Massif and the South Armorican Shear Zone; (2) in the French Massif Central; (3) in the Vosges and the Black Forest (Schwarzwald); (4) in the Mid-German Crystalline High; (5) in the allochthonous domain of the Saxothuringian Zone; (6) in the Western Sudetes; and (7) in the Moldanubian Zone. The protoliths of the infracrustal HP units were formed during the Cadomian Orogeny and/or belong to a subsequent Ordovician magmatic stage whereas the pre-metamorphic formation of the supracrustal HP units was related to post-Cadomian marine sedimentation (e.g. Mingram 1998). The diachronous exhumation of these rock complexes led to the formation of extensive nappe piles recognizable by their inverted metamorphic profiles. Pervasive ductile deformation, mylonitization, recumbent folding as well as localized crustal-scale ductile shear zones demonstrate the high strain character of these allochthonous domains. Some of the HP units were already exhumed in the Late Devonian as indicated by geochronological data as well as stratigraphic relationships (for details see below). However, as noted above, subduction processes continued until the Early Carboniferous (*c.* 340 Ma). Crustal thickening in the Variscan Internides during Early and Mid-Variscan plate convergence is considered to be non-homogeneous. For example in the Saxothuringian Zone a marine basin persisted until the end of the Lower Carboniferous.

In a later (*c.* 335–320 Ma) syn- to post-collisional stage, HT/LP metamorphism with migmatization and subsequent intrusion of voluminous granites is representative of large areas of the Internides of Central Europe. During this stage, which is generally characterized by transpressional tectonics as well as localized extensional domains, rock complexes formed at different crustal levels were juxtaposed.

Reconsidering the zones of Kossmat in terms of their internal or external position, the Westphalian and the (par)autochthonous parts of the Rhenohercynian Zone belong to the Externides whereas the Saxothuringian and Moldanubian zones form part of the Internides. The allochthonous parts of the RHZ and the MGCH constitute units of the Rheic Suture Zone, which will be described below in the context of the classical zonation of Kossmat as part of the external and the internal zones respectively. It is, however, important to note that the subdivision of the Variscan area of Europe was made by Kossmat during a time when nothing was known about the concept of plate tectonics. The astonishing recent acceptance of his ideas is reflected by the fact that different geotectonic zones of Kossmat were interpreted as individual lithospheric plates (terranes) bordered now by oceanic sutures (Franke 1992; Oncken 1997). As mentioned above, it is now considered indisputable that the continental crust affected by the Variscan Orogeny represents continent–continent collision with initial subduction of oceanic crust. However, there is no consensus about the number of independent plates or terranes and thus the number of subduction zones. Moreover, up until now there has been no satisfying Europe-wide synthesis of the orogen. For example, many authors have suggested that the geometry of the entire orogen, in relation to its present coordinates, is the result of exclusively NW–SE directed convergence (e.g. Matte 1986, 2001; Franke 1992; Oncken 1997; Schäfer *et al.* 2000). In such a view the NE–SW striking zones recognized by Kossmat represent the frontal parts of the subduction and collision zones whereas the strain transfer along NW–SE orientated domains such as the Sudetes and the South Armorican Shear Zone was accommodated by dextral shearing. However, observed NE–SW orientated *x*-axis of strain increments measured in Early Variscan as well as Mid-Variscan allochthonous units (e.g. Franke *et al.* 1995; Kroner & Hahn

2004; Leloix *et al.* 1999) within a flat-lying mylonitic foliation suggest that there was a large degree of pervasive subhorizontal shearing perpendicular to the proposed slip vectors of the plates during initial nappe emplacement. As mentioned above, the final geometry of the Variscan Orogen is dominated by the Late Variscan evolution of the external fold-and-thrust belts. At this time, however, in the Internides the Early and Mid-Variscan HP units affected by HT/LP metamorphism and intruded by post-kinematic granites were already exhumed to upper crustal levels. Thus, the majority of the strain increments of the infracrustal units pre-date this final thrusting event. It is, therefore, questionable as to what extent the relatively simple Late Variscan geometry of the Externides can be considered to be representative for the long-lasting and very complex history of the entire region. As a consequence of such objections there exist a number of alternative models which will be discussed at the end of this chapter.

Geophysical characteristics of the Variscan basement

Information concerning the recent structure of the deeper parts of the lithosphere affected by the Variscan Orogeny has been derived from geophysical investigations. Numerous seismic profiles in Central Europe combined with high resolution tomography, 3D anisotropy and receiver function analysis led to the following assumptions:

(1) The depth of the Moho varies between 24 and 38 km (Ziegler & Dézes 2006). Zones of low crustal thickness belong to Cenozoic extensional tectonics such as the rifting of the Rhine Valley Graben. Outside of this reactivated basement the average thickness amounts to *c.* 30 km. In the Bohemian and the Brabant Massifs the maximum depth of the Moho is reached at 36 km and 38 km, respectively. In the model of Hrubcová *et al.* (2005) the central part of the Moldanubian Zone is characterized by the deepest (39 km) and most pronounced Moho within the entire Bohemian Massif. If we consider that there was substantial crustal thickening during the process of Variscan collision, then the post-Variscan crust must be the result of a gravitationally driven equilibration process;

(2) A highly reflective lower crust is frequently observed within the internal part of the Variscides (Meissner 1999). In the Saxothuringian Zone, for example, this laminated lower crust is located at depths of 26–35 km. Similar values were also recorded by the MVE90 deep-reflection profile which was shot as part of the DEKORP investigation (DEKORP Research Group 1994; Bleibinhaus *et al.* 2003) and the GRANU'95 seismic refraction experiment (Enderle *et al.* 1998). The model of post-collisional delamination of dense eclogitic lower continental mafic crust (Wittenberg *et al.* 2000) provides an explanation for the felsic character with relatively little (*c.* 15%) mafic lower crustal rocks. The origin of the high reflectivity layers is considered to be the result of underplating of mantle-derived magmas during the gravitational collapse of the orogen (Costa & Rey 1995) in Late Variscan times; and

(3) There would appear to be evidence of pronounced heterogeneities within the thickness as well as the anisotropy of the lithospheric plate in the Variscan realm (albeit based on relatively sparse datasets). The depth of the lithosphere–asthenosphere boundary (LAB) is estimated at 80–90 km and 120–130 km for the Saxothuringian and the Moldanubian zones, respectively (Heuer 2006). Additionally, (Babuska & Plomerova 2001) observed divergently dipping anisotropic structures in the lithospheric mantle benath the Moldanubian and the Saxothuringian zones. Geophysical investigations within the region of the French Massif Central reveal the existence of individual lithospheric mantle domains which may be characterized by consistent large-scale crystallographic preferred orientations and separated by steeply inclined boundaries which cut through the entire lithosphere (Babuska & Plomerova 2006). These observations suggest that there was a juxtaposition of small coherent lithospheric blocks during the Variscan collision, possibly along lithospheric-scale strike-slip zones.

Structural trends in the northern Variscides: evidence from geophysical potential fields (T.P.)

Images of the magnetic and gravity fields (anomalies and their derivatives) provide powerful insights into the structure and evolution of the orogenic basement of Central Europe, particularly where it is largely concealed by thick post-orogenic cover. They reveal basement structural grain not visible at the surface and represent a dense grid of observations supplementing discrete observations derived from surface outcrops, boreholes and seismic profiles. Data gathered by the various national geoscience agencies have been standardized and compiled into colour graduated maps covering the region (for further details see Wonik 1992; Wybraniec *et al.* 1998; Banka *et al.* 2002). Interpretation of the images is of course complicated by post-Palaeozoic sources, e.g. Permian volcanic suites, and Mesozoic–Cenozoic sedimentary basins, which sometimes blur the Palaeozoic fingerprint, but occasionally enhance it, as a result of reactivation of the original basement grain. Analysis of the images is facilitated by overlaying a structural template, derived from previously published studies by the EUROPROBE Trans-European Suture Zone (TESZ) Project (Pharaoh 1999; Banka *et al.* 2002) and the PACE TMR Network (Winchester & PACE TMR Network Team 2002).

The Variscide foreland zone is heterogeneous in aspect, comprising a diversity of crustal terranes, mostly Gondwana-derived, accreted to Laurussia (or 'Old Red Continent') in several pre-Variscan orogenic phases (Ziegler 1990; Pharaoh 1999). In the west, the crust of the former Eastern Avalonia Terrane (north of the Variscan Front) is significantly less magnetic (Fig. 11.4) than that of the East European Craton (EEC). An arcuate pattern of magnetic anomalies wrapping around the Midlands Microcraton 'indenter' on its northern side, and continuing into the Brabant Massif of Belgium, spatially associated with Acadian-age slate belts (Soper *et al.* 1987; Lee *et al.* 1993), is caused by their (largely concealed) early Cambrian component, and Ordovician intrusions. Neoproterozoic meta-igneous basement rocks involved in complexly reactivated crustal lineaments (Pharaoh & Gibbons 1994) cause a number of linear magnetic highs. Permo-Carboniferous magmatic bodies also contribute to the magnetic signature (Cornwell *et al.* 1994). Beneath the southern North Sea and northern Germany, the crust of the Far Eastern Avalonia Terrane (Winchester & PACE TMR Network Team 2002) (see Krawczyk *et al.* 2008 for discussion and details), is apparently rather heterogeneous. In the north, prominent magnetic highs are attributed to magmatic complexes of either Caledonian (e.g. Williamson *et al.* 2002) or Permo-Carboniferous age (Timmerman 2004). Local magnetic highs in the northern Netherlands, north Germany and in the vicinity of the Central Graben are associated with thick accumulations of Permian volcanic rocks known from deep boreholes (Lokhorst 1998). The Variscan Foreland of central Poland comprises crust of the former Łysogóry, Małopolska and Bruno-Silesian terranes, which were also accreted during the Caledonide Orogeny (Pożaryski 1990; Dadlez *et al.* 1994; Pharaoh 1999; Winchester & PACE TMR Network Team 2002). These terranes have a much more subdued

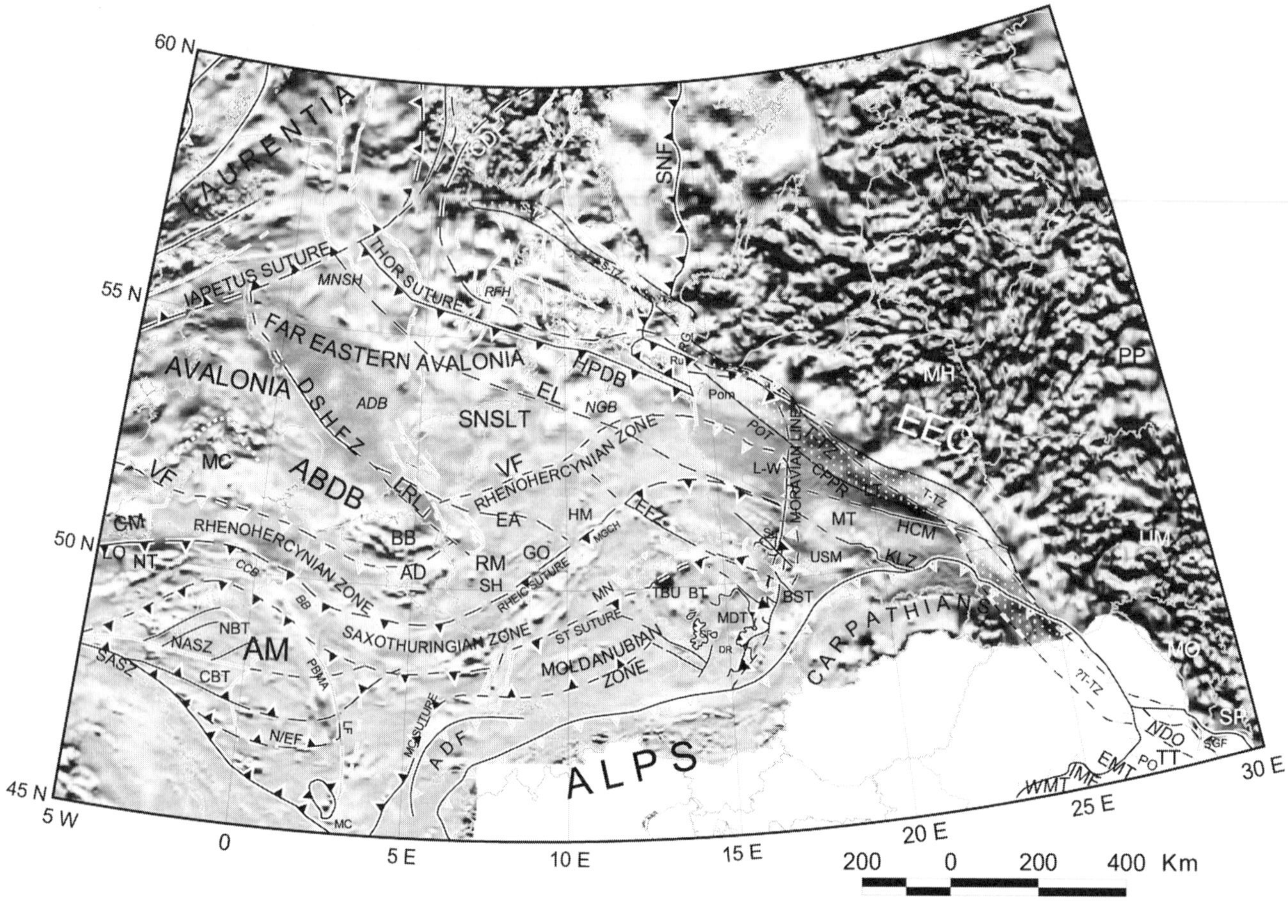

Fig. 11.4. Magnetic image (modified after Banka *et al.* (2002, and references therein), with superimposed structural template modified from Pharaoh (1999). Revisions incorporate information from Matte *et al.* (1990), Dallmeyer *et al.* (1995), Franke (1995), Chantraine *et al.* (1996) and Franke *et al.* (2000). Key to structural template. *Post-Palaeozoic basins and platforms* (indicated in italic font): ADB, Anglo-Dutch Basin; ADF, Alpine Deformation Front; BB, Bray Basin; CCB, Central Channel Basin; CD, Central Dobrogea; MNSH, Mid-North Sea High; MP, Moesian Platform; NDO, North Dobrogea Orogen; NGB, North German Basin; POT, Polish Trough; RFH, Rynkøbing-Fyn High; RG, Rønne Graben; SP, Scythian Platform. *Postulated Palaeozoic terranes and terrane/subterrane boundaries*: CBT, Central Brittany Terrane; DR, Drosendorf Unit (of BM); DSHFZ, Dowsing-South Hewett Fault Zone; EL, Elbe Lineament; GF, Gföhl Unit (of BM); KLZ, Kraków-Lubliniec Zone; LRL, Lower Rhine Lineament; ŁT, Łysogory Terrane); ML, Moravian Line; MST, Moravo-Silesian Terrane; MT, Małopolska Terrane; NASZ, North Armorican Shear Zone; NBT, North Brittany Terrane; N/EF, Nort-Erdre Fault; NT, Normannian Terrane; PCF, Peceneaga-Camena Fault; SASZ, South Armorican Shear Zone; SNSLT, Southern North Sea-Lüneburg Terrane; SGF, Sfantu Gheorghe Fault; TB, Tepla-Barrandian Unit (of BM). *Other Proterozoic–Palaeozoic tectonic elements*: ABDB, Anglo-Brabant Deformation Belt; AD, Ardennes Massifs (Stavelot, Rocroi); AM, Armorican Massif; BB, Brabant Massif; BM, Bohemian Massif; BVT, Brunovistulian Block; 'CDF', front of Caledonian deformation (see text for explanation); CM, Cornubian Massif; COF, Capidava-Ovidiu Fault; EA, Ebbe Anticline; EEC, East European Craton; EFZ, Elbe Fault Zone; GO, Giessen Ophiolite Nappe; HM, Harz Mountains; HCM, Holy Cross Mountains; HPDB, Heligoland-Pomerania Deformation Belt; LF, Loire Fault; LO, Lizard Ophiolite Nappe; L-W, Leszno-Wolsztyn Basement High; MC, Midlands Microcraton; MC (small font), Massif Central; MGCH, Mid-German Crystalline High; MH, Mazury High; MN, Münchberg Nappe (of BM); MO, Moldavian Platform; PBMA, Paris Basin Magnetic Anomaly; PP, Pripyat Trough; RM, Rhenish Massif; SNF, Sveconorwegian Front (strictly, the Protogine Zone); SH, South Hunsrück; S-TZ, Sorgenfrei-Tornquist Zone; T-TZ, Teisseyre-Tornquist Zone; UM, Ukrainian Massif; USM, Upper Silesian Massif (=MST); VF, Variscan Front. *Other geographic locations*: Pom, Pomerania; Rü, Rügen Is; Su, Sudetes Mountains (including Sowie Góry).

magnetic character than the EEC which they abut across the Teisseyre-Tornquist Zone. Most of the area of the EEC and the exposed Fennoscandian Shield is characterized by short wavelength/high amplitude magnetic anomalies, reflecting the strongly magnetic nature of the upper and middle crust (Williamson *et al.* 2002). The complex pattern of the anomalies belies the long and complex evolution of the EEC crust. The NNE-trending magnetic anomalies of the EEC are sharply truncated by the NW-trending Teisseyre-Tornquist Zone in central Poland, which defines the far-field or 'buffer zone' of the Variscan Orogen in central Europe (Berthelsen 1998). The most prominent features on the gravity map (Fig. 11.5) are the negative anomaly associated with the Polish Trough, where at least 14 km of late Palaeozoic–Cenozoic sedimentary strata may be preserved (Janik *et al.* 2002), and the positive anomaly associated with the Holy Cross Massif. To the east these anomalies are truncated by the very large gravity low associated with the foredeep of the Carpathian Orogen.

The Rhenohercynian Zone represents a part of Eastern Avalonia (and Far Eastern Avalonia), tectonically displaced by the Variscan Frontal Thrust (Cocks *et al.* 1997). Weak magnetic anomalies in Cornubia extend eastward into southern Britain, through NE France into the Ardennes massifs of Belgium (Fig. 11.4), where they exhibit a major strike swing from east–west to

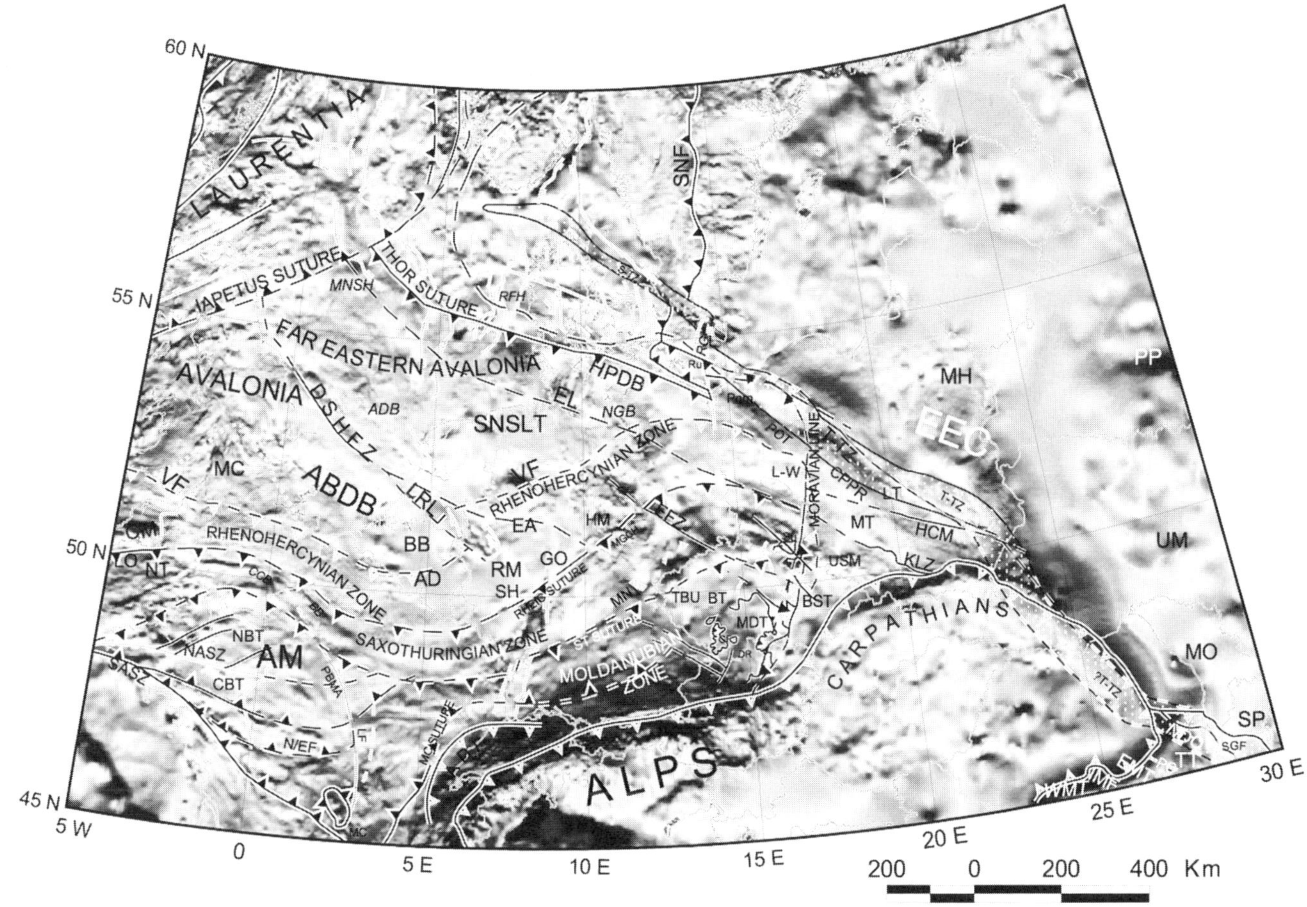

Fig. 11.5. Gravity image (modified after Banka *et al.* 2002, and references therein), with superimposed structural template modified from Pharaoh (1999). See Figure 11.4 caption for key to abbreviations.

SW–NE. In Germany, the magnetic signature of the Rhenohercynian Zone is rather featureless at this scale; nevertheless it is possible to discern a weak tectonic grain paralleling the much stronger signal from the MGCH. The magnetic images are not particularly helpful in defining the location of the Variscan Front, except in central England, where a prominent anomaly associated with igneous rocks of Precambrian, early Palaeozoic and Carboniferous age is truncated (Cornwell *et al.* 1994; Busby & Smith 2001), and in Belgium (Fig. 11.4). The gravity image (Fig. 11.5) reinforces the impression of the structural grain obtained from the magnetic images (Fig. 11.4), but in this case the source of the anomalies is frequently Mesozoic basins formed by extensional reactivation of Variscan basement structures (Whittaker 1985), e.g. the Wessex and Weald Basins. Also prominent in Figure 11.5 are anomalies discordant to the basement grain, associated with oblique structures such as the Elbe Fault Zone (Scheck *et al.* 2002), initiated as a late Variscan structure but with possible older basement inheritance, and the Rhine Graben of Cenozoic age.

The Variscide Internide Zones are associated with stronger magnetic anomalies. The MGCH, at the northern edge of the Saxothuringian Zone in Germany, is associated with a discontinuous positive magnetic anomaly, arcuate in form and broadly paralleling the Variscan Front, extending from northern France to western Poland, just to the SW of the Leszno-Wolsztyn High (Fig. 11.4), where it is sharply truncated by the inferred Moravian Line. The likely sources of the anomaly are magnetic components of a calc-alkaline magmatic arc complex, generated by southward subduction of the Rheic Ocean in Silurian–Devonian time (Dallmeyer *et al.* 1995), and/or ophiolitic relicts of ocean crust (Lefort *et al.* 1991). The magnetic anomalies associated with the MGCH enable the Rheic suture to be traced throughout the central part of the region (Fig. 11.4).

In northern France and the English Channel, linear belts of strong positive magnetic anomalies (Isle of Wight, mid-Channel anomalies) sourced in the basement have been correlated with possible ophiolitic fragments (Lefort *et al.* 1991). Some may be of Neoproterozoic (Cadomian) age, as in the St Brieuc Magmatic Arc (Ledru 2001), but a more likely source is the mid-Devonian Lizard Ophiolite (Clark *et al.* 1998; Busby & Smith 2001) which lies along strike from the mid-Channel anomalies. According to Busby & Smith (2001), a major (?oroclinal) strike-swing takes the suture into a NW orientation in the English Channel where, in the interpretation presented in Figure 11.4, it passes to the northern side of the Bray Basin and the Central English Channel Basin. The latter may reflect Mesozoic extensional reactivation of the hanging wall of the suture with appropriate expression in the gravity image (Fig. 11.5). However, this is a matter for speculation; several other authors (Leveridge *et al.* 1984; Matte *et al.* 1990; Dallmeyer *et al.* 1995) have advocated excision of the Saxothuringian Zone here altogether as a result of a major inferred strike-slip fault (the postulated Wight-Bray structure).

The Paris Basin Magnetic Anomaly (PBMA), stretching at least 400 km from central France to the English Channel, is one of the most prominent features in the magnetic image (Fig. 11.4). The Sancerre-Couy Deep Borehole terminated just above the strongly magnetic rocks which are the source of the anomaly (Million 1992), Amphibolitic relicts of a Cambro-Ordovician ophiolite are involved in a positive flower structure associated with a major north–south trending late Variscan fault (Loire Fault) on the eastern flank of the anomaly (Autran *et al.* 1992). If the PBMA links directly with the mid-Channel anomalies (Busby & Smith 2001), it is possible that the source of the latter may also contain a Saxothuringian component. Major shear zones (North and South Armorican shear zones, Nort/Erdre Fault) are inferred to partition the Armorican Massif (Matte *et al.* 1990; Chantraine *et al.* 1996) into subterranes, associated with narrow linear belts of positive magnetic anomalies, clearly visible on Figure 11.4, and also in the gravity image (Fig. 11.5). These mostly east–west trending structures show an 80 km sinistral offset on the Loire Fault.

The magnetic image (Fig. 11.4) clearly demonstrates the NE grain of the Saxothuringian Zone from the northern edge of the Vosges and Black Forest towards the NW part of the Bohemian Massif, and the strike-swing to WNW into the Sudetes Mountains, known from the surface geology (Dallmeyer *et al.* 1995). The structural grain of the Moldanubian Zone is less well imaged, but appears to parallel that of the Saxothuringian Zone. Ophiolitic rocks, associated with the intervening Saxothuringian suture, locally contribute to the magnetic signature, e.g. at the northern boundary of the Teplá-Barrandian Unit, but those associated with the Massif Central–Main Moldanubian Suture are more poorly expressed in the image. The NNE-trending Rhine Graben cuts obliquely across the Variscide structural grain, most clearly in the gravity image (Fig. 11.5). A number of NW-trending offsets to the Variscide grain are observed on the magnetic images, both within the Bohemian Massif and on its southwestern flank. These offsets are correlated with faults such as the Elbe Fault Zone, reactivated throughout the Mesozoic and Cenozoic (Scheck *et al.* 2002), and the Marginal Sudetic Fault. Where such faults have a significant amount of Mesozoic reactivation, they are more clearly expressed in the gravity image (Fig. 11.5). Also apparent is the abrupt truncation of the magnetic anomalies originating from structures of the Saxothuringian and Moldanubian zones at the NNE-trending Moravian Line, along the eastern edge of the Bohemian Massif. This boundary has recently been interpreted as a suture between the EEC (and early Cambrian accreted terranes such as the Bruno-Silesian Terrane) and the Armorican Terrane Assemblage (ATA). The original intervening ocean basin is inferred to have closed by subduction beneath the Bruno-Silesian Terrane (Winchester *et al.* 2002).

Variscan Internides: Black Forest (Schwarzwald) (H.P.H. & H.Z.)

The Black Forest (Schwarzwald) and the Vosges on the opposite side of the Rhine Graben are located within the Internides of the Variscan Orogen (Fig. 11.6). In the northern Black Forest-Vosges segment of the Variscan belt, the high-grade Moldanubian Zone is juxtaposed along an oceanic suture, the Baden-Baden Zone (BBZ), against low- to high-grade rocks of the Saxothuringian Zone (Kossmat 1927; Wickert & Eisbacher 1988; Krohe & Eisbacher 1988; Franke 1989). From north to south, the Moldanubian Zone of the Black Forest is subdivided into three units (Fig. 11.6): the Central Schwarzwald Gneiss Complex (CSGC), the Palaeozoic Badenweiler-Lenzkirch Zone (BLZ) and the South Schwarzwald Granite and Gneiss Complex (SSGC). The BLZ underwent SE-vergent thrusting and dextral wrenching during the Lower Carboniferous (Altherr & Maass 1977; Sittig 1981; Krohe & Eisbacher 1988; Eisbacher *et al.* 1989; Maass *et al.* 1990; Echtler & Chauvet 1992; Krohe 1996). Lithological differences would suggest that the CSGC and the SSGC represent different crustal blocks (Wimmenauer 1984; Lüschen *et al.* 1987; Müller 1989; Stenger *et al.* 1989; Kalt 1991) However, both basement blocks show evidence of a close relationship during the Variscan Orogeny, including an episode of HT/LP metamorphism in Visean times (Kalt *et al.* 1994*a*; Lippolt *et al.* 1994). The CSGC and the SSGC can be subdivided into internal nappes (Fig. 11.6), which reflect the Variscan collisional event (Hann & Sawatzki 2000; Kalt *et al.* 2000*a*; Hegner *et al.* 2001).

All three units within the Black Forest area, as well as the nappe boundaries, are cut by large Variscan granitoids and granite porphyry dykes with late Visean (334–332 Ma: Brewer & Lippolt 1972; Todt 1976; Schaltegger 2000) and Namurian (325 ± 7 Ma, 'Northern Schwarzwald Granite Massif'; Hess *et al.* 2000) intrusion ages.

The Baden-Baden Zone (BBZ)

According to Wickert *et al.* (1990) and Kalt *et al.* (2000*a*), the BBZ contains accreted crustal fragments with different metamorphic and tectonic histories and comprises the following rocks associations: (1) gneisses and amphibolites that equilibrated at HT/LP conditions (700–780°C, 0.40–0.45 GPa); (2) kyanite–garnet mica schists and quartzites (630–670°C, 0.9 GPa); and, (3) low-grade metasediments and metabasites (approximately 450°C, 0.2 GPa) with deformed two-mica granites. Sittig (1965) subdivided the metasedimentary successions into three series of different facies: (a) schists and quartzites formed under shallow-water conditions at the base of the succession (Lower Schindelkamm series); (b) greywackes indicating deeper-water conditions (Upper Schindelkamm series); and (c) metapelites with carbonate olistoliths indicating deeper-water conditions (Traischbach series). According to Montenari & Servais (2000) acritarchs document an Upper Cambrian to Lower Ordovician sedimentation age for the succession and these assemblages can be possibly correlated with the Villé Unit of the northern Vosges. The late Early Carboniferous intrusion of the hornblende–biotite-bearing Friesenberg Granite (Wickert *et al.* 1990) resulted in contact metamorphism of the entire succession. Undeformed sedimentary and acidic volcanic rocks of Stephanian to Early Permian age cover most of the BBZ.

The Central Schwarzwald Gneiss Complex (CSGC)

The CSGC consists of a nappe stack, which formed mainly under ductile conditions (Fig. 11.6). The gneiss complex has been interpreted as a bivergent 'pop-up structure' (Eisbacher *et al.* 1989) that was thrust over both the BBZ in the north and the BLZ to the south. The Feldberg Nappe (monotonous unit with high-pressure relicts; unit 1 after Hanel & Wimmenauer 1990; Kalt & Altherr 1996) is located in the tectonically uppermost position, while in tectonically lower positions are the Granulite Nappe (retrogressed granulites; unit 3 after Röhr 1990; Hanel *et al.* 1993), the Münstertal Nappe (variegated unit without high-pressure relicts; unit 2 after Flöttmann & Kleinschmidt 1989; Hanel & Wimmenauer 1990) and the Rand Granite Nappe (Hann *et al.* 2003*a*).

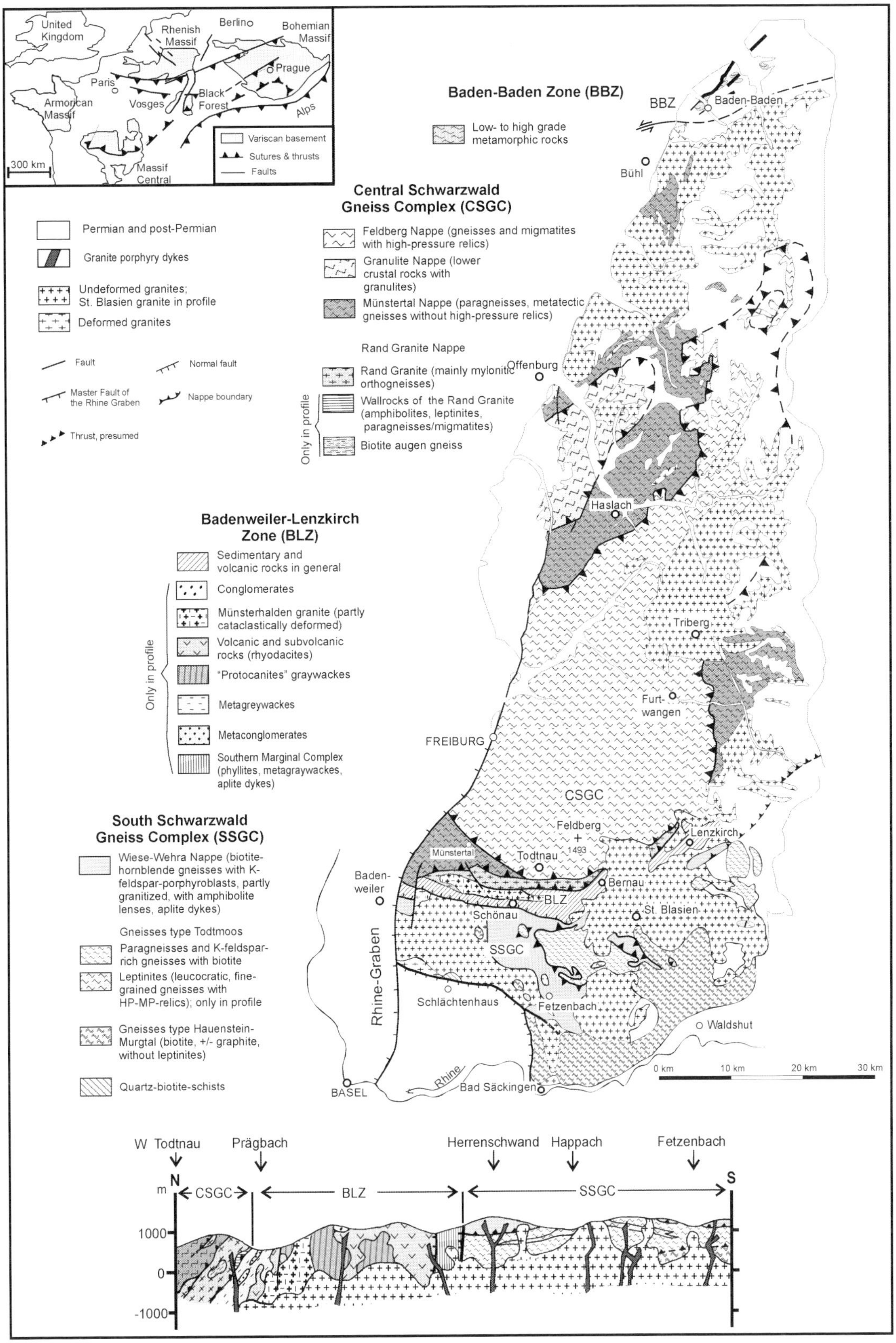

Fig. 11.6. Simplified geological map of the Black Forest area (see text for further explanation).

The Feldberg Nappe consists of paragneisses and migmatites (e.g. Mehnert 1963; Wimmenauer 1984) that record only the HT/LP metamorphic stage (700–780°C, 0.40–0.45 GPa (Kalt *et al.* 1994*a*); the mineral assemblage comprises biotite + plagioclase + quartz ± alkali feldspar ± garnet ± sillimanite ± cordierite) and contain numerous lens-shaped bodies of retrogressed eclogite (Klein & Wimmenauer 1984) and some ultramafic high-pressure rocks (Kalt & Altherr 1996). The pre-metamorphic history and the depositional conditions of the metasediments are constrained by palynological data and indicate a Late Proterozoic to Early Palaeozoic marine enviroment (Montenari 1996; Hanel *et al.* 1999). SHRIMP U–Pb dating of zircons from the paragneisses (Kober *et al.* 2004) revealed Late Proterozoic ages in inherited crystal cores.

The Granulite Nappe is characterized by granulite-facies relicts and rare eclogites and mantle-derived rocks. The gneisses are mainly of metapsammitic and metapelitic composition (Müller 1989) and bear the typical assemblage of garnet + kyanite + ternary feldspar + quartz + biotite + rutile (Kalt *et al.* 2000*a*). The metasediments contain Late Ordovician and Devonian detrital material (Kober *et al.* 2004). As in the other nappes, Archaean and Early Proterozoic zircon cores are present. The granulites required peak P-T conditions with a minimum temperature of between 950 and 1010°C at pressures of 1.4–1.8 GPa (Marschall *et al.* 2003). The metamorphism is dated at 335 ± 2 Ma (Kober *et al.* 2004).

The Münstertal Nappe consists of heterogeneous gneisses, quartzites, marbles, and amphibolites which show no relicts of a former eclogite stage. The gneisses contain rare relicts demonstrating previous MP metamorphic conditions (Rehfeld 1983). Orthogneisses reflect a Cambrian magmatic event followed by an Early Ordovician HT metamorphic phase (Chen *et al.* 2000). The results of palynological studies indicate an age of deposition in Ordovician–Silurian times (Hanel *et al.* 1999). The metasediments contain Late Ordovician detritus that probably derived from microcontinents of Gondwanan origin (Kober *et al.* 2004).

The Rand Granite Nappe forms a narrow zone, elongate in both east–west and NE–SW directions, and consisting of a heterogeneous metamorphosed granitoid rock complex with numerous wall-rock inclusions (Rand Granite association: Wimmenauer & Hanel 1997). It is situated at the southern margin of the CSGC and thrust over the BLZ. The Rand Granite Nappe comprises metagranites, metatrondhjemites and biotite augen gneisses derived from monzogranites to granodiorites. The complex developed along an active continental margin and its presence indicates long-term magmatic activity related to subduction. It is, therefore, interpreted as a remnant of the root of a Variscan magmatic arc (Hann *et al.* 2003*a*). The Rand Granite has a key position between the CSGC and the BLZ. The reconstruction of its geotectonic position during protolith formation and its tectonic and metamorphic history are important for the understanding of the geodynamic evolution of the internal Variscides of the Black Forest area.

The Badenweiler-Lenzkirch Zone (BLZ)

The BLZ is bordered by a nappe boundary to the north (Lüschen *et al.* 1987) and by a normal fault to the south. Its lithological units were imbricated and rotated into steeply inclined positions during ongoing continental collision and dextral transpression (Güldenpfennig 1997). From north to south, the BLZ can be subdivided into the following subunits (Fig. 11.6; Hann & Sawatzki 2002; Hann *et al.* 2003*b*):

(1) A subunit which comprises predominantly low-grade and subordinately medium-grade metagreywackes, metasiltstones with deformed conglomerates (Lower Palaeozoic glaciomarine diamictites; Ziegler & Wimmenauer 2001), quartzites, phyllites, and rare metavolcanic layers. Acritarchs and chitinozoans in the weakly metamorphic schists yielded Ordovician to Silurian depositional ages (Hann *et al.* 1995; Montenari *et al.* 2000; Vaida *et al.* 2004);

(2) The central subunit consists of the Tournaisian to Visean 'Protocanites greywackes' (e.g. Spiegelhalter 1910; Wilser 1933; Sittig 1963) and subordinate pelites. These experienced temperatures of high diagenesis to anchizonal metamorphism (Güldenpfennig 1997). Weyer (1962) and Kneidl *et al.* (1982) described Late Devonian conodonts from the pelitic layers. The greywacke sequence is cross-cut by subvolcanic rhyodacite and is, in places, overlain by pyroclastic rocks and rhyodacites of Visean age (340 ± 2 Ma; Schaltegger 2000). All of these lithologies are topped by Visean continental sediments (e.g. Krecher 1999). The greywacke succession has been interpreted as having been deposited in a foreland basin positioned next to a north-dipping subduction zone (Loeschke *et al.* 1998; Hegner *et al.* 2005); and

(3) The southern subunit or 'Südrandkomplex' (Southern Marginal Complex) is formed by weakly metamorphic lithologies, including cataclastically overprinted phyllites, metasiltstones and metagreywackes (Sittig 1969; Maass *et al.* 1990) penetrated by aplitic dykes. The depositional age of these rocks is Ordovician to Devonian (Vaida *et al.* 2004).

The South Schwarzwald Granite and Gneiss Complex (SSGC)

The SSGC (Fig. 11.6) consists of high-grade metamorphic gneisses of the Wiese-Wehra diatexites and gneisses of the Todtmoos and Hauenstein-Murgtal type (e.g. Hoenes 1948; Metz & Rein 1958; Mazurek 1985; Wimmenauer 1988). This assemblage also underwent Visean HT/LP metamorphism. The Wiese-Wehra diatexites with metagabbros and meta-anorthosite fragments (Sebert & Wimmenauer 1992) form a nappe structure (Hann & Sawatzki 1998). Thrusting occurred under ductile conditions between 342 and 333 Ma (Hegner *et al.* 2001). The gneisses of the Todtmoos and Hauenstein-Murgtal type constitute the autochthonous unit. The Todtmoos gneisses contain leptynites (510–500 Ma: Chen 1999) with granulite-facies relicts. Peak conditions are 1015°C at pressures of 1.5 GPa (minimum temperature, maximum pressure); no prograde relicts are preserved and isothermal decompression was less pronounced than in the CSGC granulites (Marschall *et al.* 2003). The Hauenstein-Murgtal gneisses, which are widely distributed in the southern and southeastern parts of the SSGC, predominantly consist of garnet, sillimanite and cordierite-bearing biotite gneisses containing graphite in places. Palynological studies yielded mid-Silurian protolith ages (Vaida *et al.* 2004).

Geodynamic evolution of the Black Forest (Schwarzwald) region

An overview of the geodynamic evolution of the Black Forest region has recently been presented by Hann *et al.* (2003*a*) and Hegner *et al.* (2005). The considered scenario fits the large-scale models for the evolution of the Variscan Orogen as presented by Ziegler (1989), Matte (1991), Franke *et al.* (1995), Franke (2000) and Frisch & Meschede (2005). Plate tectonic reconstructions indicate that continental blocks split off from the northern margin of Gondwana during Ordovician times, drifted towards Armorica and Laurussia and collided in the western part (Vendée, Massif Central) during the Devonian; further to the east collision took place in the Early Carboniferous (Matte 1986; Ziegler 1986, 1990; Bachtadse *et al.* 1995; Franke *et al.* 1995; Tait *et al.* 1997; von Raumer 1998; von Raumer *et al.* 2002). The CSGC and SSGC represent fragments of different microcontinents. The SSGC probably represents a separate terrane (the 'Anti-Saxothuringian' after Maass *et al.* 1990) that could be correlated with the Drosendorf Unit of the southern Bohemian Massif (Pharaoh

1999). The palynological data suggest that the sediments of the BLZ, which were deposited in an oceanic zone between these two continental realms, formed in Ordovician to Silurian times.

With regard to the lithological aspects of the succession, parts of the CSGC can be correlated with the southwestern Bohemian Massif (Kalt *et al.* 2000*a*). The BLZ probably represents the continuation of the suture zone of the southern Massif Central, situated between the 'Massif Central' and the 'Vendée-Cevennes' terranes (Fig. 11.6; Matte *et al.* 1990; Hegner *et al.* 2001) or, in a more regional context, it is the boundary between Armorica in the north and Gondwana in the south (Frisch & Meschede 2005). Correlations of the CSGC, BLZ and SSGC with the Vosges are problematic or not possible (e.g. Kalt *et al.* 2000*a*; Hegner *et al.* 2001). Convergence between the continental blocks on both sides of the BLZ occurred in Silurian to Early Carboniferous times along a north-dipping subduction zone (Fig. 11.7a). High-pressure (HP) metamorphism related to this subduction is indicated by the occurrence of eclogites in the CSGC. This Variscan HP metamorphism was constrained, by Sm–Nd dating on garnet from eclogite slivers in the CSGC, at 348–340 Ma (Chen *et al.* 2003). Similar ages – within errors – were obtained by Kalt *et al.* (1994*b*): i.e. 337 $\pm$ 6 Ma to 332 $\pm$ 13 Ma. Less precise Pb/Pb dating on garnet from the eclogites yielded isochron ages ranging between 406 and 353 Ma. All of these data suggest that subduction took place in Devonian to Early Carboniferous times. Magmatic protolith ages of the eclogites are in the range of 486 to 445 Ma (Ordovician; Kalt *et al.* 1994*b*; Chen *et al.* 2003).

Mafic granulites with MORB-like compositions and high initial εNd values (up to +9) probably indicate the formation of an oceanic basin in the Ordovician and its closure during the Devonian to Early Carboniferous (Chen 2002; Chen *et al.* 2003). Subduction of the oceanic crust in Silurian to Devonian times can be inferred from the protolith ages of the biotite augen gneisses (Silurian) and the trondhjemites (Devonian) of the Rand Granite magmatic suite. The ocean had a significant lateral dimension (probably more than 1000 km) and this is consistent with the movement of the northern margin of Gondwana from high to low palaeolatitudes during the same time interval (Frisch & Meschede 2005).

The Rand Granite represents a remnant of the plutonic level of a magmatic arc. From Silurian to Early Carboniferous times, the magmatic belt formed in an active continental margin setting located to the north of the BLZ (Fig. 11.7a). Greywackes were deposited in the trench and scraped off to build up an accretionary wedge, which now forms part of the BLZ. The Upper Devonian to Lower Carboniferous greywackes contain andesitic detritus derived from the volcanic level of the magmatic arc and clastic Cr-spinel grains, that were probably derived from ophiolites which had been incorporated into the accretionary wedge during subduction (Fig 11.7a; Güldenpfennig 1997, 1998). A thin layer of rhyolitic composition found in metagreywackes and phyllites of the northern BLZ (see Fig. 11.7a, d) has been dated at 393 $\pm$ 3 Ma (Hann *et al.* 2003*b*). This layer represents a rhyolitic ash fall of Early Devonian age derived from the magmatic arc to the north (Fig. 11.7a).

The trondhjemites from the Rand Granite have the youngest model U–Pb zircon ages (Tournaisian). If the upper-intercept ages of 436 Ma (biotite augen gneiss) to 377 Ma which were calculated from forced regression can be considered as the crystallization ages, then it can be suggested that the entire period of magmatic activity lasted from 436 Ma to 351 Ma, reflecting a long-lasting subduction scenario. A major part of the studied magmatic rocks originated from the anatexis of continental crust, as revealed by negative εNd values. Anatexis was triggered by basaltic melts generated above the subducted oceanic slab. Other parts (trondhjemites) of the plutonic complex are probably the differentiation products of mantle melts as shown by the high initial εNd values (+6.5 to +6.7) and the relatively low initial $^{87}Sr/^{86}Sr$ ratios (0.7042 to 0.7063). These isotopic characteristics indicate that a major component was derived from a mantle source. The granites and biotite augen gneisses have negative εNd values (−4.7 to −3.3) and initial $^{87}Sr/^{86}Sr$ values (0.7068 to 0.7077) indicating a mainly crustal origin. These isotopic data strongly imply that the Rand Granite forms a heterogeneous plutonic complex consisting of rocks derived from both mantle and crustal sources.

The interpretation of the Rand Granite as a heterogeneous plutonic complex is also supported by geochemical data sets. Trace element distributions are compatible with a magmatic arc to collisional setting. The zircon morphological data of the Rand Granite Complex also indicate both crustal and mantle origins. All of these results may be interpreted in terms of a subduction-related setting along an active continental margin. The presence of such a setting, and one which existed over a long period of time, led to the development of a complex magmatic arc to the north of the BLZ which, in turn, may be considered to represent the sediments of the oceanic and trench realms, which were imbricated in an accretionary wedge. Only a small part of the interpreted magmatic arc is preserved in the Rand Granite Complex.

According to Hegner *et al.* (2005), detrital zircon ages of 384–371 Ma suggest that there was a phase of igneous activity shortly before the deposition of the greywacke succession at *c.* 365 Ma. Approximate εNd values of +3 to +10 in the greywackes indicate that there was a mixing of juvenile and older crustal material. Assessment of the composition of the juvenile component shows high concentrations of light REE, and low Sm/Nd and Nb/La ratios, all of which are consistent with a subduction-related origin. In conclusion, the greywacke succession was probably derived from a Late Devonian Andean-type continental margin. Both the Rand Granite Complex and the Protocanites greywackes of the BLZ generally show the same variational trends in their εNd values. Zircons of Devonian age occur in both units. Therefore, it can be suggested that the andesitic detritus in the greywackes was derived from the eroded volcanic level of the magmatic arc.

The formation and evolution of the Rand Granite Complex as a composite and probably primarily elongate magmatic body commenced over the course of subduction and lasted until the collisional stage. Located above the subduction zone, magmatic chambers formed diachronously and with differing compositions. Mantle-derived melts penetrated the crust and assimilated crustal material or mixed with crustal melts. Such a scenario can be used to explain the observed petrographical and geochemical characteristics as well as the evolution of the zircons contained within the Rand Granite lithologies. In conclusion, in Silurian to Early Carboniferous times, subduction was accompanied by the development of a magmatic arc, from which the Rand Granite Complex is only a minor remnant.

Shortly (at least 10 Ma) after the addition of the final intrusions into the Rand Granite Complex in Tournaisian times, the entire assemblage, as a part of the CSGC, was subject to pervasive ductile deformation as a result of nappe stacking, resulting in the develolpment of widely developed mylonitic fabrics. Thrusts formed in-sequence from higher to lower structural levels under falling metamorphic conditions (amphibolite to greenschist facies). The latest movements along the

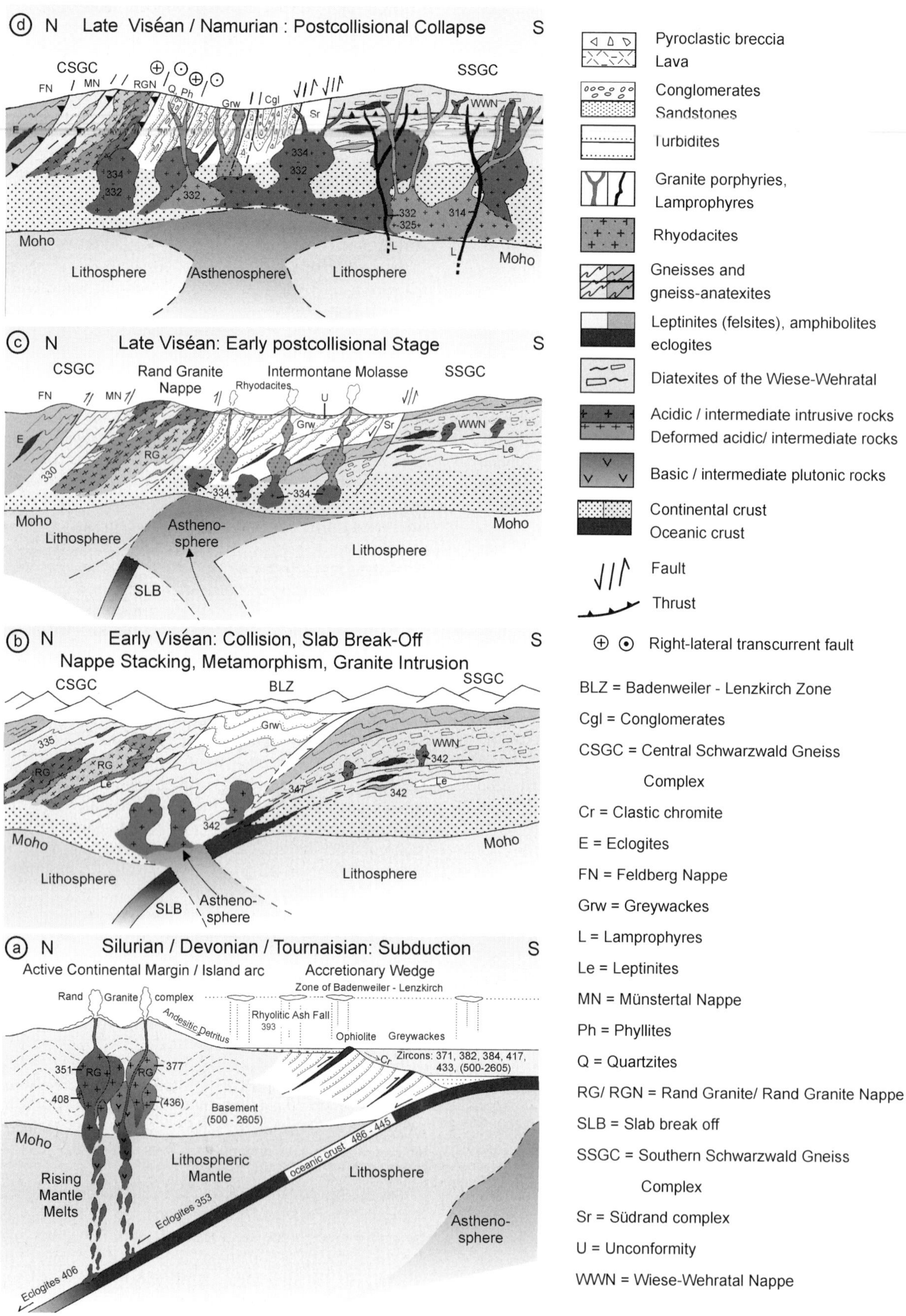

Fig. 11.7. Geodynamic model for the Variscan Orogeny in the Black Forest area (after Hann *et al.* 2003*a*).

footwall thrust of the Rand Granite Nappe occurred in the brittle state. K–Ar age data on hornblendes from a mylonitic shear zone within the Rand Granite Nappe indicate that cooling below 500–450°C occurred between 342 and 328 Ma (Werling & Altherr 1986). The Rand Granite Nappe had overthrust the BLZ and thus came into contact with the former accretionary wedge (Fig. 11.7c). The dimensions and geometric frame of the Rand Granite were determined during its emplacement in the nappe structure; thrusting separated a major part of its initial volume from the present complex.

The Early Visean evolution of the Black Forest area (Fig. 11.7b) was characterized by the collision between the northern (CSGC) and southern (SSGC) crustal blocks and the break-off of the oceanic slab. Collision was accompanied by nappe stacking, HT/LP metamorphism (*c.* 335–330 Ma) and the ascent of asthenospheric material into the crust (Fig. 11.7c). Partial melting of the deeper crust is documented by the presence of magmatic zircons in diatexites of the SSGC which provide U–Pb and Pb/Pb ages of 347 Ma (Hegner *et al.* 2001). Nappe structures with ductile deformation also formed in the SSGC south of the BLZ. Echtler & Chauvet (1992) recognized ductile shear zones with a top-to-SE directed sense of movement, active in Visean times. Nappe emplacement in the SSGC took place after 342 Ma and before the intrusion of the granites. These granites are the consequence of post-collisional extension and magmatic underplating at the beginning of the Late Visean at *c.* 334 Ma (Fig. 11.7c, d; Hegner *et al.* 2001). In the Early Visean, fragments of gabbros, which can be interpreted as the remnants of oceanic crust, were emplaced along shear zones into the SSGC (Fig. 11.7b, c).

The formation of a multiple nappe stack in the Black Forest area continued until the Late Upper Visean (Fig. 11.7c, d). Considerable shortening resulted in a crustal thickness of 50 km or more (Kalt *et al.* 2000*a*). The volcanic edifice of the magmatic arc was completely eroded, and post-collisional rhyodacitic subvolcanic plugs and lavas (Wilser 1935; Sittig 1969) erupted in an intramontane continental basin (Fig. 11.7c). The asthenosphere diapir increased in volume and resulted in further melting of the crust, thus producing large volumes of granites (Fig. 11.7d). The transition between the compressional and extensional stages may be marked by the intrusion of the deformed Klemmbach-Schlächtenhaus granites at 334 Ma (Schaltegger 2000). Right-lateral transcurrent faults were active immediately after that time (Eisbacher *et al.* 1989) and indicate the beginning of post-collisional extension and collapse of the orogen in this part of the Moldanubian Zone (Echtler & Chauvet 1992). The faults are cut by granites (from 334 to 332 Ma and 325 Ma) and finally by granite porphyries and lamprophyres (from 332 to 325 Ma; Fig. 11.7d).

Moldanubian Zone (*sensu stricto*) (S.M. & U.K.)

The Moldanubian Zone (Fig. 11.8) represents the southern part of the Bohemian Massif and comprises mostly medium- to high-grade gneisses of supracrustal origin with a late HT metamorphic overprint accompanied by voluminous intrusions of Carboniferous granites. This HT/LP metamorphism is the final regional event for the entire internal zone (i.e. Internides) of the Variscan belts of central and western Europe (Matte 1986; Zwart 1967) an area corresponding to the root of the Variscan Orogen and which continues towards the SW into the Black Forest, the Vosges and the French Massif Central. The Moldanubian Zone is tectonically

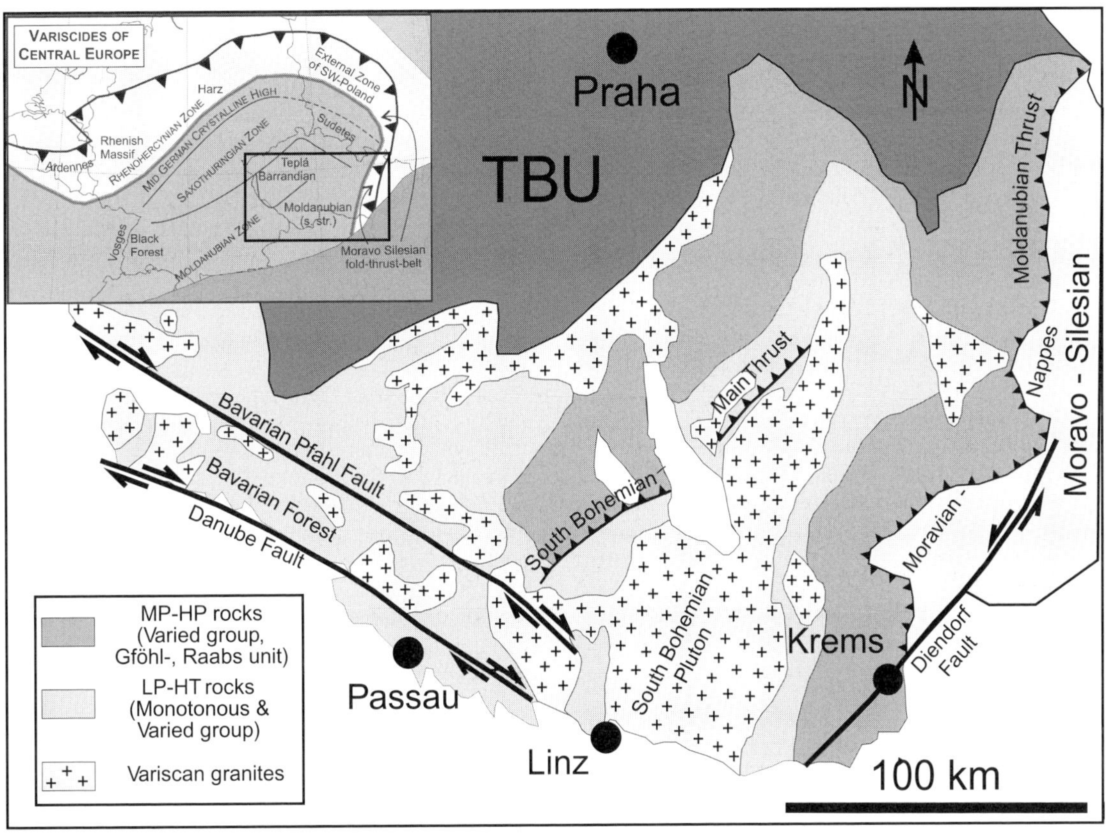

Fig. 11.8. Geological sketch map of the Moldanubian Zone (*sensu stricto*).

juxtaposed with the Teplá-Barrandian and the Saxothuringian zones in the NW and the Moravo-Silesian Zone in the SE. Despite its complex internal structure the Moldanubian Zone is subdivided into three major lithotectonic subunits: (1) the lowermost Monotonous Group that consists of biotite–plagioclase-bearing para- and orthogneisses with minor quartzites, amphibolites and rare lenses of eclogite; (2) the Varied Group formed by paragneisses intercalated with quartzites, marbles, graphitic schists, amphibolites and bodies of granitic gneisses; and (3) the structurally uppermost Gföhl Unit that consists of large bodies of HP granulites associated with garnet and spinel peridotites, pyroxenites, minor eclogites as well as high-grade partly migmatized orthogneisses, amphibolites and metagabbros. The latter two lithologies are mainly incorporated within the Raabs-Meisling Unit located at the base of the Gföhl Unit. Because the rocks of this uppermost unit show evidence of having attained the highest pressures, large-scale thrusting during the juxtaposition above the Monotonous and Varied groups must have occurred; thus the term 'Gföhl nappe' is commonly used in the literature (Fuchs & Matura 1976; Tollmann 1982)

Ages of protoliths

The South Bohemian Main Thrust is associated with a number of stretched orthogneiss bodies representing the oldest crustal components within the Moldanubian Zone which are dispersed at the base of the Varied Group and the Gföhl Unit. The largest of these, the Dobra orthogneiss, is dated at 1.38 Ga (U–Pb, zircon, SHRIMP; Gebauer & Friedl 1994) and corresponds to the basal thrust-sheet of the Varied Group. The Svetlik orthogneisses at the base of the Gföhl unit are even older, yielding Early Proterozoic ages of 2.1–2.06 Ga (U–Pb, zircon, conventional, SHRIMP; Wendt *et al.* 1993). The upper age limit for sedimentary protoliths of the Monotonous and Varied groups is constrained by the ages (U–Pb, Pb/Pb) of detrital zircons at 727 ± 106 Ma (Kröner *et al.* 1988) and 549 ± 5 Ma (Wendt 1989), respectively. The former age suggests that the sedimentary precursor to the Monotonous Group was deposited in the Late Proterozoic whereas the latter age supports the idea of an Early Palaeozoic derivation for the Varied Group. This corollary is supported by the age of 460 Ma obtained for leptynites from the Varied Group (Fiala *et al.* 1995). The orthogneiss of the Gföhl Unit yielded protolith ages of 488 ± 6 Ma (U–Pb, zircon; Friedl *et al.* 2004), comparable to the age of Lower Palaeozoic orthogneisses widespread across the Variscan belt. Interestingly, the amphibolites belonging to the Raabs-Meisling unit are significantly younger, with U–Pb zircon ages of 428 ± 6 Ma; this is taken to represent magmatic crystallization (Finger & Quadt 1995). This age is similar to that determined for a protolith of metagabbros from the Sleza ophiolite in the Sudetes (Oliver *et al.* 1993). A protolith age of 469.3 ± 3.8 Ma was obtained for granulites from the Gföhl Unit using SHRIMP II (Kröner *et al.* 2000*a*). Noticeably younger ages were acquired for some other occurrences of the granulites dated using the U–Pb zircon method at *c.* 370 Ma (Wendt *et al.* 1994). This latter age was interpreted to represent the time of an intracrustal anatectic event resulting in the formation of migmatitic to aplitic protoliths of granulites.

Metamorphism

The complexity of the Moldanubian Zone is evident from the variety of metamorphic rocks within it which preserve contrasting P-T conditions. The most striking feature of this zone is the intimate contact between rocks which underwent high pressure metamorphism at mantle depths with large volumes of migmatitic rocks (most commonly) which formed at high temperatures in mid- to upper crustal levels. For example, the SW part of the Moldanubian Zone in Bavaria (southern Germany) is characterized by the overall presence of HT/LP lithologies of the Monotonous and Varied groups. The migmatites of the Bavarian Forest are characterized by prograde biotite dehydration melting, peak metamorphic conditions of approximately 850°C and 5–7 kbar and retrograde melt crystallization at 800°C (Kalt *et al.* 2000*b*). The preservation of relicts of medium pressure paragenesis in some HT/LP samples suggests that there was an earlier metamorphic episode at least for some of the LP rocks of the Monotonous and Varied groups (Linner 1996).

The high pressure granulites of Southern Bohemia and Lower Austria, in contrast, experienced temperatures of 850–1000°C and pressures of 15–20 kbar (O'Brien 2000, and references therein). Rapid exhumation to mid-crustal levels is indicated by a nearly isothermal decompression path. It is important to note that similar metamorphic conditions have been reported from Saxon granulites of the allochthonous domain of the Saxothuringian Zone (Rötzler & Romer 2001) and the Black Forest area (Marschall *et al.* 2003). In contrast to the migmatites of the Monotonous and Varied groups, the migmatitic Gföhl gneisses preserve an early HP stage and are thus interpreted as a retrogressive product of former leucocratic granulites (Cooke & O'Brien 2001). Recently, LP granulites from the area were described which differ significantly from the HP/HT suite. The protoliths of these granulites are interpreted as the remnants of a magmatic arc, intruded between 370 and 360 Ma in the middle crust. The PT conditions of the subsequent granulite facies overprint are characterized by 4–5 kbar and 800–900°C (Janousek *et al.* 2006).

Timing

The metamorphic history of the Moldanubian Zone is intimately related to the evolution of the Variscan Orogeny. The numerous U–Pb zircon and monazite ages from the Gföhl Unit cluster around 340 Ma. Most authors interpret this time as the granulite-facies metamorphic climax (van Bremen *et al.* 1982; Schenk & Todt 1984; Kröner *et al.* 1988, 2000*a*; Aftalion *et al.* 1989; Wendt *et al.* 1994; Kotkova *et al.* 2003; Janousek *et al.* 2004, 2006). However, there is still a debate about the significance of zircon U–Pb ages in relation to the HP event, with Roberts & Finger (1997) suggesting that the majority of the zircons could only have crystallized on a retrogression path. Sm–Nd ages obtained for garnet cores with prograde zoning led Svoitka *et al.* (2002) to the conclusion that the HP/HT metamorphic peak was *c.* 10–15 Ma older than the retrograde event. This suggestion contrasts with the observation that many of the zircons which crystallized at *c.* 340 Ma were enclosed in HP phases of the granulites. The occurrence of 340 Ma old zircons in LP cordierite-bearing melt pods inside the HP granulites (Kröner *et al.* 2000*a*) could, therefore, either represent a significantly younger LP stage of the granulites or be indicative of the very rapid exhumation of the deep buried rocks. That the exhumation rate was high is evidenced by the occurrence of granulite pebbles in Upper Visean conglomerates (Vrána & Novák 2000).

Metamorphic ages obtained using a variety of geochronological methods for the Monotonous and Varied groups scatter within a timespan extending from 367 ± 20 Ma (Kröner *et al.* 1988) and 355 ± 2 Ma (Wendt *et al.* 1993) through to 328 ± 3.3 Ma (Dallmeyer *et al.* 1992) with most of the values clustering at *c.* 335 Ma. A still younger HT/LP metamorphic episode has been reported from the Bavarian part of the Moldanubian Zone with metamorphic ages ranging between 326 and 320 Ma (Kalt *et al.* 2000*b*; Grauert *et al.* 1990).

Deformation

The Lower Carboniferous exhumation of crystalline units formed at different crustal levels inside the Moldanubian Zone, and the juxtaposition of these with adjacent upper crustal units, are the most striking structural features of this zone. The sequence of events commenced with the vertical extrusion of deep-buried ultra-HP granulites into upper crustal levels, as evidenced by the occurrence of a steeply dipping mylonitic D_1 foliation followed by flat thrusting under formation of shear-zone-related subhorizontal D_2 foliations (Štipska *et al.* 2004; Tajcmanova *et al.* 2006). The exhumation process was accompanied by the development of synkinematic melts and followed by the intrusion of the post-kinematic intrusives of the South Bohemian Pluton (*c.* 335–300 Ma; Gerdes 1997; Friedl 1997; Gerdes *et al.* 2000).

In the Bavarian part of the Moldanubian Zone the final stage of LP/HT metamorphism is associated with evidence of transpressional deformation which affected the entire migmatitic rock suite (Behrmann & Tanner 1997). According to Behrmann & Tanner (1997) there are no indications of late orogenic collapse structures inside the LP/HT domain of the Moldanubian Zone. Horizontal, north–south directed shortening during the Late Variscan stage is indicated by the occurrence of a conjugate set of NW–SE orientated dextral shear zones in the 'Bavarian Pfahl' and the 'Danube fault zone' (Siebel *et al.* 2005) and of the development of NNE–SSW striking sinistral shear zones such as the Diendorf Fault (Fig. 11.8) within the southern part of the Moldanubian Zone (Wallbrecher *et al.* 1992).

Teplá-Barrandian Unit (G.Z.)

The Teplá-Barrandian Unit (TBU) forms an exceptional, largely supracrustal complex within the internal part of Variscan Europe (Fig. 11.9) and represents a shear-zone-bounded coherent block (Fig. 11.9) which is separated from the adjacent Moldanubian Zone by the Bohemian Shear Zone, the Hoher Bogen Shear Zone; and the West Bohemian Shear Zone. The North Bohemian Shear Zone (Zulauf *et al.* 2002*b*) is considered to be the boundary with the Saxothuringian Zone to the north.

In contrast to the surrounding Moldanubian and Saxothurin-

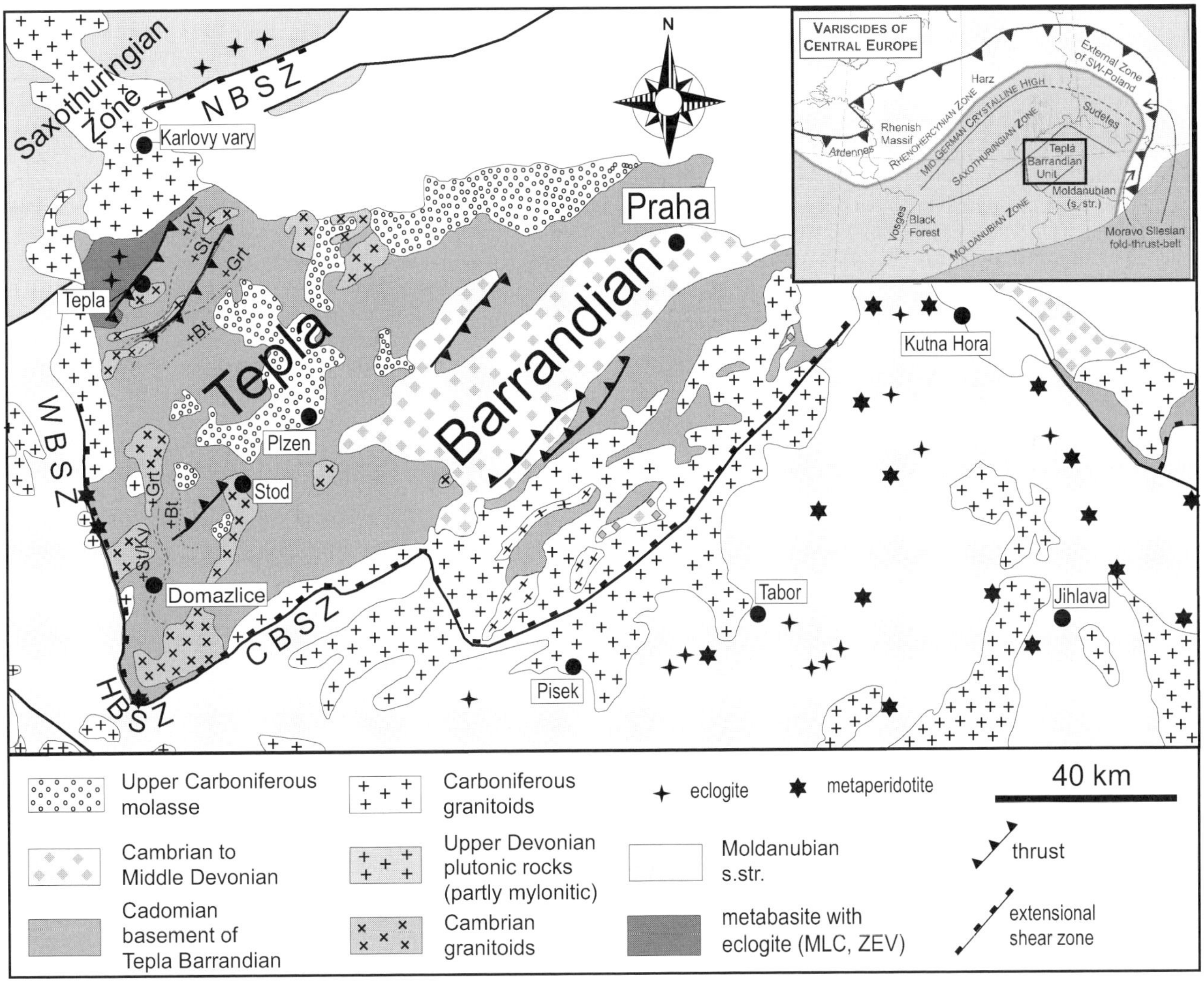

Fig. 11.9. Geological sketch map of the Teplá-Barrandian Unit and surrounding areas. Abbreviations: CBSZ, Central Bohemian Shear Zone; HBSZ, Hoher Bogen Shear Zone; NBSZ, North Bohemian Shear Zone; WBSZ, West Bohemian Shear Zone.

gian high-grade rocks, the TBU was hardly affected by Variscan deformation and metamorphism. This explains why non-metamorphic Cambrian to Middle Devonian sediments and volcanics rest unconformably on very low grade metamorphic Neoproterozoic rocks; these latter rocks preserve evidence of Neoproterozoic to Cambrian (Cadomian) orogenic activity (Chlupáč 1993*a*). It also explains why the oldest apatite fission-track ages of Central Europe have been found in this area (Glasmacher *et al.* 2002).

The metamorphic grade of the Neoproterozoic rocks increases from the Barrandian Basin (prehnite-pumpellyite facies; Cháb *et al.* 1995) towards the west and NW where amphibolite facies conditions are documented both in the Domažlice Crystalline Complex (DCC; Vejnar 1982) and in the Teplá Crystalline Complex (TCC; Žáček & Cháb 1993). Investigations of the crystallization–deformation relationships in the contact aureoles of the Cambrian plutons, described below, allowed the Cadomian and Variscan tectonometamorphic imprints within both the DCC and the TCC to be distinguished (Zulauf 1997, 2001; Zulauf *et al.* 1997; Dörr *et al.* 1998*a*).

Early Paleozoic rifting and Variscan shortening

Cambrian transtension signalled the onset of extensive Ordovician rifting that led to the separation of the TBU from Gondwana as part of the Armorican Terrane Assemblage (Tait *et al.* 2000, see also Robardet 2003). The northern drift of Armorica was associated with both the production and destruction of oceanic lithosphere. From ?Silurian to Lower Devonian times the lithosphere below the Saxothuringian Ocean was subducted towards the SE beneath the Cadomian basement of the TBU (Fig. 11.10a). This Variscan-related subduction is well documented in the eclogite-bearing Mariánské-Lázně Complex (MLC), situated along the northwestern margin of the TBU (Fig. 11.9).

The collision of the Teplá-Barrandian and Saxothuringian blocks commenced during the early Upper Devonian (Franke *et al.* 1995; Fig. 11.10b). During this time the intensity of the collapse of the overthickened wedge increased, promoting further extensional faulting and the exhumation of deeply subducted rocks. Combining the structural and kinematic results of the TCC and MLC (Zulauf 1997, 2001) with the 9HR seismic profile (Tomek *et al.* 1997), reveals that the MLC represents a metamorphic core complex, the detached roof of which is the TCC located to the SE. Extension resulted in marked 'crustal tilting' as indicated by the presence of NE–SW orientated metamorphic isograds in the TCC (Fig. 11.9). Within both the TCC and the DCC Devonian convergence occurred under medium pressure metamorphic conditions (Zulauf 1997). NE–SW trending folds and crenulations developed at the greenschist facies level (i.e. the entire DCC and the SE part of the TCC), whereas NW-directed mylonitic thrusts predominate in the amphibolite facies rocks of the TCC (Zulauf 2001).

Although Cadomian Barrovian-type mineral assemblages are present in the TCC, there is clear evidence that the major Barrovian imprint occurred during Variscan times (Zulauf 2001, and references therein). $^{39}Ar/^{40}Ar$ and K–Ar dating of hornblende and white mica from the TCC yielded ages of 383 Ma and 366–371 Ma, respectively (Dallmeyer & Urban 1998; H. Ahrendt & K. Wemmer unpublished data; H. Kreuzer unpublished data). For example, for the Variscan greenschist–subgreenschist facies transition of the Stod area (semibrittle deformation regime), a pressure interval of 4–5 kbar has been determined (Zulauf 1997). The related temperature should be just above *c.* 300°C. Geothermobarometry of the metasediments of the TCC yielded the following P-T conditions, most of which reflect the level of Variscan shortening: 6–7 kbar and 500–550°C for the garnet zone; *c.* 8 kbar and *c.* 590°C for the staurolite zone (Zulauf 1997); 5–9 kbar and 580–640°C for the kyanite zone (Cháb & Žáček 1994).

Together with the continent–continent collision which represented the main phase of Variscan activity, the convergent movements within the TBU itself migrated towards the SE affecting, in turn, the Palaeozoic cover of the Barrandian Basin. Variscan folding and top-to-the-SE thrusting in the Stod area commenced during the Famennian at *c.* 370 Ma (Zulauf 1997) and this post-dates the break in sedimentation of the Barrandian Basin at *c.* 385 Ma (i.e. the Middle–Upper Devonian boundary) (Chlupáč 1993*a*). NW–SE shortening led to the development of an Upper Devonian doubly vergent wedge, forming a megapop-up structure (Fig. 11.10b). Variscan shortening of the non-metamorphic Palaeozoic rocks of the Barrandian Basin was accommodated by top-to-the-NW thrusting in the NW and top-to the-SE thrusting in the SE. Fold axes of thrust-related folds trend NE–SW (Havlíček 1963; Svoboda 1966, fig. 52; Chlupáč 1993*b*). Early SE-directed movements in the area are documented along the Príbram, Očkov and Koda thrusts (Havlícek 1963). In some cases Cadomian basement was thrust on top of Palaeozoic rocks. A second phase of convergent movement led to the development of a series of thrusts which differ from those of the earlier phase (e.g. the Závist Thrust situated between Prague and Príbram). Reflectors imaged on the 9HR seismic profile (Tomek *et al.* 1997) suggest that the large-scale pop-up structure of the TBU includes additional, but smaller, pop-up structures.

The doubly vergent wedge and the spatiotemporal shift in thrust activity from the NW (*c.* 380 Ma) to the SE (*c.* 370 Ma) can be explained by a geometric model proposed by Martinez-Torres *et al.* (1994) that is based on a doubly vergent crustal wedge which has a counterpart in the mantle lithosphere. This model is in broad agreement with the results of numerical and analogue modelling (e.g. Willet *et al.* 1993; Brown *et al.* 1993). During the initial stage of convergence NW-directed master thrusts developed in the TCC. Subsequent SE-directed back thrusts formed in the DCC and in the area of the Barrandian Basin (Fig. 11.10a). The NW- and SE-directed thrusts represent step-up shears *sensu* Brown *et al.* (1993), along which the doubly vergent crustal wedge was lifted up. This resulted in the formation of a highly elevated plateau, which was covered with unmetamorphosed Palaeozoic rocks (Fig. 11.10c). The mantle wedge was not situated directly below the crustal wedge (Martinez-Torres *et al.* 1994) but rather developed further to the SE, below the back thrusts and in the area where recent Moldanubian rocks are exposed.

Variscan elevator tectonics and orogenic collapse

The Variscan overthickened crust of the TBU began to collapse at the Devonian–Carboniferous boundary (see below). Initial collapse was associated with the subsidence of the Barrandian Palaeozoic rocks (Fig. 11.10c). Enhanced collapse during the Lower Carboniferous is indicated by evidence of increased activity along the Bohemian Shear Zone (BSZ), which marks the boundary between the TBU and the adjacent Moldanubian and Saxothuringian high-grade rocks. Elevator-style movements along the BSZ led to the sinking of the TBU, as an upper and middle crustal block, into its hot substratum (Moldanubian and Saxothuringian lower crust, Fig. 11.10d). The BSZ is an exceptional feature, not only within the Variscides, but also worldwide.

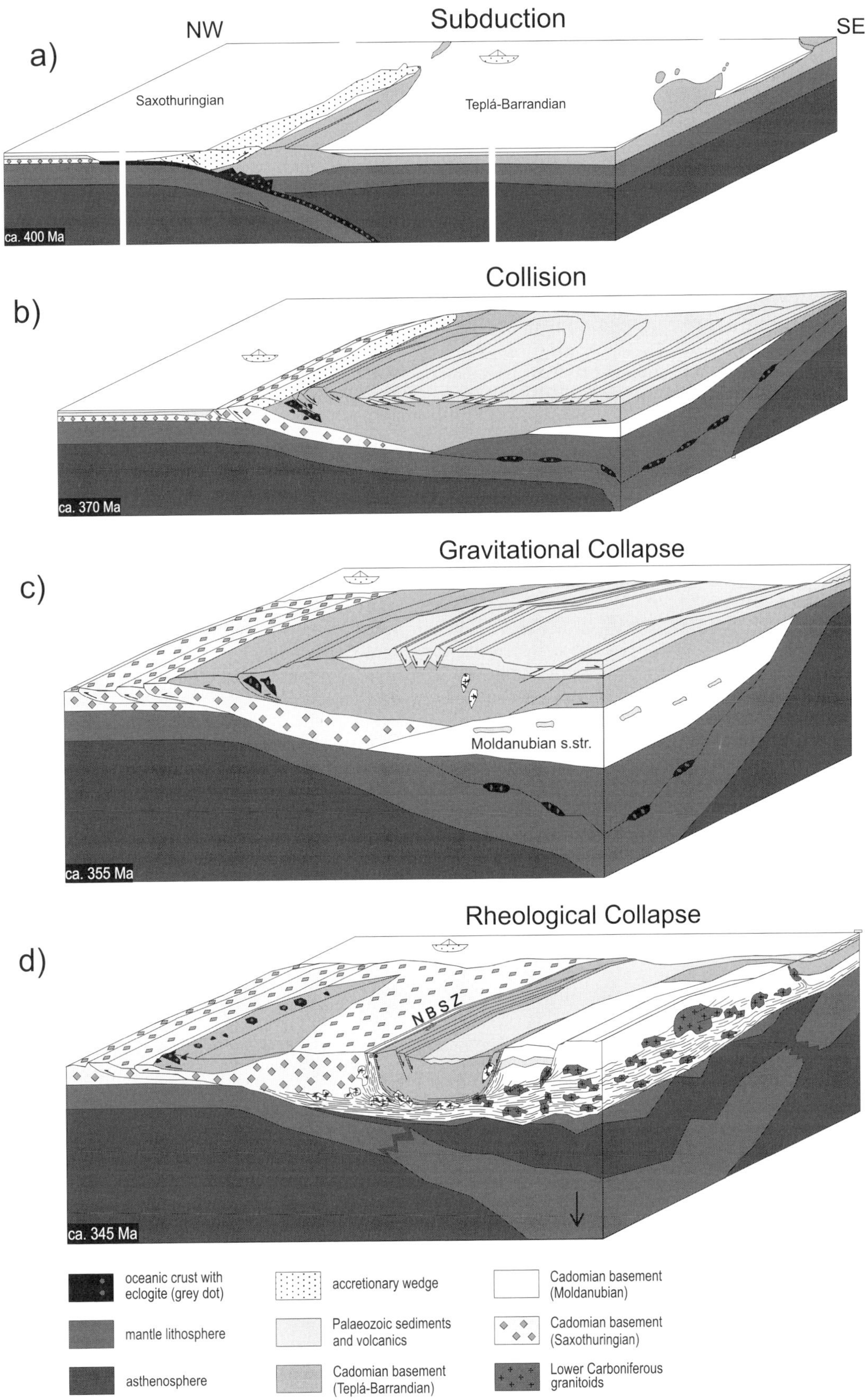

Fig. 11.10. Geodynamic model for the Variscan Orogeny in the western Bohemian Massif.

It is characterized by: (1) viscous (ductile) extensional shear zones with a high dip angle (>70°), displaying TBU-down movements; (2) retrograde metamorphic conditions during shearing, which began under amphibolite or even granulite facies conditions and ended under greenschist facies conditions; and (3) several bends in map view of up to 90° (Fig. 11.9).

The bends observed in the BSZ have been used to subdivide it into a number of segments, including: (1) the North Bohemian Shear Zone (NBSZ; Zulauf *et al.* 2002*b*); (2) the West Bohemian Shear Zone (WBSZ; Zulauf 1994, 2002*a*); (3) the Hoher Bogen Shear Zone (HBSZ; Bues & Zulauf 2000; Bues *et al.* 2002); and (4) the Central Bohemian Shear Zone (CBSZ; Scheuvens & Zulauf 2000; Artmann *et al.* 2003). There are several plutons, varying in composition and partly synkinematic, which straddle the BSZ (Fig. 11.9). U–Pb and Pb/Pb zircon ages of the plutons range from *c.* 350 to *c.* 330 Ma (Holub *et al.* 1997; Dörr *et al.* 1998*b*). Only the Teufelsberg (Čertův Kámen) diorite, situated in the Neukirchen Kdyne Complex, close to the CBSZ, yielded an Upper Devonian age (359 ± 2 Ma, concordant U–Pb zircon age; Bues *et al.* 2002). A minimum vertical throw of 10 km in the period 340–320 Ma has been determined for the elevator-style movements of the WBSZ. These dates were determined using the intrusion depth, intrusion time and cooling ages of the alkaline diorite from Mutenin and of the Babylon granite, situated in the Moldanubian Zone and in the TBU, respectively (Zulauf *et al.* 2002*a*, and references therein).

Elevator-style displacements such as those recorded from the WBSZ have also been observed in the rocks of the CBSZ (Scheuvens & Zulauf 2000), the HBSZ (Bues & Zulauf 2000) and the NBSZ (Zulauf *et al.* 2002*b*). In the case of the latter, the elevator-style movements are particularly well constrained to the period 343–337 Ma, using U–Pb dating of monazite (342 ± 1 Ma) and $^{40}Ar/^{39}Ar$ dating of muscovite (341 ± 4 Ma) (Zulauf *et al.* 2002*b*). Such an early onset of Variscan orogenic collapse has also been recorded in the Massif Central, where synorogenic extensional movements commenced at *c.* 335 Ma (Faure *et al.* 2002).

The normal movements along the WBSZ resulted in a second phase of eastward tilting of the Stod area. Based on K–Ar and $^{39}Ar/^{40}Ar$ ages of white mica (Dallmeyer & Urban 1998; Zulauf *et al.* 2002*a*), the metamorphic rocks of the DCC attained sub-greenschist facies conditions (T < *c.* 350°C) at the Devonian–Carboniferous boundary (*c.* 360 Ma), whereas K–Ar dating of white mica of the western contact aureole of the Cambrian Stod pluton yielded an age of 483 ±10 Ma (H. Ahrendt & K. Wemmer unpublished data). The latter age is compatible with an $^{39}Ar/^{40}Ar$ biotite age for the Stod granite (518 ± 8 Ma; Kreuzer *et al.* 1990). Thus, the area of Stod was located within a region subjected to subgreenschist facies metamorphism at least since Cambrian times.

In the Plzen area K–Ar dating of white mica, separated from Neoproterozic slates, also yielded high ages (496 ± 14, 489 ± 11, 499 ± 11 Ma; H. Ahrendt & K. Wemmer unpublished data). The tilting of the Domažlice-Stod-Barrandian Basin can be explained by the fact that normal slip occurred along the steeply east-dipping WBSZ, suggesting that the movements along the latter as well as the orogenic collapse in the Bohemian Massif may have commenced close to the Devonian–Carboniferous boundary.

The early collapse of the Bohemian plateau would explain why the unmetamorphosed Palaeozoic rocks of the Barrandian Basin are still preserved. An open question, however, concerns the height and extent of the plateau and the period during which it existed. Since the dramatic vertical throw (>10 km) observed along the BSZ is only possible in strongly thickened crust, with a thickness of at least 50 km, the minimum altitude of the Bohemian plateau must have been 3000 m. A similarly high altitude during the Carboniferous has also been assumed for the Massif Central of the French Variscides, based on palaeoclimate indicators (Becq-Giraudon *et al.* 1996).

According to the stratigraphic record of the Barrandian Basin, a Bohemian plateau may have been present from Frasnian to Stephanian times. Given that the Devonian and Carboniferous sediments drilled in the area of Hradec Králové are part of the TBU, this would leave only a short period of time to account for both Variscan thrusting and the presence of the Bohemian plateau (Čech *et al.* 1988). In the Jítrava-Hradec Basin, Upper Devonian (Frasnian) to Lower Carboniferous (Tournaisian) shallow-marine limestones rest unconformably on top of steeply dipping Ordovician slates (Čech *et al.* 1988). The limestones are overlain by Culm-facies clastic sediments. A link between, and a common evolution for, the TBU and the Jítrava-Hradec Basin is, however, questionable for the following reasons: (1) Famennian thrusting in the TBU (Holyšov area, see above) is not compatible with the deposition of Frasnian to Tournaisian marine limestone in the Jítrava-Hradec Basin; (2) the stratigraphy of the Jítrava-Hradec Basin is not compatible with apatite fission-track data obtained from the Barrandian Basin, which suggest that there was an initial phase of regional heating at the Devonian–Carboniferous boundary, and this has been attributed to crustal thickening (Glasmacher *et al.* 2002); and (3) the stratigraphic record of the drilled rocks of the Jítrava-Hradec Basin is similar to that of the Saxothuringian Zone, where Upper Devonian to Lower Carboniferous limestones are succeeded by Culm facies clastic sediments, and a gap in sedimentation pre-dates the deposition of the Upper Devonian limestone (Franke 1984, and references therein; Čech *et al.* 1988).

The Jítrava-Hradec Basin was situated within the Elbe Zone and this region probably underwent large-scale Variscan and post-Variscan lateral movements along NW–SE trending strike-slip faults. This would suggest that the Jítrava-Hradec Basin may have formed part of the Saxothuringian Zone and was transported dextrally towards the SE (i.e. close to the TBU).

Carboniferous cooling was probably related to the exhumation of parts of the Barrandian area. Apatite fission-track dating of an Ordovician basic tuff, located east of Rocycany, gave an age of 324 ± 8 Ma (Glasmacher *et al.* 2002). This is the oldest apatite fission-track age found in Central Europe.

Elevator-style tectonics along the BSZ can be explained by the fact that overthickened crust produced as a result of Variscan collision was weakened from below. The alkaline signature of the Mutenin diorite indicates that mantle melting, occurring at *c.* 340 Ma, was important for the thermal weakening of the crust. The steep dip of the BSZ can be explained by the fact that there was a significant rheological contrast between the extruding hot, and thus weak, Moldanubian and Saxothuringian crusts and the relatively cold, and thus rigid, crust of the TBU. The difference in thermal state between the TBU and the adjacent units is obvious from the U–Pb zircon and monazite data and from the K–Ar and $^{39}Ar/^{40}Ar$ cooling ages of the hornblendes, white micas and biotites. These data suggest that from 340 to 330 Ma the Moldanubian Zone (Teufel 1988; Kreuzer *et al.* 1989; Kalt *et al.* 2000*b*; Propach *et al.* 2000) and high pressure units of the Saxothuringian Zone (Kröner & Willner 1998; Werner & Lippolt 2000) were situated at deep crustal levels where $T > c.$ 500°C, whereas the entire TBU, including the Cadomian basement of the DCC and TCC, was situated at a supracrustal level with $T < c.$ 350°C (see references above).

The Carboniferous evolution of the TBU was probably markedly controlled by a partial removal of the overthickened mantle lithosphere (Fig. 11.10d) which explains: (1) the marked heating and rapid exhumation of the Moldanubian *sensu stricto* and Saxothuringian rocks; (2) the acceleration of the orogenic collapse and elevator-style sinking of the cold TBU into the hot and weak middle and lower crust (the recent Saxothuringian and Moldanubian units that extruded at the flanks of the TBU); and (3) the lack of a recent crustal and lithospheric root in the Variscides. Further evidence for high heat flow and igneous intrusions at deeper levels of the TBU have been obtained from the analysis of east–west trending Upper Devonian–Carboniferous tension gashes in Devonian limestones from the Barrandian Basin, which include minerals with oil-bearing fluid inclusions (Suchý *et al.* 2001).

In Lower Permian times there was a subsequent phase of increased temperature as evidenced by fission-track data. This phase was probably related to the overburden of the Upper Carboniferous to Lower Permian sediments. Most of the Upper Carboniferous coal-bearing sediments were derived from the Central Bohemian Pluton and accumulated within fault-bounded basins. The cumulative thickness of the Upper Carboniferous succession ranges from 1200 to 1700 m (Pešek 1996). Suchý *et al.* (1996) and Franců *et al.* (1998) suggested that the Lower Palaeozoic successions within the Barrandian Basin must have been buried beneath *c.* 2–3 km of overburden that was subsequently eroded. Havlíček (1998), on the other hand, has suggested that the heating of the Lower Palaeozoic rocks in the TBU was accomplished as a result of thickening related to thrusting.

A modern analogue for the Carboniferous Bohemian plateau may be the Tibetan plateau (Ménard & Molnar 1988) which is characterized by partially molten crust at depth, the latter extruding along the southern flanks of the plateau in the area of the High Himalaya (Kind *et al.* 1996; Nelson *et al.* 1996). The southern Tibet detachment system (Burchfiel *et al.* 1992) accommodated Himalayan north–south extension, whereas the north–south trending rifts of southern Tibet accommodated east–west extension (Armijo *et al.* 1986). As with the Bohemian plateau, the extensional movements in the Tibet–Himalaya orogen are operating within an overall convergent setting (Burg *et al.* 1984), and there is a marked association between extensional tectonics, high exhumation rates, steep thermal gradients, plutonism, steep topography and marked slope extremities (Molnar *et al.* 1993; Fielding 1996; Edwards & Harrison 1997; Zeitler *et al.* 2001; Chung *et al.* 2003).

Saxothuringian Zone (U.K.)

The Saxothuringian Zone (STZ) of the Variscan orogenic belt in Europe, originally defined by Kossmat (1927), is situated along the northern boundary of the Bohemian Massif. Traditionally, the fault-bounded region between the Franconian Line in the SW and the NE border of the Lusatian Block is considered to form the main part of the STZ (Fig. 11.11). The MGCH, which flanks the STZ in the NW, was originally considered to be part of the latter (Kossmat 1927), even though it contains evidence of Early Palaeozoic subduction-related magmatism (e.g. Anthes & Reischmann 2001), which is absent in other parts of the STZ (for details see below). The MGCH separates the Saxothuringian Zone from the Rhenohercynian Zone, which is located further to the NW and comprises predominantly Devonian sediments originally deposited on a passive margin of Laurussia (i.e. Old Red Continent) (Langenstrassen 1983). Sedimentary features recording post-Caledonian sedimentation are well preserved in the Rhenohercynian basins, but are absent in units of the STZ (Küstner 2000). At the SE margin of the STZ, there is a faulted contact with the Teplá-Barrandian Unit and the Moldanubian Zone (*sensu stricto*) This fault zone is now mostly concealed beneath the Tertiary age Eger rift valley.

All of the various units of the STZ were affected by the Variscan Orogeny. Indeed, the entire spectrum of regionally metamorphosed rocks is exposed, ranging from very low grade to high grade rocks, as well as rocks that experienced ultra-high temperatures and pressures. The lithological, geochemical and isotopic signatures of the protoliths are relatively uniform. Generally, all of the units of the STZ preserve features related to the Cadomian Orogeny (see Linnemann *et al.* 2008), a Cambro-Ordovician magmatic event, and Palaeozoic marine sedimentation.

The variable overprint of the Saxothuringian crust during the Variscan Orogeny is expressed in the great variety of deformational structures which are recorded. These various structures reflect the different deformation conditions which were active at the time. In general, two principal regional deformation processes can be recognized: (1) regional D_1 processes correspond to pervasive deformation during subduction and collision lasting from 370 Ma to 340 Ma, as indicated by the dating of the metamorphic complexes. D_1 processes were always associated with the development of pervasive metamorphic layering; and (2) D_2 deformation stages were associated with ductile deformation related to the final exhumation of HP and ultra-HP metamorphic rocks and their juxtaposition in the upper crust.

A subdivision of the Saxothuringian Zone into an autochthonous domain, a wrench-and-thrust zone and an allochthonous domain has been proposed by Kroner *et al.* (2007). Under this scheme the autochthonous domain and the wrench-and-thrust zone would constitute the Saxothuringian Basin (Franke 2000, and references therein) whereas the autochthonous domain would be characterized by continuous marine sedimentation extending to at least 330 Ma, and the lack of a pervasive D_1 deformation. The wrench-and-thrust zone, however, contains allochthonous complexes formed during Lower Carboniferous D_1 deformation (Kroner & Hahn 2004). The detailed structural character of this imbricated zone has been established by mapping and analysis of the hundreds of drillholes, cored for uranium exploration by the former East German SDAG Wismut Company (e.g. Russe 1995; Lange *et al.* 1999). The allochthonous units of the wrench-and-thrust zone comprise Ordovician to Lower Carboniferous rocks similar to those described from the autochthonous parts of the basin.

Geochemical signatures reported from the HP metamorphic rocks of the Western Erzgebirge, when compared to those from the autochthonous domain, reveal that the protoliths of both were broadly derived from the same crustal type (Mingram 1998). The Nd-isotope signature and TDM (depleted mantle) model ages of the Palaeozoic marine sediments within the autochthonous domain, however, indicate that the sediment source did not change from the Late Neoproterozoic until the Lower Carboniferous (pre-deep-marine deposits), suggesting either that the West African craton was continuously available as a major source supplying detritus to the basin (Linnemann *et al.* 2004) or that sediments derived from it were repeatedly recycled throughout that period. Thus, the precursors of the entire Saxothuringian Zone would appear to belong to the same crust, although there are quantitative differences. For instance, most of the subducted and exhumed HP and ultra-HP orthogneisses in the western Erzgebirge have Ordovician protoliths (Mingram *et al.* 2004). Such large volumes of Ordovician magmatic rocks, however, are

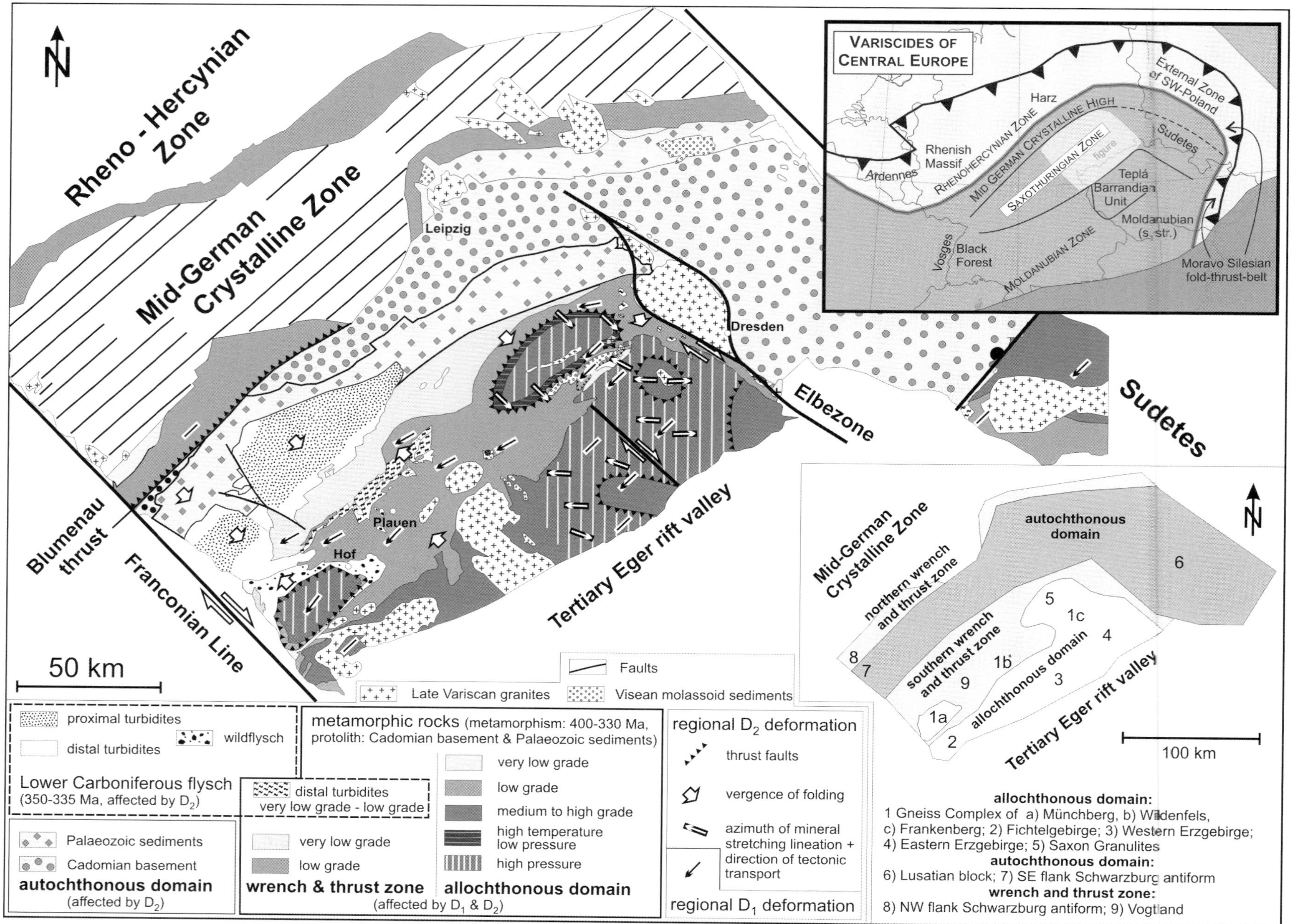

Fig. 11.11. Geological sketch map of the Saxothuringian Zone (modified after Kroner *et al.* 2007; reproduced by permission of the Geological Society of America).

not observed in the autochthonous domain of the Saxothuringian Zone.

Autochthonous domain

Locally, the autochthonous domain shows evidence for continuous basinal sedimentation, extending until *c.* 330 Ma (Fig. 11.12) and showing no evidence of contemporaneous deformation. The preserved sedimentary and volcanic record in this autochthonous unit spans the period from at least the Ordovician to the Upper Visean. Early and Middle Cambrian limestones and clastic rocks are only locally preserved, which may be due to restricted primary deposition or erosion during the Late Cambrian.

Following a sedimentation gap in the Late Cambrian, continuous marine sedimentation resumed with a transgression in the Early Ordovician (Tremadocian). The Tremadocian sediments of the inner shelf facies consist of up to 3000 m of shallow-water siliciclastic deposits; the thickness of the succession reflects the marked basin subsidence at this time. Felsic and subordinate mafic volcanic and intrusive rocks occur in the basal parts of the sequence. Arenig to Ashgillian sediments show evidence of decreasing sedimentation rates and more distal shelf facies conditions, dominated by pelites. A break in sedimentation, or a strongly condensed unit, occurred in the Llanvirn and Caradoc (oolitic iron ore horizon). Subsequently, a late Ordovician glaciomarine tillite, reflecting the Sahara glaciation, was deposited. The Cambro-Ordovician sediments indicate that this period represented a stage of widespread crustal extension which resulted in a thinned continental crust, as previously described for the adjacent areas of the Bohemian Massif (Kröner *et al.* 2000*b*; Mazur *et al.* 2004) as well as in other parts of the European Variscides (Ballevre *et al.* 2002; Sanchez-Garcia *et al.* 2003; Laumonier *et al.* 2004).

During the Silurian and Early to Middle Devonian, the inner shelf facies includes undisturbed distal shelf sediments which were deposited under conditions of low sedimentation rate. In these areas, deposition was mainly affected by global sea-level changes and variations in basin subsidence. Black shales, cherts, deep-water limestones, and pelites were deposited. Thin psammitic intercalations are occasionally present, but are restricted to the Early Devonian.

The tectonic instability associated with the early Frasnian phase of magmatism (which is preserved in the wrench-and-thrust zone) resulted in the deposition of turbiditic greywackes (partly derived from tuffaceous material) in the autochthonous domain. The association of sedimentary evidence of magmatic activity demonstrates the close spatial relationship between the autochthonous domain and the wrench-and-thrust zone. Indeed, the latter probably represents the marginal parts of the basin. The early Frasnian volcaniclastic-rich turbidites were overlain by deep-water limestones. Within these limestones there are two turbiditic sandstone intercalations of Late Fammenian age. In the Middle Tournaisian, the crenulata event resulted in the deposition of a black shale, although only the central part of the basin is overlain by Early Visean pelites. The subsequent onset of the deposition of distal turbiditic sediments is diachronous across the basin, occurring in Late Tournaisian times along the SE margin and Early Visean times in the central and northwestern parts of the basin.

During the Holkerian and the Asbian, distal deep-marine deposits passed upwards into more proximal deep-marine sediments, and this change was associated with a rapid increase in the sedimentation rate which led to the final infilling of the basin. Following the cessation of Variscan deformation-related deep-marine sedimentation, the SE part of the basin was affected by NW-directed folding (Franke 1984). Subsequent to this local D_I deformation, the basin was deformed in a SE-SSE-vergent direction (D_{II} deformation). Both phases of deformation (D_I and D_{II}) belong to the regional D_2 deformation event. D_1 is absent.

Wrench-and-thrust zone

This zone had a similar evolution to the autochthonous domain until Middle Devonian times (Fig. 11.12). The development of the wrench-and-thrust zone commenced along the SE border of the autochthonous domain in the early Frasnian, with the onset of a magmatic phase as evidenced by the presence of subvolcanic intrusions and rhyolites. Parts of the subvolcanic granites must have been rapidly exhumed, because granite-pebble conglomerates occur within Frasnian volcaniclastic units. Subsequent to the early felsic phase, voluminous mafic volcanics with pillow lavas were emplaced, followed by, and intercalated with, tuffs and volcaniclastics. The mafic lavas, as well as the associated diabase and picrite dykes, show clear within-plate geochemical signatures (Gehmlich *et al.* 2000) and are mostly related to NE–SW trending faults (Wiefel 1976; Lange *et al.* 1999). There is, however, no structural evidence for a rifting event, and thus the rapid exhumation of these felsic intrusive rocks may have been facilitated by the formation of NE–SW orientated strike-slip faults with localized transpression zones.

The intense period of fault-related volcanism led to the development of a strong submarine relief which resulted in facies variations and gaps in the Late Devonian sedimentary record. In the Late Frasnian and Fammenian, sedimentation was dominated by shallow-water to deep-water limestones. Subsequent basin subsidence is evidenced by the predominance of pelitic deposits in the Late Fammenian and the commencement, in Tournaisian times, of deep-marine sedimentation (distal facies).

The deposition of distal deep-marine siliciclastic deposits was related to the northward movement of the Variscan orogenic front. Deposition occurred in NE–SW trending basins during the Tournaisian to Early Visean. In the Arundian (*c.* 340 Ma), major parts of the area were involved in Variscan wrench-and-thrust tectonics (D_1 deformation). This deformation was related to a metamorphic overprint extending up to greenschist facies. D_1 structures comprise thrusts, recumbent folds with NNW to ESE scattering fold axes and subhorizontal foliation planes. Uniformly NE-trending stretching lineations and pressure shadows around porphyroclasts correspond to the direction of tectonic transport. Kinematic indicators such as asymmetric porphyroclasts indicate an initial top-to-the-SW transport. The NE–SW trending boundaries between the wrench-and-thrust zone and the autochthonous domain are considered to be D_1 strike-slip zones responsible for the decoupling of these two domains and the protection of the autochthonous domain from the regional D_1 deformation. Strike-slip faulting may also have been important within the wrench-and-thrust zone itself. This is evidenced by the juxtaposition of NE–SW striking units comprising similar lithologies but with different metamorphic overprints and deformational styles. Due to the subsequent compressional D_2 overprint, however, all boundaries are now represented by reverse faults.

The D_1 deformation resulted in the development of SW-directed nappes with shear lenses, duplex structures and thrusts. Palaeozoic sedimentary successions comprising Ordovician–Devonian inner-shelf facies and distal Lower Carboniferous deep-marine deposits were detached, partly metamorphosed to greenschist facies, and included in a nappe pile (Hahn 2003;

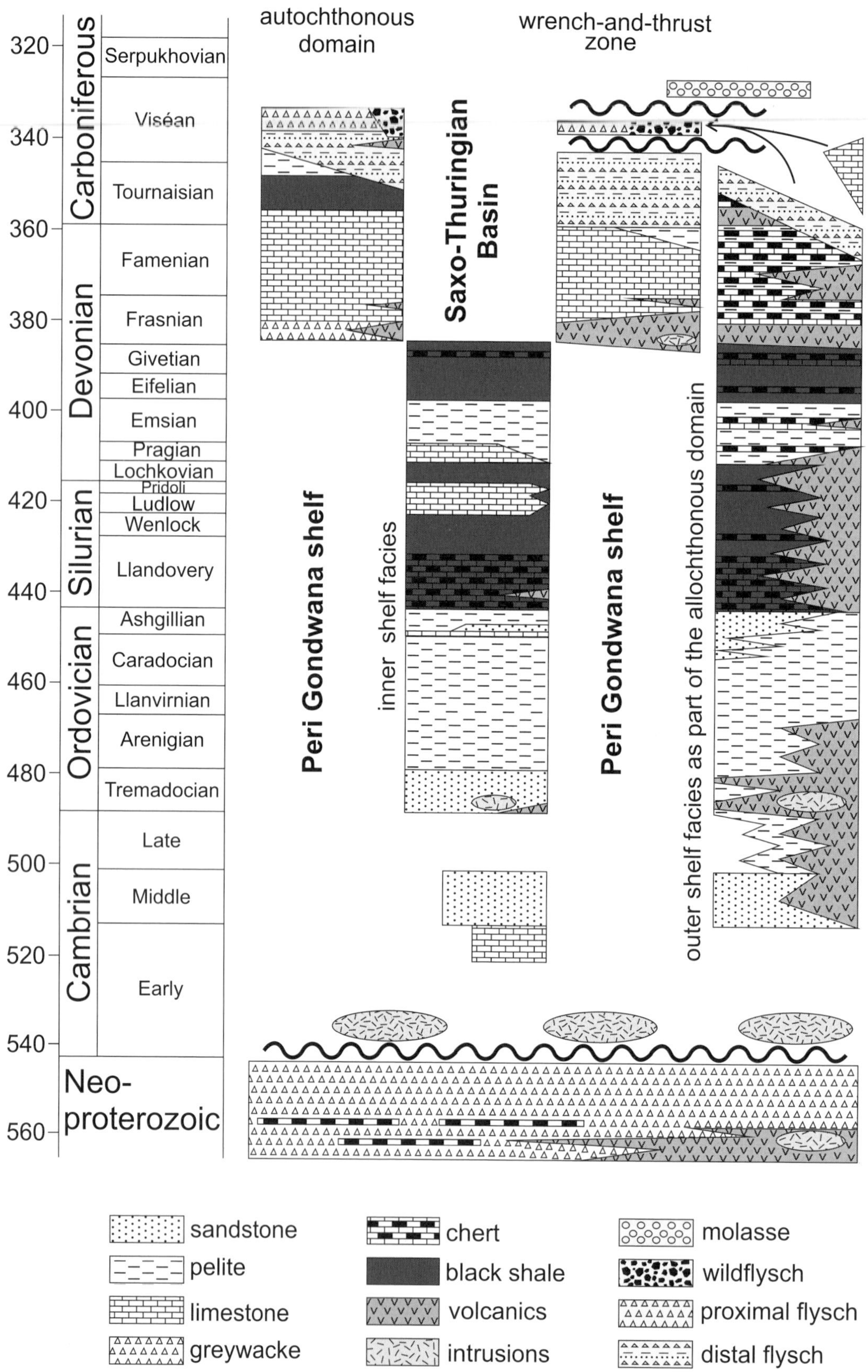

Fig. 11.12. Neoproterozoic to Lower Carboniferous stratigraphy of the unmetamorphosed to very-low-grade rocks of the Saxothuringian Zone (after Kroner *et al.* 2007, and references therein; reproduced by permission of the Geological Society of America).

Hahn & Meinhold 2005). The less competent Silurian and Middle Devonian black shales, in particular, represent preferred detachment zones. Following the end of thrusting and rapid exhumation, chaotic melange deposits (Holkerian) and proximal sandstones of the Kahmer Formation (Late Visean) were unconformably deposited on the nappe pile.

The NW border of the autochthonous domain comprises similar Lower Palaeozoic lithological successions to those described above. Here, the Blumenau Thrust subdivides the autochthonous domain of the SE flank of the Schwarzburg antiform from the northern wrench-and-thrust zone. In contrast to the SE, however, rocks reflecting the Late Devonian magmatic event are absent. The observed metamorphic layering as well as NE–SW stretched pebbles in the very low grade to low grade metamorphic sequences provide evidence for the regional D_I deformation within the NW flank of the Schwarzburg Antiform.

In contrast to the crystalline nappes described below, the metamorphic conditions experienced by the allochthonous units of the Saxothuringian Basin did not exceed medium pressures and temperatures (Fig. 11.13). The second deformation event affecting these units (D_2) corresponds to the main phases of deformation (D_I and D_{II}) as recorded in the autochthonous succession of the basin.

Allochthonous domain

The allochthonous domain of the Saxothuringian Zone includes the crystalline complexes of Münchberg, the Erzgebirge, and the Saxon granulites and low-grade metasedimentary successions. The medium- to high-grade crystalline units of the SE flank of the Saxothuringian Zone, which were partly subjected to extreme P-T conditions (Fig. 11.13), reflect the active subduction processes which were followed by rapid exhumation. From the tectonostratigraphic higher levels to the deeper ones, a series of crystalline complexes crop out, including the following: (i) eclogite-bearing Münchberg gneiss complex, Wildenfels and Frankenberg gneiss complexes; (ii) greenschist and phyllite units partly with lenses of serpentinite; (iii) Ordovician low-grade metasediments and metavolcanics, representing the classic outer shelf facies; (iv) low-grade Palaeozoic successions similar to the Lower Palaeozoic sediments and volcanic rocks of the autochthonous domain that form the upper part of the so-called Fichtelgebirge–Erzgebirge Antiform; (v) the underlying eclogite-bearing HP/LT to UHP/HT units of the Erzgebirge representing lithological equivalents to the lower Palaeozoic succession of the autochthonous domain (Mingram 1998); and (vi) the lowermost unit of the Erzgebirge includes the voluminous MP orthogneisses of the Eastern Erzgebirge, whose protoliths correspond to the Cadomian basement exposed in the Lusatian Block of the autochthonous domain (Tichomirova *et al.* 2001). The Saxon granulites are separated from this nappe pile. The granulitic core is mantled by an extremely condensed metamorphic profile. HT shear zones including migmatitic zones with garnet and cordierite gneisses at the very contact to the HP/UHT granulites are overlain by shear lenses of serpentinites, strongly deformed gabbros, gneisses, mica schists and phyllites, and very low-grade Palaeozoic metasediments. The high-grade rocks of this detachment zone experienced a D_2 isothermal decompression (Kroner 1995); the low-grade metasediments were transformed to HT schists at the contact with the hot granulitic core (Reinhard & Kleemann 1994).

The earliest HP event is recorded in eclogites of the Münchberg

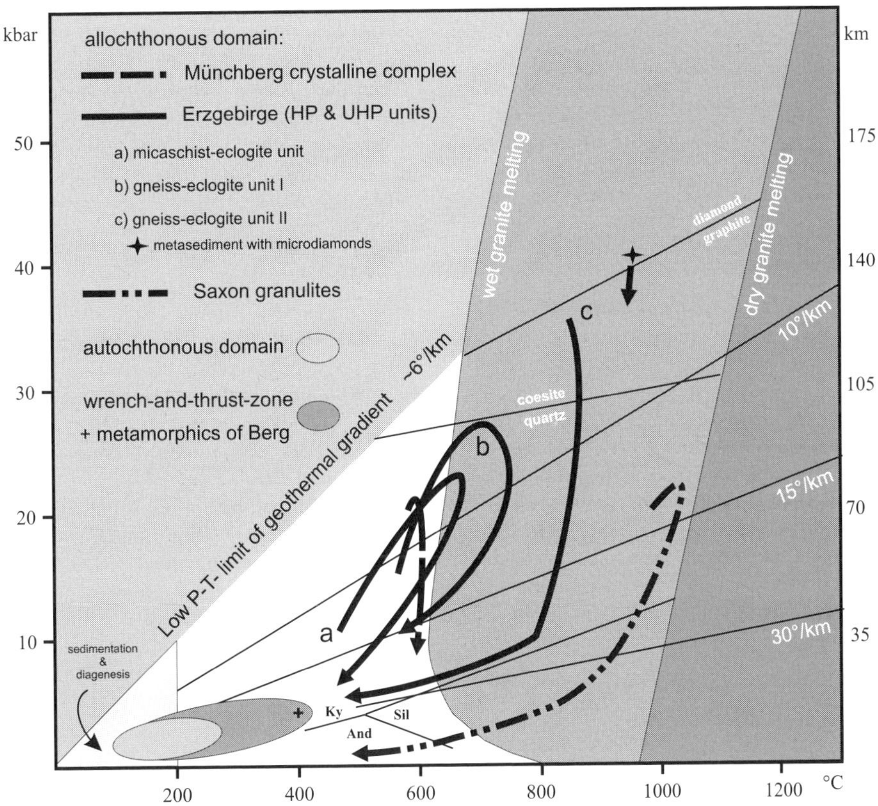

Fig. 11.13. P-T paths for selected units of the allochthonous domain of the Saxothuringian Zone (after Kroner *et al.* 2007, and references therein). The ultra-high pressure units prove subduction of continental crust (reproduced by permission of the Geological Society of America).

Massif, which had already undergone metamorphism in the Devonian (Klemd *et al.* 1991). The metamorphism of the HP complexes of the Erzgebirge took place during the Lower Carboniferous subduction of the continental crust to mantle depths (Schmädicke 1994; Willner *et al.* 1997; Klapova *et al.* 1998), as documented by the presence of microdiamonds (Massonne 1999). The felsic Saxon granulites, forming a dome structure below the wrench-and-thrust zone, experienced HP metamorphism at temperatures as high as *c.* 1050°C (Rötzler & Romer 2001; Rötzler *et al.* 2004) at 340 Ma. Tectonic lenses of lithospheric mantle in the Saxon granulites and the UHP units of the Erzgebirge (Mathé 1969, 1990) provides further evidence of deep subduction of continental lithosphere. The exhumation of the various high-grade complexes to upper crustal levels, which immediately followed the continental subduction, was diachronous within the Saxothuringian Zone and occurred over a timespan from 400 to 340 Ma (Stosch & Lugmair 1990; Quadt 1993; Kröner *et al.* 1998; Kröner & Willner 1998; Romer & Rötzler 2001, 2003). The uppermost allochthonous units (e.g. Münchberg Gneiss Complex) arrived in the upper crust at *c.* 370 Ma; the high-grade complexes of the Erzgebirge and the Saxon Granulite Massif peaked at 340 Ma and were rapidly exhumed (Romer & Rötzler 2001). At *c.* 330 Ma all high-grade units of the Saxothuringian Zone were juxtaposed in the brittle–ductile transition zone of the upper crust (Werner & Lippolt 2000).

This extensive period of tectonometamorphic activity (400 to 340 Ma) resulted in the formation of a very complex structural geometry in this part of the Saxothuringian Zone. The pervasive deformation of initial lithological contacts is indicated by the presence of inverted as well as condensed metamorphic profiles. Recumbent, isoclinal folds, partly preserved as relicts, demonstrate the transpositional character of the main foliation transecting them. The initial borders of the various inverted metamorphic sequences are orientated parallel to this flat-lying foliation suggesting that large-scale thrusting took place at this time. For example, westward thrusting of an allochthonous eclogite-bearing HP unit has been reported from the central part of the Erzgebirge (Konopasek *et al.* 2001). The evolution of this huge nappe complex and, thus, the metamorphic layering must be diachronous, as also indicated by the geochronological and stratigraphic constraints for different tectonometamorphic units (Kreuzer *et al.* 1989; Kröner & Willner 1998; Werner & Lippolt 2000).

The orientation of mineral stretching lineations are scattered over a range of 90° (Fig. 11.11). The main azimuths are orientated in NE–SW, east–west, and NW–SE directions. This strong variation of the strain increments precludes finite plane strain geometry for the allochthonous domain and demonstrates, together with the coexistence of extensional and compressional structures, the spectacular complexity of this domain. Although the complexity of the structural evolution of the tectonic uplift, as recorded by high-grade metamorphic nappes and mantled gneiss domes, is still poorly understood, the exhumation of the complexes probably occurred in two main stages. The initial increments of tectonic exhumation were related to subhorizontal top-to-the-SW thrusting (Kroner & Hahn 2004), whereas the final exhumation of the nappes was achieved by subsequent displacements along high-temperature shear zones with top-to-the-west and/or NW kinematics. The rapid exhumation of hot continental crust and its juxtaposition against cold supracrustal rocks resulted in HT/very LP metamorphism (along top-to-the-SE directed detachment zones) and the formation of a contact-metamorphic aureole in the roof phyllites enveloping the granulites (Reinhard & Kleemann 1994; Kroner 1995). There is no indication that the exhumation of the Saxon granulites took place beneath an undeformed subsiding sedimentary basin as proposed by Franke & Stein (2000). All of the rocks above the detachment zone that were affected by advective heat transfer from the hot granulitic core show evidence for an older phase of regional metamorphism with evidence of marked deformation, including recumbent folds and mylonites (Kroner 1995). The extremely condensed metamorphic profiles belong either to the upper parts of the allochthonous domain or to the phyllites of the wrench-and-thrust zone.

Rapid final exhumation and cooling of the allochthonous domain was contemporaneous with the deposition of the oldest late to post-orogenic sediments. Deposition of early Variscan continental sedimentary successions on top of the uppermost allochthon unit of the Frankenberg Zwischengebirge began at *c.* 330 Ma (late Asbian) (Gehmlich *et al.* 2000).

SW Poland Sudetes (S.M., P.A.)

The Sudetes represent the northeasternmost exposed portion of the Variscan crystalline basement in Europe. They are situated along the boundary of SW Poland and northern Bohemia and, geologically, are located on the NE margin of the Bohemian Massif. The Sudetes are a fragment of the Variscan belt, developed in Devonian and Early Carboniferous times due to the closure of at least two marine basins floored by oceanic crust and to concomitant amalgamation of Armorican terranes, followed by their accretion to the East European Platform and the adjacent collage of the Trans-European Suture Zone (cf. Matte *et al.* 1990; Cymerman *et al.* 1997; Pharaoh 1999; Franke & Żelaźniewicz 2000; Aleksandrowski & Mazur 2002; Winchester & PACE TMR Network Team 2002).

In a geological sense, the upland region of the Sudetes is considered herein together with the southern part of the Silesian-Lusatian Plain adjacent to the north, from which it is separated by the prominent Sudetic Boundary Fault (Fig. 11.14). This is a Cenozoic feature rejuvenating an older, Variscan fracture zone. Thus, the Sudetes comprise the Sudetic Mountains in the south and the Fore-Sudetic Block in the north. The entire Sudetic area extends between the WNW–ESE trending Middle Odra Fault Zone in the NE and the parallel Upper Elbe Fault Zone in the SW (Fig. 11.14). To the SE, it plunges below the Miocene Carpathian Foreland Basin and to the NW it merges with the Neoproterozoic Lusatian Block.

The geology of the Sudetes is complex, comprising a mosaic of lithologically distinct, fault-bounded pre-Permian units (Fig. 11.14) mainly affected by Devonian to Carboniferous deformation and characterized by frequent changes in the dominant structural trends. The variable geological evolution of most of the geological units in the Sudetes, combined with the occurrence of ophiolitic bodies along some of their boundaries, igneous rocks with a MORB-type geochemical signature, as well as HP metamorphic rocks (blueschists and eclogites), suggest that the area comprises a number of distinct tectonostratigraphic terranes. These are separated by tectonic sutures and major fault/shear zones. A possible terrane arrangement in this region has recently been much discussed in the literature (e.g. Matte *et al.* 1990; Aleksandrowski 1990, 1995; Oliver *et al.* 1993; Franke *et al.* 1995; Cymerman *et al.* 1997; Franke & Żelaźniewicz 2000; Aleksandrowski & Mazur 2002). The Sudetic terranes would appear to be a direct continuation of those known from the Bohemian Massif, which are generally believed to represent the Armorican Terrane Assemblage (Franke *et al.* 1995; Pharaoh 1999; Winchester *et al.* 2002; Aleksandrowski & Mazur 2002). These include the Saxothuringian, Teplá-Barrandian, Moldanubian, Moravo-Silesian and Bruno-Vistulian zones, to which local

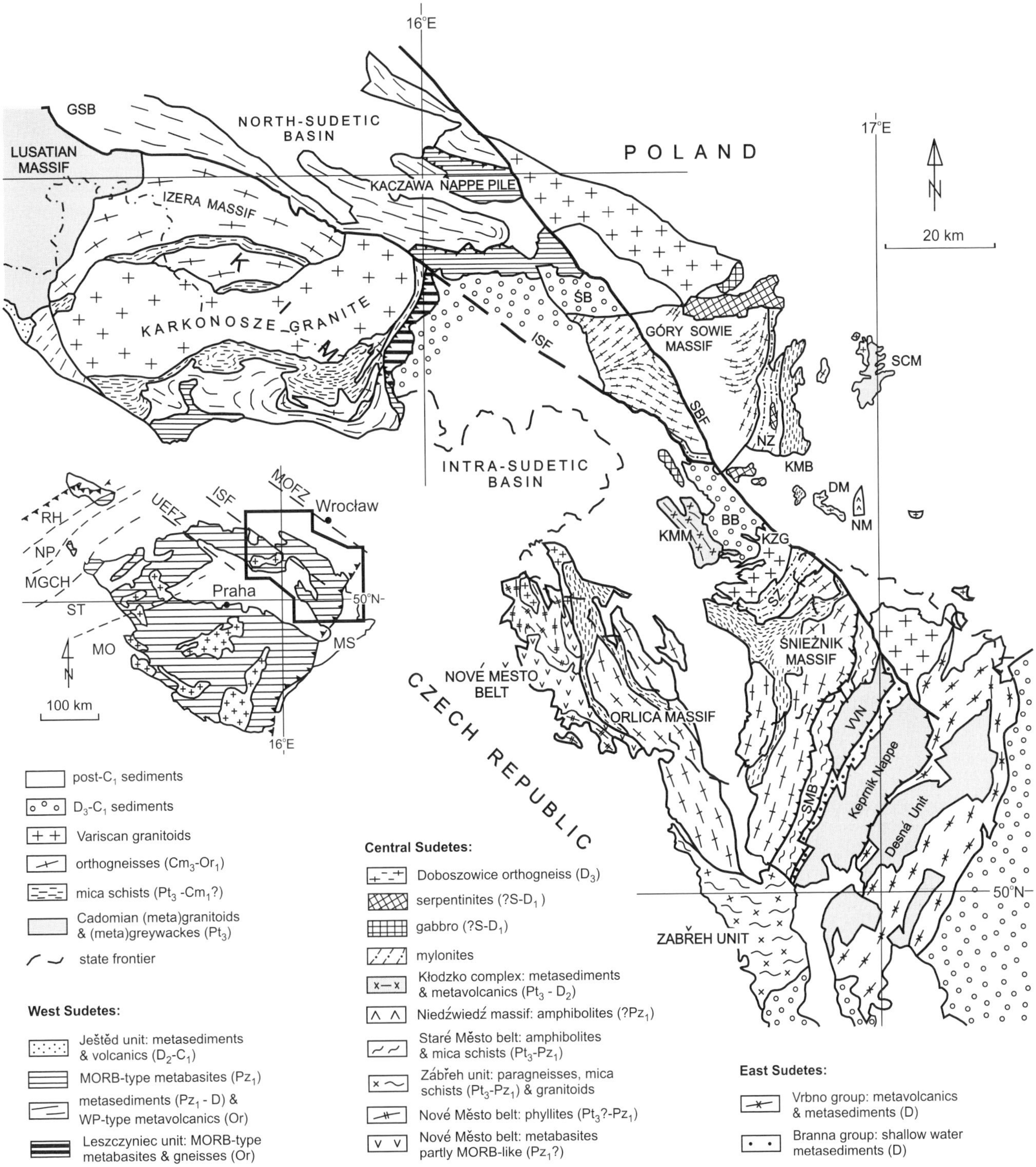

Fig. 11.14. Geological sketch map of the Sudetes. Abbreviations: BB, Bardo Basin; DM, Doboszowice Massif; GSB, Görlitz Slate Belt; ISF, Intra-Sudetic Fault; KIM, Karkonosze-Izera Massif; KMB, Kamieniec Metamorphic Belt; KMM , Kłodzko Metamorphic Massif; KZG, Kłodzko–Złoty Stok Granite; NM, Niedźwiedź Massif; NZ, Niemcza Shear Zone; ŚB, Świebodzice Basin; SBF, Sudetic Boundary Fault; SCM, Strzelin Crystalline Massif; SMB, Staré Město Belt; SZ, Skrzynka Shear Zone; VVN – Velké Vbrno Nappe. *Inset*: MGCH, Mid-German Crystalline High; MO, Moldanubian Zone; MOFZ, Middle Odra Fault Zone; NP, Northern Phyllite Zone; RH, Rhenohercynian Zone; ST, Saxothuringian Zone; UEFZ, Upper Elbe Fault Zone; hatched area corresponds to the outcrop of crystalline basement. *Age assignments*: Pt, Proterozoic; Pz, Palaeozoic; Cm, Cambrian; Or, Ordovician; D, Devonian; C, Carboniferous; *subscripts*: 1, Early; 2, Middle; 3, Late.

names are usually applied in the Sudetes (e.g. Aleksandrowski & Mazur 2002).

The Sudetes may be divided into western, central and eastern parts, showing important differences in their lithostratigraphy, structure and geological evolution (Fig. 11.15). Additionally, the Sudetic area is bisected by the major strike-slip Intra-Sudetic Fault, which trends parallel to the Middle Odra and the Upper Elbe faults (Fig. 11.14).

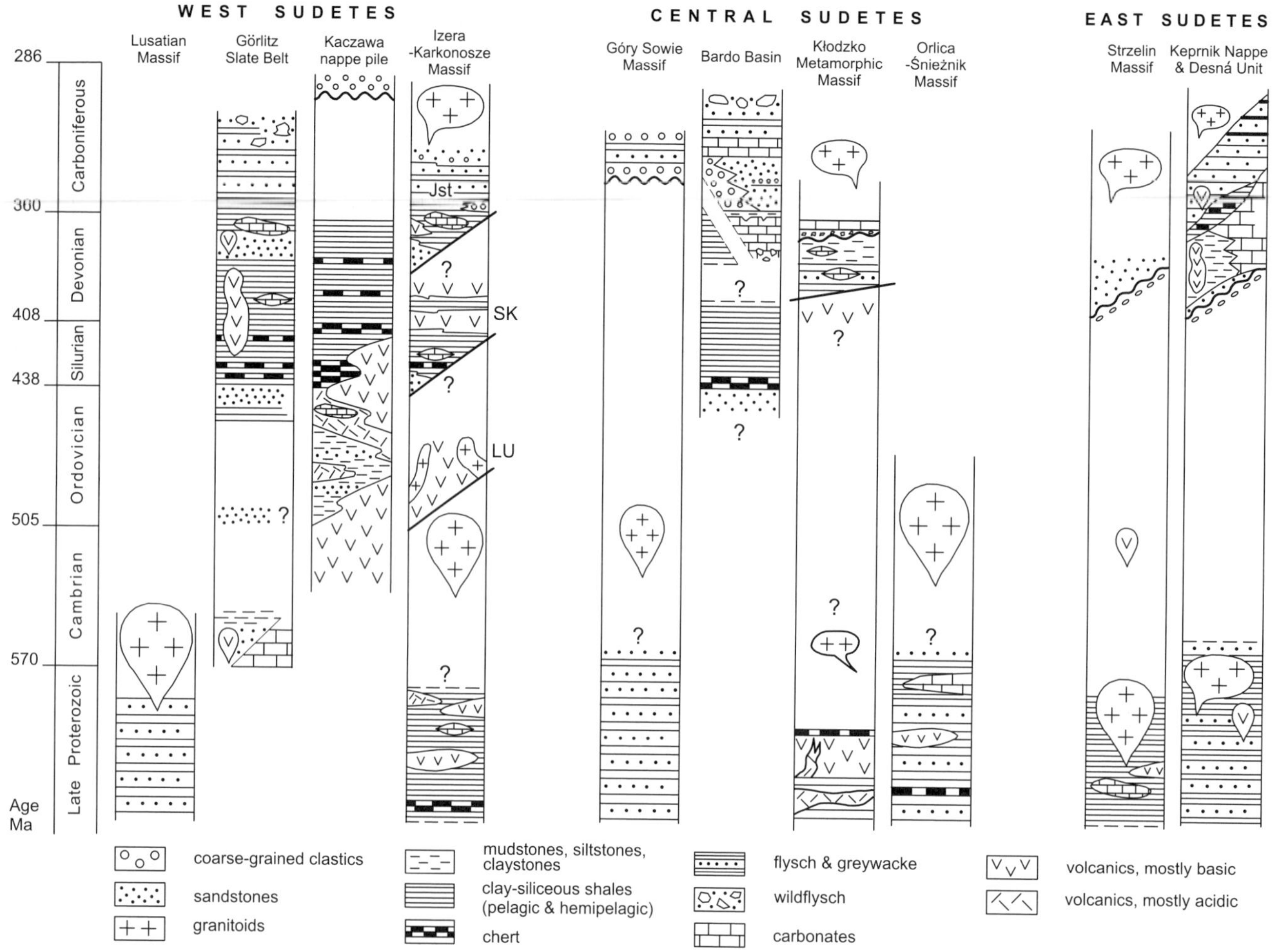

Fig. 11.15. Simplified stratigraphic columns of the main structural units of the Sudetes (modified from Aleksandrowski & Mazur 2002). Metamorphic rocks are represented by their sedimentary and igneous protoliths. Abbreviations: Jst, Ještěd Unit; LU, Leszczyniec Unit; SK, South Karkonosze Unit.

The West Sudetes comprise the eastern fragments of the Lusatian Massif, the Karkonosze-Izera Massif, the Kaczawa Nappe Pile and the Görlitz Slate Belt. All of these were deformed between the Late Devonian and the Late Carboniferous. However, the deformation climax, which was accompanied by orogenic uplift, occurred in this area during the Early Carboniferous. Subsequently, from latest Carboniferous times the region was partly covered by the North Sudetic Permo-Mesozoic Basin (Fig. 11.14).

The Central Sudetes include the Góry Sowie Massif, together with the surrounding ophiolitic bodies (the Central Sudetic Ophiolite), the Kłodzko and Orlica-Śnieżnik massifs, the Nové Město, Staré Město and Kamieniec metamorphic belts, the Niemcza and Skrzynka shear zones and the Niedźwiedź amphibolite Massif (Fig. 11.14). These are partly overlain by the intramontane Bardo and Świebodzice basins, initiated in Late Devonian times, and by the younger Intra-Sudetic Basin initiated in the middle Visean (Turnau *et al.* 2002). The main phase of deformation and orogenic uplift took place in this area in Mid–Late Devonian times, and was followed by a period of Carboniferous tectonism.

The East Sudetes form part of a collisional belt developed along the eastern margin of the Bohemian Massif. These are composed of a nappe pile of Moravo-Silesian units, the latter representing the detached Bruno-Vistulian basement, overridden from the west by the Central Sudetic Orlica-Śnieżnik Massif and the Staré Město Belt (Fig. 11.14). From top to bottom (or in map view from west to east), the East Sudetic Nappe Pile comprises the Velké Vbrno and Keprník nappes which rest on parautochthonous gneisses of the Desná Unit (Schulmann & Gayer 2000) and are covered by the allochthonous(?) Devonian metasedimentary Vrbno Group (Fig. 11.14). To the east they contact the thrust and folded Lower Carboniferous sedimentary infill of the Upper Silesian Foreland Basin, which overlies the top of a Devonian carbonate platform.

Lithostratigraphy

The Sudetes expose a variety of metamorphic complexes with Neoproterozoic, and Lower Palaeozoic to Devonian protoliths (Fig. 11.15). The complexes of Neoproterozoic protoliths most probably represent splinters from an active margin of the Gondwana continent. Indeed, they record both magmatism and metamorphism related to the Cadomian (Panafrican) Orogeny that occurred at *c.* 530–600 Ma (Kröner *et al.* 1994; Buschmann *et al.* 2001; Oberc-Dziedzic *et al.* 2003; Mazur *et al.* 2004; see also Linnemann *et al.* 2008). The Neoproterozoic complexes comprise: (1) plutonic and volcanosedimentary rocks of magmatic arc and backarc basin affinities (in the Kłodzko Metamorphic Massif); (2) granitoids intruded into greywackes (in the Lusatian Massif and the East Sudetes); and, most probably, also (3) a pelitic series enveloping Early Palaeozoic igneous intrusions (in the Izera-Karkonosze and Orlica-Śnieżnik massifs) of hitherto poorly constrained age (Gunia 1984, 1990; Kröner *et al.* 2001).

Late Cambrian–Early Ordovician granitic intrusions are widespread in the Sudetic area, and these were generally deformed into gneisses by later Variscan tectonism. The granites were emplaced in an initially continental rift environment (cf. Aleksandrowski *et al.* 2000) and, locally, preserve the record of Ordovician LP/HT metamorphism, reflecting the thinning of the continental crust (Kröner *et al.* 2000*b*; Štipska *et al.* 2001).

Late Cambrian(?)–Ordovician to Devonian volcanosedimentary basinal successions of the Kaczawa Nappe Pile and the southern Karkonosze-Izera Massif (Fig. 11.15) were initially deposited (during Late Cambrian?–Ordovician times) in a continental rift environment. The region subsequently evolved (during the Silurian–Devonian) into an oceanic basin (Furnes *et al.* 1994; Patočka & Smulikowski 2000). The successions contain Ordovician intraplate bimodal volcanics associated with probable continental rift-related sediments, Silurian metabasalts with a MORB-type geochemical signature, and Silurian–Devonian deep-marine deposits (cherts). In the eastern part of the Izera-Karkonosze Massif, the Leszczyniec Meta-igneous Unit, which comprises Ordovician mafic MORB-type rocks (of basalt/gabbro composition) associated with plagiogranite and tonalite gneisses, represents a series of plutonic and subvolcanic rocks.

Late Silurian(?)–Early Devonian (420–400 Ma; Oliver *et al.* 1993; Dubińska *et al.* 2004) oceanic crust makes up the barely metamorphosed circum-Góry Sowie (Central Sudetic) Ophiolite. This unit includes almost all of the typical ophiolitic sequence members, such as serpentinite, massive gabbros and cumulates, sheeted dykes, pillow lavas and deep-marine radiolarian cherts.

Middle Devonian to Early Carboniferous sedimentary sequences deposited on extending margins of continental blocks (continents/terranes) occur in the east Sudetes (Vrbno Unit), the Görlitz Slate Belt and the Ješted Unit. These are typified by a transition from Devonian shelf, mostly carbonate, sediments to Carboniferous turbidites and/or mélanges (e.g. in the Görlitz Slate Belt). The rocks of the Vrbno Group also contain a distinctive Middle to Late Devonian volcanosedimentary succession deposited within a backarc basin setting (Patočka & Valenta 1996).

The metamorphosed and exhumed Sudetic rock complexes subsequently formed the substratum for intramontane basins. These basins began to develop in Late Devonian (e.g. Bardo and Świebodzice basins) or in Early Carboniferous times (e.g. the Intra-Sudetic Basin; Fig. 11.15).

Structure

The vestiges of the various marine basins in the Sudetes area include the volcanosedimentary allochthonous complexes of the Kaczawa Nappe Pile and of the southern and eastern Karkonosze-Izera Massif. These form nappe piles, overthrust towards the NW (Mazur & Kryza 1996; Seston *et al.* 2000; Collins *et al.* 2000; Mazur & Aleksandrowski 2001), and they define tectonic sutures, as evidenced by meta-igneous rocks with a MORB-type geochemical signature which are associated with deep-marine sediments (Furnes *et al.* 1994; Winchester *et al.* 1995; Kryza *et al.* 1995; Patočka & Smulikowski 2000) and a record of HP, blueschist facies metamorphism (Cháb & Vrana 1979; Kryza *et al.* 1990; Smulikowski 1995; Kryza & Mazur 1995). The latter phase of metamorphism was overprinted by greenschist facies alteration. In the nappe piles, thrust sheets with MORB-type rocks tectonically overlie the thrust units that contain intraplate volcanics. An important part of the Kaczawa Nappe Pile are the mélange bodies which are often less metamorphosed than the surrounding rocks (Baranowski *et al.* 1990; Collins *et al.* 2000).

The metamorphic nappe piles of the south and east Karkonosze-Izera Massif and of the Kaczawa Nappe Pile are overthrust onto continental crust which is considered to be the pre-Variscan passive margin of the Saxothuringian Zone (e.g. Mazur & Aleksandrowski 2001; Franke & Żelaźniewicz 2002; Aleksandrowski & Mazur 2002). In the Sudetes, the continental crustal base of this passive margin is represented by the crystalline complexes of the Lusatian (Linnemann *et al.* 2000) and Karkonosze-Izera massifs (e.g. Matte *et al.* 1990; Mazur & Aleksandrowski 2001). The Lusatian Block is comprised of Neoproterozoic to Lower Cambrian Lusatian granitoids (540–530 Ma; Kröner *et al.* 1994; Tichomirova 2002) and greywackes that were deposited on an active margin of Gondwana (Kemnitz & Budzinski 1994; Gehmlich *et al.* 1997; Buschmann *et al.* 2001). To the east, within the Karkonosze-Izera Massif, the Neoproterozoic rocks were massively intruded by Early Ordovician granitoids which were later deformed into orthogneisses by Variscan tectonism. These granitoids witnessed the formation of the eastern edge of the Saxothuringian Zone as an Ordovician rifted margin. The Palaeozoic volcanosedimentary cover of this area is now only fragmentarily preserved; it comprises the Ješted Unit within the Karkonosze-Izera Massif and the Torgau-Doberlug Basin (Syncline) as well as the Görlitz Slate Belt directly adjacent to the Lusatian Massif. The Görlitz Slate Belt respresents a parautochthonous basement that was overthrust by allochthonous units of the Kaczawa Nappe pile (Aleksandrowski & Mazur 2002).

The tectonic sutures containing the allochthonous units of the Kaczawa Nappe Pile and of the southern and eastern parts of the Karkonosze-Izera Massif may be correlated with Late Palaeozoic sedimentary basins to the east. The sediments in these basins overlie the metamorphic complexes unconformably; the latter represent the upper plate, overthrusting the sutures zones. The Intra-Sudetic Basin, to the east of the Karkonosze-Izera Massif, developed in the late middle Visean (Turnau *et al.* 2002) due to top-to-the-ESE directed (Mazur 1995; Mazur & Aleksandrowski 2001) gravitational collapse. The Świebodzice Basin, east of the Kaczawa Nappe Pile complex, was initiated in the latest Devonian as a piggy-back basin.

The tectonic units located in the Central Sudetes, east of the Intra-Sudetic and Świebodzice basins, are mainly, if not entirely, allochthonous. These include the metamorphic massifs of Góry Sowie, Kłodzko, Orlica-Śnieżnik and Nové Město, as well as that of the Central Sudetic Ophiolite. The latter is the vestige of the second of the oceanic basins, whose closure resulted in the formation of the Sudetic segment of the Variscan Belt. The Central Sudetic Ophiolite underwent little metamorphism, but the Góry Sowie and Orlica-Śnieżnik massifs contain HP granulites (Kryza *et al.* 1996), which suggest that these units were derived from subducted continental crustal fragments. The granulites from the Orlica-Śnieżnik Massif also contain a record of early UHP metamorphism (Szczepański *et al.* 2004) and are associated with HP eclogites. The spatial relationships between the Góry Sowie Massif and the Central Sudetic Ophiolite remain unclear (see discussion in Aleksandrowski & Mazur 2002), whereas the Kłodzko Metamorphic Massif, and probably also the Orlica-Śnieżnik Massif, are in a tectonically higher position than the ophiolite. The Central Sudetic tectonic units are presumably overthrust to the NW, although this conclusion is based exclusively on data from the Kłodzko Metamorphic Massif (Mazur 2003; Mazur *et al.* 2004). At the boundary of the Orlica-Śnieżnik and Nové Město units, an opposite, top-to-the-SE, overthrusting polarity has been recorded (Mazur *et al.* 2005). However, the lack of metamorphic grade inversion in this area implies that there is a back-thrust contact between these two tectonic units. In general, the Central Sudetes comprise structural units that correspond to Neoproterozoic to Cambrian volcanosedimentary sequences of an active Gondwana margin (the Kłodzko Metamorphic Massif (Mazur *et al.* 2004) and the Góry Sowie Massif (Gunia 1985)), as well as those dominated by Early Ordovician granitoid intrusive bodies (Orlica-Śnieżnik Massif; e.g. Oliver *et al.* 1993; Turniak *et al.* 2000; Kröner *et al.* 2001). In contrast to the West Sudetes, the Central Sudetes do not contain volcanosedimentary successions deposited in Early Palaeozoic open-marine basins except for rare olistoliths in the Bardo Basin (Fig. 11.15).

East of the Orlica-Śnieżnik Massif, and separated from it by a suture zone (Staré Město Belt), is the East Sudetic Nappe Complex. The Staré Město Belt is composed mainly of rocks derived from an Early Ordovician initial rift that show evidence of rift-related HT/LP metamorphism (Kröner *et al.* 2000*b*; Štipska *et al.* 2001). They are intruded by synorogenic tonalites dated at *c.* 340 Ma (Parry *et al.* 1997). The East Sudetic Nappe Complex is part of the north–south striking Moravo-Silesian Zone of the Variscides, located along the eastern margin of the Bohemian Massif. This traces the suture zone between the Armorican Terrane Assemblage, represented here by the Central Sudetes, and the Bruno-Vistulian Block located further to the east. The latter extends from near Brno (Czech Republic) into Upper Silesia; to the SE it plunges below the Carpathian Frontal Thrust. The Bruno-Vistulian crystalline complex is composed of Neoproterozoic metamorphic complexes and igneous intrusions covered by Palaeozoic sedimentary strata. On the western margin of the Bruno-Vistulian Terrane, the sedimentary cover comprises Early Devonian basal conglomerates and quartzites, which are overlain by Middle–Late Devonian carbonate platform deposits and Early Carboniferous synorogenic turbidites of the Moravo-Silesian Foreland Basin.

Within the East Sudetic Nappe Complex, detached fragments of Neoproterozoic Bruno-Vistulian basement are overthrust together with metamorphosed rocks of the sedimentary cover (Schulmann & Gayer 2000). The tectonically lowermost element of the nappe pile is a 3000 m thick allochthonous sequence of Middle Devonian shales and volcanics (Vrbno Series; Cháb *et al.* 1994), representing the infill of a backarc basin (Patočka & Valenta 1996). The East Sudetic nappes were thrust to the NE in a dextral transpressional regime (Matte *et al.* 1990; Schulmann & Gayer 2000). The structural plan, which is dominated by NE–SW to NNE–SSW trending lineations, is characteristic not only of the East Sudetic Nappe Complex, but also of the adjacent part of the Central Sudetes, particularly of the Orlica-Śnieżnik Massif. At the contact zone of the latter and the Nové Město Unit, the NNE–SSW lineation is overprinted onto an older, NW–SE structural grain (Mazur *et al.* 2005).

Correlation of the Sudetes with the remaining parts of the Bohemian Massif and of particular Sudetic units with one another is hampered by two important strike-slip fault/shear zones, namely the WNW–ESE striking Upper Elbe and the Intra-Sudetic fault zones. The displacement magnitudes along these fault zones are disputed (e.g. Aleksandrowski 1990, 1995; Matte *et al.* 1990; Oliver *et al.* 1993; Aleksandrowski *et al.* 1997); since they constitute boundaries for all of the adjacent structural units, transregional correlations are difficult. Tectonic activity along both of these major fault zones was mostly terminated by Late Carboniferous sinistral displacement of, locally, up to 15–20 km, which followed an earlier, Late Devonian–Early Carboniferous(?) main phase of dextral displacement (Aleksandrowski *et al.* 1997; Mattern 2001).

Timing

Accretion of the Sudetic segment of the Variscides was a multistage process. The oldest recorded tectonometamorphic events took place at the Silurian–Devonian boundary, when HP granulites, dated at 400–395 Ma in the Góry Sowie Massif (O'Brien *et al.* 1997; Kryza & Fanning 2004) and 386 Ma in the Orlica-Śnieżnik Massif (Szczepański *et al.* 2004) show evidence of continental subduction closely following the formation of the Central Sudetic Ophiolite (420–400 Ma; Oliver *et al.* 1993; Dubińska *et al.* 2004). Uplift of fragments of the subducted continental crust resulted in the formation of a nappe pile, ophiolite formation possibly in a suprasubduction zone setting, as well as the exhumation of metamorphic complexes. The upper limit for the age of these various deformational processes is constrained by the presence of a pre-Late Devonian unconformity above the ophiolite and the nearby Kłodzko Metamorphic Massif. The unconformity is overlain by unmetamorphosed Late Frasnian to Fammenian limestones (Bederke 1924, 1929; Kryza *et al.* 1999). The Late Devonian carbonate rocks pass upwards into the Early Carboniferous clastic sequences of the Bardo Basin (Haydukiewicz 1990; Wajsprych 1986). The lower time limit for nappe emplacement is constrained by the Early Givetian fauna in the metamorphosed sediments of the Kłodzko Metamorphic Massif, below the unconformity (Hladil *et al.* 1999). Consequently, the emplacement and exhumation of the Central Sudetic nappes must have taken place within the narrow time-span of *c.* 390 to 380 Ma.

To the west, in the Kaczawa Basin, deep-water sedimentation extended until the end of the Devonian (Fig. 11.15) with the Kaczawa successions subsequently included into an accretionary prism unit. This situation is analogous to that from the adjacent Karkonosze-Izera Massif, where the age of the blueschist facies metamorphism was determined at 360 Ma (Maluski & Patočka 1997). The final thrusting of the nappes of the Kaczawa Pile and of the southern and eastern parts of the Karkonosze Massif over the eastern margin of Saxothuringian Zone, occurred not earlier than the Visean, as indicated by the age of sediments underlying the allochthonous units in the Ještěd Unit (Chlupač 1993*a*) and at the western part of the Kaczawa Nappe pile (Chorowska 1978).

The collision of the Central Sudetes with the promontory of the Bruno-Vistulian Terrane commenced as late as the earliest Carboniferous and lasted until the Early–Late Carboniferous boundary (Schulmann & Gayer 2000). At that time, in the other parts of the Sudetes the dominant types of tectonic activity were strike-slip motions localized on major fault/shear zones as well as gravitational collapse, accompanied by intense granitic plutonism. A 340 Ma collapse with a top-to-the-SE sense of shear was documented from the eastern margin of the Karkonosze-Izera Massif (Mazur 1995; Mazur & Aleksandrowski 2001). The dextral strike-slip displacements which took place at that time occurred not only along the WNW–ESE trending Upper Elbe and Intra-Sudetic fault zones (Aleksandrowski *et al.* 1997; Mattern 2001), but also along the boundary between the Nové Město Unit and the Orlica-Śnieżnik Massif (Mazur *et al.* 2005). In turn, sinistral displacements took place on the regionally significant NNE–SSW striking Niemcza Shear Zone (Mazur & Puziewicz 1995) and the NE–SW trending Złoty Stok-Skrzynka Shear Zone (Żaba & Będkowski 1995; Cymerman 1996). Granitic plutonism reached its climax at *c.* 340–310 Ma (cf. Aleksandrowski *et al.* 2000), as evidenced by the emplacement of several large plutons, such as those of Karkonosze, Strzegom-Sobótka, Złoty Stok and Żulova.

The accretion of the Sudetes took place in three main stages: (1) at the Middle–Late Devonian boundary (in the Central Sudetes); (2) at the Devonian–Carboniferous boundary (in the West Sudetes); and (3) in the Early Carboniferous (in the East Sudetes). At these times at least two marine basins underlain by oceanic crust were closed, and it is from these basins that the Sudetic ophiolite and allochthonous complexes of the Kaczawa Nappe Pile and southern and eastern Karkonosze originated. It cannot, however, be excluded that the backarc basin separating the Central Sudetes from the Bruno-Vistulian Block was underlain by oceanic crust. Vestiges of this oceanic crust may be represented by the belt of metamorphosed mafic rocks with MORB-type chemistry that are found beneath the clastic sediments of the Moravo-Silesian Foreland Basin (Přichistal 1990).

Mid-German Crystalline High (A.Z.)

The Mid-German Crystalline High (MGCH) was initially described by Brandes (1919) and later defined by Scholz (1930) and Brinkmann (1948). According to these early authors the MGCH represents a more than 350 km long SW–NE trending belt of crystalline rocks that are exposed, from SW to NE, in the northern Vosges, the Rhineland-Palatinate, the Odenwald and Spessart Mountains, the Thuringian Forest (Ruhla Crystalline Complex), and the Kyffhäuser Mountains (Fig. 11.16). In addition, rocks of the MGCH were penetrated by drillholes in the Saar-Nahe, Werra and Thuringian basins. Additionally, numerous drillholes and geophysical investigations indicate that the MGCH continues as an east–west trending belt into Saxony-Anhalt and South Brandenburg, where it can be traced to the river Neiße (Bankwitz *et al.* 2001; Kopp & Bankwitz 2003). To the east equivalents of the MGCH continue into Poland along the NE–SW trending Middle Odra Fault Zone (Don 2003). To the west, SE–NW trending equivalents of the MGCH which are parallel to the Bristol Channel-Bray Fault (Fig. 11.16) are thought to underlie the France Basin; further west MGCH-type rocks are found in the Normannian High of Cornwall (Dörr *et al.* 1999). On the Iberian Peninsula, magmatic and amphibolite facies metamorphic rocks of the Ossa Morena zone in Portugal and Spain are also considered to be equivalents of MGCH lithologies (e.g. Giese 1996).

The MGCH consists of medium- to high-grade gneisses and granitic rocks, which separate the very low- to low-grade metasediments and volcanic rocks of the Northern Phyllite Belt and the Rhenohercynian domain in the north and NW from

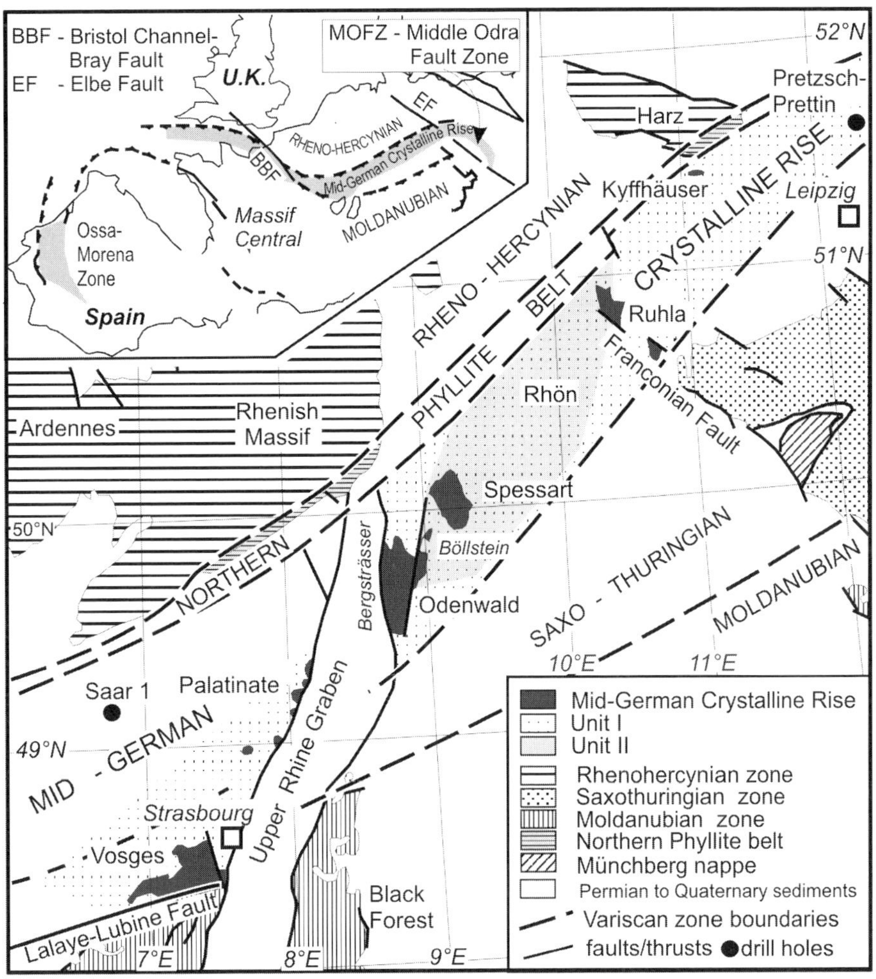

Fig. 11.16. Location of the Mid-German Crystalline High in Central Europe, and (inset) in east and west Europe.

the low- to medium-grade rocks of the Saxothuringian domain in the south and SE (Fig. 11.16). In a broader tectonic context, the MGCH is interpreted as being part of a suture zone between the Eastern Avalonia Terrane in the north and NW, and the Saxothuringian microterrane that belongs to the so-called Armorican Terrane Assemblage in the south and SE (Tait *et al.* 1997) (see also Krawczyk *et al.* 2008).

Investigations, carried out over the last ten years, indicate that the MGCH is not a coherent basement wedge, but consists of contrasting lithological units (e.g. Zeh 1996; Oncken 1997), which are presumed to have formed in different geotectonic environments and which were juxtaposed during Variscan collision (Zeh 1996; Oncken 1997; Hansch & Zeh 2000). Two main lithological units can be recognized. Rocks of unit I are interpreted to represent relicts of a Variscan magmatic arc which formed at the NW margin of the Saxothuringian domain (upper plate), whereas rocks of unit II comprise a metasediment–metabasite succession, which is assumed to have been deposited in the Rhenohercynian Ocean NE of the Saxothuringian terranes (Oncken 1997). The Rhenohercynian Ocean closed during oblique Variscan convergence between 360 and 320 Ma, and rocks of unit II were accreted to the base of the overriding Saxothuringian island arc (Oncken 1997; Altherr *et al.* 1999). The central part of the magmatic arc was eroded as a result of accretion during oblique collision, leading to the recently suggested threefold division of the MGCH. The central part of the MGCH, which comprises the Spessart Crystalline Complex, parts of the Ruhla Crystalline Complex (Ruhla and Trusetal groups), and the eastern part of the Odenwald crystalline complex (Böllstein Odenwald), now exposes underplated rocks of unit II (Fig. 11.16), whereas the SW and NE parts, exposed in the Bergsträsser Odenwald and in the Kyffhäuser Mountains and Saxony-Anhalt/South Brandenburg (east Germany), respectively, represent parts of the former Saxothuringian magmatic arc (Oncken 1997).

Saxony-Anhalt and South Brandenburg

In Saxony-Anhalt and South Brandenburg, rocks of the MGCH are completely covered by sediments. Nevertheless, their occurrence is well constrained by geophysical investigations (Lauterbach 1953; Köhler *et al.* 1989) and by numerous drillholes in the areas around Halle, Dessau, Pretzsch-Prettin-Schönewalde, Luckau and near Guben (see summary in Bankwitz *et al.* 2001). The recovered drillcore material consists of a Neoproterozoic(?)/Palaeozoic paragneiss/metabasite succession that was intruded by numerous gabbros, diorites and granites during the Variscan Orogeny between 331 and 355 Ma (Röber *et al.* 1997; Kopp *et al.* 1998, 1999; Hammer *et al.* 1999; Anthes & Reischmann 2001). A gabbro from the Züllsdorf drilling is different from the others since it yielded Late Cambrian–Early Ordovician Sm–Nd mineral isochron ages of *c.* 490 Ma (Hammerschmidt *et al.* 2003). A Late Carboniferous age of 300 Ma was determined for the Delitzsch granite (Anthes & Reischmann 2001). A K–Ar biotite age of *c.* 330 Ma indicates that paragneisses of the Hohnsdorf crystalline complex cooled to below 300°C during the Visean (Marheine 1997).

The Kyffhäuser Crystalline Complex

The Kyffhäuser Crystalline Complex (KCC) forms a small basement outcrop of only 1.2 km^2 at the northern margin of the Kyffhäuser mountains. The complex is bounded by the Kyffhäuser and Kelbra faults to the NE and NW, respectively (Fig. 11.17A). Along these faults, the KCC is uplifted by nearly 1000 m against Upper Permian to Cenozoic sediments to the north. In the south the KCC is overlain by Stephanian continental sediments, which were deposited at *c.* 300 Ma (Menning *et al.* 1996). According to Zeh *et al.* (2005), the KCC can be subdivided into three units: (1) a magmatic/metamorphic complex to the west; (2) the Borntal Intrusive Complex in the centre; and (3) the Bärenkopf granite to the east.

Geochronological data indicate that the metasedimentary rocks exposed in the western magmatic/metamorphic complex of the KCC are younger than *c.* 470 Ma (Early Ordovician) and were intruded by gabbros and diorites with island arc affinities between 345 and 340 Ma (Zeh *et al.* 2005). These *c.* 850–900°C hot intrusions caused contact metamorphism in the surrounding host rock sediments at P-T conditions of 690–750°C and 5–7 kbar, corresponding to an intrusion depth of some 19–25 km (Fig. 11.17A). Subsequently, the western magmatic/metamorphic complex was uplifted rapidly from 19–25 km to 7–11 km under high-grade metamorphic conditions (750–700°C) and was later intruded by the granites, syenites and diorites of the Borntal intrusive complex and by the Bärenkopf granite at *c.* 337 Ma (Zeh *et al.* 2005). Eventually, all of the rocks of the KCC underwent a retrograde structural/metamorphic overprint during accelerated cooling. The present-day outcrop of the KCC was exhumed prior to 300 Ma and covered by continental sediments during the Late Carboniferous.

The Ruhla Crystalline Complex

The Ruhla Crystalline Complex (RCC) forms a *c.* 100 km^2 basement outcrop situated at the northwestern end of the Thuringian Forest horst block. In this area the NE-trending MGCH is transected by several north- to NW-trending faults that belong to the Franconian fault system (Fig. 11.17B), which was active during the Late Carboniferous–Permian, Mesozoic and Upper Cretaceous (Franzke & Rauche 1991; Thomson & Zeh 2000; Zeh *et al.* 2000; Brätz 2000; Zeh & Brätz 2004). The eastern, western and northeastern part of the RCC is overlain by Late Carboniferous–Early Permian (Rotliegend) sediments and volcanic rocks, which were deposited in the Eisenach and Oberhof basins (Zeh & Brätz 2004). To the NW and south the RCC is directly overlain by Upper Permian (Zechstein) sediments and locally by Lower Triassic (Buntsandstein) sandstones. The RCC consists of four tectonometamorphic units: (1) the Trusetal Group in the SE; (2) the Ruhla Group in the west; (3) the Brotterode Group in the NE; and (4) the Central Gneiss Unit (Zimmermann 1930; Neumann 1974; Zeh 1996; Wunderlich & Zeh 2001). These units are separated from each other by the Ruhla and Trusetal granites, and/or the West Thuringian Fault (Fig. 11.17B).

The Ruhla Group forms the western part of the RCC and is subdivided into three lithostratigraphic units: the Gömigenstein, Windsberg, and Struth formations (Neumann 1974; Wunderlich & Zeh 2001). These formations consist of phyllonitic schists, quartzites and amphibolites, and are intercalated with *c.* 425 Ma Silurian orthogneisses (Erbstrom, Silbergrund and Thal gneiss); they were intruded by the Ruhla Granite at 295 Ma (Zeh *et al.* 2000). During the Variscan Orogeny rocks of the Ruhla Group underwent a prograde P-T evolution from 460°C, 2.5 kbar to 580°C, 8.5 kbar (A. Zeh unpublished work), before it was affected by a penetrative greenschist facies overprint at P-T conditions of about 400°C, 4.5 kbar (Zeh 1996).

The Trusetal Group forms the southeastern part of the RCC, east of the Trusetal Granite. In addition, 'klippen' of Trusetal

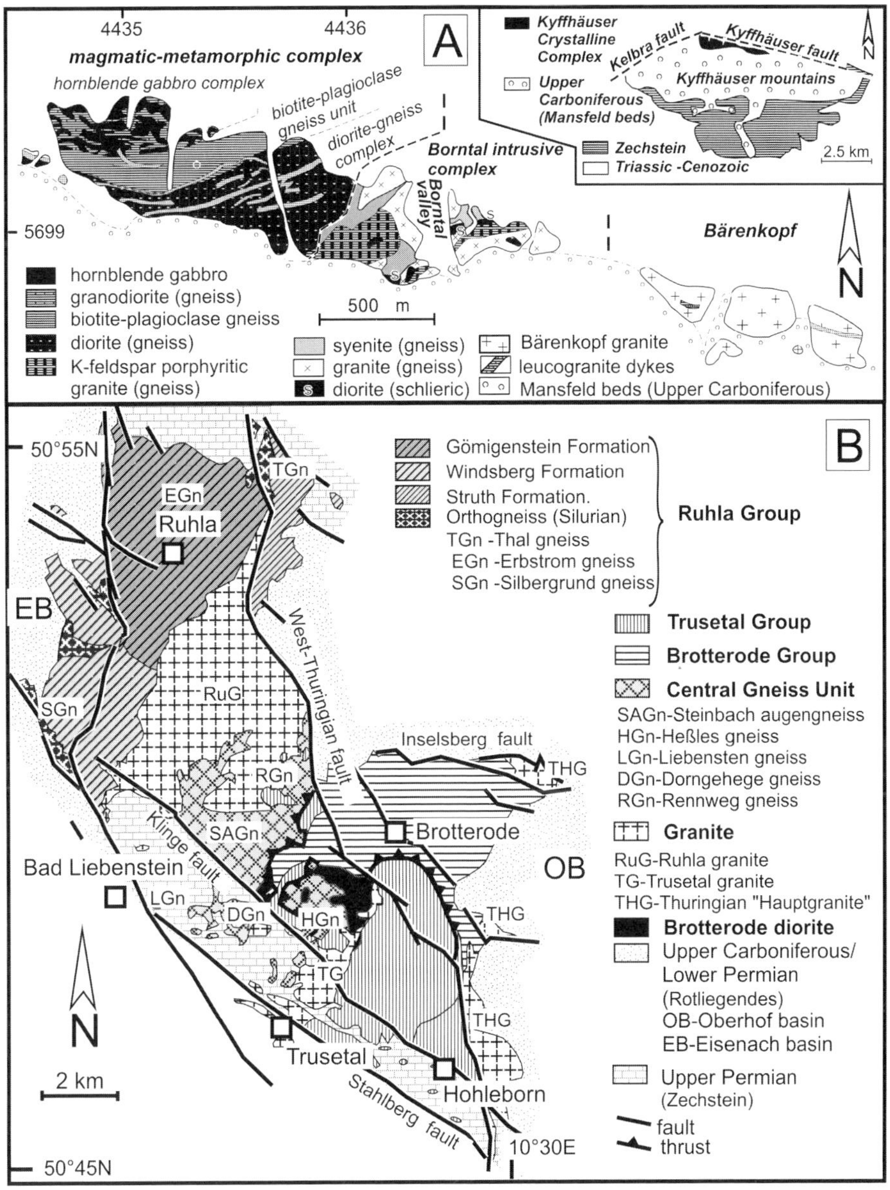

Fig. 11.17. (**A**) Geological map of the Kyffhäuser Crystalline Complex (modified after Zeh *et al.* 2005). (**B**) Geological map of the Ruhla Crystalline Complex (modified after Zeh *et al.* 2003).

Group rocks occur on top of the Central Gneiss Unit (Zeh 1996). The Trusetal Group represents a NW-vergent succession (Zeh 1996), and consists mainly of garnet-staurolite-(kyanite)-bearing mica schists, metagreywackes, quartzitic schists, quartzites and amphibolites. Most metabasalts of the Trusetal Group show typical MORB signatures (Estrada *et al.* 1992; Hansch & Zeh 2000). During the Variscan Orogeny rocks of the Trusetal Group underwent a Barrovian-type regional metamorphic overprint with a prograde P-T evolution from 400–600°C, 9–14 kbar (M_1, S_1) to 630°C, 7 kbar (M_2, S_2), followed by a retrograde overprint at 400–500°C, >4 kbar (M_3, S_3) and 400°C, >2–3 kbar (S_4) (Zeh 1996). There is no reliable evidence concerning the depositional age of the (meta)sediments and the timing of metamorphism. During the Late Carboniferous, the Trusetal Group cooled below 400°C (311 Ma: Ar/Ar muscovite: Zeh *et al.* 2000), and was intruded by the Trusetal Granite at 298 Ma (Zeh *et al.* 2000; Zeh & Brätz 2004). Subsequently, it cooled to below 300°C at *c.* 290 Ma (Ar/Ar biotite: Zeh *et al.* 2000) and, finally, was overlain by Upper Permian (Zechstein) sediments, and locally by Triassic sandstones.

The Brotterode Group is a migmatitic gneiss complex that forms the northeastern part of the RCC; it is bounded by the Thuringian Main Granite (Thüringer Hauptgranit) to the east (Fig. 11.17B). The Brotterode Group consists of garnet-(sillimanite)-(K-feldspar)-bearing paragneisses (metagreywackes, metapelites; Werner 1970) and migmatitic hornblende gneisses (metadiorites). The latter show typical island-arc signatures but there are also some massive amphibolites with MORB characteristics (Hansch & Zeh 2000). Marbles and calc-silicate rocks are very rare (Neumann 1974). *In-situ* dating of detrital zircon cores provides evidence that the metagreywackes are younger than Late Cambrian (500 Ma: Gerdes & Zeh 2006) and/or Middle Ordovician (460: Zeh *et al.* 2003), and that their source consists mainly of Proterozoic and some Archaean material, which was derived from the Avalonian–Cadomian belt, the West African and/or Amazonian Craton and the Grenvillian belt (Zeh *et al.* 2001). Metamorphic zircon overgrowths yielded a U–Pb age cluster of *c.* 355 Ma (Zeh *et al.* 2003), which is interpreted to date the high-grade metamorphism. This occurred at peak metamorphic conditions of *c.* 700°C, 5 kbar, prior to the intrusion of a diorite (340 Ma; Zeh *et al.* 2003) and the Thuringian Main Granite (337 Ma; Zeh *et al.* 2000). Subsequently, these magmatic rocks were deformed, together with their country-rocks, and cooled to below 300°C between 330 and 323 Ma (Ar/Ar biotite; Zeh *et al.* 2000; Goll & Lippolt 2001). Finally, they were covered by Rotliegend sediments at *c.* 300 Ma.

The exposure of the Central Gneiss Unit is limited by the Trusetal Granite in the east, and the Ruhla Granite in the west. The gneiss unit is characterized by a dome structure (Wunderlich 1992) and consists of granitic orthogneisses of Early Devonian ages between 413 and 398 Ma (Brätz 2000; Liebenstein Gneiss, Steinbach Augengneiss, Heßles Gneiss, Dorngehege Gneiss), and intercalated hornblende gneisses (metabasalts), which show typical IAB signatures (Hansch & Zeh 2000). Some of the orthogneisses contain 'xenoliths' with (garnet)-cordierite-sillimanite-K-feldspar-bearing assemblages indicating high-grade metamorphic peak conditions of *c.* 700°C, 4 kbar (Zeh 1996). At *c.* 290 Ma, the Central Gneiss Unit cooled below 300°C (Ar/Ar biotite; Zeh *et al.* 2000), before it was overlain by Upper Permian (Zechstein) sediments at *c.* 260 Ma.

Spessart Crystalline Complex

The Spessart Crystalline Complex (SCC) forms a *c.* 300 km^2 outcrop of crystalline rocks of the MGCH at the northwestern margin of the Spessart mountains. To the north, the crystalline basement rocks are overlain by Lower Permian (Rotliegend) sediments, and to the east and south by both Upper Permian (Zechstein) limestones and (mainly) Lower Triassic sandstones. The SCC is interpreted to form a *c.* 25 km wide, NE-trending asymmetric antiform (Weber 1995*a*; Behr & Heinrichs 1987). This antiform is assumed to consist of two main units: the Lower Tectonic Unit and the Upper Tectonic Unit. These two units are separated by a mylonitic thrust zone (including the Michelbach Fault), which itself was folded during Variscan updoming of the SCC (Weber & Juckenack 1990; Hirschmann & Okrusch 2001).

The SCC is subdivided into nine, generally NE–SW trending, lithological units (Bücking 1892; Thürach 1893; Bederke 1957; Matthes & Okrusch 1977; Okrusch 1983; Hirschmann & Okrusch 1988, 2001). These include: (1) the Alzenau Formation; (2) the Geiselbach Formation; (3) the Hörstein-Huckelheim Formation; (4) the Mömbris Formation; (5) the Rotgneiss Complex; (6) the Haibach Orthogneiss; (7) the Schweinheim Formation; (8) the Elterhof Formation; and (9) the Quartzdiorite-Granodiorite-Complex (Fig. 11.18A). The Alzenau, Elterhof and Schweinheim formations and the Haibach Orthogneiss are collectively termed the Upper Tectonic Unit while the remaining formations constitute the Lower Tectonic Unit (Hirschmann & Okrusch 1988, 2001). Based on lithostratigraphic correlations it is assumed that the rocks of the Lower Tectonic Unit represent a continuous sedimentary succession that extended from the Cambro-Ordovician through to the Silurian (e.g. Matthes 1954; Bederke 1957; Hirschmann & Okrusch 1988, 2001). The onset of deposition was marked by the greywackes of the Mömbris Formation (Smoler 1987), and ended with the deposition of the mature sandstones (now quartzites) of the Geiselbach Formation. The Silurian age of the Geiselbach Formation is well constrained by the presence of Wenlockian to Ludlowian spores (Reitz 1987). The depositional age of the other formations remains highly speculative; an upper age limit is provided by the Rotgneiss, which intruded the Mömbris Formation at 418 Ma (Dombrowski *et al.* 1995). Rocks of the Upper Tectonic Unit are assumed to be older than those of the Lower Tectonic Unit (Hirschmann & Okrusch 1988, 2001). The marble/calc silicate-bearing metasediments of the Alzenau and Elterhof formations are thought to be of Cambrian age, whereas the protoliths of the Schweinheim mica schists are thought to have formed during the Neoproterozoic (Bederke 1957; Braitsch 1957; Hirschmann & Okrusch 1988, 2001). Given the lack of reliable age information, this interpretation remains highly speculative. An upper age limit for the deposition of the Schweinheim metasediments is, however, provided by the 410 Ma protolith age of the intercalated Haibach Orthogneiss (Dombrowski *et al.* 1995).

During the Variscan Orogeny all of the rocks of the SCC underwent peak metamorphic conditions of *c.* 600–670°C at 8–10 kbar (e.g. El Shazly 1983; Nasir 1986; Okrusch *et al.* 1985; Smoler 1987; Dombrowski *et al.* 1994; Will 1998), followed by a retrograde evolution to 520–560°C at 3–4 kbar (e.g. Will 1998). Numerous age data (K–Ar, Ar/Ar and Rb–Sr) provide evidence that the SCC cooled quickly from >500°C to below 300°C between 328 and 323–315 Ma (Lippolt 1986; Nasir *et al.* 1991; Dombrowski *et al.* 1994).

Odenwald Crystalline Complex

The Odenwald Crystalline Complex (OCC) is the largest basement window of the MGCH (Fig. 11.16). It is exposed in the western part of the Odenwald mountains, which formed as a result of the *c.* 6 km uplift of the eastern Rhine Graben shoulder in Tertiary–Quaternary times (Wenzel *et al.* 1991).

It has long been recognized that the OCC consists of two major distinct tectonometamorphic/magmatic basement units: the Bergsträsser Odenwald in the west and the Böllstein Odenwald in the east (Fig. 11.18B). These units show contrasting P-T-t evolutions and are separated from each other by the Otzberg Shear Zone, which also forms one of the main boundaries of the MGCH (e.g. Schälicke 1975; Hess & Schmidt 1989; Krohe 1991, 1992; Will 2001; Stein 2001). Undeformed lamprophyres, which cross-cut the Otzberg Shear Zone, yielded an Ar/Ar biotite cooling age of 327 ± 3 Ma (Hess & Schmidt 1989).

According to Krohe (1991), the rocks of the OCC were affected by four deformational events (D_1 to D_4). The D_1 and D_2 deformation is recorded only in metamorphic rocks and occurred prior to, or during, the metamorphic peak. In contrast, the D_3 and D_4 deformation also affected Variscan granites/diorites and gabbros, which intruded the metamorphic rocks. According to Krohe & Willner (1995) and Will (2001), the D_3 sinistral strike-slip was caused by north–south directed crustal shortening, whereas the D_4 normal faulting was accompanied by north–south orientated extension.

The Böllstein Odenwald forms a SSW–NNE trending, dome-like antiform (Altenberger *et al.* 1990; Altenberger & Besch 1993) with a core of granitic to granodioritic orthogneisses and intercalated metagabbros/basalts, which is mantled by metasediments and metabasites of the so-called 'Schieferhülle' (Chatterjee 1960; Nickel 1975; Knauer *et al.* 1974). Geochronological and geochemical data (Reischmann *et al.* 2001) indicate that the orthogneiss protoliths were derived from hybrid magmas emplaced into a volcanic arc setting during the Silurian–Devonian at 405–410 Ma. The intercalated metagabbros/basalts predominantly show within-plate basalt and to a lesser extent MORB signatures (Will & Schmädicke 2001; Altenberger & Besch

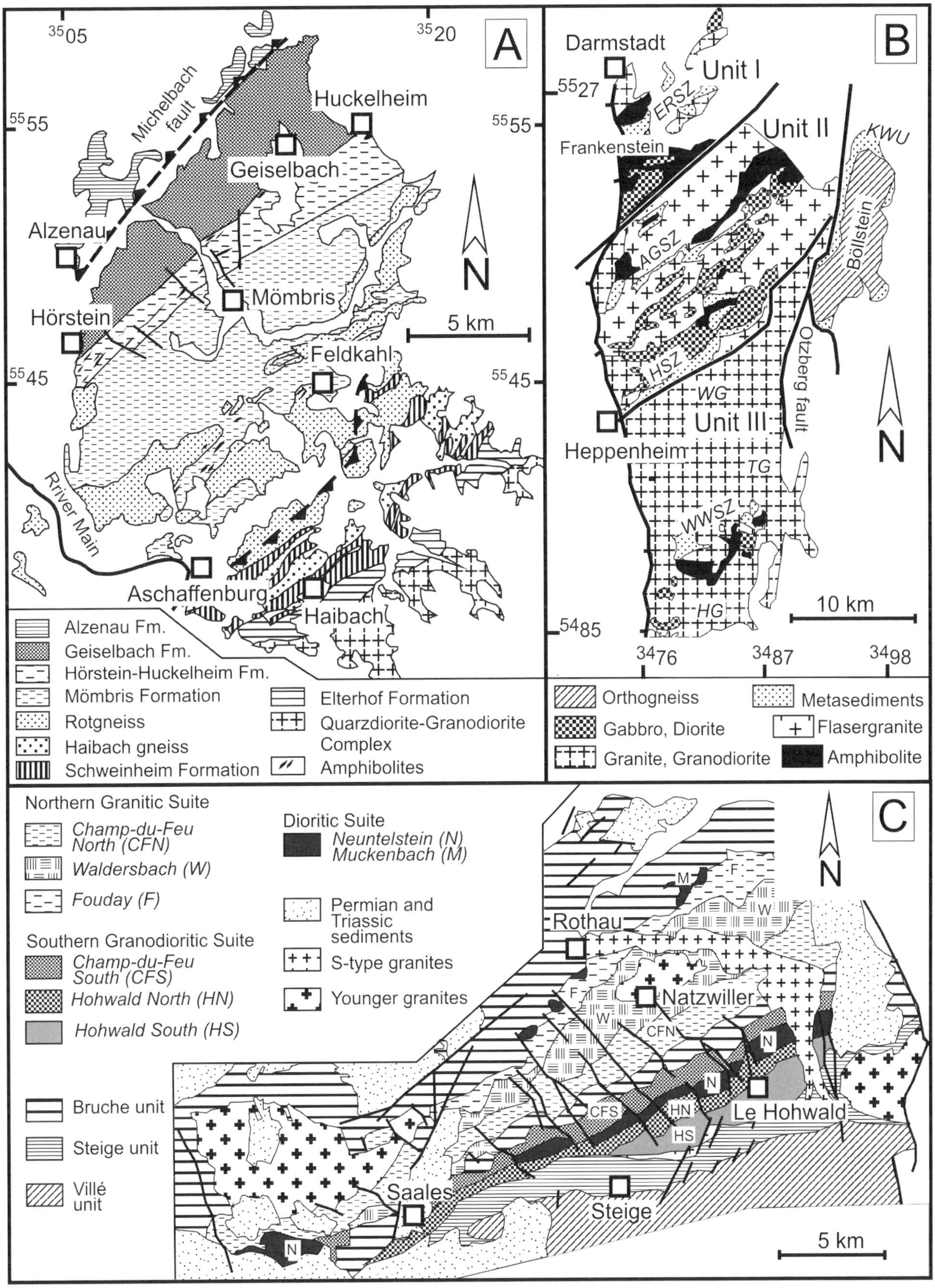

Fig. 11.18. (**A**) Geological map of the Spessart Crystalline Complex (modified after Hirschmann & Okrusch 2001). (**B**) Geological map of the Odenwald Crystalline Complex (modified after Stein 2001). Abbreviations: AGSZ, Auerbach-Großbieberach schist zone; ERSZ, Eberstadt-Roßdorf schist zone; HG, Heidelberg granite; HSZ, Heppenheim schist zone; KWU, Kainsbach-Wiebelsbach-Unit ('Schieferhülle'); TG, Tromm granite; WG, Weschnitz granite; WWSZ, Weinheim-Waldmichelbach schist zone. (**C**) Geological map of the Vosges (modified after Altherr *et al.* 2000).

1993; Scherer *et al.* 2002) and were probably emplaced during a post-Silurian rifting event (e.g. Will & Schmädicke 2001). Will & Schmädicke (2001) recognized that some of the metabasalts were metamorphosed to high-pressure eclogite facies conditions during a clockwise P-T evolution from 580°C, 8 kbar (pre-eclogitic stage) to 700 ± 50 °C, >16–17 kbar (eclogite stage), followed by a retrograde, isothermal decompression P-T path to 700°C, 15 kbar (symplectite stage) and 700–750°C, 10 kbar (amphibolite stage). Using Lu–Hf garnet–whole rock age data, Scherer *et al.* (2002) constrained the minimum age of the eclogite facies metamorphism to 357 Ma.

The schist envelope (Schieferhülle) of the gneiss core of the Böllstein Odenwald is interpreted as a metamorphosed volcano-sedimentary succession (e.g. Altenberger & Besch 1993). It consists of garnet-bearing muscovite-biotite schists, rare biotite gneisses, and minor intercalations of garnet-, graphite- and apatite-bearing quartzites, calc-silicate rocks, hornblende gneisses and biotite amphibolites (Chatterjee 1960; Altenberger & Besch 1993; Okrusch *et al.* 2000). U–Pb dating of multigrain zircon fractions from granitoid gneisses of the 'Schieferhülle' yielded a lower intercept age of 375 Ma (Todt *et al.* 1995). K–Ar dates indicate that cooling of the Böllstein rocks to temperatures below 500°C and 300°C took place at *c.* 325 Ma and 304 Ma, respectively (Kreuzer & Harre 1975; recalculated by Lippolt 1986).

The Bergsträsser Odenwald (BGO) consists of about 90% calc-alkaline magmatic rocks (Henes-Klaiber 1992; Kreher 1994; Altherr *et al.* 1999), which are divided into three magmatic complexes: the Frankenstein Gabbroic Complex (unit I), the Neunkirchen Magmatic Suite (unit II), and the Weschnitz-Heidelberg-Tromm Intrusive Complex (unit III) (Fig. 11.18B). The remaining 10% of the BGO is composed of metamorphic rocks that occur in four generally NE-trending schist tracts (the so-called 'Schieferzüge'), which are several hundred metres to a few kilometres wide: (1) the Eberstadt-Roßdorf Schist Zone; (2) the Auerbach-Groß-Bieberau Schist Zone; (3) the Heppenheim Schist Zone; and (4) the Weinheim-Waldmichelbach Schist Zone.

The magmatic rocks of the BGO have been described in detail by several authors (e.g. Magetti 1975; Nickel & Fettel 1985). Altherr *et al.* (1999) postulated that all of these magmas were formed during a progressive transition from an early subduction stage at *c.* 360 Ma to a late collision stage at *c.* 330 Ma. The oldest, subduction-related, magmatic rocks are gabbros and diorites of the Frankenstein Gabbro Complex. These intruded at shallow crustal depths (*c.* 6–9 km; Willner *et al.* 1991) into amphibolite facies metamorphic rocks at 362 Ma (Pb/Pb zircon; Kirsch *et al.* 1988). The gabbro cooled rapidly from temperatures of some 900°C (Kreher 1994) to below 200°C between 363 and 358 Ma (Ar/Ar data; Kirsch *et al.* 1988; Schubert *et al.* 2001). In contrast to the Frankenstein Gabbro Complex, the magmatic rocks of units II and III are *c.* 20–30 Ma younger (*c.* 330–340 Ma), and intruded at greater depths of 15–22 km (Altherr *et al.* 1999; Kreuzer & Harre 1975: recalculated by Hellmann *et al.* 1982; Hess & Lippolt 1996; Schubert *et al.* 2001).

The metamorphic rocks of the BGO represent a pre-Variscan volcanosedimentary succession. Amphibolites are predominant and occur in all four schist zones of the BGO (Schubert 1968; Klemm & Weber-Diefenbach 1971; von Raumer & Maggetti 1975). Geochemically, the amphibolites are similar to modern backarc basalts, and must be older than 360 Ma (Poller *et al.* 2001). These metabasalts are associated with metapelites and greywackes, quartzites, marbles and calc-silicate rocks (Matthes *et al.* 1972; von Raumer 1973). The latter are best known from the Auerbach-Groß-Bieberau schist zone (Barth 1971; Schmädicke *et al.* 2001). Extremely Al- and Fe-rich, Si-poor, corundum-bearing metamorphic rocks also occur (Meisl 1972; Krytosek & Paulitsch 1981; Altenberger *et al.* 1994; Bossdorf 1961; Adusumalli & Schubert 2001; Will & Schmädicke 2003).

Willner *et al.* (1991) suggested that the metamorphic rocks of unit I underwent a low pressure regional metamorphism of 1.5 kbar at 550°C, prior to the intrusion of the Frankenstein Gabbroic Complex at 362 Ma. Will & Schmädicke (2003) showed that rocks of the Heppenheim schist zone (unit II) underwent a hairpin-like anticlockwise P-T evolution from <660°C, 2.0 kbar to >660–680°C, 4.0 (± 0.5) kbar at the metamorphic peak, to 500°C, 4 (± 0.5) kbar. This evolution is closely related to the intrusion of voluminous magmatic rocks, which caused magmatic loading and were responsible for the 'dynamic regional contact metamorphic overprint'(Will & Schmädicke 2003).

U–Pb zircon data of Todt *et al.* (1995) and Ar/Ar hornblende data of Schubert *et al.* (2001) indicate that the metamorphic peak in the southern BGO occurred at 337–336 Ma (unit II) and 342–332 Ma (unit III), and that cooling to temperatures below 500°C took place at 335–329 Ma (unit II) and 332–320 Ma (unit III). However, there are also some K–Ar and Ar/Ar hornblende data, which record older ages between 360 and 370 Ma (Kreuzer & Harre 1975; recalculated by Schubert *et al.* 2001).

Saar-Nahe Basin

The Saar 1 borehole (Fig. 11.16) penetrates a weakly deformed Silurian granite gneiss with a zircon date of 444 ± 22 Ma (Sommermann 1993). This gneiss is overlain by unmetamorphosed middle and upper Devonian marine limestones (Zimmerle *et al.* 1976). The Upper Devonian limestones are intercalated with feldspathic sandstones of dominantly volcanic material. This sequence grades continuously into Upper Devonian–Lower Carboniferous black shales with cherts and proximal greywackes (Zimmerle *et al.* 1976), and is overlain by a basal conglomerate and a *c.* 4800 m thick succession of Westphalian and Stephanian continental sediments, which were deposited in the Saar-Nahe Basin during Upper Carboniferous rifting (Henk 1993; Weber 1995*b*) (see McCann *et al.* 2008*a*, *b*).

Rhineland-Palatinate

Along the western Rhine Graben shoulder (Fig. 11.16), the Haardt area contains very low- to medium-grade metasediments of Lower Carboniferous–Devonian and perhaps even older Neoproterozoic(?) ages (Flöttmann & Oncken 1992; Reischmann & Anthes 1996; Kuhn 2001). These rocks were intruded by several undeformed granites during the Viseán (333–325 Ma; Reischmann & Anthes 1996). In contrast, the Albersweiler granite gneiss is assumed to be of Late Devonian age (370 Ma; Reischmann & Anthes 1996). This granite gneiss is transected by basaltic (lamprophyric) dykes which yielded K–Ar and Ar/Ar biotite ages of 334 and 325 Ma (Frenzel 1971; recalculated by von Seckendorff pers. comm., 2005). In general, the metasediments of the western Rhine Graben shoulder show evidence for low-grade regional metamorphism with supposed peak metamorphic conditions of *c.* 450°C, 2 kbar (Flöttmann & Oncken 1992). Following the Variscan evolution the granites and metasediments were exhumed and the palaeolandscape was covered by Lower Permian to Triassic sediments.

Vosges

In the Vosges, the dextral Lalaye-Lubine Shear Zone separates the (very) low-grade metamorphic rocks of the Saxothuringian Zone in the north from high-grade rocks of the Moldanubian Zone in the south (Fluck 1980; Wickert & Eisbacher 1988). From south to north, the northern Vosges consists of the Villé, Steige and Bruche units (Fig. 11.18C). The Villé Unit is composed of low-grade metapelitic to metapsammitic schists and quartzites (Clauer & Bonhomme 1970) deposited during late Cambrian to early Ordovician times (Reitz & Wickert 1989). The Villé Unit was thrust over the Steige Unit, consisting of monotonous low-grade metapelites that were deposited in a shallow-marine environment during Ordovician to Silurian times (Doubinger 1963; Tobschall 1975; Müller 1989). The Bruche Unit is a sedimentary and tectonic mélange comprising middle Devonian to Early Carboniferous shelf and slope sediments and calc-alkaline volcanic rocks (Ritzki & Baroz 1988; Wickert & Eisbacher 1988). These very low-grade volcanosedimentary sequences were intruded by diorites and granites that caused a contact metamorphic overprint (e.g. de Béthune *et al.* 1968; Hess *et al.* 1995).

Based on field relationships and mineralogical and geochemical arguments, the plutonic rocks of the northern Vosges have been grouped into five suites (Fig. 11.18C) by Altherr *et al.* (2000): (1) the Dioritic Suite from the Neuntelstein and Muckenbach area; (2) the Southern Granodioritic Suite of Hohwald and the southern part of the Champ-du-Feu area; (3) the Northern Granitic Suite of the northern part of the Champ-du-Feu area, as well as the Waldersbach and Fouday areas; (4) the Younger Granites of the Andlau, Natzwiller and Seones area; and (5) the subvolcanic, S-type granites of the Kagenfels, Kreuzweg and Grendelbruch area. These five suites were intruded within a very short timespan between 335 and 328 Ma (Hess *et al.* 1995; Boutin *et al.* 1995; Altherr *et al.* 2000)

Variscan Externides: Rhenish Massif (eastern part) and Harz-Mountains (H.H.)

The Rhenohercynian Zone (RHZ) is a part of the northern, external flank of the Variscides. It resulted from a collision between the Laurussian composite plate, which included the Avalonian microcontinent already accreted to its southern margin, with the collage of Armorican terranes derived from northern Gondwana (McKerrow *et al.* 2000; see also Krawczyk *et al.* 2008). The best exposures of the Rhenohercynian Zone are in the Rhenish Massif, the Harz Mountains and SW England. The Rhenohercynian Zone extends to southern Portugal and is believed by many authors to continue to the east into the Moravo-Silesian Zone at the eastern margin of the Bohemian Massif (Engel & Franke 1983; Franke & Engel 1982, Holder & Leveridge 1986). A characteristic feature of the rocks in the Rhenohercynian Zone is their very low to low metamorphic overprint, only locally reaching greenschist facies.

Parautochthonous areas

The basement of the RHZ was consolidated during the Cadomian Orogeny and then incorporated into the Avalonian microcontinent (Franke 1995). It is mostly covered by thick Palaeozoic (mainly Devonian) sediments. The small inliers of the Ecker Gneiss (Harz Mountains) and the Wartenstein Gneiss (Hunsrück area of the Rhenish Massif) probably represent fragments of the RHZ basement. The oldest parautochthonous sediments east of the Rhine are Ordovician and Silurian shales within the Ebbe Anticline, which are conformably overlain by Devonian clastic shelf sediments. The other areas of the RHZ parautochthon consist of Devonian to Carboniferous sedimentary and volcanic successions, which allow three major evolutionary stages to be recognized (Meischner 1971; Walliser 1981; Engel *et al.* 1983*a*; Franke 1995, 2000):

(1) Clastic neritic sediments ('Rheinische Fazies' = Rhine facies) were deposited on the southern, passive margin of Laurussia during the Early Devonian (Langenstrassen 1983). The Siegenian and Emsian shelf sediments can be subdivided into three facies belts (Stets & Schäfer 2002): fluvial and deltaic to tidal and coastal marine siliciclastic sediments in the north; clay-rich, basinal sediments (up to 8000 m thick) in the central areas, and a southern belt of mostly psammitic rocks. $^{40}K/^{40}Ar$ ages of detrital muscovites within the range of 440–410 Ma suggest a common provenance area (i.e. the Caledonian orogenic belt) for the detrital material (Küstner *et al.* 1998; Huckriede *et al.* 2004);

(2) Due to continued subsidence and erosion of the northern source areas, the shelf facies retreated northwestwards over time and was overstepped from Middle Devonian through to Early Carboniferous times by hemipelagic rocks including condensed carbonate successions, Middle Devonian to Frasnian reef limestones, shales and radiolarian cherts. Crustal extension and subsidence were accentuated by felsic (Emsian and Givetian) and basic (Givetian, Late Famennian, Late Tournaisian to Early Visean) submarine intraplate volcanism (Floyd 1995) (see also Timmerman 2008); and

(3) Late Tournaisian to Namurian deep-marine sediments are overlain by the Late Namurian to Westphalian coal-bearing successions of the Ruhr District. The northwest-ward progradation of the deep-marine sediments (Sadler 1983) was controlled by an advancing active margin in the SE. Deformation and very low to low grade metamorphism prograded from the SE (325 Ma) to the NW (305 Ma) across the RHZ (Ahrendt *et al.* 1983). Deformation structures are dominated by NW-facing folds and thrusts. The thick and intensely cleaved Devonian sediments form imbricate fans separated by several prominent SE-dipping thrusts (Oncken & Weber 1995). Tectonic shortening and deformation decreases toward the NW and amounts to an average of *c.* 50% (Behrmann *et al.* 1991; Oncken *et al.* 1999).

Allochthonous units

Kossmat (1927) first introduced nappe tectonic concepts to his interpretations of the geology of the Harz Mountains. Later, most tectonostratigraphic units were interpreted as parautochthonous (e.g. Wachendorf 1986; Meischner 1991; Schwan 1991). The allochthonous nature of the Gießen-Harz Nappe is now generally acknowledged (Weber 1978; Dörr & Preiss 1982; Engel *et al.* 1983*b*; Wedepohl *et al.* 1983). Oczlon (1994) proposed that the Harzgerode Zone and some other areas of the eastern Harz Mountains are allochthonous and derived from northern Gondwana. Huckriede *et al.* (2004) demonstrated that the steeply dipping Hörre-Gommern Zone is the northern margin of the allochthonous units. The other allochthonous units form a duplex-like structure, at least 3 to 5 km thick, which is sandwiched between the Gießen-Harz Nappe and the parautochthonous rocks at its base (Figs 11.19 & 11.20). Its thick and heterogeneous roof- and floor-thrusts were previously often misinterpreted as olistostromes (e.g. Reichstein 1965; Günther & Hein 1999). Modelling of the thermal and burial history of the eastern Harz Mountains provides evidence that the thickness of

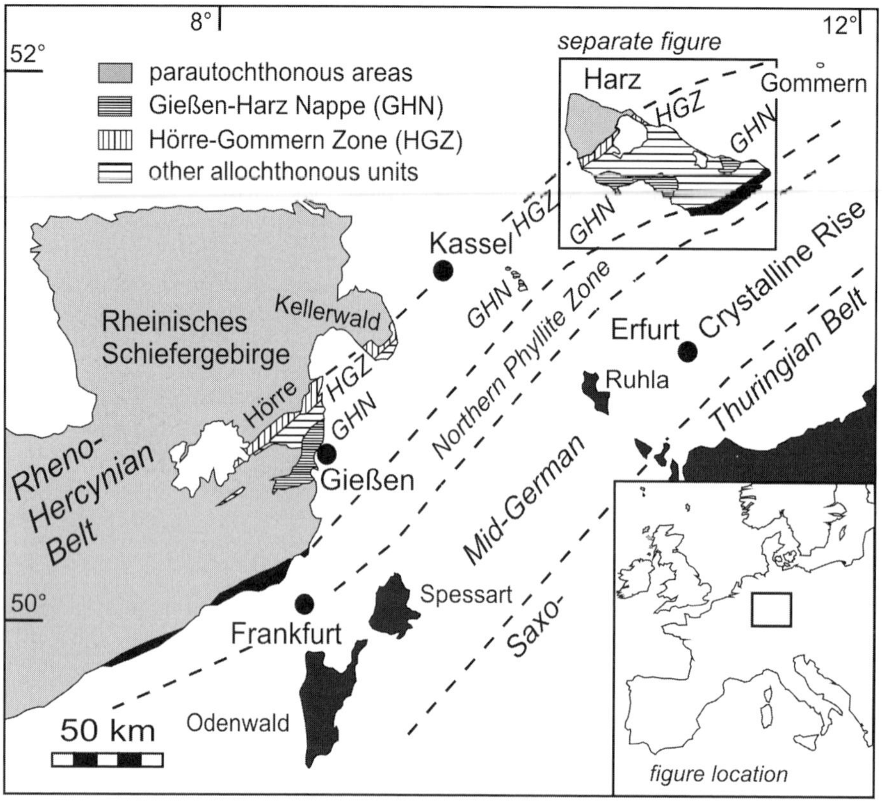

Fig. 11.19. Distribution of allochthonous units in the eastern part of the Rheinisches Schiefergebirge and in the Harz Mountains.

the allochthonous units has been significantly reduced (>8 km) by erosion (Friedel 1996).

The Hörre-Gommern Zone is a 5 to 15 km wide, NE-trending belt in the central part of the RHZ, which parallels the main tectonic trend over almost 300 km (e.g. Bender 1989; Meischner 1991; Schwan 1991; Fig. 11.19). It separates the exclusively parautochthonous areas in the NW from the allochthonous and parautochthonous areas in the SE. Stacking of the Early Carboniferous and Late Devonian sequences resulted in the formation of a tectonic wedge, which was locally backthrust to the SE. From the NW to SE, the Hörre-Gommern Zone can be subdivided into three narrow NE-trending units (Huckriede *et al.* 2004): the Bicken-Ense Unit (Eifelian through to Famennian condensed pelagic limestones), the Uebernthal-Gommern Unit (Late Devonian and Early Visean cherts, shales, turbiditic quartz-sandstones, submarine metabasalts; no greywackes), and the Hörre Unit (Frasnian siliceous shales, Famennian greywackes and limestone-turbidites, Early Tournaisian distal turbiditic quartz-sandstones followed by an alternation of cherts and limestone turbidites, and Late Tournaisian to ?Visean deep-marine sediments). The Hörre-Gommern Zone is tectonically derived from deep-water areas, directly south of the Rhenohercynian shelf margin, which were subducted beneath or overthrust by the Mid-German Crystalline High. The arrangement of tectonic units reflects gradual accretion at the front of the advancing nappe complex (Huckriede *et al.* 2004).

The Hessische Schieferserie (Franke 2000), the Gaudernbach Beds (Kegler 1967), and the Blankenburg Zone (Reichstein 1965; Wachendorf 1986) form part of a thick tectonic mélange separating the parautochthonous from the overlying allochthonous rocks (Huckriede *et al.* 2004). The dominant lithologies include Emsian sandstones, Eifelian through to Late Devonian shales with lenses of hemipelagic limestones, and Tournaisian/Early Visean radiolarian cherts. There are some rare intrusive metabasalts and no reef limestones. The sedimentary successions locally show striking similarities to those of the Hörre-Gommern Zone (Wierich & Vogt 1997). Intense deformation due to brittle shearing, boudinage, and small-scale imbrication is a common feature. The detritus of the Devonian and Early Carboniferous sandstones of these areas and from the Hörre-Gommern-Zone was derived from the Caledonian orogenic belt (Huckriede *et al.* 2004).

The Harzgerode Zone of the central and SE Harz Mountains is a chaotic assemblage of pelagic sedimentary and basic volcanic rocks of Ordovician through to Early Carboniferous age. It is separated from the Blankenburg Zone by the Tournaisian to Early Visean greywackes of the Tanne Zone. Sedimentary sequences are often transformed into a tectonic mélange possibly formed within an accretionary wedge (Oczlon 1994; Huckriede *et al.* 2004). Typical sediments within the Harzgerode Zone include Silurian graptolite shales, Devonian condensed pelagic limestones ('Hercyn-Kalke') and shales, and Lower Emsian calcareous sandstones ('Rothäuser Grauwacke'), equivalent to the 'Erbsloch-Grauwacke' of the Rhenish Massif (Bender *et al.* 1974). The eastern part of the Harzgerode Zone contains a large area of relatively undeformed Silurian graptolite shales associated with sandstones at least partly of Early Ordovician age (Huckriede *et al.* 2004). Deep drillholes (Lutzens 1972) have revealed that the graptolite shales form the top of a tectonic sequence, and are underlain by the intensely deformed, mostly Devonian part of the Harzgerode Zone, and (at the deepest level) by Early Carboniferous greywackes of the Tanne Zone. Along

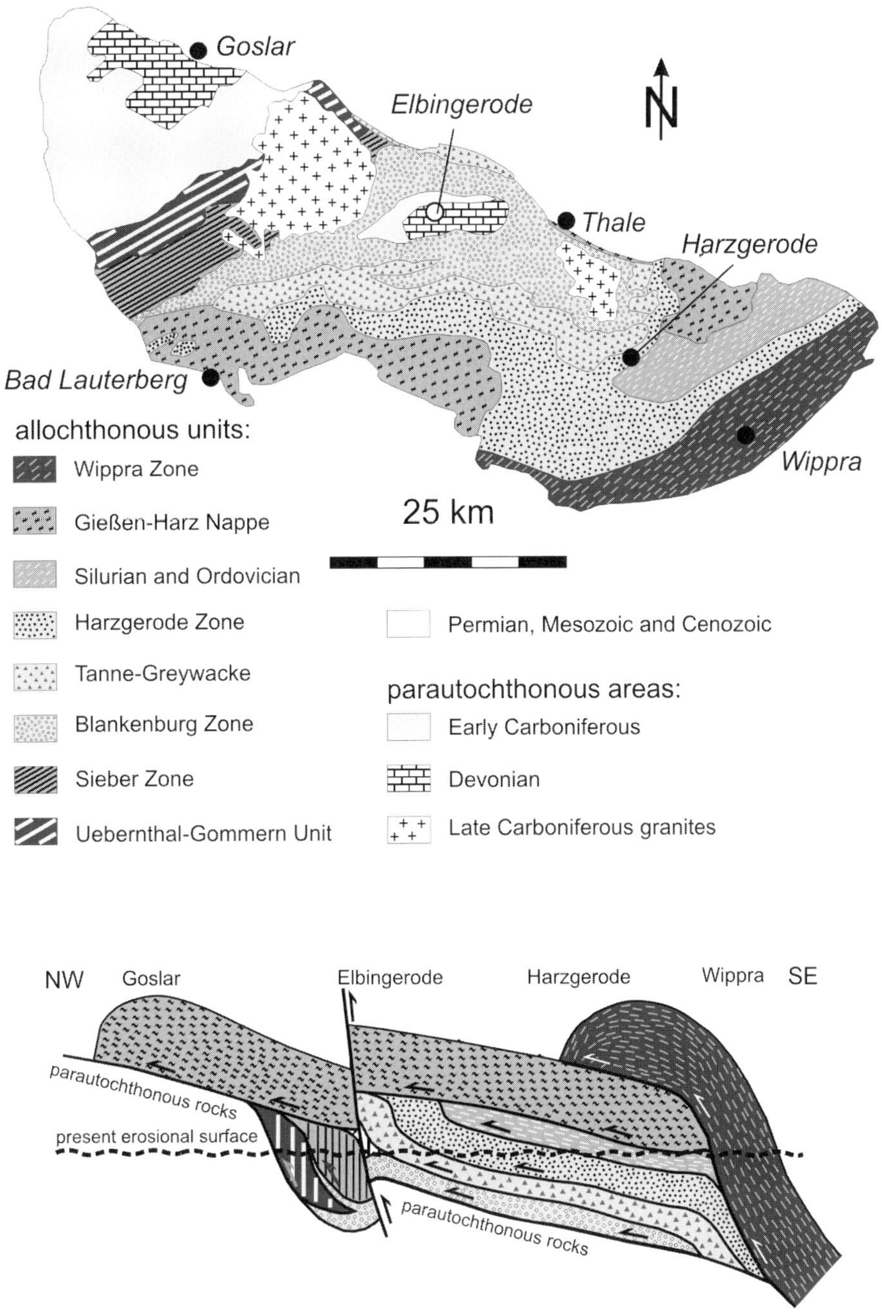

Fig. 11.20. Geological sketch map of the Harz Mountains and simplified NW–SE profile (not to scale) across the allochthonous units in the Harz Mountains.

the eastern margin of the Rhenish Massif, the 'Paläozoikum der Lindener Mark' (Kegel 1953) crops out in a narrow band between the Gießen Nappe and the parautochthon of the Lahn-Syncline. This band contains Ordovician sandstones and hemipelagic limestones of Silurian and Devonian (Early Pragian) age. Some of the sediments from the Harzgerode Zone and from the Lindener Mark show faunal and/or lithological affinities to supposed Armorican terranes and northern Gondwana (Oczlon 1994; Franke & Oncken 1995; Paris 1998; Plusquellec & Jahnke 1999). The exceptional palaeogeographic position of these rocks is confirmed by $^{40}K/^{40}Ar$ ages of detrital muscovites. Caledonian detritus is absent while the Ordovician to Emsian sandstones generally reveal Cadomian $^{40}K/^{40}Ar$ cooling ages of the detrital muscovites (Huckriede *et al.* 2004).

The Gießen-Harz Nappe (Weber 1978; Walliser & Alberti 1983) is exposed along the eastern margin of the Rhenish Massif and in the Südharz and Selke units of the Harz Mountains. NW-facing recumbent folds dominate the structure; the base of the nappe is marked by highly strained cataclasites and low temperature mylonites (Engel *et al.* 1983*b*). The nappe mainly consists of thick Frasnian to Late Famennian deep-marine successions and possible Early Carboniferous greywackes (Engel *et al.* 1983*b*; Dörr 1986; Floyd *et al.* 1990). The greywackes are locally underlain by extremely condensed units comprising Late

Emsian to Early Frasnian pelagic shales and radiolarian cherts (Dörr 1986). Tectonic slices of N-MORB-type metabasalts at the base of the Gießen Nappe (Wedepohl *et al.* 1983; Grösser & Dörr 1986) confirm the oceanic provenance of the nappe. The absence of Caledonian detritus indicates its effective separation from the Caledonian Orogen (Huckriede *et al.* 2004).

The greenschist-grade Northern Phyllite Zone (Anderle *et al.* 1995; Oncken *et al.* 1995; Klügel 1997; Meisl 1995) separates the Rhenohercynian Zone from the Mid-German Crystalline High. A prominent feature of this tectonic unit is its heterogeneity. The SW part (Hunsrück) consists of Devonian shelf sediments, the central part (Taunus) is dominated by subduction-related volcanic rocks of Silurian age, whereas the NE part (Wippra-Zone of the Harz-Mountains) is lithologically related to the adjoining Ordovician to Late Devonian sequences of the Harzgerode-Zone and the Gießen-Harz Nappe (Fischer 1929; Sehnert 1991).

Late Devonian to Namurian deep-marine sedimentation

The oldest deep-marine sediments in the Rhenohercynian area are Frasnian and Early Famennian greywackes from the Gießen-Harz Nappe and the Hörre-Gommern Zone. Detrital muscovites reveal Late Proterozoic (750 to 550 Ma) cooling ages and low to medium Si/Al ratios (low to medium pressure source rocks). Contrastingly, the main part of the deep-marine flysch sediments (parautochthonous and allochthonous units; Famennian to Early Namurian) contains Early Variscan (395 to 370 Ma) muscovites with medium to high Si/Al ratios (medium to high pressure source rocks; Huckriede *et al.* 2004). Conventionally, the source area of the Rhenohercynian deep-marine succession is considered to be the Mid-German Crystalline High directly to the south. However, the metamorphic history of this area (e.g. Lippolt 1986; Zeh *et al.* 2000; Schubert *et al.* 2001; Stein 2001) is incompatible with the data obtained from the sediments. These data are best explained by an Early Variscan collisional event involving two continental plates. Sediments from the early phase of deep-marine deposition were supplied from the upper plate, which was consolidated during the Late Proterozoic. Subsequently, MP to HP rocks of the lower plate were exhumed and stacked. Muscovite cooling ages of 395 to 370 Ma document an Eifelian to Famennian age for this orogenic event. During a period of approximately 40 Ma (Famennian to Namurian), these pressure-dominated rocks were eroded and provided the detrital material for the Rhenohercynian deep-marine succession.

Ardennes – Brabant area (F.L. & J.L.M.)

The Late Carboniferous thrusting of the Ardennes Belt onto the northern Variscan foreland along the Midi Thrust Zone provides an excellent example of the style and geometry of the structures along the North Variscan fold-and-thrust belt. The Northern Variscides in Belgium and northern France consist of three principal structural units. These are, from south to north, the Ardennes Fold-and-Thrust Belt (AFTB), the Overturned Thrust Sheets (OTS) and the Brabant Foreland (Fig. 11.21 & 11.22). All three units consist of typical Rhenohercynian Zone Palaeozoic-age sedimentary successions. Early Palaeozoic sediments with an Avalonian affinity were affected by diachronous Late Silurian to Early Devonian deformation and are unconformably overlain by Middle to Late Palaeozoic units. Early to Mid-Devonian extensional tectonics along the southern shelf of Laurussia led to the development of localized zones of strong subsidence. The phase of Late Variscan foreland deposition during the Carboniferous

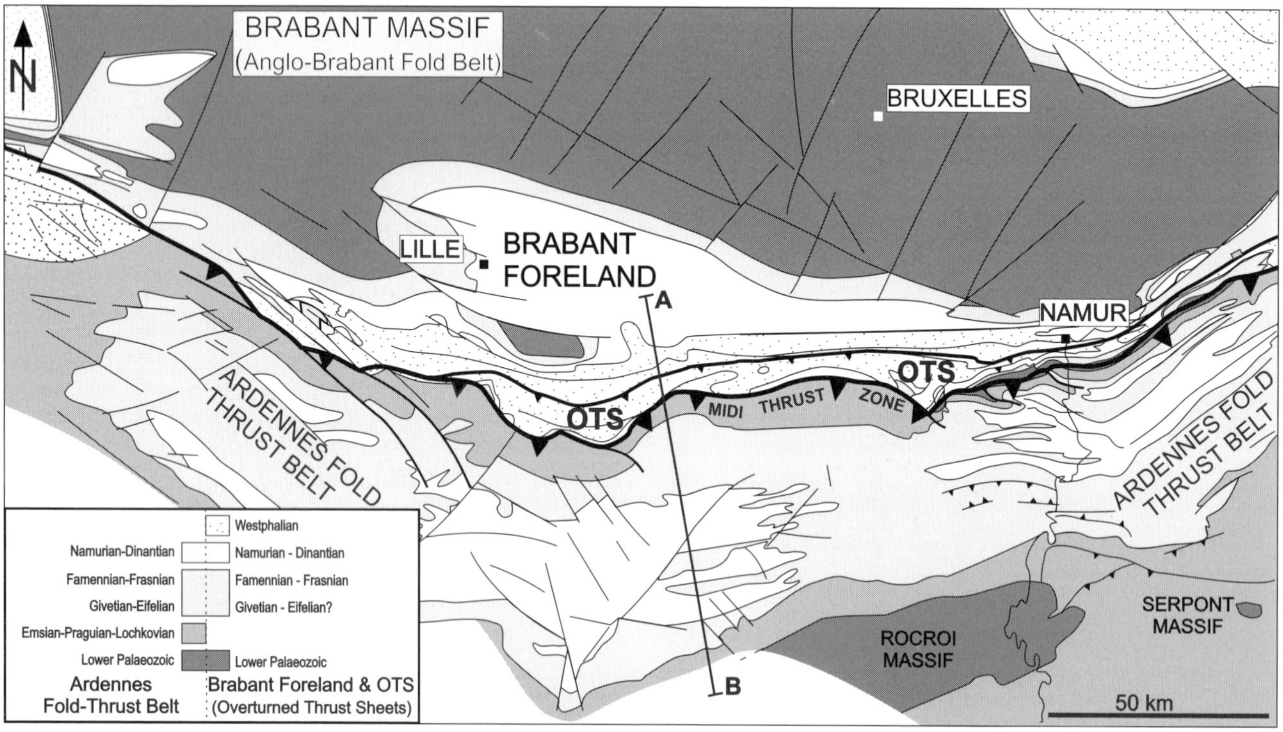

Fig. 11.21. Simplified map of the northern France–southern Belgium Variscan thrust front. Section A–B refers to the profile of Figure 11.22 (modified after Laquement *et al.* 2005).

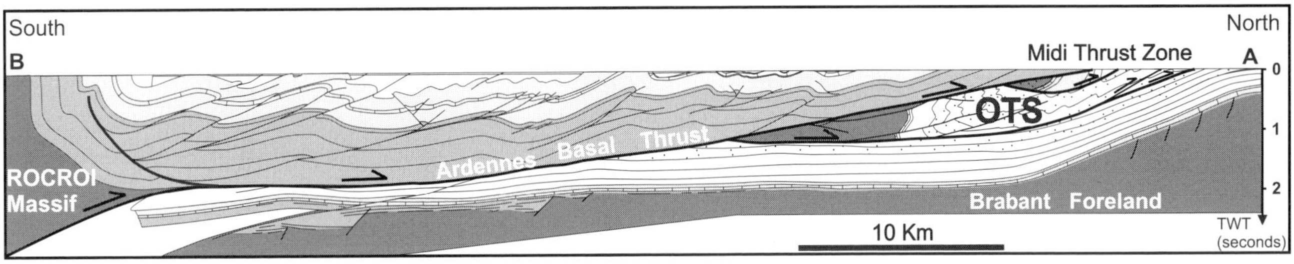

Fig. 11.22. Cross-section through the Western Ardennes thrust belt; see Figure 11.21 for location (modified after Laquement *et al.* 2005).

may be characterized by the deposition of synorogenic siliciclastic sediments and the coeval formation of a coal basin. All of these lithologies were subsequently affected by the Late Variscan NNW–SSE directed compressional tectonics.

The Early Palaeozoic substratum (Brabant Parautochthon)

The Early Palaeozoic formations are found within anticlinal domes which formed during the Variscan Orogeny. In the High Ardennes, representing the southern part of the Ardennes Allochthon, a series of massifs (Givonne, Rocroi, Serpont and Stavelot) are found. Towards the north, along the Midi Thrust Zone (Condroz), early Palaeozoic rocks are strongly deformed and constitute the basement of the Overturned Thrust Sheets. To the north, beneath the Devonian–Carboniferous cover, the Anglo-Brabant Massif represents a residual Variscan relief and forms the substratum of the Brabant Parautochthon. The age of the early Palaeozoic rocks, their geometry and the intensity of deformation vary between the various units of the area.

The angular unconformity between the folded and faulted Lower Palaeozoic succession and the overlying Lower (Ardennes) and Middle Devonian (Anglo-Brabant Massif) sediments would suggest that there was a phase of compressional tectonics prior to the onset of Devonian extension This compression would be in addition to the Late Variscan shortening of the entire NW Variscan External Zone. Evidence of the Early Devonian compressional event is found along the Anglo-Brabant Fold-Belt (Sintubin 1999); it can also be traced westwards into the Northern Appalachians where it is termed the Acadian Orogeny and is interpreted either as a Late Caledonian event (Mansy *et al.* 1999) or as the expression of the Early Variscan collisional contact of the Armorican microcontinental block south of the Rheic Ocean (Meere & Mulchrone 2006). In the Anglo-Brabant Massif, the structures linked to the Acadian phase of shortening comprise broad asymmetrical folds with the axial plane schistosity steeply inclined towards the north (Sintubin 1997). Within the OTS the intensity of Variscan deformation is such that it is virtually impossible to recognize previous tectonic activity.

The rocks of the Rocroi Massif, which have been interpreted as Cambrian in age, are characterized by the presence of a sedimentary breccia, slumps, convolutions and neptunian dykes, which suggest that there was a degree of gravitational instability at the time. This may have been linked to tectonic activity (Lacquement 2001) and as such may be related to the Early Palaeozoic rifting of Avalonia which led to the formation of the Rheic Ocean further to the south. There is, however, no evidence of significant tectonic deformation in the area at this time. The Acadian event is characterized by the formation of large folds without cleavage (Lacquement 2001) and analysis of these folds would suggest that there was a decrease in the degree of shortening from the Brabant Massif towards the south.

Post-Acadian sedimentation in an extensional setting

In Upper Silurian and Devonian times, the southern margin of Laurussia was affected by large-scale extensional tectonics in the vicinity of the Variscan Rheic suture. In the Ardennes, moderate extension, as evidenced by the rapid thickness and facies variations within the sedimentary successions, led to the development of a series of ENE–WSW and NW–SE orientated half-grabens (Meilliez 1991). In Upper Devonian times the cessation of active extension was accompanied by the emplacement of basaltic and microgranitic dykes dated at 373 +8/−9 Ma (Gofette 1989). Traces of volcanic activity were also found within the Lower Devonian sedimentary succession close to the Rocroi Massif. The geochemical characteristics of these volcanics are consistent with those of an intracontinental extensional setting.

Five principal periods of post-Acadian sedimentation have been recognized in the Ardenne-Brabant area. The initial phase (Early Devonian–Eifelian) resulted in the formation of a sedimentary wedge thinning towards the north and infilling the subsiding basin (Fig. 11.23). In Middle Devonian times, the first carbonate platform formed with the development of Givetian limestones and Frasnian reefs. These reefs were submerged in the Late Devonian at a time of renewed basin subsidence. The Early Carboniferous may be characterized by the formation of a second carbonate platform with Waulsortian reefs, in Late Visean times accompanied by slumping and the deposition of sedimentary breccias. This synsedimentary deformation is probably associated with slope steepening linked to the initial onset of Variscan shortening (De Putter *et al.* 1991). Finally, in the Namurian–Westphalian, the coal-bearing units of the northern fringe of the Variscan Front were deposited. The age of the various deposits within the basin and their relative thicknesses suggest that there was a migration of the Variscan Front from the SE towards the NW (Bless *et al.* 1977; Ahrendt *et al.* 1983; Piqué *et al.* 1984; Meilliez 1991).

Metamorphism

The metamorphism of the Ardennes and Brabant areas affects both the lower Palaeozoic units and the upper Palaeozoic cover. In the Ardennes area, the zones of maximum metamorphism are located along the southern flanks of the anticlinal domes of the Ardennes Fold-and-Thrust Belt. A summary of the estimated temperatures for these rocks was published by Fielitz & Mansy (1999), who suggested that the metapelites of the Serpont area and the Rocroi Massif experienced the highest temperatures (*c.* 500°C). However, most of the sediments in the entire region were only affected by very low grade to low grade metamorphic overprinting. Three different regional metamorphic events have been recognized (Fielitz & Mansy 1999). In the area of the Anglo-Brabant Massif a diagenetic to epizonal Early Devonian metamorphism is linked to the Acadian Orogeny and may have

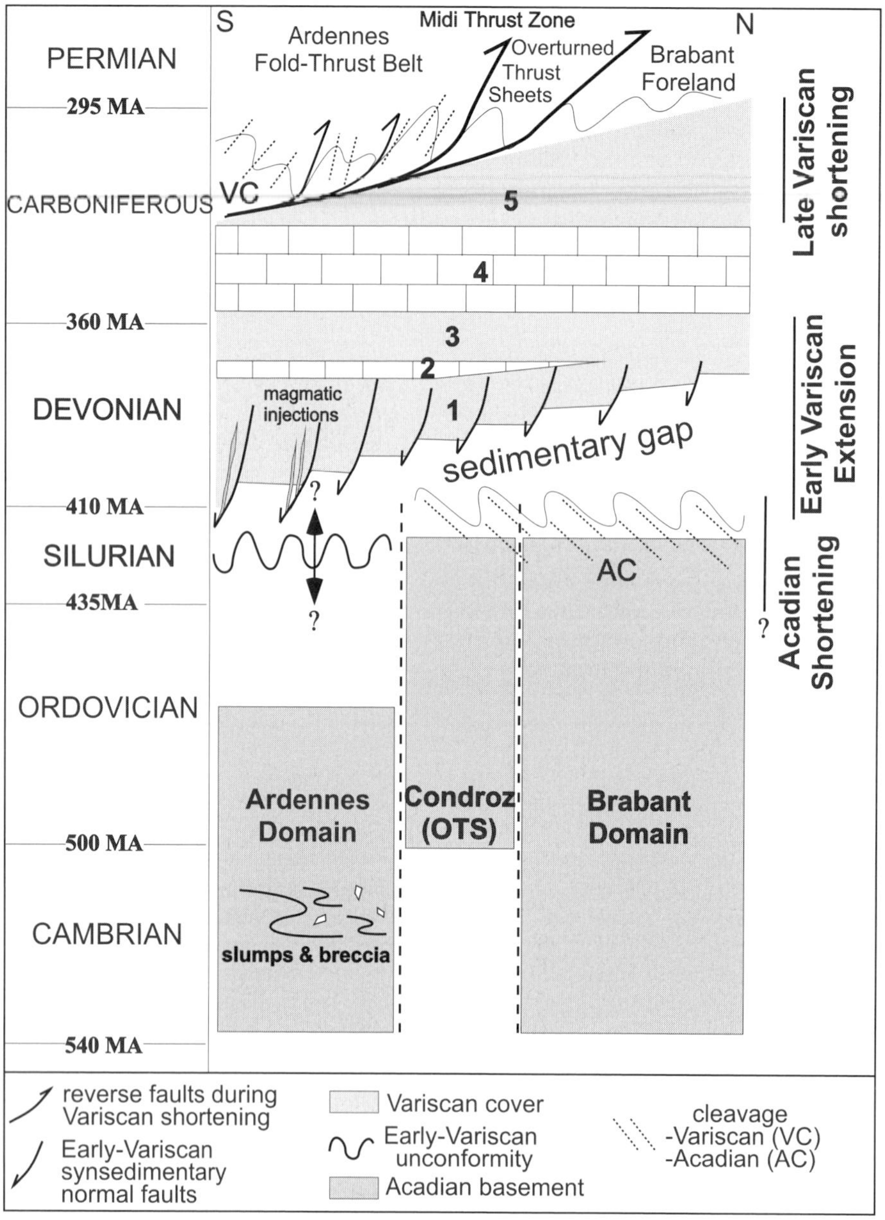

Fig. 11.23. Palaeozoic sedimentation and deformation events in the Ardennes Brabant area (modified from Meilliez 1991).

been related to the moderate thickening which occurred during this compressional event. The phase of Devonian extensional tectonics was associated with marked subsidence and sedimentation and was responsible for the development of pervasive pre-kinematic metamorphism in all of the Devonian and Lower Carboniferous rocks as well as the Lower Palaeozoic substratum of the Ardennes Fold-and-Thrust Belt. The metamorphic grade of this phase may be correlated with the age and thickness of the sedimentary cover and thus reflects the particular depth of the sediments within the basin. Such areas were subjected to warming as a result of higher heat flows due to crustal thinning. Finally, the development of the Late Variscan thrust wedge of the AFTB resulted in a very low grade to low grade metamorphic overprint in the Middle Devonian to Carboniferous sediments along the northern part of the Ardennes allochthon as well as in the sediments of the coal-rich foreland basins located to the north of the Midi Thrust Zone.

Late Variscan tectonics

Variscan shortening led to the formation of folded and faulted structures accompanied by cleavage. These structures affected both the Early Palaeozoic units and their cover. The migration of the deformation front from the SE towards the NW, as noted above, is also indicated by the rejuvenation of the age of emplacement of the schistosity across the region (Ahrendt *et al.* 1983). While older interpretations of the Ardennes suggest that the region is one of more or less pervasive folding (Fourmarier 1931), more recent interpretations based largely on seismic profiles revealed the importance of a large-scale gently south-dipping decollement zone, termed the Ardennes Basal Thrust. Generally the mechanism of deformation across the entire AFTB can be characterized by the development of thick décollement zones within the incompetent layers of the Devonian shales. The exposed frontal part the Ardennes Basal Thrust is represented by the zone of the Overturned Thrust Sheets and the Midi Thrust Zone, respectively. This latter zone separates the Ardennes allochthonous units, displaying a classic thrust-wedge geometry, from the underthrusted Namurian–Westphalian coal-bearing, foreland basin successions. These latter units are significantly deformed along the southern boundary of the Midi Thrust Zone (Fig. 11.22).

The complex deformation of the Overturned Thrust Sheets unit is characterized by the development of imbricate stacks of Early Palaeozoic sediments which are unconformably overlain by condensed Late Devonian successions and thick Carboniferous units. The base of the OTS unit is formed by a series of flat thrusts with a northern vergence that emerges on the surface within the Late Carboniferous coal-bearing formations (Delmer 1986; Raoult & Meilliez 1987). These sheets were thrust onto the Middle and Late Devonian and Carboniferous rocks (Graulich 1961; Lacquement *et al.* 1999) overlying the Brabant Foreland. Thus the classical 'Faille du Midi' is not a unique thrust plane but is rather one of the components of the faulted zone. It can be interpreted as a latest Westphalian out-of-sequence fault (Lacquement *et al.* 1999).

In general, the Variscan thrust front extending from the Channel-Boulonnais thrust belt to the Ardennes thrust belt is a composite feature characterized by a progressive change in its strike from NW–SE to NE–SW. This large-scale change in strike was accommodated by the development of second-order arcuate zones relaying in a complex way through various lateral discontinuities within the fold-and-thrust belt. Such segmentation of the belt may be well observed on the Midi Thrust Zone (Mansy *et al.* 1999). The complexity of structures reflects the fact that deformation was transpressive and that deformation was determined to an extent by the presence of pre-existing heterogeneities. These, together with the thickness variations of the Devonian sediments, resulted in oblique folding and thrust sheet rotation along lateral ramps (Lacquement *et al.* 2005).

Variscan external zone of east-central Europe (S.M., P.A., J.L.)

SW Poland

The Variscan external zone of SW Poland is almost entirely buried below the Permo-Mesozoic to Cenozoic infill of the German–Polish Basin. This has a thickness of more than 8 km. In Poland the external zone comprises a Carboniferous late- to post-orogenic clastic sedimentary succession more than 2 km thick, which was deposited in an extensive foreland basin. This basin formed part of the Variscan foreland basinal system extending from England to Poland along the northern margin of the Variscan orogenic belt. The Polish foreland basin, in its NE part (covering the areas of northern Wielkopolska, Kujavia and northern Małopolska), developed on a collage of Gondwana- (and Baltica?) derived Neoproterozoic terranes (Fig. 11.24) which were accreted to the East European Craton during Palaeozoic times (e.g. Pharaoh 1999; Winchester & PACE TMR Network Team 2002). In its marginal zone, the basin locally transgressed beyond the Teisseyre-Tornquist Line, representing the SW edge of the East European Platform (Fig. 11.25). In the SW part of the foreland basin, the Carboniferous strata onlap the

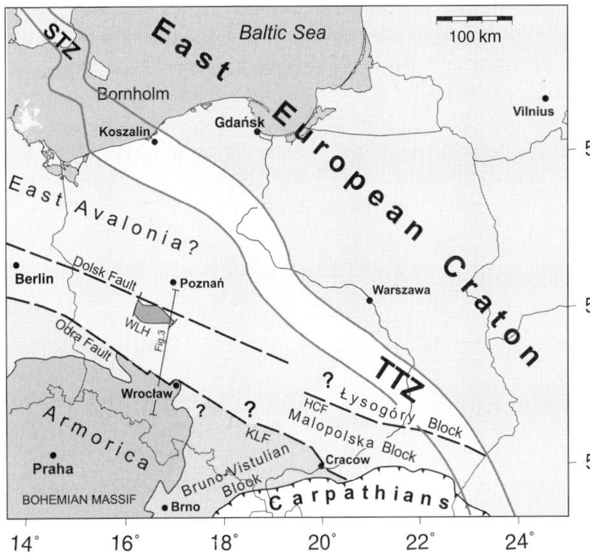

Fig. 11.24. Main basement units of the Trans-European Suture Zone in Poland. Abbreviations: HCF, Holy Cross Fault; KLF, Kraków-Lubliniec Fault; STZ, Sorgenfrei-Tornquist Zone; TTZ, Teisseyre-Tornquist Zone; USB, Upper Silesia Block; WLH, Wolsztyn-Leszno High.

Late Devonian (Haydukiewicz *et al.* 1999) pre-orogenic phyllite series of the Wolsztyn-Leszno High (Fig. 11.26) which was overthrust to the north, probably onto the southern margin of Laurussia. The age of final deformation in the basement rocks is constrained at *c.* 340 Ma by the $^{40}Ar/^{39}Ar$ cooling age of the phyllites (Żelaźniewicz *et al.* 2003). According to deep seismic refraction data (e.g. Grad *et al.* 2002), the southern part of the Variscan external zone, extending approximately up to the Dolsk Fault to the north (Fig. 11.24), is underlain by a two-layer 'Variscan-type' crust (*c.* 30–32 km thick), whereas the northern part overlies a three-layer 'Avalonian-type' crust (*c.* 30 km thick), 'transitional' to that typical of the East European Craton.

The Variscan Foreland Basin in SW Poland probably developed during late Visean–early Namurian times, directly post-dating the final, Late Devonian–Visean, collision-related events (nappe stacking, crustal thickening, widespread granitic plutonism and thermal metamorphism, gravitational collapse and orogenic uplift) in the adjacent Sudetic segment of the Variscan internal zone. The early compressional tectonism in the internal zone was apparently coeval with Late Devonian extension in the northern foreland.

The late Visean subsidence of the areas in front of the emerging Variscan Orogen resulted in foreland basin formation and the shift of late-orogenic clastic sedimentation towards the north and NE, across Wielkopolska and the Upper Silesian and Malopolska blocks, as far as the Lublin Trough in the east and Pomerania in the north (Fig. 11.25). The rapid subsidence of the foreland basin in the late Visean and Namurian A was, at least in places, interrupted during late Namurian and earliest Westphalian times by a period of non-deposition. Sedimentation resumed in the Westphalian A to B.

$^{40}Ar/^{39}Ar$ analyses of detrital white micas from Carboniferous sandstones of the Variscan foreland basin define plateau-like ages in the range of *c.* 370–355 Ma (Mazur *et al.* 2006). A comparable plateau-like age of *c.* 360 Ma was determined in neocrystalline white mica from a phyllite of the Wolsztyn-Leszno High, underlying the basin. These ages indicate that the foreland

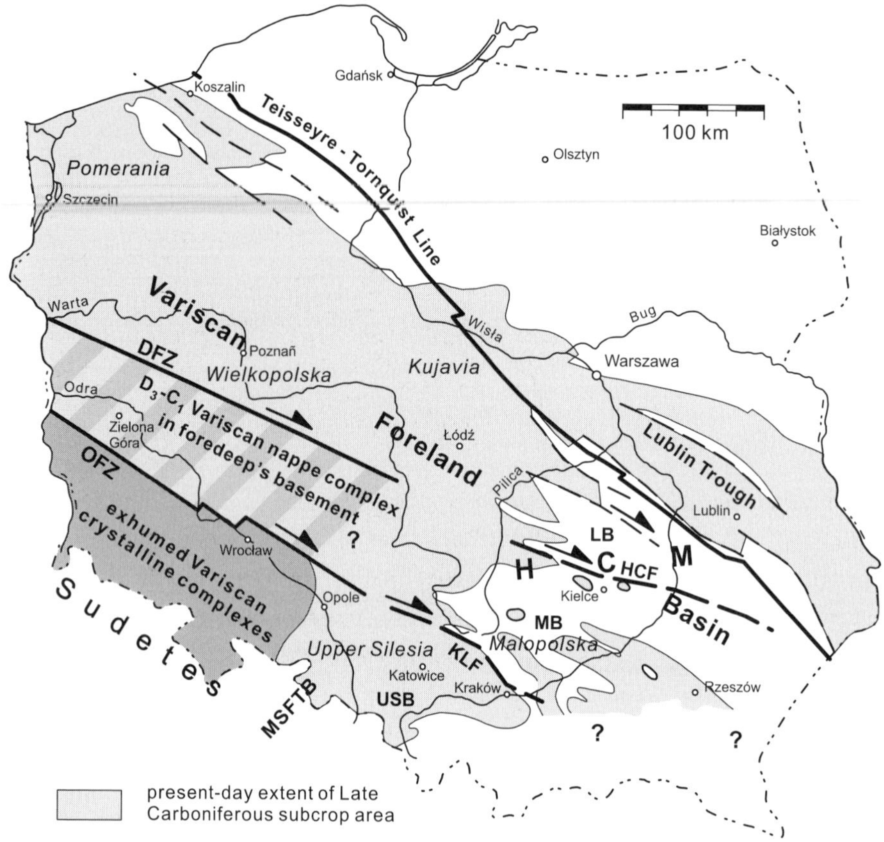

Fig. 11.25. Variscan foreland basin of Poland. Abbreviations: DFZ, Dolsk Fault Zone; HCF, Holy Cross Fault; HCM, Holy Cross Mountains; KLF, Kraków-Lubliniec Fault; LB, Łysogóry Block; MB, Małopolska Block; MSFTB, Moravo-Silesian Fold-and-Thrust Belt; OFZ, Odra Fault Zone; USB, Upper Silesian Block.

basin was supplied by source rocks that were exhumed and cooled in the Late Devonian. Since the present-day outcrop area of the adjoining Variscan belt shows only a minor component of rocks exhumed before the Carboniferous, a significant proportion of the rocks with Devonian-age cooling signatures is likely to have been removed by erosion. It is possible that such rocks once constituted a group of early Variscan nappes, occupying the uppermost position in the orogenic internal zone, and being equivalent to the phyllite units (such as the Wolsztyn-Leszno High) underlying the southern part of the foreland basin.

The inversion of the Variscan foreland basin of SW Poland took place as late as the Westphalian C, leading to localized moderate to intense folding and thrusting, and this probably occurred in an overall transpressional dextral regime partly related to major NW–SE orientated strike-slip fault activity at that time (e.g. the Teisseyre Tornquist Fault Zone, Dolsk Fault, Kraków-Lubliniec and Middle Odra faults). This deformation must have been contemporaneous with the folding of the Holy Cross Mountains, which is known to have taken place between the Namurian A and the Permian. Sedimentation in the Lublin Trough also terminated at the end of the Westphalian.

Moravo-Silesian Fold-and-Thrust Belt

The NNE–SSW trending Moravo-Silesian Fold-and-Thrust Belt (MSFTB) is located along the eastern margin of the Bohemian Massif (Fig. 11.27). It comprises synorogenic Early Carboniferous (Culm facies) and a pre-orogenic Devonian sedimentary succession of the former Moravo-Silesian Basin (MSB). This succession is up to 7.5 km thick, and was strongly deformed by subsequent thrusting and folding. The overall tectonics of the MSFTB are the result of the collisional interaction of the Bruno-Vistulian Neoproterozoic-age basement block with the Variscan collage of the central Sudetes (e.g. Matte *et al.* 1990; Aleksandrowski & Mazur 2002) and it is likely to contain a more significant NNE–SSW directed strike-slip component than has been assumed to date. The strike-slip displacements may have resulted from recurrent sinistral and dextral shearing events (cf. Rajlich 1987, 1990), whose effects are, however, thus far poorly recognized and unconstrained.

During its synorogenic stage, the MSB comprised only a relatively small component of the Variscan foreland basin system. To the north and NE, the MSB Carboniferous succession merges with that of the less deformed, Carboniferous foreland basin of SW Poland, and to the SE it plunges below the Neogene Carpathian nappe pile. The MSB succession was deposited in two sub-basins: the Jeseniky sub-basin to the north and the Drahany sub-basin to the south. These sub-basins, which are separated by the eastern extension of the upper Elbe Fault Zone (Fig. 11.27), have slightly different stratigraphic and sedimentary histories.

The clastic Carboniferous (Culm) sedimentation in the MSB commenced in the early or middle Visean and continued through to the early Namurian (Klominský 1994). The Culm succession in this area has been traditionally divided into several major formations based on litho- and biostratigraphical criteria (e.g.

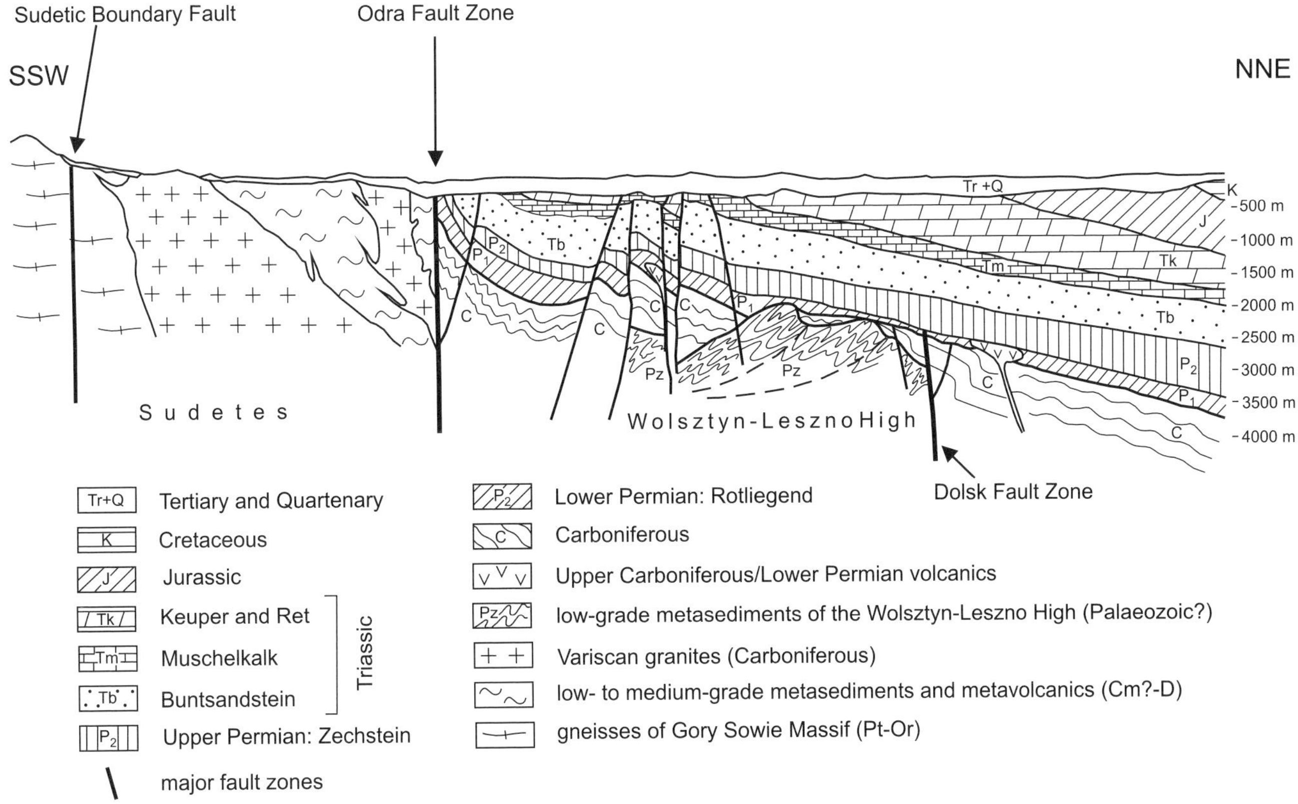

Fig. 11.26. Schematic geological section across the Variscan foreland basin in the SW Poland (modified after Grocholski 1975; vertically exaggerated).

Dvořák 1973). The contact between the Culm sedimentary rocks and the Famennian limestones is known only from boreholes close to the western border of the basin. The contact is currently tectonic, since the mechanically weak Devonian–Carboniferous boundary was reactivated by thrusting during the Carboniferous inversion of the basin (Hartley & Otava 2001). The original contact between the Devonian and the Carboniferous may have been concordant, but with a time gap (of at least 20 Ma) (Dvořák 1992). To the east, the turbiditic succession of the MSB is conformably overlain by Late Carboniferous coal-bearing deposits (Klominský 1994). These continental sediments define the succession of the Upper Silesian Coal Basin.

Sediments were supplied to the MSB from the south and SW, being transported to the north approximately parallel to the basin axis (Unrug & Dembowski 1971; Hartley & Otava 2001). During most of Visean times, the main sediment source was the low-grade metamorphic massifs as well as intrabasinal Visean limestones. This source was partly replaced during the latest Visean by high-grade metamorphic rocks. This change reflects the progressive overthrusting and stacking of the Moldanubian and Central-East Sudetic nappe pile along the eastern margin of the Bohemian Massif (Hartley & Otava 2001).

The sedimentary succession of the MSB commenced with the deposition of Pragian-age conglomerates and sandstones (Chlupac 1989), and these overlie a transgressive contact on either Neoproterozoic crystalline basement or Lower Palaeozoic sediments. The Devonian contains three main facies associations (Chlupac 1964, 1988): platform carbonates, platform slope transitional facies, and open-marine basinal facies. The carbonate platform (up to 1600 m thick) developed from late Emsian to end-Frasnian times (Hladil 1986) on top of continental and shallow-marine Emsian siliciclastics (e.g. Kalvoda *et al.* 2002). Until late Visean times, the platform underwent gradual destruction under an extensional tectonic regime, leading to the formation of localized subsidence centres (Kalvoda *et al.* 1999). In the platform slope sediments, the reduction in carbonate deposition began by the Givetian (Babek 1996), and ended during the Famennian and Tournaisian when it was replaced by pelagic basinal sedimentation. The basinal facies succession comprises Pragian to Tournaisian reef limestones, calciturbidites, pelagic carbonates and pelagic radiolarian shales deposited on the slopes of volcanic seamounts (Galle *et al.* 1995). These deposits are associated with a volcanic suite that includes alkali basalts and tholeiites with geochemical affinities to within-plate basalts and MORB. It can be inferred from this geochemical signature that the volcanics formed on a thinned continental margin and/or on oceanic crust (Souček 1981; Prichistal 1990).

The MSB succession underwent diachronous deformation after the early Namurian, resulting in a gradual eastward decrease in the age and intensity of deformation within the MSFTB (Kumpera & Martinec 1995). The inversion of the eastern part of the basin was related to the formation of an imbricate thrust system of eastern polarity, which included numerous duplex structures and east-vergent folds (Synek *et al.* 1990; Rajlich 1990). At the same time, the western part of the basin was thrust westwards (i.e. it underwent backthrusting) onto the metamorphosed Devonian succession of the east-vergent East Sudetic nappe complex (Fig. 11.28; Cháb 1990; Čižek & Tomek 1991). In this way, basin inversion resulted in the development of a divergent geometry of the MSFTB. The axial part of the belt contains Devonian-age basinal facies rocks, upthrown along high-angle, regional shear zones as a major pop-up structure (Kumpera &

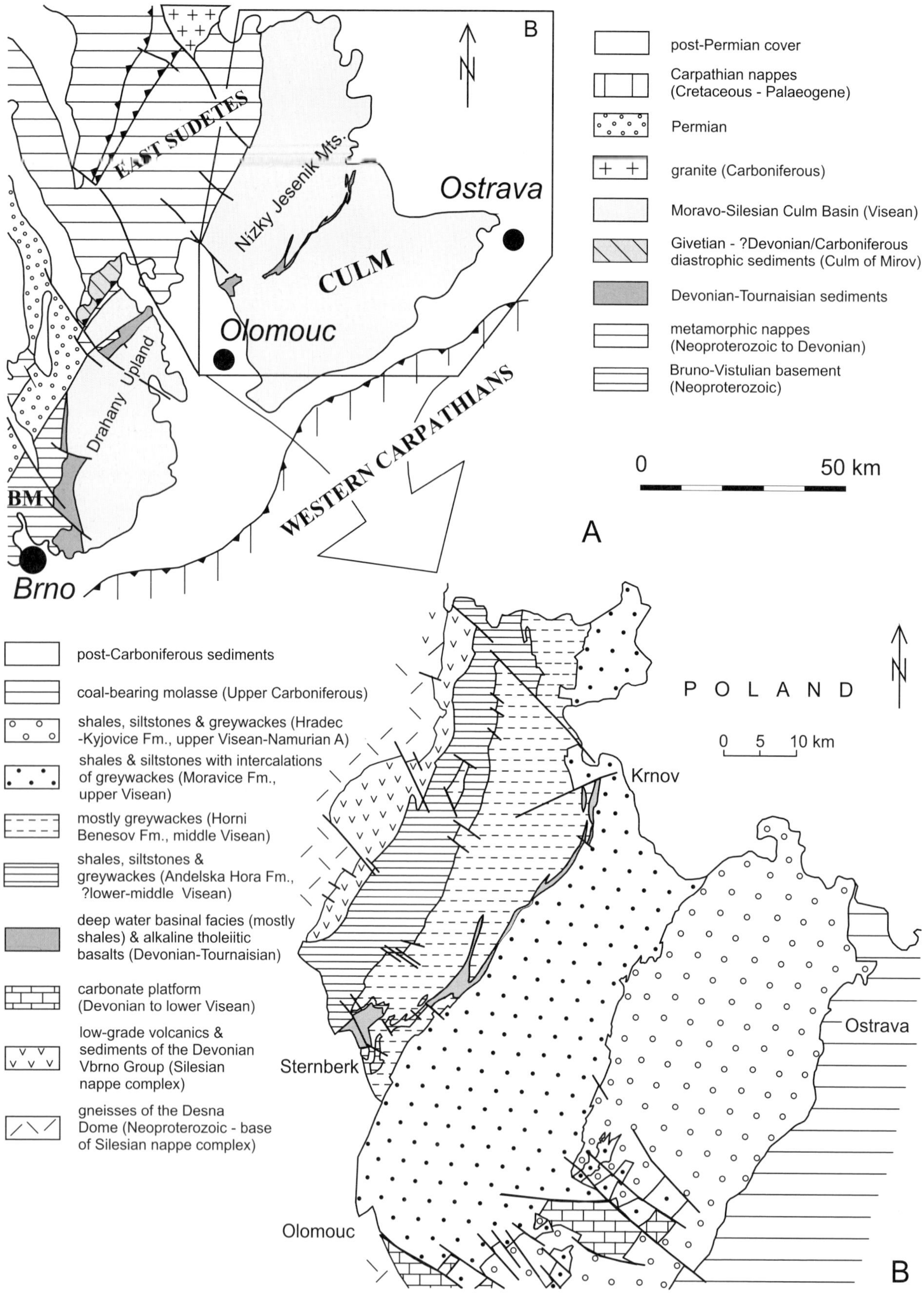

Fig. 11.27. (**A**) Location of the Moravo-Silesian fold-and-thrust belt at the eastern margin of the Bohemian Massif. BM, Neoproterozoic crystalline basement of the Brno Massif (Bruno-Vistulian Block). (**B**) Simplified geological map of the northern part of the Moravo-Silesian Fold-and-Thrust Belt (Carpathian nappes and foredeep not shown; based on Dvořák 1995).

Martinec 1995; Schulmann & Gayer 2000). The western, marginal part of the MSFTB was initially overlain by crystalline thrust sheets of the East Sudetes before the main phase of basin inversion. As a result of tectonic thickening of their overburden, the Carboniferous complexes of this part of the basin suffered greenschist facies metamorphism (Cháb 1990) and underwent a structural pattern analogous to that of the allochthonous Devonian Vrbno group adjacent to the east (Fig. 11.28). The metamorphosed Culm rocks contain a NNE–SSW trending, steeply dipping foliation and a stretching lineation plunging to the NE to north at a shallow angle. The age of the deformation presumably corresponds to the cooling of the Devonian rocks in this area, which was determined by the Ar/Ar method as *c*. 300 Ma (Maluski *et al.* 1995).

Devonian and Carboniferous of the Holy Cross Mountains

The Holy Cross Mountains region was uplifted in Miocene times (Kutek & Głazek 1972) as part of the fore-Carpathian peripheral bulge. The region contains Palaeozoic and late Neoproterozoic rocks cropping out from beneath the thick Permo-Mesozoic infill of the German–Polish Basin. The Holy Cross area straddles a boundary between two basement blocks, both components of the Trans-European Suture Zone Palaeozoic collage (e.g. Pharaoh 1999; Winchester & PACE TMR Network Team 2002). It thus includes two different tectonostratigraphic domains: that of Łysogóry to the north and of Kielce to the south (e.g. Czarnocki 1957; Szulczewski 1977). The Kielce Domain lies on the NE margin of the Małopolska Block (e.g. Pożaryski 1977), which adjoins the Łysogóry Domain and crustal block to the north, across the WNW–ESE trending, major Holy Cross Fault (Fig. 11.24). The latter thus represents a terrane boundary within the Palaeozoic basement and has been variously interpreted (e.g. Mizerski 1979; Brochwicz-Lewiński *et al.* 1983; Znosko 1983, 1996; Pożaryski *et al.* 1992; Stupnicka 1988, 1992; Lamarche *et al.* 2003*a*) in terms of multistage activity and its supposed strike-slip displacement.

The Devonian–Lower Carboniferous succession in the Holy Cross Mountains (and elswhere) overlies Early Palaeozoic, pre-Variscan complexes. In the Łysogóry Domain, Early Devonian (Pragian) clastics were deposited conformably on late Silurian rocks (e.g. Tomczyk 1988; Stupnicka 1992; Mizerski 1995, 2002; although some authors take a different view, e.g. Dadlez *et al.* 1994; Znosko 1996). In the Kielce Domain, Early Devonian (Siegenian–Emsian) sediments unconformably overlie folded Lower Palaeozoic basement (including Silurian rocks; e.g. Bednarczyk *et al.* 1970; Kowalczewski 1971; Tarnowska 1981; Głazek *et al.* 1981; Malec 1993; Szulczewski 1995). In Emsian times, detrital sediments were deposited in both domains, followed by Eifelian shallow-marine carbonates. Beginning in Givetian times, the sedimentary facies became spatially diversified (Szulczewski 1995). In both the Kielce and Łysogóry domains, the sedimentary record indicates progressively deeper-marine conditions from the Late Devonian to the Visean, reflecting the formation and development of the Holy Cross segment of the extensive Variscan foreland basin. However, some signs of basin shallowing are believed to have occurred in the Late Visean (Szulczewski 1995). The Lower Carboniferous (Tournaisian and Visean) rocks are preserved only in the central part of the Holy Cross Mountains in the Kielce Domain (Fig. 11.29). They are predominantly argillaceous and clastic sediments which were deposited in a basinal area that developed following the disintegration and submergence of the Devonian carbonate platform (Szulczewski 1995; Belka *et al.* 1996). The tectonic instability of the Holy Cross part of the Variscan foreland basin during Late Devonian and Early Carboniferous times is manifested by synsedimentary faulting and block tilting (Szulczewski 1968; Paszkowski 1988; Szulczewski *et al.* 1996; Lamarche *et al.* 2003*a*), producing numerous horsts and grabens. Their uplifted margins provided material for sedimentary breccias during the Famennian deepening of the basin (Lamarche *et al.* 2003*a*).

During the Late Carboniferous, the overall tectonic regime became compressive, leading to inversion of the Devonian–Carboniferous basin, and resulting in regional folding of Lower and Upper Palaeozoic rocks. The youngest rocks involved in the Variscan folding are of Visean age. A cross-section through the Holy Cross Mountains (e.g. Książkiewicz *et al.* 1965; Pożaryski & Tomczyk 1993; Mizerski 2002; Lamarche *et al.* 2003*a*) reveals several open synclines and anticlines involving Devonian sediments, with a wavelength of 5 to 10 km (Fig. 11.29). In most cases, the folds are slightly asymmetric and south-vergent.

The orientation of fold axial planes and of associated local fracture cleavages (e.g. Stupnicka 1992; Mizerski 1995) suggests a NNE–SSW orientated shortening direction (Lamarche *et al.* 1999) during the Late(?) Carboniferous. According to Lamarche *et al.* (2003*a*), folding was accompanied by reactivation and the

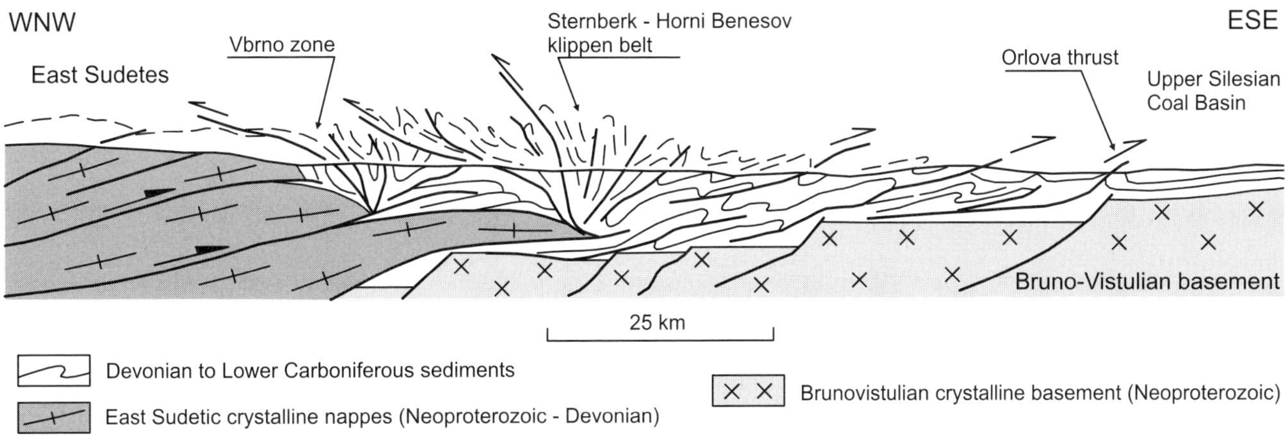

Fig. 11.28. Schematic geological cross-section across the Moravo-Silesian Fold-and-Thrust Belt (based on Grygar & Vavro 1995).

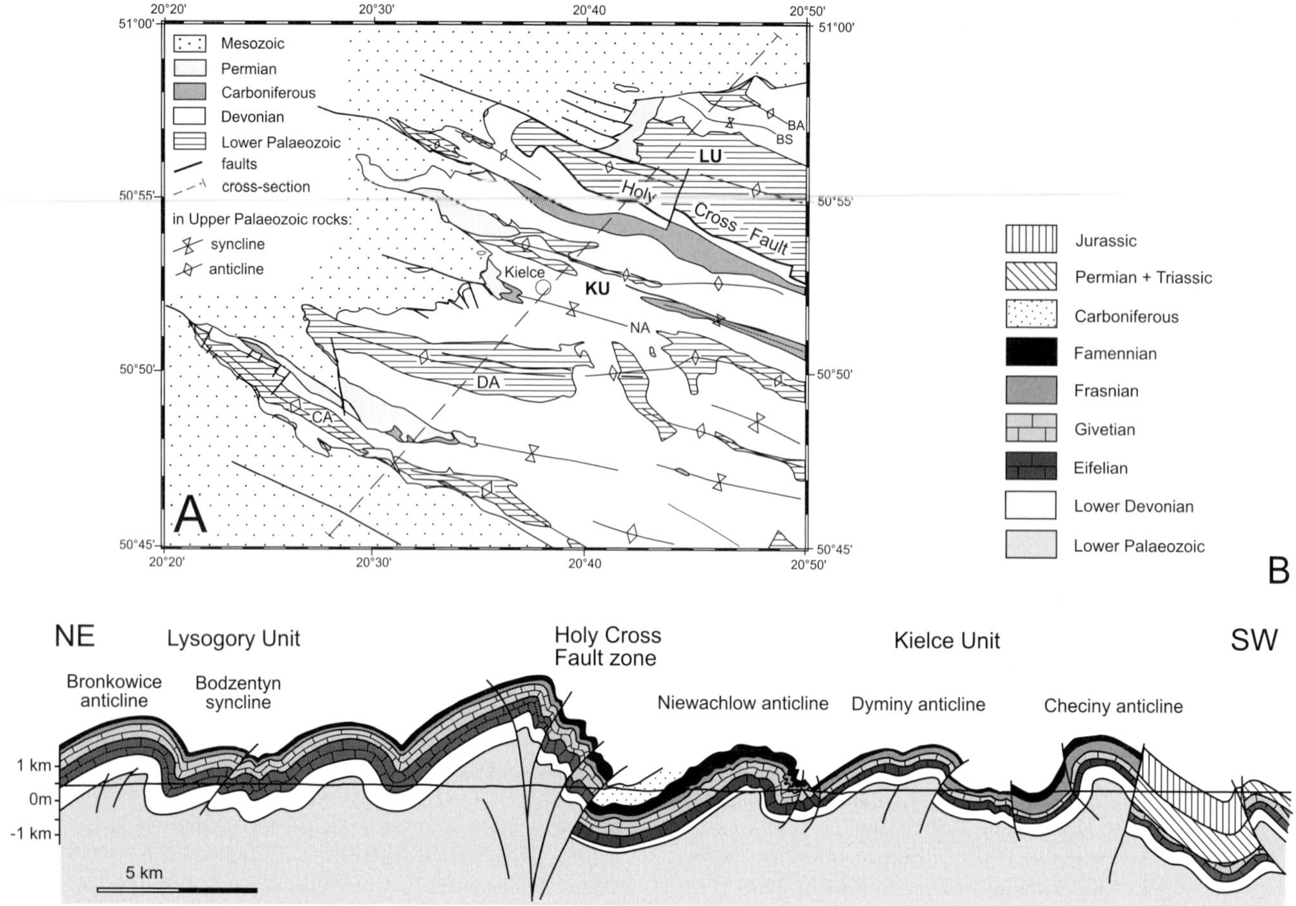

Fig. 11.29. Overview geology of the Holy Cross Mountains. (**A**) Geological sketch map of their western part (simplified after Czarnocki 1938). Abbreviations: BA, Bronkowice anticline; BS, Bodzentyn syncline; CA, Chęciny anticline; DA, Dyminy anticline; KU, Kielce Unit; LU, Łysogóry Unit; NA, Niewachlów anticline. (**B**) Schematic NE–SW cross-section across the western part of Holy Cross Mountains (after Lamarche *et al.* 2003*a*).

inversion of pre-existing normal faults. The style of the Variscan folds favours a model of thin-skinned detachment folding above a decoupling surface at depth. The main detachment may have been along the boundary between the Ordovician and the Lower–Mid-Cambrian rocks in the Kielce Domain (or maybe at an even deeper level) and must have been located below the Middle Cambrian rocks in the Łysogóry area. In the Kielce Domain, Variscan folding also affected the underlying deformed Ordovician–Silurian complex (folded during latest Silurian–Early Devonian times) and, possibly, also that of the Neoproterozoic to Early–Mid-Cambrian basement. The Variscan shortening is interpreted to have resulted in dextral transpressive reactivation of the Holy Cross Fault, resulting in the formation of a positive flower-structure around this fault zone and along major anticlines of the Holy Cross Mountains (Brochwicz-Lewiński *et al.* 1983, Pożaryski *et al.* 1992, Lamarche *et al.* 2003*b*). The low intensity of deformation in the Permo-Triassic cover at the western prolongation of the Holy Cross Fault reveals that most of the strike-slip displacement on it is Variscan, although a limited-magnitude left-lateral strike-slip reactivation may have occurred during the Alpine ('Laramide') shortening at the turn of the Cretaceous and Tertiary (Lamarche *et al.* 1999, 2003*b*). In this way, the cross-section through the Holy Cross Mountains shows their structure to be a superposition of Cambrian, Silurian to Early Devonian and Late Carboniferous deformational effects.

The structural development of the Devonian–Carboniferous complex exposed in the Holy Cross Mountains corresponds well with the general tectonic features of the buried Late Palaeozoic complex in SW Poland. These features include widespread, but mostly gentle, open folding, becoming more intense close to the WNW–ESE to NW–SE trending crustal-scale dextral, strike-slip faults, associated with major positive flower structures. These structures developed in a transpressional regime during the Late Carboniferous (in most parts not earlier than the Westphalian C), and, thus, post-date the main folding and metamorphic events in the Bohemian Massif and the Central Carpathian Variscan massifs.

Discussion and conclusion (U.K.)

The Variscan Orogeny of Europe is correlative to the final stages of the Appalachian orogenies along eastern North America, and marks the final amalgamation of Laurussia and Gondwana to form the supercontinent of Pangaea. Within Europe, a series of basement terranes form a collage of blocks with both continental and oceanic affinities. Throughout the Palaeozoic, the northern (i.e. Baltica and Laurentia) and southern (Gondwana) continents interacted through a series of rifting and collisional events involving island arcs, rifted continental fragments, and the major continents. Various terranes of Gondwanan affinity drifted northward across the palaeo-oceans that existed between Gondwana and Laurussia, and collided with both Laurentia and Baltica beginning with the accretion of Avalonia during the Ordovician–Silurian. Although the existence of Pangaea is universally accepted, the identities and precise geometries of the final colliding blocks with southern Europe are still debated (e.g. Stampfli & Borel 2002; Muttoni *et al.* 2003). Similar problems occur in unravelling the history of the Caledonian Orogeny (see Krawczyk *et al.* 2008).

Number of lithospheric plates

The complexity of the Variscan deformation event has resulted in a variety of problems in identifying the precise number of plates which were involved in orogenesis. Initial models for the evolution of the Variscan Orogen were developed prior to the paradigm of plate tectonics. Consequently, the entire orogenic process was interpreted in terms of vertical and horizontal movements of neighbouring rock units. Sedimentary basins and crystalline domains were considered as geo-synclines and geo-anticlines, respectively (e.g. Stille 1951). In terms of the Variscan Orogen the latter were explained as Precambrian crystalline complexes subsequently overprinted by an exclusively high temperature/low pressure metamorphic event followed by the intrusion of voluminous granitic batholiths (e.g. Zwart 1967).

The recognition of subduction-related high pressure domains within the Variscan belt (for details see Matte 1986) led to the assumption that the predominant and defining feature of this orogen was the final collision of continental crust during the ongoing convergence of various lithospheric plates. The idea of the closure of two oceanic domains involving the collision of three lithospheric plates (Gondwana, Armorica and Laurussia), with initial subduction beneath the intervening Armorican microplate (Matte 1986, 2001), allows the Variscides to be interpreted as a symmetrical orogen with Armorica located in the internal part. In such a view, the final folding of the various foreland basins which developed both on Gondwana as well as Laurussia, and which were located adjacent to the active plate margins, is the underlying cause for the bivergent symmetry of the external fold-and-thrust belts.

The view of a rigid and coherent Armorican microplate has been challenged in recent years by palaeomagnetic data (Tait *et al.* 1997, 2000) which suggest that the Variscan Internides consist of several microcontinental blocks (i.e. terranes or microplates). The proposed Armorican Terrane Assemblage (ATA) consists of peri-Gondwana crust south of the Rheic Ocean and north of the terranes of the Proto-Alps. In this model the collision between Gondwana, the terranes of the ATA and the Proto-Alps and Laurussia in Mid- to Late Carboniferous times led to the formation of a complex collage; this collage now comprises the Internides of the Variscan belt (e.g Franke 2000). However, the existence of a Hun-Superterrane incorporating all of the terranes of the ATA as well as of the Proto-Alps has been proposed by Stampfli *et al.* (2002). In this latter model, the Rheic Ocean is situated to the north while the Palaeotethys is in the south. It is important to note that there is no consensus about the role of Iberia in relation to Armorica, the ATA or the Hun Superterrane. The Central Iberian Zone as part of the Iberian autochthon is considered as part of Gondwana by various authors (Noblet & Lefort 1990; Ribeiro *et al.* 1990; Martinez Catalan *et al.* 1997; Matte 2001). In the model of an intervening ATA (Tait *et al.* 2000) or Hun-Superterrane (Stampfli *et al.* 2002) the entire Iberian Massif is separated from the Gondwana plate by oceanic crust which originated in Late Ordovician and Late Silurian–Early Devonian times, respectively. If Central Iberia is considered to constitute part of Gondwana then the closing of the oceanic domains to the south is precluded (Matte 2001).

The existence of microplates, terrane assemblages and a superterrane between Laurussia and Gondwana has been challenged by Robardet (2002, 2003). Comparison of the palaeobiogeographical features of the Palaeozoic sediments which cover the low-grade domains of the Iberian, French and Central European parts of the Variscan Orogen reveals that there are striking similarities in terms of sedimentological and faunal data (Robardet 2002, 2003). As mentioned above, such a view is also supported by geochronological, geochemical, and isotope-geochemical data (e.g. Linnemann *et al.* 2004) from the autochthonous domain of the Saxothuringian Zone. Thus, there is evidence that the continental crust constituting the Internides of the Variscan Orogen belonged to the same lithospheric plate prior to the collision with Laurussia. In such a two-plate model, the Rheic Ocean represents the exclusive oceanic domain between Gondwana (Gondwana *sensu stricto* + ATA + Proto-Alps) in the south and Laurussia in the north. However, the obvious juxtaposition of Cadomian blocks of the ATA during the Variscan Orogeny as well as the existence of high pressure domains which are not related to the Rheic Suture Zone requires convergence with a considerable amount of subduction inside the ATA.

Number and orientation of subduction zones

The lack of consensus on the precise number of lithospheric plates/terranes/microcontinents involved in Variscan deformation, as outlined above, is also reflected in the discussion on the number, orientation and timing of the subduction zones in the region. Consequently there exist models which explain the evolution of the Variscan Orogen, for example, for the north-western part of Iberia with ongoing convergence under initial subduction of oceanic crust (Rheic Ocean as part of the Gondwana lower plate) followed by the thrusting of allochthonous sheets on autochthonous units of Gondwana (e.g. Martinez Catalan *et al.* 2002) from Lower Devonian to Upper Carboniferous time. In such a view Laurussia occupies an upper plate position during the entire process, without the requirement of an interconnected Armorican microplate or terrane assemblage.

The existence of two subduction zones has been proposed by Matte (1986, 2001) with an Early Variscan subduction of the Massif Central Ocean and the Rheic Ocean to the south and the north of the Armorican microplate, respectively. The closing of these oceanic domains was followed by the obduction of oceanic crust along the sutures in Upper Devonian times. Ongoing convergence led to the subduction of continental crust and the formation of a series of Late Variscan fold-and-thrust belts across the region of the former Massif Central Ocean. The Armorican microplate, therefore, occupied an upper plate position during the entire process. More recently, the model of Stampfli *et al.* (2002), proposed that the so-called Hun-Superterrane (upper plate position) collided with Avalonia resulting in the formation of the Rheic suture in Mid-Devonian times. This model, therefore, would suggest that the closure of the oceanic domain in the south could not have occurred before Visean times.

If, however, Armorica is regarded as an assemblage of independent microplates then there is a geometric requirement for an increased number of active plate margins in the region. For example, Franke (2000) has suggested that the Rheic Ocean existed until Lower Devonian times. Following the first 'soft' collision (the term soft collision (Franke & Oncken 1995) has been introduced because of the lack of structural evidence along the Rhenohercynian Basin at this time) of the ATA with

Laurussia (and more precisely, the southern margin of Avalonia) the Rheic suture reopened leading to the formation of the enigmatic Rhenohercynian Ocean during Devonian times.

The more or less coeval bilateral subduction of the Saxothuringian and Moldanubian oceans beneath the Teplá-Barrandian was followed, and accompanied, by the subduction of the Rhenohercynian Ocean beneath the Saxothuringian Zone. Therefore, the final collision of the Rhenohercynian Zone (Laurussia) with the continental magmatic arc of the Mid-German Crystalline High, forming the leading edge of the ATA, is considered to be a Late Variscan process and was probably responsible for the evolution of the fold-and-thrust belt in the external zone of the Central European Variscides (Oncken 1997; Oncken *et al.* 2000).

This incomplete discussion demonstrates the recent coexistence of different plate tectonic models for the development of the region. Despite the obvious contradictions, however, there is general agreement regarding the importance of subduction processes as part of the tectonic evolution of Central Europe. The proposed upper plate position of the Armorican microplate as well as of the Hun Superterrane, however, remains problematic for obvious reasons. In such a scenario (i.e. upper plate position) extensive Andean-type magmatism, accompanied by high temperature/low pressure metamorphism in deeper parts of the upper plate, would be the dominant process. Thus Cadomian crust, which was little affected by subduction and intruded by I-type magmatites, should be the exclusive feature of the Internides. In fact, the Teplá-Barrandian Unit (TBU) of the Bohemian Massif constitutes precisely such a type of crust. As shown above, at least parts of all of the adjacent regions surrounding the TBU (i.e. Sudetes, Moldanubian Zone, allochthonous domain of the Saxothuringian Zone) represent Armorica-type crust and show features consistent with subduction-related processes. Thus, for the peri-Gondwana crust south of the Rheic suture (i.e. Armorica-type elements) both upper and lower plate positions are possible. Therefore, the existence of a coherent Armorican microplate as well as of a coherent Hun Superterrane can be ruled out and should, in fact, be abandoned.

The role of continental subduction

Subduction processes involving considerable volumes of continental crust have been considered as an important feature of many Phanerozoic collisional belts (e.g. Ernst & Liou 2000). For example, continental subduction within the allochthonous domain of the Saxothuringian Zone (Mingram 1998) at 340 Ma (Kröner & Willner 1998) underlines the importance of such processes for the European Variscides. Since the upper part of the allochthonous domain contains subducted continental crust already exhumed in Late Devonian times, multiple continental subduction events are required.

While single subduction events are explained by the pull of the leading edge of the passive continental margin following the complete consumption of oceanic lithosphere (e.g. Chemenda *et al.* 1997), multiple subduction–accretion events as described for the Scandinavian Caledonides (Brueckner & van Roermund 2004) require ongoing subduction beneath the earliest accreted continental blocks. In such a model the accreted continental blocks and the subducted crust belong to the same lithospheric plate. The application of such a subduction–accretion scenario to the Armorican Terrane Assemblage (Kroner *et al.* 2007) provides an explanation for the juxtaposition of the different Cadomian blocks of the ATA during ongoing convergence and the subduction of continental crust of the same lithospheric plate beneath early accreted blocks. Moreover, in such a model the ATA does not necessarily represent an independent microplate but can also be a part of the Gondwana plate. The reason for the contrasting material paths of the crust (i.e. subduction versus accretion) is related to differences in crustal thickness as well as the composition of the peri-Gondwana crust. Post-Cadomian inhomogeneous crustal thinning in Cambro-Ordovician times possibly led to the evolution of variously arranged Cadomian blocks surrounded by thinned continental crust (Kroner *et al.* 2007). The continuous subduction of the latter finally led to the accretion of the Armorican Terrane Assemblage during the Variscan Orogeny, i.e. the long-lasting convergence between Gondwana and Laurussia.

The role of orogenic collapse

Despite the various differences in terms of plate tectonic scenarios as outlined above, all of the models conform to the idea that plate motions were exclusively responsible for the structural geometry of the orogen. Nevertheless, there are a number of models for the development of the European Variscides which note the importance of the process of orogenic collapse in terms of the evolution of the region. Orogenic or gravitational collapse is considered to be a process which transfers gravitational potential energy from overthickened crustal regions of high potential energy to adjacent regions of lower potential energy (e.g. foreland basins) along lateral energy gradients (for details see Rey *et al.* 2001; Selverstone 2005, and references therein). The term 'late orogenic extensional collapse' has been used by some authors (e.g. Dewey 1988) reflecting the consideration that the final stage of a continent–continent collision can be characterized by large-scale normal faulting caused by the horizontal spreading of the previously thickened internal zones of an orogen. Gneiss domes or metamorphic core complexes mantled by extensional shear zones, forming condensed metamorphic profiles to adjacent upper crustal rock complexes, are the most obvious structures resulting from the process of orogenic collapse. In such a scenario the advective heat transfer, produced by the isothermal exhumation of high-grade rocks of the previously overthickened crust, result in a final phase of high temperature/low pressure metamorphism on a regional scale. In such a scenario, this final metamorphic phase is both accompanied and followed by the intrusion of voluminous granitoids.

Evidence of the activity of such processes is present in the Variscan Orogen. For example, the Internides of the Variscan belt contain features which are consistent with the process of orogenic collapse. The Montagne Noir of the French Massif Central has been interpreted as a Late Variscan metamorphic core complex (Echtler & Malavieille 1990). Two stages of late orogenic extension have also been reported from adjacent areas of the French Massif Central (Faure 1995). The first stage (Late Visean–Westphalian) was related to the emplacement of voluminous synkinematic granites, while the second (Late Carboniferous–Early Permian) resulted in the formation of intramontane coal basins and ductile normal faulting along the borders of migmatitic gneiss domes. Pervasive crustal thinning has also been proposed for the previously thickened crust of the South Armorican Zone (Cagnard *et al.* 2004). Similarly, for the Central European segment of the Variscides, the evidence of late orogenic extensional tectonics in various areas including the Odenwald (MGCH) (Krohe 1992) and the NW part of the Bohemian Massif (Krohe 1998), the presence of gneiss domes in the Erzgebirge (Sebastian 1995; Krohe & Willner 2003; Willner *et al.* 2000), the rapid isothermal unroofing of the Saxon granulites along subhorizontal shear zones (allochthonous domain of the Saxothuringian Zone) (Reinhardt & Kleemann 1994; Kroner 1995) and the juxtaposition of the upper crustal Teplá-Barrandian Unit (Zulauf 1994) with the adjacent high grade units, have all been considered as related to orogenic collapse.

Some authors explain the entire Late Variscan evolution, i.e the exhumation processes within the internal zone and the formation of the external fold-and-thrust belts, with orogenic collapse. For example, Faure *et al.* (2002) suggested that there was a phase of late orogenic extension of the entire Variscan internal zone in the Late Visean. This final extensional phase caused by horizontal spreading of previously thickened crust was accommodated by compressional tectonics in the external fold-and thrust belts. The model of diachronous (335–290 Ma) extensional collapse across the entire overthickened part of the Variscan belt with accompanying compressional compensation within the foreland basins (Praeg 2004) implies a prolonged transfer of potential energy from the internal towards the external part of the orogen.

The final stages of Variscan evolution as proposed by such a model are, however, in marked contrast to the proposed final stages of all of the various plate tectonic models outlined above. However, it does explain the termination of high pressure metamorphism in Visean times, and the fact that this was immediately followed by a regional high temperature metamorphic event. This change was related to the shift from a phase of crustal thickening to one of orogenic collapse related to equilibration processes within the continental crust.

An exclusive Late Variscan orogenic collapse scenario would negate the importance of horizontal compressional forces generated during ongoing plate convergence during the Late Variscan stage. From a structural point of view (see above), there are no indications that the Late Variscan strain increments of the entire internal region of the orogen show any evidence of exclusive lateral spreading along multiple detachment zones. Instead, this region is characterized more by its spectacular complexity in terms of the irregular juxtaposition of domains characterized by either compressional or extensional structures, or even both. For example, in the Erzgebirge region of the allochthonous domain of the Saxothuringian Zone, late orogenic extensional collapse of a complete pre-existing nappe pile has been proposed by some authors (e.g. Kroner 1995; Sebastian 1995). However, despite the indisputable existence of extensional shear zones in the area, there are certain features which cannot be explained by the process of orogenic collapse. These include the presence of lenses of coesite- and microdiamond-bearing eclogites and gneisses which were formed during the metamorphic climax at *c.* 340 Ma (depth of continental subduction >100 km) and which are now exposed as tectonic slices above huge medium-pressure orthogneiss complexes. The rapid final exhumation of the deepest parts of the allochthonous domain into mid-crustal levels (cooling at *c.* 330 Ma) would have required a large amount of thrusting at this stage. Moreover in the adjacent wrench-and-thrust zone (Saxothuringian Zone, i.e. Internides) there are no indications for regional extension in the upper crust between 340 and 330 Ma (Kroner *et al.* 2007). In place of late orogenic extension, extrusion tectonics from a subduction channel in a transpressive regime is considered responsible for the rapid exhumation of the high grade rocks of the Saxothuringian Zone (Kroner *et al.* 2007).

As noted above, there is plenty of evidence of ongoing convergence between the Central European parts of Laurussia and Gondwana through to Permian times. The Late Variscan (Upper Carboniferous–Lower Permian) evolution of the Iberian Peninsula, for example, can be characterized by initial thrusting due to east–west compression followed by oroclinal bending as a result of final north–south compression (Gutierrez Alonso *et al.* 2004; Kollmeier *et al.* 2000). The Cantabria–Asturias Arc which was intruded by voluminous granites does not contain any features of extensional tectonics. Thus, the presence of orogenic collapse structures within the internal part of the orogen reflects locally restricted, rather than orogen-wide, gravitationally driven equilibration processes of thickened crust during ongoing plate convergence.

The role of strike-slip tectonics

The existence of large-scale dextral as well as sinistral strike-slip zones has long been recognized. Strike-slip tectonics, related to oblique plate interactions, is considered to play an important role in the evolution of the Variscan orogen (e.g. Badham 1982; Shelley & Bossiere 2002). Late Variscan strike-slip tectonics along dextral shear zones affected the entire Central and Western European Variscides as well as northern Africa (Arthaud & Matte 1977). However, a proposed tectonic model which interprets the tectonic evolution of the Uralides, the Variscides and the Appalachians in terms of activity along a dextral Mega-Riedel shear system has been refuted because the plate tectonic processes active during the Uralian Orogeny (Zonenshain *et al.* 1990) were independent from those of the Variscides. Nevertheless the importance of strike-slip movements during the Variscan Orogeny has been stressed by many authors and is considered to be an essential part of all of the models outlined above. Indeed, the complex juxtaposition of different Cadomian blocks and high grade metamorphic domains requires large-scale strike-slip displacements with localized transpressional and transtensional tectonics across the entire orogen (Krohe 1996).

A generic model for the Variscides of the Central European segment

A generalized model for the evolution of the Central European segment of the Variscides has been derived from work carried

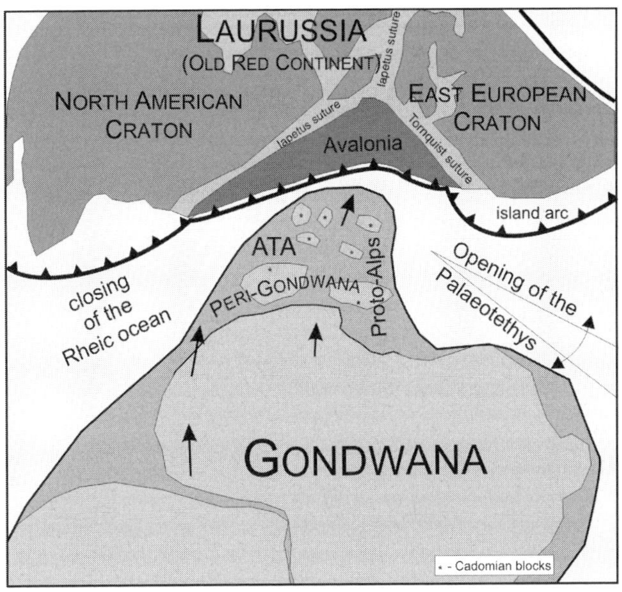

Fig. 11.30. Two-plate model for the Variscan Orogeny. The Armorican Terrane Assemblage as well as the Proto-Alps are considered as an integral part of the Gondwana plate (modified after Kroner *et al.* 2007; reproduced by permission of the Geological Society of America).

out mainly within the Saxothuringian Zone (Kroner *et al.* 2007). This model considers the ongoing convergence between Gondwana and Laurussia with initial subduction of the Rheic Ocean and implies the following constraints:

(1) The basis of the proposed evolution of the Variscides is the two-plate model (Fig. 11.30). Peri-Gondwana represents the extended northern Lower Palaeozoic continental shelf of the Gondwana plate with ongoing marine sedimentation during the Lower Palaeozoic. Cadomian crystalline blocks (e.g. Armorican Terrane Assemblage) were overlain by sedimentary units consisting of shallow-marine shelf deposits. These, in turn, are surrounded by thinned Cadomian basement, with accompanying large volumes of magmatites. The magmatic succession is Cambro-Ordovician in age, and related to the widespread extension which took place at that time and which was related to the opening of the Rheic Ocean and the separation of Avalonia from Gondwana;

(2) A period of oblique NE-directed convergence between the southern Gondwana plate and the northern plate of Laurussia along a *c.* east–west striking active plate margin commenced in the Silurian. This movement was connected with the subduction of the Rheic Ocean beneath the northern plate. Relicts of a coeval magmatic arc are exposed along the Mid-German Crystalline High (Fig. 11.31a). Associated high pressure metamorphism is documented within the uppermost allochthonous units of the ATA (e.g. Münchberg Massif, Marianske Lazne Complex, Sowie Gory Complex);

(3) Diachronous collision of the variously arranged Cadomian blocks was coeval with the subduction of the thinned continental crust of the Gondwana plate (Fig. 11.31a–c). This collision

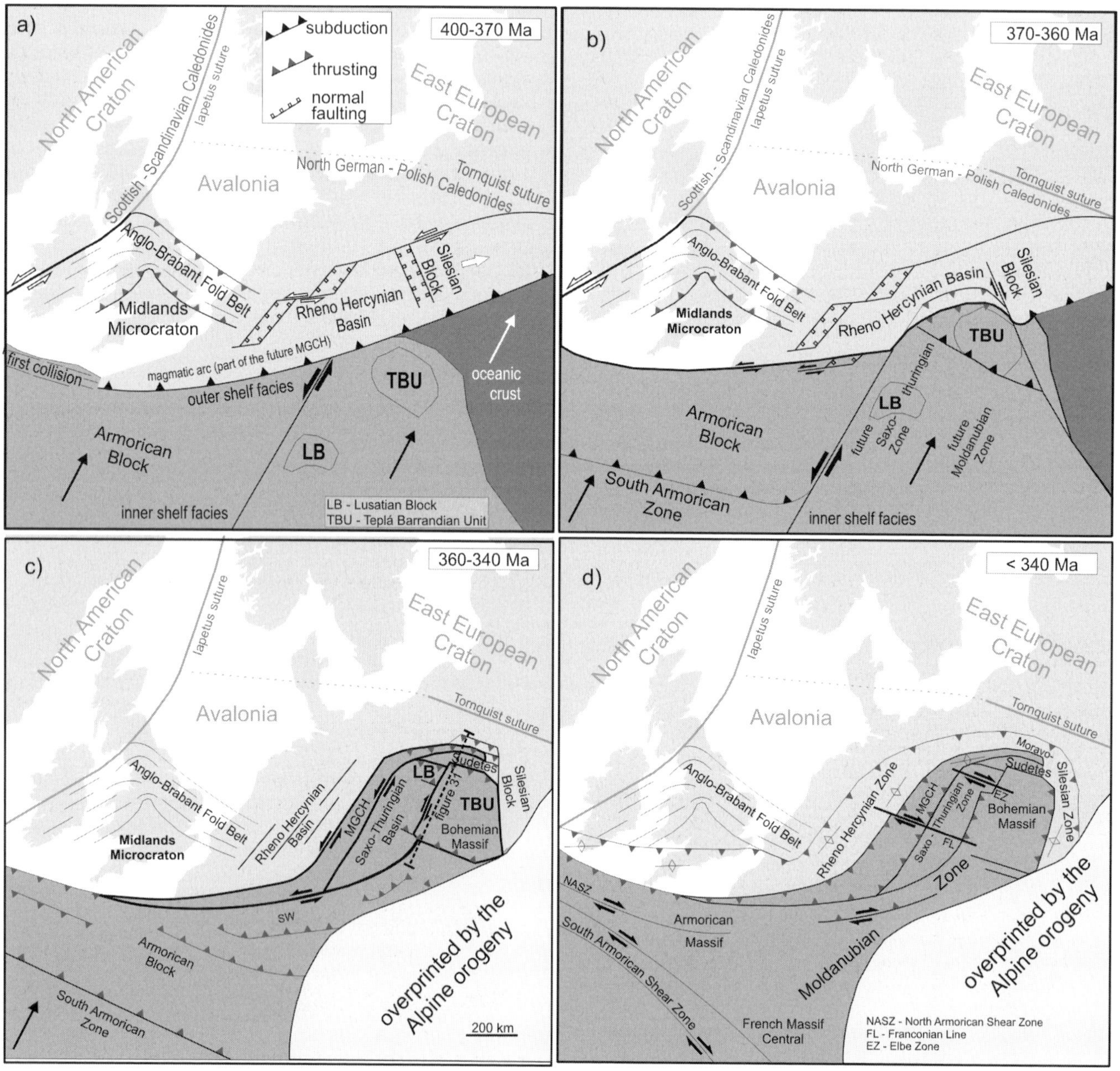

Fig. 11.31. Generic geodynamic model for the Central European Variscides (after Kroner *et al.* 2007; reproduced by permission of the Geological Society of America).

phase extended from the Early Devonian through to the Early Carboniferous (*c.* 60 Ma);

(4) Spatially restricted, fault-bounded orogenic wedges were formed, comprising the initially accreted continental blocks of the Gondwana plate together with the Early Variscan exhumed HP metamorphic units. These orogenic wedges were subsequently the sites of ongoing subduction of the Gondwana plate beneath the pro-wedge (Fig. 11.32) and coeval thrusting in the retro-wedge area of the northern plate (i.e. Laurussia). In this scenario, the Acadian Orogeny of Ireland, Britain and Belgium resulted from the collision between the Armorica Block and the Anglo-Brabant Massif of Avalonia (Fig. 11.31a);

(5) Early Variscan extensional tectonics is the main characteristic of the upper plate (i.e. the Variscan Externides) and evidence of such deformation has been observed in the Rhenohercynian as well as the Moravo-Silesian zones. Extensional tectonics, coeval with and subsequent to the Acadian Orogeny along adjacent areas of the Laurussian upper plate, were related to the ongoing oblique subduction of the Gondwana plate. This resulted in the opening of the 'Rhenohercynian Ocean' (Fig. 11.31a). Kroner *et al.* (2007) have interpreted the Rhenohercynian Basin as an area of localized transtension situated in a hinterland position, relative to the active margin, which was possibly connected with the evolution of small volumes of oceanic crust within short-lived backarc basins;

(6) NE–SW directed transform zones with localized transpressional and transtensional domains sustained strain compatibility and led finally to the juxtaposition of Cadomian blocks of the ATA as well as the tectonic contact of Cadomian blocks which belong to the Laurussian plate (Fig. 11.31a-c). This underlines the importance of strike-slip zones affecting the entire lithosphere. The Bohemian Massif, characterized by the juxtaposition of the Teplá-Barrandian Unit and the Lusatian Block representing parts of the ATA with the Moravo-Silesian Block of the Laurussian plate, is fully explained with such a scenario (Fig. 11.31c);

(7) The final exhumation of the subducted continental crust of the Gondwana plate took place during a period of regional Late

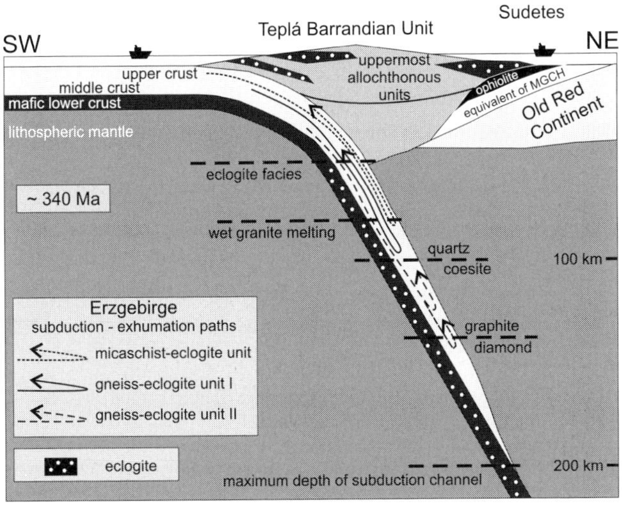

Fig. 11.32. Subduction of continental crust beneath early accreted Cadomian blocks of the Armorican Terrane Assemblage as well as beneath Early Variscan subducted and exhumed metamorphic units is considered as the essential feature of the Mid-Variscan stage of the Internides (after Kroner *et al.* 2007; reproduced by permission of the Geological Society of America).

Variscan transpressional tectonics. Associated advective heat transfer to upper crustal levels from exhumed rocks is the likely reason for this regional phase of HT/LP metamorphism. During this stage, which in addition is characterized by large-scale NW–SE directed dextral strike-slip activity and the production of voluminous granite magmatism, vertical and lateral extrusion tectonics along the former subduction channel led to the juxtaposition of units with very different evolutionary histories. This stage also represents the period of so-called 'orogenic collapse' (Fig. 11.31d); and

(8) The termination of the tectonometamorphic processes within the Internides of Central Europe is represented by the final intrusion of the post-kinematic granites which occurred in Mid-Carboniferous times (*c.* 325 Ma). The final Late Variscan (*c.* 300 Ma) thrusting in the external part of the Variscan area was considerably later. The chronological separation between these two major events precludes a causal relationship between them. Instead, ongoing convergence between the Gondwana plate and Laurussia can be considered to be the principal cause for the final compressional event affecting the entire orogen at this time, resulting in the evolution of the fold-and-thrust belt along the external zone of the Variscides (Fig. 11.31d).

References

Adusumalli, C. I. & Schubert, W. 2001. Geochemistry of metabauxites in the Bergsträsser Odenwald (Mid German Crystalline Rise) and palaeoenvironmental implications. *Mineralogy and Petrology*, **72**, 45–62.

Aftalion, M., Bowes, D. R. & Vrana, S. 1989. Early Carboniferous U–Pb zircon age for garnetiferous perpotassic granulites, Blansky les massif, Czechoslovakia. *Neues Jahrbuch für Mineralogie, Monatshefte*, **1989**(H4), 145–152.

Ahrendt, H., Clauer, N., Hunziker, J. C. & Weber, K. 1983. Migration of Folding and Metamorphism in the Rheinische Schiefergebirge: Deduced from K-Ar and Rb-Sr Age Determinations. *In:* Martin, H. & Eder, F.W. (eds) *Intracontinental Fold Belts*. Springer, Berlin, 323–338.

Aleksandrowski, P. 1990. Early Carboniferous strike-slip displacements at the northeast periphery of the Variscan Belt in Central Europe. *In: International Conference on Paleozoic Orogens in Central Europe, Göttingen-Giessen* (IGCP Program 233: Terranes in the Circum-Atlantic Paleozoic Orogens), Abstracts, 7–10.

Aleksandrowski, P. 1995. The significance of major strike-slip displacements in the development of Variscan structure of the Sudetes, SW Poland. *Przegląd Geologiczny*, **43**, 745–754 [in Polish, English summary].

Aleksandrowski, P. & Mazur, S. 2002. Collage tectonics in the northeasternmost part of the Variscan Belt: the Sudetes, Bohemian Massif. *In:* Winchester, J., Pharaoh, T. & Verniers, J. (eds) *Palaeozoic Amalgamation of Central Europe*. Geological Society, London, Special Publications, **201**, 237–277.

Aleksandrowski, P., Kryza, R., Mazur, S. & Żaba, J. 1997. Kinematic data on major Variscan strike-slip faults and shear zones in the Polish Sudetes, northeast Bohemian Massif. *Geological Magazine*, **133**, 727–739.

Aleksandrowski, P., Kryza, R., Mazur, S., Pin, C. & Zalasiewicz, J.A. 2000. The Polish Sudetes: Caledonian or Variscan? *Transactions of the Royal Society, Edinburgh*, **90**, 127–146.

Altenberger, U. & Besch, T. 1993. The Böllstein Odenwald: evidence for pre- to early-Varsican plate convergence in the Central European Variscides. *Geologische Rundschau*, **82**, 475–488.

Altenberger, U., Besch, T., Mocek, B., Zaipeng, Y. & Yong, S. 1990. Geochemie und Geodynamik des Böllsteiner Odenwaldes. *Mainzer geowissenschaftliche Mitteilungen*, **19**, 183–200.

Altenberger, U., Oberhänsli, R., Adusumalli, C. & Rasool, R. 1994. The corundum-bearing rocks around Seeheim in the Odenwald complex, Mid-German crystalline rise. *Zentralblatt für Geologie und Paläontologie, Teil I*, **1992**, 895–908.

Altherr, R. & Maass, R. 1977. Metamorphite am Südrand der

Zentralschwarzwälder Gneis-anatexitmasse zwischen Geschwend und Bernau. *Neues Jahrbuch für Geologie und Paläontologie, Abhandlungen*, **154**, 129–154.

Altherr, R., Henes-Klaiber, U., Hegner, E. & Satir, M. 1999. Plutonism in the Variscan Odenwald (Germany): from subduction to collision. *International Journal of Earth Sciences*, **88**, 422–443.

Altherr, R., Holl, A., Hegner, E., Langer, C. & Kreuzer, H. 2000. High-potasium, calc-alkaline I-type plutonism in the European Variscides: Northern Vosges (France) and northern Schwarzwald (Germany). *Lithos*, **50**, 51–73.

Anderle, H.-J., Franke, W. & Schwab, M. 1995. III.C.1 Stratigraphy. *In:* Dallmeyer, R. D., Franke, W. & Weber, K. (eds) *Pre-Permian Geology of Central and Eastern Europe*. Springer, Berlin, 99–107.

Anthes, G. & Reischmann, T. 2001. Timing of granitoid magmatism in the eastern mid-German crystalline rise. *Journal of Geodynamics*, **31**, 119–143.

Armijo, R., Tapponier, P., Mercier, J. L. & Tonglin, H. 1986. Quaternary extension in southern Tibet: Field observations and tectonic implications. *Journal of Geophysical Research*, **91**, 13803–13872.

Arthaud, F. & Matte, P. 1977. Late Paleozoic strike-slip faulting in southern Europe and northern Africa; result of a right-lateral shear zone between the Appalachians and the Urals. *Geological Society of America Bulletin*, **88**, 1305–1320.

Artmann, E., Bues, C., Scheuvens, D. & Zulauf, G. 2003. Zur tektonometamorphen Entwicklung der Zentralböhmischen Scherzone zwischen Svatá Katerina und Rittsteig unter besonderer Berücksichtigung der Forschungsbohrung Rittsteig (Böhmische Masse). *Geologica Bavarica*, **107**, 63–94.

Autran, A., Chantraine, J., Lorenz, C., Mégnion, C. & Weber, C. 1992. Conclusions générales. Forage Scientifique de Sancerre-Couy (Cher). *Mémoire Géologie Profonde de la France*, **3**, 209–212.

Babek, O. 1996. Thinning and finning upward megasequences in Middle Devonian carbonate slope deposits, Moravia (Czech Republic). *Neues Jahrbuch für Geologie und Paläontologie, Abhandlungen*, **202**, 409–432.

Babuska, V. & Plomerova, J. 2001. Subcrustal lithosphere around the Saxothuringian-Moldanubian Suture Zone – a model derived from anisotropy of seismic wave velocities. *Tectonophysics*, **332**(1–2), 185–199.

Babuska, V. & Plomerova, J. 2006. European mantle lithosphere assembled from rigid microplates with inherited seismic anisotropy. *Physics of the Earth and Planetary Interiors*, **158**(2–4), 264–280.

Bachtadse, V., Torsvik, T. H., Tait, J. A. & Soffel, H. C. 1995. X. Paleomagnetic constraints on the paleogeographic evolution of Europe during the Paleozoic. *In:* Dallmeyer, R.D., Franke, W. & Weber, K. (eds) *Pre-Permian Geology of Central and Eastern Europe*. Springer, Berlin, 567–578.

Badham, J. P. N. 1982. Strike-slip orogens – an explanation for the Hercynides. *Journal of the Geological Society*, **139**, 493–504.

Ballevre, M., Capdevila, R., Guerrot, C. & Peucat, J.-J. 2002. Discovery of an alkaline orthogneiss in the eclogite-bearing Cellier Unit (Champtoceaux Complex, Armorican Massif): a new witness of the Ordovician rifting. *Comptes Rendus Geosciences*, **334**, 303–311.

Banka, D., Pharaoh, T. C. & Williamson, J. P. 2002. Potential field imaging of Palaeozoic orogenic structure in northern and central Europe. *Tectonophysics*, **360**, 23–45.

Bankwitz, P., Kopp, J. & Ehling, B. 2001. Mitteldeutsche Kristallinzone (MKZ) im Abschnitt Halle-Guben. *In:* Steininger F. (ed.) *Stratigraphische Kommission Deutschland, Stratigraphie von Deutschland II, Ordovizium, Kambrium, Vend, Riphäikum*. Courier Forschungs Institut Senckenberg, Frankfurt, **234**, 186–196.

Baranowski, Z., Haydukiewicz, A., Kryza, R., Lorenc, S., Muszyński, A., Solecki, A. & Urbanek, Z. 1990. Outline of the geology of the Góry Kaczawskie (Sudetes, Poland). *Neues Jahrbuch für Geologie und Paläontologie, Abhandlungen*, **179**, 223–257.

Barth, H. 1971. Marmore und Kalksilikatfels von Auerbach-Hochstädten (Neue Gesichtspunkte zur Metamorphose des altbekannten Mineralfundpunktes). *Oberrheinische Geologische Abhandlungen*, **20**, 43–58.

Baumann, A., Grauert, B., Mecklenburg, S. & Vinx, R. 1991. Isotopic age determinations of crystalline rocks of the upper Harz Mountains, Germany. *Geologische Rundschau*, **80**(3), 669–690.

Becq-Giraudon, J.F., Montenat, C. & Van Den Driesche, J. 1996. Hercynian high-altitude phenomena in the French Massif Central: tectonic implications. *Palaeogeography Palaeoclimatology and Palaeoecology*, **122**, 227–241.

Bederke, E. 1924. Das Devon in Schlesien und das Alter der Sudetenfaltung. *Fortschritte der Geologie und Paläontologie*, **7**, 1–55.

Bederke, E. 1929. Die Grenze von Ost- und Westsudeten und ihre Bedeutung für die Einordnung der Sudeten in den Gebirgsbau Mitteleuropas. *Geologische Rundschau*, **20**, 186–205.

Bederke, E. 1957. Alter und Metamorphose des kristallinen Grundgebirges im Spessart. *Abhandlungen, Hessisches Landesamt für Bodenforschung*, **18**, 7–19.

Bednarczyk, W., Chlebowski, R. & Kowalczewski, Z. 1970. Geological structure of the northern limb of the Dyminy anticline in the Holy Cross Mts. *Biuletyn Geologiczny*, **12**, 197–223 [in Polish].

Behr, H.-J. & Heinrichs, T. 1987. Geological interpretation of DEKORP 2-S. A deep seismic reflection profile across Saxothuringian. *Tectonophysics*, **142**, 173–202.

Behrmann, J. H. & Tanner, D. C. 1997. Carboniferous tectonics of the Variscan basement collage in eastern Bavaria and western Bohemia. *Geologische Rundschau*, **86**, S15-S27.

Behrmann, J., Drozdzewski, G., Heinrichs, T., Huch, M., Meyer, W. & Oncken, O. 1991. Crustal-scale balanced cross sections through the Variscan fold belt, Germany: the central EGT-segment. *Tectonophysics*, **196**, 1–21.

Belka, Z., Skompski, S. & Sobon-Podgorska, J. 1996. Reconstruction of a lost carbonate platform on the shelf of Fennosarmatia: evidence from Viséan polymictic debrites, Holy Cross Mountains, Poland. *In:* Strogen, P., Somerville, I. D. & Jones, G. L. (eds) *Recent Advances in Lower Carboniferous Geology*. Geological Society, London, Special Publications, **107**, 315–329.

Bełka, Z. & Narkiewicz, M. 2008. Devonian. *In*: McCann, T. (ed.) *The Geology of Central Europe. Volume 1: Precambrian and Palaeozoic*. Geological Society, London, 383–410.

Bender, P. 1989. Die Hörre und ihre Stellung im östlichen Rheinischen Schiefergebirge. *Jahresberichte und Mitteilungen des Oberrheinischen Geologischen Vereins, Neue Folge*, **71**, 347–356.

Bender, P., Jahnke, H. & Ziegler, W. 1974. Ein Unterdevon-Profil bei Marburg a. d. Lahn. *Notizblatt des hessischen Landes-Amtes für Bodenforschung*, **102**, 25–45.

Berthelsen, A. 1998. The Tornquist Zone northwest of the Carpathians: an intraplate pseudosuture. *Geologiska Föreningen i Stockholm Förhandlingar*, **120**, 223–230.

Bleibinhaus, F., Stich, D., Simon, M. & Gebrande, H. 2003. New results from amplitude preserving prestack depth migration of the Münchberg/Vogtland segment of the MVE deep seismic survey. *Journal of Geodynamics*, **35**(1–2), 33–43.

Bless, J. M., Bouckaert, J., Calver, M. A., Graulich, J. M. & Paproth, E. 1977. Paleogeography of Upper Westphalian deposits in NW Europe with reference to the Westphalian C north of the mobile Variscan belt. *Mededelingen Rijks Geologische Dienst*, **28**, 101–147.

Bossdorf, R. H. H. 1961. Das Kristallin von Gadernheim und Laudenau im Odenwald. *Neues Jahrbuch für Mineralogie, Abhandlungen*, **95**, 370–419.

Boutin, R., Montigny, R. & Thuizat, R. 1995. Chronologie K-Ar et ^{39}Ar-^{40}Ar du métamorphisme et du magmatisme des Vosges, Comparaison avec les massifs varisques avoisinants. *Geologique de France*, **1**, 3–25.

Braitsch, O. 1957. Beitrag zur Kenntnis der kristallinen Gesteine des südlichen Spessarts und ihrer geologisch-tektonischen Geschichte. *Abhandlungen, Hessisches Landesamt für Bodenforschung*, **18**, 21–72.

Brandes, T. 1919. Die varistischen Züge im geologischen Bau Mitteldeutschlands. *Neues Jahrbuch für Mineralogie, Geologie und Paläontologie, Beilagen-Band*, **43**, 190–250.

Brätz, H. 2000. *Radiometrische Altersdatierungen und geochemischen Untersuchungen von Orthogneisen, Granite und Granitporphyren aus dem Ruhlaer Kristallin, Mitteldeutsche Kristallinzone*. Dissertation, Universität Würzburg, Germany.

Brewer, M. S. & Lippolt, H. J. 1972. Isotopische Altersbestimmungen an Schwarzwald-Gesteinen. *Fortschritte Mineralogie*, **50**, 41–50.

BRINKMANN, R. 1948. Die Mitteldeutsche Schwelle. *Geologische Rundschau*, **36**, 56–66.

BROCHWICZ-LEWIŃSKI, W., POŻARYSKI, W. & TOMCZYK, H. 1983. Palaeozoic strike-slip motions in southern Poland. *Przegląd Geologiczny*, **31**(12), 651–658 [in Polish].

BROWN, R.L., BEAUMONT, C. & WILLETT, S.D. 1993. Comparison of the Selkirk fan structure with mechanical models: Implications for interpretations of the southern Canadian Cordillera. *Geology*, **21**, 1015–1018.

BRUECKNER, H. K. & VAN ROERMUND, H. L. M. 2004. Dunk tectonics; a multiple subduction/eduction model for the evolution of the Scandinavian Caledonides. *Tectonics*, **23**, TC2004. DOI:10.1029/2003TC001502.

BÜCKING, H. 1892. Der nordwestliche Spessart. *Abhandlungen der Preussischen Geologischen Landesanstalt, Neue Folge*, **12**, 1–274.

BUES, C. & ZULAUF, G. 2000. Microstructural evolution and geologic significance of garnet pyriclasites in the Hoher-Bogen shear zone (Bohemian Massif, Germany). *International Journal of Earth Sciences*, **88**, 803–813.

BUES, C., DÖRR, W., FIALA, J., VEJNAR, Z. & ZULAUF, G. 2002. Emplacement depth and radiometric ages of Paleozoic plutons of the Neukirchen-Kdyne massif: Differential uplift and exhumation of Cadomian basement due to Carboniferous orogenic collapse (Bohemian Massif). *Tectonophysics*, **352**, 225–243.

BURCHFIEL, B.C., ZHILIANG, C., HODGES, K.V., YUPING, L., ROYDEN, L.H., CHANGRONG, D. & JIENE, X. 1992. The south Tibetan detachment system, Himalayan Orogen. Geological Society of America, Special Paper, **269**, 1–41.

BURG, J.P., BRUNEL, M., GAPAIS, D., CHEN, G.M. & LIU, G.H. 1984. Deformation of leucogranites of the crystalline Main Central Sheet in southern Tibet (China). *Journal of Structural Geology*, **6**, 535–542.

BUSBY, J.P. & SMITH, N.J.P. 2001. The nature of the Variscan basement in southeast England: evidence from integrated potential field modelling. *Geological Magazine*, **138**, 669–685.

BUSCHMANN, B., NASDALA, L., JONAS P., LINNEMANN, U. & GEHMLICH, M. 2001. SHRIMP u-Pb dating of tuff derived and detrital zircons from Cadomian marginal basin fragments (Neoproterozoic) in the northeastern Saxothuringian Zone (Germany). *Jahrbuch für Geologie und Paläontologie, Monatshefte*, **6**, 321–342.

CAGNARD, F, GAPAIS, D., BRUN, J. P, GUMIAUX, C. & VAN DEN DRIESSCHE, J. 2004. Late pervasive crustal-scale extension in the south Armorican Hercynian belt (Vendee, France). *Journal of Structural Geology*, **26**(3), 435–449.

ČECH, S., HAVLICEK, V. & ZIKMUNDOVA, J. 1988. The Upper Devonian and Lower Carboniferous in north-eastern Bohemia (based on boreholes in the Hradec Králové area). *Věstnik Ustředního Ústavu geologikého*, **64**, 65–76.

CHÁB, J. 1990. The problem of the Andělská Hora thrust in the light of new deep bores (Hrubý Jesenik Mts., Czechoslovakia). *Časopis pro Mineralogii a Geologii*, **35**, 389–401.

CHÁB, J. & VRÁNA S. 1979. Crossite-actinolite amphiboles of the Krkonoše-Jizera crystalline complex and their geological significance. *Věstník Ustředního Ústavu geologického*, **54**, 143–150.

CHÁB, J. & ŽÁČEK, V. 1994. Metamorphism of the Teplá Crystalline Complex. *KTB Report*, **94–3**, 33–37.

CHÁB, J., MIXA, P., VANECEK, M. & ŽÁČEK, V. 1994. Geology of the NW part of the Hrubý Jesenik Mts. (the Bohemian massif, Central Europe). *Věstnik Českého geologikého ústavu*, **69**(3), 17–26.

CHÁB, J., SUCHY & VEJNAR, Z. 1995. Metamorphic evolution. *In:* DALLMEYER, R.D., FRANKE, W. & WEBER, K. (eds) *Pre-Permian Geology of Central and Eastern Europe*. Springer, Berlin, 404–410.

CHANTRAINE, J., AUTRAN, A. & CAVELIER, C. 1996. *Carte Géologique de la France*. 6th edn. BRGM, Orléans.

CHATTERJEE, N. D. 1960. Geologische Untersuchungen im Kristallin des Böllsteiner Odenwaldes. *Neues Jahrbuch für Geologie und Paläontologie, Abhandlungen*, **37**, 223–256.

CHEMENDA, A., MATTE, P. & SOKOLOV, V. 1997. A model of Palaeozoic obduction and exhumation of high-pressure/low-temperature rocks in the southern Urals. *Tectonophysics*, **276**(1–4), 217–227.

CHEN, F. 1999. *Variscan and pre-Variscan geochronology of the Moldanubian zone of the Black Forest, Germany*. PhD thesis, University of Tübingen.

CHEN, F. 2002. Garnet U–Pb and Sm-Nd isotopic systems – a case study of a mafic granulite from the Moldanubian Black Forest basement. *Chinese Science Bulletin*, **47**, 864–868.

CHEN, F., HEGNER, E. & TODT, W. 2000. Zircon ages, Nd isotopic and chemical compositions of orthogneisses from the Black Forest Germany: evidence for a Cambrian magmatic arc. *International Journal of Earth Sciences*, **88**, 791–802.

CHEN, F., TODT, W. & HANN, H. P 2003. Zircon and garnet geochronology of eclogites from the Moldanubian Zone of the Black Forest Germany. *Journal of Geology*, **111**, 207–222.

CHLUPAČ, I. 1964. Towards the stratigraphic setting of the Devonian in Moravia. *Časopis pro Mineralogii a Geologii*, **9**(3), 309–316 [in Czech].

CHLUPAČ, I. 1988. The Devonian of Czechoslovakia and its stratigraphic significance. *In*: MCMILLAN, N.J., EMBRY, A.F. & GLASS, D.J. (eds) *Devonian of the World*. Canadian Society of Petroleum Geologists, Memoir, **14**(1), 481–497.

CHLUPAČ, I. 1989. Fossil communities in the metamorphic Lower Devonian of the Hrubý Jesenik Mts., Czechoslovakia. *Neues Jahrbuch für Geologie und Paläontologie, Abhandlungen*, **177**(3), 367–392.

CHLUPÁČ, I. 1993*a*. Stratigraphic evaluation of some metamorphic units in the N part of the Bohemian Massif. *Neues Jahrbuch für Geologie und Paläontologie, Abhandlungen*, **188**, 363–388.

CHLUPAČ, I. 1993*b*. *Geology of the Barrandian. A Field Trip Guide*. Waldemar-Kramer, Frankfurt a.M., 1–163.

CHOROWSKA, M. 1978. Visean limestones in the metamorphic complex of the Kaczawa Mts (Sudetes). *Annales de la Societé géologique de Pologne*, **48**, 245–261 [in Polish, English summary].

CHUNG, S.-L., LIU, D., *et al*. 2003. Adakites from continental collision zones: Melting of thickened lower crust beneath southern Tibet. *Geology*, **31**, 1021–1024.

ČIŽEK, P. & TOMEK, Č. 1991. Large-scale thin-skinned tectonics in the eastern boundary of the Bohemian Massif. *Tectonics*, **10**(2), 273–286.

CLARK, A.H., SCOTT, D.J., SANDEMAN, H.A., BROMLEY, A.V. & FARRAR, E. 1998. Siegenian generation of the Lizard ophiolite: U–Pb zircon age data for plagiogranite, Porthkerris, Cornwall. *Journal of the Geological Society, London*, **155**, 595–598.

CLAUER, N. & BONHOMME, M. 1970. Datations rubidium-strontium dans les schistes de Steige et al séries de Villé (Vosges). *Bulletin Services Carte Géologique. Alsace Lorraine*, **23**, 191–208.

COCKS, L. R. M. & FORTEY, R. A. 1982. Faunal evidence for oceanic separations in the Palaeozoic of Britain. *Journal of the Geological Society, London*, **139**, 465–478

COCKS, L.R.M., MCKERROW, W.S. & VAN STAAL, C.R. 1997. The margins of Avalonia. *Geological Magazine*, **134**, 627–636.

COLLINS, A.S., KRYZA, R. & ZALASIEWICZ, J.A. 2000. Macrofabric fingerprints of Late Devonian – Early Carboniferous subduction in the Polish Variscides, the Kaczawa complex, Sudetes. *Journal of the Geological Society, London*, **157**, 283–288.

COOKE, R. A. & O'BRIEN, P. J. 2001. Resolving the relationship between high P-T rocks and gneisses in collisional terranes: an example from the Gföhl gneiss-granulite association in the Moldanubian Zone, Austria. *Lithos*, **58**(1–2), 33–54.

CORNWELL, J.D., SMITH, N.J.P. & PHARAOH, T.C. 1994. Discussion: 'Source of the Bicester magnetic anomaly' by P.Kearey and A.M.Rabae. *Geological Journal*, **29**, 277–280.

COSTA, S. & REY, P. 1995. Lower crustal rejuvenation and growth during post-thickening collapse: insights from a crustal cross section through a Variscan metamorphic core complex. *Geology*, **23**(10), 905–908.

CYMERMAN, Z. 1996. The Złoty Stok – Trzebieszowice regional shear zone: the boundary of terranes in the Góry Złote Mts (Sudetes). *Geological Quarterly*, **40**, 89–118.

CYMERMAN, Z., PIASECKI, M.A.J. & SESTON, R. 1997. Terranes and terrane boundaries in the Sudetes, northeast Bohemian Massif. *Geological Magazine*, **134**, 717–725.

CZARNOCKI, J. 1938. Carte Géologique Générale de la Pologne 1:100000, Feuille 4: Kielce (in Polish and French). Państwowy Instytut Geologiczny, Warszawa.

CZARNOCKI, J. 1957. Geology of the Łysogóry region. *Prace Instytutu Geoologicznego*, **18**, 1–97 [in Polish].

DADLEZ, R., KOWALCZEWSKI, Z. & ZNOSKO, J. 1994. Some key problems of the pre-Permian tectonics of Poland. *Quarterly Journal,*

Polish Institute of Geology, Warsaw, **38**, 169–189.

DALLMEYER, J. R. & MARTINEZ GARCIA, E. (eds) 1990. *Pre-Mesozoic Geology of Iberia*. Springer, Berlin.

DALLMEYER, R. D. & LÉCORCHÉ, J. P. 1989. $^{40}Ar/^{39}Ar$ polyorogenic mineral age record within the central Mauritanide orogen, West Africa. *Geological Society of America Bulletin*, **101**(1), 55–70.

DALLMEYER, R. D. & URBAN, M. 1998. Variscan vs Cadomian tectonothermal activity in northwestern sectors of the Teplá-Barrandian zone, Czech Republic: constraints from $^{40}Ar/^{39}Ar$ ages. *Geologische Rundschau*, **87**, 94–106.

DALLMEYER, R. D., NEUBAUER, F. & HÖCK, V. 1992. Chronology of late Paleozoic tectonothermal activity in the southeastern Bohemian Massif, Austria (Moldanubian and Moravo-Silesian zones): $^{40}Ar/^{39}Ar$ mineral age controls. *Tectonophysics*, **210**, 135–153.

DALLMEYER, R. D., FRANKE, W. & WEBER, K. (eds) 1995. *The Pre-Permian Geology of Central and Eastern Europe*. Springer, Berlin.

DE BÉTHUNE, P., LAPANIA, E., ROUSSEAU-TURAY, M. & VON ELLER, J. P. 1968. Les diorites du Neuntelstein et du Neugrunrain prés du Hohwald (Vosges), et leurs enclaves. *Bulletin Services. Carte Géologique Alsace Lorraine*, **21**, 23–44.

DEKORP RESEARCH GROUP 1994. The deep reflection seismic profiles DEKORP 3/MVE-90. *Zeitschrift für Geologische Wissenschaften*, **22**, 623–825.

DE PUTTER, T., GROESSENS, E., & HERBOSCH, A. 1991. Le "V3a" anhydrite du sondage de Saint-Ghislain (150E387, province du Hainaut, Belgique): Description macroscopique et structures sédimentaires. *In:* SERVICE GÉOLOGIQUE DE BELGIQUE (ed.), Professional Paper, **250**(6), 38.

DELMER, A. 1986. Relations tectoniques entre le gisement houiller du Nord Français et celui du Borinage. *Annales de la Société Géologique du Nord*, CV, 111–114.

DEWEY, J. F. 1988. Extensional collapse of orogens. *Tectonics*, **7**(6), 1123–1139.

DOMBROWSKI, A., OKRUSCH, M. & HENJES-KUNST, F. 1994. Geothermobarometrie and geochronology on mineral assemblages of orthogneisses and related metapelites of the Spessart Crystalline Complex, NW Bavaria, Germany. *Chemie der Erde*, **54**, 85–101.

DOMBROWSKI, A., HENJES-KUNST, F., HÖHNDORF, A., KRÖNER, A., OKRUSCH, M. & RICHTER, P. 1995. Orthogneisses in the Spessart Crystalline Complex, Northwest Bavaria: Witnesses of Silurian granitoid magmatism at an active continental margin. *Geologische Rundschau*, **84**, 399–411.

DON, J. 2003. Herzyniden und Varisziden in Polen. *Zeitschrift für geologische Wissenschaften*, **31**, 197–204.

DÖRR, W. 1986. *Stratigraphie, Stoffbestand und Fazies der Gießener Grauwacke (E' Rheinisches Schiefergebirge)*. Dissertation, Universität Gießen.

DÖRR, W. & PREISS, R. 1982. Die Geologie des Nauborner Kopfes und seiner Umgebung. *Giessener Geologische Schriften*, **31**, 1–224.

DÖRR, W., FIALA, J., VEJNAR, Z. & ZULAUF, G. 1998*a*. U–Pb zircon ages and structural development of metagranitoids of the Teplá crystalline complex – evidence for pervasive Cambrian plutonism within the Bohemian massif (Czech Republic). *Geologische Rundschau*, **87**, 135–149.

DÖRR, W., ZULAUF, G., *et al.* 1998*b*. Cambrian transtensional and Variscan normal fault related plutons: Tectonothermal evolution within the Tepla-Barrandian (Bohemian Massif, Czech-Republic). *Terra Nostra*, **98**(2), 42–46.

DÖRR, W., FLOYD, P. A. & LEVERIDGE, B. E. 1999. U–Pb ages and geochemistry of granite pebbles from the Devonian Menaver Conglomerate, Lizard penninsula: provenance of Rhenohercynian flysch of SW England. *Sedimentary Geology*, **124**, 131–147.

DOUBINGER, J. 1963. Chitinozoaires ordoviciens et siluriens des schistes de Steige dans les Vosges. *Bulletin Services Carte Géologique Alsace Lorraine*, **16**, 125–136.

DROZDZEWSKI, G. 1993. The Ruhr coal basin (Germany): structural evolution of an autochthonous foreland basin. *International Journal of Coal Geology*, **23**(1–4), 231–250.

DUBIŃSKA, E., BYLINA, P., KOZŁOWSKI, A., DŐRR, W. & NEJBERT, K. 2004. U–Pb dating of serpentinization: hydrothermal zircon from a metasomatic rodingite shell (Sudetic ophiolite, SW Poland). *Chemical Geology*, **203**, 183–203.

DVOŘÁK, J. 1973. Synsedimentary tectonics of the Palaeozoic of the Drahany Upland (Sudeticum, Moravia, Czechoslovakia). *Tectonophysics*, **17**, 359–391.

DVOŘÁK, J. 1992. Mutual relation of pre-flysch and flysch facies in the Moravian Palaeozoic. *Vestník Českého geologického ústavu*, **67**, 363–369.

DVOŘÁK, J. 1995. VIII. Movaro-silesian zone. B. 1. Stratigraphy. *In:* DALLMEYER, R. D., FRANKE, W. & WEBER, K. (eds) *Pre-Permian Geology of Central and Eastern Europe*. Springer, Berlin, 477–489.

DVORAK, J. & PAPROTH, F. 1969. Über die Position und die Tektogenese des Rhenoherzynikums und des Sudetikums in den mitteleuropäischen Varisziden. *Neues Jahrbuch für Geologie und Paläontologie, Monatshefte*, **2**, 65–88.

ECHTLER, H. & CHAUVET, A. 1992. Carboniferous convergence and subsequent crustal extension in the southern Schwarzwald (SW Germany). *Geodinamica Acta*, **5**, 37–49.

ECHTLER, H. & MALAVIEILLE, J. 1990. Extensional tectonics, basement uplift and Stephano-Permian collapse basin in a late Variscan metamorphic core complex (Montagne Noire, Southern Massif Central). *Tectonophysics*, **177**(1–3), 125–138.

EDWARDS, M. A. & HARRISON, T. M. 1997. When did the roof collapse? Late Miocene northsouth extension in the Himalaya revealed by Th-Pb monazite dating of the Khula Kangri granite. *Geology*, **25**, 543–546.

EISBACHER, G. H, LÜSCHEN, E. & WICKERT, F. 1989. Crustal-scale thrusting and extension in the Hercynian Schwarzwald and Vosges, central Europe. *Tectonics*, **8**, 1–21.

EL SHAZLY, S. 1983. Petrography and geochemistry of the striated paragneiss series in the southwestern part of the Spessart crystalline complex. Dissertation, Technische Universität Braunschweig, Germany.

ELLENBERGER, F. & TAMAIN, A. L. G. 1980. Hercynian Europe. *Episodes*, 22–23.

ENDERLE, U., SCHUSTER, K., PRODEHL, C., SCHULZE, A., & BRIBACH, J. 1998. The refraction seismic experiment GRANU95 in the Saxothuringian belt, southeastern Germany. *Geophysical Journal International*, **133**(2), 245–259.

ENGEL, W. & FRANKE, W. 1983. Flysch sedimentation: its relations to tectonism in the European Variscides. *In:* MARTIN, H. & EDER, F. W. (eds) *Intracontinental Fold Belts*. Springer, Berlin, 289–322.

ENGEL, W., FRANKE, W. & LANGENSTRASSEN, F. 1983*a*. Palaeozoic sedimentation in the northern branch of the Mid-European Variscides; Essay of an interpretation. *In:* MARTIN, H. & EDER, F.W. (eds) *Intracontinental Fold Belts*. Springer, Berlin, 9–41.

ENGEL, W., FRANKE, W., GROTE, C., WEBER, K., AHRENDT, H. & EDER, F.W. 1983*b*. Nappe tectonics in the southeastern part of the Rheinisches Schiefergebirge. *In:* MARTIN, H. & EDER, F.W. (eds) *Intracontinental Fold Belts*. Springer, Berlin, 267–287.

ERNST, W. G. 2001. Subduction, ultrahigh-pressure metamorphism, and regurgitation of buoyant crustal slices – implications for arcs and continental growth. *Physics of the Earth and Planetary Interiors*, **127**(1–4), 253–275.

ERNST, W. G. & LIOU, J. G. 2000. Overview of UHP metamorphism and tectonics in well-studied collisional orogens. *International Book Series*, **4**, 3–19.

ESTRADA, S., GRUNEWALD, V. & WUNDERLICH, J. 1992. Zur Eduktcharakteristik und Lithostratigraphie von Truse- und Hohleborn-Serie (Ruhlaer Kristallin). *Neues Jahrbuch für Geologie und Paläontologie, Abhandlungen*, **184**, 389–419.

FAURE, M. 1995. Late orogenic carboniferous extensions in the Variscan French Massif Central. *Tectonics*, **14**(1), 132–153.

FAURE, M., MONIÉ, P, PIN, C., MALUSKI, H. & LELOIX, C. 2002. Late Visean thermal event in the northern part of the French Massif Central: new $^{40}Ar/^{39}Ar$ and Rb-Sr isotopic constraints on the Hercynian syn-orogenic extension. *International Journal of Earth Sciences*, **91**(1), 53–75.

FERNÁNDEZ-SUÁREZ J., DUNNING, G.R., JENNER, G.A. & GUTIERREZ-ALONSO, G. 2000. Variscan collisional magmatism and deformation in NW Iberia: constraints from U–Pb geochronology of granitoids. *Journal of the Geological Society, London*, **157**, 565–576.

FIALA, J., FUCHS, G. & WENDT, J. I. 1995. Moldanubian Region. C. Moldanubian Zone. 1. Stratigraphy. In: DALLMEYER, R. D., FRANKE, W. & WEBER, K. (eds) *Pre-Permian Geology of Central and Eastern Europe*, Springer, Berlin, 417–428.

FIELDING, E.J. 1996. Tibet uplift and erosion. *Tectonophysics*, **260**, 55–84.

FIELITZ, W. & MANSY, J.-L. 1999. Pre- and synorogenic burial metamorphism in the Ardenne and neighbouring areas (Rhenohercynian zone, central European Variscides). *Tectonophysics*, **309**(1–4), 227–256.

FINGER, F. & QUADT, A. VON. 1995. U/Pb ages of zircons from a plagiogranite gneiss in the south-eastern Bohemian Massif, Austria – further evidence for an important early Paleozoic rifting episode in the eastern Variszides. *Schweizer Mineralogische und Petrographische Mitteilungen*, **75**, 265–270.

FISCHER, G. 1929. Die Gesteine der metamorphen Zone von Wippra mit besonderer Berücksichtigung der Grünschiefer. *Abhandlungen der preußischen geologischen Landesanstalt, Neue Folge*, **121**, 1–64.

FLÖTTMANN, T. & KLEINSCHMIDT, G. 1989. Structural and basement evolution in the central Schwarzwald gneiss complex. *In:* EMMERMANN, R. & WOHLENBERG, J. (eds) *The German Continental Deep Drilling Program (KTB)*. Springer, Berlin, 266–275.

FLÖTTMANN, T. & ONCKEN, O. 1992. Constraints on the evolution of the Mid-German Crystalline Rise – a study of outcrops west of the river Rhine. *Geologische Rundschau*, **81**, 515–543.

FLOYD, P.A. 1995. Igneous activity. *In:* DALLMEYER, R.D., FRANKE, W. & WEBER, K. (eds) *Pre-Permian Geology of Central and Eastern Europe*. Springer, Berlin, 59–81.

FLOYD, P. A., LEVERIDGE, B. E., FRANKE, W., SHAIL, R. & DÖRR, W. 1990. Provenance and depositional environment of Rhenohercynian synorogenic greywackes from the Giessen Nappe, Germany. *Geologische Rundschau*, **79**, 611–626.

FLUCK, P. 1980. Métamorphisme et magmatisme dans les Vosges moyennes d'Alsace. *Science Géologique Bulletin*, **62**, 1–248.

FOURMARIER, P. 1931. Les plissements calédoniens et les plissements hercyniens en Belgique. *Annales de la Société Géologique de Belgique*, **34**, B364–384.

FRANCU, E., MANN, U., SUCHY, V. & VOLK, H. 1998. Model of burial and thermal history of the Tobolka-1 borehole profile in the Prague basin. *Acta Universitatis Carolinae – Geologica*, **42**, 248–249.

FRANKE, W. 1984. Variszischer Deckenbau im Raume der Münchberger Gneismasse - abgeleitet aus der Fazies, Deformation und Metamorphose im umgebenden Paläozoikum. *Geotektonische Forschungen*, **68**, 1–253.

FRANKE, W. 1989. Tectonostratigraphic units in the Variscan belt of Europe. *In:* DALLMEYER, R. D. (ed.) *Terranes in the circum-Atlantic Paleozoic Orogens*. Geological Society of America, Special Papers, **230**, 67–90.

FRANKE, W. 1992. Phanerozoic structures and events in central Europe. *In:* BLUNDELL, D., FREEMANN, R. & MÜLLER (eds) *A Continent Revealed – The European Geotraverse*. Cambridge University Press, 164–179.

FRANKE, W. 1995. Stratigraphy. *In:* DALLMEYER, R. D., FRANKE, W. & WEBER, K. (eds) *Pre-Permian Geology of Central and Eastern Europe*. Springer, Berlin, 33–49.

FRANKE, W. 2000. The mid-European segment of the Variscides: tectonostratigraphic units terrane boundaries and plate tectonic evolution. *In:* FRANKE, W., HAAK, V., ONCKEN, O. & TANNER, D. (eds) *Orogenic Processes: Quantification and Modelling in the Variscan Belt*. Geological Society, London, Special Publications, **179**, 35–61.

FRANKE, W. & ENGEL, W. 1982. Variscan Sedimentary Basins on the Continent, and relations with southwest England. *Proceedings of the Ussher Society*, **5**, 250–269.

FRANKE, W. & ONCKEN, O. 1995. Zur prädevonischen Geschichte des Rhenohercynischen Beckens. *Nova Acta Leopoldina, Neue Folge*, **71**, 53–72.

FRANKE, W. & STEIN, E. 2000. Exhumation of high-grade rocks in the Saxo-Thuringian Belt: geological constraints and geodynamic concepts, *In*: FRANKE, W., HAAK, V., ONCKEN, O. & TANNER, D. (eds) *Orogenic Processes: Quantification and Modelling in the Variscan Belt*. Geological Society, London, Special Publications, **179**, 337–354.

FRANKE, W. & ŻELAŹNIEWICZ, A. 2000. The eastern termination of the Variscides: terrane correlation and kinematic evolution. *In:* FRANKE, W., HAAK, V., ONCKEN, O. & TANNER, D. (eds) *Orogenic Processes: Quantification and Modelling in the Variscan Belt*. Geological Society, London, Special Publications, **179**, 63–86.

FRANKE, W. & ŻELAŹNIEWICZ, A. 2002. Structure and evolution of the Bohemian Arc. *In:* WINCHESTER, J. A., PHARAOH, T. C. & VERNIERS, J. (eds) *Palaeozoic Amalgamation of Central Europe*. Geological Society, London, Special Publications, **201**, 279–293.

FRANKE, W., DALLMEYER, R.D. & WEBER, K. 1995. Geodynamic Evolution. *In*: DALLMEYER, R.D., FRANKE, W. & WEBER, K. (eds) *Pre-Permian Geology of Central and Eastern Europe*. Springer, Berlin, 579–593.

FRANKE, W., HAAK, V., ONCKEN, O. & TANNER, D. 2000. Orogenic Processes: quantification and modelling in the Variscan Belt. *In:* FRANKE, W., HAAK, V., ONCKEN, O. & TANNER, D. (eds) *Orogenic Processes: Quantification and Modelling in the Variscan Belt*. Geological Society, London, Special Publications, **179**, 1–4.

FRANZKE, H.-J. & RAUCHE, H. 1991. Ablauf der rupturellen Deformationen an der Fränkischen Linie im Bereich des nordwestlichen Thüringer Waldes und Südthüringens. *KTB Report*, **91**, 21–39.

FRENZEL, G. 1971. Die Mineralparagenesen der Albersweiler Lamprophyre (Südpfalz). *Neues Jahrbuch für Mineralogie, Abhandlungen*, **115**, 164–191.

FRIEDEL, C.-H. 1996. *Tektonik und Metamorphose im östlichen Mittelharz.* Scientific Technical Report, GFZ Potsdam, **96/07**.

FRIEDL, G. 1997. *U–Pb-Datierungen an Zirkonen und Monaziten aus Gesteinen vom österreichischen Anteil der Böhmischen Masse.* PhD Thesis, University of Salzburg.

FRIEDL, G., FINGER, F., PAQUETTE, J.-L., QUADT, A. VON, MCNAUGHTON, N. J. & FLETCHER, I. R. 2004. Pre-Variscan geological events in the Austrian part of the Bohemian Massif deduced from U–Pb zircon ages. *International Journal of Earth Sciences*, **93**(5), 802–823.

FRISCH, W. & MESCHEDE, M. 2005. *Plattentektonik*. Primus, Darmstadt.

FUCHS, G. & MATURA, A. 1976. Die Geologie des Kristallins der südlichen Böhmischen Masse. *Jahrbuch der Geologischen Bundesanstalt*, **119**, 1–43.

FURNES, H., KRYZA, R., MUSZYŃSKI, A., PIN, C. & GARMANN, L. B. 1994. Geochemical evidence for progressive, rift-related early Paleozoic volcanism in the western Sudetes. *Journal of the Geological Society, London*, **151**, 91–109.

GALLE, A., HLADIL, J. & ISAACSON, P. E. 1995. Middle Devonian biogeography of closing South Laurussia-North Gondwana Variscides: examples from the Bohemian Massif (Czech Republic), with emphasis on Horní Benešov. *Palaios*, **10**, 221–239.

GEBAUER, D. & FRIEDL, G. 1994. A 1.38 Ga protolith age for the Dobra Orthogneiss (Moldanubian Zone of the southern Bohemian Massif, NE Austria): evidence from ion-microprobe (SHRIMP) dating of zircon. *Journal of the Czech Geological Society*, **39**, 34–35.

GEHMLICH, M., LINNEMANN, U., TICHOMIROWA, M., LÜTZNER, H. & BOMBACH, K. 1997. Die Bestimmung des Sedimentationsalters cadomischer Krustenfragmente im Saxothuringikum durch die Einzelzirkon – Evaporationsmethode. *Terra Nostra*, **5**, 46–49.

GEHMLICH, M., LINNEMANN, U., TICHOMIROWA, M., GAITZSCH, B., KRONER, U. & BOMBACH, K.. 2000 Geochronologie oberdevonischer bis unterkarbonischer Magmatite der Thüringischen und Bayerischen Faziesreihe sowie variszischer Deckenkomplexe und der Frühmolasse von Borna-Hainichen (Saxothuringisches Terrane). *Zeitschrift der deutschen geologischen Gesellschaft*, **151**, 337–363.

GERDES, A. 1997. *Geochemische und thermische Modelle zur Frage der spätorogenen Granitgenese am Beispiel des Südböhmischen Batholiths: Basaltische Underplating oder Krustenstapelung.* PhD thesis, Georg-August Universität, Göttingen.

GERDES, A. & ZEH, A. 2006. Combined U–Pb and Hf isotope LA-(MC-) ICP-MS analyses of detrital zircons: Comparison with SHRIMP and new constraints for the provenance and age of an Armorican metasediment in Central Germany. *Earth and Planetary Science Letters*, **249**, 47–61.

GERDES, A., WÖRNER, G. & HENK, A. 2000. Post-collisional granite generation and HT-LP metamorphism by radiogenic heating: the Variscan South Bohemian Batholith. *Journal of the Geological Society*, **157**(3), 577–587.

GHIENNE, J.-F. 2003. Late Ordovician sedimentary environments, glacial cycles, and post-glacial transgression in the Taoudeni Basin, West Africa. *Palaeogeography, Palaeoclimatology, Palaeoecology*, **189** (3–4), 117–145.

GIESE, U. 1996. Die Grenze Ossa Morena Zone/Südportugiesische Zone in SW Spanien: Geodynamik einer variszischen Sutur. *Aachener Geowissenschaftliche Beiträge (AGB)*, **20**, 1–157.

GLASMACHER, U. A., MANN, U. & WAGNER, G. A. 2002. Thermotectonic evolution of the Barrandian, Czech Republic, as revealed by apaptite fission-track analysis. *Tectonophysics*, **359**, 381–402.

GŁAZEK, J., KARWOWSKI, L., RACKI, G. & WRZOŁEK, T. 1981. The early Devonian continental/marine succession at Checiny in the Holy Cross Mountains, and its paleogeographic and tectonic significance. *Acta Geologica Polonica*, **34**, 233–254.

GOFETTE, O. 1989. Deformations caledonienne et varisque dans le massif de Rocroi; interet des felons magmatiques. *Bulletin de la Societe d'Histoire Naturelle des Ardennes*, **79**, 34–42.

GOLL, M. & LIPPOLT, H.-J. 2001. Biotit-Geochronologie ($^{40}Ar_{rad}/K$, $^{40}Ar_{rad}/^{39}Ar_K$, $^{87}Sr_{rad}/^{87}Rb$) spät-variszischer Magmatite des Thüringer Waldes. *Neues Jahrbuch für Geologie und Paläontologie, Abhandlungen*, **222**, 353–405.

GRAD, M., GUTERCH, A. & MAZUR, S. 2002. Seismic refraction evidence for crustal structure in the central part of the Trans-European Suture Zone in Poland. *In:* WINCHESTER J.A., PHAROAH T.C. & VERNIERS J. (eds) *Palaeozoic Amalgamation of Central Europe*. Geological Society, London, Special Publications, **201**, 295–309.

GRAUERT, B., GROSSE-WESTERMANN, U. & ALBAT, F. 1990. Interpretation konkordanter U-Th-Pb-Monazitalter moldanubischer Gneise – Schlussfolgerungen aufgrund vergleichbarer U-Th-Pb, Sm-Nd- und Rb-Sr-Isotopenanalysen. *Beihefte, European Journal of Mineralogy*, **2**, 82.

GRAULICH, J.-M. 1961. Le sondage de Wépion. *Service Gélogique de Belgique*, **2**, 1–86.

GROCHOLSKI, W. 1975. Variscides of southern Wielkopolska. *Przegląd Geologiczny*, **4**, 171–174 (in Polish, English summary).

GRÖSSER, J. & DÖRR, W. 1986. MOR-Basalte im östlichen Rheinischen Schiefergebirge. *Neues Jahrbuch für Geologie und Paläontologie, Monatshefte*, **12**, 705–722.

GRYGAR, R. & VAVRO, M. 1995. Evolution of Lugosilesian Orocline (North-eastern periphery of the Bohemian Massif): Kinematics of Variscan deformation. *Journal of the Czech Geological Society*, **40**(1–2), 65–90.

GÜLDENPFENNIG, M. 1997. Geologische Neuaufnahme der Zone von Badenweiler-Lenzkirch – (Südschwarzwald) unter besonderer Berücksichtigung unterkarbonischer Vulkanite und Grauwacken. *Tübinger Geowissenschaftliche Arbeiten*, A **32**, 1–120.

GÜLDENPFENNIG, M. 1998. Zur geotektonischen Stellung unterkarbonischer Grauwacken und Vulkanite der Zone von Badenweiler-Lenzkirch (Südschwarzwald). *Zeitschrift der deutschen geologischen Gesellschaft*, **149**, 213–232.

GUNIA, T. 1984. Microfossils from the quartzitic schists in vicinity of Goszów, Śnieżnik Kłodzki Massif, Central Sudetes. *Geologia Sudetica*, **18**(2), 47–57 [in Polish, English summary].

GUNIA, T. 1985. Geological position of the Sowie Góry block and its influence on the palaeography of the Palaeozoic complexes of Central Sudetes. *Geologia Sudetica*, **20**, 83–119 [in Polish, English summary].

GUNIA, T. 1990. Acritarcha and microproblematica of the crystalline limestones from the vicinity of Romanowo Górne (Central Sudetes Mts, Krowiarki). *Geologia Sudetica*, **24**(1–2), 101–137 [in Polish with English summary].

GÜNTHER, K. & HEIN, S. 1999. Die Olistostrome des Mittelharzes nordwestlich von Bad Lauterberg, eine Folge akkretionärer Prozesse. *Neues Jahrbuch für Geologie und Paläontologie, Abhandlungen*, **211**, 355–410.

GUTIERREZ ALONSO, G., FERNANDEZ-SUAREZ, J. & WEIL, A. B. 2004. Orocline triggered lithospheric delamination. *In:* SUSSMAN, A. J. & WEIL, A. B. (eds) Orogenic curvature: integrating paleomagnetic and structural analyses. Geological Society of America, Special Papers, **383**, 121–130.

HAHN, T. 2003. *Geologische Karte von Thüringen, Blatt 5339 Greiz*, scale 1:25 000. Thüringer Landesanstalt für Umwelt und Geologie, Jena.

HAHN, T. & MEINHOLD, G. 2005. Revision der Unterkarbonstratigraphie der Mehltheuerer Mulde (Vogtland, Saxothuringikum): *Geowissenschaftliche Mitteilungen von Thüringen*, **12**, 51–70.

HAMMER, J., EIDAM, J., RÖBER, B. & EHLING, B.-C. 1999. Prävariszischer und variszischer granitoider Magmatismus am NE-Rand des Böhmischen Massivs – Geochemie und Petrogenese. *Zeitschrift für geologische Wissenschaften*, **27**, 401–415.

HAMMERSCHMIDT, K., KOPP, J. & HAUCKE, L. 2003. Sm-Nd dating on minerals of a silica undersaturated, Ca-rich, crustally derived gabbro of the Mid German crystalline zone, drill hole Züllsdorf, Southern Brandenburg (Germany). *Zeitschrift für geologische Wissenschaften*, **31**, 225–238.

HANEL, M. & WIMMENAUER, W. 1990. Petrographische Indizien für einen Deckenbau im Kristallin des Schwarzwaldes. *Beihefte, European Journal of Mineralogy*, **2**, 89.

HANEL, M., LIPPOLT, H. J., KOBER, B. & WIMMENAUER, W. 1993. Lower Carboniferous granulites in the Schwarzwald basement near Hohengeroldseck (SW Germany). *Naturwissenschaften*, **80**(1), 25–28.

HANEL, M., MONTENARI, M. & KALT, A. 1999. Determing sedimentation ages of high-grade metamorphic gneisses by their palynological record: a case study in the northern Schwarzwald (Variscan Belt, Germany). *International Journal of Earth Sciences*, **88**, 49–59.

HANN, H. P & SAWATZKI, G. 1998. Deckenbau und Sedimentationsalter im Grundgebirge des Südschwarzwaldes/SW-Deutschland. *Zeitschrift der deutschen geologischen Gesellschaft*, **149**, 183–195.

HANN, H. P. & SAWATZKI, G. 2000. Neue Daten zur Tektonik des Südschwarzwaldes. *Jahrbuch der Mitteilungen des Oberrheinischen Geologischen. Vereins, Neue Folge*, **82**, 363–376.

HANN, H. P. & SAWATZKI, G. 2002. *Geologische Karte von Baden-Württemberg 1:50000, Badenweiler-Lenzkirch-Zone.* Landesamt für Geologie, Freiburg im Breisgau.

HANN, H. P, SAWATZKI, G. & VAIDA, M. 1995. Acritarchen und Chitinozoen des Ordoviziums aus metamorphen Grauwacken der Zone von Badenweiler-Lenzkirch, Schwarzwald, SW-Deutschland. *Neues Jahrbuch für Geologie und Paläontologie, Mitteilungshefte*, **6**, 375–383.

HANN, H. P., CHEN, F., ZEDLER, H., FRISCH, W. & LOESCHKE, J. 2003*a*. The Rand Granite in the southern Schwarzwald and its geodynamic significance in the Variscan belt of SW Germany. *International Journal of Earth Sciences*, **92**, 821–842.

HANN, H. P., CHEN, F., ZEDLER, H. & SAWATZKI, G. 2003*b*. Zircon ages and geochemistry of metavolcanic layers from the northern Badenweiler-Lenzkirch Zone (southern Schwarzwald, Germany). *Neues Jahrbuch für Geologie und Paläontologie, Abhandlungen*, **230**, 451–469.

HANSCH, R. & ZEH, A. 2000. Metabasites from the Ruhla Crystalline Complex: Evidence for distinct pre-Variscan, plate-tectonic environments within the Mid-German Crystalline Rise. *Chemie der Erde*, **60**, 1–25.

HARTLEY, A. J. & OTAVA, J. 2001. Sediment provenance and dispersal in a deep marine foreland basin: the Lower Carboniferous Culm Basin, Czech Republic. *Journal of the Geological Society, London*, **158**, 137–150.

HATCHER, R. D., JR. 2002. Alleghanian (Appalachian) orogeny, a product of zipper tectonics: Rotational transpressive continent-continent collision and closing of ancient oceans along irregular margins. *In:* MARTINEZ CATALAN, J. R., HATCHER, R. D., JR., ARENAS, R. & DIAZ GARCIA, F. (eds) *Variscan-Appalachian Dynamics: The Building of the Late Paleozoic Basement.* Geological Society of America, Special Papers, **364**, 199–208.

HAVLICEK, V. 1963. Tektogenetische Störungen des älteren Paläozoikums im Barrandium. *Sbornik geologickych Ved, Geologie*, **1**, 77–102.

HAVLICEK, V. 1998. Variscan folding and faulting in the Lower Paleozoic basins. *In:* CHLUPAČ, I., HAVLICEK, V., KRIZ, J., KUKAL, Z. & ŠTORCH, P. (eds) *Paleozoic of the Barrandian (Cambriam to Devonian)*. Czech Geological Survey, Prague, 165–169.

HAYDUKIEWICZ, J. 1990. Stratigraphy of Paleozoic rocks of the Góry Bardzkie and some remarks on their sedimentation. *Neues Jahrbuch für Geologie und Paläontologie, Abhandlungen*, **179**, 275–284.

HAYDUKIEWICZ J., MUSZER J. & KŁAPCIŃSKI J. 1999. Palaeontological documentation of the sub-Permian sediments in the vicinity of Zbąszyń (Fore-Sudetic Monocline). *In:* MUSZER A. (ed.) *Selected Problems of Stratigraphy, Tectonics and Ore Mineralization in Lower Silesia.* Department of Geological Sciences, University of Wrocław, 7–17 [in Polish].

HEGNER, E., CHEN, F. & HANN, H. P. 2001. Chronology of basin closure and thrusting in the internal zone of the Variscan belt in the Schwarzwald, Germany: evidence from zircon ages, trace element geochemistry, and Nd isotopic data. *Tectonophysics*, **332**, 169–184.

HEGNER, E., GRULER, M., HANN, H. P., CHEN, F. & GÜLDENPFENNIG, M. 2005. Testing tectonic models with geochemical provenance parameters in greywacke. *Journal of the Geological Society, London*, **162**, 87–96.

HELLMANN, K.-N., LIPPOLT, H. J. & TODT, W. 1982. Interpretation der Kalium-Argon-Alter eines Odenwälder Granodioritporphyritganges und seiner Nebengesteine. *Aufschluss*, **33**, 155–164.

HENES-KLAIBER, U. 1992. *Zur Geochemie der variskischen Granite des Bergsträßer Odenwaldes*. Dissertation, Universität, Karlsruhe, Germany.

HENK, A. 1993. Subsidenz und Tektonik des Saar-Nahe Beckens (SW-Deutschland). *Geologische Rundschau*, **82**, 3–19.

HESS, J. & LIPPOLT, H. J. 1996. Numerische Stratigraphie permokarbonischer Vulkanite Zentraleuropas. Teil III: Odenwald. *Geologisches Jahrbuch Hessen*, **117**, 69–77.

HESS, J. & SCHMIDT, G. 1989. Zur Altersstellung der Kataklasite im Bereich der Otzberg-Zone, Odenwald. *Geologisches Jahrbuch Hessen*, **117**, 69–77.

HESS, J., LIPPOLT, H. J. & KOBER, B. 1995. The age of the Kragenfels granite (northern Vosges) and its bearing on the intrusion scheme of late Variscan granitoids. *Geologische Rundschau*, **84**, 568–577.

HESS, J. C., HANEL, M., ARNOLD, M., GAISER, A., PROWATKE, S., STADLER, S. & KOBER, B. 2000. Variscan magmatism in the northern part of the Moldanubian Vosges and Schwarzwald. I. Ages of intrusion and cooling history. *Berichte der Deutschen Mineralogischen Gesellschaft, Beiheft 1 zum European Journal of Mineralogy*, **12**, 79.

HEUER, B. 2006. *Lithospheric and upper mantle structure beneath the western Bohemian Massif obtained from teleseismic P and S receiver functions*. Scientific Technical Report, GFZ Potsdam, **06/12**.

HIRSCHMANN, G. & OKRUSCH, M. 1988. Spessart-Kristallin und Ruhlaer Kristallin als Bestandteil der Mitteldeutschen Kristallinzone. *Neues Jahrbuch für Geologie und Paläontologie, Abhandlungen*, **177**, 1–39.

HIRSCHMANN, G. & OKRUSCH, M. 2001. Spessart und Rhön – Teil der MKZ. *In:* STEININGER, F. (ed.) *Stratigraphische Kommission Deutschland, Stratigraphie von Deutschland II, Ordovizium, Kambrium, Vend, Riphäikum*. Courier Forschungs Institut Senckenberg, Frankfurt, **234**, 93–108.

HLADIL, J. 1986. Trends in the development and cyclic patterns of Middle and Upper Devonian buildups. *Facies*, **15**, 1–34.

HLADIL, J., MAZUR, S., GALLE, A. & EBERT, J. 1999. Revised age of the Mały Bożków limestone in the Kłodzko metamorphic unit (Early Givetian, late Middle Devonian): implications for the geology of the Sudetes. *Neues Jahrbuch für Geologie und Paläontologie, Abhandlungen*, **211**, 329–353.

HOENES, D. 1948. Petrogenese im Grundgebirge des südlichen Schwarzwaldes. *Heidelberger Beiträge zur Mineralogie und Petrolologie*, **1**, 121–202.

HOLDER, M. T. & LEVERIDGE, B. 1986. Correlation of the Rhenohercynian Variscides. *Journal of the Geological Society, London*, **143**, 141–147.

HOLUB, F., COCHERIE, A. & ROSSI, P. H. 1997. Radiometric dating of granitic rocks from the Central Bohemian Plutonic Complex (Czech Republic): constraints on the chronology of thermal and tectonic events along the Moldanubian-Barrandian boundary. *Comptes Rendus Académie des Sciences Paris, Sciences de la terre et des planetes / Earth and Planetary Sciences*, **325**, 19–26.

HRUBCOVÁ, P., ŚRODA, P., *et al.* 2005. Crustal and uppermost mantle structure of the Bohemian Massif based on ELEBRATION 2000 data, *Journal of Geophysical Research*, **110**, B11305. DOI: 10.1029/2004JB003080.

HUCKRIEDE, H., WEMMER, K. & AHRENDT, H. 2004. Palaeogeography and tectonic structure of allochthonous units in the German part of the Rheno-Hercynian Belt (Central European Variscides). *International Journal of Earth Sciences*, **93**, 414–431.

JANIK, T., YLINIEMI, J., GRAD, M., THYBO, H., TIIRA, T. & POLONAISE P2 WORKING GROUP. 2002. Crustal structure across the TESZ along POLONAISE '97 seismic profile P2 in NW Poland. *Tectonophysics*, **360**, 129–152.

JANOUSEK, V, FINGER, F, ROBERTS, M. J., FRYDA, J., PIN, C. & DOLEJS, D. 2004. Deciphering the petrogenesis of deeply buried granites; whole-rock geochemical constraints on the origin of largely undepleted felsic granulites from the Moldanubian Zone of the Bohemian Massif. *Transactions of the Royal Society of Edinburgh: Earth Sciences*, **95**(1–2), 141–159.

JANOUSEK, V, GERDES, A., VRANA, S., FINGER, F, ERBAN, V, FRIEDL, G. & BRAITHWAITE, C. J. R. 2006. Low-pressure granulites of the Lisov Massif, southern Bohemia: Visean metamorphism of Late Devonian plutonic arc rocks. *Journal of Petrology*, **47**(4), 705–744.

KALT, A. 1991. Isotopengeologische Untersuchungen an Metabasiten des Schwarzwaldes und ihren Rahmengesteinen. *Freiburger geowissenschaftliche Beiträge*, **3**, 1–85.

KALT, A. & ALTHERR, R. 1996. Metamorphic evolution of garnet-spinel peridotites from the Variscan Schwarzwald (Germany). *Geologische Rundschau*, **85**, 211–224.

KALT, A., GRAUERT, B. & BAUMANN, A. 1994*a*. Rb-Sr and U–Pb isotope studies on migmatites from the Schwarzwald (Germany): constraints on isotopic resetting during Variscan high-temperature metamorphism. *Journal of Metamorphic Geology*, **12**(1), 667–680.

KALT, A., HANEL, M. & SCHLEICHER, H. 1994*b*. Petrology and geochronology of eclogites from the Variscan Schwarzwald (FRG). *Contributions to Mineralogy and Petrology*, **115**, 287–302.

KALT, A., ALTHERR, R. & HANEL, M. 2000*a*. The Variscan basement of the Schwarzwald. *European Journal of Mineralogy*, **12**, 1–43.

KALT, A., CORFU, F. & WIJBRANS, J.R. 2000*b*. Time calibration of a P-T path from a Variscan high-temperature low-pressure metamorphic complex (Bayerischer Wald, Germany), and the detection of inherited monazite. *Contributions to Mineralogy and Petrology*, **138**, 143–163.

KALVODA, J., BÁBEK, O. & MALOVANÁ, A. 1999. Sedimentary and biofacies records in calciturbidites at the Devonian-Carboniferous boundary in Moravia (Moravian-Silesian Zone, Middle Europe). *Facies*, **41**, 141–158.

KALVODA, J., MELICHAR, R., BÁBEK, O. & LEICHMANN, J. 2002. Late Proterozoic–Paleozoic tectonostratigraphic development and paleogeography of Brunovistulian terrane and comparison with other terranes at the SE margin of Baltica-Laurussia. *Journal of the Czech Geological Society*, **47**(3), 81–102.

KEGEL, W. 1953. Das Paläozoikum der Lindener Mark bei Giessen. *Abhandlungen des hessischen Landes-Amtes für Bodenforschung*, **7**, 1–55.

KEGLER, J. 1967. Die Gaudernbacher Schichten (westliche Lahnmulde, Rheinisches Schiefergebirge). Berichte der oberhessischen Gesellschaft für Natur- und Heilkunde Gießen, Neue Folge, Naturwissenschaftliche Abteilung, **35**, 23–37.

KEMNITZ, H. & BUDZINSKI, G. 1994. Die Grauwacken des Lausitz und ihre cadomische Prägung. *Abhandlungen des Staatlichen Museums für Mineralogie und Geologie Dresden*, **40**, 37–98.

KIND, R., NI, J., *et al.* 1996. Evidence from earthquake data for a partially molten crustal layer in southern Tibet. *Science*, **274**, 1692–1691.

KIRSCH, H., KOBER, B. & LIPPOLT, H. J. 1988. Age of intrusion and rapid cooling of the Frankenstein gabbro (Odenwald, SW Germany) evidenced by $^{40}Ar/^{39}Ar$ and single zircon $^{207}Pb/^{206}Pb$ measurements. *Geologische Rundschau*, **77**, 693–711.

KLAPOVA, H., KONOPASEK, J. & SCHULMANN, K. 1998. Eclogites from the Czech part of the Erzgebirge: multi-stage metamorphic and structural evolution. *Journal of the Geological Society*, **155**(3), 567–583.

KLEIN, H. & WIMMENAUER, W. 1984. Eclogites and their retrograde transformation in the Schwarzwald. *Neues Jahrbuch für Mineralogie, Monatshefte*, **1984**, 25–38.

KLEMD, R., MATHES, S., & OKRUSCH, M. 1991. High-pressure relics in meta-sediments intercaleted with the Weissenstein eclogite, Münchberg gneiss complex, Bavaria. *Contributions to Mineralogy and Petrology*, **107**, 328–342.

KLEMM, D. D. & WEBER-DIEFENBACH, K. 1971. Geochemische Untersuchungen an Amphiboliten und Dioriten des nördlichen Odenwaldes. *Neues Jahrbuch für Mineralogie, Abhandlungen*, **133**, 149–173.

KLOMINSKÝ, J. (ed.) 1994. *Geological Atlas of the Czech Republic, Stratigraphy*. Czech Geological Survey, Praha.

KLÜGEL, T. 1997. Geometrie und Kinematik einer variszischen Plattengrenze. *Geologische Abhandlungen Hessen*, **101**, 1–215.

KNAUER, E., OKRUSCH, M., RICHTER, P., SCHMIDT, K. & SCHUBERT, W. 1974. Die metamorphe Basit-Ultrabasit-Assoziation in der Böllsteiner Gneiskuppel, Odenwald. *Neues Jahrbuch für Mineralogische Abhandlungen*, **122**, 186–228.

KNEIDL, V. KREBS, W. & MAASS, R. 1982. Über Conodontenfunde im Oberdevon von Tunau (Südschwarzwald). *Neues Jahrbuch für Geologie und Paläontologie, Mitteilungshefte*, **1**, 25–35.

Kober, B., Kalt, A., Hanel, M. & Pidgeon, R. T. 2004. SHRIMP dating of zircons from high-grade metasediments of the Schwarzwald/SW-Germany and implications for the evolution of the Moldanubian basement. *Contributions to Mineralogy and Petrology*, **147**, 330–345.

Köhler, R., Becker, H., Linder, H., Schenke, G. & Schied, R. 1989. *Geophysikalische Komplexinterpretation des Zentralteils der Mitteldeutschen Schwelle im Maßstab 1: 100 000*. VEB Geophysik Leipzig. Standort: Archiv der BGR, Außenstelle Berlin.

Kollmeier, J. M., van der Pluijm, B. A. & Van der Voo, R. 2000. Analysis of Variscan dynamics: early bending of the Cantabria-Asturias Arc, northern Spain. *Earth and Planetary Science Letters*, **181**(1–2), 203–216.

Konopasek, J., Schulmann, K. & Lexa, O. 2001. Structural evolution of the central part of the Krusne hory (Erzgebirge) Mountains in the Czech Republic – evidence for changing stress regime during Variscan compression. *Journal of Structural Geology*, **23**(9), 1373–1392.

Kopp, J. & Bankwitz, P. 2003. Die Europäische Kristallinzone – eine Übersicht. *Zeitschrift für geologische Wissenschaften*, **31**, 179–196.

Kopp, J., Bankwitz, P., Bankwitz, E. & Tichomirowa, M. 1998. Variscan orthogneisses in the eastern part of the Mid-German Crystalline Zone (MGCZ). *Acta Universitatis Carolinae-Geologica, Praha*, **42**, 451–458.

Kopp, J., Bankwitz, P., Bankwitz, E., Ehling, B. & Tichomirowa, M. 1999. Ortho- und paragneisses in the eastern part of the Mid-German Crystalline Zone. *Terra Nostra*, **99**(1), 122–123.

Kossmat, F. 1927. Gliederung des varistischen Gebirgsbaues. *Abhandlungen des Sächsischen Geologischen Landesamtes*, **1**, 1–39.

Kotkova, J., Schaltegger, U. & Leichmann, J. 2003. 338–335 Ma old intrusions in the E-Bohemian massif – a relic of the orogen wide durbachitic magmatism in the European Variscides. *Journal of the Czech Geological Society*, **48**(1), 80–81.

Kowalczewski, Z., 1971. Basic geological problems of the Lower Devonian in the Holy Cross Mts. *Kwartalnik Geologiczny*, **15**(2), 263–283 [in Polish with English summary].

Krawczyk, C. M., McCann, T., Cocks, L. R. M., England, R., McBride, J. & Wybraniec, S. 2008. Caledonian tectonics. *In*: McCann, T. (ed.) *The Geology of Central Europe. Volume 1: Precambrian and Palaeozoic*. Geological Society, London, 303–381.

Krecher, M. 1999. Ein Beitrag zur Kenntnis des Kulmkonglomerates in der westlichen Badenweiler-Lenzkirch Zone (Unterkarbon, Südschwarzwald). *Berichte der Naturforschenden. Gesellschaft, Freiburg im Breisgau*, **88/89**, 277–296.

Kreher, B. 1994. Petrologie und Geochemie der Grabbrointrusionen des Frankensteins (Odenwald). *Geologisches Jahrbuch Hessen*, **122**, 81–122.

Kreuzer, H. & Harre, W. 1975. K/Ar-Altersbestimmungen an Hornblenden und Biotiten des kristallinen Odenwaldes. *Aufschluß, Sonderband*, **27**, 71–77.

Kreuzer, H., Seidel, E., Schüssler, U., Okrusch, M., Lenz, L.-L. & Raschka, H. 1989. K-Ar geochronology of different tectonic units at the northwestern margin of the Bohemian Massif. *Tectonophysics*, **157**, 149–178.

Kreuzer, H., Müller, P., *et al.* 1990. Ar-Ar conformation for Cambrian, Early Devonian, and Mid-Carboniferous tectonic units at the western margin of the Bohemian Massif. *6 Rundgespräch Geodynamik des europäischen Variszikums*, 15–18.11.1990, Clausthal-Zellerfeld (abstract).

Krohe, A. 1991. Emplacement of synkinematic plutons in the Variscan Odenwald (Germany) controlled by transtensional tectonics. *Geologische Rundschau*, **80**, 391–409.

Krohe, A. 1992. Structural evolution of intermediate-crustal rocks in a strike-slip and extensional setting (Variscan Odenwald, SW Germany): differential upward transport of metamorphic complexes and changing transformation mechanisms. *Tectonophysics*, **205**, 357–386.

Krohe, A. 1996. Variscan tectonics of central Europe: Post accretionary intraplate deformation of weak continental lithosphere. *Tectonics*, **15**, 1364–1388.

Krohe, A. 1998. Extending a thickened crustal bulge: toward a new geodynamic evolution model of the paleozoic NW Bohemian Massif, German Continental Deep Drilling site (SE Germany). *Earth-Science Reviews*, **44**(3–4), 95–145.

Krohe, A. & Eisbacher, G. H. 1988. Oblique crustal detachment in the Variscan Schwarzwald, southwestern Germany. *Geologische Rundschau*, **77**, 25–43.

Krohe, A. & Willner, A. 1995. The Odenwald Cystalline Complex. *In:* Dallmeyer, R.D., Franke, W. & Weber, K. (eds) *Pre-Permian Geology of Central and Eastern Europe*. Springer, Berlin, 174–181.

Krohe, A. & Willner, A. P. 2003. Discussion on: "Structural evolution of the central part of the Krusne Hory (Erzgebirge) Mountains in the Czech Republic – evidence for changing stress regime during the Variscan" by J. Konopasek, K. Schulmann & O. Lexa: [J. Struct. Geol. 23 (2001), 1373–1392]. *Journal of Structural Geology*, **25**(6), 1001–1004.

Kröner, A. & Willner, A. P. 1998. Time of formation and peak of Variscan HP-HT metamorphism of quartz-feldspar rocks in the central Erzgebirge, Saxony, Germany. *Contributions to Mineralogy and Petrology*, **132**, 1–20.

Kröner, A., Wendt, J. I., *et al.* 1988. U–Pb zircon and Sm-Nd model ages of high-grade Moldanubian metasediments, Bohemian Massif, Czechoslovakia. *Contributions to Mineralogy and Petrology*, **99**, 257–266.

Kröner, A., Hegner, E., Hammer, J., Haase, G., Bielicki, K. H., Krauss, M. & Eidam, J. 1994. Geochronology and Nd-Sr systematics of Lusatiann granitoids: significance for the evolution of the Variscan orogen in east-central Europe. *Geologische Rundschau*, **83**, 375–376.

Kröner, A., Jaeckel, P., Reischmann, T. & Kroner, U. 1998. Further evidence for an early Carboniferous (~340 Ma) age of highgrade metamorphism in the Saxothuringian Granulite Complex. *Geologische Rundschau*, **86**, 751–766.

Kröner, A., O'Brien, P. J. & Pidgeon, R. J. 2000*a*. Zircon ages for high-pressure granulites from south Bohemia, Czech Republic, and their connections to Carboniferous high-temperature processes. *Contributions to Mineralogy and Petrology*, **138**(2), 127–142.

Kröner, A., Štípská, P., Schulmann, K. & Jaeckel, P. 2000*b*. Chronological constraints on the pre-Variscan evolution of the northeastern margin of the Bohemian Massif, Czech Republic. *In:* Franke, W., Haak, V., Oncken, O. & Tanner, D. (eds) *Orogenic Processes: Quantification and Modelling in the Variscan Belt*. Geological Society, London, Special Publications, **179**, 175–197.

Kröner, A., Jaeckel, P., Hegner, E. & Opletal, M. 2001. Single zircon ages and whole-rock Nd isotopic systematics of early Palaeozoic granitoid gneisses from the Czech and Polish Sudetes (Jizerské hory, Krkonoše and Orlice-Sněžnik Complex). *International Journal of Earth Sciences*, **90**, 304–324.

Kroner, U. 1995. Postkollisionale Extension am Nordrand der Böhmischen Masse - Die Exhumierung des Sächsischen Granulitgebirges. *Freiberger Forschungshefte, C*, **457**, 1–114.

Kroner, U. & Hahn, T. 2004. Sedimentation, Deformation und Metamorphose im Saxothuringikum während der variszischen Metamorphose – Die komplexe Entwicklung Nord-Gondwanas während kontinentaler Subduktion und schiefer Kollision. *Geologica Saxonica*, **48/49**, 133–146.

Kroner, U., Hahn, T., Romer, R. L. & Linnemann, U. 2007. The Variscan orogeny in the Saxo-Thuringian Zone – Heterogenous overprint of Cadomian/Palaeozoic Peri-Gondwana crust. *In:* Linnemann, U., Nance, R. D., Kraft, P. & Zulauf, G. (eds) *The Evolution of the Rheic Ocean: From Avalonian-Cadomian Active Margin to Alleghenian-Variscan Collision*. Geological Society of America, Special Papers, **423**, 153–172.

Krytosek, K. & Paulitsch, P. 1981. Korund-Orientierung im Gabbrokontakt. *Geologisches. Jahrbuch Hessen*, **109**, 133–136.

Kryza, R & Fanning, M. 2004. The Góry Sowie granulites (Sudetes, SW Poland): preliminary results of SHRIMP U–Pb zircon geochronology. *International Workshop on Petrogenesis of Granulites & Related Rocks*, Náměšt 1–3.10.2004, Abstract Volume, Moravian Museum, Brno, 45–46.

Kryza, R. & Mazur, S. 1995. Contrasting metamorphic paths in the SE part of the Karkonosze-Izera block (Western Sudetes, SW Poland). *Neues Jahrbuch für Mineralogie, Abhandlungen*, **169**, 157–192.

Kryza, R., Muszyński, A. & Vielzeuf, D. 1990. Glaucophane-bearing assemblage overprinted by greenschist-facies metamorphism in the Variscan Kaczawa complex, Sudetes, Poland. *Journal of Metamorphic Geology*, **8**, 345–355.

Kryza, R., Mazur, S. & Pin, C. 1995. Leszczyniec meta-igneous complex in the eastern part of the Karkonosze-Izera Block, Western Sudetes: trace element and Nd isotope study. *Neues Jahrbuch für Geologie und Paläontologie, Abhandlungen*, **170**, 59–74.

Kryza, R., Pin, C. & Vielzeuf, D. 1996. High pressure granulites from the Sudetes (SW Poland): evidence of crustal subduction and collisional thickening in the Variscan Belt. *Journal of Metamorphic Geology*, **14**, 531–546.

Kryza, R., Mazur, S. & Aleksandrowski, P. 1999. Pre-Late Devonian unconformity in the Kłodzko area excavated: a record of Eo-Variscan metamorphism and exhumation in the Sudetes. *Geologia Sudetica*, **32**, 127–137.

Książkiewicz, M., Samsonowicz, J. & Rühle, E. 1965. *Outline of the Geology of Poland*. Wydawnictwa Geologiczne, Warszawa [in Polish].

Kuhn, W. 2001. Pfälzer Wald, Haard (19)-Teil der MKZ. *In:* Steininger, F. (ed.) *Stratigraphische Kommission Deutschland, Stratigraphie von Deutschland II, Ordovizium, Kambrium, Vend, Riphäikum*. Courier Forschungs Institut Senckenberg, Frankfurt, **234**, 109–112.

Kumpera, O. & Martinec, P. 1995. The development of the Carboniferous accretionary wedge in the Moravian-Silesian Paleozoic Basin. *Journal of the Czech Geological Society*, **40**(1–2), 47–63.

Küstner, W. 2000. *K-Ar-Datierungen an detritischen Muskoviten und Sm-Nd-Modellalter prä- und synorogener schwach metamorpher Sedimente des Rhenoherzynikum – Grundlegende Daten zur Quantifizierung orogener Prozesse am Beispiel der Varisziden*. PhD thesis, Göttingen.

Küstner, W., Ahrendt, H., Hansen, B. T. & Wemmer, K. 1998. ^{40}K/^{40}Ar-Datierungen an detritischen Muskoviten und Sm-Nd-Modellalter prä- und synorogener schwach metamorpher Sedimente im Rhenohercynikum - Grundlegende Daten zur Quantifizierung orogener Prozesse am Beispiel der Varisziden. *Terra Nostra*, **98**(2), 96–98.

Kutek, J. & Głazek, J. 1972. The Holy Cross Area, Central Poland, in the Alpine Cycle. *Acta Geologica Polonica*, **22**(4), 603–653.

Lacquement, F. 2001. *L'Ardenne Varisque. Déformation progressive d'un prisme sédimentaire pré-structuré, de l'affleurement au modèle de chaine*. Publication 29, Société géologique du Nord.

Lacquement, F., Averbuch, O. & Mansy, J. L. 1999. Geometry and strain distribution within a major lateral zone of the Variscan Ardennes fold and thrust belt (N France–S Belgium). *XXIV General Assembly of EUG*, Den Hague, 28.

Lacquement, F., Averbuch, O., Mansy, J. L., Szaniawski, R. & Lewandowski, M. 2005. Transpressional deformations at lateral boundaries of propagating thrust-sheets: the example of the Meuse Valley Recess within the Ardennes Variscan fold-and-thrust belt (N France–S Belgium). *Journal of Structural Geology*, **27**(10), 1788–1802.

Lamarche, J., Mansy, J-L, *et al.* 1999. Variscan tectonics in the Holy Cross Mountains (Poland) and role of the strutural inheritance during the Alpine tectonics. *Tectonophysics*, **313**, 171–186.

Lamarche, J., Lewandowski, M., Mansy, J-L. & Szulczewski, M., 2003*a*. Partitioning pre-, syn- and post-Variscan deformation in the Holy Cross Mts., eastern Variscan foreland. *In*: McCann, T. & Saintot, A. (eds) *Tracing Tectonic Deformation using the Sedimentary Record*. Geological Society, London, Special Publications, **208**, 159–184.

Lamarche, J., Scheck, M. & Lewerenz, B. 2003*b*. Heterogeneous tectonic inversion of the Mid-Polish Basin related to crustal architecture, sedimentary patterns and structural inheritance. *Tectonophysics*, **373**, 75–92.

Lange, G., Motz, H., Reichardt, C. & Schmidt, H. 1999. Stratigraphie und Tektonik nördlich und nordöstlich von Ronneburg. *Beiträge zur Geologie von Thüringen, neue Folge*, **6**, 139–157.

Langenstrassen, F. 1983. Neritic sedimentation of the Lower and Middle Devonian in the Rheinisches Schiefergebirge east of the river Rhine. *In:* Martin, H. & Eder, F. W. (eds) *Intracontinental Fold Belts*. Springer, Berlin, 43–77.

Laumonier, B., Autran, A., Barbey, P., Cheilletz, A., Baudin, T., Cocherie, A. & Guerrot, C. 2004. Consequences de l'absence de socle cadomien sur l'age et la signification des series pre-varisques (ante-Ordovicien superieur) du sud de la France (Pyrenees, Montagne Noire). *Bulletin de la Societe Geologique de France*, **175**(6), 643–655.

Lauterbach, R. 1953. Beiträge zur tektonischen Deutung der geomagnetischen Übersichtskarte der Deutschen Demokratischen Republik. *Wissenschaftliche Zeitschrift der Universität Leipzig, mathematisch-naturwissneschaftliche Reihe*, **3**, 271–279, Karte 11.

Ledru, P. 2001. (ed.) The Cadomian crust of Brittany (France): 3D imagery from multisource data (Géofrance 3D). *Tectonophysics*, **331**, 1–246.

Lee, M. K., Pharaoh, T. C., Williamson, J. P., Green, C. A. & De Vos, W. 1993. Evidence on the deep structure of the Anglo-Brabant Massif from gravity and magnetic data. *Geological Magazine*, **130**, 575–582.

Lefort, J. P., Bardy, P. & Poulpiquet, J. 1991. Interprétation des anomalies magnétiques reconnues en Marche occidentale et mer Celtique le long des profils SWAT. *Mémoires de la Société Géologique de France*, **159**, 133–134.

Leloix, C., Faure, M. & Feybesse, J. L. 1999. Hercynian polyphase tectonics in the northeast French Massif Central: the closure of the Brévenne Devonian–Dinantian rift. *International Journal of Earth Sciences*, **88**(3), 409–421.

Leveridge, B. E., Holder, M. T. & Day, G. A. 1984. Thrust nappe tectonics in the Devonian of south Cornwall and the western English Channel. *In:* Hutton D.W. & Sanderson D. J. (eds) *Variscan Tectonics of the North Atlantic Region*. Geological Society, London, Special Publications, **14**, 103–112.

Linnemann, U. & Heuse, T. 2000. The Ordovician of the Schwarzburg Anticline: Geotectonic setting, biostratigraphy and sequence stratigraphy (Saxo-Thuringian Terrane). *Zeitschrift der deutschen geologischen Gesellschaft*, **151**, 71–91.

Linnemann, U., Gehmlich, M., *et al.* 2000. From Cadomian subduction to Early Palaeozoic rifting: the evolution of Saxo-Thuringia at the margin of Gondwana in the light of single zircon geochronology and basin development (Central European Variscides, Germany). *In:* Franke, W, Haak, V, Oncken, O. & Tanner, D. (eds) *Orogenic Processes: Quantification and Modelling in the Variscan Belt*. Geological Society, London, Special Publications, **179**, 131–153.

Linnemann, U., McNaughton, N. J., Romer, R. L., Gehmlich, M., Drost, K. & Tonk, C. 2004. West African provenance for Saxo-Thuringia (Bohemian Massif): Did Armorica ever leave pre-Pangean Gondwana? – U/Pb-SHRIMP zircon evidence and the Nd-isotopic record. *International Journal of Earth Sciences*, **93**, 683–705.

Linnemann, U., D'Lemos, R. *et al.* 2008. Cadomian tectonics. *In*: McCann, T. (ed.) *The Geology of Central Europe. Volume 1: Precambrian and Palaeozoic*. Geological Society, London, 103–154.

Linner, M. 1996. Metamorphism and partial melting of paragneisses of the Monotonous Group, SE Moldanubicum (Austria). *Mineralogy and Petrology*, **58**, 215–234.

Lippolt, H. J. 1986. Nachweis altpaläozoischer Primäralter (Rb-Sr) und karbonischer Abkühlalter (K-Ar) der Muscovit-Biotit-Gneise des Spessarts und der Biotit-Gneise des Odenwalds. *Geologische Rundschau*, **75**, 569–583.

Lippolt, H., Hradetzky, H. & Hautmann, S. 1994. K-Ar dating of amphibole-bearing rocks in the Schwarzwald, SW Germany: I. ^{40}Ar/^{39}Ar age constraints to Hercynian HT-metamorphism. *Neues Jahrbuch für Mineralogie, Monatshefte*, **10**, 433–448.

Loeschke, J., Güldenpfennig, M., Hann, H. P. & Sawatzki, G. 1998. Die Zone von Badenweiler-Lenzkirch (Schwarzwald): Eine variskische Suturzone. *Zeitschrift der deutschen geologischen Gesellschaft*., **149**, 197–212.

Lokhorst, A. (ed.) 1998. *NW European Gas Atlas – Composition and isotope ratios of Natural Gases*. CD-ROM. British Geological Survey, Bundesantalt für Geowissenschaften und Rohstoffe, Danmarks og Grönlands Geologiske Undersögelse, Nederlands Instituut voor Toegepaste Geowetenschappen TNO, Panstwowy Instytut Geologiczny, The European Union.

Lüschen, E., Wenzel, F., *et al.* 1987. Near-vertical and wide-angle seismic surveys in the Black Forest, SW Germany. *Journal of Geophysics*, **62**, 1–30.

Lutzens, H. 1972. Stratigraphie, Faziesbildung und Baustil im Paläozoikum des Unter- und Mittelharzes. *Geologie, Beiheft*, **74**, 1–105.

McCann, T., Skompski, S., *et al.* 2008*a*. Carboniferous. *In*: McCann, T. (ed.) *The Geology of Central Europe. Volume 1: Precambrian and Palaeozoic*. Geological Society, London, 411–529.

McCann, T., Kiersnowski, H., *et al.* 2008*b*. Permian. *In*: McCann, T.

(ed.) *The Geology of Central Europe. Volume 1: Precambrian and Palaeozoic*. Geological Society, London, 531–597.

McKerrow, W. S., Mac Niocaill, C., Ahlberg, P. E., Clayton, G., Cleal, C. J. & Eagar, M. C. 2000. The Late Palaeozoic relations between Gondwana and Laurussia. *In:* Franke, W., Haak, V., Oncken, O. & Tanner, D. (eds) *Orogenic Processes: Quantification and Modelling in the Variscan Belt.* Geological Society, London, Special Publications, **179**, 9–20.

Maass, R., Prösch, T. & Schuler, D. 1990. The zone of Badenweiler-Lenzkirch: a Carboniferous accretionary wedge? *Neues Jahrbuch für Geologie und Paläontologie, Mitteilungshefte*, **1990**, 717–734.

Magetti, M. 1975. Die Tiefengesteine des Bergsträßer Odenwaldes. *Aufschluß*, **27**, 87–107.

Malec, J. 1993. Upper Silurian and Lower Devonian in the western Holy Cross Mts. *Geological Quarterly*, **37**(4), 501–536.

Malusky, H. & Patočka, F. 1997. Geochemistry and ^{40}Ar-^{39}Ar geochronology of the mafic metavolcanic rocks from the Rýchory Mountains complex (west Sudetes, Bohemian Massif): paleotectonic significance. *Geological Magazine*, **134**(5), 703–716.

Maluski, H., Rajlich, P. & Souček, J. 1995. Pre-Variscan, Variscan and early Alpine thermo-tectonic history of the north-eastern Bohemian Massif: ^{40}Ar/^{39}Ar study. *Geologische Rundschau*, **84**, 345–358.

Mansy, J. L., Everaerts, M. & De Vos, W. 1999. Structural analysis of the adjacent Acadian and Variscan fold belts in Belgium and northern France from geophysical and geological evidence. *Tectonophysics*, **309**(1–4), 99–116.

Marheine, D. 1997. Zeitmarken im variszischen Kollisionsbereich des Rhenohercynikums – Saxothuringikums zwischen Harz und Sächsischen Granulitmassiv. – Ergebnisse von K/Ar-Altersbestimmungen. *Göttinger Arbeiten Geologie Paläontologie*, **75**, 1–97.

Marschall, H. R., Kalt, A. & Hanel, M. 2003. P-T Evolution of a Variscan Lower-Crustal Segment: a Study of Granulites from the Schwarzwald, Germany. *Journal of Petrology*, **44**(2), 227–253.

Martinez Catalan, J. R., Arenas, R., Diaz Garcia, F. & Abati, J. 1997. Variscan accretionary complex of northwest Iberia: Terrane correlation and succession of tectonothermal events. *Geology*, **25**, 1103–1106.

Martínez Catalán, J. R., Díaz García, F., *et al.* 2002. Thrust and detachment systems in the Ordenes Complex (northwestern Spain): Implications for the Variscan-Appalachian geodynamics. *In:* Martinez Catalan, J. R., Hatcher, R. D., Jr., Arenas, R. & Diaz Garcia, F. (eds) *Variscan-Appalachian Dynamics: The Building of the Late Paleozoic Basement*. Geological Society of America, Special Papers, **364**, 163–182.

Martinez-Torres, L.M., Ramon-Lluch, R. & Eguiluz, L. 1994. Tectonic wedges: geometry and kinematic interpretation. *Journal of Structural Geology*, **16**, 1491–1494.

Massonne, H.-J. 1999. A new occurrence of microdiamonds in quartzofeldspatic rocks of the Saxonian Erzgebirge, Germany, and their metamorphic evolution. *Seventh International Kimberlite Conference*, Cape Town, South Africa, 533–539.

Mathé, G. 1969. Die Metabasite des Sächsischen Granulitgebirges. *Freiberger Forschungshefte*, C **251**, 1–130.

Mathé, G. 1990. Zur Geologie der Serpentinitvorkommen im sächsischen Erzgebirge. *Abhandlungen des Staatlichen Museums für Mineralogie und Geologie Dresden*, **37**, 55–72.

Matte, P. 1986. Tectonics and plate tectonics model for the Variscan belt of Europe. *Tectonophysics*, **126**(2–4), 329–332.

Matte, P. 1991. Accretionary history and crustal evolution of the Variscan belt in western Europe. *Tectonophysics*, **196**, 309–337.

Matte, P. 2001. The Variscan collage and orogeny (480–290 Ma) and the tectonic definition of the Armorica microplate: a review. *Terra Nova*, **13**(2), 122–128.

Matte, P., Maluski, H., Rajlich, P. & Franke, W. 1990. Terrane boundaries in the Bohemian Massif: result of large scale Variscan shearing. *Tectonophysics*, **177**, 151–170.

Mattern, F. 2001. Permo-Silesian movements between Baltica and western Europe: tectonics and "basin families". *Terra Nova*, **13**, 368–375.

Matthes, S. 1954. Die Paragneise im mittleren kristallinen Vor-Spessart und ihre Metamorphose. *Abhandlungen Hessisches Landesamt für Bodenforschung*, **8**, 1–86.

Matthes, S. & Okrusch, M. 1977. The Spessart crystalline complex, north-west Bavaria: rock series, metamorphism, and position within the Central German Crystalline Rise. *Colloques Internationaux du Centre National de la Recherche Scientifique* (Rennes 1974), Paris, **243**, 375–390.

Matthes, S., Okrusch, M. & Richter, S. 1972. Zur Migmatit-Bildung im Odenwald. *Neues Jahrbuch für Mineralogie, Abhandlungen*, **116**, 225–267.

Mazur, S. 1995. Structural and metamorphic evolution of the country rocks at the eastern contact of the Karkonosze granite in the southern Rudawy Janowickie Mts. and Lasocki Range. *Geologia Sudetica*, **29**, 31–98 [in Polish, English summary].

Mazur, S. 2003. *Structural evolution of the Kłodzko Metamorphic Complex and the implications for the Variscan tectonics of the Sudetes.* Acta Universitatis Wratislaviensis no. 2581, Prace Geologiczno-Mineralogiczne, **74**.

Mazur, S. & Aleksandrowski, P. 2001. The Teplá(?)/Saxothuringian suture in the Karkonosze-Izera massif, Western Sudetes, Central European Variscides. *International Journal of Earth Sciences*, **90**, 341–360.

Mazur, S. & Kryza, R. 1996. Superimposed compressional and extensional tectonics in the Karkonosze-Izera Block, NE Bohemian Massif. *In:* Oncken, O. & Janssen, C. (eds) *Basement Tectonics, Vol. 11, Europe and other Regions*. Kluwer, Dordrecht, 51–66.

Mazur, S. & Puziewicz, J. 1995. Mylonites of the Niemcza Zone. *Annales Societatis Geologorum Poloniae*, **64**, 23–52 [in Polish, English summary].

Mazur, S., Turniak, K. & Bröcker, M. 2004. Neoproterozoic and Cambro-Ordovician magmatism in the Variscan Kłodzko Metamorphic Complex (West Sudetes, Poland): new insights from U/Pb zircon dating. *International Journal of Earth Sciences*, **93**, 758–772.

Mazur, S., Aleksandrowski, P. & Szczepanski, J. 2005. The presumed Teplá-Barrandian /Moldanubian terrane boundary in the Orlica Mountains (Sudetes, Bohemian Massif): structural and petrological characteristics. *Lithos*, **82**(1–2), 85–112.

Mazur, S., Dunlap, W.J., Turniak, K. & Oberc-Dziedzic, T. 2006. Age constraints for the thermal evolution and erosional history of the central European Variscan belt: new data from the sediments and basement of the Carboniferous foreland basin in western Poland. *Journal of the Geological Society, London*, **163**(9), 1–14.

Mazurek, M. 1985. Migmatitgenese in metapelitischen Gesteinen aus Bohrungen der Nordwestschweiz: Vorläufige Resultate. *Eclogae Helvetica*, **80**, 335–343.

Meere, P. A. & Mulchrone, K. F. 2006. Timing of deformation within Old Red Sandstone lithologies from the Dingle Peninsula, SW Ireland. *Journal of the Geological Society*, **163**(3), 461–469.

Mehnert, K. R. 1963. Petrographie und Abfolge der Granitisation im Schwarzwald. *Neues Jahrbuch für Mineralogie, Abhandlungen*, **99**, 161–199.

Meilliez, F. 1991. Ardennes-Brabant. *Sciences Geologiques (Bulletin)*, **44**(1–2), 3–29.

Meischner, D. 1971. Clastic sedimentation in the Variscan Geosyncline East of the River Rhine. *In:* Müller, G. (ed.) *Sedimentology of Parts of Central Europe*. Guidebook, 8th International Sedimentologists Congress Heidelberg. Kramer, Frankfurt, 9–44.

Meischner, D. 1991. Kleine Geologie des Kellerwaldes (Exkursion F). Jahresberichte und Mitteilungen des Oberrheinischen Geologischen Vereins, Neue Folge, **73**, 115–142.

Meisl, S. 1972. Das kristalline Grundgebirge. *In:* Kupfahl, H. G., Meisel, S. & Kümmerle, G. (eds*) Erläuterungen zur Geologischen Karte von Hessen 1:25000, Blatt Nr. 6217 Zwingenberg a.d. Bergstraße*. 2. Aufl., Hessisches Landesamt für Bodenforschung, Wiebaden.

Meisl, S. 1995. Igneous activity. *In:* Dallmeyer R. D., Franke, W. & Weber, K. (eds) *Pre-Permian Geology of Central and Eastern Europe*. Springer, Berlin, 118–131.

Meissner, R. 1999. Terrane accumulation and collapse in central Europe: seismic and rheological constraints. *Tectonophysics*, **305**(1–3), 93–107.

Menard, G. & Molnar, P. 1988. Collapse of a Hercynian Tibetan Plateau into a late Paleozoic European Basin and Range province. *Nature*, **334**, 235–237.

Menning, M., van Amerom, H. W. J., Drozdzewski, G. & Weyer, D. 1996. Integrative Kalibrierung der mittel- und westeuropäischen

Karbon-Zeitskala. *Exkursionsführer und Veröffentlichung der Gesellschaft für Geologische Wissenschaften, Claustal-Zellerfeld*, **197**, 59–61.

Merschat, A. J., Hatcher, J. R. D. & Davis, T. L. 2005. The northern Inner Piedmont, southern Appalachians, USA: kinematics of transpression and SW-directed mid-crustal flow. *Journal of Structural Geology*, **27**(7), 1252–1281.

Metz, R. & Rein, G. 1958. *Erläuterungen zur geologisch-petrographischen Karte des Südschwarzwaldes 1:50000*. Lahr/Schwarzwald, Schauenburg.

Million, R. 1992. Formations magnétiques. Forage Scientifique de Sancerre-Couy (Cher). *Mémoire Géologie Profonde de la France*, **3**, 169–181.

Mingram, B. 1998. The Erzgebirge, Germany, a subducted part of northern Gondwana: geochemical evidence for repetition of early Palaeozoic metasedimentary sequences in metamorphic thrust units. *Geological Magazine*, **135**, 785–801.

Mingram, B., Kröner, A., Hegner, E. & Krentz, O. 2004. Zircon ages, geochemistry, and Nd isotopic systematics of pre-Variscan orthogneisses from the Erzgebirge, Saxony (Germany), and geodynamic interpretation. *International Journal of Earth Sciences*, **93**, 706–727.

Mizerski, W. 1979. Tectonics of the Łysogóry unit in the Holy Cross Mts. *Acta Geologica Polonica*, **29**, 1–38.

Mizerski, W. 1995. Geotectonic evolution of the Holy Cross Mts in central Europe. *Biuletyn Państwowy Institutu Geologiczny*, **372**, 5–47.

Mizerski, W. 2002. *Regional Geology of Poland*. Wydawnictwo Naukowe PWN, Warszawa [in Polish].

Möbus, G. 1966. *Abriss der Geologie des Harzes*. Teubner, Berlin.

Molnar, P., England, P. H. & Martinod, J. 1993. Mantle dynamics, uplift of the Tibetan Plateau and the Indian monsoon. *Reviews of Geophysics*, **31**, 357–396.

Montenari, M. 1996. Appearence of microfossils in high grade metamorphic rocks from SW-Germany. *IXth International Palynological Congress Meeting*, Houston, Texas, Abstracts, 110.

Montenari, M. & Servais, T. 2000. Early Palaeozoic (Late Cambrian-Early Ordovician) acritarchs from the metasedimentary Baden-Baden-Gaggenau zone (Schwarzwald), SW Germany). *Reviews of Palaeobotany & Palynology*, **113**, 73–85.

Montenari, M., Servais, T. & Paris, F. 2000. Palynological dating (acritarchs and chitinozoans) of lower Paleozoic phyllites from the Black Forest/southwestern Germany. *Earth and Planetary Sciences*, **330**, 493–499.

Müller, H. D. 1989. Geochemistry of metasediments in the Hercynian and pre-Hercynian crust of the Schwarzwald, the Vosges, and northern Switzerland. *Tectonophysics*, **157**, 97–108.

Murphy, J. B., Pisarevsky, S. A., Nance, R. D. & Keppie, J. D. 2004. Neoproterozoic–Early Paleozoic evolution of peri-Gondwanan terranes: implications for Laurentia-Gondwana connections. *International Journal of Earth Sciences*, **93**(5), 659–682.

Murphy, J. B., Gutierrez-Alonso, G., *et al.* 2006. Origin of the Rheic Ocean: Rifting along a Neoproterozoic suture? *Geology*, **34**(5), 325–328.

Muttoni, G., Kent, D. V, Garzanti, E., Brack, P, Abrahamsen, N. & Gaetani, M. 2003. Early Permian Pangea 'B' to Late Permian Pangea 'A'. *Earth and Planetary Science Letters*, **215**(3–4), 379–394.

Nasir, S. 1986. *Die Metabasite im mittleren kristallinen Vorspessart: Petrographie-Phasenpetrologie*. Dissertation, Universität Würzburg, Germany.

Nasir, S., Okrusch, M., Kreuzer, H., Lenz, H. & Höhndorf, A. 1991. Geochronology of the Spessart crystalline complex, mid-German crystalline rise. *Mineralogy and Petrology*, **43**, 39–55.

Nelson, K.D., Zhao, W., *et al.* 1996. Partially molten middle crust beneath southern Tibet: synthesis of project INDEPTH results. *Science*, **274**, 1684–1687.

Neumann, W. 1974. Mitteldeutsche Kristallinzone. *In:* Hoppe, W. & Seidel, G. (eds) *Geologie von Thüringen*. Hermann Haack, Gotha-Leipzig, 77–118.

Nickel, E. 1975. Geologische Position und Petrogenese des kristallinen Odenwaldes. *Aufschluss*, **27**, 1–25.

Nickel, E. & Fettel, M. 1985. *Odenwald – Vorderer Odenwald zwischen Darmstadt und Heidelberg*. Sammlung geologischer Führer (2. Aufl.), Borntraeger, Berlin-Stuttgart, **65**.

Noblet, Ch. & Lefort, J. P. 1990. Sedimentological evidence for a limited separation between Armorica and Gondwana during the Early Ordovician. *Geology*, **18**, 303–306.

Oberc-Dziedzic, T., Klimas, K., Kryza, R. & Fanning, C. M. 2003. SHRIMP zircon geochronology of the Strzelin gneiss, SW Poland: evidence for a Neoproterozoic thermal event in the Fore-Sudetic Block, Central European Variscides. *International Journal of Earth Sciences*, **92**, 701–711.

O'Brien, P. J. 2000. The fundamental Variscan problem: high-temperature metamorphism at different depths and high-pressure metamorphism at different temperatures. *In:* Franke, W., Haak, V., Oncken, O. & Tanner, D. (eds) *Orogenic Processes: Quantification and Modelling in the Variscan Belt*. Geological Society, London, Special Publications, **179**, 369–386.

O'Brien, P. J., Kröner, A., Jaeckel, P., Hegner, E., Żelaźniewicz, A. & Kryza, R. 1997. Petrological and isotopic studies on Palaeozoic high pressure granulites with a medium pressure overprint, Góry Sowie (Owl) Mts., Polish Sudetes. *Journal of Petrology*, **38**, 433–456.

Oczlon, M. S. 1994. North Gondwana origin for exotic Variscan rocks in the Rhenohercynian zone of Germany. *Geologische Rundschau*, **83**, 20–31.

Okrusch, M. 1983. The Spessart Crystalline Complex, northwest Bavaria. DMG SFMC Joint Meet., Excursion E4. *Fortschritte Mineralogie*, **61**, 135–169.

Okrusch, M. & Weber, K. 1996. Der Kristallinkomplex des Vorspessarts. *Zeitschrift für geologische Wissenschaften*, **24**, 141–174.

Okrusch, M., Müller, R. & Shazly, S. 1985. Die Amphibolite, Kalksilikatgesteine und Hornblendegneise der Alzenauer Gneisserie im Nordwest-Spessart. *Geologica Bavarica*, **87**, 5–37.

Okrusch, M., Schubert, W. & Stähle, V. 2000. The Odenwald, Germany: Variscan metamorphic evolution and igneous events. *Beihefte zum European Journal of Mineralogy*, **12**, 45–89.

Oliver, G. J. H., Corfu, F. & Krough, T. E. 1993. U–Pb ages from SW Poland: evidence for a Caledonian suture zone between Baltica and Gondwana. *Journal Geological Society London*, **150**, 355–369.

Oncken, O. 1997. Transformation of a magmatic arc and an orogenic root during oblique collision and its consequences for the evolution of the European Variscides (Mid-German Crystalline Rise). *Geologische Rundschau*, **86**, 2–20.

Oncken, O. & Weber, K. 1995. Structure. *In:* Dallmeyer, R.D., Franke, W. & Weber, K. (eds) *Pre-Permian Geology of Central and Eastern Europe*. Springer, Berlin, 50–58.

Oncken, O., Franzke, H. J., Dittmar, U. & Klügel, T. 1995. Structure. *In:* Dallmeyer, R. D., Franke, W. & Weber, K. (eds) *Pre-Permian Geology of Central and Eastern Europe*. Springer, Berlin, 108–117.

Oncken, O., Winterfeld, C. v. & Dittmar, U. 1999. Accretion and inversion of a rifted passive margin – the Late Paleozoic Rhenohercynian fold and thrust belt. *Tectonics*, **18**, 75–91.

Oncken, O., Plesch, A., Weber, J., Ricken, W. & Schrader, S. 2000. Passive margin detachment during arc-continental collision (Central European Variscides). *In:* Franke, W., Haak, V., Oncken, O. & Tanner, D. (eds) *Orogenic Processes: Quantification and Modelling in the Variscan Belt*. Geological Society, London, Special Publications, **179**, 199–216.

Paris, F. 1998. Palaeogeography of northern Gondwana regions in the early Palaeozoic: the verdict of the faunas. *Schriften des Staatlichen Museums für Mineralogie und Geologie zu Dresden*, **9**, 173–174.

Parry, M., Štípská, P., Schulmann, K., Hrouda, F., Ježek, J. & Kröner, A. 1997. Tonalite sill emplacement at an oblique plate boundary: northeastern margin of the Bohemian Massif. *Tectonophysics*, **280**, 61–81.

Paszkowski, M. 1988. Dinantian basin in the Cracow area: an attempt at a synthesis. *Przegląd Geologiczny*, **36**(1), 200–206.

Patočka, F. & Smulikowski, W. 2000. Early Palaeozoic intracontinental rifting and incipient oceanic spreading in the Czech/Polish East Krkonoše/Karkonosze Complex, West Sudetes (NE Bohemian Massif). *Geologia Sudetica*, **33**, 1–15.

Patočka, F. & Valenta, J. 1996. Geochemistry of the late Devonian intermediate to acid metavolcanic rocks from the southern part of the Vrbno Group, the Jeseniky Mts (Moravo-Silesian Belt, Bohemian Massif, Czech Republic): paleotectonic implications. *Geolines*,

4, 42–54.

Pesek, J. 1996. *Carboniferous of Central and Western Bohemia*. Czech Geological Survey, Prague.

Pharaoh, T. C. 1999. Palaeozoic terranes and their lithospheric boundaries within the Trans-European Suture Zone (TESZ): a review. *Tectonophysics*, **314**, 17–41.

Pharaoh, T. C. & Gibbons, W. 1994. Precambrian rocks in England and Wales south of the Menai Strait Fault System. *In:* Gibbons, W. & Harris, A. L. (eds) *A Revised Correlation of Precambrian Rocks in the British Isles*. Geological Society, London, Special Reports, **22**, 85–97.

Piqué, A., Huon, S. & Clauer, N. 1984. La schistosité hercynienne et le métamorphisme associé dans les vallées de la Meuse entré Charleville-Mézière et Namur (Ardenne Franco-Belge). *Bulletin Societé Belge de Géologie*, **93**(1–2), 55–70.

Plusquellec, Y. & Jahnke, H. 1999. Les tabulés de L'Erbslochgrauwacke (Emsien Inférieur du Kellerwald) et le problème des affinités paléogéographiques de l'allochthone "Giessen-Harz". *Abhandlungen der Geologischen Bundesanstalt in Wien*, **54**, 435–451.

Poller, U., Altenberger, U. & Schubert, W. 2001. Geochemical investigations of the Bergsträßer Odenwald amphibolites – implications for back-arc magmatism. *Mineralogy and Petrology*, **72**, 63–76.

Poole, F. G., Perry, W. J., Jr., Madrid, P. & Amaya-Martinez, R. 2005. Tectonic synthesis of the Ouachita-Marathon-Sonora orogenic margin of southern Laurentia; stratigraphic and structural implications for timing of deformational events and plate tectonic model. *In:* Anderson, J. H., Nourse, J. A., McKee, J. W. & Steiner, M. B. (eds) *The Mojave Sonora Megashear Hypothesis: development, assessment and alternatives*. Geological Society of America, Special Papers, **393**, 543–596.

Požaryski, W. 1977. The Variscan epoch in the Epi-Gothian Platform and its Border Zone. *In:* Požaryski, W. (ed) *Tectonics of Poland*. Wydawnictwa Geologiczne, Warszawa, 207–252.

Požaryski, W. 1990. The Middle European Caledonides – wrenching orogen composed of terranes. *Przegląd Geologiczny*, **38**, 1–9 [in Polish with English summary].

Požaryski, W. & Tomczyk, H. 1993. Geological cross-section through SE Poland. *Przegląd Geologiczny*, **41**(10), 687–695 [in Polish with English summary].

Požaryski, W., Grocholski, A., Tomczyk, H., Karnkowski, P. & Moryc, W. 1992. The tectonic map of Poland in the Variscan epoch. *Przegląd Geologiczny*, **40**(11), 643–651.

Praeg D. 2004. Diachronous Variscan late-orogenic collapse as a response to multiple detachments; a view from the internides in France to the foreland in the Irish Sea. *In:* Wilson M., Neumann E.-R., Davies G. R., Timmermann M. J., Heeremans M. & Larsen B. T. (eds) *Permo-Carboniferous Magmatism and Rifting in Europe*. Geological Society, London, Special Publications, **223**, 89–138.

Přichistal, A. 1990. Principal results of a study of the Paleozoic volcanism in the Šternberk-Horni Benešov belt, the Nízký Jesenik Mts. *Sborník geologických věd, Ložiskove Geolgie a Mineralogie*, **29**, 41–66.

Propach, G., Baumann, A., Schulz-Schmalschläger, M. & Grauert, B. 2000. Zircon and monazite U–Pb ages of Variscan granitoid rocks and gneisses in the Moldanubian zone of eastern Bavaria, Germany. *Neues Jahrbuch für Geologie und Paläontologie, Monatshefte*, **2000**, 345–377.

Quadt, A. von 1993. The Saxonian granulite massif: New aspects from geochronological studies. *Geologische Rundschau*, **82**, 516–530.

Quinn, D., Meere, P. A. & Wartho, J.-A. 2005. A chronology of foreland deformation: ultra-violet laser $^{40}Ar/^{39}Ar$ dating of syn/late-orogenic intrusions from the Variscides of southwest Ireland. *Journal of Structural Geology*, **27**(8), 1413–1425.

Rajlich, P. 1987. Variszische duktile Tektonik im Böhmischen Massiv. *Geologische Rundschau*, **76**, 755–786.

Rajlich, P. 1990. Strain and tectonic styles related to Variscan transpression and transtension in the Moravo-Silesian Culmian basin, Bohemian Massif, Czechoslovakia. *Tectonophysics*, **174**, 351–367.

Raoult, J. & Meilliez, F. 1987. The Variscan Front and the Midi Fault between the Channel and the Meuse River. *Journal of Structural Geology*, **9**, 473–479.

Rehfeld, U. 1983. A relictic staurolite-kyanite mineral paragenesis in paragneisses of the Central Schwarzwald. *Neues Jahrbuch für Mineralogie, Abhandlungen*, **146**, 170–181.

Reichstein, M. 1965. Motive und Probleme erneuter Deckenbauvorstellungen für den Harz. *Geologie*, **14**, 1039–1076.

Reinhard, J. & Kleemann, U. 1994. Extensional unroofing of granulitic lower crust and related low-pressure, high temperature metamorphism in the Saxonian Granulite Massif, Germany. *Tectonophysics*, **238**, 71–94.

Reischmann, T. & Anthes, G. 1996. Geochronology of the mid-German crystalline rise west of the River Rhine. *Geologische Rundschau*, **85**, 761–774.

Reischmann, T., Anthes, G., Jaeckel, P. & Altenberger, U. 2001. Age and origin of the Böllsteiner Odenwald. *Mineralology Petrology*, **72**, 29–44.

Reitz, E. 1987. Palynologie in metamorphen Serien: I. Silurische Sporen aus einem granatführenden Glimmerschiefer des Vor-Spessart. *Neues Jahrbuch für Geologie und Paläontologie, Mitteilungshefte*, **1987**, 699–704.

Reitz, E. & Wickert, F. 1989. Late Cambrian to Early Ordovician acritarchs from the Villé Unit, northern Vosges Mountains, France. *Neues Jahrbuch für Geologie und Paläontologie, Mitteilungshefte*, **1989**, 375–384.

Rey, P, Vanderhaeghe, O. & Teyssier, C. 2001. Gravitational collapse of the continental crust: definition, regimes and modes. *Tectonophysics*, **342**(3–4), 435–449.

Ribeiro, A., Pereira, E. & Dias, R. 1990. Central-Iberian Zone, allochthonous sequences: Structure in the northwest of the Iberian Peninsula. *In:* Dallmeyer, R. D. & Martinez Garcia, E. (eds) *Pre-Mesozoic Geology of Iberia*, Springer, Berlin, 220–236.

Ritzki, A. & Baroz, F. 1988. Le volcanism toléitique du massif de Schirmeck (Vosges septentrionales, France), témoin d'une zone du convergence de plagues au Paléozoiques supérieur. *Comptes rendus de l'Académie des Sciences Paris, Série II*, **307**, 511–516.

Robardet, M. 2002. Alternative approach to the Variscan Belt in southwestern Europe: Preorogenic paleobiogeographical constraints. *In:* Martínez Catalán, J.R., Hatcher, R.D. Jr., Arenas, R. & Díaz García, F. (eds) *Variscan Appalachian Dynamics: The Building of the Late Paleozoic Basement*. Geological Society of America, Special Papers, **364**, 1–15.

Robardet, M. 2003. The Armorica 'microplate': fact or fiction? Critical review of the concept and contradictory paleobiogeographical data. *Paleogeography, Paleoclimatology, Paleoecology*, **195**, 125–148.

Röber, B., Hammer, J., Eidam, J. & Ehling, B.-C. 1997. Cadomische und spätvariszische Plutonite im östlichen Teil der Mitteldeutschen Schwelle. *Terra Nostra*, **97**(5), 35–39.

Roberts, M. P. & Finger, F. 1997. Do U–Pb zircon ages from granulites reflect peak metamorphic conditions? *Geology*, **25**(4), 319–322.

Röhr, C. 1990. Die Genese der Leptinite und Paragneise zwischen Nordrach und Gengenbach im mittleren Schwarzwald. *Frankfurter Geowissenschaftliche Arbeiten, Serie C*, **11**, 1–59.

Romer, R. L. & Rötzler, J. 2001. P-T-t evolution of ultrahigh-temperature granulites from the Saxon Granulite Massif, Germany. Part II: Geochronology. *Journal of Petrology*, **42**, 2015–2032.

Romer, R. L. & Rötzler, J. 2003. Effect of metamorphic reaction history on the U–Pb dating of titanite, *In:* Vance, D., Müller, A. & Villa, I. (eds) *Geochronology: Linking the Isotopic Record with Petrology and Textures*. Geological Society, London, Special Publications, **220**, 147–158.

Rötzler, J. & Romer, R. L. 2001. P-T-t evolution of ultrahigh-temperature granulites from the Saxon Granulite Massif, Germany. Part I: Petrology. *Journal of Petrology*, **42**, 1995–2013.

Rötzler, J., Romer, R. L., Budzinski, H., & Oberhänsli, R. 2004. Ultrahigh-temperature granulites from Tirschheim, Saxon Granulite Massif, Germany: P-T-t path and geotectonic implications. *European Journal of Mineralogy*, **16**, 917–937.

Russe, B. 1995. Probleme der Überschiebungstektonik im Ronneburger Erzfeld. *Zeitschrift für geologische Wissenschaften*, **23**, 769–776.

Sadler, P. M. 1983. Depositional Models for the Carboniferous Flysch of the Eastern Rheinisches Schiefergebirge. *In:* Martin, H. & Eder, F. W. (eds) *Intracontinental Fold Belts*. Springer, Berlin, 125–143.

Sanchez-Garcia, T., Bellido, F. & Quesada, C. 2003. Geodynamic setting and geochemical signatures of Cambrian-Ordovician rift-related igneous rocks (Ossa-Morena Zone, SW Iberia). *Tectonophysics*, **365**(1–4), 233–255.

SCHÄFER, F., ONCKEN, O., KEMNITZ, H. & ROMER, R. L. 2000. Upper-plate deformation during collision orogeny: a case study from the German Variscides (Saxo-Thuringian Zone), *In:* FRANKE, W., HAAK, V., ONCKEN, O. & TANNER, D. (eds) *Orogenic Processes: Quantification and Modelling in the Variscan Belt*. Geological Society, London, Special Publications, **179**, 281–302.

SCHÄLICKE, W. 1975. Die Otzberg-Zone. *Aufschluss, Sonderband*, **27**, 47–57.

SCHALTEGGER, U. 2000. U–Pb geochronology of the Southern Black Forest Batholith (Central Variscan Belt): timing of exhumation and granite emplacement. *International Journal of Earth Sciences*, **88**, 814–828.

SCHECK, M., BAYER, U., OTTO, V., LAMARCHE, J., BANKA, D. & PHARAOH, T. C. 2002. The Elbe Fault Zone in North Central Europe – a basement controlled zone of crustal weakness. *Tectonophysics*, **360**, 281–300.

SCHENK, V. & TODT, W. 1984. U–Pb Datierungen an Zirkon und Monazit der Granulite im Moldanubikum Niederösterreichs (Waldviertel). *Fortschritte Mineralogie*, **61**, 190–191.

SCHERER, E., MEZGER, K. & MÜNKER, C. 2002. Lu-Hf ages of high pressure metamorphism in the Variscan fold belt of southern Germany. Goldschmidt Conference Abstract 2002. *Geochimica et Cosmochimica Acta Supplement*, **66**, A677.

SCHEUVENS, D. & ZULAUF, G. 2000. Exhumation, strain localization, and emplacement of granitoids along the western part of the Central Bohemian shear zone (central European Variscides, Czech Republic). *International Journal of Earth Sciences*, **89**, 617–630.

SCHMÄDICKE, E. 1994. Die Eklogite des Erzgebirges. *Freiberger Forschungshefte, C*, **456**, 1–344.

SCHMÄDICKE, E., OKRUSCH, M., SCHUBERT, W., ELWART, B. & GÖRKE, U. 2001. Phase relations of calc-silicate assemblages in the Auerbach marble, Odenwald Crystalline Complex, Germany. *Mineralogy and Petrology*, **72**, 77–111.

SCHOLTZ, H. 1930. Das varistische Bewegungsbild. *Fortschritte Geologie Paläontologie*, **25**, 235–316.

SCHÖNENBERG, R. & NEUGEBAUER, J. 1997. Einführung in die Geologie Europas. *Rombach, Freiburg*, 1–385.

SCHUBERT, W. 1968. Die Amphibolite des prävariszischen Schieferrahmens im Bergsträßer Odenwald. *Neues Jahrbuch für Mineralogie, Abhandlungen*, **108**, 69–110.

SCHUBERT, W., LIPPOLT, H. J. & SCHWARZ, W. 2001. Early to Middle Carboniferous $^{40}Ar/^{39}Ar$ ages of amphibolites and gabbros from the Bergsträsser Odenwald. *Mineralogy and Petrology*, **72**, 113–132.

SCHULMANN, K. & GAYER, R. 2000. A model for a continental accretionary wedge developed by oblique collision: the NE Bohemian Massif. *Journal of the Geological Society, London*, **157**, 401–416.

SCHWAN, W. 1991. Geologie des Acker-Bruchberg-Ilsenburg-Zuges (Oberharz) – Derzeitiger Forschungsstand und Diskussion der Probleme. *Zentralblatt für Geologie und Paläontologie, Teil 1*, **7**, 787–850.

SEBASTIAN, U. 1995. Die Strukturentwicklung des spätorogen Erzgebirgsaufstiegs und der Flöhazone – Ein weiterer Beitrag zur postkollisionalen Extension am Nordrand der Böhmischen Masse. *Freiberger Forschungshefte, C*, **461**, 1–114.

SEBERT, M. & WIMMENAUER, W. 1992. Metagabbros and meta-anorthosites in the Southern Black Forest. *Jahreshefte Geologisches Landesamt Baden-Württemberg*, **34**, 167–212.

SEHNERT, M. 1991. *Zur Biostratigraphie epizonaler Metamorphite der Wippraer Einheit/Unterharz*. Dissertation, Universität Halle.

SELVERSTONE, J. 2005. Are the Alps Collapsing? *Annual Review of Earth and Planetary Sciences*, **33**(1), 113–132.

SENGÖR, A. M. C., NATAL'IN, B. A. & BURTMAN, V. S. 1993. Evolution of the Altaid tectonic collage and Paleozoic crustal growth in Eurasia. *Nature*, **364**, 209–304.

SESTON, R., WINCHESTER, J. A., PIASECKI, M. A. J., CROWLEY, Q. G. & FLOYD, P. A. 2000. A structural model for the western-central Sudetes: a deformed stack of Variscan thrust sheets. *Journal of the Geological Society, London*, **157**, 1155–1167.

SHELLEY, D. & BOSSIERE, G. 2002. Megadisplacements and the Hercynian Orogen of Gondwanan France and Iberia. *In:* MARTINEZ CATALAN J. R., HATCHER R. D. JR., ARENAS R. & DIAZ GARCIA F. (eds) *Variscan-Appalachian Dynamics: The Building of the Late Paleozoic Basement*. Geological Society of America, Special Papers, **364**, 209–222.

SHERLOCK, S. C., KELLEY, S. P., ZALASIEWICZ, J. A., SCHOFIELD, D. I., EVANS, J. A., MERRIMAN, R. J. & KEMP, S. J. 2003. Precise dating of low-temperature deformation; strain-fringe analysis by $^{40}Ar/^{39}Ar$ laser microprobe. *Geology*, **31**(3), 219–222

SIEBEL, W., BLAHA, U., CHEN, F. & ROHRMÜLLER, J. 2005. Geochronology and geochemistry of a dyke-host rock association and implications for the formation of the Bavarian Pfahl shear zone, Bohemian Massif. *International Journal of Earth Sciences*, **94**(1), 8–23.

SINTUBIN, M. 1997. Cleavage-fold relationships in the lower Paleozoic Brabant Massif (Belgium). *Aardkundige Mededelingen*, **8**, 161–164.

SINTUBIN, M. 1999. Arcuate fold and cleavage patterns in the southeastern part of the Anglo-Brabant Fold Belt (Belgium): tectonic implications. *Tectonophysics*, **309**(1–4), 81–97.

SITTIG, E. 1963. Synsedimentäre Rutschungen im Unterkarbon des Südschwarzwaldes. *Oberrheinische Geologische Abhandlungen*, **12**, 81–94.

SITTIG, E. 1965. Der geologische Bau des variszischen Sockels nordöstlich von Baden-Baden (Nordschwarzwald). *Oberrheihnische Geologische Gesellschaft*, **14**, 167–207.

SITTIG, E. 1969. Zur geologischen Charakterisierung des Moldanubikums am Oberrhein. *Oberrheinische Geologische Abhandlungen*, **18**, 119–161.

SITTIG, E. 1981. Evidence for wrench faulting within the Paleozoic Badenweiler-Lenzkirch-zone (Southern Schwarzwald Mountains, W-Germany). *Neues Jahrbuch für Geologie und Paläontologie, Mitteilungshefte*, **7**, 431–448.

SMOLER, M. 1987. *Petrographische, geochemische und phasenpetrologische Untersuchungen an Metasedimenten des NW-Spessart/Bayern*. Dissertation, Universität Würzburg, Germany.

SMULIKOWSKI, W. 1995. Evidence for glaucophane-schist facies metamorphism in the East Karkonosze complex, West Sudetes, Poland. *Geologische Rundschau*, **84**, 720–737.

SOMMERMANN, A. E. 1993. Zirkonalter aus dem Granit der Bohrung Saar 1. *Beihefte European Journal Mineralogy*, **5**, 145.

SOPER, N. J., & WOODCOCK, N. H. 2003. The lost Lower Old Red Sandstone of England and Wales: a record of post-Iapetan flexure or Early Devonian transtension? *Geological Magazine*, **140**(6), 627–647.

SOPER, N. J., WEBB, B. C. & WOODCOCK, N. J. 1987. Late Caledonian (Acadian) transpression in North West England: timings, geometry and geotectonic significance. *Proceedings of the Yorkshire Geological Society*, **46**, 175–192.

SOUČEK, J. 1981. The geochemistry of the Devonian metabasites of the Hrubý and Nízký Jesenik Mts. *Časopis pro Mineralogii a Geologii*, **26**(2), 125–142 [in Czech].

SPEAR, F. S., KOHN, M. J., CHENEY, J. T. & FLORENCE, F. 2002. Metamorphic, thermal and tectonic evolution of Central New England. *Journal of Petrology*, **43**, 2097–2120.

SPIEGELHALTER, F. 1910. Ein Goniatit aus dem südlichen Schwarzwald. *Centralblatt für Mineralogie, Geologie und Paläontologie*, **1910**, 506–610.

STAMPFLI, G. M. & BOREL, G. D. 2002. A plate tectonic model for the Paleozoic and Mesozoic constrained by dynamic plate boundaries and restored synthetic oceanic isochrones. *Earth and Planetary Science Letters*, **196**(1–2), 17–33.

STAMPFLI, G. M., VON RAUMER, J. F. & BOREL, G. D. 2002. Paleozoic evolution of pre-Variscan terranes: from Gondwana to the Variscan collision. *In:* MARTINEZ CATALAN, J. R., HATCHER, R. D., JR., ARENAS, R. & DIAZ GARCIA, F. (eds) *Variscan-Appalachian Dynamics: The Building of the Late Paleozoic Basement*. Geological Society of America, Special Papers, **364**, 263–280.

STEIN, E. 2001. The geology of the Odenwald Crystalline Complex. *Mineralogy and Petrology*, **72**, 7–28.

STENGER, R., BAATZ, K., KLEIN, H. & WIMMENAUER, W. 1989. Metamorphic evolution of the pre-Hercynian basement of the Schwarzwald (Federal Republic of Germany). *Tectonophysics*, **157**, 117–121.

STETS, J. & SCHÄFER, A. 2002. Depositional environments in the Lower Devonian siliciclastics of the Rhenohercynian Basin (Rheinisches Schiefergebirge, W-Germany). *Contributions to Sedimentary Geol-*

ogy, **22**, 1–78.

STILLE, H. 1951. Das mitteleuropaishe varistische Grundgebirge im Bilde des gesamteuropäischen. *Geologisches Jahrbuch Beihefte*, **2**, 1–138.

ŠTIPSKA, P., SCHULMANN, K., THOMPSON, A.B., JEŽEK, J. & KRÖNER, A. 2001. Thermo-mechanical role of a Cambro-Ordovician paleorift during the Variscan collision: the NE margin of the Bohemian Massif. *Tectonophysics*, **332**, 239–253.

ŠTIPSKA, P., SCHULMANN, K. & KRÖNER, A. 2004. Vertical extrusion and middle crustal spreading of omphacite granulite: a model of syn-convergent exhumation (Bohemian Massif, Czech Republic). *Journal of Metamorphic Geology*, **22**(3), 179–198.

STOSCH, H. G. & LUGMAIR, G. W. 1990. Geochemistry and evolution of MORB-type eclogites from the Münchberg Massif, southern Germany. *Earth and Planetary Science Letters*, **99**, 230–249.

STUPNICKA, E. 1988. On the nature and origin of the Holy Cross Dislocation. *Przegląd Geologiczny*, **36**(1), 40–52 [in Polish with English summary].

STUPNICKA, E. 1992. The significance of the Variscan Orogeny in the Świętokrzyskie Mountains (Mid Polish Uplands). *Geologische Rundschau*, **81**, 561–570.

SUCHY, V., ROZKOSNY, I., ŽAK, K. & FRANCU, J. 1996. Epigenetic dolomitization of the Prídolí formation (Upper Silurian), the Barrandian basin, Czech Republic: implications for burial history of Lower Paleozoic strata. *Geologische Rundschau*, **85**, 264–277.

SUCHY, V., DOBES, P., FILIP, J., STEJSKAL, M. & ZEMAN, A. 2001. Conditions for veining in the Barrandian Basin (Lower Palaeozoic), Czech Republic: evidence from fluid inclusion and apatite fission track analysis. *Tectonophysics*, **348**, 25–50.

SUESS, E. 1885–1909. *Das Antlitz der Erde*. Tempsky & Freytag, Prag-Wien-Leipzig.

SVOBODA, J. 1966. The Barrandian Basin. The Železné hory Mountains and the metamorphic 'islets' of central Bohemia (Chrudim-Islets zone). *In:* SVOBODA, J., BENES, K., *et al.* (eds) *Regional Geology of Czechoslovakia, Part I, The Bohemian Massif.* Czechoslovak Academy of Sciences, Prague, 281–367.

SVOJTKA, M., KOSLER, J. & VENERA, Z. 2002. Dating granulite-facies structures and the exhumation of lower crust in the Moldanubian Zone of the Bohemian Massif. *International Journal of Earth Sciences*, **91**(3), 373–385.

SYNEK, J., RAJLICH, P. & URBAN, M. 1990. Transpression and transtension in the Devonian and Culm of the Nízký Jesenik Mts. *Acta Universitatis Carolinae – Geologica*, **2**, 209–234.

SZCZEPAŃSKI, J., ANCZKIEWICZ, R., MAZUR, S. & THIRLWALL, M. 2004. Timing of ultra-high pressure metamorphism in the Orlica-Śnieżnik Dome, West Sudetes. *Special Papers Mineralogical Society of Poland*, **24**, 365–368.

SZULCZEWSKI, M. 1968. Slump structures and turbidites in Upper Devonian limestones of the Holy Cross Mts. *Acta Geologica Polonica*, **18**(2), 303–330.

SZULCZEWSKI, M. 1977. Main facial regions in the Palaeozoic of the Holy Cross Mts. *Przegląd Geologiczny*, **25**(8–9), 428–432 [in Polish with English summary].

SZULCZEWSKI, M. 1995. Depositional evolution of the Holy Cross Mts. (Poland) in the Devonian and Carboniferous – a review. *Geological Quarterly*, **39**, 471–488.

SZULCZEWSKI, M., BELKA, Z. & SKOMPSKI, S. 1996. The drowning of the carbonate platform: an example from the Devonian-Carboniferous of the southwestern Holy Cross Mountains, Poland. *Sedimentary Geology*, **106**, 21–49.

TAIT, J. A., BACHTADSE, V., FRANKE, W. & SOFFEL, H. C. 1997. Geodynamic evolution of the European Variscan fold belt: paleomagnetic and geological constraints. *Geologische Rundschau*, **86**, 585–598.

TAIT, J., SCHAETZ, M., BACHTADSE, V. & SOFFEL, H. 2000. Palaeomagnetism and Palaeozoic palaeogeography of Gondwana and European terranes. *In:* FRANKE, W., HAAK, V., ONCKEN, O. & TANNER, D. (eds) *Orogenic Processes: Quantification and Modelling in the Variscan Belt.* Geological Society, London, Special Publications, **179**, 21–34.

TAJCMANOVA, L., KONOPASEK, J. & SCHULMANN, K. 2006. Thermal evolution of the orogenic lower crust during exhumation within a thickened Moldanubian root of the Variscan belt of Central Europe. *Journal of Metamorphic Geology*, **24**, 119–134.

TARNOWSKA, M. 1981. Lower Devonian in the central part of the Holy Cross Mts. *Przewodnik 53 Zjazdu Polskiego Towarzystwa Geologicznego*, 57–68 [in Polish].

TEUFEL, S. 1988. Vergleichende U–Pb- und Rb-Sr-Altersbestimmungen an Gesteinen des Übergangsbereiches Saxothuringikum/Moldanubikum, NE-Bayern. *Göttinger Arbeiten Geologie Paläontologie*, **35**, 1–87.

THOMSON, S. N. & ZEH, A. 2000. Fission-track thermochronology of the Ruhla Crystalline Complex: New constraints on the post-Variscan thermal evolution of the NW Saxo-Bohemian Massif. *Tectonophysics*, **324**, 17–35.

THÜRACH, H. 1893. Über die Gliederung des Urgebirges im Spessart. *Geognostische Jahreshefte*, **5**, 1–160.

TICHOMIROVA, M. 2002. Zircon inheritance in diatexite granodiorites and its consequence on geochronology—a case study in Lusatia and the Erzgebirge (Saxo-Thuringia, eastern Germany). *Chemical Geology*, **191**, 209–224.

TICHOMIROWA, M., BERGER, H.-J., *et al.* 2001. Zircon ages of high-grade gneisses in the Eastern Erzgebirge (Central European Variscides) – constraints on origin of the rocks and Precambrian to Ordovician magmatic events in the Variscan foldbelt. *Lithos*, **56**, 303–332.

TIMMERMAN, M. J. 2004. Timing, geodynamic setting and character of Permo-Carboniferous magmatism in the foreland of the Variscan Orogen, NW Europe. *In:* WILSON, M., NEUMANN, E.-R., DAVIES, G. R., TIMMERMANN, M. J., HEEREMANS, M. & LARSEN, B. T. (eds) *Permo-Carboniferous Magmatism and Rifting in Europe*. Geological Society, London, Special Publications, **223**, 41–74.

TIMMERMAN, M. J. 2008. Palaeozoic magmatism. *In*: MCCANN, T. (ed.) *The Geology of Central Europe. Volume 1: Precambrian and Palaeozoic.* Geological Society, London, 665–748.

TOBSCHALL, H. J. 1975. Geochemische Untersuchungen zum stofflichen Bestand und Sedimentationsmilieu paläozoischer mariner Tone: Ni, Cu, Zn, Rb, Sr, Y, Zr, Nb und Ba in den Steiger Schiefern (Vogesen). *Chemie der Erde*, **34**, 105–167.

TODT, W. 1976. Zirkon-U/Pb-Alter des Malsburg-Granits vom Südschwarzwald. *Neues Jahrbuch für Mineralogie, Monatshefte*, **12**, 532–544.

TODT, W., ALTENBERGER, U. & VON RAUMER, J. F. 1995. U–Pb data on zircons for the thermal peak of metamorphism in the Variscan Odenwald, Germany. *Geologische Rundschau*, **84**, 466–472.

TOLLMANN, A. 1982. Grossräumiger variszischer Deckenbau im Moldanubikum und neue Gedanken zum Variszikum Europas. *Geotektonische Forschungen*, **64**, 1–91.

TOMCZYK, H. 1988. Tectonics movements between margin of the East European Platform and the Holy Cross Mts. Region in the Palaeozoic. *Przegląd Geologiczny*, **36**(1), 9–16 [in Polish with English summary].

TOMEK, Č., DVORAKOVA, V. & VRANA, S. 1997. Geological interpretation of the 9HR and 503M seismic profiles in western Bohemia. *Sborník geologických ved, Geologie*, **47**, 43–50.

TURNAU, E., ŻELAŹNIEWICZ, A. & FRANKE, W. 2002. Middle to early late Viséan onset of late orogenic sedimentation in the Intra-Sudetic Basin, West Sudetes: miospore evidence and tectonic. *Geologia Sudetica*, **34**, 9–16.

TURNIAK, K., MAZUR, S. & WYSOCZANSKI, R. 2000. SHRIMP zircon geochronology and geochemistry of the Orlica-Śnieżnik gneisses (Variscan belt of Central Europe) and their tectonic implications. *Geodinamica Acta*, **13**, 1–20.

UNRUG, R. & DEMBOWSKI, Z. 1971. Diastrophic and sedimentary evolution of the Moravia-Silesia Basin. *Annales Societatis Geologorum Poloniae*, **41**(1), 119–168 [in Polish with English summary].

VAIDA, M., HANN, H. P., SAWATZKI, G. & FRISCH, W. 2004. Ordovician and Silurian protolith ages of metamorphosed clastic sedimentary rocks from the southern Schwarzwald, SW Germany: a palynological study and its bearing on the Early Palaeozoic geotectonic evolution. *Geological Magazine*, **141**(5), 629–643.

VAN BREMEN, O., AFTALION, M., BOWES, D. R., DUDEK, A., MISAR, Z., POVONDRA, P. & VRANA, S. 1982. Geochronological studies of the Bohemian Massif, Czechoslovakia, and their significance in the evolution of Central Europe. *Royal Society Edinburgh Earth Science*, **73**, 89–108.

VEJNAR, Z. 1982. Regionální metamõrfoza psamiticko-pelitických hornin domaŽlické oblasti. *Sborník geologických ved, Geologie*, **37**, 9–70 [in Czech with English abstract].

VON RAUMER, J. F. 1973. Die mineralfazielle Stellung der Metapelite und

Metagrauwacken, zwischen Heppenheim und Reichelsheim (Odenwald). *Neues Jahrbuch für Mineralogie, Abhandlungen*, **118**, 313–336.

von Raumer, J. F. 1998. The Palaeozoic evolution in the Alps: from Gondwana to Pangea. *Geologische Rundschau*, **87**(3), 407–435.

von Raumer, J. F. & Maggetti, M. 1975. Basite und Paragneise im Bereich von Heppenheim-Lindenfels, ein Vergleich geologischer Strukturen. *Aufschluss*, **27**, 39–45.

von Raumer, J. F. & Neubauer, F. 1993. *Pre-Mesozoic Geology in the Alps*. Springer, Berlin.

von Raumer, J. F., Stampfli, G. M., Borel, G. & Bussy, F. 2002. Organization of pre-Variscan basement areas at the north-Gondwanan margin. *International Journal of Earth Sciences*, **91**, 35–52.

Vrána, S. & Novák, M. 2000. Petrology and geochemistry of granulite clasts in the Visean Lulec conglomerate, Kulm in central Moravia, Czech Republic. *Bulletin of the Czech Geological Survey*, **75**, 4.

Wachendorf, H. 1986. Der Harz - variszischer Bau und geodynamische Entwicklung. *Geologisches Jahrbuch, A*, **91**, 3–67.

Wajsprych, B. 1986. Sedimentary record of tectonic activity on a Devonian–Carboniferous continental margin, Sudetes. *In:* Teisseyre A. K. (ed.) *IAS 7th Regional Meeting, Excursion Guidebook*. Ossolineum, Kraków & Wrocław, 141–164.

Wallbrecher, E., Brandmayr, M., Handler, R., Loizenbauer, J., Maderbacher, F. & Platzer, R. 1992. Konjugierende Scherzonen in der südlichen Böhmischen Masse: Variszische und alpidische kinematische Entwicklungen. *Mitteilungen der Östereichischen Mineralogischen Gesellschaft*, **137**, 237–252.

Walliser, O. H. 1981. The geosynclinal development of the Rheinisches Schiefergebirge (Rhenohercynian Zone of the Variscides; Germany). *In:* Zwart, H. J. & Dornsiepen, U. F. (eds) *The Variscan Orogen in Europe*. Geologie en Mijnbouw, **60**, 89–96.

Walliser, O. H. & Alberti, H. 1983. Flysch, olistostromes and nappes in the Harz Mountains. *In:* Martin, H. & Eder, F.W. (eds) *Intracontinental Fold Belts*. Springer, Berlin, 145–170.

Weber, K. 1978. Bewegungsbild im Rhenoherzynikum – Abbild einer varistischen Subfluenz. *Zeitschrift der deutschen geologischen Gesellschaft*, **129**, 249–281.

Weber, K. 1995*a*. Structure of the Spessart Crystalline Complex. *In:* Dallmeyer, D., Franke, W. & Weber, K. (eds) *Pre-Permian Geology of Central and Eastern Europe*. Springer, Berlin, 167–173.

Weber, K. 1995*b*. The Saar-Nahe Basin. *In:* Dallmeyer, D., Franke, W. & Weber, K. (eds) *Pre-Permian Geology of Central and Eastern Europe*, Springer, Berlin, 182–185.

Weber, K. & Juckenack, C. 1990. The structure of the Spessart Mts. Crystalline basement and its position in the frame of the European Variscides. *International Conference on Paleozoic Orogens in Central Europe, Field Guide Mid-German Crystalline Rise and Rheinisches Schiefergebirge*, 08/09/1990, Göttingen/Giessen 101–114.

Wedepohl, K.-H., Meyer, K. & Muecke, G. K. 1983. Chemical composition and genetic relations of metavolcanic rocks from the Rhenohercynian Belt of Northwest Germany. *In:* Martin, H. & Eder, F.W. (eds) *Intracontinental Fold Belts*. Springer, Berlin, 231–256.

Weil, A. B., Van der Voo, R. & van der Pluijm, B. A. 2001. Oroclinal bending and evidence against the Pangea megashear: The Cantabria-Asturias arc (northern Spain). *Geology*, **29**(11), 991–994.

Wendt, J. I. 1989. *U–Pb Zirkondatierung und Sm/Nd Systematik sowie Petrologie und Geochemie an Granuliten und Gneisen aus dem Moldanubikum Südböhmens (CSSR)*. Thesis, Johannes-Gutenberg-Universität, Mainz.

Wendt, J. I., Kröner, A., Fiala, J. & Todt, W. 1993. Evidence from zircon dating for existence of approximately 2.1 Ga old crystalline basement in southern Bohemia, Czech Republic. *International Journal of Earth Sciences*, **82**(1), 42–50.

Wendt, J. I., Kröner, A., Fiala, J. & Todt, W. 1994. U–Pb zircon and Sm-Nd dating of Moldanubian HP-HT granulites from South Bohemia, Czech Republic. *Journal of the Geological Society, London*, **151**, 83–90.

Wenzel, F., Brun, J.-P. & ECORS-DECORP working group 1991. A deep reflection seismic line across the Northern Rhine Graben. *Earth and Planetary Science Letters*, **104**, 140–150.

Werling, E. & Altherr R. 1986. Die metamorphe und tektonische Entwicklung des Südrandes der Zentralschwarzwälder Gneismasse bei Bernau. *Abstrakt Symposium für Tektonik, Strukturgeologie und Kristallingeologie, Tübingen*, 77–80.

Werner, C.-D. 1970. Geochemie der Brötteröder Serie im Ruhlaer Kristallin. *Freiberger Forschungshefte, C*, **259**, 5–82.

Werner, O. & Lippolt, H.J. 2000. White-mica $^{40}Ar/^{39}Ar$ ages of Erzgebirge metamorphic rocks: simulating the chronological results by a model of Variscan crustal imbrication. *In:* Franke, W., Haak, V., Oncken, O. & Tanner, D. (eds) *Orogenic Processes: Quantification and Modelling in the Variscan Belt*. Geological Society, London, Special Publications, **179**, 323–336.

Weyer, D. 1962. Zwei Oberdevon-Faunen von Schönau im Südschwarzwald. *Geologie*, **11**, 384–396.

Whittaker, A. (ed.) 1985. *Atlas of Onshore Sedimentary Basins in England and Wales*. Blackie, London.

Wickert, F. & Eisbacher, G. H. 1988. Two-sided thrust tectonics in the Vosges Mountains, northeastern France. *Geodinamica Acta*, **2**, 101–120.

Wickert, F., Altherr, R. & Deutsch, M. 1990. Polyphase Variscan tectonics and metamorphism along a segment of the Saxothuringian-Moldanubian boundary: the Baden-Baden Zone, northern Schwarzwald (FRG). *Geologische Rundschau*, **79**, 627–647.

Wiefel, H. 1976. Die geologische Entwicklung der Lahn-Dill-Erzlagerstätte Görkwitz bei Schleiz (Oberdevon und tiefes Dinant, Thüringisches Schiefergebirge). *Jahrbuch für Geologie*, **5/6** (für 1969/70), 451–588.

Wierich, F. & Vogt, W. 1997. Zur Verbreitung, Biostratigraphie und Petrographie unterkarbonischer Sandsteine des Hörre-Gommern-Zugs im östlichen Rhenohercynikum. *Geologica et Palaeontologica*, **31**, 97–142.

Will, T. M. 1998. Phase diagrams and their application to determine pressure–temperature paths of metamorphic rocks. *Neues Jahrbuch für Mineralogie, Abhandlungen*, **174**, 103–130.

Will, T. M. 2001. Paleostress-tensor analysis of late deformation events in the Odenwald Crystalline Complex and comparison with other units of the Mid-German Crystalline Rise, Germany. *Mineralogy and Petrology*, **72**, 229–247.

Will, T. M. & Schmädicke, E. 2001. A first report of retrogressed eclogites in the Odenwald Crystalline Complex: evidence for high-pressure metamorphism in the Mid-German Crystalline Rise, Germany. *Lithos*, **59**, 109–125.

Will, T. M. & Schmädicke, E. 2003. Isobaric cooling and anti-clockwise P-T paths in the Variscan Odenwald Crystalline Complex. *Journal of Metamorphic Geology*, **21**, 469–480.

Willet, S. D., Beaumont, C. & Fullsack, P. 1993. Mechanical model for the tectonics of doubly vergent compressional orogens. *Geology*, **21**, 371–374.

Williamson, J. P., Pharaoh, T. C., Banka, D., Thybo, H., Laigle, M., Lee & M. K. 2002. Potential field modelling of the Baltica-Avalonia (Thor-Tornquist) suture beneath the southern North Sea. *Tectonophysics*, **360**, 47–60.

Willner, A. P., Massone, H.-J. & Krohe, A. 1991. Tectonothermal evolution of a part of a Variscan magmatic arc: the Odenwald in the Mid-German Crystalline Rise. *Geologische Rundschau*, **80**, 369–389.

Willner, A. P., Rötzler, K. & Maresch, W. V. 1997. Pressure–temperature and fluid evolution of quartzo-feldspathic metamorphic rocks with a relic high-pressure, granulite-facies history from the Central Erzgebirge (Saxony, Germany). *Journal of Petrology*, **38**, 307–336.

Willner, A. P, Krohe, A. & Maresch, W. V. 2000. Interrelated P-T-t-d paths in the Variscan Erzgebirge Dome (Saxony, Germany): constraints on the rapid exhumation of high-pressure rocks from the root zone of a collisional orogen. *International Geology Review*, **42**(1), 64–85.

Wilser, J. 1933. Kulmische Schlotbrekzien und Crinoidenkalke im südschwarzwälder Paläozoicumstreifen (Geologie der Umgebung von Schönau im Wiesental). *Centralblatt für Mineralogie,Geologie und Paläontologie*, **10B**, 529–542.

Wilser, J. 1935. Südgerichteter Schuppenbau und carbonischer Vulkanismus im mittleren Schwarzwald. *Neues Jahrbuch für Mineralogie, Beilage B*, **73**, 341–381.

Wimmenauer, W. 1984. Das prävariskische Kristallin im Schwarzwald. *Fortschritte Mineralogie*, **62**, 69–86.

Wimmenauer, W. 1988. Precambrian in the horst mountains of the Rhine graben area. *In:* Zoubek, V. (ed.) *Precambrian in Younger*

Fold Belts. European Variscides, the Carpathians and Balkans. Wiley, Chichester, 381–408.

WIMMENAUER, W. & HANEL, M. 1997. Die Fortsetzung der Randgranit-Assoziation nach Nordosten und Norden. *Jahreshefte Geologisches Landesamt Baden-Württemberg*, **37**, 7–24.

WINCHESTER, J. A. & PACE TMR NETWORK TEAM 2002. Palaeozoic amalgamation of Central Europe: new results from recent geological and geophysical investigations. *Tectonophysics*, **360**, 5–21.

WINCHESTER, J. A., FLOYD, P. A., CHOCYK, M., HORBOWY, K. & KOZDRÓJ, W. 1995. Geochemistry and tectonic environment of Ordovician meta-igneous rocks in the Rudawy Janowickie Complex, SW Poland. *Journal of Geological Society London*, **152**, 105–115.

WINCHESTER, J. A., PHARAOH, T. C. & VERNIERS, J. 2002. Palaeozoic amalgamation of Central Europe: an introduction and synthesis of new results from recent geological and geophysical investigations. *In:* WINCHESTER J.A., PHARAOH T.C. & VERNIERS J. (eds) *Palaeozoic Amalgamation of Central Europe*. Geological Society, London, Special Publications, **201**, 1–18.

WINTSCH, R. P., KUNK, M. J., BOYD, J. L. & ALEINIKOFF, J. N. 2003. P-T-t paths and differential Alleghanian loading and uplift of the Bronson Hill Terrane, south central New England. *American Journal of Science*, **303**(5), 410–446.

WITTENBERG, A., VELLMER, C., KERN, H., & MENGEL, K. 2000. The Variscan lower continental crust: evidence for crustal delamination from geochemical and petrophysical investigations. *In:* FRANKE, W., HAAK, V., ONCKEN, O. & TANNER, D. (eds) *Orogenic Processes: Quantification and Modelling in the Variscan Belt.* Geological Society, London, Special Publications, **179**, 401–414.

WONIK, T. 1992. *Kompilation und Interpretation der magnetischen Anomalien der Totalintensität in Zentral- und Nordeuropa.* PhD thesis, Georg-August University, Göttingen.

WUNDERLICH, J. 1992. Petrogenese, Lagerungsverhältnisse und Gliederung der Steinbacher Augengneisfolge und ihre Position im strukturellen Bau des zentralen Ruhlaer Kristallins (Thüringer Wald). *Geologisches Jahrbuch Hessen*, **120**, 11–35.

WUNDERLICH, J. & ZEH, A. 2001. Ruhlaer Kristallin - Teil der MKZ. *In:* STEININGER, F. (ed.) *Stratigraphische Kommission Deutschland, Stratigraphie von Deutschland II, Ordovizium, Kambrium, Vend, Riphäikum.* Courier Forschungs Institut Senckenberg, Frankfurt, **234**, 135–171.

WYBRANIEC, S., ZHOU, S., *et al.* 1998. New map compiled of Europe's gravity field. *EOS Transactions of the American Geophysical Union*, **79**(37), 437–442.

ŻABA, J. & BĘDKOWSKI, Z. 1995. The sequence of shear zones in the NE part of the Śnieżnik Metamorphic Massif. *In*: MUSZER (ed.) *Góry Złote, Conference Proceedings*, University of Wrocław, 20–27 [in Polish].

ŽACEK, V. & CHAB, J. 1993. Metamorphism in the Tepla upland, Bohemian Massif, Czech Republic (preliminary report). *Vestnik ceskeho geologickeho Ustavu*, **68**(3), 33–37.

ZEH, A. 1996. Die Druck-Temperatur-Deformations-Entwicklung des Ruhlaer Kristallins (Mitteldeutsche Kristallinzone). *Geotektonische Forschungen*, **86**, 1–214.

ZEH, A. & BRÄTZ, H. 2004. Timing of Upper Carboniferous/Permian horst-basin formation and magmatism in the NW Thuringian Forest, Central Germany, a review. *In:* WILSON, M., NEUMANN, E.-R., DAVIES, G. R., TIMMERMANN, M. J., HEEREMANS, M. & LARSEN, B. T. (eds) *Permo-Carboniferous Magmatism and Rifting in Europe*. Geological Society, London, Special Publications, **223**, 319–334.

ZEH, A., COSCA, M., BRÄTZ, H., TICHOMIROWA, M. & OKRUSCH, M. 2000. Simultaneous horst-basin formation and magmatism during Late Variscan Transtension: Evidence from $^{40}Ar/^{39}Ar$ and $^{207}Pb/^{206}Pb$ geochronology in the Ruhla Crystalline Complex. *International Journal of Earth Sciences*, **89**, 52–71.

ZEH, A., BRÄTZ, H., MILLAR, I. L. & WILLIAMS, I. S. 2001. A combined zircon SHRIMP and Sm-Nd isotope study on high-grade paragneisses from the Mid-German Crystalline Rise: Evidence for northern Gondwanan and Grenvillian provenance. *Journal of the Geological Society, London*, **158**, 983–994.

ZEH, A., WILLIAMS, I.S., BRÄTZ, H. & MILLAR, I. L. 2003. Different age response of zircon and monazite during the tectono-metamorphic evolution of a high grade paragneiss from the Ruhla Crystalline Complex, Central Germany. *Contributions to Mineralogy and Petrology*, **145**, 691–706.

ZEH, A., GERDES, A., WILL, T. M. & MILLAR, I. L. 2005. Provenance and magmatic-metamorphic evolution of a Variscan island-arc complex: constraints from U–Pb dating, petrology, and geospeedometry of the Kyffhäuser Crystalline Complex, Central Germany. *Journal of Petrology*, **46**, 1393–1420.

ZEITLER, P. K., MELTZER, A. S., *et al.* 2001. Erosion, Himalayan geodynamics, and the geomorphology of metamorphism. *GSA Today*, **11**, 4–9.

ŻELAŹNIEWICZ A., MARHEINE D. & OBERC-DZIEDZIC T. 2003. A Late Tournaisian synmetamorphic folding and thrusting event in the eastern Variscan foreland: $^{40}Ar/^{39}Ar$ evidence from the phyllites of the Wolsztyn-Leszno High, western Poland. *International Journal of Earth Sciences*, **92**, 185–194.

ZIEGLER, P. A. 1986. Geodynamic model for the Palaeozoic crustal consolidation of Western and Central Europe. *Tectonophysics*, **126**, 303–328.

ZIEGLER, P. A. 1989. *Evolution of Laurussia.* Kluver Academic Publishers, Dordrecht.

ZIEGLER, P. A. 1990. *Geological Atlas of Western and Central Europe*, 2nd edn. Shell Internationale Petroleum Maatschappij B.V., Geological Society, London; Elsevier, Amsterdam.

ZIEGLER, P. A. & DÈZES, P. 2006. Crustal Evolution of Western and Central Europe. *In:* GEE, D. G. & STEPHENSON, R. A. (eds) *European Lithosphere Dynamics*. Geological Society, London, Memoirs, **32**, 43–56.

ZIEGLER, P. A. & WIMMENAUER, W. 2001. Possible glaciomarine diamictites in Lower Paleozoic series of the Southern Black Forest (Germany): implications for the Gondwana/Laurussia puzzle. *Neues Jahrbuch für Geologie und Paläontologie, Monatshefte*, **8**, 500–512.

ZIMMERLE, W., HERING, O., *et al.* 1976. Die Tiefbohrung Saar 1.- Petrographische Beschreibung und Deutung der erbohrten Schichten. *Geologisches Jahrbuch, A*, **27**, 91–305.

ZIMMERMANN, E. 1930. *Erläuterungen zur Geologischen Karte von Preußen und angrenzender Staaten 1:25000, Bl.2992 Ruhla.* Preußische Geologische Landesanstalt, Berlin.

ZNOSKO, J. 1983. Tectonics of the Palaeozoic Platform in the central-south Poland. *Kwartalnyk Geologiczny*, **27**, 73–78 [in Polish with English summary].

ZNOSKO, J. 1996. Tectonic style of the Early Palaeozoic sequences in the Holy Cross Mountains. *Geological Quarterly*, **40**(1), 1–22.

ZONENSHAIN, L.P., KUZMIN, M.I & NATAPOV, L.M. 1990. *Geology of the USSR: A Plate-tectonic Synthesis*. Geodynamics Series **21**, American Geophysical Union, Washington DC.

ZULAUF, G. 1994. Ductile normal faulting along the West-Bohemian Shear Zone (Moldanubian/Teplá-Barrandian boundary) Evidence for late Variscan extensional collapse in the Variscan internides. *Geologische Rundschau*, **83**, 276–292.

ZULAUF, G. 1997. Von der Anchizone bis zur Eklogitfazies: Angekippte Krustenprofile als Folge der cadomischen und variscischen Orogenese im Teplá-Barrandium (Böhmische Masse). *Geotektonische Forschungen*, **89**, 1–302.

ZULAUF, G. 2001. Structural style, deformation mechanisms and paleostress along an exposed crustal section: Constraints on the rheology of quartzofeldspathic rocks at supra- and infrastructural levels (Tepla-Barrandian unit, Bohemian Massif). *Tectonophysics*, **332**, 211–237.

ZULAUF, G., DÖRR, W., FIALA, J. & VEJNAR, Z. 1997. Late Cadomian crustal tilting and Cambrian transtension in the Teplá-Barrandian unit (Bohemian Massif, Central European Variscides). *Geologische Rundschau*, **86**, 571–584.

ZULAUF, G., BUES, C., DÖRR, W. & VEJNAR, Z. 2002*a*. 10 km minimum throw along the West Bohemian shear zone: Evidence for dramatic crustal thickening and high topography in the Bohemian Massif (European Variscides). *International Journal of Earth Sciences*, **91**, 850–864.

ZULAUF, G., DÖRR, W., FIALA, J., KOTKOVA, J., MALUSKI, H. & VALVERDE-VAQUERO, P. 2002*b*. Evidence for high-temperature diffusional creep preserved by rapid cooling of lower crust (North Bohemian shear zone, Czech Republic). *Terra Nova*, **14**, 343–354.

ZWART, H.J. 1967. The duality of orogenic belts. *Geologie en Mijnbouw*, **46**, 283–309.

12 Palaeozoic magmatism

MARTIN J. TIMMERMAN

This chapter summarizes the style, timing, composition and tectonic setting of the main occurrences of Cambrian to early Permian magmatic rocks in central Europe, which are here described within the framework of the Cadomian and Variscan Orogenies. In general terms, the Variscan Orogeny may be considered to be the result of Silurian to early Carboniferous accretion onto the southern margin of Laurussia of various Gondwana-derived terranes or microplates of predominantly Neoproterozoic (Cadomian/Pan-African) crust, together with their passive margin sequences and accreted island arcs (Franke 1989; Matte 1991; Ziegler 1993). These microplates originated from various parts along the northern margin of Gondwana in the Early Palaeozoic, and moved northward towards Laurentia and Baltica (see Krawczyk *et al.* 2008). These rifting, spreading, subduction, accretion and collision events occurred over a long period and were associated with magmatic activity of varying styles, compositions and volumes, of which the variously deformed and metamorphosed equivalents are found throughout Variscan Europe. Another important, late to post-Variscan phase of magmatism which occurred throughout Europe was of late Carboniferous to early Permian age.

The magmatic rocks and their metamorphosed equivalents are exposed in basement uplifts (the Variscan massifs), such as the Bohemian Massif, Odenwald, Spessart, Black Forest, Vosges, Massif Central, Iberia and the Rhenohercynian Zone (Fig. 12.1). In these internal parts of the Variscan Orogen, magmatic rocks are ubiquitous but are predominantly plutonic rocks and their metamorphosed equivalents, since mainly deep crustal levels are exposed. To the south, von Raumer (1998) estimated that over 50% of exposed rocks represent Palaeozoic basement of the Alps. Especially in the northern Variscan foreland, Palaeozoic rocks occur below thick Mesozoic to Cenozoic cover sequences and are known from deep boreholes only, or their presence is inferred from geophysical data. Magmatic rocks encountered in boreholes in, for instance, the NE German Basin, the North Sea, and NE France are mainly volcanic and of late Carboniferous to Permian age (e.g. Châteauneuf & Farjanel 1989; Aghabawa 1993; Hoth *et al.* 1993*b*; Korich & Kramer 1994; Marx *et al.* 1995; Benek *et al.* 1996; Breitkreuz & Kennedy 1999; McCann 1996, 1998; Heeremans *et al.* 2004*a*, *b*).

The early to late Palaeozoic magmatic rocks of central Europe comprise a large variety of rock types that reflect the evolution of the Caledonian and Variscan orogenies. Despite the many remaining uncertainties concerning the evolution of this orogen, and the fact that most of the earlier Palaeozoic magmatic suites have been variably deformed, metamorphosed, dismembered and eroded during later Variscan and Alpine events, several main magmatic stages can be recognized in Central Europe:

- Late Cambrian–early Ordovician mafic (-ultramafic) and mafic-felsic (i.e. bimodal) suites that were formed as a result of rifting, microplate break-off and spreading at the northern Gondwana margin. These are often associated with continental to passive margin and pelagic sediments.
- Silurian ophiolites related to the dispersal and northward drift of the microplates, and Silurian–early Devonian arc-type associations related to subduction of the oceans and seaways at the leading edge of the microcontinents.
- Early to mid-Devonian mafic and bimodal volcanic suites, in places associated with continental-shelf to deep-marine sediments, related to the opening of the Rhenohercynian Ocean.
- Late Devonian–early Carboniferous arc-type magmatic rocks related to the subduction and closure of the Rhenohercynian Ocean.
- Latest Carboniferous–early Permian, plutonic and volcanic rocks of felsic to intermediate composition that mark a period of Europe-wide extension and the collapse of the Variscan Orogen.

Data quality

Magmatic and sedimentary rocks in central Europe, especially those of early Palaeozoic age, are generally metamorphosed and strongly deformed, often during several tectonometamorphic events ranging from early Palaeozoic subduction to mid-Carboniferous collisional events and, in south and SE Europe, the Alpine Orogeny. The younger Variscan events often overprinted older structures, field relations, mineral assemblages and palaeomagnetic signatures, thereby greatly complicating the reconstruction of earlier histories (e.g. Tait *et al.* 1997; O'Brien 2000). Apart from those successions and intrusions in the basement of the Alps and Carpathians, younger, Carboniferous–early Permian magmatic rocks are much less deformed, or even undeformed as they mostly post-date the peak of the Variscan Orogeny.

The designation of, in particular, the early Palaeozoic magmatic rocks to a specific geotectonic setting and a specific Variscan event is, in many cases, not possible without ambiguity. This is due to a variety of factors, ranging from the recrystallization of the original, magmatic mineral assemblages, element mobility before, during and after metamorphism and deformation (often polyphase), dismemberment, excision and erosion of parts of the associations, unknown, imprecise or inaccurate crystallization ages of the magmatic protoliths, incomplete geochemical and isotope data, and inheritance and contamination effects. In addition, several models exist for the evolution of the Caledonian and Variscan orogens and, although broadly similar, these differ in important details. It is, therefore, often not possible to assign each and every magmatic suite to a specific terrane, microplate, intervening basin, tectonic setting, or stage of the Variscan Orogeny.

The late Devonian to mid-Carboniferous collisional history of the Variscan Orogen is well known and dated (see Belka & Narkiewicz 2008; Kroner *et al.* 2008; McCann *et al.* 2008*a*), but many aspects of the early Palaeozoic evolution of Central Europe are still unclear. In many early Palaeozoic magmatic rocks, strong reworking and metamorphic overprinting, in combination

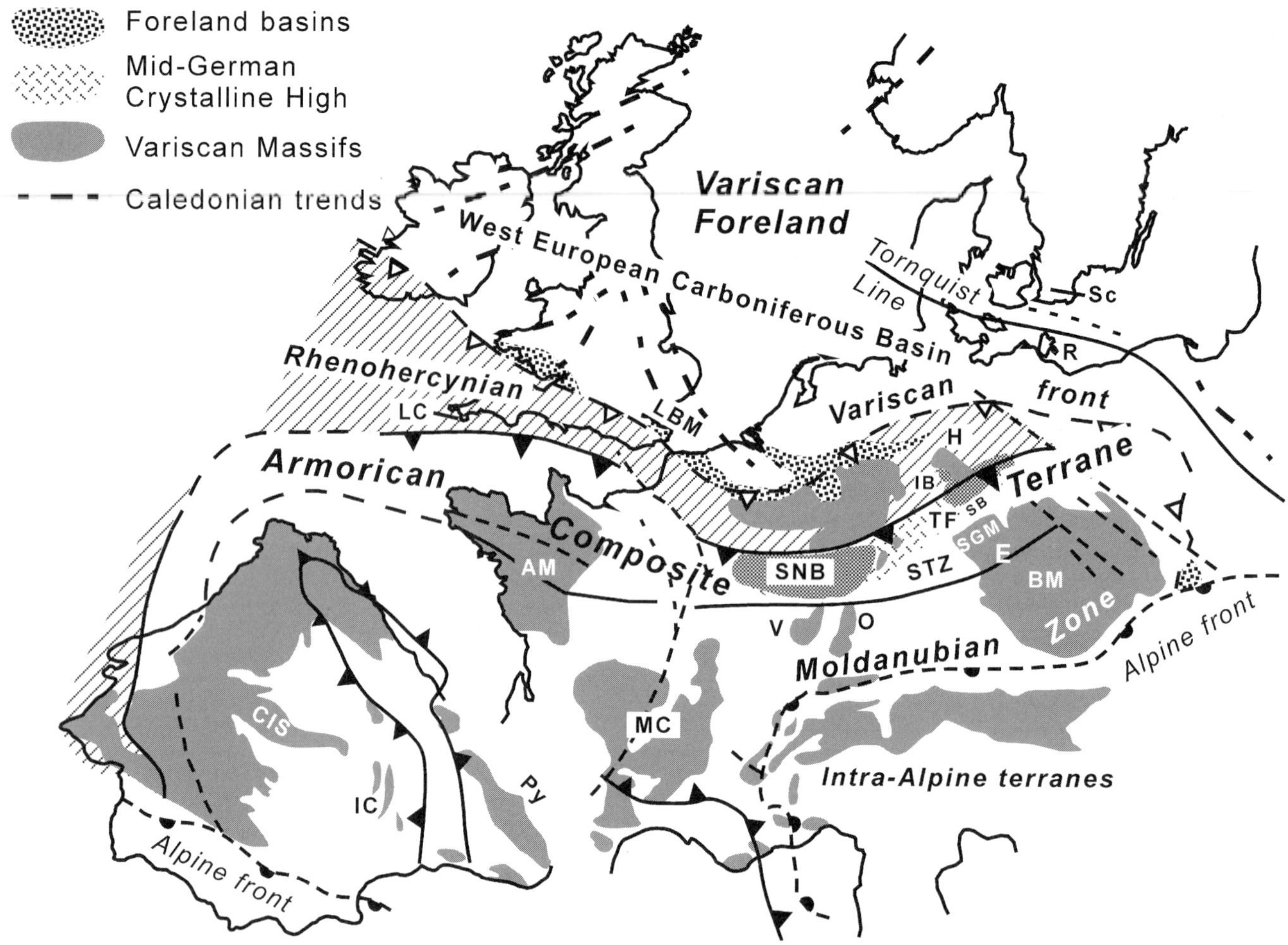

Fig. 12.1. The Variscan Orogen in Europe, with foreland, internal high-grade zones, and exposed massifs (after Franke 1989). Abbreviations: AM, Armorican Massif; BM, Bohemian Massif; CIS, Central Iberian System; E, Erzgebirge/Krušné hory; H, Harz; IB, Ilfeld Basin; IC, Iberian Chain; LBM, London-Brabant Massif; LC, Lizard Complex; MC, Massif Central; O, Odenwald; Py, Pyrenees; R, Rügen; SB, Saale Basin; Sc, Scania; SMG, Saxon Granulite Massif; SNB, Saar-Nahe Basin; STZ, Saxothuringian Zone; TF, Thuringian Forest; V, Vosges.

with pre- and post-metamorphic alteration, often caused redistribution of the mobile elements (i.e. Na, Ca, K, Mg, Fe, Rb, Sr, Ba) and disturbance of isotope systems. In most cases, the protolith character and original tectonic setting have to be inferred from their geochemical and isotope compositions using those elements that are considered to be relatively immobile during such processes. These include the high field-strength elements (HFSE) Ti, P, Nb, Ta, Hf, Zr, Y, the rare earth elements (REE), Th and, under certain conditions, Sr, Ni, Cr, Sc and V. Smooth REE patterns and a narrow range of ratios between high field-strength elements suggest these reflect the composition of the magmatic protoliths. Therefore, the quality and completeness of published geochemical, isotope and geochronological data are important factors that determine the reliability of the geological interpretations. However, geochemical differences between associated, possibly cogenetic mafic magmatic rocks will often be due to a variety of causes, such as different compositions of the mantle sources, different degrees of melting and melting depths, the effects of fractional crystallization and the degree of contamination with crustal material, or a combination of these. A clear-cut answer will not be possible in some cases.

In general terms, felsic to intermediate rocks with calc-alkaline compositions may, for instance, indicate formation in subduction settings, although such a subduction signature may also be inherited by remelting of, or contamination by, older calc-alkaline rocks. Alkaline felsic compositions indicate genesis during thinning and rifting of continental crust. Contamination of (ultra-)mafic melts with (often much older) continental crust of felsic to intermediate composition will have several effects that, in turn, are influenced by fractionation processes. Contamination with older, crustal material lowers the $\varepsilon_{Nd}(t)$ values, increases the initial $^{87}Sr/^{86}Sr$ ratios, and increases the light rare earth element (LREE) contents, and therefore steepens the slope of the REE pattern (expressed as La_N/Sm_N, La_N/Lu_N or La_N/Yb_N, chondrite-normalized ratios). The addition of LILE- (large ion lithophile element), Th- and LREE–rich, Ti- and Nb-poor crust or crust-derived felsic melts can also cause negative HFSE anomalies (Ti, Nb) with respect to the LILE and REE, resulting in high or variable LILE/HFSE ratios (a measure of the 'slope' of the multi-element pattern). It should, therefore, be kept in mind that negative Nb anomalies are not solely a characteristic of volcanic arc magmatism, but may be unrelated to subduction processes and be due to contamination by, or direct partial melting of, continental (i.e. ultimately arc-derived) crust in an ensialic setting.

In addition to geochemical character, precise and reliable

protolith crystallization ages of the Variscan and pre-Variscan magmatic rocks, and the age of metamorphism are prerequisite in order to recognize elements and events common to various units, and to correlate magmatic and metamorphic units that may have been part of a larger, contiguous terrane or microplate. In addition, reliable ages are necessary to calculate the correct initial isotope ratios. In this way, the plate tectonic environment(s) in which they were formed may be reconstructed or inferred. For instance, synchronicity or a short timespan between the time of magmatism, deformation and high-grade metamorphism is a strong indicator for a tectonically active setting at a convergent plate boundary. In addition, knowledge of the age, geochemistry and isotope composition of the pre-Variscan basement is also necessary for interpreting the origin and nature of the later Palaeozoic magmatic rocks, since the older rocks may have acted as a source reservoir for the latter's parent magmas (by direct melting, anatexis) or influenced their composition by assimilation and contamination processes.

The crystallization age of the protoliths of metamorphosed mafic rocks is difficult to determine, as many mafic melts (especially rapidly crystallized volcanic rocks) tend not to crystallize zircon that can be dated by the U–Pb method (e.g. Black *et al.* 1991; Gebauer & Grünenfelder 1979). Magmatic baddeleyite and zirconolite are not stable during high-grade, protracted metamorphism and react to form metamorphic zircon. Furthermore, Rb–Sr and Sm–Nd whole-rock isochrons are often imprecise and/or inaccurate due to element mobility, or contamination or inheritance effects. U–Pb zircon dating of intercalated felsic magmatic rocks is an alternative to indirectly date associated metamorphosed mafic rocks, assuming that these were formed coevally and not just unrelated rock associations that were juxtaposed during later tectonic events.

Due to widespread metamorphic overprinting or secondary alteration, the cited ages are mainly limited to precise and accurate U–Pb and Pb/Pb zircon and baddeleyite ages (determined by isotope dilution, ion-probe, laser ablation or the Kober evaporation technique) and $^{40}Ar/^{39}Ar$ mineral plateau ages, with uncertainties on the 2-sigma level. The time-scale used is that of Gradstein *et al.* (2004). The geochemical data taken from the literature have been plotted using the GCDkit software of Janoušek *et al.* (2006*a*) and the normalization values of Sun & McDonough (1989).

Geological framework: the Variscan terranes

In general terms, the Caledonian and Variscan orogens in Central Europe may be considered the result of the closure of several Palaeozoic oceanic basins of varying size and age, such as the Rheic, Tornquist, Saxothuringian and Rhenohercynian oceans. The destruction of oceanic crust and basin closure between the Ordovician and early Carboniferous culminated with accretion onto the southern margin of Laurussia of various Gondwana-derived terranes or microplates (some of them composite). The two main accretion and collision events in Central Europe are the Silurian Caledonian Orogeny and the Devonian to early Carboniferous Variscan Orogeny. The accreted terranes are predominantly composed of Neoproterozoic crust, Neoproterozoic to Silurian passive margin sequences, and accreted Ordovician–Devonian island arcs (Franke 1989; Pin 1990; Matte 1991; Ziegler 1993).

In the Variscan Orogen in central Europe, the recognition and delineation of individual microplates is based on the recognition of former plate margins, which can be identified by plate tectonic elements such as transcrustal faults or sutures, subduction-related eclogites and accreted arc material, ophiolites, deformed passive margin deposits, accretionary wedges, thrusts and nappes, blueschist-facies metamorphism, contrasts in age and grade of metamorphism across boundaries, in combination with geophysical data and terrane-specific (micro)fossil assemblages from Palaeozoic sediments. Correlation models between the terranes and their affiliations with Baltica, Gondwana or Laurentia depend on palaeontological and sedimentological data (e.g. Ordovician diamictites), the age of magmatism, the ages of zircons inherited from basement rocks, the age spectra of detrital zircons in (meta)sediments, and Sr–Nd isotope data. Strong deformation and metamorphism during peak-Variscan and Alpine events in the high-grade core of the orogen in Central Europe have made this a complicated task. In addition, some terranes or terrane assemblages were broken up during their northward drift or were dismembered by indentor tectonics upon collision. The fragments now occur separated and redistributed along the southern margin of Laurussia, from Central America into east Europe (e.g. Franke 2000; Winchester *et al.* 2002a, b).

Despite the limitations caused by later reworking, two main microplates have been recognized in Central Europe, namely (East) Avalonia and the Armorican Terrane Assemblage (ATA, 'Armorica'), themselves composite and comprising smaller crustal blocks as is clear from faunal similarities (Tait *et al.* 1997; Matte 2001). Separation of the dispersed terranes and the former existence of the Iapetus Ocean (between Laurentia and Baltica, and between the western part of East Avalonia and Laurentia) and the Tornquist Ocean (between Baltica and the eastern part of East Avalonia) was first inferred from the distinctive provincial faunas of Laurentia, Gondwana and Baltica (e.g. Cocks & Fortey 1982, 1990; Cocks *et al.* 1997; McKerrow & Cocks 1995; Fortey & Cocks 2003; Torsvik & Cocks 2004; and references therein). Furthermore, the occurrence of ophiolite complexes (often dismembered and metamorphosed) and eclogites suggests the nearby presence of terrane boundaries or suture zones and, thereby, the former existence of oceanic basins.

The Armorican Terrane Assemblage consisted of various fragments of continental crust, such as Iberia, Armorica (including the basement of the Paris Basin), Saxothuringia (including Lusatia/Lausitz), Teplá-Barrandia (= Perunica) and Moldanubia, that share common elements such as a basement of late Neoproterozoic rocks, Late Ordovician glacial deposits and similar Gondwanan faunas. The microplates of the ATA were separated by basins which were often underlain by oceanic crust. These basins must have had limited widths ('seaways') since there is no palaeontological or palaeomagnetic evidence for large separations (e.g. McKerrow & Cocks 1995).

The composite terranes, or terrane assemblages, dispersed at different times in the early Palaeozoic from various parts of the northern margin of Gondwana (the peri-Gondwana terranes; Tait *et al.* 1997; von Raumer 1998; Stampfli *et al.* 2002; von Raumer *et al.* 2002; Murphy *et al.* 2002; Keppie *et al.* 2003, and references therein). Gondwana itself was located at the South Pole in late Neoproterozoic to early Palaeozoic times (e.g. Torsvik & Van der Voo 2002). Peri-Laurentian terranes have not been identified in the Variscan Orogen, and the affinity of the Bruno–Silesia microplate in SW Poland is still a matter of debate (e.g. Finger *et al.* 2000; Winchester *et al.* 2002*b*; Kalvoda *et al.* 2002). Many sedimentary protoliths in Central Europe date from this period, and were deposited in accommodation space created by rifting and crustal thinning, such as Cadomian forearc and backarc basins and Cambro-Ordovician basins and passive margins. The microplates were probably situated in lateral continuity with one another and, thus, many share a common evolution prior to Cambro-Ordovician breakup and drift. Figure

12.2 shows a possible configuration of the peri-Gondwana terranes (Nance & Murphy 1996; Linnemann 2003, and references therein), but several other, slightly differing models have also been proposed (e.g. Murphy & Nance 1989; von Raumer 1998; Stampfli *et al.* 2002; von Raumer *et al.* 2002; Murphy *et al.* 2002, 2004*b*).

The cores of the terranes are composed of (late) Neoproterozoic to (locally) earliest Cambrian, generally calc-alkaline, I-type igneous and meta-igneous rocks that were formed during the Pan-African crust formation event and subsequent, latest Neoproterozoic (Cadomian), Andean-type subduction and accretion processes along the northern margin of Gondwana (Fig. 12.2; Cocherie *et al.* 2001; Compston *et al.* 2002; Guerrot & Peucat 1990; Tucker & Pharaoh 1991). These subduction-related continental arc granitoids mark the waning stages of Cadomian magmatism at the northern margin of Gondwana, which was associated with *c.* 540–550 Ma, generally low-grade metamorphism (e.g. Schaltegger *et al.* 1997; Turniak *et al.* 2000; Timmermann *et al.* 2004). During later rifting this arc was broken up and dispersed, but Cadomian magmatic arc-related rocks constitute an element common to many of the terranes of Central Europe. Their Mesoproterozoic Sm–Nd model mantle ages (t_{DM}) reflect mixing of Palaeoproterozoic crust and Neoproterozoic juvenile magmas (e.g. Friedl *et al.* 2000; Liew & Hofmann 1988).

Remnants of the Cadomian arc have been identified in different parts of the Variscides, especially of the ATA, as is clear from U–Pb zircon ages of granitoids and gabbros in the Fichtelgebirge and Erzgebirge/Krušné hory (Fig. 12.1; Kröner *et al.* 1995; Siebel *et al.* 1997; Košler *et al.* 2004; Tichomirowa *et al.* 2001; Tichomirowa 2002, 2003), North Saxony Anticline (Hammer *et al.* 1999), Lusatia/Lausitz (Kröner *et al.* 1994; Hammer *et al.* 1999; Linnemann *et al.* 2000; Dörr *et al.* 2002; Tichomirowa 2002, 2003), the Sudetes (Krkonoše/Karkonosze Complex; Kröner *et al.* 2001; Żelaźniewicz *et al.* 2004), the Orlice-Sněžník/Orlica-Śnieżnik Complex (Turniak *et al.* 2000) and the Alps (e.g. in the Silvretta nappe (Schaltegger *et al.* 1997) and Tauern Window (Schulz & Bombach 2003)).

Fragments of much older, Palaeoproterozoic (meta-)igneous rocks, such as those exposed in the Channel Islands (Samson & D'Lemos 1998; Miller *et al.* 1999; Inglis *et al.* 2004) or offshore northern Iberia (Guerrot *et al.* 1989), in combination with ages of inherited zircons in both orthogneisses and metasediments, suggest that Armorica and related terranes were situated near, and affiliated with, the West African Craton, whereas Avalonia shows affinities with the Amazon Craton (e.g. Baumann *et al.* 1991; Friedl *et al.* 2000, 2004; Gehmlich *et al.* 1997; Guerrot *et al.* 1989; Kröner *et al.* 1988, 1994, 1995; Linneman 1995; Linnemann & Romer 2002; Linneman *et al.* 2000; Murphy & Nance 2002; Murphy *et al.* 2004*a*; Nagy *et al.* 2002; Nance & Murphy 1996; Tichomirowa *et al.* 2001; Turniak *et al.* 2000; see Linnemann *et al.* 2008; McCann 2008).

The Avalonia microplate and the ATA had broken away from Gondwana during the mid-Cambrian to Ordovician period. To the north, Laurentia and Baltica collided in the Silurian, resulting in the Caledonian Orogeny in Scotland, Ireland and Norway, and the formation of the Laurussian continent (e.g. Woodcock & Strachan 2000). Further south, (East) Avalonia drifted northward, opening the Rheic Ocean behind it, and causing the closure of the Iapetus and Tornquist oceans that separated it from Laurentia and Baltica, respectively. Closure of the Iapetus Ocean was accompanied by subduction and arc magmatism in the oceanic realm and at the leading edge of (East) Avalonia (Noble *et al.* 1993; Pharaoh *et al.* 1991, 1993*a*). Accretion of Avalonia took place from the late Ordovician into the Silurian, thus forming the Caledonian Orogen of Central Europe: the low grade Anglo-

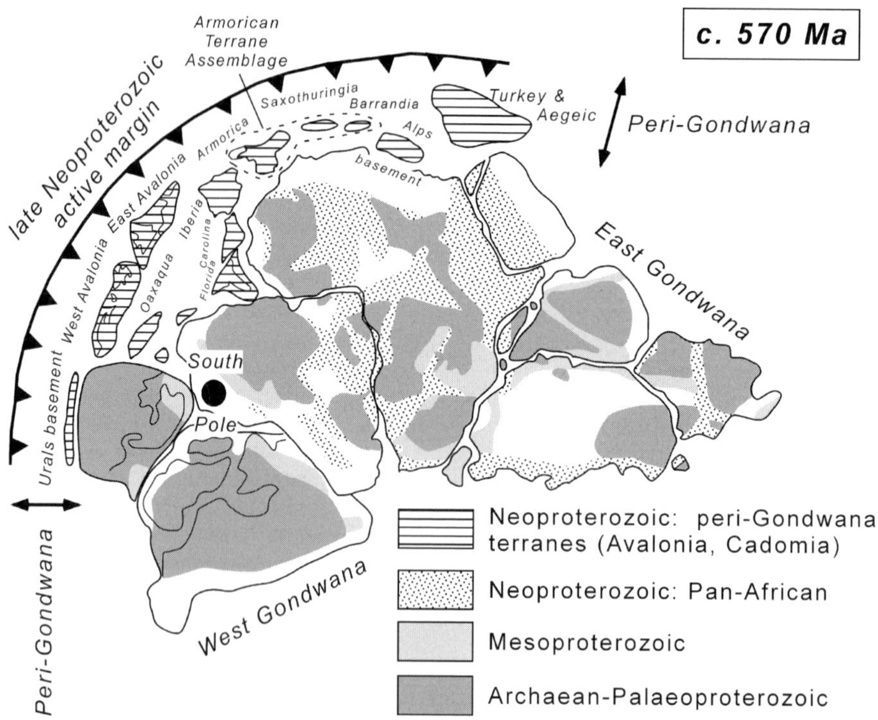

Fig. 12.2. Palaeogeographic reconstruction of the Variscan terranes at the northern margin of Gondwana in late Neoproterozoic times (peri-Gondwanan terranes), prior to fragmentation and dispersal. After Nance & Murphy (1996), Linneman (2003) and references therein.

Brabant Deformation Belt and the Trans-European Suture Zone (North German–Polish Caledonides), now largely buried (e.g. Frost *et al.* 1981; Pharaoh 1999; Pharaoh *et al.* 1993*b*; Dewaele *et al.* 2002; Verniers *et al.* 2002; Winchester *et al.* 2002*a*, *b*; Debacker *et al.* 2005).

While Avalonia docked with Laurentia and Baltica, the terranes belonging to the Armorican Terrane Assemblage had broken away from northern Gondwana, opening another oceanic basin behind it (Saxothuringian Ocean or Galicia-Southern Brittany Ocean; Matte 2001), and drifted northwards towards future Laurussia. The ATA may have been an archipelago of terranes of varying sizes, broken up during rifting and drifting, and separated by narrow seaways, some of which were underlain by oceanic crust as noted above. Closing and subduction of the Rheic Ocean generated volcanic arc-like magmatic rocks, remnants of which are found in the Mid-German Crystalline High (see below) and in the Brabant Massif of the Anglo-Brabant Deformation Belt (e.g. André *et al.* 1986; Verniers *et al.* 2002). The ATA docked with the southern, Avalonian margin of Laurussia in the late Ordovician to early Devonian. Many of the Central European terranes also contain accreted arc material, now preserved as reworked Ordovician–late Devonian orthogneisses, metavolcanics, ophiolites and subduction-related eclogites (e.g. Costa 1992; Pin & Paquette 2002).

In the late Devonian to mid-Carboniferous (Visean) the last ocean basins had closed, and microplates and subduction-related magmatic arcs had collided with and accreted to the southern margin of Laurussia (consisting of Laurentia and Baltica). In Central Europe these accretion and collision events resulted in widespread high-grade metamorphism and granitoid magmatism of predominantly Visean age (*c.* 330–350 Ma).

Cambrian–Early Ordovician

Post-Cadomian magmatism in Central Europe commenced with extension and rifting related to incipient fragmentation of the northern margin of Gondwana and the opening of the Rheic Ocean between the Avalonia microplate and Gondwana. At a later stage, the Armorican microplate and other terranes rifted away, forming other ocean basins. Rifting and terrane dispersal began in the early-mid-Cambrian and lasted into the Silurian. The existence of moving plates during this period is supported by palaeomagnetic evidence and the distribution of benthic fauna. Plate movements imply rifting and the generation and destruction of oceanic crust, but the width of the basins is more difficult to establish since palaeomagnetic data for Gondwana in the early Palaeozoic are lacking, or of poor quality. This suggests that in this period Gondwana, too, may have moved northward and that the early Palaeozoic oceanic basins were narrower than previously thought (Torsvik & Van der Voo 2002).

The magmatic rocks produced during this period are found as granitoid intrusions and granitoid orthogneisses, mafic/ultramafic assemblages, mafic to bimodal volcanosedimentary sequences and (locally) ophiolites, or parts thereof. These have been interpreted as tectonized and metamorphosed remnants of continental rifts, passive margins and ocean crust. They occur widely distributed throughout the Variscan Orogen.

In Central Europe, several early Palaeozoic volcanosedimentary complexes and ophiolites, now variably deformed, dismembered and metamorphosed, appear to form a belt of correlatives along the northern, NE and eastern margins of the Bohemian Massif. From the Vesser Complex and Schwarzburg antiform this belt extends eastwards to the Torgau-Doberlug Syncline (Basin) into the Kaczawa Mountains, the East Krkonoše (or Rudawy Janowickie) and South Krkonoše complexes, and further south- and southeastwards as the Mount Ślęża and Nowa Ruda ophiolites and parts of the Kłodzko Complex (West Sudetes). The Stronie, Nové Město and Staré Město complexes, and the Rehberg and Letovice ophiolites at the east and SE edge of the Bohemian Massif may form the southward extensions or correlatives (e.g. Finger & Steyer 1995; see Fig. 12.35). Similar, early Palaeozoic mafic, extension-related complexes occur along the NW and western margins of the Bohemian Massif, such as the Münchberg Complex, the Mariánské Lázně Complex and the Erbendorf-Vohenstrauss Zone. The Erzgebirge and Fichtelgebirge complexes also contain early Palaeozoic magmatic rocks, but these are mostly of felsic composition.

Most complexes range in age from Cambrian to late Silurian and may once have been part of a larger continental rift or passive margin system of large oceans or small basins, which once separated individual terranes or parts of composite terranes, such as the Armorican Terrane Assemblage. The present pattern of the Sudetes is that of felsic gneiss units (e.g. Jizera-Krkonoše/Izera-Karkonosze, Orlice-Sněžník/Orlica-Śnieżnik complexes) that alternate with strongly deformed and metamorphosed volcanosedimentary rift fill (South and East Krkonoše, Kaczawa, Kłodzko, Góry Sowie, Staré Město complexes). The ATA was in the early to mid-Cambrian still attached to, and part of, Gondwana, but in late Cambrian–early Ordovician times started to rift away and drift northwards. The change in sediment character and magmatic rock composition have been interpreted to reflect the progressive evolution of this process, from continental rifting and pull-apart basins, to passive margin, to full oceanic spreading (Furnes *et al.* 1994; Pin 1990; Crowley *et al.* 2000, 2002*a*, *b*; Winchester *et al.* 1995).

Erzgebirge and Fichtelgebirge complexes

The Saxothuringian Zone in the Fichtelgebirge and Erzgebirge (Germany)/Krušné hory (Czech Republic) is a composite unit that underwent a complex evolution (e.g. Franke 2000). It comprises late Neoproterozoic (Cadomian/Armorican) gneisses and granitoids, Cambro-Ordovician rift-related sediments and volcanic rocks (Kroner & Sebastian 1997; Kröner & Willner, 1998; Mingram & Rötzler 1999; Linnemann *et al.* 2000; Tichomirowa *et al.* 2001; Kemnitz *et al.* 2002), and Devonian to Carboniferous sediments (e.g. Schäfer *et al.* 2000), the latter, for instance, in the Erzgebirge basins (Schneider *et al.* 2005; Fig. 12.3).

The Erzgebirge Crystalline Complex is an antiformal stack of thrust units with distinct tectonothermal histories, separated by Variscan, retrograde shear zones and intruded by several phases of late Carboniferous granitoids (e.g. Schmädicke *et al.* 1995; Schmädicke & Evans 1997; Willner *et al.* 1997; Rötzler *et al.* 1998; Mingram 1998; Mingram *et al.* 2004). The lowermost tectonometamorphic unit is a medium pressure/medium temperature (MP/MT) gneiss unit consisting of granitic orthogneisses, minor metagreywackes and small amphibolite bodies. A shear zone separates it from the high pressure/high temperature (HP/HT) granulite-facies orthogneisses and metasediments, eclogites, microdiamond-bearing felsic gneisses, migmatitic gneisses, and kyanite-bearing mica schists (Gneiss/Eclogite Unit). A second, upper shear zone separates this unit from the overlying unit of high pressure/low temperature (HP/LT) mica schists, eclogites, granitic orthogneisses and metasediments (Micaschist/Eclogite Unit). The uppermost two units consist of low to medium pressure/low temperature garnet-bearing phyllites, quartzites and minor amphibolites, and phyllites and quartz-phyllites (Garnet-

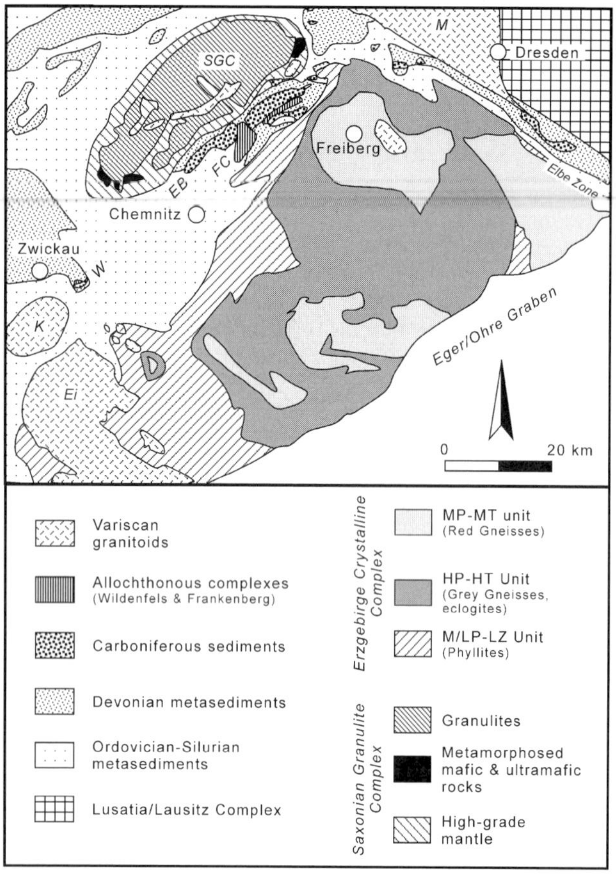

Fig. 12.3. Geological sketch map of the Erzgebirge/Krušné hory and Saxonian Granulite complexes (adapted from Linnemann 2003). Abbreviations: EB, Erzgebirge basins; Ei, Eibenstock granite; FC, Frankenberg Complex; K, Kirchberg granite; M, Meissen monzonite-diorite; SGC, Saxonian Granulite Complex; W, Wildenfels Complex.

Phyllite and Phyllite units; Mingram *et al.* 2004; Sebastian 1995; Rötzler *et al.* 1998).

The gneisses in the lower tectonometamorphic units (core of the antiform) comprise grey varieties ('inner' and 'outer grey gneisses') containing lenses of a red variety ('red gneisses'). The 'inner grey gneisses' of the lowermost MP/MT unit have strongly peraluminous, S-type compositions with relatively high Cr and Ni contents, are LREE-enriched with negative Eu anomalies, and have negative ε_{Nd}(540 Ma) values (−7.7 to −2.8). Their igneous protoliths have early Cambrian zircon (520–540 Ma) crystallization ages (Tichomirowa *et al.* 2001; Košler *et al.* 2004). The 'outer grey gneisses' of the lower shear zone have yielded Neoproterozoic zircon ages (575–580 Ma) and have peraluminous, corundum-normative, S-type compositions with negative ε_{Nd}(540 Ma) values (−7.7 to −6.9). Their precursors were probably greywackes that were derived from an active continental margin source, which is corroborated by the high abundance of detrital zircons. The maximum protolith deposition age is 576 ± 5 Ma (Tichomirowa *et al.* 2001). The 'red gneisses' occur in the MP/MT, HP/HT and HP/LT units. They have S-type granitic compositions with negative ε_{Nd} values of −7.4 to −4.1, but have more variable REE abundances with pronounced negative Eu anomalies. The igneous protolith of this gneiss variety was an S-type, crust-derived granite or rhyolite that crystallized in late Cambrian to early Ordovician times as indicated by zircon ages of 470–500 Ma (Tichomirowa *et al.* 2001; Tichomirowa 2003; Mingram *et al.* 2004). Košler *et al.* (2004) obtained mid-Cambrian (524 and 519 Ma) and early Ordovician (480 Ma) zircon crystallization ages for gneisses from the southern margin of the Erzgebirge complex.

The zircon ages indicate that the Erzgebirge contains granitoids of Cadomian and early Ordovician age, and in this respect it had a similar evolution to that of other Gondwana-derived terranes in the Variscan Orogen. The early Ordovician metagranites or metarhyolites were probably generated during a phase of extension and rifting, as similar, less deformed and metamorphosed volcanic rocks occur interbedded with passive margin sediments in lower-grade areas of the Erzgebirge (Mingram *et al.* 2004).

The Fichtelgebirge, to the west of the Erzgebirge (see Fig. 12.70), also contains Ordovician felsic metavolcanic rocks. These occur at the western end of an antiformal stack of greenschist- to lower amphibolite-facies, early Palaeozoic metasediments that are intercalated with granitic orthogneisses and intruded by several phases of late Variscan granitoids. The metavolcanic rocks are strongly deformed lavas (often porphyritic), tuffs, ignimbrites, and volcaniclastic rocks that are interbedded with late Cambrian to early Ordovician clastic metasediments (Siebel *et al.* 1997). The orthogneisses yielded a 480 ± 4 Ma Rb–Sr whole-rock isochron age, interpreted to be the crystallization age of the protolith. The metavolcanic rocks are late Ordovician as shown by 440–450 Ma zircon ages and the presence of acritarchs of Tremadoc–Arenig age in associated sediments (Siebel *et al.* 1997). The metavolcanic rocks have rhyodacitic to rhyolitic, and the orthogneisses have monzodioritic to granodioritic compositions (Fig. 12.4A). Both orthogneisses and metavolcanic have peraluminous, S-type compositions (Fig. 12.5) and are LREE-enriched with pronounced negative Eu anomalies. The orthogneisses and metavolcanic rocks have overlapping, negative ε_{Nd}(500 Ma) values of −7.7 to −3.8 and −8.2 to −4.2, respectively. In discrimination diagrams based on incompatible element ratios the metavolcanic rocks and orthogneisses plot on, or very near, the intersection of all fields, which does not permit a clear-cut interpretation (Figs 12.4B, C). However, the association with neritic to hemipelagic sediments tentatively suggests an intracratonic basin or immature rift setting for the volcanic rocks. This, in combination with the anatectic (S-type) character and the negative ε_{Nd}(t) values, suggests that the parent melts originated by melting of older (Cadomian) continental crust in an extensional setting in late Cambro-Ordovician times (Siebel *et al.* 1997).

Münchberg Complex

The Münchberg Complex to the NW of the Fichtelgebirge comprises two allochthonous, composite thrust units that overlie unmetamorphosed to very low-grade shelf sediments of the Saxothuringian Zone to the north and NW (Schätz *et al.* 2002; Fig. 12.6). The Münchberg Complex (15 km by 30 km) is the remnant of a large nappe stack, most of which has been removed by erosion, but other remnants of which are exposed further east (Wildenfels and Frankenberg complexes). The lower part is made up of thrust units of anchizone metamorphic sediments and volcanic rocks that are stratigraphically inverted, with early Carboniferous rocks forming the lowest units and (Cambro-) Ordovician (hemi)pelagic, deeper water sediments situated at the top (Reitz 1991; Schätz *et al.* 2002). Neoproterozoic elements were also incorporated into the stack, as is clear from the 609 +17/−19 Ma zircon upper intercept age for a porphyritic rhyolite

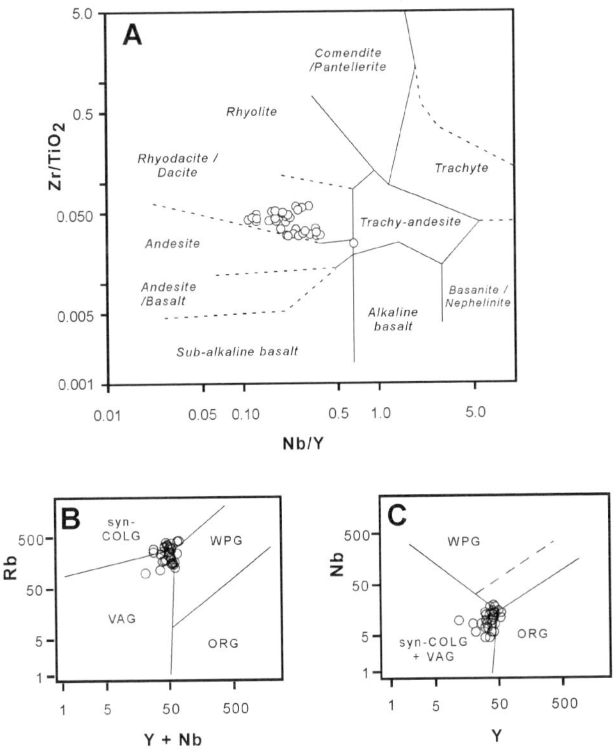

Fig. 12.4. Early Palaeozoic felsic orthogneisses and felsic metavolcanic rocks of the Fichtelgebirge. (**A**) Nb/Y versus Zr/TiO_2 classification diagram (Winchester & Floyd 1977). (**B** & **C**). Discrimination diagrams (Pearce *et al.* 1984). Field labels: ORG, ocean ridge granite; syn-COLG, syncollisional granite; VAG, volcanic arc granite; WPG, within-plate granite. Data from Siebel *et al.* (1997).

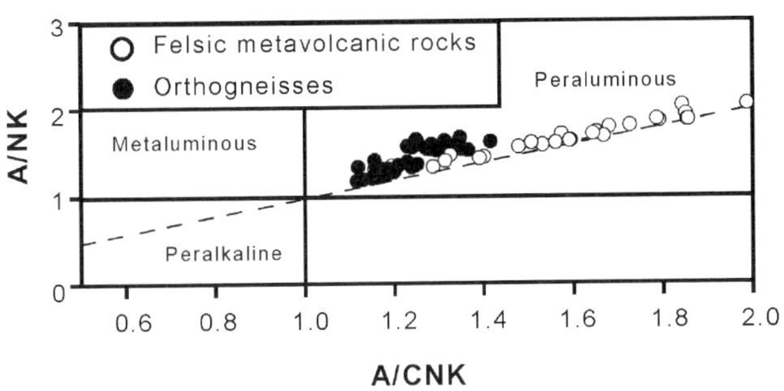

Fig. 12.5. A/CNK versus A/NK diagram for Early Palaeozoic felsic orthogneisses and felsic metavolcanic rocks of the Fichtelgebirge (Maniar & Piccoli 1989). A = Al_2O_3, C = CaO, N = Na_2O, K = K_2O. Data from Siebel *et al.* (1997).

lens NE of Stadtsteinach (Dörr *et al.* 2002), and the presence of Vendian microfossils in the SE margin of the complex (Heuse *et al.* 1994). A *c.* 80 m thick sequence of trachyandesite lavas, pillowed alkali basalts and submarine pyroclastic flows, that are interbedded with sandstones, slates and turbidites and intruded by dykes and sills, occurs near Kupferberg (Fig. 12.6). Zircons from subaqueous ignimbrites from this occurrence yielded an early Ordovician (Arenig) 478 ± 1 Ma evaporation age (Schätz *et al.* 2002; Martin *et al.* 2003), and structures in the associated

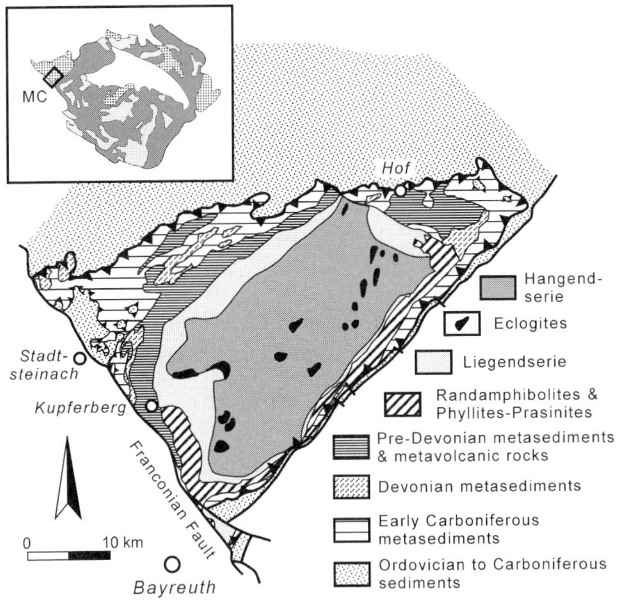

Fig. 12.6. Geological sketch map of the Münchberg Complex, NW Bohemian Massif (after Ritter *et al.* 1999).

sediments point to deposition on a continental shelf below wave base.

These low-grade units are tectonically overlain by the higher-grade, greenschist- to eclogite-facies, crystalline upper unit in which the degree of metamorphism is inverted, i.e. increases upwards. The lowermost rocks are greenschist-facies metasediments and mafic metavolcanic rocks (Prasinite-Phyllite Series) and epidote-amphibolite-facies amphibolites ('Randamphibolit' Unit). Metasediments of the Prasinite-Phyllite Series in the SE margin of the Münchberg Complex contain Vendian acritarchs indicating protolith deposition in Neoproterozoic times (Heuse *et al.* 1994). The Prasinite-Phyllite Series and 'Randamphibolit' Unit in turn are overlain by the clastic metasediments that contain sheets of granitic augen gneisses and *c.* 500 Ma old metagabbros ('Liegendserie'). The uppermost unit is the 'Hangendserie' of stratified (garnet-) amphibolites, hornblende gneisses and felsic gneisses. Eclogite lenses and layers up to 50 m thick occur at the base of this unit (Matthes *et al.* 1975).

The eclogites consist of kyanite-bearing and kyanite-free varieties that have been overprinted and retrograded to varying degrees under amphibolite- to greenschist-facies conditions. The kyanite-free eclogites yielded a Sm–Nd whole-rock isochron age of 480 ± 23 Ma, interpreted as the early Ordovician crystallization age of the magmatic protoliths. Sm–Nd mineral isochron ages of 385 ± 2 and 384 ± 3 Ma show that the eclogite-facies metamorphism occurred in mid-Devonian times (Stosch & Lugmair 1990). The varying MgO and TiO_2 contents indicate that the protoliths were derived from fractionated basalts of tholeiitic composition. Flat to LREE-depleted REE patterns, incompatible element ratios and high positive $\varepsilon_{Nd}(t)$ value (+8.7) indicate normal (N-) to enriched (E-) mid-ocean ridge basalt (MORB) compositions, suggesting that the kyanite-free eclogites represent remnants of metamorphosed early Palaeozoic oceanic crust (Stosch & Lugmair 1990). The kyanite-bearing eclogites have higher Al, Mg, Cr and Ni contents, and lower Ti, Hf and Ta contents; they are metamorphosed high-alumina basalts, gabbros or plagioclase-rich cumulates. The incompatible element ratios indicate mid-ocean ridge to island arc-like basaltic protoliths (Matthes *et al.* 1975).

The metagabbroic rocks in the underlying 'Liegendserie' range from gabbros with ophitic (cumulate) textures to foliated amphibolites. The rocks underwent prograde, amphibolite-facies metamorphism, and eclogite-facies parageneses or remnants thereof are absent. A *c.* 500 Ma U–Pb zircon age suggests an early Palaeozoic crystallization age of the magmatic protoliths (Gebauer & Grünenfelder 1979; Ducrot *et al.* 1983), similar to that of the kyanite-free eclogites (Stosch & Lugmair 1990). The low MgO and varying TiO_2 contents, in combination with positive Eu anomalies, indicate that the parent melts were fractionated to varying degrees and that some had accumulated plagioclase. Major and trace element composition (e.g. moderately LREE enrichment) and the relatively low $\varepsilon_{Nd}(t)$ values of +0.9 to +2.5 show that the gabbros were probably not cogenetic with kyanite-bearing and kyanite-free eclogites higher up in the tectonostratigraphy (Matthes & Seidel 1977; Bosbach *et al.* 1991). Petrogenetic relations, if any, have been obscured by the effects of fractionation and crustal contamination of the gabbroic melts (Bosbach *et al.* 1991).

The kyanite-free, MORB-like eclogites may represent remnants of Early Palaeozoic oceanic crust. The protoliths of the Kupferberg metavolcanic rocks, the metagabbros in the 'Liegendserie', and the eclogite varieties of the 'Hangendserie' were perhaps generated coevally in the same extensional setting, possibly during the opening of the so-called Saxothuringian Ocean that separated the Saxothuringian Terrane to the north from the Moldanubian and Teplá-Barrandian terranes to the south (Stosch & Lugmair 1990; Matte 1991; Franke *et al.* 1995; Martin *et al.* 2003). Closure of this basin was underway by mid-Devonian times, as shown by the *c.* 380 Ma, subduction-related eclogite-facies metamorphism. The fact that the amphibolite-facies metamorphism was prograde in the metagabbros, but retrograde in the eclogite varieties, indicates that these rocks had different evolutionary histories between magmatic emplacement at 480–500 Ma and subduction and basin closure at *c.* 380 Ma (Bosbach *et al.* 1991). The clockwise pressure–temperature–time paths of the eclogites point to differential, post-peak-metamorphic uplift and exhumation histories that led to various degrees of retrogradation under amphibolite- to greenschist-facies conditions (Okrusch *et al.* 1991; O'Brien 1993). The nappe units of the Münchberg Complex were thrust westwards and northwestwards onto the Palaeozoic sediments to the north. Thrusting must have taken place after the early Carboniferous, because early Carboniferous rocks are present in the lower nappe unit. Older, late Devonian (365 ± 7 Ma) $^{40}Ar/^{39}Ar$ cooling ages for phengites from the eclogites suggest that uplift and exhumation must have occurred at different stages (Hammerschmidt & Franz 1992).

Mariánské Lázně Complex

The Mariánské Lázně Complex, *c.* 230 km^2 in size, is an internally imbricated allochthonous complex that was overthrust north-westwards by the Teplá-Barrandian Terrane in the south. In the NW, the complex tectonically overlies amphibolites of the Kladská unit, interpreted as part of the Saxothuringian Zone (Jelínek *et al.* 1997; Kachlík 1997; Fig. 12.7). Eastwards it extends for *c.* 100 km under younger sediments as is clear from geophysical data and xenoliths in Tertiary volcanic rocks. The Mariánské Lázně Complex consists of a variety of magmatic rocks that were metamorphosed under amphibolite- to eclogite- and granulite-facies conditions and had different tectonometa-

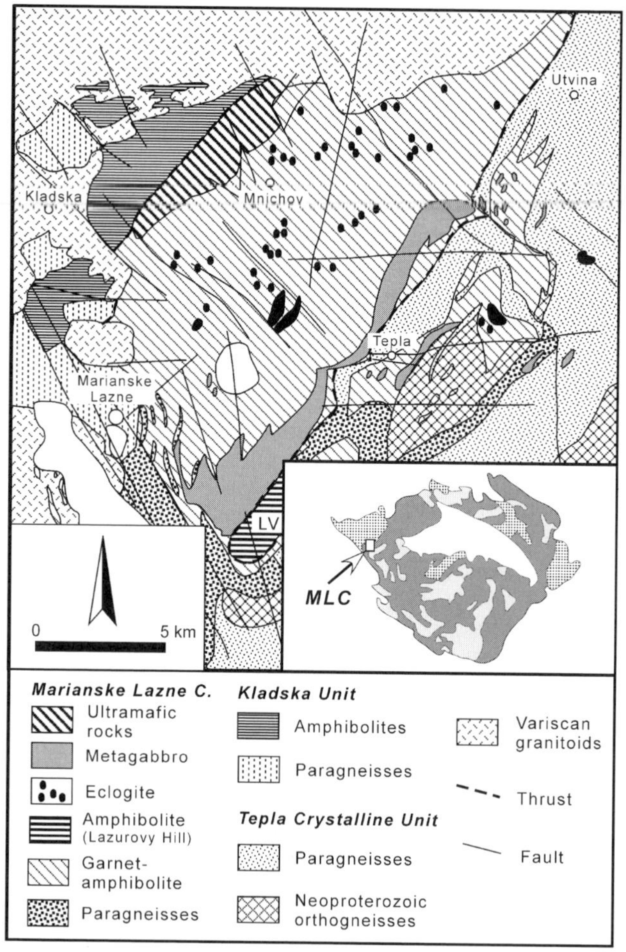

Fig. 12.7. Geological sketch map of the Mariánské Lázně Complex, NW Bohemian Massif (after Timmermann *et al.* 2004, and references therein). Abbreviation: LV, Lazurový Vrch (Lazurový Hill) amphibolites. Inset: MLC, Mariánské Lázně Complex.

morphic histories. The structural base of the complex in the NW is made up of serpentinites and tremolite-actinolite schists that were thrust onto Kladská amphibolites. Metagabbro bodies occur in the border zone with the Teplá-Barrandian Terrane, which consists of interleaved thrust sheets of Mariánské Lázně and Teplá-Barrandian lithologies. The main rock types of the complex include (garnet-)amphibolites that are associated with metagabbroic rocks, felsic gneisses, and minor volumes of eclogites and mafic granulites. The Kladská and Lazurový Hill/Lazurberg amphibolites to the north and south show no evidence of having undergone high-pressure metamorphism. Bowes & Aftalion (1991) obtained a 496 ± 1 Ma U–Pb zircon age for gabbro pegmatite from the southeastern border zone, interpreted as the crystallization age (Cambrian–Ordovician boundary). Late Neoproterozoic U–Pb zircon ages for amphibolite (540 ± 9 Ma) and eclogite (535–545 Ma), also interpreted as protolith crystallization ages, show that older, Cadomian rocks also form part of the Mariánské Lázně Complex (Timmermann *et al.* 2004). The main phase of metamorphism took place in mid-Devonian times, as indicated by 367 ± 4 and 377 ± 7 Ma Sm–Nd mineral isochron ages and a 382 ± 3 U–Pb zircon age for eclogites, a 384 ± 2 Ma zircon age for amphibolite, a 380 ± 3 Ma monazite age for felsic orthogneiss, and a 378 ± 4 Ma titanite age for leucosome (Beard *et al.* 1995; Timmermann *et al.* 2004). Hornblende from metagabbro has a 372 ± 2 Ma plateau age that is similar to hornblende and muscovite ages of 362–375 Ma from the neighbouring, northern and western marginal parts of the Teplá-Barrandian Terrane (Dallmeyer & Urban 1998).

The trace element abundances of the mafic rocks indicate a compositional range from alkaline mafic to depleted tholeiites of ocean floor character (Figs 12.8 and 12.9; Crowley *et al.* 2002*a*). The rocks have basaltic compositions, and some are enriched in Ni suggesting they are partly cumulates. The Zr enrichment trend may have been caused by fractionation or contamination with crustal material (Fig. 12.9A). Compared to modern MORB, the low-Ti tholeiites are depleted in HFSE with negative Nb

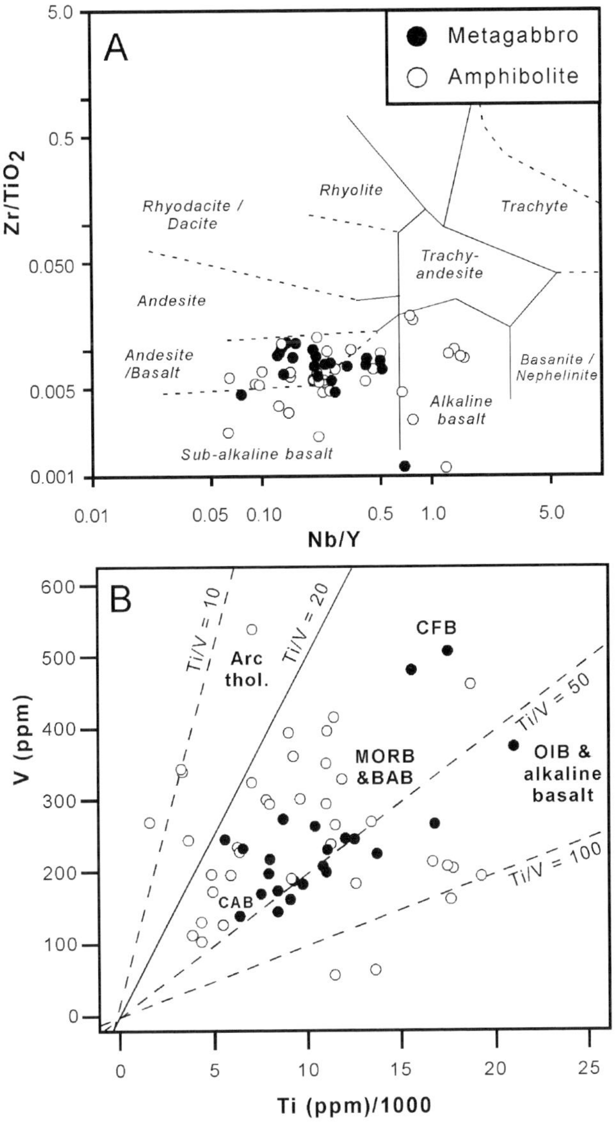

Fig. 12.8. Early Palaeozoic amphibolites and metagabbros of the Mariánské Lázně, NW Teplá-Barrandian Terrane. (**A**) Nb/Y versus Zr/TiO_2 classification diagram (Winchester & Floyd 1977). (**B**) Ti versus V tectonic discrimination diagram (Shervais 1982). Abbreviations: BAB, backarc basalt; CAB, calc-alkaline basalt; CFB, continental flood basalt; MORB, mid-ocean ridge basalt; OIB, ocean island basalt. Data from Schüssler *et al.* (1988), Crowley *et al.* (2002*a*) and Štědrá *et al.* (2002).

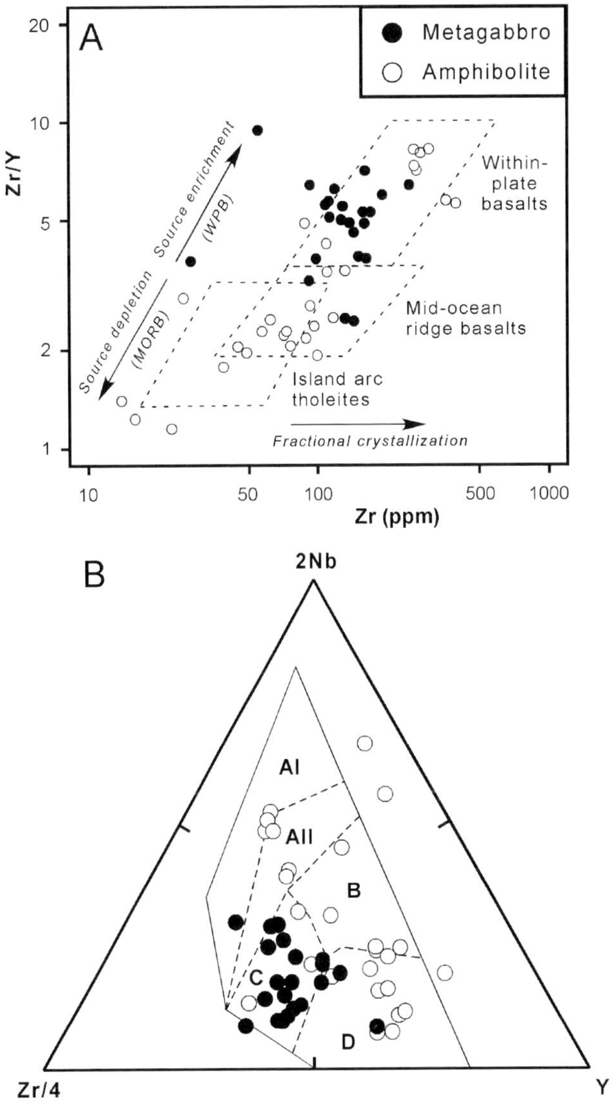

Fig. 12.9. Early Palaeozoic amphibolites and metagabbros of the Mariánské Lázně Complex, north-west Teplá-Barrandian terrane. (**A**) Zr versus Zr/Y tectonic discrimination diagram (Pearce & Norry 1979). (**B**) Nb–Zr–Y tectonic discrimination diagram (Meschede 1986). Only rocks with <52 wt% SiO_2 plotted. Data from Schüssler *et al.* (1988), Crowley *et al.* (2002*a*) and Štědrá *et al.* (2002).

anomalies (Fig. 12.10A, B) and flat to LREE-depleted REE patterns, whereas the normal tholeiites have flat to slightly LREE-enriched patterns. Alkaline metabasalts are restricted to the Lazurový Hill unit in the south and have smoother N-MORB-normalized incompatible element patterns (Fig. 12.10C) and steep LREE-enriched patterns (Fig. 12.11). The within-plate character of the latter may indicate an origin as low-degree melts of a plume-like source (e.g. higher Zr/Y ratios; Fig. 12.9A). The tholeiite varieties have MORB-like trace element compositions (Fig. 12.9) and probably represent higher degree melts from depleted subcontinental lithospheric and asthenospheric mantle (Crowley *et al.* 2002*a*). The intimate association of depleted and enriched tholeiites with alkaline metabasalts, in combination with trace element evidence for crustal contamination (i.e. presence of continental crust), suggest that the protolith formed in a Cambro-

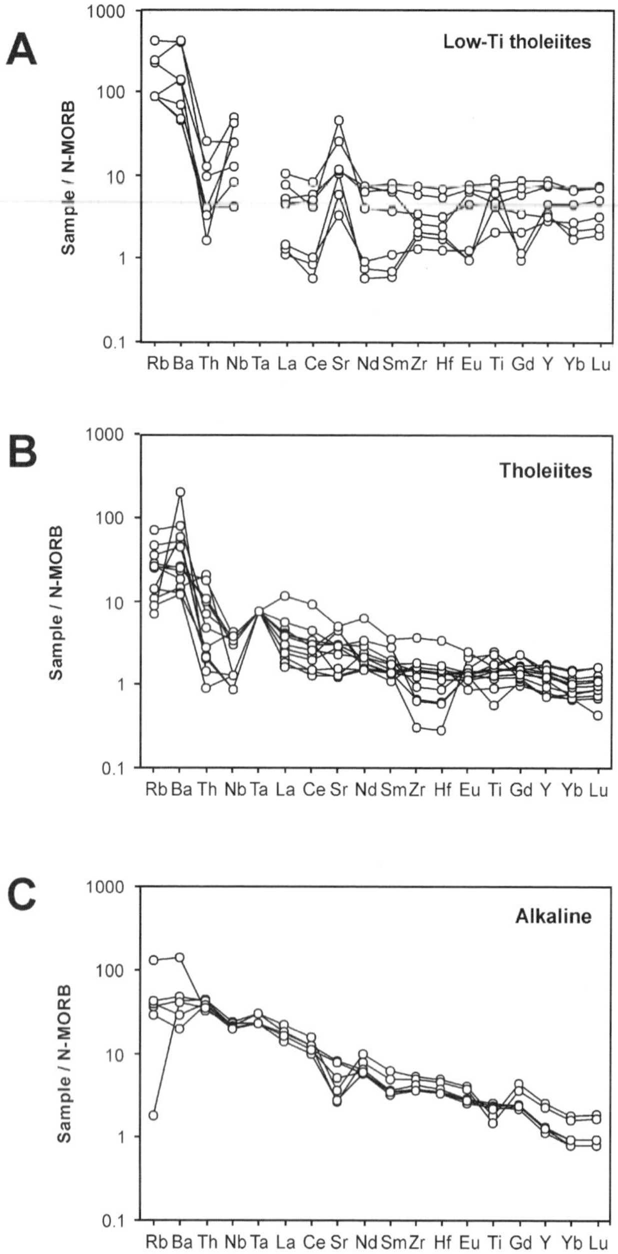

Fig. 12.10. N-MORB-normalized multi-element patterns of the amphibolite varieties of the Mariánské Lázně Complex. Data from Crowley *et al.* (2002) and Štědrá *et al.* (2002); normalization values of Sun & McDonough (1989).

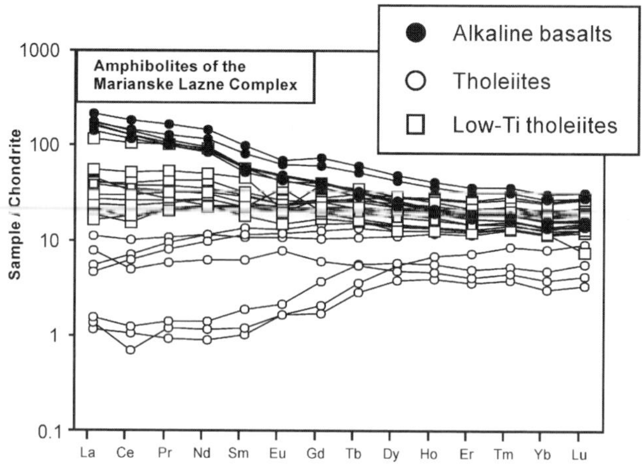

Fig. 12.11. Chondrite-normalized REE patterns of the amphibolites of the Mariánské Lázně Complex. Data from Floyd *et al.* (2002); normalization values of Sun & McDonough (1989).

Ordovician rifting and spreading environment (Crowley *et al.* 2002*a*). Basin closure, collision and high-grade metamorphism occurred in Devonian times during the early stages of the Variscan Orogeny, a scenario typical for other early Palaeozoic complexes in the Sudetes and at the western and eastern margins of the Bohemian Massif (see below).

Teplá-Barrandian Terrane

The volcanosedimentary Teplá-Barrandian Terrane constitutes the core of the Bohemian Massif (Cháb *et al.* 1997). It is separated from the high-grade rocks of the Saxothuringian and Moldanubian (*sense stricto*) zones by steep, Carboniferous shear zones and late brittle faults (Fig. 12.12; Zulauf *et al.* 2002). The Teplá-Barrandian Terrane is largely composed of a basement (Teplá Unit) of late Neoproterozoic to early Cambrian, low-grade metasediments and orthogneisses that in the south-central part are unconformably overlain by early/mid-Cambrian to mid-Devonian (Givetian) overstep sequences consisting of unmetamorphosed basinal sediments and rift-related volcanic rocks (Barrandian Basin). The NW and SW margins of the terrane are intruded by Cambrian granitoids and younger, Carboniferous granites (Chaloupský 1989). The NW parts are overlain by continental sediments and coals of the Kladno-Rakovník, Plzeň/Pilsen and Radnice basins in which sedimentation commenced around the Westphalian A–B boundary (Hess & Lippolt 1986; Opluštil & Pešek 1998, and references therein; McCann *et al.* 2008*a*).

The sequences of the Teplá Unit are up to 10 km thick and are composed of clastic sediments containing tholeiitic, alkaline and calc-alkaline volcanic rocks that were formed in a late Cadomian magmatic-arc and backarc setting at the northern margin of Gondwana. The sediments were deposited as turbidites and slump deposits in nearby peri-arc basins, later to be incorporated into an accretionary wedge (Kříbek *et al.* 2000, and references therein). Deformation was weak. Metamorphism took place under low-grade conditions but increases to amphibolite-facies, Barrovian grade towards the NW and SE. In the SW, metamorphism is directly dated at *c.* 550–540 Ma (total-Pb ages of peak-metamorphic monazites; Zulauf *et al.* 1999). Indirectly, Cadomian deformation and metamorphism are constrained by the *c.* 568 Ma zircon age of a boulder of undeformed rhyolite from Cambrian conglomerates of the Barrandian Basin, and by the *c.* 525 Ma zircon age of the oldest post-tectonic granite intrusion in the SW (see below; Zulauf *et al.* 1997; Dörr *et al.* 2002). Furthermore, *c.* 480 Ma deformed pegmatites in the SW part of the Teplá Unit distinguish between pre- and post-early Ordovician tectonometamorphic events (Glodny *et al.* 1998).

The end of (southward) subduction and tectonothermal activity at the northern margin of Gondwana is marked by a stratigraphic break and unconformity between the Neoproterozoic basement and early Cambrian conglomerates of the basins in the central part of the terrane (the Cadomian unconformity; Patočka *et al.* 1993; Zulauf *et al.* 1997; Kachlík & Patočka 1998; Drost *et al.* 2004). This reflects a change from Cadomian subduction, com-

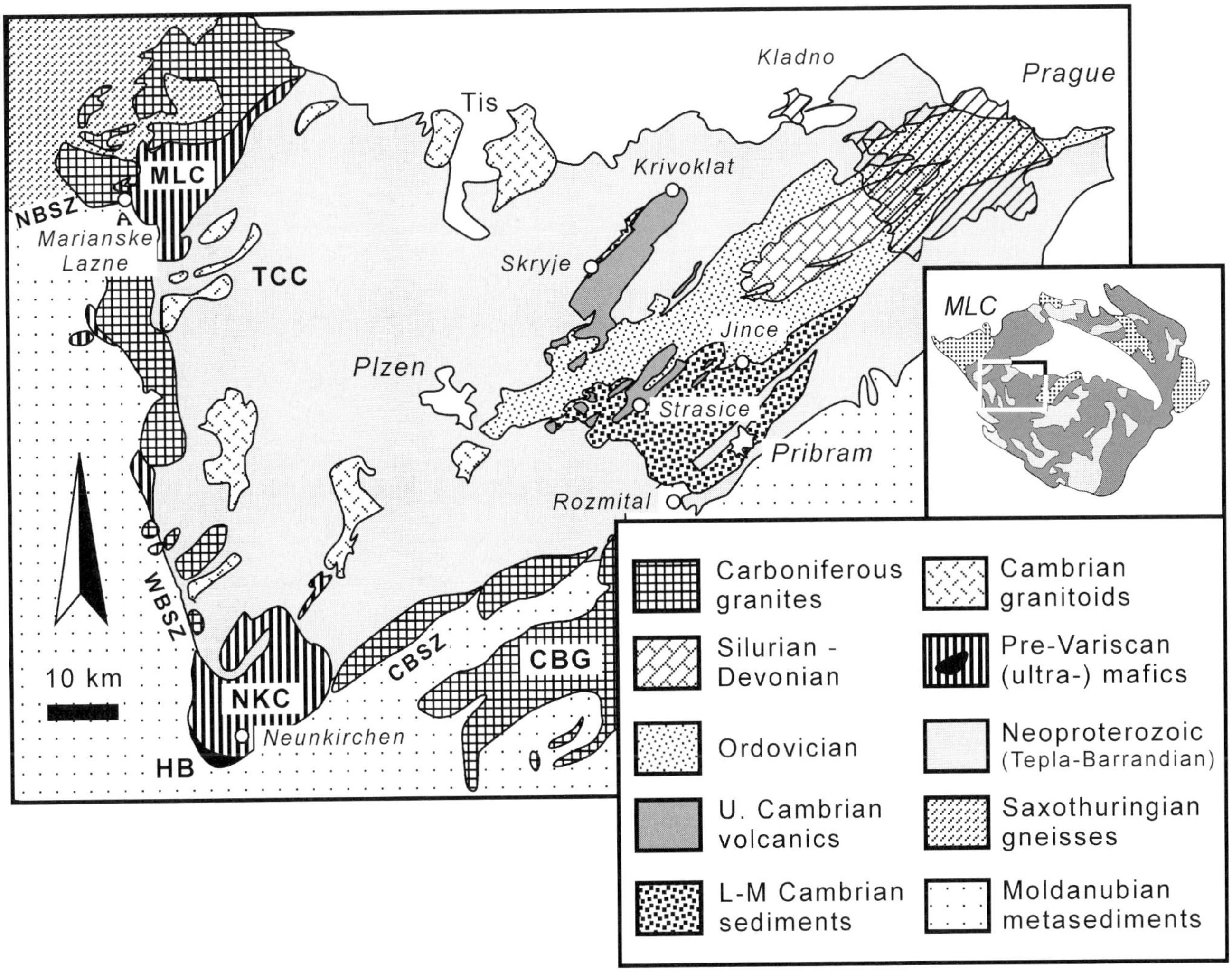

Fig. 12.12. Simplified geological map of the Teplá-Barrandian Terrane (after Dörr *et al.* 2002; Drost *et al.* 2004). Abbreviations: CBG, Central Bohemian Granite; CBSZ, Central Bohemian Shear Zone; HB, Hoher Bogen; MLC, Mariánské Lázně Complex; NBSZ, North Bohemian Shear Zone; NKC, Neukirchen-Kdyně Complex; TCC, Teplá Crystalline Complex; WBST, West Bohemian Shear Zone.

pression and arc magmatism to an extensional–transtensional setting that gave rise to the linear Barrandian Basin. Sedimentation and magmatic activity in the Barrandian Basin continued into the late Middle Devonian (Givetian) after which the basin was inverted due to Variscan compression (e.g. Patočka & Štorch 2004; Štorch 2006, and references therein; see below).

In mid-Cambrian times, trondhjemite, granodiorite and granite with *c.* 522–524 Ma U–Pb zircon ages intruded the SW part of the Teplá-Barrandian Terrane (Domažlice-Stod area; Dörr *et al.* 2002; Zulauf *et al.* 1997). Monzogranitic, granodioritic to tonalitic orthogneisses in the Teplá Crystalline Complex at the north and NW margin have similar, *c.* 505 to 517 Ma U–Pb zircon ages (Fig. 12.12; Dörr *et al.* 1998; Venera *et al.* 2000). In both areas, the intrusions cut, and therefore post-date, Cadomian structural fabrics; contact metamorphism is pronounced in many cases. The granitoids in the SW contain a late-stage magmatic fabric formed during syntectonic emplacement (Zulauf *et al.* 1997, 1999), but those in the NW were sheared during later Variscan events (Dörr *et al.* 1998). The NW suite is predominantly peraluminous in composition and incompatible element ratios indicate volcanic-arc to within-plate settings (Fig. 12.13). Dörr *et al.* (1998) suggested that the protoliths intruded in a transtensional, active continental margin setting. The SW suite is interpreted to have been emplaced in a transtensional setting during crustal thinning of the north Gondwana arc that followed slab break-off (Zulauf *et al.* 1997).

Neukirchen-Kdyně Complex

The mafic Neukirchen-Kdyně Complex in the SW part of the Teplá-Barrandian Terrane (Fig. 12.12) consists of metamorphosed (olivine) gabbros, gabbronorites and several diorite varieties that have been interpreted as the parts of a dismembered layered intrusion (Fig. 12.14; Vejnar 1986 in Propach & Pfeiffer 1998; Schüssler *et al.* 1988). The felsic end-members of this complex have yielded Cambrian (522–524 Ma) zircon crystallization ages (Zulauf *et al.* 1997; Dörr *et al.* 2002). The rocks intruded Neoproterozoic mafic metavolcanic rocks and clastic metasediments of the Domažlice Metamorphic Complex and are surrounded by contact metamorphic aureoles up to 4 km wide. The mafic plutonic rocks are slightly older than the spatially associated Cambrian granites and, furthermore, crystallized at

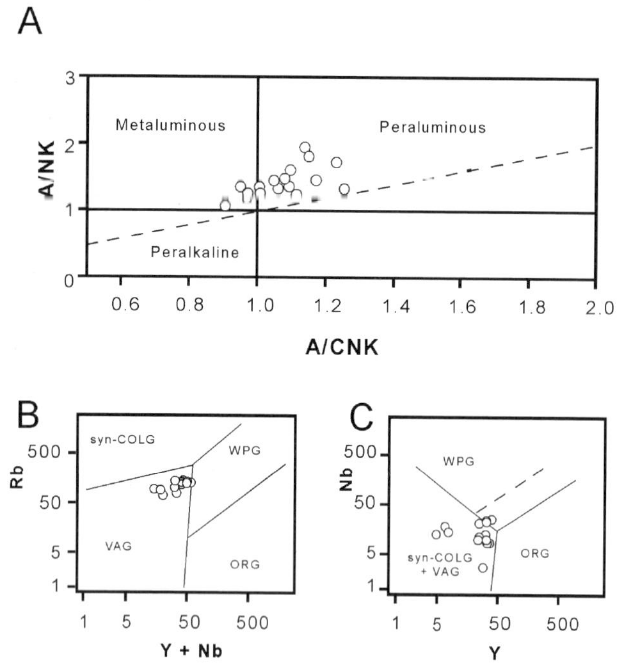

Fig. 12.13. Cambrian metagranitoids of the Teplá Crystalline Complex. (**A**) A/CNK versus A/NK diagram (Maniar & Piccoli 1989). (**B** & **C**) Tectonic discrimination diagrams (Pearce *et al.* 1984). Data from Dörr *et al.* (1998). For abbreviations see captions to Figures 12.4 & 12.5.

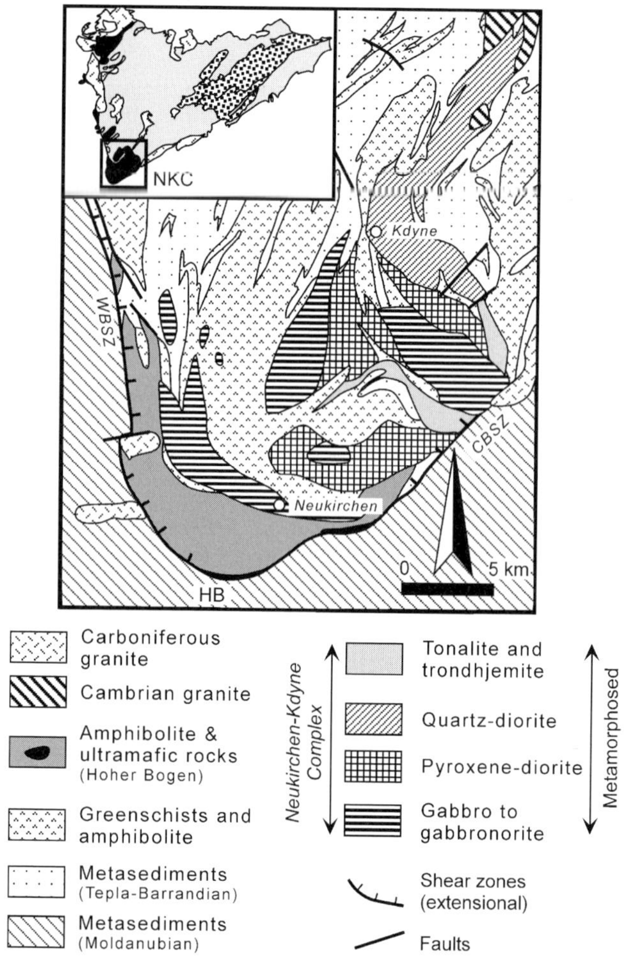

Fig. 12.14. Geological map of the Neukirchen-Kdyně Complex in the SW part of the Teplá-Barrandian Terrane, central Bohemian Massif (after Bues *et al.* 2002, and references therein). Abbreviations: CBSZ, Central Bohemian Shear Zone; HB, Hoher Bogen. Inset: NKC, Neukirchen-Kdyně Complex.

greater depths. The degree of metamorphism increases southwestwards, where the intrusive rocks are transformed into amphibolites (Schlüsser *et al.* 1988, and references therein). This amphibolite-facies metamorphism occurred during a later Variscan event, possibly thrusting the Teplá-Barrandian Terrane southeastwards onto Moldanubian units in the late Devonian (Bues *et al.* 2002). Alternatively, Pitra *et al.* (1999) suggested that the Teplá-Barrandian Terrane was juxtaposed against the Moldanubian Terrane along the West Bohemian Shear Zone in Stephanian times. The metagabbroic end-members are not primitive melts but are fractionated, as is clear from the varying Fe/Mg ratios and correlations between several elements. They have tholeiitic compositions, but the incompatible trace elements scatter across several fields in tectonic discrimination diagrams (Figs 12.15A, B & 12.16).

Hoher Bogen

The (gabbro-) amphibolites of the area around Hoher Bogen form a belt 1 to 4 km wide along the SW margin of the Teplá-Barrandian Terrane (Figs 12.12 and 12.14). They are in tectonic contact with metasediments of the Moldanubian Terrane along the extensional West and Central Bohemian shear zones. These were active during the Carboniferous (Zulauf *et al.* 2002) and host several narrow ultramafic bodies. The protoliths of the amphibolites are considered to be of (late) Cambrian age, as suggested by a 511 $\pm$ 3 Ma U–Pb zircon age for a fractionated amphibolite from the northern end of the belt (Gebauer 1993, in Propach & Pfeiffer 1998). The amphibolites are generally fine-grained and mylonitic, but grain sizes increase with increasing degree of metamorphism to the SW. Here, ortho- and/or clinopyroxene appear as part of the parageneses, and partial melting led to granulite-facies garnet + clinopyroxene-bearing hornblendites (Propach & Pfeiffer 1998). The presence of similar coarse-grained rocks with less well-developed planar fabrics have led to the view that the Hoher Bogen amphibolites represent the strongly deformed equivalents of Neukirchen-Kdyně Complex gabbros, but the different, MORB-like composition clearly contradicts this. The amphibolites are uniformly tholeiitic in composition and have elevated contents of FeO_{tot} (up to 18 wt%), TiO_2 (<3.3 wt%) and V (<600 ppm). Most trace elements are undepleted, which suggests their protoliths erupted as volcanic rocks. The trace element ratios are similar to those of modern mid-ocean ridge basalts (Figs 12.15A, B & 12.16), which is corroborated by the N-MORB-like flat and LREE-depleted REE patterns (Fig. 12.17). The ultramafic rocks in the shear zones are harzburgites and pyroxenites (i.e. cumulates) that have been altered to varying degrees. The association of MORB-like amphibolites with cumulates and harzburgites suggests they may represent the parts of a strongly dismembered, late Cambrian to early Ordovician ophiolite (Propach & Pfeiffer 1998).

Barrandian Basin

The early Palaeozoic succession in the Barrandian Basin that unconformably overlies the Neoproterozoic rocks of the Teplá

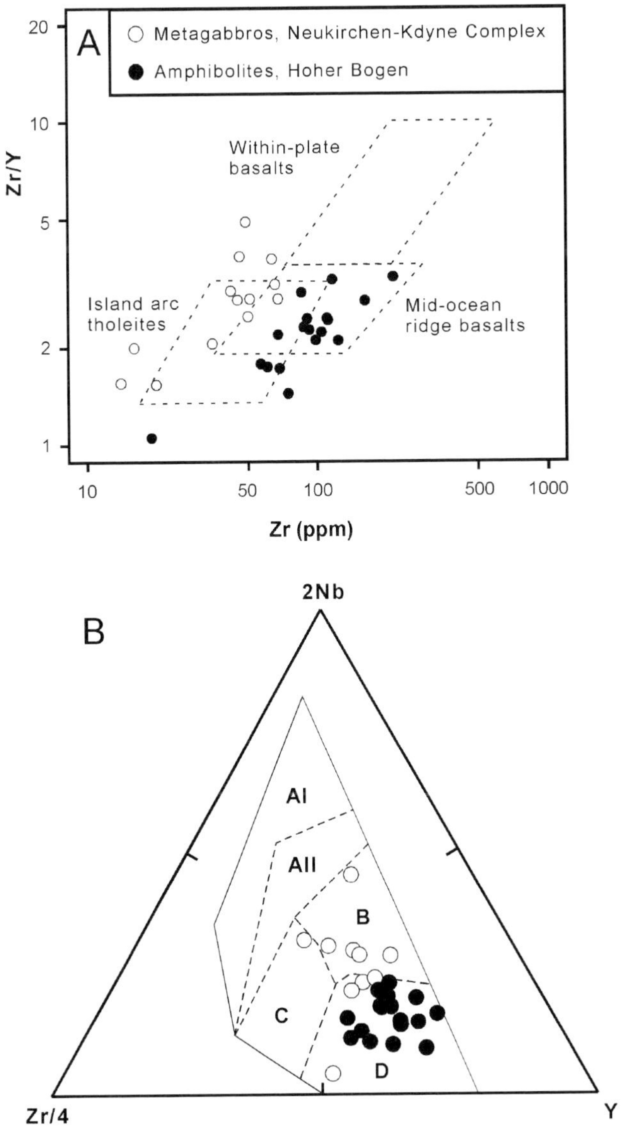

Fig. 12.15. (**A**) Zr versus Zr/Y tectonic discrimination diagram for metagabbros rocks from the late-Cambrian Neukirchen-Kdyně Complex and Hoher Bogen amphibolites, SW Teplá-Barrandian Terrane (Pearce & Norry 1979). (**B**) Nb–Zr–Y tectonic discrimination diagram (Meschede 1986). Data from Schlüsser *et al.* (1988) and Propach & Pfeiffer (1998).

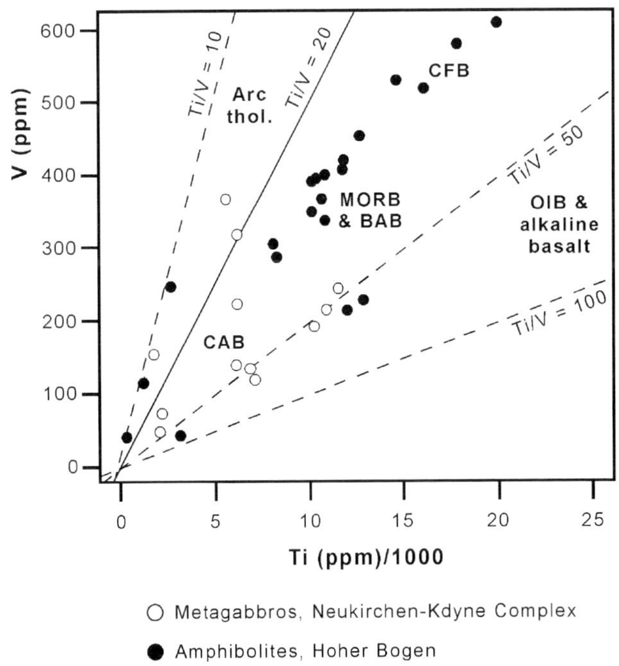

Fig. 12.16. Ti versus V tectonic discrimination diagram for metagabbros form the mid-Cambrian Neukirchen-Kdyně Complex and late Cambrian Hoher Bogen amphibolites, SW Teplá-Barrandian Terrane. Abbreviations: BAB, backarc basalt; CAB, calc-alkaline basalt: CFB, continental flood basalt; MORB, mid-ocean ridge basalt; OIB, ocean island basalt (Shervais 1982). Data from Schlüsser *et al.* (1988) and Propach & Pfeiffer (1998).

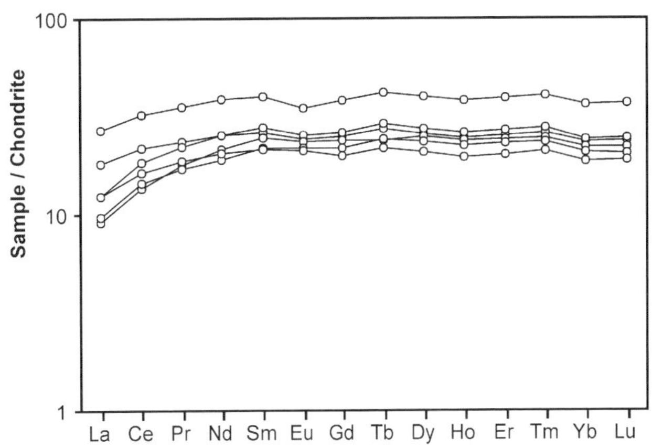

Fig. 12.17. N-MORB-like, chondrite-normalized rare-earth element diagram for the amphibolites of the Hoher Bogen area, SW Teplá-Barrandian Terrane. Data from Propach & Pfeiffer (1998); normalization values of Sun & McDonough (1989).

Unit commences with *c.* 2.5 km of Early Cambrian continental conglomerates, sandstones and greywackes that were deposited in the Příbram-Jince and Skryje-Týřovice basins. These basins were fault-bounded and show high subsidence rates. The presence of granite and rhyolite clasts within the basin successions with 594 +34/−141 and 568 ± 3 Ma U–Pb zircon ages suggests that sediments were derived from local, late Neoproterozoic arc sources (Dörr *et al.* 2002); this is corroborated by trace element evidence from the sandstones (Patočka & Štorch 2004). Furthermore, clasts of Barrovian-grade metamorphic rocks as well as the great thickness of the deposits suggest that the Cadomian basement was deeply eroded before deposition of the conglomerates (Zulauf *et al.* 1997). Locally, near Příbram and Rožmitál (Fig. 12.12), small volumes of pyroclastic trachybasalts and rhyolites occur as tuffs and tuffites with occasional unworn lapilli and magmatic crystals. Geochemical data for these rocks are limited and contradictory, and indicate deposition in an active convergent setting or, alternatively, a within-plate setting associated with contamination by continental crust. The crustal contamination may be the cause of the apparent subduction signal (Patočka *et al.* 1993; Kachlík & Patočka 1998; Drost *et al.* 2004).

Extension and crustal thinning continued into mid-Cambrian times and resulted in a transgression and deposition of marine shales; these are overlain by conglomerates and coarse-grained sandstones (Patočka & Štorch 2004, and references therein). The late Cambrian commenced with a regression, relief inversion and a depositional hiatus, followed by renewed extension, crustal thinning and the eruption of subalkaline volcanic rocks of the Křivoklát-Rokycany and Strašice volcanic complexes in the central parts of the Teplá-Barrandian Terrane (Fig. 12.12). The volcanic rocks locally unconformably overlie Middle Cambrian shales, reach thicknesses of up to 1.5 km and represent the largest volume of early Palaeozoic volcanic rocks in the Barrandian Basin. They erupted in four stages of respectively andesitic/dacitic, (basaltic) andesitic, trachydacitic/rhyolitic, and rhyolitic composition (Waldhausrová 1971; Fig. 12.18). A rhyolite from the youngest stage yielded a 499 ± 4 Ma U–Pb zircon age (Cambrian–Ordovician boundary; Drost *et al.* 2004). The mantle-derived, mafic end-members (basaltic andesites) have subalkaline compositions and moderately LREE enriched, E-MORB-like REE patterns (La_N/Yb_N = 1.8–2.3). The felsic end-members have steep LREE patterns and negative Eu anomalies that become more pronounced towards the rhyolite end of the compositional spectrum. Incompatible element ratios point to a within-plate to volcanic-arc setting. An arc setting is, however, unlikely considering their location, and the arc signature may be an inherited feature (Figs 12.19A, B; Patočka *et al.* 1993; Drost *et al.* 2004).

The early Ordovician began with the inversion of the Cambrian basins and the formation of the NE-trending Prague successor basin. This was filled with Tremadoc to early Silurian shallow-marine clastic sediments, followed by late Silurian to early Devonian shallow-marine turbiditic carbonates and limestones. Extension and sedimentation ends with deep-water siliciclastics (probably distal turbidites) in the late Middle Devonian (Givetian). The deposition of these deep-marine clastic sediments reflects the onset of Variscan tectonometamorphism in the Moldanubian, Saxothuringian and West Sudetes areas that surround the Teplá-Barrandian Terrane (Kachlík & Patočka 1998). Maturity data of organic matter from the youngest preserved, mid-Devonian rocks suggest late- to post-Devonian burial depths of *c.* 3 km (Glasmacher *et al.* 2002, and references therein). Volcanism in the early and late Ordovician occurred in the form of local pillow basalts and trachyandesites, pyroclastic rocks, and dolerite intrusions of tholeiitic to alkaline composition. The late Silurian volcanic rocks occur as subaqueous lavas,

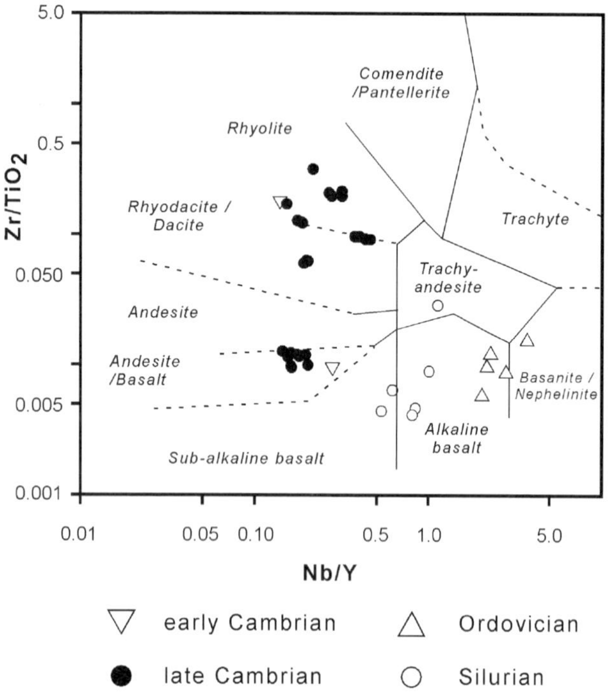

Fig. 12.18. Nb/Y versus Zr/TiO2 classification diagram for the early Palaeozoic volcanic rocks of the Barrandian Basin, Teplá-Barrandian Terrane (Winchester & Floyd 1977). Data from Patočka *et al.* (1993) and Drost *et al.* (2004).

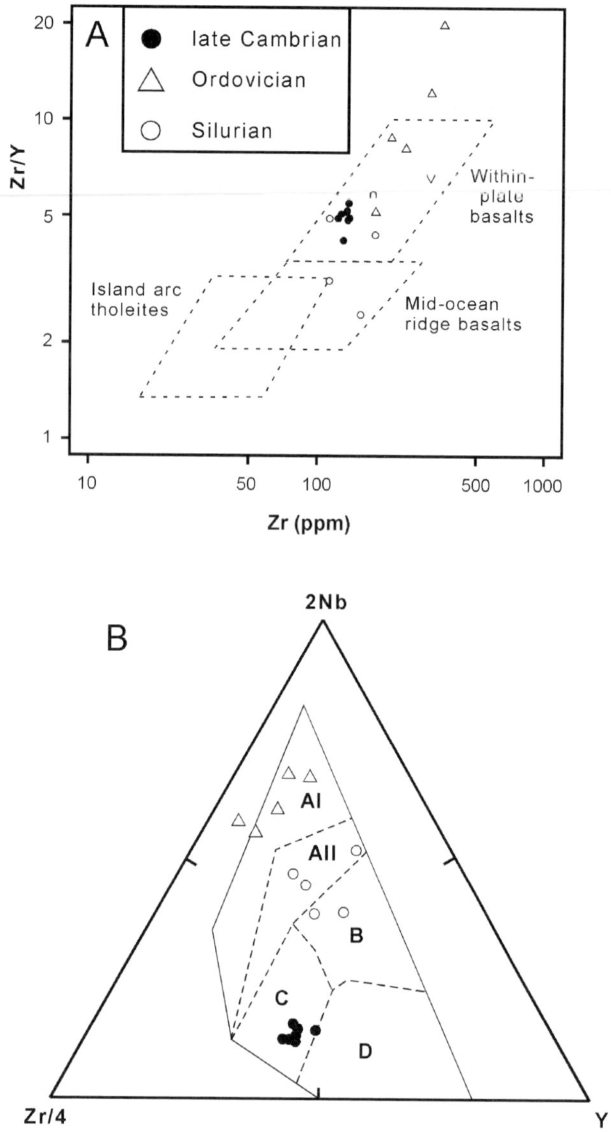

Fig. 12.19. (**A**) Zr versus Zr/Y tectonic discrimination diagram for early Palaeozoic mafic metavolcanic and -subvolcanic rocks form the central part of the Teplá-Barrandian Terrane with <52 wt% SiO_2 (Pearce & Norry 1979). (**B**) Nb–Zr–Y tectonic discrimination diagram for selected mafic metavolcanic rocks with SiO_2 <52 wt%. Field labels: AI, within-plate alkaline basalts; AII, within-plate alkaline basalts and within-plate tholeiites; B, E-MORB; C, within-plate tholeiites and volcanic-arc basalts; D, N-MORB and volcanic-arc basalts (Meschede 1986). Data from Patočka *et al.* (1993) and Drost *et al.* (2004).

doleritic sills and dykes which range in composition from basalts to basaltic trachyandesites of alkaline-transitional to tholeiitic, within-plate character (Patočka *et al.* 1993).

Faunal and palaeomagnetic data, in combination with the sedimentary record (e.g. Ordovician diamictites), indicate that the Teplá-Barrandian Terrane remained close to Gondwana during most of the early Palaeozoic. The age spectra of detrital zircons from Neoproterozoic sediments point to a West African provenance, which suggests that it was part of the Armorican Terrane Assemblage (Drost *et al.* 2004; Murphy *et al.* 2004*a*). Palaeomagnetic evidence shows that in the late Silurian, Teplá-Barrandia was situated 30° south of the equator, close to Baltica and Laurentia (Tait *et al.* 1997). This suggests that the Ordovician–Silurian volcanism may have been related to extension caused by rifting, break-off and northward drift of the terrane, a scenario that is typical for the many of the other ATA-related terranes. The presence of tholeiites (i.e. high-degree melts) suggests that extension peaked in Silurian times (Kachlík & Patočka 1998), which is supported by the presence of Silurian ophiolites in the nearby Sudetes (see below).

Western Moldanubian Terrane

The part of the Moldanubian Terrane west and SW of the West Bohemian Shear Zone consists of a nappe stack of high-pressure granulites with minor amounts of peridotites and eclogites, that overlie HT/LP migmatitic metasediments (Chaloupský 1989). Metamorphosed magmatic rocks are rare (Teipel *et al.* 2004). The sedimentary protoliths are predominantly greywackes of late Neoproterozoic age, and 323–313 Ma zircon and monazite ages for leucosomes, migmatites and metagranites show that the main phase of metamorphism took place in the mid-Carboniferous (Kalt *et al.* 2000; Propach *et al.* 2000; Teipel *et al.* 2004). The rocks are intruded by late- to post-tectonic Carboniferous granites that are also found in the West Bohemian Shear Zone and to the east of it, in the western part of the Teplá-Barrandian Terrane (see below; Siebel *et al.* 1995, 1997, 2003; Propach *et al.* 2000; Dörr *et al.* 2002; Zulauf *et al.* 2002; Chen *et al.* 2003). Ages of 321 Ma (U–Pb) for titanite, 316–322 Ma for hornblende and 313–309 Ma for biotite ($^{40}Ar/^{39}Ar$) indicate that the peak of Variscan metamorphism was followed by slow cooling (Kalt *et al.* 2000).

The subordinate volumes of meta-igneous rock comprise latest Neoproterozoic metarhyolites (555 ± 12 Ma) and metabasic rocks (549 ± 7 Ma), and early Ordovician metagranites (480 ± 6 and 486 ± 7 Ma) and metabasic rocks (481 ± 8 Ma; U–Pb zircon ages; Teipel *et al.* 2004). The ages of inherited zircons and the negative $\varepsilon_{Nd}(t)$ values (−6.3 to −0.5) of the meta-igneous rock suggest that this part of the Moldanubian is underlain by Precambrian crust. The Neoproterozoic rocks probably represent fragments of a continental magmatic arc and backarc basin situated at the north Gondwanan margin. The early Ordovician metarhyolites have negative $\varepsilon_{Nd}(t)$ values of −2.7 to −1.9, but a (possibly unrelated) metabasite has a positive value of +7.9. Teipel *et al.* (2004) suggested that the felsic rocks were formed in an Early Palaeozoic magmatic arc, a scenario that has also been proposed for the early Ordovician felsic magmatic rocks of the West Sudetes (see below).

Erbendorf-Vohenstrauss Zone

The Erbendorf-Vohenstrauss Zone is a SW-dipping, allochthonous unit situated at the NW margin of the Moldanubian Terrane, south of the boundary with the Saxothuringian Terrane. The unit is penetrated by the KTB (Kontinentaltiefbohrung) borehole of the German Continental Deep Drilling Program (Emmermann & Wohlenberg 1989; Fig. 12.20). It is composed of metasediments, amphibolites and metagabbros with minor amounts of felsic orthogneisses, eclogites and ultramafic rocks (Okrusch *et al.* 1989; Schüssler *et al.* 1989; O'Brien 1989). The rocks are polymetamorphic, being metamorphosed in Cambro-Ordovician and in mid-Devonian times as shown by the 476–496 Ma and 370–380 Ma zircon and monazite ages (Grauert *et al.* 1994; O'Brien *et al.* 1997; Söllner *et al.* 1997; von Quadt 1997; Krohe & Wawrzenitz 2000), and 371–377 Ma K–Ar ages for coarse-grained white mica (Ahrendt *et al.* 1997). The older ages are in agreement with 484–480 Ma U–Pb crystallization ages for garnet, columbite, monazite and zircon from deformed pegmatites that have been used as times markers to distinguish earlier from later stages of metamorphism and deformation (Glodny *et al.* 1998). A suite of late- to post-tectonic Carboniferous granites intruded all rock units (Siebel *et al.* 1995, 2003; see below). Apart from contact metamorphism related to the late granites, there is no evidence for high-grade, regional-scale metamorphism of Carboniferous age, and in this respect the Erbendorf-Vohenstrauss Zone differs from the surrounding Moldanubian and Saxothuringian units. The metabasic rocks can be subdivided

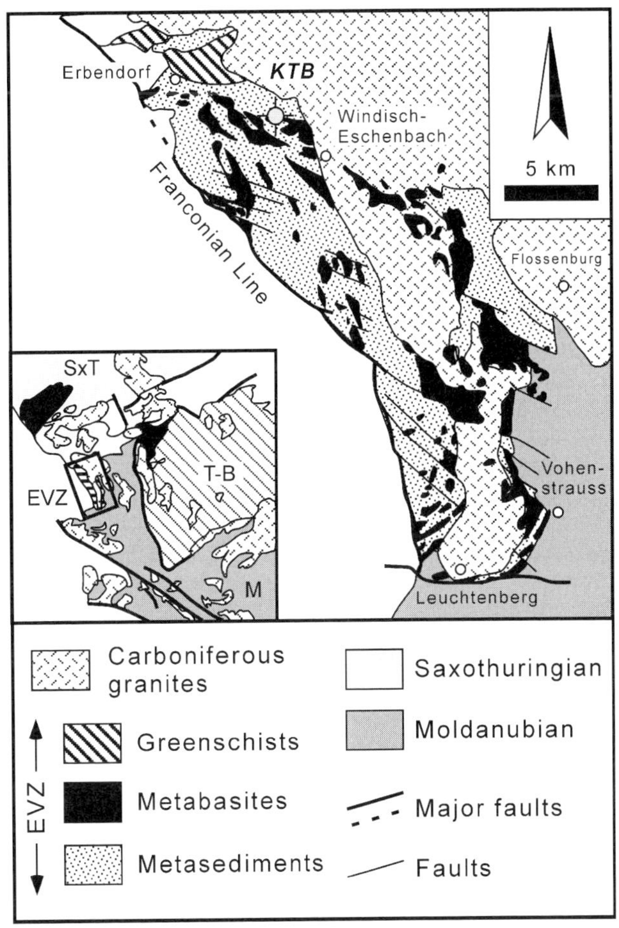

Fig. 12.20. Geological map of the Erbendorf-Vohenstrauss Zone (EVZ), NW Bohemian Massif (Bavarian Forest; after Schüssler *et al.* 1989). Inset: NW Bohemian Massif; abbreviations: KTB, Kontinentaltiefbohrung (continental deep borehole); M, Moldanubian; SxT, Saxothuringian; T-B, Teplá-Barrandian.

into lower and upper nappes of metabasic rocks and metasediments. The lower unit (Erbendorf Greenschist Zone) contains greenschist-facies metabasites, serpentinites and subordinate metasediments; the upper unit is dominated by amphibolite-facies metapelites that contain intercalations of amphibolites and metagabbros (Fig. 12.21A; Okrusch *et al.* 1989; Schüssler *et al.* 1989; Harms *et al.* 1997).

Amphibolites and metagabbros of the Erbendorf Greenschist Zone are similar in their Mg-rich tholeiitic to calc-alkaline compositions, LREE enrichment and low Ti and Nb abundances, which suggest that they were formed in a subduction-related, arc or backarc-like setting (Figs 12.21B and 12.22A, B; Schüssler *et al.* 1989). In contrast, the amphibolites from the southern margin have subalkaline compositions with flat N-MORB-normalized multi-element patterns and flat to LREE-depleted REE patterns,

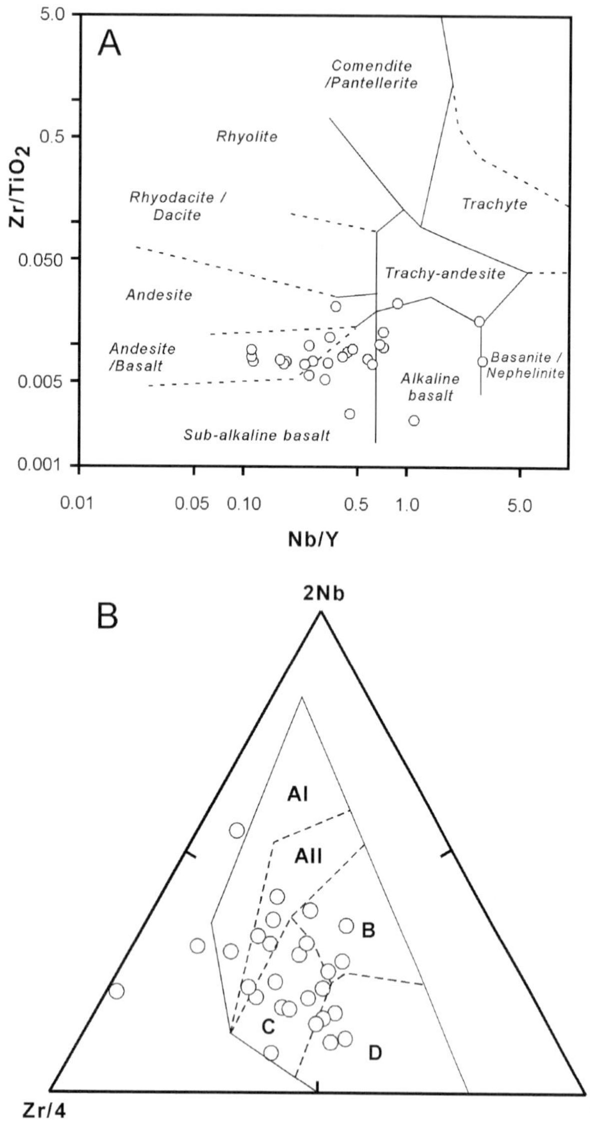

Fig. 12.21. Early Palaeozoic amphibolites and metagabbros of the Erbendorf-Vohenstrauss Zone, NW Bohemian Massif. (**A**) Nb/Y versus Zr/TiO_2 classification diagram (Winchester & Floyd 1977). (**B**) Nb–Zr–Y tectonic discrimination diagram (Meschede 1986). Data from Schüssler *et al.* (1989), von Quadt (1997) and Harms *et al.* (1997).

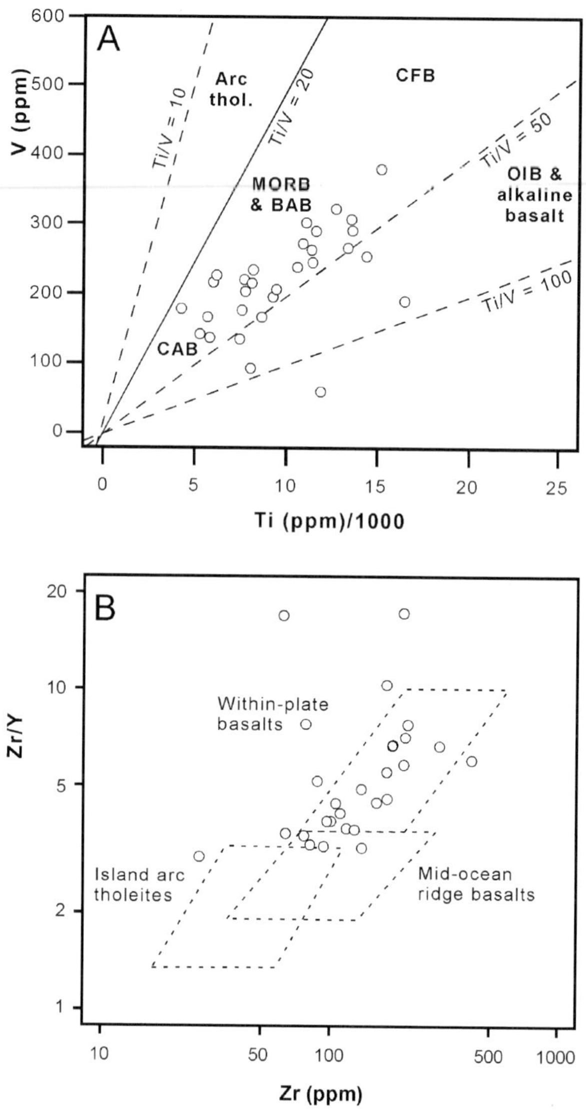

Fig. 12.22. Early Palaeozoic amphibolites and metagabbros of the Erbendorf-Vohenstrauss Zone, NW Bohemian Massif. (**A**) Ti versus V tectonic discrimination diagram (Shervais 1982). (**B**) Zr versus Zr/Y tectonic discrimination diagram (Pearce & Norry 1979). Data from Schlüsser *et al.* (1989), von Quadt (1997) and Harms *et al.* (1997).

and in this respect resemble modern N-MORB. Amphibolites from the central parts range from subalkaline to transitional, and furthermore are enriched in incompatible elements and LREE, resembling E-MORB or within-plate basalts (Fig. 12.23A). Metagabbros and amphibolites from KTB drill cores have been subdivided into an alkaline suite and a suite of various tholeiites. The alkaline mafic rocks are LREE-enriched, have $\varepsilon_{Nd}(500\ Ma)$ values of −1.1 to +4.7 and post-date the tholeiites as is clear from alkaline dykes cutting the tholeiites. The bulk of the tholeiites have normal to enriched incompatible element compositions with positive $\varepsilon_{Nd}(500\ Ma)$ values of +1.1 to +5.1 (Harms *et al.* 1997). The metagabbros have flat to LREE-enriched REE patterns and positive $\varepsilon_{Nd}(500\ Ma)$ values (+3.3 to +8.6; von Quadt 1997; Fig. 12.23B). Harms *et al.* (1997) suggested that the alkaline and tholeiitic suites were cogenetic and derived by

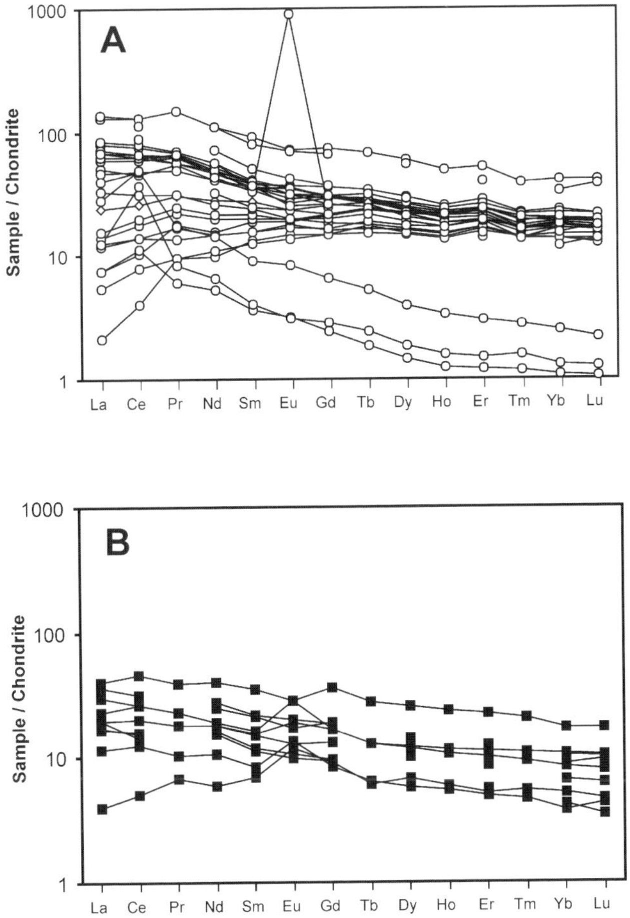

Fig. 12.23. Chondrite normalized REE patterns of Early Palaeozoic amphibolites (**A**) and metagabbros (**B**) of the Erbendorf-Vohenstrauss Zone, NW Bohemian Massif. Data from Schüssler *et al.* (1989), von Quadt (1997) and Harms *et al.* (1997); normalization values of Sun & McDonough (1989).

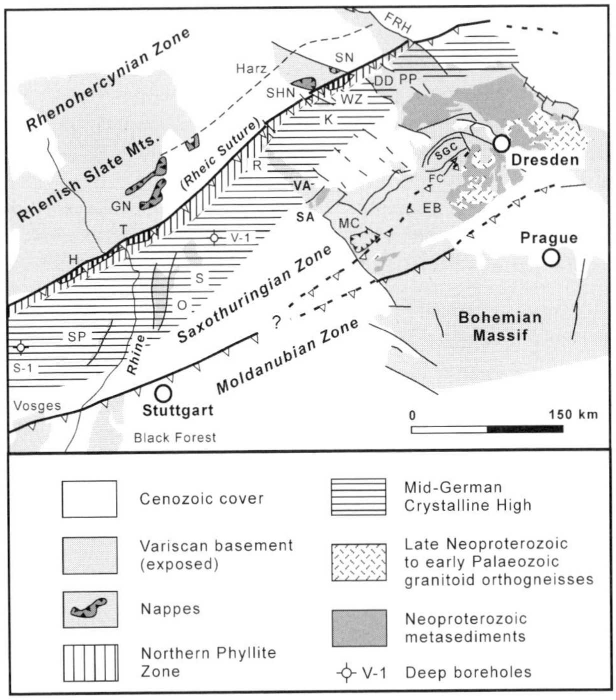

Fig. 12.24. The Mid-German Crystalline High and neighbouring units (after Anthes & Reischmann 2001; Sommer & Katzung 2004). Abbreviations: DD, Dessau-Delitzsch; EB, Erzgebirge /Krušné hory; FC, Frankenberg Complex; FRH, Flechtingen-Rosslau High; GN, Giessen nappe; H, Hunsrück; K, Kyffhäuser; MC, Münchberg Complex; O, Odenwald; PP, Pretzsch-Prettin; R, Ruhla; SGC, Saxon Granulite Complex; S, Spessart; SA, Schwarzburg Antiform; SHN, south Harz nappe; SN, Selke nappe; SP, Southern Pfalz; T, Taunus; VA, Vesser Antiform; WZ, Wippra Zone. Boreholes: S-1, Saar 1; V-1, Volkach 1.

variable degrees of melting from mantle sources with varying trace-element compositions, and possibly erupted in an oceanic basin situated near an active plate margin.

Vesser Complex and Schwarzburg antiform

The late Cambrian to early Ordovician protoliths of rift-related metasediments and metavolcanic rocks crop out in the Vesser and Schwarzburg areas in the Southern Phyllite Zone, situated between the Mid-German Crystalline High to the north and the Saxothuringian Zone to the south (Fig. 12.24; Schätz *et al.* 2002; Kemnitz *et al.* 2002; Bankwitz *et al.* 1992). The metavolcanic rocks overlie lower to middle Cambrian shallow-marine sediments that were deposited in small transtensional and pull-apart basins (Kemnitz *et al.* 2002; Linnemann 2003). Rhyodacites and a granite in the Schwarzburg area yielded middle Cambrian (*c.* 515–538 Ma) zircon ages. Rhyolitic lavas, subvolcanic intrusions and a granite intrusion have younger (490–494 Ma) late Cambrian to early Ordovician ages (Gehmlich *et al.* 1997; Kemnitz *et al.* 2002). The rocks in both areas have been deformed and metamorphosed under greenschist-facies conditions (Sommer & Katzung 2004).

The volcanism (lavas and tuffs) in the *c.* 1800 m thick volcanosedimentary Vesser Complex was bimodal and associated with gabbro and dolerite intrusions (Fig. 12.25A; Bankwitz *et al.* 1998). Kemnitz *et al.* (2002) obtained late Cambrian (502–513 Ma) U–Pb zircon ages for the magmatic rocks. The majority of the metabasalts and metagabbros in the Vesser Complex have tholeiitic compositions, but incompatible element abundances and $\varepsilon_{Nd}(t)$ values show considerable variation (Figs 12.25B, 12.26A, B). The mafic rocks range from depleted, N-MORB-like basalt with low REE contents, La-depletion and high ε_{Nd}(500 Ma) values of +7.6, to alkaline basalt that are LREE-enriched ($La_N/Lu_N = 7.2$), have low Zr/Nb ratio (7.6) and an ε_{Nd}(500 Ma) value of +3.2. The very low Th/Nb ratio of 0.07 for both enriched and depleted varieties shows that these two end-members were not contaminated by crustal material. However, the compositions of the bulk of the mafic rocks cannot be explained by simple mixing of these two end-members, since their Th/Nb ratios are higher (0.11–0.12), and they have ε_{Nd}(500 Ma) values of +4 and +5 and negative Nb anomalies; this is more likely to be the result of crustal contamination. Most of the felsic metavolcanic rocks have ε_{Nd}(500 Ma) values varying from very negative (−11.7) for anatectic melts derived from older basement, to 0 to +3.4, suggesting derivation from a mafic parent melt by assimilation and fractional crystallization processes (Bankwitz *et al.* 1992).

The close association of alkaline and MORB-like (i.e. oceanic spreading-related mafic rocks; Fig. 12.26B), mafic and crustal melts (rhyolites), in combination with the character of the

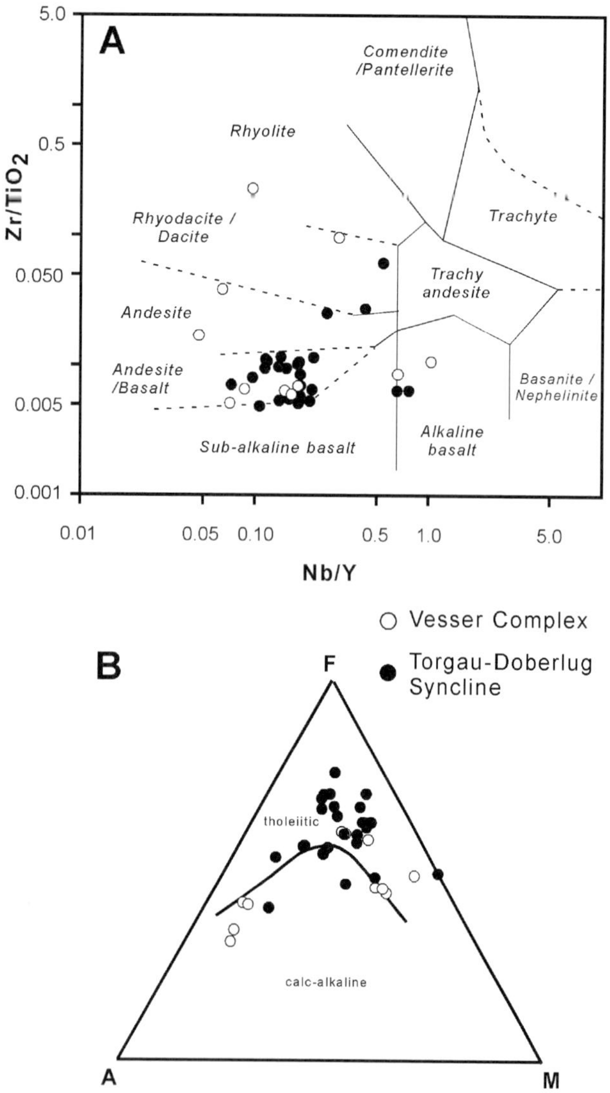

Fig. 12.25. Early Palaeozoic metavolcanic and metaplutonic rocks of the Vesser and Torgau-Doberlug complexes. (**A**) Nb/Y versus Zr/TiO_2 classification diagram (Winchester & Floyd 1977). (**B**) AFM diagram; curve separating tholeiitic from calc-alkaline field after Irvine & Baragar (1971). A = $K_2O + Na_2O$; F = FeO_{total}; M = MgO. Data from Bankwitz *et al.* (1992) and Jonas *et al.* (2000).

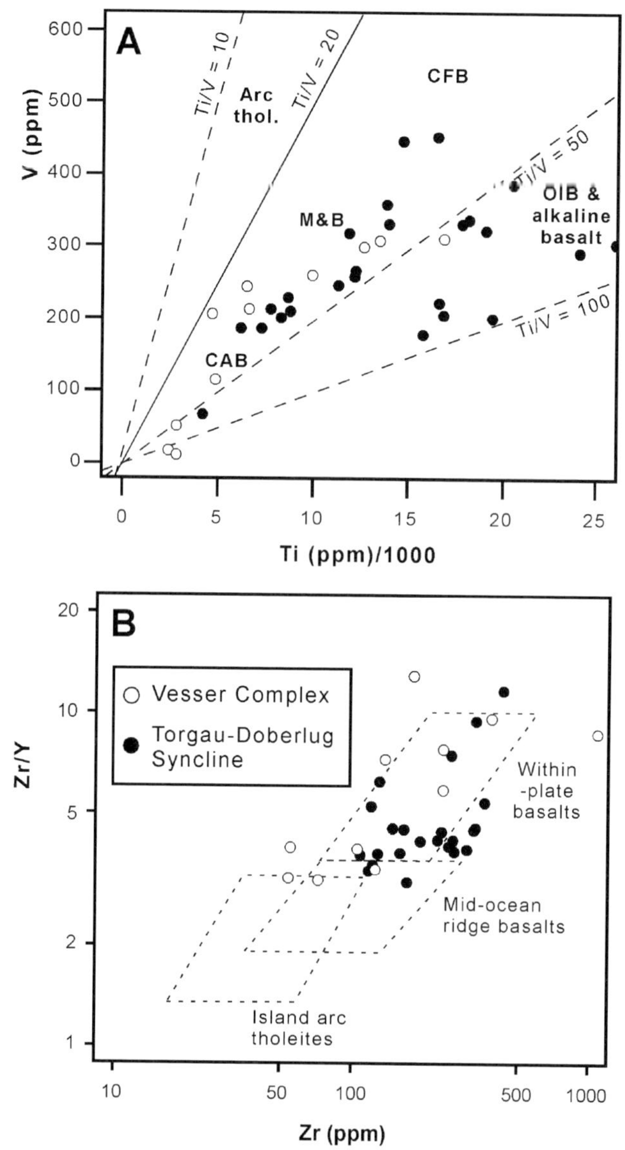

Fig. 12.26. Early Palaeozoic metavolcanic and metaplutonic rocks of basaltic composition ($SiO_2 < 52$ wt%) of the Vesser and Torgau-Doberlug complexes. (**A**) Ti versus V discrimination diagram (Shervais 1982). (**B**) Zr versus Zr/Y tectonic discrimination diagram (Pearce & Norry 1979). Data from Bankwitz *et al.* (1992) and Jonas *et al.* (2000).

sediments suggest that the magmatic activity took place in a progressively widening continental rift. This is a feature that has also been recognized in similar complexes further east in the West Sudetes (see below). The presence in the sediments of detrital white micas with early Cambrian K–Ar (cooling) ages indicates that deposition occurred not far from continental crust that was made up of Neoproterozoic (Cadomian) material (Kemnitz *et al.* 2002). This Cambrian to Ordovician-age rift was situated above attenuating crust at the northern periphery of Gondwana. The progression from continental rift to spreading is also reflected in the change of character of the sediments with time, from littoral siliciclastics, to shelf deposits (psammites, pelites) to younger black shales deposited in a deep, open-marine to pelagic environment (Bankwitz *et al.* 1992, 1998; Kemnitz *et al.* 2002).

Torgau-Doberlug Syncline (Basin)

The ENE-trending, 25 by 120 km Torgau-Doberlug Syncline (TDS) is located to the north of the Lusatia/Lausitz Complex and is mainly known from boreholes, because most of the infill is covered by Tertiary to Quaternary sediments (Fig. 12.27). The oldest rocks are deformed, anchizone metamorphic to greenschist-facies Neoproterozoic metasediments and metavolcanic rocks. The metavolcanic rocks of the basement form a suite of calc-alkaline basalts, E-MORB and andesites, and a separate suite of alkaline basalts. All have a geochemical subduction signature and probably erupted in a continental backarc, intra-arc or marginal basin. A tuff from the basement of the TDS yielded a 566 ± 10 Ma U–Pb zircon crystallization age (cited in Buschmann *et al.* 2006), and the basement rocks are an extension

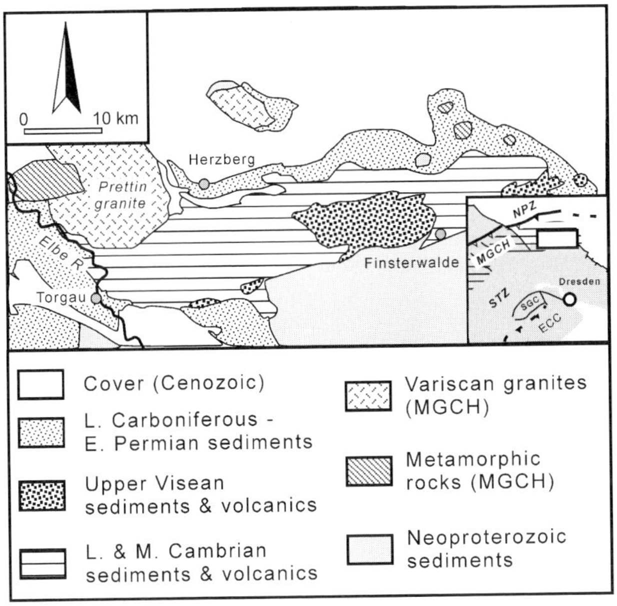

Fig. 12.27. Geological subcrop map of the Torgau-Doberlug Syncline, based on borehole data (after Jonas *et al.* 2000, and references therein). Abbreviations: ECC, Erzgebirge Crystalline Complex; MGCH, Mid-German Crystalline High; NPZ, Northern Phyllite Zone; SGC, Saxonian Granulite Complex; STZ, Saxothuringian Zone.

of, or can be correlated with, the Neoproterozoic Lusatia/Lausitz Complex to the south (Buschmann *et al.* 1995).

The Neoproterozoic basement is unconformably overlain by unmetamorphosed Lower and Middle Cambrian sediments. This Cadomian unconformity in the TDS is estimated to represent a time gap of less than 55 Ma (Buschmann *et al.* 2006). To the north, the Cambrian sediments overlie higher grade metamorphic rocks of the Southern Phyllite Zone and the Mid-German Crystalline High. In the centre of the syncline, Upper Visean sediments overlie the Cambrian with an angular unconformity, and the *c.* 330 Ma Prettin Granite that intruded the rocks in the NW is surrounded by a contact metamorphic halo. Late Carboniferous to early Permian continental sediments are present in the north and west (Buschmann *et al.* 1995; Jonas *et al.* 2000; Jonas & Buschmann 2001; Kopp *et al.* 2001, and references therein; Fig. 12.27).

The Lower Cambrian succession is up to 700 m thick and begins with basal conglomerates followed by limestones, dolostones and siliciclastic sediments of Lower to Middle Atdabian age. These were deposited in a shelf environment under arid to semiarid conditions (Buschmann *et al.* 2006). The Lower Cambrian sediments are intruded by mafic sills and dykes that are absent from the overlying turbiditic sandstones and mudstones, which have a Middle Cambrian age and were deposited in a shelf environment. Younger, Upper Visean (Asbian–Brigantian) fluviatile and paralic sediments and coals reach thicknesses of 400 to 600 m. The Carboniferous sediments are also intruded by mostly aphyric, fine-grained sills, and the limited geochemical data indicate basaltic andesitic, rhyodacitic to rhyolitic compositions. The younger intrusions may be related to the 335–328 Ma granite, diorite and gabbro intrusions in the Pretsch-Prettin area at the NW margin of the TDS (Fig. 12.27; Anthes & Reischmann 2001; Jonas *et al.* 2000; Jonas & Buschmann 2001; see below).

The dykes and sills that intruded the Lower Cambrian sediments have thicknesses ranging from a few centimetres to 60 m, and the dykes show sharp contacts against the sediment host. The sills intruded bedding-parallel and often have glassy to microcrystalline chilled margins. Locally occurring peperite-like breccias indicate shallow intrusion into unconsolidated sediments, and suggest an early Cambrian emplacement age for the intrusions (*c.* 520 Ma; Jonas & Buschmann 2001; Buschmann *et al.* 2006). The subvolcanic rocks are composed of plagioclase laths set in a groundmass of devitrified glass, clinopyroxene or amphibole, and have fine- to coarse-grained, mostly aphyric, doleritic textures. Plagioclase, olivine and amphibole phenocrysts are rare. Jonas *et al.* (2000) distinguished tholeiitic, tholeiitic andesitic and calc-alkaline andesitic basalts (Fig. 12.25A, B). Compared to N-MORB, the least fractionated basaltic rocks have flat, relatively unfractionated HREE patterns indicating that the parent melt(s) formed in a garnet-free, relatively shallow mantle source. The Nb and Ta depletion and Pb enrichment points to the presence of a subduction-related component, possibly the result of crustal contamination. The absence of an Eu anomaly suggests that plagioclase was not a fractionating phase. On the Zr versus Zr/Y diagram of Pearce & Norry (1979) the basaltic rocks plot in the within-plate field (Fig. 12.26B). Jonas *et al.* (2000) suggested that the early Cambrian magmatic activity took place in a pull-apart basin that formed on a rapidly extending continental margin, perhaps part of the Gondwanan shelf.

Jizera-Krkonoše Complex

The Jizera-Krkonoše (Czech Republic)/Izera-Karkonosze (Poland) Complex in the West Sudetes (NE Bohemian Massif; Fig. 12.28) is composed of Neoproterozoic to Cambrian-age granitic gneisses, and deformed and undeformed granitoids that intruded metasediments (mica schists). Its central part is intruded by the late Carboniferous Krkonoše Granite. The Intra-Sudetic Fault separates it from the Görlitzer Schiefergebirge to the north (Görlitz Slate Belt; Göthel 2001) and the Kaczawa Complex to the NE. To the south and east it borders the volcanosedimentary south and SE Krkonoše complexes. A granodiorite from the northern part of the Jizera-Krkonoše Complex yielded an early Cambrian (533 $\pm$ 9 Ma) U–Pb zircon age (Żelaźniewicz *et al.* 2004). This is similar to the 541 $+7/-8$ Ma age of undeformed granodiorite that intrudes Neoproterozoic greywackes near Görlitz/Zgorzelec to the north (Dörr *et al.* 2002), and a 540 $+11/-10$ Ma granite to the SW of the South Krkonoše Complex (see below; Żelaźniewicz *et al.* 2004). However, the zircon ages are up to 40 Ma younger than those of the granodiorites in the Lusatia/Lausitz Complex (542–585 Ma zircon evaporation ages, Kröner *et al.* 1994), of which the Jizera-Krkonoše Complex is regarded to be the more deformed and metamorphosed south-eastern extension or correlative. Żelaźniewicz *et al.* (2004) suggested the evaporation ages for the Lausitz granitoids may be too old due to an undetected inherited component (see also Tichomirowa 2002). Younger calc-alkaline granitoid gneisses from the northern and southern parts yielded mid- to late Cambrian zircon evaporation protolith ages in the range 502–515 Ma, and have negative $\varepsilon_{Nd}(t)$ values between -4.0 and -5.3 (Kröner *et al.* 2001).

The granite gneisses from the northern Jizera-Krkonoše Complex had fine- to coarse-grained, porphyritic granites as precursors (Oberc-Dziedzic *et al.* 2005). They have peraluminous compositions and often contain magmatic cordierite (locally with garnet $\pm$ sillimanite) which classifies them as crust-derived (S-type) melts; calc-alkaline, tonalitic and dioritic rocks are absent. Major and trace element fractionation trends and chondrite-

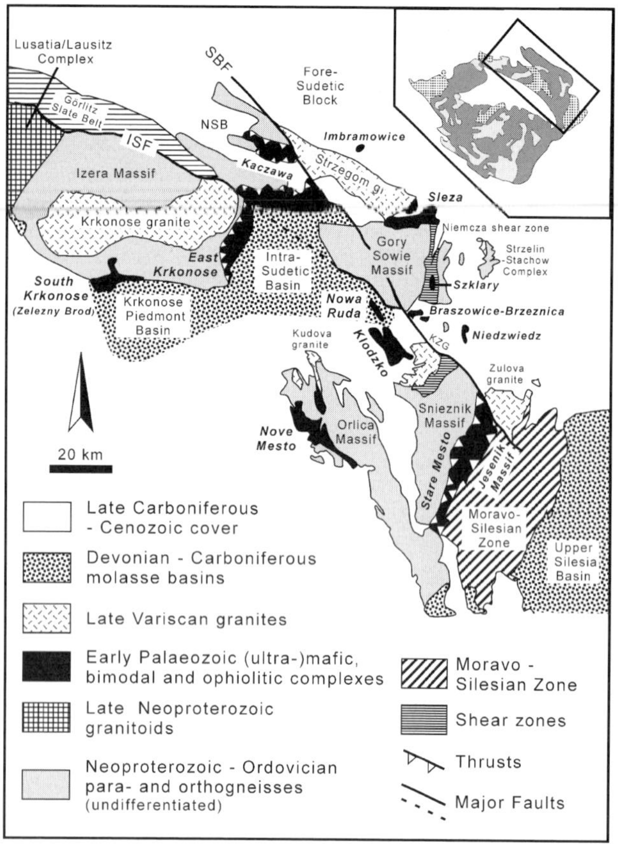

Fig. 12.28. Simplified map of the Sudetes showing the early Palaeozoic (ultra)mafic, bimodal, volcanosedimentary and ophiolitic complexes (in black: South and East Krkonoše, Kaczawa, Ślęża, Nowa Ruda, Kłzodko, Staré Město and Nové Město). Abbreviations: ISF, Intra-Sudetic faults; KZG, Kłodzko-Złoty Stok Granite; NSB, North Sudetic Basin; SBF, Sudetic Boundary Fault (after Narebski 1994; Turniak *et al.* 2000).

normalized multi-element patterns with negative Sr, P and Ti anomalies point to fractionation of feldspars, biotite, cordierite, monazite, apatite and ilmenite. Negative Nb and Ta anomalies and low Zr/Nb (18.5–4.5), Nb/Th (0.58–1.44) and Ce/Pb (1.6–3.3) ratios are in agreement with a derivation from continental crust, which is supported by the negative ε_{Nd}(500 Ma) values (−6.9 to −5.2). The granites are LREE-enriched but have flat HREE patterns, and the negative Eu anomalies point to plagioclase fractionation. In classification diagrams the granites plot in the field of volcanic-arc granites, but all other geochemical evidence is contradictory. Instead, Oberc-Dziedzic *et al.* (2005) suggested that the peraluminous granites were formed by melting of immature crustal material in a lower Palaeozoic rift setting.

The metavolcanosedimentary Kaczawa, South and East Krkonoše complexes bordering the Jizera-Krkonoše Complex have protracted and polymetamorphic histories. Most of the metavolcanic rocks underwent subseafloor alteration during and shortly after emplacement ('spilites', 'keratophyres'). This was followed by blueschist-facies metamorphism, as is evidenced by rare, but widespread occurrences of glaucophane (including crossite) and (Mg-) riebeckite in mafic rocks (Kryza *et al.* 1990; Patočka *et al.* 1996; Smulikowski 1995; Achramowicz & Wajsprych 2004). This phase of metamorphism occurred during a late Devonian subduction event, as is clear from $^{40}Ar/^{39}Ar$ ages for phengites from the southern part of the East Krkonoše Complex (Maluski & Patočka 1997). Younger, early to mid-Carboniferous $^{40}Ar/^{39}Ar$ ages for the regional greenschist-facies overprint (Maluski & Patočka 1997; Marheine *et al.* 2002) reflect peak-Variscan events that were probably due to collision between the Saxothuringian and Teplá-Barrandian terranes and the accretion of smaller terrane fragments. SE-directed extensional collapse in the early Carboniferous was accompanied by the intrusion of the Karkonosze Granite (328 ± 12 Ma Rb–Sr whole-rock isochron age; Pin *et al.* 1987) and the formation of the Intra-Sudetic Basin (Mazur & Aleksandrowski 2001).

South Krkonoše Complex

The South Krkonoše Complex (SKC) borders the Jizera-Krkonoše Complex to its south and is an internally complex belt of predominantly greenschist-facies metasediments and metavolcanic rocks (Fig. 12.29). The rocks of the SKC underwent subseafloor alteration shortly after their formation, followed by late Devonian, subduction-related blueschist-facies metamorphism, and a regional, greenschist-facies overprint in early to mid-Carboniferous times.

The complex has been subdivided into a lower and an upper unit with apparently gradual transitions (Patočka & Smulikowski 2000). The lower unit comprises terrigenous metasediments (sericite-chlorite phyllites), mafic and felsic, bimodal metavolcanic rocks, and small sheets of tectonically intercalated, strongly sheared granitic gneisses. The upper unit consists of sericite-graphite phyllites, cherts, quartzites, and calcitic and dolomitic marbles, while magmatic rocks are limited to diabase sills. Small, younger bodies of ultramafic rocks (serpentinites) and metagabbro cross-cut foliated rocks and constitute the final pulse of magmatism; they are probably Silurian to Devonian in age.

A weakly foliated granodiorite from the southwesternmost part of the SKC yielded a 540 +11/−10 Ma U–Pb upper intercept zircon crystallization age (Neoproterozoic–Cambrian boundary), but its tectonostratigraphic position with respect to the other SKC lithologies is not clear (Żelaźniewicz *et al.* 2004). Metasediments from the southern parts of the SKC contain Llandovery–Wenlock graptolite fragments, microfossils (mazuelloids–phosphatized acritarchs) and ichnofossils with ages somewhere between the

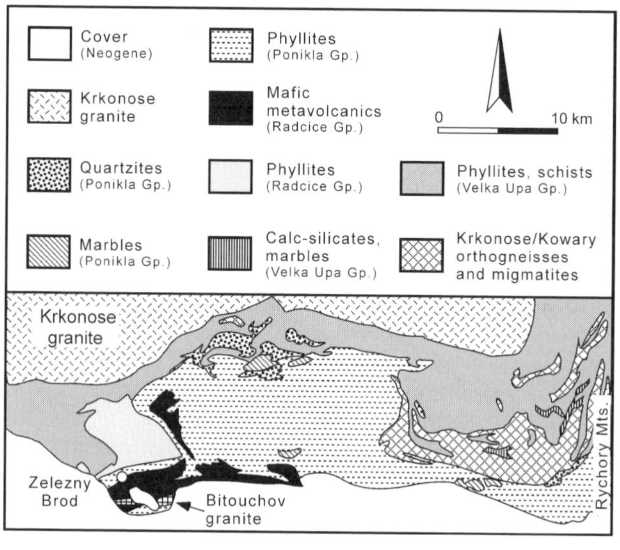

Fig. 12.29. Geological sketch map of the South Krkonoše Complex, West Sudetes (after Patočka *et al.* 1996, and references therein).

Ordovician and Devonian (Hladil *et al.* 2003, and references therein). Marbles in the SE have fragments of early Cambrian archaeocyaths and trilobites, and marbles in the south contain dacryoconarids of early to late Silurian, possibly even early Devonian (Pragian), age (Hladil *et al.* 2003).

The metavolcanic rocks of the lower unit are predominantly mafic subaqueous lavas, tuffs and tuffites; flows and pyroclastic rocks of felsic composition occur in much smaller volumes. The most common mafic rock types erupted as massive flows, pillows and pillow breccias, often vesicular. The felsic rocks were emplaced as massive flows or sills (metagranites and 'metaporphyroids'), and a fine-grained variety probably as rhyolitic glassy lavas ('metakeratophyres') or tuffs. The rare diabase intrusions in the upper unit have relict ophitic, doleritic to gabbroic magmatic textures; evidence for the blueschist-facies metamorphism in the SKC mainly comes from this rock type. The youngest metagabbros occur as coarse-grained, small lenses and pods with chilled margins, and the serpentinites are completely transformed to tremolite-actinolite schists. In general, relicts of magmatic minerals are limited to pyroxene, kaersutite and ilmenite.

The mafic metavolcanic rocks plot in the subalkaline field of the Nb/Y versus Zr/TiO_2 classification diagram of Winchester & Floyd (1977; Fig. 12.30), and have relatively flat to moderately enriched REE patterns and ocean-floor-like Cr/Ti ratios. Their protoliths were transitional (T-) to E-MORB, and true N-MORB are rare. The diabases from the upper unit have transitional to alkaline trace element patterns indicative of a within-plate setting.

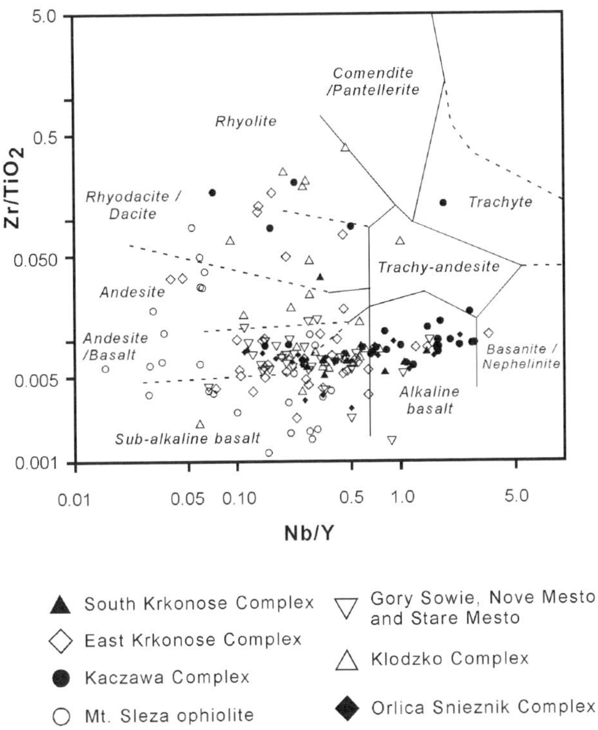

Fig. 12.30. Nb/Y versus Zr/TiO_2 discrimination diagram for the early Palaeozoic mafic to felsic metavolcanic, and mafic metaplutonic rocks of the South and East Krkonoše, Kaczawa, Mt Ślęża, Kłodzko, Góry Sowie, Nové Město, Staré Město, Rehberg and Raabs complexes (Winchester & Floyd 1977). Data from Dostal *et al.* (2001), Floyd *et al.* (1996, 2002), Fritz (1996), Furnes *et al.* (1994), Höck *et al.* (1997), Kryza & Pin (2002), Kryza *et al.* (1992, 1995, 2003), Maluski & Patočka (1997), Patočka *et al.* (2000), Patočka & Smulinowski (2000), Pin *et al.* (1988), Winchester *et al.* (1995, 1998).

The felsic metavolcanic rocks have SiO_2 contents in the 65–76 wt% range and these flows or subvolcanic intrusions classify as dacites, rhyodacites and rhyolites, and the tuffs as comendites and pantellerites. When normalized against ocean-ridge granites (ORG), the dacites, rhyodacites and rhyolites have fractionated REE patterns with a negative Eu anomaly, suggesting that their granitic parent melts were intruded or extruded in a within-plate or a volcanic arc setting. The tuffs have higher REE concentrations and steep, LREE-enriched patterns with a significant negative Eu anomaly and flat, unfractionated HREE patterns, also suggesting a within-plate setting. The narrow range in La/Nb ratios of the felsic tuffs and the mafic volcanic and subvolcanic rocks suggests that they were derived from a common parent melt, with the felsic end-members produced by extensive fractionation of plagioclase, clinopyroxene, ± amphibole. In contrast, the metagranites, metaporphyroid and metakeratophyres are strongly enriched in Th and La relative to Nb, Zr and Y, ruling out a genetic relation to the other rock types. Instead, their parent melts either originated through direct melting of (attenuated) continental crust, or from mafic melts that were extensively contaminated by continental crustal material.

The late metagabbros and ultramafic intrusions are enriched in MgO (10–20 and 20–26 wt%, respectively), have picritic to picritic basaltic compositions and may have been derived from a common parent melt, though some are cumulates. They have subalkaline compositions, are LREE-enriched and are of within-plate character.

The SKC is considered to be a Carboniferous accretionary prism, or nappe stack made up of parautochthonous and allochthonous material, the Early Palaeozoic protoliths of which were deposited on the passive margin of the Saxothuringian Terrane (part of the ATA) and in the neighbouring oceanic domain. It is possible that the SKC is the eastern extension or correlative of the Vesser Complex. Patočka & Smulikowski (2000) interpreted the decreasing REE fractionation (i.e. decreasing Ce_N/Yb_N, decreasing slope of the REE pattern) and increasing Zr/Nb and Y/Nb ratios of the mafic rocks as reflecting an evolution from early alkaline to younger, more tholeiitic melts in a widening continental rift. In this scenario, initial mantle upwelling generated the relatively small-degree melts that intruded as diabases at higher crustal levels. Further extension then developed into an early rift stage that triggered higher degrees of melting of shallower mantle, yielding more voluminous tholeiitic mafic melts. Together with crustal (anatectic) melts and felsic differentiates, these erupted as subaqueous lavas and tuffs in rift basins that were fed with terrigenous sediments. The youngest, gabbroic and ultramafic melts intruded following advanced extension and crustal thinning. No ophiolites or remnants of true ocean crust have been identified in the SKC, but remnants of blueschist-facies metamorphic minerals preserved in the coarse-grained diabases point to a subduction event of probably late Devonian age.

East Krkonoše Complex

The north–south trending East Krkonoše Complex (EKC; also termed the Rudawy Janowickie Complex) is exposed in the Rýchory Mountains, the Lasocki Ridge and the Rudawy Janowickie Mountains to the east of the Jizera-Krkonoše Complex (Fig. 12.31). Its NW part is intruded by the late Carboniferous Krkonoše Granite and to the east and south it is overlain by

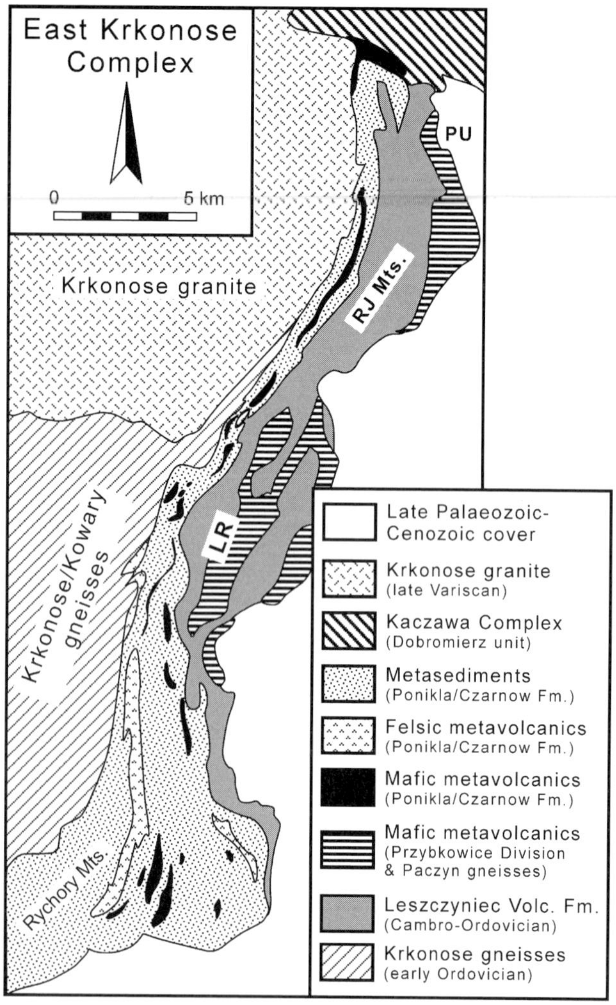

Fig. 12.31. Geological sketch map of the East Krkonoše Complex, West Sudetes (simplified after Patočka & Smulikowski 2000). Abbreviations: LR, Lasocki Ridge; RJ Mts., Rudawy Janowickie Mountains; PU, Przybkowice Unit.

Carboniferous sediments of the Intra-Sudetic Basin. To the north it is in tectonic contact with greenschists of the Kaczawa Complex.

The EKC is composed of several lithotectonic units with varying lithologies and differing metamorphic evolution, and it has been regarded as an accretionary complex (e.g. Kryza *et al.* 1995). Metasediments, amphibolites and various felsic gneisses (Karkonosze/Kowary Gneisses) are exposed in the west and SW. The felsic gneisses have early Ordovician protolith ages, as indicated by a 481–492 Ma lower intercept age range for zircons from granite gneiss (U–Pb ages: Oliver *et al.* 1993). The central part of the EKC mainly comprises greenschist-facies metasediments (phyllites, schists, quartzites) with intercalated/interbedded mafic and felsic metavolcanic rocks (Poniklá/Czarnów Formation). Graptolites from metasediments in the Poniklá/Czarnów Formation (Fm) indicate a Silurian age of deposition (Hoth 2000, and references therein). Further east, the greenschist- to epidote-amphibolite-facies Leszczyniec Volcanic Fm contains most of the magmatic rocks in the EKC, which are predominantly of mafic composition. Its southern part contains metabasalts and metadolerites, and the northern part more volcaniclastic and sedimentary rocks. The so-called Paczyn Gneisses in the eastern part of the EKC occur as variously deformed, fine- to coarse-grained, felsic to mafic (augen-)gneisses (epidote amphibolites). Metavolcanic rocks and metasediments (greenschists, phyllites) of the Przybkowice Division are restricted to the NE of the complex. U–Pb zircon ages of 505 ± 5 Ma for a felsic metavolcanic rock and 494 ± 2 Ma for Paczyn Gneiss indicate middle Cambrian to early Ordovician protolith ages for this part of the EKC (Oliver *et al.* 1993). In the Poniklá/Czarnów Fm, metavolcanic rocks of felsic composition are more voluminous than mafic rocks, but are present in only small volumes in the Leszczyniec Volcanic Fm to the east (Winchester *et al.* 1995; Patočka & Smulikowski 2000; Dostal *et al.* 2001). $^{40}Ar/^{39}Ar$ dating indicates that locally preserved blueschists-facies metamorphic mineral assemblages are Devonian in age (360–365 Ma; Patočka *et al.* 1996; Maluski & Patočka 1997).

Winchester *et al.* (1995) distinguished three geochemical provinces in the EKC with distinct incompatible element signatures. The western (Kowary) province comprises alkaline and tholeiitic metabasalts, and felsic metavolcanic rocks. The alkaline basalts range from primitive to highly evolved as shown by their widely varying Cr contents. They have steep REE patterns ($La_N/Yb_N = c.\ 34$) and high Zr/Nb ratios indicative of a within-plate setting. The tholeiitic basalts have flatter REE patterns ($La_N/Yb_N = c.\ 8$) and Ti/V ratios more typical of transitional tholeiites. Trace element ratios of the felsic metavolcanics indicate andesitic to dacitic protolith compositions (see Fig. 12.30). The mafic metavolcanic rocks that are intercalated with the metasediments of the Poniklá/Czarnów Fm to the east constitute a series of fractionated basalts of alkaline to tholeiitic composition (Dostal *et al.* 2001). The alkaline basalts have primitive mantle-normalized incompatible element patterns that resemble those of continental and ocean island basalts and are LREE-enriched ($La_N/Yb_N = c.\ 20$) and HREE-depleted, the latter suggesting residual garnet in the mantle source. The tholeiitic basalts have flatter REE patterns ($La_N/Yb_N = 1$ to 4) and have incompatible element patterns similar to E-MORB. The ε_{Nd}(500 Ma) values of the mafic rocks are positive but vary from +3.1 to +6.6, which indicates derivation from a depleted but heterogeneous mantle source, with little evidence for contamination by crustal material. The felsic metavolcanic rocks have rhyolitic compositions, are enriched in REE and Th, have negative Nb and Ti anomalies, and have negative ε_{Nd}(500 Ma) values in the range of −4.3 to −5.5; they are anatectic, crust-derived melts (Dostal *et al.* 2001).

The central geochemical province includes the Leszczyniec Volcanic Fm which is composed of tholeiitic metabasalts, locally occurring coarse-grained mafic rocks, and, as noted above, minor amounts of felsic metavolcanic rocks. The metabasalts classify as subalkaline to andesitic basalts (Fig. 12.30) and have a MORB-like character as indicated by their depleted REE patterns ($La_N/Yb_N = c.\ 0.7$), low Th and Nb abundances, high Zr/Nb (>8) and Y/Nb (>1) ratios, high ε_{Nd}(500 Ma) values (+7 to +8), and relatively flat patterns in MORB-normalized multi-element diagrams that lack negative Nb and Ta anomalies. Alkaline basalts are absent, and the wide range of Cr contents indicates that the tholeiites are fractionated to varying degrees. The coarse-grained varieties are probably cumulates as indicated by high V and low Ti and Y abundances. The felsic metavolcanic rocks differ from those in the western province in their lower Nb/Y ratios, very large negative Nb, Ti and Eu anomalies, and flat HREE patterns. The increase in Th and LREE, in combination with high, positive ε_{Nd}(500 Ma) values (+6.3 to +6.8), suggest they are related to the MORB-like mafic rocks by fractional crystallization processes (Kryza *et al.* 1995; Winchester *et al.* 1995).

The eastern geochemical province contains metagabbros and metabasalts belonging to the Paczyn Gneisses and the Przybkowice Division that are geochemically distinct from the other provinces. Metabasalts associated with the Paczyn Gneisses are LREE-depleted (La_N/Yb_N = *c.* 0.7), and the metagabbros are slightly LREE-enriched (La_N/Yb_N = *c.* 1.7); both plot in the island-arc and active continental margin fields of the Ti-V (Fig. 12.32) and Th/Yb versus Ta/Yb classification diagrams. The Przybkowice metabasalts are more LREE-enriched, having a La_N/Yb_N ratio of *c.* 2.9 and flatter MORB-normalized multi-element patterns.

No true oceanic crust or island-arc rocks have been identified in any of the EKC units. The MORB-like character of the mafic rocks of the Leszczyniec Volcanic Fm suggests that they extruded in a submarine, extensional basin that must have been underlain by (thinned) continental crust, as supported by the presence of *c.* 1.9–2.3 Ga inherited zircons (Oliver *et al.* 1993; Kryza *et al.* 1995; Winchester *et al.* 1995). For the western province, the within-plate signature of the alkaline basalts and their association with voluminous felsic volcanic rocks and large volumes of sediments also indicate an extensional setting (i.e. rift; Winchester *et al.* 1995). Dostal *et al.* (2001) interpreted the differences between the alkaline and tholeiitic basalts of the Poniklá/Czarnów Fm as being due to different degrees of melting at different depths. The tholeiites are high-degree melts from relatively shallow mantle, whereas the alkaline basalts were generated by lower degrees of melting of a deep, garnet-bearing mantle source. The melting must have been triggered by lithospheric extension since the basalts lack a subduction signature and occur in such small volumes that it would be impossible for them to have been derived from a mantle plume. Rocks of intermediate (andesitic) compositions that may suggest subduction-related magmatism are absent. Emplacement of the parent magmas of the basalts triggered anatexis of (older) crust and the formation of the felsic (rhyolitic) melts with negative $\varepsilon_{Nd}(t)$ values. The relatively large volumes of these crust-derived felsic volcanics and their association with continental sediments indicate a continental rift setting, somewhere at the northern Gondwana margin. This interpretation is corroborated by the large time gap between magmatic activity (early Ordovician) and the first, blueschist-facies, subduction-related metamorphism (late Devonian; Maluski & Patočka 1997).

The contrasting, arc-like signature of the mafic rocks of the eastern province (Paczyn Gneisses and Przybkowice Division) may be caused by crustal contamination or inheritance from a mantle source that was enriched in fluids and incompatible elements due to a previous subduction event. Magmatic rocks from this province clearly differ from those to the west, and this unit may have originated elsewhere (further east?), being subsequently juxtaposed against the other provinces during Palaeozoic subduction and accretion (Winchester *et al.* 1995).

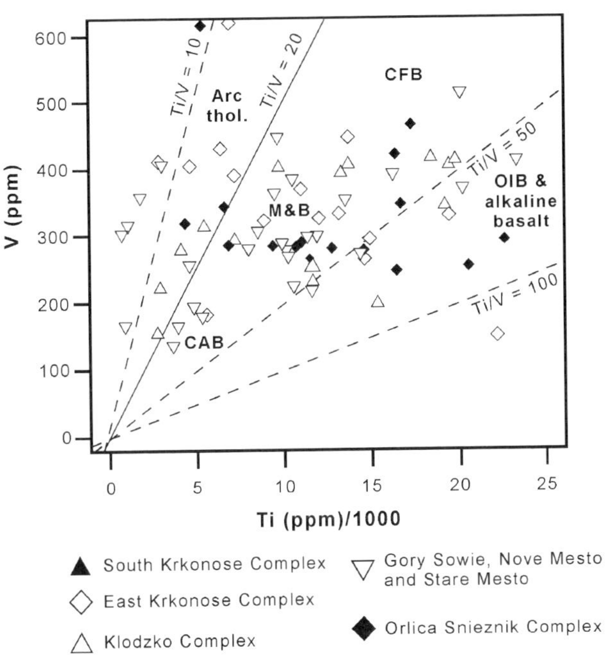

Fig. 12.32. Ti versus V discrimination diagram for metavolcanic and metaplutonic rocks from the West Sudetes area with <52 wt% SiO_2 (Shervais 1982). Abbreviations: CAB, calc-alkaline basalt; CFB, continental flood basalt; M&B, mid-ocean ridge and backarc basalt; OIB, ocean island basalt; thol., tholeiite. For data sources see caption to Figure 12.30.

Kaczawa Complex

The Kaczawa Complex is made up of low-grade volcanosedimentary suites of Cambrian to Silurian age that were thrust and stacked in Carboniferous times. Within most of these units an internal stratigraphy has been established, but precise lithostratigraphic correlations between the units is difficult to establish (Fig. 12.33; Kryza *et al.* 1992, 1994, 2004; Furnes *et al.* 1994). Three main suites have been distinguished (Fig. 12.33): the Świerzawa and Bolków units (oldest, (Cambro-) Ordovician), the Rzeszówek-Jakuszowa and Dobromierz units of Ordovician (–Silurian) age, and tectonized sedimentary breccias (mélange) of probably early Carboniferous age (Furnes *et al.* 1994; Kryza *et al.* 1994; Kozdrój *et al.* 2001). Most rock units were metamorphosed under Carboniferous, regional anchizone to greenschist-facies conditions, but locally glaucophane, crossite, phengite and jadeite relicts testify to an older, Devonian blueschist-facies metamorphic event (Kryza *et al.* 1990; Mus-

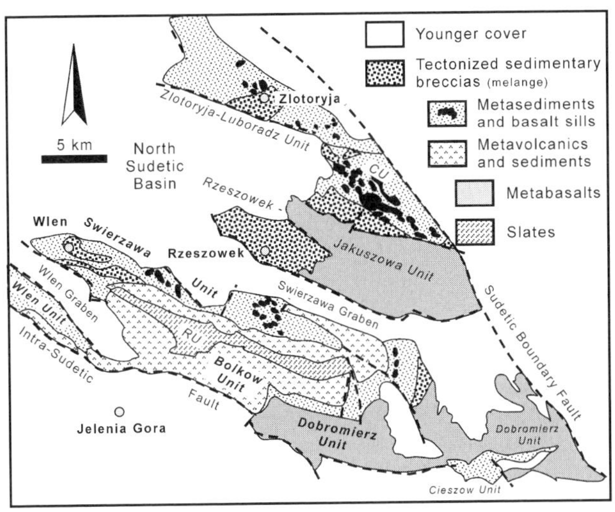

Fig. 12.33. Geological sketch map of the Kaczawa Complex, West Sudetes (after Furnes *et al.* 1994; Kryza *et al.* 1994). Abbreviations: CU, Chełmiec Unit; RU, Radzimowice Unit.

zyński & Kryza 1993; Wajsprych & Achramowicz 2003). The exceptions are the anchizone- to epizone-grade slates of the Rzeszówek and Chełmiec units (Kryza *et al.* 2002).

The oldest rocks, exposed mainly in the Świerzawa and Bolków units, are lithologically varied, metamorphosed volcano-sedimentary units. The Cambro-Ordovician Świerzawa Unit commences with up to 500 m thickness of (lower) pillow basalts that are overlain by clastic sediments of the Gackowa Fm, followed by a *c.* 100 m thick association of (upper) pillow basalts (Podgórki Volcanic Complex) and subaerial basaltic lavas. The sequence is capped by up to 200 m of bimodal metavolcanic rocks of the Lubrza association (Kryza *et al.* 1992, 1994; Furnes *et al.* 1994, and references therein).

The Radzomowice unit of low-grade metapelites (slates) separates the Świerzawa and Bolków units. The Bolków Unit starts with basalts of the Podgórki Volcanic Complex that are associated with shallow-water limestones of possibly Cambrian age. These are overlain by small volumes of volcaniclastic rocks, rhyodacitic volcanic rocks, and alkaline metavolcanic rocks of the Lubrza association. In the eastern part of the Bolków Unit these rocks are overlain by pelagic shales and cherts that yielded Silurian to Upper Devonian graptolites and conodonts (Kryza *et al.* 1994).

The sandstones of the Gackowa Fm were deposited in a shallow marine shelf setting, and were derived from rhyodacitic volcanic rocks (Kryza *et al.* 1994). The overlying Podgórki basalts comprise vesicular pillows and massive flows that erupted both subaerially and subaqueously into shallow water, as is clear from the associated volcaniclastic sandstones and shallow water limestones. Interbedded rhyodacite flows occur in the Podgórki Volcanic Fm of the Bolków Unit. The bimodal Lubrza association comprises volumetrically dominant trachyte domes and lavas that are embedded in tuffs, tuffites, and volcaniclastic sediments that were derived from rocks of felsic to intermediate composition.

Basalts from the Podgórki Volcanic Complex are enriched in incompatible elements and the LREE indicate compositions transitional between MORB and alkaline, ocean-island-like basalts, and therein resemble continental tholeiites (Figs. 12.30 and 12.34A, B). Metabasalts from the central and eastern part of the Bolków Unit have flat REE patterns and have a more MORB-like character. Both transitional basalts and tholeiites show large variations in $\varepsilon_{Nd}(t)$ values, ranging from 0 to +7.2 (at 500 and 450 Ma; Furnes *et al.* 1994). The rhyodacites are enriched in incompatible elements, but have negative Ta, Nb, P and Ti anomalies and lower, but still positive $\varepsilon_{Nd}(t)$ values (+2.7 to +3.7).

Metabasalts of probable (Cambro-)Ordovician age occur as shallow, locally porphyritic, sills in slates of the Zlotoryja-Luboradz, Chełmiec and Świerzawa units. They are moderately enriched in incompatible elements, have alkaline compositions and positive $\varepsilon_{Nd}(t)$ values of +1.8 to +3.1.

The Rzeszówek-Jakuszowa and Dobromierz units are Ordovician to Silurian in age. They comprise large volumes of pillowed and massive flows of basaltic composition, with subordinate volcanic breccias, hyaloclastites and, near the top, locally preserved black slates that contain Silurian graptolites (Furnes *et al.* 1994, and references therein). In these units, clastic sediments are rare and felsic volcanic rocks are absent. The basalts have N- to E-MORB-like REE patterns and have the highest $\varepsilon_{Nd}(t)$ values of the Kaczawa Complex (+3.2 to +8.7), indicating very little contamination by crustal material and derivation from a more depleted mantle source.

Negative Ta and Nb anomalies in some of the oldest

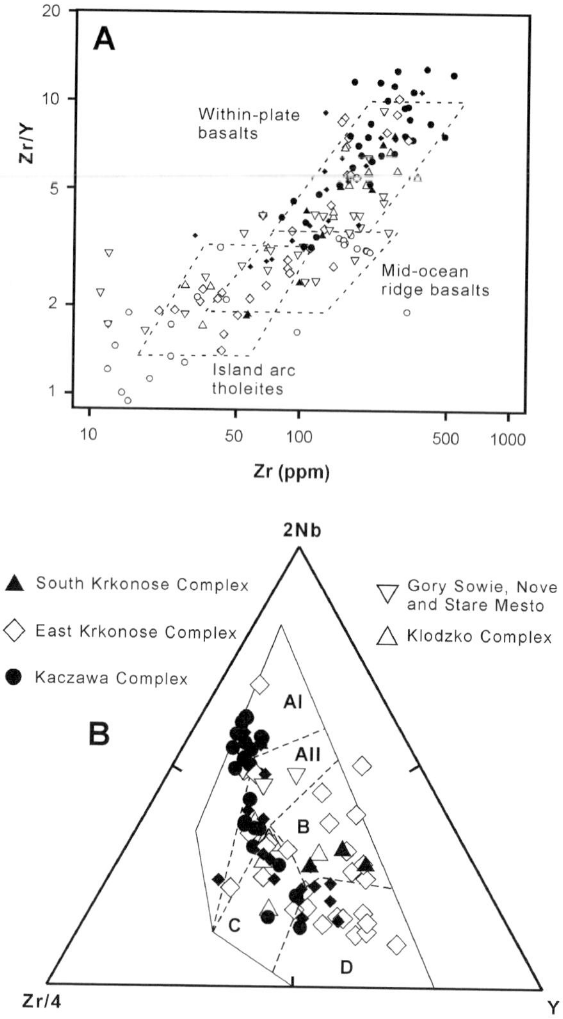

Fig. 12.34. (**A**) Zr versus Zr/Y tectonic discrimination diagram for selected mafic metavolcanic rocks from the West Sudetes with <52 wt% SiO_2 (Pearce & Norry 1979). (**B**) Nb–Zr–Y tectonic discrimination diagram for selected mafic metavolcanic rocks with SiO_2 <52 wt% (Meschede 1986). For data sources: see caption to Figure 12.30.

metabasalts (Podgórki Volcanic Complex) may suggest an origin in a subduction setting, but the close spatial association of alkaline basalts and crustal melts (rhyodacites), the lack of arc-derived greywackes and of volcaniclastic and pyroclastic rocks, and the association with shelf sediments, indicate a continental rift setting (Kryza *et al.* 1992, 1994; Furnes *et al.* 1994). The change in incompatible element and Sm–Nd isotope compositions reflects a magmatic evolution from (Cambro-)Ordovician continental rifting to a more oceanic setting (Furnes *et al.* 1994). In addition, the oldest mafic magmatic rocks may have an enriched, plume-like component that mixed with depleted mantle, and thus constituted the source of the transitional basalts. Melting of overlying crust during extension gave rise to the rhyodacite melts, which are not cogenetic with the basalts as is clear from their lower $\varepsilon_{Nd}(t)$ values. Deepening of the basin is indicated by the presence of turbidite sediments that were later intruded by sills of alkaline basalt. Further rifting and attenuation of the continental crust was accompanied by bimodal volcanism of cogenetic trachytes and alkaline basalts. The youngest volcanic rocks erupted at high rates (i.e. lack of interbedded sediments), have MORB-like trace element compositions and show little evidence of crustal contamination. This suggests that the rifting process that had begun in the (Cambro-)Ordovician ended with oceanic spreading in Silurian to Devonian times (Furnes *et al.* 1994).

Góry Sowie Complex

The Góry Sowie Complex (GSC) further to the east (Fig. 12.28) is an allochthonous unit predominantly composed of metasediments (metamorphosed sandstones, mudstones and greywackes) that are tectonically intercalated with trondhjemitic to granodioritic orthogneisses and migmatites. The metasediments contain microfossils that indicate Neoproterozoic to early Cambrian depositional ages (Gunia 1999). In addition, minor volumes of mafic and ultramafic magmatic rocks are present, and felsic to mafic granulites (metamorphosed bimodal volcanics), eclogites and garnet peridotites are locally preserved (O'Brien & Carswell 1993; Kröner & Hegner 1998; Kryza *et al.* 1996; Kryza & Pin 2002 and references therein). At its northern margin the GSC is thrust onto the mafic to ultramafic Mt Ślęża Complex. In the south it is separated from the Nowa Ruda Complex by a fault, and its relation to the Kłodzko Metamorphic Complex is obscured by the Carboniferous–Permian sediments of the Intra-Sudetic and Bardo basins (Fig. 12.28). The Mt Ślęża, Nowa Ruda and Kłodzko complexes have been interpreted as dismembered and metamorphosed ophiolites, remnants of early Palaeozoic oceanic crust, or rifted margin lithologies (e.g. Pin *et al.* 1988; Floyd *et al.* 2002). The Niemcza Shear Zone to the east of the GSC hosts the Braszowice Gabbro, which is also considered to be the middle part of a dismembered ophiolite (Dziedzic & Kramer 1992). The Stachów and Strzelin complexes to the east of the Niemcza Shear Zone are composed of felsic gneisses and metasediments that were deformed and metamorphosed during the Variscan Orogeny. Felsic orthogneisses from the Stachów Complex yielded a zircon protolith age of *c.* 500 Ma (late Cambrian). Felsic orthogneisses from the Strzelin Complex to the west yielded Neoproterozoic (600 $\pm$ 7 and 568 $\pm$ 7 Ma) zircon protolith ages and can be correlated with Neoproterozoic orthogneisses of the Moravo-Silesian Zone to the south (see below; Oberc-Dziedzic *et al.* 2003).

Kröner & Hegner (1998) obtained zircon evaporation ages of 487–482 Ma (early-mid Ordovician) for the felsic gneisses of the GSC, which they interpreted as the crystallization ages of their protoliths. The $\varepsilon_{Nd}(t)$ values range from -6.2 to $+0.2$, showing that some contain an older crustal component. The orthogneisses have calc-alkaline, I-type (subduction-related) compositions with negative to negligible Eu anomalies, but pronounced negative Nb anomalies (compared to primitive mantle). High-temperature metamorphism resulted in anatexis in Devonian times as is clear from U–Pb monazite and xenotime ages of 370–382 Ma from migmatitic gneisses in the western part (Bröcker *et al.* 1998; Timmermann *et al.* 2000). Mafic magmatic rocks are a minor component of the GSC and occur as lenses of amphibolites ($\pm$ garnet), serpentinites and corona gabbros (presumably metamorphosed lavas and shallow intrusions), and as lenses of granulite. The ages of the metamorphosed magmatic rocks are unknown, and the relation with the surrounding gneisses and migmatites is not clear (Winchester *et al.* 1998; Kryza & Pin 2002). Mafic to (predominantly) felsic granulites from the eastern part of the complex are probably metamorphosed bimodal volcanics. The mafic granulites have zircon ages of 411–414 Ma (early Devonian), perhaps the crystallization ages of the magmatic protoliths. Ages of *c.* 350 Ma for zircon overgrowths probably date late Devonian to early Carboniferous metamorphism in this part of the complex (SHRIMP U–Pb ages; Lange *et al.* 2005).

More or less independent of rock type, Kryza & Pin (2002) distinguished three geochemical groups of metamorphosed mafic rocks within the GSC: depleted amphibolites and enriched amphibolites, and a group that includes the corona gabbros ('coronitic metagabbronorites') and mafic to intermediate granulites.

The depleted amphibolites of the GSC have high Zr/Nb ratios (>30), low Nb concentrations, flat REE patterns with fractionated (convex) HREE, and negative Sr, Ti and Eu anomalies that are attributed to the removal of plagioclase and a Ti-phase during fractionation of the parent melt(s). The $\varepsilon_{Nd}(500\ Ma)$ values are high (+5 to +6), which indicates they were derived from a depleted mantle source and lack older (crustal) components. The enriched amphibolites have much lower Zr/Nb ratios (*c.* 8), are moderately enriched in LREE and have lower $\varepsilon_{Nd}(t)$ values (+0.1 to +2.9). Kryza & Pin (2002) suggested a within-plate setting for this group and derivation from a mantle source that was enriched in incompatible elements shortly before melting.

The third group includes the corona gabbros and mafic to intermediate granulites, and is geochemically more variable. The corona gabbros are medium- to coarse-grained and have preserved magmatic, ophitic textures and magmatic olivine, pyroxene and plagioclase, but their margins are foliated or faulted. Granulite-facies clinopyroxene-plagioclase rocks of mafic to intermediate composition are intercalated with felsic granulites. The rocks have Zr/Nb ratios of between 13 and 19 and are slightly LREE enriched. Some members lack a negative Nb anomaly (compared to chondrite), whereas others have pronounced negative Nb anomalies. Large, negative Ti, Eu and Sr anomalies reflect the evolved character of some of the parent melts, whereas positive Eu anomalies were probably the result of plagioclase accumulation. The $\varepsilon_{Nd}(t)$ values are also more variable, ranging from -1.0 for the evolved melts to +5.3 for cumulate-rich rocks; the low values are due to contamination with (older) continental crust. The REE patterns, variable $\varepsilon_{Nd}(t)$ values, intermediate Zr/Nb ratios, and varying Eu and Sr anomalies suggest that the parent melts were continental tholeiites derived from a depleted mantle source. During ascent and emplacement their compositions were modified by fractionation and assimilation processes under relatively low pressure conditions.

Although the rocks in the first group have a depleted character, no typical N-MORB types appear to be present, and there is no clear petrogenetic relation with the Mt Ślęża ophiolite to the north of the GSC. Compared to true N-MORB, Nb abundances are higher, $\varepsilon_{Nd}(t)$ values are positive but still lower, and the fractionated HREE point to residual garnet in the mantle source, indicating melting at greater depths. Neither do the Ti/V ratios for either depleted or enriched amphibolites support an origin in a subduction setting. Kryza & Pin (2002) proposed that the earliest melts were generated in a rift setting from deep, garnet-bearing mantle that was enriched shortly before the melting event. The resulting melts were of continental tholeiitic, within-plate affinity, but later melt generations developed towards more depleted, N-MORB-like compositions during advanced rifting and crustal thinning. During ascent and intrusion, melt compositions changed due to differentiation processes and contamination with continental crustal material.

A different, volcanic-arc setting was suggested by Kröner &

Hegner (1998) for the protoliths of the trondhjemitic to granodioritic orthogneisses and migmatites of the Góry Sowie Complex. They proposed that after rifting and break-off in earlier Palaeozoic times, an Ordovician active margin developed on the leading edge of Avalonia (or a fragment thereof) during northward drift towards Baltica, due to subduction of the Tornquist Ocean. In their view, younger U–Pb zircon ages of 463 and 473 Ma from migmatites may reflect accretion processes during drift of the microplate. However, a continental rift-related emplacement setting of the orthogneiss protoliths is suggested by others (e.g. Kryza & Pin 2002, and references therein). The granulites and garnet peridotites record ultrahigh metamorphic conditions (up to 900–1000°C, 15–20 kbar) and have 399 ± 8 and 402 ± 2 Ma U–Pb ages for metamorphic zircons (O'Brien *et al.* 1997; Kröner & Hegner 1998). This late Ordovician–early Silurian, ultrahigh-grade metamorphism was interpreted to be caused by crustal thickening during collision with Baltica, following the closure of the intervening Tornquist Ocean.

Mt Ślęża Complex

The Mt Ślęża Complex (Fig. 12.35) is the largest and best preserved of the ophiolite complexes around the Góry Sowie Complex and shows an ophiolite pseudostratigraphy comprising serpentinites, tectonized peridotites and metamorphosed ultramafic cumulates (pyroxene and amphibole), podiform chromite, gabbroic cumulates, basalts and dolerites (once possibly sheeted dykes), radiolarian cherts, locally preserved pillow basalts, plagiogranitic differentiates and rodingites (Majerowicz & Pin 1992; Pin *et al.* 1988). The metabasalts and metagabbros (Mt Ślęża Complex *sensu stricto*) occur in the north and NW, and the serpentinites in the SE (Jordanów-Gogołów Complex). To the west it is intruded by the late Palaeozoic Strzegom(-Sobótka) Granite and the southern part of the complex is overthrust by the Góry Sowie Complex. The Mt Ślęża Complex may extend northwards into the Fore-Sudetic Block as suggested by the presence in boreholes of N-MORB-like metamorphosed mafic rocks with positive ε_{Nd}(420 Ma) values (+6.7 and +9.1; Crowley *et al.* 2002*b*). Oliver *et al.* (1993) obtained a late Silurian (420 + 20/−2 Ma) single U–Pb zircon crystallization age for a Mt Ślęża gabbro, and the much younger, early Carboniferous (*c.* 353 ± 21 Ma) Sm–Nd whole-rock isochron age reported by Pin *et al.* (1988) is probably due to disturbance of the isotope system during Variscan events. The serpentinization process is dated at 400 +4/−3 Ma (early Devonian), which is the U–Pb age for zircon from rodingite in serpentinite of the Jordanów-Gogołów Complex (Dubińska *et al.* 2004).

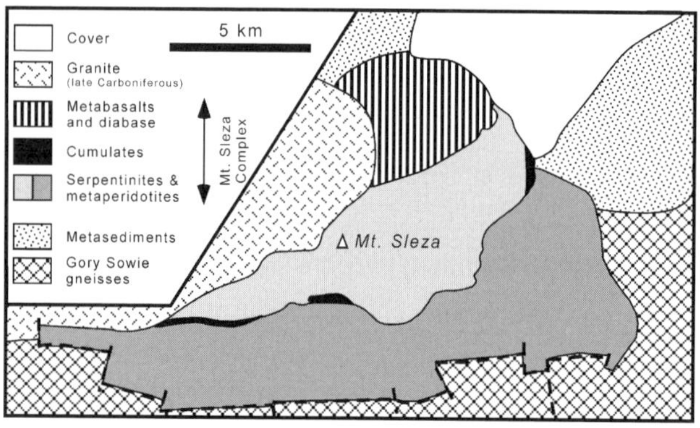

Fig. 12.35. Geological sketch map of the Mt. Ślęża ophiolite (simplified after Majerowicz & Pin 1992).

The primary mineralogy of the often strongly serpentinized peridotites is olivine, clinopyroxene, Cr-spinel and magnetite; chromitites occur as lenses. The ultramafic rocks are mostly harzburgites and lherzolites, with rare olivine-websterites, orthopyroxenites and dunites. Ultramafic cumulates, encountered in drill core in serpentinite, are composed of serpentinized olivine, relicts of ortho- and clinopyroxene, as well as serpentine, chlorite, talc, carbonates, Fe-oxides and chromite. Chromitite occurs as podiform inclusions in the ultramafic cumulates. The mafic cumulates occur in the upper parts of the complex and have a gabbroic mineralogy and show cumulate structures, such as compositional layering, cryptic layering and plagioclase-pyroxene aggregates. Hornblende, clinozoisite and epidote form secondary minerals. The dykes (locally sheeted) and volcanic rocks now occur as aphanitic to medium-grained amphibolites. Locally, fragments of diabase or metagabbro in a fine-grained to aphanitic groundmass testify to interpenetration of volcanic and plutonic members (Majerowicz 1994; Majerowicz & Pin 1992).

The metabasalts have low Ni and Cr contents, are relatively depleted in the incompatible elements Th, Nb and La, have flat REE patterns with REE concentrations 20 to 30 times that of chondrite (Pin *et al.* 1988) and have high, positive ε_{Nd}(420 Ma) values (+8.2 to +8.9; Crowley *et al.* 2002*b*). They represent fractionated melts derived from a depleted mantle source, and their MORB-like features indicate formation in an oceanic setting. Compared to the more evolved metabasalts and dolerite dykes, the metagabbros are more depleted, i.e. they have lower concentrations of incompatible elements, which is probably due to their higher cumulate content (Floyd *et al.* 2002). These authors divided the magmatic rock into high-Ti dolerites and basalts and low-Ti metagabbros that were extracted from mantle sources which differed in their degree of depletion. Although the high- and low-Ti varieties occur in close proximity, extraction did not necessarily take place at the same time. Floyd *et al.* (2002) suggested that the high-Ti melts were extracted from a MORB-type mantle source under an oceanic spreading centre, leaving a depleted mantle residue from which the low-Ti melts originated at a later time. In the Zr versus Zr/Y discrimination diagram of Pearce & Norry (1979), the dykes and lavas plot in the MORB field, whereas the (low-Ti) gabbroic rocks plot inside and below the island-arc tholeiite field due to their depleted character (Fig. 12.34A).

Braszowice Gabbro

The Braszowice Gabbro in the Niemcza Shear Zone to the east of the Góry Sowie Complex (Fig. 12.28) is an olivine-free, hornblende-rich gabbro. The absence of olivine as a fractionating phase is supported by the negative FeO versus MgO trend, which was probably caused by magnetite fractionation. MgO contents range from 5 to 13 wt%, and a change in Ti and Ca trends at *c.* 8 wt% MgO suggests that ilmenite ± Ca-pyroxene were also fractionating phases. The positive Al and Na versus MgO trends furthermore indicate plagioclase fractionation, probably under low pressure conditions. Compared to chondrite, the gabbros are depleted in Ba, La, P, Ti, Y and Yb, and enriched in Eu and Sr. In general, they are comparable to mid-ocean-ridge gabbros and cumulates, suggesting that they formed below a spreading ridge (Dziedzic & Kramer 1992; Dziedzic & Dziedzic 2000).

Kłodzko Metamorphic Complex

The Kłodzko Metamorphic Complex (KMC) south of the Góry Sowie Complex is composed of metasediments and metamorphosed magmatic rocks that are unconformably overlain by unmetamorphosed Devonian limestones in the NE, and intruded by Variscan granite in the east (Fig. 12.36). Greenschist- to (locally) epidote-amphibolite-facies metasediments and metavolcanic rocks occur in the NE, and metagabbros, amphibolites and gneisses in the SW of the KMC (Kryza *et al.* 2003; Mazur *et al.* 2004). The KMC is internally complex and several stacked thrust units of differing lithology and degree of metamorphism have been recognized. From bottom to top these are the Mały Bożków, Łączna, Bierkowice and Ścinawka units in the NW, and the Orla-Gołogłowy and Kłodzko units in the SE. The volcanosedimentary Kłodzko Unit is dominated by phyllites associated with mafic and felsic metavolcanic rocks that were deposited in an active margin environment, probably a basin bordering a volcanic arc. A deformed felsic dyke intruding the metavolcanic rocks of the Kłodzko Unit and a metagranitoid of the Orla-Gołogłowy Unit yielded protolith crystallization ages of 500 ± 5 and 506 ± 3 Ma (late Cambrian) respectively (Mazur *et al.* 2004).

The structurally lowermost Mały Bożków Unit is composed of lower greenschist-facies sediments, deposited in a shelf environment on a passive margin, and include limestones that contain

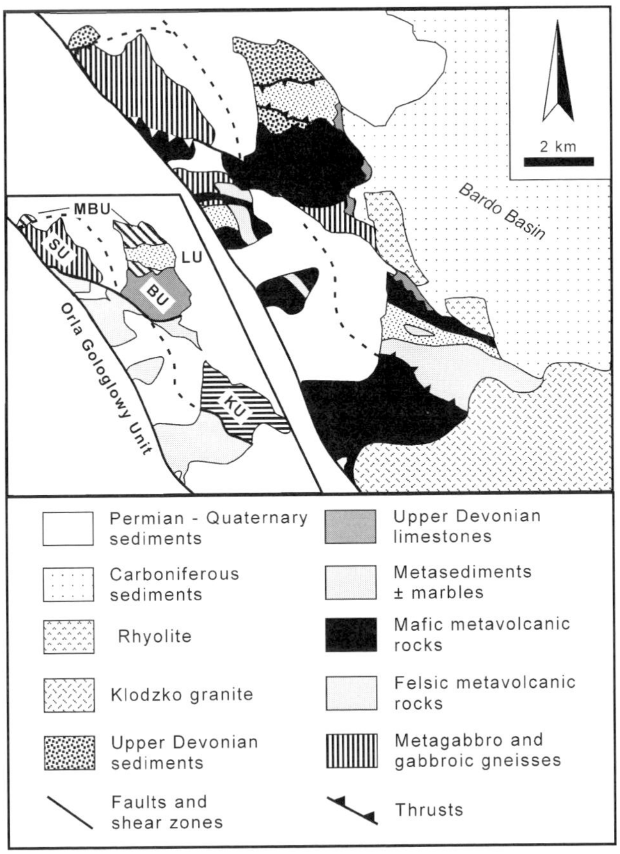

Fig. 12.36. Geological sketch map of the Kłodzko Metamorphic Complex (after Kryza *et al.* 2003; Mazur *et al.* 2004). Inset: tectonostratigraphic units of the Kłodzko Metamorphic Complex. Abbreviations: BU, Bierkowice Unit; LU, Łączna Unit; KU, Kłodzko Fortress Unit; MBU, Mały Bożków Unit; SU, Ścinawka Unit.

Givetian-age corals (late Middle Devonian); magmatic rocks are absent. The Łączna Unit to the south contains metamorphosed sedimentary breccias deposited in a continental slope environment, and includes limestones and small amphibolite bodies. The Bierkowice Unit is made up of deformed, greenschist-facies mafic rocks that erupted in a rift setting. The Ścinawka Unit contains metagabbros and amphibolites, while metamorphosed magmatic rocks are widespread in the Orla-Gołogłowy Unit, comprising intermediate to felsic subvolcanic rocks (dacite, rhyodacite and rhyolite dykes), metagabbros, and a variety of orthogneisses.

The mafic rocks in the NE units have tholeiitic basaltic compositions (Figs. 12.32, 12.34A), are moderately to strongly enriched in incompatible elements, lack a negative Nb anomaly, and have positive ε_{Nd}(500 Ma) values (+6.1 and +6.8; Kryza *et al.* 2003). Some developed a negative Sr anomaly due to plagioclase fractionation. The positive $\varepsilon_{Nd}(t)$ values, in combination with LREE enrichment, indicate derivation from a depleted mantle that was enriched in incompatible elements shortly before melting, which accounts for their within-plate tholeiitic to transitional MORB-like character (Fig. 12.32). Their composition and association with terrigenous metasediments indicate emplacement in a late Cambrian extensional setting.

The mafic rocks of the SW units have more variable compositions. Kryza *et al.* (2003) distinguished three groups of metagabbros. One group has flat to convex REE patterns, ε_{Nd}(500) values ranging from +2.2 to +4.5, and is N-MORB-like. However, fractionation and accumulation processes modified the compositions as is evident from the negative Zr and Hf, and positive Sr and P anomalies. A second group is of slightly E-MORB character. The third group is of clearly N-MORB character, having LREE-depleted to flat REE patterns, high Zr/Nb ratios (up to 64) and high, positive $\varepsilon_{Nd}(t)$ values (+7.3 to +8.8).

The felsic to intermediate rocks of the Orla-Gołogłowy Unit have calc-alkaline, peraluminous, rhyolitic to andesitic compositions (Fig. 12.30) with pronounced negative Ta and Nb anomalies (compared to N-MORB), but varying degrees of LREE enrichment. Negative Sr, P and Ti anomalies show that the melts were fractionated. The ε_{Nd}(500 Ma) values vary between +2.9 and +8.6. Kryza *et al.* (2003) suggested they formed in a volcanic arc during Cambrian subduction, but U–Pb zircon protolith ages of 590–600 Ma (late Neoproterozoic) reported by Mazur *et al.* (2004) for a metagabbro and volcaniclastic rock indicate that they are unrelated to the units to the NW. Instead, these arc-like felsic to intermediate rocks are allochthonous fragments derived from the north Gondwanan Cadomian arc, and were later thrust upon and juxtaposed against the younger, rift-related units of the KMC during the Variscan Orogeny.

Orlice-Sněžník Complex

The Orlice-Sněžník (Czech Republic)/Orlica-Śnieżnik (Poland) Complex to the south of the Góry Sowie Complex (Fig. 12.37) is composed of Neoproterozoic to early Palaeozoic amphibolite-facies granitoid orthogneisses, migmatites, metasediments, and minor amounts of marbles and amphibolites. The metasediments of the Młynowiec and Stronie formations were deposited in Neoproterozoic to Cambrian times, and were intruded by early Ordovician granitoids that were subsequently deformed into the mainly amphibolite-facies Sněžník and Gierałtów orthogneisses. The Gierałtów gneisses contain lenses of granulite and eclogite (O'Brien & Carswell 1993, and references therein). The $^{40}Ar/^{39}Ar$ argon cooling ages for biotite, muscovite and hornblende range from 328 to 338 Ma (Maluski *et al.* 1995; Steltenpohl *et*

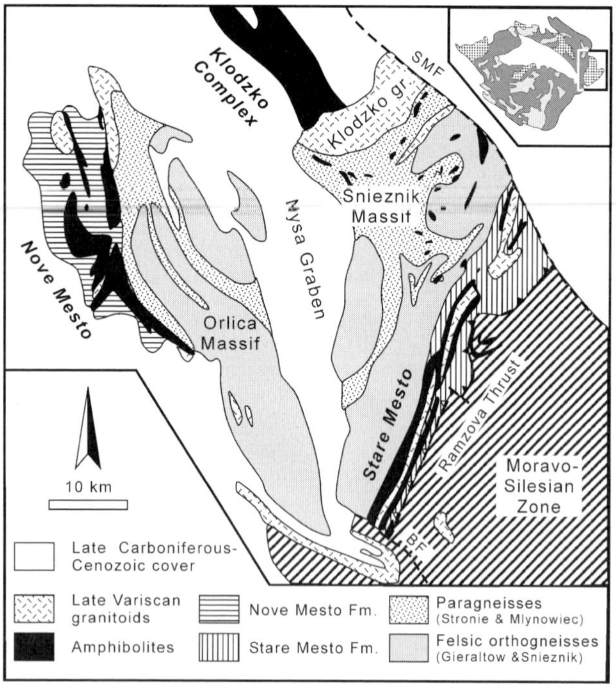

Fig. 12.37. Geological sketch map of the Orlice-Sněžník/Orlica-Śnieżnik Complex (West Sudetes) surrounded by the Stronie, Staré Město, and Nové Město volcanosedimentary complexes, West Sudetes (after Floyd *et al.* 1996; Lange *et al.* 2005). BF, Busin Fault; SMF, Sudetic Marginal Fault.

al. 1993). The (meta)granitoids are peraluminous, S-type, corundum-normative, muscovite-biotite granites with low CaO contents and high K_2O/Na_2O ratios. SHRIMP zircon dating indicates a 495 ± 7 Ma (Cambro-Ordovician) crystallization age for the protoliths, and *c.* 530–540 Ma (early Cambrian) zircon cores suggest that Cadomian rocks were the source of the protoliths (Turniak *et al.* 2000). Mid- to late Cambrian zircon protolith ages have been obtained for calc-alkaline granitoid orthogneisses from the western (Orlice/Orlica) massif (503 to 511 Ma, $\varepsilon_{Nd}(t) = -6.5$ to -3.5; Kröner *et al.* 2001), and for a granite gneiss from the eastern margin of the Sněžník Massif (511 Ma, $\varepsilon_{Nd}(t) = -6.5$ to -3; Kröner *et al.* 2000*b*; Hegner & Kröner 2000). Metamorphosed mafic magmatic rocks occur as amphibolites in the Stronie Fm, and as locally preserved high-pressure granulite- and eclogite-facies rocks within Gierałtów migmatitic orthogneisses (Kryza *et al.* 1996). In addition, amphibolites are present in the Nové Město and Staré Město belts situated to the west and east of the complex, respectively. These may have been flows or sheet-like intrusions (Floyd *et al.* 1996).

During the Variscan collision, the eastern part of the Orlice-Sněžník Complex was thrust upon the Brunia /Brunovistulia microplate to the east. This microplate is mainly composed of late Neoproterozoic (Cadomian) granitoids and high-grade metamorphic rocks, which are locally overlain by late Neoproterozoic–Cambrian, anchizone metamorphic continental sediments (e.g. Finger *et al.* 2000; Kalvoda *et al.* 2002, and references therein). The early Palaeozoic sediments in turn are disconformably overlain by early Devonian to late Carboniferous sediments that were deposited in rift, passive-margin and foreland-basin settings; they belong to the Rhenohercynian Zone and the Variscan foreland basin (e.g. Hartley & Otava 2001; Schulmann & Gayer 2000; Kalvoda *et al.* 2002). The intervening NE-trending Moravo-Silesian Zone is mostly made up of stacked nappes of Brunia-derived orthogneisses and metasediments that have yielded 685–502 Ma U–Pb zircon protolith ages (Finger *et al.* 1989, 2000; Kröner *et al.* 2000*b*).

The Staré Město Belt at the eastern margin of the Orlice-Sněžník Complex is made up of felsic and mafic, bimodal metavolcanic rocks, tonalitic gneisses, granulite-facies metasediments and, in the NE part, contains strongly retrograded eclogites (Žaček 1996). Kröner *et al.* (2000*b*) obtained a 503 Ma zircon age for a dacitic metavolcanic rock and 503–505 Ma ages for tonalitic and anorthositic gneisses. Values of $\varepsilon_{Nd}(t)$ for these Cambro-Ordovician magmatic rocks are positive (+0.5 to +6.4), indicating the presence of a juvenile component (Hegner & Kröner 2000). Rocks of the Staré Město Belt are intruded by a syntectonic tonalite sill that yielded an early Carboniferous (339 ± 7 Ma) zircon evaporation age (Parry *et al.* 1997).

The majority of the amphibolites of the Stronie Fm and the Nové and Staré Město belts have tholeiitic basaltic incompatible elements compositions; the Stronie Fm also contains minor amounts of alkaline basalt (Fig. 12.30). Based on the spread in TiO_2 contents and Nb/Y ratios, Floyd *et al.* (1996) distinguished three geochemical groups. The bulk of the mafic rocks are LREE-enriched tholeiites with La_N/Yb_N ratios between 2 and 3, high Ti/V ratios (20–50) and positive Sr anomalies due to plagioclase accumulation (Main Series; Floyd *et al.* 1996). The second group comprises low-Ti tholeiites with flat REE patterns ($La_N/Yb_N = 0.8$ to 1.0), low Ti/V ratios (<20) and lower HFSE abundances. Alkaline basalts with steep, strongly fractionated REE patterns ($La_N/Yb_N = c.$ 20), similar to ocean island basalts (OIB), constitute the third group. Metabasalts of all geochemical groups are not restricted to geographically separate suites but occur in close spatial association, and have been recognized in most of the early Palaeozoic volcanosedimentary complexes of the Sudetes (Floyd *et al.* 1996, 2002; Crowley *et al.* 2000b). The range of REE patterns, from LREE-enriched to chondritic (i.e. flat), excludes derivation from a common parental melt by means of simple fractionation processes, but indicates that the three groups originated from distinct mantle sources. Internal variations within each group, such as Fe-enrichment trends and negative Cr–Zr correlations, are due to fractional crystallization of pyroxene ± spinel. In addition, each group is made up of several melt batches as indicated by variations in Ti/V and Zr/Y ratios. The tholeiitic basalts were, furthermore, modified by assimilation of upper continental crust or sediments as indicated by enrichment in LILE, Nb and Ta, and depletion in HFSE, which is reflected in the widely varying Nb/Y, Ce/Y and Ce/Zr ratios.

Floyd *et al.* (1996) suggested the magmas derived from slightly different, depleted mantle sources during extension and crustal thinning: the high-Ti tholeiites from continental mantle lithosphere, and the low-Ti varieties from asthenosphere. The melts erupted and intruded in an early Palaeozoic continental rift setting, but extension was limited since true MORB are not present. Kröner *et al.* (2000*b*) presented a slightly different model, in which Cambro-Ordovician subduction beneath a microplate (that included the Orlice-Sněžník Complex) caused it to break away from Brunia/Brunovistulia, thus forming the Staré Město rift in between.

Rehberg and Letovice complexes

Mafic, ophiolite complexes also occur further south along the eastern margin of the Bohemian Massif (Moravo-Silesian Zone; Fig. 12.38), where the Letovice and Rehberg complexes of the

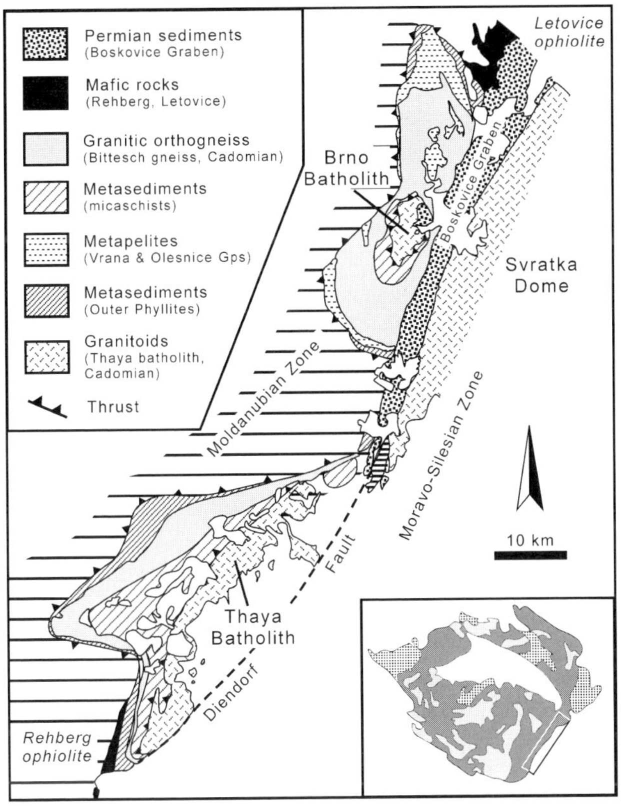

Fig. 12.38. Geological sketch map of the Moravian Zone, eastern Bohemian Massif, with Letovice and Rehberg ophiolites (after Höck 1995; Höck *et al.* 1997).

Raabs Unit/Raabs Nappe form part of a high-grade nappe stack that formed as a result of collision between the Moldanubian and Brunian/Brunovistulian terranes in mid-Carboniferous times. Thrusting occurred in an eastward direction with a northern strike-slip component, and the degree of metamorphism decreases from granulite-facies in the uppermost nappes in the west, to very low-grade metamorphism in the lower nappes and in the Devonian foreland sediments to the east (e.g. Fritz & Neubauer 1993; Finger & Steyer 1995, and references therein). The Moldanubia-derived nappes in the west are composed of Neoproterozoic passive margin sediments, calc-silicates and amphibolites that were metamorphosed to migmatitic gneisses under low- to medium-pressure conditions (the Variegated and Monotonous units). Intercalated orthogneisses, such as the Spitz and Dobra gneisses, are Neoproterozoic in age as indicated by a U–Pb zircon age of 614 ± 10 Ma for Spitz Gneiss (Finger *et al.* 1995; Friedl *et al.* 2004). The overlying Raabs Unit is made up of migmatitic gneisses and metasediments that are tectonically intercalated with serpentinites and amphibolites. The ultramafic and mafic rocks occupy a similar tectonostratigraphic position as the Rehberg and Letovice complexes (Finger & Steyer 1995; Fritz 1996). The Gföhl Unit is the uppermost of the western, Moldanubian thrust units and is composed of granulites and orthogneisses containing mantle-derived rocks. These ultra-high-pressure garnet peridotites, pyroxenite varieties, (kyanite-) eclogites, spinel peridotite and high-pressure, high-temperature leucocratic granulites occur between the Monotonous and Variegated units, and at the base of the Gföhl Unit (e.g. Beard *et al.* 1992; Becker 1996; Cooke & O'Brien 2001; Medaris *et al.* 2006, and references therein). Granulite-facies metagranites from the Gföhl Unit have yielded early Ordovician (488 ± 6 Ma) and late Devonian (366 ± 5, 373 ± 11 and 365 ± 11 Ma) zircon ages, interpreted as the crystallization age of the protoliths (Wendt *et al.* 1994; Friedl *et al.* 2004).

The nappe units were thrust northeastwards onto the imbricate nappes of the Moravian Zone, which are composed of greenschist- to amphibolite-facies I-type orthogneisses, metasediments and marbles. The lowermost Thaya Nappe is mainly made up of Cadomian orthogneisses intruded by a late-stage granite (567 ± 6 Ma; Friedl *et al.* 2004) and the Pleising Nappe higher up in the tectonostratigraphy contains very deformed metagranites. The uppermost Biteš/Bittesch Nappe comprises granodioritic to granitic orthogneisses that have yielded 578 ± 7 and 584 ± 6 Ma protolith ages (U–Pb zircon; Friedl *et al.* 2004). As in the northern part of the Moravo-Silesian Zone, the orthogneisses are mainly reworked Cadomian granitoids derived from the Brunian foreland to the east (Finger *et al.* 1989).

Mafic to granitic granulites and later, discordant anatectic melts in the Gföhl Unit have zircon and monazite ages of 338–351 Ma which show that the high-grade metamorphism was of early to mid-Carboniferous age (Aftalion *et al.* 1989; Wendt *et al.* 1994; Kröner *et al.* 2000*a*). Further to the W, in the Bohemian Massif, the Dobra gneisses and lithologies of the Variegated Unit are intruded by the undeformed *c.* 338 Ma Rastenberg Granodiorite (Klötzli & Parrish 1996). Muscovite and hornblende $^{40}Ar/^{39}Ar$ ages for both Moravian and Moldanubian units indicate cooling between 341 and 325 Ma (Dallmeyer *et al.* 1992).

The Rehberg and Letovice complexes are made up of strongly imbricated (ultra)mafic rocks and metasediments (Mísař *et al.* 1984; Jelínek *et al.* 1984; Höck *et al.* 1997). Plagiogranite gneiss from the Buschandlwand area, to the west of the Raabs Unit and correlative with the Rehberg Complex, has a U–Pb zircon upper intercept age of 428 ± 6 Ma indicating that the mafic magmatic activity occurred in mid-Silurian times (Finger & von Quadt 1995). In its structurally lower parts, the Rehberg Complex contains strongly altered harzburgite layers and lenses, followed upward by gabbro cut by basalt dykes, metagabbros and felsic to mafic metavolcanic rocks, which are capped by marbles and metagreywackes. The metavolcanic rocks range in composition from basalt to andesite, dacite and rhyolite. The Letovice Complex is made up of serpentinites, ultramafic cumulates, metagabbros (locally layered), banded amphibolites and rare mafic dykes. The ultramafic rocks are dunites, harzburgites, olivine orthopyroxenites, olivine websterites, and plagioclase peridotites with locally preserved cumulate textures. The mafic rocks have tholeiitic basaltic compositions with generally low TiO_2 contents (up to 1.5 wt%), and correlations between major elements indicate fractionation of olivine and plagioclase ± clinopyroxene. The mafic rocks of the Rehberg Complex have flat to mildly enriched (transitional) REE patterns (La_N/Lu_N between 0.8 and 2). The MORB-normalized multi-element plots show the rocks are enriched in LILE (mostly Rb and Ba), but have lower HFSE abundances (Finger & Steyer 1995; Höck *et al.* 1997). The positive Rb and Ba anomalies were not considered to be primary magmatic features, because K and Sr concentrations are not enhanced. In general, the complexes are considered to represent the volcanic, plutonic and mantle successions of ophiolites that formed near an intra-oceanic arc or in a backarc basin (Fig. 12.39; Finger & Steyer 1995; Höck *et al.* 1997). Therefore, the Raabs Unit and the Letovice and Rehberg ophiolites probably represent a suture zone that marks the former presence of a Silurian basin or rift underlain by oceanic crust (e.g. Fritz 1996). The spinel peridotites in the Gföhl nappe may be relicts of relatively shallow, suboceanic asthenospheric mantle

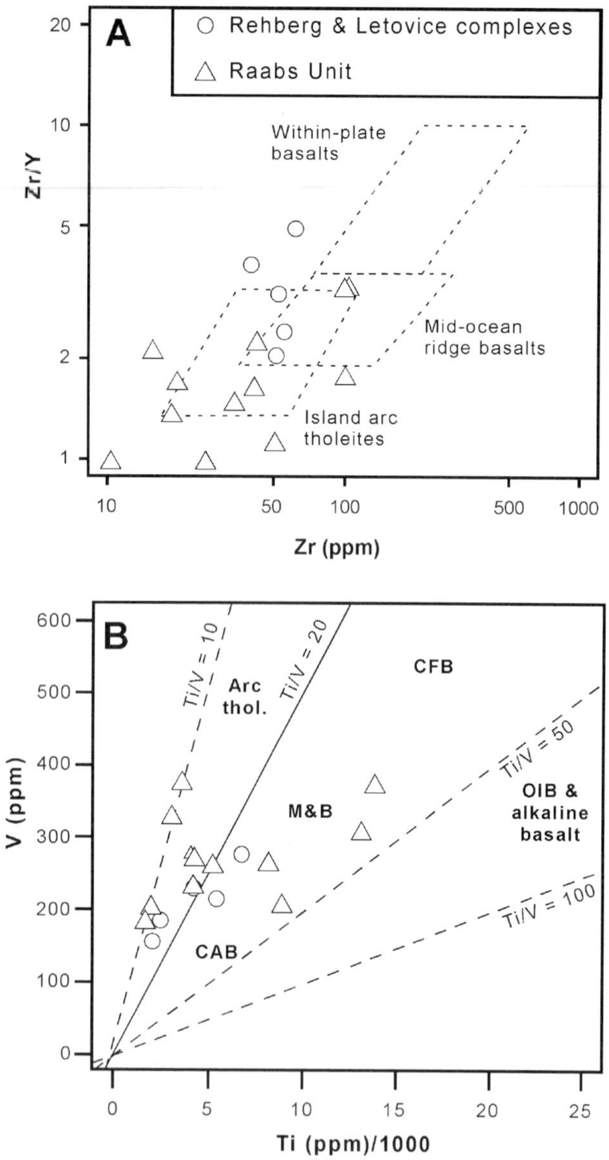

Fig. 12.39. (**A**) Zr versus Zr/Y tectonic discrimination diagram for selected mafic metavolcanic rocks from the Rehberg and Raabs complexes with <52 wt% SiO_2 (Pearce & Norry 1979). (**B**) Ti versus V discrimination diagram for metavolcanic and metaplutonic rocks from the Lettovice, Rehberg and Raabs complexes with <52 wt% SiO_2 (Shervais 1982). For abbreviations see caption to Figure 12.32. Data from Höck *et al.* (1997), Fritz (1996) and Jelínek *et al.* (1984).

(Medaris *et al.* 2006). The absence of true N-MORB-type basalts and the presence of calcareous and graphitic metasediments that occur intercalated with the mafic rocks indicate that the basin was of limited width (the Raabs Sea of Finger & Steyer 1995).

Cambrian–Early Ordovician magmatism: general

In large-scale trace element studies of the mafic and felsic magmatic rocks in the Sudetes, Floyd *et al.* (2002) and Crowley *et al.* (2002*b*) distinguished three groups of mafic rocks with differing Nb/Y and Ti/V ratios (Fig. 12.40): volumetrically dominant high-Ti tholeiites (Main Series), low-Ti tholeiites and alkaline basalts. All three groups have low MgO, Ni and Cr contents, and fractionated under low-pressure conditions (spinel fractionation). The high-Ti tholeiites show variable degrees of enrichment of LREE, Nb, Ta and Th, which were probably inherited from the mantle source. The low-Ti tholeiites generally have N-MORB-like, flat REE and HFSE patterns, and the alkaline basalts show the greatest LREE enrichment. The moderately positive $\varepsilon_{Nd}(t)$ values (0 to +4) suggest that contamination with crustal material was limited and, furthermore, restricted to the least-evolved members of the low-Ti basalts. Using ratios of immobile elements that are not significantly affected by moderate degrees of fractionation (e.g. La/Nb, Zr/Nb, Ti/Zr, Ti/Y), Floyd *et al.* (2002) suggested a lithospheric mantle source for the low-Ti tholeiites, and that the alkaline basalts originated from a deeper, ocean-island-type mantle source, perhaps a plume. The bulk of the basalts (Main Series, high-Ti group) show evidence of having formed from an asthenospheric mantle variably enriched in LREE, Nb, Ta and Th, possibly a result of mixing between depleted asthenospheric mantle and a plume component. The magmas were emplaced in rift basins formed during variable degrees of extension, ranging from continental rifts to those that were underlain by oceanic crust. These basins formed narrow seaways between continental fragments that were breaking away from northern Gondwana.

The origin of the early Palaeozoic granitoids and granitoid gneisses in the West Sudetes is still a matter of debate. An arc setting for their protoliths was suggested for example, by Oliver *et al.* (1993), Kröner & Hegner (1998) and Kröner *et al.* (2000*b*). However, the apparent arc signatures of the granitoids, granitoid gneisses, and some of the felsic metavolcanic rocks stand in contrast to the alkaline basalts and tholeiites of the nearby rift basins (Fig. 12.41). An alternative explanation may be that the arc signature was mainly inherited from their Cadomian, true arc-related crustal sources by anatexis during early Palaeozoic extension (Crowley *et al.* 2000, 2002*b*; Floyd *et al.* 2002; Oberc-Dziedzic *et al.* 2005). Crowley *et al.* (2002*b*) proposed that the felsic metavolcanic rocks in the volcanosedimentary complexes and the granitoid gneisses from the gneiss complexes were partial melts derived from older, Cadomian continental crust, or formed by extreme fractionation of mafic melts contaminated by crustal material, as suggested by their negative $\varepsilon_{Nd}(t)$ values. A similar model was proposed by Pin & Marini (1993) in order to explain the apparent subduction signature of early Ordovician bimodal magmatic suites in the Massif Central, France.

Late Ordovician–Silurian

In Central Europe, the evidence for Late Ordovician–Silurian, Caledonian events has been preserved in the Mid-German Crystalline High (MGCH) in central Germany (Fig. 12.24; Oncken 1997; Franke 2000). The MGCH is a composite, 75–100 km wide unit that can be traced for a distance of *c.* 1000 km along the northern margin of the Saxothuringian Zone and constitutes its western and northern part (Fig. 12.24; Holder & Leveridge 1986; Oncken 1997; Kopp *et al.* 2001; Kopp & Bankwitz 2003). It is exposed on the western shoulder of the Rhine Graben (southern Pfalz) and in the Odenwald, Spessart, Ruhla and Kyffhäuser complexes. East of the Kyffhäuser Complex, rocks of the MGCH have been encountered in drill cores (e.g. Alexandrov *et al.* 2001; Anthes & Reischmann 2001; Cocherie *et al.* 2004; Fig. 12.24). The eastern extension of the MGCH into southern Poland it still a matter of discussion (e.g. Kopp *et al.* 2001; Kopp & Bankwitz 2003). The tectonic contact of the MGCH with the Rhenohercynian Zone to the north is the

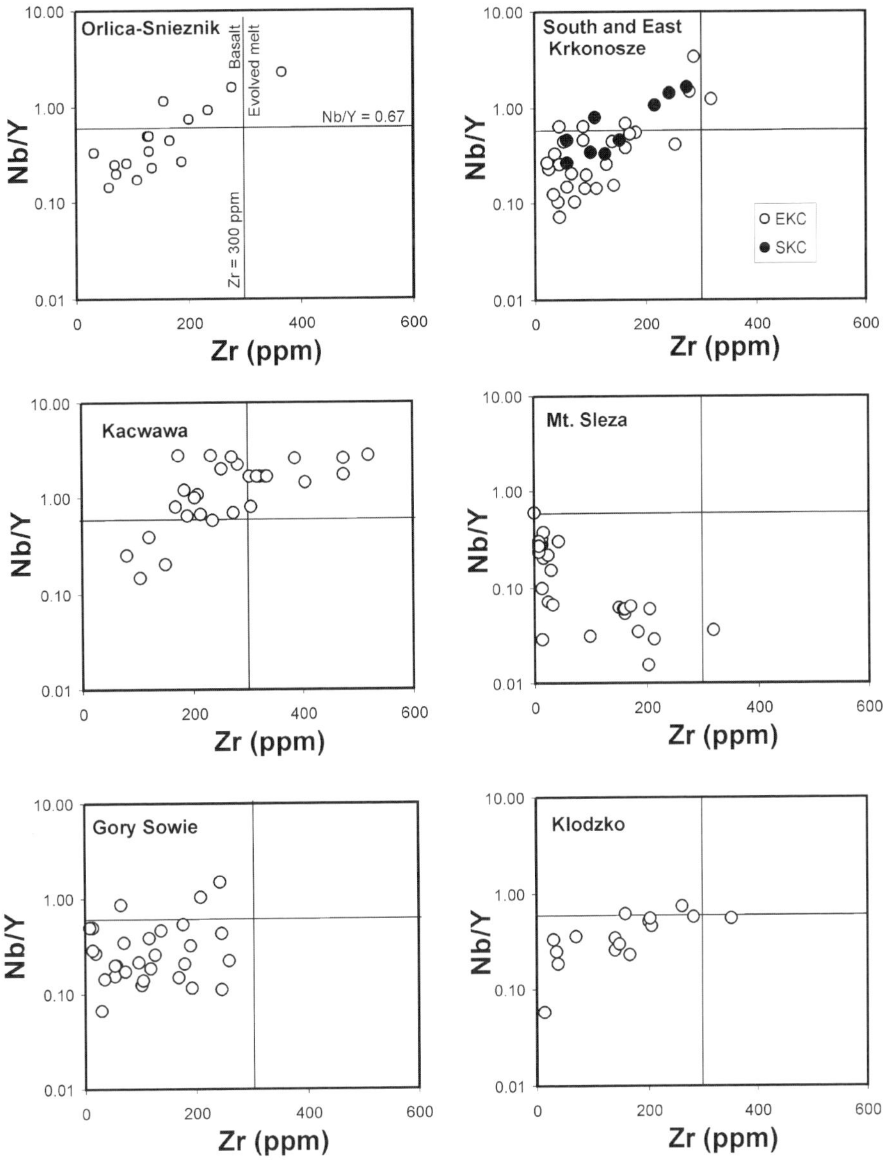

Fig. 12.40. Discrimination diagrams for the metavolcanic and metaplutonic rocks of basaltic composition (<52 wt% SiO_2) for the early Palaeozoic complexes of the West Sudetes area. Nb/Y ratio of 0.67 discriminates between alkaline and tholeiitic basalt, and a Zr concentration of 300 ppm between lesser and more fractionated melts (cf. Floyd *et al.* 2002). For data sources see caption to Figure 12.30.

Northern Phyllite Zone, exposed in the Taunus area and the Wippra Zone (SE Harz), and which is generally considered to mark the suture zone of the Rheic Ocean. Ostracods and fish fossils indicate that the suture is situated south of the Northern Phyllite Zone and south of the early Devonian fish occurrences in the Taunus area (McKerrow & Cocks 1995). The southern contact of the MGCH with the Moldanubian Zone, the Southern Phyllite Zone, is obscured by large amounts of Palaeozoic sediments, but may represent the suture zone of the Saxothuringian Ocean. The MGCH is a complex structure, containing early Palaeozoic, rift-related and passive-margin sediments, Silurian to Devonian arc-type magmatic rocks formed during the northward drift of terranes and concomitant subduction, and large amounts of granitoids generated during a later, late Devonian to early Carboniferous subduction stage. In addition, it underwent widespread, high-grade, mid-Carboniferous metamorphism related to accretion and collision processes.

Cambro-Ordovician sediments, and Silurian–early Devonian magmatic rocks occur scattered in the central parts of the MGCH (eastern Odenwald, Spessart, Ruhla), along its northern margin (Northern Phyllite Zone), and have been recognized in boreholes (e.g. Saar 1, Volkach 1). In the Pretzsch-Prettin area to the east, Cambro-Ordovician gabbroic rocks occur below cover sediments in the contact zone with the Southern Phyllite Zone (Kopp *et al.* 2001; Hammerschmidt *et al.* 2003; Fig. 12.24). Silurian-age magmatic rocks appear to be restricted to the central and western parts of the MGCH, and have not yet been recorded from the eastern part (Anthes & Reischmann 2001).

Northern Phyllite Zone

The Northern Phyllite Zone forms the contact between the Rhenohercynian Zone and the Mid-German Crystalline High. This poorly exposed and internally complex structure is exposed

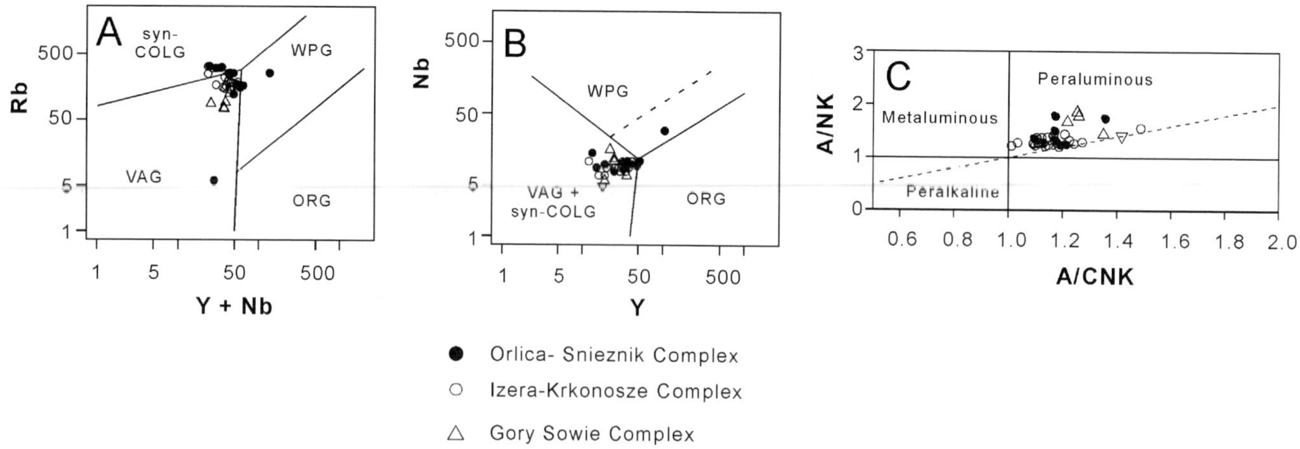

Fig. 12.41. Discrimination diagrams for early Palaeozoic granitoids and felsic gneisses from the West Sudetes (**A**, **B**: Pearce *et al.* 1984; **C**: Maniar & Piccoli 1989). For abbreviations see captions to Figures 12.4 & 12.5. Data from Kröner & Hegner (1998), Kröner *et al.* (2001), Turniak *et al.* (2000) and Oberc-Dziedzic *et al.* (2005).

in the Hunsrück, Taunus and Soonwald areas (south of the Rhenish Massif; see Fig. 12.24; Oncken 1988) and in the Wippra Zone (SE Harz; Fig. 12.24; Fig. 12.42; Schwab & Jacob 1996). Its eastern extension, the Pakendorf Zone of the Flechtingen-Rosslau High, is covered by Cenozoic sediments but is known from borehole mapping (FRH in Fig. 12.24; Borsdorf & Estrada 1991).

The Northern Phyllite Zone in the Hunsrück area comprises mainly metapelites and metabasalts with intercalated lenses of quartzite, sandstone and limestone. Conodont, acritarch and pollen evidence indicates early Devonian to early Carboniferous sedimentation ages (Bartels & Kneidl 1981; Berger *et al.* 1991, and references therein). Its tectonic contact with the Rhenohercynian Zone to the NW contains lenses of Neoproterozoic gneisses

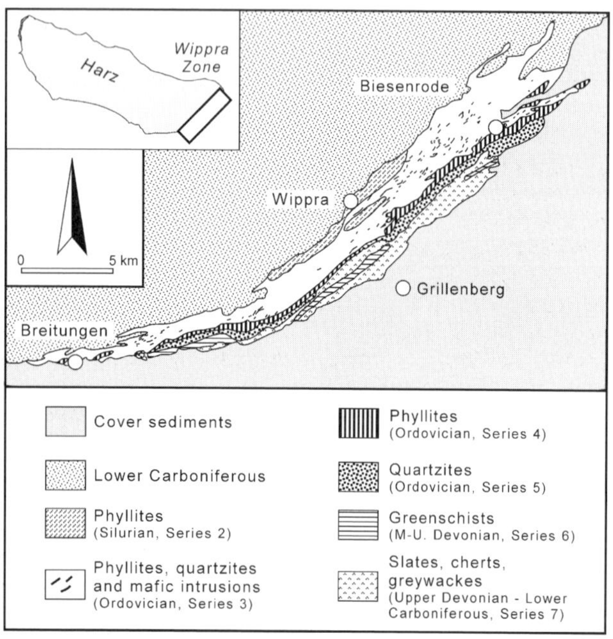

Fig. 12.42. Geological map of the Wippra Zone, SE Harz (after Löffler & Schwab 1981; Siedel & Theye 1993).

(Wartenstein Gneiss; Anderle 1987). In the Taunus area, the Northern Phyllite Zone is made up of weakly metamorphosed metavolcanic rocks at the base and phyllites and psammites that yielded early Devonian (Emsian) spores near the structural top (Anderle 1987; Reitz 1989; Meisl 1990). Metamorphosed rhyodacites and andesites (sericite gneisses and greenschists) from the south Taunus area yielded 426–442 Ma (late Ordovician to late Silurian) U–Pb zircon ages (Sommermann *et al.* 1992, 1994). Acritarchs from metasediments (phyllites) immediately to the south of the metavolcanic rocks indicate early Ordovician (Arenig) protolith sedimentation ages (Reitz *et al.* 1995). Metavolcanic rocks associated with Devonian sediments (Reitz 1989) in the SE may be younger, but the relation to the other units is not clear due to strong deformation and poor exposure. Incompatible element compositions of the metavolcanic rocks in the Taunus and Soonwald areas indicate a range of compositions, from basalt to rhyolite. The basalts comprise a group of alkaline basalts with within-plate-type trace element ratios, and a tholeiitic group of MORB-like character. The alkaline basalts are strongly enriched in LREE and often have positive Eu anomalies, suggesting plagioclase accumulation. The tholeiitic metabasalts have flat REE patterns with variable positive and negative Eu anomalies. These ocean-floor-like metabasalts perhaps correlate with MORB-type rocks in the Giessen Nappe to the north (Fig. 12.57, see below). The intimate association of within-plate and MOR-like mafic rocks is probably the result of telescoping and intercalation of passive-margin sequences and adjacent ocean floor during Carboniferous thrusting (Meisl 1990, 1995). The composition of the felsic to intermediate metavolcanics rocks ranges from rhyolite to andesite and trachyte, and these mainly erupted as porphyritic lavas; pyroclastic rocks appear to be rare. Meta-andesites and metadacites are strongly LREE enriched, and metarhyolites and metarhyodacites show very strong LREE enrichment with pronounced negative Eu anomalies. The incompatible element compositions and the presence of andesites suggest the remnant of a volcanic arc is exposed in this part of the Northern Phyllite Zone (Meisl 1995). Deformation is of late Carboniferous age, as shown by 323 ± 4 and 308 ± 4 K–Ar ages for fine-grained, foliation-defining white micas, and is related to the closing of the Devonian Rhenohercynian Ocean and the formation of the Rhenohercynian foreland fold-and-thrust belt to the north (Klügel 1997).

The Wippra Zone (SE Harz) is a *c.* 7 km wide, 35 km long, NE-trending belt of low-grade, intensely deformed metapelites, quartz sandstones, slates, greywackes, cherts, metabasalts and pyroclastic rocks, and small amounts of limestones (marbles). Seven northwestward imbricated lithostratigraphic units have been recognized, of which the Series 1 of early Carboniferous age in the NW represents the metamorphosed equivalent of the Harzgerode Zone of the SE Harz (Fig. 12.42). The other sedimentary protoliths were deposited in Ordovician (mainly Llanvirn), Silurian (Llandovery–Wenlock), and Middle and Late Devonian times (Sehnert 1991, in Schwab & Jacob 1996). Deformation and metamorphism took place under anchizone metamorphic (in the NW) to greenschist-facies conditions (in the SW; Jacob & Franzke 1992; Siedel & Theye 1993; Theye 1995). White micas and illites from metapelites have K–Ar ages in the range 350–320 Ma, showing that the deformation and metamorphism occurred in the Early Carboniferous (Ahrendt *et al.* 1996; Marheine 1997). The SE part of the Wippra Unit is unconformably overlain by Westphalian–Stephanian sediments of the Saale Basin (Jacob 1995; Schwab & Jacob 1996, and references therein).

The magmatic rocks are present as variably metamorphosed dolerite and picrite dykes, basalt lavas, and tuffs of mafic to intermediate composition. No radiometric age data are available for the magmatic rocks, but they occur within metasediments of Ordovician, Silurian and suspected Upper Devonian age (Theye 1995). Trachytic to trachyandesitic tuffs, (locally pillowed) basalt lavas, and mafic sills have been reported from Ordovician sediments (Ashgill–lower Llandovery). Picritic dolerite dykes intruded lower Ashgill sediments. Dolerite, picritic dolerite and picrite dykes that intruded Arenig–Ashgill and (suspected) Upper Devonian sediments are probably of Devonian age. Trace elements indicate that the dolerites and picrites comprise a tholeiitic and an alkaline basalt group, and that all lavas and tuffs have alkaline basaltic compositions. Most magmatic rocks classify as intraplate basalts (Borsdorf & Estrada 1991; Werner 1995).

Eastern Odenwald

The Odenwald is the largest of the exposed MGCH massifs and consists of two main parts with distinct lithologies, ages and tectonometamorphic histories (Fig. 12.43; Willner *et al.* 1991; Krohe & Willner 1995; Stein 2001). The two parts are separated by the north-trending Otzberg Fault Zone, a sinistral strike-slip zone that formed under brittle to ductile conditions between the late Devonian and the early Visean. The western (Bergsträsser) Odenwald is subdivided into three structural units, comprising mainly calc-alkaline gabbro to granite intrusions, related to late Devonian and mid-Carboniferous subduction (Units I to III, Fig. 12.43, inset). They are separated by several NE-trending, narrow zones of high-grade metasediments and other rock types of smaller volumes. The western units will be described in their proper tectonic context below.

The eastern (Böllsteiner) Odenwald differs in age and evolution from the western units. It is a NE-plunging antiformal structure cored by orthogneisses, locally containing eclogites, and mantled by high-grade metasediments (Unit IV, Fig. 12.43, inset; Altenberger *et al.* 1990; Altenberger & Besch 1993; Stein 2001; Will & Schmädicke 2001).

The antiform has a core of older granodioritic and younger granitic orthogneisses that contain inclusions of (garnet-)amphibolites, metagabbros with relict magmatic minerals and textures and, in the northern part, eclogites. The protoliths of the orthogneisses intruded amphibolite-facies metasediments and mafic metavolcanic rocks (amphibolites) that constitute the outer flanks of the antiform (Fig. 12.44; Altenberger & Besch 1993). Reischmann *et al.* (2001) report an early Devonian (410 ± 11 Ma) age for zircons from an orthogneiss. The granodiorites and granites have cotectic compositions, suggesting the parent melts were emplaced at shallow crustal levels. They are peraluminous and corundum-normative (S-type; Fig. 12.45A to C), with a narrow range of $\varepsilon_{Nd}(t)$ values (−3.2 to −2.1), but variable Sr_i ratios (0.7021–0.7124) and were probably formed by the melting of older continental crust (Altenberger *et al.* 1990). The metagabbros that post-date the granites in the core have tholeiitic compositions, are LREE-depleted, enriched in LILE, have negative Ta and Nb anomalies and resemble island-arc gabbros (Altenberger & Besch 1993). The metasediments are mainly turbiditic metagreywackes derived from a continental volcanic-arc source. They are associated with amphibolites of tholeiitic composition that are LREE-enriched, have negative Nb anomalies and, thus, resemble island-arc basalts (Altenberger & Besch 1993).

The early Devonian zircon age, in combination with the presence of eclogites and the absence of I-type granitoids, suggests that this unit had a different evolution compared to the western Odenwald, but is similar in age to parts of the Spessart to the east and the Northern Phyllite Zone to the north (Altenberger & Besch 1993; Reischmann *et al.* 2001). Altenberger & Besch (1993) interpreted the metavolcanosedimentary association as an accretionary wedge that was intruded by mafic melts, granodiorites and granites during a subduction event. Thus, the eastern Odenwald probably represents the deep part of a late Silurian to early Devonian volcanic arc that formed due to

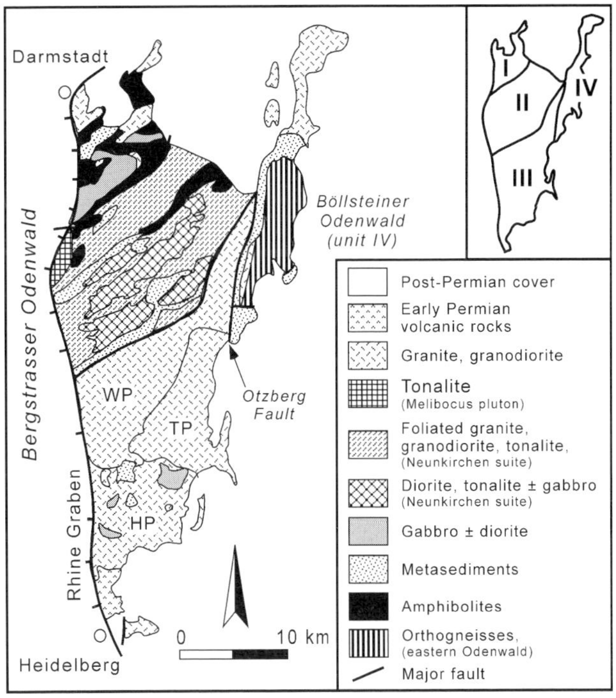

Fig. 12.43. Geological sketch map of the Odenwald (after Altenberger & Besch 1993; Altherr *et al.* 1999*a*). Abbreviations: HP, Heidelberg pluton; TP, Tromm pluton; WP, Weschnitz pluton. Inset: structural units I–III belong to the western (Bergstrasser) Odenwald, and Unit IV to the eastern (Böllsteiner) Odenwald (after Willner *et al.* 1991).

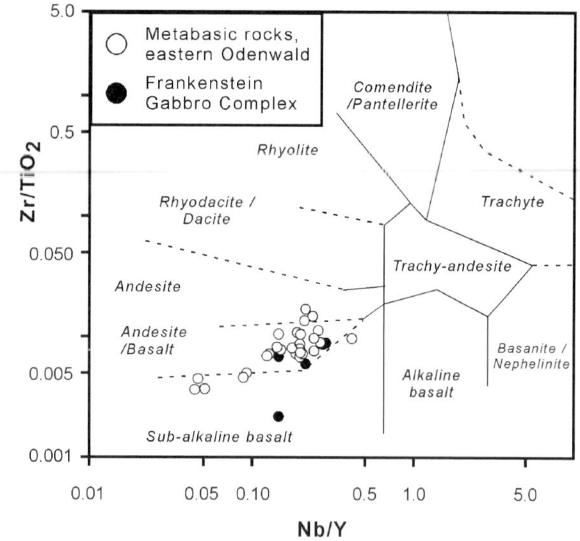

Fig. 12.44. Nb/Y versus Zr/TiO_2 discrimination diagram for the metabasic rocks of the eastern Odenwald (Unit IV) and the late Devonian Frankenstein Gabbro Complex (Unit I, western Odenwald; Winchester & Floyd 1977). Data from Altenberger & Besch (1993), Altenberger *et al.* (1990) and Altherr *et al.* (1999*a*).

subduction of the Rheic Ocean that was closing between Laurussia to the north, and an ATA-related terrane to the south. Juxtaposition of the eastern and western units of the Odenwald occurred in late Devonian to mid-Carboniferous times along the Otzberg Fault Zone (Reischmann *et al.* 2001; see McCann *et al.* 2008*a*).

Spessart Complex

The Spessart Complex to the east of the Odenwald is a NE-trending antiform cored by orthogneisses that are flanked by amphibolite- and greenschist-facies metasediments containing minor volumes of amphibolite (Fig. 12.46; Okrusch & Richter 1986; Hirschmann & Okrusch 1988; Weber 1995). The southern part of the Spessart Complex comprises undeformed, flow-foliated, heterogeneous diorites and granodiorites that yielded a 330 ± 1 Ma (Visean) zircon evaporation age (Anthes & Reischmann 2001). The orthogneisses in the core comprise muscovite-biotite granodioritic to dioritic gneisses (Rotgneiss)

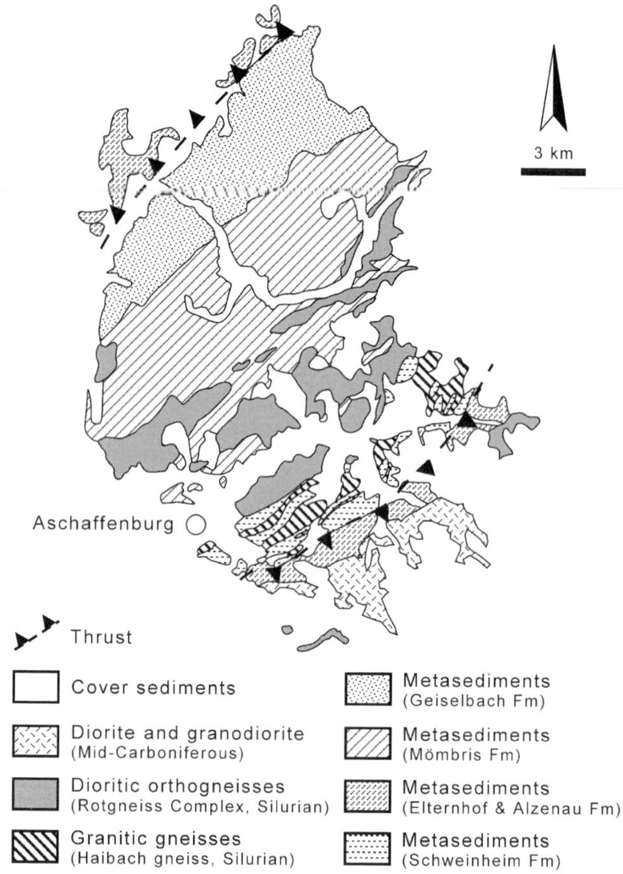

Fig. 12.46. Geological sketch map of the Spessart Complex (after Hirschmann & Okrusch 1988).

and more intensely deformed granitic gneisses (Haibach Gneiss). The latter are intercalated with metasediments. Orthogneisses from the Rotgneiss Complex and a Haibach gneiss have early Devonian to late Silurian (418 ± 18 and 410 ± 18 Ma) zircon evaporation ages, respectively (Dombrowski *et al.* 1995). Both intrusive rocks and sediments were metamorphosed under lower amphibolite-facies conditions during the Variscan Orogeny. The Rotgneiss protolith intruded the early Palaeozoic Mömbris Fm (deep marine pelites and psammites) and metasediments of the overlying Geiselbach Fm that contain spores indicating a late Silurian age of deposition (Wenlock–Ludlow; Reitz 1987). Both

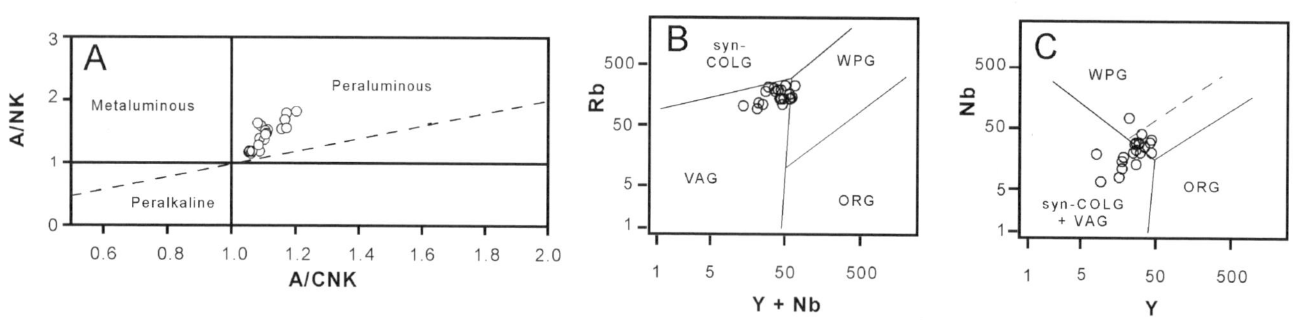

Fig. 12.45. (**A**) A/CNK versus A/NK diagram for early Palaeozoic granitoids and felsic gneisses from the eastern Odenwald (Maniar & Piccoli 1989). (**B** & **C**) Tectonic discrimination diagram (Pearce *et al.* 1984). For abbreviations see captions to Figures 12.4 and 12.5. Data from Altenberger *et al.* (1990) and Reischmann *et al.* (2001).

formations probably represent Early Palaeozoic continental-margin deposits, and the other metasedimentary formations in the Spessart Complex are probably of Neoproterozoic age (Nasir *et al.* 1991).

The Haibach gneisses and Rotgneisses are geochemically similar, since both have minimum melt compositions, are peraluminous (Figs 12.47), and have I- and S-type geochemical characteristics, though the Rotgneisses tend more towards S-type. The Haibach gneisses have variable Sr_i values ranging from 0.7203 to 0.7461, which is probably a primary feature since the Rb–Sr data define a 407 ± 14 Ma whole-rock isochron age that overlaps with the U–Pb zircon ages. One Rotgneiss sample has a rather low $\varepsilon_{Nd}(t)$ value of -8.2 indicating the presence of older crustal components (Sm–Nd data from Liew & Hofmann 1988). Incompatible element ratios indicate their protolith may have formed in a volcanic-arc or syncollisional setting (Fig. 12.48), and Dombrowski *et al.* (1995) interpreted them to be metamorphosed, anatectic two-mica leucogranites that originated in a Silurian continental arc. The southern part is composed of post-tectonic, Visean-age diorites and granodiorites (Anthes & Reischmann 2001; see below).

Devonian

Magmatic rocks of Devonian age occur in the Rhenohercynian Zone (Rhenish Massif, Harz and Jesníky Mountains), Mid-German Crystalline High, the Elbe Zone, as small intrusions in the western margin of the Teplá-Barrandian Terrane, and as roof pendants in the Carboniferous-age Central Bohemian Pluton. Those in the high-grade internal parts of the Variscan Orogen are generally attributed to subduction processes during the northward migration of the Gondwana-derived terranes.

Elbe Zone

The NW-trending Elbe Zone separates the Neoproterozoic Lusatia/Lausitz Complex to the east from the Erzgebirge Crystalline Complex and the Saxonian Granulite Complex to the west (Figs 12.3 & 12.49). Two subunits – the Nossen-Wilsdruff Schiefergebirge (Nossen-Wilsdruff Slate Belt) in the NW of the Elbe Zone and the Elbtalschiefergebirge (Elbe Valley Slate Belt) in the SE – are separated by the Early Permian Dohlen Basin and intruded by the *c.* 330 Ma Meissen monzonite-diorite. The Elbe Zone is made up of NW-trending and east-dipping fault- and thrust-bound units of varying age and lithology that were juxtaposed during early Carboniferous thrusting and dextral strike-slip movements (Kupetz 1984, 2000; Kozdrój *et al.* 2001; Linnemann 1994, 2003; Linnemann & Schauer 1999).

In the better-known Elbtalschiefergebirge these are, from the

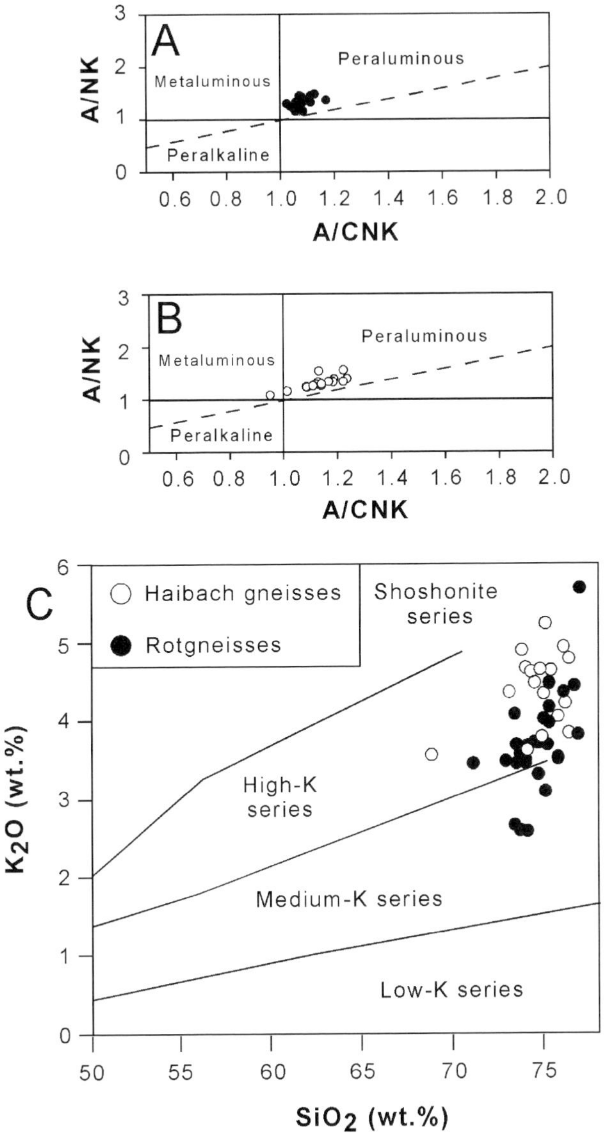

Fig. 12.47. A/CNK versus A/NK diagram for the Silurian Haibach granite gneisses (**A**) and dioritic Rotgneisses (**B**) in the Spessart, Mid-German Crystalline High (Maniar & Piccoli 1989). (**C**) SiO_2 versus K_2O diagram (Peccerillo & Taylor 1976). Data from Okrusch & Richter (1986) and Dombrowski *et al.* (1995).

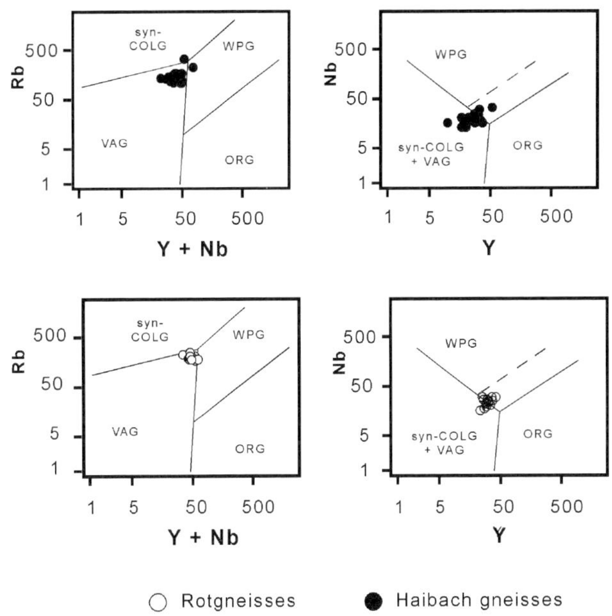

Fig. 12.48. Tectonic discrimination diagrams for the Silurian dioritic Rotgneisses and Haibach granite gneisses in the Spessart, Mid-German Crystalline High (Pearce *et al.* 1984). For field labels see caption to Figure 12.4. Data from Okrusch & Richter (1986) and Dombrowski *et al.* (1995).

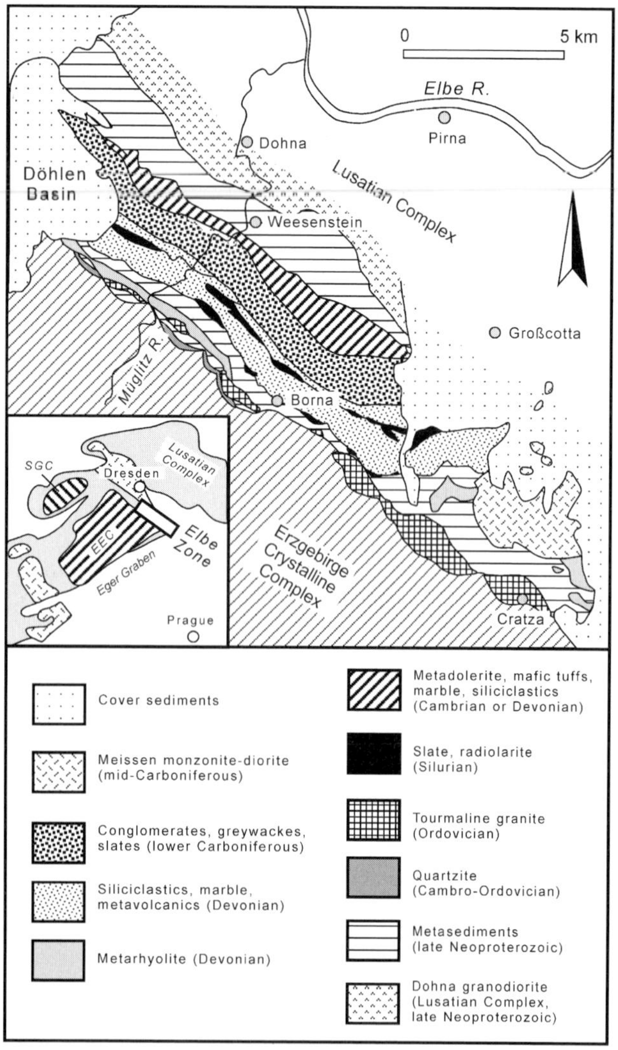

Fig. 12.49. Geological map of the Elbtalschiefergebirge, SE part of the Elbe Zone (after Gehmlich *et al.* 2000, and references therein). Abbreviations in inset: EEC, Erzgebirge Crystalline Complex; SGC, Saxonian Granulite Complex.

SW to the NE, the Phyllite Group, the Palaeozoic Unit, the Bahre Series, the Neoproterozoic Weesenstein Group, and the early Cambrian Dohna granodiorite (Linnemann 1994; Linnemann *et al.* 2000; Gehmlich *et al.* 2000; Fig. 12.49). Deformation and greenschist-facies metamorphism took place in the mid-Visean, i.e. after deposition of the Visean sediments and before intrusion of NNE-trending, post-kinematic lamprophyre dykes and granodiorite that cut the Bahre Series. The lamprophyres have 331 ± 4 and 336 ± 4 Ma K–Ar biotite ages and the granodiorite (part of the Meissen intrusion) yielded a 334 ± 7 Ma K–Ar hornblende age (Kurze *et al.* 1998).

The Weesenstein Group is made up of low-grade Neoproterozoic clastic metasediments, radiolarite/chert, and small volumes of tholeiitic metabasalt. Two granitoid clasts from a conglomerate yielded late Neoproterozoic (568 ± 4 and 559 ± 3 Ma) Pb/Pb zircon protolith ages. The rocks were deformed and metamorphosed under greenschist-facies conditions during a Cadomian event, prior to intrusion of the Dohna granodiorite which has an early Cambrian (537 ± 7 Ma) crystallization age (Pb/Pb zircon evaporation ages; Linnemann *et al.* 2000). The Weesenstein Group was probably part of the Lusatian/Lausitz Complex to the east, and may have formed the basement to the Palaeozoic sediments of the younger units in the Elbe Zone.

The Phyllite Group of the Elbtalschiefergebirge contains Upper Cambrian quartzites, Ordovician phyllites and slates, Carboniferous continental sediments (Culm facies, mid-Visean), and near the structural top, rhyodacitic metavolcanic rocks (deformed to chlorite schists), marbles, radiolarites/cherts, metabasalt lavas and tuffs, and variably mylonitized subvolcanic granite ('tourmaline granite'). The tourmaline granite in the Elbtalschiefergebirge has an Ordovician (485 ± 6 Ma age) (Linnemann *et al.* 2000). In the Nossen-Wilsdruff Schiefergebirge metarhyolites (sericite schists) occupy a similar tectonostratigraphic position and also yielded an Ordovician (486 ± 4 Ma) Pb/Pb zircon age, interpreted as the protolith crystallization age (Linnemann *et al.* 2000). A similar, *c.* 480 Ma zircon evaporation age was obtained for cataclastic granite that occurs as xenoliths in the NE part of the Carboniferous Meissen intrusion NW of Dresden (Coswig Granite; Kurze & Mohnicke 2003). Chlorite schists (strongly deformed rhyodacitic metavolcanic rocks) from the Elbe Zone yielded late Devonian (377 ± 7 and 369 ± 5 Ma) Pb/Pb zircon evaporation protolith ages (Gehmlich *et al.* 2000). Conodonts from the Phyllite Zone of the Nossen-Wilsdruff Schiefergebirge indicate a late Devonian age of deposition (Famenian; Zeidler *et al.* 2004). The Ordovician metarhyolites (sericite schists) and tourmaline granite have calc-alkaline compositions, and the low Nb, Zr, Ti and Sr concentrations and the presence of older, inherited zircons of Early Palaeozoic and Palaeoproterozoic age indicate they have an S-type geochemical signature. They represent anatectic, crustal melts (Werner 1997). The Devonian rhyodacitic rocks (chlorite schists) are rich in Zr, Nb and REE, suggesting an A-type signature; in this respect they resemble intraplate granitoids (Gehmlich *et al.* 2000). The mafic lavas and tuffs near the structural top have basaltic compositions and continental, intraplate trace element signatures (Werner 1997).

The Palaeozoic Unit is composite, comprising Upper Ordovician to Lower Carboniferous sediments separated by faults. It contains a *c.* 500 m thick sequence of metamorphosed mafic tuffs, few mafic flows, metadolerites, and marbles of late Devonian age. The flows are pillowed and vesicular and have basaltic compositions; the metadolerites have subalkaline (tholeiitic) compositions and have intraplate trace element signatures (Werner 1997; Linnemann 2003).

The Bahre Series is composed of slates, marbles and *c.* 500 m thick mafic volcaniclastic sediments, submarine lavas and tuffs of suspected late Devonian age. The lavas have tholeiitic to calc-alkaline and alkaline compositions and have intraplate trace element signatures. Werner (1997) suggested the lavas erupted in a rift setting and that the tholeiitic and alkaline rocks were derived from different mantle sources.

Rhenohercynian Zone

The magmatic rocks in the Rhenohercynian Zone (Fig. 12.1) were emplaced during Devonian to early Carboniferous extension at the southern edge of Laurussia. Progressive rifting resulted in spreading, and the formation of a broadly east–west trending oceanic or backarc basin of limited width. This basin (re-) separated Avalonia from the Armorican microplate to the south and is known as the Rhenohercynian, or Lizard-Giessen-Ostharz Ocean (Oncken 1997; Oncken *et al.* 1999; Warr 2000). U–Pb zircon dating of the Lizard Ophiolite in SW England indicates that opening and spreading was underway by 397 ± 2 Ma (Clark *et al.* 1998; Sandeman *et al.* 2000). Devonian to early Carboniferous sediments accumulated on thinned and stretched Avalonian crust of the southern passive margin, reaching thicknesses of over 10 km in the Rhenish Massif in Germany (Oncken 1997; Tait *et al.* 1997; Korsch & Schäfer 1995; Franke 2000). Zircons from altered, felsic pyroclastic rocks interbedded with clastic sediments of the Rhenohercynian Zone have yielded U–Pb zircon ages of *c.* 408 Ma, 393 Ma and 377 Ma in the Rhenish Massif (Hunsrück and Kellerwald; Kaufmann *et al.* 2004, 2005), and *c.* 385 Ma in the Munster Basin in SW Ireland (Williams *et al.* 1997, 2000). Early Devonian to early Carboniferous magmatic rocks associated with this extension event are found in SW England (Cornwall), the Ardennes, the Rhenish Massif, the Harz area and the Jeseníky Mountains (Floyd 1982, 1995; Floyd *et al.* 1993; Gofette *et al.* 1991; Kirnbauer 1991; Nesbor 2004; Wedepohl *et al.* 1983).

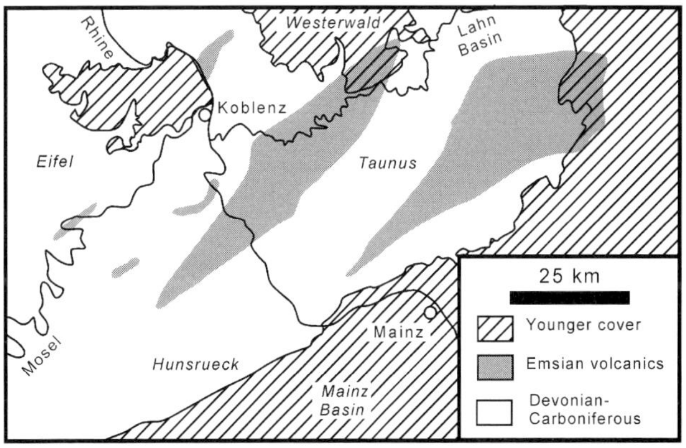

Fig. 12.50. Distribution of early Devonian pyroclastic rocks in the SE Rhenisch Slate Mountains (Rhenohercynian Zone; after Kirnbauer 1991).

Rhenish Massif (Rheinische Schiefergebirge)

In the Central European parts of the Rhenohercynian Zone, magmatic, predominantly volcanic activity was focused in the eastern Rhenish Massif, and occurred in three main stages. Early Devonian (Emsian) rhyolitic to rhyodacitic pyroclastic rocks were emplaced as debris flows that originated from an unknown (i.e. unexposed) source (Kirnbauer 1991). These were followed by a late Middle to early Late Devonian, bimodal volcanic suite with alkaline basalts (Givet-Adorf Stage) that represents the largest volume of magmatic rocks in the Rhenish Massif (Meisl *et al.* 1982; Nesbor 2004, 2006, and references therein). The final but important, rift-related phase of magmatic activity comprised early Carboniferous subalkaline and alkaline basalts (Nesbor 2004, 2006). Of all three stages, the late Devonian (Givetian–Adorf) and early Carboniferous stages are volumetrically the most important. The rocks are deformed to varying degrees and underwent low-grade alteration ('spilites', 'keratophyres') and mid-Carboniferous epi- to anchizone regional metamorphism (Ahrendt *et al.* 1983). Extension also led to the formation of sedimentary-exhalative deposits, such as the Eifelian Rammelsberg sulphide deposit in the NW Harz (Large & Walcher 1999) and the late Givetian hematitic iron ores in the eastern Rhenish Massif (Wedepohl *et al.* 1983). In contrast to the Mid-German Crystalline High to the south, rock types typical for a subduction setting, such as andesites, are not present in any of the suites. In addition, metabasalts are present in the allochthonous Giessen Nappe (SE Rhenish Massif).

Beds of early Devonian (Emsian) felsic pyroclastic rocks occur in the SE and eastern Rhenish Massif within an area of at least 3500 km^2 (Fig. 12.50). The distribution pattern is bounded by later faults, which suggests that they may once have been more widely distributed. The beds range in thickness from centimetres to several metres to tens of metres, and increase in thickness towards the NE. Some units extend over a distance of a few kilometres, but most wedge out after a few metres to tens of metres. They are interbedded with clastic sediments of early Emsian age, and range from nearly pure pyroclastic rocks to tuffites and epiclastic sediments that contain a pyroclastic component. Most of the rocks are metamorphosed and altered, but in the least-altered varieties pseudomorphs after high-temperature quartz and glass shards can still be recognized. In addition, lapilli, broken to partly rounded feldspar crystals and, in the east, rare volcanic bombs have been reported. Clastic components are detrital, rounded quartz grains, muscovite, bioclasts, and rare clasts of quartzite, siltstone and mudstone (Kirnbauer 1991). Compositionally the pyroclastic rocks plot in the rhyodacite to rhyolite fields of the incompatible element-based classification diagram of Winchester & Floyd (1977; Fig. 12.51A).

The depositional setting of the pyroclastic rocks was that of a relatively flat, deeper shelf at the southern coast of Laurussia that was fed with the erosional products of the Caledonian Orogen to the north. The (locally) great thicknesses and restricted aerial distribution of the pyroclastic rocks, in combination with the

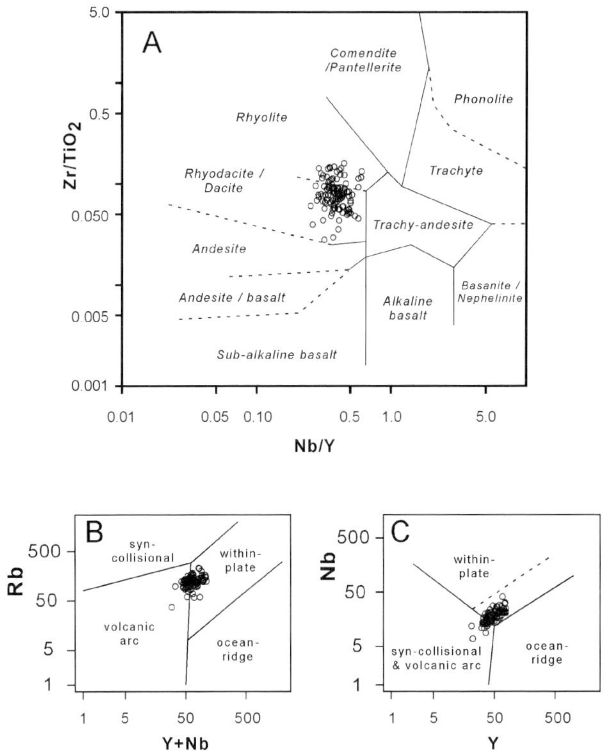

Fig. 12.51. (**A**) Nb/Y versus Zr/TiO_2 discrimination diagram for the Early Devonian pyroclastic rocks of the eastern Rhenish Slate Mountains (Rhenohercynian Zone; Winchester & Floyd 1977). (**B** & **C**) Tectonic discrimination diagrams (Pearce *et al.* 1984). Data from Kirnbauer (1991).

poor evidence for grading and sorting (i.e. absence of quartz- and feldspar-enrichment in the lower parts of beds), the absence of bioturbation, the presence of broken crystals and bioclasts, slump structures in underlying sediments and the sharp contacts with overlying beds suggest they are not submarine-deposited ashes of subaerial eruptions (i.e. true tuffs/airfall deposits). Instead, Kirnbauer (1991) pointed out that the pyroclastic rocks share similarities with submarine, possibly even welded pyroclastic flows and redeposited pyroclastic debris flows. The rapid thickness variations may be attributed to sedimentary reworking processes. The absence of similar pyroclastic rock deposits to the NW, in combination with the west- and southwestward decreasing thicknesses, and the presence of lapilli tuffs and bombs in the east, suggest they originated from (unexposed) sources located to the NE of their present location (Kirnbauer 1991).

Ongoing rifting of the southern margin of Laurussia led to the formation of several basins (mainly half-grabens). Extension was accompanied by bimodal volcanism (Nesbor 2004; Fig. 12.52). This late Middle to early Upper Devonian magmatic activity (Givet-Adorf Stage) was concentrated in the Lahn and Dill basins (Fig. 12.52, inset), where alkaline basalts, basanites and much smaller volumes of trachytic differentiates (trachybasalts, trachyandesites, peralkaline rhyolite; Fig. 12.53A) erupted from numerous overlapping, submarine volcanic centres of up to 20 km in diameter and several hundred metres high (Flick *et al.* 1991; Behnisch 1993; Nesbor *et al.* 1993; Breitkreuz & Flick 1997; Nesbor 2004). The mafic rocks erupted as submarine lavas and pyroclastic rocks and were associated with trachytes and rhyolites. The latter intruded as sills and domes into underlying sediments and formed small volcanic islands on the larger basaltic volcanoes. Locally, volcanic edifices reached elevations above sea level. Central volcanic facies are made up of (pillow) lavas, thin beds of volcaniclastic rocks and, locally, lapilli tuffs with antidune bedding deposited from base surges. Volcaniclastic

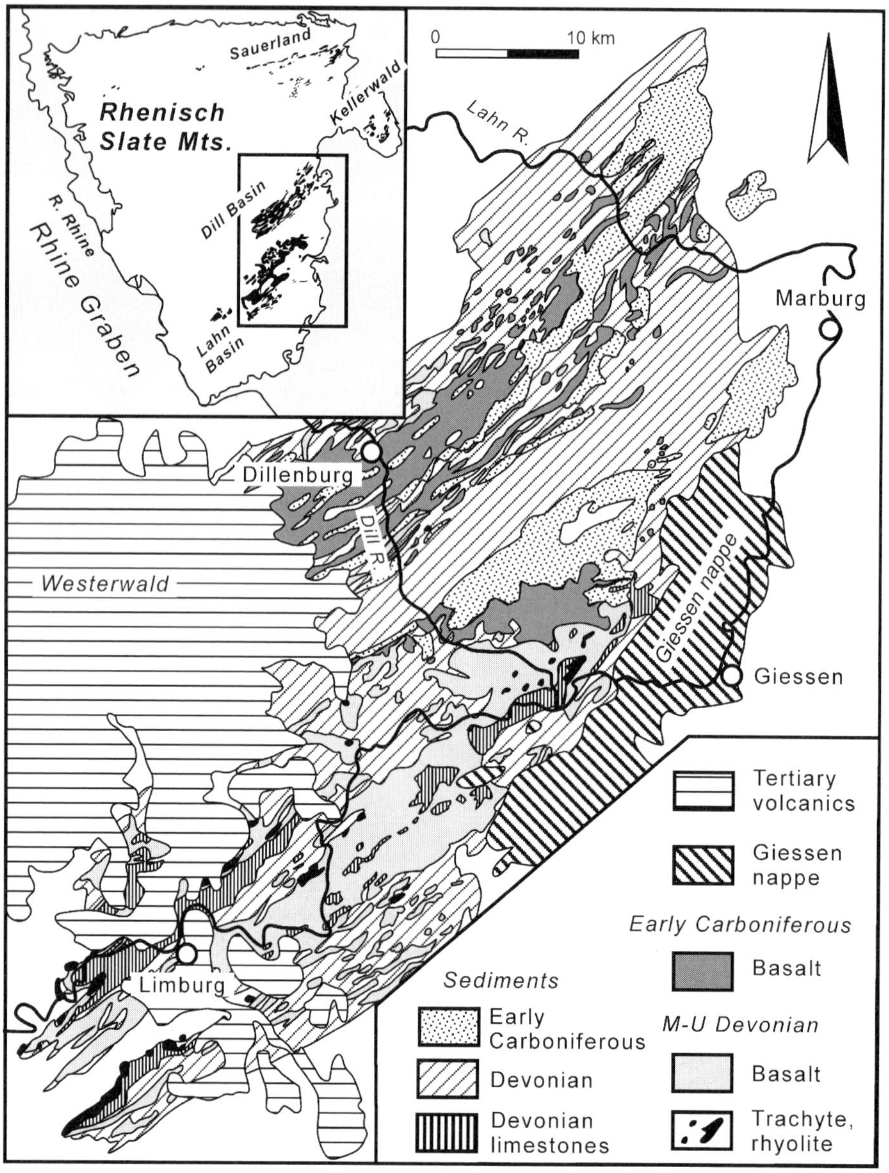

Fig. 12.52. Distribution of the late Devonian and early Carboniferous volcanic rocks in the SE Rhenish Massif (Rhenish Slate Mountains) (Rhenohercynian Zone; after Nesbor 2004).

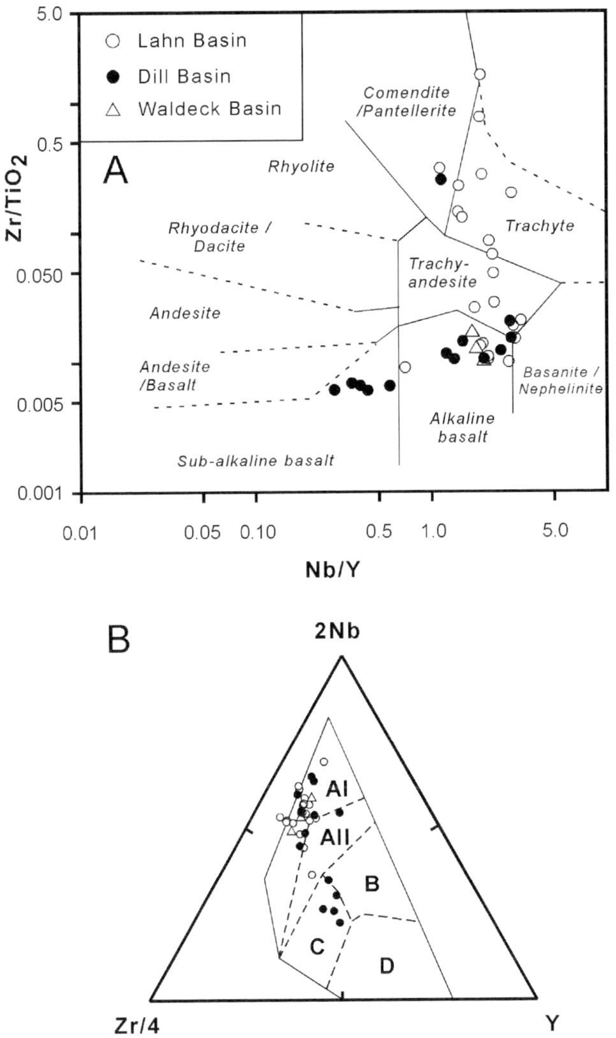

Fig. 12.53. (**A**) Nb/Y versus Zr/TiO_2 discrimination diagram for the late Devonian and early Carboniferous volcanics rocks in the SE Rhenish Slate Mountains (Winchester & Floyd 1977). (**B**) Nb–Zr–Y tectonic discrimination diagram (Meschede 1986). For field labels see caption to Figure 12.19. Data from Nesbor (2004).

rocks increase in proportion away from volcanic centres, and proximal facies contain bedded debris flows composed of vesicle-poor basaltic pillow breccias, breccias with pillow fragments and hyaloclastites, associated with polymict epiclastic rocks originating from volcanic slopes. The distal facies is made up of bedded epiclastic rocks, lapilli tuffs and pyroclastic rocks of airfall origin or deposited from submarine pyroclastic flows.

The early stage of volcanism coincided with a phase of intensified rifting, high subsidence rates and high rates of sedimentation, and was accompanied by the formation of limestone reefs on detrital material of volcanic origin (Meisl *et al.* 1982; Salamon 2003). On the northwestern margin of the Lahn Basin and the southern margin of the Dill Basin olisthostromes, submarine rafts, rock-fall deposits, side-flow and viscoplastic debrites and turbidite flows were formed. The olisthostromes are composed of locally derived sediments, such as early Emsian to Givetian sand- and mudstone, clasts of reef limestones, micritic limestones and biosparites, Emsian felsic pyroclastic rocks, and Givetian basaltic, rhyolitic and trachytic volcanic and volcaniclastic rocks. The olisthostromes are concordantly interbedded with clastic sediments and grade laterally into turbidites and debris flows. Most were deposited within a very short time interval (*c.* 1.5 Ma), lasting from the middle *varcus* until the *disparillis* conodont zones. Deposition of the olisthostromes was related to footwall uplift and block rotation during a phase of active extension; their source is situated to the west, but is not exposed (Salamon 2003).

In the Nb/Y versus Zr/TiO_2 classification diagram of Winchester & Floyd (1977) the late Devonian volcanic suite of the Rhenish Massif defines a differentiation series ranging from alkaline basalts, trachybasalts and trachyandesites to trachytes (Fig. 12.53A). The rocks are LREE-enriched ($La_N/Yb_N = 4.2$ to 20.6; Fig. 12.54A) and have primitive mantle-normalized multi-element patterns typical of alkaline mafic rocks (Fig. 12.55A). The peralkaline rhyolites are very enriched in incompatible elements, which was probably due to the assimilation of crustal material. The parent melts of this magmatic suite ascended slowly and differentiated in magma chambers prior to eruption. An intraplate setting for the mafic volcanic rocks is indicated by their Nb/Zr/Y ratios (Fig. 12.53B; Meschede 1986), and by both the character and sheer volume of the associated clastic sediments (Nesbor 2004).

Following the mid- to upper Devonian stage of magmatism, the volcanic edifices were deeply eroded in Givetian–Frasnian times. Magmatic activity resumed in late Devonian times (Famennian, Dasberg Stage) in the form of smaller volumes of rapidly ascended, primitive basalts and basanites, the latter containing mantle xenoliths. These formed submarine volcanoes built up of pillow lavas and breccias, pillow fragment breccias and hyaloclastic rocks, mantled by thick deposits of volcaniclastic rocks. Also attributed to this stage are basaltic sills that contain mantle xenoliths (spinel lherzolites, pyroxenites) that intruded into mid-Devonian sediments in the Kellerwald area (Fig. 12.52, inset). Magmatic rocks of felsic or intermediate composition are absent, and the mafic basalts and basanites are little fractionated, with high Ni and Cr abundances, and LREE enrichment ($La_N/Yb_N = 11.9$ to 15.9, Fig. 12.55B; Nesbor 2004).

The second most important magmatic phase in the Rhenish Massif is of early Carboniferous age and is preserved as a 500 to 700 m thick sequence of volcanic and sedimentary rocks in the Dill Basin (Fig. 12.52). It comprises an early stage, predominantly composed of tholeiitic basalts, and a later stage of (strongly altered) primitive alkaline basalts that contain mantle xenoliths. Related subvolcanic intrusions are more widespread, suggesting that the volcanic rocks once covered a much larger area (Nesbor 2004). The tholeiites erupted to form submarine volcanoes with a high proportion of pillow lavas and pillow breccias, and much smaller amounts of volcaniclastic rocks, suggesting deeper-water conditions. Associated dolerite sills, locally up to 100 m thick, contain olivine-rich cumulates that also intruded as individual, olivine-rich bodies. A picritic sill encountered in drill core in the Dill Basin yielded early Carboniferous (349 ± 14 and 354 ± 14 Ma) K–Ar kaersutite ages (west of Dillenburg; Fig. 12.52; Mertz 1986). The younger alkaline basalts are unfractionated and contain xenoliths of spinel-lherzolite, implicating rapid ascent. The fact that clastic sediments are interbedded with the volcanic rocks indicates moderate rates of volcanism or longer pauses between volcanic activity. The volcanic rocks are dominated by tholeiitic basalts, while alkaline basalts are rare. The tholeiites have high Ni and Cr abundances, flatter REE patterns ($La_N/Yb_N = 3.2$ to 6.6; Fig.

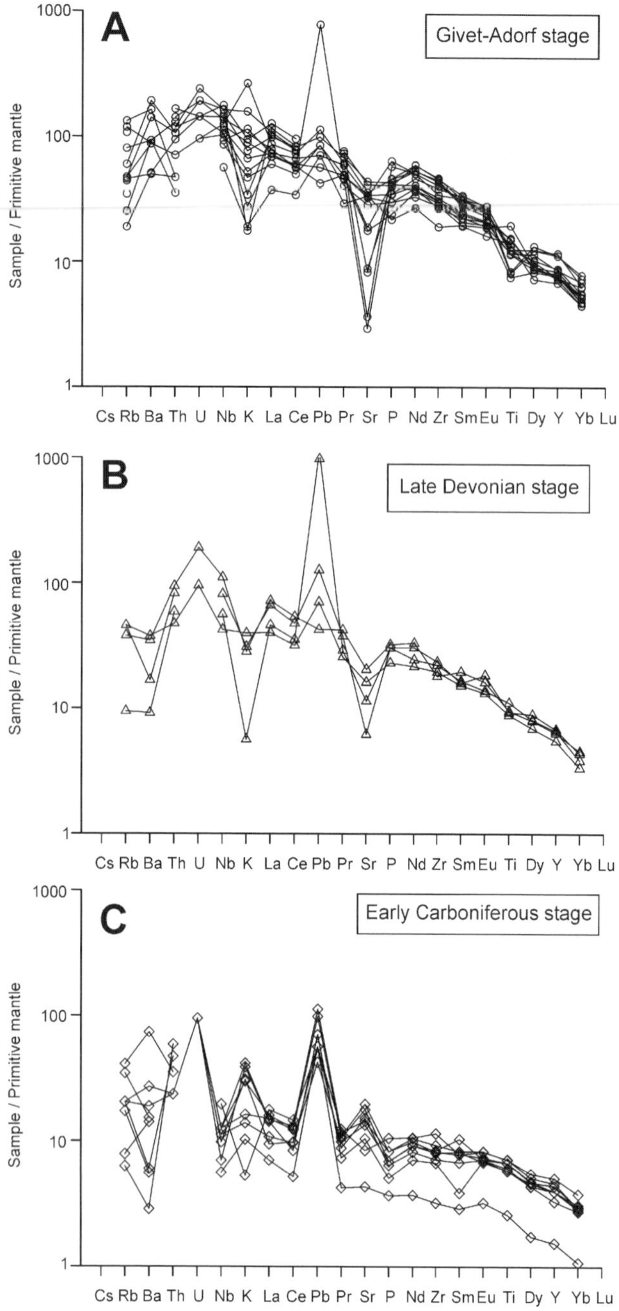

Fig. 12.54. Primitive mantle-normalized multi-element diagrams for the volcanic and subvolcanic basalts of the SE Rhenish Slate Mountains: Givet-Adorf (**A**), late Devonian (**B**) and early Carboniferous (**C**) stages. Only samples with SiO_2 <52 wt% have been plotted; the scatter in LILE is due to alteration and metamorphism. Normalizing values of Sun & McDonough (1989). Data from Nesbor (2004).

12.55C) and are little fractionated. The mantle xenoliths in the alkaline basalts are completely carbonated lherzolites, but have U-shaped platinum group element patterns typical for mantle rocks (Nesbor 2004).

The mid-Devonian to early Carboniferous magmatism on the southern margin of Laurussia took place in a shallow shelf setting, during progressive extension that lasted over a period of *c.* 30 Ma. The subalkaline composition of the early Carboniferous melts reflects higher degrees of melting during increased extension and subsidence. Airfall tuffs of (rhyo)dacitic composition, that are interbedded with the early Carboniferous sediments, are of distal origin, possibly originating from the arc magmatism that took place in the ocean basin to the south (later to become the Mid-German Crystalline High, see below).

The Giessen Nappe at the SE margin of the Rhenish Massif is composed of proximal turbiditic greywackes of late Devonian age that were overthrust onto Devonian to early Carboniferous rocks in the form of NW-vergent recumbent folds and thrust-bound units. This nappe unit is correlated with the south Harz and Selke nappes in the south Harz area further east (see Fig. 12.57). The lower *c.* 50 metres of the Giessen Nappe contains strongly deformed lenses of metamorphosed mafic rocks embedded in radiolarian cherts and early to late Devonian-age black shales (Engel *et al.* 1983; Wedepohl *et al.* 1983; Grösser & Dörr 1986). The magmatic rocks are lavas and subvolcanic intrusions which were metamorphosed under greenschist-facies conditions. They have tholeiitic basaltic incompatible elements compositions, with flat to LREE depleted REE patterns (Fig. 12.55D) and low Ti/V ratios (10 to 20), indicating an N-MORB composition that differs clearly from the enriched, within-plate type volcanic rocks in the Rhenish Massif to the north and west (Fig. 12.56). Grösser & Dörr (1986) suggested that these metabasalts represent the deformed upper parts of oceanic crust that formed due to advanced rifting and crustal thinning to the south of the passive margin. Northward transport and nappe emplacement occurred during the final closure of the Rhenohercynian Ocean in mid-Carboniferous times (see below).

Harz area

The mid-Devonian to early Carboniferous mafic magmatic rocks in the Harz area occur in both autochthonous and allochthonous units and were emplaced as submarine lavas (locally pillowed), volcaniclastic rocks (including agglomerates), sills, dykes and (locally) pyroclastic rocks of basaltic (mostly) to felsic composition. Volcanic rocks of Middle Devonian age reach thicknesses of 300–350 m in the west-central Harz near St Andreasberg, and 600–1000 m in the Elbingerode Complex in the north (Fig. 12.57). Most of the rocks underwent alteration after emplacement, followed by low-grade metamorphism dated at 309–335 Ma (K–Ar white mica ages; Marheine 1997). Magmatic activity began in mid-Devonian times and extended into the early Carboniferous in the north and NW Harz (Elbingerode and Oberharz areas). Incompatible element compositions of the basalts indicate alkaline to subalkaline compositions with within-plate signatures; E-MORB types are present in the south Harz and Selke nappes, but true MORB and volcanic-arc-type rocks have not been described from the Harz area (Gansslos er 1993; Werner 1995).

Moravo-Silesian Zone, Jeseníky Mountains

The northern, Silesian part of the Moravo-Silesian Zone is a nappe stack of Neoproterozoic orthogneisses and early to mid-Devonian clastic sediments and volcanic rocks (Dvořák 1995; Finger *et al.* 1995; Mísař 1995; Kröner *et al.* 2000*b*; and references therein). It is generally considered to be the eastern extension of the Rhenohercynian Zone, and the present northerly trend is due to *c.* 90° clockwise oroclinal bending around the NE flank of the Bohemian Massif between the late Devonian and late Carboniferous (Tait *et al.* 1996). The higher grade parts in the

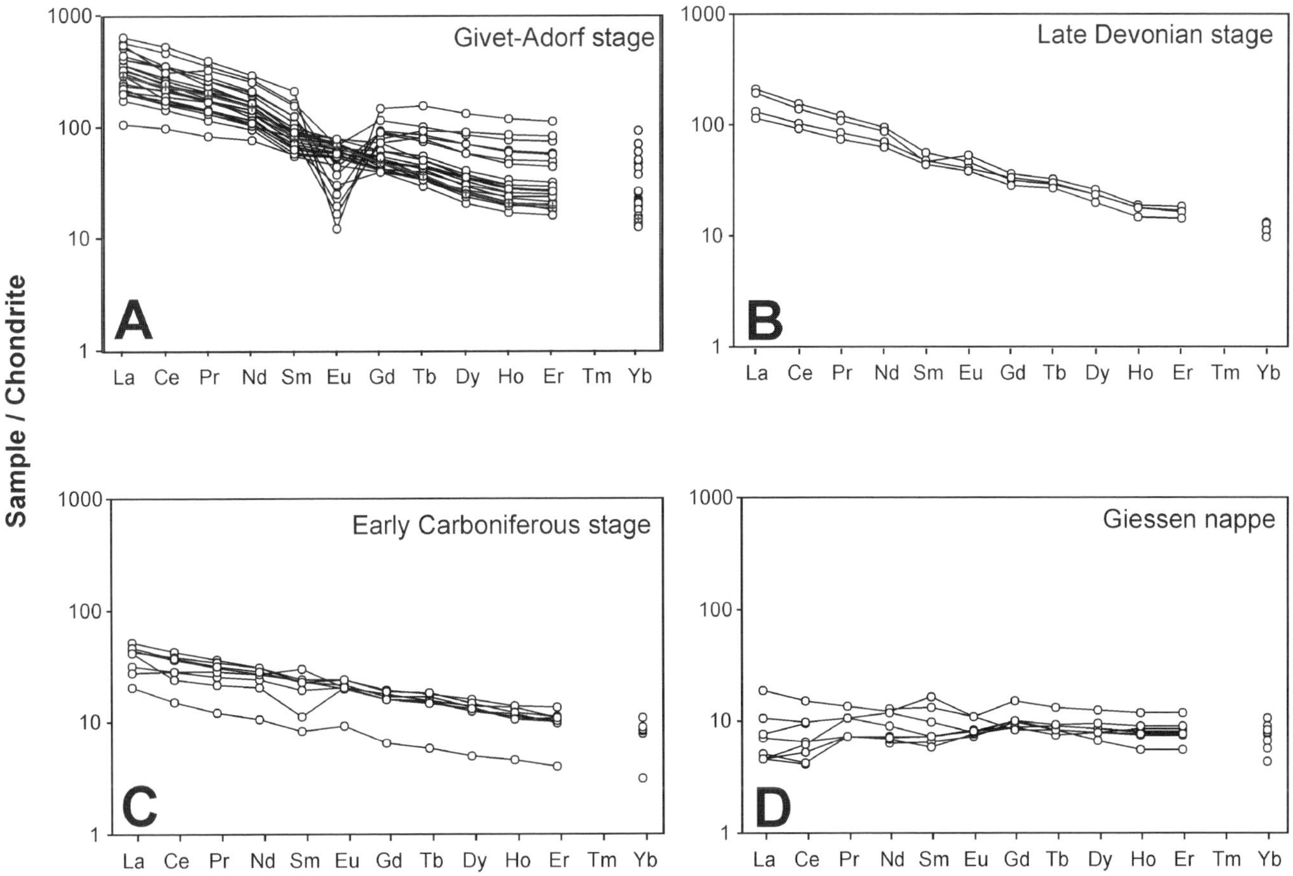

Fig. 12.55. Chondrite-normalized rare-earth element patterns for the volcanic and subvolcanic rocks of the SE Rhenish Slate Mountains: Givet-Adorf (**A**), late Devonian (**B**) and early Carboniferous (**C**) stages. (**D**) Metabasalts from the Giessen nappe. Normalizing values of Sun & McDonough (1989). Data from Grösser & Dörr (1986) and Nesbor (2004).

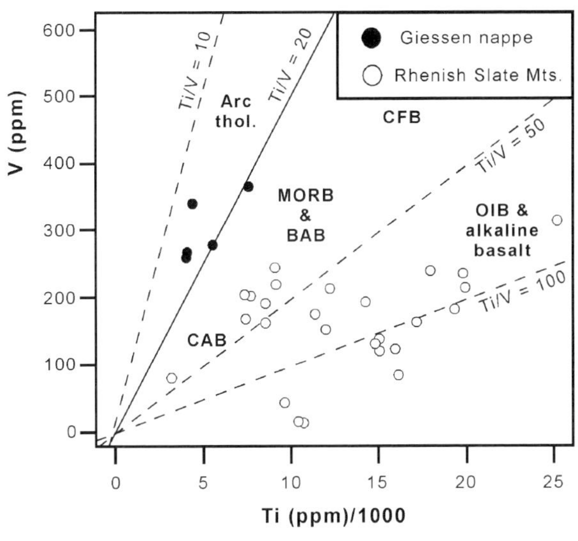

Fig. 12.56. Ti versus V tectonic discrimination diagram for N-MORB type metabasalts from the Giessen nappe and basaltic rock from the Lahn and Dill basins, SE Rhenish Slate Mountains (Shervais 1982). Only samples with $SiO_2 < 52$ wt% have been plotted. For abbreviations see caption to Figure 12.16. Data from Grösser & Dörr (1986), Nesbor (2004).

northern segment (Jeseníky Mountains; Fig. 12.58) were deformed and metamorphosed under amphibolite- to greenschist-facies conditions in early to mid-Carboniferous times during northeastward thrusting and crustal thickening (e.g. Schulmann & Gayer 2000). During this event, the Devonian sedimentary and volcanic rocks were imbricated with parautochthonous, Brunia/Brunovistulia-derived Neoproterozoic granitoid gneisses, and are presently exposed in three NNE-trending belts between the Staré Město belt to the west (eastern margin of the Orlice-Sněžník/Orlica-Śnieżnik Complex) and the Moravo-Silesian Culm Basin to the east (Fig. 12.58; Kröner *et al.* 2000*b*).

The degree of metamorphism and deformation decreases eastwards, where folded and cleaved Devonian and Carboniferous sediments of the Moravo-Silesian Culm Basin, a foreland basin, overlie Neoproterozoic basement of the Brunia/Brunovistulia microplate and (locally preserved) Cambro-Ordovician clastic sediments (Hartley & Otava 2001; Kalvoda *et al.* 2002). To the east, the Upper Silesia Coal Basin contains folded Namurian to Westphalian clastic sediments and coals associated with felsic airfall tuffs (e.g. Hess & Lippolt 1986; Kumpera 1997).

The Early Devonian (Pragian) to Early Carboniferous (Tournaisian) sediments were deposited in several environments, ranging from carbonate platform, basin, siliciclastic shelf to shallow-water shelf facies. From the latest Devonian, sedimentation is of continental, synorogenic character, including, for example, latest Visean conglomerates containing Moldanubia-

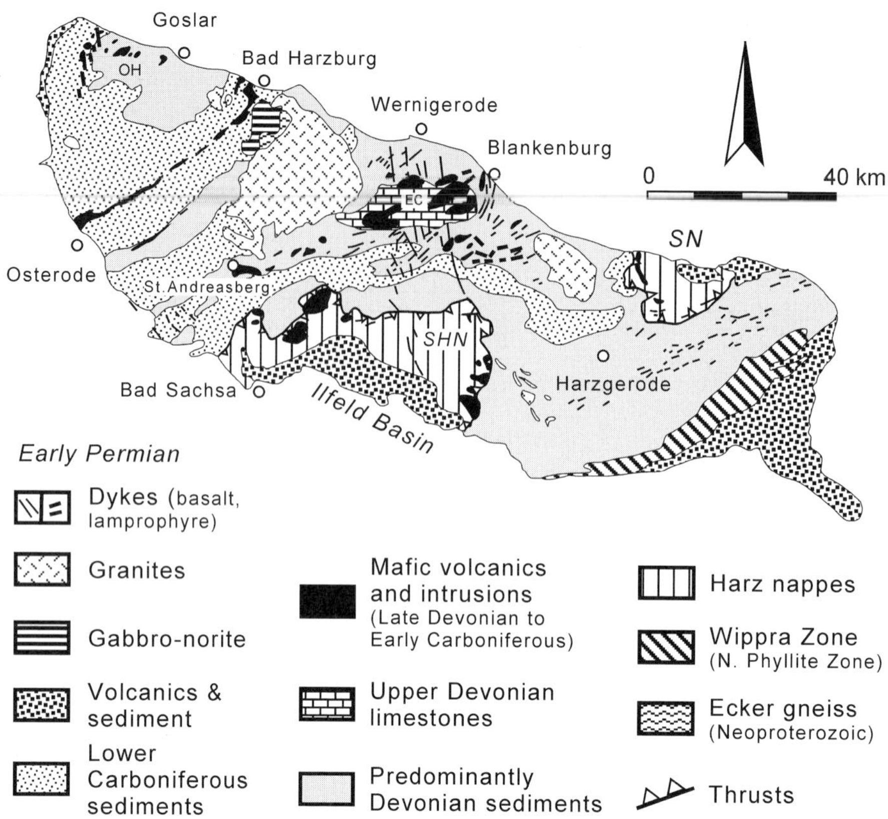

Fig. 12.57. Geological sketch map of the Harz Mountains (Rhenohercynian Zone). Abbreviations: EC, Elbingerode Complex; OH, Oberharz; SHN, South Harz Nappe; SN, Selke Nappe.

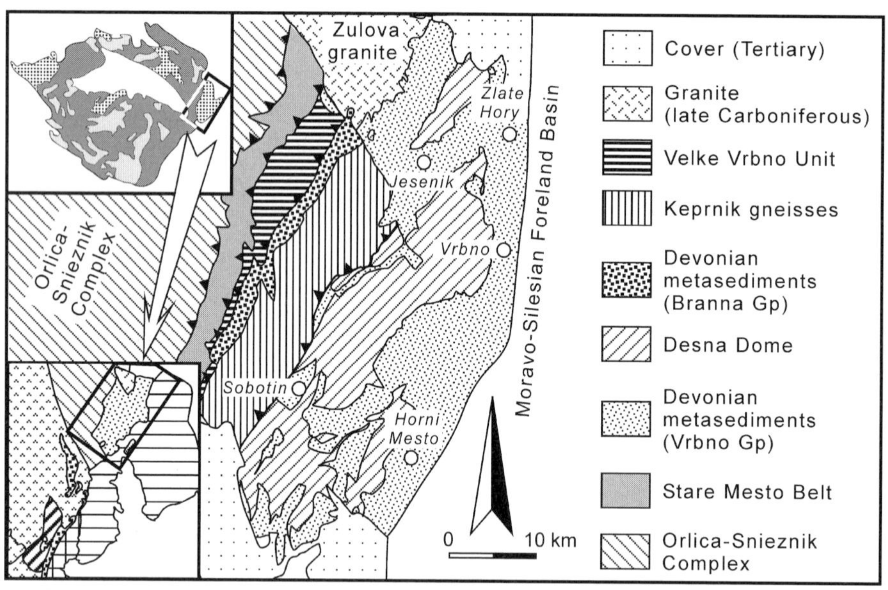

Fig. 12.58. Geological map of the Jeseníky Mountains, northern Moravo-Silesian Zone.

derived gneiss and granulite clasts (e.g. Schulmann & Gayer 2000; Hartley & Otava 2001; Kalvoda *et al.* 2002; Belka & Narkiewicz 2008; Krawczyk *et al.* 2008).

Sedimentation in the volcanosedimentary Vrbno Group (Jeseníky Mountains, Fig. 12.58) commenced in the Early Devonian (Pragian) and extended into the Late Devonian. The maximum volcanic activity took place in the late Early to Middle Devonian period. Both the granitoid gneisses and Devonian volcano-sedimentary rocks were intruded by mafic and felsic sills and dykes, probably of Devonian age. To the north, the rocks of the Vbrno Group are intruded by the latest Carboniferous–earliest Permian, post-kinematic Zulová granite (Zachalová *et al.* 2002).

The western volcanosedimentary belt contains metabasalts with pillow structures (indicating a submarine setting), meta-andesites and minor volumes of metarhyolites. The basalts and andesites have calc-alkaline compositions, negative Nb, Ti and Sr anomalies (compared to N-MORB), flat LREE patterns (La_N/Sm_N = 2.3 to 3.1) and positive ε_{Nd}(390 Ma) values (+3.3; Janoušek *et al.* 2006*b*). The metarhyolites are enriched in LREE (La_N/Sm_N = 3.3 to 4.9) and also have positive ε_{Nd}(390 Ma) values (+2.9). Field evidence, in combination with the major and trace element composition of the mafic metavolcanic rocks, reflects an origin in a volcanic arc setting. The associated felsic rocks may have originated through melting of juvenile crust or, alternatively, represent fractionation products of basic parent melts (Janoušek *et al.* 2006*b*).

In the eastern belt, mafic rocks occur in much smaller volumes and felsic to intermediate metavolcanic rocks predominate (Fig. 12.59A). These are associated with volcanosedimentary rocks and, locally, with stratabound massive sulphide deposits (Patočka 1987; Patočka & Valenta 1996). The felsic magmatic rocks comprise agglomerate tuffs and ignimbrites, indicating a subaerial setting, the latter containing fiammé and fragments of porphyritic rhyolite. The agglomerate tuffs range in composition from slightly alkaline trachybasalt and trachyandesite to dacite and rhyolite; the ignimbrites are dacitic and rhyolitic in composition (Patočka & Valenta 1996; Wilimský *et al.* 2003; Janoušek *et al.* 2006*b*). The felsic metavolcanic rocks contain alkaline, HFSE- and LREE-enriched varieties that are classified as comendite and comenditic trachyte (Fig. 12.59A–C) with rather primitive Sm–Nd isotope compositions (ε_{Nd}(390 Ma) = +2.8 to +3.8; Janoušek *et al.* 2006*b*). The mafic rocks have alkaline basaltic to trachyandesitic compositions with a within-plate trace element signature (Patočka & Valenta 1996). Field evidence and geochemistry indicate a subaerial, extensional setting for the volcanism in the eastern belt, such as a backarc or even a rift that developed on thinned continental crust (Patočka & Valenta 1996; Wilimský *et al.* 2003; Janoušek *et al.* 2006*b*).

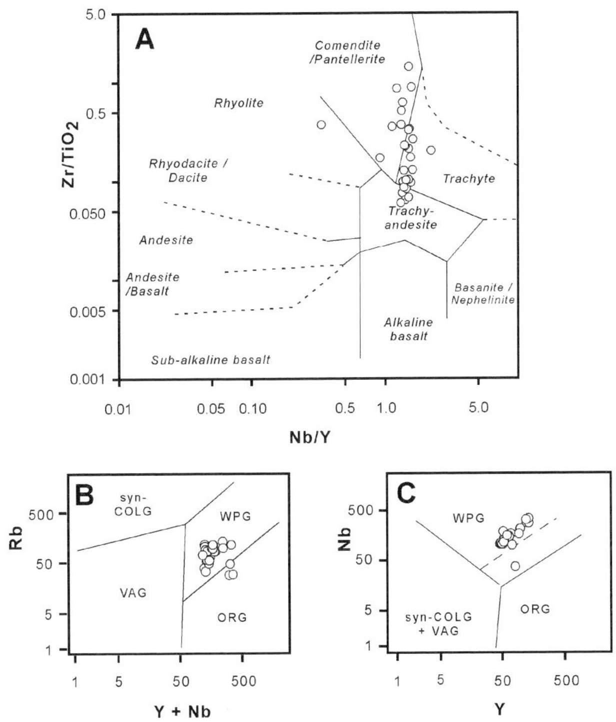

Fig. 12.59. Devonian felsic metavolcanic rocks of the Jeseníky Mountains, northern Moravo-Silesian Zone. (**A**) Nb/Y versus Zr/TiO_2 classification diagram (Winchester and Floyd 1977). (**B & C**) Discrimination diagrams (Pearce *et al.* 1984). For field labels see caption to Figure 12.4. Data from Patočka & Valenta (1996).

Most of the felsic dykes penetrating the Neoproterozoic gneisses are calc-alkaline (comenditic), closely resembling the chemistry of the volcanic rocks in the eastern belt. Dykes with an overall calc-alkaline composition and signature seem to be rarer in the western belt.

The mafic sills and dykes intruding both the Devonian and Neoproterozoic rocks are variably deformed metadolerites with ophitic textures. They have subalkaline, andesitic to basaltic compositions and have high positive ε_{Nd}(390 Ma) values (+7.8 to +8.0; Wilimský *et al.* 2003; Pecina *et al.* 2003; Janoušek *et al.* 2006*b*; Hanžl *et al.* 2007). The parental melts originated from a strongly depleted mantle source and some were contaminated by crustal material to various degrees, as indicated by positive Rb, K, Sr and Pb anomalies (Janoušek *et al.* 2006*b*).

Hladil *et al.* (1999) suggested that sedimentation took place on thinned crust at the southern margin of Laurussia in late Middle Devonian (Emsian) to early Carboniferous times. Late Devonian subduction resulted in the formation of a volcanic arc and backarc basin, the remnants of which are preserved in the western and eastern belts of the Vrbno Group, respectively. The Rehberg and Letovice ophiolites in the southern part of the Moravo-Silesian Zone may represent remnants of the subducted oceanic crust (see above).

Late Devonian–Early Carboniferous closure of ocean basins

Subduction of the Rhenohercynian Ocean commenced in the Middle to Late Devonian, as demonstrated by 376 to 396 Ma U–Pb zircon ages and 360 to 364 $^{40}Ar/^{39}Ar$ amphibole plateau ages from the obducted Lizard Ophiolite in SW England (Sandeman *et al.* 1995, 2000; Nutman *et al.* 2001). In Central Europe, SE-directed subduction was underway in the Late Devonian (Oncken *et al.*, 1999). The Lizard Ophiolite and the MORB-like metabasalts from the Giessen Nappe (SE Rhenish Massif) are probably remnants of Rhenohercynian oceanic crust. Although Devonian-age subduction-related magmatic rocks appear to be limited in volume (e.g. Frankenstein Gabbro Complex), subduction must have taken place as is clear from the occurrence of late Devonian blueschist- and eclogite-facies metamorphism in the West Sudetes, the Münchberg Complex (Fig. 12.24), and the Frankenberg Complex (Figs 12.3 & 12.24; Rötzler *et al.* 1999).

In Central Europe, late Devonian to early Carboniferous subduction and closure of the Rhenohercynian Ocean resulted in calc-alkaline, I-type arc magmatism on continental crust that probably belonged to the northward-moving Armorican terranes or elements thereof (Franke 1989; Reischmann & Anthes 1996; Oncken 1997). Early Carboniferous granitoid magmatism is widespread in the Saxothuringian (especially in the MGCH, Fig. 12.24) and Moldanubian parts of the Variscan Orogen. It is absent from the Rhenohercynian Zone to the north, where rift-related late Devonian to early Carboniferous mafic volcanic rocks are typical (see above). U–Pb zircon ages from the MGCH indicate that the magmatic activity was predominantly of Visean age (325–335 Ma; Reischmann & Anthes 1996; Anthes & Reischmann 2001; Zeh *et al.* 2000). The large volumes of magmatic material suggest their formation in Andean-type active margin environments and/or in syn- to post-collisional settings (e.g. Finger & Steyer 1990; Pin & Duthou, 1990; Finger *et al.*

1997). The MGCH represents the deep roots of a late Devonian to mid-Carboniferous continental arc.

The accretion of the magmatic arc (i.e. MGCH) and oblique collision occurred in Visean to early Namurian times resulting in 330–345 Ma metamorphism that reached amphibolite- to granulite-facies conditions in the MGCH and Moldanubian terranes to the south (Kalt *et al.* 1994*a*; Kober *et al.* 2004; Okrusch 1995; Todt *et al.* 1995; Reischmann & Anthes 1996; Altherr *et al.* 1999*a*; Rötzler & Romer 2001). However, parts of the Variscan Orogen, such as the northern Odenwald (Kirsch *et al.* 1988; Schubert *et al.* 2001), parts of the Teplá-Barrandian, and the Channel Islands (Dallmeyer *et al.* 1992), escaped reworking. In the eastern extension of the Saxothuringian Zone (Erzgebirge area, Fig. 12.3), metamorphism reached ultrahigh pressure and high temperature conditions at *c.* 341 Ma, as is clear from the presence of coesite and microdiamonds in late Neoproterozoic to early Palaeozoic felsic (crustal) gneisses and the presence of intercalated garnet peridotites (U–Pb zircon ages; Kröner & Willner 1998; Hwang *et al.* 2001; Massone 2001). In the Saxon Granulite Massif to the north, deep crustal metamorphism and deformation (*c.* 1050°C, 22 kbar) took place at *c.* 340 Ma and subsequent exhumation to mid-crustal levels occurred at very high rates (U–Pb zircon ages; von Quadt 1993; Kröner *et al.* 1998; Rötzler & Romer 2001; Romer & Rötzler 2001). This early Visean event reflects continental collision and deep subduction of continental crust that was uplifted to mid-crustal levels within a few million years after the peak of metamorphism (Kröner & Willner 1998; O'Brien 2000; Hwang *et al.* 2001).

Mid-German Crystalline High

Frankenstein Gabbro Complex, northern Odenwald

The largest outcrop of subduction-related Devonian rocks in the MGCH is the Frankenstein Gabbro Complex in the northernmost Unit I of the Odenwald (Fig. 12.43; see above). It has a late Devonian, 362 ± 7 Ma U–Pb zircon evaporation age and a 363 ± 7 Ma $^{40}Ar/^{39}Ar$ hornblende total fusion age, which show that it is a high-level intrusion that cooled rapidly (Kirsch *et al.* 1988). It intruded previously metamorphosed amphibolites and metasediments that have similar (*c.* 360 Ma) $^{40}Ar/^{39}Ar$ cooling ages (Schubert *et al.* 2001). Hornblende gabbro is the dominant rock type in the complex (Fig. 12.44), associated with smaller volumes of olivine gabbro and diorite with locally preserved, subhorizontal rhythmic layering. Variable contents of MgO (7.5–12.5 wt%), Cr (46–396 ppm), Ni (21–158 ppm) and high Al_2O_3 contents (16–21 wt%) indicate that the gabbros do not represent primitive melts, but were derived from a fractionated parent melt. The REE patterns are flat for hornblende gabbro (La_N/Yb_N = 1.9) and slightly enriched for gabbro and diorites (4.3–5.3), with unfractionated HREE abundances. Eu anomalies are positive or absent and negative Nb and Ti anomalies are pronounced. The Sr_i ratios range from 0.7035 to 0.7053 and the ε_{Nd}(360 Ma) values are positive (+3.4 to +3.8; Kreher 1994; Altherr *et al.* 1999*a*). Altherr *et al.* (1999*a*) concluded that the parent melts were generated during the late Devonian subduction event, and were extracted from garnet-free, subduction-modified mantle material (spinel-peridotite).

Rhine Graben

Carboniferous granites and granodiorites of the MGCH are exposed on the western escarpment of the Rhine Graben (Fig. 12.24), for which Reischmann & Anthes (1996) obtained U–Pb zircon crystallization ages of 325–335 Ma. The southern granitoids in this area have an I-type, subduction related character, and those exposed to the north have incompatible element signatures suggesting they intruded in a syncollisional setting (Laue & Reischmann 1994). Further to the SW, a two-mica granite from the Soultz borehole near Strasbourg yielded a 327 Ma age and a monzogranite yielded ages in the range of 331–334 Ma (U–Pb zircon; Alexandrov *et al.* 2001; Cocherie *et al.* 2004). Anthes & Reischmann (2001) interpreted these Carboniferous granitoids as having formed part of an active margin situated at the southern margin of the MGCH.

Western Odenwald

The western (Bergstrasser) Odenwald comprises three structural units consisting mainly of late Devonian to mid-Carboniferous, calc-alkaline gabbro to granite intrusions (Units II and III, Fig. 12.43). Unit I contains the late Devonian Frankenstein Gabbro Complex, and Unit IV (eastern Odenwald) differs in age and evolution from the western units; both have been described above.

The magmatic rocks that volumetrically dominate the western Odenwald range in composition from gabbro to granite (Altherr *et al.* 1999*a*). Unit II is made up of small, heterogeneous gabbro, diorite to granite intrusions (Neunkirchen Suite) which show a NE-trending, syntectonic magmatic foliation. They were emplaced during a phase of transtensional shearing (Krohe 1991) and intruded orthogneisses, amphibolites and migmatitic metasediments that were metamorphosed under amphibolite-facies conditions in Visean times (U–Pb zircon ages; Todt *et al.* 1995; Willner *et al.* 1991). The western part of Unit II is intruded by undeformed, post-tectonic tonalite to granodiorite (Melibocus Pluton; Fig. 12.43), which in turn is cut by fine-grained gabbro to diorite dykes that locally have chilled margins (Altenberger *et al.* 2001). The southern Unit III comprises gneisses and metasediments that were metamorphosed under amphibolite-facies conditions between 342 and 332 Ma (U–Pb zircon ages; Todt *et al.* 1995). These were intruded by three undeformed, homogeneous plutons: the Weschnitz Monzodiorite to granodiorite with mafic enclaves, the Tromm Granite, and the composite Heidelberg Complex of gabbro and diorite. The latter shows field evidence for magma mingling and is cut by later granites and granodiorites.

No precise crystallization ages are available for the granitoids of Units II and III, but emplacement in the period 340–330 Ma is a reasonable assumption. Their compositions range from low-K, calc-alkaline to high-K (Fig. 12.60B), and from metaluminous to peraluminous (Fig. 12.61C). The REE patterns indicate various degrees of internal fractionation, and most of the intrusive rocks of Unit III lack a negative Eu anomaly; their $\varepsilon_{Nd}(t)$ values range from −4.0 to +0.8 and Sr_i from 0.7059 to 0.7123. Fractionation trends of the diorites of the Neunkirchen Suite (Unit II) differ from those of the granodiorites and granites of the same unit, showing that they are not related by differentiation from the same parent melt. Relatively high Mg–numbers (40–69), low ε_{Nd}(340 Ma) values (−1.5 to 0.0) and elevated Sr_i ratios (0.7050–0.7053) suggest derivation from an enriched mantle source. The associated felsic end-members probably represent crustal melts modified by fractionation of plagioclase and amphibole at relatively shallow depths. The younger intrusive suites in Unit II have lower abundances of Yb (suggesting garnet was a fractionating phase) and have higher Sr abundances, limiting the role of plagioclase as a fractionating phase and implying greater depths of fractionation in thickened crust. The magmatic rocks in Unit III also have low Yb and Sc abundances and steep REE patterns that lack a Eu anomaly, pointing to fractionation of garnet and to the limited role of plagioclase. The

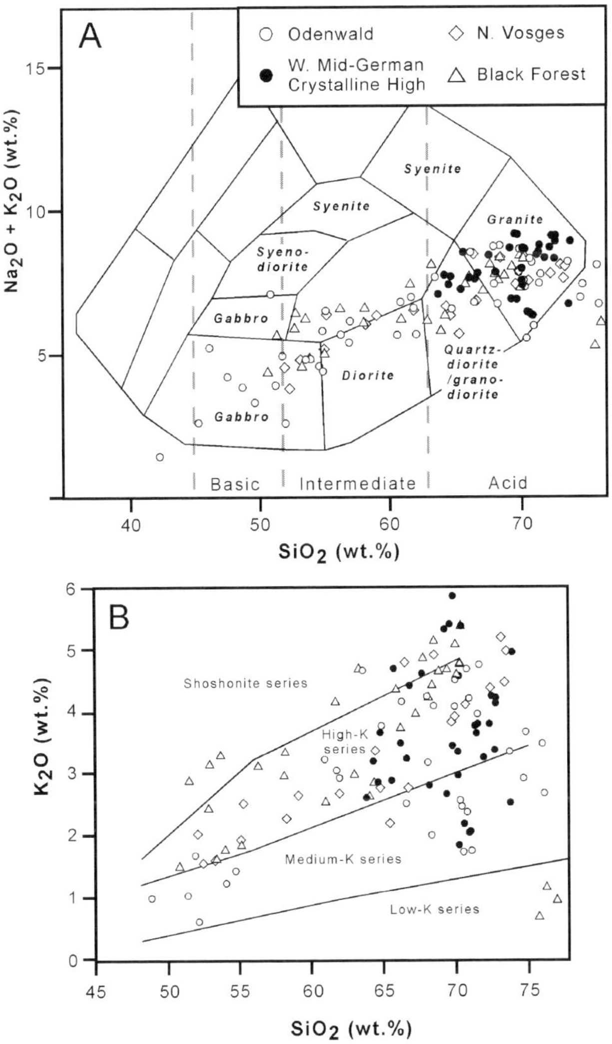

Fig. 12.60. (**A**) Total-alkalis versus silica (TAS) classification diagram for the early to mid-Carboniferous gabbro to granite intrusions in the western Odenwald, western Mid-German Crystalline High (Rhine Graben area), northern Vosges, and Black Forest. (**B**) SiO_2 versus K_2O diagram (Peccerillo & Taylor 1976). Data from Alexandrov *et al.* (2001), Altenberger *et al.* (2001), Altherr *et al.* (1999*a*, *b*, 2000), Cocherie *et al.* (2004), Hann *et al.* (2003) and Laue & Reischmann (1994).

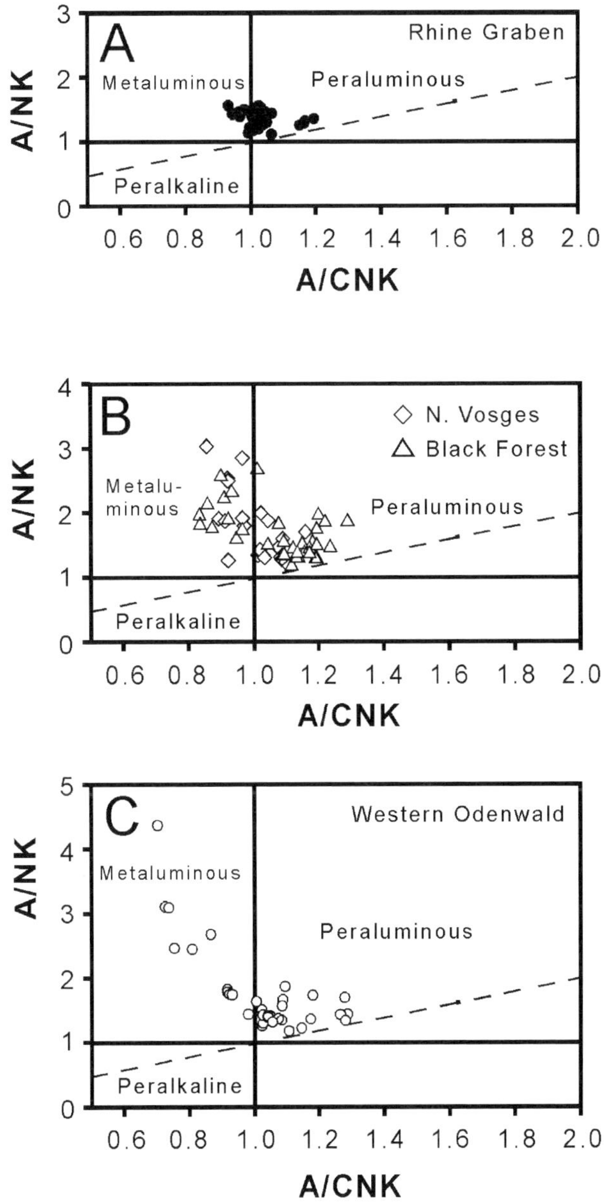

Fig. 12.61. A/CNK versus A/NK diagram for the early to mid-Carboniferous granitoid intrusions in: (**A**) the western Mid-German Crystalline High (southern Pfalz); (**B**) northern Vosges and Black Forest; and (**C**) western Odenwald (Maniar & Piccoli 1989). For field labels see caption to Figure 12.5. For data sources: see caption to Figure 12.60.

parent melts of the felsic rocks in Unit III were probably derived from metagreywacke sources (Altherr *et al.* 1999*a*).

As has been noted above, the Frankenstein Gabbro Complex (Unit I) was probably formed in a late Devonian oceanic arc setting, some time before basin closure and arc accretion in mid-Carboniferous times, and pre-dates the magmatic activity in Units II and III by *c.* 20 Ma. In tectonic discrimination diagrams, the magmatic suites in Units II and III plot in the volcanic arc and syncollisional granitoid fields, overlapping closely with granitoid rocks of similar age in the southern Pfalz, northern Vosges and northern Black Forest (Fig. 12.62). In addition, for the early Carboniferous-age rocks, Altherr *et al.* (1999*a*) noted a trend from more subduction-related older rocks (Neunkirchen suite) to the more syncollisional granitoid-like younger suites that intruded thickened crust. The western Odenwald may, therefore, represent the deeper parts of an accretionary complex or a continental magmatic arc.

Spessart Complex

The southernmost parts of the Spessart Complex are occupied by post-tectonic diorites and granodiorites (Fig. 12.46) of calc-alkaline, I-type arc character, with *c.* 330 Ma U–Pb zircon crystallization ages (Anthes & Reischmann 2001). These mark the end of Variscan tectonometamorphic events, as corroborated by the 318–324 Ma K–Ar cooling ages for hornblende, muscovite and biotite throughout the Spessart Complex (Nasir *et al.* 1991).

Ruhla Complex

The Ruhla Complex in the Thuringian Forest contains mid-Carboniferous and younger granites exposed in an early Permian

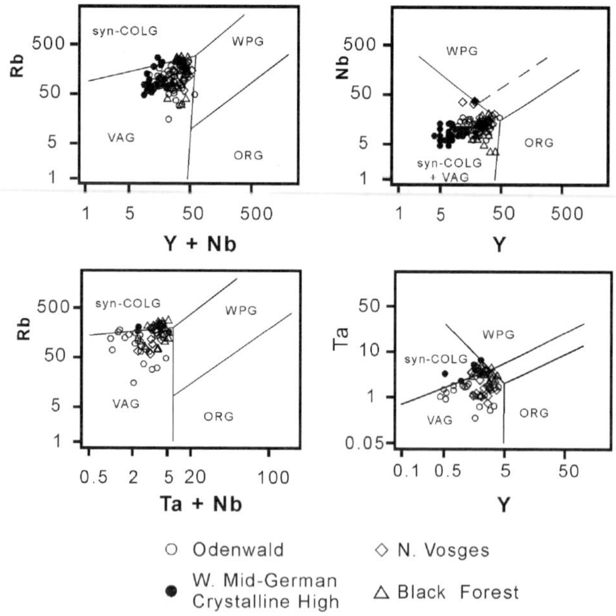

Fig. 12.62. Tectonic discrimination diagrams for the early to mid-Carboniferous granite intrusions in the Odenwald, western Mid-German Crystalline High (southern Pfalz), northern Vosges, and Black Forest (Pearce *et al.* 1984). For field labels see caption to Figure 12.4. For data sources: see caption to Figure 12.60.

horst and basin system, and surrounded by late Carboniferous to early Permian continental basins (Fig. 12.63). It is made up of a Central Gneiss Unit flanked by the metasedimentary Truse and Ruhla formations to the SE and NW that contain amphibolite-facies metasediments and amphibolites. In addition, the Ruhla Fm contains intercalated granitic orthogneisses and underwent a penetrative greenschist-facies metamorphic overprint. The Brotterode Fm to the NE comprises amphibolite-facies migmatitic metasediments and (in the west) metamorphosed mafic rocks with NE-trending planar fabrics. The gneisses of the Central Gneiss Unit were deformed into a dome-shaped structure (Fig. 12.63; Zeh *et al.* 2000). All units were intruded by I-type granites and diorites that were emplaced at shallow depths and associated with contact metamorphism. Deformation during and after intrusion resulted in varying degrees of foliation. Lamprophyre, dolerite and granite porphyry dykes cut all of the lithologies, and some of these are also foliated. The youngest ages of cores of detrital zircons from Brotterode Fm metasediments indicate protolith deposition after 460 Ma. The ages of zircon overgrowths in metasediments and zircons from a discordant granite dyke indicate that high-grade metamorphism and anatexis took place between 366 $\pm$ 6 and *c.* 359 Ma, in late Devonian to early Carboniferous times. A Visean (337 $\pm$ 4 Ma) zircon evaporation age for the foliated Thuringian Hauptgranit and the 340 $\pm$ 1 Ma zircon age for deformed metadiorite in the Brotterode gneisses show that deformation continued under lower grade conditions into Visean times. Younger ages of 289 $\pm$ 3 to 298 $\pm$ 2 Ma were obtained for the Trusetal and Ruhla granites and the Brotterode diorite (Zeh & Brätz 2000, 2004). The youngest granite porphyry dyke has a zircon age of 277 $\pm$ 7 Ma (Zeh & Brätz 2000). ^{40}Ar/^{39}Ar biotite cooling ages for metasediments and intrusions range from 337 to 323 Ma in the eastern segment to 288–298 Ma in the western segment, and reflect differential uplift histories of the Ruhla Complex (Zeh *et al.* 2000, 2003; Zeh & Brätz 2004).

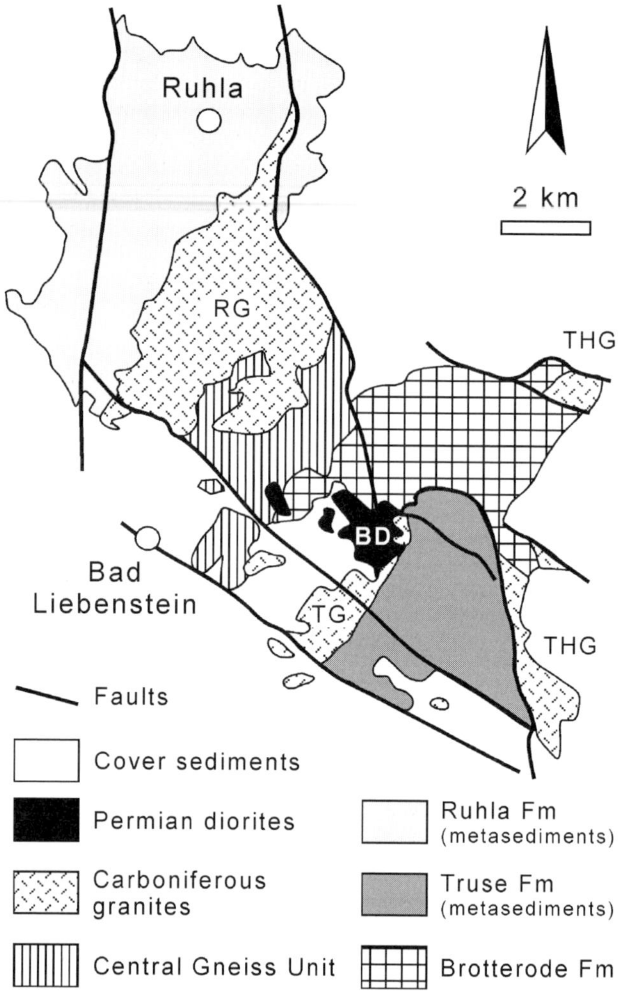

Fig. 12.63. Geological sketch map of the Ruhla Complex (after Zeh *et al.* 2003). Abbreviations: BD, Brotterode diorite; RG, Ruhla granite; THG, Thuringian Hauptgranit; TG, Trusetal granite.

Kyffhäuser and associated areas

The Kyffhäuser Complex forms a WNW-trending horst, *c.* 1 km^2 in size, in the northeasternmost part of the Mid-German Crystalline High (Fig. 12.24). Its western segment contains metasediments, hornblende gabbro (locally pegmatitic) and diorite gneisses. The centre is made up of a granite–diorite–syenite suite, and the east segment is made up of foliated granite. U–Pb zircon and monazite ages show that diorite and gabbro in the west crystallized at *c.* 341 and 345 Ma, while in the central and eastern part crystallization occurred at *c.* 337 Ma (Zeh *et al.* 2005). However, granodiorite and hornblendite have younger (326 and 329 Ma) zircon evaporation ages (Anthes & Reischmann 2001).

Pretzsch-Prettin and Dessau areas

In the Pretzsch-Prettin and Dessau areas (Fig. 12.24) to the east of the Kyffhäuser Complex and to the north and NW of the Torgau-Doberlug syncline, a large variety of granitoids were encountered in boreholes. The I-type granites have yielded 336 to 328 Ma zircon evaporation ages (Hammer *et al.* 1999; Anthes & Reischmann 2001), and gabbro from the Dessau area has a 355 $\pm$ 7 Ma zircon evaporation age (Hammer *et al.* 1999).

However, amphibole gabbro in the southern part of the Prettin Massif has yielded Sm–Nd mineral isochron ages of 491 ± 36 and 484 ± 30 Ma (Cambro-Ordovician) (Hammerschmidt *et al.* 2003). The rocks were not deformed during Variscan events and show a large variation in modal mineralogy, ranging from gabbro, granodiorite, tonalite, quartz diorite and quartz monzonite to granite. The dominant type is biotite granodiorite. Many of the granitoid intrusions contain xenoliths of biotite-, amphibole- and spinel-bearing hornfels, and locally of amphibolite, gneiss, quartzite and schist (Kopp & Röllig 1996).

Moldanubian Zone

High-grade metamorphism of Visean age (*c.* 340–350 Ma) is also typical of the Moldanubian Zone to the south of the Mid-German Crystalline High, and has been recognized in the Vosges, Black Forest and especially in the Bohemian Massif, where two of the largest intrusive complexes are situated (Fig. 12.1). The associated magmatism has a large compositional range, but is dominated by granite, (grano)diorite to syenite intrusions (e.g. Kalt *et al.* 1994*a*; Kröner *et al.* 2000*a*; Hegner *et al.* 2001; Kober *et al.* 2004).

The high-grade Moldanubian Zone contains large amounts of latest Devonian to (mainly) early Carboniferous, peraluminous (i.e. S-type, crust-derived) granites and granodiorites and subordinate volumes of peraluminous I-type (subduction-related) granitoids (e.g. Finger *et al.* 1997; Schaltegger 1997). Magmatic rocks of mafic composition occur in much smaller volumes. In the northern Vosges and northern Black Forest (Figs 12.1 & 12.24), early Carboniferous granite–granodiorite suites intruded accretionary wedges that formed due to the accretion of Armorica and related terranes to Laurussia. These are made up of stacked units of lower Cambrian (locally, uppermost Neoproterozoic) to early Carboniferous shelf to slope sediments, and are often interbedded with calc-alkaline lavas and tuffs. Older, Cadomian basement gneisses appears to be absent in the Vosges and Black Forest.

Black Forest

The Black Forest has been subdivided into several lithotectonic units: from north to south these are the Baden-Baden Zone, the Central Gneiss Complex (which is the largest unit), the Badenweiler-Lenzkirch Zone and the Southern Gneiss Complex, all of which are intruded by early Carboniferous granitoid intrusions (Fig. 12.64).

The Baden-Baden (-Gaggenau) Zone has been overthrust by the Central Gneiss Complex to the south along the Baden-Baden thrust-wrench fault. This fault, or fault zone, correlates with the Lalaye-Lubine Fault in the northern Vosges to the SW and is generally considered to separate the Saxothuringian Zone to the north from the Moldanubian Zone to the south. This fault zone comprises greenschist- to amphibolite-facies metasediments and amphibolites (metamorphosed lavas and shallow intrusions). The protoliths of the metasediments were deposited in shallow- to deep-water environments, and poorly preserved acritarchs suggest an upper Cambrian to lower Ordovician age of deposition (Montenari *et al.* 2000; Montenari & Servais 2000).

The largest unit of the Black Forest is the Central Gneiss Complex consisting of orthogneisses, migmatitic metasediments, and high-temperature mylonites. In addition, eclogites and high-pressure/high-temperature granulite-facies rocks occur in several subunits of the complex (e.g. Kalt *et al.* 1994*b*; Marschall *et al.* 2003). Tonalitic, trondhjemitic and granodiorite orthogneisses from the northern part yielded zircon protolith crystallization

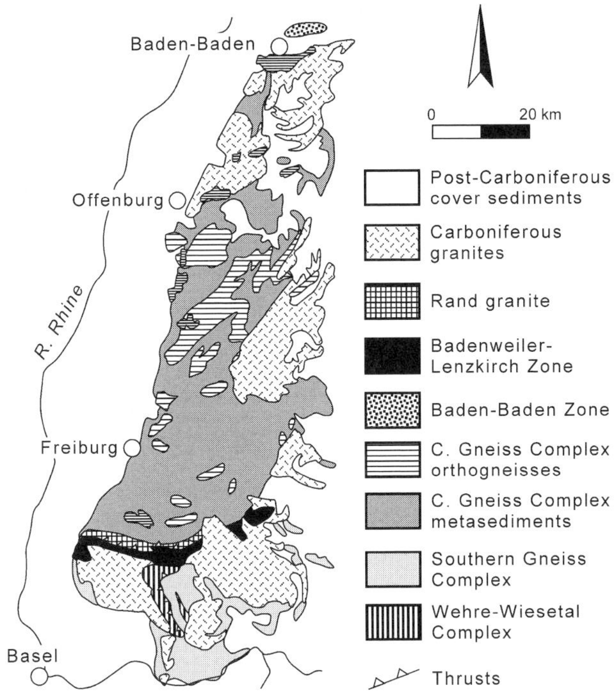

Fig. 12.64. Geological sketch map of the Black Forest (after Montenari *et al.* 2000; Schaltegger 2000; Kober *et al.* 2004).

ages of 500–510 Ma (mid- to late Cambrian). The gneisses have slightly peraluminous compositions ($Al_2O_3/(CaO + Na_2O + K_2O)$ of *c.* 1.1), K_2O/Na_2O ratios of *c.* 1, distinct negative Nb, Hf and Ti anomalies (compared to primordial mantle), are LREE-enriched (La_N/Yb_N *c.* 12) and have $\varepsilon_{Nd}(t)$ values of −3.4 to −0.1; Chen *et al.* (2000) suggested that the protoliths were formed in an early Palaeozoic magmatic arc. The eclogites with N-MORB-like compositions yielded younger (Ordovician) U–Pb zircon ages of 445 to 486 Ma, interpreted as the crystallization ages of their protoliths that were perhaps emplaced during a later, early Palaeozoic rifting stage (Chen *et al.* 2003). Metasediments from drill core from the northernmost parts of the Central Gneiss Complex yielded graphitized acritarchs and chitinozoans that suggest that their protoliths were deposited in a marine environment sometime between the lower Ordovician and upper Devonian (Hanel *et al.* 1999). Geochronological evidence from detrital zircons from the high-grade metasediments indicates deposition in early Palaeozoic times, commencing after the middle Cambrian and ending in the Ordovician. Late Ordovician detrital zircons from metasediments from the southwestern part of the complex indicate even younger depositional ages. Furthermore, the lack of Mesoproterozoic ages in the older detrital zircon populations suggests a West African affinity for this part of the Variscan basement (Kober *et al.* 2004). U–Pb ages of monazite and zircon from metasediments date the regional high-grade, low-pressure/high-temperature amphibolite-facies metamorphism and anatexis at 342 and 330 Ma (Kalt *et al.* 1994*a*; Kober *et al.* 2004).

The Central Gneiss Complex is thrust upon the narrow Badenweiler-Lenzkirch Zone to the south, which in turn is thrust upon the Southern Gneiss Complex further south (Fig. 12.64). The Badenweiler-Lenzkirch Zone comprises phyllites and Upper Devonian to Lower Carboniferous marine sediments that are interbedded with basaltic to andesitic volcanic rocks. Poorly preserved chitinozoans and acritarchs suggest Ordovician and middle to upper Silurian deposition ages of the phyllite protoliths (Hann *et al.* 1995; Montenari *et al.* 2000; Vaida *et al.* 2004). These rocks are overlain by a late Visean basin filled with continental sediments, lavas and pyroclastic rocks and, furthermore, are cut by rhyodacitic to rhyolitic dykes.

The Southern Gneiss Complex is composed of amphibolite-facies, migmatized meta-igneous and metasedimentary rocks that were metamorphosed in Visean times (341–343 Ma zircon ages; Chen *et al.* 2003). The sedimentary protoliths were deposited in Neoproterozoic to Cambrian times, and in the Ordovician, as indicated by palynological evidence (Hann & Sawatzki 1998; Vaida *et al.* 2004). The complex is overthrust by the Badenweiler-Lenzkirch Zone and the early Carboniferous Wiesetal-Wehra Nappe, the latter containing early Carboniferous arc-related gabbros and migmatites (see below; Schaltegger 2000; Hegner *et al.* 2001).

The Baden-Baden (-Gaggenau) Zone, the Central and Southern gneiss units, the Badenweiler-Lenzkirch Zone and the Wiesetal-Wehra Nappe are intruded by pre- to post-tectonic granitoid suites of early to mid-Carboniferous age (e.g. Hann *et al.* 2003; Schaltegger 2000; Figs 12.60A, B & 12.64). Those in the north Vosges and northern Black Forest were generated in a subduction-to-collision setting (Fig. 12.62), the result of southward subduction of the Rhenohercynian Ocean beneath Saxothuringian and Moldanubian crust. Subduction ended with basin closure, accretion, collision and nappe emplacement resulting in crustal thickening and regional, high-grade (low-pressure/high-temperature) metamorphism and anatexis. This occurred in Visean times, as shown by the U–Pb zircon and monazite ages of 342–330 Ma from metasediments, amphibolites and a late aplite vein (Kalt *et al.* 1994*a*; Hegner *et al.* 2001; Kober *et al.* 2004). Thrusting of the Wiesetal-Wehre Nappe onto the Southern Gneiss Complex took place sometime between 342 Ma and 332 Ma (Hegner *et al.* 2001). $^{40}Ar/^{39}Ar$ hornblende ages for the Black Forest indicate cooling below *c.* 500°C at *c.* 330 ± 30 Ma (Lippolt *et al.* 1994).

Carboniferous rocks of mafic composition are rare in the Vosges, but the Wiesetal-Wehra Nappe in the southern Black Forest contains subduction-related, metamorphosed mafic rocks that crop out in tectonic windows. Here, migmatitic granodioritic gneisses, high-grade mylonites, a layered mafic complex and locally preserved Lower Carboniferous sediments and volcanic rocks, were thrust southwards onto migmatitic gneisses, amphibolites and metarhyolites. Metagabbro and meta-anorthosite from the layered mafic complex yielded U–Pb zircon ages of 349–350 Ma, which was shortly followed by migmatization at 345–349 Ma. The gabbros and anorthosites of the layered mafic complex formed in a suprasubduction zone setting, as is clear from their negative Nb, Ce and Ti anomalies and high $\varepsilon_{Nd}(t)$ value (+6.2). The amphibolites have calc-alkaline, andesitic compositions, negative Nb, Hf and Ti-anomalies, but negative $\varepsilon_{Nd}(t)$ values (−5.9 to −3.0), indicating crustal contamination in a more continental arc setting (Hegner *et al.* 2001).

Later, S-type granites in the southern Black Forest intruded in a post-collisional and post peak-metamorphic, intracontinental setting during a Visean extensional event (Fig. 12.62; Schaltegger 2000). U–Pb zircon and monazite ages for this area range from 340 Ma for a rhyolite lava and 332 Ma for a discordant porphyry dyke in the Badenweil-Lenzkirch Zone, to 334–332 Ma granites that intrude the Southern Gneiss Complex. The ages for these post-tectonic rocks indicate that the main Variscan tectonometamorphic events had ended before the mid-Visean. In addition, the 332 Ma age for a rhyolitic ignimbrite deposited in a caldera situated in the SW part of the Central Gneiss Complex indicates that the high-grade, mid-crustal rocks were exhumed to surface conditions before the end of the Visean (Schaltegger 2000). Lamprophyre dykes intruded the Wiesetal-Wehra Nappe and the Southern Gneiss Complex during a stage of dextral transtension. They have $^{40}Ar/^{39}Ar$ plateau ages of 311–332 Ma and constitute the latest phase of Variscan magmatic activity in this area (Hegner *et al.* 1998).

Vosges

The Vosges Massif is subdivided into three units: the northern Vosges, the Southern Vosges Basin, and the central Vosges (Fig. 12.65). The northern Vosges unit is made up of low-grade, northeastward imbricated volcanosedimentary sequences of late Neoproterozoic to early Carboniferous ages that were intruded by Carboniferous granitoids (Fig. 12.65). Its northern margin is overlain by early Permian sediments and volcanic rocks and to the south it is separated from the central Vosges by the Lalaye-Lubine Fault, which is a strike-slip fault that post-dates the high-grade metamorphism in the central Vosges (Fig. 12.65; Reitz & Wickert 1989; Hess *et al.* 1995; Altherr *et al.* 2000). The northern Vosges is correlated with the Baden-Baden Zone of the northern Black Forest, and both units are considered to be

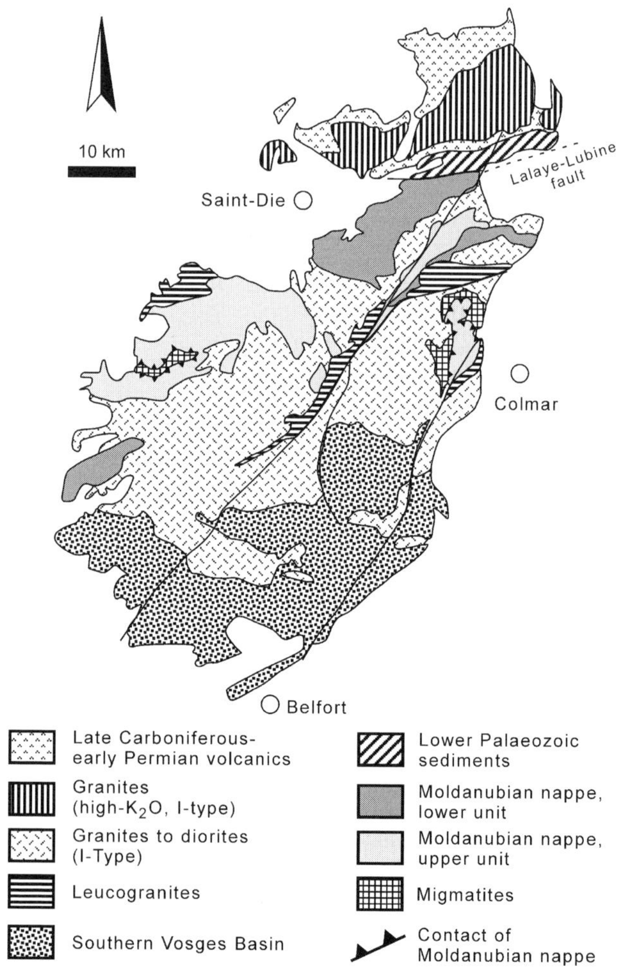

Fig. 12.65. Geological sketch map of the Vosges (after Rey *et al.* 1992; Schaltegger 2000).

fragments of a telescoped active continental margin belonging to the Saxothuringian Zone.

The central Vosges is made up of high-grade metamorphic rocks that have been subdivided into two lithotectonic units: a lower unit of amphibolite-facies, migmatitic, Devonian to early Carboniferous metasediments, separated by synkinematic granites from an upper unit of higher-grade, granulite-facies felsic gneisses containing eclogites and garnet peridotites. Rey *et al.* (1992) attributed juxtaposition of the two lithotectonic units and the intervening granite to late Visean, southwestward extension that caused near-isothermal decompression and rapid exhumation, and which was coeval with emplacement of the Carboniferous anatectic granites. Granulite-facies felsic gneiss ('leptynite') and leucosomes from migmatitic metasediments from the northern part of the central Vosges yielded zircon and monazite ages of 335 ± 4, 335 ± 4 and 337 ± 4 Ma, indicating a very short timespan for the high-grade metamorphism (Schaltegger *et al.* 1999). The amphibolite-facies, retrograde metamorphism and associated extension, uplift, decompression and anatexis were dated at 328 ± 5 Ma for a recrystallized felsic granulite, 326 ± 5 Ma for a cordierite-bearing, S-type granite, and 329 ± 2 Ma for a leucogranite intrusion from the same general area (conventional and SHRIMP zircon and monazite U–Pb ages; Schaltegger *et al.* 1999; Schulmann *et al.* 2002).

The Southern Vosges Basin comprises a succession which commences with upper Devonian (Fammenian) marine shales and turbiditic greywackes, and continues with early Carboniferous clastic sediments and volcanic rocks. This volcanosedimentary basin was intruded in the mid-Carboniferous by granites and subordinate amounts of monzodiorite and diorite of high-K, calc-alkaline character (Fig. 12.60A, B). Interbedded volcanic rocks appear at the beginning of the late Visean, during a phase of accelerated subsidence. The oldest (bimodal) volcanic rocks are submarine basalt flows and dykes of calc-alkaline, tholeiitic composition, and minor volumes of altered dacitic to rhyolitic pyroclastic rocks. Basaltic andesites to andesitic basalts occur higher up the stratigraphy within marine sediments. The youngest volcanic rocks are trachyandesites, dacites, rhyolites and trachytes associated with marine to terrestrial sediments (Lefèvre *et al.* 1994; Schaltegger *et al.* 1996). U–Pb zircon ages for rhyolite flows, dykes and ignimbrites range from 345 to 340 Ma, and the granites that intrude the sequences have 340–342 Ma zircon ages. Therefore, both the sediments and the volcanic rocks were deposited over a period of a few million years during a short-lived, Visean extension event that also exhumed the high-grade gneisses of the central Vosges (Rey *et al.* 1992; Schaltegger *et al.* 1996).

In the northern Vosges and northernmost Black Forest, several early to mid-Carboniferous granitoid suites have been distinguished based on cross-cutting relations and geochemistry, of which the older ones are dioritic, followed by granodiorites, younger high-K granites, and ending with a suite of peraluminous granites. Intrusion depths decrease with age, from *c.* 10 km for the older diorites to *c.* 3 km for the high-K granites. The older, dioritic to granodioritic suites are of I-type character and the youngest granite suite is of S-type (Altherr *et al.* 2000).

Apart from the latest, peraluminous S-type granites, all older suites in the northern Black Forest and northern Vosges have calc-alkaline compositions, and the aluminosity index indicates metaluminous to peraluminous compositions (Fig. 12.61B). The oldest diorite suite contains end-members with Mg numbers up to 62, and is richer in LILE, Nb and P than the younger suites. The diorites have relatively flat REE patterns with negative Eu anomalies, Sr initial ratios of 0.7046 to 0.7061 and negative $\varepsilon_{Nd}(t)$ values of −2.8 to −7.8. Altherr *et al.* (2000) suggested a lithospheric mantle source for this suite. The later granodiorites have higher $\varepsilon_{Nd}(t)$ values (−0.4 to +0.5), similar Sr_i ratios (0.7051–0.7053) and relatively flat HREE patterns. The granite suite that post-dates the granodiorites has similar negative $\varepsilon_{Nd}(t)$ values (−2.5 to −1.8) and similar Sr_i ratios (0.7048–0.7058) to the diorite suite, but the HREE patterns are flat. The high-K granites have increased LILE, Nb and P contents, are LREE-enriched (La_N/Yb_N = 11.8 to 38.9), but have $\varepsilon_{Nd}(t)$ values (−3.4 to −1.5) and Sr_i values (0.7046–0.7060) that overlap with those of the older suites. The REE patterns and correlations between K_2O, Na_2O, MgO and FeO(total) suggest that the older granite suite was derived by dehydration melting of meta-igneous rocks (metabasalt, metatonalite), whereas the younger suites were derived from metasediments. Thus, not all of the suites are cogenetic. Although their initial Sr isotope ratios have a limited range (0.7045 to 0.7061), the Nd isotopes indicate different source materials, probably reflecting formation in a continental arc or a post-collisional setting (Altherr *et al.* 2000). The younger granite suite belongs to the high-K to shoshonite series, as can be seen in the K_2O versus SiO_2 diagram (Fig. 12.60B) and has a more S-type geochemical character. The concave HREE patterns and Sr–Nd isotope signature of the younger granites are similar to those of the older suites, but the former have higher LILE, La, Nb and P, and lower Y and HREE abundances.

Bohemian Massif

Subsequent to the early Carboniferous high-grade Variscan events, voluminous amounts of late- to post-tectonic, Carboniferous to early Permian granitoids intruded the central, southern and western parts of the Bohemian Massif (Fig. 12.1). These range from the monzonitic Meissen intrusion in the Elbe Zone (Fig. 12.3) and large, composite batholiths such as the Central Bohemian and Moldanubian plutonic complexes, to suites of smaller-sized intrusions in the west and north parts of the Teplá-Barrandian Terrane, the NW Moldanubian Terrane, the Erbendorf-Vohenstrauss Zone and the Saxothuringian Terrane in the NW and northern parts of the Bohemian Massif (Fichtelgebirge and Erzgebirge areas; Fig. 12.3).

Central Bohemian Plutonic Complex

The Central Bohemian Plutonic Complex is a large (3200 km^2) intrusive complex that was formed by the multiple intrusion of plutons and dykes of mainly felsic to intermediate compositions (Fig. 12.66). Together with the younger Moldanubian Plutonic Complex to the south, it is one of the largest intrusive complexes in the Variscan Orogen, and has been interpreted as the exposed cross-section of an early to mid-Carboniferous, episodically constructed magmatic arc. A large variety of rock types has been recognized, ranging from gabbro, diorite, quartz-monzonite, tonalite, trondhjemite and granodiorite, to granite, melanocratic syenite and melagranite (durbachite; Fig. 12.67A). The individual intrusions differ in shape, volume, age, modal and geochemical composition, internal fabrics and their relations to the fabrics in the host rocks. The various intrusions have been grouped into several cognate suites: the Sázava, Blatná, Čertovo břemeno, and Říčany suites (Fig. 12.66). Small volumes of late, S-type granites, leucogranites (Maršovice Suite) and minette dykes are also present. High-K, shoshonitic suites constitute the largest volumes of the Central Bohemian Plutonic Complex (Janoušek *et al.* 1995; 2000*a*, *b*). The crystallization ages range from *c.* 345 Ma for the earliest intrusions (e.g. Sázava Pluton) to *c.* 337 Ma. Thus, the magmas were emplaced over a period of *c.* 20 Ma

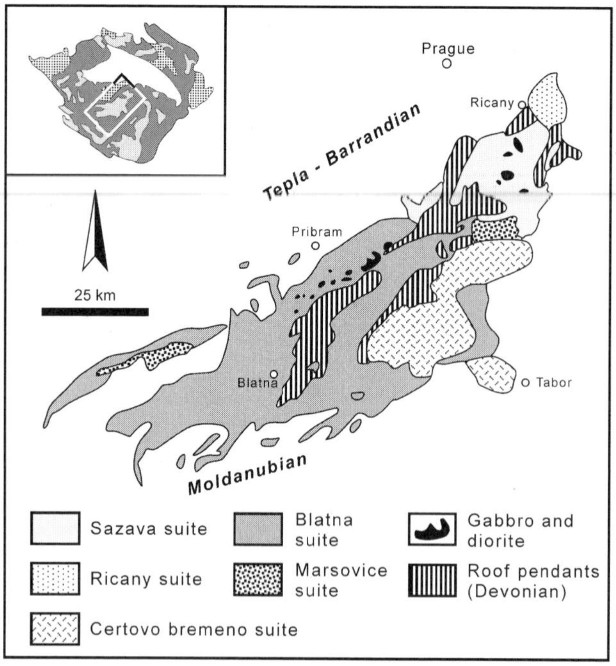

Fig. 12.66. Geological sketch map showing the magmatic suites of the Central Bohemian Plutonic Complex (after Janoušek *et al.* 2000*b*).

during, and immediately following, the peak of Variscan orogenic activity (Holub *et al.* 1997; Janoušek *et al.* 2000*b*; Žák *et al.* 2005*b*).

The complex intruded between the low-grade Teplá-Barrandian Terrane to the NW and the high-grade Moldanubian Terrane to the south. The age data suggest that the complex grew in a southwestward direction. The early intrusions occur in the north and are calc-alkaline (Figs 12.67A & 12.68B), while the later intrusions in the SE have potassic to ultrapotassic compositions (Fig. 12.67A). The earliest suite intruded upper crust during regional, arc-perpendicular compression and arc-parallel strike-slip movements (i.e. transpression), as is clear from internal fabrics of the Sázava Pluton (Žák *et al.* 2005*a*, *b*). Large, stoped blocks, kilometre-scale roof pendants and interpluton screens of host rocks, in combination with low-pressure contact metamorphism indicate emplacement at shallow crustal levels and, furthermore, show that the present level of erosion is that of the roof of the complex. The roof pendants consist of late Neoproterozoic to early Palaeozoic metasediments and metavolcanic rocks of the Teplá-Barrandian Terrane. The centre is occupied by mylonitized tonalitic to granodioritic orthogneisses of the *c.* 80 km long Staré Sedlo Complex that have zircon protolith ages of *c.* 373–365 Ma (Devonian) (Fig. 12.66; Košler *et al.* 1993). The later plutons occupy the southern part of the complex and intruded mid-crustal, Neoproterozoic metasediments of the Moldanubian Terrane. Here, emplacement during extensional movements coeval with exhumation is indicated by mineral stretching lineations in granodiorite. Intrusive contacts between the various successive stages point to multiple emplacement processes between plutons and even within individual plutons (Žák *et al.* 2005*b*). The complex was in erosional position by latest Carboniferous times, as its NE part is unconformably overlain by Stephanian–early Permian limnic sediments and conglomerates containing granitoid clasts.

The earliest Sázava and Požáry intrusions in the northern part of the Central Bohemian Plutonic Complex are high-level, syntectonic, sheet-shaped to irregularly shaped intrusions composed of quartz diorite, tonalite, granodiorite, gabbrodiorite and gabbro that contain mafic microgranular enclaves. The two intrusions yielded zircon ages of *c.* 349 and 351 Ma, respectively; they are fractionated (low Cr and Ni contents) and have calc-alkaline, metaluminous compositions (Figs 12.68A, B). The intrusions also have low K_2O, Rb, Ba and Zr contents, high Sr, Na_2O and CaO contents, and are moderately enriched in LREE ($Ce_N/Yb_N = 4.5$ to 7.2), with negative to positive Eu anomalies. Their negative Nb and Ta anomalies (compared to mid-ocean ridge granites) and low HFSE concentrations indicate formation in a subduction-related setting, or an origin from subduction-modified sources (Janoušek *et al.* 2000*b*). The Sázava Pluton is an I-type intrusion of hybrid origin (e.g. mafic enclaves), with relatively primitive major and traces element compositions, initial Sr ratios of *c.* 0.75, and bulk-earth-like ε_{Nd}(350 Ma)

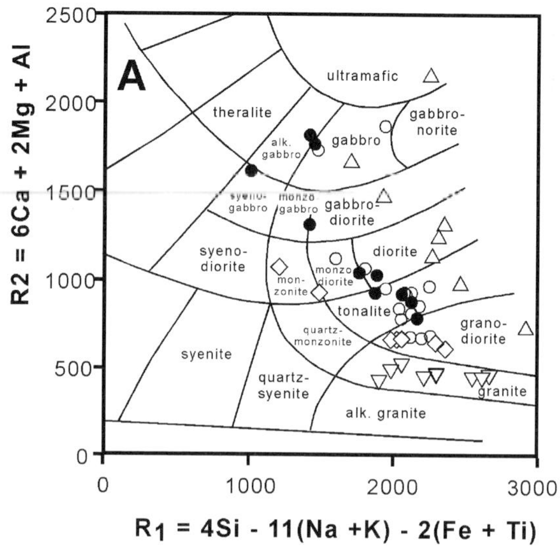

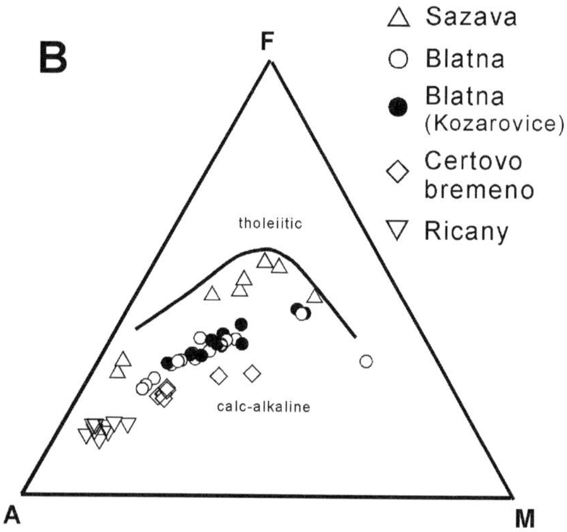

Fig. 12.67. The main intrusive suites of the Central Bohemian Plutonic Complex. (**A**) Classification diagram based on millication proportions (De la Roche *et al.* 1980). (**B**) AFM diagram; curve separating tholeiitic from calc-alkaline field after Irvine & Baragar (1971). $A = K_2O + Na_2O$; $F = FeO_{total}$; $M = MgO$. Data from Janoušek *et al.* (1997, 2000*a*, *b*).

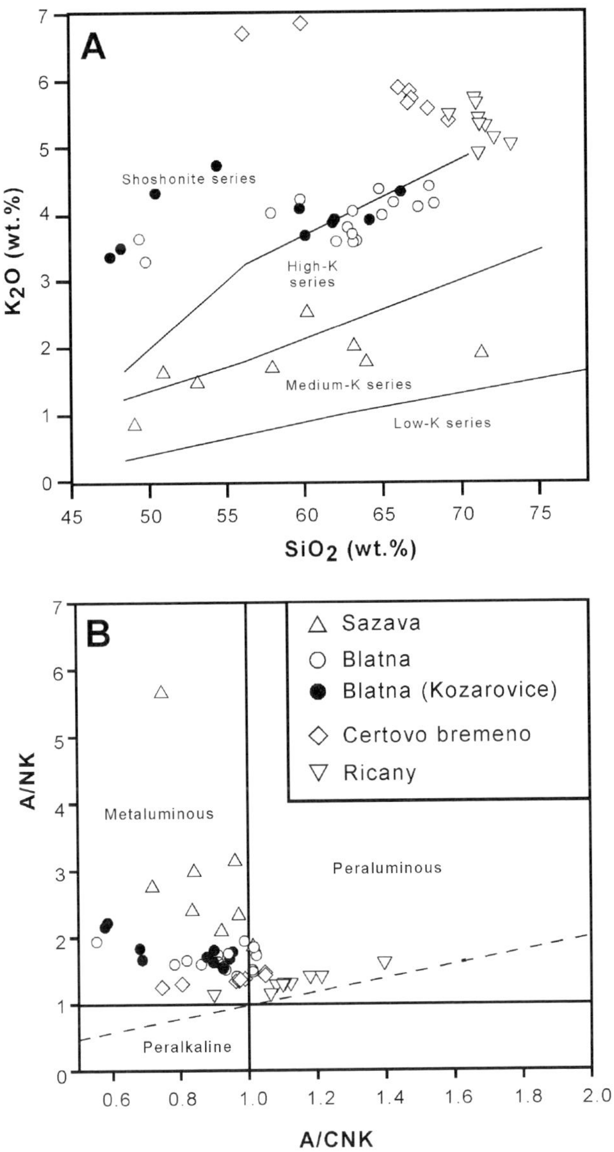

Fig. 12.68. (**A**) SiO_2 versus K_2O diagram (Peccerillo & Taylor 1976), and (**B**) A/CNK versus A/NK diagram for the main intrusive suites of the Central Bohemian Plutonic Complex (Maniar & Piccoli 1989). For abbreviations see caption to Figure 12.5. Data from Janoušek *et al.* (1997, 2000*a*, *b*).

values, and is the isotopically least evolved suite showing evidence for significant mantle input. The younger Blatná Suite to the SE is composed of granodiorite, granite and monzonite that yielded *c.* 349 to 341 Ma zircon ages and contains host rock xenoliths and surmicaceous enclaves (Janoušek *et al.* 2000*b*). The rocks contain a transpressional fabric in the NW part, which is absent from its SE margin where an extensional magmatic to subsolidus fabric developed. Therefore, the onset of exhumation of the deeper parts of the arc probably occurred during intrusion of this suite (Žák *et al.* 2005*b*). The rocks have calc-alkaline, metaluminous to peraluminous compositions (Figs 12.68A, B) and, compared with the Sázava Suite, have higher K_2O and HFSE contents, steeper REE patterns (Ce_N/Yb_N = 7.2 to 23.2) and moderately negative Eu anomalies. Mixing or contamination by Teplá-Barrandian metasediments is considered unlikely and the source of the Blatná Suite was different from that of the Sázava Suite (Janoušek *et al.* 1995). The *c.* 343 Ma Čertovo břemeno Suite comprises porphyritic granite, melagranite and melanocratic syenite that contain host rock xenoliths, and mafic microgranular and surmicaceous enclaves. The rocks are metaluminous to peraluminous, are enriched in K_2O, P_2O_5, Rb, Zr and Ba (grading to shoshonitic/ultrapotassic compositions, Fig. 12.68A), have high Cr and Ni contents and the highest HFSE concentrations of all suites. The younger Říčany Suite of porphyritic granite also contains host rock xenoliths, mafic microgranular and surmicaceous enclaves, is of peraluminous, shoshonitic composition and is strongly enriched in LREE (Ce_N/Yb_N = 15 to 20), with negative Eu anomalies of varying magnitude. Its HFSE contents are intermediate between those of the Blatná and Čertovo břemeno suites, and Moldanubian metasediments may have been its peraluminous source. The youngest intrusions comprise cordierite-bearing, peraluminous two-mica granodiorites (e.g. Maršovice Suite) and small intrusions and lamprophyre and granite porphyry dykes (Janoušek *et al.* 1995, 2000*b*).

The Sr_i ratios of the intrusions range from 0.7050 to 0.7120 and ε_{Nd}(350 Ma) values from +0.4 to −8.7, and each suite clusters around distinct Sr–Nd values. These reveal a negative correlation, from the bulk-earth-like (i.e. chondritic) composition of the Sázava Suite to the negative ε_{Nd}(350 Ma) values and elevated Sr_i ratios of the Říčany Suite. This, however, cannot be interpreted as a simple contamination trend since the geochemistry of the various suites indicates that they were not formed by a single, closed-system process but were derived from different sources and were modified by varying degrees of fractional crystallization and open-system processes, such as assimilation and magma mixing (Janoušek *et al.* 1995, 2000*b*). The earliest, Sázava Suite has typical I-type characteristics and the small volumes of granites that form the youngest phase are typically peraluminous, S-types, but the bulk of the granitoids have compositions intermediate between I- and S-type (Janoušek *et al.* 2000*b*). The K-enrichment trend from the earliest to the later stages may reflect the evolution from a continental magmatic arc (Sázava Suite) to a post-collisional setting (the shoshonites). The majority of the intrusions of the Central Bohemian Plutonic Complex appear to have been derived from lithospheric mantle sources that were variably enriched during subduction processes. Subduction may have taken place in the Devonian, as indicated by the presence of calc-alkaline Devonian orthogneisses of the Staré Sedlo Complex, and the evidence for Devonian blueschist-facies metamorphism in the West Sudetes to the NE (see above). Alternatively, subduction and mantle enrichment may have occurred much earlier, in late Neoproterozoic to early Palaeozoic times, and the arc signature in both the Devonian orthogneisses and the Sázava Suite may be an inherited feature (Janoušek *et al.* 2000*b*). The peraluminous granitoids of the Maršovice Suite contain no mantle-derived components and were probably formed through anatexis of Moldanubian metasediments (Janoušek *et al.* 1995).

Moldanubian Plutonic Complex

The composite Moldanubian Plutonic Complex (also known as South Moldanubian Batholith or Pluton) in the southern part of the Bohemian Massif forms the largest magmatic complex in the Variscan Orogen (Fig. 12.69). It comprises a variety of granitoids of mid- to late Carboniferous age that intruded Moldanubian metasediments over a period of *c.* 30 Ma (Holub *et al.* 1995, and references therein). The intrusions in the north and east have sharp contacts with their host rocks, but deeper levels are exposed in the

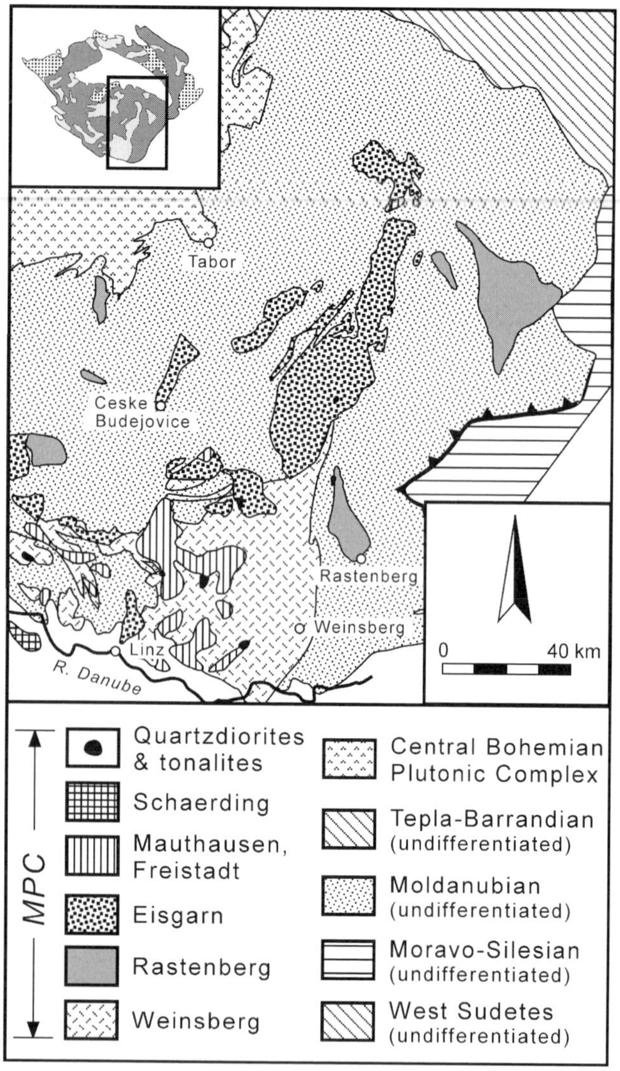

Fig. 12.69. Geological sketch map showing the magmatic suites of the Moldanubian Plutonic Complex (MPC) (after Fiala *et al.* 1995; Holub *et al.* 1995; Vrána *et al.* 1997).

west where magmatic rocks often grade into anatectic gneisses. Several types have been distinguished based on mineral contents and geochemical compositions. The Weinsberg type, which is the most widespread, is a coarse-grained, alkali feldspar megacrystic granite with accessory muscovite and, in places, garnet. Zircon ages of *c.* 355 Ma for the Weinsberg Pluton were interpreted as dating the partial melting of the source rocks, and *c.* 345 Ma zircon ages as dating emplacement (Klötzli *et al.* 2001), although Friedl *et al.* (2004) obtained a younger zircon crystallization age of 322 ± 4 Ma. The Rastenberg-type is melagranitic, but ranges in composition from granodiorite to diorite, trondhjemite and biotite granite, and contains hornblende, rare pyroxene, large to very large alkali feldspar phenocrysts and abundant gabbroic to dioritic enclaves and host rock xenoliths (Gerdes *et al.* 2000). Klötzli & Parrish (1996) obtained a 338 ± 2 Ma zircon age for the Rastenberg Pluton. Younger, post-tectonic types comprises occur as fine- to medium-grained biotite granites (Mauthausen, Schrems types) and two-mica granites with alkali feldspar phenocrysts (Eisgarn type). These were emplaced as stocks, sills and dykes and are locally associated with pegmatites, aplites, lamprophyres and may be mineralized (Vellmer & Wedepohl 1994).

The various granitoids have peraluminous to metaluminous compositions, although most have elevated Ba, Th, U, La, Ce, Rb, Nd, Sr and K contents and are LREE-enriched (La_N/Yb_N = 19 to 60). S-type granites are part of the complex, but most intrusions have both S- and I-type characteristics, although true I-types appear to be absent (Liew *et al.* 1989; Vellmer & Wedepohl 1994). The Rastenberg types have high K_2O contents (including ultrapotassic quartz monzonites), lack or have moderately negative Eu anomalies, have $^{87}Sr/^{86}Sr$(330 Ma) values between 0.7073 and 0.7082, and negative ε_{Nd}(330 Ma) values between −5.2 and −4.2 (Gerdes *et al.* 2000). The granitoid varieties of the Rastenberg Pluton were probably derived by melting of mafic sources that were previously metasomatized by subduction processes or, alternatively, represent incompletely homogenized mixtures of ultrapotassic, lithospheric mantle-derived melts and felsic, crust-derived melts (Liew *et al.* 1989; Vellmer & Wedepohl 1994; Gerdes *et al.* 2000; Gerdes 2001). In many respects they are similar to the 'redwitzites' of the NW part of the Bohemian Massif (see below). The Weinsberg type has a near-haplogranitic composition and elevated K_2O, Rb, Li, U, Th, Ba and Sr contents. Vellmer & Wedepohl (1994) suggested it was derived from a tonalitic source. Similar tonalitic sources were suggested for the younger Mauthausen and Schrems types.

The late Eisgarn-type granitoids are of S-type with high Sr_i (0.7123) and low ε_{Nd}(320 Ma) value (−7.6) and are, furthermore, strongly enriched in LREE, showing pronounced negative Eu anomalies (Vellmer & Wedepohl 1994). The fractionated HREE abundances indicate that garnet acted as a residual phase in the source, probably a metasediment. The initial Sr ratios of most types range from 0.7076 to 0.7090, and ε_{Nd}(320 Ma) values are relatively constant, falling in the range of −5.6 to −4.8. The isotopic compositions suggest that none of the parent melts were derived from mantle material, but that all had a crustal origin. Furthermore, over time, increasingly shallow sources were involved, ranging from lower crustal, incompatible element-enriched mafic sources for the Rastenberg types (such as underplated mantle-derived material), to mid-crustal tonalitic material for the Weinsberg and Mauthausen types, and metapelites for the late Eisgarn types. The heat that triggered these melting events may have come from mantle melts that underplated the Moldanubian crust (Vellmer & Wedepohl 1994).

Western and northern Bohemian Massif

In the Bavarian Forest, Oberpfalz, Fichtelgebirge and Erzgebirge areas (Fig. 12.70), granitoid emplacement took place in multiple stages as indicated by crystallization ages of *c.* 350 Ma to 300 Ma (Rb–Sr whole-isochron and U–Pb zircon) (Loth *et al.* 1999; Chen & Siebel 2004; Propach *et al.* 2000; Siebel *et al.* 1997, 2003, 2005; Kempe *et al.* 2004). Intrusions in the Oberpfalz, Fichtelgebirge and Erzgebirge areas have traditionally between subdivided into several petrographic types and subtypes that broadly correspond with their emplacement order (G1 to G4). The four distinct phases of granitoid magmatism were locally preceded by a phase of mafic plutonic rocks (e.g. Siebel *et al.* 1997). The latter are gabbros and diorites with *c.* 320–325 Ma crystallization ages that constitute the earliest post-tectonic magmatic phase (Siebel *et al.* 1997, 2003). These are restricted to the region east of the Erbendorf-Vohenstrauss Zone (Fig. 12.70) and the area near Marktredwitz, hence the local term 'redwitzites' for this rock type (Fig. 12.70). These redwitzites were cut by intrusions belonging to the 345–320 Ma 'Older Granites' (or 'Older Intrusive Complex') that often form relatively large intrusions and that may partly overlap in time with the redwitzites (Köhler & Holzl 1996; Zulauf *et al.* 2002; Chen *et al.* 2003; Siebel *et al.* 1995, 2003). Most of the G1 type granitoids belong to the Older Granites suite (e.g. B, Mu, Ba, WM, L, Zh). The 'Younger Granites' (or 'Younger Intrusive Complex') comprise a group of 315–310 Ma, smaller-sized granite intrusions (Bae, Fa, Fl, Fr, Li, M, Roz, St; Breiter & Siebel 1995; Siebel *et al.* 1995, 2003), and latest Carboniferous to early Permian (305–295 Ma) granites that are mainly restricted to the Fichtelgebirge and Erzgebirge areas (G2 to G4 types; e.g. K, P, Ei). Some of the intrusions are composite or zoned (Hecht *et al.* 1997; Chen & Siebel 2004). The Fürstenberg complex (Fig. 12.70) is an example of a composite intrusion, where 332–334 Ma diorites were followed by two suites of coarse-grained granite at 321–324 and 312–316 Ma (zircon evaporation ages; Chen & Siebel 2004).

The granite intrusions generally have sharp contacts with their country-rocks, cross-cutting the regional foliation, and some contain xenoliths of host rocks and/or enclaves of restite material. Contact metamorphic aureoles are generally present. Like the Sázava Suite of the Central Bohemian Plutonic Complex, some of the older magmatic rocks in the southern Bavarian Forest intruded synkinematically with respect to Visean to Namurian movements along the Bavarian Pfahl Shear Zone, such as the *c.* 325 to 334 Ma granitic to granodioritic 'palite' dykes (Fig. 12.70) and the Weinsberg-type granitoids further east (Büttner 1999; Siebel *et al.* 2005, 2006). In the north, magmas intruded the Saxothuringian–Moldanubian terrane boundary and are undeformed (Steinwald, Flakenberg, Friedenfels, Fig. 12.70), but those near the boundary between the Teplá-Barrandian and Moldanubian terranes are sheared by Variscan movements on the West Bohemian Shear Zone (Mutěnín, Drahotín, Babylon, Fig. 12.70; Zulauf *et al.* 2002). The intrusion shapes range from batholithic, such as the Falkenberg and Bor granites, to lens, sill or dyke-like shapes. Gravity modelling of the Oberpfalz intrusions shows that most trend north to NW, are wedge-shaped and constrained by steep faults, and extend to depths of 3 to 8 km. Deeper erosion levels are exposed in the west and east (e.g. Leuchtenberg and Bor, Fig. 12.70), whereas the granitoids in the central part are still largely buried (e.g. Flossenbürg, Rozvadov, Fig. 12.70; Trzebski *et al.* 1997).

The so-called redwitzites occur in relatively small volumes and are predominantly quartz (monzo-)diorites, but actually range in composition from gabbroic and noritic to (grano-) dioritic. Locally, they intruded as sill-like bodies (e.g. Zainhammer, Reuth-Erbendorf, Tirschentreth-Mähring, Wurz-Ilsenbach, Marktredwitz, Fig. 12.70). The rocks vary in their modal compositions, but are generally composed of olivine, orthopyroxene, biotite, plagioclase and hornblende phenocrysts set in a fine-grained matrix of clinopyroxene, andesine, oligoclase, hornblende, microcline and quartz. Hybrid rocks occur in the contact zones between the diorite and granites, reflecting the partial assimilation of granitic material by more mafic melts. Holl *et al.* (1989) suggested that the redwitzites formed by fractionation of olivine and augite, followed by post-magmatic reactions that transformed olivine and plagioclase into hornblende and clinopyroxene, and augite into hornblende. The redwitzites have calc-alkaline, metaluminous, I-type compositions; ε_{Nd}(325 Ma) values range from −4.4 to +0.2 (Holl *et al.* 1989; Siebel *et al.* 1995). Anomalously old, early Palaeozoic Rb–Sr whole-rock 'errorchrons' and Sr–Nd isotope trends indicate that the rocks are isotopically heterogeneous, and that fractionation did not significantly influence the isotopic compositions. Sr–Nd mixing trends indicate that the rocks were formed by mixing of a mafic, depleted mantle-derived melt with crustal material or crust-derived granitic melts (Holl *et al.* 1989).

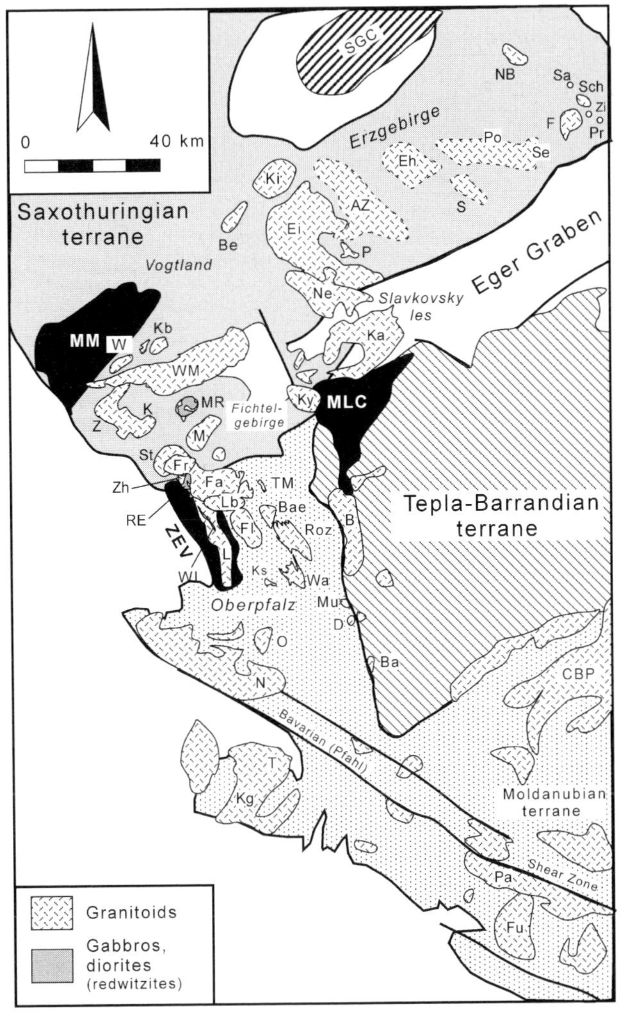

Fig. 12.70. Simplified map showing the Carboniferous to early Permian diorite (redwitzite) and granite intrusions in the Erzgebirge/Krušné hory, Fichtelgebirge, Vogtland and Oberpfalz areas (western, NW and northern Bohemian Massif). Abbreviated names of the intrusions: A, Altenberg; AZ, Aue-Schwarzenberg (comprising the Schlema-Alberoda, Lauter, Bernsbach and Beierfeld granites); B, Bor; Ba, Babylon; Bae, Bärnau; Be, Bergen; CBP, Central Bohemian Pluton; D, Drahotín; Ei, Eibenstock; Eh, Ehrenfriedersdorf; F, Fláje; Fa, Falkenberg; Fl, Flossenbürg; Fr, Friedenfels; Fu, Fürstenberg; K, Kösseine; Ka, Karlsbad; Kb, Kornberg; Kg, Kristallgranit; Ki, Kirchberg; Ks, Kreuzstein/Křížový kámen; Ky, Kynžvart; L, Leuchtenberg; Lb, Liebenstein; M, Mitterteich; Mu, Mutěnín; MR, Marktredwitz; N, Neunburg; Ne, Nejdek; NB, Niederbobritzsch; O, Oberviechtach; P, Podlesí; Pa, palite dykes; Po, Pobershau; Pr, Preisselberg; RE, Reuth-Erbendorf; Roz, Rozvadov; S, Satzung; Sa, Sadisdorf; Sch, Schellerhau; Se, Seifen; St, Steinwald; T, Trasching; TM, Tirschenreuth-Mähring; W, Waldstein; Wa, Waidhaus; WI, Wurz-Ilsenbach; WM, Weissenstadt-Marktleuthen; Z, Zentral; Zh, Zainhammer; Zi, Zinnwald. Other abbreviations: EVZ, Erbendorf-Vohenstrauss Zone; MLC, Mariánské Lázně Complex; MM, Münchberg Massif, SGC, Saxonian Granulite Complex. After Siebel *et al.* (1997, 2005), Hecht *et al.* (1997), Breiter *et al.* (1999), Förster *et al.* (1999) and Chen *et al.* (2003).

The 'Older Granites' in the Bavarian Forest, Oberpfalz, Fichtelgebirge and Erzgebirge areas form larger, often composite, internally differentiated intrusions of coarse-grained, alkali feldspar-porphyritic (monzo-)granites that are locally associated with (grano-)diorites and equigranular two-mica granitic differentiates.

The late- to post-kinematic granitoids associated with the Pfahl Shear Zone (Bavarian Forest) have ε_{Nd}(325 Ma) values ranging from −5.3 to −4.3 (Siebel *et al.* 2006). Further north, the Weissenstadt-Marktleuthen granite has ε_{Nd}(325 Ma) values ranging from −5.7 to −0.8, while those of the Zainhammer and Leuchtenberg granites range from −3.6 to −1.4 (Siebel *et al.* 1995). Their transitional S- to I-type character, intermediate Sr_I ratios and relatively high $\varepsilon_{Nd}(t)$ values suggest mixtures of partial melts derived from old crust and mantle-derived melts.

The 'Younger Granites' (G2 to G4 types) were emplaced at shallower levels as is clear from thermobarometry studies on contact metamorphic mineral parageneses and silicate melt inclusions in quartz, topaz, apatite and alkali feldspar (Siebel *et al.* 1997; Thomas & Klemm 1997). The first indicate that there was up to 6–8 km of uplift between the older and younger phases of magmatic activity, while the latter studies suggests uplifts of between 3 and 6 km (from 2 to 0.8 kbar) with temperatures concomitantly decreasing from 740 to 560°C (Thomas & Klemm 1997). The younger granitoids are equigranular to (locally) slightly porphyritic, are more differentiated and have peraluminous, S-type compositions. The $\varepsilon_{Nd}(t)$ values of the younger granitoids in the Fichtelgebirge and Oberpfalz are lower (−8.4 to −4.1; Holl *et al.* 1989; Siebel *et al.* 1995) and Sr_i ratios are higher than in the older suite; they were derived by melting of crustal material (Holl *et al.* 1989; Siebel *et al.* 1995, 2003). For example, the large amount of inherited, undissolved older zircons, peraluminous character, negative Ba, Nb, Sr, Eu and Ti anomalies, negative $\varepsilon_{Nd}(t)$ values (−6.2 to −8.2) and strong HREE fractionation (i.e. presence of restite garnet) of the 320 Ma old Neundorf and Oberviechtach granites have been interpreted by Chen *et al.* (2003) to indicate derivation from metasediments (shales, greywackes) by low-temperature melting in thickened crust (Fig. 12.70).

The Moldanubian metasediments that host the large volumes of granitoids were metamorphosed under low- to medium-pressure, high-temperature conditions between 323 and 313 Ma. This event was probably short-lived and was followed by slow cooling into late Carboniferous times (see above; Kalt *et al.* 2000; Propach *et al.* 2000; Teipel *et al.* 2004). The significant amount of heat required to generate such large amounts of granite melt was probably due to underplating of mantle melts that caused dehydration melting of meta-igneous rocks (such as mafic lower crust) and metasediments; underplating can furthermore explain the slow, post-peak cooling history of the Bavarian Forest and Oberpfalz areas (Kalt *et al.* 2000; Siebel *et al.* 1997). Subsequent differentiation processes in crustal magma chambers can partly account for the strong fractionation and compositional diversity observed both between, and within, granite intrusions. Emplacement occurred during extension along faults that were active from the early to the late Carboniferous (e.g. Behrmann & Tanner 1997; Trzebski *et al.* 1997).

Erzgebirge granites

The late Carboniferous to early Permian granites in the Erzgebirge area intruded much later than the peak of metamorphism, which here occurred in early Carboniferous times (around 340 Ma, see above). The early Visean events reflect continental collision and deep subduction of continental crust that was uplifted to mid-crustal levels within a few million years following the peak of metamorphism (Kröner & Willner 1998; O'Brien 2000; Hwang *et al.* 2001). $^{40}Ar/^{39}Ar$ ages and thermal modelling for the western Erzgebirge indicate that this event was followed by rapid exhumation around 332–325 Ma (Werner & Lippolt 2000; Franke & Stein 2000). The Erzgebirge granites intruded during and subsequent to this later stage, and were locally associated with Namurian-age lamprophyre dykes (Werner & Lippolt 2000; von Seckendorff *et al.* 2004*b*). The gneisses of the Erzgebirge Crystalline Complex and the early Carboniferous granites in the eastern Erzgebirge were in erosional position in Westphalian B/C times, because they are unconformably overlain by sediments and volcanic rocks of the Altenberg-Teplice Caldera (see below). Förster *et al.* (1999) pointed out that the presently exposed basement rocks are isotopically too evolved to have acted as the source of the biotite ± muscovite granites, but that Moldanubian lithologies exposed to the south have appropriate compositions. The Li-rich granites are not genetically related to the other granites by fractional crystallization processes, but may have originated by melting of orthogneisses and metasediments of the Erzgebirge basement (Breiter *et al.* 1999; Förster *et al.* 1999).

Compared to the 'Older Granites', the younger and more fractionated granites (G2 to G4 types) were more affected by high-temperature fluid–rock interaction causing alteration of magmatic biotite to protolithionite and zinnwaldite, and formation of secondary muscovite, sericite, chlorite, fluorite, albite, topaz and tourmaline (e.g. Irber *et al.* 1997; Förster 1998). In the Niederbrobitzsch and Markersbach granites in the SE Erzgebirge, magmatic monazite and xenotime reacted with late-magmatic fluids to form secondary REE-silicates and -carbonates (cerite and synchisite; Förster 2000, 2001).

The youngest, latest Carboniferous to early Permian granites (G4 group) tend to be strongly peraluminous S- to A-types (Fig. 12.71), often contain muscovite and/or Al-rich biotite, and may also contain aluminium-rich minerals such as garnet, cordierite, topaz, andalusite or sillimanite (e.g. Kösseine, Fig. 12.70; Hecht *et al.* 1997; Siebel *et al.* 1997, 2003; Förster *et al.* 1999). The evolved types in the Erzgebirge area form high-level stocks, cupolas and ridges that are locally associated with rare metal mineralization (Nb, Ta, Sn, U, locally W), hydrothermal alteration aureoles and breccia pipes, greisenization, tourmalinization and Li-micas (e.g. Eibenstock, Podlesí Satzung, Fig. 12.70; Štemprok 1967; Štemprok & Seltmann 1994; Breiter 2002; Breiter *et al.* 1997, 1999; Hecht *et al.* 1997; Förster *et al.* 1998, 1999; Kempe *et al.* 2004). The trend towards an increasing degree of fractionation with younger crystallization age has been corroborated by the increasing enrichment in H_2O, F and, in some cases, P in silicate melt inclusions (Thomas & Klemm 1997). In addition to the evolved S-types, transitional I- to S-type biotite and biotite-muscovite granites of similar age are also present in the Erzgebirge area, as indicated by low Sr_i ratios, elevated $\varepsilon_{Nd}(t)$ values and the presence of fine-grained mafic enclaves (Förster *et al.* 1999).

In the Erzgebirge, Bavarian Forest and Oberpfalz the gradual change from mid-Carboniferous age metaluminous/peraluminous granitoids to late Carboniferous–early Permian age, strongly peraluminous, syncollisional and post-orogenic granites emphasizes the increasing role of crustal material and the decreasing contribution of mantle-derived melts in their genesis (Figs 12.71 & 12.72). This is underlined by the increasing initial Sr ratios and decreasing $\varepsilon_{Nd}(t)$ values, which change from 0.705–0.708 and −4 to 0, respectively, for the redwitzites and the older granitoids in the Bavarian Forest and Oberpfalz, to >0.710 and <−3, respectively, for the younger granites (Holl *et al.* 1989; Siebel *et al.* 1995, 1997; Förster *et al.* 1999).

Meissen monzonite-diorite

The Meissen intrusion is situated between the eastern Erzgebirge and the Lusatia/Lausitz Complex (Fig. 12.3), and is predomi-

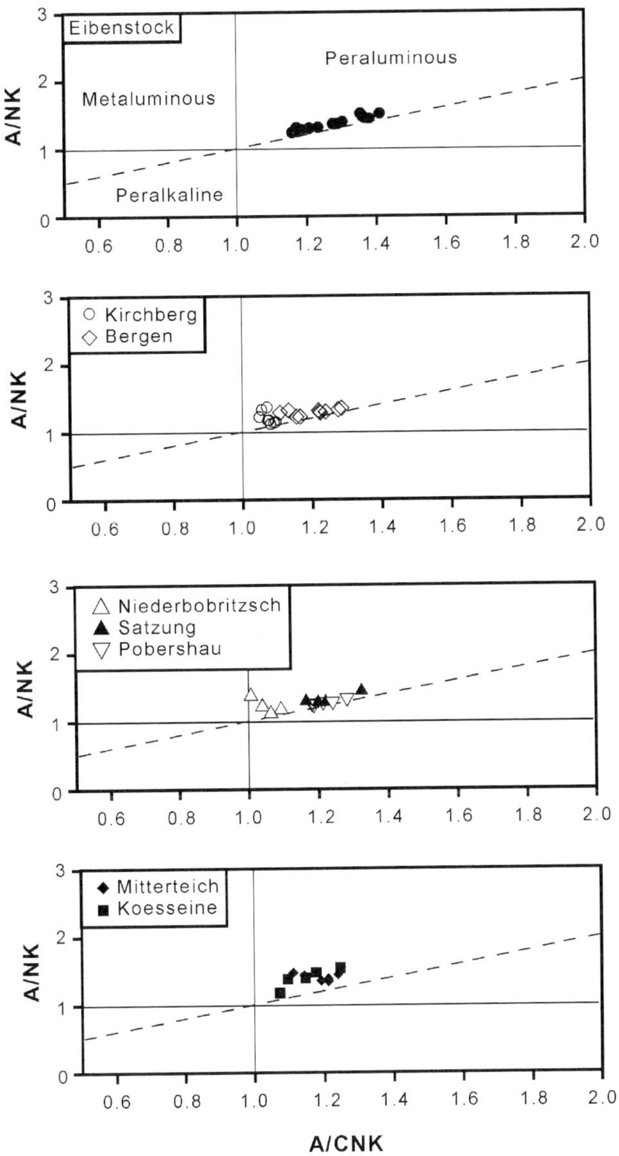

Fig. 12.71. A/CNK versus A/NK diagram for the late Carboniferous granitoid intrusions in the western Erzgebirge and Fichtelgebirge, Saxothuringian Zone (Maniar & Piccoli 1989). For abbreviations see caption to Figure 12.5. Data from Siebel (1995), Förster *et al.* (1999) and Schödlbauer *et al.* (1997).

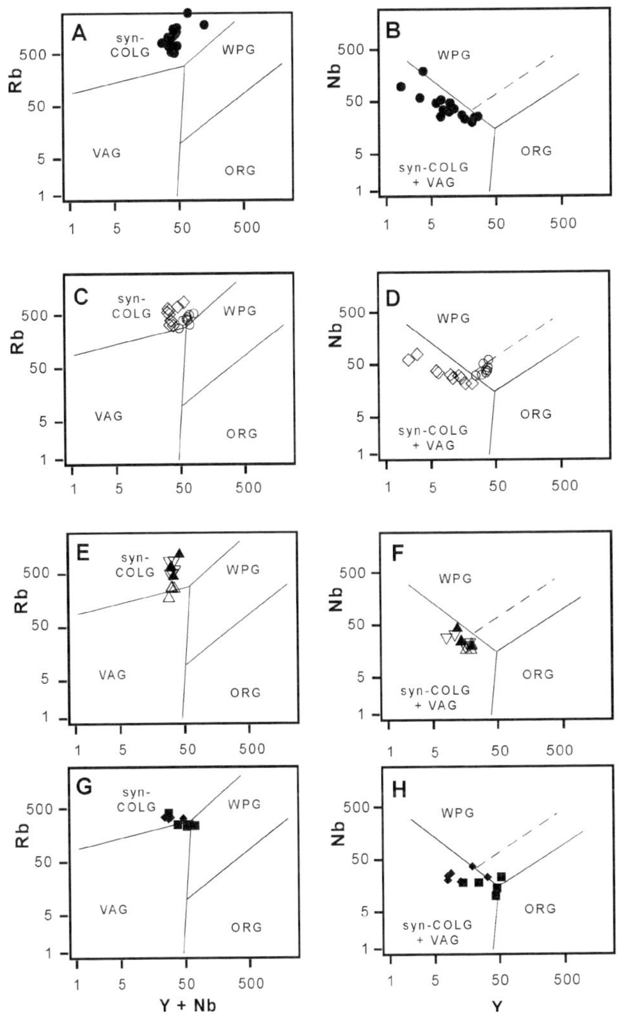

Fig. 12.72. Tectonic discrimination diagrams for the late Carboniferous granitoid intrusions in the western Erzgebirge and Fichtelgebirge, Saxothuringian Zone (Pearce *et al.* 1984). (**A** & **B**) Eibenstock; (**C** & **D**) Bergen and Kirchberg; (**E** & **F**) Niederbobritsch, Satzung and Pobershau; (**G** & **H**) Mitterteich and Kösseine granites. For symbols see Figure 12.71. For field labels see caption to Figure 12.4. For data sources see caption to Figure 12.71.

nantly composed of monzonitic to dioritic rocks that surround a centre of monzogranite, granodiorite and quartz diorite. It has a U–Pb zircon crystallization age of *c.* 326 Ma and similar $^{40}Ar/^{39}Ar$ hornblende and K–Ar muscovite and biotite cooling ages of 334–326 Ma suggesting that is was a high-level intrusion (Kurze *et al.* 1998; Wenzel *et al.* 1997, 2000). The NE part of the intrusion contains xenoliths of metasediments (chlorite and biotite ± muscovite gneisses) and cataclastic granite to granitic mylonites (Coswig Granite). One granite xenolith yielded an early Ordovician (*c.* 480 Ma) zircon evaporation age (Kurze & Mohnicke 2003). Syntectonic deformation fabrics led Mattern (1996) to suggest that the magmas were emplaced during strike-slip movements along older lineaments. The Meissen intrusion is overlain by rhyolitic volcanic rocks of probably latest Carboniferous to early Permian age (Pietsch 1959).

Gradations between the monzonites and the diorites were interpreted to be related to interactions between dioritic chamber wall or floor assemblages and later, highly differentiated melts. In addition, disequilibrium textures, such as gabbroic to monzodioritic mineral assemblages in syenitic groundmass, suggest the intrusion of multiple magma batches. The high K_2O content of the magmatic suite underlines its shoshonitic character (Fig. 12.73); the rocks are, furthermore, strongly enriched in LILE and LREE with attendant depleted Nb and Ti abundances. The high LREE concentrations of the monzonites are a feature of their enriched mantle source, which is supported by initial Sr ratios of 0.705 and ε_{Nd}(330 Ma) values of −1.9 to +0.1 for the diorites, monzonites and mafic enclaves. The most evolved monzonites have more depleted isotope compositions (Sr_i = 0.7044–0.7047, ε_{Nd}(330 Ma) = +1.7 to +2.2; Wenzel *et al.* 2000). The elevated Th, U and Zr contents have been attributed to crustal contamination. Wenzel *et al.* (2000) suggested that the mafic parent melt was derived from a subduction-modified mantle source (i.e. enriched in LILE and LREE). Subsequent fractionation of differ-

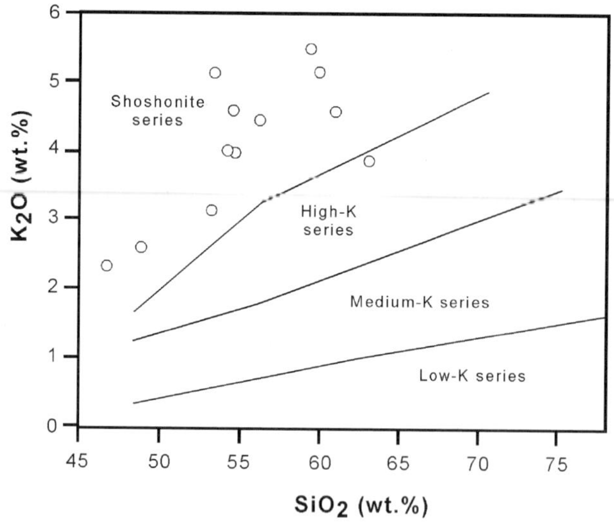

Fig. 12.73. SiO_2 versus K_2O diagram for monzonites and (monzo-) diorites of the Meissen intrusion (Peccerillo & Taylor 1976). Data from Wenzel *et al.* 2000.

ent magma batches during multiple intrusions led to the development of a gabbro–diorite–monzodiorite suite and a monzonite suite. The more depleted monzonite may have originated from less-enriched parts of a heterogeneous mantle, and its formation post-dated melting of enriched, possibly vein-like material from which the parent melt of the more enriched suite formed (Wenzel *et al.* 2000).

Late Devonian – Early Carboniferous: general

Syn- to post-convergent extension during Carboniferous collision was probably accompanied by the removal of dense lithospheric mantle through successive detachment(s) of subducted oceanic crust, and/or the removal of an orogenic root by thermal erosion or break-off (e.g. Schaltegger 1997; Arnold *et al.* 2001; Praeg 2004). Accompanying these processes, part of the lithospheric mantle may have been replaced with asthenospheric material, resulting in a thinned lithosphere of lower average density, uplift and exhumation, melting of the lower crust, and the generation of a variety of magmas that intruded higher crustal levels.

For the western extension of the Moldanubian Zone, the short timespan between the generation and intrusion of arc magmas, metamorphism, uplift, exhumation, cooling, basin formation and sediment deposition is a strong indicator for an active, convergent setting. The high-grade central Vosges probably represent the deeper parts of a magmatic arc, while the Southern Vosges Basin was a backarc basin that formed due to southward subduction of the Rhenohercynian Ocean beneath an ATA-related terrane to the south (e.g. Schulmann *et al.* 2002). The combination of extension, deposition of marine sediments (indicating lack of relief) and rapid exhumation has been explained by oblique compression and synconvergent extension during peak-Variscan accretion and collision processes (e.g. Echtler & Chauvet 1992; Schaltegger *et al.* 1996; Schulmann *et al.* 2002). In the southern Black Forest, accretion and collision occurred between 340 Ma, the age of rhyolite lava in the Badenweiler-Lenzkirch Zone, and 333–334 Ma, the age of post-tectonic granites (Schaltegger 2000). In the Vosges, these events are bracketed by the 342–345 Ma ages of the oldest granite intrusion and rhyolites in the Southern Vosges Basin, and the 326–329 Ma ages for anatectic leucosomes and leucogranite (Schaltegger *et al.* 1996, 1999; Schulmann *et al.* 2002). $^{40}Ar/^{39}Ar$ cooling ages for biotite, muscovite and amphibole mostly fall in the range 331–342 Ma (Boutin *et al.* 1995).

The short-lived and pulsed character of the Carboniferous magmatism argues against an origin by anatexis due only to radiogenic heating in thickened crust, but necessitates advective heat from the mantle. Finger *et al.* (1997) proposed that the granitic intrusions of the Moldanubian Zone were generated in a large-scale Carboniferous active margin setting at the southern margin of the Variscan Orogen. The relative paucity of true subduction-related, pre-collisional I-type granitoids may be explained by the relatively small width of the subducting ocean basins. The widespread and large volumes of high-K_2O granitoids of I- and S-type character were formed during post-collisional melting of lower and mid-crustal material in mid- to late Carboniferous times, triggered by the large heat input from mantle-derived melts. For the intrusions in the western part of the Moldanubian Zone, Schaltegger (2000) suggested that the short timespan between metamorphism and exhumation may have been caused by the removal of dense material by thermal erosion, delamination, or slab break-off after subduction. Furthermore, slab break-off or removal of a crustal root could account for the very rapid uplift and preservation of the ultrahigh-pressure granulites in the Erzgebirge and Saxon Granulite Massif (O'Brien 2000; Arnold *et al.* 2001). In contrast, the absence of decompression features and long, post-peak cooling history of the Moldanubian rocks in the Bavarian Forest area led Kalt *et al.* (2000) to suggest the presence of transient heat sources at depth, such as asthenosphere-derived mafic melts.

Namurian–Westphalian

Around *c.* 328 Ma the northward drift movements of the various Gondwana-derived terranes changed to rapid clockwise rotation by *c.* 70 degrees that finally closed the Rhenohercynian Ocean (Edel 2001). To the north of the Mid-German Crystalline High, ongoing Namurian–Westphalian oblique compression and associated dextral lateral movements caused 'thin-skinned' deformation above northward-propagating sole thrusts in the passive-margin sediments (Cooper *et al.* 1986; Oncken *et al.* 1999). The resulting Rhenohercynian Zone is a low-grade, north-vergent foreland fold-and-thrust belt that extends from SW Ireland into east Germany and the Jeseníky Mountains in the Czech Republic (Fig. 12.1; Holder & Leveridge 1986; Oncken *et al.* 1999, 2000).

As a result of lithospheric flexure due to thrust-wedge loading, a series of foreland basins developed in front of the deformation zone in which subsidence (locally quite rapid) kept pace with infilling by fluviatile/lacustrine clastic sediments and thick coal measures, punctuated by several marine incursions (see McCann *et al.* 2008*a*). This Variscan foreland basin comprises the Culm, South Wales, Namur, Ruhr, Aachen and Upper Silesia basins (e.g. Süss *et al.* 2001). Apart from airfall tuffs of distal provenance (e.g. Hess & Lippolt 1986; Königer & Lorenz, 2002; Königer *et al.* 2002), magmatic rocks of Namurian–Westphalian age are largely absent from the foreland basins and the Rhenohercynian foldbelt. This is probably due to lithospheric flexure which depressed the upper mantle, thus inhibiting decompressional melting. In the internal parts of the Variscan Orogen, small volumes of volcanic rocks are present in the Westphalian basins of western and central Bohemia, and the Altenberg-Teplice Caldera in the SE Erzgebirge; the latter probably represents the largest Westphalian magmatic complex in Central Europe (see below).

Several studies have noted that mid-Carboniferous thrusting and nappe formation in the Variscan Orogen was synchronous with the exhumation of mid-crustal metamorphic rocks and their juxtaposition with sedimentary basins in the internal parts of the orogen. A possible explanation for this is that oblique collision caused lateral escape tectonics and extension within the orogen, resulting in the 'collapse' of thickened Variscan crust (see above). This mechanism occurred along wrench faults and reactivated thrusts, leading to the development of orogen-parallel extension, basin formation and magmatic activity (e.g. Rey *et al.* 1992; Burg *et al.* 1994; Faure 1995; Schaltegger *et al.* 1996). Extension and collapse of Variscan crust may have been diachronous or occurred in several discrete stages, progressing outward (both north and south) from Visean collapse of the internal parts, to early Permian gravity-driven collapse in the Alpine region and the northern foreland (e.g. Lorenz & Nicholls 1984; Schaltegger & Corfu 1995; Schaltegger 1997; Praeg 2004). The resulting geometry is that of a large number of basins with unmetamorphosed and little-deformed infill that occur juxtaposed against metamorphic basement. This geometry has been compared to that of the Basin and Range province in the western USA (e.g. Ménard & Molnar 1988; Burg *et al.* 1994; Faure 1995; Schaltegger & Corfu 1995).

Western and central Bohemian basins

The central and western parts of the Tepla-Barrandian Terrane are overlain by the Kladno-Rakovník, Plzeň/Pilsen, Manětín, Radnice, Žihle, Mšeno-Roudnice and Kravaře basins, which once formed a single basin that drained to the NW towards the Erzgebirge basins and further into the Sub-Variscan Foredeep (Schneider *et al.* 2005). The Bohemian basins are filled with continental sediments and coals, and sedimentation started around the Westphalian A–B boundary (e.g. Hess & Lippolt 1986; Opluštil & Pešek 1998; and references therein; McCann *et al.* 2008). Jelínek *et al.* (2001) distinguished 37 volcanic horizons (tonsteins) that are generally rather altered and of reduced volume. The bulk of volcanic activity took place around the time of deposition of the Radnice Member (Westphalian C), decreasing in intensity towards the Stephanian B. Hess & Lippolt (1986) obtained a 309 ± 3.7 Ma $^{40}Ar/^{39}Ar$ sanidine plateau age for a tuff from the upper part of the Radnice Member (analytical uncertainty on 1-sigma level). Most of the tuffs and lavas have rhyolitic, dacitic to (trachy)andesitic compositions, and tuffs of basaltic to basanitic composition are rare and rather altered (Jelínek *et al.* 2001). The geochemical character ranges from subalkaline to alkaline, but the majority has a transitional character (Ulrych *et al.* 2006).

Altenberg-Teplice Caldera

The Westphalian-age Altenberg-Teplice Caldera (ATC, Fig. 12.74) in the SE Erzgebirge is *c.* 18 by 35 km in size and comprises volcanic and subvolcanic rocks of mainly rhyolitic to rhyodacitic, metaluminous to peraluminous composition (Fig. 12.75B). The southern part of the ATC is cut by the Erzgebirge/Krušné hory Fault (Fig. 12.74) but exploration boreholes show that the ATC extends further southwards under Cenozoic sediments of the Teplice Basin (Benek 1991; Seltmann & Schilka 1995; Breiter 1997; Breiter *et al.* 2001; Müller & Seltmann 2002; Štemprok *et al.* 2003). The rocks of the ATC were deposited on, and intruded, gneisses of the Erzgebirge Crystalline Complex and Carboniferous granites of the Older Intrusive Complex, such as the Fláje granite (Fig. 12.74).

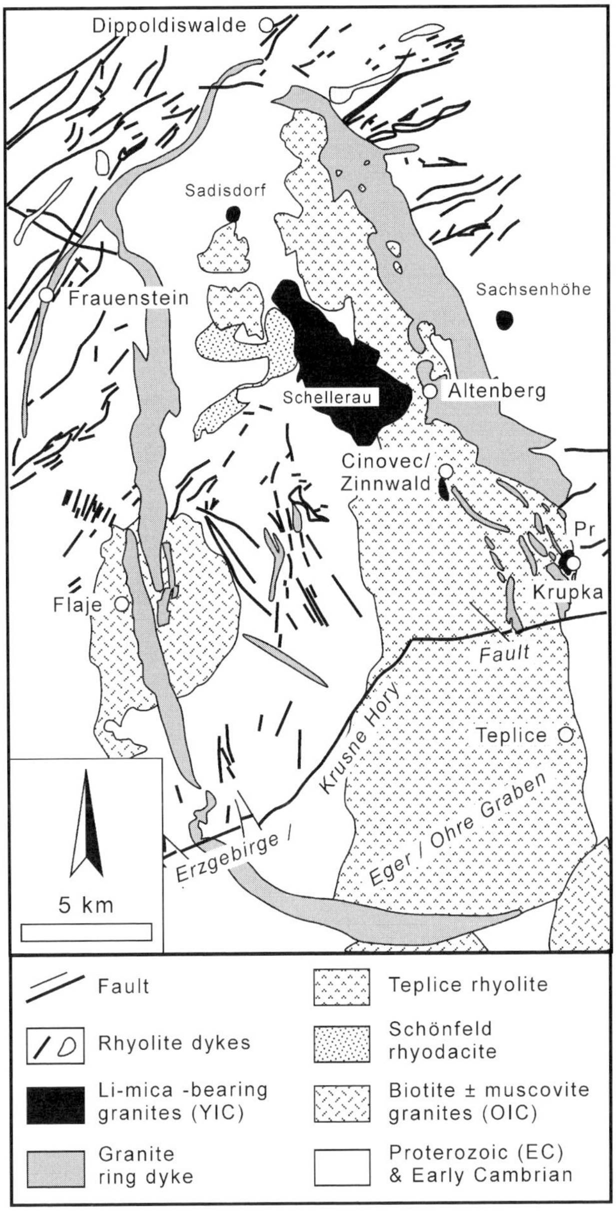

Fig. 12.74. Geological map of the Altenberg-Teplica Caldera, SE Erzgebirge/Krušné hory (after Müller & Seltmann 2002, and references therein). Abbreviations: EC, Erzgebirge Complex; OIC, Older Intrusive Complex; Pr, Preisselberg granite; YIC, Younger Intrusive Complex.

Several magmatic stages have been recognized based on field- and borehole evidence. The earliest stage (Schönfeld unit) is exposed in the NW part of the complex and is encountered in boreholes in the western part. This unit comprises dacitic to rhyodacitic tuffs, ignimbrites and dacite lavas. The volcanic rocks are interbedded with clastic sediments and coal seams containing Westphalian B–C plant remains (Breiter 1997; Breiter *et al.* 2001, and references therein).

These are overlain by the so-called Teplice Rhyolites that erupted in three main stages (TR 1 to 3) with an estimated total volume of 70 to 120 km^3 (Benek 1991). These commence with rhyolite tuffs and ignimbrites (TR 1) that are interbedded with Westphalian C–D sediments and overlain by *c.* 300 m of rhyolite lavas, tuffs, airfall tuffs and ignimbrites of the second stage (TR 2). The youngest volcanic rocks (TR 3) are rhyolite to rhyodacite

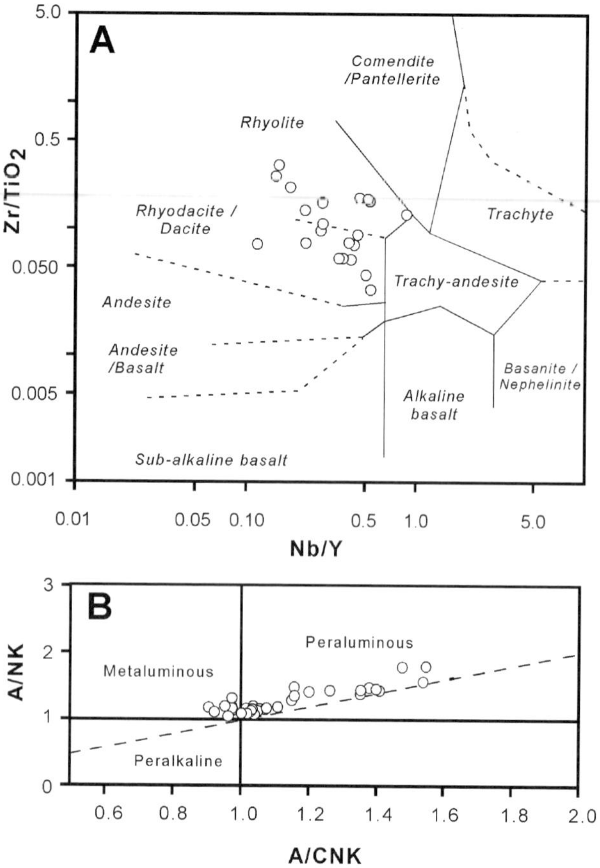

Fig. 12.75. Late Carboniferous volcanic and subvolcanic rocks of the Altenberg-Teplice Caldera (SE Erzgebirge Crystalline Complex). (**A**) Nb/Y versus Zr/TiO_2 classification diagram (Winchester & Floyd 1977). (**B**) A/CNK versus A/NK diagram (Maniar & Piccoli 1989). For abbreviations see caption to Figure 12.5. Data from Breiter (1995), Seltmann & Schilka (1995), Müller & Seltmann (2002), Štemprok *et al.* (2003) and Ulrych *et al.* (2006).

ignimbrites and subvolcanic intrusions that cover the eastern third of the outcrop area and reach cumulative thicknesses of up to 2 km, corresponding to more than 250 km^3 of lava. These represent the peak of volcanic activity and the largest volume of volcanic rocks in the ATC. The relatively thick Westphalian C–D sediments in the north suggest that volcanic activity was interrupted, but boreholes in the SW document nearly continuous volcanic activity and sedimentary intercalations here are thin (maximum 4 m thick; Breiter 1997; Breiter *et al.* 2001).

Volcanic activity ended with caldera collapse along curved, broadly north-trending ring fractures that were filled with varieties of porphyritic microgranite which also intruded the volcanic rocks (Müller & Seltmann 2002). The eastern granite porphyry ring dyke is up to 1 km wide and yielded $^{40}Ar/^{39}Ar$ biotite crystallization ages of *c.* 307–309 Ma (cited in Seltmann & Schilka 1995). Single zircon evaporation ages range from 334 $\pm$ 5 Ma to 290 $\pm$ 5 Ma (Kempe *et al.* 1999). The older, Visean zircon ages contradict the field relations and probably reflect the presence of older, inherited cores. The younger, early Permian zircon ages may be due to hydrothermal overprinting during tin mineralization (Kempe *et al.* 1999). The eruption centre of the ATC is inferred to be located to the east, in the Altenberg-Kesselhöhe area. This is suggested by the eastward thickening of the Teplice Rhyolites, and by the large, metre-scale textural variability in the porphyritic granite (GP) dykes, which are homogeneous elsewhere within the ATC.

Caldera collapse and ring dyke emplacement was followed by multiple intrusions of A-type granites of the Younger Intrusive Complex (e.g. Sadisdorf, Schellerhau, Preiselberk/Preisselberg, Cínovec/Zinnwald granites; Müller *et al.* 2000). The Sachsenhöhe and Markersbach granites are located outside the ATC (Förster 2001). As seen in boreholes and exploration mines, the Preisselberg/Preiselberk granite intruded and thus post-dates the Teplice Rhyolites (Štemprok *et al.* 2003). Lamprophyre dykes and NE-trending rhyolite dyke swarms constitute the closing stage of magmatism of the ATC.

The lowermost volcanic rocks of the Schönfeld unit have calc-alkaline compositions with relatively homogeneous Rb and Zr concentrations, but very variable Sr contents. This contrasts with the younger and more evolved, high-K calc-alkaline TR 1 to TR3 units of the Teplice Rhyolite that show evidence for internal, reverse trace element zoning. The lower parts of each TR unit have relatively high Rb, Th and HREE, and relatively low Sr, Zr and LREE concentrations, and the upper parts are depleted in Rb, Th, HREE and enriched in Sr and LREE. Breiter (1997) and Breiter *et al.* (2001) attributed the reverse zoning to repeated eruption from a magma chamber that was compositionally stratified due to the accumulation of feldspar, mafic and heavy minerals in its lower parts. Eruption produced first the felsic volcanics from the upper parts of the magma chamber, and these were followed by melts of intermediate composition from the lower parts. The parent melt of the Teplice Rhyolites probably originated by low-degree melting at high temperatures of crustal material.

The porphyritic granites of the ring dyke complex have been divided into two intrusive types (GP-I and GP-II) that have sharp intrusive contacts and different textures. The older, GP-I type volumetrically dominates the ring dyke complex and contains alkali-feldspar phenocrysts of which up to 8% are mantled by plagioclase. The GP-II type occurs in the eastern dyke close to Altenberg and is a hornblende-bearing microgranite with alkali-feldspar phenocrysts and hornblende-biotite-plagioclase-rich schlieren. Approximately 60% of the alkali-feldspar is mantled by plagioclase (Müller & Seltmann 2002). Types GP-I and GP-II are fractionated to similar degrees and are enriched in LREE with pronounced negative Eu anomalies; both types have an A-type character. The plagioclase-mantled alkali-feldspars probably formed as a result of disequilibrium processes such as mixing of mafic and felsic magmas. The change from the more fractionated, felsic GP-I granites to the less fractionated GP-II granites was explained by Müller & Seltmann (2002) to be due to continuous deflation from a compositionally zoned magma chamber, of which the middle and upper part must have consisted of non-hybrid granites. Zoning in the magma chamber may have been due to the influx of dense, mafic magma into the lower part, where mixing produced hybrid magmas of type GP-II. The most fractionated type with the most pronounced negative Eu anomaly (GP-I) originates from the silica cap of the magma chamber.

The post-caldera, A-type granites are generally highly evolved (high SiO_2 contents) syenogranites and alkali-feldspar granites of mostly peraluminous, locally metaluminous, composition. Many granites are mineralized ('tin granites') and associated with brittle faults and mineralized breccia pipes (Seltmann 1994), and often underwent strong post-magmatic hydrothermal overprinting (e.g. Seltmann 1994; Förster 2001). Often only the apical parts of buried granites are exposed, such as the Cínovec/Zinnwald intrusion (e.g. Štemprok *et al.* 2003). This is a multiple intrusion of which the upper part contains Li-rich albite granite that passes into moderately Li-enriched syenogranite. A deeper, protolithionite-bearing granite variety was encountered in boreholes. The Cínovec/Zinnwald granites extends in the subsurface southeastwards towards the Preiselberk/Preisselberg granite and a few kilometres beyond, as proven by tin exploration boreholes (Štemprok *et al.* 2003). The Preiselberk/Preisselberg intrusion is composed of biotite syenogranite (that intruded Teplice Rhyolite) and later albite granite containing Li-micas (lepidolite and/or zinnwaldite) and minor amounts of secondary fluorite and topaz. The Schellerhau intrusion forms the largest granite outcrop within the ATC and was emplaced in several stages, including porphyritic syenogranite and zinnwaldite- and protolithionite-bearing, F-rich leucogranite (Müller *et al.* 2000; Štemprok *et al.* 2003).

Stephanian–Early Permian

The youngest sediments deformed by Variscan compression are of Westphalian D age (*c.* 305 Ma) and are found in the foreland basins. These deposits mark the end of convergence in Central Europe. After a regional early Stephanian hiatus, recognized within the Variscan domain and its northern foreland, Stephanian and Early Permian sediments and volcanic rocks were deposited unconformably onto rocks that were deformed and metamorphosed during the Variscan Orogeny, indicating that by this time the orogen had already been deeply eroded (e.g. Flick 1986; Hess & Lippolt 1996). The Stephanian–Early Permian deposits mark the start of the post-orogenic evolution of Central Europe, a setting that was dominated by strike-slip faulting accompanied by alternating transpression and transtension (e.g. Arthaud & Matte 1977; Kroner *et al.* 2008; McCann *et al.* 2008*a*, *b*).

Throughout the internal Variscides and the foreland, extensive dextral wrenching in Stephanian–Early Permian times led to the development of a large number of fault-bounded, volcanosedimentary grabens, half-grabens and narrow troughs. In the internal part of the orogen, these tend to be parallel to the regional structural trend (e.g. Saar-Nahe Basin, Saale Basin, the Crediton Trough in SW England, and the Permo-Carboniferous Basin in northern Switzerland), but in the foreland area the orientations vary. The basins generally tend to be isolated and relatively small, forming in isolation from one another with their own internal drainage systems. Basin formation was associated with differential uplift, horst and graben formation on varying scales, footwall erosion, and magmatic activity. The basins were rapidly filled by (Stephanian) coal measures followed by thick successions of alluvial, fluviatile and lacustrine (i.e. locally derived) continental sediments. Early Permian sedimentation often commenced with characteristic infilling of coarse-grained clastic sediments derived from local topographic highs (i.e. alluvial fans) and deposited in semiarid to arid environments (Rotliegend). In contrast to the Visean–Westphalian basins, marine incursions appear to be absent in the Early Permian basins of Central Europe and they have been interpreted as intermontane basins.

In addition to the widespread layers of distal airfall tuffs (e.g. Hess *et al.* 1983; Hess & Lippolt 1986; Königer & Lorenz 2002; Königer *et al.* 2002), many of the Stephanian–Early Permian basins also contain volcanic rocks. The most significant occurrences are in the Southern North Sea Basin (Aghabawa 1993; Heeremans *et al.* 2004*a*, *b*; Sissingh 2004), the Oslo Graben (e.g. Neumann *et al.* 2004) and the North-German Basin (Hoth *et al.* 1993*a*, *b*; Marx *et al.* 1995; Plein 1995; Benek *et al.* 1996; Kiersnowski & Buniak 2006) to the north of, or straddling the Variscan Deformation Front, and the Saar-Nahe, Saale, North-Sudetic, Intra-Sudetic and Krkonoše Piedmont basins further south (see below). Volcanic activity in the northern Variscan foreland was furthermore accompanied by dyke and sill emplacement, such as the tholeiitic dolerites in the Midland Valley of Scotland (Macdonald *et al.* 1981), the Whin Sill Complex in northern England (Johnson & Dunham 2001), the dykes in Scania, south Sweden (Obst 1999; Obst *et al.* 2004) and the dolerite sills encountered in boreholes in Rügen, north Germany (Kramer 1995).

In many areas volcanic activity was coeval with block faulting, tilting, footwall uplift, and horst and graben formation (Cartwright 1990; Marx *et al.* 1995). The fact that lavas are found interbedded with basin sediments in the internal Variscides indicates that magmatic activity was coeval with subsidence and extension. Furthermore, extension created pathways which facilitated the passage of relatively viscous melts such as rhyolites, allowing them to escape the magma chamber and reach the surface. It has long been recognized that many Stephanian–Early Permian basins are not purely extensional, but show evidence for having formed in a dextral transtensional/transpressional setting. This is the case for basins in both the foreland (e.g. Read 1988; Ro *et al.* 1990; Ziegler 1993; Coward 1993; Mogensen 1994; Marx 1994; Plein 1995; Glennie 1997) and the internal Variscides (e.g. Stets 1990; Ziegler 1993; Faure 1995; Korsch & Schäfer 1995; Büthe & Wachendorf 1997; Mattauer & Matte 1998; Henk 1999; Zeh *et al.* 2000). A dextral strike-slip component (locally causing transtension and transpression) is inferred from basin geometry and orientation, isopach maps of synrift deposits, migration direction of depocentres, the orientations and structural data from synsedimentary faults, simultaneous horst and basin formation, and the arrangement of volcanic centres and dykes.

In the internal, central European parts of the Variscan Orogen, the volcanic rocks are often of felsic–intermediate composition and comprise a significant pyroclastic component in the form of ignimbrites and tuffs. Lavas occur interbedded with sediments showing that magmatic activity took place during subsidence and extension, and reliable U–Pb and $^{40}Ar/^{39}Ar$ ages of magmatic minerals show that this occurred in the period 305–290 Ma (summarized by Timmerman 2004, and references therein). Compositions range from granite and diorite intrusions to rhyolitic/andesitic volcanic and subvolcanic rocks, but mafic rocks are limited to diatremes, small volumes of lavas and tuffs, mafic dykes and (locally) sills. In the high-grade core of the orogen, the basins are localized and stratigraphic correlation between them is difficult (McCann *et al.* 2008*a*). Furthermore, much biogenic material that may have been present has been oxidized in Stephanian–Rotliegend primary red-beds, seriously limiting dating and correlation on biostratigraphic grounds. In contrast, deep boreholes in the NE German Basin in the northern foreland cored up to 2 km successions of latest Carboniferous–early Permian volcanic rocks that erupted in several stages and can be recognized across the region (Benek *et al.* 1996).

Plutonic rocks

Late Carboniferous to early Permian intrusive rocks are mainly of granitic composition and occur in small volumes in the Bohemian Massif, West Sudetes, Erzgebirge, Thuringian Forest, and the NW Harz areas. In the Thuringian Forest and NW Harz areas, granitoids, gabbros and rhyolite sills that are undeformed or only weakly deformed have 289–307 Ma U–Pb zircon crystallization ages (Baumann *et al.* 1991; Breitkreuz & Kennedy

1999; Anthes & Reischmann 2001; Kopp *et al.* 2002; Zeh & Brätz 2000, 2004). The leucogranites of the NW Harz are spatially associated with the *c.* 293 Ma Harzburg Gabbro (Fig. 12.57; Baumann *et al.* 1991). This intrusion ranges in composition from gabbronorite, the dominant rock type, to hybrid gabbro, norite and diorite, and harzburgite cumulates; dunite cumulates are known from drill cores (Sano *et al.* 2002). The Harzburg Gabbro is one of the larger early Permian mafic intrusions in the central part of the Variscan Orogen. Compared to primitive mantle, the dunites, harzburgite and gabbro have positive Cs, Pb, Rb, Th and U anomalies and negative Ba anomalies. The REE patterns are flat with up to five- to sevenfold chondrite concentrations. The dunites have low REE, Sr and Pb concentrations and mantle-like Sr and Nd isotopic compositions, though the Pb isotope signatures is decoupled and approaches that of upper crustal material. Hybrid gabbro has the lowest ε_{Nd}(295 Ma) values (−7.8 to −3.0), caused by a crustal component. The Sr–Nd–Pb isotope signatures suggest that, besides a crustal component, externally-derived fluids influenced the composition of the parent magma, which had an island-arc tholeiite-like composition (Sano *et al.* 2002).

Northern Germany

The North German Basin, which partly straddles the Variscan Deformation Front (Fig. 12.76), contains large volumes of Rotliegend-age felsic to intermediate volcanic rocks that occur below thick cover sequences and are only known from deep boreholes in Denmark, north Germany and east Poland. In NW Germany the volcanic rocks are mainly limited to NNW- to NNE-trending grabens and their thickness varies from a few metres to *c.* 100 m. In the NE, however, they form continuous layers up to 2 km thick and may have covered an area of *c.* 180 000 km^2 in total (Marx *et al.* 1995; Plein 1995; Benek *et al.* 1996). The rocks range from minor basalt lavas, dolerite and gabbro intrusions to much larger volumes of rhyolitic lavas, shallow intrusions and ignimbrites, basaltic andesite and andesite lavas, and tuffs that (in the east) erupted in several stages (Fig. 12.77A). Many of the volcanic rocks in the NE were emplaced as lava dome edifices and shallow mixed intrusive-extrusive complexes; true pyroclastic rocks (pumice, glass shard-rich tuffs) are restricted to local occurrences (Paulick & Breitkreuz 2005). The volcanic rocks in the North German Basin overlie and are interbedded with alluvial, fluviatile, lacustrine siliciclastic sediments and tuffites that were deposited in floodplain environments (Altmark Subgroup; Marx *et al.* 1995; Plein 1995). The total volume of volcanic rocks has been estimated at *c.* 48 000 km^3, of which rhyolitic lavas and shallow intrusions form the largest volume (*c.* 70%) and predominate in the eastern part of the basin (Benek *et al.* 1996; Paulick & Breitkreuz 2005). Basaltic andesite lavas (6 to 150 m thick) and minor volumes of gabbro and dolerite occur in the west near the Netherlands border.

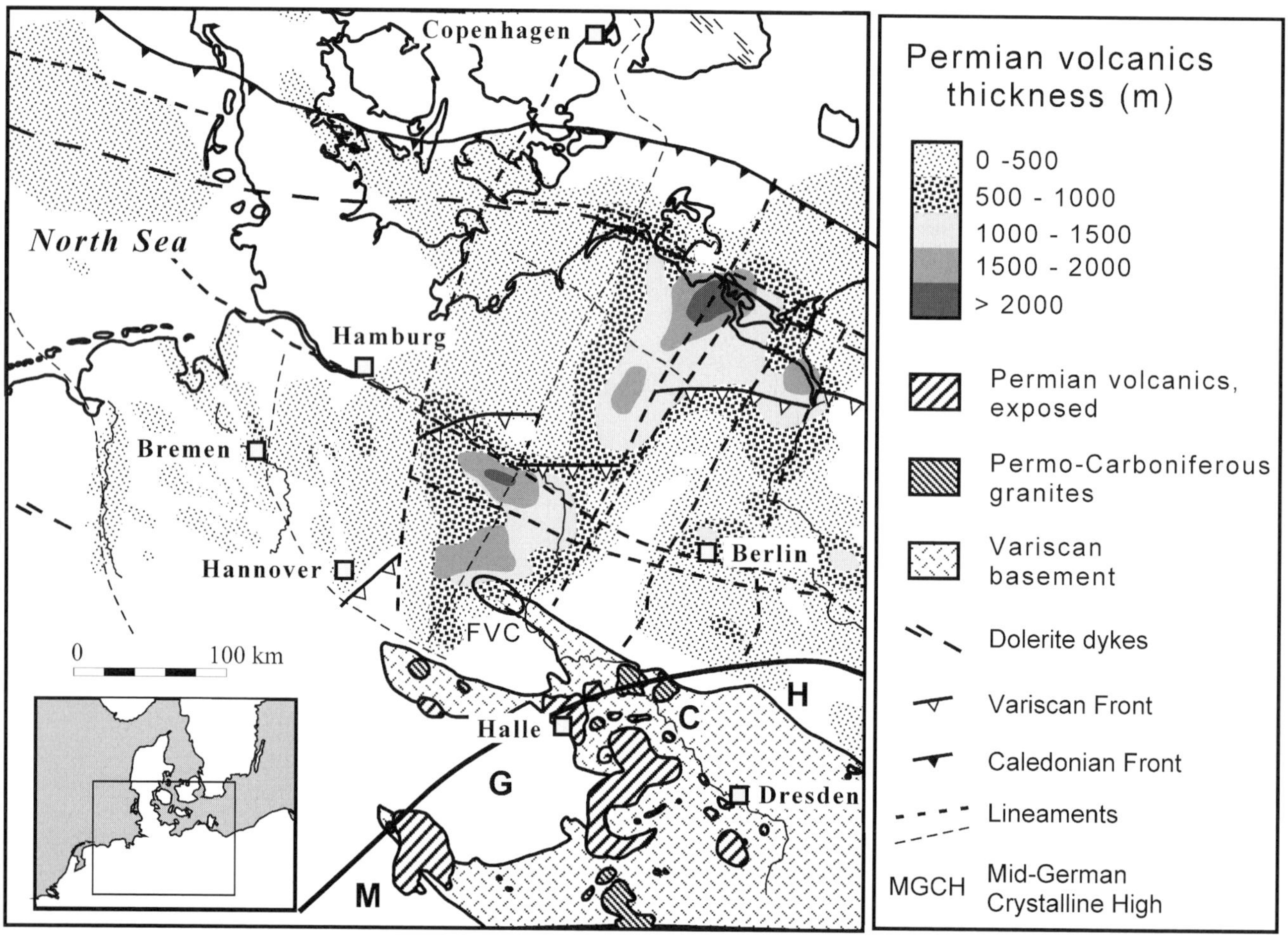

Fig. 12.76. Distribution and thickness variations of the Stephanian–Autunian volcanic rocks in northern Germany. FVC, Flechtingen Volcanic Complex. After Breitkreuz & Kennedy (1999), Marx (1994), and Marx *et al.* (1995).

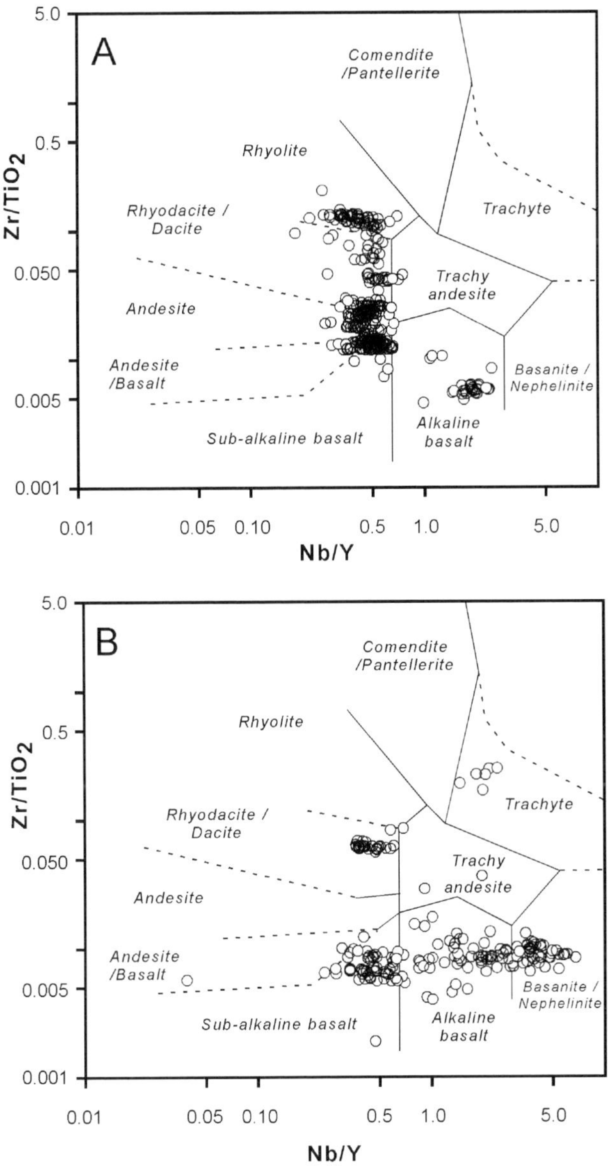

Fig. 12.77. Nb/Y versus Zr/TiO_2 classification diagram for the early Permian volcanic rocks (Winchester & Floyd 1977). (**A**) NW Germany; data from Marx (1994, and references therein). (**B**) Kattegat and central, southern and eastern North Sea; data from Aghabawa (1993, and references therein).

Basaltic andesite lavas also occur in the north-central part of the basin (between Hamburg and Bremen), and around Berlin (east Brandenburg) where they have an estimated volume of *c*. 8000 km^3 (Marx *et al.* 1995; Benek *et al.* 1996). Rhyolites, which represent the largest volume of the volcanic rocks, have U–Pb zircon crystallization ages in the range 290 ± 3 Ma to 300 ± 3 Ma (Breitkreuz & Kennedy 1999; Breitkreuz *et al.* in press).

The main period of volcanic activity took place from late Stephanian to early Permian times. East–west orientated extension caused regional-scale wrench and normal faulting along deep-seated faults (Brink *et al.* 1992; Marx *et al.* 1995). In the western part of the basin, this initial faulting caused the formation of a north-trending graben system that strongly controlled localization and thickness of the volcanic rocks. The thickest and most complete volcanic sequences are preserved in the grabens (up to 700 m thick), whereas horsts are covered by 10–200 m of the oldest volcanic formations (where not completely eroded). The erosional products of the latter are found as clasts in conglomerates that overlie the volcanic rocks in the grabens. Vertical offsets can exceed 1 km (Marx 1994; Marx *et al.* 1995; Plein 1995). Faulting and volcanism continued during deposition of thick sequences of Rotliegend-facies clastic sediments, and in late Early Permian, but pre-Zechstein times, volumetrically minor activity still took place in the form of olivine basalt lavas (Marx *et al.* 1995).

The gabbros, dolerites, basalts, andesites and rhyolites rocks of the north German Basin have subalkaline to calc-alkaline compositions, while the late Early Permian basalts are of alkaline character (Fig. 12.78A). Compared with the volcanic rocks in the central, southern and eastern North Sea, the German province

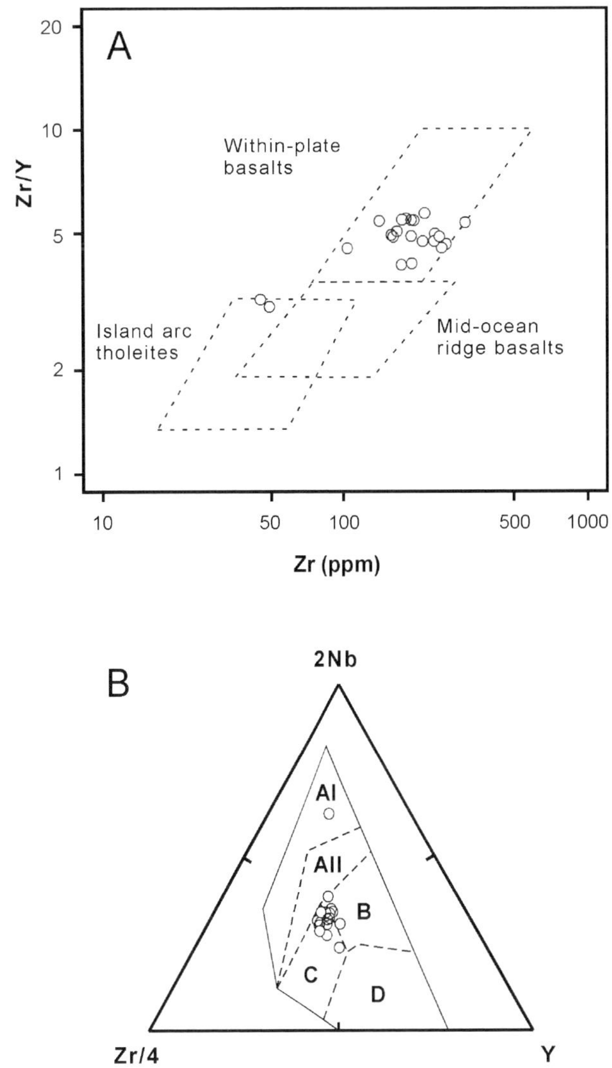

Fig. 12.78. (**A**) Zr versus Zr/Y tectonic discrimination diagram for the Permian basaltic rocks in NW Germany (Pearce & Norry 1979). (**B**) Nb–Zr–Y tectonic discrimination diagram (Meschede 1986). For field labels see caption to Figure 12.19. Data from Marx (1994, and references therein). Only samples with LOI-values <6 wt% and SiO_2 <52 wt% are plotted.

has a larger proportion of rocks of felsic to intermediate composition (Fig. 12.77A, B). The incompatible trace element composition of the least altered and least fractionated subalkaline mafic rocks suggests eruption in a within-plate setting, which is in accordance with the associated Rotliegend sediments. The early gabbro and dolerite intrusions and the later Permian alkaline basalts are not contaminated by crustal material, but the andesites and rhyolites show varying degrees of contamination. The gabbros, dolerites and late alkaline basalts are LREE-enriched and probably originated from enriched mantle sources. The REE patterns of the andesites and rhyolites from NW Germany are similar to those of continental crust. The rhyolites were probably generated by the melting of lower crustal material as further suggested by their low REE contents and the large amount of inherited zircons (Marx *et al.* 1995; Benek *et al.* 1996).

Flechtingen Volcanic Complex

The Flechtingen Volcanic Complex (Fig. 12.76) in the NW part of the Flechtingen (-Roßlau) High (Fig. 12.24) is composed of late Carboniferous, *c.* 100 m thick, lacustrine to alluvial siliciclastic sediments of Stephanian age intruded by andesite sills (Walter & Gaitsch 1988; Hoth *et al.* 1993*a*; Awdankiewicz *et al.* 2004). The sediment–sill complex overlies folded Namurian sediments (part of the Rhenohercynian Zone), and is overlain by rhyolitic ignimbrite that yielded a 302 ± 3 Ma U–Pb zircon crystallization age (Breitkreuz & Kennedy 1999). The andesite magmas intruded the siliciclastic sediments in the form of sills (the 'Older or Lower Andesites' of Benek *et al.* 1973; Flechtingen Sill Complex of Awdankiewicz *et al.* 2004) and, in the central and northern outcrop areas, cut the ignimbrites as isolated stocks and domes ('Younger or Upper Andesites').

The Flechtingen Sill Complex comprises two stacked sills up to 200 m thick and from 14 km to over 20 km in lateral (NW-SE) extension. The sills intruded the poorly lithified sediments at very shallow levels as indicated by soft-sediment deformation features and the presence of autoclastic andesite- and andesite-sediment breccias and peperites in basal and top margins of the sills. The andesites are aphyric to weakly porphyritic but strongly altered (Awdankiewicz *et al.* 2004). The upper andesites are porphyritic, containing pseudomorphs of plagioclase and (possibly) pyroxene. Samples from boreholes show flow banding, peperites and autoclastic breccias, clastic dykes and xenoliths of clastic sediments. Awdankiewicz *et al.* (2004) interpreted the younger andesites as plugs and lava domes or short flows.

Saar-Nahe Basin

Activity on many major orogen-parallel strike-slip faults was associated with opening of Namurian–Westphalian basins. Within the Mid-German Crystalline High south of the Rhenohercynian Zone, the NE-trending Saar-Nahe Basin (SNB) opened due to major strike-slip faulting and oblique extension along the Hunsrück Fault (Fig. 12.79). The SNB provides valuable information on the geological evolution in this area since it contains a stratigraphic record from the early Westphalian into the Permian (Henk 1993; Korsch & Schäfer 1995). The absence of older, Namurian sediments suggests this was an elevated area from the late Visean to the late Namurian. This period of non-deposition reflects Variscan compression, uplift, exhumation and cooling, as indicated by $^{40}Ar/^{39}Ar$ cooling ages of 320–335 Ma from MGCH basement rocks (Odenwald and Erzgebirge; Werner & Lippolt 2000; Schubert *et al.* 2001).

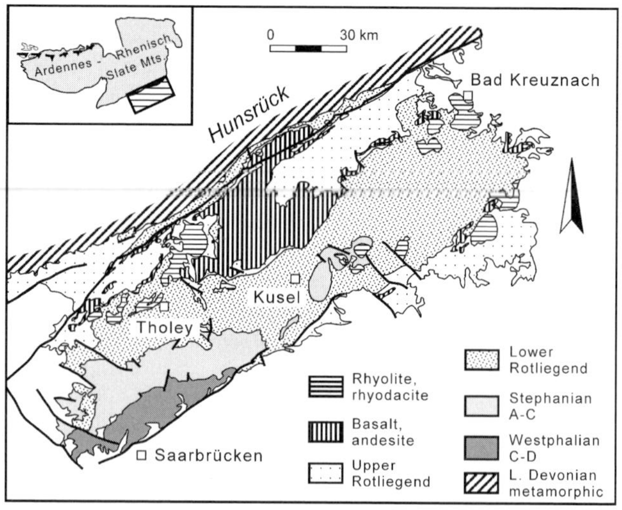

Fig. 12.79. Distribution of Westphalian–Lower Rotliegend (Autunian) sediments and volcanic rocks in the Saar-Nahe Basin, Germany (after Korsch & Schäfer 1995).

The SNB contains 6–8 km of Westphalian to Upper Rotliegend (Saxonian) alluvial, fluvial, deltaic and lacustrine sediments, coal measures, and early Permian volcanic rocks (Korsch & Schäfer 1995; von Seckendorff *et al.* 2004*a*). After the Westphalian, subsidence and deposition increased rapidly in the Stephanian and Permian. The volcanic rocks are andesite to trachyte flows and pyroclastics that are associated with subvolcanic (rhyo)dacite and rhyolite domes, sills and laccoliths, and diatremes of subalkaline basalt (Fig. 12.80A; Nicholls & Lorenz 1973; von Seckendorff 1990; von Seckendorff *et al.* 2004*a*; Schmidberger & Hegner 1999; Stollhofen & Stanistreet 1994). The several, widespread layers of altered, rhyolitic to rhyodacitic airfall tuffs have an origin outside the basin. Available U–Pb zircon and $^{40}Ar/^{39}Ar$ ages show that volcanic activity took place between *c.* 297 and 293 Ma (Lippolt & Hess 1983; Lippolt *et al.* 1989; Königer *et al.* 2002*b*). The volcanic and subvolcanic rocks (Donnersberg Fm) are lavas, ashes, lapilli tuffs, basaltic andesite, andesites and subordinate basalts and dacites that can reach cumulative thicknesses of up to 700 to 1000 m. The interaction of magmas with groundwater led to phreatomagmatic eruptions, the formation of pyroclastic rocks and the formation of maar-diatreme volcanoes on active faults or fault intersections (Lorenz & Haneke 2004). Basaltic/andesitic and rhyolite/rhyodacite sills intruded Stephanian and lower Rotliegend sediments and can be up to 12 km wide with thicknesses of over 200 m. Some intrusions are composite, and locally occurring enclaves of basaltic andesite within a rhyodacite sill indicate simultaneous emplacement and mixing of both evolved and less-evolved melts. Progressive inflation of rhyolite/rhyodacite sills and laccoliths led to the formation of extrusive domes from which block and ash flows and (locally) lavas erupted (Lorenz & Haneke 2004; von Seckendorff *et al.* 2004*a*; and references therein).

Geochemical and isotope evidence indicate that the basalts, andesites and rhyolites in the Saar-Nahe Basin form a calc-alkaline fractionation series (Fig. 12.80B). The basalts, (basaltic-) andesites and trachytes are moderately enriched in LREE and, in combination with the MORB-normalized multi-element patterns, indicate derivation from enriched mantle sources. The least evolved basalt that was analysed has the highest ε_{Nd}(295 Ma) value (+3.2) and lowest Sr_i (0.7044); the majority of the basalts and basaltic andesites cluster around a lower ε_{Nd}(295 Ma) value

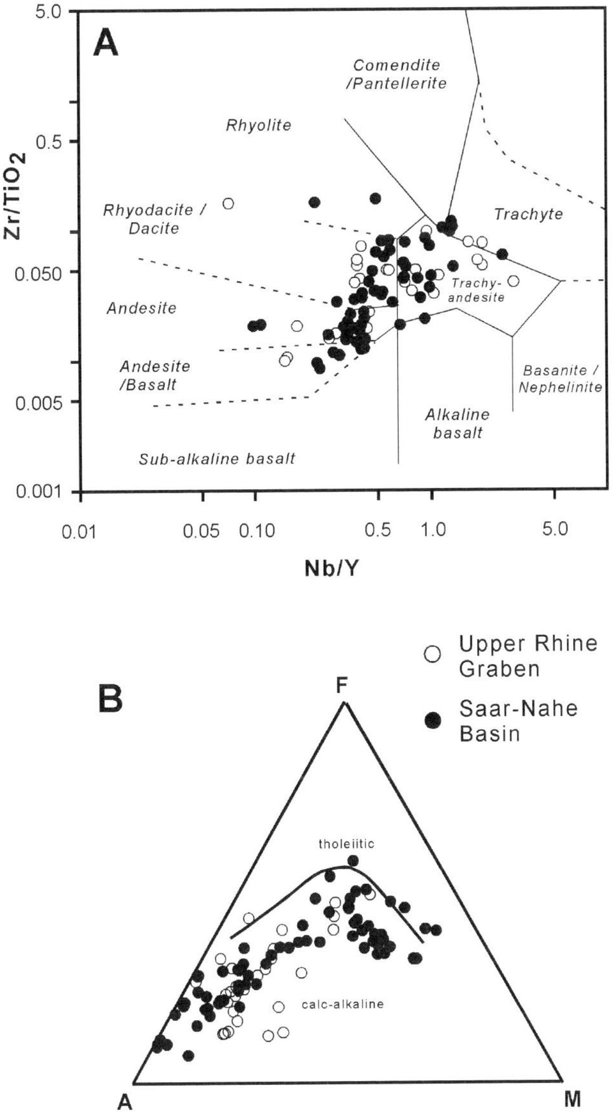

Fig. 12.80. Early Permian volcanic rocks from the Saar-Nahe Basin and boreholes in the Upper Rhine Graben. (**A**) Nb/Y versus Zr/TiO_2 classification diagram (Winchester & Floyd 1977). (**B**) AFM diagram; curve separating tholeiitic from calc-alkaline field after Irvine & Baragar (1971). A = $K_2O + Na_2O$; F = FeO_{total}; M = MgO. Data from Müller (1996), Schmidberger & Hegner (1999) and von Seckendorff *et al.* (2004a).

of −2.5 and higher Sr_i of 0.706. Rhyodacites, trachytes and rhyolites have fairly constant ε_{Nd}(295 Ma) values of *c.* −3.5 and Sr_i in the range of 0.706 down to 0.704; the latter, anomalously low ratios may be due to element mobility during secondary processes (von Seckendorff *et al.* 2004*a*). The parent melts of the mafic end-members were probably derived either from a heterogeneous mantle source that was enriched by a coeval, but distal, subduction event (e.g. Schmidberger & Hegner 1999), or from a hydrous, shallow mantle source that was moderately enriched during a previous subduction event (Nicholls & Lorenz 1973; von Seckendorff *et al.* 2004*a*, and references therein). In contrast, immobile trace element ratios and abundances of the rhyodacites and dacites indicate derivation from, or extensive contamination by, Saxothuringian crust (von Seckendorff *et al.* 2004*a*).

The SNB is part of a series of SW-trending basins that extend southwestwards into France (Lorraine), parallel to the Hunsrück and Metz faults. Most are covered by Triassic and younger sediments, but have been identified on seismic profiles and cored by several deep boreholes (e.g. Korsch & Schäfer 1995). The Metz-1 borehole contains 142 m of andesites that occur just above Stephanian-age deposits. To the south, near Nancy, 1178 m of massive rhyolites and rhyolitic breccias occur in the Foncelles-1 borehole; they overlie sediments and are probably of Permian age (Châteauneuf & Farjanel 1989). To the east, Permian-age felsic to intermediate pyroclastic rocks and basalts are known from boreholes in the Upper Rhine Graben, east of the Saar-Nahe Basin. In some boreholes they can attain thicknesses of up to 550 m and may form the eastern, sub-surface extension of the Donnersberg Fm of the Saar-Nahe Basin (Müller 1996). Lippolt *et al.* (1990) obtained a $^{40}Ar/^{39}Ar$ sanidine age of 291 ± 6 Ma for a rhyolite, and 275 ± 8 Ma for a whole-rock basalt sample from boreholes in the Upper Rhine Graben.

Thuringian Forest

In the Thuringian Forest and Harz areas undeformed, or only weakly deformed, granitoids, gabbros and rhyolite sills have 289–307 Ma U–Pb zircon ages (Baumann *et al.* 1991; Breitkreuz & Kennedy 1999; Loth *et al.* 1999; Anthes & Reischmann 2001; Zeh & Brätz 2004). Differential uplift in the Thuringian Forest was associated with 290–298 Ma granite and diorite magmatism and led to block faulting and the formation of horsts and pull-apart basins (Fig. 12.81). The basins are filled by a succession up to 2 km thick of Stephanian–Upper Rotliegend (Saxonian) continental sediments and volcanic rocks. The latter include lavas and pyroclastic rocks of mainly trachyandesitic (293 ± 2 Ma) and rhyolitic (291 ± 3 Ma) composition ($^{40}Ar/$

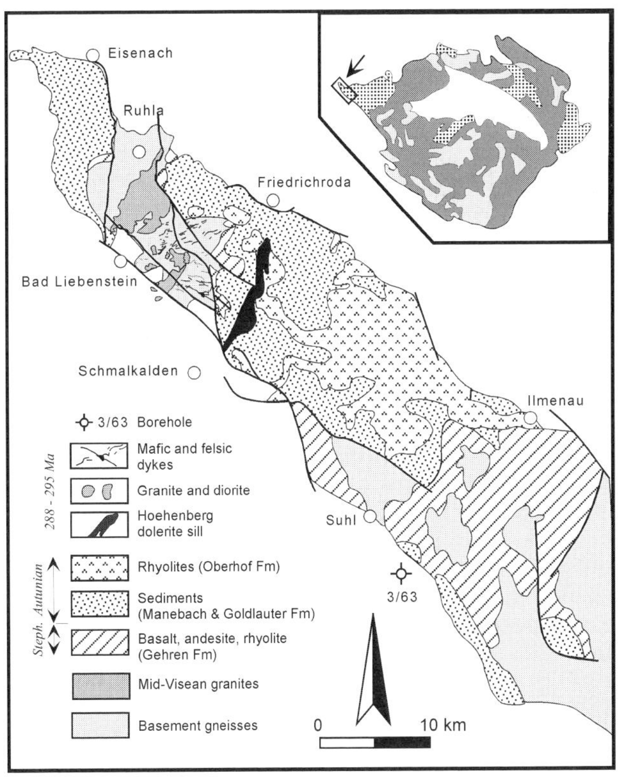

Fig. 12.81. Geological sketch map of the Thuringian Forest, with Ruhla Complex (after Zeh *et al.* 2003).

^{39}Ar biotite ages with 1-sigma errors; Goll & Lippolt 2001; Zeh & Brätz 2004). The sediments and volcanic rocks are intruded by the Höhenberg dolerite sill containing aplite dykes that yielded 294 ± 4 and 291 ± 3 Ma zircon evaporation ages (Obst *et al.* 1999).

Saale Basin and Halle Laccolith Complex

The SW-trending Saale Basin contains Stephanian to early Permian clastic sediments, volcanic rocks and subvolcanic intrusions of the Halle Laccolith Complex (also known as Halle Volcanic Complex) and the Ilfeld Basin (SE Harz area). The sediments and volcanic rocks overlie the eastern part of the Mid-German Crystalline High, the Northern Phyllite Zone and the NE part of the Saxothuringian Terrane (Fig. 12.82). Sedimentation commenced in the Stephanian with the deposition of fluviatile and lacustrine clastic sediments interbedded with coals (Rappsilber 2003; Ehling & Breitkreuz 2004, and references therein). This was followed by magmatic activity in latest Carboniferous–early Permian times which was concentrated east of the Halle Fault (Fig. 12.82). The onset of magmatic activity is indicated by the presence of porphyritic volcanic and subvolcanic rocks of basic to intermediate composition, followed by the emplacement of large rhyolitic sills of the Halle Laccolith Complex.

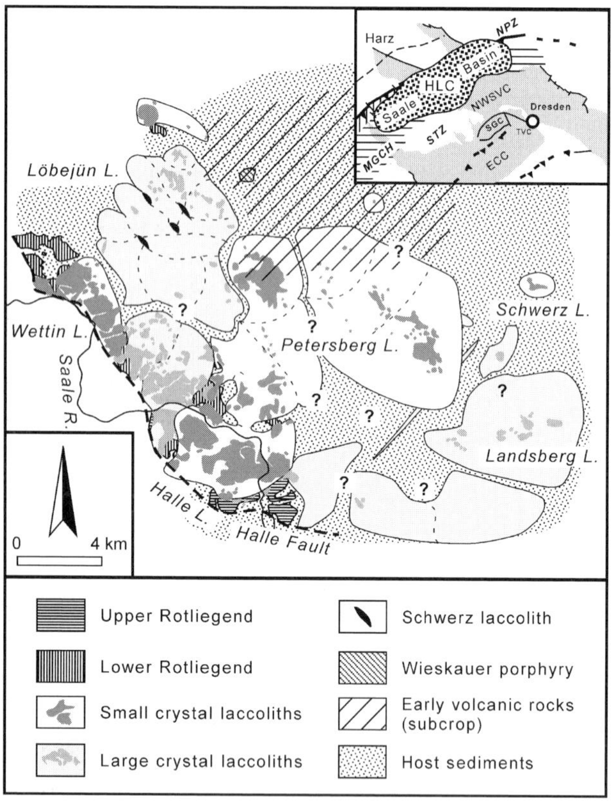

Fig. 12.82. Geological map of the Halle Laccolith Complex, Saale Basin (after Romer *et al.* 2001; Mock *et al.* 2005, and references therein). Abbreviations: L, laccolith. Inset: ECC, Erzgebirge Crystalline Complex; HLC, Halle Laccolith Complex; MGCH, Mid-German Crystalline High; NPZ, Northern Phyllite Zone; NWSVC, NW Saxony Volcanic Complex; SGC, Saxonian Granulite Complex; STZ, Saxothuringian Zone; TVC, Tharandt Volcanic Complex.

The early volcanic rocks are present as small volumes of trachybasalt, trachyandesite and trachydacite that have been divided into four units based on borehole data (Porphyrites I to IV; Fig. 12.83A; Romer *et al.* 2001). The less evolved end-members contain phenocrysts of pyroxene, olivine and Ca-rich plagioclase; locally, flow textures and miarolitic cavities are present. The volcanic rocks are enriched in LREE and have small to moderately negative Eu anomalies. The ε_{Nd}(300 Ma) values are negative (−7.4 to −0.8) and partly overlap with those of the rhyolitic laccoliths (−7.0 to −6.7; Romer *et al.* 2001).

The Halle Laccolith Complex is situated in the NE part of the Saale Basin (Figs 12.1 & 12.82) and is made up of large volumes of porphyritic rhyolite laccoliths (> 200 km^3 in total) that have yielded 294 ± 3 to 307 ± 3 Ma U–Pb zircon ages (Breitkreuz & Kennedy 1999). The laccoliths were emplaced into Stephanian to

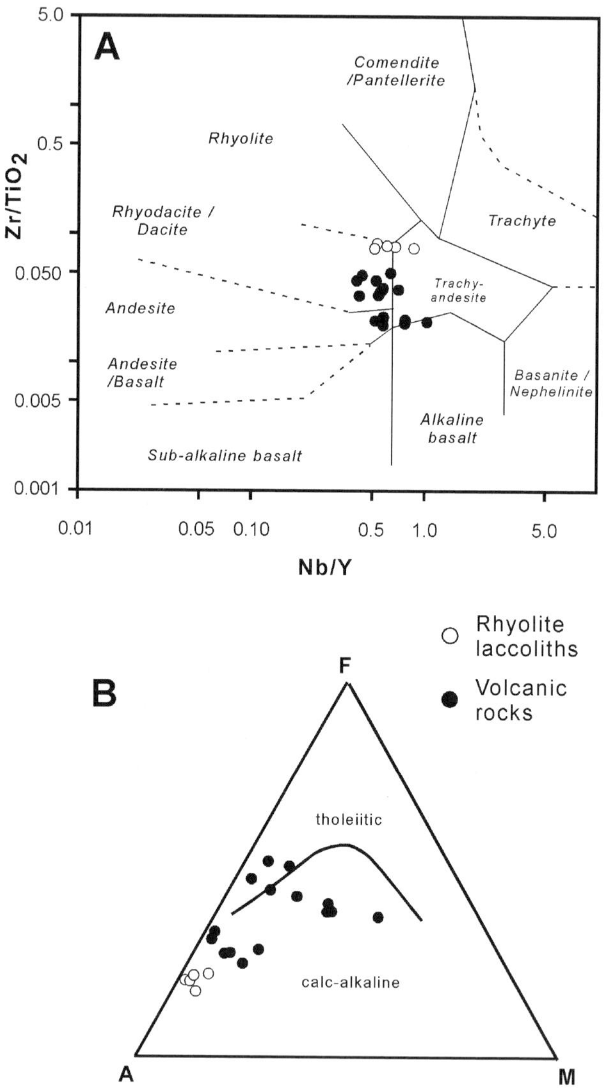

Fig. 12.83. Stephanian to early Permian volcanic and subvolcanic rocks from the Saale Basin (including the Halle Laccolith Complex (open circles). (**A**) Nb/Y versus Zr/TiO_2 classification diagram (Winchester & Floyd 1977). (**B**) AFM diagram; curve separating tholeiitic from calc-alkaline field after Irvine & Baragar (1971). A = $K_2O + Na_2O$; F = FeO_{total}; M = MgO. Data from Romer *et al.* (2001).

early Permian sediments at very shallow crustal levels, as indicated by the presence of rhyolite clasts in sediments deposited in interlaccolith depressions, chilled margins, and thin to absent contact metamorphic aureoles (Mock *et al.* 2005). Rhyolitic pyroclastic rocks and lavas occur in the southern margin of the Saale Basin (Büchner & Kunert 1997) and are probably related to the breaching of, and eruption through, the roof of the laccoliths (Ehling & Breitkreuz 2004). The rhyolites have uniform peraluminous, calc-alkaline compositions (Fig. 12.83A, B), and relatively uniform negative ε_{Nd}(300 Ma) values (−7.0 to −6.7) indicating that they were largely derived from the melting of crustal material (Romer *et al.* 2001). As in the nearby Ilfeld Basin and the Thuringian Forest area, the Halle Laccolith Complex was emplaced in an intracontinental extensional setting, and Romer *et al.* (2001) suggesting that its magmatic-arc trace element signature must have been inherited from its source, probably Saxothuringian crust.

The Ilfeld Basin at the southern margin of the Harz (Fig. 12.84) is filled with *c.* 800 m of sediments and volcanic rocks. Of these, up to 400 m comprise tuffs, ignimbrites and latitic, trachytic and rhyolitic flows that erupted at the Carboniferous–Permian boundary. Sedimentation began in the Stephanian with alluvial fan deposits that unconformably overlie folded Devonian sediments, and they were succeeded by siltstones, alluvial fans, trachytic and latitic flows, and playa carbonates. Later rhyolite domes, lavas and pyroclastic rocks cover a large area and represent an important magmatic phase within the basin (Paul *et al.* 1997). The volcanism may be related to small volumes of rhyolite lavas and subvolcanic intrusions *c.* 15 km NW of Bad Sachsa, for which Lippolt & Hess (1996) obtained 295.7 ± 2.4 Ma and 297.1 ± 5.0 Ma $^{40}Ar/^{39}Ar$ biotite ages. Büthe & Wachendorf (1997) suggested that the Ilfeld Basin was formed as a pull-apart basin during dextral transtension along older, Variscan lineaments; this setting also controlled the ascent and emplacement of the magmas (Breitkreuz & Mock 2004).

NW Saxony Volcanic Complex

The NW Saxony Volcanic Complex extends from the northern margin of the Saxonian Granulite Complex to the southern margin of the Saale Basin (inset of Fig. 12.82). It is mostly covered by younger deposits and exposure is limited. Borehole data indicate that early Permian volcanic and subvolcanic rocks cover an area of *c.* 1400 km^2, overlie lower Rotliegend conglomerates and sandstones, and are covered by upper Rotliegend conglomerates and arkoses (Pietsch 1959; Glässer 1983; Wetzel *et al.* 1995). Magmatic activity began with rhyolitic to andesitic flows and tuffs, which are limited to local occurrences in the south and east of the complex. The main phase of volcanic activity is represented by more widespread rhyolitic ignimbrites that erupted in three stages and are locally separated by continental clastic sediments. These were shortly followed by porphyritic rhyolite flows and small intrusions in the central parts of the complex. After a period of deposition of clastic continental sediments (containing tuffs), volcanism continued in the northern part of complex in the form of hybrid quartz and/or pyroxene phenocryst-bearing rhyolitic, trachytic to andesitic ignimbrites and dykes that were emplaced in three stages. Wetzel *et al.* (1995) suggested that partial melting of dioritic source rocks resulted in parent melts that subsequently fractionated and mixed in magma chambers to produce the observed range of volcanic rocks.

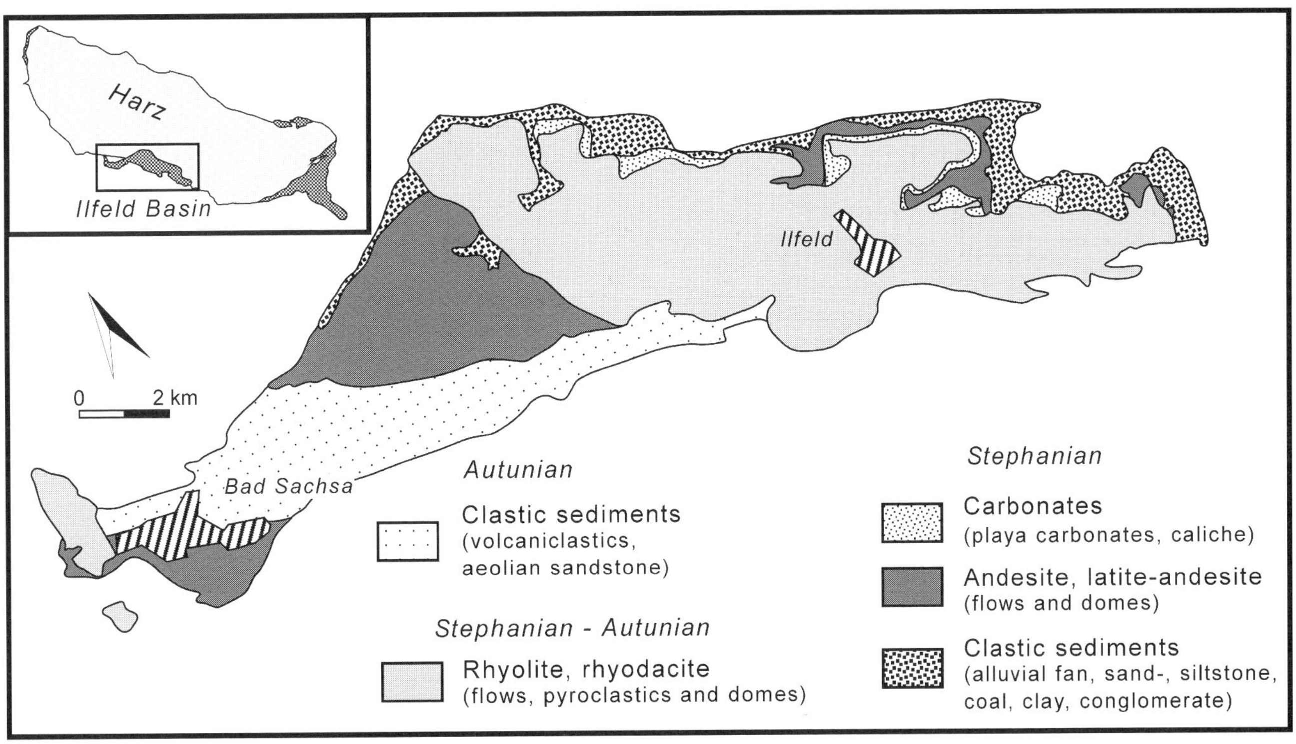

Fig. 12.84. Distribution of Stephanian–Autunian sediments and volcanic rocks in the Ilfeld Basin, south Harz (after Paul *et al.* 1997).

Tharandt Volcanic Complex

The small (*c.* 50 km^2) Tharandt Volcanic Complex SW of Dresden is composed of flow-banded and ignimbrite-like rhyolitic to rhyodacitic volcanic rocks and rhyolitic dykes (inset, Fig. 12.82; Benek 1980). The volcanic rocks erupted in two main stages (Fig. 12.85). The first stage comprises *c.* 500–600 m of rhyolite that is poor in quartz phenocrysts but contains xenoliths of basement material. The xenoliths are set in a glassy to fine-grained groundmass and comprise gneisses from the Erzgebirge Crystalline Complex in the south, and phyllites, slates and greenschists from the Nossen-Wilsdruff Zone (Elbe Zone) in the north, and locally occurring vein quartz and ignimbrite autoclasts. The size and amount of xenoliths decrease towards the younger volcanics that erupted during the second, main stage of volcanic activity and reached thicknesses of at least 300 m. This stage mainly comprises rhyolite containing quartz and alkali feldspar phenocrysts in a groundmass of glass or microcrystalline quartz and feldspar. Rhyolite dykes found outside the area of volcanism are geochemically similar to the volcanic rocks, and comprise aphyric varieties, and varieties bearing biotite, and feldspar ± quartz phenocrysts. The dykes in the east and west margin of the area show steep flow-banding that is parallel to the dyke walls, and these may represent ring and feeder dykes that were emplaced during caldera formation (Benek 1980).

Further to the NW, the Visean-age Meissen intrusion is overlain by rhyolitic volcanic rocks (devitrified and weathered porphyritic glasses, and tuffs) of probably latest Carboniferous to early Permian age (Pietsch 1959).

Döhlen Basin

The NW-trending, 22 km long, 6 km wide Döhlen Basin west of Dresden overlies rocks of the Elbe Zone, the Meissen intrusion,

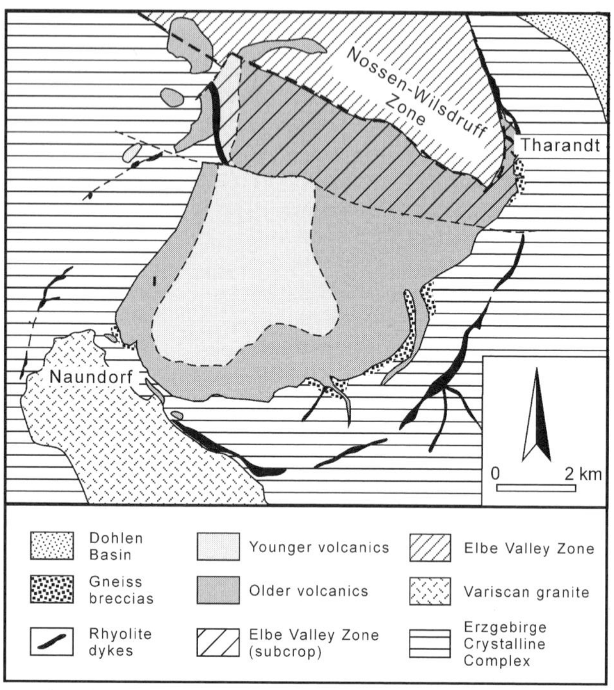

Fig. 12.85. Geological map of the Tharandt Volcanic Complex, SW of Dresden (after Benek 1980). For location see inset in Figure 12.82 (TVC).

and gneisses of the easternmost part of the Erzgebirge Crystalline Complex. It is filled with a *c.* 800 m thick sequence of continental clastic sediments and pyroclastic rocks, deposited in four megacyles. The pyroclastic rocks make up *c.* 50% of the successions. The basin has a half-graben-like symmetry and was tectonically controlled, as is clear from synsedimentary deformation structures (e.g. clastic dykes) and conglomerates and fanglomerates containing locally derived clasts. Most of the pyroclastic material appears to have been derived from the nearby volcanic centres near Tharandt and Meissen and the NW Saxony Volcanic Complex. Lavas are present only in the NW part of the basin and have rhyodacitic to trachyandesitic compositions (Reichel 1985; Hoffmann 2002).

Sedimentation began with conglomerates (containing clasts of gneisses, Meissen monzonite, rhyolite, chert and metasediments), laterally extensive pyroclastic flow horizons (probably related to those overlying the Meissen intrusion), and (locally) lapilli-, crystal- and airfall tuffs, and resedimented tuffs (Unkersdorf Fm). Locally, up to 80 m thick lavas and lava domes of porphyritic rhyodacite to trachyandesites are present. The Döhlen Fm of post-Stephanian C age is made up of fluviatile conglomerates, sandstones and siltstones, lacustrine coals, mudstones, crystal and airfall tuffs. The late upper Rotliegend-age Niederhäslich Fm commences with conglomerate containing clasts of rhyolite, followed by fluviatile and lacustrine sandstones and siltstones that are interbedded with, and overlain by, crystal- and airfall tuffs, and ends with lacustrine limestones and coals. The Bannewitz Fm of upper Rotliegend age begins with fanglomerates and conglomerates that contain clasts of rhyolite and gneisses, and reflect a period of tectonic activity. These rocks are overlain by sandstones and siltstones that are interbedded with airfall, crystal- and lapilli-tuffs (Hoffmann 2002).

To the SW of the Döhlen Basin, near Dohna, an eroded volcanic plug (<0.5 km^2) cutting the Dohna granodiorite of the Elbetalschiefergebirge (see above) consists of rhyolite composed of quartz and alkali feldspar phenocrysts (and rare biotite) set in a glassy to microcrystalline groundmass, and intrusive rhyolitic tuff. Both rhyolite and tuff shows steep to subvertically oriented flow-banding (Schwab 1959).

Intra-Sudetic Basin

The NW-trending, Intra-Sudetic Basin in the West Sudetes is one of the largest (70 by 35 km) Permo-Carboniferous basins in the internal part of the Variscan Orogen (Figs 12.28 and 12.86). It is a fault-bounded basin that contains up to 12 km of lower Carboniferous to upper Cretaceous-age sediments. The Permo-Carboniferous sediments contain three cycles of volcanic rocks ranging in composition from basaltic andesite to trachyandesites and rhyolite; true basalts appear to be absent (Fig. 12.87A; Awdankiewicz 1999*a*, *b*, 2004; Ulrych *et al.* 2004).

The sedimentary record begins with early Carboniferous continental sediments (conglomerates and sedimentary breccias) deposited as alluvial fans along the scarps of active faults. The fans prograded towards the basin centre, locally attaining thicknesses of *c.* 6.5 km. Miospore evidence from the oldest sediments in the NW indicates that sedimentation started in the Holkerian or Asbian (mid or early-late Visean; Turnau *et al.* 2002). This is in agreement with uplift-related shearing dated to *c.* 324–320 Ma by $^{40}Ar/^{39}Ar$ from basement rocks along the western flank of the basin (Marheine *et al.* 2002). The Visean sediments along the NE margin are conformably interbedded with *c.* 50 m of thick tuffs and rhyodacite flows, and cut by small felsic dykes and necks. At the northern basin margin, rare 3.5 to

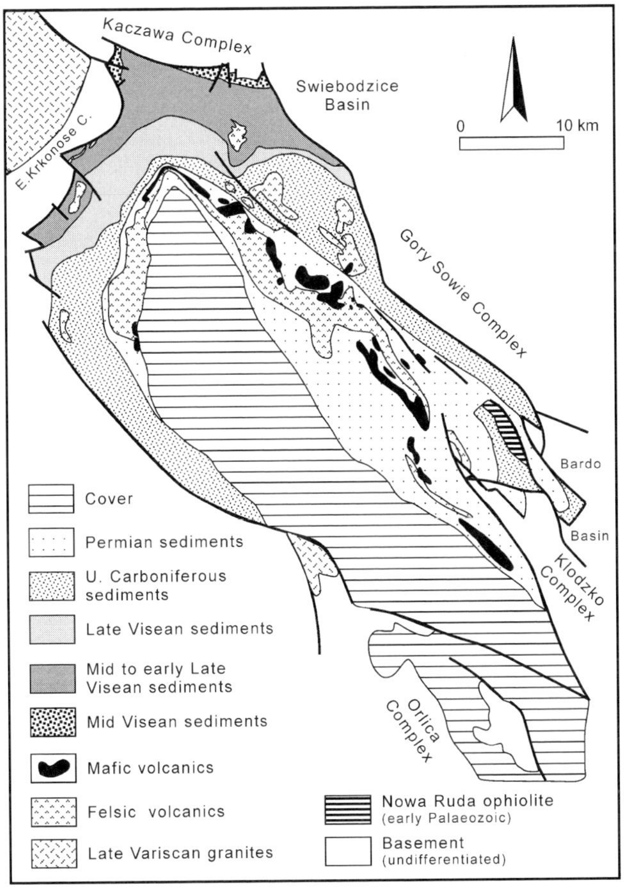

Fig. 12.86. Distribution of the mid-Carboniferous and late Carboniferous–early Permian volcanic rocks in the Intra-Sudetic Basin (after Turnau *et al.* 2002). See Figure 12.28 for location of the basin within the West Sudetes area.

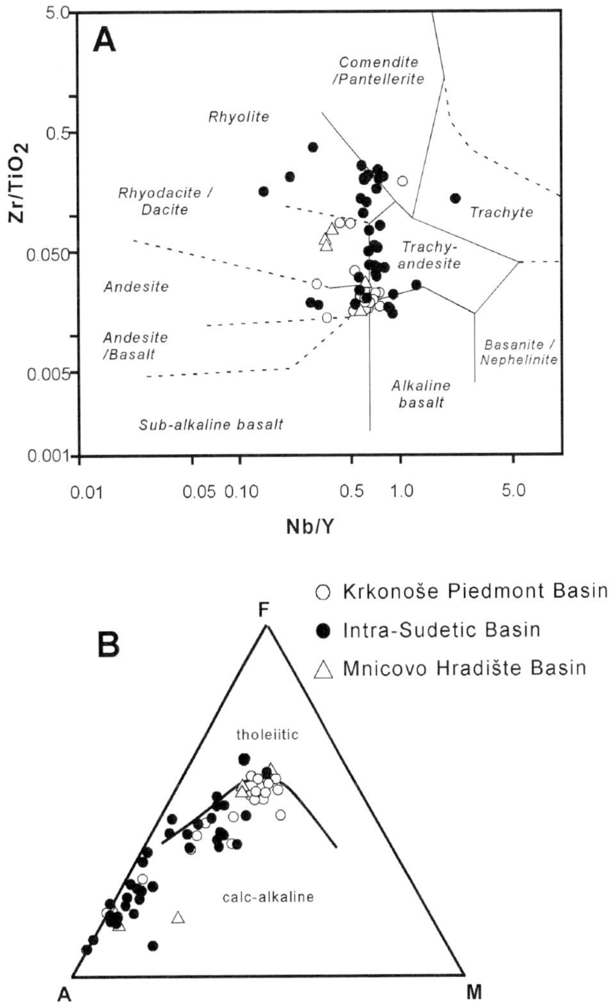

Fig. 12.87. Westphalian to early Permian volcanic rocks from the Intra-Sudetic, Krkonoše Piedmont and Mnicovo Hradiště basins (West Sudetes and NE Bohemian Massif). (**A**) Nb/Y versus Zr/TiO_2 classification diagram (Winchester & Floyd 1977). (**B**) AFM diagram; curve separating tholeiitic from calc-alkaline field after Irvine & Baragar (1971). A = $K_2O + Na_2O$; F = FeO_{total}; M = MgO. Data from Awdankiewicz (1999*a*, *b*) and Ulrych *et al.* (2002, 2004, 2006).

5 m thick andesite sills intruded poorly consolidated conglomerates (Awdankiewicz 1999*a*). Subsequent deposition of up to 600 m of late Visean marine sediments took place in the eastern part of the basin, while fluvial/deltaic sediments in the western part attain a thickness of up to 2 km.

In the mid-Carboniferous the depositional pattern in the Intra-Sudetic Basin was rearranged due to uplift in the east and south, leading to local regression and the deposition of up to 2 km of alluvial sediments and coal seams. This extended into the early Permian. Folded Upper Westphalian, and the unconformably overlying early Stephanian, sediments along the NE basin margin are cut by a suite of subvolcanic rhyolite and rhyodacite intrusions that occur as 200 to 600 m thick laccoliths and phacoliths, with associated smaller dykes and sills. Their syntectonic character suggests that they have late Westphalian–early Stephanian crystallization ages. Approximately 8 km to the SE, along the basin margin fault, late Westphalian–early Stephanian pyroclastic rocks occur infilling small, maar-like volcanoes and pipe- to funnel-shaped diatremes arranged in a NW-trending belt. These are intruded by sheets and plugs of rhyolite that can reach several tens of metres in thickness, and are in turn intruded by small volumes of late Carboniferous trachyandesite domes (Awdankiewicz 1999*a*).

Early Permian volcanic and subvolcanic associations occur in the NW, SW and south parts of the Intra-Sudetic Basin (Fig. 12.86). Basaltic trachyandesitic lavas and autoclastic lava breccias of *c.* 50–70 m thickness in the NW occur interstratified with early Permian sediments. The lavas erupted as pahoehoe-, aa- and blocky-type flows from a small shield volcano *c.* 12 km in diameter and *c.* 100 m high. The flows locally interacted with wet sediments resulting in lava/sediment breccias; interbedded laminated tuffs are pyroclastic surge deposits. Stratigraphically slightly younger rhyolites erupted as massive flows, with vesicular rhyolite and rhyolitic breccias in the upper and distal parts. In the SW parts of the basin, massive to locally brecciated basaltic andesites occur in early Permian sediments as conformable and discordant sheets up to 100 m thick. These sheets were emplaced as subvolcanic sills, lavas and lava- and breccia-filled plugs. Nearby, slightly younger, massive to weakly brecciated trachyandesite and rhyolite lava domes and flows erupted from a composite, multivent volcano. The volcanic edifice was subsequently partly eroded and covered by bedded, welded and non-welded rhyolitic tuffs that were deposited from a single, Plinian-style eruption. These pyroclastic rocks form a *c.* 50 km long belt orientated subparallel to the NE basin margin, and probably originated from a caldera of *c.* 10 km diameter to the SE. Rhyolite and trachyandesites in the south post-date the tuffs, and intruded as subvolcanic sheets up to 200 m thick, which locally pass into laccoliths and domes (Awdankiewicz 1999*a*, 2004).

The older, Carboniferous volcanic rocks of the Intra-Sudetic Basin are of subalkaline (i.e. $Nb/Y < 0.67$) to calc-alkaline character and their trace element composition indicates volcanic arc affinities. Compared to MORB, the least evolved end-members (basaltic andesites) show K, Rb and Ba enrichment relative to the other incompatible elements, suggesting active margin magmatism. In contrast, the early Permian basaltic trachyandesites are more alkaline, including shoshonitic and latitic varieties in the SE (Czech) extension of the basin (Figs. 86A, B). Their trace element compositions (enrichment of LILE, Th, Nb and LREE) point to a within-plate affinity (Awdankiewicz 1999b; Ulrych *et al.* 2004). The subduction signature of the older Carboniferous suite is likely to be a feature inherited from the mantle source, as the volcanic rocks erupted in a post-collisional, extensional setting. The late Carboniferous to early Permian rocks in the SE part of the basin have negative $\varepsilon_{Nd}(t)$ values (−6.1 to −2.7) and Sr_i values in the range of 0.705 to 0.710, indicating contamination by old continental crust (Ulrych *et al.* 2004). Adankiewicz (1999*b*) suggested that the parent magmas of both suites were derived from relatively shallow (i.e. garnet-free) mantle sources, during varying degrees of partial melting.

Krkonoše Piedmont Basin

The Krkonoše Piedmont Basin is located west of the Intra-Sudetic Basin and south of the South Krkonoše Complex (Fig. 12.28). Volcanic activity started in the Westphalian D and extended into late Early Permian times and is mostly of basaltic andesitic composition ('melaphyres' or 'andesitoids'); rhyolites appear in the upper part of the basin fill. The basaltic andesites mainly erupted as flows from several centres that occur aligned along faults. Sills, laccoliths, dykes, tuffs and agglomerates are rare. A late Carboniferous (Westphalian–early Stephanian) and an early Permian (late Autunian) magmatic suite have been distinguished. The late Carboniferous suite comprises aphyric andesites to trachyandesites, weakly porphyritic trachydacites, and moderately porphyritic and often altered trachytic dacites (Fig. 12.87A, B). The rocks are relatively evolved and have negative Nb and Ta anomalies, La_N/Yb_N ratios between 16.1 and 20.7, and are depleted in P and Ti relative to primitive mantle. The $\varepsilon_{Nd}(290\ Ma)$ values are between −3.6 and −2.6, and the $^{87}Sr/^{86}Sr$ initial ratios are between 0.7060 and 0.7072 (Ulrych *et al.* 2002). The early Permian suite comprises basaltic andesites to basaltic trachyandesites and small volumes of basalt and trachybasalts (Fig. 12.87A, B). Petrographic textures range from porphyritic, intersertal, hyalo-ophitic, ophitic to poikilo-ophitic; the lavas are often amygdaloidal and locally show flow-banding. Rhyolitic ignimbrite is fine-grained (ash) and locally contains glass shards and broken feldspar crystals. The early Permian volcanic rocks form two geochemical groups. The first group consists of relatively primitive basaltic trachyandesites and basaltic andesites with La_N/Yb_N ratios between 7.9 and 12.0, negative Nb and Ta anomalies, and depleted in P and Ti relative to primitive mantle. The second group consists of rhyolites of peraluminous, S-type geochemical character with pronounced negative Sr, Ti and Eu anomalies compared to primitive mantle. La_N/Yb_N ratios are between 4.0 and 7.4, and two samples have $\varepsilon_{Nd}(290\ Ma)$ values of −5.1 and −0.9, and $^{87}Sr/^{86}Sr$ initial ratios of 0.9563 and 0.8532, respectively. Ulrych *et al.* (2002) suggested that parent melts of the basic to intermediate rocks may have been derived from a mantle source that was enriched in incompatible elements. These melts subsequently fractionated and were contaminated with lower crustal material in a deep-seated magma chamber. The heat released from the basic magmas and the upper mantle triggered anatexis of crustal rocks, giving rise to the formation of the S-type rhyolites.

Stephanian–early Permian magmatism: summary

In Stephanian–early Permian times, a large number of fault-controlled basins were formed, accompanied by widespread volcanic and plutonic activity. The cause of basin formation may have been a combination of outward-propagating orogenic collapse, detachment of dense mantle material, and the effects of the regional extensional stress-field. Many basins show a characteristic infill of fluviatile and lacustrine (i.e. continental) sediments, and volcanic rocks. In the foreland magmatic activity resulted in the formation of large amounts of rhyolitic and andesitic volcanic rocks in northern Germany, a large variety of volcanic and plutonic rocks in the Oslo Rift (Graben), and the formation of dolerite sills and dyke swarms in northern Britain and southern Sweden. In the internal parts of the Variscan Orogen, the widespread occurrence of granitic to dioritic intrusions and andesitic to rhyolitic volcanic rocks indicates that felsic to intermediate compositions predominate. These often have a subduction-related geochemical character that does not agree with their intraplate setting. Instead, it may have been inherited from mantle sources that had been metasomatized by previous subduction events, caused by extensive assimilation of continental crust by the parent melts, and/or may have been inherited by partial melting of calc-alkaline crust.

The Stephanian–Early Permian magmatic rocks in the internal parts of the Variscan Orogen comprise a high proportion of pyroclastic rocks and are generally of intermediate to felsic composition. Their generally calc-alkaline character suggests a subduction-related origin. With the possible exception of some magmatic rocks of the Alpine basement, this contradicts their intracontinental setting and the fact that the oceanic basins had closed by mid-Carboniferous times. However, Sr–Nd isotope data and the presence of garnet and crustal xenoliths in felsic magmatic rocks indicate that many contain a significant crustal component. The parent magmas of the andesites may therefore have assimilated large amounts of crustal material, or alternatively may have been derived from mantle sources that were modified by earlier subduction events (e.g. Eckhardt 1979; Schmidberger & Hegner 1999). As in the North German Basin, the granites and rhyolites may be predominantly of crustal origin, and their calc-alkaline signature was presumably inherited by partial melting of calc-alkaline basement (Schaltegger 1997; Romer *et al.* 2001).

Geochemical evidence from the rhyolites in the North German Basin suggests that extensive crustal contamination of the parent melts has taken place (Benek *et al.* 1996; Neumann *et al.* 2004). In addition, their subduction signature contradicts the intraplate setting of the volcanism, and suggests that the parents melts may have been extracted from a mantle source that was enriched during a previous (Caledonian?) subduction event (Neumann *et al.* 2004). Assuming that 10–20% of the parent melts erupted, Breitkreuz & Kennedy (1999) suggested that the total volume of crustal melts formed (i.e. rhyolites) was of the order of 350 000–400 000 km^3, and pointed out that a similar amount of basaltic underplated melt is needed to generate this. However, there is little geophysical evidence for lower crustal mafic intrusions in this area. Instead, most of the basin is underlain by a flat Moho, and by (pre-Stephanian) crust that, below the basin centre, is thinned by *c.* 35%. Here, a high-velocity lower crust reaches a thickness of 18 km at the expense of middle crust that is thinned to a few kilometres. However, the age of this anomaly is not known (Bayer *et al.* 1999; DEKORP-BASIN Research Group 1999).

The possible mechanism for mantle melting in the internal parts of the Variscan Orogen may have been related to the break-off of subducted oceanic crust (e.g. Schaltegger 1997; Cesare *et al.* 2002). Regional extension leading to lithospheric thinning and decompressional melting of updoming asthenosphere may also have been a contributing factor in the late Carboniferous–early Permian period.

Compared to the foreland, Stephanian–Early Permian-age mafic rocks are much rarer in the internal parts of the Variscan Orogen, which suggests that the mantle-derived parent melts were unable to reach the surface, but stalled at lower to mid-crustal levels. This may have been due to the marked contrast between the density of the parent melt and the low average density of thinned Variscan crust. Only after fractionation and assimilation of sufficient amounts of crustal material did the melts attain a low enough buoyancy to be able to escape the magma chambers and erupt on the surface.

The widespread extension and thermal perturbation of the lithosphere led to an almost complete destruction of the Variscan Orogen. After convergence in the external zone of the orogen had ceased, extension and erosion removed most of the surface topography and crustal roots within a period of 20 Ma (Henk 1999). Subsequent thermal equilibration of the Central European lithosphere led to thermal subsidence leading to overstepping and amalgamation of the individual sedimentary basins and the formation of a broad depositional area in which sediments were deposited in fluviatile and playa-lake environments (the South Permian Basin). Toward the end of the Permian, the transgression of the Zechstein Sea indicates that parts of the former orogen had subsided below sea level and that the Variscan crust had attained a normal thickness.

Reviews of the manuscript or parts thereof by M. Awdankiewicz, C. Breitkreuz, K. Drost, V. Janoušek, V. Kachlík, T. McCann, J. K. Novák and J. Ulrych considerably improved the quality of text and figures, for which I am grateful.

References

Achramowicz, S. & Wajsprych, B. 2004. First Sudetic occurrence of jadeite in metabasalts. *Mineralogical Society Of Poland, Special Papers*, **24**, 53–56.

Aftalion, M., Bowes, D. P. & Vrána, S. 1989. Early Carboniferous U–Pb zircon age for garnetiferous, perpotassic granulites, Blansky les Massif, Czechoslovakia. *Neues Jahrbuch für Mineralogie, Monatshefte*, **1989**, 145–152.

Aghabawa, M. A. 1993. Petrology and geochemistry of the Rotliegendes Volcanic Rocks in Denmark and their tectonic implications. *In*: *Dynamisk/stratigrafisk analyse af Palaeozoikum i Denmark. EFP-89; Omrade 1: Olie og Naturgas*, Volume 3. Danmarks geologiske Undersøgelse Kunderapport no. 35.

Ahrendt, H., Clauer, N., Hunziker, J. C. & Weber, K. 1983. Migration of folding and metamorphism in the Rheinische Schiefergebirge deduced from K–Ar and Rb–Sr age determinations. In: Martin, H. & Eder, F. W. (eds) *Intracontinental Fold Belts*. Springer, Berlin, 323–338.

Ahrendt, H., Franzke, H.-J., Marheine, D., Schwab, M. & Wemmer, K. 1996. Zum Alter der Metamorphose in der Wippraer Zone/Harz–Ergebnisse von K/Ar-Altersdatierungen an schwachmetamorphen Sedimenten. *Zeitschrift der deutschen Geologischen Gesellschaft*, **147**, 39–56.

Ahrendt, H., Glodny, J. *et al.* 1997. Rb–Sr and K–Ar mineral data of the KTB and the surrounding area and their bearing on the tectonothermal evolution of the metamorphic basement rocks. *Geologische Rundschau*, **86**, Supplement, S251-S257.

Alexandrov, P., Royer, J. J. & Deloule, E. 2001. 331 +/- 9 Ma emplacement age of the Soultz monzogranite (Rhine Graben basement) by U/Pb ion-probe zircon dating of samples from 5 km depth. *Comptes Rendus de l'Academie des Sciences, Paris*, **332**, 747–754.

Altenberger, U. & Besch, T. 1993. The Böllstein Odenwald: evidence for pre- to early Variscan plate convergence in the Central European Variscides. *Geologische Rundschau*, **82**, 475–488.

Altenberger, U., Besch, T., Mocek, B, Zaipeng, Y & Yong, S. 1990. Geochemie und Geodynamik des Böllsteiner Odenwaldes. *Mainzer geowissenschaftliche Mitteilungen*, **19**, 183–200.

Altenberger, U., Oberhänsli, R., Stein, E. & Moghni, M. 2001. Geochemistry, tectonic setting and geodynamic significance of late orogenic dikes in the Melibocus Massif, Bergsträsser Odenwald. *Mineralogy and Petrology*, **72**, 209–228.

Altherr, R., Henes-Klaiber, U., Hegner, E., Satir, M. & Langer, C. 1999*a*. Plutonism in the Variscan Odenwald (Germany): from subduction to collision. *International Journal of Earth Sciences*, **88**, 422–443.

Altherr, R., Henjes-Kunst, F., Langer, C. & Otto, J. 1999*b*. Interaction between crustal-derived felsic and mantle-derived mafic magmas in the Oberkirch Pluton (European Variscides, Schwarzwald, Germany). *Contributions to Mineralogy and Petrology*, **137**, 304–322.

Altherr, R., Holl, A., Hegner, E., Langer, C. & Kreuzer, H. 2000. High-potassium, calc-alkaline I-type plutonism in the European Variscides: northern Vosges (France) and northern Schwarzwald (Germany). *Lithos*, **50**, 51–73.

Anderle, H.-J. 1987. The evolution of the South Hunsrück and Taunus borderzone. *Tectonophysics*, **137**, 101–114.

André, L., Hertogen, J. & Deutsch, S. 1986. Ordovician–Silurian magmatic provinces in Belgium and the Caledonian orogeny in middle Europe. *Geology*, **14**, 879–882.

Anthes, G. & Reischmann, T. 2001. Timing of granitoid magmatism in the eastern mid-German crystalline rise. *Journal of Geodynamics*, **31**, 119–143.

Arnold, J., Jacoby, W. R., Schmeling, H. & Schott, B. 2001. Continental collision and the dynamic and thermal evolution of the Variscan orogenic crustal root–numerical models. *Journal of Geodynamics*, **31**, 273–291.

Arthaud, F. & Matte, P. 1977. Late Palaeozoic strike-slip faulting in southern Europe and northern Africa: result of a right-lateral shear zone between the Appalachians and the Urals. *Bulletin of the Geological Society of America*, **88**, 1305–1320.

Awdankiewicz, M. 1999*a*. Volcanism in a late Variscan intramontane trough: Carboniferous and Permian volcanic centres of the Intra-Sudetic Basin, SW Poland. *Geologia Sudetica*, **32**, 13–47.

Awdankiewicz, M. 1999*b*. Volcanism in a late Variscan intramontane trough: the petrology and geochemistry of the Carboniferous and Permian volcanic rocks of the Intra-Sudetic Basin, SW Poland. *Geologia Sudetica*, **32**, 83–111.

Awdankiewicz, M. 2004. Sedimentation, volcanism and subvolcanic intrusions in a late Palaeozoic intramontane trough (the Intra-Sudetic Basin, SW Poland). *In*: Breitkreuz, C. & Petford, N. (eds) *Physical Geology of High-level Magmatic Systems*. Geological Society, London, Special Publications, **234**, 5–11.

Awdankiewicz, M., Breitkreuz., C. & Ehling, B.-C. 2004. Emplacement textures in Late Palaeozoic andesite sills of the Flechtingen-Roßlau Block, north of Magdeburg (Germany). *In*: Breitkreuz, C. & Petford, N. (eds) *Physical Geology of High-level Magmatic Systems*. Geological Society, London, Special Publications, **234**, 51–66.

Bankwitz, P., Bankwitz, E., Kramer, W. & Pin, C. 1992. Early Paleozoic bimodal volcanism in the Vesser area, Thuringian Forest, eastern Germany. *Zentralblatt für Geologie und Paläontologie*, Teil

I, **9/10**, 1113–1132.

Bankwitz, P., Bankwitz, E. & Kemnitz, H. 1998. The Vesser rift: a relic of Late Cambrian rifting at the peri-Gondwana margin. *Acta Universitatis Carolinae–Geologica*, **42**, 395–398.

Bartels, C. & Kneidl, V. 1981. Ein Porphyroid in der Schiefergrube Schmiedenberg bei Bundenbach (Hunsrück, Rheinisches Schiefergebirge) und seine stratigraphische Bedeutung. *Geologisches Jahrbuch Hessen*, **109**, 23–36.

Baumann, A., Grauert, B., Mecklenburg, S. & Vinx, R. 1991. Isotope age determinations of crystalline rocks of the Upper Harz Mountains, Germany. *Geologische Rundschau*, **80**, 669–690.

Bayer, U., Schleck, M., *et al.* 1999. An integrated study of the NE German Basin. *Tectonophysics*, **314**, 285–307.

Beard, B. L., Medaris, G. L., Johnson, C. M., Brueckner, H. K. & Misar, Z. 1992. Petrogenesis of Variscan high-temperature Group A eclogites from the Moldanubian Zone of the Bohemian Massif, Czechoslovakia. *Contributions to Mineralogy and Petrology*, **111**, 468–483.

Beard, B. L., Medaris, L. G., Johnson, C. M., Jelínek, E., Tonika, J. & Riciputi, L. R. 1995. Geochronology and geochemistry of eclogites from the Mariánské Lázne Complex, Czech Republic: implications for Variscan orogenesis. *Geologische Rundschau*, **84**, 552–567.

Becker, H. 1996. Crustal trace element and isotopic signatures in garnet pyroxenites from garnet peridotite massifs from Lower Austria. *Journal of Petrology*, **37**, 785–810.

Behnisch, R. 1993. Faziesabhängige Ablagerungsprozesse devonischer Vulkaniklastite im Schalstein-Hauptsattel (mittlere Lahn-Mulde). *Geologisches Jahrbuch Hessen*, **98**, 89–149.

Behrmann, J. H. & Tanner, D. C. 1997. Carboniferous tectonics of the Variscan basement collage in eastern Bavaria and western Bohemia. *Geologische Rundschau*, **86**, 15–27.

Bełka, Z. & Narkiewicz, M. 2008. Devonian *In*: McCann, T. (ed.) *The Geology of Central Europe. Volume 1: Precambrian and Palaeozoic*. Geological Society, London, 383–410.

Benek, R. 1980. Geologisch-strukturelle Untersuchungen im Tharandter Vulkanitkomplex (Südteil DDR). *Zeitschrift für geologische Wissenschaften*, **8**, 627–643.

Benek, R. 1991. Aspekte einer Volumenbilanz paläovulkanischer Förderprodukte–Beispiel Teplice-Rhyolith (Ostdeutschland). *Zeitschrift für geologische Wissenschaften*, **19**, 379–389.

Benek, R., Paech, H. J. & Schirmer, B. 1973. Zur Gliederung der permosilesischen Vulkanite der Flechtinger Scholle. *Zeitschrift für geologische Wissenschaften*, **1**, 867–878.

Benek, R., Kramer, W., *et al.* 1996. Permo-Carboniferous magmatism of the Northeast German Basin. *Tectonophysics*, **266**, 379–404.

Berger, E., Torres, P. & Wefers, J. 1991. Zur Stratigraphie der Metamorphen Zone des Hunsrücksüdrandes im Rheinischen Schiefergebirge (vorläufige Mitteilung). *Neues Jahrbuch für Geologie und Paläontologie, Monatshefte*, **1991**, 737–746.

Black, L. P., Kinny, P. D. & Sheraton, J. W. 1991. The difficulties of dating mafic dykes: an Antarctic example. *Contributions to Mineralogy and Petrology*, **109**, 183–194.

Borsdorf, K.-H. & Estrada, S. 1991. Ordovizischer Magmatismus im östlichen Rhenoherzynikum. *Zentralblatt für Geologie und Paläontologie Teil I*, **1991**(5), 1289–1296.

Bosbach, D., Stosch, H.-G. & Seidel, E. 1991. Magmatic and metamorphic evolution of metagabbros in the Münchberg Massif, NE-Bavaria. *Contributions to Mineralogy and Petrology*, **107**, 112–123.

Boutin, R., Montigny, R. & Thuizat, R. 1995. Chronologie K–Ar et $^{39}Ar/^{40}Ar$ du métamorphose et du magmatisme des Vosges. Comparaison avec les massifs varisques avoisant et détermination de l'âge de la limite Viséen inférieur–Viséen supérieur. *Géologie de la France*, **1**, 3–25.

Bowes, D. R. & Aftalion, M. 1991. U–Pb zircon isotopic evidence for early Ordovician and late Proterozoic units in the Marianske Lazne Complex, Central European Hercynides. *Neues Jahrbuch für Mineralogie, Monatshefte*, **7**, 315–326.

Breiter, K. 1995. Geology and geochemistry of the Bohemian part of the Teplice rhyolite and adjacent post-rhyolite granites. *Terra Nostra*, **7/95**, 20–24.

Breiter, K. 1997. The Teplice rhyolite (Krušné hory Mts., Czech Republic): chemical evidence of a multiply exhausted stratified magma chamber. *Bulletin of the Czech Geological Survey*, **72**, 205–213.

Breiter, K. 2002. From explosive breccia to unidirectional solidification textures: magmatic evolution of a phosphorus- and fluorine-rich granite system (Podlesí, Krušné hory Mts., Czech Republic). *Bulletin of the Czech Geological Survey*, **77**, 67–92.

Breiter, K. & Siebel, W. 1995. Granitoids in the Rozvadov Pluton, Western Bohemia and Oberpfalz. *Geologische Rundschau*, **84**, 506–519.

Breiter, K., Frýda, J., Seltmann, R. & Thomas, R. 1997. Mineralogical evidence for two magmatic stages in the evolution of an extremely fractionated P-rich rare-metal granite: the Podlesí Stock, Krušné Hory, Czech Republic. *Journal of Petrology*, **38**, 1723–1739.

Breiter, K., Förster, H.-J. & Seltmann, R. 1999. Variscan silicic magmatism and related tin-tungsten mineralization in the Erzgebirge-Slavkovský les metallogenic province. *Mineralium Deposita*, **34**, 505–521.

Breiter. K. Novák, J. K. & Chlupáčová, M. 2001. Chemical evolution of volcanic rocks in the Altenberg-Teplice Caldera (Eastern Krušné hory Mts, Czech Republic, Germany). *GeoLines*, **13**, 17–22.

Breitkreuz, C. & Flick, H. 1997. Sedimentation am trachytisch/alkalirhyolitischen Inselvulkan von Katzenelnbogen-Steinkopf (Devon/Rheinisches Schiefergebirge). *Geologisches Jahrbuch Hessen*, **125**, 5–16.

Breitkreuz, C. & Kennedy, A. 1999. Magmatic flare-up at the Carboniferous/Permian boundary in the NE German Basin revealed by SHRIMP zircon ages. *Tectonophysics*, **302**, 307–326.

Breitkreuz, C. & Mock, A. 2004. Are laccolith complexes characteristic of transtensional basin systems? Examples from the Permo-Carboniferous of Central Europe. *In*: Breitkreuz, C. & Petford, N. (eds) *Physical Geology of High-level Magmatic Systems*. Geological Society, London, Special Publications, **234**, 13–31.

Breitkreuz, C., Kennedy, A., *et al.* 2007. Far Eastern Avalonia: its chronostratigraphic structure revealed by SHRIMP zircon ages from Upper Carboniferous to Lower Permian volcanic rocks (drill core from Germany, Poland and Denmark). *In:* Linnemann, U., Nance, R.D., Kraft, P. & Zulauf, G. (eds) *The Evolution of the Rheic Ocean: from Avalonian–Cadomian active margin to Alleghanian–Variscan collision*. Geological Society of America, Special Paper, **423**, 173–190.

Brink, H.-J., Dürschner, H. & Trappe, H. 1992. Some aspects of the late and post-Variscan development of the Northwestern German Basin. *Tectonophysics*, **207**, 65–95.

Bröcker, M., Zelazniewicz, A. & Enders, M. 1998. Rb–Sr and U–Pb geochronology of migmatitic gneisses from the Góry Sowie (West Sudetes, Poland): the importance of mid-late Devonian metamorphism. *Journal of the Geological Society, London*, **155**, 1025–1036.

Büchner, C. & Kunert, R. 1997. Pyroklastische Äquivalente der intrusiven Halleschen Rhyolithe. *Mitteilungen zur Geologie von Sachsen-Anhalt*, **3**, 37–57.

Bues, C., Dörr, W., Fiala, J., Vejnar, Z. & Zulauf, G. 2002. Emplacement depths and radiometric ages of Paleozoic plutons of the Neukirchen-Kdyně massif: differential uplift and exhumation of Cadomian basement due to Carboniferous orogenic collapse (Bohemian Massif). *Tectonophysics*, **352**, 225–243.

Buschmann, B., Linnemann, U., Schneider, J. & Süss, T. 1995. Die cadomische Entwicklung im Untergrund der Torgau-Doberluger Synklinale. *Zeitschrift für geologische Wissenschaften*, **23**, 729– 749.

Buschmann, B., Elicki, O. & Jonas, P. 2006. The Cadomian unconformity in the Saxo-Thuringian Zone, Germany: Palaeogeographic affinities of Ediacaran (terminal Neoproterozoic) and Cambrian strata. *Precambrian Research*, **147**, 387–403.

Burg, J.-P., Van Den Driessche, J. & Brun, J.-P. 1994. Syn- to post-thickening extension in the Variscan Belt of Western Europe: modes and structural consequences. *Comptes Rendus de l'Académie de Sciences, Paris*, **319**, 1019–1032.

Büthe, F. & Wachendorf, H. 1997. Die Rotliegend-Entwicklung des Ilfelder Beckens und des Kyffhäusers: Pull-Apart-Becken und Rhomb-Horst. *Zeitschrift für geologische Wissenschaften*, **25**, 291–306.

Büttner, S. H. 1999. The geometric evolution of structures in granite during continous deformation from magmatic to solid-state conditions: An example from the central European Variscan Belt. *American Mineralogist*, **84**, 1781–1792.

CARTWRIGHT, J. 1990. The structural evolution of the Ringkøbing-Fyn High. *In*: BLUNDELL, D. J. & GIBBS, A. D. (eds) *Tectonic Evolution of the North Sea Rifts*. Oxford University, New York, 200–216.

CESARE, B., RUBATTO, D., HERMANN, J. & BARZI, L. 2002. Evidence for Late Carboniferous subduction-type magmatism in mafic-ultramafic cumulates of the SW Tauern window (Eastern Alps). *Contributions to Mineralogy and Petrology*, **142**, 449–464.

CHÁB, J., ŠRÁMEK, J., POKORNÝ, L., *et al.* 1997. The Teplá-Barrandian Unit. *In*: VRÁNA, S. & ŠTĚDRÁ, V. (eds) Geological model of western Bohemia related to the KTB borehole in Germany. *Sborník Geologických Věd*, **47**, 80–104.

CHALOUPSKÝ, J. 1989. Major tectonostratigraphic units of the Bohemian Massif. *In*: DALLMEYER, R. D. (ed.) *Terranes in the Circum-Atlantic Paleozoic orogens*. Geological Society of America, Special Paper, **230**, 101–114.

CHÂTEAUNEUF, J.-J. & FARJANEL, G. (eds) 1989. *Synthèse géologique des bassins permiens français*. Mémoire de Bureau de reserches géologiques et minières, **128**.

CHEN, F. & SIEBEL, W. 2004. Zircon and titanite geochronology of the Fürstenstein granite massif, Bavarian Forest, NW Bohemian Massif: pulses of the late Variscan magmatic activity. *European Journal of Mineralogy*, **16**, 777–788.

CHEN, F., HEGNER, E. & TODT, W. 2000. Zircon ages and Nd isotopic and chemical compositions of orthogneisses from the Black Forest, Germany: evidence for a Cambrian magmatic arc. *International Journal of Earth Sciences*, **88**, 791–802.

CHEN, F., TODT, W. & HANN H. P. 2003. Zircon and garnet geochronology of eclogites from the Moldanubian Zone of the Black Forest, Germany. *Journal of Geology*, **111**, 207–222.

CLARK, A. H., SCOTT, D. J., SANDEMAN, H. A., BROMLEY, A. V. & FARRAR, E. 1998. Siegenian generation of the Lizard ophiolite: U–Pb zircon age data for plagiogranite, Porthkerris, Cornwall. *Journal of the Geological Society, London*, **155**, 595–598.

COCHERIE, A., CHANTRAINE, J., FANNING, C. M., DABARD, M.-P., PARIS, F., LE HÉRISSÉD, A. & ÉGALA, E. 2001. Datation U/Pb: âge Briovérien de la série d'Erquy (Massif armoricain, France). *Comptes Rendus de l'Académie des Sciences, Paris*, **333**, 427–434.

COCHERIE, A., GUERROT, C, FANNING, C. M. & GENTER, A. 2004. Datation U–Pb des deux faciès du granite de Soultz (Fossé rhénan, France). *Comptes Rendus Geoscience*, **336**, 775–787.

COCKS, L. R. M. & FORTEY, R. A. 1982. Faunal evidence for ocean separations in the Palaeozoic of Britain. *Journal of the Geological Society, London*, **139**, 465–478.

COCKS, L. R. M. & FORTEY, R. A. 1990. Biogeography of Ordovician and Silurian faunas. *In*: MCKERROW, W. S. & SCOTESE, C. F. (eds), *Palaeozoic Palaeogeography and Biogeography*. Geological Society, London, Memoirs, **12**, 97–104.

COCKS, L. R. M., MCKERROW, W. S. & VAN STAAL, C. R. 1997. The margins of Avalonia. *Geological Magazine*, **134**, 627–636.

COMPSTON, W., WRIGHT, A. E. & TOGHILL, P. 2002. Dating the late Precambrian volcanicity of England and Wales. *Journal of the Geological Society, London*, **159**, 323–339.

COOKE, R. A. & O'BRIEN, P. J. 2001. Resolving the relationship between high P–T rocks and gneisses in collisional terranes: an example from the Gföhl gneiss–granulite association in the Moldanubian Zone, Austria. *Lithos*, **58**, 33–54.

COOPER, M. A., COLLINS, D., FORD, M., MURPHY, F. X., TRAYNER, P. M. & O'SULLIVAN M. 1986. Structural evolution of the Irish Variscides. *Journal of the Geological Society, London*, **143**, 53–61.

COSTA, S. 1992. East–west diachronism of the collisional stage in the french Massif central: implications for the European Variscan Orogen. *Geodinamica Acta*, **5**, 51–68.

COWARD, M. P. 1993. The effect of Late Caledonian and Variscan continental escape tectonics on basement structure, Paleozoic basin kinematics and subsequent Mesozoic basin development in NW Europe. *In*: PARKER, J. R. (ed.) *Petroleum Geology of Northwest Europe: Proceedings of the 4th Conference*. Geological Society, London, 1095–1108.

CROWLEY, Q. G., FLOYD P. A., WINCHESTER, J. A., FRANKE, W. & HOLLAND, J. G. 2000. Early Palaeozoic rift-related magmatism in Variscan Europe: fragmentation of the Armorican Terrane Assemblage. *Terra Nova*, **12**, 171–180.

CROWLEY, Q. G., FLOYD, P. A., ŠTĚDRÁ, V., WINCHESTER, J. A., KACHLÍK, V. & HOLLAND, J. G. 2002*a*. The Mariánské-Lázně Complex, NW Bohemian Massif: development and destruction of an early Palaeozoic seaway. *In*: WINCHESTER, J. A., PHARAOH, T. C. & VERNIERS, J. (eds), *Palaeozoic Amalgamation of Central Europe*. Geological Society, London, Special Publications, **201**, 177–195.

CROWLEY, Q. G., TIMMERMANN, H., NOBLE, S. R. & HOLLAND, J. G. 2002*b*. Palaeozoic terrane amalgamation in Central Europe: a REE and Sm–Nd isotope study of the pre-Variscan basement, NE Bohemian Massif. *In*: WINCHESTER, J. A., PHARAOH, T. C. & VERNIERS, J. (eds) *Palaeozoic Amalgamation of Central Europe*. Geological Society, London, Special Publications, **201**, 157–176.

DALLMEYER, R. D. & URBAN, M. 1998. Variscan versus Cadomian tectonothermal activity in northwestern sectors of the Teplá-Barrandian zone, Czech Republic; constraints from $^{40}Ar/^{39}Ar$ ages. *Geologische Rundschau*, **87**, 94–106.

DALLMEYER, R. D., NEUBAUER, F. & HÖCK, V. 1992. Chronology of late Paleozoic tectonothermal activity in the southeastern Bohemian massif, Austria (Moldanubian and Moravo-Silesian zones): $^{40}Ar/^{39}Ar$ mineral age controls. *Tectonophysics*, **210**, 135–153.

DEBACKER, T. N., DEWAELE, S., SINTUBIN, M., VERNIERS, J., MUCHEZ, P. & BOVEN, A. 2005. Timing and duration of the progressive deformation of the Brabant Massif, Belgium. *Geologica Belgica*, **8**, 20–34.

DEKORP-BASIN RESEARCH GROUP 1999. Deep crustal structure of the Northeast German basin: new DEKORP-BASIN '96 deep-profiling results. *Geology*, **27**, 55–58.

DE LA ROCHE, H., LETERRIER, J., GRANDE CLAUDE, P. & MARCHAL, M. 1980. A classification of volcanic and plutonic rocks using R1-R2 diagrams and major element analyses–its relationships and current nomenclature. *Chemical Geology*, **29**, 183–210.

DEWAELE, S., BOVEN, A. & MUCHEZ, P. H. 2002. $^{40}Ar/^{39}Ar$ dating of mesothermal, orogenic mineralization in a low-angle reverse shear zone in the Lower Palaeozoic of the Anglo-Brabant fold belt, Belgium. *Applied Earth Science: IMM Transactions section B*, **111**, 215–220.

DOMBROWSKI, A., HENJES-KUNST, F., HÖHNDORF, A., KRÖNER, A., OKRUSCH, M. & RICHTER, P. 1995. Orthogneisses in the Spessart Crystalline Complex, north-west Bavaria: Silurian granitoid magmatism at an active continental margin. *Geologische Rundschau*, **84**, 399–411.

DÖRR, W., FIALA, J., VEJNAR, Z. & ZULAUF, G. 1998. U–Pb zircon ages and structural development of metagranitoids of the Teplá crystalline complex: evidence for pervasive Cambrian plutonism within the Bohemian massif (Czech Republic). *Geologische Rundschau*, **87**, 135–149.

DÖRR, W., ZULAUF, G., FIALA, J., FRANKE, W. & VEJNAR, Z. 2002. Neoproterozoic to Early Cambrian history of an active plate in the Teplá–Barrandian unit—a correlation of U–Pb isotopic dilution-TIMS ages (Bohemia, Czech Republic). *Tectonophysics*, **352**, 65–85.

DOSTAL, J., PATOČKA, F. & PIN, C. 2001. Middle/late Cambrian intracontinental rifting in the central West Sudetes, NE Bohemian Massif (Czech Republic): geochemistry and petrogenesis of the bimodal metavolcanic rocks. *Geological Journal*, **36**, 1–17.

DROST, K., LINNEMANN, U., *et al.* 2004. New data on the Neoproterozoic–Cambrian geotectonic setting of the Teplá-Barrandian volcanosedimentary successions: geochemistry, U–Pb zircon ages, and provenance (Bohemian Massif, Czech Republic). *International Journal of Earth Sciences*, **93**, 742–757.

DUBIŃSKA, E., BYLINA, P., KOZŁOWSKI, A., DÖRR, W., NEJBERT, K., SCHASTOK, J. & KULICKI, C. 2004. U–Pb dating of serpentinization: hydrothermal zircon from a metasomatic rodingite shell (Sudetic ophiolite, SW Poland). *Chemical Geology*, **203**, 183–203.

DUCROT, J., LANCELOT, J. R. & MARCHAND, J. 1983. Datation U–Pb sur zircons de l'éclogite de la Boire (Haut-Allier, France) et conséquences sur l'évolution ante-hercynienne de l'Europe occidentale. *Earth and Planetary Science Letters*, **62**, 385–394.

DVOŘÁK, J. 1995. Stratigraphy. *In*: DALLMEYER, R. D., FRANKE, & W., WEBER, K. (eds) *Pre-Permian Geology of Central and Eastern Europe*. Springer-Verlag, Berlin, 477–489.

DZIEDZIC, H. & KRAMER, W. 1992. Structural and petrogenetic analysis of the Braszowice gabbro (Fore-Sudetic Block, SW Poland). *Zentralblatt für Geologie und Paläontologie, Teil I*, **9/10**, 1019–1037.

DZIEDZIC, K. & DZIEDZIC, H. 2000. Genetic relationships between metabasalts and related gabbroic rocks: an example from the Fore-Sudetic Block, SW Poland. *Geologia Sudetica*, **33**, 33–48.

Echtler, H. P. & Chauvet, A. 1992. Carboniferous convergence and subsequent crustal extension in the southern Schwarzwald (SW Germany). *Geodinamica Acta*, **5**, 39–49.

Eckhardt, F. J. 1979. Der Permische Vulkanismus Mitteleuropas. *Geologisches Jahrbuch*, **D35**, 3–84.

Edel, J. B. 2001. The rotations of the Variscides during the Carboniferous collision: paleomagnetic constraints from the Vosges and the Massif Central (France). *Tectonophysics*, **332**, 69–92.

Ehling, B.-C. & Breitkreuz, C. 2004. Spätpaläozoische Sedimentation und Vulkanismus im nordöstlichen Saale-Becken. *Schriftreihe der Deutschen Geologischen Gesellschaft*, **35**, 133–149.

Emmermann, R. & Wohlenberg, J. (eds.) 1989. The German Continental Deep Drilling Program (KTB). Site selection studies in the Oberpfalz and Schwarzwald. Springer-Verlag, Berlin.

Engel, W., Franke, W., Grote, C., Weber, K., Ahrendt, H. & Eder, F. W. 1983. Nappe tectonics in the southeastern part of the Rheinisches Schiefergebirge. *In*: Martin, H. & Eder, F. W. (eds) *Intracontinental Fold Belts*. Springer, Berlin, 267–287.

Faure, M. 1995. Late orogenic Carboniferous extensions in the Variscan French Massif Central. *Tectonics*, **14**, 132–153.

Fiala, J., Fuchs, G. & Wendt, J. I. 1995. Stratigraphy. *In*: Dallmeyer, R. D., Franke, W., & Weber, K. (eds) *Pre-Permian Geology of Central and Eastern* Europe. Springer-Verlag, Berlin, 444–452.

Finger, F. & Steyer, H.P. 1990. I-type granitoids as indicators of a late Paleozoic convergent ocean-continent margin along the southern flank of the central European Variscan orogen. *Geology*, **18**, 1207–1210.

Finger, F. & Steyer, H. P. 1995. A tectonic model for the eastern Variscides: Indications from a chemical study of amphibolites in the south-eastern Bohemian Massif. *Geologica Carpathica*, **46**, 137–150.

Finger, F. & von Quadt, A. 1995. U–Pb ages of zircons from a plagiogranite-gneiss in the southeastern Bohemian massif, Austria: further evidence for an important early Palaeozoic rifting episode in the eastern Variscides. *Schweizerische Mineralogische und Petrografische Mitteilungen*, **75**, 265–270.

Finger, F., Frasl, G., Höck, V. & Steyer, H.P. 1989. The granitoids of the Moravian Zone of northeast Austria: products of a Cadomian active continental margin? *Precambrian Research*, **45**, 234–345.

Finger, F., Frasl, G., Dudek, A., Jelínek, E. & Thöni, M. 1995. Igneous activity (Cadomian plutonism in the Moravo-Silesian Basement). *In*: Dallmeyer, R. D., Franke, W., & Weber, K. (eds) *Pre-Permian Geology of Central and Eastern Europe*. Springer-Verlag, Berlin, 495–507.

Finger, F., Roberts, M. P., Haunschmid, B., Schermaier, A. & Steyer, H. P. 1997. Variscan granitoids of central Europe: their typology, potential sources and tectonothermal relations. *Mineralogy and Petrology*, **61**, 67–96.

Finger, F., Hanžl, P., Pin, C., von Quadt, A. & Steyer, H. P. 2000. The Brunovistulian: Avalonian Precambrian sequence at the eastern end of the Central European Variscides? *In*: Franke, W., Haak, V., Oncken, O. & Tanner, D. (eds) *Orogenic Processes: Quantification and Modelling in the Variscan Belt*. Geological Society, London, Special Publications, **179**, 103–112.

Flick, H. 1986. Permokarboner Vulkanismus im südlichen Odenwald. *Heidelberger Geowissenschaflichte Abhandlungen*, **6**, 121–138.

Flick, H., Horn, M., Nesbor, H. D. & Wengert, N. 1991. Eine subvulkanische Magmenkammer des devonischen Vulkanismus (Givet/Adorf-Phase) am Nordwestrand der Dillmulde, Rheinisches Schiefergebirge. *Geologisches Jahrbuch Hessen*, **119**, 45–76.

Floyd, P. A. 1982. Chemical variations in Hercynian basalts relative to plate tectonics. *Journal of the Geological Society, London*, **139**, 505–520.

Floyd, P. A. 1995. Igneous activity. *In*: Dallmeyer, R. D., Franke, W., & Weber, K. (eds) *Pre-Permian Geology of Central and Eastern Europe*. Springer-Verlag, Berlin, 59–81.

Floyd, P. A., Exley, C. S. & Styles, M. T. 1993. *Igneous rocks of Southwest England*. Chapman & Hall, London, Geological Conservation Review, **5**.

Floyd, P. A., Winchester, J. A., Ciesielczuk, J., Lewandowska, A., Szczepanski, J. & Turniak, K. 1996. Geochemistry of early Palaeozoic amphibolites from the Orlica-Śnieżnik dome, Bohemian massif: petrogenesis and palaeotectonic aspects. *Geologische Rundschau*, **85**, 225–238.

Floyd, P. A., Kryza, R., Crowley, Q. G., Winchester, J. A. & Abdel Wahed, M. 2002. Ślęża Ophiolite: geochemical features and relationships to Lower Palaeozoic rift magmatism in the Bohemian Massif. *In*: Winchester, J. A., Pharaoh, T. C. & Verniers, J. (eds) *Palaeozoic Amalgamation of Central Europe*. Geological Society, London, Special Publications, **201**, 197–215.

Förster, H.-J. 1998. The chemical composition of REE-Y-Th-U-rich accessory minerals in peraluminous granites of the Erzgebirge-Fichtelgebirge region, Germany, Part I: The monazite-(Ce)-brabantite solid solution series. *American Mineralogist*, **83**, 259–272.

Förster, H.-J. 2000. Cerite-(Ce) and Thorian synchysite-(Ce) from the Niederbobritsch granite, Erzgebirge, Germany: implications for the differential mobility of the LREE and Th during alteration. *Canadian Mineralogist*, **38**, 67–79.

Förster, H.-J. 2001. Synchisite–(Y)-synchisite-(Ce) solid solutions from Markersbach, Erzgebirge, Germany: REE and Th mobility during high-T alteration of highly fractionated aluminous A-type granites. *Mineralogy and Petrology*, **72**, 259–280.

Förster, H.-J., Tischendorf, G., Seltmann, R. & Gottesmann, B. 1998. Die variszischen Granite des Erzgebirges: neue Aspekte aus stofflicher Sicht. *Zeitschrift für geologische Wissenschaften*, **26**, 31–60.

Förster, H.-J., Tischendorf, G., Trumbull, R. B. & Gottesmann, B. 1999. Late-collisional granites in the Variscan Erzgebirge, Germany. *Journal of Petrology*, **40**, 1613–1645.

Fortey, R. A. & Cocks, L. R. M. 2003. Palaeontological evidence bearing on global Ordovician–Silurian continental reconstructions. *Earth-Science Reviews*, **61**, 245–307.

Franke, W. 1989. Variscan plate tectonics in central Europe: current ideas and open questions. *Tectonophysics*, **169**, 221–228.

Franke, W. 2000. The mid-European segment of the Variscides: tectonostratigraphic units, terrane boundaries and plate tectonic evolution. *In*: Franke, W., Haak, V., Oncken, O. & Tanner, D. (eds) *Orogenic Processes: Quantification and Modelling in the Variscan Belt*. Geological Society, London, Special Publications, **179**, 35–61.

Franke, W. & Stein, E. 2000. Exhumation of high-grade rocks in the Saxo-Thuringian Belt: geological constraints and geodynamic concepts. *In*: Franke, W., Haak, V., Oncken, O. & Tanner, D. (eds) *Orogenic Processes: Quantification and Modelling in the Variscan Belt*. Geological Society, London, Special Publications, **179**, 337–354.

Franke, W., Dallmeyer, D. & Weber, K. 1995. Geodynamic evolution. *In*: Dallmeyer, R. D., Franke, W. & Weber, K. (eds) *Pre-Permian geology of central and eastern Europe*. Springer, Berlin, 579–593.

Friedl, G., Finger, F., McNaughton, N. J. & Fletcher, I. R. 2000. Deducing the ancestry of terranes: SHRIMP evidence for South America -derived Gondwana fragments in central Europe. *Geology*, **28**, 1035–1038.

Friedl, G., Finger, F., Paquette, J. L., von Quadt, A., McNaughton, N. J. & Fletcher, I. R. 2004. Pre-Variscan geological events in the Austrian part of the Bohemian Massif deduced from U–Pb zircon ages. *International Journal of Earth Sciences*, **93**, 802–823.

Fritz, H. 1996. Geodynamic and tectonic evolution of the southeastern Bohemian Massif: the Thaya section (Austria). *Mineralogy and Petrology*, **58**, 253–278.

Fritz, H. & Neubauer, F. 1993. Kinematics of crustal stacking and dispersion in the southeastern Bohemian Massif. *Geologische Rundschau*, **82**, 556–565.

Frost, R. T. C., Fitch, F. J. & Miller, J. A. 1981. The age and nature of the crystalline basement of the North Sea Basin. *In*: Illing, L. V. & Hobson, G. D. (eds) *Petroleum Geology of the Continental Shelf of North-west Europe*. Institute of Petroleum, London, 43–57.

Furnes, H., Kryza, R., Muszynski, A., Pin, C. & Garmann, L. B. 1994. Geochemical evidence for progressive, rift-related early Palaeozoic volcanism in the western Sudetes. *Journal of the Geological Society, London*, **151**, 91–109.

Gansslоser, M. 1993. Die Metabasalte des Harzes. *Zentralblatt für Geologie und Paläontologie, Teil I*, **9/0**, 1103–1115.

Gebauer, D. & Grünenfelder, M. 1979. U–Pb zircon and Rb–Sr mineral dating of eclogites and their country rocks. *Earth and*

Planetary Science Letters, **42**, 35–44.

Gehmlich, M., Linnemann, U., Tichomirowa, M., Lützner, H. & Bombach, K. 1997. Datierung und Korrelation neoproterozoisch-frühpaläozoischer Profile des Schwarzburger Antiklinoriums und der Elbezone auf der Basis der Geochronologie von Einzelzirkonen. *Zeitschrift für Geologische Wissenschaften*, **25**, 191–201.

Gehmlich, M., Linnemann, U., Tichomirowa, M., Todt, W. & Bombach, K. 2000. U–Pb- und Pb-Pb-Zirkondatierungen an Orthogesteinen der Elbezone: Konsequenzen für variszische Deckenüberschiebungen im Saxothuringikum. *Zeitschrift der Deutschen Gesellschaft für Geowissenschaften*, **151**, 209–230.

Gerdes, A. 2001. Magma homogenization during anatexis, ascent and/or emplacement? Constraints from the Variscan Weinsberg Granites. *Terra Nova*, **13**, 305–312.

Gerdes, A., Wörner, G. & Finger, F. 2000. Hybrids, magma mixing and enriched mantle melts in post-collisional Variscan granitoids: the Rastenberg Pluton, Austria. *In*: Franke, W., Haak, V., Oncken, O. & Tanner, D. (eds) *Orogenic Processes: Quantification and Modelling in the Variscan Belt*. Geological Society, London, Special Publications, **179**, 415–431.

Glasmacher, U. A., Mann, U. & Wagner, G. A. 2002. Thermotectonic evolution of the Barrandian, Czech Republic, as revealed by apatite fission-track analysis. *Tectonophysics*, **359**, 381–402.

Glässer, W. 1983. Beitrag zur petrologie und Vulkanologie der andesitoiden Vulkanite Nordwestsachsens. *Hallesches Jahrbuch für Geowissenschaften*, **8**, 1–30.

Glennie, K. W. 1997. Recent advances in understanding the southern North Sea Basin: a summary. *In*: Ziegler, K., Turner, P. & Daines, S. R. (eds) *Petroleum Geology of the Southern North Sea: Future Potential*. Geological Society, London, Special Publications, **123**, 17–29.

Glodny, J., Grauert, B., Fiala, J., Vejnar, Z. & Krohe, A. 1998. Metapegmatites in the western Bohemian massif: ages of crystallisation and metamorphic overprint, as constrained by U–Pb zircon, monazite, garnet, columbite and Rb–Sr muscovite data. *Geologische Rundschau*, **87**, 124–134.

Gofette, O., Liégeois, J.-P. & André, L. 1991. Age U–Pb sur zircon dévonien moyen à supérieur du magmatisme bimodal du Massif de Rocroi (Ardenne, France): implications géodynamiques. *Comptes Rendus de l'Académie des Sciences, Paris*, **312**, 1155–1161.

Goll, M. & Lippolt, H. J. 2001. Biotit-Geochronologie ($^{40}Ar_{rad}/K$, $^{40}Ar_{rad}/^{39}Ar_K$, $^{87}Sr_{rad}/^{87}Rb$) spät-variszischer Magmatite des Thüringer Waldes. *Neues Jahrbuch für Geologie und Paläontologie*, **222**, 353–405.

Göthel, M. 2001. Das autochthone und allochthone Paläozoikum des 'Görlitzer Schiefergebirges' (Mitteleuropäische Varisziden, Deutschland). *Zeitschrift für geologische Wissenschaften*, **29**, 55–73.

Gradstein, F. M., Ogg, J. G., Smith, A. G., Bleeker, W. & Lourens, L.J. 2004. A new Geologic Time Scale, with special reference to Precambrian and Neogene. *Episodes*, **27**, 33–100.

Grauert, B., Lork, A. & O'Brien, P. 1994. Altersbestimmung akzessorischer Zirkone und Monazite aus der KTB-Vorbohrung. *KTB Report*, **94-2**, B30.

Grösser, J. & Dörr, W. 1986. MOR-typ-Basalte im ostlichen Rheinischen Schiefergebirge. *Neues Jahrbuch für Geologie und Paläontologie, Monatshefte*, **12**, 705–722.

Guerrot, C. & Peucat, J. J. 1990. U–Pb geochronology of the Upper Proterozoic Cadomian orogeny in the northern Armorican Massif, France. *In*: D'Lemos, R. S., Strachan, R. A. & Topley, C. G. (eds) *The Cadomian orogeny*. Geological Society of London, Special Publications, **51**, 13–26.

Guerrot, C., Peucat, J. J., Capdevilla, R. & Dosso, L. 1989. Archean protoliths within Early Proterozoic crust of the west European Hercynian belt: possible relics of the west African craton. *Geology*, **17**, 241–244.

Gunia, T. 1999. Microfossils from the high-grade metamorphic rocks in the Gory Sowie Mts. (Sudetes area) and their stratigraphical importance. *Geological Quarterly*, **43**, 519–536.

Hammer, J., Eidam, J., Röber, B. & Ehling, B. -C. 1999. Prävariszischer und variszischer granitoider Magmatismus m NE-Rand des Böhmischen Massivs–Geochemie und Petrogenese. *Zeitschrift für geologische Wissenschaften*, **27**, 401–415.

Hammerschnidt, K. & Franz, G. 1992. Retrograde evolution of eclogites: evidences from microstructures and $^{40}Ar/^{39}Ar$ white mica dates, Münchberg Massif, northern Bavaria. *Contributions to Mineralogy and Petrology*, **111**, 113–125.

Hammerschmidt, K., Kopp, J. & Haucke, L. 2003. Sm–Nd dating on minerals of a silica undersaturated, Ca-rich, crustally derived gabbro of the Mid-German Crystalline Zone, drill hole Züllsdorf, southern Brandenburg (Germany). *Zeitschrift für geologische Wissenschaften*, **31**, 225–238.

Hanel, M., Montanari, M. & Kalt, A. 1999. Determining sedimentation ages of high-grade metamorphic gneisses by their palynological record: a case study in the northern Schwarzwald (Variscan Belt, Germany). *International Journal of Earth Sciences*, **88**, 49–59.

Hann, H. P. & Sawatzki, G. 1998. Deckenbau und Sedimentationsalter im Grundgebirge des Südschwarzwalds/SW-Deutschland. *Zeitschrift der Deutschen Geologischen Gesellschaft*, **149**, 183–195.

Hann, H. P., Sawatzki, G. & Vaida, M. 1995. Chinitozoen und Acritarchen des Ordoviziums aus metamorphen Grauwacken der Zone von Badenweiler-Lenzkirch, Schwarzwald, SW-Deutschland. *Neues Jahrbuch für Geologie und Paläeontologie, Monatshefte*, **1995**, 375–383.

Hann, H. P., Chen, F., Zedler, H., Frisch, W. & Loeschke, J. 2003. The Rand Granite in the southern Schwarzwald and its geodynamic significance in the Variscan belt of SW Germany. *International Journal of Earth Sciences*, **92**, 821–842.

Hanžl, P., Janoušek, V., *et al.* 2006. Magmatic history of granite-derived mylonites from the southern Desná Unit (Silesicum, Czech Republic). *Mineralogy and Petrology*, **89**, 45–75.

Harms, U., Cameron, K. L., Simon, K. & Brätz, H. 1997. Geochemistry and petrogenesis of metabasites from the KTB ultradeep borehole, Germany. *Geologische Rundschau*, **86**, Supplement, S155-S166.

Hartley, A. J. & Otava, J. 2001. Sediment provenance and dispersal in a deep marine foreland basin: the Lower Carboniferous Culm Basin, Czech Republic. *Journal of the Geological Society, London*, **158**, 137–150.

Hecht, L., Vigneresse, J. L. & Morteani, G. 1997. Constraints on the origin of zonation of the granite complexes in the Fichtelgebirge (Germany and Czech Republic): evidence form a gravity and geochemical study. *Geologische Rundschau*, **86**, Supplement, S93-S109.

Heeremans, M., Faleide, J. I. & Larsen, B. T. 2004*a*. Late Carboniferous–Permian of NW Europe: an introduction to a new regional map. *In*: Wilson, M., Neumann, E.-R., Davies, G.R., Timmerman, M.J., Heeremans, M. & Larsen, B. T. (eds) *Permo-Carboniferous Magmatism and Rifting in Europe*. Geological Society, London, Special Publications, **223**, 75–88.

Heeremans, M., Timmerman, M. J., Kirstein, L. A. & Faleide, J. I. 2004*b*. New constraints on the timing of late Carboniferous–early Permian volcanism in the North Sea. *In*: Wilson, M., Neumann, E.-R., Davies, G. R., Timmerman, M. J., Heeremans, M. & Larsen, B. T. (eds) *Permo-Carboniferous Magmatism and Rifting in Europe*. Geological Society, London, Special Publications, **223**, 177–194.

Hegner, E. & Kröner, A. 2000. Review of Nd isotopic data and xenocrystic and detrital zircon ages from the pre-Variscan basement in the eastern Bohemian Massif: speculations on palinspastic reconstructions. *In*: Franke, W., Haak, U., Oncken, O. & Tanner, D. (eds) *Orogenic Processes: Quantification and Modelling in the Variscan Belt*. Geological Society, London, Special Publications, **179**, 113–129.

Hegner, E., Kölbl-Ebert, M. & Loeschke, J. 1998. The origin of post-collisional Variscan lamprophyres from the Black Forest, Germany; $^{40}Ar/^{39}Ar$ phlogopite ages, compositional, and isotope characteristics. *Lithos*, **45**, 395–411.

Hegner, E., Chen, F. & Hann, H. P. 2001. Chronology of basin closure and thrusting in the internal zone of the Variscan belt in the Schwarzwald, Germany: evidence from zircon ages, trace element geochemistry, and Nd isotopic data. *Tectonophysics*, **332**, 169–184.

Henk, A. 1993. Late orogenic evolution in the Variscan Internides: the Saar-Nahe Basin, southwest Germany. *Tectonophysics*, **223**, 273–290.

Henk, A. 1999. Did the Variscides collapse or were they torn apart?: a quantitative evaluation of the driving forces for postconvergent extension in central Europe. *Tectonics*, **18**, 774–792.

Hess, J. C. & Lippolt, H. J. 1986. $^{40}Ar/^{39}Ar$ ages of tonstein and tuff sanidines: new calibration points for the improvement of the Upper Carboniferous time scale. *Chemical Geology, Isotope Geoscience section*, **59**, 143–154.

HESS, J. C. & LIPPOLT, H. J. 1996. Numerische Stratigraphie permokarbonischer Vulkanite Zentraleuropas. Teil III: Odenwald. *Geologisches Jahrbuch Hessen*, **124**, 39–45.

HESS, J. C., BACKFISCH, S. & LIPPOLT, H. J. 1983. Konkordantes Sanidinund diskordante Biotitalter eines Karbontuffs der Baden-Baden Senke, Nordschwarzwald. *Neues Jahrbuch für Geologie und Paläontologie, Monatshefte*, **5**, 277–292.

HESS, J. C., LIPPOLT, H. J. & KOBER, B. 1995. The age of the Kagenfels granite (northern Vosges) and its bearing on the intrusion scheme of late Variscan granitoids. *Geologische Rundschau*, **84**, 568–577.

HEUSE, T., KURZE, M. & REITZ, E. 1994. Paläontologischer Nachweis von Oberproterozoikum (Vendium) in der Zone Münchberg–Frankenberg. *Neues Jahrbuch für Geologie und Paläontologie, Monatshefte*, **1994**, 95–107.

HIRSCHMANN, G. & OKRUSCH, M. 1988. Spessart-Kristallin und Ruhlaer Kristallin als Bestandteile der Mitteldeutschen Kristallinzone–ein Vergleich. *Neues Jahrbuch für Geologie und Paläontologie, Abhandlungen*, **177**, 1–39.

HLADIL, J., MELICHAR, R., *et al.* 1999. The Devonian in the easternmost Variscides, Moravia: a holistic analysis directed towards comprehension of the original context. *Abhandlungen der Geologischen Bundesanstalt, Wien*, **54**, 27–47.

HLADIL, J., PATOČKA, F., KACHLÍK, V., MELICHAR, R. & HUBAČÍK, M. 2003. Metamorphosed carbonates of Krkonoše Mountains and Paleozoic evolution of Sudetic terranes (NE Bohemia, Czech Republic). *Geologica Carpathica*, **54**, 281–297.

HÖCK, V. 1995. Metamorphic evolution. *In*: DALLMEYER, R. D., FRANKE, W. & WEBER, K. (eds.) *Pre-Permian geology of Central and Eastern Europe*. Springer-Verlag, Berlin, 541–553.

HÖCK, V., MONTAG, O. & LEICHMANN, J. 1997. Ophiolite remnants at the eastern margin of the Bohemian massif and their bearing on the tectonic evolution. *Mineralogy and Petrology*, **60**, 267–287.

HOFFMANN, U. 2002. Das Rotliegend des Döhlen-Beckens (Elbezone): neue Daten zu Vulkanismus und Sedimentation. *Workshop 'Oberkarbon–Untertrias in Zentraleuropa: prozesse und ihr Timing'*. TU Bergakademie Freiberg, Geologisches Institut, 10–14.

HOLDER, M. T. & LEVERIDGE, B. E. 1986. Correlation of the Rhenohercynian Variscides. *Journal of the Geological Society, London*, **143**, 141–147.

HOLL, P. K., VON DRACH, V., MÜLLER-SOHNIUS, D. & KÖHLER, H. 1989. Caledonian ages in Variscan rocks: Rb–Sr and Sm–Nd isotope variations in dioritic intrusives from the northwestern Bohemian Massif, West Germany. *Tectonophysics*, **157**, 179–194.

HOLUB, F. V., KLECKA, M. & MATEJKA, D. 1995. Igneous activity. *In*: DALLMEYER, R. D., FRANKE, W., & WEBER, K. (eds) *Pre-Permian Geology of Central and Eastern Europe*. Springer-Verlag, Berlin, 444–452.

HOLUB, F. V, COCHERIE, A. & ROSSI, P. 1997. Radiometric dating of granitic rocks from the Central Bohemian Plutonic Complex (Czech Republic): constraints on the chronology of thermal and tectonic events along the Moldanubian-Barrandian boundary. *Comptes Rendus de l'Academie des Sciences, Paris*, **325**, 19–26.

HOTH, K. 2000. Diskussionsbeitrag zur lithostratigraphischen Gliederung der Metasedimente im südlichen uns östlichen Krkonoše/Karkonosze. *Zeitschrift für geologische Wissenschaften*, **28**, 211–277.

HOTH, K., HUEBSCHER, H.-D., KORICH, D., GABRIEL, W. & ENDERLEIN, F. 1993*a*. Die Lithostratigraphie der permokarbonischen Effusiva im Zentralabschnitt der Mitteleuropäischen Senke. *Geologisches Jahrbuch*, **A131**, 179–196.

HOTH, K., RUSBÜLT, J., ZAGORA, K., BEER, H. & HARTMANN, O. 1993*b*. Die tiefen Bohrungen im Zentralabschnitt der Mitteleuropäischen Senke–Eine Dokumentation für den Zeitabschnitt 1961–1990. *Schriftenreihe für Geowissenschaften*, **2**, 7–145.

HWANG, S.-L., SHEN, P., CHU, H.-T. YUI, T.-F. & LIN, C.-C. 2001. Genesis of microdiamonds from melt and associated multiphase inclusions in garnet of ultrahigh-pressure gneiss from Erzgebirge, Germany. *Earth and Planetary Science Letters*, **188**, 9–15.

INGLIS, J. D., SAMSON, S. D., D'LEMOS, R. S. & HAMILTON, M. 2004. U–Pb geochronological constraints on the tectonothermal evolution of the Paleoproterozoic basement of Cadomia, La Hague, NW France. *Precambrian Research*, **134**, 293–315.

IRBER, W., FÖRSTER, H. S., HECHT, L., MÖLLER, P. & MORTEANI, J. 1997. Experimental, geochemical, mineralogical and O-isotope constraints on the late-magmatic history of the Fichtelgebirge granites (Germany). *Geologische Rundschau*, **86**, 110–124.

IRVINE, T. N. & BARAGAR, W. R. A. 1971. A guide to the chemical classification of the common volcanic rocks. *Canadian Journal of Earth Sciences*, **8**, 523–548.

JACOB, G. 1995. Ergebnisse mikrogefügeanalytischer Untersuchungen an schwachmetamorphen gesteinen der Zone von Wippra (Harz). *Zentralblatt für Geologie und Paläontologie Teil I*, **1993**(9/10), 1201–1211.

JACOB, G. & FRANZKE, H.J. 1992. Die tektonische Entwicklung der Wippraer Zone im Unterharz. *Zentralblatt für Geologie und Paläontologie Teil I*, **1992**(1/2), 51–61.

JANOUŠEK, V., ROGERS, G. & BOWES, D. R. 1995. Sr–Nd isotopic constraints on the petrogenesis of the Central Bohemian Pluton, Czech Republic. *Geologische Rundschau*, **84**, 520–534.

JANOUŠEK, V., ROGERS, G., BOWES, D. R. & VAŇKOVÁ, V. 1997. Cryptic trace-element variation as an indicator of reverse zoning in a granitic pluton: the Říčany granite, Czech Republic. *Journal of the Geological Society, London*, **154**, 807–815.

JANOUŠEK, V., BOWES, D. R., BRAITHWAITE, C. J. R. & ROGERS, G. 2000*a*. Microstructural and mineralogical evidence for limited involvement of magma mixing in the petrogenesis of a Hercynian high-K calc-alkaline intrusion: the Kozárovice granodiorite, Central Bohemian Pluton, Czech Republic. *Transactions of the Royal Society of Edinburgh: Earth Sciences*, **91**, 15–26.

JANOUŠEK, V., BOWES, D. R., ROGERS, G., FARROW, C. M. & JELÍNEK, E. 2000*b*. Modelling diverse processes in the petrogenesis of a composite batholith: the Central Bohemian Pluton, Central European Hercynides. *Journal of Petrology*, **41**, 511–543.

JANOUŠEK, V., FARROW, C. M. & ERBAN, V. 2006*a*. Interpretation of whole-rock geochemical data in igneous geochemistry: introducing Geochemical Data Toolkit (GCDkit). *Journal of Petrology*, **47**, 155–1259.

JANOUŠEK, V., HANŽL, P., *et al.* 2006*b*. Contrasting petrogenesis of two volcanic suites in the Devonian Vrbno Group (Hrubý Jeseník Mts., Czech Republic). *GeoLines*, **20**, 57–59.

JELÍNEK, E., PAČESOVÁ, M., MÍSAŘ, Z., MARTINEC, P. & WEISS, Z. 1984. Geochemistry of a dismembered metaophiolite complex, Letovice, Czechoslovakia. *Transactions of the Royal Society of Edinburgh, Earth Sciences*, **75**, 37–48.

JELÍNEK, E., ŠTĚDRÁ, V. & CHÁB, J. 1997. The Mariánské Lázně Complex. *In*: VRÁNA, S. & ŠTĚDRÁ, V. (eds) Geological model of western Bohemia related to the KTB borehole in Germany. *Sborník Geologických Věd*, **47**, 61–70.

JELÍNEK, E., PEŠEK, J. & BOUŠKA, V. 2001. Late Palaeozoic volcanism in the West and Central Bohemian Basins (Czech Republic). *Acta Universitatis Carolinae, Geologica*, **45**, 33–39.

JOHNSON, G. A. L. & DUNHAM, K.C. 2001. Emplacement of the Great Whin Dolerite Complex and the Little Whin Sill in relation to the structure of northern England. *Proceedings of the Yorkshire Geological Society*, **53**, 177–186.

JONAS, P. & BUSCHMANN, B. 2001. Vorkommen und Petrographie intraformationeller Vulkanite im Unterkambrium der Torgau-Doberlug Synklinale (NE Saxothuringische Zone). *Zeitschrift für geologische Wissenschaften*, **29**, 119–147.

JONAS, P., BUSCHMANN, B. & GAITZSCH, B. 2000. Unterkambrischer und unterkarbonischer Vulkanismus der Torgau-Doberlug Synklinale (NE Saxothuringische Zone). *Zeitschrift für geologische Wissenschaften*, **28**, 157–175.

KACHLÍK, V. 1997. The Kladská Unit. *In*: VRÁNA, S. & ŠTĚDRÁ, V. (eds) Geological model of western Bohemia related to the KTB borehole in Germany. *Sborník Geologických Věd*, **47**, 70–80.

KACHLÍK, V. & PATOČKA, F. 1998. Cambrian/Ordovician Intracontinental rifting and Devonian closure of the rifting generated basins in the Bohemian Massif realms. *Acta Universitatis Carolinae–Geologica*, **42**, 433–441.

KALT, A., GRAUERT, B. & BAUMANN, A. 1994*a*. Rb–Sr and U–Pb isotope studies on migmatites from the Schwarzwald (Germany): constraints on isotopic resetting during Variscan high-temperature metamorphism. *Journal of Metamorphic Geology*, **12**, 667–680.

KALT, A., HANEL, M., SCHLEICHER, H. & KRAMM, U. 1994*b*. Petrology and geochronology of eclogites from the Variscan Schwarzwald (F.R.G.). *Contributions to Mineralogy and Petrology*, **115**, 287–302.

KALT, A., CORFU, F. & WIJBRANS, J. 2000. Time calibration of a P-T path from a Variscan high-temperature low-pressure metamorphic complex (Bayerische Wald, Germany), and the detection of inherited monazite. *Contributions to Mineralogy and Petrology*, **138**, 143–163.

KALVODA, J., MELICHAR, R., BÁBEK, O. & LEICHMANN, J. 2002. Late Proterozoic–Paleozoic tectonostratigraphic development and paleogeography of Brunovistulian Terrane and comparison with other terranes at the SE margin of Baltica-Laurussia. *Journal of the Czech Geological Society*, **47**, 81–102.

KAUFMANN, B., TRAPP, E. & MEZGER, K. 2004. The numerical age of the upper Frasnian (Upper Devonian) Kellwasser horizons: a new U–Pb zircon date from Steinbruch Schmidt (Kellerwald, Germany). *Journal of Geology*, **112**, 495–501.

KAUFMANN, B., TRAPP, E., MEZGER, K. & WEDDIGE, K. 2005. Two new Emsian (Early Devonian) U–Pb zircon ages from volcanic rocks of the Rhenish Massif (Germany): implications for the Devonian time scale. *Journal of the Geological Society, London*, **162**, 363–371.

KEMNITZ, H., ROMER, R. L. & ONCKEN, O. 2002. Gondwana break-up and the northern margin of the Saxothuringian belt (Variscides of Central Europe). *International Journal of Earth Sciences*, **91**, 246–259.

KEMPE, U., WOLF, D., EBERMANN, U. & BOMBACH, K. 1999. 330 Ma Pb/Pb single zircon evaporatrion age for the Altenberg Granite porphyry, Eastern Erzgebirge (Germany): implications for Hercynian granite magmatism and tin mineralisation. *Zeitschrift für geologische Wissenschaften*, **27**, 385–400.

KEMPE, U., BOMBACH, K., MATUKOV, D., SCHLOTHAUER, T., HUTSCHENREUTER, J., WOLF, D. & SERGEEV, S. 2004. Pb/Pb and U/Pb zircon dating of subvolcanic rhyolite as a time marker for Hercynian granite magmatism and Sn mineralisation in the Eibenstock granite, Erzgebirge, Germany: considering effects of zircon alteration. *Mineralium Deposita*, **39**, 646–669.

KEPPIE, J. D., NANCE, R. D., MURPHY, J. B. & DOSTAL, J. 2003. Tethyan, Mediterranean, and Pacific analogues for the Neoproterozoic-Paleozoic birth and development of peri-Gondwana terranes and their transfer to Laurentia and Laurussia. *Tectonophysics*, **365**, 195–219.

KIERSNOWSKI H. & BUNIAK A. 2006. Evolution of the Rotliegend Basin of northwestern Poland. *Geological Quarterly*, **50**, 119–138.

KIRNBAUER, T. 1991. Geologie, Petrographie und Geochemie der Pyroklastika des Unteren Ems/Unter-Devon (Porphyroide) im südlichen Rheinischen Schiefergebirge. *Geologische Abhandlungen Hessen*, **92**.

KIRSCH, H., KOBER, B. & LIPPOLT, H. J. 1988. Age of intrusion and rapid cooling of the Frankenstein gabbro (Odenwald, SW-Germany) evidenced by $^{40}Ar/^{39}Ar$ and single zircon $^{207}Pb/^{206}Pb$ measurements. *Geologische Rundschau*, **77**, 693–711.

KLÖTZLI, U. S. & PARRISH, R. R. 1996. Zircon U/Pb and Pb/Pb geochronology of the Rastenberg granodiorite, South Bohemian Massif, Austria. *Mineralogy and Petrology*, **58**, 197–214.

KLÖTZLI, U. S., KOLLER, F., SCHARBERT, S. & HÖCK, V. 2001. Cadomian lower-crustal contributions to Variscan granite petrogenesis (South Bohemian Pluton, Austria): constraints from zircon typology and geochronology, whole-rock, and feldspar Pb–Sr isotope systematics. *Journal of Petrology*, **42**, 1621–1642.

KLÜGEL, T. 1997. Geometrie und Kinematik einer variszischen Plattengrenze. Der Südrand des Rhenoherzynikums im Taunus. *Geologische Abhandlungen Hessen*, **101**.

KOBER, B., KALT, A., HANEL, M & PIDGEON, R. T. 2004. SHRIMP dating of zircons from high-grade metasediments of the Schwarzwald/SW-Germany and implications for the evolution of the Moldanubian basement. *Contributions to Mineralogy and Petrology*, **147**, 330–345.

KÖHLER, H. & HÖLZL, ST. 1996. The age of the Leuchtenberg granite (NE Bavaria, Germany). A revision on account of new U–Pb zircon ages. *Neues Jahrbuch für Mineralogie, Monatshefte*, **5**, 212–222.

KÖNIGER, S. & LORENZ, V. 2002. Geochemistry, tectonomagmatic origin and chemical correlation of altered Carboniferous-Permian fallout ash tuffs in southwestern Germany. *Geological Magazine*, **139**, 541–558.

KÖNIGER, S., LORENZ, V., STOLLHOFEN, H. & ARMSTRONG, R. A. 2002. Origin, age and stratigraphic significance of distal fallout ash tuffs from the Carboniferous–Permian continental Saar–Nahe Basin (SW Germany). *International Journal of Earth Sciences*, **91**, 341–356.

KOPP, J. & BANKWITZ, P. 2003. Die Europäische Kristallinzone (EKZ)– eine Übersicht. *Zeitschrift für geologische Wissenschaften*, **31**, 179–196.

KOPP, J. & RÖLLIG, G. 1996. Probleme des basischen und intermediären Magmatismus im Plutonitmassiv von Pretzsch-Prettin-Schönewalde (Zentralteil der Mitteldeutschen Kristallinscholle). *Zeitschrift für geologische Wissenschaften*, **24**, 551–568.

KOPP, J., BANKWITZ, P. & KÖHLER, R. 2001. Die Mitteldeutsche Kristallinzone (MKZ) zwischen Saale und Neiße. Teil I: Geologisches Erscheinungsbild und basischer Magmatismus. *Zeitschrift für geologische Wissenschaften*, **29**, 33–54.

KOPP, J., BANKWITZ, P. & NAUMANN, R. 2002. Magmenentwicklung im Wirkungsfeld einer transregionalen Scherzone: Geochemie und Geochronologie der Metamorphite der Bohrung Schleusingen 3/63. *Zeitschrift für geologische Wissenschaften*, **30**, 293–315.

KORICH, D. & KRAMER, W. 1994. Permosilesischer Magmatite im Untergrund von Rügen und der östlich angrenzenden Ostsee. *Zeitschrift für geologische Wissenschaften*, **20**, 249–256.

KORSCH, R. J. & SCHÄFER, A. 1995. The Permo-Carboniferous Saar-Nahe Basin, southwest Germany and north-east France: basin formation and deformation in a strike-slip regime. *Geologische Rundschau*, **84**, 293–318.

KOŠLER, J., AFTALION, M. & BOWES, D. R. 1993. Mid-late Devonian plutonic activity in the Bohemian Massif: U–Pb zircon isotopic evidence from the Staré Sedlo and Mirotice gneiss complexes, Czech Republic. *Neues Jahrbuch für Mineralogie, Monatshefte*, **1993**, 417–431.

KOŠLER, J., BOWES, D. R., KONOPÁSEK, J. & MÍKOVA, J. 2004. Laser ablation ICPMS dating of zircons in Erzgebirge orthogneisses: evidence for Early Cambrian and Early Ordovician granitic plutonism in the western Bohemian Massif. *European Journal of Mineralogy*, **16**, 15–22.

KOZDRÓJ, W., KRENTZ, O. & OPLETAL, M. E. 2001. *Comments on the Geological Map Lausitz-Jizera-Karkonosze (without Cenozoic sediments), 1:100 000*. Warsaw, Pañstwowy Instytut Geologiczny.

KRAMER, W. 1995. Phanerozoic magmatic activity in the northwestern part of the Trans-European Suture Zone. *Studia Geophysica et Geodetica*, **39**, 320–329.

KRAWCZYK, C. M., MCCANN, T., COCKS, L. R. M., ENGLAND, R., MCBRIDE, J. & WYBRANIEC, S. 2008. Caledonian Tectonics. *In*: MCCANN, T. (ed.) *The Geology of Central Europe. Volume 1: Precambrian and Palaeozoic*. Geological Society, London, 303–381.

KREHER, B. 1994. Petrologie und Geochemie der Gabbrointrusionen des Frankensteins (Odenwald). *Geologisches Jahrbuch Hessen*, **122**, 81–122.

KŘÍBEK, B., POUBA, Z., SKOČEK, V. & WALDHAUSROVÁ, J. 2000. Neoproterozoic of the Teplá-Barrandian Unit as a part of the Cadomian orogenic belt: A review and correlation aspects. *Bulletin of the Czech Geological Survey*, **75**, 175–196.

KROHE, A. 1991. Emplacement of synkinematic plutons in the Variscan Odenwald (Germany) controlled by transtensional tectonics. *Geologische Rundschau*, **80**, 391–409.

KROHE, A. & WAWRZENITZ, N. 2000. Domainal variations of U–Pb monazite ages and Rb–Sr whole-rock dates in polymetamorphic paragneisses (KTB Drill Core, Germany): influence of strain and deformation mechanisms on isotope systems. *Journal of Metamorphic Geology*, **18**, 271–291.

KROHE, A. & WILLNER, A. P. 1995. The Odenwald Crystalline Complex. *In*: DALLMEYER, R. D., FRANKE, W., & WEBER, K. (eds) *Pre-Permian Geology of Central and Eastern Europe*. Springer, Berlin, 174–181.

KRÖNER, A. & HEGNER, E. 1998. Geochemistry, single zircon ages and Sm–Nd systematics of granitoid rocks from the Góry Sowie (Owl) Mts, Polish West Sudetes: evidence for early Palaeozoic arc-related plutonism. *Journal of the Geological Society, London*, **155**, 711–724.

KRÖNER, A. & WILLNER, A. P. 1998. Time of formation and peak of Variscan HP-HT metamorphism of quartz-feldspar rocks in the central Erzgebirge, Saxony, Germany. *Contributions to Mineralogy and Petrology*, **132**, 1–20.

KRÖNER, A., WENDT, I., *et al.* 1988. U–Pb zircon and Sm–Nd model ages of high-grade Moldanubian metasediments, Bohemian Massif, Czechoslovakia. *Contributions to Mineralogy and Petrology*, **99**, 257–266.

KRÖNER, A., HEGNER, E., HAMMER, J., HAASE, G., BIELICKI, K. H., KRAUSS, M. & EIDAM, J. 1994. Geochronology and Nd-Sr systematics of Lusatian granitoids–significance for the evolution of the Variscan orogen in east-central Europe. *Geologische Rundschau*, **83**, 357–376.

KRÖNER, A., WILLNER, A. P., HEGNER, E., FRISCHBUTTER, A., HOFMANN, J. & BERGNER, R. 1995. Latest Precambrian (Cadomian) zircon ages, Nd isotopic systematics and P-T evolution of granitoid orthogneisses of the Erzgebirge, Saxony and Czech Republic. *Geologische Rundschau*, **84**, 437–456.

KRÖNER, A., JAECKEL, P., REISCHMANN, T. & KRÖNER, U. 1998. Further evidence for an early Carboniferous (similar to 340 Ma) age of high-grade metamorphism in the Saxonian granulite complex. *Geologische Rundschau*, **86**, 751–766.

KRÖNER, A., O'BRIEN, P. J., NEMCHIN, A. A. & PIDGEON, R. T. 2000*a*. Zircon ages for high pressure granulites from South Bohemia, Czech Republic, and their connection to Carboniferous high temperature processes. *Contributions to Mineralogy and Petrology*, **138**, 127–142.

KRÖNER, A., ŠTÍPSKÁ, P., SCHULMANN, K. & JAECKEL, P. 2000*b*. Chronological constraints on the pre-Variscan evolution of the northeastern margin of the Bohemian Massif, Czech Republic. *In*: FRANKE, W., HAAK, U., ONCKEN, O. & TANNER, D. (eds) *Orogenic Processes: Quantification and Modelling in the Variscan Belt*. Geological Society, London, Special Publications, **179**, 175–197.

KRÖNER, A., JAECKEL, P., HEGNER, E. & OPLETAL, M. 2001. Single zircon ages and whole-rock Nd isotopic systematics of early Palaeozoic granitoid gneisses from the Czech and Polish Sudetes (Jizerské hory, Krkonoše Mountains and Orlice-Sněžník Complex). *International Journal of Earth Sciences*, **90**, 304–324.

KRONER, U. & SEBASTIAN, U. 1997. Der Deckenbau des östlichen Saxothuringikums. *Freiberger Forschungshefte*, **C470**, 35–52.

KRONER, U., MANSY, J.-L. *et al.* 2008. Variscan Tectonics. *In*: MCCANN, T. (ed.) *The Geology of Central Europe. Volume 1: Precambrian and Palaeozoic*. Geological Society, London, 599–664.

KRYZA, P. & PIN, C. 2002. Mafic rocks in a deep-crustal segment of the Variscides (the Góry Sowie, SW Poland): evidence for crustal contamination in an extensional setting. *International Journal of Earth Sciences*, **91**, 1017–1029.

KRYZA, R., MUSZYNSKI, A. & VIELZEUF, D. 1990. Glaucophane-bearing assemblage overprinted by greenschist-facies metamorphism in the Variscan Kaczawa complex, Sudetes, Poland. *Journal of Metamorphic Geology*, **8**, 345–355. (with 1 map)

KRYZA, R., MUSZYŃSKI, A. & FURNES, H. 1992. Early Paleozoic volcanism in the Western Sudetes (Poland): A correlation between the western parts of the Świerzawa and Bolków units of the Kaczawa Mt. *Zentralblatt für Geologie und Paläontologie Teil I*, **9/10**, 1039–1052.

KRYZA, R., TURNIAK, K., MUSZYŃSKI, A. & ZALASIEWICZ, J. A. 1994. Lower Palaeozoic shallow water sequence in the eastern European Variscides (SW Poland): provenance and depositional history. *International Journal of Earth Sciences*, **83**, 5–19.

KRYZA, R., MAZUR, S. & PIN, C. 1995. The Leszczyniec meta-igneous complex in the eastern part of the Karkonosze-Izera block, Western Sudetes: trace element and Nd isotopic study. *Neues Jahrbuch für Mineralogie, Abhandlungen*, **170**, 59–74.

KRYZA, R., PIN, C. & VIELZEUF, D. 1996. High-pressure granulites from the Sudetes (south-west Poland)–evidence of crustal subduction and collisional thickening in the Variscan belt. *Journal of Metamorphic Geology*, **14**, 531–546.

KRYZA, R., ZALASIEWICZ, J., MERRIMAN, R. J., COLLINS, A. S. & KEMP, S. J. 2002. P-T paths within the Variscan accretionary prism based on illite crystallinity and dimensions of micas from metamudrocks of the Kaczawa Mountains, SW Poland. *GeoLines*, **14**, 52–54.

KRYZA, R., MAZUR, S. & PIN, C. 2003. Subduction- and non-subduction-related igneous rocks in the Central European Variscides: geochemical and Nd isotope evidence from the Kłodzko Metamorphic Complex, Polish Sudetes. *Geodinamica Acta*, **16**, 39–57.

KRYZA, R., ZALASIEWICZ, J. A., CHARNLEY, N., MILODOWSKI, A. E., KOSTYLEW, J. & TYSZKA, R. 2004. In situ growth of monazite in anchizonal to epizonal mudrocks: first record from the Variscan accretionary prism if the Kaczawa Mountains, West Sudetes, SW Poland. *Geologia Sudetica*, **36**, 39–51.

KUMPERA, O. 1997. Controls on the evolution of the Namurian paralic basin, Bohemian Massif, Czech Republic. *In*: GAYER, R. & PEŠEK, J. (eds), *European Coal Geology and Technology*. Geological Society, London, Special Publications, **125**, 13–27.

KUPETZ, M. 1984. Stratigraphie und regionale Stellung des variszisch gefalteten Paläozoikums im Nossen-Wilsdruffer Schiefergebirge. *Freiberger Forschungshefte*, **C415**, 5–40.

KUPETZ, M. 2000. Stratigraphie und Deckentektonik im variszischen Paläozoikum des Nossen-Wilsdruffer Schiefergebirges. *Zeitschrift für Geologische Wissenschaften*, **28**, 247–272.

KURZE, M. & MOHNICKE, M. 2003 Die Granite und Gneise des Elbtals bei Coswig. *Göttinger Arbeiten zur Geologie und Paläontologie*, **5**, 55–66.

KURZE, M., SEIGERT, T., WEBER, H. & HENJES-KUNST, F. 1998. Petrographie, Geochemie und Altersstellung der Lamprophyrgänge des Elbtalschiefergebirges (Sachsen). *Zeitschrift für geologische Wissenschaften*, **26**, 193–202.

LANGE, U., BRÖCKER, M., ARMSTRONG, R., TRAPP, E. & MEZGER, K. 2005. Sm–Nd and U–Pb dating of high-pressure granulites from the Zote and Rychleby Mts (Bohemian Massif, Poland and Czech Republic). *Journal of Metamorphic Geology*, **23**, 133–145.

LARGE, D. & WALCHER, E. 1999. The Rammelsberg massive sulphide Cu-Zn-Pb-Ba-deposit, Germany: an example of sediment-hosted, massive sulphide mineralization. *Mineralium Deposita*, **34**, 522–538.

LAUE, S. & REISCHMANN, T. 1994. Petrographie und Geochemie variscischer Intrusiva der westlichen Rheingraben. *Mitteilungen der Pollichia*, **81**, 195–214.

LEFÈVRE, C., LAKHRISSI, M. & SCHNEIDER, J. L. 1994. Les affinités magmatiques du volcanisme dinantien des Vosges méridonales (France); approche géochimique et interprétation. *Comptes Rendus de l'Academie des Sciences, Paris*, **319**, 79–86.

LIEW, T. C. & HOFMANN, A. W. 1988. Precambrian crustal components, plutonic associations, plate environment of the Hercynian Fold Belt of central Europe: indications from a Nd and Sr isotopic study. *Contributions to Mineralogy and Petrology*, **98**, 129–138.

LIEW, T. C., FINGER, F. & HÖCK, V. 1989. The Moldanubian granitoid plutons of Austria: chemical and isotopic studies bearing on their environmental setting. *Chemical Geology*, **76**, 41–55.

LINNEMANN, U.-G. 1994. Geologischer Bau und Strukturentwicklung der südlichen Elbezone. *Abhandlungen des Staatlichen Museums für Mineralogie und Geologie zu Dresden*, **40**, 235–250.

LINNEMAN, U. 1995. The Neoproterozoic terranes of Saxony (Germany). *Precambrian Research*, **73**, 235–250.

LINNEMANN, U. 2003. Sedimentation und geotektonischer Rahmen der Beckenentwicklung im Saxothuringikum (Neoproterozoikum–Unterkarbon). *In*: LINNEMANN, U. (ed.) Das Saxothuringikum. Abriss der präkambrischen und paläozoischen Geologie von Sachsen und Thüringen. *Geologica Saxonica*, **48/49**, 71–110.

LINNEMANN, U. & ROMER, R. L. 2002. The Cadomian Orogeny in Saxo-Thuringia, Germany: geochemical and Nd–Sr–Pb isotopic characterization of marginal basins with constraints to geotectonic setting and provenance. *Tectonophysics*, **352**, 33–64.

LINNEMANN, U. & SCHAUER, M. 1999. Die Entstehung der Elbezone vor dem Hintergrund der cadomischen und variszischen Geschichte des Saxothuringischen Terranes–Konsequenzen aus einer abgedeckten geologischen Karte. *Zeitschrift für geologische Wissenschaften*, **27**, 529–561.

LINNEMANN, U., GEHMLICH, M., *et al.* 2000. From Cadomian subduction to Early Palaeozoic rifting: the evolution of Saxo-Thuringia at the margin of Gondwana in the light of single zircon geochronology and basin development (central European Variscides, Germany). *In*: FRANKE, W., HAAK, V., ONCKEN, O. & TANNER, D. (eds) *Orogenic Processes: Quantification and Modelling in the Variscan Belt*. Geological Society, London, Special Publications, **179**, 131–153.

LINNEMANN, U., ROMER, R. L. *et al.* 2008. Precambrian. *In*: MCCANN, T. (ed.) *The Geology of Central Europe. Volume 1: Precambrian and Palaeozoic*. Geological Society, London, 21–101.

LIPPOLT, H. J. & HESS, J. C. 1983. Isotopic evidence for the stratigraphic position of the Saar-Nahe-Rotliegend volcanism. I. 40Ar/39Ar investigations. *Neues Jahrbuch für Geologie und Paläontologie, Monatshefte*, **1983**, 713–730.

LIPPOLT, H. J. & HESS, J. C. 1996. Numerische Stratigraphie permokarbonischer Vulkanite Zentraleuropas. Teil II: Westharz. *Zeitschrift der Deutschen Geologische Gesellschaft*, **147**, 1–9.

LIPPOLT, H. J., RACZEK, I. & VENZLAFF, V. 1989. Isotopic evidence for the stratigraphic position of the Saar-Nahe-Rotliegend volcanism. II. Rb–Sr investigations. *Neues Jahrbuch für Geologie und Paläontologie, Monatshefte*, **XX**, 539–552.

LIPPOLT, H. J., KIRSCH, H. & PLEIN, E. 1990. Karbonische und permische Vulkanite aus dem Untergrund des nördlichen Oberrheingrabens: Art, Altersbestimmung und Konsequenz. *Jahresberichte und Mitteilungen des Oberrheinisch geologischen Vereins*, N.F. **72**, 227–242.

LIPPOLT, H. J., HRADETZKY, H. & HAUTMANN, S. 1994. K–Ar dating of amphibole-bearing rocks in the Schwarzwald, SW Germany: I. $^{40}Ar/^{39}Ar$ age constraints to Hercynian HT-metamorphism. *Neues Jahrbuch für Mineralogie, Monatshefte*, **10**, 433–448.

LÖFFLER, H. K. & SCHWAB, M. 1981. Die Karpholite der Wippraer Einheit des Unterharzes. *Zeitschrift für Geologische Wissenschaften*, **9**, 519–539.

LORENZ, V. & HANEKE, J. 2004. Relationship between diatremes, dykes, sills, laccoliths, intrusive-extrusive domes, lava flows, and tephra deposits with unconsolidated water-saturated sediments in the late Variscan intermontane Saar-Nahe Basin, SW Germany. *In*: BREITKREUZ, C. & PETFORD, N. (eds) *Physical Geology of High-level Magmatic Systems*. Geological Society, London, Special Publications, **234**, 75–124.

LORENZ, V. & NICHOLLS, I. A. 1984. Plate and intraplate processes of Hercynian Europe during the Late Paleozoic. *Tectonophysics*, **107**, 25–56.

LOTH, G., HÖLL, R., RITTER-HÖLL, A., BARTZSCH, K. & KENNEDY, A. 1999. U–Pb SHRIMP-Alter von Zirkonen aus einem Biotit-Granit der Henneberg-Intrusion (Thüringen). *Beiträge zur Geologie von Thüringen*, N.F. **6**, 209–215.

MCCANN, T. 1996. Pre-Permian of the north-east German basin. *Geological Journal*, **31**, 159–177.

MCCANN, T. 1998. The Rotliegend of the NE German Basin: background and prospectivity. *Petroleum Geoscience*, **4**, 17–27.

MCCANN, T. 2008. Introduction and overview. *In*: MCCANN, T. (ed.) *The Geology of Central Europe. Volume 1: Precambrian and Palaeozoic*. Geological Society, London, 1–20.

MCCANN, T., SKOMPSKI, S., *et al.* 2008*a*. Carboniferous. *In*: MCCANN, T. (ed.) *The Geology of Central Europe. Volume 1: Precambrian and Palaeozoic*. Geological Society, London, 411–529.

MCCANN, T., KIERSNOWSKI, H., *et al.* 2008*b*. Permian. *In*: MCCANN, T. (ed.) *The Geology of Central Europe. Volume 1: Precambrian and Palaeozoic*. Geological Society, London, 531–597.

MACDONALD, R., GOTTFRIED, D., FARRINGTON, M. J., BROWN, F. W. & SKINNER, N. G. 1981. Geochemistry of a continental tholeiite suite: late Palaeozoic quartz dolerite dykes of Scotland. *Transactions of the Royal Society of Edinburgh, Earth Sciences*, **72**, 57–74.

MCKERROW, W. S. & COCKS, L. R. M. 1995. The use of biogeography in the terrane assembly of the Variscan belt of Europe. *Studia Geophysica et Geodaetica.*, **39**, 269–275.

MAJEROWICZ, A. 1994. Textural features and symptoms of ocean floor metamorphism in the top part of the Ślęża ophiolite (SW Poland). *Archiwum Mineralogiczne* **50**, 97–139.

MAJEROWICZ, A. & PIN, C. 1992. The main petrological problems of the Mt. Ślęża ophiolite complex, Sudetes, Poland. *Zentralblatt für Geologie und Paläontologie Teil I*, **9/10**, 989–1018.

MALUSKI, H. & PATOČKA, F. 1997. Geochemistry and ^{40}Ar-^{39}Ar geochronology of the mafic metavolcanic rocks from the Rýchory Mountains complex (west Sudetes, Bohemian Massif): palaeotectonic significance. *Geological Magazine*, **134**, 703–716.

MALUSKI, H., RAJLICH, P. & SOUČEK, J. 1995. Pre-Variscan, Variscan and Early Alpine thermo-tectonic history of the north-eastern Bohemian Massif: an $^{40}Ar/^{39}Ar$ study. *Geologische Rundschau*, **84**, 345–358.

MANIAR, P. D. & PICCOLI, P. M. 1989. Tectonic discrimination of granitoids. *Geological Society of America Bulletin*, **101**, 635–643.

MARHEINE, D. 1997. Zeitmarken im variszischen Kollisionsbereich des Rhenoherzynikums–Saxothuringikums zwischen Harz und Sächsischem Granulitmassiv–Ergebnisse von K/Ar-Altersbestimmungen. *Göttinger Arbeiten zur Geologie und Paläontologie*, **75**.

MARHEINE, D., KACHLIK, V., MALUSKI, H., PATOČKA, F. & ŻELAŹNIEWICZ, A. 2002. The Ar-Ar ages from the West Sudetes (NE Bohemian Massif): constraints on the Variscan polyphase tectonothermal development. *In*: WINCHESTER, J. A., PHARAOH, T. C. & VERNIERS, J. (eds) *Palaeozoic Amalgamation of Central Europe*. Geological Society, London, Special Publications, **201**, 133–155.

MARSCHALL, H. R., KALT, A. & HANEL, M. 2003. P-T evolution of a Variscan lower-crustal segment: a study of granulites from the Schwarzwald, Germany. *Journal of Petrology*, **44**, 227–253.

MARTIN, U., REISCHMANN, TH., BAHLBURG, H., SCHÄTZ, M., TAIT, J. & BACHTADSE, V. 2003. The Early Palaeozoic break-up of northern Gondwana: sedimentology, physical volcanology and geochemistry of a submarine volcanic complex in the Bavarian facies association, Saxothuringian Basin, Germany. *Gondwana Research*, **6**, 839–858.

MARX, J. 1994. *Die permokarbonen Magmatite Nordwestdeutschlands im Vergleich zu den magmatischen Serien angrenzender Gebiete*. PhD thesis, University of Hannover.

MARX, J., HUEBSCHER, H.-D., HOTH, K., KORICH, D. & KRAMER, W. 1995. Vulkanostratigraphie und Geochemie der Eruptivkomplexe. *In*: PLEIN, E. (ed) Norddeutsches Rotliegend-Becken – Rotliegend Monographie, Teil II. *Courier Forschungsinstitut, Senckenberg*, **183**, 54–83.

MASSONNE, H.-J. 2001. First find of coesite in the ultrahigh-pressure metamorphic area of the central Erzgebirge, Germany. *European Journal of Mineralogy*, **13**, 565–570.

MATTAUER, M. & MATTE, P. 1998. Le bassin Stéphanien de St-Etienne ne résulte pas d'une extension tardi-hercienne généralisée: c'est un bassin pull-apart en relation avec un décrochement dextre. *Geodynamica Acta*, **11**, 23–31.

MATTE, P. 1991. Accretionary history and crustal evolution of the Variscan belt in Western Europe. *Tectonophysics*, **196**, 309–337.

MATTE, P. 2001. The Variscan collage and orogeny (480–290 Ma) and the tectonic definition of the Armorica microplate: a review. *Terra Nova*, **13**, 122–128.

MATTERN, F. 1996. The Elbe zone at Dresden – a Late Paleozoic pull-apart intruded shear zone. *Zeitschrift der Deutschen Geologische Gesellschaft*, **147**, 57–80.

MATTHES, S. & SEIDEL, E. 1977. Die Eklogitvorkommen des kristallinen Grundgebirges in NE-Bayern.–X. Bestehen genetische Beziehungen zwischen Eklogit und Meta-Gabbro innerhalb des Münchberger Gneisgebietes? *Neues Jahrbuch für Mineralogie, Abhandlungen*, **129**, 269–291.

MATTHES, S., RICHTER, O. & SCHMIDT, K. 1975. Die Eklogitvorkommen des kristallinen Grundgebirges in NE-Bayern.–IX. Petrographie, Geochemie, und Petrogenese der Eklogite des Münchberger Gneisgebietes. *Neues Jahrbuch für Mineralogie, Abhandlungen*, **126**, 33–57.

MAZUR, S. & ALEKSANDROWSKI, P. 2001. The Tepla(?)/Saxuthuringian suture in the Karkonosze-Izera massif, western Sudetes, central European Variscides. *International Journal of Earth Sciences*, **90**, 341–360.

MAZUR, S., TURNIAK, K. & BRÖCKER, M. 2004. Neoproterozoic and Cambro-Ordovician magmatism in the Variscan Kłodzko Metamorphic Complex (West Sudetes, Poland): new insights from U/Pb zircon dating. *International Journal of Earth Sciences*, **93**, 758–772.

MEDARIS JR, L. G., BEARD, B. L. & JELÍNEK, E. 2006. Mantle-derived, UHP garnet pyroxenite and eclogite in the Moldanubian Gföhl Nappe, Bohemian Massif: a geochemical review, new P-T determinations, and tectonic interpretation. *International Geology Review*, **48**, 765–777.

MEISL, S. 1990. Metavolcanic rocks in the 'Northern Phyllite Zone' at the southern margin of the Rhenohercynian Belt. In: *Mid-German Crystalline Rise & Rheinisches Schiefergebirge*. Field guide of the International Conference on Paleozoic Orogens in Central Europe, Göttingen/Gießen, Aug/Sept. 1990, 25–42.

MEISL, S. 1995. Igneous activity. *In*: DALLMEYER, R. D., FRANKE, W. & WEBER, K. (eds) *Pre-Permian Geology of Central and Eastern Europe*. Springer, Berlin, 118–131.

MEISL, S., LIPPERT, H. J. & STRECKER, G. 1982. Geosynclinal vulcanism in the Lahn-Dill area. *Fortschritte der Mineralogie*, **60**, 3–15

MÉNARD, G. & MOLNAR, P. 1988. Collapse of a Hercynian Tibetan Plateau into late Paleozoic European Basin and Range province. *Nature*, **334**, 235–237.

MERTZ, D. F. 1986. Isotopische Datierung des Pikrits von Sechshelden nach der K/Ar-Methode–ein Beitrag Zur Altersstellung des variscischen Magmatismus im Rheinischen Schiefergebirge. *Geologisches Jahrbuch Hessen*, **11**, 201–207.

Meschede, M. 1986. A method of discriminating between different types of mid-ocean ridge basalts and continental tholeiites with the Nb-Zr-Y diagram. *Chemical Geology*, **56**, 207–218.

Miller, B. V., Samson, S. D. & D'Lemos, R. S. 1999. Time span of plutonism, fabric development, and cooling in a Neoproterozoic magmatic arc segment; U–Pb age constraints from syn-tectonic plutons, Sark, Channel Islands, UK. *Tectonophysics*, **312**, 79–95.

Mingram, B. 1998. The Erzgebirge, Germany, a subducted part of northern Gondwana; geochemical evidence for repetition of early Palaeozoic metasedimentary sequences in metamorphic thrust units. *Geological Magazine*, **135**, 785–801.

Mingram, B. & Rötzler, K. 1999. Geochemische, petrologische und geochronologische Untersuchungen im Erzgebirgskristallin – Rekonstruktion eines Krustenstapels. *Schriftenreihe für Geowissenschaften*, **9**, 1–80.

Mingram, B., Kröner, A., Hegner, E. & Krentz, O. 2004. Zircon ages, geochemistry, and Nd isotopic systematics of pre-Variscan orthogneisses from the Erzgebirge, Saxony (Germany), and geodynamic interpretation. *International Journal of Earth Sciences*, **93**, 706–727.

Mísař, Z. 1995. Stratigraphy. *In*: Dallmeyer, R. D., Franke, W., Weber, K. (eds) *Pre-Permian Geology of Central and Eastern Europe*. Springer-Verlag, Berlin, 521–529.

Mísař, Z., Jelínek, E. & Pačesová, M. 1984. The Letovice dismembered metaophiolites in the framework of the Saxo-Thuringian zone of the Bohemian massif. *Mineralia Slovaca*, **1**, 13–28.

Mock, A., Ehling, B.-C. & Breitkreuz, C. 2005. Anatomy of a laccolith complex – Geometry and texture of porphyritic rhyolites in the Permocarboniferous Halle Volcanic Complex (Germany). *Neues Jahrbuch für Geologie und Paläontologie, Abhandlungen*, **237**, 211–271.

Mogensen, T. E. 1994. Palaeozoic structural development along the Tornquist Zone, Kattegat area, Denmark. *Tectonophysics*, **240**, 191–214.

Montenari, M. & Servais, T. 2000. Early Paleozoic (Late Cambrian–Early Ordovician) acritarchs from the metasedimentary Baden-Baden-Gaggenau zone (Schwarzwald, SW Germany). *Review of Palaeobotany and Palynology*, **113**, 73–85.

Montenari, M., Servais, T. & Paris, F. 2000. Palynological dating (acritarchs and chitinozoans) of Lower Paleozoic phyllites from the Black Forest/southwestern Germany. *Comptes Rendus de l'Academie des Sciences, Paris*, **330**, 493–499.

Müller, H. 1996. Das permokarbon im nördlichen Oberrheingraben. *Geologische Abhandlungen Hessen*, **99**.

Müller, A. & Seltmann, R 2002. Plagioclase-mantled K-feldspar in the Carboniferous porphyritic microgranite of Altenberg-Frauenstein, eastern Erzgebirge /Krušné hory. *Bulletin of the Geological Society of Finland*, **74**, 53–78.

Müller, A., Seltmann, R, & Behr, H.-J. 2000. Application of catholodoluminescence to magmatic quartz in a tin granite–case study from the Schellerhau Granite Complex, eastern Erzgebirge. *Mineralium Deposita*, **35**, 169–189.

Murphy, J. B. & Nance, R. D. 1989. Model for the evolution of the Avalonian-Cadomian belt. *Geology*, **17**, 735–738.

Murphy, J. B. & Nance, R. D. 2002. Sm–Nd isotopic systematics as tectonic tracers: an example from West Avalonian the Canadian Appalachians. *Earth-ScienceReviews*, **59**, 77–100.

Murphy, J. B., Eguiluz, L. & Zulauf, G. 2002. Cadomian Orogens, peri-Gondwanan correlatives and Laurentia–Baltica connections. *Tectonophysics*, **352**, 1–9.

Murphy, J.B., Fernández-Suárez, J., Jeffries, T.E. & Strachan, R.A. 2004*a*. U–Pb (LA-ICP-MS) dating of detrital zircons from Cambrian clastic rocks in Avalonia: erosion of a Neoproterozoic arc along the northern Gondwanan margin. *Journal of the Geological Society, London*, **161**, 243–254.

Murphy, J. B., Pisarevsky, S. A., Nance, R. D. & Keppie, J. D. 2004*b*. Neoproterozoic–Early Paleozoic evolution of the peri-Gondwanan terranes: implications for Laurentia-Gondwana connections. *International Journal of Earth Science*, **93**, 659–682.

Muszyński, A. & Kryza, R. 1993. First finding of jadeite in the Kaczawa Mt (Sudetes). *Przegląd Geologiczny*, **41**, 24–26 [in Polish, with English summary].

Nagy, E. A., Samson, S. D. & D'Lemos, R. S. 2002. U–Pb geochronological constraints on the timing of Brioverian sedimentation and regional deformation in the St. Brieuc region of the Neoproterozoic Cadomian orogen, northern France. *Precambrian Research*, **116**, 1–17.

Nance, R. D. & Murphy, J. B. 1996. Basement isotopic signatures and Neoproterozoic palaeogeography of Avalonian–Cadomian and related terranes in the circum–North Atlantic. *In*: Nance, R. D. & Thompson, M. D. (eds) *Avalonian and Related Peri-Gondwanan Terranes of the Circum–North Atlantic*. Geological Society of America, Special Papers, **304**, 333–346.

Narebski, W. 1994. Lower to Upper Paleozoic tectonomagmatic evolution of NE part of the Bohemian Massif. *Zentralblatt für Geologie und Paläontologie, Teil I*, **9/10**, 961–972.

Nasir, S., Okrusch, M., Kreuzer, H., Lenz, H. & Höhndorf, A. 1991. Geochronology of the Spessart Crystalline Complex, Mid-German Crystalline Rise. *Mineralogy and Petrology*, **44**, 39–55.

Nesbor, H.-D. 2004. Paläozoischer Intraplattenvulkanismus im östlichen Rheinischen Schiefergebirge–Magmenentwicklung und zeitlicher Ablauf. *Geologisches Jahrbuch Hessen*, **131**, 145–182.

Nesbor, H.-D. 2006. Vulkanismus im Unterkarbon des Rheinischen Schiefergebirges. *In*: Deutsche Stratigraphische Kommission (eds) *Stratigraphie von Deutschland VI. Unterkarbon (Mississippium)*. Schriftenreihe der Deutschen Gesellschaft für Geowissenschaften, **41**, 581–584.

Nesbor, H.-D., Buggisch, W., Flick, H., Horn, M. & Lippert, H.-J. 1993. Fazielle und paläogeographische Entwicklung vulkanisch geprägter mariner Becken am Beispiel des Lahn-Dill-Gebietes. *Geologisches Jahrbuch Hessen*, **98**, 3–87.

Neumann, E.-R., Wilson, M., Heeremans, M., Dunworth, E. A., Obst, K., Timmerman, M. J. & Kirstein, L. 2004. Permo-Carboniferous rifting and magmatism in northern Europe: a review. *In*: Wilson, M., Neumann, E.-R., Davies, G. R., Timmerman, M. J., Heeremans, M. & Larsen, B. T. (eds) *Permo-Carboniferous Magmatism and Rifting in Europe*. Geological Society, London, Special Publications, **223**, 11–40.

Nicholls, I. A. & Lorenz, V. 1973. Origin and crystallization history of Permian tholeiites from the Saar-Nahe Trough, SW Germany. *Contributions to Mineralogy and Petrology*, **40**, 327–344.

Noble, S. R., Tucker, R. D. & Pharaoh, T. C. 1993. Lower Palaeozoic and Precambrian igneous rocks from eastern England, and their bearing on late Ordovician closure of the Tornquist Sea: constraints from U–Pb and Nd isotopes. *Geological Magazine*, **130**, 738–747.

Nutman, A. P., Green, D. H., Cook, C. A., Styles, M. T. & Holdsworth, R. E. 2001. SHRIMP U–Pb zircon dating of the exhumation of the Lizard Peridotite and its emplacement over crustal rocks: constraints for tectonic models. *Journal of the Geological Society, London*, **158**, 809–820.

Oberc-Dziedzic, T., Klimas, K., Kryza, R. & Fanning, C. M. 2003. SHRIMP U–Pb zircon geochronology of the Strzelin gneiss, SW Poland: evidence for a Neoproterozoic thermal event in the Fore-Sudetic Block, Central European Variscides. *International Journal of Earth Sciences*, **92**, 701–711.

Oberc-Dziedzic, T. Pin, C. & Kryza, R. 2005. Early Palaeozoic crustal melting in an extensional setting: petrological and Sm–Nd evidence from the Izera granite-gneisses, Polish Sudetes. *International Journal of Earth Sciences*, **94**, 354–368.

O'Brien, P. J. 1989. The petrology of retrograded eclogites in the Oberpfalz Forest, northeastern Bavaria, West Germany. *Tectonophysics*, **157**, 195–212.

O'Brien, P. J. 1993. Partially retrogressed eclogites of the Münchberg Massif, Germany: records of a multi-stage Variscan uplift history in the Bohemian Massif. *Journal of Metamorphic Geology*, **11**, 241–260.

O'Brien, P. J. 2000. The fundamental Variscan problem: high-temperature metamorphism at different depths and high-pressure metamorphism at different temperatures. *In*: Franke, W., Haak, V., Oncken, O. & Tanner, D. (eds) *Orogenic Processes: Quantification and Modelling in the Variscan Belt*. Geological Society, London, Special Publications, **179**, 369–386.

O'Brien, P. J. & Carswell, D. A. 1993. Tectonometamorphic evolution of the Bohemian Massif: evidence from high pressure metamorphic rocks. *Geologische Rundschau*, **82**, 531–555.

O'BRIEN, P. J., KRÖNER, A., JAECKEL, P., HEGNER, E., ŻELAŹNIEWICZ; A. & KRYZA, R. 1997. Petrological and isotopic studies on Paleozoic high pressure granulites, Góry Sowie Mts, Polish Sudetes. *Journal of Petrology*, **38**, 433–456.

OBST, K. 1999. Die permosilesischen Eruptivgänge innerhalb der Fennoskandischen Randzone (Schonen und Bornholm). Untersuchungen zum Stoffbestand, zur Struktur und zur Genese. *Greifswalder Geowissenschaftliche Beiträge*, **7/1999**, 1–121.

OBST, K., KATZUNG, G. & HAMMER, J. 1999. Dating of the late Autunian basic magmatism in the Thuringian Forest. *Neues Jahrbuch für Geologie und Paläontologie, Monatshefte*, **1999**(1), 1–10.

OBST, K., SOLYOM, Z. & JOHANSSON, L. 2004. Permo-Carboniferous extension-related magmatism at the SW margin of the Fennoscandian Shield. *In*: WILSON, M., NEUMANN, E.-R., DAVIES, G. R., TIMMERMAN, M. J., HEEREMANS, M. & LARSEN, B. T. (eds) *Permo-Carboniferous Magmatism and Rifting in Europe*. Geological Society, London, Special Publications, **223**, 259–288.

OKRUSCH, M. 1995. Mid German Crystalline High–Metamorphic evolution. *In*: DALLMEYER, R. D., FRANKE, W. & WEBER, K. (eds) *Pre-Permian Geology of Central and Eastern Europe*. IGCP 233,. Springer, Berlin, 201–215.

OKRUSCH, M. & RICHTER, P. 1986. Orthogneisses of the Spessart crystalline complex, Northwest Bavaria: indicators of the geotectonic environment? *Geologische Rundschau*, **75**, 555–568.

OKRUSCH, M., DEIDEL, E., SCHLÜSSER, U. & RICHTER, P 1989. Geochemical characteristics of metabasites in different tectonic units of the northeast Bavarian crystalline basement. *In*: EMMERMANN, R. & WOHLENBERG, J. (eds) *The German Continental Deep Drilling Program (KTB)*. Springer, Berlin, 67–79.

OKRUSCH, M., MATTHES, S., KLEMD, R., O'BRIEN, P. J. & SCHMIDT, K. 1991. Eclogites at the north-western margin of the Bohemian Massif: a review. *European Journal of Mineralogy*, **3**, 707–730.

OLIVER, G. J. H., CORFU, F. & KROGH, T. E. 1993. U–Pb ages from SW Poland: evidence for a Caledonian suture zone between Baltica and Gondwana. *Journal of the Geological Society, London*, **150**, 355–369.

ONCKEN, O. 1988. Geometrie und Kinematik der Taunuskammüberschiebung–Beitrag zur Diskussion des Deckenproblems im südlichen Schiefergebirge. *Geologische Rundschau*, **77**, 551–575.

ONCKEN, O. 1997. Transformation of a magmatic arc and an orogenic root during oblique collision and its consequences for the evolution of the European Variscides (Mid-German Crystalline Rise). *Geologische Rundschau*, **86**, 2–20.

ONCKEN, O., VON WINTERFELD, C. & DITTMAR, U. 1999. Accretion of a rifted passive margin: the Late Palaeozoic Rhenohercynian fold and thrust belt (Middle European Variscides). *Tectonics*, **18**, 75–91.

ONCKEN, O., PLESCH, A., WEBER, J., RICKEN, W. & SCHRADER, S. 2000. Passive margin detachment during arc-continent collision (Central European Variscides). *In*: FRANKE, W., HAAK, V., ONCKEN, O. & TANNER, D. (eds) *Orogenic Processes: Quantification and Modelling in the Variscan Belt*. Geological Society, London, Special Publications, **179**, 199–216.

OPLUŠTIL, S. & PEŠEK, J. 1998. Stratigraphy, palaeoclimatology and palaeogeography of the Late Palaeozoic continental deposits in the Czech Republic. *In*: CRASQUIN-SOLEAU, S., IZART, A., VASLET, D. & DEWEVER, P. (eds) Peri-tethys Stratigraphic correlations 2. *Geodiversitas*, **20**, 597–620.

PARRY, M., ŠTÍPSKÁ, P., SCHULMANN, K., HROUDA, F., JEŽEK, J. & KRÖNER A. 1997. Tonalite sill emplacement at an oblique plate boundary: north-eastern margin of the Bohemian Massif. *Tectonophysics*, **280**, 61–81.

PATOČKA, F. 1987. The geochemistry of mafic metavolcanics: implications for the origin of the Devonian massive sulphide deposits at Zlaté Hory, Czechoslovakia. *Mineralium Deposita*, **22**, 144–150.

PATOČKA, F. & SMULIKOWSKI, W. 2000. Early Palaeozoic intracontinental rifting and incipient oceanic spreading in the Czech/Polish east Krkonoše/Karkonosze complex, west Sudetes (NE Bohemian massif). *Geologia Sudetica*, **33**, 1–15.

PATOČKA, F. & ŠTORCH P. 2004. Evolution of geochemistry and depositional settings of Early Palaeozoic siliciclastics of the Barrandian (Teplá-Barrandian Unit, Bohemian Massif, Czech Republic). *International Journal of Earth Science*, **93**, 728–741.

PATOČKA, F. & VALENTA, J. 1996. Geochemistry of the Late Devonian intermediate to acid metavolcanic rocks from the southern part of the Vrbno Group, the Jeseníky Mts. (Moravo-Silesian Belt, Bohemian Massif, Czech Republic): paleotectonic implications. *GeoLines*, **4**, 42–54.

PATOČKA, F., VLAŠÍMSKÝ, P. & BLECHOVÁ, K. 1993. Geochemistry of Early Paleozoic volcanics of the Barrandian Basin (Bohemian Massif, Czech Republic): implications for paleotectonic reconstructions. *Jahrbuch Geologische Bundesanstalt, Wien*, **136**, 873–896.

PATOČKA, F., PIVEC, E. & OLIVERIOVÁ, D. 1996. Mineralogy and petrology of mafic blueschists from the Rýchory Mts. crystalline complex (West Sudetes, Bohemian Massif). *Neues Jahrbuch für Mineralogie, Abhandlungen*, **170**, 313–330.

PATOČKA, F., FAJST, M. & KACHLÍK, V. 2000. Mafic-felsic to mafic-ultramafic Early Palaeozoic magmatism of the West Sudetes (NE Bohemian Massif): the South Krkonoše Complex. *Zeitschrift für geologische Wissenschaften*, **28**, 177–210.

PAUL, J., WAGNER, K. & WESLING, C. 1997. Lithostratigraphie und Fazies des Ilfelder Beckens (Permokarbon, Harz). *Freiberger Forschungshefte*, **C466**, 109–153.

PAULICK, H. & BREITKREUZ, C. 2005. The Late Paleozoic felsic lava-dominated large igneous provinc in northeast Germany: volcanic facies analysis based on drill cores. *International Journal of Earth Sciences*, **94**, 834–850.

PEARCE, J. A. & NORRY, M. J. 1979. Petrogenetic implications of Ti, Zr, Y and Nb variations in volcanics rocks. *Contributions to Mineralogy and Petrology*, **69**, 33–47.

PEARCE, J. A., HARRIS, N. B. W. & TINDLE, A. G. 1984. Trace element discrimination diagrams for the tectonic interpretation of granitic rocks. *Journal of Petrology*, **25**, 956–983.

PECERILLO, A. & TAYLOR, S. R. 1976. Geochemistry of Eocene calc-alkaline volcanic rocks from the Kastamonu area, Northern Turkey. *Contributions to Mineralogy and Petrology*, **58**, 63–81.

PECINA, V., AICHLER, J., MIXA, P., CHLUPÁČOVÁ, M., HANŽL, P., ŽÁČEK, V. & WILIMSKÝ, D. 2003. Metadolerites of the Vrbno Group and their origin, the Jeseniky Mts. *GeoLines*, **16**, 80–82.

PHARAOH, T. C. 1999. Palaeozoic terranes and their lithospheric boundaries within the Trans-European Suture Zone (TESZ): a review. *Tectonophysics*, **314**, 17–41.

PHARAOH, T. C., MERRIMAN, R. J., EVANS, J. A., BREWER, T. S., WEBB, P. C. & SMITH, N. J. P. 1991. Early Palaeozoic arc-related volcanism in the concealed Caledonides of southern Britain. *Annales de la Société Géologique de Belgique*, **114**, 63–91.

PHARAOH, T. C., BREWER, T. S. & WEBB, P. C. 1993*a*. Subduction-related magmatism of late Ordovician age in eastern England. *Geological Magazine*, **130**, 647–656.

PHARAOH, T. C., MOLYNEUX, S. G., MERRIMAN, R. J., LEE, M. K. & VERNIERS, J. 1993*b*. The Caledonides of the Anglo-Brabant Massif reviewed. *Geological Magazine*, **130**, 561–562.

PIETSCH, K. 1959. Der subsequente Magmatismus im Sächsischen Varistikum (ein kürzer Überblick). *Geologische Rundschau*, **48**, 32–42.

PIN, C. 1990. Variscan oceans – ages, origins and geodynamic implications inferred from geochemical and radiometric data. *Tectonophysics*, **177**, 215–227.

PIN, C. & DUTHOU, J. L. 1990. Sources of Hercynian granitoids from the French Massif-Central – inferences from Nd isotopes and consequences for crustal evolution. *Chemical Geology*, **83**, 281–296.

PIN, C. & MARINI, F. 1993. Early Ordovician continental break-up in Variscan Europe: Nd-Sr isotope and trace element evidence from bimodal igneous associations of the Southern Massif Central, France. *Lithos*, **29**, 177–196.

PIN, C. & PAQUETTE, J.-L. 2002. Le magmatisme basique calcoalcalin d'age devono-dinantien du nord du Massif Central, temoin d'une marge active hercynienne: arguments geochimiques et isotopiques Sr/Nd. *Geodinamica Acta*, **15**, 63–77.

PIN, C., MIERZEJEWSKI, M. P. & DUTHOU, J. L. 1987. Isochronous age Rb/Sr of Karkonosze granite from the quarry Szklarska Poręba Huta and significance of initial ratio 87Sr/86Sr in this granite. *Przegląd Geologiczny*, **XXXV**, 512–517.

PIN, C., MAJEROWICZ, A. & WOJCIECHOWSKA, I. 1988. Upper Paleozoic oceanic-crust in the Polish Sudetes–Nd-Sr isotope and trace-element evidence. *Lithos*, **21**, 195–209.

PITRA, P., BURG, J.-P. & GUIRAUD, M. 1999. Late-Variscan strike-slip tectonics between the Teplá-Barrandian and Moldanubian terranes (Czech Bohemian Massif): petrostructural evidence. *Journal of the*

Geological Society, London, **156**, 1003–1020.

Plein, E. 1995. Norddeutsches Rotliegend-Becken. Rotliegend Monographie, Teil II. *Courier Forschungsinstitut Senckenberg*, **183**, 1–193.

Praeg, D. 2004. Diachronous Variscan late-orogenic collapse as a response to multiople detachments: a view from the internides in France to the foreland in the Irish Sea. . *In*: Wilson, M., Neumann, E.-R., Davies, G. R., Timmerman, M. J., Heeremans, M. & Larsen, B. T. (eds) *Permo-Carboniferous Magmatism and Rifting in Europe*. Geological Society, London, Special Publications, **223**, 89–138.

Propach, G. & Pfeiffer, T. 1998. Ocean floor basalt, not continental gabbro: a reinterpretation of the Hoher Bogen amphibolites, Tepla-Barrandian, Bohemian massif. *Geologische Rundschau*, **87**, 303–313.

Propach, G., Baumann, A., Schulz-Schmalschläger, M. & Grauert, B. 2000. Zircon and monazite U–Pb ages of Variscan granitoid rocks and gneisses in the Moldanubian zone of eastern Bavaria, Germany. *Neues Jahrbuch für Geologie und Paläontologie, Monatshefte*, **6**, 345–377.

Rappsilber, I. 2003. *Struktur und Entwicklung des nördlichen Saale-Beckens (Sachsen-Anhalt): Geophysik und Geologie*. Dissertation, Martin-Luther-Universität Halle-Wittenberg.

Read, W. A. 1988. Controls on Silesian sedimentation in the Midland Valley of Scotland. *In*: Besly, B. M. & Kelling, G. (eds) *Sedimentation in a Synorogenic Basin Complex; the Upper Carboniferous of NW Europe*. Blackie, Glasgow, 222–241

Reichel, W. 1985. Schichtstoerungen im unterpermischen Doehlener Becken bei Dresden; Ein Beitrag zur lithofaziellen und tektonischen Entwicklung eines intramontanen vulkanotektonischen Beckens. *Hallesches Jahrbuch für Geowissenschaften*, **10**, 21–34.

Reischmann, T. & Anthes, G. 1996. Geochronology of the Mid-German Crystalline Rise west of the river Rhine. *Geologische Rundschau*, **85**, 761–774.

Reischmann, T. Anthes, G., Jaeckel, P & Altenberger, U. 2001. Age and origin of the Böllsteiner Odenwald. *Mineralogy and Petrology*, **72**, 29–44.

Reitz, E. 1987. Silurische Sporen aus einem granatführenden Glimmerschiefer des Vor-Spessart, NW-Bayern. *Neues Jahrbuch für Geologie und Paläontologie, Monatshefte*, **1987**, 699–704.

Reitz, E. 1989. Devonische Sporen aus Phylliten vom Südrand des Rheinischen Schiefergebirges. *Geologisches Jahrbuch Hessen*, **107**, 13–25.

Reitz, E. 1991. Acritarchen des Unter-Tremadoc aus dem westlichen Frankenwald, NE-Bayern. *Neues Jahrbuch für Geologie und Paläontologie, Monatshefte*, **1991**, 97–104.

Reitz, E. & Wickert, F. 1989. Late Cambrian to Early Ordovician acritarchs from the Villé Unit, northern Vosges mountains (France). *Neues Jahrbuch für Geologie und Paläontologie, Monatshefte*, **1989**, 375–384.

Reitz, E., Anderle, H. J. & Winkelmann, M. 1995. Ein erster Nachweis von Unterordovizium (Arenig) am Südrand des Rheinischen Schiefergebirges im Vordertaunus: Der Bierstadt-Phyllit (Bl. 5915 Wiesbaden). *Geologisches Jahrbuch Hessen*, **123**, 25–38.

Rey, P., Burg, J. P. & Caron, J. M. 1992. Middle and Late Carboniferous extension in the Variscan Belt: structural and petrological evidence from the Vosges massif (Eastern France). *Geodinamica Acta*, **5**, 17–36.

Ritter, O., Haak, V., Rath, V., Stein, E. & Stiller, M. 1999. Very high electrical conductivity beneath the Münchberg Gneiss area in Southern Germany: implications for horizontal transport along shear planes. *Geophysical Journal International*, **139**, 161–170.

Ro, H. E., Stuevold, L. M., Faleide, J. I. & Myhre, A. M. 1990. Skagerrak Graben – the offshore continuation of the Oslo Graben. *Tectonophysics*, **178**, 1–10.

Romer, R. L. & Rötzler, J. 2001. P–T–t evolution of ultrahigh-temperature granulites from the Saxon Granulite Massif, Germany. Part II: geochronology. *Journal of Petrology*, **42**, 2015–2032.

Romer, R. L., Förster, H.-J. & Breitkreuz, C. 2001. Intracontinental extensional magmatism with a subduction fingerprint: the late Carboniferous Halle Volcanic Complex (Germany). *Contributions to Mineralogy and Petrology*, **141**, 201–221.

Rötzler, J. & Romer, R. L. 2001. P–T–t evolution of ultrahigh-temperature granulites from the Saxon Granulite Massif, Germany. Part I: petrology. *Journal of Petrology*, **42**, 1995–2013.

Rötzler, J., Schumacher, R., Maresch, V. W. & Wilner, A. P. 1998. Contrasts in the metamorphic evolution and exhumation history of different metapelite units in the Western Erzgebirge (Saxony, Germany). *European Journal of Mineralogy*, **10**, 261–280.

Rötzler, J., Carswell, D. A., Gerstenberger, H. & Haase, G. 1999. Transitional blueschist–epidote amphibolite facies metamorphism in the Frankenberg massif, Germany, and geotectonic implications. *Journal of Metamorphic Geology*, **16**, 109–125.

Salamon, M. 2003. Grobklastische Beckensedimente (Olisthostrome) des Oberen Mitteldevons im Lahn-Dill-Gebiet–Zeugen einer aktiven Rift-Tektonik. *Geologische Abhandlungen Hessen*, **111**.

Samson, S. D. & D'Lemos, R. S. 1998. U-P geochronology and Sm–Nd isotopic composition of Proterozoic gneisses, Channel Islands, UK. *Journal of the Geological Society, London*, **155**, 609–618.

Sandeman, H. A., Chen, Y., Clark, A. H. & Farrar, E. 1995. Constraints on the P-T conditions and age of emplacement of the Lizard ophiolite, Cornwall: amphibole-plagioclase thermobarometry and $^{40}Ar/^{39}Ar$ geochronology of basal amphibolites. *Canadian Journal of Earth Sciences*, **32**, 261–272.

Sandeman, H. A. I., Clark, A. H., Scott, D. J. & Malpas, J. G. 2000. The Kennack Gneiss of the Lizard Peninsula, Cornwall, SW England: commingling and mixing of mafic and felsic magmas accompanying Givetian continental incorporation of the Lizard ophiolite. *Journal of the Geological Society of London*, **157**, 1227–1242.

Sano, S., Oberhänsli, R., Romer, R. L. & Vinx, R. 2002. Petrological, geochemical and isotopic constraints on the origin of the Harzburg intrusion, Germany. *Journal of Petrology*, **43**, 1529–1549.

Schäfer, F., Oncken, O., Kemnitz, H. & Romer, R. L. 2000. Upper-plate deformation during collisional orogeny: a case study from the German Variscides (Saxo-Thuringian Zone). *In*: Franke, W., Haak, V., Oncken, O. & Tanner, D. (eds) *Orogenic Processes: Quantification and Modelling in the Variscan Belt*. Geological Society, London, Special Publications, **179**, 281–302.

Schaltegger, U. 1997. Magma pulses in the Central Variscan Belt: episodic melt generation and emplacement during lithospheric thinning. *Terra Nova*, **9**, 242–245.

Schaltegger, U. 2000. U–Pb geochronology of the Southern Black Forest Batholith (Central Variscan Belt): timing of exhumation and granite emplacement. *International Journal of Earth Sciences*, **88**, 814–828.

Schaltegger, U. & Corfu, F. 1995. Late Variscan 'Basin and Range' magmatism and tectonics in the Central Alps: evidence from U–Pb geochronology. *Geodinamica Acta*, **8**, 82–98.

Schaltegger, U., Schneider, J.-L., Maurin, J.-C. & Corfu, F. 1996. Precise U–Pb chronometry of 345–340 Ma old magmatism related to syn-convergence extension in the Southern Vosges (Central Variscan Belt). *Earth and Planetary Science Letters*, **144**, 403–419.

Schaltegger, U., Nägler, Th. F., Corfu, F., Maggetti, M., Galetti, G. & Stosch, H. G. 1997. A Cambrian island arc in the Silvretta nappe: constraints from geochemistry and geochronology. *Schweizerische Mineralogische und Petrographische Mitteilungen*, **77**, 337–350.

Schaltegger, U., Fanning, C. M., Günther, D., Maurin, J. C., Schulmann, K. & Gebauer, D. 1999. Growth, annealing and recrystallization of zircon and preservation of monazite in high-grade metamorphism: conventional and in-situ U–Pb isotope, cathodoluminescence and microchemical evidence. *Contributions to Mineralogy and Petrology*, **134**, 186–201.

Schätz, M., Reischmann, T., Tait, J., Bachtadse, V., Bahlburg, H. & Martin, U. 2002. The Early Palaeozoic break-up of northern Gondwana, new palaeomagnetic and geochronological data from the Saxothuringian Basin, Germany. *International Journal of Earth Science*, **91**, 838–849.

Schmädicke, E. & Evans, B. W. 1997. Garnet-bearing ultramafic rocks from the Erzgebirge, and their relation to other settings in the Bohemian Massif. *Contributions to Mineralogy and Petrology*, **127**, 57–74.

Schmädicke, E., Mezger, K., Cosca, M. A. & Okrusch, M. 1995. Variscan Sm–Nd and Ar-Ar ages of eclogite-facies rocks from the Erzgebirge and their relation to other settings in the Bohemian Massif. *Journal of Metamorphic Geology*, **13**, 537–552.

Schmidberger, S. S. & Hegner, E. 1999. Geochemistry and isotope systematics of calc-alkaline volcanic rocks from the Saar-Nahe basin (SW Germany) – implications for Late-Variscan orogenic development. *Contributions to Mineralogy and Petrology*, **135**, 373–385.

SCHNEIDER, J. W., HOTH, K., GAITZSCH, B. G., BERGER, H. J., STEINBORN, H., WALTER, H. & ZEIDLER, M. K. 2005. Carboniferous stratigraphy and development of the Erzgebirge Basin, East Germany. *Zeitschrift der deutschen Gesellschaft für Geowissenschaften*, **156**, 431–466.

SCHÖDLBAUER, S., HECHT, L., HÖHNDORF, A. & MORTEANI, G. 1997. Enclaves in the S-type granites of the Kösseine massif (Fichtelgebirge, Germany): implications for the origin of granites. *Geologische Rundschau*, **86**(Supplement), S125-S140.

SCHUBERT, W., LIPPOLT, H. J. & SCHWARZ, W. 2001. Early to Middle Carboniferous hornblende $^{40}Ar/^{39}Ar$ ages of amphibolites and gabbros from the Bergsträsser Odenwald. *Mineralogy and Petrology*, **72**, 113–132.

SCHULMANN, K. & GAYER, R. 2000. A model for an obliquely developed accretionary wedge: NE Bohemian Massif. *Journal of the Geological Society, London*, **157**, 401–416.

SCHULMANN, K., SCHALTEGGER, U., JEZEK, J., THOMPSON, A. B. & EDEL, J.-B. 2002. Rapid burial and exhumation during orogeny: thickening and synconvergent exhumation of thermally weakened and thinned crust (Variscan orogen in western Europe). *American Journal of Science*, **302**, 856–879.

SCHULZ, B. & BOMBACH, K. 2003. Single zircon Pb-Pb geochronology of the Early Palaeozoic magmatic evolution in the Austroalpine basement to the south of the Tauern Window. *Jahrbuch der Geologischen Bundesanstalt, Wien*, **143**, 303–321.

SCHÜSSLER, U., VEJNAR, Z., OKRUSCH, M., ROSE, S. & SEIDEL, E. 1988. Geochemistry of metabasites and gabbroic rocks from the Teplá-Domažlice Zone. *In*: KUKAL, Z. (ed.) *Proceedings of the 1st International Conference on the Bohemian Massif*, Prague, Czechoslovakia, Sept. 26–Oct. 3, 1988, 247–256.

SCHÜSSLER, U., RICHTER, P & OKRUSCH, M. 1989. Metabasites from the KTB Oberpfalz target area, Bavaria–geochemical characteristics and examples of mobile behaviour of 'immobile' elements. *Tectonophysics*, **157**, 135–148.

SCHWAB, M. 1959. Zur Deutung des Quarzporphyrs vom Kahlbusch bei Dohna (Sachsen) als Quellkuppe. *Geologische Rundschau*, **48**, 43–54.

SCHWAB, M. & JACOB, G. 1996. Wippraer Zone–Taunus–Hunsrück. Ein struktureller Vergleich- *Zeitschrift für geologische Wissenschaften*, **24**, 305–316.

SEBASTIAN, U. 1995. Die Strukturentwicklung des spätorogenen Erzgebirgsaufstieg in der Flöhazone–Ein weiterer Beitrag zur postkollisionalen Extension am Nordrand der Böhmischen Masse. *Freiberger Forschungshefte*, **C461**.

SEHNERT, M. 1991. *Zur Biostratigraphie epizonaler Metamorphite der Wippraer Einheit/Unterharz*. Dissertation, Martin-Luther-Universität, Halle.

SELTMANN, R. 1994. Subvolcanic minor intrusions in the Altenberg Caldera and their metallogeny. *In*: SELTMANN, R., KÄMPF, H. & MÖLLER, P. (eds) *Metallogeny of Collisional Orogens*. Czech Geological Survey, Prague, 198–206.

SELTMANN, R. & SCHILKA, W. 1995. Late-Variscan crustal evolution in the Altenberg-Teplice caldera. Evidence from new geochemical and geochronological data. *Terra Nostra*, **7/95**, 120–124.

SHERVAIS, J. W. 1982. Ti-V plots and the petrogenesis of modern and ophiolitic lavas. *Earth and Planetary Science Letters*, **59**, 101–118.

SIEBEL, W. 1995. Anticorrelated Rb–Sr and K–Ar age discordances, Leuchtenberg Granite, NE Bavaria, Germany. *Contributions to Mineralogy and Petrology*, **120**, 197–211.

SIEBEL, W., HÖHNDORF, A. & WENDT, I. 1995. Origin of late Variscan granitoids from NE Bavaria, Germany, exemplified by REE and Nd isotope systematics. *Chemical Geology*, **125**, 249–270.

SIEBEL, W., RASCHKA, H., IRBER, W., KREUZER, H., LENZ, K.-L., HÖHNDORF, A. & WENDT, I. 1997. Early Palaeozoic acid magmatism in the Saxothuringian belt: new insights from a geochemical and isotopic study of orthogneisses and metavolcanic rocks from the Fichtelgebirge, SE Germany. *Journal of Petrology*, **38**, 203–230.

SIEBEL, W., CHEN, F. & SATIR, M. 2003. Late-Variscan magmatism revisited: new implications from Pb-evaporation zircon ages on the emplacement of redwitzites and granites in NE Bavaria. *International Journal of Earth Sciences*, **92**, 36–53.

SIEBEL, W., BLAHA, U., CHEN, F. & ROHRMÜLLER, J. 2005. Geochronology and geochemistry of a dyke–host rock association and implications for the formation of the Bavarian Pfahl shear zone, Bohemian Massif. *International Journal of Earth Sciences*, **94**, 8–23.

SIEBEL, W., THIEL, M. & CHEN, F. 2006. Zircon geochronology and compositional record of late- to post-kinematic granitoids associated with the Bavarian Pfahl zone (Bavarian Forest). *Mineralogy and Petrology*, **86**, 45–62.

SIEDEL, H. & THEYE, TH. 1993. Very low-grade metamorphism of pelites in the Wippaer Zone, Harz Mountains, Germany. *Neues Jahrbuch für Mineralogie, Monatshefte*, **1993**(3), 115–132.

SISSINGH, W. 2004. Palaeozoic and Mesozoic igneous activity in the Netherlands: a tectonomagmatic review. *Netherlands Journal of Geosciences*, **82**, 113–134.

SMULIKOWSKI, W. 1995. Evidence of glaucophane-schist facies metamorphism in the East Karkonosze complex, West Sudetes, Poland. *Geologische Rundschau*, **84**, 720–737.

SÖLLNER, F., NELSON, D. R. & MILLER, H. 1997. Provenance, deposition and age of gneiss units from the KTB drill hole (Germany): evidence from SHRIMP and conventional U–Pb zircon age determinations. *Geologische Rundschau*, **86**(Supplement), S235-S250.

SOMMER, M. & KATZUNG, G. 2004. Abgrenzung, Gliederung und strukturelle Entwicklung des Neoproterozoikums im Schwarzburger Antiklinorium /Thüringisch-Fränkisch-Vogtländisches Schiefergebirge, Saxothuringische Zone). *Zeitschrift für geologische Wissenschaften*, **43**, 191–224

SOMMERMANN, A.-E., MEISL, S. & TODT, W. 1992. Zirkonalter von drei verschiedenen Metavulkaniten aus dem Südtaunus. *Geologisches Jahrbuch Hessen*, **120**, 67–76.

SOMMERMANN, A.-E., ANDERLE, H.-J. & TODT, W. 1994. Das Alter des Quarzkeratophyrs der Krausaue bei Rüdesheim am Rhein (Bl. 6013 Bingen, Rheinische Schiefergebirge). *Geologisches Jahrbuch Hessen*, **122**, 143–157.

STAMPFLI, G. M., VON RAUMER, J. F. & BOREL, G. D., 2002. Paleozoic evolution of pre-Variscan terranes: from Gondwana to the Variscan collision. *In*: MARTÍNEZ CATALÁN, J. R., HATCHER, R. D., JR., ARENAS, R. & DÍAZ GARCÍA, F. (eds) *Variscan-Appalachian dynamics: the building of the late Paleozoic basement*. Geological Society of America, Boulder, Colorado, Special Paper 364, 263–280.

ŠTĚDRÁ, V., KACHLÍK, V. & KRYZA, R. 2002. Coronitic metagabbros of the Mariánské Lázně Complex and Teplá Crystalline Unit: interferences for the tectonometamorphic evolution of the western margin of the Teplá-Barrandian Unit, Bohemian Massif. *In*: WINCHESTER, J. A., PHARAOH, T. C. & VERNIERS, J. (eds), *Palaeozoic amalgamation of Central Europe*. Geological Society, London, Special Publications **201**, 217–237.

STEIN, E. 2001. The geology of the Odenwald Crystalline Complex. *Mineralogy and Petrology*, **72**, 7–28.

STELTENPOHL, M. G., CYMERMAN, Z., KROGH, E. J. & KUNK, M. J. 1993. Exhumation of eclogitized continental basement during Variscan lithosphere delamination and gravitational collapse, Sudety Mountains, Poland. *Geology*, **21**, 1111–1114.

ŠTEMPROK, M. 1967. Genetische Probleme der Zinn-Wolfram-Vererzung im Erzgebirge. *Mineralium Deposita*, **2**, 102–118.

ŠTEMPROK, M. & SELTMANN, R. 1994. The metallogeny of the Erzgebirge (Krušné hory). *In*: SELTMANN, R., KÄMPF, H. & MÖLLER, P. (eds) *Metallogeny of Collisional Orogens*. Czech Geological Survey, Prague, 61–69.

ŠTEMPROK, M., HOLUB, F. V. & NOVÁK, J. K. 2003. Multiple magmatic pulses of the Eastern Volcano-Plutonic Complex, Krušné hory/ Erzgebirge batholith, and their phosphorus contents. *Bulletin of Geosciences*, **78**, 277–296.

STETS, J. 1990. Ist die Wittlicher Rotliegend-Senke (Rheinisches Schiefergebirge) ein 'pull-apart' Becken? *Mainzer geowissenschaftliche Mitteilungen*, **19**, 81–98.

STOLLHOFEN, H. & STANISTREET, I. G. 1994. Interaction between bimodal volcanism, fluvial sedimentation and basin development in the Permo-Carboniferous Saar-Nahe Basin southwest Germany. *Basin Research*, **6**, 245–267.

ŠTORCH, P. 2006. Facies development, depositional settings and sequence stratigraphy across the Ordovician–Silurian boundary: a new perspective from the Barrandian area of the Czech Republic. *Geological Journal*, **41**, 163–192.

STOSCH, H.-G. & LUGMAIR, G. W. 1990. Geochemistry and evolution of MORB-type eclogites from the Münchberg Massif, southern Germany. *Earth and Planetary Science Letters*, **99**, 230–249.

SUN, S.-S. & MCDONOUGH, W. F. 1989. Chemical and isotopic systematics of oceanic basalts: implications for mantle composition and processes. *In*: SAUNDERS, A. D. & NORRY, M. J. (eds) *Magmatism in the Ocean Basins*. Geological Society, London, Special Publications, **42**, 313–345.

SÜSS, M. P., DROZDZEWSKI, G. & SCHÄFER, A. 2001. The Ruhr and Aachen Basins–sedimentary environments, sequence stratigraphic model, and synsedimentary tectonics of Variscan foreland basins (Namurian B/C to Westphalian C, west Germany). *In*: HILLS, L. V. HENDERSON, C. M. & BAMBER, E. W. (eds) *Carboniferous and Permian of the World*. XIV International Congress on the Carboniferous and Permian. Canadian Society of Petroleum Geologists Memoirs, **19**, 585–598.

TAIT, J. A., BACHTADSE, V. & SOFFEL, H. C. 1996. Eastern Variscan fold belt: Paleomagnetic evidence for oroclinal bending. *Geology*, **24**, 871–874.

TAIT, J. A., BACHTADSE, V., FRANKE, W. & SOFFEL, H. C. 1997. Geodynamic evolution of the European Variscan fold belt: palaeomagnetic and geological constraints. *Geologische Rundschau*, **86**, 585–598.

TEIPEL, U., EICHHORN, R., LOTH, G., ROHRMÜLLER, J., HÖLL, R. & KENNEDY, A. 2004. U–Pb SHRIMP and Nd isotopic data from the western Bohemian Massif (Bayerischer Wald, Germany): Implications for Upper Vendian and Lower Ordovician magmatism. *International Journal of Earth Sciences*, **93**, 782–801.

THEYE, T. 1995. Metamorphosesprunge in der Wippraer Zone (Unterharz). *Zentralblatt für Geologie und Paläontologie Teil I*, **1993**(9/10), 1187–1200.

THOMAS, R. & KLEMM, W. 1997. Microthermometric study of silicate melt inclusions in Variscan granites from SE Germany: volatile contents and entrapment conditions. *Journal of Petrology*, **38**, 1753–1765.

TICHOMIROWA, M. 2002. Zircon inheritance in diatexites and its consequences on geochronology: a case study in Lusatia and the Erzgebirge (Saxo-Thuringia, eastern Germany). *Chemical Geology*, **191**, 209–224.

TICHOMIROWA, M. 2003. Die Gneise des Erzgebirges – hochmetamorphe Äquivalente von neoproterozoisch-frühpaläozoischen Grauwacken und Granitoiden der Cadomiden. *Freiberger Forschungshefte*, **C495**.

TICHOMIROWA, M., BERGER, H.-J., *et al.* 2001. Zircon ages of high-grade gneisses in the Eastern Erzgebirge (Central European Variscides)–constraints on origin of the rocks and Precambrian to Ordovician magmatic events in the Variscan foldbelt. *Lithos*, **56**, 303–332.

TIMMERMAN, M. J. 2004. Timing, geodynamic setting and character of Permo-Carboniferous magmatism in the foreland of the Variscan Orogen, NW Europe. *In*: WILSON, M., NEUMANN, E.-R., DAVIES, G. R., TIMMERMAN, M. J., HEEREMANS, M. & LARSEN, B. T. (eds) *Permo-Carboniferous Magmatism and Rifting in Europe*. Geological Society, London, Special Publications, **223**, 41–74.

TIMMERMANN, H., PARRISH, R. R., NOBLE, S. R. & KRYZA, R. 2000. New U–Pb monazite and zircon data from the Sudetes Mountains in SW Poland: evidence for a single-cycle Variscan orogeny. *Journal of the Geological Society, London*, **157**, 265–268.

TIMMERMANN, H., ŠTĚDRÁ, V., GERDES, A., NOBLE, S. R., PARRISH, R. R. & DÖRR, W. 2004. The problem of dating high-pressure metamorphism: a U–Pb isotope and geochemical study on eclogites and related rocks of the Mariánská Lázně Complex, Czech Republic. *Journal of Petrology*, **45**, 1311–1338.

TODT, W. A., ALTENBERGER, U. & VON RAUMER, J. F. 1995. U–Pb data on zircons for the thermal peak of metamorphism in the Variscan Odenwald, Germany. *Geologische Rundschau*, **84**, 466–472.

TORSVIK T. H. & COCKS L. R. M. 2004. Earth geography from 400 to 250 Ma: a palaeomagnetic, faunal and facies review. *Journal of the Geological Society of London*, **161**, 555–572.

TORSVIK, T. H. & VAN DER VOO, R. 2002. Refining Gondwana and Pangea palaeogeography: estimates of Phanerozoic non-dipole (octupole) fields. *Geophysical Journal International*, **151**, 771–794.

TRZEBSKI, R., BEHR, H. J. & CONRAD, W. 1997. Subsurface distribution and tectonic setting of the late-Variscan granites in the northwestern Bohemian Massif. *Geologische Rundschau*, **86**(Supplement), S64-S78.

TUCKER, R. D. & PHARAOH, T. C. 1991. U–Pb zircon ages for Late Precambrian igneous rocks in southern Britain. *Journal of the Geological Society of London*, **148**, 435–443.

TURNAU, E., ŻELAŹNIEWICZ, A. & FRANKE, W. 2002. Middle to early late Visean onset of late orogenic sedimentation in the Intra-Sudetic Basin, West Sudetes: miospore evidence and tectonic implication. *Geologia Sudetica*, **34**, 9–16.

TURNIAK, K., MAZUR, S. & WYSOCZAŃSKI, R. 2000. SHRIMP zircon geochronology and geochemistry of the Orla-Śnieżnik gneisses (Variscan belt of Central Europe) and their tectonic implications. *Geodynamica Acta*, **13**, 293–312.

ULRYCH, J., ŠTĚPÁNKOVÁ, NOVÁK, J. K., PIVEC, E. & PROUZA, V. 2002. Volcanic activity in Late Variscan Krkonoše Piedmont Basin: petrological and geochemical constraints. *Slovak Geological Magazine*, **8**, 219–234.

ULRYCH, J., FEDIUK, F., LANG, M. & MARTINEC, P. 2004. Late Paleozoic volcanic rocks of the Intra-Sudetic Basin, Bohemian Massif: petrological and geochemical characteristics. *Chemie der Erde*, **64**, 127–153.

ULRYCH, J., PEŠEK, J., *et al.* 2006. Permo-Carboniferous volcanism in late Variscan continental basins of the Bohemian Massif (Czech Republic): geochemical characteristic. *Chemie der Erde*, **66**, 37–56.

VAIDA, M., HANN, H. P., SAWATZKI, G. & FRISCH, W. 2004. Ordovician and Silurian ages of metamorphosed clastic sedimentary rocks from the southern Schwarzwald, SW Germany: a palynological study and its bearing on the Early Palaeozoic geotectonic evolution. *Geological Magazine*, **141**, 629–643.

VELLMER, C. & WEDEPOHL, K.H. 1994. Geochemical characterization and origin of granitoids from the South Bohemian Batholith in Lower Austria. *Contributions to Mineralogy and Petrology*, **118**, 13–32.

VENERA, Z., SCHULMANN, K. & KRÖNER, A. 2000. Intrusion within a transitional tectonic domain: the Cistá granodiorite (Bohemian Massif) – structure and rheological modelling. *Journal of Structural Geology*, **22**, 1437–1454.

VERNIERS, J., PHARAOH, T., *et al.* 2002. The Cambrian to mid-Devonian basin development and deformation history of Eastern Avalonia, east of the Midlands Microcraton: new data and a review. *In*: WINCHESTER, J.A., PHARAOH, T. C. & VERNIERS, J. (eds) *Palaeozoic Amalgamation of Central Europe*. Geological Society, London, Special Publications, **201**, 47–93.

VON QUADT, A. 1993. The Saxonian granulite massif: new aspects from geochronological studies. *Geologische Rundschau*, **82**, 516–530.

VON QUADT, A. 1997. U–Pb zircon and Sr–Nd–Pb whole-rock investigations from the continental deep drilling (KTB). *Geologische Rundschau*, **86**(Supplement), S258–S271.

VON RAUMER, J. F. 1998. The Palaeozoic evolution in the Alps: from Gondwana to Pangea. *Geologische Rundschau*, **87**, 407–435.

VON RAUMER, J. F., STAMPFLI, G. M., BOREL, G. & BUSSY, F. 2002. Organization of Pre-Variscan basement areas at the north-Gondwanan margin. *International Journal of Earth Sciences*, **91**, 35–52.

VON SECKENDORFF, V. 1990. *Geologische, petrographische und geochemische Untersuchungen an permischen Magmatiten im Saarland (Blatt 6507 Lebach)*. Geologisches–Paläontologisches Institut, Universität Kiel, Berichte–Reports, **39**.

VON SECKENDORFF, V., ARZ, C. & LORENZ, V. 2004*a*. Magmatism of the late Variscan intermontane Saar-Nahe Basin (Germany): a review. *In*: WILSON, M., NEUMANN, E.-R., DAVIES, G. R., TIMMERMAN, M. J., HEEREMANS, M. & LARSEN, B. T. (eds) *Permo-Carboniferous Magmatism and Rifting in Europe*. Geological Society, London, Special Publications, **223**, 361–391.

VON SECKENDORFF, V., TIMMERMAN; M. J., KRAMER, W. & WROBEL, P. 2004*b*. New $^{40}Ar/^{39}Ar$ ages and geochemistry of late Carboniferous to early Permian lamprophyres and related volcanic rocks in the Saxothuringian Zone of the Variscan Orogen (Germany). *In*: WILSON, M., NEUMANN, E.-R., DAVIES, G. R., TIMMERMAN, M. J., HEEREMANS, M. & LARSEN, B. T. (eds) *Permo-Carboniferous Magmatism and R ifting in Europe*. Geological Society, London, Special Publications, **223**, 335–359.

VRANÁ, S., ŠRÁMEK, J., POKORNÝ, L., MANOVÁ, M. & CHLUPÁČOVÁ, M. 1997. The Moldanubian Zone. *In*: VRÁNA, S. & ŠTĚDRÁ, V. (eds)

Geological model of western Bohemia related to the KTB borehole in Germany. *Sborník Geologických Věd*, **47**, 104–130.

WAJSPRYCH, B. & ACHRAMOWICZ, S. 2003. On the new wildflysch-to-flysch, blueschist-rich Lower Visean succession of the Western Sudetes (SW Poland). *Annales Societatis Geologorum Poloniae*, **73**, 123–137.

WALDHAUSROVÁ, J. 1971. The chemistry of the Cambrian volcanics in the Barrandian area. *Krystalinikum*, **8**, 45–75.

WALTER, H. & GAITSCH, B. 1988. Beiträge zur Ichnologie limnisch-terrestrischer Sedimentationsräume , Teil II: *Diplichnites minimus* n. ichnosp. aus dem Permosiles des Flechhtinger Höhenzuges. *Freiberger Forschungshefte*, **C427**, 73–84.

WARR, L. N. 2000. The Variscan Orogeny: the welding of Pangaea. *In*: WOODCOCK, N. & STRACHAN, R. (eds) *Geological History of Britain and Ireland*. Blackwell Science, Oxford, 271–294.

WEBER, K. 1995. The Spessart Crystalline Complex. In: DALLMEYER, R. D., FRANKE, W., & WEBER, K. (eds) *Pre-Permian Geology of Central and Eastern Europe*. Springer, Berlin-Heidelberg, 167–173.

WEDEPOHL, K. H., MEYER, K. & MUECKE, G. K. 1983. Chemical composition and genetic relations of meta-volcanic rocks from the Rhenohercynian Zone. *In:* MARTIN, H. & EDER, F.W. (eds) *Intracontinental Fold Belts*. Springer, Berlin, 231–256.

WENDT, J. I., KRÖNER, A., FIALA, J. & TODT, W. 1994. U–Pb zircon and Sm–Nd dating of Moldanubian HP/HT granulites from South Bohemia, Czech republic. *Journal of the Geological Society, London*, **151**, 83–90.

WENZEL, TH., MERTZ, D. F., OBERHÄNSLI, R., BECKER, T. & RENNE, P. R. 1997. Age, geodynamic setting, and mantle enrichment processes of a K-rich intrusion from the Meissen massif (northern Bohemian massif) and implications for related occurrences from the mid-European Hercynian. *Geologische Rundschau*, **86**, 556–570.

WENZEL, TH., OBERHÄNSLI, R. & MEZGER, K. 2000. K-rich plutonic rocks and lamprophyres from the Meissen Massif (northern Bohemian Massif): geochemical evidence for variably enriched lithosperic mantle sources. *Neues Jahrbuch für Mineralogie, Abhandlungen*, **175**, 249–293.

WERNER, C.-D. 1995. Basische Magmatite im Unter- und Mittelharz. *Zentralblatt für Geologie und Paläontologie, Teil I*, **1993**(9/10), 1257–1283.

WERNER, C.-D. 1997. Paläovulkanismus in der Elbe-Zone. *Hallesches Jahrbuch für Geowissenschaften*, **B19**, 71–82.

WERNER, O. & LIPPOLT, H. J. 2000. White mica $^{40}Ar/^{39}Ar$ ages of the Erzgebirge metamorphic rocks: simulating the chronological results by a model of Variscan crustal imbrication. *In*: FRANKE, W., HAAK, V., ONCKEN, O. & TANNER, D. (eds) *Orogenic Processes: Quantification and Modelling in the Variscan Belt*. Geological Society, London, Special Publications, **179**, 323–336.

WETZEL, K., GERSTENBERGER, H., WAND, G. & WENDT, I. 1995. Zur Geochemie der nordwestsächsischen Vulkanite. *Zeitschrift für Geologische Wissenschaften*, **23**, 371–400.

WILIMSKÝ, D., HANŽL, P., PŘICHYSTAL, A., AICHLER, J., MIXA, P., PECINA, V. & ŽÁČEK, V. 2003. Structures, textures and classification of the Devonian metavolcanics–southern part of the Vrbno group (Silesicum, Jeseníky Mts., Czech Republic). *Journal of the Czech Geological Society*, **48**, 131- 132.

WILL, T. M. & SCHMÄDICKE, E. 2001. A first find of retrogressed eclogites in the Odenwald Crystalline Complex, Mid-German Crystalline Rise, Germany: evidence for a so far unrecognized high-pressure metamorphism in the Central Variscides. *Lithos*, **59**, 109–125.

WILLIAMS, E. A., SERGEEV, S. A., STÖSSEL, I. & FORD, M. 1997. An Eifelian U–Pb zircon date for the Enagh Tuff Bed from the Old Red Sandstone of the Munster Basin in NW Iveragh, SW Ireland. *Journal of the Geological Society, London*, **154**, 189–193.

WILLIAMS, E. A., SERGEEV, S. A., STÖSSEL, I., FORD, M. & HIGGS, K. T. 2000. U–Pb zircon geochronology of silicic tuffs and chronostratigraphy of the earliest Old Red Sandstone in the Munster Basin, SW Ireland. *In*: FRIEND, P. F. & WILLIAMS, B. P. J., (eds) *New Perspectives on the Old Red Sandstone*. Geological Society, London, Special Publications, **180**, 269–302.

WILLNER, A. P., MASSONE, H. J. & KROHE, A. 1991. Tectono-thermal evolution of a part of a Variscan magmatic arc: the Odenwald in the Mid-German Crystalline Rise. *Geologische Rundschau*, **80**, 369–389.

WILLNER, A. P. RÖTZLER, K. & MARESCH, V. W. 1997. Pressure-temperature and fluid evolution of quartzo-feldspathic metamorphic rocks with a relic high-pressure, granulite-facies history from the central Erzgebirge (Saxony, Germany). *Journal of Petrology*, **38**, 307–336.

WINCHESTER, J. A. & FLOYD, P. A. 1977. Geochemical discrimination of different magma series and their differentiation products using immobile elements. *Chemical Geology*, **20**, 325–434.

WINCHESTER, J. A., FLOYD, P. A., CHOCYK, M., HORBOWY, K. & KOZDROJ, W. 1995. Geochemistry and tectonic environment of Ordovician meta-igneous rocks in the Rudawy Janowickie Complex, SW Poland. *Journal of the Geological Society, London*, **152**, 105–115.

WINCHESTER, J. A., FLOYD, P. A., AWDANKIEWICZ, M., PIASECKI, M. A. J., AWDANKIEWICZ, H., GUNIA, P. & GLIWICZ, T. 1998. Geochemistry and tectonic significance of metabasic suites in the Góry Sowie Block, SW Poland. *Journal of the Geological Society, London*, **155**, 155–164.

WINCHESTER, J. A., PHARAOH, T. C. & VERNIERS, J., 2002*a*. Palaeozoic amalgamation of Central Europe: an introduction and synthesis of new results from recent geological and geophysical investigations. *In*: WINCHESTER, J. A., PHARAOH, T. C. & VERNIERS, J. (eds) *Palaeozoic Amalgamation of Central Europe*. Geological Society, London, Special Publications, **201**, 1–18.

WINCHESTER, J. A. & THE PACE TMR NETWORK TEAM 2002*b*. Palaeozoic amalgamation of Central Europe: new results from recent geological and geophysical investigations. *Tectonophysics*, **360**, 5–21.

WOODCOCK, N. H. & STRACHAN, R. A. 2000. The Caledonian Orogeny: a multiple plate collision. *In*: WOODCOCK, N. & STRACHAN, R. (eds) *Geological History of Britain and Ireland*. Blackwell Science, Oxford, 187–206.

ŽAČEK, V. 1996. Retrograded eclogite from the Staré Město Belt, NE margin of the Bohemian Massif. *Journal of the Czech Geological Society*, **41**, 167–175.

ZACHALOVÁ, K., LEICHMANN, J. & ŠVANCARA, J. 2002. Žulovský Batholith: a post-orogenic, fractionated ilmenite-allanite I-type granite. *Journal of the Czech Geological Society*, **47**, 35–44.

ŽÁK, J., SCHULMANN, K. & HROUDA, F. 2005*a*. Multiple magmatic fabrics in the Sázava pluton (Bohemian Massif, Czech Republic): a result of superposition of wrench-dominated regional transpression on final emplacement. *Journal of Structural Geology*, **27**, 805–822.

ŽÁK, J., HOLUB, F. & VERNER, K. 2005*b*. Tectonic evolution of a continental magmatic arc from transpression in the upper crust to exhumation of mid-crustal orogenic root recorded by episodically emplaced plutons: the Central Bohemian Plutonic Complex (Bohemian Massif). *International Journal of Earth Sciences*, **94**, 385–400.

ZEH, A. & BRÄTZ, H. 2000. Radiometrische und morphologische Untersuchungen an Zirkonen aus Granitporphyren, Rhyolithen und Granitgeröllen des nordwestlichen Thüringer Waldes. *Zeitschrift der Deutschen Geologische Gesellschaft*, **151**, 187–206.

ZEH, A & BRÄTZ, H. 2004. Timing of Upper Carboniferous/Permian horst-basin formation and magmatism in the NW Thuringian Forest, Central Germany, a review. *In*: WILSON, M., NEUMANN, E.-R., DAVIES, G. R., TIMMERMAN, M. J., HEEREMANS, M. & LARSEN, B. T. (eds) *Permo-Carboniferous Magmatism and Rifting in Europe*. Geological Society, London, Special Publications, **223**, 319–334.

ZEH, A., COSCA, M. A., BRÄTZ, H., OKRUSCH, M. & TICHOMIROWA, M. 2000. Simultaneous horst-basin formation and magmatism during Late Variscan transtension: evidence from $^{40}Ar/^{39}Ar$ and $^{207}Pb/^{206}Pb$ geochronology in the Ruhla Crystalline Complex. *International Journal of Earth Sciences*, **89**, 52–71.

ZEH, A., WILLIAMS, I.S., BRÄTZ, H. & MILLAR, I. L. 2003. Different age response of zircon and monazite during the tectono-metamorphic evolution of a high grade paragneiss from the Ruhla Crystalline Complex, central Germany. *Contributions to Mineralogy and Petrology*, **145**, 691–706.

ZEH, A., GERDES, A., WILL, T. A. & MILLAR, I. L. 2005. Procenance and magmatic-metamorphic evolution of a Variscan island-arc complex: constraints from U–Pb dating, petrology, and geospeedometry of the Kyffhäuser Crystalline Complex, central Germany. *Journal of Petrology*, **46**, 1393–1420.

ZEIDLER, M., ELICKI, O., WALTER, H., KURZE, M., TRÖGER, K.-A. & KUPETZ, M. 2004. Neue Daten zur Biostratigraphie in der ''Phyllitischen Einheit'' des Nossen-Wilsdruffer Schiefergebirges

(Sachsen). *Freiberger Forschungshefte*, **C502**, 119–130.

ŻELAŹNIEWICZ, A., DÖRR, W., *et al.* 2004. The eastern continuation of the Cadomian orogen: U–Pb zircon evidence from Saxo-Thuringian granitoids in southwest Poland and the northern Czech Republic. *International Journal of Earth Sciences*, **93**, 773–781.

ZIEGLER, P. A. 1993. Late Palaeozoic – Early Mesozoic plate reorganization: evolution and demise of the Variscan Fold Belt. *In*: VON RAUMER, J. F. & NEUBAUER, F. (eds) *Pre-Mesozoic Geology in the Alps*. Springer, Berlin, 203–216.

ZULAUF, G., DÖRR, W., FIALA, J. & VEJNAR, Z. 1997. Late Cadomian crustal tilting and Cambrian transtension in the Teplá-Barrandian unit (Bohemian Massif, Central European Variscides). *Geologische Rundschau*, **86**, 571–584.

ZULAUF, G., SCHITTER, F., RIEGLER, G., FINGER, F., FIALA, J. & VEJNAR, Z. 1999. Age constraints on the Cadomian evolution of the Teplá-Barrandian unit (Bohemian Massif) through electron microprobe dating of metamorphic monazite. *Zeitschrift der Deutschen Geologischen Gesellschaft*, **150**, 627–639.

ZULAUF, G., BUES, C., DÖRR, W. & VEJNAR, Z. 2002. 10 km Minimum throw along the West Bohemian shear zone: Evidence for dramatic crustal thickening and high topography in the Bohemian Massif (European Variscides). *International Journal of Earth Sciences*, **91**, 850–864.

Index

Page numbers in *italic* denote figures. Page numbers in **bold** denote tables.